AA004550

22nd International Congress of Chemical and Process Engineering (CHISA 2016) and 19th Conference on Process Integration, Modelling and Optimisation for Energy Saving and Pollution Reduction (PRES 2016)

Prague, Czech Republic
27 - 31 August 2016

Volume 1 of 2

ISBN: 978-1-5108-5962-3

Printed from e-media with permission by:

Curran Associates, Inc.
57 Morehouse Lane
Red Hook, NY 12571

proceedings.com

Some format issues inherent in the e-media version may also appear in this print version.

Printed by Curran Associates, Inc. (2018)

Czech Society of Chemical Engineering
Novotneho Iavka 5
CZ-116 68 Praha 1
Stare Mesto, Czech Republic

Phone: +420 221 082 248
Fax: +420 221 082 366

cschi@csvts.cz

Additional copies of this publication are available from:

Curran Associates, Inc.
57 Morehouse Lane
Red Hook, NY 12571 USA
Phone: 845-758-0400
Fax: 845-758-2633
Email: curran@proceedings.com
Web: www.proceedings.com

TABLE OF CONTENTS

VOLUME 1

PLENARY LECTURES

REACTION ENGINEERING, CATALYSIS AND KINETICS

DISTILLATION AND ABSORPTION

EXTRACTION AND LEACHING

MEMBRANE SEPARATIONS

MIXING

COMPUTER AIDED PROCESS ENGINEERING

PARTICULATE SOLIDS

PHARMACEUTICAL ENGINEERING

CHEMICAL ENGINEERING EDUCATION

SYMPOSIUM ON ENVIRONMENTAL ENGINEERING

SYMPOSIUM ON SAFETY IN CHEMICAL INDUSTRY

SYMPOSIUM ON SUPERCRITICAL FLUID APPLICATIONS

SYMPOSIUM ON THERMODYNAMICS AND TRANSPORT PROPERTIES

VOLUME 2

SYMPOSIUM ON NOVEL FOOD PROCESSES AND TECHNOLOGIES

SYMPOSIUM ON POROUS MATERIALS, NANOSTRUCTURES AND NANOCOMPOSITES

SYMPOSIUM ON PROGRESS IN CHEMICAL TECHNOLOGY AND BIOTECHNOLOGY

SYMPOSIUM ON PROCESS INTENSIFICATION AND MINIATURISATION

COMMERCIAL AND TECHNICAL IMPACT FROM EU PROJECTS

PRES 2016

ADDITIONAL PAPERS

Remote control of reaction-diffusion processes with chemical microrobots

Frantisek Stepanek
Department of Chemical Engineering, University of Chemistry and Technology Prague,
Technicka 3, 166 28 Prague 6, Czech Republic
e-mail: Frantisek.Stepanek@vscht.cz

The ability to control the rates of chemical or biochemical reactions at length-scales comparable to those of single-cell organisms or their sub-cellular compartments would be desirable in a number of situations, ranging from fundamental studies of transfer phenomena to applications such as controlled delivery of actives from functional materials in the pharmaceutical, food, personal care or crop protection products. Established strategies for the control of chemical reactions known from the macroscopic world, such as control of temperature, the addition rate of reactants, or the availability of a catalyst have been applied to autonomous micro-scale systems called "chemical robots". A chemical robot is an internally structured microparticle consisting of a semi-permeable membrane that regulates molecular transport between the interior and the surroundings, a system of stimuli-responsive internal compartments that store and release reactants, immobilised enzymes or catalysts for facilitating chemical reactions, and magnetic nanoparticles that act as susceptors and enable the receiving of remotely sent radiofrequency signals.

Formulation engineering - the chemical engineering of products

J. P. K. Seville
IChemE, Davis Building, 165-189 Railway Terrace, Rugby, CV21 3HQ, UK;
tel. +44 1788 534413, e-mail: president@icheme.org

The chemical engineering of products, increasingly known as Formulation Engineering, concerns the design and manufacture of microstructured products which are often sold directly to the consumer and often designed to break down in use. Examples include foods, "instant" drinks, pharmaceutical dosage forms such as tablets, personal and household cleaning products, paints and inks – an increasingly long list. Product formulation and its subsequent processing is designed to form a specified microstructure. In use (eating, taking the dose, painting the wall, etc.) the product microstructure breaks down in a useful way, releasing flavour or active ingredients or forming a gloss layer or whatever. In manufacture, cost-efficiency and product quality are of supreme importance. In use, perception of product quality will depend on the effective breakdown or evolution of microstructure. This talk will discuss some examples and developments in teaching and research in Formulation Engineering.

Design and operations of sustainable supply chains – recent advances in life cycle optimization methods

Fengqi You
Cornell University, 318 Olin Hall, Ithaca, New York 14863, USA
Tel. 607-255-1162, email: fengqi.you@cornell.edu

As evidenced by the pope's recent encyclical on climate change and the public's interests in carbon and water footprints, sustainability issues are becoming more prevalent in the thoughts and actions of more and more people around the world. The increasing awareness of sustainability issues has driven the industry and society to pursue more economically, environmentally and socially sustainable strategies in the design and operations of supply chains. Systematic methods based on computational optimization frameworks using a life cycle perspectives play an important role in this quest. These computational frameworks integrate life cycle analysis and technoeconomic assessment with multiobjective superstructure optimization to improve overall sustainability performance of the supply chains. This talk will explore some recent results on the development of large-scale mathematical programming models and algorithms for life cycle optimization of sustainable supply chains. We will specifically focus on two computational frameworks and their applications. The first one is concerned with the functional unit-based life cycle optimization of sustainable supply chains under economic and environmental criterions. The second one is a multi-scale life cycle framework that accounts for and optimizes the direct and indirect environmental impacts from the foreground process systems scale and from the background economy scale through the integration of process systems optimization with process-based and input-output-based models. The last framework addresses the life cycle optimization in a decentralized, non-cooperative supply chains under a leader-follower game structure. We will discuss general modelling frameworks, efficient solution algorithms, and case studies on hydrocarbon biofuel and shale gas supply chains.

Chemical and process engineering: Forging a path with help from Stephen Sondheim

Cheryl I. Teich, PhD, AIChE Fellow
Process Engineering Reaction Engineering Expertise Area Leader
Engineering Solutions Technology Center
2016 AIChE Immediate Past President
The Dow Chemical Company
400 Arcola Road
Collegeville, PA 19426

In the United States, over the last thirty years, there has been a shift in industrial focus from process to product development. Over essentially the same time frame, there has been an expansion of the chemical engineering discipline to encompass areas outside the scope and the scale of the traditional CPI. These shifts in focus, combined with years of restructuring and the coincident retirement of the baby boomer population, has produced the "perfect storm" in reducing the pool of expert process practitioners.

The need to pay more attention to process engineering fundamentals is being recognized and addressed on a number of fronts. Process engineering forms the basis for ensuring a safe, secure, and economical supply of air, water, energy, food, as well as a myriad of discretionary consumer products.

The Pareto principle is valid for process development. While ever-increasingly more precise analytic tools are available, we can acquire ~ 80% of the process picture by applying chemical engineering basics: mass balance, the three laws of thermodynamics, rate laws, and economics. My talk will demonstrate the value of applying the fundamentals to frame specific "grand challenge" problems, with a goal of providing tools and guidance to newer process practitioners and speeding their own development and success.

But what does Stephen Sondheim have to do with it?

Effectiveness of capillary membrane bioreactor/reactor depending on the transport in cylindrical space

E. Nagy
Univ. Pannonia, Research Inst. Chem. Process Eng. Egyetem ut 10., 8200 Veszprém, Hungary;
tel. +36 88 603511, e-mail: nagy@mukki.richem.hu; fax: +36 88 604040.

Both enzymatic/microbial and chemical membrane reactors are widely investigated and very promising processes. Membrane can intrinsically be catalytic or the catalyst/biocatalyst can be immobilized either in the porous sponge layer (or simply on the interface of the dense layer of an asymmetric membrane in case of biocatalyst). The membrane (bio)reactor is often a capillary membrane and the reactant(s) containing phase is often fed in the lumen side of a capillary membrane Reactant(s) should reach the catalyst, thus, the chemical/biochemical reaction can take place mostly throughout the membrane matrix. Application of the convective flow through the catalytic membrane layer can significantly increase the reacted amount of the reactant. How this diffusion-convection-reaction transport process depend on the physical and the chemical properties of this membrane reactor has been analyzed e.g. by Nagy et al. [1,2]. The main question to be discussed is here how the varying cylindrical space can affect the conversion depending on the substrate transport direction. Depending on the flow direction the reaction can take place in increasing or decreasing cylindrical space, namely reactant(s) can be transported either from the shell side into the cylindrical matrix with decreasing space volume or it is transported from the lumen side into increasing membrane space volume. The varying space volume affects the concentration of the transporting reactant(s) and accordingly it can also affect the reaction rate. On the other hand the phase boundary layer can also have significant effect on the process performance [2]. A detailed analysis of mass transport through a catalytic/biocatalytic, capillary membrane is missing in the literature which analyzes the mass transfer properties depending on the cylindrical effect of a hollow fiber membrane and effect of the reaction kinetics as well as the operating modes.

It will be shown, that the cylindrical space can have significant effect on the membrane performance, depending on direction of the mass transport, when Peclet number does not differ essentially from unit. Modell results will be compared to measured data in case of both a catalytic and biocatalytic reactions.

The National Development Agency grant Hungarian Research Fund, OTKA 116727 greatly acknowledged for the financial support.

Literature

[1] NAGY E.: Mass transport through a convection flow catalytic membrane layer with dispersed nanometer-sized catalyst, *Ind. Eng. Chem Res.*, 49, 1057-1062 (2010).

[2] Nagy E., Dudás J., Mazzei, R., Drioli, E., Giorno L., Description of the diffusive-convective mass transport in hollow-fiber biphasic biocatalytic membrane reactor, *J. Membr. Sci.*, DOI: 10.1016/j.memsci.2014.11.060

Biogas upgrading via direct methanation with hydrogen from power-to-gas applications

[1]J. Witte, [1]T.J. Schildhauer, [2]A. Kunz, [3]S. Teske, [1]S.M.A. Biollaz
[1]Paul Scherrer Institute, 5232 Villigen PSI, Switzerland; tel. +41 56 310 4249, e-mail: julia.witte@psi.ch; [2]Energie 360° AG, Aargauerstrasse 182, 8010 Zurich, Switzerland; [3]Swiss Federal Laboratories for Materials Science and Technology, Überlandstrasse 129, 8600 Dübendorf, Switzerland.

Due to increasing usage of renewable energies in Europe, storage possibilities for excess electricity become more important. The Power-to-Gas (PtG) technology is a promising concept in order to store excess electricity from renewable sources, especially from summer to winter in a long term. In this project, raw biogas serves as carbon dioxide source for the PtG process and will be provided by a biogas plant of *energie 360°*, a Zurich region based utility.

Within the cooperation biogas upgrading via direct methanation is investigated. Raw biogas is produced from green and organic waste, and mainly consists of about 60 % vol. methane and 40 % vol. carbon dioxide. In order to meet the gas grid requirements further upgrading of the raw biogas is necessary, especially removing of CO_2. The conventional upgrade method via CO_2-scrubbing is cost- and energy-intensive. Instead of removing carbon dioxide, in the PtG-concept CO_2 is seen as raw material and is converted together with hydrogen into methane. Thus, not only the scrubbing unit can be saved, further more valuable product is gained from the green waste.

The project includes technical and economical modelling of the PtG- and the conventional upgrade process, as well as long duration experiments. Two possible PtG-process concepts, illustrated in Figure 1 and Figure 2, are modelled and investigated. Both concepts contain as first unit a bubbling fluidized bed (BFB) methanation, where carbon dioxide from raw biogas is converted together with hydrogen from an electrolyser into methane. The concentration of methane $\geq$ 96 % vol. must be reached to fulfil the gas grid requirements [1]. The produced water is removed by a condenser in a next step. The difference between the two concepts is the following treatment of the remaining hydrogen, whose concentration must be lower than 2 % vol. due to gas grid requirements [1]. In concept one, remaining hydrogen is converted together with the remaining carbon dioxide in a second methanation step, which is an adiabatic fixed bed (FB). After that a temperature-swing-adsorption (TSA), serving as drying unit, is implemented. The advantage of this version lies in a simple process design, but simulation results have shown, that the hydrogen fraction in the product stream is still too high in order to fulfil the hydrogen restrictions.

In concept two, after the fluidized bed methanation and the condenser, the biogas stream is compressed to the pressure required in the following H_2-membrane. Because of the higher pressure, a second condenser unit is sufficient to remove the rest of the water. The remaining hydrogen is separated by the H_2-membrane and is recycled to the methanation unit. This procedure allows a sub-stoichiometric hydrogen feed into the plant, but over-stoichiometric hydrogen-to-CO_2 ratio within the methanation reactor. Saving hydrogen is beneficial, because of the high H_2-production cost via electrolysis. On the other, hand pressure levels in the methanation and the membrane, as well as membrane area and recycle rates create an optimizing problem, which can be solved by economic considerations. The process concept two fulfils the gas grid requirements.

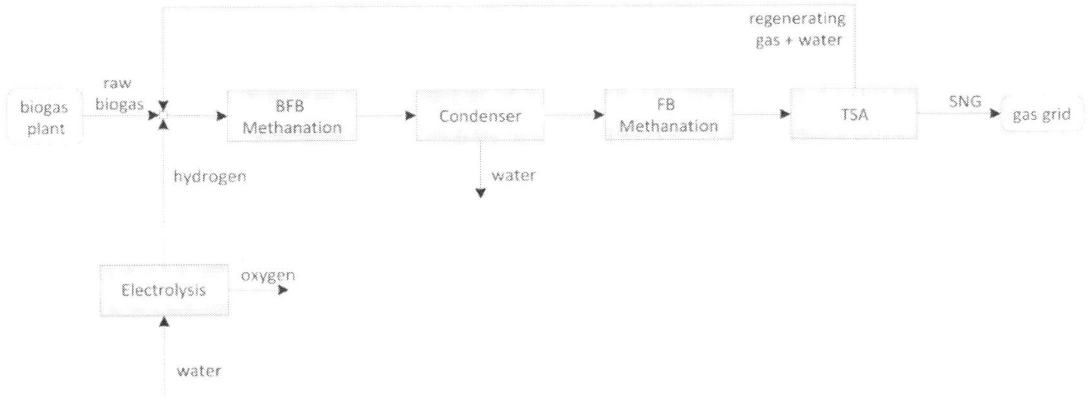

Figure 1: Schematic Flow Diagram of Concept One: Two-Stage-Methanation; BFB= Bubbling Fluidized Bed, FB=Fixed Bed, TSA=Temperature Swing Adsorption

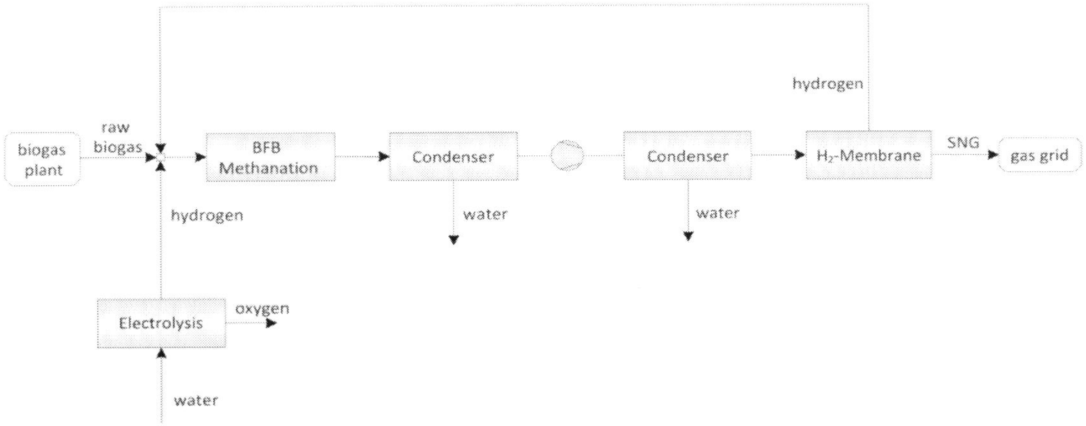

Figure 2: Schematic Flow Diagram of Concept Two: BFB-Methanation + H_2-Membrane; BFB= Bubbling Fluidized Bed

The optimising problem is solved by a detailed process chain simulation using thermodynamic models for the utilities, but a detailed rate based model for the fluidised bed [2] and the adiabatic fixed bed methanation. The hydrogen separation unit is modelled by an approach using Fick's Law, applying literature data for the permeabilities and selectivities of a commercial Imide based membrane [3]. Based on the combination of suited and sufficiently detailed models, the pareto-optima can be identified and used as input for economical modelling which compares direct methanation of biogas with more conventional biogas upgrading by CO_2 separation.

[1] Swiss Society of Gas and Water Industry - SVGW Guidelines, G13 and G18

[2] S. Teske; *Integrating rate based models into a multi-objective process design & optimisation framework using surrogate models*; Diss. EPFL No. 6302, 2015

[3] Y. Zhang, I. Musselman, J. Ferraris, K. Balkus Jr.; *Gas permeability properties of Matrimid membranes containing the metal-organic framework Cu-BPY-HFS*; Journal of Membrane Science 313 (2008) 170-181

Ionic liquid-coated transaminases for biocatalysis in organic solvents

B. Grabner, [1]M.T.Gundersen, H. Gruber-Woelfler, [2]S. Loïs, [2]S. Fantini, [3]S.Bartsch, [1]J. M. Woodley
Graz University of Technology, Institute of Process and Particle Engineering, Inffeldgasse 13/III, 8010 Graz, Austria; tel. +43 316 873 30409, e-mail: b.grabner@tugraz.at
[1]Technical University of Denmark, Dep. of chemical and biochemical Engineering, Søltofts Plads, Building 227, 2800 Kgs. Lyngby, Denmark, e-mail: jw@kt.dtu.dk
[2]Solvionic S.A., Site SNPE, Chemin de la Loge, 31078 Toulouse, France
[3]c-LEcta GmbH., Perlickstraße 5, 04103 Leipzig, Germany

Biocatalytic processing of hydrophobic substrates is of great interest in the synthesis of pharmaceutical ingredients since new APIs are become less water-soluble and more complex.[1] This is problematic since many enzymes such as transaminases can be employed in organic solvents only to a rather limited extent. In order to overcome the problem of limited enzyme stability, activity and selectivity, room-temperature solid phase ionic liquids (RTSPIL) can be used to coat the enzyme.[2-6]

Transaminases were coated with two different RTSPILs, 1-ethyl-3-methyl-imidazolium bromide [Emim]Br and 1-allyl-3-methylimidazolium chloride [Amim]Cl with different methods, melt coating, precipitation coating and colyophilization. The melt-coated enzyme was produced by heating the ionic liquid (IL) above melting point and mixing the melt with the lyophilized enzyme and cofactor. The mixture was allowed to cool to ambient temperature after the coating process. For the precipitate-coated enzyme the IL was precipitated out of a concentrated aqueous solution of the IL, enzyme and cofactor. The colyophilization was carried out by dissolving the enzyme in an aqueous buffer, addition of the coating material and lyophilisation. The prepared solid phases were then used for the enzymatic amination of various substrates in organic solvents (Figure 1). As a reference one reaction with free enzyme was carried out.

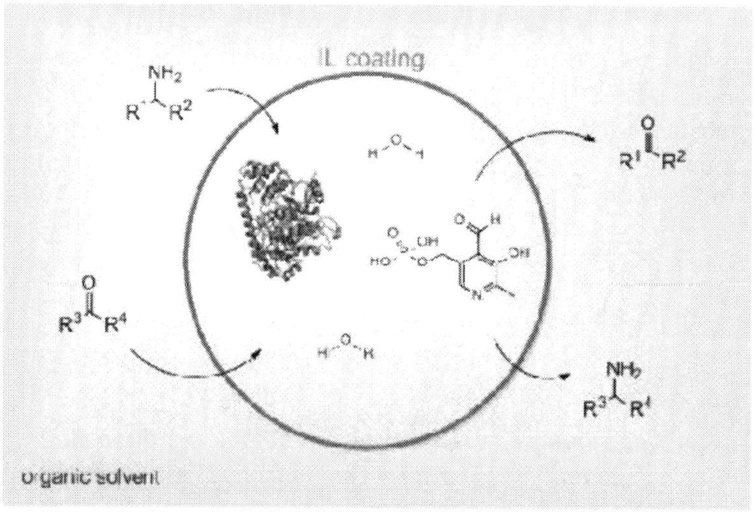

Figure 1. Reaction system for the enzymatic amination employing ionic liquid-coatied transaminase in organic solvent.

Our results show that the reactions with the coated enzymes run for over 120 h at constant reaction rate, whereas the reference reaction employing free enzyme stopped within one day. This can be attributed to cofactor decomposition, but not enzyme inactivation because further cofactor addition restarts the reaction.

In summary, the results show that transaminases immobilized via an ionic liquid show higher productivity than is soluble counterpart.

References:

[1] C. A. Lipinski, *J. Pharmacol. Toxicol. Methods* 2000, 44(1), 235-249

[2] J. K. Lee, M.-J. Kim, *J. Org. Chem.* 2002, 67, 6845-6847

[3] J. K. Lee, M.-J. Kim,, *J. Mol. Catal. B-Enzym* 2010, 68, 275-278

[4] F.-X. Dong, *J. Mol. Catal. B-Enzym* 2012, 77, 46-52

[5] J. Mutschler, *Green Chem.* 2009, 11, 1793-1800

[6] M. B. A. Rahman, *J. Mol. Catal. B-Enzym* 2012, 79, 61-65

Modelling the deposition of Fluorescent Whitening Agents (FWAs) on flat woven cotton fabrics

L. Bueno[a], C. Amador[b], S. Bakalis[a]
[a] Dep. Chemical Engineering, University of Birmingham, B15 2TT, Birmingham, UK
[b] Procter&Gamble Ltd., Newcastle Innovation Centre, Newcastle Upon Tyne, NE129BZ, UK

Fluorescent Whitening Agents (FWAs), also called optical brighteners are widely used in the textile and paper industry for improvement and maintenance of whiteness. These molecules absorb light in the UV region (350 nm) and emit on the blue region of the spectrum compensating yellowing of materials [1, 2]. In the detergent industry, FWAs are used in the formulas for whiteness maintenance of textiles; they adsorb onto the fabrics' structure during the washing cycle delivering an improved whiteness at the end of the process. Understanding and modelling the chemical and physical processes driving deposition and retention of optical brighteners as a function of washing conditions is key for optimising deposition efficiency by an optimisation of the detergent formula.

In the present work the adsorption behaviour of two stilbene derivative FWAs typically used in detergent formulas onto flat cotton fabrics has been investigated and modelled as a function of temperature, water hardness, concentration of optical brightener, fabric: liquor ratio and detergent concentration. The adsorption of the optical brighteners on textiles is tracked by a spectrophotometry based real–time methodology which allows obtaining high time resolution data of the active remaining in solution.

The resulting whiteness of textile garments is assessed by measuring their reflectance under different light conditions by spectrophotometry. In addition, the shaded component (absence of UV light absorbance) is also measured in order to understand potential yellowing of fabrics in indoor conditions. Calibration curves that correlate the adsorbed mass of optical brightener per area of flat cotton fabrics with the resulting whiteness of textiles have been developed which allow predicting the final whiteness for any adsorbed concentration of FWAs. This correlation is strongly non-linear where the whiteness benefit (delta B in LAB colorspace) plateaus as we reach high levels of optical brightener.

Based on experimental data, a mechanistic model for the deposition of optical brighteners has been developed which considers the following transformations in the system: (I) Dissolution of optical brightener, (II) convective mass transfer flow into the fabrics, (III) Fickian diffusion of brightener molecules to the fabric surface across boundary layer and stagnant layer of water in yarns and (IV) adsorption/desorption on fabrics' structure (adsorption/desorption isotherms).

The model is able to predict a) dissolution of FWAs, b) kinetics and equilibrium of brightener deposition, c) whiteness (delta B) of textile garments as a function of adsorbed concentration of FWAs under different light conditions for any temperature, hardness level, concentration of FWAs, detergent and fabric: liquor ratio.

The time to reach 95% of equilibrium as well as the equilibrium concentration, increase with increasing water to fabric ratio for a fixed brightener concentration since there is more mass of active per surface area to be adsorbed. Therefore, depending on the type of washing machine (Top Loader vs Front Loader Washing Machine) where the water to fabric ratio varies significantly (typically ~0.02 m^3:kg vs 0.004 m^3:kg), washing time plays a key role on the final brightener adsorbed on fabrics and therefore on the resulting whiteness of textiles requiring almost twice the time for actives in the higher water:fabric ratio machine to reach equilibrium. The importance of these models is that experiments can be run under laboratory conditions with less variable high water to fabric ratios and results scale up to real wash conditions.

A lack of fit is observed for both fluorescent whitening agents depositing on cotton fabrics when Fickian diffusion across a single boundary layer is considered. The hypothesis presented to explain this disagreement between experimental and predicted data is the presence of a complex dual porosity fabric structure [3]. Fabrics have two main porosities, inter-yarn porosity due to the spaces between the yarns of the textiles and intra – yarn porosity due to spaces between the fibres forming the yarns. As a consequence, optical brightener molecules diffuse across different diffusing distances before they reach the fabric surface. Initially a rapid decay of the bulk concentration is experimentally observed which is thought to correspond to the adsorption of molecules on the outer surface of the fabrics, which progressively decreases until saturation is reached leading to diffusion to deeper areas of fabrics which would promote a slower decrease of the bulk concentration due to larger diffusion distances. The approach followed for modelling the entire process is based on the consideration of two different surfaces with different surface area and diffusion distance where molecules can diffuse and get adsorbed simultaneously in parallel. Figure 1 represents experimental versus predicted values of brightener bulk concentration over time when considering a single diffusion distance (single diffusive layer) or a system with dual fabric porosity (two diffusive layers).

The approach followed to model the deposition of optical brightener on textiles as well as the key learnings can be reapplied to any other molecular active present in the washing cycle such as hueing dyes among many others.

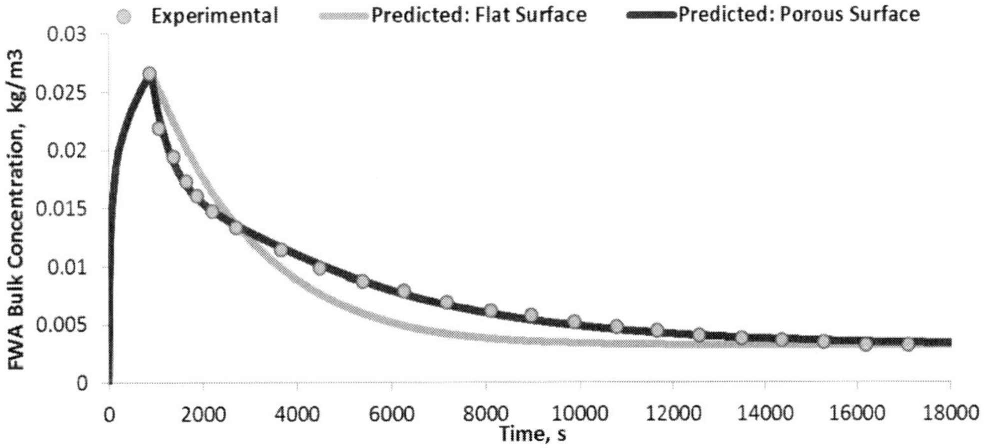

Figure 1 Experimental vs. Predicted values of brightener bulk concentration when considering a single diffusion distance or dual fabric porosity (Experimental conditions: 293K, 20gpg, 35ppm).

[1] XianNan, H. et al. (2008). Sci. China Ser. B – Chem., 51 (5), 473 – 478.
[2] Iamazaki, E.I. and Atvars, T., (2007), Langmuir, 23 (26), 12886 – 12892.
[3] Warmoeskerken, M.M.C.G., Van der Vlist, P., Moholkar, V., Nierstrasz, V., (2002), Colloids Surf A: Physiochem. Eng. Aspects 210, 277-285.

Multi-scale modelling of catalysed particulate filters for automotive exhaust gas aftertreatment

M. Leskovjan[1], M. Václavík[1], V. Novák[2], P. Kočí[1,*], D. Thompsett[2]
[1]University of Chemistry and Technology Prague, Department of Chemical Engineering, Technická 5, 16628 Praha 6, Czech Republic; *e-mail: petr.koci@vscht.cz
[2]Johnson Matthey Technology Centre, Blounts Court Road, Sonning Common, Reading RG4 9NH, United Kingdom

Despite the recent advances in development of alternative propulsion systems for automobiles, internal combustion engines will most probably remain the main technology in next few decades. During the fuel combustion, traces of pollutants are formed as undesired by-products. The emissions of toxic components are limited by legislation, focusing on four main pollutant groups: carbon monoxide (CO), hydrocarbons (HC), nitrogen oxides (NO_x) and solid particulate matter (PM). The emission limits are gradually tightened and it is not possible to meet them without sophisticated exhaust gas aftertreatment systems consisting of catalytic converters and particulate filters. Historically, particulate filters have been applied to Diesel engines that suffer from soot formation but the newly introduced limits on particulate numbers are going to enforce filters also for gasoline engines that produce small particles.

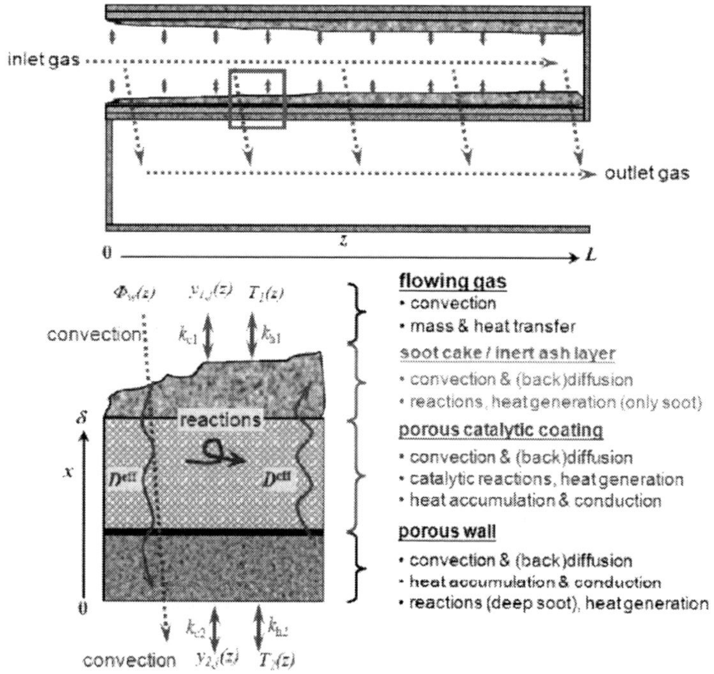

Figure 1. Schematic drawing of processes in a channel of catalysed particulate filter with catalytic layer coated on top of porous wall [1].

Both catalytic converters and filters are in the form of a cylindrical monolith with a large number of parallel channels in a honeycomb arrangement. The channels of catalytic converters are open and their walls are coated with a layer of active catalytic material where chemical reactions take place. In contrary, the filter channels are partially plugged so the exhaust gas is forced to permeate through the porous monolith walls, filtering out the soot and ash. A promising approach to make the system more compact and less expensive is to apply the catalytically active material directly on or into the porous structure of the catalysed particulate filter (Figure 1). On the other hand, there are potential negative effects such as increase of pressure loss due to deposition of catalytic material.

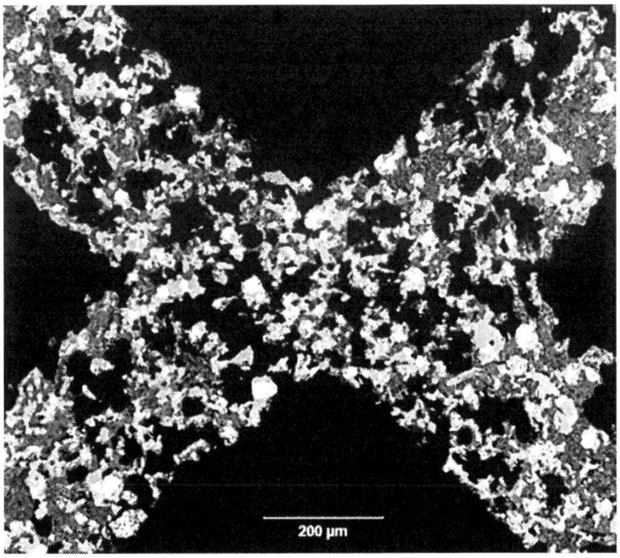

Figure 2. Cross-section SEM image of porous filter wall near the corner of adjacent square channels. White – wall material (SiC); grey – catalytic material in wall pores; black – voids.

In this contribution we present development and application of novel multi-scale models for catalysed particulate filters aiming on the description and prediction of the relations between the actual porous structure of filter (including the distribution and properties of catalytic material), and its macroscopic performance in terms of pressure loss and pollutant conversions [1,2]. Several types of filter substrates are characterized before and after the coating of active catalytic material by scanning electron microscopy (Figure 2), 3D X-ray microtomography and mercury porosimetry. The morphology of samples is then digitally reconstructed in 3D and used in micro-scale simulations of gas transport through the filter wall [2]. The obtained parameters, such as permeability and effective diffusivity in the wall, are employed in a macroscopic model of monolith channels [1], enabling to simulate the performance of the entire device. The model predictions are compared with experimental data from a lab reactor and the resulting trends are discussed.

Acknowledgements
The results were developed within the project PARTIAL-PGMs supported within the frame of EU program Horizon 2020.

References

[1] Schejbal M., Marek M., Kubíček M., Kočí P. Modelling of diesel filters for particulates removal. Chemical Engineering Journal 154 (2009), 219-230.

[2] Novák V., Dudák M., Kočí P., Marek M. Understanding the gas transport in porous catalyst layers by using digital reconstruction techniques. Current Opinion in Chemical Engineering 9 (2015), 16-26.

Real-time hybrid Monte Carlo method for modelling of 4 monomer semi-batch emulsion copolymerization

Tomáš Chaloupka, Alexandr Zubov, Juraj Kosek
Department of Chemical Engineering, University of Chemistry and Technology Prague, Technická 5, 166 28 Praha 6, Czech Republic, Tel.: 220 443 107,
E-mail: Tomas.Chaloupka@vscht.cz

Nowadays, polymerization reactors are operated according to rigidly prescribed process recipes that prohibit process optimisation and product property adjustment in real-time. Therefore, we created a model that is fast enough to enable on-line process control and optimisation, which means to reduce the batch period as much as possible while preserving the desired product properties. To do this, we implemented a hybrid Monte-Carlo approach[1,2], which provides fast predictions and, thus, enables optimisation of the production process with the controlled architecture of branched polymer, particle size distribution, polymer melt viscosity and particle morphology.

The considered reaction is carried out in a monomer-starved regime, i.e., the monomers are fed into the reactor in a manner allowing only small amount of free monomers (droplets) in the emulsion. The investigated system appears to be quite complex due to a presence of both hydrophilic and hydrophobic monomers, thermodynamics of sorption in polymer particles, etc. Moreover, we assume nonlinear polymerization with the chain transfer to polymer as a primary mechanism leading to long-chain branching and polymer cross-linking. Despite all this complexity, the simulation of the whole process trajectory must be fast enough for use in on-line control.

The hybrid Monte-Carlo approach consists of two steps. First, the monomer concentrations, temperature profiles, conversions and some other state variables are computed by a set of ODEs[3]; second, complex copolymer structures in polymer particles are simulated by Monte-Carlo repeating a random selection of physical effects that characterise the polymerisation process. The simulated copolymers are used to evaluate full molecular weight distribution, gel fraction and branching density, which directly influence polymer viscosity and particle morphology. Since the second part computes each polymer particle individually, the program can be easily parallelised to provide particle size distribution (which is also an important quality measure of the produced latex) without any significant increase in computational time.

Our model implemented in MATLAB environment simulates on standard desktop PC the whole several-hours-long batch within a few seconds and, thus, enables the on-line adjustment of the copolymerisation reaction. Implementation of our model in industry could realise significant resource and energy savings, while it would simultaneously respect high demands on the product quality.

Acknowledgement: Project RECOBA (636820-H2020-SPIRE-2014) is acknowledged.

References:

1. H. Tobita, Molecular Weight Distribution in Free Radical Polymerization with Long-Chain Branching, *J. Pol. Sci. B*, 1993, 31, 1363-1371
2. H. Tobita, Monte Carlo Simulation of Emulsion Polymerization – Linear, Branched, and Crosslinked Polymers, *Acta Polym.*, 1995, 46, 185-203
3. R. Pokorný, A. Zubov, P. Matuška, F. Lueth, W. Pauer, H.-U. Moritz, J. Kosek, Process Model for Styrene and n-Butyl Acrylate Emulsion Copolymerizaion in Smart-Scale Tubular Reactor, *Ind. Eng. Chem. Res.*, DOI: 10.1021/acs.iecr.5b02909

Substitute composition based on empirical characterization for steam-cracking modelling application

A. Karaba, P. Zámostný, Z. Bělohlav
University of Chemical Technology, Department of Organic Technology,
Technická 5, 166 28 Prague 6 – Dejvice, Czech Republic
tel. +420 22044 4301, fax. +420 22044 4340, e-mail: adam.karaba@vscht.cz; petr.zamostny@vscht.cz

In general, the feedstock characterization is very specific area in the refinery and petroleum processing industry. In some cases, exact composition can be obtained easily. In other cases, may not be available occasionally or even principally and only empirical characterization is available or even possible. But the exact compositions is important input for many supporting tools such as balance tools or predictive models of facility. In these cases, the empirical characterization can be transformed to substitute composition of fictive mixture formed by substitute components.

One of possible application is modelling of the steam-cracking process. The process transforms wide range of hydrocarbon fractions to valued products such as olefins and aromatics. Possible feedstock includes ethane, liquefied petroleum gas, gasoline, atmospheric gas oils (AGO) and many other refinery streams such as hydrocracked vacuum distillates. In cases of heavy feedstock, ranging from AGO, the exact component composition is unavailable principally. In other cases also can be unviable for relatively high analysis difficulty or human resource workload.

In this paper, we offer a detailed description of recently developed tool which is able to create substitute composition of the gasoline on the basis of usual empirical characterization: density, PIONA and distillation curve (SIMDIST). Components of substitute mixture are selected from the predefined set (about 100 compounds). But the set of compounds can be upgraded if it is needed. Components properties were collected from database but could be also estimated by Joback-type methods. Empirical characteristics of substitute mixture are determined by considering the ideal behavior of components in the mixture which is justified for a fraction of hydrocarbons. Algorithm searches for the content of individual substitute components to match the empiric criteria by an optimization procedure. Optimization parametrization is based on the selection of hydrocarbons fractions distribution in PIONA groups, so the natural molar weight distribution shape is maintained.

Paper is aimed at an analysis focused to determine the uniqueness of obtained substitute composition. The area of components content in substitute mixture satisfying selected empirical criteria will be checked. The yields of steam-cracking products of those substitute mixtures will be compared. Product yields will be determined by previously verified model of the steam-cracking PYROL [1] which is utilized in industry as a balance tool.

[1] A. Karaba, P. Zamostny, Z. Belohlav, J. Lederer and T. Herink, *Chemical Engineering & Technology*, 38, (2015) 609.

Investigation of deactivation mechanism for heterogeneous industrial hydrogenation catalyst: Accelerated aging using a special trickle bed reactor setup

[1]A. Reinsdorf, [1]W. Korth, [1]A. Jess, [2]F. Klasovsky
[1] Universität Bayreuth, Lehrstuhl für Chemische Verfahrenstechnik, Universitätsstraße 30, 95447 Bayreuth, Germany; tel. +49 2365 49-6343, e-mail: arne.reinsdorf@uni-bayreuth.de;
[2] Evonik Industries AG, Marl Germany

Oxo-alcohols are amongst the most important raw materials for the production of high volume plasticizers in chemical industry. C6 – C12 aldehydes from hydroformylation are hydrogenated on suitable heterogeneous catalysts to gain those alcohols which are further processed to the corresponding plasticizers.

During hydrogenation the formation of by-products like ester, acetal or other high boiling components has a major influence on the product quality. The influence of process control, feed quality and catalyst properties on side reactions is only sporadically described in literature.

Against the background of the high economical potential of increased alcohol yield and higher catalyst lifetime the formation of side components was investigated at industrially relevant hydrogenation conditions. In this work nonyl aldehyde was chosen as model substance feed. Reaction products were analyzed by means of GC / GC-MS measurements.

A trickle bed reactor setup with a special pre-reactor for thermal stress implementation was build up and used for accelerated aging experiments. The catalyst was stressed for at least 200 hours TOS applying harsh hydrogenation conditions as illustrated in fig 1. Subsequent soxhlet extraction with acetonitrile was applied to remove light-boiling side components. The extracted catalyst was then characterized and results were compared with catalyst from a commercial plant, prepared analogous.

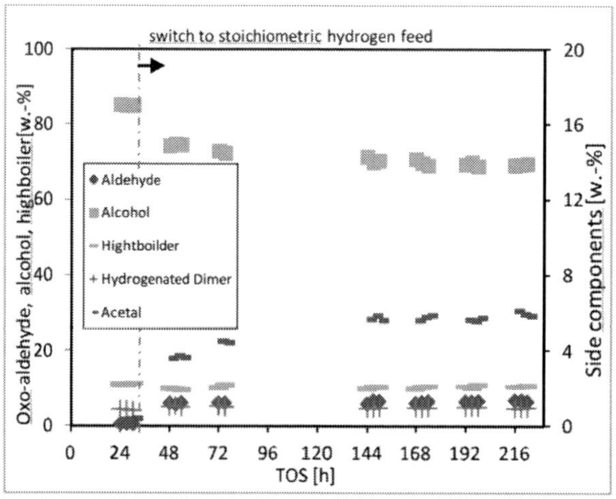

Figure 1: Continuous accelerated deactivation of Cu/Ni/Cr-Al_2O_3 catalyst in the hydrogenation of nonyl-aldehyde using a TBR

Catalyst characterization was carried out by means of TEM, EDX, XRD and BET. A fouling mechanism caused by high-boiling side components was found to be the predominant deactivation cause.

Therefore regeneration experiments for spent catalyst were conducted based on thermogravimetric analysis. The catalyst was treated at two different temperatures under either nitrogen or controlled oxygen atmosphere; subsequent hydrogenation showed a fully restored catalyst activity.

Use of Euler-Euler model and internal molecules age distribution transport theory to assess and predict trickle-bed reactors performances

Manel Fourati[1], Frederic Augier, and Romain Thomas
Affiliation : IFP Energies nouvelles
Rondpoint de l'échangeur de Solaize, Solaize, 69360, France
Corresponding author Email : manel.fourati@ifpen.fr
Phone number: +33 (0) 4 37 70 24 97

Liquid distribution in trickle bed reactors is a key hydrodynamic criterion for the reactor performance. When maldistributions exist, one dimensional classical plug-flow reactor model fails in predicting reactor performances since a more realistic description of hydrodynamics is required. 3D simulations of a lab-scale TBR using Eulerian-Eulerian model based on porous medium description have been carried out based on the model of Solomenko et al. (2014) and validated upon gamma- ray tomography data. The model has been then used to simulate hydrodynamics starting from different initial liquid maldistributions. In order to evaluate the impact of hydrodynamics on reactor performances, species transport has been solved for in adiabatic and isothermal conditions including a set of hydro-treatment chemical reactions. Simulations are carried out in the chemical as well as in the diffusion-controlled regime. The global reactor conversion is quantified and compared to plug-flow hydrodynamic regime. To suggest more realistic 1D reactor model in these conditions, the theory of age distribution transport is used to build a multi-outlet reactor model based on the molecules internal age distribution in the reactor.

. Keywords
Trickle-bed reactor, Eulerian-Eulerian model, CFD, porous medium, liquid distribution, internal ages distribution, multi-outlet reactor model

Abstract body

Trickle Bed Reactors (TBRs) have been used in refining for decades, since they are simple, cost-effective and relatively easy-to-scale-up[1].
Liquid distribution in trickle bed reactors is a key hydrodynamic criterion for the reactor performance. A poor liquid distribution can basically result in poor utilization or precocious deactivation of the catalyst and may induce local hot spots and low reactor efficiency[2]. Several studies have been done in the past to investigate experimentally mal-distributions inside TBRs[3,4], as well as numerically[5,6] . Eulerian-Eulerian CFD models are potentially interesting to simulate hydrodynamics in trickle-bed reactors since they include description of micro-scale phenomena while simulating macro-scale reactor volumes. Starting from these models, kinetics, heat balance, internal and external mass transfers can be solved for and the effect of maldistributions on reactor performances can be determined.

In this work a Eulerian-Eulerian CFD approach is used to simulate hydrodynamics in a 3D lab-scale trickle-bed reactor based on the model of Solomenko and al.(2015). The model is then validated upon Gamma-ray tomography measurements of liquid distribution.
The transport of species and energy balance are then solved for to simulate a set of reactions of hydro-treatment of a charge of hydrocarbons. Two regimes are investigated: a chemical regime and a diffusion-limited regime. It has been shown that initial liquid distribution may affect the reactor performances in a more or less moderate way depending on the maldistribution profile.

In order to build 1D reactor model that takes into account the liquid distribution profile, the transport of internal age distribution first and second moments has been simulated. This has been done by means of scalar transport equations with specific source terms developed in Liu and Tilton (2010). The profiles of first and second moments of internal age distribution allows describe the TBR by a set of parallel plug-flow reactors, each with a particular mass weight. This model is equivalent to a multi-exits plug-flow reactor and is shown to be more predictive than a classical plug-flow reactor with given axial dispersion, when the reactor is equipped with a classical chimney distribution plate. The accuracy of this reactor model is discussed for both chemical and diffusion-controlled reactor regimes.

It has been shown that the classical plug-flow reactor modeling is not relevant all time depending on hydrodynamics, especially liquid distribution inside the TBR, and the reactor controlling regime. In many cases, description via multi-exits plug-flow reactor is necessary.

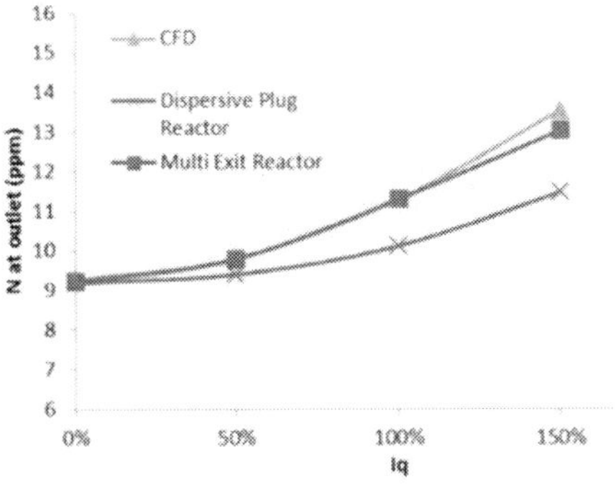

Figure 1: nitrogen concentration at reactor outlet for different maldistribution factors

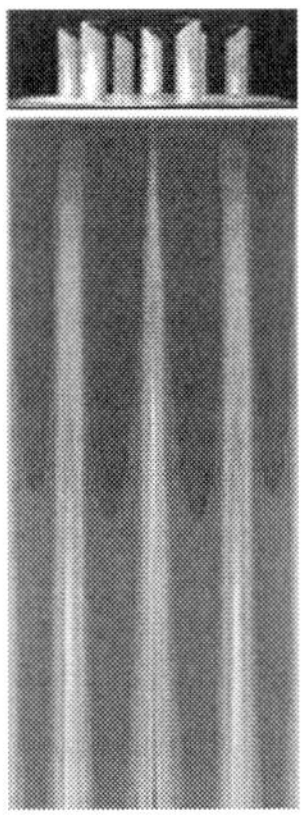

Figure 2: CFD-contours of internal mean age distribution inside a 4m-TBR below distribution plate with Chimneys

References

[1] Maiti R.N., Nigam K.D.P. Gas-Liquid Distributors for Trickle-Bed Reactors: A Review. *Ind. Eng. Chem. Res.*, 46, 6164-6182, 2007.

[2] Solomenko, Z., Haroun Y., Fourati M., Larachi R., Boyer C., Augier F., 2015. Liquid spreading in trickle-bed reactors : Experiments and numerical modeling using Eulerian–Eulerian two-fluid approach. *Chem.Eng.Sci.*, 126, 698–710.

[3] Bazer-Bachi F., Haroun Y., Augier F., Boyer C., 2013. Experimental evaluation of distributor technologies for trickle-bed reactors. *Ind.Eng.Chem.Res*, 52, 11189–11197.

[4] Marcandelli C., Lamine A.S., Bernard J.R., WILD G., 2000. Liquid distribution in trickle-bed reactor. *Oil & Gas Science and Technology*, Rev. IFP 55, 407-415.

[5] Lappaleinan K., Manninen M., Alopaeus V., 2009. CFD modeling of radial spreading of flow in trickle-bed reactors due to mechanical and capillary dispersion. *Chem. Eng. Sci*, 64, 207-218.

[6] Boyer C., Koudil A., Chen P., Dudukovic M.P., 2005. Study of liquid spreading from a point source in a trickle bed via gamma-ray tomography and CFD simulation. *Chem.Eng.Sci.* 60, 6279-6288.

[7] Liu, M., Tilton, J.N., 2010. Spatial distributions of mean age and higher moments in steady continuous flows. *AIChE J*, 56, 2561.

A TiO_2 immobilized coating prepared via spray drying for a Photo-CREC-Air Reactor

C. Lugo-Vega, H. de Lasa, [1]B. Serrano-Rosales
Chemical Reactor Engineering Centre (CREC), University of Western Ontario, Department of Chemical and Biochemical Engineering, London, Ontario, Canada N6A5B9; tel. +519 6612144, fax: +519 8502931, hdelasa@eng.uwo.ca; [1]Unidad Académica de Ciencias Químicas, Programa de Ingeniería Química, Universidad Autónoma de Zacatecas, Zacatecas, Campus UAZ Siglo XX1, 98168, México; tel. +52 492 9231006, fax: +52 492 9213990

1. Introduction

Heterogeneous photocatalysis is a potential method for the total degradation of indoor pollutants, converting VOCs into CO_2 and H_2O. However, the approaches to be used for the photocatalyst immobilization in scaled and highly efficient photoreactors are still not well established. The present study reports a method for TiO_2 immobilization on a stainless steel mesh using an Air Assisted Spray with an Automatized Spinning Coating (AAS-ASC). It is the aim of this study to report both the TiO_2-AAS-ASC preparation and its photonic performance under the scaled-up conditions of the Photo-CREC-Air Reactor.

2. Experimental Methods and Results

2.1 Immobilization Methods

TiO_2 was immobilized on a 55 x 45 cm stainless steel mesh using two methods: a) Spread Coating (TiO_2-SC) and b) Air Assisted Spray with an Automatized Spinning Coating (TiO_2-AAS-ASC). The TiO_2-SC method involves the deposition of a water Aeroxide TiO_2 P25 slurry applied using a brush on the stainless steel mesh. Concerning the TiO_2-AAS-ASC, the stainless steel mesh is set a rotating base while the nozzle sprays the mesh surface from a fixed distance. It was observed that changes of the rotating speed, distance of the nozzle to the target, air pressure and TiO_2 slurry concentration are all conditions that influence photocatalyst loadings and agglomerate size on the mesh. The Air Assisted Coating method gives uniform and reliable immobilized TiO_2, using an automatized spray nozzle with a spinning coating system.

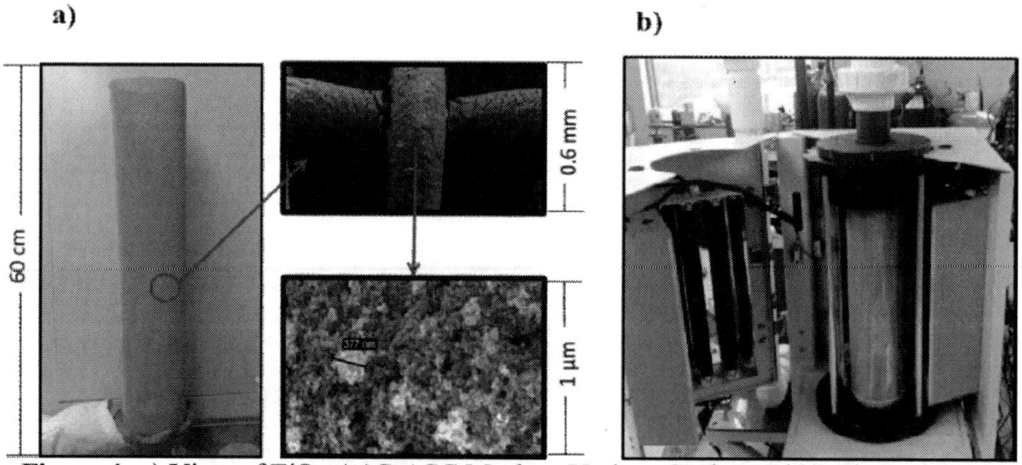

Figure 1. a) View of TiO_2-AAS-ASC Mesh at Various Scales and b) View of the TiO_2-AAS-ASC Mesh in Photo-CREC-Air Reactor.

Fig.1a reports a view of the prepared TiO_2-AAS-ASC. Mesh characterization was developed for both the TiO_2-SC and TiO_2-AAS-ASC mesh using gravimetry and SEM-EDX. The TiO_2-AAS-ASC displayed a TiO_2 homogeneous coating with a local fraction weight loading with 12% standard deviation. This is distinctively lower than the 24% standard deviation for TiO_2-SC. SEM micrographs for TiO_2-SC showed uncoated stainless steel areas and a TiO_2 accumulation on the wire joints. On the other hand, the TiO_2-AAS-ASC micrographs (refer to Fig.1a) displayed an essentially fully coated mesh surface with a 6 to 8 μm TiO_2 layer and 32 to 119 nm size particle agglomerates. Furthermore, the high adhesion of the immobilized TiO_2 was proven using the gravimetry, with only 1wt%-2wt% of the TiO_2 particles lost after 1000 hours of operation in the Photo-CREC-Air unit shown in Fig.1b.

2.2 Photo-CREC-Air Reactor

The scaled-up 55.1 L Photo-CREC-Air Reactor works as a close loop system using 8 UV lamps and a stainless steel coated mesh [1-2]. The reaction section of Photo-CREC-Air Unit consists of two concentric cylinders surrounded by UV lamps. The inner concentric cylinder is a stainless steel mesh while the exterior cylinder is made out of quartz glass. The air to be treated flows though the coated mesh from the exterior to the interior and does this with the help of a bullet nose bottom. This special design provides an excellent interaction with the photocatalyst, the air flow and the irradiation. In addition, the modified Photo-CREC Unit permits macroscopic energy balances allowing the measurement of absorbed photons, scattered photons and reflected radiation.

2.3 Photodegradation performance

The performance of the Photo-CREC-Air Unit with both TiO_2-SC and TiO_2-AAS-ASC meshes was established utilizing the photocatalytic degradation of acetone. Initial acetone concentrations ranging from 24.9 to 49 μmol/L range were employed. In this respect, different TiO_2 loadings were used to find the higher photodegradation activity as follows: 1wt% to 3wt% TiO_2 for TiO_2-SC and 0.8wt% to 2wt% for TiO_2-AAS-ASC. It was shown that the photoconversion rate of the reaction increases with photocatalyst loading and reaches a steady value at 2.5wt% and 1wt% for TiO_2-SC and TiO_2-AAS-ASC, respectively. Thus, TiO_2-AAS-ASC displays a maximum photoconversion rate with 3 times less photocatalyst than the one required for TiO_2-SC.

Quantum Yields (QYs) were evaluated for both TiO_2 immobilized meshes using the absorbed photons on the photocatalyst and the consumption of OH radicals in the acetone photodegradation. Absorbed photons were calculated through macroscopic balances that included incident, transmitted and reflected radiation measurements. Both methods reached consistent QY values of 50-60% maximum levels. A Quantum Yield per catalyst weight was also established to reflect the efficiency of the photocatalyst utilization. This yielded a QY based on unit photocatalyst weight of 25%/g and 8%/g for TiO_2-AAS-ASC and for TiO_2-SC, respectively.

3 Conclusions

a) A reliable TiO_2-AAS-ASC method for TiO_2 photocatalyst immobilization is established. The method provides good photocatalyst anchoring with the mesh, a uniform coating and a high degree of TiO_2 irradiation. b) The prepared immobilized photocatalyst coating is very active for the photoconversion of acetone in air with Quantum Yields as high as 55%.

References

[1] J.M. Garcia-Hernandez, B. Serrano-Rosales, H. de Lasa, Energy Efficiencies in a Photo-CREC-Air Reactor: Conversion of Model Organic Pollutants in Air, Industrial & Engineering Chemistry Research, 51 (2012) 5715-5727.

[2] H. Ibrahim, H. de Lasa, Kinetic modeling of the photocatalytic degradation of air-borne pollutants, AIChE Journal, 50 (2004) 1017-1027.

Photocatalysis: light, alternative energy and the environment

Hugo Ignacio de Lasa Dr.Ing., P.Eng, Ing., FCIC
Chemical Reactor Engineering Centre, University of Western Ontario, Canada

Heterogeneous photocatalysis opens new opportunities to harvest solar energy. Extensive application however, requires the availability of photocatalytic reactors such as Photo-CREC units where macroscopic radiation balances can be accurately established. Photo-CREC reactors have been invented at the Chemical Reactor Engineering Centre (CREC) of the University of Western Ontario. The high efficiency of these units has been established using quantum and thermodynamic efficiencies. Regarding the production of hydrogen as an alternative fuel, it has been proven that Photo-CREC reactors are very well suited for water splitting using Pt on TiO_2 and ethanol as an organic scavenger.

References

[1] H. de Lasa , B. Serrano, J. Moreira , P. Valades *Chem.Engng Tech.* In press (2015)DOI: 10.1002/ceat.201500305

[2] H. de Lasa, B. Serrano, M. Salaices,. *Photocatalytic Reaction Engineering*, Springer: New York, **2005**

[3] S. Escobedo, B. Serrano, H. Lasa, *Appl. Catal., B*, **2013**, 140 (141), 523 – 536. DOI: 10.1016/j.apcatb.2013.04.016.

[4] J. M. Garcia Hernandez, B. Serrano Rosales, H. de Lasa, *Ind. Eng. Chem. Res.* **2012**, 51(16), 5715- 5727. DOI: 10.1021/ie300177p

[5] B. Serrano, A. Ortiz, J. Moreira, H. I. de Lasa, *Ind. Eng. Chem. Res.* **2009**, 48, 9864-9876. DOI: 10.1021/ie900353n.

[6] M. Salaices, B. Serrano, H. de Lasa, *Chem. Eng. J.* 2002, 90, 219-229. DOI: 10.1016/S1385-8947(02)00037-2

[7] J. Moreira, B. Serrano, A. Ortiz, H. de Lasa, *Ind. Eng. Chem. Res.* 2010, 49, 10524-10534. DOI: 10.1021/ie100374f

[8] J. Moreira, B. Serrano, A. Ortiz, H. de Lasa, *Chem. Eng. Sci.* 2011, 66 (23), 5813-5821. DOI: 10.1016/j.ces.2011.07.040

[9] P.J. Valades-Pelayo, F. Guayaquil Sosa, B. Serrano b, H. de Lasa *Chem. Eng. J.* **2015**, DOI: 10.1016/j.cej.2015.03.039

[10] P. J. Valadés-Pelayo, J. Moreira, P. Solano-Flores, B. Serrano, H. de Lasa, *Chem. Eng. Sci.* **2014**, 116, 406 – 417. DOI: 10.1016/j.ces.2014.04.041

[11] M.Salaices, B.Serrano, H, de Lasa, Ind Eng.Chem.Res. 2001, 40(23) 5455-5464.DOI: 10.1021/ie0102551.

[12] B. Serrano, H. de Lasa, *Ind. Eng. Chem. Res.* **1997**, 36, 4705-4711. DOI: 10.1021/ie970104r.

[13] M. Salaices, B. Serrano; H.I. de Lasa; Chemical Engineering Science 59 (2004) 3 – 15 doi:10.1016/j.ces.2003.07.015

[14] B. Serrano, H. de Lasa, *Ind. Eng. Chem. Res.* **1997**, 36, 4705-4711. DOI: 10.1016/0927-0248(94)00230-4.

[15] B. Serrano and H.I. de Lasa, *Chem. Eng. Sci.* **1999**, 54, 3063 – 3069. DOI: 10.1016/S0009-2509(98)00478-3

[16] A.Ortiz Gomez A., B.Serrano B., H.de Lasa H. Chem.Eng.Sci, **2008**, 63, 520-557. doi:10.1016/j.ces.2007.04.053

[17] Ortiz A., Serrano B., Moreira J. de Lasa H., "Vol.36, Photocatalytic Technologies Adv. Chemical Engineering (Elsevier), eds. H.de Lasa, B. Serrano, Vol.36, 69-108 (**2009**).
[18] J. Moreira, Ph.D. Thesis, The University of Western Ontario, 2011.
[19] Garcia Hernandez J, Serrano B., de Lasa H.," Chem.Eng.J. 2010 165, 891-901 doi:10.1016/j.cej.2010.06.034
[20] H. Ibrahim H., H. de Lasa H., AIChE J., **2004** Vol.50, 1017-1027 (2004) "Kinetic Modeling of the Photocatalytic Degradation of Air Borne Pollutants"
[22] P.J. Valadés-Pelayo, J. Moreira, B. Serrano, H. de Lasa, *Chem. Eng. Sci.* **2014**, 107, 123 – 136. DOI: 10.1016/j.ces.2013.12.013
[23] B. Serrano, A. Ortiz, J. Moreira, H. I. de Lasa, *Ind. Eng. Chem. Res.* **2010**, 49, 6824-6833. DOI: 10.1021/ie9017034.

Short Resume

Hugo I. de Lasa is a Professor at the Department of Chemical and Biochemical Engineering, at the University of Western Ontario. Prof. de Lasa graduated in 1971 with a Doctoral degree from the Université de Nancy in France. Hugo de Lasa received, in 2004, an *Honoris Causa* Doctoral degree in Chemical Engineering from the Universidad Autonoma de Zacatecas, Mexico. Prof. de Lasa is the author of 310-peer reviewed articles, 6 books, and 11 U.S. and Canadian patents. He has been actively involved in the training of graduate students with the successful supervision of 42 PhD and 30 MESc. Prof. de Lasa's activities have been of key importance for furthering the understanding of the science and the technology of chemical reactors. His innovative work deals with the application of chemical reactor engineering to environmentally friendly processes and products. Examples of Hugo de Lasa's inventions are the CREC-Riser Simulator, the Pseudoadiabatic Catalytic Reactor, the CREC-Optiprobes and the Photo-CREC reactors. In 2003, de Lasa founded Recat Technologies Inc (http://www.recattechnologies.com/) with the goal of commercializing CREC innovations. In 1998, Hugo de Lasa received the Research Excellence Prize from the Faculty of Engineering Science of the University of Western Ontario. In 2000, he was designated Fellow of the Chemical Institute of Canada. That same year, he was awarded the Medal of Research and Development from the Professional Engineers of Ontario. In 2001, Hugo de Lasa received the Award in Industrial Practice and in 2004, the R.S.Jane Lecture Award, both from the Canadian Society for Chemical Engineering. In 2005, Hugo de Lasa was inducted as a Fellow of the Canadian Academy of Engineering. In 2010, The American Chemical Society published a special Industrial Engineering Chemistry Research *Festschrift* Issue in Hugo de Lasa's honour.

Application of a packed bed microreactor for determining the intrinsic kinetics of the 2-methylpropene (isobutylene) hydrogenation

Z. Vajglova, P. Stavarek, V. Jiricny, [1]J. Kolena
Institute of Chemical Process Fundamentals of the ASCR, v. v. i., EHLSP, Rozvojova 2/135 Prague 6, Czech Republic; tel. +420 220 390 235, e-mail: vajglova@icpf.cas.cz; [1]The Inorganic Chemistry Research Institute – UniCRE, Revolucni 1521/84, Usti nad Labem, Czech Republic.

This paper focused on the kinetics of the heterogeneously catalyzed reactions using a packed bed microreactor suitable for high temperature and high pressure applications. Hydrogenation of 2-methylpropene (isobutene, 2-MP) to 2-methylpropane (2-MPA) over the commercial platinum catalyst was selected as a model reaction representing total hydrogenation of the C4 fraction. The operating window of the reaction under the intrinsic kinetic control without the influence of transport phenomena was identified by the experimental diagnostic tests, dimensionless criteria and their comparison. To eliminate the transport phenomena and the transition to the kinetic control regime, the dilution of the catalyst bed by the non-porous silicon carbide exhibiting high thermal conductivity proved necessary.

The experimental data obtained under kinetic control were processed by a mathematical model of an isothermal plug flow reactor with the kinetic equation representing irreversible second-order bimolecular-type reaction valid only for low concentrations of the 2-MP. The activation energy of 131.7 $kJ.mol^{-1}$ and the frequency factor 1.6×10^{12} $m^3.mol^{-1}.s^{-1}$ were determined experimentally assuming the Arrhenius temperature dependence of kinetic constants. The study of the effect of partial pressure of the given reactant on the conversion of 2-MP determined the reaction orders with respect to H_2 and 2-MP. The experimental data were correlated by the formulated kinetic model with an accuracy of ± 10 %.

Reproducible oscillatory behaviour of the hydrogenation of 2-MP exhibiting a large length of the period (11 – 22 hours) was observed. This observed behaviour, in contrast with in literature published sources, is considered unique. Based on the character of oscillations, they were classified as thermo-kinetic oscillations resulting from the interactions of exothermic, non-linear reactions with the transport phenomena, particularly due to insufficient heat removal from the catalyst bed. The observed oscillation behaviour can be understood as a side effect of the intensification process in a microreactor, whose aim was to achieve the regime controlled solely by the intrinsic kinetics. Intensive progress of the exothermic reaction achieved in the microreactor led to the appearance of side effects that are typically expressed in normal operation.

Acknowledgement: The publication was created using the infrastructure supported by the UniCRE project, funded by the EU Structural Funds and the state budget of the Czech Republic.

A new stoichiometric and kinetic model for the substitution of polysaccharides with two reagents

Sakari Teerikoski[1], Tapio Salmi[1]

[1]Johan Gadolin Process Chemistry Centre, Laboratory of Industrial Chemistry and Reaction Engineering, Åbo Akademi, FI-20500, Turku/Åbo, Finland; tel +35822154427, e-mail:tapio.salmi@abo.fi, fax:+35822154479

Cellulose derivatives prepared by nuclophilic substitution have large applications in chemical industry and in consumer goods such as paints, detergents, cosmetics and alimentary products. Many cellulose derivatives are prepared through simultaneous substitution of the anhydroglucose units of cellulose by two reagents, such as a chloroalkane and an epoxide, which complicates the substitution stoichiometry and kinetics very much.

In this way, numerous derivatives of cellulose and other polysaccharides can be obtained, which is illustrated by Figure 1. Various derivatives can be described by changing the letter "R" to different to symbols of actual functional groups, depending on the cellulose derivative.

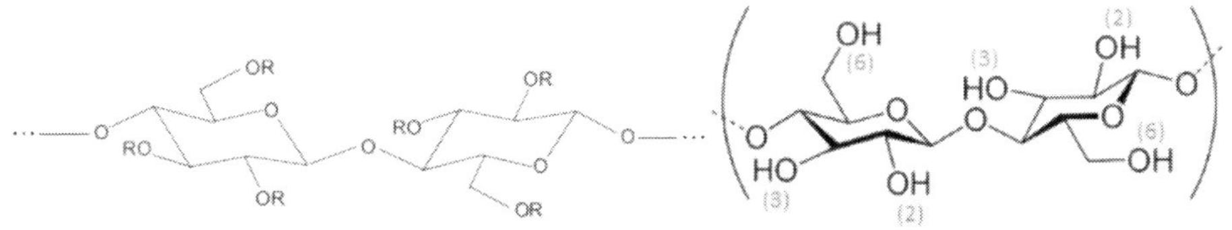

Figure 1. A cellulose derivative (left) and numbering of hydroxyl groups (right).

Current industrial practise in the production of cellulose derivatives is very much based on old experience and empirical approach. However, in order to optimize the existing processes and to develop new processes for new product, a more quantitative approach is needed, based on the reactivities of the hydroxyl groups and correct overall stoichiometry. A big shift takes place when two reagents are used instead of one: the size of the system explodes.

Mathematical models of cellulose substitution were developed for cases with two reagents. With the model it is possible to predict the substitution pattern in detail and to optimize the dosing of the reagents so that a desired product distribution can be obtained. Examples of di-substituted anhydroglucose units are given below (Figure 2).

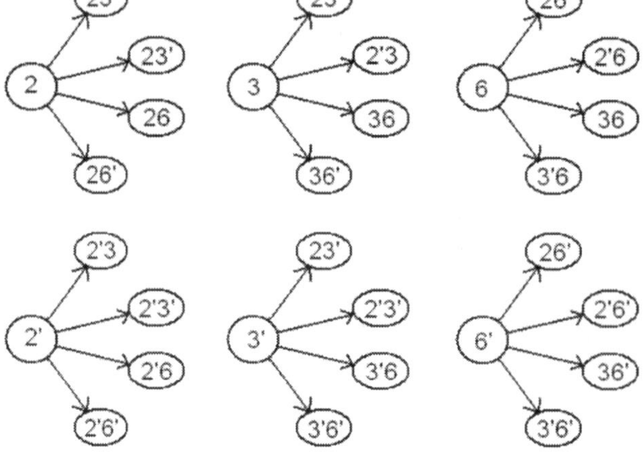

Figure 2. Di-substituted units formed from monosubstituted units when two substitution reagents are used.

When using two substituting molecules (R and R'), in total 54 reactions take place simultaneously. Leading to mono-, di- and trisubstituted products. An extensive simulation model was developed to describe the substitution pattern in isothermal batch reactors. The system consisted of 27 ordinary differential equations (ODEs) describing the mass balances of the anhydroglucose units, which were solved numerically for several cases. The results of selected numerical simulations are displayed in Figure 3, where the concentrations of mono- and disubstituted units are displayed as a function of the reaction time.

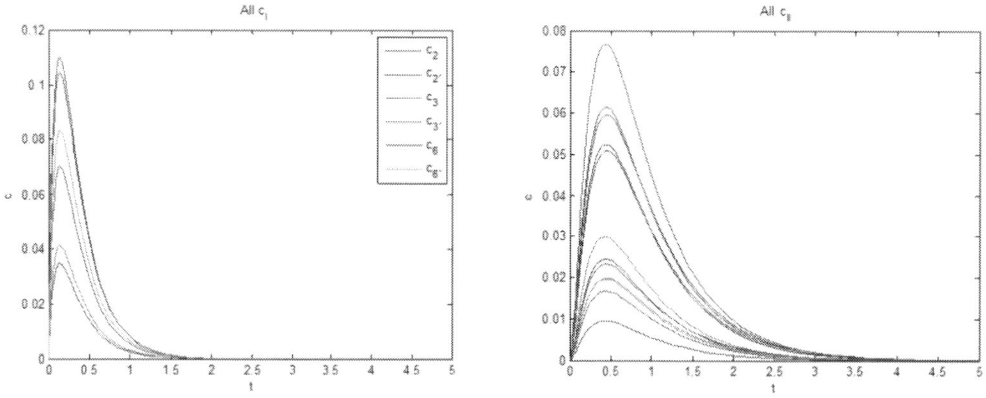

Figure 3. Concentrations of mono- (left) and disubstituted (right) anhydroglucose units as a function of reaction time in a batch reactor.

The case of dosing two reactants was investigated with numerical simulations which revealed that an optimal dosing is possible to obtain the desired product distribution. The theoretical concept developed can be applied to obtain desired degrees of polymerization (DP) and product distributions of existing and future polysaccaride derivatives.

Catalyst screening and reaction kinetics in the degradation of pharmaceuticals by ozonation

P. Tolvanen[a], S. Said[a], A. Franz[a], K. Eränen[a], N. Kumar[a], T. Salmi[a]
[a]Laboratory of Industrial Chemistry and Reaction Engineering, Johan Gadolin Process Chemistry Centre, Åbo Akademi University, Biskopsgatan 8, FI-20500 Åbo/Turku, Finland; phone: +35822154983; email: ptolvane@abo.fi

Abstract. The pollution of various pharmaceuticals into the municipal wastewaters and to rivers, lakes and sea is a major increasing concern. Due to the large water amounts in urban areas, the technology for removing efficiently all the chemicals is limited and large quantities of human pharmaceuticals are passed through the waterer purification processes. Some of the most common pharmaceuticals were studied in a semibatch reactor where the ozone was ex-situ generated continuously, and the degradation of the drug was kinetically followed by continuous sampling, O_3 quenching and HPLC analysis. The solubility of ozone as well as the degradation of pharmaceuticals was studied under various gas flow, stirring, temperature and pH conditions. An effort was put in the analysis of low-molecular weight degradation products to fully understand the reaction mechanisms of the degradation. Several catalysts screened, and the best candidates were studied for enhancing the degradation of several pharmaceuticals. The influence of the catalyst immobilization was also studied by applying a rotating bed reactor.

Keywords

ozonation, waste water, pharmaceuticals, solid catalysts, green process intensification.

Introduction. Many drugs are not fully metabolized in the human body and may therefore be excreted to the municipal sewer system. The common waste water treatment facilities are not sufficient for destruction of many of these compounds, but instead are eventually transported to the rivers and seas. The occurrence of these in wastewater effluents, in aquatic environment and even in-tap water has become a rising concern of society. The most significant of these are namely ibuprofen, diclofenac, and carbamazepine [1]. Ibuprofen is the one found in the highest concentrations of the waste waters. Indeed, the concern is tangible - toxicity effects of the aforementioned drugs on marine species such as fish, crustaceans, algae and bacteria in concentrations less than 1 mg/l have recently been reported. To improve the degradation of these kinds of substances, several treatment methods can be applied – i.e. addition of H_2O_2. However H_2O_2 is not a very efficient oxidant for the degradation of these pharmaceuticals, on the contrary, it is reported that most of them are unreacted during the purification process. Ozone (O_3), on the other hand, has recently been introduced to the water treatment as a disinfectant. Ozone is quite powerful to destroy some of the compounds; however, some of them are not degraded. Therefore a suitable catalyst should be integrated which could effectively decompose the compounds into harmless low molecular products.

This work is focused on the development of a catalytic system where different catalysts in the presence of ozone effectively destruct the target drug compound. Several compounds were tested with a range of solid catalysts. Operational parameters investigated were the reactor temperature, the pH (by NaOH), the stirring speed as well as the ozone concentration. The immobilization of the catalyst was also investigated, since it is important for the industrial process.

Experimental section. The degradation experiments were performed with the reactor system illustrated in Figure 1. The ozone was generated by a flow of a gas mixture of O_2/N_2 to produce an ozone concentration of up to 8 wt.-%, and streamed to the reactor vessel bottom, where it dispersed through a sinter to produce small gas bubbles. The volume of the double–jacket glass reactor was 1L, and the temperature was controlled by recirculating glycol-water mixture. A stirring velocity of 200-1200 rpm was achieved with a six-blade teflon impeller. The reactor was equipped with a pH-stat device which allowed controlling the pH by continuous addition of NaOH, as well to monitor the pH and reactor temperature. The outflow of unreacted ozone was neutralized by a KI-quench filter. Samples were taken with a plastic syringe and analyzed (HPLC for ibuprofen or other pharmaceutical, indigo IR-spectrometry for dissolved O_3 concentration). The pharmaceutical component was, if needed, first dissolved in a small amount of ethanol before injected to the reactor which was filled with 1L of deionized water. Typically, the waste water treatment plant conditions acted as guidance, and therefore the starting concentration of the pharmaceutical component was 10 mg/L.

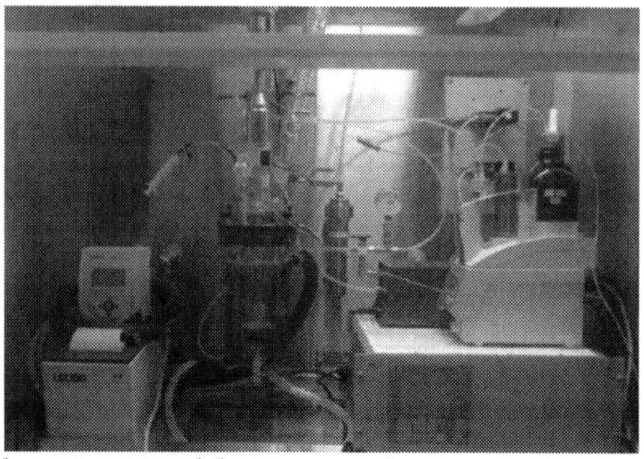

Figure 1. Schematic view of the new recycle microwave reactor system

Results. The initial experiments were devoted to the solubility of ozone in the aqueous solution at various temperatures and pH values as well as gas flow rates. The highest solubility was found at the lowest temperature (5°C) and the highest possible gas flow rate (90 v-% O_2 / 10 v-% N_2). The solubility dependency is illustrated in Figure 3.

The kinetic degradation of a typical pharmaceutical, ibuprofen, shows an interesting temperature dependency – at first, the low temperature favours the degradation rate due to the higher ozone solubility, but when inspecting the kinetics after a prolonged reaction time, the reaction becomes more complete at a higher temperature (Figure 3). The decomposition of ibuprofen was found out to be moderately fast in the absence of any catalyst, however for other pharmaceuticals, this was not the case. Therefore, an extensive catalytic screening was performed to discover which catalyst is the optimal for each pharmaceutical compound.

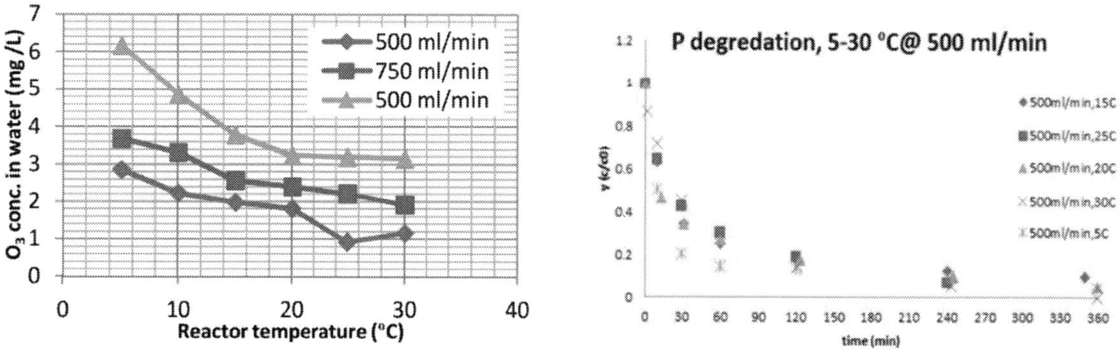

Figure 3. The ozone solubility dependency on the reactor temperature and gas flow rate (left), and the non-catalytic degradation of ibuprofen vs. the temperature with 500 ml/min (right).

Conclusions. The decomposition of pharmaceuticals by using ozone was optimized. Several catalysts were screened and studied for the reaction. A kinetic model was developed to describe the decomposition kinetics of the pharmaceutical compound.

References

[1] Antonina Kruglova, Science of The Total Environment, 09/2014, 499C(1) 394-401.

Chemical kinetics modelling of CO_2 hydrogenation to methanol over Cu/M/Al (M = Ba, Ca, Mg, Sr, Zn) heterogeneous catalysts

[1]N. Strah, [1]V. D. B. C. Dasireddy, [1]A. Pohar, [1]B. Likozar
[1]Laboratory of Catalysis and Chemical Reaction Engineering, National Institute of Chemistry, Hajdrihova 19, SI-1001 Ljubljana, Slovenia; tel. +386 1 4760281, e-mail: blaz.likozar@ki.si, neja.strah@ki.si

Carbon dioxide is a source of potential carbon-containing raw material and also one of the major greenhouse gases [1]. In the past decades, the potential use of CO_2 as an alternative feedstock replacing CO in methanol production has received attention as an effective way of CO_2 utilization [1,2].

A highly efficient catalyst is the key for methanol synthesis *via* CO_2 hydrogenation. Experiments with the aid of the isotope-labelled CO and spectroscopy have demonstrated that methanol was produced by the hydrogenation of CO_2 rather than CO, while the commonly-used ternary Cu–Zn–Al oxide catalyst for CO hydrogenation was not particularly active for the conversion of pure CO_2 at 5.0–10.0 MPa and 200–250 °C [3]. Mostly used catalyst type for CO_2 hydrogenation was the modified methanol synthesis analogue for CO hydrogenation and the existing studies addressed its chemical composition, supports, additives, different preparation methods/conditions and morphology. Notably, zinc oxide can improve the dispersion and stabilization of copper. ZnO possesses lattice oxygen vacancies, consisting of an electron pair in the lattice, which is active for methanol synthesis. The basicity of ZnO influences the catalyst activity through affecting the dispersion of copper *via* the precipitate phase of copper [4]. Studies on the precipitate of Cu-containing hydroxocarbonates revealed that some of its phases played important roles in determining catalytic performance. The mixed metal oxides, obtained by the controlled thermal decomposition of hydrotalcite-like compounds (HTlcs), with the general formula of $[M^{2+}{}_{1-x}M^{3+}{}_{x}(OH)_2]^{x+}(A^{y-})_{x/y} \times zH_2O$, also exhibited promising results. M^{2+} and M^{3+}, divalent and trivalent cation, respectively, possess homogeneous microstructure, a good dispersion of M^{2+} and M^{3+} at atomic level, an enhanced metal–oxide interaction after reduction, the stability against sintering and a high specific surface area, as well as strong basic properties [5]. Therefore, HTlcs are among the most investigated catalyst precursors considering the remarkable properties of the final catalysts. In the present study, the influence of the basicity of alkaline earth metal (*i.e.* Ba, Ca, Mg and Sr) catalysts, which were prepared *via* the hydrotalcite route on the selectivity of the methanol production in CO_2 hydrogenation, is reported. The catalysts with the molar ratio of Cu:M:Al = 6:3:1 (M = Ba, Ca, Mg, Sr, Zn) were prepared and characterised by various characterisation techniques such as scanning (SEM) and transmission (TEM) electron microscopy, X-ray diffraction (XRD), X-ray photoelectron spectroscopy (XPS), CO_2 and NH_3 temperature-programmed desorption (TPD), and N_2O chemisorption,.

Table 1: Properties of Cu/M/Al catalysts

Alkaline metal (M)	Surface area ($m^2\ g^{-1}$)	Cu dispersion (%)	Crystallite size (nm)	Total basicity (μmol $CO_2\ g^{-1}$)
Mg	85	58.2	14	2.8
Ca	69	41.1	11	3.1
Sr	63	38.5	16	3.4
Ba	48	30.7	18	4.1

Powder XRD patterns suggested that the isolated Cu^{2+} species interacted strongly with alumina, to an extent that it may have even formed the surface copper aluminate phase. However, the presence of a highly-dispersed surface CuO was present as well, distributed as smaller crystallites. The SEM and TEM images of the catalysts also supported the findings, obtained by XRD; specifically, no extensive agglomeration of CuO was observed in the respective images. Weak, strong and some moderate basic sites were attributed to the three peaks present in the CO_2 TPD profile of the prepared catalysts, while basicity was in the following order: Ba > Ca > Sr > Mg. The basic character of the catalysts was strongly dominated by the presence of the alkali metal when compared to copper and alumina.

Figure 1: Parallel reactor system.

The catalysts were consequently tested in parallel reactor system, which comprised five fixed-bed vessels (Figure 1). The catalytic performance of the alkaline earth metal-containing catalysts was compared with the commercial methanol synthesis catalysts (by Alfa Aesar and Lurgi). Methanol selectivity and CO_2 conversion were examined under the continuous flow in the temperature (T) range between 200 °C and 400 °C. The catalysts were exposed to different space velocities (GHSV) (from 2000 to 6000 h^{-1}). Prior to catalytic measurements, fresh catalysts were reduced in the stream of H_2 at 300 °C for 3 h under the atmospheric pressure (P; 1 bar). Reactors were then cooled to the ambient temperature (25 °C) introducing reactant gas (N_2:CO_2:H_2 = 1:2:6 molar ratio) flow and raising the pressure to 2.0 MPa.

The results showed that Cu/Mg/Al catalyst was at least as good as the commercial catalysts (Figure 2), while the main products detected were methanol, water and carbon monoxide. Selectivity for methanol was between 10 and 40%, depending on reaction conditions and the catalyst used. The experimental data, paralleled by model predictions showed that CO_2 conversion increased with elevated temperature, while on the contrary, methanol selectivity dropped with increasing temperature.

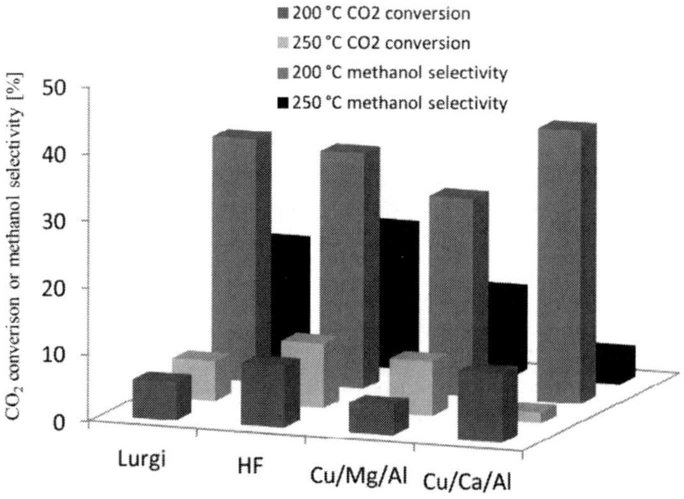

Figure 2: Methanol selectivity and CO_2 conversions over different catalysts at GHSV = 2000 h^{-1}, T = 200 and 250 °C, P = 20 bar.

Acknowledgement

The presented work was partially funded by the EU Framework Programme for Research and Innovation Horizon 2020 under the grant agreement No 637016 (MefCO2). The authors also gratefully acknowledge the financial support of the Slovenian Research Agency (ARRS) through the Programme P2–0152.

References

[1] K.-D. Jung, O.-S. Joo, Catal. Lett., 84 (2002) 21-25.
[2] Slamet, H.W. Nasution, E. Purnama, S. Kosela, J. Gunlazuardi, Catal. Commun., 6 (2005) 313-319.
[3] J. Gao, Q. Liu, F. Gu, B. Liu, Z. Zhong, F. Su, RSC Adv., 5 (2015) 22759-22776.
[4] S.G. Jadhav, P.D. Vaidya, B.M. Bhanage, J.B. Joshi, Chem. Eng. Res. Des., 92 (2014) 2557-2567.
[5] Á. Mastalir, Á. Patzkó, B. Frank, R.Schomäcker, T.Ressler, R.Schlögl, Catal.Commun., 8 (2007) 1684-1690.

Kinetic analysis of reaction via microwave dielectric measurements

L. Estel, N. Benamara, A. Ledoux, I. Polaert
LSPC-EA 4704-Normandie Université, Université et INSA de Rouen Campus Universitaire du Madrillet, 76800 Saint Etienne du Rouvray, France ;
Tel. +332 956653, e-mail: lionel.estel@insa-rouen.fr

As proposed by Astarita et al. in 1983, amines are the best absorbers for low CO_2 partial pressure in feed. The amine solutions for capturing CO_2 are currently taken as the most industrially promising option for controlling greenhouse gas emissions from power plants. In a recent paper Idem et al., claim less than 2GJ/tonne CO_2 of heat consumption for solvent regeneration. Monoethanolamine (MEA) is a typical solvent applied in continuous CO_2 capture processes, where the solvent is normally desorbed by steam at elevated temperature (typically above 373 K) in a stripper. Energy consumption was demonstrated to be the key factor reducing the efficiency of power plants by many previous researchers.

The well characterized MEA–CO_2 absorption–stripping loop has generally been utilized as a reference for evaluation of system performance with process innovations and integrations. With the goal of reducing cost and energy penalties associated with CO_2 capture, attention has been directed toward improving process efficiency. There have been theoretical studies of CO_2, absorption rates into amines or mixed amines; however, only a few authors have presented experimental data. In most of these works, the driving force (CO_2, partial pressure) was not specified for the data, prohibiting direct comparison. It is often noted that a rigorous mass transfer model is best for the interpretation of data since simplistic approximations are likely to fail. The solutions of aqueous amines loaded with CO_2, are characterized by high ionic strength, causing the activity coefficients to vary greatly over the range of conditions encountered.

When designing or simulating an absorption column for CO_2 absorption associated with chemical reactions, certain key data are required. These key data include, physical, thermal, and transport properties of the gases and the liquids involved in the system, vapour–liquid equilibrium data, configuration data for the absorber internals, and rate data for the chemical reactions. Furthermore, even though many physical properties such as density, viscosity, solubility, and diffusivity of CO_2–MEA system have recently been published in the literature (Weiland et al., 1998; Tsai et al., 2000; Ko et al., 2001), these do not include the properties for the kinetics of high CO_2-loaded, concentrated aqueous MEA systems. It is well known that precise information on the reaction rate constants as well as a detailed knowledge of the reaction mechanism for the reaction between CO_2 and MEA solutions are essential for accurate design and simulation of the CO_2 absorption process. Absorption has still some grey areas, but many researchers are devoting their efforts to provide relevant response at this process step. Yet the optimization of the overall process is mostly related to the phase of desorption and solvent regeneration. During the CO_2 capture, desorption / stripping step is known as the most energy consuming. Thus, a better knowledge of the intrinsic kinetic, hydrodynamic and thermodynamic parameters should be carried out. Especially when one finds less than 15 articles from 1995 to 2015, which can be considered as really studying this phenomenon.

In this paper, we present a new method for the understanding of the CO_2 desorption step. Among the different techniques allowing dielectric measurements, the method known as "small perturbation of a resonant cavity" is probably the most accurate and it has been used in previous works (Estel et al. 2003; Polaert et al. 2015). In this method, the perturbations caused by a sample to frequency and amplitude of cavity electromagnetic resonance are compared to standards to determine its complex permittivity. Because of the dependence of dielectric constants on the temperature, sample temperature is precisely fixed.

A small quantity of dielectric sample is placed in the center of a resonant cavity which is a TE_{013} single mode rectangular cavity. The real permittivity ε' and the dissipation factor ε" are deduced from the cavity resonance frequency and the resonance transmission by means of the measurements of the perturbation due to the sample at frequency of 2450 ±14 MHz. First of all an accurate calibration is performed with a precise volume of a standard material.

This method is available for dielectric measurements at constant and varying temperature. The cavity is equipped with a temperature control via a circulating thermal fluid in the double jacketed wall of the guide, this in order to avoid any expansion disturbance. Sample heating is operated either with a hot air stream or with a 6W regulated microwave power. When assumption of linear, homogeneous and isotropic material are available the relative complex permittivity of a mixture (ε_M' and ε_M'') can be calculated by a mixing law.

The initial reactant 30% (w/w) aqueous MEA CO_2 loaded (2.5 mol/L) and respectively the final product pure 30% (w/w) aqueous MEA have been studied separately. Figures 1 and 2, the thermal behaviour of their real permittivity and respectively their dissipation factor are presented.

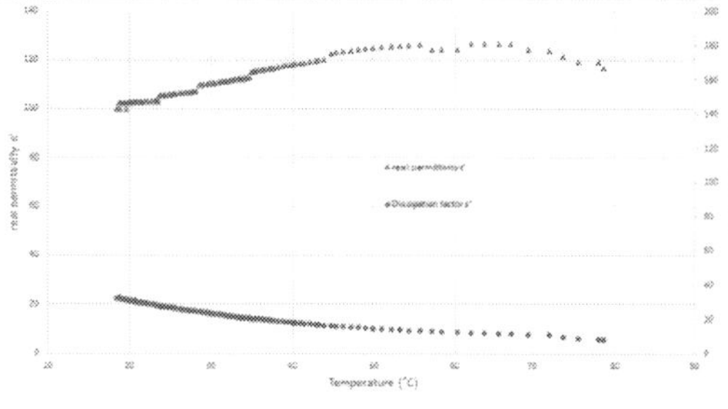

Figure 1: permittivity vs temperature
30% (w/w) aqueous MEA

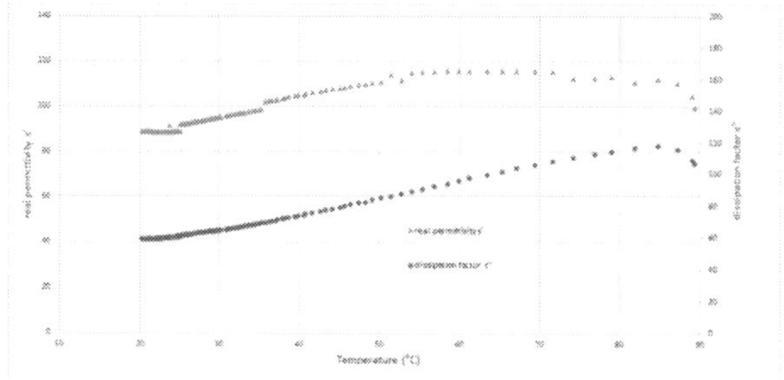

Figure 2: permittivity vs temperature
30% (w/w) aqueous MEA CO_2 loaded (2.5 mol/L)

Then the behaviour of the dielectric properties of the reaction mixture in isotherm conditions are recorded. They have been linked with the conversion rate, in order to estimate the kinetic constants. This non-intrusive method is sensitive to the chemical species concentrations and especially to the resulting ionic strength, by this way it can give information even for concentrated solutions.

References:

Astarita G., Savage D.W., Bisio A. Gas treating with chemical solvents, Wiley Interscience Publication ; 1983

Estel, L., Bonnet, C., Delmotte, M. & Cosmao, J. M. Kinetic Analysis via Microwave Dielectric Measurements. *Chem. Eng. Res. Des.* **81,** 1212–1216 (2003).

Ko J.J., Tsai T.C., Lin C.Y., Wang H.M., Li M.H. Diffusivity of nitrous oxide in aqueous alkanolamine solutions, *J. Chem. Eng. Data*, **46** (1), 160-165 (2001).

Polaert, I., Bastien, S., Legras, B., Estel, L. & Braidy, N. Dielectric and magnetic properties of NiFe2O4 at 2.45GHz and heating capacity for potential uses under microwaves. *J. Magn. Magn. Mater.* **374,** 731–739 (2015).

Tsai T.C., Ko J.J., Wang H.M., Lin C.Y., Li M.H. Solubility of nitrous oxide in alkanolamine aqueous solutions, *J. Chem. Eng. Data*, **45** (2), 341-347 (2000).

Weiland R.H., Dingman J.C., Cronin D.B., Browning G.J. Density and viscosity of some partially carbonated aqueous alkanolamine solutions and their blends, *J. Chem. Eng. Data*, **43** (3), 378-382 (1998).

Acknowledgment:

The authors gratefully acknowledge the financial support provided by "La Région Haute Normandie" through "Grand Réseaux de Recherche Sciences de l'environnement, analyse et gestion des risques - Maitrise des Risques Technologiques".

Utilization of catalytically active foams for enhancing energy efficiency in chemical reactors

F. Lali
Institute of chemical process fundamentals of the Czech academy of sciences, Rozvojova 2/135, 165 02 Prague, Czech Republic, phone: +420 220 390 233, fax:220920661,220920649, E-Mail: lali@icpf.cas.cz

About 90% of the chemical processes include catalysis, whereas the most of these process are heterogeneous. Therefore, improving the energy efficiency in chemical reactors is a very crucial issue in order to fulfill the 20-20-20 objectives of EU climate protection and energy efficiency.

Foam substrates as catalyst support have been in the focus of research in recent years especially, for multiphase reactions for instance Tourvieille et al. [1], who investigated packed millichannels with catalytically active foams reported a very high overall mass transfer when compared with conventional reactor types (Fig. 2). In the present study (Lali [2]), a tubular reactor that was packed with a Pd/Al_2O_3 foam packing with a length of 50 cm and a diameter of 1.8 cm. The hydrogenation of α-methylstyrene (AMS) to cumene was employed for characterization of the reactor performance due to availability of literature data for other reactor types that facilitated a comparison.

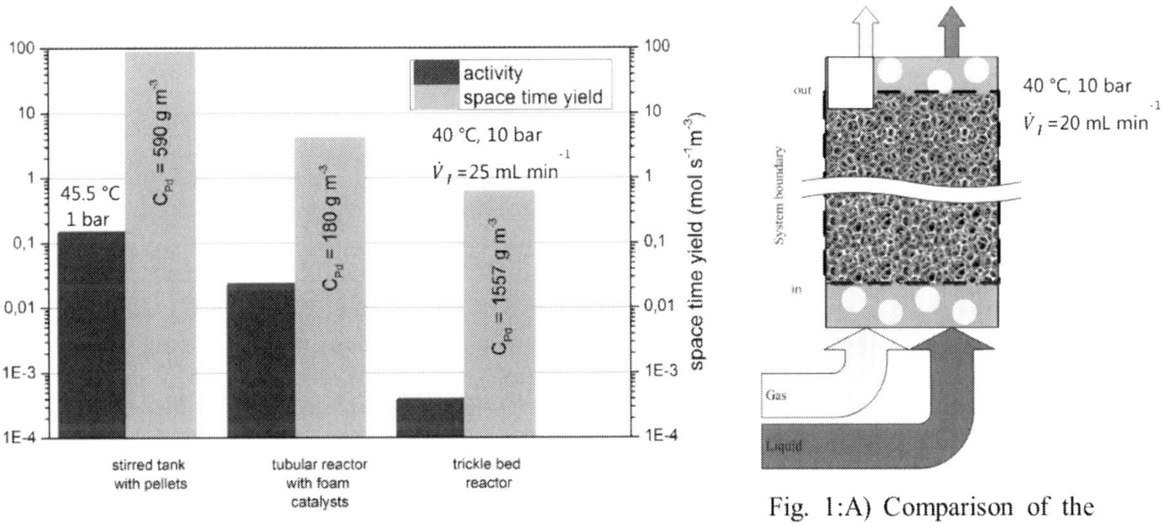

Fig. 1:A) Comparison of the reactor performance of foam catalysts with a trickle bed reactor and the intrinsic reaction rate at 40 °C B) reactor scheme in this study

Compared to a trickle bed reactor with an identical liquid flow rate and initial temperature of 40 °C and a total pressure of 10 bar, the foam catalysts show about 1000% enhancement of activity and space time yield (Fig. 1), whereas about 90% less Palladium 30% and less Hydrogen were needed for foam catalysts. The values for a stirred tank are reported by Meille et al. [3] for the intrinsic kinetic of this reaction that represent on Fig. 1 the maximum possible activity.

The experiments were carried out in a co-current upward flow. This operation mode was favoured because the downward flow was found to be very sensible to manufacturing tolerances (caused bypass flow) of the foam packing and a weak spreading of liquid in the downward flow was observed.

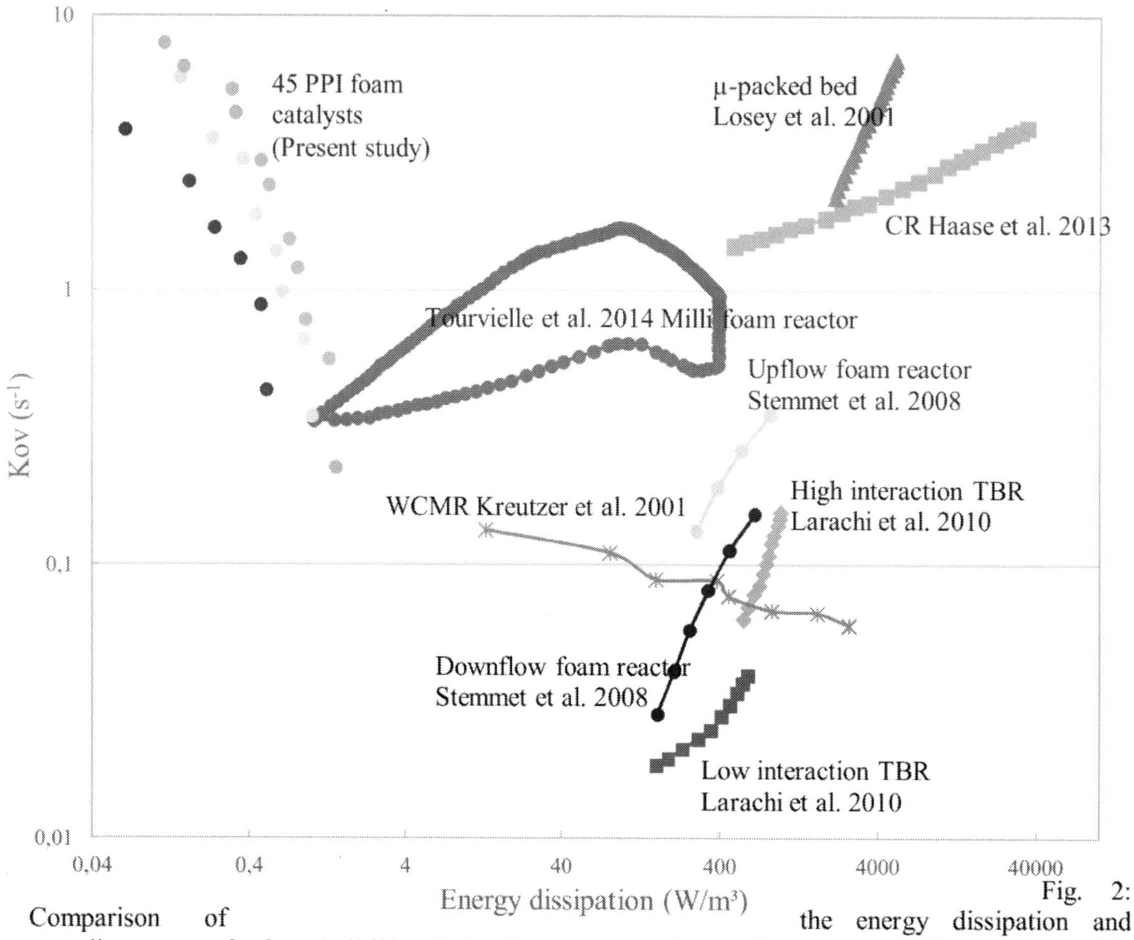

Fig. 2: Comparison of the energy dissipation and overall mass transfer from Lali [2] with the literature data; Tourvieille et al. 2015 [1], Losey et al. [4], Haase et al. [5], Kreutzer et al. [6], Stemmet et al. [7], Larachi et al. [8]

Furthermore, the overall mass transfer was measured in an externally limited condition at a temperature of 120°C and a pressure of 1 bar. The results of the measurements (Fig. 2) show a very high overall mass transfer at a very low energy dissipation. For this reason, the catalytically active foams can considerably contribute to the enhancement of energy efficiency of chemical reactors.

1. Tourvieille, J.N., Régis, P., de Bellefon, C., Chemical engineering journal, 2015. **267**: p. 332-346.
2. Lali, F., *Catalytic active foams for gas liquid solid reaction*. 2015, PhD thesis, Technische Universität Dresden: Dresden.
3. Meille, V., C. de Bellefon, and D. Schweich,. Industrial & Engineering Chemistry Research, 2002. **41**(7): p. 1711-1715.
4. Losey, M.W., M.A. Schmidt, and K.F. Jensen,. Industrial & Engineering Chemistry Research, 2001. **40**(12): p. 2555-2562.
5. Haase, S., et al.,. Chemical Engineering Science, 2013. **94**: p. 224-236.
6. Kreutzer, M.T., et al., Chemical Engineering Science, 2001. **56**(21-22): p. 6015-6023.
7. Stemmet, C.P., *Gas-liquid Solid Foam Reactors: Hydrodynamics and Mass Transfer*. 2008, Technische Universiteit Eindhoven.
8. Larachi, F., Grandjean, B. P. A. *Excel spreadsheet simulator for packed bed reactors*. 2010; Available from: http://www.gch.ulaval.ca/bgrandjean/pbrsimul/pbrsimul.html.

Verification of a macro kinetic model and development of a product distribution model for a commercial Co/Re/γ-Al_2O_3 Fischer-Tropsch catalyst in microchannel reactor

M. Ostadi, M. Hillestad[1]
Norwegian University of Science and Technology (NTNU), Department of Chemical Engineering, NO-7491, Trondheim, Norway; [1] corresponding author: e-mail magne.hillestad@ntnu.no, tel. +4773594122

In order to do a good and realistic design of a Gas-to-Liquid (GTL) plant, it is necessary to have kinetic models that correctly capture the major variations of the reaction rates and selectivities subject to changes in process parameters. There are numerous kinetic rate expressions in the literature [1], but there is little consensus to which of them is the best in terms of capturing major variations in Fischer-Tropsch kinetics. A kinetic investigation performed by Yang et al. [2] on a 40%Co/1%Re/γ-Al_2O_3 catalyst, the effect of operation temperature, pressure and gas-hourly space velocity on CO conversion, C_{5+} selectivity, methane selectivity and stability during Fischer-Tropsch synthesis in a microchannel reactor is studied. The reactor was tested at 210-240 °C, 20-40 bar and feed $H_2/CO = 2.17$. They showed that microchannel reactor could be operated at 85-90 % conversion with high C_{5+} selectivity. A kinetic model suggested by Ma et al. [1] is here implemented in a microchannel reactor model and predictions of the CO conversion are compared with the observed conversions by Yang et al. [2]. This is an empirical model structure and it is found to be able to predict the observed CO conversions in [2] quite well. However, the experimental reaction rates are higher than what is predicted by the model of Ma et al. [1]. This is mainly due to higher Co loading of the catalyst in [2] compared to the catalyst applied by Ma et al. [1]. By multiplying the rate model by a fitted parameter, the comparisons between measured and predicted CO conversions are shown in Figure 1. The model describes a positive water effect on the reaction rate.

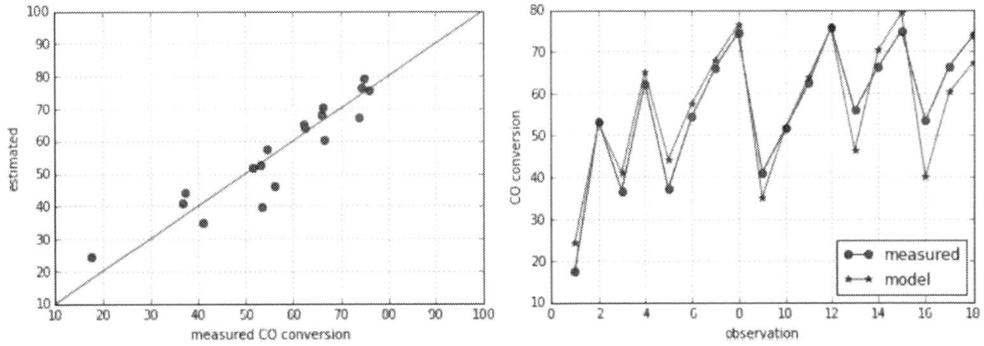

Figure 1. Comparison of measured and predicted CO conversion in a microchannel reactor.

The FT products distribution of the higher hydrocarbons follows the Anderson-Schulz-Flory (ASF) distribution, i.e. the product distribution depends only on one parameter; the chain growth probability also called the growth factor (α). If the FT reactions are lumped into one reaction and C_{5+} is a lumped component, the reaction may be written as:

$$CO + U \cdot H_2 \xrightarrow{r_{FT}} \upsilon_1 C_1 + \upsilon_2 C_2 + \upsilon_3 C_3 + \upsilon_4 C_4 + \upsilon_{5+} C_{5+} + H_2O \quad (1)$$

How to calculate the stoichiometric coefficients U, υ_i and average carbon number of the lump as function of the growth factor is shown by Hillestad [3]. In order to predict the product distribution, we have proposed a model of the growth factor and fitted parameters to experimental data. The model structure is based on the ratio between the rate of propagation and the sum of the rates of propagation and termination, though empirical exponents are introduced.

$$\alpha = \frac{1}{1 + k_2(T)\dfrac{p_{H_2}^{c}}{p_{H_2O}^{d} p_{CO}}} \quad (2)$$

Water seems to have a positive effect on the growth factor. It is well known that the selectivity of methane is higher than predicted by the ASF distribution. To account for this fact, a separate methanation reaction rate is introduced. The methanation reaction rate, r_{CH4}, is found to be proportional to the rate of methane production by ASF model; $r_{CH_4} = c \cdot r_{FT} \upsilon_1$. The model predicts C_{5+} selectivity and methane selectivity very well, as shown in Figure 2.

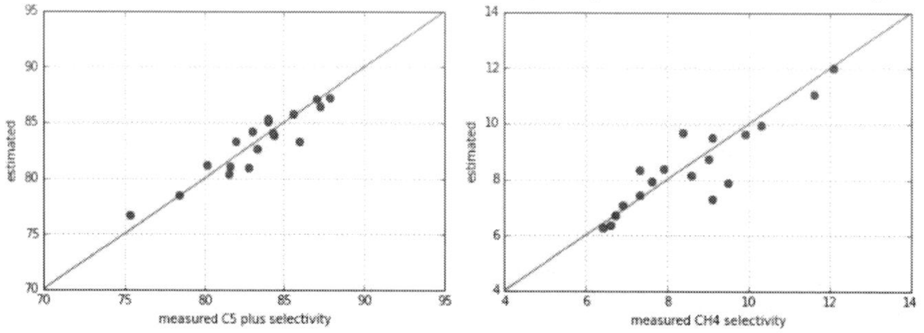

Figure 2. Comparison of measured and predicted C_{5+} and methane selectivities in a microchannel reactor.

[1] Ma W, Jacobs G, Sparks DE, Spicer RL, Davis BH, Klettlinger JLS, Yen CH, Fischer–Tropsch synthesis: Kinetics and water effect study over 25%Co/Al2O3 catalysts, *Catalysis Today*, Volume **228**, (2014), Pages 158-166.

[2] Yang J, Boullosa SE, Myrstad R, Venvik HJ, Pfeifer P, Holmen A, Fischer-Tropsch Synthesis on Co-based Catalysts in a Microchannel Reactor, effect of temperature and pressure on selectivity and stability, *Catalysts, and Catalysis: Advances and Applications*, B. H. Davis, M. L. Occelli, Eds., CRC Press, 2016, p. 223.

[3] Hillestad M, Modeling the Fischer-Tropsch product distribution and model implementation, *Chemical Product and Process Modeling*, **10** (2015), 3, 147—159

Novel mesoporous carbon from renewable sources and microporous zeolite catalysts for production of biofuel components

[1,3]L. J. Konwar, [1]P. Mäki-Arvela, [3]D. Dekka, [1]T. Riittonen, [1]N. Kumar, [1,2]J.-P. Mikkola
[1]Laboratory of Industrial Chemistry and Reaction Engineering, Johan Gadolin Process Chemistry Centre, Åbo Akademi University, 20500 Turku/Åbo, Finland, [2]Department of Chemistry, Chemical-Biological Centre, Umeå University, SE-901 87 Umeå, Sweden, [3]Department of Energy, Tezpur University, 784028 Tezpur, India

Introduction

Synthesis of biofuel components from renewable sources such as various ligno-cellulosic biomass, non-edible seeds *jatropha curcas, ricinus communis, pongamia pinnata* and different types of micro-algae are significant, taking into consideration the limited resources of fossil fuel, climate change mitigation and production of environmental friendly green fuels for stationary and transportation motor vehicles. Metal modified and solid acid catalytic materials used in the processing of biomass and production of biofuel components are of immense importance. Usually large amounts of catalytic materials are applied during the processing of biomass on industrial scale, hence it is significant that these materials are produced from renewable sources. In the present research, authors have studied the synthesis and characterization of novel mesoporous carbon catalytic materials produced from oil processing of non-edible oil seeds. The synthesized novel mesoporous carbon catalytic materials were investigated in the esterification and transesterification of acid oils to biodiesel. Further more, selective synthesis of triacetin was studied over H-Y, H-ZSM-5 and H-Beta zeolites and sulfonated novel mesoporous carbon catalysts. Conversion of bioethanol to diethyl ether was investigated over Ni-H-Y, Ni-H-Beta and H-ZSM-12 catalysts.

Experimental

The non-edible oil seeds of *mesua ferrera and jatropha curcas* were used for extraction of acid oils. The acid oil containing 43.73 % of free fatty acids were obtained by hydrolysis. The brown residual obtained after the extraction of acid oil from the oil seeds were used for synthesis of novel acidic mesoporous carbon materials. The porosity in the residual of the extracted acid oil was created by chemical pretreatment with ortho-phosphoric acid for 24 h. The chemically pretreated residual material was calcined in a muffle oven at 773 K. The porous carbon solid material was washed with distilled water and dilute HCl. The obtained carbon material was dried at 383 K and subjected to sulfonation by covalent interaction of 4-benzenediazoniumsulfate. The as synthesized novel acidic mesoporous carbon materials were characterized using SEM, TEM, IR, Raman spectroscopy, XRD, TGA, nitrogen physisorption and TPD- ammonia. The evaluation of the catalytic properties of the novel acidic mesoporous carbon material was carried out in esterification reaction of acid oil using 100 ml three neck flask, mounted on oil bath and equipped with a magnetic stirrer. The H-Y, H-ZSM-5, H-Beta zeolite catalysts were studied in batch mode using 50 ml, three neck flask for the selective synthesis of triacetin from the reaction of acetylation of glycerol with acetic anhydride. The Ni modified H-Y, H-Beta and H-ZSM-12 catalysts were investigated for synthesis of diethylether from bioethanol using 2 ml isobaric mini-reactor.

Results and discussion

The surface area of the pure mesoporous carbon material (777 m^2/g) was higher than that of acidic sulfonated mesoporous carbon (696 m^2/g), which can be attributed to the partial blocking of the pores with $-PhSO_3H$ groups. The total acid site densities determined by titration method for the sulfonated mesoporous carbon material (2.215 mmol/g) were larger than that of pure mesoporous carbon (2.032 mmol/g). The SEM exhibited the presence of irregular particle shape of pure and sulfonated mesoporous carbon. The TEM for both the materials showed the presence of aromatic/graphitic carbon sheets and high porosity for the sulfonated. The XRD patterns exhibited the amorphous nature of the sulfonated and pure mesoporous carbon materials. The thermal stability of the mesoporous carbon materials were determined by TGA. It was observed that the mesoporous carbon was stable up to 513 K under oxygen free conditions. The esterification of free fatty acid carried out over sulfonated acidic mesoporous carbon and unsulfonated mesoporous carbon. It was observed that sulfonated acidic mesoporous carbon showed higher conversion of free fatty acid in acid oil (97 %) than the unsulfonated mesoporous carbon (11 %). It is noteworthy to mention that the sulfonated mesoporous carbon catalyst exhibited higher conversion than the equivalent amount of sulphuric acid. The higher activity of the sulfonated mesoporous carbon is attributed to the presence of hydrophilic surface groups (-COOH, -OH, $-SO_3H$) present in the sulfonated carbon mesoporous material. It was also observed that the reaction of acetylation of glycerol with acetic anhydride for synthesis of triacetin over H-Y, H-ZSM-5 and H-Beta catalysts was greatly influenced by the acidity, structure and size of the pores. Further more shape selectivity property of the zeolites influenced the selectivity and product distributions. The conversion of bioethanol (72 %) and selectivity to the formation of diethylether (67 %) over the H-ZSM-12 zeolite catalyst was the highest of all the studied catalysts. Plausible explanation for enhanced catalytic property of H-ZSM-12 zeolite is attributed to the presence of large amount of Brønsted acid sites. The Ni modified H-Y and H-Beta catalysts exhibited lower conversion of bioethanol and selectivity to diethyether than the H-ZSM-12 catalyst.

References:

1. L. J. Konwar, P. Mäki-Arvela, E. Salminen, N. Kumar, A. J. Thakur, D. Deka, Applied Catalysis B: Environmental 176 (2015) 20-35.
2. L. J. Konwar, P. Mäki-Arvela, P. Begum. N. Kumar, A. J. Thakur, J.-P. Mikkola, R. C. Dekka, D. Deka, Journal of Catalysis 329 (2015) 237-247.
3. T. Riitonen, K. Eränen, P. Mäki-Arvela, A. Shchukarev, A.-R. Rautio, K. Kordas, N. Kumar, T. Salmi, J.-P. Mikkola, Renewable Energy 74 (2015) 369-378.

Modelling and experimental validation of an annular LED-based photocatalytic reactor

F. Khodadadian[1], M. de Boer[1], J.R. van Ommen[2], A.I. Stankiewicz[1] and R. Lakerveld[3]
[1] Department of Process & Energy, Delft University of Technology, 2628 CB, Delft, the Netherlands;
[2]Chemical Engineering Department, Delft University of Technology, 2628BL, Delft, the Netherlands;
[3]Department of Chemical and Biomolecular Engineering, Hong Kong University of Science & Technology, Hong Kong
tel. +3115 2783360, e-mail: f. khodadadian@tudelft.nl;

Heterogeneous photocatalysis is attracting much attention due to its large number of applications including air and water purification [1, 2], water splitting [3], and conversion of carbon dioxide to valuable hydrocarbons [4]. Nevertheless, its commercialization has been hindered by the low overall efficiency of the process [5]. One of the main challenges is to overcome the poor photon utilization within a photocatalytic reactor, which leads to high operational cost.

The rate of photocatalytic reactions depends typically on both the concentrations of the chemical species and the rate of photon absorption by a catalyst [6]. Thus, investigation of the radiation field within the reactor is crucial for improved reactor design, optimization, and scale-up. Conventional UV-lamps such as low pressure mercury lamps restrict the reactor design due to their large rigid design. Moreover, fragility, toxicity, gas leakage and disposal issues are other disadvantages of mercury lamps [7]. Alternatively, Light Emitting Diodes (LEDs) are promising light sources for photocatalysis applications [8-10]. LEDs are robust, energy-efficient, non-toxic and long lasting light sources. In addition, LEDs can be placed flexibly within each reactor configuration due to their small size and ability to deliver a range of intensities, which provides a large design space. To systematically exploit this enlarged design space, analysis and optimization of models for novel LED-based photocatalytic reactors is needed in combination with the validation of such models.

The objective of this study is to validate a model that describes the operational performance of a LED-based photocatalytic reactor. A mini-pilot plant involving an annular photocatalytic reactor using LEDs has been designed and constructed (see Figure 1). The LEDs are mounted on an inner concentric tube of the reactor. The position of the LEDs can be adjusted to modify the radiation field inside the reactor. Toluene degradation was chosen as the model reaction. The radiation field from the LED assembly was modelled and coupled with a reactor model involving mass, momentum and radiation transport balances. The model allows for kinetic parameters to be estimated from experimental data. Furthermore, the experimental setup allows for different flow rates, toluene inlet concentrations, methods for catalyst coating (hydrothermal method and atomic layer deposition) and light intensities to be applied to validate the predictive capabilities of the developed model for a broad range of operating conditions. On the long term, the validation of the proposed model advances the design, optimization and scale-up of future LED-based photocatalytic reactors for various applications.

Figure 1: a) LED-based photocatalytic reactor setup, b) LED-based photocatalytic reactor including three concentric tubes: the inner tube for mounting LEDs, the middle tube of quartz for separating reaction medium from the light sources and the stainless steel reactor shell with catalyst coating.

References

1. Herrmann, J.M., *Heterogeneous photocatalysis: state of the art and present applications In honor of Pr. R.L. Burwell Jr. (1912–2003), Former Head of Ipatieff Laboratories, Northwestern University, Evanston (Ill).* Topics in Catalysis, 2005. **34**(1-4): p. 49-65.
2. Hoffmann, M.R., et al., *Environmental Applications of Semiconductor Photocatalysis.* Chemical Reviews, 1995. **95**(1): p. 69-96.
3. Fujishima, A. and K. Honda, *Electrochemical Photolysis of Water at a Semiconductor Electrode.* Nature, 1972. **238**(5358): p. 37-38.
4. Inoue, T., et al., *Photoelectrocatalytic Reduction of Carbon-Dioxide in Aqueous Suspensions of Semiconductor Powders.* Nature, 1979. **277**(5698): p. 637-638.
5. Van Gerven, T., et al., *A review of intensification of photocatalytic processes.* Chemical Engineering and Processing: Process Intensification, 2007. **46**(9): p. 781-789.
6. Motegh, M., et al., *Photocatalytic-reactor efficiencies and simplified expressions to assess their relevance in kinetic experiments.* Chemical Engineering Journal, 2012. **207**: p. 607-615.
7. Jo, W.-K. and R.J. Tayade, *New Generation Energy-Efficient Light Source for Photocatalysis: LEDs for Environmental Applications.* Industrial & Engineering Chemistry Research, 2014. **53**(6): p. 2073-2084.
8. Chen, D.H., X. Ye, and K. Li, *Oxidation of PCE with a UV LED Photocatalytic Reactor.* Chemical Engineering & Technology, 2005. **28**(1): p. 95-97.
9. Lanfang H. Levine, J.T.R., and Janelle L. Coutts, *Feasibility ultraviloet ligth emitting diodes as an alternative ligth source for photocatalysis.* Journal of the Air & Waste Management Association, 2011. **61**: p. 932-940.
10. Kalithasan Natarajan, T.S.N., H.C. Bajaj, Rajesh J. Tayade, *Photocatalytic reactor based on UV-LED/TiO2 coated quartz tube for degradation of dyes.* Chemical Engineering Journal, 2011. **178**: p. 40-49.

A microkinetic modelling of eugenol hydrodeoxygenation over noble metal catalysts in a slurry reactor

A. Bjelić[a], M. Grilc[a], B. Likozar[a]
[a]Laboratory of Catalysis and Chemical Reaction Engineering, National Institute of Chemistry, Hajdrihova 19, 1000 Ljubljana, Slovenia, tel. +386 14760281, e-mail: blaz.likozar@ki.si;

Lignin has shown a great potential to be used as a feedstock for the fuel production and as a precursor for platform chemicals through the catalytic upgrading processes such as hydrodeoxygenation (HDO) [1, 2]. In this work HDO of eugenol (a lignin model compound, containing the same functional groups as coniferyl alcohol) was investigated in a slurry reactor over noble metal catalysts (Ru, Pt, Pd, Rh) as well as their bifunctional combinations on various supports.

All chemicals and catalysts were purchased from commercial suppliers and used as received. Catalysts were structurally characterized by scanning electron microscopy (SEM), transmission electron microscopy (TEM), energy dispersive X-ray spectroscopy (EDX), X-ray powder diffraction (XRD) analysis and temperature-programmed reduction (TPR). The specific surface area of particles was determined according to Brunauer–Emmett–Teller method (BET), while concentration of CO chemisorption was used to estimate the availability of active sites.

HDO of eugenol was performed in 300 ml autoclave (Parker/Autoclave Engineers) in the batch regime. Varying parameters were the temperature (T) 225 – 325 ^{o}C, pressure (p) 3 – 7 MPa, agitation speed (N) 100–1000 min^{-1} and catalyst to eugenol mass ratio (x) 0 – 0.35 wt%. The temperature of reaction mixture was raised from room temperature by the rate of 7.5 ^{o}C min^{-1} to the desired (final) temperature that was kept at constant value for 3 h. Liquid samples were taken in 20 min intervals, while gas phase was sampled every 30 min. Gas samples were analysed immediately after sampling by using CG (Gas Chromatography) and FTIR (Fourier Transform Infrared Spectroscopy), while liquid samples were analysed by GC–FID (Gas Chromatography with Flame Ionisation Detector) and GC–MS (Gas Chromatography with Mass Spectrometer detector).

Among the tests of eugenol HDO over various combinations of metals and supports, Ru/C results are to be presented in this abstract. Ru/C characterisation results are summarised in Table 1. Analysis of the liquid phase indicates that eugenol HDO proceeds through the following steps: eugenol → 2-methoxy-4-propylphenol → 2-methoxy-4-propylcyclohexanol → 4-propylcyclohexanol → propylcyclohexane, yielding 2-methoxy-4-propylcyclohexanol and 4-propylcyclohexanol as the dominant products, as methoxy group removal from the first is relatively slow. Results suggest that hydrogenation (allyl and aromatic ring saturation) reactions are favorised by Ru/C catalyst. Typical concentration profiles of reaction products are presented in Figure 1.

Table 1. Physicochemical properties of catalysts

wt% Ru	S_{BET}, m^{2}g^{-1}	V_{pores}, cm^{3}g^{-1}	Pore size, Å	S_{metal}, m^{2}g^{-1}	N_{AS}, µmolg^{-1}
5	648.851	0.792	48.817	0.330	220

A microkinetic model has been developed to describe experimental results based on molar balance of each component according proposed reaction pathway network. The following steps were taken into account:

$H_2 (g) \rightleftharpoons H_2 (l) \rightleftharpoons H_2 (s) + * \rightleftharpoons H_2* + * \rightleftharpoons 2H*$; (transport $g \rightleftharpoons l$, $l \rightleftharpoons s$, adsorption/desorption, dissociation/association)

$R (l) \rightleftharpoons R (s) + \gamma* \rightleftharpoons R*$; $\gamma = 1 - 8$ (transport $l \rightleftharpoons s$, adsorption/desorption)

$R* + n \cdot H* \rightarrow P* + n*$; $n = 2, 6$ (reaction of adsorbed components)

$P* \rightleftharpoons P (s) + \gamma* \rightleftharpoons P (l)$; $\gamma = 1 - 8$ (adsorption/desorption, transport $s \rightleftharpoons l$)

R represents reactants (eugenol or intermediates), *P* represents product (end product or downstream intermediates), *g* refers to gas phase, (*l*) refers to liquid phase, (*s*) refers to concentration on the catalyst surface, * refers to the adsorbed state of component. According to density function theory (DFT) calculations the number of active sites (γ) covered by each adsorbed component was estimated. (n) refers to stoichiometric factor of chemical reaction.

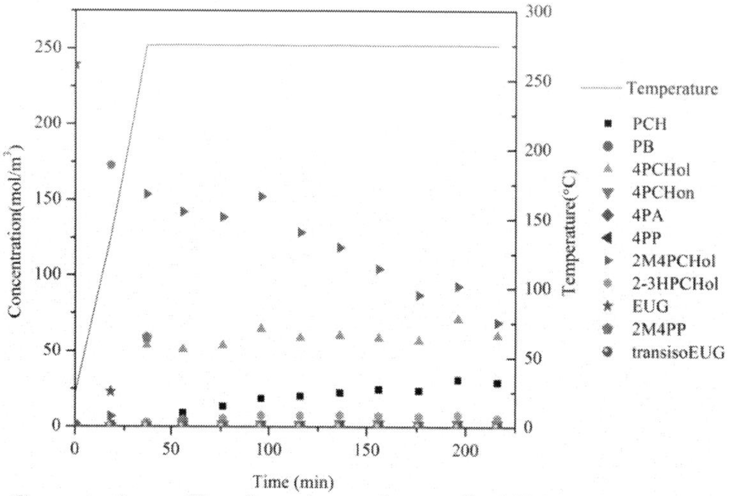

Figure 1. Concentration profiles of reaction products at $T = 275$ °C, $p = 5$ MPa, $x = 0.2$ wt%

A set of differential equations has been based on the concentration changes in all states (*l*, *s*, * and for hydrogen *g* and H*–dissociated state additionally) of all components including concentration of active sites, obtained by CO chemisorption (N_{AS}, Table 1). Equilibrium concentration of hydrogen on the bubbles' surface has been calculated according to Henry law. Modelled concentrations were obtained by solving a system of differential equations by Runge–Kutta 4–5 method, except concentrations on a *g–l* or *l–s* interphase that can be expressed with algebraic equations, as a result of negligible accumulation, according to the film theory. Adsorption constants (k^i_{ads}), desorption constants (k^i_{des}), reaction constants (k^j_r), dissociation and association constants ($k^H{}_{dis}$, $k^H{}_{ass}$) have been estimated by regression analysis, where *i* refers to the component and *j* refers to the corresponding reaction. Henry's constant as well as transport constant through the films (k_l, k_s) have been obtained by using empiric correlations. Temperature influence on the adsorption, desorption and reaction rate constants has been expressed by Arrhenius equation.

Acknowledgment

The research work was supported by Slovenian Research Agency, under the Programme P2-0152, Chemical Reaction Engineering, therefore their financial support is gratefully acknowledged.

Reference

[1] David Martin Alonso, Jesse Q. Bond, James A. Dumesic, Catalytic conversion of biomass to biofuels, Green Chemistry 12 (2010) 1493-1513

[2] Shimin Kang, Xianglan Li, Juan Fan, Jie Chang, Hydrothermal conversion of lignin: A review, Renewable and Sustainable Energy Reviews 27 (2013) 546–558

Coke deactivation pathways during the catalytic cracking of glycerol in FCC conditions

A. Errekatxo, A. Ibarra, A. Gutiérrez, J. Bilbao, J.M. Arandes, I. Hita, P. Castaño
Univ. of the Basque Country, Dept. Chemical Engineering, Barrio de Sarriena s/n, 48940 Leioa, Spain; tel. +34-946012511, e-mail: josemaria.arandes@ehu.es.

Biomass-derived feedstock is a potential source of functional fuels and chemicals, reducing the use of petroleum-derived raw materials. Glycerol is a promising renewable source, which is a by-product of bio-diesel production. From the refinery perspective, a rational and pragmatic approach is using the existing infrastructure to convert glycerol. Fluid catalytic cracking (FCC) in particular is useful to produce added value chemicals from glycerol.[1,2] The main disadvantage of co-feeding glycerol to the FCC units is the increase in catalytic deactivation, which could be correlated to the hydrogen to carbon effective ratio of glycerol.[3] In this work, the deactivation and product distribution during the cracking of aqueous glycerol has been studied, using an equilibrated (steamed) HZSM-5 zeolite as catalyst in a riser simulator.

An agglomerated catalyst based on HZSM-5 zeolite has been prepared and characterized. Reaction runs have been performed in a state-of-the-art riser simulator operating at 350-550 °C, 1 atm, feeding an aqueous solution of 50 wt% of glycerol -prepared from 99.5 wt% glycerol (Panreac Chemistry)-, catalyst/feed ratio C/O = 6 by weight and with a value of contact time of 3–20 s. Product analysis has been carried out by means of an Agilent Technologies 7890A gas-chromatograph with a FID. The products of glycerol cracking have been divided into CO, CO_2, CH_4, oxygenates (Oxg), hydrocarbons (HC) and coke. Glycerol suffers parallel and in series reactions as dehydration to form hydrocarbons or oxygenates with less oxygen moieties, decarbonylation to form CO, decarboxylation to form CO_2, olygomerization and condensation to form coke, cracking of hydrocarbons trough carbocation mechanisms (both monomolecular and bimolecular), among others. Figure 1 shows the effect of temperature on product distribution during the cracking of glycerol after 3 s on stream. The maximum yield of oxygenates is obtained at 450 °C, corresponding to the minimum yield of CO and CO_2. It is worth mentioning that the produced oxygenates and hydrocarbons (Figure 1) are mostly in the gasoline boiling point range and with a high octane number.

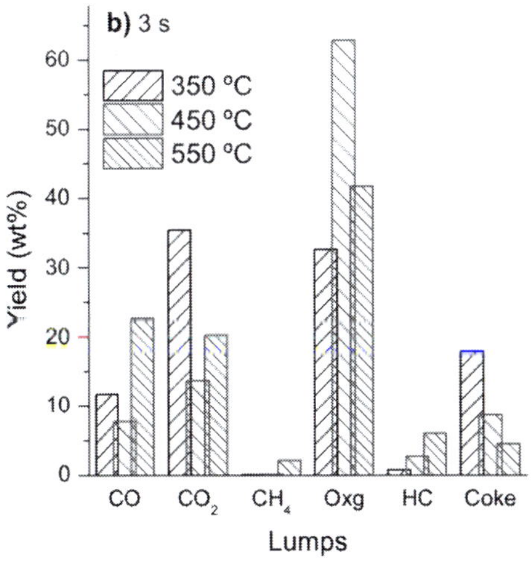

Figure 1. Product yields at different temperatures. TOS = 3 s.

The maximum yield of hydrocarbons (including CH_4) is obtained at 550 °C coinciding with the slower formation of coke. This result indicates that the transformation of glycerol into hydrocarbons compete with that into other oxygenates, and the balance between them falls into the former at higher temperatures.[4] Interestingly, coke yield decrease with temperature which is counterintuitive but has been previously observed for the transformation of other oxygenates.[5,6]

Figure 2a shows the profiles of CO_2 and H_2O signals (m/z = 44 and 18, respectively) of the exhaust generated during coke combustion using a deactivated catalyst at 450 °C and 5 s. Coke is mainly combusting at 550 °C, whereas parallel dehydrations occur at 150 and 480 °C. The deactivated catalyst has been digested with HF and extracted with CH_2Cl_2 in order to obtain the soluble coke deposited on the catalysts. The soluble coke has been analysed by ^{13}C NMR. The results of ^{13}C NMR indicate that the soluble coke deposited on the catalyst at 450 °C during the transformation of glycerol has a similar composition of that deposited in the transformation of methanol, so it is an indication that a similar hydrocarbon pool dual cycle is occurring for both feedstock,[7] with a parallel and direct coke formation from oxygenates.[4] The results demonstrate that soluble coke does not contain significant amounts of hydroxyls in alcohols (60-70 ppm) or phenols (155-160 ppm). Thus, oxygenated coke is present in H-coke, probably placed in the exterior of the zeolite crystals. The proportion of aromatics and polycondensed aromatics increases over time in detriment alkane and alkene aliphatics and carboxyls, indicating that the coke is being transformed into more recalcitrant composition.

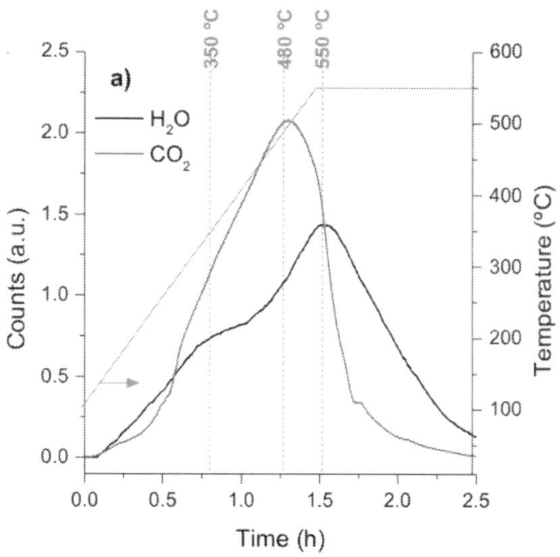

Figure 2. Evolution of the H_2O and CO_2 signals during TPO of spent catalyst at 450 °C and 20 s.

The results show that the FCC conditions of temperature (500-550 °C) and contact times (3-6 s) are optimal to transform glycerol into gasoline range oxygenates and hydrocarbons with high octane number. Indeed, using lower temperatures (350 °C) is detrimental for the process as the yield of coke is enhanced. The coke formed at these conditions contains higher proportion of hydroxyls located on the exterior of the zeolite. The use of higher temperatures favours the hydrocarbon pool mechanism, and higher proportion of hydrocarbons are obtained with lower deposition of coke. The coke formed in those conditions is more aromatic/olefinic and is located both inside and outside of the zeolite.

References

[1] A. Corma, G.W. Huber, L. Sauvanaud, P. O'Connor, *J. Catal.* **2007**, 247, 307-327.

[2] M. Niwa, N. Katada, *Chem. Rec.* **2013**, 13, 432-455.

[3] H. Zhang, Y.T. Cheng, T.P. Vispute, R. Xiao, G.W. Huber, *Energy Environ. Sci.*, **2011**, 4, 2297-2307.

[4] A. Ibarra, A. Veloso, J. Bilbao, J.M. Arandes, P. Castaño, *Appl. Catal. B: Environ.*, **2016**, 182, 336-346.

[5] B. Valle, P. Castaño, M. Olazar, J. Bilbao, A.G. Gayubo, *J. Catal.* **2012**, 285, 304-314.

[6] M. Ibáñez, B. Valle, J. Bilbao, A.G. Gayubo, P. Castaño, *Catal. Today*, **2012**, 195, 106-113.

[7] U. Olsbye, S. Svelle, M. Bjørgen, P. Beato, T.V.W. Janssens, F. Joensen, S. Bordiga, K.P. Lillerud, *Angew. Chem.-Int. Edit.*, **2012**, 51, 5810-5831.

Correlating the structure of V/SBA-15 and V/SiO_2 catalysts with their activity in the partial oxidation of ethanol

[1]J. M. Hidalgo, [1]Z. Tišler, [1]D. Kubička, [2]R. Bulánek, [2] K. Raabová
[1]Research Institute of Inorganic Chemistry / UNICRE, Areál Chempark, 436 70, Záluží 1, Czech Republic, e-mail: jose.hidalgo@vuanch.cz
[2]Department of Physical Chemistry, Faculty of Chemical Technology, University of Pardubice, Pardubice, Studentská 573, 532 10, Czech Republic

Introduction:

Vanadium-based catalysts have been used and studied for the partial oxidation or oxidative dehydrogenation reactions. [1,2] Moreover, the oxidation of ethanol to acetaldehyde is currently investigated with the aim of replacing the conventional processes based on hazardous agents such as chromate or permanganate. [3] Besides, the dispersion and nature of the vanadium over the adequate support is an important issue to obtain a high catalytic activity. [2]

Several mesoporous catalysts with some vanadium content were characterized with the aim of studying the type of vanadium present on their surface and its role in the partial oxidation of ethanol using high space velocity.

Experimental:

Two V/SBA-15 materials named V10 and V18 (10.2 and 17.8 %wt. V_2O_5 content respectively) and three V/SiO_2 fumed silica named 2VSF, 5VSF and 16VSF (2.0, 4.8 and 15.8 %wt. V_2O_5 content respectively) materials were prepared by wetness impregnation of silica by ethanol solution of appropriate amount of vanadyl acetylacetonate (Sigma-Aldrich). Impregnated samples were dried in the flow of air at 120 °C (temperature ramp: 5°C/min) overnight. Calcination was done in flow of air at 600°C for 12 hrs (temperature ramp 5°C/min). Samples were characterized using N_2 adsorption, XRF, XRD and by UV–Vis and Raman spectroscopy.

A trickle bed reactor with a length of 1000 mm made of stainless steel 316 was used. The reaction was carried out using 1 g of catalyst. The catalyst was prepared in a form of pellets which were sieved in grain size of 0.5-1 mm approximately. The catalyst and silicon carbide (0.5 mm particles) were mixed thoroughly (20 mL SiC + 1 g of catalyst) and loaded into the reactor. Finally, the remaining free volume of the reactor was filled with silicon carbide. Air flow, used for the catalytic reaction, was fed directly (activation 1 h at 400 °C; heating rate 5 °C/min) and then mixed with feedstock before entering the reactor. All catalysts were tested at 150, 200, 250 and 300 °C using 5 NL/h air flow, 1 bar of pressure and 5 g/h of ethanol. The gas output (test at 300 °C) was analysed by the method "Refinery Gas Analysis" RGA (Agilent Technologies) with a GC 7890A Agilent (Agilent Technologies, Santa Clara, CA, USA).The collected liquid samples were analysed using a GC-FID "Agilent 7890A" and GC-OFID "Agilent-Wasson-ECE Instrumentation". An ITQ 1100 GC-Ion Trap MS was used to identify the collected liquid products. Moreover, a "blank" test was carried out (with empty reactor and filled with SiC) using the same conditions than normal tests.

Results:

Blank experiment gave quite high conversion of EtOH, about 26 %wt. (the selectivity to acetaldehyde was 73 %wt.). Subsequently the results obtained over the V-doped silica should be compared to this value of blank experiment. Catalytic test showed that the activity highly depends on the concentration of vanadium (and its coordination) more than on the type of used support. Low concentrated samples (below 5%wt. of V_2O_5) gave negligible conversion of EtOH (almost the same as the empty reactor), on the other hand upon the increase of V concentration increases conversion of ethanol, reaching 38 %wt. over 16VSF and 22 %wt. over V18. These samples, together with V10, gave quite high selectivity to acetaldehyde, over 70 %wt. The main by-products were ethylene, acetaldehyde diethyl acetal and 1,3,5-trioxane.

UV-Vis together with Raman spectroscopy showed that upon the increase of vanadium concentration increases population of octahedrally coordinated bulk like VO_x units, which were not present in the samples with very low concentration of vanadium (2VSF, 5VSF). Poorly concentrated samples gave very high population of isolated monomeric VO_4 Td units together with oligomeric species. N_2 adsorption measurements showed that as the vanadium concentration approaches 10 %wt., the area of the catalysts is drastically lowered; the drop was more than 90 % thus suggesting that for this kind of reaction the important thing is not the area of the support but the vanadium concentration together with its coordination.

The whole subject will be discussed in broader concept during the presentation.

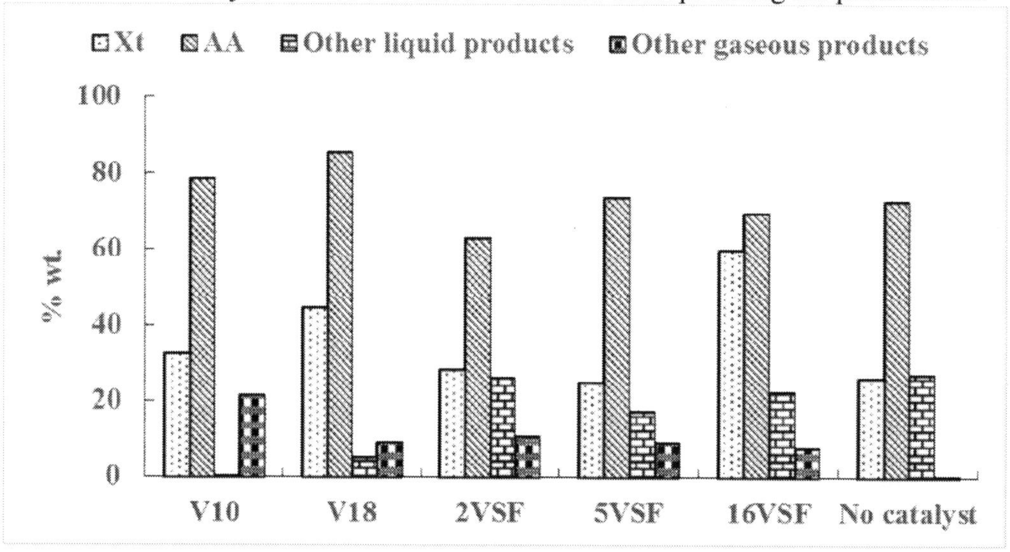

Fig. 1. Ethanol conversion (Xt; %wt.) and selectivities (%wt.) to products: AA (Acetaldehyde), "Other liquid products" (Diethyl Ether, 1,3,5 Trioxane, Acetic Acid, Acetaldehyde Diethyl Acetal) and "Other gaseous products" (Ethylene, carbon dioxide, carbon monoxide and others). The reaction was carried out at 300 °C; 5 g of EtOH/h feed flow; 5 NL/h air flow; 1 bar of pressure; 1 g of catalyst.

[1] Beck B.; Harth, M.; Hamilton, N. G.; Carrero, C.; Uhlrich, J. J.; Trunschke, A.; Shaikhutdinov, S.; Schubert, H.; Freund, H.; Schlogl, R.; Sauer, J.; Schomacker, R., J. Catal. 2012, 296, 120-131

[2] Du G.; Lim, S.; Pinault, M.; Wang, C.; Fang, F.; Pfefferle, L.; Haller, G. L.; J. Catal. 2008, 253, 74–90

[3] Redina, E. A.; Greish, A. A.; Mishin, I. V.; Kapustin, G. I.; Tkachenko, O. P.; Kirichenko, O. A.; Kustov, L. M., Catal. Today 2015, 241, 246-254

Sorption and swelling in polyethylene in gaseous and liquid media

M. Podivinská, L. Krajáková, K. Haškovcová, J. Chmelař, J. Kosek
University of Chemistry and Technology Prague, Department of Chemical Engineering, Technická 5, 166 28 Prague 6, Czech Republic,
e-mail: podivinm@vscht.cz

Sorption of monomer(s) and diluents in polyethylene (PE) influences the reaction rate in polymerization reactors, the copolymer composition and the subsequent degassing of residual organic species. The knowledge of equilibrium solubilities of low molecular weight penetrants in PE is important for the process design and optimization. Sorption of various species in polymers is accompanied by swelling, which causes the prolongation of the diffusion path and thus it has to be investigated together with the sorption itself.

Sorption and diffusion of low molecular weight penetrants is measured gravimetrically, where we measure the mass increase of polymer samples due to the gas sorption. The gravimetric data have to be corrected to the buoyancy force of the pendulous swollen polymer. Experimental measurement of swelling was carried out in video-microscopic apparatus. The method is based on the visual observation of a polymer particle by a microscope with attached digital camera.

In this contribution, we present results for the sorption and swelling of ethylene, co-monomers (1-hexene, 1-butene) and inert diluents (propane) in various grades of PE particles at conditions relevant to catalytic polymerization of olefins. Our work stands out of similar studies in the open literature due to: (i) systematic measurement for a broad range of PE densities from 0.90 to 0.96 g/cm^3 covering practically all relevant grades, (ii) reporting a comprehensive set of swelling data for several penetrants, (iii) systematic measurements for several penterants at a range of industrially relevant temperatures and pressures, (iv) reporting the sorption of both gasesous and liquid penetrants, and (v) modelling the sorption data by the advanced PC-SAFT equation of state.

Furthermore, several interesting features were observed in experimental results. The shape of the swelling isotherms follows the shape of the relevant sorption isotherms, i.e., higher swelling was observed at higher species solubilities in polyethylene. But the evaluated swelling/sorption ratios for various penetrants show a rather surprising trend.

In addition, the results are interpreted in the context of our recently proposed phase structure of PE [1]. We demonstrate that consideration of the newly proposed phase structure of PE allows for a fundamentally new and self-consistent approach to the thermodynamics of penetrant sorption in PE.

References:

[1] Chmelař J., Pokorný R., Schneider P., Smolná K., Bělský P., Kosek J.: Free and constrained amorphous phases in polyethylene: Interpretation of 1H NMR and SAXS data over a broad range of crystallinity. Polymer 2015, 58, 189-198.

[2] Chmelař J., Smolná K., Haškovcová K., Podivinská M., Maršálek J., Kosek J.: Equilibrium sorption of ethylene: Experimental study and PC-SAFT simulations. Polymer 2015, 59, 270-277.

[3] Podivinská M., Smolná K., Chmelař J., Kosek J.: Swelling of polyethylene particles and its effect on sorption equilibria under gas-phase polymerization conditions. Submitted.

Effect of oscillations in inlet gas composition on the performance of automotive three-way catalysts containing Pt, Pd and CeO_2

J. Březina, A. Arvajová, P. Kočí
University of Chemistry and Technology Prague, Department of Chemical Engineering, Technická 5, Prague 166 28, Czech Republic; tel. +420 220443293, e-mail: petr.koci@vscht.cz;

The automotive emissions are one of the most important sources of air pollution, particularly in cities and agglomerations. Due to the increasing number of cars and tightening emission limits it is necessary to develop more effective technologies for the exhaust gas treatment, which includes not only more active catalyst formulation but also an improved control of its operation.[1] One of the greatest challenges is to decrease the amount of pollutants (CO, hydrocarbons and NO_x) released after cold start, i.e., start of the engine and exhaust system from ambient temperature. Almost 80 % of the harmful emissions released during laboratory test driving cycle (and also in reality for short distance city driving) are released before the catalyst warm up and therefore before its light-off. The light-off temperature is defined as the temperature of inlet gas at which the conversion reaches 50 %.

This work is focused on systematic experimental study of the influence of inlet oxygen oscillations on the efficiency of several model catalysts containing Pt, Pd and CeO_2 in different combinations: Pt/Al_2O_3, Pd/Al_2O_3, $Pt/CeO_2/Al_2O_3$ and $Pd/CeO_2/Al_2O_3$. The measurements were performed in a heated lab reactor with temperature ramp 8 K min^{-1}, in the range of temperatures 100 – 550 °C. The inlet gas mixture simulating raw engine emissions was stoichiometric in time-average and contained propylene and propane as representative hydrocarbons. The reactivity of propane is much smaller and its conversion is not complete even at 550 °C. The composition of gas mixture was H_2 0.08 %, CO 0.166 %, C_3H_6 860 ppm, C_3H_8 860 ppm, O_2 0.775 %, NO 600 ppm, CO_2 7 %, H_2O 7 %, N_2 balance. Oscillations around the given value were applied to the inlet oxygen concentration. The forcing was symmetric – rich and lean sub-phases had the same period and amplitude.

An example of the results is shown in Fig. 1, comparing a set of experiments with oscillation frequency 0.1 Hz and amplitudes 1110 – 4440 ppm O_2. It can be seen that forcing of the inlet oxygen concentration can result in an improved NO_x conversion. The light-off temperature of NO_x was 324 °C for 4440 ppm O_2, instead of 510 °C without oscillations. It must be considered that the catalytic NO_x reduction on TWC is very limited in the presence of residual unreacted oxygen, therefore in non-oscillating stoichiometric mixture it is necessary to reach propane light-off first in order to consume all oxygen. With the oscillating inlet, the phases with lowered inlet oxygen level improve the NO_x conversion.

On the other hand, the conversion at the highest temperatures wasn't complete with the oscillating inlet, due to NO_x breakthrough in the phases with increased inlet O_2 level. The tested catalyst sample was significantly aged and its ability of oxygen storage and release on the catalytic surface was quite limited. With 0.1 Hz forcing (period 10 s), the catalyst had enough time to establish steady state during the oscillations and it was then operating in alternating steady lean conditions (characterized by NO_x breakthrough at the catalyst outlet) and steady rich conditions (characterized by hydrocarbons and CO breakthrough at the catalyst outlet). Therefore a higher oscillation frequency of 0.25 Hz was chosen for the next set of experiments (Fig. 2). This frequency is closer to that of lambda control loop in the real application and allows to achieve higher conversions while suppressing the undesired breakthrough of CO, hydrocarbons and NO_x.

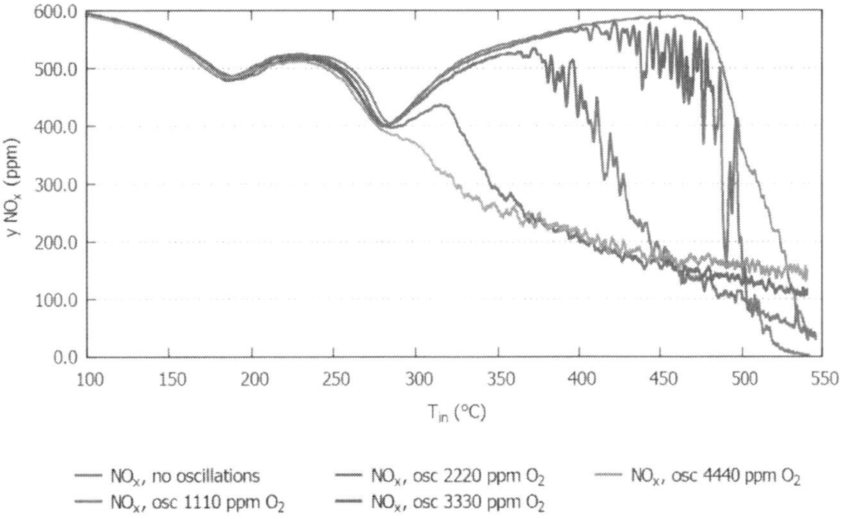

Fig. 1. Influence of inlet O_2 oscillations (frequency 0.1 Hz, varying amplitude) on outlet NO_x concentration.

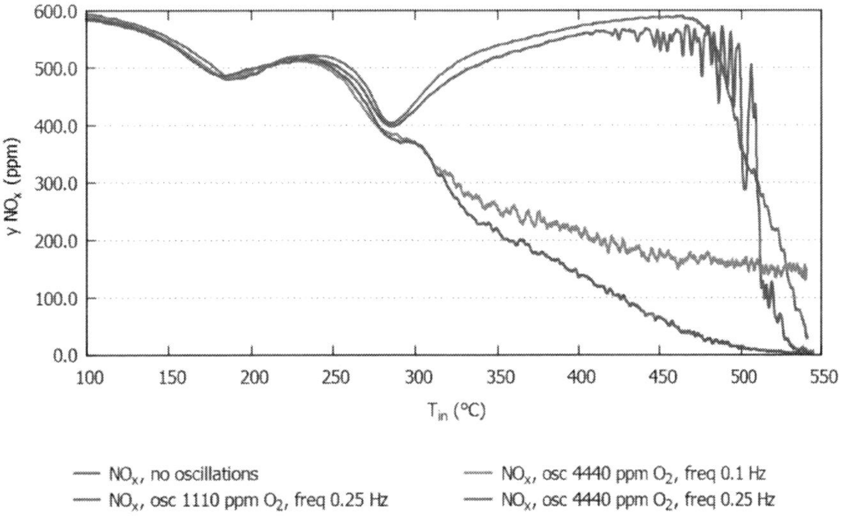

Fig. 2. Influence of inlet O_2 oscillations (varying frequency and amplitude) on outlet NO_x concentration.

It can be concluded that oscillations with proper amplitude and frequency in the exhaust gas around the stoichiometry can improve the catalysts performance. Suitable frequency depends on the actual catalyst properties, particularly on the presence of CeO_2 that provides oxygen storage capacity.[1] Similarities and differences between the performances of different catalyst formulations are discussed. The obtained results enable us to propose operating strategies leading to improved conversions of pollutants in the automotive exhaust gas.

References:

[1] Kočí P., Kubíček M., Marek M. Periodic forcing of three-way catalyst with diffusion in the washcoat. Catalysis Today 98 (2004), 345-355.

Influence of supports on catalyst activity for the reverse water–gas shift (RWGS) reaction and adsorption behavior of copper-based catalysts for CO_2 storage

[1]Andrej Pohar, [1]Damjan Lašič Jurković, [1]Venkata D.B.C. Dasireddy, [1]Blaž Likozar
[1]Laboratory of Catalysis and Chemical Reaction Engineering, National Institute of Chemistry, Hajdrihova 19, 1000 Ljubljana, Slovenia; tel. +386 1 4760281, e-mail: blaz.likozar@ki.si, andrej.pohar@ki.si

Introduction

One of the partial solutions for the issue of CO_2 emissions into the environment, and with it connected climate change, lies in the application of the reverse water–gas shift reaction. CO_2 can be captured at the point of the emissions (various industry, coal plants, *etc.*) with standard CO_2 capture procedures, after which it can be reacted with excess hydrogen to form syngas (CO + H_2), effectively reducing the carbon footprint in the process. Subsequently, syngas can be converted into liquid hydrocarbon fuels or other useful chemical products. The overall general reaction may be written as follows.

$$CO_2 + H_2 \leftrightarrow CO + H_2O \quad (1)$$

Optimally, hydrogen would be produced by the electrolysis of water from excess electrical energy at the time of its production peaks from renewable energy sources (for example from the sun or wind). For the reactions, excess hydrogen is used for obtaining syngas as the product, which also benefits the process by shifting the thermodynamic equilibrium towards the products' side. Furthermore, water can be condensed at the reactor exit and the reflux of the unreacted gases can be used for obtaining improved conversions.

On the other hand, an efficient catalyst development is also crucial for successful process operation. In the past decade, particular interest was drawn towards Au, Pt and Ir supported on Fe_2O_3 and TiO_2 due to their activity [1], while the effect of different supports and bimetallic supports was also investigated thoroughly [2].

Catalyst synthesis

10 wt% of copper oxide, supported on various substrates (Al_2O_3, SiO_2, TiO_2 and ZrO_2), were prepared using a deposition–precipitation method [3]. The catalysts were characterized by various characterization techniques such as Brunauer–Emmett–Teller (BET) surface analysis, scanning electron microscopy (SEM) coupled with energy-dispersive X-ray spectroscopy (EDS), powder X-ray diffraction (XRD), temperature-programmed reduction (TPR), CO_2 and H_2 temperature-programmed desorption (TPD) and N_2O chemisorption, while the effect of different supports on the crystallite size and the metallic dispersion of copper was also assessed.

Table 1 presents the BET surface areas and pore volumes of the prepared catalysts. The general trend observed was that a decrease in both surface area and pore volume of the catalysts, in comparison to bare supports, was due to the presence of metal oxides, and could be attributed to the clogging of the pores by the metal oxide particles [4]. The metal dispersions of copper on the surface of the catalysts could thus be paralleled to surface areas. 10 wt% Cu/Al_2O_3 catalyst showed a higher metal dispersion and smaller crystallite sizes when compared to all other catalysts. This could be due to the point of zero charge of Al_2O_3 with regard to other supports [5]. The influence of the reduction temperature, metal dispersion and crystallite size on the reverse water–gas shift reaction was subsequently investigated.

Table 1. Properties of the synthesized catalysts used for the reverse water–gas shift reaction.

Catalyst	Surface area (m^2 g^{-1})	Pore volume (cm^3 g^{-1})	Metal dispersion* (%)	Crystallite size** (nm)
10 wt% Cu/Al_2O_3	115	0.193	67	11
10 wt% Cu/SiO_2	4	0.006	23	27
10 wt% Cu/TiO_2	10	0.027	38	18
10 wt% Cu/ZrO_2	13	0.048	41	15

*Obtained by N_2O chemisorption; **Calculated using the Scherrer equation

In powder XRD profiles, all the catalysts showed the main peaks at the diffraction angles (2θ) of 35.2° (002) and 38.3° (111) with the plane spacing of 0.251 nm and 0.232 nm, respectively, and the latter were ascribed to CuO monocrystalline phase (ICDD; PDF Card 48-1548). In the H_2 TPR profiles of the catalysts, there were two H_2 TPR peaks, namely α (a low-temperature peak, below 200 °C) and β (a high-temperature peak, above 200 °C). These two observed peaks corresponded to the reduction of highly-dispersed and agglomerated CuO, respectively [3].

Modeling

A one-dimensional packed-bed convection–diffusion–reaction model was used for the description of the governing processes within the reactor. Uniform flow was assumed and the effect of flow dispersion was accounted for. The pressure drop along the reactor was calculated, as well as the pressure dependence of density and velocity. The diffusivity of the gaseous components in the mixture and its dynamic viscosity were calculated by taking into account the gas composition. The kinetic parameters were determined through regression analysis of the obtained experimental data.

Conclusions

The developed process model can be used to predict the syngas production at various operational parameters (temperature, pressure, flow rates, H_2/CO_2 ratio and reflux ratios) for different CO_2 sources (coal, biomass, *etc.*) and respective feedstock compositions, and will ultimately be correlated with process operation economics.

Acknowledgement

The presented work was partially funded by the EU Framework Programme for Research and Innovation Horizon 2020 under the grant agreement No 637016 (MefCO2). The authors also gratefully acknowledge the financial support of the Slovenian Research Agency (ARRS) through the Programme P2–0152.

References

[1] M. Aresta, A. Dibenedetto, E. Quaranta, Reaction Mechanisms in Carbon Dioxide Conversion, Springer, Berlin, 2015.
[2] M. De Falco, G. Iaquaniello, G. Centi, CO_2: A Valuable Source of Carbon, Springer, London, 2013.
[3] G.V. Sagar, P.V.R. Rao, C.S. Srikanth, K.V.R. Chary, J. Phys. Chem. B, 110 (2006) 13881-13888.
[4] J. Gao, Q. Liu, F. Gu, B. Liu, Z. Zhong, F. Su, RSC Adv., 5 (2015) 22759-22776.
[5] A.T. Bell, Science, 299 (2003) 1688-1691.

Phosphorus-containing acid activated carbon support in a bifunctional PtPd catalyst for pre-treated tire-oil hydrocracking

I. Hita[a], T. Cordero-Lanzac[a], A. Gallardo[b], J.M. Arandes[a] , J. Rodríguez-Mirasol[b], J. Bilbao[a], T. Cordero[b], P. Castaño[a].
[a]Universidad del País Vasco, Dep. Ingeniería Química, Apdo. 644-48080 Bilbao, Spain, Tel: +34 94 6015341, *e-mail: idoia.hita@ehu.es, fax: +34 94 6013500.
[b]Universidad de Málaga, Dep. Ingeniería Química, Campus de Teatinos, s/n, 29010, Málaga, Spain.

Increasingly strict environmental protection policies are exponentially pushing waste valorization strategies to move forward. Among plastic wastes, scrap tires represent for a pressing challenge, due to their predicted increasing amount and bio-hazards derived from their uncontrolled disposal [1]. As a more environmentally respectful alternative to their combustion, sequenced pyrolysis and hydroprocessing of scrap tires offers the possibility of obtaining high-quality fuels, which represents a major challenge for waste valorization at a great scale [2]. Fast pyrolysis allows for maximizing the liquid pyrolysis product (tire oil) [3], which the most interesting from both and energetic and an economic perspective, together with carbon black [4]. However, tire oil presents a series of drawbacks which need to be overcome: (i) high sulfur content, (ii) high content of aromatics and, (iii) high proportion of heavy molecules. In a conventional refinery, hydroprocessing allows for solving these barriers simultaneously, generally using Ni-Mo and Co-Mo catalysts, but these are unable of upgrading tire oil composition to the required point. To solve this, subsequent hydrotreaters with more active Pt-Pd catalysts can be used particularly for aromatics and gasoil removal [5, 6]. In this work, an activated carbon obtained from chemical activation with H_3PO_4 of olive stone has been prepared and tested as a promising acid support on the hydrocracking of low-sulfur tire oil for further composition readjustment.

The activated carbon was obtained using olive stone as precursor, impregnating it with an aqueous solution of H_3PO_4 (85 wt%, 3 g of H_3PO_4 per g of precursor) at room temperature and dried at 60 °C for 24 h. The impregnated precursor was activated in a tubular furnace under a continuous N_2 flow (150 cm^3 STP min^{-1}) at 500 °C for 2 h. Then the catalyst was cooled inside the furnace under a N_2 flow and washed with distilled water at 60 °C, dried at 100 °C and sieved to a particle size of 100-300 μm. Pt and Pd were incorporated by incipient wetting impregnation method, using $HPtCl_6 \cdot 6HCl$ and $PdCl_2$as precursors. The impregnated support was calcined at 400 °C for 4 h with a N_2 flow of 150 cm^3 STP min^{-1} in a tubular furnace to yield a supported catalyst with 0.5 wt% Pd and 1 wt% Pt. Hydrocracking runs have been carried out in a down-flow fixed bed reactor at: space time, 0-0.28 g_{cat} h g_{feed}^{-1}; 400-500 °C; 65 bar; H_2/oil ratio of 1,000 vol%; and time on stream (TOS) of 0-6 h. Liquids were collected and analysed in an Agilent Technologies 7890A GCxGC-MS System using FID and MSD.

Table 1. Properties of the bifunctional catalyst

S_{BET} (cm^2 g_{cat}^{-1})	1305
$V_{micropore}$ (cm^3 g_{cat}^{-1})	0.512
$V_{mesopore}$ (cm^3 g_{cat}^{-1})	0.829
Average pore diameter (Å)	52.3
Total acidity (mmol$_{t\text{-}BA}$ g^{-1})	0.29

Sulfur content has been measured by GC with a PFPD detector. The composition of the feed is: naphtha (BP = 35-216 °C), 27.9 wt%; diesel (BP = 216-350 °C), 50.3 wt%; and gasoil (BP > 350 °C), 21.8 wt%. As from a lump classification: paraffins and isoparaffins (P+iP), 35.4 wt%; naphthenes (N), 22.0 wt%; 1-ring aromatics (A_1), 31.9 wt%; and 2-ring aromatics (A_2), 10.7 wt%. Total sulfur in the pre-treated tire oil accounts for 1985 ppm.

During activation, a 46 wt% carbonization yield has been obtained. The main physico-chemical properties of the fresh PtPd catalyst are listed in Table 1. The catalyst has a relatively high porosity, with an average pore size of 52 Å, and high specific surface (1,305 m^2 g^{-1}), which is comparable to that of a silico-alumina support, as well as its total acidity [5]. TEM images (not shown) have proven that bimetallic Pt and Pd particles are well-dispersed on the support with homogeneous sizes.

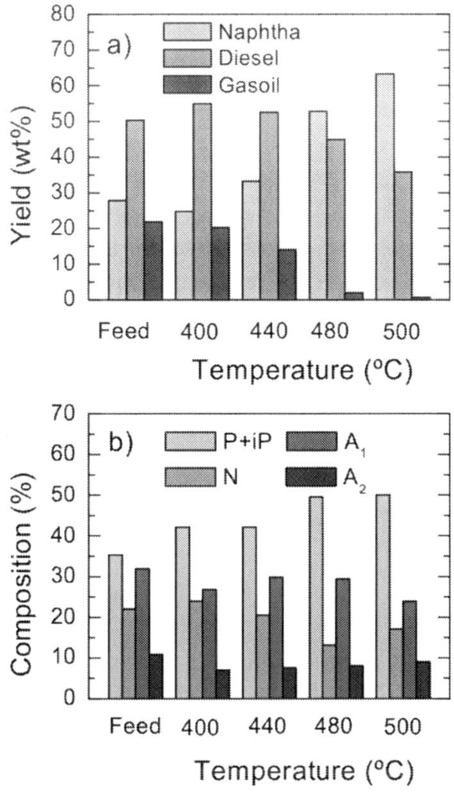

Figure 1. Temperature effect on lump distribution (a) and composition (b) of the products (65 bar; TOS, 6 h; space time, 0.28 g_{cat} h g_{feed}^{-1}).

The product lump distribution and composition are shown in Figures 1a and b, respectively. Increasing temperature allows for obtaining lighter lumps like diesel (400-440 °C) or naphtha (480-500 °C), achieving a good compromise at 480 °C with 53 wt% diesel and 45 wt% naphtha. Gasoil is removed almost completely at temperatures above 480 °C. According to a bifunctional catalyst mechanism, 1 and 2-ring aromatics hydrogenate to form naphthenes on the metallic sites to subsequently suffer ring opening to form paraffins. The first hydrogenation step is reversible and exothermic so, if the cracking is not dominant, the yield of aromatics increases with temperature (400-440 °C). At higher temperatures, paraffinic compounds are the most abundant with up to 50 wt%, particularly at 480 °C and above, when cracking and ring opening reactions are dominant.

Total sulfur (data not shown) has been reduced from 1985 ppm in the original tire oil to below 200 ppm at 500 °C, achieving a total hydrodesulfurization conversion (X_{HDS}) higher than 90 % based on total sulfur amount (in ppm) in the feed and products.

[1] A. Zabaniotou, N. Antoniou, G. Bruton (2014). *Waste Manage.*, **34**, 2335.
[2] I. Hita, A. Gutiérrez, M. Olazar, J. Bilbao, J.M. Arandes, P. Castaño (2015). *Fuel*, **145**, 158.
[3] M. Arabiourrutia, G. López, G. Elordi, M. Olazar, R. Aguado, J. Bilbao (2007). *Chem. Eng. Sci.*, **62**, 5271.
[4] P. T. Williams (2013). *Waste Manage.*, **33**, 1714.
[5] I. Hita, E. Rodríguez, M. Olazar, J. Bilbao, J.M. Arandes, P. Castaño (2015). *Energy Fuels*, **29**, 5458.
[6] I. Hita, A. T. Aguayo, M. Olazar, M. J. Azkoiti, J. Bilbao, J. M. Arandes, P. Castaño (2015). *Energy Fuels*, **29**, 7542.

$Ce_{0.495}Sn_{0.495}Pd_{0.01}O_{2-\delta}$ as highly stable and low-leaching catalyst in continuous Suzuki-Miyaura cross-coupling reactions

G.J. Lichtenegger,[1] H. Gruber-Woelfler,[1] J.G. Khinast[1]
[1] Graz University of Technology, Institute for Process and Particle Engineering, Inffeldgasse 13, A-8010 Graz, Austria, tel: +43 316 873 30409, e-mail: georg.lichtenegger@tugraz.at

In recent years, a wealth of heterogeneous catalysts for palladium catalysed cross coupling reactions has been reported in literature. While various Pd-containing organometallic complexes, anchored on different solid supports (e.g. activated carbon [1], organic polymers [2], or silica [3]), as well as ligand-free catalysts have been studied in detail, only a small number of papers reports the use of palladium substituted metal oxides for Pd-catalysed cross-coupling reactions [4,5].

Although most of the reported catalysts show high activity, good recyclability and low Pd-leaching in batch experiments, many of them suffer from rapid deactivation and considerable Pd-leaching when used in continuous flow reaction setups [6].

In this contribution we present the palladium substituted mixed oxide $Ce_{0.495}Sn_{0.495}Pd_{0.01}O_{2-\delta}$ as highly stable and low-leaching catalyst for the continuous synthesis of biphenyls using the Suzuki-Miyaura cross-coupling reaction in a fixed bed catalytic reactor.

The catalyst was synthesized out of non-toxic and inexpensive metal-precursors in a fast and convenient manner using established solution combustion methods [4]. The continuous catalytic reaction was performed in a custom made reaction setup, the so called "Plug & Play Reactor" - a versatile reactor that includes flexible modules for heating/cooling and mixing as well as an exchangeable reaction segment [7]. The Plug & Play reactor system allows not only on-line control of the flow-rates, temperatures and pressure inside the reactor, but also provides the possibility to perform an in-line analysis of the reaction progress by means of UV-Vis spectroscopy (Figure 1).

With this setup, different substituted biaryls could be synthesized (up to 10 g/d) in continuous mode at high conversion rates (up to 98%) and excellent selectivities (>99.5 %). The continuous processes proved to be stable for more than 30 hours. ICP-MS analysis of the isolated product yielded palladium contents of 1 ppm and below, confirming the high resistance of the used catalyst against Pd-leaching.

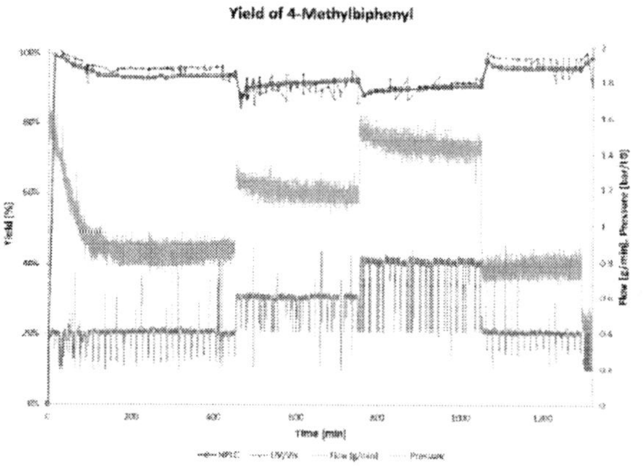

Figure 1: Synthesis of 4-Methylbihenyl via Suzuki-Miyaura coupling of p-Bromotoluene and Phenylboronic acid. Yields determined by HPLC and UV-Vis spectroscopy, flow-rates and system pressure

Furthermore, a crystallization unit for the continuous isolation of the synthesized products could be implemented.

Due to the ease of preparation from non-toxic and inexpensive precursors and their high stability, activity and selectivity in continuous Suzuki-reactions using low-cost and environmentally friendly solvents (ethanol/water mixtures), the presented cerium-tin-palladium-oxides revealed their high potential as catalysts for continuous Suzuki-Miyaura cross-coupling reactions even at larger scale.

[1] F.-X. Felpin, J. Org. Chem. 70 (2005) 8575, [2] Y. M. A. Yamada, S. M. Sarkar, Y. Uozumi, J. Am. Chem. Soc. 134 (2012) 3190, [3] H. Gruber-Woelfler, et al., J. Catal. 286 (2012) 30, [4] Sanjaykumar, S. R.; Mukri, B.; Patil, S.; Madras, G.; Hegde, M. S. J. Chem. Sci. 2011, 123 (1), 47, [5] Amoroso, F.; Colussi, S.; Del Zotto, A.; Llorca, J.; Trovarelli, A. J. Mol. Catal. A Chem. 2010, 315 (2), 197, [6] Greco, R.; Goessler, W.; Cantillo, D.; Kappe, C. O. ACS Catal. 2015, 5 (2), 1303, [7] H. Gruber-Woelfler, G. J. Lichtenegger, H. Kitzler, K. Obermaier, J.G. Khinast, ACHEMA 2015, Frankfurt, Germany

CO_2 reforming of methane over Ni-based 6%LSNZ pyrochlore catalyst

N. Kumar, [1]M. L'Abbate, [2]D. Haynes, [2]D. Shekhawat, J. J. Spivey
Louisiana State University, Department of Chemical Engineering, Baton Rouge, LA 70803, USA; tel. +1 225 578 3690, fax +1 225 578 1476, e-mail: jjspivey@lsu.edu;
[1]KTH Royal Institute of Technology, Stockholm, Sweden
[2]National Energy Technology Laboratory, U.S. Dept. of Energy, Morgantown, WV 26507

Among other options, reforming of methane and CO_2 to produce syngas can be carried out using two reforming reactions: dry-reforming and oxy-reforming:

Dry-reforming:

$$CH_4 + CO_2 \rightleftharpoons 2CO + 2H_2 \quad (1)$$

$$\Delta H_{298K} = +247\ kJ/mol,\ \ \Delta G_{298K} = +170\ kJ/mol$$

Oxy-reforming:

$$3CH_4 + CO_2 + O_2 \rightleftharpoons 4CO + 6H_2 \quad (2)$$

$$\Delta H_{298K} = +58\ kJ/mol, \Delta G_{298K} = -1\ kJ/mol$$

There are two major challenges in both these reforming reactions. The first is thermal stability, which is difficult to maintain at the demanding conditions needed to reach near equilibrium conversions, e.g., temperatures at ~900 °C. The second is deactivation by carbon deposition, which is inevitable and often quite rapid. In general, carbon deposition is much more rapid in dry-reforming than oxy-reforming, in which a small amount of oxygen in the feed helps to remove the carbon deposited from the surface.

Pyrochlores are candidate catalysts due to their thermal stability, which is attributed to its crystalline structure. For that reason, they have been used for various applications, such as gas turbine thermal barrier coatings, but relatively little is reported on pyrochlores as catalysts. Pyrochlores have the general formula of $A_2B_2O_7$, where A is a lanthanide series element and B is a transition metal and the ionic ratio of rA/rB is 1.4-1.6. The pyrochlore structure, if properly synthesized, can incorporate low levels of an active metal, such as Ni, isomorphically into the lattice at low loadings. This should provide a well dispersed, active metal that is stable, even at reforming temperatures.

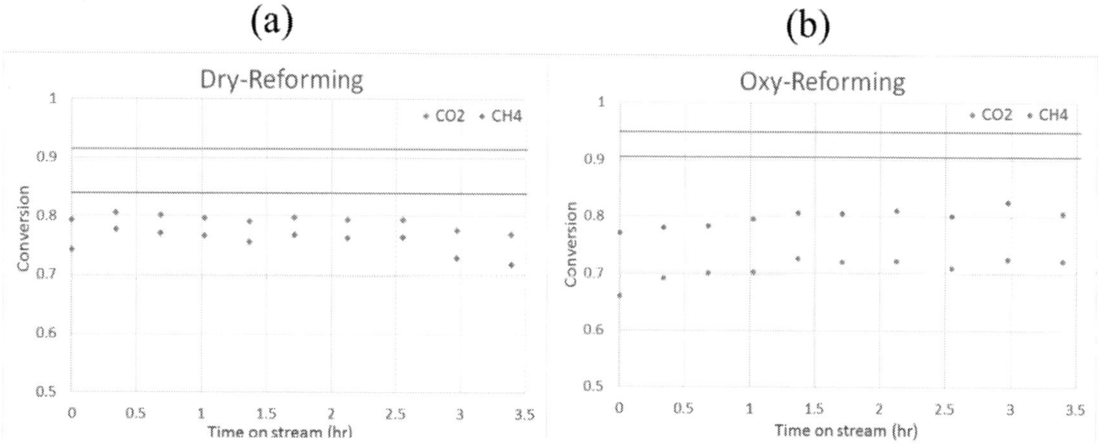

Figure 1. (a) Dry-reforming and (b) oxy-reforming over 6%LSNZ catalyst with time on stream. The solid lines represent equilibrium conversions corresponding to these conditions. Reactions: 750 °C, 1 atm, molar feed ratios: Dry-reforming: CH_4/CO_2=1/1; Oxy-reforming: $CH_4/O_2/CO_2$=3/1/1. Equilibrium CH_4 and CO_2 conversions, calculated using HSC Chemistry.

In this work, a 6 wt% Ni-based pyrochlore catalyst (6%LSNZ) was prepared by a variation of the Pechini method, which produces a uniform metal-doped crystallite catalyst. The catalyst was characterized by TPR/TPO and XRD before and after the two reforming reactions.

The time on stream results for dry- and oxy-reforming reactions is presented in Figure 1. It is seen that the CH_4 and CO_2 conversions decrease slowly but consistently with time-on-stream during the dry-reforming reaction. However, the conversions are constant for the oxy-reforming reaction, albeit just for a short 3.5-hr period tested here. Post reaction, the catalysts were characterized to determine whether there are differences in the carbon deposition at these two conditions.

Temperature program oxidation (TPO) was carried out to characterize the spent catalysts from each of the two reforming reactions and the results are presented in Figure 2. There is far more total carbon deposition on the catalyst from dry-reforming than the oxy-reforming. There is a qualitative difference in the carbon as well. The TPO shows that both catalysts contain carbon that is relatively reactive, corresponding to peaks <~410 °C, which can be attributed to amorphous carbon. A peak at ~480 °C on both catalysts is present, which can be attributed to polymeric carbon. There is far much more of this polymeric carbon after dry-reforming. Finally, there is a great deal of graphitic carbon (~650 °C), which is not present after oxy-reforming.

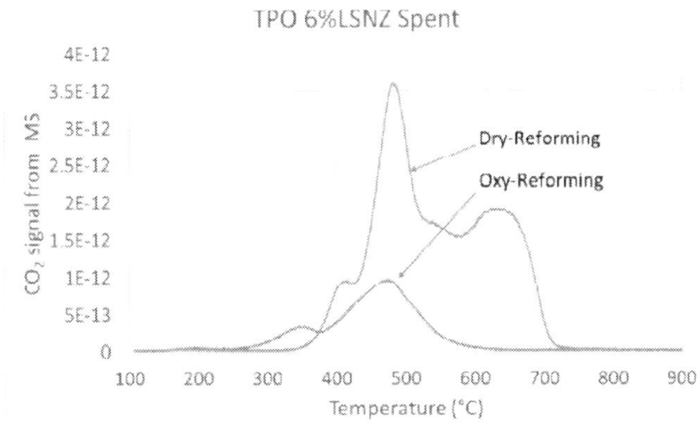

Figure 2. TPO results for spent 6%LSNZ catalyst after dry- and oxy-reforming reactions at 750 °C

Conclusions:

Ni-substituted pyrochlore catalyst has been found to be active at both dry-reforming and oxy-reforming conditions. The dry-reforming results show that the catalyst deactivates with time on stream, and a significant part of the carbon deposited on the surface of the catalyst is in the graphitic form. Under oxy-reforming conditions, however, the catalyst was found to be stable with much less carbon deposition, at least over the relatively short time-on-stream reported here.

Acknowledgements:

We acknowledge the financial support from Climate Change and Emissions Management Corporation (CCEMC), Canada. Dr. Arun Raju at UC, Riverside is gratefully acknowledged for the process development.

Effect of catalyst promotion by alkali and heavy metals on selective alkynols hydrogenation with the use of Pd-containing hypercrosslinked polystyrene

L.Zh. Nikoshvili[1,2], T.E. Hudyakova[1], E.A. Kholkina[1], A.V. Bykov[1], V.G. Matveeva[1], E.M. Sulman[1], L. Kiwi-Minsker[2,3]
[1]Tver Technical University, Dep. Biotechnology and Chemistry, A.Nikitina str. 22, 170026, Tver, Russia; tel./fax: +7 4822 789317, e-mail: nlinda@science.tver.ru
[2]Tver State University, Regional Technological Centre, 170100, Tver, Russia
[3]Ecole Polytechnique Fédérale de Lausanne, GGRC-ISIC-EPFL, CH-1015, Lausanne, Switzlerand

Selective hydrogenation of triple carbon-carbon bond using Pd nanoparticles (NPs) is of great importance as widely applicable in synthesis of fine organic synthesis. The challenge is to provide high activity, selectivity and stability of catalytic system. In our previous works [1,2] we have shown that Pd NPs stabilized by hypercrosslinked polystyrene (HPS) are promising catalysts for hydrogenation of alkynol C_5 (2-methyl-3-butyne-2-ol (MBY). However there is a room for further improvement. Recently, the Swiss group under the guidance of L. Kiwi-Minsker [3] has shown that modification of Pd NPs with second metal (Ag, Cu) allows increasing selectivity of hydrogenation of dehydroisophytol from 78% (for Pd/ZnO/SMF catalyst) up to 93% (for developed Pd-Ag/ZnO/SMF catalyst). At the same it is known that sodium can serve as a promoter of Pd-containing hydrogenation catalysts [4].

In this work we investigated the influence of promotion of 0.7%-Pd/HPS catalyst by alkaline metals (we used NaOH, KOH, Na_2CO_3, K_2CO_3, K_3PO_4 etc.) and heavy metals ($SnCl_2$, $ZnCl_2$, $Zn(NO_3)_2$, etc.). It was shown that the use of heavy metals allows increasing selectivity of MBY hydrogenation, while the use of alkaline metals results in increase of both selectivity and activity (see e.g. Fig. 1).

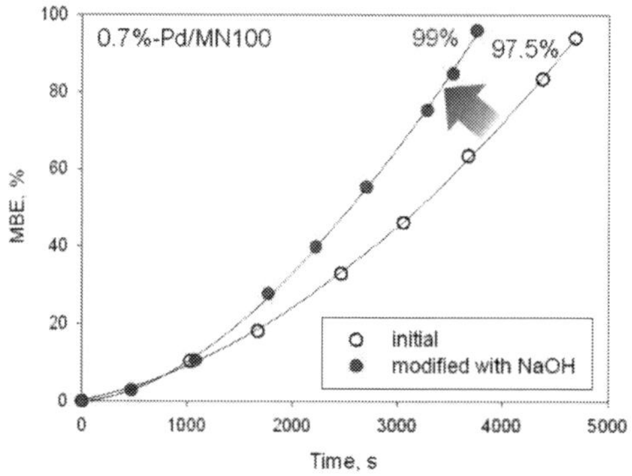

Fig. 1 – Effect of NaOH on activity and selectivity of 0.7%-Pd/HPS in hydrogenation of MBY to 2-methyl-3-butene-2-ol (MBE)

This work was funded by the Russian Science Foundation (Grant 15-19-20023).

References

[1] Nikoshvili L.Zh. et al., Catalysis Today. 256 (2015) 231-240.
[2] Nikoshvili L.Zh. et al., Catalysis Today. 241, Part B (2005) 179-188.
[3] Yarulin A. et al., Applied Catalysis A: General. 478 (2014) 186-193.
[4] Cho H.-B. et al., Bull. Korean Chem. Soc. 29 (2008) 2434-2440.

Correlating compositions of the reaction medium and coke during the steam reforming of bio-oil on a Ni/La_2O_3-αAl_2O_3 catalyst

A. Ochoa, B. Aramburu, A. Arandia, B. Valle, A.G. Gayubo, P. Castaño
Department of Chemical Engineering, University of the Basque Country (UPV/EHU), P.O. Box: 644-48080, Bilbao, Spain.Tel. +3494 601 53 61, e-mail: aitor.ochoa@ehu.eus.

1. Introduction

The depletion and environmental pollution from fossil fuels call for using new energy resources. Hydrogen has a very high energy density (~121 MJ/kg) and provides a clean energy source for the production of chemicals and is an alternative to gasoline and diesel in automotive engines. [1,2]. Catalytic steam reforming of bio-oil (produced form the pyrolysis of lignocellulose) is considered a viable and promising route for hydrogen production [3]. The process can be materialized with delocalized pyrolysis units, transport of the produced bio-oil (with higher energy density compared to lignocellulose) to bio-refineries, and subsequent steam reforming. Nickel-based catalysts have been widely used in the literature for steam reforming of bio-oil [3,4], due to the high C-C bond-breaking activity and the relatively low cost. Although high conversion and H_2 yields can be achieved, the catalyst undergoes a rapid and severe deactivation, both by sintering and, more notoriously, coke deposition [1,3,4]. The complexity and heterogeneity of the composition of bio-oil (acids, ketones, esters, alcohols, phenols and guaiacols, among others) makes it difficult to establish coke formation mechanisms during bio-oil steam reforming, which is of the utmost relevance in order to inhibit its formation routes. Consequently, the entire process would benefit by correlations predicting the amount and nature of coke deposited on the catalyst.

In this work the nature of the coke deposited on a Ni/La_2O_3-αAl_2O_3 catalyst used at different temperatures and steam to carbon (S/C) ratios in the steam reforming of bio-oil has been studied. The coke composition has been correlated with that of the reaction medium.

2. Experimental

The catalyst contains 10 wt% Ni and 9 wt% La_2O_3, and it has been prepared by incipient wetness impregnation method [4]. The bio-oil has been obtained by flash pyrolysis of pine (*pinus insignis*) sawdust in a conical spouted bed reactor at 480 °C. Bio-oil is fed to a thermal treatment stage at 500 °C to retain the pyrolytic lignin [4], and the volatile stream leaving the thermal treatment is fed, through a thermostated line, to a subsequent catalytic fluidized bed reactor: bio-oil flowrate, 0.1 mL min^{-1}; water flowrate, 0.012-0.180 mL min^{-1}; temperature, 550-700 °C; time on stream, 4 h; steam to carbon ratio, S/C=1.5-6; space time, 0.19 $g_{catalyst}$ h $g_{bio\text{-}oil}^{-1}$. The composition of the products stream leaving the reactor has been analyzed by gas chromatography combined with mass spectroscopy (Shimadzu GC/MS QP2010S). The morphology, content and composition of the deposited coke has been studied by TG/MS-TPO, SEM and TEM microscopy, and FTIR, Raman, FTIR-TPO, Raman-TPO and XPS spectroscopy. The Ni particle size of the reduced fresh and deactivated catalysts has been determined by XRD.

3. Results and discussion

Figure 1 shows the relative intensities of several IR absorbance bands in the deposited coke, as well as the mass composition of the reaction medium, at 550-700 °C and S/C = 6. IR bands in Figure 1a have been assigned to [5]: C-O bonds in acetate groups, phenolic esters and/or C=C-O-(aliphatic C) and (aromatic C)-O-(aliphatic C) type-ethers, 1240-1260 cm^{-1}; O=C-O in acetate groups, 1420-1430 cm^{-1}. IR bands in Figure 1b have been assigned to [5]: C-O bonds in groups such as ethoxy and alcohols, 1050 cm^{-1}; C=O groups in ketones and/or aldehydes, 1680-1700 cm^{-1}.

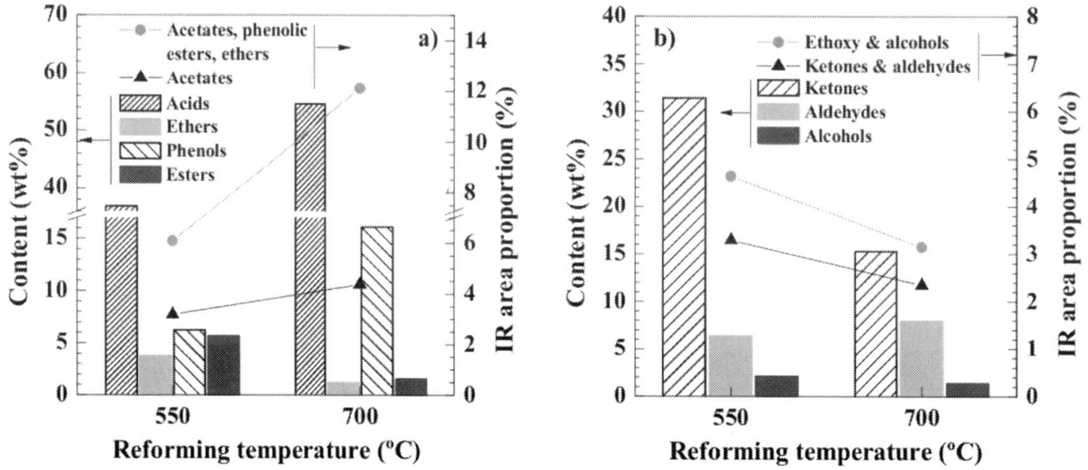

Figure 1. Composition of the reaction medium (left Y axis) and relative proportion of several IR absorbance bands (right Y axis) of coke deposited at S/C=6 of steam reforming.

Results in Figure 1a show a significant increase in the content of acids (with acetic acid as the main compound) and phenols in the reaction medium as reforming temperature is increased from 550 °C to 700 °C, whereas the content of other oxygenated groups, such as ethers and esters, is low at both temperatures and decreases as temperature is increased. These results are in accordance with those of IR bands of coke at 1240-1260 cm^{-1} and 1420-1430 cm^{-1}, as the increasing trend with temperature of acids in the reaction medium is consistent with an increase in acetate bonds in the deposited coke, in the same way that the increase in phenol groups in the reaction medium is consistent with that of phenolic bonds in the coke.

Figure 1b shows a clear decrease in the content of ketones in the reaction medium at higher temperature, whereas the content of alcohol decreases slightly and the content of aldehydes increases slightly. The decrease in the mass content of alcohols in the reaction medium is consistent with the intensity of the IR band assigned to alcohols and/or ethoxy at 1050 cm^{-1}. In addition, the combined decrease of ketones and aldehydes in the reaction medium (despite the slight increase of aldehydes) is consistent with the observed decrease in the IR band of coke assigned to ketones and aldehydes at 1680-1700 cm^{-1}.

Nevertheless, it has been observed that the S/C ratio (1.5-6) in the feed does not affect the coke composition as much as the reforming temperature.

References

[1] F.G.E. Nogueira, P.G.M. Assaf, H.W.P. Carvalho, E.M. Assaf. Appl. Catal. B., 160-1, 188-99 (2014).

[2] T. Hou, L. Yuan, T. Ye, L. Gong, J. Tu, M. Yamamoto et al. Int. J. Hydrogen Energ., 34, 9095-107 (2009).

[3] R. Trane, S. Dahl, M.S. Skjøth-Rasmussen, A.D. Jensen. Int. J. Hydrogen Energ., 37, 6447-72 (2012).

[4] B. Valle, B. Aramburu, A. Remiro, J. Bilbao, A.G. Gayubo. Appl. Catal. B., 147, 402-10 (2014).

[5] L.V. Mattos, G. Jacobs, B.H. Davis, F.B. Noronha. Chem. Rev., 112, 4094-123 (2012).

Three-phase hydrogenation of 2-methyl-3-butyn-2-ol over a Pd/ZnO catalyst: development of a kinetic model

S. Vernuccio, Ph. Rudolf von Rohr
Institute of Process Engineering, ETH Zurich, Sonneggstrasse 3, 8092 Zurich, Switzerland;
tel. +44 44 632 24 99, e-mail: svernuccio@ipe.mavt.ethz.ch; fax: +44 44 632 13 25.

The selective semi-hydrogenation of alkynes to alkenes over palladium-based catalysts are excellent examples of a catalytic three-phase processes following a consecutive reaction network.

The hydrogenation of 2-methyl-3-butyn-2-ol (MBY), in particular, is of great interest for industrial application. The partially hydrogenated product, 2-methyl-3-buten-2-ol (MBE), represents a key intermediate in the industrial synthesis of vitamins and aroma compounds.

The selectivity of this process is related to the possibility of hydrogenating MBY to MBE while preventing an over-hydrogenation reaction to 2-methyl-2-butenol (MBA) (Scheme 1). A number of catalysts have been shown to exhibit high selectivities in alkyne semi-hydrogenations including the Lindlar catalyst and more recently Pd/ZnO catalysts.

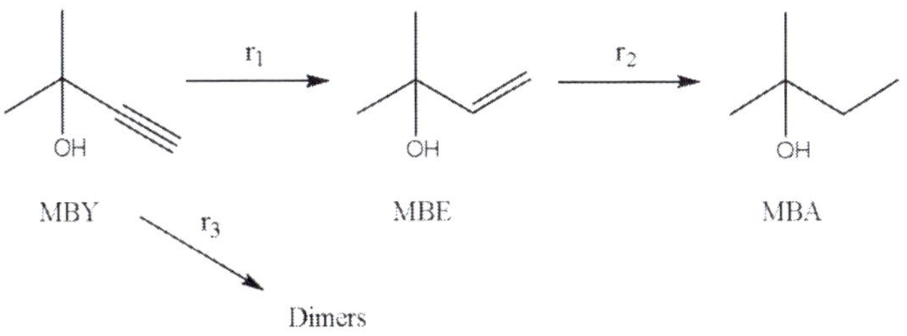

Scheme 1. Reaction network of MBY hydrogenation.

Previous studies of the selective MBY hydrogenation only led to kinetic models associated with specific operating conditions and simplified by the use of apparent kinetic constants. These models are unfortunately not able to predict the evolution of the system under a wide range of operating conditions and therefore are of limited use.

In this work, a general Langmuir-Hinshelwood-type kinetic model has been developed assuming non- competitive adsorption of the reactants/products on the catalyst active sites and dissociative adsorption of hydrogen. Initially the model was developed for the well-known Lindlar catalyst: a first order dependency with respect to hydrogen was found to describe the kinetic of the semi-hydrogenation reaction of MBY.

The kinetic regime was ensured with a quantitative analysis of the gas-liquid and liquid-solid mass transfer resistances. The gas-liquid mass transfer coefficient k_La and the hydrogen solubility in MBY were estimated by gas absorption experiments at different operating conditions. The liquid-solid mass transfer coefficients k_Sa_S were empirically evaluated.

An optimization procedure was developed to estimate the kinetic and the adsorption parameters (activation energies, pre-exponential factors and enthalpies of adsorption) governing the process. A good agreement between calculated and experimental concentration data with low standard deviations confirms the reliability of the proposed model.

The adsorption equilibrium constant for MBY was found to be significantly higher than that for MBE. This is consistent with the obtained high selectivity to MBE of the process using palladium-based catalysts. The estimated parameters were found in fair agreement with the values reported in the literature for the liquid phase hydrogenation of alkynes.

The mathematical model has been successfully validated using the best-estimated parameters to simulate the system behaviour under different operating conditions and during experiments conducted at lower initial concentration of MBY.

We now report an extension of the methodology to the recently developed Pd/ZnO catalysts which have significant advantages in terms of catalyst stability and being lead-free.

Total oxidation of ethanol over $Au/Ce_{0.5}Zr_{0.5}O_2$ catalysts

P. Topka, L. Kaluža, J. Gaálová
Institute of Chemical Process Fundamentals of the Czech Academy of Sciences, v. v. i., Rozvojová 2/135, 165 02 Praha 6, Czech Republic; tel. +420 220 390 282, e-mail: topka@icpf.cas.cz

Catalytic oxidation is a green and efficient way to handle man-made emissions of volatile organic compounds (VOC), which represent serious environmental and health issue. Despite high purchasing costs, supported noble metal catalysts are favoured over metal oxides due to their high activity and selectivity, excellent stability and superior resistance against poisoning.

Ceria attracted attention as VOC oxidation catalyst and support due to its unique property of storing and releasing oxygen. It can play the role of promoter and reducible support, provide lattice oxygen, increase the dispersion of supported metals and prevent their sintering or coke formation.

Gold supported on ceria has been reported as a good catalyst for benzene, toluene, ethanol, propanol, propene, formaldehyde and ethyl acetate oxidation. Recently, ceria-zirconia mixed oxides were proposed as VOC oxidation catalysts. It was shown that the partial substitution of cerium with zirconium leads to improved oxygen storage capacity, redox properties, thermal resistance, and better catalytic activity in oxidation reactions.

We have demonstrated that depending on the type of VOC, the introduction of gold can increase catalytic performance and/or selectivity of ceria-zirconia catalysts [1-3]. The aim of this work was to investigate in a detail the effect of Au loading on the catalytic performance of ceria-zirconia supported catalysts in the total oxidation of ethanol.

The$Ce_{0.5}Zr_{0.5}O_2$support was prepared from zirconium n-propoxide and of cerium(III)nitrate hexahydrate. The calcined support was impregnated with aqueous solution of gold(III)acetate, dried at 120 °C and calcined at 300 °C. The Au content in the catalysts according to ICP-AES analysis was 0.20, 0.37, 1.55 and 3.35 wt. %, respectively. Catalytic light-off experiments of ethanol gas-phase oxidation were carried out at 20 $m^3 \cdot kg^{-1} \cdot h^{-1}$ space velocitywith0.8 g/m^3 of ethanol in air.

The surface area S_{BET} was 45 m^2/g for the CeZr catalyst and decreased to 25 m^2/g for the 3.35Au/CeZr catalyst. With increasing gold loading, the Au particle size estimated from X-ray diffraction increased from 13 nm to 22 nm (Table 1). However, this decrease was not detrimental for the catalytic performance of gold catalysts (Table 1).

Table 1
Au particle size and catalytic performance of the catalysts in the total oxidation of ethanol

Catalyst	Au particle size* (nm)	Temperature of 50 % conversion (°C)
CeZr	-	191
0.20Au/CeZr	13	181
0.37Au/CeZr	17	182
1.55Au/CeZr	18	182
3.35Au/CeZr	22	161

* determined from X-ray diffractograms

Concluding, it was shown that the addition of gold can improve the catalytic performance of ceria-zirconia mixed oxide in ethanol oxidation. The positive effect of gold addition can be ascribed to improved redox properties of ceria-zirconia support due to the presence of Au nanoparticles.

References

[1] J. Gaálová, P. Topka, L. Kaluža, O. Šolcová, Catal. Today 175 (2011) 231-237.
[2] L. Matějová, P. Topka, L. Kaluža, S. Pitkäaho, S. Ojala, J. Gaálová, R.L. Keiski, Appl. Catal. B 142 (2013) 54-64.
[3] P. Topka, R. Delaigle, L. Kaluža, E.M. Gaigneaux, Catal. Today 253 (2015) 172-177.

Surface area-active phase reactivity correlation in scallop shells-derived solid base nanocatalysts in waste cooking oil transesterification

Frisda R. Panjaitan, [1]Shinya Yamanaka, [2]Yoshikazu Kuga
Muroran Institute of Technology, Division of Engineering, Mizumoto-cho 27-1, Muroran 050-8585, Japan
[1,2]Muroran Institute of Technology, College of Environmental Technology, Mizumoto-cho 27-1, Muroran 050-8585, Japan;
tel. +81 143 46 5747, e-mail: syama@mmm.muroran-it.ac.jp; fax: +81 143 46 5701.

A novel scallop shells-derived solid base nanocatalysts was prepared by nano-grinding method and applied to fatty acid methyl esters synthesis by transesterification waste cooking oil. Two types of solid base catalysts were obtained; CaO and Ca-glyceroxide active phases with moderate surface area. In this work, we studied: (i) prepared solid base catalysts with moderate surface area and CaO and Ca-glyceroxide active phases; (ii) compared waste cooking oil (free fatty acid content 10%) transesterification using CaO and Ca-glyceroxide active phases with moderate surface area and other catalyst (NaOH) at 40°C and 65°C, and (iii) investigated what is the true catalytic active phase because CaO drastically change its structure during transesterification reaction. Solid base nanocatalysts obtained were characterized by X-ray diffraction (XRD), thermal analysis (TGA/DTA), and BET. The results show higher surface area and active phase effect on increasing of the rate of transesterification. Large surface area of catalyst (Fig.1) was effective in reducing mass transfer limitation in the beginning of reaction that increasing the initial rate of transesterification. We also found that the real active phase that catalyzed the reaction is Ca-glyceroxide, therefore transformation CaO into Ca-glyceroxide became critical point in CaO catalyzed transesterification. With our most active prepared scallop shells-derived solid base nanocatalysts, 98.5% methyl esters yield was obtained in 60 min when 1.2wt% prepared CaO used at 65°C and 97.5 methyl esters yield in 60 min when 1.2wt% prepared CaO used at 40°C.

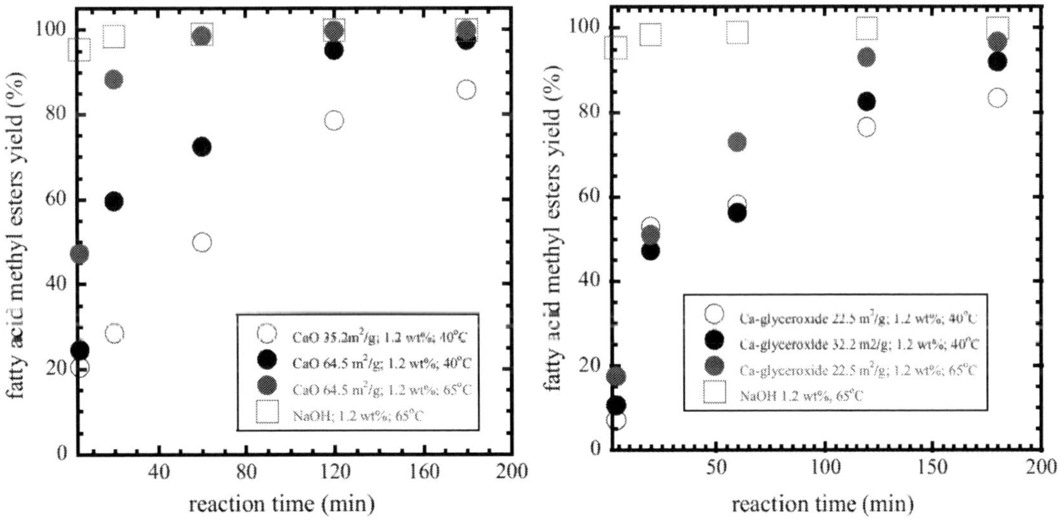

Fig.1 Yield of fatty acid methyl esters from the transesterification of waste cooking oil using scallop shells-derived solid base nanocatalysts at: molar ratio of oil:methanol (1:9); 1.2 wt% catalyst, temperature 40°C and 65°C.

A hollow core shell model for describing chemisorption of H_2S from raw syngas in a packed bed reactor of ZnO

R. Sadegh-Vaziri[1], M. U. Bäbler[1]
[1]Dept. Chemical Engineering and Technology, KTH Royal Institute of Technology, Stockholm, Sweden

Gasification of biomass produces raw syngas, or more precisely producer gas, which is a mixture of carbon monoxide, hydrogen, carbon dioxide and methane. Raw syngas contains a number of impurities such as hydrogen sulfide, ammonia, ash, and tars. These impurities cause damaging effects on downstream processes, such as Fischer-Tropsch or methanation processes which both aim at producing biofuels. Therefore, the impurities need to be removed in order for the syngas to be become a viable resource. Fig. 1 shows an overall view of biomass to fuel process and how gas cleaning fits in the process.

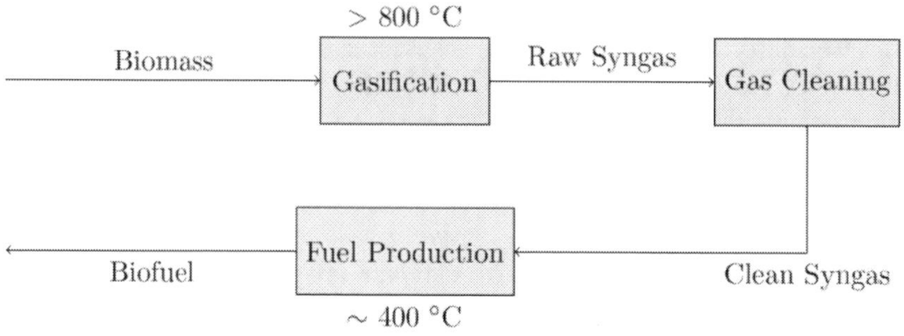

Figure 1: Schematic overview of biomass to fuel process [1].

Among the various impurities, hydrogen sulfide (and other sulfur compounds) is particularly crucial, as it has the potential to poison catalysts and also causes corrosion on process equipment. There exists several technologies for removing hydrogen sulfide, such as absorption, adsorption, and chemisorption. In chemisorption, gaseous hydrogen sulfide reacts with a solid metal oxide, which produces a solid metal sulfide and water; this removes hydrogen sulfide from the raw syngas by binding it to the solid phase. Chemisorption is typically run in a packed bed reactor at around 350-600°C [1], where solid metal oxides in form of pellets form a fixed bed through which the raw syngas is flowing. Zinc oxide is one of the popular options to be used in such chemisorption processes due to its thermodynamic characteristics and high capacity [2]. Recent studies show that nano-size particles of zinc oxide experience an outward growth when undergoing a chemical reaction with hydrogen sulfide leading to cavitation in the inside [3] [4]. This is opposite to the notion of the classical shrinking core model and grain model frequently adopted in the literature [5]. This motivated us to reconsider the modeling of the gas-solid reaction on the microscale and its implications on the process dynamics.

The chemisorption of H_2S on the packed bed of metal oxides has three distinct scales: (1) grain scale which is from few tens of nanometers to few hundred microns, (2) pellet scale which ranges from few millimeters to centimeters, and (3) bed scale which varies from few centimeters to meters. A schematic overview of theses scales is presented in Fig. 2.

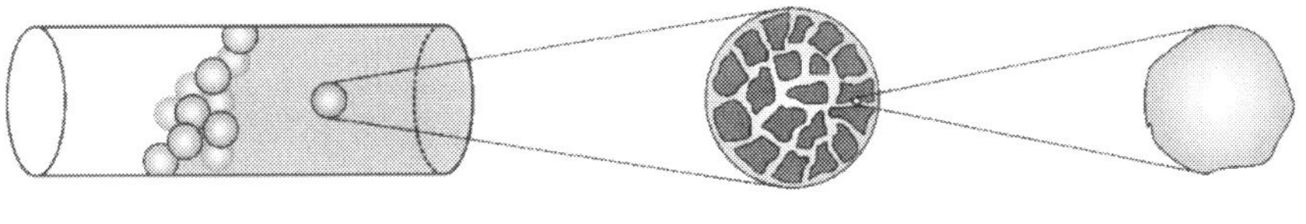

Figure 2: Different scales involved in the processs of chemisorption.

In this study a Hollow Core Shell Model (HCSM) is developed which describes the outward growth of the product layer and void formation inside the grains, and is compared with the inward growth in the classical shrinking core model. The void formation inside the ZnO grains is modelled by an Avrami-like rate expression [6]. To model the pellet scale, the developed HCSM from the grain scale is implemented in a reaction diffusion model to obtain the gas and solid concentration profile inside a pellet. The pellet model is used to show the phenomena of pore clogging inside the pellets, which translates into resistance to achieve full conversion in the pellet scale. Thereafter, a packed bed filled with pellets is modelled based on the HSCM. The effect of pore clogging on the breakthrough curves is also investigated.

References

[1] S. Cheah, D. L. Carpenter and K. A. Magrini-Bair, "Review of mid-to high-temperature sulfur sorbents for desulfurization of biomass and coal-derived syngas," *Energy Fuels,* vol. 23, no. 11, p. 5291–5307, 2009,.

[2] I. Rosso, C. Galletti, M. Bizzi, G. Saracco and V. Specchia, "Zinc oxide sorbents for the removal of hydrogen sulfide from syngas," *Ind. Eng. Chem. Res.,* vol. 42, no. 8, pp. 1688-1697, 2003.

[3] L. Neveux, D. Chiche, D. Bazer-Bachi, L. Favergeon and M. Pijolat, "New insight on the ZnO sulfidation reaction: Evidences for an outward growth process of the ZnS phase," *Chem. Eng. J.* , vol. 181, pp. 508-515, 2012.

[4] L. Neveux, D. Chiche, J. Perez-Pellitero, L. Favergeon, A. S. Gay and M. Pijolat, "New insight into the ZnO sulfidation reaction: mechanism and kinetics modeling of the ZnS outward growth," *Phys. Chem. Chem. Phys.,* vol. 15, pp. 1532-1545, 2013.

[5] Y. Fan, V. Rajagopalan, G. E. Suares and M. D. Amiridis, "Use of an "Unreacted Shrinking Core" Model in the Reaction of H2S," *Ind. Eng. Chem. Res.,* vol. 40, no. 22, pp. 4767-4770, 2001.

[6] M. C. Weinberg, D. P. Birnie and V. A. Shneidman, "Crystallization kinetics and the JMAK equation," *J. Non-Cryst. Solids* , vol. 219, pp. 89-99, 1997.

Cooking potential of oxychlorination hollow cylindrical catalyst pellet

K. R. Rout, [1] E. Fenes, [1] M. F. Baidoo, [1]T. Fuglerud, [2] De Chen [1]
Department of Chemical Engineering, Norwegian University of Science and Technology, Norway,[1] INOVYN Norge AS, Norway [2]

Oxychlorination of ethylene is a part of the process for production of vinyl chloride monomer (VCM) from ethylene [1]. VCM is the monomer for production of polyvinyl chloride (PVC) that is the third-most widely produced thermoplastics worldwide. Gaseous ethylene, HCl and air react together with $CuCl_2/\gamma\text{-}Al_2O_3$ catalyst in a fixed- or fluid bed reactor of the ethylene oxychlorination process at a temperature range of 217-257 ^{0}C and a pressure 5-6 atm following Eq. 2 [2].

$$C_2H_4 + 2HCl + \frac{1}{2}O_2 \rightarrow C_2H_4Cl_2 + H_2O \quad (1)$$

The $CuCl_2$ based catalyst is commonly used in the oxychlorination process [1], which is a typical oxidative reaction. It is generally agreed that the oxychlorination involves a redox process in which copper cycles between Cu^{2+}- and Cu^{1+} states [3]. One of the main challenges of this process is that the Cu^{1+} deposits on the surface of the catalyst during the reaction and thereby caused aggregation, and loss of Cu active materials due to its low melting temperature and volatility [3]. Furthermore, oxychlorination process is an exothermic process. Therefore, it is important to avoid high temperatures in the reactor, which will cause side reactions and catalyst deactivation. Due to deactivation in the industrail oxychlorination reactor, a carbon ring is formed in the inner part of the hollow cyclinderical pellet as shown in Fig. 1. Due to the formation of carbon ring, the mechanical strength of the oxychlorination catalyst decreases and it breaks. Hence, an attempt has been made to perform a detailed 2-dimensional hollow cyclindrical pellet modelling to find out the cooking potentiallity of the oxychlorination catalyst.

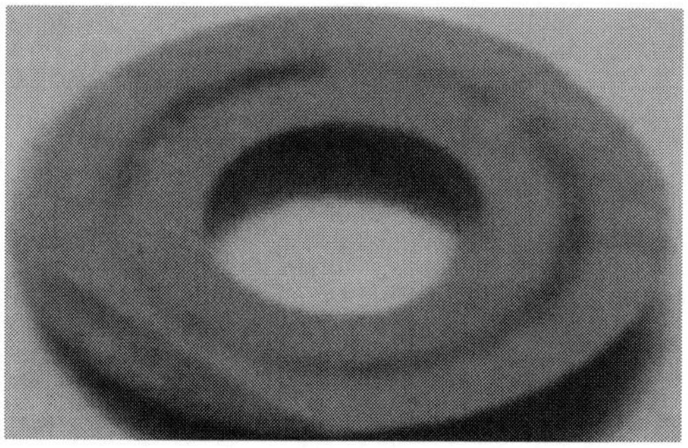

Fig. 1. Carbon ring of the hollow cyclindrical catalyst pellet

An novel hollow cylindrical 2-dimensional pellet model is derived and solved to calculate concentration profiles of different components from the center of the cylinder to the surface. Normally, pellet model is solved along the radial direction of catalyst pellet [4]. But, the flow along the inner hollow part may affect the diffusion along the radial direction. Therefore, in this study 2-dimensional (radial and axial direction of catalyst pellet) pellet model is solved. The developed model is validated qualitatively by the catalyst pellet obtained from the plant.

It is found that the place where oxygen concentration is low, the carbon starts to form. A relationship between the carbon to oxygen is generated from the simulated data and a generic mapping is done in order to show the carbon prone zone of the catalyst pellet.

Reference:

1. Lamberti, C., Prestipino, C., Bonino, F., Capello L., Bordiga, S., Spoto, G., Zecchina, A., Moreno, S., Cremaschi, B., Garilli, M., Marsella, A., Carmello, D., Vidotto S., Leofanti, G., Angew Chem Int Ed 2002. **41:2341**.

2. Laurer, P., Krome, G., Cordemans, L., Seifert, R., Danz, E., Eur. Pat., 1981. **54674**.

3. Rouco, A., Journal of Catalysis, 1995. **157:380**.

4. Rout K. R., Solsvik J., Nayak A. K., Jakobsen H. A. A numerical study of multicomponent mass diffusion and convection in porous pellets for the sorption-enhanced steam methane reforming and desorption processes. Chemical Engineering Science (2011) Volume 66 (18): 4111-4126

An industrial scale process for the direct synthesis of DMC from methanol and CO_2

H.M. Schellevis, S.A.L. Jorissen, T.D.J. Te Molder, A.G.J. van der Ham, H. van den Berg
Faculty of Science and Technology, University of Twente, P.O. Box 217, 7500 AE, Enschede, The Netherlands. h.m.schellevis@student.utwente.nl

In this research, a process is developed for the direct synthesis of dimethylcarbonate (DMC) from methanol and CO_2. DMC can be used for different purposes, mainly as anti knocking agent to replace ethanol and/or MTBE. Conventional routes are either a multiple step synthesis, involve hazardous chemicals or suffer from low single pass conversion and therefore a novel direct route is desired. Besides the benefit of a direct synthesis, this route, using CO_2 as a reactant, is also more sustainable than the conventional alternatives.

However, the equilibrium of this reaction is very unfavourable, achieving not even 1% conversion at high pressure. Therefore, a novel strategy is essential to enhance the conversion in order to develop a viable process. A literature study was conducted and the removal of water (the other reaction product) by a dehydrating agent was found to be the best method for shifting the equilibrium towards the product side. 2-cyanopyridine was selected based on the studies of Bansode et al and was the basis for this process design. Here, both reactions, the DMC synthesis and water removal, are catalyzed by a cerium oxide catalyst, in this configuration the single pass conversion can be increased to over 90% with a selectivity of 99%.

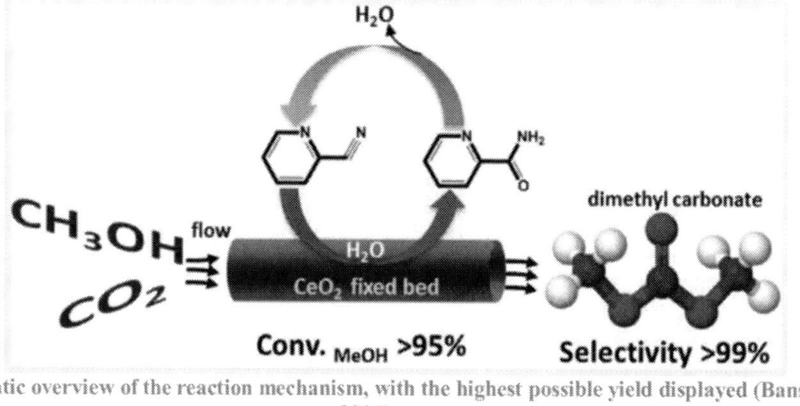

Figure 1: Schematic overview of the reaction mechanism, with the highest possible yield displayed (Bansode et al., 2014)

The process was designed in a systematic manner, starting with a conceptual design phase, where required process functions were defined, followed by assigning unit operations to the functional blocks. The process simulator UniSim Design® was used in the next level, where mass and energy flows were calculated to underpin and quantify decisions. In addition, the simulation was used to optimize the process. All major equipment was designed, which included a variety of reactors, distillation columns, flash vessels, compressors, pumps and heat exchangers. The research was finalized with an technical and economic evaluation including CAPEX and OPEX calculations.

The methanol and CO_2 react to DMC in a multi-tubular trickle bed reactor with a conversion of 92.4%. Optimal (inlet) temperature and pressure are found to be 120°C and 30 bar respectively. 97% of the CO_2, which is present in excess, can easily be recovered with two flash vessels, only suffering from a little pressure drop, minimizing compression costs. 2-picolinamide is separated by low pressure distillation (0.5 bar), regenerated and recycled, together with the unreacted 2-cyanopyridine, to the reactor. Finally, DMC is purified from the mixture containing also methanol and a small amount of CO_2. First normal distillation is applied to recover most of the DMC in the bottom stream. The mixture of DMC and methanol at near azeotropic composition, obtained in the top stream, is separated in the following step by extractive distillation with phenol as a solvent.

Little is known about the regeneration of 2-picolinamide to 2-cyanopyridine. It is known from literature that the reaction is reversible, but the time scale reported is way too long for industrial operations. Therefore this part of the process is considered a gray box. From the economics it is shown that, with the current 2-cyanopyridine price, not more than 1.5% of dehydrating agent can be lost in one cycle, in order to remain profitable. Therefore, further research on this regeneration is essential.

Overall a technically feasible process is designed for the production of DMC from methanol and CO_2. The equilibrium is successfully shifted to a single pass conversion of over 90% and an excellent selectivity is maintained. The mass balance of the designed process shows nicely that large streams are separated first. In this sequence the stream size to the following unit operation is drastically reduced, therefore reducing the equipment size. The success will strongly depend on the regeneration of the 2-picolinamide to 2-cyanopyridine in both technical and economical aspects.

An economic evaluation of the process, without taking the regeneration section into account, shows very attractive results. The payback period (0.5 yr) and return of investment (84%) are excellent. A sensitivity analysis was carried out to determine the impact of certain variables, it shows that besides the dehydrating agent recovery, also the price of methanol influences the profitability significantly. Fortunately the methanol price was relatively stable over the past years. On the other hand, the frequency of catalyst replacements and the estimated capital investment for the regeneration section do not have a large effect on the potential profitability of the plant.

In conclusion, the usage of 2-cyanopyridine as dehydrating agent potentially solves the problem of a low single pass conversion for the direct synthesis of DMC. Although, the regeneration of the 2-picolinamide to 2-cyanopyridine had to be considered as a gray box due to a lack of information. Therefore additional research is needed to investigate and improve this regeneration reaction. Overall an attractive process for the direct DMC production from CO_2 and methanol can be obtained.

Bansode, A and A. Urakawa, Continuous DMC synthesis from CO_2 and methanol over a CeO_2 catalyst in a fixed bed reactor in the presence of a dehydrating agent, ACS Catalysis, 2014, 4(11), p. 3877-3880

Comparison of oxygenated vs. chlorinated compounds in the production of light olefins from a coke deactivation perspective

J. Valecillos-Diaz, M. Ibáñez, P. Pérez-Uriarte, M. Gamero, I. Hita, A.T. Aguayo, J. Bilbao, P. Castaño
Chemical Engineering Department, University of the Basque Country UPV/EHU, P.O. Box 644, 48080 Bilbao, Spain; Telephone: +34 946015361; e-mail: jose.valecillos@ehu.eus

Introduction

Light olefins are the main building blocks of the petrochemical industry. On the other hand, the increase of natural gas reservoirs is driving attention to routes for transforming methane into chemicals and particularly light olefins. In this regard, several technologies are available and among them, industrially interesting ones [1] are (1) gasification, methanol or dimethyl ether synthesis and these to light olefins (MTO and DTO), and (2) chlorination, and chloromethane to light olefins (CTO). MTO and DTO have shown a very high activity, selectivity and stability using a HZSM-5 zeolite catalyst [2,3]. CTO has shown a higher selectivity to light olefins [4], but lower conversions in comparison to MTO under similar conditions and over the same catalyst [5]. This work compares MTO, DTO and CTO on the same HZSM-5 zeolite catalyst and identical conditions in terms of performance and, particularly, coke deactivation of the catalyst, as critical parameters for their comparison.

Experimental

The MTO, DTO and CTO were performed in a fixed bed reactor using the same conditions: 400 °C, 1.5 atm, 15 mol% of He, time on stream up to 4 h, and space time of 6.5 g h mol_{CH2}^{-1}. The catalyst was prepared agglomerating 25 wt% of HZSM-5 ($SiO_2/Al_2O_3 = 80$), with 30 wt% of bentonite and 45 wt% of α-Al_2O_3. Then, the particles were extruded, dried, crushed, sieved (0.15-0.3 mm) and calcined at 575 °C for 3 h. Both the fresh and spent catalysts were characterized by N_2 isotherms. The spent catalysts were analyzed by temperature-programmed oxidation (TPO) joint with thermogravimetry or FT-IR and mass spectroscopies, and Raman spectroscopy.

Results and conclusions

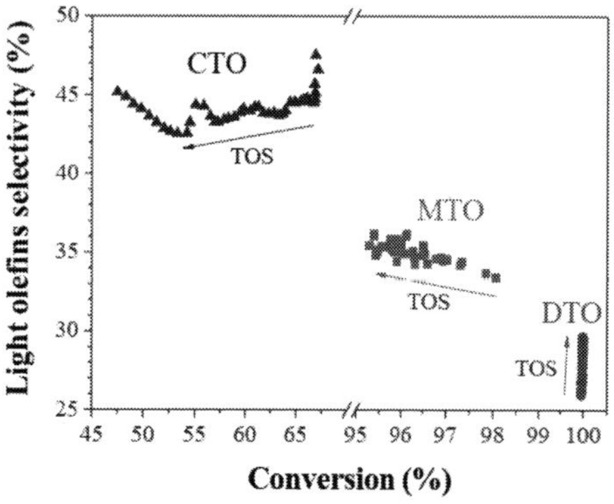

Figure 1. Evolution of conversion and C_2-C_4 olefin selectivity for MTO, DTO and CTO processes at 400 °C.

Figure 1 depicts the evolution of conversion and selectivity of light olefins in the three processes studied: MTO, DTO and CTO. Both MTO and DTO show stable conversions as the reactions progress, with an increase of light olefins selectivity. DTO exhibited c. 100% conversion during the whole reaction time while MTO conversion tends to slightly decrease with time on stream. In contrast, CTO conversions are much lower showing a faster deactivation, but light olefins selectivity is much higher during the whole reaction time. However, in terms of light olefins yield after 240 min, this parameter (in mol% of CH_2) is higher for oxygenated compounds (33.8 and 29.7 for MTO and DTO, respectively) whereas it is lower for CTO (21.5).

The pore volume (in cm^3 g^{-1}) decreased in the order: 0.338 (MTO) > 0.265 (DTO) > 0.181 (CTO). This is due to coke build up in catalyst pores, indicating a major mesopore volume loss for CTO spent catalyst and then causing a significant blocking effect of reactants to micropores (zeolite). This phenomenon is attenuated in MTO and DTO due to the presence of steam in the reaction medium. Figure 2a shows the TPO profiles of the coke deposited on spent catalysts of MTO, DTO and CTO. Coke deposited during CTO shows two combustion peaks (Figure 2a), suggesting the presence of two types of coke with different chemical nature. Coke from MTO and DTO is more homogeneous, with a predominant combustion peak. Figure 2b shows the Raman spectra of the coke deposited on the catalysts assigning D (1370 cm^{-1}) and G (1605 cm^{-1}) bands to characteristic of poorly or highly ordered aromatic structures, respectively [6]. The ratios of D/G bands, also indicative of the state of coke, are similar for the coke deposited on the catalyst used in the MTO and DTO, whereas shows two distinctive ratios for the coke deposited on the catalyst used in the CTO. The presence of two different coke structures by TPO and Raman spectroscopy formed during CTO leads us to conclude that there is a wide spatial distribution of the nature of coke deposited on the catalyst in this transformation, leading to relatively higher deposition of external (to the zeolite) and developed coke [7].

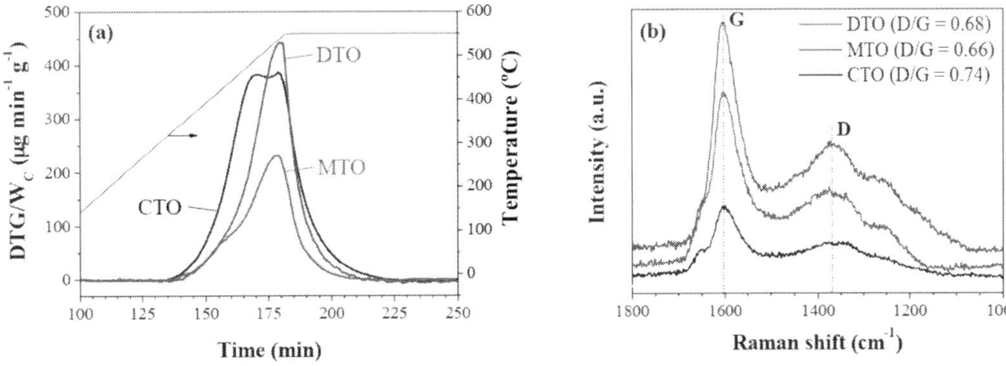

Figure 2. Coke combustion by TGA (a) and Raman spectra of the spent catalysts (b).

The selectivity of light olefins during CTO is higher in terms of the CH_2 units converted, which may reduce the costs of the downstream purification processes. The relative slower deactivation during MTO and DTO allows longer operational times, by contrast, CTO forms an additional type of external coke that blocks accessibility faster to the acid sites of the catalyst, which is related to the absence of steam in the reaction medium.

Acknowledgement: This work was financially supported by the Ministry of Economy and Competitiveness of the Spanish State General Administration (CTQ2013-46172-P and BES-2014-069980) and the Basque Government (UFI 11/39 UPV/EHU and BFI-2012 203).

References

1. U. Olsbye, S. Svelle, M. Bjørgen, P. Beato, T. Janssens, F. Joensen, S. Bordiga, K. Lillerud. Angew. Chem. Int. Ed. 51 (2012) 5810-5831.
2. A.T. Aguayo, P. Castaño, D. Mier, A. Gayubo, M.Olazar, J. Bilbao. Ind. Eng. Chem. Res. 50 (2011) 9980–9988.
3. A.S. Al-Dughaither, H. de Lasa. Fuel 138 (2014) 52-64.
4. M. Gamero, A.T. Aguayo, A. Ateka, P. Pérez-Uriarte, A. Gayubo, J. Bilbao. Ind. Eng. Chem. Res. 54 (2015) 7822-7832.
5. U. Olsbye, O. Saure, N. Muddada, S. Bordiga, C. Lamberti, M. Nilsen, K. Lillerud, S. Svelle. Catal. Today 171 (2011) 211-220.
6. J. Robertson. Mater. Sci. Eng. R 37 (2002) 129-281.
7. M. Ibáñez, M. Gamero, J. Ruiz-Martínez, B.M. Weckhuysen, A.T. Aguayo, J. Bilbao, P. Castaño. Catal. Sci. Technol.. 6 (2016) 296-306

Glycerol dehydration to acrolein in a two-zones fluidized-bed reactor

R. Meléndez[1], S. Paul[1], N. Fatah[1], B. Katryniok[1], V. Bellière-Baca[2], P. Rey[2], F. Dumeignil[1]
[1]Univ. Lille, CNRS, Centrale Lille, ENSCL, Univ. Artois, UMR 8181 - UCCS - Unité de Catalyse et Chimie du Solide, F-59000 Lille, France – Tel : 00 33 (0)3 20 33 54 57 – Mail : sebastien.paul@ec-lille.fr
[2]ADISSEO France SAS, Antony Parc 2-10, 92160 Antony, France

1. Introduction

The catalytic dehydration of glycerol to acrolein has become into the focus, due to the role of acrolein as a key-intermediate for the production of high added-value industrial chemicals such as methionine [1,2]. From the literature, three families of catalysts are widely studied for this reaction: zeolites, Keggin-type heteropolyacids (HPAs) and mixed-metal oxides. Nowadays, most of the researches are aiming at developing high performance catalysts with increased resistance to the deactivation phenomenon due to the formation of coke, whereas very few studies are focused on the development of an innovative process for continuous regeneration of the catalyst. Here within we report the use of a Two-Zones Fluidized-Bed Reactor (TZFBR) for the continuously removal of coke by means of thermal oxidation. Previously Menendez *et al.* employed this technology to mitigate the coke formation in the dehydrogenation of propane [3,4]. For the application of the TZFBR, glycerol and oxygen are injected separately. Oxygen is fed at the bottom of the reactor in the fluidizing gas and reacts with the coked catalyst, while glycerol is injected via a feed tube into the fluidized catalytic bed, allowing thus the separation of the fluidized catalytic bed into two well-defined zones: one, at the bottom, for the catalyst regeneration and the other one, at the top, for the glycerol dehydration. The circulation of the catalyst particles between the two zones due to the movement of the solid - a feature of fluidized beds - enables a continuous regeneration of the catalyst and, at the same time, the acrolein production in a single reactor. In this project, the glycerol dehydration was performed over a heteropolyacid catalyst supported on silica (20 wt.% $H_3PW_{12}O_{40}/SiO_2$) in such a TZFBR. The hydrodynamic behaviour of the fluidized bed and the influence of the operating parameters on the catalytic performance were studied.

2. Results and discussions

The fluidization of the catalyst was studied at 275 °C using 96 g of catalyst (corresponding to a 10 cm fixed bed height). Fig. 1 (left) shows that the minimum fluidization velocity (U_{mf}) of the catalyst is 8.9 10^{-3} m/s. Additionally, the results exhibit that the setup permits a good mixing and fluidization. This is illustrated by the fluidization quality, which is always superior to 80 %. The oxygen / glycerol molar ratio has a strong impact on the catalytic performance. From Fig. 1 (right), it can be observed that the increase of molar ratio (MR) from 0.3 to 2.4 favours the regeneration of the catalyst, obtaining thus a complete and stable conversion for a ratio of 2.4 with an acrolein selectivity of 60 % after 52 h.

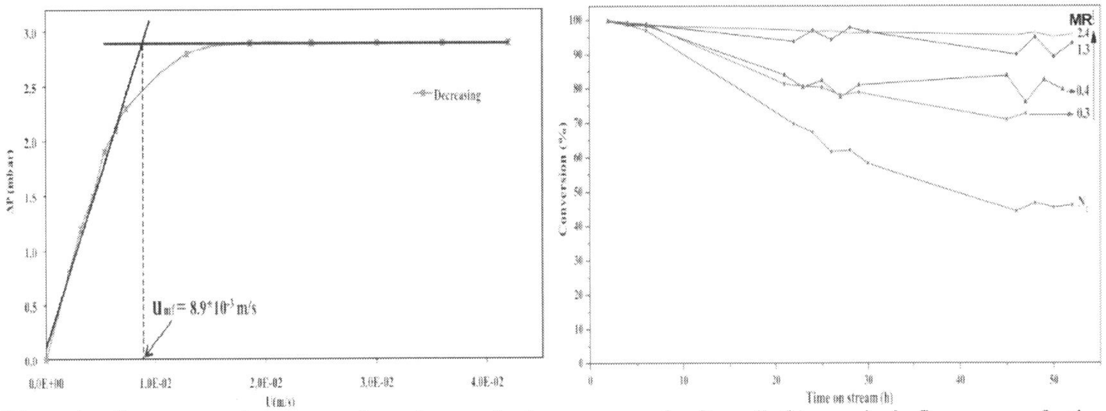

Fig. 1. Pressure drop as function of the gas velocity (left) and influence of the oxygen/glycerol molar ratio (MR) on the glycerol conversion (right) at 275 °C

3. Conclusions

The use of a TZFBR for carrying out glycerol dehydration to acrolein at 275 °C is successful using a height of 8 cm for the regeneration zone and an oxygen/glycerol molar ratio of 2.4. In these conditions the catalyst regeneration is constantly achieved and a glycerol conversion of 100 % together with a selectivity to acrolein of 60 % can be maintained during more than 50 h.

Acknowledgements

Adisseo and ANR are greatly acknowledged for having funded this work.

References

[1] B. Katryniok, S. Paul, F. Dumeignil., ACS Catal. 3, 1819–1834 (2013).

[2] A. Talebian-Kiakalaieh, N. A. Saidina Amin, H. Hezaveh., Renew. Sustain. Energy Rev. 40, 28–59 (2014).

[3] J. Gascón, C. Téllez, J. Herguido & M. Menéndez., Appl. Catal. Gen. 248, 105-116 (2003).

[4] J. Gascón, C. Téllez, J. Herguido, M. Menéndez., Chem. Eng. J. 106, 91-96 (2005).

Guidelines for proper reactor design for aldehyde-ammonias production

E. Moioli[1,2], L. Schmid[2], P. Wasserscheid[1], H. Freund[1]

[1]Friedrich-Alexander-Universität Erlangen-Nürnberg (FAU), Erlangen, Germany
[2]Lonza AG, Visp, Switzerland

The condensation of aldehydes and ammonia at high temperature is a well-known reaction in organic synthesis, which allows preparing nitrogen containing carbon chains. The reaction of aldehydes and ammonia also takes place at low temperature, spontaneously forming molecules usually referred to as aldehyde-ammonias (aldehyde-ammonia trimers). The study of the relationship between the two reactions, formation of aldehyde-ammonias and condensation to final products, can open the way for improved reactor design and operation with improved management of reaction exothermicity and separation of products. The scope of this paper is to define general guidelines that should be taken into account in the design of a reactor for aldehyde-ammonia reactions.

The first scope of this study is the fundamental understanding of the effect of the ammonia:aldehyde ratio for reactions involving aliphatic aldehydes. For aldehydes from C_2 to C_4 a common trend is found. Their common characteristic is the presence of a threshold level of ammonia excess to reach 100 % yield in the trimer product (Figure 1a). This threshold value decreases upon increasing the carbon content of aldehydes. The aldehyde not converted to trimer is lost in long chain polymers.

For the case of formaldehyde, a different behaviour is observed. This molecule forms a different structure, called urotropine (hexamethylenetetramine, $C_6H_{12}N_4$). The yield of urotropine is 100 % at a CH_2O:NH_3 molar ratio equal to 1, but tends to decrease sharply with an excess of ammonia. This represents a reversed trend as compared to the higher aldehydes. As a direct consequence, it is necessary to apply a different ammonia dosing strategy if formaldehyde is required by the reaction scheme for the desired product. When formaldehyde is used as synthetic building block together with higher aldehydes (e.g. acetaldehyde), the order of addition of raw materials added to the reaction mixture is crucial; a conceptual mismanagement in this reaction phase can cause the loss of large quantities of synthetic building blocks in the aldehyde-ammonia reactor.

In some applications, a basic pH can be undesirable because of the particular properties of co-reactants or products. For this reason, it can be necessary to add some acid to ammonia, in order to reach the desired pH. The first insight in this topic was the description of reaction of acetaldehyde with ammonia in the presence of acids. Three different acids have been chosen as reference case: hydrochloric acid, acetic acid and oxalic acid. The results in terms of trimer and acetaldehyde molar fraction are reported in Figure 2. At high pH the influence of acid is negligible; trimer can be quantitatively formed in all three cases. Decreasing pH causes a trimer concentration drop at a pH corresponding to the complete conversion of ammonia into ammonium salt. This value is different for the three acids, according to their pKa values. At lower temperature, the excess of acid tends to inhibit the reaction of acetaldehyde; this maintains low the concentration of trimer. The best results are obtained with oxalic acid. The inhibition of reactions results in the presence of high quantities of unreacted acetaldehyde. As a first screening to extend the results obtained with acetaldehyde to other aliphatic aldehydes, the three above-mentioned ammonium salts were tested in the reaction with formaldehyde, propionaldehyde and butyraldehyde. The results are shown in Figure 1b.

As observed before, formaldehyde shows a different reactivity than the higher aldehydes. Formaldehyde forms urotropine with about 100 % yield with all the three salts considered. Higher aldehydes form trimers only in limited amounts, with a minimum quantity when using ammonium acetate. The increase in the carbon atom number in the aldehydes appears to reduce the specific effect of the salt type on the reaction. This is due to the increase in the aliphatic character of the molecules, and the consequent decrease of the influence of the C=O bond on the entire molecule. Higher aldehydes show the presence of large quantities of unreacted aldehydes in the case of ammonium chloride and oxalate. Ammonium acetate, in contrast, is a catalyst for the polymerization of aldehydes, causing their irreversible loss (Figure 1c). These results confirm the analogy between acetaldehyde and higher aldehydes.

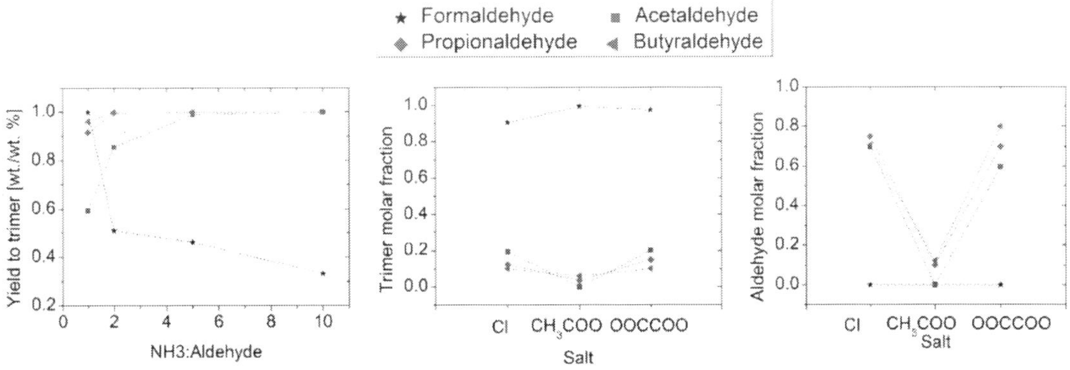

Figure 1. Reactivity of aliphatic aldehydes: a) reaction with ammonia; b) trimer molar fraction in ammonium salts; c) aldehyde molar fraction in ammonium salts.

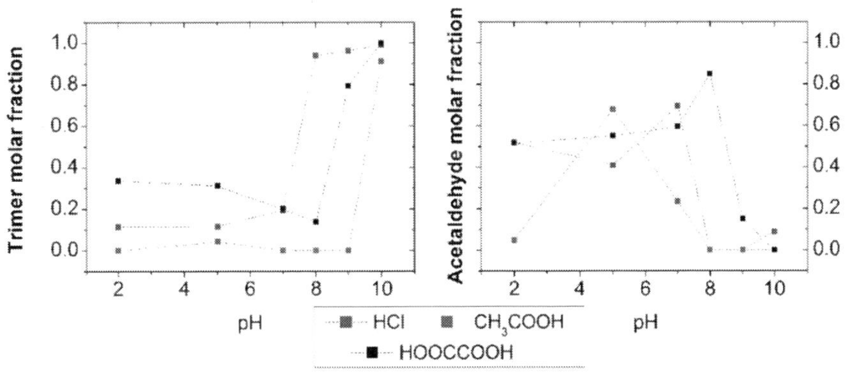

Figure 2. Trimer (a) and acetaldehyde (b) molar fraction at different pH.

These results are important to determine the minimum amount of ammonia to be used in a pre-mixer to avoid aldehyde loss. At the same time, these general guidelines allow for deriving the best dosing strategy to be applied when different aldehydes are to be used in the reaction scheme. General guidelines are also provided when using ammonium salts instead of aqueous ammonia. In this case, apart from formaldehyde which quantitatively forms urotropine, it is impossible to obtain large quantities of trimer because most of the aldehyde fed does not react with ammonium chloride and oxalate. The case of ammonium acetate is different because the reaction of C2-C4 aldehydes with the latter results in an irreversible formation of polymerization products. An extension of the results of pH dependency for acetaldehyde ammonia trimer formation will be evaluated in detail, to allow for drawing precise analogies and differences among the various aldehydes investigated in this study.

Preparation of valuable organic chemicals through Heck reaction by using Supported Ionic Liquid Catalyst (SILCA)

N. Vucetic[1], P. Virtanen[1], J.-P. Mikkola[1,2], T. Salmi[1]
1. Åbo Akademi University, Johan Gadolin Process Chemistry Centre, Laboratory of Industrial Chemistry and Reaction Engineering, Biskopsgatan 8, FI-20500, Turku, Finland.
2. Umeå University, Chemical-Biological Centre, Department of Chemistry, Technical Chemistry, SE-90187 Umeå, Sweden
e-mail: n.vucetic@abo.fi.

The palladium-catalysed reaction between aryl halides or vinyl halides and activated alkenes in the presence of a base is referred to as the "Heck Reaction". The great benefit of the Heck reaction is that it enables the creation of a carbon-carbon coupling which is a difficult task in organic synthesis.

Under homogeneous conditions ionic liquids have showed great potential in Heck reaction due to the fact that they are highly polar, thus facilitating the cationic mechanism of Heck reactions, and can contribute to the stabilization of non-associated Pd(0) species by forming anionic complexes with halide ions [1,2].

To overcome the problem which brings homogeneous catalysts associated with the handling of sensitive ligands and difficulty of catalyst recovery, it is highly desirable to develop heterogeneous catalysts for industrial applications which allow recycling of expensive Pd. Successful method includes supporting Pd complexes and an ionic liquid (IL) to various solids and avoiding the use of toxic and expensive ligands.

Direct immobilization of an IL with dissolved Pd active component onto SiO_2 was carried out by evaporation-impregnation techniques. Typically, defined mass of Pd precursor ($PdCl_2$ or $Pd(AcAc)_2$) and ionic liquid were dissolved in a solvent (acetonitrile, ethanol) with addition of THF. Then SiO_2 (125 - 250 μm particles) was added and a slurry was obtained. The slurry was mixed from 1,5 - 12 h and then dried at 30 - 40 °C under vacuum for 2 - 3 h. The yellowish powder obtained, was rinsed few times with diethyl ether and dried.

A Heck reaction of brombenzene and styrene was chosen as the model reaction. The reaction system is supposed to give trans-stilbene as the main product, and cis-stilbene and 1,1-diphenylethene as by products. The experiments were carried out in a batch reactor with a magnetic stirrer. The substrates and solvent were inserted into the reactor after which the reactor was sealed and heated up under vigorous stirring. When the temperature reached up to 110 - 150 °C, the Pd-SILCA was added and the reaction was preformed 8 - 24 h with periodical sampling for kinetic study. The samples were analysed by gas chromatography.

The contribution of many different factors during the course of the reaction was studied. The influence of different ionic liquids in terms of selectivity and yield was studied. Also, the choice of the solvent (n-decane, DMF, DMA), gas atmosphere (N_2, O_2) and the presence of bases (in reaction mixture or impregnated on SILCA) can influence the productivity as well as the selectivity of the reaction.

References

1. Li, S., Lin, Y., Xie, H., Zhang, S., Xu, J., Brønsted Guanidine Acid-Base Ionic Liquids: Novel Reaction Media for the Palladium-Catalyzed Heck Reaction, Org. Lett., 8: 391-394, 2006.

2. Petrovic, Z., Simijonovic, D., Petrovic, V., Markovic, S., Diethanolamine and N,N-diethylethanolamine ionic liquids as precatalyst-precursors and reaction media in green Heck reaction protocol, J. Mol. Cat. A: Chem., 327: 45–50, 2010.

Study of the degradation of oxygen carrier pellets in the chemical looping combustion process

E. Schaer, A. Tilland, D. Petitjean
Université de Lorraine, Laboratoire Réactions et Génie des Procédés, ENSIC, 1 rue Grandville, BP 20451, 54001 NANCY Cedex, FRANCE
tel. +33 383175304, email : Eric.Schaer@univ-lorraine.fr

The Chemical Looping Combustion (CLC) is an oxy-combustion carbon dioxide capture process using an oxygen carrier. The combustion is realized in two steps : The first one is the combustion of methane on the oxygen carrier (reduction step), the second one is the regeneration (oxidation) of the reduced material with air. Diminution of the oxygen carrier performances after some cycles of reduction and oxidation are frequently observed at industrial scale, where two fluidised beds are used for the reduction and oxidation reactions [1, 2].

The objective of this study is to understand the phenomena responsible for the decrease of the process performances. The methodology is applied to nickel oxide supported by nickel aluminate ($NiO/NiAl_2O_4$) as oxygen carrier, since it has been the subject of many studies and is known as a reference material [3, 4].

Chemical Looping Combustion experiments have been here performed in a fixed bed reactor, to determine the relation between oxygen carrier structure changes (variation of porosity, chemical composition, structural organisation…) and process performance evolutions (variation of methane conversion and oxygen carrier regeneration rate …)

The variation of the structure and physical properties of the oxygen carrier material, such as composition (measured by X-ray diffraction), specific surface area (measured by BET method), and particle size distributions (deduced from scanning electron microscopy) are related to the variations of the process performances (evolution of products composition and metal oxide reduction and oxidation) for different cycles of chemical looping combustion.

Observations shows that :

- the oxygen carrier material pellets, of several hundreds of microns, are composed of elementary particles, of micrometric sizes,
- reduction and oxidation reactions induce erosion and aggregation of the pellets, even in a fixed bed reactor, in absence of any friction between the particles,
- the proportions of nickel, nickel oxide and nickel aluminate varies against reactions cycles, reflecting some structural modification in the oxygen carrier composition during the reactions,
- the specific surface area remains constant, showing that the reactivity depends on the elementary particles which form the pellets of oxygen carrier material.

Analysis of the observations allows the development of a model accounting for the ageing of the oxygen carrier material:

- reduction of the oxygen carrier material causes a contraction of the elementary particles, which induces their erosion. This erosion phenomenon is enhanced by carbon deposition between the elementary particles, and by the reaction with the nickel aluminate support. In industrial conditions, these elementary particles may be blown out of the fluidised bed, inducing a loss of performances.

- oxidation of the elementary particles and carbon elimination cause aggregation of the oxygen carrier pellets, inducing apparition of larger and breakable pellets.

In conclusion, reduction and oxidation reactions, together with carbon deposition, have an impact on the structure of the oxygen carrier materials. Solutions to limit the evolution of the pellets, and thus prevent the efficiency loss observes at industrial scale, will finally be proposed.

References:

[1] Anders Lyngfelt, Bo Leckner and Tobias Mattisson : A fluidized-bed combustion process with inherent CO2 separation ; application of chemical-looping combustion. Chemical Engineering Science, 56:3101–3113, 2001.

[2] Juan Adánez, Alberto Abad, Francisco García-Labiano, Pilar Gayán and Luis Francisco de Diego : Progress in chemical-looping combustion and reforming technologies. Chemical Engineering Science, 38:215–282, 2012.

[3] Laihong Shen, Jiahua Wu, Zhengping Gao and Jun Xiao : Reactivity deterioration of NiO/Al_2O_3 oxygen carrier for chemical looping combustion of coal in a 10 kWth reactor. Combustion and Flame, 156:1377–1385, 2009.

[4] Lucia Blas, Sophie Dorge, Laure Michelin, Patrick Dutournié, Arnold Lambert, David Chiche and Stéphane Bertholin : Influence of the regeneration conditions on the performances and the microstructure modifications of $NiO/NiAl_2O_4$ for chemical looping combustion. Fuel, 153:284–293, 2015.

Synthesis of polymer micelle using amphiphilic macromonomer and characterization of its function

Masahiro YASUDA*, and Hiroyasu OGINO
Osaka Prefecture Univ.

1. SUMMARY

A monomer whose molecular mass is high as compared to ordinary monomer has been called as "macromonomer". In water, amphiphilic macromonomer forms a micelle composed with hydrophobic inner core and hydrophilic outer shell. In our previous study,[1] amphiphilic macromonomer which has hydrophilic poly(ethoxy) group, hydrophobic methacryloyl group, and hydrophobic alkyl group or *tert*-alkylphenoxy group were synthesized and their polymerization kinetics in chloroform were studied. These macromonomers were named as by MBC_mE_n (m: carbon number of single chain of alkyl group, n: number of poly(ethoxy) group) and MAX-n (n was the mean number of the ethylene oxide units and hydrophobic group is *tert*-alkylphenoxy group).

In this study, amphiphilic polymer micelle was synthesized with the micelle polymerization of macromonomers in water. Produced polymer micelle was characterized with its molecular mass, its distribution and micelle size.

2. EXPERIMENTAL

2.1 *Synthesis of macromonomer*

Several kinds of MAX-n and MBC_mE_n were synthesized from Triton®X or Brji® by the esterification with methacrylic acid using 1,3-dicyclohexyl carbodiimide as a dehydrating reagent.[1]

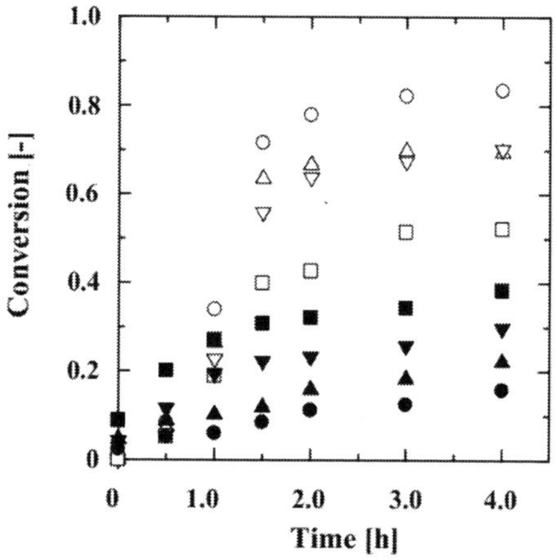

Figure 1 Time course of conversion of macromonomer in chloroform and water

Open key is polymerization in water and closed key is polymerization in chloroform. Circular: MAX-4.39, Triangle: MAX-9.2, Reverse triangle: MAX-15.8, Square: MAX-40.5

2.2 *Solution polymerization and micelle polymerization of macromonomer*

The solution polymerization of macromonomer was performed in chloroform. To determine the conversion, the reaction mixture composed of 4.0×10^{-5} mol of macromonomer, 4.8×10^{-6} mol of initiator in 0.4 cm^3 of $CDCl_3$ was poured into a test tube for NMR and was kept at 60°C. ^1H-NMR spectra were measured using a 200-MHz NMR spectrometer. In the case of micelle polymerization, owing to low solubility of macromonomer in water, 1.5×10^{-4} mol of macromonomer and 2.3×10^{-5} mol of 2,2'-azobis (2,4-dimethylvalelonitrile) were dissolved in 1 cm^3 of methanol and resulted solution was slowly poured into 29 cm^3 of distilled water in a 50 ml of poly(propylene) centrifugal tube. The reaction mixture was immersed in a water bath at 60°C for 24h and then it was quickly cooled with ice bath. The time course of conversion was also measured by ^1H-NMR spectra

To measure molecular mass of product, polymer was precipitated by *n*-hexane and was dissolved in THF. Polymer in solution was precipitated in *n*-hexane and dried again. A series of polymer purification, in which dissolution in THF, precipitation in *n*-hexane, and dry in room temperature were included, was repeated for three times. To measure molecular mass and its distribution, purified polymer was dissolved in THF and measured by GPC 8020 system (TOSOH).

3. RESULTS AND DISCUSSION

Micelle copolymerization of macromonomer

Figure 1 compares the time course of conversions of solution polymerization in chloroform and micelle polymerization in water for MAX-*n*. The rate of polymerization is high when the number *n* of the ethylene oxide units is high. The higher the number *n* of the ethylene oxide units is, the more viscous the reaction mixture is. Therefore, the termination reaction is prevented and the polymerization rate is accelerated when the number is high. This was because increase of viscosity of reaction mixture prevented the bimolecular termination reaction of polymer radicals in solution polymerization.

In the case of micelle polymerization, the conversion rapidly increased at 1-2 h. This was because macromonomer forms micelle in which vinyl group is highly accumulated and the polymerization rate is accelerated. Furthermore, the molecular mass of polymer produced by micelle polymerization is monodisperse (M_n/M_w = 1.1). This result also indicated that macromonomer forms micelle and the micelle transferred into the polymer micelle.

References

1) Yasuda et al., *J. Chem. Eng. Jpn.*, **34**, 388 (2001).

The mechanism of acetonitrile dimerization — a theoretical study

G. Zorludemir[1,2], N. McCann[2], C. Täschler[2], P. Wasserscheid[1], H. Freund[1]
[1]Friedrich-Alexander-Universität Erlangen-Nürnberg, Lehrstuhl für Chemische Reaktionstechnik, Egerlandstrasse 3, 91058 Erlangen, Germany
[2]Lonza Ltd., Research & Technology on Specialty Ingredients, Rottenstrasse 6, 3930 Visp, Switzerland

Acetonitrile (CH_3CN) is an important intermediate in organic chemistry, with an annual consumption of approximately 89,000 tonnes (2013). It is mainly used in the pharmaceutical industry as a starting material in the manufacture of vitamins A and B_1 as well as being a solvent for high-performance liquid chromatography[1]. Several pharmaceutical intermediates are produced at elevated temperature using CH_3CN in the gas phase. In these reactions, radical species play a crucial role in the reaction mechanism. Reactions of CH_3CN with halogens, hydroxyl groups and hydrogen radicals have been studied both theoretically and experimentally in the gas phase. However, self-reaction of acetonitrile species has rarely been investigated. Thus the aim of this study is to investigate acetonitrile dimerization in a mixture of CH_3CN species, in the presence and absence of chlorine radicals ($Cl^•$).

For the characterization of the transient species in a chemical reaction, density functional theory (DFT) can be used to gain insights into the electronic structure of the species, even when these species are not experimentally detectable. In this study, DFT calculations were performed using Gaussian 09[2]. The geometries of the reactants, intermediates and products were optimized at a UB3LYP/6-311+G (2d, p) level of theory. Transition states were verified by computing vibrational frequencies at the same level, where only one negative frequency indicating the saddle point on the reaction coordinate was obtained. Initially, hydrogen abstraction from CH_3CN to yield the cyanomethyl radical, $^•CH_2CN$ was investigated (Scheme 1). The standard enthalpy of reaction $\Delta rH°(298\ K)$ and the potential energy barrier ΔG^* were calculated as −35.37 and 14.20 kJ/mol, respectively. These values are in good agreement with the experimental data, in which $\Delta rH°(298\ K)$ and ΔG^* are found to be −34.77 and 17.74 kJ/mol, respectively[3].

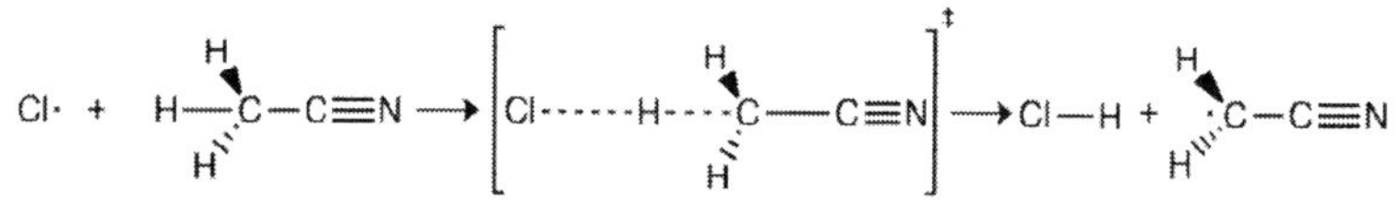

Scheme 1. Proposed reaction mechanism for hydrogen abstraction from CH_3CN

In the absence of $Cl^•$, $^•CH_2CN$ could be derived from homolytic cleavage of the C−H bond. The standard bond dissociation energy for this step was calculated to be 381 kJ/mol. Although the energy barrier for CH_2CN radical formation in the absence of a halogen radical is much higher than that in the presence of a halogen radical, the former may occur spontaneously at elevated temperature. Experimentally, trace amounts of succinonitrile

[1] Sesto, B. and Liu, S. (2014). Chemical Economics Handbook: Acetonitrile. *IHS Chemical.*
[2] Frisch, M. J., Trucks, G. W., Schlegel, H. B., Scuseria, G. E., Robb, A., Cheeseman, J.R., Scalmani, G. et al (2013). Gaussian 09, Revision D.01. *Gaussian, Inc.,* Wallingford CT.
[3] Atkinson et al. (2006). *Atmos. Chem. Phys.,6, 3625-4055.* <http://www.atmos-chem-phys.net/63625/2006/>

(which could be formed via dimerization of $^{\bullet}CH_2CN$) were observed from the reaction of CH_3CN at temperatures higher than 800 °C and in the absence of additional radicals.

Reaction of $^{\bullet}CH_2CN$ with CH_3CN may also give rise to succinonitrile formation in the presence of $Cl^{\bullet}$. The predicted mechanism depicted in Scheme 2, below, can be described in three steps:

1. $^{\bullet}CH_2CN$ binds to the cyanide carbon of the reacting CH_3CN molecule.
2. The resulting radical then undergoes isomerization via 1,3-hydrogen and carbon shifts.
3. Formation of succinonitrile occurs via $Cl^{\bullet}$ addition and hydrogen chloride elimination.

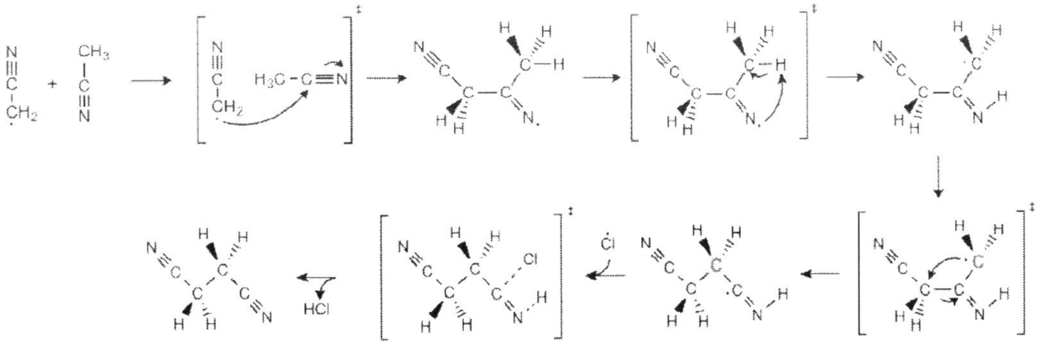

Scheme 2. Schematic representation of addition, isomerization, and elimination reactions of CH_3CN and $^{\bullet}CH_2CN$ in the presence of $Cl^{\bullet}$. The slowest step, 1,3-carbon shift, is indicated in red

The potential energy levels for this pathway are shown together with the optimized geometries in Scheme 3. The overall system is assumed to be neutral, and the presence of radical ions is neglected in the minimum energy pathway. Addition of the $^{\bullet}CH_2CN$ radical to CH_3CN appears to be thermodynamically favorable. The highest energy barrier is determined to be 285 kJ/mol for the carbon shift step, which is predicted to be the slowest step of the suggested mechanism.

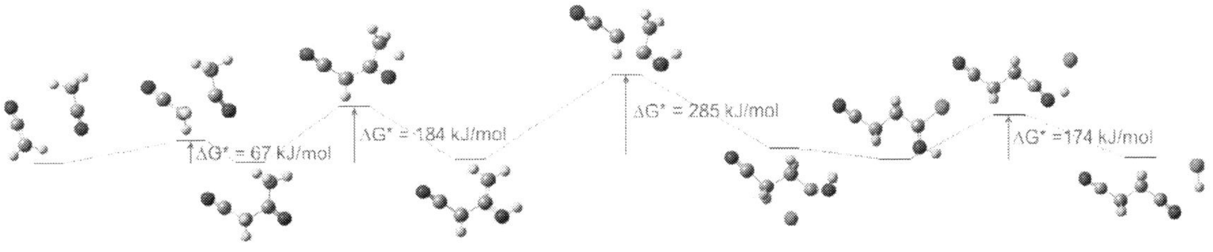

Scheme 3. Optimized geometries and potential energy surface of the suggested reaction path

These theoretical and experimental studies reveal that acetonitrile can react with its free radical to form a dicarbonitrile species in the absence or presence of halogen radicals. The insights from this mechanistic study provide a theoretical basis for the derivation of a kinetic model for two different reaction paths considered for $^{\bullet}CH_2CN$ dimerization.

Kinetic modelling of the photo-Fenton reaction and parameter fitting: a comparative study

C. Jiménez-Moreno, F. Audino, M. Graells.
Chemical Engineering Department. Universitat Politècnica de Catalunya (UPC).
Consorci Escola Industrial de Barcelona (CEIB), BARCELONATECH, EUETIB, C/ Comte d'Urgell 187, 08028-Barcelona, Spain.
tel. +34 934137275, e-mail: carlos.jimenez.moreno@estudiant.upc.edu; francesca.audino@upc.edu; moises.graells@upc.edu.

Pharmaceuticals and personal care products are continuously detected in variable, but generally very low concentrations, in wastewater treatment plants, surface and underground water and even drinking water. Hence, it is obvious that conventional wastewater treatments cannot remove this class of pollutants. Although the low concentration in which they are present, most of these compounds are considered *Persistent Organic Pollutants* (POPs), which raises the concern about their adverse effects on living organisms. Advanced Oxidation Processes (AOPs) appear to be an efficient alternative for the degradation of these compounds, being the photo-Fenton one of the processes with more potential to successfully mineralize them.

Modelling of the photo-Fenton process is a need for conveniently controlling and operating the process. In this study, a kinetic model describing the photo-Fenton process is developed and compared with a model already reported in the literatures. Both models are fit to experimental data and validated. The starting point is the model proposed by Cabrera Reina et al. (2012), which considers the following set of reactions for describing the photo-Fenton process:

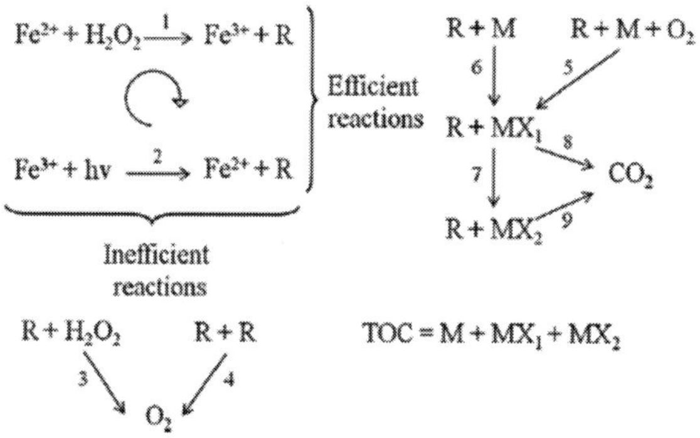

Figure 1: Reaction mechanism. (Cabrera Reina et. al, 2012)

Experiments were carried out in a pilot plant and paracetamol was selected as model pollutant. Concentrations of Total Organic Carbon (TOC) and hydrogen peroxide (H_2O_2) were measured by taking and analysing samples, while Dissolved Oxygen (DO) concentration was measured continuously using a sensor. The collected experimental data were first normalized and then a parameter fitting, searching for the kinetic constants of the model, was programmed and performed in Matlab®. The model is intended for future control and optimisation of the photo-Fenton process in the pilot plant.

Problems arise when fitting DO concentration, as also highlighted in the paper by Cabrera Reina et al. . Particularly, the measured dissolved oxygen concentration was always lower than the one predicted by the model.

Hence, with the goal of solving the problem with the DO fitting, some changes in the original model by Cabrera Reina et al. are introduced. First, an extended model is proposed, with an extra reaction, in order to take into account the ozone formation from oxygen due to UV light irradiation.. Second, a more simplified model (with only one inefficient reaction) is proposed, as shown next:

$$Fe^{2+} + H_2O_2 \xrightarrow{k1} Fe^{3+} + H_2O$$

$$Fe^{3+} + H_2O + h\nu \xrightarrow{k2} Fe^{2+} + 2H^+$$

Oxygen/Radical source

Measurement (H_2O_2)

$$2O \xrightarrow{k3} O_2 \qquad O_2 \overset{K_D}{\leftrightarrow} O_{2(g)}$$

Inefficient reaction (only one)

Measurement (DO)

$$C_n + m\,O \xrightarrow{k4} C_nO_m$$

$$C_nO_m + (2n - m)\,O \xrightarrow{k5} n\,CO_2$$

Efficient reactions (only two)

Delay (A → B →C)

$$TOC = n \cdot (\,C_nO_m + C_n\,)$$

Carbon balance

Measurement (TOC)

Figure 2: Simplified reaction mechanism proposed.

With the experimental data obtained in the pilot plant, parameter fitting is performed using the Matlab® optimisation toolbox, being the objective function to be minimized the summation of square errors (SSE) between the model and the experimental data points. Both models (the extended and the simplified) and fittings are compared based on the objective function (quantitative criteria), the adjusted coefficient of determination (R^2) and the residual plots. Once the best model is chosen, the values obtained for the parameters are checked for a range of conditions (initial concentrations of pollutant and reactants, irradiation and temperature). So, experiments are performed under different conditions and a hypothesis test is carried out through the Kruskal-Wallis technique (since the data do not follow a normal distribution), being the null hypothesis that the parameters in the whole range of conditions are the same.

Improvements it the model fitting are provided and future work is envisaged regarding the use of the photo-Fenton model for control and optimisation purposes.

Acknowledgements

Financial support from the Spanish Ministry of Economy and Competitiveness and the European Regional Development Fund, both funding the Project SIGERA (DPI2012-37154-

C02-01), and from the Generalitat de Catalunya (2014-SGR-1092-CEPEiMA) is fully appreciated.

References

Cabrera Reina, A., et al. 2012. *Modelling photo-Fenton process for organic matter mineralization, hydrogen peroxide consumption and dissolved oxygen evolution*. Applied Catalysis B: Environmental 119-120 (2012) 132-138.

Magnetically recoverable catalysts for cellulose hydrogenolysis

O.V. Manaenkov[1], L.M. Bronstein[2], V.G. Matveeva[1], O.V. Kislitza[1], E.M. Sulman[1], A.I. Sidorov[1], V.Yu. Doluda[1]
[1]Tver State Technical University, Af. Nikitina, 22, Tver, Russia, 170026. Tel./Fax: +7 4822 789317. E-mail: ovman@yandex.ru
[2] Indiana University, Department of Chemistry, 800 E. Kirkwood Av., Bloomington, IN 47405, USA

The catalyst recycling which is essential for successful industrial applications is usually not performed, because recycling of conventional supported catalysts requires an aggressive oxidative treatment which leads to deactivation via sintering. We used magnetically recoverable Ru catalysts supported on a Fe_3O_4-SiO_2 support that can be separated from the reaction mixture with a magnet. It was shown that the magnetically recoverable catalyst characterized by high selectivity to lower polyols (ethylene and propylene glycol, fig. 1). The catalyst was synthesized by a group led by Dr. L. Bronstein. For addition of Ru species, $Ru(acac)_3$ was dissolved in of THF and mixed with Fe_3O_4-SiO_2. The suspension was stirred overnight in air to allow THF evaporation. The sample was then dried in vacuum at room temperature for a minimum of 2 hours until the sample was entirely dry. The sample was heated in a quartz tube in a tube furnace under argon to 300°C for 3 hours. The catalyst was reduced in hydrogen at ambient pressure during 2 hours at 300 °C.

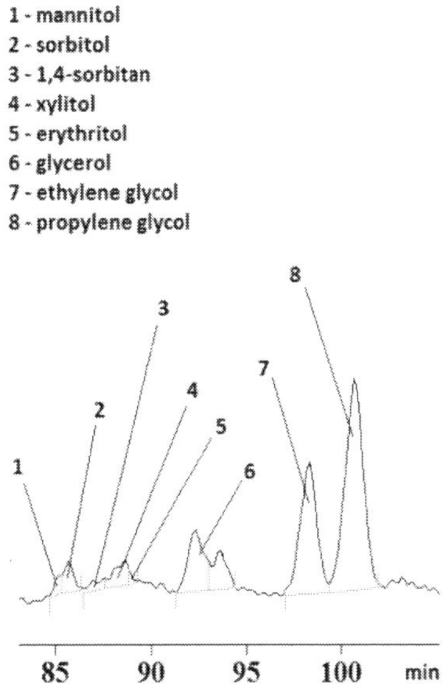

Fig. 1 –_Chromatograms of liquid phase after the reaction (255 °C, 6 MPa H_2, 30 mL water, 600 rpm, 5 % Ru-Fe_3O_4-SiO_2, 0.092 mmol Ru per 1 g cellulose, process duration 50 min.)

Cellulose conversion to polyols was carried out in subcritical water under the following conditions: temperature 205 - 260 °C, hydrogen partial pressure 4 - 106 MPa, process time 10 - 60 min. The experiments were performed in a steel reactor with volume 50 cm^3 (Parr Instrument, USA.). It was shown, that optimal reaction temperature of cellulose hydrolytic hydrogenation process was 255 °C. Optimal hydrogen pressure was found to be 6 MPa. At reaction pressures lower than 5 MPa the colored reaction solutions are formed, containing glucose destruction products. Starting form 6 MPa of overall reaction pressure the selectivity to diols decrease. Optimal cellulose hydrolytic hydrogenation time was found to be 50 minutes. The optimal conditions for cellulose hydrolytic hydrogenation over 5 % Ru-Fe_3O_4-SiO_2 catalysts sample was founded: temperature: 245 °C; hydrogen partial pressure: 6 MPa; process time: 50 minutes. The achieved results are: cellulose conversion - 100 %; propylene glycol selectivity ~ 23 %, ethylene glycol selectivity ~ 12 %.

This work was funded by the Russian Science Foundation (Grant 15-08-00455).

Methylation of toluene with methanol using modified HZSM-5 catalysts: effects of desilication and silylation

N. Peaunbida,[1,2]T. Rirksomboon
Chulalongkorn Univ., [1]The Petroleum and Petrochemical College, 254 Phayathai Rd. , Bangkok 10330, Thailand; tel. +66 881016066, [2]Center of Excellence on Petrochemical and Materials Technology,Chulalongkorn Univ., Phayathai Rd., Bangkok 10330, Thailand; tel. +662 2184141, fax: +662 6117619.
e-mail:navapass@hotmail.com;[1]tel. +66 22184136, email: t.rirksomboon@unb.ca;

p-Xylene is an important raw material in petrochemical industries for producing polyesters and plasticizers. The shape-selective alkylation of toluene with methanol is a promising way to produce *p*-xylene. To obtain a high *p*-xylene selectivity, the catalyst acidity and its proper pore size must be tailored properly. In this work, the synthesized HZSM-5 catalyst with a SiO_2 and Al_2O_3 of 160was modified by desilication, silylation, and the combination of desilication and silylation methods. The as-synthesized HZSM-5 catalyst was characterized by XRD to verify the HZSM-5 structure. The acid density of the catalysts was characterized by TPD-IPA and TPD-NH_3 techniques. The results revealed that desilication results in increased Lewis acid sites with maintaining Bronsted acid sites yielding both higher *p*-xylene and xylenes selectivities for the desilicated HZSM-5. For silylation process, the active sites on external surface was covered by TEOS resulting in lowering the activity of isomerization. However, the conversion of toluene was decreased as an increasing amount of TEOS. For the combination of desilication and silylation, the results showed that the highest *p*-xylene and xylenes selectivities are observed compared with those obtained by individual modifications because of the synergetic effect of acid strength and shape selectivity.

Photocatalytic degradation of acid orange 7 by NiO/TiO_2 and ZnO/TiO_2 and their energy storage capacity

[1]C. Wongburapachart, [1,*]P. Rangsunvigit, [2,**]P. Ngaotrakanwiwat
[1]The Petroleum and Petrochemical College, 254 Soi Chulalongkorn 12, Phayathai Rd., Pathumwan, Bangkok 10330, Thailand; tel. +662 218 4154, fax +662 215 4459, [*]e-mail: Promoch.r@chula.ac.th; [2]Department of Chemical Engineering, Burapha University, Chonburi 20131, Thailand; tel. (+66)38 10222-3351, [**]e-mail: Pailin@buu.ac.th.

Different classes of dye used in several industries include a major group of organics compound, which are toxic to both aquatic life and human beings. Among these dyes, acid orange 7 (AO7) represents up to 70% of dyes available on the world market today. The photocatalytic degradation of organic and inorganic pollutants using TiO_2 as photocatalysts has been studied, due to its relatively high photocatalytic activity, biological and chemical stability, low cost, non-toxic, and long-term stability. However, the photocatalytic activity of TiO_2 is limited by two main factors: the inability to absorb visible light and the fast recombination of photogenerated electron and hole. Therefore, many studies have been devoted to the improvement of TiO_2 photocatalytic activity. It has been reported that metal loaded on TiO_2 can trap electron to the surface, which retards the recombination of electron and hole. Moreover, the photocatalysts only function upon irradiation because the reduction and oxidation reactions are driven by photos. In this work, to keep the function of the photocatalysts in the dark, photocatalysts with energy storage ability were developed by coupling a redox-active semiconductor with TiO_2.

The photocatalysts were prepared by spin coating method on a glass slide plate as the substrate. The first layer was coated by TiO_2 and covered by a mixture of NiO and TiO_2 or ZnO and TiO_2 to obtain NiO/TiO_2 and ZnO/TiO_2 bilayer films. The activity of the bilayer photocatalysts was tested with 200 ml of a 20 ppm AO7 solution. To control the adsorption effect that could influence for the photocatalytic reaction, the bilayer film was immersed in the AO7 solution for 15 h. The experiment setup was blocked from any interference from other light sources. The solution samples were taken every hour for quantitative analysis using a UV-Vis spectrometer (UV 1800). The morphology and structure of photocatalysts were obtained by scanning electron microscope (SEM, Hitashi, S-4800).

Figure 1 shows the degradation rate of AO7 under the UV-A illumination for 8 h by the photocatalysts. The result shows that the degradation rate of AO7 increases with the operating time. The highest photocatalytic degradation rate is 89.5% by the ZnO/TiO_2 bilayer films. Figure 2 shows the degradation rate of AO7 from the performance of oxidation energy storage on the TiO_2 film. The UV-A light was turned on for 2 h and turned off 2 h for the total of 8 h operating time. The degradation rate of AO7 is about 5.0% after the first illumination and 3.1% after the second illumination. Its catalytic activity in the absence of UV illumination is virtually insignificant. For the NiO/TiO_2 and ZnO/TiO_2 catalysts, they show their energy storage capacity by active degradation of AO7 even there is no illumination.

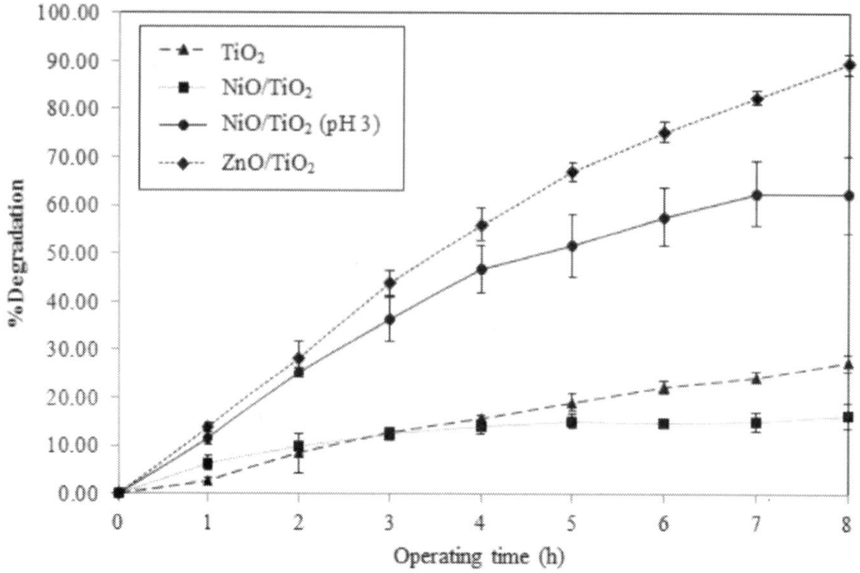

Figure 1 Photocatalytic degradation of acid orange 7 by different catalysts with the UV-A illumination for 8 h.

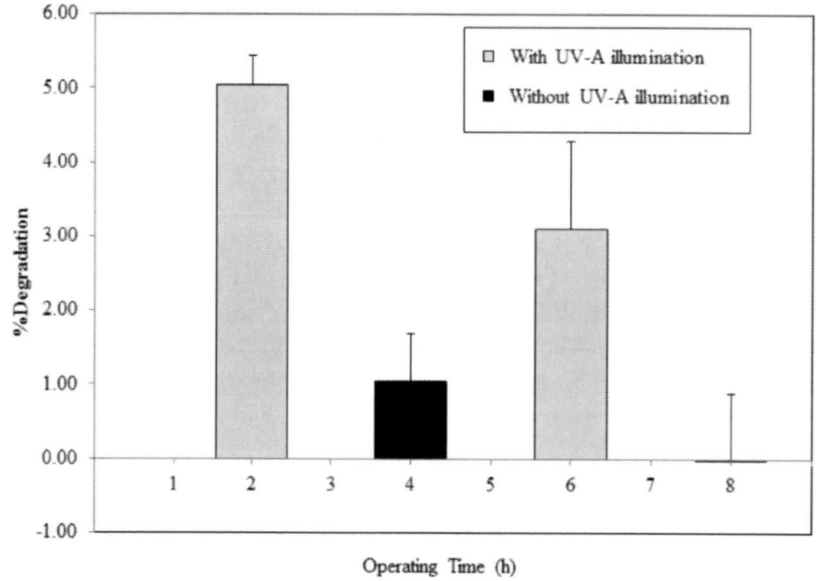

Figure 2 Photocatalytic degradation of acid orange 7 by the TiO_2 film with and without the UV-A illumination.

References:

1. Buama, S., Ngaotrakanwiwat, P., and Rangsunvigit, P. (2012). Energy storage of NiO/TiO_2 bilayer films and its effectiveness in the degradation of acid orange 7. Advanced Materials Research, 557, 2002-2007.
2. Chen, C. J., Liao, C. H., Hsu, K. C., Wu, Y. T., and Wu, J. C. (2011). P–N junction mechanism on improved NiO/TiO_2 photocatalyst. Catalysis Communications, 12(14), 1307-1310.
3. Tanaka, K., Padermpole, K., and Hisanaga, T. (2000). Photocatalytic degradation of commercial azo dyes. Water Research, 34(1), 327-333.

Production of biojet fuels from palm oil using Pt/HY^{core}-Pd/TiO_2^{shell} catalyst prepared by atomic layer deposition

[1]W. Yenying, [1,2,*]S. Jongpatiwut.
[1]The Petroleum and Petrochemical College, Chulalongkorn University, 254 Chula Soi 12 Wangmai Pathumwan Bangkok 10330, Thailand; tel. (+66)2 2184110, fax: (+66)2 2154459, *e-mail: siriporn.j@chula.ac.th; [2]Center of Excellence on Petrochemical and Materials Technology, Chulalongkorn University Research Building, Chula Soi 12 Wangmai Pathumwan Bangkok 10330, Thailand.

The depleting of petroleum reserves and increasing concern of environmental impact lead to the development of fuel production techniques from renewable resources. Biojet fuel from vegetable oil is the solution to solve the problem. Biojet fuel is obtained by hydrodeoxygenation of triglycerides followed by hydrocracking and hydroisomerization reactions. In our previous studies the Pt/HY^{core}-Pd/TiO_2^{shell} catalyst was synthesized via a combined single-step sol-gel process with surfactant-assisted templating method. In this research, the Pt/HY^{core}-Pd/TiO_2^{shell} catalyst was prepared by atomic layer deposition method for coating TiO_2 layer before covering Pt/HY^{core} with Pd/TiO_2^{shell} catalyst. The catalysts were tested for activity and selectivity for biojet fuel production in a continuous flow packed-bed reactor at 310 °C, 500 psig, LHSV of 0.5 h^{-1}, and a H_2/feed molar ratio of 60. The catalysts were characterized by XRD, XPS, TEM, BET, and TGA. Figure 1 shows XRD pattern of Pt/HY, Pt/HY^{core}-$Pd/TiO_2^{shell}$and Pd/TiO_2. The XRD pattern of Pt/HY exhibited the diffraction pattern of HY zeolite. Furthermore, the XRD pattern of Pd/TiO_2 catalyst shown the crystalline structure of the pure anatase phase of TiO_2. The XRD pattern of Pt/HY, Pt/HY^{core}-Pd/TiO_2^{shell} showed both Pt/HY and Pd/TiO_2characteristic indicating that the synthesized core-shell catalyst still remained both catalyst structure. Figure 2 shows XPS spectra of the Ti 2p peaks of synthesized Pt/HY^{core}-Pd/TiO_2^{shell} by atomic layer deposition. As shown in the XPS spectra, the Ti $2p_{1/2}$ and Ti $2p_{3/2}$ peak are located at binding energies of 465.15 and 459.35 eV respectively. These peaks indicate that the TiO_2 structure was successfully synthesized by atomic layer deposition. The conversion of refined palm oil to biojet fuel exhibited 100% which provided 47.9% of biojet yield.

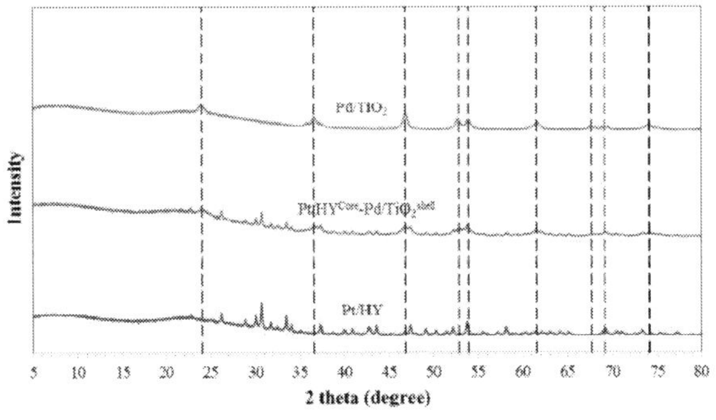

Figure 1 XRD pattern of synthesized Pt/HY, Pt/HY^{core}-Pd/TiO_2^{shell} and Pd/TiO_2 (dashed line: anatase).

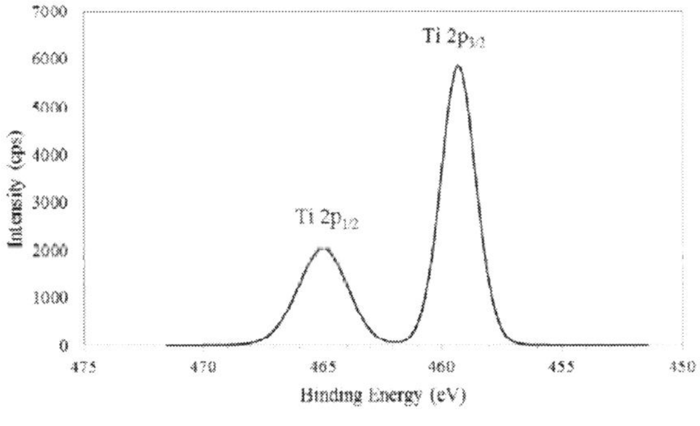

Figure 2 XPS spectra of the Ti 2p peaks of synthesized Pt/HY^{core}-$Pd/TiO_2^{shell}$by atomic layer deposition.

Synthesis and electrochemical evaluation of nano platinum supported on carbon vulcan as PEM fuel cell anode catalyst

S. Kheradmandinia, N. Khandan, M.H. Eikani
Department of Chemical Technologies, Iranian Research Organization for Science and Technology (IROST), P.O. Box: 33535111, Tehran, Iran; tel. +9821 5627 6637, e-mail: eikani@irost.ir

Platinum is the best-known catalyst for polymer exchange membrane fuel cells (PEM) that indicates highest catalytic activity in conventional PEM operation condition[1]. But main issue in these fuel cells is catalyst poisoning by carbon monoxide (CO). Thus several studies have done to synthesis and develop of catalyst with enhanced CO tolerance to be use in the anode of PEM fuel cell[2,3]. More tolerance of CO is achievable by improve the decreasing adsorption and facilitating the oxidation of CO on the catalyst surface. Since structure and morphology of the catalyst could be affected on catalyst performance[4,5,6], in this work we synthesised a nano-structural Platinum supported on carbon vulcan (Pt/C) catalyst, hiring boro hydride reduction method. Activity of our catalyst was compared with commercial one, using half-cell Cyclic Voltammetry (CV) method. CO stripping current voltammetry in half-cell was used for surveying CO electro oxidation capability of catalysts. Evaluation of catalysts using half-cell cyclic voltammetry test is less expensive, but usually this method is applicable for unsupported catalysts and shows current picks for oxidation of single layer CO, so in the case of supported catalysts, as the previous reports[7,8], there is no distinctive pick in Stripping CVs. Therefore we suggested a new test method to explore the CO electro oxidation capability of catalysts, based on recording CV current under continues CO gas flow and subtracting the blank CV current values from the that results then extra charge values attributed to CO oxidation current.

Nano-structural Pt/C catalyst contained 20 wt% of metal on carbon support was synthesised by dissolving $H_2PtCl_6.6H_2o$ in THF and then adding carbon Vulcan to solution under nitrogen atmosphere and stirring. The precursor was aged for 30 minute. Sodium borohydride added to ethanol pot with the same volume of THF and contents of pot poured in reaction flask at once. Rigorous stirring under N_2 atmosphere continued for 3-5 hours, then product isolated by centrifuge and permanently washed with distillated water and ethanol, until filtrate TDS and PH corresponded to distilled water. Finally catalyst dried in 60°C oven for one night. Product powder is analysed with XRD to check molecular structure.

Ag/AgCl and platinum rode electrodes were used as reference and counter electrode respectively in a half-cell and working eletrode with exposed catalyst surface area of 0.5 cm^2, was made by spreading catalyst ink on teflonized carbon paper. The cyclic voltammetry with 20 mv/se step from 0.0 to 1.0 v versus NHE was employed to obtain the electrochemical activity of the catalyst. CV tests were carried out once in a media of electrolyte without any gas bubbling as blank CV, and a second time under countinous bubbling of hydrogen gas in the electrolyte as H_2 CV and third time under countinous bubbling of CO gas in the electrolyte as CO CV.

H_2 CVs activate the inactive catalyst sites on the working electrod. Catalyst primary activity in electrolyte was recorded by blank CV, Then CO electrooxidation charge was recorded by 100 CO CVs. Subtracing the blank charge from CO electrooxidation charge in the positive going potentials above onset potential, resulted the charge which utilized for CO electrooxidation. As seen in Fig. 1, the synthesized catalyst indicates better activity for CO electrooxidation at the begining of CV than commercial catalyst.

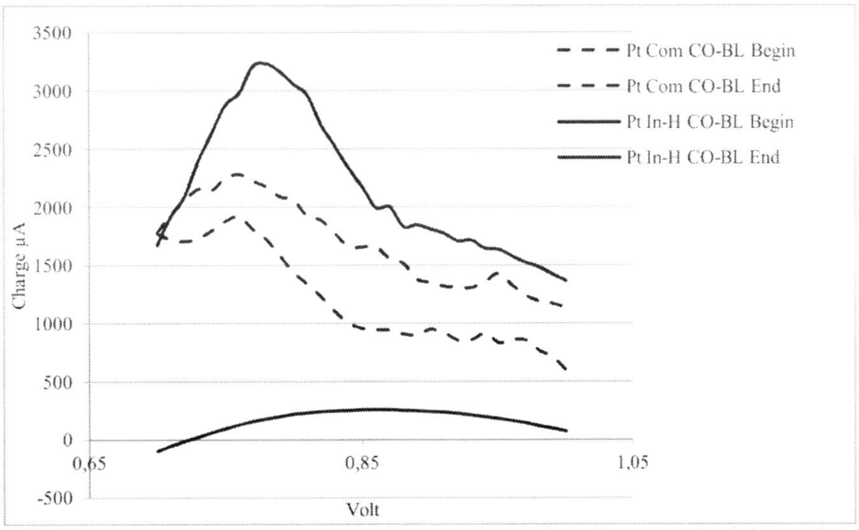

Fig. 1. Activity of Commercial and synthesized Pt/C under CO bubbling at the begining and end of CVs

Refrences

1- R. O'hayre, S.W. Cha, W. Colella- Fritz B. Prinz, "Fuel Cell Fundamentals" Second edition; John Wiley & Sons 2009.

2- S. Litster, G. McLean, 2004, Journal of Power Sources, 130:61–76.

3- S. M. Mousavi Ehteshamia, S. Hwa Chan , 2013, Electrochimica Acta, 93:334– 345

4- Jenny E. Hu, Zhufang Liu, Bryan W. Eichhorn, Gregory S. Jackson, 2012,international journal of hydrogen energy, 37:11268 -11275.

5- C. Hao. Wan, Q.H. Zhuang, 2007, Electrochimica Acta, 52:4111–4123.

6- R. Guil-Lo´pez and et all, 2010, international journal of hydrogen energy, 35:7881-7888.

7- T. Vidakovi´c, M. Christov, K. Sundmacher, 2007, Electrochimica Acta, 52 : 5606–5613.

8- P. Ochal and et all., 2011, Journal of Electroanalytical Chemistry, 655 :140–146.

TBR CFD simulation of HDS process incorporating interface momentum exchange models, and shape and the pellet porous microstructure effects

M.E. Cordero*[a,b], L.G. Zárate[a], R. Natividad[b] , J.S. Uribe[a], M. Villegas[a], José.A. Hernandez-Servin[b]
[a]Universidad Popular Autónoma del Estado de Puebla, 21 sur 1103 Barrio Santiago C.P. 72410, Puebla México, Teléfono: +52 (222) 229.94.00
[b]Centro Conjunto de Investigación en Química Sustentable UAEM-UNAM, Carretera Toluca-Atlacomulco km 14.5, Toluca, Edo. de México, México
marioedgar.cordero@upaep.mx

This work explores the effect of four different interface momentum exchange models, Attou [1], Gidaspow [2], Schiller- Naumman [3] and Wen-Yu [4], over both, the hydrodynamic, and kinetic and thermal TBR's behavior, operating in Trickle regime and where HDS reaction following LHHW kinetic type take place. Also catalyst shape effect over TBR's behavior was investigated. Spherical, four hole cylindrical, cylindrical with cross hole, trilobes, and annular cylindrical pellets were tested. Finally, with the purpose to determinates how much porous microstructure information passes from porous length scale to TBR scale, it was analyzed its geometrical representation effect over kinetic and thermal TBR's performance, the catalyst microstructure information was captured through the effective mass and thermal transport coefficients determination, as it was previously reported by Cordero [5].

The CFD simulation follows an Eulerian-Eulerian approach and includes the continuity, and momentum transport equation, (1) to (2), for both gas $(\lambda=\beta-phase)$ and liquid phases $(\lambda=\gamma-phase)$, and includes a mass and energy transport equation, (3) to (4) ,for solid (ω) and fluid (φ) pseudophases.

$$\varepsilon_\lambda \nabla\cdot(\rho_\lambda \mathbf{v}_\lambda)=0 \tag{1}$$

$$\varepsilon_\lambda \rho_\lambda(\mathbf{v}_\lambda\cdot\nabla)\mathbf{v}_\lambda=\varepsilon_\lambda\nabla\cdot\left[-P_\lambda\mathbf{I}+\mu_\lambda\left(\nabla\mathbf{v}_\lambda+(\nabla\mathbf{v}_\lambda)^T\right)\right]+\mathbf{F}_\lambda+\rho_\lambda\mathbf{g} \tag{2}$$

$$\begin{aligned}&\nabla\cdot\left(-D_\varphi\nabla C_{A_\varphi}\right)+\mathbf{v}_\varphi\cdot\nabla C_{A_\varphi}=0\\&\nabla\cdot\left(-k_\varphi\nabla T_\varphi\right)+\rho_\varphi Cp_\varphi\mathbf{v}_\varphi\cdot\nabla T_\varphi=0\end{aligned} \tag{3}$$

$$\begin{aligned}&\nabla\cdot\left(\varepsilon_\omega\mathbf{D}_{eff}\cdot\nabla\langle C_{A_\omega}\rangle^\omega\right)=\langle r_A\rangle^\omega\\&-\nabla\cdot\left(\mathbf{K}_{eff}\cdot\nabla\langle T\rangle^\omega\right)=(-\Delta H)\langle r_A\rangle^\omega\end{aligned} \tag{4}$$

The TBR geometrical model was built with a bed of 12 layer containing 124 pellets of $2\ mm$ for sphere cases and with the same $(V/A_t)_{pellet}$ value as sphere case, for other shapes (Figure 1). The TBR model has the ratio reactor to pellet diameters $N\approx 6$ value. The effective transport coefficients incorporate pellet microstructure geometrical information through its evaluation with a closure boundary value problem [6] on different representative unitary Cells (RUC's) and a realistic porous model (Figure 2).

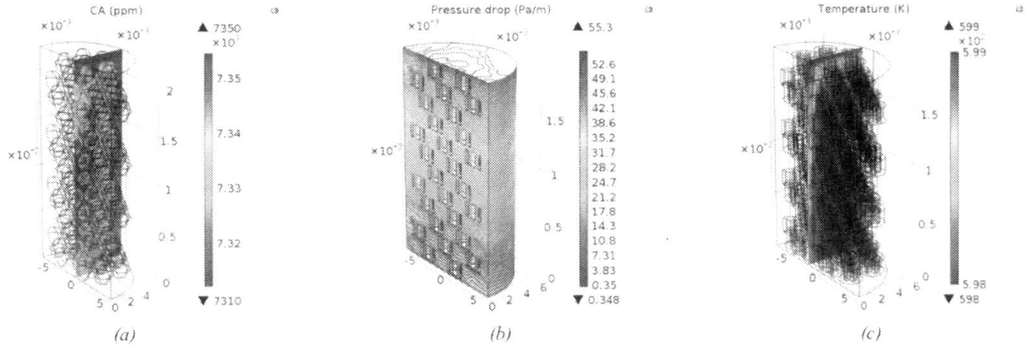

Figure 1. Geometrical TBR model with: (a) spherical pellets using attoud/attoud/attoud exchange momentum model, showing sulfurized concentration field. (b) Four hole cylindrical pellet using attoud/attoud/attoud model, showing pressure field. (c) Crus hole cylindrical pellet with attoud/attoud/gidaspow model, showing temperature field.

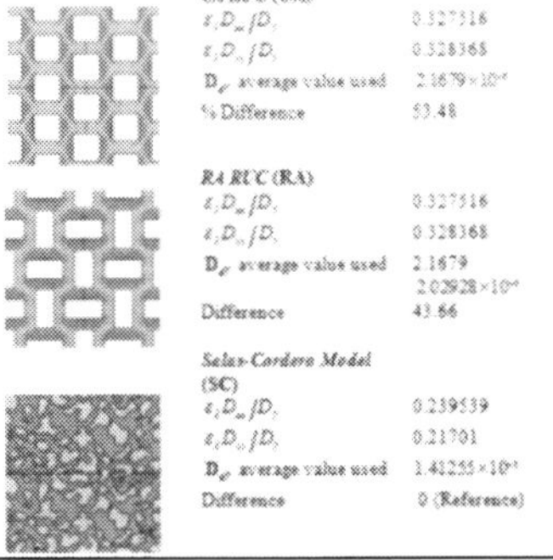

Figure 2. Different geometrical representation of microstructure of pellets, and its effect over effective transfer coefficients.

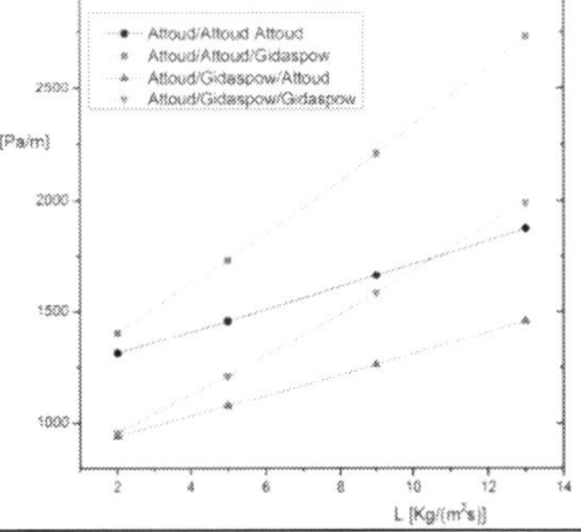

Figure 3. Effect of liquid mass flow (units) and exchange model over the pressure drop.

Our pressure drop predicted values have a tendency of experimental data found in literature [7], and have the same accuracy than other previous pressure drop prediction studies, such as [8]. It was possible to simulate a TBR model incorporating the multiphase hydrodynamics, mass and energy simultaneous transfer, and considering the effect of different interphase momentum exchange models, and of different pellet shapes. The model incorporates the multiscale analysis, through studying the influence of microstructure information over the transport phenomena at reactor length; these kind of multiscale studies are not frequently found in literature.

References

[1] A. Attou, C. Boyer and G. Ferschneider, "Modeling of the hydrodynamics of the cocurrent gas-liquid trickle flow through a trickle-bed reactor," *Chemical Engineering Science,* vol. 54, no. 6, pp. 785-802, 1999.

[2] D. Gidsapow and J. Ding, "A Bubbling Fluidization model using kinetic theory of granular flow," *AIChE Journal,* vol. 36, no. 4, pp. 523-538, 1990.

[3] L. Schiller and Z. Naumman, "A drag Coefficient correlation," *Deutsch. Ing.,* pp. 318-320, 1935.

[4] C. Y. Wen and Y. H. Yu, "Mechanics of fluidization," *Chem. Eng. Prog. Symp. Ser.,* vol. 62, pp. 3433-3441, 1966.

[5] M. Cordero, R. Natividad, L. Zárate, J. Hernández-Servin and J. Salas, "Estimation of effective diffusion coefficient and its effect on effectiveness factor for HDS catalytic process: A muti-scale approach.," *press Catalysis Today,* 2013.

[6] S. Whitaker, The Method of Volume Averaging, Netherlands: Kluwer Academic Publishers, 1999.

[7] M. H. Al-Dahhan and M. H. Dudukovic, "Pressure drop and liquid holdup in high pressure trickle bed reactors," *Chemical Engineering Science,* vol. 49, pp. 5681-5698, 1994.

[8] A. Atta, S. Roy and K. D. Nigam, "Prediction of pressure drop and liquid holdup in trickle bed reactor using relative permeability concept in CFD," *Chemical Engineering Science,* vol. 62, pp. 5870-5879, 2007.

Hydrogen production from microwave pyrolysis of spruce sawdust with char and char-doped metallic catalysts

P. Lestinsky, B. Grycova, A. Pryszcz
VSB-TU Ostrava, Institute of Environmental Technology, 17. listopadu 15, 70833 Ostrava, Czech Republic;
tel. +420 597 327 331, e-mail: pavel.lestinsky@vsb.cz

Currently, a growing demand for alternative energy sources to replace traditional fossil fuels. Although fossil fuels in the world still enough (still found new reserves or mining techniques – e.g. Shale gas) is needed looking for a solution for future substitute. The Czech Republic and other countries in the Central Europe are not self-sufficient fossil energy. However, these areas are extensively forested and wood industry in these areas have important place for several centuries. Wastes from the wood processing industry could be serving as a potential feedstock for the energy production, e.g. hydrogen.

Production of the energy from wood biomass is as old as humanity itself. In the recent 20 years there has been enormous progress in research of torrefication, pyrolysis and mainly in gasification of the biomass. Products from conventional pyrolysis of the biomass are gas known as syngas, liquid condensate known as bio-oil and char termed as bio-char. Most of the articles dealing with convection pyrolysis of biomass are focused on engaged in the production of bio-oil. Only some works deal with the possibility of producing high-quality syngas (H_2+CO) or hydrogen alone. For this purpose, it seems most appropriate technology microwave pyrolysis. Microwave is able to generate microplasmas and hop spot, which promote the heterogeneous catalytic reaction and produce greater concentration of syngas and hydrogen in the product gas than convectional pyrolysis.

In this work, experimental study of spruce sawdust microwave pyrolysis in the presence of catalysts to maximize the yield of hydrogen, or syngas was performed. Experiments were realized in the microwave oven with power up to 1 kW. The catalysts were used as the char from sawdust alone, or as the char-doped with metal ions. Ions of metals were used, which can be expected a significant effect on the yield of hydrogen (e.g. Nickel is widely used in the catalytic cracking of methane). The influence of the catalyst on the quantity of products (gas, liquid and solid), but also the amount of generated hydrogen in the syngas, or the amount of waste water in the liquid condensate were studied.

Investigation of the influence of fuel characteristics and design on pulverized fuel combustion in multifuel swirl burners

B. S. Repic, D. M. Djurovic, A. D. Marinkovic, A. M. Eric, G. S. Živkovic
University of Belgrade, Vinca Institute of Nuclear Sciences, Laboratory for Thermal Engineering and Energy, P.O. Box 522, 11001 Belgrade, Serbia; tel. +381 11 3408838, e-mail: brepic@vinca.rs

Biomass is one of key renewable energy sources, and is the fourth largest primary energy resource in the world following coal, oil and natural gas. In all its forms, biomass currently provides about 14% of the world's annual energy consumption. According to the official date Serbia is in dispose of 4.3 Mtoe of renewable energy sources, while biomass is represented with 2.7 Mtoe. 60% out of registered biomass potential are residuals from agricultural production, and the rest is wood biomass. The sources of biomass are specially grown energy crops, agricultural wastes, forestry residues and the organic fraction of municipal wastes. Its energy is converted to heat, power or chemical feedstock mainly by thermo-chemical conversion. Technologies of utilizing the biomass are determined by biomass characteristics.

Pulverized fuel firing is one of the wide modes of fuel utilization. For successful combustion of pulverized fuel it is necessary to ensure adequate conditions by the appropriate choice of burners and furnace characteristics. Combustion of pulverized fuel with swirl burners utilizing the swirl effects is very significant and widely used. Swirl flows are used extensively because of their favorable influence on stabilization and intensification of the combustion process and reduction of NOx emission. Optimal operating conditions of pulverized fuel combustion depend on burner characteristics and fuel properties.

Swirl burners are widely used in furnace technique for different types of combustion of fossil fuels. To enable detailed study of the combustion process of pulverized fuels (biomass, coal etc.) using swirl burners of different types and strengths, and compare the results of research in laboratory and real conditions fireplaces, four models of swirl burners was made with heat power of 0.1-0.7 MW. For experimental tests of burner model was used laboratory facility, which allows the implementation of pulverized fuel combustion test in a wide range of operating parameters. The following tests were performed: determination of the optimal area of the burner for the selected type of fuel, to determine the effect of fuel quality, fineness of grinding, moisture content, ash and volatile matter in pulverized fuel combustion, determining the influence of temperature on combustion and determination of additional parameters essential for the swirl burner construction and design.

In the paper are presented the results of investigation of influence that constructive and operating parameters of burner's characteristics and fuel properties (biomass and coal) exert on pulverized fuel combustion. Investigation of combustion of several types of fuels (biomass and coal) in swirl burner models and prototypes were performed. Influence of the following parameters was investigated: constructive parameters of burners (type, quarl), burner's operating parameters (velocity, temperature and quantity of air, swirl number, excess air, concentration of pulverized fuel), and fuel properties (type, quality, moisture, ash, volatile and combustible matters contents, pulverized fuel grinding fineness etc.).

SAPO-18 and SAPO-34 catalysts for the transformation of dimethyl ether to olefins (DTO process)

P. Pérez-Uriarte, M. Díaz, A. Ateka, M. Gamero, A. T. Aguayo, J. Bilbao
Department of Chemical Engineering, University of the Basque Country (UPV/EHU). P.O. Box 644, 48080. Bilbao (Spain). Tel: +34 94 601 53 61, email: paula.perez@ehu.es

The DTO (Dimethyl ether To Olefins) process is presented as an alternative to MTO process (Methanol to Olefins). The initiative aroused great interest due to the increasing availability of DME, whose synthesis in one step is considered a better option for the valorization of syngas that the Fischer Tropsch. Consequently, industrial project of DME synthesis has good prospects [1], associated with the growing availability of natural gas [2].
Various authors [3-5] have made a comparison of the kinetic behaviour (activity, selectivity of different olefins and deactivation) of different acid catalysts in the MTO process. The results have led to the proposal of SAPO-18 (AEI structure) as an alternative to SAPO-34, as it suffers lower deactivation and its lower cost.
This work contributes in the selection of an optimum catalyst for DME to olefins process in a fixed bed reactor. The study has been carried out over SAPO-n catalysts (SAPO-18 and SAPO-34) agglomerated with bentonite as a binder. The experiments were carried out under the following conditions: atmospheric pressure, 350 and 400 ºC, space time of 0.4 and 4 $g_{active\ phase}$ h $(mol_C)^{-1}$ and time on stream of 90 min. The products were continuously analyzed in a gas cromatograph (Micro-GC Agilent 490). Table 1 summarizes the physical and acid properties of the catalysts. The microporous structure has been analyzed by N_2 adsorption-desorption (ASAP Micromeritics 2000) and the total acidity and average acid strength have been determined by adsorption-desorption of NH_3, by combining termogravimetric techniques and differential scanning calorimetry.

Table 1. Physical and acid properties of the active phases and catalysts.

	S_{BET} (m^2/g)	V microp. (cm^3/g)	Total acidity ($mmol_{NH3}$/g)	Average acid strength (kJ/mol_{NH3})
SAPO-18	798	0.281	0.37	142
SAPO-34	611	0.243	0.64	153
CS-18/bentonite	236	0.072	0.10	123
CS-34/bentonite	215	0.046	0.15	127

In Figure 1, the results of selectivity at zero time on stream for SAPO-18 and SAPO-34 catalysts agglomerated with bentonite are compared. It should be noted that with the SAPO-18 catalyst a higher selectivity of propylene was obtained, which may be due to slightly smaller steric limitation for the route of methylation-dealkylation of aromatics (due to the different size of the cavities of the porous structure), which favours the propylene formation as a primary product [6].

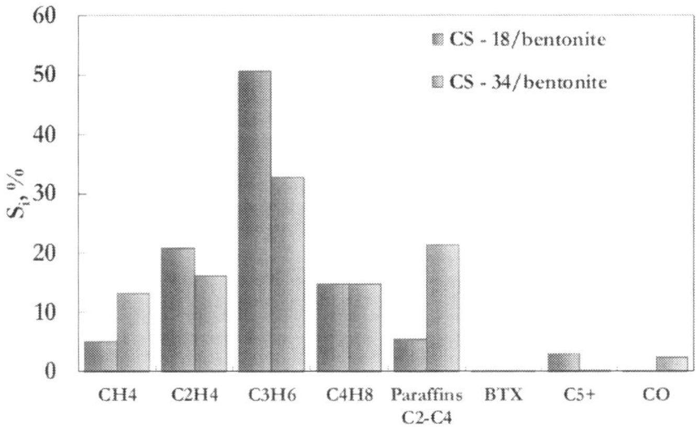

Figure 1. Comparison of the selectivity results at zero time on stream for the catalysts prepared with SAPO-18 and SAPO-34. Reaction conditions: 350 °C, 0.4 $g_{active\ phase}$ h mol_C^{-1}.

Based on these results, it is highlighted that for a reduced value of the space time (0.4 $g_{active\ phase}$ h mol_C^{-1}), and consequently, at low conversion of DME (about 5%), with the CS-18/bentonite catalyst, selectivity to light olefins is about 85%, with a propylene selectivity of 50%. These results are consistent with those obtained in the literature at zero reaction time on stream in the transformation of DME over SAPO-34 catalysts [7].

The deactivation of both catalysts is very fast due to the rapid blocking of the cavities of the microporous structure coke. Comparing both catalysts, a slower catalyst deactivation is obtained with SAPO-18, which is consistent with its lower acidity and acid strength. Furthermore, with the SAPO-34 catalyst a higher content of deposited coke is achieved, obtaining nearly twice the value of coke for high values of space time (with a space time of 4 $g_{active\ phase}$ h mol_C^{-1}, a 5.47 % is obtained for SAPO-34 and 2.92 % for SAPO-18). Consequently, the catalyst prepared with SAPO-18 has a better compromise between activity, propylene selectivity and stability than the prepared with SAPO-34.

Acknowledgements

This work was carried out with the financial support from the Ministry of Economy and Competitiveness of the Spanish Government (Project CTQ2013-46173-R), from the Basque Government (Project IT748-13) and from the University of the Basque Country (UFI 11/39).

References

[1] S. Bhattacharya, K.B. Kabir, K. Hein, Dimethyl ether synthesis from Victorian brown coal through gasification - Current status, and research and development needs, Prog. Energy Combust. Sci., 39 (2013) 577-605.

[2] M. Melikoglu, Shale gas: Analysis of its role in the global energy market, Renewable Sustainable Energy Rev., 37 (2014) 460-468.

[3] J. Li, Y. Wei, Y. Qi, P. Tian, B. Li, Y. He, F. Chang, X. Sun, Z. Liu, Conversion of methanol over H-ZSM-22: The reaction mechanism and deactivation. Catal. Today, 164 (2011), 288-292.

[4] D. Chen, K. Moljord, A. Holmen, A methanol to olefins review: Diffusion, coke formation and deactivation on SAPO type catalysts. Microp. Mesop. Mater, 164 (2012), 239-250.

[5] S. Abdollahi, M. Ghavipour, M. Nazari, R.M. Behbahani, G.R. Moradi, Effects of static and stirring aging on physiochemical properties of SAPO-18 and its performance in MTO process. J Nat Gas Sci Eng, 22 (2015), 245-251.

[6] S. Park, Y. Watanabe, Y. Nishita, T. Fukuoka, S. Inagaki, Y. Kubota, Catalytic conversion of dimethyl ether into propylene over MCM-68 zeolite. J. Catal, 319 (2014), 265-273.

[7] S.G. Lee, H.S. Kim, Y.H. Kim, E.J. Kang, D.H. Lee, C.S. Park, Dimethyl ether conversion to light olefins over the SAPO-34/ZrO_2 composite catalysts with high lifetime. J. Ind. Eng. Chem., 20 (2014), 61-67.

Benzaldehyde hydrogenation over cobalt oxide supported on silica oxide. Effect of catalyst preparation method

I. Lounas[1], H. Zazoua[1,2], A. Saadi[1], C. Rabia[1]
[1] Laboratoire de Chimie du Gaz Naturel, Faculty of Chemistry, USTHB University, PoBox 32, El-Alia, 16111 Bab-Ezzouar, Algier, Algeria. Phone: +213 21 24 73 11, fax: +213 21 24 73 11, e-mail address: asaadi@usthb.dz.
[2] Centre de Recherche Scientifique et Technique en Analyse Physico-Chimique (CRAPC Research Center), Bou-Ismail, Tipaza, Algeria

The selective hydrogenation of carbonyl compounds has growing interest since it is an important step in the preparation of fine chemicals. In the hydrogenation of benzaldehyde, the benzene ring and the carbonyl group should be hydrogenated but hydrogenolysis should also compete with hydrogenation. Benzyl alcohol is one of hydrogenation reaction products and it's intensively used as intermediates in a large number of pharmaceuticals syntheses [1, 2]. It is also used as raw material in the manufacture of various esters in the cosmetics and flavor industries [3]. The catalytic benzaldehyde reduction, in gas phase, represents an interesting alternative for the benzyl alcohol production. The obtained selectivities depended on the nature of the catalysts.

In this work, we propose the synthesis of Cobalt oxide supported on silica oxide by different methods: chemisorption-hydrolysis, ionic exchange and impregnation with two different solvents (NH_3 and NaOH). The resulting precursors were dried at 100°C and calcined in air (350°C/3 h). The solids were caracterised by their BET specific area and XRDiffraction. The catalytic reactions were carried out in a fixed bed glass reactor with 0.2g samples at atmospheric pressure and in the temperature range : 160°C-200°C (H_2 per-reduction : 350°C).

Characterization results showed that, the structural properties of all systems are similar and the characteristic XRD patterns of CoO and SiO_2 phases are observed. The pattern was that of critallized materials with well-defined bands and high degree of cristallinity of the diffraction lines. The CoO phase was of 'CFC' structure for all the oxides. Calculations showed that the mean crystallite size of metal oxide was the same, and vary in the range of 30-50 nm. The calcined simple at 350°C showed very small surface areas in the range 4.0-8.0 $m^2.g^{-1}$ and the pore volume (V_P) in the range 0.03-0.05 $cm^3.g^{-1}$.

Before testing under a dihydrogen flow, the catalysts were in-situ activated for 2h at 350°C in a current of H_2. Catalytic results showed that Co/SiO_2 is active in the benzaldehyde reduction and that the catalytic performance is sensitive to the catalyst preparation method and temperature reaction. Thus, the benzaldehyde conversion increases with increasing of reaction temperature from 160°C to 200°C. The maximum of conversion was obtained at 200°C for Co/SiO_2 prepared by chemisorption-hydrolysis method (57%), whereas it reached lower value (<40%) for Co/SiO_2 prepared by impregnation method with ammonic as solvent. At 160°C, benzyl alcohol was obtained with high selectivity (64%) on Co/SiO_2 prepared by chemisorption-hydrolysis method.

References

1. J. Scognamiglio, L. Jones, D. Vitale, C.S. Letizia, A.M. Api, Food Chem. Toxicol. 50 (2012) S140–S160.
2. Y. Zhou, J. Liu, X. Li, X. Pan, X. Bao, J. Nat. Gas Chem. 21(2012) 241–245
3. M. Li, X. Wang, N. Perret, M. A. Keane, Catal. Com. 46 (2014) 187–191

Etherification of furfuryl alcohol to butyl levulinate over ion-exchange resins, an oxygenate of second generation

J. Massanet, R. Bringué, E. Ramírez, M. Iborra, J. Tejero, R. Soto
University of Barcelona, Faculty of Chemistry, Chemical Engineering Dep., C/Martí i Franquès 1, 08028 Barcelona, Spain;
tel. +34934020155, e-mail: rogerbringue@ub.edu

Governments have established ambitious targets to replace an important fraction of fossil fuels with renewable sources within next 20 years. The EU directives of the Quality & Composition of the fuels expect that the oxygen content of automobile fuels has to be at least 15% at 2020. Consequently, the reformulation of gasoline and diesel oil is needed. Recently, alkyl levulinates, which are bio-based chemicals obtained from lignocellulose, have been proposed as gasoline and diesel additives. Butyl levulinate (BL) is an oxygenate of second generation that can be used as diesel additive because it improves the lubricity, conductivity and cold flow properties of the diesel fuel into which it is blended and leads to cleaner combustion processes with fewer smoke or NO_x emissions. It can be synthetized from furfuryl alcohol (FA) via acid-catalyzed alcoholysis. Some authors have studied this reaction using several solid acid catalysts at butanol (BuOH) to FA molar ratios ($R_{BuOH/FA}$) higher than 30 [1-3]. However, ion-exchange resins have been barely investigated. In this work, we tested five ion-exchange resins (Amberlyst 15, 35, 39, 46 and 121, from now on A15, A35, A39, A49 and A121, respectively) at 100ºC and a $R_{BuOH/FA} = 4$ in order to relate catalyst properties with activity and selectivity. Furthermore, the effect of temperature (70 to 100ºC) and $R_{BuOH/FA}$ (4 and 6) were also checked.

The experiments performed in this work prove that ion-exchange resins are promising catalysts for the etherification of FA with BuOH to obtain BL at 100ºC. The reaction has several steps (see Fig. 1): conversion of FA to an intermediate, 2-(butoxymethyl)furan, (BmF), followed by the conversion of BmF to BL. The most relevant byproducts were oligomeric insoluble substances indicating that polymerization of FA also took place.

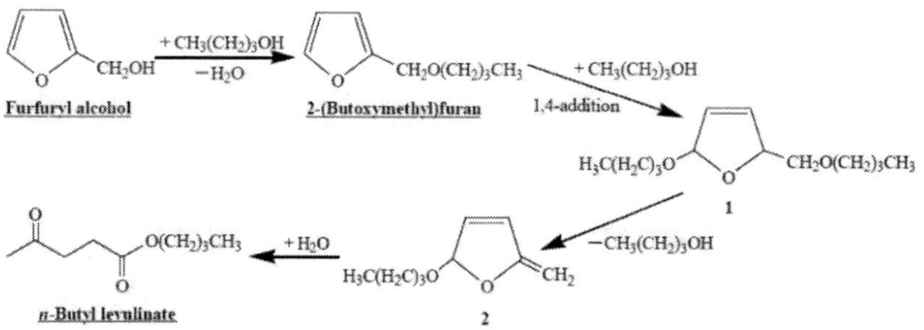

Figure 1. Reaction pathway [4]

The conversion of FA was complete for all catalysts after 4h of experiment. The catalyst with the best performance regarding selectivity and mole production was A39 (67% of selectivity after 6h), followed by A46 (59%) and A15 (58%). Besides, we found that there is an optimal between accessibility to the active sites of the resin and their acid capacity, represented by the parameter (H^+/V_{sp}, number of sulphonic groups per volume of swollen gel-phase). As Fig. 2 shows, the most selective catalyst has an intermediate value of H^+/V_{sp}, corresponding to the conventional sulphonated macroreticular resin with a low percentage of reticulant agent A39.

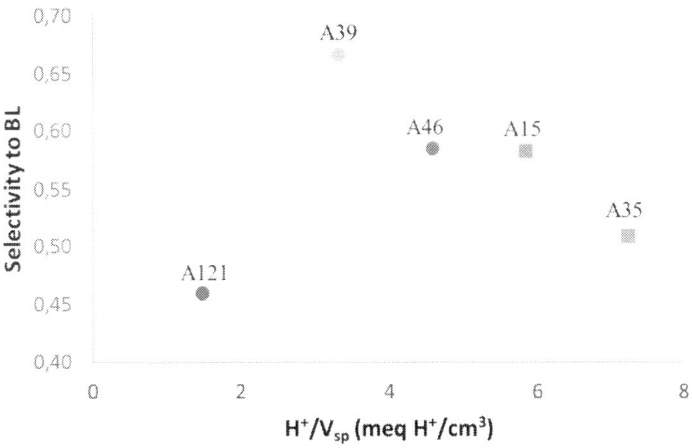

Figure 2. Selectivity to Butyl Levulinate vs H^+/V_{sp}; T= 100°C, $R_{BuOH/FA}$= 4, t= 6h.

The effect of temperature was checked on A35 at the range 70-100°C and $R_{BuOH/FA}$=4. Temperature did not affect much the conversion of FA to BmF (complete at all temperatures), but played an important role in the conversion of BmF to BL: more BL mole were produced on increasing the temperature, as Fig. 3 shows.

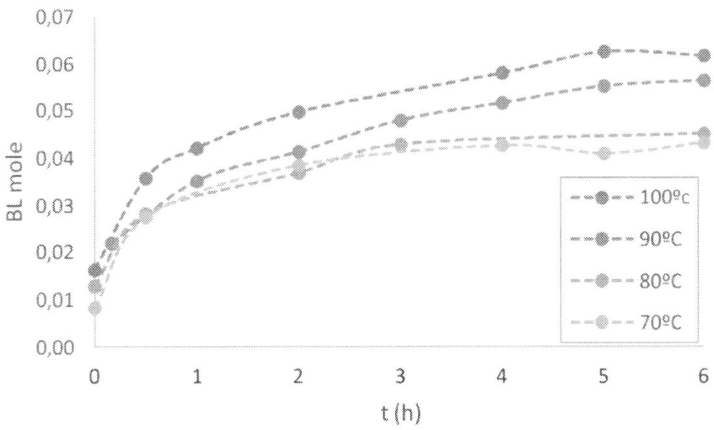

Figure 3. Influence of temperature on BL mole production after 6h, $R_{BuOH/FA}$=4

Finally, the $R_{BuOH/FA}$ was increased to 6 (T=100°C, A 35) to check the effect of FA concentration. As expected, selectivity to BL increased at the highest molar ratio, showing that a high concentration of FA promotes its polimerization to insloluble products.

[1] Demma Carà, P., Ciriminna, R., Shiju, N. R., Rothenberg, G., Pagliaro, M.; ChemSusChem, 7(3), 2014, p.835-840

[2] Zhang, Z., Dong, K., Zhao, Z. K.; ChemSusChem, 4(1), 2011, p.112-118.

[3] Neves, P., Antunes, M. M., Russo, P. A., Abrantes, J. P., Lima, S., Fernandes, A., Pillinger, M., Rocha, S.M., Ribeiro, M.F., Valente, A. A., Green Chemistry, 15(12), 2013, p.3367-3376.

[4] Peng, L., Li, H., Xi, L., Chen, K., Chen, H., BioResources, 9(3), 2014, p.3825-3834

Experimental liquid– and gas–phase study on adsorption of C_1-C_4 alcohols, C_4-C_5 isoolefins and their corresponding ethers over Amberlyst™ 35

R. Soto[1], N. Oktar[2], C. Fité[1], E. Ramírez[1], R. Bringué[1], J. Tejero[1]
[1] Chemical Engineering Department. University of Barcelona. Marti i Franqués 1-11, 08028, Barcelona. Spain. tel. +34 93 402 1304, fax. +34 93 402 1291 (email: r.soto@ub.edu)
[2] Chemical Engineering Department. Gazi University. Maltepe, 06570, Ankara. Turkey. tel: +90 312 582 3556. (email: nurayoktar@gazi.edu.tr)

From an environmental standpoint, tertiary alkyl ethers have attracted considerable attention in the last decades due to its potential application as high octane fuel additives that can reduce the emissions level of current fuels. These ethers are usually synthesized from the acid-catalyzed reaction between primary alcohols and tertiary olefins within the range of temperature 323–353K. Methyl *tert*–butyl ether (MTBE), ethyl *tert*–butyl ether (ETBE), propyl *tert*–butyl ether (PTBE), butyl *tert*–butyl ether (BTBE), *tert*–amyl ethyl ether (TAME) and *tert*–amyl ethyl ether (TAEE) are examples of such compounds. They are produced by the corresponding addition of C_1–C_4 alcohols (methanol, ethanol, propanol and butanol) to C_4–C_5 isoolefins (isobutene and isoamylenes). The most widely used heterogeneous acid catalysts for such syntheses are macroreticular ion-exchange resins, e.g. Amberlyst™ 15 and Amberlyst™ 35. Although several kinetic studies have been focused on tertiary alkyl ether synthesis over ion-exchange resins, experimental studies on adsorption and diffusion of the involved compounds on ion-exchangers are very scarce in the open literature [1-3]. Adsorption related information on these catalysts is not only important for ascertaining the actual kinetic models through which tertiary ethers are formed but also for understanding and improving the performance of new catalysts.

Experimental liquid– and gas–phase adsorption studies on the adsorption equilibria of C1–C4 primary alcohols, C4–C5 isoolefins (isobutene and isoamylenes) and the corresponding tertiary ethers over Amberlyst™ 35 have been carried out in the range of temperature from 323 to 353K. Liquid–phase adsorption experiments were performed in a bacth adsorber using very low adsorbents concentration and n-heptane as solvent. Langmuir-type isotherms were built from experiments at five different initial concentrations at each temperature and adsorption equilibrium constants were determined from linearization of such isotherms. Figure 1 shows an example of the results obtained for butanol at 323 K. Due to decomposition of tertiary ethers in liquid-phase experiments, their adsorption equilibrium constants were only determined in gas-phase. A summary of the adsorption equilibrium constants obtained in the liquid-phase runs is gathered in Table 1.

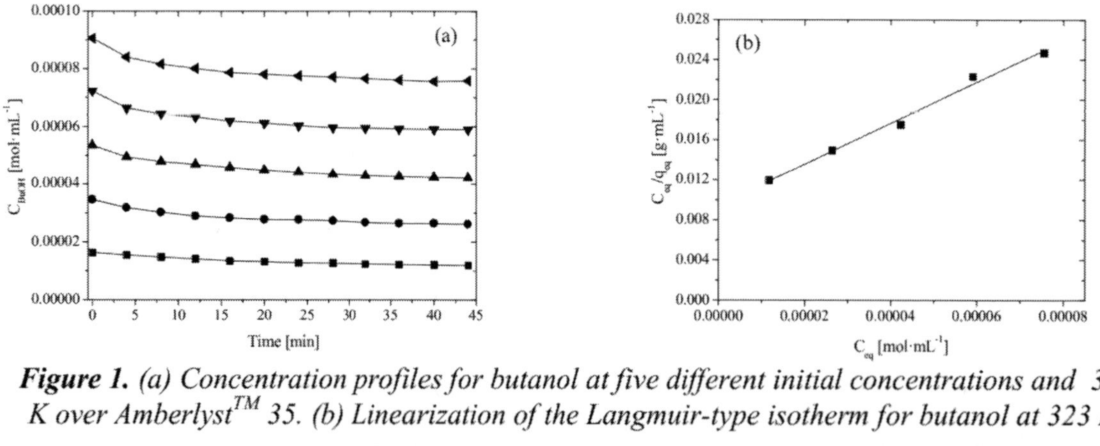

***Figure 1.** (a) Concentration profiles for butanol at five different initial concentrations and 323 K over Amberlyst™ 35. (b) Linearization of the Langmuir-type isotherm for butanol at 323 K.*

Gas–phase dynamic studies were performed in a packed-bed adsorber by carrying out pulse-response experiments. Figure 2a shows an example of the results obtained for MTBE at different temperatures and several flows of carrier gas. Adsorption equilibrium constants of alcohols have been found to be larger than those of isoolefins and ethers. Obtained heats of adsorption from van't Hoff plot of adsorption equilibrium constants confirmed the exothermicity of adsorption process in both liquid- and gas-phase studies. Among ethers from isobutene, the adsorption equilibrium constants decreased in the order: MTBE, ETBE, PTBE and BTBE. Among the isoamylenes, 2-methyl-1-butene (2M1B) showed higher value of adsorption equilibrium constants than its isomer 2-methyl-2-butene (2M2B).

Table 1. *Summary of the adsorption equilibrium constants obtained in liquid-phase.*

	$K_{a,L}$ [mL/mmol]				
T (K)	Methanol	Ethanol	Propanol	Butanol	2M2B
323	23.22	31.87	21.14	21.67	7.05
333	18.97	28.60	17.44	15.37	4.15
343	15.01	25.16	13.42	13.78	3.39
353	13.30	20.56	12.56	10.59	6.85

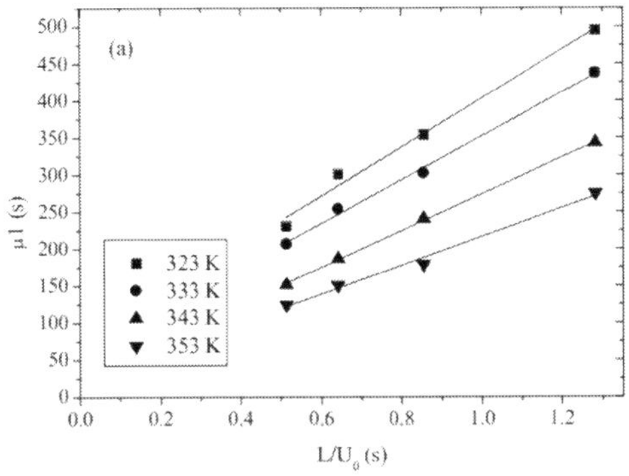

Figure 2. *(a) Example of the gas phase results obtained for MTBE. (b) SEM image for Amberlyst™ 35 at 200000X.*

Moment techniques have been applied to both set of experiments and the contributions of rate parameters concerning diffusivities of involved compounds have been estimated. Scanning electron microscopy (SEM) was performed on pretreated particles of Amberlyst™ 35 and the average micrograin radius was estimated from such images as 15.56 nm (See Figure 2b). Important contribution of micropore diffusion to the second central moments for alcohols, olefins and tertiary ethers have been found in gas-phase experiments, which suggests significant resistances toward diffusion of the involved compounds within the catalyst gel–phase. Macropore diffusion coefficients obtained varied from $1 \cdot 10^{-7}$ to $4 \cdot 10^{-7}$ cm^2s^{-1} for liquid-phase and from $9 \cdot 10^{-4}$ to $4 \cdot 10^{-3}$ cm^2s^{-1} for gas-phase. Micropore diffusion coefficients (within the catalyst gel-phase) obtained varied from $1 \cdot 10^{-15}$ to $1 \cdot 10^{-17}$ cm^2s^{-1}.

References:

- [1] N. Oktar, K. Mürtezaoğlu, G. Doğu, and T. Doğu. *Can. J. Chem. Eng.* 77 (1999) 406–412.
- [2] T. Doğu, E. Aydin, N. Boz, K. Mürtezaoğlu, and G. Doğu. *Int. J. Chem. React. Eng.* 1 (2003) 1–10.
- [3] T. Doğu, N. Boz, E. Aydin, N. Oktar, K. Mürtezaoğlu, and G. Doğu. *Ind. Eng. Chem. Res.* 15 (2001) 5044–5051.

Kinetics of the hydrogenation of 2-methyl-3-butyn-2-ol over a structured palladium-based catalyst

S. Vernuccio, Ph. Rudolf von Rohr,
Institute of Process Engineering, ETH Zurich, Sonneggstrasse 3, 8092 Zurich, Switzerland;
tel. +44 44 632 24 99, e-mail: svernuccio@ipe.mavt.ethz.ch; fax: +44 44 632 13 25.

The semi-hydrogenation of alkynes to alkenes is a key transformation for the production of fine chemical. In industry, these reactions are usually carried out with palladium-based catalysts in slurry reactors. However, these catalysts are mostly in the form of fine powders and therefore require filtration after the reaction.

To overcome this problem a novel combination of catalyst support and reactor design has been developed. A laser sintered porous structure, coated with an aluminium-oxide and zinc-oxide base layer and impregnated with palladium nanoparticles, was evaluated as a semi-hydrogenation catalyst in area of process intensification.

The advantages of this structured catalyst are a high surface area, good mixing qualities and low pressure drops. Furthermore, it is characterized by a regular geometry easily reproducible and suitable for computational simulations.

The structured reactor was integrated in a hydrogenation plant operated with recirculation of the liquid phase and continuous feeding of hydrogen.

The solvent free hydrogenation of 2-methyl-3-butyn-2-ol (MBY) to 2-methyl-3-buten-2-ol (MBE) was chosen as test reaction to evaluate its potential. This process is considered to be a key reaction for the industrial production of vitamins and aroma compounds. The selectivity of this process is related to the possibility of hydrogenating MBY to MBE while preventing its further hydrogenation to 2-methyl-2-butanol (MBA) (Scheme 1).

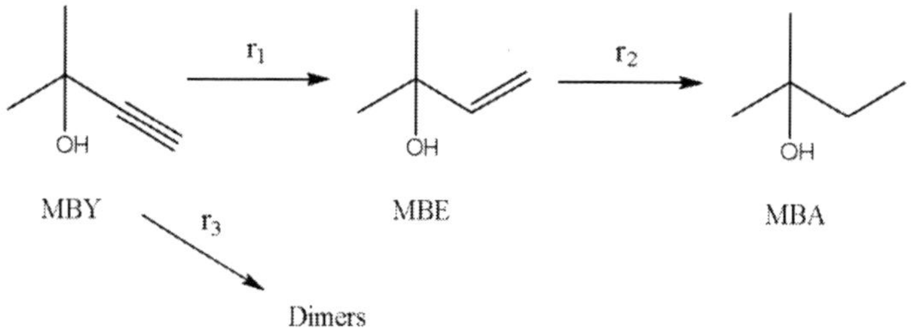

Scheme 1. Reaction network of MBY hydrogenation.

A kinetic investigation has been performed with reference to a previously developed Langmuir-Hinshelwood-type kinetic mechanism. This model assumed non-competitive adsorption of the reactants/products and dissociative adsorption of hydrogen on the catalyst surface. A first order reaction with respect to hydrogen was found to describe the kinetics of the process.

The presence of gas-liquid and liquid-solid external mass-transfer effects has also been investigated. The goal of the work was to study the kinetic behaviour of a palladium-based structured reactor to evaluate its feasibility for industrial application in the semi-hydrogenation of alkynes. This porous structure was found to be a viable alternative to increase the efficiency of the conventional batch processes, showing potential for process intensification.

Kinetics of the liquid-phase syntheses of propyl *tert*-butyl ether (PTBE) and butyl *tert*-butyl ether (BTBE) over Amberlyst^{TM} 35

J.H. Badia, R. Bringué, C. Fité, M. Iborra, R. Soto, F. Cunill
University of Barcelona, Faculty of Chemistry, Chemical Engineering Department, Martí i Franquès 1-11, 08028 Barcelona, Spain; tel. +34 93 402 0155, fax. +34 93 402 1291, e-mail: rogerbringue@ub.edu.

The kinetics of liquid-phase additions of 1-propanol and 1-butanol to isobutene to produce propyl *tert*-butyl ether (PTBE) and butyl *tert*-butyl ether (BTBE), respectively, have been studied over the ion-exchange resin Amberlyst^{TM} 35 (A-35). Both ethers have promising properties as gasoline additives. Their production from the corresponding biomass-derived alcohols allows increasing the biofuel market share of the gasoline pool. Compared to methyl *tert*-butyl ether (MTBE) and ethyl *tert*-butyl ether (ETBE), which are formed by analogous reactions, the use of heavier ethers as gasoline additives presents potential benefits, since they have higher energy content and lower Reid vapor pressure. Industrial implementation of these two syntheses would not involve noticeable modifications to the existing MTBE and ETBE production units, but their kinetics must be evaluated first since a variation of the reactive medium properties could arise, due to the molecular size of the involved compounds.

Experiments have been carried out for each studied reaction in the temperature range of 303 to 354 K using two different reaction setups. A batch reactor was used for the main body of the experimental runs and a fixed-bed tubular reactor, operating under differential regime, was used to validate the kinetic models. Also, a synthetic C_4 mixture, with a composition similar to C_4 cuts form Fluid Catalytic Cracking units, was used as the isobutene source for some of the experiments in the fixed-bed reactor. Initial alcohol to olefin molar ratio was varied from 1.0 to 2.0. According to previous works, the formation rates of MTBE and ETBE are not affected by internal mass transfer limitations at 383 K if catalyst particle sizes below 0.40 mm are used. Thus, A-35 was crushed and sieved to obtain particles below that size.

Previous experiments showed that PTBE synthesis reaction rates were not significantly affected by internal mass transfer effects when using 0.25-0.40 mm catalysts beads in the assayed temperature range. On the contrary, BTBE rates were affected by internal diffusion resistance at temperatures higher than 323 K, which could be caused by possible steric hindrances due to the size of either 1-butanol or the resulting ether. Thus, BTBE rate data obtained above that temperature were not used for the BTBE kinetic modelling. External mass transport resistances were found to be non-significant in the assayed experimental conditions.

Rate data free from mass transfer limitations were used to fit a set of kinetic equations, expressed in terms of activities (UNIFAC-Dortmund method), which stem from the LHHW kinetic formalism, in which the surface reaction is considered as the rate-determining step. The considered kinetic equations were systematically obtained by taking into account all possible combinations of adsorbed and non-adsorbed species on the catalyst. The temperature dependence of the rate coefficients and the adsorption equilibrium constants was included, and it was also considered the possibility that one or more adsorption constant values were not sensitive to temperature. Finally, when considering the possible effect of the interaction of the reaction medium with the catalyst matrix on reaction rates, a factor based on the Hildebrand solubility parameter (δ) was included. Optimal values of the kinetic equations parameters were obtained by minimization of the weighted sum of squared errors using the Levenberg-Marquardt algorithm. The best kinetic equations to represent reaction rates for the syntheses of PTBE and BTBE are presented in Equations 1 and 2, respectively:

$$r_{PTBE} = \frac{k'_{PTBE}\left(a_{IB}a_{PrOH} - {a_{PTBE}}/{K_{eq,PTBE}}\right)exp\left[\frac{\overline{V}_M\phi_P^2}{RT}(\delta_M - \delta_P)^2\right]}{\left(a_{PrOH} + k a_{PTBE}\right)^2} \quad (1)$$

$$k'_{PTBE}\left[\frac{mol}{g_{cat}\ h}\right] = exp\left[-(0.09 \pm 0.01) - (9959 \pm 73)\left(\frac{1}{T} - \frac{1}{327.7}\right)\right]$$

$$k = \frac{K_{PTBE}}{K_{PrOH}} = exp\left[-(2.53 \pm 0.06) - (10141 \pm 289)\left(\frac{1}{T} - \frac{1}{327.7}\right)\right]$$

$$\delta_P\left[MPa^{1/2}\right] = (16.37 \pm 0.35)$$

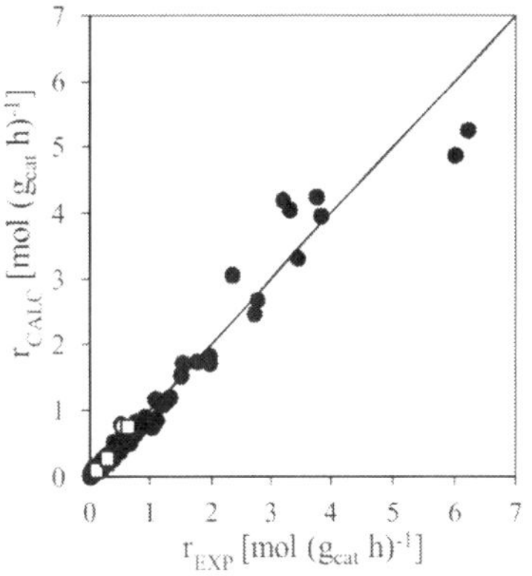

Figure 1. PTBE reaction rates estimated by Equation 1 vs. experimental ones. Batch reactor, pure IB (●); Fixed-bed reactor, pure IB (○); Fixed-bed reactor, C_4 mixture (□)

$$r_{BTBE} = \frac{k'_{BTBE}\left(a_{IB}a_{BuOH} - {a_{BTBE}}/{K_{eq,BTBE}}\right)exp\left[\frac{\overline{V}_M\phi_P^2}{RT}(\delta_M - \delta_P)^2\right]}{\left(1 + K_{IB}a_{IB} + K_{BTBE}a_{PTBE}\right)^2} \quad (2)$$

$$k'_{BTBE}\left[\frac{mol}{g_{cat}\ h}\right] = exp\left[-(2.203 \pm 0.006) - (6849 \pm 88)\left(\frac{1}{T} - \frac{1}{328.6}\right)\right]$$

$$K_{IB} = exp\left[-(1.13 \pm 0.01) + (8637 \pm 115)\left(\frac{1}{T} - \frac{1}{328.6}\right)\right]$$

$$K_{BTBE} = exp\left[-(0.72 \pm 0.05) + (8720 \pm 540)\left(\frac{1}{T} - \frac{1}{328.6}\right)\right]$$

$$\delta_P\left[MPa^{1/2}\right] = (15.0 \pm 0.1) - (0.592 \pm 0.009)(T - 328.6)$$

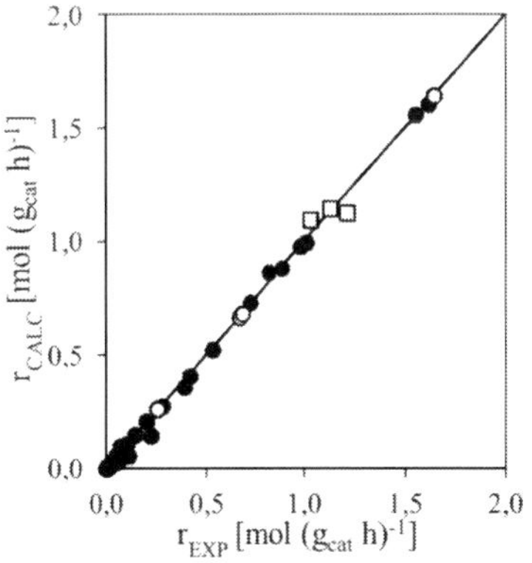

Figure 2. BTBE reaction rates estimated by Equation 1 vs. experimental ones. Batch reactor, pure IB (●); Fixed-bed reactor, pure IB (○); Fixed-bed reactor, C_4 mixture (□)

where the values of the equilibrium constants for each reaction, $K_{eq,i}$, had been previously determined in an earlier work, and K_j corresponds to the adsorption constant of the compound *j*. As seen in Figures 1 and 2, the selected kinetic expressions describe satisfactorily the rate data for the formation of both ethers.

According to the PTBE model (Equation 1), two active centers, or clusters of them, would participate in the surface reaction step, the number of vacant active sites can be regarded as non-significant, and the interaction between the reaction medium and the catalyst has a significant influence on rates. The BTBE model (Equation 2) also indicates that two active sites are needed and that the influence of the medium is significant but, for the BTBE synthesis reaction, the number of vacant active sites should be considered as significant.

From the temperature-dependent term in the apparent rate coefficients, the calculated apparent activation energies are (82.8 ± 0.6) kJ mol^{-1} for the PTBE synthesis and (57.0 ± 0.7) kJ mol^{-1} for BTBE. As seen, the values of δ_P estimated separately for the two systems are rather similar, within the margin of experimental error, which gives rise to validity of the kinetic models.

Heterogeneous catalyst study in copolymerization reactions, among propylene oxide and carbon dioxide in subcritical conditions

M. Pinilla de Dios [a], S. Pérez Nogal [a], K. Ëranen [b], T. Salmi [b] and J. García Serna [a].

[a] High Pressure Processes Group, Department of Chemical Engineering and Environmental Technology, University of Valladolid, Escuela de Ingenierías Industriales, sede Doctor Mergelina s/n, 47011 Valladolid, Spain. Tel.: +34 983184934, Tel.: +34669034365, e-mail address: mpdedios@gmail.com [b] Johan Gadolin Process Chemistry Centre at Åbo Akademi University, Turku, Finland

Carbon dioxide (CO_2) is one of the main gases that cause the Greenhouse Effect. As a way of attenuate this damage, CO_2 in these copolymerization reactions, was used as a C1 feedstock. A huge variety of heterogeneous catalysts has been used for polymerization processes. In this case, Zinc glutarate (ZnGA) was selected for copolymerization reactions. It has been studied copolymerization processes in a carbon dioxide (CO_2) atmosphere. CO_2 acted also as a monomer for the reaction. The other monomer that was involved in the reaction was Propylene oxide. Throught the mechanism of Ring Opening Polymerization, the incorporation of CO_2 in the polymer chain, can be possible. The goal of this article was to study this heterogeneous catalyst in the copolymerization reaction and determine if new green polyurethanes and polycarbonates can be synthesized with an industrial application. Several reactions have been taken place changing the amount of ZnGA (0.3, 0.6, 1, 1.5 g), the temperature of reaction (60, 80 °C), the pressure of CO_2 (6, 10, 25 barg) in the reactor, molecular weight of the final polymer (725, 3000 g/mol) and the reaction time (20, 40h). Through these results, firstly it was possible to determine which the best amount of catalyst was for the reaction and which was the effect in the polymerization reaction increasing it. Secondly, once the best amount of catalyst was determined, it was set and the temperature of the copolymerization was studied. It is possible to analyse if an increase in temperature, provides better final copolymers. Third, about the pressure in the reactor, thermodynamics and kinetics say if it has a lower impact in the polymerization than temperature. So, it was possible to determine that in subcritical conditions of CO_2, pressure was not an important variable of study. Finally, two more variables have been studied: molecular weight of the final copolymer, which showed a considerable impact in the viscosity of the final copolymer when it was increased, and the reaction time. This last variable showed if the process can be done in less time with the same quality.

Keywords: Zinc Glutarate, glycol, polypropylene carbonate, polyetherpolyols, polyurethanes, ring opening polymerization, subcritical conditions.

Oxidative conversion of light alkanes to oxygenates

S.A. Tungatarova, T.S. Baizhumanova, Z.T. Zheksenbaeva, K. Kassymkan, M. Zhumabek
D.V. Sokolsky Institute of Fuel, Catalysis and Electrochemistry, 142, D. Kunaev str., Almaty, 050010, Republic of Kazakhstan; tel. +77272916632, e-mail: tungatarova58@mail.ru;

The huge hydrocarbon reserves in Kazakhstan are consumed mainly in the form of domestic, industrial and motor fuel. The remaining gas is burnt in torches or pumped into the reservoir (petroleum gases). Therefore, processing of alkanes (the main component of natural gas and oil) to generate oxygen-containing organic compounds is one of the most urgent environmental problems in Kazakhstan. Preparation of these compounds by direct synthesis from alkanes is economically more expedient than by multistep processes.

It should be expected the formation of unsaturated hydrocarbons, aldehydes, acids and alcohols at partial oxidation of methane, ethane, propane and butane. Only the optimal selection of catalysts may be specifically carry out the process with the preferential formation of one of the listed products.

The oxidation of hydrocarbons relates to important large-capacity industrial processes. Alcohols, aldehydes, ketones and carboxylic acids, which are the main products of oxidation of light hydrocarbons, are widely used [1].

Research of activity of the developed catalysts in the oxidation of C_1-C_4 mixture into oxygenates is the aim of the present work.

Mono-, bi- and three-component catalyst systems for the oxidation of linear hydrocarbons have been developed. Their pre-treatment was carried out for the preparation of catalysts on natural carrier. Natural carriers were dried at T = 473 K for 2 h and calcined at T = 773K for 2 h, and then treated with 10% HCl and calcined again at T = 773 K for 2 h.

Catalysts based on Mo, Cr and Ga were prepared by incipient wetness capillary impregnation using a mixed aqueous solution of nitrate salts of metals. Active phase was supported on natural carriers and alumina. Then the catalysts were dried at T = 473 K for 2 hours and calcined at T = 773 K for 2 hours in air to decompose the nitrate from the catalyst surface. Research the activities of the developed catalysts were carried out at a flow installation at atmospheric pressure in a quartz tubular reactor with a fixed catalyst bed.

Analysis of the initial mixture and reaction products was performed using a "Chromos GC-1000" chromatograph with the "Chromos" software. The chromatograph is equipped with packed and capillary columns. A packed column was used for the analysis of H_2, O_2, N_2, CH_4, C2H6, C_2H_4, C_3-C_4 hydrocarbons, CO and CO_2. The capillary column was used for analysis of organic liquids such as alcohols, acids, aldehydes, ketones and aromatic hydrocarbons. The temperature of thermal conductivity detector - 473 K, the temperature of evaporator – 553 K, the column temperature - 313 K, the rate of carrier gas (Ar) - 10 ml/min.

Calculation of chromatographic peaks was carried out on calibration curves, which were built for the corresponding products using the "Chromos" software by pure substances. Exactly measured quantities of pure component or mixture of substances containing known concentrations were injected using microsyringe into the chromatograph. Calibration curve V = f (S), where V - amount of substance in ml, S - area of the peak, expressed in cm^2, was constructed based on measured peak areas corresponding to the number of injected substances. The concentrations of the resulting products were determined on the basis of the calibration curves.

Catalytic oxidation of methane of the natural gas into oxygen-containing compounds has been studied. The developed 5% Mo - 4% Cr - 1% Ga catalyst supported on the natural

Kazakhstan's clay with the structure of clinoptilolite, was investigated in oxidation of methane in flow quartz reactor under the conditions: $CH_4 : O_2 = 2 : 1$ (67% : 33%) T = 400 - 900° C and a space velocity W = 7500 h^{-1}. It was found that both liquid and gaseous products are formed during the reaction. The formation of oxygen-containing products: methanol, ethanol, propanol and acetaldehyde is observed at lower temperatures. The formation of H_2, CO and CO_2 were found in the gaseous phase.

Investigation of the activity of polyoxide catalysts containing 1, 5 and 10% of the total content of Mo-Cr-Ga as the active phase, supported on TWC (Torgai white clay), was carried out in the oxidative conversion of propane-butane mixture into oxygen-containing compositions at varying temperatures from 673 to 823 K, C_3-$C_4 : O_2 : N_2 : Ar = 5 : 1 : 4 : 5$, $\tau = 8$ s and W = 450 h^{-1}.

Research of incomplete oxidative conversion of propane-butane mixture on molybdenum-, chromium-, gallium polyoxometallates, supported on TWC, possible to determine that the preferred composition of the products is determined by the reaction temperature. It was found that 5% MoCrGa/TWC catalyst is the most optimal at changing the content of active phase of catalyst from 1 to 10%. Up to 33% acetaldehyde, 50.9% acetone and 38% methyl ethyl ketone were obtained in the test interval.

Thus, it was studied the effect of changing the catalyst and carrier composition on the composition and quantitative ratio of the major products of process.

References

1 Baizhumanova T.S., Tungatarova S.A., Zheksenbaeva Z.T., Kassymkan K., Zhumabek M. Synthesis of oxygenates by oxidation of light alkanes on modified catalysts // Chemical Engineering Transactions. – 2015. – Vol.45. – P. 1063-1068.

Modelling an integrated Resid Fluid Catalytic Cracking unit with a secondary riser

M.A. Rakib[1], M. Karakaya, S.U. Basheer, G. Singaravel, A. Taher, M. AlMusharfy
Takreer Research Centre, Abu Dhabi Oil Refining Company (TAKREER), Post Box: 3593, Abu Dhabi, UAE;
S. Spatenka, J. Aguerrevere
Process Systems Enterprise, 26-28 Hammersmith Grove, London W6 7HA, UK;
[1]tel. +971 26020851, e-mail: mrakib@takreer.com;

Fluid catalytic cracking is a major conversion process in modern refineries, and a major contributor to the commercial gasoline market. Accurate prediction of the performance of the commercial Resid Fluid Catalytic Cracking (RFCC) unit is required to enable the process engineers with guiding principles during their daily operation, which consists of several operating conditions, and also to establish an optimum operation widow to maximize refinery profits.

Low value, high molecular weight hydrocarbons like resids and vacuum gas oils from atmospheric and vacuum columns respectively are thereby converted into valuable products, thereby improving refinery profit margins. The main products are LPG and gasoline, and depending on market requirements, the process can be optimized to maximize the desired product. Coke is produced as a significant by-product, leading to a quick deactivation of the catalyst; however, this has been converted to an advantage by burning it off during regeneration step, thereby also providing the heat requirement to the otherwise endothermic process.

Atmospheric residue is fed to the main riser of the RFCC unit. Products from the riser are separated in a main fractionator, and olefin rich light cut naphtha is recycled back to the unit as feed to the secondary riser. Since the process generates significant amount of coke, catalyst regeneration by coke combustion is conducted in a 2-stage regenerator to cope up with the hydrothermal deactivation of the catalyst in the regenerator.

This work deals with the development of a comprehensive reactor model set for a dual riser and 2 stage regenerators configuration which has been optimized for processing atmospheric residue feed, while maximizing propylene production. The risers with fluidization being in the fast fluidization regime have been modelled as plug flow reactors, with an in-house generated kinetics rate expression to describe the cracking reactions and coke generation. The regenerators have been estimated to operate in a bubbling bed regime, and have been treated with the 2-phase theory of fluidization. The bubble phase has been modelled as a plug flow reactor, while the emulsion phase has been modelled as a CSTR, with consideration of interphase heat and mass transfer.

An equation-oriented modelling environment gPROMS® has been used, with the underlying fluid properties calculated from Multiflash® properties package. The model has been fine-tuned with design data for the unit, and utilized to conduct parametric studies for overall performance of the unit. Such a comprehensive reactor model for an RFCC unit gives a high level of confidence for the refinery process engineers to account for the pressure balance, heat balance of the unit, which are key for proper operation of the unit. With robust predictions, this model will be further utilized for operational optimization of the RFCC unit.

Development of composite materials by combustion synthesis for catalytic methane reforming into hydrocarbons and synthesis-gas

[1]G. Xanthopoulou, [1]K. Karanasios, [2]S.A. Tungatarova, [2]T.S. Baizhumanova, [2]Z.T. Zheksenbaeva, [2]M. Zhumabek, [1]G. Vekinis
[1]Institute of Nanoscience and Nanotechnology, NCSR Demokritos, Athens, 15310, Greece;
[2]D.V. Sokolsky Institute of Fuel, Catalysis and Electrochemistry, Almaty, 050010, Kazakhstan;
tel. +77272916632, e-mail: tungatarova58@mail.ru

Processing of natural gas into motor fuel in the XXI century has become one of the major problems of oil and gas chemistry. Currently, only steam methane reforming process is implemented in the industry for the production of synthesis gas. However, the processes of partial oxidation and carbon dioxide reforming of methane, which are not implemented in practice and are performed at high temperatures, attract a great attention of researchers in recent years. Resistant to temperature extremes and thermal shocks is one of the most important requirements for this kind of catalysts. It is desirable to have a high thermal conductivity to reduce the possibility of local overheating.

Works on research of intermetallic compounds of transition metals as a contact mass for the conversion of methane were carried out. The method of self-propagating high temperature synthesis (SHS) was used for the synthesis of catalysts. Catalysts of this type are characterized by thermal stability, mechanical strength and high thermal conductivity. SHS is a heterogeneous combustion process, which proceeds without the participation of oxygen with low energy consumption, high efficiency and rapidity. Self-propagating mode of chemical interaction is realized at local initiation, which moves along the mixture. In this case, almost no gas is released. The very high temperature develops in the reaction zone.

The catalysts of oxidative and carbon dioxide reforming of methane were prepared by SHS method from powder mixtures of metal oxides. Self-propagating combustion wave was initiated after pressing the batch and preheating to 400-600°C. The temperature of the combustion front has evolved to 830-1150°C, the specific surface area of the synthesized catalyst decreased to 0.5-0.8 m^2/g, and the temperature operating range of catalyst began from 700°C. Experiments on the catalytic reforming were carried out on flow type installation at atmospheric pressure in a tubular quartz reactor with a fixed catalyst bed.

Mixtures of salts ($Ni(NO_3)_2$ - $Co(NO_3)_2$ - $Al(NO_3)_3$ - $Mg(NO_3)_2$) are used for the synthesis of catalysts. The reactions resulting in the formation of spinels $NiCo_2O_4$, Co_2AlO_4 and $CoAl_2O_4$ and intermetallic compounds AlCo, $AlNi_3$ occur at preparation of samples by solution combustion synthesis. Figure 1 presents data on the catalysts with 50% Co, which show that the magnesium ions are almost completely replaced by cobalt ions in the $MgAl_2O_4$ spinel. As a result, the $CoAl_2O_4$ and Co_2AlO_4 spinels are formed in the reaction products.

Investigation of the activity of catalysts based on the initial mixture of $Ni(NO_3)_2$ - $Co(NO_3)_2$ - $Al(NO_3)_3$ - $Mg(NO_3)_2$ produced in the solution combustion synthesis process was carried out in the reaction of carbon dioxide conversion of methane. 100% methane conversion at 750°C was carried out on the catalyst with 47% $Ni(NO_3)_2$, whereas the conversion of CO_2 reached 81.7% at a temperature of about 900°C. Hydrogen yield reached 99.2%, yield of CO - 99.1% in the ratio of $H_2/CO = 1.2$. The activity corresponded to 0.96 mol H_2/g Cat × h and 0.78 mol CO/g Cat × h.

Tests have shown that the maximum conversion of CO_2 at 750°C was observed on the catalyst with 50% $Ni(NO_3)_2$ in the initial mixture. Catalysts with 50% $Co(NO_3)_2$ have maximum activity by CO and H_2. 100% conversion of CH_4 and CO_2, and 100% yield of H_2 and CO at a ratio of H_2/CO = 0.9 - 1.2 and the activity of 1.70 mol H_2/g Cat × h and 1.47 mole CO/g Cat × h observed with increasing of space velocity on the catalyst with 50% of $Co(NO_3)_2$ in the starting material. However, the catalysts with 50% $Co(NO_3)_2$ in the starting material have insufficient stability over time due to coking, associated with the formation of the $NiAl_2O_4$ spinel. Thus, the lower the concentration of spinel, the higher the stability of catalyst over time.

Figure 1. Electron-microscopic picture and results of chemical analysis (EDS spectrum) of catalyst 50% $Co(NO_3)_2$ + 36,15% $Al(NO_3)_3$ + 13,85% $Mg(NO_3)_2$ + 37% urea from the area indicated on the image.

Stability of the catalyst with 50% $Ni(NO_3)_2$ depending on the time at a space velocity of 3300 h^{-1} was investigated. It should be noted that the catalyst retained its high catalytic activity. At the end of the 8 h of CH_4 conversion was 92.6%, the conversion of CO_2 - 94.3%, yield of H_2 reached 91.9% and CO yield - 93.5%, the ratio of H_2/CO close to 0.9 and the activity per hydrogen was 6.05 mol H_2/g Cat × h, and the activity per CO - 6.10 mol CO/g Cat× h. This catalyst is needed regeneration after 8 h of continuous operation. The catalyst of similar composition is active and selective for the partial oxidation of methane to synthesis gas. The highest results were obtained on reduced catalyst samples.

Thus, effective catalysts for the production of synthesis gas from methane of natural gas as a result of studies have been developed. The modern method of self-propagating high temperature synthesis and solution combustion method were used for the synthesis of catalysts. XRD and scanning electron microscopy of samples with chemical analysis before and after catalysis indicated on single-phase of produced systems and the absence of significant changes in the phase composition after experiments. The activity of catalysts obtained in the solution combustion process was examined in carbon dioxide reforming of methane and partial oxidation of methane to synthesis gas.

The work was supported by the Ministry of Education and Science of the Republic of Kazakhstan (Grant No 0247/GF4).

Supported heteropoly acid catalysts of oxidative conversion of alkanes

S.A. Tungatarova, T.S. Baizhumanova, M. Zhumabek, I.S. Chanysheva, A.V. Shapovalov, V.P. Grigoriyeva
D.V. Sokolsky Institute of Fuel, Catalysis and Electrochemistry, Almaty, 050010, Kazakhstan;
tel. +77272916632, e-mail: tungatarova58@mail.ru

Physical and chemical properties of heteropoly acids (HPA) at low-temperature oxidative catalysis and action of reaction medium were investigated enough. Such investigations are insufficient for high-temperature reactions. It makes difficult exposure the nature of a stable catalytic action of supported heteropoly compounds (HPC) at high-temperature processes including partial oxidative conversion. The experimental data on investigation of thermal stability and dynamics of charge and structural transformations of compact the 12th series P-Mo HPA and supported on SiO_2 in conditions of consecutive influence of components: O_2 of air at the presence water vapor, H_2, O_2, and H_2O at the temperatures of real oxidative processes (473-1273 K) are demonstrated.

15-30% HPA supported on SiO_2 have been prepared by impregnation of the carrier by water solutions of HPA on a moisture capacity with the subsequent drying on air at 383 K within 5 h. Physical and chemical properties of catalysts were investigated by complex of methods: XRD on DRON-407 device with the cobalt anode, and IR-spectroscopy - on SPECORD-75 spectrometer in KBr tablets. Spectra of thermal programmed reduction by hydrogen (TPR) and thermal programmed oxidation by oxygen (TPO) were detected by registration of absorbed H_2 or O_2 at 293-1273 K by chromatographic method with thermal conductivity detector at passing of Ar-5%H_2 (vol.) or He-5%O_2 (vol.) mixtures (40 sm^3/min) and speed of the programmed rise of temperature - 8 K/min.

The received results allow to assume that the beginning of structural destruction of crystalline [PMo_{12}] begins from $\geq$ 773 K (disappears characteristic a.b. 950 sm^{-1} from Mo=O), MoO_3 phase is formed. Thus a.b. in the field of 1060-1150 sm^{-1} from P-O-Mo (molybdophosphates) are kept up to 1073 K. However the subsequent processing of the sample by superheated steam at 923 K leads to renewal of all a.b. in the IR-spectrum, which are characteristic for Keggin type of HPA with high intensity (780,860,940,1060 sm^{-1}). It is possible to assume that introduction of water vapor into a reaction mixture promotes "regeneration" of the Keggin structure of [PMo_{12}] from a mixture of remained fragments after calcination (Mo-O-Mo, Mo/O\Mo - bridge, molybdophosphates) as due to reoxidation, and introductions of H_2O molecules into Keggin skeleton with formation of fluidized structures. The role of water as oxidizer in the processes of partial oxidation and oxidative dehydrogenation of methane was established by a combination of isotope and kinetic methods in works. Thus, research of thermal stability of the [PMo_{12}] and [$PMo_{10}V_2$] HPA in a steam-air mixture by IR-spectroscopy has shown, that it is steady at temperatures $\leq$ 673 K. The beginning of HPA destruction on oxide components (MoO_3, V_2O_5, P_2O_5) together with intermediate oxide-like formations (P-O-Mo, V-O-Mo) is possible at the temperatures $\geq$ 773 K. Formation of MoO_3 phase and molybdophosphates with preservation of low-intensive HPA a.b. was observed at 1023 K. The subsequent processing of samples in water (superheated steam) after steam-air processing at 923, 1023 K leads to strongly pronounced "regeneration" of the Keggin structure of HPA (a.b. 780, 860, 960, 1050 sm^{-1}) at preservation of low-intensive a.b. from MoO_3 (580, 870,980 sm^{-1}). Further results of investigation on interaction of structural oxygen of catalysts with hydrogen (TPR method) on the base of compact and supported HPA are demonstrated. Thus, these stages of reduction of the $H_3PMo_{12}O_{40}/SiO_2$ are reflected in occurrence in TPR spectra the peaks of hydrogen absorption at 573-773 K (reduction and decomposition of HPA), 773-1023 K (reduction of MoO_3 into MoO_2), 1023-1273 K (reduction of MoO_2 into Mo°). Reoxidation process by oxygen of 30%$H_3PMo_{12}O_{40}$ and 30%$H_5PMo_{10}V_2O_{40}/SiO_2$ catalysts at TPO regime after their preliminary reduction by hydrogen (TPR) to formation of Mo^0 proceeds in the field of temperatures 473-1023 K with appearance of three absorption peaks of O_2: T_{Max} - 593-603, 683-703, and 743-763 K for $[PMo_{10}V_2]/SiO_2$ catalyst and T_{max} - 623-643, 703-723, and 803-823 K for $[PMo_{12}]/SiO_2$ catalyst. It is visible that reoxidation is the more low-temperature process on 100-200 K, than reduction process of supported catalytic systems (12th series Mo HPA). From comparison of data on T_{red} and T_{ox} of molybdenum in $H_5PMo_{10}V_2O_{40}/SiO_2$ and $H_3PMo_{12}O_{40}/SiO_2$ systems follows that introduction of vanadium into $H_3PMo_{12}O_{40}$ structure leads to increase of interaction temperature with hydrogen of structural oxygen of supported HPA and temperature of formation of molybdenum oxides and Mo°, but reduce reoxidation temperature of the reduced forms of molybdenum in comparison with $H_3PMo_{12}O_{40}/SiO_2$ on 30-50 K. Partial replacement of molybdenum on vanadium in $[PMo_{12}]$ HPA reduces the reduction speed (raises T_{red}) of $Mo^{6+} \rightarrow Mo^o$, but increases the reoxidation speed (reduces T_{ox}) of Mo°, formed on the basis of HPC, into Mo^{4+} and Mo^{6+} oxides. Thus, reversible transition of Mo^{6+} (MoO_3) $\leftrightarrow$ Mo° is observed in catalysts from 12th series HPA on carrier under influence of oxidation-reduction components (O_2, H_2) at the temperature interval 773-1273 K. It predetermines an opportunity of carrying out of the reaction of partial oxidation of C_1-C_4 alkanes on the oxidation-reduction mechanism. It was established that the keeping of 30%$H_3PMo_{12}O_{40}/SiO_2$ catalyst preliminary treated by TPR with hydrogen up to 923 K (there are MoO_3, MoO_2 phases) and TPO up to 873 K, or TPR with hydrogen up to 1273 K (Mo°) and TPO up to 1123 K in water leads to reduction of HPC structure on a surface of carrier. Thus we observe reversible cycle of structural and phase transformations of HPA under influence of environment and temperature at investigation of structural changes of HPA in 30%$H_5PV_2Mo_{10}O_{40}/SiO_2$ and 30%$H_3PMo_{12}O_{40}/SiO_2$ catalysts by TPR TPO, IR, and XRD methods at stages of their step-by-step reduction, subsequent oxidation at 773-1273 K and processing by water (298 K, superheated steam).

Obtained data indicate on lability of HPA/SiO_2 system in oxidation-reduction conditions and influence of water at temperatures close to temperatures of the reaction of partial oxidation and oxidative dehydrogenation of alkanes and easy regeneration of active structure of HPC from finely divided mixture of oxides under influence of O_2 and H_2O. It explains stability of the 12th series Mo HPC supported on Si-containing carrier at high temperatures (873-1073 K) at partial oxidative conversion of alkanes at the temperatures providing reversible destruction of HPA on oxides, dispersion the last under influence of O_2 and formation of HPA under influence of water vapor and temperature at some preservation of HPA elements in samples.

The work was supported by the Ministry of Education and Science of the Republic of Kazakhstan (Grant No 0247/GF4, 0330/GF4).

Comparison of CZA/S, CZZr/S and CZMn/S bifunctional catalysts for the syngas to dimethyl ether process

A. Ateka, P. Perez-Uriarte, M. Sánchez-Contador, I. Sierra, J. Ereña, A.T. Aguayo, J. Bilbao
Univ. of the Basque Country, Dep. Chemical Eng., P.O Box 644, 48080 Bilbao, Spain;
tel. +34 946015361, e-mail: ainara.ateka@ehu.eus; fax: +34 946013500.

Dimethyl ether (DME) is an environmentally friendly fuel (sulfur-free and with a low particle and NO_x emission), which has aroused a great interest in the last decades, as it is considered to be a key intermediate for producing automotive fuels and raw materials, via syngas, from alternative sources to oil (natural gas, coal or biomass). The synthesis of DME in a single step (syngas to DME process, STD) takes place over bifunctional catalysts, with a metallic function for the methanol formation, and an acid function for the selective dehydration of methanol to DME [1].

The aim of this work is to study the kinetic behaviour of $CuO\text{-}ZnO\text{-}Al_2O_3$/SAPO-18 (denoted as CZA/S), $CuO\text{-}ZnO\text{-}ZrO_2$/SAPO-18 (CZZr/S) and CuO-ZnO-MnO/SAPO-18 (CZMn/S) bifunctional catalysts on the direct synthesis of DME, as modifications by adding Zr or Mn have been reported to enhance catalyst activity and stability [2]. After catalyst screening, the effect of reaction parameters on the most suitable catalyst has been analyzed.

Runs have been carried out on a fixed bed reactor (PID Eng Tech Microactivity-Reference) under the following conditions: temperature, 250-350 °C; pressure, 20-40 bar; space time, 2.5-20 $g_{cat}h(mol_C)^{-1}$; time on stream up to 30 h. Reaction products have been analyzed continuously in a Varian CP4900 micro-gas chromatograph. The reaction indexes studied are: conversion of moles of fed CO_x (X); product yield (Y), defined as the molar ratio of DME, MeOH or paraffins in the product stream; and selectivity (S).

Figure 1a shows the yields of the reaction products (DME, methanol and paraffins) obtained with CZA/S, CZZr/S and CZMn/S bifunctional catalysts at zero time on stream under the following reaction conditions: H_2/CO ratio in the feed, 3/1; temperature, 275 °C; pressure, 30 bar; space time, 10.18 $g_{cat}h(mol_C)^{-1}$. The results evidence that bifunctional catalysts containing Zr and Mn (CZZr/S and CZMn/S) show a better kinetic behaviour, and are more active than CZA/S catalyst for the STD process. While a DME yield of 14.60 % is obtained with CZA/S catalyst, using CZZr/S or CZMn/S catalysts, the value of this reaction index is doubled. It is remarkable that a high DME selectivity is reached with the three studied catalysts, over 90 %, maintaining low methanol yields, and almost null paraffin yield in the synthesis of DME in a single step. From these results, both CZZr/S and CZMn/S catalysts seem to be suitable for use in the STD process, rather than CZA/S, the one with the most commonly used metallic function. Nevertheless, regarding to the costs related to catalyst synthesis, CZMn/S catalyst has been selected, as being noticeably more economic than CZZr/S, with a high catalytic activity, DME selectivity (Figure 1b) and stability.

Table 1 summarizes the kinetic results obtained (yield and selectivity) from the study of the effect of the reaction parameters (temperature, pressure and space time) for CZMn/S catalyst. Analyzing the effect of temperature, between 250-350 °C, a maximum DME yield (35.62 %) is observed at 275 °C, temperature reported to be the optimum for most studied catalysts. For temperatures above 300 °C, oxygenates (DME + methanol) yield decreases, and paraffin formation starts to be significant. Reaction pressure enhances DME yield, as the methanol synthesis (the limiting step of the overall process) proceeds with decrease in mole number. However, it has been reported that pressures over 40 bar do not imply a significant improvement of the process yield. On the other hand, increasing space time, CO_x conversion, and thus, DME yield increase, but does not seem to have noticeable effect on the methanol yield (despite this reaction rate also increases, its value does not exceed 1.5 % in any case) and paraffin yield.

The results are encouraging for the CZMn/S catalyst to be used in the STD process at 275 °C and under moderate reaction pressure (30 bar), obtaining DME selectivity over 95 %.

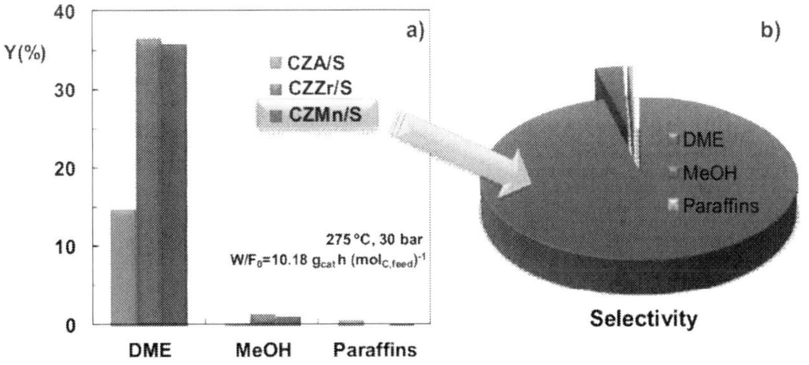

Figure 1. (a) Yields of DME, methanol and paraffins obtained with the three bifunctional catalysts: CZA/S, CZZr/S and CZMn/S; (b) Product distribution obtained with CZMn/S catalyst.

Table 1. Kinetic results (product yield and selectivity) obtained for CZMn/S bifunctional catalyst.

T	P	W/F_0	Yield (%)			Selectivity (%)		
(°C)	(bar)	($g_{cat}h(mol_C)^{-1}$)	DME	MeOH	Paraffins	DME	MeOH	Paraffins
250	30	10.18	19.22	0.78	0.07	97.43	2.04	0.53
275	30	10.18	35.62	1.25	0.25	95.37	3.39	0.84
300	30	10.18	32.91	2.15	4.27	85.50	5.38	9.11
325	30	10.18	31.16	0.96	5.93	81.35	5.56	13.09
350	30	10.18	18.58	0.84	6.43	71.52	6.12	22.36
275	20	10.18	27.69	1.11	0.39	94.86	3.81	1.33
275	40	10.18	42.65	2.00	0.23	95.02	4.46	0.52
275	30	2.50	10.02	0.87	0.35	89.11	7.76	3.13
275	30	5.00	19.90	0.71	0.34	94.97	3.41	1.62
275	30	20.00	49.50	1.70	1.00	94.58	3.27	1.92

Acknowledgements and references

This work has been carried out with the financial support of the Ministry of Economy and Competitiveness of the Spanish Government (CTQ2010-19188 and CTQ2013-46173-R), the Basque Government (Projects GIC/24-IT-220-07 and IT748-13) and the University of the Basque Country (UFI 11/39). Ainara Ateka is grateful for the Ph.D. grant from the Basque Government (BFI09.69).

[1] Z. Azizi, M. Rezaeimanesh, T. Tohidian, M.R. Rahimpour, Chemical Engineering and Processing: Process Intensification 82 (2014) 150-172.

[2] F. Meshkini, M. Taghizadeh, M. Bahmani, Fuel 89 (2010) 170-175.

Catalytic hydrogenation of D-xylose over Ru decorated carbon foam catalyst in SpinChem rotating bed reactor

Tung Ngoc Pham[1], Ajaikumar Samikannu[1], Anne-Riikka Rautio[2], Koppany L.Juhasz[3], Zoltan Konya[3], Johan Wärnå[4], Krisztian Kordas[2], Jyri-Pekka Mikkola[1,4]

[1]Technical Chemistry, Department of Chemistry, Chemical-Biological Centre, Umeå University, SE-90187 Umeå , Sweden, [2]Microelectronics and Materials Physics Laboratories, Department of Electrical Engineering, University of Oulu, P.O. Box 4500, FI-90014 University of Oulu, Finland, [3]Department of Applied and Environmental Chemistry, University of Szeged, Rerrich Béla tér 1., H-6720 Szeged, Hungary, [4]Industrial Chemistry & Reaction Engineering, Department of Chemical Engineering, Process Chemistry Centre,Åbo Akademi University, FI-20500, Åbo-Turku, Finland;

- Tung Ngoc Pham: Tel. +46 90 786 5304, Email: tung.pham@umu.se
- Ajaikumar Samikannu: Tel. +46 90 786 5304, Email: ajaikumar.samikannu@umu.se
- Anne-Riikka Rautio: Tel.: +358 50 364 5199, Email: anneriik@ee.oulu.fi
- Zoltan Konya: Tel.: +36 62 544 620, Email: konya@chem.u-szeged.hu
- Johan Wärnå: Tel.:+358 2 2154428, Email: johan.warna@abo.fi
- Krisztian Kordas: Tel.: +358 40 824 8763, Email: lapy@ee.oulu.fi
- Jyri-Pekka Mikkola: Tel: +46 90 786 5284 or +358 2 215 4574, Email: Jyri-Pekka.Mikkola@umu.se or Jyri-Pekka.Mikkola@abo.fi

In this work we introduce a novel support, flexible carbon foam (pyrolyzed melamine) on which Ru has been immobilized. Different techniques such as: scanning electron microscopy (SEM), transmission electron microscopy (TEM), X-ray photoelectron spectroscopy (XPS), inductively coupled plasma optical emission spectrometry (ICP-OES) and nitrogen adsorption–desorption measurement were used to characterize the developed catalyst. Moreover, the foam was immobilized in a SpinChem® rotating bed reactor (SRBR) (Fig. 1) during the hydrogenation D-xylose experiments. The SRBR was recently developed by SpinChem AB, Umeå, Sweden. When rotating, the SRBR takes advantage of centrifugal forces that generate a flux through a solid catalyst located at the internal parts of the reactor. This type of operations not only provides good isothermal conditions but also improves the liquid-solid and gas-liquid mass transfer to the solid phase of the catalytic sites. Through kinetic measurements which were carried out in a laboratory scale pressurized reactor (Parr®) assisted by SRBR at pre-defined conditions (40-60 bar H_2 and 100-120°C), the effect of temperature and hydrogen pressure on the conversion rate is revealed. These results have been proved by a competitive kinetics model which was found to describe the behavior of the novel system (Ru/CF catalyst used together with the SRBR) very well. Besides, it was revealed that the catalytic activity as well as the stability of our Ru/CF-SRBR is comparable with the commercial ruthenium decorated carbon catalyst (Ru/AC) under identical reaction conditions (Fig. 2). Moreover, all steps from catalyst preparation and catalyst recycling as well as catalytic testing can be performed in an easy, fast and elegant manner without any loss of materials. Briefly, the developed Ru/CF catalyst used together with the SRBR could be used an excellent alternative for the conventional Raney-Nickel catalyst in a slurry batch reactor and offers an attractive concept with obvious industrial applicability.

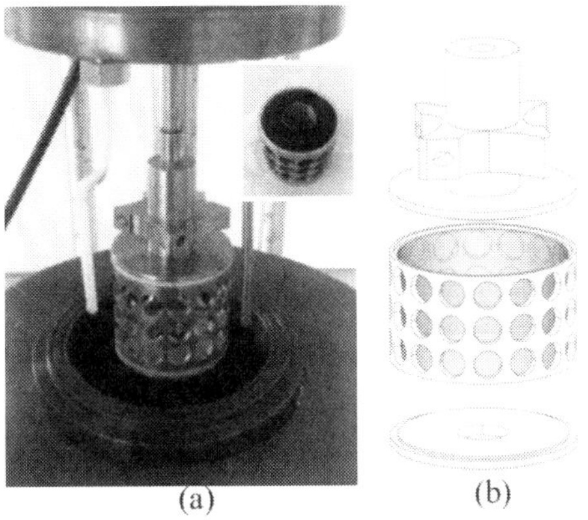

Fig. 1 (a) SRBR with Parr® reactor (big picture) and inside structure of SRBR with Ru/CF catalyst (small picture) and (b) SRBR patterns

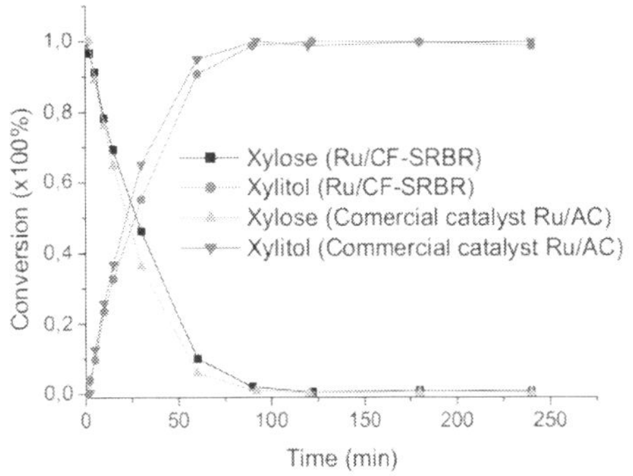

Fig. 2 Catalytic performance of Ru/CF-SRBR compared with commercial Ru/AC without SRBR

Effect of operating conditions on the one-step DME synthesis from H_2+CO+CO_2 mixtures over a CZMn/S catalyst

A. Ateka, M. Sánchez-Contador, P. Rodriguez-Vega, I. Sierra, J. Ereña, A.T. Aguayo, J. Bilbao

Univ. of the Basque Country, Dep. Chemical Engineering, P.O. Box 644, 48080 Bilbao, Spain; tel. +34 946015361, e-mail: ainara.ateka@ehu.eus; fax: +34 946013500.

Dimethyl ether (DME) is receiving a great interest, not only as clean alternative fuel, but as a key intermediate in C_1 chemistry, olefin production, or as hydrogen source. Specially, close attention has been paid to the recent advances in the bifunctional catalysts required for obtaining DME in a single reaction step from CO_2 containing syngas. Using CO_2 in place of CO to produce carbon neutral fuels has become an important task for energy and environmental sustainability [1].

Some recent papers focus on the use of Cu-ZnO-ZrO_2 metallic functions for CO_2 hydrogenation [2]. Nevertheless, more economical bifunctional catalysts composed of CuO-ZnO-MnO metallic function and SAPO-18 as acid function also offer a great catalytic performance. Thereby, the aim of this work is to study the effect of CO_2 concentration in the feed on CO_x conversion, product yield, and selectivity over a CuO-ZnO-MnO/SAPO-18 (CZMn/S) bifunctional catalyst.

For the study, a fixed bed reactor (PID Eng Tech Microactivity-Reference) connected to a Varian CP4900 micro-gas chromatograph for the continuous product analysis has been used. Reaction conditions are as follows: CO_2/CO ratio in the feed, from 0 (H_2+CO) to 1 (50 % CO_2 + 50 % CO); temperature, 250-350 °C; pressure, 10-40 bar; space time, 2.5-20 $g_{cat}h(mol_C)^{-1}$; time on stream, up to 30 h.

Figure 1 shows the effect CO_2/CO molar ratio in the feed has on the yield of the reaction products (DME, methanol and paraffins) at 275 °C, 30 bar and 10.18 g_{cat} h $(mol_C)^{-1}$ (Figure 1a), and on the evolution of DME yield with time on stream (Figure 1b). Results evidence that CO_2 rich feeds (with high CO_2/CO molar ratio), do not have a significant effect on methanol and paraffin yields, leading to a drastic decrease in DME yield. High CO_2 concentration in the reaction medium attenuates deactivation (Figure 1b), as a consequence of the water gas shift reaction. For a higher CO_2 content in the reaction medium, water formation is favoured, which reduces deactivation by coke deposition, as it has been widely reported in the literature [3].

These results, along with those obtained from the study of the effect of the reaction conditions (temperature, pressure, CO_2/CO and H_2/CO_x molar ratios in the feed, space time, and time on stream), restate that CZMn/S bifunctional catalyst is suitable for being used in the STD process cofeeding CO_2 and syngas. Moreover, DME yield and selectivity obtained with low CO_2/CO molar ratio (1/3) are similar to those obtained with the more expensive CZZr/S bifunctional catalyst.

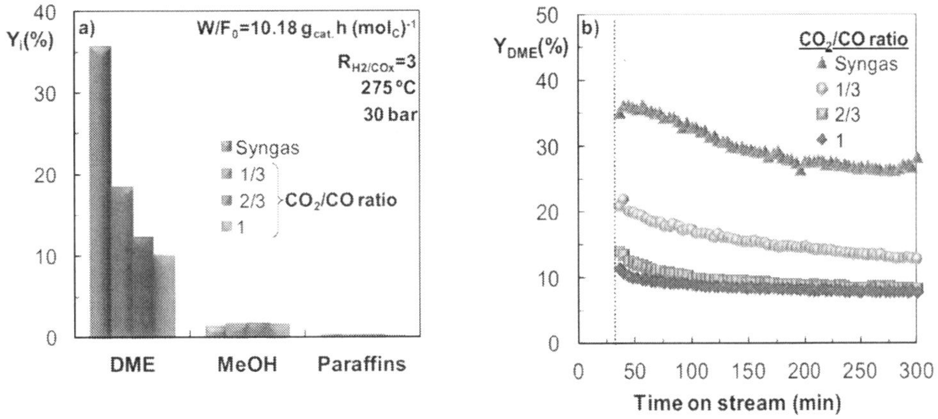

Figure 1. (a) Effect of CO_2/CO molar ratio in the feed on the yield of DME, methanol and paraffins; (b) Evolution of DME yield with time on stream for different CO_2/CO molar ratios in the feed.

Acknowledgements

This work has been carried out with the financial support of the Ministry of Economy and Competitiveness of the Spanish Government (CTQ2010-19188 and CTQ2013-46173-R), the Basque Government (Projects GIC/24-IT-220-07 and IT748-13) and the University of the Basque Country (UFI 11/39). Ainara Ateka is grateful for the Ph.D. grant from the Basque Government (BFI09.69).

References

[1] J. Sun, G. Yang, Y. Yoneyama, N. Tsubaki, ACS Catalysis 4 (2014) 3346-3356.

[2] L. Li, D. Mao, J. Yu, X. Guo, Journal of Power Sources 279 (2015) 394-404.

[3] I. Sierra, J. Ereña, A.T. Aguayo, J.M. Arandes, M. Olazar, J. Bilbao, Applied Catalysis B: Environmental 106 (2011) 167-173.

Erbium-exchanged ZSM-5 catalysed conversion of cellulose to lactic acid

Duo Li, Zhi-Ling Li, Fen-Fen Wang, Chun-Ling Liu, Wen-Sheng Dong
Shaanxi Normal University, School of Chemistry and Chemical Engineering, Westen Chang'an Rd. 620, 710062 Xi'an, China;
tel. +86 29 81530806, e-mail: wsdong@snnu.edu.cn.

The chemocatalytic conversion of cellulose to lactic acid has attracted much attention in recent years, owing to its convenient process and applicability to abundant, renewable carbohydrate biomasses. Recently, we found that $Er(OTf)_3$ and $ErCl_3$ exhibited high activity for lactic acid production from microcrystalline cellulose. As high as ~90% lactic acid yield was obtained [1]. Subsequently, we prepared erbium-exchanged montmorillonite catalysts and achieved a 67.6% lactic acid yield [2]. In the present work, erbium exchanged ZSM-5 catalysts were prepared and examined in the hydrothermal conversion of cellulose. The as-prepared catalysts exhibited excellent catalytic performance in the conversion of cellulose to lactic acid.

The erbium-exchanged ZSM-5 catalysts were prepared by a liquid-phase ion exchange method. The resulting samples were dried at 110 °C overnight and calcined in air at 550 °C for 6 h unless otherwise stated. All the catalytic reactions were performed in a 35 mL stainless-steel autoclave equipped with a mechanical stirrer. The post-reaction liquid was analyzed by a Shimadzu HPLC LC-20AT system equipped with a RID-10A detector and a Bio-Rad Aminex HPX-87H ion exclusion column.

Table 1. The compositions and textual properities of catalysts.

Catalyst	Er content (wt.%)[a]	Si/Al at. ratio[b]	S_{BET} ($m^2 g^{-1}$)	S_{micro} ($m^2 g^{-1}$)	V_p ($cm^3 g^{-1}$)	V_{micro} ($cm^3 g^{-1}$)	D_p (nm)
H-ZSM5	–	24.2	350.8	250.7	0.176	0.11	2.00
Er/ZSM5-1	3.9	19.8	334.1	236.6	0.174	0.108	2.08
Er/ZSM5-2	8.6	17.3	297.0	208.6	0.157	0.095	2.11
Er/ZSM5-3	12.4	17.9	259.4	178.4	0.141	0.081	2.17
Er/ZSM5-2-700[c]	7.9	18.0	289.8	197.3	0.152	0.093	2.17
Er/ZSM5-2-850[d]	7.1	18.4	274.5	187.2	0.149	0.101	2.22

[a] measured by ICP-OES; [b] measured by XRF; [c] calcined at 700 °C; d calcined at 850 °C.

The compositions and textual properties of various ZSM-5 catalysts are shown in Table 1. Figure 1 shows XRD patterns and NH_3-TPD profiles of the catalysts. The results of hydrothermal conversion of cellulose over the catalysts are shown in Table 2.

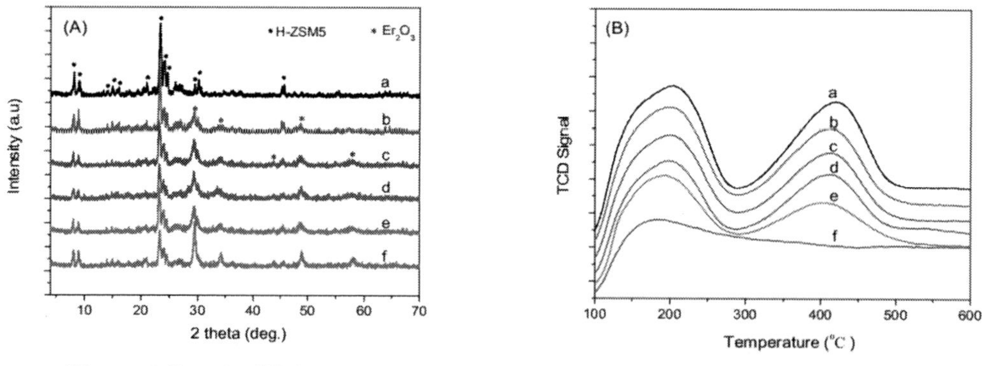

Figure 1. Powder XRD patterns (A) and NH_3-TPD profiles (B) for various catalysts:
(a) H-ZSM5, (b) Er/ZSM5-1, (c) Er/ZSM5-2, (d) Er/ZSM5-3, (e) Er/ZSM5-2-700 and (f) Er/ZSM5-2-850.

The XRD patterns of the H-ZSM5 and erbium-exchanged ZSM5 catalysts are characteristic of a typical MFI-type structure, indicating that ion exchange did not destroy the zeolite MFI topology. However the intensities of the diffraction peaks assigned to ZSM5 were decreased gradually with the introduction of erbium species, indicating partial loss of crystallinity. Moreover, some new diffraction peaks attributed to cubic Er_2O_3 crystallites appeared and their intensities increased with increasing erbium content and calcination temperatures.

NH_3-TPD measurements showed that the patent H-ZSM5 zeolite exhibited two NH_3 desorption peaks centered at 200 °C and 410 °C, respectively, implying that the catalyst contained both weak and strong acid sites. As the increase of the erbium content in the catalysts, both weak and strong acid sites decreased. In addition, higher calcination temperature also resulted in the decrease of the acid sites.

Table 2. Conversion of cellulose into lactic acid over various catalysts.

Catalysts	Conversion (%)	Yield (%)						
		Lactic acid	Formic acid	Acetic acid	Levulinic acid	Acetol	HMF	Glyceral-dehyde
Er/ZSM5-1	100	12.5	4.8	0.3	5.1	3.6	2.5	6.0
Er/ZSM5-2	100	52.3	4.6	0.4	4.5	4.9	1.3	5.1
Er/ZSM5-3	100	49.6	2.3	0.6	3.9	5.0	0.9	1.8
Er/ZSM5-2-700	100	49.1	5.7	0.5	5.0	5.1	1.2	5.3
Er/ZSM5-2-850	100	47.6	5.9	0.5	2.8	4.6	2.9	5.9

Reaction conditions: cellulose 0.3 g, catalyst 0.1 g, H_2O 30 mL, 240 °C, 2 MPa N_2, 30 min.

As the increase of erbium loading from 3.9 to 8.6 %, the lactic acid yield was increased from 12.5 to 52.3%, although the amount of acid sites on the catalyst decreased gradually. These results confirm that erbium species rather than acidic sites were active sites for the conversion of cellulose into lactic acid. As indicated previously [2,3], erbium ions can coordinate with the hydroxyl groups on the cellulose rings because of the higher electron density at these sites. This results in easier cleavage of the protonated ether bonds between the two glucosyl units in the soluble oligosaccharide intermediates that are formed by the depolymerization of cellulose catalyzed by the hydroxonium ions present in the aqueous medium at high temperatures. However, with further increases in Er content in the catalysts to 12.4%, the yield of lactic acid decreased, which may be caused by the degradation of lactic acid to undesirable products on the excessive amount of active sites.

Acknowledgments

The authors gratefully acknowledge financial support from the National Natural Science Foundation of China (Grant no. 21576161).

References

[1] F.F. Wang, C.L. Liu, W.S. Dong, Green Chem. 15 (2013) 2091–2095.
[2] F.F. Wang, J. Liu, H. Li, C.L. Liu, R.Z. Yang, W.S. Dong, Green Chem. 17 (2015) 2455–2463.
[3] X. Lei, F.F. Wang, C.L. Liu, R.Z. Yang, W.S. Dong, Appl. Catal. A 482 (2014) 78 83.

Photocatalytic CO_2 reduction by Cu loaded TiO_2 in the twin photoreactor

Bo-Ren Chen[1], Van-Huy Nguyen[1], Steven Kuan-Yu Lin[1], Jeffrey C.S. Wu[1], Kamila Kočí[2]
[1]Department of Chemical Engineering, National Taiwan University, Taipei 10617, Taiwan; tel. +886 223631994, e-mail: cswu@ntu.edu.tw
[2]Institute of Environmental Technology, VŠB-Technical University of Ostrava, 17. listopadu 15/2172, Ostrava –Poruba 708 33, Czech Republic

Currently in the carbon dioxide photocatalytic reduction, the efficiency is not high and is not always reproducible. We got high amount of products and increased the stability by adding hydrogen into the photoreactor. We studied the effects of the weight loadings and the valance states of copper on the efficiency of carbon dioxide reduction.

Sol-gel method was used to synthesize titanium dioxide. Copper was loaded on titanium dioxide by incipient-wetness impregnation method. The variations of copper valence state were obtained under different calcination atmospheres. CuO was formed under air calcination while Cu was formed under H_2 reduction. Several instruments, UV-Vis, SEM, EDS, XRD and XPS were applied to measure the light absorption, surface morphology, crystalline phase and chemical status. The valence change of copper was found from the oxidation to reduction states before and after the reaction. In the vapor-phase photoreaction, an UVC lamp was used as the light source 254 nm and the light intensity 12 mW/cm^2. The temperature was raised up to 90°C. The reactions were carried out by using TiO_2, 1%CuO/TiO_2, 2%CuO/TiO_2, 1%Cu/TiO_2 and 2%Cu/TiO_2 under 0.01 atm hydrogen. The results revealed that the reduced copper had higher activity for the CO_2 reduction under these conditions. There was not much difference in yields between 1%Cu/TiO_2 and 2%Cu/TiO_2. For these two catalysts, the highest yields were 28.72μmol/g methane, 5.74μmol/g CO and 26.13μmol/g methane, 7.23μmol/g CO, respectively, in 8 hours.

Thus, the 1%Cu/TiO2 was used in the twin reactor with Fe^{2+}/Fe^{3+} as electron mediator. We used artificial sunlight, AM1.5G Xenon lamp, to irradiate the catalysts. Pre-treated Nation cation-exchanged membrane was used for separating two different solutions that were called oxygen side and reduction side. Pt/WO_3 was put into the oxygen side solution. In the reduction side, hydrogen catalyst Pt/$SrTiO_3$:Rh, was put into the solution. The hydrogen generated by the catalyst can be used as the hydrogen source of carbon dioxide reduction. 1%Cu/TiO_2 was put on the quartz plate in the reduction side to contact with the carbon dioxide in the gas phase. The highest yields in reduction side are 0.5863μmol/g hydrogen, 0.2005μmol/g methane and 1.3008μmol/g CO.

Assessment of functional monomers for production of pressure-sensitive adhesives via emulsion copolymerization of methyl methacrylate and butyl acrylate

Natalia B. Barbosa[1], Maria M. Espinola-Colmán[1], Verónica Carranza-Oropeza[1], Dennis Chicoma-Lara[1,2], Reinaldo Giudici[1]
[1]Universidade de São Paulo, Escola Politécnica, Department of Chemical Engineering, Brazil, CEP: 05508-010
[2]Oswaldo Cruz Chemical Industry, Research and Development Department, Brazil, CEP: 07177-220
*rgiudici@usp.br

The growing demanding on the market of Pressure Sensitive Adhesives (PSA) has allowed increasing studies on this material to improve the performance of their end-use properties. This kind of adhesives are intended for applications such as removable, repositionable, low temperature, high tack, the label look and difficult bonding substrates. Also, they work fine under high speed and productivity regimes. Pressure Sensitive Adhesives form a bond simply by the application of light pressure to wed the adhesive with the adherent. PSAs are designed with a balance between flow and resistance to flow. The bond forms because the adhesive is soft enough to flow, or wet, the adherent. The bond has strength because the adhesive is hard enough to resist flow when stress is applied to the bond. Once the adhesive and the adherent are in proximity, there are also molecular interactions such as Van der Waals forces involved, which contribute significantly to the ultimate bond strength. PSAs exhibit viscoelastic properties, both of which are used for proper bonding. PSA are adhesives that are used in various types of tapes and labels. Broadly, there are three types of PSA adhesives: water-based, solvent-based and hot melt. The solvent-based adhesives have some limitations as flammability and toxicity, but they are still used because of properties such as good adhesion to certain types of materials. Thus, we can observe that the composition of the latex is closely related with the behavior of the adhesive. Then, to assess the performance of this kind of adhesives, there are mainly three properties such as loop-tack, peel adhesion and static shear, and the better understanding about them for a specific application will determine success of the product in whatever the substrates and conditions to be overcome.
In this context, this research studied the influence of two parameters on the three performance properties mentioned, in the preparation of PSAs based on polymerization of butyl acrylate and methyl methacrylate. The studied parameters are the reaction temperature (70 °C and 80 °C), and the addition of one additional functional monomer in the formulation, namely 2-hydroxyethyl acrylate (HEA), 2-hydroxyethyl methacrylate (HEMA), 2-hydroxypropyl acrylate (HPA), 2-hydroxypropyl methacrylate (HPMA). Then, reactions were obtained and characterized, such as average particle diameter by dynamic light scattering (DLS) technique, monomer conversion by gravimetry and glass transition temperature by differential scanning calorimetry (DSC). Results showed that the studied parameters such as monomer relationship and polymerization rate influenced on the polymer particle size, which PSD increase when the reaction temperature was raised. Also, a change on functional monomer influenced not only physical-chemical properties such as viscosity, pH, and superficial tension, as well in adhesive properties (loop tack, static shear, and peel adhesion). More details will discuss in this research.

Co_3O_4 supported on different supports for N_2O decomposition

A. Klyushina, K. Pacultová, K. Jiratová, L.Obalová
VSB-TU Ostrava, Institute of Environmental Technology, 17. listopadu 15, Ostrava - Poruba 708 33;
tel. +420 597 327 324, e-mail: anna.klyushina@vsb.cz.

Nitrous oxide (N_2O) is considered as an important pollutant contributing to a greenhouse effect and causing ozone layer depletion. The largest industrial sources of N_2O emissions are waste gases from nitric acid production plants. The low-temperature catalytic decomposition of N_2O (up to 450°C) to nitrogen and oxygen offers an attractive solution for decreasing of N_2O emissions in tail gas from nitric acid production plants. Important question is the choice of suitable catalytic system. Most of the studies mentioned in the literature were performed in laboratory scale in kinetic regime on the catalysts in the grain form. For industrial application, catalyst must be shaped in some way.

In this work Co_3O_4 catalysts supported on different kind of shaped supports (rings, quadrulobes) were prepared by incipient wetness impregnation method. As support materials commercial supports made of MgAl oxides, Al_2O_3 and TiO_2 were used. The catalysts were characterized by AAS, XRD, Raman, FTIR, SEM, H_2-TPR and nitrogen adsorption methods and tested for nitrous oxide decomposition. Influence of geometry and textural parameters on catalytic activity were evaluated by calculation of effectiveness factor. The influence of support composition on the obtained deN_2O catalytic activity is discussed. Samples deposited on support with a high content of Al_2O_3 (Al_2O_3 + 30 wt.% Mg) revealed lower catalytic activity for N_2O decomposition due to the negative influence of Al ions in Co_3O_4 lattice causing formation of hard reducible aluminates. The Co_3O_4 deposited on support with higher content of Mg (Al_2O_3 + 70 wt.% Mg) showed the highest catalytic activity due to favourable dissolution of Mg ions in Co_3O_4 lattice, which probably causes distortion of the spinel structure and thus enable higher oxygen desorption directly influencing N_2O decomposition activity.

This work was supported by the Czech Science Foundation (project no. 14-13750S) and by Ministry of Education, Youth and Sports of the Czech Republic in the National Feasibility Program I, project LO1208 "TEWEP" and by project of specific research SP2015/125.

Simulation study of Fischer-Tropsch reactor by combined structure with oxidative coupling of methane: increasing ethylne production

A. Ghareghashi, [1]F. Shahraki, [1]K. Razzaghi, [2,3]S. Ghader, [4]M. A. Torangi
University of Sistan and Baluchestan, Department Chemical Engineering, P.O.Box 98164-161, Zahedan, Iran Tel. 0098543113 6206, Fax: 0098543113 7092 E-mail: A_Ghareghashi@pgs.usb.ac.ir; [1] University of Sistan and Baluchestan, Department Chemical Engineering, P.O.Box 98164-161, Zahedan, Iran; [2]Department of Chemical Engineering, College of Engineering, Shahid Bahonar University of Kerman, Kerman, Iran; [3]Minerals Industries Research Center, Shahid Bahonar University of Kerman, Kerman, Iran; [4]Faculty of Engineering, University of Golestan, Gorgan, Iran

Ethylene has been viewed as one of the most commonly-used chemicals in the petrochemical industry due to its massive worldwide consumption in the production of some plastics, such as polyethylene. In this work, to obtain the heavy hydrocarbons and ethylene from methane, a two-stage reactor configuration including the oxidative coupling of methane (OCM) as first reactor and following that Fischer–Tropsch (FT) synthesis is presented (Fig. A). In this proposed structure, the FTS section is a combination of membrane-assisted fluidized bed reactor and fixed bed reactor. In the OCM reactor (first reactor), feed gas containing methane, oxygen, and nitrogen, as diluting component, enters the reactor and produces ethane, ethylene, and by-products such as synthesis gas. Eq. (i) indicate the overall pathways of OCM:

$$2CH_4 + O_2 \text{ ® } C_2H_4 + 2H_2O \quad \text{(i)}$$

$$CH_4 + O_2 \text{ ® } CO + H_2O + H_2 \quad \text{(ii)}$$

$$CH_4 + 2O_2 \text{ ® } CO_2 + 2H_2O \quad \text{(iii)}$$

Effluent synthesis gas enters the FT reactor (second reactor) and is converted into valuable hydrocarbons such as ethylene. The reactor configuration is compared with the consecutive reactors of OCM and conventional FT reactor. The set of the governing equations contain mass, energy and momentum equations are applied for both systems. The reactor model was solved numerically by applying the finite difference approximation. The set of first order ODEs equations of model in the axial direction is applied for a system of coupled equations that this obtained ODEs system was solved using the Runge-Kutta method. Effects of percent of N_2 in the feed, CH4/O2 ratio and input temperature on the yield of ethylene and other hydrocarbons are investigated. The results indicate an improvement in the ethylene and heavy hydrocarbons yield. Moreover, a significant decrease is seen in the formation of CO_2 and CH_4. As shown in Fig. B, hydrogen and carbon monoxide produce in OCM reactor and then consume in the FT reactor. Furthermore, mole flow in dual type membrane system is higher than the conventional system due to adding hydrogen to the reaction side by Pd–Ag membrane. As can be seen, the hydrogen permeable membrane has positive effect on synthesis gas conversion. Fig. C shows mole flow of ethylene in two reactors. As seen in this Fig., using of this configuration, increases the amount of produced ethylene.

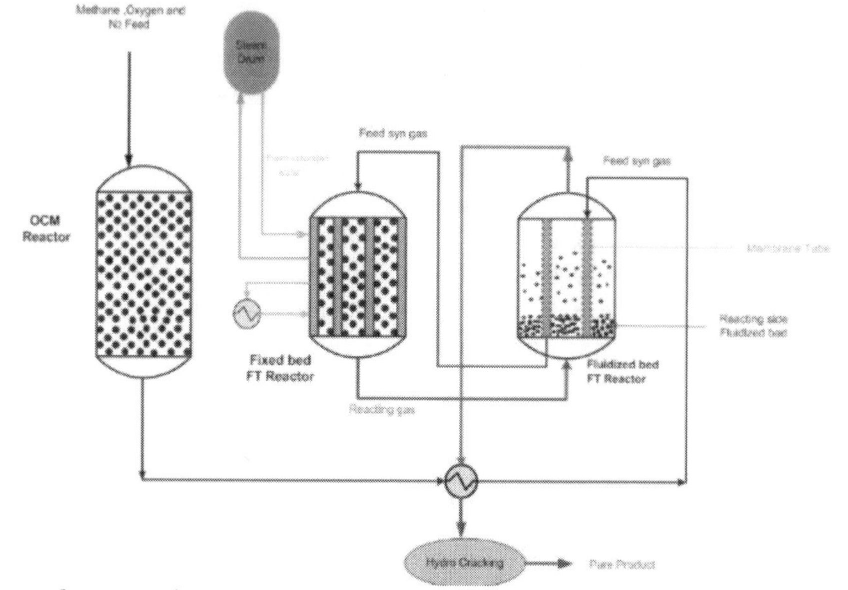

Figure A Figure of consecutive reactors: OCM and FT reactors (fixed bed and membrane fluidized bed in series)

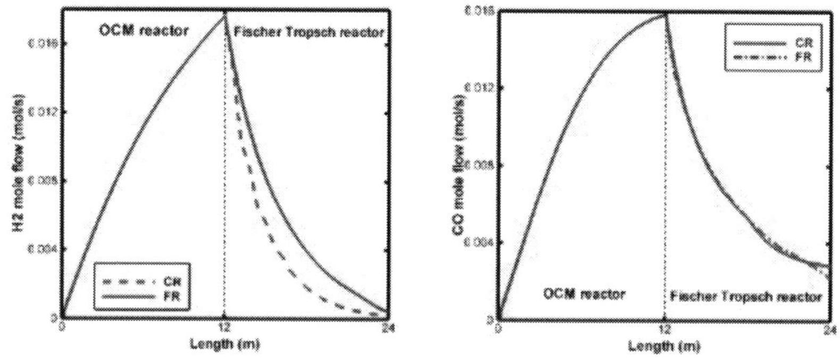

Figure B Variation of CO and H_2 mole flow in OCM and FT reactors (in CR and dual type FT in series (FR)) ($CH_4/O_2 = 12$, $T = 1103$ K).

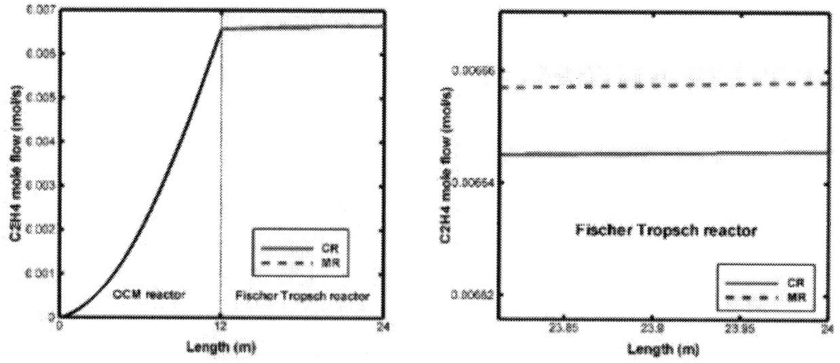

Figure C Variation of C_2H_4 mole flow in OCM and FT reactors (in CR and dual type FT in series (FR)) ($CH_4/O_2 = 12$, $T = 1103$ K).

Methanol-to-olefins : Conversion over tailor-made MFI-type zeolites

Pit Losch and Benoit Louis*
Laboratoire de Synthèse Réactivité Organiques et Catalyse, Institut de Chimie, Université de Strasbourg, UMR 7177 CNRS, 4 rue Blaise Pascal, 67000, Strasbourg Cedex, France
* corresponding blouis@unistra.fr

Actual societal and environmental challenges lead the people to revise the nature of the energy used and also the manner to use it. Recently, Nobel prize laureate, Olah suggested an attractive alternative to use more efficiently available oil and natural gas resources and, eventually, to free humankind from its dependence on fossil fuels. This approach is based on methanol, thus forming the basis of "Methanol economy" concept. Indeed, methanol is the simplest, safest, and easiest liquid oxygenated hydrocarbon to store and transport [1].

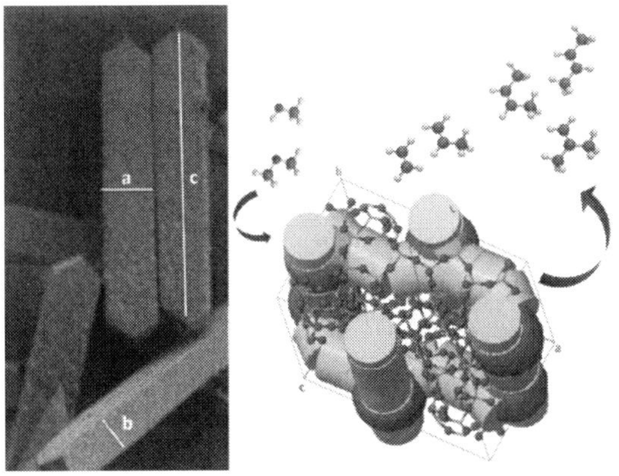

Figure 1 : SEM micrograph of a giant ZSM-5 zeolites synthesized via the fluoride route, MFI-unit cell representing in yellow the straight and sinusoidal micropores (d = 5.5 Å).

Methanol can also be conveniently converted via a single catalytic step either into gasoline (MTG process) or into light olefins, ethylene and propylene (MTO: methanol-to-olefins), which serve as building blocks in the production of synthetic hydrocarbons, polyethylene and polypropylene. This process, discovered by Mobil, is usually conducted over medium-pore zeolites with considerable acidity, rendering among others, ZSM-5 zeolite as a candidate of choice. Its defined MFI structure with peculiar pore architecture and zigzag channels, leads both to efficient shape-selectivity and high methanol conversion.

The synthesis of light olefins has attracted a huge attention during the last decades in order to fulfil the increasing worldwide demand toward ethylene and propylene.

A careful design of the zeolite catalyst is therefore warranted to keep elevated the selectivity toward light olefins. The fluoride mediated route for zeolite syntheses gained our particular interest because highly crystalline materials with a low acid site density were recently developed and proved a very interesting performance in the MTO reaction (X ≈ 100 %, $S_{C2\text{-}C4=}$ > 70 %) [2].

Zeolite	Si / Al / (P) [1]	Si / Al / P [2]	Crystal Volume (μm^3) [3]	X (%) [4]	$S_{C2\text{-}C4=}$ (%) [4]
1	100 / 3	112 / 1	1620	12	84
2	100 / 1	140 / 1	580	92	79
3	100 / 3 / 2	59 / 1 / 0.26	380	70	73
4	100 / 3 / 3	64 / 1 / 0.07	60	35	80
5	100 / 1.5 / 2	102 / 1 / 0.11	980	72	70

1 Sol-gel element ratios.
2 Measured effective element ratios for the synthesized materials (EDX, XRF or ICP).
3 Crystal volumes could be determined by SEM.
4 Methanol and dimethylether-conversion and light olefin selectivity were measured after 60 min TOS under standardized conditions; 400°C, WHSV = 1.12 h^{-1}.

Table 1: Giant zeolites synthesized under different conditions (phosphoric acid vs. hydrofluoric acid)

Those promising results led to a thorough investigation in the optimization of these giant-sized zeolite-crystals. Under standardized conditions, a specific range of zeolites were tested in terms of their MTO-activity (Table 1). The targets of this optimization process were on the one hand to further improve the fluoride mediated synthesis, and, on the other hand to gradually substitute the use of hydrofluoric acid in the synthesis by phosphoric acid. The latter study may result in the integration of phosphorus into the crystalline structure, which could be interesting considering the recent literature on post-synthetic phosphorus modification over MFI zeolites [3].

[1] G. A. Olah, A. Goeppert, G.K.S. Prakash, Beyond Oil and Gas: The Methanol Economy, Wiley VCH, Weinheim, Germany (2006).
[2] F.L. Bleken, S. Chavan, U. Olsbye, M. Boltz, F. Ocampo, B. Louis, *Appl. Catal. A* (2012) *447-448*, 178.
[3] V. Valtchev, G. Majano, S. Mintova, J. Pérez-Ramírez, *Chem. Soc. Rev.* (2013) *42*, 263.

Sol-gel catalyst $CuO/ZnO/Al_2O_3$ applied in the steam reforming of methanol and in the water-gas shift reaction

T. C. Ferrari, R. Menechini Neto, J. L. C. W. Lourenço, O. A. A. dos Santos, L. M. M Jorge
State Univ. of Maringá, Dep. Chem. Engineering, Colombo Av. 5890, 87020-900 Maringá, Paraná, Brasil;
tel. +55 044 30114778, e-mail: oneliaandreo@hotmail.com.

Hydrogen (H_2) proved to be a promising power source (1,2), especially when used in proton exchange membrane (PEM) fuel cells. However, when used directly in this type of cell has come some problems in relation to its storage and transportation (highly reactive, it requires high pressures for storage, and that when out of control, it becomes a powerful explosive). Then it suggests obtaining in situ.

The H_2 can be obtained by various processes, including from the steam reforming of hydrocarbons and alcohols, highlighting methanol (3), as show Equation 1.

$$CH_3OH + H_2O \rightarrow 3H_2 + CO_2 \quad (1)$$

The methanol steam reforming (MSR) reaction has received much attention in the past few decades as an attractive route of producing hydrogen (4). Methanol is liquid at room temperature, has a high concentration of hydrogen (4 hydrogen atoms per carbon atom), can be processed at relatively low temperatures (250 to 350°C) at atmospheric pressure and, in addition, may be obtained biomass (5,6) characterized as a simple, economical and environmentally friendly alternative.

In this process, the hydrogen stream is almost always accompanied by significant amounts of carbon monoxide (CO) poisoning that causes the electrodes of the cell PEM fuel, reducing their efficiency. Great attention has been given to using copper-based catalysts in the water-gas shift reaction (WGSR) for the removal of CO on the cell feed steam, as the Equation 2. So, the H_2-rich stream generated by reforming can be applied in fuel cell autonomous (7). In order to solve the problem, efforts have been conducted to improve the performance of catalysts for both reactions.

$$CO_2 + H_2 \rightleftharpoons CO + H_2O \quad (2)$$

Knowing the advantages that hydrogen brings as an alternative for obtaining energy in fuel cells this work aims to study the use of Cu catalysts / ZnO / Al2O3, synthesized by sol-gel method, in steam reforming reactions the methanol and shift reaction gas-water currents aimed at producing pure hydrogen. The material was characterized by TGA / DSC, BET area, pore diameter and pore volume, atomic absorption spectrophotometry, S-TPR and SEM. Experimental data was obtained in the reaction unit presented in Figure 1, which is made up of five basic parts. Gas feeding (storage cylinders); liquid feeding (water for WGSR ans water+methanol to the steam reforming), carried out using a pump; reaction unit, consisting of a microreactor (tube containing a very small amount of catalyst placed over a stainless steel screen with openings smaller than the particle size) housed inside an electrically-heated oven and a thermocouple to control the temperature; separation of condensable substances (condenser and phase separator); and chromatographic analysis, responsible for the identification and quantification of both the liquid and the gas products.

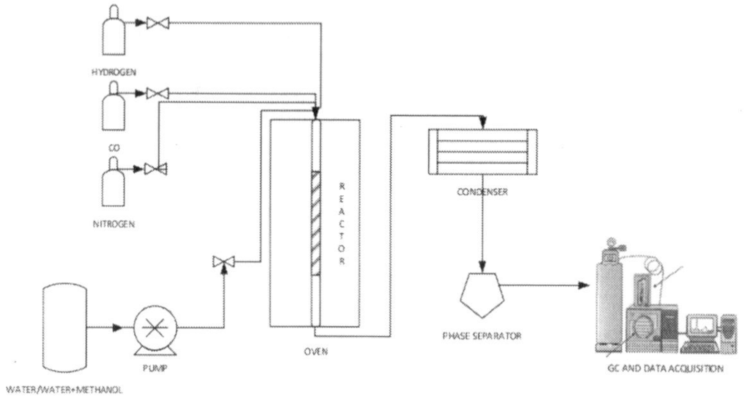

Figure 1. Schematic diagram of the reaction unit.

The results show that the sol-gel catalyst exhibits interesting physical properties, such as large metallic area, high dispersion of copper, with high crystallinity and good reducibility of Cu, which according Jeong et al. (8) is directly related to good catalytic performance . The catalyst prepared by sol-gel method shows great versatility as it featured catalytic activity for the three studied reaction systems.

References

1. G.G. Lenzi, E.K. Lenzi, C.V.B. Fávero, M.K. Lenzi, R.M.M. Jorge, O.A.A. Santos, L. M. M. Jorge. Int. J. Chem. React. Eng. **2010**, 8, A35.
2. Á. Mastalir, Á. Patzkó, B. Frank, R. Schomäcker, T. Ressler, R. Schlögl. Catal. Commun. **2007**, 8, 1684–1690.
3. S.G. Sanches, Dissertação de Mestrado, Pontifícia Universidade Católica do Rio de Janeiro, 2009.
4. C. Mateos-Pedreroa, H. Silva, D. A. P. Tanaka, S. Liguori, A. Iulianelli, A. Basile, A. Mendes. Appl. Catal. B: Environ. **2015**, 174–175, 67–76.
5. F. Hasegawa, S. Yokoyama, K. Imou. Bioresour. Technol. **2010**, 101, 109–111.
6. S.T. Yong, C.W. Ooi, S.P. Chai, X.S. Wu. Int. J. Hydrogen Energy. **2013**, 8, 9541–9552.
7. P. Guo; L. Chen; Q. Yang; M. Qiao; L. Hui; H. Li; H. Xu; K. Fan, *Int. J. Hydrogen Energy*, **2009**, 34, 2361 -2368.
8. D. W. Jeong; W. J. Jang; J. O. Shim; W. B. Han; H. S. Roh; U. H. Jung; W. L. Yooh. *Renew. Energ.*, 65, **2014**, 102-107.

The mechanism of oxidative conversion of cyclohexane on heterogeneous biomimetic catalyst

S.Aghamammadova[1], I.Nagieva[1,2], L.Gasanova[1], T.Nagiev[1,2*]
[1]Nagiev Institute of Catalysis and Inorganic Chemistry, National Academy of Sciences of Azerbaijan, 113 H.Javid av., az-1143, Baku, Azerbaijan, Tel.: (99412) 538-47-07, Fax: (99412) 538-47-67; E-mail: tnagiev@azeurotel.com
[2]Baku State University, 23 Z.Khalilov, az-1148, Baku, Azerbaijan

Heterogenized catalysts based on iron porphyrinic complexes, where the oxidation processes conducted in the gaseous phase, were developed [1,2]. Biomimetic catalyst which was a perfluorinatedtetraphenylporphyrin iron (III), applied on alumina (per-FTPhPFe (III) / Al2O3), exhibited high activity and selectivity in the oxidation processes of a large number of hydrocarbons including cyclohexane oxidation with hydrogen peroxide. The gas-phase oxidation of cyclohexane over this catalyst was carried out in the temperature range of 130-250 ^{0}C with hydrogen peroxide in an aqueous solution of 15-40% wt.

The results of studies on the effect of temperature on the cyclohexane conversion mechanism are shown in Figure 1, where it can be seen thatin the temperature up to150-160 °C the reaction that mainly proceeds is cyclohexane oxidation reaction, forming cyclohexanol and cyclohexanone. At the temperature above 180°C the reaction of oxidative dehydrogenation of cyclohexane to cyclohexene (curve 4, Figure 1) begins to accelerate, respectively, the yield of cyclohexanol and cyclohexanone decreases. The sharp increase in the yield of cyclohexene along with the decrease in the yield of formed cyclohexanol indicates that at the temperature of 230 °C or higher cyclohexanol can also be converted into cyclohexene at lower concentrations of H_2O_2 in the reaction system. This conclusion is confirmed by the results of experimental studies of the process of cyclohexanol oxidation with hydrogen peroxide at a lower concentration (20%) with the biomimetic catalyst, where at elevated temperature (200-230 °C) the cyclohexene yield is high (16-30%).

Table 1: The dependence of the reaction yield of cyclohexanol coherently synchronized oxidation with bio simulator per-FTPhPFe(III)/Al_2O_3 on the temperature: $[H_2O_2]$=20%,
$V_{H_2O_2}$ =1,41 ml/h, $V_{C_6H_{12}}$ =0,9 ml/h, $C_6H_{11}OH:H_2O_2$=1:1

	Composition of feedstock				Composition of obtained reaction yield				
t, ^{0}C	$C_6H_{11}OH$	2-methyl cyclohexanol	Hexanoic acid	Hexanoic acid cyclohexyl ester	$C_6H_{11}OH$	C_6H_{10}	1,2 cyclohexan ediol	1,3 cyclohexadi ene	2-methyl cyclohexanol
150	97,524	0,481	0,712	0,845	83,921	2,321	2,248	0,105	0,093
180	97,524	0,481	0,712	0,845	78,186	7,841	2,875	0,145	0,583
200	97,524	0,481	0,712	0,845	67,494	16,688	2,887	0,279	0,295
230	97,524	0,481	0,712	0,845	55,255	30,374	2,0	0,1	0,641

The experiments on the effect of the H_2O_2 concentration at a constant temperature (200 °C) and contact time were conducted. As shown in Figure 2, increased H_2O_2 concentration in the reaction zone contributes to the increase of oxidation reaction rate, which results with the increase of cyclohexanone yield and decrease of cyclohexene yield.

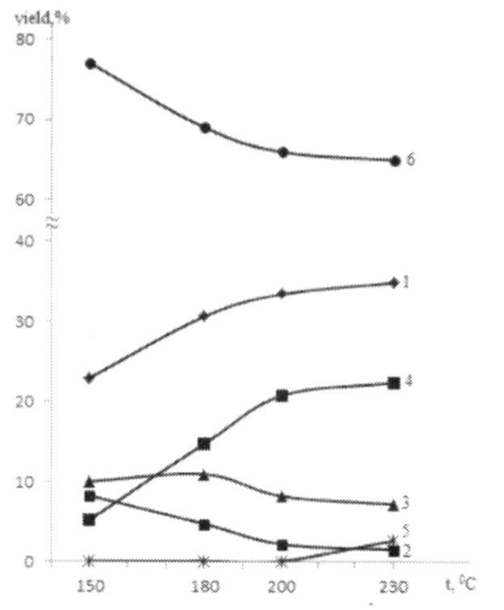

Fig.1.The dependence of the reaction yield of cyclohexanecoherently synchronized oxidation with bio simulator per-FTPhPFe(III)/Al_2O_3on the temperature: $[H_2O_2]$=25%, $V_{H_2O_2}$=1,41ml/h, $V_{C_6H_{12}}$=0,9 ml/h, C_6H_{12}:H_2O_2=1:1,7
1- conversion; 2- cyclohexanol;
3- cyclohexanone; 4- cyclohexene;
5- cyclohexadiene; 6-O_2

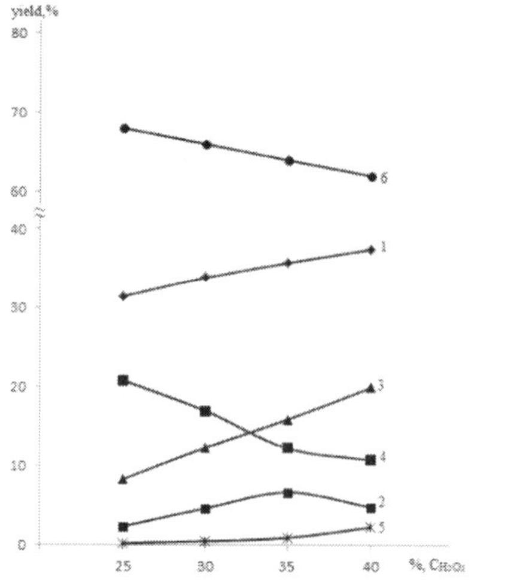

Fig.2.The dependence of the reaction yield of cyclohexanecoherently synchronized oxidation withbio simulator per-FTPhPFe(III)/Al_2O_3on concentration H_2O_2: t=200^{0}C, $V_{H_2O_2}$=1,41ml/h, $V_{C_6H_{12}}$=0,9ml/h,,
1- conversion; 2- cyclohexanol;
3- cyclohexanone; 4- cyclohexene;
5- cyclohexadiene; 6-O_2

Thus, the complex reaction, consisting of parallel-sequential oxidation and dehydrogenation reactions, which are coherently synchronized, proceeds during the process of cyclohexane oxidation with biomimetic catalyst. Depending on the reaction parameters it is possible to deliberately adjust the direction of oxidation reaction and reaction rate.

It should be noted that in terms of the mechanism of each stage of this complex reaction of cyclohexane parallel-sequential oxidation with hydrogen peroxide with biomimetics, each oxidation reaction is carried out under inducing effect of hydrogen peroxide intermediates and consists of two coherently synchronized reactions - 1) catalase (H_2O_2 decomposition reaction) and 2) monooxygenase (cyclohexane oxidation reaction) with biomimetic catalyst.

References
[1].T.M.Nagiev, *Coherent Synchronized Oxidation by Hydrogen Peroxide*, Elsevier, Amsterdam, 2007.325p.
[2].U.V.Nasirova, I.T.Nagieva, T.M.Nagiev, Journal of Materials Sciences and Engineering B2(4) (2012) pp.306-312.

Low-coordinate sites on CuPt alloy nanoclusters on TiO_2 for enhanced photocatalytic CO_2 conversion

Sooho Lee,[†] Do-Heyoung Kim[‡] and Doh C. Lee*[,†]
[†] Department of Chemical and Biomolecular Engineering (BK21+ Program), KAIST Institute for the Nanocentury, Korea Advanced Institute of Science and Technology (KAIST), Daejeon 34141, Korea; tel. +82-42-350-3977, e-mail: dclee@kaist.edu, fax: +82-42-350-3910
[‡] School of Chemical Engineering, Chonnam National University, Gwangju 61186, Korea

Central to efficient photocatalytic conversion of CO_2 to hydrocarbon is the use of metal co-catalysts that provide active sites for adsorption and protonation of CO_2 molecules. Cu nanocrystals have shown promise in this regard, as highlighted in previous studies that the chemisorption of CO_2 on Cu is relatively strong compared to the case of other transition metals. Yet, the adsorbed CO_2 molecules are reduced to produce CO, as the protonation on Cu is slow.

In this study, we use CuPt alloy nanoclusters anchored on TiO_2 as a cocatalyst for photocatalytic conversion of CO_2 to CH_4. The alloyed cocatalysts provide both active CO_2 adsorption sites and high rate of protonation, in reference to Cu and Pt nanoparticles in their mono-elemental forms. As the size of CuPt alloy decreases, the photocatalytic activity improves significantly: for example, 1.2 nm CuPt nanoclusters presented 4.5 times higher production rate than 3.6 nm CuPt nanoparticles on TiO_2. Density functional theory (DFT) analysis reveals that the higher photocatalytic rate of smaller CuPt nanoparticles is attributed to an increasing number of low-coordinated (211) edge sites that enable stronger adsorption of CO_2 molecules and intermediates (e.g., COOH, CO) than (111) planes. The alloying results in the production of CH_4 as CO_2 intermediates and hydrogen atoms are adsorbed in a proximity, facilitating protonation of the intermediate molecules.

In addition, X-ray photoelectron spectroscopy (XPS) analysis indicates that hydroxyl groups on the surface of TiO_2 also facilitate the protonation, as a result of the interaction between hydroxyl groups and intermediates adsorbed on CuPt cocatalysts. The synergistic effect of CuPt alloy and hydroxyls on TiO_2 proves to be a key to efficient CO_2 reduction, highlighting a strategic platform for the design of highly active and selective photocatalysis of CO_2 conversion.

Effect of the space time in 1-butene oligomerisation over a ZSM-5 zeolite catalyst

M. Díaz, A. Ateka, E. Epelde, P. Pérez-Uriarte, J. Bilbao, A.T.Aguayo
Department of Chemical Engineering, University of the Basque Country, P.O.Box 644, 48080 Bilbao, Spain, tel: +34 94 6015361, e-mail: marta.diaz@ehu.es

INTRODUCTION

The oligomerisation of 1-butene is a good route in order to obtain higher hydrocarbons, with the adequate composition to incorporate them to the gasoline pool (C_4-C_{11}) and fuel oil (C_{12}-C_{20}) [1-3]. The 1-butene has a low industrial interest; therefore the oligomerisation is an interesting alternative not only to supply the growing demand of fuels, but also to revalue this excess olefin. Furthermore, 1-butene can be obtained from more sustainable sources such as biomass or waste, the produced fuels will not have heteroatoms.

The zeolite HZSM-5, owing to its properties, is considered a good catalyst to promote the oligomerisation against the cracking and thus minimize the coke formation.

EXPERIMENTAL

The catalyst studied is a HZSM-5 zeolite catalyst of SiO_2/Al_2O_3= 30 ratio (Zeolyst Int.). The zeolite has been agglomerated with pseudobohemite (30 wt. %) as binder and Al_2O_3 (20 wt. %) as inert that confers a matrix with meso-and macropores to the catalyst.

The catalyst was characterized by nitrogen adsorption-desorption isotherms (ASAP Micromeritics 2000) to determine the micropore volume an external surface area (Table 1). The acid properties have been obtained by ammonia temperature desorption DSC-TPD.

The catalytic tests were performed in an isothermal fixed bed reactor at the operation conditions: 275°C; atmospheric pressure; 1-butene partial pressure, 70 %, space time 1-6 (g of catalyst) h (mol of CH_2^{-1}); time on stream, 10 h. The products obtained are divided in gases, analysed on-line in a gas chromatograph (Micro-GC Agilent 3000A), and in liquids analysed at the end of the reaction in gas chromatograph (GC-MS QP2010S Shimadzu).

Table 1: Physico-chemical properties of HZSM-5 zeolite (SiO_2/Al_2O_3= 30).

S_{BET} ($m^2\ g^{-1}$)	V_{micro} ($cm^3\ g^{-1}$)	Total Acidity (mmol $_{NH_3}$ g $_{zeolite}^{-1}$)	Acid strength (kJ mol $_{NH_3}^{-1}$)
489	0.157	0.77	125

The criteria used for assessing the performance of the catalyst were: i) conversion; ii) product yield and selectivity of each lump of products (C_3H_6, C_4H_{10}, C_5-C_7 paraffins, C_5-C_7 olefins, C_8^+), and; iii) deactivation by coke deposition.

RESULTS AND DISCUSSION

The Figure 1 (a) shows the evolution of the products selectivity with the space time. At low values of the space time the selectivity to C_8^+ reach a maximum, while the selectivity to the by-products grows when the contact time increases.

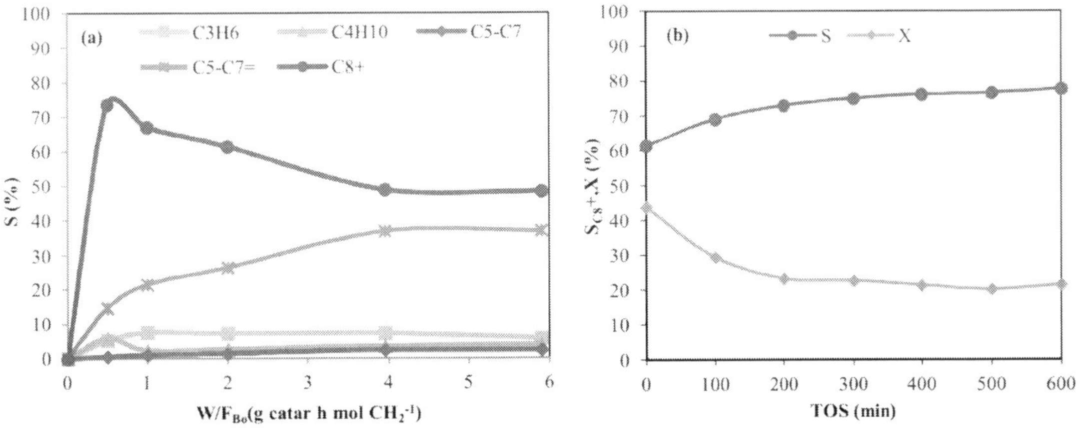

Figure 1: Evolution of the products selectivity with the space time at zero time on stream (a) and evolution of the conversion and C_8^+ selectivity, at a space time of 2 (g of catalyst) h (mol of CH_2^{-1}) (b).

In the figure 1 (b) it is observed that the conversion of 1-butene decreases after two hours of reaction and then tends to be stable during the time on stream. The selectivity of C_8^+ rises as the conversion falls with the time on stream.

These results manifest the possibility to obtain automotive fuels selectively from 1-butene in a process at atmospheric pressure and mild temperatures. The HZSM-5 zeolite catalyst regenerability allows to reuse it.

ACKNOWLEDGEMENTS

This work has been carried out with the financial support of the Minister of Science and Technology of the Spanish Government (Project CTQ2013-46173R), the Basque Government (Project PRE-2014-1-344) and the University of the Basque Country (UFI 11/39).

REFERENCES

[1] R.J. Quann, L.A. Green, S.A. Tabak, F.J. Krambeck, Ind. Eng. Chem. Res. 27 (1988) 565-570.
[2] S. Rossini, Catal. Today 77 (2003) 467-484.
[3] A. Coelho, G. Caeiro, M.A.N.D.A. Lemos, F. Lemos, F. Ramôa Ribeiro, Fuel 111 (2013) 449-460.

Modeling of NO, CO and C_3H_6 catalytic oxidation activity depending on Pt sites oxidation state

A. Arvajová, J. Březina, P. Kočí*
University of Chemistry and Technology, Prague, Department of Chemical Engineering, Technická 5, 16628, Prague, Czech Republic, *e-mail: petr.koci@vscht.cz

Oxidation of CO, unburned hydrocarbons and NO are crucial reactions in diesel exhaust gas aftertreatment systems. These reactions take place on oxidation catalysts containing platinum group metals (PGM), mostly platinum and palladium. Diesel engines are operated with lead fuel mixture (excess of air) giving rise to oxidative (lean) nature of the exhaust gas. Under such conditions, the activity of catalytic centres becomes affected by the formation of Pt oxides (PtO_x).

It is known that NO oxidation is a reaction sensitive to Pt oxidation state – higher PtO_x coverage negatively influences the NO oxidation activity [1,2]. The catalyst deactivation leads to lower NO_2 yield. However, the increased NO_2:NO_x ratio (approx. 0.5) is needed for faster NO_x reduction in an subsequent SCR catalyst. The catalyst containing PtO_x can be reactivated, for example in a reducing atmosphere returning the catalytic sites back to metallic state.

Experimental observations showed that the catalyst activity dependence on the actual PGM sites on the surface (reduced metal vs. metal oxide) should be included into a reaction kinetic model of a diesel oxidation catalyst. In this study we propose to extend the existing model for CO, hydrocarbons and NO oxidation and implement the dependence of NO oxidation on actual state of catalytic sites, represented by the fraction of the formed PGM oxides (ψ_{PtOx}), i.e.:

$$R = \Psi_{Pt}\,(1 - \psi_{PtOx})\; k(T)\; (y_{NO}\, y_{O_2}^{0.5} - y_{NO2}/K_{eq})/G_{inh},$$

where Ψ_{Pt} represents a capacity of the Pt surface sites, ψ_{PtOx} denotes platinum oxide coverage, K_{eq} stands for the equilibrium constant, G_{inh} is the inhibition term and k is a reaction kinetic constant calculated from the Arrhenius equation.

Figure 1 demonstrates experimentally observed and simulated considerable sensitivity of NO oxidation during heating and subsequent cooling to the initial state of the catalyst. In the first part of the experiment, the fully reduced catalyst (using H_2 at 400 °C and then cool down in nitrogen) exhibits much higher activity and NO_2 yield than during the cool-down ramp, when Pt oxides formation leads to a significant catalyst deactivation. At 180 °C, the catalytic activity is almost fully restored by CO pulses despite overall lean conditions. To capture these phenomena in the mathematical model, Pt oxidation by O_2 and NO_2, PtO_x reduction by NO and reductants (CO and C_3H_6) and PtO_x thermal decomposition was included [3].

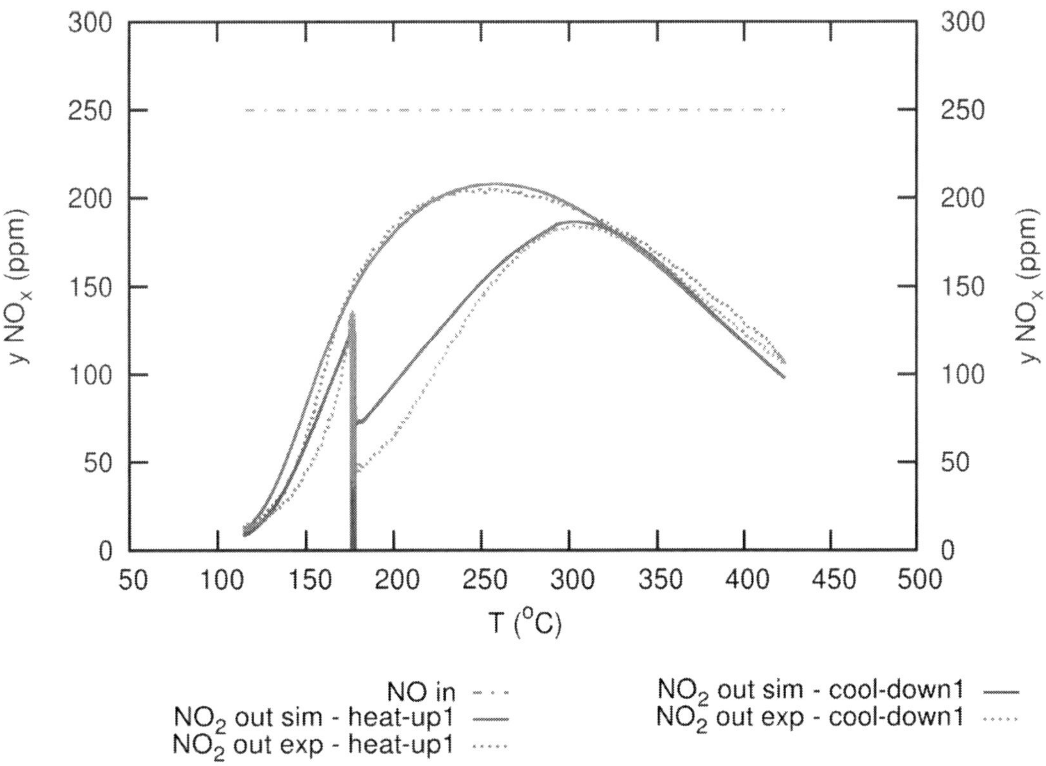

Figure 1. Inverse hysteresis during NO oxidation (heating and cooling temperature ramp) caused by PtO_x formation and the influence of CO pulses on catalyst activity - laboratory reactor experiments and simulation results [3].

References

[1] W. Hauptmann, M. Votsmeier, J. Gieshoff, A. Drochner, H. Vogel, Appl. Catal. B: Env. 93 (2009) 22.

[2] K. Hauff, U. Tuttlies, G. Eigenberger and U. Nieken, Appl. Catal. B: Env. 123-124 (2012) 107.

[3] Arvajová A., Kočí P., Schmeißer V., Weibel M. Applied Catalysis B: Environmental 181 (2016), 644-650.

Porous structure of catalysed particulate filters and its effect on pressure loss

M. Václavík[1], M. Leskovjan[1], V. Novák[2], P. Kočí[1,*], D. Thompsett[2]
[1]University of Chemistry and Technology Prague, Department of Chemical Engineering, Technická 5, 16628 Praha 6, Czech Republic; [*]e-mail: petr.koci@vscht.cz
[2]Johnson Matthey Technology Centre, Blounts Court Road, Sonning Common, Reading RG4 9NH, United Kingdom

Combustion engines produce a variety of pollutants – gaseous components (mainly CO, hydrocarbons, NO_x) are removed in catalytic converters whereas solid soot is captured in particulate filters. Both devices are in the form of a cylindrical monolith with a large number of parallel channels in a honeycomb arrangement – however, catalytic converters are flow-through while the filter channels are partially plugged so the exhaust gas is forced to permeate through the porous monolith walls, filtering out the soot. Walls of a catalytic converter are coated with a catalytically active layer, which is not the case for a standard filter.

Particulate filters were in the last decade applied mostly to Diesel engines (hence the commonly used abbreviation – DPF as diesel particulate filter). However, recent emission standards are stricter and regulate both particulate mass and number, which means gasoline engines with direct fuel injection will have to be equipped with filter (GPF) as well. Moreover, multiple catalytic converters often need to be combined, which makes the whole aftertreatment system space-demanding. A promising approach to make the system more compact is to apply the catalytically active material directly into the porous structure of the catalysed particle filters (CDPF, CGPF). There are other advantages too – overall heat-losses are reduced and the catalysed filter can be regenerated at lower temperatures when a layer of oxidation catalyst is applied.

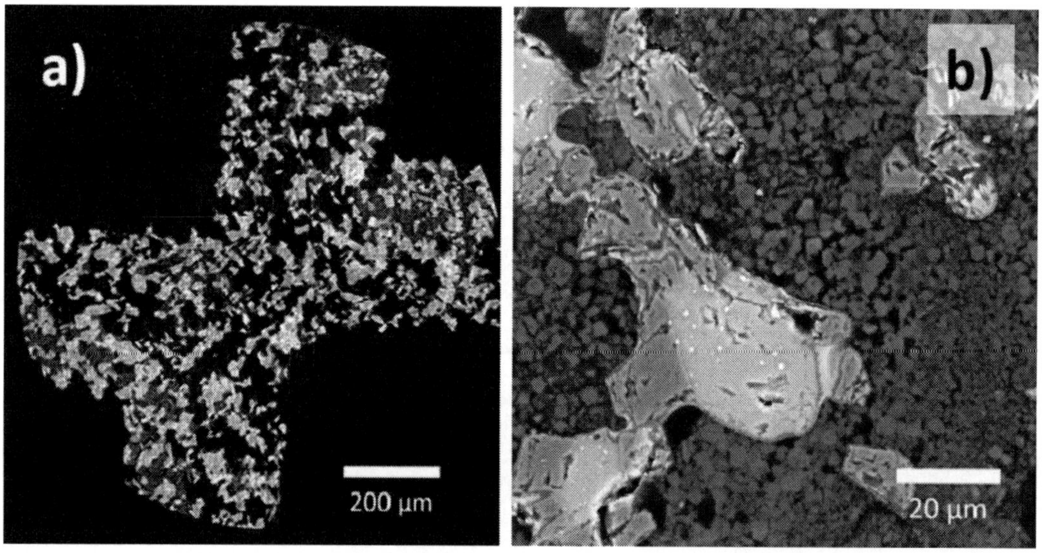

Fig. 1: A cross-section scan of a Cu-zeolite/SiC sample captured by a) X-ray microtomograph, b) SEM (light grey – SiC substrate, dark grey – Cu-zeolite coating, black – pores).

The catalysed filter performance (filtration efficiency, pressure loss, soot/ash build-up and regeneration parameters) strongly depends on the morphology of filter substrate and coated layer. Applying a thick coating may increase the pressure loss and lower the conversion of pollutants due to diffusion limitation [1-3]. In this work, novel types of filter substrate (silicon carbide, aluminum titanate) were characterized by mercury porosimetry, scanning electron microscopy (SEM) and X-ray microtomography, which allows 3D reconstruction of filter sections by combining a large number of cross-section scans (Fig. 1a). SEM on the other hand provides a more detailed view of the porous structure (Fig. 1b). The SiC filter samples were compared bare and coated with a Cu-zeolite catalyst (selective catalytic reduction of NO_x). Obtained parameters, such as macroporosities (total porosity, porosity of the substrate, porosity of the coating), pore size, location of deposited soot particles and their effect on pressure loss were studied and used for development of a mathematical model describing the catalysed particle filter [4].

Acknowledgements
The results were developed within the project PARTIAL-PGMs supported within the frame of EU program Horizon 2020. The work was further supported from specific university research (MSMT No 20-SVV/2016) and Johnson Matthey.

References
[1] Novák V., Dudák M., Kočí P., Marek M. Understanding the gas transport in porous catalyst layers by using digital reconstruction techniques. Current Opinion in Chemical Engineering 9 (2015), 16-26.

[2] Novák V., Kočí P., Marek M., Štěpánek F., Blanco-García P., Jones G. Multi-scale modelling and measurements of diffusion through porous catalytic coatings: An application to exhaust gas oxidation. Catalysis Today 188 (2012), 62-69.

[3] Guan, B., Zhan, R., Lin, H., Huang, Z. Review of the state-of-the-art of exhaust particulate filter technology in internal combustion engines. Journal of Environmental Management 154 (2015), 225-258.

[4] Schejbal M., Marek M., Kubíček M., Kočí P. Modelling of diesel filters for particulates removal. Chemical Engineering Journal 154 (2009), 219-230.

Synthesis and characterization of core- shell Ni-containing microspheres and activity test over methane dry reforming

G.Gündüz Meric, L. Degirmenci
Bilecik Seyh Edebali University, Chemical and Process Engineering Department, 11100, Gülümbe, Bilecik;
tel. +902282141765, e-mail: gamze.gunduz@bilecik.edu.tr;
Bilecik Seyh Edebali University, Chemical and Process Engineering Department, 11100, Gülümbe, Bilecik;
tel. +902282141548 , e-mail: levent.degirmenci@bilecik.edu.tr

Dry reforming of methane (DRM) with environmental and economic advantages is a promising approach with for carbon dioxide utilization and syngas generation [1]. The widespread applications of DRM resides in catalyst performance. Noble metal catalysts, e.g., platinum (Pt), palladium (Pd), rhodium (Rh), and ruthenium (Ru) are active and stable [2,3] but are highly expensive. In contrast, Ni-based catalysts are more attractive and economic [4]. Utilization of Ni based catalysts in dry reforming is limited due to sintering of active metallic particles and coke deposition on the metal surfaces [5].

Previous studies have shown that smaller Ni particles had been beneficial to the reforming reaction, with possibility of inhibiting carbon deposition [6]. To date, effective options to inhibit coke formation included chemically or physically dispersing Ni in other metal oxides, confining Ni particles in mesoporous materials and encapsulating catalyst particles in an inorganic shell [7].

In the present study, we investigated dry reforming of methane on core- shell Ni containing microspheres prepared by a modified sol- gel method. The amount of Ni was adjusted as 1.25 and 0.5 wt% based on the amount of TEOS in the synthesis solution. $NiNO_3.6H_2O$ utilized as precursor was dispersed in 20 ml deionized water, ultrasonicated for 15 min, and dispersed via homogenizator in a mixture of 50 ml ethanol, 10 ml 25 wt. % ammonia aqueous solution, followed by the addition of TEOS. Stirring rates of 300,400 and 1000 rpm was utilized in the synthesis of 1.25% Ni microspheres. Stirring rate of 1000 rpm was selected for 0.5% Ni catalysts based on the preliminary results obtained with 1.25% Ni microspheres. After stirring at room temperature for 6 h, the product was washed thrice with ethanol and deionized water, centrifuged and dried at room temperature for 24h. The structure of the samples was characterized by X- ray diffraction (XRD) and scanning electronic microscopy (SEM). Particle sizes of the samples were determined via Malvern Zeta Sizer instrument. The surface area of the samples were determined using Brunauer-Emmett-Teller (BET) method. The pore volume and pore size were derived from using Barrett- Joyner-Halanda (BJH) method. Metal loading of Ni was determined using an Inductively Coupled Plasma (ICP) Method. The catalytic activity of the microspheres were tested in dry reforming of methane.

Fig. 1 shows the SEM images of the Ni containing microspheres. The images confirmed microsphere formation. ICP results of the catalysts indicated a negligible decrease of Ni content during syntheses (Table 1). A uniform particle size was obtained for all samples and an increase in particle size with an increase in stirring rate was determined with particle size analyses (Table 2).

Figure 1. SEM images of microspheres (1.25 wt. % Ni, 300 rpm)

Table 1. ICP analyses of Ni containing microspheres

Mixing rate (rpm)	Ni amount in synthesis solution (%)	Ni amount in microsphere determined by ICP analyses (%)
300	1.25	1.23 ± 0.01
400	1.25	1.21 ± 0.02
1000	1.25	1.05 ± 0.01
1000	0.50	0.46 ± 0.02

Table 2. Particle sizes of synthesized Ni microspheres

Stirring rate (rpm)	Ni amount (%)	Particle size (nm)
300	1.25	113.4
400	1.25	126.5
1000	1.25	163.2
1000	0.5	185.7

References

[1] M.C. Bradford, M.A. Vannice, Applied Catalysis A:General 142 (1996) 73-96.
[2] D. Pakhare, J. Spivey, Chem Soc Rev 43 (2014) 7813.
[3] A. Erdöhelyi, J. Cserenyi, F. Solymosi, J. Catal. 141 (1993) 287–299.
[4] H.S. Bengaard, J.K. Norskov, J. Sehested, B.S. Clausen, L.P. Nielsen, A.M. Molenbroek, J.R. Rostrup- Nielsen, Journal of Catalysis 209, (2002), 365- 384.
[5] J. Liu, H. Peng, W. Liu, X. Xu, X. Wang, C. Li, W. Zhou, P. YYuan, X. Chen, W. Zhang, HH. Zhan, ChemCatChem 6 (2014), 2095- 2104.
[6] J. H. Kim, D.J. Suh, T.J. Park, K.L. Kim, Applied Catalysis A 197 (2000) 191-200.
[7] L. Li, P. Lu, Y. Yao, W. Ji, Catalysis Communications 36 (2012) 72-77.

Synthesis, characterization and catalytic activity of sulfonic acid functionalized silica

[1]D. Pečar, [2]A. Goršek
University of Maribor, Faculty of Chemistry and Chemical Enginering, Smetanova 17, 2000 Maribor, Slovenia; [1]tel. +385 2294 442, e-mail: darja.pecar@um.si; [2]tel. +385 2294 453, e-mail: andreja.gorsek@um.si.

Solid acid catalysts also marked as 'Green catalysts' are used to replace homogeneous acid catalysts. The use of solid acid catalyst has many advantages like: separation of products and catalyst is easy, catalyst can be reused several times without significant loss of activity, and they are environmentally and economically more acceptable because of reduced amount of produced toxic, hazardous and harmful waste. Acid catalysis is the most employed catalysis in chemical industry. For homogeneous catalysis mineral of inorganic acids are used. Those acids are hazardous and have a great corrosion impact on the plant. At the end there is a need for quenching and separation, which leads to large volumes of environmentally undesired waste.

With the aim of testing, we have synthesized the solid acid catalyst. For the support we used silica gel with a particle size (0.063-0.2) mm. Silica gel was silanized using (3-Mercaptopropyl)trimethoxysilane in toluene and further oxidized with methanesulfonic acid in H_2O_2. The catalyst was denoted as SiO_2-SO_3H.

The characterization of synthesized catalyst was carried out using nitrogen adsorption-desorption (BET), Fourier transform infrared spectroscopy (FTIR), scanning electron microscopy (SEM), thermogravimetric analysis (TGA), and X-ray diffraction analysis (XRD).

Further, the synthesized catalyst SiO_2-SO_3H was used in esterification reaction of benzoic acid with methanol. The reactions were carried out in an automated reactor system EasyMax 102 (Mettler Toledo) at temperatures ϑ = (55, 60 and 64.5) ° C, stirring rate f_S = 400 $min^{-1}$and catalyst loading m = (5, 10 and 15) g. The concentration of methylbenzoate was determined by HPLC using C18 column (Varian).

We determined kinetic parameters (reaction rate constants, activation energy and preexponential factor of Arrhenius equation) of esterification of benzoic acid with methanol. Under studied experimental conditions the reaction follows the first order kinetics. We also studied the conversion and loss of catalyst activity after repeated reactions.

Transesterefication of waste oil using biocatalysts of lipase immovilized on pillared clay

José Antonio Colín-Luna[a*], Ana Karina Medina-Mendoza[b], M. Margarita González-Brambila[a], H. Puebla-Núñez[a], Carlos R. Tapia-Medina[a], J. Ramírez-Muñoz[a]
[a] UAM-Azc, Departamento de Energía, [b] Departamento de Ciencias Básicas, Av. San Pablo 180, Col. Reynosa-Tamaulipas, 02200, DF., Méx.
*jacl@correo.azc.uam.mx

Renewable energy, also known as alternative or green, covers a range of energy sources that regenerate naturally and they are virtually inexhaustible over time. Interest in this type of energy is growing at a time when attention is focused on global warming, CO2 emissions increase, the power supply is safe and encourage reduced consumption of fossil fuels; therefore, the use of renewable energy sources is essential. Several resources from biomass are discussed in the scientific community that can deliver fuels, chemicals and energy products.

In particular the biodiesel fuel is a monoalkyl compound long chain esters derived from vegetable oils or animal fats, designated B100, obtained by the transesterification of fatty acids contained in these oils with an alcohol. The transesterification reaction can be occur without a catalyst in homogeneous phase, but requires excessive temperatures and produces a very low yield, along with numerous products. The use of a catalyst helps the reaction rate at temperatures and reduced pressures, often producing less waste and less energy consuming. The transesterification reaction of triglycerides with methanol catalyzed by acids, bases, metals or enzymes. Each class of catalyst has advantages and disadvantages often associated with the process and therefore to production costs.

Heterogeneous catalysts offer a variety of benefits, including an improved profile compared to their homogeneous counterparts environmental impact, since they are reusable, relatively easy to use, reduce the formation of byproducts and impurity levels. An example of these catalysts are natural clays such as bentonite montmorillonite, alumina, aluminum silicates and silicas, so many works have been devoted to the study of these materials as catalysts in a variety of reactions, including the transesterification . These catalytic materials are relatively inexpensive, easy to access and study corresponds to the heterogeneous catalysis.

Therefore, it is proposed to immobilize an enzyme on a modified bentonite zirconium cations to increase interlaminar space. The synthesized biocatalyst is used in the transesterification reaction of residual vegetable oils such as canola oil, using an anhydrous alcohol. The materials will be characterized by various analytical techniques through their stages of preparation and after reaction in order to determine the optimum conditions of synthesis biocalizador.

Hydrogen production by steam reforming of acetic acid in the presence of Ni containing catalysts

B. Pekmezci[1], N. Çakıryılmaz[1], H. Arbağ[1], N. Oktar[1], G. Doğu[1], T. Doğu[2]
[1]GaziUniversity, Department of Chemical Engineering, 06570 Ankara, Turkey;
tel.+90 312 5823500, e-mail: bircepekmezci@gazi.edu.tr, nurbanucakiryilmaz@gmail.com, harbag@gazi.edu.tr, oktarnuray@gmail.com, gdogu@gazi.edu.tr; fax: +90 3122308434.
[2]Middle East Technical University, Department of Chemical Engineering, Ankara, Turkey
tel.+903122102631, e-mail: tdogu@metu.edu.tr fax: +90 3122102600

Recent developments in fuel-cell technologies caused significant increase in the research and development activities for the production of hydrogen from non-fossil resources [1]. Some recent studies showed that, steam reforming of bio-oil, which was derived by flash-pyrolysis of biomass, is one of the most promising and economical ways for hydrogen production. Acetic acid is one of the major components of aqueous phase of bio-oil, which was obtained from forest/agricultural wastes[2-4]. Hydrogen production by steam reforming of biomass derived acetic acid (R-1) is also a CO_2 neutral process and attracted much attention in recent years.

$$C_2H_4O_2 + 2H_2O \Leftrightarrow 2CO_2 + 4H_2 \quad \text{(R-1)}$$

In this work, steam reforming of acetic acid (AcOH) was performed over a set of catalysts which were prepared by impregnation of different amounts of Ni (1.0, 2.5, 5.0 % by mass) into a commercial copper-zinc based reforming catalyst (Hifuel R-120 catalyst, Alfa Aesar). Catalytic activity tests were carried out in a continuous flow system. Prior to theactivity tests, catalysts were reduced with hydrogen at 750°C. Physical and structural properties of these catalysts were determined following different characterization methods, such as XRD, N_2 adsorption-desorption, XPS, DRIFTS etc. Decrease of catalyst surface area as a result of Ni incorporation is a sign of pore blocking caused by nickel. Nitrogen adsorption-desorption isotherms proved that materials were mesoporous in structure. XRD pattern of Ni particles indicated well dispersion of the metal on Hifuel R-120 surface.

Catalytic activity tests of catalysts were performed at a reaction temperature of 700°C, with a H_2O/AcOH molar ratio of 2.5.In the presence of Hifuel R-120 catalyst, the AcOH conversion was about 72 % and the amount of CO_2 was higher than H_2 in the product stream, at a space time of 0.072 s.g/cm^3 (Table 1).Also, significant amount of CH_4 was obtained, indicating occurrence of acetic acid decomposition. However, all of nickel incorporated Ni@Hifuel R-120 materials showed complete acetic acid conversion.

Table 1. Product distribution with Ni@Hifuel R-120catalysts

Products	**Hifuel R-120**	**1Ni@ Hifuel R-120**	**2.5Ni@ Hifuel R-120**	**5Ni@ Hifuel R-120**
H_2	22.02	36.77	41.37	51.78
CO	0.27	10.23	19.81	13.58
CH_4	33.45	17.57	11.85	8.17
CO_2	44.26	35.42	26.97	26.47

Moreover, activity test results showed that percent amount of H_2 increased with an increase in the amount of nickel and maximum H_2 concentration was obtained with 5Ni@Hifuel R-120. As Ni content increased, percent of methane in the product stream was also significantly decreased. While the amount of carbon monoxide increased, amount of carbon dioxide and methane decreased as a result of increase of nickel content of the catalysts. These results showed that acetic acid steam reforming and water gas shift reactions were favoured due to nickel incorporation into the catalyst structure. However, significant amount of coke formation was observed as a result of increase of Ni content. Decrease of CO mole fraction, by an increase of Ni content from 2.5% to 5%, is an indication of contribution of Boudouard reaction in coke formation.

Acknowledgements:

Financial support of Gazi University Research Fund (BAP 06/2015-05), TUBITAK 214M578 are gratefully acknowledged.

References:

[1] Gunduz S., Dogu T. 2015. "Hydrogen by steam reforming of ethanol over Co–Mg incorporated novel mesoporous alumina catalysts in tubular and microwave reactors", Applied Catalysis B: Environmental, 497-508.

[2] Hu, R., Yan, C., Zheng, X., Liu, H., Zhou, Z. 2013. "Carbon deposition on Ni/ZrO_2-CeO_2 catalyst during steam reforming of acetic acid", International Journal of Hydrogen Energy, 38, 6033-6038.

[3] Assaf, P.G.M., Nogueira, F.G.E., Assaf, E.M. 2013. "Ni and Co catalysts supported on alumina applied to steam reforming of acetic acid: representative compound for the aqueous phase of bio-oil derived from biomass", Catalysis Today, 213, 2-8.

[4] Hu, X., Zhang, L., Lu, G. 2012. "Pruning of the surface species on Ni/Al_2O_3 catalyst to selective production of hydrogen via acetone and acetic acid steam reforming", Applied Catalysis A: General, 427-428, 49-57.

Thermal stable titania nanotubes supported V_2O_5 for oxidative dehydrogenation of propane

M. Saei Moghaddam[1], J. Towfighi[1], A.H. Shahbazi Kootenaei[2], E. Goudarzi[1]
[1] Tarbiat Modares Univ., Dep. Chemical Engineering, P.O. Box 14115-143, Tehran, Iran; tel. +982182883311, e-mails: mojtabasaei@modares.ac.ir, towfighi@modares.ac.ir, e.goudarzi@modares.ac.ir.
[2] Department of Chemical Engineering, Mahshahr Branch, Islamic Azad University, Mahshahr, Iran; tel. +989127306369, e-mail: a.shahbazi@mhriau.ac.ir.

Titania nanotubes synthesized by hydrothermal treatment of crystalline titania particles with concentrated NaOH aqueous solution at 140 °C for 24 h followed by ion-exchange. The morphology and crystalline structure of nanotubes were characterized by transmission electron microscopy (TEM) and X-ray diffraction (XRD). Raman, FTIR and BET techniques were examined, too. The results confirm the presence of Sodium are located between titanate layers which it has significant role in preserving tubular structure and thermal stability. In addition it was found that the fabricated nanotubes provides higher specific surface areas (392 m^2/gr) and pore volume (1.166 cm^3/gr) more than precursor by factors of about 8 and 4 respectively. Consequently according to titania surface structure that exhibits Lewis acidity along with redox properties, titania nanotube seems to be the more practicable as a catalyst support for oxidative dehydrogenation of light olefins.

Keywords: Titania nanotubes, Hydrothermal, Sodium ions, Thermal stability, oxidative dehydrogenation.

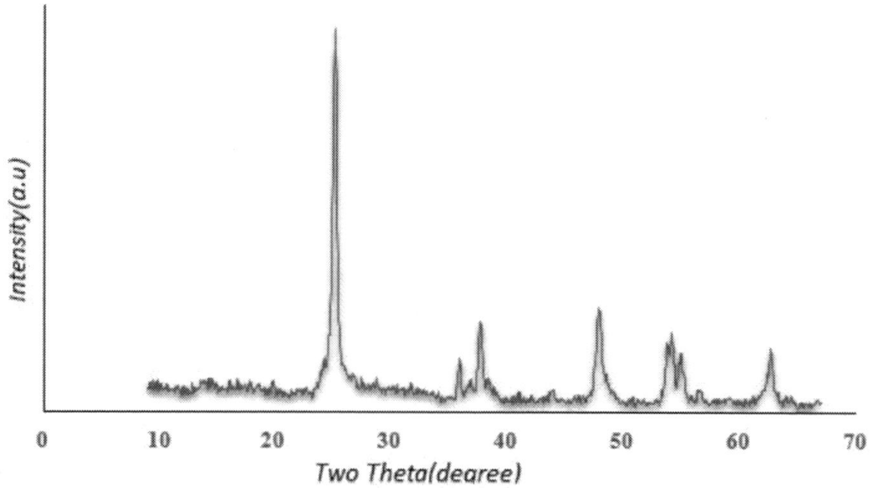

Fig.1 XRD pattern of precursor

In order to investigate phase structure of prepared sample x-ray diffraction experiment was performed. The XRD pattern is presented in Fig. 1 indicate that the raw TiO2 used as starting material for synthesis of TiNTs is fine anatase composition in the major phase diffraction peaks at 2 θ =25.3°, 36.8°, 38°, 48.1°, 53.8°, 55.1° and 62° that correspond to (1 0 1), (1 0 3), (0 0 4), (2 0 0), (1 0 5), (2 1 1) and (2 1 3) planes, respectively (JCPDS: 21-1272). Consequently, the rutile phase does not exist in precursor (JCPDS: 21-1276).

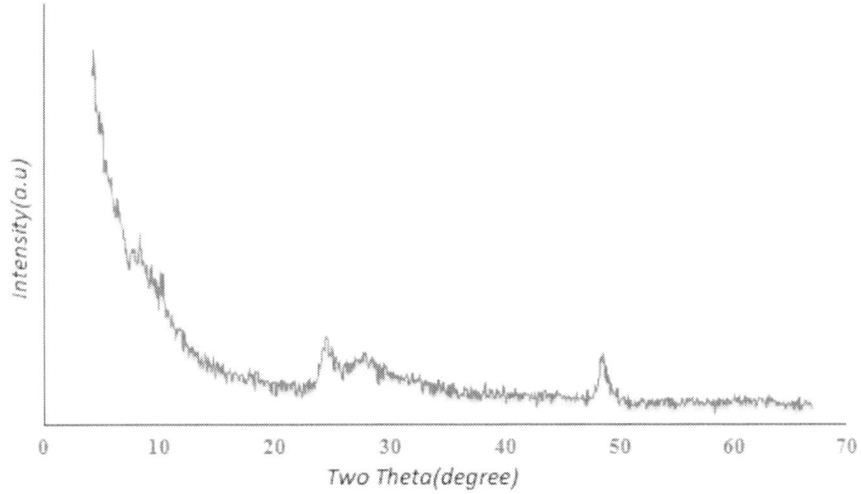

Fig.2 XRD pattern of TiNTs

After alkali hydrothermal treatment followed by ion-exchange the anatase phase of Degussa P25 transformed to TiNTs; it is according to XRD pattern indicate three broad peaks, as shown in Fig. 2 at 2θ =8.5°, 24.5°, 48.6° which correspond to (0 0 1), (0 0 3) and (0 2 0) planes, respectively may be assigned to $H_2Ti_5O_{11}.H_2O$ (JCPDS: 44-0131).
By careful of observation weak peak of $Na_2Ti_3O_7$ can be found at 10.5° that indicate the crystallization was far from complete because of low protonated degree during acid washing step. This diffraction peak around 10° corresponds to the interlayer spacing between titanate sheets. Thus, Na are located between our TiNTs layers.

Effect of temperature on hydrodeoxygenation of raw bio-oil over an activated carbon-based catalyst

R. Palos[1], T. Cordero-Lanzac[1], J. Rodríguez-Mirasol[2], P. Castaño[1], T. Cordero[2], J. Bilbao[1], J. M. Arandes[1]

[1] Universidad del País Vasco UPV/EHU, Dpto. Ingeniería Química, PO Box 644-48080, Bilbao, Spain, Tel: +34 94 6015341, *e-mail: roberto.palos@ehu.eus.

[2] Universidad de Málaga, Dpto. Ingeniería Química, Campus de Teatinos, s/n, 29010, Málaga, Spain.

Introduction. Bio-oil (liquid product of flash pyrolysis of lignocellulosic biomass) drives in particular great attention due to its prospect of becoming a sustainable source of chemicals and fuels. The bio-oil composition, with elevated concentration of oxygenated compounds (acids, aldehydes, alcohols, ketones, esters, ethers, phenols, furans, and lignin-derived oligomers) and water, makes necessary an upgrading treatment in order to valorize it as fuel [1]. Hydrodeoxygenation (HDO) offers the possibility of oxygen removal and hydrogenation of some chemical functions, obtaining a second-generation biofuel, requiring relatively harsh conditions (>200 bar, >300 °C) and expensive catalysts like noble metals supported ones (Pt, Pd, Ru…) [2]. The aim of this work is to study the oxygen removal from a raw bio-oil feedstock using a PtPd catalyst supported on an acid activated carbon (AC), and evaluate the products yields (gases, liquids and coke) obtained at different reaction temperatures. AC obtained from biomass wastes are attractive supports for HDO catalysts due to their hydrothermal resistance, lower coking tendency [3], and to their high density of acidic sites. Chemical activation with phosphoric acid allows for obtaining high thermally stable phosphates groups, which act as acid support of the required bifunctional catalyst for HDO reactions [4].

Experimental. Bio-oil was obtained by fast pyrolysis of black poplar sawdust in a conical spouted bed reactor at 450 °C and characterized in a GC-MS. The AC support was synthesized from an abundant lignocellulosic waste like olive stone (OS). The precursor was impregnated with H_3PO_4 85 wt% with an impregnation rate (wt H_3PO_4/wt OS) of 3, and carbonized up to 500 °C at 10 °C min^{-1} under a constant N_2 flow of 150 cm^3 STP min^{-1}. The metallic function (1 wt% Pt and 0.5 wt% Pd) was incorporated by simultaneous incipient wetting impregnation of a $HPtCl_6$ and $PdCl_2$ solution, calcining the impregnated support at 400 °C for 4 h with a N_2 flow of 150 cm^3 STP min^{-1}. The catalyst porous structure was characterized by N_2 adsorption–desorption isotherms, the acid function was analyzed by ammonia adsorption–desorption (TPD) and FTIR spectroscopy of pyridine adsorption, and the metallic one by XPS and CO chemisorption.

HDO reactions were carried out in a 15 cm^3 down-flow fixed bed reactor at 400–450 °C; 65 bar; WHSV, 2.6 h^{-1}; H_2/Oil ratio of 20 Nm^3 m^{-3} and time on stream (TOS), 0-8 h. Gases were analyzed on line using a MicroGC, dividing them in gas (C_1-C_2), LPG (C_3-C_4) and CO+CO_2. Liquids were collected and separated in an aqueous fraction, which was analysed in a GC-MS provided with a TCD; and in an organic fraction, analyzed in a GCxGC-MS using FID and MSD. Liquid products were grouped in two different lumps: oxygenated compounds (OX), that were soluble in the aqueous fraction, and hydrocarbons (HC), detected in the organic phase. Water content in the aqueous fraction was determined by means of Karl-Fischer titration and the aromatic and aliphatic compounds were characterized by FTIR spectroscopy. Finally, the deposited solid content and nature were evaluated by TPO analyses.

Results. The prepared AC-based catalyst (carbonization yield of the support, 46 wt%), exhibits a type IV N_2 adsorption-desorption isotherm, typical of a well-developed micro- and mesoporosity, with a BET surface area of 1305 $m^2\ g^{-1}$. Furthermore, the phosphate surface groups provide the AC a total acidity of 0.14 $mmol_{NH_3}\ g^{-1}$ and a hydrophilic character, which favours the water evacuation in the HDO process.

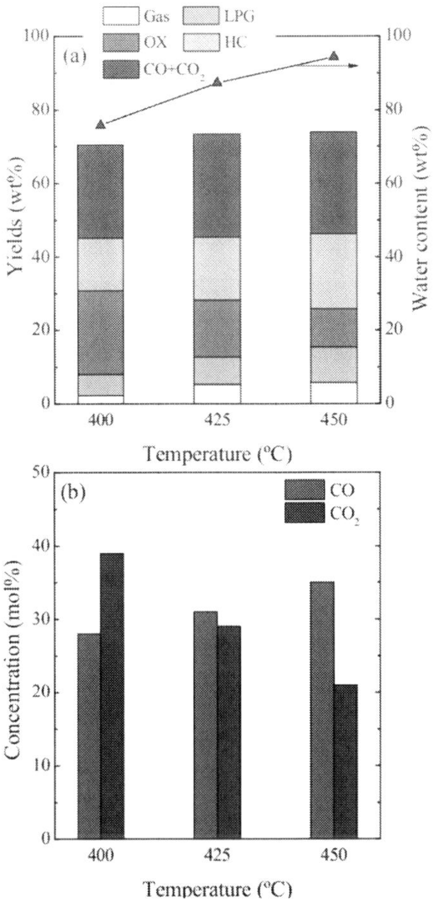

Figure 1. Temperature effect on (a) products yields and (b) CO/CO_2 ratio (65 bar; TOS, 6 h; WHSV, 2.6 h^{-1}).

As aforementioned, the reaction products are separated in an organic and an aqueous phase, Figure 1a depicts the water content in the aqueous phase at TOS = 6 h. Assuming the simplified HDO reaction $R{=}O + 2H_2 \rightarrow H\text{-}R\text{-}H + H_2O$, the water content in the reaction products is related to the HDO conversion. As can be observed, the HDO conversion increases upon increasing the reaction temperature, as well as the HC yield, obtaining a maximum HC yield of 20 wt% at 450 °C. On the other hand, the OX yield (mainly methanol, phenol, acetic acid and acetone dissolved in the aqueous phase) decreases, which is in consequence of the HDO conversion registered. Thermal cracking reactions are favoured at higher temperatures, and therefore, the gas, LPG and $CO+CO_2$ yields increase with the reaction temperature. Total product yields do not achieve a value of 75 wt% in any case due to the deposition of solid products, which includes condensed coke structures and pyrolytic lignin produced through the bio-oil compounds polymerization [5]. The evolution of CO/CO_2 molar ratio is shown in Figure 1b, where it can be seen that increasing the temperature allows for maximizing the CO/CO_2 molar ratio in the gaseous product, obtaining a maximum value of 1.7 at 450 °C. This fact, along with the highest HDO conversion and HC yields obtained, suggests that the use of activated carbon-based catalysts at elevated reaction temperatures and low pressure is a promising way to translate bio-oil into hydrocarbon fuel by directly catalytic hydrodeoxigenation, with a moderate CO_2 emission.

[1] A. Remiro, B. Valle, L. Oar-Arteta, A.T. Aguayo, J. Bilbao, A.G. Gayubo (2014). *Int. J. Hydrogen Energ.*, **39**, 6889.
[2] A.H. Zacher, M.V. Olarte, D.M. Santosa, D.C. Elliot, S.B. Jones (2014). *Green. Chem.*, **16**, 491.
[3] M. Ferrari, B. Delmond, P. Grange (2002). *Carbon*, **40**, 497.
[4] I. Hita, T. Cordero-Lanzac, A. Gallardo, J.M. Arandes, J. Rodríguez-Mirasol, J. Bilbao, T. Cordero, P. Castaño (2015). *Catal. Commun.* **DOI**: *10.1013/j.catcom.2016.01.035.*
[5] A. Ochoa, B. Aramburu, M. Ibáñez, B. Valle, J. Bilbao, A.G. Gayubo, P. Castaño (2014). *ChemSusChem*, **7**, 2597.

Acknowledgements
This work was carried out with the support of the Ministry of Economy and Competitiveness of the Spanish Government and financed by CTQ2009-12800, CTQ2012-36408 and CTQ2013-46172-P projects and the ERDF of the European Union.

Deep oxidation of ethanol over Co-Mn-Al mixed oxides supported on pelletized magnesia-alumina: role of acid-base properties

[1]K. Jirátová, [1]J. Balabánová, [2]F. Kovanda, [3]L. Obalová
[1]Institute of Chemical Process Fundamentals of the Czech Academy of Sciences, v.v.i., 165 02 Praha 6, Rozvojová 135, Czech Republic; jiratova@icpf.cas.cz; [2]University of Chemistry and Technology, Prague, Technická 5, 166 28 Prague, Czech Republic; frantisek.kovanda@vscht.cz; [3]Institute of Environmental Technology, VŠB - Technical University of Ostrava, 17. listopadu 15, Czech Republic; lucie.obalová@vsb.cz

Volatile organic compounds (VOC) emitted in industrial gases can be eliminated applying the catalytic total oxidation. Previously we found high catalytic activity of the Co-Mn-Al mixed oxides, obtained by heating of the coprecipitated layered double hydroxide (LDH) precursor, in the ethanol total oxidation. Activity of this catalyst was remarkably increased by addition of small amount of the potassium promoter. However, the catalysts prepared by pelletizing the powdered material contained high amounts of expensive cobalt, which was not fully utilized in the catalytic reaction. Therefore, we focused on preparation of supported catalysts with lower concentration of active components placed exclusively in the outer shell of the support pellets. In the present study, we applied the method of impregnation using acidic solution of metal salts in combination with a support having basic properties that could limit penetration of active components into the support pellets.

The catalysts were prepared by impregnation of magnesia-alumina pellets (5x5 mm, Sasol) with aqueous solutions containing Co, Mn and Al nitrates (molar ratio of Co:Mn:Al=4:1:1) and subsequent heating at 500 °C in air. The prepared samples were characterized by chemical analysis, powder XRD, measurement of nitrogen adsorption at -196°C, NH_3-TPD, CO_2-TPD, and TPR. Distribution of active metals in the catalyst pellets was determined by SEM. Catalytic activity in the total oxidation of ethanol, which was chosen as a model VOC, was measured under unsteady-state conditions with heating rate of 2.27 °C min^{-1} (100–400 °C), at ethanol concentration in air of 1.5 g m^{-3} and GHSV 20 l g_{cat}^{-1} h^{-1}.Temperatures T_{50}, at which 50% conversion of ethanol was observed, were chosen as a measure of the catalysts activity. In addition, specific activities of all catalysts, valid for 200 °C, were also determined. Apart from the studied catalysts, the pellets (5x5 mm) of commercial Co-Mn-Al mixed oxide catalyst (ASTIN 2-100) were also examined for comparison.

It was found out that magnesia-alumina support contained, apart from 30 wt. % of Al_2O_3, about 1 wt. % of Na. The SEM imaging of the catalysts confirmed limited penetration of active metals into the support pellets. Commercial catalyst with about 55 wt. % of (Co+Mn) exhibited uniform distribution. Sum of (Co+Mn) in the supported catalysts varied from 4 to 15 wt. %.Powder XRD patterns of all samples showed diffraction lines corresponding to spinel-like phases. Basic properties of the support as well as surface area of the samples were decreasing with gradual adding the active metal oxides, while acidity of the catalysts was increasing. Ethanol conversion over the studied catalysts showed that the most active sample, even more active than the commercial Co-Mn-Al mixed oxide catalyst with almost 55 wt.% of (Co+Mn), was the catalyst with much lower content (only about 12 wt.%) and non-uniform distribution of Co and Mn in the pellets. The main reaction byproduct, acetaldehyde, was completely transformed into CO_2 at temperatures lower than 300 °C. Formation of the other reaction byproduct, CO, was not detected.

In conclusion, impregnation of the magnesia-alumina pellets having basic properties with acidic solution of Co, Mn and Al nitrates led to limited penetration of metal cations into the support and formation of catalysts with non-uniform distribution and reduced content of active components. The formed active layer in the support resisted to attrition during the catalytic reaction. The activity of the supported catalysts slightly differed from that of the commercial catalyst with uniform distribution of Co and Mn and (Co+Mn) about 55 %. Supported catalyst with optimum active metals concentration (ca. 12 wt. %) showed better catalytic properties than the commercial one.

Acknowledgements
The authors thank the Czech Science Foundation for the financial support (project P106/14-13750S).

Modelling of the laboratory-scale packed bed reactor for partial gas-phase oxidation of alcohol using Au nanoparticles as the catalyst

[1] T. Kilpiö, [2]E. Behravech, [3]V.Russo, [4]T.Salmi
[1,2,4]ÅboAkademi, Lab. of Ind. Chem. and Reac. Eng., FI-20500 Turku, Finland. Telephone: +358 2 21531, e-mails: teuvo.kilpio@abo.fi, erfan.behravech@abo.fi tapio.salmi@abo.fi.
[3]Univ. of Naples "Federico II".Chem. Sci.Dep., IT-80126 Naples, Italy. Telephone: +39 3493290190, e-mails: v.russo@unina.it.

Packed bed reactors are commonly used for catalyst screening studies due to their simple construction and ease of operation. The use of gold nanoparticles on various supports for partial oxidation reactions has arisen general interest of the scientific community. Here a physical (phenomena based) approach was used to make an effort to generate a model to describe and explain the experimentally observed reactor performance behavior (conversion of alcohol, yields of various products). The core of the generated model for this gas-phase heterogeneously catalyzed multiple reaction system consisted of the dynamic mass and energy balances and LHHW (Langmuir-Hinshelwood-Hougen-Watson) or Tessertype of expressions to describe the reaction kinetics of each compound. Fluid flow was taken into account as the cause of generating back-mixing. Dispersion terms in combination with a semi-empiric pressure drop correlation were used for this purpose. This resulted a lightweight (fast running) model by reducing the size of the problem significantly. Momentum balances were excluded and no vector variables were consequently required. Powerful gPROMSModelbuilder Software was used as the model development, simulation and parameter estimation platform. The primary task was then to perfectly define and solve the complete set of system equations (a large set of Partial Differential Equations (PDEs), together with algebraic equations). The modelling effort was centered around the selection of the complete set of equations (to adjust the degrees of freedom down to zero) and on the simultaneous solution of the whole set in gPROMS environment. Numerical Method of Lines was used to solve the PDEs since it was known to be a method which works reasonably well scalar variables.All reactions were assumed to take place only on evenly distributed active sites of the porous catalyst particles. To take this punctually into account, the dynamic mass balances consisting of combined effects of internal diffusion and surface reaction for each gas compound within the porous particles were included and solved simultaneously with the mass and energy balances of the whole reactor. The axial and radial concentration and temperature profiles inside the reactor and the concentration profiles within the particles were then obtained as results. After the parameter estimation, the calculated profiles of conversion (ethanol) and yields of the main products (acetaldehyde and acetic acid) as a function of operating temperature followed relatively nicely the experimentally observed profiles. The estimated parameters were the rate constant and the activation energy of each reaction and the adsorption parameters for ethanol, oxygen and water with LHHW kinetics and only oxygen with Tesser type of oxidation kinetics. Finally simulation studies were done to show how operation conditions (temperature, flows and catalyst loading) influenced the results and how strongly the uncertainty in *Pe* number reflect on obtained results.

Hydrodeoxygenation of vegetable oil using microscale-based reactor

[1]Y. Sa-ngasaeng, [1,2,*]S. Jongpatiwut, [3]N. Sirimungkalakul
[1]The Petroleum and Petrochemical College, Chulalongkorn University, 254 Soi Chulalongkorn 12, Phayathai Rd., Pathumwan, Bangkok 10330, Thailand; tel. +662 2184154, fax. +662 2184154, e-mail:*siriporn.j@chula.ac.th
[2]Center of Excellence on Petrochemical and Materials Technology, Chulalongkorn University, Soi Chulalongkorn 12, Phayathai Rd., Bangkok 10330, Thailand; tel. +662 2184141, fax: +662 6117619
[3]PTT Research and Technology Institute, PTT Public Company Limited, Ayutthaya 13170 Thailand; e-mail: nichaporn.k@pttplc.com.

Because of its superior properties over biodiesel, biohydrogenated diesel (BHD) has been studied in recent years. Microscale-based reactor is considered for hydrodeoxygenation process as a result of its higher heat and mass transfer, higher surface to volume ratio and easier scale up compared with the conventional fixed bed reactor. In this study, Pd/TiO_2 catalyst was used in microscale-based reactor compared with fixed-bed reactor by using refined palm oil as the raw material. The reaction conditions were varied for each reactor to acquire the highest biohydrogenated diesel yield. The catalyst was synthesized by combined single-step sol-gel process with surfactant assisted templating method which was coated on stainless steel plate for using in the microscale-based reactor. The products will be analyzed by gas chromatography with an FID detector. The major products from fixed bed reactor were *n*-heptadecane and *n*-pentadecane produced via the decarbonylation pathway.

Temperature effect on OMW treatment by Photo-Fenton reaction

G. Hodaifa[1], C. Agabo
Pablo de Olavide Univ., Dep. Molecular Biology and Biochemical Engineering, Chemical Engineering Area, 41013 Seville, Spain; tel. +34 954 978 206, e-mail: ghodaifa@upo.es.

Management of olive oil mill wastewaters (OMW) is a difficult task because of the largeproduced volumes ($\approx 5.4\times10^6$ m^3/year on the world) and its high concentration of pollutants (such as phenolic compounds)[1]. In this sense, advancedoxidation processes (AOPs) can be the affordable optionfor the treatment of OMW if optimal conditions are achieved. Among AOPs, Photo-Fenton is one of the most efficient and economic alternatives for treating those effluents [2].

Hence, in this work different temperatures (5, 10, 15, 20, 35, 40 and 50 °C) have been tested using PhotoFenton reaction. The common operation conditions were: [$FeCl_3$] = 3g/L, [$FeCl_3$]/[H_2O_2]= 0.125, pH≈3 and agitation rate = 150 rpm.

For all studied temperatures the %$TOC_{removal}$ were varied from 94% to 98%. Also, results shown the reaction time for maximum percentage of %$TOC_{removal}$(> 90%) is equal 2 min (Fig. 1).

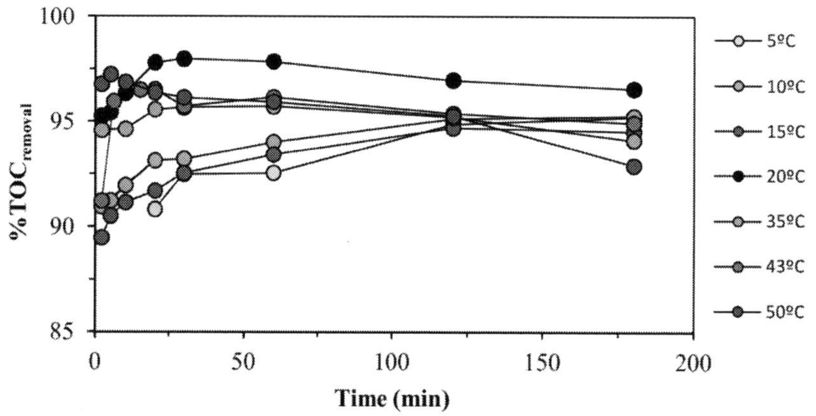

Fig. 1. OMW treatment by PhotoFenton. Temperature influence on the TOC percentages removal. Common operation conditions: [$FeCl_3$] = 3g/L, [$FeCl_3$]/[H_2O_2]= 0.125, pH ≈ 3 and agitation rate = 150 rpm.

In all experiments the phenolic compounds elimination form OMW was observed. Table 1 shows the maximum percentages of phenolic compounds removal and the reaction time necessary to reach these values. For the maximum phenolic compounds elimination temperature higher than 20 °C was determined for reaction time less than 30 min (Table 1).

According to the results obtained, the photocatalysis using PhotoFenton for OMW treatment present a good method where mild condition without energy supply (20 °C) is enough to achieved an efficient treatment because constituting a fair compromise between high degradation (%$TOC_{removal}$ = 97.2%, %$TC_{removal}$ = 96.3%, %$COD_{removal}$ = 95.7% and%Phenolic $compounds_{removal}$ = 97.2%) and reaction time. All that imply a little reactor volume.

Table 1. Phenolic removal percentage in OMW treatment by photoFenton.Common operation conditions: $[FeCl_3]$ = 3g/L, $[FeCl_3]/[H_2O_2]$= 0.125, pH ≈ 3 and agitation rate = 150 rpm.

Temperature (°C)	%Phenolic Compounds$_{removal}$	
	Maximum	Reaction Time (min)
5	96,9	180
10	99,5	180
15	97,8	120
20	97,2	30
35	97,4	20
40	96,4	6
50	97,1	5

References

[1] Sobhi B., Isam S., Ahmad Y., Jacob H., 2007. Reducing the environmental impact of olive mill wastewater in Jordan, Palestine and Israel. *Water Resource in the middle East.* s.l., Springer, pp. 409-421.

[2] Michael I., Panagi A., Ioannou L.A., Frontistis Z., Fatta-Kassinos D.,2014.Utilizing solar energy for the purification of olive mill wastewater using a pilot-scale photocatalytic reactor after coagulation-flocculation. Water Research 60, 28-40.

Kinetic study of nitration a biofuel to synthesis of bio-additive fuels

I. Ben Talouba, H. M. Vélez Reina, L. Balland, N. Mouhab,
Université de Rouen Normandie/INSA-Rouen ; Laboratoire de Sécurité des Procédés Chimiques (LSPC)
BP 8 F-76131 Mont St Aignan Cedex FRANCE
Tel : +33.2.35.52.84.55, e-mail: imed.bentalouba@univ-rouen.fr, fax: +33.2.35.52.84.52.

Diversification of raw materials to the use of renewable resources is a real interest for the development of innovative functional products [1-4]. The development of green processes and the use of new environmentally friendly products (less waste, low toxicity of renewable raw materials, carbon balance...) must therefore be designed in a viable economic context for companies but also for a safety chemistry to prevent accidents, explosions, fires and releases.

2-EthylHexy Nitrate (EHN) is an energetic material used in particular as additive to raise cetane number of diesel fuels and to improve its combustion properties. This additive is usually synthesized though nitration of iso-octanol coming from fossil resource with mixed acid. Research works are trying to find alternative additive derived from renewable resources to substitute EHN. The most commonly used oils to produce this biofuel are extracted from sunflower, soybean, palm, canola, peanut [5] ... Canoira et al. [6] showed that the production of nitrated biodiesel from waste oil in fuming nitric acid or 70 % w/w of nitric acid can increase the diesel cetane number from 54.7 to 60.5.

Despite nitration reaction are widely carried out on industrial scale and studied in the past, there is no literature on kinetic parameters of biofuel nitration. The knowledge of these parameters, as well as associated energetic parameters, is necessary for the prediction of the reactive medium behavior as well for normal and degraded operating modes.

In this work, we are interested in nitration of biodiesel from canola to produce an additive to improve the cetane number of diesel from a biofuel. The nitration reaction by a mixed sulfonitric acid ($HNO_3/H_2SO_4/H_2O$) was carried out in a 500mL capacity RC1-RTCal reaction calorimeter and a stirring speed of 600 rpm under isothermal conditions. Monitoring of reaction conversion over time is identified by combining the thermal conversion and chemical analysis. Temperature and mass ratio of mixed sulfonitric acid (HNO_3 / H_2SO_4) were highlighted on the nitration performance conversion.

The methodology for estimating kinetic parameters and associated energies is reached by comparison of power profile building by the model to that obtained by reaction calorimeter. The model adopted in this study takes account: power released by the physical transformations, linked to mixing phenomena due of acid dosing, power released by the nitration and power released by the degradation of bio-additive. The reaction medium is considered as a pseudo-homogeneous mixture.

Kinetic parameter estimation of model is achieved thanks to a minimization method of a criterion. This method is based on a genetic algorithm coupled with Rosenbrock's local convergent method.

References:

1. Mohanty A. K., Misra M., Drzal L. T., Sustainable Bio-Composites from Renewable Resources: Opportunities and Challenges in the Green Materials World, J. of Polymers and the Environment, Vol. 10, no.1, 19-26, 2002

2. White, A. A.; Platz, M. S.; Aruguete, D. M.; Jones, S. L.; Madsen, L. D.; Wesson, R. D.The National Science Foundation's Investment in Sustainable Chemistry, Engineering, and Materials ACS Sustainable Chem. Eng. 1, 871- 877, 2013.

3. Canoira L., Alcantara R., Torcal S., Tsiouvaras N., Evripidis L., Dimitrios M.K., Nitratoion of biodiesel of water oil : nitration biodiesel as a cetane number enhancer, Fuel, 965-971, 2007.

4.Abdullah, Wicakso D.R., Junaidi A.B., Badruzsaufari, Production of cetane improver from Jathropa curcas oil, Indo.J. Chem., Vol. 10, no. 3, 396-400, 2010.

5. Likozar B., Levec J., transesterification of canola, palm, peanut, soybean and sunflower oil with methanol, ethanol, isopropanol, butanol and tert-butanol to biodiesel: Modelling of chemical equilibrium, reaction kinetics and mass transfer based on fatty acid composition, Applied Energy; 123, 108-120, 2014 .

6. Canoira L., Alcantara R, Tsiouvara N., Lois E., Korres D., Nitration of biodiesel of waste oil : Nitrated biodiesel as a cetane number enhancer, Fuel, 86, 965-971, 2007.

Evaluation of the Nonlinear Frequency Response Method for the Hydrolysis of Acetic Anhydride

M. Felischak[1], D. Nikolic[2], M. Petkovska[2], C. Hamel[1,3], A. Seidel-Morgenstern[1,4]
[1]Otto-von-Guericke University Magdeburg, Dep. Chemical Process Engineering, Universitätsplatz 2, 39106 Magdeburg, Germany tel. +49 3916758128, e-mail: matthias.felischak@ovgu.de; [2]University of Belgrade, Serbia; [3]Anhalt University of Applied Sciences, Köthen, Germany; [4]Max Planck Institute for Dynamics of Complex Technical Systems, Magdeburg, Germany.

Dynamic operation can provide favorable equilibrium and kinetics, mass and heat transfer in chemical reactors, by an optimization of performance parameters. Usually fluctuations of inlet parameters are unwanted and can be even dangerous. Therefore detailed analysis of the system is needed to fully predict possible outcomes [1]. The method of nonlinear frequency response (NFR) attempts to predict the outcome of perturbation for the instructive input variables using nonlinear functions within the model equations. The aim of the derived models is the estimation of optimal frequencies and amplitudes for sinusoidal changes of the total inlet flow, concentration, temperature and the cooling temperature, as well as combinations of several of these inputs [2].

The NFR is an analytical method for the evaluation of periodically perturbed chemical reactors. Based on the concept of high order frequency response functions (FRFs) weakly nonlinear systems can be described. As an indicator for the performance of the forced periodic operation the value and sign of the non-periodic (DC) component of the outlet have to be analyzed, as shown in Fig. 1. This gives insight of possible improvements that can be implied by the input changes. By applying the NFR method the non-periodic term can be estimated for the modulation of a single and for multiple input parameters.

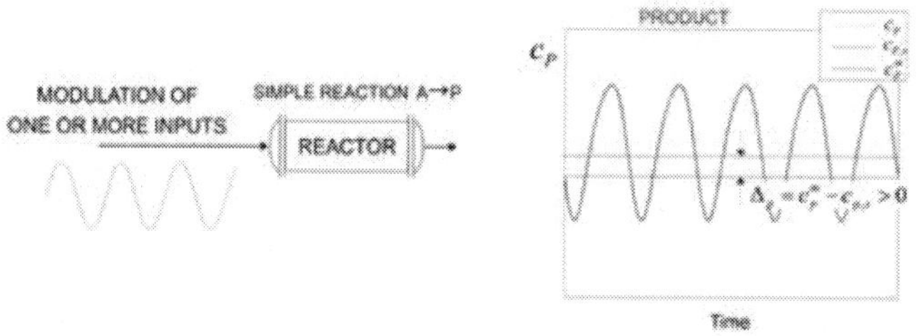

Fig 1: General Approach of the Nonlinear Frequency Response Function with sinusoidal change of the inlet parameters, resulting in a possible process intensification [3].

As a model reaction the hydrolysis of acetic anhydride in a continuously stirred tank reactor (CSTR) has been chosen. This exothermal reaction is investigated for the cases of separately and simultaneously modulated inlet concentrations and feed flow rates. For the experimental validation of predicted model-based results at first a verification and adjustment of the kinetic parameters is performed. Based on these results an optimal steady-state is calculated, which is used as an initial state of periodic operations. To estimate the performance of the reaction system mean values of the conversion of acetic anhydride and the yield of acetic acid are compared to the corresponding steady-state values of the reactor. Different forcing frequencies, input amplitudes and, for multiple input parameter, phase shift between them are investigated. The theoretical foundation as well as systematic exploration of the experimental possibilities will be presented.

[1] J.E. Bailey, Chemical Engineering Communications, 1 (1973) 111-124.
[2] P.L. Silveston, R.R. Hudgins, Periodic Operation of Chemical Reactors, ISBN: 978-0-12-391854-3.
[3] M. Petkovska, D. Nikolic, A. Markovic, A. Seidel-Morgenstern, Chemical Engineering Science, 65 (2010) 3632-3637.

The conversion of fructose into lactic acid catalysed by La-modified phosphomolybdic acid

Bo-Wen Zhao, Hao Li, Chun-Ling Liu, Wen-Sheng Dong
Shaanxi Normal University, School of Chemistry and Chemical Engineering, Westen Chang'an Street 620, 710119 Xi'an, China;
tel. +86-29-81530806, e-mail: clliutt@snnu.edu.cn.

Lactic acid is regarded as a high-potential platform chemical for the production of a wide range of useful intermediates. It also has extensive applications in pharmaceuticals, the food industry, cosmetics and biodegradable plastics *etc.*. Currently, lactic acid is typically produced via biotechnological fermentation of sugars. However, this process suffers from drawbacks e.g. expensive enzymes and a complex purification process. Hence, it is important to explore alternative chemo-catalytic routes for the production of lactic acid from renewable resources.

In the present work, lanthanum-exchanged phosphomolybdic acid homogenuous catalysts were prepared and examined in the conversion of fructose. The as-prepared catalysts exhibited excellent catalytic performance in the production of lactic acid from fructose.

The lanthanum-exchanged phosphomolybdic acid catalysts were prepared by a liquid-phase ion exchange method using $H_3PMo_{12}O_{40}$ and $La(NO_3)_3$ as raw materials. The mixed solution was cast onto a glass substrate and then transferred into an oven for evaporation of water at 110 °C overnight. All the catalytic reactions were performed in a 35 mL stainless-steel autoclave equipped with a mechanical stirrer.

Table 1. The compositions of catalysts determined by X-ray fluorescence

Catalyst	La (%)	P (%)	Mo (%)	Mo/La atomic ratio
HPMo	–	2.1	97.9	–
La-HPMo-1	6.3	1.7	92.0	21.2
La-HPMo-2	12.6	1.8	85.6	9.8
La-HPMo-3	15.2	1.9	82.9	7.9
La-HPMo-4	25.1	1.6	73.3	4.2

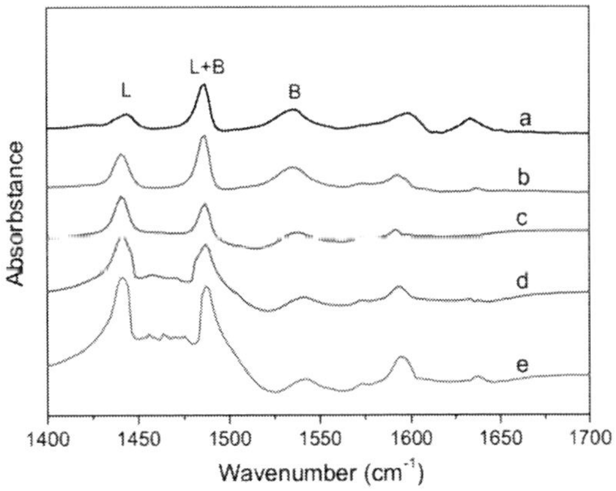

Figure 1. FTIR spectra of pyridine adsorption on various catalysts: (a) HPMo, (b) La-HPMo-1 (c) La-HPMo-2, (d) La-HPMo-3, (e) La-HPMo-4.

The compositions of various catalysts are shown in Table 1. Figure 1 shows pyridine adsorption infrared spectra of the catalysts. As shown in Figure 1, the parent phosphomolybdic acid ($H_3PMo_{12}O_{40}$, HPMo) contained mainly Brønsted acidic sites together with lesser amount of Lewis acid sites. After the introduction of La ions into HPMo, the amount of Brønsted acidic sites in the samples decreased with increasing the La content; whereas, the amount of Lewis acid sites increased. This result indicated that exchanging of H^+ with La cations modified the acidity of HPMo and increased Lewis acidity.

Table 2. The hydrothermal conversion of fructose catalysed by various catalysts

Catalyst	Temp (°C)	Fructose conversion (%)	Yield (%) Lactic acid	Formic acid	Acetic acid	Levulinic acid	Acetol	HMF
HPMo	130	81.5	6.3	4.8	–	2.4	2.4	16.2
	170	100	25.4	7.6	2.6	4.1	4.3	8.9
La-HPMo-1	130	81.2	13.4	1.6	0.8	0.5	1.9	10.7
	170	100	27.3	3.1	2.4	1.8	2.7	6.1
La-HPMo-2	130	82.1	19.0	1.7	0.8	0.6	2.3	10.1
	170	100	49.2	4.1	1.2	1.1	2.3	4.5
La-HPMo-3	130	84.5	28.7	1.5	0.1	0.5	1.4	9.2
	170	100	65.0	4.5	0.3	1.0	2.5	3.7
La-HPMo-4	130	93.7	23.7	1.9	1.0	0.5	1.8	8.1
	170	100	56.8	4.6	3.2	1.2	1.4	3.8
$La(NO_3)_3$	130	56.1	5.0	–	1.2	–	1.6	1.0
	170	100	40.7	7.1	4.3	2.1	4.9	4.4

Reaction conditions: 0.1 g fructose, 0.07 g catalyst, 30 ml H_2O, 2 MPa N_2, 4 h, 500 rpm.

As shown in Table 2, the main liquid products in the conversion of fructose were lactic acid, 5-hydromethylfurfural (HMF), and small amounts of levulinic acid, acetic acid, formic acid, and acetol. During reaction, two possible reaction pathways for fructose conversion may be involved. Pathway A involves dehydration of fructose to HMF, and rehydration of HMF to levulinic acid and formic acid [1]. Pathway B involves the retro-aldol condensation of fructose to trioses and conversion of trioses to lactic acid, and probably acetol, acetic acid and formic acid [2,3]. Pathway A is catalysed by Brønsted acid. The retro-aldol condensation of fructose in pathway B is catalysed by Lewis acid; whereas the conversion of trioses can be catalysed by both Brønsted and Lewis acidic sites. Clearly, the increased Lewis acidic sites can promote the conversion of fructose to lactic acid. This is confirmed by the fact that with increasing the content of La in the catalysts the yield of lactic yield was enhanced. La-HPMo-4 showed lower yield of lactic acid than La-HPMo-3, although the former had higher content of Lewis acid sites. The result suggests that fine balance between Brønsted and Lewis acidic sites may be also important.

We also examined the recycling of La-HPMo-3 in the reaction, and found that the catalyst can be used repeatedly at least five times without loss of lactic acid yield.

Acknowledgments

The authors gratefully acknowledge financial support from the National Natural Science Foundation of China (Grant no. 21576161).

References

[1] M. Dusselier, P. Van Wouwe, A. Dewaele, E. Makshina, B.F. Sels, Energy Environ. Sci. 6 (2013) 1415–1442.
[2] Y.L. Wang, W.P. Deng, B.J. Wang, Q.H. Zhang, X.Y. Wan, Z.C. Tang, Y. Wang, C. Zhu, Z.X. Cao, G.C. Wang, H.L. Wan, Nat. Commun. 4 (2013) 2141–2148.
[3] H.B. Zhao, J.E. Holladay, H. Brown, Z.C. Zhang, Science 316 (2007) 1597–1600.

The role of external mass transfer on corn stover saccharification: fractal kinetic modeling

M. J. Wojtusik[1], Mauricio Zurita[2], Juan Carlos Villar[3], Miguel Ladero[1,*], Félix García-Ochoa[1]

[1]Ingeniería Química, Universidad Complutense de Madrid, Madrid, [2]Abengoa Research, Abengoa, Sevilla, [3]Productos Forestales, INIA-CIFOR, Madrid, Spain mladerog@ucm.es

Introduction

Enzymatic saccharification of pre-treated lignocellulosic biomass is a very complex process, involvinga broad net of reactions. Due to this complexity, it has been difficult to model the kinetics of the reaction system. A novel approach to modelling this reacting system considers it as a fractal solid evolving with time[1]. Because of the mass transfer problems, it can be said that in this type of heterogeneous reactions, the enzymatic saccharification of cellulose is limited on the solid cellulose with a space non-integer dimension less than three. In this situation, is possible to apply a first order potential model modified with fractal assumptions[2]. The rate coefficient, which corresponds to rate constant in the traditional kinetic analysis, is time-dependent in the fractal kinetics. This dependence is correlated to the fractal exponent, h [1]. For the determination of the parameters of this model, several experiments were carried out at different stirring rates.

Experimental

Pre-treated corn stover (PCS) was used as the raw material, provided by Abengoa. The pre-treatment employed was a dilute acid pre-treatment, under usual operational conditions. All of the experiments were carried out in 1L stirred tank bio-reactors, with an initial mass of 550 g and with an initial total solid concentration of 20 %. The enzymatic cocktail employed in this work was provided by Abengoa. Temperature and pH values were those usually employed, while stirring speed was studied up to 500 rpm. Several samples were taken at different times, from 0 to 30 hours. The liquid fraction was analysed by means of HPLC, using a Phenomenex Rezex RHM-Monosaccharide (H+). An aqueous solution of 0.05 M H_2SO_4 was used as mobile phase. Aspen Custom Modeler v8.4 was used for the mathematical fitting of the experimental results to the fractal model.

Results

Several values of stirring rate were proved, from 50 to 500 rpm. For each experiment, the fractal model previously described was applied. The initial value of the rate constant (k), and the fractal factor (h) was determined for each run. The results shown that the higher the stirring speed is, the lower is the value of h. Also, it can be said that h is related to the accessibility of the solid for the enzymes. If this parameter value is high, the diffusional hindrances are important, at a constant reactivity of the solid. However, at medium to high stirring speeds, where (based on final glucose concentration values and initial overall reaction rates) no mass transfer hindrance is observed, h is very low: 0,15 (h equal to 0 means global first order, as expected for the elemental hydrolysis reaction). On the other side, the evolution of the first-order kinetic constant is hyperbolic with N, as expected when external mass transfer is removed by increasing stirring speed, and moving towards a turbulent fluid dynamic regime.

So, according to the results, it can be said that external mass transfer has an important effect on the final yield obtained during the enzymatic saccharification of lignocellulosic biomass. Reynolds number estimation suggests that laminar regime is to be avoided if high productivity in high-solid content systems is desired. The stepwise depolymerisation due to glucanases reduces dramatically overall viscosity, and increases, in parallel, Re values, but it seems that an critical initial Re number is needed.

Table. Statistical fitting parameters and kinetic constants (with their standard errors) for the fractal model when applied in the fitting of each experiment data

N (rpm)	k'	h	F-value	AICc	RMSE
500	0.054±0.002	0.19±0.02	5083	-92.11	0.0118
400	0.053±0.003	0.18±0.04	2449	-85.19	0.0162
300	0.056±0.002	0.21±0.02	4864	-92.31	0.0117
250	0.052±0.002	0.25±0.02	6552	-109.95	0.0121
200	0.044±0.001	0.26±0.02	9174	-128.02	0.0086
150	0.052±0.002	0.25±0.02	9547	-114.43	0.0109
100	0.035±0.001	0.48±0.03	1496	-105.88	0.0173
75	0.036±0.001	0.54±0.03	1611	-107.06	0.0145
50	0.021±0.001	0.61±0.03	1518	-111.75	0.0111

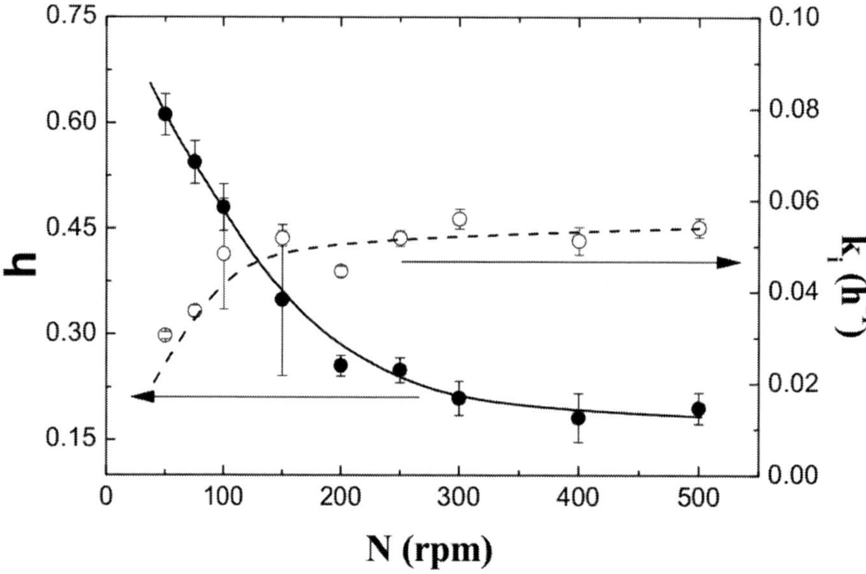

Figure. Evolution with agitation speed of kinetic parameters of fractal model (black filled circles with error bar for the fractal exponent h and black empty circles with error bar for the first-order kinetic constant k_a)

References

[1] R. Kopelman, Fractal Reaction-Kinetics, Science, 241 (1988) 1620-1626.
[2] Z. Wang, H. Feng, Fractal kinetic analysis of the enzymatic saccharification of cellulose under different conditions, Bioresource Technology, 101 (2010) 7995-8000.

Optimization of isomerization activity and aromatization activity in catalytic naphtha reforming via reforming of MO over tri-metallic modified catalyst

Fawzi.M. Elfghi[1], N. A. S. Amin.
[1]Department of Chemical and Petrochemical engineering, The College of Engineering & Architecture, Initial Campus, Birkat Al Mouz Nizwa, Sultanate of Oman

In this work, the estimation capacity of the response surface methodology (RSM), in the catalytic naphtha reforming to enhance the octane number of reformats via isomerization reaction pathway and minimize the aromatization activity over tri-metallic modified Pt-Re/Al_2O_3 catalyst were investigated by applying Design of experiment (DOE).

The parent bimetallic catalyst were modified using a relatively inactive metal (Sn) by means of employing non-conventional method of anchoring technique called controlled surface reaction method in order to favor the intimate contact of Sn with the active Pt–Re. The correlation between RON, aromatization and ismerization activiries with three reaction variables namely temperature (480-510°C), pressure (10-30 bar) and space velocity LHSV (1.2-1.8 h^{-1}) were presented as empirical mathematical models via reforming of complex mixture (80°C -185°C).

Numerical results indicated that the minimum aromatization activity is 20% when reaction temperature is 460°C and pressure of 35 bar. Results also show that maximum ismerization activity of 58% was achieved when pressure is 30 bar and space velocity is 1.8 h. it has been found that optimum value of RON = 89 was attained at 449.9°C, 32 bar and 1.7 h^{-1}. However, high operating pressure and low reaction temperature significantly decrease the total aromatics coupled with substantial decrease in RON which can be enhanced by producing high yield of siomers.

Keywords: catalytic naphtha reforming; central composite design; response surface methodology;

Flow reactor study of hydrazine decomposition in sub- and supercritical water

I. Svishchev, A. Plugatyr
Trent Univ., Chemistry Dept., 1600 West Bank Dr., Peterborough, Canada, K9L0G2;
tel. +1 705 748 1011 x 7063, e-mail: isvishchev@trentu.ca.

The unique properties of high-temperature and supercritical water make it a very attractive medium for a number of environmentally friendly technological applications such as supercritical water oxidation, hydrogen production, as well as for power generation. The goal of this study is two-fold. First, examine the hydrodynamic performance of the flow-through reactor from ambient to supercritical water conditions by conducting residence time distribution measurements (RTD). The results of our recent studies demonstrate that hydrothermal reactors may exhibit non-ideal hydrodynamic behavior. Consequently, experimental determination of the mean residence time of the fluid in the "hot zone" of the flow-through system is crucial for obtaining reliable kinetics data, particularly at supercritical water conditions. The second goal of this study is to examine thermal decomposition of hydrazine hydrate under operating conditions relevant to a supercritical water-cooled nuclear reactor (SCWR). SCWR concept is a direct cycle operating at the pressure of 25 MPa with core inlet and outlet temperatures of about 550 K and up to 900 K, respectively. Hydrazine is one of the potential candidates for chemistry control for both in-core and out-of-core components of a SCWR. Above room temperature, hydrazine decomposes primarily into ammonia and nitrogen, both acceptable products for the nuclear plant chemistry.

The supercritical water test facility is described. The hydrodynamics of employed reactor configuration is examined from ambient to supercritical water conditions by conducting RTD measurements. The RTD results indicate that at 25 MPa, the system exhibits plug flow behavior with small extents of dispersion over the examined temperature and flow rate ranges. Experimental data obtained from the RTD measurements are used to determine the volume of the "hot zone" for the flow-through system. Thermal decomposition of hydrazine in aqueous solution is investigated from 473 to 723 K with pressure fixed at 25 MPa. The first-order rate constant for the decomposition reaction increases from 3.73×10^{-4} s^{-1} at 473 K to about 0.31 s^{-1} at 725 K and follows Arrhenius behavior with effective pre-exponential factor and activation energy being $10^{4.275}$ s^{-1} and 70.24 kJ/mol, respectively.

High conversion of phenolic compounds in hydrodeoxygenation process over Ni-Mo nanoparticles supported on SBA-15

R.W. S. Lima[1], T. L. R. Hewer[1], M. Schmal[2], R. M. B. Alves[1].
[1]University of São Paulo, Polytechnic School, Dep. Of Chemical Engineering, Av Prof Luciano Gualberto, tv 3, no. 380, São Paulo, SP, CEP: 05508-970, Brazil; tel. +55 11 30912237, email: rubens.lima@usp.br; [2]Federal University of Rio de Janeiro, COPPE, Chem.Eng. Program , P.O. Box: 68502, 21941-972, Rio de Janeiro, RJ; tel.+55 21 2562-8349.

The limitation of fossil energy resources and the increasing need for chemicals and liquid fuels require strong efforts in utilizing renewable energy sources. In this way, biomass is the main renewable carbon source and can be converted into an energy-dense oily mixture (bio-oil) by fast pyrolysis, a type of combustion at high temperatures and in absence of oxygen. However, these bio-oils contain considerable amounts of oxygenated molecules, phenolic compounds and water, which makes necessary an upgrading to use them in the chemical industry or fuel production. This can be achieved via hydrodeoxygenation (HDO), in which bio-oil is treated with hydrogen at moderate temperatures (250-450 °C), high pressure (15–300 bar) and in presence of a catalyst, eliminating the oxygen in the oil. HDO of phenolic derivatives, typical representatives of the oxygenated compounds in bio-oils, has great relevance on the process and its mechanism is still not fully understood. Sulfided CoMo- and NiMo-based catalysts, are well suited for this kind of reaction, typically, they are supported on alumina, silica and carbon. In this way, mesoporous silica materials, i.e. SBA-15, have generated interest as supports due to their high structure regularity, low-cost and non-toxicity.

In this study, we present the performance of a Ni-Mo catalyst supported on SBA-15 for the conversion of anisole and guaiacol, either separated or in mixture, in a continuous flow reactor at high pressure (15 kgf cm^{-2}). The influence of the temperature on activity, selectivity to HDO and product distribution is also analyzed. Further, the behavior of the model compounds on the reaction mechanism of the HDO treating is discussed.

The catalyst was characterized using the following methods: N_2 adsorption isotherms, powder X-Ray diffraction, temperature-programmed desorption of ammonia, scanning electron microscopy (SEM), temperature-programmed reduction and infrared spectroscopy. The catalytic tests were carried out using approximately 250 mg of catalyst and at 200, 250 and 300 °C, depending on the test. Previously to the experiments, the catalyst was calcinated and reduced in-situ, under a flow of 50 cm³ min^{-1} of H_2 (AGA) for 4 hours. In a typical experiment, the feed mixture was injected at 14.8 cm³ h^{-1} and vaporized with a flow of 100 cm³ min^{-1} before entering the reactor. Three different reactant solutions were tested: 2% vol. guaiacol (VETEC), 2% vol. anisole (MERCK) and a mixture of guaiacol and anisole 1% vol. each, all employing n-heptane (Aldrich) as solvent. The feed and the products were analyzed inline by gas chromatography. Conversion, products selectivity and HDO were determined in accordance with the equations 1, 2 and 3, respectively:

$$X\,(\%) = \frac{(Reactant)_{in} - (Reactant)_{out}}{(Reactant)_{out}}.100 \quad (1)$$

$$S_i\,(\%) = \frac{Product\ Conc_i}{\sum Product\ Conc_i} = \frac{Product_i/\sum Product_i}{\sum(Product_i/\sum Product_i)}.100 \quad (2)$$

$$HDO\,(\%) = \frac{n_0.a_0.X - \sum n_i.a_i}{n_0.a_0.X}.100 \quad (3)$$

where n_0 is the molar flow of reactant converted, a_0 is the number of oxygen atoms in the reactant (1 for anisole and 2 for guaiacol), n_i is the molar flow of the product i and a_i is the number of oxygen atoms in its molecule. The stoichiometric ratio between anisole or guaiacol and the main products was taken as the unit.

The catalyst characterization tests showed a pore blocking after metal deposition and nanometric particles on its surface. The BET area was found to be 332 m^2 g^{-1}; the total pore volume was 0.68 cm^3 g^{-1}; and the pore diameter was 7.4 nm. Before the metal deposition on the support, the BET area was 630 m^2 g^{-1} and the total pore volume 1.17 cm^3 g^{-1}, which suggests that the metal particles blocked part of the pores present in the SBA-15. Figure 1a shows the SEM micrograph, in which length wise-organized regular micrometric structures are observed, typical of the SBA-15 mesoporous silica. Also, the deposited Ni and Mo nanoparticles can be seen and their size distribution is presented in figure 1b. The mean particle size is 39±12 nm, greater than the pore size, supporting the hypothesis of pore blocking by the metal particles.

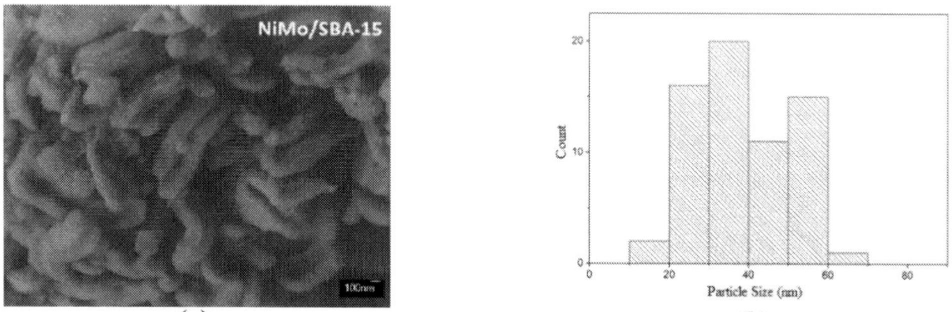

(a) (b)

Figure 1 – (a) SEM micrograph and (b) Particle size distribution of the Ni-Mo/SBA.

Figure 2 shows the results of the catalytic tests. On one side, guaiacol and anisole conversions increased with temperature, reaching 84% (guaiacol only), 75% (anisole) and 50% (guaiacol in mixture) at 300 °C. Guaiacol conversions were higher than anisole possibly due to the effect of the extra oxygen atom in the guaiacol molecule, which can act as adsorption site on the catalyst surface. On the other side, HDO percentage decreased with temperature for all reactants. At 200°C, guaiacol HDO was 97%, anisole was 41% and guaiacol (mixture) was 81%. At low temperatures, non-oxygenated compounds were found to be the main products, mainly benzene and cyclohexane, which increased the HDO results. In contrast, high temperatures favor reactions routes that produce molecules with 1 or more oxygen atoms, reducing thus HDO extension. In light of these results, we could understand the effects of temperature and chemical properties of the model compounds on conversion, selectivity and HDO percentage over a nanoparticled Ni-Mo catalyst supported on SBA-15.

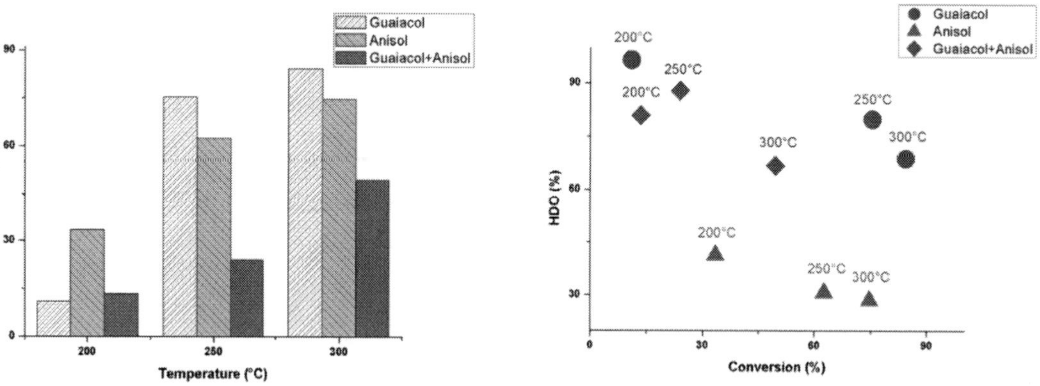

(a) (b)

Figure 2 – (a) Conversion and (b) HDO results on the NiMo/SBA-15 catalyst

Synthesis of iron oxide nanoparticles and application for liquid phase catalytic oxidation of propanol

D. Lokhat[1], M. Carsky[1]
School of Engineering, University of KwaZulu-Natal, Mazisi Kunene Road, Durban 4041, South Africa; tel. +27312601897, email: lokhat@ukzn.ac.za; fax: +27312601118.

A series of iron oxide nanoparticles were prepared using a co-precipitation technique. An appropriate mixture of Fe^{3+} and Fe^{2+} salts were precipitated with an aqueous solution of NaOH or KOH under ultrasonic irradiation. The resulting hydroxide slurry was further oxidized using air or CO_2before filtering and drying to yield nanosized iron oxide particles. All samples were analyzed using XRD to determine the mineralogical composition. The samples oxidized in air yielded a mixed metal oxide composed of magnetite, goethite and wuestite. These samples were also found to be magnetic. There was no observable difference in the mineralogical composition of the mixed metal oxide when KOH or NaOH was used as the precipitating agent. The samples prepared using CO_2 were almost exclusively composed of Fe_2O_3 and did not have any magnetic properties.

The prepared nanoparticles were applied for the liquid phase catalytic oxidation of 1-propanol. Approximately 100 ml of 1-propanol and 7.7 g of nanoparticles were placed in a 250 ml round bottom flask equipped with an overhead stirrer, gas sparger, temperature probe and reflux condenser. The reactor was heated to 50 °C using a heating mantle and air was introduced at the bottom of the liquid via the gas sparger at a flowrate of 100 $cm^3 \cdot min^{-1}$. The reaction was allowed to proceed for 1 hour.

The Fe_2O_3 catalyst prepared under CO_2 showed no activity for oxidation of 1-propanol. In the experiments conducted using the magnetite catalysts small quantities of 2-propanone were obtained as a product. The yields of 2-propanone ranged between 1.23 and 2.15 %. The presence of 2-propanone was confirmed via GCMS and the yields of 2-propanone were determined using conventional gas-liquid chromatography. Oxidation of primary alcohols usually gives aldehydes and carboxylic acids as products, however no propanal or propanoic acid was observed in the product liquid. The conversion of 1-propanol to 2-propanone is suggested to proceed via a dehydration of 1-propanol and hydrolysis of the resulting intermediate to first give 2-propanol, which is subsequently oxidized to the corresponding ketone.

Development of structured reactors for transformation of biomass components to high-value products

V. Shumilov[1], K. Eränen, N. Kumar, D. Yu. Murzin, T. Salmi
Laboratory of Industrial Chemistry and Reaction Engineering
Abo Akademi University, Biskopsgatan 8, 20500, Turku, Finland
tel. +358449175938, e-mail: vshumilo@abo.fi

In the mechanism of heterogeneous catalysis a crucial role plays catalyst structure. The objective of this project is to develop structured reactor for the transformation of biomass components to valuable products by heterogeneously catalyzed reductive amination reaction. Porous materials are used as catalytic carriers in this work.

Porous materials with porosity within the range from 20% to 97% and different pore type and size made of ceramics, polymers or metals are widely applied in chemical industry. The unique properties of porous materials allow to carry out a spectrum of well-established and recent applications in catalysis.

In this work polyurethane replica technique used in the producing of porous ceramic catalytic carriers, optimizing such parameters as strengths, surface area, pore size. Alumina in the α phase was used as a material for this purpose. α-Al_2O_3 the strongest and the hardest of oxide ceramics, its wearability resistance properties, refractoriness, dielectric and good thermal properties make it suitable for catalysis. Different techniques were implemented to increase surface area of the carrier, they consist of adding the additional layer of γ-alumina slurry on the surface of the foam. γ-Al_2O_3 has much bigger surface area, according to nitrogen adsorption surface area of the carrier reaches 250 m^2/g.

Wet deposition methods were used to add an active metal component (Platinum, Ruthenium) on the surface of the obtained catalytic carrier. The catalyst samples were characterized by following analysis: transmission electron microscopy (HR-TEM) to examine metal nanoparticles size (see Fig. 1), X-ray photoelectron spectroscopy (XPS-ESCA), inductively coupled plasma mass spectroscopy (ICP-MS) to estimate the total amount of metal.

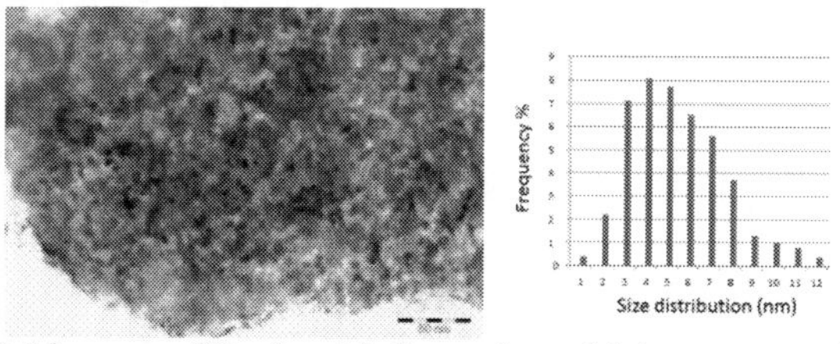

Figure 1. TEM image of Ru/ γ-Al_2O_3 and, and histogram of size distribution of Ru nanoparticles

Extensive reductive amination experiments are planned to be carried out with ethyl levulinate in mixture with 3-aminopropanol at 10 to 25 bar H_2 pressure, temperature between 100 and 250°C. The work includes preparation and testing of reactor structures and kinetic studies, as well as investigation of heat and mass transfer effects and flow studies carried out with classical methods and tomography.

The main parameters of the experiments are initial composition of the mixture, temperature, hydrogen pressure and residence time. Long-term experiments will also be performed in order to confirm the catalyst durability. The products will be analyzed with high performance gas chromatography.

Improved *p*-xylene selectivity on the conversion of light paraffins to aromatics using modified HZSM-5 catalysts

[1]T. Chuntachat, [1,2]S. Jongpatiwut, [1,2]T. Rirksomboon, [1,2]B. Kitiyanan
[1]Chulalongkorn Univ., The Petroleum and Petrochemical College, Phayathai Rd., 10330 Bangkok, Thailand; tel. +66 22184154, fax. +66 22184154, e-mail: siriporn.j@chula.ac.th
[2]Center of Excellence on Petrochemical and Advanced Materials Chulalongkorn University, Phayathai Rd., 10330 Bangkok, Thailand; tel. +662 2184141, fax: +662 6117619.

Presently, the abundance and lower cost of light paraffins especially *n*-pentane have created interest in converting them into aromatics because of its useful for increasing product value. Among the aromatics, *p*-xylene is industrially more important than the others because it is the key chemical for the production of polyethylene terephthalate (PET). In particular, ZSM-5 zeolites are widely used because their suitable pore sizes favor *p*-xylene production. However, their stability was not satisfactory under high *p*-xylene selectivity conditions. Therefore, the modification of ZSM-5 zeolite catalyst was required. In this work, modifications of HZSM-5 zeolite catalysts were desilication, Ga ion-exchange, and silylation, respectively. For desilication condition, various concentrations of NH_4OH, temperature and time used in treatment process will be studied for the conversion of *n*-pentane to aromatics maximizing *p*-xylene yield. Moreover, the desilication under microwave and conventional heating will also be studied. The catalyst activity testing was carried out in a continuous flow fixed-bed reactor at 500 °C, atmospheric pressure, and WHSV of 5 h^{-1}. The catalysts were characterized by several techniques including ^{27}Al MAS NMR, BET, TPD, TPO, XRD, and XRF.

Influence of alloying components salts on the catalytic hydrolysis kinetics of tetraethoxysilane in production of high purity silicate materials

A. Bessarabov, A. Kvasyuk, V. Trokhin, A. Vendilo
R&D Centre "Fine Chemicals", Krasnobogatyrskaya st. 42, 107564 Moscow, Russian Federation; tel. +7 495 9835888, e-mail: bessarabov@nc-mtc.ru

Polycomponent SiO_2-based materials containing various oxides have been extensively used. A promising method for their synthesis is the liquid-phase hydrolysis of tetraethoxysilane (TEOS) in the presence of water-soluble salts of alloying elements. Kinetic regularities of the process should be known for developing technologies for the synthesis of polycomponent silicate materials.

TEOS hydrolysis under acidic conditions is a complicated process involving consecutive reactions of ethoxide hydrolysis (whose degree is 75%) and silanol condensation. Final products of the hydrolysis are high-molecular polyethoxysiloxanols with reactive silanol groups (Shalumov et al., 1984).

At the first stage kinetics of autocatalytic hydrolysis of tetraethoxysilane in the presence of aluminium nitrate has been studied experimentally. According to the suggested reaction mechanism and the proposed mathematical model, kinetic constants have been determined.

In the presence of soluble aluminum nitrate the process is complicated by exchange reaction between free silanol groups of polyethoxysiloxanol with aluminum nitrate. This reaction leads to the increase of the acidity. The acid formed catalyzes TEOS hydrolysis, hence in the presence of aluminum nitrate it becomes autocatalytic.

Kinetic studies were carried out in an isothermal, propeller-stirred reactor at different temperatures. The process was monitored according to changes in the TEOS concentration and the aqueous phase acidity determined with an ionometer. Hydrodynamic conditions were chosen so that all experimental studies were performed in the automodel region. The initial catalyst concentration depends on the hydrolysis of aluminium nitrate whose initial equilibrium concentration was 0.12 mol/l.

At the second stage the effect of neutral salt concentrations on the hydrolysis kinetics in three-component systems containing TEOS and aqueous solutions of Na, Nd and Pb nitrates was shown. A generalized model for the synthesis of polycomponent hydrolytes is suggested, which permits to predict the conditions to prepare compositions with specified characteristics.

The mechanism of TEOS hydrolysis has been described as:

$$\equiv Si-O-C_2H_5 + H_3O^+ \overset{K}{\Leftrightarrow} \equiv \overset{\overset{HOH}{\vdots}}{Si} - \overset{\overset{H^+}{\vdots}}{O}-C_2H_5 \qquad (1)$$

$$\equiv \underset{\underset{HOH}{\vdots}}{Si} - \underset{\underset{H^+}{\vdots}}{O}-C_2H_5 \overset{k}{\rightarrow} \equiv Si-OH+H^+ + C_2H_5OH \qquad (2)$$

where K is the equilibrium constant and k is the reaction rate constant.

When water-soluble hydrolyzing salts are used for alloying, the formation of metal-siloxane bonds can be described as:

$$\equiv Si-OH+Me^{n+} \Leftrightarrow [\equiv Si-O-Me]^{(n+1)^+} + H^+ \qquad (3)$$

Hence, proton extraction, i.e. changes in solution acidity during co-condensation, can be a criterion of interaction. Sharp decrease in the pH of final hydrolyte in the presence of aluminum nitrate has been established previously (Bessarabov and Kvasyuk, 2015).

890 and 815 cm bands in the IR spectra of alumina silica gels that are due to the vibrations of Si-O-Al bond, confirm the formation of a metal-siloxane bond during TEOS cohydrolysis with an alloying element salt. On the other hand, the bands of amorphous silica and NO_3^-, in the IR spectra of gels containing sodium, neodymium and lead nitrates, indicates that samples have individual phases of their components. This fact and the unchanged acidity of product hydrolytes suggest that no metal-siloxane bond is formed during the liquid-phase hydrolysis in the presence of unhydrolyzed compounds. Neutral salts, which do not react with hydrolytic polycondensation products, however, exert a pronounced effect on the kinetics.

Kinetic modelling of hydrolysis in the presence of an alloying additive has revealed that the hydrolysis reaction is zero and first order with respect to TEOS and catalyst, respectively. It has been established that sodium and neodymium nitrates influence the energy component, with a linear dependence of the activation energy on the concentration of alloying compounds. Taking into account the limitations imposed by reaction (2), the kinetic model is:

$$\frac{dX}{d\tau} = -KkX_K = -K_O X_K \exp\left(-\frac{E_O + \beta_{Me} X_{Me}}{RT}\right) \qquad (4)$$

where X, X_K and X_{Me} are the concentrations of TEOS, catalyst and alloying compound, respectively; K_O = 18416.7 s^{-1} is the pre-exponential factor; R is the universal gas constant; T is temperature; E_O =27.57 kJ/mol is the activation energy for acidic TEOS hydrolysis; β_{Na} =3.38 and β_{Nd} = 20.12 kJ/mol^2 are the alloying factors accounting for the effect of sodium and neodymium nitrates on the kinetics.

From experimental investigations and mathematical processing the alloying factor for lead nitrate is β_{Pb} = 9.27 kJ/mol^2. This value is between the parameters for sodium nitrate and neodymium nitrate. Growth of the alloying factor in this sequence leads to an increase in the activation energy, i.e. to the retardation of TEOS hydrolysis, which is likely to be due to variations in the ionic strength of solution.

Contribution of the concentration and nature of alloying components to the rate of TEOS hydrolysis can also be treated in terms of the Brönsted effect in electrolysis. This theory accounts for the influence of neutral salt concentration on the rate in aqueous solutions.

Investigations of hydrolysis in three-component systems involved the analysis of paired interaction of alloying additives and its influence on the kinetics. We have examined two three-component systems: TEOS - $Pb(NO_3)_2$ - $Nd(NO_3)_3$ and TEOS - $NaNO_3$ - $Pb(NO_3)_2$. To supplement the previously studied influence of nitrates on activation energy, we have plotted three kinetic dependences for each system in the presence of alloying additives.

From the mathematical analysis we have derived the generalized kinetic equation:

$$\frac{dX}{d\tau} = K_O X_K \exp\left(-\frac{E_O + \sum_{i=1}^{n} \beta_{iX_i}}{RT}\right) \qquad (5)$$

where n is the number of alloying compounds.

Taking into account the dispersion of reproducibility, we have applied the Fisher criterion to prove the adequacy of model and experiment for three binary, two ternary and one quaternary systems.

References

Shalumov B Z, Bessarabov A M, Mikheev M V, Gornostaeva L A, Polyakov B I (1984) Kinetics of autocatalytic hydrolysis of tetraethyl orthosilicate in the presence of aluminium nitrate. Reaction Kinetics and Catalysis Letters 25(3-4):177-179. doi: 10.1007/BF02064396

Bessarabov A, Kvasyuk A (2015) Cybernetic researches in technology of chemical reagents and high-purity substances. Clean Technologies and Environmental Policy 17(5):1365-1371. doi: 10.1007/s10098-015-0985-y

Influence of presence of heavy metals on the kinetics of pyrolysis of almond shell

A. Ronda[1], M. Calero, G. Blázquez, A. Pérez, M.A. Martín-Lara
University of Granada, Dep. Chemical Engineering, Avda. Fuentenueva, s/n, 18071 Granada, Spain; tel. +34958244075; e-mail: alirg@ugr.es; fax: +34958248992.

This work presents a comparative study on thermal degradation behaviour of pyrolysis of raw and metal-loaded almond shell after its use as material for the removal of copper and lead from aqueous solutions in a fixed bed column. First, breakthrough curves of biosorption process were obtained and main dynamic biosorption parameters were determined and analyzed. Then, non-isothermal thermogravimetric experiments were carried out with raw and metal-loaded biomass in a thermobalance under nitrogen atmosphere at different heating rates (5, 10 and 20 °C/min). Although, results show the same shape of the TG curves, an increase in the thermal stability of metal-loaded samples in the early stages is observed. It is evidenced by a slight displacement of the curves to the right in the first half of the TG curve. Also, at higher temperatures, a displacement of TG curves downward is observed, modifying the temperature ranges to which the main thermal degradations take place. Furthermore, the residual mass fraction of metal-loaded samples is lower than that observed for raw almond shell. This behaviour could be due to the inhibitory and/or catalyst effect that copper and lead performs on the primary and secondary reactions taking place during the thermal degradation of the main constituents of the raw almond shell, which could decrease the proportion of the resulting solid product. Next, activation energies were determined by Flynn-Wall-Ozawa, Kissinger-Akahira-Sinose and Friedman methods. A significant variation on activation energies in function of conversion degree was obtained. This fluctuation of activation energy in model-free methods can be considered the result of thermal degradation reactions of different pseudo-components of the lignocellulosic material (hemicellulose, cellulose and lignin). Therefore, thermal degradation step cannot be represented by a single kinetic model or a single type of degradation for the whole process. However, three stages can be distinguished, the first stage at low conversion degrees, another stage at conversion degrees between 0.4 and 0.7 that has higher activation energies and can be associated mainly to cellulose degradation and a third stage with lower values of the activation energy for conversion degrees of 0.8 and 0.9. Also, Coast-Redfern method was applied to different $g(\alpha)$ functions for describing possible thermal decomposition mechanisms. The results of the Coast-Redfern showed that nth-order mechanism is the effective mechanism for majority of regions of raw and metal-loaded almond shell samples. Finally, a rigorous mechanistic model that considers three independent nth-order parallel reactions was analysed. Copper and lead contained in metal-loaded samples not modified remarkably kinetic parameters values determined for describing pyrolysis process.

Kinetic analysis of the oil shale pyrolysis using thermogravimetry and differential scanning calorimetry

J. P. Foltin, A. C. L. Lisbôa, G. N. Prado
Univ. of Campinas (UNICAMP), Dep. Process Engineering, Cidade Universitaria Zeferino Vaz, Av. Albert Einstein, 500 – Zip Code: 13083-852 - Campinas - SP – Brazil;
e-mail: juliana.p.foltin@gmail.com.

The organic matter in oil shale is tightly bound to the mineral matter and can only be liberated by pyrolysis. The oil shale has kerogen, a polymer with complex structure in three dimensions, the pyrolysis process of shale becomes very complex, with, many reactions. This study aims to evaluate the kinetic behavior of shale pyrolysis through thermogravimetry analysis (TGA) and Differential Scanning Calorimetry (DSC) non-isothermal. Oil Shale comes from Basin Irati located in the South of Brazil and is provided by Petrobras/ SIX located in the city of Sao Mateus do Sul, Parana, Brazil. Considering the methods ASTM E-698 and Kissinger, were performed an analysis and a comparison of the parameters activation energy and factor pre-exponential using the DSC and TGA data. Experimentally, the heating rates used were 2, 5, 10, 15, 20, 25, 40, 50 ° C / min under constant nitrogen flow of 50 ml / min. The results indicates that the temperature peak increases logarithmically with increasing heating rate values and the kinetic parameters calculated using the different methods are similar when compared to the literature.

Keywords: DSC, oil shale, kinetic, model-free, TGA.

Development of the novel rate-based mass-transfer model for the packed absorption and distillation columns

F. Rejl, J. Haidl, L. Valenz
University of Chemistry and Technology, Prague, Technická 5, 166 28 Praha 6, Czech Republic; tel. +420 22044 3299, e-mail: frantisek.rejl@vscht.cz;

The paper refers about the development of the new absorption/distillation rate-based mass-transfer model which will govern important phenomena neglected in previous and contemporary models (namely liquid phase axial mixing and thermal distillation). The effect of the process conditions will be quantified by standard experiments performed not only under absorption but also under distillation conditions. The profile method enabling determination of the volumetric mass-transfer coefficients under distillation conditions will be used for evaluation of the distillation experiments.

The basic comparison of the absorption and distillation processes has been verified on the absorption and distillation wetted wall columns. The physical properties of the involved absorption systems have been chosen to be close to the methanol/n-propanol distillation system. While the transfer of the mass-transfer coefficients between absorption and distillation has been found generally possible in the device with known interfacial area and with minimal effect of the phase longitudinal mixing, the mean error of transferred coefficients was found to be rather high - 13 % (k_G) and 25 % (k_L). The poor quality of the diffusion coefficient correlations as well as some other unrevealed phenomena is suspected to be a cause of the deviations.

The axial mixing of the liquid phase was found to be an important phenomenon under absorption as well as distillation conditions and separation of its effect from the mass-transfer phenomena seems highly desirable. It was revealed, that the Bodenstein numbers evaluated from the distillation experiments are for structured packings generally lower (axial mixing is higher) than those evaluated under absorption conditions, while for the random packing the opposite is true. k_La and k_Ga evaluated from the distillation experiments –when axial mixing is taken into account- are substantially higher than those with the plug flow assumption. The direct transfer of the Bo_L from absorption to distillation was found inaccurate.

The mass-transfer coefficients, interfacial area and Bodenstein numbers were measured using standard systems and procedures in the packed absorption column with two different diameters. In order to quantify the effect of the liquid phase physical properties on individual transport/hydrodynamic quantities the properties of the liquid phase were altered by addition of the surfactants or directly by replacing of the aqueous system with organic one. The gas-phase properties were heavily modified by introduction of helium and sulphur hexafluoride as a carrier gas and results have been correlated in dimensionless form.

Preliminary analyses of optimal control of batch distillation of non-ideal binary mixtures

Marija Stojkovic [a,b*], Vincent Gerbaud [a,b], Nataliya Shcherbakova [a,b]
[a] Université de Toulouse, INP, UPS, LGC (Laboratoire de Génie Chimique), 4 allée Emile Monso, F31432 Toulouse Cedex 04, France
[b] CNRS, LGC (Laboratoire de Génie Chimique), F31432 Toulouse Cedex 04, France
[*] corresponding author: marija.stojkovic@ensiacet.fr

1. Motivation

Our study is motivated by the problem of industrial solvents regeneration. In France, the solvent regeneration has a strong potential: according to a 2011 French national environmental agency ADEME, only 18% of waste solvent are partially regenerated, while 82% are incinerated [2]. Typically, the concerned mixtures are non-ideal, and a suitable processes such as azeotropic and extractive batch distillation is required. In solvent regeneration industry the batch operation mode gives the flexibility enabling to handle various solvent lots over the year. In practice, batch operation reflux and heating policies are mostly set empirically, based on a valuable know-how.

2. Interest for optimal control

In the context improving energy and separation efficiency, there is a steady interest in the finding of the best control configuration and policy,[6] but much less is known about the best way to control the batch distillation process using the optimal control technique. Hence, the aim of our research is to propose the best open loop control strategy that improves the process operation.

3. System, Model and Optimal control problem.

We present preliminary results of the application of an optimal control strategy for the separation of zeotropic and azeotropic binary mixtures by batch rectification. The non-ideal thermodynamics of the binary mixture is modelled according to the equations given in Doherty and Malone, [4]. We consider an N-stage batch distillation column under total condensation with constant vapour flow rate and constant tray liquid hold-up. Such a system is describes by a system of ordinary differential equations and the distillate rate plays the role of the control variable [3]. The main objective is to find the best control policy in order to maximize the final product output of desired purity over a fixed time period. We initialize the column in the steady state mode. The purity condition is treated as a differential constraint for an additional state variable with prescribed final value.

4. Method

To solve the problem we use the so-called direct method of optimal control. The method is based on a full discretization of the dynamics of the system using an appropriate numerical scheme. This allows to transfom the original optimal control problem into a standard optimisation problem. This method is used in the recently developped optimal control solver BOCOP [5] which uses the IPOPT as the non-linear solver.

5. Our first results

We present the first results of optimal control strategies for batch distillation obtained using the BOCOP solver. In particular, we are interested to describe the structure of the optimal control for different types of mixtures. Figure 1 displays the optimal evolution for inputs along with the optimal evolution of manipulated variable for the P-xylene-hexane. On the upper side, we display liquid molar compositions evolution with time; on the down side we display distillate rate evolution with time.

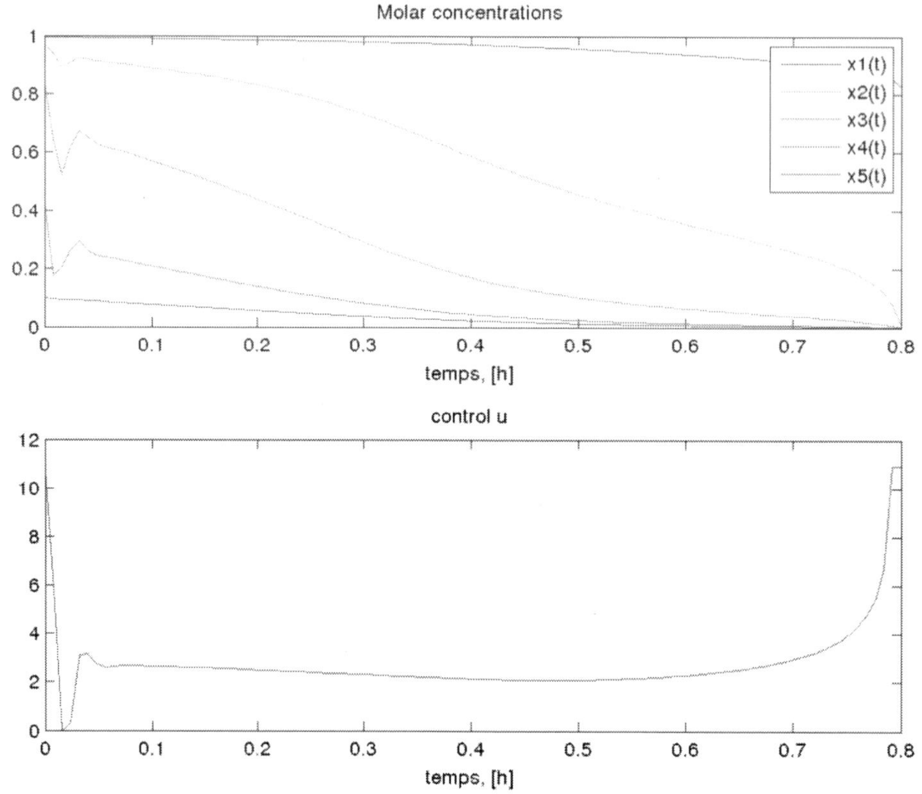

Figure 1. a) Liquid molar concentration vs time, b) Control parameter (D-distillate rate) vs time.

[1] Simant Ranjan Upreti. Optimal Control for Chemical Engineers. CRC PRES, 2011

[2] V. Gerbaud, I. Rodriguez-Donis, Distillation de melanges non-ideaux : Distillation azeotropique et distillation extractive. Choix de l'entraineur. Technique de l'ingenieur, J2-612, 12-2010.

[3] E. Trelat Controle optimal : Theorie et application. Vuibert; IInd edition, 2008

[4] M.F Doherty and M.D. Malone, Conceptual Design of Distillation Systems, Ist editions, MCGraw Hill chemical engineering series ISBN 0-07-017423-7, 2001.

[5] Pierre Martinon, INRIA-Saclay, Institute of Mathematics and Informatics, Bocop Documentation, www.bocop.org

[6] S. Skogestad, I. Postlethwaite, Multivariable Feedback Control: Analyses and Design, Iind edition, Wiley , ISBN 13 978- 0-470-01167-6 (H/B), 2010.

Simultaneous mass and heat transfer model of distillation column

J. Haidl, F. J. Rejl, L. Valenz
UCT Prague, Dept. of Chemical Engineering, Technicka 3, 166 28 Prague; tel. +420 739665286, e-mail: haidlj@vscht.cz.

The rate-based modeling of the packed absorption and distillation columns is an advanced approach to design these high-capacity devices. The model of the column should, in principle, describe all the phenomena occurring on the packing, and thus it should provide reliable predictions of the column behavior. The complexity of such general model, however, requires vast number of parameters which are hard to obtain. To enable the usage of this approach, the model of the column is therefore seriously simplified by the omitting of the phenomena which are assumed to have only minor impact on the process. The interphase heat transfer in distillation performed in packed column is an example of such phenomenon. In common practice, it is assumed that both the phases in distillation, liquid and vapor, are saturated, i.e. they are on its bubble and dew point, respectively. For these conditions the values of the liquid and vapor heat-transfer (as well as mass-transfer) coefficients can be evaluated out of the experimental concentration profiles measured along the distillation column using the profile method developed on our workplace. When applied on the distillation data measured on the ethanol-water, methanol-ethanol and methanol-propanol systems the profile method provided the values of the heat transfer coefficients significantly different than values predicted on basis of the mass- and heat transfer analogy (the difference of 25 – 75 % has been found). This finding suggests either that the phases are not saturated or that the description of the mass- and heat-transfer mechanism is incomplete. The simultaneous mass and heat transfer (SMHT) model of the distillation has been developed to enable the study of the influence of the heat transfer on the distillation column performance. In contrast to the ordinary model of the distillation column the SMHT model enables the subcooling of the liquid and overheating of the vapor according to the intensity of the heat flux given by the heat-transfer relations. The oversaturation of the phases is prevented by two aditional transfer mechanisms – bubbles and mist formation. The SMHT model was validated on the distillation data measured in the wetted wall column; the observed behavior of the phases in the column – liquid subcooling and mist formation – has been predicted by the model. In this paper the influence of the heat transfer on the mass-transfer performance of the packed distillation column is presented. The predicted concentration and temperature profiles for the distillation of methanol-propanol and cyclohexane-heptane mixtures are compared to the experimental data and the heat-transfer induced mass transfer is evaluated.

Transferability of hydraulic and mass transfer characteristics from absorption to the distillation conditions. Study of column diameter effect

L. Valenz, F.J.Rejl, J. Haidl
University of Chemistry and Technology Prague
Technická 5, 166 28 Prague, Czech Republic
lukas.valenz@vscht.cz

Our paper assesses success of the recalculation of the mass-transfer data from absorption to distillation conditions for the Mellapak 250Y and 452Y structured packings. The tests of the recalculation procedure are namely focused on the effect of column diameter and of taking into the account of the axial mixing of the liquid phase or acceptance of the usual assumption of the plug flow. The absorption data in the form of the volumetric mass-transfer coefficients were measured in the absorption column of inner diameter of 0.15m and 0.29m using proven aqueous test systems. The distillation data in the form of the HETP were measured under the total reflux in the 0.15m atmospheric column using alcohol systems.
The recalculation of the acquired absorption mass-transfer coefficients to the distillation conditions was performed through utilization of the model proposed by Billet and Schultes (1999). The raw absorption data were used for optimization of the all six packing specific constants of the model. The hydraulic/hydrodynamic data necessary for the evaluation were also measured and the effect of the column diameter is assessed.
The HETP prediction based on the recalculated mass-transfer coefficients led to poor results for all packings. However, the disagreement is not addressed to the fundamental impropriety of the test systems or the inability of the BS model to transfer the data from absorption conditions to distillation ones. We demonstrate that reasonably accurate values of HETP can be obtained when the influence of the liquid phase axial mixing is involved into the HETP prediction. Therefore we suggest as a reasonable further development step of the process simulation software incorporation of the more sophisticated hydrodynamic models.

Glycol dehydration: evaluation of volumetric mass transfer coefficients in the case of absorption and stripping structured packing towers

T. Todinca, [1]A.D. Filep, [2]G. Vescan
Univ. Politehnica of Timişoara, CAICON Dep., 6, Vasile Pârvan B-vd, 300223, Timişoara, România, tel. +40256403078, teodor.todinca@upt.ro; [1]Univ. Politehnica Timisoara, CAICON Dept., [2]Armax Gaz.S.A., 35A, Aurel Vlaicu St., 551041, Mediaş, Romania

Between the well known methods for the drying of natural gas (compression and cooling, absorption in selective liquids, adsorption on solid desiccants, hydration of deliquescent compounds, etc.), one of the most widely used is the counter-current absorption in triethylene glycol. In its simplest configuration, the TEG dehydration unit contains a contactor (counter-current absorber), a flash tank acting as a phase separator, rich/lean TEG heat exchangers (lean TEG going to contactor is cooled down by the rich TEG), filters and, finally, a regeneration column where water is removed from TEG. In order to obtain lean TEG regeneration up to 99.9 w%, a stripping gas is injected in counter current with the TEG leaving the reboiler in a short column (the "stripping" column) placed between a surge tank and the reboiler.

In order to perform process simulation, because of its dedicated property packages (Peng Robinson, Glycol packages) and for the possibility of integrating new equipment models through Aspen Custom Modeler, Aspen HYSYS is utilized in this work. A typical simulation scheme is shown in the figure 1: in both absorption and regeneration zones, recycle units are implemented in order to analyze different parts and to facilitate the convergence of the simulation scheme. For an accurate evaluation of the water Dew point, a component splitter is used for extracting TEG vapor from the flux leaving gas/TEG heat exchanger.

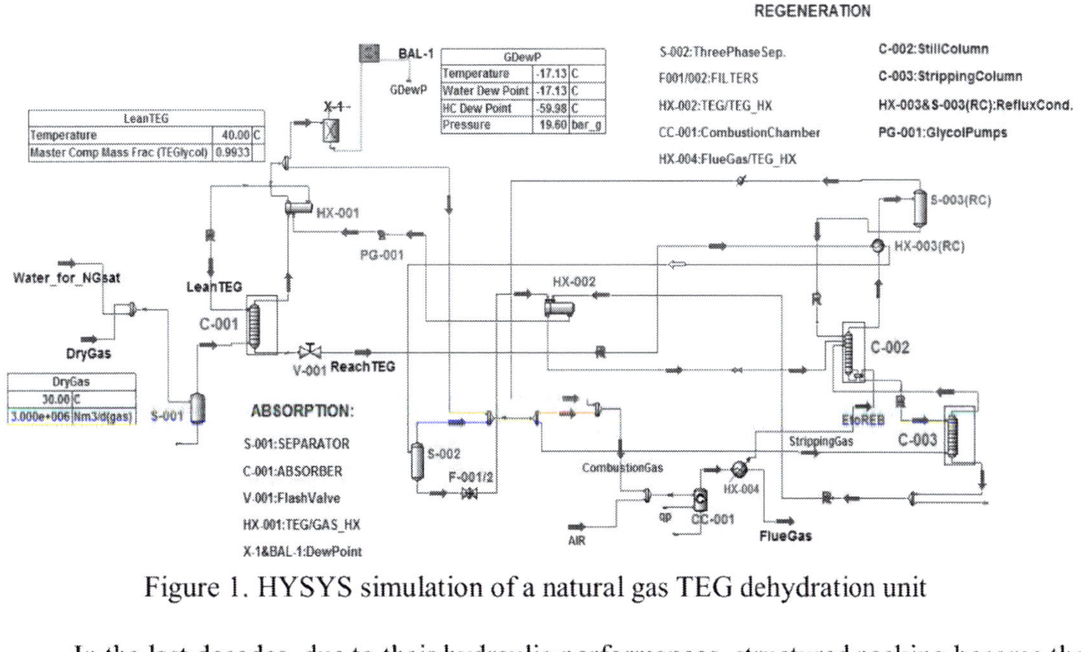

Figure 1. HYSYS simulation of a natural gas TEG dehydration unit

In the last decades, due to their hydraulic performances, structured packing became the first option for internals in TEG dehydration units. However, the available open literature correlations for the evaluation of mass and heat transfer coefficients in packed beds (Larachi et al, 2005, Billet et al, 1999, etc.), while checked against industrial plant data, where found not enough accurate for the case of water absorption/stripping in TEG process. For this reason, the values of the mass transfer coefficients were estimated on a volumetric basis using data from three different industrial units by minimizing the square sum of the differences between models predictions and plant data. For both absorption and stripping, Parrish et al. methodology (see Twu et al, 2005) was used for the calculation of the activities in the liquid phase.

In the case of the **contactor**, a non-isothermal plug flow model describing phases composition and temperature profiles along tower adimensional axial coordinate was used for the evaluation of the volumetric mass transfer coefficient *Kya* and the volumetric heat transfer coefficient. For *Kya*, the following correlation is giving the best results:

$$Kya = 8.424 \cdot (F_f)^{0.85} \cdot \left(\frac{Re_L}{h_L}\right)^{0.41} \cdot Sc_L^{0.33} \quad (1) \qquad \text{where:} \quad h_L = 0.41 \cdot \left(\frac{\eta_L}{\rho_L} \cdot w_L \cdot a_P^2\right) \quad (2)$$

In the equations 1 and 2, F_f is the F factor, Re_L is the Reynolds number, Sc_L is liquid Schmidt number, h_L, η_L, ρ_L and w_L are the hold up, viscosity, density and superficial velocity of the liquid phase while a_P is the packing specific surface.

Regarding the volumetric heat transfer coefficient, its value was found as being 5 time higher than the value calculated on the basis of Chilton – Colburn analogy.

In the case of the **stripping column**, because for all experimental data temperature profile has been found as close to isothermal, a plug flow model describing only gas and liquid phase compositions along axial coordinate was used. The volumetric mass transfer coefficient *Kya* was estimated again on the basis of the minimization of the square sum of the differences between model predictions and plant data. In order to express the variation of volumetric mass transfer coefficient with stripping gas and liquid stream characteristics, a quite simple dependence was finally validated:

$$Kya = 30.1 \cdot F_f^{0.407} \cdot L_0^{0.26} \quad (3)$$

Such simple correlation is justified by the specificity of the process and by the limited range of the experimental data which are covering liquid loads (L_0) from 20 to 40 $m^3/(m^2{*}hr)$ and F-factors from 0.02 to 0.15 $Pa^{0.5}$.

The correlations presented in this paper can be used for a quick evaluation of the structured packing height, facilitating a preliminary evaluation of the investment costs and the estimation of the optimal process parameters.

REFERENCES (selection)

[1] *Larachi, F., Lévesque, S., Grandjean, B.P.A.*, Industrial & Engineering Chemistry Research 47.9, 3274-3284, 2008

[2] *Billet, R., Schultes*, M., Chemical Engineering Research and Design 77.6, 498-504, 1999

[3] *Twu, C. H., Tasone, V., Sim, W.D., Watanasiri, S.*, Fluid Phase Equilibria 228, 213-221, 2005

CFD simulation on desulfurization and decarbonization processes of natural gas in a packed column

W. Su, X. Su, Y. Wang, X. Lan, J. Gao
State Key Laboratory of Heavy Oil Processing, China University of Petroleum, Beijing 102249, China
tel. +86 18810986165, e-mail: suwucup@163.com; [1] China University of Petroleum, Beijing, China, 102249

The following processes and applications are closely related to the process of natural gas desulfurization and decarbonization, which can improve the safety of natural gas storage, transportation and utility. The amine solvents are widely used in this field due to its high efficiency in removing acid gas from the natural gas. Methyldiethanolamine (MDEA) is one of the most commonly used amine solvents. With the increasing demand for natural gas, it has already been a focus that how to improve the safety of natural gas by enhancing the absorption process of acid gas in the packed column. Therefore, we applied Computational Fluid Dynamics (CFD) to analyze the flow, mass transfer and reaction behavior in this process. Firstly, gas and liquid phases are under countercurrent operation in column and they can contact with each other in the filler area. Euler-Euler models are generally applied to describe multiphase flow, and for random packed filler, porous medium model is commonly adopted. Secondly, gas and liquid phases have an apparent interface, and two-film theory can be used to describe mass transfer between these two phases. Thirdly, two reactions are included in this process. According to the special properties of MDEA as tertiary amine, the desulfurization reaction is instantaneous which can be treated as a first-order reaction, and for the decarbonization reaction, which has a fast reaction rate, it is often treated as a pseudo-first reaction. Regarding their sufficiently fast reaction rate, mass transfer and chemical reaction equilibrium theory is suitable to describe the absorption process. In addition, an enhancement factor was introduced in order to correct the rate of these two reactions and to describe their interactions. Finally, in this paper an Euler-Euler based 3D numerical flow-reaction model was developed to predict the absorption of H_2S and CO_2 into MDEA in a random packed column. The distributions of pressure, velocity, phase volume fraction and components concentration within the packed column were obtained with the present model. In addition, these results can provide necessary information for the design and scale-up of the column to enhance the absorption process and maximize the capacity of treatment.

Fatigue cracking of distillation and absorption trays induced by flow oscillations

R. C. Darton[1], W. Zhang[2] and P. H. Taylor[1]
[1] Department of Engineering Science, University of Oxford, Parks Road, Oxford; email: Richard.darton@eng.ox.ac.uk
[2] Centre for Offshore Foundation Systems, University of Western Australia, Australia

Fatigue cracking and failure of column trays can occur when they vibrate in resonance with pressure pulsations in the inter-tray space. Such pulsations are caused by synchronous oscillation of gas flow through the tray perforations on sieve or valve trays, when the perforations are evenly spaced, and the velocity through them is not too high. The controlling mechanism is that of a Helmholtz resonator, in which a mass of liquid near to each perforation oscillates radially, exchanging kinetic energy with potential energy (head) in the bulk liquid across the tray. The rate of exchange is determined by the wave speed in the bulk liquid. A correlation scheme is suggested for the oscillation frequency based on literature data for a range of perforated plate layouts installed in an air-water simulator. The oscillation frequency increases as hole velocity increases, as hole-spacing decreases and at greater liquid heights on the tray. The oscillation frequency is not affected by the volume of the inter-tray space.

We show that effective elastic constants can be calculated for perforated tray panels, and draw attention to the stress concentration around holes where fatigue cracking can be initiated. Vibration analysis of an idealized tray structure identified the first 30 modes with resonant frequencies between 25 and 70Hz, the range where excitation by flow-induced pressure pulsations has been observed in practice. Based on this analysis we make recommendations both for design and operation, to avoid vibrational damage of distillation and absorption trays.

How to use carbonic anhydrase enzymes in carbon capture technology

Arne Gladis[1], Maria T Gundersen, Philip L Fosbøl, John M Woodley, Nicolas von Solms
Center for Energy Resources Engineering, Chemical Engineering Department, Technical University Denmark, Søltofts Plads, 2800 Kgs. Lynby, tel.+4545252978, email: arngkt.dtu.dk,

The Carbon capture process is a technology that could provide an immediate impact on the greenhouse gas emission. The solventbased chemical absorption/desorption process is the most mature technology within and could be retrofitted to large single point sources such as fossilfuel burning power plants. In an ideal process a solvent is used fast reaction kinetics andwith low heat of reaction, to minimize the equipment size and energy requirement.

Solvent with low heat of absorption and favourable desorption performance show often very poor reaction kinetics and slow mass transfer. This would result in much bigger equipment sizes and push up the capital costs and make the application not feasible. The use of catalysts can speed up the reaction rate and enhance the absorption/desorption. One novel catalyst is the enzyme carbonic anhydrase which catalyses the reversible reaction of carbon dioxide hydration. It is known as a very robust enzyme with one of the fastest turn over reaction rate in the world. Enzymes behave very different from conventional catalysts; several factors, such as solvent type, solvent concentration, temperature, pH and CO_2 concentration affect the enzyme performance.

The outline of this presentation is to give an overview of how the enzyme acts in the technical solvent media under absorption and desorption, and which of the factors have the biggest influence on the performance. It comprises of thorough experimental study with different solvent types on a wetted wall column apparatus and deriving a kinetic model for the enzyme. These enzyme enhanced solvents will then be compared to the benchmark solvent 30 wt% monoethanolamine. Process conditions will be determined under which the enzyme works best and an estimation will be derived on how much the absorption performance can be enhance with enzyme addition.

Calorimetric study of absorption of CO_2 in amine-based solvents

H. Yamada
Research Institute of Innovative Technology for the Earth, 9-2 Kizugawadai, Kizugawa, Kyoto 619-0292, Japan;
tel. +81 774 75 2305; fax: +81 774 75 2318; e-mail: hyamada@rite.or.jp.

CO_2 capture technologies are of potential importance for the reduction of anthropogenic CO_2 emissions in the atmosphere. Although amine scrubbing is the currently applicable technology used in an industrial scale for the capture of CO_2 from a gas stream, it still has high energy penalty [1]. Thus the CO_2 absorption properties of the amine solutions have been extensively studied to reduce energy costs in the capture process [2-4]. We have studied a wide variety of amine compounds and developed novel solvents by rational molecular design [5-8].

Primary and secondary amines react with CO_2 to form carbamate anions and protonated amines:

$$2R^1R^2NH + CO_2 \rightleftarrows R^1R^2NCOO^- + R^1R^2NH_2^+ \quad (1)$$

where R^n represents the substituent group. In the presence of water, the bicarbonate anion forms as another product of CO_2 absorption [2,9].

$$R^1R^2R^3N + CO_2 + H_2O \rightleftarrows HCO_3^- + R^1R^2R^3NH^+ \quad (2)$$

Because these absorption reactions are exothermic, the carbamate and bicarbonate dissociate to release CO_2 on heating in the regeneration process. The energy cost in the regeneration process represents more than half the operating cost of a capture plant. Therefore the heat of absorption of CO_2 in the solvent is a major factor in determining the thermal effects in the process [10,11]. Here, we study the heat of absorption of CO_2 in the amine solutions, focusing on several factors including molecular structure of amine, concentration of amine, and CO_2 loading.

The heat of absorption of CO_2 was acquired using a differential reaction calorimeter (DRC Evolution, Setaram) [5]. An amine solution was placed into two glass flasks (150 mL each), with one serving as the reaction cell and the other as the reference cell. The flasks were connected in parallel to a water bath (313 K). Then, 99.9 % CO_2 gas was added to the reaction flask at a flow rate of 120 mL/min at atmospheric pressure. The amount of absorbed CO_2 in a solution was determined using a total organic carbon analyzer (TOC-VCSH, Shimadzu). Some of amines used in the present study are listed in Table 1.

Table 1. CO_2 loading and heat of absorption for aqueous amines [a].

Amine (pK_a) [b] Formula	$\Delta\alpha$ [b]	$-\Delta H$ [d]	Amine (pK_a) [b] Formula	$\Delta\alpha$ [c]	$-\Delta H$ [d]
MEA (9.5)	0.0 → 0.2	88.2	AMP (9.8)	0.0 → 0.2	88.0
$NH_2CH_2CH_2OH$	0.0 → 0.6	79.3	$NH_2C(CH_3)_2CH_2OH$	0.0 → 0.6	75.3
DEA (9.0)	0.0 → 0.2	75.7	AHPD (8.2)	0.0 → 0.2	63.5
$NH(CH_2CH_2OH)_2$	0.0 → 0.6	65.7	$NH_2C(CH_2OH)_3$	0.0 → 0.6	57.5
MDEA (8.5)	0.0 → 0.2	55.0			
$CH_3N(CH_2CH_2OH)_2$	0.0 → 0.6	52.2			

[a] 2 mol/L; [b] Dissociation constant measured at ca. 298 K [12]; [c] Delta loading (mol CO_2/mol amine); [d] Heat (kJ/mol CO_2) at 313 K.

Figure 1 shows the temperature differential between the reaction and reference cells during CO_2 absorption in aqueous amine solutions. Apparently, MEA (pK_a = 9.5) absorbs CO_2 with dissipation of large amounts of heat compared with DEA (pK_a = 9.0). This is attributed to the difference in basicity between the amines, which also determines the branching ratio between reactions (1) and (2) [12].

The heat of absorption depends on the CO_2 loading as well as the molecular structure of amine (Table 1). As can be seen Figure 1, the heat of absorption of CO_2 in aqueous MEA solutions abruptly decreases in the CO_2 loading region above 0.5 mol/mol amine. Recently, we studied CO_2 capture mechanisms in aqueous MEA solutions based on ab initio molecular orbital calculations and proposed pathways for the conversion between carbamate and bicarbonate [13], which explains the decreasing heat observed in Figure 1. As a tertiary amine, MDEA is a weaker base and absorbs CO_2 exclusively by reaction (2) with dissipation of lower amounts of heat.

Table 1 suggests that the heat of absorption of CO_2 is predominantly dependent upon the pK_a value, when the type of product, either carbamate or bicarbonate, is same. As an alternative to MEA, sterically hindered amines such as AMP have attracted attention because of their low regeneration energies [9]. However, it should be noted that the heat of absorption in AMP solvents is comparable to that of MEA in the low-CO_2 loading region (Table 1).

The heat of absorption is also governed by solvation. Because there is an abundance of amino lone pairs, hydroxyl groups and ions in the amine–H_2O–CO_2 system, the solvation energies are significantly affected by the polarity of the solvent [9]. Water, being the most polar solvent, stabilizes the ion products, which largely contributes to the enthalpy change of the CO_2 absorption. Hence, the heat of absorption of CO_2 generally depends on amine concentration through such solute-solvent interactions. As shown in Figure 2, however, it was hardly affected by concentration of aqueous MEA solutions in the low-CO_2 loading region. This result indicates that for the CO_2 absorption reaction, an enthalpic stabilization effect by MEA as a surrounding solvent is almost equal to that of water solvent.

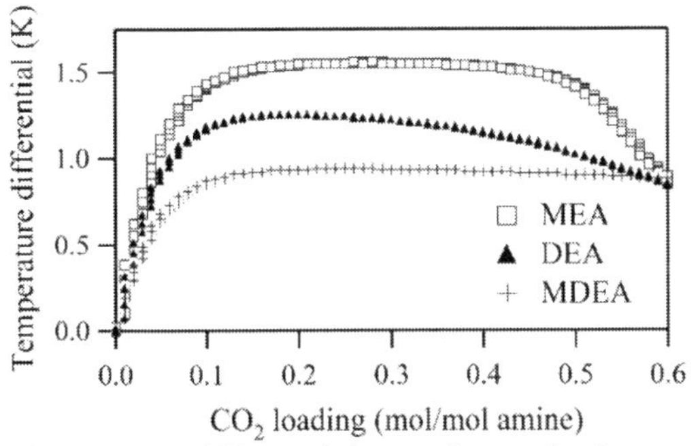

Figure 1. Differential reaction calorimetry for CO_2 absorption in aqueous amine solutions (2 mol/L) at 313 K.

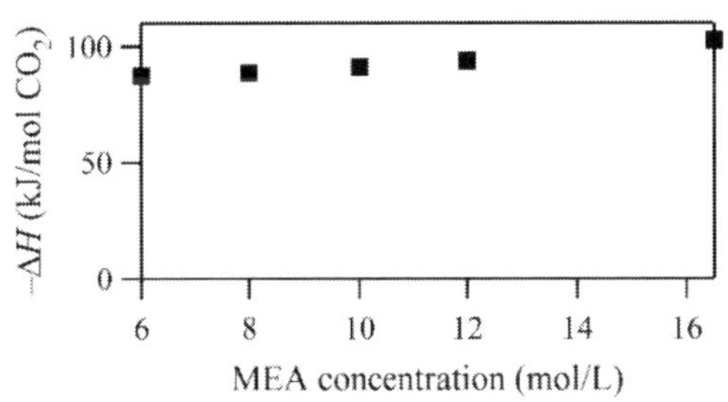

Figure 2. Heat of absorption of CO_2 in aqueous MEA solutions (6.0–16.5 mol/L) in the low-CO_2 loading region (0.0–0.2 mol CO_2/mol amine) at 313 K.

[1] G. T. Rochelle, *Science* **2009**, *325*, 1652. [2] P. D. Vaidya, E. Y. Kenig, *Chem. Eng. Technol.* **2007**, *30*, 1467. [3] L. Dubois, D. Thomas, *Chem. Eng. Technol.* **2009**, *32*, 710. [4] L. Dubois, D. Thomas, *Chem. Eng. Technol.* **2012**, *35*, 513. [5] F. A. Chowdhury et al., *Ind. Eng. Chem. Res.* **2013**, *52*, 8323. [6] H. Yamada et al., *Int. J. Greenhouse Gas Control* **2013**, *17*, 99. [7] H. Yamada et al., *J. Mol. Model.* **2013**, *19*, 4147. [8] H. Machida et al., *J. Chem. Eng. Data* **2015**, *60*, 814. [9] H. Yamada et al., *J. Phys. Chem. A* **2011**, *115*, 3079. [10] I. Kim, H. F. Svendsen, *Ind. Eng. Chem. Res.* **2007**, *46*, 5803. [11] A. V. Rayer, A. Henni, *Ind. Eng. Chem. Res.* **2014**, *53*, 4953. [12] H. Yamada et al., *Ind. Eng. Chem. Res.* **2010**, *49*, 2449. [13] Y. Matsuzaki et al., *J. Phys. Chem. A* **2013**, *117*, 9274.

Improved reactive distillation columns with a double reactive section by feed splitting and external recycle

H. Chen, K. Huang
Beijing University of Chemical Technology, Dep. Information Science and Technology, Beisanhuan East Rd. 15, 100029 Beijing China; tel. +86 10 64437805, e-mail: chenhs@mail.buct.edu.cn.

Reactive distillation column (RDC) is a promising technology of the process intensification and receives considerable attentions over the past 20 years. However, its applications for the separations of the reacting mixtures with the most unfavorable ranking of relative volatilities (i.e., two reactants are the lightest and heaviest components and two products are the light and heavy ones) have rarely been studied so far. The reason is that the most unfavorable ranking of relative volatilities makes the lightest and heaviest reactants extremely difficult in close contact with each other and poses thus great challenge to process synthesis and design. In order to overcome this difficulty, Tung and Huang [1] proposed a kind of reactive distillation columns with a double reactive section (RDC-DRS) in 2007 as shown in Fig. 1a. The two reactive sections were placed, respectively, onto the top and bottom of the RDC-DRS and so were the feeding locations for the heaviest and lightest reactants. With this unique arrangement, the heaviest and lightest reactants could contact each other relatively easily. In 2010, we proposed to add a top-bottom external recycle to a reactive distillation column with a single reactive section [2]. The top-bottom external recycle could transfer a certain amount of the unconverted reactants from one end where it accumulated to the other end where the reactive section was located and helped consequently to boost reaction conversion rate. Detailed studies revealed that with the careful determination of the external recycle the RDC-TBER could be considerably superior to the RDC-DRS. Very recently, we advocated to reinforce the RDC-DRS with feed splitting (RDC-DRS-FS), as shown in Fig. 1c [3, 4]. The lightest and heaviest reactants were introduced simultaneously into the condenser and reboiler and the RDC-DRS-FS was found to have a better economic performance than the RDC-TBER. Stimulated by these encouraging outcomes, we anticipated that the RDC-DRS-FS could be reinforced further with a top-bottom external recycle because the heaviest and lightest reactants still showed high presences at the bottom and top, respectively. The idea gives rise to a RDC-DRS-FS with a top-bottom external recycle (RDC-DRS-FS-ER) as shown in Figs. 1d and 2d. The external recycle should be arranged from the top to bottom for exothermic reactions and from the bottom to top for endothermic reactions. The heaviest and lightest reactants are expected to be in close contact, thus leaving a great potential of improvement in steady-state performance.

This work is to evaluate the RDC-DRS-FS-ER with two examples. For example I, the esterification of lactic acid with methanol (an exothermic reaction), the external recycle of the RDC-DRS-FS-ER is directed from top to bottom. Through the careful searches of catalyst distribution, feed stages, the stage for withdrawing the intermediate product, the external recycle flow rate (for the RDC-DRS-ER and RDC-DRS-FS-ER only), and the feed splitting ratios of the reactants (for the RDC-DRS-FS and RDC-DRS-FS-ER only), the detailed process designs are obtained and sketched in Fig. 1. The RDC-DRS-FS-ER lowers the reboiler heat duty by 98.72 % and 10.28 %, respectively, as compared with the RDC-DRS and RDC-DRS-FS. For example II, the esterification of palmitic acid with isopropanol (an endothermic reaction), the external recycle of the RDC-DRS-FS-ER is directed from bottom to top. Fig. 2 sketches the detailed designs of the four process schemes (i.e., the RDC-DRS,

RDC-DRS-ER, RDC-DRS-FS, and RDC-DRS-FS-ER). The RDC-DRS-FS-ER reduces the reboiler heat duty by 26.06 % and 13.9 %, respectively, as compared to the RDC-DRS and RDC-DRS-FS. An in-depth comparison is included in Table 1 between the RDC-DRS, RDC-DRS-ER, RDC-DRS-FS, and RDC-DRS-FS-ER. The RDC-DRS-FS-ER is found to be the best process design, followed by the RDC-DRS-FS, RDC-DRS-ER, and RDC-DRS. These outcomes indicate the effectiveness of employing a top-bottom external recycle to the RDC-DRS-FS. Therefore, the RDC-DRS-FS-ER should be the most competitive alternative for the separations of the reacting mixtures with the most unfavorable ranking of relative volatilities.

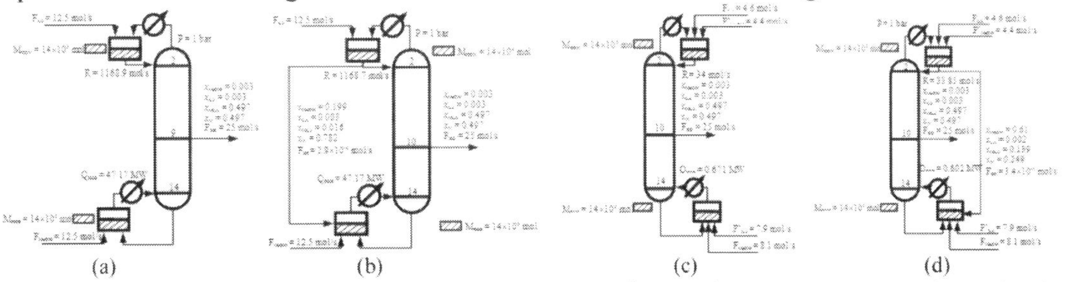

Fig. 1 Process designs for Example I: (a) RDC-DRS, (b) RDC-DRS-ER, (c) RDC-DRS-FS, (d) RDC-DRS-FS-ER.

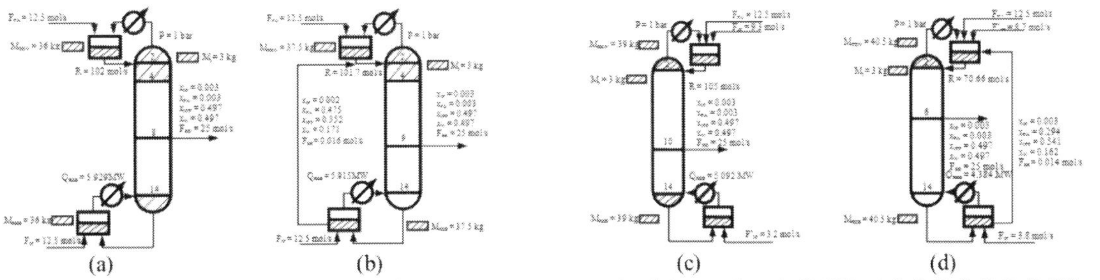

Fig. 2 Process designs for Example II: (a) RDC-DRS, (b) RDC-DRS-ER, (c) RDC-DRS-FS, (d) RDC-DRS-FS-ER.

Table 1. Comparison between all process designs for Examples I and II

Example	Process design	Q_{REB} (MW)	Comparison
Example I	RDC-DRS	47.17	100 %
	RDC-DRS-ER	47.17	100 %
	RDC-DRS-FS	0.671	1.42 %
	RDC-DRS-ER-FS	0.602	1.276 %
Example II	RDC-DRS	5.929	100 %
	RDC-DRS-ER	5.915	99.8 %
	RDC-DRS-FS	5.092	85.9 %
	RDC-DRS-ER-FS	4.384	73.9 %

References:

[1] S. T. Tung, C. C. Yu, Effects of relative volatility ranking to the design of reactive distillation, AIChE J. 53 (2007) 1278–1297.

[2] H. Chen, K. Huang, W. Liu, L. Zhang, S. Wang, S. J. Wang, Enhancing mass and energy integration by external recycle in reactive distillation columns, AIChE J. 59 (2013) 2015–2032.

[3] L. Zhang, H. S. Chen, Y. Yuan, S. Wang, K. Huang, Adopting feed splitting in design of reactive distillation columns with two reactive sections, Chem. Eng. Process. 89 (2015) 9–18.

[4] L. Zhang, H. Chen, Y. Yuan, S. Wang, K. Huang, Synthesis and design of reactive distillation columns with two reactive sections, Chem. Eng. Res. Des. 100 (2015) 311–322.

Scale-up of a packed bed column for wastewater treatment

A. Ronda[1], M.A. Martín-Lara, V. Castillo, G. Blázquez, M. Calero
University of Granada, Dep. Chemical Engineering, Avda. Fuentenueva, s/n, 18071 Granada, Spain; tel. +34958244075; e-mail: alirg@ugr.es; fax: +34958248992.

Wastewaters with a high content in heavy metals are a problem from industry. Wastewater treatment is a process in which the heavy metals in wastewater are partially removed until levels are under legal limits. Adsorption process is among the most effective techniques for heavy metal removal from wastewater and olive tree pruning is an effective and cheap solid widely used as an adsorbent.

The main aim of this work is to carry out a study of the effect of scaling in the process of biosorption of Pb(II) with olive tree pruning (OTP) in fixed bed column. Carried out experimental tests are classified in two groups: laboratory scale (internal diameter of 1.5 cm and a length of 23 cm) and pilot plant (internal diameter of 5 cm and a length of 56 cm packed). The scaling was performed according to similarity criteria proposed by Inglezakis et al., 2006. A total of 8 experiments were performed keeping constant the residence time, bed depth and linear and space velocities (subcriteria 1) and 4 experiments keeping constant the residence time, relation z/d and space velocity (subriteria 2).

Firstly, the effect of bed depth and the flow rate on biosorption of lead by OTP was studied in both scales. It is noted that the increasing of bed depth moved the breakthrough curve to right, maintaining a similar shape. Thus, the breakthrough time increases from 101 min to 241 for a 4.7 cm and 11.3 of bed depth respectively for the laboratory scale and from 70 min to 195 for the same change on bed depth for the pilot scale. It is also observed that the increasing of flow rate decreased the breakthrough and exhausted times.

Finally, to study the scaling of the process, equivalent breakthrough curves in both scales were studied and compared according to two defined subcriteria.

Table 1: Characteristic parameters from experimental breakthough curves.

			Flow rate, mL/min	Bed depth, cm	t_b, min	t_{ex}, min	q_r, mg/g	q_s, mg/g	% R
SC 1	1	Laboratory scale	4	5.7	101	420	10.99	21.91	99.5
		Pilot scale	44.44	5.7	70	375	7.59	20.76	99.4
	2	Laboratory scale	8	11.3	105	495	11.75	23.84	99.6
		Pilot scale	88.89	11.3	75	345	8.60	19.49	99.6
	3	Laboratory scale	4	11.3	241	660	12.11	17.61	99.8
		Pilot scale	44.4	11.3	195	840	11.13	22.54	99.6
	4	Laboratory scale	8	22.6	210	720	11.79	23.05	99.7
		Pilot scale	88.89	22.6	252.5	615	14.87	22.39	99.8
SC 2	1	Laboratory scale	4	5.7	101	420	10.99	21.91	99.5
		Pilot scale	148.15	18.9	81	345	9.71	19.41	99.4
	2	Laboratory scale	4	11.3	241	660	12.11	17.61	99.8
		Pilot scale	148.15	37.7	212.5	630	12.77	24.04	99.8

It is noted that following the subcriteria 1, breakthrough curves parameters in equivalent tests were similar in both scales, although almost experiments the curve of pilot column was advanced to laboratory one. Thus, breakthrough times and capacities were lower in the laboratory scale. Obtained results according subcriteria 2 were also consistent. Anyway, in all experiments the % of Pb removal was higher than 99%. Experimental data were fitted by the Thomas model and results were used to design an industrial scale column.

Modeling of Cr (VI) and Ni (II) biosorption in a fixed-bed column. Estimation of mass transfer parameters

I.L. Rodríguez Rico[1], M. Calero de Hoces[2], R.J. Cabrera Carrazana[3], Y. Cruz Bermúdez[4]
[1] Department of Chem. Eng., Central University "Marta Abreu" of Las Villas, Carretera a Camajuaní Km 5[1/2], Santa Clara, Villa Clara, Cuba. Tel. 5342211825, e-mail: ivanl@uclv.edu.cu.
[2] Department of Chem. Eng., University of Granada, 18071-Granada (SPAIN), Telf.: +34 958 243315, Fax: +34 958 248992, e-mail:mcaleroh@ugr.es
[3] Department of Chem. Eng., Central University "Marta Abreu" of Las Villas, Carretera a Camajuaní Km 5[1/2], Santa Clara, Villa Clara, Cuba., Tel. 5342211825, e-mail: rjcabrera@nauta.cu
[4] Department of Chemistry, Central University "Marta Abreu" of Las Villas, Carretera a Camajuaní Km 5[1/2], Santa Clara, Villa Clara, Cuba. Tel. 5342211825, e-mail: yennierc@uclv.edu.cu

The first part of this work is directed to the analysis of data obtained in the mechanism that describes the biosorption equilibrium of Cr (VI) and Ni(II) in fixed bed columns filled with natural sugar cane bagasse under isothermal conditions. The breakthrough curves along with the characteristic removal parameters like retention capacity and removal percent were experimentally determined at different operating conditions.

Thomas´s model (Thomas 1944) is one of the most general models that are used to describe the biosorption behaviour in the fixed bed columns. This model is based on the second order kinetics and considers that the mass transfer in the interface controls the sorption process.

$$\frac{C}{C_o} = \frac{1}{1 + \exp\left(\frac{k_{Th}}{F}\left(q_o m - C_o V_{ef}\right)\right)}$$

Where is the k_{th} Thomas rate constant (mL/min·mg); q_o is the maximum concentration of solute in the solid phase (mg/g); m is the mass of biosorbent (g).

Linearizing the above equation and plotting ln (C_o/C-1) versus V_{ef} or versus t at a given flow rate, k_{TH} and q_o are obtained from the slope and intercept respectively.

In the fixed-bed system, the main design criterion is to predict the service time of the bed, which is the time required for the removal a specified amount of metal from solution before regeneration is required. This period of time is called the service time of the bed.

$$t = \frac{N_o H}{C_o u_o} - \frac{1}{k_a C_o}\ln\left(\frac{C_o}{C_t} - 1\right)$$

The model is known as the bed depth service time (BDST) model, where t is the service time, N_o is the dynamic bed capacity (mg/L). H is the bed height of the column (cm), u_o is the influent linear velocity (cm/min) defined as the ratio of the volumetric flow rate to the cross-sectional area of the bed, C_t is the effluent concentration of solute in the phase liquid (mg/L); C_o is the initial concentration of solute in the liquid phase (mg/L) and k_a is the rate constant in BDST model (L/mg-min), See table 1.

Table 1: Thomas's model parameters for the dynamic runs.

Metal	C_o	m	H	k_{TH}	q_o	R^2
Cr (VI)	10	1.0	15	5.876	1.324	0.830
	10	1.5	20	3.792	1.113	0.968
	15	1.0	15	2.497	1.912	0.949
	15	1.5	20	2.999	2.093	0.966
Ni (II)	15	1.0	15	2.468	2.893	0.975
	15	1.5	20	2.414	3.428	0.964
	25	1.0	15	2.439	3.210	0.979
	25	1.5	20	2.347	3.666	0.977

Mass transfer parameters like unused bed length and overall mass transfer coefficient were investigated. Additionally, Thomas and Bed Depth Service Time models were used to describe and simulate the columns behavior. High removal percentages of Cr (VI) and Ni(II) were obtained, showing that sorption of these metals is irreversible and is controlled by the mass transfer at the interface.

The concentration in the fluid and the solid phase change with time as well as with position in a fixed-bed. The transfer process is described by the overall volumetric mass transfer coefficient ($K_c a$) obtained from a solute material balance in the column considering irreversible isotherms

$$K_c a = \frac{u_o N}{H}$$

Where, u_o (cm/min) is the superficial velocity of fluid, H (cm) is the bed length and N is the overall number of transfer units defined as:

$$N(\tau - 1) = 1 + \ln\left(\frac{C}{C_o}\right)$$

The parameter $\tau$is dimensionless time defined as:

$$\tau = \frac{u_o C_o \left(t - H\frac{\varepsilon}{u_o}\right)}{\rho_p (1-\varepsilon) H_t W_{sat}}$$

Where, $H_t\varepsilon/u_o$is the time to displace fluid from the bed voids (normally negligible); $u_o C_o t$ is the total solute fed to a unit cross section of bed up to time t; and $\rho_p(1-\varepsilon)H_t W_{sat}$ is the capacity of the bed, or the amount of the solute exchanged if the entire bed came to equilibrium with the feed (ε is the bed porosity and ρ_p is the bulk density).

Calculation of the unused bed surface is a method to evaluate the adsorption capacity of biosorbents in continuous-flow packed columns.

Hence it, for a full bed length (H), the length of unused bed HUNB (cm) is:

$$H_{UNB} = \left(1 - \frac{t_b}{t_s}\right) H$$

Where, t_b(min) is the break-point time and t_s(min) is the saturated point. The results are shown in Table 2: Mass transfer parameters for the dynamic runs.

Metal	C_o	m	H	$K_c a$	N	H_{UNB}
Cr (VI)	10	1.0	15	0.72	4.24	7.17
	10	1.5	20	0.72	5.68	15.00
	15	1.0	15	1.42	8.38	12.79
	15	1.5	20	2.30	18.03	13.33
Ni (II)	15	1.0	15	1.94	11.41	12.82
	15	1.5	20	2.42	19.01	11.61
	25	1.0	15	5.06	29.79	13.00
	25	1.5	20	9.89	77.40	13.33

Operational cost minimization of distillation columns using metamodeling techniques

V.M. Martín Barrios, A. Shokry, M. Graells.
Chemical Engineering Department. Universitat Politècnica de Catalunya (UPC).
Consorci Escola Industrial de Barcelona (CEIB), BARCELONATECH, EUETIB, C/ Comte d'Urgell 187, 08028-Barcelona, Spain.
tel. +34 934137275, e-mail: victor.manuel.martin@estudiant.upc.edu; ahmed.shokry@upc.edu; moises.graells@upc.edu.

Operation optimization of distillation columns is usually a complex engineering problem, due to the complexity of the rigorous distillation models to be considered. This complexity is due the high nonlinearity which stems from the embedded relations, as reaction kinetics, thermodynamics, heat and mass transfer...etc. Additionally, such models often require high convergence time, and include several convergence problems, lack of transparency and unavailability of derivative (i.e. black box models. e.g. modular process simulators) [1],[2],[3],[4]. On the other hand, the use of simplified or short cut models that consider/describe only the main features of system in the optimization problem, leads to deteriorated and unrealistic optimization results [1],[2],[3],[4].

A new efficient way receiving considerable research interests for the rigorous optimization of distillation columns consists on the use of data driven surrogate models or metamodels, which are built with data generated form the rigorous complex model. They can be used in the optimization problem instead of the complex model, providing both high accuracy and computational simplicity [1],[2]. In this line, Quirante, et al. [4] have developed an optimization algorithm which replaces complex black box distillation model (ASPEN model) with kriging metamodels, which are then optimized in a sequential/adaptive way using derivative based (local) optimizer. Their approach showed robust performance with very high accurate/deterministic results; however, one of its major shortages is that it can be easily trapped in a local optima or separated feasible region (in case of non-continuous search space). Ochoa-Estopier, et al.[3] have presented an optimization algorithm based on the replacement of a complex distillation system (ASPEN model) with Artificial Neural Networks metamodels, which are then optimized by derivative free optimizer (Simulated Annealing); their algorithm showed ability of global convergence, due to the stochastic nature of the optimizer, although this characteristic also resulted in the inability of providing deterministic solutions.

In this work, a two stage optimization algorithm is presented, which is based on the use of metamodels (e.g. Radial Basis Functions), which replace the complex rigorous model: in the first stage, the metamodels are explored/optimized with a derivative free optimization algorithm (Genetic Algorithm (GA)) in order to find the best candidate global solution/area. In the second stage, a deterministic (derivative based) optimization algorithm (*fmincon matlab* optimizer) is used to refine/enhance the optimization search/results around this candidate global solution/area obtained from the first stage. The proposed method is applied to a mathematical example and a case study adapted from [4], of a distillation column modeled by ASPEN HYSYS. The results show that the proposed algorithm is able to provide very accurate results, and it likely converges to global optimal solution. The results also illustrate that the proposed algorithm couples the advantages of both works [3] [4] and avoids their limitations: the first stage of the algorithm guarantees a global convergence, as the stochastic optimizer (GA) is able to globally explore the whole search space, avoiding the entrapment in a local optima or a separated feasible region. And the second stage enhances and refines the optimization results, taking advantage of the derivative based deterministic optimizer.

Acknowledgements

Financial support received from the Spanish Ministry of Economy and Competitiveness and the European Regional Development Fund (research Project SIGERA, DPI2012-37154-C02-01), the 'Agencia de Gestió d'Ajuts Universitaris i de Recerca-AGAUR' (2014 FI00305), and the Research Group CEPEiMA (2014SGR1092), is fully appreciated.

References

[1]. Ahmed Shokry , Antonio Espuña,(2012). Using surrogate models for process design and optimization.Uncertainty Modeling in Knowledge Engineering and Decision Making, 483–488.

[2]. Ahmed Shokry , Antonio Espuña, (2014). Applying Metamodels and Sequential Sampling for Constrained Optimization of Process Operations, Lecture Notes in Computer Science ,8468, 396-407.

[3]. Ochoa-Estopier, Lluvia M., Megan Jobson, and Robin Smith. 2013. "Operational Optimization of Crude Oil Distillation Systems Using Artificial Neural Networks." *Computers & Chemical Engineering* 59 (December): 178–85.

[4]. Quirante, Natalia, Juan Javaloyes, and José A. Caballero. 2015. "Rigorous Design of Distillation Columns Using Surrogate Models Based on Kriging Interpolation." *AIChE Journal* 61 (7)

From microextraction experiments to apparatus design

S. Willersinn, H.-J. Bart
University of Kaiserslautern, Chair of Separation Science and Technology, Gottlieb-Daimler-Street 44, 67663 Kaiserslautern, Germany; tel. +49 6312052414, e-mail: bart@mv.uni-kl.de

Liquid-liquid extraction is a well-known separation technique in hydrometallurgical metal recovery processes such as copper, iron or zinc [1]. Recently, the recovery and recycling of other valuable metals from low concentrated (secondary) resources in ppm range has become increasingly important, which is also addressed and funded by European research initiatives (Horizon 2020 etc.).

Usually, industrial metal extraction is conducted in large mixer-settler batteries, where the residence time is adjusted to reach the desired extraction performance and the extraction kinetics only plays an underpart. Extraction columns for example have a much lower footprint and multiple theoretical stages can be achieved within one single apparatus in counter-current mode.

Therefore, the extraction kinetics plays an important role in process designing, where the residence time is mainly determined and restricted by the hydrodynamics (e.g. flow rates, energy dissipation etc.). For metal extraction, the mass transfer by diffusion is additionally coupled with a chemical reaction of the metal ion with an ion exchanger molecule, usually at the interface (reactive extraction). Consequently, the mass transfer is limited either by diffusion or chemical reaction or by both respectively. In literature there are several methods for investigating the reactive mass transfer [2] including Lewis cells, single drop experiments or rotating diffusion cells etc. in order to determine the kinetic regime. The result of such investigations is quite often, depending on the system, a "lumped parameter" (i.e. an overall mass transfer coefficient), which consist of a diffusional and a reaction part.

The aim of this work is to get a detailed view into this coupled mass transfer, i.e. the quantification of individual mass transfer resistances. For this, a microstructured contactor was developed (s. Fig. 1). Here, the extraction process is conducted non-dispersively, both phases are separated by a metal-woven membrane and the interface is pinned at the pore edges depending on the wetting characteristics. This apparatus design allows the experimental investigation of the extraction kinetics in a well-defined system (i.e. laminar flow, constant and flow-independent interfacial area) at very low sample volumes. The obtained kinetic constants for a specific test system and experimental conditions can be used for process and apparatus designing. Additionally, the higher specific interfacial area in the microchannels leads to a process intensification, where for example dilute feed streams are treated more efficiently.

In a first step, a mass transfer model was developed [3] using the standard test system zinc/di-(2-ethylhexyl)phosphoric acid (D2EHPA)/isododecane, proposed by the European Federation of Chemical Engineering (EFCE). In the next step, it is used to quantify the chemical reaction resistance for various experimental conditions, which depends mainly on the acidity of the feed. From these results, it is possible to derive kinetic constants such as the forward reaction rate constant. The comparison with Lewis cell experiments shows a good agreement, when taking a stagnant film layer thickness of 10^{-3} – 10^{-4} cm [4] into account, which cannot be neglected due to the fast chemical reaction.

In order to expand the methodology, it is validated with a test system, where the overall mass transfer is limited by chemical reaction, the extraction of germanium(IV) with the extractants LIX 63 and KELX 100 is investigated. Usually, for metal extraction, underlying a cation exchange mechanism, the extraction efficiency increases with an increase in pH. The germanium ion behaves in the opposite direction, feed acidities of > 200 g/l H_2SO_4 (for LIX 63) and < pH 1 (for Kelex 100) are necessary to efficiently extract germanium. It was found that the extraction mechanism depends on the germanium species in the aqueous phase, which is pH-dependent as well [5]. From equilibrium and stripping experiments the two following two reactions are proposed to occur at the interface for (1) the extraction with LIX 63 and (2) the extraction with Kelex 100, where R represents the extractant molecule:

$$(1)\quad [Ge(SO_4)_2 + H_2O] + \overline{[RH]} \rightleftarrows \overline{[GeR(OH)(SO_4)]} + [H_2SO_4]$$

$$(2)\quad [Ge(SO_4)_2 + 5H_2SO_4] + 3\overline{[RH]} \rightleftarrows \overline{[GeR_3(HSO_4) + 6H_2SO_4]}$$

The presence of sulphuric acid in the aqueous phase overcompensates the proton release by ion exchange (sulphuric acid is co-extracted), which means that at higher sulphate concentrations the equilibrium is shifted to the right-hand side.

Compared to the zinc extraction, the chemical reaction of the germanium extraction is by orders of magnitude slower, as shown in Fig. 2 for a typical Lewis cell experiment, where the extraction rate is governed only by chemical reaction.

The forward reaction rate constant is calculated by the "F"-method [6] from micro-contactor experiments, as it is usually done in Lewis cell experiments too. Here, the molar flux is plotted consecutively against the sulphuric acid concentration, the extractant concentration and the germanium concentration, which yields the forward reaction rate constant.

Additionally, the extraction kinetics is monitored online, by implementing the setup into a confocal Raman spectroscopy system. The microchannels are optically accessible and local concentration gradients in all three dimensions are obtained.

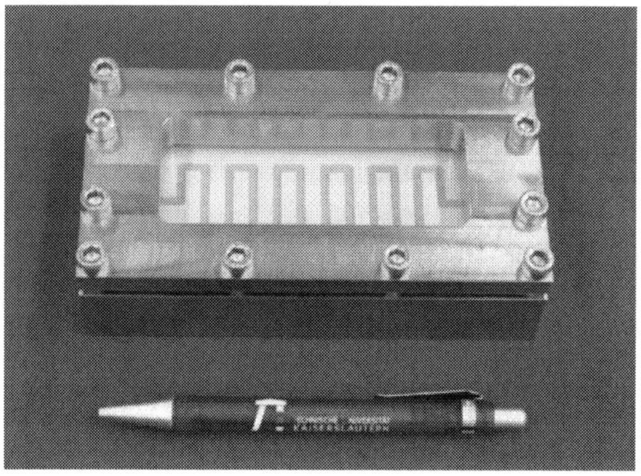

Fig. 1: Modular microcontactor.

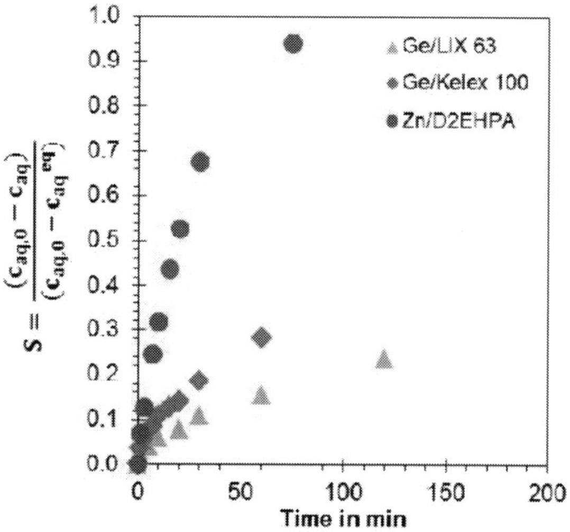

Fig. 2: Extraction kinetics of Ge and Zn in a Lewis cell.

References

[1] Bart, H.-J (2001), Reactive Extraction, Heidelberg, Springer.

[2] Stevens, G.W. & Perera, J.M. (1997), *Miner. Process. Extr. Metall. Rev.* 17, (1-4), 205-226.

[3] Willersinn, S. & Bart, H.-J. (2015), *Chem. Eng. Process. Process Intensif.* 95, 186-194.

[4] Rydberg, J. et al. (2004), Principles and Practices of Solvent Extraction, Dekker, New York.

[5] Cote, G. & Bauer, D. (1980), *Hydrometallurgy* 5, (2-3), 149-160.

[6] Biswas, R.K. & Mondal M.G.K. (2002), Proceedings ISSE, Bhubaneswar, Allied Publishers, New Delhi.

Effect of sieve tray hole diameter on the efficiency of a vibrating plate extractor

N.D. Sincuba, S.Rathilal, M. Carsky
Durban University of Technology, Department of Chemical Engineering, PO Box 1334, Durban, 4000, Republic of South Africa, tel. +27833079955, email: carskym@ukzn.ac.za

The acetone- toluene-water system was studied in a 4.77 cm ID vibrating plate liquid-liquid extraction column (VPE) with a total number of 31 sieve trays. The effect of agitation level (product of frequency and amplitude of vibration) and sieve tray hole diameter on the dispersed phase holdup, drop size distribution and Sauter mean diameter was studied. Normally, the mean drop diameter decreases more rapidly with an increase in the agitation levels and as the sieve tray hole diameter decreases. The dispersed phase hold up in a mixer settler decreased as the agitation level increased until a minimum dispersed phase holdup was reached at the agitation level of 3.75 mm/s. During the dispersion regime, as the agitation level increased the dispersed phase holdup was increased. The drop size and Sauter mean diameter were correlated to the agitation level, whereas, the dispersed phase holdup was correlated to the operating regimes (mixer settler regime and dispersion regime) of the VPE. Sieve trays with a hole diameter of 3.00 mm was the most effective from the ones studied.

Effect of sub-critical water treatment on oil extraction and sugar hydrolysis in palm oil milling processes

R. Yunus, [1]Y. Li[2], C. Chang[2], D. Jiang[2]
[1]Department of Chemical and Environmental Engineering, Universiti Putra Malaysia
43400 UPM, Serdang, Selangor, Malaysia
tel. +6012 2306009, e-mail: robiah@upm.edu.my;
[2] Ohio Agricultural Research and Development Center, 1680 Madison Ave, Wooster, OH 44691, USA
Tel:+330-263-3855, email: li.851@osu.edu and chang.1509@osu.edu

Biomass is an inexpensive alternative feedstock for biofuels and fine chemicals. In recent years, increasing research effort has been spent on developing an economically viable and environmentally safe process for producing green chemicals like levulinate esters from the biomass feedstock. Levulinate esters, like methyl levulinate, ethyl levulinate, and butyl levulinate, are important intermediates and chemicals having widespread applications in many fields. Recently, levulinate ester has been identified as a renewable, environmental friendly alternative to methyl tertiary-butyl ether (MTBE). MTBE is mainly used as an oxygenated fuel additive for diesel fuels, gasoline fuels, biofuels, and blends. However, the use of MTBE as fuel additive has raised environmental concerns due to the detection of MTBE in surface and ground water. Up to now, there are four developed potential pathways for the synthesis of levulinate esters from biomass conversion, including the direct acid-catalyzed alcoholysis of biomass, esterification of levulinic acid derived from the hydrolysis of biomass, alcoholysis of 5-(chloromethyl)furfural derived from biomass, and alcoholysis of furfuryl alcohol from the hydrogenation of furfural. An advantage of the reaction in alcohol is the wastewater is minimized and products are purified easily by distillation of the esters. Methanol and ethanol, which give the highest yields of levulinate esters, can be obtained at a lower production cost. The resulting keto esters are substrates for a variety of condensation and addition reactions at the ester and keto groups. However, a market for these higher-value products is needed to justify the process cost.

Even though, there are many processes available for the synthesis of levulinate esters from biomass conversion, there is still a need for a process for producing levulinic acid and its esters from biomass that is cost effective, yet providing the desired products with high product yield and purity. In this study, it is proposed that the production of levulinic acid is integrated into an existing subcritical water process in palm oil mill that is being developed mainly to extract oil from the mesocarp fiber of palm oil. To achieve this, the subcritical water process must be designed in such a way that it will produce levulinic acid in its aqueous solution for the subsequent synthesis of levulinate esters. There are 5 steps involved in this process; Sub-critical water (SCW) process to separate oil from the digested palm oil-mesocarp fiber, separation of oil, aqueous solution and sludge via centrifugation, separation of levulinic acid and formic acid from water by distillation, esterification levulinic acid with a polyol (glycerol) under vacuum condition to produce levulinate polyol ester and the final transesterification of levulinate polyol ester with methanol to produce levulinic ester.

In this study, the effect of subcritical water (SCW) as a pre-treatment process on sterilized fruits prior to palm oil extraction was investigated. The influences of process parameters namely temperature and degree of maceration on oil recovery and fiber hydrolysis are reported. The oil recovery from the dried pre-treated fiber was nearly 90% in all cases. The macerated fibers only affect the hydrolysis of fibers into sugar but not the oil recovery. The free fatty acid content of the oil after pre-treatment remained below 5% which nullify the claim that high temperature process would affect the quality of oil. Maintaining the temperature at 200°C or below will not affect the quality of oil. Since the process took place at temperature between 180 to 210°C, some of the palm mesocarp fibers were hydrolyzed and degraded into sugars and acids. The HPLC analysis on the aqueous layer from the SCW pre-treatment showed that the hydrolysis of fibers did take place to C5 and C6 sugars as well to acids. However, subsequent acid hydrolysis of fibers using sulfuric acid revealed that the remaining fibers contained 50% glucose, 5 % xylose, 40% lignin and 5% ash. Consequently, the fibers can be hydrolyzed into HMF and subsequently to levulinic acids for further processing to methyl levulinate.

Effect of the diluent on the solvent extraction of molybdenum (VI) with ionic liquids

E. Quijada-Maldonado, M.J. Torres, J. Romero
University of Santiago de Chile, Dep. of Chemical Engineering, Av. Lib. B. O'Higgins 3363. Estación Central, Santiago, Chile;
tel. +56 2 2718 1838, e-mail: esteban.quijada@usach.cl;

Molybdenum(VI) is commonly recovered from leach liquors by means of solvent extraction. The extraction media consists of bis(2-ethylhexyl) phosphoric acid (D2EHPA) as the extractant and kerosene as the diluent. Also, an aliphatic alcohol (1-dodecanol) is required as a phase modifier to dissolve the organometallic complex in the extraction media. Kerosene is highly toxic, flammable and volatile which brings serious environmental issues. Recently, ionic liquids have been proposed as a green alternative to replace the organic diluents due to the excellent properties such us hydrophobicity, non-flammability and negligible vapour pressure. Besides that, a previous work have determined that the use of ionic liquids increases the extractability of lanthanides when compared to the organic diluent hexane [1].

Ionic liquids have been already used as diluent for the extraction of many metal ions such as copper, vanadium, some rare earths, etc. From these previous studies, 1-butyl-3-methylimidazolium bis(trifluoromethylsulfonyl)imide, [bmim][Tf_2N], is the most studied ionic liquid because it is highly hydrophobic. The main goal of this work is to compare the common organic diluent kerosene with [bmim][Tf_2N] on the solvent extraction performance of molybdenum using D2EHPA as extractant. Furthermore, it has been pointed out that this ionic liquid is not capable to dissolve some organometallic complexes [2] and it would require an organic modifier. Therefore, the performance of the ionic liquid 1-octyl-3-methylimidazolium bis(trifluoromethylsulfonyl)imide, [omim][Tf_2N], was also studied since this ionic liquid is more aliphatic than [bmim][Tf_2N] and could dissolve organometallic complexes without the need of a phase modifier.

The solvent extraction experiments were carried out by mixing an aqueous solution (pH = 1) of sodium molybdate with an organic solution composed by D2EHPA and the diluent in an open flask with stirring for 40 minutes to assure the equilibrium. Meanwhile, initial and equilibrium pH were measured. The concentration of molybdenum (VI) in the aqueous phase was determined by AAS.

The results indicate that the use of [bmim][Tf_2N] as the diluent in the solvent extraction of molybdenum required the phase modifier tributylphosphate (TBP) to dissolve the organometallic complex and the amount of this modifier depends on the initial concentration of molybdenum (VI) in the aqueous phase. For the case of [omim][Tf_2N], the phase modifier was not required.

On the other hand, Figure 1 shows that when increasing the D2EHPA concentration the extraction percentage increase for all the diluents. However, the best extraction performance was observed when using [omim][Tf_2N] providing better extraction capacity than the other diluents.

In Figure 2, for all the aqueous-organic phase ratios (A/O) it can be observed that the diluents [bmim][Tf_2N] and [omim][Tf_2N] show better extraction percentages than those obtained with kerosene. This behaviour is due to that, these ionic liquids have higher dielectric constants than kerosene resulting in high extraction capacities. However, when comparing both ionic liquids, [omim]Tf_2N] is the diluent that showed the best results. Although, [bmim][Tf_2N] exhibits higher dielectric constant than [omim]Tf_2N], the longer aliphatic chain of the latter ionic liquid would provide a more nonpolar phase to dissolve the organometallic complex.

These promising results could allow the replacing of the conventional organic diluent with a green alternative solvent and the simultaneous significant decreasing of the amount of diluent required to achieve the desired separation.

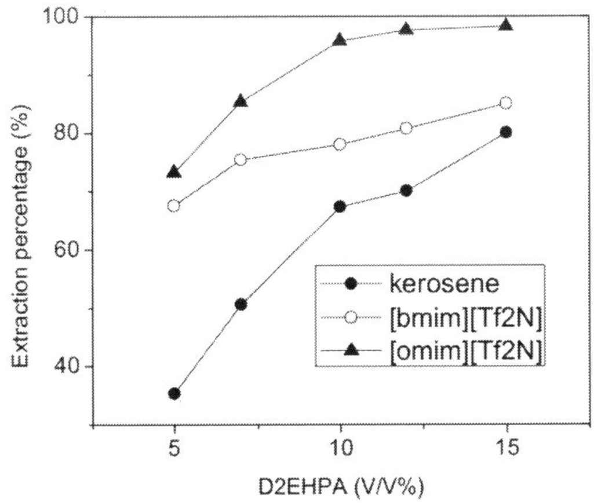

Figure 1. Extraction percentages as a function of the extractant concentration in the diluent. Initial [Mo (VI)] = 6 g/L, initial pH = 1, A/O = 1.

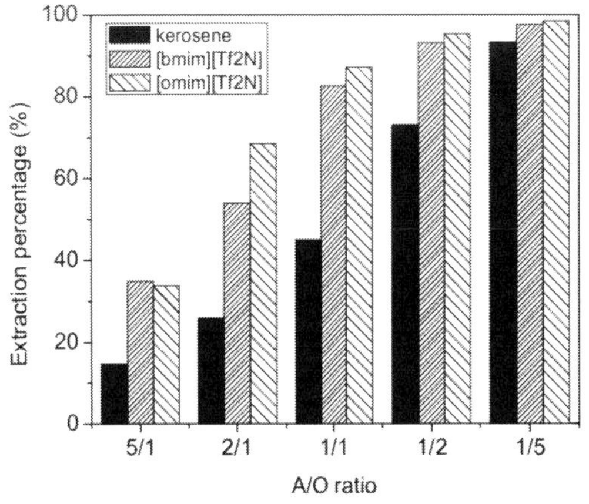

Figure 2. Extraction percentages as function a of aqueous/organic (A/O) ratios for all the diluents. Initial [Mo (VI)] = 4.5 g/L, initial pH = 1, [D2EHPA] = 5 (%V/V)

Reference List

[1] S.J. Yoon, J.G. Lee, H. Tajima, A. Yamasaki, F. Kiyono, T. Nakazato, and H. Tao, Extraction of lanthanide ions from aqueous solution by bis(2-ethylhexyl)phosphoric acid with room-temperature ionic liquids, Journal of Industrial and Engineering Chemistry, 16 (2010) 350-354.

[2] H. Mehdi, K. Binnemans, K. Van Hecke, L. Van Meervelt, and P. Nockemann, Hydrophobic ionic liquids with strongly coordinating anions, Chem. Commun., 46 (2010) 234-236.

Prediction of dispersed phase hold-up in rotating disc contactor liquid extraction column

A. M. I. Al-Rahawi, Department of Engineering, (GUtech,) P.O. Box: 1816,PC 130, Sultanate of Oman. ayham.alrahawi@gutech.edu.om,
Tel: +(968) 2206 1125, Fax: +(968) 2206 1002.

The factors affecting the dispersed phase hold up measured in Rotating Disc Contactor liquid–liquid extraction columns were studied under various operating conditions. The experiments were designed to generate data required to determine the main variables control the hold up in RDC. The results from around 80 runs were analyzed to derive precise correlations for predicting dispersed phase hold up in liquid-liquid RDC extraction column.

Two precise correlations were proposed for predicting dispersed phase holdup in RDC extraction column. The first was based on; distributor hole diameter, operating variables, includes the dispersed and continuous phases flow, the disc speed, system physical properties and column geometry.

The second one considered the same variables, except for the column geometry. This could be used for design purposes.

The proposed correlations were tested with four different distributors, three different systems and with two different columns.

The two correlations are the first of their kinds to consider the distributor hole inlet diameter in RDC column. These have been neglected by previous investigators.

The accuracy of the correlation was high and the standard deviation for all data is 2.5%.

Recovery of lithium and cobalt from spent Li-ion batteries using organic acids as leaching agents

Rabeeh Golmohammadzadeh[a], Fereshteh Rashchi[a] , Mahdi Karimi Jafari[a], Ehsan Vahidi[b]
[a] School of Metallurgy and Materials Engineering, College of Engineering, University of Tehran, Tehran, Iran; tel. 9821-8801 2999, e-mail: rashchi@ut.ac.ir; fax: 9821-8800 6076.
[b] Division of Environmental and Ecological Engineering, Purdue University, West Lafayette, IN 47907, USA; tel. 813-466 2471, e-mail: evahidim@purdue.edu; fax: 765-494 4482.

There is a valid assumption that 30,000 metric tons of Li-ion batteries (LIBs) will be disposed each year by 2030. The requirments for developing sustainable mobility and achieveing cuts in oil consumption will lead the automotive industry to provide a fresh perspective on implementing a policy of mass-production of electric vehicles in which the lithium (Li) consumption will be dramatically high. Furthermore, considering production of digital cameras, laptops, and mobile phones, it is anticipated that Li resources will be completely consumed even before 2020 unless an efficient recycling process can satisfy this growing demand in the near future. Therefore, it is vital to develop processes to recover Li and cobalt (Co) from spent batteries to answer future need.

In this investigation, the goal is to optimize parameters in leaching of spent LIBs in the presence of different organic acids as leaching agents and applying ultrasonic agitation. The LIBs were dismantled and after separation of cathodes and anodes, a hydrometallurgical route was utilized. In this study, five parameters were surveyed which include different organic acids as leaching reagent (citric acid, acetic acid, DL-malic acid and oxalic acid), different concentrations of acids (1-6 mol.L^{-1}), hydrogen peroxide (H_2O_2) concentration (0-5Vol%), solid to liquid (S/L) ratio (15-50 gr.L^{-1}) and leaching temperature (25 °C - 90 °C). Characterisation techniques which used in this study are inductively coupled plasma optical emission spectrometry (ICP-OES) and X-Ray diffraction (XRD). The XRD patterns before and after calcination indicate that the major phase is lithium cobalt oxide ($LiCoO_2$). To optimize the major parameters, design of experiments (DOE) method and the subdivision of the Taguchi were applied. The responses were recovery of Li and Co. The developed model represents the effect of 5 parameters of different organic acids as leaching reagent, different concentrations of acids, H_2O_2 concentration, S/L ratio and leaching temperature. Applying the optimum conditions, leaching efficiencies of more than 70% for Co and more than 90% for Li were achieved.

Key words: spent Li-ion batteries, cobalt, lithium, organic acid.

Rhenium extraction by aqueous biphasic systems formed with PEG/PPG block copolymers

[1]K. Hirayama, [2]K. Sawada, [1]Y. Enokida
Nagoya Univ., [1]Graduate School of Engineering and [2]IMaSS, Furo-cho, Chikusa-ku, Nagoya, 464-8601 Japan;
tel. +81 789 3602, e-mail: hirayama.kazuki@b.mbox.nagoya-u.ac.jp; fax: +81 789 5936.

Aqueous biphasic systems (ABS) formed with polymers, *e.g.*, polyethylene glycol (PEG), poly propylene glycol (PPG), polyvinylpyrrolidone *etc.* and ammonium sulfate, are known to be effective for the extraction of technetium and rhenium. [1][2] As the polymer, PEG / PPG block copolymers (Pluronic-L64) can form ABS which show high distributions of technetium, while its workable concentration range for the formation of ABS is narrow.[1] In this study, we investigated on ABS using other Pluronics, Pluronic-L31 and L61, and compared with those with Pluronic-L64 in terms of the workable range and the rhenium distribution.

Pluronics (Sigma-Aldrich Co. LLC.) and ammonium sulfate (Wako Pure Chemical Ind., Ltd) were dissolved with pure water and adjusted their concentrations at 10-50 wt% and 1.0-2.5 mol dm^{-3}, respectively. For ammonium sulfate solutions, small amount of ammonium perrhenate was spiked to adjust their rhenium concentrations at 2.0 mmol dm^{-3}. A 4 cm^{3} of ammonium sulfate solution was mixed with the same volume of the Pluronic solution in a tube by a shaker for 10 min, and it was centrifuged at 10^{3} G for 1 h. After visual observation of the ABS formation, the rhenium concentration in each phase was measured by ICP-AES.

Pluronic-L31 and L61 could form ABS with lower concentrations of ammonium sulfate in the initial solution than that needed for Pluronic-L64, *e.g.*, they were approximately 0.5 mol dm^{-3} for 30 wt% Pluronic-L31 and L61, while it was 1.0 mol dm^{-3} for L64. In the ABS with Pluronic-L64 and high concentrations of ammonium sulfate, *e.g.*, 1.8 mol dm^{-3} for 30 wt% Pluronic-L64, the upper phase had high viscosities, resulting in the narrow workable range. On the other hand, Pluronic-L31 and L61 could form ABS with more than 2.5 mol dm^{-3} ammonium sulfate solution.

The rhenium distribution ratios in ABS formed from 30 wt% Pluronics are shown in Fig. 1. From the viewpoints of the workable range and the distribution ratio, Pluronic-L31 was suitable for extraction of rhenium among the three Pluronics.

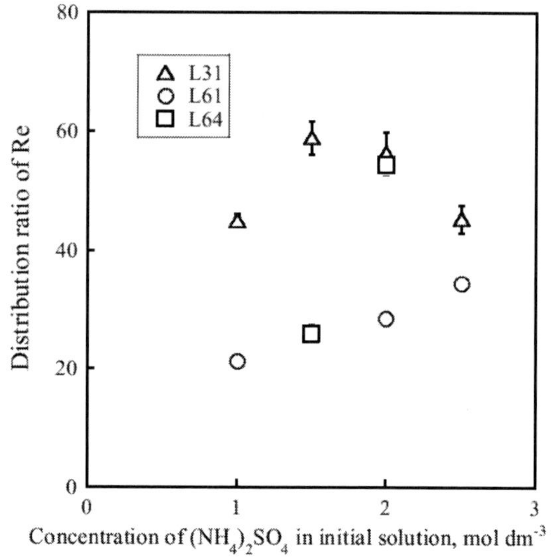

Fig. 1 Distribution ratio of rhenium

[1] R. D. Rogers, Jinhua Zhang, *Journal of Chromatography B*, **680**, 231 (1996).

[2] T. D. Batueva *et al.*, *Russian Journal of Inorganic Chemistry*, **55,** 804 (2010).

Solvent extraction of uranium(VI) ions from aqueous solutions by the 2-butyl-imidazolium di-(2-ethylhexyl) phosphate as ionic liquid

M. Bouhoun Ali, [1,2]A. Y. Badjah Hadj Ahmed, M. Attou, [3]A. Elias, [4]M. A. Didi, S. Chegrouche
Nuclear Research Center of Draria, Box 43, Draria, Algiers, Algeria; e-mail: mbouhounali@gmail.com; [1]University of Sciences and Technology, Houari Boumediene, Faculty of Chemistry, Laboratory of Functional Organic Analysis, Box 32, El Alia, Bab Ezzouar, Algiers, Algeria, [2]University of King Saud, College of Science, Department of Chemistry, Box 245 Riyadh 11451, Saudi Arabia; e-mail: ybadjah@hotmail.com; [3]University of Mouloud Mammeri, Faculty of Sciences, Department of Chemistry, Box 17, Tizi-Ouzou, Algeria; e-mail: hamghil@yahoo.fr; [4]University of Tlemcen, Department of Chemistry, Laboratory of Catalysis, Box 119, Tlemcen, Algeria; e-mail: madidi13@yahoo.fr

In this work, the extraction of uranium(VI) from aqueous solutions using room temperature ionic liquid 2-butyl-imidazolium di-(2-ethylhexyl) phosphate, [C_4mim][D2EHPA], in chloroform has been investigated. Ionic liquid was synthesized and characterized by elemental analysis and by FT-IR and NMR spectroscopy. The effects contact time, pH, aqueous/organic phase ratio, initial uranium(VI) concentration, ionic liquid concentration and temperature on the extraction process have been studied. An empirical modelling was performed by using a 2^4 full factorial design, and regression equation for extraction uranium(VI) was determined from the data. The ionic liquid concentration has the most significant parameter affecting uranium(VI) extraction, whereas the pH and the initial uranium(VI) concentration has the most negative effect on the process, followed by the temperature. The experimental results showed that the optimum conditions obtained were: contact time = 30 min, aqueous/ organic phase ratio = 1, pH value = 2.5, initial uranium(VI) concentration = $4.2x10^{-5}$M, ionic liquid concentration = $5x10^{-4}$M and temperature of 293.15 K. The slopes method showed that the ratio of [C_4mim][D2EHPA] to extracted uranium(VI) (ionic liquid/metal) is 2:1. For the extraction of uranium(VI), $UO_2(NO_3)_2$.2[C_4mim][R.HR] complex was formed (D2EHPA is denoted as HR). Thermodynamic parameters showed that the extraction process was exothermic heat and spontaneous. The obtained optimal parameters were applied to real effluents containing uranium(VI) ions.

Keywords: Solvent extraction; uranium(VI) ions; 2-butyl-imidazolium di-(2-ethylhexyl) phosphate, modelling.

Oil extraction from sunflower seed using green solvent

S. S. de Jesus,[1] A. Santana[2], R. Maciel Filho[1]
[1]Laboratory of Optimization, Design and Advanced Control/School of Chemical Engineering University of Campinas – 13083-970, Campinas, SP, Brazil. [2] Laboratory of High Pressure Process, Department of Chemical Engineering and Environmental Technology, University of Valladolid, Zip 47011, Valladolid, Spain. e-mail: ssjesus@gmail.com

Currently million tons of toxic solvents are used in the oil extraction industry, in which we can mention the n-hexane, representing a major environmental problem[1]. Many solvents can be obtained from renewable feed stocks. Brazil is the world's largest producer of sugarcane and its residue can be a valuable source of raw material for the production of green solvents[2]. Among these solvents we can mention 2-methyltetrahydrofuran obtained from the sugarcane bagasse and isoamyl alcohol that is a residue from alcohol distillation. The objective of this study was to choose and determine a green solvent system with similar extraction efficiencies as the traditional Bligh and Dyer method[3]. Sunflower seed oil was extracted using Bligh and Dyer method[3] (Figure 1) with 2-methyltetrahydrofuran and isoamyl using alcohol ratios of 1:1; 2:1; 3:1; 1:2; 3:1 and the water volume was calculated by Equation 1[4]. At the same time comparative experiments was performed with chloroform and methanol ratios of 1:1; 2:1; 3:1; 1:2; 3:1. Comparison study was done using 5 replicates (n=5). Statistical analysis was performed using Microsoft Office Excel (Microsoft, USA) to determine means and Tukey's Honestly Significant Difference test for comparison between treatments ($\alpha = 0.05$).

$$\varepsilon_{mix} = \sum \varepsilon_i \phi_i + \ldots + \varepsilon_n \phi_n \qquad (1)$$

The results showed that using classic method with methanol and chloroform presented the extraction oil yield with the values of 31-44% (w/w) and values of 36-45% (w/w) using green solvents for extractions. Among the two extraction methods, 2 methyltetrahydrofuran and isoamyl alcohol ratio 2:1 provided the best results (45% w/w), while the classic method using chloroform and methanol with ratio of 3:1 presented a extraction oil yield of 44% (w/w). It was concluded that the proposed extraction method using 2-methyltetrahydrofuran and isoamyl alcohol in this work allowed the same efficiency level as chloroform and methanol.

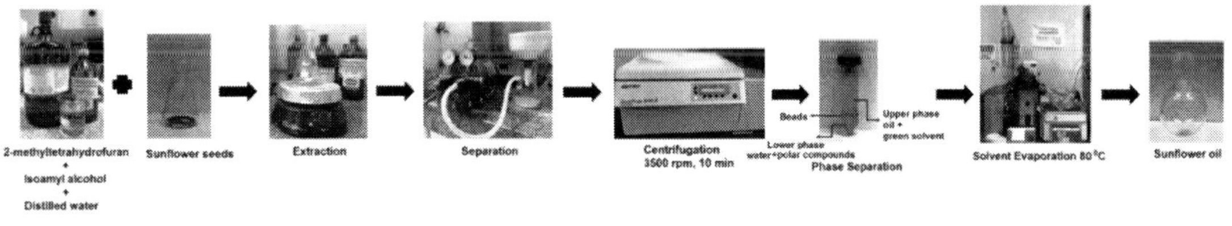

Figure 1: Schematic of lipid extraction for green solvent system.

References

[1] Capello C., Fischer U., Hungerbuehler K. *Green Chem.* **2007**, *9*, 927-34.

[2] Gerbens-Leenes, P.W., Hoekstra, A.Y. UNESCO-IHE Institute for Water Education. **2009**, Delft, the Netherlands.

[3] Bligh, E.G., Dyer, W.J. *Can. J. Biochem. Physiol.* **1959**, *37*, 911-917.

[4] Probst, K.V. Ph.D. Thesis. **2014**, Kansas State University.

Aqueous two-phase system for recovery of building blocks using a green deep eutectic solvent

I. Domínguez[1],* E.A. Macedo[1], A. Domínguez[2], B. González[2]
[1]Department of Chemical Engineering, Faculty of Engineering, University of Porto, Rua Dr.
[2]Advanced Separation Processes Group, Edif. Isaac Newton, Department of Chemical Engineering, University of Vigo, Lagoas-Marcosende, 36310 Vigo, Spain
*Corresponding/Presenting author: irene@uvigo.es.

ATPS based on ionic liquids (ILs) are potential separation systems for biomolecules and added-value compounds. Many novel IL-based ATPS were characterized, by measuring and modeling their liquid-liquid equilibria and studying the partitioning of antibiotics, amino acids, alkanoids and proteins[1]. ILs can form ATPS with salts, polymers, amino acids, and carbohydrates. Those based on ILs and carbohydrates combine the tunable properties of ILs with the chemical functionality of sugar-based molecules, providing an "eco-friendly" separation media. However, the synthesis process of ILs is complex and expensive and ILs are difficult to purify. In addition, pyridinium or imidazolium based ionic liquids are not completely "green". Their toxicity is no less than traditional organic solvents. To overcome the limitations of ILs, deep eutectic solvents (DESs) have emerged. DESs based on choline represent a class of biodegradable and non-toxic ILs, which together with carbohydrates can form a green and mild environment for the separation and purification of biomolecules.

Biomass-derived building blocks [2] such as furan derivatives (furfural), or dicarboxylic acids (levulinic acid) present similarities with the mentioned biomolecules. Since ILs have been applied for biorefining purposes [3], extraction with IL-based ATPS can be seen as a reliable separation process for biorefineries. Moreover, as these biomass-derivatives are produced from sugars, the ATPS (DESs+carbohydrates+water) can provide an useful simultaneous reaction and separation system for production and purification of such building blocks.

In this work, the synthesis of two green deep eutectic solvent (DES) together with the Liquid-liquid equilibrium data and the phase diagrams of aqueous biphasic systems at 25°C are presented. In addition, the partitioning of levulinic acid and furfural is investigated at 25°C and atmospheric pressure.

[1] Freire, M.G., A.F.M. Claudio, J.M.M. Araujo, J.A.P. Coutinho, I.M. Marrucho, J.N.C. Lopes, and L.P.N. Rebelo, Aqueous biphasic systems: a boost brought about by using ionic liquids. Chemical Society Reviews, 2012. 41(14): p. 4966-4995.
[2] Werpy, T.A. and G. Petersen, Top Value Added Chemicals From Biomass: I. Results of Screening for Potential Candidates from Sugars and Synthesis Gas. 2004.
[3] Tan, S.S.Y. and D.R. MacFarlane, Ionic Liquids in Biomass Processing. Topics in Current Chemistry, 2009. 290: p. 311-339.

Acknowledgements
This work was partially supported by FCT/MEC and FEDER under Programe PT2020 (Project UID/EQU/50020/2013), and co-financed by QREN, ON2 and FEDER (Projects NORTE-07-0162-FEDER-000050 and NORTE-07-0124-FEDER-0000011), for which the authors are thankful.
The authors are also grateful to the Comisión Interministerial de Ciencia y Tecnología (Spain) for the project CTM2013- 46093-P and Xunta de Galicia (Spain) for the project EM2014/039. I. Domínguez is grateful to the Xunta de Galicia for her scholarship (ED481B 2014/104-O)..

The solvent extraction of uranium, plutonium and thorium from aqueous solution by 1-hydroxyhexadecylidene-1,1-diphosphonic acid

M. Bouhoun Ali, [1,2]A. Y. Badjah Hadj Ahmed, M. Attou, [3]A. Elias, [4]M. A. Didi

Nuclear Research Centre of Draria, Box 43, Draria, Algiers, Algeria; e-mail: mbouhounali@gmail.com; [1]University of Sciences and Technology, Houari Boumediene, Faculty of Chemistry, Laboratory of Functional Organic Analysis, Box 32, El Alia, Bab Ezzouar, Algiers, Algeria, [2] University of King Saud, College of Science, Department of Chemistry, Box 245 Riyadh 11451, Saudi Arabia; e-mail: ybadjah@hotmail.com; [3] University of Mouloud Mammeri, Faculty of Sciences, Department of Chemistry, Box 17, Tizi-Ouzou, Algeria; e-mail: hamghil@yahoo.fr; [4] University of Tlemcen, Department of Chemistry, Laboratory of Catalysis, Box 119, Tlemcen, Algeria; e-mail: madidi13@yahoo.fr

In this paper, the solvent extraction of uranium(VI), plutonium(IV) and thorium(IV) from aqueous solutions using 1-hydroxyhexadecylidene-1,1-diphosphonic acid (HHDPA) in carbon tetrachloride has been investigated. The HHDPA was previously synthesized and characterized by FT-IR, ^{1}H NMR, ^{31}P NMR spectroscopy and elemental analysis. At organic-to-aqueous phase ratio (1/1) and at room temperature, 95% of uranium(VI), 92% of plutonium(IV) and 83% of thorium(IV) were recovered in 30 min. The conventional log-log analysis of the extraction data reveals that ratios of extractant to extracted U(VI), Pu(IV) and Th(IV) are 1:2.

The solvent extraction of uranyl nitrate from aqueous solution using synergistic mixture of 1-hydroxyalkylidene-1,1-diphosphonic acids and trioctylphosphine oxide

M. Bouhoun Ali, [1,2]A. Y. Badjah Hadj Ahmed, M. Attou, [3]A. Elias, [4]M. A. Didi

Nuclear Research Centre of Draria, Box 43, Draria, Algiers, Algeria; e-mail: mbouhounali@gmail.com; [1]University of Sciences and Technology, Houari Boumediene, Faculty of Chemistry, Laboratory of Functional Organic Analysis, Box 32, El Alia, Bab Ezzouar, Algiers, Algeria, [2]University of King Saud, College of Science, Department of Chemistry, Box 245 Riyadh 11451, Saudi Arabia; e-mail: ybadjah@hotmail.com; [3]University of Mouloud Mammeri, Faculty of Sciences, Department of Chemistry, Box 17, Tizi-Ouzou, Algeria; e-mail: hamghil@yahoo.fr; [4]University of Tlemcen, Department of Chemistry, Laboratory of Catalysis, Box 119, Tlemcen, Algeria; e-mail: madidi13@yahoo.fr

The solvent extraction of uranyl nitrate by 1-hydroxy-1-hydroxyalkylidene-1,1-diphosphonic acids and 1-hydroxyalkylidene-1,1-diphosphonic acids - trioctylphosphine oxide (TOPO) mixture from aqueous solution was investigated. The 1-hydroxy-1-hydroxyalkylidene-1,1-diphosphonic acids tested for extraction were : 1-hydroxyhexadecylidene-1,1-diphosphonic acid (HHDPA) and 1-hydroxydodecylidene-1,1-diphosphonic acid (HDDPA), which were previously synthesized and characterized. The extraction test out have revealed the synergistic effect arising from use of 1-hydroxyalkylidene-1,1-diphosphonic acids and TOPO mixture for extracting uranyl nitrate.

Coalescence and electrocoalescence: extended chemical engineering

A.M. Spasic, [1]J.M. Jovanovic, V. Manojlovic, [2]M. Jovanovic, [3]S. Alexandrova, [4]A. Saboni
Institute for Technology of Nuclear and Other Mineral Raw Materials, Dept. of Chem.Engng., 86 F. d'Esperey St., 11000 Belgrade, Serbia, E-mail: a.spasic@itnms.ac.rs; [1]Inovacioni Center, Dept. of Met. & Tech., University of Belgrade, 4 Karnegijeva St., 11000 Belgrade, Serbia;
[2] Dept. of Met. & Tech., University of Belgrade, 4 Karnegijeva St., 11000 Belgrade, Serbia
[3]Laboratoire de Thermique et Procedes, ENSGTI, rue Jules Ferry, BP 7511, 64 075 Pau, France; [4]Laboratoire SIAME-IUT, UPPA, Avenue de l'Universite, 64 000 Pau, France

Introduction; Developed liquid-liquid interfaces play an important role in some physical, chemical, and biochemical processes, for example strongly related are the processes of formation, existence and destruction of emulsions and double emulsions, life of biological cells, and various membrane separation processes. Since the particular events and phenomena that occure in these processes are not sufficiently well or deeply understood the introduction of theoretical models known in classical and quantum electrodynamics is needed. Hence, the objective of this research is a try to provide an approach to elaborate the complex phenomena that occure at developed liquid-liquid interfaces or at small separations. The representative events that are to be discussed are coalescence and electrocoalescence using electrohydrodynamics (EHD) principles.

History and motivation: The material present results obtained during the last two decades long engineering work, in the beginning analytic and synthetic, and later on scientific work, anticipative and contemplative. So, this material is a kind of experimental and theoretical engineering and multidisciplinary scientific research.

This story begins in the pilot plant for uranium extraction from wet phosphoric acid by D2EHPA-TOPO process. The light phase loop for solution of the entrainment problems was developed and constructed. Equipment, of particular interest for this research, used for a breaking of double emulsions, was the lamellar coalescer. Double emulsion passing through the lamellar coalescer involves the sequences: approach, rest, disturbance, rupture on an inclined plate, coalescence/electrocoalescence, and flow up. Physical chemical system, which has been used to corroborate the validity of theoretical predictions, was a heavy-phase droplet/light-phase film structure immersed in a heavy-phase continuum, double emulsion. This system was the heavy-phase output from a "pump-mix" mixer-settler battery together with its entrained light phase. The heavy liquid was 5.6 M phosphoric acid, and the light liquid was the synergistic mixture of 0.5 M di(2-ethylhexyl) phosphoric acid and 0.125 M tri-n-octylphosphine oxide in dearomatized kerosene.

Theoretical concepts: It was necessary to develop two concepts for easier understanding of the physical events at developed liquid-liquid interfaces. The first is a concept of an entity, and the corresponding classification of finely dispersed systems; this concept permit consideration of the electron transfer phenomenon beside the heat, mass, and momentum transfer phenomena commonly used in *Chemical Engineering*. An entity is derived from Reynolds magnetic criteria as a limit value of characteristic velocity over characteristic resistance. Entities may be differential elements of mass/demons, ions, the phonons as quanta of acoustic energy, the infons as quanta of information, the photons, or the electrons. The second concept is an introduction of almost forgotten basic electrodynamics element the memdiode or memristor as a current controlled device, and the corresponding memristive systems. This concept is needed because the fundamental electrodynamics elements, the resistor, the inductor, and the capacitor are not sufficient for a satisfactory explanation of complex phenomena at developed liquid-liquid interfaces. Now, the electrodynamics elements are: the capacitor and memristor as electroelastic elements, and the inductor and resistor as electroviscous elements. Also, the terms electrical forces and electrical interfacial potential have to be introduced beside the terms mechanical forces and interfacial tension because at small separations the forces of electrical origin become dominant compared to the forces of mechanical origin.

Coalescence and electrocoalescence: Here, nanorheology and electromechanical analogies enter into the game; a few among more than forty constitutive models of liquids, developed during the last two centuries, are mentioned. A corresponding constitutive model of liquids was developed and applied to the selected emulsions and double emulsions. Considering Marangoni instabilities of the first and second order, and possible electrical analogues the coalescence and electrocoalescence phenomena, in some extent, are elucidated.

Theory: As a central part of this research theory of electroviscoelasticity is postulated. The terms electroviscosity and electroviscoelasticity may be broadly defined as dealing with fluid flow effects on physical, chemical, and biochemical processes.

Experimental: In order to justify and corroborate more general approach the obtained calculated results were compared to those experimentally measured using the representative liquid-liquid system. Theory of electroviscoelasticity was experimentally corroborated using the developed and constructed liquid-liquid contact cell for measurements of the electrical interfacial potential, and the nuclear magnetic resonance spectrometer was used for the energetic analysis.

Implications: The term *Extended Chemical Engineering* seems to be appropriate when the electron transfer phenomenon is added to the heat, mass, and momentum transfer phenomena. Now, applying EHD principles deeper understanding of the events at smaller separations (micro, nano, atto) become possible; among these events are coalescence and electrocoalescence.

These concepts, models, and theory discussed may facilitate the understanding of, for example, entrainment problems in solvent extraction, very developed interfaces in colloid and interface science, chemical and biological sensors, electro-analytical methods, biology or biomedicine, in particular hematology (non-Newtonian fluid flow), genetics (DNA models, electron transfer) and electroneurophysiology (quantum spin transport). Furthermore, both, the model and theory may be implemented in studies of structure: interface barriers or symmetries,-surface examples-bilipid membrane cells, free bubbles of surfactants, Langmuir Blodget films,-line examples-genes, liquid crystals, microtubules,-point examples-fullerenes, micro-emulsions-overall surface, line, and point, examples-dry foams, polymer elastic and rigid foams.

References: [1] A.M. Spasic et al. CHISA 1987-2014 All Congresses, Prague, CZ. [2] J-P Hsu, A.M. Spasic, *Interfacial Electroviscoelasticity and Electrophoresis*, CRC Press-Taylor & Francis, Boca Raton London New York, 2010. [3] A.M. Spasic, J.M. Jovanovic, M. Jovanovic, Adv. Colloid Interface Sci. 2012; 31-49. [4] A.M. Spasic, M.P. Lazarevic, M.V. Mitrovic and D.N. Krstic In A.M. Spasic and J.P. Hsu (Eds.) *Finely Dispersed Particles: Micro-, Nano-, and Atto-Engineering*, CRC Press-Taylor and Francis, Boca Raton London New York, 2006. [5] A.M. Spasic In A.V. Delgado (Ed.) *Interfacial Electrokinetics and Electrophoresis*, Marcel Dekker Inc.: New York, 2002. [6] A.M. Spasic, *Sci. Rew., Special Issue:* Nonlinear Dynamics, Dedicated to: Milutin Milankovic (1879-1958) Serbian Scientific Society, Belgrade, 2013; *S2,* 269-289. [7] A. M. Spasic, M.P. Lazarevic, M.V. Mitrovic, D.N. Krstic, *Chem.Ind.* 2009**;** *63-5a,* 511-526.

Effect of dispersed phase inlet distributor on drop sizes and dispersed phase hold-up in RDC liquid extraction column

A. M. I. Al-Rahawi, (1) Department of Engineering, German University of Technology (GUtech), P.O. Box: 1816,PC 130, Muscat - Sultanate of Oman.
ayham.alrahawi@gutech.edu.om, Tel: +(968) 2206 1125, Fax: +(968) 2206 1002.
Salam Kadhim Al Dawery, College of Engineering, Department of Petroleum and Chemical Engineering, Nizwa – Oman

Hydrodynamic characteristics like hold-up and mean drop diameter were measured in an extraction RDC column for different distributor holes diameters and under different operating conditions.

Change in behavior of hold-up was observed in experimental data when the distributor holes diameter changed. Hold-up increases with decreasing dispersed phase holes inlet distributor. This behavior, however, can very well be explained by the small drop sizes generated in the lower part of the RDC.

Runs have been conducted to determine the hold-up and Sauter mean drop diameter as function of the flow rate and rotation disc speed. Experiments were carried out in the presence and absence of mass transfer. The drop size was measured by analysis of images and hold-up by the shut-off method. The factors affecting the dispersed phase drop size and hold up were investigated using two different columns size. Three systems were tested.

Effects of electric current pulse on the destruction of inorganic water solution

D.U.Alshimbayeva*, B.T.Utelbaev*, E.N.Suleimenov *
*Kazakh-British Technical University, Tole bi str., 59, Almaty, 050000, Republic of Kazakhstan
Phone +7 727-266- 8314, Fax +7 727-333-4175, e-mail: d.alshimbaeva@gmail.com

Experimentally have been established the possibility of formation of inorganic macromolecules in aqueous solutions . [1] It was shown that their structure depends on all physicochemical factors affecting to solution (concentration, temperature, electric current, etc.). The influence of the shape and parameters of the electrical signal into inorganic aqueous solutions were investigated. As the object of research various concentrations of sodium molybdate solution were chosen. In the scientific and technical literature can be found the assertion that "the true conductivity of" inorganic aqueous solution can be installed in the application of a unidirectional electric current with a very short pulse [2,3]. The laboratory facility was collected for determination of the conductivity of these solutions for the duration of electric pulse current of 1 nanosecond. The shape and characteristics of the signal is shown in Figure 1. During the measurment of electrical conductivity of an aqueous solution of sodium molybdate, it was decomposed and sedimented. Figure 2 shows the change of electrical conductivity of water solutions of sodium molybdate (the voltage across the electrodes, which characterizes the change in the electrical conductivity) over time measurement. As seen in Figure 2, in 5 hours electrical conductivity decreased almost eight times because of the sedimentation of sodium molybdate. Decomposition of the solution began immediately after the supply of current to the electrodes (indication devices), but visually determine the start of decomposition can be an hour after the start of measurement. Of course, not all inorganic aqueous solutions may be decomposed by a pulsed electrical current. However, we can clearly argue that the supplied pulsed energy affects the microstructure of aqueous solutions.

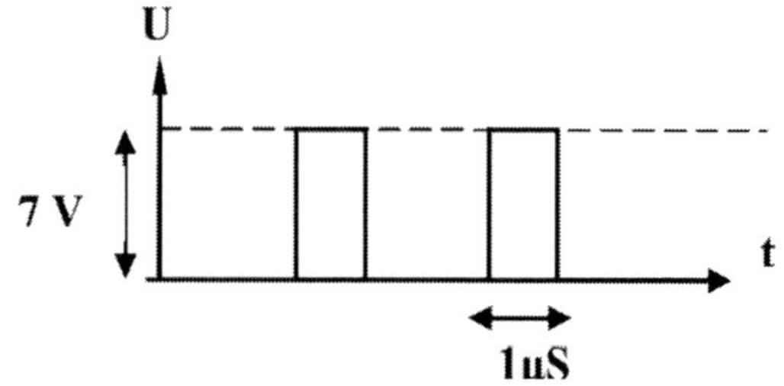

Figure 1 - The parameters and the waveform of electric signal

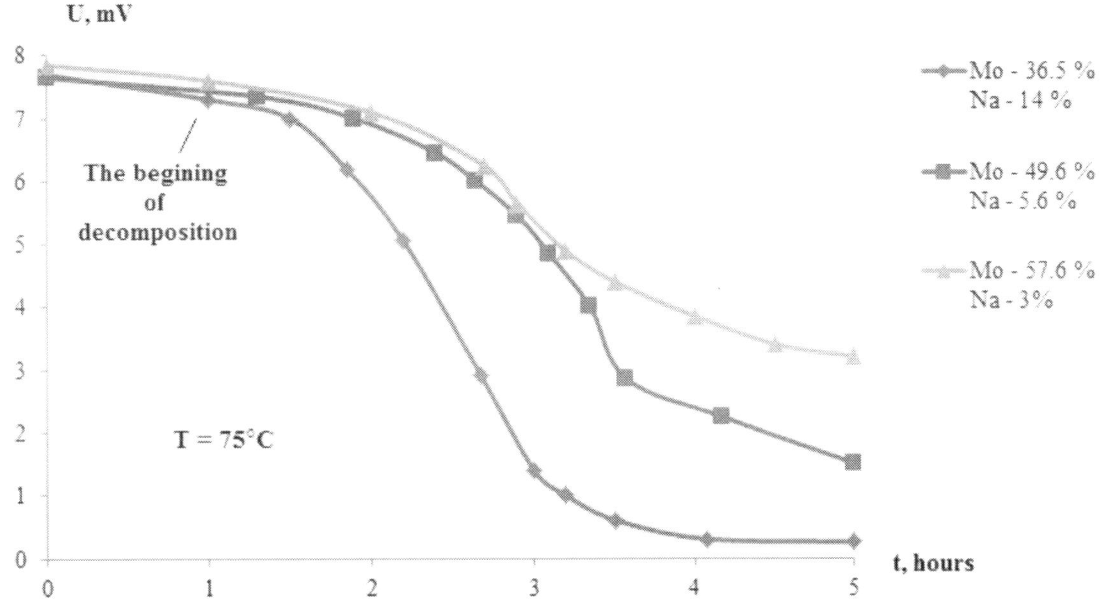

The dependence of sodium molybdate solution conductivity on time

Figure 2 - Changes in conductivity of an aqueous solution of sodium molybdate (the voltage at the electrodes, which characterizes the electrical conductivity) as a function of time due to the decomposition solution

References

1. Suleimenov EN, Romanteev YP, Suleimenov IE Ctrukturoobrazovanie in aqueous solutions of vanadium oxide - 5. European Science and Technology.Materials of the III International Research and Practice Conference. Munich, Germany. October 30th-31st, 2012, vol. 1, p.75 -79.

2.Zaykov YP Molostov OG Batukhtin VP, L. Ivanovsky Using a pulsed current in the preparation of the copper-calcium alloy by electrolysis of molten melts // 2000, №1, S.43-48.

3.Lopatin BA Conductometry. RISO SB RAS, Novosibirsk, 1964, 3080 p.

Essential oils extraction from cumin seeds using subcritical HFC-R134a

M.H. Eikani, N. Khandan, E. Feizi
Department of Chemical Technologies, Iranian Research Organization for Science and Technology (IROST), P.O. Box: 33535111, Tehran, Iran; tel. +9821 5627 6637, e-mail: eikani@irost.ir

Among extractable components of different plants, essential oils are an important and valuable part with increasing applications in food and pharmaceutical industries. They are usually heat sensitive and it is preferable that extraction from plant materials be carried out at low temperatures. Application of non-CFC solvents, named phytosol or florasol (based on 1,1,1,2-tetrafluoroethane; HFC-R134a) is a new technique for extraction of essential oils at room temperature and low pressure[1,2]. Such materials have a very interesting potential to be used as a solvent in some special areas of chemical industries. HFC-R134a is an ozone-friendly product; however, probable influences on global warming will remain a concern. It is possible practically to apply them in a dead-end cycle, like a refrigeration cycle, and using a well-designed equipment, reduce its emission to the atmosphere. A HFC-R134a based extraction system can be an alternative and competitive method for similar extraction technologies like supercritical CO_2 extraction. The main advantage is lower needed extraction pressure, e.g. less than 10 bar.

The aim of this research work was to extract essential oils from cumin seeds (*Cuminum cyminum* L.) as an essential oil-bearing material by subcritical HFC-R134a in a dynamic mode for the first time in Iran. Cumin essential oil is employed in many types of flavoring compounds. As mentioned, the main advantage is suitable extraction conditions which give rise to better quality of the product. The essential oils are mostly pure and contain little to no foreign substances. The comparative chemical composition of cumin seed oil produced in different ways such as steam distillation, supercritical CO_2 extraction and superheated water extraction has been reported[3,4].

The extraction yield was defined as the amount of cuminaldehyde as the major component of cumin seeds. Design of experiments was based on Taguchi method and the effect of selected factors at different levels was studied. Some important operating conditions including pressure (P) from 7 to 10 bar, mean particle sizes (D) from 0.30 to 1.00 mm, and extraction time from 60 to 120 min on essential oil yield was selected for this research work. The response was extraction yield (Y). The model was built based on the variables with confidence levels of 95%. The calculations were carried out by Minitab software. The optimal extraction condition was obtained and comparison between the subcritical HFC-R134a extraction and hydrodistillation was carried out.

References

1. Dean, J.R., Liu, B., Price, R. (1998) *J. of Chromatography A*, 799, 343-348.
2. Dean, J.R., Liu, B., Price, R. (1998) *Phytochemical Analysis*, 9, 248-252.
3. Eikani, M.H., Goodarznia, I., Mirza, M. (1999) *Flavour and Fragrance J.*, 14, 29-31.
4. Eikani, M.H., Golmohammad, F., Rowshanzamir, S., Mirza, M. (2007) *J. of Food Process Engineering*, 30, 255-266.

Extraction of phenolic compounds using [BMpyr][Nf_2] ionic liquid

B. González, O. G. Sás, R. Fidalgo,[1]E. J. González
Advanced Separation Processes Group, Edif. Isaac Newton, Department of Chemical Engineering, University of Vigo, Lagoas-Marcosende, 36310 Vigo, Spain; tel. +34986812292, e-mail: bgp@uvigo.es; [1]Departamento de Ingeniería Química Industrial y Del Medio Ambiente, Escuela Técnica Superior de Ingenieros Industriales, Universidad Politécnica de Madrid, E-28006 Madrid, Spain

Introduction:

Phenolic compounds are frequently found in wastewaters, which are a kind of organic compounds with specific toxicity can be considered as a serious pollutant of water and soil[1]. The presence of phenolic compounds into waterways may adversely affect human health and flora and fauna [2]. No evidence exists to indicate that phenols have any carcinogenic potential [3]. Due to these reasons phenols are considered priority pollutants by the European Decision 2455/2001/CE [4]. The World Health Organization (WHO) was fixed the amount limit of Chlorophenol in drinking waters by their toxicity in 0.1 μg/L [5] and for phenolic compounds in 1 μg/L6. The legal limit for phenols in industrial wastewater is 1 mg/L [7].

The development and design of new and "greener" separation technologies has been a hot topic of research in the past decades. Meanwhile liquid-liquid extraction since include various volatile organic compounds (COVs), which have high volatility and toxicity, therefore, this work present the idea of replacing these traditional solvent by new green solvents: ionic liquids (ILs).

The ionic liquid 1-butyl-3-methylpyrrolidinium bis(fluorosulfonyl)imide [BMpyr][Nf_2] has been tested as solvents for the extraction of 2-chlorophenol and o-phenol from water using liquid-liquid equilibrium.

Experimental:

Various extraction tests were conducted for removal of phenolic compounds from aqueous solutions using [BMpyr][Nf_2]. Liquid-liquid equilibrium (LLE) experimental data of the ternary systems water + 2-chlorophenol + [BMpyr][Nf_2]. and water + o-phenol + [BMpyr][Nf_2] were determined at atmospheric pressure. The experiments were carried out using a known ratio of solvent and aqueous 2-chlorophenol or o-phenol solutions. Mixtures were vigorously stirred, and then, samples were left to settle. After equilibrium was reached, the concentrations of 2-chlorophenol in the aqueous phase were determined by the standardized 4-aminoantipyrine method using a UV-spectrophotometer [1].

Results and conclusions:

The tests carried out with o-phenol and 2-chlorophenol have demonstrated the same behaviour in both cases, so, complete extraction process with the optimal parameters was applied reached a high extraction yield. The legal limit for industrial wastewater is not exceeded at initial concentrations below 1 g/L.

Acknoledgements

O. G. Sas is grateful to the Comisión Interministerial de Ciencia y Tecnología (Spain) for her FPI grant (BES-2014-067694). All authors are grateful to the Comisión Interministerial de Ciencia y Tecnología (Spain) for the project CTM2013- 46093-P.

References

(1) Eaton, A. D.; Clesceri, L. S.; Greenberg, A. E. In Standard Methods for the examination of water and wastewater. Apha: 1995.

(2) Mohanty, K.; Jha, M.; Meikap, B. C.; Biswas, M. N. Industrial and Engeneering Chemical Research. 2005, 44, 4128-4138.

(3) Busca, G.; Berardinelli, S.; Resini, C.; Arrighi, L. Journal of Hazardous Materials 2008, 160, 265-288.

(4) EC Decision 2455/2001/CE L331 of 15-12-2001.

(5) World Health Organization 2004.

(6) World Health Organization (WHO) Geneva, Switzerland 1996, Vol. II.

(7) European Comission 1991.

High pressure extraction techniques of chamomile and determination of antioxidant activity and umbelliferone content in obtained extracts

M. Molnar, N. Mendešević, D. Šubarić, S. Jokić
Josip Juraj Strossmayer University of Osijek, Faculty of Food Technology Osijek, Franje Kuhača 20, 31000 Osijek, Croatia; tel: +38531/224-320, fax: +38531/207-115, e-mail: stela.jokic@ptfos.hr

Cultivation of medicinal and aromatic plants has increased in recent years and more and more arable land in Croatia is designed specifically for this type of farming. Medicinal and aromatic plants are grown primarily because they contain biologically active components that make up the basic raw material for the production of drugs, cosmetics and flavors for foodstuff. Chamomile (*Matricaria chamomilla*) has been known as a medicinal plant for a long time, mainly due to its anti-inflammatory, antiseptic and antispasmodic activity. For this reason, chamomile is used as the starting material for consideration in this paper. Among many biologically active compounds found in chamomile, a coumarin derivative, umbelliferone, was also found.

The aim of this study was to compare chamomile extracts, produced with various extraction techniques (soxhlet, hydrodestillation, macaretion, supercritical CO_2 extraction, high pressure maceration system) based on their antioxidant activity and umbelliferone content determined by high performance liquid chromatography (HPLC). The following samples of chamomile were analyzed: unprocessed chamomile flowers first class, processed chamomile flowers first class, pulvis and processing waste obtained from the company Matricia Ltd.

The greatest extraction yield (22.3%) was obtained with 50% ethanol in high pressure maceration system and the highest yield of umbelliferone (41.03 mg/100g) was obtained using high pressure maceration system. Extracts have also proven to possess antioxidant activity (45.4 - 73.7% DPPH scavenging activity). The significance of this study offers a possibility of potential utilization of waste from chamomile processing, applying different extraction techniques and attaining the final natural product by using supercritical CO_2 extraction without any residues of organic solvents and with potential use in the cosmetic industry. Furthermore, a new developed system of maceration under pressure significantly reduces the time of extraction providing extracts with higher antioxidant activity.

Keywords: chamomile, extraction, high pressure, umbelliferone, antioxidant activity

Liquid-liquid extraction of uranium(VI) from aqueous solution using 1-hydroxyalkylidene-1,1-diphosphonic acids

M. Bouhoun Ali, [1,2]A. Y. Badjah Hadj Ahmed, M. Attou, [3]A. Elias, [4]M. A. Didi
Nuclear Research Centre of Draria, Box 43, Draria, Algiers, Algeria; e-mail: mbouhounali@gmail.com; [1]University of Sciences and Technology, Houari Boumediene, Faculty of Chemistry, Laboratory of Functional Organic Analysis, Box 32, El Alia, Bab Ezzouar, Algiers, Algeria, [2]University of King Saud, College of Science, Department of Chemistry, Box 245 Riyadh 11451, Saudi Arabia; e-mail: ybadjah@hotmail.com; [3]University of Mouloud Mammeri, Faculty of Sciences, Department of Chemistry, Box 17, Tizi-Ouzou, Algeria; e-mail: hamghil@yahoo.fr; [4]University of Tlemcen, Department of Chemistry, Laboratory of Catalysis, Box 119, Tlemcen, Algeria; e-mail: madidi13@yahoo.fr

The extraction of uranium(VI) from aqueous solutions has been investigated using 1-hydroxyhexadecylidene-1,1-diphosphonic acid (HHDPA) and 1-hydroxydodecylidene-1,1-diphosphonic acid (HDDPA), which were synthesized and characterized by elemental analysis and by FT-IR, ^{1}H NMR, ^{31}P NMR spectroscopy. In this paper, we propose a tentative assignment for the shifts of those two ligands and their specific complexes with uranium(VI). We carried out the extraction of uranium(VI) by HHDPA and HDDPA from [carbon tetrachloride + 2-octanol (v/v: 90%/10%)] solutions. Various factors such as contact time, pH, organic/aqueous phase ratio and extractant concentration were considered. The optimum conditions obtained were: contact time = 20 min, organic/aqueous phase ratio = 1, pH value = 3.0 and extractant concentration = 0.3M. The extraction yields are more significant in the case of the HHDPA which is equipped with a hydrocarbon chain, longer than that of the HDDPA. Logarithmic plots of the uranium(VI) distribution ratio vs. pHeq and the extractant concentration showed that the ratio of extractant to extracted uranium(VI) (ligand/metal) is 2:1. The formula of the complex of uranium(VI) with the HHDPA and the DHDPA is $UO_2(H_3L)_2$ (HHDPA and DHDPA are denoted as H_4L). A spectroscopic analysis has showed that coordination of uranium(VI) takes place via oxygen atoms.

The solvent extraction of uranium(VI) using alkyliminodimethylenediphosphonic acids

M. Bouhoun Ali, [1,2]A. Y. Badjah Hadj Ahmed, M. Attou, [3]A. Elias, [4]M. A. Didi

Nuclear Research Centre of Draria, Box 43, Draria, Algiers, Algeria; e-mail: mbouhounali@gmail.com; [1]University of Sciences and Technology, Houari Boumediene, Faculty of Chemistry, Laboratory of Functional Organic Analysis, Box 32, El Alia, Bab Ezzouar, Algiers, Algeria, [2]University of King Saud, College of Science, Department of Chemistry, Box 245 Riyadh 11451, Saudi Arabia; e-mail: ybadjah@hotmail.com; [3]University of Mouloud Mammeri, Faculty of Sciences, Department of Chemistry, Box 17, Tizi-Ouzou, Algeria; e-mail: hamghil@yahoo.fr; [4]University of Tlemcen, Department of Chemistry, Laboratory of Catalysis, Box 119, Tlemcen, Algeria; e-mail: madidi13@yahoo.fr

The solvent extraction of uranium(VI) from aqueous solutions has been investigated using hexadecyliminodimethylenediphosphonic acid (HIDMPA), and decyliminodimethylenediphosphonic acid (DIDMPA), which were synthesized and characterized by elemental analysis and by FT-IR, ^{1}H NMR, ^{13}C NMR, ^{31}P NMR spectroscopy. In this paper, we propose a tentative assignment for the shifts of those two ligands and their specific complexes with uranium(VI). We carried out the extraction of uranium(VI) by HIDMPA and DIDMPA from [carbon tetrachloride + 2-octanol (v/v: 90%/10%)] solutions. The extraction yields are more significant in the case of the HIDMPA which is equipped with a hydrocarbon chain, longer than that of the DIDMPA. Logarithmic plots of the uranium(VI) distribution ratio vs. pH_{eq} and the extractant concentration showed that the ratios of extractant to extracted uranium(VI) is 2:1. The formula of the complex of uranium(VI) with the HIDMPA and the DIDMPA is $UO_2(H_3L)_2$ (HIDMPA and DIDMPA are denoted as H_4L).

Sorption of gases, vapours and their mixtures in glassy and rubbery polymers: Unified modelling with GAB model of multilayer adsorption

O. Vopička, D. Radotínský, K. Pilnáček, M. Lanč, P. Číhal, A. Randová, K. Friess
University of Chemistry and Technology, Prague, Department of Physical Chemistry, Technická 5, Prague 6, 166 28, Czech Republic, tel. +420220444029,
e-mail: vopickao@vscht.cz

As a matter of fact, polymers are capable of sorbing various gases, vapours and liquids. The parameterization of the isothermal sorption data (sorption isotherms) however remains a challenging task. This is naturally so because the sorptives (gases, vapours, liquids) and also the polymers can have miscellaneous chemical structures, which alter the intermolecular cohesive interactions and which can lead to the cooperative and competitive sorption effects in the case of mixed sorbates. Furthermore, the ability of polymers to sorb compounds changes with respect to their physical state. A polymer can naturally be in the glassy or rubbery state. The transition between these two states occurs at a polymer-specific glass transition temperature. All this diversity obviously restrains the attempts of the researchers to develop rather universal models of sorption in polymers.

The GAB model of multilayer adsorption, named after Guggenheim, Anderson and De Boer, has recently been shown to effectively parameterize the sorption of pure gases in glassy polymers [1,2], single-compound vapours in glassy polymers [2,3], see Fig. 1, and single-compound vapours and liquids in rubbery polymers [4]. Based on this rather wide applicability of GAB model, it can be concluded that the sorption in polymers might have the nature of multilayer adsorption on the inner adsorption centres placed in the bulk polymers. It further follows from the analyses of the monolayer sorption capacities that one sorption centre is formed by approx. three mer units for a number of systems [3]. Apart from the applicability of the GAB model to the sorption of pure gases, vapours and liquids in polymers, this model has recently been extended [5] to the sorption of binary mixtures in rubbery polymers [5] as well as in glassy polymers [6], see Fig. 1.

In this work, the capability of the GAB model to effectively parameterize various types of sorption isotherms of pure compounds and their temperature dependences will be discussed on examples. The extension of the model to the sorption of mixed sorptives will be addressed as well. Historically, GAB model was published 50-70 years ago [7-9] and was firstly formulated as the Langmuir's case IV adsorption even 98 years ago [10]. Despite this long history, the usability of the GAB model for polymers remained somewhat overseen, although it enables the unified modelling of the sorption phenomena of various compounds in numerous polymers.

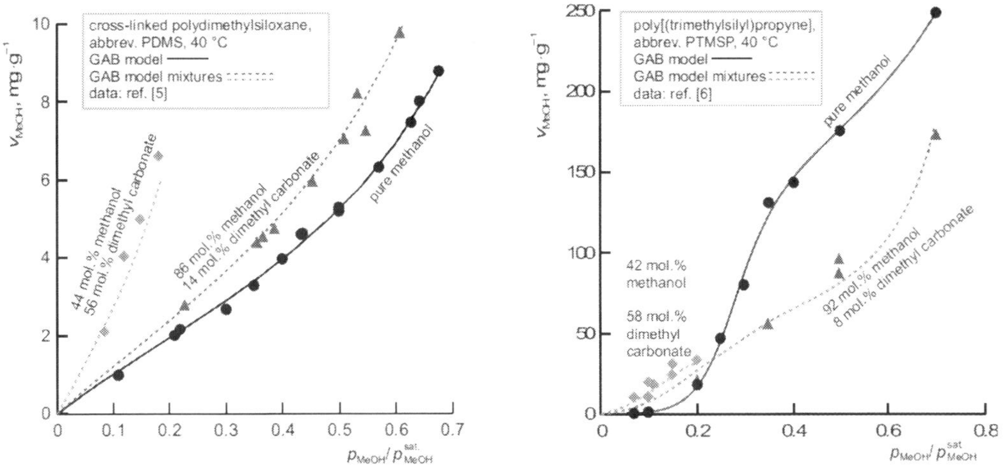

Fig. 1. Sorption of pure methanol and methanol together with dimethyl carbonate at two molar fractions in PDMS (left) and PTMSP (right). GAB model for the sorption of pure compounds (solid curves) and of mixtures (dashed curves) are displayed.

Literature

[1] O. Vopička, K. Friess, Analysis of Gas Sorption in Glassy Polymers with the GAB Model: An Alternative to the Dual Mode Sorption Model, *J. Polym. Sci., Part B: Polym. Phys.* 52 (2014) 1490-1495.

[2] O. Vopička, K. Friess, V. Hynek, P. Sysel, M. Zgažar, M. Šípek, K. Pilnáček, M. Lanč, J.C. Jansen, C.R. Mason, P.M. Budd, Equilibrium and transient sorption of vapours and gases in the polymer of intrinsic microporosity PIM-1, *J. Membr. Sci.* 434 (2013) 148-160.

[3] O. Vopička, K. Pilnáček, P. Číhal, K. Friess, Sorption of Methanol, Dimethyl Carbonate, Methyl Acetate, and Acetone Vapors in CTA and PTMSP: General Findings From the GAB Analysis, *J. Polym. Sci., Part B: Polym. Phys., in press*, http://dx.doi.org/10.1002/polb.23945

[4] O. Vopička, A. Randová, K. Friess, Sorption of vapours and liquids in PDMS: novel data and analysis with the GAB model of multilayer adsorption, *Eur. Polym. J.* 60 (2014) 49-57.

[5] O. Vopička, D. Radotínský, K. Friess, Sorption of vapour mixtures of methanol and dimethyl carbonate in PDMS: Experimental study, *Eur. Polym. J.* 73 (2015) 480-486.

[6] O. Vopička, D. Radotínský, K. Friess, Sorption of vapour mixtures of methanol and dimethyl carbonate in PTMSP: Cooperative and competitive sorption in one system, *Eur. Polym. J.* 73 (2016) 480-486.

[7] E.A. Guggenheim, Application of Statistical Mechanics, Clarendon Press, Oxford, UK, 1966.

[8] R.B. Anderson, Modifications of the Brunauer, Emmett and Teller Equation, *J. Am. Chem.Soc.* 68 (1946) 686-691.

[9] J.H. de Boer, The Dynamical Character of Adsorption, Clarendon Press, Oxford, UK, 1953.

[10] I. Langmuir, The adsorption of gases on plane surface of glass, mica and platinum, *J. Am. Chem. Soc.* 40 (1918) 1361-1403.

Acknowledgements

Results were obtained under the financial support of the Czech Science Foundation within the Grant Projects 13-32829P and P106/10/1194. Partial financial support was obtained from the Specific University Research Project (MSMT No 20/2015).

Modelling the membrane formation dynamics using the Cahn-Hilliard equation, a way for controlling the membrane morphology

H. Manzanarez*, J.P. Mericq*, P. Guenoun**, J. Chikina**, D. Bouyer*
* IEM (Institut Européen des Membranes), UMR 5635 (CNRS-ENSCM-UM2), Université Montpellier II, Place Eugène Bataillon, F-34095 Montpellier, France
** CEA Saclay, DSM-IRAMIS, UMR 3299 (CEA-CNRS), LIONS, F-91191 Gif-sur-Yvette, France
e-mail : herve.manzanarez@univ-montp2.fr; Denis.Bouyer@umontpellier.fr;

In this paper, we developed a numerical model coupling Cahn-Hilliard and Navier-Stokes equations [1, 2] to simulate the membrane formation dynamics by phase inversion. The phase separation process is widely used to produce porous polymeric membranes, whose morphology depends on several factors such as the polymer concentration, the solvent quality, or the mass transfer involved during the process [3]. The phase inversion is a de-mixing process involving the formation of two phases: a rich polymer phase (the final membrane after the solidification of the membrane) and a lean polymer phase (the pores of the membranes after the solvent extraction).

One of the most important challenges in the membrane formation field concerns the control of the formation mechanisms, which will have a great impact on the final membrane structure and thus on the functional properties such as the permeability and the selectivity. The elementary formation mechanisms can be optimized and monitored using computational simulation to describe the structuration dynamics once the phase inversion starts.

The phase separation dynamics can be described by the fourth order Cahn-Hilliard equation (Equation 2), where the relaxation of the order parameter is driven by local minimization of the Ginzburg-Landau free energy functional (Equation 1).

$$F_{GL} = \int_V \left(F_{HF} + \frac{1}{2}C(\phi)|\nabla\phi|^2\right) dV \tag{1}$$

$$\frac{\partial\phi}{\partial t} + v\nabla\phi = \nabla \cdot \left(M(\phi)\nabla\frac{\partial F_{GL}}{\partial\phi}\right) \tag{2}$$

In this work, we developed 2D and 3D models (figures 1 and 2) using the Flory-Huggins-De Gennes theory [4, 5] to describe the thermodynamic potential involving also hydrodynamics effects induced by the phase inversion. We first verified the linear model of the spinodal decomposition, i.e. the Cahn model involved during the first demixing period. This work investigated the effects of the polymer concentration and the temperature quench on the patterns evolution and hence on the final membrane porosity and pore connectivity. The growth laws were also determined depending on the temperature quench and the initial formulation.

This work shows great potential for better understanding the most important parameters and mechanisms involved in the phase inversion and to predict the final membrane morphology from initial conditions of formulation and process parameters.

Figure 1: 2D model of spinodal decomposition of polymer/solvent

Figure 2: 3D model of one shape of concentrated phase during spinodal decomposition polymer/solvent system

References

[1] J. W. Cahn and J. E. Hilliard, *Free Energy of a Nonuniform System. I. Interfacial Free Energy*, J. Chem. Phys. 28, 258 (1958).

[2] V.E. Badalassi, H.D. Ceniceros, S. Banerjee, *Computation of multiphase systems with phase field models*, J. Comput. Phys. 190 (2003).

[3] G.T. Caneba, D.S. Soong, *Polymer membrane formation through the thermalinversion process. 1. Experimental study of membrane structure formation*, Macromolecules 18 (1985) 2538.

[4] P.J. Flory, *Principles of Polymer Chemistry* (Cornel lUniversity Press, Ithaca, NewYork, 1986).

[5] P.G. De Gennes, *Dynamics of fluctuations and spinodal decomposition in polymer blends*, J. Chem. Phys. 72, 4756 (1980).

Heterogeneous ion-exchange membranes and the mechanisms for the occurrence of the overlimiting current

L. Vobecká[1], M. Svoboda[1,2], L. Chroust[1], N. Kurospayeva[1], Z. Slouka[1,2]
[1] University of Chemistry and Technology Prague, Dept. of Chemical Engineering, Technická 3, Prague 6, 16628, Czech Republic
2 University of West Bohemia, New Technologies - Research Centre, Univerzitni 8, 306 14 Pilsen, Czech Republic
e-mail: sloukaz@vscht.cz; tel: +420220443251

Heterogenous ion-exchange membranes form an inherent functional component of electrodeionization and electrodialysis units and present a subject of intensive scientific research that mainly focuses on synthesis of membranes with optimal performance. Besides aforementioned industrial applications for electroseparation, these membranes have recently been employed as functional elements in a point-of-care diagnostic platform where they serve as biosensors, biomolecular filters and biomolecular preconcentrators. These functionalities are given by phenomena occurring at the heterogeneous ion-exchange membranes on application of DC electrical field and their interactions with biomolecules.

These phenomena result from concentration polarization of membranes in DC electrical field. Concentration polarization is characterized by generation of ion depleted and ion enriched regions on either side of the membrane that correspond to the diluate and concentrate sides, respectively. The ion depleted side is responsible for the occurrence of limiting and overlimiting regions clearly distinguishable on current-voltage characteristics. The ion depleted side, specifically the interface between the membrane and the adjacent electrolyte, is the place where interesting electrokinetic (vortex formation) and electrochemical (water splitting) phenomena take place. These phenomena are considered as critical mechanisms responsible for the overlimiting region (OR).

To study the mechanisms standing behind the occurrence of the overlimiting region and relate the behavior to the structure of the membrane, we developed an experimental fluidic system that allows to analyse small pieces of heterogeneous ion-exchange membranes, measure their electrochemical properties and observe phenomena taking place at membrane-electrolyte interface through fluorescent microscopy. MicroComputed Tomography was employed to analyse 3D structure and composition of the membranes swelled in electrolytes of various ionic strengths and evaluate the effect of ionic strength on structural changes in the membrane (Figure 1 right). The following experimental analysis by fluorescent microscopy (Figure 1 left) and electrochemical characterization provided fundamental insight into the mechanisms that govern the behavior of membranes in OR.

In this paper we will present selected experimental results for cation and anion exchange membranes provided by Membrain a.s.

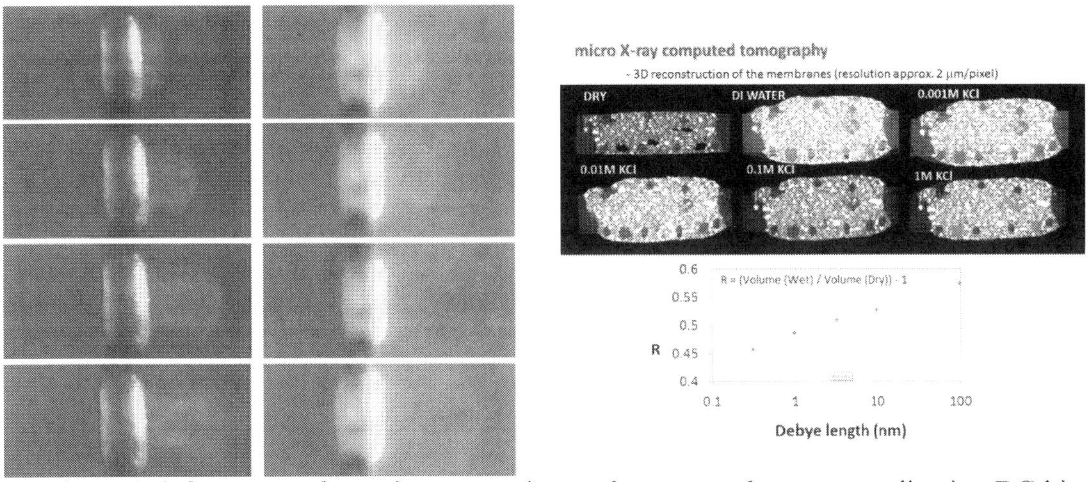

Figure 1: left – vortex formation at a cation exchange membrane on application DC bias for fluorescein and fluorescently tagged particles; right – microCT sections of a cation exchange membrane swelled in KCl solutions of various concentrations.

Structure and transport properties of polymer membranes modified by hybrid fullerene C_{60}-containing stars

A Pulyalina, [1,2]G. Polotskaya, [2]L. Vinogradova, [3]Z. Pientka, [1]A. Toikka
Saint Petersburg State University, Institute of Chemistry, Universitetsky pr. 26, 198504 Saint Petersburg, Russia; tel. +78124284805, e-mail:alexandra.pulyalina@gmail.com; [1]Saint Petersburg State University, Institute of Chemistry, Universitetsky pr. 26, 198504 Saint Petersburg, Russia; tel. +78124284805, [2]Institute of Macromolecular Compounds, Russian Academy of Sciences, Bolshoy pr. 31, 199004, Saint Petersburg, Russia; tel. +78123237407; [3]Institute of Macromolecular Chemistry, Czech Academy of Sciences, Heyrovskeho nam. 2, 162 06 Praha 6, Czech Republic; tel. +420296809280

The combination of individual properties of polymers and its modifier can create synergic characteristics for the promising materials production for various applications. Among the important ones is the membrane technology, in particular, pervaporation and gas separation techniques.

In the present work the novel membranes were developed based on poly (2,6-dimethyl-1,4-phenylene oxide) (PPO) modified by an inclusion of hybrid star-shaped macromolecules (HSM) as a filler up to 5 wt%. The HSM consists of a small branching center, which is a molecule of fullerene C_{60} (diameter 7.1 A), and two types of tethered arms of different chemical nature: non-polar polystyrene (PS) and polar copolymer of poly (2-vinylpyridine-*block*-poly-*tert*-butylmetacrylate (P2VP-*block*-PTBMA)). It is known that blends of PPO and PS are completely inter-soluble and the phase separation does not occur until the temperature of their thermal destruction. Therefore, one can expect that the PS arms will be good compatible with PPO matrix and will promote the uniform distribution of HSM in PPO matrix. Also HSM containing polymer chains of different chemical nature is capable to self-organization in solutions.

Hybrid star-shaped macromolecules (HSM) were prepared by a multistage synthesis:

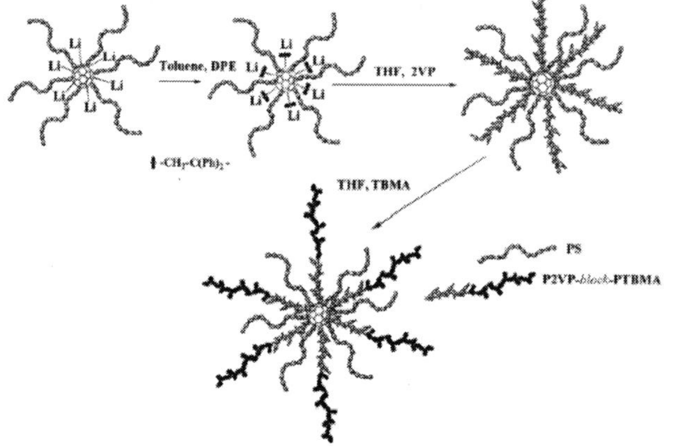

PS arms were characterized by size exclusion chromatography, $M_n = 6.9 \cdot 10^3$ and $M_w/M_n = 1.04$. Molecular weight of the diblock copolymer arm was higher to double than that of PS arm at the given equal length of both block P2VP and PTBMA.

The PPO/HSM composites containing 1, 3 and 5 wt% HSM were prepared by mixing solutions of 3 wt% PPO in chloroform and 3 wt% HSM in chloroform. Polymer membranes of 40 μm thick were obtained by casting of the polymer solution in chloroform on a cellophane surface followed by solvent evaporation at 40 °C and drying in vacuum oven at 60 °C up to the constant weight.

The thermal stability and glass transition temperature of membranes were estimated by thermogravimetric analysis and differential scanning calorimetry. Peculiarity of membrane structure was established by scanning electron microscopy and small-angles neutron scattering. Some physicochemical properties were determined such as density, fractional free volume, contact angles. Sorption and diffusion parameters were estimated by tests in liquid medium. The mass transfer through the membrane was investigated in organophilic pervaporation and gas separation.

The inclusion of star self-organized macromolecules significantly affects on the morphology of the membranes. Data of small-angles neutron scattering indicates qualitatively on the change in the geometry of diffusion channels in PPO/HSM as comparison with that of PPO membranes. It was established that star-shaped macromolecules trend to internal and external segregation with PPO which led to appearance of small disturbances in the polymer package. The overlap of these areas creates an elongated nanoscale and submicron channels. Introduction of hybrid stars in the polymer matrix transforms highly branched canal system to structure with linear channels. This fact significantly influenced on transport properties of membranes.

The effect of star macromolecules inclusion into PPO membranes on pervaporation properties was studied in the process of methanol–ethylene glycol mixture separation (this problem arises from the joint involvement of these alcohols in a number of syntheses or technological processes). In the pervaporation of methanol–ethylene glycol mixture over the concentration range of 5 ÷ 30 wt% methanol in feed, all membranes showed high affinity to methanol and produced permeate concentrated in methanol. The total flux through membrane increased with the rise both the HSP content in membrane and the methanol concentration in the feed. The separation factor achieved maximal value at modifier content in membranes equal to 3 wt% HSP.

The gas separation properties were measured using a laboratory high vacuum apparatus with a constant-volume permeation cell for five representative gases: N_2, O_2, H_2, CO_2, and CH_4. Introduction of the hybrid stars significantly changes morphology and influence on physical and chemical properties of the PPO membrane. These facts are influence on the results of gas separation. Transport of small molecules is a method that is sensitive to change of polymer density, so increase of PPO/HSM density led to decreasing permeability of nitrogen molecules resulting in the increase of selectivity in separation of H_2/N_2 and O_2/N_2 gas pairs.

Acknowledgment. A.M.T. and G.A.P. acknowledge the Russian Foundation for Basic Research (project 15-03-02131). A.Yu.P. is also grateful for the support by Fellowship of the President of Russian Federation (project № 184, SP-1469.2015.1) and Saint Petersburg State University (event 6, project № 12.42.1452.2015).

Analysis of some engineering parameters relevant to the performance and reliability of hollow fiber spinning system

Shadia R. Tewfik[a,b], Abdel Ghani M.G.Abulnour[a,b], , Hayam F. Shaalan[a,b] ,Mohamed I. El-Anwar[a,c], SaharS.Ali[a,b], Kamal I. Abed[a,c], Mervat A. Badr[a,c] and Mohamed H. Sorour[a,b]
National Research Centre, Engineering Research Division ,ElBohous St. Dokki, Giza, Egypt .
POB. 12622,Tel:0020233370933 - Fax 0020233370931[a] email:shadiatewfik@yahoo.com
Chemical Engineering and Pilot Plant Department[b], Mechanical Engineering Department[c]

Hollow Fiber (HF) membranes manifest wider applications on numerous environmental and industrial fields. The peculiar characteristics of surface to volume ratio, standalone characteristics and lack of structurally separate turbulence promoters motivate the rapid advancement of HF technology and engineering. The spinning system is subject to numerous parameters and variables that critically influence the characteristics and performance of the produced HF.

This paper outlines theendeavors to develop a unique cellulose acetate (CA) HF membrane under varying dope rheological setting, spinning nozzle dimensions and post curing temperatures. Characterization results using scanning electron microscope (SEM), atomic force microscopy (AFM) in addition to changes in mechanical properties are outlined. Furthermore, the effect of the abovementioned parameters on water flux is addressed.

Moreover, factors governing the fabrication of polysulfone/polyether sulfone (PS/PES) hollow fibers have been investigated with emphasis on the effect of polyvinyl pyrrolidone additives and air gap on the morphology of the prepared fibers. Characterization studies comprising (SEM), (AFM) and mechanical characteristics reveal the milestones affecting dimensional features and average roughness parameters. The paper is concluded with recommendations pertinent to fabrication of CA and PS/PES hollow fibers with preferred morphology.

Desalination using RO and NF: New insights, limitations and improvements

V. Freger
Wolfson Department of Chemical Engineering, Technion – IIT, Haifa, Israel
E-mail: vfreger@technion.ac.il

Polyamide composite membranes for reverse osmosis (RO) and nanofiltration (NF) are available today in a wide range of permeabilities and selectivities for various types of feedwater. In general, they show a good rejection of salts and most pollutants and reasonably high fluxes. However, a need for improvement may still be identified. For instance, rejection of boron, a micropollutant in seawater desalination, is insufficient and costly solutions are required. The polyamide membranes are also too prone to fouling and biofouling due to unfavourable surface characteristics. Another currently unresolved challenge is the very high rejection of multivalent ions as compared to NaCl that poses serious limitations in waste water desalination for re-use in agriculture.

I will first present the new insights into the membrane structure and selectivity mechanism obtained theoretically, using molecular dynamics, and experimentally, using impedance spectroscopy of RO membranes and appropriate model materials. This analysis shows that the solute rejection takes place within a dense polymer film; however, molecular defects in the membrane structure might provide the main path for non-selective transport of solutes. Appropriate "sealing" of such defects may then improve selectivity with only moderate loss of permeability, which will be illustrated using two different sets of results: (a) a "2D sealing or caulking" of the membrane with a surface layer that blocks major defects and (b) a 3D sealing, i.e., filling the cavities within the polyamide layer, with appropriately selected small molecules. The last part of the talk will address the challenge overcoming limitations on water recovery and scaling in waste water desalination for irrigation purposes using NF membrane with novel selectivity developed in our lab.

Preparation and characterization of polyethersulfone mixed matrix ultrafiltration membranes embedded with functionalized SBA-15 nanoparticles

Jing Guo[1], Arcadio Sotto[2], Antonio Martín[2] and Jeonghwan Kim[1]*
[1]Department of Environmental Engineering, [1]INHA University, Inha-ro. 100, Incheon, Republic of Korea, 22122
[2]Chemical and Environmental Engineering Group, ESCET, Rey Juan Carlos University C/ Tulipán s/n 28933-Móstoles- Madrid-Spain
*Corresponding author: tel. +082-32-860-7502, e-mail: jeonghwankim@inha.ac.kr; fax: +082-32-865-1425

Polyethersulfone (PES) membrane suffers from severe membrane fouling in water and wastewater treatment applications due to its instinct hydrophobic property. In this study, we conducted preparation, characterization and evaluation of the performance of the PES mixed matrix membranes (MMMs) embedded with Ti- and Zr- incorporated SBA-15 nanoparticles. The casting solution was prepared by adding 0, 0.3wt%, 0.6wt%, 0.8 wt % of nanoparticles, which corresponds to the MMMs named as PES, PES-Ti0.3, PES-Ti0.6, PES-Ti0.8, PES-Zr0.3, PES-Zr0.6 and PES-Zr0.8, respectively. Ti-SBA-15 and Zr-SBA-15 nanoparticles with size around 1 μm were prepared by directly hydrothermal procedure using Pluronic P123 triblock polymer as structure-directing agent, tetraethylorthosilicate as silicon precursor and titanium isopropoxide as Ti source in the case of Ti-SBA-15 sample or zirconocene dichloride as Zr precursor in the case of Zr-SBA-15 material . The functionalized silica materials were dispersed into PES polymer casting solution. The membranes were then fabricated by performing Nonsolvent Phase Inversion Method. The SBA-15 nanoparticles and MMMs fabricated were characterized by TEM, SEM, FTIR, EDX, AFM, TGA, and contact angle measurement. The membrane fouling of the MMMs fabricated was investigated by filtering protein compound of bovine serum albumin (BSA). Static adsorption tests were also performed with the MMMs fabricated to observe the extent of adsorptive fouling with BSA compound. The SEM results of the MMMs showed that the embedded nanoparticles did not change the finger like structure of the membranes. The membrane pore size increased when the nanoparticle concentration increased from 0 to 0.3 wt%, but decreased with increasing further dosage of nanoparticles into casting solution. Due to the changes of membrane pore size, the pure water permeability of the composite membranes increased with lowering nanoparticle concentration (0.3 wt%). However, the pure water permeability decreased as the concentration of nanoparticles increased from 0.3 to 0.8 wt%. After allowing membrane fouling, the flux recovery rate of PES-Ti0.8 and PES-Zr0.8 membrane was higher than that observed with pure PES membrane. Moreover, the membranes fabricated with highest dosage of nanoparticles resulted in lowest static adsorption with BSA compound, indicating 74.23 μg BSA equilibrium concentration/cm^2 of membrane surface area, 49.82μg/cm^2 and 58.19μg/cm^2 for pure PES membrane, PES-Ti0.8 and PES-Zr0.8, respectively. The results indicated that the antifouling property of PES membrane should be improved by embedding SBA-15 nanoparticles.

Comparison of feed pretreatment by organic and inorganic coagulants on membrane microfiltration of titanium dioxide dispersions

M. Grulich and P. Mikulášek

Institute of Environmental and Chemical Engineering, Faculty of Chemical Technology, University of Pardubice, Studentská 573, 532 10 Pardubice, Czech Republic, Miroslav.Grulich@upce.cz,
Petr.Mikulasek@upce.cz, Tel.: +420 466 037 503

The influence of coagulant type and operating parameters on crossflow microfiltration of aqueous dispersions of titanium dioxide has been examined. The experiments were carried out with a tubular ceramic microfiltration membrane with a nominal pore size of 0.1 μm at various operating parameters. Two types of coagulants were used for a series of crossflow microfiltration experiments: organic (polyacrylamide, poly(diallyldimethylammonium chloride), poly(acrylamide-co-acrylic acid) partial sodium salt, and inorganic (iron chloride, aluminium potassium sulfate dodecahydrate). The values of steady-state permeate flux for the crossflow microfiltration with and without pretreatment have been experimentally evaluated.

The results of the experiments with coagulants showed that the flux initially declines rapidly and then stabilizes. The results also suggested that some organic coagulants were a better for this system than inorganic coagulants. The better coagulant for this system was poly(diallyldimethylammonium chloride) and its optimal concentration was 30 mg l^{-1}. Finally it was shown that pretreatment of the feed by poly(diallyldimethylammonium chloride) resulted in a permeate flux that was more than three times higher than that obtained without any pretreatment. Moreover, there has been a very positive effect of this coagulant on the particle size. Pretreatment by 30 mg l^{-1} poly(diallyldimethylammonium chloride) led to an average particle size that was almost 18 times higher than that obtained without pretreatment. The other two organic coagulants did not produce such improvements; pretreatment of the feed by polyacrylamide increased the permeate flux by only 10%, whilst pretreatment by poly(acrylamide-co-acrylic acid) showed lower permeate flux than in the case without pretreatment. Pretreatment of the feed by two inorganic coagulants resulted in a permeate flux that was higher than without pretreatment, but showed lower permeate flux than in the case of the use poly(diallyldimethylammonium chloride).

From the above experimental results it is evident that the results indicate the need for careful selection of the coagulant due to their differing influences on the permeate flux.

Keywords: Microfiltration; Coagulation; Particle size; Titanium dioxide; Ceramic membrane

Peculiarity of membrane separation of lower hydrocarbons: basic aspects and practice

V.V.Zhmakin, V.V. Teplyakov
A.V.Topchiev Institute of Petrochemical Synthesis, Russian Academy of Sciences (TIPS RAS), 119991, GSP-1, Leninsky pr. 29, Moscow, Russia, e-mail tepl@ips.ac.ru, tel. +74952585304

Despite recent dynamic development of the membrane science and technology in general, practical application of membrane processes for separation of lower hydrocarbons is rather limited due to low selectivity of membranes. Nevertheless according to the dependencies of lower hydrocarbons permeability on temperature for certain membrane materials the selectivity of particular gas pairs becomes higher in the area of lower temperatures but lower than for individual components. For the moment there is a deficiency of data for lower hydrocarbons transfer in membranes at lower temperature especially for propane and butane containing mixtures. Paper considers theory of "hard spheres" for estimation of solubility and heat of sorption for C1-C14 linear hydrocarbons as an explanation of thermodynamic contribution in selective transfer through rubber-like polymers (e.g. polydimethylsiloxane). As a result, an estimation of hydrocarbons permeability of real commercial membranes can be calculated as negative or positive dependences on temperature.
Non-isothermal method of gas permeability measurement as for single gases as for gas mixtures was suggested and applied in this work for quick and reliable evaluation of commercial membranes properties. Experiments were carried out with MDK-1 membranes (produced by ZAO STC "Vladipor", Russia), C1-C4 hydrocarbons and mixtures og their in the temperature range from -20 to +40 °C. MDK membranes consist of three layers: (1) macroporous support layer based on polyethylene terephtalate (PET), (2) mesoporous support layer based on fluoroplast and (3) dense selective layer based on silicon-containing polymer. Small thickness of selective layer (1-3 μm) provide very short time-lag ($\sim 10^{-2}$ s) that in combination with slow heating rate (0.3 °C/min) allows to consider steady-state gas transfer at each moment. Permeability measurements were carried out using differential method with vacuum-pumping of volume under membrane. Heating (cooling) of membrane cell was provided by thermocryostat Huber CC-410, gas concentration measurements were carried out by Shimadzu GC-17 equipped with GS-Tek 8253-5015 column and TCD, helium was used as a carrier gas.
It was demonstrated that methane permeability decreases with temperature decrease however propane and butane permeability increases as temperature decreases. Ethane has intermediate properties. Obtained temperature dependencies were built in Arrhenius plot coordinates and corresponding values of apparent activation energy of permeability (E_p) for C1-C4 were found. Combination of experiments with theoretical approach expands the permeability parameters estimation up to C14 for linear saturated hydrocarbons.

Acknowledgments

Selected results presented in this paper were obtained with financial support from TIPS RAS State Program (Registration number 01200902404).

Key words: Theory of absorption, Lower hydrocarbons, Membrane, Separation

Highlights:

Non-isothermal method of membrane permeability quick measurement is suggested.

It was found that propane/methane selectivity increases with temperature decrease

It is estimated that good methane purification can be obtained at -20 °C at low stage-cut.

Superior molecular size screening of calcium alginate membranes composited with agar

S. Kinoshita, M. Imai*

Course in Bioresourse Utilization Sciences, Graduate School of Bioresourse Sciences, Nihon University. 1866 Kameino, Fujisawa, Kanagawa-pref. 252-0880 JAPAN

Tel.: +81-466-84-3978; Fax: +81-466-84-3978, E-mail address: XLT05104@nifty.com

Superior molecular size screening of calcium alginate membranes composited with agar was demonstrated. Sodium alginate and agar mixed aqueous solution was spread on glass-plate using applicator at 250 μm gap. It was gelated by cooling and then cross-linked by immersed into the calcium chloride aqueous solution to obtain the calcium alginate-agar composite membrane. Mass fraction of agar in the mixed polymer aqueous solution (f_{agar}) was defined by Eq. 1.

$$f_{agar}[-] = \frac{\text{mass of Agar [g]}}{\text{mass of Alginate [g] + mass of Agar [g]}} \quad (1)$$

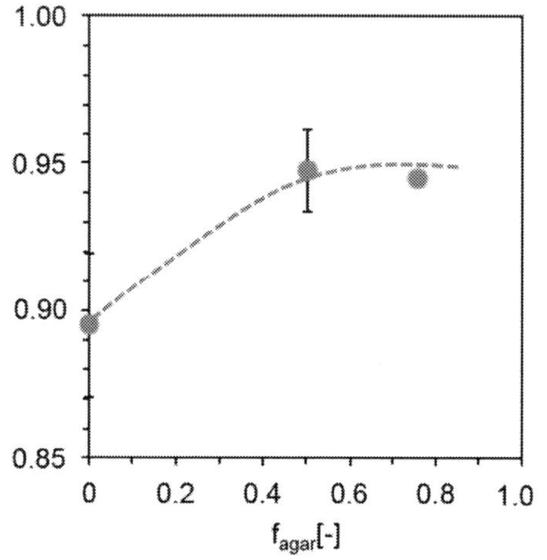

Fig. 1. Effect of mass fraction of agar on water content in the membrane.

Figure 1 shows the change of water content (H_m) of composite membrane in wet state. It was determined by gravimetric method based on mass of wet membrane. The water content was increased with increasing of mass fraction of agar and then leveled off. Stable membrane was obtained less than 0.75 of mass fraction of agar. Maximum water content was measured as 0.95 due to hydrophilic character of agar.

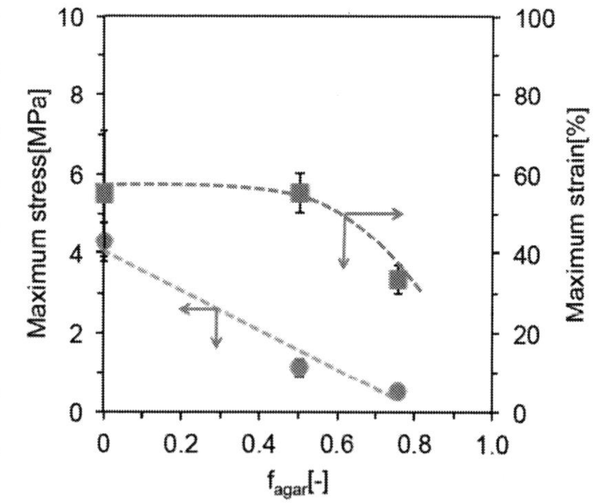

Fig. 2. Effect of mass fraction of agar on ●: maximum stress, and ■: maximum strain at membrane rupture.

Figure 2 shows the effect of mass fraction of agar in the membrane on the mechanical strength of wet membrane. Maximum stress and maximum strain at membrane rupture were examined with various mass fraction of agar. Maximum stress was decayed by increasing of mass fraction of agar. Maximum strain of membrane remained constant at 0.5 of mass fraction of agar and then gradually decayed with increasing of mass fraction of agar. Sufficient mechanical strength for practical application of alginate-agar composite membrane was expected less than 0.5 of mass fraction of agar in the casting solution.

Table 1 presents component characte- rization for mass transfer examination. Mass transfer character of the composite membrane was evaluated from the effective diffusion coefficient based on mass transfer flux.

Figure 3 shows the promising molecular size screening ranging from 60 to 1017 Da. Effective diffusion coefficient of water-soluble components was strongly depended on molecular size. In the case of alginate solo membrane (f_{agar}=0), it was changed 694 folds indicating Line 1. In constant, the composite membrane (f_{agar}=0.5), it was changed 141 folds indicating Line 2.

Mass fraction of agar of memrane (f_{agar}) was a dominant factor to regulate effective diffusion coefficient. Mass fraction of agar rolled as enlargement factor of mass transfer channel in the composite membrane.

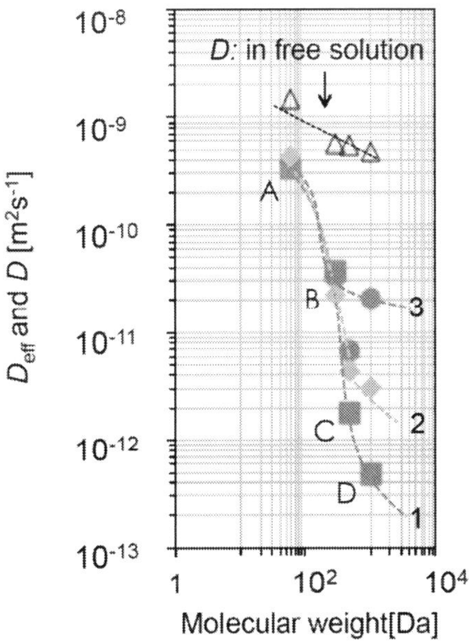

Fig. 3. Effective diffusion coefficient of water-soluble components in the membrane.

Line 1 ■: f_{agar}=0

Line 2 ◆: f_{agar}=0.5

Line 3 ●: f_{agar}=0.76

Table 1. Water-soluble components as the indicator of mass transfer of membrane.

		Molecular formula	Molecular weight[Da]	Estimated diameter[Å]
A	Urea	NH_2CONH_2	60	6
B	Neutral Red	$C_{15}H_{17}ClN_4$	288	10
C	Indigo Carmine	$C_{16}H_8Na_2O_8S_2$	466	12
D	Rose Bengal	$C_{20}H_2Cl_4I_4Na_2O_5$	1017	16

Reference

1. K. Kashima, M. Imai, I. Suzuki, Superior molecular size screening and mass-transfer characterization of calcium alginate membrane, Desalination and Water Treatment 17, 143-149 (2010)

H_2SO_4 and NaOH reclamation from industrial wastewater by electrodialysis with heterogeneous bipolar membrane

J. Kinčl[1], T. Jiříček[1], D. Neděla[1], N. Václavíková[1], M. Amrich[1], A. Ashrafi[1], J. Kroupa[2], J. Cakl[2], P. Doleček[2], B. Šiška[2], B. Velen[3], F. Toman[3], D. Černínová[4]
[1] MemBrain s.r.o., Pod Vinicí 87, CZ-47127 Stráž pod Ralskem, e-mail: jan.kincl@membrain.cz, +420 487 805 238
[2] University Pardubice, Department of Environmental and Chemical Engineering, Studentská 95, CZ-53210 Pardubice
[3] DIAMO a.s., o.z. GEAM, Máchova 201, CZ-47127 Stráž pod Ralskem
[4] Česká membránová platforma o.s., Mánesova 1580, CZ-47001 Česká Lípa

Pure sodium sulphate is produced at uranium mine DIAMO s.p. o.z. GEAM located in Dolní Rožínka, Czech Republic. Industrial technology consist of precipitation of Ca and Mg by lime and soda ash, sedimentation, sand filtration, sorption of heavy metals and uranium in ion exchange column, oxidation, sorption of oxidation products on activated carbon and acidification by H_2SO_4. 65 m^3/h of pretreated water with total dissolved solids (TDS) 35 g/l is partially desalted by electrodialysis (ED) to 12 g/l. ED diluate is further desalted by reverse osmosis (RO) reaching TDS 0.18 g/l and discharged into the local river meeting effluent restrictions. RO retentate is mixed with feed water of ED. The final step of full scale technology (not tested in pilot scale) lies in concentration of ED concentrate having TDS about 110 g/l by evaporation and Na_2SO_4 crystallization.

Figure 1: Industrial electrodialysis (left) and reverse osmosis (right) in GEAM (Toman 2009)

New technologies were considered because of low price and low market demand for Na_2SO_4 product. Electrodialysis with bipolar membrane (EDBM) was chosen because both NaOH and H_2SO_4 can be reused on site. About 9% of Na_2SO_4 salt in form of 4% water solution (RO retentate) can be split into 4% NaOH solution and 1.5% H_2SO_4 solution, which can be further concentrated to 10% H_2SO_4 by existing ED technology. EDBM diluate

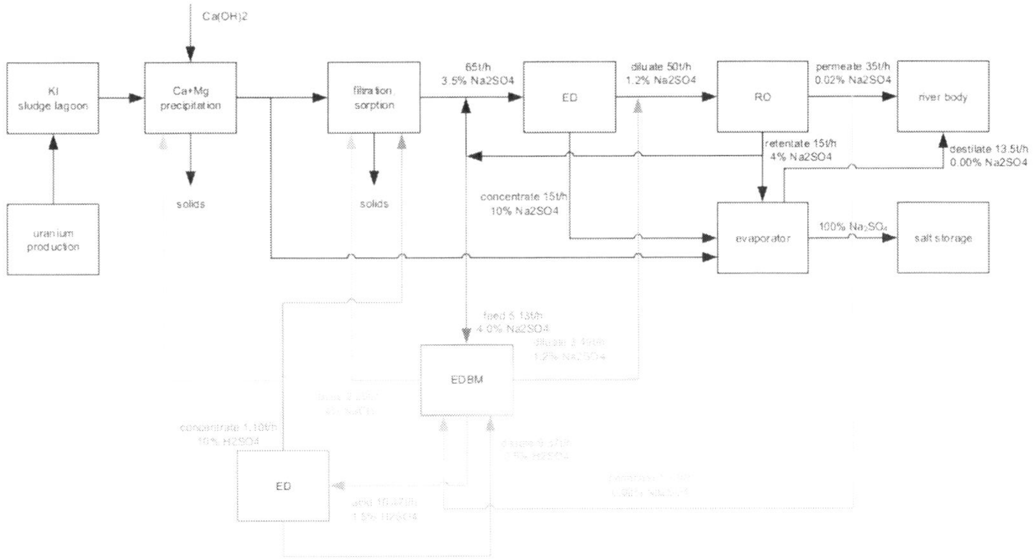

Figure 2: EDBM integration (in colour) into existing technology (in black)

Next version of heterogeneous bipolar membrane was developed and produced for pilot scale testing. New four compartment pilot module (salt, acid, base, electrode solution) was developed and produced. Pilot unit was developed and produced too and applied for treatment of wastewater coming from uranium production sludge water at GEAM. Industrial size EDBM module and unit is being developed for this application.

Figure 3: Pilot module EDBM-Y/3x25-0.8 (left), pilot unit P1 EDR-Y/4x (middle), industrial module EDBM-II/3x150-0.8 (right)

Addition of industrial EDBM would decrease operating costs by decreasing NaOH and H_2SO_4 purchase. Value of NaOH and H_2SO_4 produced is 317 034 €/year at unit prices of 50% NaOH 210 €/t and 94% H_2SO_4 55 €/t. Operation costs (energy, chemicals, spare parts, wastes) were estimated at 197 624 €/year making operation profit 119 410 €/year. Technology would be also much more flexible with EDBM installed. For example only EDBM can be used in winter when low rain and snow is increasing water level in sludge lagoon only slightly, no need to run expensive evaporator.

Toman F.: Membrane processes in a water treatment at the uranium chemical mill, Book of abstracts PERMEA Prague, 2009

Recovery of valuable substances by treatment of electroplating liquids with membrane distillation

C. Brunner, [1]J. Buchmaier [1], C. Platzer [1]
[1]AEE – INTEC Dep. Industrial Processes and Energy Systems Feldgasse 19, 8200 Gleisdorf Austria +43 3112 5886470 e-mail: c.brunner@aee.at;.

The process of applying metallic coatings to an article by passing an electric current through an electrolyte is called electroplating. Many different industries use the electroplating process in order to achieve changes of physical, chemical and mechanical properties. Various pre-treatment steps such as polishing, grinding and degreasing are conducted. During the process of galvanizing the work piece runs through various basins which provide specific bathes, which have to be exchanged frequently. This is realized by either rinsing directly above the bath or in a separate basin. Thereby the concentration of the chemicals in the baths gets reduced and a huge amount of expensive electroplating liquids is discharged during the process. Since conventional technologies are relatively energy intensive, recovery strategies for the chemical solutions are rarely applied. The membrane distillation (MD) offers a promising alternative since available waste heat from other processes can be used as the driving force for the separation technology.

In this study the applicability of the MD in the electroplating industry is examined by treating three different kinds of bathes, thereby concentrating the electroplating liquids for the reuse and providing process water for recycling.

In a first step, the stability, in terms of the chemical resistance, of a poly tetra fluor ethylene (PTFE) membrane was proven by conducting liquid entry pressure (LEP) tests such as contact angle (CA) measurements with membranes that have been exposed to the respective electroplating liquid before. Small scale pre-experiments provided information for further operational parameters in the pilot scale facility with a capacity of 200l per hour. A custom-made spiral wound direct contact module (DCMD) was used in a pilot plant, suitable for treating several m3/d. Experiments were carried out on-site for a few month. A cleaning strategy was developed and the performance of MD for the applications was proven. An integration concept of the MD in the electroplating process including the company-internal energy supply and economic feasibility of the approach is a result of this study.

The paper will present results from field tests at the electroplating company on the long-term behaviour of the whole system. In the scope of this paper a concept for the integration of the membrane distillation facility into the electro-plating plant will be presented and the potential for the integration of waste heat or solar thermal energy will be pointed out. An essential part is also the calculation of the economic feasibility of this new technology.

The result will include detailed knowledge about the operation parameters such as flow rates, temperatures, energy consumption etc. Based on these information, a scale up for large-scale facilities with an optimized membrane module and optimized operation parameters will be presented.

Improvement of energy generation by pressure retarded osmosis process

E. Nagy[1], I. Hegedüs[1], B. Kakasi[1]
[1]Univ. Pannonia, Research Inst. Chem. Process Eng. Egyetem ut 10., 8200 Veszprém, Hungary;
tel. +36 88 603511, e-mail: nagy@mukki.richem.hu; fax: +36 88 604040.

A unique area of forward osmosis research is the pressure retarded osmosis (PRO) for generation of electricity from mixing of saline and fresh water. The basic idea is the mixing of seawater and river water for extraction of the energy of the mixing of the two phases by means of a semipermeable membrane. The latest investigations are concluded, that both the operating conditions and/or even the salt concentration of the high salinity and the low salinity solutions should essentially be improved in order to get economically efficient energy generation [1]. Crucially important is the exact description of the salt and water transport in order to be able to predict the effect of operating conditions. Applying the solute fluxes given for every single transport layer, namely external boundary layers, selective-, and the support layer, new expressions were developed to define the overall mass transfer coefficient and the interface solute concentrations. These equations make possible much deeper investigation of the mass transport process and process efficiency in pressure retarded osmosis system. It was stated that the effect of the feed side boundary layer on the energy generation must not be neglected without checking its effect. The extractable energy is dominated mainly by the membrane selectivity, structural parameter and also by the solute concentrations. Essential improvement of the membrane selectivity and/or decrease of the value of the structural parameter is needed to get more efficient pressure retarded osmosis process for energy extraction. Accordingly, not only the operating conditions but the change of the salt concentration of the draw and feed phases should also be regarded as an alternative process instead of seawater-river water pair. Thus, one must seek processes where the liquid phases can be regenerated and repeatedly applied. Obviously this raises perfectly different problems to be investigated and solved. Such an alternative process can be the so called closed-loop realization where both phases are continuously regenerated and reused.

How the efficiency can be improved by the improvements of the membrane selectivity and the decrease of the value of structural parameter and the feed phase salt concentration change will be discussed. It will also show how the closed-loop process can be realized, what kind of technological problems, and how should be solved for it.

The National Development Agency grant, Hungarian Research Fund (OTKA 116727) greatly acknowledged for the financial support.

Reference

[1] Straub, A.P., Deshmukh A., Elimeléech, M., Pressure-retarded osmosis for power generation from salinity gradients: it is viable? Energy & Environ. Sci., DOI:10.1039/cSee02985f

[2] Nagy E., A general, resistance-in-series salt- and water flux models for forward osmosis and pressure-retarded osmosis for energy generation, *J. Membr. Sci.*, 460 (2014) 71-81

Continuous in-line mixed gas permeability and selectivity measurements of polymeric membranes by mass spectrometric analysis

A. Fuoco, M. Monteleone, E. Esposito, P. Bernardo, [1]C. Brazinha, [1]S.C. Fraga, [1]J.G. Crespo, [2]M. Carta, [2]N.B. McKeown, [3]M. Lanc, [3]K. Friess, J.C. Jansen

Institute on Membrane Technology, ITM-CNR, Via P. Bucci, Cubo 17/C, 87036 Rende (CS), Italy, Ph.: +39 0984-492031, e-mail: johannescarolus.jansen@cnr.it; [1] REQUIMTE/CQFB, Chemistry Department, Universidade Nova de Lisboa, Campus de Caparica, 2829-516 Caparica, Portugal, Ph. +351 21 294 8385; [2] School of Chemistry, University of Edinburgh, David Brewster Road, EH9 3FJ, UK, [3]University of Chemistry and Technology Prague, Department of Physical Chemistry, Technická 5, 166 28 Prague 6, Czech Republic

Membrane-based gas separation processes are becoming increasingly important in various industrial sectors, where they are gradually replacing some of the traditional gas separation processes. Many studies have been devoted to the development of novel membrane materials. Detailed studies of the transport properties in polymeric membranes are often carried out by sorption studies, to analyse the gas solubility and diffusivity, and by permeation studies in a so-called time lag instrument, to measure the permeability and diffusivity. With the traditional time lag studies in a fixed volume/pressure increase instrument, only permeation data on pure gases can be obtained. In mixed gas permeation experiments, the gas composition of the feed, permeate and retentate streams is most commonly analysed by gas chromatography. Even in the case of relatively fast micro-GC, the analysis times are relatively long and multiple injections must be performed to follow the membrane performance as a function of time.

The present paper proposes the in-line analysis of the feed, permeate and retentate gas flux and composition in mixed gas permeation experiments by the use of a quadrupole Mass Spectrometer (MS). Recently, Crespo *et al.* proposed this technique for the analysis of the permeate in relatively slow pervaporation experiments [1,2]. In this work will optimize this method to make it suitable for the much faster mixed gas separation experiments using polymeric membranes[3] (Fig. 1).

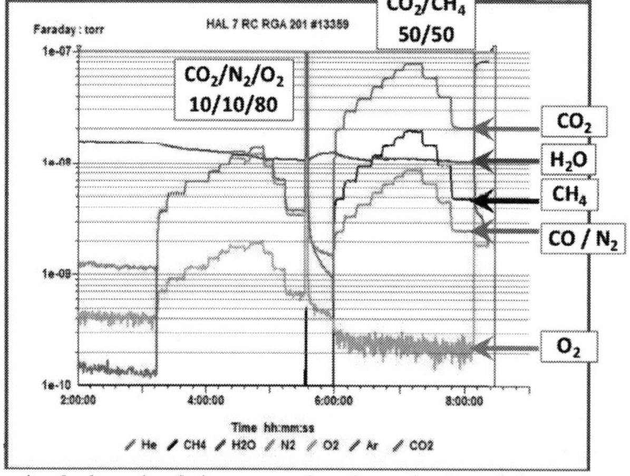

Figure 1. Typical signal of time-resolved analysis of the permeate by Mass

Spectrometry during mixed gas permeation at different feed pressures.

The aim of the work is to analyse the possibility to follow time dependent phenomena taking place in the polymer membranes during gas permeation. One feature will be the analysis of the transient behaviour, similar to the time-lag method, in order to evaluate the effect of the composition on the diffusivity of the individual species in a gas mixture. The work will focus on two different membrane types, based on polymer/ionic liquid gels and on polymers of intrinsic microporosity, respectively.

Two different experimental setups will be studied: the first one with a large-bore connection between the membrane and mass spectrometer for vacuum operation, and the second one with a narrow-bore capillary for use with a sweeping gas. Both systems will be optimized in terms of quick response times and high sensitivity. Besides the method development, the effect of membrane operation conditions on the separation performance will be reported.

Acknowledgements: Part of the research leading to these results received funding from the European Union's Seventh Framework Program (*FP7/2007-2013*) under *grant agreement* n° 608490, project M^4CO_2. Financial support was also received from the CNR/FCT bilateral collaboration program 2015-2016.

References

(1) Fraga, S. C.; Trabucho, L.; Brazinha, C.; Crespo, J. G. *J. Memb. Sci.* **2015**, *479*, 213–222.
(2) Brazinha, C.; Fonseca, A. P.; Teodoro, O. M. N. D.; Crespo, J. G. *J. Memb. Sci.* **2010**, *347*, 83–92.
(3) Pilnacek, K.; Jansen, J. C.; Bernardo, P.; Clarizia, G.; Bazzarelli, F.; Tasselli, F. *Procedia Eng.* **2012**, *44*, 1027–1029.

Pure versus mixed gas permeability of novel polymers of intrinsic microporosity

E. Esposito, P. Bernardo, A. Fuoco, M. Monteleone, G. Clarizia, [1]M. Carta, [1]N.B. McKeown, J.C. Jansen*
Institute on Membrane Technology, ITM-CNR, Via P. Bucci, Cubo 17/C, 87036 Rende (CS), Italy, Ph.: +39 0984-492031, *e-mail: johannescarolus.jansen@cnr.it; [1] School of Chemistry, University of Edinburgh, David Brewster Road, EH9 3FJ, UK

From their first introduction in 2004 as potential membrane materials,[1] the research on polymers of intrinsic microporosity (PIMs) has undergone an explosive growth because of a number of unique properties. A typical feature of PIMs is their extremely rigid and highly contorted polymer backbone, which prevents efficient packing. Therefore, most PIMs have a high BET surface area and a high free volume, arranged as continuous interconnected channels in the polymer matrix. The large free volume fraction and the interconnected void structure, to which they owe the classification of "intrinsically microporous" polymers, provides PIMs with a very high permeability and a high diffusion coefficient for most gases. Besides fast diffusion, PIMs also exhibit very high solubility of easily condensable gases like CO_2, and of organic vapours. This combination of a strong size sieving character because of the polymer stiffness, and the high sorption due to the high surface area and microporosity, is quite unique for PIMs and provides them with exceptional performance.[2]

Most studies on novel PIM-based membrane materials focus on pure gas permeation, and relatively few on mixed gases. Depending on their structure, either the size sieving behaviour or the strong sorption selectivity will dominate and this may cause significant differences between pure and mixed gas permeability. It was previously found that properties may even be opposite to what can be expected intuitively, for instance, when the affinity of amino groups[3] for CO_2 yields a decrease rather than the expected increase in CO_2 permeability.

The present paper will study in detail the difference between pure gas permeation and mixed gas separation for a number of different polymers. The effect of mixture composition and other operational conditions (temperature, pressure) will be analysed. From one side the pure gas permeation measurements, especially in the time lag mode, offer valuable information on the basic transport phenomena, such as permeability, diffusivity and solubility. From the other side, only mixed gas permeation experiments can show the behaviour of the membranes under representative operating conditions. The attention will be focused on a number of separations of great industrial interest, such as CO_2/N_2 (CO_2 capture from flue gas), CO_2/CH_4 (biogas upgrading) or O_2/N_2 (nitrogen production or oxygen enrichment)

Acknowledgements: Part of the research leading to these results received funding from the European Union's Seventh Framework Program (*FP7/2007-2013*) under *grant agreement* n° 608490, project M^4CO_2.

References

(1) Budd, P. M.; Elabas, E. S.; Ghanem, B. S.; Makhseed, S.; McKeown, N. B.; Msayib, K. J.; Tattershall, C. E.; Wang, D. *Adv. Mater.* **2004**, *16*, 456–459.
(2) Carta M., Malpass-Evans R., Croad M., Rogan Y., Jansen J.C., Bernardo P., Bazzarelli F., M.

N. B. *Science* **2013**, *339*, 303–308.
(3) Mason, C. R.; Maynard-Atem, L.; Heard, K. W. J.; Satilmis, B.; Budd, P. M.; Friess, K.; Lanci, M.; Bernardo, P.; Clarizia, G.; Jansen, J. C. *Macromolecules* **2014**, *47*, 1021–1029.

New hyperbranched polyimide membranes for targeted gas separations

Karel Friess[1], Petr Sysel[2], David Malý[2], Ondřej Vopička[1], Marek Lanč[1], Kryštof Pilnáček[1], Daniel Radotínský[1]
1Departments of Physical Chemistry, University of Chemistry and Technology, Technická 5, Prague 6 -Dejvice, 166 28, Czech Republic;
2Department of Polymers, University of Chemistry and Technology, Technická 3, Prague 6 - Dejvice, 166 28, Czech Republic;
karel.friess@vscht.cz

Hyperbranched polyimides (HBPI) represent the promising materials for gas separation membranes. Their specific inner architecture of highly branched structure creates the accessible cavities of atomic sizes [1]. Thus, such size-sieving properties are desired for efficient gas separation.

In this work, new gas separation membrane materials based on hyperbranched polyimides were prepared *via* solvent evaporation procedure from (i) 4,4'-oxydiphthalic anhydride and 4,4',4''-triaminotriphenylemethane, (ii) 4,4' -(hexafluoroisopropylidene)diphtalic anhydride and 4,4',4''-triaminotriphenylemethane and (iii) 4,4'- (hexafluoroisopropylidene)diphtalic anhydride and 4,4',4''-triaminotriphenylmethane with linear polyimide Matrimid 9725. New materials were characterized using thermogravimetry, IR and dynamic-mechanical analyses.

Permeation properties were tested using the self-developed manometric apparatus [1]. The time-lag permeation experiments of the pure gases (hydrogen, nitrogen, oxygen, methane, carbon dioxide) were performed at 1.1 bar (feed) and at 25°C. The permeability coefficients and diffusion coefficients of these gases were determined. The gas transport properties were correlated with the composition and morphology of the membranes. Determined values (permeability, ideal selectivity) for H_2 / CH_4 and CO_2 / CH_4 gas pairs lies close to the Robeson 2008 upper bound [2].

Acknowledgements
The financial support of the Czech Science Foundation (grant GACR No. 15-06479S) is gratefully acknowledged.

[1] Friess K., Sysel P., Minko E., Hauf M., Vopicka O., Hynek V., Pilnacek K., Sipek M.: Desalination and Water Treatment 14, 165 (2010).
[2] Robeson L., Journal of Membrane Science 320, 390 (2008).

Role of Ti in different titanosilicates (TS-1 and ETS-10) for the separation of CO_2/CH_4 using mixed matrix membranes

V. Martin-Gil[1], A. Lopez-Garcia[2], M. Z. Ahmad[1], P. Hrabanek[3], R. Mallada[2], I.F.J. Vankelecom[4], V. Fila[1].
[1]Department of Inorganic Technology[a], Institute of Chemical Technology of Prague, Prague, 16628, Czech Republic.
[2]Department of Chemical Engineering, Nanoscience Institute of Aragon (INA), Universidad de Zaragoza, E-50018-Zaragoza, Spain.
[3]Heyrovský Institute of Physical Chemistry of the ASCR, v.v.i.; Dolejškova 2155/3, 18223 Prague 8, Czech Republic
[4]Centre for Surface Chemistry and Catalysis, KU Leuven, Kasteelpark Arenberg 23, 3001 Heverlee, Belgium

Polymeric membranes are limited by Robeson's upper bound [1] which establishes when permeability increases, selectivity decreases and vice versa. Mixed Matrix Membranes (MMM) have caught the attention of many researchers in last years in order to overcome this limitation and improve the gas separation performance of polymeric membranes in order to achieve the necessary features for industrial applications. Although, studies using silicate particles are widely spread [2–6], only a few studies using titanium silicates have been done for gas separation processes [7,8]. In this work, three titanium silicates with different content of titanium were synthesized for using them as fillers for MMM's: (i) ETS-10, (ii) TS-1 using Si/Ti = 100 and (iii) TS-1 using Si/Ti = 25.

Titanosilicate samples were characterized by XRD, XPS and SEM. TS-1 particles showed a good size distribution and the particle size ranges from 200 nm to 400 nm. In the case of ETS-10, the size distribution was broader and the particle size ranges from 600 nm to 800 nm. Mixed Matrix Membranes for CO_2/CH_4 separation applications were prepared using Matrimid® as continuous phase and different inorganic loads (10, 20, and 30 wt%.). The membranes were measured at 8 bars of transmembrane pressure for CO_2/CH_4 mixed gases system at 50/50 vol./vol. concentration.

MMMs using TS-1 (Si/Ti = 25) as filler showed a maximum increase of 89.1% of CO_2 permeability for the MMM with 30 wt% of filler and 23.9% increase in separation factor for 20 wt% MMM. In the case of TS-1 (Si/Ti = 100) only permeability increased significantly with a maximum increase of 90.1% in the case of 30 wt% loading. These results showed that content of Ti inside TS-1 plays an important role in the sorption of carbon dioxide. High content of Ti inside the zeolite lead to increase CO_2 adsorption. Regarding ETS-10 MMMs, both permeability and separation factor increased slightly with respect to the unfilled membranes, In conclusion, it can be said that TS-1 (Si/Ti=25) is the most suitable filler to use in MMMs for gas separation applications.

References

[1] L.M. Robeson, The upper bound revisited, J. Memb. Sci. 320 (2008) 390–400.

[2] P. Burmann, B. Zornoza, C. Téllez, J. Coronas, Mixed matrix membranes comprising MOFs and porous silicate fillers prepared via spin coating for gas separation, Chem. Eng. Sci. 107 (2014) 66–75.

[3] B. Zornoza, O. Esekhile, W.J. Koros, C. Téllez, J. Coronas, Hollow silicalite-1 sphere-polymer mixed matrix membranes for gas separation, Sep. Purif. Technol. 77 (2011) 137–145.

[4] B. Zornoza, C. Téllez, J. Coronas, Mixed matrix membranes comprising glassy polymers and dispersed mesoporous silica spheres for gas separation, J. Memb. Sci. 368 (2011) 100–109.
[5] T. Suzuki, Y. Yamada, Characterization of 6FDA-based hyperbranched and linear polyimide-silica hybrid membranes by gas permeation and129Xe NMR measurements, J. Polym. Sci. Part B Polym. Phys. 44 (2006) 291–298.
[6] A.B. Beltran, G.M. Nisola, E. Cho, E.E.D. Lee, W.-J. Chung, Organosilane modified silica/polydimethylsiloxane mixed matrix membranes for enhanced propylene/nitrogen separation, Appl. Surf. Sci. 258 (2011) 337–345.
[7] A. Galve, D. Sieffert, C. Staudt, M. Ferrando, C. Güell, C. Téllez, et al., Combination of ordered mesoporous silica MCM-41 and layered titanosilicate JDF-L1 fillers for 6FDA-based copolyimide mixed matrix membranes, J. Memb. Sci. 431 (2013) 163–170.
[8] A. Galve, D. Sieffert, E. Vispe, C. Téllez, J. Coronas, C. Staudt, Copolyimide mixed matrix membranes with oriented microporous titanosilicate JDF-L1 sheet particles, J. Memb. Sci. 370 (2011) 131–140.

Pervaporation and gas separation with supported liquid membranes

Pavel Izák, Michal Žák, Zuzana Sedláková, Magda Kárászová, Marie Kačírková, Jiří Vejražka
Institute of Chemical Process Fundamentals, Rozvojová 135, 165 02 Prague 6, Czech Republic, Fax: +420-220920661, Tel: +420-220390204, E-mail: izak@icpf.cas.cz

Membrane separation still continues to be one of the major challenges not only in chemical and biotechnological production processes, but also in waste water treatment. The goal of this work is to combine the recent developments in the field of membrane technology, e.g. supported liquid membranes [1,2], pervaporation [3,4], and gas separation [5,6] with the use of ionic liquids (ILs) for providing novel solutions in downstream processing or process intensification. ILs are currently explored as new reaction media for chemical synthesis or electrochemical applications, have emerged as a new class of solvents in the last decade. ILs seem to have a large potential in downstream processing (pervaporation or gas separation), especially when applied in a form that requires only a small amount of them, e.g. in supported liquid membranes. The special property of ILs is their non-measurable vapour pressure at ambient temperature that makes their application in liquid membranes attractive for pervaporation or gas separation. Pervaporation and gas separation is considered as a perspective and modern membrane process for the separation of various liquids or vapour mixtures. The major problem associated with solute recovery by membrane processes is the restrictive compromise between selectivity and flux: High solute selectivity involves the use of conditions that lead to relatively low mass fluxes in the membrane. Supported ionic liquid membranes offer a range of possible advantages: 1. Molecular permeation is much higher in liquids than in polymers, as liquids allow higher fluxes; 2. The selectivity of the separation can be influenced by variation of the liquid - especially ILs offer the advantage of a wide variety of properties; 3. Thanks to their special mixing behaviour ILs as liquid membranes easily allow three-phase systems; 4. Contrary to the extraction, only small amount of liquid is necessary to form the liquid membrane, thus allowing also the use of more expensive materials. Practical examples of utilization of ILs in downstream processes are given.

Acknowledgements
The Czech Science Foundation Grant No. 14-12695S is acknowledged for financial support.

References

1. P. Izák, K. Schwarz, W. Ruth, H. Bahl, P. J. Dyson, U. Kragl, Applied Microbiology and Biotechnology, 78, 597 (2008).
2. M. Kohoutová, A. Sikora, Š. Hovorka, A. Randová, J. Schauer, M. Tišma, K. Setničková, R. Petričkovič, N. Greenspoon, P. Izák, Eur. Polym. J. 45, 813 (2009).
3. P. Izák, N.M.M. Mateus, C.A.M. Afonso, J.G. Crespo, Sep. Pur., 41, 141 (2005).
4. P. Izák, U. Kragl, M. Köckerling, Green Chem., 8, 947 (2006).
5. J. Jansen, K. Friess, G.Clarizia, J. Schauer, P. Izák, Macromolecules 44, 39 (2011).
6. M. Kárászová, J. Vejrazka, V. Veselý, K. Friess, A. Randová, V. Hejtmánek, L. Brabec, P. Izák, Sep. Pur., 89, 212 (2012).

Pure and mixed gas transport of permanent gas mixtures through polysulfone membrane

M. O. Midda, S. S. K. Jampa, A. K. Suresh
Indian Institute of Technology Bombay, Dept. of Chemical Engg., Mumbai-400076, India;
tel. (+91)9920130052, e-mail: oayesmidda@gmail.com; aksuresh@iitb.ac.in;

Introduction

An important issue in the separation of gases using membranes is the prediction of mixed gas permeability using pure gas data. But owing to the presence of coupling effects, the models normally used for mixed gas permeation often lead to erroneous results by such predictions. Gas-gas and/or gas-polymer interactions are believed to exist in the separation of real gas mixture through polymeric membranes, which cause these discrepancies.

The gas transport through glassy polymeric membranes is mostly described by solution-diffusion mechanism. According to this mechanism, gas molecules first sorb into the polymer matrix and then diffuse through the film owing to the presence of a concentration gradient, in turn caused by a pressure difference driving force. Therefore, the gas permeation flux depends on the relative affinity of gases towards the polymer sorption sites. Among the sorption models presented in literature so far, the dual-mode sorption and partial immobilization model is the most successful model for explaining pure gas transport through glassy polymeric membranes, because of its versatility and applicability to numerous gas-polymer systems. Whereas, a very limited number of mixed gas study has been reported in literature till date[1–7] using this model with few experimental data points. Because of this lack of experimental data, there is a gap in the fundamental understanding the mixed gas permeation process through dense glassy membranes, and especially on the question of whether the permeability parameters determined from pure gas experiments can be used to predict mixed gas results. This motivates the present study. The study includes the separation of permanent gas mixture with rigorous number of data points. Separation of carbon dioxide from its mixtures with oxygen and nitrogen has been chosen as the model system for this study.

Materials & Methods

Flat sheet PSf membranes were synthesized by solution casting method described elsewhere[8–10]. The polymer (Udel-3500), kindly gifted by Advanced Solvay Polymers (USA), was degassed under vacuum conditions at 80°C for 24 h to remove adsorbed moisture. Then, a viscous solution was prepared by dissolving 1.2 g polymer in 6 ml chloroform (13 wt% of polymer), followed by magnetic stirring for 24 h. The solution was cast on a glass plate by Doctor's blade (Sheen instruments, UK) and maintained at room temperature (30°C) in a partially closed desiccator for 24 h to ensure a slow evaporation of the solvent. Afterwards, the formed films were completely dried in vacuum conditions at 80°C for 24 h.

The gas permeation set-up was comprised with mass flow controller and back pressure regulator to maintain flow rate, mixed gas composition and pressure in the feed side. Sweep gas was used to carry the permeate volume into an on-line GC to measure exact composition; and Eq 1 was used to calculate the gas permeability.

$$P = \frac{Q\,.\delta}{A.\Delta p} \tag{1}$$

Results & Discussions

The experimental set-up for obtaining gas permeability is able to measure both the pure and mixed gas permeability data. Each experimental data point was collected after ensuring that the system had reached steady state. The pure gas permeability data reveals that the permeability of all the studied gases followed a non-linear decreasing trend with increase in feed pressure. This behaviour is normally attributed to the variation of gas solubility in the polymer matrix; the solubility decreases exponentially with pressure and leads to a decrease in the absorption uptake at high pressure and hence in reduced permeability. The permeability data were fitted to the dual mode (Henry+Langmuir) sorption model, and the parameters calculated by non-linear least square fitting.

The mixed gas data were collected very accurately to study the effect of each gas on the permeation of the other. The permeation data were collected for the binary mixture of CO_2/O_2 and CO_2/N_2 system by varying the feed pressure and feed composition. The variation of CO_2 permeability with CO_2 partial pressure is presented in Fig 1. Mixed gas data were collected for four different feed pressure (2, 4, 6 and 8 bar absolute pressure) and three different feed composition (CO_2/O_2 and CO_2/N_2 in the ratio of 10/90, 30/70 and 50/50). The results show the non-linear decrease in permeability with pressure for a particular gas mixture as seen for pure gas also. Mixed gas permeability is lower than pure gas permeability at a particular CO_2 partial pressure which indicates that the mixed gas transport is affected in presence of other gas component. The variation in CO_2/N_2 selectivity is shown in Fig 2. It shows that the mixed gas selectivity is higher than pure gas selectivity in all the cases. The solid lines in Fig 1 and Fig 2 represent the values predicted by the dual mode sorption theory, assuming competition. It is seen that the theory is only partially successful in predicting the data, implying that there are other factors (such as interaction between the gas molecules or additional mass transfer resistance in the feed side) because of which the experimental and calculated values show deviation. Therefore, the dual-mode sorption model itself is only partially able to explain the experimental data for the studied gas mixture. A mass transfer resistance can improve the agreement to a great extent, suggesting methods for extending pure gas data to mixed gases.

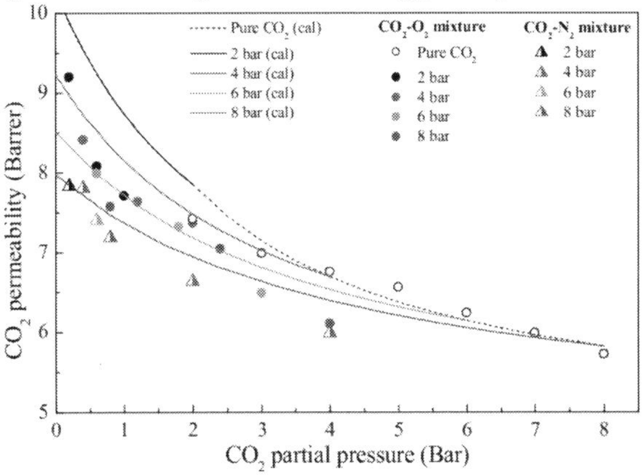

Fig 1. CO_2 permeability with CO_2 partial pressure for CO_2/O_2 and CO_2/N_2 mixture.

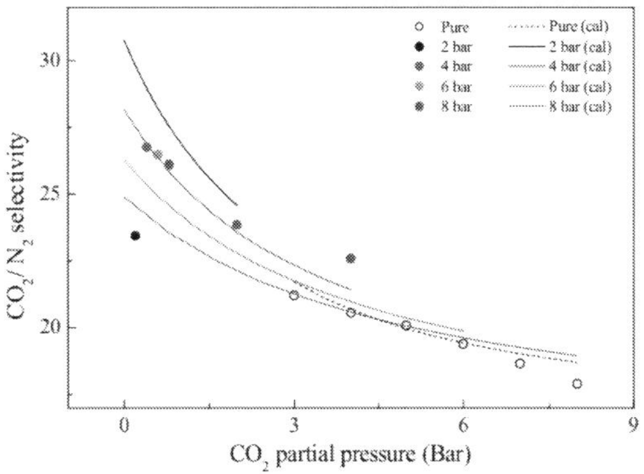

Fig 2. CO_2/N_2 selectivity with CO_2 partial pressure.

Conclusions

The general conclusions from the study are – (a) gas permeability decreases non-linearly with feed pressure, (b) experimental pure gas selectivity also decreases gradually with feed pressure, (c) mixed gas permeability is lower than pure gas permeability at the same partial pressure, (d) mixed gas selectivity is higher than pure gas selectivity for a particular partial pressure, and (e) extending pure gas data to mixed gas permeation requires accounting for factors such as mass transfer resistence which come into play only in gas mixtures.

References

(1) David, O. C.; Gorri, D.; Urtiaga, A.; Ortiz, I. *J. Memb. Sci.* **2011**, *378*, 359.
(2) Raharjo, R.; Freeman, B.; Sanders, E. *J. Memb. Sci.* **2007**, *292*, 45.
(3) Yeom, C. K.; Lee, S. H.; Lee, J. M. *J. Appl. Phys.* **2000**, *78*, 179.
(4) Dhingra, S. S.; Marand, E. *J. Memb. Sci.* **1998**, *141*, 45.
(5) O'Brien, K. C.; Koros, W. J.; Barbari, T. a.; Sanders, E. S. *J. Memb. Sci.* **1986**, *29*, 229.
(6) Sanders, E. S.; Koros, W. J.; Hopfenberg, H. B.; Stannett, V. T. *J. Memb. Sci.* **1984**, *18*, 53.
(7) Chern, R. T.; Koros, W. J.; Sanders, E. S.; Yui, R. *J. Memb. Sci.* **1983**, *15*, 157.
(8) Kim, S.; Chen, L.; Johnson, J. K.; Marand, E. *J. Memb. Sci.* **2007**, *294*, 147.
(9) Zornoza, B.; Esekhile, O.; Koros, W. J.; Téllez, C.; Coronas, J. *Sep. Purif. Technol.* **2011**, *77*, 137.
(10) Zornoza, B.; Seoane, B.; Zamaro, J. M.; Téllez, C.; Coronas, J. *Chemphyschem* **2011**, *12*, 2781.

High permeable zeolite W membranes for CO_2/CH_4 separation: synthesis and characterization

Mojtaba Mirfendereski, [1]Toraj Mohammadi
Faculty of Mechanical Engineering and Energy, Shahid Beheshti University, Tehran, Iran; tel. +98 912 203 7345, email: mirfendereski@alumni.iust.ac.ir
[1]Research Centre for Membrane Separation processes, Faculty of Chemical Engineering, Iran University of Science and Technology, Narmak, Tehran, Iran; tel. +98 21 77240051, email: torajmohammadi@iust.ac.ir

In this paper high permeable zeolite W membranes were synthesized for separation of CO_2 from CH4. The effects of synthesis parameters (temperature, time and coating repetition) on permeability and selectivity of synthesized membranes were investigated. Experiments were carried out at three levels of synthesis temperature: 165, 185 and 200 °C; synthesis time: 6, 12 and 18 h and two levels of coating: 1 and 2 layers. The CO_2 and CH4 permeabilities versus synthesis temperature first decreased with increasing synthesis temperature from 165 °C up to 185 °C, then increased with further increasing the values of synthesis temperature. These permeabilities decreased with increasing synthesis time from 6 h up to 12 h and continued decreasing up to 18 h. In the case of CO_2/CH_4 selectivity, reverse trends were observed. This behavior was attributed to the dual effect of increasing synthesis temperature and time on membrane permeability and selectivity. Repetition of zeolite layer coating strongly increased membrane selectivity and decreased permeability for CO_2/CH_4 mixture. Zeolite layers were thickened and larger crystals were formed at higher synthesis temperatures and times. Formation of larger crystals accelerated the rate of zeolite layer integration, which was responsible for gas separation, in one hand and reduced the density of deposited zeolite layer on the support, due to void increase, on the other hand. Gas permeances through the membranes were in the range of 10-8 to 10-7 (mol/m2 s Pa) which showed high permeable zeolite membranes were synthesized. In terms of maximizing the CO_2/CH_4 selectivity (=20.1), medium synthesis temperature with high synthesis time (185 °C and 18 h) and two layer coating were selected. High permeable zeolite W membranes with moderate selectivities were successfully synthesized in this work. It found as good choices for the separation of CO_2 from CH_4 in commercial applications.

Novel chemistry routes to prepare solvent-resistant nanofiltration membranes

Ivo F.J. Vankelecom
Centrum voor Oppervlaktechemie en Katalyse (COK), Dept. M²S, Faculty Bio-science engineering, KU Leuven
Leuven Chem&Tech, P.O. Box 2461, Celestijnenlaan 200F
3001 Leuven, Belgium
Tel: + 32-16-32.15.94
http://www.biw.kuleuven.be/m2s/cok/Groups/membrane-technology
ivo.vankelecom@kuleuven.be

Nanofiltration (NF) and reverse osmosis (RO) are pressure-driven membrane processes in which low molecular weight components can be retained on a molecular level. Many large-scale aqueous applications exist, e.g. in saltwater desalination and wastewater treatment. Nowadays, RO represents 53% of the total worldwide desalination capacity, showing a three to eight times lower specific energy consumption than distillation technologies. Solvent resistant NF (SRNF) has been emerging more recently as a potential energy-saving replacement for distillation and a waste-free alternative for extractions and chromatographic separations in (petro)chemical, pharmaceutical and food industry.

Thin film composite (TFC) membranes, consisting of a very thin, mostly polyamide selective layer on top of a porous support, are a widespread membrane type in RO and aqueous NF, and are now also being explored for use in SRNF. The porous support, made via phase inversion, typically consists of poly(ether)sulfone for aqueous applications. The polyamide film is prepared via interfacial polymerization, by impregnating the support with an aqueous amine solution and subsequently contacting its surface with an immiscible organic (typically hexane) acyl chloride solution. Common monomers are m-phenylenediamine (MPD) or piperazine and trimesoylchloride (TMC). Several novel concept of interfacial polymerization to form TFC membranes will be presented. They will include the introduction of ionic liquids as one of the reaction phases, the formation of the interfacial polymerised layer before creating the porous support and of the introduction of novel types of chemistries for interfacial polymerisation for membranes, such as Click chemistry.

A second type of membranes commonly applied in SRNF is prepared via the process of phase inversion in which a cast polymer solution in converted into a solid porous material by immersion in a non-solvent. By mixing acrylates in such mixtures and irradiating the solidified membranes by UV or electron beam, highly stable SRNF-membranes and support layers could be obtained.

Finally, a third concept will be shown in which the 2 methods of phase inversion and interfacial polymerisation will be combined to obtain TFC-membranes on very stable, crosslinked PI-membranes in a drastically reduced number of processing steps.

Seawater desalination by combined nanofiltration and ionic exchange

R. Bórquez., Chemical Engineering Department, University of Concepción, Concepción, P.O. Box 160-C, Concepción, Chile. Tel.: 56 - 41- 2204534; fax: 56- 41- 2247491.
rborquez@udec.cl

Desalination is considered as an attractive and reliable technology for making new water available from saline sources that forms 97.5% of the planet's water. However, the conventional desalination technologies including the reverse osmosis process are often considered an energy intensive process. Reverse osmosis (RO) desalination operates at a very high hydraulic pressure in order to overcome the osmotic pressure of the saline water and therefore requires significant energy. More energy is consumed since RO process is prone to membrane scaling and fouling problems.

This paper presents experimental results on the use of nanofiltration and ionic exchange to obtain drinking water from seawater.

Ionic exchange experiments were conducted in columns using both anionic and cationic resins to remove chloride and sodium ions.

Nanofiltration at trans-membrane pressures (TMP) in the range 20-40 bars, temperatures between 10-22°C, and feed flow between 707-1730 L/h was able to reduce total dissolved solids (TDS) from 33.5 g/L to 0.01 g/L, with 48 L/m^2h permeate flux. This was much greater that 16 L/m^2h obtained with typical reverse osmosis (RO) membranes operating at the same TMP.

Results show that 99.9% total dissolved solids (TDS) and Ca^{2+},Cl^-, Mg^{2+}, Na^+ and SO_4^{2-} ions could be removed by the microfiltration and two stage of nanofiltration, operating at 40 bar and 20 bar, respectively. Resulting permeates meet the Chilean drinking water regulation.

Nanofiltration membranes produce a much higher permeate flow (over 3 times permeate flow) than reverse osmosis membranes at same operation conditions.

However, one-step of nanofiltration at 40 bar was unable to remove the required Cl^- and Na^+ ions concentration to meet the Chilean and WHO drinking water regulations. Further treatment with ionic exchange resins (anionic Purolite A-300 (1.35 meq/g), and cationic Purolite C-100 (1.53 meq/g) yielded acceptable quality water, with breakthrough times around 115 min, and 150 min, respectively.

Hydrophobic post-treatment of PVDF-SAPO34 membrane in membrane gas absorption for CO_2 removal

Nor Aini, Ahmad, Choe Peng, Leo*, Abdul Latif, Ahmad.
School of Chemical Engineering, Engineering Campus, Universiti Sains Malaysia, Seri Ampangan,14300 Nibong Tebal, S.P.S., Pulau Pinang, Malaysia.
*Corresponding author. tel.: +60 4 5996425, fax: +60 4 5941013, email address: chcpleo@ usm.my.

Biomethane (CH_4) or biohydrogen (H_2) are of great importance as they are expected to become the major fuels in near future. However, the presence of excessive CO_2 in the raw gas mixtures reduced the calorific value greatly. Membrane gas absorption (MGA) offers an effective method to remove CO_2 at low pressure and low cost, matching with the main requirements in fuel gas purification. Microporous membranes are commonly used as the contactors in MGA for CO_2 absorption into liquid absorbent. The specific gas-liquid interfacial area in MGA is considerably high, but membrane wetting should be prevented to minimize the mass transfer resistance in the pores [1]. Among the hydrophobic polymers, PVDF appears to be the easiest processing polymer because it is soluble in solvent. The microporous structure can be easily formed via phase inversion. In recent years, the membrane researchers continued to develop PVDF mixed matrix membranes (MMM). This is because that the incorporation of inorganic fillers in PVDF membranes resulted in great hydrophobicity, porosity and mechanical stability for other applications. Up to date, MMMs were rarely tested in MGA. The filler in MMMs should be designed with special character to enhance CO_2 absorption through the membrane barrier. Compared to other porous materials, small pore zeolites especially SAPO-34 appealed to be an interesting filler due to its high adsorption affinity to CO_2 [2, 3]. The incorporation of zeolite improved the thermal stability of the MMMs as depicted by an increase in the glass transition temperature (T_g) and TGA results [4]. However, the hydrophilic nature of SAPO-34 zeolite may reduce the hydrophobicity of MMMs. Pre-fluorinated treatment was conducted to produce hydrophobic SAPO-34 zeolite in our previous work [5]. Although the hydrophobicity of MMMs increased, the presence of pre-fluorinated SAPO-34 induced the formation of dense membrane structure. In this work, post-fluorinated on PVDF/SAPO-34 is proposed to improve the membrane hydrophobicity for CO_2 removal.

SAPO-34 zeolite was prepared from the solution of Al_2O_3: P_2O_5:$0.6SiO_2$:1.07TEAOH:$56H_2O$ using hydrothermal method under microwave. SAPO-34 zeolite loadings in MMMs were varied from 2 to 8 wt%. PVDF from Solef Company (Solef 6010/1001) was used as the major polymer to prepare the hydrophobic and microporous membranes via phase inversion. N-methyl-2-pyrrolidone (NMP) (Emplura, Merck Chemical) was used as the solvent. The SAPO-34 zeolite was dispersed into the polymer solution as explained by Shen and Lua [6]. The PVDF/SAP-34 membranes were prepared using two-stage soft coagulation bath system [7]. The polymer solution was casted on glass plate using a blade with a gap of 40 µm. (Heptadecafluoro-1,1,2,2-tetra hydrodecyl) triethoxysilane (FAS) (*Gulf Chemical* Pte. Ltd., Singapore) was used as an agent to fluorinate the SAPO-34. Volume ratio of 1 ml of ethanol: 50 ml of FAS was used to modify PVDF/SAPO-34 membranes. The PVDF/SAPO-34 membranes were immersed into a grafting solution for 24 hours and rinsed with ethanol. The modified PVDF/SAPO-34 membranes were dried at 60 °C for 24 hours. The membranes were characterized with the goniometer, porometer, scanning electron microscopy (SEM), and *differential scanning calorimetry (DSC)*. The membrane performance was further tested using a MGA system with water and diethanolamine (DEA) as absorbents. The membrane sample with a diameter of 19.5 cm was placed in the membrane permeation cell. Feed gas at 200 ml/min contains 60 % of CO_2 in H_2 gas. The system was operated for 15 min to achieve a steady state condition prior to sample measurement. The concentration of CO_2 in the outlet water from the absorbent was measured using chemical titration method to determine CO_2 flux. Using DEA as absorbent, the retentate was analysed using a gas chromatography (*Agilent 7890AGC)*.

The PVDF/SAPO-34 membrane pore size reduced proportional with the SAPO-34 zeolite loading. High SAPO-34 zeolite loading at 8 wt% formed denser membrane structure and might introduce mass transfer resistance. The water contact angle is representing surface hydrophobicity. The water contact angle of PVDF/SAPO-34 increased rapidly than the neat PVDF membrane. Improvement on membrane hydrophobicity was due to fluorination effect by the FAS on SAPO-34 zeolite. PVDF/SAPO-34 membrane with 8 wt% of zeolite showed the highest CO_2 absorption using water and DEA. Although the membrane structure was much compact, but the high zeolite loading enhanced the CO_2 absorption via the MMMs after post-treatment.

References

1. Zhang Y, Wang R: Novel method for incorporating hydrophobic silica nanoparticles on polyetherimide hollow fiber membranes for CO2 absorption in a gas–liquid membrane contactor. *Journal of Membrane Science* 2014, 452:379-389.
2. Junaidi MUM, Khoo CP, Leo CP, Ahmad AL: The effects of solvents on the modification of SAPO-34 zeolite using 3-aminopropyl trimethoxy silane for the preparation of asymmetric polysulfone mixed matrix membrane in the application of CO2 separation. *Microporous and Mesoporous Materials* 2014, 192(0):52-59.
3. Junaidi MUM, Leo CP, Ahmad AL, Kamal SNM, Chew TL: Carbon dioxide separation using asymmetric polysulfone mixed matrix membranes incorporated with SAPO-34 zeolite. *Fuel Processing Technology* 2014, 118(0):125-132.
4. Ahmad J, Hägg M-B: Preparation and characterization of polyvinyl acetate/zeolite 4A mixed matrix membrane for gas separation. *Journal of Membrane Science* 2013, 427(0):73-84.
5. Junaidi MUM, Leo CP, Ahmad AL, Ahmad NA: Fluorocarbon functionalized SAPO-34 zeolite incorporated in asymmetric mixed matrix membranes for carbon dioxide separation in wet gases. *Microporous and Mesoporous Materials* 2015, 206:23-33.
6. Shen Y, Lua AC: Preparation and characterization of mixed matrix membranes based on PVDF and three inorganic fillers (fumed nonporous silica, zeolite 4A and mesoporous MCM-41) for gas separation. *Chemical Engineering Journal* 2012, 192(0):201-210.
7. Ahmad AL, Ramli WKW: Hydrophobic PVDF membrane via two-stage soft coagulation bath system for Membrane Gas Absorption of CO2. *Separation and Purification Technology* 2013, 103(0):230-240.

Influence of operational conditions and membrane type on the performance of degassing contactors for CH_4 recovery from anaerobic bioreactor effluents

M. Henares, [1]M. Izquierdo, C. Gabaldón and V. Martínez-Soria
Chemical Engineering Dep., Univ. Valencia. Research Group on Environmental Engineering (GI^2AM), Av. Universidad s/n, 46100 Burjassot (Spain). E-mail: maria.henares@uv.es

Anaerobic bioreactors are considered as high-strength treatments of industrial wastewaters, with the main advantage of production methane, which can be used as biofuel. Low wastewater temperature processes involve a significant quantity of residual dissolved methane (D-CH_4) present in the water effluent of the reactor, since the methane solubility raises with the decrease of temperature. The recovery of the residual D-CH_4 in the anaerobic effluent is essential, since D-CH_4 represents a loss of potential energy source and also can difficult the reuse of water in a close loop due to the possible oversaturation of methane. In addition, D-CH_4 causes a negative impact to the environment due to fugitive emissions of CH_4 to the atmosphere. Despite Degassing Membrane (DM) contactors, specially hollow fiber modules, are commonly used to remove dissolved gases such as CO_2 and O_2in several processes, information and/or studies about the recovery of D-CH_4 are very scarce(Bandara et al. 2011; Cookney et al. 2012).

Two types of hollow fibers membrane contactors were employed in this study. The first module is the PDMSXA-250, a dense membrane contactor with fibers of polydimethylsiloxane (PDMS) with an internal surface area (A_i) of 159 cm^2, supplied by PermSelect®, MedArray Inc. (USA). The second module is the 1x5.5 MiniModule® supplied by Liqui-Cel, Membrana GmbH (Germany). This contactor was a microporous membrane module made of polypropylene (PP) fibers with an A_i of 1800 cm^2.

DM modules were coupled to a laboratory scale Expanded Granular Sludge Bed (EGSB) anaerobic reactor, which treated a synthetic wastewater polluted with ethanol at 25°C. Contactors were feed with the flow from the recirculation stream, in which the concentration of the D-CH_4 was similar to the effluent stream. Experiments with both module configurations, A) liquid in lumen-side (vacuum in shell-side) and B) liquid in shell-side and vacuum in lumen-side were carried out. The filtered (40 μm) anaerobic effluent was introduced into the lumen or shell side using a peristaltic pump, resulting in fluxes (Q_L/A_i) from 6 to 190mL $s^{-1}m^{-2}$.Two operational modes were also studied: sweep gas mode with N_2 and vacuum mode. Sweep gas flow rate was fixed in the range allowed by the manufacturer specifications. The vacuum pressure (P_{vac}) was set from 140 to 800 mbar. The measurement of D-CH_4 was carried out with a headspace method(Henares et al. 2014).

As was expected, the performance in methane removal of both DM modules decreased with flux and increased with vacuum pressure. The maximum RE reached for lumen side configuration experiments was of 80% and 93% for PDMS and PP contactors, respectively, obtained at the lowest flow-rate (6 mL s^{-1} m^{-2}) and the highest vacuum pressure (800 mbar). Nevertheless, the influence of flux was higher in experiments with PP modules, and RE decreased more sharply with the flux with this contactor. This phenomenon can be observed clearly in Figure 1, where the performance of both degassing modules are compared at experiments carried out at 140 mbar shell side vacuum pressure. As can be seen in this figure for fluxes lower than 25mL s^{-1} m^{-2} PP module showed higher RE, but for higher values of flux PDMS results more convenient to remove methane. Similar results were obtained at different vacuum pressures. These results can be explained taking into account the potential wetting of hydrophobic membranes, such as PP. In Figure 2 results obtained for PDMS module at both configurations are shown at a vacuum pressure of 800 mbar. As can be seen best performance was obtained when the liquid is flowing in lumen-side. Similar results were obtained for PP module. Although these results are in agreement with the manufacturer's suggestions, the usual configuration of these devices for industrial operations is the shell-side mode, as this configuration facilitates maintenance and cleaning of the membrane.

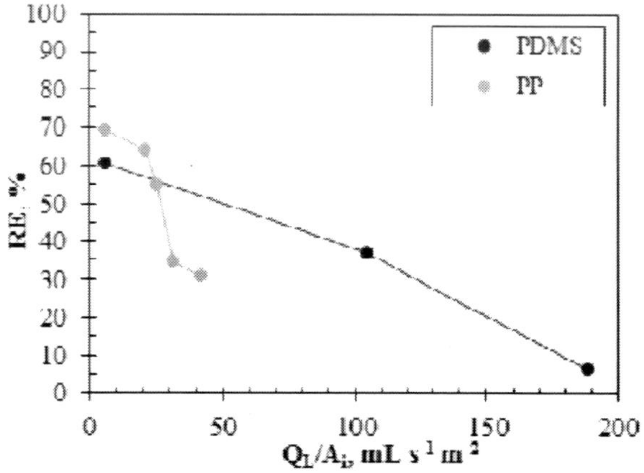

Figure 1. Evaluation of RE in PDMS and PP modules at 140 mbar of shell side vacuum pressure

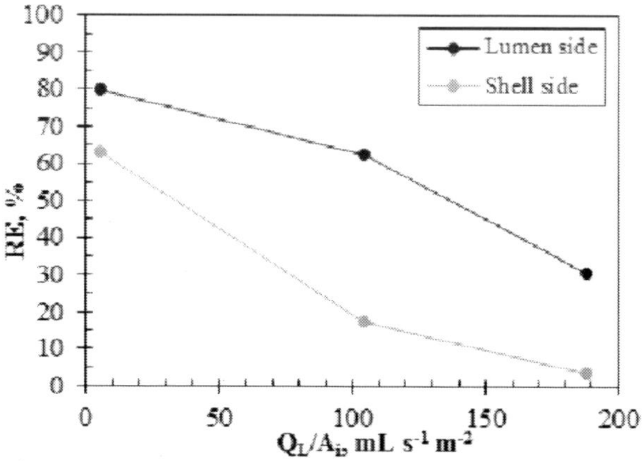

Figure 2. Evaluation of RE in PDMS module at lumen-side (A) and shell-side (B) configurations. Vacuum pressure 800 mbar.

Experiments using N_2 as sweep gas resulted in higher efficiencies than in the experiments with vacuum pressure. Nevertheless, vacuum pressure operation uses to be preferred for the recovery of $CH_4$and its usage as a fuel due to the difficulty of recovery of methane from sweep gas. Experimental mass transfer coefficients were calculated and the results corroborate the occurrence of the wetting phenomenon. Satisfactory results were achieved for the evaluation and modelling of mass transfer coefficients with the experimental data.

This study shows degassing membrane units as a promising technology to recover the D-CH_4 from anaerobic effluents.

Acknowledgements: Financial support from Ministerio de Economía y Competitividad (Spain, Project CTM2014-54517-R co-financed with FEDER funds) is acknowledged. Financial support from Generalitat Valenciana (Spain, Prometeo project 2013/53) is also acknowledged. M. Henares acknowledges Generalitat Valenciana for financing support of her PhD.

References:

Bandara, W.M.K.R.T.W. et al., 2011. Removal of residual dissolved methane gas in an upflow anaerobic sludge blanket reactor treating low-strength wastewater at low temperature with degassing membrane. *Water research*, 45(11), pp.3533–40.

Cookney, J. et al., 2012. Recovery of methane from anaerobic process effluent using poly-di-methyl-siloxane membrane contactors. *Water science and technology*, 65(4), pp.604–10.

Henares, M. et al., 2014. Methane recovery from anaerobic bioreactor effluents by degassing membrane technology. In Proceedings of *13th Mediterranean Congress of Chemical Engineering*. Barcelona (Spain), p. 262.

Investigation of effects and interactions of reaction mixture composition ratios on the characteristics of NaA zeolite nanoparticles

Mojtaba Mirfendereski, [1]Toraj Mohammadi
Faculty of Mechanical Engineering and Energy, Shahid Beheshti University, Tehran, Iran; tel. +98 912 203 7345, email: mirfendereski@alumni.iust.ac.ir
[1]Research Centre for Membrane Separation processes, Faculty of Chemical Engineering, Iran University of Science and Technology, Narmak, Tehran, Iran; tel. +98 21 77240051, email: torajmohammadi@iust.ac.ir

Summary

The objectives of present study were to investigate the effects of synthesis parameters (a=SiO_2/Al_2O_3, b=Na_2O/Al_2O_3, c=H_2O/Al_2O_3 ratios, crystallization temperature (T) and time (t)) on formation of zeolite-NaA nanocrystals (crystallinity and crystal size). The synthesized zeolite-NaA nanocrystals were characterized using XRD and SEM. The hydrothermal method was used to prepare zeolite-NaA nanocrystals. Composition ratios of the gel were Al_2O_3: (1.92-5.0) SiO_2: (3.16-50.0) Na_2O: (128-1000) H_2O.

Effects of synthesis parameters

SiO_2/Al_2O_3 ratio; the results showed that slightly increasing SiO_2/Al_2O_3 ratio reduces crystallinity significantly. Thus, it can be concluded that at these conditions (low alkalinity), zeolite-NaA crystals can be only produced if SiO_2/Al_2O_3 ratio is the lowest (1.92).

Na_2O/Al_2O_3 ratio; the results showed that increasing Na_2O/Al_2O_3 ratio terminates formation of zeolite-NaA crystals. Thus, zeolite-NaA crystals can be only produced with medium crystallinity if alkalinity is the lowest (0.003). Also, it can be said that increasing alkalinity reduces nucleation time and accelerates crystal formation. Thus, during experiments in which alkalinity increases within a constant period, crystallinity increases first, however the synthesized zeolite crystals are then dissolve again in the alkali mixture and crystallinity finally decreases.

H_2O/Al_2O_3 ratio; the results showed that increasing H_2O content (H_2O/Al_2O_3 ratio) has the weakest effect on crystallinity and crystal size of zeolite-NaA. Which decreased crystallinity and crystal size of zeolite-NaA.

Synthesis temperature; increasing synthesis temperature from 80 to 100 °C significantly increases crystallinity of zeolite-NaA. A similar trend can be observed for crystal size that increases with increasing synthesis temperature. Thus, larger zeolite-NaA crystals can be produced with higher crystallinity at higher crystallization temperature.

Synthesis time; According to the results, increasing synthesis time terminates formation of zeolite-NaA crystals. Crystallinity and mean crystal size of zeolite-NaA decreases with increasing synthesis time from 3 to 5 h. As known, zeolites are thermodynamically metastable phases. In general, Ostwald's law of successive reactions is followed in zeolite synthesis, i.e., a metastable phase appears first and then successively more stable phases are replaced with each other.

Interactions of SiO_2/Al_2O_3 and Na_2O/Al_2O_3 ratios; as mentioned, increasing SiO_2/Al_2O_3 and Na_2O/Al_2O_3 ratios reduce crystallinity and crystal size of zeolite-NaA significantly. However, increasing both the ratios simultaneously neutralizes their individual effects and caused no significant effect. Thus, it can be concluded that increasing Si and Na contents in zeolite-NaA synthesis gel have a reciprocal effects making these two parameters have approximately the same minor effects on crystallinity and crystal size of zeolite-NaA.

Interactions of SiO_2/Al_2O_3 and H_2O/Al_2O_3 ratios; Increasing both the ratios simultaneously reduces crystallinity and crystal size of zeolite-NaA significantly. Thus, it can be concluded that the effect of SiO_2/Al_2O_3 ratio is more important than that of H_2O/Al_2O_3 and crystallization of zeolite-NaA cannot be performed when there is a high-silica content and alkalinity is weak ($Na_2O/Al_2O_3 = 3.16$).

Interactions of Na_2O/Al_2O_3 and H2O/Al_2O_3 ratios; increasing both the ratios simultaneously reduces crystallinity and crystal size of zeolite-NaA significantly. Thus, it can be concluded that crystallization of zeolite-NaA in a mixture with strong alkalinity (0.05) is not performed. While, at weak alkalinity (0.024), crystallinity and crystal size of zeolite-NaA increase significantly.

Interactions of synthesis temperature and time; increasing both temperature and time simultaneously has no significant effect. In other words, increasing the two parameters simultaneously neutralizes the individual effects of each other. Thus, it can be concluded that increasing synthesis temperature increases nucleation rate and thus crystal size of zeolite-NaA. On the other hand, increasing synthesis time causes synthesized crystals to be dissolved in alkaline solution and this decreases crystallinity and crystal size. Thus, these two effects are canceled with each other and therefore crystallinity of zeolite-NaA does not considerably change.

Conclusion

The results indicated that the zeolite crystals synthesized with composition ratios of $1.92SiO_2$: Al_2O_3: $3.16Na_2O$: $128H_2O$ at 100 °C for 3 h had the highest degree of crystallinity and the lowest crystal size. In the next step, the effects of five synthesis parameters of SiO_2/Al_2O_3, Na_2O/Al_2O_3, H_2O/Al_2O_3, synthesis temperature and synthesis time on crystallinity of zeolite-NaA were evaluated. The results indicated that increasing each parameters of SiO_2/Al_2O_3, Na_2O/Al_2O_3, H_2O/Al_2O_3 and synthesis time alone while the other parameters remained at constant levels, reduced the crystallinity and crystal size of zeolite-NaA. For synthesis temperature the opposite effect was observed which both the crystallinity and crystal size increased with increasing temperature.

The evaluation of interactions between these parameters indicated completely different results. For example increasing both SiO_2/Al_2O_3 and Na_2O/Al_2O_3 simultaneously neutralized the above explained effects for each one singly and approximately kept the crystallinity constant. In other words, the high silica zeolite-NaA could be synthesized only in case of increasing alkalinity of reaction mixture. In other case, the effects of increasing SiO_2/Al_2O_3 and H_2O/Al_2O_3 simultaneously reinforced above explained effects for each other and significantly affected crystallinity and crystal size. It was found that alkalinity is the most important parameter in zeolite-NaA synthesis so that increasing alkalinity accelerates the crystallization rate and as a result shortened the required synthesis time. In cases which the synthesis time remain at initial level, the synthesized crystal grains are dissolved again in the alkali mixture and reduce crystallinity again.

The investigation of effects of Na/Si ratio on zeolite-NaA synthesis indicated that in samples with high Na/Si ratio, both crystallinity and crystal size increased even at low alkalinity. It could be explained that increasing Na^+ cations content in the reaction mixture, compensate the effects of decreasing alkalinity and keep the crystallinity constant.

As a final conclusion, it can be said that the five considered synthesis parameters have different effects on zeolite-NaA synthesis. The signification of these parameters in order of influences on crystallinity and crystal size of zeolite-NaA can be arranged as follows:

- $SiO_2/Al_2O_3 \approx$ Alkalinity > H_2O/Al_2O_3
- Synthesis temperature > Synthesis time

A standardized experimental approach for organic solvent nanofiltration by ceramic membranes

K. Lechner, R. Haseneder, J.-U. Repke
TU Bergakademie Freiberg, Institute of Thermal, Environmental and Natural Products Process Engineering, Leipziger Str. 28, 09599 Freiberg; tel. +49 3731 392377, e-mail: kai.lechner@tun.tu-freiberg.de

Presently, liquid mixtures of organic solvents are separated by applying proven techniques such as extraction or distillation. The latter is by far the most widely employed unit operation for solvent regeneration. Separation is achieved by evaporating more volatile components, which results in a comparatively high energy requirement in order to attain the separation objectives. Hence, organic solvent nanofiltration (OSN) with ceramic or polymeric membranes is considered a promising, energy and cost saving alternative since separation takes place at molecular scale, e.g. by size exclusion and electrostatic interactions without the demand for heat transfer.

However, due to the absence of entire thermodynamic knowledge and process understanding and the missing of appropriate predictive tools to select nanofiltration membranes, extensive experimental screening is still required to assess the feasibility of a process within the scope of process synthesis or retrofitting.

Therefore, the German Federal Ministry of Economic Affairs and Energy funds the research project ESIMEM (03ET1279G) to reveal and investigate mass transfer principles within polymeric and ceramic membranes with the objective to develop heuristics and short-cut methods to establish nanofiltration as a widely accepted unit operation in chemical engineering.

In this study, permeation of polar and apolar organic solvents like toluene and ethanol through hydrophobic inorganic membranes, which consist of ZrO_2 and TiO_2 with pore sizes of 0.9 and 3.0 nm respectively, is systematically investigated. Therefore, a standardized method for membrane testing has been established to ensure reliable and comparable results for both polymeric and ceramic membranes under the same experimental conditions. In addition, retention measurements have been carried out for the systems polystyrene-toluene and polystyrene-ethanol. The first results obtained from experimental work will be presented and discussed with regard to the research program's objectives.

Further research concentrates on expanding the experimental range to less polar, protic and aprotic solvents like methyl ethyl ketone, ethyl acetate and n-heptane. Additionally, retention measurements will be conducted using solutes of different molecular size, polarity and structure such as sucrose octaacetate and poly-(2-vinylpyridine). Thus, a broad database of experimental results is obtained allowing a first categorization of membrane types suitable for a certain application with respect to the given solvent-solute system.

Furthermore, solvent permeation will be described using the pore-flow model as first modelling approach.

Air filtration using polymeric hollow-fibre membranes

P. Bulejko, T. Svěrák, [1]M. Dohnal, P. Kejík, O. Krištof, K. Mayerová
Heat Transfer and Fluid Flow Laboratory, Faculty of Mechanical Engineering, Brno University of Technology, Technická 2896/2, 616 69, Brno, Czech Republic; tel. +420 541 144 912, email: bulejko@lptap.fme.vutbr.cz; [1]Department of Economics, Faculty of Business and Management, Brno University of Technology, Kolejní 2906/4, Královo Pole, 612 00, Brno, Czech Republic

Air filters perform an important function in commercial, institutional as well as industrial facilities. As indoor air contains two to five times higher concentrations of pollutants than outdoor, air filtration systems are needed. Their main function is to remove respirable particles such as microorganisms, dust and allergens from the air, which is important due to associated health concerns. It has been found that there is a direct relationship between increased concentrations of airborne particles and human health disorders.

The use of membranes for air filtration has significantly increased over the past decade. Their unique performance, chemical, surface and physical properties are preferred in a number of air filtration applications. The effort of this work is to introduce hollow fibre membranes to air filtration which, to the best of authors' knowledge, have not been used, so far. This type of membranes can be used for microfiltration of air and to provide high efficiencies in particulate matter removal down to 0.3 μm sizes which is comparable with high efficiency particulate air (HEPA) filters. However, next issues must be solved. What are the pressure losses and thus energy demands for operation of these membranes? Is it advantageous in terms of the acquisition cost? And what about the operation costs? Together with particle capturing capability, is it able to accomplish the requirements for air filtration? To find out the answers to these questions is the main aim of this work.

Ammonia removal from reject water by membrane contactor

D. Podstawczyk, A. Witek-Krowiak
Wrocław University of Technology., Dep. Chemistry, Wybrzeże Wyspiańskiego 27, 50-370 Wrocław, Poland;
tel. +48713203437, e-mail: daria.podstawczyk@pwr.edu.pl; anna.witek@pwr.edu.pl

One of the major environmental problems in conventional municipal wastewater treatment plant is reject waters generated during digested sewage sludge dewatering on press. The reject waters treatment is very expensive and often difficult due to significant fluctuations in the composition over time as well as high concentrations of specific biogenic substances such as phosphates and ammonia.

Wastewater from sludge dewatering process may contain more than 500 ppm ammonium nitrogen and other contaminants such as phosphates and chemical oxygen demand (COD). Treatment of reject water is very expensive and often difficult to implement due to considerable time fluctuations in the composition of wastewater and high concentrations of phosphates and ammonia. Although the flow of effluents from the press is low (below 5% of the total influent wastewater stream), it may increase the load on the plant resulting in occasional overloading. Consequently, it affects the efficiency of wastewater treatment processes and results in an increase of costs of nitrification/denitrification systems (Ocansey, 2005).

The aim of the study was to investigate the feasibility of a membrane contactor system for ammonia removal from actual reject water. The membrane contactor system consisting of low pressure gas-liquid mode was evaluated in terms of ammonia removal efficiency at different operating parameters (pH, feed temperature, permeate cooling temperature). Equipment photo is shown in Fig. 1.

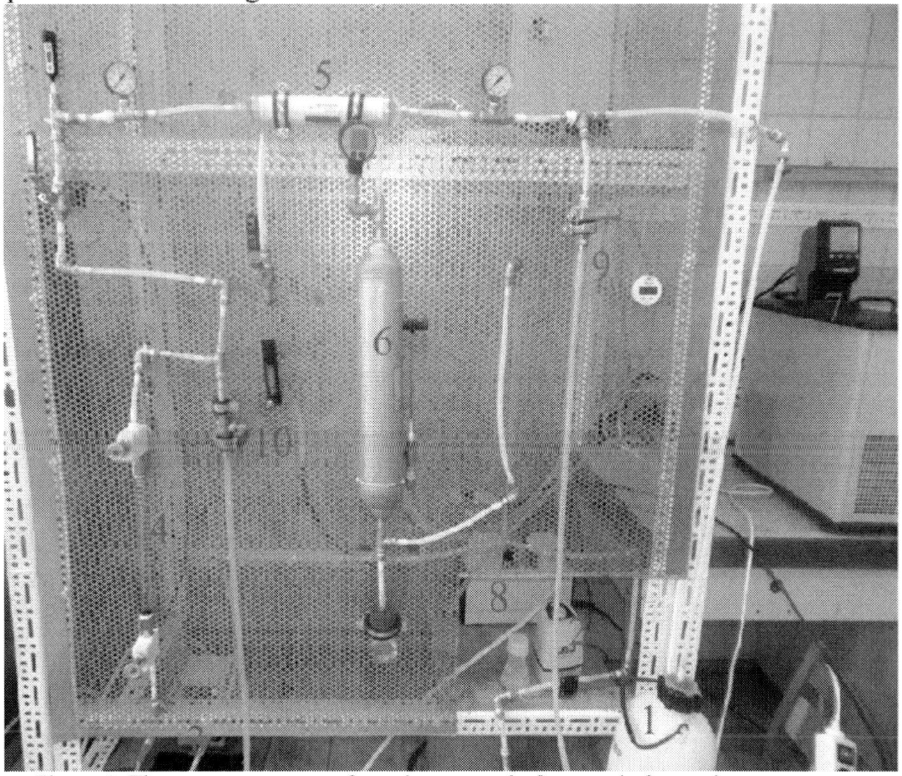

Figure 1. The apparatus system form the removal of ammonia from reject water.

The wastewater was pumped from a feed tank (**1**) by means of a diaphragm pump (**2**). The feed was heated to 60°C in coils placed in a thermostat (**3**). The flow rate was controlled by a rotameter (**4**). Liquid stream was introduced to the lumenside (inside) of the hollow fibre polypropylene membrane module Liqui-cel (1,7x8,75 MiniModule^{TM}) (**5**), while the permeate (concentrated ammonia solution) was cooled in a heat exchanger (**6**) outside of the fibre using a cryostat (**7**). The underpressure outside of the membrane was mantained at a value of approx. 0.8 bar by a vacuum pump (**8**). The reject water was circulated in the system for 5h, until the ammonia concentration was approx. 40 ppm. NH_3 concentration was measured by means of ammonia electrode detector on the basis of a calibration curve.

The highest removal efficiency (77%) was obtained for pH above 12, temperature of feed 60°C, cooling temperature -2°C and lumenside underpressure of 0.8 bar. The concentration of ammonia in dewatering effluent decrease from 668 ppm before removal process, to 40 ppm after the treatment. The pH decreased from 12.5 to 10.2, confirming that ammonia was successfully removed from reject water.

References

Ocansey, F.N. 2005. New trends in treatment of reject water from dewatering of sludge. Master's thesis number: 2005-01. Department of Water and Environmental Engineering, Lunds Institute of Technology University of Lund, Sweden.

Application of upper critical solution temperature systems embedded with carbon nanotubes for preparation of high flux ultrafiltration membranes

T. V. Plisko[1], I. V. Silaeva[1], A. V. Bildyukevich[1] , Y. A. Isaichykava[1], M. E. Dmitrenko[2], A.V. Penkova[2]
[1]Institute of Physical Organic Chemistry, National Academy of Sciences of Belarus, Surganov str., 13, 220072, Minsk, Belarus; tel. +375172922822, e-mail:plisko.v.tatiana@gmail.com;
[2] St. Petersburg State University, 7/9 Universitetskaya nab., St. Petersburg, 199034 Russia, tel.: +7(812)4284805

Polyphenylsulfone (PPSU) is an aromatic high performance amorphous thermoplastic polymer [1]. The excellent chemical and physical properties such as high stability under acidic conditions, resistance to organic solvents, high heat and impact resistance as well as durability, mechanical stability and relatively low cost make it a considerably attractive membrane material for solvent resistant nanofiltration, pervaporation, fuel cell membranes, membranes for hydrogen separation and ultrafiltration [1-5].

The nonsolvent induced phase separation (NIPS) method is widely applied for the preparation of anisotropic ultrafiltration membranes with dense selective skin-layer and porous support. However, studies on development of the PPSU membranes reveal that a very low pure water flux (PWF) was achieved for ultrafiltration membranes prepared via NIPS which can be attributed to limitations of the variation of casting solution composition due to low miscibility of PPSU with commonly used pore-forming additives (polyethylene glycol (PEG), polyvinypyrrolidone (PVP)) in aprotic amide solvents ((N-methy-2-pyrrolidinone (NMP), N,N-dimethylformamide etc.).

We have found earlier that polymer systems PPSU-PEG-PVP-NMP have upper (T=38°C) and lower (LCST) (T=130°C) critical solution temperatures. Cloud points were determined by visual observation of the turbidity change of polymer solution. At the T<38°C the PPSU solutions are turbid thermotropic gels. It was found that one-phase homogeneous solution exists at the temperature region 38-130°C, at T≥130°C phase separation occurs and two phases are formed. To overcome the processing restrictions of these systems an effective approach to preparation of high flux ultrafiltration membranes via upper critical solution temperature thermally induced phase separation method (UCST-TIPS) was developed. This method involves application of the dope solution at the temperature region where homogeneous polymer solution exists and using the coagulation bath at the T<UCST. Our studies revealed that using PPSU-PEG-PVP-NMP systems and UCST-TIPS method allows enhancing PWF of flat sheet membranes from 5-20 $l \cdot m^{-2} \cdot h^{-1}$ (NIPS) [5] to 200 $l \cdot m^{-2} \cdot h^{-1}$.

This study focuses on the modification of PPSU flat–sheet membranes for ultrafiltration, prepared via UCST-TIPS, using multiwalled carbon nanotubes (MWCNT). MWCNT were carboxylated using the mixture of sulfuric and nitric acids (1:1) by standart technique. The effect of MWCNT loading to the PPSU-PEG-PVP-NMP systems on the phase state, viscosity and thermodynamic parameters of viscous flow was studied. The temperature region of component miscibility was shown to narrow when the MWCNT concentration exceeds 0.06 mass.%. When the solutions loaded with MWCNT undergo phase separation at T≥130°C MWCNT were found to precipitate in the most viscous phase with the highest PPSU concentration which is the evidence of the high dispersion degree of the MWCNT in the polymer solution due to PPSU-MWCNT contacts formation. The viscosity of casting solutions was found to rise slightly when the MWCNT concentration increases.

Flat-sheet ultrafiltration composite PPSU/MWCNT membranes were prepared via UCST-TIPS and the impact of MWCNT loading on membrane structure, flux, human serum albumin (HSA) rejection, and hydrophilic/hydrophobic properties of membrane selective layer was investigated. PWF was shown to pass through the maximum when the MWCNT concentration increases, herewith the maximum PWF (480 $l \cdot m^{-2} \cdot h^{-1}$) is achieved at the MWCNT concentration 0.08 mass. % in the cassting solution (1 on figure). The formation of more permeable structure at 0.06-0.08 mass.% of MWCNT is attributed to the increase of demixing rate of the casting solution upon the contact with coagulant (water) during the phase inversion process due to the lower thermodynamic stability of these casting solutions compared to the solutions with lower MWCNT concentration. When the MWCNT concentration increases up to 0.15% the viscosity of the casting solution rises which results in decreasing the precipitation rate and less permeable membrane structure formation.
HSA feed solution was prepared in the phosphate buffer at pH=7,2 and HSA concentration 0.05 mass.%. HSA feed solution flux was shown to increase upon MWCNT introduction to the casting solution and pass through the maximum at the MWCNT concentration 0.08 mass.% (2 on figure). It was shown that HSA rejection increases from 85% to 87-91% when MWCNT are introducted to PPSU casting solution.

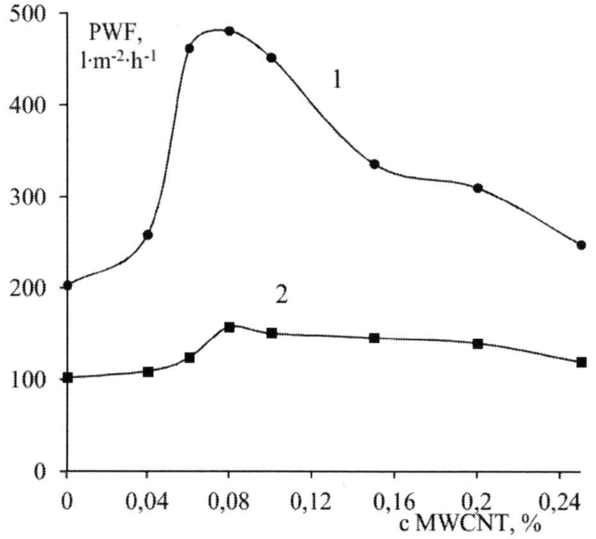

Figure. Pure water flux (1) and human serum albumin solution flux (2) of the composite PPSU/MWCNT membranes vs the MWCNT concentration in the casting solution.

Modification of PPSU membranes using MWCNT allows enhancing PWF without sacrificing protein rejection. Novel composite PPSU/MWCNT membranes prepared via UCST-TIPS can be utilized for water treatment, separation in biotechnology as well as a support for composite membrane preparation for solvent resistant nanofiltration and gas separation applications.

1. N. Jullok, S. Darvishmanesh, P. Luis // Chemical Engineering Journal, 175:306– 315, 2011.
2. Darvishmanesh S. et al Journal of Membrane Science, 379:60–68, 2011.
3. B. Decker et al // Chemistry of Materials, 22:942–948, 2009
4. T.H. Weng et al // International Journal of Hydrogen Energy, 33:4178–4182, 2008.
5. Moideen I. K. et al // Desalination and Water Treatment, 2015, doi:10.1080/19443994.2015.1106985

This work was carried out with a financial support from Belarussian Republican Foundation for Fundamental Research (project №X15PM-061).

Cascade filtration system to recover heavy metals from industrial wastewater

A.C. Macarie, I. Balcu, N. Mirica
National Institute for Research and Development in Electrochemistry and Condensed Matter, 144 Dr. A. Paunescu Podeanu, 300569, Timisoara, Romania, Tel.: +40 256 222119,

Industrial waste water treatment and processing is a problem of continuous interest.The development of the recover technology for heavy-metals and the use of reverse osmosis at the end of the waterfall process is leading to a mix of synergy between characteristics related to functionality,usability,durability and cost.In addition,the proposed solutions show a high degree of novelty and complexity,providing the basis for developing original technologies,with remarkable results in terms of wastewater treatment and water recycling solutions for various industrial processes using Zn,Cr,Cu and As.Optimizing of a complex cascade filter allows the recovery of a variety of heavy metals from industrial wastewater,and their final treatment and disposal of pathogenic elements through reverse osmosis process.

Clarification of the phosphoric acid by used organic membranes

Khaoula Khaless[a], Said AMAL,[a]
[a] Laboratoire "Acide, Procédés et Technologies de Séparation", Direction Recherche & Développement, Groupe OCP, El Jadida, Morocco

Various treatments have been used to clarify phosphoric acid: precipitation, flotation and other. The precipitation method is considered the most economical. However, this technique produces a large amount of precipitate that requires further treatment. No studies have been conducted for the regenerated membranes. So far, There is only one study reported for nanofiltration, she is currently employed as a downstream process (finishing process), by cons, no study has been conducted for the regenerated membranes.
The regeneration of membranes was performed on a set of 12 samples of 0.15 m^2 active area. These samples were used to study the 29% P_2O_5 phosphoric acid clarification laboratory scale. They were extracted from membrane sheets from 2 modules 8" and 2, 5", each batch was immersed in an oxidizer for a period not exceeding seven days. The samples have been characterized systematically before the immersion in the oxidant. In this study, we were able to regenerate used membranes and clarify the 29% P_2O_5 phosphoric acid by these ones.
We show, by tests, that the NF and RO regenerated membranes allow to achieve 70% for solids matter and 65% for organic matter abatement. These encouraging results will pave the way for implementing these membranes in the treatment of phosphoric acid. In addition to economic gain, the use of the used membrane allow to respect the environment too (no waste, no organic solvent and effluent to be regenerated).

KEYWORDS
Used Membranes, Recycling, Phosphoric Acid, Clarification

Dehydration of acetic acid by pervaporation using PVA-fullerenol mixed matrix membranes

[1]M. Dmitrenko, [1]A. Penkova, [2]D. Roizard
[1]St. Petersburg State University, 7/9 Universitetskaya nab., 199034 St. Petersburg, Russia, e-mail: m.dmitrienko@spbu.ru, a.penkova@chem.spbu.ru;
[2]Laboratoire Réactions et Génie des Procédés, CNRS, Université de Lorraine, ENSIC, 1 rue Grandville, 54000 Nancy, France, e-mail: denis.roizard@univ-lorraine.fr.

Membrane processes are developing very quickly and the materials with improved transport characteristics are needed. One of the perspective membrane technologies is pervaporation which can be used for various industrial separations. The potential of this process is evaluated for majority applications, especially for the dehydration of different mixture compositions. Among the variety of membrane materials, the polymers take an important place as they have good mechanical characteristics, film forming properties and selectivity. However, the majority of known polymers have low water permeability including the commonly used polymer polyvinyl alcohol (PVA). One of the most promising ways to improve the permeability and selectivity of membranes is the preparation of mixed matrix structures to combine advantages of a priori non compatible structures, using for instance carbon nanoparticles. In the present work a low-hydroxylated fullerene has been used as a modifier and a cross-linking agent for polyvinyl alcohol.
The aim was to study the effect of fullerenol $C_{60}(OH)_{12}$ on transport and physico-chemical properties of polyvinyl alcohol (PVA). The conditions of preparation and crosslinking of membranes based on nanocomposite PVA – fullerenol were developed. Structural and physico-chemical characteristics of the membranes were studied by various physical and chemical methods of investigation (spectroscopic methods, differential scanning calorimetry, thermogravimetric analysis, scanning electron microscopy and sorption experiments).
The transport properties of membranes were investigated for separation of acetic acid - water mixtures at different temperatures by pervaporation. It has been shown that the addition of fullerenol particles to PVA matrix leads to the significant change and the improvement of membrane properties due to the changes in the structure and morphology.

This work was supported by Fellowship of President of Russian Federation СП-1153.2015.1 (Penkova A.V.) and the support of Région Lorraine and CNRS. The experimental work was facilitated by equipment from Resource Centers: for Nano technology, Research Centre for Physical Methods Surface Investigation, Thermal Analysis and Calorimetry, Chemical Analysis and Materials Research Centre and GEOMODEL at St. Petersburg State University.

Desalination of shallow brackish waters by electrodialysis and photovoltaic solar energy in isolated locations of Atacama Desert

J. R. Quispe,[1] F. Remonsellez,[1] P.G. Toledo
Universidad Católica del Norte, Dep. of Chemical Engineering, Av. Angamos 0610, Antofagasta, Chile; tel. +56 55 2355987, fax: +56 55 2355917, e-mail: jquispe@ucn.cl;[1] Universidad de Concepción, Dep. of Chemical Engineering, Casilla 160-C, Correo 3, Concepción, Chile, Fax: + 56 41 224 74 91

The desalination of brackish water is a mean of obtaining low cost drinking water. The method of desalination used in this work is based on an electrodialysis system powered by photovoltaic panels. The advantages of the use of the photovoltaic power are that is non-polluting; it is also silent; the fuel-sunlight- is free, abundant and a renewable resource, decentralized; and low maintenance cost. A system of these characteristics would be especially useful in isolated locations such as Quillagua (Antofagasta, Chile) with access to scarce and high salinity water resource like the Loa River where connection to the electric grid is not possible.

Antofagasta is located in the Atacama Desert (North of Chile) with the highest solar radiation in the world. Water desalination by solar energy can be achieved effectively by electrodialysis (ED) process powered by photovoltaic (PV) panels. In this paper, the experimental work involved the use of a lab scale ED system. The photovoltaic system consisted of 3 PV panels, 2 batteries and 1 inverter capable to generate 540p watts per hour, enough to run the ED system.

The aim of this paper is to demonstrate the feasibility of the desalination of brackish water from the Loa River (total dissolved solids = 8020 - 15100 mg/L) in Quillagua (Chile) by means of an electrodialysis system powered directly by photovoltaic solar panels in order to generate water for agricultural irrigation. This experimental work involved a small-scale commercial-type electrodialysis stack powered by photovoltaic cells. The stack consisted of 25 cell pairs. The feed water was fed from river water of high salinity. The experiments were done at product flow rates ranging from 10 to 50 L/h. Increasing the flow rate resulted in lower product quality given in terms of percent salt removal. The average feed water concentrations of chemical species that exceed the maximum levels permitted by the Chilean's standard 1333 for agricultural irrigation water are chlorides 2399 mg/L, sulphates 814 mg/L, arsenic 1,49 mg/L, boron 15 mg/L , lithium 8.33 mg/L, sodium 1733 mg/L, calcium 396 mg/L among others. The approximate energy consumption ranged from 2 to 3.5 kWh / m^3 of treated water depending on the required quality: drinking water or irrigation water. The percentage of pollutant removal was over 95%. This allowed even get drinking water but with more energy consumption. The experimental results (Figure 1) allowed to demonstrate the feasibility of using electrodialysis process for desalinating brackish water in a remote areas like Quillagua where there are no connection to the electric grid.

Figure 1. Electrodialysis pilot system for experimental tests developed in Quillagua, Antofagasta, Chile.

Desirable preparation of metal ion assist alginate membrane to develop bio-friendly separation technology for food and pharmaceutical industry

K. Kashima[1], M. Imai[2]
[1] Department of Materials Chemistry and Bioengineering, National Institute of Technology, Oyama College, 771, Nakakuki, Oyama, Tochigi-pref. 323-0806, Japan.
Phone: +81-285-20-2808
[2] Graduate School of Bioresource Sciences, Nihon University, 1866, Kameino, Fujisawa, Kanagawa-pref. 252-0880, Japan
* Corresponding author: keitakashima@oyama-ct.ac.jp

Desirable preparation of novel alginate membrane combining attractive molecular size recognition and sufficient mechanical strength were successfully achieved to select various metal ions as a cross-linker, Ca^{2+}, Sr^{2+} and Ba^{2+}. Those membranes were expected practical applications for food engineering fields such as selective separation of monosaccharides and oligosaccharides. The membranes have not only molecular size recognition in the range of those saccharides, especially under 1000Da but also biocompatible property.

Alginate is abundantly and sustainably produced by marine biological resources, especially brown seaweed. It has been widely applied in the food industry. In addition, alginate was continuously developed as useful materials for biomedical applications, especially for controlled delivery of drugs and other biologically active compounds and for the encapsulation of cells. The molecular chain of alginate is constructed of a block co-polymer of β-D-mannuronate and α-L-guluronate. Sodium alginate easily forms a cross-link in the presence of divalent cations such as Ca^{2+}, resulting in a highly compacted gel network. Homopolymeric blocks of α-L-guluronate (GG blocks) are constructed mainly of a cross-linked zone with divalent cations. GG blocks chelate alkaline earth-metal ions because of the spatial arrangement of the pyranose ring and the hydroxyl oxygen atoms. GG blocks are constructed mainly of a cross-linked zone with Ca^{2+}. This is called an "Egg-Box Junction", where the ions represent the "Eggs". Authors had previously reported preparation of stable calcium alginate membrane and molecular size recognition in nanofiltiration range ($60 < MW < 604$) [1-2]. Desirable preparation of alginate membrane was expected paying attention to kind of metal cations as a cross-linker.

A water permeation coefficient (L_P) [$m^3\ m^{-2}\ s^{-1}\ Pa^{-1}$] was evaluated based on the volumetric pure water permeation flux (J_V) [$m^3\ m^{-2}\ s^{-1}$] using membrane permeation apparatus driven N_2 pressure. The permeation flux was expressed conventionally according to Kedem-Katchalsky equation.

$$J_v = L_p(\Delta P - \sigma \Delta \pi) \quad (1)$$

Where ΔP is operational pressure [Pa], σ is reflection coefficient [-], $\Delta \pi$ is osmotic pressure [Pa]. In pure water permeation, osmotic pressure is ignored between feed and stripping phase ($\Delta\pi$=0). Eq. (1) can be rearranged as followed:

$$J_v = L_p \Delta P \quad (2)$$

The water permeation coefficient is this work was calculated using Eq. (2). The water permeation

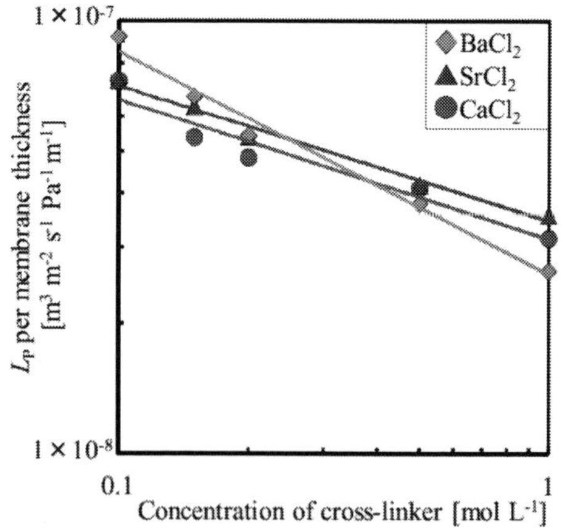

Fig. 1 Effect of concentration of cross-linker on permeability coefficient (L_P)

coefficient decreased logarithmically with increasing concentration of cross-linker. This suggests that the number of permeation channels and/or the size of the permeation channels decreased with increasing concentration of cross-linking ions. The effect of cross-linker concentration on water permeability appeared strongly in the $BaCl_2$ used, suggesting that Ba^{2+} have a much stronger interaction in α-L-guluronate block. Binding strength of metal ions to alginate molecular chains is differed depending on kind of cross-linked metal ions [3]. Hence, the alginate membrane cross-linked with $BaCl_2$ constructed a dense polymer network due to strong bonding between alginate molecular chain and Ba^{2+}.

Molecular size recognition was evaluated based on rejection rate of some molecular size indicator (glucose 180Da; Methyl Orange 327Da; Rose Bengal; 1017Da; bovine serum albumin, BSA 66000 Da). The rejection (R [-]) was determined from following formula Eq. (3).

$$R = 1 - (C_s / C_f) \quad (3)$$

Where C_s and C_f are the concentration of stripping solution [mol L^{-1}] and concentration of feed solution [mol L^{-1}], respectively. In higher concentration of cross-linker, the membranes performed high rejection even if glucose (180Da) regardless kind of cross-linker.

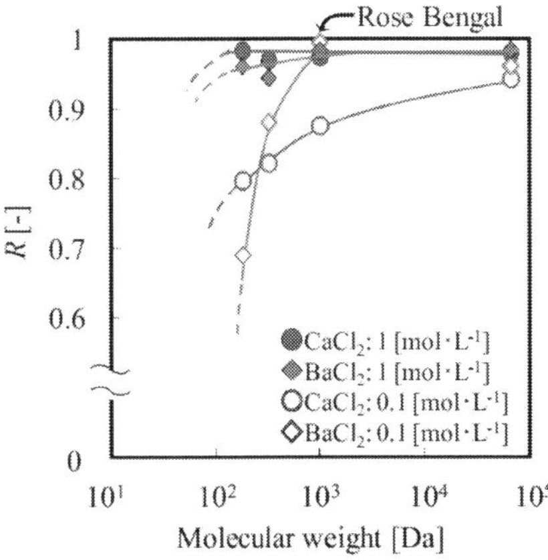

Fig. 2 Molecular size recognition of alginate membranes cross-linked with various divalent cations

In the membrane prepared with lower concentration of cross-linker, rejection was decreased with declining molecular weight of model size indicators. Especially, alginate membrane prepared 0.1mol L^{-1} $BaCl_2$ performed remarkable molecular size recognition ranging less than 1017Da. It indicates that mass transfer channels in the barium alginate membrane are mono-disperse at nano-meter range less than molecular size of Rose Bengal.

The volumetric water content of a swollen membrane can be assumed to an indicator of the void fraction of the membrane structure. The void fraction decreased with increasing concentration of the cross-linkers. The alginate membrane cross-linked with $BaCl_2$ was highly densified by the provided Ba^{2+}.

Desirable preparation of metal ion assist alginate membrane is promise technology to create rapid and bio-frendly separation engineering for forthcoming food and pharmacutical inovations.

Acknowledgment

The authors are sincerely grateful to "Grant-in-Aid for World-Class Technological Research and Development" from Tochigi Industrial Promotion Center in Japan for financial support.

References

[1] Kashima K., M. Imai, Impact factors to regulate mass transfer characteristics of stable alginate membrane performed superior sensitivity on various organic chemicals. Procedia Enginnering. 42 (2012) 964-977.

[2] Kashima K., M. Imai, I. Suzuki, Superior molecular size screening and mass-transfer characterization of calcium alginate membrane. Desalination and Water Treatment. 17 (2010) 143-149.

[3] Donati I., S. Paoletti, Material Properties of Alginates. Alginates: Biology and Applications. Springer (2009) 1-53.

Determination of the sorption of an azeotropic mixture of organic compounds in PDMS

Ondřej Vopička, Daniel Radotínský and Karel Friess
University of Chemistry and Technology Prague, Department of Physical Chemistry, Technická 5, 166 28 Prague 6
daniel.radotinsky@vscht.cz

A knowledge of sorption properties of mixed organic compounds in polymers is useful for the prediction of separation performance of membranes used for permeation or pervaporation applications. In this work, we present an extended method based on earlier works of Clement et al. [1] and Rios et al. [2] for determination of sorption of liquid mixtures in polymers together with work of Schabel et al. [3] for determination of vapour mixture sorption in polymers.

The main goal of this work was to determine sorption of an azeotrope-forming mixture of methanol and dimethyl carbonate in PDMS [4]. For that purpose, we developed a new apparatus that consists of two independent measuring cells, each containing a polymer sample of known mass. Both cells were washed continuously gaseous mixture of studied vapours at atmospheric pressure and at 40 °C in an inert carrier gas. The first cell is optimized for weighing to determine the total weight of the sorption of organic vapours. The second cell was designed for analysis of vapours which have been adsorbed by polymer sample. The mixture of vapours after desorption and freezing using liquid nitrogen were analyzed using the FTIR spectrometer Nicolet iS10.

Taking into account the fact that the dimethyl carbonate was ranked among the so-called green solvents, and that it is produced from mixtures with methanol, obtained results can be interesting for applications in chemical industry.

Acknowledgement
Results were obtained with the financial support of GACR grants (no. 13-32829P, P106/10/1194. The work also was partly funded by the specific university research (Ministry of Education No.20 / 2015) and the internal UCT grant (project A2_FCHI_2015_013).

[1] Clément, R.; Kanaan, Ch.; Brulé, B.; Lenda, H.; Lochon, P.; Jonquières, A. An original automated desorption apparatus for measuring multi-component sorption properties of barrier polymer films. *J. Mem. Sci.* 2007, *302*, 95–101.
[2] Rios De Anda, A.; Fillot, L.; Preda, F.; Rossi, S.; Long, D.; Sotta, P. Sorption and plasticization effects of ethanol–toluene–isooctane ternary mixtures in polyamide 6,6 and induced plasticization effects. *Eur. Polym. J.* 2014, *55*, 199–209.
[3] Schabel, W.; Scharfer, P.; Kind, M.; Mamaliga, I. Sorption and diffusion measurements in ternary polymer–solvent–solvent systems by means of a magnetic suspension balance - Experimental methods and correlations with a modified Flory–Huggins and free-volume theory. *Chem. Eng. Sci.* 2007, *62*, 2254–2266.
[4] Vopička, O.; Radotínský, D.; Friess, K. Sorption of vapour mixtures of methanol and dimethyl carbonate in PDMS: Experimental study. *Eur. Polym. J.* 2015, *73*, 480–486.

Development of membranes for CO_2/N_2 separation: Comparison of homogeneous flat, asymmetric disk-type and composite hollow fiber morphologies

A. Fernández-Barquín[1], C. Casado-Coterillo[1], M. Etxeberría-Benavides[2], J.Zúñiga[2], A. Irabien[1]

[1]Universidad de Cantabria, Department of Chemical and Biomolecular Engineering, 39005 Santander, Cantabria, Spain; tel.+34942206777, e-mail: fbarquina@unican.es

[2] Tecnalia Research and Innovation, Energy and Environmental Division, 20009 Donostia-San Sebastián, Spain

The atmospheric concentration of greenhouse gases (GHGs) has increased significantly over the last years. The ability of membranes to separate efficiently CO_2 from other light gases, such as N_2, has received a great deal of attention due to its importance as alternative energy-efficient process reducing greenhouse gas emissions [1].

Polymer membranes are the most mature candidates for membrane-based post-combustion carbon capture. However, the lack of thermal and mechanical stability, as well as water resistance limit their use at large scale power production. Besides, polymer membranes usually present a trade-off between permeability and selectivity in a gas pair separation. This is the reason why we developed matrix membranes (MMM) from highly CO_2 permeable and thermally robust polymers and compatible inorganic and organic fillers to enhance the CO_2/N_2 permselectivity and mechanical resistance of the pure polymer materials [2, 3].

The separation of CO_2/N_2 requires of high productivity and performance membranes in order to treat the huge flow rate of flue gases. On one hand, composite membranes allow reducing the thickness of the selective layer by casting on a macroporous support that provides mechanical strength. On the other hand, hollow fiber configuration presents elevated membrane area/module volume ratio, good operating flexibility and handling in the manufacturing of modules. These factors affect the efficiency of the membrane in the separation process and facilitate the scaling up of the permeation unit [4].

In this work, we measured the CO_2/N_2 separation performance of asymmetric disk-type and composite hollow fiber membranes where the selective layer is made from those materials that overcome the existent material capacity in the CO_2/N_2 separation [2], in terms of single gas permeation tests under different temperatures to evaluate the influence of morphology and geometry on membrane behavior. Table 1 shows the results for one of these CO_2 selective membrane materials in different morphologies.

Table 1. CO_2/N_2 separation using a 5 wt.% HKUST-1/Ionic Liquid-Chitosan MMM as selective layer.

Membrane	Selective layer thickness [μm]	CO_2 permeance [GPU] (*)	$\alpha(CO_2/N_2)$
Homogenous flat	91.8	52	19
Asymmetric disk	48	105	4.79
Composite hollow fiber	0.34	7229	4.83

(*) 1 GPU = 10^{-6} $cm^3(STP)\ cm^{-2}\ cmHg^{-1}$

Financial support from MINECO under project CTQ2012-31229, Ramón y Cajal programme (RYC-2011-08550) and the Early Stage Researcher grant (BES2013-064266) at the University of Cantabria are gratefully acknowledged.

References:

[1] R. S. Haszeldine, Science, 325, (2009), 1647-1652

[2] C. Casado-Coterillo, A. Fernández-Barquín, B. Zornoza, C. Téllez, J. Coronas, A. Irabien, RSC. Adv. 5, (2015), 102350-102361.
[3] E. Santos, E. Rodríguez-Fernández, C. Casado-Coterillo, A. Irabien, Int. J. Chem. React. Eng. 13 (2015)
[4] M. Sandru, S.H. Haukebo, M.-B. Hägg, J. Membr. Sci., 346 (2010) 172-186.

Development of PDMS/[omim][Tf_2N]/PDMS membranes for separation of biobutanol by pervaporation

R. Cabezas[1], C. Maldonado[1], C. Araya[1], G. Merlet[1], A. Hasanoglu[2], J. Romero[1*]

[(1)] Laboratory of Membrane Separation Processes (LabProSeM) University of Santiago de Chile, Av. Lib. B. O'Higgins 3363. Estación Central. Santiago. Chile.

[(2)] Yıldız Technical University, Department of Chemical Engineering, Davutpaşa, 34210 Istanbul, Turkey.

E-mail: julio.romero@usach.cl

In this work, the separation of butanol as the main product of the ABE fermentation process was carried out by pervaporation using membranes based on a specific ionic liquid. These membranes were synthesized by means of the formation of a gelled [omim][Tf_2N] layer, which was placed between two polydimethylsiloxane (PDMS) films. These membranes were assessed in an experimental setup where the pervaporation process of model ABE fermentation mixtures was implemented, obtaining values of transmembrane flux and extraction efficiency at different operating conditions.

The performance of the ionic liquid-based membranes was compared to the obtained one for single PDMS layer membranes. From the experimental assays, improved butanol/water selectivity was verified for ionic liquid membranes when these were compared to the PDMS membranes.

Transmembrane fluxes of butanol, acetone, ethanol and water obtained with PDMS/[omim][Tf_2N]/PDMS were compared to the observed ones with single PDMS layer membranes. The main differences between both membranes is represented by the higher permeability of butanol (80% higher) and lower permeability of water (88% lower) exhibited by PDMS/[omim]/PDMS membrane. These values could be explained by the high solubility of butanol in the gelled ionic liquid layer, which shows a highly hydrophobic character. Furthermore, the selectivity of the ionic liquid-based membrane has been estimated from these experiments and its values are reported in **table 1**.

Table 1. Selectivity of pervaporation membranes tested in this work. Values calculated as solute/water selectivity.

Compound	Single PDMS layer membrane	PDMS/[omim][Tf_2N]/PDMS membrane
Butanol	4.6	62.3
Acetone	8.0	45.7
Ethanol	1.9	13.1

On the other hand, a mass transfer model has been proposed in order to describe the mass transfer of each species through the ionic liquid-based membrane. This model has been developed on a resistances-in-series approach, which describes individual mass transport steps through the aqueous feed boundary layer, the selective layer of ionic liquid and both PDMS layers that support the selective one. Classical phenomenological equations were solved using a script developed in Matlab® applying an iterative algorithm based on the Regula Falsi method in pseudo-steady state condition. The model was used to correlate the diffusion coefficient value in each membrane layer: both PDMS support films and the gelled [omim][Tf_2N] layer placed in the middle.

From the transfer model, it can be estimated that both PDMS support layers represent the main mass transfer resistances with a percentage close to 90% of the overall value; meanwhile, the gelled ionic liquid layer represents a lower percentage of this overall resistance. The percentage of the mass transfer resistance represented by the gelled [omim][Tf_2N] layer is negligible for butanol, but it represents close to 17% of the overall resistance for the ethanol transfer.

Effects of pressure oscillation on permeate flux of a micro-channel cross-flow ultrafiltration module

[1]S. Yoshikawa, [1]Y. Takahashi and [1,2]S. Ookawara
[1]Tokyo Institute of Technology, Dep. Chemical Engineering, 2-12-1-S4-2 O-okayama, Meguro-ku Tokyo, 152-8551, Japan
tel. +81-3-5734-3278, e-mail: syoshika@chemeng.titech.ac.jp; [2]Egypt-Japan University of Science and Technology, Alexandria, Egypt, New Borg El-Arab, Alexandria, Egypt

A cross flow ultrafiltration module with micro channels was designed by authors in previous studies [1], [2], [3]. The module consisted of feed liquid side and permeate liquid side channels. Membrane was put between two channels. The channel height, width were 0.1 mm, 1.5mm, respectively. The module with such small sizes of micro channels has the advantages of a high ratio of membrane surface area to feed liquid volume and high shear stress, which is expected to reduce the degree of membrane fouling. Taking the advantages into account, such kinds of ultrafiltration modules are expected to be used for separation and refining of macromolecules and as membrane reactors in micro-chemical processes. Authors have investigated the transport phenomena in the module. In previous studies [1], [2], a micro-channel filtration module described above was fabricated and the experiments of the cross-flow ultrafiltration of PolyVinylPyrrolidone (PVP) aqueous solution were carried out. PVP was taken up as a model solute of macromolecules such as enzymes. In the experiments, changes in permeate flux with time were measured under various conditions of transmembrane pressure, feed liquid concentration, feed flow rate and length of channels. Permeate flux reached a constant value after enough time passed. It appeared that the constant permeate flux increased with increase of feed flow rate and decreased with increase of concentration. In addition, the average flux over the whole module in a longer channel was smaller than that in a shorter one. It was considered that the effects of convective transport of feed liquid in the axial direction, concentration polarization and osmotic pressure would be significant. On the basis of the consideration, a model on the transport phenomena including the effects described above was proposed. Estimated results by means of the model agreed appropriately with the experimental results. The estimated results suggested that the permeate flux decreased in the downstream region of the channel because of the effects of concentration polarization and high osmotic pressure. This would be the reason why the fluxes decreased with increase of the length of the channel. In order to increase the flux, it would be necessary to reduce the concentration close to the surface of the membrane.

In the further study of authors [3], improved modules were proposed. It might be effective to put a mixing device e.g. a static mixer, in the micro-channel. But, pressure loss through the device would become large. Another idea is to reverse the direction of the permeation. In the previous module, the membrane was put on the bottom side of the feed liquid channel. As a result, the feed liquid close to the membrane was concentrated extremely in the downstream of the micro-channel. An improved module with a staggered arrangement of the membrane on the top and the bottom sides of the feed liquid channel was designed. The permeation direction changes at the axial position in the channel, where the side of the membrane changes from the bottom to the top, and vice versa. As results of the experiment, the permeate flux of the improved module was larger than that of the previous one. It appeared that the staggered arrangement was effective to cancel the concentration polarization.

Though the effect of the concentration polarization was reduced by means of the improved module described above, fabrication and operation would be very difficult because of the complicated design. In this study, instead of change in the module design, pressure oscillation was tried to be applied for cross-flow ultrafiltration process with an original micro-channel module, in which the membrane was put on the bottom side of the feed liquid channel. In the operation, pressure in the feed liquid channel was oscillated between 100kPa and 0kPa under various time intervals. Inverse fluxes were observed just after the pressure was reduced to 0kPa. After 100kPa was applied again, fluxes were recovered to larger values than smaller values caused by the osmotic pressure before pressure reduction. It was considered that the concentration polarization was canceled by the inverse fluxes. In addition, the relationship between the time averaged fluxes and time intervals of the oscillation was discussed. As a result, the possibility that the pressure oscillation would be effective on the recovery of the permeate flux was suggested.

References

[1] A. Nishimoto, S. Yoshikawa, Characteristics of Cross-Flow Ultrafiltration Module with Micro-Channels Based on a Model of Transport Phenomena, International Workshop on Process Intensification 2008, Tokyo (2008) PA28

[2] A. Nishimoto, S. Yoshikawa, S. Ookawara, A Model for Transport Phenomena in a Cross-Flow Ultrafiltration Module with Micro-channels, Membranes (2011) **1**:13-24

[3] S. Yoshikawa, A. Nishimoto, S. Ookawara: A New Design of a Micro-Channel Cross-Flow Ultrafiltration Module with a Staggered Arrangement of Membranes, International Membrane Science and Technology Conference 2013, Melbourne (2013) 25206

Gas transport properties of polyphenylene oxide membranes modified by polystyrene stars with fullerene C_{60} cores

[1]Z. Pientka, [2]A. Pulyalina, [3]L. Vinogradova, [2,3]G. Polotskaya
[1]Institute of Macromolecular Chemistry, Czech Academy of Sciences, Heyrovskeho nam. 2, 162 06 Praha 6, Czech Republic; tel. +420296809280 e-mail: pientka@imc.cas.cz; [2]Saint Petersburg State University, Institute of Chemistry, Universitetsky pr. 26, 198504 Saint Petersburg, Russia; tel. +78124284805, e-mail:alexandra.pulyalina@gmail.com; [3]Institute of Macromolecular Compounds, Russian Academy of Sciences, Bolshoy pr. 31, 199004, Saint Petersburg, Russia; tel. +78123235963, e-mail: polotskaya@hq.macro.ru

Membrane separation processes are an essential part of scientific and technical progress in many areas. Currently, the significant stage on the way of membrane processes application is the design of membranes with specific parameters as well as high permeability and selectivity. To improve the membrane transport properties, a novel approach based on physical modification of polymer matrix by including macromolecules of complex architecture such as polyhedral oligomeric silsesquioxane (POSS), different dendrimers and stars have been proposed.

In the present work star-shaped macromolecules with six arms of polystyrene grafted onto fullerene C_{60} core or fullerene-containing polystyrene (FPS) were used for modification of poly(2,6-dimethyl-1,4-phenylene oxide) (PPO). Preparation of PPO/FPS dense thin film membranes of ~40 μm thick was made by casting 3 wt% polymer solutions in chloroform on a cellophane surface. The solvent was removed by evaporation at 40 °C; the membrane was separated from the substrate and dried to the constant weight. Optimal concentration of FPS was determined experimentally and did not exceed 5 wt%.

Thermogravimetric analysis, study on physical properties and contact angles as surface tension measure were used for membrane characterization. Peculiarity of membrane structure was studied by atomic force microscopy (AFM) with tapping mode method. AFM shows that incorporation of FPS particles in PPO matrix leads to a significant modify of surface of polymer films and increase in the roughness of membranes. The gas permeability of five representative gases: N_2, O_2, H_2, CO_2, CH_4 was measured using a laboratory high vacuum apparatus with a constant-volume permeation cell. Gas transport properties of PPO/FPS membranes change in permeability and selectivity as compared with PPO membranes. Addition of FPS decrease the permeability, however, change selectivity. In case of CO_2/H_2 causes inversion. Thus, it is established that polystyrene stars with fullerene C_{60} cores profoundly affect on the morphology and transport properties of PPO/FPS mixed matrix membranes.

Acknowledgment. The authors acknowledge the Russian Foundation for Basic Research (project 15-03-02131). A.Yu.P. is also grateful for the support by Fellowship of the President of Russian Federation (project № 184, SP-1469.2015.1).

High pressure gas separation performance of PEI / HNT mixed matrix membranes: A pilot study

M. Z. Ahmad[1,*], V. Fila[1],
Tel: +420 777 322 292, Email: ahmadm@vscht.cz
[1] Department of Inorganic Technology, Faculty of Chemical Technology, UCT Prague, Technická 5, 166 28 Praha 6 – Dejvice, Czech Republic

Gas separation by polymeric membranes has been proven to be a feasible technology for the harsh operating conditions of an offshore natural gas pre-treatment system. Currently, more advanced mixed matrix membranes (MMMs) are being widely studied due to their superior performance. Polyetherimide (PEI) known to possess high mechanical stability under pressure.

We investigated gas separation properties of asymmetric hollow fiber PEI/modified halloysite nanotube MMM. *N*-*β*-(aminoethyl)-*γ*-aminopropyltrimethoxysilane (AEAPTMS) was used to modify the filler. The fabrication of hollow fiber membrane was first optimized for five fabrication factors: die extrusion rate, pump pack temperature, polymer concentration, quench bath and leach bath temperature while considering CO_2/CH_2 selectivity, CO_2 permeance and tensile strength as the primary responses. We used a pilot scale membrane testing unit which simulated high pressure processing conditions to study the performance stability and CO_2-induced plasticization of this MMM in the presence of aggressive CO_2 feeds, over time. FESEM, XRD, TGA and tensile strength test were conducted to characterize the mechanical and physical properties of the samples.

Our results indicate the hollow fiber was produced as intended at 0.3/0.6 (ID/OD) with a good filler dispersion. The CO_2/CH_4 separation selectivity improved by up to 110% compared to pristine hollow fiber PEI. The pilot study suggested this type of MMM is stable at pressure as high as 50 bar with an acceptable performance reduction over time, at different testing parameters. Overall, this findings suggest that the MMM to be a potential candidate for the use in an offshore natural gas pre-treatment system.

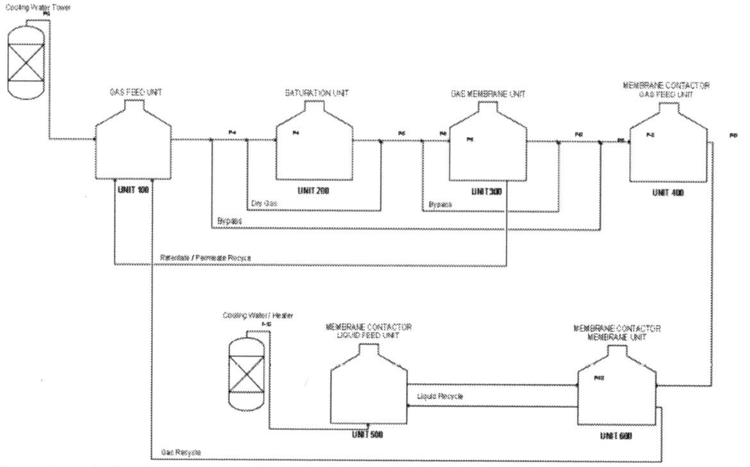

Figure 1: The simplified process flow diagram of the pilot scale membrane testing unit

Improvement of permeate recovery rate in brackish groundwater reverse osmosis plants by pre-treatment and recovery of discharged brine

J. R. Quispe,[1] F. Remonsellez,[1] P.G. Toledo
Universidad Católica del Norte, Dep. of Chemical Engineering, Av. Angamos 0610, Antofagasta, Chile; tel. +56 55 2355987, fax: +56 55 2355917, e-mail: jquispe@ucn.cl;[1]
Universidad de Concepción, Dep. of Chemical Engineering, Casilla 160-C, Correo 3, Concepción, Chile, Fax: + 56 41 224 74 91

Desalination is a widely used technology in the world to meet the increasing demand for fresh and clean water. The main operation used in the conventional desalination process is reverse osmosis (RO). However, a key challenge imposed by desalination by RO is the brine management and disposal issue because the current membranes reach a water recovery and produce liquid brine reject. For inland desalination, the standard approach to manage this reject is the use of evaporation ponds. However, wasting water through evaporation in one of the driest deserts in the world is not the best solution for sustainable management of scarce water available in the north of Chile. Especially if the recovery rate of fresh water in brackish water reverse osmosis plants (BWRO) varies between 50-75%. In this paper it is proposed an alternative to help resolve issues associated with RO brine handling for BWRO plants by brine treatment using a low cost, less footprint and low energy consumption pre-treatment method of inline coagulation-sand filtration in order to increase the permeate recovery. The permeate recovery rate in RO process is generally defined as the permeate flow rate divided by the membrane feed flow rate. A high recovery rate means a high process yield. However, in a desalination process, operation at high recovery results in higher average concentrate salinities in the membrane elements, higher osmotic pressures and higher membrane feed pressures compared to operation at low recovery. In addition, supersaturation of the concentrate can result in more scaling and high membrane flux can result in more fouling. This is the reason why RO plants do not operate at high recovery rates (55%) like CAPRA BWRO plant (Figure 1) located in San Pedro de Atacama, Chile.

The aim of this study was to increase the overall permeate recovery rate from 55% to 75% through recovering water from the discharged brine (RO rejection). For this, a chemical treatment was performed to partially remove silica from 175 to 96 ppm, arsenic from 1.33 to 0.62 ppm, calcium from 453 to 140 ppm, among others. This process consisted in two phases: 1) the pH was raised up to 8 to promote precipitation of silica, 2) sodium hypochlorite used for the oxidation of arsenite to arsenate and subsequent coagulation with ferric chloride was added, then a sand filter was used to retain the precipitates. The treated brine water was very similar to the well water in terms of the main compounds promoting membrane fouling (scaling). So this treated water can be mixed with the feed water coming from the wells in order to increase the availability of feed water for CAPRA plant instead of pumping water from the wells. The tests were conducted in a pilot plant in situ and the module 7 from CAPRA RO plant that process 24 m^3/h of brackish water with a permeate recovery rate of 55%, achieving an overall permeate recovery rate of 75%. The results of this work will increase the availability of water for further desalination and thus reduce the consumption of water from the wells for a sustainable management of scarce water available in San Pedro de Atacama, Chile.

Figure 1. CAPRA BWRO Plant and RO pilot plant in San Pedro de Atacama, Chile.

Mass transfer characterization of calcium alginate membrane containing hydrophobic domain involving with hydrophilic/hydrophobic effect on molecular size selectivity.

Y.Ichikawa, M.Imai*
Course in Bioresourse Utilization Sciences, Graduate School of Bioresourse Sciences, Nihon University, 1866 Kameino, Fujisawa-city, Kanagawa-pref., 252-0880, JAPAN.
TEL.: +81-466-84-3978, E-mail address: XLT05104@nifty.com

Abstract

A stable calcium alginate membrane containing hydrophobic domain was successfully fabricated by the casting method. Sodium alginate and dispersed heptane mixed solution was prepared high speed agitation (7000 min^{-1}). Surfactant Span 80 was added 0.1% in heptane. The emulsion polymer solution was spread on glass-plate using applicator at 150 μm gap. Spread solution was cross-linked by immersion into calcium alginate aqueous solution (0.5M) for 5mim at room temperature (298K). The volumetric ratio of hydrophobic domain was calculated by Eq. (1).

$$\varphi_h = \frac{\text{heptane[ml]}}{\text{Solution of sodium alginate[ml]} + \text{heptane[ml]}} \quad (1)$$

The membrane contained hydrophobic domain was a little optically clouded by dispersed hydrophobic domain.

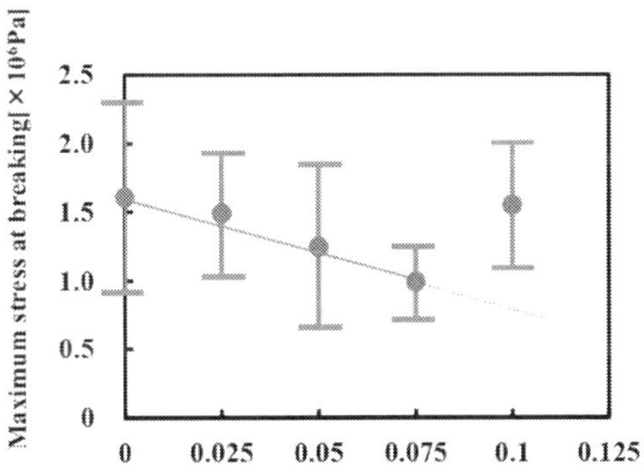

Fig.1: Changing of the maximum stress at breaking.

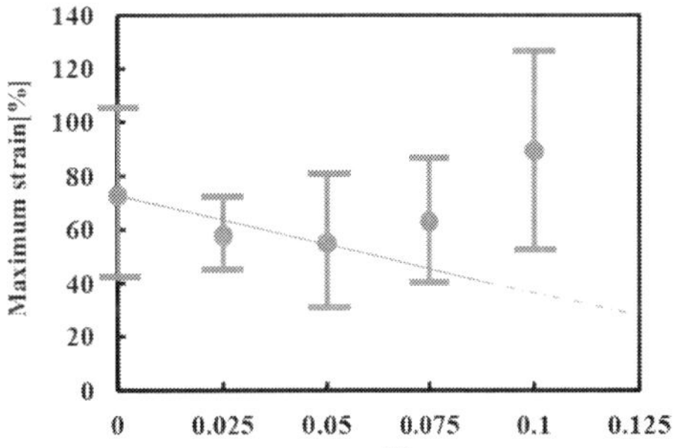

Fig.2: Changing of the maximum strain.

The mechanical strength of the membrane was evaluated by the maximum stress (Figure 1) and the maximum strain (Figure 2). The maximum stress at membrane rupture was decayed with increasing of volumetric fraction of hydrophobic domain. The maximum strain of the membrane was decreased with incresing volumetoric fraction of hydrophobic domain. The data in higher range of volumetric fraction of hydrophobic domain (over 0.075) was unstable.

The water content were demonstrated with the volumetric fraction of hydrophobic domain (Figure 3). The water content was the ratio of mass of contained water to the mass of membrane in swollen state. It was slightly decreased with increasing volumetric fraction of hydrophobic domain.

Figure 4 shows the effective diffusion coefficient (*Deff*) of water-soluble markers (Methyl Orange 327Da, Indigo carmine 466Da, Bordeaux S 604Da, Rose bengal 1017Da, Sirius red 1373Da) in the membrane measured from mass transfer flux.

Effective diffusion coefficient of calcium alginate solo membrane was strongly depended on the molecular weight of water –soluble markers.It was changed 13 folds even if the molecular size of tested component 4 folds.

Previous investigation on molecular size recognition of calcium alginate was almost 20000 fold [1]. In contrast, the dependency of effective diffusion coefficient on the molecuar size in this work was decilned compared with previous study [1]. That was depended on the membrane preparation procedure and internal structure of membrane.

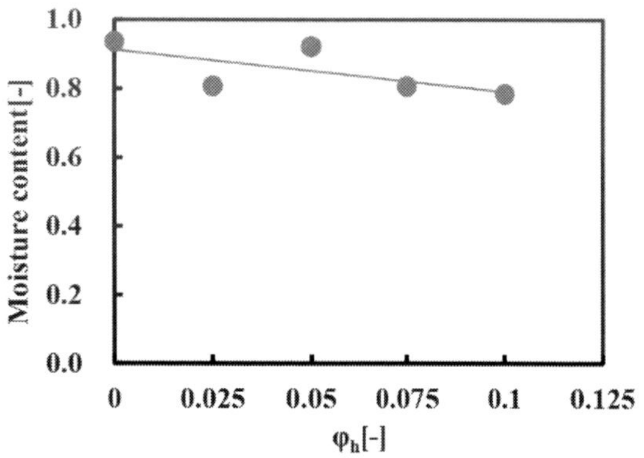

Fig.3: Effect of mass fraction of heptane on the water content of calcium alginate membrane.
($CaCl_2$ con.=0.5 M)

▲: Bulk aqueous phase.(estimated by Wilke-Chang's eq.)
■: Calcium alginate solo membrane.
●: Calcium alginate-hydrophobic composite membrane.

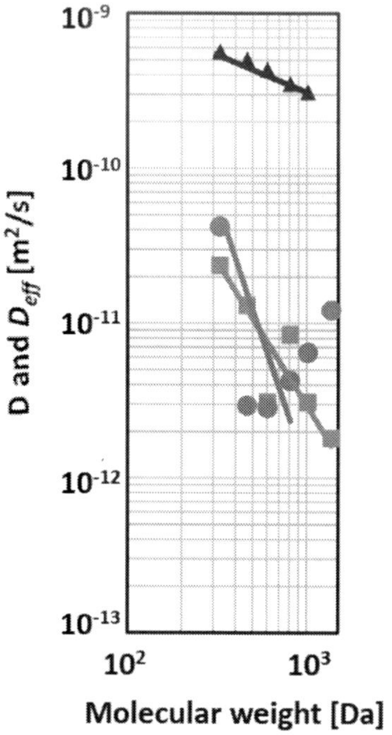

Fig.4: Changing of effective diffusion coefficient.

Reference

[1] K.Kashima, M.Imai, I.Suzuki, Desalin. Water Treat. 17, 143-149 (2010).

Membrane contactor for CO_2 capture with aqueous ammonia: filling a gap of knowledge by getting closer to industrial conditions

K. Villeneuve, D. Roizard, S. Rode
Laboratoire de Réaction et Génie des Procédés, 1 rue Grandville 54001 Nancy, France ;
tel. +33383175289, e-mail: kevin.villeneuve@univ-lorraine.fr; email : denis.roizard@univ-lorraine.fr; sabine.rode@univ-lorraine.fr

Over the past three decades, membrane contactors have been considered as an innovative alternative to conventional gas-liquid contactors such as packed columns. In this technology, a membrane separates the gas and the liquid streams, leading to a non-dispersive and modular character, thus avoiding liquid entrainment or flooding. A technology commonly proposed is the use of hollow fibre membrane contactor (HFMC), consisting of a fibre bundle with an outer diameter of between 0.1 and 1 mm. The small diameter allows this contactor to develop sizeable interfacial areas, which could potentially lead to significant reduction in equipment volume.

HFMC technology has been widely investigated for its use in CO2 absorption with chemical solvents, mainly monoethalonamine. Aqueous ammonia appears to be another promising candidate for CO_2 capture, because of its lower energy demand for regeneration. HFMC involving composite membranes are of particular interest, as they can decrease the ammonia slip, which is a key issue in ammonia-based carbon capture technology. Composite membranes make use of a thin, dense polymer skin coated on a microporous support, avoiding fluid breakthrough and retaining selectively ammonia. Due to the high permeance of the dense film, the overall mass-transfer coefficient of these membranes is reasonably close to that measured for microporous membranes [1].The interest of the technology has been demonstrated in our research group [2] with a contactor volume reduction ratio of about nine, compared to classical gas-liquid columns.

Despite the potential advantages of membrane contactors, very few investigations have studied implementing this technology for CO2 capture within an industrial framework. The performance of CO2 absorption and liquid absorbent regeneration using this technology under industrial conditions, i.e., wet dusty gas, loaded absorption solutions, continuous operation, is still unknown. In order fill up this lack, experiments closer to industry were performed, using commercial membranes and wet inlet gas.

In order to study and to model multi-component mass-transfer in the system, the membrane mass-transfer coefficient of all volatile species, i.e. CO2, ammonia and water must be known. Moreover, in reactive systems characterised by important absorption heats, the estimation of the water exchange flux is crucial, as it strongly influences the heat balances due to the high latent heat involved. Accordingly, experiments, with water and nitrogen were performed, in order to determine the mass-transfer coefficient of water vapour through three commercial membranes of different types, i.e. porous, dense and thin skin. Experimental results are shown in Figure 1. As could be expected, the dense membrane is characterized by a low mass-transfer coefficient, whereas the porous and the dense-skin membranes are both characterized by reasonably high mass-transfer coefficients. All measured mass-transfer coefficients were in accordance with values estimated using literature analogies.

The experiments with inlet humidified flue gas intend to evaluate the relative impact of gas humidity on CO2 capture performances. These results will be described and discussed in the presentation in the light of previous published [2, 3].

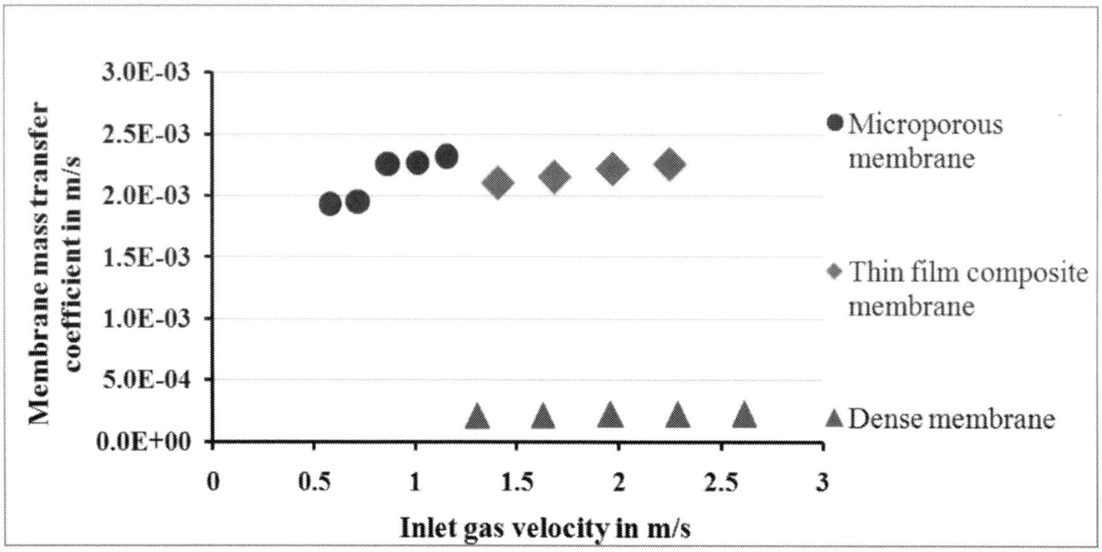

Figure 1 : Mass transfer coefficient of water vapor for different types of membrane at 40 °C

Acknowledgment: This project is sponsored by “Agence Nationale pour la Recherche grant” ANR. The LGC (Toulouse, FRANCE) is gratefully acknowledged for the preparation of the membrane contactor.

References:

[1] P.T. Nguyen, E. Lasseuguette, Y. Medina, J.C. Remigy, D. Roizard, E. Favre, A dense membrane contactor for intensified CO2 gas/liquid absorption in post-combustion capture, J MembrSci, (2011) 377 (1-2), pp. 261-27.

[2] C. Makhloufi, E. Lasseuguette, J.C. Remigy, B. Belaissaoui, D. Roizard, E. Favre, Ammonia based CO2 capture process using hollow fiber membrane contactors, J. Memb. Sci. 455 (2014) 236–246. doi:10.1016/j.memsci.2013.12.063.

[3] McLeod, P. Buzatu, O. Autin, B. Jefferson, E. McAdam, Controlling shell-side crystal nucleation in a gas–liquid membrane contactor for simultaneous ammonium bicarbonate recovery and biogas upgrading, J. Memb. Sci. 473 (2015) 146–156. doi:10.1016/j.memsci.2014.07.063.

Membrane separation of dimethyl carbonate and methanol mixtures using cellulose triacetate - poly(butylene succinate) blend membranes

P. Čihal, O. Vopička, K. Friess
University of Chemistry and Technology Prague, Department of Physical Chemistry, Technická 5, 166 28 Prague, Czech Republic;
tel. +420 220 444 266, e-mail: petr.cihal@vscht.cz.

Dimethyl carbonate (DMC) became a very popular chemical precursor, solvent and anti-knock additive in fuels due to its low environmental toxicity [1]. DMC is usually produced from methanol, with which it forms minimum-boiling-point-azeotrope with the composition of 82 mol.% of methanol at 40 °C [2]. This azeotropic mixture can be classically separated by high pressure distillation [3] or extractive distillation [4][5]. An alternative way is the utilization of low-cost, low-temperature and low-pressure membrane separation.

Recently, new polymeric blends composing of cellulose triacetate (CTA) and poly(butylene succinate) (PBS) have been developed [6]. Significantly, these blends are homogenous at concentrations lower than 40 wt.% of poly(butylene succinate). We report the separation properties of such membranes that were tested using a flow vapour permeameter with mass-spectrometric detection. Permeation of vapour mixtures containing 82 mol.% (i.e. composition of the azeotrope) and 62 mol.% of methanol through CTA/PBS membranes was measured for blends containing 0, 5, 10, 20 and 30 wt.% of poly(butylene succinate).

Our results demonstrated that the separation factor of the blend membranes significantly increased compared to pure cellulose triacetate membrane. This was caused by the decrease of dimethyl carbonate flux at the azeotropic composition of the feed; methanol flux decreased as well. The X-ray diffraction analyses revealed that the addition of poly(butylene succinate) blocks inner microporous structure of cellulose triacetate and thus erases specific pathways through which dimethyl carbonate can diffuse through the membrane.

The best ratio of separation factor and permeation coefficient was shown by the CTA/PBS blend containing 10 wt.% of poly(butylene succinate).

Acknowledgements

Results were obtained under the financial support of the Czech Science Foundation within the Grant Project 13-32829P. Partial financial support was obtained from the Specific University Research Project (MSMT 20-SVV/2016).

References

[1] Zhao, T.; Han, Y.; Sun, Y., Novel reaction route for dimethyl carbonate synthesis from CO2 and methanol. *Fuel Processing Technology* **2000,** *62* (2–3), 187-194.

[2] Comelli, F., sothermal Vapor−Liquid Equilibria Measurements, Excess Molar Enthalpies, and Excess Molar Volumes of Dimethyl Carbonate + Methanol, + Ethanol, and + Propan-1-ol at 313.15 K. *J. Chem. Eng. Data* **1997,** *42* (4), 705-709.

[3] Shah, V. M.; Bartels, C. R.; Pasternak, M.; Reale, J., *Opportunities for membranes in the production of octane enchancers.* 1989.

[4] Hsu, K.-Y.; Hsiao, Y.-C.; Chien, I. L., Design and Control of Dimethyl Carbonate−Methanol Separation via Extractive Distillation in the Dimethyl Carbonate Reactive-Distillation Process. *Industrial & Engineering Chemistry Research* **2010,** *49* (2), 735-749.

[5] Blahut, A.; Dohnal, V., Interactions of volatile organic compounds with the ionic liquids 1-butyl-1-methylpyrrolidinium tetracyanoborate and 1-butyl-1-methylpyrrolidinium bis(oxalato)borate. *The Journal of Chemical Thermodynamics* **2013,** *57*, 344-354.

[6] Uesaka, T.; Nakane, K.; Maeda, S.; Ogihara, T.; Ogata, N., Structure and physical properties of poly(butylene succinate)/cellulose acetate blends. *Polymer* **2000,** *41* (23), 8449-8454.

Morphology and gas separation performance of polyimide / polysulfone asymmetric membranes synthesized by dual bath method

B. Haider, [1]M. Kaspereit
University of Erlangen-Nuremberg, Institute of Separation Science & Technology, Egerlandstrasse 3, 91058 Erlangen, Germany; tel. +49-9131/85-27454, e-mail: bilal.haider@fau.de; [1]University of Erlangen-Nuremberg, Institute of Separation Science & Technology, Egerlandstrasse 3, 91058 Erlangen, Germany; tel. +49-9131/85-27 441, e-mail: malte.kaspereit@fau.de; fax: +49-9131-85-27441.

Membrane morphology plays a key role in defining gas permeability and selectivity in membranes. In the current study polyimide and polysulfone asymmetric membranes are synthesized by dual bath method to promote a delayed demixing of solvent. Polyimide and polysulfone solutions of 25% (w/w) are prepared in N-Methyl-2-pyrrolidone (NMP) as solvent. Afterwards these solutions are spin casted on glass plates in N_2 environment using a spin coater. Immediately after spin casting the glass plates are first transferred to a 50/50 water / NMP coagulation bath for 30 seconds. Subsequently, the membranes are transferred to a deionized water coagulation bath for overnight to achieve complete removal of solvent. After completion of drying procedure, finally the membranes are coated with 5% PDMS solution in n-hexane to seal any potential micro defects.

Fig. a) Polysulfone asymmetric membrane coated with PDMS

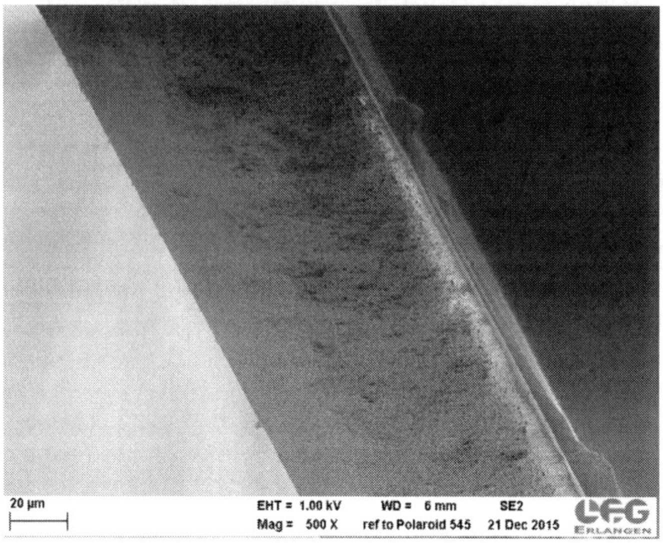

Fig. b) Polyimide asymmetric membrane coated with PDMS

As shown in the SEM analysis in the figure, the basic morphology of both, polyimide and polysulfone membranes prepared by the mentioned technique is the same, i.e. the pore size at the top is smaller, while it gradually increases towards the bottom. Importantly, a skinned dense layer is absent in the both cases. This can be seen as an attractive feature to enhance membrane permeability. This absence of the skinned layer may be attributed to the delayed demixing of polymer and solvent which causes the cellular asymmetric pores in the membrane.

The synthesized membranes are characterized regarding their permeability and selectivity for 5 different gases (H_2, He, CO_2, CH_4 and N_2). The constant pressure method is used to evaluate permeability at room temperature and a pressure between 1 and 20 bar. In the case of both polyimide and polysulfone, the order of permeability is as follows: $H_2 > He > CO_2 > CH_4 > N_2$. The ideal selectivity is also calculated using pure gas permeability data. Uncoated polyimide showed an ideal selectivity of 25 at 10 bar for the gas pair H_2/N_2 while uncoated polysulfone achieves a value of 20 for the same pair. A coating with PDMS may be of interest for the sake of enhancing performance of these membranes by improving selectivity and sealing micro defects. For the two mentioned membrane types, such coating leads to almost doubled selectivity for above mentioned gas pair, while permeability decreased, but remained in the same order of magnitude.

Nanofiltration purification of olive mill wastewater: assessment of the isoelectric point

J.M. Ochando-Pulido*[1], M.D. Victor-Ortega[1], A. Martínez-Férez[1]
[1] Chemical Engineering Department, University of Granada, 18071 Granada, Spain; jmochandop@ugr.es

Membrane fouling still remains nowadays one of the main challenges of the broad applied membrane technology, especially in liquid-liquid separation processes [1]. Membrane fouling may lead to dramatically shorten the life time of membrane modules and alters its selectivity and rejection performance [2]. In other words, the lack of knowledge and control of membrane fouling is an additional cost for the industry which should be minimized to permit successful competiveness of the technology. This applies especially on wastewater purification processes.

In the present work, a thin-film composite polymeric nanofiltration (NF) membrane is examined for the tertiary treatment of secondary-treated two-phase olive mill wastewater (OMW2-ST). Overcoming the deleterious fouling phenomena persistently encountered in membrane processes managing wastewater streams was indeed pursued. With this purpose, the isoelectric point (PZC) of the membrane was determined as a key for the operating conditions. This will determine the intensity of the electrostatic interactions between the organic and inorganic solutes present in the OMW2-ST stream and the active surface of the NF membrane, further to the steric effects.

Table 1: NF performance as a function of the feedstream pH

Feedstream pH_0	P_{TM}, bar	J_{th}, L/hm^2	$-\Delta J_p$, %
5	13	26.9	57.3
5.5	13	52.6	16.7
6.5	13	57.8	8.1
7.5	13	59.0	5.5

P_{TM}: operating pressure; pH_0: feedstream inlet pH; J_{p0}: initial permeate flux; J_{th}: threshold flux; $-\Delta J_p$: flux loss.

Working above the isoelectric (pH > 5.8) of the membrane ensured high steady-state permeate productivity (~ 60 L/hm^2), also economically sustainable in time owed to minimization of the fouling-build up rate (0.91 h^{-1}). Moreover, these conditions also provided high feed recovery (90%) and significant rejection efficiencies for the electroconductivity (58.1%) and organic matter (76.1%). This led to a purified permeate stream exiting the NF membrane operation exhibiting average EC and COD values that permit complying with the water quality parameters established by different regulations for discharge public waterways and irrigation purposes.

References

[1] M. Stoller, M. Bravi, Desalination 250 (2010) 578-582.
[2] M. Stoller, A. Chianese, Desalination 191 (2006) 62-70.

Performance evaluation of a catalytic membrane reactor for hydrogen separation in methane steam reforming reaction

M. Amanipour, J. Towfighi, M. Heidari, E. Ganji Babakhani, A. Zamaniyan
Chemical Engineering dep., Tarbiat modares University, Gisha, Tehran, Iran.
Tel: +989125304394, e-mail: m.amanipour@modares.ac.ir

Cylindrical alumina microfiltration membrane (GMITM Corporation, i.d=9 mm, o.d.=13 mm, length= 50 mm) with an average pore size of 0.5-0.6 μm and porosity of about 0.3 was used as the support for membrane reactor. This support was soaked in boehmite sols, the mean particle size of which was adjusted in the range of 50 to 500 nm by carefully controlling hydrolysis time, and calcined at 927 K for two hours. This process was repeated with different boehmite solutions in order to achieve an intermediate layer with an average pore size of about 50 nm. The resulting substrate was then coated with a thin and dense layer of silica+alumina by counter current chemical vapour deposition (CVD) method.

A boehmite sol which was prepared by a standard procedure was used to prepare the catalytic layer. The catalytic membrane reactor was placed in an experimental set-up to evaluate the permeation and hydrogen separation performance for a steam reforming reaction. The set-up consisted of a tubular module in which the membrane was fixed and the reforming reaction occurred at the inner side of the membrane. Methane stream, diluted with nitrogen, and deionized water with a steam to carbon (S/C) ratio of 3.0 entered the reactor after the reactor heated up to 873 K and the catalytic layer was reduced at presence of hydrogen for 2 hours. A counter-current nitrogen flow was used as sweep gas through the outer side of the reactor. Any liquid produced was separated at reactor exit by a cold trap and the produced gases were analysed by an on-line gas chromatograph (Agilent 7890A) to measure total CH_4 conversion and H_2 permeation.

BET analysis indicated uniform size distribution with average pore size of 280 nm and average surface area of 275 $m^2.gr^{-1}$. Single-component permeation tests were carried out for hydrogen, methane and carbon dioxide at temperature range of 650-800°C and the results showed almost the same permeation and hydrogen selectivity values for hydrogen as the composite membrane without catalytic layer. Performance of the catalytic membrane was evaluated by applying it as a membrane reactor for methane steam reforming reaction at gas hourly space velocity (GHSV) of 100,000 h^{-1} and 3 bar. CH_4 conversion increased from 50% to 90% with increasing reaction temperature from 600°C to 750°C, which is sufficiently above equilibrium curve at reaction conditions, but slightly lower than membrane reactor with packed nickel catalytic bed because of its higher surface area compared to the catalytic layer. Hydrothermal stability of the catalytic membrane reactor was evaluated in a reforming reaction at 650°C. Hydrogen permeation dropped by only 1.85 times from 5.0×10^{-7} $mol.m^{-2}.S^{-1}.Pa^{-1}$ to 2.8×10^{-7} $mol.m^{-2}S^{-1}.Pa^{-1}$ after exposure to steam for two successive days, which means stability in morphological changes in catalytic membrane structure .

Separation of aroma compounds from binary model solutions by pervaporation; effect of permeate cooling temperature

A. Dawiec, A. Witek-Krowiak
Wroclaw University of Technology, Faculty of Chemistry, Department of Chemical Engineering, ul. Norwida 4/6, 50-373 Wroclaw, Poland, e-mail: anna.dawiec@pwr.edu.pl
tel. +48 71 320 34 37

Pervaporation is a membrane separation process where some components from a liquid mixture permeate selectively through a dense membrane and evaporate on a downstream side of the membrane. Pervaporation due to gentle process conditions, high selectivity, modular design and low energy consumption, is a promising alternative to currently used conventional separation techniques. This process is most frequently used in dehydration of solutions containing small amounts of water (special interest for industry application is related with separation of azeotropes or close-boiling mixtures)[1] also, in separation of organic-organic solutions [2] or in the recovery of organic compounds even from very dilute solutions [3]. Organophilic pervaporation has a high potential for separation of natural aroma compounds from e.g. fruit juices, highly diluted in complex aqueous media, due to its high selectivity towards aroma compounds, modular design and its ability to operate under mild conditions. However, pervaporation is a complex process, dependent on many factors whose total effect has a direct impact on the process performance.

The main goal of this study was to determine the effect of permeate cooling temperature on efficiency of PV in binary aroma solutions separation. In the literature there is no accessible work related with the effect of cooling temperature on PV performance. Two different aroma compounds (butyl acetate (BA) and ethanol (ET)) were used to prepare binary (BA/ET+water) model solutions.

In this study, the partial and total fluxes as well as separation factors from pervaporation of binary aroma compounds solutions were compared, to investigate how permeate cooling temperature affect the performance of whole PV process. In first step, pervaporation of binary ET-water model solution was conducted. The results are presented in Fig.1. As can be seen, a partial ethanol flux increase gently with the temperature in the range of -20 to -5^{0}C. For the permeat cooling temperature equal 0 and 10^{0}C the ethanol partial flux slightly decreases. The highest values of aroma stream and enrichment factor, which were 657 g/(m^2•h) and 4.37 respectively, were obtained for permeate cooling temperature equals -5^{0}C.

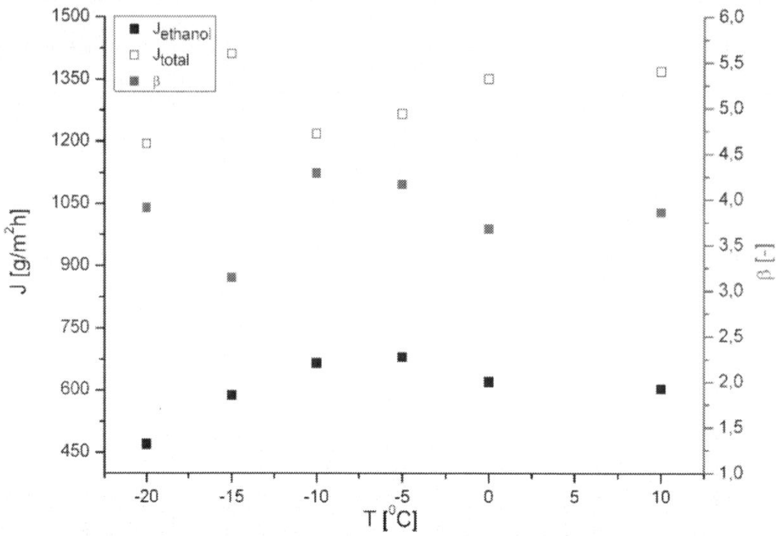

Fig. 1. Effect of permeate cooling temperature on the ethanol permeate flux (total flux, J_{total}; partial ethanol flux, $J_{ethanol}$) and enrichment factor (β).

In the second step, the performance of pervaporation of butyl acetate-water model solution was determined. As in the former case, BA partial fluxes and enrichment factors obtained in different permeate cooling temperature were compared to estimate optimal temperature value. The highest values of BA partial flux and enrichment factor, which were 620 g/(m^2•h) and 174 respectively (cooling temperature -5^{0}C).

On the basis of the experimental results, the permeate cooling temperature in the sweeping gas pervaporation has influence on pervaporation performance, like flux and separation factor. The optimal cooling temperature could improve PV efficiency and reduce energy consumption. The determined optimum permeate cooling temperature was -5^{0}C.

The work was financed by grant number NN209096740 (National Science Centre, Poland).

References:

[1] S. Xia et al., Dehydration of ethyl acetate–water mixtures using PVA/ceramic composite pervaporation membrane, Separation and Purification Technology 77 (2011) 53–59

[2] B. Smitha et al., Separation of organic–organic mixtures by pervaporation—a review, Journal of Membrane Science 241 (2004) 1–21

[3] M. Khayet et al., Studies on pervaporation separation of acetone, acetonitrile and ethanol from aqueous solutions, Separation and Purification Technology 63 (2008) 303–310

Separation of biobutanol from fermentation solutions by perstraction using [omim][Tf_2N] as receiving phase

G. Merlet[1], C. Aravena[1], R. Cabezas[1], E. Quijada-Maldonado[1], J. Romero[1]
[1] University of Santiago de Chile, Department of Chemical Engineering, Av. Lib. B. O'Higgins 3363, Estación Central, Santiago, Chile.
E-mail: julio.romero@usach.cl

Perstraction or solvent extraction assisted by membranes is a separation technique that combines membrane separation with liquid-liquid extraction in a single step, where the driving force is the concentration difference between the aqueous phase and the extractant; meanwhile, the membrane represents a selective interface [1].

On the other hand, ionic liquids (ILs) are a novel class of green solvents, which show several advantages over organic solvents, such as negligible volatility, high thermal and chemical stability [2, 3]. Thus, ILs become an alternative to conventional solvents used as stripping phase in perstraction processes.

In this work, the selective separation of acetone-butanol-ethanol (ABE) from aqueous mixtures was studied simulating the production of biobutanol from ABE fermentation process [4]. In this way, biobutanol involves a high commercial interest as a second-generation biofuel.

A perstraction system was implemented for the selective removal of butanol from ABE model solutions. In this experimental setup, the membrane and the receiving phase were a flat sheet PDMS film and a specific imidazolium-based ionic liquid, respectively. **Table 1** summarizes the values of operation parameters. Two different ionic liquids were tested in this work: [bmim][Tf_2N] and [omim][Tf_2N].

Table 1. Operating conditions of perstraction experiments.

Variable	Values / Range	Units
Ionic Liquid	[bmim][Tf_2N] [omim][Tf_2N]	
Aqueous feed flow rate	0.48 – 0.90 – 1.25	L min^{-1}
Ionic liquid flow rate	0.48 – 0.90 – 1.25	L min^{-1}
Acetone concentration **Butanol concentration** **Ethanol concentration**	6000 12000 2000	ppm ppm ppm
Temperature	30	°C
Operating time	6	h

From the perstraction experiments, transmembrane fluxes of butanol, acetone, ethanol and water were obtained and its values were compared with the obtained ones in vacuum pervaporation tests with the same membrane.

From the perstraction experiments with [bmim][Tf_2N], it can be observed that average transmembrane fluxes of acetone and butanol were $3.29x10^{-3}$ [kg m^{-2} h^{-1}] and $3.61x10^{-3}$ [kg m^{-2} h^{-1}], respectively. Meanwhile, the transmembrane flux of ethanol was considered negligible. These values of flux are lower than the obtained ones for pervaporation. However, the butanol/water selectivity is significantly increased respect to pervaporation. The best separation was achieved with perstraction with [omim][Tf_2N]. Simultaneously, the water transmembrane flux could be obtained by mass balance and compared with its value obtained in pervaporation. The perstraction system using [omim][Tf_2N] shows again the best separation response. All these results are summarized in figures 1 and 2.

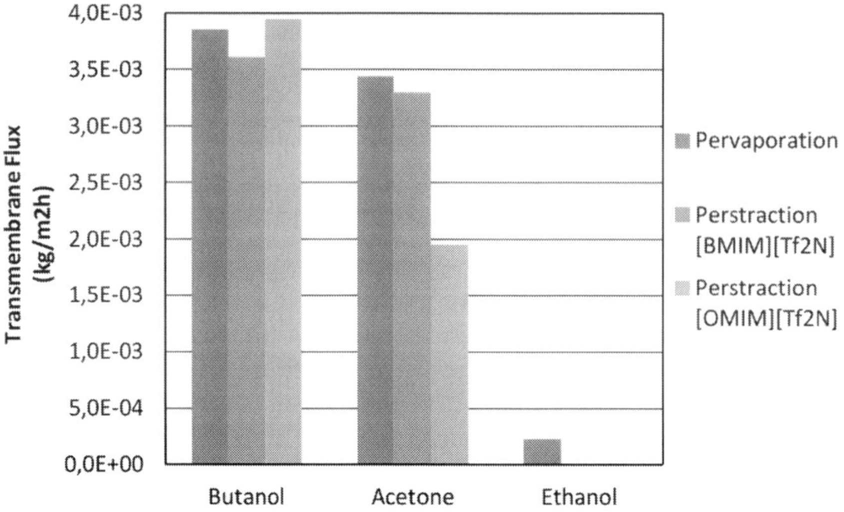

Figure 1. Experimental transmembrane fluxes for perstraction and pervaporation. These values were calculated from the average ones of each experiment.

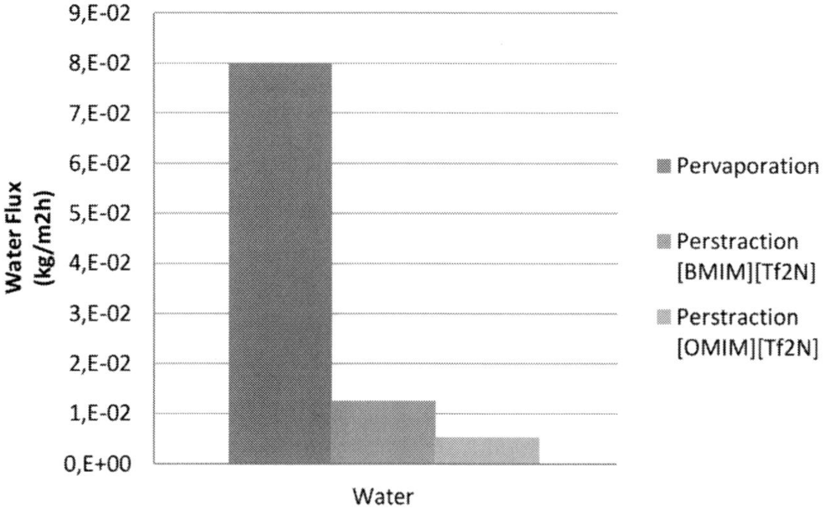

Figure 2. Average transmembrane fluxes of water obtained by perstraction and pervaporation assays.

Thus, perstraction of ABE mixtures using a PDMS membrane and [omim][Tf_2N] as extractant represents a suitable alternative for effective butanol separation from fermentation solutions, obtaining transmembrane fluxes slightly higher than the obtained ones in pervaporation with similar membranes. Simultaneously, perstraction with [omim][Tf_2N] was highly selective to butanol. Future experiments would consider assays with real fermentation broths in order to assess membrane fouling.

1. Huang, H.-J., S. Ramaswamy, and Y. Liu, *Separation and purification of biobutanol during bioconversion of biomass.* Separation and Purification Technology, 2014. **132**(0): p. 513-540.
2. Plaza, A., et al., *Separation of butanol from ABE mixtures by sweep gas pervaporation using a supported gelled ionic liquid membrane: Analysis of transport phenomena and selectivity.* Journal of Membrane Science, 2013. **444**: p. 201-212.
3. Keskin, S., et al., *A review of ionic liquids towards supercritical fluid applications.* The Journal of Supercritical Fluids, 2007. **43**(1): p. 150-180.
4. García, V., et al., *Challenges in biobutanol production: How to improve the efficiency?* Renewable and Sustainable Energy Reviews, 2011. **15**(2): p. 964-980.

Simultaneous tuning of ionic conductivity and oxidative resistance of SPEEK membrane electrolyte by incorporation of sulfated zirconium titanate

Taghiyeh Saririchi Aboudzadeh[a,b], Soosan Rowshanzamir[a,b]
[a] School of Chemical Engineering, Iran University of Science and Technology, Narmak, Tehran 16846-13114, Iran
[b] Fuel Cell Laboratory, Green Research Centre, Iran University of Science and Technology, Tehran, Iran. Tel.: 98 2177491223; fax: 98217491242.
E-mail address: rowshanzamir@iust.ac.ir (S. Rowshanzamir).

Nowadays, proton exchange membrane fuel cells (PEMFCs) have emerged as promising power supply systems for stationary, vehicular, and portable applications. Traditionally, these devices were based on perfluoro-sulfonic acid membrane electrolytes, given the commercial name Nafion. However, Nafion membranes are restricted to operating temperature below 100°C [1]. Various studies in the open literature confirmed that Sulfonated PEEK (SPEEK) is a promising PEM because of its low cost, and good mechanical and thermal stabilities. Nevertheless, depending on the variation of the sulfonation degree (DS), the mechanical properties of S-PEEK membranes could deteriorate progressively (at higher DS). To solve this key issue, S-PEEK has been used as a major component in the synthesis of hybrid/blend membranes [2]. The introduced nanoparticles to polymer membranes might be silica [3, 4], aluminum phosphate [5], clays [6], zirconia [7, 8], titania [9, 10], and polymeric nanoparticles [11].
Our work has been focused on sulfated zirconium titanate as an additive to SPEEK for electrolyte membranes. RSM design was used to optimize the two independent variables: the sulfonation time and the content (wt%) of sulfated zirconium titanate. The dependent variables were the proton conductivity and oxidative stability. The experimental design array made from central composite set to achieve high oxidative stability, together with maximum proton conductivity. The optimum parameters were found 6.9 h sulfonation time and 11.11 wt. % of sulfated zirconium titanate that represented proton conductivity of 22.44 $mS.cm^{-1}$ (at 100°C and 100 %RH) and oxidative stability of 152.8 min. The sulfonation time and content (wt%) of sulfated zirconium titanate had the same effect on conductivity, while, the content (wt%) of sulfated zirconium titanate is more crucial to oxidative stability (Fig. 1). Moreover, the addition of sulfated zirconium titanate nanoparticles improved both oxidative stability and proton conductivity. The physic-chemical properties of the membranes were examined by FESEM, TGA, DSC and tensile strength test. Also, the XRD, EDX and TEM analysis were conducted to characterize the nanoparticles.

a b

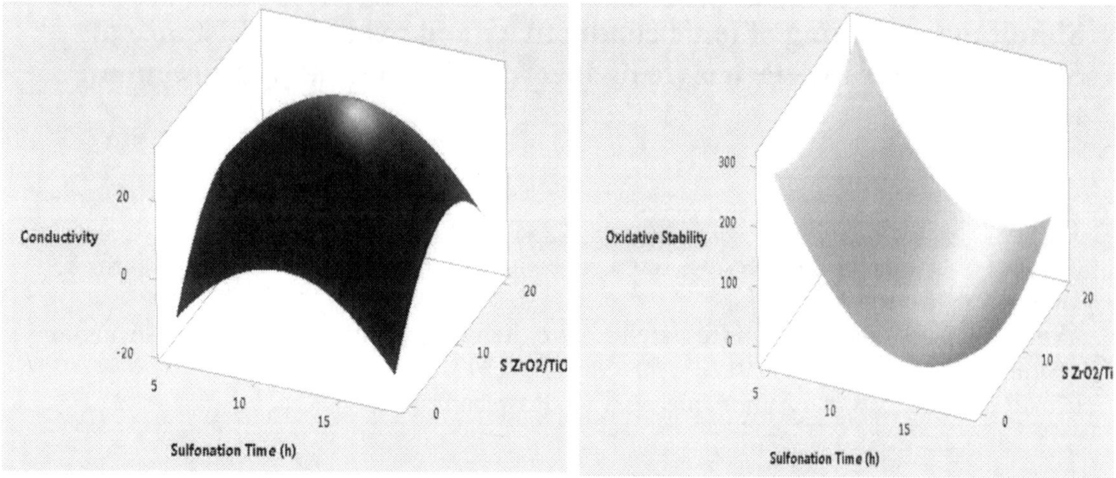

Fig. 1. Surface plot of (a) conductivity vs sulfonation time; $SZrO_2TiO_2$, (b) oxidative stability vs sulfonation time; $SZrO_2TiO_2$

References

1. Jun, Y., et al., *Functionalized titania nanotube composite membranes for high temperature proton exchange membrane fuel cells.* International Journal of Hydrogen Energy, 2011. **36**: p. 6073-6081.
2. Iulianelli, A. and A. Basile, *Sulfonated PEEK-based polymers in PEMFC and DMFC applications: A review.* International Journal of Hydrogen Energy, 2012. **37**: p. 15241-15255.
3. Mishra, A.K., et al., *Silicate-based polymer-nanocomposite membranes for polymer electrolyte membrane fuel cells.* Progress in Polymer Science, 2012. **37**(6): p. 842-869.
4. Du, L., et al., *SPEEK proton exchange membranes modified with silica sulfuric acid nanoparticles.* International Journal of Hydrogen Energy, 2012. **37**: p. 11853-11861.
5. Rangasamy, V.S., et al., *Preparation and characterization of composite membranes based on sulfonated PEEK and AlPO4 for PEMFCs.* Solid State Ionics 2012. **216**: p. 83–89.
6. Ilbeygi, H., et al., *Transport properties of SPEEK nanocomposite proton conducting membranes: Optimization of additives content by response surface methodology.* Journal of the Taiwan Institute of Chemical Engineers 2014. **45**: p. 2265-2279.
7. Zhang, Y., et al., *Promotion of PEM Self-Humidifying Effect by Nanometer-Sized Sulfated Zirconia-Supported Pt Catalyst Hybrid with Sulfonated Poly(Ether Ether Ketone).* J. Phys. Chem. , 2007. **111**: p. 6391-6399.
8. Silva, V.S., et al., *Zirconium oxide hybrid membranes for direct methanol fuel cells—Evaluation of transport properties.* Journal of Membrane Science 2006. **284**: p. 137–144.
9. Aslan, A. and A. Bozkurt, *Nanocomposite membranes based on sulfonated polysulfone and sulfated nano-titania/NMPA for proton exchange membrane fuel cells.* Solid State Ionics, 2014. **255**: p. 89-95.
10. Cozzi, D., et al., *Organically functionalized titanium oxide/Nafion composite proton exchange membranes for fuel cells applications.* Journal of Power Sources, 2014. **248**: p. 1127-1132.
11. Ng, L.Y., et al., *Polymeric membranes incorporated with metal/metal oxide nanoparticles: A comprehensive review.* Desalination 2013. **308**: p. 15–33.

Solid-polyimide mixed matrix membranes for gas separation: inorganic and organic fillers

[1,a]C. Khajornpob, [1,b]T. Rirksomboon, [2,c]S. Kulprathipanja, [2,d]C. Liu
[1]Chulalong Univ., The petroleum and petrochemical college, Phayathai Rd. 254, 10330 Bangkok, Thailand; [a]tel. +66 881016066, e-mail: khanjae.chtt@gmail.com;
[b]tel. +66 22184136, email: t.rirksomboon@unb.ca; [2]UOP LLC, Algoquin Rd. 25 E, 60017-5017 Illinois, USA; [c]tel. +1 8473913138, email: santi.kulprathipanja@honeywell.com
[d]tel. +1 8473912657, email: chunqing.liu@honeywell.com.

A mixed matrix membrane is a hybrid composite membrane containing dispersed organic or inorganic fillers in a continuous organic polymer matrix. The mixed matrix membranes (MMMs) provide the advantages of polymer membranes such as low cost and ease of processibility and molecular sieve membranes such as high selectivity and permeability. In this study, MMMs were fabricated using different inorganic (AlPO-14, Nanosilicalite, SAPO-34, and UZM-54) and organic (β-Cyclodextrin) fillers as dispersed phase in Matrimid 5218 polymer as continuous phase. The solution-casting method was used to fabricate the MMMs with 10, 20, and 30 wt% fillers. The MMM gas separation performance was investigated by the permeability and selectivity via single gas measurements of CO_2 and CH_4 using the time-lag method at 50 °C and a pressure of 100 psig. The results showed that Nanosilicalite, SAPO-34 and UZM-54 were aggregated after casting. These results could be because of the incompatibility between the polymer and inorganic phase properties. For β-Cyclodextrin/Matrimid, its permeability of both CO_2 and CH_4 decreased with increasing filler loading. This result can be explained in terms of inclusion in the cavity of β-CD. However, the selectivity of CO_2/CH_4 was slightly increased at 15 wt% and highest at 20 wt% β-CDs loaded in Matrimid polymer when compared with pure Matrimid membrane.

Speciation, analysis and extraction of lanthanum element from Algerian's natural and industrial ressources using ion exchange process

Nawel RAHMOUNE, Mourad AMARA
Laboratoire d'hydrométallurgie et chimie inorganique moléculaire, Faculté de Chimie, USTHB, Algiers, ALGERIA, Email : mamara@usthb.dz

Rare earth elments became during the last two decades an alternative ressources to the developpement of sustainable materials. An important demand of these elements and the high cost of extraction from ores leads researchers and industrial to investigate recycling this ressource.

This study focuses on the speciation using x-ray diffraction analysis and extraction by ion exchange procedure using chelating resin modified to a better selectivity for lanthanum element. Modification of commercial amberlite resin by a phosphorus group was achieved by immersion in an appropriate solution in particular experimental conditions such as illumination by uv light allowing creation of radical functional groups with high reactivity. Interesting results have been obtained and high efficiency was confirmed by analysis using induced coupled plasma technique. The same study has been conducted using membrane separation procedure in the presence and in the absence of chelating agent.

Keywords : lanthanum, ion exchange, extraction, separation

The synthesis and characteristic of an anion conductive polymer membrane for alkaline anion exchange fuel cell

M. Iravaninia[1], S. Azizi[1], S. Rowshanzamir[1,2]
[1] School of Chemical Engineering, Iran University of Science and Technology, Narmak, Tehran 16846-13114, Iran; [2] Fuel Cell Laboratory, Green Research Center, Iran University of Science and Technology, Narmak, Tehran, Iran; Tel./fax: +98 2177491242; e-mail: rowshanzamir@iust.ac.ir.

Abstract: In this study, anion exchange membrane based on polysulfone was prepared via chloromethylation, amination and alkalization. In amination step, trimethylamine and N,N,N',N'-tetramethyl-1,6-hexanediamine were used as amination and crosslinking agents, respectively. The chloromethylated polysulfone was characterized by ^{1}HNMR spectroscopy. Ionic conductivity at different temperatures, ion exchange capacity, water uptake and swelling ratio were measured for the prepared anion exchange membrane. The ionic conductivity is in the range of 12-29 mS/cm at 25-80°C and the ion exchange capacity is 1.4 meq/gr. Furthermore, the water uptake and swelling ratio values are 38% and 22%, respectively.

Keywords: Anion Exchange Membrane, Ionic Conductivity, Polysulfone, Ammonium groups, Solid Alkaline Fuel Cell.

Introduction: In recent decade, anion exchange membrane fuel cells (AEMFCs) attract more attention among different types of fuel cells. AEMFCs are new technologies that utilize anion exchange membranes (AEMs) as solid polymer electrolytes. These types of fuel cells have some advantages over proton exchange membrane fuel cells such as: employing non-precious metal catalysts (Ag, Ni, or Pd), effective reduction of oxygen in cathode, using various liquid fuels, minimizing corrosion problems under alkaline media and reducing fuel crossovering due to the counter-current migration of fuel and hydroxide ion [1-2]. AEM is the key component of AEMFCs and has an important effect on the fuel cell performance. In this article the homogeneous AEM was prepared via chloromethylation, amination and alkalization. This membrane is based on polysulfone and functionalized with amine groups, trimethylammonium (TMA) and 1,6-tetra hexanediamine (TMHDA). Degree of chloromethylation (DC) for chloromethylated PSF (CMPSF), ionic conductivity, ion exchange capacity (IEC), water uptake (WU) and swelling ratio (SR) were measured for the AEM.

Methods: All chemicals used for membrane preparation were supplied from Merck and Sigma Aldrich companies. The chloromethylation step was performed as discussed in other reference [3] at 55°C for 72 hr and the CMPSF was obtained. In amination step, 1 g of CMPSf was dissolved in 10 cc dimethylacetamide (DMAc). Then 0.18 cc TMHDA and 1.41 cc TMA were added to the solution and the reaction continued for 48 hr at 80°C. The resulted solution was casted and the anion exchange membrane in chloride form was obtained. In alkalization step, the membrane was soaked in 1 M KOH solution to provide the membrane in hydroxide form. The hydroxide conductivity was calculated by electrochemical impedance spectroscopy (EIS). The IEC of the membranes was determined by back titration method as discussed in other work [4], WU and SR were obtained by weight and length difference between the wet and dry AEMs as reported in elsewhere [5] and the DC of CMPSF was determined by proton NMR (^{1}HNMR) spectroscopy. The DC is average number of chloromethylated groups per PSF monomer unit and was calculated as discussed in other work [5].

Results and discussion: The chloromethylation of PSF was confirmed by ^{1}HNMR spectra. In Figure 1, the ^{1}HNMR spectra of pristine PSF (a) and CMPSF (b) were illustrated. As clearly seen in Figure 1 (b), the CMPSFs show a new peak at 4.4-4.6 ppm (designated as '5') which is assigned to the hydrogen proton of chloromethyl groups ($-CH_2Cl$). According to this figure the DC was 1.43.

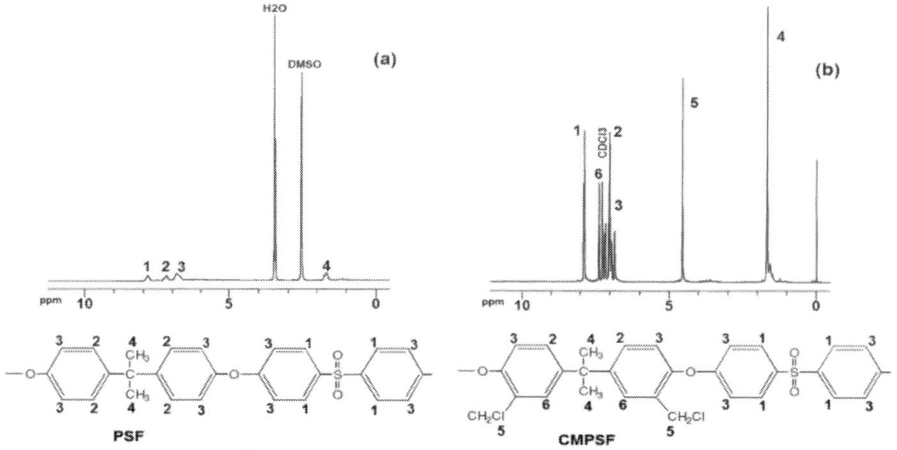

Figure 1- ^{1}HNMR spectra for pristine PSF (a) and CMPSF (b)

Hydroxide conductivity values at 25-80 °C were calculated using resistance obtained from Nyquist plot. As shown in Figure 2, by increasing the temperature, the ionic conductivity values were improved. Obviously, as the temperature increases, the free volume expands and the anion transferring channels become wider, which improves overall mobility of ions and polymer chains and consequently the ionic conductivity increases [6]. The IEC, WU and SR values for the prepared AEM are reported in Table 1.

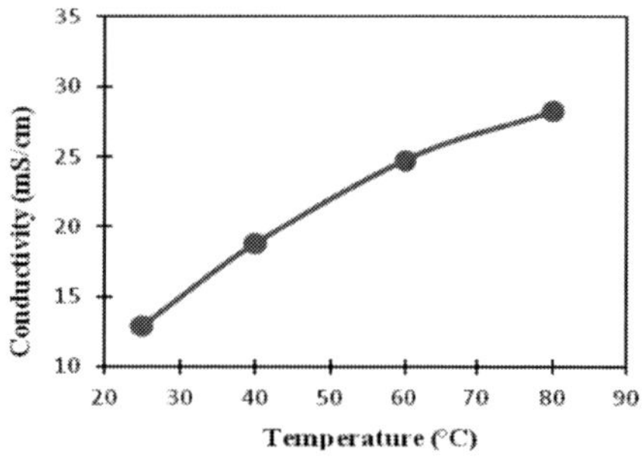

Figure 2- Temperature dependency of ionic conductivity for prepared

Table 1- IEC, WU and SR values for the prepared AEM

IEC (meq/gr)	WU (%)	SR (%)
1.43	38.24	21.95

Conclusions: AEM based on polysulfone containing ammonium groups was prepared by chemical modification method using chloromethylation, amination and alkalization. In amination step, TMHDA was used as both amination and crosslinking agent. The properties of resulted AEM including ionic conductivity, DC, IEC and swelling ratio were measured. According to the ionic transport measurements and water content characterizations, the prepared AEM can be denoted as good candidates for AEMFC application.

References:

[1] Merle G., Wessling M., Nijmeijer K., J. Membr. Sci., 377, 1 (2011).
[2] Couture G., Alaaeddine A., Boschet F., Ameduri B., Prog. Polym. Sci., 36, 1521 (2011).
[3] Avram E., Butuc E., Luca C., Druta I., Sci., Part A: Pure Appl. Chem., 34, 1701 (1997).
[4] Arges C., Parrondo J., Johnson, A., Ramani, V., J. Mater. Chem. 22, 3733 (2012)
[5] Iravaninia M., Rowshanzamir S., Fuel Cells, DOI: 10.1002/fuce.201500154, (2015).
[6] Li X., Wang L., Cheng S., J. Appl. Polym. Sci., 132 (2015).

Effect of polyether sulfone spinning parameters on morphological and mechanical properties

Shadia R. Tewfik [a,b],Mohamed H. Sorour[a,b] , Hayam F. Shaalan[a,b] and Heba A. Hani[a,b]
National Research Centre, Engineering Research Division ,ElBohous St. Dokki, Giza, Egypt . POB. 12622,Tel 0020233370933 - Fax 0020233370931 [a]
email: hayam64@hotmail.com
Chemical Engineering and Pilot Plant Department[b]

Hollow fiber (HF) membranes are regaining wide interest over flat membranes due to its compaction and high area to surface volume ratio. Typical polymers used for hollow fiber fabrication include cellulose acetate, polysulfone, polyether sulfone, polyamide…etc. This work addresses the fabrication of HF from PES using n-methyl pyrrolidone (NMP) as solvent in addition to other components to achieve desired characteristics. The spinning system includes, jacketed reaction vessel, four spinneret block, coagulation and washing baths in addition to dryer and winder. The dope viscosity has been in the range from12000 to 30000 centi-poise. Air gap has been adjusted from 7 to 35 cm and coagulation bath temperature from 20 to 50 °C. The effect of several other parameters has been explored including bore fluid composition, take-up roller speed, dryer temperature……etc.The fabricated fiber has been subjected to post treatment using glycerol solution. The HF membranes have been characterized by scanning electron microscope (SEM), atomic force microscope (AFM) and mechanical properties.Pure water permeability has been evaluated for the fibers under varying conditions.
Results indicate prevalence of finger like structure. AFM study indicated average surface roughness from about 15 to over 70 nm. Profile of stress strain characteristics reveals suitability of the fibers for downstream interventions for fabrication of thin film composite membrane.

Mass transfer investigation of membranes based novel aromatic copolyamides

A Pulyalina, [1,2]G. Polotskaya, [2]M. Goikhman, [2]I. Podeshvo, [1]A. Toikka
Saint Petersburg State University, Institute of Chemistry, Universitetsky pr. 26, 198504 Saint Petersburg, Russia; tel. +78124284805, e-mail:alexandra.pulyalina@gmail.com; [1]Saint Petersburg State University, Institute of Chemistry, Universitetsky pr. 26, 198504 Saint Petersburg, Russia; tel. +78124284805, [2]Institute of Macromolecular Compounds, Russian Academy of Sciences, Bolshoy pr. 31, 199004, Saint Petersburg, Russia; tel. +78123237407;

Polymers of heteroatomic structure are perspective class for membrane fabrication because of their high mechanical strength and heat resistance properties. Structural ordering, fixed free volume and physicochemical properties of polyheteroarylenes make them especially useful as efficient membrane materials for the separation of liquid mixtures. Chemical structure of monomer unit play dominant role in determining physical and chemical properties of polyheteroarylenes including membrane transport parameters. Studies on polymers with anthrazoline units in the backbone have found that advantages of polymers (high-strain-strength properties, heat resistance, processability) and a number of anthrazoline characteristics (optical properties, semiconductor and others) are combined in polymers with anthrazoline units in the backbone.

Thus the aims of this work are the development of the membranes based on novel polyheteroarylenes - aromatic copolyamides containing fragments with anthrazoline units in the backbone and the investigation of its physicochemical and transport characteristics. The effect of anthrazoline fragments inclusion in aromatic polyamide was estimated for dense nonporous films prepared from polyamide (PA) and two copolyamides (coPA) containing 20 and 30 mol.% anthrazoline units in the backbone.

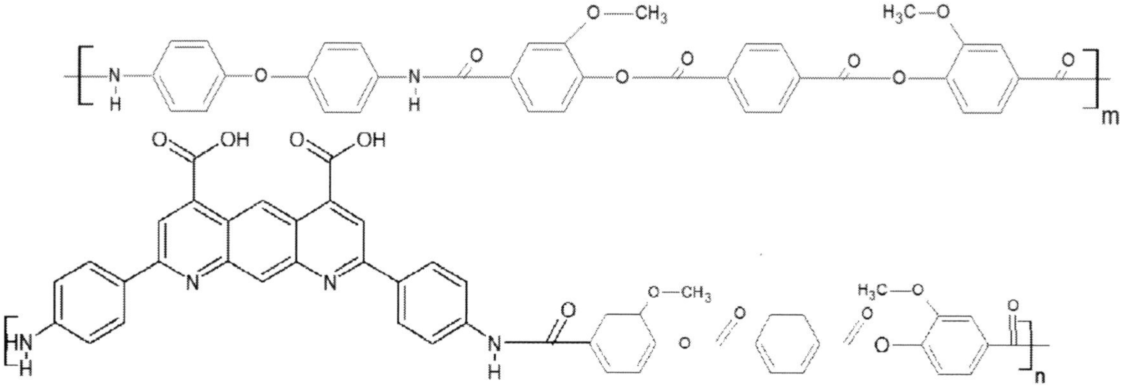

Dense polymer films (30–40 μm thick) were obtained by casting a polymer solutionsin NMP on a glass plate, followed by evaporation of the solvent at 100°C in air. Films fixed on the glass plate were dried to a constant weight at 50°C in vacuum for 5 days.

Scanning electron microscopy and thermogravimetric analysis were carried out for investigation the structure of the membrane and its thermal stability. The mass transfer through membrane was estimated by sorption and diffusion experiments.

The performance of the membranes was studied in pervaporation of commercial significant methanol–toluene mixtures. For the evaluation of intrinsic properties of penetrant-membrane system, permeability and selectivity were calculated. It was found that membranes under study are thermal, mechanical and chemical stable. Glass transition temperature and the density of films such as the value of fraction free volume increases with the introduction of anthrazoline units that indicates on the structure transformation in the polymer films. PA is characterized by the less separation factor and total flux as compared with both copolymers for the entire range of feed concentration. Permeance of both penetrants methanol and toluene goes up with increasing concentration of anthrazoline fragments.

Acknowledgment. The work was supported by the Russian Science Foundation (grant 16-13-10164).
Equipment of Resource Centers of St. Petersburg State University, namely "Centre for X-ray Diffraction Studies", "Thermogravimetric and calorimetric methods of investigation", "Chemical Analysis and Materials Research Centre", Interdisciplinary Resource Center "Nanotechnologies" and Education Resource Centre in the direction of chemistry were used for membranes investigation.

Fluid-solid adsorption from batch to continuous

V. Russo, [1]R. Tesser, M. Trifuoggi, M. Di Serio
[1]University of Naples "Federico II", Chemical Sciences Department, via Cintia, 4, 80126 Napoli, Italia; tel. +39 081674036, e-mail: v.russo@unina.it.

Adsorption is a very versatile separation technique that can be used both for product separation and waste purification, consisting in the selective interaction of one or more compounds from the liquid phase on a solid called adsorbent.

Even if adsorption is a classical separation technique, there is still a lot of confusion and lack of robustness in interpreting correctly the physics of the system. Thus, the development of a robust mathematical model, that describes the kinetics of this phenomenon, is demanding to design adsorption columns. An Adsorption Dynamic Intraparticle Model (ADIM) has been recently proposed and tested [1-2], obtaining good results in terms of adsorption kinetics data fitting. The model is based on the assumption that the mass balances are solved both in the liquid bulk phase and in the particle, considering in this last case both the diffusion in the liquid pore and the migration on the solid surface. Both the mentioned papers are based on the data collected in the literature, with no possibility to check its high flexibility being the literature lacking of complete datasets where all the possible operative variables are investigated.

In the present work, an experimental effort has been made performing an adequate experimental study chosing methylene blue adsorption over silica as a case study. It has been possible to demonstrate that starting from the mass transfer parameters obtained from batch experiments, it is possible to correctly predict the behavior of adsorption columns, by installing a lab-scale equipment. The adsorption column simulations have been performed by modifying the mass balances equations, obtaining a three-dimension model, where concentration profiles are solved in both dimensionless residence time (τ), reactor axial position (z) and particle radius (r_p) coordinate; see Figure 1 as example of dimensionless concentrations in the bulk (C/C_0) and liquid inside particle (C_L/C_L^{MAX}) phases. With this model it has been possible to forecast the breakthrough curves of the continuous columns, obtaining good results in terms of simulations.

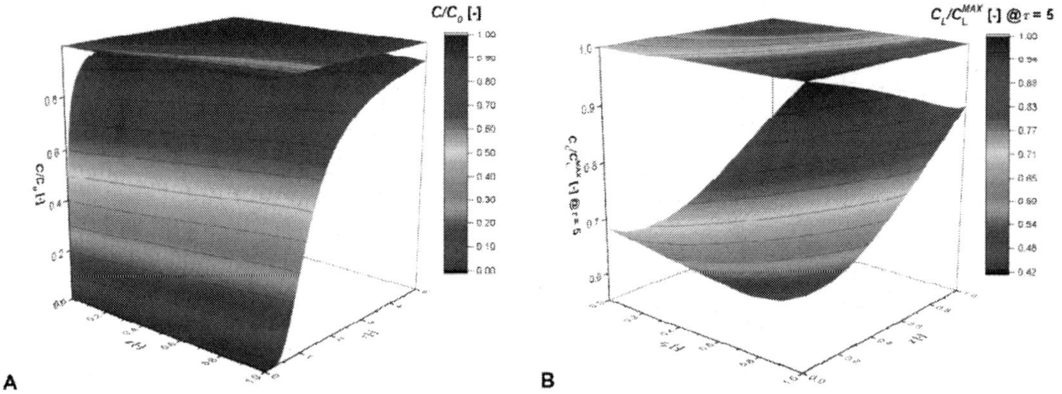

Figure 1 – A. bulk dimensionless concentration; B. liquid inside particles at τ=5.

References

[1] V. Russo, R. Tesser, M. Trifuoggi, M. Giugni, M. Di Serio, Computers and Chemical Engineering 74 (2015) 66-74.

[2] V. Russo, R. Tesser, D. Masiello, M. Trifuoggi, M. Di Serio, Chemical Engineering Journal 283 (2015) 1197-1202.

Experimental and numerical analysis of a ribbed hydrocyclone for separation of particles

Gayatree Patra[1], Sudipto Chakraborty[1], B. C. Meikap[1, 2]
[1]Department of Chemical Engineering, Indian Institute of Technology (IIT) Kharagpur, West Bengal- 721 302, India; e-mail: gayatree@che.iitkgp.ernet.in;
Tel. +91 -3222 283942, e-mail: sc@che.iitkgp.ernet.in;
Tel. +91 -3222 283958, e-mail: bcmeikap@che.iitkgp.ernet.in;
[2]Chemical Engineering, School of Engineering, Howard College, University of Kwazulu-Natal, Durban, South Africa

The effects of ribs on hydrocyclone performance are investigated by experimental and as well as FLUENT software. Hydrocyclone operation is simple but the multiphase flow behaviour inside a hydrocyclone is complicated. Computational fluid dynamics (CFD) codes are useful to study the flow behaviour, but because of the complex flow structure, numerical simulation is at present not able to completely substitute experiments but can reduce, to a certain extent, experimental cost for design and optimization. The Reynolds averaged Navier- Stokes equations are solved by using finite volume method based on pressure correction SIMPLE algorithm.

Experimental set-up: According to Rietema's optimum geometry, a conventional type of hydrocyclone was designed and a spiral rib of 8 mm width and 1 mm thick was inserted inside the cylindrical portion of the hydrocyclone. The hydrocyclone was made of Perspex material having a cyclone diameter of 15.2 cm. The length of the cylindrical portion is 22.8 cm and that of the conical portion is 38 cm. The liquid is injected tangentially by a rectangular inlet of dimension of 5.08×2.54 cm^2 fitted tangentially to the cylindrical portion of the hydrocyclone. The hydrocyclone was operated from 4 kPa to 13 kPa. A large tank having capacity of 500 l is provided to collect recycled liquid from apex pipe and overflow pipe and recycled back to inlet section of the hydrocyclone. The slurry is pumped into the hydrocyclone by a 0.5 H.P. centrifugal pump. Sand particles having density 2500 kg/m^3 was used for the experimental study of the hydrocyclone. Fine sand particles having size ranges from 50 μm to 200 μm and tap water was used for the preparation of slurry. A number of samples having different concentrations were prepared, like 2 kg of sand is mixed with 100 l of water. The centrifugal pump injected the slurry at a very high velocity to the hydrocyclone. The slurry volumetric flow rate was 1.2-3 m^3/h. The efficiency of separation is calculated as

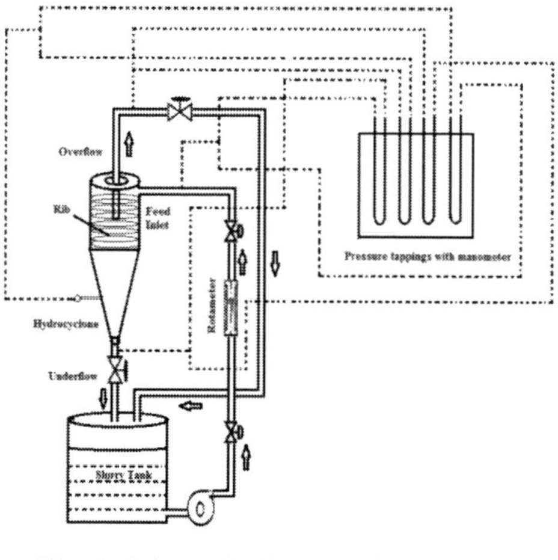

Fig. 1: Schematic diagram of experiment set up

$$\eta = \frac{\text{Inlet particle(wt\%)} - \text{Overflow particle (wt\%)}}{\text{Inlet particle (wt\%)}} \times 100$$

CFD simulations: The geometry of helical ribs is created in AUTOCAD and is inserted in hydrocyclone. The flow field was modelled using commercial CFD code fluent version 15.0.

Reynolds stress model (RSM) for highly swirling flow and volume of fluid (VOF) for multiphase interactions are used. At the inlet 'velocity-inlet' and at both outlets 'pressure-outlet' boundary conditions are defined. For discretization of partial differential equations, finite volume method has been used.

Results and discussion: Fig. 2.shows that the separation efficiency increases with increase in flow rate and the efficiency is higher for with ribs than without ribs. This may be due to insertion of ribs, the hydrocyclone guide the fluids better and the particles lose their kinetic energy faster. The tangential velocity of a ribbed hydrocyclone is higher as compared with a conventional hydrocyclone under similar geometry and operational conditions. This result is compatible with the results of collection efficiency. From the CFD result, the pressure contour shows that the pressure drop of a ribbed hydrocyclone is lower than that of without ribs. Similar results are found from the experiment also.

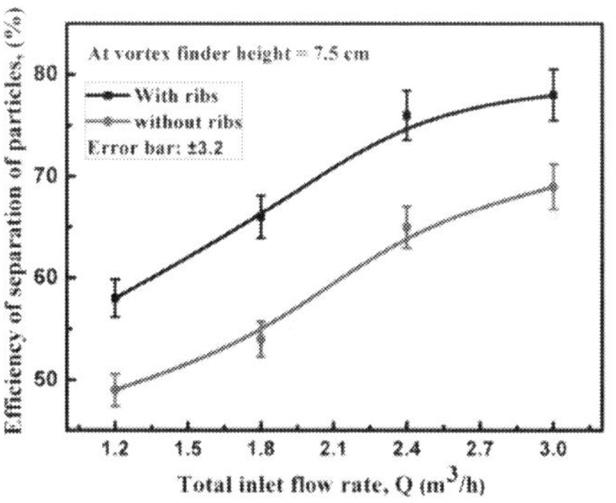

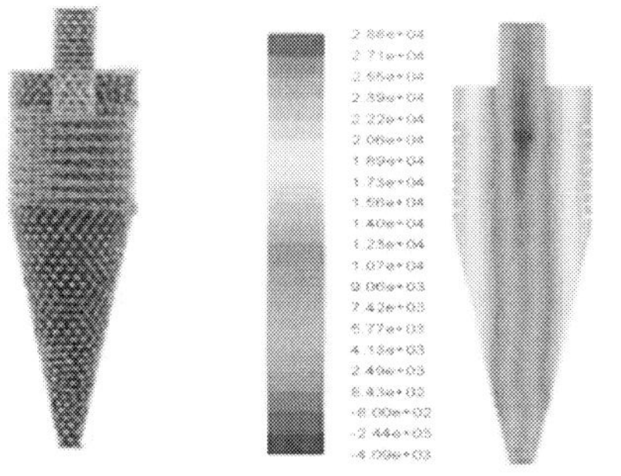

(a) (b)

Fig. 2: Effect of total flow rate on separation efficiency

Fig. 3. CFD grid generation (a) and pressure contour (b)

Conclusion: Pressure drop in a ribbed hydrocyclone is lower than that of conventional hydrocyclone. Separation efficiency of a ribbed hydrocyclone is better than that of without ribs under the same geometry and operating conditions.

References:

1. S. M. Noode Farahani, V. Tahmasbi, H. Safikhani and A. Abbasi, Eng.Appl.comp.Fluid 5(2), 180-187 (2011).
2. R. Sripriya, M. D. Kaulaskar, S. Chakraborty, B. C. Meikap, Chem. Eng. Sci. 62(22), 6391-6402 (2007).
3. Ghodrat, S. B. Kuang, A. B. Yu, A. Vince, G. D. Barnett, P. J. Barnett, Miner. Eng. 63, 125-128 (2014).

Microalgae biomass harvesting from anaerobic effluent: pH effects

S.Cassini, [1] S. Francisco, R. Keller, P.W.P. Antunes
Univ. Federal Esp. Santo (UFES), Dept. Environmental Engineering, Av. Fernando Ferrari 541 Campus Goiabeiras, 29075-910 Vitoria, ES., Brazil.

The recovery of the biomass microalgae (BMA) is a critical step in the separation process biomass for various applications such as bioproducts and biofuels (Milledge & Heaven 2013). Among the existing processes for biomass, harvesting from the aqueous medium, the coagulation / flocculation and precipitation is the preferred choice as the most economically advantageous process (Uduman et al. 2010, Zeng et al. 2011). In the case of microalgae recovery biomass, the pH values observed in the growth environment of the microalgae biomass is always greater than 9-10, turning it a true alkaline medium for solid-liquid separation. At this pH range, most coagulants used for biomass recovery require higher doses or simply do not work. The objective of this study was to evaluate alternative coagulants for grown BMA recovery in anaerobic effluent and characterize effluent quality after recovery. Tanfloc Pop®, cationic starch SUPERION® 300 and Aluminum Sulfate promoted removal of more than 90% of biomass concentrations greater than 80 mg / L, at pH 7 to 10, wherein the cationic starch which coagulant was promoted most biomass removal at various pH values evaluated. The seeds of *Moringa oleifera* and the *Hibiscus esculentus* polymers used for coagulation flocculation worked only on average in the acid pH range. The liquid final effluent generated after sedimentation of the biomass showed higher removal than 80% for phosphorus and nitrogen and greater than 50% BOD and COD, using the coagulant of aluminum sulfate, cationic starch and Tanfloc Pop® SUPERION® 300. In conclusion, natural polymers as coagulants in a wide pH range can replace the most common coagulants used in water treatment and for microalgae harvesting, without decreasing the biomass removal efficiency and with increasing water quality of final effluent by nitrogen and phosphorus removal, turning it high quality for wastewater treatment.

References

Milledge, J. J.; Heaven, S. A review of the harvesting of microalgae for biofuel production. Reviews In Environmental Science and Biotechnology, Springer, v. 12, n. 2, p.165-178, 2013.

Uduman, N. et al. Dewatering of microalgal cultures: A major bottleneck to algae-based fuels. Journal of Renewable and Sustainable Energy, American Institute of Physics, v. 2, p.1-15, 2010.

Zeng, X. et al. Microalgae bioengineering: from CO_2 fixation to biofuel production. Renewable and Sustainable Energy Reviews, Elsevier, v. 15, p.3252-3260, abr. 2011.

Structural characterization of chemically and physically separated asphaltene

S.Kananpanah[1, 2*] , M. Bayat[2], M.A.Moosavian[1],R.kheirkhah[3], A.Sadr[1]
1. Department of Chemical Engineering, Faculty of Engineering, University of Tehran, Iran;
2. Faculty of Research and Development in Downstream Petroleum Industry, Research Institute of Petroleum Industry (RIPI), Tehran, Iran;
3. Nargan Company, Tehran, Iran
*kananpanahs@ripi.ir

One of the approaches to prevent asphaltene deposition is removing aggregated asphaltene from crude oil by de-asphaltene methods in downstream process. Changing in temperature, pressure and composition can cause asphaltene flocculate and then these flocculated asphaltene can separate from crude oil. The method of de-asphaltene procedure can affect the structural properties of asphaltene. The parameters that affect the separation of the asphaltene are directly enhanced by their physicochemical properties, for this reason the experimental separation techniques are extremely important.

In this study, asphaltene are isolated chemically and physically from an Iranian heavy crude oil under three different experimental conditions by using heat (thermal de-asphaltene), heat and toluene (thermal-toluene de-asphaltene) and n-heptane (n-heptane de-asphaltene) as precipitation issue. The effect of isolation method on the crystallite structure and surface morphology of asphaltene is characterized by X-ray diffraction (XRD) and scanning electron microscope (SEM) technique, respectively.

Iranian dead crude oil samples were chosen for this study. The API of crude oil is 18.9 (by using ASTM D 1298). The Standard IP-143 is used for asphaltene content measurement; the wt% content of asphaltene is 11 for studied crude oil. N-heptane and toluene was supplied by Merck Chemical Co., Germany.

In the first de-asphaltene method, physical thermal de-asphaltene procedure was used. A high-pressure, jacket reactor, equipped with a stirrer and temperature-controlled oil bath, was used for heating the crude oil at 60°C temperature. The crude oil was heated for 30 minutes before being filtered through a sinter glass filter using a vacuum pump. Asphaltene particles at fist trapped and accumulated through sinter glass pores and then gradually the asphaltene accumulated on the surface of the sintered glass filter. Actually two types of asphaltene particles presents; the aggregated asphaltene on the surface of sinter glass filter (type A) and aggregated asphaltene which trapped in sinter glass pores (type B). Firstly, Type A asphaltene remove directly from the surface of sintered glassed filter. Secondly, type B asphaltene was dissolved in toluene and separate from sintered glassed filter. Then all sample asphaltene (type A and B) oven dried till constant weight percent of asphaltene aggregates was achieved. In the third de-asphaltene method modified IP-143 procedure was used to achieve n-heptane asphaltene in de-asphaltene method as type C asphaltene. Extracted asphaltene sample was characterized to compare the effect of separation method on asphaltene morphology and structural parameter by using XRD and SEM analyzer. The X-ray patterns for all extracted asphaltene in present paper are in general very similar, and each contains the γ and graphene layer stacking peak at respected angles of asphaltene particles. The positions of the γ band is located around 24(°2θ), and the position of gamma band is around 29(°2θ) for all isolation technique with Cobalt anode XRD. The distance between the saturated portions of the molecules $d\gamma$ was estimated from the maximum of the gamma band and layer distance between aromatic sheets d_m was estimated from the maximum of the graphene band which is reported in Table1. The average height of the stack of aromatic sheets perpendicular to sheet surfaces L_C was between 27 and 42 Å. Finally, number of aromatic sheets in the stack of aromatic sheets (M) was in the range of 8 to 13. The results confirm the more tendencies of condensed aromatic sheets to form stacks in the type C asphaltene structure. Resulted particles of N-heptane de-asphaltene have the lowest value of dm and consequently the highest value of L_C (42 Å) and M (13) as reported in table.1.

Table 1.XRD structural parameters of asphaltene aggregated by three different methods

Asphaltene Isolation method	dm(Å)	dγ(Å)	L_C(Å)	M
Thermal de-asphaltene at 60°C	3.49	4.30	27	8
Thermal-Toluene de-asphaltene at 60°C	3.50	4.28	34	10
N-heptane de-asphaltene	3.48	4.13	42	13

Based on IP143 the relaxation time for asphaltene precipitation is the prevalent factor for stacking molecules. The precipitation times in n-heptane de-asphaltene can cause asphaltene particles preferably stack to generate asphaltene clusters while in thermal de-asphaltene just 30 minutes of relation time was observed. Consequently the number of aromatic sheets in stacked cluster has its maximum value of 13 for extracted asphaltene by IP143 test method. On the other hand the number of aromatic sheets in stacked cluster of toluene washing of thermal extracted asphaltene is more than thermal aggregated asphaltene. Actually by adding extra precipitation time to asphaltene particles by dissolving in toluene, asphaltene particles aromatic sheets steadily stacked to grow around 34 Å.

Figures.1A, B and C show the images of SEM, used for investigation of asphaltene surface morphology related to asphaltene type A, B and C respectively.

Figure1. SEM images of asphaltene type A, B and C from left to right

Surface morphology study show pore and cavities and also agglomerated and smooth surfaces of asphaltene. SEM show smoother and more semi-solid surfaces for toluene washed asphaltene particles while more micellar formation was observed for thermal isolated asphaltene particles. It is also found that by adding n-heptane to the oil sample with longer precipitation time in compare with thermal de-asphaltene precipitation time, tighter surface morphology achieved. Appearance of pores in the SEM images was used as an indicator of the resin detachment.

As results shows (table 1 and figure 1) extraction procedure has a strong influence on the physicochemical properties of the isolated asphaltene. The results showed that at thermal de-asphaltene solid like micellar shape was the dominant morphology of asphaltene particles which cause the least height of the crystallites and number of aromatic sheets in a stacked cluster. This morphology change to semi-solid smooth surface and agglomerated asphaltene with cavities by altering the separation method to thermal-toluene de-asphaltene and n-heptane de-asphaltene. The maximum heights of the crystallites belong to n-heptane de-asphaltene.

Drying of sludge from waste water treatment plant using solar energy

Awni Al-Otoom[1], Fahmi Abu Al-Rub[1], Hasan Mousa[1,2], Mays Shadeed[1]
[1]Dept. of Chemical Engineering, Jordan University of Science and Technology, P. O. Box 3030, Irbid 22110, Jordan
[2]Department of Petroleum and Chemical Engineering, Sultan Qaboos University, P. O. Box 33, P. C. 123, Muscat, Oman

The work in this study presents a newly developed system to utilize solar energy to dewater the sludge in a semi continuous mode. The semi continuous dryer mainly consists of a moving belt on which aluminum bars are fixed to hold the sludge and a fan used to drive the air at the required speed. The effects of operating conditions, such as air velocity, belt speed, and mode of contact between the sludge and air, on the drying rate of water were investigated. The drying rate of water was measured under the effects of different operating conditions including air velocity, belt speed, and mode of contact between the sludge and air. The experimental results proved the technical feasibility of the developed solar drying system to efficiently evaporate water. The achieved drying rates were 2.02 kg water/m^2. hr and 0.49 kg water/m^2.hr in September and February, respectively. Under a given insolation rate a high evaporation rate is achieved at large air velocity and slow motor speed.

An empirical relationship to predict the mass transfer coefficient (drying flux constant) was derived by conducting a well-designed experiment using a tray dryer. The data obtained successfully described the mass transfer coefficient expressed as $J_D=15.1Re^{-0.72}$.

[1] Correspondence author; email: awni_otoom@just.edu.jo

Hydrodynamic study of a fluidized bed reactor used for the wastewater treatment

Khaled Ghozlane, [1]Kechroud Nassima, [1]Bourouina-Bacha Saliha, [2]Bourouina Mustapha

Department of Chemical Engineering, Faculté de technologie, Université de Bejaia
Department of Chemistry, Faculté des Sciences Exactes, Université de Béjaia
Email : kghouzlane@yahoo.fr
Phone : +213 793344277

In order to study the performances of a fluidized bed reactor of active carbon for wastewater treatment, preliminary essays were conducted to characterize the flow thanks to the residence time distribution experiments (RTD).
The use of the packed bed reactors is generalized in the wastewater treatment due to the ease and the reliability of this operation. Nevertheless, several problems (such as excessive pressure drop and fouling) are significantly reduced in fluidized-bed reactors (FBR).
The response surface methodology (SRM) is widely used in order to improve the process yield, reduce development time and overall process costs. The aim of our paper is to apply this methodology in order to model and optimise the flow in the fluidised bed column. To investigate the flow through the column, RTD tests are performed with NaCl tracer. A statistical experimental design based on half central composite design (CCD) was employed.
The two chosen independant variables, found to influence significantly the flow in this reactor, were temperature and flow rate. The responses were the pressure drop and the RTD.
The results were analyzed statistically using Minitab 16 (ANOVA) to define most important process variables affecting the parameters of flow model (Peclet number, axial dispersion coefficient, Number of stirred tank in series)..
These RTD experiments form the basis for the study of mass transfer in fluidized bed reactors because it allows us to deduce the performance of the adsorption operation whether in open or closed circuit.

Key words : Residence time distribution, fluidised bed, NaCl, modelling. Central Composite Design

Modeling and optimization of adsorption processes on activated carbon in fixed bed reactors. Application to the treatment of phenolic wastewater

H. Guenounou, [1]S. Bourouina-Bacha , [2]M. Bourouina, [3]F. Oughlis-Hammache
University of Bejaia, Department of Chemical Engineering, faculty of technology, Department of Chemistry, Bejaia, Algeria. tel: +213 778855701, email: guenounou_h82@yahoo.fr

The objectif of this study is the dynamic modeling of phenol adsorption process on activated carbon in a fixed bed reactor operating in closed circuit. On the basis of surface diffusion model, a simplified model has been successfully applied to the simulation of this process. the evolution of simulated concentrations in adsorbed phase and in a liquid phase along the bed and into the reservoir are compared to those obtained experimentally based on various operating conditions, such as the flow rate (from 0.67 to 1.34ml/s), the temperature of the phenol solution (from 25 to 35°C), the hight of the bed (from 8 to 14cm) or initial phenol concentration (from 35 to 85mg/L). Adjustement errors are all less than 1%. The estimated values of the overall mass transfer coefficient Kg vary between 0.21 and 0.48 cm/min. The overall coefficient always increases with the flow rate, whatever the temperature or the operating conditions, which is a very clear indication that turbulence promotes adsorption by eliminating the film resistors on the activated carbon surface.

Keywords: Adsorption, Mass transfer, Modelling, Simulation

High shear impellers use effect on the leaching rate of silver from refractory mineral

R. Juarez-Mata[1], A. Alonso[2], J. Ramírez-Muñoz[*,1], R.M. Luna-Sanchez[1]
[1]Universidad Autónoma Metropolitana-Azcapotzalco, Departamento de Energía. Av. San Pablo 180, C.P. 02200, Mexico City, Mexico.
[2]CoNaCyT-Universidad Autonóma Metropolitana-Azcapotzalco, Departamento de Energía. Av. San Pablo 180, C.P. 02200, Mexico City, Mexico. *jrm@correo.azc.uam.mx

Abstract

It has been showed in previous works that is possible to reach silver recoveries from some mineral concentrates using thiourea system. It has been calculated that the leaching rates are more than three times faster than those results obtained from cyanide process. In this work, a typical high shear impeller (HSI), Norstone® (NT), was used in the agitation of a leaching tank for the recovery of silver from a refractory mineral. The NT performance was compared with others agitators, such as Rushton turbine (RT) and stirring solutions with magnetic bar with and without ultrasound waves (USW), in thiourea solutions. Results shows that both, the USW and the NT, break the preg-robbing effect, which inhibits silver recovery due to the presence of naturally occurring carbonaceous matter, reaching silver recoveries near to 70% in both cases.

Introduction

Metal recovery from minerals and concentrates, is performed according to values content on it. Pyrometallurgy processes is commonly used to obtain several base metals like copper, lead or tin, due to its high concentration in most of the minerals treated. Gold and silver are often contained in concentrations below to 0.01 and 0.1 %, respectively. Pyrometallurgical processes has low energetic efficiencies, which leads to the use of other alternatives [1].

Thiourea is one of the most promising solution, capable to be applied in the refractory ore treatment where cyanide has been failed [2]. In further works, also has been demonstrated that silver electrodeposition can be used as a process separation [3]. Despite these achievements in the solution chemistry, the hydrodynamic of agitation process during the leaching stage has received little attention. The present work was focused on the study and comparison of a NT and RT impellers used in the agitation of leaching tank for the recovery of silver from a refractory mineral, in with thiourea solutions.

Methodology

A Mexican ore named Fresnillo-Zn was used. Its silver content was evaluated in approximatively 20 kg/t using a wet analysis. Mineral was milled and screened, the size between -100+200 mesh was used in all the experiments. Thiourea solutions were prepared with 0.4 mol/l thiourea, 0.1 mol/l H_2O_2 as oxidant agent and H_2SO_4 to adjust pH value on 1.

In order to stablish a comparison point, two more experiments were carried out. In the first, the stirrer media was a magnetic bar and an USW was added, in order to improve the leach rates; 100 ml of a thiourea solution with the same composition as in the NT and RT experiments was used. In a second experiment, the leaching agent was cyanide; 100 ml of 0.1 mol/l NaCN solution was prepared, at pH 10.5. All reagents used were reactive grade and deionized water was employed. In all cases the solid-liquid ratio was 1:5.

NT and RT impellers were attached to an equipment with power and rotation speed control in order to ensure the same power input into the leaching solution, which was set in 25 W for all experiments. Besides, a solution total volume of 1.5 liters was used with the NT and

RT impellers. All tests were carried out in three replicates in order to ensure reproducibility. The metal analysis was carried out in a Spectrophotomer.

Results and discussion

The silver extraction percentage as a function of leaching time is plotted in Fig. 1. The cyanidation reached only a 20% of silver extraction (X marker) according to the refractory character of the ore. The use of a magnetic bar with 100 ml of solution (◆marker) gives 30% extraction after six hours; in this case it is important to note that after the first sample (15 min) where the extraction was more than 40%, silver seems to leave the solution. This effect has been reported by Rong-yu [4] and is known as preg-robbing, caused by the adsorption of the silver thiourea complexes on the mineral surface. This phenomenon appears also when RT and NT were used, but in the last one, looks like the preg-robbing effect is broken after two hours reaching similar values to that obtained with the USW inclusion, which curve is ascending for all times. It is possible that the high shear effect induced by USW and NT into the solution produce the effect of releasing the silver complex adsorbed in the mineral surface.

The residual solid was analyzed after the leaching, on a SUPRA VP 55 microscope applying energy dispersive X-ray spectroscopy (EDS). Results on Table I shows only the atomic percentage for lead, zinc and silver in order to stablish the phases that were liberated by the effect of the stirring system used. While magnetic bar and RT presents a marginal outcome on the exposed phases with metallic sulfides, the use of NT and USW allows higher amounts, mainly for silver. On the other hand, in the cyanide system there is an important liberation of Zn but not Ag, maybe zinc leaching is the cause of the refractory for this ore. This explains why the high shear stirring methods leads to a significant extraction improvement.

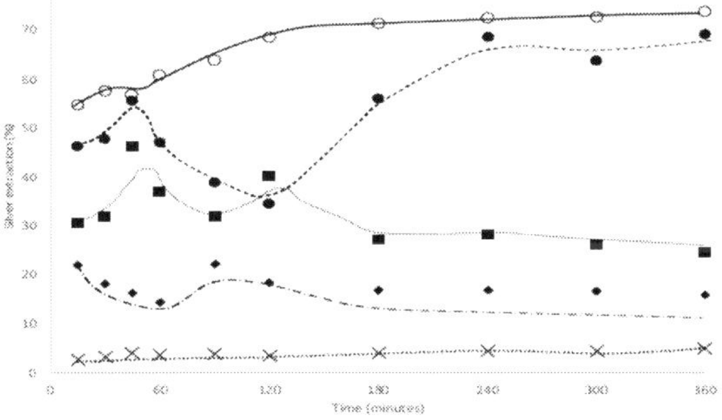

Figure 1. % Silver extraction percentage as a function of leaching time for the systems CN (X), magnetic bar ◆,RT ■,NT ●, USW O.

Table I. Atomic percentage in the residual solid

	Atomic %				
	CN	Magnetic bar	RT	NT	USW
Pb	0.98	0.41	0.56	1.26	2.63
Zn	7.00	0.30	0.17	1.81	1.91
Ag	--	0.05	0.06	1.21	1.62

Conclusions

Silver extraction from a refractory mineral can be improved by using an alternative chemical system like thiourea. The combination of thiourea with a high shear stirring, like USW or HSI increases the silver extraction outstandingly. Nevertheless, the USW presents problems for implementation in industrial processes batch. On the other hand, a Norstone HSI gives similar results than USW and tangible possibility for its adopting in any industrial facility.

References

[1] Habashi, F., 1999. A textbook of hydrometallurgy. Métallurgie Extractive Québec, Canadá, ISBN 2-980-3247-7-9.

[2] Poisot, M.E., Martínez, I., Lapidus, G.T., 2008. Electrodeposition of a Silver-Gold Alloy (DORÉ) from Thiourea Solutions in the Presence of Other Metallic Ion Impurities, Hydrometallurgy 93(23-29).

[3] Poisot, M.E., Alonso, A., Martínez, I., Lapidus, G.T., 2012. Influence of oxygen reduction and hydrogen evolution in the gold and silver direct electrodeposition process from thiourea solutions in a filter press type reactor, Hydrometallurgy 129-130 (90-96).
[4] Rong-Yu W., LeVier, K. M., Clayton, R. B., 1994. Hydrometallurgical process for the recovery of precious metal values from precious metal ores with thiosulfate lixiviant. US Patent US5354359 A.

Effect of various polycarboxylates on the rheological and physicomechanical properties, of different local cements

M. Samar , A.Benmounah
samarmadjid2000@yahoo.fr., benmounah2000@yahoo.fr
Research Unit: Materials, Processes and Environment (UR/MPE), Boumerdès University, Algeria

R.Kheribet , M. Saidi
k_rabia2007@yahoo.fr, saidimoh@yahoo.fr.
Research Unit: Materials, Processes and Environment (UR/MPE), Boumerdès University, Algeria

Polycarboxylate superplasticizers (PCP) can significantly reduce the amount of mixing water and can produce concrete of low water / binder ratio.
However, it is well known that hydration, setting or curing concrete may be affected by the addition of dispersing additives. This can lead to undesirable phenomena such as the loss of workability, a long delay of cement setting or premature of stiffening of the cement paste. This is called incompatibility cement / adjuvant. To contribute to the understanding of these phenomena, we studied the influence of three polycarboxylate superplasticizers, from different manufacturers, (A3, T12 and M145), on the rheological behavior , the physicomechanical properties and the hydration process of four different Algerian cements: sulphate resistant cement (CRS), pozzolanic cement (CPZ) blast furnace slag cement (CPL), and a cement with addition of limestone (CPC). To this end, we used the AR2000 rheometer, analysis (DTA-TGA) and the (XRD) in order to understand the influence of each polycarboxylate and of the cement composition .
The results of mechanical strength test according to the age of adjuvanted mortars with W / C = 0.35 showed that:
- A3 has given the best mechanical resistances of the four cements and at all ages. This reflects its good interaction with cements.
- T12, also gave good resistances with all cements used.
- M145, has even decreases the resistance of CRS and CPL, this is due to its bad compatibility with these cements.
These results are confirmed by the DTA-TGA analysis.
Spreading measures with mini-cone of cement slurries were also made at 5, 60 and 120 minutes after mixing and at temperatures of 20 ° C and 35 ° C.
In conclusion, we can say that the action of polycarboxylates on hydration, physico-mechanical and rheological properties of cement varies from one polycarboxylate, to another and from one type of cement to another.

Keywords: *cement, polycarboxylate, DTA-TGA, XRD, shear stress, plastic viscosity, mechanical strength.*

Optimization of thermal de-asphalting process

S. Kananpanah[a], R. Kheirkhah[b*], M.A. Moosavian[c] ,M. Bayata, M. Bahari[a]
[a] Research institute of petroleum industry, Tehran, Iran
[b] Nargan Company, Tehran, Iran(r.kheirkhah@nargan.com)
[c] University of Tehran, Chemical engineering department, Tehran, Iran

In the present study, the optimal dissociation temperatures of asphaltene separated from crude oil in thermal de-asphaltene at elevated temperatures (100-170°C) has been investigated. The dependence of asphaltene particle size and aggregation on temperature has been inquired by dynamic light scattering technique to investigate the best temperature for maximum asphaltene separation in thermal de-asphalting, in the absence of n-alkane for Iranian heavy crude oil.

Iranian dead crude oil samples were chosen for this study. The API of crude oil is 18.9 (by using ASTM D 1298). The Standard IP-143 is used for asphaltene content measurement; the wt% content of asphaltene is 11 for studied crude oil. N-heptane and toluene was supplied by Merck Chemical Co., Germany. Thermal de-asphalting procedure in which asphaltene particles are aggregated via heating crude oil at an elevated temperature is used for separation of asphaltene from crude oil. A high-pressure, double layer reactor, equipped with a stirrer and temperature-controlled oil bath, was used for heating the crude oil at various temperatures in the range of 100-170°C. All extracted asphaltene samples were characterized using Dynamic Light Scattering (DLS) analyzer.

Figures 1 show average diameter of particles/aggregations by intensity (intensity (%) versus size (nm)) at various isolation temperatures (100-170°C). As observed in Figure.1, increasing temperature from 100 to 150 °C increases the size of asphaltene particles. In addition, by increasing the isolation temperature above 150°C, the average particle sizes decreases by increasing the temperature up to 170°C. According to Figure.1, by increasing the isolation temperature above 100°C, the average size of particles increases from 142 to 865 nm. Based on colloidal aspects asphaltene is partly colloidal and partly dissolved in crude oil, stabilized by resin adsorbed on their surfaces[1]. Changing temperature alters resin concentration in the oil and then the fraction of asphaltene active surface sites [1-3]. Kinetic energy of asphaltene also increases by increasing number of empty active sites on the asphaltene surface have increased due to resin leaving the asphaltene surface.

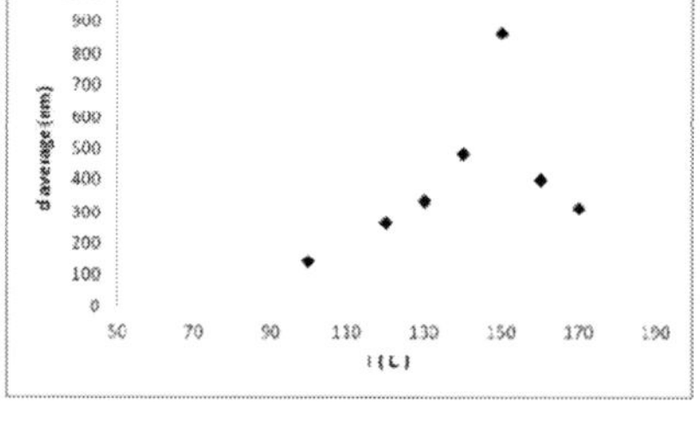

Fig1. The average diameter of asphaltene aggregates versus temperature in thermal de-asphalting process

The reversibility of asphaltene dissociation mechanism is proven by this phenomenon. The kinetic energy of asphaltene particles will increase, in other words, the large colloidal state asphaltene and asphaltene particles are broken and re-dissolved in crude oil by partial reversible growing mechanism. The asphaltene precipitation is partially reversible under changing temperature. The size of asphaltene particles has decreased from 865 (at 150°C) to 315 nm (at 170°C). At lower temperatures (100-150°C), larger aggregates form, which appear to break into lower mass at higher temperatures (150-170°C). As far as separation process is concerned, the temperature at which the maximum particle size has occurred can be selected as the optimal separation temperature due to the maximum size of asphaltene aggregates. Actually this maximum temperature should provide proper size separation for thermal de-asphaltene filtration by appropriate filtration rate and consequently the maximum yield of asphaltene separation. This claim is evaluated in the next section. The yield (figure.2) at different temperatures was measured through thermal de-asphaltene technology. The highest yield of asphaltene separation was achieved at 150°C, as observed in Figure.2.

The maximum yield of asphaltene at 150°C was decreases by increasing temperature up to 170°C. Maximum yield of asphaltene separation is the main objective of de-asphaltene process. The maximum size of asphaltene particles at optimal temperature (150°C), leading to maximum efficiency of asphaltene separation.

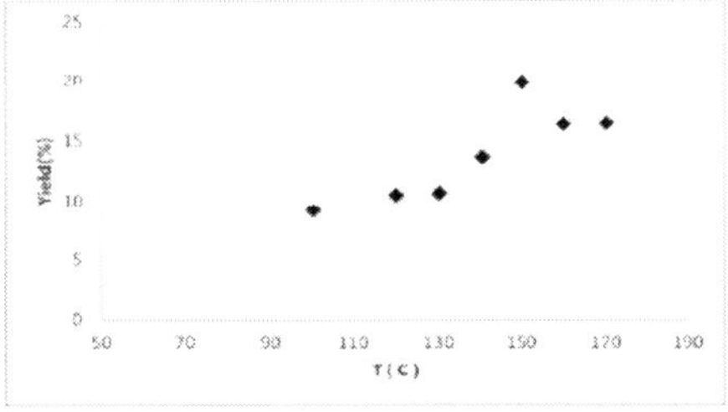

Fig.2 Yield of asphaltene separation in thermal de-asphalting process

Reaching and finding the optimal temperature for maximum separation efficiency is an important part in separation process studies. Moreover, the reversibility of aggregation has occurred by increased temperature over the optimal value. This observation also relates reversibility of agglomeration of asphaltene particles to separation temperature. In other words, given the type of crude oil studied, observation of reversibility conditions should lead to a specific temperature.

References

1. Park, S.J. and G. Ali Mansoori, Aggregation and Deposition of Heavy Organics in Petroleum Crudes. Energy Sources, 1988. 10(2): p. 109-125.

2. Gawrys, K.L., P. Matthew Spiecker, and P.K. Kilpatrick, The Role of Asphaltene Solubility and Chemical Composition on Asphaltene Aggregation. Petroleum Science and Technology, 2003. 21(3-4): p. 461-489.

3. Hirschberg, A., et al., Influence of Temperature and Pressure on Asphaltene Flocculation. SOCIETY OF PETROLEUM ENGINEERS JOURNAL, 1984. 24(3): p. 283-293.

Role of flocculation-sedimentation on OMW treatment processes

G. Hodaifa, J.A. Páez, C. Agabo, E. Ramos
Pablo de Olavide Univ., Dep. Molecular Biology and Biochemical Engineering, Chemical Engineering Area, 41013 Seville, Spain; tel. +34 954 978 206, e-mail: ghodaifa@upo.es.

The high olive-oil production entail large volumes of washing wastewaters (OMW) with high organic matter (COD ≈ 2-35 g O_2/L and a pH ≈ 5-9). The usualsolution for these wastewaters is the storage/evaporation in rafts. Nowadays,the disposal of OMW is one of the most serious environmental problemsdue to its high organic matter and the presence of highly phytotoxic compounds such as phenolic compounds, pesticides, etc.

The application of Advanced Oxidation Processes (AOPs) isone of the most efficient OMW treatments [1,2]. AOPs are very promising method producing high quality effluent but their operating costs normally exceed other treatments, e.g. biological treatments [3].In this sense, the combination of a physico-chemical pre-treatment with an Advanced Oxidation treatment is gaining importance as an interesting alternative.

In this work a pre-treatment by flocculation-sedimentation is proposed owing to its successful ability to removing suspended and colloidal fraction from OMW. Three coagulant-flocculants have been analyzed with and without pH control: Bentonite as a mineral coagulant, Flocudex CS-51 as a cationic synthetic flocculant and QG-2001 as an anionic synthetic flocculant.

Fig. 1 illustrates the rate of sludge formation volume (volume of solid occupied), V = f(t), during the flocculation operation for different concentrations of Flocudex CS-51flocculant.A small addition of Flocudex CS/51 (5 mg/L) significantly improves the solid phase separation, increasing the sludge volume to 14 cm^3/L. Increasing flocculant doses from 5 to 400 mg/L does not imply an increase in the volume of sludge formed (Fig. 1). Only, in the dose of 100 mg/L this sludge volume increasedup to 20 cm^3/L.

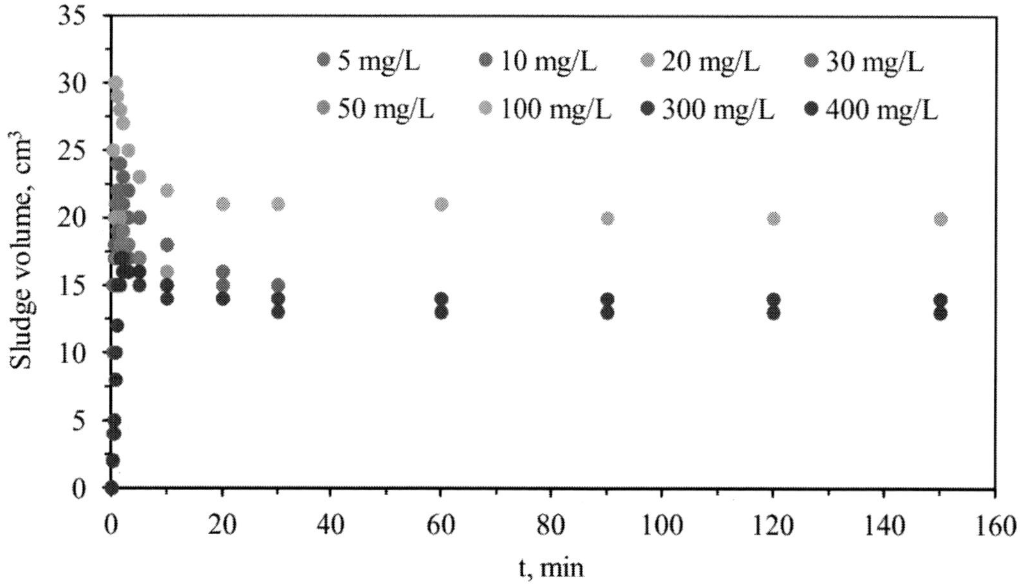

Fig. 1. Dynamic behaviour of sludge volume formed during flocculation with different concentration of Flocudex CS-51.

Also, from Fig. 1 it can be observed that the phase's separation (solid-liquid) was need less than 2 min. In addition, in all cases it can be detected that the compression zonefinished after 20 min.Similar behaviour was registered in the case of Bentonite but the compression zone only detected when the coagulant dose was higher 5 g/L.In the case of QG-2001 no compression zone was detected and the flocculant need more than 40 min to finish the sedimentation.

For all coagulants-flocculants (Bentonite,Flocudex CS-51 and QG-2001) used, higher removal percentages of total carbon ($\%TC_{removal}$ = 29%, 41% and 42%, respectively) at pH-OMW = 3 were registered. However, lower$\%TC_{removal}$ (11%, 7% and 14%, respectively) were determined at pH-OMW without control. On the other hand and as expected, the highest percentages of total organic carbon removal ($\%TOC_{removal}$ = 27%, 35% and 33%, respectively) were obtained at pH 3.

The removal percentage of total phenolic compounds ($\%TPC_{removal}$) is higher at pH-OMW = 3 in the case of Flocudex CS-51 ($\%TPC_{removal}$ = 15%) and QG-2001 ($\%TPC_{removal}$= 61%) flocculants. However, higher $\%TPC_{removal}$ = 32% was registered at basic pH-OMW =10 in the case of Bentonite.On the other hand, higher percentages of turbidity removal have been reported in experiments at pH-OMW without control ($\%Turbidity_{removal}$ = 51%, 54% and 38% for Bentonite, Flocudex CS-51 and QG-2001, respectively). These facts imply that the election of the optimum flocculent/coagulant and operational pH depend on the type of the subsequent process to use in the OMW treatment.

References

[1] L.M. Nieto, G. Hodaifa, S.R. Vives, J.A. Casares, and J. Ochando, "Degradation of organic matter in olive-oil mill wastewater through homogeneous Fenton-like reaction," *Chem. Eng. J.* 173, 2011, 503-510.

[2] B. Bianco, I.D. Michelis, and F. Vegliò, "Fenton treatment of complex industrial wastewater: Optimization of process conditions by surface response method,"*J. Hazard. Mater.* 186, 2011, 1733–1738.

[3] S.E. Garrido Hoyos, L.M. Nieto, F.C. Rubio, and A.R. Cormenzana, "Kinetics of aerobic treatment of olive mill wastewater (OMW) with *Aspergillus terreus*,"*Process Biochem.* 37(10), 2002, 1169–1176.

Empirical modelling of a 100 mm micro-hydrocyclone capacity

J. Izquierdo[1], A. Portillo[1], J. Vicente[2], A. Pablos[1], R. Aguado[1], M. Olazar[1]
[1]Department of Chemical Engineering, University of the Basque Country (UPV/EHU), SPAIN;tel. +349 46012510, e-mail:javier.izquierdo@ehu.eus; [2]Development Department, Novattia Desarrollos Ltd., Spain; tel. +349 46444421, e-mail: novattia@novattia.com

Small diameter hydrocyclones, or micro-hydrocyclones (below 100 mm), are commonly used in the industry due to their ability to achieve cut sizes lower than 40 μm. While some empirical correlations have been proposed for hydrocyclones bigger than 200 mm, few studies have been done for modelling the capacity of micro-hydrocyclones.

The work presented is part of one of the first stages of a R&D project performed between the Catalytic Processes & Waste Valorization Research group, (University of the Basque Country, UPV/EHU) and Novattia Desarrollos Ltd., technology development company, with the objective of achieving a more flexible, compact and efficient hydrocyclone technology than the actual one.

In this first stage of the work, experiments have been done with a 100 mm hydrocyclone in a pilot plant using different feeds, concentrations and geometrical configurations, in order to study the behaviour of the hydrocyclone in different conditions. The aim of the study presented, consisted in finding a correlation that fits correctly the experimental results of pressure drop and feed flowrate obtained in the pilot plant. In a following stage, similar experiments will be done with a smaller hydrocyclone and it will be checked if the correlation fits well that results too.

To achieve the objective, it was selected a correlation proposed by Yang et al [1] that links the Euler number (pressure drop) with the Reynolds number (feed inlet velocity). This correlation was chosen because this two adimensional numbers considered the fluid properties, and it is thought that for this reason the experimental data will fit better. However, it was seen that the experimental results did not fit well, so it was decided to add new parameters to the correlation. It was tried to add parameters that considered the dimensions of the most important hydrocyclone design parameters, like the overflow and underflow outlets diameter.

After some attempts and a statistical analysis, the results obtained proved that the correlation (Eq 1) that fits best the experimental results is the one that consider together with the Euler and Reynolds numbers, the vortex finder and spigot diameters and the pressure drop. In Figure 1 can be seen how well fit the experimental results with the proposed correlation.

$$\mathrm{Eu} = k \cdot Re^{a} \cdot \left(\frac{D_0}{D}\right)^{b} \cdot \left(\frac{D_u}{D}\right)^{c} \cdot \left(\frac{\Delta P}{P_o}\right)^{d} \qquad \text{Eq 1}$$

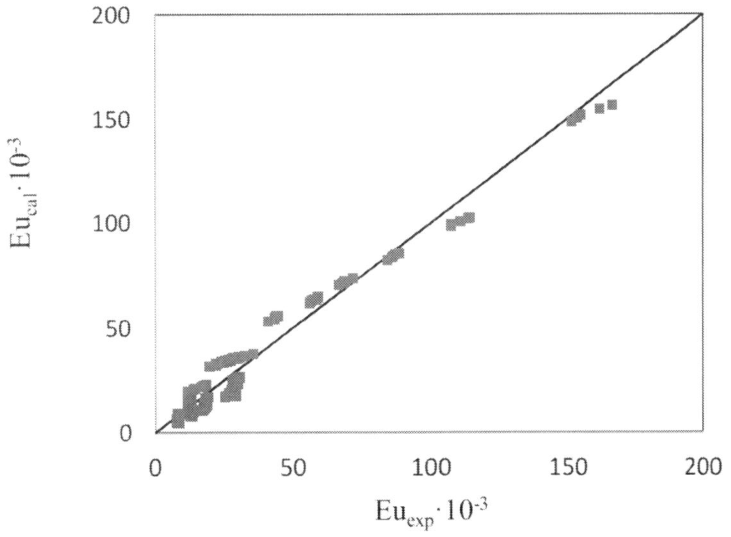

Fig. 1: Comparison of the experimental values of Eu number with those calculated using Eq 1.

REFERENCES

[1] Q. Yang, Z. Li, W. Lv, H. Wang, On the laboratory and field studies of removing fine particles suspended in wastewater using mini-hydrocyclone, Sep. Purif. Technol. 110 (2013) 93.

Development of a 1-dimensional boiling model: Part I – A two-phase flow pattern map for a heavy hydrocarbon feedstock

P. Verhees, [1]A.R. Akhras, K.M. Van Geem, G.J. Heynderickx
Laboratory for Chemical Technology, Ghent University, Technologiepark 914, 9052 Ghent, Belgium, tel. +329 264 56 77, e-mail: Kevin.VanGeem@UGent.be [1]Saudi Aramco, R&D Center, Dhahran, KSA, tel. +966 13 876-8459, e-mail: abdulrahman.akhras@aramco.com

Nowadays, the use of heavy fossil fractions as steam cracker feed is growing. Consequently, fouling problems are extended to heat exchangers in the steam cracker convection section. In order to limit this fouling, the design and operating conditions of the different heat exchangers in the steam cracker convection section need to be optimized. One such heat exchanger in the convection section is the evaporator. Optimization of the evaporator based on CFD simulations is not possible in an acceptable time period. Therefore, the development of a 1-Dimensional boiling model will result in considerable speedup of evaporator simulations.

The numerous available 1D models to simulate heat transfer in a boiling flow are developed for refrigerants. Since, the operating conditions, geometry and feed properties in a convection section significantly differ from cooling applications, a new flow boiling model should be constructed. The boiling model is developed following the mechanistic method[1]. Parameters characterizing the flow pattern, e.g. liquid holdup, film thickness, etc. can be determined based on a flow regime map. Based on these parameters, the heat transfer coefficient can be calculated. In literature it is stated that such models are valid in the complete range of vapor qualities (0 to 1)[2]. The first step developing a mechanistic model is the construction of a flow regime map.

CFD modelling is done using the Volume-Of-Fluid model with a piecewise linear interface interpolation scheme. Combined with an evaporation model, an accurate description of the flow characteristics is obtained. In a first step regime maps for a heavy hydrocarbon component are constructed. Several CFD simulations for different combinations of wall temperature, vapor quality and mass flux are performed. The obtained regime maps show a good agreement with regime maps available in literature, thus validating the applied models. Similar regimes are observed, e.g. wavy flow illustrated in Figure 1.

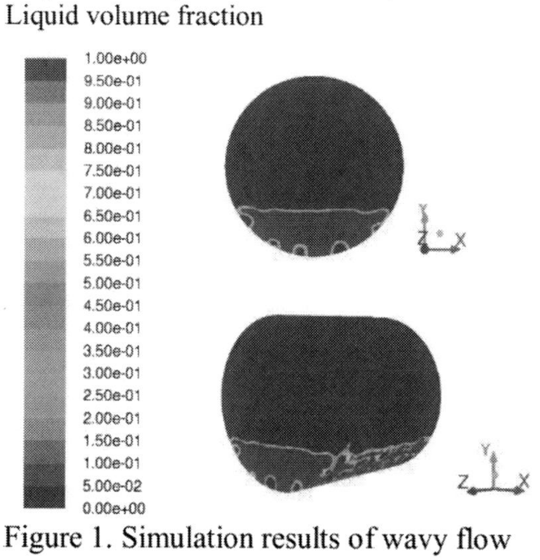

Figure 1. Simulation results of wavy flow

After the construction of the flow regime map, the parameters characterizing the flow patterns are determined based on the results of the CFD simulations. These parameters are detrimental to calculate the required heat transfer coefficient.

References

1. Wojtan L.; Ursenbacher T.; Thome J. R., Int. J. Heat Mass Transfer, 2005, 48, 2955-2969
2. Boissieux X., Heikal M.R., Johns, R.A., Int. J. of Refrigeration, 2000, 23, 269-283

Binary nucleation of H_2SO_4-H_2O in laminar co-flow tube

[1]T. Trávníčková, [1]L. Škrabalová, [1,2]J. Havlica, [3]J. Hrubý, [1]V. Ždímal
[1]Institute of Chemical Process Fundamentals of the CAS, v. v. i., Rozvojová 2/135, 165 02 Prague 6, Czech Republic; phone: +420 220 390 262, fax: +420 220 920 661
e-mail: travnickovat@icpf.cas.cz; [2]University of Jan Evangelista Purkinje, České mládeže 8, Ústí nad Labem, 400 96, Czech Republic; [3]Institute of Thermomechanics of the CAS, v. v. i., Dolejškova 1402/5, 182 00 Prague 8, Czech Republic.

Atmospheric aerosols significantly affect many areas of human being, whether it be on the global climate, local formation of clouds, visibility and signal transmission, or human health. One of the main mechanisms of fine aerosols formation is a binary nucleation of sulphuric acid with water. Because accurate description of all processes occurring in the atmosphere and conditions under which these phenomena occur is very difficult, the mechanism of nucleation of sulphuric acid with water remains still insufficiently described. In order to overcome this problem several types of experimental devices have been proposed. For this purpose flow chambers with turbulent mixing are most commonly used. However, these chambers encountered problems, how to determine the position of the nucleation zone and the concentration of sulphuric acid in this zone.

We measure binary nucleation of sulphuric acid with water using Laminar Co-Flow Tube, which overcomes both of these problems. In this paper we present the first results of measurements on this device. It appears that with this type of chamber equivalent results as measured on the flow chambers with turbulent mixing can be obtained. Moreover, the use of super-clean gases in our experiment helps to eliminate the impact of amines and other compounds to stabilize the clusters.

Laminar Co Flow Tube is actually a tube within tube, where the inner tube is fed with a mixture of saturated vapour of sulphuric acid in a nitrogen stream, and the packing stream is at approximately equal gradual rate fed with a mixture of saturated water vapour in nitrogen. The central tube ends after several tens of cm and on its edge, it comes to diffusion contact of these two mixtures, which then further flow throw the chamber together. Laminar parallel flow can be modelled and gain by this way a position of nucleation zone in which stable nuclei of a new phase are formed. These nuclei in the supersaturated mixture continue to grow and are carried out of the chamber. Thanks to the placement of acid into axial flow, and because the diffusion coefficient of acid in nitrogen is much lower than the diffusion coefficient of water in the nitrogen, losses of H2SO4 on device walls are effectively suppressed. At the outlet of the chamber, the particles are detected by a condensation particle counter with enlarger of clusters (PSM AirModus).

Mass and momentum balances were solved using commercial software tool FLUENT 14.5. Composition of the inlet stream was obtained by calculation from the saturated vapour pressure of water (Katz 1967) and sulphuric acid (Kulmala 1990) at the saturation temperature. Based on knowledge of temperature, these pressures and flow rates composition of the mixture were calculated at every point of the apparatus and the profiles of relative humidity *RH* and the relative acidity *Ra* were obtained. The theoretical nucleation rate was calculated according to Wyslouzil al. (1991) and experimental nucleation rate J_{exp} as a function of T and *Ra* was determined by Anisimov and Wagner (1993) parametrization from the particle flow detected at the outlet from the chamber using the PSM (see Fig. 1).

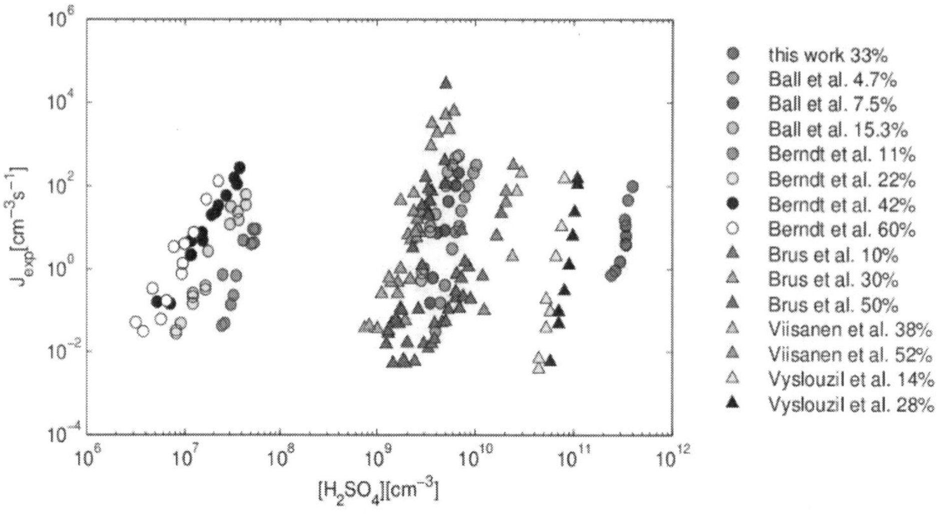

Fig. 1: Comparison of our data with previous studies - Dependence exp. nucleation rate on the concentration of H2SO4 at $T = 25\ °$ C.

The obtained nucleation rates are roughly two orders of magnitude lower than those obtained by other authors on flow chambers with turbulent mixing. A shift of nucleation isotherm towards higher concentrations of H2SO4 may be caused by the high purity of the used substances, which are free of amines and other ingredients stabilizing critical clusters. We have shown that LCFT is a relevant tool for the measurement of nucleation rate of binary mixtures.

Acknowledgements:

The authors wish to thank for the support the work of the grant P105 / 12/0664 of Grant Agency of the Czech Republic and the EU project 7. RP EU Marie-Curie ITN HEXACOMM č. 315760.

Literature:

Reiss, H., Margolese, D.I., Schelling, F.J., Experimental study of nucleation in vapor mixtures of sulfuric acid and water. *J. Colloid Interface Sci.*, 56, 511–526, (1976).

Katz, J.L., Ostermier, B.J., Diffusion cloud-chamber investigation of homogeneous nucleation, *J. Chem. Phys.*, 47, 478, (1967).

Kulmala, M. and Laaksonen, A.: Binary nucleation of water-sulfuric acid system: Comparison of classical theories with different H2S04 saturation vapor pressures, *J. Chem. Phys.*, 93, 696–701, (1990).

Wyslouzil, B.E., Seinfeld, J.H., Flagan, R.C., Okuyama, K., Binary nucleation in acid water-systems.2. sulfuric-acid water and a comparison with methanesulfonic-acid water, *J. Chem. Phys.* 94, 6827 (1991).

Wagner, P.E., Anisimov, M.P., Evaluation of nucleation rates from gas flow diffusion chamber experiments, *J. Aerosol Sci.* 24, S103-S104 (1993).

Coupling of a macro-scale CFD model with a bubble-scale model for simulation of polyurethane foams

M. Karimi,[1] P. Ferkl[2], D. L. Marchisio[3], J. Kosek[4]
[1]Politecnico di Torino, Dep. Applied Science and Technology, Institute of Chemical Engineering, C.so Duca degli Abrizzi 24, 10129 Torino, Italy;
tel. +39-0110904758, e-mail: mohsen.karimi@polito.it;
[2]University of Chemistry and Technology Prague, Dept. Chemical Engineering, Czech Republic; e-mail: ferklp@vscht.cz;
[3]Politecnico di Torino, Dep. Applied Science and Technology, Institute of Chemical Engineering, C.so Duca degli Abrizzi 24, 10129 Torino, Italy;
tel. +39- 0110904622 / 4622, e-mail: daniele.marchisio@polito.it;
[4]University of Chemistry and Technology Prague, Dept. Chemical Engineering, Czech Republic; e-mail: Juraj.Kosek@vscht.cz;

The market demands and the wide range of applications have distinguished the polyurethane foam (PU) modelling methods as profitable, yet challenging subject. The complexity of the physical phenomena, occurring during the foaming process, such as the evolution of bubble/cell size distribution (BSD) due to bubble growth and coalescence, confines the modelling techniques to mostly kinetics-based models, for example the modelling procedures developed by Baser and Khakhar [1-2]. Although these modelling methods can reasonably describe the principal of foaming process (i.e., the gelling and blowing reactions), the lack of information on the bubble scale hinders obtaining a-priori knowledge of the final product characteristics. This leads us to developing a methodology for coupling both macro-scale and bubble-scale physical phenomena taking place during the foaming process of PU.

The novelty of the proposed modelling scheme is that one can simulate the macro-scale evolution of the PU foam using a CFD model, while a detailed bubble-scale model accounts for the growth of bubbles due to the molecular diffusion and chemical reactions. Moreover, the CFD solver not only tracks the interface between the expanding foam and the surrounding air (beneficial for mold filling applications), it solves a population balance equation, using QMOM approach [3], to determine the evolution of bubble size distribution.

The macro-scale model takes into account the polymerization reactions and the presence of physical and chemical blowing agents, while a non-Newtonian formulation evaluates the rheological behaviour of the foam. The bubble growth model, on the other hand, simulates the growth of a single spherical bubble due to the exchange of blowing agent between the bubble and its shell. For a case with two blowing agents, it solves 2 partial and 3 ordinary differential equations as well as three additional ODEs for the kinetics of the reactions. The outcome of the bubble-scale model identifies the continuous rate of change of the number of moles of each blowing agent which will be converted and fed into the macro-scale tool to evaluate the growth source terms for solving the population balance equation.

One of the main contribution of this work is the coupling between the two modelling tools using MoDeNa interface library [4]. Using the MoDeNa framework the macro-scale model evaluates the bubble growth rates only at the quadrature nodes, however since the full simulation of the detailed bubble growth model could be expensive a "surrogate model" is derived and applied instead, providing an accurate approximation of the bubble growth rate.

As far as the implementation is concerned, the bubble-scale model is written in Fortran language, while the macro-scale tool is founded by developing a new Volume-of-Fluid based solver within open-source CFD code OpenFOAM.

The developed modelling strategy is applied to simulate a 2D planar "mixing-cup" experiment for a cup of 10 cm × 30 cm. Two blowing agents including n-pentane as the physical blowing agent and water as the chemical blowing agent are used and 2% of the cup is filled. The initial bubble size distribution is assumed to be a log-normal distribution with the mean size of 1 μm. Figure 1 shows the time evolution of foam density and temperature rise during the foaming process. The numerical predictions (solid lines) are compared with the experimental measurements (symbols) reported in [5]. As seen, the general trend of temperature rise is captured reasonably well, even though the final temperature reduction due to the external cooling is not described. The foam density is, however, predicted with high level of agreement with the measurements. The results show the successful coupling of the two modelling tools for the simulation of PU foam. The next step is to investigate the predicting capability of the developed framework for the simulation of industrial scale mold filling applications.

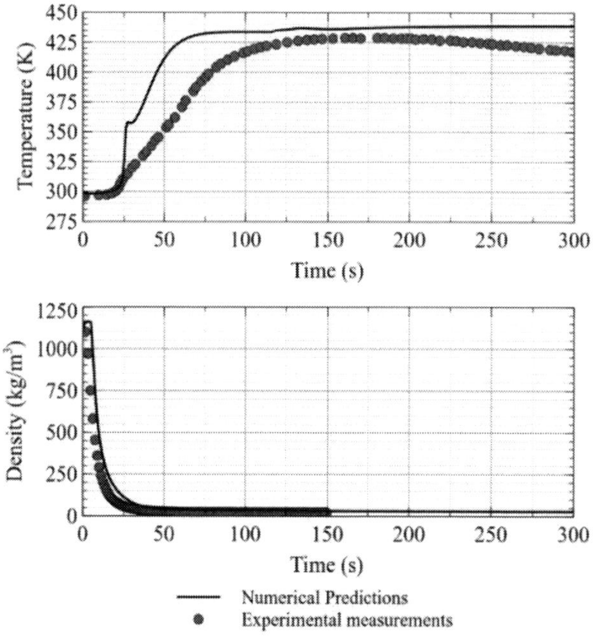

Figure 1. The time evolution of foam density and foam temperate; comparison of the experiments (symbols) and numerical predictions (solid lines)

References.

[1]Baser S. A. and Khakhar D. V. Modelling of the dynamics of R-11blown polyurethane foam formation. *Polymer engineering and science*, 34(8): 632-641, 1994.

[2]Baser S. A. and Khakhar D. V. Modelling of the dynamics of water and R-11 blown polyurethane foam formation. *Polymer engineering and science*, 34(8): 642-649, 1994.

[3] Marchisio D. and Fox R. Computational Models for Polydisperse Particulate and Multiphase Systems. Cambridge Series in Chemical Engineering.

Cambridge University Press, 2013.

[4]MoDeNa EU-Project. Modelling of morphology development micro-and nanostructures 2015. [Online. accessed: 26-November-2015]

[5] Winkler C.A. Numerische und experimentelle Untersuchungen zu Formfüllvorgängen mit Polyurethanschäumen unter komplexen Randbedingungen. Ph.D. thesis, University of Stuttgart, 2009.

Viscosity of concentrated dispersions: Modeling by discrete element method

Martin Kroupa, Michal Vonka, Miroslav Soos, Juraj Kosek
University of Chemistry and Technology Prague, Department of Chemical Engineering, Technicka 5, 166 28 Praha 6, Czech Republic;
tel. +420 220 443 296, e-mail: Juraj.Kosek@vscht.cz

The viscosity of a dispersion is a crucial quantity affecting its flow and handling properties. The current theoretical understanding of colloidal dispersion rheology is restricted to simplified hard-sphere systems in the limit of diluted or moderately concentrated dispersions. However, in real industrial processes (e.g., emulsion polymerization) one often needs to handle particles with complex surface properties (stabilization, adhesion) reaching volume fractions up to 50 %. There is therefore a large need to broaden the understanding of these systems in terms of their viscosity and related phenomena such as the stability and cluster formation. Mathematical modeling can be a useful tool to achieve this goal.

We use the Discrete Element Method (DEM) to model the dynamic behavior of a waterborne dispersion subjected to a high shear. The interaction of the particles with the fluid is accounted for using the two-way coupling technique. The interaction between the particles that are stabilized, adhesive and elastic is described by the combination of the Derjaguin-Landau-Verwey-Overbeek (DLVO) and Johnson-Kendall-Roberts (JKR) theories. We developed a method for the evaluation of the dispersion viscosity that is based on the balance of forces acting on the solid wall in the simple shear model. We also address the slippage of solvent molecules on the particle surface (characterized by the slip distance) and its effect on the dispersion viscosity using a novel approach inspired by Atomic Force Microscopy measurements.

At first, we validated our model with literature experimental data and established theories for a simplified hard-sphere model. In this system, the viscosity is a strong function of particle volume fraction (see Fig. 1). The validation resulted in the determination of the slip distance in the order of units to tens of nanometers, which are perfectly realistic values for this quantity. Having this only uncertain parameter specified, we proceeded to simulations of systems with more complex inter-particle interactions, for which there is no established description. Our model predicts the viscosity to rise for more adhesive coagulating particles and provides a clear link between the dispersion viscosity and cluster size. Moreover, the model is also able to capture the viscosity increase during the autocatalytic coagulation of stabilized latex providing a tool for the determination of the characteristic coagulation time. Finally, our results suggest that the dispersion viscosity is primarily controlled by the lubrication forces and that other dissipative forces play only a minor role.

The results of this work bring a theoretical insight into the rheology of concentrated dispersions. By predicting the viscosity, a directly measurable quantity, the model can serve as a basis for the operation of real processes that handle colloidal latexes.

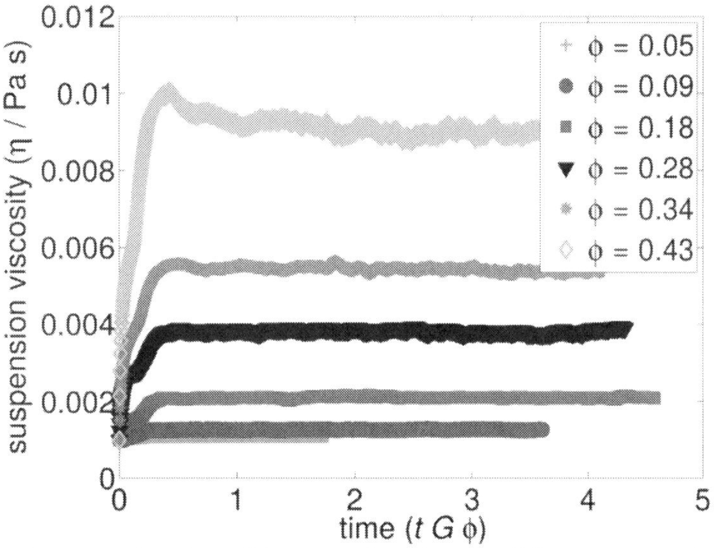

Figure 1: Viscosity η as a function of dimensionless time for different values of particle volume fraction ϕ. The symbol G represents the shear rate.

References

[1] Kroupa, M.; Vonka, M.; Kosek, J. Modeling the Mechanism of Coagulum Formation in Dispersions. *Langmuir* **2014**, 30, 2693–2702.

[2] Kroupa, M.; Vonka, M.; Soos, M.; Kosek, J. Size and Structure of Clusters Formed by Shear Induced Coagulation: Modeling by Discrete Element Method. *Langmuir* **2015**, 31, 7727–7737.

Hydrodynamics of nano-confined fluid: fluid inhomogeneities, slip and flow enhancement compared with H-P relation

J.B. Xu, Y.Y. Wang, Y.M. Yong, C. Yang
Key Laboratory of Green Process and Engineering, Institute of Process Engineering, Chinese Academy of Sciences, 1 North 2nd Street, Zhongguancun, Haidian District, 100190 Beijing, China ;
tel. +86-10 82544928, e-mail: jbxu@ipe.ac.cn; fax: +86-10 62551257.

The fluidic transport at nanometer scales shows great potential in many fields, such as membrane separation, nano-confined catalysis, energy conversion and DNA sequencing, benefiting from its specific phenomena occurring at this small scale. The solid/liquid interactions in nanochannels are more significant and non-negligible than those in larger scales. To be specific, phenomena as fluid inhomogeneity, hydrodynamic slip, abnormal phase behavior and electrical double layer can be initiated by the interactions.

In this work, the hydrodynamics of fluid flow within nanoslit or nanotube was investigated by the dissipative particle dynamics (DPD) simulation and theoretical analysis. The influences of wall/fluid interactions and the dimensions of the channels were discussed respectively by examining the density profiles, velocity profiles and the fluxes. The results show density and viscosity inhomogeneities near the wall/fluid interfaces, which are determined by the wall/fluid interactions in spite of the dimensions or the flow rates. The apparent slip length of shear-driven flow, body force-driven flow and flow driven by external force acting on adsorbed solutes was acquired by introducing the fluid inhomogeneities into Stokes equation.

For the body force-driven flow, the flow pattern varies with the size of the nanochannels, and in particular, plug flow can be observed in a highly confined condition. The enhancement, defined as the ratio of the simulated flux to that calculated by the Hagen-Poiseuille (HP) relation, keeps increasing with the decrease of the slit size and the wall wettability. The flux models were theoretically analyzed according to the proportion of the inhomogeneous region in the flow system, and four flow patterns were introduced. In the flow with negligible inhomogeneity occurring when the characteristic dimension D is greater, the continuum based method as the HP relation is valid. In the flow with non-dominant inhomogeneity, the equation based on the Navier's condition can be used to compensate the slight deviation near the walls. In the flow with dominant inhomogeneity where D is comparable with molecular size, an inhomogeneous model derived from the inhomogeneous equation of motion can calculate the flux precisely. And finally, in the flow with overlapped inhomogeneity when plug flow can be observed, the friction model for plug flow is the easiest way for flux calculation. The four models together can provide a reasonable description of flow characteristics in all dimensional ranges. DPD simulations have been perfomed and the consistency between our theory and the simulation has been verified.

This work was supported by the National Natural Science Foundation of China (21490584, 91534105).

Spatial cross-correlation of optical probes for measurement of horizontal bubble diameters, validated in bubble colums up to 1 m

[1]A. Forret, [1]F. Augier, [1]P. Raimundo, [2]D. Beneventi, [3]A. Cartellier
[1]IFP Énergies Nouvelles, Rond-point de l'échangeur de Solaize, BP 3, 69360 Solaize, France tel. +4 37 70 24 57, e-mail: pedro-felipe.maximiano-raimundo@ifpen.fr; [2]Laboratoire Génie des Procédés Papetiers, 461 rue de la Papeterie, 38402 Saint-Martin d'Hères Cedex, France; [3]Laboratoire des Ecoulements Géophysiques et Industriels, BP 53, 38041, Grenoble cedex, France, tel. +4 76 82 50 48 , e-mail:alain.cartellier@legi.grenoble-inp.fr

Bubble column reactors are widely used in industrial biological and chemical processes. This simple reactor configuration is attractive due to its efficiency in liquid mixing and gas-liquid mass transfer. However their scale-up is still a challenge as it impacts many physical variables as phase velocities, turbulence and mixing properties. The use of CFD models in this goal is still limited by the lack of physical closure laws concerning turbulence and momentum phase exchange terms, even more when considering bubble sizes and their velocity distributions in heterogeneous regimes. At these conditions, there is no optical access to the interior of the column due to the high bubble concentration, excluding the use of most of the available methods.

The goal of this study is to develop measurement techniques that can be used in 3-D columns in a wide range of superficial gas velocities (up to 35 cm/s). Experiments were carried out in different diameters cold flow air-water bubble column (D=0.15; 0.40 and 1 m).

Bubbles horizontal diameters are measured by an innovating method that uses the cross-correlation of signals from two monofiber probes, placed at the same elevation, with a given distance between them. A theoretical approach of the spatial cross-correlation of the signals, based in a probability approach, was developed. This method was used to study the impact of the considered bubble shape (oblate, prolate), eccentricity and inclination on the measurement of the cross-correlation. As an outcome of this study, an expression to calculate a horizontal bubble diameter *Dh* as a function of the cross-correlation value *CC* and probes distance *d* was found. In order to validate it, an endoscopic imaging methodology has been used to obtain in-situ images of the flow. The images were treated in order to obtain handmade measurements of bubble size, eccentricity, angle and shape. Comparison between cross-correlation measurements and endoscopic imaging shows a good agreement, even in the challenging conditions encountered in bubble columns at large gas fraction (up to 40%). An example of the horizontal bubble diameter radial profile measured by endoscopy and deduced by cross-correlation (considering an oblate or a prolate shape) in a 40 cm bubble column at a gas velocity of 16 cm/s is given in the following figure.

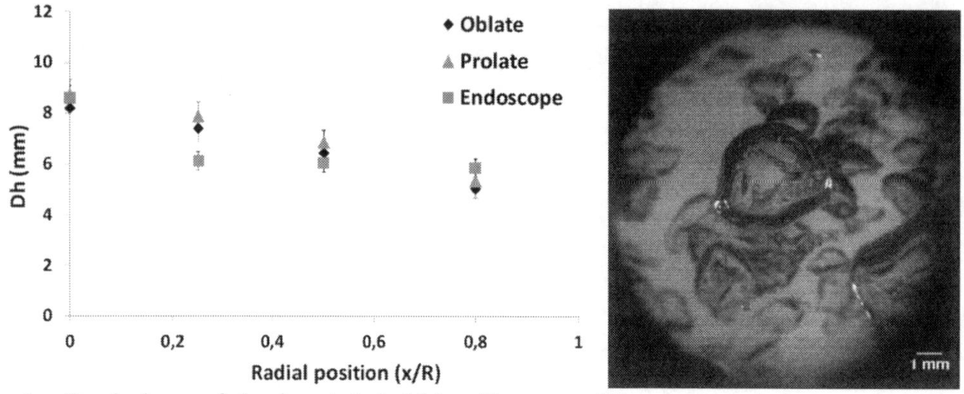

Figure 1: Evolution of horizontal bubble diameter Dh measured by cross-correlation (considering an oblate or a prolate shape) and endoscopic imaging along the column radius (D=40cm, h/D=3.75,vsg=16cm/s)

Investigation of the multiple holdup problem in developing laminar-laminar stratified flow in inclined channels

[1,2,3]D. Thibault, [4]R. Belt, [1,2,3]A. Liné
[1]Université de Toulouse ; INSA, UPS, INP ; LISBP, 135 Avenue de Rangueil, F-31077 Toulouse, France ; tel. +33 (0) 5 61 55 92 75, e-mail : dthibaul@insa-toulouse.fr; [2]INRA, UMR792 Ingénierie des Systèmes Biologiques et des Procédés, F-31400 Toulouse, France ; [3]CNRS, UMR5504, F-31400 Toulouse, France ; [4]TOTAL E&P, DSO/ED/ECP, 2 place Jean Millier, La Défense 6 – Cedex 45, 92078 Paris, France ; e-mail : roel.belt@total.com.

Stratified gas-liquid or liquid-liquid patterns are frequently encountered in many applications like production pipelines, nuclear or process industries under steady, developing or transient conditions, where because of the gravity, the heavier fluid is flowing in the bottom part of the pipe with the lighter one above it. The two-fluid 1D model is generally used to predict the pressure drop and the holdup values. It is known since Baker et al. (1988) that for certain operating conditions, this model can lead to non-unique holdup solutions. This typically occurs for upward inclined pipes and low or negative flow rates of the heavier phase and also for downward inclined pipes with high flow rates of the heavy fluid. Such conditions are unfortunately sometimes corresponding to operating conditions and the question that arises is to determine which of these solutions will actually occur. Such information is of practical interest in gas-condensate pipelines for instance, where the liquid content in the pipeline dictates the choice of the pipe diameter and the minimum gas flow rate at which the pipeline can be operated.

Landman (1991) showed through theoretical investigations that this issue is not a mathematical artefact of the model since the multiple holdup occurrences persist when considering the exact solutions of the Navier-Stokes equations for laminar-laminar flow in a channel. This was also confirmed experimentally by Ullmann et al. (2003) who observed that part of the predicted solutions is feasible. Barnea and Taitel (1992) performed structural stability analysis in order to pick out the right holdup solution. In their study, they assumed a transient solution in a uniformly filled pipe and used shear stress closure laws derived from the single-phase case. At the opposite, in the present study a stationary but developing flow is assumed and the exact expressions of the shear stresses are used in the set of equations of motion since the flow is assumed to be laminar in both phases.

A simple model based on a first order ODE (Ordinary Differential Equation) is then derived and used to follow the evolution of the holdup from the inlet of the pipe. The system will converge from a non developed solution at the inlet of the channel to a fully developed solution of the holdup downstream. When 3 solutions are predicted by the two-fluid model (co-current flow), the intermediate one is found to be unstable while the two other equilibria are stable. These results are discussed regarding the approach of a potential that the system tends to minimize. This approach provides the same conclusions that the structural stability analysis and offers a qualitative interpretation for occurrence of multiple stable or unstable modes (or equilibria), sudden jumps of solution, hysteresis phenomenon, sensitivity, ... that stratified two-fluid flow systems are subject to. In this approach, a local peak of potential is associated to an unstable equilibrium, as it is the case for the intermediate solution when 3 holdup are predicted. The two others holdup solutions are to be relied on local wells of potential and thus considered as stable equilibria.

Critical condition or hydraulic shock is a limitation of the first order model used previously and thus supercritical flow condition is assumed first in the framework of this study. Nevertheless, when critical flow condition is encountered during the longitudinal evolution of the holdup, a model based on a second order ODE considering the viscous stress diffusion is used. This model is able to handle situation where a hydraulic shock is predicted before finally reaching the downstream supercritical steady-state solution. In addition, maps can be drawn to know whether a hydraulic shock can occur. The first one represents the critical condition curves along the dimensionless superficial velocities and allows the user to know if a hydraulic shock may occur and the holdup values the subcritical flow zone stand between. The second map interprets the longitudinal evolution of the holdup as a Froude number trajectory. The steady-state solution(s) of a system can thus be placed on such a map in order to determine if supercritical or subcritical flow regime is expected for the downstream solution(s).

References

- Baker, A., Nielsen, K., Gabb, A., 1988. Pressure loss, liquid-holdup calculations developed. Oil & Gas Journal 86, 55–59.
- Landman, M., 1991. Non-unique holdup and pressure drop in two-phase stratified inclined pipe flow. International Journal of Multiphase Flow 17, 377 – 394.
- Ullmann, A., Zamir, M., Gat, S., Brauner, N., 2003. Multi-holdups in co-current stratified flow in inclined tubes. International Journal of Multiphase Flow 29, 1565 – 1581.
- Barnea, D., Taitel, Y., 1992. Structural and interfacial stability of multiple solutions for stratified flow. International Journal of Multiphase Flow 18, 821 – 830.
- Thibault, D., Munoz, J.M., Liné, A., 2015. Multiple holdup solutions in laminar stratified flow in inclined channels. International Journal of Multiphase Flow 73, 275 – 288.

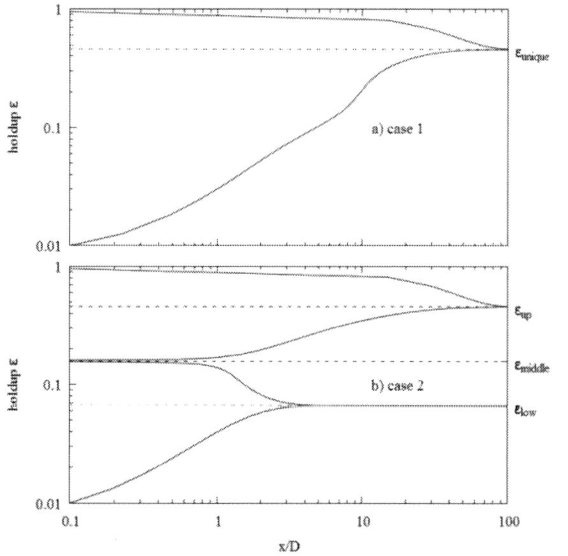

Fig. 1: Longitudinal evolution of holdup.

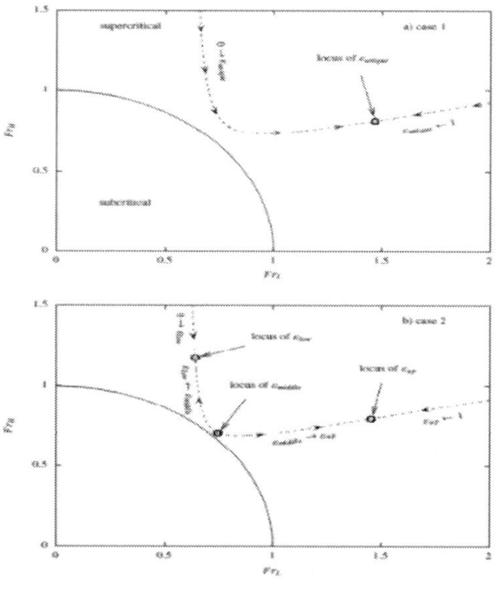
Fig. 2: Froude numbers trajectories and critical condition line.

Hydrodynamic behaviour and pressure drop of an upwardly two-phase flow through single- and multi-hole orifices

A. Zeghloul, A. Azzi, F. Saidj, Barry J. Azzopardi[1]
University of Sciences and Technology Houari Boumedien (USTHB), FGMGP/LTPMP, Bab Ezzouar, 16111, Algiers, Algeria, e-mail: aazzi@mail.com, [1] Faculty of Engineering, University of Nottingham, University Park, NottinghamNG7 2RD, United Kingdom.

Multi-hole orifices are used as chokes in petroleum production. The fundamental flow through them in their full open position is considered in this paper. An experimental investigation of the hydrodynamic behaviour and pressure drop of a two-phase mixture flowing upwardly in a pipe containing single- and/or multi-hole orifice is presented. Time series of cross-sectionally averaged void fractions have been measured at nine axial positions by using a conductance probe technique. The measurements of upward single- and two-phase gas-liquid flow pressure drop across the orifices were achieved using accurate differential pressure transmitters.

The 34 mm internal diameter pipe test section was made of transparent acrylic resin to permit visual observation. A series of four orifices with the same open area ratio (area or total area of the holes divided by the area of the pipe); and same thickness are employed. The superficial velocities of the gas was varied from 0 to 4 m/s and for the liquid from 0.3 to 0.91 m/s covering thus, single-phase liquid, bubbly, slug and churn two-phase flows.

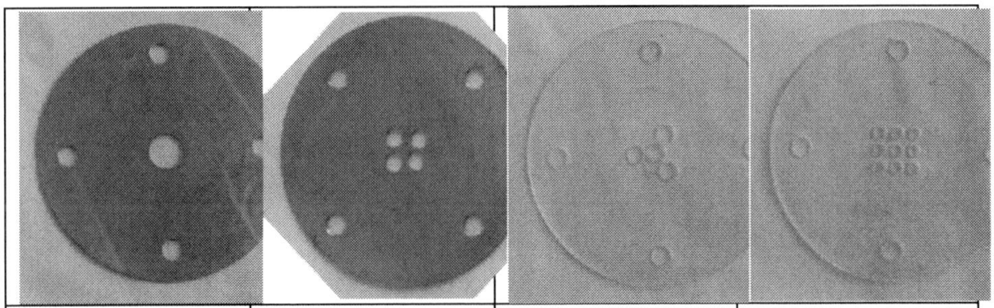

Standard orifice	4-hole square layout	4-hole trianglular layout	9-hole square layout

Figure 1: Test plates employed in test.

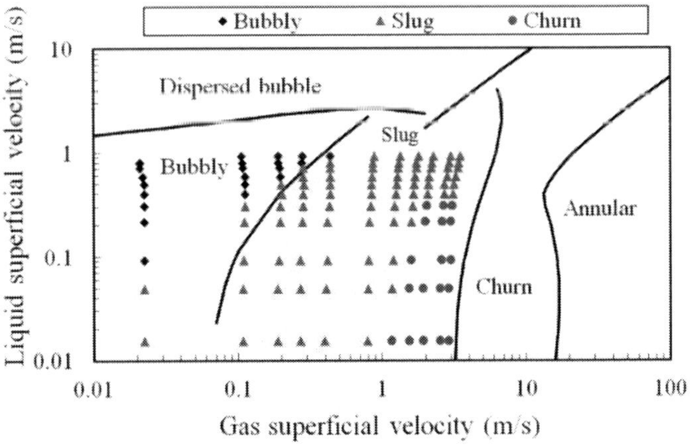

Figure 2: Flow pattern map for the vertical pipe before the orifice illustrating the flow rates studied.

The example time traces shown in Figure 3 illustrate how the different orifices affect the flow through them. The multi-hole orifice (9 holes) shows more evidence of larger bubbles cap bubbles than the single hole orifice and that these persist through the orifice and downstream.

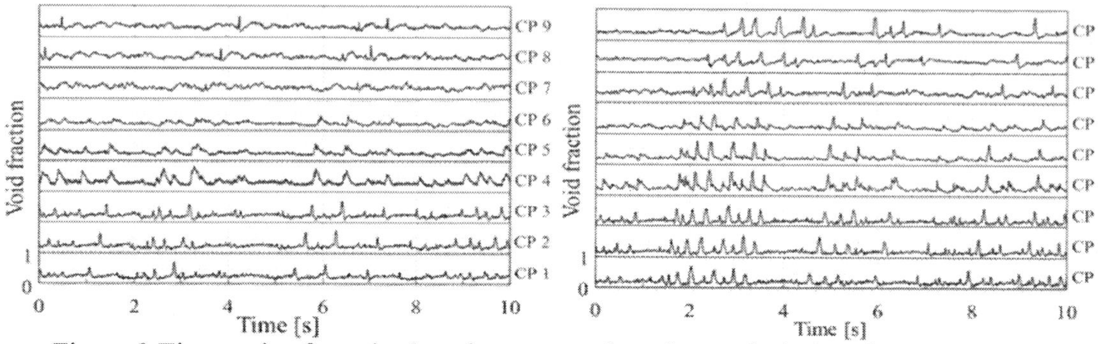

Figure 3 Time series from the 9 probes mounted up the vertical pipe, for a gas superficial velocity of 0.1 m/s and a liquid superficial velocity of 0.91 m/s Top standard orifice, bottom 9 hole orifice

Figure 4 shows how the different orifices affect the orifice pressure drop. For each orifice the pressure drop goes through a maximum as the gas velocity is increased and then begins to drop again. The effect of the different orifices is much clearer as gas velocities below the maximum than above it. Reference to Figure 2 shows that the maximum is linked to the transition between bubbly and slug flow.

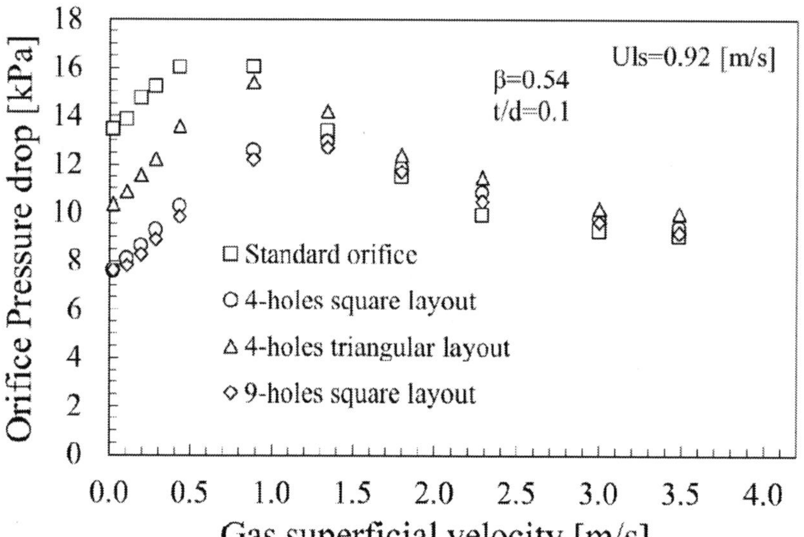

Figure 4: Two-phase flow pressure drop function of the gas superficial velocities for orifices.

Probabilistic method of calculation of processing of water in the spiactor

A. Grechushkin, V. Lvov, V. Zakharova, A. Fatova, L. Janzen.
Bauman State Technical University, Power Engineering Faculty,
105005 Moscow, Russia; tel. +7 499 2636893, fax: +7 499 2636707.
e-mail: wasser2004@bk.ru;

In this research, we proposed a novel modelling method of processes in the spiraktor. The spiractor, for example, are used for water softening, processing of a concentrate of systems of the revese osmosis and processing of a drain of ion-exchange filters. The spiractor is usually conical tank where the bottom is tangentially fed raw water and alkali spiral movement is brought to the sublime limestone and quartz sand or marble grains with 0,1-0,3 mm, which is to be added at the beginning spiractor operation. On the grain settles excluded carbonate and increasing in size from 1 to 2 mm.

Process of change of mass of grains in the spiractor has pronounced stochastic character. We proposed of change of mass of grains in the spiractor as casual Markov process of the birth and death of grains. Process of change of mass of grains in the reactor has the determined and stochastic parts and this process can be described by the equation Fokker-Planck-Kolmogorov.

$$\frac{\partial f(m,t)}{\partial t} = \frac{\partial [w(m)\cdot f(m,t)]}{\partial m} + \frac{\partial^2 [D(m)\cdot f(m,t)]}{\partial m^2}$$

$f(m,t)$ - density of distribution of concentration of grains with m mass on coordinate m and t time.

Stationary distribution on the mass of grains:

$$f_s(m) = \frac{1}{m_{cp}} \exp\left(-\frac{m}{m_{cp}}\right)$$

m_{cp} - the average mass of grains in the reactor.

Change of average mass of grains in the reactor can be determined by phenomenological methods authentically.

Including the mass of grains $m \sim a(\rho-\rho_0)r^3$, ρ and ρ_0 - density of grains and water.

$$f_s(r) = \frac{3r^2}{r^3{}_{mcp}} \exp\left(-\frac{r^3}{r^3{}_{mcp}}\right)$$

r_{cp} - the average radius of grains in the reactor.

Having empirical these qualities of the conditioned water from the size of grains of loading it is possible to define the main indicators of quality of the conditioned water and frequency of replacement of grains. This method of calculation allows to define an optimum profile of the spiractor also.

For check of theoretical model the experimental stand was created (Fig. 1). The spiractor consists of a pipe with a diameter of 75 mm and has wide top part. The wide top part interferes with ablation of grains. We applied grains with a size from 0,20 to 0,25 mm. Optimum time of contact in the spiractor made 6 min. The increase in the average size of grains in 36 hours made 40%.

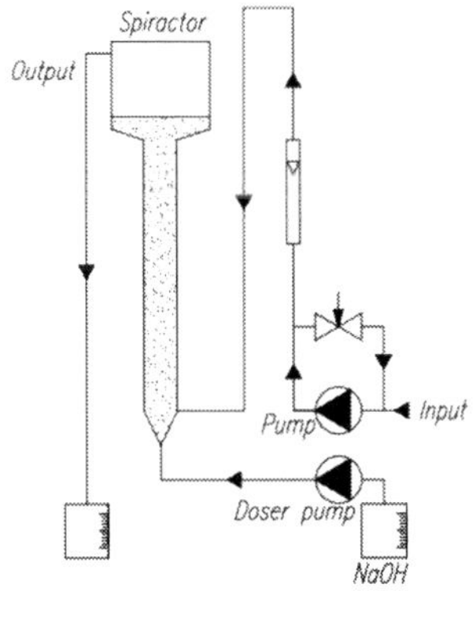

Fig. 1. Experimental stand.

A probability and statistical model of particle classification process in hydrocyclones classifiers of small sizes

A.V. Krokhina, V.A. Lvov, G.P. Pavlikhin
Bauman Moscow State Technical University, Dep. Environmental engineering,
2-nd Baumanskaya 5, 105005 Moscow, Russian Federation;
tel. +7 926 592 49 49; e-mail: alina@krokhina.com.ru.

Hydrocyclones are widely used in chemical engineering and technology for processing polydisperse systems of variable composition and features. Processes of "liquid-solid" polydisperse systems have a special place among them. Modern technical requirements have prompted a detailed study of solid fine fractions classification process with particle size less than 5 μm. Here the specific features of using small cylinder-conic hydrocyclones-classifiers should be emphasized.

At the same time, implementation of this apparatus is constrained because there are no universal calculation procedures and the general theory of transfer processes in hydrocyclone-classifiers is incomplete. The above-mentioned circumstances are related to the complex behavior of dispersed particles in the hydrocyclones, and it is currently impossible to describe it quantitatively using deterministic methods. In this case, an integrated approach based on the probability and statistical method of computing should be considered the most promising.

A theoretical model of suspension classification process in hydrocyclones

Proposed scheme of the dispersed phase motion in small hydrocyclone is shown in fig. 1. The specialty of the particle motion is the secondary flow circuit of the dispersed phase.

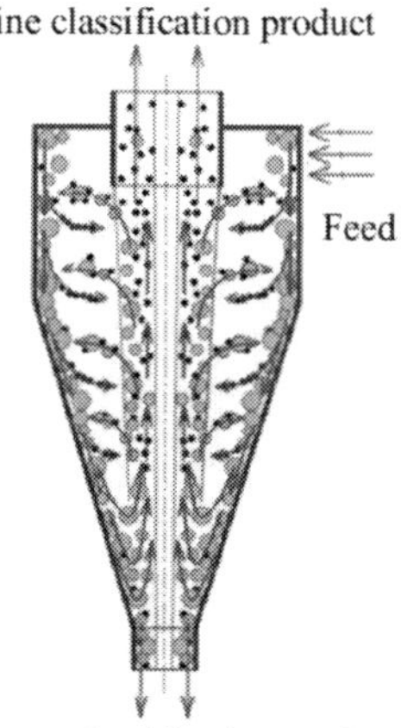

Fig. 1 Scheme of the dispersed phase motion in the hydrocyclone

Determining the stochastic components of the process, we will assume that random influences on a dispersed system particle at different moments are statistically independent (δ-time-correlated) and change much faster than the changes in observable values induced by them. Also let us assume that quasi-steady stochastic motion of dispersed particles conforming to the theory of Markov processes takes place in the cylindrical working part of a hydrocyclone.

Regarding a disperse system as having low or medium concentration C, let us consider a cylindrical coordinate system connected with the hydrocyclone's axis of symmetry. We will also assume that the tangential velocities of both the dispersing medium and dispersed particles are nearly identical.

Given curvilinear motion with radius R and angular velocity ω of a Stokes spherical particle with mass m and diameter d in a moving coordinate system, we calculate the particle's velocity:

$$\vec{v} = \vec{w} + \vec{v}_c, \qquad (1)$$

where $\vec{w}$ is the velocity of the dispersing medium, and $\vec{v}_c$ is the particle's relative velocity.

The equation of motion for a single particle may be written as:

$$m_ч \frac{d\vec{v}_c}{dt} + \frac{\pi d^3}{6}(\rho_s - \rho_m)\vec{\omega}\times(\vec{\omega}\times\vec{R}) + m(\frac{d\vec{\omega}}{dt}\times\vec{R}) + 2m(\vec{\omega}\times\vec{v}_c) + \sum = -3\pi\mu d\vec{v}_c, \qquad (2)$$

where t is time, ρ_s and ρ_m are the densities of the solid particles and the dispersing medium, R is the current rotation radius, ω is the angular velocity, μ is the dynamic viscosity of the dispersing medium, and Σ are other forces, including non stationary and random components of all forces.

At a first approximation for radial motion of the particles with size d we obtain:

$$u = u' + \frac{d^2}{18\mu}(\rho_s - \rho_m)\cdot \omega^2(R)\cdot R \ , \tag{3}$$

where $u' = u'(Q,R)$ is the mean velocity of radial movement of the dispersing medium.

Denoting $R^{n+1} = S$ as a new generalized coordinate for new mean particle velocity W in space S we obtain:

$$W = W' + \widetilde{W} = \frac{dR^{n+1}}{dt} = \frac{dS}{dt} = -\left(\frac{A_0(n+1)}{S^{(1-n)/(n+1)}} - \frac{C_p(n+1)}{S}D^2\right) + \widetilde{W} = -(A_\Sigma S^{\frac{n-1}{n+1}} - \frac{C_\Sigma}{S}) + \widetilde{W} \ , \tag{4}$$

where n is constant index, $n \in [0,5,1]$, $A_0 = Q/k_0(Q)h$, $A_\Sigma = (n+1)A_0$ and $C_p = k_f^2 d^2(\rho_s - \rho_m)/18\mu$, $C_\Sigma = (n+1)C_p D^2$ - are some constants, where $k_0 = k_0(Q) = k_0(w'_0)$ - is a constant taking into account the height variation of the working area in the apparatus, h is the length of the hydrocyclone's vertical section between the vortex finders and upper end of the apex, $k_f = k_f(d)$ is a fractional shape factor of particles with size d, $w' = w'(R)$ is the tangential velocity in the hydrocyclone, $D = w'R^n = const$. Here, together with the mean process velocity W', the random component of this velocity $\widetilde{W}$ is still present, and its effect cannot be neglected.

As a result, it is clear that the radial distribution of dispersed particles may be characterized only statistically, for example, using density function $f = f(S,t)$ normalized to unity.

Then, in order to describe the hydrodynamic step of the system's evolution in space S, we use the known Fokker-Planck-Kolmogorov kinetic equation in the form:

$$\frac{\partial f(S,t)}{\partial t} = -\frac{\partial}{\partial S}(W'(S)f(S,t)) + \frac{1}{2}\frac{\partial^2}{\partial S^2}(B(S)f(S,t)) \ . \tag{5}$$

Then a general solution of this equation in the limiting case $n=1$ was obtained in the form:

$$f(x,t) = \left\{ x^{\frac{\Theta}{2}} e^{-\frac{mx}{2}} \sum_{i=0}^{i=\infty} \left(\frac{\Theta\cdot m}{(\Theta+2i)} x \right)^{\frac{\Theta}{2}} e^{-\frac{\Theta m}{2(\Theta+2i)}x} C_i L_i^{(\Theta-1)}\left(\frac{\Theta\cdot m}{(\Theta+2i)}x\right) \exp\left(-\frac{k^2(\Theta+i)}{b(\Theta+2i)^2} i\cdot t \right) \right\}, \tag{6}$$

where $k = A_\Sigma / S_0^{\frac{2}{n+1}}$, $c = C_\Sigma / S_0^2$, $b = b_0 / S_0^2$ are constant coefficients characterizing the intensity of the classification effect, centrifugal forces and random perturbations, respectively and they make complexes $\Theta = c/b$ and $m = k/b$.

Considering that all particles reached the wall of apparatus are completely trapped and removed through the apex, classification efficiency curve function can be written as:

$$T = T(d,t) = \int_1^\infty f_n(x,d,t)dx \ . \tag{7}$$

Results and discussions

Comparison of experimental and theoretical studies (Fig. 2) shows that the suspensions classification process in the hydrocyclone-classifiers of small sizes can be satisfactorily described by an equation of diffusion type, in particular, by the Fokker-Planck-Kolmogorov equation in the space of the new generalized coordinate S.

Here the relative error of approximation of calculated curve does not exceed the relative error of experimental determination of the classification efficiency curve *T(d)* with a probability of 0,95 was about 15%.

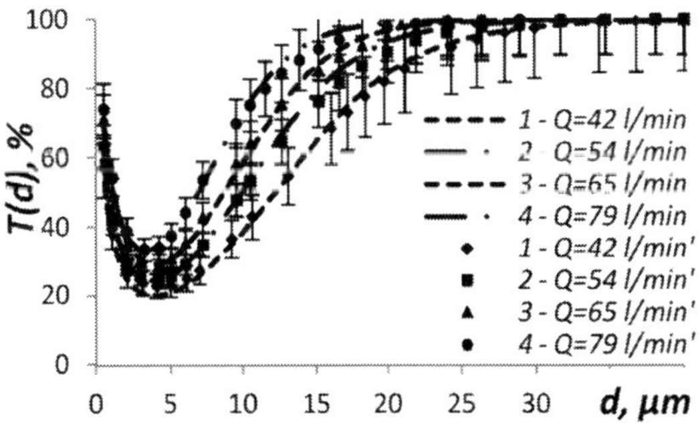

Fig.2 Classification efficiency curves *T(d)*: lines - calculations, points - experiments for hydrocyclone *d*=50mm

Investigation of influence of channel vibration on the dissolution rate of single bubbles in milli-channels

Mohammadreza Haghnegahdar[1], Stephan Boden[1], Uwe Hampel[1,2]
[1]Institute of Fluid Dynamics, Helmholtz-Zentrum Dresden-Rossendorf, Bautzner Landstr. 400, 01328 Dresden, Germany; tel. +49 351 260 3767, Fax: +49 351 260 2383, Email: m.haghnegahdar@hzdr.de; [2]AREVA Endowed Chair of Imaging Techniques in Energy and Process Engineering, Technische Universität Dresden, 01062 Dresden, Germany.

In the past decades, milli- and microreaction technology had intensive development with the numerous advantages such as intensification of heat and mass transfer, easy scale-up, reducing energy and resource consumption. Optimization of the chemical processes is one of main requirements put forward by modern industry with the aim to achieve technological efficiency and environmental safety. A new design and method for measurement of mass transfer rate of single bubbles in vibrating milli-channels is being developed with the claimed goal of process intensification [1],[2].

In this work, the dissolution rate of a single Taylor bubble of carbon dioxide in water is investigated using high resolution X-ray radiography technique in a vibrating vertical channel. The liquid-side mass transfer coefficient is calculated by measuring the changes in the size of the bubble at constant pressure. The experiments cover a large range of initial Taylor bubble length varying from 8 to 24 mm. The channel is a glass pipe with 6 mm inside diameter and circular cross section. The bubble is unceasingly monitored by holding the bubble stationary using the technique of Schulze and Schluender [3]. The glass channel is vibrated using a calibrated vibrator in horizontal direction. The amplitude and frequency of vibration is controlled by a wave generator accurately. The method which is used to measure the variation of the bubble size is X-ray radiography. This technique was qualified to disclose the three-dimensional shape of Taylor bubbles in capillary and enabled the acquisition of a series of high-resolution radiographic images of nearly stationary Taylor bubbles. The processed images which give volume (and also the interfacial area) of the bubble with high accuracy as a function of time, are used to evaluate the liquid side mass transfer coefficient between bubble and liquid using the mass conservation equation. The liquid phase is filtered-deionized water and the gas phase is CO_2.

The results for the short term dissolution of single CO_2 bubbles show that the channel vibration with high frequency (50, 100, 1000 and 10000 Hz) does not have a detectable influence on the rate of mass transfer for stationary single bubbles however, for channel vibrations lower than 50 Hz, the liquid-side mass transfer coefficient increases by more than 32%. In addition, it is shown that the measured mass transfer coefficients do not have intensive dependency on the bubble length and also equivalent diameter (diameter of the sphere having the same volume).

References

[1] H. Su, S. Wang, H. Niu, L. Pan, A. Wang, and Y. Hu, "Mass transfer characteristics of H2S absorption from gaseous mixture into methyldiethanolamine solution in a T-junction microchannel," *Sep. Purif. Technol.* , vol. 72, no. 3, pp. 326–334, 2010.

[2] T. Bauer, M. Schubert, R. Lange, and R. S. Abiev, "Intensification of heterogeneous catalytic gas-fluid interactions in reactors with a multichannel monolithic catalyst," *Russ. J. Appl. Chem.*, vol. 79, no. 7, pp. 1047–1056, 2006.

[3] G. Schulze and E. U. Schlünder, "Physical absorption of single gas bubbles in degassed and preloaded water," *Chem. Eng. Process.*, vol. 19, no. 1, pp. 27–37, 1985.

On the dynamic measurements of dissolved oxygen concentration in viscous liquids

L. Labík, R. Petříček, T. Moucha
University of Chemistry and Technology, Prague, Chemical Engineering Dep., Technická 5, 166 28 Praha 6, Czech Republic; tel: +420220443297, e-mail: libor.labik@vscht.cz

A dynamic measurement of dissolved oxygen concentration *DOC* is important for wide range of industrial, medical and environmental applications. For example, in recent years attempts were made to measure fast *DOC* oscillation in arterial blood, *DOC* profiles in oceans on moving gliders or *DOC* in absorption columns and fermenters with dynamic pressure method. All above mentioned experiments need a reliable description of transient characteristics of dissolved oxygen (*DO*) probe for correct data interpretation. Various models, based on Fickian diffusion, are available in literature to describe transient characteristics of *DO* probes, which are successfully used for low-viscosity liquids like water. On other hand, there is a lack of studies, which confirm the usability of the models in viscous liquids. Our goal is to test the reliability of current models when they are used to describe *DO* probes transient characteristics in viscous liquids. We also suggest a reliable and repeatable technique for transient characteristics measurement.

To measure transient characteristics of *DO* probe it is necessary to make a fast step change in *DOC* in front of the probe. First possibility is to change *DOC* in flowing liquid or gas, for example with a three way valve. Another possibility is to move mechanically the probe from one environment to another one with a different *DOC*. Three unique apparatuses were constructed to enable transient characteristics measurement. First was based on two stirred cells with different *DOC*. A *DO* probe was mechanically transferred between them. The second apparatus enabled to measure transient characteristics in gas, where the concentration change was very rapid. The last one contained a measuring cell with tangential flow around the probe, which enabled to measure transient characteristics in flowing liquid. Optical and polarographical probes were used in our experiments.

The technique with *DO* probe displacement did not exhibit a good reproducibility. Conversely, the *DOC* step change realization using a three way valve gave repeatable and reproducible results both in gas and in liquid phase, even using higher viscosity liquids. Transient characteristics were measured using various liquid velocities. The time constants in liquid were strongly dependent on superficial velocity, especially in the case of optical probes because of high oxygen solubility in their membrane. Therefore, the optical probes suffer from the exhaustion of the dissolved oxygen in liquid film in front of the membrane. In dynamic measurements, this exhaustion of dissolved oxygen in diffusion layer leads to the deceleration of its concentration change in the vicinity of membrane compared to the oxygen concentration change in liquid bulk.

The reliable technique was developed to measure *DO* probes transient characteristics in liquid. This technique, together with a correct physical model, provides the precise prediction of *DO* probe behaviour during dynamic *DOC* measurement, even in higher viscosity liquids.

Measurement of volume, voidage and porosity: a novel application of the material balance

M.J Heslop and [1]A. McDonald
University of Sheffield, Department of Multidisciplinary Education, 32 Leavygreave Road, Sheffield S3 7RD, United Kingdom; tel. +44 114 222 7601, m.j.heslop@sheffield.ac.uk;
[1]University of Strathclyde, Department of Chemical Engineering, 75, Montrose Street, Glasgow, United Kingdom

The measurement of volume, voidage and porosity is important in a range of scientific, industrial and domestic situations. First, consider the simple example of a regular shape such as a rectangular box, cylinder or sphere. The dimensions can be measured and from the appropriate equations, the volume can be determined subject to the standard errors. This approach is not possible if the shape is irregular. Second, an important characteristic of soils is the voidage, that is, that fraction of the total volume that is empty space. In the case of construction and foundations, too low a voidage can compromise the safety of buildings. In the case of the topical subject of drainage, too low a voidage can result in there not being enough capacity to contain water and result in flooding.

In order to justify a new experimental technique, it is helpful to make reference to conventional and established experimental techniques. It is also helpful to categorise the measurement of volume as either internal (for example, the volume of an empty chamber), external (the volume of a solid object) or voidage/porosity (the volume *fraction* of a material that consists of empty space accessible to a gas). One such method is the displacement method – in which the (external) volume to be measured is added to an incompressible fluid – the resulting total system volume can be used to determine the sample volume. This has been used to determine the volumes of seeds and potato chips. The hydrostatic method involves measuring the weight of an object both in air and in a fluid of known density: the difference can be used to infer the (external) volume of the object. The permeation method is based on measuring the flow of gas out of a pressurised container through a membrane. Finally, the method of acoustic volumetry requires a loudspeaker and two sealed chambers (one being the reference). The loudspeaker imposes a sinusoidal variation in the two chamber volumes which gives similar variations in pressure.

There are a number of reasons why we feel that an alternative method for the measurement of volume is worthy of publication. First, the new method is distinctive in that it is gas-based, is not destructive and can be applied to a range of applications. Second, the two stages of the experiment (flow-addition and flow-removal) provide a self-check on the results. Third, the method is robust and cheap, capable of detecting very small changes in volume. Fifth, the continuous flow of gas through the chamber helps ensure that the process is carried out at constant temperature.

The schematic experimental arrangement is shown below and consists of: upstream and downstream resistances, the volume under investigation, a valve to add a perturbation flow and a pressure transducer to measure the pressure between the two resistances. Initially, the system is at steady-state whereby the values of M_1 and M_2 are equal, and the pressure P_1 is time invariant (at $P_1(0)$). An experiment is conducted by switching the valve to add a small perturbation flow (m) to the main flow upstream of the volume. This causes a slow increase in the pressure P_1 until a new steady-state is increased, at which $P_1 = P_1(\infty)$. The step change in pressure at steady state is ΔP_1. The return (perturbation removal) experiment is then conducted once the system has reached steady state. From a consideration of the material balance for such an unsteady-state process, as well as the assumption of laminar flow through the resistances, it can be shown that the area bounded by the dashed lines and the variation in pressure depends on the flow resistances as well as the initial pressure and the value of the perturbation flow.

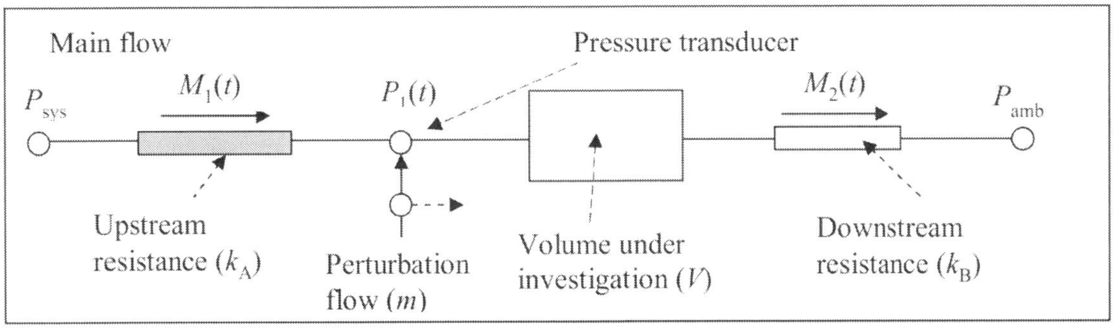

A typical experimantal result is shown below. This was obtained by adding a nitrogen perturbation flow of 20 ml/min to a nitrogen main flow of 70 ml/min. The resistances are formed from lengths of 1.6 mm stainless-steel tubing. The volume is approximately 300 ml. Such an experiment would allow the system to be calibrated, and this calibration would then be used to investigate unknown volumes.

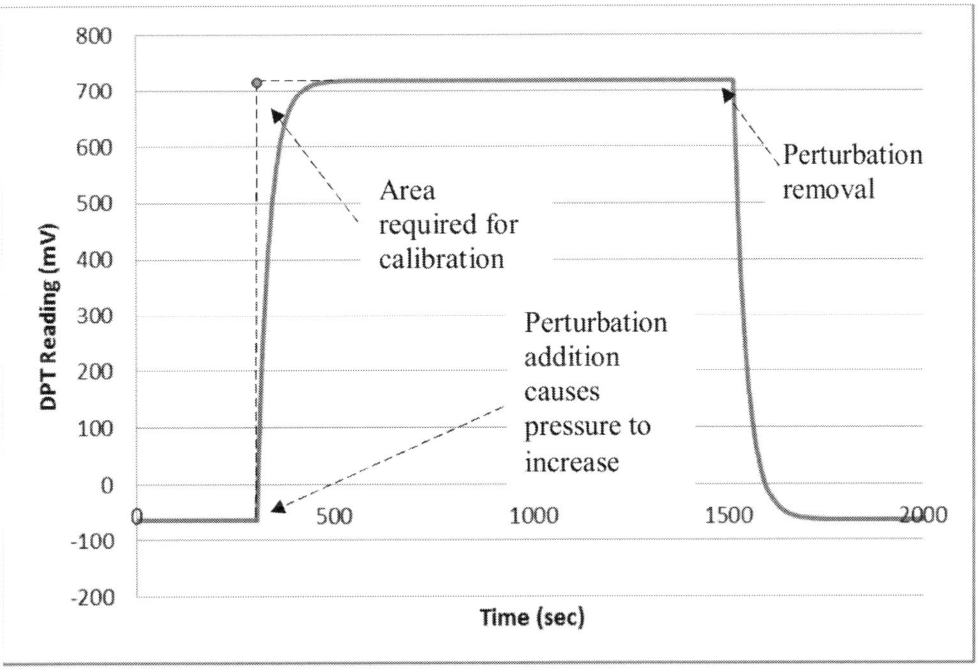

One advantage of this technique is that the sensitivity can be increased significantly by increasing the flow resistances, so that this simple arrangement is capable of measuring very small changes in volume. Other advantages are the quick response time, simple construction and the use of gases – which make the method non-destructive. Furthermore, the perturbation-removal experiment (return stage) provides a check on whether there has been any drifting in the baseline – and allows this to be corrected. The technique above is shown applied to simple empty volumes. In order to apply this method to say packed beds and porous solids, a different experimental arrangement would be required – one option of which is to locate the sample in a side-chamber. The voidage would then be characterised by both the area and shape of the response curve.

Circulation of catalyst particles in a fluidized bed in the reactor with internal elements and their effect on the reaction of gas dehydrogenation

S.A. Solovev, [1]A.A. Lamberov, S.R. Egorova, [2]O.V. Soloveva
Kazan (Volga region) Federal Univ., Inst. of Mathematics and Mechanics, Kremlevskaya St. 35, 420008 Kazan, Russia; e-mail: serguei_s349@mail.ru; [1]Kazan (Volga region) Federal Univ., Inst. of Chemistry, Kremlevskaya St. 29, 420111 Kazan, Russia; [2]Kazan State Power Engineering Univ., Inst. of Heat Power Engineering, Krasnoselskaya St. 51, 420066 Kazan, Russia.

Fluidized bed apparatus are widely used in the chemical industry. The advantage is the high speed of heat and mass transfer between components of the reaction, which are resided in different aggregation states. Such apparatus are used in industry for more than fifty years, but interest in the study of properties and applications of the fluidized bed was not lost. New schemes and modernizations of existing technologies are continuously appeared.

One of the available ways to get the data of interest on any processes is a numerical simulation, where the physical phenomenon is associated with the mathematical model. Numerical calculations of fluidization are usually solved by Eulerian-Eulerian approach, when the carrier (gas, liquid) and discrete (solid) phase are considered as continuous. To account for the interaction of particles in the fluidized bed, by analogy with the kinetic theory of gases, was added an equation describing change in the kinetic energy of particles due to their collision (e.g., [1]). The possibility of numerical solution of equations allows to carry out calculations of different variants of the fluidization process, both in terms of basic research and practical application.

A downward movement of the solid particles along the walls with high concentration is observed in fluidization in cylindrical columns. This effect ensures continuous circulation of the particles in the apparatus. Each structural element and the gas feeder in particular make a contribution to the efficiency of the apparatus operability. Location of internal elements on the way of the main flow stream of particles may affect the operation of the apparatus as a whole. This may improve the efficiency of chemical reactions by optimum distribution of the catalyst in gas flow in the study of chemical reactors catalytic processes. Changes in the character of the movement and concentration of particles along the wall with deflectors are shown in [2]. Papers [3, 4] are devoted to experimental and numerical study of effects on the hydrodynamics of the fluidized bed with ring baffles located on the walls of cylindrical apparatus. Experimental studies with internal elements in a liquid-solid fluidized bed are presented in [5]. Investigations of fluidized bed processes in the column with an inner cylinder were experimentally made in [6].

Fluidized bed apparatus of cylindrical shape with a uniform gas feeding are considered in this paper. Ring baffles, deflectors and grids were investigated as internal components. Ring baffles and deflectors on the walls provide a deviation in the downstream zone to the central area. Grids provide breaking of rising bubbles. Both cases with and without clearance between walls and grids are considered. The first aim of present paper was to determine the influence of internal elements on the circulation flow of gas and particles in the apparatus.

The temperature field plays an important role in catalytic chemical reactions. Due to the temperature field is organized continuous circulation of the heated catalyst from the regeneration chamber. The second aim of the paper is to determine effect of circulation flow to the chemical reaction on the example of isobutane to isobutylene dehydrogenation. Isobutane fraction was used as a gas raw. Microspherical aluminum-chrome catalyst particles of 20-200 micrometers in diameters are located in the reactor. The dehydrogenation of isobutane is accompanied with the release of hydrogen and the absorption of heat. Dehydrogenation reaction of C_3-C_5 paraffins is thermodynamically feasible only at temperatures 300-900 ^oC [7]. Therefore, to maintain the temperature of the reaction gas is supplied at a temperature of 550 ^oC, and the regenerated catalyst is fed at a temperature of 650 ^oC. Examples of [8, 9] describe the kinetics of the dehydrogenation of paraffins in fixed and slow movement beds of catalyst. Following them was composed the algorithm that takes into account the conversion of gas, the gas temperature, the volume fraction of catalyst, an elementary volume of reaction area, and the time step.

Differential equations that describe hydrodynamic processes and the processes of heat and mass transfer in the field of computational model of the reactor were solved in CFD package ANSYS Fluent. A full simulation of the reactor unit in differential equations for components mass of the gas mixture is necessary to consider changes associated with chemical reactions. For this problem were written users defined functions (UDF), which added terms of mass transfer and absorption of heat, depending primarily on gas temperature and catalyst concentration.

Hydrodynamic and thermal characteristic operating parameters, such as the field of catalyst concentration and the temperature of gas, were obtained from calculations. Pictures of the main circulating catalyst stream were created and results of influence on the concentration of the catalyst were determined. The effect of internal elements on the behavior of the catalyst in reactor and the efficiency of chemical reaction in the dehydrogenation of isobutane example were investigated.

[1] Gidaspow D. *Multiphase Flow and Fluidization*. Boston: Academic Press, 1994.

[2] Gorliz M.R., Grace J.R. *Augmentation of heat transfer by deflectors on circulating fluidized bed membrane walls* // International Journal of Heat and Mass Transfer. 2002. Vol. 45. P 1149.

[3] Samruamphianskun T., Piumsomboon P., Chalermsinsuwan B. *Effect of ring baffle configurations in a circulating fluidized bed riser using CFD simulation and experimental design analysis* // Chemical Engeneering Journal. 2012. Vol. 210. P. 237–251.

[4] Chalermsinsuwana B, Samruamphianskun T., Piumsomboon P. *Effect of operating parameters inside circulating fluidized bed reactor riser with ring baffles using CFD simulation and experimental design analysis* // Chemical Engineering Research and Design. 2014. Vol. 92. P. 2479.

[5] Piovano S., Salierno G.L., Montmany E., D'Agostino M., Maestri M., Cassanello M. *Bed Expansion and Particle Classification in Liquid Fluidized Beds with Structured Internals* // Chemical Engeneering and Technology. 2015. Vol. 38. No. 3. P. 423.

[6] Gujjula R., Mangadoddy N. *Experimental Investigation of Hydrodynamics of Gas-Solid Flow in an Internally Circulating Fluidized Bed* // Canadian Journal of Chemical Engineering. 2015. Vol. 93. P. 1380.

[7] Beccari M., Romano U. *Encyclopaedia of Hydrocarbons*. Rome: ENI & Istituto della Enciclopedia Italiana G.Treccani. 2006. Vol. 2. P. 687.

[8] Gyngazova M.S., Kravtsov A.V., Ivanchina E.D., Korolenko M.V., Chekantsev N.V. *Reactor modeling and simulation of moving-bed catalytic reforming process* // Chemical Engineering Journal. 2011. Vol. 176-177. P. 134.

[9] Frantsina E.V., Ivashkina E.N., Ivanchina E.D., Romanovskii R.V. *Developing of the mathematical model for controlling the operation of alkane dehydrogenation catalyst in production of linear alkyl benzene* // Chemical Engineering Journal. 2014. Vol. 238. P. 129.

A spatially coupled two-fluid model and discrete particle model for the simulation of gas-solid fluidized beds

Xizhong Chen[a,b], Junwu Wang[a]*
[a]State Key Laboratory of Multiphase Complex Systems, Institute of Process Engineering, Chinese Academy of Sciences, Beijing 100190, P. R. China
[b] University of Chinese Academy of Sciences, Beijing, 100049, P. R. China
*T: +86 10 82544842; F: +86 10 62558065; E: jwwang@ipe.ac.cn

The Gas-solid two-phase flows are widely found in chemical engineering, such as fluid catalytic cracking and coal combustion fluidized bed. Computational fluid dynamics has become a powerful and useful tool for design and scale-up these kinds of equipment. However, accurate and efficient description of gas-solid flows remains a great challenge. At present, two most popular approaches are usually used to simulate gas-solid flow: two-fluid model and discrete particle model. The major difference between these two methods is the way how the particulate phase is treated and each method has its own strengths and weaknesses. Two-fluid model is computationally efficient but is insufficient in describing the constitute relation and boundary condition for solid phase. On the other hand, the discrete particle model has a good resolution for solid phase but requires extremely high computational cost for industrial processes. In order to make full use of the strengths but discard the weaknesses of these two methods, we have developed a hybrid multi-scale method, by spatial coupling of two-fluid model and discrete particle model, for gas-solid flows, as shown in figure 1. The strategy is to utilize discrete particle model in critical part of the simulation domain to improve the accuracy of the simulation, while utilizing two-fluid model in the majority of the simulation domain to keep the efficiency of the simulation. The hybrid multi-scale method is firstly validated in simulation of a granular channel flow and a gas-solid channel flow without gravity. For the simulation of a typical gas-solid flow in fluidized bed riser, the key of the proposed method, the discrete-to-continuum and continuum-to-discrete mapping method that are used in the spatially coupled regions, is further developed to consider the effect of mesoscale clustering structures inside risers.

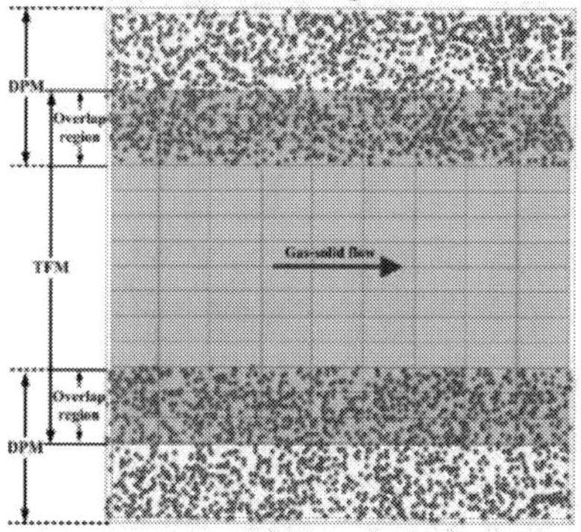

Figure 1 Schematic of the coupling strategy in the proposed hybrid multi-scale model

Investigation of gas-solid fluidized system with side injections using non-invasive based techniques and CFD simulation

J. S. Rawat, R. K. Upadhyay and P. Tiwari
Department of Chemical Engineering, Indian Institute of Technology Guwahati, Guwahati, Assam, 781039, India
*Corresponding author: Tel.: +91 361 2582263; fax: +91 361 2582291
Email address: pankaj.tiwari@iitg.ernet.in (Pankaj Tiwari)

Nozzle assisted gas-solid fluidized bed is usually preferred in the polypropylene industries, drying and pharmaceutical industries. The distributions of the solids inside the system play an important role in the designing of such type of system. The presence of the jet(s) contributes to turbulence locally, due to the action of high fluid velocity and, hence affects the size distribution of the solids (catalyst particles) inside the reactor. Sometimes the surface area increases drastically which is the key parameter to be controlled in any reactive system. It is clear from the literature that the design of the nozzle, type of the nozzle, number of nozzle(s) and its location and orientation is still under investigation. Although few research articles predicted the quality of the fluidization with and without side injection(s), the application of multiple injection(s) provides an open challenge for experimentation as well as scale up. For this purpose a cold flow model with cylindrical column of diameter 0.21 m and height 2.25 m was used as the fluidized system. The glass beads and air were used as a fluidized material and fluidizing medium respectively. In addition, the compressed air was fed to the system with a nozzle assembly (of nine nozzles) which is having a flexibility to change the orientation with the side wall. The present research work focussed on the computation fluid dynamics (CFD) simulations in Euler-Euler framework and radioactive based non-invasive techniques to quantify the performance of gas-solid fluidized system with side injection(s). Various parameters such as particle size, particle density, fluidization gas velocity, nozzle flow rates, its location and orientation were investigated. Gamma ray tomography and radioactive particle tracking (RPT) techniques provided the insight of solid distribution and other flow parameters such as instantaneous velocity, fluctuations and shear stresses respectively. Further the experimental finding can be validated using the Euler-Euler simulations in Ansys 14.5.

Theoretical and experimental investigation of ultrasound propagation through porous medium

Z. Karakaş, Y. Uludağ
Middle East Technical University, Department of Chemical Engineering, 06800 Çankaya/Ankara
tel. +90 03122104354, e-mail: zkarakas@metu.edu.tr, fax:

Ultrasound propagation through porous medium is investigated theoretically and experimentally. Allowing measurements with opaque medium, giving fast measurements, requiring low cost and being non-invasive make the use of ultrasound based techniques convenient[1]. The propagation of sound is directly related to the physical properties of a porous medium such as porosity, bulk modulus of medium, density of medium, particle size *etc.* Therefore, it becomes possible to characterize a porous medium by using ultrasonic techniques provided that useful models capturing the sound porpagation through such medium are readily available.

In order to model sound propagation through porous medium Newton - Laplace equation, which gives the sound velocity in a medium, can be applied. Physical properties of the porous medium, e.g. density and bulk modulus, are obtained by volume averaging of the relavent properties of the individual phases. According to the Newton-Laplace equation, as porosity increases, sound velocity shows a negative dispersion as expected. Since speed of sound in solids higher than that of liquids, as liquid amount increases, means as porosity increases, sound velocity decreases. Since the equation includes only the bulk modulus, density and porosity, dependence of the sound speed on different characteristics of porous medium, such as particle size, can not be captured. Therefore, more detailed investigation is done by employing Biot Model that also includes the frequency of sound wave as a parameter enabling to investigate particle size as well.

In Biot model, the medium is assumed to be fully saturated with liquid, porous medium is assumed as elastic. Also, particles are assumed as spherical and particle size and distribution are uniform inside the medium. Moreover, the most important assumption behind this model is that, the wavelength is much larger when compared to the size of grains. For the model, cubic control volume of porous medium is selected[2]. As sound propagates through medium, it exerts a force on solid and liquid phases. Forces acting on a solid body can be written in terms of stresses and on a liquid body in terms of pressure. Moreover, as sound propagates through the medium, it disturbs the particles and particles starts to move. Here, the important criteria is the movement of liquid particles. In porous medium as solid particles move, some of liquid particles are stuck on solid particles and they move with them. Also, because of the movement, liquid particles have velocity relative to solid medium. In model, the velocity terms are given in terms of displacement of solid and liquid particles. The movement of

---

[1] Mihoubi, M., Bélorgey, M., & Kettab, A. (2008). Analysis, in a free surface steady flow, of the interstitial velocity field inside a sedimentary bed. *C. R. Geoscience,* (340), 858-864.

[2] Biot, M. (1955). Theory of Propagation of Elastic Waves in a Fluid - Saturated Porous Solid. I. Low - Frequency Range. *Acousticel Society of America, 28*(2), 168-178

particles translates into kinetic energy. Biot defined this kinetic energy according to displacement of solid particle, displacement of liquid particle and relative velocity of liquid particle[3]. According to the defined kinetic energy, Biot defined the force acting on each body by Lagrangian method. From the equality of forces and elastic moduli definitions, dynamic equations for porous medium are obtained as[4];

$$\mu\nabla^2\underline{u} + (H-\mu)\nabla(\nabla\cdot\underline{u}) + C\nabla(\nabla\cdot\underline{w}) = \frac{\partial^2}{\partial t^2}(\rho\underline{u} + \rho_f\underline{w}) \quad (1)$$

$$C\nabla(\nabla\cdot\underline{u}) + M\nabla(\nabla\cdot\underline{w}) = \frac{\partial^2}{\partial t^2}\left(\rho_f\underline{u} + \frac{(1+\alpha)\rho_f}{\beta}\underline{w}\right) + \rho_f\frac{v_{0s}}{k_s}\frac{\partial\underline{w}}{\partial t} \quad (2)$$

According to this model, as frequency increases, sound velocity shows a negative dispersion while attenuation, which is the decreases in the energy of sound, exhibits a positive dispersion. However, in this model, attenuation mechanism is taken only as absorption. In absorption mechanism, mechanical energy is converted into heat. However, as the grain size and wavelenght values become closer to each other, then the attenuation mechanism changes. In this stuation, scattering becomes dominant. When the effects are compared, scattering has more impact than absorption on energy loss. For that reason in order to observe the investigation of sound velocity in a porous medium having grains larger than wavelength, experimental studies are conducted.

For the experimental studies, three sets of media are prepared with glass beads having grain size as 300 microns, 700 microns and 1 mm. Furthermore, frequency values are 1 MHz, 2 MHz and 4 MHz. Therefore the size of grains are close to the ultrasound wavelenght. The experiments are conducted by DOP 2000, Pulsed Ultrasonic Doppler Velocimetry. According to the experimental results, it is observed that as frequency increases, sound velocity shows a negative dispersion. Also, deviation in phase velocity for water saturated glass beads media with 1mm, 700 microns and 300 microns are 458 m/s, 741 m/s and 813, respectively. Furthermore, as frequency increases, attenuation increases. Different from Biot model, in this state, the attenuation coefficients have much higher values. For water saturated glass beads medium with grains 1 mm, 700 microns and 300 microns attenuation show $f^{1.7}$, f^2 and $f^{1.1}$ dependency, respectively. The difference in sound velocity and attenuation from Biot model suggests a different attenuation mechanism than that assumed in the model. In order to use the Biot model for the calculation of physical properties of such medium, the effect of scattering mechanism is needed to be integrated into the model. By this way Biot model may become not only applicable for the medium having smaller grain sizes but also for the medium having higher grain sizes.

---

[3] Biot, M. (1962). Generalized Theory of Acouzstic Propagation in Porous Dissipative Media. *The Journal of the Acoustical Society of America,* 1254-1264.

[4] Biot, M. (1956). Theory of Propagation of Elastic Waves in a Fluid - Saturated Porous Solid. I. Higher Frequency Range. *Acoustical Society of America, 28*(2), 179-191.

Analysis of oil recovery by injection of polymer solution in sandpack and microfluidic chip

N. M. Lima[1], R. P. Dias[2], D. Maza[3], M. S. Carvalho[4]
Pontifical Catholic University of Rio de Janeiro (PUC-Rio), Department of Mechanical Engineering - Laboratory of Microhydrodynamics and Flow in Porous Media, Rua Marquês de São Vicente, 225, Gávea, Rio de Janeiro, 22453-900, Brazil
Phone: +55 21 3527 1164, Fax: +55 21 3527 1165, nicollelimarj@hotmail.com[1]; ricardod@ipb.pt[2]; danmerm@gmail.com[3]; msc@puc-rio.br[4]

Waterflooding is the most commonly used oil recovery method in the industry. However, the high mobility ratio limits the amount of oil displaced by the water phase. An effective alternative to minimize this limitation is the application of technologies that act as mobility control agents. Polymer solution is used in many cases as a way to increase the water phase viscosity and consequently reduce the mobility ratio.

Experimental evidences have shown that the elastic behavior of some polymer solution may contribute to a better pore level oil displacement, reducing the residual oil saturation. This pore level behavior is not clearly understood.

Here, a glass microfluidic chip is used as a two-dimensional model porous space that is most useful in order to understand the results of the oil displacement experiments performed with silica sandpack.

The microfluidic device has the principal features of a porous media and provides means for pore level flow visualization. The evolution of the water phase as it displaces oil is recorded.

Using sandpack or the microfluidic device, four different water phases were used: pure water, two high molecular weight polymer (PEO or HPAM) solutions and a glycerol-water mixture with the same viscosity of the polymer solutions.

Visualization of the flows in the microfluidic chip provides specific information about the presence of the trapped oil phase and the movement of the oil/water interface in the network.

Results show that the polymer solution can modify the liquid distribution in the porous media, improving the displacement efficiency at pore scale and consequently the final oil recovery factor in both the microfluidic device and sandpack.

Water-in-oil emulsion formation for preventing wax deposition of heavy crude oil through pipeline

[1]K. Kruethong, [1,*]A. Chareonsang, [1,**]U. Suriyapraphadilok, [2]D. A. Sabatini
1The Petroleum and Petrochemical College, 254 Soi Chulalongkorn 12, Phayathai Rd, Pathumwan, Bangkok 10330, Thailand, Tel (662)2184110-11, Fax (662)2154459,
*e-mail: Ampira.c@chula.ac.th, **e-mail: Uthaiporn.s@chula.ac.th; [2]School of Civil Engineering and Environmental Science, University of Oklahoma, 202 W. Boyd St., Rm 334 Norman, Oklahoma 73019-1024; Tel. 405-325-4273, Fax: 405-325-4217

An emerging challenge of waxy crude oil transportation in pipelining system is wax deposition such as paraffin wax deposition on wall surfaces of pipeline in oil production and refinery systems. Solid and/or semi-solid wax builds up when the temperature falls below its cloud point or wax appearance temperature (WAT), causing the deposition of a layer of paraffin molecules. In this work the water-in-oil (W/O) emulsion is formulated with the aim to prevent wax crystal formation and wax precipitation using nonionic alcohol ethoxylated as a surfactant. This study, the appropriate surfactant system and its emulsion characteristics were investigated. The affecting parameters on emulsion formation including surfactant structures and concentrations, brine conditions (NaCl + $CaCl_2$), water content and wax content were also investigated at various conditions. The emulsion phase behavior and its stability was observed by visual observation at room temperature (25 ± 2 °C). For the simulating model water-in-oil emulsion study, the appropriate surfactant system with different structures, nonionic alcohol ethoxylated series, anionic extended with propoxylate group and conventional surfactant, dodecyl benzene sulfonate are investigated. The result showed that nonionic alcohol ethoxylated surfactant with 3EO groups can form an emulsion with high stability comparing with those of anionic based surfactant systems. The emulsion characterizations was determined by viscometer and polarization microscope. Figure 1 demonstrates the effect of surfactant concentration with 5 %wt. brine concentration on the emulsion viscosity and stability. The result indicated that the correlation between viscosity and separated water content where the minimum viscosity presented at surfactant concentration in the range of 5 to 10 %wt. The micro-observation of emulsion characteristic as shown in Figure 2. It can be seen that the emulsion solution is more homogeneously at the surfactant concentration of 7% wt.

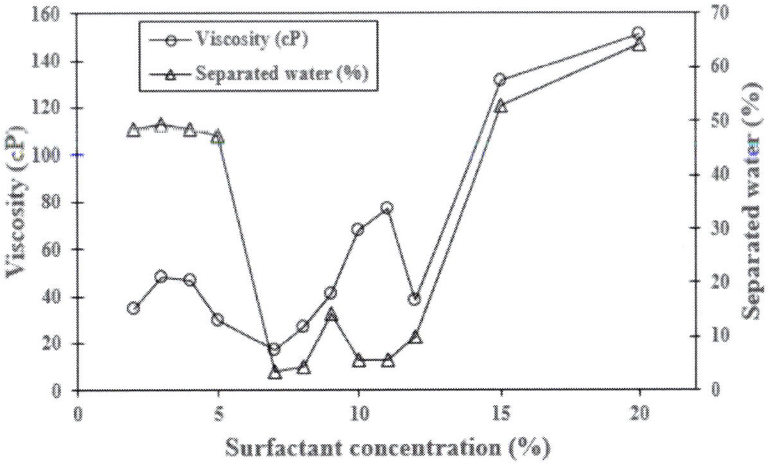

Figure 1 The effect of surfactant concentration with 5 %wt. brine concentration on the emulsion viscosity and stability.

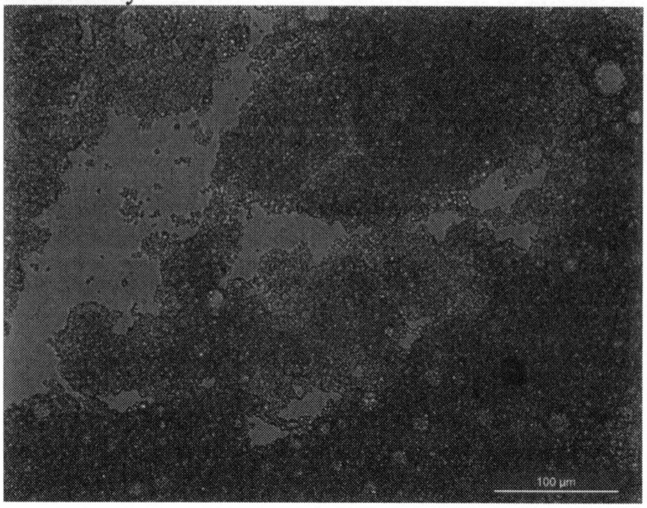

Figure 2 The microscopic photo of the dispersed phase particle size distribution with the water content 50%, 7% surfactant concentration and 5% brine concentration.

References:
1. Zheng, S. and Fogler, H.S. (2014). Fundamental Investigation of Wax Diffusion Characteristics in Water-in-Oil Emulsion. Industrial & Engineering Chemistry Research.
2. Wang, P., Wang, W., Gong, J., Zhou, Y. and Yang, W. (2013). Effects of the Dispersed Phase on Oil/Water Wax Deposition. Journal of energy resources technology 135(4), 042902.
3. Abdurahman, N., Azhari, N. and Yunus, Y. (2008). Formulation and Evaluation of Water-Continuous Emulsion of Heavy Crude Oil Prepared for Pipeline Transportation.

The modifications of the rheological behaviour of Na-bentonite suspensions with chitosan biopolymer effect

B. Abu-Jdayil, K. B. Alsayyed Ahmed, M. Djama, M. Ghannam
Chemical & Petroleum Engineering Department, U.A.E. University, Al-Ain, P.O. Box: 15551, UAE, tel:+97137135317, fax:+9713, e-mail: babujdayil@uaeu.ac.ae

Recently, the use of natural minerals in different industrial applications increasingly attracts attention because of their abundance and low cost. One of the important natural minerals is the bentonite, which consists mainly of montmorillonite and contains varying amounts of other minerals like quartz and feldspar.

With exceptional rheological properties, bentonites are used in different branches of industry including drilling fluids, dyes, pharmaceuticals, paper, cement, nanocomposites, polymer composites, and ceramics. In the use of bentonite efficiently in these industrial applications, different additives are recommended to optimize the rheological behaviour of the formulation.

One of the most important applications of bentonite dispersions is their use as drilling muds in the oil industry. Bentonite's usefulness in drilling comes from its unique rheological properties: relatively small quantities of bentonite in water form a viscous, shear-thinning material. Suspensions of bentonite are most often thixotropic and at high enough concentrations, characteristics of a gel (a fluid with a yield stress) can be observed. Different polymers have been used in bentonite dispersions to control the rheological behavior of the dispersions that meet the specification of the desired applications, such as poly vinyl pyrolidone, poly vinyl alcohol, xanthan, polyanionic cellulose, polyethyleneimine, poly (ethylene glycol), and carboxymethyl cellulose sodium salt.

Two grand challenges for this century are renewable energy and the utilization of green resources. Recently, much attention has been paid to develop low cost, sustainable, environmentally friendly, and high performance water-based drilling muds. In this study, the effect of chitosan biopolymer on the rheological properties of Na-bentonite suspensions was investigated in the polymer concentration range of 0.1 – 3.0 wt%.

Chitosan is a derivative of chitin (a natural polysaccharide) which may be prepared by deacetylation of chitin in an alkaline medium. Chitin itself is the second most abundant natural polymer in the world and both chitin and chitosan being biorenewable, biocompatible, environmentally friendly, biodegradable, and bio functional these are natural resources waiting for a market.

Two concentrations (4.0 and 8.0 wt%) of bentonite suspensions were prepared in this study. These two concentrations have been chosen to represent two different rheological behaviours of bentonite dispersions. Figure 1 shows the flow curves of the pure bentonite dispersions for 4.0 and 8.0 wt% solid concentrations. Bentonite dispersions examined in this investigation showed non-Newtonian behavior at both concentrations. Modeling analysis was carried out to predict the most relevant model that represents the rheogram behavior. From the investigation of this analysis, the well-known Herschel-Bulkley model was found to fit all the measured data of the bentonite suspensions. The regressed parameters of Herschel-Bulkley model revealed that the 8 wt% bentonite suspension exhibited shear thinning behaviour with a remarkable yield stress. On the other hand, the 4 wt% sample behaved as Bingham material at low shear rate, and as a shear thickening fluid at high shear rate. In addition, the rheological tests revealed that that the 8.0 wt% bentonite dispersion exhibited a clear thixotropic behaviour, while that for 4.0 wt% dispersion was not clear.

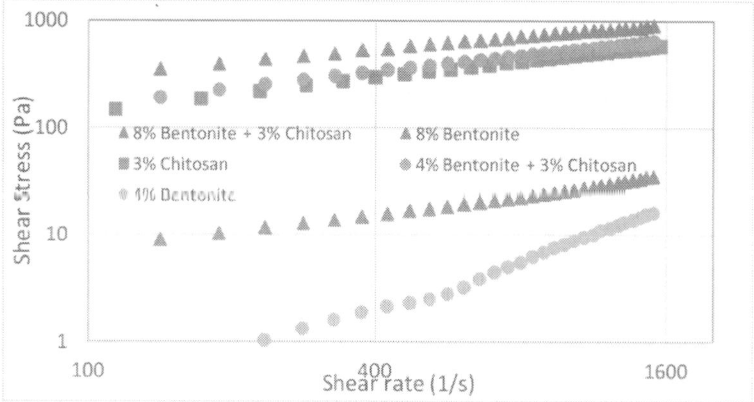

Figure 1: Flow curves of pure bentonite, pure chitosan (3.0 wt%) and suspensions of bentonite and chitosan.

Figure 2 shows the flow curves of pure chitosan solutions at different concentrations. It is clear that the shear stress is a strong function of chitosan concentration and shear rate. Shear stress gradually increases by increasing the shear rate and chitosan concentration. The effect of chitosan concentration is more pronounced at high shear rate. Modeling analysis according to the power-law model was carried out for all the examined concentrations of chitosan solutions. This model (soild line in Fig. 2) fits very well all the tested solutions of chitosan. Low concentrations of chitosan solutions up to 0.5 wt% show Newtonian behavior. Increasing the chitosan concentration beyond 1.0 wt% causes a significant decrease of the flow behavior index from unity (i.e. Newtonian behavior) to 0.53 (i.e. shear thinning behavior).

Addition of chitosan to the bentonite suspension increased significantly the shear stress and apparent viscosity of suspension (see Fig. 1). This significant increase of shear stress and viscosity of the bentonite dispersion can be attributed to the adhering of chitosan molecules on the clay particles surfaces and absorb water by swelling in slurry. The increase of the suspension viscosity is an indication of forming a network structure within the suspension. The advantage of chitosan that the desired rheological properties can be obtained with less bentonite by adding chitosan polymer and the undesirable effects of high bentonite concentrations can be avoided. The viscosity of 8.0 wt% pure bentonite dispersions was achieved by adding less than 1.0 wt% chitosan to 4.0 wt% bentonite dispersions. The combined flow behaviors of Bingham and shear thickening observed for the 4.0 wt % pure bentonite dispersion was shifted toward the shear thinning profile with a yield stress in the presence of the chitosan. This yield stress increased substantially with chitosan concentration. In addition, the thixotropic behavior of bentonite dispersions was significantly enhanced by adding chitosan polymer.

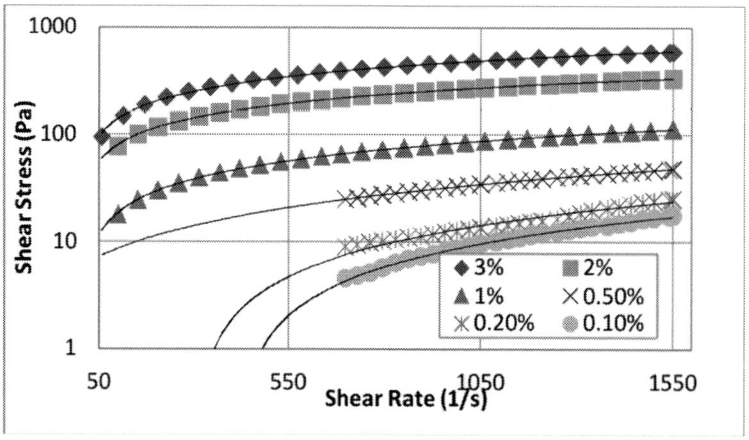

Figure 2: Flow curves of chitosan at different concentrations.

Study of the factors which control the behavior of suspensions of the activated bentonite. Case study of Meghnia's commercial product- Algéria

M. A. BELGHOUL[1], K. DERAMCHI[1], M.OULD HAMOU[1], , E.MOUSSAOUI[1], B.BOURBALA[1], D. BENKACI[1], S. BOUTRIA [1], S.BENTAALLA [1]
Ecole Nationale Polytchnique, 10, avenue Hassen Badi, El Harrach Alger, Algérie Tel. : +213 23 82 85 39 e-mail: karimaderamchi@yahoo.fr [(1)]Laboratoire Génie Minier, Ecole Nationale Polytchnique, 10, avenue Hassen Badi, El Harrach Alger, Algérie.
Fax : 00 213 23 82 85 29

The aim of this study is the characterization of the rheological properties, essentially the viscosity, of MEGHNIA's Bentonite based on water/Bentonite ratio. This, in order to underscore the viscosity variation problem of this bentonite.
we run several viscosity measurements' tests on several samples of bentonite with different activation degrees, using Na_2Co_3 , at different concentrations of Bentonite

MEGHNIA's main product is the drilling mud Bentonite, which is conditioned by the strict API standards, applied by most petroleum actors. During its production, MEGHNIA's bentonite, does not meet the minimum viscosity requirements, which is about 20 Cp, while its viscosity do not stop increasing with time.

Once the product is stored from 15 days to two months, the viscosity of Bentonite meets perfectly the requirements of the API standard, which is around 30 Cp.

This viscosity, increasing with time, presents a serious chemical commercialization issue of the product..

Clays are defined from a size side as materials corresponding to the fraction smaller than 2μm of a geological formation. They are defined also from a mineralogical side as aluminosilicate minerals most often from the family of "phyllosilicates" which consist of very thin particles which are composed of very thin layers (7, 10 or 14 Å).

The Bentonite is classified as a member of the silicates clays family. The biggest one is a smectite. The clays of this family are composed of sheets of minerals. The one sheet of clay is formed by octahedral layer between two tetrahedral layers.

The elementary layers are separated by water molecules and exchangeable cations. These cations restore the neutrality of the system and also provide the connection between the adjacent sheets.

Montmorillonites swelling properties are due to the hydrophilic nature of their surfaces and the presence of hydratable cations in the interlayer gallery, the degree of hydration of the clay depends on the nature of the hydrated cation. When the hydration continues by the swelling of the particles, in the presence of water, the bentonite can form a hydrated solid, a suspension or a gel.

Bentonite concentration and the structures adopted by the sheets are the parameters influencing the rheological properties of aqueous Bentonite suspensions.

A fluid is thixotropic if its viscosity, at constant shear rate, decreases over time with a condition of reversibility, if it increases the fluid is anti-thixotropic. The thixotropic behavior is relates to the fluidizing materials.

For this study, after reducing the Bentonite original samples using rotate divider, we prepare various Bentonite/water fluids: under several mass concentrations of 3%, 6% and 9%

of Bentonite. For each concentration, we activate the Bentonite under three degrees: 2%, 4% and 8% of Na_2Co_3, so we do have 9 samples. We measure the viscosity of each sample in several timing, respecting an interval of 3 days between each measurement.

We do measure for all the samples the 75 µm residue, Humidity and the swelling indicator. We do realize that all the samples are conform to API standards, so we can measure the viscosity using a BROOKFIELD viscometer at 100RPM speed (constant speed).

C%	Activation degree %	Temperature C°	PH	Viscosity Cp	Measure
3%	2%	26,60	9,60	2,30	Mes 1
	4%	24,90	7,30	4,50	Mes 1
	8%	25,20	7,80	2,00	Mes 1
6%	2%	27,10	10,55	3,00	Mes 1
	4%	23,60	10,00	7,00	Mes 1
	8%	25,30	7,60	2,50	Mes 1
9%	2%	27,80	10,42	4,00	Mes 1
	4%	24,00	7,10	7,70	Mes 1
	8%	24,90	8,00	3,50	Mes 1

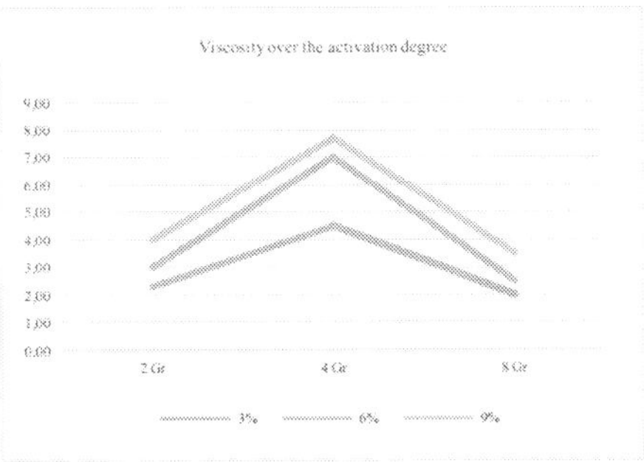

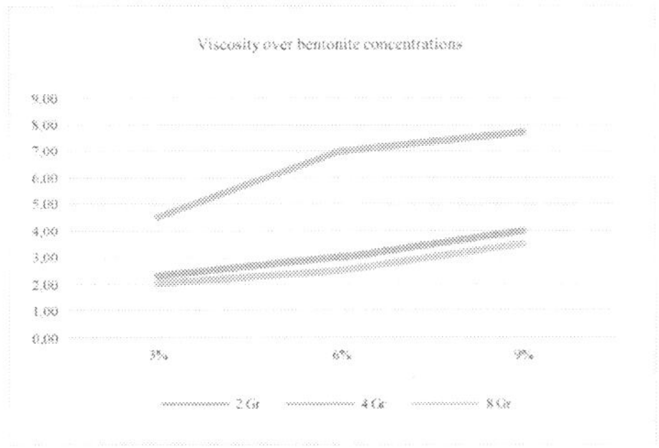

In the first graph, we observe that the viscosity increase until the optimal bentonite activation degree of 4% and then it decrease.

In the second one, we do observe an increase in viscosity by increasing the concentration, with the fact that the 8% activated Bentonite show the smallest viscosity values.

The add of Na_2Co_3, Sodium Carbonate produce an exchange between the Na^{2+} and the clay's Ca^{2+}which adsorb the water and increase the fluid's viscosity. Once the solution is saturated of Ca^{2+}, in our case 4%, the reaction between the excess of Na^{2+} and the OH^-will release H_2O decreasing the fluid's viscosity.

Thus, the optimal Na_2Co_3 activation degree of MEGHNIA bentonite is 4%.

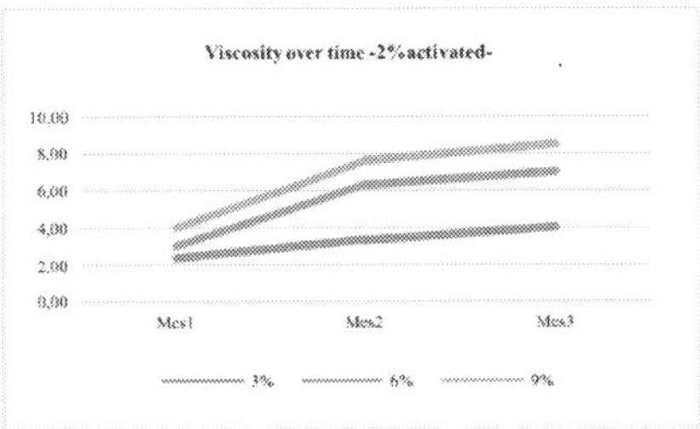

For all the fluids, different concentrations and activation degrees, the viscosity increase under the same shear rate with the same spindle speed: 100RPM. The 4% activated Bentonite present the highest viscosity results, since it represents the optimal activation degree.

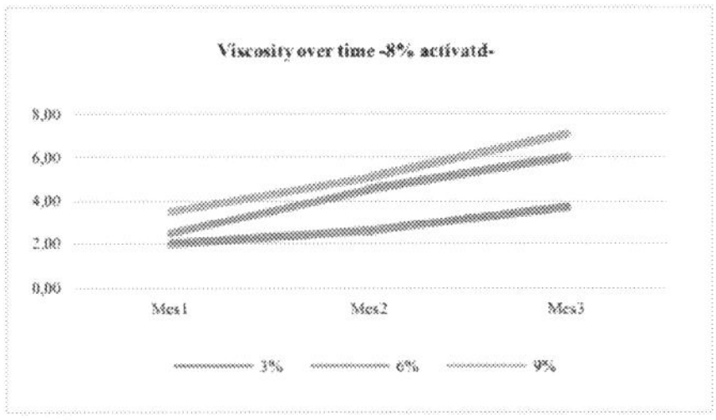

Thus, we can confirm that the Bentonite/water fluid is an anti-thixotropic solution, the viscosity decreased with the same shear rate over time. The phenomenon is explained by rheofluidifiency of the system, usually, it is due to the reconstruction of the physical tridimensional network of the structure which means the return of the particles to the minimum energy position is progressive under the Brownien movement.

Evaluation of produced water pipes mineralogical scaling in a central Tunisian oilfield

S. Barbouchi, [1]L. Monser
ETAP, Petroleum Services Department, 54 Avenue Mohamed V-1002 Montplaisir , Tunis, Tunisia; tel: +216 71 285 300, e-mail: sonia.barbouchi@etap.com.tn;
[1]INSAT, Inorganic Chemistry Department, Centre Urbain Nord BP 676 - 1080 Tunis; tel.(+216) 71 703 829, e-mail: lotfi.monser@insat.rnu.tn

In oilfields, every barrel of produced oil is associated with the production of 3 to 5 barrels of water at global level. One of the major flow assurance problems related to this coproduction of water is the inorganic scale deposition in water pipes

In order to evaluate the scaling mechanism in a central Tunisian oilfield, the chemical compositions of produced water and scale deposits samples have been identified by ICP-OES analysis. All samples have been pre-treated by a microwave assisted acid digestion.

The scale mineralogical characterisation has been realized using XRD and SEM; and hydrocarbons traces have been detected by Rock-Eval pyrolysis.

The physic-chemical produced water characterization has been used to calculate the Langlelier Saturation Index values, helping to determine the scaling and corrosive potential and to understand the scale deposition mechanism.

The future scale inhibition and removal strategies to be implemented in the concerned oilfield are being derived in a large part from the results of the present study.

Characterization of paraffin from different Tunisian crude oils

M.Samcha[1], A.Boukadi[2]
[1] ETAP, petroleum services department, 4 rue des entrepreneurs Charguia II-2035 Tunisie, tel: +21670838440, fax: +21670838988, e-mail: mariem.samcha@etap.com.tn;
[2] ETAP, department of petroleum services, 4 rue des entrepreneurs Charguia II-2035 Tunisie, tel: +21670838440, fax: +21670838988, e-mail: amel.boukadi@etap.com.tn;

Wax deposition is one of the serious problem in oil industry. Indeed, paraffinic waxes present in petroleum crude can precipitate when temperature decreases during oil production, transport or storage.

These deposits are mainly composed by n-paraffins (linear alkanes) and branched paraffins.

The precipitation of these paraffin is responsible for changes in the waxy crude oil properties, including gelation of oil and an increase in viscosity.

In this context we are interested in the characterization of paraffins, after isolation from different crude oils using precipitation method and purification by column chromatography.

Important information about the structural composition of these waxes were obtained from Nuclear Magnetic Resonance Spectroscopy: 1H NMR, 13C NMR, DEPT 135 and Heteronuclear Single-Quantum Correlation (HSQC).

Flow assurance of a Libyan waxy crude oil

F. Mohamed[1], H. Benkreira[2], S. Al-hengari[1], R. Patel[2] and M. Elbousiffi[1]

[1]Libyan Petroleum Institute Tripoli-Libya
[2] Faculty of Engineering and Informatics, University of Bradford, UK

In general, Libyan crude oils are classified among the best crude oils producing in the world that is for their high API and low sulfur contents which makes them friendly to the environment and yield more light distillates. However, some Libyan crude oils contain moderate to high contents of paraffin wax (2-29 wt. %) and pour point up to51°C.
Paraffin waxes are considered valuable hydrocarbons however; solid paraffin's precipitate out from waxy crude oils can complicate the low temperature flow properties causing major difficulties in pipeline transportation and storage. In this paper, some experimental results describing the flow behavior of waxy Libyan crude oils are presented including the effect of temperature on crude oil viscosity. Other related parameters such as thixotropy, yield stress, thixotropy recovery and yield stress recovery are discussed. Multiphase simulation is used to describe the pipeline cool down carrying the crude because of emergency shut down. The results presented show the importance of rheological data in assessing the flow assurance strategies.

Keywords: Rheology, viscosity, paraffin wax, multiphase simulation, flow assurance

The alternative solution for the pipeline transport of heavy crude using the heaters in pumping stations

A. Benmounah, [1]A. Chebouba
Univ. UMBB, Dep. Transport FHC, Indep. Street, 35000 Boumerdes, Algeria;
tel. +213 773377601, e-mail: amarbenmounah@yahoo.fr; [1]Bat.19G8, 35000 Boumerdes,
tel. +213 561851167, e-mail: chebouba@yahoo.fr; fax: +213 24795162.

There are several possibilities to transport the viscous crude by pipeline. The best known are the transportation of the product mixed with a diluent and the transport with heating of the product. In the 1st case the disadvantages are:
1. Reduction of the efficiency of transport caused by the diluent product,
2. Harm to the environment in the event of loss of the product
3. Increase in the cost of transport.

In the 2nd case the major drawback lies in the selection and dimensioning of stations pumping and heating of the viscous oil.

The original idea of this paper is to propose a method of reduction in viscosity of oil by heating. The difference lies in the type of proposed heating. Indeed it is to practice heating inside pumping stations. The distances and the type of pump with motor drive will be chosen on the basis of thermal and hydraulic calculations that will determine the optimal distance of reheating, i.e. the distances travelled by the heavy oil in turbulent regime.

It is proposed to install heaters in tank of storage in pumping stations at the optimal distance of flow according to the turbulent regime. The advantage of this process is to save heating energy and pumping. So the product will retain its initial properties. We propose a theory based on the thermal and hydraulic calculation of oil pipeline to determine reheating range, the temperature corresponding of the flow regime change for a given oil type (each type of product has its own limit temperature of heating), the generated load losses, the number of the heating and the pumping stations and the necessary thickness of insulation to maintain, as far as possible, the heat in the oil pipeline.

Keywords: Viscosity, Heating, Energy, Environment, Economy.

Hydrodynamic behaviour of an upward two-phase slug flow

F. Saidj, A. Zeghloul, H. Bouyahiaoui, A. Azzi, B J. Azzopardi[1]
University of Sciences and Technology Houari Boumedien (USTHB), FGMGP/LTPMP, Bab Ezzouar, 16111, Algiers, Algeria; e-mail: faiza78saidj@yahoo.fr , [1]Faculty of Engineering, University of Nottingham, University Park, NottinghamNG7 2RD, United Kingdom.

Flows of gas-liquid mixtures are frequently encountered such as in chemical, petroleum and energy production industries.

In vertical flow, several flow patterns are observed and may be classified by gas proportion increasing as follows: bubbly, slug, churn and annular flow. Depending on the flowrates, properties of each phase as well as on pipe size (diameter and length of the pipe), one of these flow pattern may take place. Among these, slug flow is worth to investigate. Indeed, in oil and gas pipelines, an understanding of flow structure as well as its behaviour is necessary for a safe and optimal design of separators situated downstream the riser. These separators can be exposed to periodical impacts caused by the arrivals of intermittent slugs. Slug flow exhibits clearly period structure. It is composed of a gas bubble often called Taylor bubble, which occupies almost the whole of the pipe cross-section surrounded by a thin film of liquid. This latter will either upward or downward flow. Taylor bubbles are followed by liquid slugs, which often contain a dispersion of small bubbles.

In the present study, characteristics of the slug flow, mainly average void fraction, Taylor bubble and liquid slug lengths, and slug frequency were extracted from the time series of cross-sectionally averaged void fraction. This parameter has been measured using a conductance probe technique at nine positions along a transparent acrylic pipe of 34 mm diameter and about 6 m length. The superficial velocity of the air was varied between 0.3 and 4 m/s and that for the water between 0.21 and 0.91 m/s.

Spherical cap bubbles, slug and churn flows have been observed in the test section for the ranges of gas and liquid superficial velocities investigated. These observations have been confirmed by using the method of Costigan and Whalley for the recognition of the flow pattern configuration. The data have been plotted on the mechanistic flow pattern map of Shoham.

The analysis of the average void fraction time series and their corresponding PDFs, show that the effect of the mixer on the flow essentially disappears once the seventh probe is reached. This corresponds to a distance of 95 pipe diameters.

It was found that the average void fraction increases with the gas superficial velocity.

Correlations proposed for the calculation of the velocity of the structure, i.e., Nicklin et al. and Azzopardi et al., give good prediction of these velocities.

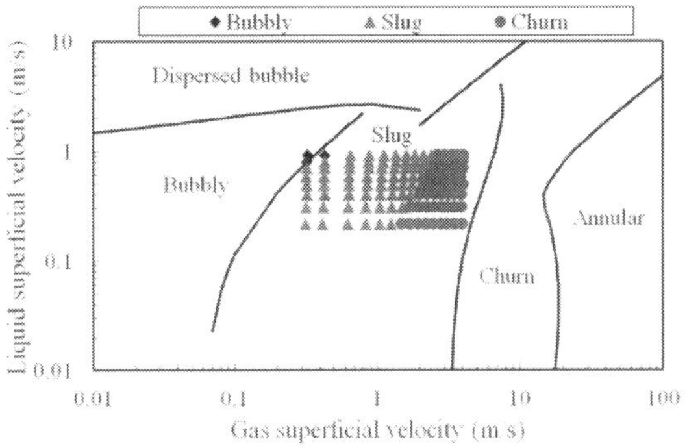

Figure 1: Flow pattern map in the vertical pipe

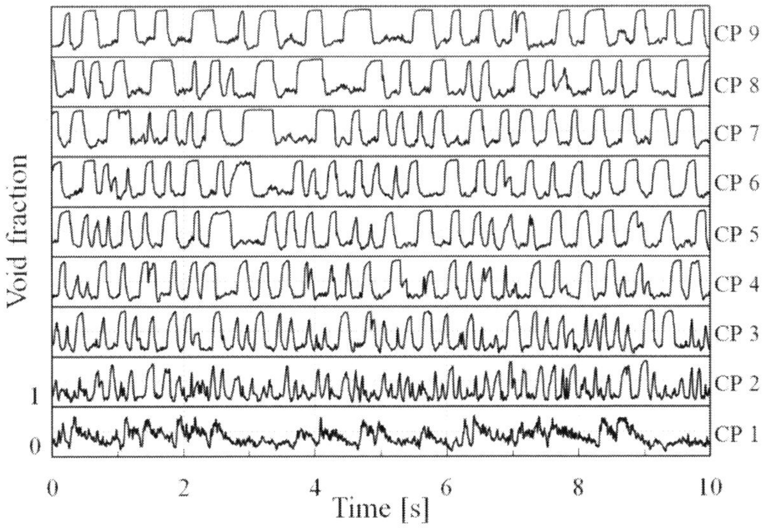

Figure 2: Time series void fraction along the test section, Uls=0.58m/s and Ugs=0.64m/s

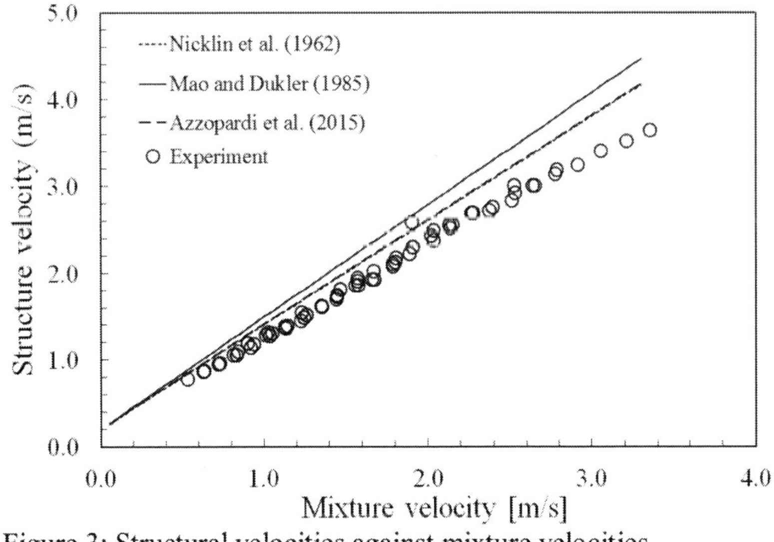

Figure 3: Structural velocities against mixture velocities

Influence of aerators on heat transfer in multiphase contactors

M. Žižka, R. Šulc, P. Ditl
Czech Technical University in Prague, Faculty of Mechanical Engineering 166 07 Praha 6, Technická 4, tel: +420 224 352 715, fax: +420 224 310 292, e-mail: martin.zizka@fs.cvut.cz.

In chemical and biochemical processes gas is often dispersed into liquid through aerators. The aerator design differ mutually n in process plants implementation. The aerator design affects the gas dispersion into the liquid and severally two-phase flow creation. This flow has an impact on heat and mass transfer between liquid and gas dispersion. The aerator design should be of the main designer's interest.

The main aim of this paper is to determine the effect of an aerator for the heat transfer between gas and liquid and to compare three different aerators.

Our experimental study was performed on one single orifice aerator and two different aerators with different orifices pattern shown in Fig. 1. The gas bubbles are dispersed into the liquid through this aerator. Entering the gas bubbles have a lower temperature than the liquid. From the viewpoint of the gas, heat transfer occurs from the liquid to the gas. The gas is warmed up, and the liquid is cooled, so the liquid temperature decreases with time. The inlet gas temperature is practically unchanged and remains constant. The outlet gas temperature changes with time, depending on heat transfer. Heat transfer can be also accompanied by simultaneous evaporation of the liquid, which has to be covered into the evaluation.

This study was performed on the bubble column 0,15 m in diameter with various water levels within the range of superficial gas velocity varying from 0,01 m/s to 0,1 m/s.

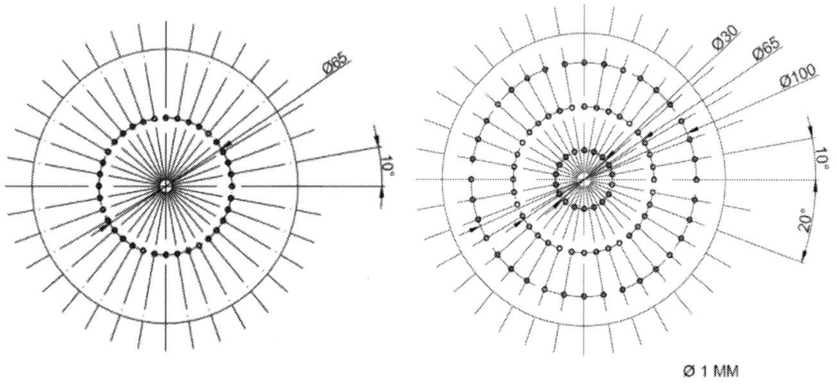

Fig. 1. Aerators orifices pattern

CFD simulation of the two phase flow ejector by Eulerian AUSM method

M. Zolfaghari[1], S. Sadeghpour Galooyak[2], B. Dabir[3]
1-Amirkabir University of Technology, Dep. Chemical Engineering, 424 Hafez Ave, Tehran, Iran, 15875-4413, Tel: (+98) 21-48255301, E-mail: mm.zolfaghari.r@gmail.com
2-Refining Technology Development Division, Research Institute of Petroleum Industry, Tehran, Iran, 14665-137, Tel: (+98) 21-48255301, E-mail: sadeghpours@ripi.ir
3-Amirkabir University of Technology, Dep. Chemical Engineering, 424 Hafez Ave, Tehran, Iran, 15875-4413, Tel: (+98) 21-48255301, E-mail:drbdabir@aut.ac.ir

Ejectors are widely used for creating vacuum and increasing the pressure of low pressure stream in industrial applications. In the brief description, Ejector is made from three different parts: inlet, mixing and exhaust section. In the entrance section hi-momentum stream (Primary Flow) drives low momentum flow (Secondary Flow) to the mixing zone, where two streams reach the same momentum with the supersonic flow; finally, in the exhaust section, as a result of the shock phenomena, supersonic stream reaches the higher exhaust pressure comparing to the secondary flow. Simulation of single phase flow Ejectors was done by many investigators; in this project, Two phase flow ejector is simulated where gas exists as the continuous phase and water droplets present the disperse flow. Subsequently valuation of the CFD simulation is carried out by the experimental data to certify accuracy of model. AUSM+-up method has been used for simulation of the supersonic flow and distinguishing exact point of shock in the ejector. AUSM$^+$-up, the latest version of AUSM family, was proposed by Liou(2006) to be applicable in all speed and also be stable in a wide range of applications. In a this project, simulation is done by Eulerian approach, developed by H.Paillere(2003) and J.R.Garcia(2006) for two phase flow AUSM method. Furthermore, because of cylindrical geometry of ejector and also in order to saving in a time and cost of the calculation, simulation is done in 2D axisymmetric model. For viscous parts of the H.Paillere(2003) formulation, turbulent k-ε as robust and accurate model, used extensively in the industrial application and also in other previous ejector simulation, was employed.

According to Y.Bartosiewicz(2009) et al. article experimental data was provided for a wide range operation condition. In this work, the main focus is simulation of two phase flow ejector by FORTRAN90 for two different pressures off primary flow (4, 5bar) in which droplets were considered just in the primary flow according to the experimental data.

Simulation results for P_I=4, 5bar is depicted in Fig.1. In the on-design simulation, average entrainments ratio was 3.2% lower than experimental data and also in the off-design operation condition 4.5% more than experimental data. By comparing the results with the other works, the agreement between model and experiment almost is acceptable over the range of the operation condition. In the couture, disperse phase distribution is shown for mass fraction α=1% and droplet diameter 17 micron.

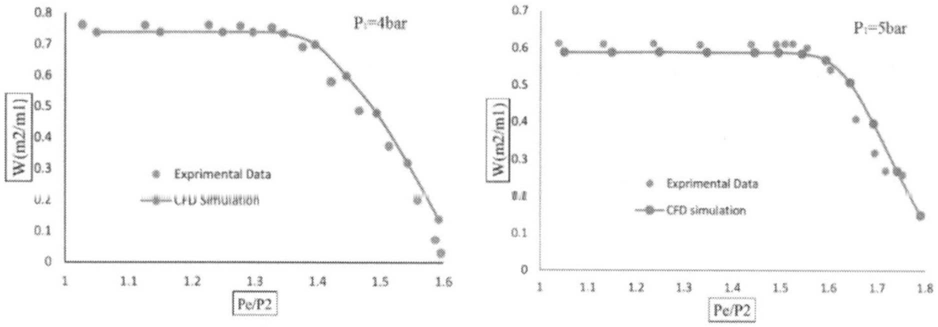

Fig.1. Entrainment Ratio to ratio of the Backpressure/Secondary pressure

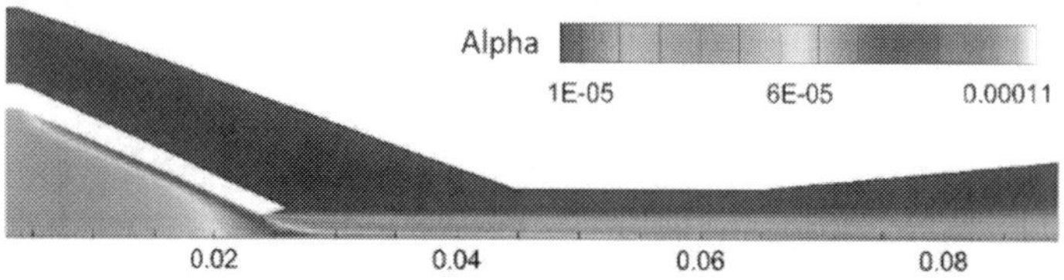

Fig.2.Volume fraction of disperse phase

Experimental investigation of single fluid particle breakage due to turbulence

E. H. Herø, J. Solsvik, H. Knuutila, H. A. Jakobsen
Department of Chemical Engineering, Norwegian University of Science and Technology (NTNU), Norway; tel. +47 4849244, e-mail: eirik.h.hero@ntnu.no;

Simulation of multiphase flow is a common industrial challenge, examples comprise dispersed flows in chemical reactor and in phase separation equipment. The dispersed phase may be described by a population balance equation, PBE, which can be given as[1]:

$$\frac{\partial f_n(\xi,r,t)}{\partial t} + \nabla_r \cdot [f_n(\boldsymbol{\xi},\boldsymbol{r},t)\boldsymbol{v}_r] + \nabla_\xi \cdot [f_n(\boldsymbol{\xi},\boldsymbol{r},t)\boldsymbol{v}_\xi] = -B_D(\boldsymbol{\xi},\boldsymbol{r},t) + B_B(\boldsymbol{\xi},\boldsymbol{r},t) - C_D(\boldsymbol{\xi},\boldsymbol{r},t) + C_B(\boldsymbol{\xi},\boldsymbol{r},t) \quad (1)$$

where $\boldsymbol{r}$ is the space coordinate vector, $\boldsymbol{\xi}$ is the internal coordinate vector and t is the time. $\boldsymbol{v}_r$ is the velocity transporting the particle in physical space and $\boldsymbol{v}_\xi$ is the velocity in the internal coordinates[1]. The terms on the RHS of eq. (1) represent death and birth of fluid particles of size due to breakage and coalescence events. The terms for the breakage events can further be given, for death and birth respectively, as[1];

$$B_D(\boldsymbol{\xi},\boldsymbol{r},t) = b(\boldsymbol{\xi})f_n(\boldsymbol{\xi},\boldsymbol{r},t) \quad (2)$$

$$B_B(\boldsymbol{\xi},\boldsymbol{r},t) = \int_{\xi}^{\xi_{max}} \nu P(\boldsymbol{\xi},\boldsymbol{\xi}')b(\boldsymbol{\xi}')f_n(\boldsymbol{\xi}',\boldsymbol{r},t)\, d\boldsymbol{\xi}' \quad (3)$$

The breakage frequency, $b(\boldsymbol{\xi})$, represents the fraction of particles of size $\boldsymbol{\xi}$ breaking per unit time[1]. The breakage frequency may further be divided into breakage time, t_b, and breakage probability, P_b. ν denotes the average number of particles formed from the breakage of a particle of size $\boldsymbol{\xi}'$[1]. Finally, the size distribution probability, $P(\boldsymbol{\xi},\boldsymbol{\xi}')$, denotes the probability for creating a particle of size $\boldsymbol{\xi}$ from the breakage of a particle of size $\boldsymbol{\xi}'$[1].

Several models of turbulence breakage exist[2]. To discriminate between these models, data from single fluid particle breakage experiments is needed[3]. In particular, the physics of the kernels in Eqs. (2) and (3) must be determined experimentally. Such an experiment must identify mother drop size, breakage time, breakage probability, number and size distribution of daughters. It is important that the flow is characterized so that turbulent kinetic energy dissipation level in the breakup region is known. To adequately identify the breakage events, single particle experiments must be performed instead of dense dispersed phase experiments in which each breakage event cannot be observed.

An apparatus for investigating turbulent breakage of single droplets should meet a broad range of requirements. Mainly, single fluid particle breakage should occur in a region with even and sufficiently high turbulence level and no shear-flow, so that the breakage probability and breakage time may be related to the turbulent energy dissipation level. To that end it is also important that the fluid particle should enter this region as spherical and undisturbed by the flow as possible. In addition it must be possible to film the entire breakup event and in such a way that the fluid particle deformation, breakage and all daughters are captured. Lastly, for obtaining enough data for statistically relevant results within a reasonable timeframe, the image analysis and the restarting of the experiment should be automated.

Several experimental setups for dispersed phase breakage investigations exist in the literature. Mostly they focus on dense dispersions, i.e. large amount of dispersed entities. Grid flow is common, e.g. introducing the dispersed phase downstream of a grid[4], upstream of several grids[5] or let buoyancy transport the dispersed phase upstream towards a rotating grid[6]. Some singe drop experiments are known; the stirrer tank is an industry relevant experiment which may be performed as a single drop experiment[7], which can be adapted[8]. Introducing single droplets and bubbles inside a static mixer is also previously attempted[3]. In addition, experiments that focus on homogenous turbulence, e.g. [9], may be adaptable for single fluid particle experiments.

The major advantages associated to the previously proposed set-up designs can generally be listed as; 1) the experimental runs can easily be fully automated, 2) the magnitude of the turbulent energy dissipation rate is fairly homogeneous in the measuring volume (and measured), 3) the turbulent flow is close to homogeneous without high viscous shear flows or jet flows to identify the main breakage mechanism, 4) the entire fluid particle deformation dynamics and breakage event can be captured in the measuring volume, 5) the costs of building the apparatus, running the fluid particle analysis and performing the flow characteristics are moderate.

Likewise, the previously proposed set-up designs are also associated to major disadvantages generally listed as; 1) the experimental runs can not easily be fully automated, 2) the magnitude of the turbulent energy dissipation rate is far from being homogeneous in the measuring volume (and often not measured), 3) the turbulent flow is very inhomogeneous containing high viscous shear flows or jet flows locally thus the main breakage mechanisms are not identified, 4) the entire fluid particle deformation dynamics and breakage event can not be captured in the measuring volume, 5) the costs of building the apparatus, running the fluid particle analysis and performing the flow characteristics are very high.

We intend to construct a new apparatus that adopts the major advantages of the previous experimental set-up designs and at the same time removes the main disadvantages of some of the previous experimental designs. The presentation will point out the main advantages and disadvantages of the different experimental set-ups found in the literature. Followed by an evaluation of which of the advantages identified for the previous set-ups are most important to keep in a new apparatus and which of the major disadvantages identified for the previous set-ups are most important to remove. From this analysis we did construct a new set-up which are more optimal for single fluid particle breakage investigations. Finally, sample sets of experimental data will be presented.

References:

[1] Jakobsen, H.A., Chemical Reactor Modeling. Berlin: Springer; 2014
[2] Solsvik, J., Tangen, S., Jakobsen, H.A. Rev. Chem. Eng. 2013; 29(5): 241-356.
[3] Andersson, R., Andersson, B. AIChE J, 2006; 52: 2020-2030.
[4] Lance, M., Bataille, J. J. Fluid Mech., 1991; 222: 95-118.
[5] Azizi, F., Al Taweel, A.M. Chem. Eng. J. 2011; 166: 715-725.
[6] Ravelet, F., Colin, C., Risso, F., Physics of Fluids, AIP, 2011; 23: 103301.
[7] Solsvik, J., Jakobsen, H.A. Chem. Eng. Sci. 2015; 131: 219-234.
[8] Maaß, S., Gaebler, A., Zaccone, A., Paschedag, A.R., Kraume, M. Experimental Chem. Eng. Res. Des., 2007; 85(A5): 703-709.
[9] Shy, S.S., Tang, C.Y., Fann, S.Y. Exp. Thermal and Fluid Science. 1997; 41: 251-262.

Unsteady flow inside reversing chamber

R. Kłosowiak
Poznań University of Technology, Faculty of Machines and Transport, Chair of Thermal Engineering, Piotrowo 3, 60965 Poznań, Poland;
tel. +48 6652046, fax: +48 6652281, e-mail: robert.klosowiak@put.poznan.pl;

The presented results concern on axisymmetric jet flowing to the cylindrical vessel. Axisymmetric jet are widely used in various technological processes which involve drying, mixing, heating and cooling. Conducted experimental studies and numerical analysis on the axisymmetric jet flow are used to intensify the phenomena of mass and energy transport. Cylindrical vessel walls are a barrier for entering into the jet. Accordingly, the axial free jet bounded by lateral exhibits the characteristics of a free jet and limited. The research includes measurements on the actual position and numerical analysis. Experimental measurements were carried out by Thermal Anemometry Systems - TSI. While the numerical analyses were performed using ANSYS - CFX. Numerical analysis contains the results carried out using RANS and URANS. In both of the numerical analysis was used the SST model of turbulence. The next step of the analysis was to perform numerical analysis LES. The results of analyses are shown in graphs and lists them together in order to compare results. Utilitarian feature of the chamber is the possibility of its application to the intensification of heat exchange in the combustion chamber. In addition, the results obtained can find the optimal operating point of the chamber with the current heat load and low emissions.

a)

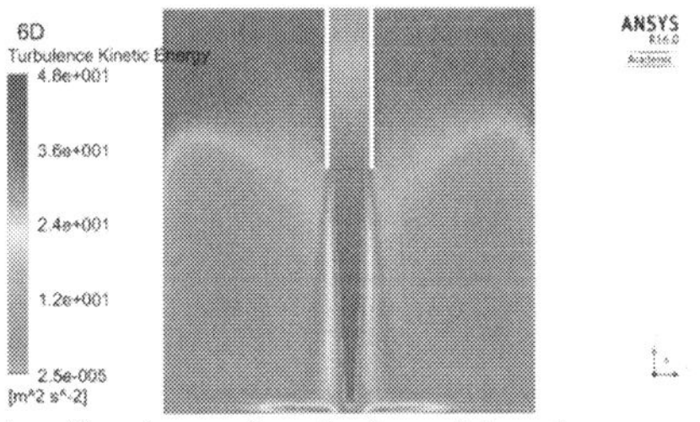

Distribution of turbulent kinetic energy

b)

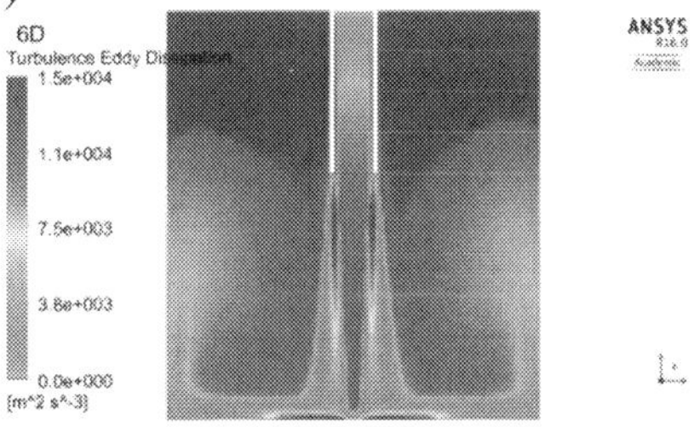

Distribution of dissipation rate

c)

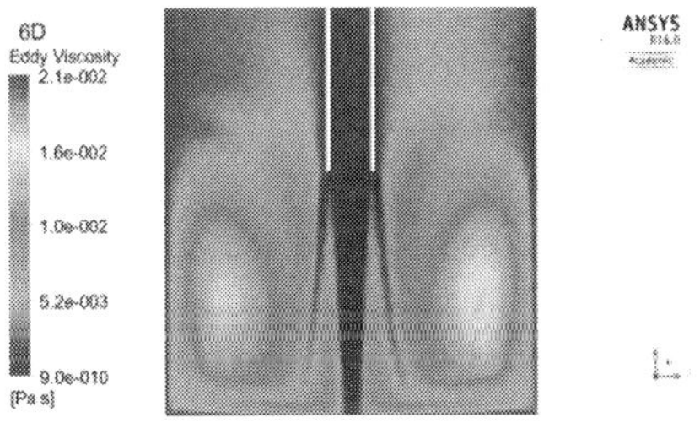

Distribution of turbulent viscosity

d)

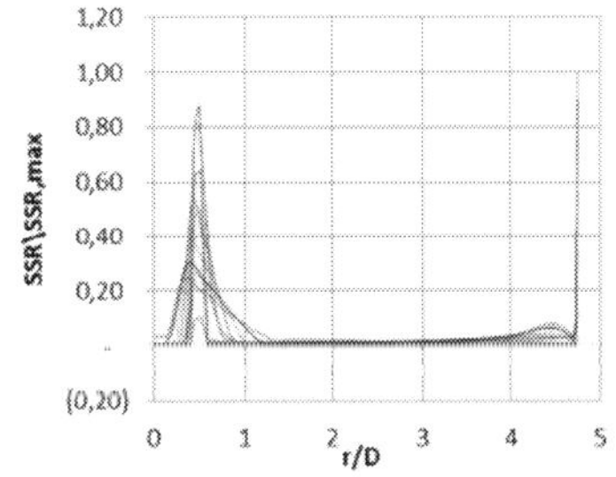

Shear stress distribution

Figure 1. Results of RANS simulation for 6D distance and 30m/s.

The flow of liquid around the bubbles generated from two nozzles

P.Dzienis[1)], D. Tomaszuk[1)], W. Suchecki[2)], R. Mosdorf [1)]
[1)] Bialystok University of Technology, Faculty of Mechanical Engineering
Address: Wiejska 45C, 15-351 Bialystok, Poland
tel. +48 85 746 92 00, fax. +48 85 746 92 10
[2)] Płock Campus of Warsaw University of Technology, Department of Process Equipment,
Address: Jachowicza 2/4, 09-402 Płock, Poland
tel. +48 24 262 26 10, fax. +48 24 262 65 42
*e-mail: p.dzienis@pb.edu.pl

The dynamics of bubble departure and bubble chain formation is important to control a gas – liquid mass transfer process. The bubbles flow and bubbles interactions change the liquid flow around them and around the nozzle. The liquid flow around the nozzle affects alternative bubble departure from two neighbouring nozzles. During alternative bubble departures the bubbles do not coalesce. In other cases chaotic bubble departures or bubble coalescence are observed.
In the present study the bubbles were generated from two bras nozzles (with the inner diameter equal to 1 mm) to distilled water. The nozzles were placed at the bottom of a glass tank. During the experimental investigation a distance between the nozzles and an air volume flow rate were changed. It was shown that for some air volume flow rates and some distances between the nozzles alternative bubble departures occurred. For the cases in which the bubbles departed alternatively, the bubble paths have been analysed. The results showed that the bubbles paths were repetitive in the distance around the nozzle equal to 13.5 mm. In order to study the liquid flow around the bubbles which depart alternatively and simultaneously from twin nozzles, the PIV method was used.
It was shown that during the simultaneous bubble departures, the bubbles repel each other immediately above the nozzles. The bubbles generate temporary liquid flow above the nozzles and grow up in the common liquid stream. The localization of the temporary streams, over the nozzles outlets, is not repetitive for respectively departing bubbles. The movement of the previous bubbles generates steady liquid stream from the distance of 50mm over the nozzle outlet to the end of the tank. In the final stage of the flow, the bubbles are drawn to the steady stream and move one over another. This stream is wider than the stream generated during the alternative bubble departures. During the alternative bubble departures, growing bubbles generate their own liquid flow around themselves. The bubble's own stream causes that the bubbles departing from the neighbouring nozzles do not coalesce. The bubbles move in a straight line until they are drawn into the main steady stream and start to flow one over another.
The study has been accomplished under the research project S/WM/1/2013 financed resources to study MNiSW.

Effect of liquid properties on interaction of a pair of rising bubbles – statistic evaluation

M. Vecer, K. Wichterle
VSB-Technical University of Ostrava, FMME, dept. Chemistry, tr. 17. listopadu 15, 70833, Ostrava-Poruba, Czech Republic; tel. +420 596324230, e-mail: marek.vecer@vsb.cz;

Bubble rising in a liquid, that flows downward in a diverging conical canal, should occupy a specific position in the canal; moreover, slight rotation of the liquid brings about better probability of bubble position in the centre of the channel. As consistent with this conception, both of identical-volume bubbles should tend to the same position, where coalescence would take place. However, visual observations recorded by high-speed camera indicate that bubbles on short distance stop to exhibit independent behaviour. Most common ellipsoidal bubbles show no collisions in practice. Experimentally found probability density of relative position was treated for bubbles of volume 25-200 mm3 in liquids of various viscosity, surface tension and conductivity. Relevant hydrodynamic forces of mutual bubble interactions were estimated. Vortex formation around moving bubble can explain the fact that the latter bubble is forced out from wake area. That is why formation of in line bubble configuration is not probable in that cases. Most usual configuration of pair of bubbles is parallel position with the same height. The pair of parallel rising bubbles maintain wobbling motion and sinusoidal trajectory, whereas in phase oscillations proceed and collisions do not take place. Statistical evaluation of the relative bubble position and the relative bubble velocity will be presented here as relevant arguments.

Keyword: bubble interactions, relative position, pair of bubbles

Acknowledgements: Presentation of this study was supported by "Institute of clean technologies for mining and utilization of raw materials for energy use - Sustainability program" (reg. No. LO1406).

Distribution of the flow turbulence behind sphere for different inflow turbulence

L. Bogusławski
Poznań University of Technology, Faculty of Machines and Transport, Chair of Thermal Engineering, Piotrowo 3, 60965 Poznań, Poland;
tel. +48 6652046, fax: +48 6652281, e-mail: leon.boguslawski@put.poznan.pl;

The flow turbulence influence on momentum and heat transfer processes and modify local transport characteristics. In many cases this phenomena are used for intensification of mass and heat transport. The strong velocity gradient in shear layer of free jet is able to generate high turbulent flow which can intensify mixing processes. The flow turbulence is transported downstream mainly due to convection. What happened if in such flow is additionally generated turbulence by any obstacle which can change flow turbulence. To verified relation between convection of turbulence and its local production the experimental measurements were curried out

In jet flow the sphere was located and average velocity and "rms" of turbulent fluctuations were measured. Measurements were done for different inflow turbulence levels from about 1.5% till about 17.3%.

Open wind tunnel was used to perform this experimental investigations. Turbulent flow was generated by a free round jet issued from round nozzle diameter of 0.15 m . Free jet generated flow with different turbulence level dependent on distance from jet outlet section. Jet velocity at nozzle outlet was change to get constant value of average velocity at any distance from nozzle outlet. This gives possibility to get turbulent flow with the some velocity but with different level of turbulence. The Constant Temperature Anemometer TSI IFA 300 was used to measure average flow velocity and "rms" of turbulent fluctuations of velocity. Measured signal was digitally collected and processed by used Thermal Pro 5.0 computer program. It was also possible to get the power spectrum of turbulent fluctuations of velocity in chosen locations.

The analysis of the turbulent flow velocity fluctuations before and behind the sphere have been performed.

For low level of inflow turbulence one can observe increasing of the turbulent fluctuations behind sphere. The turbulence increase visible in the region where the average velocity profile strongly change as is shown in Figure 1. Increasing of the turbulence is good correlated with gradient of average velocity changes. This means that increase of turbulence is the results of local production of turbulence proportional to the gradient of average velocity changes.

For high level inflow turbulence any vital change of turbulence profiles are observed in Figure 2. Local production of turbulence is not observed. Due to convection of high level turbulence of order 17% the local production do not influence on turbulence profile. This means that models of turbulence applied in numerical simulations of turbulent flows ought to be adapted dependent on the level of turbulence in the flow.

Analysis of the power spectrum of turbulent fluctuations will be presented too as a rough information as far as turbulent flow structure concern.

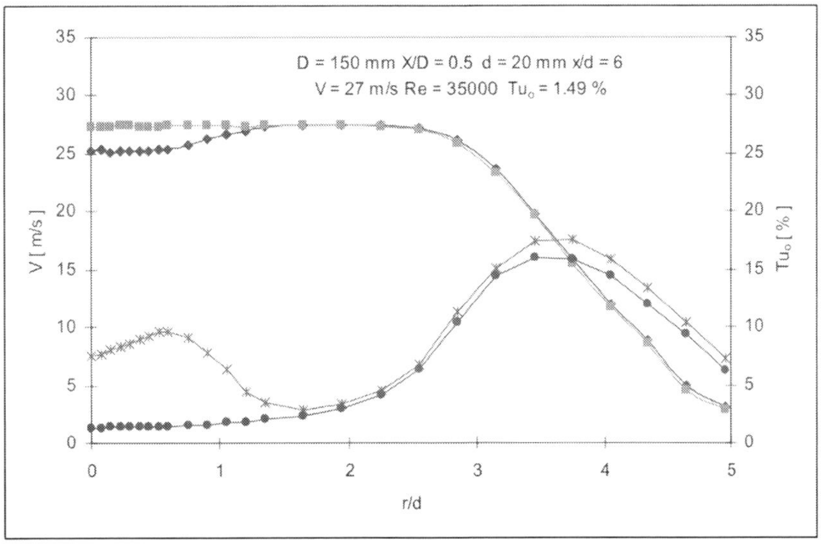

Figure 1. Radial profiles of the average velocity and turbulent velocity fluctuations in the free jet and behind sphere for the inflow turbulence level Tu_{oo} =1.5 %.

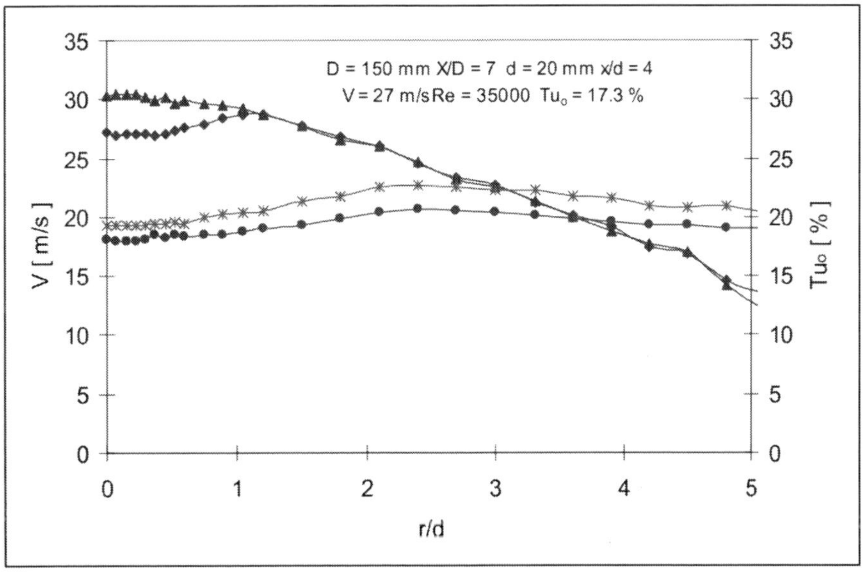

Figure 2. Radial profiles of the average velocity and turbulent velocity fluctuations in the free jet and behind sphere for the inflow turbulence level Tu_{oo} =17.3 %.

The role of near-wall drag effect on the dynamics of a tethered polymer under shear flow

G. Nir[1], E. Chetrite[2], A. Vivante[3], Y. Garini[4] and R. Berkovich[5]
[1] Harvard Medical School, Dep. of Genetics, Boston, MA 02115, USA; tel. +1-617-432-4432, e-mail: guy_nir@hms.harvard.edu.
[2] Ben-Gurion University of the Negev, Dep. of Chemical-Engineering, Beer Sheva 84105, Israel.
[3] Bar Ilan Univ., Dep. of Physics and Institute of Nanotechnology, Ramat Gan 52900, Israel.
[4] Bar Ilan Univ., Dep. of Physics and Institute of Nanotechnology, Ramat Gan 52900, Israel; tel. +972-3-531-7433, email: yuval.garini@biu.ac.il.
[5] Ben-Gurion University of the Negev, Dep. of Chemical-Engineering, Beer Sheva 84105, Israel; tel. +972-8-646-1489, e-mail: berkovir@bgu.ac.il.

Local interactions at the polymer-surface interface under external shear flow conditions govern the dialogue that ultimately affects the dynamical macroscopic behavior. This kind of interaction is Interactions of such kind are associated with numerous phenomena, ranging from melt fracture to DNA interactions with the nucleus envelope, all of which are manifestation of visco-elastic instabilities. Simplified descriptions of such systems, in terms of force acting on a polymer in the vicinity of a surface have proven to be valuable for modeling. However, crucial features of such systems still remain elusive, partially due to challenges in decoupling the intrinsic diffusion of a polymer from surrounding affects, such as, surface proximity. Here we used a Tethered Particle motion setup with a microfluidic device to monitor the time evolution of single dsDNA molecules, as a model polymer, under shear flow, for probing near-wall effects. The proximity of the tethered spherical body under flow parallel to a wall results in an increased drag force, and reduction of velocity. We introduce an explicit model that incorporates near–wall hydrodynamic effects (*Faxén* corrections) with the polymer nonlinear elasticity. Applying this model to the relaxation dynamics of the chain, we probe through the internal diffusion of the system and the near–wall drag effect on the *Faxén* correction factors. Together with characteristic-time analysis subjected to various flow rates, we show through the interplay between the external flow and the near-wall viscous restraining effects, that the later is prominent to the tension along the chain until a turnover that takes place at a relative extension of ~0.8 of the chain.

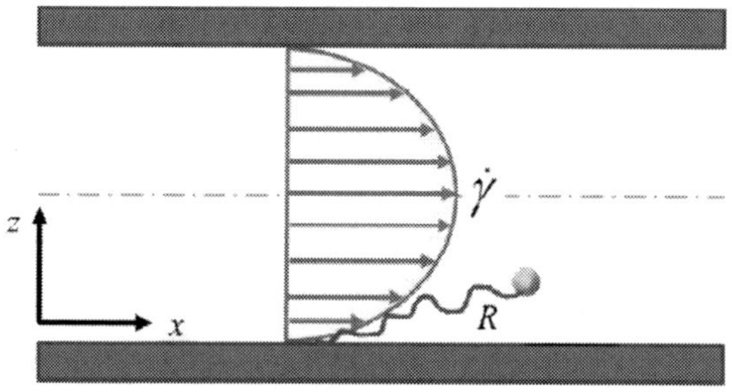

The role of ionic surfactants in electrophoretic addressing of oil-in-water droplets

L. Vobecká[1], J. Tuček[1], Z. Slouka[1,2], M. Přibyl[1,2]
[1] University of Chemistry and Technology Prague, Department of Chemical Engineering, Technická 5, 166 28 Prague 6, Czech Republic
[2] University of West Bohemia, New Technologies - Research Centre, Univerzitní 8, 306 14 Pilsen, Czech Republic
e-mail: lucie.vobecka@vscht.cz; tel: +420220443239

Controlled spatial delivery of discrete fluid volumes (mostly droplets) dispersed in a continuous immiscible phase to a specific location e. g. on a microfluidic chip is a very attractive technique for dosing of required reagents in low-volume synthesis of chemical substances or in separation of proteins, nucleic acids and other biomolecules. One of the means how to manipulate these small discrete volumes in space (and also in time) is electrophoresis that occurs on application of electrical field provided the controlled fluid packages contain charge. Since bulk solutions are electroneutral the only place where spatial charge can be found is at the interface between the continuous and discrete phase, which depends on the type of the two-phase system. It is thus of broad interest to control the surface charge of the discrete phase. In this contribution, we will present experimental study focused on ionic surfactants as potential charge sources in two phase systems comprising water as continuous phase and organic liquid as discrete phase. Ionic surfactants are organic molecules composed of hydrophilic polar groups bearing charge and hydrophobic carbon chain(-s). While hydrophobic carbon chains should easily penetrate into the organic phase, the polar head should remain on the other side of the interface and provide the discrete volume with charge which could be used in electrophoretic transport of the droplets. To test this hypothesis, we fabricated a simple millifluidic chamber with separate compartments serving as electrode reservoirs and studied experimentally the effect of ionic and also non-ionic surfactants on electrophoretic movement of kerosene droplets dispersed on water.

The initial experiments were aimed at the effect of concentration of surfactants and the strength of the electrical field on the electrophoretic movement of droplet. Interestingly, we observed no directed or controlled movement of droplets in the electrical field. However, dynamic gradients in concentrations of both ionic and non-ionic surfactants generated in the absence of electrical field resulted in the movement of kerosene droplets always away from the source of the surfactant. This observation can be explained by the Marangoni effect that occurs in systems with gradients of surface tension. To study mutual interaction of diffusion and migration of surfactants and their resultant effect on kerosene droplets we equipped our millifluidic chamber with a dialysis-like tubing as a source of diffusing surfactant. The following parametrical study showed that (i) the electrical field enhanced the transport of surfactants from the tubing into the millifluidic chamber, and (ii) the kerosene droplets always moved away from the dialysis tubing independent of type of ionic surfactant and the polarity of the electrical field. The experimental results showed Marangoni phenomenon dictates the behavior of droplets in our system.

Transient mass transfer from a spherical bubble in homogeneous shear liquid flow: A three-dimensional numerical simulation

A. Saboni[1], S. Alexandrova[2], M. Karsheva[3], and C. Gourdon[4]
[1]MAT-IUT-UPPA, Avenue de l'Université, 64000 Pau, France, e-mail : abdellah.saboni@univ-pau.fr, [2]LeTEP-ENSGTI, rue Jules Ferry, BP 7511, 64075 Pau cedex, France, [3]Dep. Chemical Engineering, UCTM, Kl.Ohridski blvd, 1756 Sofia, Bulgaria, [4]LGC, ENSIACET, 4 allée Emile Monso, 31432 Toulouse cedex 1, France

Knowledge of the basic phenomena occurring between a bubble and an external flow is important to the understanding of many gas-liquid systems in industrial plants such bubble columns and distillation columns as well as environmental flows such as air bubble entrainment which drives the exchange of nitrogen, oxygen and carbon dioxide between the atmosphere and the sea.

In the present study the model of mass transfer from a bubble to a continuous phase was developed. The transient mass transfer from a spherical bubble in homogeneous shear liquid flow is examined by means a three-dimensional numerical simulation. First we solve the Navier-Stokes equations and obtain the flow fields around a bubble in homogenous shear flow. Then we use the velocity components for solving the diffusion-convection equation and derive the rate of mass transfer from the bubble.

The effect of the fluid shear on drag coefficients, on the concentration profiles and on Sherwood numbers, is investigated for different ranges of Reynolds numbers ($0<Re<200$) and Peclet numbers ($0<Pe<10^5$).

References

Auton T. R., 1987, The lift force on a spherical body in a rotational flow. J. Fluid Mech. 183,199–218.

Bagchi P., Balanchandar S., 2002, Effect of rotation on the motion of a solid sphere in linear shear flow at moderate Re. Phys. Fluids. 14, 2719–2737.

Clift R, Grace J.R., Weber ME., Bubbles, Drops and Particles, New York, Academic Press, 1978.

Legendre D., Magnaudet J., 1998 The lift force on a spherical bubble in a viscous linear shear flow. J. Fluid Mech. 368, 81–126.

Sugioka K., Komori S., 2009, Drag and lift forces acting on a spherical gas bubble in homogeneous shear liquid flow. J. Fluid Mech. 629, 173–163.

Saboni A., Alexandrova S., Karsheva M., Gourdon C., 2011, Mass transfer from a contaminated fluid sphere, AIChE J, 57: pp 1684–1692

Saboni A., Alexandrova S., Mory M., 2010, Flow around a contaminated fluid sphere, International Journal of Multiphase Flow, 36, Issue 6, Pages 503-512

Numerical model of microdroplet charging and oscillatory movement between electrodes

R. Flittner, M. Přibyl, Z. Slouka
Univ. of Chemistry and Technology, Chemical Engineering Dep., Technická 5,
166 00 Praha 6 - Dejvice, Czech Rep.; tel. +420 22044 3251, e-mail: sloukaz@vscht.cz

Charged microdroplets can be useful as sample containers in lab-on-a-chip systems. They act as protection against sample loss due to diffusion or against mixing with other samples or substances, while their charge enables precise control of their movements using the electric field. Although, droplet addressing has been used widely in literature, the nature of charge and the precise mechanism of the charging have yet to be described.[1] Nevertheless, qualitative descriptions of the charging process have been given.[2] These descriptions incorporate electric field strength and droplet size as parameters. But there are more parameters that could have an influence. To identify these influential parameters and because physical experiments are limited by property values of the incorporated materials, we would like to offer our numerical model as a research tool.

Our numerical model describes the oscillatory movement of a microdroplet that bounces between 2 electrodes exchanging charge when the droplet is in contact with an electrode. Our model uses all relevant physical parameters such as electric field strength, dielectric constants and the droplet conductivity to scale the oscillatory movement and the exchange of charge. Our model also supports droplet deformation according to surface tension and wall contact angle as it was built on COMSOL 4.4 Microfluidics software.

We have identified key parameters to evaluate the charge exchange and the oscillatory movement and we have run case studies on the scaling parameters of our model. From the result data, we have obtained and discussed fitted curves. Our model is the first qualitative description of the charging process taking into account droplet dynamics and deformation in a closed two-phase system as well as charge transfer between an electrode and the attached droplet.

[1] D. J. Im, "Next generation digital microfluidic technology: Electrophoresis of charged droplets", Korean J. Chem. Eng., 2015 (Vol. 32, Iss. 6)

[2] D. J. Im et al., "Electrophoresis of a Cherged Droplet in a Dielectric Liquid for Droplet Actuation", J. Anal. Chem., 2011 (Vol. 83)

Mass transfer enhancement of liquid-liquid system by droplet deformation in a microfluidic expansion device

Xiaoda Wang, Ting Qiu
School of Chemical Engineering, Fuzhou University, Fuzhou 350116, China; tel. +86 13075819265, e-mail: xdwang@fzu.edu.cn.

This contribution presents experimentally the mass transfer enhancement of liquid-liquid system induced by droplet deformation in a microfluidic expansion device. In this microfluidic device sketched in Fig. 1, after generated in a microfluidic T-junction, droplets flow downstream along the microchannel to a square expansion of 1200 μm wide, where the deformation of droplet interface occur. The droplets continue to flow downstream to the exit of microchannel, after leaving the expansion. Except the expansion, the cross-sections of the microchannels are 400 μm (wide) ×400 μm (height). The microchannel was fabricated on a polymethyl methacrylate (PMMA) plate by a precision machine and sealed by another PMMA plate of equivalent size.

Methylbenzene and acetic acid-saturated water, which were used as dispersed and continuous phases, were pumped into the microchannel by two syringe pumps (Harvard Apparatus, PHD 22/2000, USA), respectively. The glycerol (0-60wt%) and surfactant Tween 20 (1wt%) were added into the continuous phase to adjust the viscosity of continuous phase and stabilize the droplet formation process, respectively. The interfacial tension was measured with a tensiometer on a Tracker apparatus (Dataphysics, Germany), by the pendant drop technique. An Ubbelohde capillary viscometer (Shanghai Qihang Glass Instruments Factory, China) was used to characterize the viscosities of the continuous and dispersed phases. The liquid density was surveyed with a vibrating tuber density meter (Anton Paar DMA-4500-M, Austria).

In the experiment of mass transfer, the liquid-liquid mixture collected at the exit of the microchannel was send into an oil-water separator to be separated. Both the concentrations of acetic acid in continuous and dispersed phases were determined by the method of acid-base titration. We systematically investigated the effects of droplet size, droplet velocity of movement, physical property of fluid and concentration of solute on mass transfer coefficient. In addition, how the geometric construction of the expansion affected the mass transfer efficiency was also examined in detail. The experimental results suggested that the mass transfer in the droplet-based liquid-liquid system was enhanced by the microfluidic expansion device.

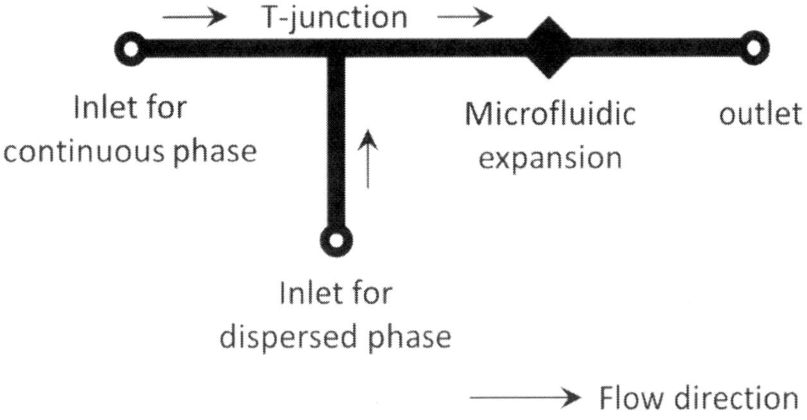

Figure 1 Schematic diagram of the microfluidic chip.

Evaluation of residence time distribution in millireactor with laminar flow: comparison of experiments, CFD and laminar flow model

S. Hejda[1], T. Trávníčková[1], J. Havlica[1,2], P. Stavárek[1], V. Staněk[1]
[1] Institute of Chemical Process Fundamentals of the ASCR, v. v. i.; Rozvojová 2/135, 165 02 Prague, Czech Republic, hejda@icpf.cas.cz
[2] Jan Evangelista Purkyně University, České mládeže 8, Ústí nad Labem, 400 96, Czech Republic

Knowledge about the residence time distribution (RTD) of a reactor is essential to the assessment of its performance. For an optimal control of the chemical reactions, one key parameter for the design of such reactors is just the RTD of the fluid. Many models have been suggested to characterize the RTD through pipes, one of the most commonly known being the axial dispersed plug flow developed by Taylor in 1953.

We present a study of RTD in the case of small straight channels. RTD measurements in liquid phase were performed by a dye injection and its absorbance measurements with a spectrometer at the inlet and the outlet of the reactor (Figure 1). The method consistency was confirmed by a dye mass balance between inlet and outlet which was satisfied with an error of 5% rel. The RTD in the measured section was characterized by so called e-curve E (t), mean residence time and Peclet number for different flow conditions (Reynold´s numbers). The RTD data for various channel lengths and liquid flow rates were compared.

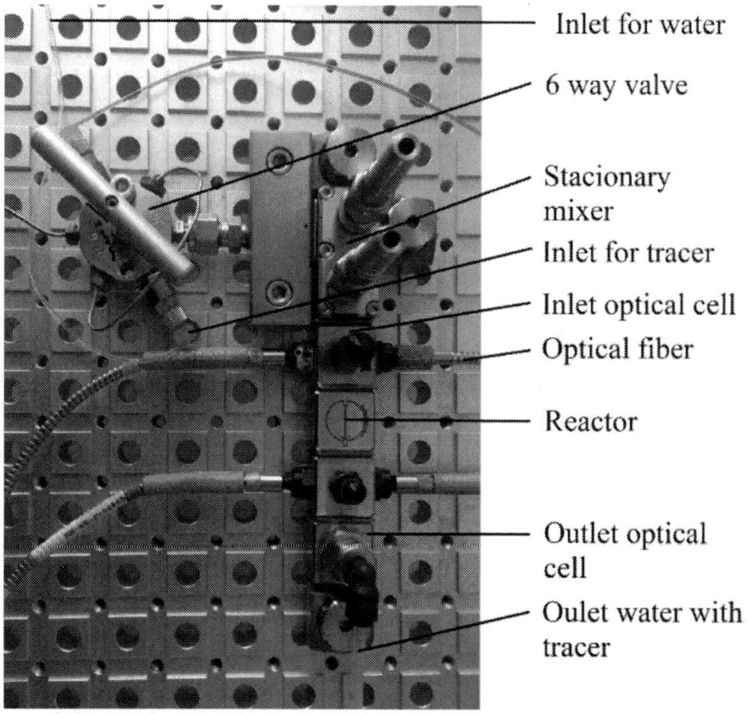

Fig.1 Exploded view of the equipment

To evaluate a contribution of a laminar flow with a parabolic profile to overall back-mixing in the channel, we have formulated a model for a general dye impulse deviation under the conditions of pure laminar flow.

The final formula of the model predicts a mean outlet concentration profile of the tracer C_{mean}^{outlet} for generally variable inlet tracer concentration profile:

$$C_{mean}^{outlet} = \int_0^1 f\left(t - \frac{L}{v}\right)\left[1 - \left(\frac{r}{R}\right)^2\right] 4\left(\frac{r}{R}\right) d\left(\frac{r}{R}\right) \qquad (1)$$

In this formula *f(t)* designates the concentration of the tracer at the inlet as a function of time, *L* is long of inner radius *R* fed at the inlet by liquid carrying a tracer, *v* is flow velocity and *r* is radial coordination. For a simple inlet concentration functions the outlet concentration may be obtained analytically.

In the last step, CFD simulations were conducted. At the inlet of the reactor a pulse of the tracer was generated, its deviation was then simulated along the whole reactor.

The experimental data and model results from purely laminar model and CFD simulations were compared in terms of E-curves, mean residence time and Peclet number.

Thorough evaluation of laminar flow experiments and application of mathematical models and CFD simulation brings new insight to individual flow phenomena contributions to back-mixing in small channels.

Acknowledgements:
This study was supported by GACR through projects no. P105/12/0664

No adiabatic flow of fluid through the channel filled with metal foam

M. Płaczek, [1]R. Dyga, [2]S. Witczak
Opole University of Technology, Department of Process Engineering,
Mikołajczyka 5, 45-271 Opole, Poland, tel. (+4877) 4498374, fax. (+4877) 4498366,
e-mail: m.placzek@po.opole.pl; [1] tel. (+4877) 4498440, e-mail: r.dyga@po.opole.pl;
[1] tel. (+4877) 4498370, e-mail: s.witczak@po.opole.pl

Over the last few years, the interest in metal foams has rapidly increased when the production technology of these materials was improved, making it possible to produce significant quantities of high quality foams from various metal alloys. Metal foaming technologies allow to obtain material with specific properties, which are suitable for difficult exploitation, i.e. in such conditions as high temperature, chemically aggressive environment or high mechanical loads so they are used in many different areas e.g. as thermal insulations, or as filling of heat accumulators which help their charging and discharging cycles, especially in the case when the heat accumulation medium undergoes a phase change [1-3].

More than 90-percent porosity, completely open cell structure and high specific surface area of metal foams makes these materials are increasingly seen as an alternative to other types of structured packings in process equipment. They are known cases of use of open-cell metal foam as packing of heat exchangers to improve heat transfer [4], in packed columns or chemical reactors, i.e. to enhance the mixing process by change in turbulence of the flow, as a catalyst support and in some cases as a stirrer used in order to mix the gas and liquid phases and create better condition for mass transfer [5-8].

Due to the relatively high thermal conductivity of metal foams there are wide possibilities of their use in heat transfer and no adiabatic mass transfer processes. However, even in cases where the metal foam has a different role than the intensification of heat exchange, eg. in catalytic reactors in engineering practice it is necessary to know the conditions of heat transfer.

The paper describes experimental tests with the heating of various kinds of fluids (air, water and oil) which flowed through a channel filled with open-cell metal foams. In the research three types of aluminum foams were investigated. Changes of channel wall temperature as well as fluid temperature during flow were measured. The frictional pressure drop was measured too.

Based on the literature data and analysis of experimental results, the evaluation of the influence of flow conditions, including geometrical parameters of foams, velocity and fluid properties on the nature of the fluid flow in the cellular space and heat transfer in channel was done.

Despite of laminarization of fluid flow by porous packing the fluid flow in the cellular metal foams may be the transitional or even turbulent flow. The character of the flow can be determined using a modified Reynolds number, wherein the characteristic linear dimension was defined on the basis of parameters which describe the geometrical structure of the foam.

The results of the study indicate that the fluid forced through a heated channel filled with metal foam takes heat primarily from the wall of the channel and to a lesser extent through its skeleton. The intensity of heat transfer depends mainly on hydrodynamic flow conditions, especially in a turbulent flow. The influence of foam structural parameters such as cell size and porosity is visible in a greater extent in laminar flow. However, the convective heat transfer coefficient during heating of each of the fluids used in research, has different values depending on the type of foam filling the channel. Change in heat transfer coefficient values depends on the character of the flow. In a turbulent flow convective heat transfer coefficient is approximately proportional to the velocity of the fluid.

As a result of heat transfer from the channel wall to foam by conduction, the temperature of the wall is lower (in comparison to the empty channel), which in high-temperature apparatuses may be of high importance from the viewpoint of structural strength.

The paper presents the developed methods for calculating the frictional pressure drop and convective heat transfer coefficient for the flow of liquids and gases through pipes filled with metal foams.

REFERENCES

[1] Stephani G. et al., Cellular metals for functional applications – an overview, in: Proc. Powder Metallurgy World Congress & Exhibition. PM2010, Florence, Italy, 10–14 October (2010) vol. 4, 95–102.

[2] Quadbeck P. et al., Open cell metal foams – application-oriented structure and material selection, in: G., T. Hipke, Scheffler M., Adler J. (Eds.), Proceedings CELLMAT 2010 (2010) 279–288.

[3] Xiao X. et al., Effective thermal conductivity of open-cell metal foams impregnated with pure paraffin for latent heat storage, Int. J. Therm. Sci. 81 (2014) 94–105

[4] Dyga R., Płaczek M., Heat transfer through metal foam–fluid system, Experimental Thermal and Fluid Science 65 (2015) 1–12

[5] Tschentscher R. et al., Gas-liquid mass transfer in rotating solid foam reactors, Chem. Eng. Sci. 65 (2010) 472-479

[6] Stemmet C.P et al.: Gas-liquid mass transfer and axial dispersion in solid foam packings, Chem. Eng. Sci. 62 (2007) 5444-5450

[7] Leon M.A. et al., Residence time distribution and reaction rate in the horizontal rotating foam stirrer reactor, Chem. Eng. Sci. 117 (2014) 8-17

[8] Leon M.A. et al., Mass transfer modeling of a consecutive reaction in rotating foam stirrer reactors: Selective hydrogenation of a functionalized alkyne, Chem. Eng. Sci. 73 (2012) 412-420

Void fraction predictive method in mini-channel two-phase flows

M. Pietrzak, M. Płaczek, R. Dyga
Opole University of Technology, Department of Chemical and Process Engineering, Mikołajczyka 5, 45-271 Opole;
tel. +48 77 449 83 71, e-mail: m.pietrzak@po.opole.pl

Gas-liquid two-phase flow is one of the most frequently observed flow in practical applications. Although most of the research concerning two-phase flow in the last decades were related to nuclear reactor cooling applications, this topic is currently experiencing a renewed interest in connection with the design of mini and micro-thermal systems, such as mini and micro-heat exchangers and micro-cooling elements for microelectronic components. Two-phase flow in min-channels has been the flow configuration of choice for many cooling applications demanding very high rates of heat dissipation per unit volume.

To date, research aimed at predicting the void fraction in mini-channels are still limited in comparison to studies that rely on correlations for conventional channel or empirical correlations. But as the number of published studies continues to rise, design engineers are confronted with tremendous confusion when selecting a suitable model or correlation. The primary reason behind this confusion is limited validity of most published methods to a few working fluids and narrow ranges of operating conditions.

The present study addresses this limitation by discussing the development of two consolidated mini-channel databases. The first is for air-water two-phase flows, and consists of 83 void fraction data points. The second is for refrigerants flow (ammonia and carbon dioxide), and consists of 171 void fraction data points. These consolidated databases can used to assess the accuracy of presented models and correlations as well as to develop universal correlations that may apply to a large number of fluids and very broad ranges of operating conditions.

Experimental determination and estimation of the pressure drop through the column, gas holdup and bed expansion in Type II-TBC

B. Bensebia[1], F. Bentahar[2], O. Bensebia[3]

[1]Laboratoire "Chimie Verte-Eau", Département Génie des Procédés, Faculté de Technologie, Université Hassiba Benbouali-Chlef, B.P. 151, Chlef-02000, e-mail : bensebiab@yahoo.fr, Tel: +213554091433; Fax: 21327721794,
[2]Laboratoire "Phénomènes de Transfert" Faculté de Génie Mécanique et de Génie des Procédés, Université des Sciences et de la Technologie-Houari Boumediene, BP 32, El Alia-Bab Ezzouar 16111, Alger.
[3]Laboratoire "Laboratoire des Sciences du Génie des Procédés Industriels (LSGPI)", Faculté de Génie Mécanique et de Génie des Procédés, Université des Sciences et de la Technologie-Houari Boumediene, BP 32, El Alia-Bab Ezzouar 16111, Alger.

The expanded bed height was studied because it is essential to the design of TBCs. The bed expansion, which determines the height of the fluidized bed and therefore that of the column, is one of the key design parameters for the fluidized bed of TBC. Knowledge ofthis factor would allow calculation of the expanded volume of the bed and providea basis for the determination of liquid and gas holdups and pressure drop.Thus, the aim of this work is to establish the response to the dependent expanded bed height as the independent, hydrodynamic process factors.

For this purpose, experimental data on pressure drop and bed expansion were collected over a wide range of operating variables in a Type II-TBC.

The process parameters considered are gas and liquid velocities, free-open area of the distributor grid, static bed height, particle diameter. The existence of three regions for the expansion of the bed has been shown for all experimental systems. The corresponding hydrodynamic regimes in these zones were analysed.

The bed expansion was correlated with independent parameters and the average deviation AARD was 10.75 %. The estimated values were compared to the values predicted by many other correlations.

This correlation was compared with those in the literature. The dependence of hydrodynamic variables on process parameters predicted by simulation agrees well with experimental data.

Characteristics of coal fluidised bed

R. Kłosowiak, J. Bartoszewicz, R. Urbaniak
Poznań University of Technology, Faculty of Machines and Transport, Chair of Thermal Engineering, Piotrowo 3, 60965 Poznań, Poland;
tel. +48 6652046, fax: +48 6652281, e-mail: robert.klosowiak@put.poznan.pl;

This paper presents results of experimental research aimed at determining the minimum level of fuel in a storage bunker influencing on the amount of dust being released above the fuel during transition of the pressure wave arising as a result of an explosion in the pipe between the fuel bunker and the coal pulveriser. Tests were conducted in a fluidisation chamber with gas flow velocities lower than critical ones. Results allowed us to determine the level of fuel bed in the boiler's bunker, at which the lowest amount of dust was emitted in the region of oxidizer occurrence after the pressure wave had passed. Presented analysis enabled developing efficient passive method of explosion protection for grinding and dust storage systems.

A polycarbonate column of the internal diameter of 5 cm was used for tests. End caps with nozzles was placed on its bottom. Oxygen under pressure was delivered to the column from a gas cylinder through those nozzles as well as through the reduction valve and the control valve of 0.2 s closing time. Figure 1 presents end caps used in tests.

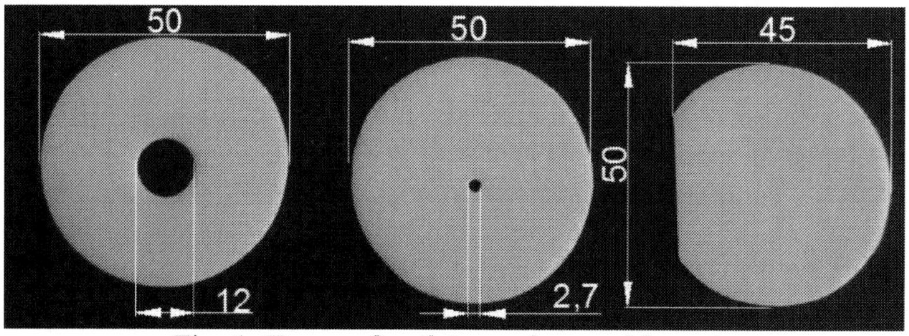

Fig. 1. Types of outlet nozzles used in tests

End caps were made as flat lids with nozzles of a diameter of 12 mm and 2.7 mm, respectively, and the aperture of 45/50 mm. A grid with meshes of a dimension of 0.3 mm was placed under end caps to homogenize the flow from the bottom of bed. Change in the end cap diameter enables conducting tests for various forces acting under the bed, and the aperture enables observation of edge effects, that is oxygen flow in the neighbourhood of the test chamber wall. Test were registered with the speed of 50 frames per second by a HD camera installed on the stand. Bed mass consists of mixtures of a density of 770 and 1580 kg/m3.

Presented results of observation (fig. 2) were obtained for the smallest outlet nozzle in the cap end, with gradually increased supply pressure. The sand layer raised for approx. 5–7 mm, and then it settled leaving a distinct trace on the column. Bed was raising at a uniform rate and occurring dust came from the external layer. Force acting on the column's cross-section equalled to 29.4 N. Further supply pressure increase resulted in forming the fluidised bed. The layer raised gradually and fluffed (figures from 2a to 2c), and next it decomposed forming the fluidised bed (figures 2d and 2e). However, at this pressure value fluidization was not stable and the bed subsided periodically and next raised again (figures from 2f to 2h). Force acting on the column's cross-section equalled to 196 N.

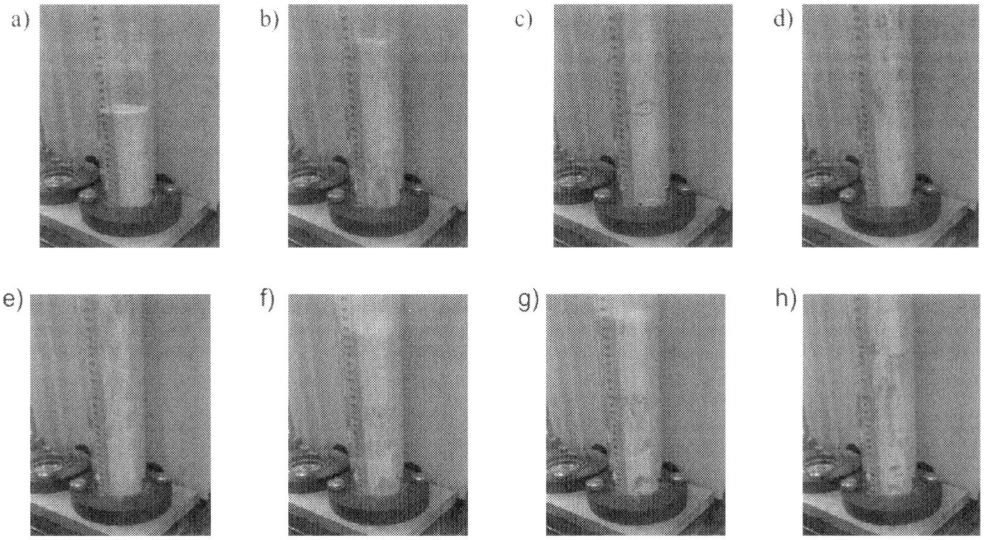

Fig. 2. Nozzle diameter Ø2.7, pressure $\Delta p = 0.1$ MPa, height $h = 15.5$ cm

As the result of tests it was stated that the location of the outlet nozzle on the cap end in the column is measureable. For great height of the bed and near the aperture the gas flows along the wall and escapes from the bed ejecting significant amount of dust, and with the pressure increase the bed stratified, what is the reason of great dust emission.

Effect of particle cluster size on flow, heat transfer behavior and the reaction process in FCC riser

M. Wang, L. Lv, N. He, X. Lan, J. Gao.
State Key Laboratory of Heavy Oil Processing, China University of Petroleum, Beijing 102249, China
tel. +86 15201414457, e-mail: cupwangmin@163.com; State Key Laboratory of Heavy Oil Processing, China University of Petroleum, Beijing 102249, China.

Fluid Catalytic Cracking (FCC) is an important process to convert heavy petroleum hydrocarbons into light oils in the presence of catalysts. In FCC unit (FCCU), the riser is the core part, in which the distribution and quality of the products are affected by the flow hydrodynamics, mass transfer and heat transfer. Recently, Computational Fluid Dynamics (CFD) method has been widely used to simulate and analyze the catalytic cracking process in the riser reactors. However, the influence of particle clusters was ignored by most of the traditional simulation methods which were based on the average method, leading to the difference between simulation results and actual industrial process. Our previous research showed that the particles clustering hindered the sufficient contact and the heat transfer between the oil gas and the catalyst particles, which affected the catalytic cracking reactions ofVGO. Furthermore, the different size of clusters had different impacts on the flow pattern, mass and heat transfer, and the reaction behaviors. Therefore, the effect of cluster size on the flow, heat transfer and the reaction process in FCC risers is studied.

Based on the preliminary research, the behaviors of flow, heat transfer and cracking reactions in two common types of clusters (ellipsoidal and spherical) in the reactors are simulated taking the catalytic cracking reactions of the VGO as an example. The distribution of gas velocity, temperature, species concentrations and the reaction rates are obtained. The results show that with an increase of cluster size, the oil gas is harder to pass through the clusters, and the resistance of clusters increases. Besides, the increasing trend of spherical clusters is higher than that of ellipsoidal clusters in the same longitudinal area. This suggests that the increasing of both lateral and longitudinal area will cause the increase of the force in the same gas velocity.

Moreover, the heat transfer between oil gas and particles is affected by flow hydrodynamics. With an increase of cluster size, the oil gas is harder to renew in clusters, leading to the decrease of heat transfer in which both heat transfer coefficient and heat flux decrease. Furthermore, the heat transfer coefficient and heat flux in spherical cluster surface is obviously higher than that in ellipsoidal cluster surface, which shows that the increase of both lateral area and longitudinal area of cluster will cause the decrease of the heat transfer between cluster and gas in the same gas velocity.

In addition, the bigger the cluster size, the higher the temperature of gas around the cluster surface, which will have an influence on the cracking reaction on the cluster surface that both primary reaction and secondary reaction rates decrease with the increase of cluster size. Besides, there is stronger impediment of clusters to primary reaction than secondary reaction. Affected by all the factors, the concentration of primary reaction product gasoline decreases slightly, while the concentration of secondary reaction products such as coke and gas increases.

Therefore, the generation of large particle clusters is supposed to be avoided in the FCC riser to reduce the yields of gas and coke.

Assessment of permeability of coal chars under conditions of multi directional flow of gas

G. Wałowski[1], G. Filipczak[2], E. Krause[3]
[1]Institute of Technology and Life Sciences, Biskupińska Street 67, 60-463 Poznań, Poland, g.walowski@itp.edu.pl; [2]Faculty of Mechanical Engineering, Chemical Engineering Department, Opole University of Technology, Mikołajczyka Street 5, 45-271 Opole, Poland, g.filipczak@po.opole.pl; [3]Central Mining Institute – Experimental Mine of „Barbara", Podleska Street 72, 43-190 Mikołow, Poland, ekrause@gig.eu

The results of hydrodynamic of gas flow through various type of porous material kind of coal char (carbonizers) are presented. For each type of these materials have been assessed process parameters such as apparent density, porosity and rate of anisotropic. With respect to hydrodynamics the studies have focused on the evaluation of the permeability of such materials and resulting from the gas flow pressure drop allowed to determine the permeability coefficient and the coefficient pressure drop, too. For the measuring of permeability the gas volume stream, as a result from differential of reference pressure was accepted – both for unspecified shaped samples (Fig. 1a) and of uniquely defined shape in the form of a cube (Fig. 1b). In both cases, an independent assessment of the physical characteristics of chars was the subject of analysis. Experimental studies in the evaluation of the permeability of the porous material was carried out under conditions of water bubbling.

Due to anisotropic structure of chars the studies were performed with respect to the flow direction, i.e. directivity permeability, as a result of three-dimensional orientation for each sample of tested materials. The data results indication of strongly diversified hydrodynamic parameters for investigated types of materials, and comparison of test results leads to a conclusions that the main reason of this is an anisotropic structure of carbonizers – Fig. 2.

Fig. 1. Char-coal samples: a) shapeless sample (piece of the deposit), b) cubic specimen with a directional flow of gas (dimensions of 20×20×20 mm)

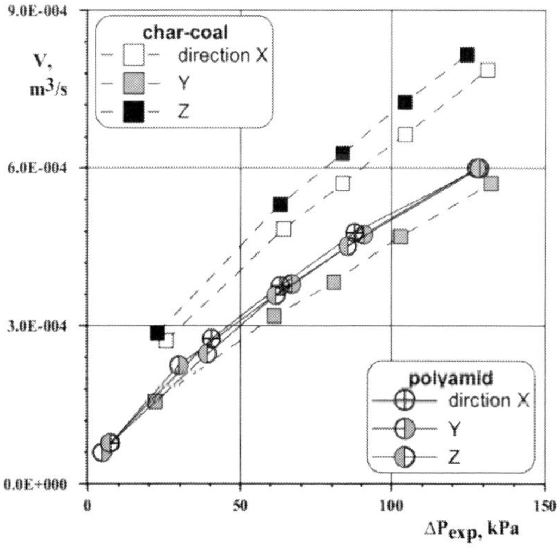

Fig. 2. Air permeability of anisotropic cube char-coal sample and for comparison the isotropic cubic sample (the same size) of polyamide material

With the purpose of obtain detailed description of microstructure process conditions as the hydrodynamic result of gas flow through porous materials, attempted to use the numerical method to assessment of this process. Numerical modeling indicated the concept of gas flow through porous a tortuous structure of the microchannels and micropores, together forming a porous structure with a specific geometry – both an elementary unit constituting the porous layer-bed. Numerical analysis was performing by using the finite volume package by means of ANSYS Fluent Programming. The particular analysis of the process conditions has allowed to develop a method of create computational grid for the parallel and complex network of microchannels with their different geometrical configurations. The calculation parameters of indispensable to numerical analysis, such as material porosity, anisotropic permeability, shape, length and diameter of microchannels, together with particular to take into account tortuous of microchannels were evaluated relative to really material characteristics – Fig. 3.

Fig. 3. The numerical image of the porous structure of tested char
(the cubic sample)

Based on the analysis it can be concluded that the way of numerical modelling in parallel with an experimental measurement of gas permeability, creates the possibility to comprehensive assessment of the hydrodynamics of gas flow through porous materials with anisotropic structure. This also provides possibilities to use of numerical modelling for the estimation of gas flow hydrodynamic through such materials in larger scale process.

Cycle times study off different solids in draft tube conical spouted beds

I. Estiati, M. Tellabide, H. Altzibar, M. Olazar
Univ. of the Basque Country, Dep. of Chemical Engineering, P.O. Box 644, 48080 Bilbao, Spain; tel. +34 946015994, e-mail: idoia.estiati@ehu.es; fax: 946013500.

The spouted bed is a fluid-particle contact technique that has been successfully applied to systems where fluidization has yielded unsatisfactory results. Accordingly, the main difference with fixed and fluidized beds lies in their cyclic movement.

Conical spouted beds are highly versatile in the gas flow rate, allowing operating with particles of irregular texture, fine particles and those with a wide size distribution and sticky solids, whose treatment is difficult using other gas-solid contact regimes. These features of the conical spouted beds make them attractive for different applications.

The ratio between the inlet diameter and particle diameter limits the scaling-up of spouted beds (the inlet diameter should be smaller than 20-30 times the particle diameter). The insertion of a draft tube is the way of overcoming this limitation.

Particle cycle time is defined as the time the particle takes to travel from the top of the annulus downward and back again to its starting point. Since the proportion of time spent by a particle in the spout is insignificant compared with that spent in the annulus, particle cycle times can be deduced from solid flow patterns in the annulus [1].

Knowledge of particle cycle time is very useful to ascertain the bases of the spouted bed technique. Furthermore, information on this parameter and particle trajectories is essential for spouted bed applications, given that the average cycle time regulates energy and mass transfer, and influences chemical reactions [2].

The main aim of this work is to gather information about solid cycle times in conical spouted beds provided with different internal devices and without devices. Based on this information, the factors of major influence on particle cycle time and solid circulation are determined. According to a statistical analysis, the effect of the most significant variables on the cycle times and solid circulation have been established. The particle diameters used are 0.01 m for chickpeas and 0.006 m for glass beads.

The results show that in the experimental systems with open-sided draft tubes, the contactor angle is the factor of greatest influence on the cycle times and solid circulation rate using glass beads and chickpeas. However, the most important factor using glass beads is the stagnant bed height and using chickpeas the height of the entrainment zone of the nonporous draft tube. Furthermore, inserting the draft tube as a factor in the statistical analysis, the results show a great influence on the particle cycle time of different experimental systems, so the use of this device modify the hydrodynamic of spouted beds.

[1] N. Epstein, J.R. Grace. Eds. Cambridge University Press: New York, 2011.
[2] J. Makibar, A.R. Fernandez-Akarregi, I. Alava, F. Cueva, G. Lopez, M. Olazar. Chem. Eng. Process 2011, 50, 790-798.

3D DEM simulation of coal gasification in gas-solid bubbling bed

Y. Wu, L. Peng, X. Lan, and J. Gao
State Key Laboratory of Heavy Oil Processing, China University of Petroleum, Beijing 102249, China
tel. +86 89731773, e-mail: yaya0224@foxmail.com; State Key Laboratory of Heavy Oil Processing, China University of Petroleum, Beijing 102249, China; fax: +86 89731773.

Computational fluid dynamic (CFD) simulation has widely used in the investigation of gas-solid flow dynamics and chemical reactions. Because Eulerian–Eulerian approach does not recognize the characteristics of heat and reactions from particle level, in this paper, the discrete element method (DEM) is used to simulate the coal gasification at the particle level in a 3D gas-solid bubbling bed. The gas phase is treated as a continuum and the solid phase is modeled by DEM. The reactions of coal gasification involve the heterogeneous reactions of coke pyrolysis, char oxidation and gasification, and the homogeneous reactions of gas phase combustion. The heat transfer in the bubbling bed consists of the conduction and convective heat transfer of particle-fluid, the heat transfer resulting by particle collision and radiative transfer by hot solid phase. The model predicts the effects of the particle heterogeneous flow structure on the thermal characteristics of coal particles when heating and burning, and the gaseous emissions from a fluidized sand-coal binary mixture. The collision heat transfer is weak in the bubbling bed less than 5% during this process. The results of product outlet and temperature field in the bubbling bed of gas and solid phase are agreement with experimental data from literature. The results show that the higher temperature is favorable for the products in H_2 and CO reactions. The carbon conversion decreases with the increase of the height of injection point owing to the increase of solid entrainment and the decrease of particle residence time and particle temperature.

Modeling and optimization of biomass combustion process

D. M. Djurović, B. S. Repić, D. V. Dakić[1], A. D. Marinković, M. J. Paprika
University of Belgrade, Vinca Institute of Nuclear Sciences, Laboratory for Thermal Engineering and Energy, P.O. Box 522, 11001 Belgrade, Serbia; tel. +381 11 3408837, e-mail: dejan2004@vinca.rs; [1]Innovation Center, Faculty of Mechanical Engineering, University of Belgrade, Kraljice Marije 16, 11120 Belgrade, Serbia, tel. +381 11 3370364.

Biomass is one of key renewable energy sources in the Republic of Serbia. This is the reason for the development of cheap thermal devices burning biomass from agricultural production as quite available and cheap energy source. Researches in this area are very complex and in order to obtain reliable data it is necessary to carry out theoretical and experimental research of the process. Therefore, efforts have been made in the Laboratory for Thermal Engineering and Energy of the "Vinča" Institute to develop a technology for utilizing bales of various sizes and shapes for energy production. In this paper has been presented optimization of the existing furnace for biomass combustion in the way of improving energy efficiency and environmental protection. By proper dimensioning of furnace also could be avoided unnecessary investment costs. Two-dimensional turbulent flow model with homogeneous chemical reactions has been developed. Turbulent flow is considered using time averaging *Navier-Stokes* equations that are closed by k-ε turbulence model. Calculations based on the proposed models were conducted using commercial CFD package *FLUENT*. Comparative analysis of the results of modeling existing and proposed (optimized) furnace has shown lower CO emission at the proposed furnace outlet.

Parametric analyses of numerical simulations of specific processes or so called numerical experiments are very useful in engineering practice, because it is cheaper than experiments which may need costly changes to the plant. Also, it is the faster way to get the desired results and the data about the modified plant. In addition, it is possible to obtain a detailed picture of the relevant field sizes, all over the considered area, while experiment can be difficult, or almost impossible to realize.

Mathematical model of the process in the adiabatic combustion chamber for combustion of biomass was analyzed for the case of furnace volume change. Since the previous analyses confirmed the efficient operation of the furnace, the question arises whether the same effect could be achieved with some similar furnace of smaller volume. Also any reduction in combustion chamber volume will lead to reduction in investment cost and accordingly to cheaper energy obtained from this facility.

Keywords: Biomass, combustion, CO emission, mathematical modeling.

Numerical simulation on gas-solid flow and mass transfer in FCC packing strippers

Y. Wang, G. Zhao, W. Su, X. Lan, J. Gao
State Key Laboratory of Heavy Oil Processing, China University of Petroleum, Beijing 102249, China
tel. +86 15650796889, e-mail: wyipingsunshine@sina.com; [1] China University of Petroleum, Beijing, China, 102249

Stripper is an important part of the fluid catalytic cracking (FCC) unit. Stripping techniques for economic and stable operation of the FCC are of great significance, which can effectively reduce the coke yield and save energy. For the sake of improving the stripping performance, efficient stripping techniques have been proposed in this work. With stripping technology, a new packing stripper was presented to enhance gas-solid contact, suppress solid back-mixing and eliminate some operating problem such as flooding, bridging and mal-distribution, compared with various types of traditional stripping internals. Based on adsorption/desorption and diffusion properties, light components and heavy components of the oil gas were treated differently. The gas-solid flow in the packing stripper was simulated by CFD method, and then the gas-solid mixing behaviour were deeply analysed. The parameters indicating the back-mixing degree were obtained. The influence of bubble size and gas/solid flow distribution on the mass transfer between the bubble and emulsion phase was also analysed. The effect of the five key operation parameters (stripping temperature, stripping time, gas velocity, crude oil properties, catalyst structure) was investigated. The effect of the four key structural parameters (height-diameter ratio, interval distance between two packed bed, tilt angle, the number of packed segments) was also investigated. The optimized parameters were obtained and the stripping efficiency was extremely improved. The results showed that the packing stripper had a better gas-solid mixing performance and operating flexibility. The study revealed the hydrodynamic behaviour, to provide necessary information for assisting the packing stripper engineering design and optimization.

CFD simulation on the influence of particle clusters and bubbles in FCC regenerator

N. He, Y. Wu, M. Wang, X. Lan, J. Gao
State Key Laboratory of Heavy Oil Processing, China University of Petroleum, Beijing 102249, China
tel. +86 18810987273, e-mail: he_ningning@163.com; [1] China University of Petroleum, Beijing, China, 102249

Fluid catalytic cracking (FCC) is considered to be an important heavy oil upgrading process. The regenerator, where catalyst activity is recovered, is an important part of the fluid catalytic cracking unit (FCCU). The catalyst regeneration is a complex process that involves a gas-solid two-phase flow, heat and mass transfer and coke combustion reaction, which determines heat balance and production capacity of the entire FCCU. At present, the operation regime of most FCC regenerators is taken as turbulent flow, which is a transition state between bubbly flow and fast fluidization regime, and it contains bubble characteristics of bubbling fluidized bed and particle cluster characteristics of fast fluidized bed. There have been extensive researches in regenerators and many regenerator models have been established by CFD simulation. However, most of the traditional simulation methods were based on average methods within the Eulerian frame, ignoring the presence of bubbles and the influence of particle clusters, thus making the simulation results quite different from actual industrial burning efficiency.

In the present work, the behaviors of flow, heat transfer and reaction process in typical clusters (spherical and ellipsoid cluster) were firstly simulated by CFD. The results reveal that the particle clusters hinder the sufficient contact and heat and mass transfer between gas and solid, resulting non-uniform distributions of temperature and components concentration in the clusters, which absolutely affects heat and mass transfer, and consequently, influences coke combustion reaction.

Secondly, the presence and properties of bubbles also affect heat and mass transfer. Existing experiments on bubbles indicate that bubbles have effects in two aspects. In positive aspect, the coalescence and breakup of the bubbles aggravate gas-solid mixing and constantly update the particle surface, which is beneficial to reaction. In negative aspect, bubbles occupy certain space, and the conductivity of bubbles is poor, and they are unfavorable to heat and mass transfer. In addition, if the bubbles rise too fast, gas embraced bubbles has not enough time to react with particles before leaving the bed. Which one of the two aspects is more dominant depends on operation conditions.

In this work, a CFD model based on particle clusters and bubbles of FCC regenerators is established. Gas-solid flow behavior, heat and mass transfer, and coke combustion process are investigated. Results show that the reaction rates of the coke combustion reaction in the clusters are greatly lower than those in the isolated single particle. And the existence of bubbles affects heat and mass transfer in different aspects. The simulation results indicate that the burning efficiency is in better agreement with industry production than previous models, which validates the rationality of the present model for regenerators and its feasibility in describing a commercial regenerator.

Electrospraying: mathematical modelling of nanoparticle deposition

A. Zítková, M. Kroupa, J. Kosek
Department of Chemical Engineering, University of Chemistry and Technology Prague, Technická 5, 166 28 Praha 6 – Dejvice, Czech Republic
Phone: +420 220 443 296, Fax: +420 220 444 320, e-mail: Juraj.Kosek@vscht.cz

Few years ago the prefix "nano" used in the relation with a product guaranteed a better market position (often despite the price). As time goes, the products taking advantages of the special properties of nanostructured systems are becoming more common and producers have to concentrate on the reduction of the product's final price. Ideal manufacturing equipment works with high efficiency and low costs.

An example of equipment with these properties is the electrospray. It is based on a simple principle: liquid evaporates from highly charged droplets during their flight that is caused by electric field between charged nozzle and grounded substrate while the repulsive force of positive charges on the droplet surfaces causes their disintegration. The broad utilization of this technology is supported by many advantages, i.e., high efficiency, low-cost, simple principle, but also by the possibility of the deposition of various materials including inorganic, organic and biological substances and even living microorganisms. But there is one crucial disadvantage – the morphology of the final layer of deposited nanoparticles strongly depends on the working conditions and is hardly predictable.

We decided to take the advantages of the mathematical modelling and aim at clarifying the influence of various effects taking place during the final part of the droplet's motion on the final morphology of the deposited layer. Our model is based on the Discrete Element Method (DEM) which uses the second Newton's law and the summation of the forces affecting each particle to simulate its motion. Besides the forces between particles (i.e., non-contact van der Waals and contact elastic adhesive forces) and between particles and the surrounding fluid (drag force) the most important force in the system is the electric force. The final magnitude and the direction of electric force depends on many phenomena: various charge on different particles, the permittivity of the particles and the surrounding environment, the electric field between the nozzle and the substrate etc. Therefore, the electric force is computed from the electric potential distribution obtained by solving of Poisson's equation on the discretized domain using the Finite Volume Method (FVM).

The model was used to conduct several parametric studies with different types of simulations focused on different effects. The obtained results brought more insight into the processes taking place during the deposition and they also helped to further develop the model. Although the model was able to simulate the particle deposition as well as the creation of the final layer, the results showed that the extension of the model about the description of the discharging process is the most important direction for the next development. Due to the absence of the established description of the nanoparticles discharging, we came up with a model based on a phenomenological basis. It computes the dynamics of the particle discharging according to the charge difference between the particle and the objects in the contact with it. The suitability of this model extension was proved by testing simulations, where the dependence of the particle deposition on the discharging of other particles can be clearly observed.

Although mathematical modelling is a very useful and powerful tool, it is always important to compare the results with experimental data. The morphologies obtained from the multiparticle simulations resemble the morphologies known from the experiments (see the following figure), but clearly, there is still a long and exciting way to completely understand the complicated phenomena involved in the morphology evolution of the layer.

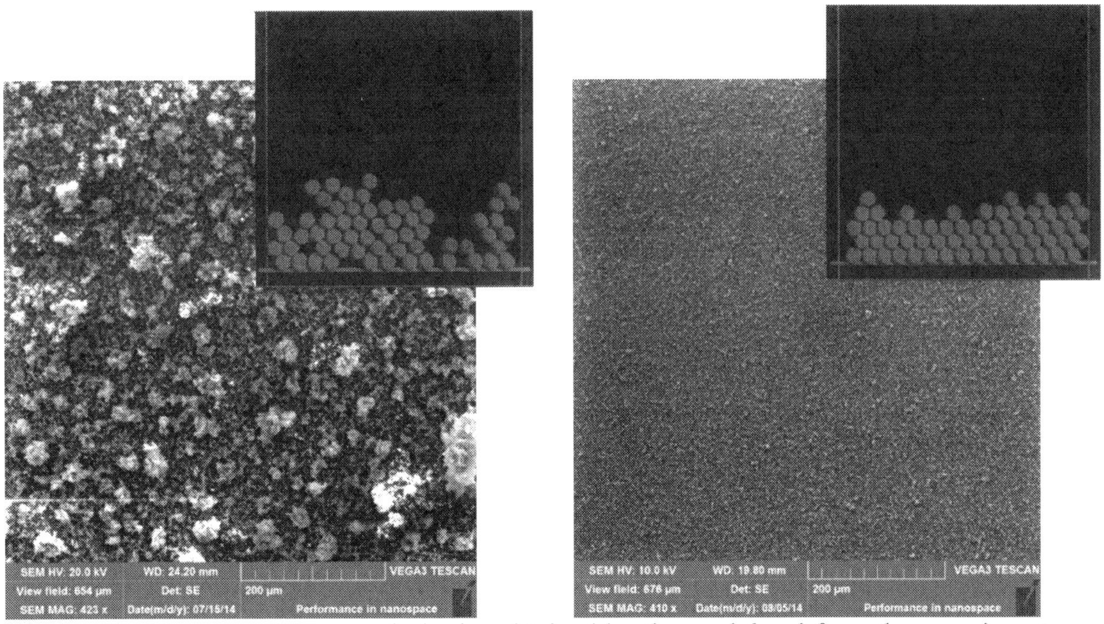

Figure: Comparison of the morphologies obtained by the model and from the experiment.

Experimental investigation of the breakup of spiral-nozzle-produced liquid curtain for absorption processes

O. Krištof, T. Svěrák, P. Kejík, P. Bulejko, K. Mayerová, J. Kalivoda
Transfer and Fluid Flow Laboratory, Faculty of Mechanical Engineering, Brno University of Technology. Technická 2896/2, 616 69 Brno, Czech Republic; e-mail: kristof@LPTaP.fme.vutbr.cz, tel. +420 541 144 908

Absorption processes as a chemical-engineering technology for gaseous pollution prevention are appropriate and effective solution in terms of their financial and constructional advantages. In order to understand the behavior of fluid flow and in order to design and evaluate absorption devices the study of liquid atomization, droplet size, droplet velocity, spray pattern and fluid distribution is essential.

This study deals with the experimental investigation of liquid curtain evolving inside a sprinkler head of the pilot-plant wet shower scrubber. The experiments were carried out in a transparent enlarged model of the sprinkle head. The liquid was atomized by a spiral nozzle which produced full cone pattern consisting of three liquid curtains in the shape concentrically ordered hollow cones. The investigation was focused on determination of the relevant liquid atomization zones, droplet size distribution along the liquid sheet, droplets radial velocity and effective spraying angle of each cone. The LDA and PIV methods were used for the observation of the evolving liquid curtain. An Nd:YAG laser-pulse light with a 5 ns pulse length and a Canon D70camera with an EF 100 mm f/2.8 USM Macro objective were used. It is necessary to compare the results obtained from the experiments with the CFD simulation in the future.

Static vortex contactor for intensification of interphase processes between liquids

A. O. Kuzmin[1], M. Kh. Pravdina[2]
[1] Boreskov Institute of Catalysis, pr. Lavrentieva 5, 630090 Novosibirsk, Russia; tel. +7(383)326573, e-mail: kuzmin@catalysis.ru;
[2]Institute of Thermophysics SB RAS, pr. Lavrentieva 1, 630090 Novosibirsk, Russia; e-mail: model@itp.nsc.ru

Liquid-liquid contacting for solute extraction or carrying out chemical reactions is widely employed in chemical industry. Various types of contactors used for liquids mixing are available: mechanically stirred vessels, rotor–stator and centrifugal apparatuses, static mixers, sieve-tray columns, etc. Conventional contactors are known to have both advantages and disadvantages. For example, static mixers cannot provide long residence time, their pressure drop increases with length. Centrifugal and stirred contactors are efficient, but they may result in emulsification, their rotating parts and sealing need to be maintained. Vertical sieve-tray columns cannot provide fast mass-transfer.

It is known that structured confined vortex flows can provide some benefits for multiphase processes intensification purposes [1], e.g. for gas-liquid contacting [2]. As an example of a liquid-liquid contactor, the efficient contacting can be achieved by using of two-fluid Taylor-Couette flow where two immiscible liquids are stratified and contacted without emulsification in the annulus between spinning coaxial cylinders [3].

The present work is devoted to the development of an approach to liquids contacting in a strained confined swirled flow produced in a static vortex chamber (VC). Here VC is a substantially cylindrical shell having the top and the bottom end walls with at least one of them providing for liquid outlet. Fluid flow enters the body of VC via a swirler located at the periphery. The swirler consists of a ring with multiple tangential speeding up slots designed to convert the pressure energy to rotational motion. Fluid enters the VC body with a great angular momentum, which is slowed down by flow friction on the end wall surfaces to provide a spatial shear stress field within totally turbulent vortex flow with effective dissipation of incoming pressure energy inside the flow. The developed VC displays also sufficient values of centrifugal gravity to sustain the vortex motion and of residence time to complete solute extraction.

LLVC is intended initially for realization of efficient liquids contacting for the extraction purposes of sour impurities, such as mercaptans, from liquid hydrocarbon stream to an aqueous caustic solution in the MEROX process [4,5]. Current industry practice is to use a sieve-tray column with its large mass and overall dimensions. Developed approach can serve to facilitate contacting and settling while minimizing the size of contactor.

For example (see figure below), the water addition into the vortex flow of hexane leads to formation of homogenous liquid emulsion inside a VC, which is immediately partitioned after leaving the VC. The one of the most important advantages of the developed approach is the possibility to carry out effective mixing and contacting of liquids with possible gas addition and/or removal of the gaseous reaction products. Tested configurations of LLVC provided residence times of 3÷6 sec and energy dissipation rates of 7÷15 W/L at relatively low pressure drops of 30÷50 kPa.

Developed simplified theoretical description of the flow in a VC is made with an axisymmetrical flow assumption. It is shown that friction stresses on the side and end walls of VC is responsible in establishing a desired flow pattern inside VC. The angular momentum balance conservation relation leads to the law for angular velocity distribution, which can be used then for calculation of other vortex flow parameters.

Liquid-liquid vortex contactor has some attractive characteristics. Experimental evidences were obtained that the shear stresses inside the vortex flow result in very intensive liquids contacting with fast renovation of interfacial surface. Contactor provides sufficient yet not excessive mixing facilitating the subsequent phase separation and does not contain any moving or spinning parts. It was shown, that almost 100% of a theoretical extraction may be achieved in few seconds of residence time, as measured for a process of a dye extraction from water into pentanol stream.

Various designs of LLVC may be suggested to satisfy different requirements such as intensity of mixing, residence time value, number of stages and throughput.

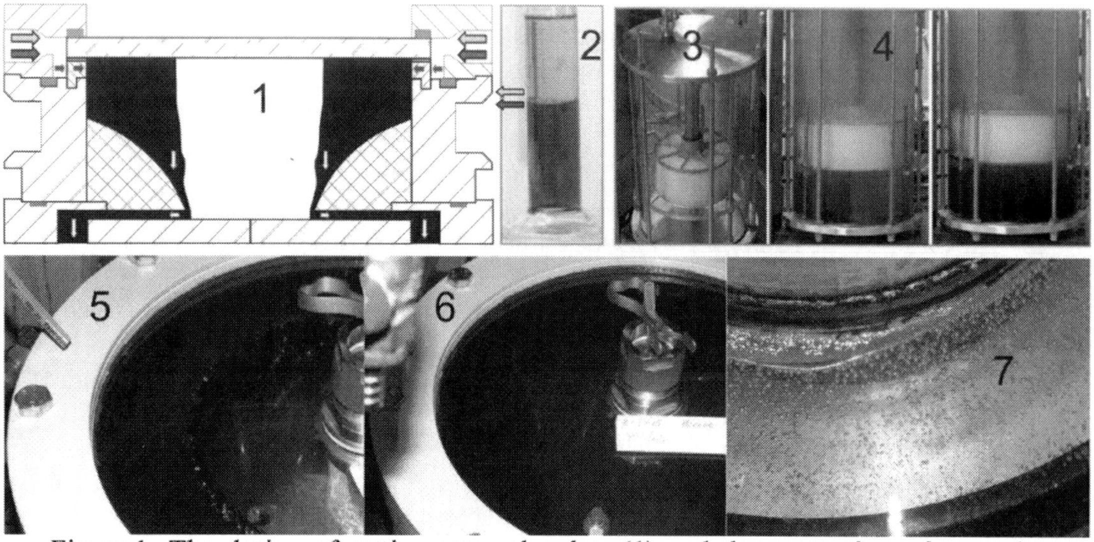

Figure 1. The design of static vortex chamber (1) and demonstration of water-hexane system (2) mixing process at flowrates for hexane - 40 L/min, for water - 1.7 L/min. The separation in a vortex breaker device (3) of water-hexane mixture (4) after contacting in VC. Pictures represent the comparison of water-hexane mixing and contacting in non-submerged (5) and submerged (6) flow operations of VC Air injection into vortex water flow through one of the swirler nozzles (7).

1. M. Goldshtik, 1981. Vortex flows. Novosibirsk. NAUKA (Russian)
2. Kuzmin, A. O., et al. (2005). Vortex centrifugal bubbling reactor. Chemical Engineering Journal 107(1-3): 55-62.
3. G. Baier, M.D. Graham, E.N. Lightfoot, 2000. Mass transport in a novel two-fluid Taylor vortex extractor. AIChE Journal. 46(12). 2395-2407
4. Andrey Kuzmin, Aziz Sattar, Lev Davydov, 2016. Apparatus and process for contacting liquids, US 9233319 B2
5. Lev Davydov, Daniel K. Aiken, Andrey Kuzmin, 2014. Process and apparatus for extracting, US 8813976 B2

Comparison of two high pressure emulsification processes; high pressure jet and microfluidizer

S. Sadeghpour Galooyak[1], M. Zolfaghari[2], N. Fatourehchi[3]
1-Refining Technology Development Division, Research Institute of Petroleum Industry, Tehran, Iran, 14665-137, Tel: (+98) 21-48255301, E-mail: sadeghpours@ripi.ir
2-Amirkabir University of Technology, Dep. Chemical Engineering, 424 Hafez Ave, Tehran, Iran, 15875-4413, Tel: (+98) 21-48255301, E-mail: mm.zolfaghari.r@gmail.com
3-Department of Chemical Engineering, Robat Karim Branch, Islamic Azad University, Tehran, Iran, Tel: (+98) 21-56887901, E-mail: n.fatourehchi@gmail.com

In this study, preparation of O/W nano-emulsion using a high pressure jet apparatus (new emulsification device) has been studied and the results were compared with those fabricated by a microfluidizer. Pressure (408, 680 and 952 bar) and oil content (5, 9 and 13%) were the affecting parameters on the high pressure homogenizer. Also, pressure (408, 680 and 952 bar), oil content (5, 9 and 13%) and number of passes (2, 3 and 4) have effect on the final droplet size distribution. The schematic for these two apparatus have been shown in figures 1-a and b. A response surface method (central composite design, CCD) was used to determine the effects of specified parameters on the performance of the microfluidization device and to specify the affecting mechanisms on the nano-emulsion fabrication. Since the number of parameters in high pressure jet are low, the full factorial method has been selected to study the emulsification process in this device. Sauter mean diameter (SMD) and Polydispersity index (PdI) are two available responses. The SMD and PdI decreased by increasing the homogenization pressure, while oil content has negative effect on the disruption of emulsion droplets for both of apparatuses. Furthermore, both of responses improved by increasing the number of passes in microfluidizer. Comparison of two emulsification process depicted that SMD and PdI in microfluidizer is less than high pressure jet. So, microfluidization is more efficient process than high pressure jet, especially at higher number of passes. While temperature rise and production of contaminant material due to contact of high pressure liquid to the nozzle wall in high pressure jet was less than microfluidizer. Furthermore, the results implied that laminar elongational flow is an alternative disruption mechanism in addition to inertia in turbulence flow.

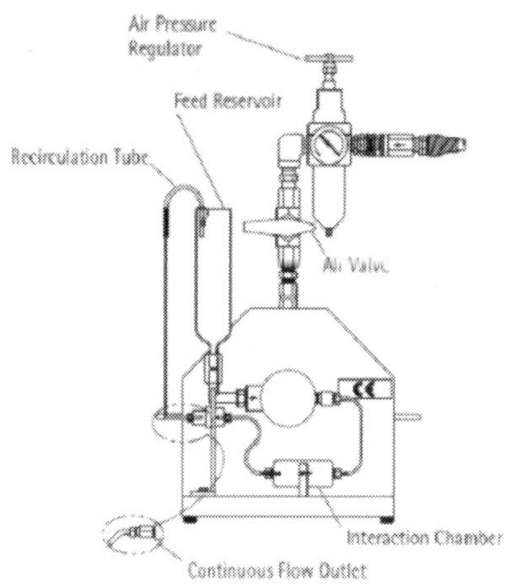

(a)

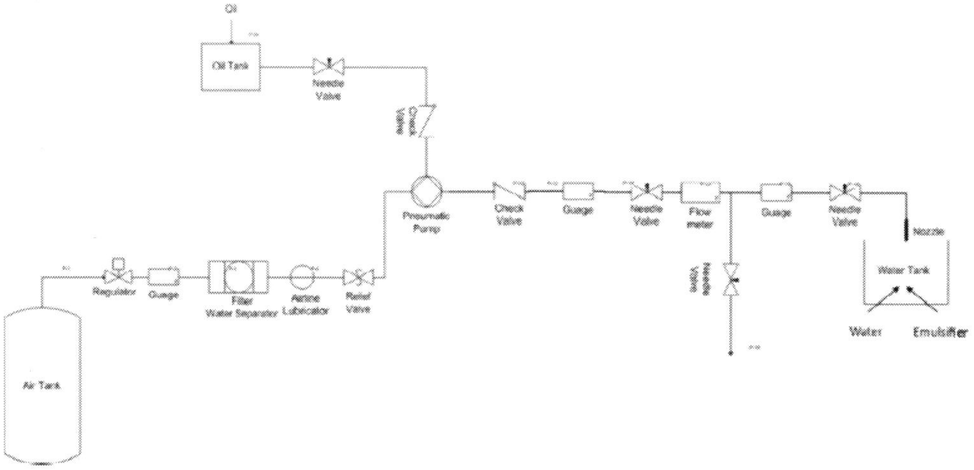

(b)

Figure 1. Schematic for different emulsification process a) Microfluidization device, b) high pressure jet.

Hydrodynamics of mixing: processing of Particle Image Velocimetry data

A. Liné, [1]& Coauthors
Toulouse Univ., INSA-LISBP, Chemical Engineering Dep., 135 Avenue de Rangueil, 31077 Toulouse cedex, France;
tel. +33561559786, e-mail: alain.line@insa-toulouse.fr.

This paper presents post-processing techniques and results of 2D Particle Image Velocimetry (PIV) data acquired in a mixing tank and in single phase flow. The goal of the research is two-fold:

- Better understanding of hydrodynamics phenomena,
- More efficient processing and extraction of information from huge amount of data created by PIV technique.

Three kinds of post-processing of PIV data are implemented:

- Proper Orthogonal Decomposition is performed in order to rank 2D velocity fields in a series of modes related to mean flow, organised motions and turbulence. The contributions of each mode to the total kinetic energy and to the total dissipation rate of kinetic energy are presented. The number of modes needed to reconstruct mean flow, organised motions (trailing vortices), kinetic energy and its dissipation rate are discussed.
- Spectral analysis of velocity (Energy Density Spectrum) is performed in terms of wave numbers, in the case of classical PIV data, and in terms of frequencies, in the case of time resolved PIV data. Post-processing issues and refinements are highlighted and characteristic time and length scales are determined. Taylor macro and micro-scales as well as Kolmogorov scales are addressed.
- Direct estimation of instantaneous dissipation rate and quantification of intermittency are issued from processing of time resolved PIV data.

The experiments were carried out in a transparent vessel of diameter T=450 mm filled with tap water, stirred by a Rushton turbine of diameter equal to D=T/3 with six blades of thickness T/150. A standard configuration was employed with liquid height H equal to vessel diameter (H = T) and four equi-spaced baffles of width B = T/10. The clearance, C, was equal to the impeller diameter and was measured between the flat bottom of the mixing vessel and the impeller-disk plane. The blade height was equal to 0.2 D and the disk thickness was equal to blade thickness.

The time-resolved PIV data, in terms of instantaneous velocity without filtering the high frequency fluctuations, were measured by PIV equipment provided by LaVision; it includes a double-pulsed Nd-YLF Newwave laser Pegasus (2×10 mJ at 1 KHz) and a Photron CMOS RS3000 camera (up to 3 KHz at a resolution of 1024 × 1024 pixels). Rhodamines particles were used as seeding particles and their average sizes were less than 20 μm (density=1.1). The dedicated processor, Davis 7.1, was used to acquire data of the flow fields. For each set of 2-D PIV measurements, several thousand pictures were taken at a frequency of 2 KHz and the measurements were repeated several times to establish reproducibility.

Different PIV techniques (time unresolved ones) also used in this studies consisted of lasers and image acquisition systems provided by LaVision and Dantec. The LaVision system included a laser Nd-Yag (Quantel, 10 Hz, 200 × 2 mJ), a synchronization system and a

Charge-Coupled-Device (CCD) camera (Imager Intense, 12 bits, 1376×1040 pixels^2). The Dantec system consisted in a laser Nd-Yag (Quantel, 15 Hz, 30 × 2 mJ), a synchronization system and cameras (Flowsense, 2M or 4M pixels).

The main results of the data processing are the following:

- Automatic processing of instantaneous PIV data is proposed, enabling to differentiate organised and turbulent motions; it is based on POD analysis and phase resolved measurement may thus be avoided;
- Eigen values issued from POD express the contribution of each mode to the total kinetic energy measured in the 2D PIV plane; eigen modes analysis enables to determine the range of length-scales associated to each mode;
- Energy density spectrum versus wave numbers are reconstructed highlighting the contribution of the different modes;
- Energy density spectrum versus frequency are available from time resolved PIV data;
- Characteristic time and length scales of mixing phenomena are then extracted from huge amount of PIV data.
-

The perspectives of the data processing are two-fold:

- Make progress on velocity data reconstruction based on POD,
- Propose original methodology of reduced order modelling in a mixing tank that could replace subjective partition of the vessel volume.

Liquid mixing time in dense solid-liquid stirred tanks

A. Paglianti, [2]C. Carletti, [3]G. Montante
Department of Civil, Chemical, Environmental and Materials Engineering, Università di Bologna, via Terracini 28, 40131 Bologna, Italy; [2]Process Design and Systems Engineering Laboratory, Åbo Akademi University, Biskopsgatan 8, FIN-20500 Åbo, Finland; [3]Department of Industrial Chemistry, Università di Bologna, via Terracini 28, 40131 Bologna, Italy; tel: +390512090406 e-mail: giuseppina.montante@unibo.it

Optimized design and successful operation of solid-liquid stirred tanks depend on the capability of predicting their complex fluid dynamic characteristics, which are affected by the solid loading. The understanding gained so far on single phase and dilute solid-liquid systems cannot be directly extended to dense suspensions, since important features, including solids suspension and distribution, power consumption, liquid mixing, vary greatly due to the different interactions between the continuous and the dispersed phases. Among the different parameters of interest for the estimation of the overall performances of slurry stirred tanks, the liquid mixing time is the aspect addressed in this paper.

Based on the current knowledge, the overall time for the achievement of a suitable level of liquid mixedness in dense suspension is not predictable with confidence, as is the case of single-phase or dilute two-phase flows. The complex dependency of the liquid mixing time on the solid content and on the clear liquid layer height has been already highlighted in previous works (Kraume, 1992; Bujalski et al., 1999). Either important or negligible deviations with respect to single phase flows have been found depending on the specific conditions, being the mixing time about equal or much longer depending on the impeller speed and on the solid content. In this work, we aim to contribute to the characterization of the homogenization dynamics of a slurry stirred tank, adopting Electrical Resistance Tomography (ERT) as the investigation tool.

The suspensions of glass particles of mean size d_p.=385μm and mass ratio X=0.43 (kg solids/kg liquid) in a very dilute solution of sodium chloride in demineralised water in a fully baffled stirred tank is considered. The laboratory stirred tank of diameter T=0.232 m and height H=0.280 m is of a quite standard geometry (flat bottomed, cylindrical, fully baffled, open on top). Agitation is provided by a single centrally-mounted Lightning A310 impeller of diameter D=0.096m and placed at the off-bottom clearance of C=T/3. The local conductivity before and after the almost instantaneous injection of a small amount of saturated NaCl aqueous solution in the solid-liquid suspension of lower conductivity is measured on horizontal vessel sections. To this end, equally spaced squared stainless steel electrodes of 20 mm side and 1 mm thickness are fixed circumferentially on the vessel lateral wall (16 electrodes per plane). The voltage difference is measured instantaneously by the circular adjacent strategy, the local conductivity (316 values per plane) is obtained from the voltage measurements by the linearized back projection algorithm on a mesh with square cells of 11.5 mm side.

For a detailed analysis of the homogenization dynamics, the time evolution of the normalized liquid conductivity on two measurement planes, located one below and one above the impeller, is considered. The effect of different injection locations is analysed on a local basis. The experimental results are discussed taking into account the suspension conditions with respect to the experimentally determined just suspended impeller speed, N_{JS}, and the local solid concentration distribution determined in a previous investigation (Carletti et al.,

2014). The comparison with the results obtained in single phase conditions and stirring with a Pitched Blade turbine is also addressed.

As for the mixing time estimation, the dimensionless conductivity in the cell i, $C_i(t)$, is calculated as the conductivity of the slurry during the tracer dispersion divided by the conductivity of the medium before the tracer injection. Afterwards, the measured dimensionless conductivity in the cell i, Ci(t), is adopted for the estimation of the local tracer homogenization dynamics after a suitable normalization, being it proportional to the tracer concentration. The normalized local conductivity, χ_i, is calculated as:

$$\chi_i(t) = \frac{C_i(t) - C_i(0)}{C_i(\infty) - C_i(0)} \tag{1}$$

where Ci(0) is the conductivity before the tracer addition and Ci(∞) is the final conductivity after the complete homogenization.

The normalized average conductivity obtained on two measurement planes for two different injection locations is shown in Figure 1, as an example.

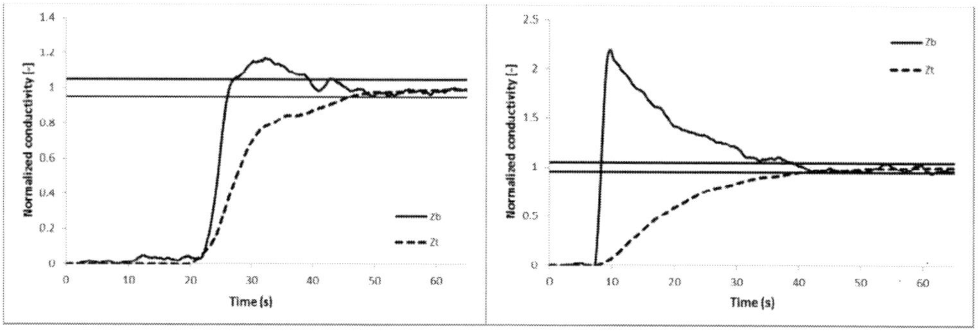

Figure 1. Average conductivity on horizontal planes located below (zb) and above (zt) the impeller, after the tracer injection from top (right) and bottom (left).

The mixing time is estimated from the time evolution of the normalized local conductivity averaged over each measurement plane and significant differences with respect to the corresponding single phase conditions are obtained, as expected.

Following Kukukova et al. (2009), the intensity, the scale and the rate of change of mixing are also calculate, for gaining insight into the process dynamics by exploiting the detailed ERT data.

References

Bujalski, W., Takenaka, K., Paolini, S., Jahoda, M., Paglianti, A., Takahashi, K., Nienow, A.W. and Etchells, A.W. (1999). Suspension and liquid homogenization in high solids concentration stirred chemical reactors. Chem. Eng. Res. Des., 77, 241-247.

Carletti, C., Montante, G., Westerlund, T., Paglianti, A., 2014. Analysis of solid concentration distribution in dense solid-liquid stirred tanks by electrical resistance tomography. Chem. Eng. Sci., 119, 53-64.

Kraume, M. (1992). Mixing times in stirred suspensions. Chem. Eng. Technol., 15, 313-318.

Kukukova, A., Aubin, J., Kresta, S.M. A new definition of mixing and segregation: Three dimensions of a key process variable. Chemical Engineering Research and Design, 87 (4), 633-647, 2009.

Scaling of velocity gradients in a vessel agitated by a Rushton turbine

Radek Šulc, Pavel Ditl
Czech Technical University in Prague, Faculty of Mechanical Engineering, Department of Process Engineering, Technická 4, 166 07 Prague, Czech Republic;
Phone: +420 22435 2549, e-mail: Radek.Sulc@fs.cvut.cz;

One of indirect methods used for the dissipation rate estimation is based on fluctuation velocity gradient measurements. In PIV studies, the spatial distributions of measured velocity gradients are often presented. However, inspection analysis shows that the validity of the spatial distribution of any dimensionless property for arbitrary process conditions in geometrically similar agitated vessels requires independence of a given dimensionless property on the impeller Reynolds number. The spatial distribution obtained is than in general valid only when this independence is both theoretically predicted and experimentally verified. Unfortunately, in many studies the results are presented only for one impeller speed and, in addition, when a Rushton turbine is used as an impeller, the experiments are often carried out in the transient flow regime.

In this work, the scaling of fluctuation velocity gradients have therefore been tested at a given spot under eleven different conditions, representing: i) two different vessel sizes, ii) three different viscosities, and iii) different impeller rotational speeds. Further, the scaling of isotropy ratios calculated for known measured velocity gradients will be also presented.

The hydrodynamics and the flow field in an agitated vessel were measured using 2-D Time Resolved Particle Image Velocimetry (2-D TR PIV). The experiments were carried out in fully baffled cylindrical flat bottom vessels 300 and 400 mm in the inner diameter. The dimensionless impeller diameter D/T was 1/3. The dimensionless impeller clearance H2/D taken from the lower impeller edge was 0.85. The tank was filled by distilled water and dimensionless liquid height H/T was 1. The dimensionless baffle width B/D was 1/10.

Three liquids of different viscosities were used as the model liquids, namely: i) distilled water ($\nu = 9.35 x 10^{-7}$ m2/s), ii) a 28% vol. aqueous solution of glycol ($\nu = 2 x 10^{-6}$ m2/s), and iii) a 43% vol. aqueous solution of glycol ($\nu = 3 x 10^{-6}$ m2/s). The velocity fields were measured at an impeller rotation speed in the range from 300 rpm to 850 rpm, that covers the Reynolds number range from 50 000 to 189 000. This Re range secures the fully-developed turbulent flow.

CFD study of the hydrodynamic induced by a high shear impeller Norstone® type in an unbaffled stirred tank

G. Martínez-de Jesús*[,1], A. De la Concha[1], J. Ramírez-Muñoz[1]
[1]Departamento de Energía, Universidad Autónoma Metropolitana- Azcapotzalco. Av. San Pablo 180, Azcapotzalco, Mexico City, 02200, Mexico. *g.martinez.dj@gmail.com.

Abstract

The purpose of this work was to evaluate the hydrodynamic produced by the Polyblade® Norblade® polymer high shear impeller (Norstone®, inc.) –namely in this work PNN- at $25 \leq Re \leq 75$, and to compare its performance with the standard Rushton turbine (RT). The study was carried out by using the Computational Fluid Dynamics software ANSYS FLUENT 15.0. The dependence of power (N_P) (numerical and experimental) and pumping (N_Q) numbers on *Re* were obtained. In addition, the flow structure and the contours of the viscous dissipation induced by both, the PNN and RT impellers are reported.

Introducction

Dispersion of nanoparticle powders into a liquid in stirred tanks is commonly carried out by using mixing devices well-known as high shear impellers (HSI's). Because HSI's are operated at relatively high angular velocities, a zone of high shear force is created, which provides the energy necessary to break up particle agglomerates. PNN is a HSI commonly used in the major USA paint industries to incorporate abrasive powdered pigments into a liquid and break down particle agglomerates to produce a fine dispersion [1]. Nevertheless, in the scientific literature there is a lack of information about its performance. Therefore, evaluating the hydrodynamics induced by this mixing device could be of great use in the assessment of its performance. To characterize an impeller is important to know as the energy is spent. A common practice to evaluate the performance of different impellers is by comparing its values of N_P and N_Q as function of *Re* [2], which are a measurement of its shear and pumping capacities, respectively.

Methodology

The fluid flow was modeled as laminar and a multiple reference frame approach was used to solve the discretized equation of motion as suggested by [2]. The fluid used in the experiments was a mixture of food grade glucose (45° Bx) and distilled water. Solutions of 36 and 37.8 % by weight of glucose were prepared. The fluid properties were: μ=0.95, 1.7 Pas and ρ=1376.46, 1386.26 kg m^{-3}, respectively. Experiments were performed at 23°C with a cylindrical vessel with dished bottom. The diameter *T* of tank was equal to 0.132 m. The dimensionless geometric ratios corresponding to the impeller off-bottom clearance (*C*), liquid height (*Z*) and impeller diameter were: C/T, D/T=0.3848, Z/T=1, respectively. Agitation and power measurements were carried out by using a Dispermat® equipment type AE01.

Results

The dependence of N_P on *Re* is shown in Fig. 1a. N_P values were obtained by integrating the viscous dissipation function from its definition [2]. It can be observed that the experimental values of N_P for PNN are in agreement with the numerical data and in general are lower than the corresponding RT values. The values of N_Q are shown in Fig 1b, as can be seen, N_Q values for RT are higher than those for PNN. Note that increasing *Re* for PNN (i.e. the power draw), N_Q values remains almost constant. These results suggest that the energy delivered by this impeller is mainly dissipated in the fluid instead of being spent in pumping.

In Fig. 1c it can be observed that, according to the relation N_P/N_Q, the PNN have higher capacity to dissipate energy and its pumping capacity is limited. The opposite is observed for the RT.

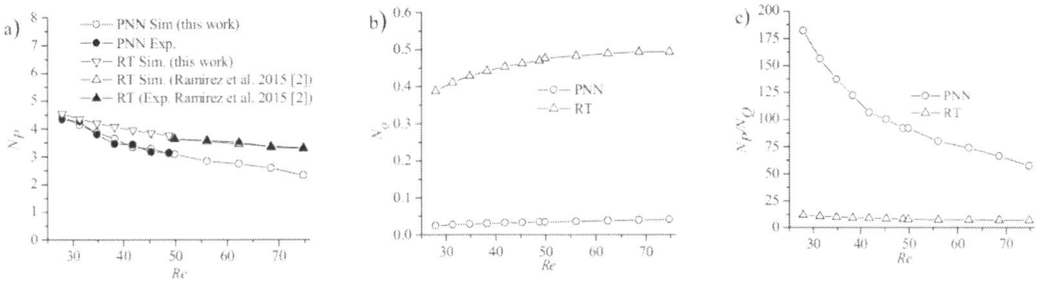

Figure 1. N_P and N_Q values and its ratio N_P/N_Q for PNN and RT.

In order to visualize the induced hydrodynamic of both impellers, in Fig 2a are shown the velocity fields. Note that for both impellers, circulation loops are generated above and below the impeller, which is characteristic of radial flow impellers, as it is for RT [3]. In Fig. 2b are depicted the normalized contours of viscous dissipation ($\varepsilon/\varepsilon_m$), where ε_m is the average dissipation in the whole fluid volume. As can be seen, the highest values of $\varepsilon/\varepsilon_m$ are located in the tip of the impellers. Note that PNN impeller exhibits broader surfaces than RT where values of $\varepsilon/\varepsilon_m$ are maximum. Thus, PNN impeller might give better results for pigment dispersion processes.

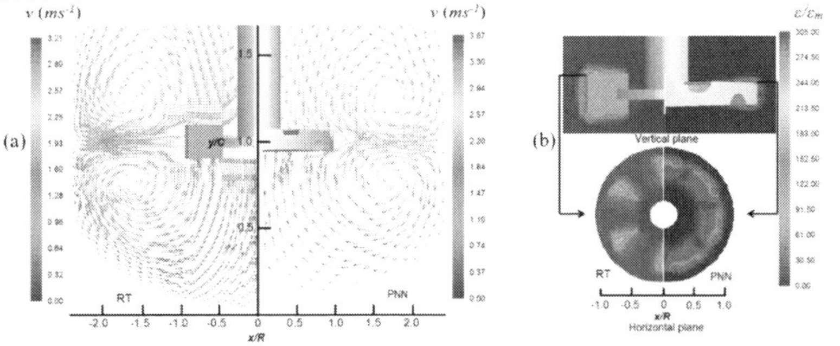

Figure 2. (a) Projected velocity vectors and (b) contours of viscous energy dissipation, for RT and PNN, at power input of 12.92 *W*. The arrows in black indicates the position of the horizontal planes. $R=D/2=25.4$ mm.

Conclusions

The PNN impeller has a large ratio N_P/N_Q, this means that the energy delivered to the fluid is spent mainly in viscous dissipation, and in contrast, its pumping capacities are low. In addition, the contours of viscous dissipation suggested that the PNN impeller concentrates the energy dissipation near the impeller, and therefore, this would be more appropriate for dispersion processes. However, its pumping capacity is low and the processing time could be affected.

References

[1] Norstone® Inc (2010). Polyblade®- Norstone® Inc (online): Available at: http://www.norstoneinc.com/our-products/polyblade/ [accessed 29 Nov. 2015].

[2] R. Ramírez-Gómez, D. García-Cortés, G. Martínez-de Jesús, M. González-Brambila, A. Alonso, J. Ramírez-Muñoz. 2015. Performance evaluation of two high shear impellers in an unbaffled stirred tank. *Chem. Eng. Technol.* 38, 1519-1529.

[3] Rice, M., Hall, J., Papadakis, G. and Yianneskis, M. 2006. Investigation of laminar flow in a stirred vessel at low Reynolds numbers. *Chem. Eng. Sci.* 61, 2762–2770.

Heat transfer performance of the NETmix – a novel micro-meso structured mixer/reactor applied for Carbon Storage and Capture (CCS)

M. F. Costa, M. M. Dias, [1]J. C. Lopes
Laboratory of Separation and Reaction Engineering, Faculty of Engineering, University of Porto, Rua Dr. Roberto Frias s/n, 4200-465 Porto, Portugal;
tel. +351 225 081 167, e-mail: lopes@fe.up.pt

It is accepted worldwide that climate changes are occurring, and many associate these changes to the human use of fossil fuels. Emissions of carbon dioxide is a serious problem in present electricity and oil refinery industries, as well as steel industries, that have to pay to either treat or reduce CO_2 emissions. One of the most interesting strategies to solve this environmental problem is the capture and storage of CO_2. However, this process is costly, contributing by 70-90% of the total operating cost. Continuous efforts have been made to search for alternative methods for CO_2 capture, so that the overall operating cost can be reduced.

Hydrate-based CO_2 capture (or HBCC) technology is a novel process that has gained large interest by the scientific community. It consists in the precipitation of CO_2 hydrates, concentrating large amounts of gas in solid crystals. Gas hydrates are structurally similar to ice, consist on a lattice of water that traps one or more hydrocarbon or non-hydrocarbon gas molecules, and are thermodynamically stable at low temperatures (below 10°C) and moderate pressures (20-30 bar). Several studies of this process are described in literature. In all of them, an important problem of CO_2 hydrate precipitation is the large amounts of energy that the reaction releases to the reaction medium that inhibits further precipitation of hydrates. In particular, the use of stirred tanks or tubular reactors has been shown to be insufficient to remove the reaction heat in an adequate period of time. Also, it has been shown that once the heat transfer limitations are overcome, precipitation is fast as long intense mixing is achieved in the reaction medium, something that is not observed in typical stirred tanks, especially at large scale operation.

Static mixers have been used in several industrial applications, to perform continuous process operations, and are standard equipment for the mixing of miscible or immiscible fluids. In particular, the NETmix reactor, a novel micro-meso structured mixer/reactor has the potential for being used in the production of hydrates for CO_2 capture. The NETmix reactor consists of a network of mixing chambers interconnected by transport channels. Above a critical channel's Reynolds number, the system evolves to a self-sustained oscillatory laminar flow regime inside the mixing chambers, inducing local strong laminar mixing. Past works haven shown that strong laminar mixing is achieved at the cost of the flow energy, without requiring mechanical parts and with small associated pressure drop. The use of this technology in the precipitation of nanocrystalline materials, where mixing quality is a key factor, has been previously studied and applied in industrial products. One of the great advantages of the NETmix structured the reactor is the simplicity of scale-up to industrial scale, by simply adding several networks in parallel, without compromising the flow patterns the are observed, being capable to work at both small and large scale with the same performance.

The objective of the present study is to demonstrate that besides its mixing capacity, the NETmix reactor greatly enhances the heat transfer rates, overcoming the heat transfer limitations, allowing the use of a continuous process of CO_2 capture. In general, the NETmix can be used as a chemical reactor for reactions where the mass and heat exchange are important for the viability of the process.

Heat transfer coefficients were computed and presented in their dimensionless form, using the Nusselt number. A CFD model (NETmix Unit Block model or NUB) was developed to simulate the laminar flow inside a significant portion of the NETmix domain. Due to the periodic flow patterns observed in the NETmix, a model comprising 2 central chambers and 6 half-chambers was found to be sufficient to predict heat transfer rates. With the purpose of studying and quantifying the heat transfer in the NETmix, simulations at various Reynolds numbers were performed for constant wall temperature and constant wall heat flux.

In order to better evaluate the performance of the NETmix capacity of removing heat and to validate the CFD simulations, the obtained Nusselt numbers were compared with the Nusselt numbers for the well-known case of laminar flow between parallel plates. The comparison is made using the asymptotic solutions for two different reference heat transfer conditions: constant wall temperature and constant wall heat flux, derived for Newtonian fluids with constant physical properties and fully-developed velocity profiles.

CFD simulations of the NETmix domain have shown that the NETmix is capable to remove 3-5 times more heat than the case of flow between parallel plates. Heat transfer enhancement is observed when the oscillatory behaviour is observed, and increases with the increase of the Reynolds number.

To understand the real capacity of the NETmix technology and its use in the CO_2-hydrate based process, a comparison between the NETmix and several other technologies that widely used in the industry is made (Figure 1). As it can be seen, the NETmix reactor presents heat removal capacity similar to microreactors, but with much larger mixing capacity, qualities that are essential for the continuous production of CO_2 hydrates, without increasing the process costs.

Thus, the NETmix mixer/reactor is shown to be a very efficient equipment to remove/supply large amounts of heat, at small and large scale operations, particularly suited in highly exo/endothermal reactions, and in particular in the precipitation of hydrates for CO2 capture and storage.

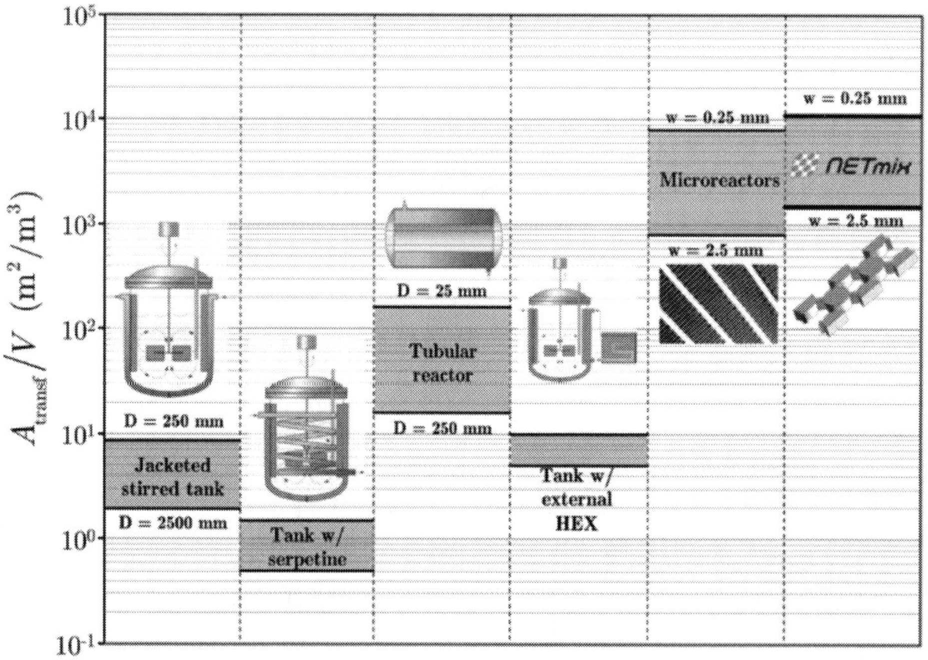

Figure 1 - Heat exchange capacity of several industrial equipments.

Application of Gaussian cubature to model chemical engineering processes

J. Bałdyga[1], G. Tyl[1], M. Bouaifi[2]
[1.] Warsaw University of Technology, Faculty of Chemical and Process Engineering,
ul. Waryńskiego 1, 00 645 Warsaw, Poland;
tel. +48 22 234 63 22, e-mail: g.tyl@ichip.pw.edu.pl, fax: +48 22 8256037.
[2.] Centre de Recherche et d'Innovation de Lyon, Solvay, 85 Avenue des Frères Perret, BP 62,
69192 Saint-Fons Cedex,France. email : Mounir.Bouaifi@Solvay.com

One of the commonly used approaches for solving population balance equations in chemical engineering applications is the quadrature method of moments, that has been introduced by McGraw (1997) to simulate the course of particulate processes. It is based on the approximation of the density function in the source term by the Gaussian quadrature so that it preserves the moments of the original distribution. In this method one determines weights and abscissas that are used in the quadrature approximation. To this end one uses so called inversion algorithms. McGraw (1997) applied the product-difference algorithm by Gordon (1968). More such algorithms are available in the literature.

Wright, McGraw and Rosner (2001) extended the QMOM method to the multivariate population problem. In this case there are no such algorithms available. Several attempts have been made to apply the above mentioned algorithms also for the multidimensional case, which resulted for example in a multiple 3-point quadrature technique and a 12-point quadrature technique (McGraw et al., 2001). The DQMOM equations (Marchisio and Fox, 2005) are derived via the moment transfer method, in a similar way to QMOM.

In this work we propose another method to be applied to the multivariate population problem, namely a Gaussian cubature (GC) technique that applies linear programming for the approximation of the multivariate distribution. This approach has been presented by DeVuyst and Preckel (2007) for the modeling of economic processes. In this method the first step is to set up a grid over each variable. Then a lattice is created by the Cartesian product of those grids. Linear programming approach is then used on the lattice using the system of equations for the moments as the constraints and objective function set to zero. Linear program returns weights (restricted to be positive) for each of the lattice points. Maximum number of the points with strictly positive weights is equal to the number of constraints used.

Examples of application of the Gaussian cubature (GC) are presented for two processes typical for chemical engineering applications. The first one represents drop dispersion accompanied by mass transfer in the liquid-liquid dispersions, the second one is devoted to crystallization modeling with direction-dependent growth rates, which gives rise to multi-dimensional population balance equations (PBE).

DeVuyst E. A. and Preckel P.V, 2007, "Gaussian cubature: A practitioner's guide. Mathematical and Computer Modelling, 45 (2007) 787–794.

Gordon, R.G., 1968. Error bounds in equilibrium statistical mechanics. *J. Math. Phys.* ***9***, 655-663.

Marchisio D. L. and Fox R. O., 2005, "Solution of population balance equations using the direct quadrature method of moments. Journal of Aerosol Science, 36(1), 43-73.

McGraw R., 1997, "Description of aerosol dynamics by the quadrature method of moments." *Aerosol Science and Technology* 27, 255–265.

Wright D. L., McGraw R. and Rosner D. E., 2001, "Bivariate extension of the quadrature method of Moments for modeling simultaneous coagulation and sintering of particle population." *Journal of Colloid and Interface Science* 236, 242–251.

Mixing time in a lab-scale stirred reactor with a catalytic basket

V. Santos-Moreau[1], L. Brunet-Errard[1], M. Braga[1], C. Pereira da Fonte[1], I. Pitault[2], M. Rolland[1], T. Danzo[1] and C. de Bellefon[2]

1- IFPEN, Rondpoint de l'échangeur de Solaize. BP 3. 69360 Solaize, France
Tel : 0033 4 37 70 22 64, vania.santos@ifpen.fr

2- Laboratoire de Génie des Procédés Catalytiques, Université de Lyon, CNRS-CPE;

Stirred reactors with catalytic basket (SRCB) are often used at the laboratory scale (volumes smaller than 1 ℓ) to test and evaluate millimetre sized shaped catalysts performances and measure kinetics in Solid/Gas/Liquid systems. In most cases, the SRCB are operated in batch mode on the liquid and solid phases while the gas (reactant) is continually introduced in the system in order to maintain the system pressure. In SRCB reactors, the catalyst (extrudates or beads) is placed in a basket located around the impeller. The impeller is usually radial, in order to push the gas and liquid through the catalyst bed, increasing the liquid/solid velocities. The gas is pumped from the gas headspace to the liquid through a self-inducing impeller, whose advantage is promoting an "internal" recycling (cheap and suitable for any gas). The gas localised in the headspace of the reactor is sucked trough the hollow shaft through inlet orifices located at the top of the shaft and exits at orifices placed near the impeller blades by the pressure gradient generated by the stirring.

The hydrodynamics and mixing in the SRCB's are quite complex and a significant impact of the basket on the flow patterns, and consequentially on the mass transfer and on the quality of the blending, can be expected. Many studies have described experimental measurements and modelling to obtain mixing times of the liquid phase agitated by turbines in vessels or reactors. The majority concern large size tanks for chemical and biochemical industrial applications, wherein various types of turbines and configurations (single or multiple turbines) have been used with ungassed or gassed agitation (aerated vessel) and loading or flooding regimes. Regarding lab-scale stirred reactors with a catalytic basket, some results of mixing time measurements are given in the works of Fongarland (2003), Mitrovic (2001) and Schweitzer (1998). Different invasive and non-invasive techniques have been used to determine the mixing time in stirred vessels.. In the case of small reactors, intrusive techniques should be avoided since the presence of probes can disrupt the flow in the reactor.

In the present work, the mixing time in a SRCB was determined by a colorimetric technique in hydrocarbon media in absence of chemical reaction. This method is particularly interesting since there is the possibility to estimate local mixing times in different zones of the reactor. The principle of this technique is based in the introduction of a colorant in a transparent mock-up and follow the tracer dispersion in the reactor until homogenisation. The variation of the tracer concentration with time in different parts of the reactor (regions of interest, ROI) is followed by a camera (Figure 1). The analysis of the images pixel variation allows to estimate the mixing time. Image processing has been carried out using the image processing library *Halcon HDevelop* software (version 10), which allowed an automated image processing. $t_{95\%}$ mixing time was chosen as criteria to define homogenisation. To each ROI and each image a relative C_r concentration was defined. Having all ROI $t_{95\%}$ mixing times (local mixing times), a global mixing time was defined as the longest of all ROI mixing times. Different configuration of the reactor were studied: with and without basket, different turbine diameters, 20 and 25 mm (respectively, R20 and R25), different basket fillings and different of liquid heights (Fig 2 and 3). For the reactor without the basket the effect of the presence of gas in mixing time was also studied.

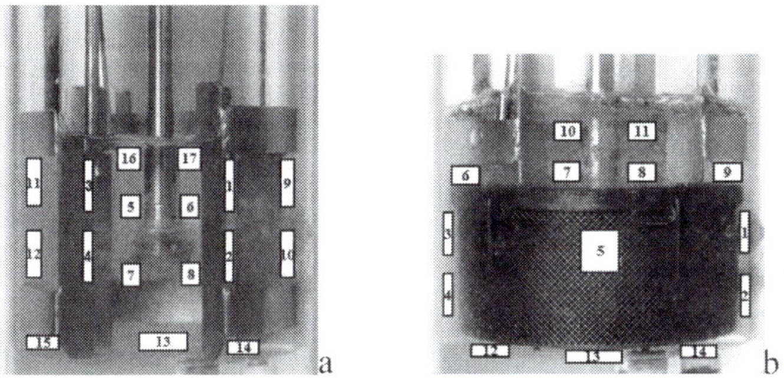

Figure.1- Dispersion of colorant in the reactor (PMMA mock-up) and definition of different ROI. a) Reactor without the basket and b) Reactor with the basket.

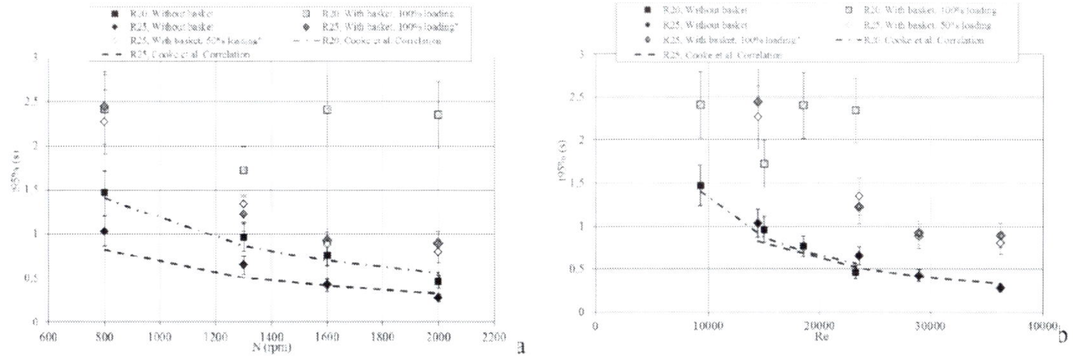

Figure 2 – Mixing time in the reactor at different configurations : R20 or R25, with and without basket at different fillings in function of the a) stirring rates and b) Reynolds numbers.

For the reactor without the basket the global mixing times are low and close to 0.5-1.5 s and the results closely agree with the correlation of Cooke and Middleton, validating the experimental method of determination of the mixing time (see Figure 2). At the same stirring rate the mixing time are slightly higher in the case of the reactor agitated with the R20. However, at the same Reynolds numbers the mixing times curves are identical meaning that the mixing time is not highly influenced by the stirrer diameter but only by the hydrodynamic conditions of the system and by the ratio between inertial and viscous forces.

The introduction of the basket in the reactor results in a significant increase of the mixing time. Results shows (Figure 2) that the mixing time is influenced by the reactor internal configuration and not only by the impeller agitation speed. However, the impact of the basket filling in the mixing time is insignificant comparing with the effect of the basket grid. The presence of the basket grid is the main responsible by the increase of the mixing time.

In the basket configuration, the effect of the R20 and R25 turbines in the flow diferent. In the case of the R25 turbine the mixing time decreases with the stirring and Reynolds number, with similar behaviour to the reactor without basket. In the case of the R20 a changing in the flow regime appears. The analysis of the local mixing times allowed to identify a flooding regime characterized by the accumulation of gas near the stirrer, hindering the flow of liquid though the basket.

Cooke, M., J. C. Middleton, and J. R. Bush (1988). Mixing and mass transfer in filamentous fermentations, in: E.L. Paul, V.A. Atiemo-Obeng, S.M. Kresta (Eds.), Handbook of Industrial Mixing: Science and Practice, Wiley-Interscience, Hoboken, NJ, 2004, pp. 605–607. Chapter 11.

P. Fongarland, Etude cinétique de l'hydrodésulfuration de composés soufrés de gazoles en réacteur continu parfaitement agité, PhD thesis, University Claude Bernard Lyon 1, (2003)

M. Mitrovic, Étude des transferts de matière dans un réacteur triphasique gaz-liquide-solide, d'investigation cinétique (réacteur Robinson-Mahoney), PhD thesis, University Claude Bernard Lyon 1, (2001) 52-53

J.M. Schweitzer, Modélisation cinétique des réactions catalytiques d'hydrocraquage par la théorie des événements constitutifs, PhD thesis, University Claude Bernard Lyon 1, (1998)

Mixing in CIJS: insight into highly efficient 2D laminar micro / mesomixers

N. D. Gonçalves, C.P. Fonte, M. M. Dias, R. J. Santos, J. C. Lopes
Laboratory of Separation and Reaction Engineering, Faculdade de Engenharia da Universidade do Porto, Rua Dr. Roberto Frias, 4200-465 Porto, Portugal
e-mail: lopes@fe.up.pt

Confined Impinging Jets (CIJs) are highly efficient mixing devices composed by a semi-confined mixing chamber, two opposed injectors near the closed top of the chamber and an outlet in the open end from where the fluid leaves the reactor (Fonte, et al. 2015). The fluids enter into the mixing chamber through two injectors as high-speed opposed jets that decelerate in the short distance from the injector to the chamber axis causing strong energy dissipation, which promotes very effective mixing. Two main flow regimes have been identified in CIJs, one at low Reynolds number, where the flow is steady and the fluids injected from opposed jets are kept almost segregated, and a chaotic or dynamic flow field where a street of vortices is formed upon the impingement of the opposed jets.

The CIJs have length scales ranging from 1 mm, for the inlet jets, to 10 mm for the mixing chamber. Within this range of scales the CIJs produce very homogenous mixtures with mixing times ranging from 10 to 100 ms at throughput rates that can go up to 100kg/min for the case of reactive polymerization. The short mixing times are promoted at low Reynolds number, around 200, thus at laminar flow regime. The intense mixing in CIJs at laminar flow regimes is mainly described in the framework of 2D mixing mechanisms. The expansion of the jets lengths scale in the mixing chamber produces an inverse energy cascade of the different flow scales, where the energy is injected from the small scales and feeds the larger eddies. This energy flow from smaller to larger scales is the typical cascade of energy in 2D turbulent flows (Boffetta, and Ecke, 2012). The marked 2D features of the flow in CIJs have been the subject of previous works (Santos, et al. 2010). The scale at which the energy is injected in the mixing chamber is the jets diameter, and this jets feed vortices with a diameter that is approximately half the mixing chamber width, these vortices engulf into larger vortices that extend throughout the entire chamber width. The distribution of energy into the different scales shows a great concentration of energy in a limited range of vortices sizes, with the inertial sub-range of energy dissipation being confined to scales ranging from the mixing chamber width (D) to the injectors width (d), where D/d= 10/1.5. The width of the jets is clearly shown from the kinetic turbulent energy power spectrum to be the scale where energy is injected into the system. The flow scales grow and from the turbulent kinetic energy spectrum is also clear the concentration of energy in larger scales where the size of the vortices is defined from the mixing chamber geometry. Figure 1 shows the mixing of a clear and dyed fluid in CIJs with the identification of the typical length scales.

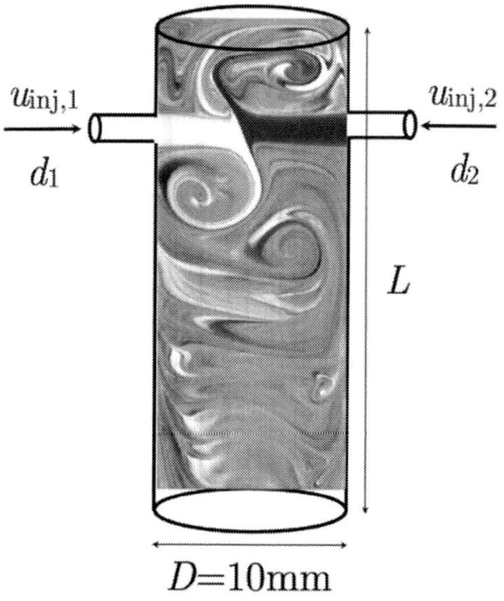

Fig 1. – Plannar Laser Induced Fluoresence in a CIJ at Re = 200.

The hydrodynamic scales are growing from the jets inlet to the scale of the mixing chamber width, but the mass transfer scale is being reduced from the inlet scale that is the largest mixing scale identified in Figure 1. The jets are then subject to intense shear rates by the larger vortices in the mixing chambers, which cause a fast decrease of the mixing scales to smaller sizes than 100 μm.

The strong flow dynamics in CIJs is associated with typical frequencies of the flow that are associated to mixing scales. The flow frequencies can be controlled from an external stimulus, namely the modulation of the flow rate of the opposed jets – active mixing. Mixing dynamics is analysed for several modulation frequencies. The external modulation can promote states of resonance throughout the whole mixing chamber, i.e. the flow fluctuation take on a single frequency for some specific frequencies of the external stimuli. This generalised resonance of the flow in the mixing chamber causes a larger concentration of energy in a single wavelength while increasing the local values of turbulent kinetic energy and decreasing its dissipation along the mixing chamber. Several modulation frequencies were tested, but the largest enhancement of turbulent kinetic energy was consistently observed only in the range of frequency values that induced the resonance of the flow field.

The control of mixing scales and of the turbulent energy spectrum opens several prospects on the application of active mixing. In the particular case of CIJs, the larger vortices were observed from previous works to be the major mixing mechanism in CIJs, and so their induction has the potential to enhance mixing, particularly on steady state flow regimes.

References

Boffetta, G. and Ecke, R.E., *Two-Dimensional Turbulence.* Annual Review of Fluid Mechanics, 2012. **44**(1): p. 427-451.

Fonte, C.P., Sultan, M.A., Santos, R.J., Dias, M.M., and Lopes, J.C.B., *Flow imbalance and Reynolds number impact on mixing in Confined Impinging Jets.* Chemical Engineering Journal, 2015. **260**: p. 316-330.

Santos, R.J., Teixeira, A.M., Erkoç, E., Sultan, M.A., Karpinska, A., Dias, M.M., and Lopes, J.C.B., *Validation of a 2D CFD Model for Hydrodynamics' Studies in CIJ Mixers.* International Journal of Chemical Reactor Engineering, 2010. **8**: p. A32.

Effects of mass transfer and micromixing on product distribution of complex reactions

J. Bałdyga, M. Kotowicz
Warsaw University of Technology, Faculty of Chemical and Process Engineering,
ul. Waryńskiego 1, 00 645 Warsaw, Poland;
tel. +48 22 2346376, e-mail: J.Baldyga@ichip.pw.edu.pl, fax: +48 22 8256037.

Many chemical reactions leading to desirable intermediate and end-products (such as pharmaceutical intermediates and other fine chemicals) are accompanied by side reactions producing undesired byproducts. Creation of byproducts decreases the yield of desired reactions and requires costly product separation. Even when the desired reaction is very fast comparing to undesired ones, its rate can be controlled by mass transfer, and the final selectivity can result then from competition between the desired, mass transfer controlled chemical reaction and the other, slower reactions. Hence, to control selectivity of complex chemical reactions one needs not only to have detailed information on kinetics and thermodynamics of chemical reactions but also understand and interpret effects of the flow, mass transfer and following it micro-mixing on the course of chemical reactions.

In this work we consider the case of fast mass transfer from droplets shrinking due to fast dissolution in the continuous aqueous phase. The dispersed organic phase consists of diisopropyl ether and p-Toluenesulfonic acid (pTsOH) as acidic reactant in the system of competitive test reactions being modification of the set of test reactions presented in [1].

$$NaOH + pTsOH \rightarrow H_2O + pTsONa$$

$$H_3CC(OCH_3)_2CH_3 + H_2O \xrightarrow{H^+} (CH_3)_2CO + 2CH_3OH$$

The first reaction can be treated as instantaneous, the second reaction is catalyzed by hydrogen ions H^+ and can be treated as fast. Ethanol is used to increase solubility of pTsOH in the ether. The continuous, aqueous phase contains 2,2-dimethoxypropane (DMP), NaOH and ethanol that is used as an internal standard for chromatographic analysis (GC). Experiments were carried out in the rotor-stator mixer. The rotor-stator mixers are characterized by high energetic efficiency due to a focused delivery of energy to the active high-shear regions that occupy very small fraction of internal mixer space [2,3].

To interpret experimental data a model combining the mass transfer model with the micromixing engulfment model is applied.

The authors acknowledge the financial support from Polish National Science Centre (Grant agreement number: DEC-2013/11/B/ST8/00258).

Literature:

1. Baldyga J., J.R. Bourne J.R., and Walker B., (1998) Non-isothermal micromixing in turbulent liquids: Theory and experiment. The Canadian Journal of Chemical Engineering, 76(3):641{649, 1998.
2. Jasińska M., Bałdyga J., Cooke M., Kowalski A., Mass transfer and chemical test reactions in the continuous-flow rotor-stator mixer. Proceedings of the 15th European Conference on Mixing, June 28-July 3, 2015, Saint Petersburg, Russia, 169-174.
3. Jasińska M., Bałdyga J., Cooke M., Kowalski A., (2013) Investigations of mass transfer with chemical reactions in two-phase liquid-liquid systems, Chem. Eng. Res. Des., 91, 2169-2178.

Influence of buffer and acid choice in the Villermaux-Dushman method

C. Baqueiro[1], N. Ibaseta[1], P. Guichardon[1], , L. Falk[2]
[1]Aix-Marseille Université, CNRS, Centrale Marseille, M2P2 UMR 7340, Technopole de Chateau-Gombert, 38 rue Frédéric Joliot-Curie, 13451, Marseille, Cedex 20, France; pierrette.guichardon@centrale-marseille.fr
[2]Université de Lorraine, Laboratoire Réactions et Génie des Procédés (LRGP), UMR 7274 du CNRS, 1, rue Grandville BP 451 F-54001 Nancy, Cedex, France

The Villermaux-Dushman method, originally developped in Nancy (Fournier et al, 1996; Guichardon and Falk, 2000), has become the reference method for micromixing characterization. It is based on a competitive parallel reaction system. The first reaction is the instantaneous neutralisation of dihydrogenoborate ions, whereas the second one is a fast (but slower) reaction: the comproportionation of iodide and iodate ions to iodine (the so-called Dushman reaction). In perfect micromixing conditions, only the neutralisation reaction will take place, if the acid is the limiting reagent. On the other side, in poor micromixing conditions, the dihydrogenoborate will be locally consumed without completely consumming the acid, and the second reaction can take place. Hence, the micromixing can be characterized by the iodine generation. A micromixedness ratio can be then calculated (Guichardon and Falk, 2000), with values included between 0 (no micromixing) and $+\infty$ (perfect micromixing).

Since then, questions have arised concercing the choice of the acid and the buffer solution. The original method used sulfuric acid. The micromixedness ratio was calculated by considering that the sulfuric acid was completely dissociated into sulfate anions and protons. Kölbl et al (2010) pointed out that sulfuric acid concentrations were too high to have a complete acid dissociation, and proposed the replacement of sulfuric acid by perchloric acid. Moreover, Pinot et al (2014), pointed out the health risk associated with boric acid handling, and proposed the use of a phosphate buffer instead of the borate buffer.
The present work concerns the comparison of the results obtained depending on the choice of both the acid and the buffer solution. Experiments have been carried out in a standard vessel of 1 L with two baffles, stirred by an inclined blade turbine. 2×10^{-3} L of an acid has been added to 1 L of a mixture of iodate, iodide and buffer ions at injection point <2> as described in Guichardon and Falk 2000. The choosen feed flow rate (equal to 40×10^{-3} L/s) was slow enough for reaching perfect macromixing conditions. The micromixedness ratio, α, has been calculated from spectrophotometric absorption at 353 nm. Table 1 shows the reactants concentrations for each set of experiments.

Table 1 : Reactants concentrations

Experiment set	$[H_2SO_4]$ mol/L	$[HClO_4]$ mol/L	$[H_2BO_3^-]$ mol/L	$[HPO_4^{-2}]$ mol/L	$[Na_2SO_4]$ mol/L	[KI] mol/L	$[KIO_3]$ mol/L
1	0.5	-	0.1	-	-	0.011	0.00233
2	0.5	-	-	0.1	-		
3	-	1	0.1	-	-		
4	-	0.5	0.1	-	-		
5	-	1	-	0.1	-		
6	-	1	0.1	-	0.5		
7	-	1	-	0.1	0.5		

Figure 1 shows the results that have been obtained. When comparing the results obtained by using the phosphate and the borate buffer with the same acid (experiments 1 and 2 or experiments 3 and 5), it can be obviously seen that phosphate buffer leads to higher values of the micromixedness ratio, for a same stirring speed, injection time and injection position. If these two last parameters are kept constant, the micromixing time should also be constant. Hence, any difference in the micromixedness ratio may be attributed to a different reaction kinetics. It may then be deduced that the Dushman reaction may be faster in the presence of the borate buffer. Estimation of the micromixing time by using a phenomenological model (such as the incorporation model, Fournier et al 1996b), would then require a kinetic model which takes into account an eventual buffer catalytical effect (Schmitz, 1999). The influence of acid concentration is well brought out comparing experiments 1 and 3 or experiments 2 and 5 where different results are obtained because of a partial acid sulfuric dissociation. In contrary, adding sodium sulfate salt to perchloric acid allows to obtain same micromixedness ratio as those obtained with sulfuric acid (comparing experiments 1 and 6 or experiments 2 and 7). Indeed, in a such acid medium (pH<2) sulfate ion behaves as a weak base.

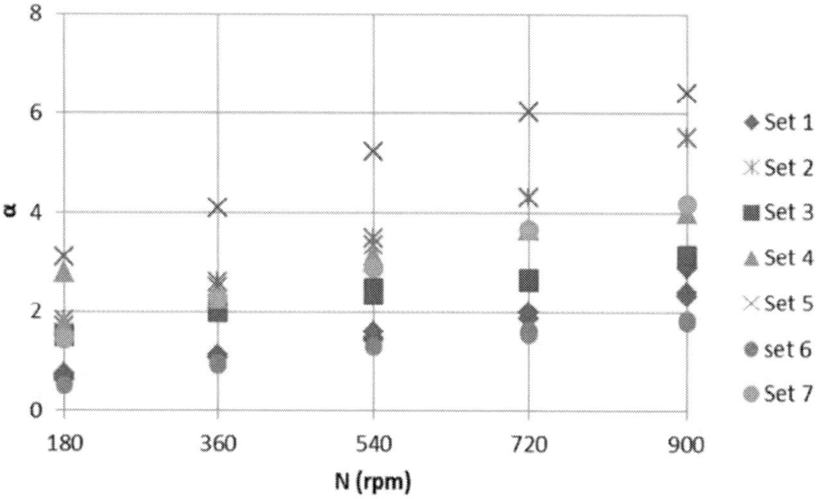

Figure 1: Micromixedness ratio as a function of the stirring speed

References

Fournier, M.-C., Falk, L., Villermaux, J. (1996a). A New Parallel Competing Reaction System for Assessing Micromixing Efficiency – Experimental Approach. Chem. Eng. Sci., 51, 5053

Fournier, M.-C., Falk, L., Villermaux, J. (1996b). A New Parallel Competing Reaction System for Assesing Micromixing Efficiency – Determination of Micromixing Time by a Simple Mixing Model. Chem. Eng. Sci., 51, 5187.

Guichardon, P., Falk, L. (2000). Characterisation of Micromixing Efficiency by the Iodide-iodate ReactionSystem. Part I: Experimental Procedure. Chem. Eng. Sci., 55, 4233

Kölbl, A., Schmidt-Lehr, S. (2010). The iodide iodate reaction method: The choice of the acid, Chem. Eng. Sci., 65, 1897

Pinot, J, Commenge, J.-M., Portha, J.-F., Falk, L. (2014). New protocol of the Villermaux–Dushman reaction system to characterize micromixing effect in viscous media, Chem. Eng. Sci., 118, 94

Schmitz, G. (1999). Kinetic and Mechanism of the Iodate-Iodide Reaction and Other Related Reactions. Phys. Chem. Chem. Phys., 1, 1909

Study of mixing performance of gas-liquid fluid intensified by flexible shaft stirring reactor

Zuohua Liu[1 *], Na Chen[1], You He[1], Haixian He[1], Shuang Li[1], Renlong Liu[1], Changyuan Tao [1]
Sha Zheng Street, No. 174, Shapingba District, Chongqing,China;
tel. 15922926287, e-mail: liuzuohua@cqu.edu.cn, [1] School of Chemistry and Chemical Engineering, Chongqing University, Chongqing 400044, China.

Gas liquid dispersion mixing operation is to pass into a mixing tank of gas dispersion, increase in unit volume of liquid and two phase's contact area, which can improve the mass transfer rate and accelerate reaction rate. The reasonable design of the mixing device can help the fluid to form suitable flow state, gas evenly and fine dispersion, improve the contact area of the two phase, and improve the mixing efficiency. The largest Lyapunov exponent (LLE) was calculated with Matlab and combined with image contrast analysis to analyze the chaotic characteristic of the fluid in stirred tank. Results showed that the maximum Lyapunov exponent of the flexible shaft stirred reactor is higher than that of the rigid shaft stirred, and the gas liquid mixing effect of the flexible shaft stirred reactor is better. The effect of the flexible shaft mixer in different air inlet positions on the gas liquid chaotic mixing behavior in the slot was compared with the image contrast experiment. And the flexible shaft stirred reactor with the center air inlet enhanced the macro mixing behavior of the gas liquid greatly. When the flexible axis system of A/B in 2:1 is more favorable for the hybrid of gas liquid chaotic mixture. With the increase of air intake, the maximum Lyapunov index increased. However, when the air intake was too large, flooding phenomena was heavily, which reduced the effect of gas liquid mixing greatly.

Volumetric mass transfer coefficient prediction in viscous liquid in mechanically agitated fermenters

R. Petříček, L. Labík, T. Moucha
University of Chemistry and Technology, Prague, Chemical Engineering Dep., Technická 5, 166 28 Praha 6, Czech Republic; tel: +420220443297, e-mail: libor.labik@vscht.cz

In the industrial fermentation processes, most liquids are non-coalescent ones, which often exhibit increased viscosity. However, due to the limitations of most measurement methods, there is a lack of reliable data for predicting volumetric mass transfer coefficients (k_La) for non-coalescent viscous batches, especially under high dissipated energies. Our goal is to supply accurate k_La data measured using reliable experimental technique - Dynamic Pressure Method (DPM). This method was modified for the measurement in viscous liquids. Dissolved oxygen (DO) probes were placed in a bypass measuring cells. This set up brings well defined transient characteristics of DO probes, which is crucial for the correct k_La evaluation. Measurements were conducted in a two phase multiple-impeller fermenter with a non-coalescent viscous Newtonian batch under a wide range of experimental conditions and in the apparatuses of two scales. Two types of DO probes (polarographic and optical) were used for measurements. Using pure oxygen as gas phase, it was confirmed that DPM is independent of driving force of absorption. The improved set up of DPM enabled to use optical DO probes. It was confirmed that optical DO probe can be used for k_La values up to 0.4 s^{-1}. Based on the experimental data, reliable correlations were developed to predict k_La in industrial fermenters. Classic correlation, based on gassed power input P_g and superficial gas velocity v_s, has low standard deviation but it is not scale specific, see separate zones of the data in fig. 1 for the vessels of i.d. 30 cm (T30) and 60 cm (T60), respectively. When the term of impeller tip speed (ND) is used instead of P_g, predicted data do not exhibit neither over- nor under-estimation of k_La for particular apparatus scale. The aggregation of the data even if they were measured using different impeller types shows the impeller power number P_o as a reliable predictor of k_La in a common correlation for most impeller types, when used together with the impeller tip speed term, see fig. 2. The correlations suggested above can be used for fairly accurate design of industrial fermenters or for the optimization of their performance under a wide range of operational conditions.

$k_La = 2.04\cdot10^{-3}(P_g/V_L + v_s\rho_L g)^{0.935} v_s^{0.553}$, $SD = 28\%$

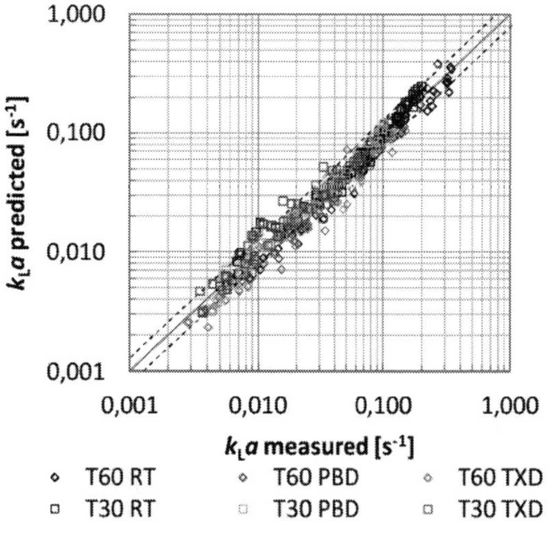

$k_La = 0.402(ND)^{2.08} v_s^{0.521} P_o^{0.689}$, $SD = 31\%$

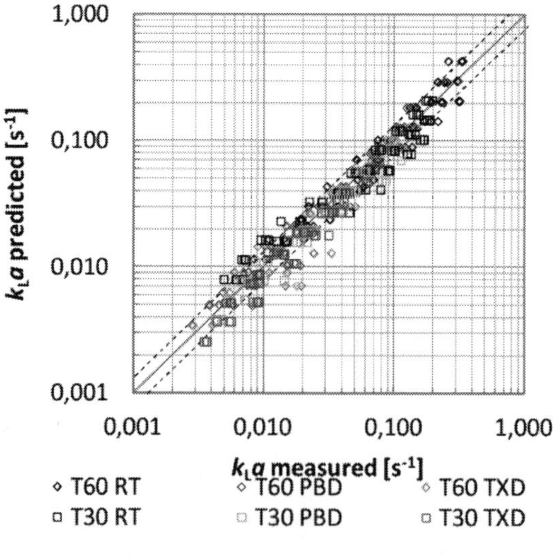

Mass transfer from small droplets suspended in turbulent flow

M. Jasińska, J. Bałdyga
Warsaw University of Technology, Faculty of Chemical and Process Engineering,
ul. Waryńskiego 1, 00 645 Warsaw, Poland;
tel. +48 22 2346319, e-mail: M.Jasinska@ichip.pw.edu.pl, fax: +48 22 8256037.

Objectives of this paper are related to identification of mass transfer mechanism between continuous and dispersed phase in two-phase liquid-liquid systems as well as development of related mathematical models. There are of course many models describing mass transfer between droplets and the continuous phase, starting from the classical models, such as the film and penetration models. The models based on physics of mass transfer phenomena were developed by Levich [7] and Batchelor [2] in sixties and seventies years of the 20th century, and were further developed by other researchers, see for example [3]. In these models in the case of mobile surface the Sherwood number is proportional to the 1/2 power of the Pe'clet number even for infinitely high viscosity of the dispersed phase, whereas for the solid particles the power on Pe'clet number is equal to 1/3. This problem was discussed in previous publications of present authors [1, 6] and some first models were developed.

The model presented in this work describes relations between fluid deformation, surface mobility, molecular diffusion and mass transfer. The developed model has a form that enables its implementation to the CFD code to account for mass transfer between the drops and continuous phase in turbulent flows, which can lead to practical applications described briefly below.

The problem of practical importance considered in this work is modelling of effects of mass transfer on the course of simple and complex chemical reactions. Usually the classical film and penetration models are applied for this purpose. Hence, the second objective of this work is application of the model described above to investigate effects of model structure including fluid deformation and surface mobility on selectivity of complex reactions. Selectivity can be considered as a good indicator of mass transfer mechanism; on the other hand the problem of selectivity is very important from industrial and environmental points of view.

Model predictions will be compared with experimental data published recently by the authors of this work [4, 5]. In experiments a system of test reactions based on a set of two parallel reactions was used, the first of them being instantaneous and the second one fast relative to mixing and mass transfer. Two of reactants, benzoic acid (B) and ethyl chloroacetate (C), initially dissolved in toluene were transferred from a dispersed, organic phase to the continuous aqueous phase, where they react with the same third reactant sodium hydroxide (A). The product distribution of this set of parallel chemical reactions expressed by reaction selectivity is a good measure of a competition between reactions, mixing and mass transfer and can be used to identify energetic efficiency of mass transfer [4, 5].

The authors acknowledge the financial support from Polish National Science Centre (Grant agreement number: DEC-2013/11/B/ST8/00258).

Literature:

1. Bałdyga J., Jasińska M., Effect of model structure on complex liquid-liquid heterogeneous reactions, Proceedings of the third European Process intensification Conference, EPIC 2011, 20-23 June 2011, Manchester, UK, CD, 175-181.
2. Batchelor G. K., (1980). Mass transfer from small particles suspended in turbulent flow, J. Fluid. Mech., 98 (3), 609-623.
3. Favelukis M., (2010), Mass transfer around oblate spheroidal drops at low Reynolds numbers. Chemical Engineering Science 65, 3808–3813.
4. Jasińska M., Bałdyga J., Cooke M., Kowalski A., Mass transfer and chemical test reactions in the continuous-flow rotor-stator mixer. Proceedings of the 15th European Conference on Mixing, June 28-July 3, 2015, Saint Petersburg, Russia, 169-174.
5. Jasińska M., Bałdyga J., Cooke M., Kowalski A., (2013) Investigations of mass transfer with chemical reactions in two-phase liquid-liquid systems, Chem. Eng. Res. Des., 91, 2169-2178.
6. Jasińska M., Lewandowski P., Bałdyga J., (2013) Nowy model wnikania masy z reakcją chemiczną w układach heterofazowych ciecz- ciecz, Inżynieria i Aparatura Chemiczna, 52(4), (2013) 325-327.
7. Levich V.G., Physical hydrodynamics, Prentice-Hall, Englewood Cliffs, N.J., 1962.

CFD study of the maximum effective viscous dissipation for powder dispersion processes induced by a ring-style high shear impeller

J. Ramirez-Muñoz[*,1], G. Martínez de Jesús[1], A. D. De la Concha[1], A. Alonso[2]
[1]Universidad Autonoma Metropolitana-Azcapotzalco, Departamento de Energia. Av. San Pablo 180, C.P. 02200, Mexico City, Mexico. [2]CONACYT-Universidad Autónoma Metropolitana-Azcapotzalco. Av. San Pablo 180, Col. Reynosa Tamaulipas, Azcapotzalco, D.F. 02200. *jrm@correo.azc.uam.mx.

Abstract

A ring-style high shear impeller (HSI) Hockmeyer® type of two rings (HockD-2R) of practical importance (Hockmeyer® Equipment Corp. D-Blade), was investigated by using the computational fluid dynamics (CFD) Software Ansys-Fluent® 15.0. The mesh volume inside and near the impeller swept volume was divided in six volume regions. Since the maximum viscous dissipation values induced by the impeller are closely related with the ultimate particle size in actual pigment dispersion processes, these values were calculated from simulations. In the range of operating conditions evaluated in this work, it was found that, the maximum averaged viscous dissipation values are located outside the ring surfaces and in the volume swept by the impeller blades. Whilst, at the same power input, the viscous dissipation in the impeller blades swept volume increases drastically as the fluid viscosity decreases.

Introduction

HSIs are mixing devices typically mounted in a single shaft and are commonly used in the coating industries for dispersing and de-agglomerating pigment solids into liquids [1]. Small pigment particles may agglomerate due to their high surface contact area resulting in high surface energy, which causes particles form agglomerates. For the mechanical breakdown of these associated particles, it is necessary to create a zone of intense shear. To this purpose, the HockD-2R HSI is commonly used in order to rapidly break lumps of powdery material. The HockD-2R is claimed to be a HSI since it creates a high shear stress zone (or high energy dissipation) with intense flow in the gap between the rings [2,3] (see Fig. 1). Despite the practical importance of high shear impellers (HSIs), operating in the laminar flow regime, literature information on their performance is scarce and the knowledge of their hydrodynamic from fundamental principles is limited. In this work, the viscous dissipation induced by a HockD-2R HSI in an unbaffled tank was studied at *Re* of 36 and 110.

Methodology

Fig. 1 shows the ring-style Hockmeyer® impeller and Fig. 2 depicts the computational grid. It can be seen that the computational grid near the impeller was segmented in several volumes. Grid refinement in the region near the walls and the impeller can be appreciated. Due to its rotationally periodic nature, only one-sixth of the tank containing the impeller was considered as a solution domain and periodic boundary conditions were applied. Other details on simulations can be found in a previous work from our team group [3]. Simulations were performed in a cylindrical tank with dished bottom and inner diameter T= 132 mm. The dimensionless geometric ratios corresponding to the impeller off-bottom clearance (C), liquid height (Z) and impeller diameter (D) were C/T=0.3848, Z/T=1 and D/T=0.3848, respectively. The fluid viscosity and density were μ = 0.945 Pa s and ρ = 1346 kg m^{-3} (for Re=110), and μ = 2.47 Pa s and ρ = 1366 kg m^{-3} (for Re=36), respectively.

Fig. 1. High shear impeller of two rings Hockmeyer® type.

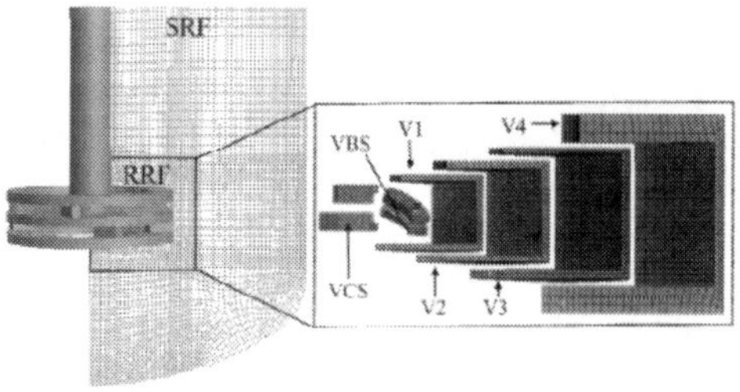

Fig. 2. Computational grid and different volume regions used in CFD simulations.

Results and discussions

Averaged viscous dissipation values normalized with the mean viscous dissipation in the entire fluid volume, were calculated from its definition for two fluid viscosity values [3] at a power input of 36 *W*. These values were evaluated in different regions near the impeller and are shown in Fig. 3. As can be seen, at μ=2.47 Pa s, the maximum averages viscous dissipations are located in V1 and V2, however, at μ=0.945 Pa s, the viscous dissipation in VBS increases drastically and the maximum induced values are located in VBS and V1, being these values are very alike. Considering that the maximum viscous dissipation is a key mixing parameter which has been related experimentally to the breakage of nanoparticle clusters [3], it is likely that for μ=0.945 Pa, the ultimate particle size due to the breakage of nanoparticle agglomerates is carried out in the VBS volume region, and in de V1 region, as well.

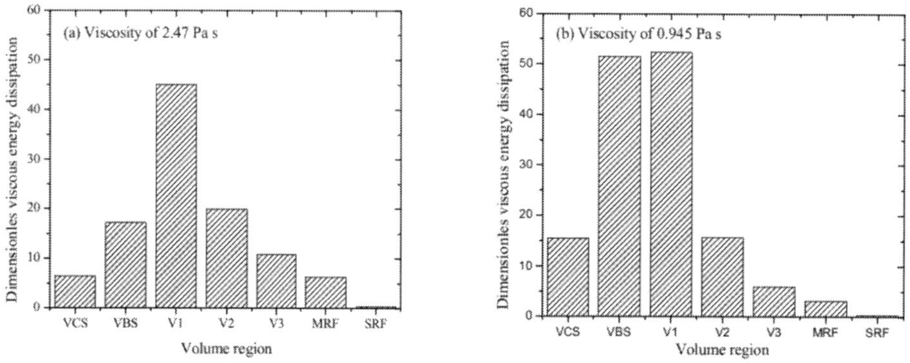

Fig. 3. Viscous dissipation in different volumes shown in Fig. 2, at power input of 36 *W*, (a) μ = 2.47 Pa s and (b) μ = 0.945 Pa s

Conclusions

Averaged viscous dissipation values obtained from simulations, were used as indicators of a ring style HSI impeller performance. It was found that at power inputs of 36 *W* or higher and fluid viscosity of 0.945 Pa s, the impeller blades swept volume might be considered as effective region for pigment dispersion processes. This is due to its higher effective viscous dissipation produced inside the impeller blades swept volume.

References

[1] T.C. Patton, Paint Flow and Pigment Dispersion, 2nd ed., John Wiley & Sons, New Jersey, (1979).

[2] H. Hockmeyer, A Practical Guide to High-Speed Dispersion, Paint Coat. Ind., 26 (2010) 32 - 36.

[3] R. Ramírez-Gómez, D. García-Cortés, G. Martínez-de Jesús, M.M. González- Brambila, A. Alonso, S.A. Martínez, J. Ramírez-Muñoz, Performance evaluation of two high-shear impellers in an unbaffled stirred tank, Chem. Eng. Technol., 38 (2015) 1519 - 1529.

Experimental investigation of impeller position effect on nucleation kinetics of borax decahydrate in dual impeller cooling crystallizer

A. Kaćunić, M. Ćosić, N. Kuzmanić
University of Split, Faculty of Chemistry and Technology, Department of Chemical Engineering, Ruđera Boškovića 35, 21000 Split, Croatia;
tel. +385 21 329-453, e-mail: akrap@ktf-split.hr.

The production process of crystalline products is governed by two kinetic steps - nucleation and crystal growth. Nucleation kinetics is highly correlated to the width of the metastable zone and with regard to product characteristics it usually has a strong influence on crystal growth. Considering the fact that batch crystallization is usually carried out in suspension, mixing can affect the width of the metastability region, nucleation kinetics and, finally, the properties of the product which include crystal size distribution, purity, morphology, etc. Thus, for a complete understanding of the crystallization process, besides a rigorous description of the crystallization phenomena, the hydrodynamics of the system needs to be analyzed in detail [1,2]. The majority of the researches dealing with the influence of mixing on crystallization are conducted in a single impeller crystallizer, which is common in laboratory scale investigations, but on an industrial scale, multiple impeller crystallizers are usually employed. It is well known that hydrodynamic conditions in stirred reactor depend on the impeller and crystallizer geometry. The flow fields in these systems are very complex, with vortical structures that are not always well defined and may significantly change with impeller position. The flow complexity greatly increases if a configuration with two impellers is used, especially when applied impellers are of the different type, each having its own flow characteristics.

The aim of this work was to investigate the influence of impeller position on the nucleation mechanism and nucleation rate of disodium tetraborate decahydrate (borax) in a dual impeller batch cooling crystallizer. Examinations were performed in a crystallizer of 15 dm^3 with a constant impeller to tank diameter ratio (D/d_T) of 0.33 and an aspect ratio (H/d_T) of 1.3 which required the addition of a second impeller. In this research, two different types of impellers were mounted on the same shaft: axial down pumping pitched blade turbine (PBT) was used as a lower impeller and radial straight blade turbine (SBT) was used as an upper impeller. Mixing was performed at the speed which ensured the state of complete suspension in the system (N=N_{JS}). In order to examine different impeller positions, experiments were carried out at different impeller off-bottom clearances - c/D, in the range of 0.2 to1.5, as well as at different spacing between the impellers - s/D, in the range 0 to1.5. Crystallization was carried out by a controlled cooling of the solution, which was saturated at 30°C, at the rate of 6 °C h^{-1}. Concentration of the solution was measured in line by the means of *Na* ion-selective electrode, while the onset of nucleation, i.e. the metastable zone width, was detected from the change in concentration values measured over process time. In order to gain a complete insight into the intensity and structure of overall fluid flow in the crystallizer the mixing time was determined by potentiometric method and the photographs of the flows were taken. For all examined positions, during the experiments, torque measurements were performed and power consumption per unit mass was calculated.

From the obtained results of metastable zone width, at all examined mixing conditions according to the Mersmann's nucleation criterion, nucleation in the crystallizer took place by a heterogeneous nucleation mechanism. The values of heterogeneous nucleation rate were calculated by the expression proposed by Mersmann [3].

With regards to the impeller position in the crystallizer, results show that when impeller off-bottom clearance increases in examined range, nucleation rate increases almost by three times. Although the nucleation rate changes significantly, it is not a direct consequence of the impeller clearance but rather of the value of Reynolds number (Re_{JS}). As stated earlier, mother liquor mixing was performed at just suspended impeller speed, which drastically increased with an impeller off-bottom clearance due to the change in the overall fluid flow pattern. A higher value of N_{JS}, resulted in an increase of turbulence in the system, what reflected on the metastable zone width and finally on nucleation rate.

When mixing was performed at different impeller spacing, a regular correlation between the values of the nucleation rate and impeller spacing was not found. The lowest value of nucleation rate was obtained at the lowest and the highest examined value of impeller spacing. At minimum impeller spacing, impellers acted as a single impeller but with a changed geometry, with twice the width of impeller blade. On the other hand, when impellers were set apart at position s/D=1.5, completely different hydrodynamic conditions, when compared to the ones at s/D=0, resulted in almost the same value of nucleation rate. To gain insight into intensity of the overall fluid flow at examined mixing conditions systems mixing time was determined as well. It was interesting to notice that the values of mixing time corresponded with the values of metastable zone widths, thus indicating their dependence on the overall fluid flow pattern. Obviously, the obtained results could only be explained if interactions between the flows generated by individual impellers are clearly understood. For that reason the photographs of the flows were taken. This way, it was possible to identify the region in the overall fluid flow structure, in which an interaction of flows, generated by each of the impellers, occurred. By taking into consideration the values of nucleation rate, mixing time, power consumption and overall fluid flow pattern, it is possible to define the impact of impeller position on nucleation kinetics in examined dual impeller system.

Acknowledgments
This research work is supported by the Croatian Science Foundation and is a part of the HETMIX Project.

REFERENCES:

J. Urlich, C. Striege, Some Aspects of Importance of Metastable Zone Width and Nucleation in Industrial Crystallizers, J. Crystal Growth 237-239 (2002) 2130-2135.

E. L. Paul, V. A. Atiemo-Obeng, S. Kresta, Handbook of Industrial Mixing, John Wiley and Sons, Inc., Hoboken, New Jersey, 2004.

A. Mersmann, Supersaturation and Nucleation, Trans IChemE 74 (Part A) (1996) 812-820.

Distribution of the turbulent kinetic dissipation rate in an agitated vessel

B. Kysela[1], R. Sulc[2], J. Konfrst[1], Z. Chara[1], I. Fort[2], P. Ditl[2]
[1]Institute of Hydrodynamics, AS CR, v.v.i., Pod Patankou 30/5, 166 12 Prague 6, Czech Republic; [2]Department of Process Engineering, Faculty of Mechanical Enginnering, Czech Technical University in Prague, Technicka 4, 166 07 Prague 6, Czech Republic.

Introduction

The design of the agitated tanks depends on the proposed operating conditions and processes for that they are used for. Namely dissipation rate of the turbulent kinetic energy is important parameter for the scale-up modelling. The dissipation rate is commonly determined as integral value based on power input of the impeller, but without information about distribution inside the agitated volume. The cumulative distributions of the dissipation rate within an agitated vessel are estimated by evaluations of the CFD (Computational Fluid Dynamics) results, where the data was obtained from RANS (Reynolds Averaged Navier-Stokes equations) and LES (Large Eddy Simulations). The simulations were performed for an agitated vessel equipped with four baffles and stirred by a standard Rushton turbine (tank diameter 0.3 m, impeller diameter 0.1 m, off-bottom clearance half of tank diameter, impeller speed 200 rpm). The values of the dissipation rate from the LES calculations were approximated by computing the SGS (Sub Grid Scale) dissipation rate.

Calculations

The pressure-based solver pimpleDyMFoam from OpenFOAM 2.3.1 package (for incompressible flow) was used for calculations in agitated vessel. The impeller movement was simulated by SM (Sliding Mesh) method. Turbulence was modelled by LES (Large Eddy Simulation) approach, with Smagorinsky SGS model and standard k-epsilon model. The final non-structured consist of 3.3 mil. (mainly hexagonal) cells. Agitated tank walls with bottom, baffles, impeller, shaft and a lit eliminating aeration effects were set to the no slip condition. The part of the impeller shaft outside the sliding region was defined as a moving wall with impeller speed velocity. The time step was 0.0005 s that corresponds to 0.6° of the impeller movement.

Fig.1. ISO volumes of dissipation rate of TKE > 8 m^2/s^3 evaluated from k-epsilon (left) and LES (right).

Results

The dissipated energy was evaluated in agitated vessel, where from cell values and their volumes the energy was calculated and sorted by epsilon values and consequently the cumulative distribution was calculated.

The results show that the energy is dissipated mainly in impeller vicinity and it also depends on formation of trilling vortices as it is shown in fig 1. The depicted iso-surfaces also indicate more smooth results for RANS calculations because the averaged velocity field probably affects the calculation of other derivatives.

Fig. 2. Comparison of the cumulative distributions of the dissipated energy in agitated vessel evaluated from RANS k-epsilon model and from LES approach.

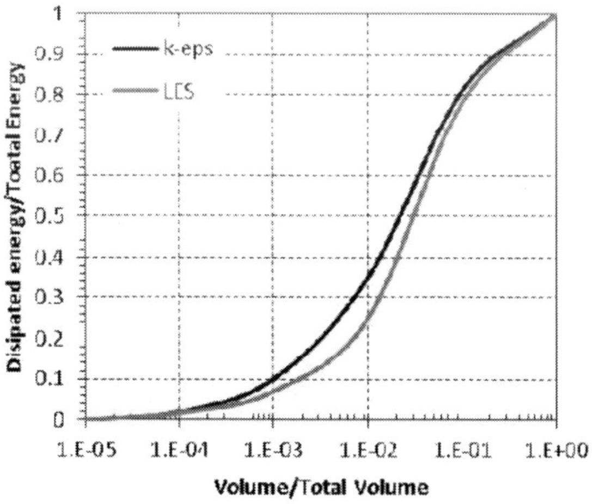

The distribution curves fig 2 are in agreement with experimental results where the main working volume is determined as 1/10 of the total volume and in vicinity of impeller. In our case the dissipated energy is there around 80 % and around 30 % for 1/100 of the total volume.

Conclusions

The cumulative distributions of dissipation rate of turbulent kinetic energy were evaluated from results of k-epsilon turbulence model and LES. The results indicate the distortion of results probably affected by using of RANS models.

The evaluated distribution curve indicates that around 80 % of energy is dissipated in 1/10 of vessel volume.

Acknowledgements: This research has been subsidized by the research project No. GA CR 16-20175S and RVO: 67985874.

Driss, Z., Bouzgarrou, G., Chtourou, W., Kchaou, H., Abid, M.S.: Computational studies of the pitched blade turbines design effect on the stirred tank flow characteristics, European Journal of Mechanics - B/Fluids 29(3), 236–245, 2010.

Ducci, A., Yianneskis, M.: Turbulence kinetic energy transport processes in the impeller stream of stirred vessels, Chemical Engineering Science 61(9), 2780–2790, 2006.

Huchet, F., Liné, A., Morchain, J.: Evaluation of local kinetic energy dissipation rate in the impeller stream of a Rushton turbine by time-resolved PIV, Chemical Engineering Research and Design 87(4), 369–376, 2009.

Sheng, J., Meng, H., Fox, R.O.: A large eddy PIV method for turbulence dissipation rate estimation, Chemical Engineering Science 55(20), 4423–4434, 2000.

Soos, M., Kaufmann, R., Winteler, R., Kroupa, M., Lüthi, B.: Determination of maximum turbulent energy dissipation rate generated by a Rushton impeller through large eddy simulation, AIChE Journal 59(10), 3642–3658, 2013.

Rotary sloshing instabilities in unbaffled stirred vessels

Antonio Busciglio[1], Tomáš Kracic[2], Tomáš Moucha[2], Giuseppina Montante[1], Alessandro Paglianti[3]
[1]Dipartimento di Chimica Industriale "Toso Montanari", Alma Mater Studiorum – Università di Bologna, via Terracini 34, 40131 Bologna, Italy
[2]Depatment of Chemical Engineering, University of Chemistry and Technology in Prague Technická 3, 166 28 Praha 6 – Dejvice, Czech Republic
[3]Dipartimento di Ingegneria Civile, Chimica, Ambientale e dei Materiali, Alma Mater Studiorum – Università di Bologna, via Terracini 34, 40131 Bologna, Italy

Unbaffled stirred tanks are less often employed in the process industry with respect to baffled tanks due to their poorer mixing performances. Nevertheless, they may bring advantages in several applications, such as pharmaceutical and food processes, in the production of biofuels, in bioreactors for the growth of shear sensitive cells among the others (Scargiali et al., 2013). They are always characterized by a highly swirling fluid motion and, unless they are completely filled and closed with a lid, a surface vortex, whose depth depends on the impeller rotational speed.

It was recently shown that rotating waves may spontaneously develop in unbaffled partially filled stirred vessels, regardless of the system axial symmetry (Busciglio et al., 2016; Pieralisi et al., 2016). These rotating waves give rise to periodic stresses that may lead in turn to mechanical failures or to spills and overflows.

In this contribution, an experimental study on the onset and the characterization of free-surface periodic oscillations in an unbaffled stirred tank is carried out. In particular, the investigated geometry resembles an industrial digester at lab scale. It corresponds to a partially filled tank of low height-to tank diameter ratio, due to the requirement of a free space for the biogas accumulation over the top. The oscillation frequency of the free liquid surface is experimentally evaluated by both visual observations and measurements of the local dynamic pressure time traces. The oscillation frequency is identified from the time traces by means of simple Fast Fourier Transform.

Different modes of oscillation of the liquid free surface are observed, depending on impeller type, fill ratio and fluid viscosity. In all cases, the oscillation frequency is found to increase linearly with the impeller speed, with a slope that depends on the impeller geometry and on the oscillation mode under analysis.

An interpretative model based on the theory of liquid sloshing dynamics is proposed.

References

A. Busciglio, F. Scargiali, F. Grisafi, A. Brucato, Oscillation dynamics of free vortex surface in uncovered unbaffled stirred vessels, Chem. Eng. J., 285 (2016) 477-486.

F. Scargiali, A. Busciglio, F. Grisafi, A. Tamburini, G. Micale, A. Brucato, A. Power consumption in uncovered unbaffled stirred tanks: Influence of the viscosity and flow regime, Ind. Eng. Chem. Res., 52 (2013) 14998-15005.

I. Pieralisi, G. Montante, A. Paglianti, Prediction of fluid dynamic instabilities of low liquid height-to-tank diameter ratio stirred tanks Chem. Eng. J. 295 (2016) 336-346.

Axial solid concetration profile in an agitated vessel

M. Ayas, T.Jirout
Czech Technical University in Prague Faculty of Mechanical Engineering Department of Process Engineering Technická 4 166 07 Praha 6 – Dejvice Czech Republic. Phone: ++420-2-2435 2551, e-mail: Mehmet.Ayas@fs.cvut.cz

Abstract
This study deals with investigation of axial solid-liquid concentration profile in an agitated vessel. First, solid concentration distribution was obtained experimentally by using sampling method and then numerical model was created in Fluent by using multiphase (two-phase *k*-e turbulence) model. Experiment was performed in a cylindrical vessel with four baffles and with two different solid loadings .Flow was generated by using single axial flow impeller (six blade turbines, 45°). The simulated particle axial concentration profiles were compared with the experimental data.

Figure 1. Mixing Vessel

Gas hold-up prediction in viscous liquid in mechanically agitated fermenters

T. Kracík, J. Šmidrkal, L. Labík, T. Moucha,
University of Chemistry and Technology, Prague, Chemical Engineering Dep., Technická 5, 166 28 Praha 6, Czech Republic; tel: +420220443297, e-mail: libor.labik@vscht.cz

In chemical, biochemical and food processes, mechanically stirred tank reactors, fermenters are frequently used to ensure gas-liquid mass transfer. Often, a batch used in these apparatuses exhibits increased viscosity in comparison with water. Knowledge of transport characteristics is essential for their proper design, operating and optimization. One of the transport characteristics is a gas hold-up. The equations commonly used for gas hold-up estimation do not reflect the effect of increased viscosity. However, a viscous batch exhibits specific behaviour during mass transfer process. Our goal is to propose a reliable correlation for gas hold-up under a wide experimental condition for batches with increased viscosity, based on our experimental data obtained both in pilot-plant and in laboratory vessel.

The measurements were carried out in the vessels of inner diameter 0.29 and 0.59m, respectively. Different types of impellers and their combinations were used, including radial, axial, and combined liquid flow impellers. The gas hold-up was measured in a multiple-impeller vessel at different impeller frequencies and at several gas flow rates. The position of a float on the batch level was recorded.

Literature correlation equations were tested for gas hold-up. On the one hand, classic correlation equation, based on total gassed power input and superficial velocity, have a standard deviation (SD) 18% in average, when each impeller data treated separately. On the other hand, the correlation common for all impeller types give the SD = 27%. Another literature correlation (Abrardi et al., 1988) $\varepsilon_g = D_1\left[P_g(1-\varepsilon_g)\right]^{D_2} v_S^{D_3}$ shows low reliability in the case of processing data sets from two vessel sizes. We proposed a new correlation based on the impeller tip speed and superficial gas velocity, which took into account better the size of vessels. For individual impellers the SD were 13% in average. The correlation for all impeller types together gave the SD = 32%. When the impeller power number was included into the correlation, the SD decreased to 21% as is shown in the figure.

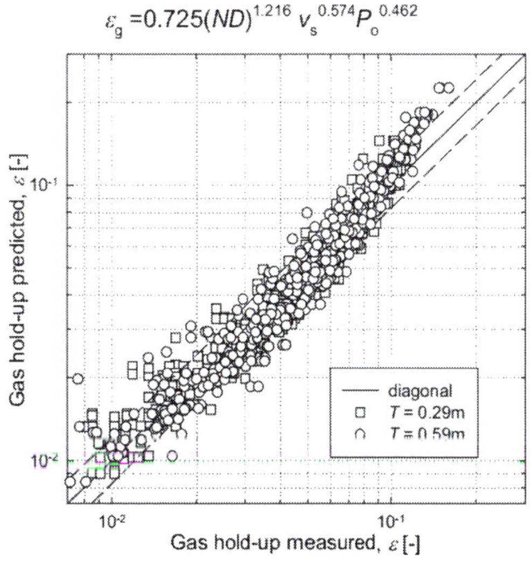

The proposed correlations can be used for impeller gas hold-up prediction in laboratory, pilot-plant and industrial scale fermenters under a wide range of operational conditions.

Abrardi, V., Rovero, G., Sicardi, S., Baldi, G., Conti, R., 1988. Sparged vessels agitated by multiple turbines. Proc. Eur. Conf. Mixing 6th, 329-336.

Gas-liquid mass transfer rates in unbaffled stirred tank

L. Labík[1], A. Brucato[2], F. Scargiali[2], T. Moucha[1]
1) University of Chemistry and Technology, Prague, Chemical Engineering Dep., Technická 5, 166 28 Praha 6, Czech Republic; tel: +420220443297, e-mail: libor.labik@vscht.cz
2) Università di Palermo, Dipartimento di Ingegneria Chimica, Gestionale, Informatica e Meccanica, Viale delle Scienze, Ed. 6, 90128 Palermo, Italy

Unbaffled stirred tanks are increasingly recognized as a viable alternative to common baffled tanks for a range of processes where the presence of baffles is undesirable for some reason. For instance, in the case of shear sensitive cell cultivation (e.g. human cells), unbaffled tanks have been recently found to be able to provide sufficient mass transfer through the free surface vortex. As a consequence the need for bubble formation and subsequent bursting, along with relevant cells damage, is conveniently avoided.

In this work the influence of impeller geometry, vessel size and viscosity of the batch on mass transfer performance an unbaffled stirred vessel operating both in sub-critical conditions (the free surface vortex has not yet reached the impeller) and in super-critical conditions (the free surface vortex has reached the impeller and a gas phase is ingested and dispersed inside the reactor) is presented.

The mass transfer performance of unbaffled systems is mainly affected by degree of turbulence rather than specific power consumption. Although the division for sub and super critical condition is suitable for description of specific power input and impeller Power number, the experimental results showed that the mass transfer intensity can be divided into four regimes. This makes the description of mass transfer intensity more complicated than for conventional baffled vessels. First regime is characterized by practically flat liquid surface and not well mixed liquid. Second regime (reg. I) is characteristic for formation of central vortex, well mixed liquid but no gas ingestion. The third regime (reg. II), is transitional regime where the gas is started to be ingested into the impeller blades area. This regime is characteristic for sharp increase of volumetric mass transfer coefficient (k_La). The fourth regime (reg III) is starting when the impeller frequency is high enough to make full gas dispergation. Then the k_La is starts to behave like in baffled stirred tanks.

Among the stirrer geometries investigated a simple PBT was found to provide the most interesting oxygen transfer performance in the sub-critical regime, and can therefore be regarded as a particularly suitable stirrer for shear sensitive cultures. In the super-critical regime, unbaffled tanks are found to provide a performance comparable with that of the standard (baffled) bioreactors, hence resulting in a viable alternative also for fermentations involving robust cells.

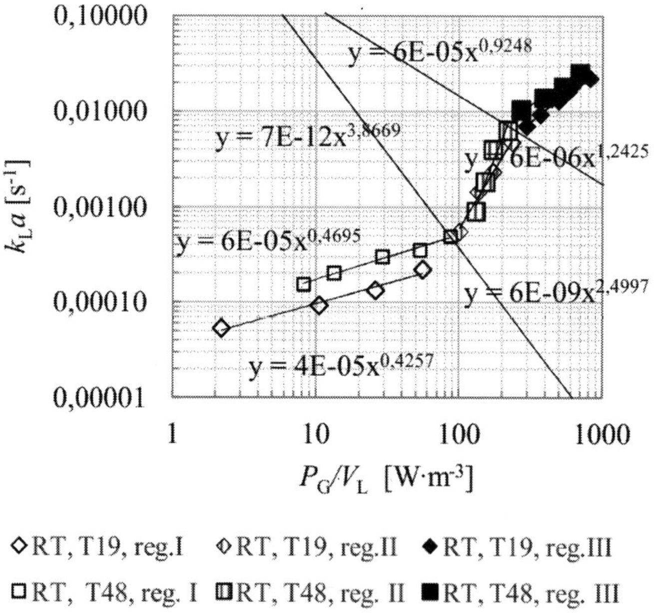

Effect of blade rake angle on dynamics and homogenization of granular system in a vertical bladed mixer

[1]T. Trávníčková, [1,2]J. Havlica, [3]M. Kohout
[1]Institute of Chemical Process Fundamentals of the CAS, v. v. i., Rozvojová 2/135, 165 02 Prague 6, Czech Republic; phone: +420 220 390 262, fax: +420 220 920 661
e-mail: travnickovat@icpf.cas.cz; [2]University of Jan Evangelista Purkinje, České mládeže 8, Ústí nad Labem, 400 96, Czech Republic; [3]Department of Chemical Engineering, University of Chemistry and Technology; Technická 5, 166 28 Prague, Czech Republic.

The granular materials are found in many areas of human exposure (food, pharmaceuticals, cosmetics, chemicals, building materials ...) and their mixing is one of the most spread industrial operations. Despite that, or perhaps because of that, granular systems appear in so many variants, there is still not a solid theoretical foundation for describing the processes taking place inside the mixing equipment. Design of industrial devices for mixing of granular systems is mostly the result of long experience or the method of trial and error. The complex dynamics of large number of particles is very difficult to monitor by using experimental methods and until recent time its description was time consuming also for mathematical simulations. With the development of computer technology, modelling of mixing granular systems using mathematical methods such as DEM (discrete element method), CFD (computational flow dynamics), or Monte Carlo method gets back to the forefront.

The mixing process was studied by using Discrete Element Method (DEM) with varying blade rake angle (30°, 45°, 60°, 75°and 90°) and blade rotational speed (from 0.1 to 960 rpm). With change of blade rake angle, the ratio between the height of the blade and bed depth changes for constant number of particles in the granular material. This ratio has a significant impact on the dynamics of the mixing process. For this reason, in addition to basic numerical experiments with constant bed depth (approximately 42 thousand particles, bed depth = 4cm), simulations were also carried out with held fixed ratio (bed depth/height of the stirrer blade = 2).

Depending on the blade rake angle and blade rotational speed the primary and secondary flows arising in a stirred granular layer, volume fraction, concentration patterns and global transport characteristics have been described. Subject of our interest is to point out the connection between the dynamics and the homogenization process that has been investigated and quantified by the Lacey mixing index (see Figure 1) for two different initial packing configurations: side-by-side and bottom-up.

In general, the side-by-side arrangement is better homogenized than axial bottom-up arrangement for all blade rakes and all blade rotational speeds. At low blade rotational speeds, the difference in homogenization by differently inclined blades is most significant. For blade rotational speed 15 RPM even agitators with 60°, 75° and 90° blade rake does not homogenize at all. The whole system rotates together with the agitator as one solid block. For higher blade rotational speed (from 150 to 960 RPM), the differences among homogenization Lacey's curves are reduced. For very high blade rotational speeds (960 RPM) the efficiency of stirring decreases and Lacey index is less than one for all blade rakes.

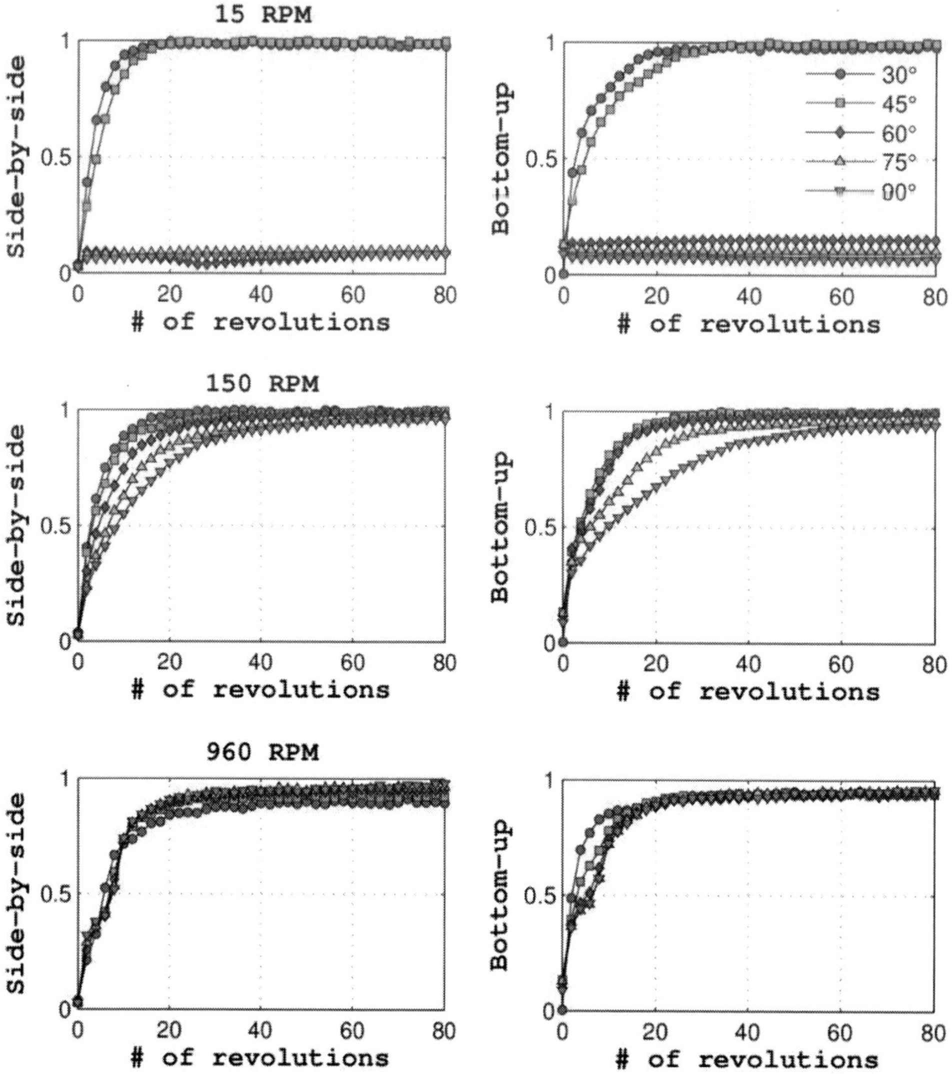

Figure 1: Lacey indexes for side-by-side and bottom-up initial packing arrangements computed for 30°, 45°, 60°, 75° and 90° blade rake and different blade rotational speeds. Fixed bed depth/height ratio of the stirrer blade used.

In our contribution we will explain the relationship between the dynamics of the mixing process and the process of homogenization. We believe that results of this comprehensive study will be helpful in optimizing the design and operation of the mixing devices.

Acknowledgements:

This research was supported by the grant GACR P105/12/0664 from the Czech Science foundation.

Assessment of the Newtonian viscosity effect on the viscous dissipation of a Hockmeyer® type impeller in an unbaffled stirred tank

[1]G. Martínez de Jesús, [1]A. De la Concha, [2]A. Alonso, [3]D. García-Cortés, [4]A. Soria, [1,*]J. Ramírez-Muñoz
[1]Departamento de Energía, Universidad Autónoma Metropolitana- Azcapotzalco. Av. San Pablo 180, Col. Reynosa Tamaulipas, Azcapotzalco, D.F. 02200, Mexico. [2]CONACYT-Universidad Autónoma Metropolitana-Azcapotzalco. Av. San Pablo 180, Col. Reynosa Tamaulipas, Azcapotzalco, D.F. 02200. [3]Departamento de Medio Ambiente, Instituto Superior de Tecnologías y Ciencias Aplicadas, Avenida Salvador Allende, esq. Luaces, La Habana, 10400, Cuba. [4]Departamento de IPH, Universidad Autónoma Metropolitana-Iztapalapa, San Rafael Atlixco #186, Col. Vicentina, Iztapalapa, D.F. 09340, Mexico.
[*]jrm@correo.azc.uam.mx.

The ring-style high shear impeller (HSI), also known as *disperser blade*, Hockmeyer® type, is a mixing device frequently used in industries of paint, paper, inks, cosmetics and pharmaceutical to disperse and deagglomerate lumps of solid particles into a liquid. As indicated by the supplier, a ring style HSI is a powerful tool for optimizing disperser performance due to its patented Hockmeyer innovation [1] with double-cutting combination, in the teeth and in the external surface of the outer rings [2]. Because it runs at relatively high tip speeds, in principle, a zone of high dissipation exists around the impeller swept volume. In this zone are localized maximum energy dissipation values, which are necessary to break apart lumps of powdery material for solid-liquid dispersion processes [3]. Is supposed that the passage of the slurry through existing slots between the rings creates considerable shear on the agglomerates. Therefore, among the various applications of HSIs, a common practice in coating and paint industries is dispersing pigment slurries in unbaffled tanks. Nonetheless, in comparison with radial, axial and mixed flow impellers, the hydrodynamics induced by the high shear dispersers has been relatively little studied in the literature [4, 5].

In order to gain understanding on the performance of a HSI Hockmeyer® type of two rings (HockD-2R) of practical importance (Hockmeyer® Equipment Corp. D-Blade), its hydrodynamic performance was investigated. The study was carried out by using the Computational Fluid Dynamics software ANSYS FLUENT 15.0. Simulations were performed in an unbaffled tank with dished bottom and inner diameter T= 132 mm. The dimensionless geometric ratios corresponding to the impeller off-bottom clearance (C), liquid height (Z) and impeller diameter (D) were: C/T=0.3848, Z/T=1 and D/T=0.3848, respectively.

The study focused into evaluate the viscous dissipation, ε, as a function of Newtonian viscosity at different power input at Reynolds number (Re) between 20 and 110. The values of ε were obtained from simulations and the ratio $\overline{\varepsilon_i}/\overline{\varepsilon}$ was calculated. Where $\overline{\varepsilon_i}$ is the averaged viscous dissipation in different regions around the impeller as shown in Fig. 1, and $\overline{\varepsilon}$ is the mean viscous dissipation in the fluid whole domain. The obtained results are shown in Fig. 2. It can be observed that at the same power input and with the lowest viscosity evaluated in this study, 0.95 Pa·s (closed symbols), the regions VBS (the impeller blades swept volume) and V1 (the first region around the impeller) exhibit the highest values of $\overline{\varepsilon_i}/\overline{\varepsilon}$. At the highest power input (36 W), the values of $\overline{\varepsilon_i}/\overline{\varepsilon}$ in VBS and V1 regions are very similar. On the other hand, it can be seen that beyond of V1, the values of $\overline{\varepsilon_i}/\overline{\varepsilon}$ decrease drastically, being very similar and regardless of viscosity and power input. The results shown in this work for Newtonian viscous fluids confirm that in the region nearest to the impeller, i.e. V1 and VBS, are generated the maximum viscous dissipation, which is in agreement with the provider indicates on your website [2].

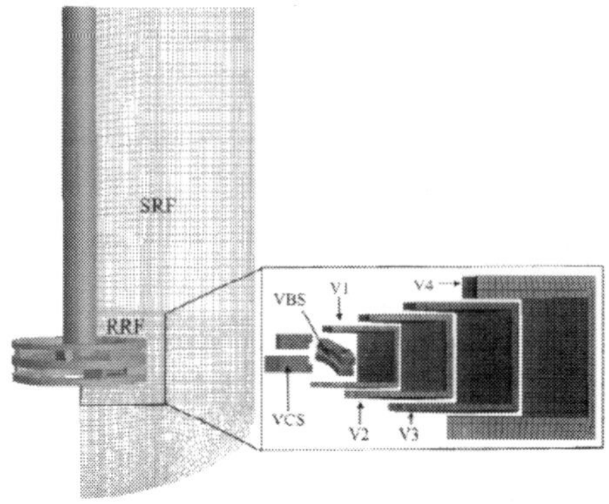

Fig. 1. Computational grid and different volume regions used in CFD simulations.

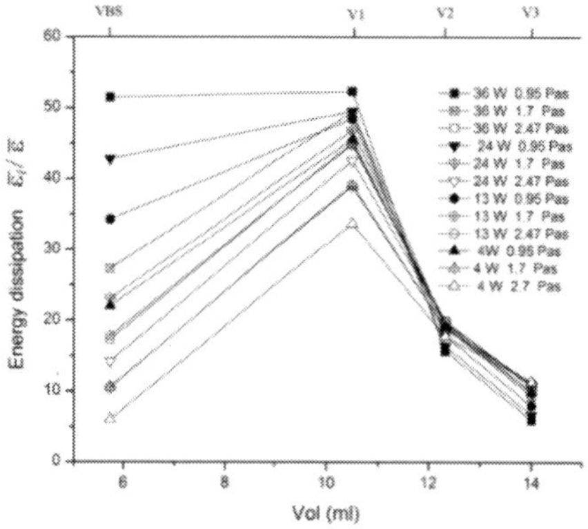

Fig. 2 Normalized averaged viscous dissipation ($\varepsilon_i / \bar{\varepsilon}$) in the regions shown in Fig. 1 as function of the Newtonian viscosity and power input.

References

[1] Klein, S., Hockmeyer, H., (Sidney Klein, 484 Duryea Terrace, West Hempstead, N.Y., and Herman Hockemeyer, 33 Hudson Terrace, Englewood Cliffs, NJ, USA), US19633071353A, 1963.

[2] Hockmeyer (2012). 'D' Style "Ring" Blade (online): Available at: http://www.hockmeyer.com/products/dispersers-agitators/disperser-blades/d-style-ring-blade-detail.html [accessed 29 Nov. 2015].

[3] Hockmeyer, H., 2010. Practical guide to high-speed dispersion. *Paint & Coatings Ind.*, 26, 32-36.

[4] Unadkat, H., Rielly, C. D., and Nagy, Z. K. 2011. PIV study of the flow field generated by a sawtooth impeller. *Chemical Engineering Science*, 66, 5374–5387.

[5] R. Ramírez-Gómez, D. García-Cortés, G. Martínez-de Jesús, M. González-Brambila, A. Alonso, J. Ramírez-Muñoz. Performance evaluation of two high shear impellers in an unbaffled stirred tank. *Chem. Eng. Technol.* 38, 1-12.

On the dynamics of gas adsorption processes for systems governed by different phenomena

J. Vargas Schmitz, [1]M. Hovestadt, M. [1]Kriesten, [1]M. Hartmann, M. Kaspereit.
Institute of Separation Science and Technology, Friedrich-Alexander University of Erlangen-Nuremberg, Egerlandstr. 3, 91058 Erlangen, Germany, tel. +49 9131 8527453, e-mail: juergen.vargas@fau.de; [1]Erlangen Catalysis Resource Center, Friedrich-Alexander University of Erlangen-Nuremberg, Egerlandstr. 3, 91058 Erlangen, Germany.

A relevant prerequisite for analysis and design of adsorption processes is the understanding of the dynamic behaviour of the process. Depending on the system and operating concept under consideration, various effects can play a significant role, in particular competitive equilibria, mass transfer resistance, axial dispersion, and heat transfer – but also control devices. The relevant effects have to be accounted for by corresponding mathematical process models.

We present a discussion of the mentioned effects on the adsorber dynamics for three experimental systems that are each governed by different phenomena. Experimental breakthrough curves measured for a variety of conditions are compared to simulation results obtained by different modelling approaches.

As a model system, the separation of N_2/CO_2 on activated carbon is studied. Adsorption isotherms are determined using a magnetic suspension balance. Breakthrough curves are measured on a single bed of a pressure swing adsorption plant (250 mL, PSA-1000) for varying composition, pressure, flow rate, and external temperature control. Figure 1 reveals that a fair to excellent agreement can be achieved depending on the modelling approach. Heat transfer effects are highly relevant given the non isothermal conditions of the adsorber bed. Importantly, it can be shown that competitive effects observed are related not only to competitive adsorption, but also to the interdependence of the partial pressures. In addition, the back pressure regulator must be modelled to achieve a realistic representation of the system dynamics at high volume flow rates. Based on the results, PSA experiments are designed, which deliver N_2 of high purity.

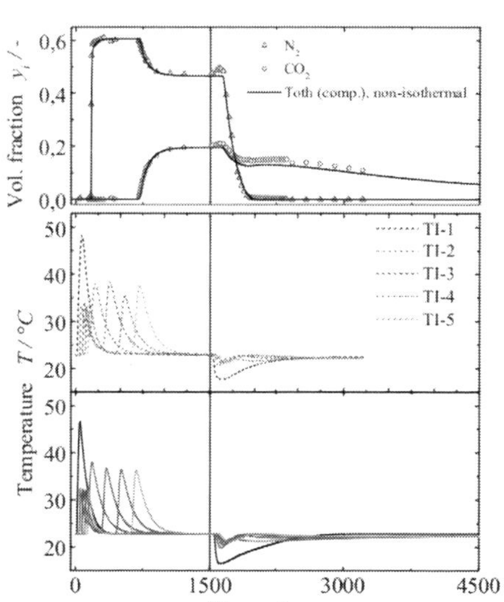

Figure 1. Comparison of experimental and simulated CO_2/N_2 breakthrough curves and temperature profiles.

More complex process dynamics will be found for systems that exhibit adsorption isotherms with inflection points or pronounced adsorption/desorption hysteresis. As an example of both (Figure 2), we discuss experimental results for CO_2 adsorption on the metal organic framework MIL-53 and a modelling approach for the breathing effect shown by this adsorbent.

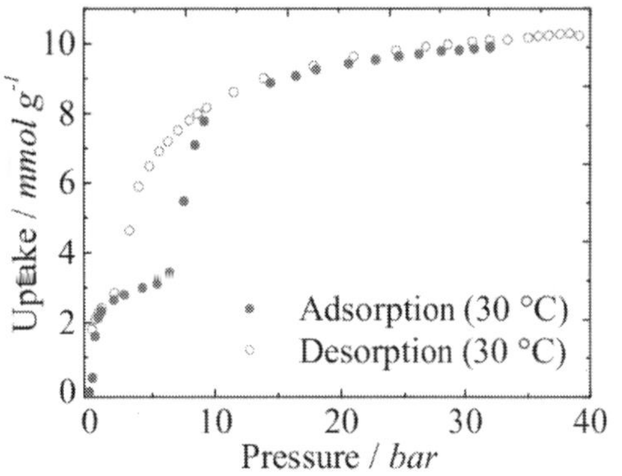

Figure 2. Adsorption and desorption isotherms of CO_2 on MIL-53 at 30 °C

Finally, an economically interesting application could be the separation of paraffins and olefins on zeolitic imidazolate frameworks (ZIF). In this context, we discuss an example where the separation is controlled by both, the adsorption equilibrium and adsorption kinetics.

The relevant findings are summarized and conclusions are given regarding the choice and parameterization of corresponding process models.

Development of a 1D simulation model for a steam cracker convection section

P. Verhees, I. Amghizar, J. Goemaere, [1]A.R. Akhras, K.M. Van Geem, G.B. Marin, G.J. Heynderickx
Laboratory for Chemical Technology, Ghent University, Technologiepark 914, 9052 Ghent, Belgium, tel. +329 264 56 77, e-mail: Kevin.VanGeem@UGent.be [1]Saudi Aramco, R&D Center, Dhahran, KSA, tel. +966 13 876-8459, e-mail: abdulrahman.akhras@aramco.com

The radiant and the convection section of a steam cracker are thermally coupled by process gas and flue gas flow, illustrated by Figure 1. A correct simulation of the convection section is one of the essential requirements to fully optimize the cracker's operation, in particular when heavy liquid feeds are used. One of the consequences is an incomplete evaporation of the heavy tail leading to fouling problems. To locate the fouling problems an accurate model of the convection section describing evaporation and predicting inner tube wall temperatures is needed. In this work a 1D simulation model describing flow boiling and convective heat transfer is developed. Modeling both flue gas and process gas side allows a coupled simulation of the convection section.

At the process gas side both single and two phase (feed evaporation) flow need to be simulated. For the former, it is observed that the three most popular heat transfer correlations[1-3] give equal results. Heat transfer in the evaporating feed is more challenging to model. Models using empirical correlations, and mechanistic methods using feed-specific flow pattern maps considering flow regimes are applied. Modeling results are observed to differ, especially in the low and high vapor fraction range. In this work, these mechanistic models, for single component evaporation, are extended to model evaporation of a multicomponent mixture, based on its true boiling point curve. Since the models are not yet applied to complex mixtures, validation data is obtained by performing CFD simulations. CFD modeling is done using the VOF model combined with an evaporation model.

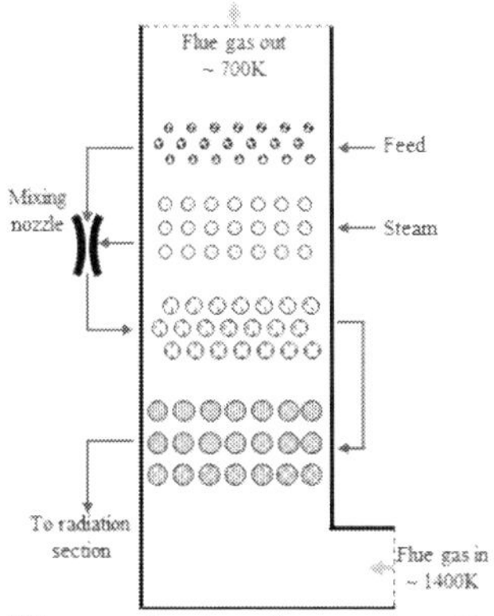

Figure 1: A schematic of the convection section

Heat transfer at the flue gas side is mainly due to convection. The convective heat transfer is calculated using the Zukauskas correlation[4] and an analytical method[5]. Given the typical layout of tube banks, Prandtl and Reynolds numbers are within the validity range of the correlation. No differences in simulation results are observed when using the two methods. In the last step, an iterative procedure is developed to couple the process gas and the flue gas side simulations.

The developed 1D model for the coupled simulation of a steam cracker convection section is validated using plant data. The results show that the applied heat transfer models are applicable over a wide range of operating conditions. A correct modeling of heat transfer in the evaporating hydrocarbon mixture is the main challenge to develop a reliable 1D simulation model for the convection section.

References

1. Dittus, F. W.; Boelter, L. M. K. Univ. California Publ. Eng. 1930, 2, 443.
2. Gnielinski, V. Int. J. Chem. Eng. 1976, 16, 359-368.
3. Sieder, E. N.; Tate, G. E. Ind. Eng. Chem. 1936, 28, 1429-1435.
4. Zukauskas, A. In Adv. Heat Transfer, 1972, p 93-160.
5. Khan, W. A.; Culham, J. R.; Int. J. Heat Mass Transfer 2006, 49, 4831-4838.

Modelling of chemical reactors: from systemic approach to compartmental modelling

J. Haag[1], C. Lemaitre[1], C. Gentric[2], J.P. Leclerc[1]
[1]Laboratoire Réactions et Génie des Procédés, UMR CNRS 7274, Université de Lorraine,
1, rue Grandville, BP 20451, 54001 Nancy cedex, France
tel. +33383175042, e-mail: jeremie.haag@univ-lorraine.fr
tel. +33383175083, e-mail: cecile.lemaitre@univ-lorraine.fr
tel. +33383175056, e-mail: jean-pierre.leclerc@univ-lorraine.fr
[2]Laboratoire Génie des Procédés Environnement et Agroalimentaire, UMR CNRS 6144,
37, Bd de l'Université, BP 406 44602 Saint-Nazaire cedex, France
tel. +33240172632, e-mail: caroline.gentric@univ-nantes.fr

Modeling of chemical reactors consists in dividing the reactor volume into smaller volumes of assumed behavior (such as perfect mixing or plug flow) in which mass and thermal balances are implemented. These volumes are defined based on the hydrodynamics, and phenomena governing the considered process such as heat and mass transfer and/or chemical reaction. The approaches start from the basic one in which the whole reactor is assimilated to an ideal reactor (plug flow or perfect mixer) up to a fine discretization using CFD. The aim of this study is to describe the different methodologies commonly used during the last 40 years in chemical engineering and to present the recent developments of an intermediate approach: compartmental modeling.

The oldest approach is the systemic method which is based on an interconnected network of a maximum of a dozen of elementary reactors (plug flow, perfectly mixed reactor, dead volume…) representative of the global hydrodynamics behavior. It is a global approach that has been developed by (Levenspiel, 1999) and progressively improved to obtain much more realistic models. For instance, (Claudel et al., 2003) have developed a useful general tool, which is able to give the most appropriate structure by transforming qualitative data into quantitative data using the possibility theory. In every case, the model has to be able to represent properly the residence time distribution (RTD) curves. More recently, (Ben Gaida et al., 2012) have studied a gas-liquid bioreactor with a complex geometry in order to product ethanol from yeast. The study focuses only on hydrodynamics (abiotic) and allows deeper investigation of productivity and conversion yield when biotic conditions occur. Residence time distribution (RTD) and internal age distribution (IAD) of different operating set-ups was necessary to develop the model structure for each phase. (Leon et al., 2014) have modelled a horizontal rotating foam stirrer reactor in order to study a hydrogenation reaction. The global structure of the model is directly defined by the structure of the reactor which is divided into several similar stages. Conversion and yield are well predicted. Influence of the stage number is studied in order to find the highest selectivity toward the desired product.

The development of reactive computational fluid dynamics (CFD) method based on the resolution of the Navier-Stokes equations in a large number of computing cells (typically hundreds of thousands of cells) led to a drastic reduction of systemic approach applications. Indeed, it allows to simulate all together complex hydrodynamics, heat and mass transfer and chemical reactions problems with high precision. For example (Aghbolaghy and Karimi, 2014) have simulated enzymatic hydrogen peroxide production within a continuous stirred tank reactor. Concentration of hydrogen peroxide has been optimized in function of inlet water flow rate, glucose molar rate injection and impeller speed.

Based on systemic approach and CFD modelling, an intermediate approach, named the "compartmental modeling", has been developed since the last ten years. It relies on the description of the reactor by a structural and functional subdivision of a few hundreds of compartments. These compartments are created using information deduced from tracer experiments, technical information about the studied process and CFD flow simulations. The introduction of mass and thermal balances in each compartment allows faster simulations of the reactor behavior than CFD with more physical meaning. For example (Gresch et al., 2009) have studied a pilot-scale ozonation reactor in continuous flow configuration. A prediction of reactor performance was established by calculation of conversion rates for different orders and rates of reaction. (Le Moullec et al., 2010) have studied a biological wastewater treatment reactor and they predicted successfully oxygen and COD profiles along the reactor. (Nauha and Alopaeus, 2015) have simulated algae culture in a bubble column using several sub models representing bubble size distribution, bubble coalescence and breakup, drag force, light intensity, and zenith angle or length of day.

These three different approaches have emerged at different times but they are still employed nowadays. This paper will present several applications of the different approaches and will guide toward an approach rather than the others. The most recent approach, which is based on the two oldest, has been applied only to specific reactors and cannot be straightforwardly transposed to other ones. A generalization of the compartmental approach will be presented and applied to some applications.

References:

Aghbolaghy, M., Karimi, A., 2014. Simulation and optimization of enzymatic hydrogen peroxide production in a continuous stirred tank reactor using CFD–RSM combined method. J. Taiwan Inst. Chem. Eng. 45, 101–107. doi:10.1016/j.jtice.2013.05.009

Ben Gaida, L., Andre, C., Bideaux, C., Alfenore, S., Cameleyre, X., Molina-Jouve, C., Fillaudeau, L., 2012. Modelling of hydrodynamic behaviour of a two-stage bioreactor with cell recycling dedicated to intensive microbial production. Chem. Eng. J. 183, 222–230. doi:10.1016/j.cej.2011.12.065

Claudel, S., Fonteix, C., Leclerc, J.-P., Lintz, H.-G., 2003. Application of the possibility theory to the compartment modelling of flow pattern in industrial processes. Chem. Eng. Sci. 58, 4005–4016. doi:10.1016/S0009-2509(03)00269-0

Gresch, M., Brügger, R., Meyer, A., Gujer, W., 2009. Compartmental Models for Continuous Flow Reactors Derived from CFD Simulations. Environ. Sci. Technol. 43, 2381–2387. doi:10.1021/es801651j

Le Moullec, Y., Gentric, C., Potier, O., Leclerc, J.P., 2010. Comparison of systemic, compartmental and CFD modelling approaches: Application to the simulation of a biological reactor of wastewater treatment. Chem. Eng. Sci., 20th International Symposium in Chemical Reaction Engineering—Green Chemical Reaction Engineering for a Sustainable Future 65, 343–350. doi:10.1016/j.ces.2009.06.035

Leon, M.A., Nijhuis, T.A., van der Schaaf, J., Schouten, J.C., 2014. Residence time distribution and reaction rate in the horizontal rotating foam stirrer reactor. Chem. Eng. Sci. 117, 8–17. doi:10.1016/j.ces.2014.06.016

Levenspiel, O., 1999. Chemical Reaction Engineering, third. ed. John. Wiley & Sons Inc.

Nauha, E.K., Alopaeus, V., 2015. Modeling outdoors algal cultivation with compartmental approach. Chem. Eng. J. 259, 945–960. doi:10.1016/j.cej.2014.08.073

The algorithm for modeling and calculation of bioreactors for mammalian cells cultivation

E. Vorobeva, A. Smirnov, E. Guseva, N. Menshutina
D.I. Mendeleyev University of Chemical Technology of Russia, Computer Aided Process Engineering Faculty, Geroev Panfilovtsev Str. 20 bld.1, 125480 Moscow, Russia;
tel./fax: +7(495)495-00-29, e-mail: emulsja@rambler.ru.

One of the most important process in biopharmaceutical industry is a cultivation of mammalian cells. These cells are capable to produce metabolites (e.g. proteins, monoclonal antibodies, etc.) that can be used for the medicines applications - vaccines, drug substances and others. Mammalian cells are very sensitive to the parameters of cultivation (e.g. temperature, pH and pO_2 concentrations, etc.), to the shear stress and other hydrodynamic conditions. Thereby the process of the cells viability maintaining at the high level is very difficult to control.

The algorithm of calculation of the volumetric flow rate of nutrient medium at every stage of a cultivation process depending on the type of a cultivation system (type of cells and bioreactor) has been developed. Computational fluid dynamics (CFD) has been used to analyze the fluid flows and different engineering parameters of bioreactors.

At present work as one of the research objects the process of cultivation of mammalian cells has been investigated in hollow-fibre membrane bioreactor as one of the most popular systems for cell lines cultivations. CHO cells have been chosen as a model cell line due to their wide popularity in many biopharmaceutical productions. The cells grow on the outer fiber surface, affecting the nutrient medium flow throughput of the porous membrane material. According to the developed algorithm the volumetric flowrate of nutrients has been calculated. The CFD model has been developed for one membrane fiber for evaluating the nutrient medium flows inside the membrane, on its surface, and the model of the whole bioreactor has been composed. At this stage of modeling it has been established that the calculated volumetric flow rate corresponded to the laminar flow that could promote the low shear stress.

Calculation procedure has also held for suspension type of bioreactors with the using of microcarriers for cells growth. As microcarriers have been used porous material spherical particles. CHO cells have grown on the outer surface of carriers and in its pores, completely filling the free area. The presence of the carriers in the growth medium has influenced the hydrodynamic environment within the bioreactor and made the fluid more viscous. Therefore it has also been needed to hold the CFD modeling to calculate the optimum volume flow of the nutrient medium to achieve the required number of cells in the bioreactor.

Consequently, the developed algorithm including CFD modeling has been approved for the calculation of the optimum nutrients flows for different types of bioreactors, which is testimony its wide application for different biopharmaceutical processes.

Steady state modelling of a Kühni liquid extraction column using the spatially mixed sectional quadrature method of moments

S. Alzyod, [2]M. Attarakih,[1,2] H-J. Bart[2]
[1]The University of Jordan, Faculty of Engineering & Technology, Chemical Engineering Department, 11942-Amman, Jordan; tel. +49 (0)631 205-2414; email: attarakih@yahoo.com
[2]University of Kaiserslautern, Chair of Separation Sciences and Technology, 67637-Kaiserslautern, Germany; tel. +49 (0)631 205-2414, email: samer.alzyood@yahoo.com, bart@mv.uni-kl.de; fax: +49 (0) 631 205-2119

In spite of decades of research, the extraction equipment design procedures are still faced by the non-ideal behaviour of the dispersed phase. In such equipment, the droplet size distribution plays a decisive role in the determination of the physicochemical product properties of the dispersed phase. In this regard, the dispersed fluid phases can be described using the general Boltzmann-type equation which is called the population balance equation (PBE). The PBE is considered as an efficient framework which takes all of the coupled hydrodynamics and mass transfer phenomena into account to model the real behaviour of the dispersed phase. These phenomena include: Breakage, coalescence, growth, and inter-phase mass transfer. The PBE consists of highly nonlinear transport equations and hence the solution methods rely either on stochastic Monte Carlo methods, sectional methods or the Quadrature Method of Moments (QMOM). In contrast to the classical Quadrature Method of Moments (QMOM), the Sectional QMOM (SQMOM) reconstructs the distribution of the droplet density function and still keeps the advantages of the classical QMOM (Attarakih et al., 2009). As a result of this, the holdup and the mass transfer profiles can be accurately predicted. In this regard, Attarakih et al., (2009) proved that most of the available moments based methods are special cases of the general SQMOM framework by varying the number of primary and secondary particles. For example, if only one primary particle is used, the QMOM is obtained, while if only one secondary particle is used the Moving Pivot Technique (MPT) and the Multi Primary One Secondary Particle Method (MPOSPM) are obtained.

In this contribution, we extended the SQMOM to the physical domain to model the steady state hydrodynamics and mass behaviour of a Kühni liquid-liquid extraction column. To facilitate the model implementation, the resulting integral spatial numerical flux is closed using the MPOSPM, while the hydrodynamics source terms are closed using the Two-Equal Weight Quadrature (TEqWQ) formula derived by Attarakih et al., (2009). In addition to this, the resulting column hydrodynamics equations are coupled with the One Primary One Secondary Particle Method (OPOSPM) to model the mass transfer behaviour inside the column. The present numerical scheme is fifth order accurate in space and fully implemented using MATLAB software.

As a numerical case study, the mixed-SQMOM model is validated using PPBLab software which is based on the extended fixed pivot technique (Attarakih et al., 2012) where a multi-sectional space-time population balance equation solver is used. In addition to this, the mixed model prediction is validated against the published steady state experimental data (Garthe, 2006) for both EFCE chemical test systems: water-acetone-toluene and water-acetone-butyl acetate. The required coalescence model parameters are obtained from Attarakih et al., (2015) and Alzyod (2014), where both authors implemented the OPOSPM as a reduced population balance model using MATLAB SIMULINK environment to simulate the same column geometry under the same operating conditions. The spatially mixed SQMOM model was found flexible to predict the column hydrodynamics as well as the mass transfer profiles. A sample of results is shown in Fig.(1) where the mean droplet velocity and steady state mass transfer profiles are shown for a Kühni extraction column of DN80. The comparisons of SM-SQMOM with PPBLab software shows that the two predicted mass transfer profiles, as compared to the experimental data of Garth (2006), are in excellent agreement.

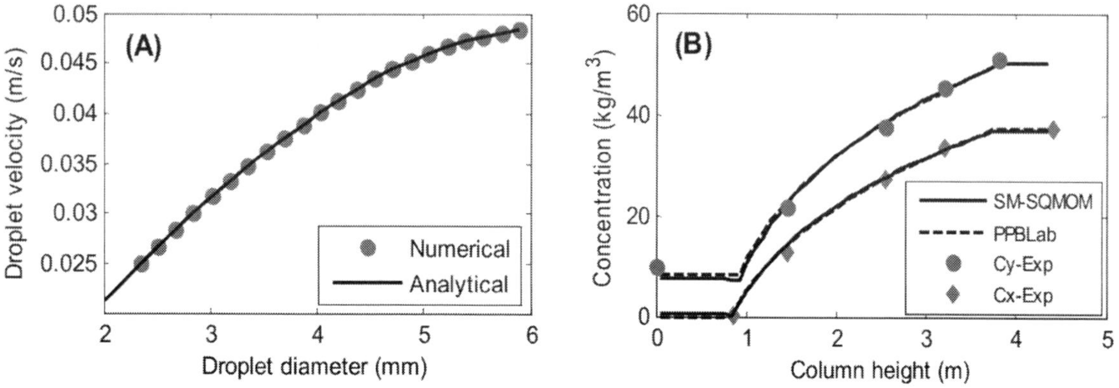

Figure (1): Comparison between the simulated and reference solution profiles inside a Kuhni DN80 liquid-liquid extraction column: A) Dispersed phase velocity. B) Solute mass transfer profiles.

References:

1- M. Attarakih, S. Al-Zyod, M. Abu-Khader, H.-J. Bart, "PPBLAB: A New Multivariate Population Balance Environment for Particulate System Modelling and Simulation," Procedia Eng, vol. 42, pp. 1445-1462, 2012.

2- M. Attarakih, M., C. Drumm, H-J. Bart, "Solution of the population balance equation using the Sectional Quadrature Method of Moments (SQMOM)," Chemical Engineering Science, vol. 64, pp. 42-752, 2009.

3- D. Garthe, "Fluid Dynamics and Mass Transfer of Single Particles and Swarms of Particles in Extraction Column," Dissertation, Technischen Universität München, Germany, April 2006.

4- Alzyod, S. (2013). One primary one secondary particle method SIMULINK module for the steady state and dynamic modelling of Kühni extraction column. Master thesis, The University of Jordan, Amman, Jordan.

5- Attarakih, M., Al-Zyod, S., Hlawitschke, M., & Bart, H-J. "OPOSSIM: A Population Balance-SIMULINK Module for Modelling Coupled Hydrodynamics and Mass Transfer in Liquid Extraction Equipment." Computer Aided Chemical Engineering, vol. 37, pp. 257-262, 2015.

Equilibrium theory based comparative assessment of dual-reflux pressure swing adsorption process cycle configurations

[1]Tushar S. Bhatt, [2]Giuseppe Storti, [1]Joeri F.M. Denayer, [3]Renato Rota
[1]Vrije Universiteit Brussel, Department of Chemical Engineering, Pleinlaan 2 , 1050 Brussels, Belgium; [2]ETH Zürich, Department of Chemistry and Applied Biosciences, Vladimir-Prelog-Weg 1, HCI F 125, 8093 Zürich, Switzerland; [3]Politecnico di Milano, Chemistry, Materials and Chemical Engineering Department "Giulio Natta", Via Mancinelli 7, 20131 Milan, Italy.

Pressure swing adsorption (PSA) is a well-established technology for the separation and/or purification of many gaseous mixtures[1]. Some of its most popular industrial applications include: hydrocarbon separation, hydrogen purification, air drying, and air separation[2]. PSA systems employed for the processing of binary feed gas mixtures can be categorized[3] into three main types: (i) stripping, (ii) rectifying and (iii) dual-reflux. Stripping PSA systems are capable of producing only the light product (weakly adsorbed; species B) at high purity[4]. On the other hand, rectifying PSA systems possess the capability of producing only the heavy product (strongly adsorbed; species A) at high purities[4].

Dual-reflux pressure swing adsorption (DR-PSA) systems combine the features of stripping and rectifying PSA and are capable of achieving complete separation of binary feed gas mixture[1]. In a DR-PSA unit, binary feed gas can be delivered to the bed operational at either constant high (P_H) or low pressure (P_L) and pressure swing can be carried out either with pure A or B species. This flexibility allows for four distinct DR-PSA cycle configurations[4]: (i) DR-PH-A: Feed to P_H and pressure swing with A; (ii) DR-PL-A: Feed to P_L and pressure swing with A; (iii) DR-PH-B: Feed to P_H and pressure swing with B; (iv) DR-PL-B: Feed to P_L and pressure swing with B.

Separation and/or purification of gaseous mixtures in laboratory-scale DR-PSA process units has been reported in literature[3],[5-7]. Several modeling techniques, ranging from simple to complex, have also been reported in literature for design, optimization and/or simulation of DR-PSA process[1-4],[8-10]. The aforementioned literature survey revealed that: majority of the researchers used either of the DR-PSA process cycle configurations (for their experimental and/or modeling analysis) without clearly stipulating the motive for such selection. Kearns and Webley[11] provide the only available guideline for such a choice based on their productivity and energy consumption criterion.

The simplest modeling approach (usually indicated as the equilibrium theory), involves simplifying assumptions such as: instantaneous linear equilibrium throughout the columns, isothermal conditions, negligible mass transport resistances and axial mixing, negligible pressure drop, and ideal gas behavior. Although approximate, equilibrium theory offers vital insights into the system behavior and frequently represents a very effective design tool to elucidate process performance[1]. It was previously employed[1] to propose an optimal design strategy (that identified a triangular operating region inside which, complete separation at cyclic steady state conditions can be achieved) for a specific cycle configuration: DR-PH-A. In this work, we extended the same design strategy to other cycle configurations: DR-PL-A, DR-PH-B and, DR-PL-B.

For each cycle configuration, the equilibrium theory allowed meticulous tracking of the characteristic curves and shock transitions during constant and non-constant pressure steps of the cyclic process. Additionally, an in-depth analysis of the impact of process variables on the design parameters was performed for each cycle configuration, and the results were compared with one another. With respect to previous literature, following major accomplishments can be mentioned:

(i) the proposed design strategy provides novel insights into the separation behavior of individual process cycle configuration;
(ii) for each cycle configuration, correlation between the single specific feed injection position and its minimum to maximum range has been established by identifying a triangular operating region, inside which, complete separation can be achieved;
(iii) distinct design constraints and boundary conditions of the complete separation region for each process cycle configuration have been determined;
(iv) a selection criteria, that facilitates the selection between DR-PSA process cycle configurations has been proposed;
(v) the optimal high to low operating pressure ratio range for all of the DR-PSA process cycle configurations has been identified.

[1] Bhatt, Tushar S., Giuseppe Storti, and Renato Rota. "Optimal design of dual-reflux pressure swing adsorption units via equilibrium theory." *Chemical Engineering Science* 102 (2013): 42-55.

[2] Bhatt, Tushar S., Giuseppe Storti, and Renato Rota. "Detailed simulation of dual-reflux pressure swing adsorption process." *Chemical Engineering Science* 122 (2015): 34-52.

[3] Bhatt, Tushar S., et al. "Experimental and Modeling Analysis of Dual-Reflux Pressure Swing Adsorption Process." *Industrial & Engineering Chemistry Research* 53.34 (2014): 13448-13458.

[4] Kearns, David T., and Paul A. Webley. "Modelling and evaluation of dual-reflux pressure swing adsorption cycles: Part I. Mathematical models." *Chemical engineering science* 61.22 (2006): 7223-7233.

[5] Diagne, Doudou, Motonobu Goto, and Tsutomu Hirose. "Parametric studies on CO2 separation and recovery by a dual reflux PSA process consisting of both rectifying and stripping sections." *Industrial & Engineering Chemistry Research* 34.9 (1995): 3083-3089.

[6] McIntyre, J. A., Armin D. Ebner, and James A. Ritter. "Experimental study of a dual reflux enriching pressure swing adsorption process for concentrating dilute feed streams." *Industrial & Engineering Chemistry Research* 49.4 (2010): 1848-1858.

[7] Saleman, Thomas L., et al. "Capture of low grade methane from nitrogen gas using dual-reflux pressure swing adsorption." *Chemical Engineering Journal*281 (2015): 739-748.

[8] Diagne, Doudou, Motonobu Goto, and Tsutomu Hirose. "Numerical analysis of a dual refluxed PSA process during simultaneous removal and concentration of carbon dioxide dilute gas from air." *Journal of Chemical Technology and Biotechnology* 65.1 (1996): 29-38.

[9] Thakur, R. S., Nitin Kaistha, and D. P. Rao. "Process intensification in duplex pressure swing adsorption." *Computers & Chemical Engineering* 35.5 (2011): 973-983.

[10] Ebner, Armin D., and James A. Ritter. "Equilibrium theory analysis of dual reflux PSA for separation of a binary mixture." *AIChE journal* 50.10 (2004): 2418-2429.

[11] Kearns, David T., and Paul A. Webley. "Modelling and evaluation of dual reflux pressure swing adsorption cycles: Part II. Productivity and energy consumption." *Chemical engineering science* 61.22 (2006): 7234-7239.

Transient model of a fluorine electrolysis cell coupling electric currents, heat transfer, two-phase flow and transport of species

J. Vukasin[(1) (2)], I. Crassous[(2)], B. Morel[(2)], J. Sanchez-Marcano[(1)].
(1) Institut Européen des Membranes, Université Montpellier, 2 Place Eugène Bataillon, 34095 Montpellier cedex 5, France.
E-mail: julien.vukasin@areva.com.
(2) HRP, AREVA NC, BP 16, 26701 Pierrelatte cedex, France.

In the nuclear fuel cycle, fluorine is produced by the electrolysis of the molten salt KF-2HF according to the following equation:
$2\ HF \longrightarrow H_2 + F_2$.

Hydrogen is produced at the cathode and fluorine is produced at the anode. The latter is then added to UF_4 to obtain UF_6. This electrolysis is a complex process to study since hydrofluoric acid and fluorine are hazardous and highly corrosive. Also it must be performed in anhydrous conditions. The reaction is exothermic and a heat exchanger inside the cell is used to maintain the electrolyte at the working temperature. For more than 50 years, many studies have been done to improve the electrolysis but some phenomena are still not understood.

The purpose of this work is to couple experimentation and modeling in order to understand better the interactions between the different phenomena involved during the electrolysis. Recently, a lab-scale electrolysis cell has been designed with a particular cooling system consisting in two concentric tubes in which water is flowing. A 3D-model of this cell has been developed in Comsol Multyphysics™ coupling electrokinetics, two-phase flow, heat transfer and species transport. The comparison between experimental and simulated data allowed the validation of the model at the laboratory scale. It is important to notice that this model gives access to some additional data that are difficult to obtain experimentally such as the hydrodynamics of the system or the local HF concentration.

One particular feature of this model is to take into account the solidification of the electrolyte occurring when the temperature and/or the HF concentration drop. Since very few studies on KF-2HF had been published, only data for liquid KF-2HF were available. The diffusion coefficient of HF was derived from molecular dynamics. Some key parameters for the solid phase were missing, such as heat capacity or thermal conductivity. The first one was determined using theoretical approach. To get the last crucial parameter, a reactor filled with solid KF-2HF was designed and experiments with heating-cooling cycles were carried out. Heat fluxes were then measured and enabled determining the value of the thermal conductivity of solid KF-2HF.

Electrolysis experiments were carried out using a lab-scale cell. Operating parameters of the process were varied in order to assess their impact. Those parameters include the entrance temperature of the cooling system, the HF concentration and the current density at the electrode. The potential of the electrodes were measured as well as the output temperature of the cooling system. It was also possible to detect and quantify the solidification of the electrolyte around the heat exchanger and on the cathode.

Transient simulations of several minutes were carried out, using a segregated solver with a mesh composed of 269000 elements. Comparison between simulated and experimental results were drawn in order to evaluate the accuracy of the model. Special attention was given to the output temperature of the heat exchanger since a small gap between the simulations and the experiments would mean that the heat transfer between the electrolyte and the cooling system is well simulated. The solidification process of the electrolyte on the cathode was also carefully studied because it gives a lot of information about the HF consumption at the electrodes. Figure 1 illustrates the solidification of the electrolyte around the heat exchanger. One may notice that a thin layer of electrolyte remains partly liquid between the tube and the cathode. Figure 2 shows the active side of the cathode on which the solidification of the electrolyte occurs. This phenomenon is called cathodic polarization which is triggered by the HF depletion on the active side of the cathode.

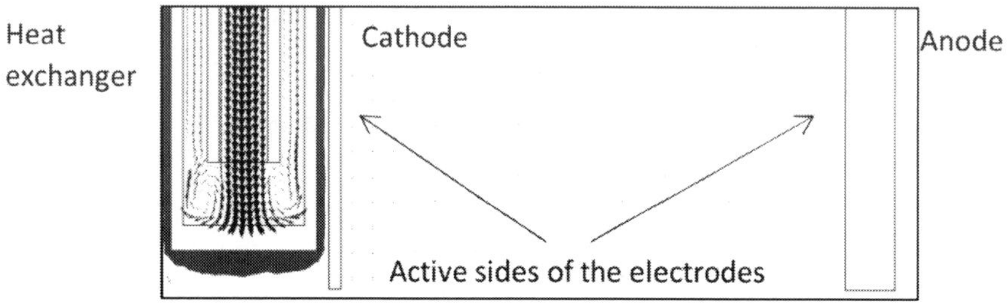

Figure 1: Solidification of the electrolyte around the heat exchanger. In red: solidified electrolyte; arrows: speed of the cooling water. Conditions: $T_{cooling\ water}$ = 10 °C, initial mass fraction of HF = 40 %.

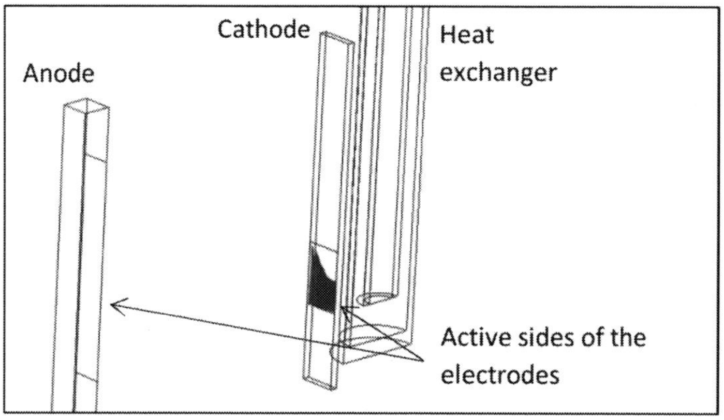

Figure 2: Solidification of the electrolyte on the active side of the cathode. In red: solidified electrolyte. Conditions: $T_{cooling\ water}$ = 10 °C, initial mass fraction of HF = 40 %.

The comparison between the model and the experiment will be discussed in this presentation. A parametric study will also be carried out to explain the interactions between the solidification of the electrolyte and the electrolysis.

Application of stepwise regression with global search for parameter estimation in complex dynamic systems

M. Shacham, [1]N. Brauner
Dept. Chem. Eng., Ben-Gurion Univ. of the Negev, Beer-Sheva 8415, Israel;
tel. +972 86461481, e-mail: shacham@exchange.bgu.ac.il; [1] School of Mech. Eng., Tel-Aviv Univ., Tel-Aviv 69978, Israel; tel. +972 36408127, e-mail: brauner@eng.tau.ac.il.

Reliable mechanistic kinetic models of chemical/biological processes are essential for the understanding, design, optimization and control of such processes. Such models are often described by systems of ordinary differential equations (ODEs). The models usually contain unknown physical parameters that need to be determined using a set of measurements (experimental data). The complexity of parameter estimation problems, whose solution is attempted, has been increasing continuously. The so-called "Dow challenge problem" (Biegler et al. 1986), which was of complexity typical of the nineteen eighties, could be formulated so as to contain about 10 unknown parameters. Recent benchmark problems recommended by Villaverde et al. (2015) contain more than 100 parameters (with measured data available for parameter estimation) and up to 1759 parameters (with simulated data). The increase of the problem complexity may cause difficulties in converging to the globally optimal solution and may render some of the parameters unidentifiable. Sin et al. (2010), for example, presents an example based on a dynamic cellulose hydrolysis model, where only 6 out of 26 parameters are identifiable based on the available data. Attempting to identify more parameters leads to large errors in the estimated parameter values.

Recently Shacham and Brauner (2014) proposed a stepwise regression algorithm to identify the most influential parameters and separate them from the unidentifiable ones in unconstrained parameter estimation problems. In this presentation the extension of our stepwise regression algorithm is presented. It enables handling constrained parameters, uses global search methods and revised stopping criteria for addition of parameters to the model.

A typical dynamic parameter estimation model is shown in Eq. A1 and A2 (available in Appendix A of the full abstract that presents the example problem and can be found at the ftp site: ftp://ftp.bgu.ac.il/shacham/Chisa_16/ParmEstFull.pdf). This example (Liu and Wang 2009) describes a fermentation process that uses a high ethanol tolerance yeast, to produce ethanol. The process model includes 4 differential equations that represent the state variables x, s, p_1 and p_2, and 19 unknown parameter values that have to be estimated based on experimental data. Two parameters are constrained ($Y_{p1/s}$ = [0.1, 0.51], $Y_{p2/s}$ = [0.1, 0.2]), and the rest of the parameters are constrained to positive values.

Let us denote the parameters that are included in the regression model (and the associated terms of the explanatory variables) at any stage of the stepwise regression process as basic parameters, and the remaining parameters as non-basic parameters. At every step the objective function is calculated while adding each of the remaining non-basic parameter (and the associated variable) into the basic set, in turn. The parameter whose addition causes the greatest reduction of the objective function value is selected for inclusion in the basic parameter set at this particular stage. The addition of parameters continues until some stopping criterion is satisfied.

In the nonlinear case (such as the example discussed here) the stepwise regression must be preceded with an initial stage of model analysis and establishment of an initial basic set. Stepwise regression requires setting parameters not included in the basis at zero. In a nonlinear expression some parameters often cannot be set at zero for various reasons (e.g., model singularity). Consequently the model analysis and modification must start by re-definition of the model parameters as to allow a zero parameter value if necessary. There is no need to modify parameters constrained to nonzero values (i.e., parameters constrained to intervals that do not include the zero value). In addition, if a term contains a product of several parameters, this product must be replaced by a new parameter in order to enable consideration of a single parameter at a time. The model equations containing the reformulated parameters are shown in Eq. A3 and Eq. A4.

The next stage is of the problem analysis includes assignment of hierarchy levels to the parameters. The model equations can usually be divided to terms (i.e., groups of expressions linked by multiplication and division operators). Zeroing one (or more) of the parameter values may cause removal of the whole term from the model. Such parameters are categorized as of first hierarchy ($H = 1$), as they must be included in the basis prior to considering the inclusion of additional parameters of lower hierarchy that are associated with the same term in the basis. For the model equations A3 and A4, the $H = 1$ parameters include $Y_{p1/s}$, $Y_{p2/s}$, π_1, π_2 and π_3 the other parameters are of lower hierarchy level (higher H).

The initial basic set of parameters must include all the nonzero constrained parameters and the ones from amongst the $H = 1$ parameters, which are sufficient for representing the initial trend of the state variables and ensure that none of the state variables will remain constant throughout the integration. For the nonzero constrained variables, the constrains may provide good initial estimates, while for the rest of the H=1 parameters the initial trend of the state variables can be used for obtaining such estimates.

The calculation of the parameter values that minimize the objective function (comprising of the sum of squares of the differences between the experimental and calculated values of the state variables) is carried out in an outer loop using the globally convergent "Scatter Search" (Ugray et al., 2007) algorithm as implemented in the *GlobalSearch* library function ensemble of MATLAB. In an inner loop, an integration routine is used to determine the state variable values at time intervals where experimental data are available. In this study, the integration is carried out either by the MATLAB *ode45* function, which is based on an explicit Runge-Kutta (4,5) formula, or for stiff systems the *ode15s* function is used, which utilizes the backward difference formulas (BDF) of Gear (1971). The addition of the parameters to the model in a stepwise fashion continues as long as all the parameters are significantly different from zero (thus their values are larger, in absolute value, than the respective confidence intervals, stopping criteria 1) or until reaching a minimum of the objective function of a "validation" data set. The use of the globally convergent algorithm at each step of the stepwise regression assures that the solution obtained is unbiased to the order of the parameter selection and their initial estimates.

In the conference the globally optimal solution of the example problem with the full set of parameters will be presented and compared with the optimal solution obtained by the proposed stepwise regression algorithm, pointing out the advantages of the latter.

References

Biegler, L. T., Damiano, J. J., Blau, G. E. AIChE Journal,1986; 32: 29-45.

Gear CW. Numerical initial value problems in ordinary differential equations. Englewood Cliffs: Prentice Hall; 1971.

Liu P-K, Wang F-S. Comput. Chem. Eng. 2009;33:1851–60.

Shacham M., Brauner N., Comput. Chem. Eng. 2014;69:26–38.

Sin G., Meyer A.S., Gernaey K.V., Comput.Chem.Eng., 2010;34:1385–92.

Ugray Z.,, Lasdon L., Plummer J., Glover F., Kelly J., Martí R., INFORMS Journal on Computing, 2007; 19: 328–340.

Villavarde A. F., Henriques, D., Smallbone, K., BMC Systems Biology, 2015; 9:8: 1-15.

On the combined grassroots-retrofit integration of first and second generation biorefineries

[1,2]A.D. Mountraki, [1]A B.Benjelloun. Mlayah, [2]A.C. Kokossis
[1]Compagnie Industrielle de la Matière Végétale (CIMV), 109 rue Jean Bart - Diapason A, 31670, Toulouse, France; tel. +33(0)534 318 241, e-mail: a.mountraki@cimv.fr; [2]National Technical University of Athens (NTUA), 9 Heroon Politechniou, Politechnioupoli, Zografos, 15780, Greece; tel. +302107724275, e-mail: mountrak@central.ntua.gr.

The time has come for second generation biorefineries to pass to the industrial scale. Although lignocellulosic biorefineries can, in principle, be developed as new plants, there is an equal interest to upgrade first generation technology. As they process different feedstock, the integration of different generation biorefineries has attracted attention. Existing work is focused on the potential to reduce energy consumption, demonstrating that the integrated scheme can contribute to a more economical design with lower environmental impacts than the stand-alone. The upgrade of first generation plants constitutes, however, a retrofit problem. The modified and improved process has to make the best use of the invested equipment of the existing plant. The most economical design is searched among numerous scenarios, which multiply when also other portfolios of products need to be investigated. While first generation installations are nascent plants, still possible to amend and modify, second generation upgrades account for new technology and investment very flexible to adjust so that the upgrades are best.

To date, first and second generation processes have been integrated by the adoption of grassroots design methodologies, while, in reality, their coupling is far more complex. The integration of first and second generation biorefineries is a combined grassroots–retrofit design problem that hides inner trade-offs and proves the deficiency of present-day practices in proposing realistic designs for such mixed design problems. There is significant potential to exchange (by-)products and not only to integrate energy. For example, bagasse can be used as feedstock in the second generation process, and, similarly, lignin can be used as fuel for the co-generation unit. Moreover, the processes may utilize common sections, like sharing the same downstream processes. Finally, the co-generation (CHP) unit, after modifications, can support both plants. That implies a retrofit problem that is much larger and complex than most conventional applications. The size and the complexity of the problem impose the need for a systematic approach in the form of mathematical models.

This work studies the different symbiotic options in the integrated pair of first and second generation biorefineries, scoping for the development of a new systematic methodology, which can combine different process systems engineering tools for the integration of grassroots-retrofitted problems. Detailed designs are developed in Aspen Plus®, validated by the industrial partner (CIMV Process™). The detailed models feed the mathematical models, developed in GAMS®. The two processes are integrated following a mixed mathematical model. The model applies the methodology of grassroots integration on second and retrofit integration on first generation technology, following the transhipment analysis. Preliminary results indicate a significant potential in internal trade-offs savings, both in operating and capital cost.

Acknowledgment: The financial support from Marie Curie European Research Program, RENESENG (FP-607415) is gratefully acknowledged.

Optimal design for removal pharmaceutical contaminants from wastewater streams

M. Hidalgo-Muñoz[1], J. M. Zamora-Mata[2], J. Ramírez-Muñoz[1,*]
[1]Universidad Autónoma Metropolitana-Azcapotzalco, Departamento de Energía. Av. San Pablo 180, C.P. 02200, Mexico City, Mexico. [2]Universidad Autónoma Metropolitana-Iztapalapa, Departamento de IPH, Av. San Rafael Atlixco No. 186, Col. Vicentina, Iztapalapa D.F. 09340, Mexico. email:*hidalgo.mmg@gmail.com.

The occurrence of pharmaceutical contaminants in aquatic environment or groundwater is of continued concern for the health and safety of consuming public due to the development of antibiotic resistant bacteria. The existing conventional water treatment plants were not designed for the removal of these contaminants, and therefore antibiotics are being regularly released into the aquatic environment and groundwater. Regulation, treatment and emission of these substances have received little attention in most countries around the world. Its inclusion as dangerous materials depend upon the damage to human health, as well as its presence and dispersion in water resources (Petrovic et al., 2003). Pharmaceutical drugs and antibiotics are very widely used and its natural metabolism inside the human body makes that its concentration in excretions retain important levels, mainly in the health care facilities effluents, e.g. hospital sewage. Therefore, its remotion from wastewater has become a major issue recently (Homem et al., 2010; Padilla et al., 2014).

This paper presents an evaluation of two different alternatives for the selective treatment of wastewater streams carrying paracetamol. The considered treatment technologies are: 1) electrochemical oxidation, and 2) electrocoagulation. For each technology, a mathematical programming model is developed on the basis of a basic treatment superstructure that includes various wastewater treatment configurations. The basic superstructure of the network presented by Zamora et al. (2004) includes splitters (S), mixers (M), a treatment unit (UT) and the interconnections. Figure 1 shows the network superstructure which covers all the treatment alternatives. The paracetamol outlet concentration of both electrochemical processes depends on several variables that include the treatment unit wastewater flowrate and inlet concentration. Treatment unit performance is modelled with a variable order kinetic, with different constant values for each of the analyzed technologies. The problem can be stated as follows: Given is a set of effluent streams, with known flow-rates that contain certain concentration levels of paracetamol (Table 1). The objective is to determine the corresponding optimum intermediate flowrates so that discharge composition regulations for the pollutant is met, with a minimum total annual cost of the system. For each treatment unit, a nonlinear programming model (NLP) is formulated and solved to optimality. The NLP problem consists of an objective function (e.g. cost), mass balances (e.g. UT, S, M), and input-output relationships for the treatment unit.

Table 1. Wastewater process data.

I	S_i (L/min)	$C_{i,j}$ (mg/L)	$m_{i,j}=S_{i,j}C_{i,j}x10^{-3}$ (g/min)
1	11	60	0.66
2	9	43	0.387
3	7.5	32	0.24
4	4	21	0.084
5	3	10	0.03
	$m_j = \Sigma m_{i,j}$		1.401

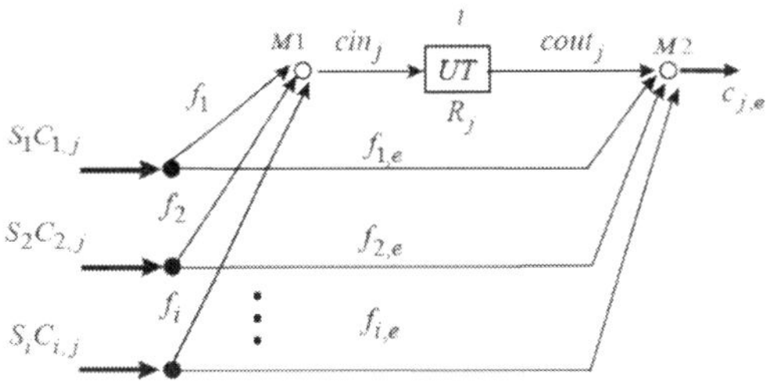

Fig. 1. Basic superstructure (Zamora et al., 2004).

References

Homem, V., Alves, A., Santos, L., 2010. Amoxicillin degradation at ppb levels by Fenton oxidation using design of experiments, Sci. Total Environ., 408, 6272– 6280 (2010).

Padilla, B.G., Alonso, A., Martínez, S.A., González, M., Jaúregui U.J., Ramírez-Muñoz, J., 2015. Electrochemical degradation of amoxicillin in aqueous media, Chem. Eng. Process, 94, 93–98.

Petrovic, M., Gonzalez, S., Barceló, D., 2003. Analysis and removal of emerging contaminants in wastewater and drinking water, Trends Analyt. Chem. 22, 685–696.

Zamora, J., Hernández, R., Castellanos, J. 2004. Linear programming model to assist the design of distributed wastewater treatment systems (In Spanish), Rev. Mex. Ing. Quím. 3, 121–134.

Integration of process design and hierarchical structure-based control

M. Zhou, Y. Cai, F. Tao, H. Pan
Zhejiang University of Technology, Dep. Chemical Engineering, Chaowang Rd. 18, 310014, Hangzhou, China;
tel. +86-571-88320329, e-mail: htpan@zjut.edu.cn;

Traditionally, process design and control system design are carried out sequentially. That is, in the first step, process design is preformed that is based on steady state economic calculations; subsequently, a suitable control structure is synthesized that is generally based on heuristic controllability measures. Thus, the process design and process control aspects have been generally studied independently. However, Process design and control are closely related in chemical processes. The integration of process design and control (IPDC) may result in numerous economic and operability benefits over the traditional sequential design approach and has become increasingly important and received much attention from both academic and industrial interests over the recent decades.

The conventional IPDC problem is solved by dynamic optimization, which results in computational complexity. In addition, it should be pointed out that most of the works in the field of IPDC utilize conventional PID controllers to manipulate the process. In this work, advanced process control is embedded into the framework of integrated design and control process based on a robust modeling approach that could avoid the solution of dynamic optimization. And then the designed chemical processes are characterized by economic and dynamic performance, and can be operated close to design boundaries and specification limits without design constraint violations. The key idea of the employed approach is that the system's closed-loop dynamic behavior is represented as a nominal linear state space model with uncertain model parameters. Then, the uncertain closed-loop process model obtained is used to compute a bound of the worst-case scenario, process stability and process feasibility. Specifically, a hierarchical control structure, which consists of lower-level conventional PID regulators and a higher-level advanced process control (APC), is researched in the IPDC framework.

The structure of the IPDC strategy is given in Figure 1.

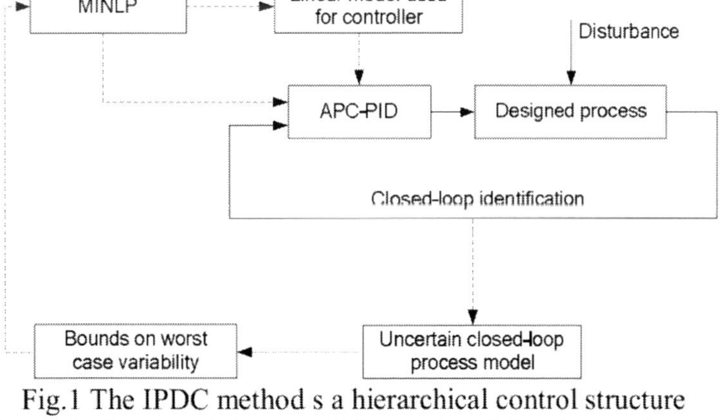

Fig.1 The IPDC method s a hierarchical control structure

As shown in figure 1, the closed loops which consist of the designed processes and PID control are treated as generalized object and are cascaded with APC. At each optimization step, uncertain models from the disturbance to the process variables require to be estimated and are used to obtain the bounds on worse case variability. The optimization variables considered in the IPDC framework are the best process structure, dimensioning and operating conditions together with control structures and controller parameters in the presence of external perturbations and uncertainties in the process parameters.

In the paper, from the APC applications point of view, the hierarchical control structure has been taken into account within the IPDC framework. The key for solving the resulting IPDC problem is to utilize robust tools and formulates IPDC problem as a nonlinear programming problem. A simulation example is used to illustrate the proposed IPDC scheme.

Nonlinear generalized predictive control based on narimax model of a reactive packed dıstillation column

S. Karacan[1], I. Bayram[1] and A. Giwa[1]
[1]Ankara University, Engineering Faculty, Department of Chemical Engineering, Tandoğan 06100, Ankara, TURKEY. Tel:+90(312) 2033523, Fax: +90 (312) 212 15 46 (E-mail: karacan@eng.ankara.edu.tr)

Esters are of great importance to chemical process industries. Among them, acetate esters are important organic solvents widely used in the production of varnishes, ink, synthetic resins, and adhesive agents. They are produced from the reactions of acid and alcohols under an acidic condition. A key issue in the production of these esters is the low conversion from the reactions. As a result, heavy capital investments and high energy costs are inevitable. The reactive distillation is a very attractive way to reduce these investments and energy costs.
Reactive distillation is a process that combines both separation and chemical reaction into one unit. It has a lot of advantages for those reactions occurring at temperatures and pressures suitable for the distillation of the resulting components. Its main advantages are derived from the elimination of equipment and the constant removal of products because it is known that increased overall conversion can be achieved in equilibrium reactions if the products are continuously removed from the reaction zone. The combination of both chemical reaction and separation in a single unit results in the complexity of this process which makes its modeling a very challenging task to chemical engineers because the unavailability of process models capable of reliably describing the several complexes and interrelated phenomena including simultaneous reaction and separation has made the safe scale-up from laboratory experiments to industrial plants quite challenging, especially when a packed column is the type involved in the reactive distillation.
Over the past two decades, nonlinear model predictive control (NMPC) has emerged as a promising tool for solving complex and moderately large-scale industrial control problems With the goals of control performance, energy conservation and cost effectiveness of process operations in industries, the design of optimal controllers and controller performance assessment have received great attention in both industries and academia. The main objective of control is to maintain a product purity within the desired range. This study is aimed at controlling a process used for ethyl acetate production from the esterification reaction between ethanol and acetic acid in a reactive distillation column. Experimental studies are carried out in a laboratory scale reactive packed distillation column having a height of 1.5 m. A control module connected on-line to a computer is used to manipulate the reflux ratio and the feed ratio of the process. Reflux ratio is chosen as the manipulated variable to control the distillate temperature of the column. Improving the nonlinear NARIMAX model of the system in batch time, model parameters are obtained by System Identification Toolbox of MATLAB. This model is used in generalized predictive control algorithm. Distillate temperature of the reactive packed distillation column is controlled for set-point tracking and disturbance rejection. The comparison of the results of NLGPC with those of PID revealed that better control performance are obtained using NLGPC.

Integrated disturbance observer based control and inferential control of a binary distillation column

Simon William Jeffries, Jie Zhang
School of Chemical Engineering and Advanced Materials, Newcastle University, Newcastle upon Tyne, England, UK, e-mail: jie.zhang@newcaslte.ac.uk

This paper presents an integrated disturbance observer based control and inferential control of a binary distillation column. Disturbance observers have been successfully used in previous research providing excellent disturbance attenuation. Research into the application of inferential control methods in distillation columns have achieved a much higher quality of control in comparison to the feedback control using composition analysers which possess substantial measurement delays. This paper takes the novel approach of integrating two frequency domain disturbance observers with multiple tray temperature based inferential estimations of top and bottom product compositions. In order to cope with colinearities in tray temperature measurements, principal component regression (PCR) and partial least square (PLS) regression are used in the soft sensor development. The number of principal components in the PCR model and the number of latent variables in the PLS model are determined using cross validation. The frequency domain disturbance observers have been designed by optimising the value of the filter time constants within a first order filter, in order to maximise disturbance rejection.

The proposed method is applied to a simulated methanol-water column where the simulation is based on a rigorous mechanistic model based on detailed tray by tray mass and energy balances. The control performance is measured using sum of squared errors. The following control schemes are compared: compositional analyser based feedback scheme, multiple tray temperature based inferential control, and multiple tray temperature inferential methods integrated with disturbance observers. A series step disturbances with magnitudes of +/-15% in feed flow rate and feed composition are tested. It is shown that the integration of multiple tray temperature based inferential methods with the disturbance observer gives the best control performance among the compare control schemes. The effect of the filter time constant used in the disturbance observer filter on control performance is also investigated.

Modelling, optimisation, and cyclic steady-state behaviour of an adsorptive reactor for the reverse water gas shift reaction

A. A. Munera Parra, J.L. Cadavid Cardenas, C. Asmanoglo, D. W. Agar
Laboratory for Chemical Reaction Engineering, Technische Universität Dortmund, Emil-Figge Straße 66, 44227, Germany
Tel: +49(0)2317552350, e-mail: alejandro-augusto.munera-parra@bci.tu-dortmund.de ; fax: +49 (0) 231 755-2698.

Adsorptive reactors represent an attractive multifunctional alternative for process intensification. An ideal case study for such reactor is the low-temperature reverse Water Gas Shift reaction (LT-rWGS):

$$CO_2 + H_2 \leftrightarrow CO + H_2O \qquad \Delta H^0_{R,298K} = 41.1 kJmol^{-1}$$

In this reaction, the limited equilibrium conversion at low temperatures (~250°C) can be displaced to more favourable conversions by adsorbing the water produced by the reaction, by integrating the separation of the water into the reactor operation. In addition, the endothermic heat of reaction is comparable to the heat of adsorption, for instance the heat of adsorption of water on 3A Zeolite has been reported to have values ranging from -43 to -72 $kJmol^{-1}$ [1], which allows for the heat effects to be coupled for nearly isothermal operation. Furthermore, for a pressure swing operation, a degree of freedom is added as to where the heat of desorption can be extracted from depending on the available heat sources in either the upstream or downstream processes.

The modelling of adsorptive reactors presents some challenges due to the presence of steep fronts. A preliminary study on the numerical methods available to simulate adsorptive reactors for systems with steep fronts was performed by the authors [2]. In this work, the results obtained are applied to the real case of the LT-rWGS. The set of PDAEs describing the species concentration, adsorbent loading as well as the change in velocity due to the adsorption of the water with respect to time and position are derived. A typical result of the unsteady-state simulation for the adsorptive reactor is presented in Fig. 1.

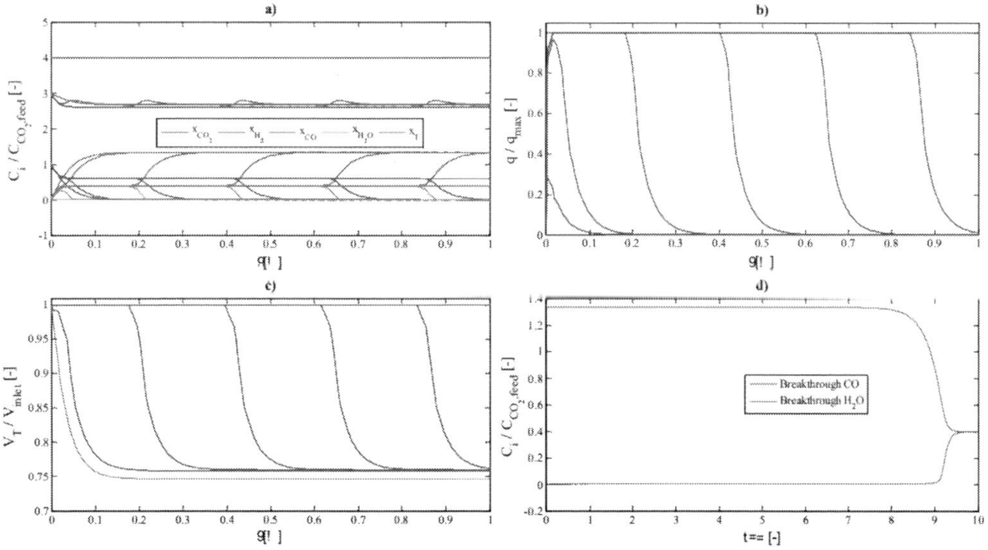

Figure 1: a) Product concentration profiles as a function of time. b) Adsorbent loading as a function of time. c) Flow rate profiles as a function of time d) Breakthrough of water and concentration of product at the outlet as a function of time

As can be seen in Fig. 1, the reactor operates at full conversion for a certain time before water finally breaks through. In order to prolong this period and thus maximise the product throughput, the adsorptive reactor offers several degrees of freedom for optimisation: the catalyst and adsorbent distribution, their respective fractions and the length of each region as depicted in Fig. 2. A dynamic optimisation problem was set up and solved to maximise the production time.

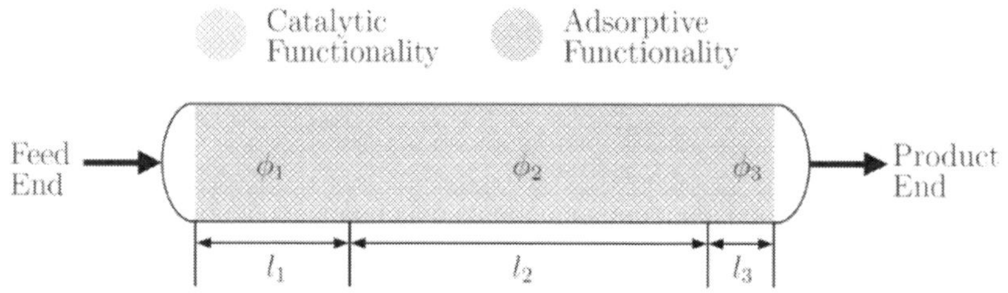

Figure 2: Adsorptive reactor with its degrees of freedom for optimisation

Finally using the optimised catalyst and adsorbent distribution, the cyclic steady state behaviour of the full operation , i.e. reaction, desorption, purge and pressurisation, as shown in Fig. 3, has been calculated to establish the full cycle time, and therefore the necessary number of units needed to guarantee continuous operation.

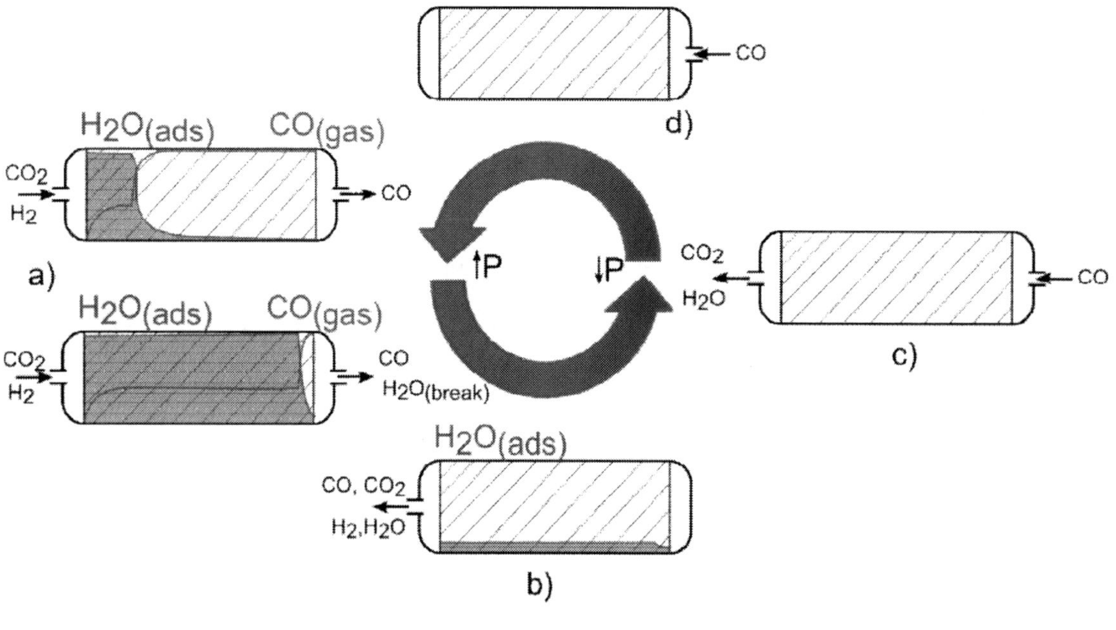

Figure 3: Cyclic operation. a) Reaction/Adsorption. b) Depressurisation/Desorption. c) Purge. d) Feed pressurisation.

References

[1] Simo, M., Sivashanmugam, S., Brown, C.J., Hlavacek, V. (2009) Adsorption/Desorption of water and Ethanol on 3A Zeolite in Near-Adiabatic Fixed bed. *Ind. Eng. Chem. Res.*, 48, 9247-9260.

[2] Munera Parra, A.A., Cadavid Cardenas, J.L., Agar, D.W., (2015) Modelling and Simulation of Adsorptive Reactors under Advective Dominated Conditions, In Preparation.

Value chain synthesis and optimization in microalgae biorefineries

M. Psycha, A. C. Kokossis
National Technical University of Athens, School of Chemical Engineering, Zografou Campus, 9, Iroon Polytechniou Str., GR-15780 Athens, Greece; tel. +302107723129, e-mail: melpsycha@mail.ntua.gr.

Microalgae as a source of biomass have been recently gaining ground especially due to the wide range of products they can offer. Thus, biorefineries capitalizing on this particular biomass are worth exploring not only from an experimental but also from a conceptual point of view. Most of the focus is concentrated on the design of integrated processes for the production of biofuels and respective approaches have been developed to tackle the issue [1, 2]. The study pays attention to the valorization of specialty chemicals and attempts to address the limitations involved in the field of microalgae biorefineries. The selection of suitable technologies in addition to the identification of the product portfolio compose the elementary challenges amplified by the presence of competitive chemistries and the need for multi-product plants to ensure sustainability [3, 4].

The methodology developed follows the combinatorial use of a superstructure and mathematical techniques to simultaneously screen and evaluate potential paths for the extraction of desired products. The problem is formulated starting with a set of feedstocks f that can be converted either to intermediates i or to final products p by the use of processes t. For feedstocks, intermediates and products the prices are known, whereas for the technologies yields and capital costs are known. The selection of processes and product portfolios are treated as degrees of freedom. In order to form all the available options, a superstructure is developed involving conceptual feedstocks, processes, intermediate mixtures or fractions and products. This is a crucial point, especially in the case of microalgae biorefineries, since establishing the feasible connectivities is a great challenge due to lack of industrial applications. Therefore, information exchange with practitioners is of high significance. For the construction of the synthesis model, units are linked to each other via connections. Units comprise blocks which represent the processes, and nodes which can be feedstocks, intermediates and products. The connections between units reflect the input/output relations between blocks and nodes. One block can have more than one output connection, either with intermediates or products, but only one input connection, either with feedstocks or intermediates. The "activation" of a connection depends on the formulation of the model.

The abovementioned synthesis model is optimized taking into consideration economic data and results to the optimal paths based on the developed superstructure. For the units, variables and parameters are used to describe them. Parameters refer to the input data, such as prices, yields and economic factors. The model employs both continuous and binary variables. Inlet and outlet flows as well as variables for the economic estimation are considered continuous, whereas binary variables play the role of decision making regarding the selection or deselection of units. Additionally, a number of constraints are incorporated into the model in the form of mass balances, costing equations and logical constraints.

For the purpose of this study, a case for the extraction of high-value chemicals from microalga Dunaliella is analyzed as part of the D-Factory, an FP7 collaborative project. The extraordinary tolerance to both salt stress and environmental contaminants render Dunaliella a promising candidate for large scale outdoor cultivations. For the construction of the model, Dunaliella paste is considered the single feedstock. The blocks (processes) involved are: ultrafiltration, nanofiltration, ethanol extraction, hexane extraction, supercritical CO_2

extraction, H_2O extraction, solvent extraction and chromatography. In order to extract multiple products from microalgae, the biomass is subjected to two processing stages involving the same technologies. Furthermore, chromatography is used for the separation of three different groups of products. Thus, the number of synthesis blocks is 18. Intermediates can be microalgae cells after the 1st processing stage, mixtures of fractions (proteins/glycerol, glycerol/carotenoids/lipids, carotenoids/lipids) or single fractions (proteins, lipids, carotenoids) pinpointing the number of intermediate nodes to 11. The products are 15 in total: glycerol, residue, group 1 (trans-β-carotene, cis-β-carotene, other carotenes, zeaxanthin, violaxanthin, lutein, phytoene), group 2 (chlorophyll a and b, pufas, EDK vitamins) and group 3 (glycerol enzymes, carbonic anhydrase). The study takes into account the stages after cultivation and harvesting of the microalgae, where screening paths of candidate technologies and product portfolios is a matter of necessity due to the numerous available options. The selected paths resulting from the model are presented in Figure 1 reporting the processes and product portfolios for each selection. It is noteworthy that even though path II has less products in its portfolio and thus lower revenues, it is considered a more profitable solution than path III, which has a bigger product portfolio.

The combination of a superstructure design and an optimization model presented in this study tackles the designing problem in the field of microalgae biorefineries and offers guidelines for the development of more advanced and detailed design analysis. This methodology is setting the baseline for a future multicriteria model for the selection of optimal paths regarding the capitalization of microalgae for high-value chemicals.

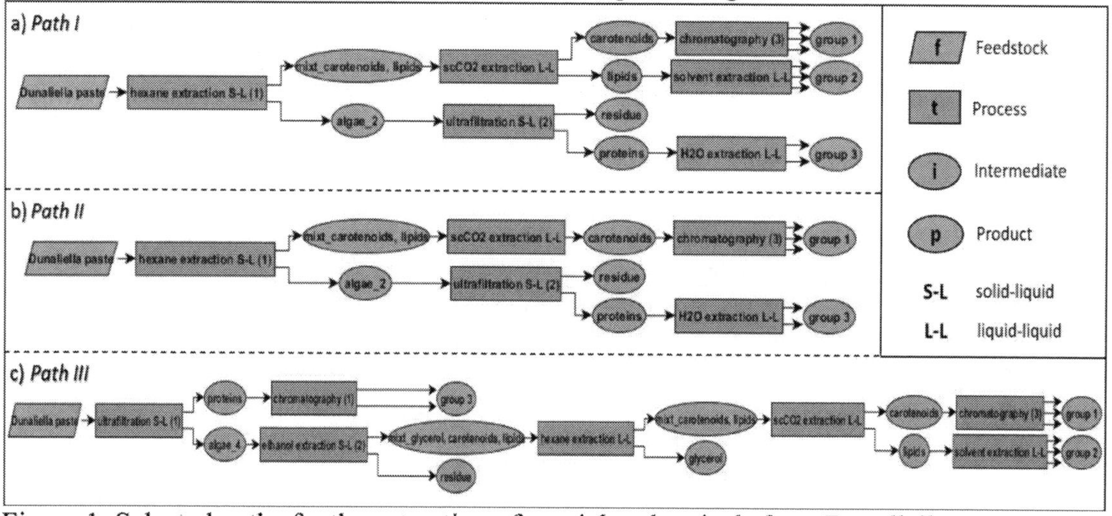

Figure 1. Selected paths for the extraction of specialty chemicals from Dunaliella

References

[1]. M. Martín, I.E. Grossmann, 2014, Design of an Optimal Process for Enhanced Production of Bioethanol and Biodiesel from Algae Oil via Glycerol Fermentation, Applied Energy, 135, 108-114

[2]. M. Rizwan, J.H. Lee, R. Gani, 2013, Optimal Processing Pathway for the Production of Biodiesel from Microalgal Biomass: A Superstructure Based Approach, Computers and Chemical Engineering, 58, 305-314

[3]. M. Psycha, K. Pyrgakis, A.C. Kokossis, 2014, Process Design Analysis for the Valorisation and Selection of Integrated Micro-algae Biorefineries, Computer Aided Chemical Engineering, 33, 1543-1548

[4]. M. Psycha, K. Pyrgakis, P.J. Harvey, A. Ben-Amotz, A.K. Cowan, A.C. Kokossis, 2014, Design Analysis of Integrated Microalgae Biorefineries, Computer Aided Chemical Engineering, 34, 591-596

Operational optimisation of industrial cooling water systems

F. Song[1], N. Zhang[2], R. Smith[3]
the University of Manchester, the Center for Process Integration, Oxford Road, M139PL, Manchester, United Kingdom; tel. [1]+447425813381;
email:[1]fei.song-2@postgrad.manchester.ac.uk;[2]nan.zhang@manchester.ac.uk;[3]Robin.Smith@manchester.ac.uk.

Cooling water systems are used to remove waste heat from processes, in which there are three main components namely cooling towers, cooler networks and water distribution networks. In process industries, the cooling water system is an important part of the utility system in terms of utility consumption and the impact on both processes and other systems. Cooling water systems keep processes running properly and steadily. Without cooling water systems, many production processes are unable to proceed effectively and efficiently. Due to a huge amount of cooling duty, cooling water systems consumes a large amount of water and power, resulting in high operating cost. Besides, cooling water systems can affect the operation of other systems like refrigeration systems and chilled water system. Cooling water systems are very complex, as they are affected by processes and climate as well as that the components in them are strongly interacted with each other. In a cooling water system, the cooling water from cooler network enters cooling towers and the cooling water from cooling towers is distributed to cooler network. Therefore, the flowrate and temperature of cooling water into individual coolers are confined by cooling towers, and in turn the flowrate and temperature of cooling water into individual cooling towers are confined by cooler network. Processes determine the cooling duty to individual coolers. The air conditions like temperature, pressure and humidity have an effect on the performance of cooling towers. Besides, water distribution network restrict water distribution in cooler network. When all of these factors are considered at the same time, the operating conditions obtained allow cooling water system work properly and steadily. In order to minimise the operating cost of cooling water systems or to optimise the impact of cooling water systems on processes or other systems, operational optimisation of cooling water systems are one way to achieve these purposes. The aim of this work is to develop a complete model for cooling water system and a solution method for operational optimisation of cooling water systems.

Castro et al. [1] integrated cooling towers, pumps and cooler network for operational optimization. However, the model for cooling towers cannot predict water evaporation rate accurately, and the model for coolers was expressed by energy balance without consider the structure of coolers. Cortinovis et al. [2] developed a model for cooling water system for operational optimization, including a cooling tower model, a hydraulic model and a thermal model for coolers. The phenomenological model of cooling towers used by Cortinovis et al. [2] contained differential equations which are unable to be used in an optimization program directly. The methods mentioned above were developed for a cooling tower and cooler networks in parallel arrangement. Besides, the piping network was not considered to confine the water distribution in cooling water systems.

In this work, a mathematic model of overall cooling water systems is developed, which includes models for cooling towers, coolers and the network. In cooling tower modelling, a simplified cooling tower model is regressed out based on a rigorous cooling tower model developed by Poppe [3] for forced draft counter flow wet cooling towers. The cooling tower model is validated for reliable results of evaporation rate and water outlet temperature of cooling towers. In the cooler modelling, the effects of water flowrate on overall heat transfer coefficient and pressure drop are considered. Cooler networks are modelled for both parallel and the combination of parallel and series configurations. In the network modelling, a hydraulic model is used to confine water distribution in cooling water systems. As the model of cooling water systems is formulated as an NLP problem, the CONOPT solver is then applied for optimisation. Case studies are presented to illustrate the application of the models developed in this work.

References

[1] M.M. Castro, T.W. Song, J.M. Pinto, "Minimization of operational cost in cooling water systems," Trans IChemE , vol. 78 (part A), pp. 192-201, 2000.

[2] G.F. Cortinovis, J.L. Paiva, T.W. Song, J.M. Pinto, "A systemic approach for optimal cooling tower operation," Energy Conversion and Management , vol. 50, pp. 2200-2209, 2009.

[3] J.C. Kloppers, D.G. Kröger, "A Critical Investigation into the Heat and Mass Transfer Analysis of Counterflow Wet-Cooling Towers," International Journal of Heat and Mass Transfer, vol. 48, pp. 765–777, 2005.

Steady state modelling and parametric optimization of drying oil process using COCO as a CAPE-OPEN flowsheet simulator

M. Attarakih, [1,2] S. Alzyod[2], O. Aldmour[1], O. Markarian[1], H-J. Bart[2]
[1]The University of Jordan, Faculty of Engineering & Technology, Chemical Engineering Department, 11942-Amman, Jordan; tel. +49 (0)631 205-2414; email: attarakih@yahoo.com;
[2]University of Kaiserslautern, Chair of Separation Sciences and Technology, 67637-Kaiserslautern, Germany; tel. +49 (0)631 205-2414, email: samer.alzyood@yahoo.com

Acetylated Castor Oil (ACO) is the main raw material used to produce Drying Oil (DO), which is one of the major chemical components used in the paint industry. DO is produced by the thermal decomposition of the Acetylated Castor Oil (ACO) with acetic acid as a byproduct. As a side reaction, DO dimerizes to form gum as a solid phase, which reduces the yield of DO. Due to the complex composition of ACO, DO and the gum, they are modeled as palmitic acid, tetradecene and 1-octacosene respectively [1]. In this contribution, the steady state DO flowsheet (Reactor & Distillation sections) is modeled using the CAPE-OPEN flowsheet simulator (COCO), where ACO, DO and gum are added as new components to the thermodynamics package (TEA). The plug flow reactor model, along with the reaction package CORN, are used to implement the reactor model and reaction kinetics [1] (main and side reactions). The distillation section, which is composed of a sequence of two distillation columns is implemented using the ChemSep distillation module. To improve the economics of the DO process, parametric optimization of the distillation section is performed by minimizing the Equivalent Annualized Operating Cost (EAOC) with column operating pressure and reflux ratio as decision variables. Moreover, preliminary economic evaluation (profitability analysis) based on discounted and non-discounted methods are performed based on a DO heat integrated flowsheet with an optimized distillation section. This results in an average rate of return on investment (ROROI) of about 60 percent, and a non-discounted payback period of 0.7 years.

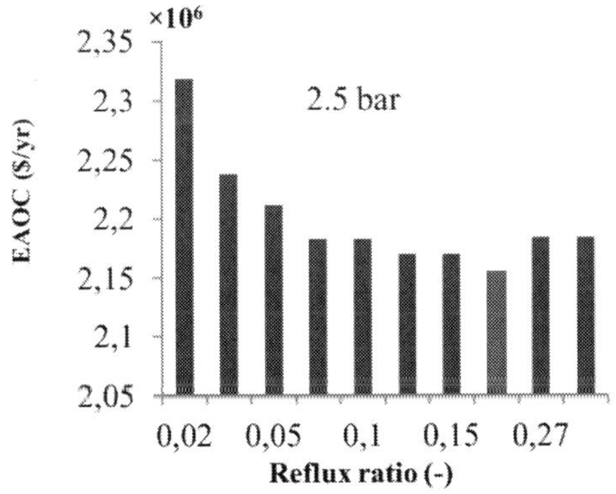

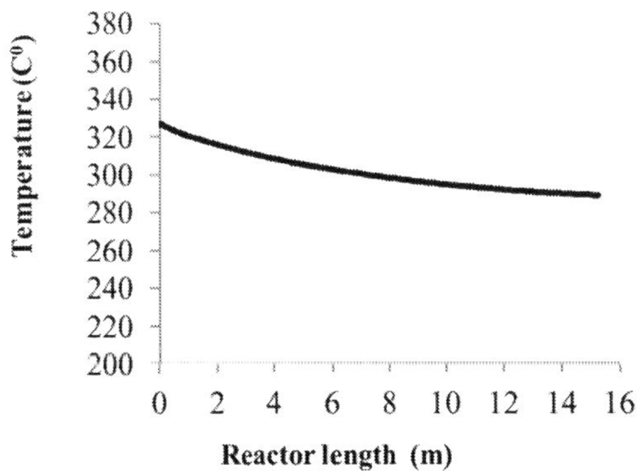

Figure (1): a) ACO recovery tower optimization b) DO Reactor Temperature profile.

Fig.(1) shows a sample of the distillation section optimization with column operating pressure as a parameter. On the right hand side of Fig. (1) is the DO reactor temperature profile which is optimized to reduce the reactor size and eliminate heat input by controlling the design inlet reactor feed temperature.

References:

1- N. A. Amenaghawon, C. O. Okieimen, C. O. (2012). Modelling and simulation of thermal decomposition of acetylated castor oil in a non-isothermal reactor, Journal of Engineering and Applied Sciences, 8, pp. 45-52, 2012.

Determination of the optimum vapor bleeding arrangement of the quintuple-effect evaporator in the evaporation process for raw sugar manufacturing

S. Chantasiriwan
Thammasat Univ., Fac. of Engr., Khlong Luang District, Pathum Thani 12121, Thailand;
tel. +66-(0)-25643001, fax: +66-(0)-25643010, e-mail: somchart@engr.tu.ac.th.

Raw sugar manufacturing process may be considered as consisting of two processes: the juice extraction process and the evaporation process. The first process extracts juice from sugar cane, and the second process evaporates water from the diluted juice and produces granulated sugar as the final product. Main components of this process are juice heater, quintuple-effect evaporator, and crystallizer.

The schematic representation of the evaporation process is shown in the following figure. The juice heater is of the indirect type consisting of 4 heat exchangers (HC, H1, H2, H3, and H4). It receives diluted juice at the flow rate of $m_{f,i}$ from the juice extraction process. After passing successively through H4, H3, H2, and H1, the juice temperature increases from $T_{h,4}$ to $T_{h,0}$. It is important that $T_{h,0}$ must be above the boiling point at the atmospheric pressure so that dissolved gases in the juice are got rid of by passing the juice through the flash tank (FC). This means that pressure of the diluted juice at the exit of H1 is a little above the atmospheric pressure. Finally, the juice pressure is raised from the atmospheric pressure to the pressure in the first effect of the evaporator (p_1), and the juice is passed through HC to increase its temperature to the boiling point. High-pressure steam is used to heat the juice.

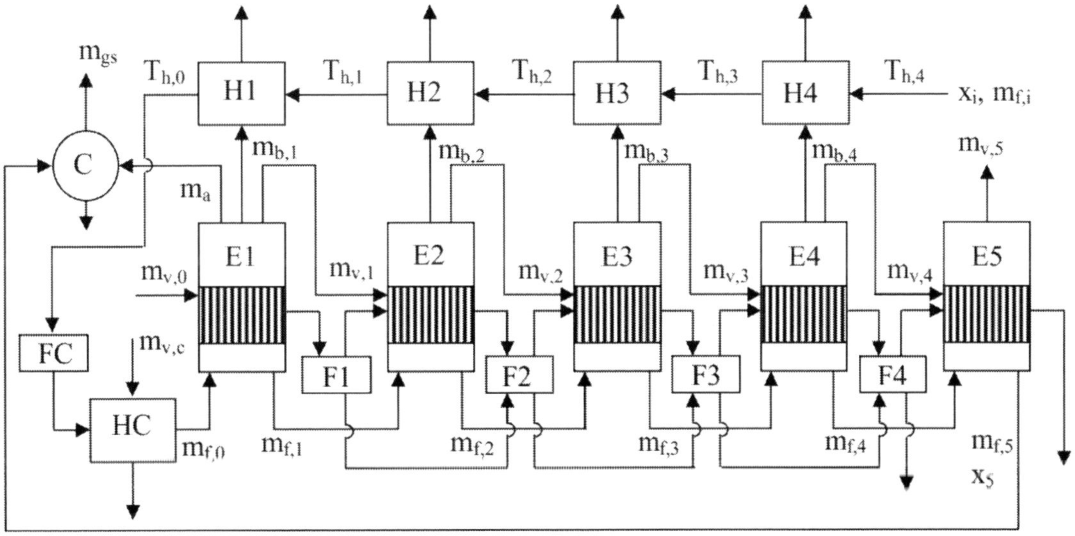

The quintuple-effect evaporator requires a supply of high-pressure steam from the turbine exhaust. The thermal energy released by the condensation of the steam causes the evaporation of sugar juice at a lower pressure p_1 in the first vessel (E1), resulting in vapor and more concentrated sugar juice. The vapor leaving all vessels except the last vessel (E5) is used to evaporate sugar juice in the succeeding vessel. The arrangement in the above figure makes use of condensate flash recovery in order to improve the efficiency of the evaporator. A flash tank is placed after each effect except the last one. The first flash tank (F1) receives condensate from the first vessel at pressure p_0 to produce vapor and condensate at pressure p_1. Each of the other flash tanks (F2, F3, and F4) receives condensate from the preceding vessel and the preceding flash tank at pressure p_i to produce vapor and condensate at pressure p_{i+1}.

Vapor is bled from all vessels of the evaporator except the last one. Vapor bled from the first, second, third, and fourth vessels are used to increase juice temperature in H1, H2, H3, and H4, respectively. Additional vapor is bled from the first vessel, and used in the crystallizer (C) to evaporate the water content in the concentrated juice leaving the evaporator. The output of the crystallizer is raw sugar.

The mathematical model of the evaporation process indicates that the performance of the process is affeted by the vapor bleeding scheme, which depends on the distribution of juice heater surface among H1, H2, H3, and H4. If the total heating surface (A_h) and the concentration of juice leaving the evaporator are specified (x_4), only two of the surfaces are free parameters. Simulation result is obtained for the variation with the surfaces of H2 and H3 ($A_{h,2}$ and $A_{h,3}$) of the steam economy, defined as the amount of water evaporated as the diluted sugar juice becomes granulated sugar divided by the amount of high-pressure steam required for the evaporation process. Parameters used to obtain this result are p_0 = 200 kPa, p_5 = 16 kPa, x_i = 12%, $T_{h,0}$ = 103°C, $T_{h,4}$ = 30°C, $A_{e,1} = 1.3 \times 10^4$ m^2, $A_{e,2} = 7.0 \times 10^3$ m^2, $A_{e,3} = A_{e,4} = A_{e,5} = 4.0 \times 10^3$ m^2., $A_h = 5.0 \times 10^3$ m^2, and x_5 = 60%. It can be seen that there exists the optimum distribution ($A_{h,1}$ = 1025.73 m^2, $A_{h,2}$ = 1082.97 m^2, $A_{h,3}$ = 1011.57 m^2, and $A_{h,4}$ = 1879.73 m^2) that yields the maximum steam economy (SE = 2.789).

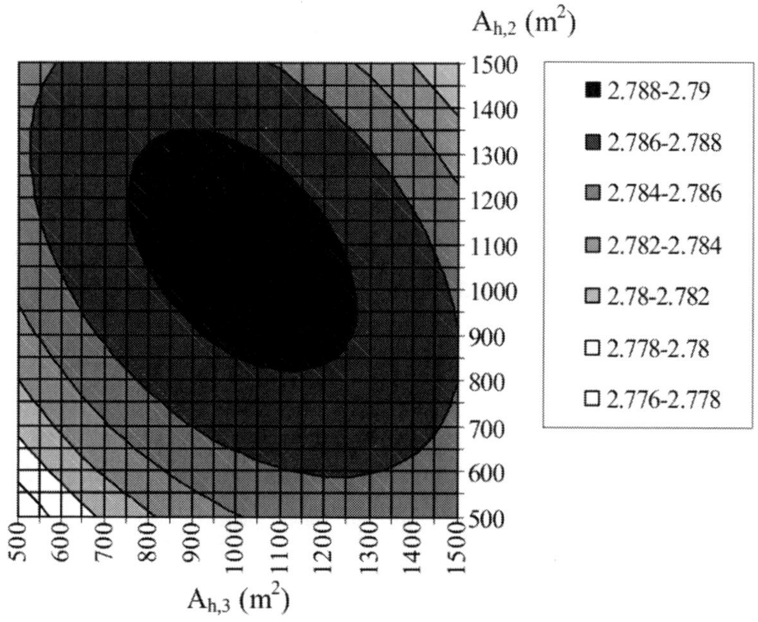

Further simulation shows that there is a different optimum disribution if the amount of granulated sugar produced by the process is to be maximized. The designer of the evaporation process must, therefore, select one of the two optimum distributions, depending on the design objective.

A holistic approach for energy optimization of highly cross-linked plants, using the example of a refinery

[1]F. Pöllabauer, [2]G. Bachmann, [1]T.Wallek
[1]Institute of Chemical Engineering and Environmental Technology, Graz University of Technology, NAWI Graz, Austria; tel. +43 316 873 4988, florian.poellabauer@tugraz.at;
[2]OMV Refining & Marketing GmbH, Vienna, Austria; tel. +43 1 40440 42427, gerald.bachmann@omv.com

An oil refinery is a typical example of a complex compound system consisting of several, usually highly cross-linked plants, in which the interactions among these plants in terms of advantageous energy use are posing a challenge, particularly when considering a variety of crudes as a feedstock and varying utilization of capacity.

For single plants, the use of commercial process simulation tools is state of the art as a basis for tuning of operating points, conducting energetic optimization on column-level, trouble management, case studies and even dynamic simulations for emulation of step-tests to determine control-parameters. While for these applications a modeling-depth down to each single equilibrium stage is desirable, the usually huge number of streams, recycles, apparatuses and chemical compounds makes it difficult to combine such models for simultaneous optimization of several plants, and maintain model stability while spending reasonable amounts of simulation time.

This is one reason why the energetic optimization of a complex compound system of plants is usually still based on empirical considerations.

In contrast, this paper suggests a comprehensible methodology to reduce the modeling-depth of a compound system in such a way that the resulting model is reduced to the main product streams while providing links with the facilities like steam network, electric supply network, hydrogen network, fuels, as well as the associated power plants. Notably in view of using the model for cross-plant energetic optimization tasks, the implementation of an integrated static model covering plants and utilities is required because the power plants are interlinked with all process units via steam- and powers networks, and residues are used as power plant fuels.

Reduction of the modeling-depth is achieved by breaking down the real system into an aggregation of appropriate sub-systems for which the fulfilment of mass balances is a crucial aspect of the model. Based on measured data from the plants and/or detailed process simulation, each plant is simplified to a deterministic black box as a function of a few variables by a fit of linear functions. In this way, the real system is stepwise reduced to a linearized system of splitters. The resulting set of functions describing the main characteristics of each process step in terms of product yield and energy consumption can be used to formulate one target criterion or a weighted sum of several target criteria that can be influenced by both production and utilities. Due to increasing economic interests, an efficient operation of the power plants also with respect to smart interaction with the power grid and the use of combined heat and power (CHP) technology in addition to optimal conditions in the production plants is gaining importance as a key factor for sustainable operation.

The effective use of the model is demonstrated by an exemplary optimization task. For the treatment of recycle streams, a best-practice approach is demonstrated on the basis of three examples. We point out that the modular structure of the model provides possibilities for further detailing, and the methodology presented in this paper can smoothly be transferred to other compound systems of plants in refining or chemical industry.

Artificial intelligent tools to analyze thermographic images. Assisting in early diagnosis of illnesses

E. S. Pariente, J. C. Cancilla, G. Matute, R. Aroca-Santos, J. S. Torrecilla*
Departamento de Ingeniería Química, Facultad de Ciencias Químicas, Universidad Complutense de Madrid, 28040 Madrid, Spain; Tel. +34913944244, email: jstorre@ucm.es.

Abstract

All types of objects at room temperature emit infrared radiation. This is an invisible electromagnetic radiation (ER) which possesses longer wavelengths than the visible spectrum (700 nm). Therefore, this ER transmits lower quantities of energy than the visible radiation. Given that infrared radiation is emitted by every object because of its temperature, their measurement does not imply and hazard to the environment or to the human being and, therefore, it can be successfully used in very different sectors.

It is important to note that with the measurement of the infrared radiation, the temperature variation, with the position and time of a given object, can be determined very precisely. That is, anomalous sources of heat can be accurately located thus enabling the identification of malfunctions in equipment that originate a variation of their temperature, even during the earliest stages. This is why this technique is widely used to estimate the performance of industrial applications, such as maintenance of equipment, monitoring electric installations or chemical engineering process, and so on **[1,2]**. The most relevant aspect is the selection of the most adequate sensor to determine this heat or temperature variation in the most reliable way possible. It is important to select an equipment with a proper temperature sensitivity, as well as spatial resolution. **[3]**

Since 400 BC, temperature has been proven to be a very good indicator of health and it has been used for clinical diagnosis **[3]**. Currently, following this important historical concept, this technique can be applied to detect illnesses early or to control the postsurgical period. It is important to highlight that due to the low ER used and the need of locating the point at an anomalous temperature in the body, this thermographic technique is only reliable to capture heat generation which is close to the human being's surface. This is why this technique is more than adequate to assist in the diagnosis different types of cancer, diabetic disorders, rheumatic diseases, gynecological illnesses, dental problems, and so on. **[3]** In this line, the main objective of this work is to mathematically treat the thermographic images for the early diagnosis of skin cancer and to evaluate the possibility to extrapolate these algorithms to other types of illnesses and to control the evolution of the postsurgical status of patients.

Given the mathematical nature of these thermal images, after statistical pre-treatment of the matrixes that come from the thermal images, these values are introduced as independent variables in the supervised algorithm to classify the illness as a skin cancer or not. The target samples used have been classified as skin cancer or not by medical experts who works in several Spanish Hospitals. The mathematical algorithms used here are mainly based on supervised neural networks in combination with other linear and non-linear algorithms **[4]**. In the preliminary studies carried out up to now, and using published thermographic images, the ratio of correct diagnosis was higher than 80%. In addition to this, currently, the application of these algorithms to head and neck cancer and to control the postsurgical status of patients is being successfully done as well.

In conclusion, these promising preliminary results open the door to the diagnosis of skin cancer with thermographic images, giving more information to the oncologist to reach reliable early diagnosis of this cancer in combination with other commonly employed techniques. Infrared thermal imaging is becoming a medical diagnostic tool to measure abnormal temperature patterns originated by illnesses. It is important to note that the integration of the aforementioned mathematical algorithms and thermographic images will result in a fast and inexpensive methodology which is easy to employ and implement.

References

[1] A.S.N. Huda, S. Taib, K.H. Ghazali, M.S. Jadin. A new thermographic NDT for condition monitoring of electrical components using ANN with confidence level analysis. ISA Transactions 53 (2014) 717–724

[2] A.S.N Huda, Soib Taib. Application of infrared thermography for predictive/preventive maintenance of thermal defect in electrical equipment. Applied Thermal Engineering 61 (2013) 220-227.

[3] B.B. Lahiri, S. Bagavathiappan, T. Jayakumar, John Philip. Medical applications of infrared thermography: A review. Infrared Physics & Technology 55 (2012) 221–235.

[4] D. Hernandez-Contreras, H. Peregrina-Barreto, J. Rangel-Magdaleno, J. Ramirez-Cortes, F. Renero-Carrillo. Automatic classification of thermal patterns in diabetic foot based on morphological pattern spectrum. Infrared Physics & Technology 73 (2015) 149–157.

Industry 4.0 – Reluctant Communication?

Armin Fricke, Jan C. Schöneberger
Chemstations Europe GmbH, Lindencorso, Unter den Linden 21, 10117 Berlin / Germany;
tel. +49-30-20 200 600, e-mail: info@chemstations.eu.

Industry 4.0 is the computerization of manufacturing processes in the industry, thus making plant engineers and operators of chemical, pharmaceutical and life science industries to both parties as well as to those affected. While plant engineers focus on the project earnings and project objectives (time, cost, quality), the operating profit and the operating flexibility are relevant to operators. Industry 4.0 requires the use of software and the exchange of data. Software leads to interfaces between different systems, which are often applied across companies and thus bring about integration on the value chain. Industry 4.0 is therefore very concrete cooperation and exchange of information between different companies. This gives rise to opportunities and risks. Data cannot be conceived without a data architecture, data management and data exchange formats. Industry 4.0 is therefore very concrete cooperation and exchange of information of software and hardware. This results in numerous options in terms of architecture and the exchange formats of data interfaces.

This paper presents the co-operation and exchange of information of people in departments and companies as well as machinery and equipment from the perspective of a process simulator and the example of an ethylene glycol process. A special focus is on the technical aspects of data exchange and data security ("Security by Design") and on the organizational aspect of data ownership ("Business Value Added"). The examples include the process design and operation with the use of a process simulator in the foreground.

Resistors are identified that hinder the implementation of industry 4.0. To designate them in a first step makes sense because it reveals conflicting interests of EPCs and operators. To overcome the resistors brings significant opportunities for further growth of these industries.

A minimal canonical representation of physical models

T. Haug-Warberg
NTNU, Dep. Chem. Eng., Sem Saelands v. 4, 7094 Trondheim, Norway;
tel. +47 73594108, e-mail: haugwarb@nt.ntnu.no

Mathematical modelling is becoming increasingly important to the modern world. With the advances of physical theories, ranging from microscopic quantum mechanics to classical thermodynamics; there is a wide range of applications that can now be attacked by computer software. On the flip side of the coin, advances in computer science have been of an equal magnitude. However, a cross-breed between computer science and physical science is still not easy because the scientist must master, not only one, but two advanced subjects.

In an attempt to bridge the gap between computer science and engineering disciplines this paper addresses the problem of formulating physically based models in a canonical way. With canonicity we mean a form that is related to a general rule or standard formula. Needless to say, a canonical form cannot embrace all conceivable models without becoming so general that there is hardly any common structure left at all. Hence, we are going to restrict ourselves to the discussion of the *minimal* canonical form. It is to a lesser degree generic, while it represents the model with the smallest set of programming entities. The class of models we can handle is surprisingly large still.

It should be stressed that the minimal canonical representation is an abstraction that it is not tied to a particular programming language or to a particular software platform – not even to a particular programming paradigm. This makes it ideal as a backend for automatic exportation of code from graphical user interfaces.

In order to define the minimal canonical model representation there must exist a set of tokens having a precise mathematical meaning without being tied to a particular programming language. In addition there must exist a set of rules for how the model is being calculated. This outset recalls us about the lambda calculus that was founded by Church (1936) and Turing (1937). But, whereas they researched the general theory we shall be seek a much slimmer variant of the realm. In particular, we shall be content with limiting ourselves to a subset of all existing physical models, knowing that we cannot implement them all.

The canonical representation must capture discrete state models frequently found in mechanics, thermodynamics, and engineering; while maybe steering away from distributed systems and the concept of fields. The handling of distributed systems require that we discretize the model spatially thus forming a set of ordinary differential equations, which definitely is within our scope. Finally, we want the model formulation to be explicit to avoid recursive calls within the model definition itself. Borrowing a term from graph theory we describe the model graph as acyclic, but possibly disconnected. In short, the proposed representation of the model is on the following form:

$$y_{k<n} = c_k$$
$$y_{k=n} = x$$
$$y_{k>n} = f_k(y_1, y_2, \cdots, y_{k-1})$$

Here, c is a set of universal constants, model parameters, or any other objects known a priori to the definition of the model; x is a list of free variables defining the model state; y is the corresponding set of calculated state properties; and f is an (immutable) list of model expressions connecting x and y. Based on this simplification, the instantiation of a new model object can be written:

$$a = A(x, y, f)$$

A is the data type (class) which is responsible for creating object a. Once the model object has been created it can calculate its state properties like any mathematical function:

$$a(x) \rightarrow a$$

The next step is to utilize several model objects in order to build a complete model framework. Bringing in an example from thermodynamics we might for example want to formulate a two-phase separator:

$$separator \supset equations \supset (vap, liq)$$

All four objects taking part in this equation have the same canonical form, and since they are isomorphic we can start applying operators defined with a single function interface. Basically, we only need to specify two sets of active x and y elements:

$$opr(\{x_i\} \in x, \{y_j\} \in y)$$

This interface is sufficient for calculating Jacobian matrices, for solving sets of non-linear equations, and for integrating sets of ordinary differential equations:

$$jac(\{x_i\}, \{y_j\}) \rightarrow \frac{\partial\{y_j\}}{\partial\{x_i\}}$$

$$solve(\{x_i\}, \{y_j\}) \rightarrow \{y_j\}(\{x_i\}) = 0$$

$$odeint(\{x_i\}, \{y_j\}) \rightarrow \int \frac{d\{x_i\}}{dt} = \int \{y_j\}$$

The operators cannot be implemented directly into the model object where they apply. This would make the object definition recursive, which is forbidden. The trick is to let the parent object do the operation on behalf of the child. Hence, to complete the separator example, we should make the operators visible in the model graph:

$$separator \supset eqn \cup solve \supset (vap \cup jac, liq \cup jac)$$

In order to calculate two-phase equilibrium we only have to specify the separator variables x:

$$separator(x) \rightarrow \left(vap(x^v), liq(x^l)\right)_{eq}$$

The equilibrium conditions, and the constraint specification, are taken care of by eqn, which is called from $separator$ to iterate on the states of vap and liq until all thermodynamic requirements are fulfilled. It makes the formalism completely general from a thermodynamic point of view. A more advanced example discussed in this paper shows a dynamic flash tank under the influence of a PI-controller.

A new population balance module for pulsed sieve plate liquid extraction column using PPBLab software

M. Attarakih,[1,2] S. Alzyod,[2] H-J. Bart[2]
[1]The University of Jordan, Faculty of Engineering & Technology, Chemical Engineering Department, 11942-Amman, Jordan; tel. +49 (0)631 205-2414; email: attarakih@yahoo.com;
[2]University of Kaiserslautern, Chair of Separation Sciences and Technology, 67637-Kaiserslautern, Germany; tel. +49 (0)631 205-2414, email: samer.alzyood@yahoo.com

Liquid-liquid extraction process is one of the most widely used separation processes after distillation which is based on the intensive contact of two immiscible (or partially miscible) phases to ensure high mass transfer rates. This separation process uses many types of equipment which can be categorized into two agitated and non-agitated columns. The non-agitated columns have many advantages when compared to the agitated ones. These include high throughput, high separation efficiency and less maintenance. The presence of the column internals in these columns enhances the droplet breakage rates, which facilitates the control of the available interfacial area for mass, momentum and energy transport. The non-agitated columns can be operated with mechanical pulsation which results in high turbulent and shear forces and enhances the solute mass transfer rate. As a results of this, the dispersed phase inside the column undergoes instantaneous (due droplet breakage and coalescence) and continuous (because of mass transport) changes and loses its identity through a finite residence time distribution. Accordingly, detailed modelling at the discrete level is an obvious need using detailed population balances as transport equations [1, 2].

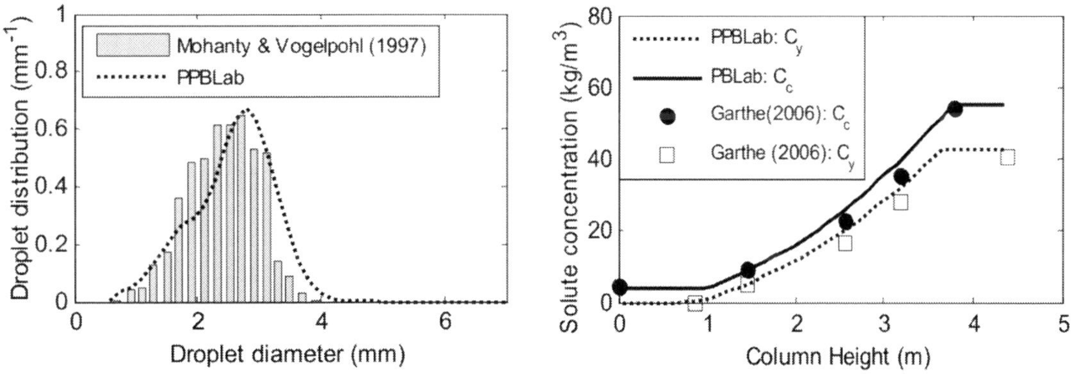

Figure (1): Comparison between the experimental and the simulated hydrodynamics and mass transfer profiles using two different sieve plate columns: A) Droplet size distribution in a pulsed sieve plate DN225 column. B) Mass transfer profiles in a pulsed sieve plate DN80 column.

In this contribution, we present a new comprehensive population balance based module for modelling the coupled hydrodynamics and mass transfer in pulsed sieve plate liquid extraction columns. The new module is fully implemented using PPBLab software which utilizes the recent population balance model solution algorithms and solvers [1,2]. In this regard, the PPBLab detailed and reduced extended fixed pivot solvers are used to discretize the internal coordinates (droplet diameter and droplet concentration), while the PPBLab built-in space-time solver is used to discretize the physical spatial domain which is a multi-sectional space-time solver. The software PPBLab provides a robust finite volume numerical scheme (at the space level) which can be also coupled with the MATLAB standard ordinary differential stiff time solvers. In addition to this, a user friendly interface is designed to facilitate the user inputs and outputs and to allow a full access to the CAPE-OPEN thermodynamics package (TEA) databank. As a case study, the new module is validated using the published steady state experimental [1, 2] data for both water-acetone-butyl acetate and water-toluene chemical test systems as shown in Fig.(1). The columns used in the simulations are the pulsed sieve plate columns DN80 and DN225 respectively, where the droplet size distribution and the steady state mass transfer profiles are compared with the experimental data [3, 4]. The new PPBLab module is found flexible and accurate in predicting the experimental data and is able to follow the liquid extraction column behaviour during scale-up. As a conclusion, the developed sieve plate PPBLab module is considered as efficient software for modelling and simulation purposes of liquid-liquid extraction columns in general and pulsed sieve columns in particular.

References:

1- M. Attarakih, S. Al-Zyod, M. Abu-Khader, H.-J. Bart, "PPBLAB: A New Multivariate Population Balance Environment for Particulate System Modelling and Simulation," Procedia Eng, vol. 42, pp. 1445-1462, 2012.

2- M. Attarakih, M. W. Hlawitschka, M. Abu-Khader, S. Al-Zyod & H.-J. Bart, "CFD-population balance modeling and simulation of coupled hydrodynamics and mass transfer in liquid extraction columns". Applied Mathematical Modelling, 39, pp. 5105-5120, 2015.

3- S. Mohanty, A. Vogelpohl, "A simplified hydrodynamic model for a pulsed sieve-plate extraction column," Chemical Engineering and Processing, vol. 36, pp. 385-395, 1997.

4- D. Garthe, "Fluid Dynamics and Mass Transfer of Single Particles and Swarms of Particles in Extraction Column," Dissertation, Technischen Universität München, Germany, April 2006.

Predictive calculations of the transport properties of alkylamines

A. Rendón-Calle[1], G. Orozco[2+], S. Builes[1*]
1: Grupo de Investigación DDP, Universidad EAFIT. Medellín, Colombia
2: Grupo de Modelado y Computación Científica, Facultad de Ciencias, Universidad Antonio Nariño. Bogotá, Colombia
+ gustavo.orozco@uan.edu.co
* sbuiles@eafit.edu.co

The use of fossil fuels for energy generation has led to a continuous increase of CO_2 concentration on the atmosphere. This rise in CO_2 concentration increases the human-driven greenhouse effect. Economic reasons make it necessary to find means to keep relying on fossil fuels for energy, but separating and capturing CO_2 emissions.

It is possible to separate CO_2 from flue gases using an acid-base reaction. This process, called acid gas removal, consists on reacting aqueous amines with CO_2 forming carbonates in an exothermic reaction, which can be reversed by increasing the temperature. Thus, in a different process unit the carbonates are heated in order to desorb the CO_2, separating it, and recovering the amines, for its subsequent use in the absorption process. One of the main drawbacks of the process is the high viscosity of the amines, which have to be diluted in order to be pumped around the absorption-desorption system. Effective selection of the amines requires a method that can relate the structure of the adsorbent to its performance. This can be achieved by using molecular simulations, as they relate the microscopic behavior of the molecules during the absorption to their macroscopic properties, such as viscosity and density. One of the objectives in this research consists on exploring systematically the properties of different amines using molecular simulations.

A systematic experimental study of the CO_2 capture potential of amines is an expensive process. An alternative to measurement of experimental properties, is the calculation with predictive methods. In this work, we studied the behavior of properties such as viscosity and density of kinds of amines at varying operating conditions in order to generate new data that could be used in the design of equipment for the CO_2 removal process. First, we validated molecular dynamics (MD) simulations using experimental data of density, viscosity and self-diffusion coefficients of primary, secondary, and tertiary alkylamines. These properties are involved in the calculation and design of the process units required for CO_2 separation and capture. In order to determine the most accurate or appropriated model to make the predictions of thes properties, the simulations were performed using two different interaction potentials TraPPE [1] and OPLSaa [2]. In addition the properties calculated with the recently developed AUA4 [3-5] were included for comparison. The results were compared with available experimental data for validation of the applicability of these three different potentials as accurate predictive means of generating absorption data for applications of CO_2 separation and capture.

Properties of primary, secondary, tertiary and bifunctional alkylamines were calculated using the GROMACS software for molecular dynamics calculations. Equilibration runs in the NVT and NPT ensembles, of 1 ns each, were used. Production simulations of 10 ns were performed in the NPT ensemble using a Berendsen thermostat and a Noose-Hoover barostat. The integration time step was 2 fs for equilibration and 1 fs for production. The intermolecular potential parameters are derived from the work of Wick et al [1] and Rizzo et al [2]. The potential parameters for unlike site pairs are expressed via the Lorentz-Berthelot mixing rules. The equations of motion were integrated using the velocity Verlet algorithm with constrained bonds and using the Lincs algorithm for both force fields TraPPE and OPLSaa. In all cases, 343 molecules were placed in a cubic box with periodic boundary conditions. The shear viscosity coefficient was estimated using the Green-Kubo formalism.

$$\eta = \frac{V}{K_B T} \int_0^{\infty} \langle P_{xz}(t_0) P_{xz}(t_0 + t) \rangle_{t_0} \, dt \qquad (1)$$

The self-diffusion coefficient was calculated as the time integral of the velocity autocorrelation function (VACF).

$$D = \frac{1}{3N} \int_0^{\infty} dt \left\langle \sum_{k=1}^{N} v_\alpha^k(t_0 - t)\, v_\alpha^k(t_0) \right\rangle \qquad (2)$$

Acknowledgements

Support for this work was provided by Universidad EAFIT (project 621-000003). Computational time was provided by Apolo Centro de Computación Científica.

References:

1. Wick, C.D., et al., *Transferable Potentials for Phase Equilibria. 7. Primary, Secondary, and Tertiary Amines, Nitroalkanes and Nitrobenzene, Nitriles, Amides, Pyridine, and Pyrimidine.* The Journal of Physical Chemistry B, 2005. **109**(40): p. 18974-18982.
2. Rizzo, R.C. and W.L. Jorgensen, *OPLS all-atom model for amines: resolution of the amine hydration problem.* Journal of the American chemical society, 1999. **121**(20): p. 4827-4836.
3. Orozco, G.A., et al., *A Transferable Force Field for Primary, Secondary, and Tertiary Alkanolamines.* Journal of Chemical Theory and Computation, 2013. **9**(4): p. 2097-2103.
4. Orozco, G.A., et al., *Transferable Force Field for Equilibrium and Transport Properties in Linear, Branched, and Bifunctional Amines I. Primary Amines.* The Journal of Physical Chemistry B, 2011. **115**(49): p. 14617-14625.
5. Orozco, G.A., et al., *Transferable Force Field for Equilibrium and Transport Properties in Linear and Branched Monofunctional and Multifunctional Amines. II. Secondary and Tertiary Amines.* The Journal of Physical Chemistry B, 2012. **116**(21): p. 6193-6202.

An inexpensive approach to quantify binary and ternary mixtures of monovarietal extra virgin olive oils

R. Aroca-Santos,[1] J. C. Cancilla,[1] G. Matute,[1] A. Pérez-Pérez,[1] E. S. Pariente,[1] K. W. Wierzchos,[2] A. Moral,[3] A. Tijero,[1] J. S. Torrecilla*[,1]
[1] Departamento de Ingeniería Química, Facultad de Ciencias Químicas, Universidad Complutense de Madrid, 28040 Madrid, Spain; Tel. +34913944244, email: jstorre@ucm.es; [2] Applied Physics Concentration in Astrophysics. University of South Florida e-mail: kwierzchos@mail.usf.edu [3] Departamento de Biología Molecular e Ingeniería Bioquímica, Área de Ingeniería Química, Universidad Pablo de Olavide, 41013 Sevilla, Spain; email: amoram@upo.es.

Abstract
Olive oil, and particularly extra virgin olive oil (EVOO), is the leading source of fat in the well-known Mediterranean diet [1]. It is extracted from olives through mechanical processes at low temperatures, and its consumption is becoming more and more extended, among other reasons, due to the abundant scientific studies regarding its organoleptic properties and demonstrating its favorable effects on human health [2-4]. Its complex composition, which turns EVOO into an economically and nutritionally valuable product, has been conventionally split in two major parts: the greater one, the saponifiable fraction, which is mostly formed by fatty acids and triacylglycerides (TAGs), and the minor one, the insaponifiable fraction, which possesses a wide-ranging amount of compounds regarding chemical nature such as polyphenols, sterols, and pigments, such as carotenoids and chlorophylls (mainly lutein and pheophytin-a, respectively) [3,5]. Due to the importance of EVOO in the market, the legal protection of such a high-quality product is required, and it must be backed up by various chemical processes to monitor, assess, and certify the quality and denomination of origin of EVOOs, and to avoid adulterations or illegal mixtures with low-grade or -quality oils [6-8].
In this sense, monovarietal EVOOs (gathered from a single variety of olive) have been thoroughly analyzed, as they maintain the organoleptic properties which are intrinsically linked to the fruit variety chosen, such as flavor, odor, phenolic compounds and amount, and so on [9]. These oils have been studied and characterized during multiple analyses in order to try and understand the relation between the organoleptic profile and the composition of a particular EVOO, and even with the roles they may play in human health [6]. Locating the chemical differences between different monovarietal EVOOs can lead to the creation of methodologies which are able to provide the estimation of the quality of EVOOs.
Three distinct Spanish monovarietal EVOOs (Cornicabra, Picual, and Hojiblanca), as well as their binary and ternary mixtures, have been studied using visible spectroscopy. The information gathered from the spectra was mathematically treated and modeled following two chemometric approaches: multiple linear regression (MLR) models, and non-linear artificial neural networks (ANNs). All the models were validated employing k-fold cross-validations, and the largest mean absolute errors for the olive varietal quantifications were 10% in volume for the MLR and 2.8% for the ANN. This validation assures that the mathematical tools can generalize well, proving they can accurately quantify EVOO varietals in mixtures only using visible spectroscopy information.
These results manifest that ANNs can be employed to very accurately quantify EVOO mixtures (better than MLRs), not only to gather information of the different monovarietals or to classify them, but also to control the quality of these valuable products following an

inexpensive and straightforward procedure which, in addition, could be used for real-time applications.

To sum up, a reliable and accurate approach to quantify the composition of monovarietal EVOO mixtures relying on visible spectroscopy and non-linear chemometrics has been successfully developed, without the need of any previous complex sample treatment or chemical use. This could lead to the development of inexpensive quality control systems which would be able to directly inform about the composition of blends of EVOOs in real time, without the need of specialized personnel.

References

[1] Franco, M.N.; Galeano-Díaz, T.; Sánchez, J.; De Miguel, C.; Martín-Vertedor, D. Total phenolic compounds and tocopherols profiles of seven olive oil varieties grown in the south-west of Spain. J. Oleo Sci. 2014, 63, 115-125.

[2] Grosso, G.; Mistretta, A.; Frigiola, A.; Gruttadauria, S.; Biondi, A.; Basile, F.; Vitaglione, P.; D'Orazio, N.; Galvano, F. Mediterranean diet and cardiovascular risk factors: a systematic review. Crit. Rev. Food Sci. 2014, 54, 593-610.

[3] Dalen, J.E.; Devries, S. Diets to prevent coronary heart disease 1957-2013: what have we learned? The American Journal of Medicine, 2014, 127, 364-369.

[4] Lozano-Sánchez, J.; Cerretani, L.; Bendini, A.; Gallina-Toschi, T.; Segura-Carretero, A.; Fernández-Gutiérrez, A. New filtration systems for extra-virgin olive oil: effect on antioxidant compounds, oxidative stability, and physicochemical and sensory properties. J. Agric. Food Chem. 2012, 60, 3754-3762.

[5] Sinelli, N.; Casale, N.; Di Egidio, V.; Oliveri, P.; Bassi, D.; Tura, D.; Casiraghi, E. Varietal discrimination of extra virgin olive oils by near and mid-infrared spectroscopy. Food Res. Int. 2010, 43, 2126–2131.

[6] Cancilla, J.C.; Wang, S.C.; Díaz-Rodríguez, P.; Matute, G.; Cancilla, J.D.; Flynn, D.; Torrecilla, J.S. Linking chemical parameters to sensory panel results through neural networks to distinguish olive oil quality. J. Agric. Food Chem. 2014, 62, 10661-10665.

[7] Reboredo-Rodríguez, P.; González-Barreiro, C.; Cancho-Grande, B.; Simal-Gándara, J. Quality of extra virgin olive oils produced in an emerging olive growing area in north-western Spain. Food Chem. 2014, 164, 418-426.

[8] International Olive Council. (2013). Trade standard applying to olive oils and olive pomace oils; Madrid, Spain. T.15/NC No.3/Rev.7.

[9] Longobardi, F.; Ventrella, A.; Casiello, G.; Sacco, D.; Tasioula-Margari, M.; Kiritsakis, A.K.; Kontominas, M.G. Characterisation of the geographical origin of Western Greek virgin olive oils based on instrumental and multivariate statistical analysis. Food Chem. 2012, 133, 169–175.

Comparison between the predicted coke yields of the FCC unit with different ANN models

X.Su, Y. Wu, H. Pei, J. Gao, X. Lan.
State Key Laboratory of Heavy Oil Processing, China University of Petroleum, Beijing 102249, China
tel. +8613426259855, e-mail: suxin_0806@163.com; State Key Laboratory of Heavy Oil Processing, China University of Petroleum, Beijing 102249, China.

Fluid catalytic cracking (FCC) is one of the most popular processes to convert the heavy petroleum hydrocarbons into light oil products like dry gas, liquefied gas, gasoline and diesel and into by-products like slurry oil and coke in the presence of high temperature and catalyst. The coke is a more important by-product because it relates to the heart balance of the FCC unit. The increasing of coke yield not only reduces the total liquid yield, but also affects the heat balance of the plant and increases the device operation difficulty. It is of great significance to control the coke yield of the catalytic cracking unit at lower levels on a small scale fluctuation to improve the total liquid yield as well as the operation stability and to increase economic benefits. FCC coke formation and burning process is a continuous process with a complex reaction mechanism which is affected by many parameters which influence each other. Therefore, it is difficult to establish the prediction model of many parameters using traditional models.

The Artificial Neural Network (ANN) has a strong self-learning and adaptive ability, which shows obvious superiority in nonlinear forecasting. To establish a good method to predict the coke yield, this paper combines the industrial production data and respectively establishes the generalized regression neural network (GRNN) forecasting model and the back propagation (BP) neural network prediction model by using the artificial neural network, in which 28 key parameters of influencing the catalytic cracking coke yield are regarded as the input and the coke yield as the output. The comparison and analysis results show that the accuracy and stability of the BP neural network prediction results are better than that of the GRNN. In this study, we respectively use the particle swarm optimization (PSO) and genetic algorithm (GA) to optimize the BP neural network in order to further improve the prediction precision. The analysis and comparison of these two kinds of optimization algorithm show that they both improve the prediction precision of the BP neural network. Considering the accuracy and stability of the prediction results, the BP neural network optimized by the genetic algorithm forecasting model is better than that of the BP neural network optimized by the particle swarm algorithm prediction model.

Design and synthesis of regulation system for controlled release of CO_2 during the probiotic beverage fermentation

A. Goršek, [1]J. Ritonja, D. Pečar
University of Maribor, Faculty of Chemistry and Chemical Enginering, Smetanova 17, 2000 Maribor, Slovenia; tel. +385 2 2294 453, e-mail: andreja.gorsek@um.si; [1]Jožef Ritonja, University of Maribor, Faculty of electrical engineering and computer science, Smetanova 17, 2000 Maribor, Slovenia; tel. +386 2 2207 074, e-mail: jozef.ritonja @um.si; tel. +385 2 2294 453, e-mail: darja.pecar@um.si.

Milk is an excellent growth medium for microorganisms, which means that fermentations could be performed with different starter cultures yielding a range of dairy products. One of the basic parameters, which describe the starter culture, is its ability for production of aromatic products. Therefore, yogurt and similar products have characteristic smell and taste, which is the consequence of volatile/non-volatile acids and carbonyl products formed during fermentation. Because of its unique smell and taste kefir is very well-known among milk drinks.

It has been known for a long time that one of the growth and metabolism products of the microorganisms is carbon dioxide (CO_2). To understand the impact and use of this gas in the microbial digestion it is necessary to monitor its level when it is dissolved in the fermentation medium. Too low concentration means atypical flavor of the drink, too high acting inhibitory to microorganisms. Fluctuations of CO_2 concentration may occur due to technological errors, or due to unforeseen action of microorganisms. Such fluctuations are not desirable on the industrial scale of milk-based drinks production. There are several known methods of measuring the CO_2 content in the fermentation medium, but there is still a lack of control systems.

For this reason we developed the simplified mathematical model of CO_2 release system. An intuitive approach based on experimental analysis was used for the modeling. The studied chemical system was dealed as two-inputs (temperature, mixing frequency) one-output (CO_2 concentration) dynamical system.

First the time response of the output quantity on the constant input quantities by different initial conditions was meassured. It was shown that the ouput response could be described by a simple exponential function. For the different constant values of the input variables the different numerical parameters of the exponential function were obtained.

After that we determined the effect of the input variables' dynamics to the output variable's response. During the fermentation process we generated a step changes of the input variables. Based on a comparison of CO_2 concentration responses in the case of constant input quantities and in the case of the step changed input quantities, we determined the transfer functions which connect the input- and output variables. The experimentally obtained model is consistent with the theoretical model of the process. Comparison of simulation and experimental results showed a high degree of correlation. The model is suitable for further analysis of this process and for the design and synthesis of the control for CO_2 release systems.

Development and validation of a 3D transient heterogeneous model of a structured plate reactor-exchanger for the methanation of CO_2

R. Try[1, 2,*], A. Bengaouer[1], C. Jallut[2]
[1]French Alternative Energies and Atomic Energy Commission (CEA), Laboratory for Innovation in New Energy Technologies and Nanomaterials (LITEN), 17 Rue des martyrs, 38000 Grenoble, France.
[2]Université de Lyon, F-69622, Lyon, France; Université Lyon 1, Villeurbanne ; LAGEP, UMR 5007, CNRS, CPE, 43 bd du 11 novembre, 69100 Villeurbanne, France.
*Corresponding author: tel. (+33)4.38.78.20.42, email: rasmey.try@cea.fr

The integration of renewable energy into the electric sources grid underlines the issue of the storage of an excess of electricity production. The Power-to-gas is one of the solutions to store this energy by converting the electric surplus into fuel gas (hydrogen, SNG: Substitute Natural Gas).SNG is preferred due to the available gas infrastructure, but it needs a step of hydrogenation of carbon dioxide into methane, or methanation. The intermittent nature of the electricity production implies that the methanation step has to support a certain level of flexibility of working condition. Consequently, the dynamic behavior of the reactors that are used to perform this step has to be studied.

Methanation reaction is balanced, thermodynamically favored at low temperature (250°C to 300°C) and highly exothermic. Its kinetic follows the Arrhenius law leading to strongly non-linear phenomena both in stationary and intransient regime. In the frame of our studies, the methanation step is performed in an intensified milli-structured plate reactor-exchanger. The reactor is made of a parallelepiped piece of stainless steel. Several tens of millimetric channels with a rectangular section are placed in the axial direction and are filled with fixed beds of nickel-alumina catalyst powder. A cross flow of thermal oil allows to cool the system (Figure 1). The study of this particular reactor has been done in stationary so far [1]. My work is devoted to investigate its dynamic behavior and stability under varying operating conditions.

In literature, numerous papers on dynamic behavior of fixed bed reactors are available but most of them deal with classical tubular fixed-bed adiabatic reactors. For example, it has been shown that unexpected behavior could happen when changing the operating point such as wrong way behavior, travelling hot spot, overshoot and bifurcation. These phenomena are undesirable if they are not controlled because they can modify the reactor's performance or they can lead to reduce the catalyst lifetime. These phenomena have been observed in our milli-structured reactor.

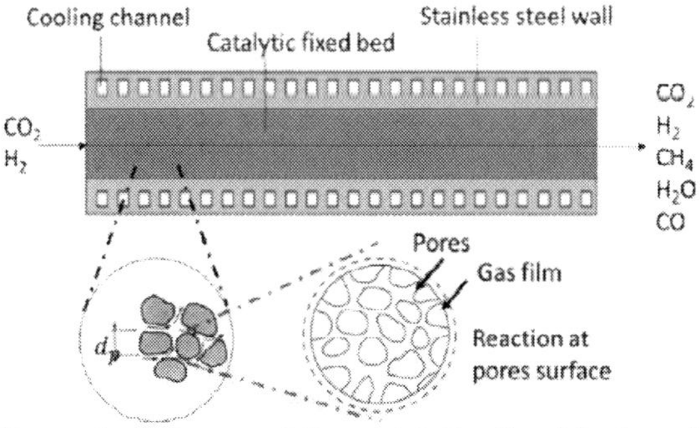

Figure 1: Scheme of the catalytic fixed bed methanation reactor cooled by a cross-flow thermal oil

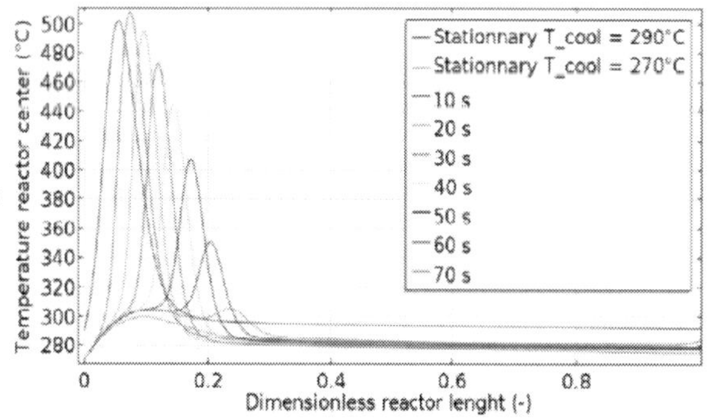

Figure 2: Simulation of the transient behavior of the reactor after a step down of the cooling temperature

In order to study the dynamic behavior of our reactor, a dynamic heterogeneous model has been developed by using the COMSOL Multiphysics software. As each channel of our reactor has a rectangular section, phenomena are not radially uniform, thus it is a 3D model based on coupled energy, material and momentum balances (Figure 1) is used for the gaseous phase. Furthermore, diffusion limitations take place in the catalyst and between the catalyst powder and the reacting gas. Under such conditions, the dynamic model has to describe these limitations, this is done by solving heat and mass transport equations in the catalyst powder. For the moment, we assume that no mal-distribution occurs between the channels so that a single channel is represented. Firstly, the steady-state simulations obtained from this model are compared with stationary experimental data so that some intrinsic parameters of the reactor's model, such as permeability, heat transfer coefficient or kinetics are fitted to have a good agreement. Secondly, the model is used to simulate over time the evolution of the reactor performance (temperature, concentrations, pressure drop) when the operating point changes.

Simulation results predict wrong way behaviors and traveling hot spots inside the reactor. For instance, if the cooling fluid temperature suddenly drops from 300°C to 270°C (Figure 2),the maximum reactor temperature increases at first (curve at t= 10s) then the temperature peak decreases and moves toward the outlet of the reactor (curves for $t \geq 10s$).The amplitude of these phenomena highly depends on the thermal properties of the catalyst as density, heat capacity and thermal conductivity that are not very well-known.

These phenomena are experimentally observed at the reactor scale and will be investigated at the channel scale in order to avoid potential interactions between channels. In this way, a one-channel reactor pilot plant highly instrumented by thermocouples is developed and will be helpful to identify parameters and mechanisms governing the reactor behavior under non-stationary working conditions.

[1] Ducamp, J. (2016). *Conception et optimisation d'un réacteur-échangeur structuré pour l'hydrogénation du dioxyde de carbone en méthane de synthèse dédié à la filière de stockage d'énergie électrique renouvelable*. Université de Strasbourg.

Discovering waste integrated valorisation routes using an ontology engineering approach

F. Barla, F. Lykokanellos, A. C. Kokossis
National Technical University of Athens, School of Chemical Engineering, 9 Heroon Polytechniou Rd., 15780 Zografou, Greece.

The ever increasing global waste production in combination to the scarcity of materials creates an enormous strain on our resource system [1]. The needs to reduce the use of virgin resources give rise in research areas such as waste Industrial Symbiosis (IS) [2], waste biorefineries, waste valorisation and circular approach economies. Waste resources can considered abundant generated by both industrial and municipal activities. The chemistries or else known as "synthesis paths" to extract materials from waste can also be considered abundant.

Waste is a multicomponent material, containing valuable chemical compounds. If recovered, they can become standalone feedstock or be also integrated with alike feedstocks to be processed by existing industrial facilities. Such sustainable, retrofitting practice will enlarge the process capacity, benefiting from economies of scale, improve LCA performance and cost-effectiveness originating from waste's off-market prices [2].

To achieve such task, a) to appropriately integrate waste resources with virgin feedstocks and 2) to identify suitable alternative synthesis paths, producing diverse products, requires the use of two well-established research fields: ontology engineering and mathematical optimisation. Therefore, it is firstly required a large well-structured database to organise knowledge, achieving interoperability of information across various manufacturing organisations and processing facilities. Secondly, an optimisation based decision support platform is necessary to take into consideration all three decision resolution levels: strategic, tactical and operational [3].

This work provides a framework that combines an ontological based platform with an optimisation process systems tool to first organise and manage tacit knowledge such as chemistries, process models, paths, technology specifications, feedstock characteristics etc. for knowledge sharing across cooperative processing units [3] and secondly to discover the optimum synthesis path according to set environmental, economic and technical objectives [4]. The semantic algorithm is used to establish layers of interactions and translate process models into networks that are further fathomed using optimisation and superstructure technology.

Acknowledgements
The financial support from Marie Curie European Research Program, RENESENG (FP-607415) is gratefully acknowledged.

References
[1] Lin, C. S. K., Pfaltzgraff, L. A., Herrero-Davila, L., Mubofu, E. B., Abderrahim, S., Clark, J. H., Koutinas, A. A., Kopsahelis, N., Stamatelatou, K., Dickson, F., Thankappan, S., Mohamed, Z., Brocklesby, R., Luque, R.: Food Waste as a valuable resource for the production of chemicals, materials and fuels. Current situation and global perspective, Energy Environ. Sci. 6, 426-464 (2013)

[2] Cecelja, F., Trokanas, N., Raafat, T., Yu, M.,: Semantic algorithm for Industrial Symbiosis network synthesis, Comp. Chem. Eng. (2015) http://dx.doi.org/10.1016/j.compchemeng.2015.04.031
[3] Munoz, E., Capon-Garcia, E., Lainez, J. M., Espuna, A.: Integration of enterprise levels based on an ontological framework, Chem. Eng. Res. Des. 91, 1542-1556 (2013)
[4] Tsakalova, M., Lin T.-C., Yang A., Kokossis, A. C.: A decision support environment for the high-throughput model-based screening and integration of biomass processing paths. Ind. Crops Prod. (2015), http://dx.doi.org/10.1016/j.indcrop.2015.05.035

Evaluating the applicability of reactive distillation to quaternary reversible reaction systems

Rahma Muthia,[1] Arjan Reijneveld,[1] Louis van der Ham,[1] Gerrald Bargeman,[2] Antoon ten Kate,[2] Anton A. Kiss[1,2]
[1] Sustainable Process Technology, Faculty of Science and Technology, University of Twente, PO Box 217, 7500 AE Enschede, The Netherlands
[2] AkzoNobel Research, Development and Innovation, Process Technology Strategic Research Group, Zutphenseweg 10, 7418 AJ Deventer, The Netherlands
E-mail: a.a.kiss@utwente.nl, tony.kiss@akzonobel.com

Reactive distillation (RD) is as an attractive unit operation combining simultaneous reaction and separation in a single column. Since the early 20th century, scientific literature and patents related to design methods of RD (e.g. graphical methods, optimization techniques, and heuristic approaches) have been published [1-3]. However, the lack of knowledge for the applicability range of RD in the early stage of design (conceptual phase) is hindering the selection and use of RD in chemical processes [3]. To overcome such limitation, this study presents a fast screening method to check the applicability of RD for quaternary reversible reactions. For this assessment only minimal basic data is required, such as the chemical equilibrium constant (K_{eq}), kinetic rate, and the relative volatilities of the key compound pairs.

The quaternary reversible reaction system 2R-2P ($A + B \leftrightarrow C + D$) was investigated by specifying both reactants as mid-boiling components ($T_{b,C} < T_{b,A} < T_{b,B} < T_{b,D}$) and assuming ideal VL behaviour, i.e. constant relative volatility (α). The boiling point ranking is the most favourable order, since good separation of products is attainable. The feed streams of pure A and B are fed in stoichiometric ratio (as saturated liquids) to a RD column operating at atmospheric pressure. The light reactant is fed at the bottom part of the reactive zone and the heavy reactant is fed at the top of reactive zone, in order to obtain a counter current flow in the reactive zone (Figure 1). As design variables, the value of the equilibrium constant (K_{eq} = 0.1-10), column configuration (number of theoretical stages NTS ≤ 100, location and length of reactive zone), and reflux ratio (RR ≤ 100) were varied in the sensitivity analysis. Generic cases were investigated and simulated using the process simulator Aspen Plus (v8.6) in order to cover an extended range of equilibrium constants and relative volatility combinations.

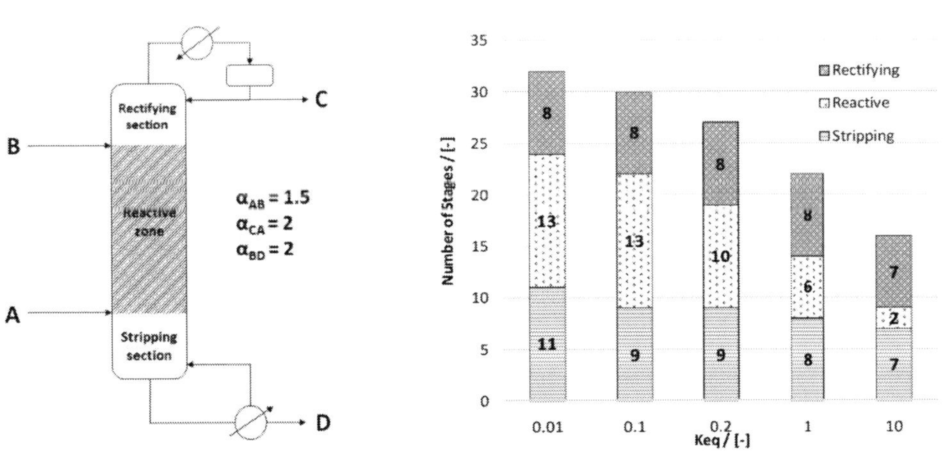

Figure 1. RD column configuration (left) and optimal RD column configurations for an equilibrium limited system (right)

For convenience and considering space limitations, this abstract highlights the results of the base case only, which considers relative volatility values α_{A-B}, α_{C-A}, α_{B-D} being equal to 1.5-2-2, respectively. By setting the product purity to 99% as an important constraint, and fixing the bottom to feed (B/F) ratio to 0.5, a sensitivity analysis was performed in each scenario considering a variable number of stages in order to obtain the lowest energy requirements. For each value of K_{eq}, the optimal configurations were determined based on the combination of number of theoretical stages (NTS) and reflux ratio (RR) that resulted in the minimum value of the objective function, which is directly linked to the total annual cost:

$$F_{ob} = NTS \times (RR + 1) \qquad (1)$$

The effect of reaction kinetics was investigated next, since slow reaction rates indicate a kinetically controlled system [3]. When the system was assumed to reach equilibrium, this gave the best performance of the RD column as the system is not limited by the reaction rate but only by chemical equilibrium. In the kinetically controlled model, the residence time per stage was specified at a reasonable value of 30 s. By changing the kinetic rate constant the Damköhler number ($Da = k_f \times V/F$) number was varied in the range 0.01-10 to show the effect of kinetics. The RR vs NTS plot and the optimal column configurations for various K_{eq} values are shown in Figure 1 and Figure 2. Additionally, for a specific value of $K_{eq} = 0.1$, the effect of kinetics is shown in Figure 2, at various Damköhler numbers. In the range of K_{eq} =10-0.1 the minimum reflux ratio hardly changes, being in contrast with the significant increase of RR when $K_{eq} < 0.01$. This is necessary to increase the reactant concentrations in the reactive zone to deal with the unfavourable K_{eq} value. The increase in the minimum number of stages is about constant (3-4 stages) for each factor 10 decrease in K_{eq} value. This increase fully covers the increase in the number of reactive stages. The kinetically controlled RD simulations show that above Da=0.1 the results are close to the equilibrium results. In contrast, below Da=0.05 the optimal reflux ratio and number of stages increases significantly.

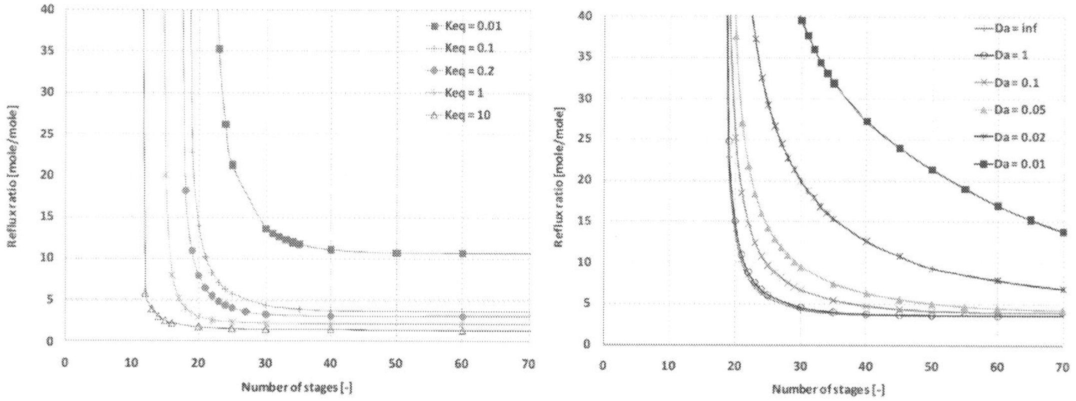

Figure 2. Design parameters for an equilibrium limited RD column (left) and influence of reaction rate (expressed as Da) at $K_{eq} = 0.1$ for a kinetically controlled RD column (right)

The results of this work ultimately allow a quick estimation and sanity check of the RD column configuration, prior to performing any rigorous process simulations, and based only on minimal basic data (equilibrium constant, kinetic rates and relative order of volatilities).

References

1. Hoyme C. A., *Doctoral dissertation*, University of Tennessee, Knoxville, 2004.
2. Taylor, R., Khrisna, R., *Chem. Eng. Sci.*, **2000**, 22, 5183-5229.
3. Shah M., Kiss A. A., Zondervan E., de Haan A. B., *Chem. Eng. Process.*, **2012**, 60, 55-64.

Evaluation and optimal design of two electrochemical units for amoxicillin removal from wastewaters streams

A. López-Yáñez[1], J. M. Zamora-Mata[2], A. Alonso[3], J. Ramírez-Muñoz[1,*]
[1]Universidad Autónoma Metropolitana-Azcapotzalco, Departamento de Energía. Av. San Pablo 180, C.P. 02200, CDMX, Mexico. [2]Universidad Autónoma Metropolitana-Iztapalapa, Departamento de IPH, Av. San Rafael Atlixco No. 186, Col. Vicentina, Iztapalapa, CDMX. 09340, Mexico. [3]CONACYT-Universidad Autónoma Metropolitana-Azcapotzalco. Av. San Pablo 180, C.P. 02200, CDMX, Mexico. *email: jrm@correo.azc.uam.mx.

Abstract

In this work, a nonlinear programming (NLP) model, incorporating the electrocoagulation kinetic is developed based on a basic treatment superstructure, which covers many treatment alternatives. The objective function minimizes the flow through the reactor, which is considered proportional to the treatment cost. It was found that by using a NLP model it is possible to obtain an optimal design for an effluent treatment network contaminated with amoxicillin. The use of the proposed NLP model is depicted with a study case which shows its versatility to attain optimal results for a set of effluent streams.

Introduction

The existing conventional water treatment plants were not designed for antibiotics removal and therefore, they are being regularly released into the aquatic environment and groundwater. Regulation, treatment and emission of these substances have received little attention in most countries around the world. Its inclusion as dangerous materials depend upon the damage to human health, as well as its presence and dispersion in water resources [1]. Amoxicillin is a very widely used antibiotic and its natural metabolism inside the human body makes that its concentration in excretions retain important levels, mainly in the health care facilities effluents, e.g. hospital sewage. Therefore, its removal from wastewater has become a major issue recently [2,3]. This paper presents an evaluation of one alternative for the selective treatment of wastewater streams carrying amoxicillin. The considered treatment technology is electrocoagulation.

Methodology

The basic superstructure of the network presented by Zamora et al. [4] includes splitters (S), mixers (M), a treatment unit (UT) and the interconnections. The problem studied in this work for two treatment units can be stated as follows: Given a set of effluent streams, with known flow-rates at certain concentrations level of amoxicillin (Table 1). The objective is to determine the corresponding optimum intermediate flowrates so that discharge composition regulations for the pollutant is met. For the treatment units, a NLP model is formulated and solved. The NLP problem consists of an objective function (e.g. cost), mass balances (e.g. UT, S, M), and input-output relationships for the treatment units.

Table 1. Wastewater process data.

I	S_i (L/min)	$C_{i,j}$ (mg/L)	$m_{i,j}=S_{i,j}C_{i,j}x10^{-3}$ (g/min)
1	6	28	0.168
2	5.5	21	0.1155
3	4.5	19	0.0855
4	4	17	0.068
5	3	16	0.048
$m_j = \Sigma m_{i,j}$			0.485

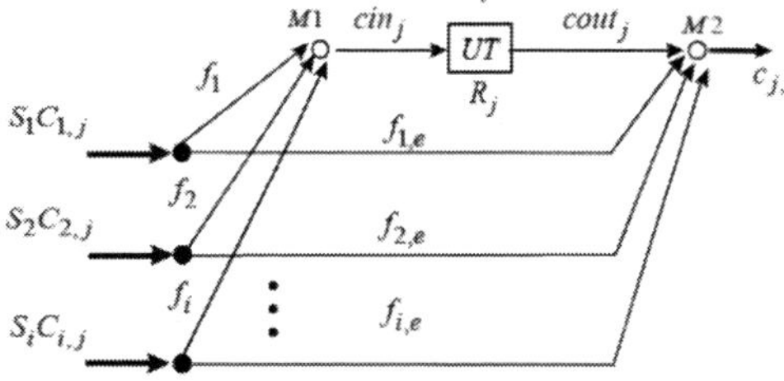

Fig. 1. Basic superstructure [4].

Results and Discussion

The NLP model [5] and basic superstructure [4] were adapted for two electrochemical treatment units. This approach is explained with the solution of an illustrative example. Because the amoxicillin is considered an emergent contaminant [3], a discharge limit of 10 mg/L was assumed. In the example, an electrochemical technology model [3] was used for the removal of the contaminant. GAMS has been used as the equation modeling system, by using GAMS/MINOS as the NLP solver. The electrochemical treatment cost is optimized by minimizing the treatment flowrate. To obtain a global optimal design, a stochastic multi-start optimization algorithm was utilized for the solution of the NLP model.

The final design for the wastewater treatment network is presented in Figure 2. As can be seen, the stream 1 and a fraction of flowrate 2 are treated in the UT_1, while the untreated fraction of flowrate 2 is treated in the UT_2. Fraction of flowrate 3 is treated in the UT_2, while the untreated fraction of flowrate 3 is carried directly to mixer 3 (M_3). The streams 4 and 5 were bypassed directly to mixer 3. The total flowrate treated for this case is 14.378 L/min.

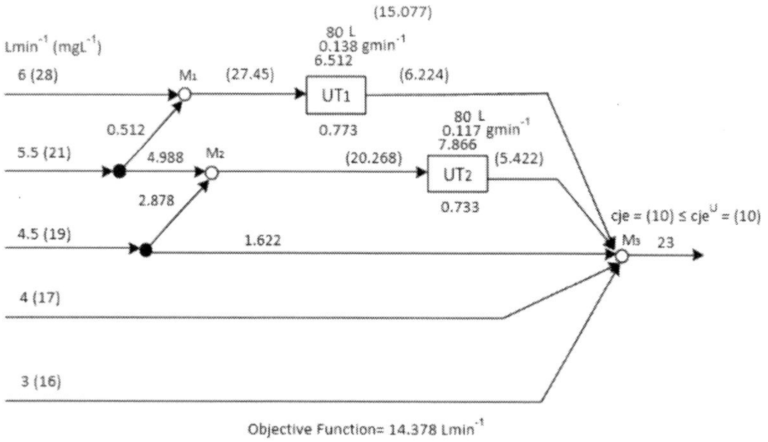

Fig. 2: Topological design for a wastewater treatment network.

Conclusions

The results for the illustrative design example presented in this work show that the use of a versatile design approach is important when the utilized removal process exhibits a relatively complex kinetic model; such is the case of an electrocoagulation treatment process. The developed NLP design approach represents a methodology that allows the design of a minimum treatment flowrate system based on an electrochemical reactor. This could be greatly helpful in order to explore the use of this type of treatment technology, which has been considered traditionally as a non-efficient process.

References

[1] Petrovic, M., Gonzalez, S., Barceló, D., 2003. Analysis and removal of emerging contaminants in wastewater and drinking water, Trends Analyt. Chem. 22, 685–696.

[2] Homem, V., Alves, A., Santos, L., 2010. Amoxicillin degradation at ppb levels by Fenton oxidation using design of experiments, Sci. Total Environ., 408, 6272– 6280.

[3] Padilla, G., Alonso, A., Martínez, S.A., González, M., Jaúregui U.J., Ramírez-Muñoz, J., 2015. Electrochemical degradation of amoxicillin in aqueous media, Chem. Eng. Process, 94, 93–98.

[4] Zamora, J., Hernández, R., Castellanos, J. 2004. Linear programming model to assist the design of distributed wastewater treatment systems (In Spanish), Rev. Mex. Ing. Quím. 3, 121–134.

[5] López-Yáñez, A., Zamora-Mata, J.M. Alonso, A., Ramírez-Muñoz, J., 2016. Determination of an optimal structure for electrochemical treatment of wastewater streams contaminated with amoxicillin traces (In Spanish), Inf. Tecnol., 27 (5) In press.

Fuzzy control design for a biochemical process

A. Vasičkaninová, M. Bakošová, J. Oravec, A. Mészáros
Slovak University of Technology in Bratislava, Faculty of Chemical and Food Technology, Institute of Information Engineering, Automation, and Mathematics, Radlinského 9, 812 37 Bratislava, Slovak Republic; e-mail: anna.vasickaninova@stuba.sk.

Biochemical processes are very difficult to control due to their strong nonlinear dynamics and time varying parameters. Fermentation process, wherein the cultivation medium consists of living microorganisms, is vital and important in many biotechnological applications and it is necessary to eliminate such negative effects as substrate inhibition, catabolite repression, product inhibition, glucose effect and so on, many variables are of difficult measurement, and usually involve laboratory analyses.

The nonlinear model describing the response of Saccharomyces cerevisiae, known as baker's yeast, has been used for the non-linear process control simulation. This mathematical model is based on limited oxidation capacity of yeast leading to a switch-over from oxidative to oxide-reductive metabolism.

The main goal is to maintain a desired profile of dissolved oxygen concentration in the fermenter by manipulating the process inlet air flow rate.

Conventional methods of controller design require detailed information about the controlled system, especially time constants, time delays and steady-state gains. The nonlinearity and time-dependence variation of plant parameters can cause incorrect identification of system as well as controller parameters. This parameters mismatch usually leads to degradation of control performance resulting in oscillations, overshoot or long regulation times.

Fuzzy logic control based on the Takagi–Sugeno inference method has been applied for the control of the baker's yeast fermentation. The simulation results show that the fuzzy logic controller improves the performance of the conventional scheme. In spite of everything, the disadvantage of the fuzzy controller is in the tuning. The fuzzy controller demands a deeper knowledge of the process in study and of the own operation of the controller.

Lattice Boltzmann simulation of blood and comparison with real results

D.A. Roa Romero[1], G.A. Orozco Alvarado[2]
Departamento de Física, Facultad de Ciencias, Universidad Antonio Nariño
Carrera 3 Este No. 47A – 15. Bogotá, Colombia
Tel. +57-316 3064800. e-mail: [1]dieroa@uan.edu.co, [2]gustavo.orozco@uan.edu.co

Thrombosis and atherosclerosis are pathologies produced by the combination of several hemodynamic factors that can be difficult to identify. Furthermore, the treatments are mainly based on the use of substances that affect the blood components and functions with dangerous effects, for instance hemorrhage [1]. Therefore, the diagnosis must be as accurate as possible since it is essential in the treatment selection.

In order to propose new diagnose techniques and treatments, it is necessary to deeply understand the main physical and chemical blood features as well as its flow in the vascular system. Some studies have reported the experimental velocity profile of blood flowing in a life vessel by way of techniques based on fluorescent markers, e.g., dyeing red blood cells or nanoparticles. Those markers are inserted into the veins of life specimens such as rats or rabbits. The velocity distribution of blood can be obtained recording several images in a video and then reconstructing the markers trajectories in time [2, 3].

Another approach consists in studying the blood flow using computational algorithms. From this point of view, blood is modeled through the Navier-Stokes equations (NSE). As known, several methods have been proposed in order to find numerical solutions of the NSE. One of them, based on the transport Boltzmann equation has recently been widely used and is called Lattice-Boltzmann (LB) [4, 5]. LB is an explicit method in which the dynamics of a fluid is modeled using a set of particles interacting between them. This algorithm discretizes the fluid space in a mesh made up of cells. Each cell has associated a probability to find a certain particle at a certain point and with a certain velocity. In the specific case of blood flow, some works have been already studied features such as the non-newtonian blood behavior including different rheological models [6, 7].

The aim of this work is to study the blood flow in a rat mesenteric vessel using the LB algorithm and to compare it with the available experimental information in the open literature. Blood will be study as non-newtonian and homogenous fluid using several well established blood rheological models [6]. The algorithm was completely implemented in C++ and parallelized using the OpenMP libraries.

References

[1] F. Lavorini, B. Di Bello, et. al., "Diagnosis and treatment of pulmonary embolism: a multidisciplinary approach", Multidisciplinary Respiratory Medicine Vol 8. 2013

[2] G. J. Tangelder, *et. al.* "Velocity profiles of blood platelets and red blood cells flowing in arterioles of the rabbit mesentery", Circulation Research. Vol. 59, 505, 1986

[3] H. Ha, *et. al.*,"Hybrid PIV-PTV technique for measuring blood flow in rat mesenteric vessels". Microvascular Research, Vol. 84, 242, 2012.

[4] X. He and L. S. Luo, "Theory of the lattice Boltzmann method: From the Boltzmann equation to the lattice Boltzmann equation", Phys. Rev. E Vol. 56. 1997

[5] S. Succi, "The lattice Boltzmann equation for fluid dynamics and beyond". Oxford University Press, 2001. ISBN 0198503989

[6] M. Ashrafizaadeh, H. Bakhshaei "A comparison of non newtonian models for Lattice-Boltzmann blood flow simulations". Computer and mathematics with Applications,Vol 58. 2009

[7] Boyd.J, Buick.J.M & Green.S. "Application of the lattice Boltzmann method to non-newtonian flow carotid artery model". Australian Institute of physics 17th national congress 2006

Modelling of a dividing-wall distillation column and comparisonwith conventional 2-columns distillation system using EMSO software

J. P. Henrique, [1]R. Sousa Jr,[2]A. R. Secchi,[3]M. A. S. S. Ravagnani,[3]C. B. B. Costa
Federal University of São Carlos, Chemical Engineering GraduateProgram, São Carlos - SP- Brazil, e-mail: joaopaulohenrique@yahoo.com.br; [1]Federal University ofSão Carlos, Chemical Engineering Department, São Carlos - SP- Brazil; [2]Federal University ofRio deJaneiro, Chemical Engineering Program - PEQ/COPPE, Rio de Janeiro, RJ - Brazil; [3]State University of Maringá, Chemical Engineering Department, Maringá, PR - Brazil.

Conventional distillation columns models are composed of a large number of nonlinear differential-algebraic equations.Divided wall column (DWC) design presents an additional challenge for their modeling and optimization because much more equations are needed. However, the DWC can save up to 40% of the reboiler energy consumption [1]. Despite this advantage, DWC models are not readily available in commercial software, such as Aspen Plus© and HYSYS Process©. Hence, the usual representation of DWC in analysis and simulation reports in the literature is made with coupling of conventional distillation column models.

The aim of this work was to create a DWC model (Figure 1b) in an equation-oriented process simulator, free of charge to academic research and educational activities. This software, EMSO, has also the advantage that all developed models are open for inspection and extension by any user, which makes it extremely flexible for research use [2, 3]. A sensitivity analysis of streams composition and heat duties is then performed concerning design (total number of trays and number of divided trays) and operating (reflux ratio) parameters. Results are also compared to those obtained by a model of conventional 2-columns distillation system (Figure 1a) in the same software.

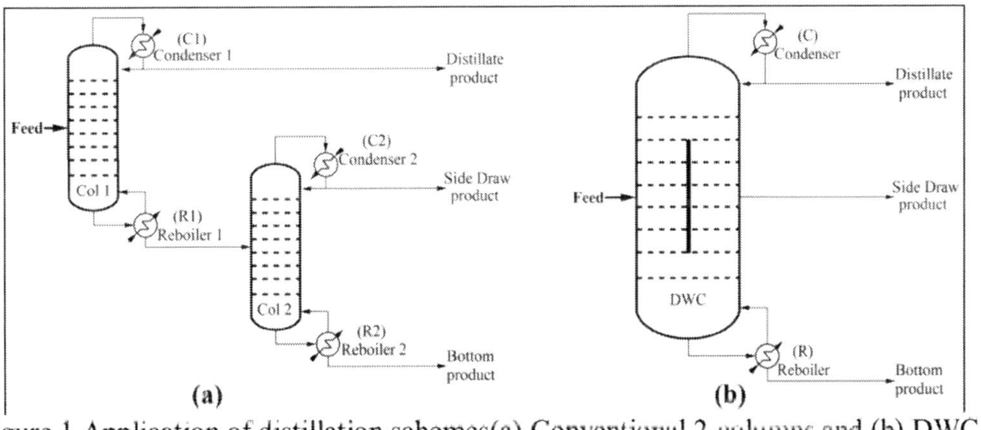

Figure 1.Application of distillation schemes(a) Conventional 2-columns and (b) DWC.

The models wereused to separate 480 kmol/h of a ternary equimolar mixture (n-pentane, n-hexane, n-heptane)using the Peng-Robinson state equation to predict the thermodynamic properties,and the boiling point of the mixture to calculate the temperature of each tray.In both schemes, each output stream has a flowrate of 160.00 kmol/h and the columns are operating at 510 kPa. Steady state results were obtained using the dynamic models with 8.3 h of simulation to generate good initial guess for the Newton method to obtain the steady state.

In the conventional 2-columns scheme, Column 1 (Col 1) has 20 trays with the feed in the 9th tray.Its bottom product feeds the 7^{th}tray of Column 2 (Col 2), which has16 trays. A reflux ratio (RR) of 3.00 is used in both columns.

According to Table 2, the DWC was defined with 27, 47 or 107 trays. The RR as well as the number of divided trays were evaluated to maintain the same separation (molar fractions in the outlet streams) performed by the conventional 2-columns scheme (Table 1). The downward liquid flow is equally divided at the inlet ofthe divided trays, and the same happens to the upwardvapor.The inlet stream and outlet side draw product are positioned in the central tray of the DWC, in opposite sides of the dividing wall.

Table1–Molar fractions products and heat duties for the conventional distillation scheme.

		Distillate product	Side draw product[1]	Bottom Product[2]	Total condenser duty (kW)	Total reboiler duty (kW)
Conventional system	n-C5	0.983	0.036	0.000	8419.24	9510.11
	n-C6	0.017	0.897	0.077		
	n-C7	0.000	0.067	0.923		

[1]Distillate and [2]Bottom product of Col 2 in the conventional 2-column scheme.

Table 2 – Molar fractions products and heat duties for the DWC distillation scheme.

	Total number of trays	Divided trays	Reflux Ratio		Distillate Product	Side draw product	Bottom Product	Total condenser duty (kW)	Total reboiler duty (kW)
DWC	27	10	7.40	n-C5	0.983	0.037	0.000	8324.94	10034.8
				n-C6	0.017	0.898	0.075		
				n-C7	0.000	0.065	0.925		
	47	31	6.11	n-C5	0.983	0.037	0.000	7044.27	8495.86
				n-C6	0.017	0.896	0.077		
				n-C7	0.000	0.067	0.923		
	107	91	6.10	n-C5	0.983	0.037	0.000	7035.92	8482.29
				n-C6	0.017	0.898	0.075		
				n-C7	0.000	0.065	0.925		

The results show that DWC is able to obtain separation equivalent to those of the conventional 2-columns distillation system, even with smaller number of trays. Though the capital investment is lower due to the smaller number of trays, with few trays it is necessary a much larger RR, increasing the operating cost.With RR approximately twice that of the conventional 2-columns system, it is possible to obtain the same separation with 10.66% less of total reboiler duty by a column 30.5% higher, if the comparison is made to the sum of the number of trays of the two conventional.

In this way, a proper DWC model, easily editable by any user,was developed and is available inanopen-source software.

Acknowledgement: This study was conducted with the support of CNPq (National Council for Scientific and Technological Development) and CAPES (Coordination for the Improvement of Higher Education Personnel), Brazil.

References:

[1] Sangal, V. K.; Kumar, V.;Mishira, I. M. Optimization of a dived wall column for the separation of C4-C6 normal paraffin mixture using box-behnken design.*Chem. Ind. & Chem. Eng. Q. (Online)*, v. 19, n. 1, p. 107-119, 2013.

[2] Soares, R. P.; Secchi, A.R. EMSO: A new environment for modelling, simulation and optimisation. *Comp. Aid. Che.Engng*.p. 947-952, 2003.

[3] Rodrigues, R.; Soares, R.P.; Secchi, A.R. Teaching chemical reaction engineering using EMSO simulator. *Comput.Appl. Eng. Educ.* v. 18, n. 4, p. 607–618, 2010.

Molecular precursors for electrochemical machines

[1]M.I. Iorga, [1]N. Mirica, [1]A. Tudoran, [1,2]M.V. Putz
[1]National Institute for Research and Development in Electrochemistry and Condensed Matter –INCEMC– Timisoara, Laboratory of Renewable Energies-Photovoltaics, St. Dr. A. Paunescu Podeanu, No.144, 300569 Timisoara, Romania, tel. 00 40 256 222119, fax 00 40 256 201382, www.incemc.ro, e-mail: mirela_iorga@yahoo.com; [2]West University of Timisoara, Laboratory of Structural and Computational Physical-Chemistry for Nanosciences and QSAR, Biology-Chemistry Department, Str. Pestalozzi No. 16, 300115 Timisoara, ROMANIA

Most of the artificial molecular machines designed are based on interlocked molecular species as rotaxanes, catenanes, and related species. However, the main interactions present in the case of rotaxanes and catenanes are: charge transfer capacity, hydrogen bonding, hydrophobic-hydrophilic character, π-π stacking, electrostatic forces, metal-ligand bonding. Yet, the extension of the machine concept at molecular level is very important both for fundamental research and for the practical applications of nanoscience and nanotechnology. They may lead to bottom-up construction of nanoscale devices and machines powered by light, as well as to molecular electronics, such as miniaturized electric circuits, much smaller than the conventional digital circuits fabricated on semiconductor chips. These two conceptual tendencies may be combined by the so called Threading–dethreading process, e.g. the thread-like species trans-1, containing a π-electron rich azobiphenoxy unit, and the π-electron deficient macrocycle, by self assembling gave a pseudorotaxane, stabilized by electron donor-acceptor interactions, as presented in figure.

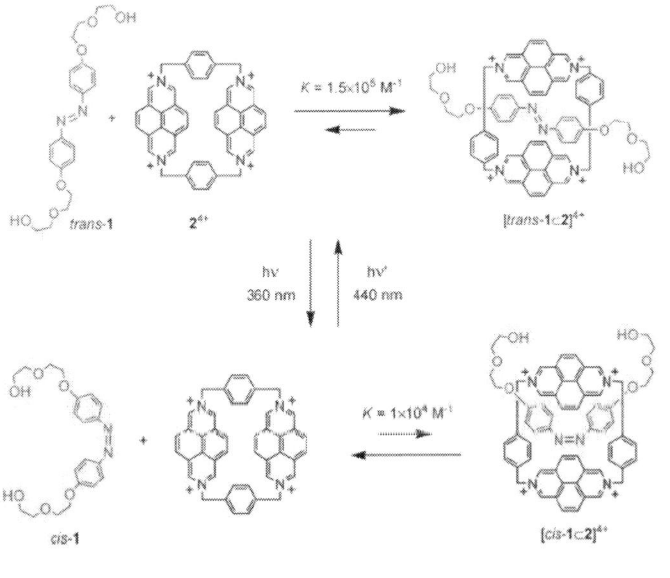

The present work explores the possibility to repeat the dethreading-rethreading cycle of the above-mentioned case by computational modeling and predictions to fluorescence signal; the results may be used for designing a precursor cycle for molecular photoelectrochemcial machine with threading-dethreading controlled by light energy and no waste products.

Acknowledgements: Grant PN16140301, and Prof. Margherita Venturi from Bologna University.

REFERENCES:

Balzani, V.; Credi, A.; Venturi, M. (2007), Molecular devices and machines, *Nanotoday*, 2, 2, 18-25.

Balzani, V.; Credi, A.; Venturi, M. (2009), Light powered molecular machines, *Chem.Soc.Rev*, 38, 1542-1550.

Cao, D.; Amelia, M.; Klivansky, L.M.; Koshkakaryan G.; Khan S.I.; Semeraro, M.; Silvi, S.; Venturi, M.; Credi, A.; Liu, Y (2010), Probing Donor-Acceptor Interactions and *Co*-Conformational Changes in Redox Active Desymmetrized [2]Catenanes, *J. AM. CHEM. SOC.*, *132*, 3, 1110–1122.

Venturi, M.; Credi, A. (2014), Electroactive [2]catenanes, *Electrochimica Acta*, 140, 467–475.

Multivariable controllers for anaerobic digestion in UASB reactors

J.G. Pérez-Vian[1], F. Caballero Echeverria[1], E. Torres-Quintanilla[1], H. Puebla[1], and R. Aguilar-López[2]

[1] Universidad Autónoma Metropolitana Azcapotzalco, Posgrado en Ing. de Procesos, Av. San Pablo 180, Col. Reynosa-Tamaulipas CP 02200 Mexico City; tel. +52 53182146, e-mail: hpuebla@correo.azc.uam.mx

[2] Centro de Investigación y de Estudios Avanzados del Instituto Politécnico Nacional. Av. Instituto Politécnico Nacional 2508, Col. San Pedro Zacatenco CP 07340, Mexico City. E-mail: raguilar@cinvestav.mx

Anaerobic digestion (AD) processes have been proved to be effective for the degradation wastewater with high content of organic matter. AD degrades the organic matter into a biogas, composed mainly by carbon dioxide and methane. The simultaneous production of biogas and substrate degradation has increased the attention of AD processes. Thus, the AD process can be framed as multivariable control problem. However, controlling AD units is a challenging task due to model complexity, nonlinear dynamics, and constrained variables and cross coupling interaction between inputs and outputs.

In this work we have addressed a general design of multivariable feedback control configurations for CSTR and tubular upflow anaerobic sludge blanket (UASB) reactors. Numerical simulations on case studies are used to show the effectiveness of several multivariable control configurations under disturbances and uncertainty parameters.

Quantifying refined olive oil present in extra virgin samples through a non-linear chemometric method

J. C. Cancilla,[1] R. Aroca-Santos,[1] E. S. Pariente,[1] G. Matute,[1] A. Pérez-Pérez,[1] K. W. Wierzchos,[2] A. Moral,[3] A. Tijero,[1] J. S. Torrecilla*[,1]
[1] Departamento de Ingeniería Química, Facultad de Ciencias Químicas, Universidad Complutense de Madrid, 28040 Madrid, Spain; Tel. +34913944244, email: jstorre@ucm.es; [2] Applied Physics Concentration in Astrophysics. University of South Florida, e-mail: kwierzchos@mail.usf.edu; [3] Departamento de Biología Molecular e Ingeniería Bioquímica, Área de Ingeniería Química, Universidad Pablo de Olavide, 41013 Sevilla, Spain.; email: amoram@upo.es.

Abstract

Olive oil is the juice gathered from olives, and extra virgin olive oil (EVOO) is its high quality form, which is obtained by pure mechanical processes. The use of EVOO is on the increasing side because of the many studies concerning its valuable organoleptic properties and the continuous evidence that its routine consumption is beneficial for human health [1-3]. Due to the relevance of EVOO in the global market, its legal protection is required, as its high-quality and complex composition may become compromised by intentional illegal mixtures or unintentional contaminations with lower quality oils such as refined olive oils [4-6].

In this research, binary blends of refined olive oil with four different EVOO varietals (Picual, Cornicabra, Hojiblanca and Arbequina) have been characterized through a straightforward method based on the combination of visible spectroscopy and non-linear artificial neural networks (ANNs). The spectra obtained from the spectroscopic analysis was treated, and the independent variables that were used for a multilayer perceptron (MLP) model were extracted. The non-linear model was capable of perfectly identifying the EVOO varietal (100% correct classification rate), while the error during the quantification of EVOO in mixtures possessing from 0% to 20% refined olive oil was 2.1% mean prediction error. These results reveal that the combination of visible spectroscopy and MLP models are a trustworthy, user-friendly, and inexpensive approach which can potentially be implemented to for the online detection of EVOO samples which are contaminated with refined olive oils.

Therefore, the suggested methodology is straightforward and efficient in terms of time and economy. Minor sample treatments required, while the spectrophotometer utilized to measure is easy to operate and the technique is non-destructive (samples do not suffer any kind of alteration). As a last advantage, this methodology enables the real-time monitoring of the production of EVOO, which is useful for both producers and distributors.

References

[1] Grosso, G.; Mistretta, A.; Frigiola, A.; Gruttadauria, S.; Biondi, A.; Basile, F.; Vitaglione, P.; D'Orazio, N.; Galvano, F. Mediterranean diet and cardiovascular risk factors: a systematic review. Crit. Rev. Food Sci. 2014, 54, 593-610.

[2] Dalen, J.E.; Devries, S. Diets to prevent coronary heart disease 1957-2013: what have we learned? Am. J. Med. 2014, 127, 364-369.
[3] Lozano-Sánchez, J.; Cerretani, L.; Bendini, A.; Gallina-Toschi, T.; Segura-Carretero, A.; Fernández-Gutiérrez, A. New filtration systems for extra-virgin olive oil: effect on antioxidant compounds, oxidative stability, and physicochemical and sensory properties. J. Agric. Food Chem. 2012, 60, 3754-3762.
[4] Cancilla, J.C.; Wang, S.C.; Díaz-Rodríguez, P.; Matute, G.; Cancilla, J.D.; Flynn, D.; Torrecilla, J.S. Linking chemical parameters to sensory panel results through neural networks to distinguish olive oil quality. J. Agric. Food Chem. 2014, 62, 10661-10665.
[5] Reboredo-Rodríguez, P.; González-Barreiro, C.; Cancho-Grande, B.; Simal-Gándara, J. Quality of extra virgin olive oils produced in an emerging olive growing area in north-western Spain. Food Chem. 2014, 164, 418-426.
[6] International Olive Council. (2013). Trade standard applying to olive oils and olive pomace oils; Madrid, Spain. T.15/NC No.3/Rev.7.

Robust control of biobutanol production in a CSTR

E. Torres-Quintanilla[1], H. Puebla[1], H.I. Velázquez-Sánchez[2] and R. Aguilar-López[2]
[1] Universidad Autónoma Metropolitana Azcapotzalco, Posgrado en Ing. de Procesos, Av. San Pablo 180, Col. Reynosa-Tamaulipas CP 02200 Mexico City; tel. +52 53182146, e-mail: hpuebla@correo.azc.uam.mx
[2] Centro de Investigación y de Estudios Avanzados del Instituto Politécnico Nacional. Av. Instituto Politécnico Nacional 2508, Col. San Pedro Zacatenco CP 07340, Mexico City. E-mail: raguilar@cinvestav.mx

Nowadays the production of biofuels is relevant due its potential use as substitute of non-renewable energy resources. Biobutanol is one of the most important biofuels due to its fuel properties. The most important biobutanol production method is the ABE fermentation. ABE fermentation has been carried out mainly in batch operation mode. However, the continuous operation for biobutanol production is desirable in order to increase its production and quality. Moreover, it is well known that the operation of a process can be improved through systematic optimization and control studies. Continuous production of biobutanol process through ABE fermentation could be more attractive via the introduction of control schemes that guarantee the process operation in a given operation point. Since biological processes are subject to important model uncertainties, control designs must be design with good robustness capabilities.

In this work, we have designed a couple of robust controllers for regulation of biobutanol production using as manipulable variables the dilution rate and the substrate glucose input concentration. Our control design is based on a low-order step response model enhanced with estimation of model uncertainties due model reduction and uncertainties in model parameters. Step response models and the control performance evaluation are based on a non-structured model of seven states, including substrate, acetate, butyrate, and biomass. The results show that it is possible to control the production of biobutanol using the proposed robust control.

Sensitivity analysis of a cryogenic separation unit of the plant of natural gas of Malvinas for the recovery of ethane

E. Jara-Morante, [1]A. Mallqui-Antialón, A. Bueno-Lazo, Co-authors
National University of San Marcos, Department of Chemical and Chemical Engineering, A. Amezaga Av. s/n, Lima, Peru; tel. +-51-1-6797000, e-mail: elianejm@hotmail.com;[1]
National University of Engineering, Department of Chemical Engineering, Tupac Amaru Av. s/n, Lima, Peru.

The natural gas is a source of energy that plays an important and strategic role in the energetic development of Peru. The source is located in the south-east jungle region of the country named Camisea, where the Malvinas plant is placed. Malvinas is a plant that process the natural gas extracted in two cryogenic units, each one processing a nominal capacity of 2×9.65 MMNm^3/day. The gas is dry and rich in methane, ethane and propane; its composition in ethane is up to 10% molar.

Nowadays, Malvinas plant is focussed on separate propane as the main component in the heavy fraction and a mixture of methane, ethane and propane as the light fraction. This mixture is commercialized as a whole mixture and used in the fertilizer industry, dynamite production or burned, no special treatment is made for the separation of ethane. As we know, ethane is the base component for the production of ethylene and other by products, its processing will promote the development of the petrochemical industry in Peru.

The objective of our study is to determine the mode and operating conditions that the cryogenic units should operate in order to increase the recovery of ethane in the light fraction. We focused in two cryogenic units with a 5.89 MMNm^3/day design capacity each and able to operate in two modes: SCORE mode and GSP mode, each unit can alternatively work with the Joule-Thompson valve device or the expander. Four operating modes were evaluated: SCORE/JT, SCORE/Expander, GSP/JT and GSP/Expander modes, we used ASPEN Plus V7.3 to simulate the complete process and analyse the effect of process variables through a sensitive analysis.

The results show that in the actual mode of operation the recovery of ethane in the light fraction is 9.41% molar and 9.37% molar in the SCORE/JT and SCORE/Expander while is lower with the GSP/JT and GSP/Expander modes. The operating variables that exert a significant effect in the process and increase the ethane recovered are: temperature in the accumulation tank and the distillate fraction in the top of the column. The actual temperature in the accumulation tank is -137°F for both SCORE/JT and SCORE/Expander and when the temperature is increased to -8°F and -18°F the recovery of the ethane is increased in a 12.44% and 12.05% respectively. It can be concluded that increasing the temperature in the accumulation tank the recovery of ethane is increased, therefore the requirements of energy to reach cryogenic conditions are lower than in the actual operation. It is possible to implement these new operation conditions in the SCORE/JT or SCORE/Expander mode since no additional equipment or significant physical modifications are need. The change in the production objective of the plant will give an aggregated value to the product that can be used as a raw material in the petrochemical industry.

Study of system-wide and structural properties of the pre-polyamidation tank

E.A. Alekseev, B.A. Golovushkin, A.N. Labutin
Ivanovo state university of chemistry and technology, Sheremetevskiy str. 7, 153000, Ivanovo, Russia; tel. +7 910 6866514, e-mail: eugene-17@mail.ru.

At present time researchers widely use simulation methods of technological processes of synthesis synthetic polymers for solving different optimization and control tasks. It's necessary to create effective control system for performing process under economic or technical point of view optimal behaviors. Processes are studied as control object for solving this problem.

In this work object is the pre-polyamidation tank in which stage of pre-polyamidation of polyamide-6 in solid phase is existing. Earlier mathematical model of this stage was derived, simulation was performed and adequacy of derived model was confirmed.

The aim of this work is to study system-wide and structural properties (connectivity, controllability and observability) of tank by using mathematic and simulation methods.

Method using Bristol matrix which characterize connectivity degree in static was used for evaluating degree of system connectivity. It has been shown that our object is connected. Hence it's necessary to compensate cross links for controlling of tank.

Presentation dynamic object in the state space are used for studying such properties as stability of the free movement, controllability and observability. Original nonlinear model of object with distributed parameters was presented by using her discrete analog (cell's model). Linearization of object in the neighborhood of working point was performed in Simulink app «Linear Analysis Tool», MATLAB. Matrix of state A (dimension 490×490), matrix of control B (dimension 490×3) and matrix of observe C (dimension 3×490) were derived.

Study of stability of unperturbed system in the state space was performed. Result shows that our object is stable in the neighborhood of working point. In this way the one has property of stabilizability.

Controllability N_c and observability N_o matrixes ($N_c = [B{:}AB{:}A^2B{:}\ldots{:}A^{n-1}B]$, $N_o = [C^T{:}A^TC^T{:}(A^T)^2C^T{:}\ldots{:}(A^T)^{n-1}C^T]$, where n is order of system) and their rank were derived and calculated by using inbuilt MATLAB functions. Rank of controllability matrix is 17. It is small than order of system that equals 490, hence our pre-polyamidation tank is not fully controllable. In this way there are such initial conditions in the phase state when object cannot be transferred to specified final condition. Rank of observability matrix is 18. It shows that our object is not fully observable.

Controllability and observability of object on outputs were estimated also. Matrixes of state, controllability and observability has dimensions 7×7, 7×3, 3×7 respectively. The received results show that our object is fully controllable and observable in the space of outputs. Object is stable and stabilizable in the neighborhood of working point.

Evaluation of system-wide and structural properties of pre-polyamidation tank carried out allowed us to choose control actions and measured state variables. Problem of structural, algorithmic and parametric synthesis of control system minimize energy consumption for performing process synthesis of polyamide-6 will be solved next time.

The optimization for vinyl chloride polymerization process based on modelling

H. T. Pan[1], P. F. ZHU[2], M. F. Zhou[1], L. Y. Xia[1]
1 College of Chemical Engineering, Zhejiang University of Technology, Hangzhou, Zhejiang, China;
Tel. + 86-571-88320329, e-mail: htpan@zjut.edu.cn;
2 Zhejiang Longsheng Chemical Research Co. Ltd., Shaoxing, Zhejiang, China;
e-mail: pfzhu1987@126.com;

The Polymerization of Vinyl Chloride (PVC) is a typical exothermic and batch process. On the premise of meeting the requirements of product quality indicators, reducing agents consumption and shortening the reaction time are effective ways to save energy and improve the productivity. This paper focuses on the process optimization strategy based upon multiple models, which include quality models, thermodynamic model, kinetic model and optimization models. The structure of the optimization strategy is given in Figure 1.

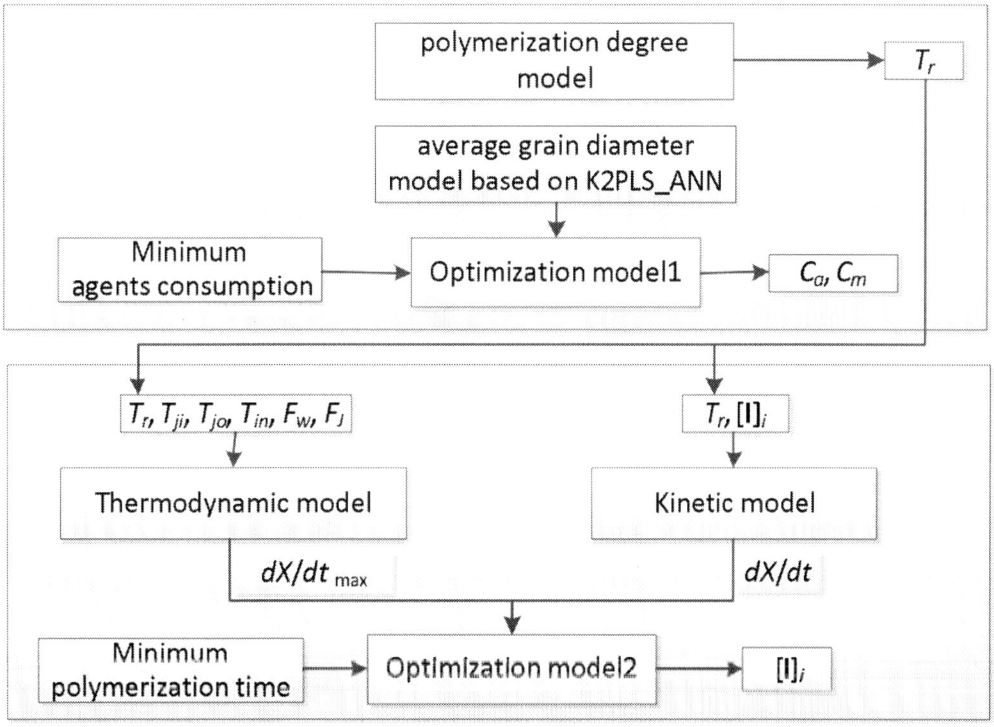

Figure 1 The structure of the process optimization strategy for vinyl chloride polymerization

PVC has two main quality indicators: polymerization degree and average particle size. Without the chain transfer agent, the polymerization degree depends only on the polymerization temperature T, thus a regression model method is constructed based on the measured data. The average particle size is affected by an amount of factors such as the concentration of dispersant (C_m), the concentration of nano micro emulsion of $CaCO_3$ (C_a)

and the polymerization temperature T. The increasing of average particle size is determined by the breakage rate and coalescence rate of vinyl chloride droplets. However, it is difficult to identify the parameters for the mechanism model. Therefore, a data-driven soft-sensor modelling method is proposed by combining kernels partial least squares (K_2PLS) with artificial neural network (ANN). The K_2PLS-ANN based average particle size model is designed to predict the quantitative nonlinear information between the impact variables and the average particle diameter. The polymerization temperature T is obtained by the polymerization degree model. The optimal concentration of dispersant (C_m) and the nano micro emulsion of $CaCO_3$ (C_a) are obtained by nonlinear programming with the average size model and the cost optimization model, aimed at the lowest agents consumption.

For the safety of the polymerization process, the reaction rate is strictly limited. Without the chain transfer agent, the reaction temperature T, the initiator type and concentration are key elements affecting the polymerization rate. In this work, a kinetic model of composite initiators is referenced to predict the polymerization rate. A thermodynamic model based on the heat balance equations is presented to estimate the heat transfer characteristics of the reactor and compute the maximum limiting polymerization rate. An optimization strategy of composite initiators based on multi model integration (i.e. the thermodynamic, the kinetic model and a polymerization time optimization model) is proposed, aimed at minimizing the polymerization time. With the optimal concentration of composite initiators, the polymerization proceeds smoothly and swiftly, and the heat transfer capacity of the polymerization reactor is fully utilized.

The research results show that the multi-model integration based optimization strategy, with the quality models as constraints in the optimization models, can not only meet the requirements of quality indicators but also reduce the consumption of agents and achieve a significant reduction in the total processing time.

The use of cellular automata with varying sizes of the cells for modeling dissolution of solid dosage forms

I.A.Tiptsova, S.I.Ivanov, E.S Vorobeva, N.V.Menshutina
D.I. Mendeleyev University of Chemical Technology of Russia, Computer Aided Process Engineering Faculty, Geroev Panfilovtsev Str. 20 bld.1, 125480 Moscow, Russia;м
tel. +7(495)495-00-29, e-mail: yirish@yandex.ru

The dissolution of the pharmaceutical industry is an essential process in the development stage of new original drugs as well as for the selection of generic compositions. In order to reduce time and finances for carrying out multiple experiments on the dissolution of the different composition of solid dosage forms model dissolution based on cellular automata with changing sizes of cells was developed.

The dissolution process in the cellular automaton model with changing sizes of the cells is described as follows. The substance from the cells with the state named "solid" goes into cells with the state named "liquid" in the process of dissolution thus the amount of the substance in the cells changes. On this basis, we can assume that the volume of cells change and thus their sizes change. There is a shift of the cells border in the direction of the cell with the "solid" state in all pairs of cells "liquid - solid ", i.e in the cell with the "solid" state amount of substance decreases, and in cells with the "liquid" state amount of solute increases.

Based on this cellular automaton model of the dissolution process software system for calculating the kinetics of dissolution was developed. The software product is written in C #, Microsoft Visual Studio 2013.

The dissolution objects in this model can be different pharmaceutical forms (tablets, granules). Objects may be monocomponent (crystalline active substance), multicomponent, containing dispersed inclusion of various shapes, as well as insoluble components.

Fig. 1 shows schematic illustration of granule dissolution. Granule represents the distribution of inclusions "labrasol-API" (Fig. 1 are marked in green) in a mixture of solutes - maltodextrin (marked in blue) and gum arabic (marked in red).

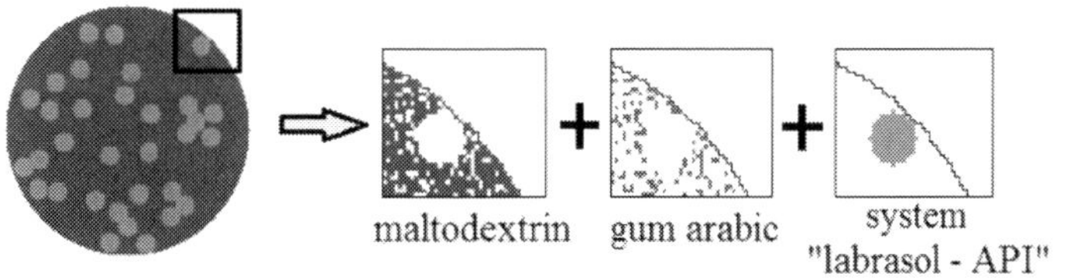

Figure 1. Schematic illustration of granule dissolution

Visualization states granules in dissolution process at various times is shown in Figure 2.

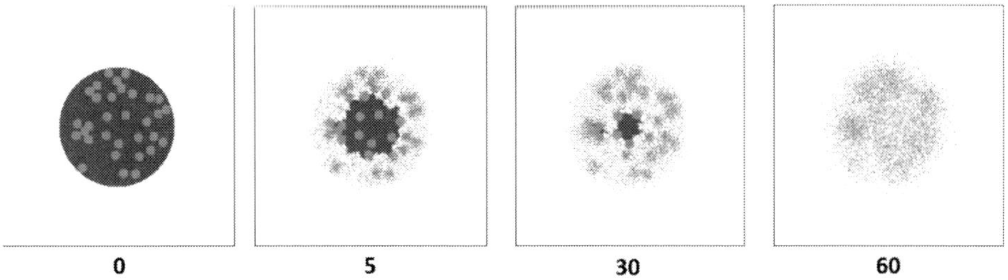

Figure 2. Visualization states granules in dissolution process at various times (0, 5, 30, 60 minutes) with content ratio of maltodextrin (MD) and gum arabic (GA) MD: GA= 2: 1, with content of ibuprofen in the "labrasol-API" =20%

Changing sizes of cells in the cellular automaton model gives several advantages:

- changing sizes of cells allows calculating surface area changing in the dissolution process with high accuracy;
- cell with "solid" state is removed, when the dissolution is complete. Gradual removal of cells with "solid" state reduces the number of cells in the cellular automaton, which accelerates the process of calculation.

Thus, the developed model allows predicting the kinetics of dissolution of solid dosage forms (tablets, granules) by varying compositions (monocomponent, multicomponent, containing dispersed inclusion of various shapes as well as insoluble components) and the geometric form of the drug.

Dynamic evolution of morphological characteristics during particulate processes through on-line experimental analysis and population balance modelling

[1]C. Frances, [1]C. Coufort-Saudejaud, [2]A. Liné
[1]Université de Toulouse, Laboratoire de Génie Chimique, INPT, CNRS, UPS, 4, allée Emile Monso 31432 Toulouse, France, tel. +33 534323639, e-mail: christine.frances@ensiacet.fr ;
[1]Université de Toulouse, Laboratoire d'Ingénierie des Systèmes Biologiques et Procédés, INSA, CNRS, INRA, 135 Avenue de Rangueil, 31077 Toulouse, France, tel. +33 561559786, alain.line@insa-toulouse.fr.

Processing of suspensions is an important operation in many industrial applications, as for the synthesis of particulate or composite materials, the separation of pollutants in the water treatment or the production of biological systems. In order to optimise, control or predict these processes, the change of the dispersed solid properties in relation with the process conditions has to be modelled. The dynamic evolution of the characteristics of the particles is commonly obtained by the solution of the population balance equation (PBE) [1]. The population of particles often undergoes several phenomena, such as nucleation, growth, aggregation, breakage, restructuring, …. In the PBE, particles are usually distributed with respect to one intrinsic property, such as the particle volume or size. However, other internal parameters, in relation with the particle shape or structure may be of interest. In other applications, it can be important to recover information on an active agent concentration, or the moisture content of particles. In such situations, bivariate or multidimensionnal population balance models incorporating all the relevant internal solid properties must be addressed.

Among the different methods used to solve population balance models, (classes or sectional methods, Monte Carlo methods or moment-based methods), the quadrature-based moments methods present the advantage of requiring low computational costs while allowing acceptable accuracy of the predicted distribution moments [2]. The objective of this presentation is to show how such methods can be used to model particulate processes in which it is needed to track the dynamic evolution of at least two internal solid properties. It will be illustrated on several examples in order to emphasize the advantages and the limitations of these methods.

The first studied case concerns a flocculation process as it is encountered in water treatment. The flocs morphology, related to the particle size and shape, affects their ability to settle down which is important to ensure a correct separation after the flocculation process. Flocculation experiments were performed under turbulent conditions in a Taylor Couette apparatus associated with an image processing system enabling to determine various morphological characteristics of flocs [3]. The perimeter and the surface area of flocs, which can be obtained from image analysis, can be first chosen as internal coordinates in a bivariate PBE to track the evolution of the floc structure versus time. The challenges are then to choose relevant aggregation and breakage kernels based on these two properties and to solve the PBE using DQMOM. The validation is done comparing experimental and predicted moments.

Flocculation runs were also performed in a stirred tank associated with an on-line scattering analysis giving access to the floc size and the fractal dimension versus time. The fractal dimension was then included in the PBE for the time evolution of the size distribution moments as well as in the agglomeration and breakage kernels. By this way, the PBE, remains monovariate and can be easily solved using QMOM. Fig. 1 gives an illustration of the agreement between the experimental and the predicted moments.

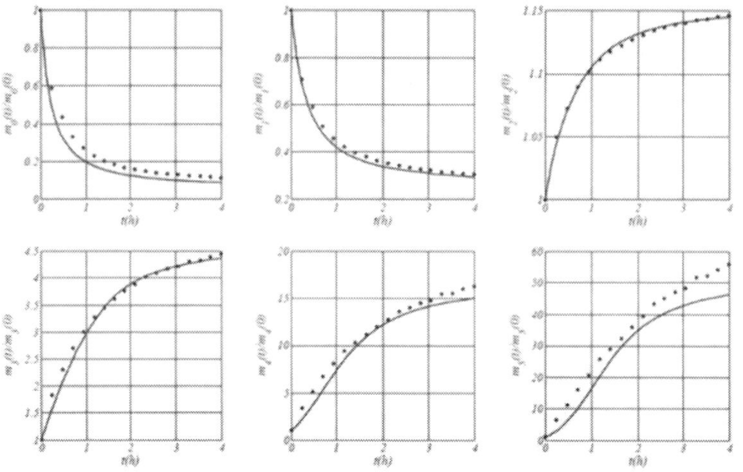

Fig. 1 : Comparison of the evolution of the predicted and experimental moments during a flocculation process using QMOM taking into account of the fractal dimension of aggregates in the PBE and in the aggregation efficiency and the breakage kernel

The second studied case is related to a comminution process. Although the particle size is the only solid property usually considered in grinding processes modelling, the particle strength is another key parameter needed to be taken into account into realistic analysis of these processes. A bivariate population balance equation applied to a grinding process was implemented in which the particles were simultaneously characterized by their size and their mechanical strength, expressed by the minimum energy needed to break them. The benefit of of a bivariate approach can be shown analysing the sensitivity of some parameters and comparing the results of the monovariate and bivariate cases [4]. The effect of the parameters of the quadrature-based method (the number of nodes and the selected set of moments needed to solve the system) on the accuracy of the results is also discussed (Fig. 2).

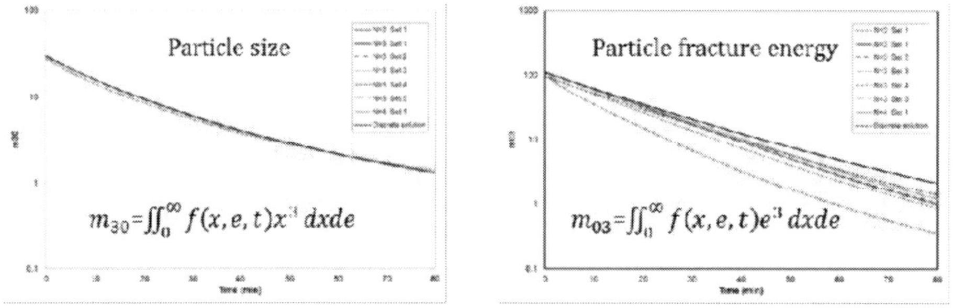

Fig. 2 : Effect of the number of nodes (N) and of the selected set of moments on the pure mixed moments related to the particle size and the particle fracture energy during a batch grinding process using DQMOM to solve the population balance model.

[1] D. Ramkrishna (2000). Population Balances, Academic Press, San Diego
[2] D.L. Marchisio, R.O. Fox (2013) Computational models for polydisperse particulate and multiphase systems, Cambridge Univ. Press, Cambridge
[3] M. Vlieghe, C. Saudejaud, C. Frances, A. Liné (2014). AIChE J., 60, 2389-2403
[4] C. Frances, A. Liné (2014). AIChE J., 60, 5, 1621-1631

Flowability alteration of surface modified cohesive glass particles

Z. Todorova, W. Hintz, J. Tomas
Otto von Guericke University Magdeburg, Institute for Process Engineering, Mechanical Process Engineering, Universitätsplatz 2, 39106 Magdeburg, Germany
tel.: +49 (0)391 6754912, fax: +49 (0)391 6711160, e-mail: zinaida.todorova@ovgu.de

The fine and ultrafine particles find broad application in many industrial branches due to their special properties, such as small size or high specific surface area. However, handling of this kind of particles causes a number of problems due to their high adhesion potential. It occurs because the adhesion forces, and the van der Waals forces in particular, exceed the gravitational forces by several orders of magnitude. However, the morphology and chemical composition of the particle surface strongly influence the interparticular interactions and, therefore, it is advisable to formulate the solid surface properties by thin layers of functional materials. These layers may be adsorbed, bound to, or coated on the surface.

In this study a cohesive micro glass powders ($d_{50,3}$ = 7 µm) are used as a model system because of their smoothness, high sphericity and wide availability. A chemical surface modification with hydrophobic silanes was carried out to alter the van der Waals interactions of the glass particles, which in case of dry and fine dispersed powders are of large importance. Five different silanes that possess three different hydrophobic rests - fluoroalkyl, phenyl and alkyl- were applied. The grafting with fluoroalkyl silanes proved to be the most effective hydrophobization process for the reduction of the adhesion force: it not only reduced and in some cases completely eliminated the effect of capillary forces, but also led to reduction of the London dispersion forces. A better powder flowability for almost all silanized powders was measured using a Schulze ring shear tester.

Another way to influence the interaction between the particles is to modify the surface roughness by adding of nanoscaled particles which play the role of so called flow additives. As nanoscale flow additives, in the present work, were used TiO_2 and SiO_2 nano particles, which were coated on the glass surface by means of hetero coagulation coating process in aqueous media. Using again the shear testing method the macromechanical consolidation function of the coated powder was measured and an increased flowability was observed as expected. The effect of the particle size used for the coating and the porosity of the coating is also discussed.

Using a constitutive contact model 'stiff particles with soft contacts' [1], important micromechanical contact properties as: adhesion force, contact consolidation and adhesion force intensification were back-calculated from the shear tests and thoroughly discussed.

[1] J. Tomas, Adhesion of ultrafine particles - a micromechanical approach, Chemical Engineering Science 62 (2007) 1997-2010

Population balance model for dissolution of particle agglomerates

[1]W. L. Zhao, [1]A. Buffo,[2]B. Han, [2]M. Louhi-Kultanen,[1]V. Alopaeus
1. Aalto University., Department of Biotechnology and Chemical Technology, Kemistintie 1, 02150 Espoo, Finland; tel. +358 451674316, e-mail: wenli.zhao@aalto.fi; 2. Lappeenranta University of Technology, Department of Chemical Technology, FI-53851, Finland

Dissolution is often the rate-limiting step in leaching and aqueous mineral carbonation processes. Shrinking particle model usually adopted to describe the dissolution of non-porous powder particles is based on the assumption that particles are monodispersed, i.e., all the particles have the same diameter. In reality, particles are polydispersed and particle size distribution has strong influence on the dissolution rate. For this reason, a population balance model, accounting for the evolution of particle size distribution, can give an insight of the behaviour of dissolution process [1]. Such population balance model must contain different sub-models for all the relevant phenomena involved in the process: the dissolution rate can be expressed as a shrinking term due to the mass transfer from the solid phase to the liquid. Another important phenomenon, usually neglected in these cases, is the breakage of the particles due to high shear rates present in the equipment (e.g., stirred tanks) where the dissolution process occurs. In addition, the raw materials are not always perfect sphere or cubic, but rather agglomerates, and the breakage of such large agglomerates may alter the total mass transfer area and increase the dissolution rate significantly in the stirred tank reactor. Furthermore, the electroneutrality must be satisfied during the dissolution of particles in the electrolyte solution. This aspect can be considered by including the Nernst-Planck equation into the mass transfer model to confirm the electrically neutral constraint as well.

The dissolution of $Mg(OH)_2$ solids into a HCl solution is chosen as our study case. The reaction between $Mg(OH)_2$ and HCl solution at acidic region (pH<7) is a instantaneous reaction which is much faster than the diffusion rate. Therefore, the external mass transfer is the rate-controlling step. The mass balance model including electroneutrality can be written as:

$$\frac{dc_i}{dt} = N_i a = k_{s,i} a \Delta c_i - k_{s,i} a z_i \overline{c_i} \left[\sum_i (z_i k_{s,i} \Delta c_i) / \sum_i (z_i^2 k_{s,i} \overline{c_i}) \right] \quad (1)$$

where N_i is the mass transfer flux, $mol \cdot m^{-2} \cdot s^{-1}$;$k_{s,i}$ is the solid-liquid mass transfer coefficient, $m \cdot s^{-1}$; c_i is the concentration of components in the bulk solution, $mol \cdot m^{-3}$; Δc_i is the driving force of dissolution, $mol \cdot m^{-3}$; z_i is the charge of cationic and anionic.

As previously mentioned, the physical phenomena considered here are the particle breakage and particle shrinking due to mass transfer; all other mechanisms such as nucleation and aggregation are neglected. The population balance equation can be written for diameter as the internal coordinate as:

$$\frac{d(n(L))}{dt} = \int_L^{\infty} \beta(L,\lambda) g(\lambda) n(\lambda) d\lambda - g(L) n(L) + \frac{\partial (G(L) n(L))}{\partial L} \quad (2)$$

where $n(L)$is the density of particle size distribution, $\# \cdot m^{-4}$; $\beta(L, \lambda)$ is the daughter particle size distribution, which tell us the number of particles and their size produced after the breakage event; $g(\lambda)$ is the breakage rate, $\# \cdot s^{-1}$, function of different physical parameter of the investigated system, e.g., turbulent dissipation rate, liquid viscosity, size of the particles [3]; $G(L_j)$ is the shrinking rate, $m \cdot s^{-1}$, estimated from the dissolution model. The specific mass transfer area (a) is obtained from the solution of Eq. (2), carried out with the High-order Moment Method of Classes (HMMC) [2].

In this work, the initial size distribution and the morphology of solids were investigated experimentally by the laser diffraction analyzer and the scanning electronic microscope respectively.

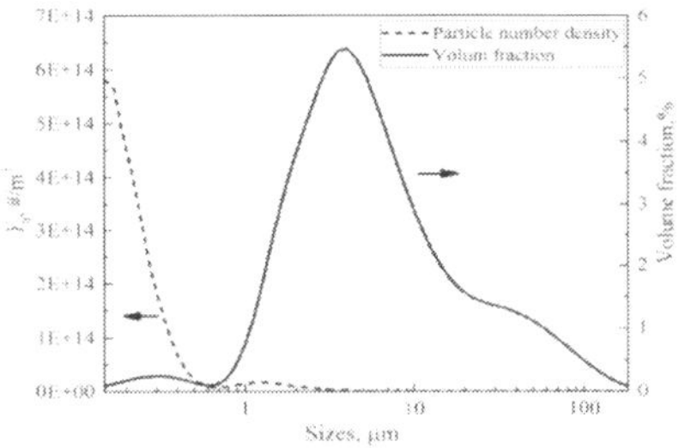

Figure 1. The size distribution of raw material

Figure 2. The SEM image of raw material

The initial distributions of particle number density and volume fraction obtained from our sample are shown in Fig.1. As it is possible to see, the large agglomerates with small number density contributes the main peak of volume distribution. The phenomenon can be confirmed qualitatively by looking at the SEM image of the same sample in Fig.2.

Then the following numerical case was considered: an initial weight of $Mg(OH)_2$, 0.0075kg, was included into a stirred tank containing a 2.64L HCl solution at pH=1. The rotating speed of impeller is 500 rpm. This system was assumed spatially homogeneous in first approximation. The dissolution profile and size distribution of solids after 15 s of stirring are shown in Fig.3 and Fig.4.

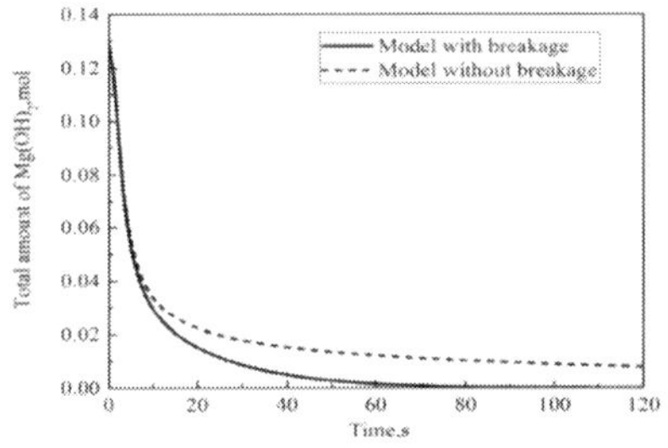

Figure 3. The dissolution profile

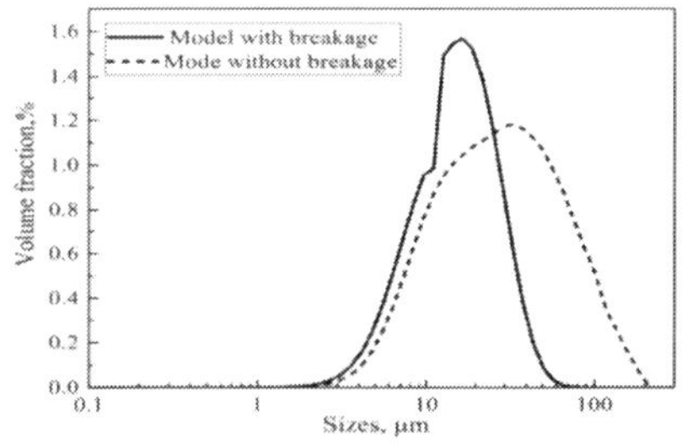

Figure 4. The size distribution of particle at 15 s

As it is shown in Fig.3, the dash line is the dissolution profile calculated without breakage model while the solid line is the dissolution profile calculated with breakage model. As expected, the dissolution rate increases significantly by considering the agglomerates breakage [3], since the interfacial area is enhanced by the presence of small particles produced. This is also confirmed by the size distribution at 15 s reported in Fig.4. However, it seems that the breakage does not affect the dissolution rate of particles smaller than 10μm. Future developments of this work include the evaluation of the influence of the actual fluid field on the dissolution rate through CFD calculation, the model validation by choosing larger agglomerates and the experimental evaluation of the dissolution rate and time evolution of particle size distribution.

References

[1] M. Hänchena, S. Krevorb, M. Mazzottia, K. S. Lacknerb. Validation of a population balance model for olivine dissolution. Chem. Eng. Sci. **2004**, 62, 6413-6422.

[2] V. Alopaeus, M. Laakkonen, J. Aittamaa. Solution of population balances with growth and nucleation by high order moment-conserving method of classes. Chem. Eng. Sci. **2006**, 62, 2277-2289.

[3] J. A. Wojcik, A. G. Jones. Particle disruption of precipitated CaCO3 crystal agglomerates in turbulently agitated suspensions. Chem. Eng. Sci. **1998**, 53,1097-1101.

Nanoparticulate zeolite production by the combination of mechanochemical and hydrothermal procedure

A. Muramatsu, K. Kanie, M. Sakaguchi, H. Kobayashi, and M. Nakaya
Tohoku Univ., IMRAM, Katahira 2-1-1, Aoba-ku, Sendai 980-8577, Japan;
tel. +81-22-217-5163, e-mail: mura@tagen.tohoku.ac.jp.

We have already reported the development of a new synthesis method for TS-1 utilizing mechanochemical reaction[1]. This mechanochemical synthesis method employs titania and silica powders as starting materials, which are allowed to react mechanochemically to be a silica-titania composite as a precursor to TS-1. Figure 1 illustrates the precursor preparation process using the mechanochemical reaction occurring at the grinding using a planetary ball mill. By the rotation and the revolution of a milling pot, balls installed inside the pot hit each other to produce powerful impact and/or friction energy according to operational conditions such as the milling time and the disk rotation speed[2]. This mechanical energy promotes the degradation of crystalline structure, the cleavage of bonds, and the formation of new bonds

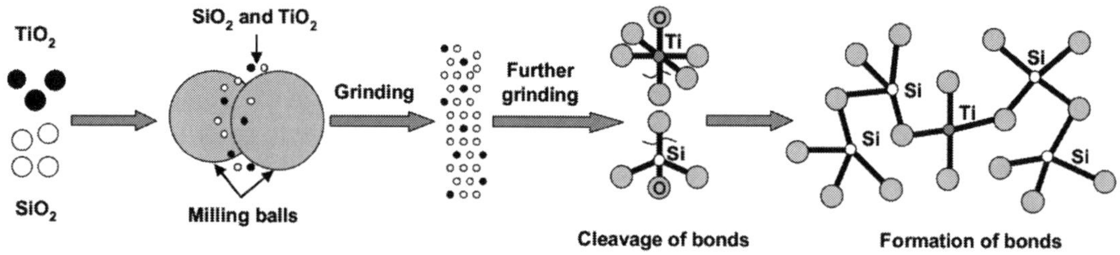

Fig. 1 Schematic illustration of the precursor preparation via the mechanochemical process.

between silicon and titanium species, resulting in the formation of a uniform silica-titania mixture. Thus prepared silica-titania mixture is hydrothermally treated in the presence of organic structure-directing agent (SDA) to be crystallized into TS-1 materials.

We have also investigated their application to the synthesis of the other zeolite. Silica-titania composite is prepared by the mechanochemical reaction of bulk silica and titania and crystallized into a *BEA-type zeolite through the hydrothermal treatment[3]. By using this silica-titania composite as a silicon and titanium sources, the crystallization time to *BEA-type titanosilicate can be greatly shortened. Thus obtained Ti-Beta material shows high catalytic activity in the oxidation of

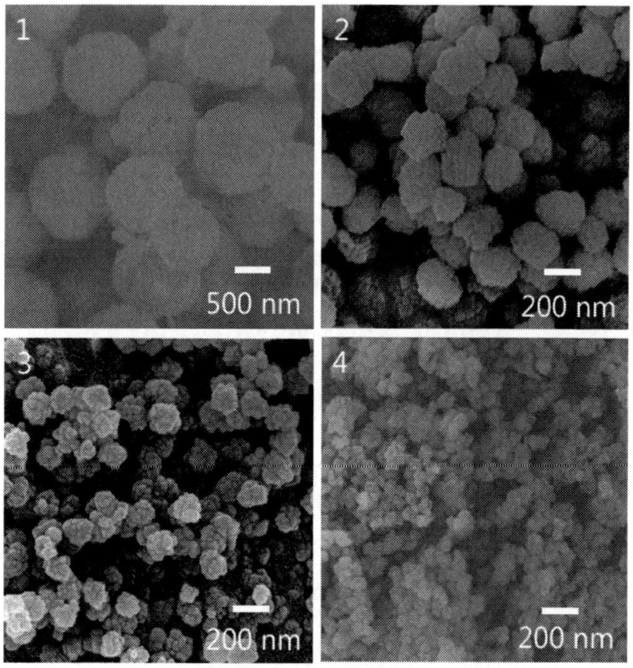

Fig. 2 Size-controled synthesis of Fe-silicalite in the same manner composed of the mechanochemical and hydrothermal process.

cyclohexene using hydrogen peroxide as an oxidant. The contamination of zirconia from a milling media into a silica-titania composite not only disturbs the crystallization of the *BEA-type structure but also depresses the catalytic activity of the final Ti-Beta product.

Figure 2 shows the synthesis of Fe-silicalite by the combination of mechanochemical and hydrothermal processes, where the starting materials were fumed silica and hematite. The size of Fe-silicalite was controlled by stirring time and water content in the hydrothermal step. We also succeeded in the synthesis of Fe-ZSM-5 in the same procedure, but AlOOH addition to the precursors in the grinding process.

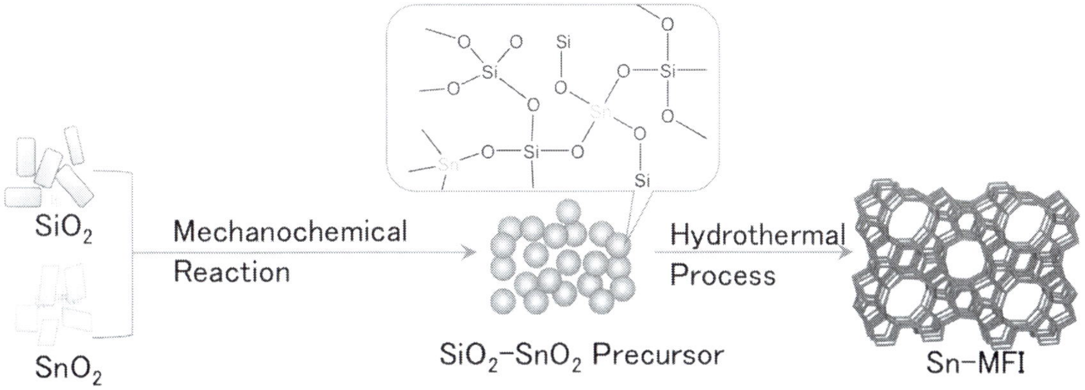

Fig. 3 Schematic illustration of the novel synthesis method of Sn-MFI zeolite in the same manner composed of the mechanochemical and hydrothermal process.

Nowadays, we have been developing the novel synthesis method of MFI zeolite as a general procedure of the mechanochemical and hydrothermal methods. Sn substituted silicates of MFI topology (Sn-MFI) have attracted our attention to the catalysts for industrially reactions, for example the isomerization of glyoxal containing in bio-oil as by-products to glycolic acid. However, it is difficult to increase Sn content by conventional hydrothermal synthesis. In order to introduce Sn to the MFI framework, using SnO_2-SiO_2 precursor is effective. We have focused on a mechanochemical reaction to prepare a SnO_2-SiO_2 precursor. This mechanochemical route may allow to mix Sn and Si well uniformly at atomic level. In this study, Sn-MFI materials are synthesized by combination of a mechanochemical and hydrothermal treatment (Fig. 3).

The precursor SnO_2-SiO_2 was prepared by milling the mixture of tin dioxide and fumed silica using planetary ball mill. The obtained SnO_2-SiO_2, amorphous silica, H_2O, NaCl aq., and TPAOH aq. were added into a Teflon vessel. The solution was treated at 160 °C for 120 h and then calcined at 540 °C for 12 h.

UV-vis spectra and XRD pattern of SnO_2-SiO_2 precursor showed that Sn was uniformly diffused into silica by the mechanochemical reaction. From UV-vis spectra and FT-IR spectra of Sn-MFI prepared with using the obtained SnO_2-SiO_2 precursor, Sn species were substituted into the MFI framework. The Si/Sn ratio of the Sn-MFI could be controlled by the feed ratio of Si/Sn.

References

1) Yamamoto, K., Borjas Garcia, S.E., Saito, F., Muramatsu, A., Chem. Lett. 35, 570 (2006).
2) Mio, H., Kano, J., Saito, F., Kaneko, K., Mater. Sci. Eng. A 332, 75 (2002). 3) Borjas Garcia, S.E., Yamamoto, K., Muramatsu, A, J. Mater. Sci. 43, 2367 (2008).

Solvent Effects of Zinc Sulfide Quantum Dots Classified by Size Selective Precipitation

Y. Mori, K. Yamamoto, [1]D. Segets, [1]W. Peukert, K. Tuchiya
Doshisha Univ., Dep. Chem. Eng. & Mater. Sci., Tatara Miyakodani 1-3, 610-0321 Kyotanabe, Japan; tel. +81 774 65 6626, e-mail: ymori@mail.doshisha.ac.jp;
[1]Insitute of Particle Technology (LFG), Friedrich-Alexander-Univ. Erlangen-Nürnberg (FAU), Erlangen, Germany

The mean particle size and the size distribution (PSD) are very important factors for functional characteristics of semiconductor nanoparticles (NPs). NPs with relatively high monodispersity can be synthesized in wet chemical system when the reaction field is controlled, such as a hot soap method or a hydrothermal synthesis. Even using such synthetic-controlled techniques, however, particle size is often distributed. Hence, classification or separation operation is strongly needed not only due to PSD but also from the view of the purification of product after synthesis. The classification technique based on size-selective precipitation (SSP) could be effective. We demonstrated how effective SSP technique is for the classification of NPs to use 1-thioglycerol-capped ZnS NPs, which was performed by gradual addition of 2-propanol (poor solvent) to well-dispersed aqueous colloidal solution (good solvent) [1, 2].

In this paper, in order to elucidate solvent dependency for SSP process, the separation efficiency of ZnS NPs was investigated using three kinds of good solvents (water, dimethyl sulfoxide, dimethylformamide) and five kinds of poor solvents (methanol, ethanol, 2-propanaol, acetone, acetonitrile) with various poor solvent ratios. As-prepared ZnS NPs were dispersed in good solvent. After gradual addition of poor solvent, the suspension was centrifuged to separate the flocculated NPs from primary particles in solution. The flocculated NPs were dried in vacuum and redispersed in water. PSD of redispersed NPs was calculated from UV-vis adsorption spectra using the algorithm reported by Segets et al. [3], based on which mean volume diameter, coefficient of variance of PSD and cut size by SSP were evaluated. Figure 1 shows the relationship between the recovery ratio by SSP of ZnS NPs and the mean volume diameter after SSP. When the recovery ratio was low, the mean volume diameter of redispersed NPs was large, because the larger NPs tended to be aggregated by SSP. The mean volume diameter of redispersed NPs did not depend on the combination of good and poor solvents in this figure. And the coefficient of variance of PSD and the cut size of the classification also did not depend on the combination of good and poor solvents, only depended on the recovery ratio of NPs. This means that a certain volumetric good solvent/poor solvent combination always leads to a defined distribution of particles into courses and fines.

We tried to apply the DLVO theory together with the Flory-Huggins theory for polymer to explain this phenomenon, because the recovery ratio of NPs depended on dispersion-aggregation phenomena among NPs, that is, the inter-particle interaction potentials between two particles during SSP process. The interaction potentials could be expressed by the sum of van der Waals, electrostatic, Born, osmotic and elastic potentials. The last two potentials were contributed by the ligand of 1-thioglycerol, and the Flory-Huggins parameter, χ, was found as the most dominant parameter on dispersion-aggregation phenomena. χ can be expressed by the following equation using Hansen solubility parameters $(\delta_d, \delta_p, \delta_h)$, which reflect the atomic dispersion interaction, polar interaction, and hydrogen bonding interaction of molecules, respectively [4].

$$\chi = \alpha \frac{v_l}{RT}\left[\left(\delta_{d1}-\delta_{d2}\right)^2 + 0.25\left(\delta_{p1}-\delta_{p2}\right)^2 + 0.25\left(\delta_{h1}-\delta_{h2}\right)^2 \right], \qquad (1)$$

where v_l is molar volume of the mixture solvent, and α is constant. The subscript of "1" and "2" means the Hansen solubility parameter (HSP) coordinates of the ZnS NPs and the HSP coordinates of the solvent mixture, respectively. HSPs of the solvent mixture were estimated based on the values of the pure solvents according to their volumetric mixing ratio [5, 6].

For the molecules covered on ZnS NPs, we assume that good solvent molecules are preferentially adsorbed to the ligands, and then cover on molecules of the ligands. Equation 1 indicates this case, where the subscript "1", means the good solvent. The recovery ratio of NPs could be expressed by χ parameter for any poor solvent, although the data were scattered somewhat. When changing a good solvent from water to dimethyl sulfoxide or dimethylformamide, the tendency of the relationship between the recovery ratio and χ parameter was same as Figure 2. We could conclude that Flory-Huggins parameter is the most dominant parameter on dispersion-aggregation phenomena, and that the separation properties of SSP process can be estimated by Flory-Huggins parameter.

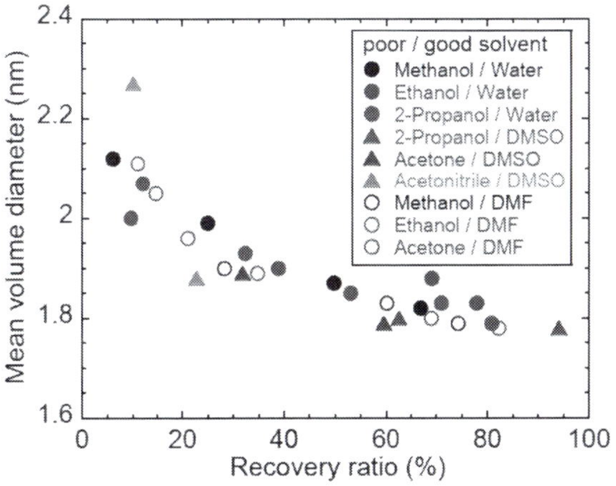

Figure 1. Mean volume diameter after SSP process for different combination of mixture solvents system as a function of recovery ratio.

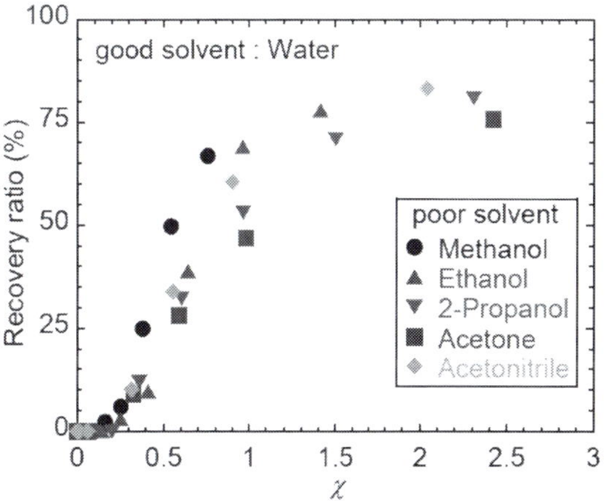

Figure 2. Recovery ratio for ZnS NPs against χ parameters for different mixtures of a poor solvent and a good solvent (water).

Reference

[1] D. Segets, *et al.*; *J. Nanoparticle Res.*, 2013, **15**, 1486-1499.
[2] D. Segets, *et al.*; *J. Phys. Chem. C*, 2015, **119**, 4009-4022.
[3] D. Segets, *et al.*; *ACS nano*, 2012, **6**, 9021-9032.
[4] T. Lindvig *et al.*, *Fluid Phase Equilibria*, 2002, **203**, 247-260.
[5] L. Niu *et al.*, *Soft Matter*, 2013, **9**, 7780-7786.
[5] B. J. Immergut, *et al.*; *Polymer handbook*, fourth edition. John Wiley & Sons, 1990.

Control of the chemical composition of particulate solid solutions used as cobalt and nickel mixed oxide precursors

A. Teyssier[1], B. El Goundali[2], M. Kaddami[2], J.J. Counioux[1], C. Goutaudier[1]
[1]Lyon Univ., Claude Bernard Lyon 1 Univ., Laboratoire des Multimatériaux et Interfaces UMR CNRS 5615, 43 bd du 11 novembre, 69622 Villeurbanne Cedex, France;
tel. +33(0)4 72 43 12 05, fax: +33(0)4 72 44 06 18, e-mail: angelique.teyssier@outlook.com;
[2]Hassan 1st Univ, Faculté des Sciences et Techniques de Settat, B.P.: 577 route de Casablanca, Settat, Morocco.

Nickel and cobalt oxides, pure or in the form of mixed crystals, are known for their widespread properties and applications. Considering their growing interest and the significance of their synthesis methods to reach desirable properties (purity, composition, morphology), a large number of papers dealing with the mixed oxide preparation by the metal salt precursor decomposition were published recently [1-3]. Furthermore, the particulate solids present specific features, such as size, polymorphic form, texture, that must be controlled to obtain the required properties; in addition, for catalytic applications, the chemical composition of the precursors and their homogeneity need to be known as well as their morphology and their structure.

In the case of cobalt and nickel mixed oxides, the solid solutions of the hydrates of Co and Ni nitrates are highly relevant as precursors for catalysis or for the preparation of mixed ceramics. They present important qualities as precursor since nitrates can be obtained easily, they are highly soluble in water and so the cationic nickel/cobalt ratio is controllable both in the liquid and solid solutions, and their thermal decomposition (loss of H, O, N) occurres at relatively low temperature (< 400°C) [4].

In aqueous solutions, these nitrates form multicomponent systems, which can be exploited if the H_2O-$Co(NO_3)_2$-$Ni(NO_3)_2$ ternary system is known. This is the reason why the solid-liquid equilibria of this system have been investigated, at atmospheric pressure, within the temperature range [-25, +30°C]. The solubility curves were first determined by synthetic methods: the conductivity of the solution was followed depending on the quantity of added water, each break in the conductivity curve representing a change of domain. Then strategic mixtures were prepared in each domain and the mixtures were stirred at the chosen temperature until the equilibrium was reached. After separation, the compositions of the solid and liquid phases were determined by UV-visible spectroscopy. Then, the solid phases were characterised both by powder and single-crystal X-ray diffraction.

The chemical and structural analysis of the crystals show that the solid phases are the solid solution $Co_{1-x}Ni_x(NO_3)_2.6H_2O$ and reciprocally the solid solution $Ni_{1-y}Co_y(NO_3)_2.6H_2O$ (or $Ni_{1-y}Co_y(NO_3)_2.9H_2O$), within a large composition area and stable in the entire studied temperature range.

During the first stage, the solid liquid equilibria in the H_2O-$Co(NO_3)_2$-$Ni(NO_3)_2$ ternary system were delimited and so the stability domain of the solid solutions depending on the temperature and the composition was defined.

Moreover, in the ternary systems involving concentrated electrolyte aqueous solutions, liquidus curve modeling is usually complex because of the very high activity coefficient values of ions. As a consequence, a large number of experimental mixtures must be done in the chosen temperature range. However, in these systems the ice liquidus curves often follow the linear laws [5-6] as predicted by the ideal solution model and then can be calculated by using the Quasi Ideal Model [7] from the both equations of the ice liquidus curves and the limit aqueous binary systems. In a similar way, it appears that the liquidus curves of the nitrates follow an identical scheme. The solid solutions, in which a few nickel atoms are substituted by cobalt and reciprocally, as $Co_{1-x}Ni_x(NO_3)_2.6H_2O$, $Ni_{1-y}Co_y(NO_3)_2.6H_2O$ and $Ni_{1-y}Co_y(NO_3)_2.9H_2O$, and their corresponding liquidus curves also follow linear laws, thus can be determined from the same model in the all temperature range without the need of multiplying experiments.

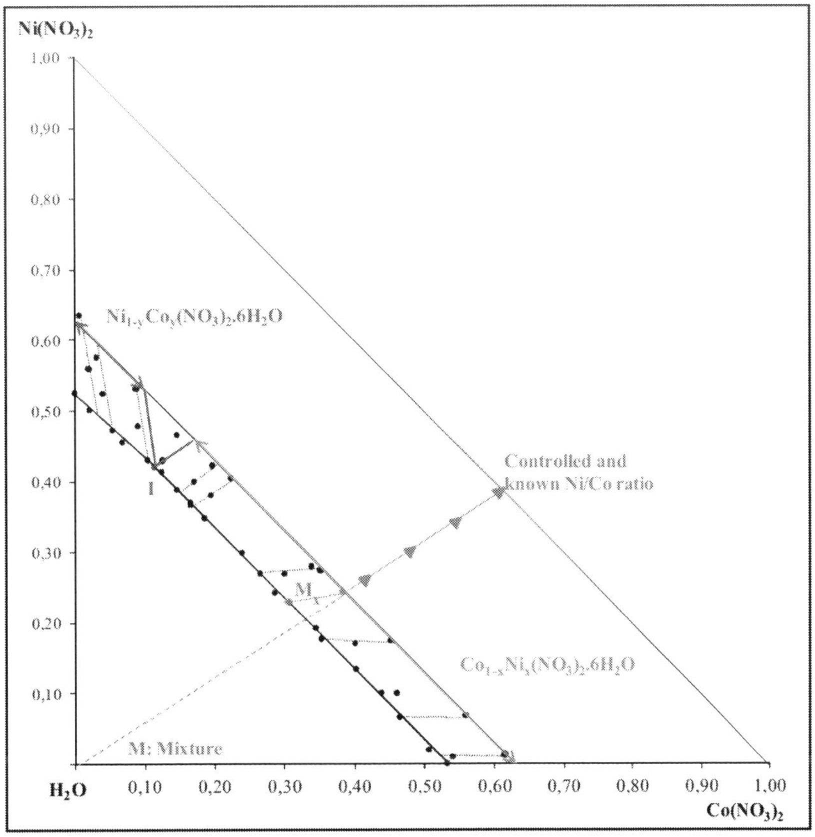

Fig 1. The H_2O-$Co(NO_3)_2$-$Ni(NO_3)_2$ ternary system at 30°C: the two first steps to obtain homogeneous mixed oxide.

The second stage is the design of one homogeneous mixed oxide powder in three main steps. First, a solution of hexahydrated nitrate solid solution with a known nickel/cobalt ratio is prepared. At a defined temperature, the composition of the liquid phase can be found in the ternary phase diagram. In the second step, the solid solution used as precursor is dehydrated to obtain the anhydrous solid. Then, the nitrate is decomposed into the corresponding oxide at low temperature to keep the homogeneity and the chosen nickel cation ratio, which is accurately known.

[1] L.A. De Faria, M. Prestat, J.F. Koening, P. Chartrier, S. Trasatti, *Electrochim. Acta*, **1998**, 44(8-9), 1481-1489.
[2] C. Bocca, A. Barbucci, M. Delucchi, G. Cerisola, *Int. J. Hydrogen. Energ.*, **1999**, 24(1), 21-26.
[3] E. A. Brocchi, D. W. Macedo, G. Solorzono, F. J. Moura, *Mater. Charact*, **2012**, 63, 70-76.
[4] W. M. Keely, H.W. Maynor, *J. Chem. Eng. Data*, **1963**, 8(3), 297-300.
[5] L. Misane, S. El Allali, M. Kaddami, A. Zrineh, R. Tenu, J. Berthet, J.J. Counioux, *Thermochim. Acta*, **2000**, 356, 117-126.
[6] A. Ibnlfassi, M. Kaddami, K. El Kacemi, *J. Therm. Anal. Calorim.*, **2003**, 74, 341-347.
[7] R. Tenu, C. Goutaudier, B. El Goundali, M. Kaddami, J.J. Counioux, *J. Therm. Anal. Calorim.*, **2013**, 112, 263-270.

Fabrication of compartmentalized microparticles by extractive gelation in a microfluidic device

A. Pittermannová[1,2], A. Zadražil[1], N. Bremond[2], J.Bibette[2], F.Štěpánek[1]
[1]Department of Chemical Engineering, University of Chemistry and Technology, Prague, Technicka 3, 166 28 Prague 6, Czech Republic; e-mail: pitterma@vscht.cz
[2]Laboratoire Colloïdes et Matériaux Divisés, ESPCI ParisTech, 10 rue Vauquelin, 75005 Paris, France

For successful application of compartmentalized microparticles in the therapeutic treatment several requirements have to be met, the size of the microparticles is definitely one of them. Therefore this work is focusing on the reduction of a characteristic dimension of the composite hydrogel microparticles to that of the red blood cells.

In the present work, compartmentalized microparticles were synthesized from a calcium alginate gel matrix and contain several functional components, namely iron oxide nanoparticles for radiofrequency heating and MRI visualisation, liposomes for the storage and release of active molecules of their precursors. Liposomes are spherical structures formed by a phospholipid bilayer, which are able to release the active matter in response to temperature. The permeability of liposomes can be significantly increased once they are heated above its phase transition temperature. The needed temperature change is achieved by iron nanoparticles upon exposure to external radio-frequency magnetic field.

Monodisperse microparticles were prepared by a novel extractive gelation process on a custom-designed microfluidic chip (Figure 1.a). First, droplets of aqueous solution of alginate in the continuous oil phase are formed in a flow-focusing channel. Second, gelation of the droplets is established in a long wavy channel by diffusion of calcium ions from the continuous phase. At the same time, the volume of the droplets is reduced by partial extraction of water into the continuous phase. Depending on the volumetric flow-rates of the aqueous and the oil phase, different conditions of the production were mapped and the effect of extraction on the achievable size of the microparticles was evaluated. The ideal conditions to produce monodisperse microparticles and to encapsulate iron oxide nanoparticles (Figure 1.b) and liposomes (Figure 1.c) were found. The stability of liposome in the alginate microparticles was investigated and the first remotely triggered released experiments were conducted.

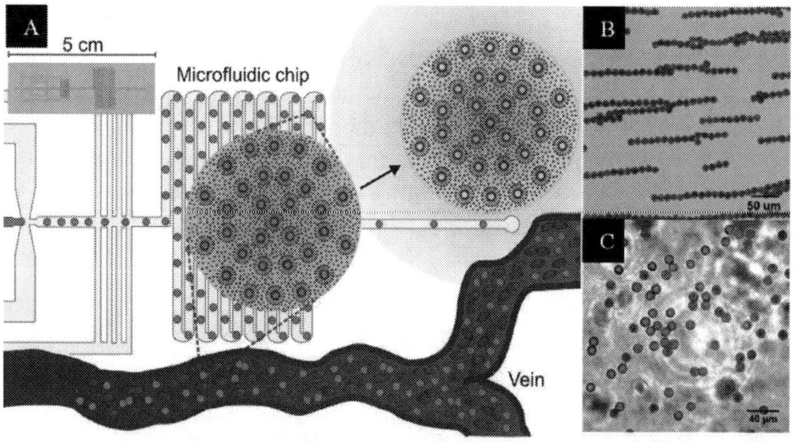

Figure 1: A) Production of compartmentalized alginate microparticles in a microfluidic device; B) Orientation of the alginate microparticles in the magnetic field; C) Microparticles with encapsulated liposomes visualized by confocal microscopy

DEM simulation of the dynamic response to discharge initiation of the model grain silo

R. Kobyłka, J. Horabik, M. Molenda
Institute of Agrophysics Polish Academy of Sciences
20-290 Lublin, Doświadczalna 4, Poland
e-mail: j.horabik@ipan.lublin.pl; tel: +48 81 74 450 61; fax: +48 81 74 450 67

Evolution of stress distribution inside the storage silo is an inherent effect in dynamic operations of handling of granular materials. Sudden increase in lateral pressure with simultaneous ramp down of vertical pressure at the opening of discharge gate may generate severe pulsations of bin structures and result in damage. Due to nonlinearity and heterogeneity of granular systems, mechanism of generation and propagation of stress waves is very complex and not yet completely understood. With an increase in availability of affordable computing power, numerical methods became promising tool for examination of mechanical behavior of granular systems.

In the reported study, numerical simulations of the discharge initiation of the grain silo were performed. Discrete Element Method (DEM) modeling of start of discharge of a flat-bottomed cylindrical container (0.12 m in diameter and 0.5 m high) was carried out. 118 000 polydisperse spherical particles, having diameter of 3.79±0.05 mm and material parameters of wheat, were used. The silo was discharged through the orifice or through the entire bottom of the container. Two cases of gravity-driven discharge were studied: 1) controlled velocity of discharge initiation slowly building up to 0.05 m/s (e.g. Wensrich, 2003), and 2) immediate removing of the orifice or the entire bottom. Based on DEM modeling, the propagation of pressure waves at the silo wall and bottom were analyzed.

Start of the discharge initiates propagation of rarefaction wave up the silo height: plane in the case of discharge through the entire bottom, and spherical in the case of discharge through the orifice. The rarefaction wave may be followed by the compaction wave. Speed of the wave front ranged from 70 to 80 m/s (Kobyłka et al, 2015).

The first step in the dynamic pressure switch is the decrease in the wall pressure along the width of the rarefaction wave. The lateral pressure along the rarefaction wave have shape of a parabolic envelope with the amplitude decreasing as the wave travel up the silo. At the position of switch from the rarefaction to compaction the lateral pressure decreases to zero and immediately increases along the width of the compaction wave to values exceeding the static pressure. In the case of discharge through the orifice no dynamic overpressure was observed but vibrations of the lateral pressure with the frequency of about 420 Hz took place.

Passage of the rarefaction wave results in increase in mobilization of particle-particle and particle-wall friction. Location of the wall friction increase coincides with the point of the start of particle velocity increase while mobilization of particle-particle friction is broad and coincides with the entire wave front.

DEM simulations were found closely following the results from an experimental and numerical study already published in literature, but allows for greater inside into the granular bed. Therefore DEM might be considered useful tool for explaining such phenomena.

Kobyłka R., Horabik J., Molenda M., 2015. Development of a rarefaction wave at the discharge initiation in a storage silo – DEM simulations. Powder Technology (submitted).
Wensrich C., 2003. Numerical modeling of quaking in tall silos. International Journal of Mechanical Sciences, 45, 541-551.

A systematic study on air guided discharge of cohesive powders

Authors: J. Necas, S. Zigan, P. Garcia-Trinanes, M.S.A. Bradley
[1] VSB – Technical University of Ostrava, Laboratory of bulk materials, 17. Listopadu 15, Ostrava – Poruba, 708 33 Czech Republic (Jan.Necas@vsb.cz)
[2] The Wolfson Centre for Bulk Solids Handling Technology, University of Greenwich, Central Avenue, Kent ME4 4TB, United Kingdom (P.GarciaTrinanes@greenwich.ac.uk ; S.Zigan@greenwich.ac.uk)

KEYWORDS
Silo discharge, shear cell, fluidisation

Companies handling a range of cohesive powders with a particle size smaller than 100 microns face often the challenge to use their facilities effectively. The bulk loading and storage systems at ports are one example were a wide range of fine powders such as cement, barite are handled and stored in large storage silos. The discharge of these powders is sometimes problematic and causes long discharge times and additional costs for the plant/ facility operators. Poor discharge of powders is caused by rat holing and channelling of material in a silo. To control poor discharge behaviour of fine powders from silos, industry uses air to reduce the strength of the bulk material and initiate bulk flow. The air is injected into the bulk by air pads in the conical section of the silo. One reason why the addition of air supports the discharge is the enhanced flowability of the powders. The flowability depends on bulk/ particle properties such as particle size, shape and density as well as surface moisture. Controlling the airflow helps to control the flowability of cohesive powders.
Understanding the process parameter influencing the flowability of powders can help to optimise the process of discharging cohesive powders. Cohesive materials such as cement, barite were compacted and fluidised in a cylindrical fluidised apparatus. The fluidisation velocity was well below minimal fluidisation velocity of the material i.e. well in the incipient flow region of the material. A rheometer was utilised in the fluidised bed to estimate the shear stress at failure of the material at constant torque and changing air velocities. In another test the air velocity was kept constant and the torque was varied. Results of both experiments revealed that varying the air flow rate and varying the torque gave similar shear stress at failure of the material.

These are very promising results because they allow plant operators to estimate with a bench scale test the amount of air required to reduce the strength of the cohesive material. These results will also feed into an empirical model predicting the amount of air required for any consolidated powder.

Assisted fluidization of cohesive powders

P. Garcia-Trinanes[1], S. Zigan[1], J.Nečas[2], M.S.A Bradley[1]
[1] The Wolfson Centre for Bulk Solids Handling Technology, University of Greenwich Central Avenue, Kent ME4 4TB, United Kingdom (P.GarciaTrinanes@greenwich.ac.uk; S.Zigan@greenwich.ac.uk)
[2]VSB – Technical University of Ostrava, Laboratory of bulk materials, 17. Listopadu 15, Ostrava – Poruba, 708 33 Czech Republic (Jan.Necas@vsb.cz)

This article highlights the application of mechanical stirring to be used during the fluidization process applied to cohesive powders. The quality of fluidization is improved breaking the stable channels and dissociating the forces that maintain the agglomerates together.In addition the agglomerate size of these group C powders is estimated using the logarithm mean density difference.

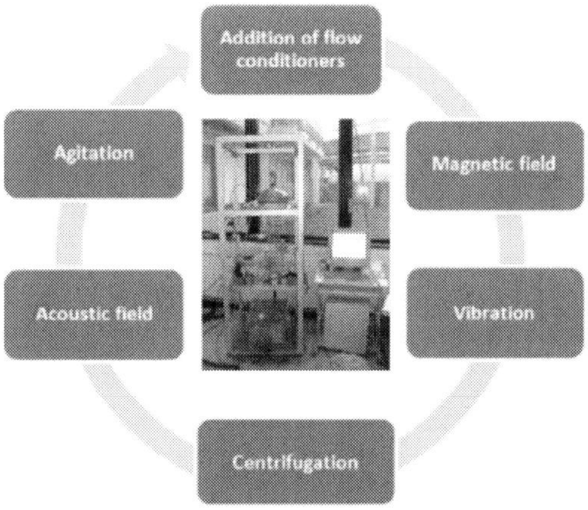

Fig.1.Aid techniques used as assisting methods for improving fluidization

The flow properties of two different calcium carbonates are investigated through experimental measurements: i.e. Sevilla Powder Tester and Schulze Ring Shear Tester. These tests make possible to characterize the flow properties through an estimation of intrinsic parameters at different stress loads (consolidation states) which simulate processing operations such as feeding, transportation, and storage of these materials in multiphase systems.

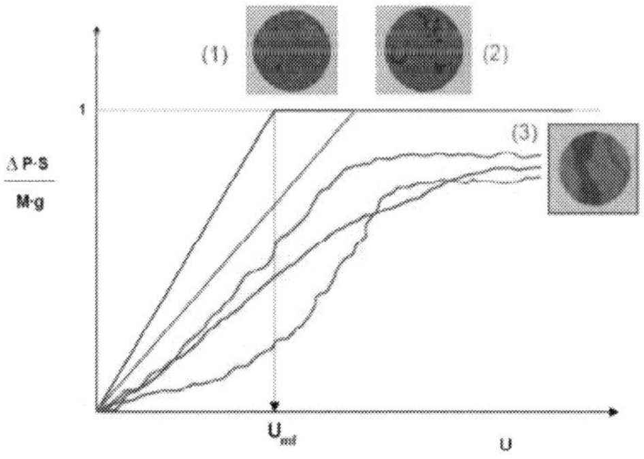

Fig.2.Profile of the pressure drop as a function of the superficial gas velocity.

The most obvious characteristic of Type C powders is their intractability to fluidise. In practice, some phenomena are observed which differ from the ideal behavior of a fluidized bed[1], for instance; particle bed fragmentation, slugging or channelling. The interparticle friction (small particle sizes and high cohesion forces amongst the particles) makes the fluidization very heterogeneous and the size of the bubbles formed can be considerable, therefore well differentiated zones can be observed in the particle bed. When this happens, the fluctuations are important, and as a direct consequence the operation results erratic[2]. This paper delves into the analysis of the fluidization conditions of several well-differentiated group C powders and how the quality of fluidization can be improved by using agitation which has a direct effect and permits to destroy the formation of agglomerates by exceeding the forces that allow the particles to stay together and form dynamic aggregates[3,4,5].

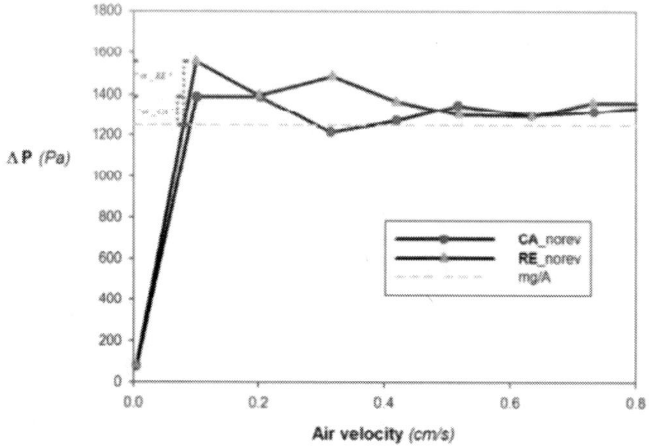

Fig.3.Experimental values of tensile stress

The results obtained indicate that, when mechanical stirring is introduced into the column, the powders performed into a fluidized state. Therefore the use of agitation eliminates the contraventions associated with the non-uniform fluidization of group C particles and makes possible to achieve a fluidized state allowing a non- bubbling fluid-like fluidization at a relatively lower gas velocity computed with the minimum fluid velocity(U_{mf}) necessary to initiate the fluidization without agitation. Furthermore it is possible to determine the U_{mf} and using a RS Rheometer allowing the homogenisation of aggregate fluidization for improved gas-solid contacting. A semi-quantitative mechanism is proposed that replicates trends between torque measurements and pressure difference across the bed.

1. D. Kunii, O. Levenspiel; Fluidization Engineering, 2nd Ed. Butterworth-Heinemann (1991)
2. K. Rietema; The Dynamics of Fine Powders, Elsevier 1220 Science Publishers Ltd, (1991)
3. J.P.K. Seville, U. Tuzun, R. Clift; Processing of Particulate Solids, Blackie Academic&Professional (1997)
4. K.Saleh, X.B.Cami, A.Thomas, P.Guigon; An Experimental Study on the Fluidisation Behaviour of Geldart C Glass Powders; KONA No.24 (2006)
5. J.M. ValverdeMillán; Fluidization of Fine Powders, Particle Technology Series Vol. 18, pp 13-22 (2013)

Compression tests of fine, surface modified soda lime glass particles for the determination of their micromechanical material and contact properties

K. Mader-Arndt, J. Tomas
Institute for Process Engineering, Mechanical Process Engineering, Otto von Guericke University Magdeburg, Universitätsplatz 2, 39106 Magdeburg, Germany,
tel.: +49 391 6751866, fax: +49 391 6711160, e-mail: katja.mader@ovgu.de

The flow behavior of cohesive bulk powders is very important in many industrial branches. With the decrease of particle size, the interparticular adhesion force between the particles increases significantly. The result of this are increased handling problems. Especially for fine to nano-dispersed particle sizes the van der Waals attraction forces exceed the weight forces, which can lead, for example, to undesired (silo outlet) or desired (tablet press) adhesion depending on the process.

Using reference micro soda lime glass particles ($d_{1,50}$ = 284.3 µm and $d_{2,50}$ = 513.2 µm) the single contact of smooth spherical particles is investigated. In order to determine the micromechanical product properties of the particles an in-house designed compression apparatus is used in the present work. With this compression apparatus normal force-displacement curves are recorded and the physical material properties are model-based back-calculated. Therefore the model "stiff particles with soft contacts" is used to quantify and compare the elastic-plastic contact properties.

To investigate the influence of the van der Waals attraction force, the particles are functionalized to get hydrophilic and hydrophobic surfaces. To hydrophilize the particle surfaces a very strong oxidizing agent is used – the peroxymonosulfuric acid. In order to generate a hydrophobic glass surface, the process of silanization is applied.

Electric charge limits on settled powders

[1] J. Pérez-Vaquero, M.A.S. Quintanilla, A. Castellanos
Universidad de Sevilla, Dep. Electronics and Electromagnetism, Avda. Reina Mercedes s/n, 41012 Seville, Spain;
tel. +34 954 559551, e-mail: jperez11@us.es; [1]Fac. Física, Avda. Reina Mercedes s/n, 41012, Seville.

In flows of dry particulate systems, electric charge is generated on particle surface by their collision with walls and with other particles. Charge build-up on single particles can yield local charge values high enough to surpass the limiting electric field for corona discharge into the surrounding gas. Then, local charge is decreased to a lower value that becomes stabilized when flows stop and particles deposit in a container. In this paper, we have used a Faraday pail system to measure the residual particle charge after using two different devices -tribochargers- for particle charging. One of the tribochargers allowed us to directly measure the total charge that was transferred from the walls to the particles, and this can be compared to the final values in the bulk powder once it is collected in the Faraday pail. Tests with different materials, dispersing gas and powder storage humidity were done by using the two tribochargers. The results show bigger influence of particle mean size on the charging levels than the storage relative humidities and dispersing gases we have used. In addition, a simple model based on electrostatic considerations is developed to understand the discharge processes behind the results obtained when the powders are collected in the Faraday pail system.

Triboelectric charging in powder production and processing

Ladislav Konopka, Simon Jantač, Juraj Kosek
University of Chemistry and Technology Prague, Dept. of Chemical Engineering, Technická 5, 166 28 Praha 6, Czech Republic, Tel.: 220 443 107, E-mail: Ladislav.Konopka@vscht.cz

The presence of electrostatic charge is the ubiquitous feature of our everyday life. We generate the charge unintentionally simply by combing our hair, driving our cars, walking in shoes and by other daily activities. Thus a friction of objects is sufficient for asymmetrical electron transfer and the subsequent charging. This type of charging is called triboelectric charging or triboelectrification. Oppositely to charging, it is remarkably difficult to discharge objects, mainly those that are good insulators. As a result of this disproportion, we are surrounded by an excess electrostatic charge, concentrated mostly on the surface of insulators. Since triboelectrification is solely a surface-related phenomenon, it is most significant in systems with relatively high surface-to-mass ratios like powders.

During the production and processing of industrially important powders (e.g. polymers, drugs, flour), the powder particles frequently undergo collisions with other particles or with device walls, thus an excess electrostatic charge is generated on their surfaces. Both the repulsion of likely charged particles and the attraction of oppositely charged particles severely complicate the manufacturing and storage of the powders. For example, in fluidized bed polymerization reactors the charged polymer particles form aggregates and wall sheets, which at least negatively affect the optimal flow regime and at worst choke up the reactor, resulting thus in the reactor shut-down. Complications associated with unwanted aggregation/segregation of powder particles appear also, for example, during micronization and tableting of pharma powders. Plus, tonner and pigment industries exploit triboelectrification and strongly benefit from any progress in the understanding of the phenomenon. Nevertheless, the triboelectrification is still insufficiently understood, regardless of the industrial needs and the fact that this phenomenon has been observed for centuries.

Historically, there are two approaches attempting to describe triboelectric charging. The first one is the so called triboelectric series, which is a list of materials sorted according to experiments in a way that a material from the upper part of the list earns positive charge after a friction contact with a material from the lower part of the list. The differences are caused by different material work functions (i.e., the different capability of capturing electrons). Regrettably, the triboelectric series prediction can quantify the charging process only in a limiting case when materials are always in the same type of friction (e.g., rubbing of two parallel planes). The second approach considers the asymmetric electron transfer when evaluating the charging process, stating that each object in contact can gain either positive or negative charge depending on the type of the friction contact rather than on the object material. Again, the quantitative prediction of the charging is limited and is applicable well for identical materials. A proper description must thus correspond to both of the aforementioned approaches. In the case of particle-wall contact, a capacitor model can be utilized to combine both the asymmetric friction of objects and the difference in material work functions in a specific particulate system. The applicability of the model predictions is limited by the accuracy of included charging function and by the amount of experimental measurements providing us with charging function parameters.

In order to better understand the charging processes, we performed a systematic series of experiments accompanied by mathematical modelling in the exemplary system of polyethylene (PE) powder particles and a metal wall. We combine cascade method apparatus (a slide followed by the Faraday's pail) with the Discrete Element Method (DEM) model of sliding polyethylene particles describing thus the charging of PE particles caused by particle-wall collisions. Using this apparatus, we measure not only the saturation charge of PE particles, but also the charging dynamics as a function of industrially relevant parameters: temperature, sorption equilibria, low humidity atmosphere, PE density and molecular weight. Our DEM model is complemented by the capacitor model of particle-wall charging and validated by the measured data. Additionally, we address the general question of whether the charging depends only on the contact area of colliding objects (which correlates with the spontaneous release of electrons) or also on the amount of dissipated mechanical energy during the contact (which correlates with the excitation energy necessary for the electron transfer). Our study thus reveals the fundamental understanding of triboelectric charging while providing valuable information for polyolefin industry.

An electrochemical technique to measure the degree of dispersion of solid particles in aqueous medium

A. D. De La Concha[1], C. Haro Pérez[1], A. Alonso[2], J. Ramírez-Muñoz[1], G. Martínez de Jesus[1]
[1]Universidad Autónoma Metropolitana-Azcapotzalco, División de Ciencias Básicas e Ingeniería. Av. San Pablo 180, Azcapotzalco, C.P. 02200, Mexico City, Mexico.
[2]CONACYT-Universidad Autónoma Metropolitana-Azcapotzalco. Av. San Pablo 180, Azcapotzalco, C.P. 02200, Mexico City, Mexico. Email:*g.martinez.dj@gmail.com.

Abstract

In actual pigment dispersing processes, the design objective is to achieve maximum dispersion efficiency, which ideally is attained when agglomerate particles are reduced to its primary size. Nowadays, most methods used to measure the degree of pigment dispersion are based on optical measurements, however those might become inefficient when they are used in high reflectivity pigments (e.g. titanium dioxide, silicon dioxide). In this work, novel methods to measure wet pigment dispersion, which are based on the use of two electrochemical techniques known as cyclic Voltammetry and Chronoamperometry are explored. It was found a strong relationship between the dispersion degree and the way that the electrical current is passing through the system under analysis.

Introduction

In measurements pigment dispersion techniques, the majority of methods use light as their main source. Those using DLS [1] (Direct Light Scattering, also known as PCS, Photon Correlation Spectroscopy), measure Brownian motion and relate it to the actual size of the particle. The problem here is that the sample under analysis must have such motion in order to perform measuring. Turbidimetry [2] could be another option, but it needs differences between refraction indexes from the particles and the medium where they are contained.

To avoid aforementioned problems, in the current work two well-known electrochemical techniques [3] are presented and they rely solely on the current flux (I) when a potential (v) is applied over a thin layer called interface. Cyclic Voltammetry uses a potential sweep as the input signal and records the electrical current passing over two electrodes, whilst Chronoamperometry uses a constant potential pulse and records also the electrical current as response.

Methodology

Experiment one consisted of five 80 mL samples of a solution of KCl at 0.01 M within deionized water (18 MΩ cm^{-1}), first sample consisted with no SiO_2 diluted within, while the remaining four consisted each one of 1 gr of SiO_2 diluted within the solution. To achieve different particles dispersion degrees for the samples containing SiO_2, a stirrer and ultrasonic waves (usw) were used. For the first sample with SiO_2, a stirrer was used for 3 min, and for the remaining three of them, it was used full amplitude ultrasonic waves (Hielscher UP200S, 200W, 24kHz) at various time intervals: 3, 12 and 15 min. Voltammetry analysis was used at a scan rate sr=50mV/s, and the first inversion potential V1 can be set to V1=-1.0V, while for the second inversion potential V2, that can be set to V2=0.8V.

Experiment two consisted of three agglomerates samples of 200 μL of Sulfate latex diluted within 15 mL of deionized water (18 MΩ cm^{-1}). In order to attain different degrees of dispersion, ultrasonic waves were used only on the last two samples: at 60% wave amplitude in the first one, and full 100% wave amplitude over the last one for a time period of 2.5 min each. Voltammetry analysis was applied all over these three samples, fixing sr=50mV/s with V1=-1.0 V and V2=0.5 V for this case in particular. Finally, double pulse Chronoamperometry was used over the same samples applying a first pulse E1=-1.0 V for a time period T1=30 s, and a second pulse E2=0.5 V for a time period of T2=45 s.

Results and Discussion

For the experiment one (Figure 1), it can be seen from the first signal coming out from sample one without SiO_2 and without agitation, that the maximum peak for the current flow in this case is -243.1 μA when potential reached -1.0 V, whereas for the second signal that represents SiO_2 inside the solution KCl-H_2O with 3 min of agitation using only a stirrer, the maximum peak is found to be at -196.7 μA. Signal 3 coming out from sample 3 on which it was used usw for 3 min, shows a maximum peak of -187.5 μA. Signal 4 and 5 coming out from sample 4 and 5 respectively on which usw was used at times of 12 and 15 min, respectively, there exists an obvious difference over the previous samples, for these two signals recorded maximum peaks of -143.5 μA and -135 μA.

In the experiment two, Voltammetry shows a similar behavior to those obtained from experiment 1 about the current intensity, that is, it seems that the more intensive the agitation procedure is used, the less flow is passing through the interface. It was decided to verify this results with Chronoamperometry for which it was used the same settings as in experiment one (figure 2). Chronoamperometry shows that when steady-state behavior is reached (flat region) a mean value for the current-flow passing through the electrodes can be computed, in which case is -4.5 μA for the signal that represents sample 1, -2.32 μA for the signal in sample 2, and -1.63 μA for signal representing sample 3.

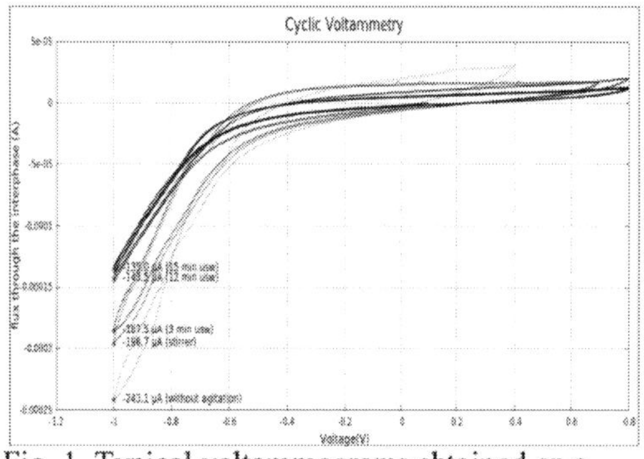

Fig. 1. Typical voltammograms obtained on a vitreous carbon electrode.

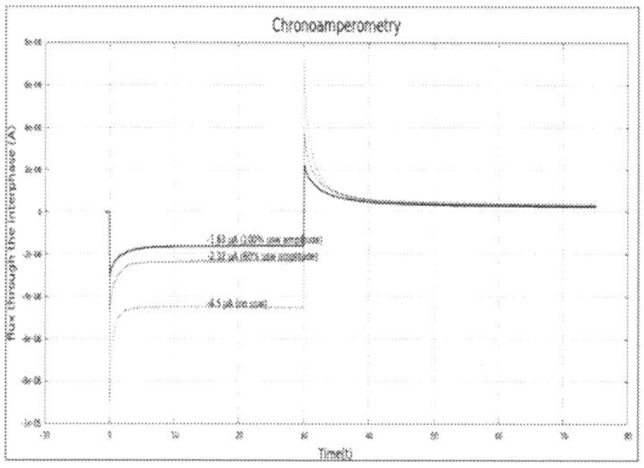

Fig. 2. Typical Chronoamperometry obtained on a vitreous carbon electrode.

Conclusions

In this preliminary work it was found statistical data that suggest a strong relationship between the agitation method used to disperse solid particles into a liquid and the current magnitude passing over the electrodes. These results suggest that Voltammetry and Chronoamperometry might be used as a promising technique to measure the dispersion degree of solid particles. This novel technique might be applied in wet systems where traditional ones have poor reproducibility.

References

[1] Zetasizer Nano Series User Manual. England. July 2004.
[2] Michael J. Sadar, Turbiditry Science, Technical Information Series, Booklet No. 11, p. 4.
[3] Allen J. Bard, Larry R. Faulkner, Electrochemical methods: fundamentals and applications, Wiley, 1980.

Analysis of effect of particle density on mixing behaviour in a rotating drum by DEM

Michinori Yamamoto[1,2], Shingo Ishihara[3], Junya Kano[3]
1) SUMITOMO BAKELITE CO.,LTD., Japan
2) Graduate School of Environmental Studies, Tohoku University, Japan
3) Institute of Multidisciplinary Research for Advanced Materials, Tohoku University, Japan

The mixing of particulate materials is one of crucial unit operations in many industries such as chemicals, pharmaceutical, food and plastics. It is important to give the best homogenization by understanding material properties in terms of quality and process control. However, the influences of material properties (e.g., density, restitution coefficient, frictional coefficient and so on) on the mixing behaviour are still not completely understood. In this study, the mixing behavior of particles in a rotating drum was analized by experiment and Distinct Element Method (DEM) in order to discuss the effect of particle density.

In both the experiment and the DEM simulation, mixing behavior of alumina and stainless steel particles in the rotating drum was observed and evaluated with an in-house image processing. The diameter of alumina and stainless steel particles were 3.0 mm and the number of each particles was 753. Rotational speed was adopted 25 rpm. Alumina and stainless steel particles initially set in a side-by-side arrangement.

In the experiment, most of the alumina particles with lower density exist close to the container wall and in the surface layer of particle bed. While, most of stainless steel particles with higher density concentrate to the center of particle bed. The segregation of alumina and stainless steel particles was obviously observed. The quite similar mixing behaviour of particles was also observed in the DEM simulation, and it was confirmed that the DEM simulation has high reliability.

The DEM simulation was used to analyze the effect of particle density on mixing behaviour and to clear the segregation mechanism in a binary system. These results indicated that the particle density ratio is higher, the mixing behaviour in a rotating drum is strongly affected by it. The particles with lower density could move easily in comparison with higher density particles and the segragation was caused. In order to improve the mixing state of particles, the rotational speed increased. As a result, the particles motion is comparatively uniform and the random mixture is leaded.

Assessment of floc morphology and operating conditions interactions of a pilot-scale activated sludge process using image analysis technique

[1]P. Oliveira, [1]C. Saudejaud, [1]M. Alliet, [1]C. Frances
[1]Laboratoire de Génie Chimique, Université de Toulouse, CNRS, INPT, UPS, 4, Allée Emile Monso, 31432 Toulouse Cedex 4, France, tel. +33 (0)5 34 32 36 39, e-mail: christine.frances@ensiacet.fr

The improvement of wastewater treatment processes has become gradually more required to guarantee the strict environmental standards regarding wastewater discharge [1,2]. For this purpose, activated sludge (AS) systems are a regular biological process used in both municipal and industrial plants. This process occurs in two steps: (1) biodegradation of pollutants inside an aeration tank and (2) bioflocs separation by gravitational force through a settler. Although AS systems are quite efficient, a number of problems can arise due to malfunctions which are generally associated to the operating conditions, for example: oxygen dissolved (OD) concentration, pH, solids retention time or organic loading. Such problems can greatly compromise sludge structure (size and shape) and consequently the overall process performance [3,4]. With the objective of intensifying wastewater treatment control, some innovations have been recently implemented as image analysis technique which has proven to be a suitable tool for morphological characterization of the AS flocs [5, 6].

The aim of the current work is to get a deeper insight into activated sludge behavior and identify relevant relationships between microscopic variables (morphological parameters) and the macroscopic conditions (operational and physiochemical parameters). To this end, samples from an AS pilot plant were analysed using a precise automated microscope.

This experimental pilot (Figure 1) consists in a conventional AS process, i.e. it comprises a cylindrical aeration tank with a total volume of 30 L (Ø 25.5 cm) followed by a 6.6 L secondary clarifier (Ø 22 cm). Sludge provided from a large-scale wastewater treatment plant was first inoculated into the pilot to initiate the process. The system was then fed with a synthetic effluent having a predefined composition (Table 1) and it was equally fed with tap water in order to achieve a COD of 800 mg/L. Measurements of pH, dissolved oxygen and temperature were provided thanks to probes introduced inside the aeration tank. An aeration controller allowed setting the duration of the anoxic and oxic cycles. MLSS (Mixed Liquor Suspended Solids) concentration and Sludge Volume Index (SVI) were measured every weeks.

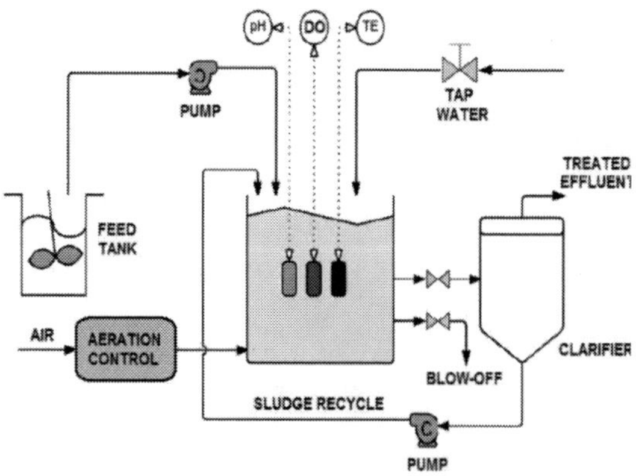

Fig. 1 : Scheme of the pilot plant used

Table 1 : Composition of the synthetic wastewater

Composition	Concentration (mg/L)
Glucose ($C_6H_{12}O_6$)	750
Ammonium chloride (NH_4Cl)	56
Calcium chloride ($CaCl_2$)	10
Ferric chloride ($FeCl_3$)	0,3
Magnesium sulphate ($MgSO_4$)	25
Potassium phosphate monobasic (KH_2PO)	12
Total COD	800

Morphological characteristics of AS flocs were assessed using an automated microscope (Morphologi G3 from Malvern®) combining an optical microscope (with a FireWire™ digital camera) and a software that enables image acquisition, processing and analysis. Preliminary tests were first done to define the morphology analysis procedure and the following conclusions were drawn: (1) a wet cell carrier allows getting a good image quality; (2) among the dilutions 1:10 - 1:20 - 1:40, the dilution of 1:20 provides reproducible results and enables a good isolation of the AS flocs; (3) a threshold of 183 (grey level) allows an efficient image binarization; (4) measures must be accomplished under highest contrast and employing a magnification of 10, in this scale of magnification, the level of visible details is significantly improved and the macroflocs are contained and (5) as stop scanning criteria, choosing the number of particles yields more repeatable results than imposing the scanned area. It means 100,000 individual flocs.

With respect to the tests conducted in the pilot-scale plant, Figure 2 presents the changes of biomass concentration (Mixed Liquor Suspended Solids) and Sludge Volume Index (SVI) over acclimation phase (30 days). During this phase, biomass seems stabilized at a mean MLSS concentration of 8.0 g/L while SVI values have settled down at 97.4 ± 6 mL/g. The SVI value is indeed contained in an optimal range (80 – 150 mL/g) where the sludge is considered having good settle ability.

Changes of the operating parameters (solids concentration, aeration cycles and rates...) were then applied to analyze the response of the system in terms of both morphological parameters and COD removal efficiency.

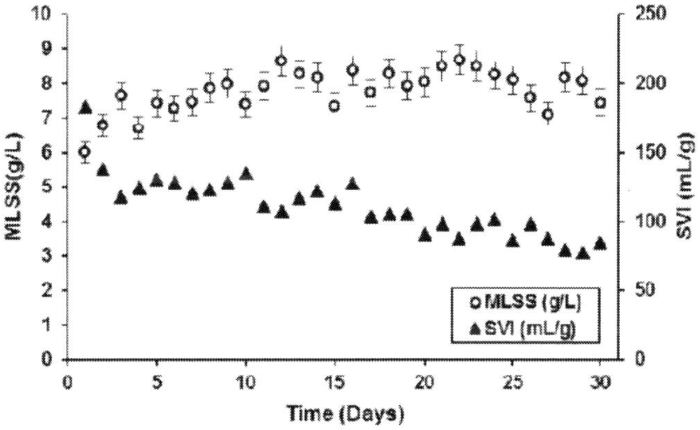

Fig. 2: Changes of MLSS and SVI values over acclimation phase in the pilot plant.

[1] R. Jenné, E. N. Banadda, N. Philips, et J. F. V. Impe, *J. Environ. Sci. Health Part A*, vol. 38, n° 10, p. 2009-2018, oct. 2003.
[2] R. Jenné, E. N. Banadda, G. Gins, J. Deurinck, I. Y. Smets, A. H. Geeraerd, et J. F. V. Impe, *Water Sci. Technol.*, vol. 54, n° 1, p. 167-174, juill. 2006.
[3] E. Koivuranta, J. Keskitalo, A. Haapala, T. Stoor, M. Sarén, et J. Niinimäki, *Environ. Technol.*, vol. 34, n° 5, p. 679-686, mars 2013.
[4] E. Koivuranta, J. Keskitalo, T. Stoor, J. Hattuniemi, M. Sarén, et J. Niinimäki, *Environ. Technol.*, vol. 35, n° 13, p. 1605-1610, juillet 2014.
[5] D. P. Mesquita, A. L. Amaral, et E. C. Ferreira, *Anal. Chim. Acta*, vol. 802, p. 14-28, nov. 2013.
[6] D. P. Mesquita, A. L. Amaral, et E. C. Ferreira, *Chem. Eng. J.*, vol. 285, p. 349-357, févr. 2016.

Caspase-3 siRNA functionalized poly(butylcyanoacrylate) nanocarrier for anti-apoptosis in brain

XW. Zhang[1], W. Hintz[1], P. Henrich-Noack[2], J. Tomas[1], B. Sabel[2]
[1] Otto von Guericke University Magdeburg, Institute for Process Engineering, Mechanical Process Engineering, Universitätsplatz 2, 39106 Magdeburg, Germany
tel.: +49 (0)391 6752001, fax: +49 (0)391 6711160, e-mail: xiwei.zhang@med.ovgu.de
[2] Otto von Guericke University Magdeburg, Institute for Medical Psychology, Leipziger Str.44, 39120 Magdeburg, Germany

Apoptosis which caused by damage or disease could be shut down by several therapeutics but rarely in central nervous system due to the protection from the blood brain barriers which resist the penetration of approximately 98% of the compounds tested for treatment of damage and diseases of the brain. siRNA which is one of dramatic drugs of anti-apoptosis also failed to penetrate the barriers. However, we developed a caspase-3 siRNA capsuled poly(butylcyanoacrylate) (PBCA) nanocarrier which could cross the blood brain barrier and protect the neurons from death.

In this case, the PBCA nanocarrier, which had a water core with caspase-3 siRNA, was synthesized by interfacial polymerization in water in oil (w/o) emulsions and composed of non-ionic surfactants which have excellence permeability to the barrier. PBCA nanocarriers were characterized via polydisperse index, diameter and size distribution. After synthesis, the siRNA capsuled PBCA nanocarrier transfected HEK cell directly. Caspase-3 expression was detected by westernblot.

As a result, the expression of caspase-3 was blocked after 48 hours of the addition of siRNA loaded PBCA nanocarrier, which means the siRNA loaded PBCA nanocarrier was degraded and released siRNA intracellular and a down-regulate affection of caspase-3 had been found as the increasing of the concentration of siRNA in nanocarrier.

For in vivo, a brain tumor model will be involved in this case. The tumor will be measure after the injection of siRNA loaded PBCA nanocarrier in proper time. And the nanocarrier will be modified with Rhodamine 6G and supervised by in vivo confocal microscopy after injection.

Effect of activated carbons particle size and surface chemistry on the slow release of urea

M. Galhetas[1,2], S. Ventura[2], W. Souza[1], A.P. Carvalho[2], Figueiredo, M.A.G[1]
[1]Universidade do Estado do Rio de Janeiro, Departamento de operações e Projetos Industriais, Rua São Francisco Xavier 524, 20550-900, Maracanã, Rio de Janeiro, Brasil; tel. +55 21 23340563; e-mail: wallace.souza@uerj.br; mgaya@uerj.br;
[2]Centro de Química e Bioquímica, Faculdade de Ciências, Universidade de Lisboa, 1749-016 Lisboa, Portugal; tel. +351 217500897; e-mail: amgalhetas@fc.ul.pt; ana.carvalho@fc.ul.pt; fax. +351 217500888

The agricultural sector has been using high quantities of fertilizers to meet the increasing food demands, which results from the exponential growth of the global population (around 7.2 billion according to United Nations). The use of high quantities of fertilizers have far demonstrated undesirable environmental impacts, as is the case of surface and groundwater contaminations, leaching of, for example, nitrates, and in last instance the eutrophication of water resources [1]. The sustainable use of fertilizers may be attained by using systems suitable to slowly release the nutrients, according to the needs of the plants. These are named controlled release fertilizers (CRF) and the advantages of using these systems over the conventional ones are losses reduction, sustainable supply of nutrients, and decrease of application and of negative effects derived from overdose. The most common CRF are those coated with a natural or semi-natural, environmentally friendly macromolecular material [2] and the superabsorbent-fertilizer system [3]. Another type of CRF that, to the best of our knowledge, have not been yet addressed in the literature, is the one that considers the fertilizer supported on activated carbon. In this sense, the aim of the present work is to study the slow release of urea adsorbed on activated carbons.

The materials used in this study will be two commercial carbons, supplied by Norit, with different particle size: (i) a powdered activated carbon (PAC) and (ii) a granular activated carbon (GAC).

The influence of the particle size and surface chemistry of both commercial PAC and GAC on the adsorption and desorption (through kinetic and equilibrium assays) of urea will be studied through batch experiments. For that, the GAC will be sieved to $20 < \phi < 30$ mesh and $\phi < 170$ mesh and, the surface of all materials will be acidified with concentrated HNO_3.

The modified and non-modified materials will be characterized according to the texture (N_2 adsorption isotherms), surface chemistry (pH_{PZC}) and morphology (SEM).

The results obtained so far, reveal that the acid treatments promoted a decrease in the pH_{PZC} of around 50 % (from 9 to 4) for NS carbon, which proves the presence of acid groups in the materials surface.

A breakthrough curve regarding the impregnation of a selected carbon with urea is also foreseen.

References

[1] W. Yuan, Solihin, Q. Zhang, J. Kano, F. Saito. Powder Technology 260 (2014) 22-26.
[2] N.N.R. Ahmad, W.J.N. Fernando, M.H. Uzir. Biosystems Engineering 129 (2015) 78-86.
[3] A. Rashidzadeh, A. Olad. Carbohydrate Polymers 114 (2014) 269-278.

Effect of synthesis process parameters on stoichiometric hydroxyapatite properties

[1]B. Galey, [1]M. Tourbin, [2]P. Gras, [2]N. Abi Chebel, [3]N. Rouquet, [2]F. Brouillet, [2]D. Grossin, [1]C. Frances
[1]Laboratoire de Génie Chimique, Université de Toulouse, CNRS, INPT, UPS, 31432 Toulouse Cedex 4, France, tel. +33 (0)5 34 32 36 39, e-mail: christine.frances@ensiacet.fr
[2]CIRIMAT, Université de Toulouse, CNRS, INPT, UPS, Toulouse, France,
[3]URODELIA Company, Saint-Lys, France

Hydroxyapatite (HAP) is a prominent biomaterial. It is mainly used as a coating of metal prosthesis, owing to its strong chemical and structural similarities with the natural bone mineral component. It is usually synthesized as a powder, and applied on the metal surface by means of plasma spray. Many properties, including the chemical composition of the hydroxyapatite powder, as well as the particle size, morphology and density, affect the characteristics of the final coating [1]. Industrial processes used to synthesize hydroxyapatite for coating applications often include successive steps of precipitation, maturation, washing, spray drying and calcination/sintering. The properties of the final powder result from the agglomeration of hydroxyapatite particles initially formed by precipitation and occurring during these different successive steps [2]. Nevertheless, actual processes generate very low yields, due to the lack of control of the agglomerate size. On one hand, the finest particles are lost during process steps of washing and spray drying. On the other hand, oversized particles are removed by sieving, leading to an additional loss of material.

This work focuses on the hydroxyapatite agglomeration during the precipitation and maturation steps in order to control the aggregate size. As the synthesis conditions play a key role in the agglomeration of HAP particles, several semi-continuous synthesis experiments were conducted to investigate the effects of the main operating parameters (stirring rate, reagent flow rates, supplementary ammonia addition, ...) on the properties of the precipitated product. In order to characterize the effects of process parameters on the physicochemical properties of the slurry, characterization techniques that investigate the chemical composition (such as XRD or FTIR) were coupled to physical analyzes that examine the shape and size distribution of the HAP agglomerates in suspension (laser diffraction, scanning electronic microscopy).

The precipitation of hydroxyapatite was performed in a stirred tank and results from the reaction at a high temperature of phosphoric acid and calcium nitrate in an aqueous ammoniacal medium. The purity of the product is controlled by X-ray diffraction after its calcination at 1000°C for 15h (standard norm ISO 13779-3). In such conditions, only the products with a stoichiometric ratio Ca/P=1.67 are stable; the other compounds being decomposed into HAP and CaO (Ca/P>1.67) or HAP and β-TCP(Ca/P<1.67). Fig. 1 shows the diffractogram which is obtained in each case. During the runs, a special attention was paid on the pH evolution over process time, since it seems to be a key parameter for the synthesis of stoichiometric HAP and is strongly dependant on the process conditions. As an example, the pH evolution during the synthesis at different stirring rates is reported on Fig. 2.

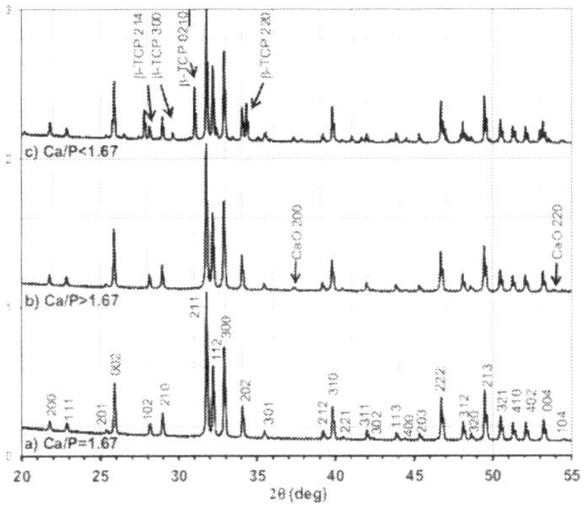

Fig. 1: X-ray diffractogram after calcination

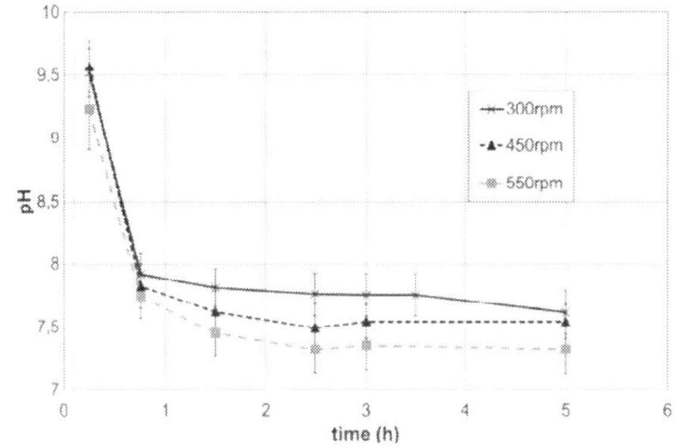

Fig.2: Change over time of the pH value

Hydrodynamic conditions also affect the size distribution of HAP agglomerates and its evolution over process time. Under given process conditions, it was observed that the initial agglomerates had a rather large and dispersed size (see Fig. 3). The size then decreases during the synthesis probably thanks to breakage and restructuring phenomena. The size distribution of the product at the end of the synthesis step presents a main proportion of particles of a few microns and eventually another population constituted by larger agglomerates. It was shown that the maximum size of these agglomerates was calibrated by the Kolmogorov micro-scale depending on the applied shear rate [3].

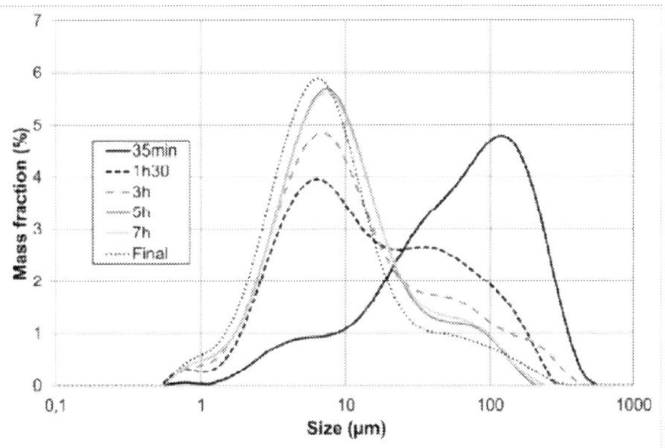

Fig. 3: Change over the process time of the particle size distribution

The reagents flow rate variation and the addition of supplementary ammonia were also considered in order to improve the process yield while keeping a good quality of the produced HAP, expressed in terms of purity as well as physical properties (size distribution, density, ..).

[1] I. Demnati, D. Grossin, O. Marsan, G. Bertrand, G. Collonges, C. Combes, M. Parco, et al., Comparison of Physical-Chemical and Mechanical Properties of Chlorapatite and Hydroxyapatite Plasma Sprayed Coatings, The Open Biomedical Engineering Journal 9, no. suppl. 1–M3 : 26–39, 2015.

[2] R. Rodríguez-Clemente, A. López-Macipe, J. Gómez-Morales, J. Torrent-Burgués, V.M. Castaño, Hydroxyapatite precipitation: A case of nucleation-aggregation-agglomeration-growth mechanism. J. Eur. Ceram. Soc., 18, 1351–1356, 1998.

[3] C. Coufort, C. Dumas, D. Bouyer, A. Liné. Analysis of floc size distributions in a mixing tank. Chemical Engineering and Processing: Process Intensification, 47(3): 287–294, 2008.

Experimental characterization of electrostatic charging of polymer powders

S. Jantač, L. Konopka, J. Kosek
UCT Prague, Department of chemical engineering, Technická 5, 166 28 Praha 6,
E-mail: jantacsi@vscht.cz

Charging of objects caused by their mutual friction is called triboelectric charging and is still poorly understood. Triboelectric charging is responsible for severe industrial problems, occurring mainly during the production and storage of insulating powders like polyethylene (PE), flour, drugs or pigments. The accumulated energy can be released in from of spark discharge. This discharge may cause an ignition of dispersed powders or an ignition of flammable gases. In order to better understand the triboelectrification, we complemented our previously built apparatus (slide and Faraday's pail) with high speed camera and measured particle-wall charging dynamics. We observed more extensive charging at higher concentration of organic solvents. This increase can be explained by a change in elastic properties of particles (bigger contact area) due to absorption of solvent in to the particle. We observed more extensive charging at higher temperatures. This increase can be explained by temperature-dependent elastic properties of particles (bigger contact area) or by the excitation of electrons on the particle surface. We measured the particle elasticity (restitution coefficient) as a function of temperature. Our results of temperature dependent charging provide not only the important information for industry, but when combined with the measurements of particle elasticity represented by restitution coefficient they broaden our insight into the charging mechanism itself.

Characterization of banana waste (pseudo stem and leaves) by thermogravimetric analysis for potential use as biofuel

J. F. Saldarriaga[1], I. Estiati[2], A. Atxutegi[2], A. Aristizabal[1], R. Aguado[2], M. Olazar[2]
Environmental Engineering Program. Faculty of Engineering. University of Medellin, Carrera 87 N° 30 – 65, Medellín. Colombia, tel +5743405817, e-mail: juanfelorza@gmail.com
Department of Chemical Engineering, University of the Basque Country, PO Box 644 - E48080 Bilbao. Spain.

For use of biomass as a fuel it is required, determining physicochemical parameters such as moisture content, volatile matter, fixed carbon and ash, and the content of the three natural polymers constituents (hemicellulose, cellulose and lignin), which condition degradation kinetics. These polymers can decompose at a temperature range of 225-325, 305-375 and 250-500 °C respectively (White, Catallo, and Legendre 2011; Carrier et al 2011.). Similarly, it has been determined that cellulose reactivity is much higher than that of lignin, which makes a greater number of cellulose occur more volatile (Shen et al. 2009; Carrier et al. 2011). This leads to the greater the lignin composition, the greater the amount of char produced or carbonaceous residues, because there is less thermal degradation.

This work aims to design a tool that is versatile and reliable to calculate, from a simple and quick analysis as a thermogram, related to the quality of solids, such as lignocellulosic composition, moisture content, ash biofuels properties, volatile matter and fixed carbon. This tool will subsequently be integrated into a rigorous model that will allow predicting the behavior of alternative biomass combustion plant waste.

The methodology of thermogravimetric analysis was implemented allowed from one run to determine the moisture content, volatile matter, fixed carbon, ash, and adequate separation of the peaks corresponding to the degradation of the three natural polymers constituents (hemicellulose, cellulose and lignin).

The results of chemical composition, have been reported by many authors, showing some similarities or differences in chemical characteristics for different biomasses, in Table 1 shows the results of fixed carbon, volatile material, ash and moisture content of the 8 types of biomasses, in which you can see high ash content for fern very similar results to those obtained by other authors (Demirbas 2009; Raveendran, Ganesh, and Khilar 1995).

A high heating value indicates higher energy potential (Lopez & Llera, 2010), and according to different authors the optimum values of calorific power biomass for direct use as fuel in combustion processes vary between 16 to 18 MJ/kg (Marangoni et al. 2012), range in which biomass are studied leaf and pseudo stem of bananas, and the mixture of these residues, which are values of 17.45 MJ/kg, 17.87 MJ/kg and 17.33 MJ/kg, respectively. Therefore, the samples analysed allow inferring that waste banana (leaf and stem) are favourable for power generation raw materials.

The volatile compounds derived from both organic part of the biomass, and the inorganic part; a high content of these compounds indicates the presence of an organic filler when in 74-89% range, which provides a susceptibility to thermal degradation in processes such as pyrolysis and combustion (Lopez & Llera, 2010; Sellin et to 2013; Marangoni et al 2012).

Thus the results in the analysed samples, the present values for the leaf, stem and the mixture, 74.70%, 85.30% and 80.78%, respectively, are located in the range, allowing affirm the feasibility resource for fuel production using a thermochemical process. The percentages of fixed carbon for leaf, stem and mixing are 15.68%, 9.85% and 11.01%, the values of the sheet and the mixture are similar to the results found by Fernandes et al. (2013) and Marangoni et al. (2012) which are between 12-15%.

Table 1. Characterization of banana waste

Properties	Sawdust	Banana leaves	Banana pseudo stem	Pseudo stem and leaves mixing
Particle size (mm)	0.76	0.55	0.67	0.62
Density (kg/m^3)	496	421	456	433
Moisture content (% p/p, b.s.)	9.53	8.41	7.62	7.83
HHV (MJ/kg)	18.84	17.45	17.87	17.33
Volatile Matter (wt%)	83.96	74.7	85.3	80.78
Carbon Fixed (wt%)	15.75	15.68	9.85	11.01
Ash content (wt%)	0.29	9.61	4.85	8.21
Carbon (wt%)	46.47	42.69	41.77	42.66
Hydrogen (wt%)	5.95	5.44	5.37	5.41
Nitrogen (wt%)	0.2	1.46	1.02	1.48
Oxygen (wt%)	29.78	47.35	41.8	43.73
hemicellulose	21.94	21.88	23.28	22.71
Cellulose	35.67	30.51	34.15	30.28
Lignin	33.07	29.61	30.07	31.94

TGA analysis allows better study of the raw material, with a ramp optimum heating for sufficient degradation of the three components (hemicellulose, cellulose and lignin) with a previous drying step, which allows moisture loss, generating better conditions of combustion. Moreover, high temperatures with an optimal isothermal period, allows the degradation of lignin content and then passes through an oxidation step at high temperatures to cause combustion of the char formed by the pyrolysis process.

This tool allows the lignocellulosic analysis reliably from a thermobalance analysis. This may be of particular interest in the context of more sophisticated and expensive measure equipment for experimental analysis as is hemicellulose, cellulose and lignin, also volatile matter, fixed carbon and ash.

Modelling and simulation for elastic-plastic materials using ADEM

S. Ishihara, J. Kano
Tohoku Univ., Institute of Multidisciplinary Research for Advanced Materials, Katahira 2-1-1, Aobaku, Sendai, Japan; tel. +81 222175136, e-mail: ishihara@tagen.tohoku.ac.jp

Grinding is one of the most effective operations to produce fine particles, and the significance of grinding grows with each passing year due to the development of nanotechnology. Grinding has been used in many fields, for example, foods, cosmetics, medicines, mining, and so on. However, in general, the efficiency of grinding is quite low because the mechanism of grinding have not elucidated. Therefore, it is necessary to clarify the grinding mechanism and also to control the grinding process. In this work, it has been attempted to analyse the grinding phenomena using computer simulation. A new simulation model, ADEM (Advanced Distinct Element Method)[1)] has been developed and it can represent both non-spherical particles motion and particle breakage behaviour. In ADEM, single particle is represented by agglomerate of primary particles, and primary particles are connected by joint spring. To express the breakage and plastic deformation, each joint spring has two kind of thresholds. Single particle breakage was analysed by ADEM, the methods and results is shown as follows.

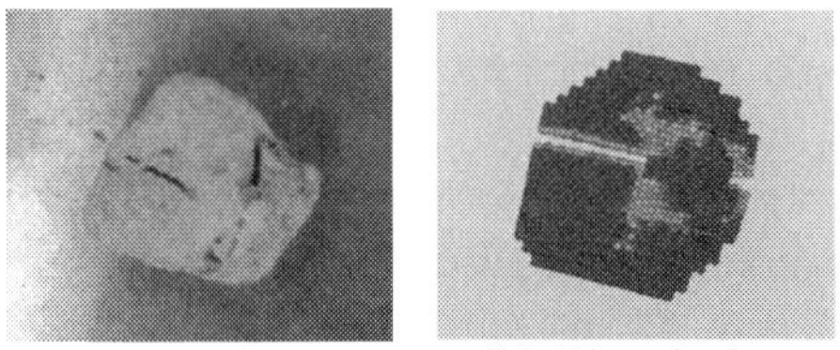

(a) Experiment (b) Simulation

Fig.1 Particle shape and breakage behavior of compression test.

Compression test and impact test have been done to observe the breakage behavior at various load velocities. Both experiment and simulation have been carried out to confirm the validity of simulation model by comparisons of each results. **Table 1** shows the conditions of simulation. **Fig.1** and **Fig.2** shows the results of compression test and impact test, respectively. In compression test, simulation results have good agreement with experimental results from the view point of load-displacement curve and the shape of fragments. In impact test, plastic deformation behaviour was observed. At the initial state, single particle shape is spherical, however, the shape gradually changed into flat plate by continuous impact. These results suggest that ADEM would be a useful tool to analyse the particle breakage behaviour.

Table 1 Conditions of simulation

Primary particle diameter [mm]	1.0
Time step [sec]	1×10^{-6}
Hegiht of falling ball [mm]	300
Joint-spring coefficient [kN/m]	1.5
Plastic strain [-]	0.05
Maximum strain [-]	0.09

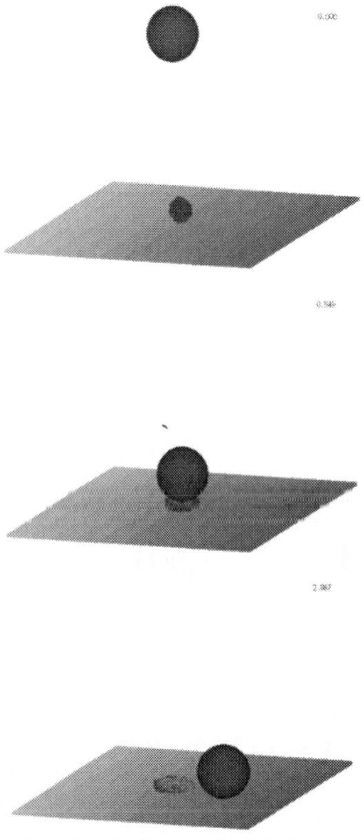

Fig.2 Plastic deformation behavior after impact.

Natural–calcium based residues for carbon dioxide capture

N. Chalermwat, R. Rattanaprapanporn, B. Chalermsinsuwan, S. Poompradub
Department of Chemical Technology, Faculty of Science, Chulalongkorn University, 254 Phayathai Rd., Patumwan, Bangkok 10330, Thailand;
tel. +662 2187523, fax: +662 2555831, e-mail: sirilux.p@chula.ac.th

In recent years, various solid sorbent materials such as calcium oxide (CaO), activated carbon, zeolite, or silica, had been proposed for carbon dioxide (CO_2) capture from flue gases, industrial process and deforestation due to its low cost and no corrosion problem. In this work, clam shell was introduced as a precursor of CaO sorbents for CO_2 adsorption in a bubbling fluidized bed reactor. Benefits of using natural calcium-based materials are low price, abundant, easy to find and non-toxicity. The 2^k experimental design was used to find the operating conditions for the calcination process. The experimental parameters were including particle sizes (150 - 250 μm and 250 - 425 μm), heating rates (10 °C/min and 50 °C/min) and duration time (15 min and 30 min) at 950 °C under N_2 atmosphere. In addition, the characteristics of clam shell before and after calcination were characterized by energy dispersive X-ray (EDX), thermogravimetric analyzer (TGA), X-ray diffraction (XRD), scanning electron microscope (SEM), and Brunauer-Emmett-Teller (BET). The results showed that the CO_2 adsorption capacity was significantly depending on the heating rate. The optimum conditions for calcination process were the use of CaO particle sizes of 250 - 425 μm under heating rate of 10 °C/min for 15 min. The maximum CO_2 adsorption capacity of CaO obtained from clam shell was 31.1 mg CO_2/g CaO, which exhibited higher capacity than that of commercial CaO due to the higher surface area and surface properties. The morphology transformation of calcined clam shells comparing with commercial CaO after carbonation confirmed that the sorbent materials from clam shells had the ability to capture CO_2. Accordingly, clam shell can be used as a CaO sorbents in a fluidized bed reactor to capture CO_2 from the combustion gases in the power plants and industries at high temperature.

Figure 1. (a) Clam shell and (b) Clam shell powder after grinding.

Numerical investigation of agitated beds homogenization in dependence on bed depth via DEM

[1]T. Barczi, [2]T. Trávníčková, [1]M. Kohout, [2,3]J. Havlica
[1] Department of Chemical Engineering, University of Chemistry and Technology; Technická 5, 166 28 Prague, Czech Republic; tel. +420 2 2044 3168, e-mail: martin.kohout@vscht.cz
[2] Institute of Chemical Process Fundamentals of the ASCR, v. v. i.; Rozvojova 2/135, 165 02 Prague, Czech Republic
[3] University of Jan Evangelista Purkinje, České mládeže 8, Ústí nad Labem, 400 96, Czech Republic

Granular materials such as packed beds present a unique state of matter and are currently in the urgent need of developing the proper theoretical basis for its basic aspects. The particle mixing is therefore studied comprehensively and a lot of equipment with uncertain internal behaviour is used. In contrast to mixing of continuous liquid phase where behaviour is relatively well known, in case of description of discrete phase the present state of affairs is rather unsatisfactory. The process of homogenization is mainly studied at macroscopic (equipment) scale and, therefore, results can be used for describing and understanding only the mixer type of particular interest. The studying of homogenization from the first principles, which can be taken for general description, remains still poorly understood due to its complexity.

One of the possible ways for mathematical modelling of mechanical interactions between particles in granular systems is based on discrete element methods (DEM). In DEM models each particle is represented as a sphere with defined properties and is tracked individually. The motion of each sphere is calculated as a sum of external (gravitational force) and collision forces (caused by the collision of particles between each other or by the contact of the particle with the wall). Resulting translational and angular velocities than characterize the movement of the particles. In this contribution open source DEM particle simulation software LIGGGHTS was used. Collision forces are there modelled by the soft-sphere approach with the Hertzian collision model.

As an application we studied agitated particle beds (packed from glass spherical particles of two colours – packed vertically and horizontally) mixed in a vertical cylindrical mixer with two opposed flat blades. Dependence of spatiotemporal dynamics on geometrical aspects (blade rotational speed and packed-bed depth) was analysed at microscopic scale by the complex flow patterns with different nature. At macroscopic scale degree of homogenization was characterised by commonly used global mixing characteristics such as Lacey and Poole mixing indexes. These indexes represent overall degree of homogenization and therefore should be used for prediction of suitability of chosen geometrical aspects. In addition macroscopic and microscopic approaches were connected and from flow pattern observation optimal conditions for the best homogenization were predicted.

On the use of natural limestone and dolomite for thermochemical energy storage of concentrated solar power

B. Sarrion[1], J.M. Valverde[3], A. Perejon[1,2], L.A. Perez-Maqueda[1], P.E. Sanchez-Jimenez[1], J. Perez-Vaquero[3]

[1]Instituto de Ciencia de Materiales de Sevilla (C.S.I.C.-Universidad de Sevilla). C/Americo Vespucio 49, Sevilla 41092. Spain. [2]Departamento de Química Inorgánica, Universidad de Sevilla. C/Profesor Gonzalez s/n, Sevilla 41071, Spain. [3]Facultad de Fisica, Universidad de Sevilla. Avda. Reina Mercedes s/n, 41012 Sevilla, Spain, tel. +34 954 559 551, e-mail: jmillan@us.es, jperez11@us.es.

CaO has been long proposed as an energy intensive material for the storage of energy in a chemical form by means of carbonation/calcination cycles. This Calcium-looping process (CaL) is the basis of a pilot-scale proven technology for CO_2 capture, which is accomplished by carbonation of CaO and its regeneration in a calciner reactor operated under high CO_2 partial pressure and high temperature. The wide availability of low cost limestone and dolomite to be used as CaO precursors (~ 10€/ton) is a key factor for the feasibility of the CaL process. However, the huge potential for lime (CaO) also to store energy has been inhibited by its propensity to sinter at the high temperatures of the standard CaL cycle for CO_2 capture, which reduces dramatically its multicycle conversion. This work shows that limestone and dolomite exhibit an optimum performance for their use to store thermochemical energy by integrating the CaL process into Concentrated Solar Power (CSP) tower plants. The CSP-CaL integration yields high storage energy density (3.2 GJ/m^3) with possible long time gaps between load and discharge. Multicycle carbonation/calcination tests show that the residual conversion exhibited by CaO derived from calcination of limestone and dolomite is rather high at new CaL conditions that maximize the efficiency of its integration into CSP tower plants using already mature and inexpensive technologies.

Preparation of Pickering emulsion stabilized by stearate microparticles

N. Higashide[1], K. Naoe[1*], N. Matsuda[2], and M. Imai[3]
[1]Advanced Chem. Eng. Course, Faculty of Advanced Eng., National Institute of Technology, Nara College, Yamato-Koriyama, Nara 639-1080, Japan; +81 743 55 6163, e-mail: naoe@chem.nara-k.ac.jp; [2] R&D Division, Taihei Chemical Industrial Co. Ltd., Ikaruga, Nara 636-0104, Japan; [3]Graduate School of Bioresource Sci., Nihon University, Fujisawa, Kanagawa 252-0880, Japan

Pickering emulsions are emulsions stabilized by solid adsorbed at liquid-liquid interface, and expected as food, pharmaceutical, and cosmetic media because this emulsion system can reduce the use of artificial surfactant. In this study, preparation of Pickering emulsions by food additive stearate microparticles was investigated. Magnesium stearate was used as a model stearate microparticles. Stearate magnesium powder was added to a mixture of aqueous solution containing 0.01 M $CaCl_2$ and n-heptane solution (aq. soln./n-heptane = 1:1 vol.), and mixed by using magnetic stirrer for 3 min. After mixing the emulsion phase was formed at the bottom of oil phase. In the emulsion phase, emulsions with diameters less than 1 mm were successfully prepared without any surfactant. After preparing the emulsions in the presence of aqueous fluorescence dye calcein under the same condition, fluorescence microscopy revealed the formation of water-in-oil (W/O) emulsion. The average diameter of emulsion decreased with increasing rotation speed of magnetic stirrer in the range of 1,600-2,800 rpm. The minimum average size of 0.273 mm was obtained at 2,800 rpm. Thus, it was found that the diameter of emulsions could be controlled by changing rotation speed. The prepared emulsions were stable for more than 3 months. Magnesium stearate microparticles can also stabilize water droplets in olive oil. The W/O emulsions were successfully prepared via moderate stirring. The average diameter of the obtained W/O emulsion was as small as ca. 60 μm. However, the emulsions exhibited low stability. These demonstrate that food additive stearate microparticles have the ability to form emulsions in edible oils, leading to biocompatible emulsion systems.

Production and characterisation of superparamagnetic magnetite nanoparticles using a continuous co-precipitation process in a T-mixing system

W. Hintz, M. Khalid, J. Tomas
Otto von Guericke University Magdeburg, Institute for Process Engineering, Mechanical Process Engineering, Universitätsplatz 2, 39106 Magdeburg, Germany
tel.: +49 (0)391 6752295, fax: +49 (0)391 6711160, e-mail: werner.hintz@ovgu.de

Superparamagnetic iron oxide nanoparticles (SPIO-NP), such as magnetite Fe_3O_4 are among the most frequently used biocompatible particle system in the biomedicine. Based on their physical properties (e.g. particle size), these nanoparticles have numerous advantages in diagnostics and therapeutically applications in medicine (selective release of drugs in the human body, a better pharmaceutical kinetics and dynamics, local medication). Although a huge progress has been made, there is still the need to solve fundamental aspects of the production of magnetic fluids (controlled nucleation and growth processes; narrow particle size distribution; stability of the nanoparticle system against agglomeration and chemical oxidation processes; surface modification and coating).

Today, single domain particles with a high magnetizability have a primary particle size between 3 to 15 nm, only. Multi domain agglomerates have particle sizes up to 100 nm. In fact, these particles have a high magnetizability (the integral magnetizability can be calculated by multiplication of each primary particle of the aggregate). In contrast, larger single domain particles have the attraction that the magnetizability scales with d^3 (volume proportional magnetizability!), but they still keep the switchable superparamagnetic properties as well as the stability of suspension. These not comprehensively explored processes are built up by complex connected micro processes.

In the present work, a magnetite continuous co-precipitation process has been investigated using a T-mixing system starting from Fe^{2+}/Fe^{3+} in alkaline medium. There has been a main focus on the influence of several process parameters (mixing and flow rate, educt concentrations, molar of Fe^{2+}/Fe^{3+}, pH value (addition of NaOH), concentration of stabilizer (sodium citrate)) on particle size, and zeta-potential.

In the continuous co-precipitation process, particle sizes have been obtained between 9.5 nm and 390 nm; a higher flow rate (0.06 m/s … 0.33 m/s) shows a slight increase in particle sizes; pH range was varied between 8.0 to 13.0, while a higher pH range gives larger agglomerates; a decreasing concentration of sodium citrate led to more agglomerated particles; the molar ratio of Fe^{2+}/Fe^{3+} has no significant influence on particle sizes. Optimal process parameters (flow rate 0.03 m/s, stoichiometric concentration ratio, Fe^{2+} 0.31 mol/L, Fe^{3+} 0.62 mol/L, concentration NaOH 2.4 mol/L, pH 8.0, concentration sodium citrate 4.29 g/g magnetite) leads to 10.3 nm superparamagnetic magnetite nanoparticles. Measured zeta-potentials are around - 30 mV.

Production and characterization of fluorescent polybutylcyanoacrylate nanoparticles to investigate the transport mechanisms of the blood-brain-barrier

[1]T. Hopf, [2]Y. Qing, [1]W. Hintz, [2] P. Henrich-Noack, [1]J. Tomas, [2] B. Sabel
[1] Otto-von-Guericke-University Magdeburg, Institute for Process Engineering, Mechanical Process Engineering, Universitätsplatz 2, 39106 Magdeburg, Germany; tel.: +49 (0)391 6752001,fax: +49 (0)391 6711160 e-mail: Talea.Hopf@ovgu.de
[2] Otto-von-Guericke-University Magdeburg, Institute of Medical Psychology, Leipziger Str. 44, 39120 Magdeburg, Germany

In previous studies it has already been shown that polybutylcyanoacrylate (PBCA) - nanoparticles are capable of crossing the blood-brain barrier. The blood-brain barrier is a natural, highly selective filter that protects the brain from in the blood circulating messengers, toxins and pathogens and at the same time it supplies the brain with nutrients [1]. However, the blood-brain barrier is an insurmountable barrier for many drugs. At this point PBCA nanoparticles should be used as drug carriers to transport the active ingredients directly into the central nervous system. For a systematic investigation of the unexplained transport mechanism fluorescent PBCA nanoparticles are required to visualize the nanoparticles inside the cell tissue.

The aim is to modify the particles with special focus to their loading with fluorescent dye. For the preparation of the particles the established process of miniemulsion polymerization was used. To modify the properties of the nanoparticles like size, surface structure and electro kinetic potential (zeta-potential), different surfactants (Tween 80, Lutrol F68) and surface modifier (DEAE-Dextrane, Dextrane 70.000) were used.

For analysing the particles, the particle size distributions, zeta-potential, mass content and amount of unbounded fluorescence dye were measured. For the migration mechanism In Vivo Confocal Neuroimaging (ICON) was used. Therefore nanoparticles were injected into anesthetized rats and the cell tissue of the retina was focused to see the fluorescent nanoparticles crossing the blood-retina-barrier.

Thus, it was found that Rhodamine 123 and Propidium Iodide can be used as fluorescence dyes. For both fluorescence dyes it was proven, that incorporation inside the nanoparticles is much more effective then adsorption on one particle surface. Furthermore nanoparticles with a double labelling (Rhodamine 123 and Propidium Iodide) were successfully produced and tested.

[1] J. Kreuter, „Transport of drugs across the blood-brain barrier by nanoparticles", Current Medicinal Chemistry - Central Nervous System Agents, 2002

Relaxation of polymers and polymers emulsion using method TD-NMR

P.Schneider, T. Chaloupka, J. Kosek
University of chemistry and technology Prague, Department of chemical engineering, Technicka 5, 166 28 Prague 6,
E-mail: schneidp@vscht.cz

Polymers are one of the most common materials and have a broad range of applications (constructions, car industry, medicine, packaging, etc.), but their inner morphology on meso-scale is often not fully understood. Final properties of polymer materials (sorption capability, viscosity, density, melting point, etc.) heavily depend on the morphology. Thus, for optimization of industrial production and innovation of new polymer products, it is necessary to investigate proper methods for morphology characterization and determine morphology changes during sorption processes. Moreover, these data are essential for programs, which perform on-line control of industrial reactors and, thus, realise significant resource and energy savings.

We employ a time-domain nuclear magnetic resonance (TD-NMR) for the morphology characterization of polyolefins and polymer emulsions. Using several pulse sequences methods (Free induction decay, Solid-Echo sequence, Saturation recovery measurement, etc.), we obtain T1 and T2 characteristic relaxation time constants of components in the polymer samples. For better understanding of chain relaxation in the samples, we measured temperature dependency of the relaxation constant. From these constants, we identify present phases and determine their content in the samples by evaluating the contribution of each component to the total TD-NMR signal. Additionally, using the TD-NMR data, we are able to distinguish between several types of polymer morphologies. Thus, TD-NMR can be further used for the polymer morphology characterization and enable on-line control of industrial reactors.

Acknowledgement: Project RECOBA (636820-H2020-SPIRE-2014) is acknowledged.

Segregation of fine sands in a conical spouted bed

A. Pablos, [1]J. Vicente[2], J. Izquierdo[1], M. Olazar[1], R. Aguado[1]
[1]Department of Chemical Engineering, University of the Basque Country (UPV/EHU), SPAIN;tel. +349 46012510, e-mail: aitor.pablos@ehu.eus; [2]Development Department, Novattia Desarrollos Ltd., Spain; tel. +349 46444421, e-mail: novattia@novattia.com

Segregation has been studied in a conical spouted bed with open-sided draft- tube and confinement system, being the design parameters the optimal obtained in a previous hydrodynamic study. This research is framed in the joint project of Catalytic Process & Waste Valorization Research group, (University of the Basque Country, UPV/EHU) and Novattia Desarrollos Ltd., technology development company, with the objective of designing a new dryer for fine and ultrafine sand based on spouted bed technology.

For a good design of the continuous process equipment is crucial to know the behaviour of the material over time, especially to know whether the type of contactor and process conditions generates segregation in the bed. This phenomenon determines, for example, the optimal place to put the solids outlet or if it will be necessary to perform a purge for coarse particles.

The material used in the assay were about 5 kg of fine sand with a particle size distribution between 0 and 800 μm which have been divided by size into 5 groups. To determine the bed segregation, after 8 hours of fluidization with 80 Nm^3/h, samples of sand have been taken in six points along the bed and have been granulometrically contrasted with the results for the entire bed.

The most important result of the study can be seen in the Figure 1. It shows that the bed has not a significant segregation with these working conditions. This good behaviour could be attributed to the benefit of using the confinement system and reinforces the conviction to achieve significant advances in the treatment of fine particles with this type of technology.

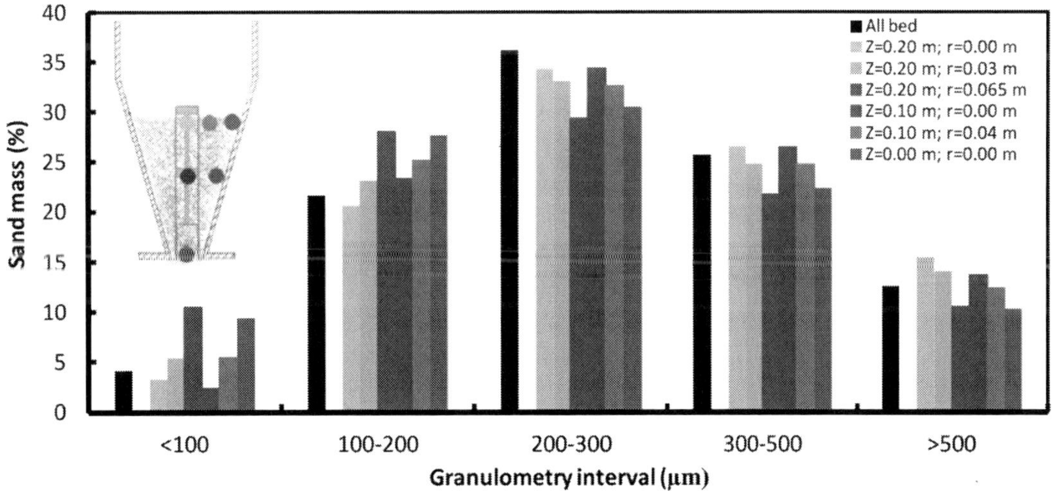

Fig. 1: Mass percentages of the different particle size ranges for entire bed compared with the sand samples collected in several point along the height and the radius of the bed.

Selective surface-modification in main-chain polymer-metal complex nanocrystals

R. Suzuki, T. Onodera, H. Kasai, and H. Oikawa
Tohoku University, IMRAM, 2-1-1 Katahira, Aoba-ku, 980-8577 Sendai, Japan;
tel. +81-22-217-5645, e-mail: hkasai@tagen.tohoku.ac.jp

Metal-organic complex crystals exhibit unique optical, electronic and magnetic properties. Especially, polymer-metal complex crystals have coordinatively-unsaturated center metal or ligand molecules exposed on the surface of crystals, where we would perform sophisticated surface modification so as to modulate and/or tune the unique properties. On the other hand, nanocrystals have large specific surface area, compared to bulk crystal. However, it is frequently difficult to nanocrystallize polymer-metal complexes, due to their poor solubility. In addition, since well-control of inorganic/organic interface structure has not still been achieved enough, research on metal-organic complex nanomaterials has been far behind from both viewpoints of foundation and application.

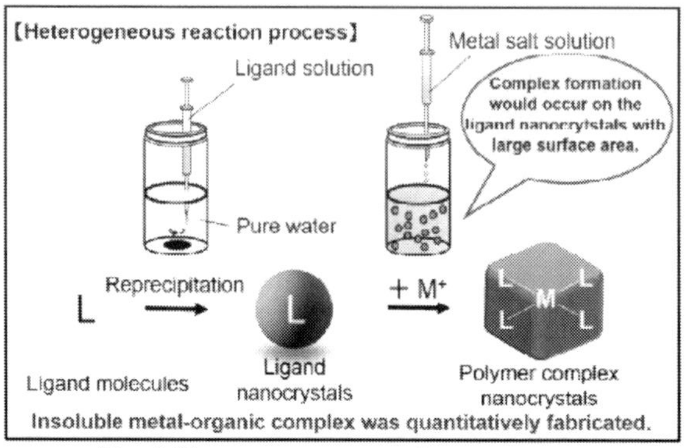

Fig. 1 Heterogeneous reaction process

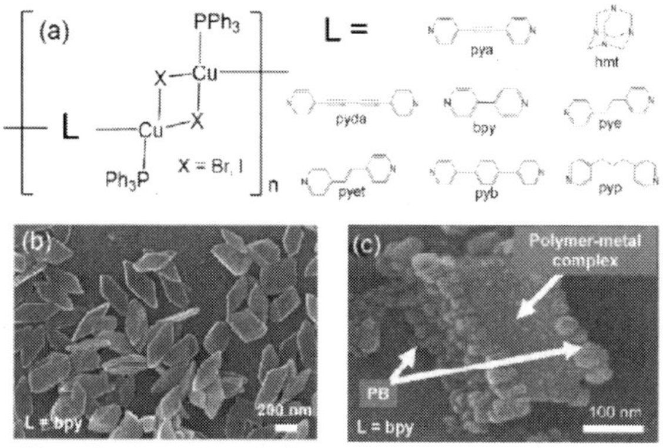

Fig. 2 (a) Chemical structure of polymer-metal complexes, SEM image of (b) as-prepared and (c) PB-modified polymer-metal complex nanocrystals

In the present study, we have developed a useful method to prepare a well-defined metal-organic complex nanocrystals as shown in Fig. 1 [1]. In this method, complex formation would occur on the previously-prepared organic ligand nanocrystals with large surface area. As a result, insoluble main-chain polymer-metal complexes (Fig. 2(a)), which show luminescence in solid state, have been nanocrystallized successfully. Especially, bipyridine complex (L = bpy) [2] nanocrysltas were obtained as well-defined pallarerogram plate (Fig. 2(b)). From TEM observation and analysis of electron diffraction pattern, it is found that {010} plane of bipyridine complex nanocrystals has grown well, that is, the terminal chain would be located on the {010} plane of nanocrystals as a dangling bond. So, we have tried to modify the surface of nanocrystals by end-capping the terminal chain of polymer complex. Actually, Prussian blue (PB) nanoparticles were selectively adsorbed on the {010} plane of polymer-metal complex nanocrystals (Fig. 2(c)). Ligands having N-terminal located on the surface would bind to PB nanoparticles. On the other hand, polymer-metal complex nanocrystals have been also modified with phenylazopyridine (PAZ) showing photochromism. Interestingly, PAZ would also serve as a crystal-growth inhibitor, so that the size of nanocrystals was controlled with the concentration of added PAZ. Moreover, optical properties of nanocrystals were changed by photoisomerization. This fact would open the pathway to novel photoresponsive materials induced by surface effect.

[1] R. Suzuki *et al.*, *Jpn. J. Appl. Phys.*, **2014**, *53*, 06JH03.
[2] Araki *et al.*, *Inorg. Chem.*, **2005**, *44*, 9667.

DEM simulation of powder caking process under humid conditions

Homayra TAVARES[1], Mohamed GUESSASMA[1]
Zahra AFRASSIABIAN[2], Mikel LETURIA[2], Khashayar SALEH[2]

[1]Université de Picardie Jules Verne, LTI-EA 3899, France
[2]Sorbonne Universités, Université de technologie de Compiegne, TIMR-EA 4297, France

Abstract

In this work we present the results of numerical simulation of powder caking process in granular media exposed to variations of ambient relative humidity. Caking consists in unwanted agglomeration of powder particles which is caused by reinforcement of interparticle forces. This process concerns a large number of industrial sectors (food, pharmaceutical, petrochemical ...) and induces every year considerable economic losses. Caking is often irreversible and not easy to predict. In this study, Discret Elements Method (DEM) was used to simulate the caking phenomenon. This numerical method is an alternative to the continuous media approach, dealing with a set of particles in interaction. In DEM, the second law of Newton is solved for each particle and the contact forces between particles are calculated by specific constitutive laws. This allows depicting the behavior of granular media at a macroscopic scale from binary interactions at particles scale. Several type of interactions could exist in a granular media namely van der Waals forces, electrostatic, magnetic and capillary forces. In the present work, we mainly focus on the role of capillary condensation and subsequent liquid bridge formation within a granular medium exposed to ambient relative humidity. The capillary condensation phenomenon which is responsible for the appearance of liquid bridges in binary contacts can be caused by variation of the local relative humidity and ambient temperature. Such bridges cause an attractive force between particles, leading to the formation of a cake with modified physicochemical and mechanical properties [1]. By considering a REV, the DEM is then performed by means of a MULTICOR3D [2] software to characterize the characteristics of the cake (degree of saturation and tensile strength) in order to establish relationships between the microscopic parameters and the macroscopic properties. In fact, the numerical simulations allow a better understanding of the influence of some parameters such as the grain size, the volume of liquid and the variation of relative humidity on the macroscopic response of the media [3]. So far, we have validated the MULTICOR3D software for regular packing, the following steps will emphasize on random packing, sensitivity analysis of the parameters and macromechanical behavior of the cake under varying loadings (compression test, shear test…).

Keywords: Caking of powders, capillary condensation, Discret Elements Method

References

[1] H. Rumpf, The strength of granules and agglomerates. In: Agglomeration, *Ed. KNEPPER, (W.A.)*, John Wiley, New York, p. 379-418, (1962).
[2] S. Martin, M. Guessasma, J. Léchelle, J Fortin, K. Saleh, and F. Adenot, Simulation of sintering using a non smooth discrete element method. Application to the study of rearrangement. Computational Materials Science, 84 :31–39, 2014.

[3] Z. Afrassiabian, M. Leturia, M. Benali, M. Guessasma, K. Saleh, An overview of the role of capillary condensation in wet caking of powder, *Chimical Ingeneering and Disign*, http://dx.doi.org/doi:10.1016/j.cherd.2016.03.020, 2016.

Hot melt coating for QbD oriented patient-centric product development: An overview

[1]S. Salar-Behzadi, [2]D. Lopes, [3]Karin Becker, [4]Roland Hohl, [5]M. Stehr, [6]D. Lochmann, [7]D. Haack, [8]M. Bresciani, [9]J. Khinast, [10]A. Zimmer
[1]RCPE GmbH, Inffeldgasse 13, 8010 Graz, Austria; tel.+4331687330948, e-mail: sharareh.salar-behzadi@tugraz.at;[2]RCPE GmbH, Graz, Austria; e-mail: diogo.lopes@tugraz.at;[3]Karl-Franzens University, Inst. of Pharm. Sciences, Dept. of Pharmaceutical Technology, Universitätsplatz 1, 8010 Graz, Austria, e.mail: karin.d.f.becker@gmail.com; [4]RCPE GmbH, Graz, e.mail: roland.hohl@tugraz.at; [5]Cremer Oleo GmbH & Co. KG, Witten, Germany,e.mail: michael.stehr@cremer.de;[7] Hermes Arzneimittel GmbH, Großhesselohe, Germany, e-mail:haack@hermes-pharma.com; [6]Cremer Oleo GmbH & Co. KG, Witten, Germany,e.mail: dirk.lochmann@cremer.de; ; [8]RCPE GmbH, Inffeldgasse 13, 8010 Graz, Austria; e-mail: massimo.bresciani@tugraz.at; [9]RCPE GmbH, Inffeldgasse 13, 8010 Graz, Austria; e-mail:Khinast@tugraz.at; [10]Karl-Franzens University, Inst. of Pharm. Sciences, Dept. of Pharmaceutical Technology, 8010 Graz, Austria, Member of BioTechMed, email:andreas.zimmer@uni-graz.at

Introduction:

Multiparticulate systems as "Direct to Mouth" dosage forms are a recent strategy to meet the special needs of children, elderly people and patients suffering from dysphagia. This summary presents a novel and cost-efficient strategy for the manufacturing of taste-masked multiparticulate systems with stable performance using hot melt coating (HMC). A quality by design (QbD) approach was undertaken including quality risk management (QRM) tools and process analytical technology (PAT) for production of taste masked multiparticulate systems.

Product engineering:

The product development is described in this summary in the following sections:

1) Providing multiparticulate systems by direct using of active pharmaceutical ingredient (API) crystals or by roller compaction of API for engineering high API-loaded particles with optimal strength and size distribution for further HMC processing. The roller compaction of Ibuprofen Na dihydrate (Ibu Na 2H) (BASF, Germany) is presented in this summary. Figure 1 shows the granular friability using different excipients.

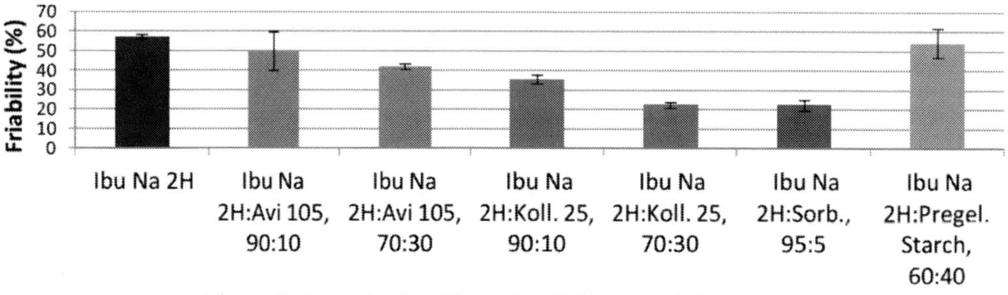

Figure 1: Granular friability using different excipients

Selecting sorbitol (Parteck SI150, Merck, Germany) as binder, a statistical model was used to investigate the impact of different process parameters and sorbitol concentration on the granular friability (Figure 2). The surface of granules is shown in Figure 3.

2) Engineering of lipid-based formulations for using as coating material in a solvent-free HMC process for taste masking of multiparticulate systems.

HMC technology provides the avoidance of organic and aqueous solvents. The process is fast and cost-effective, since evaporation steps or solvent recovery and disposal are not required. Additionally, the frequently applied lipid excipients, which exist in nature, are comparatively cheaper and "generally recognized as safe" (GRAS). However, taste-masking using lipid-based materials poses challenges in attaining the desired immediate release profile. The molten lipid can be re-solidified in different polymorphic structures and stabilities, affecting the stability of end product. In this study tripalmitin (Dynasan 116, Cremer Oleo, Germany) was used as lipid. Polysorbate 65 (Tween 65, Croda, Germany) was used to induce the stable polymorph of lipid during the manufacturing process. A statistical design was conducted that revealed the critical process parameters affecting the taste masking efficiency and drug release. The impact of the emulsifier concentration on the transformation of thermodynamically less stable α-polymorph to stable β-form of lipid is shown in Figure 4.

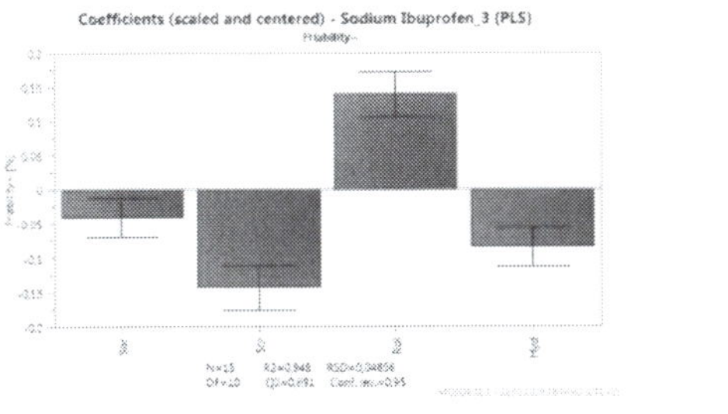

Figure 2: Impact of critical process parameters and excipient on the granular friability

Figure 3: Surface of produced granules

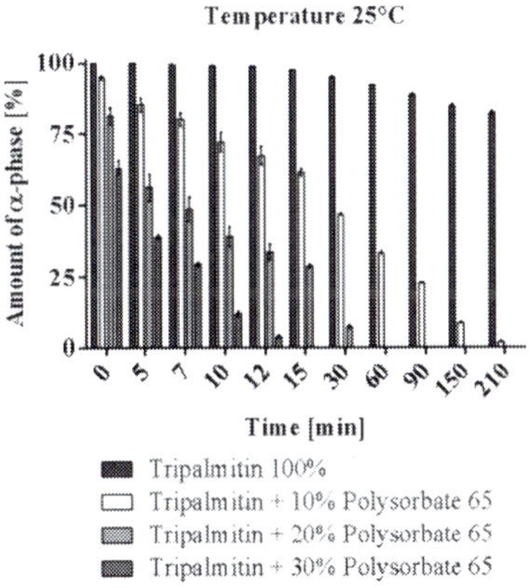

Figure 4: Impact of emulsifier concentration on the solid state of tripalmitin

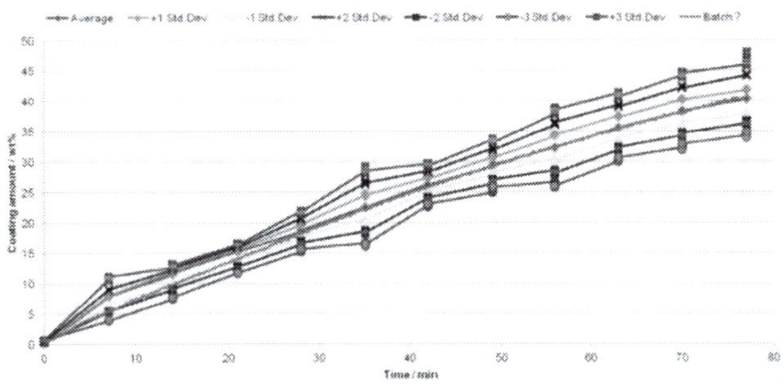

Figure 5: End-point determination of HMC process using inline-NIR spectroscopy

3) QRM tools were applied during the early phases of product design based on the ICH Q9. The product design was evaluated with failure mode and effects analysis, focusing on the prioritization of CQAs. Thereafter, flowcharting and criticality assessment were used for further risk assessment of the HMC process. The defined risk was reviewed and controlled by hazard and critical control points analysis, and fault tree analysis. Near infrared (NIR) spectroscopy was implemented for in-line monitoring of the process and controlling the applied coating amount to ensure specified release profile and taste masking. The end-point of the HMC process was determined using in-line NIR-spectroscopy (Figure 5).

This abstract is a short overview on the successful science-based engineering of the population-driven products such as "direct to mouth" dosage forms.

Effect of tabletting on the microstructure and dissolution characteristics of pharmaceutical granules

[1]D. Smrcka, [2]J. Dohnal, [1]F. Stepanek
[1]UCT Prague, Technická 3, 166 28, Praha 6 – Dejvice; tel. +420220443048, e-mail: David.Smrcka@vscht.cz; [2]Zentiva k.s., U Kabelovny 130/22, 102 00 Prague 10.

Granulation is an agglomeration process, where the primary particles create bigger grains where original particles are still recognizable. As a foregoing step to tabletting could have a significant influence on a final dosage form (i.e. a tablet) properties. Granule structure is determined by the conditions under which the granulation is conducted (e.g. main impeller and chopper speed). Basically the higher power input is greater degree of a granule consolidation is observed, which affects the structure as well as a granulometry. How strongly could granule properties affect the properties of the final dosage form, is a question not easy to answer. Granule strength and the tabletting force are contradictory parameters deciding the degree of the inner granule structure preservation. In our recent work [1,2,3], we have shown how the manipulation of the granulation process parameters can affect the granule microstructure, and in turn how the granule microstructure affects the disintegration and dissolution kinetics.

The aim of this work is to carry out parametric study considering qualitatively varied batches of the granules in combination with different applied tabletting force. Granules produced at very low (Fr = 0.26) and high (Fr = 2.04) main impeller speed were subjected to low and high tabletting force (5 and 18 kN respectively). This pilot study showed that low applied force produced the tablets of various strengths, while the tablets pressed by high force exhibit the same strength. Structure of used granules were analysed by computational micro tomography. The kinetics of active pharmaceutical ingredient (API) release and a granule disintegration were determined by methodology combining an online measurement of a particle size distribution by a static light scattering method and offline determination of API concentration using uv-vis spectrophotometry. In next stage of study the coupled tablets and granules characteristics (i.e. strength and dissolution profiles) will be compared to provide information about transfer of properties or not in the particular cases. The main contribution of this study is in a deeper insight into a non-trivial problematic often challenging in pharmaceutical industry.

[1] Kašpar O., Tokárová V., Oka S., Sowrirajan K., Ramachandran R., Štěpánek F., "Combined UV/Vis and micro-tomography investigation of acetaminophen dissolution from granules", Int. J. Pharm. 458, 272-281 (2013).

[2] Smrčka D., Dohnal J., Štěpánek F., "Effect of process scale-up on the dissolution of granules with a high content of active pharmaceutical ingredient", Powder Technol. 285, 88-95 (2015)

[3] Smrčka D., Dohnal J., Štěpánek F., "Dissolution and disintegration kinetics of high-active pharmaceutical granules produced at laboratory and manufacturing scale", European Journal of Pharmaceutics and Biopharmaceutics, submitted 2015.

Effect of colloidal silica dioxide on rheological properties of common pharmaceutical excipients

D. Majerová, D. T. Tran, [1]L. Kulaviak, [1]M. Růžička, [2]F. Štěpánek, P. Zámostný
University of Chemistry and Technology in Prague, Dep. Organic Technology, Technická 5, 166 28, Prague, Czech Republic; tel. +420 220 444 301, e-mail: majerovd@vscht.cz;
[1]Institute of Chemical Process Fundamentals of the ASCR, Rozvojová 2/135, Prague, 165 02, Czech Republic; tel. +420 220 390 260, e-mail: kulaviak@icpf.cas.cz; [2] Department of Chemical Engineering, Institute of Chemical Technology, Technická 3, Prague, 160 00, Czech Republic; tel. +420 220 443 236, e-mail: Frantisek.Stepanek@vscht.cz

Manufacturing of pharmaceutical solid dosage forms involves several processing steps including flow through hoppers, sieving, pouring, blending, die-filling, and compaction. These steps are highly sensitive to powder properties such as flowability and bulk density, which are to some extent inter-related. The knowledge of flow properties of raw excipients and their mixtures is essential for predicting their behaviour during blending, compression, handling or leaving in storage and these properties are critical for manufacturing since they can affect the final product quality and the efficiency of the processes.

Very effective tool for assessing properties such as flowability and powder cohesion is powder rheology. There are various commercially available instruments; one of them is the FT4 Universal Powder Rheometer (Freeman technology, UK). It allows the measurement of the powder response to various stress fields, thus simulating the range of processing conditions more closely. During the testing, materials are subjected to dynamic and shear tests which provide set of basic rheological characteristics. Moreover there is a conditioning procedure between each measurement which allows generation of a stable consolidation state and ensures the reproducibility of the data.

The flowability of powders can be improved significantly by the addition of small amounts of substances described as 'glidants' in the area of tableting technology. The primary particles of glidants exist in the form of agglomerates which are broken down into smaller aggregates during the blending process. These small glidant particles adsorb at the surface of the other solid component grains, increase the roughness of the host surface and thus improve the powder flow.

In this work, the rheological properties of silicified microcrystalline cellulose and pregelatinized starch was investigated in comparison to the characteristics of pure excipients. At first, pure materials were analysed on scanning electron microscopy (SEM) to obtain their particle size distribution and particle shape. Also the rheological properties of all three substances were examined by the Freeman FT4 rheometer. The observed properties were in accordance with known characteristics and typical behaviour of the excipients.

Then the binary mixtures of these excipients and different concentration of anhydrous colloidal silicon dioxide (Aerosil 200®) were prepared by blending in a Turbula® mixer for a defined period of time. The Freeman FT4 rheometer was used to measure flow properties of the mixtures. The bulk (conditioned and compressed densities, compressibility index), dynamic (basic flow energy) and shear (friction coefficient, flow factor) properties were determined. The optimum flowability was exhibited by the mixtures comprising 0.25 - 1 % of Aerosil, but the results revealed also additional more complex phenomena affecting the flowability of binary mixtures (blends). In order to support these findings and investigate the blend behaviour in more detail, the mixtures of Aerosil and microcrystalline cellulose were selected for further experiments. A compressibility test on Gamlen Tablet Press GTP-1 (Gamlen Tabletting, UK) was performed, confirming the binary mixture compressibility being affected by the composition, which allowed to explain poor flowability in Aerosil-rich blendss. The samples were also subjected to Energy-dispersive X-ray spectroscopy to reveal the relationship between the rheological properties and the inter-particle structure of the materials. The images confirmed the theory that glidant fills cavities in the grains at first, and then covers the surface. In order to clarify the effect of particle size on the glidant action, the excipients were pre-sieved and selected fraction were mixed with Aerosil and re-tested by the rheometer. It was found that compressibility is mainly determined by the properties of the excipient itself and that the fraction with the larger particles are less compressible in general. Also the kinetics were studied with different time of mixing selected when preparing mixtures in the blender. The first results show that with the increasing time of mixing the compressibility decrease. Also, the basic flow energy is changing rapidly during prolonged mixing and it is lower at shorter mixing.

3D-printing of pharmaceutical tablets

[1]M. Novák, [2]A. Zadražil, [3]P. Kovačík, [4]F. Štěpánek
[1,2,4]University of Chemistry and Technology, Laboratory of Chemical Robotics, Technická 5, 16628 Prague, Czech Republic; tel. +420 220 443 292, e-mail: matej.novak@vscht.cz; [2]tel. +420 220 443 048, e-mail: ales.zadrazil@vscht.cz; [4]tel. +420 220 443 236, e-mail: Frantisek.Stepanek@vscht.cz; [3]Zentiva Group, a.s., U Kabelovny 130, 102 37 Prague, Czech Republic; tel. +420 220 443 048, e-mail: Pavel.Kovacik@zentiva.cz;

Production of tablets using 3D printers is a novel method of pharmaceutical formulation. Although still in the research&development phase, since recently it already has its first commercial application. The main advantage of this approach is, that every single tablet can be printed to meet the needs of a specific patient, both the dosage and dissolution rate can be controlled by the composition and porosity of the tablet, furthermore, two or more drugs can be present in one tablet. The aim of this project is to print pharmaceutical tablets, using Fused Deposition Modeling, a commercially available 3D printing technology. To do this, the feed material must be produced first in the form of a solid filament, made from biocompatible polymer matrices, plasticizers and active pharmaceutical ingredients (drugs). This is achieved by using hot-melt extrusion technique. The first model filaments contain Hydroxypropylcellulose, triacetin as a plasticizer and Ibuprofen as a model drug. After the fabrication, filaments of varying compositions are analyzed by UV spectroscopy in order to determine remaining drug contents. Dissolution properties and mechanical characteristics are investigated in order to evaluate printability and biocompatibility of the filaments. Furthermore, 3D models of tablets are prepared using modeling software and first printing attempts are conducted. Results of filament and tablet analysis are utilized to determine optimal composition of the feed material and to evaluate, which pharmaceutical ingredients are suitable for this formulation method.

Protein crystallization in a novel meso oscillatory flow reactor

[1,2]F. Castro, [1,2]A. Ferreira, [2]J.A. Teixeira and [1]F. Rocha
[1]LEPABE-Laboratory for Process Engineering, Environment, Biotechnology and Energy, Faculty of Engineering of Porto, Dep. of Chemical Engineering, Univ. of Porto, Rua Roberto Frias, s/n, 4200-465 Porto, Portugal; tel. (+351) 225 081 884. [2]Centre of Biological Engineering, Dep. of Biological Engineering, Univ. of Minho, Campus de Gualtar, 4710-057 Braga, Portugal; tel. (+351) 253 604 400, fax: (+351) 253 604 429, e-mail: fcastro@deb.uminho.pt;

Protein crystallization is best known for its use in protein three-dimensional (3D) structure determination by X-ray diffraction. On the other hand, little systematic know-how exists for protein crystallization as a process step [1]. However, protein crystallization offers great potential to be used for separation and formulation applications on an industrial scale [1, 2]. Purification of proteins via crystallization may offer significant cost reductions compared to conventional downstream processing techniques such as chromatography. Moreover, protein crystals have a longer storage life and greater purity compared to the dissolved form.

Protein crystallization as a process step requires rapid and quantitative crystallization. However, most of the studies in protein crystallization had been focused on producing high quality crystals obtained by slow crystallization from small volumes under static conditions. But, crystallization under static conditions is often diffusion-limited, leading to rather long process durations and/or low yields. Hence, protein crystallization under flow conditions has drawn attention with the capability to improve the convective protein transport, reducing the crystallization time and improving yield [3]. In contrast to structural analysis, large single crystal shape is not as important, but crystals with a certain size, shape and mechanical strength are desirable so as to be easily separated from their mother liquor. Optimization with respect to all these criteria has to be based on a fundamental understanding of the process, that is, knowledge of the phase diagram. However, complete phase diagrams for prediction of protein crystallization are rarely available and crystallization conditions are usually found by a trial-and-error approach.

In this work, we propose to investigate protein crystallization in a novel meso oscillatory flow reactor (meso-OFR) designed by members of the present research team (Figure 1) [4]. It consists of a glass jacketed tube provided with smooth periodic cavities (SPC). The reactor has an approximate volume of 4 mL and is operated under oscillatory flow mixing, controlled by the oscillation frequency and amplitude. The present work aims at investigating for the first time the applicability of this novel meso-OFR on protein crystallization for purification purposes. Lysozyme was used in this work as a protein crystallization model. Different experimental conditions were screened by varying protein and precipitating agent (sodium chloride) concentration at pH 4.7 and T = 20°C (Figure 1) in order to derive a sufficiently quantitative crystallization phase diagram of the protein under study. The resulting crystals were characterized by optical microscopy and their activity was measured. Induction time was also measured for all the experimental conditions studied by online monitoring of the turbidity of the crystallization solution (Figure 1) in order to study the nucleation phenomenon.

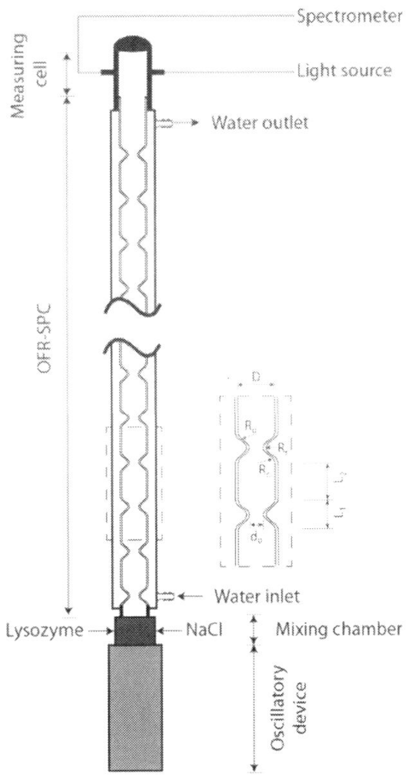

Figure 1. Schematic representation of the experimental set-up for lysozyme crystallization experiments

References:

[1]. S. Schmidt, D. Havekost, K. Kaiser, J. Kauling, and H.-J. Henzler, "Crystallization for the Downstream Processing of Proteins," Engineering in Life Sciences, vol. 5, pp. 273–276, 2005.

[2]. J. V. Parambil, M. Schaepertoens, D. R. Williams, and J. Y. Y. Heng, "Effects of Oscillatory Flow on the Nucleation and Crystallization of Insulin," Crystal Growth & Design, vol. 11, pp. 4353–4359, 2011.

[3]. B. Smejkal, B. Helk, J.M. Rondeau, S. Anton, A. Wilke, P. Scheyerer, J. Fries, D. Hekmat, D. Weuster-Botz, "Protein crystallization in stirred systems-scale-up via the maximum local energy dissipation", Biotechnol. Bioeng, vol. 110, pp. 1956-63, 2013.

[4]. A. Ferreira, F. Rocha, J.A. Teixeira, A. Vicente, Apparatus for mixing improvement based on oscillatory flow reactors provided with smooth periodic constrictions, WO/2015/056156, 2014.

Controlling the dissolution rate of APIs by formulation on porous nano- and microcarriers

[1]M. Šoltys, [2]P. Kovačík, [2]J. Beránek, [1]F. Štěpánek
[1]University of Chemistry and Technology in Prague, Department of Chemical Engineering, Technická 5, 166 28 Praha 6, Czech republic, *E-mail: marek.soltys@vscht.cz;
[2]Zentiva, k.s., a Sanofi Group company, U Kabelovny 130, 102 00 Praha 10, Czech republic.

High number of newly discovered active pharmaceutical ingredients (APIs) are poorly soluble in water which decreases their bioavailability and makes their formulation into a functional pharmaceutical product quite a challenge. One of the possible solutions is amorphization of the API. The amorphous form is thermodynamically less stable and thus generally provides better dissolution kinetics than more stable crystalline states of the APIs [1]. However the amorphous state tends to spontaneously crystallize over time and needs to be stabilized somehow in order to be viably formulated.

There are several methods of achieving stable amorphous formulations. One of them is formulation on porous carriers. The confined space of mesopores prevents the API from crystallizing and in combination with high surface area leads to faster release kinetics [2]. However the previous experiments were all done with microparticles. The first half of this work studies the use of hollow or solid mesoporous sub-micron silica particles loaded with poorly soluble API and its release kinetics behavior.

High amount of API was loaded using only simple sorption process. Using the sub-micron particles in combination with poorly soluble API resulted in significant increase in dissolution rate compared to dissolution rate of crystalline API.

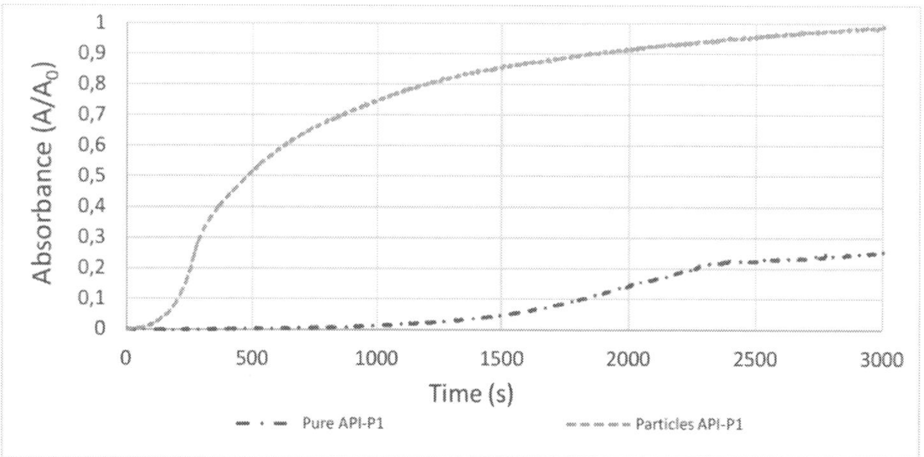

Figure 1: Dissolution profiles of pure poorly soluble API and release kinetics of the same API from porous particles. All dissolution experiments were conducted in 100 ml of deionised water.

References

[1] J. Willart and M. Descamps, "Solid state amorphization of pharmaceuticals," *Mol. Pharm.*, vol. 5, no. 6, pp. 905-920, 2008.

[2] K. Qian and R. Bogner, "Application of Mesoporous Silicon Dioxide and Silicate in Oral Amorphous Drug Delivery Systems," *J. Pharm. Sci.*, vol. 101, no. 2, pp. 444-463, 2011.

Stealth pHPMA-modified nanocarriers for antibody-targeted drug delivery systems

[1]D. Lizonova, [1]J. Hanus, [1]J. Hasa, [1,2]V. Kral, [3]M. Pechar, [3]R. Pola, [1]F. Stepanek
[1]University of Chemistry and Technology Prague, Department of Chemical Engineering, Technicka 3, Prague 6, 16628, Czech Republic; tel. +420 220 443 048, e-mail: denisa.lizonova@vscht.cz, fax: +420 220 444 320;
[2]Laboratory of Structural Biology, Institute of Molecular Genetics of the ASCR, v.v.i, Flemingovo nam. 2, Praha 6, 16637, Czech Republic
[3]Laboratory of Biomedical Polymers, Institute of Macromolecular Chemistry of the ASCR, v.v.i., Heyrovskeho nam. 2, Prague 6, 16206, Czech Republic

The efficient treatment of malignant tumors requires relatively high doses of cytotoxic drugs, which are usually toxic also to normal, healthy cells. This non-specific toxicity results in many adverse effects on the human body. The fundamental principle of Targeted Drug Delivery (TDD) is to reduce the unwanted toxicity against the healthy organs and tissues and, at the same time, to increase the concentration of the drug in the tumor using tumor-specific targeting ligands.

Monoclonal antibody IgG M75 specific for the antigen of Carbon Anhydrase IX (CA IX) overexpressed on the surface of hypoxic tumor cells (for example colorectal carcinoma) was covalently bound to the model fluorescent silica nanoparticles and to the liposome-aggregates (liposome-Poly-L-lysine-SPION (SuperParamagnetic Iron Oxide Nanoparticles)) via biocompatible hydrophilic (*N*-(2-hydroxypropyl)methacrylamide)-based copolymers (pHPMA). The pHPMA coverage improves the physical parameters of the particles, almost zero zeta potential facilitates the stabilization of the suspension. The pHPMA covered particles have also lower affinity to the plasma proteins, which results in the longer circulation time.

The ability of such particles to bind the antigen was examined using ELISA (Enzyme-Linked ImmunoSorbent Assay)-like test. *In vitro* tests with human blood serum were carried out to determine the nanoparticle-protein affinity. Size and zeta potential of the particles were determined using dynamic light scattering (DLS). The successful pHPMA modification of the liposome aggregates was confirmed via confocal microscopy using the fluorescently labeled pHPMA surface layer.

Teaching efficiency in chemical engineering

J. Glassey[1]*, N. Kockmann[2], V. Meshko[3], A. Porjazoska Kujundziski[3], M. Polakovic[4], L.M. Madeira[5], E. Schaer[6]
(1) Newcastle University, Newcastle upon Tyne, United Kingdom,
(2) TU Dortmund University, , Germany,
(3) International Balkan University, Skopje, R. Macedonia,
(4) Slovak Technical Univerity, Bratislava, Slovakia,
(5) FEUP, Porto, Portugal;
(6) ENSIC-LRGP, UL Nancy, France,
*Eric.Schaer@univ-lorraine.fr

The iTeach European project (http://www.iteach-chemeng.eu/) consortium aims at developing a robust and objective framework for assessing the effectiveness of various pedagogical methodologies in chemical engineering education, in particular those aimed at promoting important core competencies related to employability, in a range of geographical and educational contexts.

A review of the learning outcomes of a chemical engineering higher education formation (as set out by the Bologna recommendations, the review carried out by the Working Party on Education of EFCE and various professional accreditation requirements) has been developed in consultation between industrial and academic partners.

It indicated a significant alignment of the requirements world-wide with some differences in the emphasis on fundamental science, related engineering disciplines and information technology applications.

State-of-the art in assessing the effectiveness of teaching of core (chemical) engineering knowledge and of the development of professional skills and competencies required to increase the employability of the graduates was also reviewed. The differences in the perception of the effectiveness of various pedagogical methodologies between the stakeholder groups (employers, academics, graduate and students) provided a valuable input into defining various indicators of the effectiveness of teaching in chemical engineering higher education.

Decision making technology and multiobjective optimisation were subsequently used to identify the most appropriate evaluation methods and to develop 2 frameworks for supporting effective delivery of core knowledge and employability competencies.

The first one assesses the effectiveness of a whole formation, based on more than 160 parameters, and allocated in 7 indicators such as Pedagogy, Learning Outcomes, Attractiveness, Relations with Research and with Industry, Employment and Quality.

The second one, devoted to the effectiveness of a single teaching unit, is based on the assessment of 6 indicators such as Strategic Nature of the Teaching Unit, Relevance of the proposed formation, Pedagogical relevance of the teaching approach, Perception of relevance of the pedagogical approach by the students, Evaluation of the acquisitions and Evaluation of Transfer by different stakeholders.

The proposed frameworks will be detailed, and the first results, for the evaluation of teaching effectiveness in chemical reaction engineering in different countries (Great Britain, France, Germany, Portugal, Slovakia, Macedonia) and using different teaching methodologies (traditional lectures, work-based learning, recorded lectures, problem based learning, practical instruction via labs and problem based learning) will be presented.

Research and development by chemical engineering students: increasing motivation and cross course knowledge

E. Székely, A. Deák, M. Tőrincsi, Z. Hórvölgyi
Budapest University of Technology and Economics, Faculty of Chemical Technology and Biotechnology, Műegyetemrkp. 3., H-1111 Budapest, Hungary; tel: +3614631111, e-mail: sz-edit@mail.bme.hu

Chemical engineering, as it is now, is interdisciplinary. As a result of this interdisciplinarity, chemical engineering curricula contain a large variety of courses strongly or only marginally connected to each other. One of the greatest challenge in the education of chemical engineers (and related, like biochemical engineers, environmental engineers, pharmaceutical engineers etc.) is how to help our students to synthetize their knowledge and abilities gained at different courses.

Our aim is to demonstrate how real individual projects prior thesis work may help undergraduates to understand and accept the role of different disciplines within chemical engineering. The traditional way in our faculty is to invite interested students as early even as their first year to participate in various research or development projects. Their work is supervised by a staff member, and in most of the cases the undergraduates have the possibility to join a research team of 5-10 persons for a short training first. Right after having the skills to do valuable individual work (either experimental, modelling or else) they become responsible for a segment of a bigger project. Often the final outcome of such a volunteer work is a so called scientific student contribution, combination of written and oral presentation of the work done. On the other hand, and in our opinion this is the greatest advantage of such a possibility for the undergraduates, during their involvement into a research team they face a number of different practical and theoretical problems; do synthesis or use chemicals, use different equipments, analytical techniques and/or data, try to understand their observations by using all their knowledge; observe different problem solving techniques; learn efficient communication and information sharing by being responsible for smaller tasks within a bigger project.

Most of the students participating in such a project is very satisfied, and they are motivated to continue for several years. After a year or more they can be involved in the training of the groups' newcomers, which deepens their understanding in the field and also gives the chance to develop a professional network across the years as well. On the other hand supervisors also feel motivated, it is fascinating to show to new and new young colleagues how exciting our profession is. We will present that the extra time and work of these students do not decrease their study results, on the contrary, they also perform excellently in their regular studies.

Education and management in using especially hazardous chemicals in a research and teaching institution

Sebastian Brückner,[1] Jean-Luc Marendaz[2] and Thierry Meyer[1,3]
[1] Safety Competence Center, Ecole Polytechnique Fédérale de Lausanne, 1015 Lausanne, Switzerland; tel. +41 216930315, sebastian.bruckner@epfl.ch
[2] Section of Chemistry and Chemical Engineering, Institute of Chemical Sciences and Engineering, Ecole Polytechnique Fédérale de Lausanne, 1015 Lausanne, Switzerland; tel. +41 216939851, jean-luc.marendaz@epfl.ch
[3] Group of Chemical and Physical Safety, Institute of Chemical Sciences and Engineering, Ecole Polytechnique Fédérale de Lausanne, 1015 Lausanne, Switzerland; tel. +41 216933614, thierry.meyer@epfl.ch

Wrong manipulation, storage or disposal of chemicals can cause great damage whether it occurs on industrial plants, in academia or at home. Amongst the numerous reasons, lack of knowledge and haste are the most common ones. Universities are preparing the future professionals with the required skills. As safety becomes more and more important in the daily life it is important instructing the latter by actively responding with innovating curricula and learning experiences. To make researchers and students aware of the hazards and risks associated to chemicals, and protect them accordingly, the Safety Competence Center (SCC) has decided to submit to authorization the purchase and manipulation of very hazardous chemicals. The person willing to use one of these chemicals needs to collaborate with this occupational safety and health team to set up the experiment.

The process starts by a thoughtful discussion with the requester about the substance and whether a less hazardous alternative could be used instead. If the request is validated, the work procedure is analysed. This also includes how the waste shall be treated. Eventually, the laboratory's infra-structure is improved accordingly. Finally, the adapted personal protective equipment is also investigated with the requester. Consequently, the SCC authorizes or not the requesting person to order the compound. This concept will be illustrated with osmium tetroxide, as an example of a lethal compound used in chemistry, biology and electronic microscopy. The management of the entire lifecycle of this chemical will be presented.

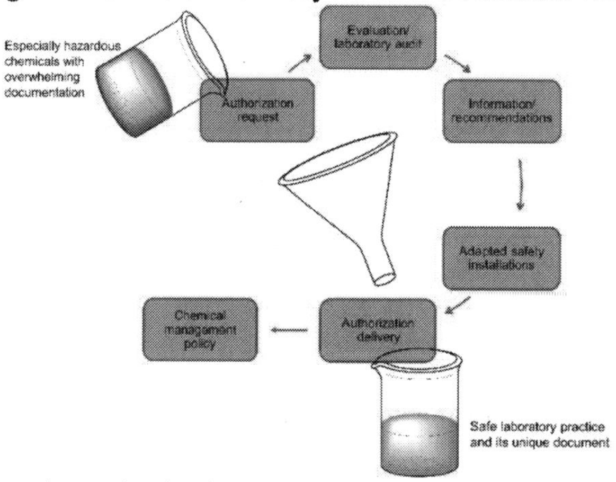

The data resulting from the authorization process is recorded in a database and the need to review the associated procedure and/ or the authorization itself is performed twice a year during safety audits of the workplaces.

In the interest of educating and protecting researchers and students, the SCC team has put in place a clear and straightforward authorization concept for highly hazardous chemicals. The implementation of this collaborative authorization process has simplified and encouraged discussing safety aspects at all levels amongst all research groups. Moreover, following this implementation, no accident involving such compounds has been reported on site. Finally, by rising the future professionals safety awareness, this innovative concept may help industries in their continuous occupational health and safety efforts.

Pre-lab tests, integrated experiments and targeted reports: approaches to increase deeper learning and student satisfaction in practical work

M.J Heslop
University of Sheffield, Department of Multidisciplinary Education, 32 Leavygreave Road, Sheffield S3 7RD, United Kingdom; tel. +44 114 222 7601, m.j.heslop@sheffield.ac.uk.

There is an interesting analogy between the discipline of chemical engineering (CE) and the subject of education: where the flow of material through pipes is replaced by the transfer of information between educator and learner. Consider the following pictorial representation of this process:

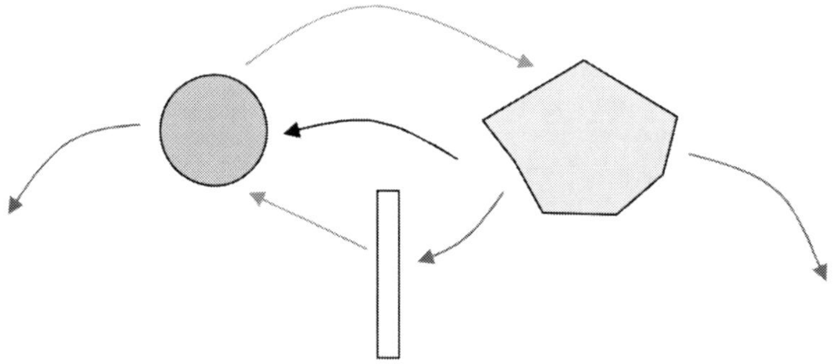

The gold arrow at the top represents the transfer of information from lecturer to student. The dark arrow in the middle represents the "unedited" feedback and ideas the lecturer receives from the students and the blue represents the information the lecturer receives via evaluation surveys. The red and green arrows represent the information that is transferred that is simply discarded and forgotten. Ideally these arrows should be as short as possible, certainly I would hope that the green arrow might be close to zero. The obvious analogy here would be with process leaks and emissions. The lecturer is represented as a circle because in order to represent expected conformity. The group of students is represented as irregular to represent the inevitable range of attitude and ability, although the size and shape will vary from year to year. The narrow rectangle is intended to represent a filter, perhaps changing the amount and type of information, which can then lead to misinterpretation. The list of stakeholders is by no means complete: other additions might be the university, faculty, professional body, government and society.

Practical exercises and lab experiments are an important part of any CE curriculum for a number of reasons. First, they enable the student to put into practice what they learn in taught classes – and to have an appreciation that it is often important to make assumptions to theory in order to obtain a workable solution. Second, they allow students to gain experience of operating equipment, taking measurements and appreciating that measurements are subject to error. Third, they allow students to develop the key transferable skill of group working – ideal preparation for many career paths. A typical lab experiment might involve the following: reading the instructions, carrying out the experiment, recording data, calculating the results and writing a report. A typical experiment might last three hours.

The evidence from student evaluations would suggest that there are a number of issues with the way that such practical work is carried out. First, despite the importance of practical work, only approximately 10% of the student contact time is spent engaged in practical work. Second, because of the typically broad CE curriculum, the wide range of experiments means that students are required to carry out experiments on a "round-robin" basis. For exmaple, Group1 might carry out an experiment on radiation heat transfer in Week 1 and then an experiment on distillation in Week 3. However, the relevant subject material might not be covered until Week 5. Furthermore, very few departments have the space or resources to have multiple copies of the same experimental apparatus. This disjoint between timetabling and availability of equipment then leads to the third issue: that the students are not prepared for the experiment, and so do not make the best use of the time available. Finally, with the increasing student numbers in CE courses, it is clear that different approaches are required to try and make the best of the contact time available.

At Sheffield, we have taken three approaches to try and alleviate these issues. First, in order that the students are well prepared for the practical work, they are required to carry out a pre-lab test. This is short, and involves questions both on health and safety as well as the principles of the experiments. The questions are typically multi-choice and multi-answer. The students are allowed to take the pre-lab as many times as possible in order to determine the correct answers. The tests are marked automatically through the virtual learning environment (VLE). Second, rather than implement long experiments, we have looked again at how the experiment might be re-designed so that it takes less time. The advantage of this re-design is that with say three experimental apparatus and an experimental time of say 1 hour, it is possible to process the entire cohort within a week. For example, the subject of radiation heat transfer might be covered in Week 5, and then in Week 6 the whole cohort carry out an experiment based on the said subject. This is what is termed an "integrated experiment": directed to the time in the semester at which the taught material is covered. Third, rather than ask the students to complete full laboratory reports, we only ask for one section at a time, so that they are only required to take on board targeted feedback before applying this to a complete report for their final submission.

The evidence from student evaluations and anecdotal comments is in general positive. The students welcome the opportunity for a structured preparation – and that is not too time-consuming. Furthermore, considering the various issues in group work, the students appreciate the fact that their lab partners will also be prepared. The overall preparedness helps to improve group cohesion. Another positive point is the variation and breadth that comes from carrying out more short experiments rather than fewer longer experiments – thus making better use of the time available. Finally, report writing can be a laborious exercise, especially when it involves submitting many reports – often making the same mistakes. Targeted feedback, one report section at a time, allows key information to be taken on board – and also reduces the workload for teaching staff. This work will include the following contributions: both student evaluation information as well as examples of resources that have been produced.

Teaching of process simulation in chemical engineering program

R. Tuunila, M. Huhtanen, E. Lahdenperä
LUT School of Engineering Science, Lappeenranta University of Technology, Lappeenranta, Finland; tel. +358 40 5296855, e-mail: ritva.tuunila@lut.fi

Abstract

Simulation-based teaching has become a common method in addition to a traditional class room teaching and hands-on laboratory exercises in engineering education. Especially in chemical engineering, where teaching of process and plant design is essential part of curricula, utilizing simulation as a tool is very important. In this paper we present how teaching of process simulation is integrated to curricula of chemical engineering program at Lappeenranta University of Technology.

Introduction

Chemical engineering is a profession that deals with materials and processes to produce reactants and chemicals or convert raw materials or chemicals into useful forms. To do that successfully, engineers must have theoretical knowledge of different phenomena that they gain in the classroom, but engineers also need some practical knowledge and skills. That knowledge is traditionally studied in the laboratories with hands-on experiments, which have always been essential part in chemical engineering programs (Feisel and Rosa 2005, Müller et al. 2007). In the laboratory a student has a possibility to use real, even often a laboratory scale, equipment, in chemical engineering also with real raw materials or chemicals, and deal with real data, which improves his/her practical skills (Ferrero et al. 2003).

Nowadays, the hands-on laboratories are seen quite costly due to needs of often large laboratory space, chemicals and equipment to perform experiment, and also instructor time. Also many lab experiments may be rather complex and even dangerous (Müller et al. 2007). When at the same time development of computer technology has enabled using simulation in teaching more effectively, simulation laboratories are replacing and already have replaced part of hands-on laboratories in engineering education (Balakrishnan and Woods 2013).

Simulation is today a crucial part of process and plant design, especially due to growing importance of environmental and safety aspects and reliable operation of the processes (Seider et al. 1999, Towler and Sinnot 2013). When the target of higher education is to give best possible knowledge and skills for university students required in work life, utilizing process simulation as a tool in chemical engineering teaching is an essential part of curricula.

Integrating simulation to curricula

In the Bachelor level chemical engineering students study basics of chemistry, mechanical and chemical unit operations and basics of process and plant design and process simulation. In the Master level the students study a course Advanced process design, in which they combine knowledge from previous courses to learn to use simulation tool in process design and development tasks. Schematic description of the timeline of course delivery in the curricula is given in Figure 1.

Course objectives

The main objectives of the course *Introduction to process simulation* are that after the course a student know basics and common applications of process simulation, can simulate simple chemical processes with the chosen program (ASPEN Plus), and is able to analyze the achieved results on certain level. The course includes an individual simulation assignment, which main purpose is to show students that changing the operating parameters of one processing unit has a profound effect on the operation environment of the other processing units. Another target is to emphasize for students the importance of careful analyzing of the results by demonstrating that results from process simulation can be seemingly correct but closer look may reveal results that are not realistic.

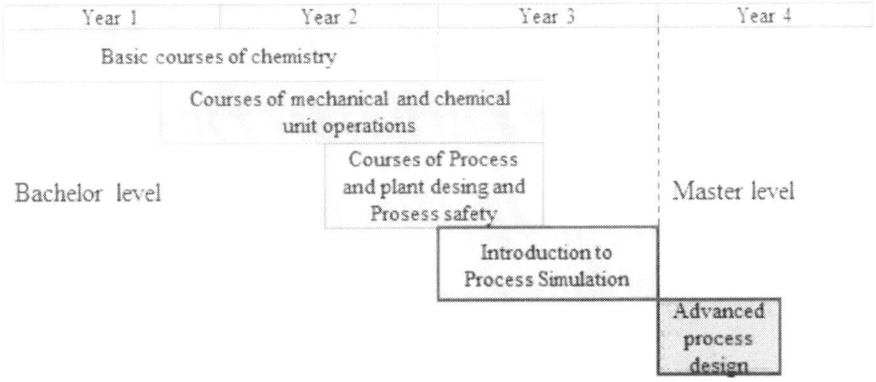

Figure 1. Schematic timeline of course delivery

In the course *Advanced process design* the students are expected to be able to apply process simulation in design of chemical processes. The final goal is make students understand and to gain skills to perform process conceptual design supported by process simulation in order to perform physical and chemical property estimation required to support design, know how to go from process requirements into actual process flowsheet (process synthesis), and perform process analysis, evaluation and optimization including material and energy balances, economy, safety, flexibility.

Conclusions

Curricula of chemical engineering program has been developed to give students basic skills, including skills for process simulation, which allow students enter widely to different duties and being straightly useful for industry after graduation. Teaching of process simulation instruct students to standard methodologies define problems, construct and evaluate alternative solutions in process design and optimization.

References

Balakrishnan, B., Woods, P.C. (2013), A comparative study on real lab and simulation lab in communication engineering from students' perspective, *European Journal of Engineering Education* **38**(2), pp. 159-171

Feisel, L.D., Rosa, A.J. (2005), The Role of the Laboratory in Undergraduate Engineering Education, *Journal of Engineering Education* **94**, pp. 121-130

Ferrero, A., Salicone, S., Bonora, C., Parmiagiani, M. (2003), ReMLab: a Java-based remote, didactic measurement laboratory, *IEEE Transactions on Instrumentation and Measurement* **52**, pp. 710-715

Müller, D., Bruns, F.W., Erbe, H-H., Robben, B., Yoo, Y-H. (2007), Mixed reality learning spaces for collaborative experimentatation: A challenge for engineering education and training, *International Journal of Online Engineering* 3(4), pp. 15-19

Seider, W.D., Seader, J.D., Lewin, D.R. (1999), *Process Design Principles: Synthesis, Analysis, and Evaluation*, John Wiley & Sons, Inc., New York, USA

Towler, G., Sinnot, R. (2013), *Chemical Engineering Design: Principles, Practice and Economics of Plant and Process Design*, 2 ed., Butterworth-Heinemann, Oxford, UK

The software ASPEN Plus® for graduate and undergraduate courses in chemical and environmental engineering

E. Ruiz, C. Cara, F. Espínola, M. Moya, I. Romero, J.C. López-Linares, J.M. Romero-García, E. Castro
Department of Chemical, Environmental and Materials Engineering, University of Jaén, Campus Las Lagunillas s/n. 23071 Jaén, Spain; tel. +34953212163, e-mail: ecastro@ujaen.es

ASPEN Plus® is a powerful simulation software that can also be used as an excellent learning tool in both graduate and undergraduate courses of chemical and environmental engineering.

This work summarizes the preliminary results of the authors' experiences for connecting the main chemical and environmental engineering concepts with their own research work in the context of several undergraduate and Master courses taught in the University of Jaén, Spain, e.g., Management and Treatment of Residues and Soils, Chemical Technology, Chemical Process Engineering or Industrial Biotechnology.

As a prominent example, the production of biofuels from agricultural residues was selected to apply the simulation process because this matches well with the content of the above mentioned courses. In addition, students perform lab works on the same topic, particularly on the production of bioethanol and biodiesel form residual materials, so that the comparison between real and simulated operations can represent an interesting integration of theoretical and actual results.

The work is part of a learning innovation program establish by the University of Jaén whose financial support is greatly appreciated.

Operator training in industrial biotechnology using simulators

V. C. Hass, [1]I. Gerlach, [1]C. F. Mandenius
Hochschule Furtwangen University, Faculty Medical and Life Sciences, Jakob-Kienzle-Straße 17, D - 78056 Villingen-Schwenningen, Germany; tel. +49 7720 307 4235, e-mail: volker.hass@hs-furtwangen.de; [1]Linköping University, Division of Biotechnology, Department of Physics, Chemistry and Biology, SE 581 83 Linköping, Sweden.

Due to their increasing complexity industrial biotechnological processes are carried out using advanced and demanding process control strategies in order to achieve maximum product yields and optimal utilization of resources and energy. This leads to high demands on the operational skills of plant staff and operators.

Operator training simulators (OTS) are computer based training tools representing a real plant using an interactive virtual environment. They are currently applied in aviation, navigation, power plants and other process industries (Gerlach, 2015). So far, only very few OTS applications have been reported for processes in industrial biotechnology, which might be due to the complex kinetics and dynamics of these processes. However, OTS provide an efficient means of industrial operator training as well as to support academic education and laboratory training in the field of biotechnology (Hass, 2016).

Using a recombinant protein production process and an industrial bioethanol process as examples we will show, how a systematic OTS development and build up leads to a powerful tool for efficient operator training and successful academic education.

According to Gerlach (2015) the OTS development may be structured into seven phases. In the preparatory phase (1) the process under consideration and training targets are defined. The development tools and mathematical models are specified. In a preliminary development phase (2) the functional process description of the training process and possible design configurations are investigated according to biomechatronic design principles (Mandenius, 2010). Phase (3) is concerned with mathematical process modelling, followed by the OTS development phase (4), including the development of graphical user interfaces and the process control system. In a testing phase (5) not only the technical functionality of the OTS is tested, but also the fulfilment of the training requirements must be investigated. The testing phase is followed by a correction phase (6) and finally an evaluation (7) phase.

The recombinant protein production process simulator in this work was mainly used for academic training and education, whereas the bioethanol process simulator was developed for industrial training. Industrial recombinant protein production processes comprise several upstream and downstream steps. The academic process under consideration consists of a complex recombinant *E.coli* cultivation and a cell disruption step using a high pressure homogenizer (HPH). The industrial bioethanol process mainly consists of a starch hydrolysis unit, a bioethanol fermentation, centrifugation and distillation processes but also includes auxiliary equipment such as different heat exchangers. Both simulators have been equipped with several holding and product tanks, pumps, valves etc. as well as process controllers equivalent to the real systems.

Five key industrial and academic training targets were identified from discussions with experienced plant staff and academic teachers: (1) improve fundamental biotechnological understanding and skills, (2) development of plant operational skills in critical situations, (3) improve capability to design and plan plant operation and process experiments, (4) train handling of complex process control systems, (5) support of modelling education (academia). From these demands, it becomes clear that an OTS must include models that realistically describe the processes as well as the control actions, even under suboptimal operational conditions, throughout start up and shut down procedures and in case of selected incidents. Accelerated and real time simulation modes support training scenarios in different scales of time. A realistic graphical user interface very similar to the one of the real process control system further enhances training efficiency.

The mathematical unit operation models in the OTS were formulated in form of systems of nonlinear differential equations with embedded systems of nonlinear equations, resulting from dynamic mass balances and taking into account biological reaction kinetics as well as kinetics of heat and mass transfer. They were solved numerically using the Runge-Kutta-algorithm and embedded into the process control system WinErs (Ingenieurbüro Schoop GmbH, Hamburg), which also allows for the individual design of control and automation schemes as well as graphical user interfaces. Model calibration and validation was performed using data from laboratory processes (e.g. recombinant protein cultivation with *E.coli*, high pressure homogenisation) or industrial data.

Our studies on the training efficiency were performed with different groups of students at Bremen University of Applied Sciences and Linköping University. Different techniques to investigate the training efficiency, such as multiple choice tests prior and after OTS training and comparative observations of groups of trained and untrained students during laboratory work were used. With these studies it could be shown that not only the operational skills of the trainees were improved by the training with the developed OTS but also their process understanding related to basic principles as well as their capability to interpret process data and thus the capability to identify critical process situations.

From the findings of this study, it may be concluded, that systematically developed OTS can have significant impacts on the biotech industry. The developed OTS for the bioethanol process as well as the recombinant protein production process allow for trouble shooting training without potential operational risk and provide hands-on control system training to get operators acquainted with the behaviour of applied controllers. Due to interactive hands-on training process understanding was taught very efficiently. Operational skills were trained in a cost effective way without disturbing the real process flow, even for critical process steps such as start up and shut down procedures.

<u>Literature</u>

Gerlach, I. (2015): Operator Training Simulators towards Industrial Biotechnology. Linköping Studies in Science and Technology, Dissertation No. 1683; ISBN: 978-91-7519-031-0

Hass, V.C. (2016): Operator training simulators for bioreactors. In: Mandenius C.F. (Ed.) (2015). Bioreactors: design, operation and novel applications. Wiley-VCH, Weinheim, Germany; in press.

Mandenius, C.F. and Björkman, M. (2010): Mechatronic design principles for biotechnology product development. Trends Biotechnol. 2010, 28, 230-236.

Chemical engineering education in a developing country: where are we heading?

J.H. Knoetze
Faculty ofEngineering, Stellenbosch University, Private Bag X1, Matieland, 7602, South Africa. Tel +27-21-8084204
e-mail: jhk@sun.ac.za

Most developing countries need more engineers as the number of engineers per 1000 of the workforce is an important indicator of economic activity in many sectors of the economy. Furthermore, chemical engineering is an important discipline in these countries as it provides the technical expertise for adding value to raw materials (e.g. from mining and agriculture). However, when looking at chemical engineering one needs to take the broader engineering education picture into account. In most developing countries there is a great need to improve the infrastructure, technical abilities of the workforce and increase the economic growth rate. This usually has to be achieved under conditions of high unemployment, non-optimal secondary school systems and severe budgetary constraints for higher education. In addition, there are further differences between developed and developing countries such as poverty, children not exposed to high technology, shortage of qualified mathematics and science teachers, huge differences between rich and poor within these countries, no tradition of higher education, language, etc. A number of similarities are however also identified, namely children tend not to read, undeveloped communication skills, disciplinary problems, etc. In South Africa elements of both the developed and developing world are present and to succeed as a nation more and more of the previously disadvantaged students have to be attracted into technical careers (and thus also chemical engineering).

The following factors were identified as the main factors that are having an effect on chemical engineering education in South Africa, namely: changes in economic activity patterns, globalisation of the economy, longer technical careers, emerging technologies (e.g. biotechnology, nanotechnology, green processes, etc), the environment and sustainability, evolution of IT technology (computing skills), shift from process engineering to product engineering, employment of chemical engineers, outcomes based education, standard of first year entrants, legislation, life-long learning, alternative energy sources and feedstocks, the increase in knowledge, skills and capabilities that have to be developed within a four-year degree programme and the conflicting requirements between academic excellence and industry requirements. The emerging technologies most important to South Africa are identified and discussed.

These factors and many more have to be accounted for in programme specifications, curricula, assessment methodology and personnel and personnel management. The effect of outcomes-based educational programmes as well as accreditation processes on the management of engineering programmes is another area that needs consideration. In developing countries the management of the interface between teaching and research is even more critical to ensure that the needs of the country are addressed, often leading to trade-off between local relevance and international ranking positions. The last major factor is the high failure rates, especially amongst first generation students.

Engineering academics therefore need to be innovative, motivated and look at multi-disciplinary approaches to develop new solutions applicable to the local conditions, but stay internationally relevant and competitive.

Self-training module and serious game for a flipped classroom about liquid-liquid extraction for future engineers in lifelong learning

M. Debacq
Cnam-GPIP, 2 rue Conté, case 2D1P20, 75003 Paris - France
tel. +33.1.58.80.87.07, e-mail: marie.debacq@cnam.fr

The Conservatoire National des Arts et Métiers (Cnam) is a public education institution dedicated to professional lifelong higher education established under the Convention in 1794 on a proposal by Abbe Henri Grégoire "to perfect national industry" and "illuminate the ignorance of he who does not know and the poverty of he who has no means of knowing". Its mission is to provide continuing education open to all, anywhere.

The Cnam chemical engineering team started to offer its distance learning program in 2003; since 2007 [1], 90% of the credits towards the engineering degree can be obtained at a distance; presence is compulsory only for experimental labs. Over the years, many tools have been tested [2], but until 2014 only traditional teaching practices were used. The first experiment of flipped classroom took place in fall 2014 and was about applied fluid mechanics [3,4].

Flipped classrooms are the subject of numerous studies and research; including those of the famous professor Marcel Lebrun [5].

The experiment described here took place in fall 2015 and relates to a liquid-liquid extraction course for engineers, using not only the flipped classroom but also a serious game (for learning the use of ternary diagrams).

The self-training module and the serious game were developed with the publishing chain Scenari (Opale for the module and Topaze for the serious game). The module is inserted in the learning environment Moodle which allows the use of discussion forums and the development of rich quizzes. Virtual classrooms organized with AdobeConnect complete the system.

The self-training module is precisely scripted and richly illustrated; this is achieved thanks to the highly structured Scenari-Opale publishing chain.

Learners emphasize a more in-depth acquisition of knowledge and really appreciate doing the most difficult exercises with the teacher. Through such pedagogical methods, they acquire new skills, such as autonomy, cooperation, creativity and communication.

Flipped classrooms require a true change of posture: the teacher becomes a guide and is no more the only one who knows; the learner becomes much more active and open.

[1] Debacq M., 2007, "Vers la mise en place d'une offre complète de formation à distance en génie des procédés au Cnam : pragmatisme, efficacité et conséquences sur le présentiel", Récents Progrès en Génie des Procédés, n°96, article 514

[2] Debacq M., 2013, "Pédagogie 2.0 : outils efficaces ou gadgets pour une formation en génie des procédés au Cnam ?", Récents Progrès en Génie des Procédés, n°104, article 120421

[3] Debacq M., 2015, "A Flipped Classroom Experiment In The Context Of Lifelong Learning In Chemical Engineering", ECCE2015, Nice - France

[4] Debacq M., 2015, "Scenari + Moodle = Self-training module for a flipped classroom in distance lifelong learning or in traditional teaching", AIChE2015, Salt Lake City - USA

[5] Lebrun M., décembre 2014, "Essai de modélisation et de systémisation du concept de Classes inversées" Blog de Marcel - http://bit.ly/ML-Classes-inversées

Electrocatalytic approach to carbon dioxide utilization

A. Irabien, M. Alvarez-Guerra, J. Albo[1], A. Dominguez-Ramos, E. Alvarez-Guerra, A, del Castillo, I. Merino-Garcia
Departamento de Ingenierías Química y Biomolecular. Universidad de Cantabria, Santander (Spain); [1]Departamento de Ingeniería Quimica, UPV-EHU, Leioa (Spain)
E-mail: angel.irabien@unican.es

Climate Change is a great challenge of the 21st century, Carbon Capture and Utilization (CCU) explores the ways to produce useful chemicals from CO2. Electrochemical reduction of CO2 is considered as a promising reaction for the storage of intermittent renewable energy in the form of different chemicals such as CO, formic acid, methanol and methane, which can be redirected to the market. There are different chemical ways, to activate CO2 to obtain useful products at low temperatures:

- Heterogeneous Catalysis: it is based on the CO2 catalytic hydrogenation. Methane, high hydrocarbons, oxygenates and syngas could be the main products obtained in this way. The limitation of this process is associated to the hydrogen demand of the process.
- Photocatalysis: semiconductor materials such as TiO2, p-GaP, pSi and pGaAs are used as electron source to produce CO2 reduction. Co-catalyst such as pyridine could be used to increase the process efficiency.
- Electrocatalysis: CO2 reduction takes place at the cathode. Different reactions at the cathode can be found as a function of number of electrons.

The electrochemical reduction of carbon dioxide to useful products represents a great alternative, which permits reducing greenhouse gas emissions, due to their associated environmental and economic benefits. Electrochemical CO2 reduction may also potentially control the anthropogenic carbon cycle by recycling CO2 from various sources into feedstock materials for fuels and chemical manufacturing. However, some issues such as the renewable energy supply of CO2 capture and electrochemical conversion, and the limitations of the carbon trade market make difficult the practical application of CO2 capture and utilization in short time.

The products depend on the number of electrons involved in the red-ox reaction: two, four, six, eight, twelve-electron reduction paths can be found to produce useful products such as carbon monoxide (CO), formic acid (HCOOH) or formate (HCOO-), formaldehyde (CH2O), methanol (CH3OH), methane (CH4), ethylene (C2H4), ethanol (CH3CH2OH), as well as others different products. However, a mixture of different products such as hydrocarbons and carbon monoxide is usually observed.

Experimental results of the carbon dioxide electroreduction to useful products will be reported leading to different products: Albo et al. [1] studied the electrochemical reduction of carbon dioxide into methanol using Cu2O and Cu2O/ZnO-catalyzed carbon papers as electrodes at the cathode. A Nafion® 117 membrane was also used to separate the cathode and anode compartments. When a Cu2O-carbon paper electrode was used, a FE to methanol of 45.7 % was achieved, whereas the highest FE to methanol in Cu2O/ZnO (1:1)-based electrodes was 17.7 % at -1.3 V vs Ag/AgCl, so the use of Cu2O-ZnO mixtures for the continuous electrochemical CO2 reduction to methanol seems to be promising. Recently, a review have been developed in order to understand the different possibilities in the CO2 electrochemical reduction into methanol [2].

In order to compare the influence of the catalyst for the electrochemical reduction of CO2 to formate, the use of a lead or a tin cathode electrode has been studied [3,4,5] All the experiments in these works used Nafion® membranes in order to separate the two compartments. Alvarez-Guerra et al. [3] utilized a lead plate as cathode and a solution of 0.45 M KHCO3 + 0.5 M KCl as catholyte. The highest formate FE achieved was 94.7 % at low current density values (2.5 mA/cm2) for which the measured concentration was 14.4 mg/L. At the same conditions, a tin plate was used as working electrode to compare the performance of the process in terms of rate of formate production and Faradaic efficiencies observed [4]. With a Sn cathode, the rates of formate production were 25 % higher with efficiencies about to 70 %. The value obtained with a Pb plate was lower in comparison with the Sn cathode. Considering these results, Del Castillo et al. [5] changed the cathode material compared with the experimental system used by Alvarez-Guerra et al. [3,4], so a GDE loaded with Sn particles with the aim of operating at higher current densities than the Sn plate electrode was used. In fact, higher concentrations were also obtained with the Sn-GDE (885 and 1348 mg/L) with an efficiency of 70 %. The GDEs compared to plate electrodes were better in terms of high rate, concentration and efficiency.
The review and discussion of the carbon dioxide electroreduction and its technical challenges is focused in the electrocatalytic materials and the electrochemical reactor design.

References

[1] J. Albo, A. Sáez, J. Solla-Gullón, V. Montiel, A. Irabien, Production of methanol from CO2 electroreduction at Cu2O and Cu2O/ZnO-based electrodes in aqueous solution, Appl. Catal. B-Environ. 176-177 (2015) 709-717.

[2] J. Albo, M. Alvarez-Guerra, P. Castaño, A. Irabien, Towards the electrochemical conversion of carbon dioxide into methanol, Green Chem. 17 (2015) 2304-2324.

[3] M. Alvarez-Guerra, S. Quintanilla, A. Irabien, Conversion of carbon dioxide into formate using a continuous electrochemical reduction process in a lead cathode, Chem. Eng. J. 207-208 (2012) 278-284.

[4] M. Alvarez-Guerra, A. Del Castillo, A. Irabien, Continuous electrochemical reduction of carbon dioxide into formate using a tin cathode: Comparison with lead cathode, Chem. Eng. Res. Des. 92 (2014) 692-701.

[5] A. Del Castillo, M. Alvarez-Guerra, A. Irabien, Continuous electroreduction of CO2 to formate using Sn gas diffusion electrodes, AIChE J. 60 (2014) 3557-3564.

Techno-economic evaluation of a low-temperature CO_2 capture unit for IGCC plants

David Berstad[1], Simon Roussanaly[1], Rahul Anantharaman[1], Petter Nekså[1], Jana Jakobsen[1]
[1]SINTEF Energy Research, PO Box 4761 Sluppen, NO-7465 Trondheim, NORWAY
Corresponding author. Tel.: +47 735 93 767; fax: +47 735 97 250. E-mail address: simon.roussanaly@sintef.no.

CO_2 capture conditions vary greatly between large point sources within power generation and industries [1]. The CO_2 concentration ranges from roughly 1% for diluted off-gas from aluminium smelters, up to 80–90% for dried oxy-fuel flue gas. The differences become even more distinct when considering the CO_2 partial pressure as this varies by 3–4 orders of magnitude, from approximately 1 kPa to 2 MPa. The higher range can be found in shifted syngas compositions from coal gasifiers. With high CO_2 partial pressure high energy and cost efficiency for CO_2 capture processes can be achieved.

This work presents a high-efficiency low-temperature CO_2 capture unit for integrated gasification combined cycle (IGCC) power plants [2]. The capture process is based on partial condensation and phase separation of liquid CO_2 from non-condensables. The shifted syngas assumed in this work consists close to 40 % CO_2, while the remainder is manly made up by hydrogen.

Different process designs are possible for the low-temperature CO_2 capture unit. The baseline design generates high-pressure supercritical CO_2 for pipeline transport. The syngas feed is first dehydrated and compressed to specified separation pressure and subsequently cooled to separation temperature in a heat exchanger network. After the first separation stage the liquid CO_2 is further purified in a secondary flash separator and subsequently pumped and heated to transport state. The hydrogen-rich non-condensable fuel product is heated and expanded to the combustor feed state. Auxiliary refrigeration for the capture unit can is provided by vapour-compression cascade refrigeration units.
The design and performance of heat exchangers has been verified through simulations of geometric multi-stream plate-fin heat exchanger models. The assumed performance of compressors and expanders has also been shown to be achievable with state-of-the-art technology. This indicates that all process components may be available within a relatively short time frame. The low-temperature capture unit should therefore provide a relatively low-risk technology option for pre-combustion CO_2 capture.

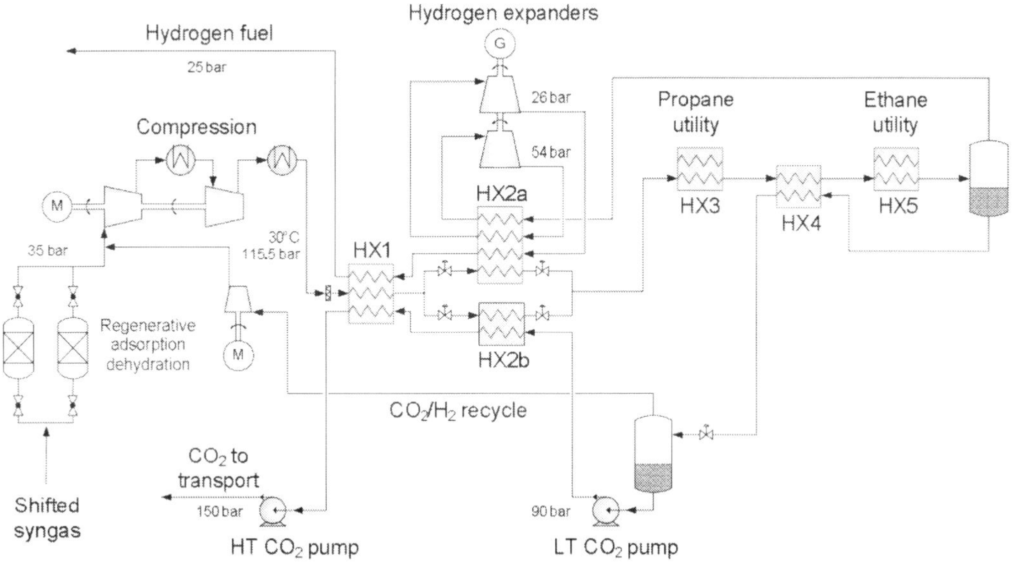

Figure 1: Process design for the low-temperature CO_2 capture unit [3]

The net specific power requirement for the base case has been estimated to 273 kJe/kg$_{CO2}$ at 85% CO_2 capture ratio [2, 3]. Based on the electricity output and the plant costs, the cost of electricity (LCOE) of the process has been evaluated and benchmarked to Selexol process as the reference case . The evaluation shows that the cost of electricity with the base case low-temperature CO_2 capture is 9% less expensive than the reference plant with Selexol CO_2 capture [3]. When looking directly at the cost of capturing CO_2, the low-temperature concept for pipeline transport, with a cost of only 22 €/t$_{CO2,avoided}$, is 35% more cost-efficient than the Selexol process.

Acknowledgements

This publication has been produced with support from the BIGCCS Centre, performed under the Norwegian research program Centres for Environment-friendly Energy Research (FME) and Gassnova. The authors acknowledge the following partners for their contributions: Gassco, Shell, Statoil, TOTAL, GDF SUEZ and the Research Council of Norway (193816/S60).

References

[1] Berstad D, Anantharaman R, Nekså P. Low-temperature CO2 capture technologies – Applications and potential. International Journal of Refrigeration. 2013;36:1403-16.
[2] Berstad D, Anantharaman R, Nekså P. Low-temperature CCS from an IGCC Power Plant and Comparison with Physical Solvents. Energy Procedia. 2013;37:2204-11.
[3] Berstad D, Roussanaly S, Skaugen G, Anantharaman R, Nekså P, Jordal K. Energy and Cost Evaluation of A Low-temperature CO2 Capture Unit for IGCC plants. Energy Procedia. 2014;63:2031-6.

Electro-Coagulation: a step towards chemical-free water treatment

D. Tito, P. Krystynik, P. Kluson
Institute of Chemical Process Fundamentals, Academy of Sciences of the Czech Republic, Rozvojova 135/1, 165 02 Prague 6, Czech Republic, Tel: +420 220390149; email: tito@icpf.cas.cz.

There is a well established need for more efficient and effective technologies to deal with contamination in wastewater treatment. The challenges result not only from a greater amount and a larger variety of contaminants appearing in effluent streams but also novel contaminants (e.g. EDC's) that show relevant activity at trace levels (ppb), requiring new treatment processes.

Precipitation followed by physical separation is a strategy often used to remove a broad range of contaminants from solution through a process called coagulation. Although this is typically carried out by chemical precipitation, an electrochemical alternative is possible which can reduce chemical usage and footprint: the process is called electro-coagulation. This controlled corrosion process has the potential to substantially reduce the amount of chemicals used in chemical coagulation and offers the capability of taking basic wastewater treatment to remote locations, this way improving health but also tackling pollution at source.

As a result of this great potential, this technique has received substantial attention over the past decade but gaps in the understanding of the principles still exist and ambiguity in reporting of results can be found in the literature. This work will focus on presenting the principles of electro-coagulation, clarifying what are the key process parameters and demonstrating the applicability of the technology by showcasing large-scale case studies and process performance data.

The investigation of removal of hexavalent chromium from wastewater in fixed bed column

N. Sivri, [1]N. D. Aksoy
Department of Chemical Engineering, Faculty of Engineering, Istanbul University, Avcilar, 34160, Istanbul, Turkey; tel. +90-212-4737000/17669, e-mail: nurcan.sivri@istanbul.edu.tr, fax: +90-212-4737180; [1]Department of Chemical Engineering, Faculty of Chemical-Metallurgy, Istanbul Technical University, Ayazaga, 34469, Turkey; tel. +90-212-2853527, e-mail: nuran@itu.edu.tr, fax: +90-212-2852925.

Chromium is widely used in textile, metallurgy, electroplating and several other industrial applications which leads to danger for human health and environment. The aim of this study is to investigate the chromate adsorption performance of surfactant modified Bigadiç clinoptilolite. Optimum conditions are also investigated for continuous system experiments, namely; flow rate, initial Cr(VI) concentration and bed height. Adsorption is also characterized by using Thomas and BDST model.

By utilization of surfactant (oleylamine) it was aimed to improve Cr(VI) removal capacity of clinoptilolite. For modification of clinoptilolite with oleylamine, the ECEC and CEC of clinoptilolite were found by using a method studied by Ming and Dixon [1]. CEC and ECEC were calculated as 1.02 meq/g and 0.296 meq/g respectively. Modified clinoptilolite was characterized by FTIR (Fourier Transform Infrared Spectroscopy) and TGA (Thermal Gravimetric Analysis).

After the oleylamine modified clinoptilolite samples characterization, Cr(VI) uptake capacity of clinoptilolite was investigated in fixed bed column. For this purpose, experiments were conducted under different operational conditions by using bench-scale fixed bed column operated under continuous down-flow mode. Firstly, experiments were conducted in order to evaluate the effect of flow rate on removal of Cr(VI) in fixed bed column operation. As a result of these experiments, the lowest flow rate was resulted in the most effective process with the highest Cr(VI) adsorption capacity. Results obtained from these experiments indicate that lower flow rate end up with increase in the retention time required for the adsorption process. Secondly, effect of initial concentration on removal of Cr(VI) was investigated in fixed bed column operations. It was observed that adsorption arrived saturation faster with increasing influent Cr(VI) concentration. Thirdly, effect of bed height on removal of Cr(VI) was investigated in fixed bed column operations. It was clearly that as the bed depth increased, the exhaust time increased due to the increased binding sites of the column. On the other hand, more contact time of adsorbent/ Cr(VI) resulted in higher adsorption capacity. Finally, the Thomas and BDST model was fitted to investigate the breakthrough behavior of Cr(VI) onto clinoptilolite modified oleylamine. The results clearly indicated that predicted curves proposed by Thomas model were in good agreement with the experimental curves at all experimental conditions.

[1] Ming, D.W., Dixon, J. B. *Clays and Clay Minerals,* 35, 463-468, 1987.

Getting to grips with the water-energy-food Nexus? – A *Chemical Engineering Matters* Green Paper

Institution of Chemical Engineers, [1]A. Howe
Institution of Chemical Engineers, Davis Building, Railway Terrace, Rugby, CV21 3HQ, UK; tel. +44 1788578214, e-mail: chemengmatters@icheme.org; [1]

Global demand for food, energy and water is increasing and these resources cannot be considered in isolation. Chemical engineers play a major role in delivering sustainable solutions at the water-energy-food Nexus. This paper presents a range of case studies highlighting examples where the application of a systems thinking approach has improved our understanding of Nexus.

Water, energy and food are essential to sustain human life. Nevertheless, around a billion of the world's poorest people lack satisfactory access to one or more of these basic needs. All are traded internationally and the demand for each is growing. Regional availability varies and each is heavily regulated. But most importantly water, energy, food production and climate change are intrinsically linked.

Chemical engineers have a central role to play in shaping the understanding of the connections between water, energy and food – and, more importantly, in devising practical solutions to address this dilemma through:

1. Applying life cycle analysis (LCA)
2. Working across Nexus boundaries
3. Thinking globally but acting locally

IChemE asserts that applying systems thinking to the interconnectedness of water, energy and food is required to improve the sustainability of human wellbeing and the world in which we live. The following ideas are put forward by chemical engineers for consideration to support the development of the water-energy-food Nexus and to help all those who need to work more sustainably, including decision-makers, manufacturers, companies and regulators:

1. Greater investment is required from funding agencies for systems thinking Nexus research. Stronger collaboration between disciplines is imperative.
2. Governments must work together to solve this problem by developing international platforms to promote Nexus-thinking through the circular economy and other novel solutions.
3. The role of the Nexus in the new 17 Sustainable Development Goals needs to be explored, and the consequences for policy and commercial activity clarified.
4. Operating companies and manufacturers should refrain from making decisions on sustainability and efficiency until they have studied their impact on the water-energy-food Nexus.
5. Industry should publicise case studies of effective, good practice chemical engineering solutions to Nexus challenges.
6. New tools are required for investigating the interconnectedness of the Nexus at different scales, and developing solutions to Nexus challenges.

7. Better understanding is required of the trade-offs between Nexus security and ecosystem health.

Chemical engineers are working to inform the research surrounding the Nexus by using systems thinking to examine the whole supply and value chains of natural resources. Analysis of this nature is essential to using a Nexus approach to process solutions. Chemical engineers will continue to make a key contribution towards ensuring that all processes are efficient, cost-effective and sustainable. By applying systems thinking to the water-energy-food Nexus, chemical engineers offer a unique insight into creating, sustaining and improving wellbeing for all.

An assessment of the potential of solar photovoltaic (PV) application in South Africa

Silas Mulaudzi, Steve Bull

School of Chemical Engineering and Advanced Materials
Newcastle University
Newcastle upon Tyne NE1 7RU
E-mail: s.k.mulaudzi@newcastle.ac.uk

Keywords: solar PV, hybrid energy system

Abstract

The world benefits greatly from fossil fuel resources with more than 80% of world's energy demand coming from fossil fuels. Proven coal reserves are sufficient for the next 113 years, while natural gas reserve could last up to 55 years. However, given the finite life of these resources and the need to reduce emissions alternative energy resources such as solar, biomass, hydro and wind are suggested as they are clean, abundant and effectively infinite. Solar PV is best known as a method for generating electric power by using solar cells to convert energy from the sun into a flow of electrons. In practice, it can be used in several variants depending on the demand and technical development of the user communities; grid connected, off-grid and hybrid energy systems are in use worldwide.

This study was aimed at investigating and establishing the potential of solar PV energy for both off-grid and grid connected application across South Africa. Moreover, it aimed to explore the viability and suitability of a hybrid renewable energy system (PV plus wind turbines) in coastal areas of South Africa. An *Optioneering* approach was used to determine the potential of solar PV and the hybrid energy system. This is a structured evaluation of the options in support of decision-making. The options (in this case the provinces) were modelled in 3 scenarios with different weights assigned to different key parameters. The parameters used in the modelling process were solar radiation intensity, electricity tariff, population, land availability, electricity consumption and electrification backlog.

The Northern Cape Province has the highest solar PV potential for scenarios 1 and 2, followed by the Eastern Cape and Kwazulu Natal Provinces. Gauteng was found to have the highest solar PV potential for scenario 3. The parameters and their weighting are therefore significant in deciding the location where solar PV plant is to be established.

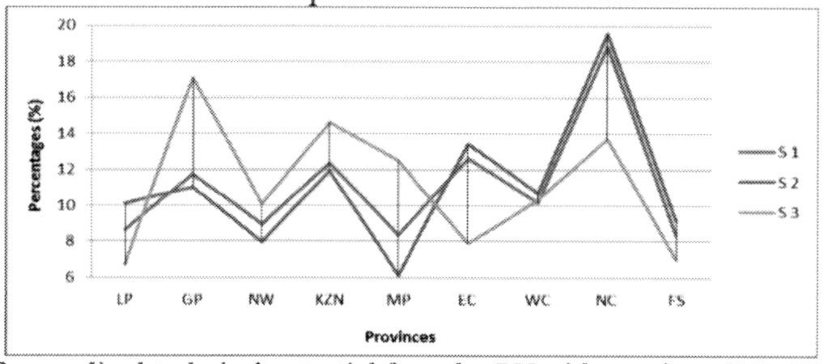

Figure 1: Variation of normalised technical potential for solar PV with province (LP, GP, etc.) in South Africa based on three different selection scenarios (s1-s3).

The PV module efficiency is critical in quantifying the potential power output over a certain period of time. The PV efficiency has shown a tendency of declining over time and it has been noted that PV degradation occurs at a faster rate than anticipated. This would have dire consequences for the solar PV investment in South Africa. The social dimension, variability and intermittency, regulatory framework, technical infrastructure and skills, training & capacity are the other main barriers for the development and growth of solar PV energy.

Biomass gasification in air-blown fluidized bed: Measurements and modelling at pilot scale

O. Authier[1], J.F. Brau[1], M. Debal[2], P. Girods[2], G. Mauviel[3], Y. Rogaume[2]
[1]EDF R&D, Fluid Dynamics, Power Generation and Environment Department, 6 quai Watier, 78400 Chatou, France;
tel. +33130878028, e-mail: olivier.authier@edf.fr;
[2]LERMAB, Université de Lorraine, 27 rue Philippe Séguin, 88000 Epinal, France
[3]LRGP CNRS, Université de Lorraine, 1 rue Grandville, 54000 Nancy, France

Biomass gasification is an efficient technology for generation of combined heat and power (CHP), substitute natural gas, hydrogen, liquid fuels and chemicals. For CHP application, the main lignocellulosic biomass conversion technologies are atmospheric direct-fired and gasification system. Air-blown gasification systems convert biomass into a gas with low heating value by partial oxidation at elevated temperature (700-900°C). After treatment and cooling to produce a commercial-quality product, the product gas can be burned in an internal combustion engine for CHP. Heat can be recovered from the system to improve the global energy efficiency.

Air-blown gasification reaches a high technology readiness level in operational environment and has the advantage of an attractive electrical efficiency (25-35$\%_{LHV}$) especially for decentralized power generation using locally supplied biomass to better control the issue of use conflict. Among the available technologies, bubbling fluidized bed (BFB) gasifiers are often preferred in medium scale applications (1-10 MW_e). Indeed, BFB gasifiers are flexible given that different types of fuels can be treated in the same reactor at high efficiency. They also have the advantage of efficient solid mixing and high heat transfer, leading to nearly isothermal bed conditions and fast solid conversion in comparison with fixed bed gasifiers.

To ensure the development of the technology for challenging biomass fuels, a new biomass gasification pilot plant (50 kg/h) has been built in LERMAB (Figure 1) to test the continuous production (> 24h) of a high-quality product gas in an environment representative of industrial conditions. Experiments have been achieved with pellets of softwood and pellets of straw in an olivine fluidized bed at low equivalence ratio (0.3 to 0.4), high bed temperature (800 to 850°C) and high gas velocity (10 U_{mf}). Gas composition at the exit of the gasifier and after gas cleaning is determined in terms of (i) N_2, O_2, H_2, CO, CO_2, CH_4 and C_2 hydrocarbons (μGC-TCD) (ii) tar compounds (tar protocol, GC-FID and HPLC-fluorescence detector) (iii) H_2O (Karl Fischer).

Figure 1. Biomass gasification pilot plant (50 kg/h) at LERMAB

Accurate models are needed to predict cold gas efficiency and product gas composition profiles inside the fluidized bed. A steady-state gasifier model has been recently developed and validated at lab-scale (Authier et al. 2015) using Aspen Plus software. The Aspen Plus (8.6) fluidized bed model describes one-dimensional isothermal fluidized bed and entrainment of particles in the freeboard. Particle size and density, vessel geometry, additional gas supply, chemical reactions (i.e. biomass drying and devolatilization, tar cracking, char and gas oxidation, methane reforming, water-gas shift and tertiary tar conversion) and their impact on fluid-mechanics and bed expansion are considered. Different correlations provided in Aspen Plus are used in order to determine transport disengagement height and entrainment of solids from the bed. The simulated gas composition profiles are finally compared with experimental data obtained thanks to the new pilot plant.

Reference

O. Authier, B. Cluet, A. Delebarre, A. Dufour, G. Mauviel, Biomass gasification in bubbling fluidized bed: Measurements and modeling of gas composition profiles, 22nd International Conference on Fluidized Bed Conversion, 2015, Turku, Finland

Investigation of the balled biomass combustion in experimental furnace

B. S. Repic, A. M. Eric, D. V. Dakic[1], D. M. Djurovic, M. R. Mladenovic, S. Dj. Nemoda
University of Belgrade, Vinca Institute of Nuclear Sciences, Laboratory for Thermal Engineering and Energy, P.O. Box 522, 11001 Belgrade, Serbia; tel. +381 11 3408838, e-mail: brepic@vinca.rs; [1]Innovation Center, Faculty of Mechanical Engineering., University of Belgrade, Kraljice Marije 16, 11120 Belgrade, Serbia, tel. +381 11 3408836

Development of the sector of heat and power generation represents a strategic goal of the Republic of Serbia primarily for the reason of faster economic development and provision of energy security. Since Serbia is quite dependent on the imported fuels, special attention should be paid to the more widespread use of domestic resources, primarily renewable energy sources (RES). The use of RES is promoted by market conditions and requirements of the EU Directives. Prices of fossil fuels are increasing at the same or higher rate than the rate of reserves depletion. EU Directives require increased use of RES. By signing several international conventions Serbia has accepted an obligation to reduce GHG emissions, which can be achieved through increased use of RES.

The use of RES significantly contributes to more efficient valorization of domestic potentials in energy production, CO_2 and GHG emission reduction, reduced import of fossil fuels, country's industrial development and new job creation, primarily in rural areas. Energy potential of RES in Serbia is quite significant, estimated at about 4.2 Mtoe. Biomass contributes with about 2.7 Mtoe and represents the most important RES in Serbia (63% of total renewable potential). A share of agricultural biomass equals 60%, while forest biomass contributes with 40% to the overall biomass potential. While forest biomass is extensively used, agricultural biomass is practically unused due to the lack of appropriate combustion technologies. Due to unfavorable temperature characteristics of ash and milling properties of agricultural biomass special combustion technologies need to be used. Agricultural biomass should be used in the same form as collected, e.g. in bales, with minimum energy needed for its preparation. The best economic effects are achieved if biomass is used close to where it is being collected, without being transported long distances.

Bearing in mind the diversity of characteristics of residues from agricultural production, which can be burnt in the same combustion chamber in the form of bales, the optimization of process is an important link in a long and successful exploitation. Thereby, consideration must include all sections of the combustion process, from drying phase to the phase of burning out products of incomplete combustion. One of the critical phases is process of burning coke residue, which lasts the longest, and it is taking place in combustion zone where the highest temperature are formed. High temperatures are unfavorable for several reasons, but two are the most important. The first reason is the process of melting ash biomass in some species (e.g. wheat straw) which is particularly pronounced and starts at temperatures below 800°C. Another reason is the formation of nitrogen oxides, because at the high temperatures dominant mechanism is thermal NOx formation. Both problems require extensive research that is not easy to perform in large boilers, and therefore experimental combustion chamber that uses biomass in the form of small bales is very suitable to perform the required tests.

Biomass combustion on cigarette principle is a new and undeveloped technology. For studying the combustion of baled biomass in the Laboratory for Thermal Engineering and Energy, Institute "Vinca" is designed, constructed and built several experimental systems with cigarette combustion. The biggest and most important facility is the industrial demonstration boiler of 1.5 MW for combustion waste baled biomass for heating 1 ha of greenhouses. In order to analyze the process of biomass combustion and to assess the impact of recirculation on the maximum temperature in the combustion chamber a special experimental apparatus was designed and constructed. In addition the corresponding mathematical model was established and verified by comparing the results with experiment. A mathematical model of the combustion chamber is simulated work with varying degrees of recirculation in terms reducing the maximum temperature. The results are of great importance for optimizing of the combustion process in furnaces that operate on cigarette principle.

Experimental apparatus for studying the combustion of baled agricultural biomass is shown in Figure 1. Experimental furnace adapted for burning bales of dimensions 0.35×0.45 ×0.80 cm. Baled biomass using a ball screw is introduced in isolation furnace, where it relies on the grid. Combustion air is entered into combustion chamber using distributive chamber, and its distribution is done by flaps. Ratio of the amount of air and recalculated flue gas is controlled by control valves. Combustion fluid is transported by an electric fan. For determination of the temperature profile in bale, which participating in the combustion process, a certain thermocouple is attached to the bale and with it introduced in the combustion chamber. The resulting temperature profile was used in the verification of the developed mathematical model of combustion.

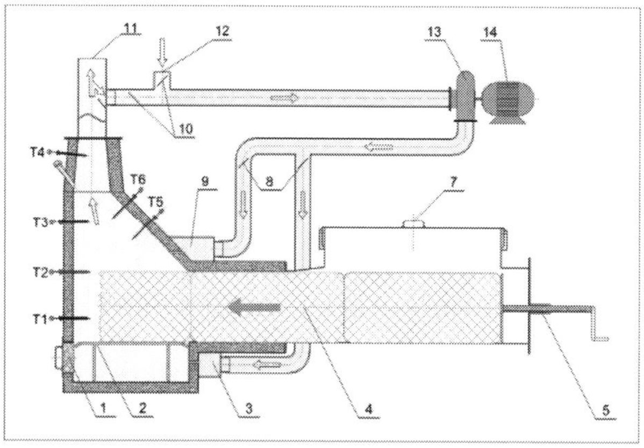

Fig. 1. The scheme of the experimental facility

The results of experimental tests that included measurement of characteristic parameters necessary to determine material and energy balance as well as the quality of the combustion process of biomass is presented in the paper. Model results show satisfactory agreement with experimental results both in terms of the calculated temperature and the composition of gases at the outlet cross section and in terms of temperature profiles in the central plane of balled biomass. It follows that the developed model can be successfully used to simulate a series of the furnace working regimes, as well as to simulate the combustion of various biomasses in baled form. Conducted a parametric analysis showed that the recirculation of cold flue gas has a positive effect on the reduction of the maximum temperature in the combustion chamber and the width of the zone of maximum temperature. This is very important for solving the problems of ash melting. At the same time, a lower content of NOx is achieved by in the flue gas, which is very important and from an ecological point of view.

Assessment of the effect of fuel and type of measuring for the Hg emissions from combustion of coal

K. Borovec, [1]T. Ochodek
VSB-TU Ostrava, Energy Research Center, 17. Listopadu 15/2172, Ostrava – Poruba, Czech Republic; tel.: +420 597 323 868, fax: +420 597 324 295, e-mail: karel.borovec@vsb.cz,
[1] VSB-TU Ostrava, Energy Research Center, 17. Listopadu 15/2172, Ostrava – Poruba, Czech Republic; tel.: +420 597 324 265

The EU is prepared pursuant to Directive 2010/75 / EU since 2018, a strict limit on emissions of mercury in large combustion sources. It is possible to reach the limit concentration of mercury by optimizing the combustion process or the FGT system so without being forced to invest in other technologies?

The aim of the project was to determine the effect of mercury content in the fuel concentration of mercury in emissions, and operation stability of the entire combustion process and system for flue gas cleaning. During the 5-day measurements were taken and subsequently conducted an analysis of samples of fuels, solid residues from incineration, limestone and gypsum, limestone and gypsum slurry and process water. Simultaneously were monitored all major operating systems are compared and methodologies for measurement of emissions of two accredited groups. Test results indicate the effect of operating states of various components of power unit of the amount of flue gas Hg concentrations at relatively the same concentration of mercury in the fuel. It is especially noticeable affect stability DeSOx devices. During the test were confirmed (validated) the accuracy of the use of accredited methods for periodic measurements of total Hg simultaneous measurement of two groups. To determine the effect of stability depending detailed deoxy emission hg mobile laboratory will be built using KME Hg CV using UV absorption, HM 1400 TRX

In paper will be a lot of information about results of tests and other parts of methodologies for measurement of mercury emissions, type of analyzers, measurement results and other related information.

Sampling and analysis of trace metal elements in natural gas by high pressure bubbling sampler and ICP-MS analysis

[1, 2, 3]M. Cachia, [2]H. Carrier, [1]B. Bouyssière, [3]G. Caumette, [3]L. Mahenc, [1]I. Le Hécho.
[1]IPREM-LCABIE, Université de Pau et des Pays de l'Adour, 2 avenue du président Pierre Angot, 64053 Pau cedex, France ; [2]LFC-R, Université de Pau et des Pays de l'Adour, BP1155-64013 PAU cedex, France ; [3]TIGF, espace Volta, 40 avenue de l'Europe, 64010 Pau cedex, France ; [1, 2, 3]tel. +33 559407667, email: maxime.cachia@uni-pau.fr ; [1]email : isabelle.lehecho@univ-pau.fr ; [2]email : herve.carrier@univ-pau.fr ; [3]email: guilhem.caumette@tigf.fr .

Natural gas is a fossil energy commonly used in developed countries. Natural gas is mainly composed of methane (> 90% vol.) and contains trace elements such as heavy metals (< 0,01% vol.) [1, 2]. Heavy metals can have an impact on the environment in case of gas storage in geological formations, even with low concentrations. It is thus necessary to know the trace elements composition of the gas in order to prevent any environmental contamination from storage activities.

A system has been developed to sample trace metals in natural gas, directly on the transportation grid with a pressure of 70 bars. Three successive vials are filled with an adequate solutions to trap metals from the gas. One solution is nitric acid with hydrogen peroxide for arsenic, cadmium, copper, chrome, lead, tin, vanadium and zinc. For the mercury, the solution is permanganate with sulphuric acid. The connected gas flows with a rate of 20 L/min, through these three 150 mL vials. The vials are then closed and brought to laboratory for analysis by inductively coupled plasma mass spectrometry (ICP-MS). On a regular sampling campaign, the High Pressure Bubbling Sampler is connected to the gas network for 24 hours. Thus 30 Nm^3 of gas goes through the 150 mL of trapping solution, which makes a trace elements pre-concentration factor of 2.105.

In order to test the efficiency of the system for metal trapping, the three vials have been tested in laboratory with a controlled gas (argon) spiked with mercury at around 0,1 µg/L concentration. Argon was chosen instead of methane for security reason, and mercury for its high volatility and its difficulty to be trapped. Pressure, temperature and flow rate were controlled during the experiment to be at 6 or 50 bars, at room temperature and at 120-150 ml/min. The solutions tested were permanganate and sulphuric acid (as the sampling of natural gas, described in the norm EN 13211) or nitric acid with hydrogen peroxide. The gas with mercury crosses the vials and mercury concentration in the flowing gas was measured at the entrance in the vials and at its exit using gold traps Amasil®, and Sir galahad® analyser (PSAnalytical) in order to guaranty the mass balance during the experiment. Solutions are analysed at the end of the experiment with ICP-MS (Agilent 7500). A scheme which describes these experiments is presented figure 1.

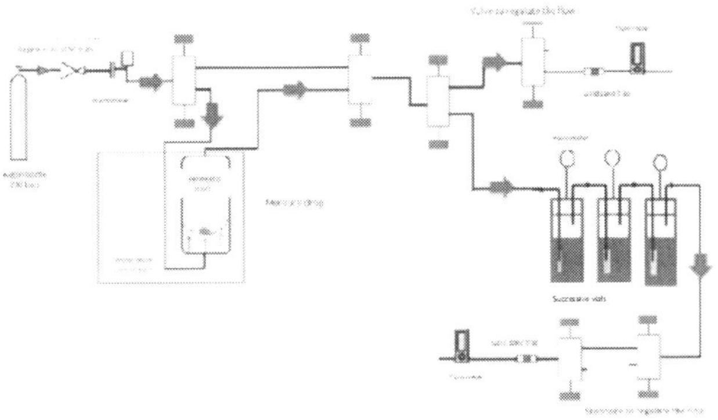

Spargers : filled with bubbling solution

Permeation cell : gas mercury concentration

Figure 1 : scheme of the experiment to test the three spargers

The experiments showed that bubbling works well to trap mercury in gas with the permanganate and acid sulphuric solution. It is very efficient: about 96% of the mercury present in the gas is removed and trapped in the solution. Moreover, about 94% of mercury is found in the first vial indicating that no leak of mercury occurs during sampling. Figure 2 shows an example of result of the bubbling efficiency with the mercury concentration in gas before and after the crossing of the vials.

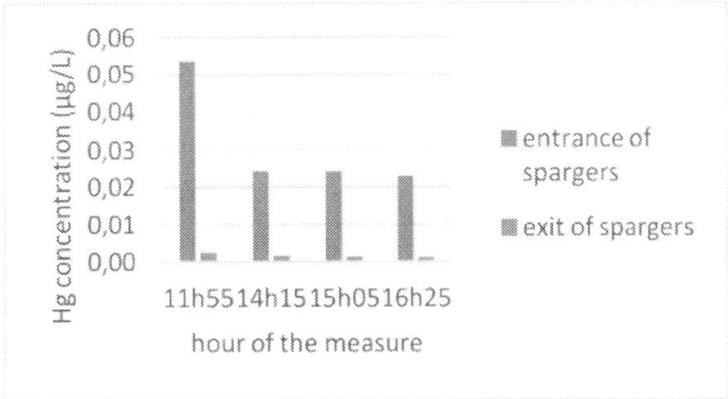

Figure 2 : mercury concentration in gas at different time of measurement (argon 6 bars, 17°C)

To go further, the same experiment was realised with nitric acid and hydrogen peroxide instead of permanganate: this is easier to analyse. However, in this case, mercury is not trapped in the bubbling solution. This solution is inefficient to remove the mercury from the gas. We can find mercury in the three spargers and also in the gas at the exit of the bubbling system. In conclusion, the bubbling system in efficient for metals and especially mercury analysis in gas at high pressure but it is important to optimise the trapping.

The different steps of development and optimisation will be presented, together with the results from several analysis campaigns on TIGF gas network.

The equipment can therefore be used to monitore quality of such gas in such environment.

References:

1. Duoyi et al. , *Discovery of the metal trace elements in natural gas and its ecological environment significance.* Earth Science Frontiers 15(6) (2008) 124-132.
2. Institute of gas technology, *Characterization and measurement of natural gas trace constituents. Volume 1. Arsenic.* (1995).

A dynamic model of carbon dioxide absorption in a bubble column reactor

Mahshid Shaaker Dushanlu and Leila Vafajoo
Islamic Azad University, South Tehran Branch, Chemical and Environmental Engineering Group, Deh haqi Ave., Ahang Blvd., Tehran, Iran, Zip code:1777613651,
Tel.:+98 21 33010400, Fax: +98 21 3371714
Corresponding Author's E-mail: vafajoo@azad.ac.ir

IPCC fifth assessment has showed a globally averaged warming of 0.85 °C over the period of 1880 to 2012 which is projected to rise to 6.5°C by 2100. Emissions of carbon dioxide from fossil fuels and industrial process contributed about 80 percent of this increase. Hence, a delay in mitigation efforts or the limited availability of low carbon technologies further increases this challenge [1]. The CO_2 capturing using dry sorbents such as alkaline oxides has been recently investigated as an alternative method so as to emit major drawbacks of conventional technologies such as Amine based solutions toxicity and high energy cost. [3] A bubble column reactor has been used as the contractor in this study in which the gas is dispersed in the aqueous $Mg(OH)_2$ solution. Bubble columns provide several advantages in chemical industry such as high heat and mass transfer rates, compactness, high selectivity, low maintenance, low operating cost and being the main considerations in three phase systems. [4]

The purpose of this study was developing a slurry bubble column model to assist in understanding of the reaction of CO_2 removal within a gas-liquid-solid system, discussing the effects of reaction parameters such as inlet gas velocity (0.145 and 0.0145 m/s) in term of absorption efficiency ($Mg(OH)_2$ mass fractions) [Fig. 1]. A dynamic model of a bubble column reactor with a diameter of 10 and height of 130 centimeters has been unsteadily developed for 60 minutes accorded with operating conditions as shown in Table 1. It is found that the overall $Mg(OH)_2$ mass fraction participated in the reaction increases by more than 10^2 in lower CO_2 velocity. Experimental data and modeled results in term of CO_2 removal have been compared so as to validate the modified model with an average error of 18% which indicates the accepted validity of model. [Fig. 2]

Table1. Operational Parameters

Value	Parameter
≈ 52 °C	Inlet gas temperature
25-54 °C	Liquid temperature
0.5 L/min	Gas flow rate
≈ 12% vol.	Inlet CO2 concentration
93	Mg(OH)2 concentration
1 M	Liquid volume
2.1 L	Mixing rate
25-50 micron	Porosity of gas bubbler

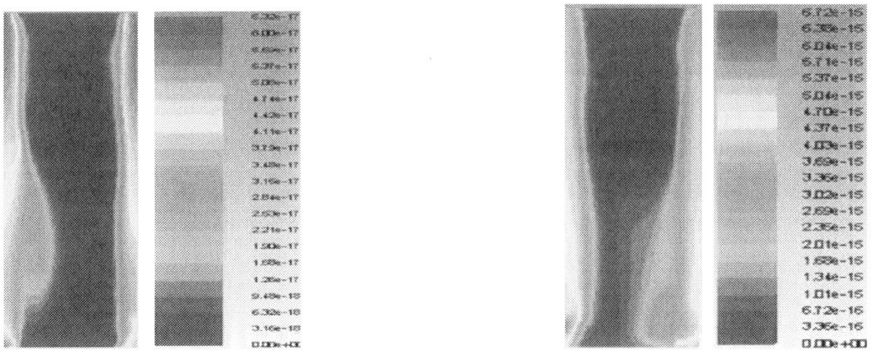

v= 0.0145 m/s   v= 0.145 m/s

Figure1. $Mg(OH)_2$ mass fraction profile in 2 different inlet gas velocities

Figure2. Experimental and modeled CO_2 removal percentage

Resources

[1] "IPCC, 2014. Intergovernmental Panel on Climate Change Fifth Report. ".

[2] L. Cheng, T. Li, T. C. Keener , J. Y. Lee, "A mass transfer model of absorption of carbon dioxide in a bubble column reactor by using magnesium hydroxide slurry" International Journal of Greenhouse Gas Control, pp. 240-249, 2013.

[3] T. Li, C. Tim ,L. Cheng, "Carbon dioxide removal by using Mg(OH)2 in a bubble column: Effects of various operating parameters" ,Greenhouse Gas Control, pp. 67-76 , 2014.

[4] E. A. Lvarez, J. M. Navaza , B. Sanjurjo "Continuous removal of carbon dioxide by absorption employing a bubble column" , Chemical Engineering, pp. 251-256, 2008.

[5] EIA, "World Energy outlook, Energy Information Administrartion" 2015.

Analysis of the degradation of multiple organic micropollutants using laccase enzymes

R. Pype, C. Hautphenne, J. Blavier, F. Debaste
Université Libre de Bruxelles, Dep. Transfers, Interfaces and Processes, Av. F.D. Roosevelt 50, ULB CP165/67, 1000 Brussels, Belgium; tel +3226506756, e-mail: fdebaste@ulb.ac.be

Laccase enzymes have been tested in many studies for their ability to degrade a large spectrum of phenolic compounds that are associated to pollution even at really small concentrations, the so called *micropollutants* [1]. The most efficient laccases, produced by various fungi such as *Trametes Versicolor,* are able over short time scales (minutes to hours) to significantly degrade numerous xenobiotics that might cause a variety of threats to the environment such as Estradiols, Bisphenol A, Diclophenac [2], Triclosan [1], …

Most of the studies that are realized in this promising field are dedicated to the elimination of a single micropollutant. Yet, the main source of micropollutants in the environment is due to water coming out of wastewater treatment plants (WWTP) that are aggregating highly variable pollution sources. At the exit of these WWTP, mixes of micropollutants are encountered and should be treated [3]. A few studies have begun to address this problem and showed that it was indeed possible to degrade multiple micropollutants simultaneously up to the pilot scale [4]. However, no study detailed the impact of treating mixes on the kinetics of degradations. Yet, interactions between the different micropollutants, their degradation products and the enzymes might drastically change the observed reaction rate by mechanisms such as competition for the enzyme (decreasing the reaction rate) or reaction of an enzymatically generated radical with another micropollutant (thus increasing the global reaction rate). Characterizing, understanding and modelling these kinetic interactions is thus of high interest to design efficient treatment solutions for organic micropollutants elimination.

This study aims at a first analysis of the interactions between different micropollutants in a general framework of process development. Considering the already large spectrum of micropollutants that might be addressed, testing combinations will lead to an even larger number of tests. Key model systems, with few micropollutants but allowing to highlight the main possible mechanisms, should be identified. This work propose to make a large literature analysis of existing study of mixes of micropollutants degraded by laccases and to try to extract trends and quantitative informations about the kinetics (reaction rate, impact of temperature and pH;…). These informations are then confronted to new experimental data.

The first step of this study is the identification of existing works on mix of micropollutants degradation using enzymes. This excludes studies where a mediator is added to speed up the reaction of a single micropollutants. In final, 18 articles were isolated.

General methodology trends can be extracted from these 18 studies. First, only 3 different laccases were tested with mixes: *Trametes Versicolor*, *Corriolopsis Gallica* or *Myceliophthora thermophile.* Up to 31 different micropollutants were mixed [4]. Yet, only a few, summazired in Table 1, were found repeatedly, including oestrogenes (17α-ethinylestradiol, 17β-estradiol, estrone, estriol), bisphenol A and diclofenac.

Various type of reactors (batch, continuous membrane reactor, fixed bed) with free or immobilized enzymes were found. Most were realized in synthetic water with spiked micropollutants. Only one study [2] did explicitly mention the observation of a coupling mechanism between the micropollutants.

Based on all these studies, it was chosen to focus on oestrogenes, bisphenol A and diclofenac. The set of data from literature was then completed with new batch experiment for 17β-estradiol and bisphenol A mixes in presence of free laccases from *Trametes Versicolor*. With all these data, a systematic analysis of degradation rate was realized by expressing, for each experiment, a kinetic constant, independent of the pH, the temperature, enzyme concentration and initial pollutant concentration. A similar constant was also evaluated based on studies with micropollutants taken one by one. The comparison of these constants from one experiment to another for a given strain allows to evaluate the potential impact of the presence of multiple micropollutants on the global kinetics.

As all the different studies taken into account vary in methodology, only trends can be considered as significant in the comparison of the different experiments. The main findings are:

- The global order of degradation of given micropollutants change when mixes are considered, underlying the possible coupling between the reactions
- Diclofenac can be degraded up to 10 time faster in the presence of other micropollutants but only at pH higher than its pKa.
- Dependence of the reaction rate to the pH is not identical for all the micropollutant, highlighting the possible importance of their pKa values.
- Both *Trametes Versicolor* and *Myceliophthora thermophile* show similar ability to degrade the considered micropollutants

Globally, this works allows to confirm that reaction of mixes of micropollutants show significant difference compared to the degradation of single micropollutants. Further work is required to harness the potential interest of these interaction as some slowly degraded micropollutants seems to be eliminated faster in presence of others. This could leads to more efficient and cost effective micropollutants treatments.

Table 1: Number of study including the micropollutants

Micropollutant	Number of studies
17β-estradiol	10
Bisphenol A	9
Diclofenac	9
Carbamazepine	7
Ketoprofen and naproxen	6

References

[1]F. Debaste, G. Songulashvili, M.J. Penninckx, "The potential of Cerrena unicolor laccase immobilized on mesoporous silica beads for removal of organic micropollutants in wastewaters", Desalination and Water Treatment, vol 52.,no. 10-12, pp. 2344-2347, 2014

[2] R. R. Nair, P. Demarche, and S. N. Agathos, "Formulation and characterization of an immobilized laccase biocatalyst and its application to eliminate organic micropollutants in wastewater," *New Biotechnol.*, vol. 30, no. 6, pp. 814–823, Sep. 2013.

[3] S. Snyder and M. Benotti. "Endocrine disruptors and pharmaceuticals: Implications for water sustainability", *Water Science and Technology*, vol. 61, no. 1, pp. 145–154, 2010.

[4] L. N. Nguyen, J. P. van de Merwe, F. I. Hai, F. D. L. Leusch, J. Kang, W. E. Price, F. Roddick, S. F. Magram, and L. D. Nghiem, "Laccase – syringaldehyde (SA) – mediated degradation of trace organic contaminants (TrOCs) in an enzymatic membrane reactor (EMR): Removal efficiency and effluent toxicity," *Bioresour. Technol.*, 2015.

Modeling and optimization of dye biodegradation process in batch reactor

H. Bouchène[1], N. Imadalou-Idres[2], M. Bourouina[3*], S. Bacha-Bourouina[1], S.Benallaoua[2]
[1] Department of Chemical Engineering, Faculté de Technologie, Université de Béjaia
[2] Department of Microbiology, Faculté des Sciences de la Nature et de la Vie, Université de Béjaia
[3]Department de Chimie, Faculté des Sciences Exactes, Université de Béjaia
*Email : bouryas@yahoo.fr
Phone : + 213 66 120 40 70

The aim of this work is the modeling and optimization of wastewater biodegradation batch processes using archea salt requiring bacteria.

The degradation yield and kinetic parameters were estimated for three conditions : Temperature, pH and dye initial concentration. To determine the responses patterns, central composite design was followed.

Primary models based on the techniques of predictive microbiology were used to estimate the parameters of microbial growth needed to study the process. The estimation of the model parameters was performed using the non-linear regression based on Levenberg Marquart algorithm.

A rotatable orthogonal central composite design at five levels allowed us to acquire a second-order model with the terms of interaction between the three influential factors chosen.

The statistical significance of the model was evaluated by analysis of variance (ANOVA) using Minitab 16. Application of the response surface methodology in secondary model and evaluation of growth parameters through the primary model, enabled a rapid and simultaneous exploration of the influence of the 3 factors (T, pH and MB concentration) on the growth and antagonistic activity of the haloarcheon and degradation performance.

The main and interaction effects of these independent variables were evaluated and their influences on bacterial growth and on degradation performance are characterized. These models are then applied to predict the time and optimal operating conditions for the best responses: depollution yield, growth and microbial activity.

Key words: Methylen Blue, Halobacteria, central composite design, degradation, optimization

Screening of green microalgae from stressed environments for antibacterial activity

[1]G.N.A., Senhorinho; [2]G. M. Ross; [3]J. A. Scott
[1]Biomolecular Sciences Program, Laurentian University, 935 Ramsey Lake Rd, P3E2C6 Sudbury, Ontario, Canada; tel: +1 705 6627288, e-mail: gsenhorinho@laurentian.ca, fax: +1705 6754858.
[2]Northern Ontario School of Medicine, 935 Ramsey Lake Rd, P3E2C6 Sudbury, Ontario, Canada; tel: +1 705-662-7288, e-mail: gross@nosm.ca, fax: 705-6754858.
[3]Bharti School of Engineering, Laurentian University, 935 Ramsey Lake Rd, P3E2C6 Sudbury, Ontario, Canada; tel: +1 705-6574858, e-mail: jascott@laurentian.ca, fax: 705-6754858.

The discovery of penicillin in the late 1920s and other antibiotics that followed represented a major scientific achievement against deadly bacterial infections. Natural sources of antibiotics, such as from certain soil bacteria and fungi, have provided many of the available antibacterial drugs. These antibiotics are derived from secondary metabolites, which appear to help microorganisms survive stress in their environments (Leflaive and Ten-Hage, 2007).

However, increasing resistance to existing antibiotics has become a significant worldwide concern (Stanton, 2013), and combined with the lack of new synthetic compounds is encouraging the search for antibiotics from previously unexplored natural sources (Ling *et al.*, 2015). This includes relatively little investigated microalgae (Mudimu *et al.*, 2014; Senhorinho *et al.*, 2015).

Microalgae are photosynthetic microorganisms found in almost every ecosystem, including fresh and marine water, desert soils, rocks, snow and hot springs (Lee, 2008). Despite their enormous ecological diversity, there have been to date only limited studies that have shown that certain microalgae have an ability to produce compounds with antibacterial activity (Ordog *et al.*, 2004; Rodriguez-Meizoso *et al*, 2010; Kokou *et al.*, 2012). The active metabolites appear to be produced by a variety of genera and species, and are dependent on environmental stress factors (Senhorinho *et al.*, 2015).

A systematic and broader screening of microalgae, particularly those collected from stressed environments may, therefore, provide a new source of unique compounds with antibacterial activity. As a consequence, samples were collected from previously unexplored water bodies around abandoned mine sites in Northern Ontario Canada. Microalgae at these sites would have been exposed to various stresses, such as a wide-range of annual temperatures, low pH and/or high metal levels.

Our aim was to develop a rapid and reliable methodology that provides efficient preliminary screening for microalgal antibacterial activity. This was achieved through cultivating green microalgae from various stressed environments on Bold's basal medium agar plates (Fig. 1) prior to obtaining cellular extracts for testing for antibacterial activity.

Once isolated and identified, 40 microalgal strains were cultivated on solid medium for 28 days (Fig. 1). Intracellular extracts were then obtained and screened for antibacterial activity by putting drops on plates streaked with active human pathogens. From the 40 extracts investigated, 37% showed an ability to inhibit bacterial growth, in particular *Staphylococcus aureus* (Fig. 2). This bacterial species causes serious worldwide hospital and community-acquired infections and is resistant to almost all commercially available antibiotics (Pantosti *et al*, 2007).

In conclusion, the results of this study show that certain green microalgal strains that survive in a range of stressed conditions offer significant promise as a potential new source of natural antibacterial compounds.

Figure 1: Isolation of green microalga on Bold's basal medium agar

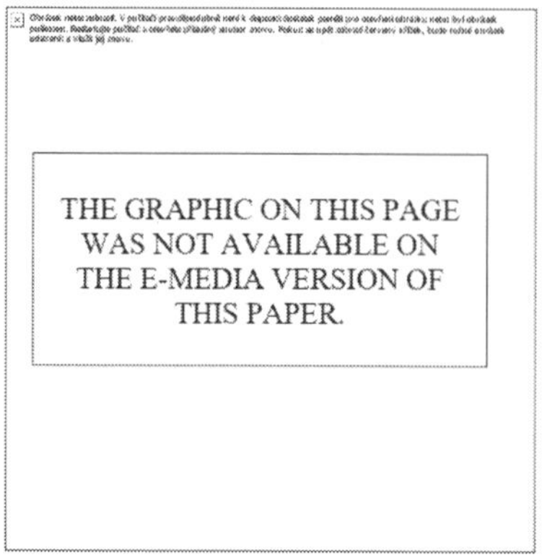

Figure 2: Inhibition of *S. aureus* growth (clear zones) by microalgal extracts.

References

Kokou, F., Makridis, P., Kentouri, M., Divanach, P. 2012. Antibacterial activity in microalgae cultures. Aquac. Res. 43, 1520–1527.

Lee, R. E. 2008. Phycology. Basic characteristics of the algae, 4th edition, chapter 1. Cambridge University Press, New York, USA, 547 p.

Leflaive, J.P., Ten-Hage, L. 2007. Algal and cyanobacterial secondary metabolites in freshwaters: a comparison of allelopathic compounds and toxins. Freshwater Biol. 52, 199–214.

Ling, L.L., Schneider, T., Peoples, A.J., Spoering, A.L., Engels, I., Conlon, B.P., Mueller, A., Schaberle, T.F., Hughes, D.E., Epstein, S., Jones, M., Lazarides, L., Steadman, V.A., Cohen, D.R., Felix, C.R., Fetterman, K.A., Millett, W.P., Nitti, A.G., Zullo, A.M., Chen, C., Lewis, K. 2015. A new antibiotic kills pathogens without detectable resistance. Nature 7535, 455–459.

Mudimu O., Rybalka, N., Bauersachs, T., Born, J., Friedl, T., Schulz, R. 2014. Biotechnological screening of microalgal and cyanobacterial strains for biogas production and antibacterial and antifungal effects. Metabolites 4, 373–39.

Ordog, V., Stirk, W.A., Lenobel, R., Bancirova, M., Strnad, M., Van Staden, J., Szigeti, J., Nemeth, L. 2004. Screening microalgae for some potentially useful agricultural and pharmaceutical secondary metabolites. J. Appl. Phycol. 16, 309–314.

Pantosti, A., Sanchini, A., and Monaco, M. (2007) Mechanisms of resistance in *Staphylococcus aureus*. Future Microbiol., 2, 323-334.

Rodriguez-Meizoso, I., Jaime, L., Santoyo, S., Senorans, F. J., Cifuentes, A., Ibanez, E. 2010. Subcritical water extraction and characterization of bioactive compounds from haematococcus pluvialis microalga. J. Pharm. Biom. Anal. 51, 456–463.

Senhorinho, G. N. A., Ross, G. M., Scott, J. A. 2015. Cyanobacteria and eukaryotic microalgae as potential sources of antibiotics. Phycologia 54, 271–282.

Stanton, T. B., 2013. A call for antibiotic alternatives research. Trends in Microbiol. 21, 111-113.

Impacts of main parameters affecting the regeneration process of Amberlyst A26 anion exchange resin previously used for final treatment of olive mill effluent

M.D. Víctor-Ortega[1*], J.M. Ochando-Pulido[1], A. Martínez-Ferez[1]
[1]Chemical Engineering Department, University of Granada, 18071 Granada, Spain
email: mdvictor@ugr.es

The key for the cost-efficient implementation of an ion exchange (IE) process is to ensure that the IE resins can be easily and effectively regenerated, in order to guarantee their operation for the maximum possible working cycles [1]. In this sense, IE processes must be stopped periodically and the IE resin regenerated. Concretely, anionic resins are usually regenerated by using concentrated base solutions.

In former research works, the final purification of olive mil effluent coming from the two-phase extraction system (OMW-2) was carried out through IE process comprising a combination of two resins in serial connection: one cationic resin (Dowex Marathon C) and one anionic resin (Amberlyst A26) [2]. With the aim of reducing its high organic load, previous to the final treatment by means of IE technology, OMW-2 was conducted to a secondary treatment based on Fenton reaction-advanced oxidation process followed by a flocculation–sedimentation step and filtration-in-series through different filtration medium [3,4].

The aim of this research study was to optimize the regeneration process of Amberlyst A26 strong-weak anion exchange resin, previously used for the purification of the secondary-treated olive mill wastewater stream (OMW-2ST), by investigating the impact of the main parameters affecting the regeneration efficiency. For this purpose, the effects on the regeneration and rinse times as well as the overall regeneration efficiency of the IE resin were addressed as a function of the regeneration operation factors, comprising the regenerant concentration and type, the operating temperature and the flow rate.

The results demonstrate that regeneration time decreased and rinse time increased with increasing regenerant concentration (**Fig. 1**). In this sense, 0.7 L and 1 L NaOH 4% solution was used for complete regeneration of Amberlyst A26 resin. Furthermore, the optimal temperature for regeneration of Dowex Marathon C was 298 K (regeneration efficiencies higher than 96 %). The impact of flow rate on the selected IE resins regeneration process efficiency revealed that, regeneration efficiency decreased with increasing flow rates (**Fig. 2**). In this sense, optimum flow rate was found to be 2.5 L h^{-1}, which yielded regeneration efficiencies close to 100 %.

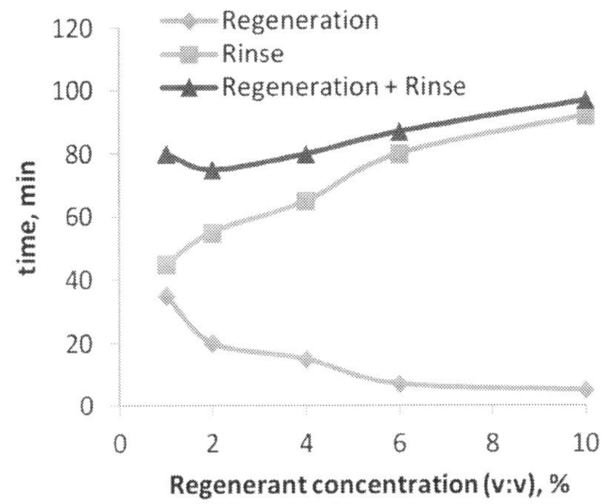

Fig. 1. Impact of regenerant solution concentration on regeneration and rinse time of Amberlyst A26 strong-base anion exchange resin. Regenerant: NaOH; Temperature: 298 K; Flow rate: 2.5 L h^{-1}.

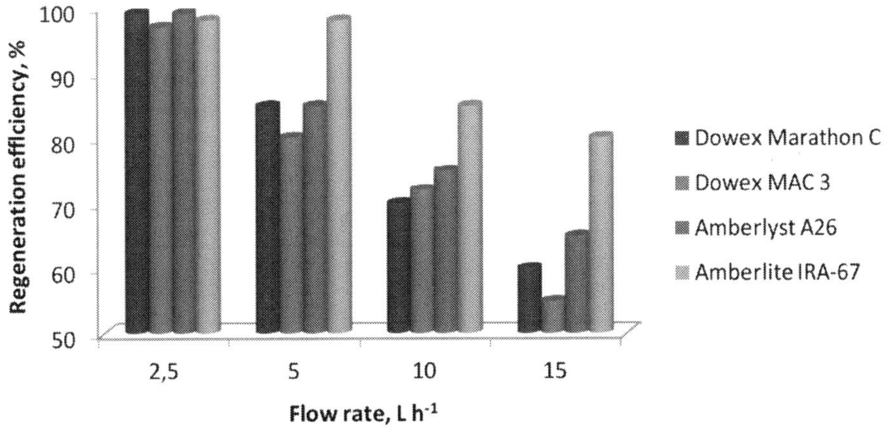

Fig. 2. Impact of flow rate on regeneration efficiency of the selected resins. Regenerant: 4% NaOH; Temperature: 298 K.

References

[1] Simon, G.P. 1991. Ion Exchange Training Manual. Chapter 6. Springer Science. New York.

[2] Víctor-Ortega, M.D., Ochando-Pulido, J.M., Airado-Rodríguez, D., Antonio Martínez-Ferez, A. 2015. Process Safety and Environmental Protection. Under review.

[3] Martínez Nieto, L., Hodaifa, G., Rodríguez Vives, S., Giménez Casares, J.A., Ochando, J. 2011a. Chem. Eng. J. 173, 503-510.

[4] Martínez Nieto, L., Hodaifa, G., Rodríguez Vives, S., Giménez Casares, J.A., Ochando, J. 2011b. Clean - Soil, Air and Water 39(10), 949–955.

Leaky solution-difussion modelization of membrane performance during reverse osmosis of pretreated olive mill wastewater

J.M. Ochando-Pulido*[1], M.D. Victor-Ortega[1], A. Martínez-Férez[1]
[1] Chemical Engineering Department, University of Granada, 18071 Granada, Spain; jmochandop@ugr.es

The increasing demand of olive oil worldwide required replacing the discontinuous olive oil extraction process by the more efficient continuous centrifugation-based one. This brought as a drawback a significant increase of the generated wastewater, called olive mill wastewater (OMW). In this work, OMW from a two-phase olive oil extraction process (OMW-2) was treated by means of batch membrane operations in sequence comprising UF followed by NF and finally RO. This treatment is capable of successfully removing the organic matter and other pollutants from the wastewater, but membrane fouling can limit drastically the operation and longevity of the membranes.

Membrane fouling still remains nowadays one of the main challenges of the broad applied membrane technology, especially in liquid-liquid separation processes [1]. Membrane fouling may lead to dramatically shorten the life time of membrane modules and alters its selectivity and rejection performance [2]. In other words, the lack of knowledge and control of membrane fouling is an additional cost for the industry which should be minimized to permit successful competiveness of the technology. This applies especially on wastewater purification processes.

In this work, a leaky solution-diffusion model was applied to predict and modelize the rejection behaviour of a spiral-wound thin film composite reverse osmosis membrane in the purification of OMW-2:

$$Ri = (P_{TM} \cdot \sigma_i)/(P_{TM} + \beta_i)$$

where the rejection of the solute i depends primarily on the applied net operating pressure ($P_{TM} = \Delta P - \Delta\pi$) and two parameters: σ_i which is a reflection coefficient that indicates the rejection capability of the membrane ($0 < \sigma_i < 1$) and βi which is a fitting parameter [3].

An accurate prediction of the experimental results was attained by the applied leaky solution-diffusion model in all cases (R^2 above 0.99). The reflection coefficients with respect to COD rejection (σ_{COD}) fitted for the RO membrane were very next to a value of 1. These results are in very good agreement with the salt-rejecting nature of RO membranes.

Finally, the COD values measured in the purified water samples exiting the RO membrane pilot plant permitted complying with irrigation quality standards in all cases (COD values below 1000 mg L^{-1} according to Spanish legislation).

References

[1] M. Stoller, M. Bravi, Desalination 250 (2010) 578-582.
[2] M. Stoller, A. Chianese, Desalination 191 (2006) 62-70.
[3] S. Jain, S.K. Gupta, , J. Membr. Sci. 232 (2004) 45–62.

Removal of alkylphenols from industrial and municipal wastewater

J. Derco, J. Dudáš, L. Sumegová, M. Valičková, [1]S. Murínová
Institut of Chemical and Environmental Engineering, Faculty of Chemical and Food Technology, Slovak University of Technology, Radlinského 9, 812 37 Bratislava, SK;
tel.: +421 2 59325231, e-mail: jan.derco@stuba.sk
[1]Water Research Institute, Nábrežie L. Svobodu 5, 812 49 Bratislava, Slovak Republic

Alkylphenols (APs) are used as antioxidants and plasticisizers, and also in the production of alkylphenol polyethoxylates (APEOs). Approximately 80% of all the APEOs production is released to the market as nonylphenol polyethoxylates (NPEOs) (Ying et al., 2002), which are widely used in the production of plastics, textiles and agricultural chemicals, and in household goods such as detergents, paints, pesticides and cosmetics. These consist of a very complex mixture of isomers (Boehme et al., 2010). 4-nonylphenol (NP) is used as antioxidant, in plastic additives and mainly for synthesizing NPEOs as non-ionic surfactants in detergents, emulsifying agents, pesticide formulations, herbicides, and dispersing agents in the paper and pulp industry.

NPEOs are degraded primarily to the corresponding NPs, and thus ubiquitous in the environment. Nonylphenol etoxylates are oxidised to 4-nonylphenol (NP) under anaerobic conditions (i.e. in sewer system, sediments and during anaerobic processes at WWTP), they are known to be unstable and toxic in water environment and exhibit estrogenic activities (Gültekin and Ince, 2007). NPEOs are transformed by microorganisms to more toxic compounds, e.g. NP and NPEOs ethoxylates with shorter chains (Vazquez-Duhalt et al., 2005). Nonylphenyl and nonylphenyl ethoxylates are designed as priority hazardous substances in Water Framework Directive (Directive 2000/60/EC, 2000) and most of their use are currently regulated Directive 2003/53/EC, 2003; Soarez et al., 2008). Accordingly, a large majority of the research related to APEOs and APs destruction in the water environment is focused on NP, which enter the environment through wastewater streams with NPEOs.

NPs often occur in wastewater as degradation product of NPEOs (Soarez et al., 2008). The principal pathway for nonylphenol removal in waste water treatment plants is sorption to the sludge solids (Shao et al., 2003). Ning et al. (2007) investigated the destruction of 4-tert-octylphenol (OP) and NP by direct reaction with ozone (ozonolysis). They found that ozone reacts easily with both alkylphenols. Tertial ozonation contributed also to decrease of water estrogenic activity (Bertanza et al., 2011).

The results of the study case of removal of NP, OP and APEOs from a real industrial and municipal wastewater (WW) are presented. Industrial wastewater was pre-treated by coagulation with $FeCl_3$ and adsorption on zeolite (ZEO), before discharging into municipal sewer system.

The aim of the work was to investigate an alternative WW treatment processes in a case of extending of production and consequent increase of alkylphenols release. Evaluation of municipal WWTP capacity located closest to the industrial plant discharging alkylphenols, including alternative process tertiary secondary treatment of municipal wastewater, were another goals. Biotic and abiotic removal processes with activated sludge, adsorption on granular activated carbon (GAC), ozonation and combined O_3/GAC processes were investigated.

Removal efficiencies of investigated APs and APEOs in primary sedimentation tank were very low. From practical point of view, the highest and the most significant removal efficiencies within the whole WWTP were observed for NP and NPEOs. Dominancy of abiotic mechanisms of APs removal follows from adsorption measurements. Activated sludge cultivated in lab-scale extended aeration tank accounted for relatively high adsorption affinity to APs and APEOs. The removal efficiencies were in the range from 34.7% (OP) to 94.9% (NPEOs). Activated sludge from municipal WWTP receiving industrial wastewater (WW) containing Aps, accounted for very low adsorption affinity to APs.

Ozone selectivity of the examined APs and their ethoxylates depends on WW composition. High removal efficiencies of NP and NPEOs were measured during ozonation of industrial WW. Mean values of removal efficiency varied from 6.2% for OP to 93.2% in case of OPEOs. Ozonation of municipal WWTP effluent resulted in low residual APs and APEOs concentrations. Similar removal rates of OP and their ethoxylates published Ning et al (2007) based on results of single component (NP, NPEOs) ozonation. Mean values of treatment efficiencies in the range from 9.2% NP to 47.9% NPEOs were observed during ozonation of municipal WW after receiving an industrial WW with APs content. The highest removal efficiency 33.0% was observed for nonylphenol etoxylates.

High variability in treatment efficiency during ozonation of both, municipal and industrial WW were observed when treating WW samples taken in different time (mean values 70.5% for NP and 48.9% for NPEOs). Negligible differences were observed for removal of NP and NPEOs when comparing O_3 and O_3/GAC processes performed under the same conditions. On the other hand, significantly higher removal/transformation rate of OPEOs was observed with O_3/GAC process when comparing with O3 process. NP and NPEOs residual concentrations in all analyzed samples of municipal WW complied with the limits of Environmental Quality standards (Directive 2008/105/EC).

Significantly higher volumetric ozone consumption was observed for industrial WW treatment in comparison with municipal WW ozonation, due to higher organic pollution. The lowest toxicity impact of intermediates and products of ozonation treatment on *Vibrio fisheri* was measured after 10 minutes of ozonation which was lower in comparison to products of O_3/GAC process. Significantly higher removal efficiency of octylphenol etoxylates was observed with O3/GAC process when comparing with ozonation process alone. Lower toxicity impact of intermediates and products of ozonation treatment on *Vibrio fisheri* was measured in comparison to O3/GAC process. It can be concluded that removal of NPs from industrial WW before its discharging to the municipal sewer system is not efficient. Actually, the MWWTP effluent concentration complies with EQS values, including NPs.

Acknowledgements
This work was supported by the Slovak Research and Development Agency under the contract No APVV-0656-12. The authors would like to thank also for the support from the VEGA Grant 1/0859/14.

References

Bertanza G, Pedrazani R, Dal Grande M, Papa M, Zambarda V, Mantani C, Steimberg N, Mazzoleni G, Di Lorenzo D. *Water Research* 45 (2011) 2473-2484.
Boehme RM, Andries T, Dötz KH, Thiele B, Guenther K. *Chemosphere* 80 (2010) 813-821.
Gültekin I and Ince NH. *Journal of Environmental Management* 85 (2007) 816-832.
Ning B, Graham NJD, Zhang Y. *Chemosphere* 68 (2007) 1163-1172.
Ying GG, Williams B, Kookana R. Environ. Int. 28 (2002) 215-226.
Shao B, Hu J, Yang M. *Bull Environ Contam Toxicol* 70 (2003) 527-532.
Soarez A, Guieysse B, Jefferson B, Cartmel E, Lester JN. *Environmental International* 34 (2008) 1033-1049.
Vazquez-Duhalt R, Marquez-Rocha F, Ponce E, Licea AF, Viana MT. *Applied Ecology and Environmental* Research 4 (2005) 1-25.

Investigating the Microbiologically Induced Corrosion (MIC) and the associated integrity risks in underground gas pipeline systems

P.- M. P. Spanidis ([1]), C. Siontorou ([2])
([1]) *ASPROFOS Engineering S.A.*, 40 Helidonous str., 145 64, Athens, Greece, tel.: +00 30 2106253349, email: spanidisp@gmail.com
([2]) *Department of Industrial Management and Technology, University of Piraeus*, 80 Karaoli and Dimitriou str., 18534, Piraeus, Greece, tel.:+00 30 2104142368, email:csiontor@webmail.unipi.gr

Corrosion is an electrochemical process inducing damages in the lattice of metals. The underground steel gas pipelines are often subjected to corrosion attack. The most common effects of corrosion are metal loss, mechanical fatigue and premature aging of the corroded pipeline segments. In several cases, these effects may create major problems, such as presence of holes onto the pipe body, gas release to atmosphere in the form of a jet, ejection of the overlaying soil and, subsequently, environmental, social and safety impacts.

Literature offers interesting theories and numerous publications on approaching the mechanisms through which the corrosion phenomenology is initiated and developed. Some of those theories are focused on the Microbiologically Induced Corrosion (MIC), which is a mechanism explaining how various bacterial communities may enhance the manifestation of corrosion, under certain subsoil conditions. MIC is significant for gas supplying and operating organizations since, once it has been diagnosed, except the inspection costs, MIC inserts also rehabilitation costs and expenses for compensation of the consumers due to the capital loss they are suffering, due to unpredicted interruptions of the contracted gas deliveries.

For the underground gas pipelines, research reveals areas where MIC investigation may be further developed from a different viewpoints, especially under the consideration that MIC can be seen as a multidisciplinary biogeochemical phenomenon generated by three (3) discrete factors stepping synergistically: (a) the structural imperfections of pipe material and deterioration of the external polyethylene coating, (b) the under soil bacterial communities which may be developed in anaerobic conditions and close to areas with identified pipe imperfections, and, (c) the subsoil inorganic matter and its properties (limestone compounds, hydrogen sulfide, nitrogen oxides, soil moisture, acid producing free ions, pH, etc.).

This paper presents a contextual analysis of the synergistic MIC phenomenology and also examination and identification of the risks that MIC may insert to the pipelines integrity from the moment that MIC symptoms or defects are diagnosed onto the body of operating gas pipelines. The paper is organized in two parts: the first deals with the MIC phenomenology representation by means of Fault Tree Analysis (FTA) ontology, based on literature references and collection of information from findings of the empirical research. The analysis represents a decomposition and reorganization of the multidisciplinary MIC knowledge considering biogeochemical conditions and factors acting synergistically to the manifestation of the corrosion process; the second part deals with an analysis and evaluation of risks that gas operating organizations have to manage against MIC, in order to secure the integrity of the gas networks they are operating. Rehabilitation and remedial methods specific to identified risks are also recommended, on the basis of an integrity risk response planning. Finally, techno-economic aspects related to MIC management are also drawn, while conclusions and proposals for further research are offered.

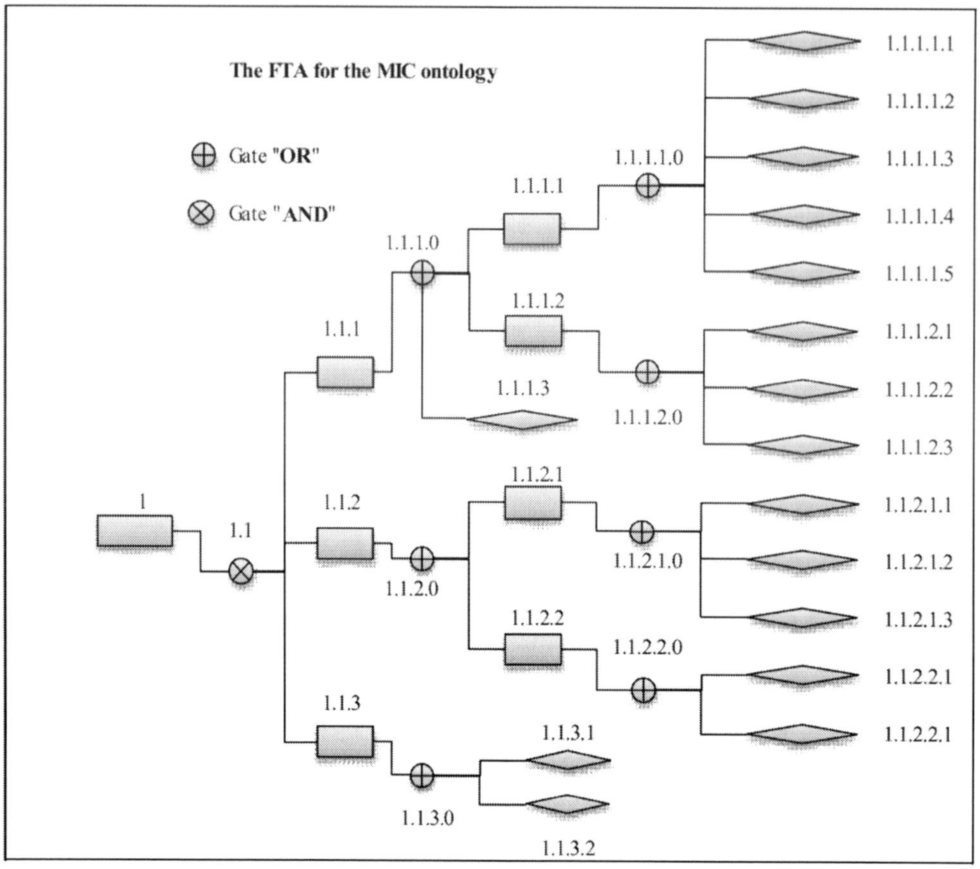

Codification		Description of the FTA ontology content
N	1	MIC of Underground Gas Pipelines
N	1.1.1	Disbonding of external Polyethylene (PE) coating
N	1.1.2	Pipe Fabrication Irregularities
N	1.1.3	Action of Corrosive Bacterial colonies
N	1.1.1.1	PE material fabrication failure(s)
N	1.1.1.2	Failure(s) in field application of coating sheets (pipe welding sections)
L	1.1.1.3	Coating damages during transportation, storage and laying of the pipes
N	1.1.2.1	Insufficient pipe wall thickness
N	1.1.2.2	Corrosion caused by atomic hydrogen $\|H^+\|$ action
L	1.1.3.1	Action of Acid Producing Bacteria (APB) to produce lactic, formic or acetic acids and generate of corrosion pits
L	1.1.3.2	Action of Sulfide Reducing Bacteria (SRB) influencing the pipe substrate to produce $Fe(OH)_2$ and/or FeS scalings
L	1.1.1.1.1	Melting point <120^0 C
L	1.1.1.1.2	Melt flow index> 0.35 g/10 min
L	1.1.1.1.3	Density < 0.94 gr/cm^3
L	1.1.1.1.4	Strength at yield point < 15 Mpa on 23^0 C
L	1.1.1.1.5	Elongation at break <600% on 23^0 C
L	1.1.1.2.1	Presence of soil, concrete, oil and grease residues on the uncoated pipe surface
L	1.1.1.2.2	Air incisions beneath the insulation surfaces
L	1.1.1.2.3	Coating openings due to insufficient overlapping between the PE sheets (<30 mm)
L	1.1.2.1.1	Imperfections (lamination, dents) with depth >12.5% of the nominal wall thickness
L	1.1.2.1.2	Wall thickness tolerances <-8% of the nominal wall thickness for welded seam pipes
L	1.1.2.1.3	Wall thickness tolerances <-10% of the nominal wall thickness for seamless pipes
L	1.1.2.2.1	H_2 absorption from atmospheric moisture in pipe storage yards
L	1.1.2.2.2	H_2 absorption from additives used in pipe manufacturing processes (forging, welding)

Strategies for enhancing hydrophobic VOCs oxidation by biofiltration

A. Zehraoui[1], G. A. Sorial[2]
Environmental Engineering Program, School of Energy, Environmental, Biological and Medical Engineering, College of Engineering and Applied Science, University of Cincinnati, 701 Engineering Research Center, 2624 Clifton Ave. Cincinnati, OH 45221-0012, USA;
Tel[1]. +1 (513) 748-7237, e-mail: zehraoan@mail.uc.edu;
Tel[2]: +1 (513) 556-2987, e-mail: george.sorial@uc.edu;
Fax: (513) 556-2599.

The passage of the 1990 Amendments to the Clean Air Act have led to the development of more stringent regulations, standards, guidelines and codes of volatile organic compounds (VOCs) emissions. It also paved the way to the development of processes aimed at reducing air pollution. Since VOCs contribute greatly to air pollution, several techniques were applied for VOCs abatement such as physico-chemical, adsorption, scrubbing, condensation, oxidation processes, and biofiltration.

Recently, biofiltration systems have emerged as an effective technique for non-methane hydrocarbon removal from air operating under dynamic VOCs loading rates and stressed conditions. Combined by its cost effectiveness, it is an attractive option for controlling VOCs emissions from various industrial processes. More interestingly, Trickle–Bed-Air-Biofilters (TBABs) offer more optimal and controllable operations which result in low maintenance costs over traditional biofilters. Thus, they are more efficient in treating VOCs that do not engender acidic by-products. Combined by its cost effectiveness, compared to the traditional and energy-intensive methods, and its higher and consistent removal efficiency performance and harmless by-products generated it is an attractive option for controlling VOCs emissions from various industrial processes.

The performance of TBABs is significantly affected by the VOCs solubility in water and their Henry's law constant. Hydrophilic compounds, water soluble ketones and esters, are easily biodegradable than hydrophobic ones, mostly aromatic hydrocarbons. The latter are more resilient to degradation, as VOC mass transfer from gas to liquid phase is rate limiting in TBABs. Therefore, the TBAB's challenge is to biodegrade hydrophobic VOCs and prevent their conveyance into waste streams. Hence, an increase in bioavailability of VOCs in the liquid biofilm phase will overcome both mass transfer and reaction rates. In this aim, several processes were tested for addressing the bioavailability of hydrophobic contaminants, like introducing surfactants, providing favorable conditions for fungi, and most recently the utilization of two-phase reactors.

For this purpose, and in order to increase the bioavailability of n-hexane, a bench of experiments was conducted in Trickle Bed Air Biofilters (TBABs) to overcome the mass transfer rate limiting by either increasing n-hexane solubility or augmenting surface area of the media. The following scenarios were tested for enhancing n-hexane biodegradation:

- ✓ Biodegradation of n-hexane in mixture with methanol
- ✓ Operating condition at pH 4, the TBABs were examined at low pH buffered nutrient solution stimulating fungi growth over the traditional bacteria working at neutral pH
- ✓ Presence/absence of methanol on n-hexane biofiltration

- ✓ Use of methanol as an initial acclimation period for enhancing n-hexane biodegradation in TBABs
- ✓ Installation of an integrated 2-bed adsorption/desorption system before the biofilter to buffer the transient operating condition.

In general, n-hexane removal efficiencies obtained in these studies are extensively higher as compared to other published values in treating n-hexane alone, with surfactant or in mixture with benzene either at pH 4 or 7. This implies that optimizing the introduction of methanol or its alternate use in TBAB could be a promising means for advance oxidation of hydrophobic odours and hence fulfilling the regulatory requirements.

Fly ash treatment and metal recovery

H. van den Berg[1], K.A. Kruders[2], A.G.J. van der Ham[1], H.P.M. Oudenhoven[3]
[1]Faculty of Science and Technology, University of Twente, P.O. Box 217, 7500 AE Enschede, The Netherlands, h.vandenberg@utwente.nl
[2]Zeton B.V., P.O. Box 9, 7500 AA Enschede, The Netherlands
[3]Twence B.V., P.O. Box 870, 7550 AW Hengelo, The Netherlands

Metal recovery from fly ash will contribute to a circular economy. Since several years recovery of metals in the form of metal hydroxides has been practiced in Switzerland in a number of plants (1). Since a few years the recovery of pure zinc is in operation (2). Our project deals with the investigation of the process steps leaching and sedimentation to prepare the application of the recovery processes in the Netherlands. Next to recovering the metals, an important economic goal is to clean the fly ash so it can be disposed of as inert or non-hazardous waste.

One of the first steps currently used to obtain the metals from the fly ash is solid-liquid extraction. Acidic scrub water resulting from wet flue gas cleaning is used for this. Pourbaix diagrams can assist in the prediction of the most suitable conditions for extraction. Pourbaix diagrams give a graphical presentation of the most stable thermodynamic state of solids in an aqueous environment for a potential range and pH range. Acidic conditions favor leaching of certain metals whereas additional oxidative conditions offer an even higher extraction rate. For pure zinc, a Pourbaix diagram predicts that the zinc is extractable in an acidic solution. However, the zinc can be found in many forms in the fly ash. One of those forms is K_2ZnCl_4 which actually can be extracted. Some zinc silicates seem to be not extractable. Therefore, the exact extraction ratio cannot be predicted for fly ashes produced in incinerators. Pourbaix diagrams provide only a partial understanding of the leaching mechanism. Nevertheless, they indicate feasible conditions for extraction. Furthermore, together with electrochemistry basics, they offer insight into the precipitation behavior when the metals are actually in solution and the alkalinity is increased. They provide also at least indications for process operation.
Lime water can be used to precipitate $CaSO_4.2H_20$ and $Fe(OH)_3$.
Secondly, we work on simulation of the solid-liquid extraction using the ESP OLI program. Also using this tool, we cannot unambiguous predict in which form the metals are present. E.g. some actually detected components cannot be predicted by the simulation program. A zinc chloride component is known in the component database of OLI, but K_2ZnCl_4, however, is not. It does, nonetheless, simulate precipitation of already dissolved metals satisfyingly.
We will show — in as much — how the different tools can give contributions to the understanding of the mechanism of metal leaching from fly ash.
Leaching of metals by washing of fly ash by acidic scrub water has to be improved before the residual sediment can be classified as "non-hazardous".
An economic evaluation shows that the project is close to break-even. Sensitivities have been determined for e.g. disposal costs, zinc metal recovery fraction.
Electrochemistry forms the basis for this urban mining project.

Reference 1. A. Bühler and S. Schlumberger, "Recovering heavy metals from filter ash «acidic filter ash leaching - fluwa process» a future-oriented process in waste incineration," BSH Umweltservice AG, Sursee, Switzerland.

FlyashChisa2016

Reference 2. S. Schlumberger, " New technologies and options for the treatment of flue gas cleaning residues to promote sustainable resource management", BSH Umweltservice AG, Sursee, Switzerland.

Application of water vapour for reactivation of limestones in carbonate looping technology

M. Staf, K. Ciahotný
UCT Prague, Department of Gas, Coke and Air Protection, Technická 5, 166 28 Prague 6 – Dejvice;
tel. +420 220444458, e-mail: marek.staf@vscht.cz; fax +420 220445010.

High temperature carbonate looping technology represents one of the promising methods for capturing carbon dioxide from flue gases, especially in power industry. Its principle is reaction of gaseous CO_2 with solid calcium oxide at the temperatures in range of 870 – 970 K. This reaction is reversible, where decarbonation (calcination) is achievable by reaching the temperature to 1200 K. While the calcination is endothermic carbonation process is exothermic with reaction enthalpy -178.2 $kJ.mol^{-1}$ CO_2.

This heat can be technically utilised, which makes the overall energetic balance of the process more economical. Besides the raw materials on the basis of calcium carbonate there were also different materials tested by numerous researchers. All these materials, including especially raw or modified lithium zirconates, allow reversible chemical sorption of carbon dioxide at elevated temperatures. As regards design of the CO_2 capturing unit there are also other systems similar to the carbonate looping and working on the principle of adsorption onto active surface of suitable solid substrates. Among such adsorbents being tested on a long-term basis it is especially hydrotalcite-like compounds (HTlcs).

Each of the above mentioned materials has its advantages, but it also shows certain weaknesses, which justify further research. The fundamental problem of calcium carbonate, and materials derived from it, is a gradual decline of the initial sorption capacity. This undesired effect is caused by the high temperature sintering of particles and also by blocking of pores due to structural changes. Decrease of the initial capacity during cyclic repeating of calcination and carbonation leads to devaluation of the material. The main disadvantage of zirconate-based sorbents and HTlcs is their purchase cost. Moreover zirconates in their initial form do not offer sorption capacity comparable to calcium oxide.

The study hereby presented is aimed at utilisation of the natural limestones for the technology of post combustion capture of CO_2, based on carbonate looping. Czech Republic disposes with relatively big exploitable quantities of the limestones, which some of them have been already successfully used in the desulfurization technology in the coal-fired power plants since 1990s. Uncomplicated availability and the related low financial costs were the main arguments why natural limestones were considered preferentially for CO_2 capture instead of artificial or modified sorbents.

The main objective of this study is research of possibilities of re-activation of the sintering devalued limestone using water vapour. Within the experiments a comparison of the sorption capacities of several selected natural limestones with representatives of HTlcs and lithium zirconate was realised. Basic characterisation of all samples was achieved using following set of analytical methods: measurement of BET surface, XRF analysis of the elemental composition, XRD analysis and thermogravimetric analysis.

The experiments with cyclic CO_2 sorption followed by decarbonation were carried out in a special laboratory apparatus, based on a vertical quartz reactor, allowing testing about 90 - 100 g of the sample at temperatures up to 1273 K. Temperature was measured using a thermocouple Ni-CrNi, placed in the reactor axis and in the middle of the layer of the sample. Data were stored continuously during whole experiment. Gas was introduced at the bottom of the reactor and after passing the sample it exited the top of the reactor and after being cooled down to the laboratory temperature it flowed into the IR spectrometer, which provided online analysis of the outlet CO_2 concentration. All tests were carried out at the atmospheric pressure inside the apparatus. The gas flow was set up and measured using mass flow meter/controller Bronkhorst EL-FLOW Select installed before the gas inlet into the heated space of the reactor. For the purpose of verification of the gas flow measurements the wet drum gas meter was installed at the end of the whole apparatus. Two different gas mixtures simulated desulfurised and ash-free flue gas. The molar composition of these gases was 14 % CO_2 in nitrogen and ternary mixture of 12 % CO_2 and 12 % O_2 in nitrogen respectively.

Water steam regeneration unit consisted of a heating mantle with glass flask, where distilled water was boiled. Generated vapour was stripped using nitrogen flow, which outlet humidity and temperature was measured. Nitrogen enriched by water vapour was transported through electrically heated pipe into the sample layer.

First series of the measurements was intended for verification the changes in sorption capacities occurring when all cycles were realised without activation of the water steam unit. Calcination was conducted with heating rate 10 K/min until reaching final temperature 1273 K whereas carbonation was always realised under isothermal conditions, concretely at the temperature 923 K. Within the whole sample base the average sorption capacity after the first cycle was 26 g (CO_2)/100 g (calculated for initial limestone). After 10 cycles the capacity decreased to 8.5 g/100 g in average, which represents relative decrease by 67.3 %.

The principle of steam regeneration is as follows. After calcination all calcium carbonate is converted to CaO. Calcium oxide reacts with the water molecules strongly exothermically to form $Ca(OH)_2$. Within the hydration process the sintered particles are disintegrated, which leads primarily to increase of the BET surface, resulting in improved CO_2 sorption kinetics and also technically useful sorption capacity. The hydration process should be performed in each cycle, after completion of the calcination stage. Realisation of the experimental hydration in the above described apparatus was following. After termination of each calcination phase the reactor was cooled down to the temperature 375 – 378 K. Subsequently the nitrogen stream was switched on and enriched by water vapour to 95 % relative humidity at the given temperature. After reaching thorough saturation of the tested sample the water vapour generation was stopped and the reactor was heated up until reaching the carbonation temperature. Then the sorption phase of the cycle started. Experiments with water saturation at higher sample temperatures were realised in the same way.

Since the dehydration temperature of $Ca(OH)_2$ is 785 K this value should never be reached during regeneration stage. Within this study there were numerous tests performed showing the quantity of water vapour as well as reaction times, which are necessary for achieving maximum effect of sorption properties regeneration. Generally it can be stated that this method of regeneration is able to improve sorption capacities in each cycle by 4 – 6 g/100 g.

Within the experiments also samples of hydrotalcite and Li_2ZrO_3 were tested in the above described apparatus. After 10 cycles the achievable sorption capacity was approximately half if compared with the average value of the natural limestones. However, both HTlcs and modified zirconates have still big research potential, which may lead to better performance in the future.

Acknowledgements: This research was supported by grant from Norway, project number: NF-CZ08-OV-1-005-2015.

Effective absorption of nitrogen oxides in water using new absorption equipment having glass fiber filter and its absorption mechanism

Daiyasu A.[1], Yamauchi S.[2], Taniguti S.[3], Asano H.[3], Asano M.[3], Ogino H[1]., Yasuda M[1].
*[1]Osaka Prefecture Univ., Department of Chemical Engineering, 1-1 Gakuen-cho, Naka-ku, Sakai, 5998531Osaka, Japan; [2] Tada Electric co. LTD, 488 Shimokasaka, Oku-cho, Setouchi, Okayama 7014247, Japan; [3] Pollution Prevention Apparatus Laboratory Co., Ltd., Reosu otoshi, 12-13, Otoshi-cho, Neyagawa, Osaka 5720048, Japan
tel. +81-722-254-9299, e-mail: yasuda@chemeng.osakahu-u.ac.jp;

Introduction

Automobile and chemical plant emissions of nitrogen oxides (NOx) is one of the result in photochemical smog, air pollution and acid rain. Therefore, NOx emission regulation are adopted by many governments and countries. Economical and high efficient NOx emission have not been still complete while the treatment technique of other emission gases have been established. These are because water solubility of nitrogen oxide is quite low and ordinary NOx removal equipment requires expensive catalyst or produces a huge amount of chemical waste, and so on.

We propose a smart and efficient NOx elimination process in which nitrogen oxides absorb in water using new absorption equipment having glass fiber filter. When only water was used as an absorbent, a good NOx removal efficiency comparing ordinary common packing materials was attained [1]. Using our absorption equipment, 95-98% of NOx (from 10,000 ppm to 400,000ppm) can absorb in water and absorbent water becomes up to 30wt% of HNO_3. When the mass balance of NOx in gas and nitric acid in water was taken, the detected amount of nitric acid in water is quite small as compared with the decreased amount of NOx in gas. Furthermore, even if concentration of HNO_3 become high, NOx removing efficiency doesn't decrease.

To clarify this phenomenon, we measured the residence time distribution of water and gas in the absorption equipment. A schematic illustration of gas and liquid flows in the NO_X absorption tower is shown in Fig.1.

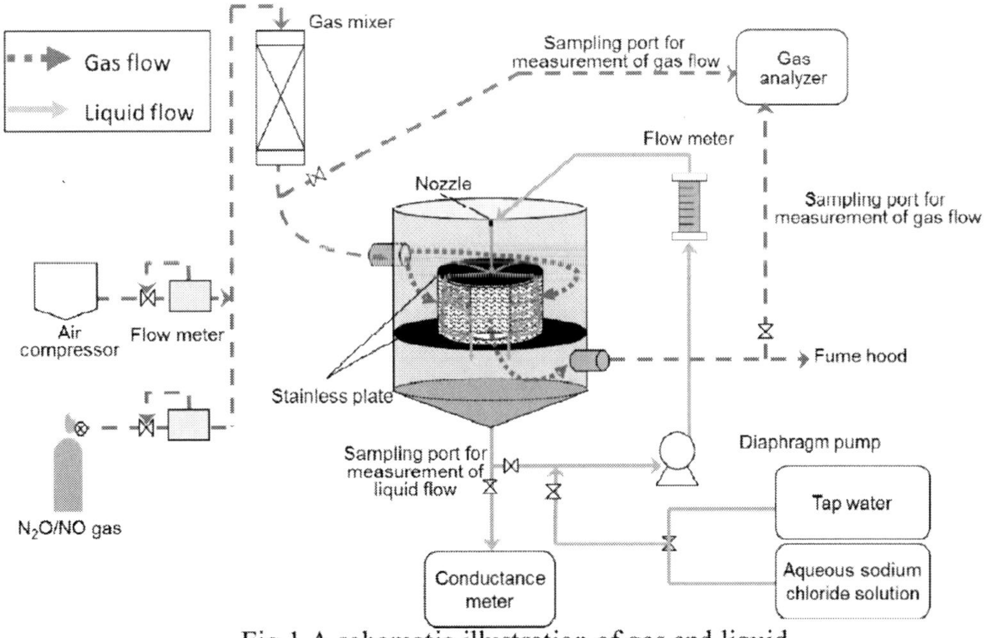

Fig.1 A schematic illustration of gas and liquid flows in the NO_X absorption tower

Experimental method

Experiment were conducted by step response method. We measure the residence time distribution of water by using 0.01 mol·L^{-1} aqueous sodium chloride solution as a tracer and the residence time distribution of gas by using nitrous oxide as a tracer.

Results and discussion

The result of experiment about residence time distribution of water is shown in Figs.

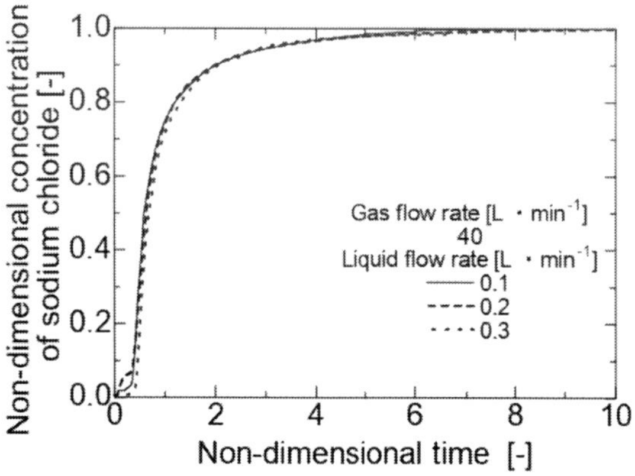

Fig.2 Effect of liquid flow rate on residence time distrubution of water

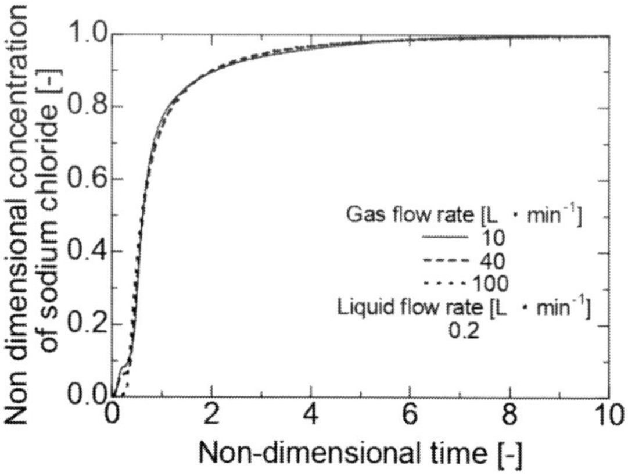

Fig.3 Effect of gas flow rate on residence time distrubution of water

We calculated non-dimensional time by dividing time by average residence time. We also calculated non-dimensional concentration of sodium chloride by dividing concentration of sodium chloride in outlet by concentration of sodium chloride in tracer. As shown in Figs.1, non-dimensional residence time distribution of liquid doesn't change depending on liquid flow rate and gas flow rate. It is seen same trend when gas flow rate fixed 10 or 100 L · min^{-1} and water flow rate fixed 0.1 or 0.3 L · min^{-1}. To verify this result, we weighed the glass fiber filter when it took 2 days after starting operation. Under every operating condition, weight of the filter is 22.5 kg. As a result, weight of the filter didn't changed depending on liquid and gas flow rate.

Conclusion

In this study, we measured the residence time distribution of water and gas in order to clarify the phenomenon that when the mass balance of NOx in gas and nitric acid in water was taken, the detected amount of nitric acid in water is quite small as compared with the decreased amount of NOx in gas. Furthermore, even if concentration of HNO_3 become high, NOx removing efficiency doesn't decrease. The result shows residence time distribution and weight of the filter doesn't change depending on liquid and gas flow rate. This result means liquid phase ratio in glass fiber filter is a constant and its ratio is over 70 %. Since liquid phase ratio in the filter is quite large, it takes so long time that all of water in the filter replace HNO_3.

Moreover, we measure NOx absorption efficiency by 200 ppm of NOx (about 90 % of nitrogen monoxide). At the conference, change of HNO_3 concentration by time, HNO_3 concentration ratio, and the relationship between this absorption equipment's operating condition and pressure drop in this equipment will be introduced.

Reference

[1]Yasuda et al., *Environ.Sci.Technol.*, 2011. **45**(5), pp1840-1846

A new approach of electrode modification for enhancement capacitive deionization (CDI) cell

P. Shahmohammadi[a,b], S. Rowshanzamir[a,b]
[a] School of Chemical Engineering, Iran University of Science and Technology, Narmak, Tehran 16846-13114, Iran
[b] Fuel Cell Laboratory, Green Research Centre, Iran University of Science and Technology, Tehran, Iran. Tel.: 98 2177491223; fax: 98217491242.
E-mail address: rowshanzamir@iust.ac.ir (S. Rowshanzamir).

According to the growing world population and limited fresh water reserves in the world, the quicker and cheaper desalination of sea-water in order to provide any water consumption has been importantly highlighted. Capacitive deionization technology (CDI) is the new promising process for desalination of brackish water [1]. As can be seen from Fig. 1 low cost estimation associated with CDI rather that other conventional methods such as stage flash distillation (MSF), multi effect distillation, vapour compression distillation (VCD), membrane process such as reverse osmosis, makes this technology as an optimistic method to meet fresh water. In this method the salt ions in brackish water is removed by passing through between porous electrodes. In fact, by applying voltage difference between electrodes the negatively and positively charged ions move towards the desired electrodes, assemble in electric double layer and temporarily remain there without any movement. Under such condition crossing brackish water convert to water without and salt.

Several effective structure of material such as the composite of activated carbon cloth [2] and carbon-fiber veil has been used widely that has been improved performance of such like systems in many cases. Nevertheless, surface non-uniformity, low surface area, presence of metal catalyst [3] and pore size mostly affect problems of clogging and demise of microorganism in long term operation. As regards good conductivity and electrosorption to the surface plays an important role on CDI. Hence, research trends are concentrated to find new structures, which have good conductivity and low resistance for transferring of electron, high-quality toughness and excellent activity. So electrode structure and materials which used in their framework play a crucial role in performance of such like system. Hence, in order to enhance electrical conductivity and provision of compatible bed, a complex electrode structure based on stainless steel-304 was prepared.

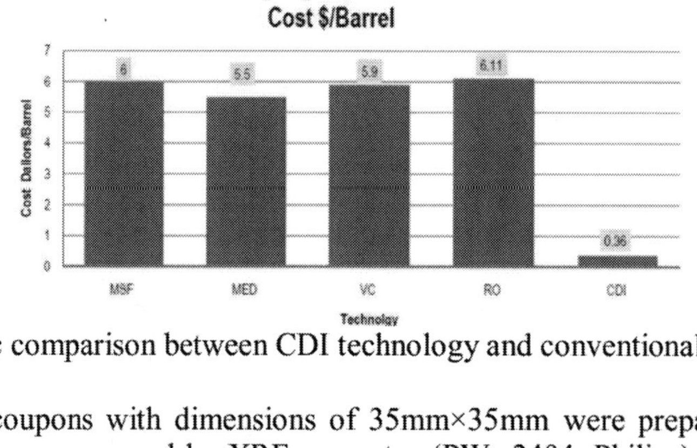

Fig. 1 Economic comparison between CDI technology and conventional method

Stainless steel coupons with dimensions of 35mm×35mm were prepared. The percent weight composition was measured by XRF apparatus (PW- 2404, Philips) that was 12.6745 Cr, 1.375 Mn, 66.373 Fe, 10.609 Ni, and 8.968 Cu. The coupons (pore size 250 μm) were rinsed with acetone 100% in order to eliminate any grease on it. Then, composite of polyvinyl alcohol (10% w) + activated carbon + Deionized water (35% v/v) was casted onto the electrode, then the electrodes put into oven at 120°C 2h for drying and vacuum oven at 80°C for 4h in order to vaporize organic solvent as well. Fig 2 presents metallography test for two sides of SsAc electrode coupon (A, A′ and B, B′ are the upper and down surface respectively) and carbon paper (CP), F and graphite rode GR (G)). Uniformity and ordered meso-porous arrangement has been clearly found out for SsAc rather than others.

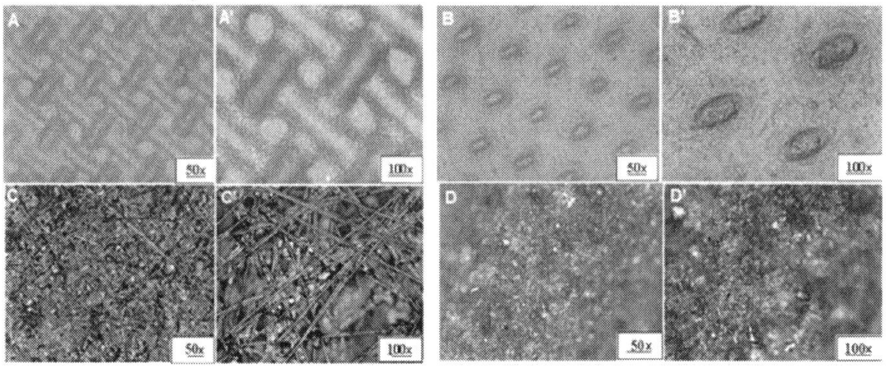

Fig. 2 Metalography of the surface of the SsCG electrode; **A, A′)** outer side and, **B, B′)** inner side. **C, C′** and **D, D′** respresent the metalography of the surfaces of the CP and GR electrodes

Two parallel probes were performed for electronically assessment of electrodes. As it is shown in **table 1**, electrical conductivity of SsAc is nearly ~2 times greater than activated carbon cloth. Stainless-steel 304 as a base of the electrode which it has the superior capability to facilitate electron transfer because of high conductivity ($\sigma = 1.38 \times 10^4$ S cm^{-1}) has a decisive effect on it. Moreover, a linear increase in conductivity was observed up to a certain amount of active carbon dose of coating. The lower resistance of the electrode reduces the internal resistance of the cell, one of the major limits in electron transfer across the load.

Conductivity (S cm^{-1})	Electrode
0.1 – 0.15	Stainless steel-304 plus Activated carbon (SsAc)
0.075	Carbon Paper (CP)
0.055	Graphite Rod (GR)
0.070[4]	Activated carbon cloth

CDI further improvements in the near future towards a comprehensive technology will be beyond water desalination usage alone. Many applications such as lithium from sea water, phosphate ions from agricultural sources will provide valuable metallurgical operations which might be considered as an alternative advantage of this technology.

[1] K. Laxman, M.T.Z. Myint, H. Bourdoucen, J. Dutta, Enhancement in ion adsorption rate and desalination efficiency in a capacitive deionization cell through improved electric field distribution using electrodes composed of activated carbon cloth coated with zinc oxide nanorods, ACS Appl. Mater. Interfaces 6 (2014) 10113–10120.
[2] C. Yan, L. Zou, R. Short, Polyaniline-modified activated carbon electrodes for capacitive deionisation, Desalination 333 (2014) 101–106.
[3] S. Porada R. Zhao A. van der Wal V. Review on the science and technology of water desalination by capacitive deionization. Progress in Materials Science 58 (2013) 1388–1442.
[4] Zhao F, Rahunen N, Varcoe J, Roberts A, Avignone-Rossa C, Thumser A, and Slade R. Factors affecting the performance of microbial fuel cells for sulfur pollutants removal. Biosens Bioelectron 2009;24:1931–1936.

A process simulation model for anaerobic bioscrubber using Aspen Plus

D.Bravo, F.J. Álvarez-Hornos, J.M. Penya-roja and C.Gabaldón
Univ. Valencia, Dep. Chemical Eng., Avd. de la Universitat, s/n, 46100 Burjassot, Valencia, Spain; tel.+34963543437, e-mail: daniel.bravo@uv.es

The emission of tropospheric ozone precursors such as Volatile Organic Compounds (VOCs) is a public health concern worldwide (Genc et al., 2012) being the flexographic industry an important sources of VOC. The use of solvent inks in flexographic companies is increasing in the last decades in Europe. Air emissions in this facilities are characterized by high air flow rates and low VOC concentrations of high-water soluble compounds (Rothenbuhler et al., 1995; Sempere et al., 2012), thus, bioscrubbing could be a feasible biological alternative versus conventional physico-chemical processes. In this sense, the anaerobic bioscrubber technology is a novel option where the scrubbing liquid is regenerated anaerobically. The development of a model for the anaerobic bioscrubber can provide a useful tool for researchers and/or industries in order to design the system and/or predict the influence of the operational parameters in the performance of the process.

The process simulation model has been developed in Aspen Plus® v8.0, which implement meticulous thermodynamic and rigorous property methods calculations and a wide range of unit operations (Rajendran et al., 2014). The model has been linked with Matlab® vR2015a and calibrated by using experimental data obtained in an industrial anaerobic bioscrubber prototype. The experimental set-up of 18 m^3 of water volume was tested at a real flexograpohic facility (Altacel Transparant Verpakkingsindustrie B.V, The Netherlands). The prototype (BONCUS^{TM}) was supplied by Pure Air Solutions and mainly consisted in a packed scrubber, a high-rate anaerobic granular sludge bed reactor and a buffer tank. The unit was connected to the exhaust gases of the facility and operated during two years to demonstrate the proof of concept of the technology. The system was provided with instrumentation that continuously monitored the inlet and outlet VOC concentration in the gas phase, temperature, pressure drop in the scrubber and the pH and the biogas production of the anaerobic reactor. Chemical Oxygen Demand (COD), Volatile Fatty Acids (VFA) and nutrient concentration (N and P) in the effluent water of the anaerobic reactor were weekly measured.

The scrubber was simulated using the thermodynamic model Peng-Robinson with Wong-Sandler mixing rules (PRWS) and the RadFrac Aspen Model. The experimental data used for the calibration of the scrubber were the water and air flow rate, total VOC concentration in the inlet and outlet stream and the average inlet composition measured by GC-MS (65.5 % of ethanol, 25.3% of ethyl acetate and 9.2% of 1-ethoxy-2-propanol). The number of stages for each compound were calibrated (specifying the individual Murphee efficiencies) comparing the individual removal efficiency per component with its pertinent experimental value. Equations correlating the liquid-to-gas flow ratio with the High Equivalent to a Theoretical Plate (HETP) was obtained per compound. The HETP for ethyl acetate ranged in 1.5 and 2.5 m for liquid-to-gas flow ratios between $3.5 \cdot 10^{-3}$ and $9.1 \cdot 10^{-3}$. The HETP for ethanol and 1-ethoxy-2-propanol were less variable (around 0.78 m for both compounds), showing that soluble compounds are less influenced by the air and liquid loads. This observation could be explained by the very high solubility of this compounds whose Henry's law constant in the case of ethanol is $5.2 \cdot 10^{-9}$ atm L mol^{-1} at 20°C, same results were obtained by other researchers Le Cloriec et al., (2001).

The anaerobic reactor and the gas liquid separator (GLS) were simulated with rigorous continuous stirred tank reactor and Flash2 Aspen Model, respectively, using electrolyte non-random two-liquid (ELECNRTL) as a thermodynamical model. Monod kinetic equations were used for the acetogenesis (solvents to acetate) and methanogenis steps and were introduced in Aspen using Langmuir-Hinshelwood Hougen-Watson (LHHW) reaction class (same mathematical expression as Monod model). The model input was the composition, temperature and pH of the water coming from the scrubber model. The volume of the reactor was set as the volume of the granular sludge bed of the industrial prototype. The biomass concentration (20 kg_{VSS} m^{-3}) and yield biomass coefficient (0.1 mg_{VSS} mg_{COD}^{-1} for acetogenesis and 0.023 mg_{VSS} mg_{COD}^{-1} for methanogenesis) were taken from literature (Tchobanoglous et al., 2003). From experimental data, calibration of mean parameters is being developed. The preliminary results of these parameters were: $\mu_{ethanol}$ and $\mu_{ethyl\ acetate}$ ($\sim$1.5 d^{-1}), $\mu_{1\text{-}ethoxy\text{-}2\text{-}propanol}$ ($\sim$0.1 d^{-1}), $\mu_{methanogenesis}$ ($\sim$0.03 d^{-1}), Ks for ethanol and ethyl acetate ($\sim$150 g_{COD} m^{-3}), Ks for 1-ethoxy-2-propanol ($\sim$1500 g_{COD} m^{-3}) and Ks for methanogenesis ($\sim$160 g_{COD} m^{-3}).

Process simulation model for anaerobic bioscrubber was developed with Aspen Plus, being able to estimate the experimental data accurately. This model will be used as a design tool and will be able to predict the influence of the operational parameters in the performance of the process.

Acknowledgements:

The research leading to these results has received funding from the People Programme (Marie Curie Actions-ITN) of the European Union's Seventh Framework Programme FP7/2007-2013/ under REA grant agreement n°606942 (TrainonSEC). Financial support from Ministerio de Economía y Competitividad (Spain, Project CTM2014-54517-R co-financed with FEDER funds) is acknowledged. Financial support from Generalitat Valenciana (Spain, Prometeo project 2013/53) is also acknowledged. The authors would like to give special thanks to Wim Schaap (Heineken, The Netherlands) for providing the granular sludge.

References:

Genc, S., Zadeoglulari, Z., Fuss, S. H. and Genc, K. (2012). The adverse effects of air pollution on the nervous system. *Journal of Toxicology*, *2012*, 782462.

Le Cloriec, P., Humeau, P. and Ramirez-Lopez, E. M. (2001). Biotreatments of Odours Control and performances of a biofilter and a bioscrubber.pdf. *Water Science and Technology : A Journal of the International Association on Water Pollution Research*, *44*(9), 219–226.

Rajendran, K., Kankanala, H. R., Lundin, M. and Taherzadeh, M. J. (2014). A novel process simulation model (PSM) for anaerobic digestion using Aspen Plus. *Bioresource Technology*, *168*, 7 13.

Rothenbuhler, M., Heitz, M., Beerli, M. and Marcos, B. (1995). Biofiltration of organic volatile organic emissions in reference to flexographic printing processes. *Water, Air, and Soil Pollution*, *83*(1), 37–50.

Sempere, F., Martínez-Soria, V., Penya-Roja, J. M., Waalkens, a and Gabaldón, C. (2012). Control of VOC emissions from a flexographic printing facility using an industrial biotrickling filter. *Water Science and Technology : A Journal of the International Association on Water Pollution Research*, *65*(1), 177–82.

Tchobanoglous, G., L.Burton, F. and Stensel, H. D. (2003). *Wastewater Engineering: Treatment and reuse.* (I. Metcalf & Eddy, Ed.) (4th Editio.). Mc Graw Hill.

A study on removal of reactive blue 19 from aqueous solutions by using potato shell powder

E. Demirhan, B. Ozbek, A.E. Gemikoz,
Yildiz Technical University, Department of Chemical Engineering, Davutpasa Campus, 34210, Esenler/Istanbul, Turkey;
Tel. +90 212 383 4769, fax: +90 212 383 4725;
e-mail: demirhan@yildiz.edu.tr

Dyes are widely used in textiles, paper, rubber, plastics, leather, cosmetics, pharmaceutical and food industries in order to colour the products. Basic and reactive dyes are extensively used in the textile industry because of their favorable characteristics of bright color, high solubility. Considerable amounts of dyes can cause pollution problems in the form of colored wastewater discharged into environment. So, the removal of dyestuffs has received considerable attention in the textile industry in order to find appropriate treatment technologies. Among the methods of dye removal, adsorption is one of the preferred techniques because of its easiness and the availability of a wide range of adsorbents [1,2].

Activated carbon due to its extended surface area is commonly used as adsorbent for dye removal; but it's hard to regenerate and it's also expensive, resulting in high costs. Thus, in order to decrease the cost of treatment, some attempts have been made to find low cost alternative adsorbents. For this purpose, in this present study, potato shell powder which is an agricultural residue was used to investigate the adsorption performance on reactive blue 19 removal. The effects of pH, temperature, and adsorbent concentration were examined. Moreover, the most suitable adsorption kinetic models were examined according to the data obtained from experiments.

Keywords: Potato shell, reactive dye removal, adsorption, modelling, kinetic

REFERENCES

[1] Sharma, P., Kaur, R., Baskar, C., Chung, W.J., (2010), Removal of methylene blue from aqueous waste using rice husk and rice husk ash, Desalination 259, 249–257.
[2] Hao, G.P., Li, W.C., Wang, S., Zhang, S., Lu, A.L, (2010), Tubular structured ordered mesoporous carbon as an efficient sorbent for the removal of dyes from aqueous solutions, Carbon, 48: 330-3339.

Abatement of toluene in air by non-thermal plasma: Effects of plasma reactor types and added catalysts

Siwada Deepracha[a], Sumaeth Chavadej[*,a,b], Thitiporn Suttikul [c]
[a] Chulalongkorn Univ., The Petroleum and Petrochemical College, Phayathai Rd., 10330 Bangkok, Thailand; tel. +66 22184139, fax. +66 22184139, e-mail: sumaeth.c@chula.ac.th;
[b] Center of Excellence on Petrochemical and Advanced Materials Chulalongkorn University, Phayathai Rd., 10330 Bangkok, Thailand; tel. +66 22184141, fax. +66 26117619.
[c] King Mongkut's University of Technology North Bangkok., Department of Chemical Process Engineering Technology, Tambon Nonglalok, Amphur Bankhai., 21120 Rayong, Thailand; tel. +66 25552000, fax. +66 25874350.

In this work, the oxidation of toluene in various low-temperature plasma reactors including corona discharge, parallel dielectric barrier discharge (DBD), cylindrical DBD and gliding arc discharge was investigated. The effects of operational parameters of applied voltage, input frequency and residence time were investigated. The parallel DBD reactor showed the best performance in terms of the highest toluene conversion of 92% and the highest CO_2 selectivity of 55.2% with a CO sectivity of 1.5% when compared with the other reactors. However, the corona discharge reactor was selected to investigate the effects of added catalysts (Mn, Cu and Ag) because it gave the lowest power consumption and the shortest residence time. Another reason was that the corona discharge was the only plasma reactor could be loaded and operated with added catalysts. The 3 wt% Mn/TiO_2 calcined 500°c was the best catalyst to provide the highest toluene conversion (97.4%) and the highest CO_2 selectivity (55.1%) with the lowest CO selectivity of 0.8%.

Adsorption of oxytetracyclin on orange peels

F. Mohellebi, [1]F. Kies, I. Ait Amer Meziane
Laboratoire de Valorisation des Energies Fossiles, Departement Génie Chimique, Ecole Nationale Polytechnique, 10 Avenue Hassen BADI, El-Harrach, Alger, Algérie;
tel. +21356132 11 69, e-mail: ferroudja.iddir@g.enp.edu.dz
tel.+ 213555 97 14 46, e-mail: fairouz.kies@g.enp.edu.dz

It is impossible to imagine our society without effective modern drugs. They help us to prevent and treat disease. Nowadays, large quantities of various pharmaceutical active substances are manufactured to protect humans and animals. Following the improvement of medical care, longer life expectancy and progressive industrialisation of agriculture, the amounts of drugs consumed have increased. However, in many cases, these products are not fully absorbed and metabolised by patients and are partially excreted. So, traces of these products join the water cycle. With significant advances in chemical analysis technology, it is now possible to measure many pharmaceutical residues in water at extremely low concentrations, often several times lower than those that were possible to measure a few years ago. Indeed, studies have highlighted the presence of more than 80 pharmaceutical products in water. We do not know exactly what effects these residues have on the aquatic environment for example on microorganisms. Which implications these drugs have on the food chain, on the biodiversity and on the entire ecosystem, is a frequently cited issue.

At first glance, it would seem reasonable to completely avoid the rejection of pharmaceutical residues in order to protect our water system. The abandonment of production and medication is not a realistic or desirable scenario. Another option is to take technical measures to treat contaminated water.

In order to achieve the removal of a pharmaceutical micro pollutant (oxytetracyclin) from the aquatic environment, its adsorption was conducted on orange peels, a food waste largely abundant in Algeria. Our study has three main objectives: reduce pollution levels and in the same time manage and valorise an agricultural waste.

Adsorption of oxytetracyclin on grinded orange peels (particle size of 100 μm) was studied in batch reactor. This study investigated the influence of various parameters on the oxytetracyclin removal efficiency. These include contact time, initial concentration of oxytetracyclin, mass and particle size of orange peel powder and pH of the medium. Results show that for a solid/liquid ratio of 8 g/L, the removal efficiencies of oxytetracyclin are of 100% and 94 % for antibiotic concentrations of 5 and 10 ppm, respectively. The adsorption kinetics are well described by the pseudo second-order model. The adsorption isotherms obtained are in accordance with the Langmuir model. In addition, the adsorbed amounts decrease with increasing particle size and the adsorption of oxytetracyclin is favoured in the acidic reaction medium, which is the natural pH of orange peels in solution.

Keywords: Antibiotics, Oxytetracyclin, Orange peels, Toxicology, Adsorption, Water pollution control.

Advantages of the two-compartment reactor for biodiesel production with ultrasound irradiation

R. Nakayama[a †], M. Imai[a]*, J.M Woodley[b]
[a] Course in Bioresource Utilization Sciences, Graduate School of Bioresource Sciences, Nihon University, 1866 Kameino, Fujisawa, Kanagawa. 252-0880, Japan
[b]CAPEC-PROCESS Research Center, Department of Chemical and Biochemical Engineering, Technical University of Denmark, DK-2800 Kgs. Lyngby, Denmark

*Phone /Fax: +81 466 84 3978, E-mail: XLT05104@nifty.com

Biodiesel (fatty acid methyl ester: FAME) was successfully produced using a two-compartment reactor with ultrasound irradiation. The reactor was composed of a mechanically stirred compartment (ST) and ultrasound irradiation compartment (US). The reaction solution was circulated between the ST and the US. The enzyme was exposed to ultrasound irradiation solely in the US. In the ST, the reaction solution was mixed without ultrasound irradiation. Enzymatic transesterification to produce FAME from rapeseed oil using Callera Trans L™ (a liquid formulation of a modified *Thermomyces lanuginosus* lipase) was enhanced by the aid of ultrasound irradiation. The reaction rate was evidently increased. A shorter mean residence time in US and higher ultrasound power performed higher initial reaction rate of FAME production.

† Present Address :
Department of Environmental Chemistry and Chemical Engineering, School of Advanced Engineering, Kogakuin University, 2665-1 Nakano-machi, Hachioji, Tokyo, 192-0015, Japan

Aerobic sequencing batch reactor for the treatment of industrial wastewater from the brewery

[2]K. P. Shabangu, [1]Dr B.F. Bakare, [2]Dr. M. Chetty
[1]Faculty of Engineering, Department of Chemical Engineering, Mangosuthu University of Technology, P.O. Box 12363 Jacobs, Durban 4026, South Africa, Email: shabanguk@mut.ac.za ;Tel: +27 31 819 9288; Fax: +27 31 907 7301
[2]Faculty of Engineering, Department of Chemical Engineering, Durban University of Technology, P.O. Box 1334, Durban,4000, South Africa.

One of the major effects of social and economic change due to industrialization is the generation of industrial wastewater which requires treatment before being released into the environment. Previously, sequencing batch reactor (SBR) has been engaged as an efficient technology for wastewater treatment, principally for domestic wastewater because of its simple configuration and high treatment efficiency in terms of organic compounds and solid removal. In this project a laboratory scale sequencing batch reactor was constructed and used for the treatment of wastewater generated from a brewery process.

The overall objective of the project was to investigate operating conditions for a laboratory scale SBR for treating brewery wastewater and to determine the effects of continuous low oxygen concentration versus cyclic aeration schemes on SBR performance. It was envisaged that characterization of the following parameters had to be implemented, pH, conductivity, COD, BOD_5, ammonia, solids to analyze the performance of the SBR.A series of laboratory analysis were carried out in Mangosuthu University of Technology research laboratory. All analysis was based on standards methods (Association, 1998).

The data showed distinctive characteristics of brewery wastewater which coincided with previously done research studies on characterization of brewery wastewater. To mention a few crucial parameters; COD value was: 13 209 ± 6583 mg/l, BOD_5: 3017 ± 2713 mg/l, l and turbidity were around 390 ± 84 NTU. This data clearly verifies that brewery wastewater had high strengths organics that required a biological treatment for efficient solids removal and organics strength stabilization. CON SBR1 had effectively presented a great aptitude to attain prime treatment of COD and BOD_5 with 5 days yielding above 90% COD removal and an average of 85 % BOD_5 stabilization as well agreeing to previously done work from other authors (Huige, 1995, Irvine et al., 1989, Ling, 1998). The Total dissolved solids and total suspended solids displayed an average percentage removal of about 60% for both CON SBR1 and CYC SBR2.CON SBR1 can confidently be engaged for proficient brewery waste water treatment according to the obtained results in this study. This reactor occurred sustainable on COD and BOD_5 stabilization in most batches as compared to the CYC SBR2. Convincingly a continuous low oxygen dosing concentrations favors a high activity rate of the microorganism for the optimum treatment of brewery waste water.

Keywords: Brewery Wastewater, Chemical Oxygen Demand, 5-days Biological Oxygen Demand, Sequencing Batch Reactor, Hydraulic Retention Time, Continuous (CON), Cyclic (CYC)

ASSOCIATION, A. P. H. 1998. APHA. 1998. *Standard methods for the examination of water and wastewater,* 20.

HUIGE, N. 1995. Brewery by-products and effluents. *FOOD SCIENCE AND TECHNOLOGY-NEW YORK-MARCEL DEKKER-*, 501-501.

IRVINE, R. L., KETCHUM JR, L. H. & ASANO, T. 1989. Sequencing batch reactors for biological wastewater treatment. *Critical Reviews in Environmental Science and Technology,* 18, 255-294.

LING, L. 1998. *Brewery Wastewater Treatment Using Aerobic Sequencing Batch Reactors with Mixed Culture Activated Sludge.* University of British Columbia.

Available techniques applicable for efficient removal of antibacterial agent Triclosan from water

J. Pérko, T. Weidlich
Institute of Environmental and Chemical Engineering, Faculty of Chemical Technology, University of Pardubice, Studentská 95, 532 10 Pardubice, Czech Republic; e-mail: jan.perko@upce.cz.

Triclosan (2,4,4'-trichloro-2'-hydroxydiphenyl ether) is a widely used antibacterial and antifungal agent. It can be found in many personal care products and cosmetics (creams, toothpastes, shampoos, shower gels etc). Triclosan (Fig. 1) is commonly detected in the aquatic ecosystems, as it is only partially removed during wastewater treatment process. It is not proven to be harmful to humans but as a strong antimicrobial agent may cause damage in the environment to many aquatic organisms (e.g. algae) and to increase toxicity by formation of toxic degradation products such as 2,4-dichlorophenol, chlorodioxins etc.

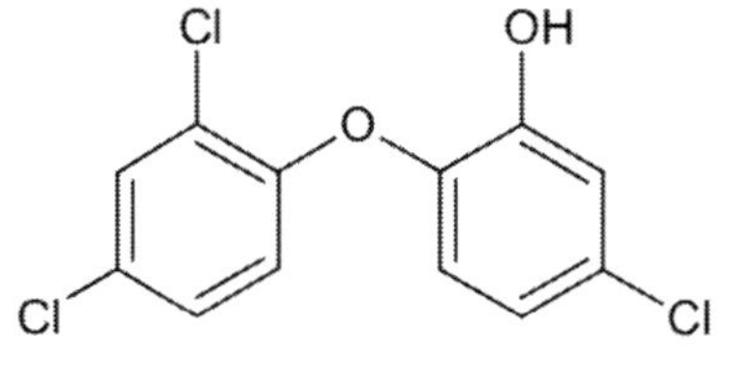

Fig. 1 – Chemical structure of triclosan

There had been tried out three different techniques of triclosan breakdown by reductive dechlorination using zero-valent metals, such as Al-Ni alloy, Devarda´s alloy (Cu, Al, Zn) and zinc powder (Fig. 2).

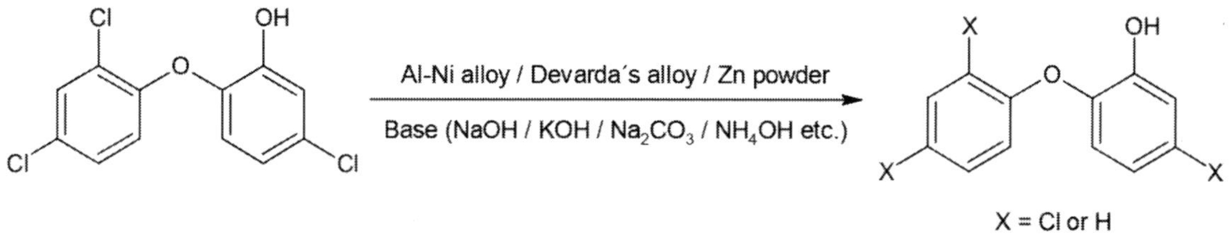

Fig. 2 – Triclosan reductive dechlorination using zero-valent metals and various bases

Procedure is as follows: first triclosan is transferred to phenolate for higher triclosan concentrations (triclosan solubility in aqueous solution is very low, up to 10 mg/l; with base it could be increased to millimoles per liter). Then the various bases and various types of reductive agents (zero-valent metals) are added. Reaction is conducted under ambient conditions. According to results, the most of dehalogenation product is provided by using Al-Ni alloy which seems to be working efficiently. Various results were obtained by using various bases. Usage of e.g. NaOH, KOH, NaF seems to be more efficient than the usage of NH_4OH or CH_3COONH_4. It has been proven that this process works however it needs to be conducted more experiments for optimization and possible application on the industrial scale.

Biomarkers distribution in crude oil feedstock compared with refined products

[1]M. Kerntong, [2,*]S. Homklin, [3,**]S. Narumon, [1,4,***]S. Jongpatiwut.
[1]The Petroleum and Petrochemical College, Chulalongkorn University, 254 Chula Soi 12, Wangmai, Pathumwan, Bangkok 10330, Thailand; tel. (+66)2 2184110, fax: (+66)2 2154459; [2]University of Phayao, 19 Village No. 2, Maeka, Muang Phayao, Phayao 56000, Thailand; tel. (+66)54 466666, fax: (+66)54 466690, [*]e-mail: supreeda.ho@up.ac.th; [3]PTT Global Chemical Public Company Limited, 59, Ratniyom Road, Noenphra, Mueang Rayong, Rayong 21150 Thailand; tel. (+66)38 994000, fax: (+66)38 994111, [**]e-mail: narumon.s@pttgcgroup.com; [4]Center of Excellence on Petrochemical and Materials Technology, 7th floor, Chulalongkorn University Research Building, 254 Chula Soi 12, Wangmai, Pathumwan, Bangkok 10330, Thailand, [***]e-mail: Siriporn.J@chula.ac.th

In recent years, there have been spill oils appearing on the beach in Thailand which affect the environment and ecology of the living animals in the areas, however the origins of the spill oils are still implied. Due to the physical properties of the spill oils are similar to fuel oil. Consequently, it cannot confirm that these spill oils have the origin from refinery product or crude oil. As the differences of the forming conditions and its compositions result in unique chemical, the fraction of the high resistant compounds can be used as a fingerprinting to identify the crude oil origins. In petroleum industry, crude oil was processed and converted to other petroleum products. The mixed crude oil was used in the process as feedstock for the economic and product yield purpose.

In this study, attention is focused on the appearing of hopane biomarker in refinery products and compared with the feedstock by chemical fingerprinting analysis. The detection of hopane biomarkers in crude oil feedstock (COF) and its refined products (RP01 and RP02) is shown in Table 1. Hopane biomarkers at 5816, 5951, 6029, 6041, 6053, 6062, 6128, 6143, 6329, and 6782s were disappeared due to the refinery process (atmospheric, vacuum distillation and visbreaking). Hopane with retention time 6104s only appeared in RP01 and RP02 samples, indicating that the hydrocracking and visbreaking process may generate the new biomarkers (Russell et al., 2004). From the hydrocracking unit, there are two main chemical reactions occurring in the process: catalytic cracking of heavy hydrocarbons into lighter unsaturated hydrocarbons that damaged the biomarkers and the saturation of these newly formed hydrocarbons with hydrogen which is the reason for the generation of new biomarkers. From the result, 15 biomarkers that appear in RP01 is 58% of the total biomarkers that exist in COF and RP02 have 20 types of hopanes which is 77% of total hopanes that appeared in COF.

Table 1 Hopane biomarkers in crude oil feedstock and refined product

RT	5714	5804	5816	5951	5963	5975	6029	6041	6053	6062	6083	6104	6128	6143	6161	6224	6245	6326	6329	6353	6365	6455	6491	6593	6635	6734	6782
COF	✓	✓	✓	✓	✓	✓	✓	✓	✓	✓	✓		✓	✓	✓	✓	✓	✓	✓	✓	✓	✓	✓	✓	✓	✓	✓
RP01	✓	✓			✓	✓					✓	✓			✓	✓	✓	✓		✓	✓	✓	✓	✓		✓	
RP02	✓	✓		✓	✓	✓			✓		✓	✓	✓	✓	✓	✓	✓	✓		✓	✓	✓	✓	✓	✓	✓	

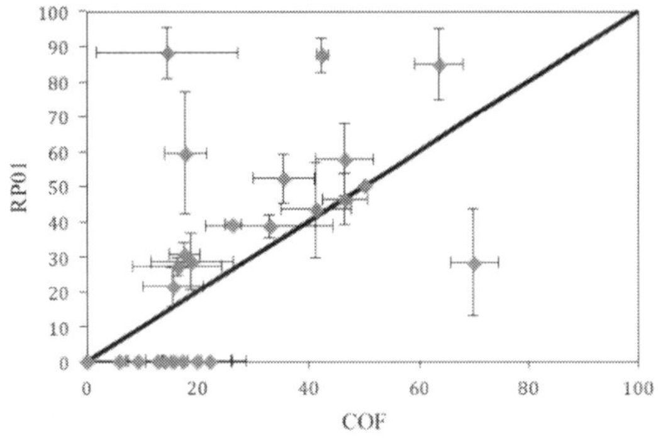

(a)

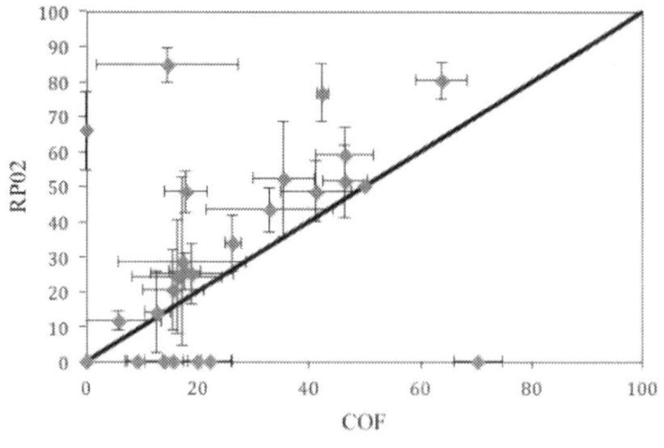

(b)

Figure 1 The correlation plot of diagnostic ratios of hopanes biomarkers between (a) COF and RP01 and (b) COF and RP02.

Figure 1 shows the correlation plots of diagnostic ratio of hopanes biomarkers between COF and refined products while using the 95 % confidence limit. It is clearly seen that there were several keys not overlapping the line x=y, indicating non-match because some biomarkers disappeared during the process and some new biomarkers were generated. Therefore, results have shown that there are significant differences in distribution of hopanes biomarkers in refined product produced from COF as raw material.

Bisphenol-A and 2,4-dichlorophenol photocatalytic degradation by B-doped TiO_2 catalysts under simulated sunlight irradiation

E.Bilgin Simsek
Yalova Univ., Dep. Chemical and Process Engineering, Yalova, 77100 Yalova, Turkey;
tel. +90 226 8155403, e-mail: ebilgin.simsek@yalova.edu.tr, esrabilgin622@gmail.com

The environmental pollution of endocrine disrupting chemicals (EDCs) in water sources has received scientific and public attention over the past decade as EDCs are capable of interfering with normal hormone function through mimicking or blocking endogenous hormones [1]. Bisphenol A (2,2-bis (4-hydroxyphenyl) propane, BPA) is a representative endocrine disrupter. On the other hand, chlorophenols (CPs) are organic pollutants and are most widespread in the aqueous environment. The heterogeneous photocatalysis has many advantages compared to other advanced oxidation processes in water treatment and purification, including avoiding the use of hazardous chemicals, minimum waste streams, energy efficient compared to photolysis and sonolysis, and operating under ambient condition [2-Wang-lim-2010]. Due to its good photoactivity, high chemical and biological stability, relatively low cost, and low toxicity, TiO_2 has attracted much attention for applications in water treatment. Doping of non-metals (C, N, B, F, S) into TiO_2 can extend the absorption band-edge of TiO_2 and could result in higher visible-light responses for the degradation of organic pollutants. In the current study, highly visible-light photoactive boron doped TiO_2 catalyst was synthesized via a solvothermal method, which showed efficient visible light photo-degradation of BPA and 2,4-dichlorophenol (DCP) compared with raw TiO_2.

The derived photocatalysts are predominantly of anatase phase, with well-developed mesoporosity. XPS spectrum for B 1s of B-TiO_2 catalyst (B content is 8%) revealed the occurrence of interstitial boron atoms (Ti–O–B) in the anatase structure. The band gaps of samples were also estimated from the intercept of UV–vis spectra. The absorption edge position of all B-doped TiO_2 moved toward longer wavelength which results a decrease in the bandgap with B-doping. In the FT-IR spectrum of B doped catalysts, new peaks appeared at 1200 cm^{-1} and 1393 cm^{-1} which can be ascribed to B–O and Ti–O–B bonds.

The photocatalytic activities of the obtained B-doped TiO_2 catalysts were investigated by the degradation of BPA and 2,4-Dichlorophenol (DCP) under both UV-A light and simulated solar light irradiation, and the results are shown in Figure 1 and 2.

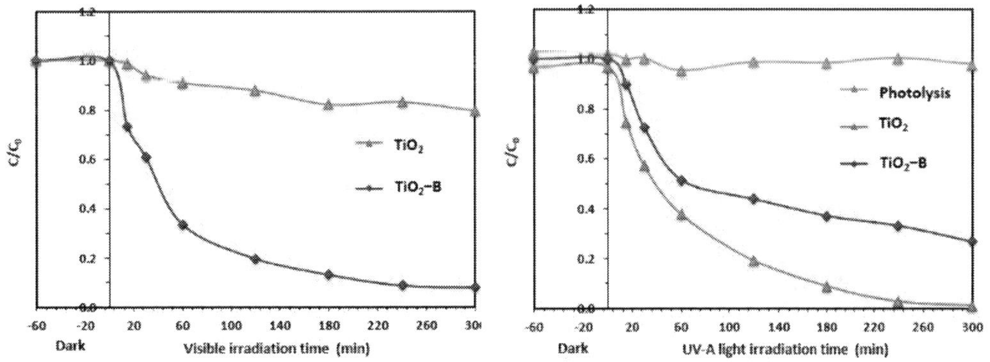

Figure 1. The degradation curves of DCP by the TiO_2 and TiO_2-B

As can be seen from Figure 2 that the photocatalytic degradation of BPA also depends on the nature of the catalyst and the kind of the light irradiation (UV-A or visible). After 300 min of photocatalysis under UV-A light and visible irradiation, TiO_2 degraded BPA by 73% and 98%, respectively. Under 5 h of visible, the extents of BPA degradation of pure TiO_2 and B-doped TiO_2 were calculated as 51%, and 89%, while the corresponding percentages for DCP were 20% and 92%, respectively.

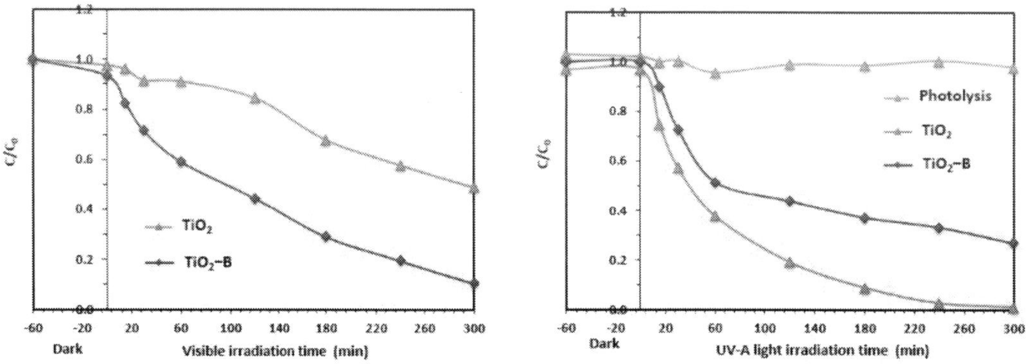

Figure 2. The degradation curves of BPA by the TiO_2 and TiO_2-B

BPA and DCP removal was found most favorable under circumneutral pHs. In order to identify the optimum dosage of photocatalyst, B-TiO_2 catalyst were varied from 0.25 to 2 g/L in 100 mL of DCP and BPA (20 mg/L) at pH value of 6.0 up to 5 h. As the photocatalyst dosage was increased from 0.25 to 2 g/L, the photocatalytic removal efficiency of BPA increased from 26.1 to 93.5% and that of for DCP it increased from 66.2 to 95.9%.

B-doped TiO_2 catalyst was synthesized by a direct hydrolyzation of n-tetrabutyl titanate in a solution of boric acid, and was treated by solvo-thermal synthesis. The photocatalytic activity of as-prepared B-doped TiO_2 was determined and compared by photodegradation of BPA and DCP under UV-A light and visible light. Under visible light boron doped catalyst showed an excellent photocatalytic performance towards demonstrating that B-doping could enhance the photocatalytic activity of TiO_2. The current work may provide new insights into the preparation of effective visible light photoactive TiO_2 catalysts.

References

[1] Luo, L., Yang, Y., Zhang, A., Wang M. et al. A. *App Surf Sci* 353 (2015) 469–479.
[2] Wang, X., Lim, T.-T. *App Catal B: Environ* 100 (2010) 355–364.

Column performance of PAD 910 resin packed bed for BTEX compounds removal in industrial waste water

T.P. Makhathini and [1]S. Rathilal
Mangosuthu University of Technology, Faculty of Engineering, Department of Chemical Engineering, P.O. Box 12363 Jacobs, Durban – Umlazi 4026, South Africa, Email: thobeka@mut.ac.za ; [1]Durban University of Technology, Faculty of Engineering and the Built Environment, Department of Chemical Engineering, P.O. Box 1334, Durban 4000, South Africa, Email: rathilals@dut.ac.za

Industrial wastewater containing organic compounds and/or substances is an increasing problem due to its increasing toxic threat to humans and the environment. The removal of organic compounds has become an imperative issue due to stringent measures that are introduced by the Department of Environmental Affairs in South Africa to enforce regulations concerning wastes that emanate from petrochemical industries. Thus, wastewater containing the BTEX compounds must be well understood so as to device adequate treatment processes. Hence, this study seeks to evaluate the PAD 910 polystyrenic resin as an alternative adsorbent for BTEX compounds.

Fixed bed column was used to evaluate the dynamic adsorption behaviour of BTEX/PAD 910 polystyrenic resin through a dynamic column approach. The performance of small-scale fixed bed column, containing PAD 910 polystyrenic resin was evaluated using 14.47 mg/L of BTEX concentration as synthesised wastewater feed. The column with 32 mm diameter, studied bed depths of 40, 80 and 120 mm and flow rate of 6 ml/min were used in order to obtain experimental breakthrough curves. The bed depth service time (BDST) model was used to analyze the experimental data and design parameters like adsorption capacity, adsorption rate and service time at 20% and 60% breakthrough. BDST was also used to predict the service times of the column operated under different influent concentrations and flow rates to produce theoretical values that were compared to the experimental values.

The experimental breakthrough curves for the adsorption of BTEX in a column packed with PAD 910 polystyrenic resin at different bed heights were obtained. The adsorption capacity per unit bed volume (N_0) and adsorption rate constant (K) are estimated by making use of the Bed Depth Service Time (BDST) equation. Breakthrough is defined as the point where a specified amount of the influent is detected in the effluent (Lagrega et al., 1994). A breakthrough of 20% on the column occurs when the concentration of the effluent is 20% of the influent concentration (~15mg/L) and 60% breakthrough occurs when the effluent is 60% of the influent on PAD 910 polystyrenic resin column.

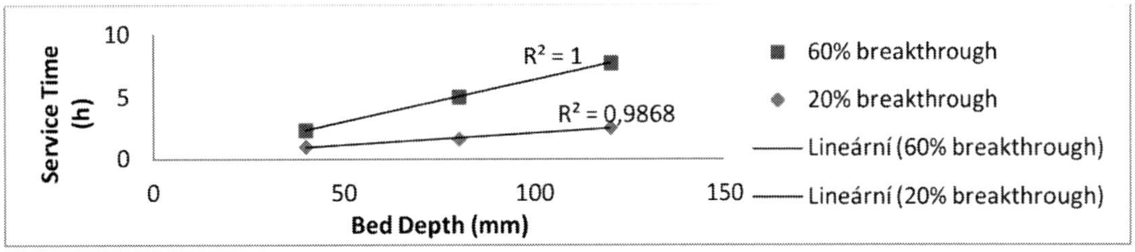

Figure 1.1: BDST plot for BTEX adsorption onto PAD 910 polystyrenic resin determined from a theoretical approach.

For this study, it was confirmed from the 20% and 60% breakthrough that the increase in the bed depth increases breakthrough time t_b and that the use of a single column is not applicable in a case where the influent concentration that is to be treated is high. In addition, the slope of the S-curve from t_b to the point of exhaustion is found to decrease as the bed depth increases from 80 mm to 120 mm, which means that the breakthrough curve becomes steeper as the bed depth decreases. Therefore there is a higher percentage of BTEX removed from the wastewater due to the increase in the bed depth of the packed column. The adsorption rate constant (K_a); which characterizes the rate transfer of the adsorbates from the liquid phase onto the solid was found to be 1.02 X 10^{-2} l/mg-h for benzene, 0.98 X10^{-2} l/mg-h for toluene, 1.23 X10^{-2} l/mg-h for ethylbenzene and 1.11 X 10^{-2} l/mg-h for xylene.

Table 1.1: Comparison of theoretical and experimental adsorption capacity at 20% breakthrough of initial adsorbate concentration

Adsorbates	N_0, (mg/l)	Theoretical Capacity, (mg/g)	Actual Capacity, (mg/g)
Benzene	2.45	0.52	0.58
Toluene	3.33	0.44	0.49
Ethylbenzene	0.23	0.22	0.23
m-; p-Xylene	1.23	0.12	0.18
o-Xylene	0.45	0.09	0.11

The adsorption capacity (N_0) was calculated for both 20% and 60% breakthrough and from the actual experimental data obtained. Table 1.1 is evident that PAD 910 polystyrenic had higher adsorption capacity per unit bed volume for 60% breakthrough than 20% breakthrough for BTEX. The same table also compares the theoretical adsorption capacity determined using the BDST parameters to actual experimental adsorption capacity obtained during continuous packed column operation.

The column design parameters obtained from the earlier test with 6 ml/min, initial concentration of 14.47 ppm and bed depths of 40, 80 and 120 mm can be used for the design of the new adsorption column with a new influent concentration of ~30 ppm. A new BDST equation was developed with new slope and intercept due to the changes in the concentration. This new equation for a change in concentration is developed from the equation below instead of running additional experiments:

$$New\ Slope = Old\ slope\left(\frac{V_{old}}{V_{new}}\right) \quad (1\text{-}1)$$

$$New\ b = Old\ b\left(\frac{C_{0,old}}{C_{0,new}}\right)\frac{\ln\left[\left({C_{0,new}}/{C_b}\right)-1\right]}{\ln\left[\left({C_{0,old}}/{C_b}\right)-1\right]} \quad (1\text{-}2)$$

The equation was based on the new slope and the new intercept derived from the BDST equation for 14.47 mg/l feed concentration. The increase in the influent concentration of the adsorbates from 14.47 mg/l to 30 mg/l also leads to a decrease in the breakthrough times, which is to be expected. The predicted point for the 30 mg/l concentration was seen to match the three different bed heights obtained from the 14.47 mg/l concentration. The breakthrough times corresponding to 20% actual concentration of 14.47 mg/l and the calculated 30 mg/l were obtained and presented.

Development of a phenomenological model to describe the kinetics of the laccase-catalyzed degradation of micropollutants

R. Pype, S. Flahaut, F. Debaste
Université Libre de Bruxelles, Dep. Transfers, Interfaces and Processes, avenue F.D. Roosevelt 50, 1050 Brussels, Belgium; tel. +322 6502916, fax: +322 6502910, e-mail: rosapype@ulb.ac.be; Université Livre de Bruxelles, Institut de Recherches Microbiologiques Jean-Marie Wiame, avenue E. Gryson 1, 1070 Brussels, Belgium; tel. +322 5267715, fax: +322 5267273, e-mail: sflahaut@ulb.ac.be; Université Libre de Bruxelles, Dep. Transfers, Interfaces and Processes, avenue F.D. Roosevelt 50, 1050 Brussels, Belgium; tel. +322 6502916, fax: +322 6502910, e-mail: fdebaste@ulb.ac.be.

In recent years, pollution of surface and drinking water by organic micropollutants has become a growing environmental issue. Micropollutants are substances which have a negative impact on the organisms of the environment, even at very low concentrations (ng/L to µg/L). Their presence in the environment is often associated with negative impacts as long and short term toxic effects, endocrine perturbation effects or microorganisms developing antibiotics' resistance. Thus, their release in the environment should be avoided. However, current wastewater treatment plants (WWTPs) are not specifically designed to remove micropollutants, especially because of the cost, the energy consumption and the difficulty of treatment. Therefore, numerous micropollutants are not sufficiently removed by WWTPs and end up in the natural environment at potentially problematic concentration levels. Presently, micropollutants can be found in wastewaters but also in surface waters, sediments, underground waters or drink waters where they are free to induce their adverse effects on the organisms.

A way to improve the micropollutants removal is the use of biodegradation processes such as enzymatic methods. Laccases are lignin-modifying enzymes found in white rote fungi (among other organisms) that show an interesting potential for the oxidation of highly recalcitrant organic micropollutants. Laccases can oxidize, polymerize or transform phenolic or anthropogenic compounds to less toxic derivatives. Laccases can oxidize a broad range of molecules, including phenols, dyes, pesticides, endocrine disruptors and polycyclic aromatic hydrocarbons. Moreover, to catalyze the substrate oxidation, laccases only need molecular oxygen which is reduced in water simultaneously with the substrate oxidation.

Among the micropollutants oxidized by laccases, there is the Remazol Brilliant Blue R (RBBR). RBBR is a typical anthraquinone dye used in textile industry. As all dyes, when released in the aquatic environment, it reduces the amount of sunlight to photosynthetic organisms resulting in decreased oxygen levels in aquatic ecosystems. Moreover, it represents an important class of toxic and recalcitrant organopollutants.

The goal of this work is to develop a phenomenological model describing the kinetics of micropollutant degradation by laccases. The development of this model is based on the experimental data obtained for the degradation of RBBR, used as a model-micropollutant. Indeed, the fact that it is a colorant makes RBBR a compound whose concentration can easily be followed using a spectrophotometer. Then, RBBR is a good model-compound to study the kinetics of the degradation of a micropollutant by laccases. This model development is a first step before the more general framework of the study of the degradation of mixtures of micropollutants by laccases.

The model developed so far is based on a modified Briggs-Haldane kinetics where the classical assumption of steady state quantity of enzymatic complex is relaxed and a low substrate concentration is assumed. The full mathematical development can be found in [1]. The evolution of the micropollutant concentration during the laccases treatment is given by:

$$\frac{A}{A_0} = exp\left\{-k_{cat}\left[t + \left(\frac{B}{k_i} - \frac{B}{k_i}\, exp[-k_i\, t]\right)\right]\right\}$$

where A is the micropollutant concentration at any time (M), A_0 is the initial micropollutant concentration (M), t is the reaction time (min) and k_{cat} (min^{-1}), B (no units) and k_i (min^{-1}) are kinetic constants.

Experimental enzymatic degradation of RBBR was realized at 20.1°C and pH 5 (citrate-phosphate buffer) with an initial RBBR concentration of 3 10^{-4} M for different activities of *Trametes versicolor* laccases. The enzymatic activities were determined as described in [2]. The evolution of RBBR concentration was monitored with a spectrophotometer.

Figure 1 presents typical results of this model compared to a Michaelis-Menten (also assuming a low substrate concentration) for three different initial enzymatic activities. This shows that the classical Michaelis-Menten is not suitable but that relaxing an assumption leads to a good fitting. Yet, no single values of the kinetics constants allow a good fitting for the 3 enzymatic activities. This underlies a more complex behavior that will be studied in the coming months.

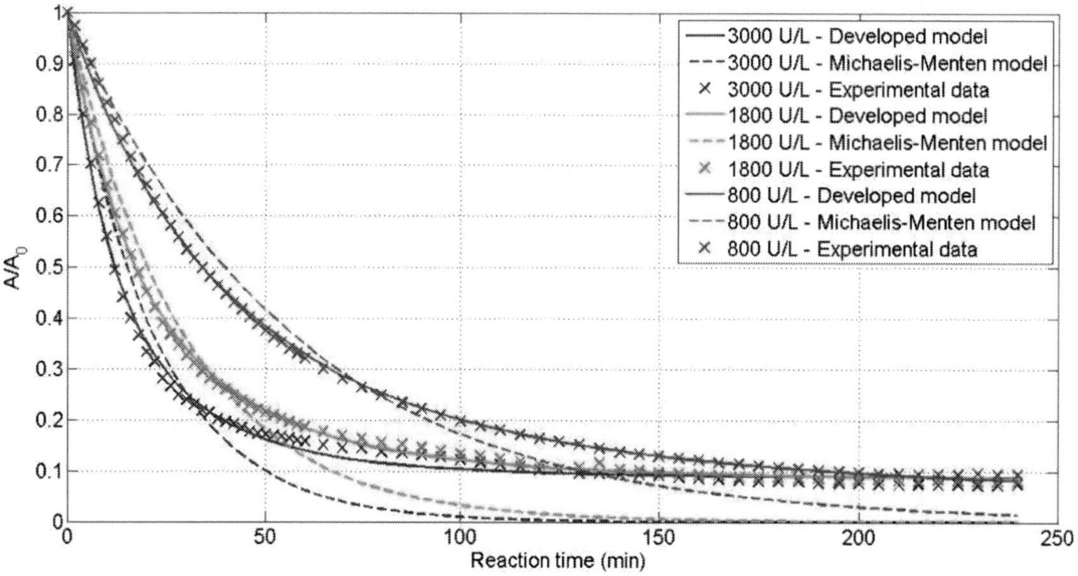

Figure 1: Degradation kinetics of RBBR (developed model, Michaelis-Menten model and experimental data) for different enzymatic activities.

Bibliography

[1] C. Hautphenne, "Laccase-catalized oxidation of bisphenol A: kinetic study and fate of reaction products," Thesis presented in fulfillment of the requirements for the degree of Docteur en Sciences de l'Ingénieur, Université Libre de Bruxelles, 2016.

[2] C. Hautphenne and F. Debaste, "Harnessing Laccases for the Synthesis of Bisphenol A Biopolymers," *Chem. Eng. Technol.*, vol. 38, no. 7, pp. 1223–1228, Jul. 2015.

Electrochemical reactor for advanced removal of metal ions from solution

M.I. Iorga, N. Mirica, M.C. Mirica, D. Buzatu
National Institute for Research and Development in Electrochemistry and Condensed Matter – INCEMC– Timisoara, Dept. of Applied Electrochemistry, St. Dr. A. Paunescu Podeanu, No.144, 300569 Timisoara, Romania, tel. 00 40 256 222119, fax 00 40 256 201382, www.incemc.ro,
e-mail: mirela_iorga@yahoo.com

The paper presents an original electrochemical reactor for advanced removal of metal ions from solution used in environmental protection, for the treatment of water containing metals, with the possibility of metal recovery (may apply to copper, silver, zinc, gallium ions, etc.).
It is well known that electrochemical methods for metal ions recovery from diluted solution come into attention especially in recent years, due to the increasing difficulties encountered in the treatment of wastewater contaminated with extremely nocive polutants, which often can't be effective destroyed or removed by classical methods. They refer to advantageous treatment or pre-treatment of wastewater containing heavy metals. If at the same time it is possible to recover those metals (especially in the case of valuable metals) the cost of the treatment process is significantly reduced. The recuperative electrochemical methods are based on metal ions discharge at the cathode, under the reaction conditions assured by an appropriate design of the electrochemical reactor.
Regardless of the reactor type used, there are four main operating modes: simple bath, recycled bath through a loop with a stirring reactor, one pass through the reactor, electrolyte flow through n series reactors. In industrial processes if finally is intended to obtain solutions very low metal concentration, the last alternative is preferred.
The electrochemical reactor proposed is a single reactor, with several compartments, each one for every electrolysis stage, disposed circular, concentric, as shown in figure 1, where the electrolyte flows gravitationally. The number of compartments, which is the same with the number of electrolysis cells, is established taken into account the initial concentration of the solution, the global degree imposed for depletion and the intermediary depletion degree for each electrolysis cell.
Copper is recovered from a solution with an initial concentration of 20 g/l copper ions. A global degree of depletion of 90 % is imposed; in this case the electrolysis is done in three stages, and the intermediary depletion degree is 60 %. For the first electrolysis stage the anode and cathode radius are chosen, and subsequently the anode and cathode radius for the second and third electrolysis stage and the interelectrodic distances for each electrolysis stage are calculated. Finally, an overall depletion degree of 93.6 % was obtained, which is higher than initially imposed (90 %).

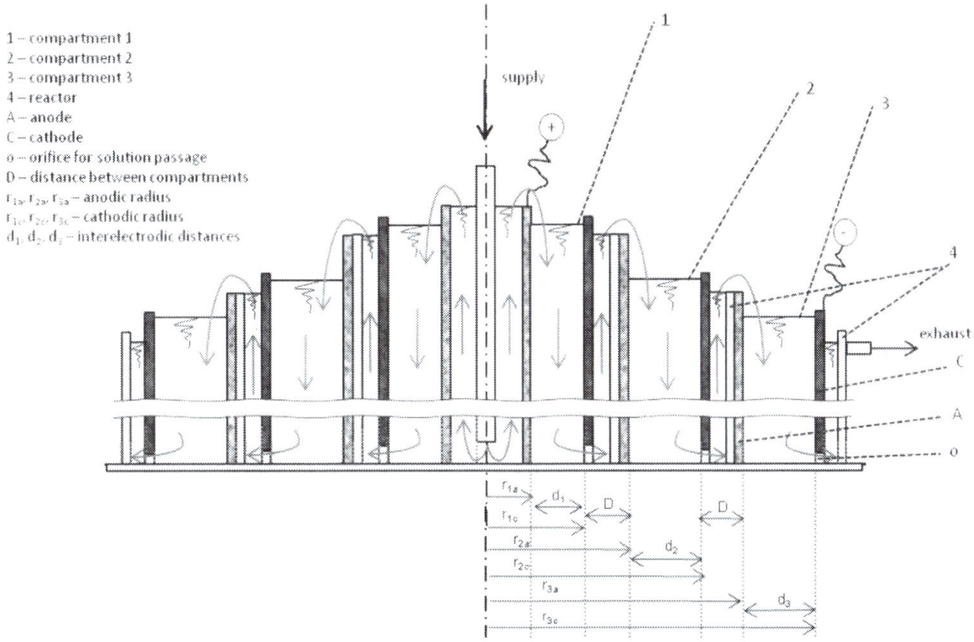

Figure 1. Electrochemical reactor for advanced removal of metal ions from solution in case of a three stages process (vertical profile)

REFERENCES:

Davis, M.L.; Cornwell, D.A. (2008). *Introduction to Environmental Engineering*, Fourth Edition, McGraw-Hill NY.

European Commission (2001). *Integrated Pollution Prevention and Control (IPPC). Reference Document on Best Available Techniques in the Nonferrous Metals Industries*.

Kuhn, A. T. (1971). *Industrial Electrochemical Processes*, Elsevier, Amsterdam.

Iorga, M.I.; Pop, R.; Mirica, M.C.; Buzatu, D. (2012). Methods for the Enhancement of Mass Transport at the Recovery of Metal Ions from Hydroelectrometallurgical Processes, in *Recent Researches in Metallurgical Engineering – From Extraction to Forming*, Dr Mohammad Nusheh (Ed.), ISBN: 978-953-51-0356-1, InTech, DOI:10,5772/38717.

Iorga, M.I., Mirica, M.C., Mirica, N., Balcu, I. (2016). *Electrochemical Reactor and Method for Metal Ions Removal from Solutions*, Patent no.127174, OSIM Bucuresti, Romania.

Mirica, N., Dragoş, A., Mirica, M. C., Iorga, M.I., Macarie, C. (2006). *Electrochemical Methods for Metal Ions Recovery from Solutions*, Ed.Mirton, Timişoara, Romania.

Pletcher, D., Walsh, F.C. (2000). *Industrial Electrochemistry*, 2nd Edition, Springer.

Rajeshwar, K., Ibanez, J.G. (1997). *Environmental Electrochemistry. Fundamentals and Application in Pollution Abatement*, Academic Press, Inc., San Diego.

Walsh, F. C., Reade, G. W. (1994). Electrochemical techniques for the treatment of dilute metal-ion solutions, in *Studies in Environmental Science 59. Environmental Oriented Electrochemistry*, Edited by C.A.C. Sequeira, Elsevier, Amsterdam.

*** Formatting Issue - Best Available Paper/Graphic ***

Energy and exergy analysis of hydrogen network of refinery: Case study on Imam Khomeini Shazand refinery

M. M. Shalchi[1], O. Tavakoli[2], A. A. Sharafi[3], Y. Saboohi[4]
[1]School of Chemical Engineering, College of Engineering, University of Tehran, 16 Azar St., Tehran 14176, Iran; tel. +989122971285, e-mail: m.m.shalchi@ut.ac.ir; [2]School of Chemical Engineering, College of Engineering, University of Tehran, 16 Azar St., Tehran 14176, Iran; tel. +982166498984, e-mail: otavakoli@ut.ac.ir (corresponding author); [3]Imam Khomeini Shazand Refinery, 20th km of Arak-Borujerd Road, Iran, Iran; tel. +988633674541, e-mail: amir_abbas_sharafi@yahoo.com; [4]Department of Energy Engineering, Sharif University of Technology, Taymouri Alley, Ghasemi St., Habibollahi St., Azadi Ave., Tehran, Iran; tel. +982166085185, e-mail: saboohi@sarif.edu.

Todays because of changing the crude oil into heavy and high sulfur oil, refineries use more the hydrocracking, hydrotreating and hydroprocessing processes. These processes reduce sulfur and aromatics content and increase light components that can decrease the emission of Green House Gases(GHG) and grow the profit of refinery [1]. These processes consume hydrogen, so supply of hydrogen in refineries very important and Hydrogen Network (HN) has a key rule in refinery.

Anlysis of HN enegy and exergy can determine the portion of energy consumption of HN, amount of energy losses and exergy destruction. Optimization of HN with considering the energy and exergy can diminish energy losses and exergy destruction and in general, dwindle the operational cost of refinery. In this paper a model for HN is developed, then energy and exergy analysis is done and HN is optimized based on the exergy destruction, energy losses and economy parameters.

To do that, at first the Reference Energy System (RES) of HN should be formed. The RES determines the interaction between units in HN and distribution of energy carrier streams like electricity, steam etc. in HN. The RES is composed of several layer, each of which has multiple units. The Fig. 1 is represented the HN RES of Imam Khomeini Shazand Refinery. This RES is consist of 8 layer and 19 carrier energy.

After that the mass and energy balance of HN RES is calculated. The Fig. 2 shows the energy losses of RES. The most Energy losses is in "Hydrogen Treatment" layer by 35%. This layer is the first consumer of High Pressure (HP) steam, second consumer of fuel gas in HN and has the highest capacity between other layers of RES. These reasons have caused this layer to become the first energy loser.

The next step is calculation of exergy balance. Total exergy destruction of HN is 95,800 GJ/Day. The most amount of exergy destruction of a unit in HN is 25,900 GJ/Day belongs to Reduced-Crude Desulfurization (RCD) unit and it is 27% of total exergy destruction. The next phase is optimization of HN by considering of energy and exergy analysis.

*** Formatting Issue - Best Available Paper/Graphic ***

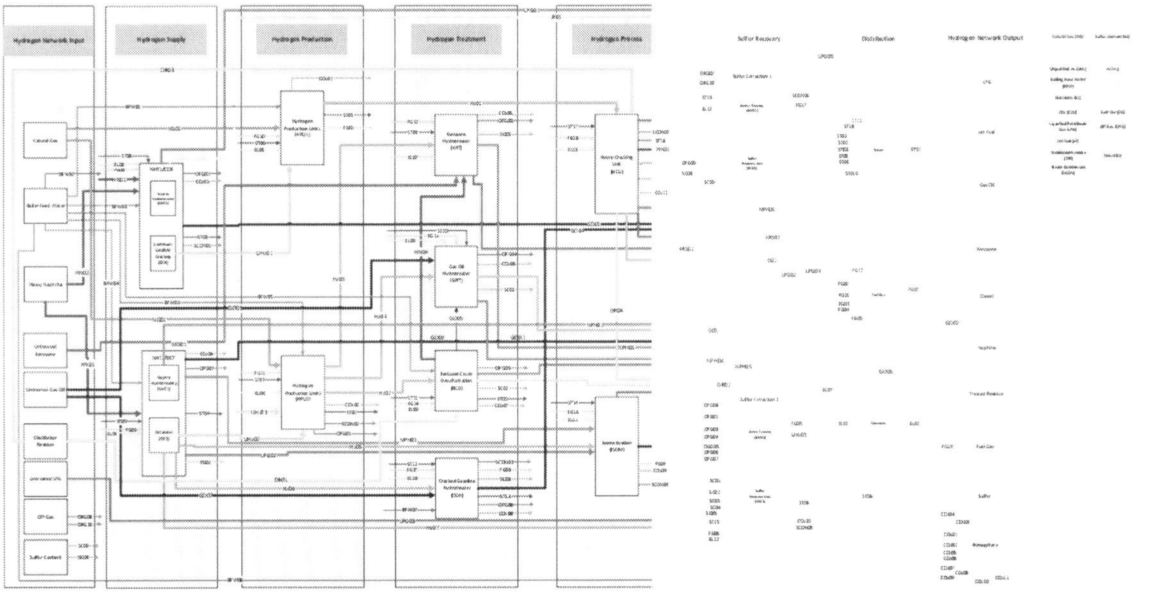

Fig. 1 Reference Energy System of Hydrogen Network of Shazand Refinery

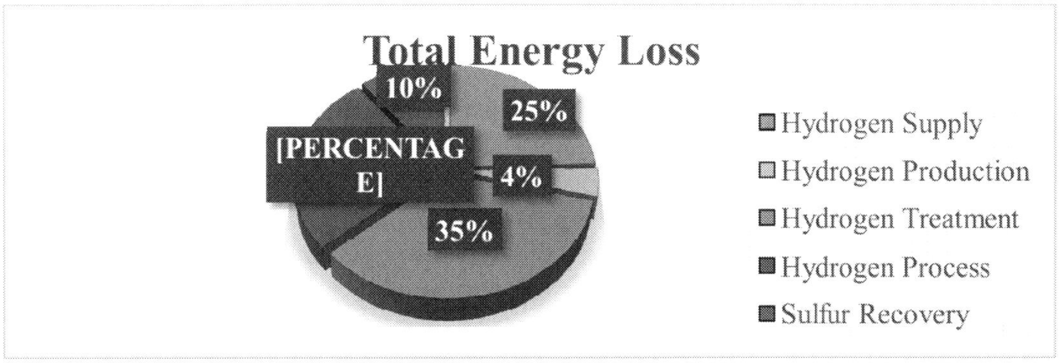

Fig. 2 Total energy losses of HN RES

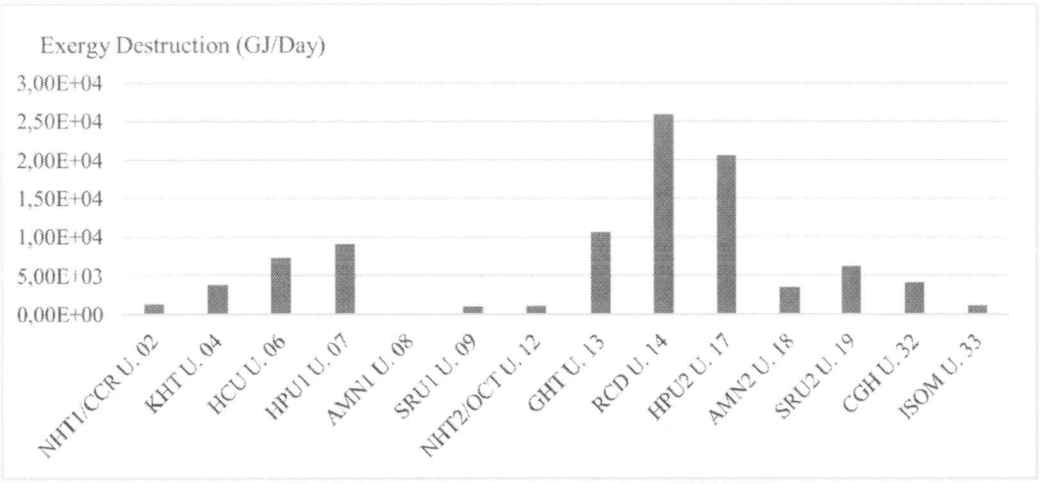

Fig. 3 Diagram of exergy destruction of HN RES

[1] C. Deng et al., Comparative analysis of different scenarios for the synthesis of refinery hydrogen network, Applied Thermal Engineering (2014), http://dx.doi.org/10.1016/j.applthermaleng.2014.04.036.

Evaluation of wastewater treatment process based on ecotoxicological bioassays

B. Urminska[1], H. Zlámalová-Gargošová[2]
[1]Slovak University of Technology, Faculty of Chemical and Food Technology, Department of Environmental Engineering, Radlinského 9, 812 37 Bratislava, Slovakia;
tel. +421 949 441 129, e-mail: barbora.urminska@stuba.sk;
[2]Brno University of Technology, Faculty of Chemistry, Institute of Chemistry and Technology of Environmental Protection, Brno, Czech Republic.

Wastewater treatment processes have a key role in sustainable development as a tool of protection of surface waters from anthropogenic pollution. The possibilities of environmental pollution control are increasing and more advanced wastewater treatment methods are being emphasized, as well as more specific control of contaminants presence in treated waters. It is obvious that none of the wastewater treatment plants (WWTP) is able to eliminate residues completely – it can only decrease the pollution level to an acceptable value. Chemical analysis can detect many of the pollutants but it is hardly possible to identify all of the compounds that could be present in the wastewater, often in very low concentrations. Furthermore, the mere identification of the compounds does not provide sufficient information on their effects on the environment. Therefore, apart from routine chemical analysis it is useful to perform also ecotoxicological bioassays as the means of evaluation of biological effects caused by wastewater discharges to water ecosystems and to find out whether the treatment process is sufficient to provide a satisfactory decrease of environmental impact (ecotoxicity) [1, 2].

The acute toxicity depends on the treatment level of the WWTP and the species tested. The aim of this study was to examine the impact of treatment technology on the ecotoxicity of effluents and to evaluate the efficiency of wastewater treatment process using selected ecotoxicological bioassays. Wastewater samples were collected at three municipal wastewater treatment plants in the Czech Republic: Brno-Modřice (500 000 PE), Mikulov (25 000 PE) and Lednice (12 000 PE). The wastewater treatment plants had various values of population equivalent (PE) and different treatment technologies.

The samples were taken from influent, after the primary treatment and from the effluent of WWTPs. Ecotoxicity testing of wastewaters is not yet common and required by the Czech legislation. However, the general rule is that the test battery for the ecotoxicological evaluation should include organisms of all three trophic levels. Therefore, the ecotoxicity tests were conducted on following organisms: *Thamnocephalus platyurus* and *Daphnia magna* represented the consumers, an aquatic plant *Lemna minor* and a terrestrial plant *Allium cepa* represented the producers and *Vibrio fischeri* represented the decomposers [3-7]. Obtained results expressed as EC50, IC50 or LC50 were also compared to available results from other similar studies [1, 8, 9].

Based on the obtained results, the performed tests can be set up into the order from the most sensitive to the least sensitive: *V. fischeri* > *T. platyurus* > *D. magna* > *A. cepa* > *L. minor*. That corresponds to the results of the studies conducted by Mendonça et al. [1, 9].

The results demonstrated that secondary (biological) treatment is sufficient to reduce wastewater ecotoxicity toward the selected test organisms to the minimum. On the other hand, acute toxicity tests used in the study were not sensitive enough to distinguish one treatment technology from another. The ecotoxicological approach can contribute to the environmental management of the treatment plant but it is necessary to focus also on the use of tests of chronic toxicity, mutagenicity and genotoxicity which can detect the long-term and sublethal effects of effluents on target organisms. Several studies have shown that these tests have high information value and should be the norm also in testing and implementing new technologies for wastewater treatment and sludge processing regarding its further use [10-12].

Acknowledgements

This study was supported by the project No. FCH-S-15-2869 from the Ministry of Education, Youth and Sports of the Czech Republic.

Resources

[1] Mendonça, E., Picado, A.,Paixão, S., Silva, L., Barbora, M. and Cunha, M. A. (2011). The role of ecotoxicological evaluation in changing the environmental paradigm of wastewater treatment management. The 6th Dubrovnik Conference on Sustainable Development of Energy, Water and Environment Systems (SDEWES'2011). DOI: 10.13140/2.1.2168.3205.

[2] Bundschuh, M. (2014). The Challenge: Chemical and ecotoxicological characterization of wastewater treatment plant effluents. Environmental Toxicology and Chemistry. Vol. 33, Issue 11, p. 2407.

[3] MicroBioTests Inc. (2015). Thamnotoxkit F^{TM}: Standard Operational Procedure.

[4] MicroBioTests Inc. (2009). Daphtoxkit F^{TM} – Crustacean toxicity Screening Test for Freshwater: Standard Operational Procedure.

[5] ISO 20079:2005. Water quality - Determination of the toxic effect of water constituents and wastewater on duckweed (*Lemna minor*) - Duckweed growth inhibition test.

[6] Fiskesjó, G. (1993). *Allium* test I: A 2–3 day plant test for toxicity assessment by measuring the mean root growth of onions (*Allium cepa* L.). Environmental Toxicology and Water Quality. Vol. 8, Issue 4, p. 461–470.

[7] ISO 11348-3:2007. Waterquality – Determination of the inhibitory effect of water samples on the light emission of *Vibrio fischeri* (Luminescent bacteria test) – Part 3: Method using freeze-dried bacteria.

[8] Morbey, M. A. et al (2008). Ecotoxicological assessment of wastewaters: a Portuguese case study (Trancao river basin). Environmental toxicology. Vol. 23, Issue 4, p. 466–472.

[9] Mendonça, E., Picado, A.,Cunha, M. A. and Catarino, J. (2011). Environmental Management of Wastewater Treatment Plants – the Added Value of the Ecotoxicological Approach. In: Elzbieta Broniewicz (Ed) Environmental Management in Practice. ISBN: 978-953-307-358-3, InTech, DOI: 10.5772/18389.

[10] Englert, D., Zubrod, J. P., Schulz, R. and Bundschuh, M. (2013). Effects of municipal wastewater on aquatic ecosystem structure and function in the receiving stream. Science of the Total Environment. Vol. 454–455, p. 401–410.

[11] Magdeburg, A., Stalter, D., Schlüsener, M., Ternes, T. and Oehlmann, J. (2014). Evaluating the efficiency of advanced wastewater treatment: Target analysis of organic contaminants and (geno-)toxicity assessment tell a different story. Water Research. Vol. 50, p. 35–47.

[12] Tilton, F., Benson, W. H. and Schlenk, D. (2002). Evaluation of estrogenic activity from a municipal wastewater treatment plant with predominantly domestic input. Aquatic Toxicology. Vol. 61, Issue 3-4, p. 211–224.

Influence of Fe^{3+} on removing of nitrate ions by autotrophic denitrification using *Thiobacillus denitrificans*

Z. Blažková, V. Trousil, E. Slehová, J. Palarčík, M. Slezák, J. Cakl
University of Pardubice, Faculty of Chemical Technology, Institute of Environmental and Chemical Engineering, Studentská 573, 532 10, Pardubice, Czech Republic

Thiobacillus (T.) denitrificans is widely distributed microorganism. It is not toxic and pathogenic bacterium and belongs to group of gram-negative, chemolithoautotrophic and facultative anaerobic microorganism. It can obtain energy by the oxidation of elemental sulfur and other sulfur-containing inorganic compounds, while electrons are released and used for reduction of nitrate and nitrite ions to dinitrogen gas (under anoxic conditions).

T. denitrificans is found in soil, mud, marine and freshwater sediments, domestic sewage, industrial waste-treatment lagoons and digestion tanks. From the viewpoint of environment, it is very important organism - *T. denitrificans* can be used in processes of alternative methods of removing nitrates from contaminated water.

Autotrophic denitrification can be affected by many factors as pH, temperature and the ration C/N and S/N etc. This work is focused on influence of ferric iron on removing nitrate ions, because it belongs to compounds accelerating denitrification. The aim of the work was to determine optimal amount of ferric iron for specific conditions. The best added concentration of ferric iron was 0.1 mg/L for removing of nitrates using bacteria *T. denitrificans*. All experiments were in batch reactors and elemental sulfur was used as electron donor.

Nutrient removal in an intermittently aerated bioreactor

B. Urminská, J. Derco, Z. Škottová, K. Šimkovič
Slovak University of Technology, Faculty of Chemical and Food Technology, Department of Environmental Engineering, Radlinského 9, 812 37 Bratislava, Slovakia;
tel. +421 949 441 129, e-mail: barbora.urminska@stuba.sk.

Different and even conflicting requirements of biological processes used to removal of organic pollution, nitrogen and phosphorus have resulted in significantly more complex technologies in terms of their construction as well as operation. Not only the system configuration and its operation increased in complexity but also more strict standards for effluent quality have to be satisfied [1]. Intensification of wastewater treatment plants (WWTP) often increases growth of filamentous microorganisms which leads to worse sludge separation properties and sludge bulking.

In smaller WWTPs, activated sludge process is often low-loaded with use of completely mixed reactor. In this case conditions necessary to nutrients removal even increase the risk of filamentous sludge bulking.

The aim of this study was to examine the influence of operating conditions in an intermittently aerated bioreactor (cycle length, length of mixing and aeration) on nitrogen removal and nitrogen and phosphorus removal, respectively. The main advantage of this type of bioreactor is high flexibility in maintenance and control of biochemical environments in the reactor. Intermittently aerated bioreactors allow maintaining both oxic and anoxic conditions in one reactor, depending on time. This type of bioreactor is a convenient solution for smaller WWTPs where there is not enough space for spatial segregation of oxic and anoxic processes. This type of bioreactor is a convenient solution for smaller WWTPs where there is not enough space for spatial segregation of different biochemical environments [2, 3].

The study was performed using lab-scale completely mixed and intermittently aerated bioreactor with volume of 4.2 dm^3. We examined impact of various time cycles on nitrogen removal (nitrification and denitrification process) and luxury uptake of phosphorus. The bioreactor was operated as low-loaded activated sludge process with model wastewater flow 2.1 $L.d^{-1}$, i.e. hydraulic retention time (Θ) of the wastewater in the bioreactor was approximately 48 hours. The secondary settling tank had volume of 0.9 dm^3.

Concerning the nitrogen removal, the operation cycles 15-45 and 30-60 were examined. The former cycle (15 minutes of aeration, 30 minutes mixing only, i. e. length of the whole cycle 45 minutes,) was not satisfactory, as the aerobic conditions prevailed substantially over anoxic and the concentration of nitrate nitrogen at the effluent was high. The cycle 30-60 was better; concentrations of inorganic nitrogen at the effluent were very low. On the other hand, the sedimentation properties of the sludge were bad and excessive growth of filamentous microorganisms was observed.

For the experiments with luxury uptake processes we used operation cycles 15-60, 15-120, 30-90 and 15-90. The cycle 15-60 was unsatisfactory because effluent concentrations of nitrate nitrogen were high and luxury uptake of phosphorus did not occur. The same applies to the cycle 15-120 but also the amount of filamentous microorganisms was too high and the biomass concentration was continuously decreasing. The cycle 30-90 was also unsuitable - relatively short anoxic/anaerobic period was sufficient for denitrification but not for the luxury uptake process. The cycle 15-90 was the most satisfactory with convenient lengths of aerobic, anoxic and anaerobic period, high efficiency of nitrification and denitrification process and significant decrease of phosphorus concentration. Concentration of ammoniacal nitrogen did not exceed 2 mg·L^{-1} and nitrite concentration was bellow 1.0 mg L^{-1}. Increased efficiency of biological enhanced removal of phosphorus and very low effluent concentration values of ammonium, nitrite and nitrate concentration values were obtained after changing of operational conditions.

It can be concluded that the intermittently aerated bioreactors are suitable for nitrogen removal as well as for enhanced biological removal of phosphorus. The main advantage is the flexibility of possible changes of conditions in the bioreactor. However, the hydrodynamic regime of the reactor, alternating oxic, anoxic and anaerobic conditions and occurrence of low concentration of dissolved oxygen, organic substrate and nutrients can lead to excessive growth of the filamentous microorganisms in the bioreactor. This fact increases the requirements for control and process management.

Acknowledgements

This work was supported by the Slovak Research and Development Agency under the contract No. APVV-0656-12. The authors also wish to thank for the support from VEGA Grant 1/0859/14.

Resources

[1] European Commission (2000). Directive 2000/60/EC of the European Parliament and of The Council of 23 October 2000 establishing a framework for community action in the field of water policy, Off. J. Eur. Communities, L 327/1

[2] Derco, J., Králik, M., Kovács, A., Berešíková, Z., Žarnovský, Ľ. (1994). Modelling of Carrousel Plant by na Intermittently Aerated Activated Sludge Process. Polish J. Environ. Studies 1, 25-30.

[3] Nakajima, J., Kaneko, M. (1991) Practical Performance of Nitrogen Removal in Small-Scale Sewerage Treatment Plants Operated in Intermittent Aeration Mode. Wat. Sci. Tech., 23 pp. 709-718.

Optimization of final treatment of olive mill wastewater through a combination of strong-acid cation exchange and strong-base anion exchange resins

M.D. Víctor-Ortega[1*], J.M. Ochando-Pulido[1], D. Airado-Rodríguez[2], A. Martínez-Ferez[1]
[1]Chemical Engineering Department, University of Granada, 18071 Granada, Spain
email: mdvictor@ugr.es
[2] Nofima AS - the Norwegian Institute of Food, Fisheries and Aquaculture Research, Osloveien 1, 1430 Ås, Norway

Water scarcity specially concerns agricultural irrigation, which represents the largest demand of fresh water worldwide, that is, more than 70% of the total water consumption [1]. However, the possibility of using regenerated wastewater for irrigation purposes would solve this problem and bring very positive environmental and economic impacts.

On the other hand, olive oil industry, which is currently one of the main agro-industrial activities in the Mediterranean Basin countries, generates one of the most highly polluted agro-industrial effluents: olive mill wastewater (OMW). The disposal of this effluent into the environment causes severe deteriorations such as colouring of natural waters, serious threat to the aquatic life, pollution in surface and ground waters, alterations in soil quality, phytotoxicity and odour nuisance [2,3].

In former works, a treatment process based on Fenton's reaction was optimized with the aim of removing the organic and phenolic pollutants of OMW coming from the 2-phase extraction process (OMW-2). The treatment process comprises natural precipitation to remove coarse particles followed by Fenton-like advanced oxidation, flocculation–sedimentation and filtration in series (secondary treatment, ST) [4,5]. However, high levels of monovalent ions, mainly sodium and chloride, as well as iron, are found in OMW-2ST, derived from the dosage of the catalyst and neutralizing agents during the secondary treatment [6]. In addition, a significant phenols amount above maximum legislated standard level (0.005 mg L^{-1}) [7] was present in this secondary-treated OMW-2ST effluent and iron concentration exceeded the maximum levels allowed by the European legislation for drinking water (0.2 mg L^{-1}) [8].

Therefore, a tertiary treatment is necessary for final purification of OMW-2ST. In this scenario, novel IE resins developed over the last decades have promoted this technology as a suitable purification process for wastewater treatment. In this sense, IE is considered a green and simple technology, capable of achieving high removal efficiencies and selectivity [9].

In this research study, the final purification of OMW-2ST is addressed through a combination of two IE resins: a strong-acid cation exchange resin and a strong-base anion exchange resin. The main objective was the optimization of the final purification of OMW-2 by means of the IE system comprising the mentioned resins in serial connection.

Firstly, recirculation mode experiments were carried out to examine the impacts of the main operating parameters on the IE process performance - operating temperature, initial pH and flow rate - as well as the interactions among them. For this purpose, the IE processes base on the different IE resins were modelled by means of central composite design (CCD).

Subsequently, continuous experiments were performed under the optimum operating conditions found for this pair of resins in order to find the best IE resins combination for its application in the purification of OMW-2ST at industrial scale. Finally, water quality standards for reusing the purified olive mill effluent in the olive oil production process and thus reducing the environmental impact were checked.

Results reveal that, the best resins disposition was: Dowex Marathon C resin followed by Amberlyst A26 (see **Fig. 1**). The proposed IE process yielded up to 89 % removal of these pollutants when the initial pH, operating temperature and flow rate were set to 4.2, 20.6 ° C and 13.8 L h^{-1}, respectively. Under these conditions, the achieved concentrations for all contaminants in the final stream were maintained below the maximum established limits. Finally, high treated effluent volume recovery (31 L) could be achieved in continuous mode under the optimal operating conditions found, complying with standards for reuse in the production process.

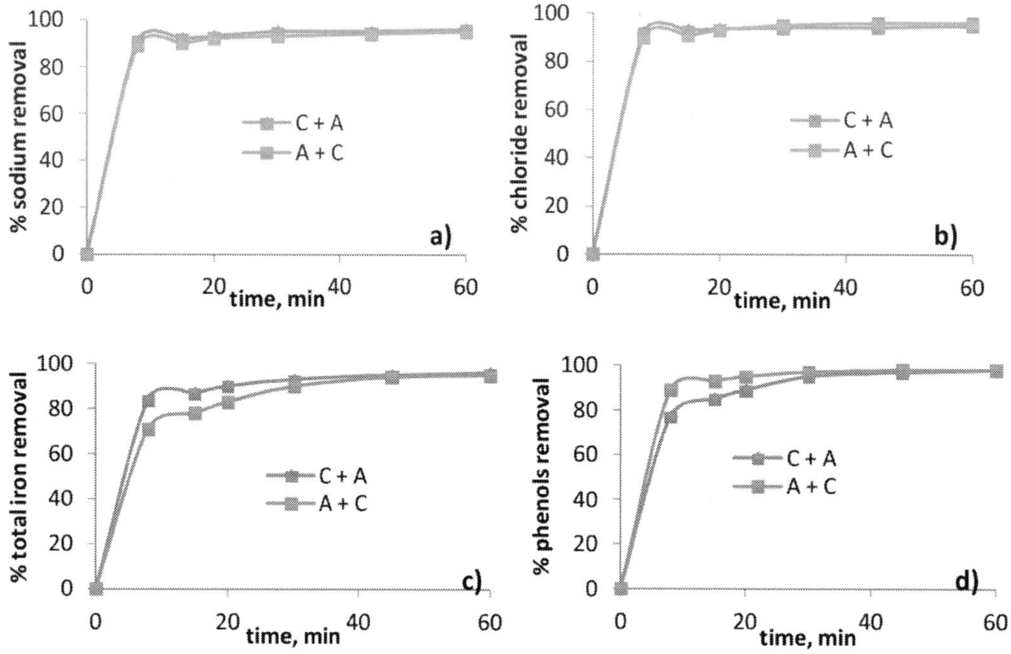

Fig. 1. Effect of resins disposition (A + C: cationic resin + anionic resin; A + C: anionic resin + cationic resin) on a) sodium, b) chloride, c) total iron, and d) phenols IE removal efficiencies as a function of recirculation time.

References

[1] Food and Agricultural Organization (FAO). Review of world water resources by country. Water Reports 23, Rome, Italy (2003).
[2] A. Dhouib, F. Aloui, N. Hamad, S. Sayadi. Process Biochem. 41 (2006) 159-167.
[3] K. Kestioğlu, T. Yonar, N. Azbar. Process Biochem. 40 (2005) 2409-2416.
[4] L. Martínez Nieto, G. Hodaifa, S. Rodríguez Vives, J.A. Giménez Casares, J. Ochando. Clean - Soil, Air, Water 39(10) (2011a) 949-955.
[5] L. Martínez Nieto, G. Hodaifa, S. Rodríguez, J.A. Giménez, J. Ochando. Chem. Eng. J. 173(2) (2011b) 503-510.
[6] G. Hodaifa, J.M. Ochando-Pulido, S. Rodriguez-Vives, A. Martinez-Ferez. Chem. Eng. J. 220 (2013b) 117-124.
[7] European Commission, Council Directive 80/778/EEC of 15 July 1980 on the quality of water intended for human consumption (1980).
[8] European Commission, Council Directive 98/83/EC of 3 November 1998 on the quality of water intended for human consumption (1998).
[9] A. Demirbas, E. Pehlivan, F. Gode, T. Altun, G. Arslan. J. Colloid and Interface Sci. 282 (2005) 20-25.

Performance of catalysts TiO_2 modified with iron in the photocatalytic reduction of Hg (II) in different reaction systems

A.L.S Coelho, [1]A. F. Almeida Neto; F. F. Ivashita; C. M. Oliveira; G. G. Lenzi; L. M. M. Jorge; O. A. A. dos Santos
State Univ. of Maringá, Dep. Chem. Engineering, Colombo Av. 5890, 87020-900 Maringá, Paraná, Brasil; tel. +55 044 30114778, e-mail: oneliandreo@hotmail.com; [1]State Univ. of Campinas, Univ.,College Chem. Enginnering, Av. Albert Einstein, 500, 13083-852 Campinas, São Paulo, Brasil.

Mercury is severely hazardous water contaminants, both organic and inorganic forms are cumulatively and highly toxic (1). There are currently in Brazil record of 80 documents, between decrees, laws, ordinances, regulatory instructions and resolutions, addressing the issue mercury, and its applications in different ways – products, industry, agricultural, industrial wastewaters, etc. generating waste activities containing Hg in Brazil mainly produce liquid effluents. In this context, photocatalytic reduction of Hg(II) to Hg(0) from aqueous solutions has been developed in environmental clean-up and remediation. (2,3). This method is based on the irradiation of semiconductor particles and reactive features of electron-holes pairs generated when catalyst is illuminated with energy greater than the semiconductor bandgap (4).

TiO_2 has been widely applied photocatalytic reactions, however, its wide band gap energy (3,2 eV for anatase) means that only 5% of the solar spectrum is used (3). In order to improving the photo-property and expand the photo response of titanium dioxide, it can be used selective doping with transition metals into matrix TiO_2. The Fe^{3+} can be considered the most appropriate dopant due to the ionic radius (0,64 Å) be similar to Ti^{4+} (0,68 Å), which induces the incorporation of the cation in the lattice structure of TiO_2, reducing bandgap (5,6).

Thus, recent studies were focused on the production of iron-doped TiO_2 prepared by wet impregnation with different Fe contents (5, 8, 10 e 15% wt. Fe) and calcined at 400 °C. Characterization of materials was performed by TGA / DSC, BET area, pore diameter and pore volume, PZC, EDX, SEM, DRX, FT-IR, spectroscopy photoacoustic and spectroscopy Mossbauer. The photocatalytic reduction of Hg^{2+} to Hg^0 was performed in batch stirred tank reactor (BSTR), in the presence of catalysts, under four different conditions reactions.

The results show that addition of iron modify the electronic structure of TiO_2 by the incorporation of Fe^{3+} in matrix of semiconductor. Although only 8 wt% showed improving the photo-activity in reduction, with a total Hg^{2+} removal (100%, 90 minutes) ratio for Hg^0.

For the iron content of 5wt% the ratio of reduction of Hg^{2+} were 67% in 180 minutes of reaction, this is probably due to the low specific area surface ($7 m^2 g^{-1}$) and pore volume ($0,0003 cm^3 g^{-1}$) when compared to TiO_2 8 wt% Fe, that displays values of specific area surface and pore volume of 14 $m^2 g^{-1}$ and 0,0092 $cm^3 g^{-1}$, respectively.

For the iron content of 10 and 15wt% a large amount of iron ions dose not substitute for Ti sites, but forms hematite phase (α-Fe_2O_3), formation of hematite pellets, disadvantaging the insertion of Fe^{3+} in the crystalline structure of TiO_2, besides reducing the photocatalytic activity (5).

References

1. K. Tennakone, U.S. Ketipearachchi. Applied Catalysis B: Environmental, **1995**, 5, 343-349.

2. X. Wang, S.O. Pehkonen, A.K. Ray.Electrochimica Acta, **2004**, 49, 1435-1444.
3. G.G. Lenzi, C.V.B. Fávero, L.M.S. Colpini, H. Bernabe, M.L. Baesso, S. Specchia, O.A.A. Santos.Desalination, **2011**, 270, 241-247.
4. M.J. López-Muñoz, J.Aguado, A. Arencibia, R. Pascual. Applied Catalysis B: Environmental, **2011**, 104, 220-228.
5. S. Zhu, Y. Li, C. Fan, D. Zhang, W. Liu, Z. Sun, S. Wei. Physica B, **2005**, 364, 199-205.
6. R.M. Mohamed, M.A. Salam. Material Research Bulletin, **2014**, 50, 85-90.

Performance of sulfaquinoxaline degradation by Fenton and photo-Fenton processes

V. R. Urbano[1, 2], M. C. S. Teran[1], M. G. Maniero[2], J. R. Guimarães[2], M. Pérez-Moya[1]
[1]Escola Universitaria d'Enginyeria Tècnica Industrial de Barcelona, Universitat Politècnica de Catalunya, BARCELONA TECH;
tel. +34 93 4137275, e-mail: montserrat.perez-moya@upc.edu; EUETIB, C/Comte d'Urgell 187, 08028-Barcelona, Spain.
[2]School of Civil Engineering, Architecture and Urban Design, University of Campinas, Unicamp;
tel. +19 3521 2378, e-mail: jorober@fec.unicamp.br; 951 Albert Einstein Avenue, 13084-971, Campinas, Brazil.

The presence and effects of contaminants of emerging concern on environment has been largely discussed on literature. The scientific community agrees in the lack of technologies for a properly degradation of these compounds and the needs to implement alternative treatments of the conventional ones. Sulfaquinoxaline is an antimicrobial from the sulfonamides class, used mainly in veterinary medicine. Scientific studies show the presence of sulfaquinoxaline on environment at concentrations ranging from 0.01 to 112.1 ng L^{-1} (García-Galán et al., 2011 a; 2011 b; Iglesias et al., 2012).

Recently, sulfaquinoxaline (SQX) degradation by a combination of UV radiation with hydrogen peroxide (Liao et al., 2016) and solar irradiation (Fur et al., 2012; Hoff et al., 2014) were reported. Fenton and photo-Fenton are advanced oxidation processes based on hydroxyl radical generation that has been proved being an efficient technique for sulfonamides degradation (Pérez-Moya et al., 2011; Yamal-Turbay et al., 2014).This work proposes study the efficiency of Fenton and photo-Fenton processes for sulfaquinoxaline degradation.

For preliminary studies the experimental setup consisted in use hydrogen peroxide (H_2O_2) corresponding to the stoichiometric concentration to get complete SQX mineralization (25 mg L^{-1} of SQX to 118.68 mg L^{-1} of H_2O_2). Experiments were conducted in a batch reactor of 2 liters. The experimental conditions adopted were as follows: 118.68 mg L^{-1} of H_2O_2, SQX 25 mg L^{-1}, 10 mg L^{-1} of iron, pH 2.8 and temperature of 26 °C. A sunlight lamp (Ultra-Vitalux®, Osram, 300W) was used. The mineralization and degradation of the compound were evaluated monitoring the total organic carbon concentration using a TOC analyzer and concentration of SQX using a HPLC with diode array detector.

Photo-Fenton process was slightly more effective to degrade SQX when compared with Fenton process (Figure 1). However, the photo-Fenton process improved the mineralization of the compound of 0% to 41% after 240 minutes of reaction. High mineralization of SQX is as important as high degradation to evaluate the efficiency of the process studied, since the products of degradation can be more harmful than the original compound.

Sulfaquinoxaline reduction was higher than 80% after 180 minutes of reaction on photo-Fenton process, so probably using a higher H_2O_2 concentration even better performance of the process would be reached, decreasing the time of reaction and/or increasing the degradation.

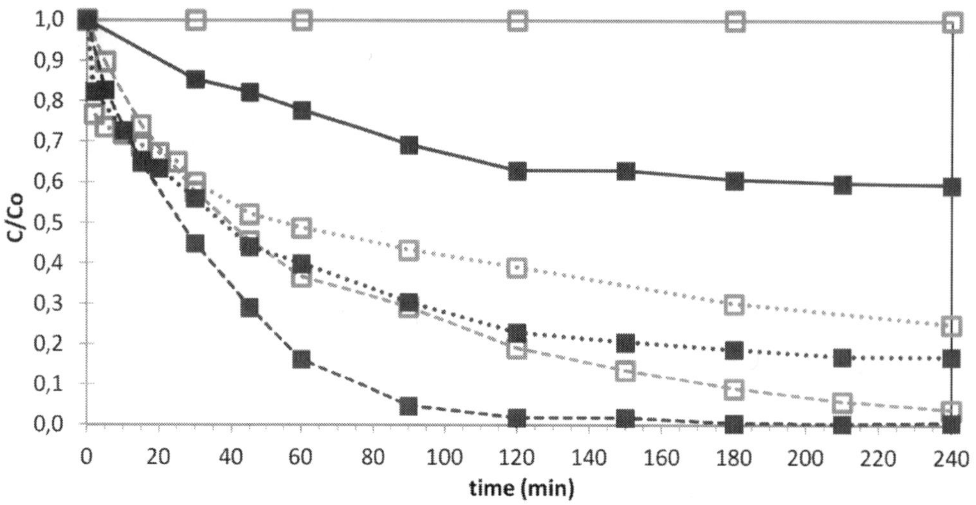

Figure 1. Fenton (grey color) and photo-Fenton (blue color) performance of sulfaquinoxaline and its byproducts were evaluated: SQX concentration (dotted line), TOC evolution (solid line) and residual H_2O_2 (discontinuous line).

After this preliminary screening a 2^2 design of experiments with star points and four centre points for statistical validity are planned in order to determine enhanced conditions for the photo-Fenton mineralization of SQX solutions, the dependence of the treatment output on factors such as hydrogen peroxide amount and the presence or absence of source of light.

Moreover a more detailed evaluation of the mineralization efficiency and toxicity reduction obtained in this study will be carried out using Gram-positive (*Staphylococcus aureus*) and Gram-negative (*Escherichia coli*) microorganisms.

Acknowledgements

The authors gratefully acknowledge the financial support received from the Spanish Ministry of Economy and Sustainability, the European Regional Development Fund and São Paulo Research Foundation through projects SIGERA (DPI2012-37154-C02-01), Generalitat de Catalunya (2009SGR-861) and FAPESP (2013/04656-8 and 2013/09543-7). Authors acknowledge CAPES for providing scholarship to V. R. Urbano and FAPESP for the research grant of M. G. Maniero (2013/-7817-2).

References

C. Le Fur, B. Legeret, P. De Sainte Claire, P. Wong-Wah-Chung, M. Sarakha, Rapid Communications in Mass Specrometry, 27 (2013) 722.

M.J. García-Galán, T. Garrido, J. Fraile, A. Ginebreda, M.S. Díaz-Cruz, D. Barceló, Analytical and Bioanalytical Chemistry, 399 (2011 a) 795.

M.J. García-Galán, M.S. Díaz-Cruz, D. Barceló, Environment International, 37 (2011 b) 462.

R. B. Hoff, L. Meneghini, T. M. Pizzolato, M. C. R. Peralba, M. S. Díaz-Cruz, D. Barceló, Analytical Chemistry, 86 (2014) 5579.

A. Iglesias, C. Nebot, J. M. Miranda, B. I. Vázquez, A. Cepeda, Environmental Science and Pollution Research, 19 (2012) 3235.

Q. Liao, F. Ji, J. Li, X. Zhan, Z. Hu, Chemical Engineering Journal, 284, (2016) 494.

M. Pérez-Moya, H. D. Mansilla, M. Graells, Journal of Chemical Technology and Biotechnology, 86 (2011) 826.

E. Yamal-Turbay, L. P. González, M. Graells, M. Perez-Moya, Environmental Technology, 35 (2014) 1695.

Removal of nonsteroidal anti-inflammatory drugs by photocatalytic processes from surface waters

V. Trousil[1], Z. Blažková[1], E. Slehová[1], J. Palarčík[1], O. Machalický[2], J. Cakl[1]
[1]University of Pardubice, Faculty of Chemical Technology, Institute of Environmental and Chemical Engineering, Studentská 95, 53210, Pardubice, Czech Republic
[2]University of Pardubice, Faculty of Chemical Technology, Institute of Organic Chemistry and Technology, Studentská 95, 53210, Pardubice, Czech Republic

Pharmaceuticals belong to a class of substances, which are very often found in outlets of wastewater treatment plants or in surface waters. These substances have by their nature a strong biological effect, so non-target organisms are affected. Diclofenac and naproxen are pharmaceuticals from the group of Nonsteroidal anti-inflammatory drugs (NSAIDs). These two pharmaceuticals are ubiquitously presented in surveyed surface waters. These substances are able to affect organism across the food chain at environmental concentrations. Also diclofenac and naproxen were found in bile of wild species *A. brama* and *R. Rutilus*.

Diclofenac and naproxen (500 µg/L) were removed by heterogeneous photocatalysis from model samples. As a source of UV, the UV-LED diodes Luminus with a peak emmision of 365 nm were used. 0.5 g/L of TiO_2 (AV-01, Precheza) was used as a catalyst. Also the influence of hydrogen peroxide H_2O_2 or pH adjustment were explored.

Photocatalytic processes showed good results in diclofenac and naproxen removal. These values varied from 80 to almost 98 % for naproxen and diclofenac. Removal rate was also enhanced by H_2O_2 addition or acidic pH conditions.

Although good results were obtained in photocatalytic processes, the feasibility of this method on real waters was investigated. Samples from surface river water were obtained from Labe, Pardubice, during March 2016. Samples were purified and preconcentrated because of detection of pharmaceutical residues in these types of waters. For preconcentration, SPE manifold (Macherey-Nagel) and OASIS HLB (Waters) 6 mg sorbents was used. Spiked surface water samples were treated by heterogeneous photocatalysis. Also, the influence of hydrogen peroxide H_2O_2 and pH adjustment were explored as well. Also, basic water quality parameters were done before and after heterogeneous photocatalysis.

Heterogeneous photocatalysis of surface water samples showed lower efficiency than in case of model samples due to non-selective character of these photocatalytic processes. Diclofenac was removed by almost 94 %, but naproxen only by 64 %. However, H_2O_2 addition or acidic pH conditions enhanced the photocatalytic processes as well. Photocatalytic processes also affected parameters as a COD (decreased by 27 %). Sufficient results were also obtained in bacteria removal (85 %).

Removal of selected organochlorine compounds by ozone based processes

K. Šimkovič, J. Derco, J. Dudáš, B. Urminská
Faculty of Chemical and Food Technology, Slovak University of Technology, Radlinského 9, 812 37 Bratislava, Slovakia; tel. +421917515472, e-mail: xsimkovic@stuba.sk

Discharging various micropollutants into aquatic environment in general and in particular substances classified as priority, hazardous and persistent generate serious concerns due to their potential adverse effects on human health and the living organisms in the environment[1,2]. Mostly these are synthetic and non-synthetic substances and their effective removal requires non-traditional processes, novel approach and technologies[3]. Ozonation and ozone based processes are considered the prospective ones[4].

The results of examination of the ozonation process and combined process of ozone with zero-valent iron nanoparticles (nZVI) are presented. Five selected organochlorine pesticides, i.e. hexachlorobenzene (HCHB), hexachlorobutadiene (HCHBD), pentachlorobenzene (PCHB), (1r,2R,3S,4r,5R,6S)-1,2,3,4,5,6- hexachlorocyclohexane (common name lindane, γ-HCH; abbreviation LIN) and 1,4,5,6,7,8,8-hepta-chloro-3a,4,7,7a-tetrahydro-4,7-methano-1H-indene (common name heptachlor, abbreviation HCH). First three of these compounds are specified as priority hazardous substances while all pesticides except for HCHBD are listed in the Stockholm Convention of Persistent Organic Pollutants[2,4,5].

We investigated the effectiveness of 5 $ml.l^{-1}$ nZVI NANOFER 25 suspension for the optimization of micropollutants removal from model water. This suspension contains 14–18 wt. % of Fe, 2 - 6% of Fe_3O_4, 0-1% of carbon and 80% of water. Specific surface area of these nanoparticles is larger than 25 $m^2.g^{-1}$ and specific gravity is in the range 1.15 – 1.25 $g.cm^{-3}$ (20°C)[6]. Experiments were carried out in a reactor with effective volume of 2 dm^3 under laboratory conditions. A Life Tech ozone generator with the maximum ozone production of 5 $g.h^{-1}$ was used. Flow rate of oxygen was Q_{O2} = 60 $l.h^{-1}$ (T = 21 °C, p = 107 650 Pa) and ozone generator was operated at 50% of the maximum power. Ozone was prepared from pure oxygen. The mixture of ozone and oxygen was transported to the reactor via a frit for fine bubble transfer between gas and liquid phase. The effectiveness of 5 $ml.l^{-1}$ NANOFER 25 nZVI suspension (equal to 1g Fe^0) including combined O_3/nZVI processes in model water was also investigated. In the process of using nZVI without ozonation, the reaction mixture of model water and nZVI was homogenized by inert gas – nitrogen N_2. The suspension used in the experiments was added into the reactor at the beginning of the process.

These results indicate that nano zero-valent iron type NANOFER 25 is capable of removing organochlorine pesticides from wastewater to varying extents. NZVI shows very promising results in regard to removal efficiency of the majority of substances. High removal efficiencies were observed for all pesticides during the first 2 min of the process using nZVI (more than 92%). Higher degradation efficiencies of LIN were measured in single nZVI process compared to the ozonation process. By combination of both processes - O_3/nZVI removal efficiency of organochlorinated compounds reached slightly higher values in comparison to single O_3 process. On the other hand, the process efficiency was lower than in the case of single nZVI. Lindane was better removed in the absence of ozone. It can be concluded that the use of nZVI for the removal of synthetic organic substances is a promising option due to its easy application, high reactivity and efficiency.

This work was supported by the Slovak Research and Development Agency under the contract No. APVV-0656-12. The authors also wish to thank for the support from VEGA Grant 1/0859/14.

1. Directive 2000/60/EC of the European Parliament and of The Council of 23 October 2000 establishing a framework for community action in the field of water policy, Off. *J. Eur. Communities*, L 327/1 **2000**.
2. Decision No. 2455/2001/EC of the European Parliament and of the Council establishing the list of priority substances in the field of water policy and amending Directive 2000/60/EC, 2001. *Off. J. Eur. Union*, OJ L **2001**, 44, 1–5.
3. Hollender J, Saskia G, Zimmermann SG, Koepke S, Krauss M, McArdell CHS, Ort Ch,
4. Stockholm Convention on Persistent Organic Pollutants, 23 May 2001, Stockholm. **2001**.
5. Stockholm Convention on Persistent Organic Pollutants, 9 May 2009, Geneva. **2009**.
6. www.nanoiron.cz

The assessment of operation of oil refinery wastewater treatment plant by dynamic simulations

I. Balážová Pijáková, J. Derco
Institute of Chemical and Environmental Engineering, Faculty of Chemical and Food Technology, Slovak University of Technology, Radlinského 9, 812 37 Bratislava, SK, e-mail: ivana.pijakova@stuba.sk

Modelling is an effective tool to investigate ecological and chemical quality of water resources. In this paper, we present the models that were applied for calculation of WWTP effluent quality for quantifying the impact of discharged wastewater from this WWTP on receiving water. Dynamic simulations were performed using a real experimental data.
The models for similar wastewater treatment plant and receiving water were developed. The ASM1 model was used for modelling industrial WWTP in this study. A simplified model was applied due to very similar biodegradability, i.e. BOD_5/COD ratio of the studied industrial wastewater in comparison with municipal wastewater. Simple Water Quality Model was used for river water quality modelling. Dissolved Oxygen Sag Curve was also applied as an alternative model to calculate dissolved oxygen concentration profile in the river.
The main goals of this study were:

1. to investigate the effect of changed operating conditions on the performance of studied WWTP
2. to study the effect of hydraulic loads due to operational problems on the performance of studied WWTP
3. assessment of WWTP capacity at oil refinery for additional industrial wastewater treatment
4. to investigate the impact of the current WWTP operation on recipient quality

The results of simulation show that the WWTP will be able to meet effluent standard limits after increase in temperature and loads. The results of simulation aimed at the impact assessment of oil refinery WWTP effluent on receiving water quality show that the values of monitored parameters in the river are complying with the Environmental Quality Standards.

Acknowledgements

This work was supported by the Slovak Research and Development Agency under the contract No. APVV-0656-12. The authors wish to thank also for the support of the VEGA Grant 1/0859/14.

Thermodynamic limits for dehalogenation of producer gas by Na_2CO_3, $NaAlO_2$, $CaO/CaCO_3$, $BaO/BaCO_3$ and mixed CaO-$BaO/BaCO_3$ sorbents

[1]K. Svoboda, [2]J. Leitner, [1]J. Havlica, [1,3]J. Brynda, [3]S. Skoblia, [1]M. Hartman, [1]M. Pohořelý, [4]H.-T. Hsu
[1]Institute of chemical process fundamentals of the CAS, v.v.i., Rozvojová 135, 165 02 Praha 6, CZ, E-mail: svoboda@icpf.cas.cz Tel. + 420 220 390 160 Fax: +420 220 920 661
[2]Dep. of Solid State Engineering, Faculty of Chemical Technology, University of Chemistry and Technology Prague, Technická 3, 166 28 Prague 6, CZ, E-mail: jindrich.leitner@vscht.cz
[3]Department of Gas, Coke and Air Protection, University of Chemistry and Technology Prague, Technická 5, 166 28 Prague 6, CZ, E-mail: Sergej.Skoblja@vscht.cz
[4]Chemistry Division, Institute of Nuclear Energy Research, No. 1000, Wenhua Rd., Jiaan Village, Longtan Township, Taoyuan County 32546, Taiwan (R.O.C.)
E-mail: dukeininer@iner.gov.tw Tel.: 886-3-4711400 ext. 5346

The advanced applications (e.g. catalytic processes and fueling of high temperature fuel cells) of producer gas from gasification processes are extremely sensitive to inorganic and organic contaminants in the gas and on limits of cleaning processes. Among the gas contaminants, stable halogen containing compounds, HCl, HF and HBr produced in the gasification process should be considered as clear candidates for deep removal. Producer gas cleaning for attaining hydrogen-halides concentrations approx. below 1 ppm_v requires either application of low temperature wet absorption processes or, particularly for fueling of high temperature fuel cells (SOFC), deployment of medium and high temperature (600 – 1200 K) cleaning methods with solid sorbents. Our study is focused on thermodynamic analyses and minimization of Gibbs energy for gas de-halogenation (removal of HCl, HF and HBr) by selected solid sorbents (Na_2CO_3, $NaAlO_2$, $CaO/CaCO_3$, $BaO/BaCO_3$, and mixtures containing CaO-$BaO/BaCO_3$). The cheap Na_2CO_3 and $NaAlO_2$ based sorbents are suitable for deep removal of HCl and particularly HBr at temperatures below 800 K, but they are less suitable for HF removal. For attaining low HF equilibrium concentrations rather the $CaO/CaCO_3$ based sorbents are the best choice. The promising $BaO/BaCO_3$ based sorbents and mixed CaO-BaO/BaCO3 sorbents for removal of HCl, HBr and HF (or even their mixtures with H_2S) are dependent on appropriate solid solution form stabilizing BaO for its sorption applications, because BaO has significantly more favorable equilibria than $BaCO_3$ with all studied hydrogen-halides. Unfortunately stabilization of BaO in mixtures with CaO and similar compounds requires temperatures over 1150 K – 1200 K.

Tell and sell safety knowledge

N. Schmersal-Hoekstra, [1]
Fluor B.V. Taurusavenue 155, 2132 LS Hoofddorp, the Netherlands
Tel 0031235432532, e-mail: nynke.schmersal@fluor.com.

Late September 2015 in Nice (France), during the 10th European Congress of Chemical Engineering there was a special two-day event on process safety. During a round table discussion, a thesis was introduced by Shakeel Kadri, the executive director of the Center for Chemical Process safety. The thesis entitled: 'process safety needs to be part of the curricula at universities'. After the introduction a lively discussion started and was joined by all the safety experts and engineers present. Is this correct, is process safety not taught at universities; have graduates not been exposed to safety or operational design aspects in their education? If this is indeed the case, what is the reason for this; should actions be taken? What can we do, as process safety engineers, to enhance safety knowledge and create a culture of process safety from the educational level upwards? The thesis and questions are discussed in this paper, as view in the Dutch context.

In the Netherlands, safety is part of the curricula of the bachelor phase and the master programs of the three technical universities. However, the curricula do not show enough details to verify if there is sufficient focus on process safety.

Vapro has investigated if process safety is part of the curriculum of vocational universities (HBO in the Netherlands, which results in a Bachelor Degree) (report by A. van der Velden et al, 2015). (Process) safety is part of the curricula, however to improve education in safety in chemical engineering it is recommended to develop a curriculum (or to improve the existing curricula) for safety in Chemical Engineering.

In the mean time, how does industry cope with new employees with limited process safety knowledge? Employers fill this gap in knowledge by training on the job and by providing training to their employees. That is how it has been done mainly up to now. But if we want to improve the process safety knowledge in industry, it starts with education of the next generation in an earlier phase, to create the right mindset.

Is it possible to create a focus on process safety? For that question, we can take a look at the initiatives for a focus at technical studies in general. To counter the trend of the reduction in the number of technical students, all parties involved, employees, educational institutes and government both in Europe and in the Netherlands have started initiatives. These initiatives show a positive result for technical studies in general. This approach could be translated to an approach to create focus in education for a process safety.

To improve the level of process safety knowledge in industry, we as safety engineers, as professionals, should contribute to create a focus on process safety. We should enhance knowledge in safety and specific in process safety, by sharing knowledge with our colleagues and within our networks. We have to act as ambassadors, we have to 'tell and sell' process safety.

Enhancement of pipeline safety by early leak detection using external Vapor Sensing Tubes (VST)

J. Zach, Ph.D., [1] Dr. Ing. Walter Knoblach [2]
AREVA GmbH, Dep. Monitoring and Diagnostics, Paul-Gossen-Str. 100, 91052 Erlangen, Germany;
tel. +49 9131 900 911 44, e-mail: jan.zach@areva.com; [1] tel. +49 9131 900 92367, e-mail: walter.knoblach@areva.com; [2] fax: +49 9131 900 941 69.

The risk of hydrocarbon and toxic spills increases with the aging of chemical plant related infrastructure. The need for early detection of such kinds of pollutions is essential, particularly in view of potential environmental damage, cleanup costs, and the loss of public confidence in industry's ability to quickly respond to leaks.

The first line of defense in the event of a leak is early detection and operational response in order to limit product loss from the pipe. Sophisticated mathematical modeling of flow regimes coupled with multiple instruments at pipelines has increased the sensitivity of the conventional leak detection methods. However, despite these technological improvements significant leaks have occurred recently on major pipeline systems with damaging consequences.

Leak detection from pipelines or vessels can be significantly enhanced by the application of a leak detection system using so-called external vapor sensing tube technology. This publication describes the LEOS® leak detection system. Its application and operational experience in high sensitivity locations is discussed with selected case studies. The system does not compete with conventional leak detection techniques (e.g. mass balance based methods) but is a complementary measure ensuring extremely low detection thresholds independent of throughput and flow conditions of the pipeline. As it is designed for weeping and pinhole leaks, it reliably detects leakages at very early stage, long before they lead to severe environmental damage or public hazard.

An escape pathfinding algorithm in gas fire of large petrochemical industry

Zhenxing Cheng, [1]Qunxiong Zhu, Yuan Xu, Yanqing Wang
[1]College of information science & technology, Beijing University of Chemical Technology, No.15, Beisanhuandong Road, Chaoyang District, Beijing, China; zip-code: 100029; phone: (+86)-010-64426960; fax: (+86)-10-64437805;e-mail: zhuqx_buctielab@163.com;

In the field of industrial production, due to the lack of scientific guidance, the incident of large petrochemical industrial leads to huge casualties and major economic losses. Proposing an effective pathfinding algorithm in gas fire for escaping plays a significant role in human's survival.

This paper provides an escape route for individuals in large petrochemical industry. In preliminary, wireless sensors will be set to detect fire in real time. Tanks and roads are surrounded by sensors. Based on the inference, the gas emitting velocity $V(t)$ has linear correlation with time. Here, ignoring the affection by resistance decreasing, the gas fire expansion model can satisfy accelerating expansion: $S = a*t^2 + b*t + c$, these parameters will be ensure by the data from sensor network and the fire expansion model can be built.

After that, the escape pathfinding algorithm in gas fire will be provided in detail by the following three steps.

Firstly, this process will be divided into two sub procedures. In beginning, based on fire expansion model above, calculate the in-fire time of each tank by hierarchical traversal, this value should be updated if the outcome calculated in this process is smaller than it was. Add the tank into set if it is in the state of fire increasing while individual arrives, by using this method, some tanks firing completely or not in fire can be excluded, and the tank which will be taken into next step is reduced. At last, the tanks with light color as figure 1in the state of fire increasing can be searched.

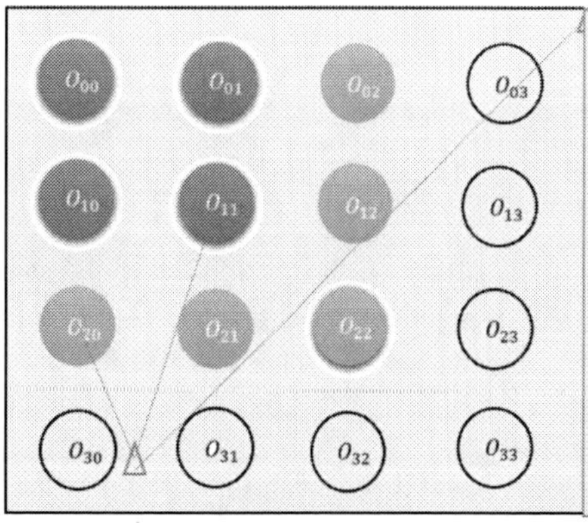

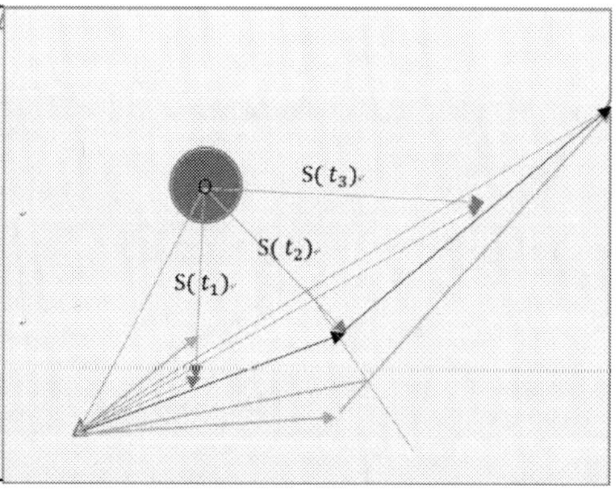

Fig.1 Firespreading among tanks in our scene   Fig.2 Calculate critical point in each direction

Additionally, search for the critical points in fire's moving margin. To each tank got by the former process, cosine theorem will be used in each direction with the fire expansion model. As shown in figure 2, we can infer $S_h(t_2)^2=S(t_2)^2+OH_2-2*OH*S(t_2)*cos\theta$, in which the only variable t_2 can be solved. With this method, all critical positions can be ensured and stored in the critical point set.

Thirdly, use greedy idea as the strategy to find the next insertion. Firstly, based on the real time positions of both human and exit, a straight line can be drawn. Through the points in critical set above, select the max angle direction by greedy algorithm for keeping the path safe and short. Then get the next insertion as intermediate point in the path. The greedy idea can be proved by figure 3.

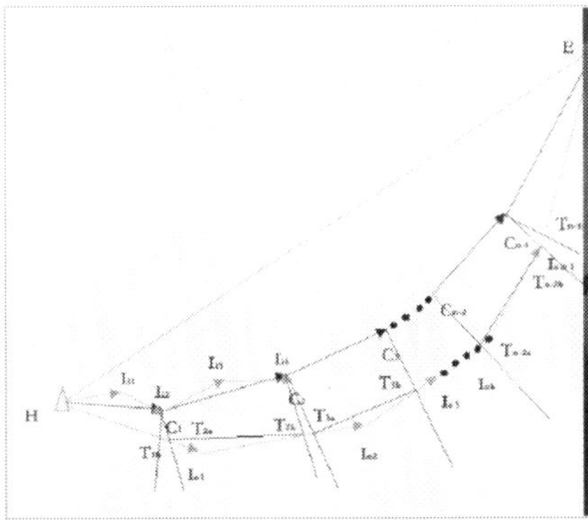

Fig. 3 Choose shorter safety path

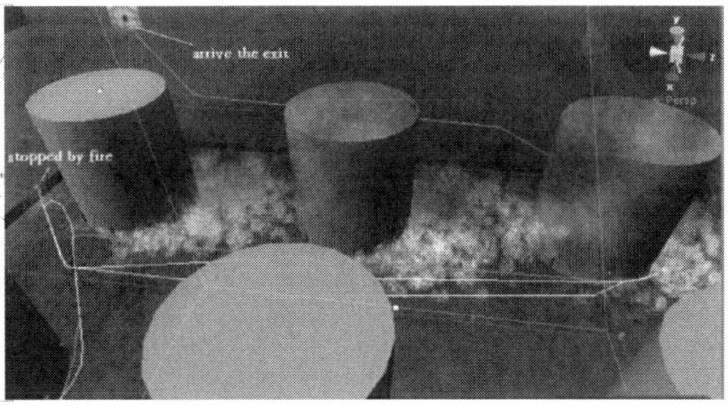

Fig. 4 The comparison of two pathfinding algorithm

The former process is a sub process until arriving the exit, and three dimension technology will be applied for experiment. From figure 4, it is clear that the individual with the proposed algorithm can escape from the gas fire while the individual without predicting the spread of fire's moving is stopped. In table 1, it is a comparison between two algorithms which can keep individual in safety. From the table, we know that the path by the proposed algorithm is shorter. Therefore, the algorithm we proposed can provide individual a safe and shorter path in emergency escape system of large petrochemical industry.

Table 1. The comparison between two pathfinding algorithms

Start point of individuals	Destination point of individuals	path without predicting		path we proposed	
		Access	Distance	Access	Distance
(37.5, 5.0, 9.6)	(-50.0, 0.0, -30.0)	N	——	Y	181.5398
(47.3, 5.0, 12.8)	(-50.0, 0.0, -30.0)	Y	175.2583	Y	136.0492
(32.0, 5.0, 48.8)	(-50.0, 0.0, -30.0)	Y	130.1546	Y	130.3085
(44.6, 5.0, 32.0)	(-50.0, 0.0, -30.0)	N	——	Y	147.9065
(27.5, 5.0, 12.7)	(-50.0, 0.0, -30.0)	Y	137.5653	Y	115.9371

ACKNOWLEDGEMENTS

This research has been supported by National Natural Science Foundation of China under Grant No. 61573051, No. 61472021, Beijing Natural Science Foundation under Grant No. 4142039, Open Fund of the State Key Laboratory of Software Development Environment under grant SKLSDE-2015KF-01, and Fundamental Research Funds for Central Universities of China under Grant No. YS1404.

Autoignition temperature data for binary mixtures of alcohols with water

J. Skřínský[*], J. Vereš, J. Trávníčková, A. Dalecká
Energy Research Centre, VŠB-TU Ostrava, 17. listopadu 15/2172, 708 33 Ostrava, Czech Republic; tel. +420 597 324 948, e-mail: jan.skrinsky@vsb.cz

Mixtures of water with alcohols are important in numerous engineering applications. AIT is an important variable used to characterize fire and explosion hazard of liquids and is needed for safe handling, storage and transportation.

This paper describes application of a model for predicting the AIT of binary alcohol mixtures. To confirm the predictive efficiency of the derived AIT values, the model was verified by comparing the predicted values with the experimental data for the studied selected mixtures: water + methanol; water + ethanol; water + 1-propanol; water + 1-butanol; and, water + 1-pentanol.

Measurements of the autoignition temperatures were obtained from the AIT-12-1 autoignition tester (SN: 321-OZM-13-01, OZM Research s.r.o., Czech Republic). The experimental set-up consists of an electrically heated crucible furnace capable of attaining a temperature of 600°C or higher, commercial 500-ml borosilicate round-bottom short-necked boiling flask closely wrapped in reflective aluminium foil and fine chromel-alumel thermocouple (36 B and S gage) for measuring the gas temperature inside the flask. The apparatus involved control devices that program the instrument to heat the sample at a specific heating rate within a temperature close to the expected autoignition temperature. Autoignition temperatures of the pure liquid flammable substances were determined experimentally according to the EN 14522:2005 and ASTM E659-78:2005 standards with an accuracy of 1.5 °C.

Dimensionless experimental formulae based on a rational reciprocal function for correlation of AIT data of binary mixtures of two flammable components have been applied. The formulae are based on data obtained from AIT experiments. The proposed approach requires only two coefficients, molar fraction of components and AIT temperatures of the pure flammable components to be known in advance.

Results reveal that immiscibility in the two liquid phases should not be ignored in the prediction of AIT. Potential application for the model concerns the assessment of fire and explosion hazards, and the development of inherently safer designs for chemical processes containing binary partially miscible mixtures of aqueous-organic system.

Development of new ammonia sensor for industrial wastewater effluents

*Melaschenko I.A.[1] and Moozyckine A.U.[2] 1. Kharkiv National University, 21 Frunze St., Kharkiv, 61002, Ukraine, mobile: +380508663535, e-mail: iam001@meta.ua, 2. Swansea University, Singleton Park, Swansea, SA2 8PP, UK.

The large amounts of wastewater effluents are produced during manufacturing processes in ammonia fertilizers industry; this ammonia containing wastewater is routinely directed to biological water cleaning facilities.

The concentration of ammonia in wastewater effluents strongly affects the efficiency of their biological treatment, and when it reaches levels above 0.05 kg×m^{-3} this can result in high mortality rate of bacteria in aeration tanks. Therefore, continuous control of ammonia levels in wastewater effluents presents an important challenge for the industrial water treatment pilot plants.

Many well developed and efficient analytical solutions (e.g. use of glass and liquid ammonia-selective membranes, enzyme-coated electrodes, etc.) are deemed unsuitable for the conditions existent in wastewater technological streams containing high amounts of organic solvents, synthetic oils and surfactants.

As a possible solution to this long-standing problem, we have recently developed solid-state ammonia-selective sensor based on insoluble iridium salts of the following stoichiometry: $\{IrS_2\}_x{*}\{(NH_4)_2\{IrCl_6)\}_y$.

Iridium sulfide provides a perfect inert retaining matrix for ammonium hexachloro-iridate, due to its extremely low solubility (solubility product for IrS_2 $K_{sp}=10^{-75}$), low electrical resistance and high chemical inertness.

Selectivity coefficients of the developed sensor are as follows:

$$k_{NH4^+/Cu^{2+}} = 10^{-5}; \quad k_{NH4^+/Pb^{2+}} = 10^{-6}; \quad k_{NH4^+/Zn^{2+}} = 10^{-5}$$

The developed sensor shows stable work within an extremely wide range of pH values: 2 – 11, and has an operating range of NH_4^+ concentrations 0.005 – 0.05 kg×m^{-3}.

The impedance R of the developed sensor is 15 MΩ.

The dependence of the electrode potential on ammonia concentration is liner within the range of indicated pH values, and the quotient has been determined as follows:

$$dE/d(lg\, a_{NH4^+}) = 54 \text{ mV}$$

From the body of experimental data, the optimal composition of ion-selective membrane has been determined, allowing for improved sensitivity, robustness and longevity. An advanced method for the manufacturing of the ammonia-sensitive membranes has been developed and tested.

We have found that, in the presence of high chloride ion concentrations in wastewater effluents, the sensitivity of the developed sensor is reduced substantially, and this creates the scope for the future research and developmental work.

Reliability approach to the study of resistance of the oil transport pipelines rehabilitated by composite materials

N. Abdelbaki [1*], M. Bettayeb [2*], M. Gaceb [2*], E. Bouali [2*], R. Bouzid [2*]
[1]University Akli Mohand Oulhadj of Bouira,10000, Algeria
[2]University M'hamed Bougara Boumerdes, Faculty of Hydrocarbons and Chemistry
* Laboratory of Petroleum Equipments Reliability and Materials, Independence Street 35000, Boumerdes,

lfep@umbb.dz

The management of the maintenance as well as the rehabilitation of the older sections of a network of transport by pipeline are fundamental and vital tasks for the safety and continuity of hydrocarbons delivery operations, from the production wells, to the transformation and consumption centres. They contribute effectively in maintaining and even, in the improvement of the performance of the installations and their availability, particularly following renovation operations. Indeed, and to ensure the availability of the network, each line of the latter, shall be inspected periodically and a repair plan is drawn up on the basis of the results obtained from inspections, thus, defining defects to be repaired and the forms of reparation to be implemented.

Repair of hydrocarbons transport pipelines using sleeves made of composite materials, is a widespread technique, with the advantage to intervene on the pipeline in full delivery of the product, without operation stoppage and without welding, significantly reducing the impact, risks, the duration and the cost of the operation. In this context, a study of resistance of these pipelines, reinforced with composite materials was made based on the reliability approach.

Calculation and experimental validation of maximum explosion pressure for C1-C5 alcohol mixtures in a 1 m^3 spherical closed volume

J. Skřínský*, J. Vereš, J. Trávníčková, A. Dalecká
Energy Research Centre, VŠB-TU Ostrava, 17. listopadu 15/2172, 708 33 Ostrava, Czech Republic; tel. +420 597 324 948, e-mail: jan.skrinsky@vsb.cz

Gas explosion severity parameters are the explosion pressure, defined as the highest pressure reached during an explosion in a closed volume at a given fuel concentration, and the maximum explosion pressure, found as the highest explosion pressure over the flammable range. These are needed for risk assessment, pressure vessel design for explosion containment and design of pressure relief systems.

In the present contribution we calculated the maximum pressure for gaseous alcohol-oxidiser mixtures in a spherical closed volume at different initial temperatures up to 250 °C. Calculation of the maximum rates of pressure rise is based on the assumption of chemical equilibrium and combustion in constant volume. The minimum free energy method is used to calculate the composition and properties of products.

The chemical equilibrium model assumes adiabatic conditions in constant volume, and formation of equilibrium-defined concentrations of post explosion compounds and their expansion due to the temperature rise caused by the liberated heat assuming ideal gas behaviour and ideal condensed phase products.

To confirm the predictive efficiency of the derived maximum pressure, the model was verified by comparing the predicted values with the experimental data for the studied mixtures: air + methanol; air + ethanol; air + 1-propanol; air + 1-butanol; and, air + 1-pentanol. Explosion pressures are calculated compared to the experimental ones and differences explained.

The shape of the explosion pressure curves with varying concentration is discussed at all investigated initial temperatures. The increase in the initial temperature lowers the explosion pressure, and increases the flammability range. The upper explosion limits increases and the lower decreases. When the mixture compositions approaches the flammability limits the explosion pressure drops sharply to zero.

Calculated explosion pressures show a reasonable agreement at the near stoichiometric concentrations. Outside this range the agreement becomes poor, the deviations from the measured values increase when the flammability limits are approached. Based on the pressure time traces different regimes of explosion development or combustion conversion have been identified. The regimes depend on the initial mixture composition, at given conditions.

Three conclusion have been made: (1) the calculated adiabatic flame temperature was found to be a weak function of the initial pressure and initial temperature; (2) the increase in the initial temperature and the initial pressure has no significant effect on the post explosion mixture composition; (3) other factors, like change of flow characteristics (turbulence level) during the explosion at elevated conditions, have no significant effect on the course of the explosion.

Flash point of organic ternary mixtures containing alcohols: experiment and prediction

[1]P. Dolníček, [2]J. Skřínský
Faculty of Safety Engineering, VŠB- TU Ostrava, Lumírova 13/630, 700 30 Ostrava, Czech Republic, tel. +420 597 322 835, e-mail: petr.dolnicek@vsb.cz; [2]Energy Research Centre, VŠB-TU Ostrava, 17. listopadu 15/2172, 708 33 Ostrava, Czech Republic, tel. +420 597 324 931, e-mail: jan.skrinsky@vsb.cz

In a given liquid, the flash point is the temperature determined experimentally at which the substance emits sufficient vapour to form a combustible mixture with air, with a lower flash-point value indicating relatively greater fire and explosion hazard. The fire and explosion hazard of liquids, such as essential oils, is primarily characterized by their flash point. In addition to the usage and accumulation of flammable liquids, such as is outlined above, the transportation requirements for these mixtures are primarily related to their flash-point values.

The predicted and experimental flash point value for the flammable mixtures of alcohols was obtained. As flammable mixtures in this contribution are considered ternary mixtures in which all three components are flammable. All of the predictions are done assuming ideal vapours above ideal and non-ideal liquid solutions. One of the purposes of the contribution is to show the erroneous prediction that could be made if an ideal solution is assumed when the solution is non-ideal. The vapour pressure as a function of T of each mixture component is obtained from the extended Antoine equation.

To confirm the predictive efficacy of the derived flash points, the model was verified by comparing the predicted values with the experimental data for the studied mixtures: methanol + ethanol + 1-propanol; ethanol + 1-propanol + 1-butanol; 1-propanol + 1-butanol + 1-pentanol.

An PM-4 Pensky-Martens Flash Point Analyzer (Petrotest, Germany), which meets the requirement of ČSN EN ISO 2719 standard, was used to measure the flash points for a aqueous-alcohol solutions at different compositions. The basic system configuration of the close cup tester consists of an external cooling system, test cup, heating block, electric igniter, measuring module, thermometer and indicator/operating display.

The predictive results of applied model describe the experimental data well. Based on this evidence, therefore, it appears reasonable to suggest potential application for our model in assessment of fire and explosion hazards, and development of inherently safer designs for chemical processes containing binary mixtures of flammable solvents.

Dimensionless experimental formulae based on a rational reciprocal function for correlation of flashpoint data of binary mixtures of two flammable components have been applied. The formulae are based on data obtained from flash-point experiments. The proposed approach requires only two coefficients, molar fraction of components and flashpoint temperatures of the pure flammable components to be known in advance.

Our results reveal that immiscibility in the three liquid phases should not be ignored in the prediction of flash point. Based on this evidence, therefore, it appears reasonable to suggest potential application for our model in assessment of fire and explosion hazards, and development of inherently safer designs for chemical processes containing binary partially miscible mixtures of flammable solvents.

Flash-point prediction for ternary mixtures of alcohols with water in order to improve their safety

[1]P. Dolníček, [2]J. Skřínský
Faculty of Safety Engineering, VŠB- TU Ostrava, Lumírova 13/630, 700 30 Ostrava, Czech Republic, tel. +420 597 322 835, e-mail: petr.dolnicek@vsb.cz; [2]Energy Research Centre, VŠB-TU Ostrava, 17. listopadu 15/2172, 708 33 Ostrava, Czech Republic, tel. +420 597 324 931, e-mail: jan.skrinsky@vsb.cz

Aqueous solutions of alcohols are often employed in the extraction and manipulation of labile materials such as proteins. Flashpoint is an important variable used to characterize the fire and explosion hazard of liquids and is necessary to know for safe handling, storage and transportation.

The predicted and experimental flash point value for the flammable mixtures of alcohols was obtained. As flammable mixtures in this contribution are considered ternary mixtures in which two components are flammable and one non-flammable. All of the predictions are done assuming ideal vapours above ideal and non-ideal liquid solutions. One of the purposes of the contribution is to show the erroneous prediction that could be made if an ideal solution is assumed when the solution is non-ideal. The vapour pressure as a function of T of each mixture component is obtained from the extended Antoine equation.

The flash point for a solution was estimated by equation of Le Chatelier, an extended Antoine equation for estimating the saturated vapour pressure, and the Wilson, NTRL and UNIQUAC models for estimating activity coefficients. To confirm the predictive efficiency of the derived flash points values, the model was verified by comparing the predicted values with the experimental data for the studied selected mixtures: water + methanol + ethanol; water + ethanol + 1-propanol; water + 1-propanol + 1-butanol; water + 1-butanol + 1-pentanol.

PM-4 Pensky-Martens and AB-5 Abel semi-automatic flash point analyzers (Petrotest, Germany), which meets the requirement of EN ISO 2719:2002 and EN ISO 13736:2013 European standards, were used to measure the flash points for a aqueous-alcohol solutions at different compositions. The basic system configuration of the used close cup tester consists of an external cooling system, test cup, heating block, electric/gas igniter, measuring module, thermometer and indicator display.

Dimensionless experimental formulae based on a polynomial and rational reciprocal functions for correlation of flashpoint data of ternary mixtures of two flammable components have been applied. The formulae are based on data obtained from flash-point experiments. The proposed approach requires only two coefficients, molar fraction of components and flashpoint temperatures of the pure flammable components to be known in advance.

Results reveal that immiscibility in the two liquid phases should not be ignored in the prediction of flash point. Expected application for the combination of the dimensionless correlations and the thermodynamic models concerns the assessment of fire and explosion hazards, and the development of inherently safer designs for chemical processes containing ternary partially miscible mixtures of aqueous-organic system.

Influence of the vent duct design in case of gas explosion in 20 litre sphere apparatus

V. Mikaczo[1], Z. Simenfalvi[2], G. Szepesi[3]
University of Miskolc, Faculty of Mechanical Engineering and Informatics, Institute of Energy Engineering and Chemical Machinery, Department of Chemical Machinery
Address: H-3515 Miskolc-Egyetemváros, Hungary. mikaczo.viktoria@uni-miskolc.hu

The main explosion values (K_{ST}, K_G) were determined in 20 litre sphere where the measuring must be executed according to standards. Based on this standardized method, this paper shows some experiments with different static activation overpressures without vent duct and with different vent duct design. Temperatures and pressures were measured by sensors with high frequency in the vessel and in different positions in the vent duct. The activation pressure induced by rupture discs with 3 bar_g burst pressures and in some lower pressures aluminium and plastic membranes was used. In every cases the membranes was installed in special holder to achieve a reproducible activation pressures.

However numerous studies was being regarded to vented explosions this paper's main goal to prepare a measurement method to describe the overpressure increment in vessel and vent duct using several vent duct design. CFD simulation were used to investigate the description of the pressure changes in the vent duct. This CFD simulation was validated by experimental measurements. Further aim was to use this numerical simulation of large volumes (>20L) loaded by gas explosion.

The formulas in the standard to determine the pressure increment for vented explosion have limited validations. This limitation relates to the diameter/length ratio of vent duct, the range of the static activation pressure, the volume of the vessel and the maximum reduced pressure. Based on experimental measurements and CFD simulations this paper gives opportunities to review those limitations.

---

[1] PhD, Student;
[2] Associate Professor
[3] Associate Professor

Towards pure enantiomers with supercritical carbon dioxide based crystallizations

E. Székely, M. Zsemberi, L. Lőrincz, M. Kőrösi, A. D. Zodge, [1]T. Sohajda, [2]J. Madarász
Budapest University of Technology and Economics, Department of Chemical and Environmental Process Engineering, Műegyetem rkp. 3., H-1111 Budapest, Hungary; tel: +3614632202, e-mail: sz-edit@mail.bme.hu [1]Cyclolab LtD., llatos út 7, H-1097 Budapest, Hungary; tel: +361347606 [2]Budapest University of Technology and Economics, Department of Inorganic and Analytical Chemistry, Műegyetem rkp. 3., H-1111 Budapest, Hungary; tel: +3614634047, e-mail: madarasz@mail.bme.hu

Pharmaceutical, food and flavour industries need large amounts of chemically pure, solvent free enantiomers of a large variety. Enantiopure chemicals are typically those which have an enantiomeric excess >99%, often even >99.9% purity is required. Optically active molecules can be obtained by several techniques from isolation from natural source towards total synthesis, the most common is still the optical resolution. Typically the procedures starting from a prochiral or racemate chemical do not provide a purity high enough, thus further separation and purification steps are necessary. [1] As further purification recrystallization and chromatographic methods are the most frequently applied, but the later one requires a large amount of solvent and also energy for the evaporation afterwards. On the other hand recrystallization (or if it is not possible crystallization with some chiral or non-chiral reagents) is time and also solvent consuming.

Supercritical techniques, namely supercritical fluid extraction [2-3], antisolvent precipitation [4-5] and enzymatic catalysis [6-7], efficiently applied to produce enantiopure compounds. Besides, supercritical carbon dioxide based techniques offer lower organic solvent consumption, variable solvent properties via controlling pressure, temperature and composition, thus a possibility for optimisation. Gas or supercritical antisolvent methods require short processing time and lower organic solvent residues in the crystals can be achieved compared to conventional crystallization methods.

By antisolvent precipitation with carbon dioxide we obtained pure enantiomers of some organic acids, namely ibuprofen, mandelic acid, *p*-chloro-mandelic acid, metoxy-phenylacetic acid by recrystallization of their diastereomeric salts and by repeated resolutions. While the first resolution step was more efficient with the half-equivalent method (resolving agent added in half molar equivalent to the racemate), further purification lead to higher selectivity with molar equivalent method (recrystallization).

Initial enantiomeric excess – enantiomeric excess of the product correlations were systematically compared to the melting point diagrams and results of conventional recrystallizations. The results comparing antisolvent (13 MPa, 45 °C, methanol) and atmospheric recrystallisations of ibuprofen – α-methyl-benzylamine salts are presented in Fig. 1. Not only the enantiomeric excess values of the raffinates were superior, to the conventionally obtained crystalline phases, but also higher yields were obtained by antisolvent precipitation.

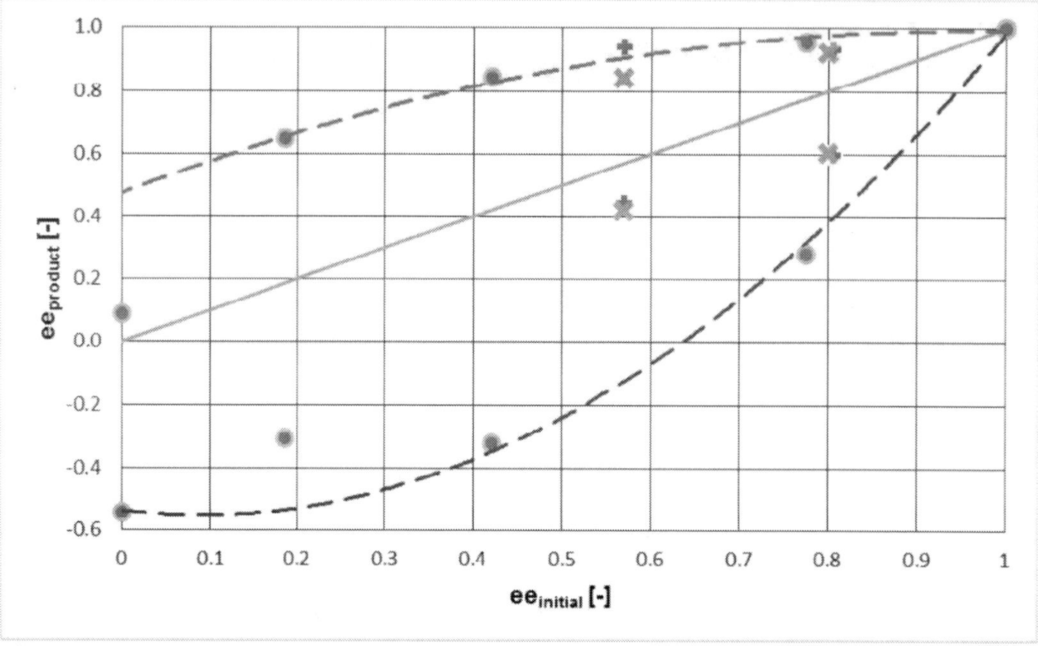

Figure 1. Comparison of the enantiomeric excess of the product as a function of initial enantiomeric excess (ibuprofen – α-methyl-benzylamine; dots: supercritical antisolvent, x atmospheric from ethanol, + atmospheric from methanol). Enantiomeric excess value of extracts (mother liquor) are above the reference line, enantiomeric excess value of raffinates (crystalline product) are under the reference line.

The research work was supported by the Hungarian Academy of Science via János Bólyai Research Grant (E.Sz.), DoHip Network, European Community's Seventh Framework Programme (FP7/2013-2016) under grant agreement no: 316959 (A.D.Z) and the Hungarian National Research Fund (K108979).

[1] Faigl F, Fogassy E, Nográdi M, Pálovics E, Schindler J: Separation of non-racemic mixtures of enantiomers: an essential part of optical resolution, Org. Biomol. Chem., 8:(5) 947–959 (2010).

[2] Szekely E, Simandi B, Laszlo K, Fogassy E, Pokol G, Kmecz I: Effect of achiral support on the resolution of tetramisole by supercritical fluid extraction, Tetrahedron-Asymmetry 13:(13) 1429-1434 (2002).

[3] Székely E, Bánsághi G, Thorey P, Molnár P, Madarász J, Vida L, Simándi B: Environmentally benign chiral resolution of trans -1,2-Cyclohexanediol by two-step supercritical fluid extraction, Ind. Eng. Chem. Res. 49:(19) 9349-9354 (2010).

[4] Kordikowski A, York P, Latham D, Resolution of ephedrine in supercritical CO2: A novel technique for the separation of chiral drugs, J. Pharm. Sci. 88(8):786-91 (1999).

[5] Bánsághi Gy, PhD thesis, Budapest University of Technology and Economics (2015).

[6] Utczás M, Székely É, Tasnádi G, Monek E, Vida L, Forró E, Fülöp F, Simándi B: Kinetic resolution of 4-phenyl-2-azetidinone in supercritical carbon dioxide, J. Supercrit. Fluids 55:(3) 1019-1022 (2011).

[7] Székely E, Utczás M, Simándi B, Kinetic enzymatic resolution in scCO2 – Design of continuous reactor based on batch experiments, J. Supercrit. Fluids, 79: 127-132 (2012).

Supercritical water gasification of microalgae and their constituents

G. Caputo, P. Rubio Blazquez, M. Dispenza, F. Scargiali, G. Marotta, A. Brucato
University of Palermo, Dep. of chemical, managment, informatics and mechanical engineering, viale delle scienze, ed. 6, 90128 Palermo, Italy.
tel. 09123896761, giuseppe.caputo01@unipa.it

Microalgae offer several advantages relative to lignocellulose biomasses to convert CO_2 into fuels. Microalgae are produced as wet biomass for which conversion technologies like pyrolysis and dry gasification are unfavorable due to high-energy demand for drying. Using water as a reaction medium would eliminate the need for the drying process. In this context, supercritical water gasification (SCWG) is a sound technology for the conversion of microalgae to produce high heating value product gases [1].
The main idea of SCWG is to take benefit from the special properties of SCW as a solvent and as a reaction partner that allow relatively fast hydrolysis of biomass and rapid degradation of the polymeric structure of cells. As a result, pressurized gas mainly rich in H_2, CO, CO_2 and CH_4 can be obtained [2] with limited amount of tar and char, thanks to high solubility of the intermediates in the reaction medium. This leads to high gas yields at relatively low temperature (≤700 °C).
To date, a limited number of studies have been reported on the SCWG of algae [3]. Due to the lack of exhaustive data, more experimental work is necessary to develop SCWG at industrial scale. A laboratory scale SCWG plant complete of heat recovery section and pre-heat of the biomass slurry to simulate more closely an industrial design was employed in this work. The reactor has a down-flow configuration with a volume of 20 dm^3. The plant operates continuously with a residence time up to 120 s. Detailed description of the plant can be found in a previous work [4].
A description of the fluid flow into the reactor has been obtained by means of computational fluid dynamics (CFD) simulations of the reactor. The simulations were conducted by the ANSYS CFX 15 code which uses a finite volume approach. Simulation allowed not only to understand the complex fluid dynamic phenomena that describe the system, but also to optimize the operational parameters to attain the best performance for the process. In order to obtain the distributions of the species concentration, the kinetic simulation have been implemented in the second part of the CFD work.
The CFD simulation showed that when the microalga feed is introduced through the top of the reactor by a nozzle, a jet stream is formed that causes a back mixing in the reactor chamber, contributing to rapid initiation of reaction and down flow of species through the chamber. The reactor thus has the top section that behaves like a mixed reactor and the bottom section that behaves like a plug flow reactor.
For what concerns experimental results, various experiments have been performed in order to investigate the effect of microalga concentration, residence time and pre-heating temperature of the feed on the gasification yield and gas composition. Also the effect of inorganic salt catalysts has been studied. Operating at 25 MPa and 650°C in the presence of K_2CO_3, gasification efficiency (GE) of nannochloropsis gaditana microalga has been obtained up to 80%. GE was measured as the ratio between the mass of gas produced to respect of the mass of alga fed to the reactor. The produced gas is composed of hydrogen, methane, carbon dioxide and carbon monoxide. Hydrogen concentration up to 55% by mole has been obtained at the best operating conditions. This work was also aimed at obtain information of the role of individual constituents of microalga with respect to resistance to gasification. Indeed,

microalgae are a mixture of various components such as oil, proteins, carbohydrates and fatty acids which concentration changes from species to species. The role played by various kinds of proteins in the gasification mechanism is particularly important in the case of applying gasification to different kind of microalgae or after the removal of the oil content (defatted microalgae). Preliminary experiments showed that GE of individual components of microalga (oil and proteins) could be significantly higher or lower than that of raw microalga. Therefore, the initial composition of microalgae is fundamental to assess a feasible process for the treatment of microalgae.

1. Haidue, A. G.; Brandenberger, S. S.; Vogel, F.; Bernier-Latmani, R.; Ludwig C.; SunChem: an integrated process for the hydrothermal production of methane from microalgae and CO2 mitigation, J. Appl. Phycol., 21, 529-541, 2009.
2. Chakinala, A. G.; Brilman, D. W.F.; van Swaaji, W.P.M.; Kersten, S.R.A.; Catalytic and non catalytic supercritical water gasification of microalgae and glycerol, Ind. Eng. Chem. Res., 48(3), 1113-1122, 2010.
3. Guan, Q.; Savage, P. E.; Wei, C.; Gasification of alga nannochloropsis sp. In supercritical water, Journal of Supercritical Fluids, 61, 139-145, 2012.
4. Caputo, G.; Rubio, P.; Scargiali, F.; Marotta, G.; Brucato, A.; Experimental and fluid dynamic study of continuous supercritical water gasification of glucose, Journal of Supercritical Fluids, 107, 450-461, 2016.

High yield immobilization of phospholipid A_1 and quick initiation of hydrolysis of phospholipid in supercritical carbon dioxide

T .Tateno and M .Imai*
Course in Bioresource Utilization Sciences, Graduate School of Bioresource Sciences, Nihon University, 1866 Kameino, Fujisawa, Kanagawa-pref., 252-0880 Japan
*Phone /Fax: +81 466 84 3978, E-mail: XLT05104@nifty.com

High yield immobilization of phospholipid A_1 (PLA_1) and quick initiation of hydrolysis of phospholipid in supercritical carbon dioxide were demonstrated. Immobilized phospholipase A_1 was prepared using porous polypropylene carrier, commercially called Accurel MP100 (Membrana GmbH, Germany). Phospholipase A_1 was adsorbed onto the carrier surface in acetate buffer (pH5.0). Adsorbed lipase was successfully immobilized by cross-linking with the aid of glutaraldehyde (GA). Adsorbed amount of PLA_1 was depended on the equilibrium concentration of feed aqueous phase. The immobilized amount was almost equivalent to the adsorbed amount. The immobilized yield was stably exceeded 98% in our experimental condition (Figure1) [1].

Rate study of enzyme reaction of immobilized PLA_1 in W/O microemulsion was carried out to determine rate parameters. The reaction substrate (phospholipid from soy bean) was dissolved in a mixed organic phase composed by isooctane and 1-butanol which volumetric ratio was 10 to 1. The range of substrate concentration was regulated from 1.25×10^{-3} to 9.0×10^{-2}M. Acetate buffer aqueous solution (pH5.0) was employed as an additive water to form a W/O microemulsion.

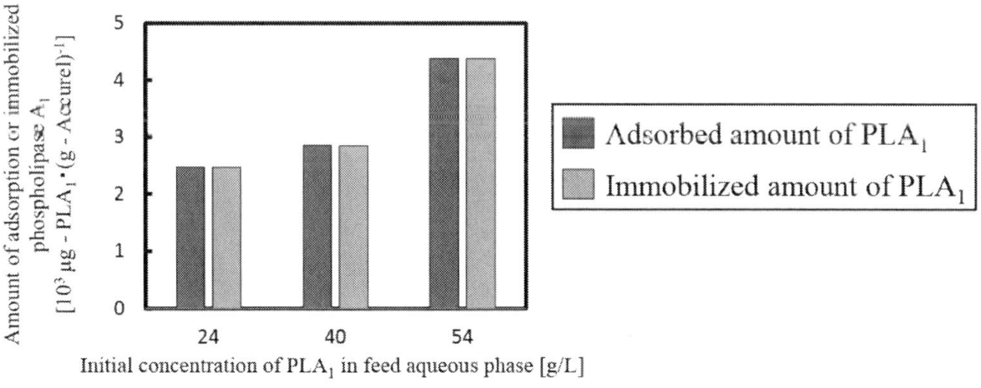

Fig. 1 Immobilized amount of phospholipase A_1 in Accurel MP100

Hydrolysis reaction was initiated when immobilized PLA_1 was introduced into the W/O microemulsion phase (40 ℃). Reaction progress was monitored by the concentration increase of produced palmitic acid. Reaction parameters Vmax 0.85mM/min and Km 0.03mM were determined by Lineweaver – Burk plot.

Schematic illustration of $SCCO_2$ reaction system was presented in Figure 2. Immobilized PLA_1 (1g, $5cm^3$) was installed in a reaction vessel (cylindrical free volume $9.3cm^3$, inside diameter 1.3cm, axial length 7.0cm) made of stainless steel (SUS 316). Lecithin (2g, purity ca.50%) was introduced in the vessel. Reaction was initiated when $SCCO_2$ (313K, 10MPa) was introduced and stored in the vessel. The reaction was carried out in batch wise.

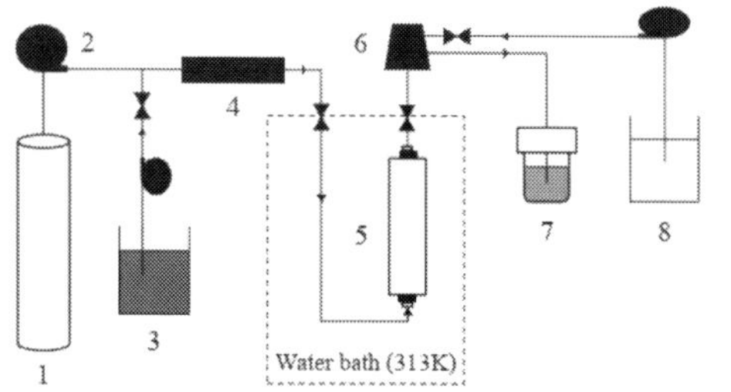

(1) CO_2 gas cylinder (2) Compressor
(3) Substrate (4) Mixing CO_2 and substrate
(5) Reactor vessel (6) Switching valve
(7) Sampling trap using isooctane
(8) Washing fluid (isooctane)

Fig. 2 Schematic illustration for immobilized PLA_1 reaction in $SCCO_2$

The reaction was monitored by concentration change of produced palmitic acid. Reaction was quickly initiated (Figure 3). Initial reaction rate obtained 0.7mM/min was equivalent match to the calculated reaction rate (0.77mM/min) that was based on the rate parameters in W/O microemulsion system.

The results depicts that $SCCO_2$ was a promising reaction solvent for production of fatty food component by enzymatic reaction.

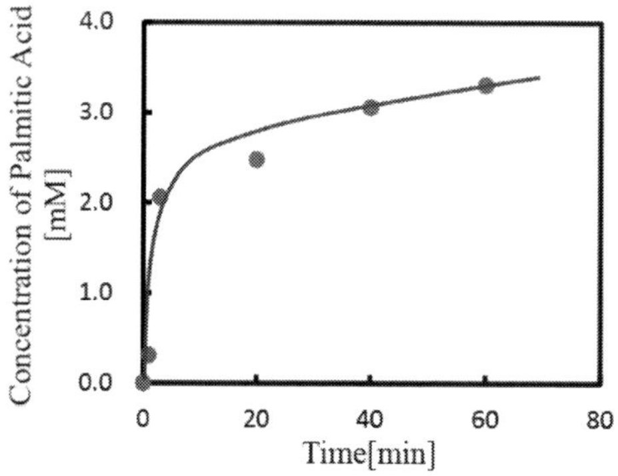

Fig. 3 Production of palmitic acid in $SCCO_2$ reaction system

[Reference]

1. Masakazu Naya, Masanao Imai. Regulation of the hydrolysis reactivity of immobilized *Candida rugosa* lipase with the aid of hydrophobic porous carrier, Asia-Pacific J. of Chemical Engineering, 7, S157 – S165, (2012).

Thermodynamics methodology for ternary chelatant/cationic exchanger/$SCCO_2$ system applied for the remediation of ^{137}Cs contaminated soil

A. Leybros, [1]A. Dartiguelongue, A. Grandjean
CEA, DEN, DTCD, SPDE, LPSD, F-30207 Bagnols-sur-Cèze, France;
[1] tel. +334 66391641, e-mail: antoine.leybros@cea.fr;

Countries worldwide have become increasingly concerned about the presence of pollution, results of years of industrial development and activity. In March 2011, Fukushima Daiichi nuclear power plant accident led to the release of a large quantity of volatile fission products such as ^{137}Cs in the environment. Despite the numerous existing soil decontamination technologies, there is a need for environmentally friendly process that will limit the production of unsuitable secondary waste and population exposure. Extraction in supercritical CO_2 ($SCCO_2$) solvent is one of these promising, smooth technologies to be applied to soil remediation [1].

$SCCO_2$ exhibits gas-like mass transfer rates and yet have liquid-like solvating abilities. Their high diffusivity and low viscosity enable them to penetrate and transport solutes from porous solid matrices to the $SCCO_2$ phase. Furthermore, its flexibility in terms of ability to adjust solvating power by tuning operating pressure and temperature is quite interesting. Direct extraction of metal ions by $SCCO_2$ is inefficient due to charge neutralization requirement and weak ionic solute-solvent interactions. Hence, organic ligands are needed in order to chelate these ions and make them soluble in $SCCO_2$. Crown ethers and calixarenes have been intensively studied for their selectivity in the extraction for alkali metal cations such as cesium in liquid-liquid processes. Crown ethers have also been successfully used to extract cesium in $SCCO_2$ with perfluorocarboxylic acid as a cationic exchanger.

A general methodology to study ternary system extractant/cationic exchanger/ $SCCO_2$ is needed to improve the efficiency of the process. The aim of this study is to optimize amounts of extractant and exchanger, considering decontamination yield, energetical and economic process optimization in the aim of scale-up. This work is divided into the following steps: determination of pure component properties, acquisition of data for each binary system, model implementation by binary interaction parameter fitting and validation with typical ternary data. Solubilities of considered extractant and cationic exchanger in $SCCO_2$ data have never been published.

Feasibility of $SCCO_2$ extraction processing for decontamination of contaminated soils has been studied using simulated sand soil and non-radioactive ^{133}Cs as contaminant. Extractant system considered was a mixture of DB18C6 crown ether and perfluorooctanoic acid. Extraction yields up to 95% have been obtained. Process optimization in terms of extraction yield was carried out by setting suitable operating conditions such as pressure, temperature, extractant system concentration and soil moisture.

From experimental results and ternary system thermodynamics data, extraction process scale-up is in progress in the aim of transportation on contaminated site. CO_2 and extractant fluxes management, expansion/compression cycles have to be considered carefully in order to minimize financial and energy costs. Feasibility of back extraction step is also studied.

As a conclusion, $SCCO_2$ soil remediation cleaning remains a promising technique that could be combined with other pre or post treatment processes.

[1]G. Anitescu et al., J. of Supercritical Fluids 38(2006)167-180.

The story of the 1978's Essen symposium that got started our research field on supercritical fluid applications

José M. del Valle
Department of Chemical and Bioprocess Engineering, Pontificia Universidad Católica de Chile, Avda. Vicuña Mackenna 4860, Macul, Santiago, Chile.Tel.: +562 23544418; Fax: +562 23545803; email:delvalle@ing.puc.cl.

Günther Wilke, an influential German organometallic chemist and the director of the Max Planck Institute (MPI) for Coal Research from 1967 (when he was 42 years old) to 1992, and the vice-president of the Max Planck Society, was astonished that, given the importance of separations on the chemical engineering industry, long time had to elapse prior to the first symposium on a "novel" method of separation such as "extraction with supercritical gases" (currently supercritical fluid extraction and allied processes relaying of the use of supercritical fluids). This symposium was held in Essen in 1978, and he wrote the preface of a special issue of *Angewandte Chemie* that was devoted to it. The then "novel" supercritical fluid extraction process dissolved nonvolatile solutes (sometimes even selectively) to form a charged supercritical phase, tuned the properties (particularly the density) of the supercritical fluid within wide ranges with small changes in temperature and specially pressure, applied distillation and liquid-liquid extraction principles simultaneously to develop the process, and allowed continuous separation of (the) dissolved solute(s) by decreasing pressure and/or increasing temperature.

The astonishment of Director Wilke was due to the long time elapsed since the first observation of solvation phenomena in pressurized gases, which raised limited academic interest but gave no rise to practical applications in the chemical engineering industry. However, chance individual observations in the intervening 80-year period from the first observation led to the fundamental studies started by Kurt Zosel in MPI for Coal Research in 1962. These fundamental studies produced his very influential first patent (USP 3,969,196) that, for many, signaled the birth of supercritical fluid extraction. According to Director Wilke, it was no coincidence that all the authors of the progress report that was subsequently published in the October issue of the International Edition in English of *Angewandte Chemie* were from then West Germany, because it was there that supercritical fluid extraction was discovered and developed. The single exception was the supercritical toluene extraction of coal developed in the UK (USP 3,970,541) whose economics were somewhat suspects.

The Essen symposium and subsequent special issue of *Angewandte Chemie* included contributions of Kurt Zosel; of Peter Hubert and Otto Vitzthum from HAG AG, who showed the application of supercritical CO_2 extraction to plant material, including hop cones, for food and related applications; of Gerhardt Schneider from Ruhr University of Bochum, who unveiled the physicochemical principles of this novel separation method; of Ernst-Ulrich Franck and a coauthor from University of Karslruhe, on phase equilibria of binary mixtures at supercritical conditions; of Egon Stahl and co-authors from University of Saarbrücken, on their elegant micro-analytical technique combining an extraction cell and a thin-layer-chromatography to explore the effect of extraction conditions on the selective extraction of the constituents of biological matrices, mainly plant material, using supercritical fluids, mainly supercritical CO_2; of Ernst Klesper from the University of Freiburg on chromatographic separations using supercritical fluids; of Siegfried Peter and disciple Gerd Brunner of University of Erlangen-Nuremberg on the countercurrent separation of nonvolatile substances in a high-pressure column using supercritical fluids; and of Rudolf Eggers from

Thyssen Industrie AG on the mechanical design of large-scale industrial plants for the supercritical fluid extraction of solid substrates.

The participants in the Essen symposium were personally responsible for the industrial application of supercritical fluid extraction. Indeed HAG AG was the first company to commercially implement supercritical CO_2 extraction to decaffeinate coffee (Kaffee HAG) soon after the symposium, in 1978.The company was set up by Ludwig Roselius who is the co-author of some patent applications with Peter Hubert and Otto Vitzthum. Kaffee HAG was purchased General Foods in 1979, and is now owned by Kraft Foods (General Foods and Kraft fused as Kraft General Foods under the ownership of Phillip Morris). Two of Prof. Stahl's former graduates (Karl-Werner Quirin and Dieter Gerard), and his co-authors in a very influential book on supercritical fluid extraction, founded Flavex Naturextrakte, which to this day specializes in the production of high-value plant ingredients using supercritical CO_2 extraction.

The Essen symposium also assembled several generations of extremely talented researchers that influenced academy for many years to come. Although Dr. Zosel retired from MPI the year after the Essen symposium and died ten years later, other participants had a more lasting influence. In addition to his contributions to supercritical fluid extraction, Prof. Stahl revolutionized analytics by inventing thin-layer chromatography. Unfortunately, he died prematurely (at age 62, in 1986), but this did not precluded the most prestigious medal of the Society for Medicinal Plant Research being named after him. Soon after the meeting, Prof. Klesper moved to Aachem University of Technology. He had a special issue of the *Fresenius Journal of Analytical Chemistry* being devoted to him in 1992, as the father of supercritical fluid chromatography. Prof. Franck pioneered work on physical and chemical properties of supercritical fluids and fluid mixtures, particularly water and aqueous solutions at high temperatures and pressures. He kept active following retirement, authoring 63 publications as emeritus professor. He died at age 84 in 2004. His research interests were shared by Prof. Schneider who focused more on supercritical CO_2. He had a special issue of the *Journal of Chemical and Engineering Data* being dedicated to him in 2009, on the occasion of his 77th birthday. Prof. Peter, the same as Prof. Franck, also had very prolific post-retirement career. He died in 2007 when he was 91 years old. Among the youngest participants of the Essen symposium were Prof Brunner and Prof. Eggers, who moved to the young Technical University of Hamburg at Harburg and were responsible for the development of the research unit on thermal separation processes. Prof. Brunner had a special issue of *The Journal of Supercritical Fluids* (where he holds as position as Regional editor for Europe) dedicated to him in 2012 on the occasion of his 70th birthday. Prof. Eggers passed away prematurely last year, shortly after his retirement.

This presentation will review 1978's Essen symposium on supercritical fluid extraction and the persons that made it so influential.

Supercritical impregnation of thymol in LDPE nanocomposites for active food packaging: assessment of kinetic release

A. Rojas[1,2], F. Martínez[1], A. Torres[1], A. Guarda[1], M.J. Galotto[1], J. Romero[2*]

[1]Center for the Development of Nanoscience and Nanotechnology (CEDENNA), Laboratory of Food Packaging, University of Santiago de Chile, Av. Ecuador 3493, Santiago, Chile.
[2]Laboratory of Membrane Separation Processes (LabProSeM), University of Santiago de Chile, Av. Lib. B. O'Higgins 3363, Santiago, Chile; tel. +5627181834, e-mail: julio.romero@usach.cl

In the last decades, there has been a change in the paradigm of food packaging. Nowadays it is expected that the package will have a positive effect on quality, food safety and shelf life of the food that it contains. This characteristic defines the active food packaging. Among the different type of active food packaging systems, the antimicrobial ones have gained great attention, especially when these contain natural antimicrobial agents. However, these agents are often thermolabile being degradated during traditional processing, which generates important losses of antimicrobial agent during the packaging preparation. In this way, the impregnation of active compounds in plastic materials using supercritical fluids represents a green technology, which could overcome these drawbacks.

On the other hand, the release kinetic of the antimicrobial agent could represent other issue in the design of an active package. In this case, nanotechnology could allow the development of solutions in terms of release rate control.

In this work, the technologies mentioned above are combined to study the supercritical impregnation of thymol in nanocomposites formed by LDPE and nanoclay Cloisite 20A (C20A). After preparation of LDPE/C20A films by extrusion, these were impregnated with thymol using supercritical CO2 at pressures varying from 9 to 15 MPa and depressurization rates between 1 and 10 MPa min^{-1} at 40°C.

The thermal properties of the impregnated nanocomposite films were obtained from DSC and thermogravimetric analysis. Meanwhile, its structural characteristics were assessed by FTIR spectroscopy.

The incorporation of thymol was improved by the presence of nanoclay Cloisite 20A in the LDPE films. Thus, this procedure allowed the nanocomposites preparation with thymol concentrations ranged between 0.98 and 1.26% and LDPE films without Cloisite 20A with thymol concentrations ranged between 0.38 and 0.76%. The incorporation of thymol was improved using the lowest rate of depressurization. Thymol incorporation has been verified without significant thymol losses. Furthermore, the incorporation of nanoclay and thymol in LDPE films does not involve a significant modification of its thermal properties. Future experiments will consider release kinetic assays with two different food simulants to verify the modification of transport properties during the thymol transfer from the nanocomposite to the food.

The effect of experimental design on crystal size of TiO_2 thin films prepared by using modified supercritical carbon dioxide

M. Cerhová, M. Sajfrtová, [1]L. Matějová, V. Dřínek, [2]S. Daniš, V. Jandová
Institute of Chemical Process Fundamentals of the ACS, v.v.i., Rozvojová 135, 165 02 Prague 6, Czech Republic, cerhova@icpf.cas.cz; [1]Institute of Environmental Technology, VŠB-Technical University of Ostrava, 17.listopadu 15/2172, 708 33 Ostrava – Poruba, Czech Republic; [2]Department of Condensed Matter Physics, Faculty of Mathematics and Physics, Charles University in Prague, Ke Karlovu 5, 121 16 Prague 2, Czech Republic

TiO_2 is known as an efficient photocatalyst for its good ability to produce the pairs electron-hole under UV light [1]. Moreover, titania exhibits high chemical stability, strong oxidative activity and non-toxicity. It has been investigated in a form of thin films in many application areas; as sensing films of gas sensors [2], coatings for self-cleaning surfaces [3], photocatalyst promising in water- and air-purification [4] or degradation of organic pollutants [5].

Properties as phase composition, crystallinity or crystallite-size influencing the photocatalytic activity and are size dependent. That means the controlled preparation of nanocrystallites is important for their applications. Calcination is a commonly used method for the purification and to crystallize TiO_2 thin films [6]. Unfortunately, this thermal treatment is accompanied by excessive sintration, crystallite growth or recrystallization. To reduce the process temperature and to increase the thermal stability and the photocatalytic activity the TiO_2 thin films can be also extracted by $scCO_2$, subcritical water or pressurized organic solvents [7].

In this work, the method using pure and modified supercritical carbon dioxide was tested to prepare a set of crystalline TiO_2 thin films deposited on soda-lime glass. Precursor solution for TiO_2 films was prepared by a templated sol-gel method in the reverse micelle environment. Simple approach or combination of several steps performed at different temperature (40-150 °C) and solvent composition was applied at pressure 30 MPa. Water, ethanol and water : ethanol (1:1) mixture were added into supercritical CO_2 as modifiers at concentration 10-30 wt.%. The effect of extraction design on microstructure and purity of TiO_2 thin films were examined.

The thin films were characterized with respect to the (micro)structural properties by Raman spectroscopy. The most promising thin films samples were analyzed by means of X-ray diffraction to determine the crystallite size and the phase composition.

TiO_2 thin films treated by water modified $scCO_2$ exhibited the crystalline structure anatase. Subsequent modification of $scCO_2$ by 10 wt. % ethanol led to the removal of the organic precursor used in the sol-gel method and thus to increase the purity of TiO_2 thin films. The additional drying with 200 g of pure $scCO_2$ had no positive effect on crystal growth but the mixture of anatase and brookite crystals was formed. At the optimal conditions of SFC by water : ethanol (1:1) mixture modified $scCO_2$ (30 MPa, 150 °C, 100 g of $scCO_2$) the anatase crystals with the size of 13 nm were obtained.

Acknowledgements. The financial support of the Grant Agency of the Czech Republic (projectno.14-23274S) is gratefully acknowledged.

[1] Prieto O. et al. *Sol. Energy*. **2005**, *79*, 376-383. [2] Savage N. O. et al. *Sens. Act. B*. **2001**, *72*, 239-248. [3] Fujishima A. et al. *Electrochim. Acta.* **2000**, *45*, 4683–4690. [4] Hsien Y.-H. et al. *Appl. Catal. B: Environ*. **2001**, *31,* 241–249. [5] Nakata K. et al. *Electrochim. Acta*. **2012**, *84*, 103. [6] Zhou M. et al. *J. Hazard. Mat.* **2008**, *154*, 1141–1148. [7] Matejova L. et al., EMSF 2014, Marseille, France, p.161 (Full text on Flash disc).

Scale analysis of the generalized supercritical fluid extraction model for diffusive mass transfer in polydisperse packs

A.A. Salamatin
Kazan Federal Univ., Dept. Aerohydromechanics, Kremlyovskaya Str. 18, 420008 Kazan, Russia; fax. +7 (843) 292-72-79, e-mail: arthouse131@rambler.ru

Elaborate supercritical fluid extraction (SFE) models conventionally distinguish between different mechanisms of the solvent mass transfer on the marco-scale level of porous packing bed in the vessel and micro-scale level of single particles. While macro-scale modelling is traditionally based on spatially averaged convective mass balance equations, generally accounting for axial dispersion effects [1-4], the principles of particle-scale models are not so well established or, even, understood. As a rule, two limiting approaches are discussed and employed in SFE data interpretation; they are the shrinking core (SC) model [1, 2] and broken and intact cells (BIC) model [3, 4]. The latter one assumes high longitudinal permeability of the internal microscopic channels (cell walls), resulting in lumped parameter system approximation with average solute concentration in the ground material. On the other hand, the SC model considers the case of high transversal permeability of the cell membranes and assigns all the mass transfer resistance to diffusion along the transport channels. Consequently, the narrow diffusive front moves inside each particle and separates the shrinking oil-filled core from the outer depleted transport zone. Obviously, there exists a wide intermediate range of the ground-plant-material properties when typical rates of solute diffusion across the cell membranes are on the same order as those through the particle, along its transport channels, and the two above (SC- and BIC-) limits fail to describe the SFE process on the micro-scale level.

A generalized SFE model for diffusive solvent-mass transfer in particles is proposed and analysed in this study on the basis of scale analysis. Each (spherical) particle is considered as an agglomerate of cells (Fig. 1). Every cell is considered as a lumped system containing solvent (1) with dissolved-solute concentration θ_s and droplets (4) of undissolved oil with volume-averaged density x (initial value x_0). The cells are bounded by cell membranes (6) surrounded by cell walls (2) which together with intercellular voids (3) compose the so-called apoplast transport system with solute concentration θ_a and apparent diffusion coefficient D_a. For finite solute saturation concentration θ_*, the relationship between x and θ_s is written as

$$\theta_s = \min\{\theta_*; x/(1-\varepsilon_0)\}, \quad (1)$$

where ε_0 is the cell volume fraction, occupied by insoluble substances (5). Eq. (1) implies that undissolved oil (if any) instantaneously dissolves in the solvent inside the cell up to saturation level θ_*.

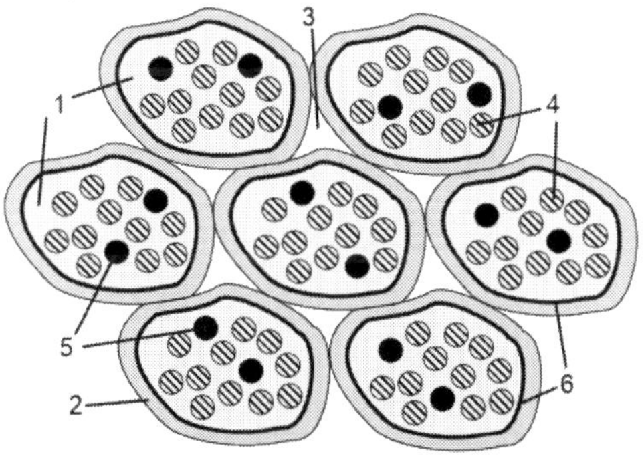

Fig. 1. Schematic of internal structure of ground particles. 1 – solute, dissolved in the solvent, 2 – cell wall, 3 – intercellular space (voids), 4 – undissolved oil, 5 – insoluble substances (proteins, etc.), 6 – cell membrane (bold solid line).

Eventually, mass-balance master equations of solute diffusion inside a particle can be presented in terms of dimensionless time τ, radial coordinate r, and concentrations x, and θ_s, θ_a normalized by their respective scales τ_{sc}, a, x_0, and θ_*

$$\delta_{cw}\frac{\mathrm{M}}{1+\mathrm{M}}\frac{\partial\theta_a}{\partial\tau}=(1-\varepsilon)\mathrm{M}\left(\theta_s-\theta_a\right)+\frac{1}{6}\Delta\theta_a,\qquad \frac{\mathrm{M}}{1+\mathrm{M}}\left((1-\varepsilon)\frac{\partial x_s}{\partial\tau}+\delta_{cw}\frac{\partial\theta_a}{\partial\tau}\right)=\frac{1}{6}\Delta\theta_a, \tag{2a}$$

$$\theta_s=\min\left\{1;\frac{x}{\Theta}\right\},\quad \mathrm{M}=\frac{a^2\beta_c}{2D_a},\quad \delta_{cw}=\frac{\varepsilon\Theta}{1-\varepsilon_0},\quad \Theta=(1-\varepsilon_0)\frac{\theta_*}{x_0} \tag{2b}$$

$$\tau_{sc}=\frac{x_0}{\theta_*}\left(\frac{1}{3\beta_c}+\frac{a^2}{6D_a}\right).$$

Here, β_c is the reduced mass transfer coefficient across the cell membranes (units – s^{-1}), ε (~0.01) is the apoplact volume fraction, a – particle radius, and τ_{sc} is the timescale of particle full depletion defined as the sum of two typical times of diffusion across the cell membrane and along the cell wall. Parameter δ_{cw}, as well as ε, is usually negligible if compared to 1, and only two dimensionless complexes, similarity numbers, Θ and M entirely specify the SFE regime.

For $\Theta \ll 1$, i.e. for plant material with high initial oil content such as seeds, two limiting cases of small (I) and high (II) values of M can be formally defined. Asymptotic expansion of Eqs. (2) in case I (Θ, M $\ll 1$) shows that the principal-order approximation corresponds to BIC model. In case II ($\Theta \ll 1$, M $\gg 1$), the SC model is obtained with the width of moving diffusive front estimated as $a\mathrm{M}^{-0.5}$. At the same time, it must be emphasized that the above scale analysis performed for polydisperse packed beds with typical range of particle size, $a \sim 10^{-4}$-10^{-3} m, reveals at D_a / β_c-ratios $\sim 10^{-7}$ m^2 a wide spectrum of various SFE scenarios from BIC- to SC-limits simultaneously existing in different particle-size fractions with different M-values from M ~ 0.1 (for smaller particles) to M ~ 10 (for bigger particles). Consequently, only BIC- or SC-type limiting mass transfer models alone may not be applicable for kinetic data description on micro-scale level in case of polydisperse packed beds.

The work was supported by the Russian Foundation for Basic Research and the Republic of Tatarstan, grant No. 15-41-02542 r_povolzhe_a, and grant No. 16-31-00007 mol_a.

References:

[1] *Goto M., Roy B.C., Hirose T.* Shrinking core leaching model for supercritical fluid extraction // **J. Supercrit. Fluids.** 1996. V. 9. № 2. P. 128-133.

[2] *Egorov A.G., Salamatin A.A.* Bidisperse shrinking core model for supercritical fluid extraction // **Chem. Eng. Technol.** 2015, V. 38, № 7, P. 1203-1211.

[3] *Sovova H.* Rate of the vegetable oil extraction with supercritical CO_2 – I. Modelling of extraction curves // **Chem. Eng. Sci.** 1994. V. 49. № 3. P. 409-414.

[4] *Sovova H.* Steps of supercritical fluid extraction of natural products and their characteristic times // **J. Supercrit. Fluids.** 2012. V. 66. P. 73-79.

Mass transfer induced particle swelling during fixed-bed extraction – A model study

A. Rieger[1], R. Haseneder[1], J.-U. Repke[1]
TU BA Freiberg Institute for Thermal, Environmental and Natural Products Process Engineering, Leipziger Str. 28, D – 09599 Freiberg, Germany
tel. +49 3731 392377, e-mail: andre.rieger@tun.tu-freiberg.de

Swelling or shrinking of particles during the course of solid-fluid extraction (SFE) is often observed experimentally, but only scarcely incorporated in mathematical process models, that are often employed for process optimization and scale-up. Only recently, Meyer et al. (2012) presented his research on the swelling of natural matter exposed to $scCO_2$ during SFE of differently pre-treated rapeseed and included his findings in a mathematical model by assuming that bed voidage, solid density and particle size are linear functions of solute (rapeseed oil) mass fraction in the solid. While quite effective in modelling, the parameters lag physical meaning and therefore do not provide insight in the link between mass transfer and transient changes in the particle size nor their connection to the bed void.

To overcome this obstacle, we treated the solid phase as a bed of spherical particles and derived mass balance equations from first principles including the transient change of the particle size due to mass transfer of solvent and solute components. In order to achieve coupling of particle size and bed voidage we introduced a relationship between particle size and bed voidage that is applicable to a moderate extent of swelling, so that severe shape deformation does not occur. The derived balance equations, namely overall mass and species continuity equations for the fluid and solid phase as well as interface motion equations for the solid phase, were non-dimensionalized in order to reduce the combinations of sets of parameters in an overall parameter study. The derived systems of equations accompanied with an appropriate set of initial and boundary conditions has been discretized using the finte volume method in combination with the method of lines and subsequently solved employing the Mathworks Matlab R2015b DAE solver ode15s. The use of this DAE solver was required since an overall steady state mass balance has been added to the set of differential equations in order to account for the increase/decrease in flow velocity in the bed due to mass transfer and the dynamics of the bed void.

Fig. 1 shows computational results for extraction curves over time for some example calculations on a semi-log time scale, where it has been assumed that the ratio of superficial residence time to external mass transfer resistance of the solvent is 10^{-6} and solvent internal mass transfer resistance is in the same order, respectively. Solute to solvent density ratio in the solid has been assumed 100 for curves with the effect of swelling present (continuous lines in Fig. 1), and internal mass transfer resistance is 100-fold the value of the external resistance. Computed extraction curves in Fig. 1 are shown for different ratios of external mass transfer resistance of solvent to solute, $\tau = (10^{-5}, 1)^T$.

The presented extraction curves in Fig. 1 clearly show the effect of swelling for the scenario $\tau = 1$, where external mass transfer resistances of solvent and solute are in the same order of magnitude. While in the beginning of the extraction solvent absorption and subsequent swelling is insignificant to extraction of the solute, the gap between the curves increases to a significant difference in extraction degree of up to 10%, caused by an increase in the internal mass transfer resistance due to longer diffusion paths caused by the swelling. In the final section of the extraction however, the difference between both curves reduces again due to a decrease of the diffusion resistance caused by solute desorption induced shrinking of the particle. The curve for $\tau = 10^{-5}$ shows the insignificance of swelling effects caused by solvent sorption since solute extraction occurs on a much lower time scale than solvent absorption.

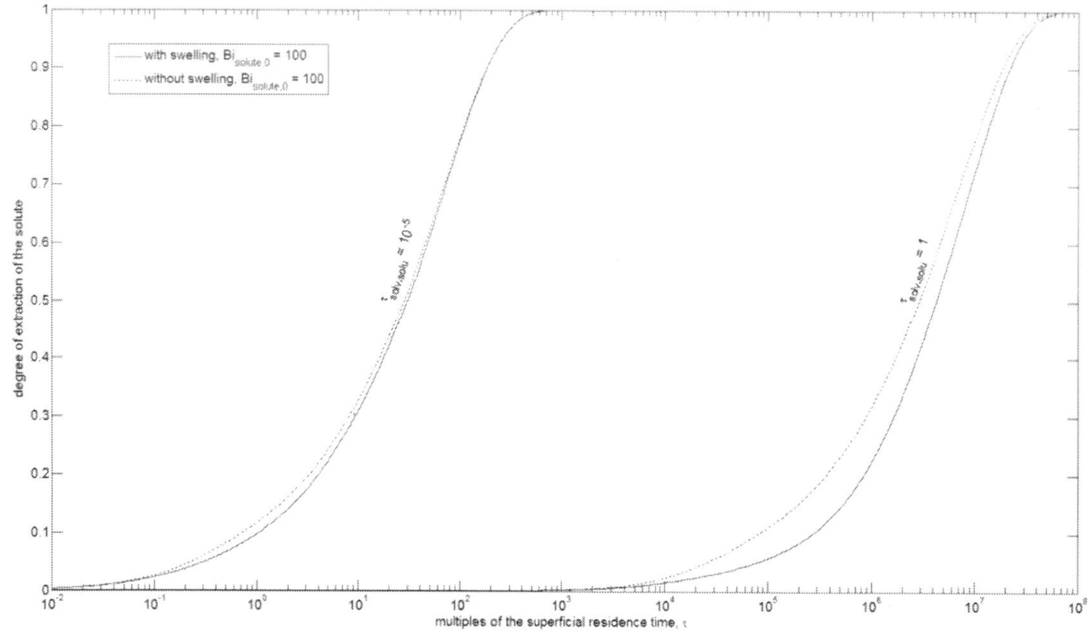

Fig. 1: Degree of extraction over log-scale time for different ratios of time scales of solute to solvent external mass transfer and solute internal mass transfer being 100-fold its external mass transfer (Bi = 100).

The proposed model has been employed to simulate a number of mass transfer scenarios in terms of different commonly met time scales of mass transfer mechanisms for both solvent and solute, repectively. All computed results produced physically meaningful extraction curves based on the chosen set of mass transfer time scales and provided us with a simple yet powerful tool for the analysis of factors influencing the significance of swelling on solid-fluid fixed-bed extraction processes. Future work is going to comprise of an analysis of the effect of swelling on cost factors such as solvent recycle, solvent-to-solute ratio in connection with solvent residence time in the bed and furthermore a computational study on a feedback control strategy for solvent flow rate in the bed as a function of swelling state based on the dynamics of the pressure drop in the bed.

Meyer, F.; Stamenic, M.; Zizovic, I.; Eggers, R.: Fixed bed property changes during scCO2 extraction of natural materials – Experiments and modelling. J Supercritical Fluids 72 (2012), pp. 140 – 149.

Supercritical fractionation of turmerones from turmeric rhizomes by sorption – comparison of different types of silica gel

M. Topiar, M. Sajfrtova, H. Sovova
E. Hála Laboratory of Separation Processes, Institute of Chemical Process Fundamentals of the CAS, v.v.i., Rozvojova 135, 16500 Prague, Czech Republic;
tel. +420 220390241, e-mail: topiar@icpf.cas.cz;

In the recent years, natural products from plants became more often used in many fields. Supercritical fluid extraction with carbon dioxide (SFE) is considered as more selective method for obtaining plant isolates than traditional separation techniques, i.e. hydrodistillation or oxhlet extraction. Due to the nonpolar character of carbon dioxide, the SFE isolates mostly contain a mixture of volatile and non-volatile compounds such as terpenes and cuticular waxes. Fortunately, we can use several fractionation techniques by which we are able to separate the SFE isolate to several fractions and increase this way the concentration of desired compounds. Beside simple ones, more sophisticated techniques such as using several separators or fractionation of isolates with the help of sorbent can be applied.

The powder obtained from turmeric (*Curcuma longa* L.) rhizomes is commonly used as a natural dye or spice. The best known chemical components of turmeric are curcuminoids, with the most famous compound – curcumin. Other important compounds extractable by SFE are turmerones, β-sesquiphellandrene or zingiberene. According to the scientific studies, ar-turmerone shows hepatoprotective[1] and anticancer[2] effects, among others. Another study indicates that β-sesquiphellandrene has anticancer potential comparable with curcumin[3].

Our previous research was focused on the optimization of turmerones extraction using supercritical CO_2 (sc-CO_2) and their sorption on different sorbents with particular interest in determination of a suitable sorbent for their separation from the SFE isolate.[4] Eventually, silica gel was determined as the most appropriate sorbent for this purpose. Nevertheless, we can find on the market a large amount of silica gel types with different chemical and physical properties.

Aim of this study was to find the optimal type of silica gel for the separation of turmerones from the turmeric SFE isolate and their concentration in eluted fractions.

The chemical composition of isolates and their yields were determined by GC-MS and GC-FID. The sorption experiments were done with turmeric isolate obtained by SFE at optimal conditions for the extraction of turmerones – 9 MPa, 50 °C. Main components of the initial isolate were turmerone (22.7 wt.%), ar-turmerone (13.3 wt.%), curlone (13.8 wt.%) and (1.7 wt.%).

Several types of silica gels with different pore (Å) and particle (μm) size were used. The elution of the turmeric isolate from silica gel was performed with sc-CO_2 at 66 MPa and 50 °C. (Lower pressures were not sufficient to start elution because of the weak solvent power of sc-CO_2.) The flow rate of CO_2 was adjusted to 1.8 $g.min^{-1}$. Several fractions were collected during experimental run to determine the changes in the isolate composition. The concentration of β-sesquiphellandrene in isolate increased from 1.7 wt.% to 28.8 wt.% by using of supercritical sorption technique. The increase in concentration of turmerones, especially ar-turmerone in isolate was observed, when silica gel with smaller pore size and narrowed range of particles was used.

The type of silica gel showing the best results for the observed parameters was used in an experiment where the extraction from plant material and extract sorption on silica gel were integrated into one process.

1. Hail, N., Jr.; Lotan, R., Cancer chemoprevention and mitochondria: targeting apoptosis in transformed cells via the disruption of mitochondrial bioenergetics/redox state. Mol. Nutr. Food Res. 2009, 53 (1), 49-67.

2. Li, Y.-H.; Chung, H.-C.; Liu, S.-L.; Chao, T.-H.; Chen, J.-C., A novel inhibitory effect of Antrodia camphorata extract on vascular smooth muscle cell migration and neointima formation in mice. Int Heart J 2009, 50 (2), 207-209.

3. Tyagi, A. K.; Prasad, S.; Yuan, W.; Li, S.; Aggarwal, B. B., Identification of a novel compound (β-sesquiphellandrene) from turmeric (*Curcuma longa*) with anticancer potential: comparison with curcumin. Invest. New Drugs 2015, 33 (6), 1175-1186.

4. Topiar M., Sajfrtova M., Karban J.: Fractionation of turmerones from turmeric rhizomes (Curcuma longa L.) by methods using supercritical carbon dioxide. (Czech). 62. Konference chemického a procesního inženýrství - CHISA 2015, Proceedings p. 64, hotel Jezerka, přehrada Seč, Czech Republic, 09-12 November 2015.

Acknowledgments: The financial support from the Institute of Chemical Process Fundamentals of the CAS, v.v.i. via internal Ph.D. grant is gratefully acknowledged.

Porosity and cell size of CO_2 blown polystyrene foams: experiments and modelling

H. Sovová, [1]A. Nistor, M. Topiař, [1]J. Kosek
Institute of Chemical Process Fundamentals of the CAS, Rozvojová 135, 16502 Prague 6, Czech Republic, tel. +420 220390234, e-mail: Sovova@icpf.cas.cz; [1]University of Chemistry and Technology Prague, Dept. of Chemical Engineering, Technická 5, 16628 Prague 6, Czech Republic, e-mail: juraj.kosek@vscht.cz.

Polymer foaming with CO_2 is a promising area of research with applications already commercialized and more on the horizon. In batch experiments, polymer absorbs supercritical CO_2 at the temperature adjusted above its glass transition temperature, Tg, until saturation is reached. Rapid depressurization is then applied, causing gas supersaturation and allowing cell formation. In our experiments, polystyrene films of the thickness of hundreds of micrometers up to a few milimeters were exposed to supercritical CO_2 at saturation conditions in the range of 40-100 °C and 18-28 MPa. Additional experiments were conducted with polystyrene modified with light hydrocarbons. The cell size distribution and porosity of foams were measured and related to the saturation conditions and depressurization time.

The polymer foaming is an interdisciplinary field reaching to polymer rheology, mass transport, thermodynamics, and nucleation. Computational approaches enable insight into foaming phenomena, while experimental techniques are coming that can provide reliable data. Most of theoretical effort is focused on phenomena of cell nucleation and cell growth.[1-3] Less attention has been paid to foam vitrification.[4] As the temperature of CO_2 decreases by its expansion, the polymer in tight contact with growing cells is cooled down until Tg is reached and polymer vitrification occurs. The cell size and foam porosity are fixed at this moment and can be measured after the depressurization is finished. CO_2 solubility in the polymer and Tg depression under the effect of CO_2 play an important role in the process.

While the cell size is related to the solvent solubility in the polymer, in respect to the foam porosity we propose a novel approach considering the changes in the temperature of the mixture in the course of CO_2 expansion due to the Joule-Thomson effect. Heat balance equations including heat transfer over the film surface were integrated to the time when the decreases to Tg. (The necessary properties of CO_2 were available on the NIST web pages and the properties of polystyrene mixtures with CO_2 necessary for the calculations were published in recent research papers.) The estimates were compared with experimental results.

References

1. E. Reverchon and S. Candea, Production of controlled polymeric foams by supercritical CO2. J. Supercrit. Fluids 40 (2007) 144.
2. D.L. Tomasco et al., Development of CO2 for polymer foam applications. J. Supercrit. Fluids 47 (2009) 493.
3. R. Liao et al., Rheological control in foaming polymeric materials: I. Amorphous polymers. Polymer 51 (2010) 568.
4. R. Li et al., Glass transition temperature in microcellular foaming process with supercritical carbon dioxide: A review. Polymer-Plastics Technology and Engineering 54 (2015) 119.

Hydrothermal degradation of rutin in subcritical water: Effect of reaction temperature and analysis of degradation products

M. Ravber, Ž. Knez, M. Škerget
Faculty of Chemistry and Chemical Engineering, University of Maribor, Smetanova 17, SI-2000, Maribor, Slovenia; tel. +386 22294463, e-mail: mojca.skerget@um.si; fax: +386 22527774.

Subcritical water, which is defined as pressurized liquid water in the temperature range from 100 °C to 374 °C, is recently gaining more and more popularity as a green and environmentally friendly solvent with many beneficial properties compared to conventional organic solvents. A high number of studies have shown its potentials, particularly being interesting as an extraction solvent for bioactive compounds from natural materials, which are applied as suitable and safer alternatives to synthetic antioxidants in the food industry. Nevertheless, the high applied temperatures in these extractions tend to degrade these thermally labile compounds into other, often non-edible or even toxic compounds, which have an unpleasant smell and taste. Therefore, studies on the stability of the bioactive compounds need to be performed at these extreme conditions. Furthermore, the identification and quantification of the degradation products must be assessed, in order to better understand when and how these unwanted products are formed. In that way the optimization of the extraction parameters can be performed with less difficulties as well as more efficiently.

In this study, the quercetin glycoside rutin, a common bioactive compound found in many natural materials, was subjected to degradation under subcritical water conditions. The study was performed in the temperature range of 120-220 °C in a batch-type reactor. Aqueous rutin solutions were subjected to subcritical water for two hours at steady-state conditions and the reaction was monitored by collecting an aliquout of sample every 10 min. The aliquots were diluted accordingly and analyzed for different degradation products using high performance liquid chromatography coupled to a variable wavelenght detector (HPLC-VWD).

Results have shown that rutin begins to hydrolyze into its free aglycone form – quercetin already at the lowest temperature (120 °C). The formation of quercetin at this temperature is slow but this quickly changes when the temperature is increased to 140 °C, where more than 50 % of the rutin is degraded into quercetin already after 1 h of reaction. At 140 °C however, quercetin begins to slowly degrade mostly into two products, namely protocatechuic acid (3,4-dihidroxybenzoic acid) and 4,6-dihydroxy-coumaran-3-one. At higher temperatures (≥160 °C) the formation of these products is further promoted. Simultaneously, at these temperatures the presence of 5-hydroxymethylfurfural, a common degradation product formed from hexose sugars, could be confirmed. At 200 °C and 220 °C, practically all rutin is degraded into quercetin and its degradation products already after 10 min. At these temperatures, protocatechuic acid could no longer be detected but instead higher amounts of catechol could be observed.

Keywords: Rutin, Subcritical water, Degradation, Quercetin, Analysis.

Supercritical fluid extraction of eucalyptus volatile oil in view of thermodynamic modelling

H. Sovova, [1]R.P. Stateva, M. Topiar
E. Hála Laboratory of Separation Processes, Institute of Chemical Process Fundamentals of the CAS, v.v.i., Rozvojova 135, 16500 Prague, Czech Republic; tel. +420 220390234, e-mail: sovova@icpf.cas.cz; [1]Institute of Chemical Engineering, Bulgarian Academy of Sciences, Georgi Bontchev str., bl. 103, Sofia 1113, Bulgaria; tel. +359 29793481, e-mail: thermod@bas.bg.

The extraction with supercritical carbon dioxide (scCO2) is very well suited for obtaining plant extracts rich in essential oils of authentic composition. The main components of essential oil are usually terpenoids: monoterpenes, oxygenated monoterpenes, sesquiterpenes, and oxygenated sesquiterpenes. It is well known that the solubility of these substances in scCO2 decreases with increasing molecular size and number of hydroxyl groups in a molecule. Thus, under mild extraction conditions, the mixture of terpenoids in extract is initially richer in the most soluble monoterpenes and contains almost no sesquiterpenes, and only gradually the mixture composition approaches that which was originally in the plant and which can be determined by extraction with n-hexane. In contrast to *n*-hexane, $scCO_2$ does not leave any traces of foreign substances in extracts, which can be directly used in various cosmetic or food products. Evidently, a good knowledge of the relationship between the composition of the mixture in the plant and solubility of its components in scCO2 under different pressures and temperature would enable standardization of plant extracts by the control of extraction conditions.

The aim of this contribution is to determine this relationship in the case of eucalyptus oil. The oil from the leaves of *Eucalyptus grandis* L., which was selected for experiments, is rich in a monoterpene α-pinene, a monoterpene oxide 1,8-cineole, a sesquiterpene aromadendrene, and a sesquiterpene alcohol globulol. The composition of saturated vapour phase in volatile oil+CO_2 system was measured by sampling the extract from either milled dry leaves or essential oil on inert carrier, using a sufficiently large solvent residence time. Gas chromatography was used to analyse the composition of the samples, leaves, and essential oil. The extraction pressure and temperature were varied in the region of $scCO_2$ with relatively low density, where the solvent selectivity is more pronounced. The experimental results were used to determine the temperature dependence of the parameters of thermodynamic model for the studied system.

Finally, the calculation of mixture solubility in $scCO_2$ based on thermodynamic modelling was implemented in a model for extraction kinetics, and optimum extraction conditions were calculated for several examples of eucalyptus leaves composition where a fixed composition of CO_2 extract is the optimization criterion.

Composition and antioxidant activity of pumpkin seeds grown in different production systems

[1]T. Perko, [1]M. Škerget, [1]M. Knez Hrnčič, [1]Ž. Knez, [2]M. Jakop, [2]F. Bavec
[1]University of Maribor, Faculty of Chemistry and Chemical Engineering, Laboratory for Separation Processes and Product Design, Smetanova 17, SI-2000 Maribor, Slovenia; tel. +386 222 94 463, fax: +386 225 27 774, e-mail: mojca.skerget@um.si
[2]University of Maribor, Faculty of Agriculture and Life Sciences, Chair of Organic Agriculture, Field Crops, Vegetable and Ornamental Plants, Pivola 10, SI-2311 Hoče, Slovenia, +386 232 09 001, e-mail: manfred.jakop@um.si

Pumpkins (*Cucurbita pepo L.*) are widely cultivated vegetables used for human consumption. The Styria region in north-eastern Slovenia and the adjacent Styria region in southern Austria have a long tradition of growing pumpkins as an oil crop. Pumpkin oil contains biologically active components which show antidiabetic, antibacterial, hypocholesterolemic, antioxidant, anticancer, anti-mutagenic, immunomodulatory, antihelmintic, and anti-bladderstone activity, and other miscellaneous effects. It is rich in phenolic, flavonoids, polysaccharides, mineral salts, vitamins and other substances beneficial to health.

In our work the chemical composition of pumpkin seeds and antioxidant activity of their oils were determined. The recommended methods of the Association of Official Analytical Chemists (AOAC) were used to determine the chemical composition of the pumpkin seeds including the contents of moisture, ash, and total carbohydrates. The moisture content was determined by drying the seeds in an oven at 105 ± 1 °C to a constant weight. Ash was determined by incinerating the sample at 550 °C in a muffle furnace. Total carbohydrate concentration of pumpkin seeds was determined with the phenol/sulfuric acid colorimetric method. Pumpkin seed oil was extracted by using subcritical propane. The extraction was performed at 40 °C and 50 °C and pressures of 50 and 100 bar. All the analyses were performed in triplicate. The obtained extracts were analyzed for radical scavenging activity.

The aim of this study was to establish how different ways of production (conventional, integrated, control and organic) influence the chemical composition and antioxidant activity in pumpkin seeds.

Key words: pumpkin seed, subcritical extraction, propane, antioxidant activity

Extraction of glycyrrhetinic acid from licorice (*Glycyrrhiza uralensis*) using organic solvents and supercritical carbon dioxide with additive water

Y.Saotome and M.Imai*
Course in Bioresource Utilization Sciences, Graduate School of Bioresource Sciences, Nihon University, 1866 Kameino, Fujisawa, Kanagawa-pref. 252-0880, Japan
*Phone /Fax: +81 466 84 3978, E-mail: XLT05104@nifty.com

Licorice (*Glycyrrhiza uralensis*) glows mainly in Asia. It is extensively used in herbal medicine for its strongly anti-inflammatory action. Glycyrrhetinic acid (La) included in licorice brought anti-inflammatory action for human [1]. glycyrrhizin (Rg) is a glycoside of glycyrrhetinic acid (La). Chemical structure of glycyrrhetinic acid and glycyrrhizin presents in Figure 1.

In this study, glycyrrhetinic acid (La) was dominantly extracted by supercritical carbon dioxide at 10~20MPa, 304.4~318K and organic solvents (butanol, ethanol, and propernol) at 353K. The amount of glycyrrhetinic acid (La) was increased with increasing pressure and decreasing temperature. The highest extracted amount was 6.91×10^{-6} [mol-glycyrrhetinic acid/g-dried sample/mol-$SCCO_2$] at 20MPa, 308K.

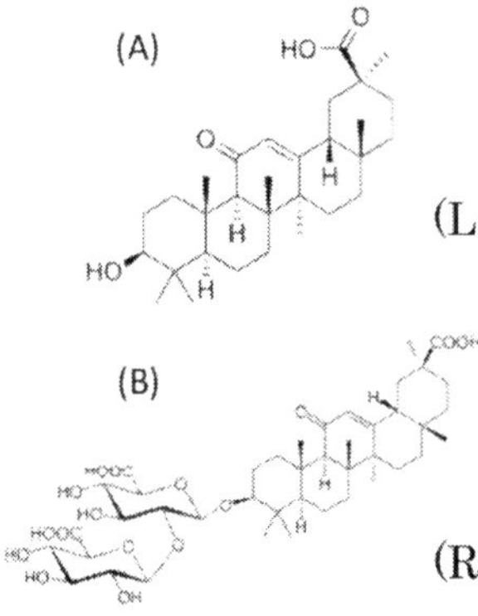

Fig.1 Chemical structure of Glycyrrhetinic acid (La) and Glycyrrhizin (Rg).

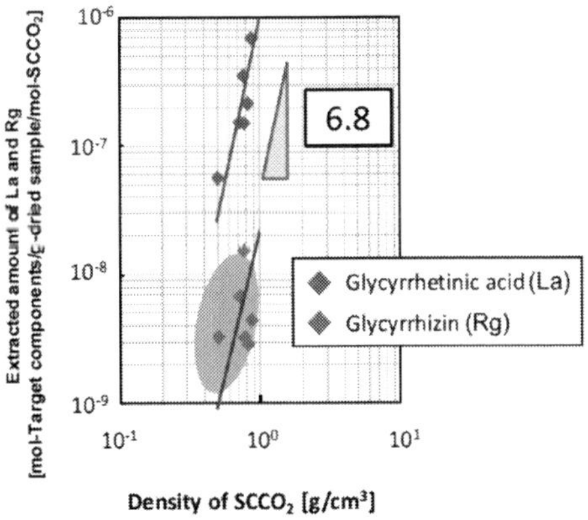

Fig.2 Extracted amount of glycyrrhetinic acid and glycyrrhizin vs. density of $SCCO_2$.

The density of $SCCO_2$ was a predominant factor including the effect of pressure and temperature in extraction using $SCCO_2$. Figure 2 depicts the correlation of the extracted amount of glycyrrhetinic acid (La) and glycyrrhizin (Rg) with the density of $SCCO_2$. The density of $SCCO_2$ was calculated using Bender's equation [2]. They were proportional to 6.8^{-th} power of density of $SCCO_2$.

The solubility parameter is a representative factor indicating the physicochemical properties of solvent. The solubility parameter of $SCCO_2$ was calculated by Gidding's study [3], as following Eq. (1). The solubility parameter of glycyrrhetinic acid (La) and glycyrrhizin (Rg) according to Fedor's method [4] were 23.9[(J/cm^3)]$^{0.5}$ and 30.5[(J/cm^3)]$^{0.5}$, respectively. The difference in solubility parameter between extract and solvent was a predominant factor for required energy for solubilisation as Eq. (2).

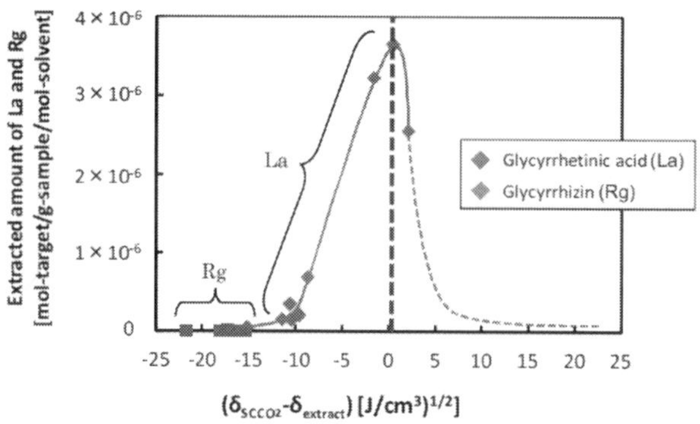

Fig.3 Correlation of extracted amount of target components and solvents

$$\delta_{SCCO_2} = 1.25Pc^{1/2}\left(\frac{\rho_{rSCF}}{\rho_{r\,liq}}\right) \quad (1) \qquad \Delta E = V_A\ \varphi_{solvent}{}^2\left(\delta_{SCCO_2} - \delta_{extract}\right)^2 \quad (2)$$

Figure 3 indicates the correlation of extracted amount and the difference of solubility parameter between solute and solvent. The parabolic curve was appeared and well accorded with Eq. (2). Small difference in solubility parameters between solute and solvent is an attractive condition for higher yield of target component (glycyrrhetinic acid (La)). $SCCO_2$ rolls excellent solvent for glycyrrhetinic acid (La). In contrast, glycyrrhetinic acid (La) was extracted in $SCCO_2$ with additive water, which was over the critical amount of water.

[1] J. Kim, ShinAnti-inflammatory effect of roasted licorice extracts on lipopolysaccharide-induced inflammatory responses in murine macrophages, Biochemical and Biophysical Research Communications 345 (2006) 1215–1223.

[2] E. Bender, Equations of state exactly representing the phase behavior of pure substances, Proc. 5th Symposium on Thermophysical Properties; Amer. Soc. Mech. Engr. (1970) New York.

[3] J. C. Giddings, M. N. Myers, L. Mclaren, R. A. Keller, High-pressure gas chromatography of non-volatile species, Science 162 (1968) 67–73.

[4] R. F. Fedors. A method for estimating both the solubility parameter and molar volume of liquids, Polymer Engineering and Science 14 (1974) 147-154.

Supercritical CO_2 extraction of bioactive triterpenes from rosemary leaves: kinetics and modeling

O. Bensebia[1], B. Bensebia[2], Kh. Allia[1], D. Barth[3]
[1] Université des Sciences et de la technologie Houari Boumediene (USTHB), Faculté Génie des Procédés, Laboratoire des Sciences du Génie des Procédés Industriels (LSGPI), BP. 32, El Alia, Bab Ezzouar 16111, Algiers, Algeria. phone:+213 21 24 79 19, e-mail: bensebiao@yahoo.fr
[2]Université Hassiba Ben Bouali, Faculté de Technologie, Département de Génie des Procédés. Laboratoire Chimie verte-Eau-Energie, BP 151, Chlef 02000, Algeria
[3] Université de Lorraine, Laboratoire Réactions et Génie des Procédés, UMR 7274, 1, Rue Grandville BP 20451, 54001-Nancy, France.

Supercritical fluids (SCFs) are alternative solvents in the field of Green Chemistry that are being developed as advanced separation techniques. When using conventional solvent extraction methods is it very difficult to control the selectivity of the extraction process. For this reason, supercritical fluid extraction with fractionation is a very attractive option for extracting selective compounds from herbs and plants.
In this work, Supercritical carbon dioxide (SC-CO2) was applied to fractionate rosemary supercritical extracts in order to produce fractions with high concentration of pentacyclic triterpenes, mainly α,β-amyrin, these triterpenes are promising leading compounds for the development of new multi-targeting bioactive agents. Supercritical rosemary extraction was carried out by operating at 180 bar and 50 °C in the extraction vessel, and on-line fractionation in a two-step depressurization system to selectively precipitate the pentacyclic triterpenes. Phytochemical composition of extracts was analyzed by means of GC–MS and HPLC. The overall extraction curve was represented using three models.

Key words: Supercritical fluid extraction, Fractionation, Pentacyclic triterpenes, rosemary

Supercritical CO_2 fractionation of sage leaves

O. Bensebia[1], B. Bensebia[2], Kh. Allia[1], D. Barth[3]
[1]Université des Sciences et de la technologie (USTHB), Faculté Génie des Procédés, Laboratoire des Sciences du Génie des Procédés Industriels (LSGPI), tel. +213 21247919, e-mail: bensebiao@yahoo.fr;[1] BP. 32, El Alia, Bab Ezzouar 16111, Algiers, Algeria; fax +213 21247919.
[2]Université Hassiba Ben Bouali, Faculté de Technologie, Département de Génie des Procédés Laboratoire Chimie verte-Eau-Energie, BP 151, Chlef 02000, Algeria
[3] Université de Lorraine, Laboratoire Réactions et Génie des Procédés, UMR 7274, 1, Rue Grandville BP 20451 54001-Nancy, France

The use of supercritical fluid technology as an innovative technology to extract bioactive compounds has grown considerably in recent decades. In the past few decades however, sage has been the subject of an intensive study for its phenolic antioxidant components. In this study, sage leaf oil was isolated by a supercritical fluid extraction using CO2 in which the extraction was followed by a two-stage fractional separation.
Experiments were conducted at 50 °C and 180 bar and fractionation of the extract was accomplished maintaining separator (S1) at 0 °C and separator (S2) at 20 °C. The sage leaves were also extracted with Soxhlet extraction, using hexane solvent. The chemical compositions of the extracted oils were analysed using a gas chromatography-mass spectrometer (GC–MS). The SFE yield was 2.59 % comparable in magnitude to that of the hexane Soxhlet extraction of 3.89 %. The chemical compositions of the extracts were quite different for the two extraction methods. The extract from the SFE and fractionation was characterized by the occurrence of labdane type diterpenes (35.37%) in high amounts with *epi*-13-manool (34.95%) as main compounds, while the oil extracted by soxhlet contained a high amount of waxes compounds (38.19 %) with octacosane (13.86%) as main compounds.

Key words: Supercritical extraction, Fractionation, Labdane diterpene, Sage leaves

Creation of new functional materials: "Aerogel - CNTs" for the use as supercapacitors

P. Tsygankov, I. Khudeev, A. Tyrtyshnikov, E. Vorobeva, S. Ivanov, N. Menshutina
D.I. Mendeleyev University of Chemical Technology of Russia, Computer Aided Process Engineering Faculty, Geroev Panfilovtsev Str. 20 bld.1, 125480 Moscow, Russia;
tel./fax: +7(495)495-00-29, e-mail: pavel.yur.tsygankov@gmail.com.

Development of the technologies connected with power supplies and energy storing is in great progress. Currently portable electronic devices become more multifunctional and have a tendency to reduce the size, thickness and weight. A possible solution to reducing the weight and increasing the thermal capacity is the development of a functional material having a low density, high strength and at the same time providing a large amount of energy.

There are two types of supercapacitors. The first type are supercapacitors whose capacity of a double electric layer is concentrated at the electrode / electrolyte interface. Carbon materials are usually used as material. The second type is the so-called pseudocapacitance supercapacitors where accumulation of energy is based on Faraday reaction at the electrode / electrolyte interface. Oxides of metals or electroconductive polymers are usually used as material.

Our work deals with creation of new materials: "aerogel – CNT". The work describes two ways of impregnation of carbon nanotubes (CNT) and their structuring. The first way is based on impregnation of CNTs into the sol stage and subsequent forming a gel. Structuring was carried out in an electromagnetic field. The second way is based on impregnation of the CNT's growth centers into gel, carrying out supercritical drying and subsequent growing CNTs in the structure of aerogel.

Increase in the specific capacitance and stability of the electrode electroactive material of the supercapacitor is achieved by a combination of a redox reaction (pseudocapacitance effect) and the electrical capacitance component of the carbon material.

The last part of the work describes the developed mathematical model. The mathematical model is based on the modified method of diffusion-limited aggregation (diffusion-limited aggregation - DLA). At the first stage the set quantity of nanotubes is randomly placed in the field. Particles of gel arrive from a remote source and make Brownian motion thus modeling the diffusion. The source can be of various forms: circle, sphere, surface. After reaching the cluster (originally – the center of crystallization) or nanotubes particles "adhere" to it, forming a branchy structure. The undoubted advantage of the DLA algorithm is an opportunity to generate both two-dimensional and three-dimensional structures. Comparison of calculated and experimental data was carried out, the adequacy of model was evaluated.

Use of supercritical carbon dioxide for direct preparation of crystalline monolithic TiO_2 aerogels

M. Cerhová, M. Sajfrtová, [1]L. Matějová, V. Dřínek, [2]S. Daniš, V. Jandová
Institute of Chemical Process Fundamentals of the ACS, v.v.i., Rozvojová 135, 165 02 Prague 6, Czech Republic, sajfrtova@icpf.cas.cz; [1]Institute of Environmental Technology, VŠB-Technical University of Ostrava, 17.listopadu 15/2172, 708 33 Ostrava – Poruba, Czech Republic; [2]Department of Condensed Matter Physics, Faculty of Mathematics and Physics, Charles University in Prague, Ke Karlovu 5, 121 16 Prague 2, Czech Republic

Titania aerogels are special highly-porous materials often used in the form of monoliths [1]. Their very high specific surface area makes them attractive candidates in many applications: solar energy conversion [2], photocatalysis [3], electronic devices [4], or as supporting material in the catalytic reactions [5]. The (photo)catalytic activity of monoliths critically depends on the structural properties such as crystal phase composition and crystallite size. Crystalline and pure TiO_2 aerogels are commonly prepared by calcination. However, this thermal treatment results in uncontrollable destruction of porous structure of titania aerogels, commonly accompanied by the decrease of the surface area as well as the porosity [6]. Therefore, gentler methods for the preparation of TiO_2 aerogels are desirable to optimize their morphology and functionality. Use of supercritical carbon dioxide ($scCO_2$) seems to be perspective way to obtain crystalline and pure TiO_2 monoliths without any subsequent thermal treatment [7, 8].

The present work deals with the crystallization of monolithic TiO_2 aerogels in environment of pure and water modified $scCO_2$. To remove organic precursors used during sol-gel synthesis and to crystallize TiO_2 the combination of several steps with pure and water-modified $scCO_2$ was tested with respect the high porosity and the specific surface area of titania blocks.

The effect of temperature (40-100 °C) and water concentration in $scCO_2$ (5-30 wt. %) on structural and textural properties was investigated at constant pressure of 30 MPa. The structural properties of the prepared monolithic TiO_2 were characterized by Raman spectroscopy. The textural properties such as specific surface area, content of mesoporous and volume of microporous were determined from nitrogen physisorption and skeletal density by helium pycnometry measurements.

The pure $scCO_2$ caused any crystallization. When 5 or 15 wt. % of water was added as a $scCO_2$ modifier a crystalline structure was obtained. The specific surface areas of mesoporous titania changed between 200-634 m^2/g. Increasing the temperature and the concentration of water in $scCO_2$ resulted into the decrease of specific surface areas and change of the monoliths into the more fragile blocks.

Acknowledgements. The financial support of the Grant Agency of the Czech Republic (project no.14-23274S) is gratefully acknowledged.

[1] Dagan G. et al. *J. Phys. Chem.* **1993**, *97*, 12651-12655. [2] Li X. et al. *Electrochim. Acta.* **2014**, *145*, 281-285. [3] Meng Nan Chong et al. *Water res.* **2010**, *44*, 2997-3027. [4] Baia L. et al. *Thin Solid Films.* **2006**, *511-512*, 512-516. [5] Mariscal R. et al. *Catal. Today.* **2002**, *75*, 385-391. [6] Pourmand M. et al. *Current nanoscience.* **2008**, *4*, 151–156. [7] Shimoyama Y. et al. *Chem. Eng. Res. Des.* **2010**, *88*, 1427-1431. [8] Matejova L. et al. *Mater. Res. Bull.* **2012**, *47*, 3573–3579.

Preparation of microcellular polystyrene foams with supercritical CO_2 and co-solvent

M. Topiar, [1]A. Nistor, [1]Déborah de Masi, [1]J. Kosek
Institute of Chemical Process Fundamentals of the CAS, Rozvojova 135, 165 02 Prague 6, Czech Republic;
[1]University of Chemistry and Technology Prague, Dept. of Chemical Engineering, Technická 5, 166 28 Prague 6, Czech Republic;
tel. +420 220 443 296, fax: +420 220 444 320, e-mail: Juraj.Kosek@vscht.cz

Polystyrene (PS) foams are used for packaging and mainly as thermal insulators. PS microcellular foams have the potential to replace common commercial PS foams (with cell sizes above 100 μm) because they can offer improved thermal insulation properties [1, 2]. This improvement is reached by decreasing the cell size below 10 μm. Such microcellular foams can be prepared using supercritical CO_2 ($scCO_2$) as the blowing agent. However, the prepared microcellular foams often have an insufficient porosity that is lower than 90 %.

In this work, we use co-solvents to increase the foam porosity. Microcellular PS foams were prepared by the pressure induced foaming method using $scCO_2$ and various co-solvents. The foam morphology was characterized via SEM. The effect of co-solvents on the foam morphology was the following: the use of co-solvents caused an increase in porosity by the occurrence of larger cells and the formation of interconnected cells, which made the morphology less homogeneous. But at higher co-solvent concentrations (e.g., > 6 wt.%, at 70°C) the morphology became homogeneous and a very high porosity was reached (Figure 1). Additionally, the use of co-solvents enabled us to prepare homogenous microcellular PS foams at lower impregnation temperatures, which has not been possible before just with the use of $scCO_2$.

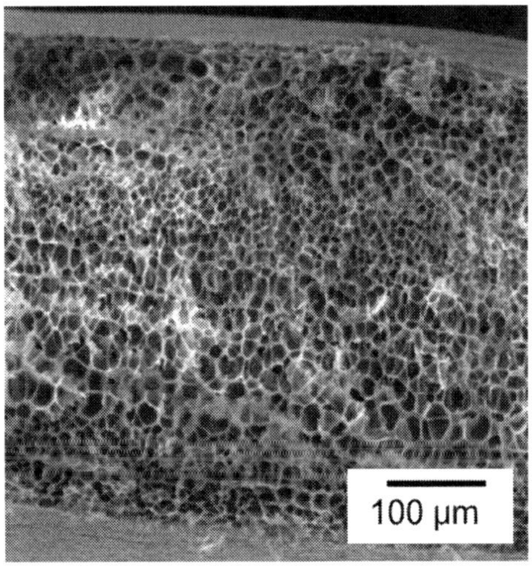

Figure 1: Cross-section cut of a microcellular PS foam.

References:

1. Ferkl, P., R. Pokorny, M. Bobak, and J. Kosek, *Heat transfer in one-dimensional micro- and nano-cellular foams.* Chemical Engineering Science, 2013. **97**: p. 50-58.
2. Notario, B., J. Pinto, E. Solorzano, J.A. de Saja, M. Dumon, and M.A. Rodríguez-Pérez, *Experimental validation of the Knudsen effect in nanocellular polymeric foams.* Polymer, 2015. **56**: p. 57-67.

Incorporation of natural compounds in biopolymers by supercritical impregnation for active food packaging

A. Torres[1*], E. Ilabaca[1,2], A. Rojas[2], C. Villegas[1], J. Romero[2], F. Rodríguez[1], A. Guarda[1], M.J. Galotto[1]

[1]Food Packaging Laboratory. Food Science and technology Department. Center for the Development of Nanoscience and Nanotechnology (CEDENNA), University of Santiago de Chile, Av. Manuel Umaña 050, Santiago, Chile; tel. +56227184522, email: alejandra.torresm@usach.cl (Alejandra Torres).
[2]Laboratory of Membrane Separation Processes (LabProSeM), University of Santiago de Chile, Av. Lib. B. O`Higgins 3363, Santiago, Chile.

Most of the petroleum based plastics used for food packaging are non-degradable and cause many environmental problems associated with their disposal, including damage to the ecosystems, water supplies, and sewer systems. The use of biodegradable plastics obtained from renewable sources is seen as one of the many strategies to minimize the environmental impact of petroleum-based products, leading to a renewed interest in packaging materials based on biopolymers, which offer favorable environmental advantages in terms of biodegradability compared with conventional synthetic polymeric materials. Among all these biodegradable polymers, polylactic acid (PLA) can be considered as a promising alternative to the petroleum-based commodity materials, because it can be produced from corn, potato, and other agricultural products. However, the efficiency of bio-based polymers in food packaging is limited by its high moisture sensibility (hygroscopicity), poor mechanical properties and high permeability to gases. Today, new trends in food packaging involve the use of biopolymers in order to develop eco-friendly materials and simultaneously propose new techniques to give specific properties to these materials. Antimicrobial (AM) packaging is one of the most interesting and promising components of Active Packaging (AP), which potentially allows industry to combine the preservative functions of AM agents with protective function for foods. On the other hand, the demand for the use of natural additives has produced a clear increase in the number of studies based on natural extracts such as essential oils (EOs), which are categorized as Generally Recognized as Safe (GRAS) by US Food and Drug Administration. Among the EOs, thymol (major component of thyme and oregano essential oils) and cinnamaldehyde (from cinnamon essential oil) have been widely used as active components in antimicrobial packaging.

Additionally, one of the problems that currently occur in the preparation of active packaging with essential oils by the traditional methods is related to the volatilization or degradation of the active agent(s) because of the high temperature values used during the plastic processing. In these cases the supercritical impregnation method can represent an efficient and advantageous alternative. This technique has been largely used in the development of materials for applications in the biomedical field mostly to prepare improved sustained drug delivery devices. Because of this, the aim of this work is the supercritical impregnation of thymol and cinnamaldehyde in PLA films at different impregnation pressures (9 and 12 MPa) and depressurization rates (0.1, 1 and 10 MPa min^{-1}) at a constant temperature (313 K), in order to obtain polymeric materials with potential to be used as active food packaging. Thus, this procedure allowed the active compounds incorporation with concentrations ranged between 11 and 21%. The impregnated materials were characterized in function of thermal, mechanical and structural properties. An important modification of the thermal and mechanical properties were obtained.

Mathematical constraints for an optimal design of temperature-dependent attractive parameter expressions in cubic equations of state

J.N. Jaubert, R. Privat, S. Lasala, Y. Le Guennec
Université de Lorraine, ENSIC-LRGP, 1 rue Grandville, 54000 Nancy, France.
tel. +33 383175128, e-mail: romain.privat@univ-lorraine.fr

Numerous modifications to the Van der Waals model have been presented over the years with the aim of representing with increasing accuracy the thermodynamics of complex systems. As a matter of facts, the most of those do not incorporate a substantial difference in the functional combination of attractive and repulsive forces, with respect to the original formulation introduced by Van der Waals.

Although the analytical expressions of the repulsive and attractive terms proposed in literature do not correctly quantify the actual repulsive and attractive contribution to pressure, their sum results in a quantitative representation of fluid properties being sufficiently accurate to make their combination the "cornerstone of the generalized van der Waals theory".

From the Van der Waals proposal, even the most successful engineering two-parameters cubic equations of state (EoS) still express their attractive and repulsive term by introducing a parameter a, which is a measure of the attractive forces ("energy parameter") between molecules, and the parameter b, which is a measure of the size ("intrinsic volume" or "co-volume") of the molecules. The pure-component a energy parameter is directly proportional to the temperature-dependent alpha function $\alpha(T)$ which is a measure of how the a parameter deviates from its critical value.

A wide variety of α-functions has been proposed over the past years. Those contain parameters which are adjusted according to different criteria: either over a whole set of components or for each specie ("generalized" versus "component-specific" α-function); either considering the whole temperature domain or specifically in the subcritical or supercritical regions ("overall" versus "domain-specific" α-function). Furthermore, among the mostly applied formulations, two different functional forms are identifiable: polynomial and exponential forms.

In this work, the popular component-specific Twu (1991) alpha function (denoted Twu91), which exhibits an exponential form has been selected to model Vapour-Liquid Equilibrium (VLE) behaviour of mixtures containing supercritical components.

A key outcome of our work is that the use of the Twu91 α-function together with some sets of component-dependent parameters has resulted in the inaccurate VLE modelling of such mixtures, in spite of the use of advanced EoS/g^E mixing rules. A detailed examination of this issue has shown that the application of such sets of parameters entails the generation of abnormal deviations of the α-functions, when extrapolating at supercritical conditions. In parallel, the use of the simple generalized and polynomial Soave alpha function incorporated in the same advanced EoS/g^E mixing rules as previous, has led to very accurate results.

In the basis of these counterintuitive results, it was attempted to understand how the use of an EoS coupled with a generalized alpha function (with no adjustable parameter) could induce better modelling results than an EoS coupled with a more flexible alpha function involving 3 adjustable parameters. It has been shown that the calibration of α-functions requires the imposition of constraints to the parameters under regression in order to guarantee consistent predictions of the derived thermodynamic properties (residual enthalpy, entropy, heat capacity …) at all temperatures (sub- or super-critical).

On the basis of such conclusions, the Twu91 alpha-function has been re-parameterized for a large series of pure components (more than 400). In the light of the newly regressed parameters, a generalized Twu-type α function, performing better than Soave's α function, has been proposed. The use of this new generalized Twu alpha function has been used to model the VLE data of binary systems containing a supercritical component and has led to improved modelling results with respect to the use of the simple Soave alpha-function.

Effect of pressure, temperature and molecular weight of polymer on the interfacial tension and gas solubility of binary systems at elevated pressures

[1]M. Knez Hrnčič, [1]G. Kravanja, [1]Ž. Knez
[1]University of Maribor, Faculty of Chemistry and Chemical Engineering,
Laboratory for separation processes and product design, Smetanova ulica 17, 2000 Maribor, Slovenia;
tel. +386 2 2294 477, e-mail: masa.knez@um.si

Knowledge of basic thermodynamic and transport data like phase equilibria, density, viscosity, dielectric constant, diffusion coefficient and interfacial tension is essential when designing a process in order to satisfy economical requirements. Literature data, reporting behavior of binary systems at elevated pressures and temperatures are still relatively scarce and comprise methods that are either expensive either time consuming.

Interfacial tension of binary systems at elevated pressures and temperatures was taken under investigation in the scope of the present research. Since interfacial tension plays a key role in understanding many polymer engineering processes and applications at high pressure, a special highlight should be dedicated to the development of new techniques for the determination of the interfacial tension at the conditions pertaining to the processing at elevated pressures by using supercritical fluids as a processing tool. Supercritical fluids are promising media for reducing interfacial tension and increasing the solubility of gas into polymers melts since the use of supercritical fluids enables processing of different materials without the use of liquid organic solvents. This will certainly bring both economic and environmental benefits thus absence of organic solvents means no need of additional, expensive steps for removal, recovery or disposal of toxic solvent; there will be little environmental issues related to the energy consuming, high polluting processes of solvent purification or disposal, limiting the risk of pollution of air, soil, surface or underground water sources.

Effect of pressure, temperature and molecular weight of polymer on the interfacial tension was investigated by a capillary rise method adapted to the measurement conditions. Experiments were performed in a glass-windowed equilibrium cell. Binary systems of CO_2 + PEG (polyethylene glycol) and argon + PEG were taken as model systems to investigate the interactions at the gas + polymer interface as a function of pressure and temperature. The molecular masses of PEGs differed from Mw = 400 g mol^{-1} up to Mw = 10 000 g mol^{-1} up to a pressure of 30.0 MPa at a constant temperature of 343 K. The reliability of the technique was confirmed by conducting measurements of the interfacial tension between CO_2 and pure water at a temperature of 318 K. The accuracy of the obtained results compared to the ones found in the literature was better than 4 %. Additionally, the effect of temperature was also investigated by measuring the solubilities and the interfacial tensions in the systems CO_2 + PEG 1500 and argon + PEG 1500 within the range of temperatures from 328 K up to 363 K.

The required densities were obtained under the same working conditions by a gravimetric method involving accurate magnetic suspension balance.

It was observed that interfacial tension behaves as a strong function of pressure for all of the investigated systems. In the case of argon saturated solutions of PEGs, the effect of pressure on interfacial tension was less pronounced in comparison with CO_2 saturated solutions of PEGs which resulted in up to 3 times higher values at pressures above 10 MPa. A possible reason for that is the difference in solubilities of both gases in polyethylene glycol: the solubility of carbon dioxide in polymer was found to be up to 10 times higher than the solubility of argon, most probably as a consequence of different molecular structures of both gases. The highest solubilities and consequently densities of CO_2 and argon were determined in PEGs of lower molecular masses due to the fact that a PEG with a low molecular mass has a matrix with more empty spaces and can accept more gas molecules.

Key words: Diffusivity, solubility, density, binary systems, argon, carbon dioxide.

Property prediction for complex hydrocarbon mixtures by means of real-component surrogates

K. Kos, T. Wallek
Graz University of Technology, Institute of Chemical Engineering and Environmental Technology, NAWI Graz, Inffeldgasse 25C/I, 8010 Graz, Austria;
tel. +43 316 873 4989, e-mail: kathrin.kos@tugraz.at.

Conventional petroleum-derived fuels are complex, heterogeneous mixtures of several hundreds of individual hydrocarbon species of uncertain composition. In order to emulate fossil fuels as well as other miscellaneous refinery streams by means of simulation technology, simplified substitute mixtures referred to as surrogates are required. The ultimate goal in surrogate design is to specify a blend of few representative compounds that is able to mimic various selected target properties (design criteria) of the respective target fuel or product. Regarding the two existing approaches for surrogate formulation, the traditionally applied pseudo-component approach has limited applicability compared to the more promising real-component approach, due to the following reasons: Using real chemical species allows application of rigorous thermodynamic models such as equations of state or activity coefficients for equilibrium calculations, property estimation and experimental validation. Another advantage is the broader application area comprising elevated temperatures and pressures as well as non-fossil polar components such as the biogenic energy carriers ethanol and biodiesel.
Real-component surrogates can be easily used for experimental work as they can be exactly reproduced at any time and in simulations. Thus, inaccuracies arising from oversimplified or imperfect product characterization within simulations can be avoided. In the case of gasoline and diesel fuel surrogates, such real component mixtures can be used for all stages of internal combustion engine CFD simulation.
For the generation of such multi-component surrogates an algorithm has been developed and successfully applied to various diesel fuels [1]. A comprehensive data bank containing pure-component property data of over seventy hydrocarbon species as well as interaction parameters serves as a data basis for mixture properties calculation. Surrogates are generated based on the simultaneous fitting of selected fuel properties of the substitute mixture to given data of the target fuel, using a least-squares approach. Design criteria of first choice are: true boiling point curve, liquid density, cetane number, C/H ratio, lower heating value and biodiesel content. By means of this optimization method, the optimal composition of the respective surrogate is obtained according to the selected target properties, their wheighting and the chosen hydrocarbon species. The selection of an appropriate set of real components is in fact an important criterion for the successful representation of the target fuel. For this purpose, an algorithm for automated component selection based on the 'F-to-enter' and 'F-to-remove' approach was developed and implemented. Depending on the application area (spray vs. combustion CFD simulation) the resulting surrogates typically contain four to ten real chemical components and allow reliable emulation of diesel fuel both in process simulations and engine combustion simulations.
The developed surrogate calculation approach can easily be adapted for application to gasoline or jet fuel and to various other refinery streams as well.

References

[1] A. M. Reiter, T. Wallek, A. Pfennig, M. Zeymer "Surrogate Generation and Evaluation for Diesel Fuel" Energy & Fuels 2015, 29: 4181-4192

VLE description of ternary systems containing ionic liquids

E. Graczová, P. Steltenpohl
Department of Chemical and Biochemical Engineering, Faculty of Chemical and Food Technology, Slovak University of Technology Bratislava , Radlinského 9, 812 37 Bratislava, Slovakia; tel. +421 2 59325760, e-mail: elena.graczova@stuba.sk; fax: +421 2 52493198.

The aim of this study is to show the thermodynamic description of isobaric ternary vapor – liquid equilibrium data of systems containing ionic liquids using G^E equations extended by the universal ternary contribution. These G^E equations, such as NRTL, UNIQUAC and Redlich–Kister, consist of original binary parameters and ternary contribution parameters. The original binary parameters are evaluated independently based only on the binary equilibrium data of binary subsystems; ternary parameters of ternary contributions are evaluated from experimental ternary vapor–liquid equilibrium (VLE) data. Due to the ternary contribution, the extended G^E model equations significantly improve the correlation of the ternary VLE data compared to the prediction using the original binary G^E model. This method allows for correct description of the VLE of ternary systems even if the amount of experimental equilibrium points is insufficient or just incomplete VLE data are available (e.g. P–x or y–x data).

The procedure was applied to 10 ternary VLE systems containing ionic liquids. Assuming isobaric VLE, parameters of the ternary contribution were treated as temperature independent and also as temperature dependent similarly to the temperature dependence of binary parameters.

Results of the thermodynamic description of ternary VLE of the studied systems containing ionic liquids are presented in form of mean deviations as well as root mean squared deviations of variables such as temperature and mole fractions of the components in the vapor phase. The correlation results obtained using extended NRTL equation show that the use of a universal ternary contribution improves significantly the thermodynamic description of the ternary VLE in regard to the prediction using binary parameters only. Mean deviation of the temperature decreased on average about twice for all the chosen ternary systems; mean deviation of the component mole fractions decreased by approximately 1.5 times. In comparison to the direct fitting of ternary VLE data, we obtained better results in case of the average mean deviation of the temperature; less important effect was observed in case of the average mean deviation of the component mole fractions. Results of the correlations are presented in form of 3D diagrams.

Acknowledgement
The authors acknowledge the Research and Development Assistance Agency APVV (Grant APVV-0858-12) for financial support.

Discrete modeling: the concept of Shannon entropy used for thermodynamic model development

[1]T. Wallek, [2]A. Pfennig
[1]Institute of Chemical Engineering and Environmental Technology, Graz University of Technology, NAWI Graz, Austria; Tel. +43 316 873 7966, e-mail: Thomas.wallek@tugraz.at;
[2] Department of Chemical Engineering, Université de Liège - Sart-Tilman, Belgium; +32 (0) 4 366-3521, e-mail: andreas.pfennig@ulg.ac.be

Thermodynamic models such as equations of state and g^E-approaches are being challenged by the need to describe complex and/or oxygenated molecules, which are characterized by multiple, spatially distributed interaction sites. In this context, established models often use local compositions for representation of molecular properties. However, the fact that these local compositions represent *averaged* system values, limits the flexibility of such approaches to describe complex mixtures. To incorporate a more detailed molecular picture into thermodynamics, previous papers introduced 'discrete modeling' as a novel approach. As a proof of concept, the thermal and caloric equations of state, heat capacity and Maxwell-Boltzmann distribution for ideal gas were derived on the basis of discrete states of individual molecules [1-2].

In this paper, the discretized states of a lattice system of finite size are used as the basis for the application of discrete modeling to spherical molecules in binary condensed-phase mixtures, with the medium-term goal of g^E-model development. The discrete model is based on the definition of discrete system states, where states of equal energy are combined in discrete energy classes j, represented by discrete values of local composition, $x_{12,j}$, in contrast to averaged local compositions. In this way, the model uses the same level of discretization as classical statistical thermodynamics in terms of its partition functions, yet avoids 1.) *a-priori* averaging of local compositions by utilizing a distribution, and 2.) the confinement to systems of infinite size. Considering that each energy class can be realized by several system states, the possible arrangements of the system at a given energy are characterized by degeneracy functions. The formalism of discrete modeling then uses probabilities of discrete local compositions, $p_{12,j}$, along with degeneracies to express the Shannon entropy of the system together with all relevant constraints, like energy due to molecular interactions. Using Shannon entropy equivalently to thermodynamic entropy, constrained maximation of entropy then yields the equilibrium distribution for the probabilities of discrete local compositions, $p_{12,j}$, from which all thermodynamic functions of interest can be derived. The figure illustrates the resulting equilibrium distribution for a binary system of global composition $x_1 = 0.3$ at three different dimensionless interchange energies, ω_{12}^*/k_BT, representing repulsion, random mixing and attraction. The results of this approach are compared to those derived from Monte-Carlo simulations and by applying the well-known Guggenheim model of chemical theory. As an outlook, a further refinement of the discretization level from discrete local compositions towards discrete states of individual molecules is discussed.

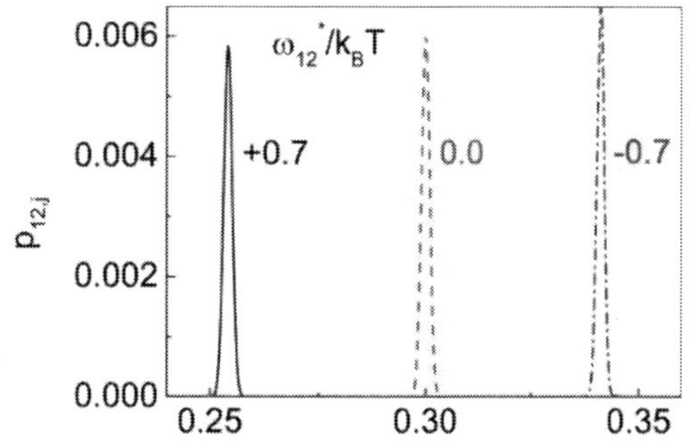

[1] M. Pfleger, T. Wallek, A. Pfennig "Constraints of Compound Systems: Prerequisites for Thermodynamic Modeling Based on Shannon Entropy" Entropy 2014, 16: 2990-3008.

[2] M. Pfleger, T. Wallek, A. Pfennig "Discrete Modeling: Thermodynamics Based on Shannon Entropy and Discrete States of Molecules" Ind. Eng. Chem. Res. 2015, 54: 4643-4654.

10 years with the PPR78 model: capabilities and limitations of predictive cubic equations of state involving classical mixing rules

R. Privat, J.N. Jaubert
Université de Lorraine, ENSIC-LRGP, 1 rue Grandville, 54000 Nancy, France.
tel. +33 383175128, e-mail: romain.privat@univ-lorraine.fr

Today, the synthesis design and optimization of the processes are carried out with the help of process simulators (ProSim, PRO/II, ASPEN ...). It is however well known that the accuracy of the simulated results mainly depends on the quality of the thermodynamic model. In most cases, the phase behaviour and the energetic properties (enthalpy, heat capacity ...) of multi-component systems for which nearly no data are available has to be known. The phase behaviour can obviously be measured, but measurements are very time consuming. This is why modern process design requires models:

(i) capable of predicting the equilibrium properties without the preliminary use of experimental data,
(ii) yielding accurate results in both the sub- and super-critical regions,
(iii) ensuring an acceptable level of accuracy in energy, entropy and exergy balances.

Simultaneous fulfilment of these requirements is a very difficult and challenging task for a thermodynamic model.
To this end, the PPR78 model is a predictive cubic equation of state (EoS) developed since the early 2000. It is based on the Peng-Robinson equation of state coupled with classical Van der Waals mixing rules. A cubic EoS has been chosen because in process design, due to their low complexity and their high accuracy for non-polar compounds, such EoS allow for fast screening of a large number of design alternatives and pre-selection of the most favourable candidate structures.
Chemical intermolecular interactions are quantified through the use of the so-called binary interaction parameters (BIPs) that are assumed to be temperature-dependent. A proper estimation of these interaction parameters from data regression procedures is increasingly difficult if not impossible (due to a lack of appropriate data) as the number of molecules involved in the studied blend is growing. A group-contribution based method, able to estimate BIPs from the mere knowledge of the chemical structures of the molecules, was thus derived to make the model predictive.
The PPR78 model can nowadays be used to predict thermodynamic properties of mixtures the molecules of which contain any of the following elementary structural groups:

Alkane groups: CH_3, CH_2, CH, C
Alkene groups: $=CH_2$, =CH, =C, ethylene, $=CH_{cycle}$, $=C_{cycle}$
Aromatic compounds groups: CH_{aro}, C_{aro}, $CH_{double\ ring}$
Naphthene groups: $CH_{2,cycle}$, $CH_{cycle} \Leftrightarrow C_{cycle}$.
Permanent gases (molecules are single groups): CO_2, N_2, H_2S, H_2, He, Ar, CO, SO_2, O_2, NO
Mercaptan group: SH
Water: H_2O
Refrigerant groups: C_2F_6, CF_3, $CF_2 \Leftrightarrow$ CF, R152a, R1234a.

This presentation will offer the opportunity to take a global view on the real advantages and limitations of the PPR78 model. In particular, it will be shown how this approach was successfully applied to:

- aqueous systems,
- refrigerant mixtures,
- synthetic petroleum fluids,
- real petroleum fluids (involving ill-defined compounds).

The possibility to define a predictive SRK-based EoS from the PPR78 model through the use of the PR_2SRK theory will be also mentioned.

Finally, whereas it is often claimed that cubic-based models using classical mixing rules:
- lead to worse results at low pressure than the so called $\gamma - \phi$ approach,
- are not able to properly predict the phase behaviour of size-asymmetric systems,
- poorly predict the critical locus of binary systems,
- lead to a disaster in terms of aqueous system phase behaviour prediction.

it will be attempted to convince that the PPR78 model invalidate such assertions.

Reduction methods for phase equilibrium calculations: which independent variables?

Dan Vladimir Nichita
Université de Pau et des Pays de l'Adour, CNRS UMR 5150, Laboratoire des Fluides Complexes, BP 1155, 64013 Pau, France;
tel. +33 5 59407699, e-mail: dnichita@univ-pau.fr

Phase equilibrium calculations are in the core of chemical and petroleum engineering simulations. In some simulations, a very large amount of phase equilibrium calculations is required, representing the most important part of the computational effort. It is imperative that calculation algorithms are not only efficient, but also highly robust since a single failure may cause significant error propagations leading to false solutions.

An interesting alternative to conventional methods (seeking for the solution in the compositional space) for phase equilibrium calculations is given by the so called reduction methods, in which the reduced number of independent variables does not depend on the number of components. Since the reduction methods were formalized by Hendriks [1], all implementations used as variables either the reduction parameters, which are linear forms in composition (methods denoted Red-Q) or the Lagrange multipliers associated with the constrained minimization of the Gibbs free energy with respect to the reduction parameters subject to a set of specific constraints (methods denoted Red-h). Red-h methods were found to be superior to Red-Q methods in terms of robustness and convergence speed [2-4].

Hendriks and van Bergen [5] introduced the generalized distribution coefficients, defined as the ratio of the reduction parameters of equilibrium phases (method denoted Red-K). Surprisingly, even though an important number of scientific papers on reduction methods were published the last decades, the generalized distribution coefficients were not used, except in one paper [6] on multiphase equilibrium calculations (with excellent results, but no comparison with other formulations is given). In this work, a reduction method based on the generalized distribution coefficients is presented and compared with existing formulations (conventional and reduced).

Numerical experiments were carried out for several hydrocarbon mixtures containing non-hydrocarbon components which require relatively large binary interaction parameters, with emphasis to the convergence behavior near phase boundaries and in the vicinity of critical points.

It was found that: i) Red-K requires almost the same number of iterations as Red-h and the conventional method using the logarithm of equilibrium constants (lnK) as independent variables; ii) Red-K exhibits almost the same convergence path in the compositional space as Red-h and lnK; iii) The Jacobian matrix in Red-K Newton iterations has the best condition number as compared to *all* formulations in the reduced and compositional spaces; iv) all values of the generalized distribution coefficients are very close one to each other (at the limit, they could be identical for mixtures containing components belonging to homologous series for which the two EoS parameters are linear dependent; in practice the relationship between these two parameters is almost linear); v) the values of the generalized distribution coefficients are relatively few changing during iterations. The last two features positively influence conditioning and convergence properties. Property iii) is maybe the most important: round-off errors are smaller than for other methods, mainly in the critical region and near the convergence locus for negative flash calculations. It influences properties i) ensuring a faster convergence and ii) ensuring monotonic convergence where other methods did not.

Summarizing, the Red-K method behaves similarly to Red-h and lnK for most conditions, except for difficult conditions where it exhibits superior convergence behavior. In conclusion, it seems that the independent variables proposed by Hendriks and van Bergen, that is, the forgotten generalized distribution coefficients are the most suitable variables in the reduction methods applied to phase equilibrium calculations, and lead to the fastest and more robust formulation.

References

[1] E.M. Hendriks, Reduction Theorem for Phase Equilibria Problems, Ind. Eng. Chem. Res. 27 (1988) 1728-1732.

[2] D.V. Nichita, A. Graciaa, A New Reduction Method for Phase Equilibrium Calculations, Fluid Phase Equilib. 302 (2011) 226-233.

[3] M. Petitfrere, D.V. Nichita, A comparison of conventional and reduction approaches for phase equilibrium calculations, Fluid Phase Equilib. 386 (2015) 30-46.

[4] M. Petitfrere, D.V. Nichita, Multiphase equilibrium calculations using a reduction method, Fluid Phase Equilib. 401 (2015) 110-126.

[5] E.M. Hendriks, A.R.D. van Bergen, Application of a Reduction Method to Phase Equilib. Calculations, Fluid Phase Equilib. 74 (1992) 17-34.

[6] D.V. Nichita, D. Broseta, J.-C. de Hemptinne, Multiphase Equilibrium Calculation Using Reduced Variables, Fluid Phase Equilib. 246 (2006) 15-27.

New measurements and modeling of electrical conductivity and pH of n-propanethiol and n-butanethiol in *N*-methyldiethanolamine + piperazine (MDEA-Piperazine) aqueous solution at 306 K and 1 atm

Muhammad Yaqoob[1], Gull Zaib[1], Mudasir Hussain[1], Safwan Ali[1], Christophe Coquelet[2], Javeed A. Awan[1]*,
[1]Institute of Chemical Engineering and Technology, Faculty of Engineering and Technology University of the Punjab, Lahore, Pakistan. [2]MINES ParisTech, PSL Research University, CTP - Centre of Thermodynamic of Processes, 35 rue Saint Honoré 77305 Fontainebleau, France.

In this work new measurements of electrical conductivity (μS.cm^{-1}) and pH of n-propanethiol and n-butanethiol in *N*-methyldiethanolamine (MDEA) + piperazine aqueous solution at 306 K and atmospheric pressure. The experiments were performed at three 2.5 ml, 3.0 ml and 5.0 ml volumes of n-propanethiol and n-butanethiol in 45 wt % and 18 wt % MDEA aqueous solutions. The piperazine concentration has been kept constant by 4 wt % in all the experiments. The changes in electrical conductivity (EC) with time have been related to the change in concentration of ionic species and change of pH has been related to the acid-base neutralization reaction.

Keywords: electrical conductivity, kinetics of thiols, acid-base neutralization, thiol removal, organic sulfur species

Chemical equilibrium and liquid–liquid equilibrium for the system acetic acid – n-butanol – n-butyl acetate – water

A.A. Samarov, P.V. Naumkin, M.A. Toikka, A.M. Toikka
Saint-Petersburg State Univ., Dep. Chem. Thermodynamics & Kinetics, Universitetskiy pr. 26, Peterhof, 198504 Saint-Petersburg, Russia; tel. +7 812 4284052, e-mail: samarov@yandex.ru

In this paper we consider the phase behavior of reactive system in the case of liquid phase splitting. The systems with simultaneous processes are of special interest both for the development of basic thermodynamic theory and for chemical engineering applications. The liquid phase splitting in the run of chemical reaction could be often considered as unwanted phenomena because it leads to the unfavorable changes in reaction kinetics and hydrodynamic conditions in chemical reactor. In other cases the reactions in heterogeneous media may be useful for the design of coupled industrial processes such as reactive extraction, reactive distillation or extractive distillation. In our work we consider some singularities of reactive fluid systems with liquid phase splitting from thermodynamic point of view and on the base of new experimental data on liquid–liquid equilibrium (LLE) including our recent results [1-3].

n-Butyl acetate is one of the most abundant chemical solvents because of a low environment impact in comparison with other solvents. The system with n-butyl acetate synthesis reaction is one of the most investigated systems. Due to a low environment impact n-butyl acetate is considered as convenient chemical solvent. It is also used in various industrial fields. The design of n-butyl acetate production demands special knowledge on phase and chemical equilibria (CE). Unfortunately, most researches have focused on the solubility and phase equilibrium (liquid–vapor and liquid–liquid).

In this paper we present the results of experimental study on LLE, CE and modeling of the LLE in the acetic acid – n-butanol – n-butyl acetate – water system at 308.15 K and atmospheric pressure (Fig. 1). Due to the limited miscibility of water and ester the liquid phase splitting also occurs in quaternary system.

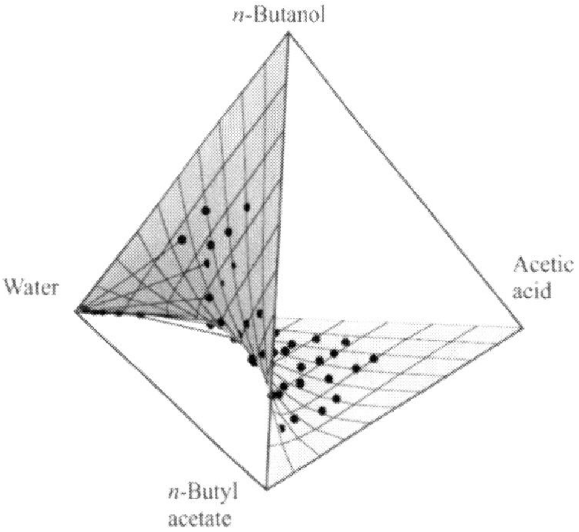

Fig. 1. The diagram of CE surface in acetic acid – *n*-butanol – *n*-butyl acetate – water system

Acknowledgement: Authors are grateful to SPbSU for Research Fellow grant (12.50.1564.2013).

References

[1] Trofimova M., Toikka M., Toikka A. Solubility, liquid–liquid equilibrium and critical states for the quaternary system acetic acid – ethanol – ethyl acetate – water at 293.15 K. *Fluid Phase Equilib.*, **313** (2012) 46-51.

[2] Toikka A., Toikka M. Solubility and Critical Phenomena in Reactive Liquid–Liquid Systems. *Pure Appl. Chem.*, **81** (2009) 1591-1602.

[3] Toikka M.A., Gorovits B.I., Toikka A.M. Solubility in the System Constituted by Acetic Acid, *n*-Propanol, Water, and *n*-Propyl Acetate. *Russ. J. Appl. Chem.*, **81** (2008) 223-230.

Electronic database of ternary contribution parameters to G^E equations

E. Graczová, P. Steltenpohl
Department of Chemical and Environmental Engineering, Faculty of Chemical and Food Technology, Slovak University of Technology Bratislava , Radlinského 9, 812 37 Bratislava, Slovakia; tel. +421 2 59325760, e-mail: elena.graczovai@stuba.sk; fax: +421 2 52493198.

Electronic database of ternary contribution parameters to common GE equations (specifically: NRTL, UNIQUAC and Wilson equations, Redlich – Kister expression) extended by the ternary contribution is presented. The database offers sets of model parameters of GE equations that are proper for the prediction of vapour–liquid, liquid-liquid and vapour-liquid–liquid equilibrium data of more component systems. The aim of this database is to provide access to binary and ternary parameters and to improve the prediction of multi-phase more component equilibriums essential for the separation system design.

The database contains 81 isothermal and 46 isobaric ternary systems; as well as 306 isothermal and isobaric birnary subsystems containing various kinds of organic constituents: alkanes, alkenes, cycloalkanes, aromatics, ethers, esters, halogen derivatives, nitrogen compounds, water... Also systems with ionic liquids are included. The database is continuously extended.

The main source of chosen equilibrium data and original binary parameters of GE equations (NRTL, Wilson, UNIQUAC) is the Vapor–Liquid Equilibrium Data Collection DECHEMA. Binary parameters of the Redlich – Kister expression were evaluated from binary equilibrium data. Parameters of the ternary contribution to the chosen excess molar Gibbs energy models were calculated from the available ternary VLE data using the corresponding original binary model parameters. Results of thermodynamic evaluations are presented in form of statistical parameters, as well as 2D and 3D diagrams of ternary systems.

A demo-version of the electronic database is publicly available at the web of the Institute of Chemical and Environmental Engineering, FCHPT-STU Bratislava (http:/kchbi.chtf.stuba.sk/TCDB).

Acknowledgement

The authors acknowledge the Research and Development Assistance Agency APVV (Grant APVV-0858-12) for financial support.

A new method for determining average boiling points of narrow boiling range oil fractions using a thermogravimetric analyzer

R. Rannaveski, O. Järvik, V. Oja
Tallinn Univ. of Technology, Dep. Chemical Engineering, Ehitajate tee 5, 19086 Tallinn, Estonia; e-mail: vahur.oja@ttu.ee

Here we propose an alternative experimental method to evaluate weight average boiling points of oil narrow boiling range fractions (per-prepared) in a fast manner. An average boiling point is a commonly used average input parameter in empirical determination methods (correlations) to predict thermodynamic properties. In practice, there are five different average boiling points used (volume average boiling point, weight average boiling point, molar average boiling point, cubic average boiling point and mean average boiling point); however, for fractions with narrow boiling ranges these different boiling points are considered equal.

The new experimental method proposed was developed to determine atmospheric boiling points with only a small amount of sample, usually less than 20 mg. The application of the method was aimed to support the development of thermodynamic property estimation correlations for unconventional oils with a chemical character substantially different than that of conventional petroleum. When creating the method, the principle behind the ASTM E1782 standard (Standard Test Method for Determining Vapor Pressure by Thermal Analysis) was considered as a starting point. Although the ASTM E1782 is for determining the vapor pressures of pure substances, it has lately been extended to determine the vapor pressure and initial boiling points of oil fractions with narrow boiling ranges [1,2]. Here, it is also useful to note that most ASTM E1782 applications are based on using a differential scanning calorimeter (see the latest summary as of 2014 [1]) and only few are based on thermogravimetric analyzer.

The new thermogravimetric analyzer based method relies on measuring a differential mass loss curve (i.e. the vaporization curve of the progressively less volatile oil remaining in the sample pan) when the oil vaporizes from a hermetically sealed sample pan through a pinhole with a diameter of 50 μm. And this under an assumption that, under certain experimental conditions, the differential mass loss curve roughly parallels the boiling point distribution of the fraction (continuous mixture). The vaporization of the sample through a pinhole was measured at a constant heating rate of 10 °C/min (or 20 °C/min in few cases) in a nitrogen flow of 200 ml/min. Then the differential mass loss curve was converted into a boiling (or condensation) curve of the vaporized species (narrow fractions as pseudocomponents). From the latter the weight average boiling point is then calculated. Further details can be found from [3].

The performance of the method was evaluated by comparing average boiling points calculated from this thermogravimetric analyzer based method with those obtained from rectification results (calculated as the average of the initial and final temperatures of the cut). For evaluating the method's performance about 35 fractions, including fractions with narrow boiling ranges (boiling ranges of up to 20 °C) and wider fractions that were obtained by combining the closest narrow fractions (boiling ranges of up to 60 °C), were used. The narrow boiling range oil cuts were prepared by distillation of a shale oil middle fraction (technical fraction with a boiling range of about 160 to 470 °C, manufactured from Kukersite oil shale by an industrial retorting unit based on the Galoter process). The middle oil was distilled into fractions by a packed rectification column with 24 theoretical plates and a reflux ratio of 6:1(in accordance with the ASTM D2892 standard). The absolute average deviation of the average boiling points determined using the new average boiling point experimental determination method was of 2 °C (note that with the thermogravimetric analyzer applied, the accuracy of measuring the temperature was ±2 °C) and the maximum deviation was of 8 °C (with only 6 points deviating from the TBP values more than 3 °C). Figure 1 presents graphically the comparison of determined average boiling points, obtained by the thermogravimetric analyzer, with average boiling points from rectification data for fractions studied.

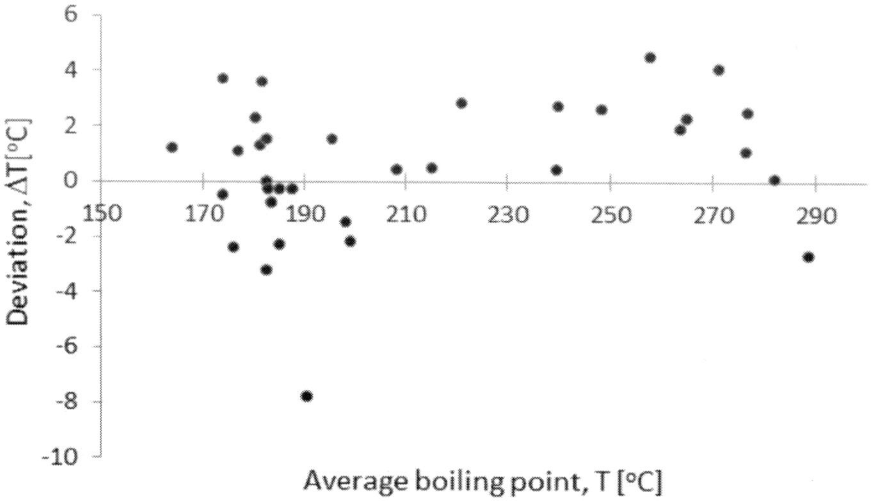

Figure 1. Comparison of the average boiling points determined by the thermogravimetric analysis method, proposed in this study, and the traditional true boiling point distillation (calculated as the average of the initial and final temperatures of the cut).

Acknowledgements:
The authors gratefully acknowledge financial support provided by Estonian Ministry of Education and Research, under target financing SF0140022s10.

References:
[1] Siitsman C., Oja V., Extension of the DSC method to measuring vapor pressures of narrow boiling range oil cuts, Thermochim. Acta 622 (2015) 31-37.
[2] Siitsman C., Oja V., Application of a DSC based vapor pressure method for examining the extent of ideality in associating binary mixtures with narrow boiling range oil cuts as a mixture component, Thermochim. Acta (2016), 10.1016/j.tca.2016.05.11
[3] Rannaveski R., Järvik O., Oja V., A new method for determining average boiling points of oils using a thermogravimetric analyzer: application to unconventional oil fractions, submitted to J. Therm. Anal. Calorim. (2016).

Improved solution windows for the Rachford-Rice equation

Dan Vladimir Nichita, [1] Claude F. Leibovici
Université de Pau et des Pays de l'Adour, CNRS UMR 5150, Laboratoire des Fluides Complexes, BP 1155, 64013 Pau, France; tel. +33 5 59407699, e-mail: dnichita@univ-pau.fr;
[1] CFL Consultant, Hélioparc, 2 Avenue Pierre Angot, 64053 Pau Cedex, France

The Rachford-Rice equation is solved in an inner loop of flash calculations. It consists in a sum of hyperbolas, and the unique root of interest is located between two adjacent asymptotes which define the negative flash window [1] (the mathematical domain which is wider than the physical one). In petroleum and gas compositional reservoir simulators, the Rachford-Rice equation must to be solved a very large number of times (often billion of times). Due to the shape of the function, hybrid Newton-bisection methods are usually used to solve the nonlinear equation; the number of iteration can be important (dozens) if the root is located very close to a pole.

A recent study [2] showed that using appropriate changes of variables the original function can be transformed either in convex functions (by removing one asymptote) or in quasi-convex functions (by removing both asymptotes) very close to linearity in a wide range [2,3]. Since the sufficient condition of monotonic convergence is fulfilled for the convex transformations, (second- and certain higher-order) Newton methods can be used without interval control (it is guaranteed that an overshoot of the solution cannot occur). For all the algorithms proposed in Ref. [2], the quality of the initial guess, thus the convergence speed, are directly related to the existence of narrow solution windows.

In this work, new, extremely simple solution windows are proposed, which depends only on feed compositions. It is proved that novel solution windows are always smaller than previously proposed ones [2-4]. First, several pathological cases are discussed, then numerical experiments are carried out for a very large number of randomly generated flashes, showing that the use of new solution windows systematically decreases the number of iterations required for convergence. In the range of number of components usually encountered in compositional reservoir simulations, the reduction in number of iterations (and thus in computer time) is up to 20 % as compared to the fastest previous formulations [2,4].

References

[1] Whitson, C.H., Michelsen, M.L., The negative flash. Fluid Phase Equilibria 53, 51-72, 1990.

[2] Nichita, D.V., Leibovici, C.F., A Rapid and Robust Method for Solving the Rachford-Rice Equation Using Convex Transformations, Fluid Phase Equilibria, 353, 38-49, 2013.

[3] Nichita, D.V., Leibovici, C.F., Rapid and robust resolution of Underwood equations using convex transformations, Computers & Chemical Engineering, 71, 574-590, 2014.

[4] Leibovici, C.F., Neoschil, J., A new look at the Rachford-Rice equation. Fluid Phase Equilibria 74, 303-308, 1992.

Simple equipment for measurement of total pressure tested on poly(ethylene glycol) + water system

Jan Pavlíček, Grozdana Bogdanić, Ivan Wichterle
Institute of Chemical Process Fundamentals, The Czech Academy of Sciences, Rozvojová 135
165 02 Prague 6, Czech Republic. E-mail: wichterle@icpf.cas.cz

This study belongs to a series of continuing studies of vapour–liquid equilibria (VLE) in diluted polymer + solvent systems. We investigated polystyrene + butan-2-one [1] or + toluene [2], poly(methyl methacrylate) + acetone [3] or + butan-2-one [4], poly(acrylic acid) + water [3], including earlier measurements of VLE in solutions of co- and terpolymers of octadecyl acrylate and octadecyl methacrylate with styrene and 1-vinyl-2-pyrrolidone [5–6]. All those experiments have been carried out by ebulliometric method, which prevents from use of more concentrated mixtures.

In this work, new equipment designed for total pressure determination in mixtures where one of component is not volatile, such as polymer + solvent systems, is presented. It enables measurement of pressure above solutions in a wider concentration range. The equipment has been tested and calibrated using vapour pressure of pure water. It consist of thermostat with temperature measurement (AΣΛ, model F250), 100 ml vessel for solution with magnetic stirrer with a very strong driver, and of heated manifold for charging, degassing and for pressure measurement. Pressure is measured with use of absolute heated capacitance unit MKS Baratron (pressure range 1000 Torr = 133 kPa). Mixtures were prepared by weighing to obtain volume of 50 ml, approx.

New experimental data on VLE in the poly(ethylene glycol) [PEG, M_n = 6000 g/mol] + water were determined under isothermal conditions at 308.15 and 313.15 K. It can be seen that it was possible to cover concentration region up to 50 wt. % of solvent.

The experimental data were correlated using the UNIQUAC-FV model [7]. In order to compare our data with other available sources, we correlated all published data for the PEG + water system determined by different methods [8–10] using the same model. In addition, we predicted the activities of water in mixtures with PEG using the Entropic-FV model [11].

It can be concluded that the set-up enables to determine vapour pressure data of mixtures with higher amount of polymer in comparison with the ebulliometric method, which provides the data at lower concentrations only due to problems with smooth boiling, usually occurring above 15 % of polymer. The pressure measurement unit must be calibrated for application at higher temperatures (up to 450 K). Another unit of MKS Baratron (with pressure range 10 Torr = 1.33 kPa), heatable to 500 K will be used later on as zero indicator. Data for the PEG + water system were successfully correlated using the UNIQUAC-FV model and they complete published data sets. The Entropic-FV model was tested against the obtained experimental data and can be used to predict VLE in the studied system.

References

[1] J. Pavlíček, G. Bogdanić, I. Wichterle, *Fluid Phase Equilib., in press.*
[2] J. Pavlíček, G. Bogdanić, I. Wichterle, *Chem. & Biochem. Eng. Q.,* 29 (2015) 1–4.
[3] J. Pavlíček, G. Bogdanić, I. Wichterle, *Fluid Phase Equilib.,* 358 (2013) 301–303.
[4] J. Pavlíček, G. Bogdanić, I. Wichterle, *Chem. Biochem. Eng. Q.*, 28 (2014) 447–450.
[5] J. Pavlíček, G. Bogdanić, I. Wichterle, *Fluid Phase Equilib.*, 297 (2010) 142–148.

[6] G. Bogdanić, I. Wichterle, *J. Chem. Eng. Data*, 56 (2011) 1080–1083.
[7] A. Jonquières, L. Perrin, S. Arnold, P. Lochon, *J. Membrane Sci.*, 150 (1998) 125–141.
[8] R. Sadeghi, Y. Shahebrahimi, *J. Chem. Eng. Data*, 56 (2011) 789.
[9] R. Sadeghi, R. Hosseini, B. Jamehbozorg, *J. Chem. Thermodyn.*, 40 (2008) 1364.
[10] M. Herskowitz, M. Gottlieb, *J. Chem. Eng. Data*, 30 (1985) 233.
[11] G.M. Kontogeorgis, A. Fredenslund, D.P. Tassios, *Ind. Eng. Chem. Res.*, 32 (1993) 362–372.

This study was partly supported by the Czech Science Foundation (Grant No. 15-19542S).

AUTHOR INDEX

AUTHOR INDEX

AUTHOR INDEX

AUTHOR INDEX

AUTHOR INDEX

AUTHOR INDEX

AUTHOR INDEX

AUTHOR INDEX

AUTHOR INDEX

AUTHOR INDEX

AUTHOR INDEX

AUTHOR INDEX

AUTHOR INDEX

AUTHOR INDEX

AUTHOR INDEX

AUTHOR INDEX

AUTHOR INDEX

AUTHOR INDEX

AUTHOR INDEX

AUTHOR INDEX

AUTHOR INDEX

AUTHOR INDEX

AUTHOR INDEX

22nd International Congress of Chemical and Process Engineering (CHISA 2016) and 19th Conference on Process Integration, Modelling and Optimisation for Energy Saving and Pollution Reduction (PRES 2016)

Prague, Czech Republic
27 - 31 August 2016

Volume 2 of 2

ISBN: 978-1-5108-5962-3

Printed from e-media with permission by:

Curran Associates, Inc.
57 Morehouse Lane
Red Hook, NY 12571

Some format issues inherent in the e-media version may also appear in this print version.

Printed by Curran Associates, Inc. (2018)

For permission requests, please contact Czech Society of Chemical Engineering
at the address below.

Czech Society of Chemical Engineering
Novotneho Iavka 5
CZ-116 68 Praha 1
Stare Mesto, Czech Republic

Phone: +420 221 082 248
Fax: +420 221 082 366

cschi@csvts.cz

Additional copies of this publication are available from:

Curran Associates, Inc.
57 Morehouse Lane
Red Hook, NY 12571 USA
Phone: 845-758-0400
Fax: 845-758-2633
Email: curran@proceedings.com
Web: www.proceedings.com

TABLE OF CONTENTS

VOLUME 1

PLENARY LECTURES

REACTION ENGINEERING, CATALYSIS AND KINETICS

DISTILLATION AND ABSORPTION

EXTRACTION AND LEACHING

MEMBRANE SEPARATIONS

SOLID-LIQUID SEPARATIONS

FLUID FLOW AND MULTIPHASE SYSTEMS

MIXING

COMPUTER AIDED PROCESS ENGINEERING

PARTICULATE SOLIDS

PHARMACEUTICAL ENGINEERING

CHEMICAL ENGINEERING EDUCATION

SYMPOSIUM ON ENVIRONMENTAL ENGINEERING

SYMPOSIUM ON SAFETY IN CHEMICAL INDUSTRY

SYMPOSIUM ON SUPERCRITICAL FLUID APPLICATIONS

SYMPOSIUM ON THERMODYNAMICS AND TRANSPORT PROPERTIES

VOLUME 2

SYMPOSIUM ON NOVEL FOOD PROCESSES AND TECHNOLOGIES

SYMPOSIUM ON POROUS MATERIALS, NANOSTRUCTURES AND NANOCOMPOSITES

SYMPOSIUM ON PROGRESS IN CHEMICAL TECHNOLOGY AND BIOTECHNOLOGY

SYMPOSIUM ON PROCESS INTENSIFICATION AND MINIATURISATION

COMMERCIAL AND TECHNICAL IMPACT FROM EU PROJECTS

PRES 2016

ADDITIONAL PAPERS

Property prediction subject to insufficient reliable data for the training set

N. Brauner , [1]M. Shacham
School of Mech. Eng., Tel-Aviv Univ., Tel-Aviv 69978, Israel; tel. +972 36408127, e-mail: brauner@eng.tau.ac.il; [1] Dept. Chem. Eng., Ben-Gurion Univ. of the Negev, Beer-Sheva 8415, Israel; tel. +972 86461481, e-mail: shacham@exchange.bgu.ac.il.

Current methods used to predict physical and thermodynamic properties can be classified into group contribution methods (GC, e.g., Marrero and Gani, 2001), methods that use some combination of measured properties and GC (e.g. Nannoolal et al. 2007) , use of reference series (RS, e. g. Shacham et. al. 2013) and quantitative-structure-property relationships (QSPRs, e.g., Wakeham et al. 2002, Brauner et al, 2006). The GS methods usually use a "training set" of compounds from various chemical groups with available experimental data for derivation of the group contribution values. The QSPR methods use similar training sets for identification of descriptors included in a QSPR model and calculation of the corresponding parameter values. Another set of "evaluation set" compounds, also containing only compounds with available experimental data, are used to validate the prediction accuracy of the resultant GC and QSPR models. There has been extensive effort in developing more accurate and more widely applicable correlations as more experimental data with higher precision becomes available. Nannoolal et al (2007) , for example, have shown that the GC methods that they have developed are able to predict the critical pressure and temperature, P_c, T_c with significantly higher accuracy than previously developed GC methods.

However, the prediction of P_c and T_c for groups of compounds that are not sufficiently stable at (or near) T_c to enable accurate measurement of the critical constant represent a special challenge. Consequently, attempt to predict these properties for a member of such a group, (the "target compound") may rely on a training set where very few compounds that are structurally similar to the target are included, because of the lack of similar compounds for which experimental critical data are available. Thus, the published prediction error statistics which rely on evaluation sets of compounds with reliable experimental data may not be applicable to compounds that are unstable at the critical point.

We have checked this hypothesis using recently measured P_c and T_c data for two groups compounds of current industrial interest. Such data was published for the first group, that includes: acenaphthene, fluorene, anthracene, phenanthrene and pyrene (cancerogeneous arenes) by Nikitin and Popov (2015). The similarity group selection algorithm of the Targeted QSPR method (Brauner et al. 2006) was used for identifying the compounds most similar to the target compound in the DIPPR (www.aiche.org/dippr) database. Amongst the 10 most similar compounds, experimental T_c data are listed only for 2 members of the acenaphthene's training set, 4 for fluorine, 3 for anthracene, 2 for phenanthrene and 2 for pyrene. Experimental P_c data are listed only for 2 members of the acenaphthene's training set, 3 for fluorine, 1 for anthracene, 1 for phenanthrene and none for pyrene.

Nikitin and Popov (2015) compared their experimental results with several prediction techniques including methods published after the year 2000: Marrero and Ganni (MG, 2001), Wakeham et al. (W, 2002) and Nannoolal et al(N, 2007). In some cases these most recent state of the art methods yielded predictions with larger errors than older methods that are considered less accurate. For example, predicting the T_c of fluorene with the method of Nokay (1959) resulted in prediction error of 0.48%, while the more recent methods yielded prediction errors of 0.99% for MG and 4.57% for N. The P_c of anthracene predicted by the method of Thodos (1957) resulted in prediction error of 2.52%, while the more recent methods yielded prediction errors of 8.52 % for MG, 7.89% for W and 6.94 % for N. Thus, the selection of the most appropriate prediction method to be used for a particular property of a target compound should consider the size of the available training set.

The second group tested included the methyl ester biodiesel compounds: methyl tetradecanoate, hexadecanoate, octadecanoate and (Z)-9 methyl octadecanoate for which recently measured P_c and T_c data are available (Nikitin and Popov, 2014). These compounds contain long methylic chains (with 15, 17, 19 and 19 carbon atoms respectively). For such compounds we (Shacham et al. 2013) recommended the use of the RS method for property prediction. Using this method only a few (3-4 or more) measured property values from the methyl ester homologous series are required for prediction. Such information is available in the DIPPR database with T_c data uncertainty values of 1% to 3% and P_c data uncertainty values of 3% to 10%. We compared our reference series results with the experimental data of Nikitin and Popov (2014) identified the compounds similar to these target compounds, as was done for the aromatic hydrocarbon groups, and found very few closely similar compounds with experimental critical property data available. Nikitin and Popov (2014) compared their experimental results with the prediction techniques of Marrero and Ganni(2001) and and Wilson and Jasperson (as described by Poling et al., 2000).

In the presentation the detailed results of the study will be described and the impact of the lack of sufficient reliable data on the various prediction techniques will be discussed.

References

1. Brauner, N., Stateva, R. P., Cholakov, G.St. and M. Shacham, *Ind. Eng. Chem. Res.*, 45, 8430-8437 (2006).
2. DIPPR Data Compilation of Pure Chemical Properties, http//www.aiche.org/dippr.
3. Marrero, J. and Gani, R. *Fluid Phase Equilibria.* 183, SI 183-208 (2001)
4. Nannoolal, Y., Rarey, J. and D. Rajugernath, *Fluid Phase Equilib.* 252, 1–27 (2007)
5. Nikitin, E. D. and A. P. Popov, *Fluid Phase Equilibria.* 380, 11-17 (2014)
6. Nikitin, E. D. and A. P. Popov, *J. Chem. Thermodynamics* 80, 124–127 (2015)
7. Nokay, R., "Estimate Petrochemical Properties"; *Chem. Eng.* 66, 147-148 (1959)
8. Shacham, M., Paster, I. and N. Brauner, *AIChE J.*, 59(2), 420–428 (2013)
9. Thodos, G., *AIChE J.* 3, 428-430 (1957)
10. W.A. Wakeham, G. St Cholakov, R.P. Stateva, J. Chem. Eng. Data 47 (2002) 559

Vapor-liquid equilibria for mercaptans + hydrocarbons binary mixtures. Experimental results and DISQUAC or modified UNIFAC (Do) analysis

Farid Brahim Belaribi, [1] Rachid Sahki, [2]Jacques Jose
University of Sciences and Technology Houari Boumédiene, Faculty of Chemistry, BP 32 El-Alia, 16110 Alger, Algérie. Tel. +213 21248008, e-mail: bbelaribi@usthb.dz (2nd e-mail: faridbrahimbelaribi@yahoo.com); [1]University UMMTO, Tizi-ouzou, Algerie; [2]University of Lyon 1, France, e-mail: Jacques.jose@univ-lyon1.fr

Vapor pressures, for mercaptans (ethanethiol, propanethiol) + hydrocarbons (cyclohexane, benzene) binary mixtures, were measured in the temperature range from [283.15 to 363.15] K using a static apparatus. Isoplethal *T–P* data, for pure liquids and mixtures, were correlated with Antoine's equation. Interpolated isothermal VLE data were reduced using Barker's method.

The excess molar Gibbs energies, G^E, were calculated for several constant temperatures, taking into account the vapor phase imperfection in terms of the second virial coefficients and the variation of the Gibbs energy of the pure liquids components with pressure. G^E data were fitted to the Redlich–Kister empirical equation.

The excess molar enthalpies, H^E, were estimated from the temperature dependence of the G^E, using the Gibbs-Helmoltz equation.

Finally, the H^E and G^E results were analyzed in terms of DISQUAC and Modified UNIFAC (Do) models.

Solubility coefficient effect on chemical demulsification process for a Brazilian crude oil

Filipe Xavier Feitosa[1], Marina Dias Marques,[1]Sameque de Oliveira Nascimento[3], Sandra Soares Aguiar[2], Hosiberto Batista de Sant'Ana [1], Maria Carmen Bezerra Moreira[4], Rilvia Saraiva Santiago Aguiar [1]*.
[1]Universidade Federal do Ceará – Grupo de Pesquisa em Termofluidodinâmica Aplicada (GPTA), [2]Universidade Federal do Ceará
[3]Secretaria de Educação do Estado do Ceará – SEDUC
[4]Petrobras - CENPES
*Corresponding author e-mail: rilvia@ufc.br
Keywords: Solvent, Nonylphenol, Ethoxylated, Emulsions

The increasing concerns about the environmental impact of petroleum industry has been a driven force in the search of the reduction of use of aromatics compounds, e.g., BTX (benzene, toluene, and xylene) in different processes. During crude oil production, emulsion formation is an important issue, resulting in several highly difficult problems, as complex multiphase flow regime, corrosion and viscosity limitation problems. For this reason, it is frequent the use of water-soluble and/or oil-soluble demulsifiers in order to separate water from crude oil emulsion. Solvents such as toluene, xylenes and co-solvents as methanol and ethylene glycol monobutyl ether are usually used in order to improve cold flow properties and compatibility. This study investigated the possibility of improving the utilization of eco-friendly compounds as a carrier soluble additive solvent during demulsification process by studying the influence of solubility coefficient of different solvents, such as different alcohols, along with acetone and toluene. To do this, an experiment was conducted by changing the solvents (methanol, diethylene glycol, buthanol, triethylene glycol, ethanol, 1-propanol, acetone or toluene), at a constant concentration (200 ppm) of nonylphenol polyethoxylated surfactant (commercial IGEPAL CO 720) and pH (3, 7, and 10), at constants temperature (T = 60°C). These variables have been tested for a crude oil emulsion produced with a 30 % (v/v) water cut, and 240 g/L NaCl of salinity, und 3,200 rpm. Bottle test was carried out at 60 °C in ASTM tubes for water separability tests. As a result, it has been observed an important effect of solubility coefficient of solvents and an erratic behaviour of pH for solvents with a Hildebrandt solubility coefficient (δ) above 26 $MPa^{1/2}$. It has been observed that for higher δ the demulsification efficiency decreases. This behaviour could be attributed to intramolecular forces undisclosed by Hildebrandt coefficient, such as hydrogen bonds, and dispersive interactions. Probably, higher hydrogen bonds interactions lead to water solubility into solvents. For this reason, the interface of emulsions droplets keeping an asphaltene rigid film, promoting a stability of emulsions.

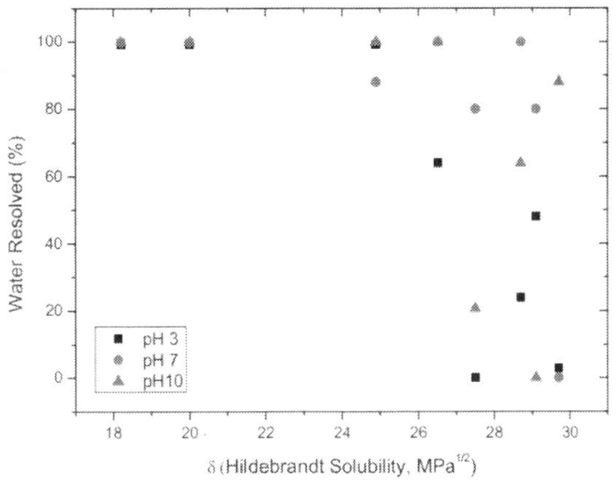

Figure 1 – Bottle test for an ethoxylated compound in differents pH's for a Brazilian crude oil emulsion (30 % water cut, 240 g/l NaCl) and differents solvents (methanol, diethylene glycol, buthanol, triethylene glycol, ethanol, 1-propanol, acetone or toluene)

Chemical modifications on alkyl-phenolic compound to enhance demulsification performance in Brazilian crude oils

F. X. Feitosa[1], M. D. Marques,[1] R. S. Alves[1], S. de O. Nascimento[3], J. R. M. de Andrade[2], S. de A. Soares[2], H. B. de Sant'Ana [1], M. C. B. Moreira[4], R. S. Santiago-Aguiar [1]*.
[1]Universidade Federal do Ceará – Grupo de Pesquisa em Termofluidodinâmica Aplicada (GPTA), [2]Universidade Federal do Ceará
[3]Secretaria de Educação do Estado do Ceará – SEDUC
[4]Petrobras - CENPES
*Corresponding author e-mail: rilvia@ufc.br

Emulsion formation is one of the most important problems faced by petroleum companies in order to guarantee a flow assurance during petroleum production. This formation is function oil reservoir fluid composition, along with operational conditions (pressure and temperature). For this reason, chemical additives should be frequently used in order to reduce or prevent emulsion formation. Unfortunately, it does not exist a general chemical compound that could be used for different reservoirs. Consequently, scanning and screening analysis for new compounds should be performed with the purpose of knowing their dosage and synergic effects on the prevention of emulsion formation. Three different route reactions (ethoxylation, sulfonation, and nitration) have been carried out from an alkyl-phenolic compound in order to know effective are different "head-group" size as demulsifier agent. The additives were characterized by RMN analysis. The demulsification activity were studied in crude oil emulsions produced with a 30 % (v/v) water cut, 240 g/L NaCl of salinity, at different pH (range from 3 to 10), under agitation (3200 rpm). Bottle test was carried out at 60 °C in ASTM tubes for water separability tests, by adding 200 ppm of each chemical tested. The results (Figure 1) show that pH increases phase separation for all chemicals synthetized. It seems that demulsification is more significant for ethoxylated compounds that present four main groups in their head. This behaviour could be attributed to the ability of ethoxylated compound to stabilize asphaltene in the emulsion.

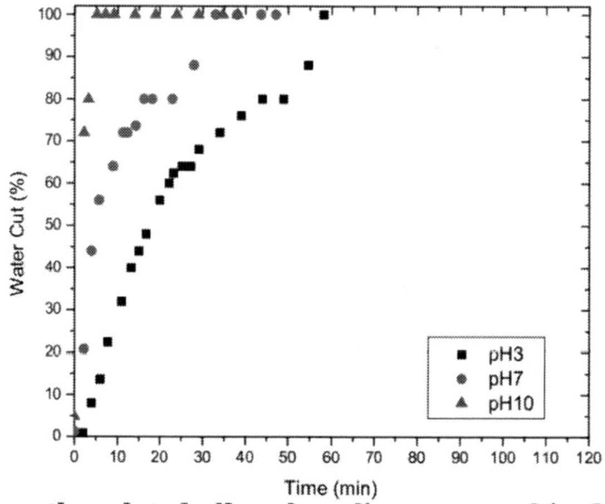

Figure 1 – Bottle test fo an ethoxylated alky-phenolic compound in differents pH. Brazilian crude oil emulsion formulation uses 30 % water cut, 240 g/l NaCl.

Unconventional use of a DSC based vapor pressure method for examining the extent of ideality in associating binary mixtures composed of narrow boiling range oil cuts

C. Siitsman, V. Oja
Tallinn Univ. of Technology, Dept. of Chemical Engineering, Ehitajate tee 5, 19086, Tallinn, Estonia; email: vahur.oja@ttu.ee

The current preliminary study presents some experimentally measured data aimed at gaining preliminary insight into the degree of ideality of binary systems, where at least one of the mixture components is a narrow boiling range distillation fraction of oil rich in phenolic moieties. The degree of ideality was evaluated by looking at deviations from Raoult's law at the selected liquid phase concentration. A vapor pressure determination technique for pure compounds, based on a differential scanning calorimeter (DSC), was used together with the underlying goal of evaluating the suitability of the DSC technique for determining the vapor pressures of binary mixtures of two narrow boiling range distillation fractions or of a narrow boiling range fraction and a pure compound. In this method the vapor pressure curve is determined by measuring, via separate experiments, a number of boiling points at various specified pressures (as an onset from a DSC thermogram, resulted from a constant heating rate heated sample evaporation through a pinhole from a hermetically sealed container at a specified pressure). For a list of previous applications of the DSC technique on pure compounds, binary mixtures of pure compounds and narrow boiling range oil fractions see the latest summary of DSC based vapor pressure experiments, as of 2014 [1].

The current data obtained suggest that the DSC based vapor pressure method could be a useful tool for preliminary examining the extent of ideality of these kinds of multicomponent systems. To illustrate the performance, Figure 1 graphically presents illustrative experimental data for an equimolar binary mixture of an oil fraction and a pure compound (nicotine), whereas Figure 2 presents illustrative data for an equimolar binary mixture of two practically non-overlapping narrow boiling point oil fractions. A total vapor pressure curve (solid line in these figures) for each of the binary mixtures, as if they were ideal solutions, was calculated from Raoult's law as $\sum P_i\, x_i$, where x_i is the component's (or fraction's) mole fraction in the solution and P_i is the vapor pressure of the component (or fraction). For calculation of Raoult's law curves shown in these figures the narrow boiling point oil fractions were taken as single pseudocomponents characterized by average properties such as an average molecular weight. Deviation from ideality was evaluated using average absolute relative deviation (ARD %). Further experiment and calculation details can be found from reference [2].

In regards to Figure 1, it can be seen that the presence of a compound containing a nitrogen base in the binary mixture with the oil fraction, with phenolic OH content of 7.9 wt% and practically no nitrogen, exhibits exchange of molecular interactions, and as a result, has a lower total vapor pressure than expected from Raoult's law. This is consistent with, although not as pronounced as, the trend seen with the Illinois No. 6 coal tar and phenanthridine equimolar mixture determined by the Knudsen effusion technique in references [3,4]. In regards to the Figure 2, it can be seen that the binary mixture of fractions with difference in average boiling point of about 150 K indicated considerable positive deviation from ideality. However, depending on narrow boiling range fractions mixed and their binary mixture compositions either positive or negative deviations from Raoult's law can be observed. Furthermore, it is also to mention that for binary mixtures of fractions taken closer together during the distillation, but still with boiling point ranges essentially not overlapping (difference in average boiling point of about 50 K), the ideal mixture behavior could be a fair approximation (for more details see the reference [2]).

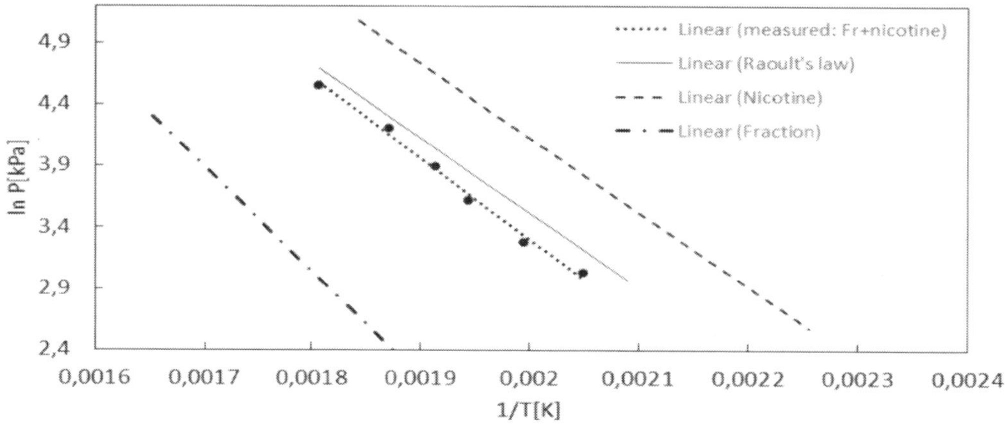

Figure 1. Vapor pressure behavior of an equimolar mixture of oil Fraction (average boiling point 639 K, MW 238 g/mol, H/C 1.29, phenolic OH content 7.9 wt%) and nicotine (boiling point 531 K). Average deviation from ideality as ARD% is 14 %.

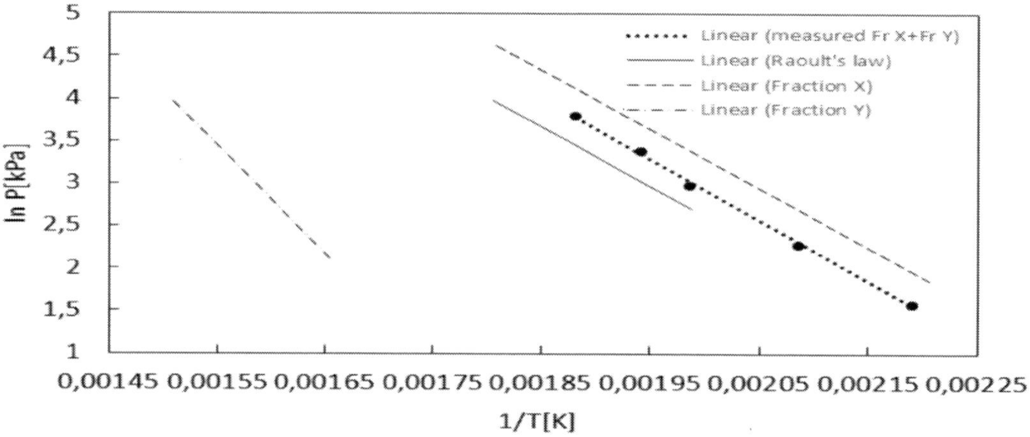

Figure 2. Vapor pressure behavior of an equimolar mixture of oil Fractions X (average boiling point 569 K, MW 191 g/mol, H/C 1.41, phenolic OH content 5.9 wt%) and Y (average boiling point 712 K, MW 383 g/mol, H/C 1.26, phenolic OH content 5.3 wt%). Average deviation from ideality as ARD% is 40%.

Acknowledgements:
The authors gratefully acknowledge financial support provided by Estonian Ministry of Education and Research, under target financing SF0140022s10.

References:

[1] Siitsman C., Oja V., Extension of the DSC method to measuring vapor pressures of narrow boiling range oil cuts, Thermochim. Acta 622 (2015) 31-37.
[2] Siitsman C., Oja V., Application of a DSC based vapor pressure method for examining the extent of ideality in associating binary mixtures with narrow boiling range oil cuts as a mixture component, accepted by Thermochim. Acta (2016), 10.1016/j.tca.2016.05.11
[3] Oja V., Suuberg E.M., Vapor liquid equilibrium in polycyclic aromatic compound mixtures and in coal tars; M. Nomura, P.M. Rahimi, O.R. Koseoglu (editors), Heavy Hydrocarbon Resources, Oxford University Press, Oxford, 2005, pp. 113-122.
[4] Oja V., Suuberg E.M., Vapor liquid equilibrium in polycyclic aromatic compound mixtures and in coal tars, Prepr. Pap.-Am. Chem. Soc., Div. Fuel Chem. 48 (2003) 50-51.

Solid-liquid phase diagrams of binary systems containing chlorobenzene and chlorocyclohexane with n-alkanes: experimental data and correlations

S. Mounsi, [1]O. Dahmani
Université des Sciences et de la Technologie Houari Boumediene, Faculté de Chimie, BP 32, El Alia, Bab Ezzouar, Alger, Algeria.

Experimental solid – liquid phase diagrams [1] were determined for binary systems consisting in chlorobenzene and chlorocyclohexane with n-hexane, n-heptane, n-octane, n-nonane and n-decane.

DISQUAC group contribution model [2] [3] was applied to predict the studied solid – liquid phase diagrams.

The interaction parameters (g_{ad}/RT) and (h_{ad}/RT) between alkane group (a) and chlorine group (d) in chlorocyclohexane are the same as those used in 1- chloroalkanes + n-alkanes binary systems.

The interaction parameters (g_{ad}/RT) and (h_{ad}/RT) between alkane group (a) and chlorine group (d) in chlorobenzene are adjusted on experimental excess Gibbs energy and excess enthalpy of chlorobenzene + n-alkane binary system.

Calculated solid-liquid phase diagrams are in good agreements with experimental data.

References

[1] W.M. Smit, Rec. Trav. Chim., 75(1956) 1309
[2] H.V. Kehiaian, JPE Grolier et G.C. Benson, J. Chem. Phys. , 11-12 (1978)75
[3] A.Bondi, Physical Properties of Molecular Crystals, Liquids and Glasses. John Wiley, New York 1968. P. 502

Volumetric properties of binary mixtures of N-methyl piperidine with alkanols at 298.15 and 313.15 K. Experimental measurements and application of the Prigogine-Flory-Patterson theory

A. Boussebissi, G. Boukais-Belaribi, F. Belaribi.
Laboratoire de Thermodynamique et Modélisation Moléculaire, Faculté de Chimie, Université des Sciences et de la Technologie Houari Boumediene, BP32 El Alia,16111 Bab Ezzouar, Alger,Algérie. Tel :+213 0555 81 54 38. E-mail address:boussebissi@yahoo.fr

Densities, ρ, of binary liquid mixtures of N-Methyl piperidine with methanol, ethanol, 1-propanol, 1-butanol, 1-pentanol, 1-hexanol and 1-octanol, were measured using an Anton Paar vibrating-tube densimeter, at 298.15 and 313.15 K and at atmospheric pressure, over the whole composition range. The excess molar volumes, V^E, for the mixtures, were determined from the experimental density data. All the mixtures exhibit large negative V^E values suggesting the existence of specific interactions between unlike molecules and indicating that the strong cross-association between amine and hydroxyl groups is the dominant effect. As the temperature increases the V^E magnitude slightly increases. Experimental V^E values were fitted with the Redlich-Kister equation. Partial molar volumes at infinite dilution were determined. Finally, the applicability of the Prigogine-Flory-Patterson theory was tested and discussed.

Modeling the possibility of wax formation in gas refinery units: Case study (LNG UNIT)

Mozhgan Hosseini
Shiraz University of Technology, Iran

The paraffin wax deposit can cause clogging in oil and gas pipelines and it leads to low efficiency and sometimes causes shut down in production units. Generally, temperature in hydrocarbon solution when the first solid hydrocarbon is formed, measured as the temperature of Wax phase formation. This temperature is normally measured in the lab. There are two thermodynamic approaches for calculating this temperature. In the first model, the paraffin wax phase is assumed to be a solid solution and in the second model, the wax phase is considered as a set of pure solid components. In the solid solution model, an equation of state is applied for liquid-vapor equilibrium calculation and so an activity model is used for solid-liquid equilibrium. In the multi-solid solution model, due to assuming that solid phases are pure, the activity coefficients are equal to unity for the solid phases. Thus only one equation of state is applied for calculation of the vapor-liquid equilibrium. In this study, the multi-solid model of *Nikita et al.* is used for calculating temperature and the amount of deposit wax. The *Ghanaei et al.* method was used to predict the pointing correction factor in high pressure conditions. An innovative method based on the Katz method is applied to separate additional heavy component. Al-Sahhaf, Won and Ghanaei mathematical equations were used to calculate critical properties, acentric factor and melting point properties. Furthermore, improved Peng-Robinson equation of state -introduced by Feyzi *et al.*- was used to calculate fugacity of pure and mixed components. The results of the present model was compared with several previous models for evaluation of the model. Then using the present model, the amount of wax formation for the mixture of natural gas in the train 1 of the phase 4,5 in the South Pars Gas Complex (SPGC) was calculated for different temperatures.

1- Prediction of Wax Formation in Gas Refinery LNG Unit

After validating the model in the previous sections, a sample of entering natural gas to gas refinery process unit of South Pars Gas Complex (SPGC) was simulated to predict wax formation in operational conditions. Feed compositions to unit have over than 50 components.

This operational unit works in pressure of 60 bar and temperature range of -30 to -40 °C. Therefore, wax formation for compounds was also simulated in this pressure and 200-270 K temperature range. By this modeling, wax formation temperature in this pressure is 260 K.

2- Conclusion

According to the obtained results from simulation of various samples and their comparison to experimental data and studies of the other researchers, the following results from this study are obtained:

1- Multi-solid (MS) model in comparison to solid solution (SS) model has better description and prediction of wax formation in hydrocarbon compounds.

2- Adaptation of simulation results in comparison to experimental data showed that the introduced model is valid in prediction the wax formation amount and temperature.

3- MS model was improved using the correct Peng-Robinson state model by considering Poyuting correction term for high pressure and Ghanaei et al. relation for Poyuting correction term totally improves results of the present research in comparison to results of other researchers' models.

4- Since in triple phase systems of wax formation, the variables numbers, equilibrium equation, and balance are more in comparison to double phase systems, selecting a proper algorithm is so important to accelerate solving computing software and system convergence. In this study, a new method was introduced based on simultaneous using of flash technics and repetition-replacement-correction method that is quick and secured method to solve triple phase systems for wax formation.

5- As it is clear from table (4), the heaviest element of studied gas compound was C_{16} and molar fraction of heavier compounds than C_{15} are few. Therefore, wax formation for previous parts with much more heavier compounds was negligible. However, this low wax formation accumulates solid particles in the processes installations in high rates and makes serious operational problems such as increasing pressure drop, reducing operational yield of systems, and also pipes and other installations clogging.

Food application of RS4 resistant starch

[1]E. Šárka, [1]M. Kubová, [1]I.Wiege, [2]P. Horák, [1]P. Smrčková,
[3]V. Dvořáček, [1]D. A. Chena Aldao, [1]Z. Bubník
[1]UCT Prague, Department of Carbohydrates and Cereals, Technicka 5, 166 28 Prague 6, Czech Republic; tel. +420 22044 3115, e-mail: evzen.sarka@vscht.cz; [2]Institute of Macromolecular Chemistry AS CR, v. v. i., Heyrovsky Sq. 2, Prague 6, Czech Republic; [3]Crop Research Institute, Drnovská 507, 161 06 Prague, Czech Republic; .

Introduction

Resistant starch (RS) escapes digestion by human enzymes within the small intestine and passes into the colon, where it is metabolised into secondary products by colon microbiota. It is assumed to have positive health benefits. RS is classified into several categories; the group RS4 includes particular chemically modified starches (Jane et al., 2010). Starch acetates of degree of substitution (DS) 0.01 ÷ 0.2 provide thickness, body, and texture in foods. On the other hand, high DS derivatives (AS) have different properties such as hydrophobicity and melt processibility; they are used as tablet binders, hot melt adhesives, biodegradable plastic materials and pharmaceutical tablets (Chi et al., 2008). The aim of the presentation is the introduction of RS4 as a potential food additive and the recording of experimental results of AS impacts on bread, specifically on properties of dough and on nutritional, physical and sensory attributes of the bread itself.

Material and methods

A commercial wheat A-starch and acetic anhydride were weighted into a two-neck round flask. As a catalyst, a 50 % (w/w) solution of NaOH was added. The reaction was performed in a microwave reactor with a controlled temperature of 118 °C. The mixture was then cooled down, washed and dried at 43 °C. DS 0.82 was determined by ^{1}H NMR.

The fermentation and baking were done in a Kenwood BM 450 bread machine (program No. 1, weight 500 g). Part of the dough flour was substituted with acetylated starch: see the ratios in the following table:

	m_m (g)	ratio (%)
flour	300,0	100,0
water	200,0	66,7
sugar	7,0	2,7
oil	5,6	2,3
salt	8,0	1,9
yeast	5,0	1,7

The water absorption capacity and the dough properties such as dough development time, dough stability, protein weakening, starch gelatinization, amylase activity and starch gelling were determined by Mixolab (Chopin). The rheological properties of the starch/flour mixtures were tested by an RVA analyser as well. The volume of the loaf was measured according to the AACC (2000) method 10-05.01, the shelf-life by PNR 10 (Petrotest Instruments). Determination of RS was carried out using the AACC 32-40 (2002) method with a Megazyme kit. Additionally, the appearance, elasticity, taste and smell of the bread were tested.

Results and discussion

Fig. 1 illustrates the negative influence of AS addition on the volume of the loaf; it led to disruption of the starch-gluten matrix. Fig. 2 demonstrates the results of rheological measurements by Mixolab. The fall of the starch viscosity (point C3) is connected to very limited dissolution of AS.

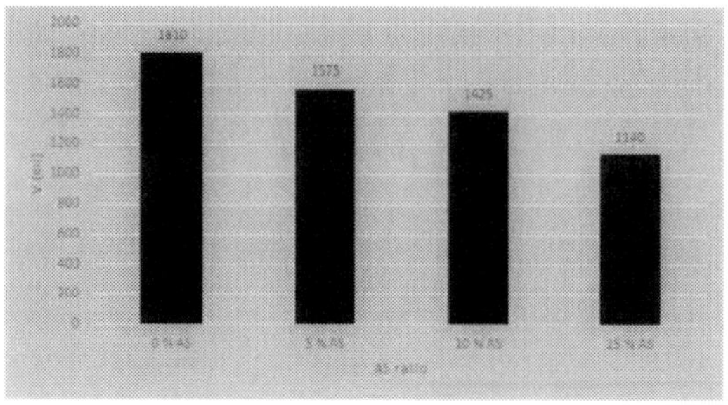

Fig. 1

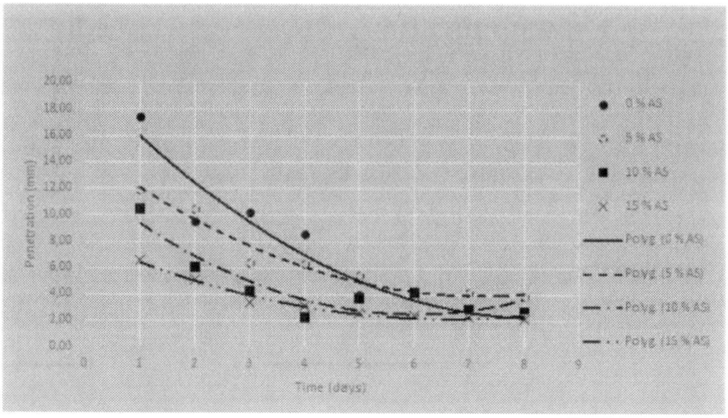

Fig. 3

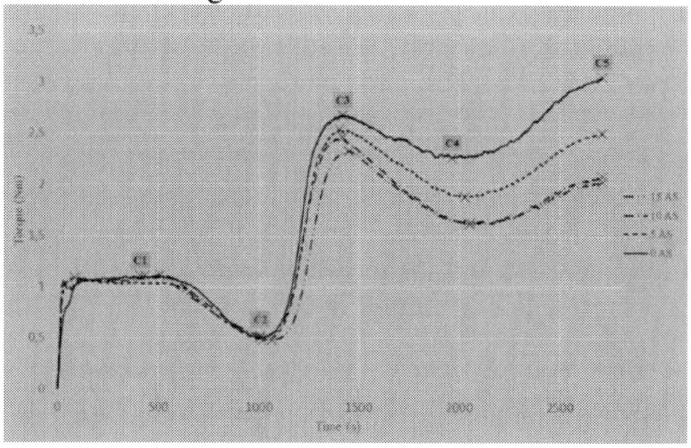

Fig. 2

Fig. 3 shows the ageing of the prepared bread. The added starch decreased the penetration of the bread, but after 5 days the penetration was practically the same when compared to the bread without AS addition.

Conclusions

RS4 offers exciting new potential as a hydrophobic food ingredient since it remains unchanged during heat processes such as baking or extrusion and survives in most food processes.

Prepared breads were fortified by various AS additions. AS influenced the porosity of the bread while the elasticity was unchanged. Additions of 10 % and 15 % caused worse appearance, lower volume and a crust light in colour. The substitution of 5 % wheat flower by AS resulted in better sensory properties and decrease in starch retrogradation, and RS content was 2.4 % - approx. twice more than that of the bread without the AS addition.

References:

1. Chi H., Xu K., Wu X., Chen Q., Song Ch., Zhang, W.; Wang, P.: Food Chemistry 106, 923–928 (2008).

2. Jane J.-L., Jiang H., Hasjim J.: Resistant starch and its health benefits. Proceedings of the 6th International Conference on Polysaccharides-Glycoscience, Prague, p. 9 (2010).

Acknowledgment

This research was supported by the "Wheat with specific starch composition and features for food and non-food purposes" QJ1310219 research grant of the Ministry of Agriculture.

Dextran molecular mass effect on raw juice purification during processing of frost damaged sugar beets

M. Wojtczak, R. Gruska, P. Bąk, A. Antczak-Chrobot
Lodz University of Technology, Faculty of Biotechnology and Food Sciences,
Institute of Food Technology and Analysis, Stefanowskiego 4/10,
90-924 Lodz, Poland, e-mail: maciej.wojtczak@p.lodz.pl, tel. +48426313453

Since European sugar market regime results in longer campaigns, the problem of processing deteriorated beets is still an important topical problem. Deterioration of beets is mainly a result of storage of defrosted beets. Frost-damaged beets are very susceptible to microbiological infections. The growth of microbiological infections in frost-damaged beets leads to several changes in the chemical composition of the beets. These changes concern mainly the hydrolysis of sucrose by microorganisms and the production of various metabolites (e.g. glucose, fructose, raffinose and polysaccharides like dextran or levan).

The most harmful result of deterioration of beets is the appearance of dextran. Dextran is polymeric chains of glucosyl units, synthesized by dextransucrase. Dextrans contain substantial number of consecutive α-(1-6) linkages in their main chain, usually comprising more than 50% of the total linkages. Dextran also has side chains, stemming mainly from α-(1-3) and occasionally from α-(1-4) or α-(1-2)-branched linkages. Microbiological dextran is not a pure chemical substance but it consists of a mixture of polymeric chains of glucose of different degrees of polimerisation. Usually, we divided dextran into 3 groups: high molecular weight dextran, that is, above 1000 kDa, medium molecular weight, that is, from 100 to 1000 kDa, and low molecular weight, that is, below 100 kDa. It may be expected that the disadvantageous influence of dextran on sugar production will depend on the degree of dextran polymerization.

The presence of dextran in the raw juice leads to a disruption of normal processing operations, particularly first and second carbonatation filtration. The blockage of filters is mainly caused by increasing juice viscosity and decreasing the average size of a particle of calcium carbonate precipitated during carbonatation.

The material for analyses was low quality raw juice obtained from deteriorated beets, but free of dextran. In order to analyse the influence of dextran on the purification process dextran standards of different molecular weight were added to the raw juice. The dextran standards of the following molecular weight: 100, 200, 500 and 2000 kDa were used. The concentration of dextran added to the raw juice was from 250 to 1000 ppm. The samples of juice without dextran and with dextran underwent the process of purification in laboratory conditions.

The particle size distribution of sludge in juice after first and second carbonatation was analysed by means of Laser Particle Sizer ANALYSETTE 22. At the same time the viscosity and the filtration coefficient were determined in juices during purification process of juice with different content of dextran.

The presence of dextran in the raw juice leads to a disruption of filtration after the first and the second carbonatation by blockage of filters and decreasing the viscosity of juice. The blockage of filters in the purification process is mainly caused by increasing the content of fine crystals and increasing the standard deviation of size distribution of a particle of calcium carbonate precipitated during carbonatation. The negative influence of dextran on the filterability and the viscosity of juice after carbonatation is depend on average molecular mass of dextran.

Effects of the asparaginase on the reduction of the acrylamide levels in biscuits

A. Budelli[1], M. C. Parnoffi[2], R. Nigro[2], C. Schioppa[3]
[1]Heinz Innovation Center, NieuweDukenburgseweg 19, 6534, Nijmegen, Netherlands
[2]Università degli Studi di Napoli Federico II, DICMAPI, P.leTecchio 80, 80125 Naples, Italy
[3]I.T.P S.r.l., Via Bisignano a Chiaia, 68 – 80121 Naples, Italy

Recent data dealing with acrylamide content in heated food prove the concrete exposure by diet to the toxin for people and especially for infants and young children. European Union has considered necessary to recommend maximum levels of the toxin for each food category. Biscuits for young children and infant are respectively 500 and 200 μg/kg.

The primary mechanism of acrylamide formation in food is associated with the Maillard reaction between asparagine aminoacid and a reactive carbonyl in form of reducing sugars during thermal treatment over 120°C and at low water content.

Based on recent studies, there is tangible evidence that, in foodstuff such as biscuits, the acrylamide concentration level is uneven in the whole product, and more precisely, it is higher on the surface and lower in the centre.

Asparaginase is an enzyme able to hydrolyse asparagine into aspartic acid and ammonia, thereby removing a key precursor for toxin formation. Asparaginase addition in doughs for bakery products has been claimed to effectively reduce acrylamide levels without altering the colour and taste of the final product.

Despite the large relevant literature, there seem not to exist many studies dealing with specific strategies aimed at reducing the acrylamide levels in food. This is the main focus of the present work, facing the efficiency of two different techniques for the inhibition of acrylamide for biscuits: 1)the addition of the powdered enzyme in the bulk during the dough preparation and 2)the deposition by water solution on the surface of the biscuits before cooking. Distributions of acrylamide in cooked biscuits have also been investigated measuring toxin levels in the whole product, in the centre and on the surface.

The biscuits were prepared according to a classified recipe, accounting for specific steps and temperature levels of single ingredients and doughs. The main ingredients are wheat flour, sucrose, vegetable oils, milk powder, barley malt and raising agents (ammonium bicarbonate, sodium bicarbonate, disodium diphosphate). The dough was laminated down to a thickness of 2 mm and cut in rectangular slabs of about 70x25 mm.

Cooking was performed using a multifunction domestic electric oven (Hotpoint-Ariston 2800W) suitably equipped by a PID temperature controller to maintain cooking temperature at 170°C. The baking time were respectively set to 7 min.

The Acrylamide content was measured by HPLC with MS/MS according to AOAC Recommandation.

Different amounts of asparaginase (Novoenzymes, 3500 U/g) were added to the above mentioned recipe. The doses used for tests did not modify dough rheology and physical properties. The enzyme was increased up to 0,5 g/(kg of dry solids) as suggested by enzyme producer for semi-sweet biscuit doughs. The key results, for bulk treatmentent, are summarized in figure 1 in which the adimensional concentration of acrylamide (normalised at 100 ppb level) is analysed at increasing enzyme content. A deep decrease of acrylamide can be observed at a level of asparaginase content of 0,125 g/(kg of dry solids). At higher enzyme content the level of toxin is almost constant at levels between 100-40 ppb. From the figure, it is also possible to evaluate the distribution of acrylamide in the different zones of the biscuit.

As expected the level of acrylamide detected in the crust (1 mm in deep from surface) is sensibly higher than in the centre.

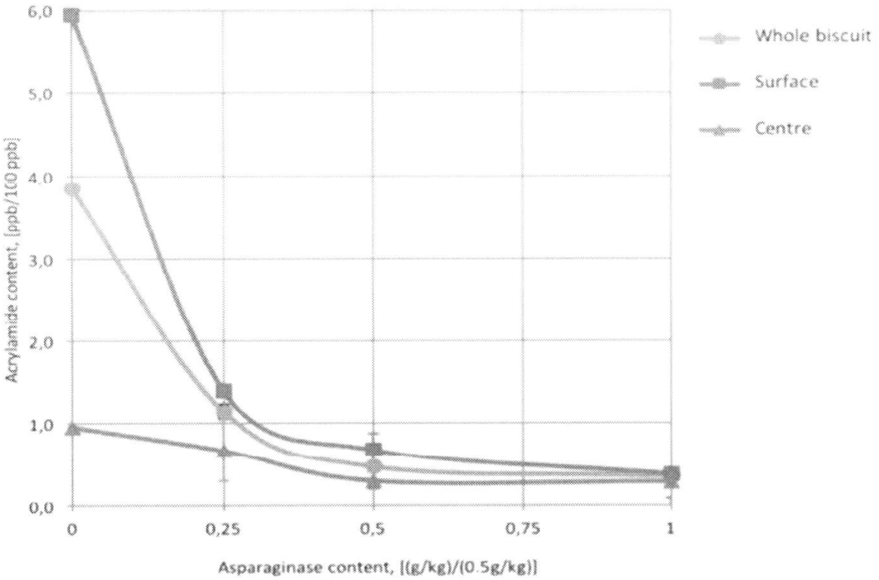

Figure 1: Effect of Asparaginase addition in bulk for semi-sweet dough biscuits.

Tests performed adding asparaginase as water solution on the surface on biscuits at two different temperatures, namely 26°C e 40°C, evidenced no relevant effect on acrylamide concentrations in the differet zones of the biscuit.

Hydrolyzed galactomannans fractionation using automated membrane filtration unit

S. Henke, S. Gillarová, A. Hinková, V. Pour, Z. Bubník
University of Chemical Technology, Prague, Department of Carbohydrates and Cereals, Technická 5, 166 28 Praha 6 - Dejvice, Czech Republic; tel. +420 220 443 111, e-mail: svatopluk.henke@vscht.cz.

The presented paper is a continuation of our previous work, where the automation of existing membrane filtration unit TIA (Bollene, France) has been done. The unit TIA is a laboratory and pilot-plant scale membrane filtration separator with two tubular ceramic membranes (Membralox, Pall) having the filtration area 0.24 m^2 each, covering the micro- and ultrafiltration range. The station is now equipped with Siemens Simatic S7-1200 controller system, allowing controlling the pumps, pressure gradient, temperature and back washing unit. This work deals with separation of hydrolyzed galactomannans, which were prepared using different ways – acidic and enzymatic hydrolysis of guar gum. The fine membrane ultrafiltration is suitable separation step before chromatographic treatment and necessary method for isolation of desired monosaccharides from the rest of higher saccharides in the solution.

Increase of oil yield by enzymatic degumming of sunflower oil

S. Lassen[1], O. Wenzel[1], K. Suck[2], U. Sohling[2], B. Niemeyer[1]
[1] Helmut Schmidt University / University of the Bundeswehr Hamburg, Holstenhofweg 85, D-22043 Hamburg, Germany, bernd.niemeyer@hsu-hh.de
[2] Clariant Produkte (Deutschland) GmbH, Ostenriederstr. 15, D-85368 Moosburg, Germany, Ulrich.Sohling@clariant.com

Degumming of oils from natural plants is an essential processing step for a later application as nutrition, chemical feedstock as well as biodiesel. In this step an essential loss of oil as the major value compound is recognized. Considering an increasing worldwide production of around 200 million tons of natural oils per annum until 2022 [1], even a small improvement in oil recovery results in a significant amount of oil available. This would contribute to an improved process economy as well as to ecology, as the demand on further soil for agriculture (mostly gained from natural forests in Asia, e.g. Malaysia, Indonesia) would be lowered. Actually in the industrial degumming process the oil is heated in acidified water in order to make hardly water soluble compounds like phospho- and glycolipids water soluble and remove them together with other water soluble oil compounds from the oil.

The enzymatic degumming data with phospholipase A1 (PLA1) (3rd column in Figure 1) were compared with the results obtained with water and acidified water, like in the industrial process (1st and 2nd column). PLA1 treatment reduces the phosphor content from 145 ppm to 3.1 ppm (3rd column), which is below the permissible limiting value of 4 ppm for biodiesel according to DIN EN 14214. An economically relevant result was obtained for the yield of the aimed free fatty acids (FFA) which increased by 19 % for a processing time of 30 min (data not shown). Proposed mechanisms of the loss versus the recovery of oil, influenced by the different process conditions are optically proved by a series of images taken by in-situ video microscopy [2]. The presentation overviews these results, valuable for further process development. This cooperative project is funded by the German Federal Ministry of Education and Research (BMBF) under the support code 0316063A within the scope of the frame project of the BioCatalysis2021.

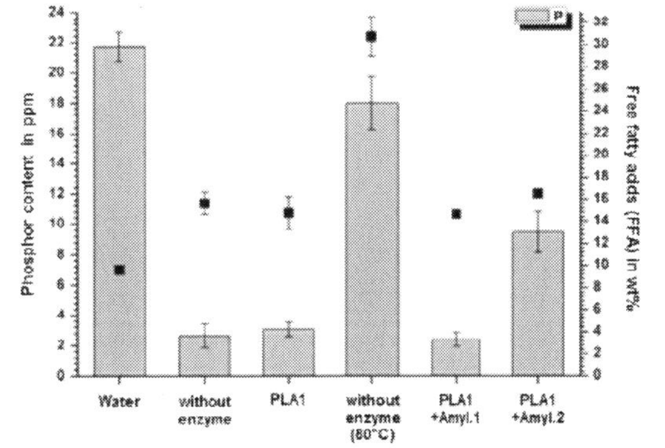

Figure 1 Final content of phosphor, and residual oil after a processing time of 45 min [1].

[1] O. Wenzel, 2015, Enzym-katalytische Entschleimung von Sonnenblumenöl zur Steigerung der Ölausbeute, Dr.-Thesis, Helmut-Schmidt-University Hamburg, Germany.

[2] U. Sohling et al., 2015, Inorganic Adsorbents in Enzymatic Processes: Adsorption of Substrates, Reaction Products, Enzymes, Surface Modification, and Enzyme Immobilization; in: L. Hilterhaus, A. Liese, U. Ketting, G. Antranikian (eds.), Applied Biocatalysis – From Fundamental Science to Industrial Applications, VCH-Wiley, Weinheim, FRG, 2016.

Isolation of mannose from natural polysaccharides

S. Gillarová, S. Henke, Z. Bubník
University of Chemical Technology, Prague, Department of Carbohydrates nad Cereals, Technická 5, 16628 Prague 6, Czech Republic; tel. +420 22044 3111, e-mail: simona.gillarova@vscht.cz

The aim of this work was chromatographic isolation of mannose bounded in plant polysaccharides. Such materials are in nature, e.g. guar gum (*Cyamopsis tetragonoloba*) and locust bean gum (*Ceratonia siliqua*). As source material for experiments, a commercially available Guar gum 5000 CPS (Vega Provita s.r.o., Czech Republic) has been used. The polysaccharide was subjected to acid hydrolysis in methanolic medium with HCl, subsequently pH balanced with NaOH, or subjected to enzymatic hydrolysis. Solution containing hydrolyzed polysaccharides were passed for membrane filtration to remove rest of higher saccharides. Permeate from this step is then used as a feed for discontinuous anex and catex chromatography, aiming mannose isolation. The analysis for each step was performed using HPAEC/PAD. Operational parameters for continuous chromatography were calculated from obtained concentrations courses of discontinuous chromatography and fuzzy controlled simulation of continuous mode at 8-column system in the SMB mode has been done to obtain optimized operational parameters, and continuous separation of hydrolyzed mixture has been run.

Microencapsulation of autochthonous probiotics bacteria's and natural antioxidants compounds to increase thermal resistance during and after drying

R. Bórquez., and J. De Bruijn, Chemical Engineering Department, University of Concepción, Concepción, P.O. Box 160-C, Concepción, Chile. Tel.: 56 - 41- 2204534; fax: 56- 41- 2247491. rborquez@udec.cl

Use of probiotic microorganisms and natural antioxidants during the formulation of functional foods receives every day more attention. However, these species are very unstable without encapsulation, being a complication to use and incorporate them in food products. Stability of bioactive compounds can be improved by microencapsulation (**ME**) using biopolymers, which protect them from oxygen, moisture and other stresses. Modification of the rate or place at which these active substances will be released, can be realized by using materials with specific barrier properties for manipulating their release and to provide unique sensory, health and / or functional benefits.

Microencapsulated material can easily be transported, stored and incorporated; it competes with freezing as a conservation method for advanced applications. However, development of these technologies is still going on, particularly, when high thermal resistance in the microencapsulated product is required.

This study about microencapsulation of probiotic microorganisms and bioactive compounds proposed the following steps:

1. The evaluation of the effect of microencapsulation on the survival and the properties of native probiotic bacterias, as well as on antioxidant capacity and composition of bioactive compounds from berries extracts, using techniques such as emulsification in gels (alginates, chitosan, gums, fatty acids, natural polymers), followed by atmospheric freeze drying or spray drying using solid protectors to reduce thermal conductivity; the main goal was to improve thermal resistance during drying and other high temperature processes when it used as an ingredient for food preparation.
2. The evaluation of thermal stability of encapsulated products, simulating their performance at high temperature process conditions used in food processing such as pasteurization, cooking and extrusion.

Native lactic acid bacteria were used that are available after screening and characterization by our R&D group. Different encapsulation materials were selected for microencapsulation by emulsification, atmospheric freeze drying and spray drying. Thermal resistance and stability of microcapsules during storage were studied. A concentration about 10^{10} CFU/g of encapsulated microorganisms was achieved for functional foods with a maximum retention of bioactive compounds and antioxidant activity, combined with high stability of the formulated products at storage.

This research is a contribution to the development of technological innovation in microencapsulation processes of biomaterials, as well as to the generation of fundamental knowledge about spray drying and atmospheric freeze drying process applied to raw materials sensitive to heat, mass and momentum transfer mechanisms combined with quality changes of processed products.

Model for drying and inactivation of baker's yeast in fluidized beds operated under reduced pressure

S. Zarekar, A. Bück, [1]M. Jacob, E. Tsotsas
Thermal Process Engineering, Otto von Guericke University, Universitaetsplatz 2, Magdeburg, Germany, 39106; tel. +49-(0)391-67-12322, e-mail: sayali.zarekar@ovgu.de;
[1]Glatt Ingenieurtechnik GmbH, Nordstrasse 12, Weimar, Germany, 99427.

Fluidized bed drying processes are widely used in the formulation of bioactive materials such as foodstuff, dairy products, and drugs. Such materials are often temperature sensitive, and it is essential to minimize their deactivation during the course of the drying process. While the deactivation can be lowered by reducing the drying temperature, it can also lead to a low process throughput due to the reduced drying capacity of the gas medium. Other low temperature methods such as freeze-drying can be used, but they are costly and time consuming for bulk production as compared to high temperature drying processes. An alternate approach is to operate the fluidized bed at moderate vacuum conditions which can result in considerable reduction in the product temperature as well as equipment costs.

To minimize the inactivation of thermo-sensitive substances, the optimization of drying conditions is extremely crucial. In this study we investigate the influence of sub-atmospheric pressure on drying and inactivation of baker's yeast in a batch fluidized bed operation using mathematical modeling. The kinetics of drying is modeled using van Meel's concept of a modified normalized drying curve (NDC) of a single particle. Single particle drying rates for yeast are obtained from published experimental data [1]. Inactivation, which is a function of the rate of change of temperature and water activity of yeast, is modeled using first order kinetics.

At constant inactivation, a kinetically controlled drying process leads to higher drying rates at low operating pressures. This is due to the increase in specific capacity of air for water vapor at low pressures resulting in an increased driving force for evaporation. However, a consideration of inactivation kinetics in the model results in a larger reduction in cell viability at higher drying rates, as can be seen in Figure 1. We, therefore, develop a coupled model for drying and inactivation of yeast to account for the competing kinetically and equilibrium (capacitive) controlled aspects of the process. This model is used to optimize the process path for drying of baker's yeast at partial vacuum conditions.

Reference

[1] F. Debaste, V. Halloin, L. Bossart, B. Haut. A new modeling approach for the prediction of yeast drying rates in fluidized beds, Journal of Food Engineering, 84 (2008) 335-347.

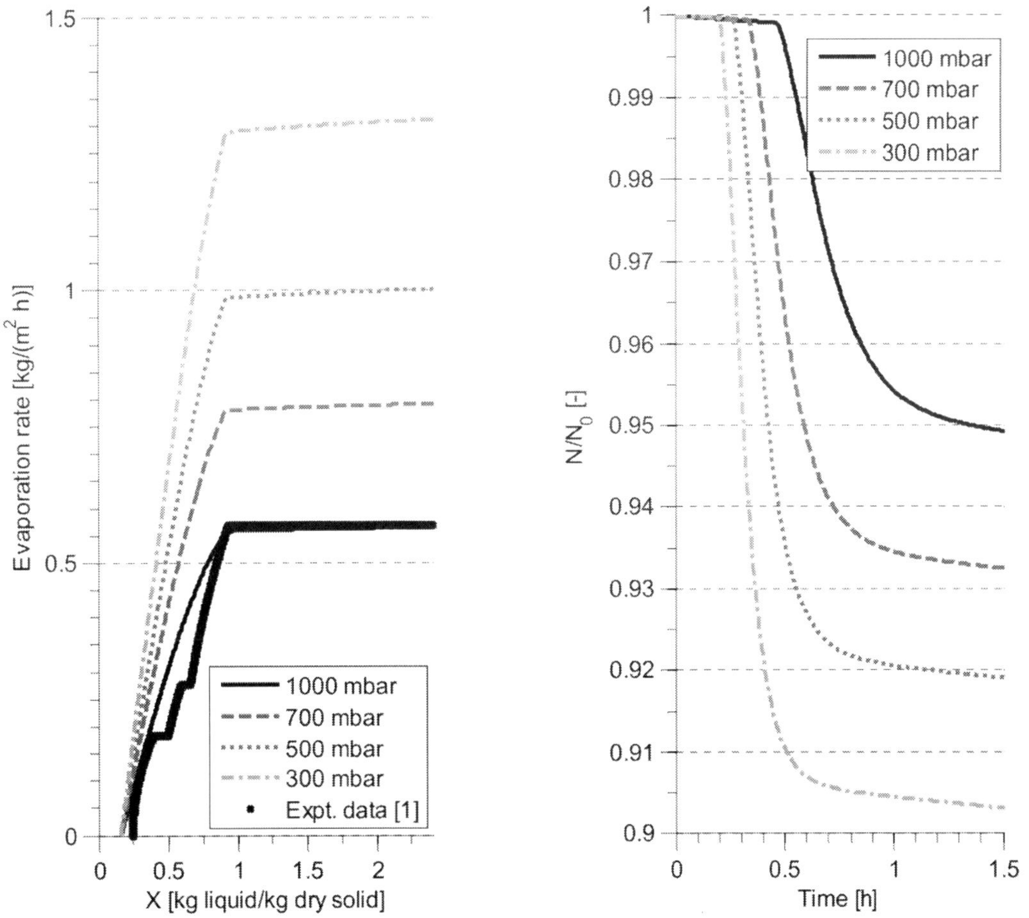

Figure 1. (Left): Evaporation rate of water versus the moisture content of the yeast particles for different operating pressures in the fluidized bed; (Right): Change in cell viability (live cell count ratio) over drying time. The inlet mass flow rate of air is kept constant for all values of operating pressure.

RVA profiles of non-traditional plants and composite with flour – statistical evaluation

I. Švec, M. Hrušková
University of Chemistry and Technology Prague, Department of Carbohydrates and Cereals, Technicka 5. 166 28 Prague 6, Czech Republic;
tel. +420 220 443 206, e-mail: Ivan.Svec@vscht.cz; fax: +420 220 445 130

In food industry, non-traditional plants usage became very popular within the last ten years. Re-discovering of plant staple foods eaten by ancient civilisation as well as renaissance of forgotten cereals count two advantages for final consumers – enlargement of bakery products offer, and mainly a substantial improvement of nutritional value of wheat-flour based foods. To such plants, seeds of chia, teff, fonio or flour from nopal, chestnut and acorn as well as barley could be ranked.

Majority of those raw materials are known as non-gluten and/or non-starch ones, and so their application in common processes aimed to food manufacturing requires usually partial modifications. For this reason, exploration of thermal and rheological properties and behaviour of alternative materials in a laboratory scale could save time and costs. Within the cereal chemistry field, apparatuses developed by Brabender GmbH. (farinograph, extensigraph, amylograph), Chopin (mixolab) or Newport Scientific Inc. (Rapid Visco Analyser) could be used. The last mentioned one represents a relatively new technique in Europe, user-friendly for operator – low consumption of tested materials, necessity of any chemicals and middle long duration of the test. The method approved by the American Association of Cereals Chemists (AACC Method 76-21.01) describes pasting behaviour of flour-water suspension during heating and cooling, simulating processes in leavened dough and bread during baking and after it. Selection of the RVA equipment was also conditioned by uneasy accessibility of some of tested plant materials in limited amount.

Samples tested within the work could be sorted into a four groups:

a) 12 pure raw-plant materials:
   - Wheat flour (WF), barley flour (BF);
   - Wholemeal from white and dark chia and teff seeds (CH1, CH2, T1, T2);
   - Commercial and laboratory prepared fonio flour (F1, F2);
   - Commercial nopal and chestnut flour (NF, CF);
   - Golden flax fibre of granulation 0-300 μm and 500-700 μm (GF1, GF2);
   - Brown flax fibre of granulation 300-500 μm and 500-700 μm (BF1, BF2);

b) Wheat-barley flour premix 70:30 (w/w, abbreviation WBF) and 12 tri-composite blends containing 5% or 10% CH1, CH2, T1, T2, NF, CF;

c) WBF control and 8 tetra-composite blends containing 5% or 10% chia and teff wholemeals combined together (i.e. 2.5%/5.0% CH1 + 2.5%/5.0% T1, and similarly for pairs CH1-T2, CH2-T1, and CH2-T2);

d) WF and bi-composite blends containing 2.5% or 5.0% GF1, GF2, BF1, BF2, F1 and F2.

Samples gained in form of seeds were disintegrated to wholemeal by using blade grinder Concept KM 5000 (sample dosage 50 g, operation time 1 min).

For the RVA test, predefined procedure "Flour.rvp" was used; water amount was kept constant (25.00 g), and basic sample weight 3.5 g was adjusted to actual moisture. Measurement of RVA profiles was performed in duplicate, and for statistical evaluation, median of standard deviations in the data pairs (i.e. joint deviation) was calculated. Using Statistica 7.1 software (StatSoft Tulsa, USA), verifiable differences between tested samples were depicted by analysis of variance (P = 95%). Impact of factors plant species/type and dosage level was explored by principal component analysis.

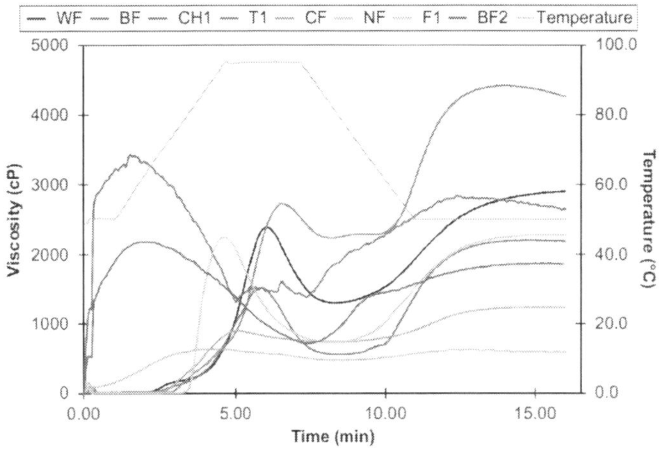

Fig. 1. RVA profiles of selected non-traditional plants. WF – wheat flour, BF – barley flour, CH1 – wholemeal from white chia seeds, T1 – wholemeal from white teff seeds, CF – chestnut flour, F1 – fonio flour, BF2 – brown flax fibre of granulation 500-700 μm

Fig. 1 shows some similarity rate between RVA profiles of WF, BF, T1 and F1, demonstrating their botanical classification within the *Gramineae* family, and thus similar pasting behaviour of starch as the main polysaccharide. Besides very low peak viscosity for chestnut flour, course of its RVA profile appears wheat-like one. Chia and brown flax fibre demonstrated quick beginning of gelatinization, i.e. clearly at lower temperature than other samples. Unambiguously diverse pasting behaviour was recorded for NF, reflecting modified representation of polysaccharide structures.

By the RVA test, single botanical species could be statistically distinguished mainly according to the parameters Peak time, Final viscosity and Setback (Tab. 1). Values of the Peak viscosity were less dependent on tested samples, but WF, T2 and F1 as cereals reached comparable level.

Acknowledgement

The work was supported by the grant QI 151027 of the NAZV, Czech Republic.

Table 1. Statistical differences of RVA curve characteristics between tested plant species and variants (ANOVA, P = 95%)

Plant species	Pasting temperature (°C)	Peak time (min)	Peak viscosity (cP)	Hold time (min)	Hold viscosity (cP)	Final viscosity (cP)	Setback (cP)
Wheat	67.8 c	6.07 j	2292 e	8.20 e	1327 c	2870 g	1543 h
Barley	82.3 f	5.80 i	1520 c	8.33 e	560 a	2184 de	1624 hi
Chia white	50.4 a	1.53 b	3432 g	7.33 cd	1388 c	2643 f	1255 f
Chia dark	50.0 a	1.20 ab	3799 h	7.00 bc	1285 c	2682 fg	1397 g
Teff white	74.3 d	6.53 k	2721 f	8.20 e	2235 f	4261 j	2026 j
Teff brown	75.2 d	6.73 k	2361 e	9.80 f	1920 e	3642 h	1722 i
Chestnut	71.0 c	5.00 h	895 b	8.13 e	736 b	1233 b	497 c
Nopal	57.4 b	4.13 f	634 a	8.13 e	478 a	590 a	112 a
Golden flax 1	50.0 a	1.33 bc	2720 f	7.27 c	1275 c	2009 cd	734 d
Golden flax 2	50.1 a	2.93 e	3335 g	6.80 b	1890 e	3926 i	2036 j
Brown flax 1	85.5 g	1.07 a	1760 d	7.67 d	474 a	1241 b	767 d
Brown flax 2	50.0 a	2.07 d	2183 e	7.33 cd	718 b	1860 c	1142 e
Fonio 1	78.3 e	4.67 g	2245 e	8.20 e	737 b	2273 e	1536 h
Fonio 2	94.5 h	7.00 l	548 a	6.07 a	483 a	759 a	276 b
Repeatability	0.6	0.05	41	0.07	27	35	18

Golden flax 1, 2: flax fibre of granulation 0-300 μm and 500-700 μm, respectively.
Brown flax fibre of granulation 300-500 μm and 500-700 μm, respectively.
Fonio 1, 2: commercial flour and laboratory prepared wholemeal.

The advantage of modern agarose based chromatography sorbents for milk proteins separation and purification

V. Skalka, S. Remišová, L. Čurda
University of Chemistry and Technology, Department of Dairy, Fat and Cosmetics, Technická 3, 166 28, Prague, The Czech Republic; tel. +4207764127, e-mail: volodymyr.skalka@vscht.cz.

There are several milk proteins, which are produced in industrial scale. Among them are α-lactalbumin, lactoferrin and lactoperoxidase. α-Lactalbumin is one of the dominant whey protein, which has calcium binding properties. It has molecular weight 14178 Da, and the isoelectric point between 4.2 and 4.5. Concentration of these proteins in milk varies from 0.7 to 4.5 g per litter of milk. α-Lactalbumin above all is used in production of infant formula, sport food and food supplements. This protein is separated mainly by complex of ultrafiltration and precipitation techniques. Lactoferrin and lactoperoxidase are minor whey iron-containing glycoproteins, which have molecular weight about 80 kDa. Isoelectric point of lactoferrin is 7.9- 8.0, while isoelectric point of lactoperoxidase is 9.2–9.9. Concentration of these minor proteins in milk varies from 200 to 500 mg per litter of milk. Lactoferrin is used mainly in production of infant formula, food supplements and cosmetic. Lactoperoxidase is used as alternative preserving agent in production of dairy products and cosmetic. Both these glycoproteins can be separated simultaneously by chromatography with using cation exchange sorbents and cross flow filtration.

Main criteria for choosing of chromatography media for milk proteins separation are cost of sorbents, binding capacity and work flow rate. For separation and purification of milk proteins previously are used sorbents with 200-500 μm beads size. Over past five year has been made significant improvements of chromatography resins. The driver of this improvement was immunoglobulins, using of which is common in science and medicine. As result nowadays sorbents with binding capacity over 100 mg of immunoglobulins per ml of sorbent are present on the market. Some of these sorbents could be effectively used for milk proteins separation and purification. By using of such sorbent is possible to increase amount and purity of target proteins from one separation batch and decrease expenses for time, water and reagents.

During our experiments were used two types of agarose cation exchange sorbent Bio-Works WorksBead 40S(Sweden) and Bio-Works WorksBead 100S (Sweden) Target proteins were α-lactalbumin, lactoferrin and lactoperoxidase. For separation of these proteins were used sweet and acid milk whey. For determination of target proteins concentration and purity was used C8 RP-HPLC. Additionally for determination of purity was used electrophoresis in polyacrylamide gel.

First step of lactoferrin and lactoperoxidase separation conducted by direct capturing with using of chromatography sorbents. After capture step elution was processed by using linear gradient of sodium chloride in phosphate buffer. For each type of sorbents were selected optimal conditions of feeding and elution and calculated dynamic binding capacity. For both sorbent dynamic binding capacity was over 70 mg/ml and was depended on feeding conditions and performance of column cleaning.

Purification of α-lactalbumin included several steps. The first one was precipitation technique using citric acid, centrifugation of precipitated apo α-lactalbumin and final reconstitution of α-lactalbumin by diluting in solution with calcium chloride. By using of this method, it was possible to get up to 75% pure α-lactalbumin and main contaminant component was β-lactoglobulin. For last step purification was used chromatography. In this case sorbent was used to bind β-lactoglobulin, which has higher isoelectric point than α-lactalbumin. For this separation was used acetic acid – sodium acetate buffer system without any gradient elution. The final purity of α-lactalbumin after chromatographic purification was over 85%.

After comparing literature dates and obtained results, we can argue that the using of tested modern chromatography sorbents can improve efficiency and reduce the cost of milk protein's separations.

Application of fungicidal microorganisms in malting process

[1]K. Zítková, [1,2]M. Poštulková, [1]J. Řezanina, [1]P. Dostálek, [2]M. C. Růžička, [1]T. Brányik
[1]University of Chemistry and Technology Prague, Department of Biotechnology, Technická 5, 166 28 Prague 6, Czech Republic; tel. +420 220444126, e-mail: tomas.branyik@vscht.cy;
[2]Academy of Sciences of the Czech Republic, Institute of Chemical Process Fundamentals, Rozvjova 2/135, 165 02 Prague 6, Czech Republic

Currently, one of the biggest problems in the malting and brewing industry is fungal contamination of raw materials. In particular, the infection of barley by filamentous fungi (molds) causes serious problems (Fusarium head blight, mycotoxin and hydrophobin production) during and after the malting process. To suppress the undesirable mold growth, chemical, physical, and biological methods can be used. One such biological method is the use of *Pythium oligandrum* as "starter culture" during malting process. In this study, we compare the effectiveness of *P. oligandrum* with *Geotrichum candidum*, which is currently used as starter culture.

For the malting of barley samples a computer controlled micromalting device was used (Ravoz, Czech Republic). *G. candidum* and *P. oligandrum* were inoculated onto the barley and after the malting process their presence was monitored by cultivation methods and qRT-PCR. Mycotoxin production was determined by LC-MS and the treated malts were tested for their gushing potential.

The results strongly suggest that *P. oligandrum* is a more efficient starter culture than *G. candidum*, which is currently used as the starter culture in malt-houses. This observation is supported by physiological characteristics and mode of action of the fungicidal microorganisms. While *G. candidum* is able to overgrow other molds present on barley, *P. oligandrum* actively suppresses the growth of undesirable molds by its mycoparasitic lifestyle. Lower mycotoxin levels and gushing potentials were detected for malt treated by *P. oligandrum* as compared with control samples.

The application of *P. oligandrum* on barley contaminated by *Fusarium* sp. during the malting process was confirmed as a highly efficient method against fungal contamination. Oomycete *P. oligandrum* possesses a high anti-fungal potential.

Automatic control of sugar crystallization with the use of measuring crystal mass by means of near infrared spectrometry

M. Wojtczak
Lodz University of Technology, Faculty of Biotechnology and Food Sciences, Institute of Food Technology and Analysis, Stefanowskiego 4/10, 90-924 Lodz, Poland, e-mail: maciej.wojtczak@p.lodz.pl, tel. +48426313453

The process of batch crystallization resulting in obtaining right quality sugar crystals requires particular care of measurement and control of crystal growth. An ideal method of controlling the crystallization process should be based on measuring the supersaturation. Controlling supersaturation within the metastable zone during the crystallization guarantees orderly growth of the crystals. Unfortunately, at present there are no instruments for direct measurements of supersaturation.

Theoretical calculations of supersaturation during crystallization is possible by means of on-line measurements of mother liquor and massecuite concentration measurements. However, in order to accurately measure and control supersaturation during crystallization, a number process inputs are required. Among the more important inputs are the purity and temperature of the mother liquor as well as the concentration of the solution.

Proper controlling crystallization requires maximum growth rate of the crystals without the formation of secondary grain. Thus, the process may be controlled by managing crystal growth rate without taking into consideration the real saturation of the mother liquor. This may be realized by modelling the crystals content in massecuite on the basis of measuring the concentration of massecuite and mother liquor.

Another way of controlling the process of crystallization on the basis of the content of crystals in massecuite is a direct measurement of crystals content by means of a near infrared spectrometer. In our study we used PROM OPEKS crystal analyser for measuring the crystals content and the concentration of dry substance in the process of boiling the seed magma. The analysis was made in on of Polish sugar factories during 2014/2015 campaign.

The analysis embraced measuring the accuracy of the PROM OPEKS instrument, and a comparison of the quality of the seed magma boiled according to refractometric and microwave control with that by means of the near infrared instrument. The poster present the usefulness of NIR spectrometry for controlling of crystallization process.

ACKNOWLEDGMENT. The authors thank the companies: Krajowa Spółka Cukrowa S.A. POLIMEX-CEKOP-MODER Sp. z o.o., AP SYSTEM Automatyzacja Procesów Przemysłowych, PROM-OPEKS for valuable cooperation and support of this study.

Centrifugation as an assisted technique to freeze concentration of commercial orange juice

P. Orellana, N. Troncoso, G. Petzold
Department of Food Engineering, Universidad del Bío-Bío, Casilla 447, Chillán, Chile;
tel. +56422463136, e-mail: patricio.o.palma@gmail.com

Concentration of liquid foods such as a fruit juice is an important unit operation, because the concentrated products occupy less volume and weight. This unit operation can be made through evaporation, membrane separation and freeze concentration. Freeze concentration as compared to other technologies has some significant potential advantages for producing a concentrate with high quality because the process occurs at low temperatures, resulting in minimal loss of volatiles (Morison & Hartel, 2007). The industrial future of freeze concentration has been associated more with developments in the configuration of one-step systems (block freeze concentration or progressive freeze concentration) than conventional freeze concentration systems (suspension crystallization), because of the simpler separation step (Miyawaki et al., 2012; Petzold & Aguilera, 2009; Sánchez et al., 2010). Other advantage of these one-step systems is their simplicity in terms of both the construction and operation of the equipment (Sánchez et al., 2009). The alternatives of assisted techniques applied to block freeze concentration are external forces such as vacuum or centrifugation. In this way, centrifugation has been proposed by Bonilla-Zavaleta et al. (2006) in frozen pineapple juice to separate ice from concentrated juice, while Luo et al. (2010) obtained ice crystals of high purity during the freezing concentration of brackish water. Thus, centrifugation is an alternative assisted technique to improve the efficiency of block freeze concentration applied to solutions and fruit juices (Petzold & Aguilera, 2013; Petzold et al., 2015). In this study, centrifugation was used to enhance the separation of concentrated solution from frozen commercial orange juice in three cryoconcentration cycles. The orange juice was unidirectionally frozen at -20° C for 12 h and subjected to centrifugation (2400 RCF for 10 min at 20 °C), achieving the separation of ice and concentrated. This procedure was performed in three cycles, in each cycle collecting the concentrate being used as initial concentrate for the next stage of cryoconcentration. Physicochemical parameters were determined as soluble solids (°Brix), pH, density, color and ascorbic acid, for fresh juice, ice and concentrated by centrifugation and rotary evaporator. For rotary evaporator, the concentrated extract (°Brix) to obtain the equivalent freeze-centrifugation processing each cycle. Thus, compare the concentrate between cryoconcentration and high temperature. The results indicate increase of the measured parameters, except pH, which is maintained constant in all samples. There is increased to 11.4 to 32.2 ° Brix, 0.9 to 1.6 g / ml for soluble solids and density, respectively. The color changed from light orange to dark orange, and reported an important retention of ascorbic acid content with a yield close to 50% in final cryoconcentrate samples, compared to concentrate by rotary evaporator with less to 40% of retention. It is concluded that centrifugation is an effective technique to remove the concentrated solution of the matrix ice. The performance of this technique applied to orange juice was attributed to ice matrix acting as a porous solid through which the concentrated solution percolates through drainage channels of the ice improved by the centrifugal force.

Acknowledgments

Authors are grateful for the financial support provided by CONICYT through FONDECYT project number 11140747.

References

- Bonilla-Zavaleta, E., Vernon-Carter, E.J. & Beristain, C.I. (2006). Thermophysical properties of freeze-concentrated pineapple juice. Italian Journal of Food Science, 18, 367–376.
- Luo, C., Chen, W. & Han, W. (2010). Experimental study on factors affecting the quality of ice crystal during the freezing concentration for the brackish water. Desalination, 260, 231–238.
- Morison, K.R. & Hartel, R.W. (2007). Evaporation and freeze concentration. Eds. Heldman, D.R. Lund, D.B. & Sabliov, C. Handbook of Food Engineering, CRC Press, New York, pp. 495–552.
- Miyawaki, O., Kato, S. & Watabe, K. (2012). Yield improvement in progressive freeze-concentration by partial melting of ice. Journal of Food Engineering, 108, 377–382.
- Petzold, G. & Aguilera, J.M. (2009). Ice morphology: Fundamentals and technological applications in foods. Food Biophysics, 4, 378–396.
- Petzold, G. & Aguilera, J.M. (2013). Centrifugal freeze concentration. Innovative Food Science and Emerging Technologies, 20, 253–258.
- Petzold, G., Moreno, J., Lastra, P., Rojas, K., Orellana, P. (2015). Block freeze concentration assisted by centrifugation applied to blueberry and pineapple juices. Innovative Food Science and Emerging Technologies, 30, 192-197.
- Sánchez, J., Ruiz, Y., Auleda, J.M., Hernández, E. & Raventós, M. (2009). Review. Freeze concentration in the fruit juices industry. Food Science and Technology International, 15, 303–315.
- Sánchez, J., Ruiz, Y., Raventós, M., Auleda, J.M. & Hernández, E. (2010). Progressive freeze concentration of orange juice in a pilot plant falling film. Innovative Food Science and Emerging Technologies, 11, 644–651.

Combined process and energy modelling for sustainable production in breweries

B. Muster-Slawitsch, [1]C. Brunner[1], H.Schnitzer [2]
[1]AEE – INTEC Dep. Industrial Processes and Energy Systems Feldgasse 19, 8200 Gleisdorf Austria +43 3112 5886470 e-mail: b.muster@aee.at;.

To reach the targets for the reduction in fossil greenhouse gas emissions that have been set by the European Union, industry will have to radically change its current practices. While intensification of processes for maximising process yields at minimal resources has been promoted over the past years, the effects of these new intensified technologies on the overall production site and its energy consumption have not yet been evaluated. In this paper a methodology that holistically optimizes process technologies and overall energy efficiency has been applied to 5 breweries.

The applied methodology combines detailed process modelling including selected qualitative criteria with heat integration for real-life process data. For process modelling, a brewery model is applied [1] that allows the comparison of different technologies and operating conditions regarding thermal energy requirement, hot water management and product quality. For evaluating heat integration strategies the heat integration tool SOCO has been used [2] including algorithms for generating heat exchanger proposals based on variable process streams and a simulation model to simulate interaction between heat exchangers and storages. This poster presents the results applying the combined approach of process modelling and heat integration for five different brewing sites (see Table 1).

Table 1: Specification of brewing sites

	brewing capacity	packaging capacity		brewing technology					Brew volume		
		bottles	kegs	mashing	mash separation	kettle	Evap. rate	vapour recovery	cold cast wort	brews /week	Gravity
	[hl]	[%]	[%]						hl/brew		[g/100ml]
site 1	659979	100%	0%	infusion	lauter tun	Dynamic low pressure boiling	6%	energy storage + wort preheating	454	30	13.7
site 2	915195	66%	34%	decoction	lauter tun	internal heater	2.8%	energy storage + wort preheating	629	30	12.8
site 3	327713	57%	43%	decoction	lauter tun	external heater	8.6%	mechanical vapour compression	223	30	12.6
site 4	695890	60%	40%	infusion	mash filter	internal heater	2.8%	energy storage + wort preheating	238	60	13.7
Site 5	696113	60%	40%	infusion	mash filter	Schonkochverfahren	5.2%	warm water generation	238	60	13.7

Based on existing research results on kinetic modelling and mashing experiments, site 4 and 5 are using infusion mashings with gradual heating at low heating rates [1]. Mashing times can be reduced by 15% and power demand by 50%. This allows for low temperature heat supply of mash tuns without extensive increase of heat exchange surfaces.

The Brewery Model allows fast calculation of minimal thermal energy demand requirement per technology and useful process heat requirement. Parametric studies effectively help to analyse the impacts of different technologies and operating conditions. Hot water management is again proven to be a key for efficient brewhouse operations

.

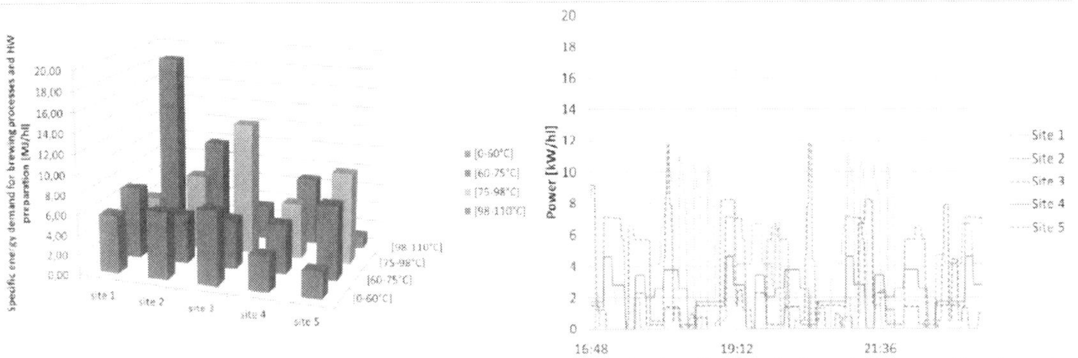

Figure 1: Specific energy demand on different temperature levels(right) and variation of energy demand over time (left) for the five brewing sites

The algorithms of the heat integration tool SOCO can design and optimize practical heat integration concepts. For brewhouses a promising heat integration concept combines the heat of vapour condensation and wort cooling to charge an energy storage which is discharged by wort preheating and mashing. An optimized design of this concept may cover 95% of the thermal energy demand of these processes. A brewery with an optimized technology set and more continuous process profile (site 4) can realise this heat integration strategy with 40% less storage volume and up to 80% smaller heat exchangers in comparison to a conventional site (site 1). Peaks in power demand can be reduced by 80%.

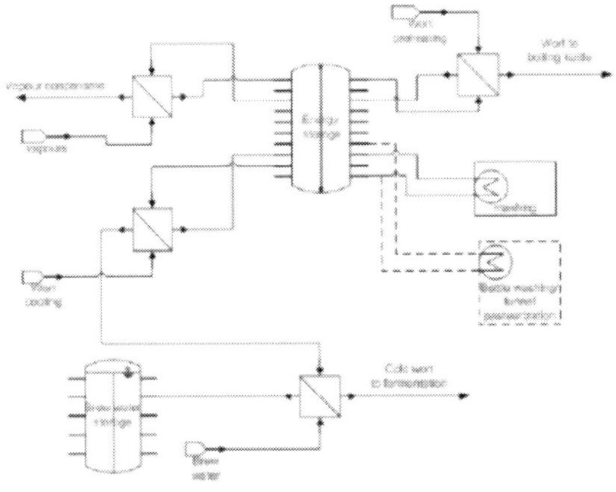

Figure 2: Proposed heat integration scheme

This work shows the significant effects of technology choices on energy efficiency, heat integration and the overall thermal energy management in industry. The combined approach of process modelling that generates variable process profiles and heat integration analysis that allows planning of heat exchanger and storage networks based on these profiles, enables to design optimized technology sets in industry. Detailed models of those processes posing bottlenecks to efficient thermal energy supply can reveal new processing strategies.

1. Muster-Slawitsch, B., et al., Process modelling and technology evaluation in brewing, *Chemical Engineering and Processing: Process Intensification* **84**, 98-108 (2014).
2. Muster-Slawitsch, B., et al. Methodology for Batch Heat Integration and Storage System Design for Ideal Integration of Solar Process Heat. in *Chemical Engineering Transactions*. Prague: AIDIC Servizi S.r.l., 2011.

Determination of acoustic fields in suspensions of orange waste during pretreatment with high-intensity ultrasound

T. C. Polachini, [1] R. S. Siqueira, L.M. Laurindo, J. Telis-Romero
Instituto de Biociências, Letras e Ciências Exatas, DETA, Universidade Estadual Paulista, São Jose do Rio Preto, CEP 15054000, SP, Brasil; tel. + 17 32212251, e-mail: javier@ibilce.unesp.br.

The increasing in CO_2 emissions allied to the search of nonpetroleum-based sources of energy has strengthened studies involving lignocellulosic wastes as renewable source to biofuel. As seen that biomass is less expensive and more available in large quantities than sucrose, for example, its use represents a great potential for bioethanol industry.

Concentrated orange juice is one of the most important commodities worldwide and Brazil is the major producer. A great example of lignocellulosic byproduct is the orange waste, obtained from its processing. Although orange waste offers big advantages, a viable cellulose treatment is still a challenge. New technologies have been developed to enhance the process of biomass accessibility. Among them, there are included process involving supercritical fluids, acid or alkaline treatments, steam explosion, high-intensity ultrasound and others.

The benefits of using high-intensity ultrasound in diverse processes stimulated researches based on biomass pretreatment. But, the correct design and operations depends on how the input energy behaves in the volume of interest until its conversion to acoustic energy. For this, calorimetric method is widely used to measure the real power absorbed, and so, the cavitation effects. Orange waste, an available byproduct of orange juice processing, were used in acidified suspensions to determine the absorbed power, intensity and power yield. For this, specific heat was obtained and modeled in different conditions. These values ranged between 3.22 and 4.28 $kJ \cdot kg^{-1} \cdot K^{-1}$, with low values for more concentrated suspensions. Acoustic power and acoustic intensity varied linearly with applied power and quadratically with solids concentration. Maximum acoustic power of 279.6 W was found in higher input power and less concentrated suspensions. Similar behavior was observed for intensity varying from 1.33 to 69.11 $W \cdot cm^{-2}$. A power yield of 78.8% was encountered for 4% of solids at 320 W while 7.1 % of efficiency was observed when suspensions are more concentrated at low input energy (80 W).

Evaluation of composite flour containing flax fibre

M. Hrušková, I. Švec
University of Chemistry and Technology Prague, Department of Carbohydrates and Cereals, Technicka 5. 166 28 Prague 6, Czech Republic;
tel. +420 220 443 206, e-mail: Marie Hruskova@vscht.cz; fax: +420 220 445 130

Flax (*Linum usitatissimum* L.) is old utility plant originated in Asia. Slim stalk and light-blue flowers could distinguish the plant. Its fruit is boll containing tiny brown seeds. There exist flax varieties (e.g. Amon, Raciol) with lighter seed, named as "golden flax".

Flax seed is typical by high oil content (40%), dietary fibre (28%) and proteins (21%). Further known constituents are minerals (4%) and non-starch polysaccharides (6%) as lignans, hemicelluloses and phenolics.

Flax fibre is commercial food supplement gained as by-product during oil pressing or extraction, sold in a dry powder form. Walramcom company, flax fibre producer from the New Zeeland, present mean nutritional values for the supplement from brown and golden flax: saccharides 2.4%, proteins 32.0% and total fat 16.6% (of which 13% unsaturated fatty acids). According to production method, majority mass portion presents dietary fibre (TDF 45.2%, IDF 37.9% and SDF 7.3%). Proteins have a non-gluten nature, and thus they are safe for coeliac patients. Further, flax fibre is a good source of constituents with high anti-oxidant activity, especially lignans (e.g. secoisolariciresinol diglucoside) and vitamin E.

The pilot study was aimed at characteristics comparison of several commercial types of flax fibre, using them in form of composite flour during non-fermented properties testing. Verifying technological potential and consumer quality of these flour composites, cookies were manufactured in a laboratory scale.

Samples tested within the work could be sorted according to four flax groups (FF):

- Golden flax fibre of granulation 0-300 µm and 500-700 µm (GF1, GF2);
- Brown flax fibre of granulation 300-500 µm and 500-700 µm (BF1, BF2);

Technological features of wheat flour (WF) and four flour composites are described by Zeleny test (ČSN ISO 5529) and Falling number (ČSN ISO 3093). Non-fermented dough properties were determined by using of farinograph and extensigraph Brabender (Germany), following the international norms (ČSN ISO 55 30-1, 55 30-2, respectively). From prepared wheat-flax fibre composites, cut-off biscuits were manufactured, following internal methods ended by quality evaluation.

Evaluation of technological quality of flour composites

According to Zeleny test values, bakery quality of proteins could be considered as lower compared to WF. Addition of GF caused a decrease in a higher extent (about 30%) in relation to BF influence (-18%) and WF control. As could be noticed in Figure 1, neither addition level nor FF granulation did not trigger significant differences. In wheat flour sample, amylases activity as the Falling number was estimated about ca 25% lower than optimum for standard bakery processing is (250 s). With respect to variability in the FF type (GF vs. BF), granulation (2 types) as well as addition level (2.5% vs. 5.0%), flax fibre contributed to Falling number change softly. An insignificant lowering and increase was observed for GF and BF, respectively, with weak impact of addition level.

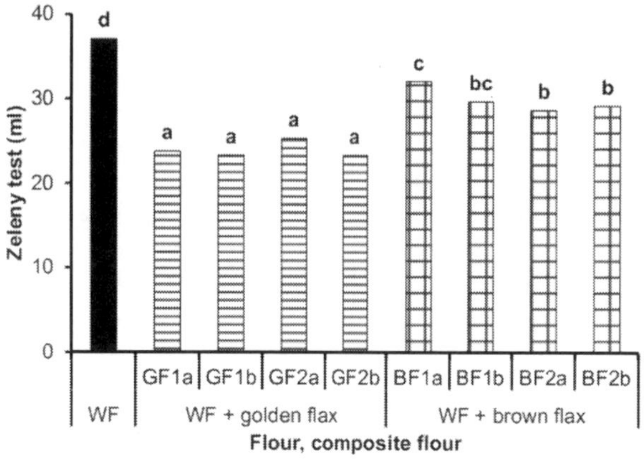

Fig 1. Characteristics of protein quality (Zeleny test value) - wheat flour and flour composites

Evaluation of non-fermented dough features

In case of standard farinograph proof, rheological properties of non-fermented dough prepared from flour composites are characterised by water absorption increase about 3%-7%. Higher values were determined for 5% addition and BF of coarser granulation. Dough development time was shortened by the GF effect (from 3 min to 2 min), and reversal trend was observed for the BF (prolongation about 25%). According to extensigraph characteristics, change in dough viscoelastic properties implies the elasticity-to-extensibility ratio determined after 60 min of dough resting, which simulates II. phase of fermentation process. The ratio increased from 2.18 (WF control) about 10% – 70% reflecting FF type and granulation of GF and BF. In composite dough, diluted gluten protein contributed to lowering of bakery quality of tested dough samples; extensigraph energy decreased seriously by incorporation of finer GF (about 15%), while impact of coarser GF addition was softer (decrease ca 10%).

Evaluation of cut-off biscuit features

Cut-off biscuits are characteristic by manufacturing technique, i.e. by cutting-off from dough plate of calibrated thickness. The internal method operates with seven quality characteristics, of which three (specific volume of baked biscuits, shape as spread ratio (diameter-to-height ratio) and sensory profile allow compare different recipe variants. For biscuits containing GF, 2.5% such fibre had a positive influence on the evaluated features; the impact of GF granulation was less provable. Brown flax fibre caused a soft specific volume decrease (about approx. 10%), and biscuits shape spread into approx. a half scale. Sensorial profiles of GF and BF biscuits did not deviate from the wheat control one. Golden flax contributed to attractive yellow shade of biscuit surface, and brownish one in case of BF. Coarser flax fibre was visually detectable as darker dots in biscuit crust, especially in case of biscuits containing brown seeds FF.

Conclusion

Owing to chemical composition, flax fibre has a potential to be used in food industry. Added into wheat flour, flax fibre changed its technological and rheological behaviour in correspondence with type (golden/brown flax), granulation as well as addition level. Compared to wheat flour, technological tests demonstrated the lowering of protein quality. Towards to cut-off biscuit attributes, flax fibre from golden seeds had stronger positive effect than the brown one.

Acknowledgement

The work was supported by the grant OI 151027 of the NAZV, Czech Republic.

Experimental study on drying kinetics of rocket leaves undergoing microwave drying

E. Demirhan, B. Özbek, Y. Ayata
Department of Chemical Engineering, Yıldız Technical University, Davutpaşa Campus Esenler/İstanbul, 34210, Turkey;
tel. +90 212 383 4769, fax: +90 212 383 4725.
e-mail: demirhan@yildiz.edu.tr

Drying, in general, means removal of water from the material. The purpose of drying food products is to allow longer periods of storage with minimized packaging requirements, reduce shipping weights and preserve seasonal plants and make them available to consumers during the whole year. Besides these advantages, some important changes take place, as structural and physicochemical modifications that affect the final product quality during dehydration. Microwave drying is an alternative method because of its uniform energy and high thermal conductivity to the inner sides of the material, space utilization, sanitation, energy savings, precise process control and fast startup and shutdown conditions [1-4].

Rocket leaves have been used in salads and they have strong and peppery flavor. In the present study, the effects of microwave output powers and sample amounts on moisture content, moisture ratio and drying time of rocket leaves were investigated. A domestic microwave oven was used for the drying experiments. Experiments showed that drying time was proportional to microwave output powers and sample amounts for rocket leaves. The mathematical models depending on process conditions were also developed. Also, the minimum energy to evaporate one gram of water was calculated by using the drying constants for each sample.

REFERENCES

[1] G. V. Barbosa-Canovas and H. Vega-Mercado, Dehydration of Foods, 1st Ed., Chapman & Hall, New York (1996).
[2] K. J. Park, Z. Vohnikova and F. P. R. Brod, J. Food Eng., 51(3), 193 (2002).
[3] M. Maskan, J. Food Eng., 44(2), 71 (2000).
[4] Y. Soysal and S. Oztekin, J. Agric. Eng. Res., 78(1), 57 (2000).

Impact of physicochemical properties of additive aqueous phase on rheological properties of capillary suspension and application for fluid food

M. Imai and H. Sato
Course in Bioresouce Utilization Sciences, Graduate School of Bioresource Sciences,Nihon University,
1866 Kameino, Fujisawa, Kanagawa-pref. 252-0880, Japan.
Corresponding author's E-mail (M.Imai): XLT05104@nifty.com

Novel capillary suspension composed from green tea powder and coconut oil was successfully provided with small additive amount of aqueous phase.
Green tea powder was suspended in coconut oil with volumetric fraction of green tea particle was 0.45. When the volumetric fraction of additive pure water more than 0.032, the fluid dynamics of suspension was drastically changed to gel-like capillary suspension. It presents attractive potential of capillary suspension for forthcoming novel fluid food products.
The capillary suspension can be obtained by additive small amount of additional fluid into normal suspension. Additive fluid caused the change of network structure between particles based on capillary force. Accordingly, the internal structure of suspension is physically intensified. The rheological properties of suspension were dramatically changed and the yield stress can be controlled by the additive fluid.

Key word: Capillary suspension; Rheological properties; Additive fluid; Novel fluid food products; Yield stress

Optimization of galactose and mannose preparative continuous SMB separation

S. Gillarová, S. Henke, Z. Bubník
University of Chemical Technology, Prague, Department of Carbohydrates and Cereals, Technická 5, 16628 Prague 6, Czech Republic; tel. +420 22044 3111, e-mail: simona.gillarova@vscht.cz, fax: +420220445130.

The work deals with modelling and optimization of continuous SMB chromatography using of software package Aspen Chromatography. The goal of this work is to calculate output concentration of separated compounds in both streams, extract and raffinate. The data from discontinuous separation of glucose and mannose aqueous solution has been used to obtain input equilibrium and kinetic parameters for given model. Optimization was then used for calculating operational parameters to get optimal purity, yield and production of each monosaccharide on 8-columns preparative SMB pilot plant.

Physical and sensory properties of corn snack products enriched with food industry by-products

A. Jozinović, Đ. Ačkar, J. Babić, B. Miličević, J. Panak Balentić, [1]M. Jašić, D. Šubarić
Josip Juraj Strossmayer University of Osijek, Faculty of Food Technology Osijek, Franje Kuhača 20, HR-31000 Osijek, Croatia; tel: +38531/224-337, fax: +38531/207-115, e-mail: dackar@ptfos.hr; University of Tuzla, Faculty of Pharmacy and Technology, Univerzitetska 8, 75000 Tuzla, Bosnia and Herzegovina; tel/fax: +38735/320-745

By-products of plant food processing, which are rich in fibre and bioactive compounds, have attracted much attention for their functionality and potential as food ingredients in recent years. Extrusion, as a continuous process with high versatility and productivity, offers a good avenue to incorporate different types of by-products in ready-to-eat snacks and breakfast cereals. In addition, incorporating of these by-products can improve nutritional value of extruded products that are otherwise carbohydrate-rich, high glycaemic products.

The aim of this study was to investigate the possibility of application brewer's spent grain (BSG), sugar beet pulp (SBP) and apple pomace (AP) in production of corn snack products. By-products were mixed with corn grits at different ratios (0/100; 5/95; 10/90 and 15/85), where in the mixtures with BSG and SBP pectin was added at level 1% d. m., with the aim to produce products with properties of ready-to-eat snacks. Prepared samples with 15% of moisture content were extruded in the laboratory single screw extruder at temperature profile 135/170/170 °C, using a screw with compression ratio 4:1 and die with 4 mm diameter. Obtained extrudates were air-dried overnight, and physical and sensory properties of extrudates were determined.

The obtained results showed that addition of by-products to corn grits resulted in decrease of expansion ratio (ER) and fracturability, whereas bulk density (BD) and hardness of extrudates increased. Acceptable scores in sensory evaluation received all samples with brewer's spent grain and sugar beet pulp (with addition of 1% pectin), and samples with 5 and 10% of apple pomace. It can be concluded that these three by-products can be successfully incorporated in corn grits with the aim of production acceptable expanded corn snack products.

Keywords: corn grits, by-products, extrusion, physical properties, sensory properties

Rheological and sensorial analysis of tomato ketchup

A. Budelli[1], G. D. De Prisco[2], M. Gallo[2], N. Grizzuti[2], R. Nigro[2]
[1]Heinz Innovation Center, NieuweDukenburgseweg 19, 6534, Nijmegen, Netherlands
[2]Università degli Studi di Napoli Federico II, DICMAPI, P.leTecchio 80, 80125 Naples, Italy

Reduction of the amounts of sugar and salt in the diet is part of the new health care WHO/FDA reform and its emphasis on preventive health care measures. The high concentration of salt and sugar in processed foods and beverages is widely considered a significant contributor to the obesity epidemic and other nutrition-related diseases, like diabetes, hypertension, heart disease, stroke and cancer.

The reduction of these ingredients has a great influence on sensorial perception, which is essential for determining product quality and consumer acceptance, resulting in a decrease of flavour and taste intensity. Studies show that it is related to an increase of viscosity. Modification of the product composition, that lower its viscosity, can so induce consequences on aroma/taste perceptions.

It is thus useful to understand the relationship between food structure and sensory properties.

Tomato paste can be considered from a rheological point of view as a weak gel, with complex rheological behaviour: it is a non-Newtonian, shear-thinning and a thixotropic fluid that shows an apparent yield stress.

Studies have shown that different salt content induces large differences in rheological properties and texture perception of model cheeses [1]. There are no other similar studies for tomato ketchup.

Aim of this work is to analyse by means rheological procedures and sensory profiling characteristics of five tomato ketchup that differ in salt, sugar, vinegar and tomato solid content; a sample characterized by a zero amount of added salt is used as reference in the sensorial analysis.

A Physica MCR501 stress-controlled rheometer with parallel plate geometry, 25 mm in diameter and 1 mm gap distance, was used to study the viscoelastic characteristics of tomato ketchup.

Different protocols are used, they differ in probe geometries, gap, frequency and stress utilized.

The oscillatory frequency sweep has shown that, under linear viscoelastic conditions, these ketchups behaved as gels and has the typical non-linear behaviour of a suspension. The trends in all ketchups were very similar.

For the sensorial analysis a Quantitative Descriptive Analysis (QDA) was chosen as technique, for all samples at disposal were all very similar too. This proved the skills of the company of manufacturing products whose tastes are very difficult to differentiate by consumers.

The research of possible relationships between the rheological and sensorial variables brought to relate the apparent viscosity to the saltiness of ketchup. Yet, it was not possible to find out a clear model that could help to predict the perceived quality of ketchup from the rheological characterization.

References:

[1] Saint-Eve A., Lauverjat C., Magnan C., Déléris I., Souchon I., 2009, Reducing salt and fat content: Impact of composition, texture and cognitive interactions on the perception of flavoured model cheeses, Food Chemistry 116, 167-175

Separation properties of commercially available spiral wound membranes in fractionation of organic acids in sweet cheese whey

A. Hinková, V. Pour, S. Gillarová, S. Henke, L. Jandová, Z. Bubník
University of Chemical Technology, Prague, Department of Carbohydrates and Cereals, Technická 5, 16628 Prague 6, Czech Republic; tel. +420 220443111, e-mail: hinkovaa@vscht.cz, fax: + 420 220445130.

The separation of molecules using nanofiltration and reverse osmosis membranes is affected by many factors, especially in complex solutions containing several components of different size, shape, degree of dissociation, solvation or charge.

This paper studies the effect of process conditions (pH and pressure) on separation of organic acids (acetic, propionic, lactic and citric acids) in a solution of sweet cheese whey using two commercially available spiral wound membranes (SWM) with different cut-offs. Desal SC2540F1457 membrane is intended for reverse osmosis, whereas Filmtec NF 270 has lower rejection and is mostly used for nanofiltration. A solution prepared from dried whey was purified by two-step ultrafiltration (500 nm and 20 nm) to remove proteins and permeate was used as a feed for filtration on SWM. The pH of feeds for SWM filtration was adjusted to values from 5 to 6.5 and pressures applied were 1.5, 2 and 2.5 MPa.

The rejections of acetic and propionic acids were higher with rising pH on both membranes used. Due to their smaller molecule size (60 and 74 g/mol; respectively) the sieving effect was less significant mechanism of separation and the separation was mostly affected by the degree of dissociation, which increases with higher pH values. With higher dissociation, the repulsion of negatively charged membrane is also higher. On the other hand, lactic and citric acids (molecular weight of 92 and 190 g/mol; respectively) rejections on both membranes were steady regardless the initial pH, showing that the size and steric effect is a dominant mechanism in a separation in this range of pH. The rejections of acids on the Filmtec membrane were slightly declining with higher pressure applied which means that the concentration polarisation layer repulsion is reduced at higher pressure. The same effect was not observed for the Desal membrane where the rejections of acid were not dependent on the driving pressure.

The effect of diafiltration on recovery of organic acid and sugars during cheese whey membrane ultrafiltration

A. Hinková, V. Pour, S. Henke, M. Sluková, S. Gillarová, Z. Bubník
University of Chemical Technology, Prague, Department of Carbohydrates and Cereals, Technická 5, 16628 Prague 6, Czech Republic; tel. +420 220443111, e-mail: hinkovaa@vscht.cz, fax: + 420 220445130.

Whey, produced in large quantities in cheese manufacture, is a valuable source of many compounds, which can be further utilised. For instance, whey proteins are used in special nutritional products and lactose can serve as a source of carbon for fermentation to produce organic acids (e.g. lactate, propionate). Ultrafiltration (UF) represents an efficient technique for separation of whey proteins however a small part of lactose and organic acids is always retained with the protein fraction. From these reasons, sweet cheese whey was treated by a single-step ultrafiltration (UF) using 50 kDa tubular inorganic membrane followed by three diafiltration steps. The aim was to enhance the transport of present sugars and organic acids into permeate to minimise their losses in the retentate protein fraction. Obtained UF permeates were collected and consequently concentrated by nanofiltration (NF) on a spiral wound polymeric membrane NF 270-2540 (Dow, Filmtec).

The recovery of nitrogen compounds was 81% in the UF retentate, both for single-step ultrafiltration and diafiltrations. In the single step UF, lactose rejection was 13.3% and the relative recovery of lactose in retentate was 37%. The subsequent three diafiltrations reduced the loss of lactose to 14%. Similar effect was seen in organic acids rejection where the recoveries in permeate after the three diafiltrations were 80 – 90%. Hence, the loss of acetic and citric acid was reduced by 31 and 23%, respectively.

Lactose and galactose rejection factor on the nanofiltration membrane was 93.3% and 77.1%, respectively. Rejection of organic acids on the same membrane was 71 – 76%, which provides good concentration of carbohydrates and organic acids using relatively low pressure of 0.6 MPa.

The effect of temperature and sucrose concentration on the viscosity of lime milk

R. Gruska, A. Antczak-Chrobot, P. Bąk, M. Wojtczak
Lodz University of Technology, Faculty of Biotechnology and Food Sciences,
Institute of Food Technology and Analysis, Stefanowskiego 4/10,
90-924 Lodz, Poland, e-mail: maciej.wojtczak@p.lodz.pl, tel. +48426313453

In sugar factories the lime milk in purification is more often prepared by using sweet solutions (sweet water). Since sucrose has the ability to form coordinate bonds with calcium ions, a significant change in viscosity of this lime milk was observed This may cause difficulties in its transport or storage.

Therefore, the aim of our study was to show how temperature and sucrose content in sweet solutions affect on the viscosity of lime milk. The studies were performed by using fresh burned lime, sweet solutions taken from sugar factory and model solutions. The viscosity measurements were performed using a rotational viscometer. The lime slaking was carried out in temperature from 70°C to 90°C in a ratio: 1 part of burned lime to 5.5 parts of solution. In the next part of the experiment, the lime milk was cooled and heated, which was a simulation of temperature changes during transport and storage.

The results have shown that the formation of high viscosity colloid (or coordinate bonds of sucrose and calcium) is facilitated when the initial slaking temperature is higher than 80°C and the sucrose content is below 10%. This observation applies to both solutions: model and sweet water. The highest viscosity (above 200 cPa) had lime milk prepared by using sweet solutions at 90°C and sucrose content equal 6%. The increase of sugar content in the solutions caused a decrease of initial viscosity of milk. In the case of model solutions, the sucrose content higher than 10% at temperature 90°C produced a lime milk with viscosity less than 80 cP. It was more than three times lower than the viscosity of milk prepared from solutions with a 6% sugar content (250 cP). Similar correlations were observed in technical solutions (sweet water). It was also observed that if the lime slaking is carried out using solutions at low temperature (70°C), in each case, the lime milk has a low viscosity (less than 50 cP). In this temperature the sucrose content in the solution is irrelevant.

In the next stage of the study the lime milk was cooled and heated. In the process of cooling the fresh lime milk any substantial change in viscosity was observed. Significant changes of the viscosity were noted during the heating of the cooled lime milk. This applies both to the model sweet solutions and to the sweet water. If the sucrose content was higher than 10%, in temperature exceeding 45 – 50°C (model solutions) and 55 – 60°C (sweet water) we observed an increase of viscosity to about 370 cPa (model solutions) or to 200 cP (sweet water). The highest viscosity (even more than nine times higher) remained relatively short. When the temperature reached about 75 – 80°C, the viscosity rapidly decreased and its value was equal almost to the viscosity at the beginning of the process (about 30 - 50 cP).

It can be concluded that the viscosity of the lime milk depends on the temperature and the content of sucrose in sweet solutions. In sugar factories it is recommended to avoid cooling the lime milk below 60°C and then re-heating it. This may cause an increase of its viscosity and problems during transportation and storage (for example overgrowing pipes and tanks). The increase in viscosity may be probably caused by the colloidal properties of the calcium and the ability of sucrose to form coordinate bonds with calcium.

ACKNOWLEDGMENT. The authors thank the Sugar Company "Krajowa Spółka Cukrowa S.A" for valuable cooperation and support of this study

Novel strategies to design zeolite catalysts with tailored acidities and porosities

Benoît Louis,[1] Pit Losch,[1] Alessandra Vieira,[2] Elisa Silva Gomes,[2] Ana Belen Pinar,[3] Marcelo Maciel Pereira,[2]

1. Laboratoire de Synthèse Réactivité Organiques et Catalyse, LASYROC, Institut de Chimie, UMR 7177 CNRS, Université de Strasbourg, 1 rue Blaise Pascal, 67000 Strasbourg cedex, France

2. LACES, Universidade Federal do Rio de Janeiro, Centro de Tecnologia, Departamento de Química Inorgânica, Avenida Athos da Silveira Ramos, 149, Ilha do Fundão, 21941-909, Rio de Janeiro, RJ, Brazil

3. Laboratory of Crystallography, Department of Materials, ETH Zurich, Vladimir-Prelog Weg 5, CH-8093 Zurich, Switzerland

Spontaneous self-organization of matter demonstrates the capacity of nature to structure itself when specific conditions are present. The design and control of chemical systems over multiple-length scales from the molecule to the crystal thus represent one of the greatest challenges in science [1,2]. In particular, the ability to understand how molecules organize, thus build supramolecular architectures to produce functional nanomaterials. Supramolecular chemistry deals with the association of two or more chemical species, based on molecular non-covalent interactions [2]. Heterogeneous catalysis can also take advantage from the peculiar properties of self-assembly processes. Herein, we report the synthesis of zeolites (LTA, FAU, MFI, BEA) with tailor-made silicon / aluminum ratios (SAR) as well as controlled porosities through different synthetic approaches: in the presence of biomass-derivate compounds, via solid-solid recrystallization, seeding procedure. For instance, the use of biomass extracted molecules in the gel for preparing allowed to produce either solely microporous or micro / mesoporous MFI zeolite crystals [3,4].

A ZSM-5 zeolite (MFI structure) was synthesized with an exceptional low SAR value = 8 (Figure 1), representing the lowest value reported so far [3]. As *'a priori'* expected, this low SAR ZSM-5 exhibited a higher catalytic activity in hydrocarbon cracking compared with regular ZSM-5 catalyst (Figure 2).

Figure 1. SEM image of SAR=8 ZSM-5 zeolite

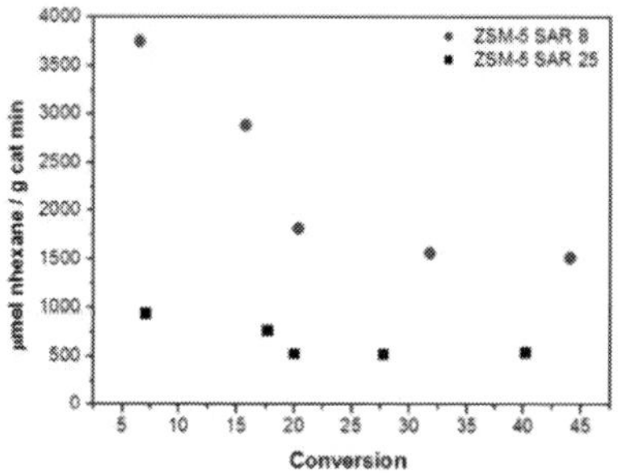

Figure 2. Rate of n-hexane cracking at 773 K versus conversion over ZSM-5 (SAR 8) and reference ZSM-5 (SAR 25)

Depending the methodology chosen, morphological and textural properties of zeolites were largely affected [5]. This simple and inexpensive method was recently extrapolated for improving textural properties of γ-alumina [6]. Both surface areas and pore volumes increased up to 2.5-fold and 1.5-fold, respectively.

In the extended presentation, we will tempt to rationalize how the preparation procedure influences the zeolite crystals formation (morphology, size, porosity, acidity) and hence modifies their catalytic properties in acid catalysis.

References

[1] A. Stein, S.W. Keller, T.E. Mallouk, Science 259 (1993) 1558.

[2] J.M. Lehn, Science 227 (1985) 849.

[3] B. Louis, A. Vieira, M.M. Pereira, *Br. Patent* (2016) *deposited.*

[4] F. Ocampo, J.A. Cunha, M.R.L. Santos, J.P. Tessonnier, M.M. Pereira, B. Louis Appl. Catal. A 390 (2010) 102.

[5] C.C. Rocha, A. Balanqueux, M. Boltz, P. Losch, C. Bernardon, V. Bénéteau, P. Pale, M.M. Pereira, B. Louis L'Actualité Chimique 393-394 (2015).

[6] C.S. Cardoso, Y.E. Licea, X. Huang, M. Willinger, B. Louis, M.M. Pereira, Micro. Meso. Mater. 207 (2015) 134.

Micro-mesoporous polystyrene matrices – a base for effective Pd nanocatalyst development of Suzuki cross-coupling and selective alkynols hydrogenation

L.Zh. Nikoshvili[1,2], N.A. Nemygina[1], E.A. Kholkina[1], T.E. Hudyakova[1], V.G. Matveeva[1], M.G. Sulman[1,3], E.M. Sulman[1], L. Kiwi-Minsker[2,4]
[1]Tver Technical University, Dep. Biotechnology and Chemistry, A.Nikitina str. 22, 170026, Tver, Russia; tel./fax: +7 4822 789317, e-mail: nlinda@science.tver.ru
[2]Tver State University, Regional Technological Centre, 170100, Tver, Russia
[3]A.N. Nesmeyanov Institute of Organoelement Compounds of RAS, 119991, Moscow, Russia
[4]Ecole Polytechnique Fédérale de Lausanne, GGRC-ISIC-EPFL, CH-1015, Lausanne, Switzlerand

Pd-catalyzed reactions are one of the most widespread in fine organic synthesis. For example, selective hydrogenation of alkynols is used for production of a large variety of terpenoids and also fat-soluble vitamins A, E and K. Cross-coupling reactions are also of high importance in fine organic synthesis. Among the existing cross-coupling methods, Suzuki cross-coupling between aryl halides and arylboronic acids is one of the most widespread and effective methods of synthesis of biaryls, which are important semi-products in synthesis of a number of pharmaceuticals, ligands and polymers.

This work summarizes our experience in the development of novel composite Pd-containing polymer-based catalysts for selective hydrogenation of alkynols and Suzuki cross-coupling. As a polymeric support for the catalyst synthesis, different types (functionalized and without functional groups) of hypercrosslinked polystyrene (HPS), which possess both high porosity and possibility for internal mass transfer, were used.

While analyzing all the possible factors influencing the catalytic properties of Pd/HPS composites, the size of Pd nanoparticles (NPs) seems to be the key factor. Through careful choice of palladium precursor ($PdCl_2(CH_3CN)_2$, $(PhCN)_2PdCl_2$, $(StyPdCl_2)_2$, $Pd(OAc)_2$, $PdCl_2$) in combination with the type of HPS, the possibility for control of NP growth and dispersion can be achieved (Fig. 1), and thus the tuning of catalytic properties is possible.

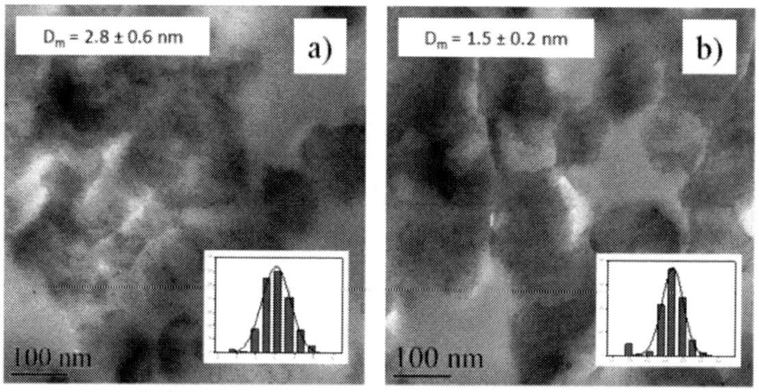

Fig. 1 – Examples of TEM images for 1.5%-Pd/HPS catalysts obtained while using $PdCl_2(CH_3CN)_2$ (a) and $Pd(OAc)_2$ (b) as precursors

Synthesized Pd/HPS catalysts were shown to be active and selective (selectivity and conversion >98%) in chosen hydrogenation and cross-coupling reactions.

This work was funded by the Ministry of Education and Science of the Russian Federation (RFMEFI57414X0121) and the Russian Science Foundation (Grant 15-19-20023).

Performances of MOF vs. zeolite coatings in adsorption cooling applications

M. Tatlier
Department of Chemical Engineering, Istanbul Technical University, Maslak, 34469 Istanbul, Turkey; tel.: +90 2122856234, e-mail: tatlierm@itu.edu.tr, fax: +90 2122852925.

Adsorption heat pumps/chillers may be good alternatives for heating/cooling applications whenever suitable sources of energy, such as waste heat and solar energy are available. Significant improvements have been demonstrated to take place in these systems when the adsorbent is directly crystallized on metal heat exchanger surfaces, instead of being used in powder or pellet form. Elimination of the limitations originating from inefficient heat transfer inside the adsorber as well as the easy manipulation of the adsorbent thickness, in a manner avoiding mass transfer limitations, may be achieved by using coatings.

Materials such as zeolite, silica gel and activated carbon have been proposed to be used as adsorbents in adsorption heat pumps/chillers. Zeolites have unique properties suitable for these systems. The possibility of preparing their coatings on various metal surfaces is another advantage for zeolites leading to the improvements mentioned above. Metal Organic Frameworks (MOFs) have been developed quite recently and have been studied in the last two decades for their application especially in gas sorption and separation. However, water sorption in MOFs has not been much explored, perhaps due to the fact that most of them are not stable in water. The water sorption studies have generally focused on well known carboxylate based MOFs, such as HKUST-1, MIL-100/101, MIL-53, CPO-27, UiO-66, MOF-801 and MOF-841. It still seems to be a challenge to find materials which have relatively high water uptake in a narrow relative humidity range while maintaining high structural integrity. The attempts to increase the water stability of MOFs have led to decreases in their water uptake capacity due to enhanced hydrophobicity and reduction in surface areas. Investigations on the preparation of coatings of MOFs on metal surfaces have also started recently.

In this study, a mathematical model [1] was used to determine the performances of a new hydrophylic triazolyl phosphonate MOF [La3L4(H2O)6]Cl·x H2O(1,L(2-)=4-(4H-1,2,4-triazol-4-yl)phenyl phosphonate), and a zeolite, NaX, which were envisioned to be grown as coatings on stainless steel heat exchanger tubes. It has been reported that the mentioned MOF is water stable and has a water adsorption capacity comparable to zeolite NaX [2]. The duration of a single cycle of an adsorption cooling system was obtained for different thicknesses of coatings and the optimum thickness, maximizing the power, was determined under various operating conditions. The results obtained were also compared to those determined previously for zeolite A coatings theoretically [3] and experimentally [4].

In the mathematical model used, both the temperature and concentration gradients across the adsorbent and the temperature gradient across the metal wall of heat exchanger tubes were determined. The differential equations were solved by a computer program using a finite difference scheme to obtain the duration of a single cycle of an adsorption heat pump/chiller during the distinctive periods of heating, desorption, cooling and adsorption. The average cooling power of the chiller may be defined as

$$P = Q_u / \Delta t \qquad (1)$$

where Q_u represents the useful effect that may be obtained from the evaporator in a single cycle completed in the time interval Δt. The useful effect may be defined as follows

$$Q_u = \Delta W\, M_z\, L - C_{pw}\, M_z\, \Delta W\, (T_{in} - T_{out}) \quad (2)$$

C_{pw} denotes the specific heat capacity of water while ΔW is the amount of adsorbate circulating in the adsorption system and L is the latent heat of vaporization of water. T_{in} and T_{out} represent the condenser and evaporator temperatures and M_z is the zeolite mass. The power was estimated for a special design of the adsorber, where numerous heat exchanger tubes coated by the adsorbent were placed. The evaporator and condenser temperatures were selected as 2°C and 20°C, representing the operation of a refrigerator. The mass of adsorbent varied with the coating thickness. The parameters in the mathematical model pertaining to the adsorbents, namely the effective diffusivity of water, equilibrium concentration of water on the surface, density, specific heat capacity, heat of adsorption, thermal conductivity were obtained from the literature.

The cycle duration of the adsorption cooling system as well as its power was estimated for various coating thicknesses of the MOF and zeolite NaX adsorbents. The optimum coating thickness was determined to be much higher for the metal organic framework when compared to the zeolite. An interesting observation was the relatively high difference between the durations of adsorption and desorption cycles for zeolite NaX. This seemed to originate from the strong temperature dependency of the effective diffusivity of the mentioned zeolite, leading to rather low diffusivity values at the lower end of the temperature range used. The cycle durations were considerably shorter for the MOF investigated in this study and the power was higher. For the metal organic frameworks with relatively low densities, the advantage of having shorter cycle durations may be reduced and much higher coating thicknesses should be used for the same coating mass. The results indicated that zeolite NaA may also be considered in the adsorption cooling systems as a good alternative to zeolite NaX under some operating conditions, in spite of its lower adsorption capacity.

The modeling studies revealed that the selection of a suitable adsorbent for adsorption heat pump/cooling systems is not a straightforward task. Different operating conditions may favor the use of different adsorbents. The determination of the optimum mass equivalent coating thickness, maximizing the power of the adsorption devices, helps to avoid using thicknesses for which severe mass/heat transfer limitations exist. It should also be remembered that the optimum actual coating thickness may be higher than the optimum mass equivalent thickness in case of inhomogeneous coatings prepared by different synthesis methods.

References

[1] Tatlier, M., Tantekin-Ersolmaz, Ş.B., Erdem-Şenatalar, A., Micropor. Mesopor. Mater., 27, 1-10 (1999).

[2] Begum S, Wang Z, Donnadio A, Costantino F, Casciola M, Valiullin R, Chmelik C, Bertmer M, Kärger J, Haase J, Krautscheid H., Chemistry, 20, 8862-8866 (2014).

[3] Tatlier, M., Erdem-Şenatalar, A., Int. J. Refrig., 23, 260-268 (2000).

[4] Tatlier, M., Munz, G., Fueldner, G., Henninger, S., Micropor. Mesopor. Mater., 193, 115-121 (2014).

Assessment of adsorption isotherms of hydrogen above the critical point on activated carbons from agricultural waste materials

K. Soukup, [1]G.J.F. Cruz, V. Hejtmánek, O. Šolcová
Institute of Chemical Process Fundamentals of the CAS, v. v. i., Rozvojová 135, CZ-165 02 Prague 6, Czech Republic; e-mail: soukup@icpf.cas.cz, tel.: +420 390 283; [1]National University of Tumbes, Department of Forestry Engineering and Environmental Management, Av. Universitaria s/n, Campus Universitario, Pampa Grande-Tumbes, Peru

The design of the adsorption-based hydrogen storage systems requires a fundamental understanding of hydrogen physisorption and/or chemisorption processes and adsorption/desorption kinetics. Modeling of hydrogen adsorption isotherms represents a fundamental issue for finding the best operating parameters and it can also be utilized for the design of new materials in terms of their hydrogen uptake capacity. Additionally, the knowledge of the model adsorption isotherms can reduce a number of the requested experiments due to the possibility to predict adsorption properties over a wide range of temperatures and pressures.

Evaluation of the adsorption isotherms over a wide range both of temperatures and pressures was performed to optimize hydrogen uptake as well as release capacity. A model for the adsorption isotherms of microporous adsorbents must rely on a minimum number of parameters, which should have a clear physical interpretation. In addition, it must predict the high-pressure maximum typical for supercritical conditions. We found the Dubinin-Astakhov equation as an appropriate model that meets all aforementioned conditions.

In the framework of the presented work, the low-cost activated carbons from various tropical agricultural waste biomass including mango pits, red mombin seeds, coffee husks and cocoa pod husks were prepared. The raw materials collected in Tumbes region in Peru were mixed with activating agent based on $ZnCl_2$ in the weight ratio of 1/1. Subsequently, the mixture was put in a ceramic reactor and afterwards in a horizontal oven to be directly carbonized under nitrogen atmosphere. The temperature increased with the rate of 10 min/min until a final carbonization temperature of 600 °C was reached and it was kept during 2 h.

The hydrogen sorption capacities of the prepared microporous samples were measured at a temperature range corresponding to normal boiling point of nitrogen (77 K), 273 K, and 298 K and equilibrium pressure up to 1000 kPa. It is to be expected that the hydrogen physisorption of all samples is strongly correlated with their inner surface area. Moreover, their hydrogen capacity decreases with increasing equilibrium temperature. It was found that sample based on mango pits reveals the highest adsorption capacity (2.3 wt.%, 0.06 wt.%, and 0.05 wt.%, at 77 K, 273 K, and 298 K respectively) towards hydrogen. Furthermore in the case of temperatures of 273 K and 298 K, hydrogen storage capacity increases linearly with increasing hydrogen equilibrium pressure. On the other hand, the maximum on adsorption isotherms of hydrogen occurs at 77 K as soon as the bulk density of H_2 increases as a higher rate than the adsorbed density of H_2.

Acknowledgement

The financial support of the Czech Science Foundation (project no. 15-14228S) and of the National University of Tumbes (Canon project – Res. N° 0722-2014/UNT-R) is gratefully acknowledged.

Aging of thermal-insulation properties of polyurethane foams investigated by multiscale modelling

P. Ferkl[1], A.Nistor[1], M. Podivinska[1], M. Vonka[1], M. Toulec[2], J.Kosek[1]
[1]University of Chemistry and Technology Prague, Department of Chemical Engineering, Technicka 5, 166 28 Prague 6, Czech Republic, www.kosekgroup.cz, email: juraj.kosek@vscht.cz;
[2]Robert Bosch, spol. s.r.o., Roberta Bosche 2678, 370 04 České Budějovice, Czech Republic

Polyurethane foams are extensively used in many branches of industry. To name a few applications, we can mention furniture, bedding, interior car parts and packaging. Our contribution is focused on the thermal insulation properties of rigid polyurethane foams.

The polyurethane is formed by the reaction of isocyanates with polyols in the presence of catalysts. Polyurethane foams are created by the so-called reactive foaming. In this process the foam is formed during the exothermic polymerization reaction. The combination of chemical and physical blowing agents is commonly used. Chemical blowing agent is in this case water, which reacts with isocyanate and gradually releases carbon dioxide into the system. The physical blowing agent is usually pentane or cyclopentane, which is dissolved in the initial mixture and evaporates when the temperature is increased by the chemical reaction. By the adjustment of polyurethane chemistry (i.e., reactants) and the type and amount of the blowing agents it is possible to produce foams with a broad range of porosities and cell size distributions, and thus with different application properties.

Unfortunately, the thermal insulation properties of polyurethane foams degrade over the years. The blowing agents spontaneously diffuse out of the foam whereas the air diffuses into the foam. Because the air has generally higher thermal conductivity than blowing agents, this phenomenon causes the unwanted increase in the foam conductivity. Thus the general understanding of this process would help to produce foams, which don't have good insulation properties only at the beginning but also over the lifespan of the product.

The mathematical modelling of the aging of thermal insulation properties of polyurethane foams is challenging for several reasons. First, there are very little data available on solubility and permeability of blowing agents in polyurethanes in the open literature. Second, the thermal conductivity of gas mixtures must be carefully evaluated using complex mixing rules to avoid unwanted discrepancies. Third, the models are requiring advanced image analysis algorithms to reliably characterize the foam morphology. Fourth, the heat in polyurethane foams is transferred by both conduction and radiation. Thus, a model for the prediction of the radiation through and in the foams is needed[2]. Fifth, a model, which simulates the diffusion of gases and considers the real foam morphology was developed. And finally, since we are dealing with many models and a multiscale problem, an interface allowing the interaction of all the deployed models has to be employed.

We used a newly developed MoDeNa[1] orchestrator for the multiscale interaction of the models. It provides a systematic way for the higher-scale model to call the lower-scale model. Moreover, to reduce the computational time, it provides simplified surrogate models as the representation of the complex lower-scale models and these surrogate models are fitted automatically by the design of experiments. Thus, the complex lower-scale models are executed only several times during the simulation. The orchestrator stores these results and uses them to fit the surrogate model, which can be then used by the higher-scale model.

We developed a new model, which is able to simulate the diffusion of gases through a series of successive polymer walls and gas cells. This model estimates the time evolution of the concentration profiles of individual gases inside the foam. Once the composition of gases in the foam is known, the gas conductivity and then the foam conductivity are calculated.

The mathematical modelling was complemented by experimental work. The solubility and the diffusivity of gases in polyurethane were measured on gravimetric and pressure decay apparatuses. Polyurethane foam samples were prepared using carbon dioxide and cyclopentane as blowing agents. The evolution of foam conductivity during accelerated aging of the foam samples at elevated temperatures was used to validate the mathematical model.

[1] MoDeNa-EUProject, 2015. Modelling of morphology development of micro- and nanostructures. [Online]. URL https://github.com/MoDeNa-EUProject/MoDeNa

[2] Ferkl P., Pokorny R., Kosek J. (2014). "Multiphase approach to coupled conduction–radiation heat transfer in reconstructed polymeric foams." International Journal of Thermal Sciences 83:68-79.

Morphogenesis of micro-structured polystyrene foams prepared by thermally induced phase separation: a theoretical and experimental study

A. Nistor, M. Vonka, A. Rygl, M. Voclova, M. Minichova, J. Kosek
University of Chemistry and Technology Prague, Dept. of Chemical Engineering, Technická 5, 166 28 Prague 6, Czech Republic;
tel. +420 220 443 296, e-mail: Juraj.Kosek@vscht.cz; fax: +420 220 444 320

Polymer foams are suitable for a wide range of applications depending on polymer properties and especially on the foam morphology. Currently produced polystyrene (PS) foams have a closed cell structure and they are used as thermal insulators. Generally, the preparation of foams with various morphologies by using of one production method would open new application possibilities. Such a method could be the thermally induced phase separation (TIPS). However, a deeper understanding of the TIPS method is still missing.

In this work, we present a validated model for the morphogenesis of micro-structured PS foams formed by TIPS, in particular by the spinodal decomposition [1]. The in-home designed and constructed apparatus enabled us to prepare PS foams at various process conditions and with different solvents via the TIPS method. Then, the foam morphology was characterized by SEM. Moreover, we measured on another in-home made device the binodal curves (cloud-points) by a thermo-optical method. The binodal curves and foam morphologies were compared with the thermodymic model and the morphology evolution model predictions, respectively. For the thermodynamic model we used the Flory-Huggins theory and for the morphology predictions we used the Cahn-Hilliard model. The Cahn-Hilliard approach is suitable for systems, where the phase separation takes place by the mechanism of spinodal decomposition. The model predicts correct length scales of closed-cell morphologies for various system compositions and temperatures in 2D and 3D (Figure 1). Moreover, the model enables to predict different morphologies: (i) aggregated particles, (ii) open-cell structures, (iii) combination of open- and closed-cell structures, and (iv) closed-cell foams.

In addition, we present our experimental and theoretical results of the TIPS morphogenesis dynamics, which will enable us to fully optimize the process and, thus, to prepare foams with desired structures.

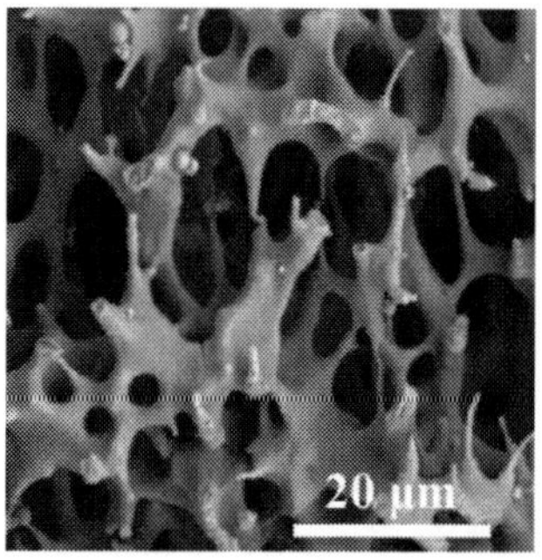

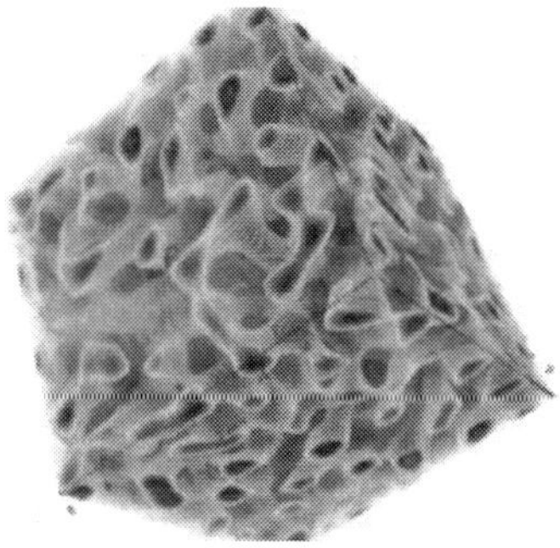

Figure 1: SEM image of a microcellular PS foam and its 3D morphology prediction by Cahn-Hilliard model.

Reference:

1. M. Vonka, A. Nistor, A. Rygl, M. Toulec, and J. Kosek, *Morphology model for polymer foams formed by thermally induced phase separation.* Chemical Engineering Journal, 2016. **284**: p. 357-371.

Simultaneous removal of antimonate and arsenate from tap water by using cellulosic carbon fibers decorated with iron (oxy)hydroxide

E. Bilgin Simsek[1], I. Novak[2], D. Berek[2]
[1]Yalova Univ., Dep. Chemical and Process Engineering, 77100 Yalova, Turkey;
tel. +90 226 8155403, e-mail: ebilgin.simsek@yalova.edu.tr, esrabilgin622@gmail.com;
[2]Polymer Ins., Slovak Academy of Sciences, 84541 Bratislava, Slovakia; e-mail: i.novak@seznam.cz; dusan.berek@savba.sk,

Carbonaceous materials, such as activated carbon, carbon nanotubes (CNTs), carbon nanofibers (CFs), and graphene nanosheets, have emerged as promising adsorbents for the removal of contaminants due to their surface functional and outstanding mechanical and electrical properties. CFs have become one of the most important advanced composite materials produced today owing to its light weight, high strength and elasticity, high electrical/thermal conductivity, and excellent corrosion resistance [1]. On the other hand, antimony and arsenic pollutions are mainly found in the ground water on global levels, affecting large populations where antimony is a toxic bioaccumulative element with similar chemical and toxicological properties to arsenic, and moderate levels of them may lead to bioavailability and harmful environmental effects [2]. The dominant forms of arsenic and antimony in water environment are arsenate [As(V)] and antimonate [Sb(V)] under oxidized conditions [3]. As arsenic and antimony can be found to co-occur in the environment and as antimony may well compete with arsenic for adsorption on adsorbents in aqueous solutions, it is essential to examine the competitive sorption characteristics of them. The objective of this work is to develop an efficient functional material for the adsorption of coexistent As(V) and Sb(V) in water, and investigate the possibility of simultaneous removal of both As(V) and Sb(V) from tap water samples.

Cellulosic carbon fibers (CF) decorated with iron (oxy)hydroxide materials (CF-Fe) were synthesized with iron weight ratios of 3%, 7%, 11% and 20%. Figure 1. shows the SEM images of raw and iron modified carbon fibers.

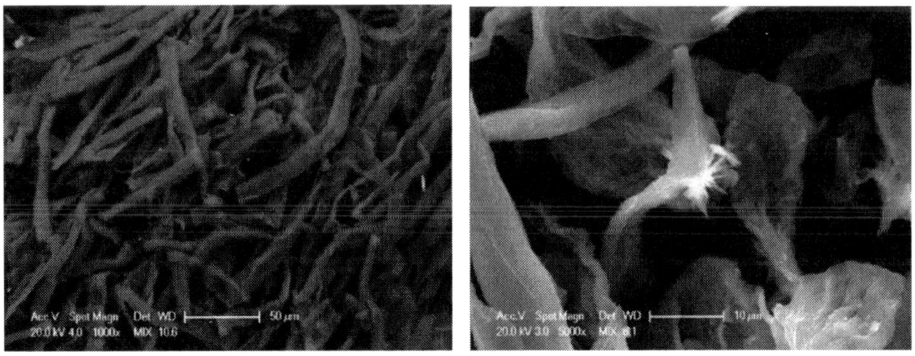

(a) (b)

Figure 1. SEM images of (a) pristine and (b) iron modified carbon fibers

Batch adsorption experiments were conducted by transferring 25 ml of each As(V) and Sb(V) solution (in tap water) with 100 µg/L concentration.. The effect of initial solution pH, temperature, competing anions, adsorption kinetics and adsorption isotherm on As(V) and Sb(V) adsorption. Equilibrium concentrations were determined by atomic absorption spectrophotometer by the graphite furnace system. The binary adsorption experiments for mixed As(V) and Sb(V) solution were carried out to study the isotherm of competitive system.

Figure 2. presents the adsorption performances of both CFs in unitary and binary system. When compared with pristine CNF, the iron loaded samples presented high efficiency. CFs decorated with Fe_2O_3 surface functional groups have maximum As(V) and Sb(V) adsorption capacity of 134.3 and 132.4 μg/g, while in binary system, it decreased to 128.0 and 125.8, respectively.

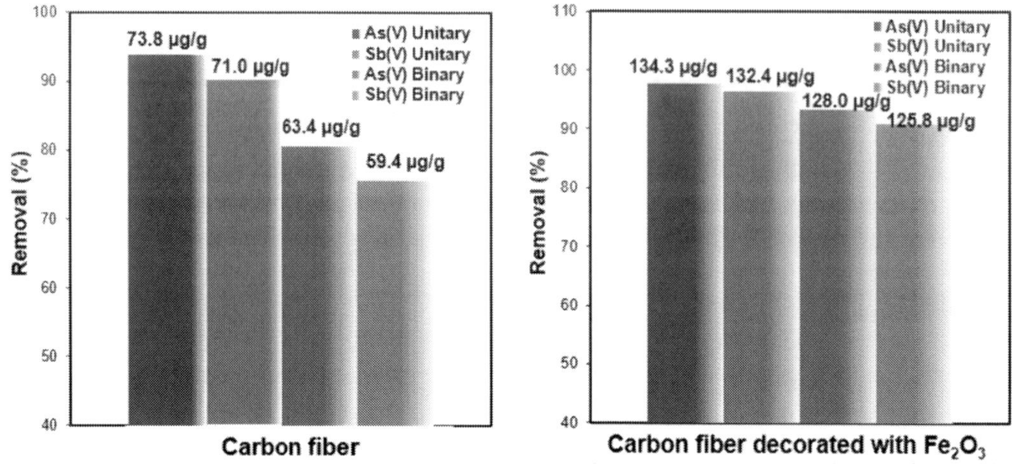

Figure 2. Comparison of As(V) and Sb(V) removal performance of carbon fibers

In order to examine Sb(V) and As(V) adsorption characteristics of CFs, Langmuir, Freundlich and Dubinin–Radushkevish (D–R) isotherm models models were used to analyze the adsorption experimental data. Considering the determination coefficient values (R^2) ranging between 0.95-0.99, the adsorption data of both raw and iron loaded CFs were well simulated with both the Freundlich and D-R models implying heterogeneous nature of adsorption. The adsorption process between CF and Sb(V) & As(V) was identified to be an exothermic while positive enthalpy values of CF-Fe samples supported the endothermic nature of system

The adsorption kinetics of CFs decorated with iron hydroxides was investigated and compared with raw CF. It was found that both Sb(V) and As(V) adsorption rate increased with increasing amount of iron loading. With the adsorbent dosage of 0.5 g/L, within 2 h, CNF-Fe (3%), CNF-Fe (7%) and CNF-Fe (11%) samples were able to reduce As(V) concentration below its maximum level (10 μg/L) regulated by the US EPA while CNF-Fe (20%) sample decreased this level within 20 min. The kinetic parameters revealed that the equilibrium data was described better by the pseudo-second-order model for As(V) and Sb(V), as the correlation coefficient (R^2) was closer to 1.0.

The results can contribute to the decontamination technology of As(V)/Sb(V) polluted water and further understanding the competitive adsorption behaviour of As(V) and Sb(V) on carbon fiber based materials.

References

[1] Youe, W.-J., Lee, S.-M., Lee, S.-S., Lee, S.-H., Kim, Y.-S. *Int J Biol Macromol* 82 (2016) 497–504.

[2] Duan, L-Q., Song, J.-M., Li, X.-G., Yuan, H.-M. *Cont Shelf Res* 30 (2010) 1522–1534.

[3] Lu, H., Zhu, Z., Zhang, H., Zhu, J., Qiu, Y. *Chem Eng J* 276 (2015) 365–375.

Ice templating based approaches for porous materials and nanostructures

Aled D Roberts, Ulrike Wais, Adham Ahmed, Lei Qian, Richard Hayes, Haifei Zhang*
Department of Chemistry, University of Liverpool, Oxford Street, Liverpool, United Kingdom. Tel: +44 151 7943545. Email: zhanghf@liv.ac.uk.

Ice templating has been widely used to prepare a wide range of porous materials and nanostructures. The process is highly straightforward. By freezing a solution or suspension and removing the frozen solvent crystals, ice-templated structures can be produced. It is important to adjust the solution compositions and control the freezing conditions in order to generate the materials with desired structures or morphologies. A directional freezing process has been developed by our group to prepare a wide range of aligned porous structures. This method works by orientating crystal growth during the directional freezing process and removal of ice crystals during the freeze-drying process. This method can be used for solutions and suspensions (including microspheres and nanoparticles) with solvents such as water, organic solvents, and compressed CO_2. Quite importantly, by combining ice templating and emulsion templating, hierarchically porous materials with systematically tuned porosity can be produced.

We will show in this lecture how aligned porous polymers, organic cages, composite materials, and nanofibers can be prepared via the versatile ice-templating approach. Our group is also specialized in preparing monodisperse porous silica microspheres for high performance liquid chromatography. Particularly, a one-pot synthesis method was developed to prepare a unique type of spheres-on-sphere silica particles which have shown fast and effective chromatographic separation of both small molecules and large biomolecules. The suspension of such silica microspheres could be processed by ice templating to produce hierarchically silica porous materials with controlled pore morphology. Importantly, by combining ice templating and emulsion templating, hydrophobic organic nanoparticles could be formed *in situ* in interconnected porous polymer. This has proved to be an effective and patentable method to prepare stable aqueous nanoparticle dispersion, which has very interesting applications for poorly water-soluble drug compounds.

To address the solubility and low mechanical stability of the freeze-dried materials, a UV frozen polymerization has been developed to produce crosslinked aligned porous polymers which showed improved stability and chromatographic separation. We also reported for the first time the preparation of aligned macroporous metal-organic framework (MOF) monoliths with intrinsic microporosity.

In conclusion, the ice templating based approaches can be used to prepare a wide range of porous materials or nanostructures which have great potential in supported catalysis, separation, battery electrodes, and pharmaceutical fields.

Synthesis of zeolite fibers by ice-templating and steam-assisted crystallization of hydrogels

H. Tamon, T. Iwaya, H. Mori, N. Sano
Kyoto Univ., Dep. Chemical Engineering, Kyoto 615-8510, Japan; tel.: +81-75-383-2664, e-mail: tamon@cheme.kyoto-u.ac.jp.

Zeolite has been used as a catalyst or an adsorbent in various industries. Mass transfer is one of the most important factors controlling the performance of microporous materials. The short diffusion length is desired to decrease the resistance of mass transfer. Although extremely fine particles of zeolite are useful for the short diffusion length, the pressure drop is extremely large in the case where the particles are used as a packed bed. Hence, the fabrication is important for the effective application of zeolite. Especially, monolith or fiber morphology has recently attracted great attention from its low pressure drop and short diffusion length. However, it is extremely difficult to fabricate zeolite in these morphologies without any binder.

The authors applied ice-templating and steam-assisted crystallization (SAC) to the synthesis of zeolite monolith [1, 2]. Here tetrapropylammonium hydroxide (TPAOH) was used as a structure-directing agent (SDA). In the present work, zeolite fibers are synthesized by ice-templating and steam-assisted crystallization, and their characteristics are discussed.

Silica gel fibers used as the precursor of zeolite fibers were prepared by applying ice-templating to silica hydrogels obtained by sol-gel polymerization of commercial sodium silicate solutions. The silica hydrogels were unidirectionally frozen in a liquid nitrogen bath at a constant rate of 30 cm/h. The hydrogels were freeze-dried at -10°C for 72 h, and then silica gel fibers were obtained. The fibers were subsequently heated at 800°C for 2 h in air atmosphere to give alkali-proof properties to the samples. Aluminum nitrate ($Al(NO_3)_3$) was supported in the precursor by the impregnation method to prepare zeolite fibers. The dry silica gels containing SDA (TPAOH) and $Al(NO_3)_3$ were crystallized by SAC in an autoclave. The samples after crystallization were calcined at 600°C for 4 h in air atmosphere to remove SDA from crystallized samples, and finally zeolite fibers were obtained.

Figure 1 shows SEM images of the precursor silica gel and the sample after crystallization. Here SAC was conducted under the following conditions: the SDA concentration; 3.0 wt%, the SAC temperature; 130°C, the SAC time; 12 h. The fibrous morphology can be kept for the crystallized sample as shown in Figure 1. Figure 1 (c) and (d) show that the particulate crystals are synthesized by SAC and the surface of fiber is densely covered with the synthesized crystals. Figure 1 (d) also indicates that the hollow fibers having inside voids can be obtained by SAC. It is considered that the unique morphology is formed because the precursor is dissolved out by high-pressure steam during SAC.

Figure 1. SEM images of precursor silica gel (a. b) and silicalite fiber after crystallization (c, d).

The XRD patterns of the precursor silica

gel and the samples after crystallization were measured. Only the broad halo centered at around $2\theta = 23^{\circ}$ was observed in the XRD pattern of the precursor silica gel, and no crystalline phase was detected. In the sample after crystallization, sharp peaks were observed at around $2\theta = 8^{\circ}$ and 23°. All of the observed diffraction peaks corresponded to those of MFI-type zeolites. Since the prepared samples contained no aluminum atom, the crystalline phase of the synthesized sample was identified as silicalite-1.

Figure 2 shows the influence of SAC time and SAC temperature on the degree of crystallinity of silicalite fibers. The crystallinity increases with the increase in SAC temperature and SAC time, and reaches the values above 80 %. Here the SDA concentration for impregnation is 3.0 wt%. The influence of SAC time and SDA concentration on the degree of crystallinity has also been certified. The crystallinity increases with the increase in SAC time and SDA concentration. These results imply that silicalite fibers with high crystallinity can be synthesized without any binders by ice-templating and SAC.

Figure 3 shows the relation of crystal size, SDA concentration, and SAC time. It is found that the crystal size decreases with the increase in SDA concentration for arbitrary SAC time and SAC temperature. The data indicated by four-sided figures in Figure 1 indicate that the degree of crystallinity is around 80%. Hence, the silicalite fibers with high crystallinity and small crystal size can be prepared by selecting a proper SDA concentration.

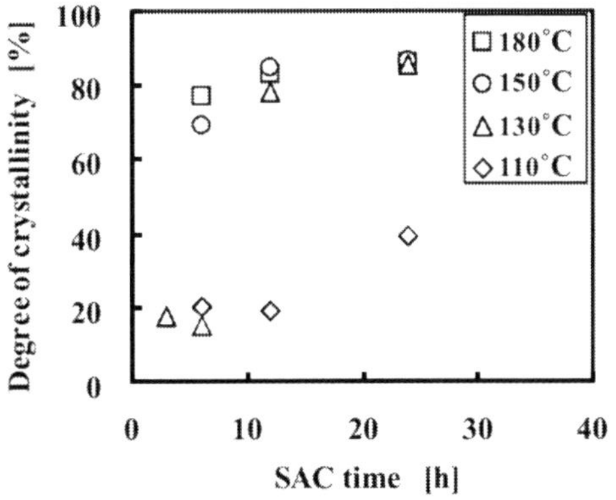

Figure 2. Influence of SAC temperature and SAC time on degree of crystallinity of silicalite fibers (SDA concentration: 3.0 wt%).

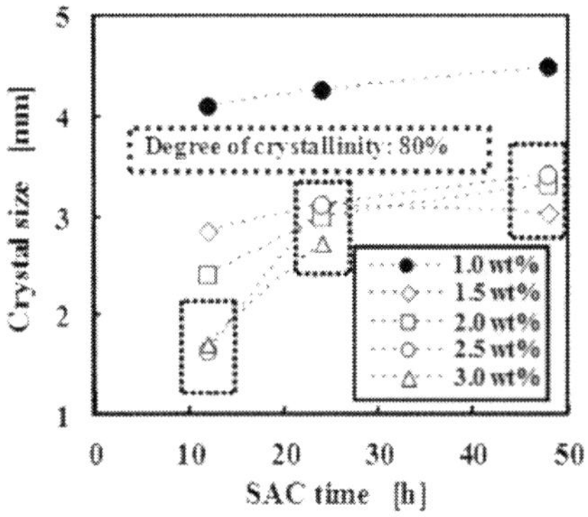

Figure 3. Influence of SAC time and SDA concentration on crystal size of silicalite fibers (SAC temperature: 130°C).

Zeolite fibers were synthesized for Si/Al = 100, the SDA concentration =3.0 wt%, the SAC temperature = 150°C, and the SAC time = 96 h. SEM images of cross section of the fibers showed that hollow fibers having inside voids were obtained by SAC as same as silicalite fibers shown in Figure 1. The surface of fibers was densely covered by particulate zeolite crystals. The XRD pattern suggested that the fibers prepared were MFI-type zeolites.

The existence of aluminium in zeolite was identified by NMR. The ^{27}Al MAS NMR spectrum suggested that the sharp peak of 4-coordinated Al^{3+} was observed at chemical shift of $\delta = 54$ ppm. The result certified that the sample had a framework structure composed of AlO_4 tetrahedra. Therefore, it can be concluded that not only silicalite fibers but also zeolite fibers can be synthesized by ice-templating and SAC.

[1] H. Mori, K. Aotani, N. Sano, H. Tamon: J. Mater. Chem., 21, 5677-5681 (2011).

[2] H. Tamon, T. Akatsuka, H. Mori, N. Sano: Chem. Eng. Trans., 32, 2059-2064 (2013).

Magnesium silicide for energy storage and recycling technology

J. Bumba, P. Dytrych, F. Kaštánek, O. Šolcová
Institute of Chemical Process Fundamentals of the ASCR, v. v. i., Rozvojová 135, 16502, Prague, Czech Republic, fax: +420220920661, tel: +420220390278, bumba@icpf.cas.cz

The silicides as binary compounds of silicon with less electronegative metals can be utilized in many industrial fields owing their mechanical, physical, chemical and thermoelectric properties as a function of their composition and structure. For example magnesium silicide can be used for energy storage. One of many routes for preparation of magnesium silicide involves mixing of powdered components and their heating under various conditions; e.g. a thermal treatment under vacuum or inert gas. As a possible source of raw material, silicon obtained by reutilization of photovoltaic panels and magnesium scraps could be used. Prepared magnesium silicide is then hydrolysed with diluted phosphoric acid to produce silanes and magnesium phosphate as a by-product. Thermal decomposition of silanes leads to hydrogen and silicon with a ultra-high purity. Formed magnesium phosphate can be with advantage utilized as a very effective fertilizer.

A series of magnesium silicide samples using starting materials with a various level of surface oxidation were prepared. The reaction was carried out under various conditions, e. g. temperature, reaction time, and gaseous environment. The prepared samples were characterized by a scanning electron microscopy, an energy dispersive X-ray spectroscopy, a photoelectron and Raman spectroscopy. The samples were then hydrolysed with a series of dilute acids solutions. The resulting gaseous products were characterized by a gas chromatography-mass spectrometry. The prepared samples of magnesium silicide and the products of hydrolysis were compared with the commercial standard. Silane, as one of the hydrolysis products, was thermally decomposed in a reaction cell using platinum wire and its loss and the formation of nano-silicon was observed.

The aim of this study is to optimize the reaction conditions that would ensure efficient and economical process. This could lead up to new design for photovoltaic panel recycling technologies, hydrogen production and storage.

The financial support of the Grant Agency of the Czech Republic, projects No.: GA15-14228S, is gratefully acknowledged.

Developing of a preparative chromatography process for resolving enantiomers of chiral anaesthetics using modified porous glass beads

T. Munkelt[1], I. Mutavdzin[1], C. Hamel[1], A. Seidel-Morgenstern[1], C. Küster[2], D. Enke[2], 1 Inst. of Chemical Engineering, Otto-von-Guericke Univ., Universitätsplatz 2, 39106 Magdeburg, Germany, tel.: +49391 6754653, email: thomas.munkelt@ovgu.de; 2 Inst. of Chemical Technology[2] Univ. of Leipzig, Leipzig, Germany

For maintaining general anaesthesia during the clinical surgeries volatile chiral halogenated ethers are used. These chiral substances are currently applied as racemic mixtures (50:50 mixtures of two enantiomers) although several studies indicated specific differences in the anaesthetic potential and side actions to human body [1]. Hence, there is interest in developing a separation process for chiral anaesthetics.

To achieve the objective of separating chiral anaesthetics the model substances isoflurane and desflurane were investigated. Tailor-made glass beads were used as the carrier material for a modified enantioselective γ-cyclodextrine as reported in [2, 3]. Two different ways for immobilising this selector, via chemical bonding and coating onto the surface with polysiloxane, were studied.

Porous glass beads were chosen as carrier material for the gas chromatographic separation because of their unique control regarding texture properties, particle size distribution and geometrical shape [4]. The texture and chemical properties of the glass beads were optimized based on the results of standard physical (measurements mercury intrusion, nitrogen adsorption) and classical dynamic chromatographic experiments using unmodified and modified porous glass beads.

Successful separation of the two inhalation anaesthetics was achieved with the prepared enantioselective glass beads. The chromatographic parameters were optimized to improve the resolution potential of the chromatographic column. Afterwards data were gained for parameterising the multi-Langmuir isothermal equations, as a basis for optimization and scale up of the preparative separation processes. Aiming to produce 1 g of pure enantiomers per day in the laboratory, the results of the optimization were used to calculate the dimensions of the column needed exploiting simple scale up rules. In an intermediate experimental step these scale rules proved to be viable.

Currently capture columns are designed, which collect the different fractions after the chromatographic separation, the two anaesthetic enantiomers and an unresolved mixture. When reaching the maximum capacity of these capture columns, they will be regenerated by heating and pressure reduction respectively.

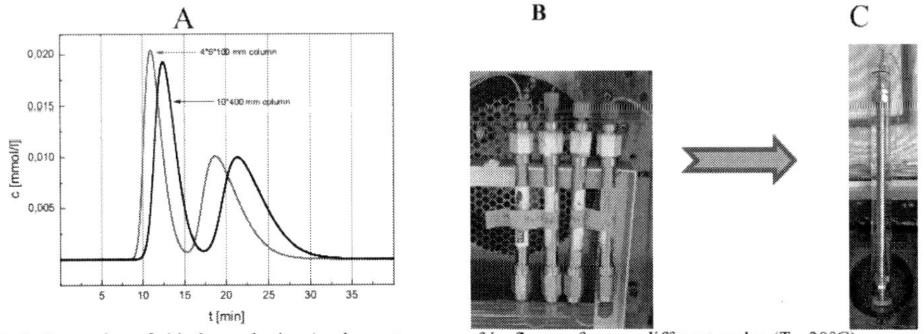

Fig 1: Separation of chiral anesthetics **A:** chromatograms of isoflurane for two different scales (T= 20°C) corresponding to columns shown in **B** (4 connected columns of di= 6 mm and l= 100 mm) **C** (di= 10 mm, l= 400 mm)

[1] **H. A.** Haeberle, H. G. Wahl, G. Aigner, K. Unertl, H.-J. Dieterich, Eur. J. of Anaesthesiology 21 (2004) 144.
[2] W. A. König, R. Krebber, P. Mischnick, Journal of High Resolution Chromatography,12 (1989) 732.
[3] Schurig, Salient Features of Enentioselective Gas Chromatography: The Enantiomeric Differentietin of Chriral Inhalation Anasthetics as a Representative Methodological Case in Point, Springer , 2013
[4] A. Inayat, B. Reinhardt, H. Uhlig, W.-D. Einicke, D. Enke, Chem. Soc. Rev. 42 (2013) 3753.

Liquid-crystalline organic-inorganic hybrid dendrimer with a CdS nano-core: The self-organized structure-dependent photoluminescence behavior

K. Kanie,[1,*] M. Matsubara,[1] W. Stevenson,[2] Y. Yamazaki, J. Yabuki,[1] K. Kojima,[1] M. Nakaya,[1] S. F. Chichibu,[1] A. Muramatsu,[1] X. Zeng,[2] G. Ungar[2,3]

[1]Institute of Multidisciplinary Research for Adv. Mater., Tohoku Univ., 2-1-1, Katahira, Aoba-ku, Sendai 980-8577, Tel. +81-22-217-5165, e-mail: kanie@tagen.tohoku.ac.jp. [2]Dept. of Materials Science and Engineering, Univ. of Sheffield, Sheffield S1 3JD, UK, [3]Dept. of Physics, Zhejiang Sci-Tech Univ., Xiasha College Park, 310018 Hangzhou, China

Semiconductor quantum dots (QDs) have attracted a great deal of attention in material science due to their size-dependent and tuneable absorption and emission properties. In addition, periodic arrays of QDs have a considerable potential in applications in novel nano-electronic and optical devices. Recently, we developed liquid-crystalline (LC) organic-inorganic hybrid dendrimers,[1] where the dendrimer has a spherical gold nanoparticle (NP) at its core. Dense surface modification of spherical gold NPs with LC **G2** dendron molecules was found to result in a hybrid exhibiting a thermotropic LC phase with simple cubic structure.

The chemical structure of **G2** is shown in Figure 1. In the present study we focus on imparting such ability to self-organize onto monodisperse QD nanospheres. As a result, dendron-modified QDs were obtained, displaying ordered superlattices. Such structures were found to exhibit unusual photoluminescence (PL) behaviour.

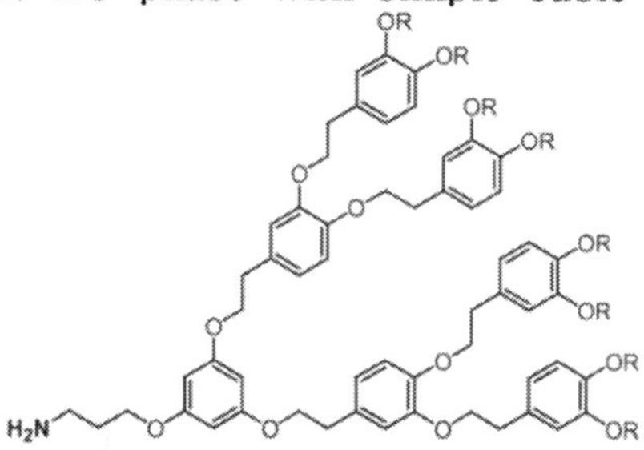

Figure 1. A chemical structure of dendron **G2.** R = $C_{12}H_{25}$-.

As the internal core of the dendrimer, CdS NPs **C1-C3** with different degrees of CO_2H modification amount was prepared. Surface modification of **C1-C3** by **G2** was carried out by amidation reaction. Average particle size of core part of **C1**, **C2**, and **C3** were calculated as 3.9, 3.9, and 3.8 nm, respectively. Figure 2 shows TEM images of **C1** NPs and **G2** modified **C1** (**G2/C1**). The interparticle distance of **G2/C1** expanded into *ca.* 9 nm after modification with **G2**. The numbers of **G2** molecules per one particle for **G2/C1**, **G2/C2**, and **G2/C3** were assigned as 69, 54, and 47 molecules, respectively, from TG analysis. SAXS and DSC measurements showed that **G2/C1** exhibited thermotropic LC phases in a wide range of temperatures. At 150 °C, **G2/C1** formed a thermotropic cubic LC phase with a novel structure. The cubic structure was retained at room temperature after cooling. On the other hand, only random structure was seen for **G2/C2** and **G2/C3**. The **G2/C1** dendrimer as deposited on a glass substrate, also formed a disordered structure . In such a state, it showed strong photoluminescence when UV irradiated at 365 nm (Figure 3a).

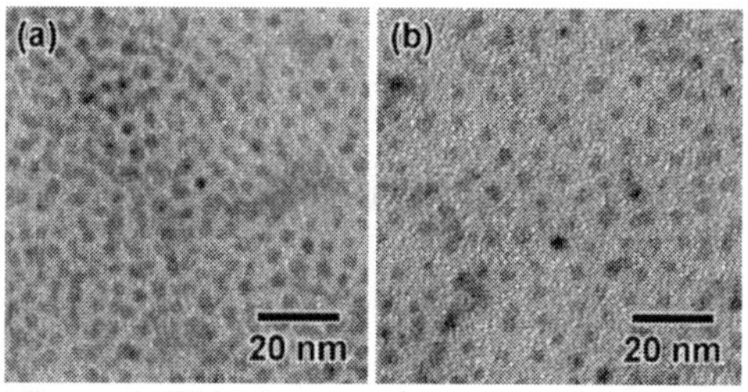

Figure 2. TEM images of CO_2H-modified CdS NPs and **G2**-modified CdS NPs: (a) **C1**; (b) **G2/C1**.

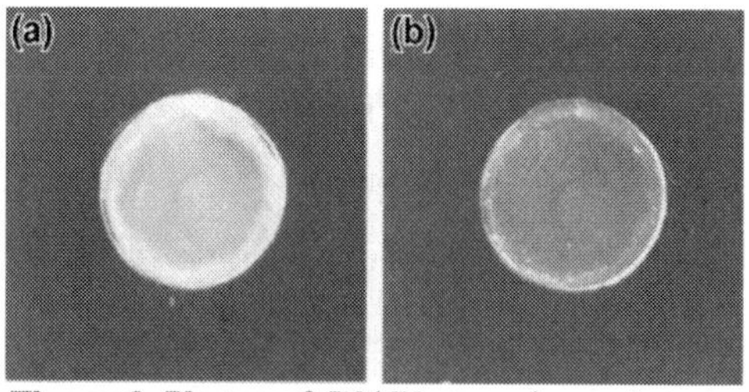

Figure 3. Photos of **G2/C1** on a glass substrate under UV irradiation (365 nm) at r.t.: (a) Before annealing; (b) After annealing at 150 °C and cooled to r.t.

However the PL was quenched when the **G2/C1**
dendrimer
self-organized in the cubic phase after annealing at 150 °C followed by cooling (Figure 3b). Such PL quenching behaviour was totally reversible, and appears to be derived from the periodic structure of **G2/C1**. The schematic illustration of our present study is shown in Figure 4. The detailed analysis on PL quenching by femto-second laser is now in progress. Such unusual PL behavior might be a powerful tool to develop future functional devices.

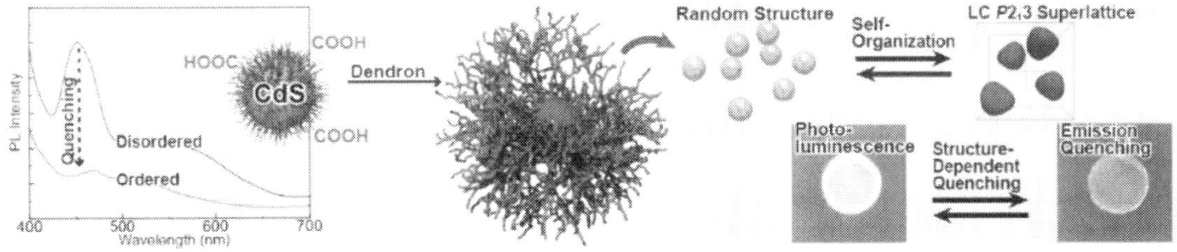

Figure 4. A schematic illustration of structure-dependent photoluminesence behavior of self-organizing organic-inorganic hybrid dendrimer with a CdS nano-core.

Reference

1) K. Kanie, M. Matsubara, X. Zeng, F. Liu, G. Ungar, H. Nakamura, A. Muramatsu, *J. Am. Chem. Soc.* **134**, 808 (2012).

Hydrophilic modification of porous expanded perlite particles by plasma polymerization for soilless agriculture

M. Karaman, M. Gürsoy
Selcuk University, Department of Chemical Engineering, Kampus, 42030, Konya, Turkey
tel. +90 332 223 2108, e-mail: karamanm@selcuk.edu.tr; fax: +90 332 241 0635.

Climate change and population growth cause the scarcity of arable land and water. Therefore, recently the ways of growing more plants in less space with the less resource-consumption have been investigated. Soilless agriculture, namely hydroponic agriculture is one of these methods. Hydroponic agriculture differs from traditional agriculture in that, the plants are produced in growing media in place of soil. Expanded perlite has become an important "soil alternative" material because of its superior features. Expanded perlite absorbs the nutrients and water, and allowing the plants to use them as needed. In addition to water, roots need oxygen for plant growth. Due to its porous structure, expanded perlite has excellent oxygen retention. Furthermore, expanded perlite is chemically inert, non-toxic, non-water-soluble, easy to obtain and reusable.

The main objective of this study is to make expanded perlite surfaces more hydrophilic through an all-dry encapsulation process, namely rotating-bed plasma enhanced chemical vapor deposition (rPECVD) method. An 13.56 MHz radio frequency plasma discharge was employed to deposit thin poly(2-hydroxyethyl methacrylate) (PHEMA) polymeric films on expanded perlite surfaces . A rotating plasma reactor was used to provide effective agitation and mixing of the particles during the depositions. Chemical and morphological properties of uncoated and coated expanded perlites were determined by FTIR, XPS, SEM, and contact angle measurements. Based on results of this study, it can be concluded that rotating-bed PECVD method can be used effectively for conformal coatings around porous particles having complex geometries like expanded perlite. PECVD avoided many problems associated with existing liquid-based particle coating techniques such as particle agglomeration, toxic solvents, and poor uniformity. The important plasma deposition parameters, specifically plasma power and operation mode (continuous vs. pulsed) were found to be effective in providing control of the surface hydrophilicity and water retention of expanded perlite particles. The water holding capacity of expanded perlite was increased from 70% up to over 80% with PHEMA coating. These features can widen the area of use of expanded perlite especially in agriculture. The method of this study can be extended to functionalize many other types of particles having dimensions down to micro and even nano scale.

Removal of nickel and cadmium from aqueous solution using Chitosan/ *Chlorella vulgaris* nanocomposite

Mehrzad Dadban[1], Omid Tavakoli[2], Mohammad Ali Faramarzi[3]
[1]School of Chemical Engineering, College of Engineering, University of Tehran, 16 Azar St., Tehran 14176, Iran; tel. +989122971285, e-mail: m.dadban@ut.ac.ir; [2]School of Chemical Engineering, College of Engineering, University of Tehran, 16 Azar St., Tehran 14176, Iran; tel. +982166498984, e-mail: otavakoli@ut.ac.ir; [3]Pharmaceutical Biotechnology Laboratory, Faculty of Pharmacy, Tehran University of Medical Sciences, Tehran 14176, Iran; tel. +982161112187, e-mail: ffaramarzi@yahoo.com

Removal of heavy metals from wastewater is a very important action with respect to human's health and environmental impacts. Adsorption method can be considered as an effective and widely used process due to its simplicity, moderate operational conditions and economical feasibility. In this study a nanofiber was prepared by the electrospinnig of chitosan and then *Chlorella vulgaris (C. vulgaris)* microalgae's cells was immobilized on chitosan electrospun nanofiber mat during *C. vulgaris* cultivation. The potential of the prepared bio-nanocomposite for adsorption of nickel (Ni) and cadmium (Cd) from aqueous solution was investigated. The adsorbent was characterized by Scanning Electron Microscopy (SEM) and Fourier Transform Infrared (FTIR). The influence of variable parameters such as adsorbent dose, pH, contact time, initial concentration of nickel (Ni), cadmium (Cd) ions and temperature were studied in batch experiments.. The kinetic data of metal ions were analyzed by pseudo-first-order, pseudo-second-order kinetic models. Three isotherm models of Freundlich, Langmuir and Dubinin-Radushkevich were applied to describe the equilibrium data of metal ions. Based on the obtained results the adsorbent showed remarkable sorption capacities of both heavy metal ions. The sorption capacity was in order of Ni>Cd. The calculated thermodynamic parameters (ΔG°, ΔH° and ΔS°) showed that the nature of both metal ions sorption onto the adsorbent was endothermic and spontaneous.

Keywords: Chitosan, *Chlorella vulgaris*, Nanofiber, Heavy metal, Adsorption

Numerical simulation of flow and convective diffusion in porous media with pore variables by a REV-scale lattice Boltzmann method

Yumei Yong, [1]Xiaodan Zhang, Weicheng Xu, Chao Yang
Key Laboratory of Green Process and Engineering, Institute of Process Engineering, Chinese Academy of Sciences, Beijing 100190, China; tel: +86 010 82544928, email: ymyong@ipe.ac.cn; [1]School of chemical & environmental engineering, China University of mining & technology, Beijing 100083,China; tel: 86 010-82544929, email: xdzhang@ipe.ac.cn

Based on the lattice Boltzmann method at representative elementary volume(REV) scale, in this paper the flow and convective diffusion processes in porous media coupling with convection diffusion equation are numerically simulated. A two-dimensional nine-speed incompressible lattice Boltzmann model D2G9 is built up and couples with a double-distribution-function of two-dimensional D2Q5 to simulate the convective diffusion process. According to the geometry variables, porosity and permeability which are from numerical results at pore scale, different porous media structure types, variables of the fracture and pores are introduced into the REV mathematical models in order to analyse and explore how the porous structure and typical geometric variables influence the characteristics of flow and convective diffusion in porous media. Numerical results show that the responding convective diffusion coefficient in porous media packed by the irregular granular is larger than that in the tubular porous media because the medium resistance in porous media packed by the irregular granular is smaller than that in the tubular porous media (Fig.1). The convective diffusion coefficient in porous media packed by the irregular granular increases with the grain diameter. The fractures parallel to the flow direction intensify the convection diffusion process but the fractures perpendicular to the flow direction contribute little to the convection diffusion process. Pores are so little that they have no influence on convective diffusion coefficients.

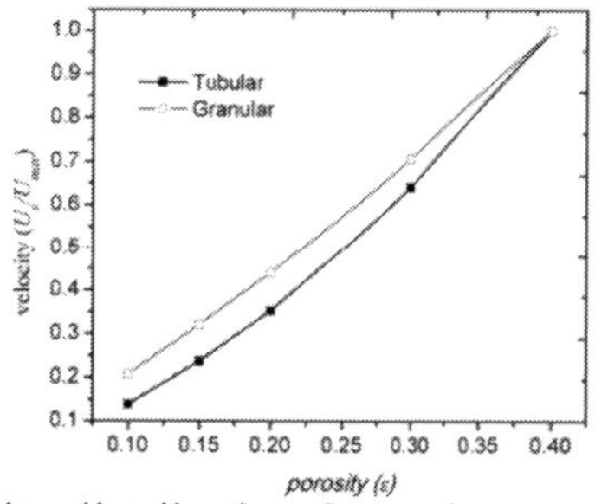

a) Velocity distribution for various porosity

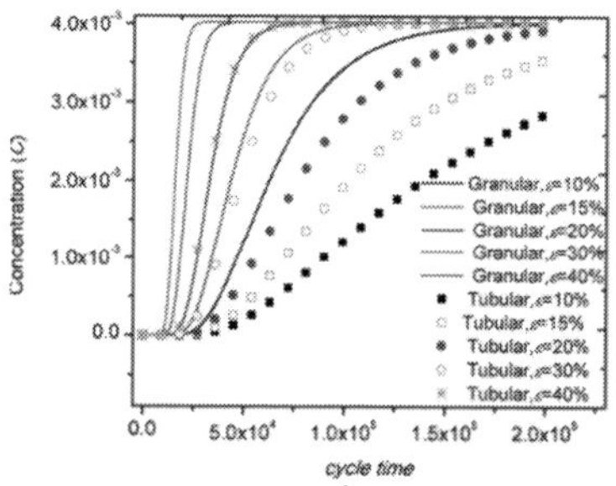

b) Average concentration versus cycle time

Fig.1 Flow and convective diffusion rules in the granular and tubular porous media.

Investigation of physical conditions on thermally synthesized carbon nanoparticles from *Nerium oleander* extracts

S. Simsek[1], M.O. Alas[2], B. Ozbek[1], R. Genc[2]
[1]Yıldız Technical University, Chemical Engineering Department, 34210, Esenler, İstanbul; tel. +90 212 383 4755, e-mail: ssimsek@yildiz.edu.tr; tel. +90 212 383 4747, e-mail: bozbek@yildiz.edu.tr; [2]Mersin University, Chemical Engineering Department, 33343, Yenisehir, Mersin; tel. +90 324 361 00 01/7393, e-mail: rukangnc@gmail.com; e-mail: melis_alas@hotmail.com

Nanoparticles have been used as carriers for drug delivery systems because of their designing potential such as target-specific, non-toxic, biodegradable, high drug loading capacity and etc. Based on these processable features, especially the nanoparticles originated from natural products are so sensational for cancer therapy and diagnosis.

Nerium oleander (Oleander), a member of *Apocynaceae* family, is an evergreen medium sized shrub to 6 m high with very narrowly elliptic green leaves reaching to 20 cm [1, 2]. Its biological preparations generally show anticancer, antimicrobial, anti-inflammatory and neuro-protective activities [2, 3]. Because of Oleander's anticancer activity, it is thought that the carbon nanoparticles originated from *N. oleander* extracts can be potential drug carrier systems for cancer treatment.

Therefore, in the present study, it is aimed to synthesize carbon nanoparticles from leaf extracts of *Nerium oleander* by using thermal synthesize method. Within the scope of this consideration, primarily, ethanol and water extracts of *N. oleander* dried leaves were prepared by ultrasonic extraction. Using these extracts, the formulations were prepared with or without polyethylene glycol. The thermal synthesize of *N. oleander* originated carbon nanoparticles were carried out at 250°C for 45 minutes in an oven [4]. By centrifugation, large particles were removed and clear solutions were obtained. Then, for the characterization of synthesized carbon nanoparticles, the studies were performed by UV-Vis spectrophotometry, fluorescence spectrophotometry, Fourier transform infrared spectroscopy, zeta particle size and zeta potential analysis.

Following the studies above, the effects of physical conditions on the particle characteristics were examined since the particle feature has a key role for drug delivery and controlled release, especially, for cancer treatment. Particle size and/or charge can change the pathways of cellular ingestion, the blood circulation time and intratumoral distribution of the molecules [5]. Therefore, size, fluorescent luminescence, surface charge of the particles were investigated at various physical conditions such as temperature, pH and ionic charge.

References:

1. Davis P.H., 1978, *Edingburgh University Press*, 6th ed.:158-159.
2. Zibbu G., Batra A., 2010, *J. Chem. Pharm. Res.*, 2(6):351-358.
3. Sharma P., Choudhary A. S., Parashar P., Sharma M. C., Dobhal M. P., 2010, *Chemistry&Biodiversity*, *7*:1198-1207.
4. Mukherjee P., Misra K. S., Grkya M. C., Chang H.-H., Tiwari S., Wilson W. L., Scott J. W., Bhargava R., Pan D., 2015, *Small*, 11(36):4691-4703.
5. Bae Y. H., Park K., 2011, *Journal of Controlled Release*, 153:198-205.

A promising pathway to fabricate microporous mixed matrix (ZeoTIPS) membranes of ideal structure for gas separations

[1]Meisam Dabiri Havigh, [1,2]Sobhan Neyrizi, [1]Reza Yegani
[1]Faculty of Chemical Engineering, Membrane Technology Research Center, Sahand University of Technology, P. O. Box: 51335-1996, Tabriz, Iran. [2]MESA+ Institute for Nanotechnology, Faculty of Science and Technology, University of Twente, P.O.Box 217, 7500 AE, The Netherlands, E-mail: s.neyrizi@student.utwente.nl

Microporous mixed matrix (ZeoTIPS) membranes, comprising highly selective zeolite particles suspended in a cellular and microporous polymer matrix, offer an attractive route to the development of high performance and efficiency membranes required for refined size-based separations. These membranes which are typically synthesized by thermally induced phase separation method, in an ideal state, combine the separation capability of zeolite films with the processability of polymer matrix, while yielding high permeability of microporous membranes. The development of ZeoTIPS structure,however, has not been without its challenges. Non-ideal morphologies at the interface between the polymer matrix and dispersed phase, which typically appear as undesirable voids or varying degrees of polymer coating, are major hurdles impeding the realization of microporous mixed matrix paradigm .

The work presented here focuses on defining procedures and parameters to surmount these challenges and form a microporous mixed matrix membrane displaying ideal interface morphology.Polymer-zeolite adhesion (interface morphology) is controlled through engineered selection of membrane materials. Liquid-solid interaction parameter (ε^{o}) for polymer-zeolite pair is compared with that of polymer-diluent pair to achieve Relative affinity parameter R_{AP} (equation 1), which can be used to predict the interaction strength between polymer/diluent and 4A zeolite. As shown in Fig.1, R_{AP} values for different polymer-diluent systems are associated with varying degrees of polymer coating. From SEM images it is clear that for a particular zeolite there should exist a point where polymer-zeolite affinity transitions from weak to strong as R_{AP} value is increased. This transition point is where moderate polymer-zeolite affinity can hopefully be achieved to form ZeoTIPS membranes of ideal structure.

$$R_{AP} = \frac{\varepsilon^{o}{}_{P-Z}}{\varepsilon^{o}{}_{D-Z}} = \frac{-0.519 + 0.043\ \delta t_{polymer}\left(\frac{1}{MPa^{\frac{1}{2}}}\right)}{-0.519 + 0.043\ \delta t_{Diluent}\left(\frac{1}{MPa^{\frac{1}{2}}}\right)} \quad (1)$$

δt: Hildebrand solubility parameter

ε^{o}: Liquid-solid interaction parameter

R_{AP}: Relative affinity parameter

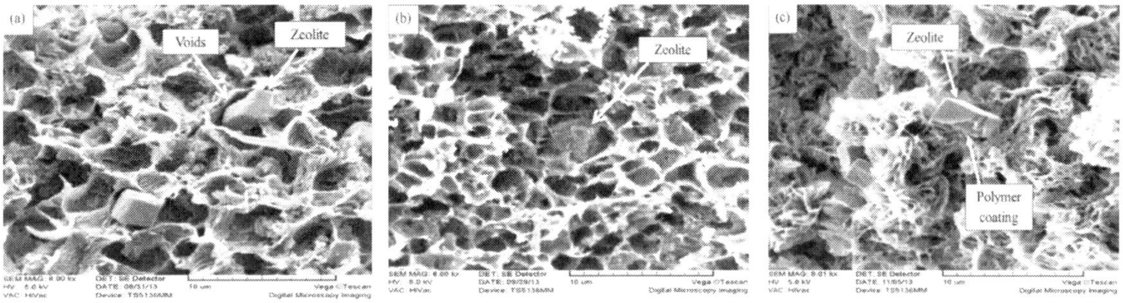

Figure.1. Microporous mixed matrix membranes with different types of interface morphology (a) High density polyethylene-Diphenylether-4A Zeolite (R_{AP}=0.58) (b) Isotactic polypropylene-Diphenylether-4A Zeolite (R_{AP}=0.77) (c) High density polyethylene -Mineral Oil-4A Zeolite (R_{AP}=1.44).

Four types of polymer- diluent systems including High density polyethylene (HDPE) - Diphenyl ether (DPE), Isotactic polypropylene (iPP) - Diphenyl ether (DPE), High density polyethylene (HDPE) - Mineral oil and Poly(methyl methacrylate) (PMMA)-Cyclohexanol incorporated with 18 wt.% 4A zeolite are examined through CO_2 and CH_4 single gas permeation tests (Table.1, results are shown for the first three systems). Gas permeation results coupled with SEM images are used to investigate that range of R_{AP} values over which moderate polymer-zeolite affinity can occur.The results suggest that for 4A zeolite the transition point between weak and strong polymer-zeolite affinities should occur at R_{AP} values ranging from ~ 0.6 to 0.85. Therefore, to prepare an ideal ZeoTIPS membrane with 4A zeolite the appropriate polymer-diluent pair should be selected over this range.

Table.1. Pure gas performance of ZeoTIPS membranes incorporated with 4A zeolite with different R_{AP} values at ambient temperature and 2bar.

Membrane	R_{AP}	$P_{CO2}\times10^{-3}$	$P_{CH4}\times10^{-3}$	P_{CO2}/P_{CH4}
HDPE-DPE	0.58	642.9	43.41	14.81
iPP-DPE	0.77	614.7	22.53	27.28
HDPE-Mineral oil	1.44	472.9	19.83	23.85

Combination of various nanomaterials for effective water purification

M. Morozova, P. Krystynik, J. Bumba, O. Solcova
Institute of Chemical Process Fundamentals of the CAS, v. v. i., Rozvojova 2/135, 165 02 Prague 6 – Suchdol, Czech Republic
tel. +420 220 390 282, e-mail: morozova@icpf.cas.cz

Water protection is one of the most important tasks of the environment and also the most challenging areas. The quality of surface water is affected by a series of contaminated sources; primarily by industry. Industrial waste water is generally contaminated by various organic and inorganic more or less toxic compounds. The important indicators of water contamination are pH, temperature, chemical oxygen demand, suspended solids, ammonia nitrogen, total nitrogen, phosphorus, cyanides, hydrocarbons, heavy metals etc. Total removal of these pollutants is not easy and it is necessary to combine the advanced oxidation techniques with the classical separation processes. Waste water can be cleaned by various chemical, physico-chemical and biological methods. Nevertheless, these techniques are not able to clean waste water sufficiently. There are plenty of water-soluble substances, based on hormones, which cannot be removed by conventional techniques. Thus, the final purification is an urgent task, nowadays. This work is focused on preparation and characterization of various nanomaterials applied in sorption, oxidation, electrocoagulation and photocatalysis techniques. The main aim is the study of their effective combination and activity testing in purification processes.

The financial support of the Technology Grant Agency of the Czech Republic, projects No.: TA04020700, is gratefully acknowledged.

Diffusiophoresis of an isolated, charged toroidal polyelectrolyte

Y. R. Hsu,[1] J P. Hsu,[1] S. Tseng[2]
Natl. Taiwan Univ., Dept. Chem. Eng., 10617 Taipei, Taiwan; tel. +886 2 33663022, e-mail: jphsu@ntu.edu.tw;[1] Tamkang Univ., Dept. Math., 25137 Tamsui, New Taipei City, Taiwan fax: +886 2 26209916, e-mail: topology@mail.tku.edu.tw[2]

Diffusiophoresis, a solute concentration gradient driven phenomenon, has been applied in many areas of practical significance including, for example, separation, coating, purification, and scavenging. It also plays a role in modern nanotechnology based applications such as DNA sensing, drawing the attention of researchers in various fields, recently. Since particles are driven by a solute concentration gradient in diffusiophoresis, the Joule heating effect in the applied electric field driven case (i.e., electrophoresis) is absent, and there is no particle-electrode contact. However, the mechanisms involved in diffusiophoresis can be more complicated than that in electrophoresis. For a charged particle in a salt solution, these include chemiphoresis, where particles are driven by an unbalanced ionic distribution and double layer polarization (DLP), and electrophoresis, where they are driven by a local electric field arising from the difference in ionic diffusivities. Due to DLP, the particle behavior is also influenced hydrodynamically by the flow of the ion-rich fluid inside (outside) the double layer towards the low (high) salt concentration direction (i.e., osmotic flow).

Toroidal shaped entities are not uncommon in biological science. For instance, bacterial DNA, plasmid DNA of eukaryotic cells, and most viral DNA can all be isolated as a single intact toroidal chromosome. While free DNA in solution is of linear shape, the DNA in living cells is highly condensed into toroidal arrays, especially in sperm cells and viruses where gene transcription is inactive. *Escherichia* coli also has a toroidal structure with a hole large enough to encircle double-stranded DNA. Anabaenopsis is another example of toroidal biocolloids.

Often, biocolloids such as DNAs, RNAs, and proteins are simulated by polyelectrolyte (PE), a porous, high molecular weight entity bearing dissociable functional groups, the dissociation of which in an aqueous medium yields fixed charge. Although the behavior of a PE in an applied field has been studied extensively, the diffusiophoresis of a toroidal PE has not been considered so far. Considering the potential applications of diffusiophoresis in modern nanotechnology and biotechnology, extending previous electrokinetic analyses to the case of toroidal PE diffusiophoresis is highly desirable and necessary. This is done in the present study.

Ionic transport in a polyelectrolyte-modified conical nanopore

H. H. Wu,[1] C. Y. Lin,[1] J P. Hsu,[1] S. Tseng[2]
Natl. Taiwan Univ., Dept. Chem. Eng., 10617 Taipei, Taiwan; tel. +886 2 33663022, e-mail: jphsu@ntu.edu.tw;[1] Tamkang Univ., Dept. Math., 25137 Tamsui, New Taipei City, Taiwan fax: +886 2 26209916, e-mail: topology@mail.tku.edu.tw[2]

Due to its potential in mimicking ion channels in living organisms, the ionic transport in artificial nanopores has drawn the attention of researchers in various fields. Recent advancement in nanopore fabrication technology also makes relevant applications versatile. As the size of a nanochannel is down to the order of Debye length (or thickness of electric double layer, EDL), the overlapping of EDLs yields distinctive electrokinetic phenomena, including, for example, ion concentration polarization (ICP), ion selectivity, and ionic current rectification (ICR). ICR is the phenomenon that a nanopore shows a diode-like current-voltage behavior, or the preference for ionic current in a certain direction when an electric potential is applied. For charged nanopores, ICR might arise from several reasons, such as asymmetric pore geometry, imposed salt gradient, and bipolar surface properties.

Polyelectrolyte (PE) is often used to modify the surface of a solid-state nanopore, rendering it to have versatile characteristic and, therefore, ICR nature. The coating of materials on a solid-state nanopore also provides a way to estimate the pKa of the coating materials through experimentally measured pH-ionic current relationship. In this study, we analyze the ICR in a PE-modified conical nanopore, focusing on the influence of the way that a nanopore is modified. A complete set of electrokinetic equations is solved, comprising Poisson-Nernst-Planck (PNP) equations for ionic transport and electric field, and modified Stokes-Brinkman equations for flow field. The behaviors of the system considered under various conditions are simulated by varying applied voltage, bulk salt concentration, and the thickness of the PE layer.

New carbon-containing materials such as electrodes for electrochemical processes

A.A. Atchabarova, M.S. Lepikhin, R.R. Tokpayev, A.T. Kabulov, M.K. Nauryzbayev
The Center of Physical-Chemical Methods of Research and Analysis (CPCMRA)
al-Farabi Kazakh National University, 96a Tole bi street, 050012, Almaty, Kazakhstan
e-mail: azhar05@mail.ru

The composite electrodes have advantages against the electrode composed of one conductive phase. And that's why they are widely used and their manufacturing methods are intensively developing. The advantages of composite electrodes are: lower cost, lower weight, varieties of shapes and designs of the electrodes, higher signal / noise ratio, the possibility of volume or surface modification as the conductive and insulating phases. The volume and surface modification opens the great perspectives for creating electrodes with the given electrochemical properties, including microelectrode ensembles [1].

Apricot pits, walnut shells which are plant raw materials and shungite concentrate which is carbon-mineral raw material from "Bakyrchik" deposit of East Kazakhstan were used as a material to obtain carbon-containing electrodes.

The flotation process was carried out to stabilize the structure of shungite materials. The concentrate with carbon content 40 ± 2% by weight was obtained [2].

The thermic carbonization at the temperature 800 ° C (in an inert atmosphere of argon) and activation with superheated steam at the temperature of 850-950 ° C were carried out to increase the specific surface area and mechanical strength of the materials. Increasing the specific surface area happens because of the burnout of unstructured carbon, organic compounds and the opening of "blocked" pores. The studied materials were obtained with the following parameters:

- the sample based on shungite concentrate - specific surface area is 153.36 m^2/g; mechanical strength is 53%;
- the sample based on apricot pits - specific surface area is 505.68 m^2/g; mechanical strength is 91%;
- the sample based on walnut shell - specific surface area is 423.45 m^2/g; mechanical strength is 86%.

The electrodes were obtained based on the given materials. The electrodes were obtained by the method of spreading paste from the test material and PVDF (Polyvinylidene fluoride) on copper carrying base.

The intercalation capacity is the most important characteristic of the electrodes. The intercalation capacity is quantity of electricity that is generated at the electrode when full charged, per unit mass.

The results were obtained on Neware battery testing systems. The results showed that the average value of the discharge and charge capacities of samples based on carbonizate of walnut shell exceed significantly the capacity of samples based on shungite concentrate (Q charge - 31.4 mA · h/g; Q discharge - 2.1 mA · h/g) and carbonizate apricot pits (Q charge - 6.5 mA · h/g; Q discharge - 0.3 mA · h/g). It may be connected with the structure of walnut shell. The initial charge and discharge capacities of this sample are equal to 566.0 mA · h/g and 18.9 mA · h/g respectively.

The given results let us to make a conclusion that the samples based on carbonizate of walnut shell can be used as the main material in the further experiments for sorption of metals (Au^+, Ag^+, Ca^{2+}, Mg^{2+}, etc.).

References

1 G.N. Noskov Solid carbon-containing composite electrodes to determine the elements by voltammetric methods / author's abstract of the Doctor of Chemistry: 02.00.02. - Tomsk, 2012.

2 S.V. Nechipurenko, V.N. Duhnitsky, S.A. Yefremov, M.K. Nauryzbayev Production of carbon-mineral sorbents based on shungite / Sorbents as a factor of life quality and health: Materials of the 2nd All- Russian Scientific Conference. - Belgorod, 2006. - P. 170-173.

Preparation of dextran modified magnetic nanoparticles

Katja Vasić, Željko Knez, Maja Leitgeb
University of Maribor, Faculty of Chemistry and Chemical Engineering, Laboratory for separation processes and product design, Smetanova 17, SI-2000 Maribor, Slovenia; tel. +386 2294 462, e-mail: maja.leitgeb@um.si.

Advantages and applicability of magnetic nanoparticles has opened up a large scope of possibilities in biomedicine and biotechnology. Comparing to larger sized particles, the main advantage of magnetic nanoparticles is in their high surface-to-volume ratio, which contributes to a higher surface energy and to excellent magnetic properties.

Magnetic nanoparticles are typically crystals of inorganic elements for which the largest characteristic dimension is approx. 1-100 nm. Their inorganic magnetic core is usually surrounded by layers of functional coatings. Magnetic nanoparticles are therefore used as carriers to which different active substances can bind. They also have a high coercivity and high magnetic susceptibility and are often bound together in aggregates, which gives them many new possibilities in properties and functionalization.

Dextran is a biocompatible hydrophilic polymer, made of many glucose molecules consisting of mostly α-1, 6-glycosidic linkages, which in alkaline solutions are physically absorbed with non-covalent interactions onto magnetic nanoparticles. Hydroxylic groups can therefore easily be cross-linked to dextran and also be functionalised with primary amines to bind to targeted ligands and peptides. Dextran is also used in medicine as an antithrombotic agent to reduce blood viscosity and as a volume expander in hypovolaemia.

Maghemite nanoparticles were prepared with co-precipitation of iron (II) and iron (III) ions. Dextran was dissolved in distilled water at different concentrations (1 mg/mL, 2 mg/mL, 3 mg/mL). Afterwards, dissolved dextran was added to the solution of iron (II) and iron (III) ions, which was heated to 60°C and mechanically stirred for 20 min. After stirring, the pH of the solution was adjusted to 12, using ammonium hydroxide. Later on, dextran coated nanoparticles were functionalised with oleic or citric acid in order to control particle size and to prevent agglomeration of the nanoparticles. SEM analysis of prepared dextran coated nanoparticles was performed. Functionalized dextran magnetic nanoparticles prepared as described before are being further researched and studied using different cross-linking agents, including epichlorohydrin and glutaraldehyde.

Preparation of thermoresponsive Pickering emulsions formulated with gold nanoparticles by biopolymer addition

S. Onishi[1], K. Naoe[1]*, and M. Imai[2]

[1]Advanced Chem. Eng. Course, Faculty of Advanced Eng., National Institute of Technology, Nara College, Yamato-Koriyama, Nara 639-1080, Japan; tel. +81 743 55 6163, e-mail: naoe@chem.nara-k.ac.jp;[2]Graduate School of Bioresource Sci., Nihon University, Fujisawa, Kanagawa 252-0880, Japan

Preparation of thermoresponsive Pickering emulsions formulated with gold(Au) nanoparticles was investigated. Au nanoparticle Pickering emulsions were prepared by using mercaptocarboxylated Au nanoparticles. Thermoresponsive Pickering emulsions were prepared by adding biopolymer gelatin to the system. The Au nanoparticles were prepared by phase transfer method with11-mercaptoundecanoic acid(MUA) as a stabilizing agent[1)].From the TEM images the average size of the metal core of MUA-Au nanoparticles were 2.1 nm. A mixture of the Au nanoparticle aqueous solution containing NaCl as an electrolyte and toluene solution (aq. soln./toluene = 1:1 vol.) was sonicated for 2 min. After ultrasonication, the upper phase became clouded. Fluorescence microscopy of the emulsions with fluorescence dye calcein revealed the formation of water-in-oil (W/O) emulsions. The emulsion size was measured by optical microscopy. The average size of emulsion was 0.548 μm with a size distribution of 20.1%. The stability of the emulsion was examined by measuring the turbidity change of the emulsion phase with time. Incorporating gelatin to the emulsion was successfully achieved by adding gelatin powder to the emulsion phase. In the emulsion system containing gelatin the decrease in the turbidity with time was effectively suppressed at 293 K while not suppressed at 313 K, indicating that the gelation of gelatin at lower temperatures would make the emulsions more stable against coalescence. From optical microscopic measurement the size distribution of emulsions with gelatin was changed with the temperature. Thus, the thermoresponsive behavior of the emulsion was obtained owing to the gelatin addition.

[1] Naoe, K., Yamanaka, K., Takeuchi, S., Kawagoe, M., and Imai, M., *J. Chem. Eng. Jpn.*, **45**, 789-792 (2012)

Sorption of heavy metal ions by cellulose filled with fullerene

O.V. Alekseeva, N.A. Bagrovskaya, A.V. Noskov
G.A. Krestov Institute of Solution Chemistry of Russian Academy of Sciences, Akademicheskaya Str, 1, Ivanovo, 153045, Russia,
tel. +74932 351859, e-mail: ova@isc-ras.ru

Cellulose is one of the most abundant biopolymers having a permanently renewable raw material resources base. Cellulose is widely used in manufacture of paper, fibers, membranes, sorbents, as well as in medicine and pharmaceutics. Natural cellulose materials effectively recover the heavy metal ions from aqueous electrolyte solutions, exhibit selectivity, easily regenerate, and remain operable in aqueous-organic media. Introduction of inorganic nanoparticles into biopolymers enables purposeful control over their structure and physicochemical properties, which imparts novel properties to a composite material. Nanoparticles of metal oxides, carbon nanotubes, and fullerenes can be used as modifying fillers.

The goal of our study was to obtain cellulose/C_{60} nanocomposites and examine the sorption-kinetic properties with respect to the heavy metal ions.

Cellulose was modified by impregnation of the polymer with a solution of C_{60} nanoparticles in *o*-xylene. Further, the polymer was separated from the solution and dried in air until the solvent was fully removed. The sorption kinetics of Cu(II), Ni(II), and Cd(II) ions was performed from aqueous solutions of sulfates of these metals by the obtained cellulose/C_{60} composite films. The initial and the residual concentrations of metal ions in the filtrate were determined by atomic-absorption spectroscopy.

In case of unmodified cellulose, the process in which Cu(II) and Ni(II) ions pass into the sorbent phase has a high rate. The sorption equilibrium is attained in 40÷60 min after the beginning of sorption on cellulose. For Cd(II) ions, the equilibrium is attained more slowly, in approximately 5 h. Introduction of 0.38 wt. % of fullerene into cellulose leads to an increase in the rate of sorption process on the composite, compared with the unmodified sample. This is indicated by the decrease in the time of equilibration to 40 min for Cu(II) ions and to 90 min for Cd(II) ions.

To describe the sorption kinetics it is necessary to determine the reaction order and the equation that can be used to calculate the quantitative parameters of process. In the present study, we processed experimental data on adsorption of metal ions in terms of models of pseudo-first-order and pseudo-second-order reactions. Values of the equilibrium concentration of ions sorbed and the kinetic constants for the pseudo-first- and pseudo-second-order reactions were found.

It was found that the sorption kinetics of copper and nickel ions by both original and fullerene-containing cellulose are described by the equation of the pseudo-first-order reaction. The composite film recovers copper more effectively, compared with the unmodified cellulose. But, in the case of nickel sorption, introduction of C_{60} into cellulose hardly affects the rate constant. Sorption of cadmium ions by the modified polymer are described by the equation of the pseudo-first-order reaction too. .

The reported study was supported by RFBR, research project No. 15-43-03034-a.

Sorption on clay for transformer oil regeneration

M. Matějková, [1]V. Kužílek, [1]L. Košanová, F. Kaštánek, O. Šolcová
Institute of Chemical Process Fundamentals of the ASCR, v. v. i., Rozvojova 135, 16502, Prague, Czech Republic, fax: +420220920661, tel: +420220390280, matejkova.martina@icpf.cas.cz
[1] ORGREZ, a.s., Division for electrical laboratories, Vítkova 17, 186 00, Prague 8, Czech Republic

New clay sorbents originated in the Czech Republic were applied for regeneration of the used transformer insulation oil to prevent the transformer failures. The cause of failure has been a short circuit in the dielectric isolation. This effect is attributed to the so-called "corrosive sulphur". Corrosive sulphur is defined as various forms of organic sulphur compounds (often thermally unstable) such as Cu_2S, which can cause corrosion of metal parts of the transformer, in particular copper and silver. Dibenzyl disulfide (DBDS) has been found to be the leading corrosive sulphur compound in the insulation oil

The key physico-chemical and chemical properties of transformer oils containing corrosive sulfur were defined. Thus, viscosity at 40°C, density at 20°C, group composed of transformer oils and gas chromatograph with chemiluminescence sulfur detector (GC SCD) were determined in the real transformer oil.

This work is focused on finding an effective way to decontaminate such oils. For this purpose sorption was chosen as a cheap and simple decontamination technique. As sorbents two types of Czech clays were applied and their sorption efficiencies were compared with commercial sorbent usually used for treatment of an insulation oil. The real used transformer electroinsulation oil was applied in two differently sized column filled by sorbent.

Moreover, new sorbents were thoroughly characterized by physical adsorption of nitrogen, mercury porosimetry, helium pycnometry as well as by XRD.

Acknowledgement: The financial support of TA CR (Grant No. TA04020151) is gratefully acknowledged.

Comparison of hexane vapour permeation indifferent membranes

Z. Sedláková, L. Morávková, J. Vejražka, P. Izák
Inst. Chem. Proc. Fund., CAS, Rozvojova 135, 16502Prague, Czech Republic;
tel. +420220390133, e-mail: sedlakova@icpf.cas.cz.

Nowadays, the industrial development causes a rapid increase of anthropogenic pollution of environment. Therefore, there is a need of an intensive search for sustainable and "green" technologies. One of the current issue is to minimalize the emissions of hydrocarbons into atmosphere. The emission of organic volatile compounds (VOCs) means not only an ecological problem but also an economic loss. Therefore, it is necessary to avoid a further increasing of the emissions as well as to solve the capture of the pollutants.

The separation of VOCs can be realised by various methods such as condensation, absorption, adsorption, etc. In recent decades, membrane technology has attracted attention as an alternative and versatile tool. Its advantages consist in its low operating cost, simple operation, compact design, and modular equipment.

The aim of this work was the VOC removal from a polluted air stream. The model mixture contained nitrogen saturated by various concentrations of hexane. The mixture was chosen as a representative binary mixture of the polluted air by gasoline vapours. The nitrogen represented the major component of air and hexane represented the gasoline vapours, respectively.

Vapour permeation experiments were carried out at various conditions using the test stand apparatus [1, 2]. The apparatus allow to measure at different pressures above and under the membrane, cell temperatures, saturation temperatures, two flow arrangements (co- or counter-current flow) and two modes with either sweeping gas or vacuum applied at the permeate side of the membrane.

The transport and separation properties were compared for three different kind of membranes, namely supported ionic liquid membrane [1], low density polyethylene and polyamide thin film composite membrane. Further, the influence of the membrane material on the transport and separation properties was discussed. The highest flux was obtained for the supported ionic liquid membrane while the permeability of hexane through the polyethylene membrane was the lowest. The differences between the fluxes in different membranes were almost two orders of magnitudes.

Acknowledgement

The financial support from the Ministry of Education, Youth and Sports of the Czech Republic (project LD14094) and from the Technology Agency of the Czech Republic (project TE01020080) is gratefully acknowledged, respectively.

Literature

1. Morávková L., et al. (2014). Chemical Papers 68(12): 1739-1746.
2. Vopička O., et al. (2015). Chemical Engineering and Processing: Process Intensification 94: 72-77.

Comparison of membrane materials for biogas purification

Z. Sedláková, M. Kárászová, L. Morávková, J. Vejražka, P. Izák
Inst. Chem. Proc. Fund., CAS, Rozvojova 135, 16502 Prague, Czech Republic;
tel. +420 220390133, e-mail: sedlakova@icpf.cas.cz.

The bio-methane seems to be very promising source of energy. This alternative source of energy can be obtained from the biomass after the removal of unwanted impurities. The methane concentration in biogas varies usually from 50 vol. % to 70 vol. % depending on its origin. The major impurity of biogas is carbon dioxide. Furthermore, the biogas contains traces of hydrogen sulphide, ammonia, siloxanes, and other compounds. It is usually not possible to remove the impurities in one or two technological steps due to their different chemical nature. The methods utilised for the biogas purification are for example physical or chemical absorption, water scrubbing, pressure swing adsorption, and membrane separation.

Among the possible biogas purification techniques, membrane processes present the technology with many advantages such are operation simplicity, easy maintenance, safety and low operation cost. The main attention has to be paid to the choice of a membrane material. Therefore, the choice of membrane for gas separation is based on specific physical and chemical properties, since these materials should be tailored to separate specific gas mixtures.

The aim of this study is the comparison of different membrane material for biogas separation. The comparison is based on the both our experimental results and found literature data for the binary mixtures containing methane and carbon dioxide. The compared membrane materials cover both flat sheet membranes (ethylene-octene copolymer membranes [1], elastomeric block co-polyamide membranes [2], water swollen thin film composite membranes [3,4], ionic liquid membranes [2,5,6] and mixed membrane matrix [1,7]) and hollow fibre modules (polyacrylonitrile PAN hollow fibre coated by Pebax [8]).

The most selective membranes were found to be ionic liquid membranes (selectivity up to 25) and very good selectivity were obtained for both water swollen membrane (selectivity up to 20) and PAN/Pebax (selectivity up to 18).

Acknowledgement

The financial support from the Technology Agency of the Czech Republic (project TE01020080) is gratefully acknowledged.

Literature

1. Sedlakova Z., et al. (2012). Membranes 4(1): 20-39.
2. Bernardo, P., et al. (2012). Separation and Purification Technology 97: 73-82.
3. Poloncarzova, M., et al. (2011). Angewandte Chemie-International Edition 50(3): 669-671.
4. Dolejš, P., et al. (2014). Separation and Purification Technology 131(0): 108-116.
5. Kárászová, M., et al. (2014). Separation and Purification Technology 132(0): 93-101.
6. Kárászová, M., et al. (2015). Separation and Purification Technology 153: 14-18.
7. Habibzare S. ,et al. (2014). International Journal of Nano Dimension 5(3): 247-254.
8. Esposito, E., et al. (2015). Chemical Engineering and Processing: Process Intensification 94: 53-61.

Diffusiophoresis of a pH-regulated toroidal polyelectrolyte in a solution containing multiple ionic species

Shang-Hung Hsieh, Jyh-Ping Hsu
National Taiwan University, Department of Chemical Engineering, No. 1 Sec.4 Roosevelt Rd., 10617, Taipei, Taiwan;
Tel. +886-2-23635230, Fax: +886-2-23623040, e-mail:r04549027@ntu.edu.tw

Considering recent application of diffusiophoresis in biosensing technology, we extend previous analysis for the diffusiophoresis of a toroidal PE at a fixed charge density to that at a pH-regulated condition in the presence of multiple ionic species so that the underlying assumptions are more realistic. The diffusiophoretic behaviors of the toroidal PE considered under various conditions are examined by varying its functional group density, size, and softness, and the solution pH and bulk salt concentration. The results of numerical simulation reveal that due to the competition between the hydrodynamic force and the electric force acting on a particle, the behavior of a toroidal PE can be different both quantitatively and qualitatively from the corresponding spherical PE. This implies that using an equivalent sphere to simulate the actual entity can yield significant deviation and, therefore, knowing the exact shape of an entity is necessary both in interpreting experimental data and in designing relevant devices.

Generation of zeolite coatings on glass and Al_2O_3 beads in a fluidized bed

E. Roydeva, [1]F. Scheffler
Otto von Guericke University Magdeburg, Laboratory of Industrial Chemistry, Universitaetsplatz 2, 39106 Magdeburg, Germany, tel.: +49-(0)391-67-51060, [1]Otto von Guericke University Magdeburg, Laboratory of Industrial Chemistry, Universitaetsplatz 2, 39106 Magdeburg, Germany, tel.: +49-(0)391-67-58824

Introduction

Zeolites are crystalline, microporous aluminosilicates. The framework is usually formed by SiO_4 and AlO_4tetrahedra, which are 3D-linked via oxygen atoms. They can be found as natural materials in a broad range of structure types as well as synthesized in a hydrothermal crystallization process [1]. These porous materials own a very well defined pore system and a narrow pore size distribution. Based on their specific properties, zeolites find applications as catalysts, molecular sieves and also adsorbents [2].

For certain applications, it is necessary to generate a thin zeolite coating layer on an inert support material, e.g. to avoid undesired consecutive reactions or high costs [3,4]. Various methods for the generation of zeolite coatings on a support material are available, for example: hydrothermal synthesis, wash coating or partial transformation of the support [5,6].

The current work focuses on the deposition of zeolite coatings on spherical support materials in a fluidized bed granulator.

Experimental

To investigate the influences of the process parameters and suspension composition on the deposition of the zeolite layer on the support materials, a coating process was carried out in a lab scale fluidized bed granulator. Compressed air was used as atomization and also as fluidization gas. For the atomization was used a two fluid nozzle, configured in top spray position. An inlet temperature of 80°C was chosen for all experiments. The atomization pressure was varied between 1 and 2 bar and the gas mass flow rate in the range between 60 and 80 kg/h. Other parameter, which was varied was the spray rate. Furthermore, different suspensions were prepared, containing 5, 10 or 15 wt% zeolite and different binder components, like methyl cellulose or colloidal silica. Two types of support materials were applied: non-porous glass beads with a diameter between 300 and 400 µm and alumina beads with a diameter of 1 mm.

The coated particles were characterized by scanning electron microscopy, x-ray diffraction, light microscopy and dynamic image analysis.

Results

Porous layers were successfully generated on spherical support materials. Depending on the process parameters different layer thickness and yield are available.

SEM images of glass and alumina beads, coated at 2 bar atomization pressure with a slurrycontaining5 wt % of zeolite are shown in fig.1. It could be observed, that the particle size has an influence on the homogeneity of the coating layer. By the glass particles, which have a smaller mean diameter in comparison to the alumina beads, could be seen an inhomogeneous coating layer.

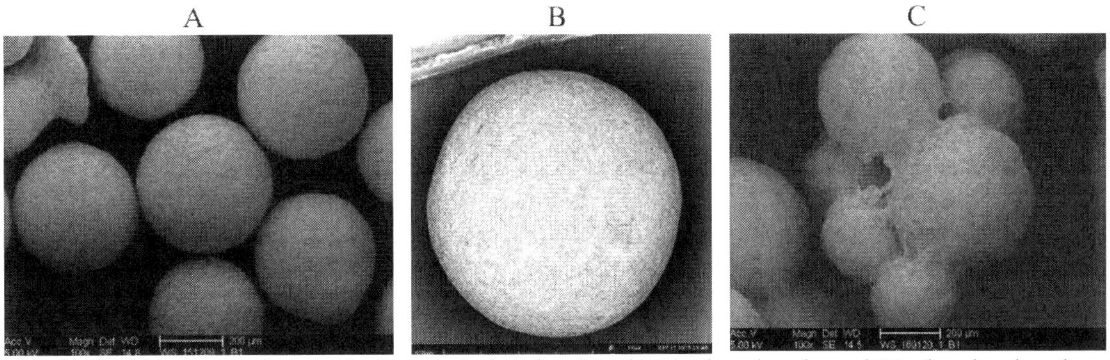

Fig. 1: SEM images of coated glass and alumina beads:A) glass beads and B) alumina beads coated at 2 bar atomization pressure and 5 wt % of zeolite in the suspension, C) glass beads coated at 1 bar atomization pressure and 10 wt% of zeolite in the suspension.

Furthermore, the influence of the atomization pressure was investigated. The dominant process at lower atomization pressure by the experiments with glass beads was agglomeration (see fig.1 C). Some agglomerates were observed also by the coating of alumina beads under the same experimental conditions.

References

[1] G. Busca, Heterogeneous Catalytic Materials: Solid State Chemistry, Surface Chemistry and Catalytic Behaviour, Elsevier, Amsterdam, Oxford, (**2014**)

[2] A. H. Sulaymon, A. S. Mahdi, Spherical Zeolite-Binder Agglomerates, *Chem. Eng. Res. Des.* 77 (4) (**1999**) 342-350

[3] A. Erdem-Senatalar, M. Tatlier, M. Ürgen, Preparation of zeolite coatings by direct heating of the substrates, *Micropor. Mesopor. Mater.* (**1999**) 331-343

[4] J. C. Jansen, W. Nugroho, H. van Bekkum, in: R. von Ballmoos, J. B. Higgins, M. M. J. Treacy. Proc. 9th Int. Zeolite Conf. Monreal, Canada, Butterworth-Heinemann, Boston, MA, (**1992**), Vol.1, 247

[5] J.M. Zamaro, M.A. Ulla, E.E. Miro, Zeolite washcoating onto cordierite honeycomb reactors for environmental applications, *Chem. Eng. J.* 106 (1), (**2005**)25-33

[6] F. Scheffler, R. Herrmann, W. Schwieger, M. Scheffler, *Micropor. Mesopor. Mater.* 67 (**2003**) 53.

Green synthesize of carbon nanoparticles from *Nerium oleander* extracts via two different microwave assisted methods

S. Simsek[1], M.O. Alas[2], B. Ozbek[1], R. Genc[2]
[1]Yıldız Technical University, Chemical Engineering Department, 34210, Esenler, İstanbul; tel. +90 212 383 4755, e-mail: ssimsek@yildiz.edu.tr; tel. +90 212 383 4747, e-mail: bozbek@yildiz.edu.tr; [2]Mersin University, Chemical Engineering Department, 33343, Yenisehir, Mersin; tel. +90 324 361 00 01/7393, e-mail: rukangnc@gmail.com; e-mail: melis_alas@hotmail.com

Natural products are the potential and wealthy sources of new drugs, new drug leads and new chemical entities. Therefore, plant originated molecules and/or agents inspired interest for the cancer treatment [1].

Nerium oleander (Oleander), belongs to *Apocynaceae* family, is extremely toxic [2]. It is an evergreen plant with long, narrow green leaves and white or pink flowers [3].The genus has various biological activities such as cytotoxic, antibacterial, anti-inflammatory and anticancer etc. [2]. Since, the fractions and/or compounds of *N. oleander* show anticancer activity, it was planned the preparation of carbon nanoparticles from this plant to design potential drug carrier systems for cancer treatment.

Consequently, in this present study, it was aimed to use the leaf extracts of *Nerium oleander* as carbon source for the microwave assisted synthesize of carbon nanoparticles. For this purpose, ethanol and water extracts from *N. oleander* dried leaves were prepared by using ultrasonic extraction. Then, by use of these extracts, the formulations with or without polyethylene glycol were prepared. The green synthesize of *N. oleander* based carbon nanoparticles were fulfilled by two different microwave irradiation methods. In the first procedure, the prepared solutions were treated in a domestic microwave oven until the samples were dried out[4,5]. Then, the residues were dissolved in distilled water and large particles were removed till clear solutions obtained by centrifugation. In the second method, the solutions under the same conditions with the first one were placed at microwave synthesizers and the reactions were carried out under the microwave irradiation[6]. Large particles were again removed from the solution by centrifugation.

Finally, for the characterization of synthesized carbon nanoparticles, the studies were performed by UV-Vis spectrophotometry, fluorescence spectrophotometry, Fourier transform infrared spectroscopy, zeta particle size and zeta potential analysis.

References:

1. Bhanot A., Sharma R., Noolvi M. N., 2011, *International Journal of Phytomedicine*, 3: 9-26.
2. Sharma P., Choudhary A. S., Parashar P., Sharma M. C., Dobhal M. P., 2010, *Chemistry&Biodiversity*, 7:1198-1207.
3. Zibbu G., Batra A., 2010, *J. Chem. Pharm. Res.*, 2(6):351-358.
4. Duan J., Yu J., Feng S., Su Li, 2016, *Talanta*, 153: 332-339.
5. Jiang J., He Y., Li S., Cui H., 2012, Chem. Commun, 48:9634-9636.
6. Hongying L., He Z., Jiang L.-P., Zhu J.-J., 2015, ACS Appl. Mater. Interfaces, 7:4913-4920.

Ionic current rectification in a conical nanopore for various types of aqueous salt solution

Shu-TuanYang, Chih-Yuan Lin, Jyh-Ping Hsu
National Taiwan University, Department of Chemical Engineering, No. 1 Sec.4 Roosevelt Rd., 10617, Taipei, Taiwan;
Tel. +886-2-23635230, Fax: +886-2-23623040, e-mail:r04524038@ntu.edu.tw

The influence of electroosmotic flow (EOF) on the behavior of the ion current rectification (ICR) in a conical nanopore connecting two large identical reservoirs is investigated. In particular, the effect of the type of salt is examined by considering LiCl, NaCl, and KCl. We show that neglecting EOF is capable of influencing ICR both quantitatively and qualitatively. If EOF is neglected, the rectification factor at each level of the applied electric potential bias across the two reservoirs for the case of LiCl (KCl) is always the largest (smallest). However, if EOF is taking into account, the relative magnitude of the rectification factors for various salt depends upon the level of the applied electric potential bias. This behavior is consistent with the experimental observation in the literature, and can be explained by the degree of ion deplection/enrichment in a nanopore.

Cross-linked polymers as products of lignin biotreatment

E. Kozliak, Y. Ji, A. Kubátová, F. Asina, I. Brzonova, K. Voeller
University of North Dakota, SUNRISE/Dakota BioCon Research Group (Chemistry and Chemical Engineering Depts.), Grand Forks, ND, 58202 U.S.A., tel. +1 701 7772145, email evguenii.kozliak@und.edu

Lignin is an abundant phenolic biopolymer accounting for approximately 30% of the Earth's renewable non-fossil organic carbon. Based on the lignin structure, a number of valuable chemicals could be produced from lignin to replace petrochemicals by breaking down its structure to smaller-MW fragments. Yet, none of the currently explored protocols presents an efficient and economical way of lignin utilization. In this study, modern methods of chemical analysis were applied to the complex water-soluble product of microbial lignin degradation to achieve mass balance closure and assess the organic carbon fractionation. Efficient fungal lignin degraders, e.g., *Coriolus versicolor*, were compared to both bacterial degraders, *Microbacterium* and *Streptomyces sp.*, and pure laccase in long-term cultivations, 54d, along with abiotic thermal treatments, to determine the source of the problem.

When lignin was degraded in liquid media with glucose added as a co-substrate, the carbon mass balance closed successfully not only for laccase, as expected, but also for bacterial treatments. Only the fungal treatments yielded a sizable loss of lignin carbon (as assessed by gravimetric analyses), presumably, as CO_2, along with a partial conversion of aromatic protons to aliphatic as shown by NMR. The newly developed thermal carbon analysis (TCA) showed that all three treatments increased the non-volatilizable (up to 850 ^oC) carbon fraction, which elutes only in the presence of oxygen as CO_2. Thus, microbial growth is linked not only to lignin catabolism but also to the formation of apparently highly cross-linked polymers.

Formation of such polymers becomes the main pathway if no co-substrate is added to the medium. The process may be accelerated by conducting it with fungi grown on agar, particularly in the presence of 2-10% dimethyl sulfoxide as a lignin solubilizing agent. Another way of forming oligomers/polymers as the main products is the enzymatic treatment with laccase as opposed to bacteria or fungi. These high-MW products were lightly colored and remained water-soluble unlike the coked lignin re-polymerization products formed as a result of thermal (abiotic) treatment in subcritical water. However, no biotreatment appears to occur without formation of such polymers. Lignin solubilization in neutral aqueous media as a result of biotreatment, which had been perceived to be a result of depolymerization, was shown to occur even when the polymerization prevails, as the polymers formed appear to be more polar than the original lignin.

To obtain further insights into the newly formed polymers' chemical structure, the biotreatment products were characterized by HPLC-DAD using a reverse phase column and by Pyr-GC-MS. HPLC-DAD showed an increase of polarity and decrease of polydispersity as a result of fungal treatment. Pyr-GC-MS demonstrated that the thermal breakdown of polymer products does not result in the formation of new phenolic products. Thus, cross-linking appears to be the main path of lignin biopolymerization as opposed to oxygenation and other structural changes. The formation of modified and more recalcitrant polymers appears to be an inherent feature of lignin biodegradation. Thus, the economically sustainable processes of biomass utilization envisioned in the future should address the utilization of such polymer products.

Comparison of lab scale single biofilter, biotrickling filter and bubble column and combined biofilter/biotrickling filter treating styrene/acetone mixture

M. Halecky, [1]L. Zapotocky, J. Paca, [2]E. Kozliak, [3]K. Jones
Institute of Chemical Technology, Department of Biotechnology, Prague, Czech Republic, e-mail: Martin.Halecky@vscht.cz, [1]Dekonta a.s. Prague, Czech Republic. [2]University of North Dakota, Department of Chemistry, Grand Forks, ND, USA, [3]Texas A&M University-Kingsville, South Texas Environmental Institute, Kingsville, TX, USA

Three most common bioreactors, single biofilter, biotrickling filter and bubble column, and combined biofilter/biotrickling filter were tested under the same conditions for cleaning of the waste gas containing mixture of styrene and acetone. Acetone and styrene significantly differ in both their biodegradability and mass transfer rates. As the opposite of styrene, the gas-liquid of acetone mass transfer is facile as shows in its Henry's law constant value, which is two orders of magnitude lower than that of styrene.
Used bubble column (pH control and dissolved oxygen measurement) and biotrickling filter (pH control) allowed much better process control in comparison to a biofilter. On the other hand, biofilter is simple and cheap reactor with great long-term stability.
Biotrickling filter was able to remove acetone with higher efficiency, however, significant accumulation of acetone under high loadings led to drop of the degradation efficiency of both pollutants. Tests performed in bubble column shown high oxygen demand and limitation of biodegradation process by oxygen especially under high loadings. Similarly to the biotrickling filter, significant accumulation of acetone leading to the decrease of degradation efficiency was observed in bubble column. Degradation efficiency of combined biofilter/biotrickling filter was much greater than both single biofilter and biotrickling filter, thus, the synergic effect of such bioreactor combination was improved.

The work was financially supported by the Czech Technology Agency, Project TA03020395 and by the Ministry of Education of the Czech Republic, Project KONTAKT II LH13073. North Dakota researcher participation was sponsored by the North Dakota SUNRISE BioProducts Center of Excellence.

Verification of pilot combined biotrickling filter/biofilter treating styrene/acetone mixture in real operating conditions

L. Zapotocky, [1]M. Halecky, [1]J. Paca,
Dekonta a.s., Prague, Czech Republic, e-mail: zapotocky@dekonta.cz, [1]Institute of Chemical Technology, Department of Biotechnology, Prague, Czech Republic.

The pilot biotrickling filter, biofilter and the combined biotrickling filter/biofilter were tested in real operating conditions for cleaning of the waste gas containing mixture of styrene and acetone. Acetone and styrene are volatile organic compounds used for production of laminates. Both substances have very different physico-chemical properties and different biodegradability. Used pilot scale biotrickling filter with pH control allowed much better process control in comparison to the pilot biofilter. On the other hand, biofilter is simple and cheap reactor with great long-term stability.
During long-term tests were obtained basic operating parameters (elimination capacity and removal efficiency) of all three bioreactors. Operating conditions allowed control only the waste gas flow. Concentration of both pollutants fluctuated depending on the stage of production. It was also verified the effect of single-shift operation and operational shutdowns to eliminate efficiency organic solvents combined biotrickling filter/biofilter. Synergy of combination of biofilter and biotrickling filter was proven when much greater elimination capacity of combined biofilter/biotrickling filter than single biofilter or biotrickling filter was observed.

The work was financially supported by the Czech Technology Agency, Project TA03020395.

Anaerobic bioreactor treating a solvents-containing wastewater: biodegradability studies and enhancement via micronutrients dosing optimization

N. Vermorel[2], M. Izquierdo[1], J. M. Penya-Roja[1], A.Waalkens[2], C. Gabaldón[1]
[1] University of Valencia., Dep. of Chemical Engineering, Av. de la Universidad s/n, 46100 Burjassot, Spain; e-mail: nadine.vermorel@uv.es
[2] Pure Air Solutions, P.O. Box 135, 8440 AC Heerenveen, The Netherlands.

Complex mixtures of solvents can be found in industrial wastewaters and should be treated to avoid negative environmental impacts and comply with legislations such as the European Industrial Emissions Directive (2010/75/EU). Laboratory and pilot-scale studies have been carried out to assess and improve the feasibility of anaerobic treatment of solvents found in the food packaging industry. The solvents commonly emitted in this sector are either ethanol or 2-propanol (isopropanol), as the main solvent, and ethyl acetate, isopropanol, 1-ethoxy-2-propanol and 1-methoxy-2-propanol as secondary solvents.

Ethanol is known for being easily biodegradable; some of the other compounds have either not been studied yet or have been reported to be less easily degradable, presenting low inhibitory concentrations in the case of isopropanol and ethyl acetate (Tonouchi, 2004; Ince et al., 2011; Schwartz, 1991). Hence, further studies are required to investigate if anaerobic treatment can be used to treat these solvents' mixtures.

Furthermore, optimization of the nutrient supplementation has also been tested for enhancing the secondary solvents degradation rate. Indeed, when synthetic wastewaters are treated, the microorganisms responsible for the solvent degradation – a carbon source for bacterial growth – need additional nutrients for their metabolisms. If macronutrients dosing has been throughfully investigated, micronutrients supplementation studies are scarce and the ranges of recommanded values are broad, due among other factors, to the diversity of both the subtrates and experimental conditions. Thus, specific studies on the dosing of trace elements for the anaerobic digestion of synthetic wastewaters are needed.

Batch experiments using the AMPTSII by BioProcess Control®, operated at 25 °C, were run for inhibition studies of each of the secondary solvents on the main solvent- namely ethanol. Reactors of 500 mL were seeded with an anaerobic granular sludge previously acclimatized to the mixture of solvents .The assays were done according to the EPA guidelines (EPA, 2008). The biodegradability (biochemical methane potentials and specific methanogenic rates) of the secondary solvents was assessed for COD concentrations from 1 to 25 g L^{-1}.

The resulting findings were further investigated at pilot-scale with an 8 m^3 Upflow Anaerobic Sludge Bed (UASB) reactor seeded with 3 m^3 of granular sludge originating from a brewery wastewater treatment plant.

At laboratory scale, a 2^{4-2} fractional factorial experimental plan has been designed for the optimization of micronutrients supplementation, in order to study six trace elements (Co, Fe, Mn, Mo, Ni, Zn) at two different dosing concentrations. This plan allows a first screening of the importance of each trace element for the solvents' anaerobic degradation. In this case, a mixture of ethanol and 1-methoxy-2-propanol was taken as a case study.

The results of the batch experiments indicate that there is no inhibition of the secondary solvents on the degradation of ethanol, in the experimental conditions of the tests.

Furthermore, each of the secondary solvent could be successfully degraded, after acclimatization periods ranging from 3.6 to 126 days. Removal efficiencies higher than 90% were achieved, based on TOC analysis. All the secondary solvents presented a lower degradation rate than ethanol- the main solvent-. These degradation rates, characterized by the Specific Methanogenic Activity (SMA), increased for initial solvent concentrations up to 10 g COD L^{-1} before decreasing for high substrate concentrations (25 g COD L^{-1}).

Pilot-scale experiments confirmed that ethanol could be treated quickly and easily at ambient temperatures (in average around 21 °C) for Organic Loading Rates (OLR) around 15 kg COD m^{-3} d^{-1}. Ethanol spikes of 6.5 kg COD were degraded within a few hours. Isopropanol was also tested as the main solvent. Stable substrate removal for OLR up to 6 kg COD m^{-3} d^{-1} were achieved within 2 weeks, with Volatile Fatty Acids values kept under 100 mg HAc L^{-1} and residual COD values in the water around 2000 mg O_2 L^{-1}. Further acclimatization is undertaken in order to reach higher OLR.

Concerning the micronutrients study, the trace elements playing the most important roles in the anaerobic degradation of the solvents were found to be: iron in the first position, followed by cobalt and manganese and with a lower impact zinc. The highest positive interaction was found between Ni and Mo, followed by Fe-Co and Mo-Zn. The main negative interactions were observed for Ni-Fe and Mo-Mn. When trace elements were supplemented in an optimized combination, the solvents removal rate was increased several times, in comparison to the basic dosing corresponding to the methanogens bacteria composition (Rajeshwari et al., 2000) and common growth yield. These trace metals requirements, higher than expected, can be explained through the partial bioavailability of the compounds and the additional enzymatic needs.

This study indicates that anaerobic biotreatment could be applied successfully to the treatment of solvent emissions from the food-packaging industry, with ethanol as the main solvent – allowing simultaneous pollutants removal and biogas production. When 2-propanol is the main solvent instead of ethanol, first results show that a longer acclimatization period will be required. Further studies are being undertaken to determine the maximal OLR applicable. Additionally, optimization of the micronutrients supplementation is shown to clearly enhance the degradation rate of secondary solvents present in the wastewaters considered.

Acknowledgments: The research leading to these results has received funding from the People Programme (Marie Curie Actions-ITN) of the European Union's Seventh Framework Programme FP7/2007-2013/ under REA grant agreement n°606942 (TrainonSEC). Financial support was also obtained from Generalitat Valenciana (Spain, Prometeo project 2013/53).

References:

Agency, E. P. (n.d.). Fate, Transport and Transformation Test Guidelines OPPTS 835.3420 Anaerobic Biodegradability of Organic Compounds in Digested Sludge: By Measurement of Gas Production.

Ince, B., et al. (2011). Inhibition effect of isopropanol on acetyl-CoA synthetase expression level of acetoclastic methanogen, Methanosaeta concilii. *Journal of Biotechnology*, 156(2), 95–9.

Rajeshwari, K., et al. (2000). State-of-the-art of anaerobic digestion technology for industrial wastewater treatment. *Renewable and Sustainable Energy Reviews*, 4(2), 135–156.

Tonouchi, A. (2004). Anaerobic 2-propanol degradation in anoxic paddy soil and the possible role of methanogens in its degradation. *Current Microbiology*, 49(2), 75–8.

Schwartz, L. J. (1991). Toxic effects of selected industrial solvents in batch and continuous anaerobic reactors. *Applied Biochemistry and Biotechnology*, 28-29(1), 297–305.

Novelty in reactor design and optimization for biotechnology applications using a systems approach

G. P. PANAYIOTOU[1], A. D. MOUNTRAKI[1,2], A. C. KOKOSSIS[1*]

[1]School of Chemical Engineering, National Technical University of Athens (NTUA), Heroon Polytechniou 9, Polytechnioupoli Zografos 15780, Athens, Greece
e-mail: akokossis@mail.ntua.gr
[2]Compagnie Industrielle de la Matière Végétale (CIMV), 109, Rue Jean Bart - Diapason A - F-31670 Labège, France

Reactors in biotechnology applications are mainly heterogeneous systems that are strongly non-ideal and operate at mild temperatures with several reversible paths in place. Therefore, mass transfer aspects may become crucial as desired products are possible to remove in the course of reaction, eventually improving the overall yields. In contrast with the homogeneous systems, in the heterogeneous schemes, apart from the kinetics and the heat transfer effects, the mass transfer and the general hydrodynamic behavior of the system becomes important. The overall reaction rate depends on kinetics and on the system's ability to exchange mass between different phases. Usually, mass transfer and kinetics are balanced. There is a huge potential for innovation and there is strong evidence that non-conventional reactors are capable to bring significant merits. Still, the design of bioreactors is mainly based on experience and heuristics as they are developed for conventional reactors. Moreover, the diversity of feedstock, along with the complicated activity of the microorganisms used, introduce several design uncertainties to consider.

Significant systems technology is available to cope with multiple paths but a focus on reaction kinetics much less on mass transfer. A systematic approach is proposed in the paper that capitalizes on past knowledge and methods as it is available for multi-phase reactors. The method is intended to support decisions towards unconventional reactors, account for reaction and multi-phase, systems but also consider options for mass transfer aspects and a better link of models with industrial reactors. A superstructure approach has been presented using shadow reactor compartments and following Kokossis & Floudas (1990). The extension of this work will be described considering mass transfer links that are are now modelled taking into consideration the implicit arrangement for each shadow compartment (e.g. differentiation between packed bed reactors, fluidized bed reactors, segregated reactors stirred tank bioreactors, tubular bioreactor, airlift bioreactors, etc.). The work capitalizes on the work from a separate community that has provided libraries of correlations that relate mass transfer parameters and holdups with different types of industrial reactors. The methodology uses a superstructure approach that is optimized using evolutionary methods. Results suggest novel sections, novel/non-conventional reactors, and provide for a natural interpretation of optimization results into industrial units. Design parameters include the type of micro and macromixing and the type and size of the industrial reactor. Examples include gas-liquid, solid-liquid, liquid-liquid reaction systems; the methodology could extend for any number of phases.

Acknowledgements: The financial support from Marie Curie European Research Program, RENESENG (FP-607415) is gratefully acknowledged.

References

Kokossis, A.C. and Floudas, C.A., 1990, Optimisation of complex reactor networks—I. Isothermal operation, Chem Eng Sci, 45(3): 595–614.

A new six compartment model for submerse cultivations

V. C. Hass, [1]S. Brüning, D. Kanzler, [2]R. Pörtner
Hochschule Furtwangen University, Faculty Medical and Life Sciences, Jakob-Kienzle-Straße 17, D-78056 Villingen-Schwenningen, Germany; tel. +49 7720 307 4235, e-mail: volker.hass@hs-furtwangen.de; [1]Institute for Environmental and Bio-Technology, University of Applied Sciences Bremen, Neustadtswall 30, D - 28199 Bremen, Germany; [2]Institute for Bioprocess and Biosystems Engineering, Hamburg University of Technology, 21071 Hamburg, Germany

Complex bioprocesses require effective process control strategies. Especially model based control strategies show high potential for the optimization of bioprocesses. The mathematical models in these algorithms must fulfill two major requirements. First, the model must be able to describe the mechanistic relationship between actuating variables like feeding rates etc. and the main state variables, like biomass and product concentrations. Second, the parameterized model must be able to predict the time course of the state variables of a cultivation process. Model development appears to be a time consuming and laborious task. In order to reduce the time necessary for model development, a new structured compartment model was developed, which is easy to adapt to different cultivation processes, as will be shown in this contribution.

The model developed is a new six compartment model (SCM), which was used to describe the time course of batch and fed-batch cultivations of the bacteria *Escherichia coli* and *Lactobacillus delbrueckii*, the yeast *Saccharomyces cerevisiae*, the fungus *Cyathus striatus* (see fig. 2) and two different animal cell lines (hybridoma and CHO).

The model is divided into the compartments primary biomass X_{pri}, product forming biomass X_p, structural biomass X_s, inactive primary biomass X_i, inactive structural biomass X_{si} and dead biomass X_d (Figure 1). The primary biomass represents all reactions related to cell growth. This compartment is reproduced autocatalytically by carbon and nitrogen source uptake. The product producing biomass is responsible for secondary metabolite or induced protein (e.g. GFP) production. The formation of this compartment is catalyzed by the primary biomass. The growth of the compartment of structural biomass is also catalyzed by the primary biomass. X_s consists of membranes, cell wall and complex polysaccha-

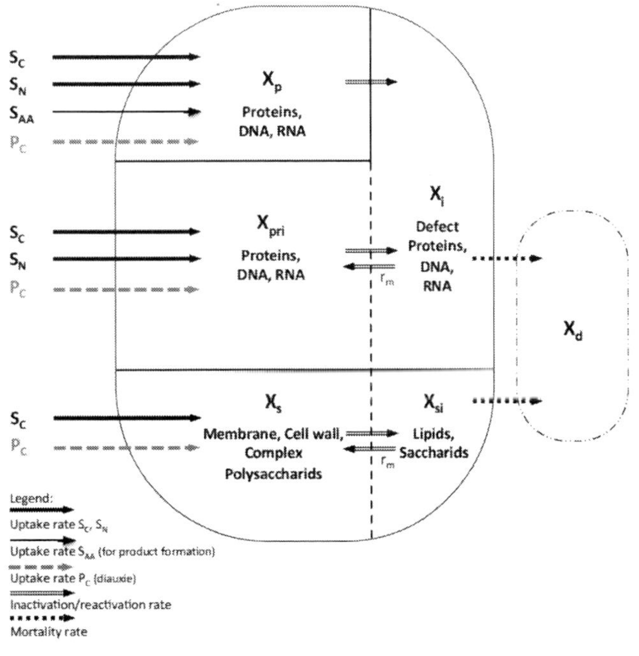

Figure 1: Simplified model scheme for SCM. PC – Metabolite, SAA – Amino acid, SC – C-Source, SN – N-Source, Xd – Dead biomass, Xi – Inactive biomass, Xp – Product forming biomass, Xpri – Primary biomass, Xs – Structural biomass, Xsi – Structural inactive biomass.

rides. It is formed from carbon sources. Primary biomass and structural biomass may be inactivated to form inactive primary and structural biomass and may be reactivated to represent maintenance metabolism. Under certain process conditions the inactive biomass compartments will be converted into dead biomass. Major properties of the model are the following:

1. The substrate concentration influences its uptake rate and therefore the rates of all further reactions.
2. Substrates are used for biomass and for product formation as well as for energy generation via fermentation or energy generation via cellular respiration.
3. Uptake rates, inactivation and death rates as well as yield coefficients are functions of the state variables and are described by double sigmoidal functions.

In order to adapt the model to a bioprocess a four-step modeling procedure was applied. First, a detailed process description was derived from experimental data and literature. Second, stoichiometric equations were formulated with respect to the metabolism under investigation. In the third step cause-effect relationship hypothesis, which were derived from process analysis, were used to identify the parameters of interest. Finally, these model parameters were estimated manually to adjust the model to the measured data.

The model is able to describe the time courses of important state variables (Figure 2) such as carbon and nitrogen sources, intermediate products, and products of primary and secondary metabolism as well as induced products. It is capable of describing O_2 uptake und CO_2 production via exhaust gas composition (aerobic processes) or gas flow rates (anaerobic processes). The model was adapted to each cultivation process only by choosing different values for the model parameters. This reduces the modelling effort for a new process significantly. In future work the model will be tested for prediction and process development.

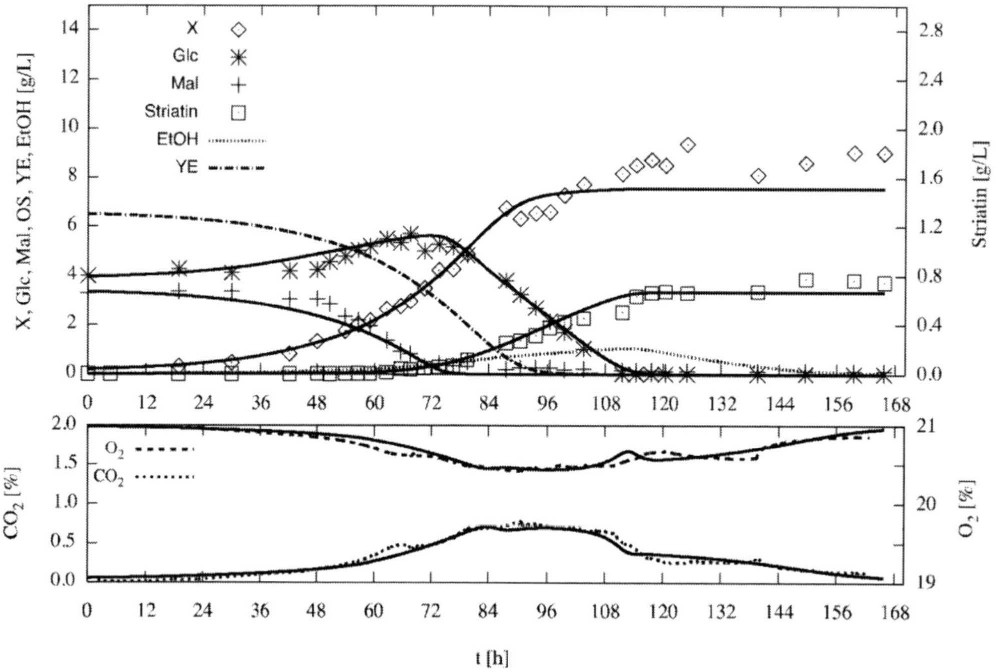

Figure 2: Comparison between experimental data (symbols and dots) and simulated data (continuous lines) of a *C. striatus* cultivation. Offline data: EtOH – Ethanol; Glc – Glucose; Mal – Maltose; X – Biomass; YE – Yeast extract. Online data: CO_2 (vol% - exhaust gas) –; O_2 (vol% - exhaust gas).

We gratefully acknowledge the financial support of the BMBF (FKZ: 031A124 A-C) enabling us to perform this research.

Black yeast cells of *Phaeotheca triangularis* disruption by a mechanical method

Maja Čolnik, Željko Knez, Maja Leitgeb
University of Maribor, Faculty of Chemistry and Chemical Engineering, Laboratory for Separation Processes and Product Design, Smetanova ulica 17, 2000 Maribor, Slovenia;
tel. +386 2 2294462, e-mail: maja.leitgeb@um.si

The cells of *Phaeotheca triangularis* were disrupted by homogenizer in order to use the enzymes from this black yeast for biotransformations.

Black yeasts represent an important form of microorganisms that are able to survive in extreme habitats such as salt, hot and cold deserts and deep water. They have been isolated from a wide range of extreme environments, oligotrophic stone base, highly saline habitats and Arctic glacier ice. Black yeasts are characterized by slowly expanding colonies. *P. triangularis* is a black yeast which belongs to a single order, the Dothideales, within the Ascomycetes. *P. triangularis* is an obligate halophile.

Intracellular biological molecules can be released from the cells by cell disruption (lysis). Cell disruption is a sensitive process because of the cell wall's resistance to the high osmotic pressure inside them. The very convenient way to achieve cell disruption are mechanical cell disruption methods. The results of these methods are often evaluated in terms of the activity level of a cellular enzyme released to the disrupted suspension, combining the efficiency of the disrupting process with an estimate of the degree of cell disruption.

Homogenizer is an efficient way to apply shear to cell suspension. It may be used to disrupt bacteria and yeast on both laboratory and industrial scale.

For this purpose, *P. triangularis* cell suspension was treated by homogenizer at 35 °C and at different treatment times. The viability of *P. triangularis* cells was monitored spectrophotometrically by measuring optical density of the cultures. Additionally, the total protein concentration in the suspension was also determined. Our interest was to excrete the intracellular enzymes α-amylase and celullase from *P. triangularis* and to keep the extracellular enzymes protease and β-glucosidase in the active form after the treatment of the cell suspension culture by homogenizer. A high amount of proteins was released from *P. triangularis* cells during the mechanical method. The residual activity of all examined enzymes was sufficient and they could be used as biocatalysts in this medium.

Desirable regulation of surface hydrophobicity of polypropylene for lipase immobilization for lipid modification in supercritical carbon dioxide

M. Naya[1], A. Shono[1], M. Imai[2]
[1]Department of Industrial Chemistry, Tokyo University of Science, 1-3 Kagurazaka, Shinjuku-ku, Tokyo. 162-8601, Japan. Phone: +81-3-5228-8310;
[2]Course in Bioresouce Utilization Sciences, Graduate School of Bioresource Sciences, Nihon University, 1866 Kameino, Fujisawa, Kanagawa-pref. 252-0880, Japan.
Corresponding author's E-mail (M. Naya): m.naya@ci.kagu.tus.ac.jp

Desirable regulation of surface hydrophobicity of polypropylene carrier for lipase immobilization was demonstrated. It was applied for higher reproducible lipid modification in supercritical carbon dioxide. Surface hydrophobicity of carrier for immobilized lipase was a key factor to improve reactivity and reproducible of fatty acid in supercritical carbon dioxide ($SCCO_2$). Supercritical carbon dioxide has been expected to be a suitable solvent for industrial production of biochemical components by enzymatic processing [1]. Lipases, *Candida cylindracea*, *Candida rugosa*, *Rhizopus arrhizus* and wheat germ were employed. Immobilized lipase was prepared using polypropylene-based hydrophobic porous particles (Accurel MP100) [2]. In our study, Accurel MP100 was shaken in ethanol (99.5%) for 2h (298 K) as a pre-treatment to regulate the surface before adsorption of lipase. Contact angle and critical surface tension were measured by droplet shape analysis (DMs-401, KYOWA Co. Ltd, Tokyo) at room temperature (298 K). They were representative indicators to understand hydrophilic character of solid surface. To evaluate hydrophilic character of solid surface of carrier, the critical surface tension was determined according to Zisman plot (Fig.1). The critical surface tension of original polypropylene was 26 mN/m. After pretreatment by ethanol, it was obtained as 37 mN/m. The cos θ was increased by pretreatment. The change of critical surface tension and cos θ indicated hydrophilic regulation of carrier surface by adsorbed polar head group of ethanol. Pretreatment of polypropylene particle by ethanol was effective to enhance adsorption of lipase.

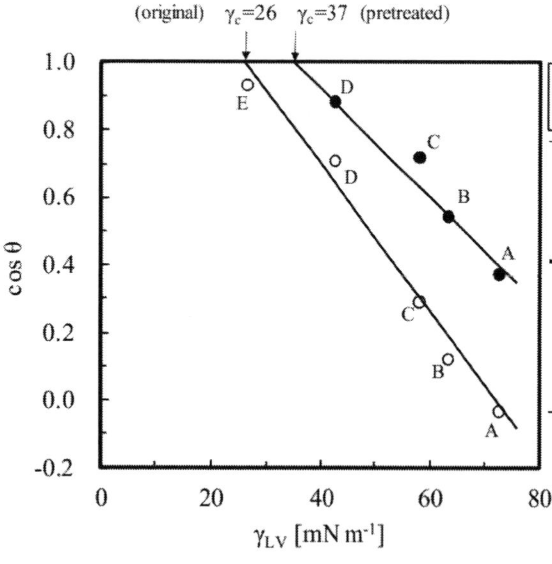

keys	Liquid	γ_{LV} [mN m^{-1}]*,**	Contact angle θ [°] on Polypropylene (cos θ) pretreated	original
A	Water	72.8*	92 (-0.0349)	68 (0.375)
B	Glycerol	63.4*	83 (0.122)	57 (0.545)
C	Formamide	58.2*	73 (0.292)	44 (0.719)
D	Aniline	42.5**	45 (0.707)	28 (0.883)
E	1-Octanol	26.7**	21 (0.934)	ND

The data indicated in parenthesis were cos θ.

Critical surface tension was refered from Fox, H. W. and Zisman, W. A., *J Colloid Science* 7: 428-442 (1952)* and Quayle, O.R., *Chem Reviews*, **53**: 439 (1953).**

Fig.1 Zisman plot to evaluate critical surface tension of polypropylene. Close circles: polypropylene pretreated by ethanol. Open circles: polypropylene original. The critical surface tension γ_c was obtained from extraplated value on the upper abscissa.

After adsorption, lipase was immobilized by cross-linking in soaking in aqueous glutaraldehyde (3%). The immobilized yield of lipase based on the adsorbed amount exceeded 98% for every tested source. Cross-linking of lipase by glutaraldehyde was very beneficial for achieving high yield of immobilization. Triolein was adopted as a typical lipid substrate. Oleic acid was assumed to be a main reaction product in this study. Immobilized *R. arrhizus* lipase performed outstanding rapid initiation and higher production of oleic acid in $SCCO_2$. In five repeated uses of immobilized lipase for every tested source, highly reproducible and a satisfactory reaction yield were well preserved. Reactivity of immobilized *R. arrhizus* lipase in $SCCO_2$ was equally matched in reaction conversion of non-immobilized lipase in W/O microemulsion (Fig.2). The excellent advantages of $SCCO_2$ for lipid modification by immobilized lipase are expected to be applied for industrial use.

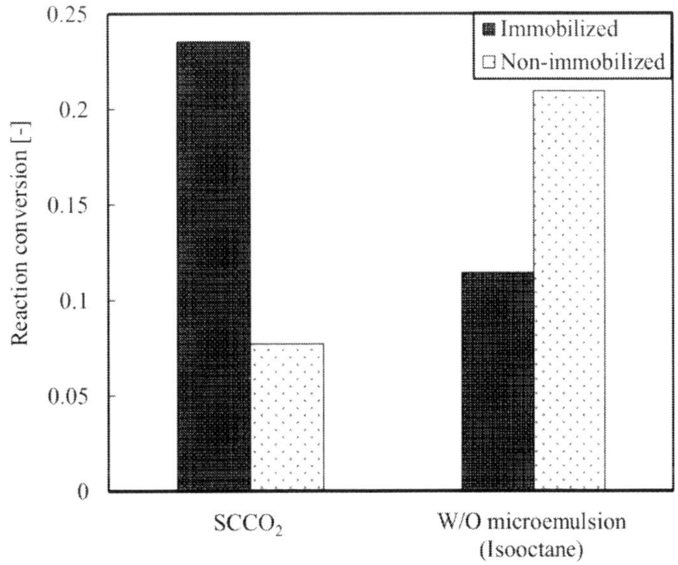

Fig. 2 Comparison of reaction conversion between supercritical carbon dioxide (10MPa, 313K) and W/O microemulsion (3.5%(v/v) 1-butanol/isooctane, 310K). Hydrolysis reaction using immobilized *Rhizopus arrhizus* lipases on Accurel was performed. Reaction conversion of non-immobilized lipase was presented as a reference. The reaction conversion was defined as the molar ratio of the amount of oleic acid produced from triolein at 60 min after initiation to the maximum molar number of oleic acid, that is twice of the initial moles of triolein, becouse of *Rhizopus arrhizus* perform 1,3 specifisity to triolein molecule.

Reference

[1] H. Taher, S. Al-Zuhair, A. H. Al-Marzouqi, Y. Haik, M. Farid, “Enzymatic biodiesel production of microalgae lipids under supercritical carbon dioxide: Process optimization and integration”, *Biochemical Engineering Journal*, 90 (2014) 103–113

[2] M. Naya and M. Imai, “Advantages of supercritical carbon dioxide for lipid hydrolysis by immobilized lipase with higher reaction rate and reproducible of repeated use”, *Journal of Chemical Technology and Biotechnology*, In press (2016)

Design of an industrial process for the production of aniline by direct amination

R.T. Driessen[1], P. Kamphuis[1], L. Mathijssen[1], R. Zhang[1], A.G.J. van der Ham[1], H. van den Berg[1], A.J. Zeeuw[2]
[1]University of Twente, Faculty of Science and Technology, PO box 217, 7500 AE Enschede, The Netherlands, e-mail: r.t.driessen@student.utwente.nl; [2]Huntsman Belgium BVBA, Everslaan 45, B-3078 Everberg, Belgium

The scope of this work is to develop and design a chemical process, which is able to produce aniline (250 kton/year) from benzene by direct amination with a nitrogenous compound. At the moment aniline is used to produce 4,4-methylenedianiline (MDA), which is a precursor for producing methylene diphenyl di-isocyanate (MDI). MDI accounts for 75% (volume based) of the global aniline market. Conventional techniques use a two-step process to produce aniline from benzene. The drawbacks of these techniques are: formation of acids, low atomic efficiency and expensive reactants. Our intention is to design a one-step process, i.e. via the direct amination of benzene to aniline, to reduce the number of reaction steps and to achieve a higher atomic efficiency. We expect that this design will reduce the number of unit operations, lower the cost on reactants and lower the energy consumption and therefore is expected to be more economical profitable.

A number of routes for the direct amination have been investigated. The direct amination of benzene by hydroxylamine is found to be the most optimal route because this route results in the highest aniline yield with reasonable reaction conditions.

Hydroxylamine is an instable and reactive component, which is expensive, and therefore the production of hydroxylamine is included in the process design. The most feasible route of producing hydroxylamine is via the hydrogenation of nitric oxide because it gives the highest hydroxylamine yield, has reasonable reaction conditions and it has already been applied in industry. Nitric oxide is not available in large quantities on the market and besides it is toxic, which means it is difficult to purchase due to strict regulations. Therefore the production of nitric oxide is also included in the process design. This reactant can be produced via the oxidation of ammonia – a technology applied for the production of nitric acid.

The methodology for chemical process design is based on Douglas' method for conceptual design [1]. After the process overall consideration, the first phase of the process design is the identification of the required reactor systems and recycle flows which resulted in a number of conceptual design alternatives. The second phase of the process design is the design of the separation system, again resulting in a number of conceptual designs. Eventually, based on (a) the magnitude of the calculated recycle flows, (b) the number of unit operations and (c) the early removal of inerts and corrosive components, the most promising conceptual design is developed up to a basic engineering level. This includes a process flow diagram and corresponding mass and energy balances calculated in UniSim Design.

The proposed process design can be divided into three major sections; the production of nitric oxide, the production of hydroxylamine and the production of aniline. A schematic overview is given in Figure 1 in which the challenge of each section is given in red. The production process of nitric oxide deals with the lower explosion limit of ammonia and the production of a significant amount of water. Part of this water is recycled to avoid that the ammonia concentration becomes larger than its explosion limit. The second section, which contains the production of hydroxylamine, has the challenge that an excess of hydrogen is required in the reactor because this excess enhances the reaction rate and thus the conversion. This also induces a significant recycle of hydrogen. Since other gases are present, hydrogen must be recovered using pressure swing adsorption (PSA). Furthermore, an acidic environment is required to increase the stability of the hydroxylamine salt. The challenge of the aniline production section is caused by the acidic environment, which is required for the direct amination of benzene by hydroxylamine. This means that the produced aniline is protonated. Since protonated aniline is also used as reactant in the MDA process, neutralization of protonated aniline is not needed.

The fraction of aniline in the aqueous phase must be larger than 75 mass percent to satisfy the needs for the consecutive MDA process. Furthermore, benzene is recycled in a large extent to ensure complete consumption of hydroxylamine.

The simulated process shows a very good carbon atomic efficiency: it is almost equal to unity. The atomic efficiency of nitrogen is 72%. The main loss of nitrogen is caused by the purge after the PSA due to ammonia, which is not converted in the nitric oxide production section. From the hydrogenous components 80% will end up in the products, the main loss (19%) is the purge of produced water during the oxidation of ammonia.

From a heat integration study it is concluded that there is an excess of high temperature heat available. From this excess heat, a considerable amount of high and medium pressure steam can be produced. External utilities are needed to cool all streams. Model investigation shows that the external cooling utilities can be substantially reduced by operating the process at slightly higher temperatures.

Furthermore, the results from an economic evaluation shows that the payback period of this process design is 8 years and the return on investment (ROI) is 6% assuming a depreciation period of 20 years.

To conclude, the proposed process is technically and economically feasible. The process can be even improved by (a) investigating if enriched air can be used instead of air, (b) separating aniline from the organic recycle by extraction, adsorption or membrane technology and (c) reducing the purge of nitric oxide. Most of these recommendations can be handled by available technologies so this would increase the potential of this process.

[1] J. M. Douglas, Conceptual design of chemical processes. McGraw-Hill, 1988.

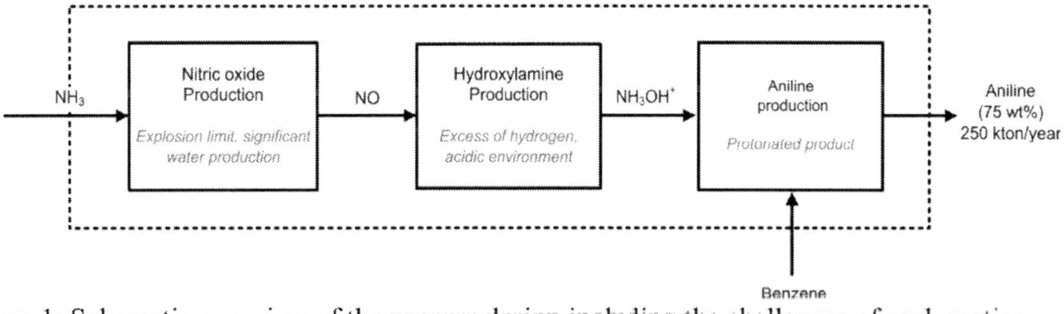

Figure 1: Schematic overview of the process design including the challenges of each section

Peroxidase immobilized on magnetic nanoparticles for the oxidation of 2,3,6-trimethylphenol

O.V.Grebennikova, V.Yu. Doluda, E.M. Sulman
Tver State Technical University, Department of biotechnology and chemistry, 170026, Tver nab. A. Nikitina 22, e-mail: sulman@online.tver.ru

Recent years magnetic nanoparticles are a promising support for enzymes in biotechnology. The use of magnetic nanoparticles as an enzyme support allows easily and quickly separating the biocatalyst from the reaction mixture via an external magnetic field.

In this paper the immobilization of horseradish peroxidase (HRP) on magnetic nanoparticles Fe_3O_4 synthesized by polyol method is proposed. Fe_3O_4 nanoparticles were treated with 3-aminopropyltriethoxysilane (APTS) which was attached to free OH groups on the surface of the iron oxide. Enzymes can easily be attached to the reactive amino groups of APTS. Peroxidase was immobilized on the modified support of iron oxide. The resulting biocatalyst can be successfully applied in the oxidation of 2,3,6-trimethilphenol (TMP) to produce 2,3,5-trimethylhydroquinone (precursor of vitamin E).

Optimal conditions of synthesis of the biocatalyst and the optimal conditions of the TMP oxidation were chosen. It was shown that pH 6.5 was optimum for the oxidation of TMP in the presence of native and immobilized peroxidase. When studying the dependence of the activity of peroxidase-based biocatalyst immobilized to magnetic nanoparticles on temperature (Figure 1) it was found that catalyst activity increased with increasing of temperature, in that case the enzyme became thermostable. However, the optimum temperature was set to 40 C, as this temperature was optimal for the native peroxidase. The further temperature increase leads the impact of support nanoparticles on TMP oxidation.

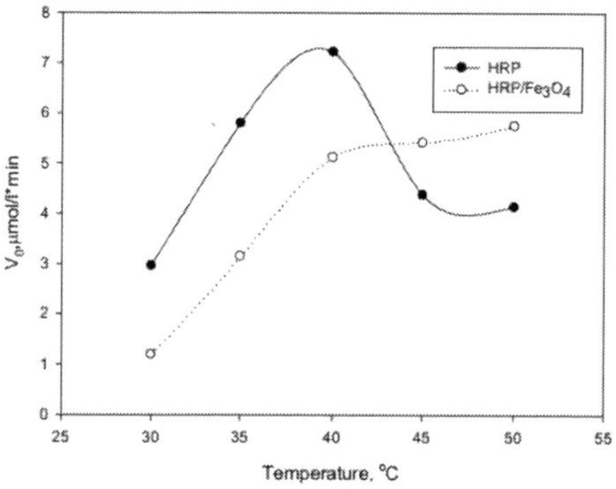

Figure 1 – Temperature influence on TMP oxidation
Initial reaction volume 30 mL, biocatalyst concentration 0.2 g/L, substrate (TMP) concentration 1.5 mmol/L, pH=6.5, $C(H_2O_2)$=1.5 mol/L

Synthesized biocatalyst was characterized by physical and chemical methods: transmission electron microscopy, vibrating magntitometry and low-temperature nitrogen adsorption. It was shown that the biocatalyst was stable for 5 cycles. The main advantage of this biocatalyst is its easy separating from the reaction mixture.

Application of Acti-zyme in wastewater treatment recovering biogas

M. M. Manyuchi,[1]D. I. O. Ikhu-Omoregbe, O. O. Oyekola
[1]Department of Chemical and Process Systems Engineering, Harare Institute of Technology, P O Box BE 277, Belvedere, Harare, Zimbabwe (mmanyuchi@hit.ac.zw)
[2]Department of Chemical Engineering, Cape Peninsula University of Technology, Bellville, Western Cape, Cape Town 7530, South Africa

Water scarcity is a global problem hence the need for sustainable wastewater management technologies. Acti-zyme a bio-catalyst has been applied for wastewater treatment with the aim to recover value added products mainly biogas and bio-solids. Wastewater treated with Acti-zyme has resulted in more than 45% removal of bio-contaminants through the Acti-zyme kinetics. Biogas harnessed from the wastewater treatment was 30% more compared to compare to Acti-zyme catalyzed systems.
Keywords: Acti-zyme, biogas, bio-nutrients, wastewater treatment

Anaerobic treatment of wastewater using Acti-zyme

Anaerobic treatment of wastewater is becoming increasingly popular since it provides process stability and control for wastewater treatment plants. In anaerobic treatment of wastewater there is no additional nutrient requirement and valuable by-products such as biogas are produced. Therefore, anaerobic treatment of wastewater using biocatalysts such as Acti-zyme is an attractive treatment option. Acti-zyme is a bacteria enzyme that biodegrades organic waste anaerobically (Duncan, 1970; Powell and Lundy, 2007; Tshuma, 2010; Dzvene, 2013; Manyuchi *et al.*, 2015a). During the digestion process, Acti-zyme multiplies and reproduces due to the availability of nutrients in the wastewater. Acti-zyme can reproduce as much as 2 billion bacteria colonies per gram added to water in 48 hours (Tshuma, 2010). Anaerobic Acti-zyme action on wastewater results in enhanced water treatment due to reduced BOD_5, nitrates, phosphates and TSS by more than 40% according to studies that were conducted on Mid-Mupfuure Dam water,on piggery wastewater in Zimbabwe, on wool scouring wastewater and sewage (Cail *et al.*, 1986; Tshuma, 2010; Dzvene, 2013; Manyuchi *et al.*, 2015b). Acti-zyme neutralises the wastewater pH and increase the wastewater effluent DO by more than 100% (Tshuma, 2010; Manyuchi *et al*; 2015b). Acti-zyme can also be used to treat blocked sewer pipes (Tshuma, 2010). Acti-zyme also helps in eliminating odour, which is a challenge in sewage treatment and can also be used for treating municipal, agricultural, commercial and food industries wastewater (Powell and Lundy, 2007). Acti-zyme loading in wastewater ranges between 3-50 g/m^3 at stirring rates of 40-60 rpm have been employed during wastwater treatment (Powell and Lundy, 2007).

Bio-nutrient removal in wastewater

The COD/BOD_5, BOD_5/TKN, COD/TKN and COD/TP ratios are used as good indicators in bio-nutrient removal through biodegradability and de-nitrification during biological sewage treatment (Tas *et al.*, 2009). Muserere *et al.* (2014) reported COD/BOD_5 ratios of 1.5-3-5 and COD/TP ratios of 20-60 for City of Harare in Zimbabwe wastewater when they were determining their treatability, and the values obtained indicated the wastewater was easily biodegradable. The COD/BOD_5, BOD_5/TKN, COD/TKN and the COD/TP ratios obtained for Acti-zyme treated wastewater were > 1.2, 4.0, 8.0 and 15.0 respectively (Manyuchi *et al.*, 2015c).

Biogas from wastewater

Biogas was generated from wastewater using Acti-zyme as bio-catalyst by Cail *et al.* (1986) from wool scouring wastewater, Duncan (1970) in hog wastes and from sewage

(Manyuchi *et al.*, 2015d). The amount of bio-methane generated was also higher as compared to previous studies with Acti-zyme by Duncan (1970) whereby they obtained a biogas with a composition of 60% CH_4 for digesting hog waste using Acti-zyme with 0.00625% of Acti-zyme for 50 days at 35 °C for the hog waste loading of 0.75-0.99 kg/kg VSS. Earlier studies by Cail *et al.* (1986) also reported a biogas composition with 68% CH_4 after digesting wool scouring wastewater at 35 °C using 1% (v/w) of Acti-zyme for 207 days for the wastewater loading of 0.5-1.5 L/day. Manyuchi *et al.* (2015d) also had a high bio-methane quality in this attributed to higher Acti-zyme loading which assisted in catalyzed digestion of the sewage sludge.

Conclusion

Acti-zyme can be used as a bio-catalyst in wastewater treatment for value added resource recovery mainly biogas. The properties of Acti-zyme can be exploited to achieve high quality biogas.

References

Cail, R. G., Barford, J. P., and Linchacz, R. (1986) Anaerobic Digestion of Wool Scouring Wastewater in Digester Operated Semi-Continously for Biomass Retention, Agricultural Wastes, 18, p. 27-38.

Duncan, A. C. (1970) Two Stage Anaerobic Digestion of Hog Wastes, Master of Science Thesis, University of British Columbia, 1970.

Dzvene, D. K. (2013) An Investigation on the Effectiveness of Acti-zyme Bacteria to Treat Organic Waste at Tripple C Pigs (Colcom), Bachelor of Science Honours Degree in Environmental Sciences and Technology, School of Agricultural Sciences and Technology, November 2013.

Muserere, S. T., Hoko, Z. and. Nhapi, I. (2014) Characterisation of Raw Sewage and Performance Assessment of Primary Settling Tanks at Firle Sewage Treatment Works, Harare, Zimbabwe, Physics and Chemistry of the Earth, 67-69, p. 226-235.

Powell, B. and Lundy, J. (2007). Acti-zyme Agricultural and Municipal Wastewater Treatment. Environment Depot Canada, p. 1-23.

Tas, D. O., Karahan, O., Insel, G., Ovez, S., Orhon, D. and Spanjers, H. (2009) Biodegradability and Denitrification Potential of Settleable Chemical Oxygen Demand in Domestic Wastewater. Water Environment Research, 81 (7), p. 715-727.

Tshuma, A. C. (2010) Impact of Acti-zyme Compound on Water Quality along Mid-Mupfuure Catchment-Chegutu, BSc Honours Thesis, Geography and Environmental Studies.

Manyuchi, M. M., Ikhu-Omoregbe, D. I. O. and Oyekola, O. O. (2015a) Acti-zyme Biochemical Properties: Potential for Use in Anaerobic Sewage Treatment Co-Generating Biogas, Asian Journal of Science and Technology, 6 (3), p. 1152-1154, 2015.

Manyuchi, M. M., Ikhu-Omoregbe, D. I. O. and Oyekola, O. O. (2015b) Effect of Acti-zyme Loading, Retention Time and their Interaction on Sewage Physicochemical Properties during Anaerobic Treatment. Chemical Technology: An Indian Journal, 10 (6), p. 259-265.

Manyuchi, M. M., Ikhu-Omoregbe, D. I. O., Oyekola, O. O., Zvarevashe, W and Makumbe, Z. (2015c) Mathematical Modelling for Bio-nutrient Removal in Sewage Treatment Using Acti-zyme as Bio-catalyst. BioMaths Communications, 2015.

Manyuchi, M. M., Ikhu-Omoregbe, D. I. O. and Oyekola, O. O. (2015d) Innovative Biogas and Bio-solids Generation from Anaerobic Sewage Sludge Digestion Using Acti-zyme as Biocatalyst as a Sustainable Measure for Developing Countries, Micro Energy Supply, Practitioner Proceedings, p. 46-49.

Silica gels of different densities: the kinetics of formation and several properties

B. G. Pokusaev, S. P. Karlov, A. V. Vyazmin, D. A. Nekrasov, V. V. Reznik
Moscow State Univ. Mech. Engin., ul. Staraya Basmannaya 21/4, 105066 Moscow, Russia;
tel. +7 495 2671051, e-mail: av1958@list.ru.

Gels are specific disperse systems with liquid dispersing medium, and the dispersion phase makes up a spatial structured mesh due to the intermolecular interaction at the contact sites [1]. Gels are micro- or nanostructured media whose internal structure depends on the material and on the method of preparation. The characteristics of gel do not remain constant over time, and can change due to the internal recondensation and recrystallization processes.

The presence of microchannels makes gels an exceptional tool for use in regenerative medicine for growth of biological issues, including the problem of culturing stem cells in vitro [2]. The capillary network of gel matrix with ordered spatial system of nanochannels is intended for the delivery of nutrients to individual cells, and to remove products from the metabolism. It seems promising the idea that gels can be used as a carrier during the formation of bio-structured elements (based on the 3D printing) for additive technology.

For experimental study we used gels on the basis of silicates. Such gels are well investigated (see, for example [3]), simply prepared, not toxic. Such gels are appropriate for additive technologies used for the creation of skeleton matrix (retaining biologically active gels, but transmissive to liquid nutrient) of bioreactors by 3D-printing. Gels of different density were investigated. The stock solution of sodium silicate was diluted with distilled water to the levels of density 1.04 – 1.08 g/cm^3. The gel-generating mixture was prepared by adding the hydrochloric acid. The density of silicate solution was controlled by measuring the refractive index by the Abbe refractometer.

For the investigation of the kinetics of gel formation the spectroscopic method is used. This method analyses the form of spectrum of transparent, reflected and scattered light of the sample. The experimental setup diagram is shown in Figure 1. The experimental setup provides the instantaneous registration of 2048 channels in spectral range of 200 – 1100 nm. The optical resolution is in average 1.0 nm. The registration system is able to output and to analyze the spectra during the gel formation.

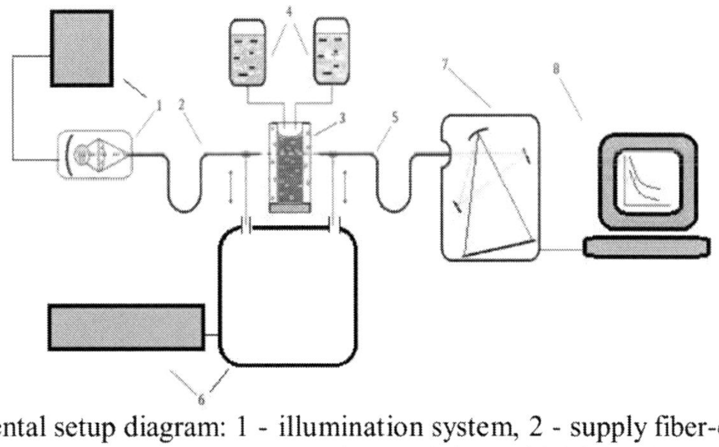

Figure 1. Experimental setup diagram: 1 - illumination system, 2 - supply fiber-optic cable, 3 - sample of the gel,4 - device preparation and submission of initial solutions for gel, 5 - receiving fiber-optic cable, 6 - device for scanning the sample gel, 7 - fiber-optic spectrophotometer USB2000+, 8 - the block of registration and processing of spectra.

The data obtained from the spectrum of the passing light through the simple are analyzed. Characteristic shapes of spectra during the gel formation at different times are shown in Figure 2. The reference time corresponds to the moment of adding in initial mix of the hydrochloric acid. The intensity of the passing light through the sample during the gel formation decreases for all wavelengths, meaning that the optical density of the media increases. This demonstrates the process of the formation of a new dispersed phase. The maximum of wavelength of the passing light shifts to infrared range during the gel formation.

Experimental data of rates of formation of gels are shown in Figure 3. Relations are obtained for the light wavelength of 680 nm, corresponding to the maximum of the lighting system spectrum. During the gel formation the intensity of light passing through the sample is reduced, which corresponds to the appearance of the structured dispersed phase. With the increasing density of the mixture the gel is also becoming more tightly structured. The time of the gel formation measured at the inflection points is almost unchanged.

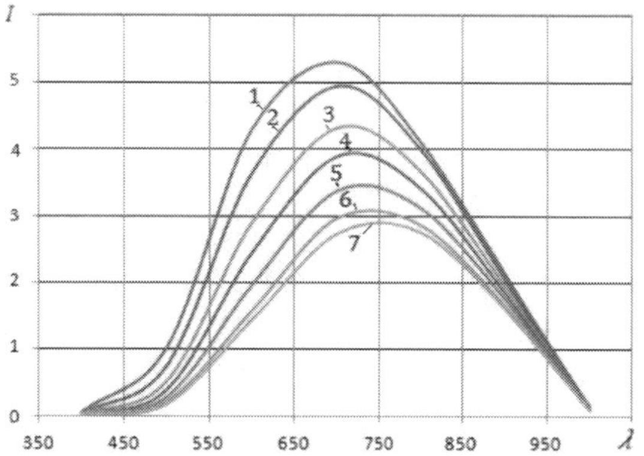

Figure 2. The spectra of transmitted light: I (conv. units.) - conventional light intensity, λ (nm) - the length of the light wave. Time of the gel formation: 1 - 0 s; 2 - 20 s; 3 - 40 s; 4 - 60 s; 5 - 120 s; 6 - 300 s; 7 - 600 s. The density of the gel 1.07 g/cm^3.

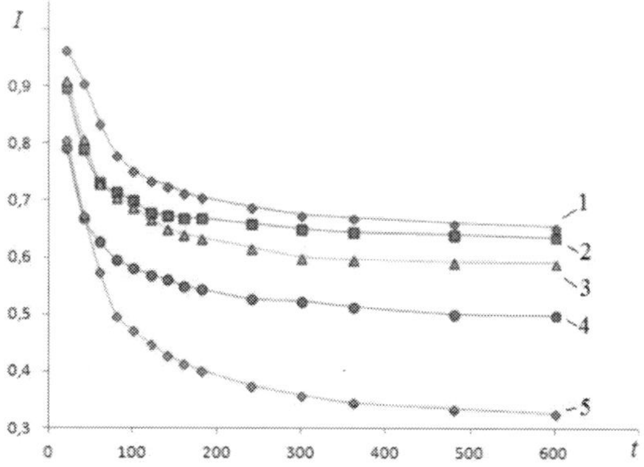

Figure 3. Intensities of transmitted light I (relat. units) vs. time t (s) for the gels of varying density: 1 - 1.04 g /cm^3; 2 - 1.05 g/cm^3; 3 - 1.06 g/cm^3; 4 - 1.07 g/cm^3; 5 - 1.08 g/cm^3.

The study of changes in the structure of the aging gel (long-term storage) is performed by spectral method. It is found that the sealing of the structure of the gel continues, though much slower than during the gel formation. Probing of the structure of the gel showed that it is not uniform in the thickness. Depending on the density of the initial solution, its optical density increases by 20 – 50% with respect to the surface layers with the distance from the surface.

Studies of silicate gel formed from solutions containing silver nanoparticles showed that such gels are optically denser. The time of their formation increases with the increase of the concentrations of nanoparticles. However, they are spatially more uniform and less prone to aging. The diffusion rate of true solutions in gels distinctly decreases with the increase of density of the gel and the concentration of nanoparticles. Herewith the size and the structure of the diffusing molecules significantly affect to reduction of the diffusion rate.

This work was supported by the Russian Science Foundation (project no. 15-19-00177).

Reference

1. Tokita M., Nishinary K., Eds., 2009, Gels: structures, properties, and functions. Fundamentals and applications, Progress of Colloid and Polymer Science, vol. 136.

2. Westrin B.A., Axelsson A., 1991, Diffusion in gels containing immobilized cells: a critical review, Biotechnology and Bioengineering, vol. 38, pp. 439-446.

3. Shabanova N.A., Sarkisov P.D., 2012, Sol-gel Technology. Nanodispersed Silica. Binom. Moscow, Russia (in Russian).

Contribution to the modelling of matting waterborne polyurethane dispersions for the tannery industry

J. Uribe[1], M. Graells[1], J. Salgado[2]
[1] Escola Universitària d'Enginyeria Tècnica Industrial de Barcelona, Consorci Escola Industrial de Barcelona, Universitat Politècnica de Catalunya, BARCELONATECH, EUETIB, C/ Comte d'Urgell 187, 08028-Barcelona, Spain; tel. +34 93 4137275, e-mail: julio.urpad.upc@gmail.com , moises.graells@ upc.edu .
[2]BASF Española S.L. , Cúrtex Division, C/ del Mig 121, 08907-Hospitalet de Llobregat, Spain; tel. +34 93 2616206, e-mail: juan.salgado-valle@basf.com

Chemical engineering has traditionally been a discipline constrained to the design and operation of chemical plants. However, product design is growing in importance due to a new paradigm targeting the implementation of holistic solutions. Applications on different fields have been addressed (Ng et al., 2009), but small attention has been paid to the modelling of electromagnetic properties such as gloss, which has an undeniable economic importance in the painting and coating industries.

In this work, final gloss has been studied for mixtures of a new hybrid matting polyurethane dispersions (PUD), composed out of conventional silica matting agent (product A), and a novel self-matting agent (product B). The main objective of the work was to develop a mathematical model for predicting the final gloss of thickened-for-application mixtures. Such a mathematical model is intended for both quality control and the design of tailored coating products (i.e. meeting specified gloss).

The work has been developed in BASF, at Cúrtex division, where experimental work has been carried out. Samples have been prepared by mixing both matting agents and incorporating the required amounts of thickener. Measurements have been performed using a rheometer and measuring the gloss of product-coated drawdown cardboards.

A two-step model was proposed with the following main assumptions:

a) The system can be satisfactorily described by means of the gloss of product A, the gloss of product B, the dynamic viscosity of the unthickened mixture, and the final viscosity of the product.
b) There is no effect of other minor additives in the final gloss or viscosity
c) The gloss of the thickened mixture is a fraction of the unthickened mixtures gloss.

The first part of the model predicts the result of mixing both matting agents and certain additives. In this model, an empirical approach was used by means of an excess gloss function that accounted for the deviations of the gloss behaviour from ideal mixing rules.

The second part of the model consists of a modification of the first principle model proposed by Fletcher (2002). This model accounts for the gloss depletion due to thickening that the matting agent experiences when its viscosity rises.

Different models were proposed using different mixing rules and functions. All these models were tested with several techniques including residual plots, normality tests and also benchmarked towards untrained data sets. Finally, a model has been found that accurately correlates the different properties. The goodness of fit is high in regard of the different statistical tests performed. Experimental validation with new mixtures has finally proved the practical use of the model developed.

References:

Ka M. Ng, Rafiqul Gani and Warren D. Seider (Editors). Chemical Products: From conceptualization to commercialization. Computers & Chemical Engineering, Volume 33, Issue 5, Pages 929-1114 (21 May 2009)

Fletcher, T. E. 2002. "A Simple Model to Describe Relationships between Gloss Behaviour, Matting Agent Concentration and the Rheology of Matted Paints and Coatings." Progress in Organic Coatings 44 (1): 25–36. doi:10.1016/S0300-9440(01)00206-5.

Microfluidic microbial fuel cell as a bioenergy generator for implantable medical devices

Mohammad Mahdi Mardanpour, Soheila Yaghmaei
Sharif University of Technology, Department of Chemical and Petroleum Engineering, Azadi Avenue, P.O. Box 11365-9465, Tehran, Iran, Tel.: +982166166430; Email: yaghmaei@sharif.edu, fax:+982166166430.

The present study evaluated nickel as a novel alternative for conventional electrodes in microfluidic microbial fuel cells (MFCs) to improve power density and reduce the cost of materials for medical usage. The feasibility of a microfluidic MFC as an efficient power generator for production of bioelectricity from glucose and urea as organic substrates in human blood and urine for IMDs was investigated for the first time. *Escherichia coli*, which normally live in the lower intestine, were used as biocatalysts in the microfluidic MFC.

The microfluidic MFC was constructed using poly methyl meth acrylate (PMMA) plates as the main body; a single microchannel (1 mm in depth; 1 mm in width; 8 cm in length) was cut using a laser beam as an anodic compartment. A nickel plate (0.5 mm in thickness; 1 mm in width; 5 cm in length) was glued to one side of the anodic compartment.

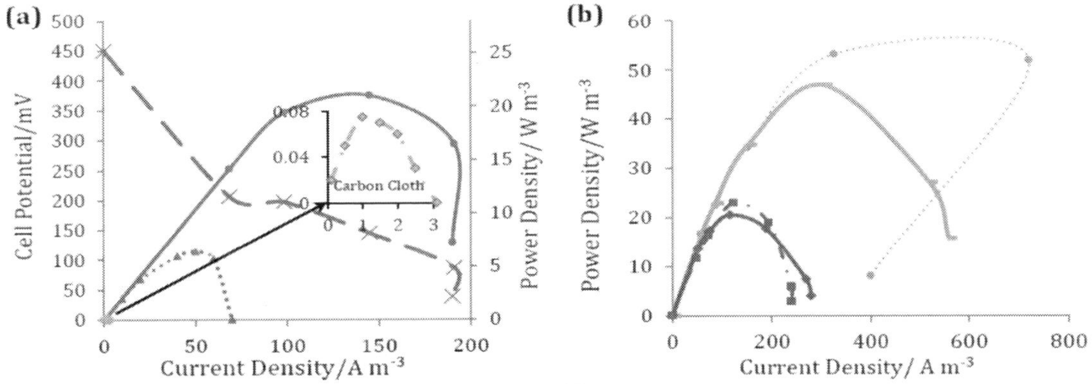

Fig. 1. (a) polarization curve of nickel electrode ( )and power density curves of nickel ( ), graphene ( )and carbon cloth ( ) electrodes in microfluidic MFC, (b) power density curves in flow rate of 30 μl h^{-1}( ),100 μl h^{-1}( ),300 μl h^{-1}( ), and 1 ml h^{-1}( ) at glucose concentration of 1200 mg l^{-1}.

Fig. 1a compares polarization and power density curves for nickel, graphene, and carbon cloth as reported by Mink et al. (2014) for batch-fed microfluidic MFCs. A maximum power density of 21 Wm^{-3} was obtained by the microfluidic MFC using nickel as the anode electrode with 10 kΩ external resistance. Fig. 1a shows that the use of a nickel electrode in batch mode for a microfluidic MFC increased the power density for carbon cloth 280-fold and for graphene 3-fold [1].

In continuous mode, the maximum power density (about 104 Wm^{-3}) increased more than 4.9-fold over that obtained during batch mode operation. The maximum power density of 53.1 Wm^{-3} was obtained for glucose oxidation in microfluidic MFC under condition of 30 μl h^{-1} flow rate and glucose concentration of 1200 mg l^{-1}. The significant point in Fig. 1b is the variation of microfluidic internal resistance with changing glucose flow rate and occurrence of overshoot phenomena in low substrate flow rates.

A comparison of maximum power densities for glucose fuel cells (as power supplies for IMDS) with modified microfluidic MFCs introduced in the present work are shown in Table 1. Comparing the maximum power density of the modified microfluidic MFC in the present study with microfluidic MFCs fed with different substrates and biocatalysts show that *E. coli* biofilm on nickel electrodes performed adequately for bioelectricity production. The glucose-fed microfluidic MFC recorded a power density of 5.2 μW cm^{-2}, which is more than 13-fold the maximum power density reported in microsized MFCs and 18% of that for a glucose fuel cell.

Table 1. Comparison of maximum power density generated in microfluidic MFCs and microsized glucose fuel cells.

Anolyte volume (μl)	Anode	Substrate	Biocatalyst	Power Density	Reference
50	Nickel	Glucose	*Escherichia coli*	52 ± 2 μW cm^{-3} 5.2 ± 0.2 μW cm^{-2}	Present study
16	Cr/Au	Glucose	*S. cerevisiae*	2.3 nW cm^{-2}	[2]
15	Gold	Glucose	*S.cerevisiae*	0.4 μW cm^{-2}	[3]
Glucose fuel cell	Si & Pt	Glucose	-	2 μW cm^{-2}	[4]
Glucose fuel cell	Zn & Pt	Glucose	-	4.4 μW cm^{-2}	[5]

The advantages of the modified microfluidic MFC over the microsized glucose fuel cell are the significant power density of the microfluidic MFC, low production cost (<$1 per device), the significant decrease in consumption of expensive materials such as platinum, simple construction, self-regeneration, management of human excreta, and elimination of oxygen injections.

Acknowledgments

This research was supported by Sharif University of Technology, Vice President for Research Grant G930111.

Reference

[1] J. E. Mink, R. M. Qaisi, B. E. Logan, and M. M. Hussain, *NPG Asia Materials,* vol. 6, p. e89, 2014.

[2] M. Chiao, K. B. Lam, and L. Lin, *Journal of Micromechanics and Microengineering,* vol. 16, p. 2547, 2006.

[3] C.-P. Siu and M. Chiao, *Journal of Microelectromechanical Systems,* vol. 17, pp. 1329-1341, 2008.

[4] V. Oncescu and D. Erickson, *Journal of Power Sources,* vol. 196, pp. 9169-9175, 2011.

[5] S. Kerzenmacher, U. Kräling, T. Metz, R. Zengerle, and F. Von Stetten, *Journal of Power Sources,* vol. 196, pp. 1264-1272, 2011.

Essential oils extraction from *Bunium Persicum* (Boiss.) by instant controlled pressure drop

E. Feizi, M.H. Eikani, F. Golmohammad, [1]B. Tafaghodinia,
Department of Chemical Technologies, Iranian Research Organization for Science and Technology (IROST), PO.Box 3353111, Tehran, Iran; tel: +98218838324, e-mail: eikani@irost.ir; [1]Department of Agriculture, Iranian Research Organization for Science and Technology (IROST), Tehran, Iran.

Bunium persicum (Boiss.), a grassy plant, is indigenous to Iran. This plant belongs to the *Apiaceae* family and is predominantly found in the mountainous regions in Kerman province of Iran which is commonly used as antispasmodic, carminative, anti-obesity and lactogage[1]. It has essential oil content 1-4% (v/w). In this research work, essential oil of *B. persicum* was extracted by instant controlled pressure drop as a new technology (*Détente Instantanée Contrôlée*, DIC). The extraction yield was compared with hydrodistillation and ultrasound.

DIC is a process introduced and developed by Allaf *et al.* in 1988 at university of La Rochelle, France[2]. It can be used for several purposes such as Pretreatment for extraction of natural compounds and non-volatile molecules from plants, Drying of raw material, Decontamination of Solid and Powder Foodstuffs, Extraction volatile organic compounds (VOCs)[3]. Nowadays, there is an interest to use green extraction methods to achieve better quality and quantity, e.g. microwave-assisted extraction, ultrasound-assisted extraction, supercritical fluid extraction, pulsed electric field and DIC methods[4]. DIC apparatus comprises of 3 main parts: (1) The autoclave with heating jacket also named the processing vessel, (2) the vacuum tank linked to a vacuum pump, and (3) the instant valve enabling the abrupt connection between (1) and (2). There is a double jacket around the vacuum tank in order to condensate the extracts[5]. The schematic diagram of the DIC apparatus is presented in Fig. 1. Other aspect is the effect of DIC on the Plant tissue that makes a porous structure[6]. There are many different compounds in *B. persicum* essential oil, incl. α- pinene, myrcene, ρ –cymene α-terpinene, limonene, α- terpineolene, terpinene-4-ol, cuminaldehyde, α-terpinen-7-al, γ-terpinen-7-al. The last three compounds, were selected to be the major compounds for measuring the amount of yield oil. They are oxygenated and desirable compounds.

Applying response surface methodology (RSM), the factors are pressure (1-4 bar), time (20-60 s), cycle (1-12), and mean particle diameter (0.3-1.18 mm) and statistics was carried out by Design Expert software. Separation and identification of the compounds were done by GC-FID and GC-MS. The results showed that the optimum pressure, mean particle size, time and cycle were 4 bar, 0.3-0.7mm, 60s and 12 respectively. The order of factors based on its impact on the essential oil yield were cycle>particle diameter >pressure> time. Also, DIC extract had more amount of oxygenated compound with respect to the other methods.

References

1. F. Sharififar, N. Yassa, and V. Mozaffarian (2010) *Pak J Pharm Sci,* 23, 300-304.
2. M. Fatima-Zohra, A. Abdelfattah, K. Ibtisam, L. Badiaa, A. Tamara, and A. Karim (2014) *J. of Advances in Chemistry,* 7, 1287-1295.
3. M. A. Vian, T. Allaf, E. Vorobiev, F. Chemat "Solvent-Free Extraction: Myth or Reality?" in *Alternative Solvents for Natural Products Extraction*, ed: Springer, 2014, 25-38.

4. N. Rombaut, A. S. Tixier, A. Bily, and F. Chemat (2014) *Biofuels, Bioproducts and Biorefining,* 8, 530-544.
5. T. Allaf, V. Tomao, K. Ruiz, K. Bachari, M. ElMaataoui, and F. Chemat, (2013) *LWT-Food Science and Technology,* 51, 111-119.
6. B. Berka-Zougali, A. Hassani, C. Besombes, and K. Allaf (2010) Extraction of essential oils from Algerian myrtle leaves using instant controlled pressure drop technology," *J. of Chromatography A,* 1217, 6134-6142.

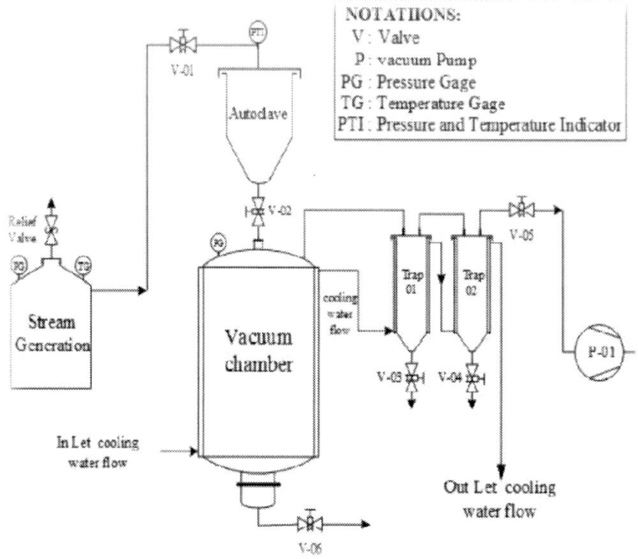

Fig 1. DIC laboratory system

Isolation and purification of laccase producing fungi and determination of optimize condition

[1] Z. Ghobadi Nejad, [2]S. M. Borghei, [3]S. Yaghmaei, [3]S. Morshedi, [3]H. Sharifi
Graduate School of the Environment and Energy, Science and Research Branch, Islamic Azad University, Tehran, P.O. Box: 14515-775, Tehran, Iran
z_ghobadi@mehr.sharif.edu
[2]Biochemical and Bioenvironmental Research Center, Sharif University of Technology, Tel: +982166164103, Fax: +982166005417, P. O. Box 11155-1399 Tehran, I.R. IRAN, e-mail: mborghei@sharif.edu
[3]Department of Chemical and Petroleum Engineering, Sharif University of Technology, Tel: +982166166430, Fax: +982166005417, P. O. Box 11365-6891 Tehran, IRAN, Yaghmaei@sharif.edu
[3] Department of Chemical and Petroleum Engineering, Sharif University of Technology, Tel: +989357563108, P. O. Box 11365-6891 Tehran, IRAN, shayanmorshedi@gmail.com
Department of Chemical and Petroleum Engineering, Sharif University of Technology, Tel: +9899118427708, P. O. Box 11365-6891 Tehran, IRAN, h2.sharifi1994@gmail.com

Enzyme production is an expanding field of biotechnology. Laccase is a copper containing enzyme, catalyzes theoxidation of various aromatic compounds specifically phenols and anilines while concomitantly reducing molecular oxygen to water. Promising industrial and environmental applications of laccase include textile-dye decolourization, pulp bleaching, food industry, bioremediation of soils and water, polymer synthesis, the development of biosensors and biofuel cells, pharmaceuticals & cosmetic industry. Solid state fermentation holds tremendous potential for large scale production of enzymes. Although the enzyme is present in plants, insects and bacteria, the most important source are fungi and particularly basidiomycetes.

SSf is generally defined as the growth of microorganisms on solid materials in the absence or near absence of free water. SSF has marked advantages over SmF in terms of productivity, concentration of the product and effluent generation.

In this study, samples include soil and rotting wood were collected from different forest and gardens localities and universities campus in Tehran, Iran. Fungal isolates were obtained from these samples using serial dilution method and point inoculation method on PDA agar plates. Pure cultures were maintained on PDA slants and were screened for laccase production. For all of these plates, guaiacol (1%) was added as an indicator. Formation of reddish brown zone around the colony on agar plates indicates presence of laccase. The reddish- brown zone appears due to oxidation of guaiacol by laccase. After screening several fungal strains, some of them were chosen due to their relatively high laccase activity compared to other laccase producing fungi.

Selection of an appropriate substrate is a key factor in SSF, which determines the success of the process. It has been a practice to use lignocellulosic agroindustrial residues for the production of laccase. Wheat bran and rice bran were screened for the production of laccase. 5 gm. of them were autoclaved in petri plates. Buffer solution of pH=5.5 (10 m M sodium-acetate buffer) were used as moistening medium, and to maintain desirable pH.

It was found that under relative humidity 60 to 70%, neutral pH, 30°C temperature, and 5 days incubation time, and no additional nitrogen supplement, rice bran as a substrate give maximum enzyme activity. Enzyme activity is measured in U/gfs (where gfs is defined as the number of units of enzyme produced from 5 g fermented substrate).

The present isolate H1 is found to be a very efficient producer of laccase under solid state fermentation. The present investigation has confirmed and evaluated the use of rice bran as an inexpensive and easily available agroindustrial waste for laccase production under solid state fermentation.

Measurement and prediction of condensate gas hydrates formation conditions – a study case

Mimoun Hadj[2] ,Samer Said[1], Otmanine Ghazi[2] and Belloum Mohamed[1]
1. Laboratory of Materials Science (LCMVAR), Dpt of Chemistry, Faculty of Science, University Hadj Lakhdar of Batna,, Algeria
2. Laboratory of Hydrocarbons technology, Dpt of Chemistry, Faculty of Hydrocarbons & Chemistry, University Mhamed Bouguerra of Boumerdes, Algeria.
mimounhadj@yahoo.fr

Abstract: A new field of gas condensate recently discovered in south of Algeria caused a lot of difficulties and even accidents during its operation due to unexpected hydrates plug formation. To prevent such damage, we carried out an experimental study using a visual high pressure PVT cell to investigate the thermodynamic conditions of gas hydrates formation and stability, then we established hydrate formation curves, identifying safety zones, allowing operators to work safely. The influence of inhibitors on gas hydrate formation was also investigated. This study was gathered with a numerical analysis using PVTsim commercial software to confirm our experimental results.

Key words: hydrates, gas condensate, plug, PVT, safety

Methane production from ethanol wastewater using a two-stage UASB system: Granular sludge formation with high performance

[1]S. Yimthachote, [1,2]S. Chavadej
[1]Chulalongkorn Univ., The Petroleum and Petrochemical College, Phayathai Rd., 10330 Bangkok, Thailand; tel. +6622184139, fax. +66 22184139, e-mail: sumaeth.c@chula.ac.th;
[2]Center of Excellence on Petrochemical and Advanced Materials Chulalongkorn University, Phayathai Rd., 10330 Bangkok, Thailand; tel. +662 2184141, fax:+662 6117619.

A two-stage upflow anaerobic sludge blanket (UASB) system is one of high rate anaerobic processes for methane production. The aim of this study was to investigate the effects of COD loading rate on methane production and to correlate the system performance to the formation of granular sludge. The UASB system consisted of a 4-L hydrogen UASB unit and a 24-L methane UASB unit connected in series. The system was operated at different COD loading rates of ethanol wastewater under mesophilic temperature (37°C) with a constant recycle ratio of 1:1 (effluent flow rate: feed flow rate). The pH in the hydrogen UASB unit was controlled at 5.5 while the pH in the methane UASB unit was not controlled. The system could be operated up to the highest COD loading rate of 28 kg/m^{3}d based on methane working volume with good process stability, indicating by a very low VFA of 34 mg/L as acetic acid. The extremely high activity of the studied two-stage UASB system (about 4 times of that of a typical UASB) resulted from the formation of granular sludge in the system with the size range of 0.18-2.64 mm and a very high microbial concentration of 14780 mg/L. Under the highest COD loading rate, the system had highest SMPR of 7363 mL/L_Rd (or 343 mL/g MLVSS·d) and the highest methane yield of 309 mL/g COD removed (or 298 mL/g COD applied).

On-site cellulolytic enzyme production using solid-state fermentation of mixed filamentous fungi without the extraction step

L. Maehara[1,2], S.C. Pereira[1], A.J. Silva[2], C.S. Farinas[1,2]
[1]Embrapa Instrumentação, Rua XV de Novembro, 1452, São Carlos, SP, Brazil; [2]Federal Universityof São Carlos, Graduate Program of Chemical Engineering, Rd. Washington Luís, km 235, São Carlos, SP, Brazil; tel. +551621072908, e-mail: cristiane.farinas@embrapa.br

The development of novel and efficient processes for the production of cellulolytic enzymes is a key step to ensure the economic viability of the enzymatic route for the production of biofuels and other bioproducts from biomass, within the biorefinery concept. To address this technological challenge, this work demonstrates the feasibility of using the whole solid-state fermentation(SSF) medium of mixed filamentous fungi containing enzymes, fungal cells, and residual solid substrate for the saccharification of a lignocellulosic biomass and biofuels production. The use of SSF for industrial enzyme production has received increasing attention over the past years due to the inherent advantages of this cultivation system. However, in traditional SSF processes, after the cultivation period the enzymes synthesized by the microorganisms are extracted from the solid substrate by conventional solid-liquid extraction, with the final products after filtration being a liquid supernatant containing the enzymes of interest, which can be further concentrated, together with a solid residue. A potential way of avoiding production of this solid residue is to use the whole fermentation medium, containing the enzymes, mycelia, and the residual solid substrate, for the saccharification of lignocellulosic biomass such as sugarcane bagasse. By using the whole SSF medium, the steps of extraction and filtration can be excluded from the overall process, providing advantages in terms of cost reduction and also avoiding the generation of waste streams, hence being compatible with the biorefinery concept. Here, the selection of culture conditions to maximize the production of enzymes directly in sugarcane bagasse as a substrate of SSF as well as the hydrolysis of biomass with the whole SSF medium of isolated and mixed fungal strains were evaluated.

The use of a combination of sugarcane bagasse and wheat bran (1:1, on a dry mass base) plus 0.075 g/g of lactose favored the production of cellulolytic enzymes, leading to an increase of enzyme activity of 73% for β-glucosidase, 67% for endoglucanase, and 72% for xylanase, when compared to the control condition using steam-exploded pretreated sugarcane bagasse as the substrate, without the presence of lactose or wheat bran.This selected medium was then used for SSF of ten different combinations of fungal strains (Figure 1).

The mixed cultivation of *Trichodermareesei* and *Arpergillusoryzae* resulted in an enzymatic cocktail more efficient on the saccharification of pretreated sugarcane lignocellulosic biomass. Comparing the individual fungal cultivations with the mixed cultivations, the results showed a significant difference between them in terms of the glucose released during the hydrolysis reaction (Figure 1). The mixed cultivation of *T. reesei* with *A. oryzae*, resulted in levels of glucose released of 6.2 g/L and 14.1 g/L of reducing sugars after 36 hours of enzymatic hydrolysis. This concentration of glucose released compared to the best individual cultivation (*A. niger*) corresponds to a gain of about 50%.

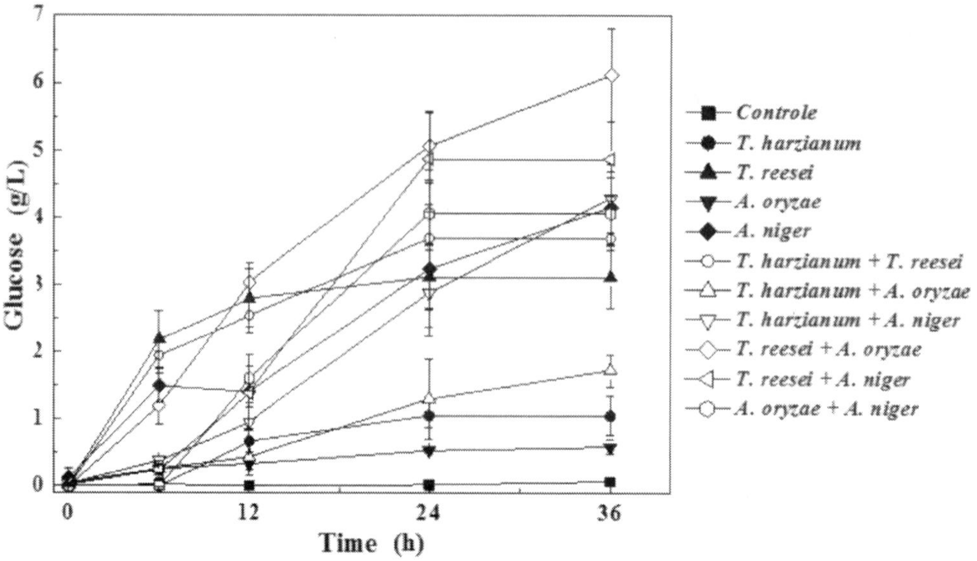

Figure 1. Amount of glucose released during enzymatic hydrolysis using the whole fermentation medium of solid-state fermentation (SSF) for ten combinations of fungal strains cultivated isolated and in a mixed culture (combined two by two) and the control sample containing only the steam-exploded pretreated sugarcane bagasse. Hydrolysis reactions were carried out with 5% (w/v) solids loading at 50 °C for 36 h at 200 rpm of agitation.

These results are very promising when compared to our previous reports using isolated fungal SSF cultivations without the extraction step[1, 2]. Moreover, it indicates that the enzymes that remains adsorbed to the solid medium contribute significantly to the hydrolysis step. These findings may contribute to the development of process for on-site production of the enzymes needed for biomass conversion into renewable energy and bioproducts.

References

1 Pirota, R., Delabona, P. S. and Farinas, C. S. (2014) Simplification of the biomass to ethanol conversion process by using the whole medium of filamentous fungi cultivated under solid-state fermentation. *Bioenergy Research*. **7**, 744-752

2 Pirota, R., Baleeiro, F. C. F. and Farinas, C. S. (2013) Saccharification of biomass using whole solid-state fermentation medium to avoid additional separation steps. *Biotechnology Progress*. **29**, 1430-1440

Processing of Sidi Bouali's sand – Algeria, for glass industry

M.OULD HAMOU[1], K. DERAMCHI[1], A. MERCHICHI [1], L. OUBRAHIM[1], S. BOUTRIA [1]
Ecole Nationale Polytechnique, 10, avenue Hassen Badi, El Harrach Alger, Algérie Tel. : +213 23 82 85 39 e-mail: ouldhamoumalek@yahoo.fr [(1)]Laboratoire Génie Minier, Ecole Nationale Polytechnique, 10, avenue Hassen Badi, El Harrach Alger, Algérie.
Fax : 00 213 23 82 85 29

The sand as a raw material requires one or more processing operations, such as sizing, washing, desliming, mixing, leaching and by using gravimetric, electrostatic, magnetic and flotation methods, before being used in various industrial applications.
As a useful substance, sand has found a very large application in the glass industry, after being successfully recovered in order to meet the quality required by users. In this paper are given the results obtained after the processing of the sand of Sidi Bouali (Algeria) for glass industry. The material is located in a sandy foundation composed of three varieties: light, yellow and brown. The gray sand layer located in the gray-white light can be operated separately. Characterization, particle size, mineralogical and chemical studies have been undertaken. The processing of a raw material with 96.26 % SiO_2 and 1.13 % Fe_2O_3 has yielded to a purification of this sand, which reached a suitable concentrate of 98.7 - 99.30 % SiO_2 and 0.06-0.08 % Fe_2O.

Keywords: *sand, iron oxide, silica, flotation, glass.*

Production of ethanol from sugarcane bagasse: real data and its bottlenecks

C. K. Yamakawa[1,2], S. C. Rabelo[1], M. B. W. Saad[1], C. M. de Carli[1], S. T. Rojas[3], E. L. M. Arias[1,2], C. E. V. Rossell[1,2], R. M. Filho[1,2]
[1]Laboratório Nacional de Ciência e Tecnologia do Bioetanol (CTBE), Centro Nacional de Pesquisa em Energia e Materiais (CNPEM), Caixa Postal 6192, CEP 13083-970, Campinas, São Paulo, Brasil, tel. +55 19 3518 3139, e-mail: celina.yamakawa@bioetanol.org.br; [2]School of Chemical Engineering, State University of Campinas (UNICAMP), Campinas, SP, Brazil. [3]School of Food Engineering, State University of Campinas (UNICAMP), Campinas, SP, Brazil

The production of fuel ethanol from sugarcane bagasse was evaluated technically starting with the bagasse reception until obtaining ethanol and vinasse. This study focuses on the conversion of hexoses stream to ethanol. The reception of sugarcane bagasse from Granelli Mill (Charqueada, state of São Paulo, Brazil) occurred in the vibratory sieve multideck (MultiVibro, Brazil) from CTBE's Pilot Plant to remove impurities such as stone, sand, and lumpy material. Then, the cleaned bagasse was pretreated from hydrothermal treatment conditions at 190°C/ 10 minutes with 10% (m/v) of solids in the mixture reactor of 350 L (Pope Scientific, USA) from CTBE's Pilot Plant. The discharge of reacted slurry succeeded in the nutsche filter (Pope Scientific, USA) from CTBE's Pilot Plant. The solid stream (cellulignin) was washed with water in a basket centrifuge (Ferrum, Switzerland) from CTBE's Pilot Plant.

The whashed cellulignin was hydrolysed with a commercial enzyme cocktail (10 FPU/ g cellulignin) in the same reactor used to pretreatment. The hydrolysis conditions were 50°C, 60 hours, and the pH was adjusted to 4.8 with NaOH and H_2SO_4. After that, the liquid stream rich in glucose called cellulosic hydrolysate or C6 liquor was recovered by centrifugation in the basket centrifuge. The unreacted phase was discarded appropriately. The raw C6 liquor with 2.3 °brix was concentrated until 26.5 °brix in the wiped film evaporator (Pope Scientific, USA) from CTBE's Pilot Plant. The conditions of evaporation were 80°C and 475 mmbar of vacuum. The concentrated cellulosic hydrolysate was clarified according to Copersucar (1987). Then, the material was divided in erlenmeyer flasks for sterilization in the autoclave during 15 minutes at 121°C. Nutrient supplementation was assessed previously to conduct an ethanol fermentation with 100% of cellulosic hydrolysate with *Saccharomyces cerevisiae* and showed the requirement of molasses for successful conversion to ethanol. Therefore, molasses diluted with potable water was sterilized separately and then after blended with the sterilized concentrated C6 liquor to compose the substrate for ethanol fermentations. The substrate was prepared for 10 successive ethanol fermentation with cell recycle in a typical current industry configuration. The fermentations were conducted in a bioreactor Bioflo 115 of 2 L (New Brunswick Scientific, USA) from CTBE's Bioprocess Lab, at 33°C, 6 hours of substrate feeding and 2 hours to ensure the uptake all fermentable sugars. After each fermentation, the cells were recovered through centrifugation in the Avanti J-26 XP high performance centrifuge (Beckman Coulter, USA). The pellet was diluted with water and pumped back to the bioreactor for acid treatment then followed by new fermentation. The fermented media from all fermentations, wine, were joined for distillation stage in a batch distillation system autodest® 800 AC (Fischer, Germany) from Unicamp's Lopca Lab. Two hydro alcoholic fractions with 77% (v/v) and 40% (v/v) of ethanol were characterized in terms of physical properties and composition. The BOD, COD, nitrogen, potassium, sulfates and phenols of vinasse were analyzed. The preliminary results demonstrated a requirement to the development process to achieve expectations productivity according pointed by Milanez *et al.* (2015) to boost second-generation ethanol.

COPERSUCAR, Cooperativa de produtores de cana, açúcar e álcool do estado de São Paulo, Fermentação. São Paulo, 1987.

Milanez, Artur Yabe *et al.* De promessa a realidade: como o etanol celulósico pode revolucionar a indústria da cana-de-açúcar: uma avaliação do potencial competitivo e sugestões de política pública. BNDES Setorial, Rio de Janeiro, n.41, p. [237]-294, 2015.

Production of lipolytic enzymes using agro-industrial residues

L. M. Pinotti [1], J. X. Lacerda [1], M. M. Oliveira [1], R. D. Teixeira[1], C. Rodrigues [2], S.T.A. Cassini [2].

[1] Department of Engineering, Federal University of Espírito Santo, Rodovia BR 101 Norte Km 60 São Mateus – ES – Brazil, tel. +552733121592, e-mail: pinotti2008@hotmail.com

[2] Department of Environmental Engineering, Federal University of Espírito Santo, 29075-910 Vitoria –ES- Brazil

Lipase (E.C.3.1.1.3) is a versatile and key enzyme in various bioprocesses involving the esterification and transesterification reactions for biodiesel generation. However, the production and recovery of enzymes is very expensive and often constitute an obstacle to wide use in bioprocesses. In this context, the solid state fermentation (SSF) could be an low-cost alternative, since it allows the use of agro-industrial wastes with low added value and promoting the production of more concentrated biocatalyst. The utilization these residues in the fermentations, not only minimize the quantity off these residues in the environmental, but also add value to raw material, trough the production of economical interesting substances. The objective of this work is to verify the lipase production using sugarcane bagasse as substrate inoculated with lipolytic microorganism. The sugarcane bagasse is an agro industrial residue with high availability in Brazil and can have multiple uses, representing 25 to 30% of the total weight of the sugar cane. The microorganisms used were *Penicillium sp.* and *Rhizomucor sp.* combined with temperature operating conditions of 28, 33 and 38 ° C, moisture content of 60, 70 and 80% and olive oil concentration 5, 7,5 and 10% as inducer. The study was performed by using factorial design of type (3^3) with central points. The results shown that best microorganism for lipase production was the *Rhizomucor sp.* (0.58 IU / g substrate), although there is very slight difference when compared to the *B. megaterium* (0.47 IU / g substrate). These results were found in the conditions of 33 °C, 80% moisture content and 10% of inducer for both microorganisms. After analyses statistic was verified that the temperature and moisture content of the medium interfered in the enzyme production for *Penicillium* sp. For the *Rhizomucor* sp. the moisture content and the concentration of the inductor olive oil interfered in the enzyme production.

Keywords: agro-industrial residues, microorganisms, lipases, sugarcane bagasse, solid state fermentation.

Removal of oxygen from wine pipelines transportation

V. Báleš[1], P. Timár[1], P. Timár[1], M. Ševčík[2]
[1]Slovak University of Technology, Department of Chemical and Food Technology, Radlinského 9, 812 37 Bratislava, Slovakia, e-mail: vladimir.bales@stuba.sk, pavol.timar@stuba.sk, pavel.timar@stuba.sk;[2]Winery Chateau Topoľčianky

This work is focused on the use of nitrogen as an inert gas in the desorption of oxygen from wine during his transportation in the pipeline. Preferably, measurements are made to determine the optimal flow rate of the nitrogen gas injected into the flow of the wine to reach the desired oxygen concentration in wine. The results of the individual measurements and their analysis are made. The measurements were realized in the real factory conditions. Next figure shows one example of measurement:

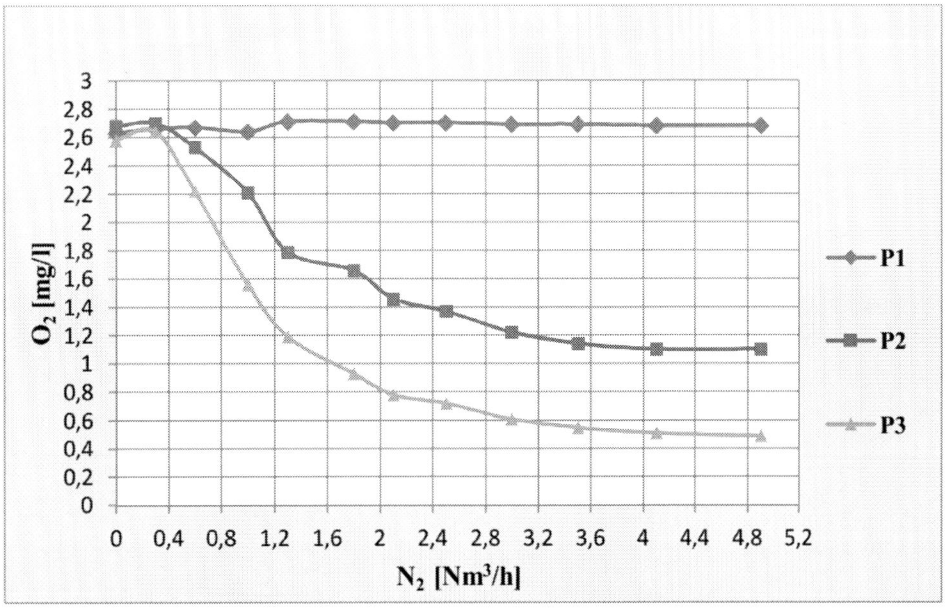

P1 - Exit from wine tank, P2 - 4 m after the place of nitrogen injection, P3 - Inlet to the second tank, 34 m after the place of nitrogen injection

The optimal nitrogen flow injection were determined for reaching of oxygen concentration under 0,6 mg/l.

Rheological study and application of asphalt modified with tatajuba's wood

[1]J. G. M. Coelho, A. L. A. Mesquita, C. G. B. T. Dias
Federal Univ. of Pará, Graduate Program in Engineering of Natural Resources of the Amazon, Augusto Corrêa St. 01, Belém, Pará, Brazil;
tel. +05191982528879, e-mail: Johnny@ufpa.br; [1]Ecocomposite Laboratory, Augusto Corrêa St., Belém, Pará, Brazil; fax: +559132017000.

In pavement is commonly used coating consisting of mixed ligand based oil and asphalt cement aggregate as pebble, sand, stone materials and cements. The main objective of this study is the evaluation of asphaltic layer with waste wood. In which follow the following steps: (1) shear mixture of sawdust, wood type tatajuba (bagassa guianensis), red quartz sand (RQ) and ground tire rubber (CR), (2) to prepare nine specimens of hot mix asphalt (HMA) second dosage Marshall in small scale, with composite tatajuba/RQ/CR dosage 20/2/10 by weight, (3) rheological characterization by Haake RS 6000 of parallel plates which was analyzing the viscosity ($\eta(\dot{\gamma})$) as a function of shear rate and the complex modulus ($G^*(\omega)$) as a function of angular velocity of the sample with 7%, 7,5% and 8% of asphalt and 2%, 3% and 4% of pre-mixes with Tatajuba, scanning electron microscopy (SEM) and strength tests the maximum radial compression of the specimens to obtain the parameters of Marshall. The results showed that wood type tatajuba was inserted into the polymeric structure of rubber modifying the structure and rheological properties of the material in the composite increase of stability in mechanical testing, the intermediate values of relationship bitumen/voids in sample 3% of tatajuba wood and 8% of asphalt. Thus, the applicability of the composite studied is satisfactory and environmentally friendly when it comes to the reuse of materials that replace natural fields. Thus, the composite application viability study is satisfactory and environmentally friendly when it is reusing materials that can replace natural materials deposits.

Semiconductor biomimetic electrode – TPhPFe(III)/Al_2O_3/Si, for catalase and peroxidase types

N.N.Malikova, N.I.Ali-zadeh, T.M.Nagiev
Nagiev Institute of Catalysis and Inorganic Chemistry, National Academy of Sciences of Azerbaijan,
113 H.Javid av., az-1143, Baku, Azerbaijan, Tel.: (99412)538-47-07, Fax: (99412)538-47-67;
E-mail: tnagiev@azeurotel.com

Biosensors allow to continuously monitor biochemical processes in biotechnology, todetermine the quality of food products, their composition, the content of toxins, antibiotics, and to perform environmental monitoring.

In this respect, the works in the field of simulation catalysis[1,2]with the help of which it is possible to replace the active part of biosensors with their chemical analogues –biomimics, are of the greatest scientific value.

Nowadays, semiconductor biosensors characterized with good reproducibility, high sensitivity, availability and cheapness, arewidely used.

For this purpose ironporphyrin biomimetic electrode based on silicon (Si)was examined.Silicon (Si) is one of the basic materials used for the creation of bio- and chemical sensors, which is due to manufacturability and low cost. Tetraphenylporphyrin iron adsorbed on alumina (TPhPFe(III)/Al_2O_3) was applied on silicon (Si) substrateby gluing.

Results of potentiometric studies of TPhPFe(III)/Al_2O_3 biomimetic electrode catalase activity are shown in Figure 1.

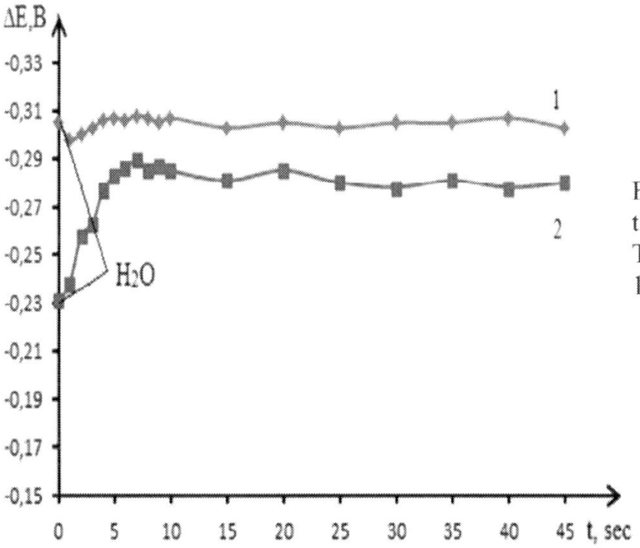

Fig.1. Change of e.m.f. system depending on the time at low concentrations of H_2O_2 for TPhPFe (III)/Al_2O_3/Si electrode.
1. 1×10^{-6} wt.%; 2. 1×10^{-4} wt.%.

Considering the curves 1 and 2 of Figure 1.it can be seen that the ironporphyrin biomimetic sensor of catalase type exhibits high sensitivity for a few seconds.

The electrode potential of the carriersolution is ΔE = (-0,231 mV) for 1×10^{-4} wt. % and ΔE = (- 0,305 mV) for 1×10^{-6} wt. %.[3].

It can be seen from the figure that there is a great leap of electrode potentialfor a few seconds when adding H_2O_2 (in both cases).

Ethyl alcoholis one of the most important technical products and raw material. As we know a considerable amount of ethanol produced from food raw materials is consumed for the preparation of alcoholic beverages, which requires a clear and precise control over the content of ethyl alcohol in these products. So it was interesting to explore catalase biomimetic sensor on the peroxidase activity.

Figure 2 shows the experimental data obtained during the study of peroxidase activity of tetrafenilporphyrin biomimetic electrode at a concentration of 0.5 wt. % C_2H_5OH and various concentrations of H_2O_2. It was necessary to measure the electrochemical potential of the 0.5 % - aqueous solution of ethyl alcohol while analyzing curves of Figure 2, to study the peroxidase reaction.

As can be seen from the fig.2 the value of 0.5 % - aqueous solution of ethyl alcohol is identical with potentials of redistilled water (Fig.2 curve 1).Thus, the water-alcohol solution remains nearly inert with respect to the electrode and the concentration of $H_2O_2$is ranged from 1×10^{-6}wt % to 0,5wt %.[4].

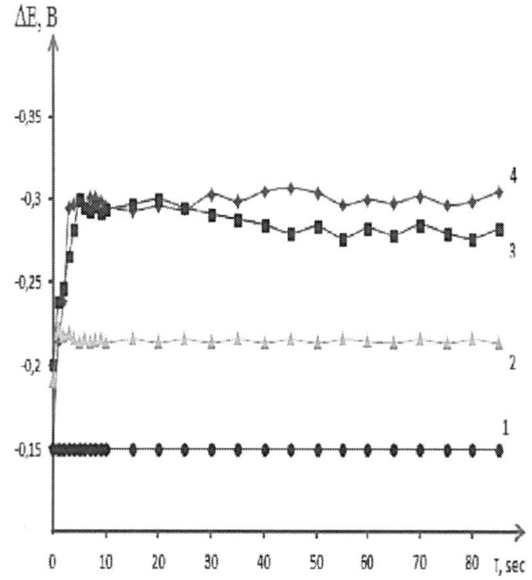

Fig 2. Change of e.m.f. system depending on the time of peroxidase sensor.
1. 99.0 wt % H_2O, and 0.5 wt % C_2H_5OH
2. H_2O 99.0 wt %, 0.5 wt % C_2H_5OH + H_2O_2 1×10^{-4} % by weight;
3. 99.0 wt % H_2O, 0.5 wt % C_2H_5OH + H_2O_2 and 0.5 % by weight;
4. H_2O 99.0 wt% , 0.5 wt % C_2H_5OH + H_2O_2 0.1 wt %;

Considering the curves (2,3,4) of Figure 2 we can see that biomimetic sensor of peroxidase type exhibits high sensitivity.

The electrode potential of carrier solution is: forH_2O_2concentration of0.5 wt % ΔE = (-0,191 mV); for H_2O_2concentration of1×10^{-4}wt % ; ΔE = (-0,2 mV); for H_2O_2concentration of0.1 wt % ΔE = (-0,1 mV), where the water-alcohol solution remained constant and was 0.5 wt %.in all cases.

Reference

[1] Nagiev T.M. Coherent Synchronized Oxidation by Hydrogen Peroxide. Amsterdam: Elsevier, 2007. 325 p.

[2] Nagiev T.M. "Biomimetic Based Application". Preface IX. Chapter 4. Croatia. INTECH. 2011. P.105.

[3] N.N.Malikova, N.I.Ali-zadeh, T.M.Nagiev Catalase-biomimetic sensor on base of electrochemical electrode TPhPFe(III)/Al_2O_3/Pb and TPhPFe(III)/Al_2O_3/Si. //Journal of Chemistry and Chemical Engineering (USA), www.davidpublishing.com Vol. 9, Number 1, January 2015 (Serial Number 86) P.67-70

[4] N.N.Malikova, N.I.Ali-zadeh, T.M.Nagiev Peroxidase-biomimetic sensor on base of silicon electrode–TPhPFe(III)/Al_2O_3/Si //ECCE10+ECAB3+EPIC5 September 27th – October 1th 2015, Nice, France, p.1039

Separate production of hydrogen and methane from cassava wastewater by three-stage anaerobic sequencing batch reactor (ASBR)

[1]A. Jiraprasertwong, [1,2]S. Chavadej
[1]Chulalongkorn Univ., The Petroleum and Petrochemical College, Phayathai Rd., 10330 Bangkok, Thailand; tel. +662 2184139, fax:+662 2184139, e-mail: sumaeth.c@chula.ac.th;
[2]Center of Excellence on Petrochemical and Advanced Materials Chulalongkorn University, Phayathai Rd., 10330 Bangkok, Thailand; tel. +662 2184141, fax: +662 6117619.

Biological hydrogen and methane production from cassava wastewater by mixed cultures was investigated by using a three-stage anaerobic sequencing batch reactor (ASBR) system. The feed wastewater contained around 21,000 mg/L total COD while the three-stage comprising dark, photo, and methane reactors with 4, 10, and 20 working volumes, respectively were operated under mesophilic temperature (37±1 °C) and 6 cycles per day. The photo bioreactor with the illumination intensity approximately 6,000 lux (led lamp) and the pH in the hydrogen reactor was controlled at around 5.5. A recycle ratio of 1:1 was used to minimize a NaOH addition in the hydrogen reactor for pH adjustment. The three-stage ASBR system was carried out at different COD loading rate (ranging from 4 to 14 kg/m^{3}d based on methane reactor) in order to study the effect of organic loading rates (OLR) to obtain the optimum condition that provide maximum performance on anaerobic digestion system of the three-stage ASBR.

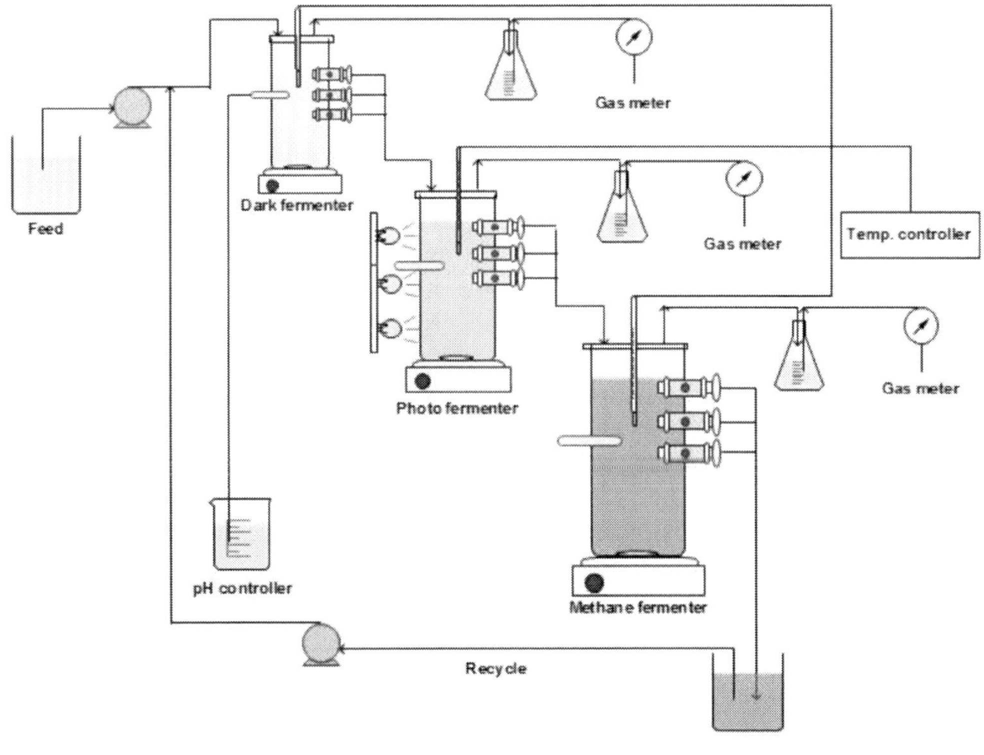

Figure 1: A three-stage anaerobic sequencing batch reactor (ASBR) system.

Small and high scale biorefineries from lignocellulosic biomass in Colombia

[1]V. Hernández, [2]Eulogio Castro, [1]C. A. Cardona
[1]Tecnoparque, Servicio Nacional de Aprendizaje SENA, Km 10 vía al Magdalena, Manizales-Colombia
[2]Jaen University, Department of Chemical, Environmental and Material Engineering Campus Las Lagunillas, tel. +34 953212163.
[3]Universidad Nacional de Colombia Sede Manizales, Research Group in Chemical, Catalytic and Biotechnological Processes, Km 9 vía al Aeropuerto, Manizales, Colombia, tel. +57 68879300 ext 55880, e-mail: vhernandezp@unal.edu.co;

Because of its geographic location and its wide range of climates, Colombia is the third-ranked country in the world in terms of biodiversity, housing approximately the 10% of the world species, and the fourth-ranked in terms of hydrographic resources. According to the Departamento Administrativo Nacional de Estadística (DANE), from 2000 to 2009, agriculture contributed an increase of 27.4% to the Colombian Gross Domestic Product (GDP). Furthermore, the World Trade Organization reported that the agroindustrial exports represented for Colombia approximately 30% of its total exports. It is estimated that approximately 44.77% of the continental surface of the country (11.174.800 hectare) is destined to agricultural activities. Although Colombia counts with approximately 21,5 million ha that could be dedicated to agriculture, just 4,9 million ha are actually used. In contrast, only 20 million ha are suitable for livestock, but almost 38.6 million ha are currently used for this activity.

By 2005, 4.058.470 ha were dedicated to perennial and annual crops, which correspond to the 98% of the agriculture production while the remaining 2% corresponds to fallow areas. In Colombia, the cutting and transformation processes of the different crops lead to the generation of more than 27 million tons of wastes per year. These agricultural crop residues can be classified as field and processing residues. Residues left in an agricultural field after harvesting the crop such as empty fruit bunches from palm plantations; discarded fruits, leaves and stems, straw and stalk and seed pods are considered field residues. On the other hand, materials obtained as byproducts after processing the crops such as husks, seeds, bagasse and roots are classified as processing residues. According to the Ministerio de Minas y Energía, 75% of the total biomass produced by the agricultural sector in Colombia corresponds to crop residues, while processing residues represent the remaining 25%. Although 100% of the processing residues can be recovered to obtain valuable products such as energy and biofuels, a percentage of the field residues should be recycled onto the land. Field residues protects the soil from erosion, helps build organic matter and contributes nutrients back to the soil. The percentage of residue removal depends on the specific local climatic and soil conditions. However, it is estimated that the residue removal can reach 35% without adverse effects on future yields. Therefore, Colombia has a great production of lignocellulosic biomass with no currently applications.

In this work, three of the most important agroindustrial residues in the country have been considered as feedstocks for small and high scale biorefineries to obtain valuable products: Sugar cane bagasse, empty fruit bunches and coffee husk. The raw materials were firstly characterized following the NREL Laboratory methods. Then the commercial software ASPEN Plus was implemented to obtain the mass and energy balances for both small and high scale biorefineries. After the commercial software ASPEN Economic Evaluation was used to develop the economic assessment. According to the results, to obtain high value added products from residues in the Colombian context using the above mentioned residues as feedstock could improve the rural and industrial sectors in the country, although high scale biorefineries offer higher profit margins.

Synthesis and characterization of fatty acids using microwave assisted in alkaline hydrolysis *Jatropha curcas* oil

[1]N. R. Yusuf, [1]R. N. M. Kamil, [2]S. Yusup,
[1]Fundamental and Applied Sciences Department, Universiti Teknologi PETRONAS, 32610 Bandar Seri Iskandar, Perak;
[2] Chemical Engineering Department, Universiti Teknologi PETRONAS, 32610 Bandar Seri Iskandar, Perak;
tel. +605 3687681, e-mail: ruzaimah_nikmkamil@petronas.com.my; [1]Universiti Teknologi PETRONAS, 32610, Bandar Seri Iskandar, Perak, Malaysia; fax: +605 365 4075.

Fatty acids materials are the most important preliminary materials in biochemical industrial process. Alkaline hydrolysis process is known to be the easiest and fastest way to achieve maximum fatty acids yield. In this study impact of microwave irradiation assistance applied in production of fatty acids from *Jatropha curcas* oil and their chemical characterization was verified using NMR analysis and compared with commercial fatty acids. Other physical properties such as % FFA, density, viscosity, cloud point and flash point were studied. ^{1}H-NMR and ^{13}C-NMR spectra showed similar structure of both synthesized fatty acids and commercial fatty acids. The physical properties are also in good agreement for both fatty acids. These findings emphasize that alkaline hydrolysis process under the assistance of microwave irradiation does not change the properties of fatty acids but enhance the fatty acids production to about 97 wt % with shorter reaction time of 15 minutes.

Table 1 : Comparison between physical properties of synthesized and commercial fatty acids

Properties	Units	Fatty Acids	
		This work	Commercial
Free Fatty Acids	wt %	97	99
Viscosity @ 40 °C	mPa·s	16.79	20
Density @ 40 °C	g/cm^3	0.88	0.91
Cloud Point	°C	-5.1	14
Pour Point	°C	12	10
Flash Point	°C	204	190

Technology of ceramic gravel for various concretes

B.R. Khaidarov[1], R.A. Khaidarov[1], R.T. Akhmetova[1]
[1]Department of Chemical Engineering, Kazan National Reseach Technological University (KNRTU), Phone (+7) 9173971127, K.Marks str., 68, Kazan - Russia. Corresponding author: E-mail address: [1]rachel13@list.ru

A cost-effective technology for producing ceramic gravel is developed, which in its physical, mechanical and chemical parameters is an alternative to the classical aggregate obtained from natural minerals (granite, diabase). The main advantage of ceramic gravel is the possibility of production it from low-quality raw clay.

It was found that the low bulk density of ceramic gravel reduces the load on the foundation of roads, reduces the thermal conductivity and significantly increases the service life of asphalt concrete. Ceramic rubble can be used as aggregate for concrete structures, thereby reducing the load on the structure.

According to technology the duration of the firing and pre-drying of materials is not more than 6-7 hours. The introduction of these technologies is possible in conventional brick kilns.

Resulting in vitro ceramic rubble was tested in industrial laboratories. For all materials were obtained high quality parameters. Ceramic rubble strength is not inferior to granite, and a number of indicators (water absorption, resistance to decay and bulk density) even surpasses it.

The main consumers of the materials developed are road construction companies, and concrete factories.

Promising to get new types of products and materials based on the production of ceramic rubble:

- lightweight aggregate (expanded clay MPE);
-easy blocks with high thermal insulation performance and durability;
- ceramic gravel with low thermal conductivity and high strength;
- climb a high-quality abrasive material;
- paving slabs and curbs and others.

Temperature effect on biogas production from cassava residue

Natsiri Sukthomya [a*], Sumaeth Chavadej [a,b], Malinee Leethochawalit[c]
[a)] The Petroleum and Petrochemical College, Chulalongkorn University, *Bangkok*, Thailand
[b)] Center of Excellence on Petrochemical and Materials Technology, *Bangkok*, Thailand
[c)] Innovative Learning Center, Srinakharinwirot University
[*)] Corresponding author: n.sukthomya@gmail.com

The main objective of this study was to compare the biogas production performance from cassava residue operated at mesophilic (37 °C) and thermophilic (55 °C) temperatures using a continuous stirred tank reactor (CSTR). A feed solution was the UASB effluent treating cassava wastewater mixed with 2% cassava residue which was fed into two identical 4-L CSTR units at different TS loading rates (1.5 to 3.0 kg/m^3 d). The results showed that the optimum TS loading rates of mesophilic and thermophilic operation for the maximum TS removal, maximum methane production and maximum methane yield were 2.0 and 2.5 kg/m^3 d, respectively. Under the optimum TS loading rate of 2.5 kg/m^{3}d at 55 °C, the system provided a methane yield of 231.1 mL CH_4/g TS applied (483.8 mL CH_4/g TS removed) and a specific methane production rate of 601 mL CH_4/L d or 154.7 mL CH_4/g MLVSS d which were significantly higher than those at mesophilic temperature. The degradation performances of cassava residue under thermophilic temperature were 62.1, 60.9 and 2.0% for cellulose, hemicellulose and lignin, respectively whereas the digestibilities under mesophilic were 59.7, 50.9 and 0.95%, respectively.

The components activation in technology of sulphidic materials

R.T. Akhmetova[1], A.A. Yusupova[1], L.R. Baraeva[1], A.I. Khatsrinov[1], A.Yu. Akhmetova[1], B.R.Khaidarov[1]
[1]Department of Chemical Engineering, Kazan National Reseach Technological University (KNRTU), Phone (+7) 9173971127, K.Marks str., 68, Kazan - Russia. Corresponding author: E-mail address: [1]rachel13@list.ru

Sulfur is an important and large-tonnage raw material in chemical industry. A sulfur overproduction is observed recently, so options of expansion of its use in nonconventional material-intensive spheres are offered. One of nonconventional scopes of sulfur is its use as the knitting component for receiving the sulfuric concrete possessing a number of positive properties in comparison with cement concrete: fast set and preservation of high durability, low water absorption and frost resistance. However release of toxic dioxide of SO_2 sulfur by production of sulfuric concrete possible. For prevention of this process most often by researchers offer introduction to composition of sulfuric concrete of various functional additives. However there is also other way of the solution of this problem: to connect element sulfur in the received material in sulfides with use the elektrophilic activators, for example chlorides of zinc, iron or aluminum. In our opinion, it will allow to improve also operational properties of the received sulphidic materials (SM).

For receiving SM widespread components are used: sulfur – the by-product of an oil and gas complex which is formed at purification of crude oil, natural silicon containing raw materials (filler) with the content of amorphous dioxide of silicon not less than 65% and quartz sand. As the activating additive of process of synthesis of sulfides – one of the chlorides.

Formation of sulfides at interaction of chlorides of zinc, iron or aluminum both with a sulfuric component, and with amorphous dioxide of silicon was confirmed with theoretical (quantum and chemical) methods of research and modern experimental methods of the analysis (IR-spectroscopy, EPR, XRF). For chemical binding of components and getting dense and high-strength samples the technology can be realized due to modernization of the existing production base and is carried out in some stages: crushing and drying of a filler; modifying of amorphous dioxide of the silicon which is a part of a filler, the activating elektrofilic additive; mixtion of the modified filler with sulfuric melt; laying or filling of the received weight in forms with the subsequent pressing under pressure. The received SM were tested by standard techniques with an assessment of their mechanical properties: durability on compression, water absorption, density, frost resistance and resistance to hostile environment. During researches the optimum structure of SM answering to the following characteristics is received: durability on compression of 70-80 MPa, density is 2,23-2,53 g/cm^3, frost resistance of 200-240 cycles and water absorption of 3,9-7,8 masses. %, resistance to 5% to solutions of salts and acids from 0,94 to 0,98.

SM can be applied in designs of buildings and constructions which have increased requirements on resistance to hostile environment and impermeability. Such designs are applied in industrial, agricultural, warehouse constructions both for loose, and liquid materials. SM are most perspective for production: elements of pavings (plates, paving slabs, face checkers, onboard stones, road protections); designs subject to salt aggression (floors, drain trays, etc); engineering constructions (collector rings, sewer pipes, treatment facilities).

The study of the influence of load's chemical composition in the coking process on the efficiency and the quality of coke

Dr. H. MIMOUN[*,a],Dr.N.BENTAHAR[a] .F. HENNAOUI[b], M. KERBOUA[c],
[a] Maitre de conferences, Boumerdes, Algerie ; [b] Magister, Boumerdes, Algerie ;
[c] Ingénieur, Boumerdes, Algerie.
Laboratoire de recherche Technologie des hydrocarbures ; Faculté des hydrocarbures et de la chimie ; Université de Boumerdès
Mimounhadj@yahoo.fr

This research was performed in order to allow the study of the chemical composition influence of the coking process load on the efficiency and the quality of coke. For this reason, the coking of the following loads was realized: Atmospheric residue (RAT), vacuum Residue (RSV) and catalytic Residue of cracking (RCC). (The residues are obtained from an Algerian crude oil).As the oil residues are rich for their strongly polar composition, such as the asphaltene resins, and complex structures units (SCU), which has a role in the formation of coke, and as the dispersion of these latter improves the quality of coke, a study on the stability of aggregation was carried out by the addition of one stabilizer (oil Extract) in the coking process load. The Compounding (Extracted from /RCC oil) has been drived to the best efficiency of coke. The study consists of the influence…., this is characterized by the analyses Infra-red (IR) and x-ray diffraction (XRD).

Keywords: *Coking; oil residue; dispersant;* ***aggregation*** *stability*

Three-stage ASBR system for separate production of H_2 and CH_4 from ethanol wastewater

[1]T. Watcharanurak, [1,2]S. Chavadej
[1]Chulalongkorn Univ., The Petroleum and Petrochemical College, Phayathai Rd., 10330 Bangkok, Thailand; tel. +66 22184139, fax. +66 22184139, e-mail: sumaeth.c@chula.ac.th;
[2]Center of Excellence on Petrochemical and Advanced Materials Chulalongkorn University, Phayathai Rd., 10330 Bangkok, Thailand; tel. +662 2184141, fax: +662 6117619.

The objective of this work was to develop a new anaerobic process, three-stage anaerobic sequencing batch reactor (ASBR), and to determine the optimum COD loading rate for maximum hydrogen and methane production from ethanol wastewater under mesophilic operation (37 °C) and at a constant pH of 5.5 for the first ASBR unit. The three-stage ASBR system consisted of three units of ASBR connected in series which had liquid working volumes of 4, 8, and 24 L, respectively. The recycle ratio of effluent flow rate to feed flow rate of 1:1 was applied for both of first and second ASBR units. Under the optimum COD loading rate of 15 kg/m3d (based on feed COD load and total ASBR volume), the system provided the highest production performance in terms of yields and the specific production rates of both gases. Under the optimum COD loading rate, the produced gas contained 48.9% H2, 4.3% CH4 and 46.8% CO2 in the first ASBR unit; 54.1% CH4, 45.2% CO2 and 0.7% H2 in the second ASBR unit; and 81.2% CH4 and 18.8% CO2 without H2 in the third ASBR unit, respectively.

Upgrading acetic acid to acrylic acid – A process design

J. Pieper, M.P. van Gaalen, J.H.R. Lok, L.J. Noll, A.G.J. van der Ham, J.P. Lange, H. van den Berg.
Faculty of Science and Technology, University of Twente, P.O. Box 217, 7500 AE Enschede, The Netherlands, j.pieper-1@student.utwente.nl

As the world's population increases, so does the demand for end consumer products. Acrylic acid is an important chemical intermediate for the industrial production of plastics, coatings, adhesives and superabsorbent polymers. The current market is 5.6 Mt/a and is expected to increase to 8 Mt/a by 2020.

The current production route is based on the oxidation of propylene, however, there is a renewed interest in using acetic acid as a precursor. This work shows the feasibility of a novel process based on the aldol condensation of acetic acid with formaldehyde to acrylic acid. The formaldehyde production by methanol oxidation is integrated into the design. A plant capable of producing 200 kt/a (4% of the global market) glacial acrylic acid (99.5%) is designed. The plant is to be located in the US to take advantage of the abundancy of shale gas, as the feedstock of the novel process consists entirely of shale gas derivatives (methanol, acetic acid). Following, a stable or even decreasing price of methanol and acetic acid is expected in the next year, which will boost the economic attraction of the process.

Initially the economic potential of the desired reaction was calculated, then a black box was designed to evaluate the process based on the conversions and selectivities collected from literature. Conceptual design methods have been applied. Here, the different concepts are evaluated by a process overall mass balance, assuming the empirical values stated in literature. The conceptual process design is divided into a reactor block and a separation block, where the design variables are first evaluated separately because of the large degree of interplay between them.

Key variables identified early on for the reactor block were the oxygen feed (pure oxygen vs. air), reactor setup (single vs. dual reactor system), separation between reactors (no recovery or H_2O and/or MeOH recovery) and feed ratio (stoichiometric or excess acetic acid). Based on literature data and engineering heuristics, the following conceptual process design was formulated.

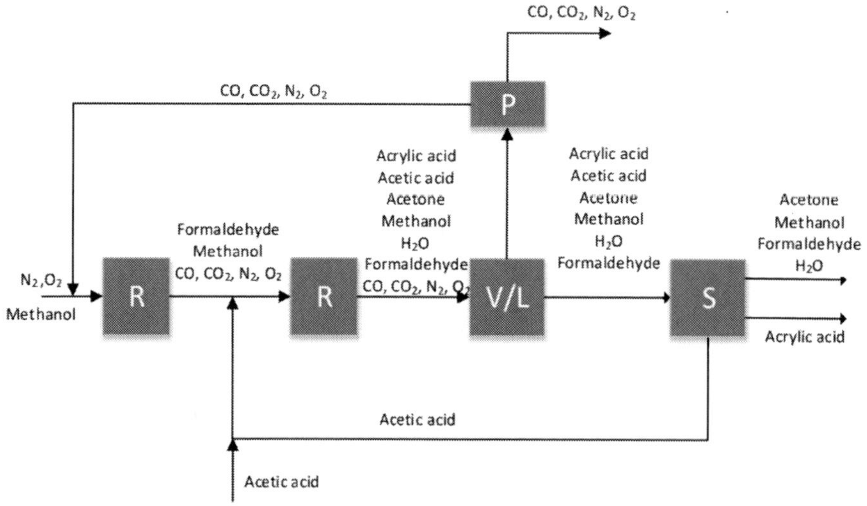

The novel acrylic acid process is based on the aldol condensation of formaldehyde and acetic acid.

It is a 2-step process in which the 1st step is the integrated formaldehyde production and the 2nd step is the acrylic acid synthesis. A cost analysis indicated, that an air feed with its purge off gas treatment in the first reactor is cheaper than using a pure O_2 feed. Due to the adiabatic temperature rise of the methanol oxidation based on thermodynamic calculations, a large inert recycle is necessary to dilute the contents of the 1st reactor.

The main challenge of the process is the recovery of unreacted feedstock, formaldehyde and acetic aid, out of the reactor effluent. Due to the interaction of formaldehyde with water, a full conversion of formaldehyde was desired. Hence, an excess of acetic acid was inserted to the second reactor. Driven by economic and ecologic incentives, additional separation of the unconverted feedstock and by products has to be applied. In order to recover the excess of acetic acid multiple separation stages were required, a recovery up to 85% were achieved. However, energy consumption showed that recovery of 70% of acetic acid is more profitable.

The separation train was designed in UniSim to determine the most optimal sequence of separation steps. The two main options were either drying the crude product with subsequent recovery of acrylic acid from an acetic acid/acrylic acid mixture, or first recovering acrylic acid from the crude product with subsequent recovery of acetic acid from the acetic acid/water mixture. Key parameters were the obtained purity of the acrylic acid (>99.5 wt.%) and the energy expenditure of the separation train.

From the model followed that glacial acrylic is readily recovered by a series of distillations in which water is first separated from the crude product. Acetic acid has to be recovered by a combination of distillation and extraction due to difficulties with separation from water. In order to push the by-product acetone and leftover methanol are also recovered with distillation and extraction.

The data obtained in UniSim are used to construct a piping and instrumentation diagram (P&ID). Finally, the P&ID is completed by the addition of a process control scheme of all components.

Once the design of the reactor and the chosen separation sequence were finalized, equipment sizing was performed for the process units. The total capital investment cost (CAPEX) of the plant is 160 M$ and the cost of manufacture (COM) is 230 M$/a. A linear depreciation period of 10 years is assumed. The payback period is determined to be 3.3 years. Yearly profits are calculated to be 30 M$/a, which results in a return on investment of 20%. The major cost component is the feedstock costs (70%). Heat integration analysis showed energy savings of 30%, where heat from the process streams could be converted to steam as utilities.

Based on industrial standards the process is both technically and economically deemed feasible.

Key areas for further improvement are the large gas recycle and the composition of the 2nd reactor. The large compressor costs are due to the high dilution in the 2nd reactor, which is not necessary for the operation. In order to achieve reduction in equipment sizing and compressor costs, lowering the total inert recycle is essential.

The investigation of waxes effect on the formed crystal in the demethanization unit in the phases 4,5 of South Pars gas complex

Mozhgan Hosseini, Shiraz University Of Technology
Iran, Fars Province, Noorabad, Taleghani Av,

The paraffin wax deposit can cause clogging in oil and gas pipelines and it leads to low efficiency and sometimes causes shut down in production units. Generally, temperature in hydrocarbon solution when the first solid hydrocarbon is formed, measured as the temperature of Wax phase formation. This temperature is normally measured in the lab. There are two Thermodynamic approaches for calculating this temperature. In the first model, the paraffin wax phase is assumed to be a solid solution and in the second model, the wax phase is considered as a set of pure solid components. In the solid solution model, an equation of state is applied for liquid-vapor equilibrium calculation and so an activity model is used for solid-liquid equilibrium. In the multi-solid solution model, due to assuming that solid phases are pure, the activity coefficients are equal to unity for the solid phases. Thus only one equation of state is applied for calculation of the vapor-liquid equilibrium. In this study, the multi-solid model of *Nikita et al.* is used for calculating temperature and the amount of deposite wax. The *Ghanaei et al.* method was used to prredict the pointing correction factor in high pressure conditions. An innovative method based on the Katz method is applied to separate additional heavy component. Al-Sahaf, Won and Ghanaei mathematical equations were used to calculate critical properties, acentric factor and melting point properties. Furthermore, improved Peng-Robinson equation of state -introduced by Faizi *et al.*- was used to calculate fugacity of pure and mixed components. The results of the present model was compared with several previous models for evaluation of the model. Then using the present model, the amount of wax formation for the mixture of natural gas in the train 1 of the phase 4,5 in the South Pars Gas Complex (SPGC) was calculated for different temperatures.

Key Words: Wax, Gas stream, Pressure drop, LNG unit.

An experimental study on the optimization of subtilisin immobilization using eggshell waste

B. Özbek, G. Bacak, Ş. Ünal
Yildiz Technical University, Department of Chemical Engineering, Davutpasa Campus, 34210, Esenler/Istanbul, Turkey;
Tel +90 212 383 47 47, fax +90 212 383 47 25
e-mail: bozbek@yildiz.edu.tr

Alcalase is a protease that generally used in the production of detergents and cosmetic goods. High costs and unstable production conditions of the enzymes cause the low efficiency in industrial applications. Enzyme immobilization is a widely studied technique to produce more economical and efficient enzymes. Immobilized enzyme increases its stability and that is an advantage to expand the application areas of the enzymes [1]. Eggshell waste generated by the food processing industry has great potential for enzyme immobilization support as it is biodegradable material and nontoxic for human bodies. Eggshell and the eggshell membrane consist of 94% to 97% calcium carbonate, 3% to 6% protein and %1 various minerals, such as magnesium, potassium and traces of iron, sulphur and phosphorus. Immobilization process requires porous and strong support structure [2]. Eggshell has pores ranging from 7000 to 17000 and strong structure. Therefore, eggshell waste is unique material which has been studied in various field, e.g. biochemical and medical applications, catalytic usages in chemical processes, adsorption of some materials such as dyes and inorganic ions [3]. Therefore, in the present study, immobilization of subtilisin enzyme on the eggshell was investigated using with glutaraldehyde as cross-linking agent. The immobilization conditions (e.g. immobilization time, optimum pH, enzyme concentration effect) were studied in detail.

Keywords: Eggshell wastes; enzyme immobilization; subtilisin

References:

1. Gupta, M. N., Kaloti, M., Kapoor, M., and Solanki, K., (2011). "Nanomaterials as matrices for enzyme immobilization.", Artificial cells, blood substitutes, and immobilization biotechnology, 39(2):98–109.
2. Tsai, W.T., Yang, J.M. ,Lai, C.W. , Cheng, Y.H. , Lin, C.C. , Yeh, C.W. , (2006). "Characterization and adsorption properties of eggshells and eggshell membrane", Bioresource Technology, 97:488-493.
3. Guru, S.P., Dash, S., (2014). "Sorption on eggshell waste – A review on ultrastructure, biomineralization and other applications", Advances in Colloid and Interface Science, 209:49-67.

Assessment of global potential of biohydrogen production from agricultural residues and its application in nitrogen fertilizer production

M. Karimi Alavijeh[1,2], S. Yaghmaei[1]
[1]Department of Chemical and Petroleum Engineering, Sharif University of Technology, Azadi Avenue, P.O. Box 11365-9465, Tehran, Iran, tel.: +982166166430; fax:+982166166430, e-mail: yaghmaei@sharif.edu.
[2]Center of Advanced Research and Development of Elite Affairs, ETKA organization, Tehran, Iran, e-mail: masih.karimi@chmail.ir.

Natural gas is one of the primary sources to supply global energy demand. From 1980 to 2013, the global natural gas consumption has increased from 1499 to 3437 billion cubic meters, i.e., about 129% enhancement. A wide variety of natural gas applications has been developed, especially in the industrial section. The main application of natural gas in agriculture is to produce hydrogen via the steam reforming process to manufacture ammonia and its derivatives as the foremost nitrogen fertilizers. However, natural gas is non-renewable and will run out by 60 years later. Hence, in the future its usage for the ammonia production will be limited, due to the price rising and other dominant applications in providing electricity and fuel. On the other hand, agricultural biomass is a highly potential energy resource and a large amount of crop residues is annually left on the field. The biohydrogen production from these materials can be used as a supplement in parallel with hydrogen obtained by natural gas.

In the present study, a method has been applied to estimate collectable crop residues including wheat, corn, barley, rice, sugarcane, potato, alfalfa and oat as the sustainable major crops in the Africa, Asia, Central America, Europe, Northern America, Oceania and South America. Hydrogen yields from the pretreated and enzymatic hydrolysed residues were estimated and the potential biochemical production of biohydrogen from these residues has been calculated. Moreover, the global potential of biohydrogen production from agricultural residues by 2030 has been prognosticated. The production capacity of ammonia from biohydrogen was estimated. In addition, the application of biohydrogen from cereals residues to provide their nitrogen requirement was calculated based on the cultivable area. Eventually, a discussion has been made on the amount of potentially saveable natural gas and greenhouse gas (GHG) emissions decline as well as the social cost of carbon dioxide (SCC) reduction by using biohydrogen.

The results reveals that about 1.18 billion tonnes of 3.37 billion tonnes of the aforementioned residues can be collected for the production of 10.25 million tonnes of biohydrogen, i.e., 77% of the world's total hydrogen production in 2013. The amount of collectable agricultural residues and biohydrogen might increase to 2.09 billion tonnes and 17.66 million tonnes, respectively, by 2030. Approximately 58.06 million tonnes of ammonia can be obtained by biochemically produced hydrogen and, nitrogen requirements of near 17.19%, 32.53%, 11.52%, and 33.47% of the global cultivation area of wheat, corn, barley and rice would be possibly supplied by the collected residues of the same crop, respectively. Results demonstrated that for higher crop yields, a further area could be fertilized, e.g., nitrogen could be provided for 93.75% and 34.34% of the rice total cultivated area in Oceania and Asia, respectively. The rice production in Asia is about 573 times more than Oceania, but the rice yield in Oceania is 3.79 times higher than Asia.

About 34.32 million tonnes of natural gas could be saved and in comparison with the steam reforming method, 73.26 million tonnes CO_2 e of GHG emissions and 2.93 billion dollars of SCC could be reduced by the biochemical production of hydrogen from agricultural residues.

Basic design and optimization of an NGL recovery plant

R.D. Tweehuysen[1], K.D. van Munster[1], A.G.T Reijneveld[1], J.A. Eertink[1], A.G.J. van der Ham[1], H. van den Berg[1], H van der Valk[2], D. Hoogendoorn[2]
[1]University of Twente, P.O. Box 217, 7500AE Enschede, The Netherlands. Faculty of Science and Technology, r.d.tweehuysen@student.utwente.nl
[2]CB&I, P.O. Box 93252, 2509AG,The Hague, The Netherlands

Natural gas liquids (NGL, $\geq C_3$) recovery has been around for quite some time. Recent plant designs are based on heuristics and proven technology. With a fresh mind a typical NGL recovery plant was systematically redesigned based on state of the art technology. Starting from a black-box evaluation, different functions and options were evaluated. The resulting basic design was optimized for minimal energy requirements, while maintaining high NGL recoveries.

The incoming feed gas was collected from a multi-well system and consists of: a sales gas fraction (C_1-C_2), a LPG fraction (C_3-C_4), a condensates fraction (C_5-C_{12}), an inert fraction (CO_2 and N_2) and contaminants H_2S and water. For our case study a raw gas capacity of 5 mega ton per annum was assumed, with an NGL recovery of at least 95%. Based on typical product specifications and a black-box evaluation the NGL plant should have at least four functions: removal of H_2S, dehydration, separation of the feed gas into a: sales gas, LPG and condensates fraction. Lastly, the sales gas has to be compressed to reach pipeline pressure. In principle the sequence of these functions was free and should be optimized.

To identify possible unit operations a literature study was conducted for each function. The literature study resulted into a plethora of possible combinations of unit operations. Selecting the best combination is not a straightforward task. Each unit operation was evaluated based on literature and process simulations. Simultaneously evaluating the unit and plant performance, in the process simulator UniSim®, the most optimal combination of unit operations was selected.

The removal of H_2S can be done in several ways: membranes, absorption and adsorption. The limiting factor for membranes and adsorption is hydrocarbon loss and regeneration duty respectively. Therefore absorption is selected to remove H_2S. To prevent hydrate formation the water concentration needs to be as low as 1 ppmv. Only adsorption can reach these low water concentrations. However, bulk removal of water can be done using membranes, absorption and cooling. Comparing duties and equipment count, condensation followed by adsorption is the best option to dehydrate the natural gas. The schematic overview of the NGL recovery plant was:

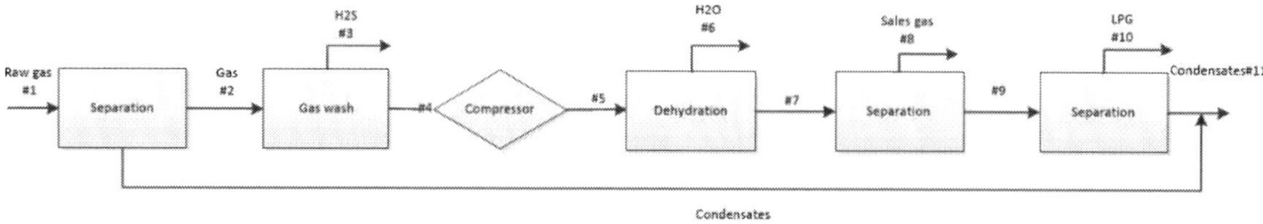

The condensates are removed in the first separation, to prevent two phase behavior and to limit the load on consecutive equipment. The gas wash was placed in front of the dehydration step, as aqueous amine solutions were used to remove H_2S. Placing the compressor before the dehydration step facilitated easy condensation for bulk water removal followed by deep water removal by adsorption. Lastly the sales gas, LPG and NGL separation took place.

The NGL recovery unit was the last function of the NGL plant. Here membranes, swirl inducing devices, Joule Thompson valves, Turbo-expanders and absorption were compared on recovery, energy duty and equipment count. Literature studies and simulations showed that a NGL recovery higher than 95% was not feasible in single unit operation. Therefore combinations of unit operations were considered. Membranes were not suitable as the state of the art membranes did not combine a high selectivity with a high flux, resulting in low NGL recoveries and large membrane areas. Lean oil absorption and adsorption were not favorable since NGL recovery techniques based on dewpointing outperformed ad- and absorption. In dewpointing processes with a high NGL recovery (>95%) temperatures as low -80°C were required. Conventional cooling could not reach these low temperatures, without an expensive external cooling cycle. As gas was cooled to recover NGL, heat or in this case cold integration was necessary to enhance the recovery, and limit energy demands.

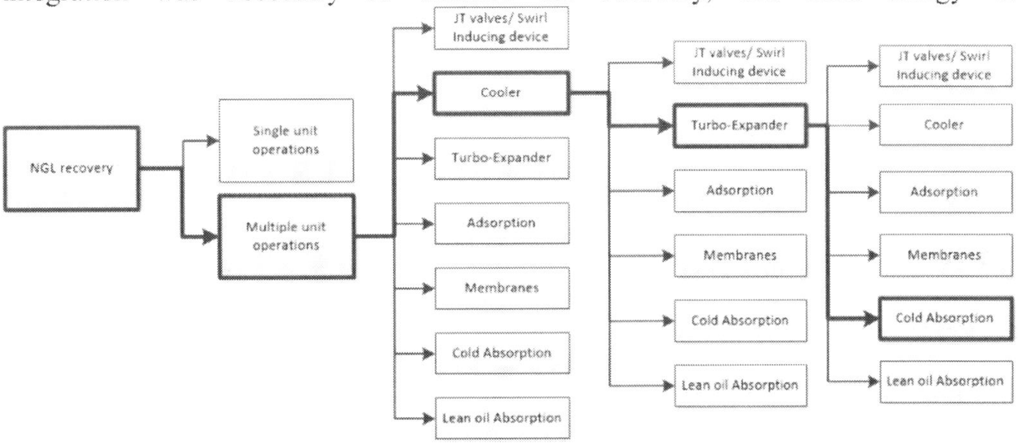

Evaluating all the options at a plant level (see scheme above) cooling first by heat integration was favorable. This reduced the pressure drop over the downstream turbo-expander. The turbo-expander was selected over the JT-valve, as it was more efficient and some of the expansion duty can be recovered. To reach a recovery of 95% cooling and a turbo-expander were not sufficient. Therefore, an extra unit operation was required. Considering all the options and evaluating them on a plant level yielded cold absorption as the most energetically favorable option. In cold absorption, the top product of the de-ethanizer is partially condensed. The liquefied fraction was used to wash the vapor at the outlet of the turbo-expander in a wash column. This scheme was then optimized for minimum compressor duty, with a closed cold cycle. Therefore no external cooling cycle was needed.

Compared with currently built processes this design is similar, with slight differences. Therefore we conclude that in the past years the state of the art for NGL recovery has evolved instead of revolutionized. For the future we expect that swirl inducing devices and membranes can play a role in new NGL recovery plant designs.

Biorefinery: An economic feasible waste transformation technology?

A. Kutsay, L. Krátký
Czech Technical University in Prague, Faculty of Mechanical Engineering,
Department of Process Engineering, Technická 4, Prague 6, Czech Republic,
phone +420 773 883 884, e-mail: Andrey.Kutsay@fs.cvut.cz

As it appears, economic analysis is the reason of why renewable energy plants are not so famous on world market. However, biorefinery concept open new doors for spreading it on world commercial production.

As the motivation the previous studies were taken into account [1]. It was concluded that the profit from electricity selling is simply not enough. The Fig. 4 shows the proposed biorefinery concept with implemented of previously studied thermal expansionary pretreatment (TEP).

Briefly, after TEP it is possible to separate treated suspension (wheat straw, water) into liquid, which is reach of sugars, and fibers. The fibers could be used for production of biobased polymers, e.g. bioplastics. The separated liquid could be used for biofuels production, e.g. biogas or bioethanol. The biofuel could be burned in CHP unit in order to produce electricity and heat. After combusting of biofuel the carbon dioxide is produced. The carbon dioxide can be mixed with water and with help of Sun it is possible to get algae. Algae by itself consist of biomass and oil, the ratio is 50:50. The separated biomass can be sent back to TEP, in order to enrich the production, and oil can be used for biodiesel production. The complexity of proposed concept is going to increase energy efficiency, reduce wastes and emissions and finally will contribute to the overall economic and environmental sustainability of biorefinery.

Only after such a proposed solutions it is possible to create a biorefinery plant, which could compete among other energy producing plants on the world market.

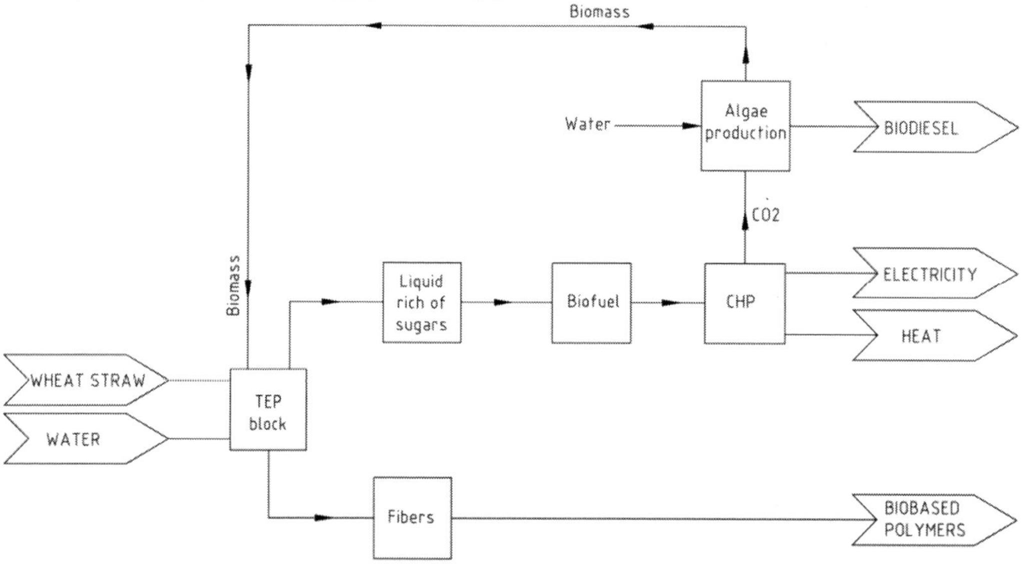

Figure 1. Biorefinery with TEP implementation

[1] A. Kutsay, L. Kratky, T. Jirout. Energy-economic study of thermal-expansionary pretreatment for its implementation at biogas plant.

Detoxification of hemicellulose hydrolysate of the sugarcane bagasse by liquid-liquid extraction

L. R. Roque[1],G. P. Morgado[1], R. R. Pinto[2], J. L.Ienczak[1], E. B. Moraes[1], M. Aznar[2*], S. C. Rabelo[1]

[1] Laboratório Nacional de Ciência e Tecnologia do Bioetanol (CTBE), Centro Nacional de Pesquisa em Energia e Materiais (CNPEM), Caixa Postal 6192, CEP 13083-970, Campinas, São Paulo, Brasil, tel.+55 19 35183191, e-mail: sarita.rabelo@bioetanol.org.br

[2] School of Chemical Engineering, State University of Campinas (FEQ-UNICAMP), Av. Albert Einstein, 500, CEP 13083-582, Campinas, São Paulo, Brazil.

in memorian

Lignocellulosic biomass is a renewable source of energy production on a scale sufficient to play a significant role in developing vital programs for renewable energy and creating a more environmentally conscious society. Among the lignocellulosic biomasses, bagasse and straw from sugarcane in Brazil stand out as being economically viable for the production of "environmentally friendly" fuel.

In the process of ethanol production from sugarcane bagasse (Ethanol 2G), a step of pretreatment is required, mainly to reduce the recalcitrance of the material and maximizing the subsequent hydrolysis and fermentation of sugars. During this step, there may be formation of some by-products, which interfere negatively in the fermentation process, such as acetic acid, which is formed by hydrolysis of the acetyl group present in hemicellulose; formic and levulinic acids, degradation products of sugars; phenolic compounds, mainly formed by the partial degradation of lignin; and furan aldehydes, particularly furfural and hydroxymethylfurfural (HMF) formed by the degradation of pentoses and hexoses, respectively.

Furfural and HMF are known to inhibit the growth rate and ethanol yield, and carboxylic acids, such as acetic acid that decreases the driving force of protons across the cell membrane, especially in the presence of ethanol. Furthermore, the presence of acetic acid in the fermentation medium has been reported as the cause of reduced enzymatic activity of xylose reductase. Phenolic compounds are mentioned as major inhibitors in the fermentation process, significantly affecting the yield of ethanol.

An alternative to minimize inhibitory effects of the fermentation medium is the removal of these constituents prior to the fermentation step. Thus, the aim of this work was studied the removal of the inhibitors by liquid-liquid through of the solvents known commercially and the evaluation of the process proposed through the results obtained after the fermentation step.

In order to evaluate the liquid-liquid extraction in this work, it is necessary to know the behaviour of inhibitors that are present in hemicellulose hydrolysate, with the chosen solvents, in terms of distribution coefficients and environment concerns, such as ethyl acetate, n-butyl acetate, n-heptyl acetate, isobutyl acetate and isoamyl propionate. To represent the non-idealities present in the liquid–liquid equilibria, the most used thermodynamic models NRTL and UNIQUAC, and group contribution method, UNIFAC and their modifications were considered in the simulations. These models/methods were used to generate liquid–liquid equilibrium diagrams using the Commercial Simulator Aspen Plus®, considering a ternary system, with water/solvent/inhibitors. It was verified, through the simulations, that acetate solvents present a reasonable water solubility in the organic phase (separation region) that can introduce complications when the extract phase is to be treated. However, increasing the acetate molecular weight, this negative effect decreases.

The solvents evaluated by simulation were employed in liquid liquid extraction assays and after phases separation (extract and raffinate), the detoxified liquors were used in the fermentability evaluation by *Sheffersomyces stipitis* NRRL Y-7124 yeast.

In summary, the use of solvents in liquid-liquid extraction showed that it is possible to reduce inhibitors effect in the hemicellulose hydrolysate in order to maximize the fermentation of the sugars to produce ethanol 2G.

Effect of different substrates and inducers on production of laccase by a new isolated fungi in solid state fermentation

[1] Z. Ghobadi Nejad, [2]S. M. Borghei,[3]S. Yaghmaei
[1] Graduate School of the Environment and Energy, Science and Research Branch, Islamic Azad University, Tehran, P.O. Box: 14515-775, Tehran, Iran
z_ghobadi@mehr.sharif.edu
[2,3]Biochemical and Bioenvironmental Research Center, Sharif University of Technology, Tel: +982166164103, Fax: +982166005417, P. O. Box 11155-1399 Tehran, I.R. IRAN, e-mail: mborghei@sharif.edu
[3]Department of Chemical and Petroleum Engineering, Sharif University of Technology, Tel: +982166166430, Fax: +982166005417, P. O. Box 11365-6891 Tehran, IRAN, Yaghmaei@sharif.edu

Laccases (benzenediol: oxygen oxidoreductases, EC1.10.3.2) are an interesting group of multi-copper enzymes, which have received much attention of researchers in the last decades due to their ability to oxidize both phenolic and non-phenolic lignin related compounds as well as highly recalcitrant environmental pollutants. The use of enzymes can offer many possibilities in biotechnological applications. Such applications include the detoxification of industrial effluents, mostly from the paper and pulp, textile and petrochemical industries, polymer synthesis, bioremediation of contaminated soils, wine and beverage stabilization. Laccases are also used as catalysts for the manufacture of anti-cancer drugs and even as ingredients in cosmetics. Recently, the utility of laccases has also been applied toNanobiotechnology.

In this study, the production of laccase by an indigenous strain of H1 was studied on solid state fermentation (SSF). Culture parameters such as type of Substrate and Relative Humidity (RH) and the length of incubation time were optimized.

Selection of an appropriate substrate is a key factor in SSF, which determines the success of process. It has been a practice to use of lignocellulosic agroindustrial residues for the production of laccase. Different lignocellulosic substrates such as Wheat straw, wheat bran, rice straw, rice bran, banana peelings, and corn germ were screened for production of laccase. Among the various substrates, rice bran was found to be the best substrate. In addition, the present investigation has confirmed and evaluated the use of rice bran as an inexpensive and easily available agroindustrial waste for laccase production under SSF.

Optimization of initial moisture content is essential in SSF, as it affects the substrate utilization and laccase production. The moisture content levels on SSF processes vary between 60 and 90% and the optimum moisture content for growth and substrate utilization is between 60 and 70%.

The enzyme activity were assayed in different incubation times (from 3 to 12 days) with the different moisture content level (60%, 70%, 80%, 90%) and It was observed that maximum laccase activity was obtained on 5th day of incubation with 66% of RH. In this study, the laccase activity was measured by monitoring the oxidation of 1ml of 10mM Guaiacol buffered with 3 ml 10mM Sodium acetate buffer at 470 nm. Enzyme activity is measured in U/gfs (where gfs is defined as the number of units of enzyme produced from 5 g fermented substrate). The production of laccase by H1 fungi was increased or induced when an aromatic compound (such as vanilla or guaiacol) or different metals ions(such as Cu^{2+})was added to the growth medium. In this research all inducer compounds were filter-strilized and added at the time of incubation. Various inducers such as $CuSO_4$, $FeSO_4$, vanilla, $MnSO_4$, Na_2SO_4, $CaCl_2$ and Yeast extract were compared and $CuSO_4$ was obtained as the best inducer.

Effects of biomass particle size on ultrasound pretreatment of cassava waste

T. C. Polachini, L.F.L. Betiol, J. Telis-Romero, A. Mulet
Instituto de Biociências, Letras e Ciências Exatas, DETA, Universidade Estadual Paulista, São Jose do Rio Preto, CEP 15054000, SP, Brasil; tel.+ 17 32212251, e-mail: javier@ibilce.unesp.br.

In order to satisfy the growing demand for renewable energy without affecting the food sector, raw materials, such as lignocellulosic bio-resources, should be utilised. Technologies for the conversion of biomass resources into fuels and chemicals such as ethanol have sparked growing interest as a means of reducing dependency on fossil fuels, and of developing environmentally friendly and renewable energy sources. Such biomass resources include energy crops, agricultural crops, forestry residues, aquatic crops and residues from processing industries. While it is well known that ethanol can be produced from lignocellulosics through enzymatic hydrolysis and fermentation, in order to reach high yields and high ethanol concentrations the cellulosic biomass must be pre-treated. Pre-treatment, this is crucial to improve the enzymatic digestibility of biomass, is projected to be among the most expensive process steps, and thus a considerable bottleneck in the production of ethanol. Biomass conversion in the second-generation ethanol industry comprises the pre-treatment to the biomass, conversion of the cellulose and hemicelluloses into fermentable sugars and fermentation of the sugars into bio-fuels, physicochemical, chemical and biological pre-treatment methods, or a combination of these. Pre-treatment processes for biomass hydrolysis generally refer to breaking of the naturally resistant carbohydrate lignin shield that limits accessibility of the enzymes or chemicals to the cellulose and hemicellulose. Thus the application of power ultrasound is an alternative to the above-mentioned methods, potentiating the production of fermentable sugar in the biomass pre-treatment. Various pre-treatment techniques can change the physical and chemical structure of lignocellulosic biomass and improve the hydrolysis rates. High-intensity ultrasound could be a promising technique in the biomass pre-treatment process. The objective of this work was to study the effect of biomass particle size, biomass concentration, ultrasonic power level and sonication time on the production yield in total sugars (ST) and reducing sugars (SR) during the pre-treatment of cassava waste biomass.

Fuzzy modeling for properties prediction of green polymeric blends from renewable resources: a novel approach for biomedical purposes

N. M. N. Lima,[1] L. Zuniga Linan,[1] F. Manenti,[2] R. Maciel Filho,[1] M. Embiruçu,[3] M. R. Wolf Maciel[1]
[1]University of Campinas (UNICAMP), Department of Processes and Products Development (DDPP), 13083-852, Campinas/SP, Brazil
[2]Politecnico di Milano, CMIC Dept. "Giulio Natta", Piazza Leonardo da Vinci 32, 20133, Milano, Italy
[3]Federal University of Bahia (UFBA), Polytechnic Institute, 40210-630, Salvador/BA, Brazil.

Biomaterials and their blends have experienced enormous growth in latter years and studies show that in future they will expand further more. Such compounds has been used in several clinical applications to repair, rebuild, replace or even regenerate damaged areas of the body. Therefore, there is a huge need and interest by researchers all over the world to develop scientific and technological extensive research in this field. It seeks to improve people's health and to reduce production costs of involved materials.

Poly(methyl methacrylate) (PMMA) and Poly(Lactic Acid) (PLA) are polymers widely used in the biomedical area.

PMMA is used in several medical applications, such as for the fabrication of bone cement, contact and intraocular lens, screw fixation in bone, filler for bone cavities and skull defects, and vertebrae stabilization in osteoporotic patients. This polymer has interesting qualities of biocompatibility, ease of manipulation and low toxicity.

PLA can be directly synthesized from lactic acid and it has been proved to be a biodegradable polymer, absorbable, non-toxic, hydrolysable, and with high mechanical performance when compared to other materials. Furthermore, the lactic acid could be obtained from sugar cane bagasse, being an upgrading of the lignocellulosic residue from the sugar industry. In these regards, the PLA production has attracted considerable attention recently.

However, the isolated use of both PMMA and PLA like biomaterials presents restrictions and it is necessary to further enhance the properties of them with the formulation of blends each other or with other materials, which must be suitable for due medical application. The blending of different polymers is a simple and economical method to enhance the properties of a particular polymer matrix [1]. Also, the use of co-materials from renewable resources is very important by environmental and economic issues.

This paper describes the development of fuzzy models to predict properties of green polymeric blends according to preparation techniques and laboratory scale operational conditions. The proposed fuzzy methodology allows the formulation of global nonlinear long-range prediction models from the conjunction of a number of local linear fuzzy dynamic models. Data sets from experimental measurements is used to obtain identification data of the models. Then, appropriate techniques for fuzzy models generation were applied and the formulations were built in Fortran 90 language. Lignin from sugar cane bagasse was adopted to modify the properties of both PMMA and PLA. Several proportions for the polymer blends were evaluated. According to previous studies, compatible blends of PMMA with lignin are expected to exhibit higher biodegradability compared to the pure PMMA, while PLA/lignin blends could have substantially greater structural strength than the pure PLA. Satisfactory results were achieved and this supposes that the proposed methodology can be useful to obtain appropriate estimates of systems of difficult predictions.

References

1. X. Wang, F. Du, J. Jiao, Y. Meng, and R. Li, "Preparation and properties of biodegradable polymeric blends from poly(propylene carbonate) and poly(ethylene-co-vinyl alcohol), "*Journal of Biomedical Materials Research Part B: Applied Biomaterilas*", vol. 83, pp. 373-379, November 2007.

Improved rock phosphate solubilization by using a combined mechanical-biological approach

R. Klaic[1,2], F. Plotegher[1], C. Ribeiro[1], T. C. Zangirolami[2], C. S. Farinas[1,2]
[1]Embrapa Instrumentation, Rua XV de Novembro, 1452, São Carlos, SP, Brazil; [2]Federal University of São Carlos, Graduate Program of Chemical Engineering, Rd. Washington Luís, km 235, São Carlos, SP, Brazil; tel. +551621072908, e-mail: cristiane.farinas@embrapa.br

Chemical fertilizers are used worldwide to provide the nutrients necessary for plant growth and agricultural development. Phosphorus (P) is one of the key required nutrients which is usually supplied in the form of P-fertilizers obtained from the chemical processing of rock phosphate (RP). However, this process involves high costs and the generation of undesirable waste streams [1]. Thus, there is a growing interest in the development of biotechnological processes to allow RP application using P-solubilizing microorganisms (PSMs). The P-solubilizing activity of PSMs is determined by their ability to produce and release metabolites such as organic acids that through their hydroxyl and carboxyl groups chelate the cations (mainly calcium) bound to phosphate, being the latter converted into soluble forms [2]. Although there are reports on bacterial and fungal sources of PSMs, the aerobic fungi are recognized for their high growth rates and production of organic acids [3]. Others important aspects to consider on this bioprocess development include the adequate cultivation method to access all the available phosphate present in the RP. These microbial cultivations can be conducted using a liquid (submerged fermentation, SmF) or a solid medium (solid-state fermentation, SSF).

In this paper it is proposed a novel combined mechanical-biological strategy to produce soluble phosphate from RP by using a selected strain of the filamentous fungi *Aspergillus niger*. An alternative to accelerate the P-solubilization process employed here was to reduce the particle size of the RP to the nanometric range, so the phosphate solubility could be increased as a consequence of the higher surface area [4]. Such combined strategy would allow the P-solubilization driven by the reduced particle size to be favored by other factors, such as the local change of pH caused by the organic acid released during *A. niger* cultivation. To address this proposal, the effect of particles size on P-solubilization by *A. niger* cultivated under SSF and SmF processes were evaluated using Itafós phosphate rock (IRP) as a model material, since this rock contains a significant amount of phosphatic phase (P_2O_5, 20%).

The IRP was initially characterized using different physical-chemical methods. The X-ray diffraction results showed that the main observed phases were quartz [SiO_2] and a phosphate phase corresponding to fluorapatite [$Ca_{10}(PO_4).F_2$]. The X-ray fluorescence analysis showed that IPR is composed by 10 traces of surface oxides (SiO_2, 36.4%; Al_2O_3, 4.6%; Fe_2O_3, 2.03%; CaO, 30.4%; MgO, 0.52%; TiO, 0.22%; P_2O_5, 20.29%; Na_2O, 0.3%; K_2O, 0.88%; MnO, 0.1%; LOI, 4.33%). These results confirm the identification of the mineralogical phase of quartz observed in the X-ray diffraction and the presence of a phosphate phase of about 50% (w/w) in the IRP. The phase of quartz present in relatively large amount evidences the igneous origin this rock.

The effect of the IRP particle size on P-solubilization was then investigated by cultivating *A. niger* under both SmF and SSF using the IRP that was subjected to milling under different periods (Figure 1). It was observed a significant increase in P-solubilitization for both cultivation processes with the reduction of IRP particle size. For instance, there was a 56% increase in P-solubilization using IRP subjected to 10 min milling for SmF and a 53% increase for the SSF, when compared to the natural rock without milling. Therefore, a particle size after a 10 min milling time (which resulted in a surface area of 5.13 m^2 g_{rock}^{-1}) was choosen to be further evaluated, since this condition could be considered as a relatively short milling time that will not add a significant cost and energy demand to this process.

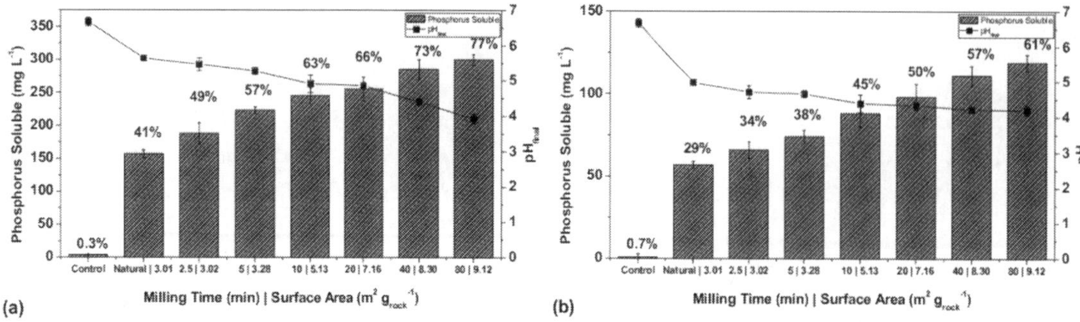

Figure 1: Phosphorus solubilization and pH_{final} using different particles sizes of IRP for *A. niger* cultivation **a)** Submerged fermentation using Pikovskaya medium, 96 h, 30 °C, 220 rpm, $pH_{initial}$ 7 and 5 g L^{-1} IRP. **b)** Solid-state fermentation using of sugarcane bagasse, 96 h, 30 °C, moisture 80%, $pH_{initial}$ 7 and 5 g L^{-1} IRP.

It is also important to consider that the acidification is a continuous process, and the kinetics of P-solubilization during fungal cultivation is important for bioprocess development. Thus, a kinetic study of P-solubilization was carried out using IRP at the selected particle size to evaluate the time profile of P-solubilization under both SmF and SSF cultivation systems. In both processes, maximum P-solubilization were achieved after 96 h of cultivation, reaching values of 63% of P-solubilization using SmF and 45% using SSF. However, these values were not significantly different from the ones achieved after 72 h. These findings show that *A. niger* strain used in this study can potentially solubilize a relatively large amount of phosphorus in a short period of time. It is also important to mention that the culture medium used in this study was not optimized yet. Therefore, the production of organic acids can still be optimized to further increase P-solubilization, since one of the main P-solubilization mechanisms is the reduction of pH by the production of organic acids [3].

Our results show that the mechanical treatment combined with the biological cultivation process is a potential strategy for P-solubilization and that even short periods of milling can be very effective to improve the solubilization of RP. Such proposed strategy contributes to future developments to implement large-scale industrial production of biofertilizers.

References

[1] Vassilev, N., Vassileva, M. 2003. Biotechnological solubilization of rock phosphate on media containing agro-industrial wastes. *Appl. Microbiol Biotechnol*, 61, 435-440.

[2] Kpomblekou, K. Tabatabai, MA. 1994. Effect of organic acids on release of phosphorus from phosphate rocks. *Soil Sci*, 158(6), 442–453.

[3] Mendes, G.O., et al. 2014. Mechanisms of phosphate solubilization by fungal isolates when exposed to different P sources. *Ann Microbial*, 64, 239-249.

[4] Chien, S.H., Menon, R.G. 2003. Factors affecting the agronomic effectiveness of fertilisers by mechano-milling. *Nutr. Cycl. Agroecosys.*, 67-177.

Influence of presence of asphaltenes on organic deposits in the petroleum process

A. Hemissi Boukadi, [1] M. Samcha
Entreprise Tunisienne d'Activités Pétrolières, rue des entrepreneurs 2037 la Charguia II Ariana, Tunisie;
tel. +216 70838302, e-mail: amel.boukadi@etap.com.tn;
[1] Entreprise Tunisienne d'Activités Pétrolières, rue des entrepreneurs 2037 la Charguia II Ariana, Tunisie; tel. +216 70838844, e-mail: meriam.samcha@etap.com.tn;

Organic solids that precipitate in installation petroleum lines during crude oil production could be paraffins, asphaltenes or a mixture of both in varying proportions, depending on the chemical and physical quality of the produced effluent. These precipitations are troublesome because they slow the flow, reduce the flow rates and can cause serious situations such as blockage of pipelines and in critical conditions, stop the flow.

Several studies have been conducted to determine the nature of this precipitations and responsible factors in their occurrence (1).These factors may be related to, thermodynamic changes between the tank and surface production lines; the quality of produced fluid, light, medium or heavy, and even the presence of water may cause or not precipitation of organic products.

In this context, we propose to present the main factors responsible for the breakdown of the natural balance that existed in reservoir before production, the different phases of transition between liquid and solid and interdependence between the asphaltenes and paraffins during the formation of deposits in the pipelines. At the end, we will present an optimization model for better assurance of flow, in oil production.

(1): Ariza-León, E., Molina-Velasco, D. R. & Chaves-Guerrero, A. (2014). Review of studies on asphaltene - wax interaction and the effect thereof on crystallization. CT&F - Ciencia, Tecnología y Futuro, 5(5), 39-53.

Keys words: Asphaltenes, organic deposits, flow assurance

Iron-based nanoparticles production by flame spray pyrolysis

C. Allouis[1] A. Budelli[2], S. Cimino[1], R. Nigro[3], P. Schiattarella[4]
[1] Istituto Ricerche sulla Combustione - CNR– Naples (IT)
[2]Heinz Innovation Center, NieuweDukenburgseweg 19, 6534, Nijmegen, Netherlands
[3]Università degli Studi di Napoli Federico II, DICMAPI, P.leTecchio 80, 80125 Naples, Italy
[4]I.T.P S.r.l., Via Bisignano a Chiaia, 68 – 80121 Naples, Italy

Iron is a component of heme in hemoglobin and myoglobin in which it plays an important role in the transport, storage and utilization of oxygen. Iron deficiency is one of the primary risk factors for health worldwide (especially woman and children). This deficiency induce anemia, alteration in mental development and decrease in immunity. Nutritional iron deficiency arises when physiological requirements cannot be met by iron absorption from diet.

Particle size is a determinant of iron (Fe) absorption from poorly soluble Fe compounds. Decreasing the particle size of metallic Fe and ferric pyrophosphate added to foods increases Fe absorption and bioavailability. The compounds most suitable for this purpose are iron salts as FePO4 e FeSO4.

Iron phosphate is commonly used as a human nutritional supplement and as a fertilizer ingredient.

Flame Spray Pyrolysis (FSP) is an industrial consolidated process used to obtain nanosized powders in large scale. However the application of nanotechnology to the field of medical food is the subject of controversial opinions due to the shortage of available toxicological data.

The aim of this study was to produce and characterize nanoparticles of FePO4 made by Flame spray pyrolysis (FSP). The lab-scale combustion system was a diffusion flame reactor with a stainless-steel gas-assisted atomizer to feed liquid precursor solution (Fig. 1)

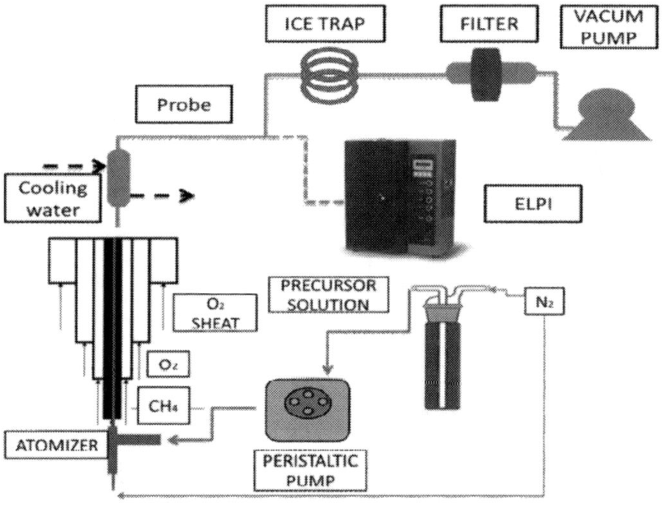

Figure 1. Experimental set-up for nanoparticles synthesis, in line sampling and powder collection.

The effect of precursor flow rate and concentration on particles size and powder production rate was investigated.

The particle size distribution functions were in-line characterized in the range 7 nm-10 μm by a low pressure impactor, ELPI^{TM}. Furthermore qualitative and quantitative analysis was made by gravimetric measurement, Scanning Electron Microscope (SEM) and nitrogen adsorption BET.

To characterize the reactive environment before precursor flame inclusion, also the CH4/xylene/O2 flame were studied by ELPI in-line analysis, SEM analysis and microanalysis Spherical particles were produced with average size ranging between 80-100nm: the concentration of iron was 67% wt%, while that of the phosphorous was 15% wt%. An increase in the numerical concentration of the particulate was observed with increasing the flow of precursor.

A microfluidic tool box for emulsification and separation research

K. Schroën, [1]K. Muijlwijk, [1]T, van der Laar, [1,2] S. Sahin, [1]J. Sprakel, [2]C. Berton-Carabin [1]
[1]Wageningen University, Dept. Agrotechnology and Food Science, Food Process Engineering group, Bornse Weilanden 9, 6708 WG Wageningen, The Netherlands, +31-317 4833896, karin.schroen@wur.nl, kelly.muijlwijk@wur.nl, sami.sahin@wur.nl, claire.carabin-berton@wur.nl.
[2]Wageningen University, Dept. Agrotechnology and Food Science, Physical Chemistry and Soft Matter group, Dreijenplein 6, 6703 HB Wageningen, The Netherlands, ties.vandelaar@wur.nl, joris.sprakel@wur.nl.

Microfluidic devices are known for their accurate control on liquid flows, reactions, temperature effects, etc. However, it is less known that these findings can also be used to develop novel process technology. In this contribution, we will show microfluidic tools that are applied in the emulsification and separation field, and illustrate how we applied them to broaden our knowledge base beyond the current state of the art, and design new processes based on it.

In microfluidic devices, it is no problem to make nice uniform droplets, but making a lot of them is not that simple. We took the so-called EDGE device as a starting point; it is unique in its ability to produce multiple droplets from one droplet formation unit. We have improved its' performance dramatically (> 100x) by designing the substructure such that also at high applied pressure the droplets remain the same size[1]. By placing these devices in parallel, productivities can be reached that are industrially relevant, although it should be mentioned that construction wise, there are considerable challenges.

After the droplets are formed, they also should stabilise fast enough to prevent coalescence. To monitor this effect on relevant size and time scales, and with a statistically relevant number of droplets, the coalescence chamber[2] and the micro-centrifuge[3] were developed. These techniques allow us to monitor stability both under flow conditions and under enhanced gravity, and that leads to a window of operation in which emulsion can be formed and kept in a stable way. Examples from both devices are shown in Figure 1.

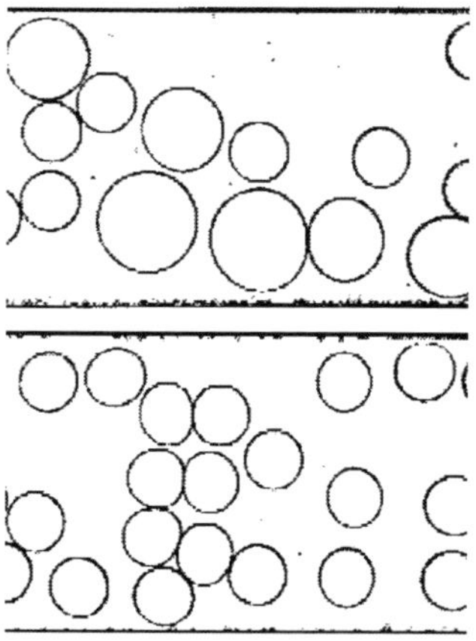

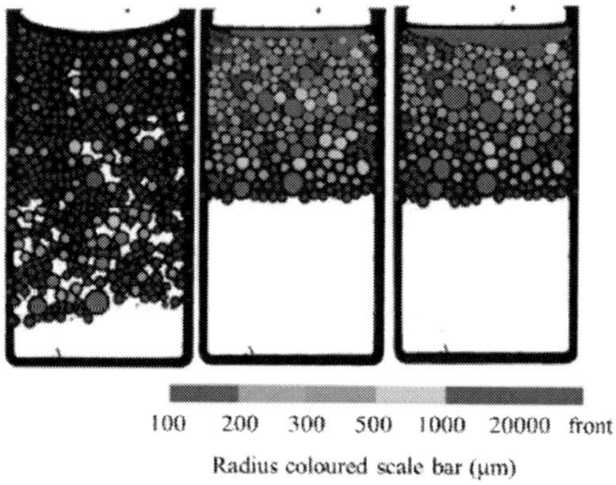

Figure 1. Examples of emulsions in the coalescence channel (left) top image unstable emulsion, bottom image stable emulsion, in the micro-centrifuge (right) in which the emulsion is compressed (going from left to right) and the oil layer on top (red) increases as function of time.

Also in the field of separation, microfluidic tools have helped us understand behaviour during filtration in great detail[5]. For example, we monitored particles in flow, and found that larger ones are mostly found in the centre of the channel, while small ones are found more closely to the wall. This facilitated filtration greatly; due to the segregation in flow the actual pore size did not matter that much anymore, only the ratio between transmembrane flow and cross-flow velocity determined the transmission of particles as illustrated in Figure 2. When chosen appropriately, large particle could be kept out of the permeate completely[6].

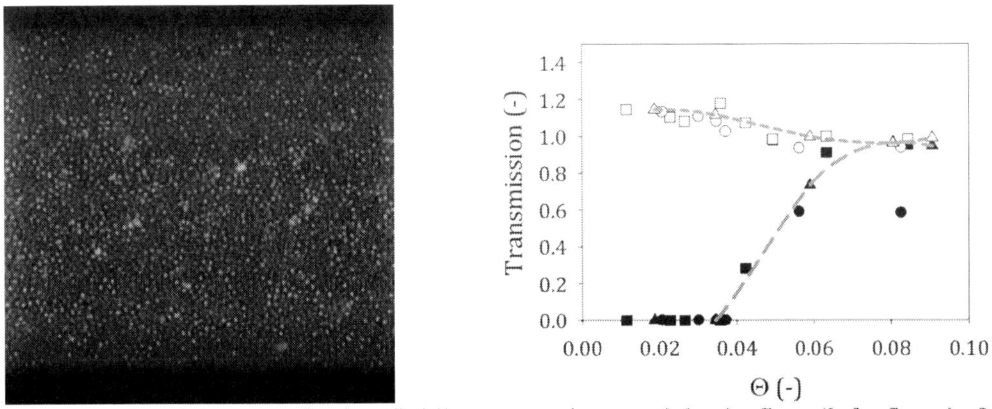

Figure 2. Example of the use of microfluidics to monitor particles in flow (left; flow is from left to right, large particles are green and concentrated in the middle) and the resulting process characteristic (right). If the ratio of transmembrane flow and cross-flow (Θ) is chosen properly, only small particles can be found in the permeate (open symbols) while the large ones (filled symbols) are completely retained.

In summary, microfluidic tools have helped us tremendously in understanding processes at the micrometre scale (and millisecond range), and this has allowed us to develop not only tools that can probe these scales, but also to take it one step further and develop new process technology based on our findings.

1. Sahin S, Schroën, K: Partitioned EDGE devices for high throughput production of monodisperse emulsion droplets with two distinct sizes. Lab on a Chip 2015, 15(11): 2486-2495.
2. Krebs T, Schroen K, Boom R: A microfluidic method to study demulsification kinetics. Lab Chip 2012, 12(6):1060-1070.
3. Krebs T, Ershov D, Schroen CGPH, Boom RM: Coalescence and compression in centrifuged emulsions studied with in situ optical microscopy. Soft Matter 2013, 9(15):4026-4035.
4. Steegmans MLJ, Warmerdam A, Schroën CGPH, Boom RM: Dynamic interfacial tension measurements with microfluidic Y-junctions. Langmuir 2009, 25(17):9751–9758.
5. Van De Laar T, Schroën K, Sprakel J: Cooperativity and segregation in confined flows of soft binary glasses. Physical Review E - Statistical, Nonlinear, and Soft Matter Physics 2015, 92(2):022308.
6. Van Dinther AMC, Schroën CGPH, Boom RM: Particle migration leads to deposition-free fractionation. Journal of Membrane Science 2013, 440:58-66.

Advanced millireactor technology for the kinetic investigation of very rapid reactions: preparation of epichlorohydrin by dehydrochlorination

Cesar A. de Araujo Filho[1], Shuyana Heredia[1], Kari Eränen[1], Tapio Salmi[1]
[1]Johan Gadolin Process Chemistry Centre, Laboratory of Industrial Chemistry and Reaction Engineering, Åbo Akademi, FI-20500, Turku/Åbo, Finland; tel +35822154427, e-mail:caraujo@abo.fi, fax:+35822154479

In recent years, increasing attention has been paid on the use of biomass-derived raw materials for the production of fuels and chemicals. Following this trend, glycerol has emerged as a potential platform molecule for the production of several value-added chemicals. Glycerol is a stoichiometric by-product of biodiesel production via trans-esterification reaction; in fact, it constitutes ca. 10% wt. of the crude biodiesel produced. Epichlorohydrin is a valuable chemical intermediate which can be used in the production of epoxy resins and plasticizers. It can be obtained from glycerol originating from biofuel production instead of fossil raw materials.

The first step in the production of bio-based epichlorohydrin involves the hydro-chlorination of glycerol in the presence of a weak organic acid as a homogeneous catalyst, such as acetic, propionic, malonic or adipic acid. The reaction scheme of glycerol hydro-chlorination gives 3-chloro-1,2-propanediol (α-MCP) and 2-chloro-1,3-propanediol (β-MCP) as primary products and 1,3-dichloro-2-propanol (αγ-DCP) and 1,2-dichloro-3-propanol (αβ-DCP) as secondary products. Typically α-MCP and αγ-DCP are very dominant among the products. Both αγ-DCP and αβ-DCP can be transformed to epichlorohydrin by dehydrochlorination in the presence of alkali (NaOH, KOH or $Ca(OH)_2$).

The generally accepted mechanism for the dehydrochlorination of αγ-DCP involves a base catalyzed dissociation of the αγ-DCP molecule, followed by an internal nucleophilic substitution as depicted below,

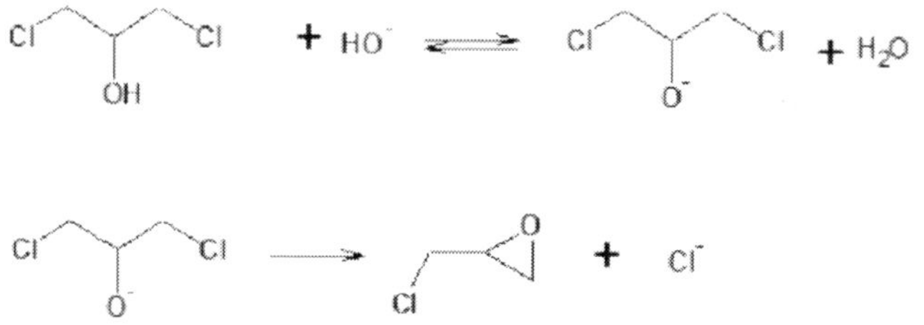

The experiments were conducted in a continuous millireactor constructed by us. The equipment consisted of four stainless steel tubes (á 1 m) of different internal diameters (0.5, 0.7, 1.0 and 1.7 mm) connected in series. At the very beginning of the reaction zone and at the intersection of each tube, sampling valves were placed, allowing sampling at different residence times. Due to the small volume of the reactor, short residence times were achieved and samples could be collected at residence times as short as 0.98 s. Before the first sampling valve, the reactants (αγ-DCP and NaOH) travelled through separate 1 meter long pre-heating tubes. The pre-heating and reaction zones were immersed in a water bath. HPLC pumps provided a precise flow of the liquids. Isothermal experiments were conducted at 30-70°C with various reactant ratios (1:1, 1:3 and 3:1). The concentrations of αγ-DCP and epichlorohydrin were analyzed by using a gas chromatograph (capillary column JW Scientific DB-WAX, 30 m x 0.250 mm x 0.25 μm; flame ionization detector).

The reaction rate for the dehydrochlorination of αγ-DCP can be written as

$$r = kc_{\alpha\gamma}c_{OH^-}$$

The flow characteristics in the millireactor was checked by calculating the Reynolds number in the different parts of the reactor at the reaction temperatures. It turned out that the reactor can be described with a simple plug flow model. The steady-state mass balance equation is

$$\frac{dc_i}{d\tau} = r_i$$

where τ represents the mean residence time and i=αγ-DCP and epichlorohydrin.

The fit of the experimental data at different temperatures and reactants ratio is depicted in Figure 1, indicating a very good description of the experimental data by the model.

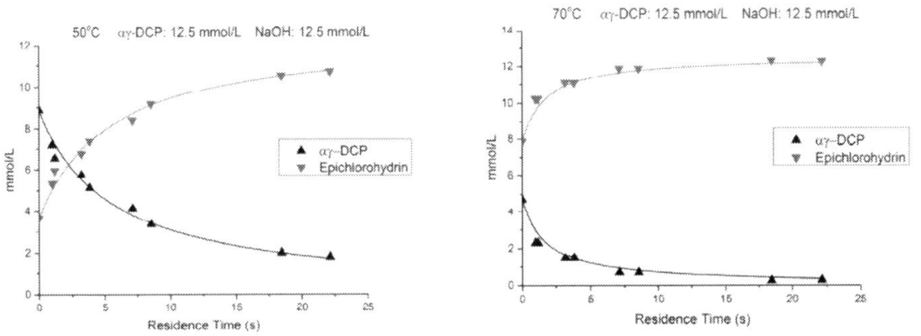

Figure 1. Examples of experimental data (dots) and model predictions (continuous lines).

The kinetic parameters were obtained by nonlinear regression analysis: $k(T_{ref})=0.0212m^3/mol/s$ and Ea=82.6kJ/mol. The errors of the kinetic parameters were 2.0 and 1.5%, respectively (T_{ref}=50°C). A contour plot showing the sensitivity analysis of the objective function is depicted in Figure 2.

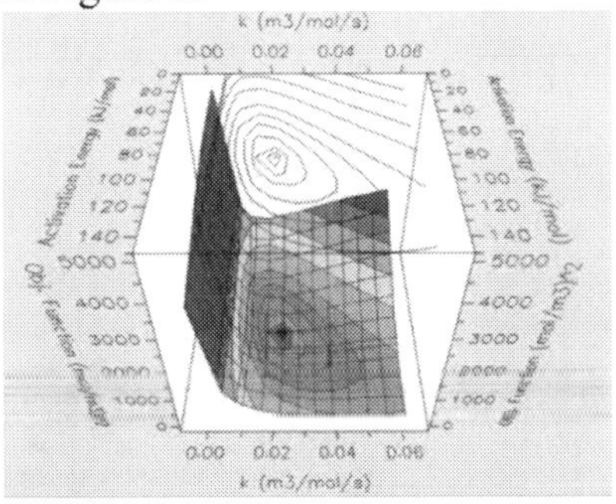

Figure 2. 3D contour plot depicting objective function surface (Q) as a function of the two estimated parameters.

The contour plot displayed in Figure 2 shows a single well-defined minimum of the objective function surface, determined by varying the two kinetic parameters adjusted by the model. It attests the excellent accuracy of the parameter estimation. The work shows that a continuous millireactor is a strong tool in the investigation of rapid liquid-phase reaction kinetics with a high precision. The millireactor construction is flexible and inexpensive.

Carbon dioxide hydrogenation by means of plasmonic resonance activation

S. Muñoz[a], A. Navarrete[a], L.M. Sanz-Moral[a], A. Martín[a], R. Dittmeyer[b], M.J. Cocero[a].
a) University of Valladolid, Department of Chemical Engineering and Enviromental Technology, High Pressure Processes Group, Calle Dr Mergelina S/N 47011 Valladolid, Spain. Tel: +34 639677958, e-mail: sergiomp@iq.uva.es.
b) Institute for Micro Process Engineering, Karlsruhe Institute of Technology, Karlsruhe, Germany.

COP21 has been an important achievement to fight climate change at a global scale[1]. It is going to limit the use of conventional fuels and to impulse renewable energies. However, one of the main problems of renewable energies is their intermittence and therefore we need of efficient storage methods which allow their use at any moment.

The use of solar energy for the synthesis of products such methanol provide new economic opportunities and an alternative to transform CO_2 into clean fuels. In fact, current production of methanol would store more than 60 million tons of CO_2.

Direct hydrogenation of CO_2 shows low conversion, but the industrial CAMERE process improves the conversion via an intermediate RWGS reaction.

$$CO_2 + H_2 \Leftrightarrow CO + H_2O;\ \Delta H^0 = +41.19\ kJ/mol \quad (1)$$

$$CO + 2H_2 \Leftrightarrow CH_3OH;\ \ \Delta H^0 = -90.70\ kJ/mol \quad (2)$$

The RWGS is endothermic, and performing it with renewable energies is still a challenge.

In this work, the plasmonic catalytic activation is proposed as a new method to carry out the hydrogenation of CO_2. Visual light is going to be used to activate by plasmon resonance in a Cu/ZnO catalyst, which has shown good results in conventional hydrogenation of CO_2[2].The catalyst will be impregnated in glass microreactors for efficient energy transference and silica transparent aerogels will be used as a support for the catalyst as they combine both high surface areas and good light transmission.

The synthesis of a plasmon tuneable catalyst was made using the method reported by Tan et al (2013)[3]. This method allows the synthesis of Cu/ZnO nanocomposites whose visual light absorption have been tested and their morphology and size was analyzed with TEM (Figure 1).

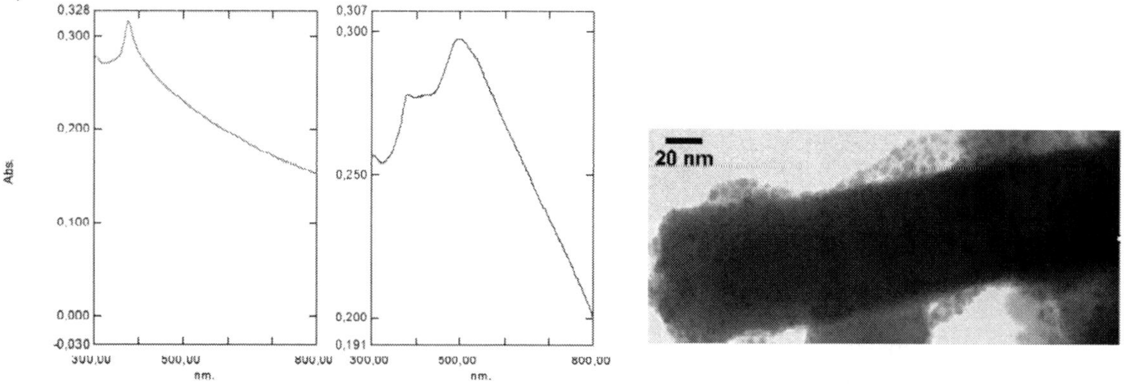

Figure 1. Visual light and TEM image of Cu/ZnO nanocomposites.

The nanocomposites were then supported by direct synthesis on silica aerogels and dried with superctitical CO_2 as reported by Sanz- Moral et al (2014)[4]. A new simple method has been developed in this research to facilitate the impregnation of the aerogel on the walls of the microreactors creating a sole entity for the transmission of light. With this method, the aerogel containing the metal catalyst is impregnated when it is still liquid (hydrogel), and the aging and supercritical drying is performed directly inside the microchannels (Figure 2).

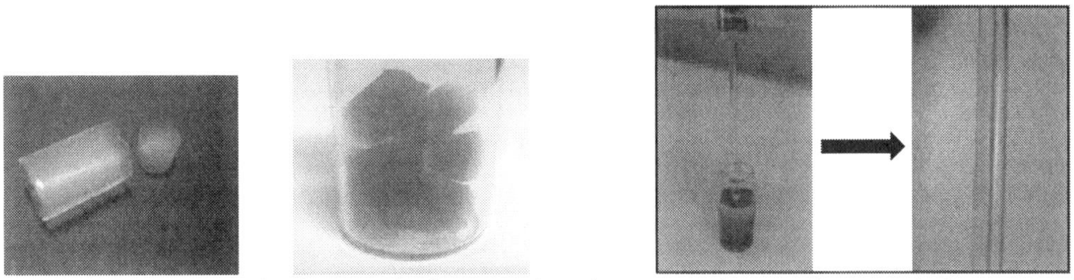

Figure 2. Aerogels before and after catalyst impregnation, and aerogels impregnation in microchannels

An experimental setup for performing the CO_2 hydrogenation has been built up. Hydrogen and carbon dioxide are mixed in a 3:1 proportion and heated in an oven together with the glass microreactor. Visual light is provided by 36 LEDs which provide a nominal power of 9780 W/m^2. The setup allows a precise control of gas inlet flows, pressure and temperature, to study their influence in the process. The oulet gases are analyzed by in-line gas chromatography analysis.

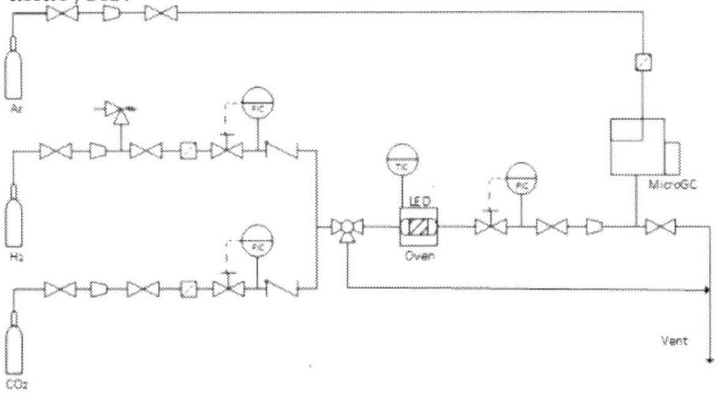

Figure 3. Experimental setup for CO_2 hydrogenation.

First experiences working under low temperatures (50 – 70 °C) and 20 bar have shown encouraging results, achieving CO_2 conversion rates similar to other works at higher temperatures, with an average CO_2 conversion rate of 381 mmol / (gcat·min).

The work is going to continue studying the influence of reaction variables, such as temperature, pressure, inlet flows and amount of catalyst. Optimization routes will be proposed to achieve the highest CO_2 conversion rate as a way to high methanol yields.

This work presents a new route to use renewable energy to create "solar commodities". And in this way we propose a new family of microreactors which integrate in a single entity, plasmonic activation, high light transmittance and continuous reaction evolution. The proof-of-concept is made for the CO_2 hydrogenation with visual light.

1- Cozier, M. The un COP21 Climate Change Conference and the role of CCS. *Greenhouse Gases: Science and Technology*, **2015**, vol. 5, p. 697-700.
2- Jadhav, S.G.; Vaidya, P.D.; Bhanage, B.M.; Joshi, J.B. Catalytic carbon dioxide hydrogenation to methanol: A review of recent studies. *Chemical Engineering Research and Design*, **2014**, vol. 92, p. 2557-2567.
3- Tan, Z. Y.; Yong, D.; W. Y.; Zhang, Z.; Low, H. Y.; Chen, L.; Chin, W. S. Nanostructured Cu/ZnO Coupled Composites: Toward Tunable Cu Nanoparticle Sizes and Plasmon Absorption. *The Journal of Physical Chemistry C,* **2013**, vol. 117, no 20, p. 10780-10787.
4- Sanz-Moral, L.M.; Rueda, M.; Mato, R.; Martín, A. View cell investigation of silica aerogels during supercritical drying: Analysis of size variation and mass transfer mechanisms. *The Journal of Supercritical Fluids*, **2014**, 92, 24-30.

Debottlenecking of chemical processing through Process Intensification: ethylene glycol case study

M. Barecka[ab], M. Jaberi[a], M. Skiborowski[a], A. Górak[ab],
[a]Laboratory of Fluid Separations, TU Dortmund University, Emil Figge Str 70, 44227 Dortmund, Germany, Tel.: +49 231-755 26 82, e-mail: magda.barecka@bci.tu-dortmund.de
[b]Lodz University of Technology, ul. Wólczańska 213, 90-924 Łódź, Poland, Tel. +49 426313784

The economic and environmental challenges that the chemical industry is facing have led to a rising interest in Process Intensification (PI). PI has been defined as the development of innovative apparatuses and techniques, which offer drastic improvements in chemical manufacturing and processing [1], by enhancing of phenomena through integration of unit operations, integration of functions, integration of phenomena and/or targeted enhancement of a phenomenon within an operation [2]. Overall, the idea of PI is to replace large, expensive and energy-intensive process parts with ones that are smaller, more efficient and economical beneficial [3]. As a matter of fact there is already a wide variety of intensified equipment available, which can tackle many of the most difficult process bottlenecks [1]. However, up till now, only a limited number of PI technologies have achieved implementation [4]. This can mainly be attributed to the absence of means for a systematic identification of potential PI options and a quick evaluation of the potential improvement. Due to missing data and models, detailed modeling and simulation can rarely be performed for PI options. Even if the required information is available, a detailed simulation can only be performed for a few selected options due to time and resource limitations. As a result, a systematic identification of the most promising PI equipment for a given bottleneck is a complex task. To this end, this method is proposed, that enables a fast identification of the relevant process bottlenecks and the identification of feasible PI options in combination with a quantitative evaluation of the possible improvement concerning the given bottleneck (e. g. reaction selectivity).

Building up on the approach of Lutze et al. [2], numerous PI units for tackling a given bottleneck are identified on the basis of a broad PI database. The database includes more than 150 intensified equipment options and technologies previously reported in the literature. All of them are described by numerous parameters characterising the feasibility of their application and the specific phenomena that the equipment is reported to improve (e.g. mass transfer). Furtherly, the database contains characteristic parameters describing the equipment, as e.g. surface to volume ratio for microreactors. The technologies included in the database are referenced to the literature data on its experimental demonstration and applications.

The PI database enables to screen out the infeasible equipment or technologies for improvement of the target phenomena by considering such criteria as e.g. feed and product stream phases, constraints on process operating conditions or feasibility indicators derived from thermodynamic insights [5]. The remaining debottlenecking options are further evaluated by means of shortcut models to confirm their feasibility and provide a quantitative assessment of the intensification potential. Based on these estimates a preselection of the most promising intensified equipment is achieved in an early design phase, at which available data is commonly sparse.

The debottlenecking methodology and associated models are illustrated for the case study of ethylene glycol production. Within the framework of proposed debottlenecking approach, at first the limitations of the given process are analysed based on a detailed simulation of a base case design for the ethylene glycol production taken from literature [6].

The major bottlenecks of the process are identified in respect to the specified objective of improvement, which is the operational cost per unit of product. The selectivity of the reactor towards production of the ethylene glycol is determined as the main process bottleneck. Taking into account the given bottleneck, a stepwise screening of the PI database is performed in order to identify feasible PI options. Thereby more than 150 reported technologies are taken into account. After the stepwise screening 5 remaining PI options are further evaluated and a quantification of the potential improvement of feasible options is performed by application of shortcut methods. By means of this screening procedure reactive distillation is determined as the most promising option for improving the ethylene glycol production. In order to validate the potential improvement suggested by the shortcut evaluation, the process flowsheet is modified with a reactive distillation column and a rigorous simulation of the whole flowsheet is performed. While the rigorous simulation confirms the potential improvement of the selectivity by means of the reactive distillation, it also shows that application of the reactive distillation does not result in a significant reduction of operational costs for the case study. Nevertheless, the flowsheet modification results in a simplified flowsheet with a reduced number of distillation columns needed to achieve the desired product quality.

The presented approach provides a systematic analysis for identifying the most relevant process improvement and the determination of the most promising intensification options by means of a quantitative assessment. The whole method can be applied as a support in the early conceptual design phase, when the available data is sparse and detailed simulation for numerous equipment options is infeasible. The method is therefore meant as support tool to identify the PI equipment which should be investigated in more detail.

Keywords: Process Intensification, shortcut modelling, reactive distillation, debottlenecking

References:

[1] Stankiewicz, A.; Moulijn, J.; Process intensification: transforming chemical engineering, Chem. Eng. Process 2003, 42, p. 137-444

[2] Lutze, P.; Gani, R.; Woodley, J.; Process intensification: A perspective on process synthesis, Chem. Eng. Processing: Process Intensification 2010, 49, p.547-558

[3] Tsouris C, Porcelli JV. Process Intensification - Has its time finally come?, Chem. Eng. Prog. 2003, 99(10), p. 50-55.

[4] Becht, S.; Franke, R.; Geißelmann, A.; Hanh, H.; An industrial view of process intensification, Chem. Eng. Process 2009, 48, p. 774-780

[5]Jaksland, C.; Gani, R.; Lien, K.; Separation process design and synthesis based on thermodynamic insights, Chem. Eng. Science 1995, 50, p. 511-530

[6] Dye, R., Ethylene Glycols Technology, Korean J. Chem. Eng., 2001, 18(5), p. 571-579

Microscale technology and biocatalytic processes: from lab on a chip to production scale

Igor Plazl
Faculty of Chemistry and Chemical Technology, University of Ljubljana, Večna pot 113, 1000 Ljubljana, Slovenia; tel. +38614798580, e-mail: igor.plazl@fkkt.uni-lj.si

Numerous excellent publications in the last two decades demonstrated the advantages of microfluidic devices for very efficient performing of chemical and biochemical processes at controlled and repeatable conditions. New concepts such as continuous processing, flow chemistry, high-throughput screening, and process intensification, have been established in order to open novel pathways in process design and engineering. Besides, a number of highly innovative and systematic approaches, protocols, tools, and strategies has currently being developed in both industry and academia, all to minimize the gap between research and industry, and to define a smooth transfer of (bio)chemical processes to the industrial environment. To meet these challenges, we must advance the field from (bio)catalyst discovery to (bio)catalytic microprocess design. This will require not only a new level of understanding of reaction mechanisms and transport phenomena at the micro scale, but also the development of relevant computational tools [1].

In this work the microscale bioprocess development based on scale-up/scale-out/numbering-up concept in combination with modeling-based optimization is presented (Fig. 1). Theoretical description of transport phenomena and the kinetics at the micro scale is discussed together with the presentation of multiscale modeling concept. Multiscale modeling combines existing and emerging methods from diverse scientific disciplines to bridge the wide range of time and length scales that are inherent in a number of essential phenomena and processes in materials science and engineering. Multiscale modeling may help to design and optimize microsystems technology for chemical and biochemical processes from a microscopic point of view [1].

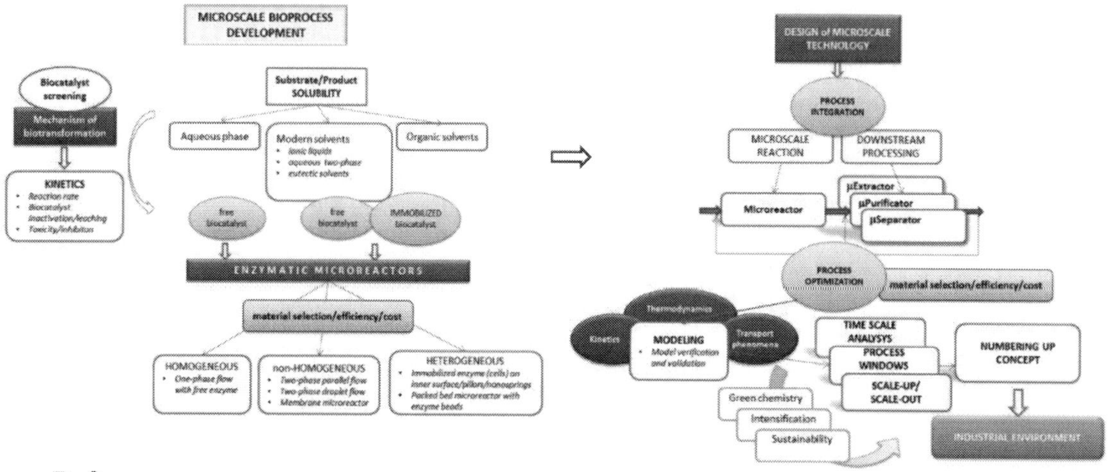

References:
1. Wohlgemuth, R., Plazl, I., Žnidaršič-Plazl, P., Gernaey, K.V. and Woodley, J. M. Microscale Technology and Biocatalytic Processes – Opportunities and Challenges for Synthesis, *Trends Biotechnol.*, **2015**, 33: 302-314

Forced convective transport of alumina-water nanofluid in micro-channels

C. Yang, K. L. Peng, T. Qiu
Fuzhou Univ., School of Chemical Engineering, Xue Yuan Rd. 2, 350116 Fuzhou, China;
tel. +591 22866241, e-mail: cyang@fzu.edu.cn;

Mathematical and physical models

In the modified Buongiorno's model proposed by Yang et al.[1], the nanofluid density variation resulting from the non-uniform distribution nanoparticles was considered in the governing equations. However, the influence of shear-induced and viscosity-gradient induced diffusions on the migration of nanoparticles in the base fluid have been neglected. In micro-channels, viscous dissipation could play a significant role in the temperature field solution. Hence, this model should be further modified to comprise the effects of shear-induced, viscosity-gradient induced diffusions, and viscous dissipation as

$$\nabla\cdot(\rho\mathbf{u})=0 \tag{1}$$

$$\frac{\partial\rho\mathbf{u}}{\partial t}+\rho\mathbf{u}\cdot\nabla\mathbf{u}=-\nabla p+\nabla\cdot(\mu\nabla\mathbf{u}) \tag{2}$$

$$\begin{aligned}\frac{\partial\rho cT}{\partial t}+\nabla\cdot(\rho c\mathbf{u}T)=\nabla\cdot(k\nabla T)+\mu\nabla\mathbf{u}:\nabla\mathbf{u}\\ +\rho_p c_p\left(D_B\nabla\phi+\frac{D_T}{T}\nabla T+\frac{D_S}{\dot\gamma}\left(\nabla\dot\gamma+\frac{\dot\gamma}{\phi}\nabla\phi\right)+\frac{D_\mu}{\mu}\nabla\mu\right)\cdot\nabla T\end{aligned} \tag{3}$$

$$\frac{\partial\phi}{\partial t}+\nabla\cdot(\phi\mathbf{u})=\nabla\cdot\left(D_B\nabla\phi+\frac{D_T}{T}\nabla T+\frac{D_S}{\dot\gamma}\left(\nabla\dot\gamma+\frac{\dot\gamma}{\phi}\nabla\phi\right)+\frac{D_\mu}{\mu}\nabla\mu\right) \tag{4}$$

where $\mathbf{u}$, p, T, and ϕ are the velocity vector, pressure, temperature and nanoparticles volume fraction of the nanofluid, respectively. Moreover, ρ, c, μ and k are the density, heat capacity, dynamic viscosity and thermal conductivity of the nanofluid, respectively.

The steady-state, hydrodynamically and thermally fully developed nanofluid flows into a horizontal micro-channel with its wall subject to constant and uniform heat flux, as illustrated in Fig. 1. In this study, the nanofluid in consideration is a suspension with alumina nanoparticles in distilled water. In the cross section of micro-channel, nanoparticles distribution is controlled by four driving mechanisms, which are Brownian diffusion, thermophoresis, shear-induced diffusion and viscosity gradient-induced diffusion, respectively.

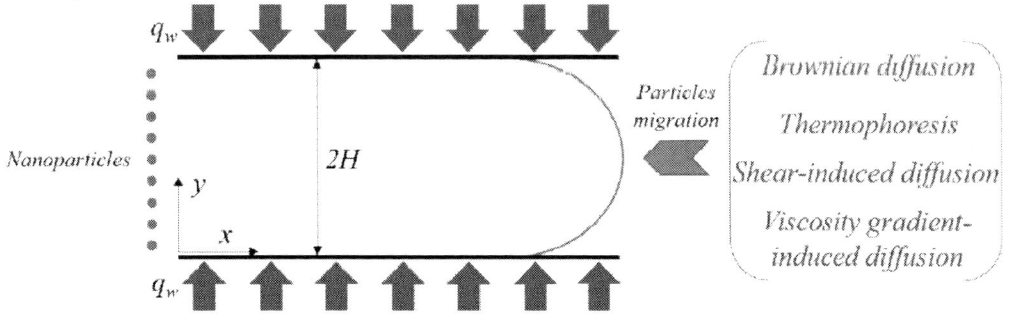

Fig. 1 Schematic diagram of alumina-water nanofluid in a horizontal channel

In order to consider the effects of slip velocity and temperature jump conditions at the wall, the following boundary conditions are used in this study:

$y=0$ (on the wall)

$$u_N = \lambda \frac{\partial u}{\partial y}\bigg|_w \tag{5a}$$

$$-k\frac{\partial T}{\partial y} = q_w \tag{5b}$$

$$D_B\frac{d\phi}{dy}+\frac{D_T}{T}\frac{dT}{dy}+\frac{D_S}{du/dy}\left(\frac{d^2u}{dy^2}+\frac{1}{\phi}\frac{du}{dy}\frac{d\phi}{dy}\right)+\frac{D_\mu}{\mu}\frac{d\mu}{dy}=0 \tag{5c}$$

$y=H$ (on the centerline)

$$\frac{du}{dy}=\frac{\partial T}{\partial y}=\frac{d\phi}{dy}=0 \tag{6a, b, c}$$

Effect of viscous dissipation

From the physical point of view, the Brinkman number Br is the ratio of the heat production due to viscous force to heat transferred from the wall to the fluid, and defined as follow

$$Br \equiv \frac{\mu_w u_B^2}{q_w H} \tag{7}$$

Hence, zero Br number means that there is no viscous dissipation. Moreover, positive or negative Br number corresponds to heating of fluid or its cooling, respectively. In Fig. 2, heat transfer characteristic of alumina-water nanofluid in a channel is investigated under the variation of Brinkman number. For the heat transfer characteristic of alumina-water nanofluid, the increase of Br weakens the heat transfer performance for the heating cases as shown in Fig. 2. The reason for this is that the rise of Br not only reduces the temperature gradient near the wall, but also forces more nanoparticles to migrate from the wall to the centerline, which decrease the two elements of Nu_B, respectively. Based on the discussion above, the heat transfer enhancement resulted from the increase of Br for the cooling cases can also be explained in the same way. Moreover, the role of Br is better appreciated by considering the case of fluid cooled in micro-channel.

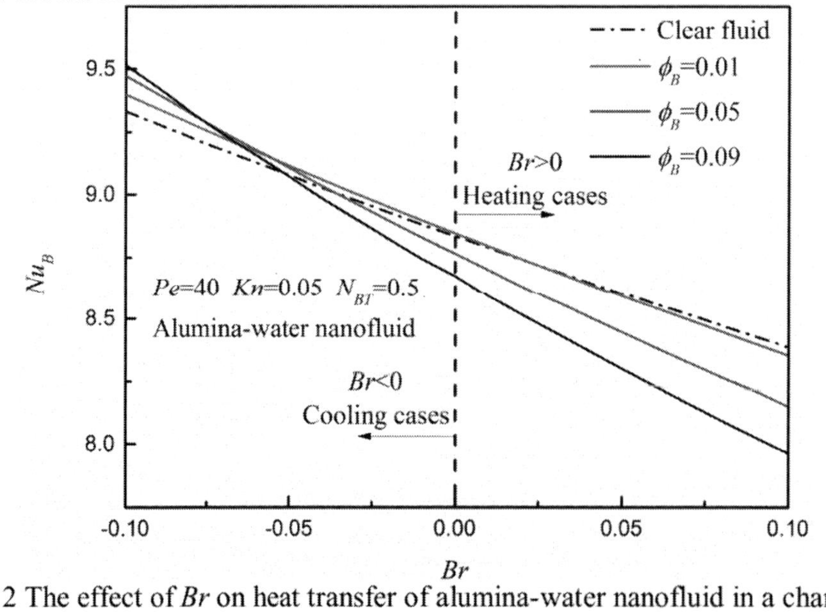

Fig. 2 The effect of Br on heat transfer of alumina-water nanofluid in a channel

References

1. Yang C, Wang Q, Nakayama A, Qiu T. Effect of temperature jump on forced convective transport if nanofluids in the continuum flow and slip flow regimes. Chem. Eng. Sci. 2015; 137: 730-739.

Lab-in-a-particle microchemical systems: remote control of polymerase chain reaction (PCR) by radiofrequency stimulation

Ayse Beyza Aysan[1], Zdeněk Knejzlík[2], František Štěpánek[1]
[1]University of Chemistry and Technology, Faculty of Chemical Engineering Department, Prague, Technicka 5, 166 28 Prague 6 – Dejvice;
tel. +420 220444320, e-mail: aysana@vscht.cz;
[2]University of Chemistry and Technology, Faculty of Food and Biochemical Technology Department, Prague, Technicka 5, 166 28 Prague 6 – Dejvice

Nanomaterials have gained considerable attention owing to their exceptional physical, chemical and biomedical properties including optical, mechanic, magnetic; reaction activity and catalytic properties; curing, delivery and controlled release properties. These nanomaterials can be synthesized and carried inside the spatially confined volume of individual micro and submicroreactors. The modelling and the mimicing of biochemical processes that combined nanomaterials with living cells is a new and interseting area to light the hindrances.[1,2,3]

The aim of this work is to design and implement microscopic biochemical reactors, in the form of cell-sized microcapsules, that can perform biochemical reactions such as the reproduction and amplification of nucleic acids *in vitro* through the polymerase chain reaction (PCR) and be controllable remotely in a contactless way. To achieve this, three technologies have to be combined: (i) magnetic iron oxide nanoparticles (SPIONs), which act as the local source of heating by dissipating heat upon exposure to a radiofrequency (RF) alternating magnetic field; (ii) PCR mixture (primers, enzyme, etc.), and (iii) core-shell microcapsules, in which all the components are embedded. In this specific contribution, we present an investigation of the compatibility of the SPIONs and the PCR reaction components.

SPIONs were fabricated succesfully and characterized in terms of concentration, size & size distribution, surface morphology, radiofrequency heating measurement. The concentration of the SPIONs was calculated as 22.6 mg/ml Fe (4.517 μg/μl in the PCR mixture). The particle size range was about 10-15 nm (from TEM images, Fig 1a). Magnetic induction heating studies was carried out by calculating Specific Absorption Rate number and determined the heating capability of nanoparticles with this concentration and size. Measurements showed that nanoparticles increase the temperature from 30°C to 100°C within 28 seconds (Fig 1b). PCR reaction was carried out in the presence of SPIONs. In the beginning, the concentration was 4.517 μg/μl in the PCR mixture then it was decreased to see the effect. Agarose gel electrophoresis results clearly indicated that the amplification efficiency decreased with higher concentration that was seen in the Figure 2.

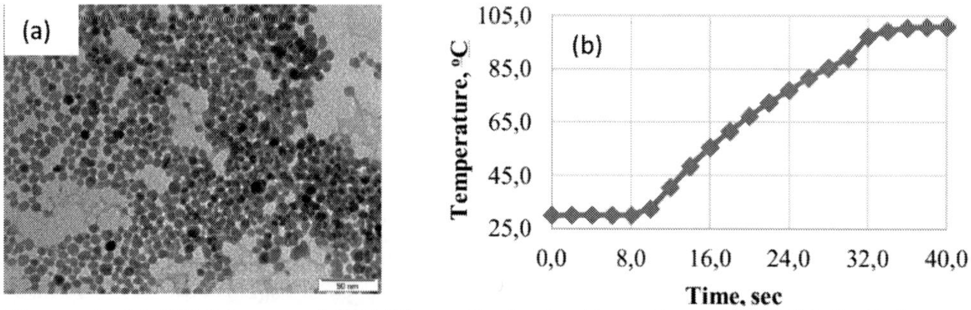

Figure 1. (a) TEM image of SPIONs synthesized with sodium citrate (b) Radiofrequency Heating Induction curve of SPIONs synthesized with sodium citrate

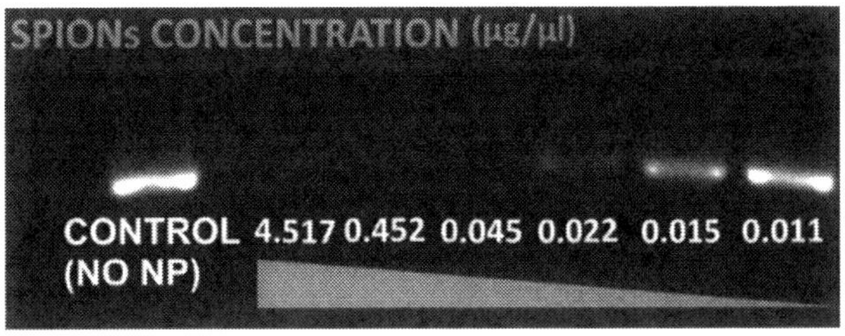

Figure 2. Agarose gel electrophoresis of PCR products with SPIONs synthesized with sodium citrate

Also, it was observed that nanoparticles were not stable in the PCR solution. They aggregated while storing the reaction buffer in empty spaces and resulted in less amount of reaction mixture for amplification. Different stabilizing agents including citrate, dextran, 3-aminopropyltriethoxysilane (APTES) and polyethylene glycol (PEG) was used to improve the stability and compatibility of SPIONs in PCR reagents. All experiments were done in the same way and optimized conditions were obtained for the further experiments.

Acknowledgement
This study is supported by Specific University Research (MSMT No. 29/2015).

References

1. Shchukin D.G., and Sukhorukov G.B., 2004, Nanoparticle Synthesis in Engineered Organic Nanoscale Reactors, Advanced Materials, 16 (8), 671-682.
2. Nune S.K., Gunda P., Thallapally P.K., Lin Y.Y., Forrest M.L., Berkland C.J., 2009, Nanoparticles for biomedical imaging, Expert Opin Drug Deliv., 6(11), 1175-1194.
3. Mahmoudi M., Sant S., Wang B., Laurent S., Sen T., 2011, Superparamagnetic iron oxide nanoparticles (SPIONs): Development, surface modification and applications in chemotherapy, Advanced Drug Delivery Reviews, 63, 24-46.

Liquid–liquid micromixing in a large spinning disc reactor: Application of smooth disc surface and single-point distributor

S. R. Al-hengari[1, 2] , K.V. K. Boodhoo[2]
[1]Libyan Petroleum Institute, P.O. Box: 6413, Tripoli – Libya
Tel. +218 92 4538805; Email: salah_alhengari@yahoo.co.uk ; Fax +218 21 4830031
[2]School of Chemical Engineering and Advanced Materials, Newcastle University, Newcastle upon Tyne, NE1 7RU, UK

Micromixing phenomena (i.e. mixing at molecular level) plays a very important role in the chemical industry and the study of micromixing improves the understanding of important processes such as combustion, polymerization, crystallization and competing fast chemical reactions. The (SDR) is designed to create highly sheared thin liquid films which have excellent heat/mass transfer properties. The liquid is spread out on the rotating surface in the form of very thin films (50-500 μm) having high shear rates, a large number of surface waves and intense surface ripples which are known to play an important role in improving the selectivity, yield, and quality of final products of a chemical reaction. This paper concerns the characterisation of the micromixing process within a thin film flowing on the 30cm SDR using a smooth disc surface and Single-Point Distributor. A parallel-competitive reaction test scheme (i.e. the iodide-iodate test reaction) was adopted to quantify micromixing in terms of the segregation index (Xs), micromixdess ratio (α), power dissipation (ε) and micromixing time(t_m). Various parameters where investigated including: effects of disc rotation rates, feed flowrates, and effect of total flow rate ratio, liquid feed concentrations and viscosities. A comparative analysis of the micromixing efficiency in 30cm SDR and 1.37l semi-batch stirred reactor, (SBR), was carried out. The SDR is shown to give significantly better micromixing performance which highlights its potential as an alternative device for processes where a high degree of mixing is critically important.

Kay words: Micromixing characterisation; Spinning disc Reactor; iodide-iodate reaction; segregation index (Xs); micromixdess ratio (α); power dissipation (ε); micromixing time(t_m); smooth disc surface; Single-Point Distributor.

Scale-up of oscillatory helically baffled reactors (OHBRs)

Safaa M.R. Ahmed, Anh N. Phan, Adam P. Harvey
School of Chemical Engineering and Advanced Materials, Newcastle University
Newcastle upon Tyne - NE1 7RU
Mobile:+44 7774101215
E-mail: s.m.r.ahmed@newcastle.ac.uk
Keywords: Oscillatory baffled reactor, scale-up, residence time distribution

An Oscillatory Baffled Reactor (OBR) is an intensified design of continuous plug flow reactor (PFR) in which plug flow behaviour can be achieved at very low net flows (laminar flow regime). OBRs consist of tubes with periodically spaced baffles of various designs (orifice, helical, integral etc. baffles). There is a net flow through the reactor, and a superimposed oscillatory flow. The oscillatory flow interacts with the baffles to produce flow structures (usually vortices) that provide mixing. The mixing in the OBR is therefore independent of the net flow. As a result, the OBR's niche application is to operate "long" reactions in continuous mode. This is usually impractical in conventional tubular reactors.

Scale-up in conventional reactors, i.e. stirred tank reactors, is unpredictable due to the non-uniform mixing at large scale, leading to large variations in concentration, temperatures etc. This means that optimum conditions obtained from laboratory scales cannot be directly used at large scales, therefore process development/ product-to-market time would increase. However, scale-up of OBRs should be more predictable, as the flow structures can be reproduced across length scales.

Recent studies on oscillatory helically baffled reactors (OHBRs) at small scales (millilitre volume) found that the helical baffled design could provide high degree of plug flow across a wide operating window due to the combined effect of vortex formation and swirl flow. It was found that the behaviour of residence time distribution remained the same at all tested scales (10mm diameter to 25mm diameter, corresponding to volume 0.078L to volume 0.834L) when the similarity in geometric and dynamic parameters was maintained. The degree of plug flow was quantified in terms of number of tanks-in-series (N). At a fixed geometry, a scale-up correlation was established and validated over a range of operating conditions such as oscillatory conditions (Strouhal number, *St*, and oscillatory Reynolds number, Re_o) and the velocity ratio of oscillatory flow and net flow (ψ).

$$N = 25\, Str^{-0.3}\ \psi e^{-0.13\psi} + 30.2$$

Selective adsorption behaviors of para-xylene, Co2+ and Mn2+ ions from pure terephthalic acid wastewater on zeolites

T. Qiu, C. S. Ye, Z. Y. Qi, M. M. Zhou
Fuzhou Univ., School of Chemical Engineering, Xue Yuan Rd. 2, 350116 Fuzhou, China;
tel. +591 22866241, e-mail: tingqiu@fzu.edu.cn;

As reported, the quantity of effluent produced in the refined unit of Pure Terephthalic Acid (PTA) has surpassed 180 million tons per year in China. It is estimated that the production of 1 ton PTA will approximately generate 3 tons PTA wastewater, which consist of high concentration components, such as Para-Toluylic acid and some metal pollutants. Para-Toluylic acid can be reacted to product PTA and the Co2+ and Mn2+ ions can be used as catalysts in PTA industry. In order to lower the cost and protect the environment, all kinds of organic matter and Co2+ and Mn2+ ions must be recycled from the wastewater. Authors reported a process to deal with PTA wastewater, which includes Para-xylene (PX) liquid-liquid extraction, ultra-filtration and reverse osmosis. And it can effectively recycle organics and metal ions. However, it was found that the life of reverse osmosis membrane is very short for the reason that its base material is polyamide, which could be decomposed by residual PX. So it is necessary to treat the trace of PX before ultra-filtration operation. The wastewater after extraction technology still contains PX (40-50ppm), Co2+ and Mn2+ ions (30ppm, 30ppm). Selective adsorption PX technology is considered to be one of the most promising approaches to achieve the aim. In this work, the great interest is in ion exchange with zeolite to selectively adsorb PX from PTA wastewater.

A series of different zeolites were synthesized and characterized by SEM analysis, X-ray microanalysis and N2 adsorption, and the ZSM-5 with H+ or Fe3+ exchange zeolite was confirmed to have the maximum equilibrium adsorption capacity and the biggest selectivity factor. The selectivity factors of PX to Co2+, PX to Mn2+ can attain 1878.992 and 1995.242, respectively, and the saturation adsorption capacity of PX even increase to 126.10 $mg \cdot g^{-1}$. Besides, effects of various important parameters such as zeolite dosage, initial PH and temperature were determined in batch experiments. The adsorption thermodynamics and kinetics of PX on H-ZSM-5 zeolites were investigated completely and systematically. Compared with the Freundlich isotherm model, the Langmuir isotherm model agreed better with the experimental data obtained from batch studies in a liquid phase at temperatures between 25 and 55℃, and the adsorption parameters of thermodynamics are $\Delta H = -12.33$ $kJ.mol^{-1}$, $\Delta S = -17.83\ J \cdot mol^{-1} \cdot K^{-1}$ and $\Delta G < 0$. It demonstrated that the adsorption process is a spontaneously exothermic physisorption process. On the other hand, the results indicated that the adsorption of PX on H-ZSM-5 zeolites followed the pseudo-second-order rate expression, and the calculated equilibrium adsorption capacity matched well with the actual equilibrium adsorption capacity. Therefore, H+ exchange ZSM-5 zeolite can be used as a highly efficient adsorbent to selectively remove PX from PTA wastewater.

Distributed mini-plant scale production of fuels and chemicals from renewable energy

A. Laari, T. Koiranen
Lappeenranta University of Technology, School of Engineering Science, PO Box 20, 53851 Lappeenranta, Finland; tel. +358 294 462 111, e-mail: arto.laari@lut.fi

Energy markets are currently experiencing a rapid transition from fossil raw materials to renewable bio-, solar and wind energy. It is estimated that even 60 % of the new built energy capacity is at present time based on renewable resources. However, some challenges in the wider use of renewable energy still remains. Renewable wind and solar energy is intermittent having large temporal and regional variations. It is therefore necessary to develop methods to store energy in large scale to be used in times when and in locations where renewable energy is not available.

There are many ways to store renewable wind and solar energy. Some of the proposed storage methods are mechanical or potential, such as compressed air, spinning masses or hydraulic energy storage. In these cases the stored energy can be easily converted back to electricity. However, these storage systems have generally poor efficiency, they require large volumes, and the storage is regional without possibilities to transport energy from one location to another. Chemical energy storage, on the other hand, is often considered as an efficient way to store energy. Chemicals can be easily transported to other locations were energy is needed, transformed back to electricity in power plants or fuels cells, or used for heating or transportation whenever needed. To avoid energy transfer costs, it is economically cost-effective to consume the produced electricity close to the site where it is procuded. This limits the optimal scale of the chemical production plants using renewable energy and leads to the idea of chemical production in distributed mini-plants. This is contrary to the prevailing economy of scale thinking and might cause changes in the ways chemical industry operates.

The objective of this research is to search for ways to store renewable wind and solar energy chemically. The research is carried out by searching existing knowledge from literature and by applying screening matrix methods common in chemical product design to compare different options. The purpose is to evaluate all the relevant aspects of different products, i.e. chemicals and their production technologies, to look for the best products for the purpose. The factors that should be considered are such as costs, maturity of technology, engineering complexity, energy efficiency, safety, and customer acceptability etc. These criteria are often contradictory and difficult to consider by other means. The screening matrix method allows taking quantitatively into account important criteria that influence the selection of the best products. Not much of this kind of quantitative comparison and analysis of renewable fuels can be found from literature. Some studies have been, however, reported recently. Tremel et al. (2015) have compared different aspects of the most important renewable fuels, such as methanol, Fischer-Tropsch (FT) diesel, dimethyl ether (DME), synthetic natural gas (SNG), and ammonia. They conclude in their publication that methanol production may be identified as the synthesis technology that achieves the highest score considering all the aspects that influence the product selection. The price of the produced methanol is reported to be 980 €/t including all the costs. Pérez-Fortes et al. (2015) have compared the technological, economic and environmental aspects of the production of methanol and formic acid from captured CO_2 and H_2 produced by electrolysis. According to their analysis the price of the produced methanol (1380 €/t) or formic acid (2540 €/t) is currently not competitive to the products produced from fossil sources.

Methanol may be hence considered a promising candidate for renewable energy storage. Olah (2005, 2013) has proposed a methanol economy as a reasonable alternative to hydrogen economy. Methanol is generally considered an excellent fuel and it can also be easily blended with gasoline. Methanol can be also considered as a versatile platform chemical and a starting point for further synthesis of other chemicals, for instance gasoline or monomers. Traditionally, methanol is synthesized from syngas ($CO/CO_2/H_2$) in gas phase at a temperature range of 523-573 K and a pressure range of 50-100 bar. The reaction is thermodynamically limited and the product yield in one-pass of gases is only around 15 to 25 %. Therefore, large gas circulation rates are necessary which increases the operational and capital costs of the process. To overcome the thermodynamic limitations of the gas-phase process, a liquid phase methanol synthesis has been also considered. One potential methanol production technology is to use alcohols as a reaction promoter (Fan et al., 1999; Tsubaki et al., 2001). This methanol process proceeds in liquid phase catalyzed by heterogeneous catalysts, such as Cu/ZnO and $Cu/Zn/Al_2O_3$. It consumes directly carbon dioxide and hydrogen and is expected to proceed through the formation of formates (or formic acid) at catalyst surface. Formic acid then reacts with alcohol to form ester which is further reacted with hydrogen to produce methanol. It is claimed that this process works well at low temperature and pressure (443 K and 30 bar) with high selectivity and yield of methanol with low gas circulation rates. As an exothermic reaction, the equilibrium of methanol synthesis shifts towards the products at low temperatures which increases the reaction yield. From the point of view of distributed mini-scale processing, this liquid-phase methanol process, which converts directly CO_2 and hydrogen to methanol using alcohols as a promoter, is a very attractive process. The process is relatively simple consisting of only a few processing steps, CO_2 capture, H_2 production from electrolysis, methanol synthesis, and methanol separation from water. Another remarkable benefit is that large gas circulation rates can be avoided.

As a result of the study, it is conluded by comparing different options that methanol is probably the most potential chemical than can be produced from carbon dioxide and hydrogen. The current price of renewable methanol is not competitive with methanol produced from fosil resources. The high costs are mostly influenced by the cost of renewable electricity and the high capital investment of electrolyzer. However, it is expected that these costs come down rapidly when the renewable energy production expanses. It is also concluded that the liquid-phase methanol process which uses heterogeneous catalysts and alcohols as a reaction promoter is the most potential technology for distributed mini-scale production of methanol. To continue further in process development, the synthesis technology of the liquid-phase methanol process in mini-plant process environment is to be studied in details. Novel catalysts, including supported gold and copper catalysts, are to be developed to improve methanol production rate and product yield.

Tremel, A., Wasserscheid, P., Baldauf, M., Hammer, T., Int. J., Hydrogen Energy, 40 (2015), 11457-11464.
Pérez-Fortes, M., Schönenberger, J. C., Boulamanti, A., Harrison, G., Tzimas, E., 4th Conference on Carbon Dioxide as Feedstock for Fuels, Chemistry and Polymers, 29 - 30 September 2015, Essen, Germany.
Olah, G. A., Angew. Chem. Int. Ed., 44 (2005), 2636-2639.
Fan, L., Sakaiya, Y., Fujimoto, K., Applied Catalysis A: General, 180 (1999), L11-L13.
Tsubaki, N., Ito, M., Fujimoto, K., 197 (2001), 224-227.

Process intensification by equilibrium shift during ibuprofen-glycerol enzymatic esterification: kinetic modelling

M. Ladero, M. Ravelo, F. Garcia-Ochoa
Department of Chemical Engineering, College of Chemical Sciences, Complutense University of Madrid, 28040, Madrid (Spain). Phone number: 0034-913945139
e-mail address: fgochoa@ucm.es and mladero@ucm.es

Introduction

Bio-diesel production has resulted in higher glycerol amounts, being this compound a by-product obtained at a 10% mass ratio refered to the biofuel. Therefore, glycerol prices has dropped and the chemical is now looked upon as a platform chemical (C3) in the biorefinery context [1].Glycerol esterification with carboxylic acids leads to several products of interest in the chemical, food, cosmetic and pharmaceutic industries, including the production on profen prodrugs [2]. Well-known lipases as isozyme B of *Candida Antarctica* (CALB)present an opportunity to develop new processes and products based on glycerol, while Fisher esterification is well-known to be limited by equilibrium position, so hybrid operations shifting it toward the desired products are commonly employed, if possible [4]. In this work, a liquid industrial preparation of CALB (Lipozyme CALB L, Novozymes Ltd.) is employed for the esterification of ibuprofen and glycerol in the presence of toluene in reaction-only and reactive distillation devices and kinetic models are used to compare their performance.

Experimental

Lipozyme CALB L was kindly provided by Novozymes Ltd., while all other chemicals and solvents were of analytical grade and purchased to Sigma-Aldrich. Preliminar reaction tests were performed in open batch reactors to study ibuprofen solubility in glycerol and toluene (and other apolar solvents), ibuprofen partition between these two solvents, and compositional and temperature effects on the activity and stability of the enzyme. Further runs in closed batch reactors and batch reactors coupled to Dean-Stark apparatus at 50-80ºC and several concentrations of ibuprofen in glycerol-toluene were the basis of the selection of adequate kinetic models by statistical non-linear regression coupled to numerical integration of the differential equations. Several goodness-of-fit parameters were taken into account for this selection.

Results

Several solvents were considered for the solubilisation of ibuprofen, being toluene the most appropriated one, both in terms of solubility and of mass transfer balance between glycerol and the apolar organic solvent in contact. Water content higher than the one added with the enzymatic preparation resulted in lower reactivities and productivities, while mass transfer hindrances were avoided at stirring rates higher than 750 rpm. Enzyme concentration effects on initial reaction rate showed a saturation behaviour typical in interfacial kinetics, being an amount of 2 g/L enough to reach maximum initial reaction rate.

In the best conditions for maximum initial rate, batch operations were performed in closed tank reactors and in tank reactors coupled Dean-Stark apparatus at long reaction time and several temperatures and initial ibuprofen concentration values so to observed final equilibrium ibuprofen conversion. As shown in the Figure, the final value of X changed from 0.6 to 0.8 as temperature changed from 50 to 80 ºC if closed systems were employed. On the contrary, conversion was total is water was continuously withdrawn from the reacting liquid during operation with the aid of toluene (azeotropic distillation). Data obtained at shorter time values were used for kinetic model selection. Application of goodness-of-fit criteria (Fischer's F, Akaike and Bayesian information criteria, sum of squared residuals and percentage of explained variation –VE-), allowed for the selection of the following kinetic models:

System A: simple reactive distillator (Dean-Stark apparatus). Equation 1.

$$r = \frac{\left[(15.09 \pm 0.39) - \frac{6812 \pm 133}{T}\right] \cdot C_I}{1 + (0.68 \pm 0.14) \cdot C_I}$$

System B: closed batch reactor. Equation 2.

$$r = \frac{\left[(11.42 \pm 0.44) - \frac{5274.14 \pm 148.63}{T}\right] \cdot C_I - \left[(2.30 \pm 0,75) - \frac{2468.39 \pm 255.28}{T}\right] \cdot C_{MG}}{1 + (0.86 \pm 0,29) \cdot C_I + (16.92 \pm 1.26) \cdot C_{MG}}$$

Where the concentrations of water and glycerol are much higher than those of ibuprofen and its monoglycerides and can be taken as constants during the reaction progress.

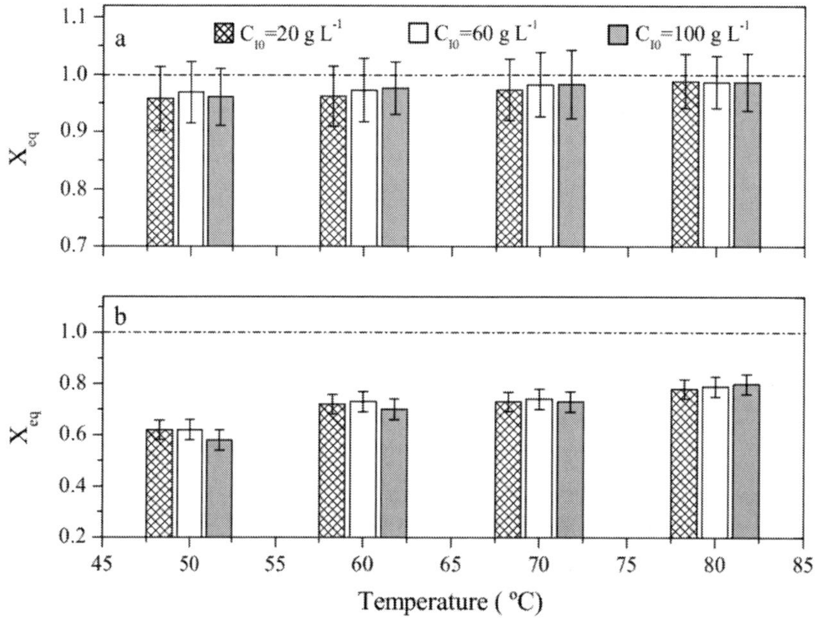

References

[1] Wolfson, A., C. Dlugy, and Y. Shotland, Glycerol as a green solvent for high product yields and selectivities. Environmental Chemistry Letters, 2007. 5(2): p. 67-71

[2] Jean-Ching, C. and T. Shau-Wei, Enantioselective Synthesis of (S)-Ibuprofen Ester Prodrug in Cyclohexane by Candida rugosa Lipase Immobilized on Accurel MP1000. . Biotechnology Progress, 2000. 16: p. 986-992

[3] Liu, Y., F. Wang, and T. Tan, Effects of alcohol and solvent on the performance of lipase from Candida sp. in enantioselective esterification of racemic ibuprofen. Journal of Molecular Catalysis B: Enzymatic, 2009. 56(2–3): p. 126-130

[4] Elgue, S., Conté, A., Marty, A. and Condoret, J.-S. Continuous lipase esterification using process intensification technologies. Journal of Chemical Technology and Biotechnology, 2014. 89: p. 1590–1598

Harnessing the value from scientific and technical collaborations – a pharmaceutical industry perspective

H.-J. Federsel
Chemical Development, AstraZeneca, Macclesfield, SK10 2NA Cheshire, UK;
tel.+44 7884 735 300, email: hans-jurgen.federsel@astrazeneca.com

The past two to three decades have seen a profound change in the way pharmaceutical companies are operating. Admittedly, discovering and developing medicines is a complex, tedious, and long-lasting endeavour circumvented by many hurdles and risks. At the end of this "journey" covering in most cases 10 years or more only very few of the projects initiated will result in a novel pharmaceutical product – the global average figure is somewhere around 4-5% of successful outcomes.[1]Notwithstanding, the challenges in the medical field are enormous and we are facing many therapeutic areas where there are diseases with a high unmet need for better treatment paradigms. Studies aimed at finding statistically validated factors which play a significant role in ensuring an increased likelihood of success in drug projects have unambiguously identified the quality of the science applied and the level of innovation as main contributors.[2]

Historically, pharmaceutical companies (especially those that often are referred to as Big Pharma) have been rather self-sufficient in the sense that in house capabilities existed to cover the whole range of activities and the associated skills required along the R&D supply chain - from early discovery through to full development and registration. This business model also took account of the need for more basic research with the aim to develop platform technologies and methodologies that could be applied across wide parts of the drug project portfolio. Thus, the prevailing philosophy was that due to the high level of confidentiality which was the traditional modus operandi across the pharma business, most if not all of this work had to be performed internally with the option that the key results and potential IP (intellectual properties) were owned and controlled by the company itself.

Moving into the 21st century, these principles have largely been abandoned – if nothing else than for resource restrictions and lack of funding as a consequence of putting more emphasis on the core activities such as clinical trials underpinning the delivery of a new medicine to patients.Hence, in this lean environment many of the so called enabling sciences, e.g. compound synthesis, process R&D, manufacture, and the development of scientific tools and technologies are either completely outsourced or conducted in a collaborative manner with external partners. Whilst the former model is often arranged on the basis of a fee-for-service agreement, the latter relies on the insight that a technology or capability doesn't have to be owned as long as the freedom to operate is guaranteed. With this new paradigm in mind it is pleasing to note the change in mind-set that has taken place over the recent past, opening up for collaborations of various shape and form on a wide range of topics; albeit, always with the aim to create a win-win situation for everyone involved.[3] A brilliant example of this is offered by aplethora of EU funded research projects over recent years and the experiences from this and the conclusions to be drawn will be discussed in more detail.

1. (a) Paul, S. M.; Mytelka, D. S. et al*Nature Rev. Drug Discov.* **2010**, *9*(3), 203-214. (b) Calcoen, D.; Elias, L.; Yu, X. *Nature Rev. Drug Discov.* **2015**, *14*(3), 161-162.

2. Ringel, M.; Tollman, P.; Hersch, G.; Schulze, U. *Nature Rev. Drug Discov.* **2013**, *12*(12), 901-902.
3. Federsel, H.-J. *Org. Process Res*. Dev. **2012**, *16*(2), 260-261.

Green electricity-based processing: the inevitable future

A. Stankiewicz
Chair of Intensified Reaction and Separation Systems, TU Delft Process Technology Institute, Delft University of Technology, Leeghwaterstraat 39 2628 CB Delft, The Netherlands; e-mail: a.i.stankiewicz@tudelft.nl, tel. +31 15 2786678, fax: + 31 15 2786975

The long-term vision on the future intensified chemical processes encompasses the utilization of renewable ("green") electricity as primary energy source for those processes. With electricity becoming the most widely available, versatile energy form on Earth, the electricity-based methods will play an important role in the development of flexible, distributed production units for clean manufacturing of fuels and chemicals in various environments.

In this context the current lecture addresses the thermodynamic domain of Process Intensification and reviews recent advances in the utilization of various electricity-based forms of energy to intensify chemical reactions and separations at the molecular level. The focus is on using electric, electromagnetic and acoustic fields to influence the orientation, alignment and activation of molecules. The challenges of implementation of the resulting novel reactor concepts are discussed.

The lecture is illustrated with numerous examples including the new "ADREM" project (Adaptable Reactors for Resource- and Energy-Efficient Methane Valorization) realized within the EU SPIRE program. ADREM will provide highly innovative, economically attractive and environmentally friendly processes and equipment for efficient transformation of methane into useful chemicals and liquid fuels to be run in the long term on green electricity. Among other things, novel reactor systems developed in the project will provide means for the conversion of methane in cases when it presently has to be flared or released because of lacking connections to gas pipelines. This presents a strong positive environmental impact (anti global warming), which will still be increased when using renewable electricity as a source for operation of the new reactors.

To realize the above-described vision, more symbiotic research between catalysis, physics and chemical engineering is postulated. Better understanding of the underlying phenomena and interrelations between the catalysis and various energy forms is a prerequisite here. The concurrent development of tailored, energy-responsive catalysts and novel reactor concepts with enhanced field control will lead to a new quality in chemical processing.

Chemical engineering potential for methane and carbon dioxide activation and conversion within sustainable process industry through resource and energy efficiency (SPIRE)

[1]Blaž Likozar, [1]Andrej Pohar, [1]Miha Grilc, [1]Venkata D.B.C. Dasireddy, [1]Matej Huš, [1]Mitja Ogrizek, [1]Andraž Pavlišič, [1]Jing-Lin Liu, [1]Neja Strah, [1]Damjan Lašič Jurković, [1]Matic Grom, [1]Janez Levec

[1]Laboratory of Catalysis and Chemical Reaction Engineering, National Institute of Chemistry, Hajdrihova 19, SI-1001 Ljubljana, Slovenia; tel. +386 1 4760281, e-mail: blaz.likozar@ki.si

Introduction

The Laboratory of Catalysis and Chemical Reaction Engineering (LCCRE) within the National Institute of Chemistry, Slovenia, is alongside several European academic and industrial partners involved in the development of innovative green chemical production technologies in an industrially-relevant environment by increasing the overall resource efficiency and energy intensity, and by decreasing the greenhouse gas emissions into the environment. The mentioned research and innovation efforts are joined under the SPIRE initiative within the Horizon 2020 Framework Programme.

CO_2 Conversion to Methanol: MefCO2

The MefCO2 (Synthesis of Methanol from Captured Carbon Dioxide Using Surplus Electricity) project is aimed at an increase of renewable energy usage and a decrease of CO_2 emissions, which are important European strategic objectives. Carbon dioxide, a greenhouse gas, which would normally be emitted into the atmosphere, is captured and converted to methanol with hydrogen, which is produced by water electrolysis. The electricity, used for electrolysis, is obtained from the surplus electricity at the peaks of its generation from renewable resources (wind, solar, *etc.*), simultaneously stabilizing the electrical grid. The product, methanol, is a widely-useable platform chemical and a precursor for further synthesis. The project will conclude with the construction and test runs of an operational pilot plant in Lünen, Germany.

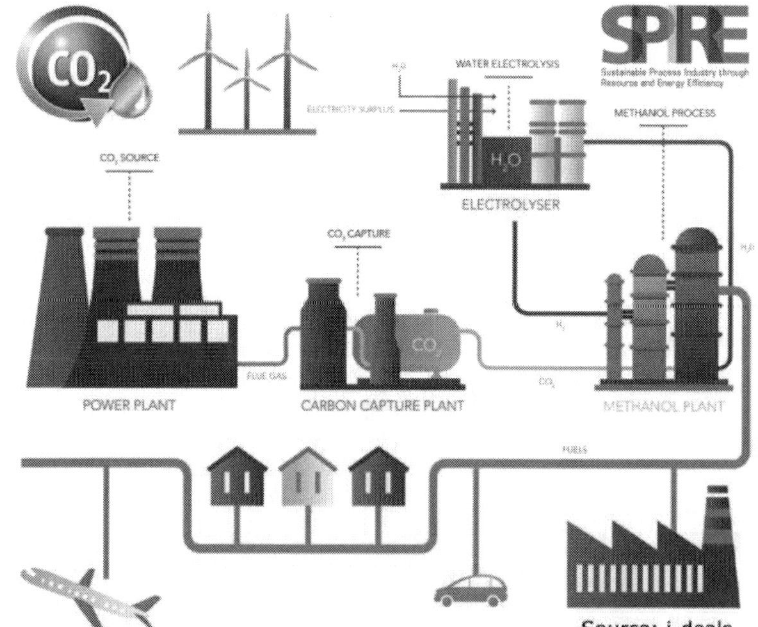

Source: i-deals

Results

The development of an active, selective and stable catalyst for the methanol production from CO_2 is essential for achieving a favorable economic balance. Thus, the methanol synthesis from CO_2 over current industrial catalysts suffers from a lack of selectivity, since large amounts of CO are formed as a by-product. The understanding of the Cu–ZnO synergy in the creation of the active sites

for methanol and CO formation therefore holds the key for the development of the catalysts for the low-CO methanol synthesis. Several copper-based catalysts with alkali earth metals or transition metals, and alumina support were synthesized and tested.
For catalyst performance and operational parameter screening, 5 parallel high-pressure, high-throughput cartridge-type packed-bed reactor systems with a swift gas chromatography (GC) composition analysis, were designed and built. The system was pressure-tested up to 200 bar with single-pass and recycle capabilities.

Methane and CO_2 Activation: ADREM
The Horizon 2020 ADREM project (Adaptable Reactors for Resource- and Energy-Efficient Methane Valorization) is aimed at the development and validation of a scalable and adaptable reactor technology that will boost the selectivity of the methane conversion reactions for an efficient (co-)transformation into useful chemicals or liquid fuels (*e.g.* C_{2+} hydrocarbons). The designed reactors will be able to operate with changing feedstock composition and will deliver the desired products on-demand.

Results
LCCRE is involved in the transformation of methane to liquid chemicals, assisted by plasma and augmented by the product removal *via* condensation within a high-temperature gradient (HTG) reactor. The general aim is to design and develop an energy-efficient plasma and catalyst system for the methane conversion to value-added chemicals. The reactor design is considered, based on the following two concepts:
- Single-step: plasma and catalytic reactor for (co-)converting CH_4 directly to chemicals.
- Two-step: plasma and catalytic reactor for (co-)converting CH_4 to syngas, combined with the HTG reactor for product removal.

The effects of the experimental conditions (plasma power, reactor construction, gas composition, gas flow, *etc.*) on the performance of the plasma reactors will be assessed. Plasma sources will be adopted for the CH_4-to-liquid chemicals reactor, and for the CH_4-to-syngas reactor, respectively. Further process intensification will be achieved with the proposed HTG technology.

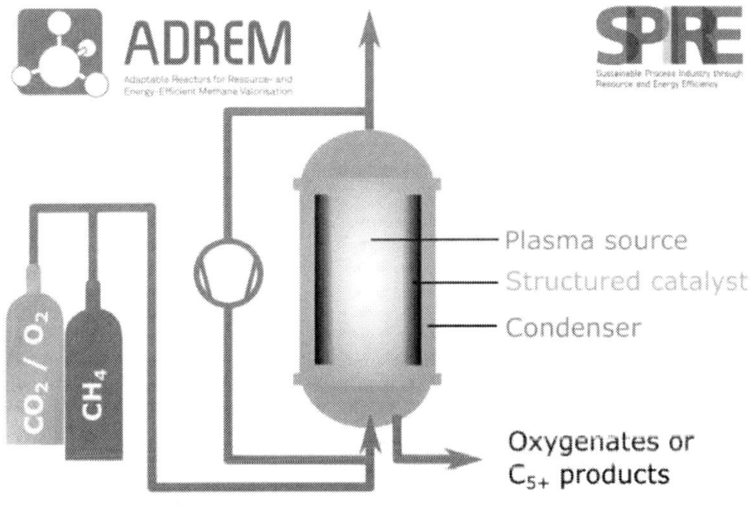

Source: National Institute of Chemistry

Acknowledgement
The presented work was partially funded by the EU Framework Programme for Research and Innovation Horizon 2020 under the grant agreements No. 637016 (MefCO2) and No. 636820 (ADREM). The authors also gratefully acknowledge the financial support of the Slovenian Research Agency (ARRS) through the Programme P2–0152, Chemical Reaction Engineering.

Process design and life-cycle assessment of energy-efficient routes for the production of gasoline from biogas and pyrolysis oil

S. Sundaram[1], G. Kolb[2], V. Hessel[1], Q. Wang[1]
[1] Department of Chemical Engineering & Chemistry, Eindhoven University of Technology, The Netherlands
[2] Fraunhofer ICT-IMM, Decentralized and Mobile Energy Technology Department, Mainz, Germany

Introduction

Most of the world's gasoline today is produced via crude oil distillation and subsequent refining.[1] Some alternate routes also exist, such as natural gas-to-liquid fuels (GTL) and coal-to-liquid fuels (CTL) techniques. These methods have been investigated in detail, and have even been implemented in the industry.[2,3] Even so, coal and natural gas are non-renewable resources, and have a negative impact on the environment in their production and use phases.[4] Thus, of late, there has been an increased focus within the European Union on the development of cleaner ways to produce liquid fuels, by using biomass as the starting material.[5]

With this background, this work looks at the ex-ante design of two new processes (termed ATR and CR), currently under experimental investigation in the EU Large-Scale NMP project BIOGO, for the production of gasoline from a mixture of biogas and pyrolysis oil. The processes are designed for an industrial capacity of 30 ktons/year, followed by a life-cycle assessment (LCA), both benchmarked against a previously commercialised process producing gasoline from natural gas.

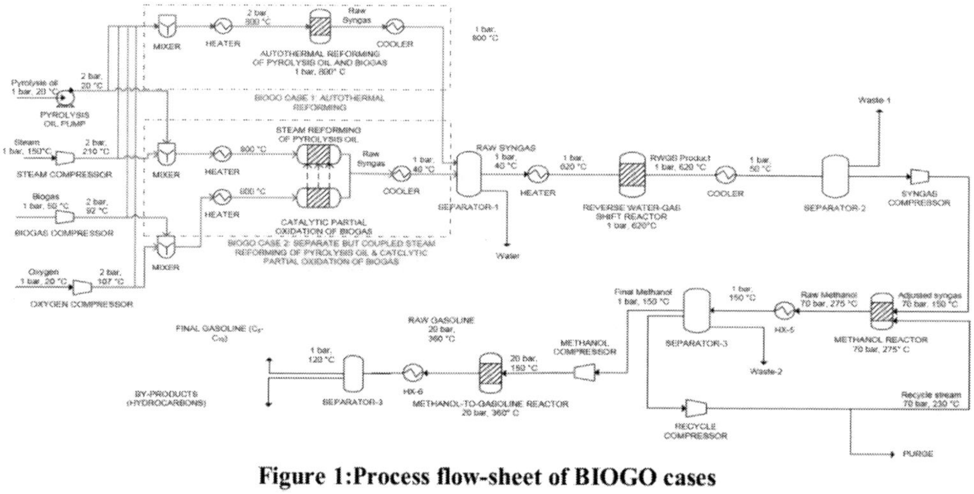

Figure 1:Process flow-sheet of BIOGO cases

Methodology

In the conventional process (Base case), natural gas, steam and oxygen in a specified ratio are fed into an autothermal reactor. The syngas produced is then adjusted, sent for methanol synthesis and then further to the gasoline production step. On the other hand, the two BIOGO routes use pyrolysis oil and biogas as the starting material. In the ATR case (Autothermal reforming of pyrolysis oil and biogas), the pyrolysis oil and biogas are subjected to autothermal reforming in a single reactor, with the addition of steam and oxygen. In the CR case (Steam reforming of pyrolysis oil and catalytic partial oxidation of biogas –

separate but coupled reactions), the steam reforming of pyrolysis oil (an endothermic reaction) and catalytic partial oxidation of biogas (an exothermic reaction) are coupled, such that they balance each other thermally. In both cases, the syngas generation step will need no external energy input. The processes downstream of this step remain the same for both routes, as in the Base case. The three cases are then subjected to a pinch analysis using the Aspen Energy Analyzer, and the most optimum utility consumption is found. Lastly, the mass and energy flows from the Aspen flowsheets are fed into the LCA for all cases, and their environmental performance is measured for certain impact categories.

Results

The Aspen simulations showed that the Base case has a carbon efficiency of 0.48, while the ATR and CR scenarios yielded 0.40 and 0.53, respectively. The LCA showed that for the Global Warming Potential (Figure 2), the ATR and CR scenarios are 90% and 47.5% lower than the Base case respectively, while for the Primary Energy Demand, they are 8.4% and 43% lower than the Base case, respectively. Moreover, the amount of oxygen consumed in the ATR and CR scenarios is lower than that in the Base case, which has lower associated impacts.

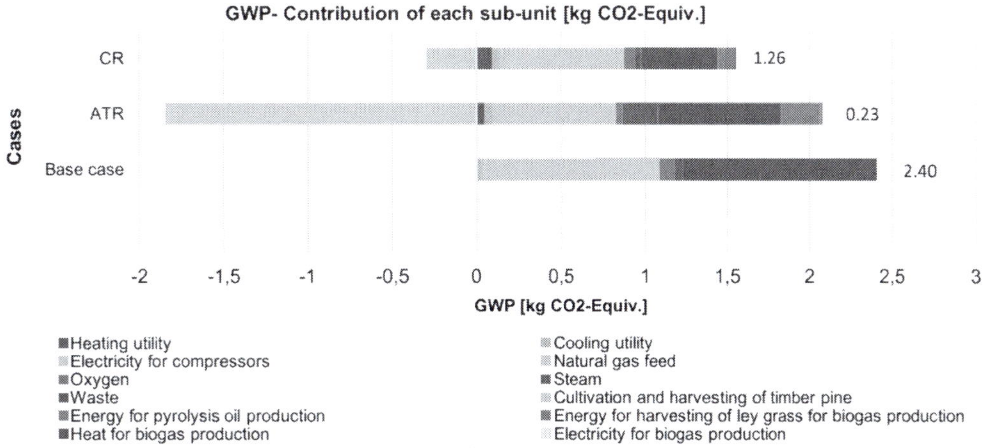

Figure 2. Comparison of Global Warming Potential for the three cases

However, in the categories of Terrestrial Ecotoxicity Potential (TETP), Photochemical Ozone Creation Potential (POCP), and Freshwater Aquatic Ecotoxicity Potential (FAETP), the Base case does better than both ATR and CR scenarios. The major contributor to almost all impact categories is the electricity; this is due to the fact that the electricity is sourced from the power grid, which is typically a mix of different kinds of power plants (dominated by nuclear and fossil-fuel based plants). The impact of the electricity is also reflected in other sub-processes like oxygen, and cultivation and harvesting timber pine, since these units also consume electricity from the grid. Thus, a conclusion has to be made based on a multi-criteria decision-making using a weighting of the impact categories.

References

(1) Schmer, R. (2000) The Motor Gasoline Industry: Past, Present & Future. DIANE Publishing.

(2) Glebova, O. (2013) Gas to Liquids : Historical Development and Future Prospects.

(3) Jacobs, T. (2013) Gas-to-Liquids Comes of Age in a World Full of Gas (SPE-0813-0068). *J. Pet. Technol.* 68–73.

(4) Burnham, A., Han, J., Clark, C. E., Wang, M., Dunn, J. B., and Palou-Rivera, I. (2011) Life-Cycle

Greenhouse Gas Emissions of Shale Gas, Natural Gas, Coal, and Petroleum. *Environ. Sci. Technol.* 619–627.
(5) European Biofuels Technology Platform. (2006) The EBTP Strategic Research Agenda. *EBTP*.

Engineering of photocatalytic reactions in microreactors: reaction development and scale up

Timothy Noël, Eindhoven University of Technology, Micro Flow Chemistry & Process Technology, Building 14, Helix, room STW 1.48, De Rondom 70, 5612 AP Eindhoven, The Netherlands. Phone: +31-40-2473623; E-mail: t.noel@tue.nl; Twitter: @NoelGroupTUE

Photoredox catalysis is emerging as a new and powerful tool in synthetic organic chemistry to facilitate such photochemical reactions by means of visible light.[1] Notable examples involve the use of ruthenium(II)polypyridine complexes which upon irradiation produce a photoexcited state. This photoexcited state gives rise to a single electron transfer process (SET) with organic substrates which can undergo subsequently a synthetic transformation. Whereas a myriad of different reactions have recently succumbed to this mode of catalysis, it is important to note that several challenges still remain with respect to high catalyst loadings, extended reaction times, scalability, quantum yield and generality of the catalyst system.

In this oral communication, we will report on the acceleration of gas-liquid photocatalytic reactions in continuous photomicroreactors.[2] We will go into detail on both engineering and chemical/catalytic aspects of continuous-flow photochemistry. Relevant examples from our group, such as the photocatalytic aerobic disulfide formation will be detailed on during the presentation.[3,4,5] Furthermore, we will discuss the engineering aspects concerned with photocatalytic gas liquid reaction in flow. This includes the potential (i) to extract efficiently kinetics,[6] (ii) to increase the energy efficiency of the photomicroreactor[7] and (iii) to scale-up the chemistry with a numbering-up strategy.[8] Insight information will be provided about the discovery process and the different parameters that led to success.

[1] C. K. Prier, D. A. Rankic, D. W. C. MacMillan, Chem. Rev. 2013, 113, 5322-5363.

[2] (a) Y. Su, N. Straathof, V. Hessel, T. Noël, Chem. Eur. J. 2014, 20, 10562-10589. (b) N. J. W. Straathof, Y. Su, V. Hessel, T. Noël, Nature Protocols 2016, 11, 10-21.

[3] For a photocatalytic Stadler-Ziegler reaction: X. Wang, G. Cuny, T. Noël, Angew. Chem. Int. Ed. 2013, 52, 7860-7864.

[4] For photocatalytic trifluoromethylation reactions: (a) N. Straathof, H. Gemoets, X. Wang, J. C. Schouten, V. Hessel, T. Noël, ChemSusChem 2014, 7, 1612-1617. (b) N. Straathof, D. van Osch, A. Schouten, X. Wang, J. C. Schouten, V. Hessel, T. Noël, J. Flow Chem. 2014, 4, 12- 17. (c) N. J. W. Straathof, B. J. P. Tegelbeckers, V. Hessel, X. Wang, T. Noël, Chem. Sci. 2014, 5, 4768-4773.

[5] For photocatalytic disulfide formation: A. Talla, B. Driessen, N. J. W. Straathof, L. G. Milroy, L. Brunsveld, V. Hessel, T. Noël, Adv. Synth. Catal. 2015, 357, 2180-2186.

[6] Y. Su, V. Hessel, T. Noël, AIChE J. 2015, 61, 2215-2227.

[7] Y. Su, A. Talla, V. Hessel, T. Noël, Chem. Eng. Technol, 2015, 38, 1733-1742.

[8] Y. Su, K. Kuijpers, V. Hessel, T. Noël, Reaction Engineering & Chemistry, 2016, DOI: 10.1039/c5re00021a.

Biocatalytic process design and reaction engineering

R. Wohlgemuth
Sigma-Aldrich, Member of MERCK Group, Industriestrasse 25, CH-9470 Buchs, Switzerland, e-mail: roland.wohlgemuth@sial.com

Biocatalytic processes occurring in nature provide a wealth of inspiration for manufacturing processes with high molecular economy. The molecular and engineering aspects of bioprocesses converting available raw materials into valuable products are therefore of much industrial interest. Modular reaction platforms and straightforward working paths from the fundamental understanding of biocatalytic systems in nature to the design and reaction engineering of novel biocatalytic processes have been important for shortening development times. Building on broadly applicable reaction platforms and tools for designing biocatalytic processes and their reaction engineering are key success factors. Process integration and intensification aspects are illustrated with biocatalytic processeses to numerous small-molecular weight compounds, which have been prepared by novel and highly selective routes, for applications in the life sciences and biomedical sciences.

References:

[1] M.A.K.Vogel, H.Burger, N.Schläger, R.Meier, B.Schönenberger, T.Bisschops, R. Wohlgemuth, React. Chem. Eng. DOI: 10.1039/C5RE00009B (2016).
[2] E.E.Ferrandi, C.Sayer, M.N.Isupov, C.Annovazzi, C.Marchesi, G.Iacobone, X. Peng, E.Bonch-Osmolovskaya, R.Wohlgemuth, J.A.Littlechild, D.Monti, FEBS Journal 282, 2879-2894 (2015).
[3] E.E.Ferrandi, C.Marchesi, C.Annovazzi, S.Riva, D.Monti, R.Wohlgemuth, ChemCatChem 7, 3171-3178 (2015).
[4] P.Tufvesson, M.Nordblad, U.Krühne, M.Schürmann, A.Vogel, R.Wohlgemuth, J.M.Woodley, Org. Process Res. Dev. 19, 652-660 (2015).
[5] R.Wohlgemuth, I.Plazl, P.Žnidaršič-Plazl, K.V.Gernaey, J.M.Woodley, Trends in Biotechnology, 33, 302-314 (2015).
[6] V.Hélaine, R.Mahdi, G.V.Sudhir Babu, V.de Berardinis, R.Wohlgemuth, M. Lemaire, C.Guérard-Hélaine, Adv. Synth. Catal. 357, 1703-1708 (2015).
[7] G.S.Molla, R.Wohlgemuth, A.Liese, J. Mol. Catal. B: Enzymatic, 124, 77-82 (2016).
[8] D.Gauss, B.Schönenberger, G.S. Molla, B.M.Kinfu, J.Chow, A.Liese, W.Streit, R.Wohlgemuth, in: Applied Biocatalysis - From Fundamental Science to Industrial Applications (Eds.: A.Liese, L.Hilterhaus, U.Kettling, G. Antranikian), Wiley-VCH, Weinheim, Germany, in press (2016).

Biocatalytic process development using miniaturized reactors with ω-transaminase immobilized in LentiKats®

M. Bajić[1], I. Plazl[1], R. Stloukal[2], P. Žnidaršič-Plazl[1]
[1]Faculty of Chemistry and Chemical Technology, University of Ljubljana, Večna pot 113, 1000 Ljubljana, Slovenia; tel. +38614798572, e-mail: polona.znidarsic@fkkt.uni-lj.si;
[2]LentiKat's a.s., Pod Vinicí 83, 471 27 Stráž pod Ralskem, Czech Republic

Production of bioactive, enantiomerically pure compounds either by direct asymmetric synthesis from prochiral substrates or by kinetic resolution of racemic compounds using biocatalysis as highly selective, diverse and environment-friendly approach, is gaining increased attention. ω-transaminases, which are pyridoxal-5′-phosphate-dependent enzymes that catalyze transfer of amino functionalities from amine donors to ketone groups, were shown to be very effective catalysts in the production of chiral amines, widely used as highly valuable products or key intermediates in the production of optically pure compounds [1-2].

In order to develop economically relevant biocatalytic process, continuous operation with immobilized biocatalyst is typically beneficial. Packed bed columns with enzymes or cells entrapped within porous particles are often a matter of choice, although problems with transfer limitations within particles and fluid channelling in real reactors might prevent efficient exploitation of biocatalysts. Polyvinyl alcohol lens-shaped particles prepared by LentiKats' methodology enable efficient immobilization of biocatalytic material in a porous matrix which possesses excellent physico-mechanical characteristics and at the same time provide very good biocatalyst activity and stability [3]. Further operational stability must be ensured by proper reactor design and selection of operational conditions.

In this study, LentiKats® with immobilized ω-transaminase were used within a robust and reusable miniaturized packed bed reactor (MPBR) consisted of a laminated system of polymeric materials creating microchannels. A systematic scale-out of the MPBR was achieved by increasing individual dimensions of the rectangular- and rhomboid-shaped (Figure 1) channels and reactors were evaluated regarding their efficiency and further compared with the performance of a small batch reactor using model transamination reaction. Developed MPBR was also used to establish temperature dependence of the reaction catalyzed with immobilized ω-transaminase yielding 55 °C as an optimal temperature. Furthermore, operational stability of the MPBR with ω-transaminase immobilized in LentiKats® was tested under the continuous flow during the period of 21 days revealing over 80% retained initial productivity. The benefit of using miniaturized reactors in process development study which could be performed in a very short time and at low chemicals consumption was again confirmed.

Figure 1. One of the tested miniaturized packed bed reactors with LentiKats.

Acknowledgement

Financial support from the EU FP7 Project BIOINTENSE – Mastering Bioprocess Integration and Intensification across Scales (Grant No. 312148) is acknowledged, while M. Bajić was supported through Ad futura scholarship program provided by the Slovene Human Resources Development and Scholarship Fund (Grant No. 11011-73/2013). ω-transaminase was kindly provided by c-LEcta (Leipzig, Germany). Authors acknowledge Dr. U. Novak and A. Belovič for their generous help in experimental work.

References:

1. Koszelewski, D., Tauber, K., Faber, K., Kroutil, W. ω-Transaminases for the synthesis of non-racemic α-chiral primary amines. *Trends Biotechnol.* **2010**, 28:324–332
2. Malik, M.S., Park, E.S., Shin, J.S. Features and technical applications of ω-transaminases. *Appl. Microbiol. Biotechnol.* **2012**, 94:1163–1171
3. Wittlich, P., Capan, E., Schlieker, M., Vorlop, K.D., Jahnz, U. Entrapment in Lentikats®, in: Nedovic, V., Willaert, R. (Eds.) Fundamentals of Cell Immobilisation Biotechnology, Springer Science+Business Media, Dordrecht, 2004, pp. 53–63.
4. Wohlgemuth, R., Plazl, I., Žnidaršič-Plazl, P., Gernaey, K.V., Woodley, J.M. Microscale technology and biocatalytic processes: opportunities and challenges for synthesis. *Trends Biotechnol.* **2015**, *33* (5), 302–314

Multi-scale simulation framework applied to morphology evolution and properties of polyurethane foams

H. Preisig[1], S. Pricl[2], A. Daiss[3], B. Eling[4], D. L. Marchisio[5], H. Rusche[6], P. Andreson[7], J. Llorca[8], J. Gross[9], J. Kosek[10], P. Albrecht[11]
[1]Norwegian University of Science and Technology, Trondheim, Norway
[2]University of Trieste, MOSE-DEA, Trieste, Italy
[3]BASF SE, Polymer Processing & Engineering, Ludwigshafen, Germany
[4]BASF Polyurethanes GmbH, PU Research - Novel PU Precursors, Lemfoerde, Germany
[5]Politecnico di Torino, Institute of Chemical Engineering, Torino, Italy
[6]WIKKI Limited, London, United Kingdom
[7]Technische Universiteit Eindhoven, Polymer Technology, Eindhoven, Netherlands
[8]IMDEA Materials Institute, Getafe, Spain
[9]Universität Stuttgart, Institut für Thermodynamik, Stuttgart, Germany
[10]University of Chemistry and Technology, Prague, Czech Republic
[11]DIN German Institute for Standardization, Berlin, Germany

The MoDeNa project [1] aims at developing, demonstrating and assessing the easy-to-use multi-scale software framework application under an open-source licensing scheme that delivers models with feasible computational loads for process and product design of complex materials. The concept of MoDeNa is an interconnected multi-scale software framework. Polyurethane (PU) foams and thermoplastic polyurethanes (TPU) were selected to demonstrate the capabilities of MoDeNa framework because the complexity of involved physico-chemical phenomena practically prevented attempts for computer-aided design and optimization of PU material structure and production processes in the past.

Polyurethanes are an excellent example of a large turnover product produced in a large variety of grades and of which the properties are the result of designing and controlling the material structure on all levels of scale, from the molecule to the final product. The scales considered in MoDeNa are identified either classically as nano-, micro-, meso- and macro-scales (with somewhat arbitrary definition of scales) or by the types of involved models as quantum-chemical, molecular dynamics, population balance, particle-based or continuum models. The physico-chemical processes considered involve reaction kinetics, thermodynamics, diffusion and permeation transport, colloid chemistry, complex evolving rheology of polymer and foam, heat transfer including conduction and radiation, mechanical properties and fluid dynamics. The morphology of PU foams and TPU is described either by statistical descriptors or a full spatially 3D morphology is considered. The project is complemented by experimental activities providing data for parameter estimation and model validation. Moreover, project MoDeNa initiated efforts in the standardization of multi-scale modeling.

The contribution will first introduce the MoDeNa framework enabling the multi-scale coupling between models developed for specific purposes and scales. The framework consists from an orchestrator, a database and an interface library. The information exchange is governed by protocols and may occur in two ways, namely:

- forward mapping, i.e., passing information from a fine to a coarse scale in the upward direction,
- backward mapping, i.e., passing information from a coarse to a fine scale in the downward direction.

Forward mapping is relatively straightforward, while backward mapping inevitably required iterations since changing the operating conditions at the fine level changes the feedback to the coarse level. Backward mapping can be realized either by two-way coupling or by automatic fitting the surrogate models. The first approach is usually computationally expensive either due to the data exchange or the complexity of the microscopic-scale simulation. Thus replacing the microscopic-scale simulation with a surrogate model presents usually the only viable alternative. This operation inherently constitutes a transfer of data across scales and MoDeNa is unique in that it focuses on this approach.

Two case studies of MoDeNa framework will be presented:

- **PU foaming model** based on coupled Computational Fluid Dynamics (CFD) and Population Balance Equations (PBE) approach employing by means of surrogate models information about reaction kinetics, heat and mass transfer, thermodynamic driving forces, polymer and foam rheology and material properties.
- **Heat-insulation model** of PU foam considering the spatially 3D structure of foam, heat conduction and scattering of heat radiation and deterioration of heat insulation properties due to gradual permeation of gases out/into the foam.

Coming to the final stage of MoDeNa project we realized, that it could have a much broader impact on the modeling of multi-scale problems than thought initially:

- Multi-scale modeling is usually also multi-disciplinary in nature and requires expertise from distant areas of science. The MoDeNa framework is a reusable tool allowing for the fast cooperation of multi-disciplinary teams addressing complex problems that haven't been efficiently addressed in the past.
- MoDeNa is a step towards the future plug-and-play integration of software tools for multi-scale simulations. Moreover, it is based on open-source code, thus eliminating the hurdles of proprietary solutions.
- MoDeNa has also become a role model for the collaboration in multi-scale modeling and the organization of the previously scattered efforts. The testing of MoDeNa on the complex case studies of industrial interest means that a framework allowing for fast innovation and optimization of materials and process has been developed.

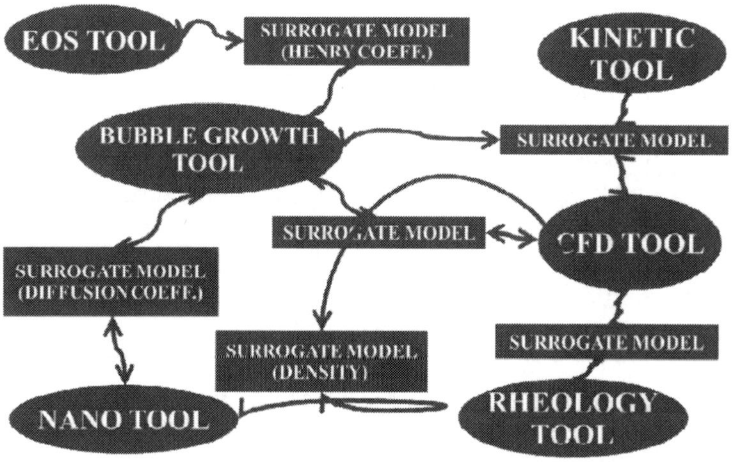

Figure: Example of tools and connections developed for the PU foaming case study.

Acknowledgment: EU FP7 project "MoDeNa" (nr. NMP4-SL-2013-604271, 2014-2016) is acknowledged.

References

[1] MoDeNa project web-site http://www.modenaproject.eu/

EU project RECOBA: Expected impact and an overview about progress in advanced process control of emulsion copolymerization

L. Šeda[1], E. Jahns[1], A. Mhamdi[2], A. Mitsos[2], P. Joy[2], S. Hamzehlou[3], J. R. Leiza[3], J. M. Asua[3], N. Rajabalinia[3], J. Kosek[4], T. Chaloupka[4], M. Kroupa[4], M. DeMiguel-Ramos[5], A. J. Flewitt[5], A. Lapkin[6], J. Suberu[6], F. Gjertsen[7], P. Singstad[7]
[1]BASF SE, Group of Polymer Reaction Engineering, Carl-Bosch Strasse 38, 67056 Ludwigshafen am Rhein, Germany, tel. +49 621 60-43250, e-mail: libor.seda@basf.com; fax: +49 621 60-41766; [2]Chair of Process Systems Engineering, AVT.SVT, RWTH Aachen University, Turmstraße 46, 52056 Aachen, Germany; [3]POLYMAT, University of Basque Country UPV/EHU, Joxe Mari Korta Center, Avda. Tolosa, 72, 20018 Donostia-San Sebastián, Spain; [4]Department of Chemical Engineering, University of Chemistry and Technology Prague, Technická 5, 166 28 Praha 6, Czech Republic; [5]Electrical Engineering Division, Department of Engineering, University of Cambridge, 9 JJ Thomson Avenue, Cambridge CB3 0FA, UK; [6]Department of Chemical Engineering and Biotechnology, New Museum Site, University of Cambridge, Cambridge CB2 3RA, UK; [7]Cybernetica, Leirfossveien 27, 7038 Trondheim, Norway.

The annual production of emulsion polymers in the EU is about 2 million tons. Usually, the production process follows a fixed schedule with predefined temperature and reactant feeding profiles in order to produce products with desired properties. Between properties of importance in case of emulsion polymers belong, e.g., molecular weight distribution, solid content, copolymer composition, size distribution of latex particles or their morphology. The disadvantage of the above process control is, however, its inability to react on various disturbances and, if necessary, to make corrective actions leading to desired product. Such a disturbance could be, for instance, an impurity in the reactants leading to lower efficiency of the utilized initiator. Then, the product properties may be negatively affected and in the worst case the whole batch can be lost. With the advanced process control such scenarios could be avoided. Due to large production amount of emulsion polymers the processing industry aims for the highly optimised processes as such a technological advance brings both significant environmental and economic benefits.

The presented work regards to the overview of industrial and scientific goals and common partner achievements in the frame of the EU funded project RECOBA (Cross-sectorial REal time sensing, advanced COntrol and optimization of BAtch processes, saving energy and raw materials). In general, the main goal of the RECOBA project is to replace the current process control based on repetition of fixed schedule by a model predictive control (MPC) and thus optimise the process in the real time and also to ensure opimised product quality without batch to batch quality variations. In one of the example processes within the RECOBA project the partners aim for the MPC of an emulsion copolymerization of two water soluble and two water insoluble monomers which are commonly used, for example, for painting applications. The optimisation of the emulsion polymerization process leads also to an increase of the resource and energy efficiencies of the process, which are besides the minimal product properties variations another very important ecological and economic outcomes of the project. The overview about RECOBA project advances on the sensors and process models fields for online control and their applicability for emulsion polymerization will be presented.

For more information about the RECOBA project, please, visit:
www.spire2030.eu/recoba

This project has received funding from the *European Union's Horizon 2020 research and innovation programme* under grant agreement No 636820.

EU project PrintCR3DIT: Potential of 3D printing technology for process intensification in chemical industries

P. Stavárek, F. Lali, Z. Vajglová, M. Ujčič, [1]C. Grande
Institute of Chemical Process Fundamentals of the ASCR, v. v. i.; Rozvojová 2/135, 165 02 Prague, Czech Republic,tel. +420 220 390 142, e-mail:stavarek@icpf.cas.cz,[1] SINTEF Materials and Chemistry, Forskningsveien 1, 0373, Oslo, Norway

Catalytic reactors account for production of 90% of chemicals we use in everyday life. To achieve the decarbonisation of European economy and comply with the 20-20-20 target, resource utilization and energy efficiency will play a major role in all industrial processes.

The concept of PRINTCR3DIT is to employ 3D printing to boost process intensification in the chemical industries by adapting reactors and structured catalysts to the requirements of the reaction. This manufacturing technique is particularly useful in reactions where diffusion, mixing and/or heat transfer are limitations against reaching higher performance. The utilization of the concept of 3D printing will also reduce the resource utilization of reactor and catalyst manufacture, energy consumed (< 15%) and transportation.

The methodology will be applied to three markets of fine chemicals, specialty chemicals and fertilizers, ranging from few tons to millions of tons of production per year. This demonstrates the enormous versatility of 3D printing for reactor and catalyst designs that cannot be improved with traditional building and design tools. For all these processes, the challenges to be solved are thermal management, innovative reactor design and flow distribution. These examples will provide realistic data in different markets to delineate business case scenarios with the options of new integrated plants or retrofitting for large-scale applications.

Application of cutting-edge 3D printing to catalytic reactors will foster higher productivity, a more competitive industrial sector and higher value jobs in Europe - keeping leadership in such a challenging arena. PRINTCR3DIT is a joint effort between world-leading industries (4), innovative SMEs (4), R&D institutes (4) and a university that aim to accelerate deployment of a set of products to the market.

New developments in heat exchanger network retrofit

Robin Smith
Centre for Process Integration,
University of Manchester, UK.

The goal for heat exchanger network retrofit can be to reduce energy consumption, increase capacity or modify to allow for process changes. Academic attention to the problem has focussed on the reduction of energy consumption. Industrially, this problem is of paramount importance. Yet, despite a huge academic effort, the identification of cost effective retrofit remains one of our greatest challenges. The fundamental problem is usually not excessive capital cost incurred by the installation of new units in the network, but the capital cost incurred by changes in duty to the remaining units in the network that have their connections maintained, together with the cost of piping and civil engineering. These capital costs often make all but the simplest retrofit projects uneconomic. Thus, the challenge is not so much finding design changes that improve the energy consumption, but to ensure that the overall retrofit is economically viable. Minimising changes to the units that have their connections maintained and minimizing piping and civil engineering costs is just as important as identifying structural changes to improve the network.

The first systematic retrofit methods proposed were based on pinch analysis. However, pinch analysis is fundamentally not suited to heat exchanger network retrofit, as it analyses networks from the perspective of a new design and attempts to turn the existing network into the ideal grassroot design. This does not recognise the constraints imposed by the current design and the difficulties of changing it. Mathematical programming methods have also been developed, but were not capable of addressing realistic industrial problems. The quest to improve on both these methods led to the development of hybrid methods that combine aspects of both pinch-based and mathematical programming methods. Yet, despite much research, none of these methods have been able to consistently deliver practical and economic retrofit designs.

The use of heat transfer enhancement can in principle provide economic retrofit by enhancing the performance of existing heat exchangers and avoiding the installation of new heat transfer area. However, it has limited capabilities if used on its own without modifying the network structure.

This presentation will review new insights into network retrofit that allow effective modifications to be identified through the application of simple rules. It will also demonstrate how network changes can become more effective when systematically combined with heat transfer enhancement. The automation of retrofit will also be discussed.

Regional and total site CO_2 integration considering purification and pressure drop

Wan Norlinda Roshana Mohd Nawi [1], Sharifah Rafidah Wan Alwi [1], Zainuddin Abdul Manan [1], Jiří Jaromír Klemeš [2], Petar Sabev Varbanov [2]
[1] Process Systems Engineering Centre (PROSPECT), Research Institute on Sustainable Environment, Universiti Teknologi Malaysia, 81310 UTM Johor Bahru, Malaysia
[2] Pázmány Péter Catholic University (PPKE), Faculty of Information Technology and Bionics, 1083 Budapest, Hungary, e-mail: varbanov.petar@itk.ppke.hu

The application of Pinch Analysis (PA) targeting method has been recently explored to the design of CO_2 emission reduction in Total Site planning of CO_2 capture, utilisation and storage. The algebraic method based on Problem Table Algorithm (PTA) for Total Site CO_2 Integration (TSCI) provides the designer with integrated CO_2 capture, utilisation and storage (CCUS) for the optimal CO_2 emission reduction. In TSCI, CO_2 is captured with certain quality from various plants and supply into system header pipeline. The CO_2 header could satisfy various CO_2 demands for various industry located in the header, and only surplus CO_2 is to be sent to storage. The extended methodology with consideration of purification and pressure drop in TSCI planning is proposed. The CO_2 supply from the header could satisfy the purity demand through a process of purification. Purification is a process to upgrade the purity level have not been considered in the previously used in TSCI method. In addition, pressure drop during CO_2 transportation in the pipeline system has been included to identify the implication of pressure drop in TSCI design and has resulted about 29.01 MPa of the total pressure drop (ΔP_d). Therefore, 68.2 t/h of flow rate (FT) of 81 % purity level header is supplied and purified to satisfy the 50 t/h of demand, at 99 % purity level. The improved methodology of TSCI network has been further developed and provides a more realistic scenario for CCUS implementation.

A P-graph model for multi-period optimization of isolated energy systems

K.B. Aviso[1,2], J.-Y. Lee[3], R.R. Tan[1,2]
[1]Chemical Engineering Department, [2]Center for Engineering and Sustainable Development Research, De La Salle University, 2401 Taft Avenue Manila, Philippines, 0922, [3]Department of Chemical Engineering and Biotechnology National Taipei University of Technology, 1, Sec 3, Zhongxiao E Rd, Taipei 10608, Taiwan, ROC
tel. +632 5244611, e-mail: kathleen.aviso@dlsu.edu.ph

Reliable isolated energy systems are necessary for supplying energy to remote areas where grid extension is not feasible. It may also be required to harness renewable energy to reduce the use of conventional fuel that involves potentially high transportation cost and carbon emissions. Polygeneration systems are suitable for distributed energy supply in remote areas with the advantages of compactness and operational flexibility. Also, the simultaneous production of multiple utilities and products provides the opportunity for process integration, hence increased fuel efficiency and reduced carbon emissions. The challenge is in identifying the optimal design of the processes such that the system remains in operation regardless of anticipated changes in raw material supply and product demand during multi-period operations. Graph theoretic models in the form of P-graph (process graph) have previously been developed for synthesizing single period polygeneration systems. This work develops a P-graph model to handle multi-period operations of isolated energy systems. The resulting network is thus more robust since it is able to operate amidst changes in the availability of raw material supply and/or variations in product demand. In addition, the P-graph model is also capable of generating sub-optimal solutions which provide insights on other intangible parameters which may be significant to decision-makers. A case study will be presented to demonstrate the proposed approach.

Modular without containers – the Lonza multi-purpose plant concept

B. C. Seyfang, T. Gruetzner
Research & Technolgy; Lonza AG; Rottenstr. 6; CH-3930 Visp
tel. +41(0)27 948 7034, e-mail: bernhard.seyfang@lonza. Com ; Lonza.com

Chemical and biotechnological production processes are exposed to a strong cost pressure, especially in countries with high labour cost such as Germany or Switzerland. Chemical industries try to avoid this pressure by reduction of research and development cost and reduction of investment. Initiatives such as the ‚50% idea' or modular plant systems such as ‚F^3 factory' or ‚Invite' are the direct results.

In order to maintain competitiveness for products with a short life cycle, it is essential that the technology transfer of processes into production plants is realised as fast as possible. Efficient process development, resulting in time periods of 6-12 months between a first experiment in the lab and production start, is an important element of the concept. Automation, 24/7 experiments and PAT are key to a successful tech transfer.

Lonza uses its modularized, but fix installed multi-purpose plants to manufacture products with annual amounts of 10-700 mt. A central aspect is a high flexibility of the plant that allows the implementation of highly different processes. Both factors must not hamper the quality or security of supply of the product. The plant staff must be highly qualified to work in these surroundings.

Being back-integrated to a cracker, as well as coordinated by a detailed and flexible asset planning, multi-purpose plants are the backbone of an efficient production of intermediates, additives and actives.

The interaction of these factors results in a competitive advantage that has proven its success over the years and that is still lacking for other modular concepts. In our presentation we are going to present our accelerated development process as well as elementary technical solutions to utilize the full flexibility of Lonza's multi-purpose plants without containers.

A risk-based criticality analysis in bioenergy parks using P-graph method

M. F. D. Benjamin, [1]C. D. Cayamanda, [1]B. A. Belmonte, [1]R. R. Tan, [1]L. F. Razon
Research Center for the Natural and Applied Sciences, University of Santo Tomas, España Blvd., 1006 Manila, Philippines, tel. +632 7314041, e-mail: mdbenjamin@mnl.ust.edu.ph;
[1]Chemical Engineering Department, De La Salle University, 2401 Taft Avenue, 0922 Manila, Philippines

The adoption of bioenergy parks is a prospective solution to increase the sustainability of stand-alone biomass processing plants (Martin and Eklund, 2011). Production and resource efficiency, lower carbon emissions, and economic sustainability are achieved by synergistic exchanges of material and energy resources between components plants. However, such increased plant interdependency and the resulting tightly integrated energy system is vulnerable to capacity disruptions (i.e., plant inoperability). Cascading failure due to such disruptive event is an inherent risk in bioenergy parks and may pose as a barrier in implementing such system (Benjamin et al., 2015b). The extent of risk originating from disrupted critical plants in the network is demonstrated to be higher. A previous study by Benjamin et al. (2015a) developed a novel risk-based criticality index, based on input-output (I-O) models, to quantify the effect of a component plant's disruption within a bioenergy park. *Criticality* is defined as the measure to which a disrupted component plant is the source of network vulnerability. This index is used to rank the plant's relative risk in the network based on its disruption consequence.

In addition to I-O approach, graph theory methods such as P-graph (process graph) can be used in analyzing potential risks in bioenergy parks (Klemeš and Varbanov, 2015). P-graph is a graph theory and combinatorial algorithm-based framework developed by Friedler et al. (1992a) for process network synthesis (PNS). This method is based on axioms and theorems that develops maximal structure and determines feasible structures of a given system (Friedler et al., 1992b; Friedler et al., 1993).

Recently, the P-graph method is used to optimize disrupted integrated systems at different scales. This approach is applied to polygeneration systems under process inoperability conditions (Tan et al., 2014), allocation of electricity in economic sectors during climate change-induced crisis conditions (Aviso et al., 2015), allocation of inoperability in urban infrastructures (Tan et al., 2015), and used to determine optimal "damage control" during crisis conditions in industrial complexes (Tan et al., 2015). However, to date, P-graph method has not been applied to develop risk-based metrics to determine the criticality of component plants within a bioenergy park.

In this work, a P-graph approach is proposed as an alternative methodology for criticality analysis of component plants in a bioenergy park. Such risk-based metric can also be used for developing risk management measures to protect critical infrastructures, thereby increasing the robustness of bioenergy parks against disruptions. A case study is then presented to demonstrate the effectiveness of this method. Figure 1 shows the general framework for the risk-based criticality analysis using P-graph.

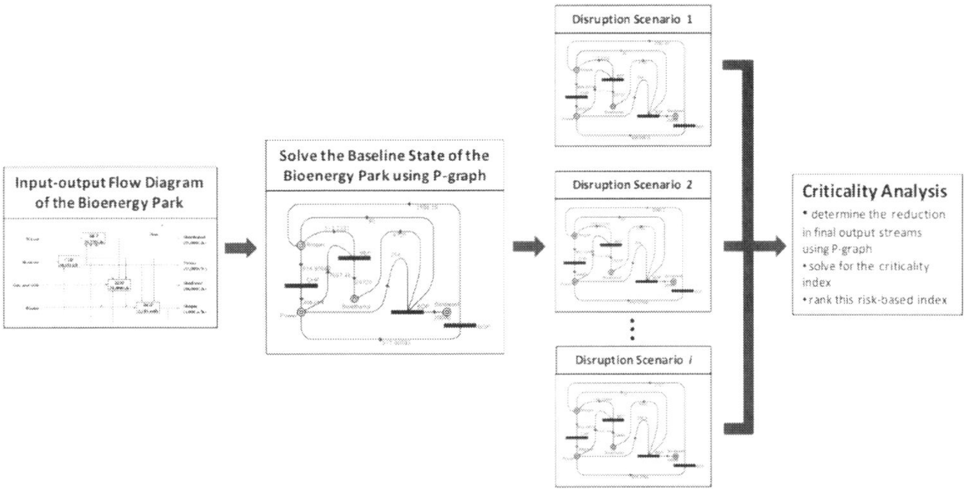

Figure 1: General framework for criticality analysis using P-graph.

References

Aviso, K. B., Cayamanda, C. D., Solis, F. D. B., Danga, A. M. R., Promentilla, M. A. B., Yu, K. D. S., Santos, J. R., Tan, R. R. (2015). P-Graph approach for GDP-optimal allocation of resources, commodities and capital in economic systems under climate change-induced crisis conditions. *Journal of Cleaner Production*, 92, 308-317.

Benjamin, M. F. D., Tan, R. R., Razon, L. F. (2015a). A methodology for criticality analysis in integrated energy systems. *Clean Technologies and Environmental Policy*, 17, 935-946.

Benjamin, M. F. D., Ubando, A.T., Razon, L. F., and Tan, R. R. (2015b). Analyzing the disruption resilience of microalgal multi-functional bioenergy systems using dynamic inoperability input-output modeling. *Chemical Engineering Transactions*, 45, 1579-1584.

Friedler, F., Tarjan, K., Huang, Y. W., Fan, L. T. (1992a). Graph-theoretic approach to process synthesis: axioms and theorems. *Chemical Engineering Science*, 47, 1973-1988.

Friedler, F., Tarjan, K., Huang, Y. W., Fan, L. T. (1992b). Combinatorial algorithms for process synthesis. *Computers & Chemical Engineering*, 16, 313-320.

Friedler, F., Tarjan, K., Huang, Y. W., Fan, L. T. (1993). Graph-theoretic approach to process synthesis: polynomial algorithm for maximal structure generation. *Computers & Chemical Engineering*, 17, 929-942.

Klemeš, J. J., Varbanov, P. S. (2015). Spreading the Message: P-graph enhancements: implementations and applications. *Chemical Engineering Transactions*, 45, 1333-1338.

Martin, M., Eklund, M. (2011) Improving the environmental performance of biofuels with industrial symbiosis. *Biomass and Bioenergy*, 35, 1747-1755.

Tan, R. R., Cayamanda, C. D., Aviso, K. B. (2014) P-graph approach to optimal operational adjustment in polygeneration plants under conditions of process inoperability. Applied Energy, 135, 402-406.

Tan, R. R.; Aviso, K. B.; Yu, K. D. S; Promentilla, M. A. B; Santos, J. R. (2015). P-Graph Approach to Allocation of Inoperability in Urban Infrastructure Systems. Chemical Engineering Transactions, 45, 1339-1344.

Tan, R. R., Benjamin, M. F. D., Cayamanda, C. D., Aviso, K. B., and Razon, L. F. (2015). P-graph approach to optimizing crisis operations in an industrial complex. *Industrial & Engineering Chemistry Research*. (in press, doi:10.1021/acs.iecr.5b03205).

Control of solid mass flow rate in circulating fluidized bed by a pulsed gas flow

M. Ishizuka, H. Mizuno, Y. Kansha, A. Tsutsumi
Institute of Industrial Science, The University of Tokyo, 4-6-1 Komaba, Meguro-ku, Tokyo 153-8505, Japan;
tel. +81 3 5452 6727, e-mail: a-tsu2mi@iis.u-tokyo.ac.jp; fax: +81 3 5452 6728.

As a coal-fired power generation technology for further improvement of power generation efficiency of coal-fired power generation, exergy regeneration type coal gasification power generation technology (Tsutsumi, 2004), a triple-bed circulating fluidized bed (Guan et al., 2010), has been proposed.

Authors analyzed flow characteristics of the riser and the downer in the triple-bed circulating fluidized bed as an equivalent circuit model. The equivalent circuit model of the riser and downer are shown in Figure 1 (Ishizuka et al., 2015). It can be seen from figure 1, this equivalent circuit has a nature of a low-pass filter. A combination of a low-pass filter and a pulse voltage is generally used as a switching power supply because of output power control. Solid mass flow rate in the circulating fluidized bed represented by the switching power supply in the model. Thus, we tried to control the solid mass flow rate of the triple-bed circulating fluidized bed by using the pulsed gas supply to the riser with the low-pass filter characteristics of this model in this research.

The input-output characteristics of the equivalent circuit of the riser inputting a pulse voltage were investigated. We used the SPICE electric circuit simulation software to analyze the circuit behavior. Circuit constant is to use the value of the reference (Ishizuka et al., 2015), the input pulse height is set to 80V. When the input pulse width is changed, the output current is changed depending on the pulse width. Moreover, the output current is changing depending on the pulse density. This result by giving a pulsed gas supply to the riser, it shows the possibility controlling the particle circulation rate of the triple-bed circulating fluidized bed.

References
A. Tsutsumi. Future generation IGCC using exergy recuperation technology. Clean Coal Technol. J., 11:17-22, 2004.
G. Guan, C. Fushimi and A. Tsutsumi. Downward gas-solids flow characteristics in a high-density downer reactor. Chem. Eng. J., 164:221-229, 2010.
M. Ishizuka, H. Mizuno, Y.Kotani, Y. Kansha, and A. Tsutsumi. Modelling of flow behavior at downer and riser in triple bed circulating fluidized bed using equivalent circuit. Chem. Eng. Trans., 45:889-894, 2015.

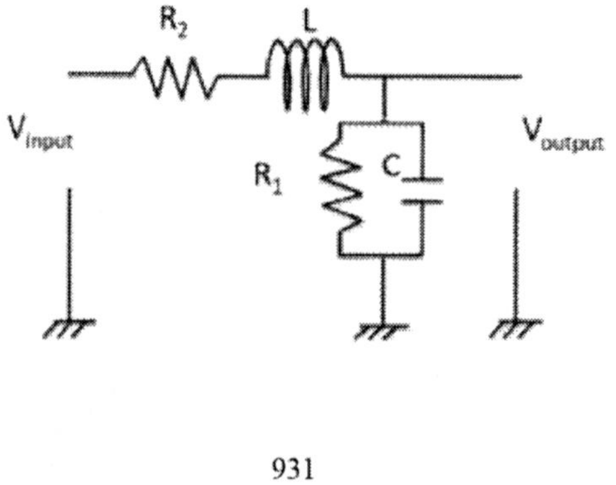

Figure 1. Equivalent circuit model of the downer/riser flow.

Flexible and efficient power production in a grate boiler though model predictive control and a novel mechanistic model of biomass combustion

A. Boriouchkine, S.-L. Jämsä-Jounela
Aalto Univ., Dep. Biotech. and Chem. Tech., P.O. Box 00076, Aalto, Finland;
e-mail: alexandre.boriouchkine@aalto.fi;

Increasing share of renewable energy resources (RES) of intermittent nature, for instance, wind and solar energy impose challenges on the power production from more conventional technologies such as grate boilers. In addition, the trend observable in the development of renewable energy production points towards an emergence of grids comprising several independent power producers. Thus, a grate boiler, as a part of such a grid, has to be able to operate on low loads and under fluctuating power demand while maintaining efficient and clean combustion. In this paper, we explore the power production flexibility of a grate boiler which utilizes biomass and which is regulated by a model predictive controller (MPC) for improved plant performance. In the study, a special emphasis is set on the factors neglected in the majority of the studies, namely combustion cleanliness and efficiency of a boiler under various loads. The analysis of these factors is enabled by a novel mechanistic model of biomass combustion upon which the MPC is built.

The model achieves faster-than-real-time computational times yet describes the most important phenomena occurring in a burning fuel layer, including fuel drying due to heat radiation and convection, chemical reactions: pyrolysis of cellulose, hemicellulose and lignin as well as char oxidation to carbon oxides. With this model, which was successfully validated against industrial data, we are also able to monitor temperatures, fuel composition and quantity inside the fuel layer which ensures stable boiler operation also during rapid changes in power demand. The results obtained from this study are presented, analysed, discussed and the guidelines for the successful MPC implementation in a grate boiler to achieve flexible and efficient power production are proposed.

Decision support for optimised cooling tower operation using weather forecasts

D. Ackerschott, [1]B. Beisheim, [2]S. Engell
TU Dortmund University, Emil-Figge-Straße 70, 44227 Dortmund, Germany; tel. +49 231 755 7378, e-mail: daniel.ackerschott@bci.tu-dortmund.de; [1]INEOS Köln GmbH, Alte Straße 201, 50769 Köln, Germany; tel. +49 221 3555 1415, e-mail: benedikt.beisheim@ineos.com; [2]TU Dortmund University, Emil-Figge-Straße 70, 44227 Dortmund, Germany; tel. +49 231 755 5126, e-mail: sebastian.engell@bci.tu-dortmund.de

Cooling towers have a significant share in the consumption of electrical energy at chemical production sites. Thus the optimal operation of this equipment has a large potential for energy savings. Because the cooling water provided by the towers strongly influences the plants to which the water is delivered, a quality (temperature and pressure) is usually specified. To fulfil these requirements even under difficult conditions, like hot days in the summer, the cooling towers are overdesigned in most operational regimes. As a consequence, the number of active towers is frequently higher than necessary resulting in an immoderate energy demand.

To overcome this problem a model based on lumped mass and energy balances was developed. Since only measurements of the ambient conditions as well as of the temperature and flow rates of the inlet and outlet cooling water are needed, the model can easily be implemented with the given instrumentation. An optimisation uses this model and sets up a baseline which takes the ambient conditions and the desired cooling water quality into account. To deal with changes in the external factors the baseline is updated frequently. The current tower performance is rated via this baseline and reveals potential for energy savings.

The first step of decision support is the evaluation of the trade-off between cooling water temperature and the electrical energy consumption, whereof the results are used to provide timely feedback to the operational personnel. The second step is a prediction of the cooling tower performance using an open online weather forecast. This enables the operators to act in anticipation and to foresee limitations in the cooling capacity. Consequently, other plants can shift production capacity to times where the shortage does not exist, e.g. to the night.

This decision support system is set up as a web service at the production site of INEOS Cologne, Germany, accessible for the operators and plant managers with standard web browsers, which allows a convenient utilisation.

A genetic algorithm based plant layout design methodology considering piping and safety

Yan Wu, Yufei Wang*, Xiao Feng, Shuo Feng
State Key Laboratory of Heavy Oil Processing, China University of Petroleum(Beijing)
18 Fuxue Road, Changping, Beijing 102249, China.
wangyufei@cup.edu.cn

Plant layout design is the key way to help enterprise improving production efficiency, operation safety and energy saving. At the level of facilities in a plant, the constraints of layout design have been gradually improved. However, the main constraints will change when the problem of design extends to the level of plants in a factory or an industrial park. Because the long distance between the plants cannot be neglected. Pipeline network and safety issues for the whole factory are deemed to be the more remarkable constraints. In this paper, the pipeline network is specified into two types: material pipeline and steam pipeline. The investments for both types are evaluated through temperature, pressure and the property of material individually. For the length of steam pipelines, the minimum length of steam pipelines is optimized by Kruskal algorithm. Additionally, the likelihood of safety accidents, such as explosion, is taken into account to make the plant layout design more practical. A genetic algorithm is used to solve the problem. The objective function is to minimize the total pipeline cost and the economic property damage which is caused by safety accidents simultaneously. This paper presents a mathematical model for the plant layout design and a case study reveals the effectiveness of proposed methodology.

Procedure for the simultaneous synthesis of Heat Exchanger Networks at process and total site levels

A. Nemet, L.Čuček, Z. Kravanja
University of Maribor, Faculty of Chemistry and Chemical Engineering, Smetanova ulica 17, 2000 Maribor, Slovenia,
email:zdravko.kravanja@um.si

Although Heat Exchanger Network (HEN) synthesis methods have already been in existence for the last forty years, several issues still need to be resolved in order to obtain more realistic networks. Moreover, the complexity of the problem further increases when Heat Integration is performed at Total Site level, compared to that at process level. The usual approach is first to perform Heat Integration within processes, and later to consider the non-integrated parts for Total Site Heat Integration. This sequential approach can omit some promising solutions and thus generally lead to worse results compared to the approach where the entire Total Site is synthesized simultaneously (Nemet et al, 2014). In order to obtain most realistic results, several practical constraints should be accounted for such as transport between processes, optimal intermediate utility temperature level(s), pipeline design, heat losses and pressure drop, etc. (Nemet et al, 2015).

In this work a synthesis of the entire Total Site is performed which simultaneously considers integration within and between plants (at plant and at Total Site level). For this purpose, the superstructure optimization approach is used. The superstructure contains all the possible matches for heat exchange within and between processes. Heat exchange between processes can be performed i) directly by exchanging heat between hot stream of one process with cold stream of another process or ii) indirectly by utilizing an intermediate utility at optimal temperature levels (these are optimization variables). In both cases the transport of the heat carrier, either process stream or intermediate utility stream, is considered.

Due to numerous options for heat recovery, especially when the number of streams is large, the model could become computationally too extensive to be solved. However, when evaluating the possible matches, it is usually found that most of the matches are either infeasible due to heat transfer limitation or unviable due to economics. A two-step approach is, therefore, proposed, where in the first step match alternatives are pre-screened with respect to infeasibility and unviability, and in the second step more detailed design is synthesized/optimized on the basis of the reduced superstructure, obtained from the first step. The basic idea of the pre-screening is to apply energy targeting simultaneously with rough area targeting in order to identify a reduced set of feasible and economically efficient matches - a reduced HEN superstructure for the entire Total Site. An extended MILP transhipment model has been developed, which alongside minimization of external utilities accounts also for a rough estimation of heat exchanger areas. However, some simplifications are applied within the model to make it linear, namely the mentioned rough area are estimated, the temperatures of intermediate utilities are fixed, and heat losses and pipeline costs are estimated regarding distances, proportionally to the transferred heat. In the second step a more detailed MINLP model (Nemet, 2015) is applied, which accounts for variable intermediate utilities' temperatures, more exact heat losses, pressure drops, and pipeline cost calculations, preheating requirements and, moreover, condensate return and its preheating requirement. Optimal operating and area costs and the HEN structure are determined on the basis of the stage-wise superstructure model (Yee and Grossmann, 1990), now extended for different exchanger types (Soršak and Kravanja, 2002) and applied to the reduced superstructure obtained in the first step.

The first and the second step have been implemented in a new version of TransGen and HENSYN, respectively, initially developed for the retrofitting of large-scale Total Site HENs (Čuček and Kravanja, 2014; Čuček and Kravanja, 2015). The solutions obtained by the proposed two-step procedure are more realistic and can handle larger numbers of streams involved in the heat integration.

Acknowledgement

The authors acknowledge the financial supports from the Slovenian Research Agency (programs P2-0032 and P2-0377) and from SCOPES joint research project CAPE-EWWR 'Computer Aided Process Engineering applied to energy, water and waste reduction during process design and operation'.

References

Čuček L., Kravanja Z., 2014, Efficient Transhipment-Based Framework for Energy Targeting and Retrofitting the Industrial Total Sites, Chemical Engineering Transactions, 39, 1813-1818.

Čuček L., Kravanja Z., 2015, Retrofitting of Large-Scale Heat Exchanger Networks within Total Sites Under Uncertainty by Considering Trade-Offs Between Investment and Operating Cost, Chemical Engineering Transactions, 45, 1723-1728.

Nemet A., 2015, Synthesis of Processes and Process Subsystems for Entire Lifetime, PhD thesis, University of Maribor, Slovenia.

Nemet A., Klemeš J.J., Kravanja Z., 2014, Mathematical Programming Approach to Total Site Heat Integration, Computer Aided Chemical Engineering, 33, 1795-1800.

Nemet A., Klemeš J.J., Kravanja Z., 2015, Designing a Total Site for Entire Lifetime under Fluctuating Utility Prices, Computers and Chemical Engineering, 72, 159-182.

Input-output based system perturbation analysis of the environmental implications of rice hull utilization for power generation in the Philippines

K.D.S. Yu[1], K.B. Aviso[2,3], R.C. Eusebio[3], C. Felix[4], A.P. Mayol[4], M.A.B. Promentilla[2,3], J.R. Santos[5], A.T. Ubando[3,4] and R.R. Tan[2,3]
[1]School of Economics De La Salle University, Manila, Philippines, 0922
[2]Chemical Engineering Department, De La Salle University, Manila, Philippines, 0922
[3]Center for Engineering and Sustainable Development Research, De La Salle University Manila, Philippines, 0922
[4]Mechanical Engineering Department, De La Salle University, Manila, Philippines, 0922
[5]Engineering Management and Systems Engineering Department, The George Washington University, Washington D.C., U.S.A. ;
tel. +632 5244611, e-mail: krista.yu@dlsu.edu.ph

Climate change has influenced economic policies, as governments encourage electricity distributors to utilize renewable energy resources whenever possible. This has caused a shift from traditional to wind, water, solar and biomass as sources of energy. Agriculture dependent economies commonly produce low-value products relative to industrialized economies. Converting agricultural waste into electricity will increase the value of engaging such activities. However, climate change also contributes immensely to the shifts in seasons thereby effecting agricultural yield and biomass production. This study analyzes the impact of using rice hull waste into electricity generation to economic activities and carbon emissions using an input-output (IO) approach. The IO model is widely used for illustrating the interdependent relationship between various economic sectors. This study extends the traditional IO model such that it links the IO table to another table composed of further subsectors not found in the original database, providing a more detailed snapshot of activities. The Philippine case is considered, given that rice is the main staple crop. However, the volume of rice production is insufficient which causes the Philippines to import rice from neighboring countries. Despite the high level of demand for rice in the Philippines, farmers tend to shift their production to higher valued crops. Alternatively, the by-product of rice production, rice hull, is an underutilized crop residue available in large quantities which may be used for power generation. The potential electricity contribution of rice hull can either reduce the power shortage that has been going on for the last decade or displace fossil fuel based electricity, at the same time, increase the value of rice production. Hence, government can reduce rice imports and service the utility demand of its constituents.

Process and sustainability assessment of gas liquefaction pathways

T.-V. Nguyen, F. Bühler, B. Elmegaard
Technical University of Denmark, Dep. of Mechanical Engineering, Nils Koppels Allé, 2800 Kgs. Lyngby, Denmark;
tel. +45 254102, e-mail: tungu@mek.dtu.dk;

Maritime traffic is responsible for significant SOx and NOx emissions, as well as large amounts of discharged wastewater. Heavy fuel oils (HFOs) have been extensively used, because of their low prices and wide availability, and despite their high viscosity and high emission factors. In the last years, new regulations were put in place to make sea traffic more environmental-friendly. In an effort to switch to cleaner fuels such as marine gas oil (MGO), marine diesel oil (MDO) and liquefied natural gas (LNG), the International Maritime Organization (IMO) has limited the fuel sulphur content to a maximum of 0.1% m/m in 2015 and the NOx emissions to 3.4 g/kWh in 2016 within the Baltic Sea.

The present work deals with the possible integration of small-scale LNG production facilities in Danish harbours. This would promote an increased usage of this fuel in the shipping industry, using natural gas from the grid, or, if possible, biogas. Liquefaction plants are based on refrigeration cycles, which can be categorised into the cascade, mixed-refrigerant and expansion-based processes. They differ by their equipment inventory, working fluids and operating conditions, which results in different design configurations, system performances, capital costs and environmental impacts.

The study is therefore divided into two steps and builds on the comparison of six gas liquefaction pathways. It focuses at first on the differences in the process design and performances, and then on the environmental impact of these systems. The latter considers (i) the local emissions involved with the production of LNG from biogas and natural gas, (ii) the global benefits of introducing LNG as marine fuel instead of HFO, and (iii) the life cycle impact of the complete system, from cradle to grave.

Total site analysis as a synthesis model (MILP) to select, optimize and integrate processes in multiple-product plants

Konstantinos A. Pyrgakis, Antonis C. Kokossis
National Technical University of Athens, Department of Chemical Engineering, Iroon Polytechneiou 9, 15780 Greece; tel. +307723129, e-mail: kpyrg@mail.ntua.gr

The industrialization of biochemicals production leads to the development of biorefineries. The biorefinery concept includes a wide range of bio-sources and bio-/thermo-chemical routes that convert biomass into bulk chemicals and specialties, polymers, plastics, surfactants, pharmaceuticals, cosmetics and fuels. The integration of multiple production lines in the same plant yields in significant savings by exchanging materials and energy among them. Due to the high impact of energy cost on sustainability, energy integration plays crucial role in building efficiencies in the upcoming industry. Energy integration among multiple processes naturally relates to Total Site Analysis (TSA). TSA involves (i) direct integration between hot/cold streams of each process and (ii) indirect integration between streams of different processes via steam production/reuse. The conventional methodology uses the graphical tool of Site Sources and Sinks Profiles (SSSPs) to estimate energy targets in Total Sites. Though TSA has been extensively used to address cases of known/installed processes, the existence of numerous alternative routes raises questions about which processes should be selected, collocated and integrated together to maximize steams savings and minimize energy cost of the (understudied) plant. Under these conditions, TSA should be used as a synthesis tool to systematically evaluate all potential process combinations.

The candidate products and processes take the form of value chain paths (Figure 1.a), which incorporate all possible transitions from raw materials to intermediates and final products. The energy efficiencies of the plant vary according to the type and number of integrated processes. The conventional TSA approach requires the exhaustive use of graphical tool to evaluate all possible combinations that emerge from value chains(Figure 1.a); the conventional approach is counter-productive and incapable to cope with multiplicity and complexity of synthesis problem. The problem requires the combination of thermodynamics with mathematical programming. Two novel representations are proposed to mathematically formulate (i) the mass balances across value chains (selection of paths) and (ii) the energy balances in Total Site integration (evaluation). The first, Biomass Bipartite graph Representation (BBR), is similar to a superstructure approach that represents all candidate routes; the second, Total Site Representation (TSR), uses a new concept of heat cascading.

The BBR uses three building blocks: product nodes (chemicals), process units (paths) and interconnections to (i) transform value chains into comprehensive representations (Figure 1.b);(ii) translate interconnections into chemicals flow rates; (iii) record mass balances along the BBR. The selection of processes and their capacities is implied by the flow rates of interconnections that feed them. The maximum capacities and energy needs (per unit mass)of candidates are estimated in advance, given the conversion rates of each path (by experimental groups) and the capacity of raw materials. Since the mass and energy flows inside a process are linearly depended on its capacity, the energy contents of hot/cold streams are linearly expressed by the flow rates (variables) and the energy needs per unit mass (parameters). The selection of processes (BBR) implies the contribution of their streams in integration (TSR).

The construction of SSSPs requires the development of all Grand Composite Curves assuming fixed/known capacities and energy content for each candidate; namely, processes are considered as binary options. However, sharing upstream chemicals among competitive processes and properly adjusting their capacities – and thus, the energy contents of involved hot/cold streams – steam savings of the overall plant may be improved. Those benefits cannot be detected by conventional TSA methodology. Treating process capacities as continuous variables (via flow rates), the optimal matches of steam-source/sink processes are detected. The proposed concept could be interpreted as if SSSPs were using variable (moving) curves.

Total Site integration is conducted by a super-interval representation, Total Site Cascade (TSC), which is composed by intervals of individual cascades of all candidate processes. TSC incorporates all heat exchange options between hot/cold steams and utilities. A properly modified transshipment model is used to mathematically formulate heat flows along the TSC. Heat exchange between streams of different processes is not allowed and direct integration of contributed processes is secured by TSC; the overall utility needs of selected processes are estimated. A parallel set of blocks, utility operators, is applied together with TSC to gather the utilities of individuals and: (i) translate excess heat (high temperature cold utility levels are additionally incorporated to extract this heat from TSC) into steam generation, (ii) estimate the maximum overlap(savings)between generated and demanded steam, (iii) call for external utilities to cover the remaining heating and cooling needs. The utility operators actually detect the optimal matching of steam-sources/sinks, subsequently minimizing Total Site energy cost. The system of TSC and utility operators composes the TSR (Figure 1.c), which includes the necessary energy balances that mathematically formulate Total Site Analysis avoiding the use of the graphical tool. Mass (BBR) and energy (TSR) balances are constructed as an optimization model (MILP) to select the optimal processes and products portfolio that minimizes energy cost and secures maximum energy savings.

The conventional and proposed TSA approaches were used to assessa real-life biorefinery with 7 candidate processes and 9 chemicals. The optimal portfolio that obtained by proposed model includes two more processes and one more chemical compared to solution that obtained by conventional TSA. Moreover, the optimal solution yields in higher energy efficiencies (compared to conventional TSA) in any level: 9% (hot) and 14% (cold) lower utilities consumption; 8% higher steam production; 2% lower energy cost. Improvements are due to the selection of a mix of competitive steam-source/sink processes that optimally – qualitatively (steam levels) and quantitatively (steam generation≥demand) – integrate each other. The model detects processes that increase steam production and adjusts their capacities to optimally valorise produced steam across site, thus, utilities cost is minimized.

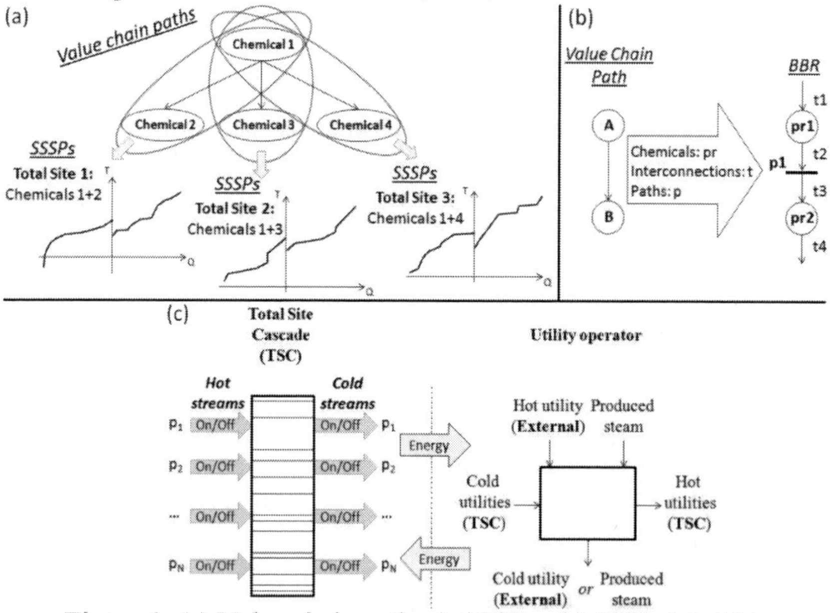

Figure 1: (a) Value chain paths & SSSPs, (b) BBR, (c) TSR

Simultaneous optimization and heat integration of effluent recovery process in polyolefin plants

Tu Gaonv, Liao Zuwei, Huang Zhengliang, Jiang Binbo, Wang Jingdai, Yang Yongrong
State Key Laboratory of Chemical Engineering, College of Chemical and Biological Engineering, Zhejiang University, Zheda road 38#, Hangzhou, Zhejiang 310027, P.R. China;
Tel. +86 13605709456, e-mail: yangyr@zju.edu.cn.

In the worldwide, the demand of energy like crude oil, natural gas and coal is increasing sharply. As we all know, the industries are the major energy contributors to operating costs, and also, major waste and effluent manufacturers to contaminate the environment. Polyolefin is one of the most widely applied polymer products for its lower cost and better processing performance. During the reaction of polymerization, the powder of polyolefin will be generated and usually adsorb certain amount of unreacted hydrocarbons. After purging by nitrogen, the desorbed gas will be burnt with the purging nitrogen [1]. However, the tightening environmental regulations on CO_2 emissions aimed at alleviating global warming are among the many driving forces behind the need for energy conservation and efficiency [2,3]. Therefore, the effluent gas should be recycled to realize cleaner production of polyolefin.

This article presents a novel effluent gas recovery process that separates components by self-refrigeration which consumes no additional cold utility as shown in Figure 1. Since the recovery process is a multicomponent phase separation procedure, the process streams exhibit nonlinear temperature-enthalpy relations (*T-H* curve). Therefore, the accuracy of state equations are critical. To balance the nonlinearity and accuracy of the state equations, we regressed the enthalpy (H) as a cubic polynomial function of temperature (T), pressure (P) and compositions (m) of the process streams. To find the optimal flowsheet of the recovery process, a superstructure that contains compressor, expander, expansion valve, multistream heat exchanger and gas-liquid separation unit is proposed. Based on this superstructure, a MINLP model is developed. In the model, the phase change of streams is allowed inside the multistream heat exchanger. The object is to maximize total annualized revenue by regulating connecting structures as well as operating parameters. The model is applied in an industry example, and a large total annualized revenue is obtained.

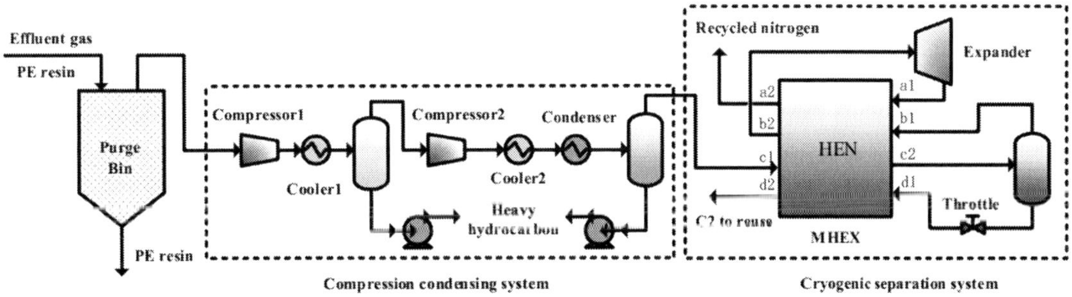

Figure 1. The flow sheet of compression and condensation, cryogenic separation combined process

References

1. Baker, R.; Jacobs, M., Improve monomer recovery from polyolefin resin degassing. *Gas* **1996,** 700, 4.000.
2. Hasan, M.; Jayaraman, G.; Karimi, I.; Alfadala, H., Synthesis of heat exchanger networks with nonisothermal phase changes. *AIChE journal* **2010,** 56, (4), 930-945.
3. Gharaie, M.; Zhang, N.; Jobson, M.; Smith, R.; Panjeshahi, M. H., Simultaneous optimization of CO_2 emissions reduction strategies for effective carbon control in the process industries. *Chemical Engineering Research and Design* **2013,** 91, (8), 1483-1498.

Activity and stability of novel silica-based catalysts for hydrogen production via oxidative steam reforming of ethanol

V. Palma, C. Ruocco, E. Meloni, A. Ricca
Salerno Univ., Dep. Industrial Engineering, Via Giovanni Paolo II, 132 84084 Fisciano (SA), Italy; tel. +39 089 964027, fax +39 089 964057 email: cruocco@unisa.it

Introduction

The global concern due to the depletion as well as the environmental damages of oil reserves has increased the demand of clean energy. In particular, the employment of sustainable sources (i.e. biomass derived liquids) as feedstock for H_2 production to be employed in fuel cells represents an excellent chance. Among the different available technologies, oxidative steam reforming (i.e. a combination of steam reforming and partial oxidation) is a promising approach: the exothermic contribution of oxidation reactions may reduce the external heat supply and air co-feeding tends to decrease carbon deposits on catalysts [1].
The objective of this work is to study the activity and stability of a novel catalyst supported over a CeO_2/SiO_2 system for steam and oxidative reforming of ethanol. Activity tests were performed in the low-temperature range and the impact of space velocity on product gas distribution was investigated. The influence of O_2 co-feeding on stability was also studied.

Experimental

The Pt-Ni/CeO_2/SiO_2 catalysts were prepared by wet impregnation. Power silica gel, calcined at T=600°C for 3 h (specific surface area of 400 m^2/g), was added to an aqueous solution containing CeO_2 precursor. The next step involved active species deposition, carried out by impregnating Ni earlier than Pt on CeO_2/SiO_2 surface: the selected salt precursors, supplied by Strem, were $Ce(NO_3)_3*6H_2O$, $Ni(NO_3)_2*6H_2O$ and $PtCl_4$ respectively for CeO_2, Ni and Pt deposition. The impregnation-drying-calcination cycles were repeated to reach 3wt% Pt-10wt% Ni/CeO_2; the CeO_2 content was 20wt % with respect to SiO_2. The specific surface area of the catalyst was 232 m^2/g. Fresh and spent catalysts were characterized by means of several techniques, including specific surface area (SSA) measurements, X-ray diffraction and fluorescence and temperature programmed reduction.

Catalytic tests were carried out in a fixed-bed stainless steel reactor, placed in an electrical oven. Prior to the reaction, the sample was reduced in situ under a 5%H_2 in N_2 stream at 600°C for 1 h (heating rate 10°C/min). The catalytic bed is located in the reactor annular section and process temperature is monitored by a thermocouple placed at the middle of the catalytic outer section. The liquid ethanol/water mixture, having a fixed molar ratio, is stored in a tank and sent to the boiler after premixing with a N_2 dilution stream. For oxidative tests, air can be feed directly to the reactor through an independent line. The products at the reactor outlet can be detected through an online FT-IR multi-gas analyzer: as diatomic molecules do not adsorb infrared radiation, the H_2 and O_2 concentrations are measured by a thermoconductivity and paramagnetic analysers.
Catalytic tests between 300 and 600°C were performed at 10000, 20000 and 30000 h^{-1} at H_2O/C_2H_5OH (r.a.) molar ratio equal to 4 and O_2/C_2H_5OH (r.o.) molar ratio equal to 0 or 0.5 under a total follow rate of 550 Ncm^3/min and the results were also compared in terms of products selectivity. The catalysts stability for ESR (steam reforming) and OSR (oxidative steam reforming) was investigated at T=500°C and GHSV=20000 h^{-1}.

Results & Discussion

Total ethanol conversion was recorded at T>430 °C for ESR as well as OSR tests and the results were not affected by GHSV variations. However, for ethanol steam reforming reaction, the reduction of contact time from 360 to 180 ms was responsible for lessened conversions of C_2H_5OH, which at 340°C dropped to 73% and 47%, respectively at 20000 h^{-1} and 30000 h^{-1}. On the contrary, a conversion of almost 100% was observed at the lowest value of space velocity even at T<340°C during OSR. In fact, CeO_2presence increases ethanol conversion due to the higher C–C bond cleavage capacity promoted by O_2 [2]. It's also interesting to note that $O_2$was completely converted between 300 and 600°C and this parameter was not affected by contact time decrease from 360 and 120 ms. Moreover, whatever the contact time selected, the system reached thermodynamic equilibrium at almost 480°C under both oxidative and non-oxidative conditions. The impact of oxygen co-feeding on catalyst stability was investigated at 500 °C and 20000 h^{-1}. Products gas distributions versus time-on-stream (TOS) are shown in comparison with equilibrium data in Figure 1.

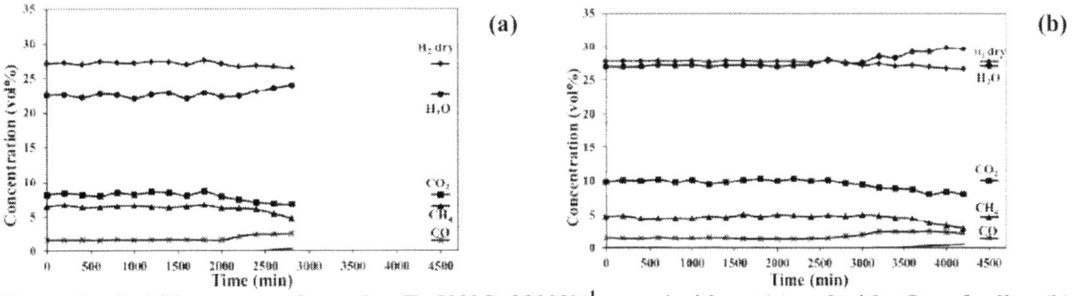

Figura 1. . Stability tests performed at T=500°C, 20000h^{-1}, r.a.=4 without (a) and with oO_2co-feeding (b).

Concerning ESR, products distribution was not affected during TOS for almost 34 h. However, at t>2000 min, a higher extent of CH_4 decomposition reaction was observed, leading to lessened C_2H_5OH conversions after 2500 min. A very similar trend was recorded during OSR test. However, in the latter case, ethanol was completely converted for almost 60 h: O_2 availability enhanced coke removal and resulted in H_2yields close to the ESR case.

Conclusions

The experimental study has been conducted over a wide range of T (300-600°C) and GHSV (10000-30000 h^{-1}). At T>480°C the system reached equilibrium independently from the contact time during both ESR and OSR. However, at lower T, ethanol conversion as well as reforming activity decreased with space velocity. The catalyst was also very stable, as total conversion was recorded for almost 60 h in the oxidative steam reforming condition.

Acknowledgments

The presented work is funded within the FluidCELL project as part of the European Union's Seventh Framework Programme (FP7/2007-2013) for the Fuel Cells and Hydrogen Joint Technology Initiative under grant agreement n°621196. Note: "The present publication reflects only the authors' views and the FCH JU and the Union are not liable for any use that may be made of the information contained therein".

References

[1] Maia T.A. et al., 2014, Fuel Process. Technol., 128, 134-145.
[2] de Lima S.M. et al., 2008, Catal. Today, 138, 162-168.

Hierarchical method for prioritization of sustainable products' characteristics

O. Canciglieri Junior[1*], Suewellyn Krüger[1], A. M.O. Sant'Anna[2]
[1]Production and System Engineering Graduate Program at Pontifical Catholic University of Parana (PPGEPS/PUCPR), Rua Imaculada Conceicão, 1155, Curitiba, Brazil;
[2]Federal University of Bahia, Department of Industrial Engineering, Aristides Novis street 2, Salvador, Brazil;
[*]tel. +55 41 3271 2579, e-mail: osiris.canciglieri@pucpr.br

The conception of the manufacturing process to be initially applied in the parts rapid production targeting a first embodiment of an idea (prototypes) and without many demands on strength and accuracy is called rapid prototyping. New rapid prototyping technologies are more accessible under the aspects of cost, accuracy, usability, environmental and prototypes started to be increasingly explored in the sustainable product development process. This paper proposes the application of hierarchical analysis method to support the decision-making in Rapid Prototyping processes based on the analysis of product and RP technology and prototype characteristics. The application of the developed method consists of assigning numeric values to the product characteristics and to the important characteristics for users in order to build the comparison matrix to perform the assessment and the prioritization steps of the relevant features. Performing the prioritization of features considered most important along with the characteristics of the chosen prototyping technology make possible the analysis of the effect of the characteristics prioritization. The concepts of product development, rapid prototyping and analytic hierarchy process for the sustainable product was used in a structure (lightweight dual) to evaluate the proposed strategy. The lightweight dual product from the 3D printing technology has been successful to show the product's purpose and functionality due to its lower cost, short execution time, good surface finishing and dimensional accuracy. The results show that the analysis of the sustainable product for rapid prototyping technology based on characteristics prioritization strategy improves the product design.

Keywords: **Rapid prototyping, analytic hierarchy process, decision making; sustainable products**

Pinch analysis approach to energy planning using weighted composite quality index

Mayuresh Patole,[1] Raymond R. Tan,[2] Dominic C. Y. Foo,[3] Santanu Bandyopadhyay[1]
[1]Department of Energy Science and Engineering, Indian Institute of Technology, Bombay Powai, Mumbai, 400076, India; fax: +91 22 25726875; [2]Chemical Engineering Department, De La Salle University, 2401 Taft Avenue, 0922 Manila, Philippines; fax: +632 536 4226; [3]Department of Chemical and Environmental Engineering/Centre of Excellence for Green Technologies, The University of Nottingham Malaysia, Broga Road, 43500 Semenyih, Selangor, Malaysia; fax: +6 03 8924 8001

Pinch Analysis has evolved over the past four decades from a methodology originally developed for optimising energy efficiency of industrial plants. Diverse extensions are documented in Process Integration literature, such as Mass Integration, Hydrogen Integration, Financial Pinch Analysis, etc. In all cases, Pinch Analysis applications are based on common principles of using stream quantity (e.g., enthalpy) and quality (e.g., temperature) to determine optimal system targets. This targeting step identifies the Pinch Point, which facilitates problem decomposition for subsequent network design.

One important class of Pinch Analysis problems is energy planning with footprint constraints. This area of work began with the development of Carbon Emissions Pinch Analysis (CEPA), where energy sources and demands are characterized by carbon footprint as the quality index. This methodology has been extended by using alternative quality indexes, such as water footprint, land footprint, emergy transformity, inoperability risk, energy return on investment (EROI) and human fatalities. Despite such developments, these Pinch Analysis variants have the limitation of being able to use one quality index at a time. To date, attempts at developing multiple-index Pinch Analysis methods have only been partially successful.

In this work, a multiple-index Pinch Analysis method is developed by using a composite quality index; the latter is assumed to be a weighted linear function of different quality indexes normally used in energy planning, as discussed previously. The weights used to compute the composite index are determined via the Analytic Hierarchy Process (AHP). A case study adapted from a literature example is solved to illustrate this approach.

Mixing pressurization and chemical absorption based carbon capture and hydrogen recovery from low pressure refinery off-gases

Qiao Zhang, Huachao Song and Guilian Liu
Affiliation: School of Chemical Engineering & Technology, Xi'an Jiaotong University, Xi'an, Shaanxi, China.
Zip code: 710049
Telephone: 86-15389294760, 86-13259736312, 86-13032944525
E-mail: q.zhang1986@stu.xjtu.edu.cn, songhuachao009@stu.xjtu.edu.cn, guilianliui@xjtu.edu.cn

Refinery off-gases always encompass remarkable amount of hydrogen and meanwhile a large quantity of carbon dioxide. Because of the low pressure, such off-gases are neither easy nor economical to separate by physical process, such as PSA(Pressure Swing Adsorption), MS(Membrane Separation) etc. As a result, they are usually burnt to generate heat, simultaneously devaluating hydrogen resource and emitting undesired carbon dioxide.

Aiming to recover hydrogen and capture carbon dioxide, this paper will put forward a refinery off-gases reusing system, and then set up a mathematical model to perform the tech-economic assessment, optimizing the process design. In this system, gas pressurization and carbon dioxide capture are two key issues. In consideration of the low pressure is even not enough to enter the compressor, we introduce another high pressure hydrogen source to lift the low one by mixing them together in Venturi tube. Afterwards, the carbon dioxide in the mixture, an acid gas, can be chemically absorbed by alkaline liquor while hydrogen is insoluble. In such a way, hydrogen is concentrated and recovered as fresh resource and carbon dioxide is also sequestrated to get both economic and environmental profits. A practical refinery case is employed to demonstrate the applicability of the proposed method.

Key words: Mixing pressurization; chemical absorption; carbon capture; hydrogen recovery; optimization

Data reconciliation for total site integration

Jun Yow Yong[a], Andreja Nemet[b], Petar Sabev Varbanov[c], Zdravko Kravanja[b], Jiří Jaromír Klemeš[c]
[a]Faculty of Information Technology, University of Pannonia, Egyetem utca 10, 8200 Veszprém, Hungary
[b]Faculty of Chemistry and Chemical Engineering, University of Maribor, Smetanova ul. 17, SI-2000, Maribor, Slovenia
[c]Pázmány Péter Catholic University, Faculty of Information Technology and Bionics, Práter u. 50/A, 1083 Budapest, Hungary
junyow.yong@cpi.uni-pannon.hu

Data extraction is a critical step performed before any Heat Integration study can commence (Klemeš, 2013). Particularly for existing plants, data reconciliation is a must in data extraction to obtain representative data. The main Heat Integration analysis method – Pinch Analysis (PA), requires only the heat capacity flowrates (CP) and temperatures (T) of process streams. While there are already some publications on reconciliation of heat exchanger network data (Nemet et al., 2015), works presenting data extraction and reconciliation for PA are scarce. The models proposed in the literature reconcile CP and T simultaneously. The previous work by the current authors proposed an iterative method (Yong et al., 2016). Iterating between two sub-models, the method keeps one type of parameters constant (e.g. T) while reconciling the parameters of another type (e.g. CP), until satisfactory convergence is achieved. Although the iterative method features some inaccuracy, compared to the simultaneous method, it is less computationally intensive and is based on a linear model.

However, that specific work has performed reconciliation only on a single plant from a site. In this work, the scope is expended into including Total Site Heat Integration (TSHI). The scope does not only include all streams for each plant, but also includes other energy system components (boilers, steam turbines) as well as energy carriers (steam, refrigerants, hot oil). The necessary data collection and analysis steps are formulated, providing criteria and tools for evaluating the reconciliation results.

References

Klemeš J.J. (ed) (2013) Handbook of Process Integration (PI): Minimisation of Energy and Water Use, Waste and Emissions, Woodhead/Elsevier, Cambridge, UK, 1184 ps. ISBN – 987-0-85709-0.

Nemet, A., Klemeš, J.J., Varbanov, P.S., Mantelli, W, Heat Integration retrofit analysis an oil refinery case study by Retrofit Tracing Grid Diagram, Frontiers of Chemical Science Eng. 2015, 9(2): 163–182, doi: 10.1007/s11705-015-1520-8.

Yong J.Y., Nemet A., Bogataj M., Zore Ž., Varbanov P.S., Kravanja Z., Klemeš J.J., 2016. Data Reconciliation for Energy System Flowsheets. Proceedings of the 26th European Symposium on Computer Aided Process Engineering – ESCAPE 26, June 12th -15th, 2016, Portorož, Slovenia (Submitted).

Cost effective heat exchangers network of total site heat integration

S. Boldyryev, G. Krajačić, N. Duić
Department of Energy, Power Engineering and Environment, Faculty of Mechanical Engineering and Naval Architecture, University of Zagreb, Ivana Lučića 5, 10000 Zagreb, Croatia, E-mail: stas.boldyryev@fsb.hr.

The energy efficiency in reprocessing industries is one of main objectives in coming future. It is connected with some important issue that have to take into account building sustainable future. From one side, more efficient process plants consume less fuel and produce less emission, and from the other side it guarantees additional employment to operate the advanced industrial clusters. Nevertheless, even modern process industry has a waste heat that have to be utilised and fuel requirements for process operation. The industrial zone are becoming more and more interesting as potential source of heat for district heating systems, municipal heat needs, interplant heat usage, heat to power generation etc. However, the reduction of the energy prices is decreasing the investment attractiveness of retrofit projects, this fact forces finding the solutions with minimised investments to get appropriate money payback.

Last time there are a lot research were published solving problem of energy efficiency of industrial systems. Wide scale of different methodologies have been used to get a solution or, in some case, even to reduce the problem. One of the most effective method that provide the solutions in systematic way for industrial systems and regional energy efficiency is Total Site Analysis (TSA). This methodology still has been developed due to a lot of local tasks and constraints as well as wide potential applications.

The present work is focused on finding the cost effective solution for implementation of heat exchangers network (HEN) during Total Site Integration (TSI). The authors discussed the solution that proposes the compromise between number of heat exchangers and heat transfer area to reduce the investments of retrofit of HEN. The feasibility of intermediate streams number is also analysed for better plants interconnection taking into account optimal amount of heat recovered.

The presented case study shows the pathway selection for creation of site recovery system with certain number of heat exchangers and interconnection between production processes, residential customers, heat generation facilities and other members of regional units. The results of this work can be used for further development of Total Site methodologies as well as for decision making during interplant heat integration and improvement of side wide heat recovery.

Acknowledgements. The financial support by the EC and Croatian Ministry of Science Education and Sports project "CARBEN" (NEWFELPRO Grant Agreement No. 39).

Relative concentration based integration of hydrogen networks with multiple impurities

Qiao Zhang, Huachao Song and Renjie Shen
School of Chemical Engineering & Technology, Xi'an Jiaotong University, Xi'an, Shaanxi, China
Zip code: 710049
Telephone: 86-15389294760, 86-13259736312, 86-18392522122
E-mail: q.zhang1986@stu.xjtu.edu.cn, songhuachao009@stu.xjtu.edu.cn, shenrenjie@xjtu.edu.cn

Hydrogen is an important and expensive resource for petroleum industries, especially for refineries. Hydrogen network integration is an effective way to make full use and conserve hydrogen resource. This paper will present a new mathematical programming method based on relative concentration to improve the resource utilization efficiency of hydrogen networks with multiple impurities.

Traditional approaches, either pinch based conceptual analysis or mathematical programming methods, are based on total flow rate and hydrogen/impurity absolute concentration, indicating hydrogen merged with impurities. However, refinery hydroprocessing operation requires hydrogen and restrict impurities separately, namely hydrogen to oil ratio, and impurities upper limit. Therefore, merging together is tighter constraint than dividing them off on conserving hydrogen resource. Zhang et al.[1] proposed a new pinch analysis based on relative concentration for hydrogen network with single impurity. For two cases without and with purification reuse, the conclusions reveal that the latter case based on relative concentration can save more fresh hydrogen that its counterparts on absolute concentration. Consequently, the relative concentration analysis on hydrogen networks with multiple impurities is more promising to save fresh hydrogen that on the case of single impurity because of the former is higher in problem complexity. Thus, such a relative concentration based mathematical model is set up to optimize the resource allocation. A practical refinery case is used to demonstrate the advantages of the proposed method.

Key words: Hydrogen network with multiple impurities; relative concentration; Mathematical model; hydrogen resource utilization efficiency; optimization

On the systematic allocation of natural gas under footprint constraints in industrial clusters

[1,2]D. Al-Mohannadi, [1,2]K.Abdulaziz, [1]S. Alnouri, [1,3]P. Linke,
[1]Department of Chemical Engineering, Texas A&M University at Qatar, Doha, Qatar [2]Qatar Foundation Research Division, Doha, Qatar [3]Qatar Energy and Environment Research Institute (QEERI), PO Box 5825, Doha, Qatar

Natural gas can be monetized through a number of different options. For instance, it can be sold as natural gas, or converted into a diverse set of alternative products. Hence, the development of effective strategies for natural resource allocation is vital, as a result of natural resource depletion, as well as the decrease in oil/gas prices. In the meantime, concerns for increased carbon dioxide emissions into the atmosphere, have caused the global society to seek large-scale carbon dioxide emission reduction targets. Since the majority of CO2 emissions are from large industrial stationary sources and fuel consumption, this often presents the industrial sector with many difficulties. Many industries would need to be able to tolerate strict emission regulations, whilst balancing profit-making activities. Carbon reduction methods in the past have been limited in technology, energy integration or geographical proximity to apply carbon capture and sequestration (CCS). More recently, a carbon integration approach has been introduced to determine the most efficient carbon dioxide reduction options in industrial parks [1]. The methodology considers multiple carbon dioxide sources, potential carbon sinks, the layout of the city, and the associated costs of CO2 conditioning and transmission.

Carbon integration considers the various conversion routes that take carbon dioxide into value added products, which in turn many be converted chemically, biologically or even through geographical utilization (such as enhanced oil recovery). This often creates many incentives for carbon emission reduction. Moreover, possibilities for exploring synergies between different industries, so as to diversify a certain product portfolio within an industrial cluster, while adhering to emission mitigation targets, may also be considered. A key design challenge is to combine the most promising configurations of alternative natural gas monetization processes from the many alternatives that could potentially exist, in which carbon management options could also be applied. Hence, the development of a methodology which could assist in exploiting synergies between natural gas conversion and carbon management is essential.

Many previous work have focused on maximizing profit of a single process while managing emission was reduced to energy integration. However, this work is a first attempt to systematically allocate natural resource under footprint constraints. The outcome is to design natural gas and carbon dioxide networks for industrial cities that will yield the maximum profit. The work explores different reduction scenarios, expansion plans and determines most profitable product mix from an industrial cluster. It takes into consideration growth demands and understands the trade-off between environmental performance and potential profitability of natural resource allocation. It strives to provide policy makers and regulators with an effective tool for planning of industrial cities and in developing strategies both nationally and regionally. The work will be illustrated using a case study to demonstrate the application of the method on parks similar to structure adopted in oil and gas centred economies.

[1] Al-Mohannadi D.M., Linke P., On the Systematic Carbon Integration of Industrial Parks for Climate Footprint Reduction. Journal of Cleaner Production. (2016), pp. 4053-4064

Valorization of low grade heat with process integration methods

[1]S.Y. Oh, [1]D. H. Kwak,-[1]J.K. Kim
[1]Department of Chemical Engineering, Hanyang University, Wangsimni-ro 222, Seongdong-gu, Seoul 04763, Republic of Korea; tel. +82(0)2-2220-2331, email: jinkukkim@hanyang.ac.kr

Low grade heat in process industries is not fully recovered, due to marginal economic gains and difficulties in retrofitting of existing processes. Various technologies and options are available for improving energy efficiency based on utilization of low grade heat. In this presentation, we would like to compare different design options available in industries and evaluate their techno-economic impacts.

Process Integration methodology is highlighted to assess potential for the recovery of low grade heat within the site in a systematic and holistic manner. Also difficulties related to valorization of low grade heat is to be addressed to provide practical guidelines.

Case study will be presented about how Process Integration methods can be effectively applied for improving the value of waste heat as well as for identifying the optimal options, subject to constraints and limitation.

Acknowledgement
This research was supported by the International Research & Development Program of the National Research Foundation of Korea (NRF) funded by the Ministry of Science, ICT & Future Planning of Korea (Grant Number: 2015K1A3A708074117).

Synthesis and targeting of total water networks using two cascade transhipment models coupled with recycle flows

A. Nikolakopoulos, [1]A. Kokossis
National Technical University of Athens, Department of Chemical Engineering, Heroon Polytechneiou, 9, Athens, 15780, Greece;
tel. +302107723236, e-mail: nikolako@mail.ntua.gr; [1]tel. +302107724275, e-mail: akokossis@mail.ntua.gr.

The chemical industry is facing an increasingly stringent legislative framework on discharge limits, while fresh water resources are rapidly depleting. On the other hand industries are struggling to operate in the most economic way. Overall costs depend on the fresh water requirements, the selection of the appropriate technologies for waste treatment, the flowrates of the treated effluents and the possibilities for water reuse and recycle. Prior to detailed designs it is useful to have an overall picture of the maximum potential of the system in terms of clear water consumption, treatment processes to integrate, the minimum wastewater flowrate and maximum reuse and recycle [1]. The possibility of reuse and recycle increases significantly the potential for water savings. Graphical targeting methods [2] do not offer a holistic approach to the water integration problem that could take into account the recycle option for the minimization of fresh water intake and synchronously deal with the selection of the most appropriate treatment technologies among alternative options and the production of targets for multiple treatment flowrates. Furthermore, shortcut models for synchronized water targeting and selection of treatment processes are absent from literature [3]. The present work integrates the extension of the transhipment models of [4], presented in [5], into a closed loop (see Figure 1) iterative procedure that selects treatment processes, targets the minimum treatment flowrates and at the same time calculates the optimal recycle rates and fresh water requirements ahead of detailed design.

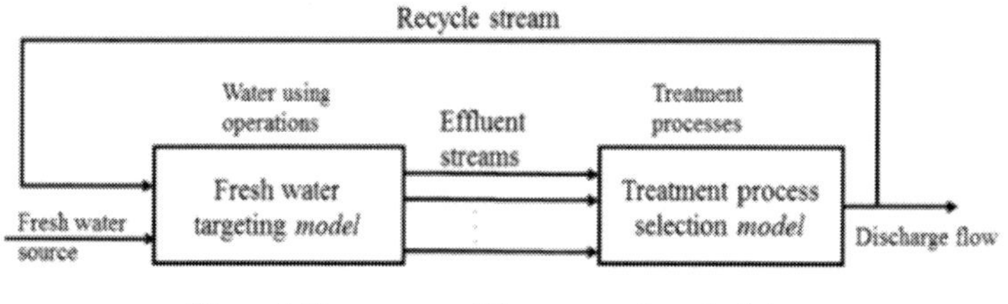

Figure 1 The concept of the proposed methodology

The novelty and advantage of the proposed method is that it assesses types and capacities of treatment processes simultaneously to calculating optimal recycle flows, further ensuring that any process configuration within the approved set of design features will satisfy the system's constraints. The proposed models act complementary to the water reuse targeting transhipment methods proposed [6], the integrated water targeting and network design approach of [7], and the transhipment mathematical model of the Total Site approach introduced by [8] for assisting the optimal synthesis of biorefinery processing paths

The proposed approach adopts the concept of concentration interval diagrams proposed for mass exchange networks [9], and the transhipment logistics model adjusted to heat exchange network design [10]. In the iterative procedure (see Figure 2) the first model calculates the minimum fresh water flow at *Cost 1* and the resulting wastewater flows. In Step 2 the second model uses the features of the wastewater flows to select treatment technologies at *Cost 2* and to calculate minimum treatment flows. *Total Cost* is then compared to the current cost (*CTC*), and the procedure continues with Step 1 and iterates until it satisfies the stopping criterion at Step 3. The output of the procedure is the selection of treatment processes and the optimal fresh water, recycle and treatment flows.

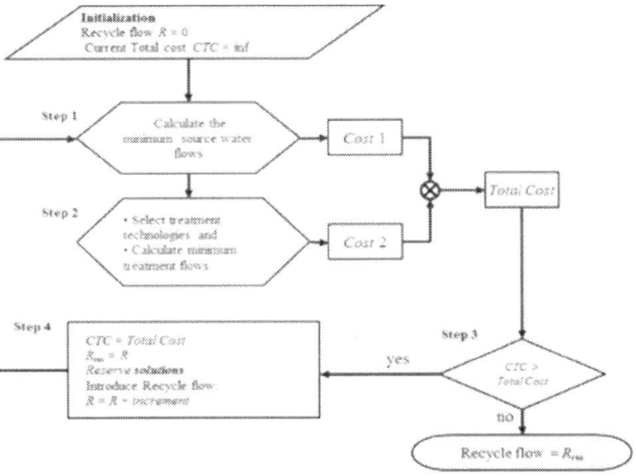

Figure 2 Steps of the iterative procedure

The methodology has been tested successfully on problems of systems with various water using operations and treatment technologies producing valid selection of treatment processes and treatment flows together with optimal targets for fresh water and recycle flows.

References

1. W.C.J. Kuo and R. Smith, "Effluent treatment system design," Chemical Engineering Science, 52, 23, 4273-4290 1997.
2. Y.P. Wang and R. Smith, "Design of distributed effluent treatment systems," Chemical Engineering Science, 49, 18, 3127-3145 1994.
3. D.C.Y. Foo, "State of the art review of pinch analysis techniques for Water network synthesis," Industrial and Engineering Chemistry Research, 48, 11, 5125-5159 2009.
4. A. Nikolakopoulos, L. Thomaidis and A. Kokossis, "Mathematical Programming Shortcut Screening Models for the Design of Integrated Waste Treatment Systems," Computer Aided Chemical Engineering,vol. 34, pp. 381-386, 2014.
5. A. Nikolakopoulos and A. Kokossis, "A Mathematical Programming Targeting Method to Select Treatment Technologies Ahead of Design,"Computer Aided Chemical Engineering, vol. 37, pp. 1091-1096, 2015.
6. A. Nikolakopoulos, P. Karagiannakis, A. Galanis and A. Kokossis, "A water saving methodology for the efficient development of biorefineries," Computer Aided Chemical Engineering,vol. 30, pp. 7-10, 2012.
7. A. Nikolakopoulos, A. Galanis, P. Karagiannakis and A. Kokossis, "An Integrated Targeting and Design Method for Saving Water in 2nd Generation Biorefineries," Proceedings of the 2013 AIChE Annual Conference, 2013.
8. D. Koufolioulios, A. Nikolakopoulos, K. Pyrgakis and A. Kokossis, "A Mathematical Decomposition for the Synthesis and the Application of Total Site Analysis on Multi-product Biorefineries," Computer Aided Chemical Engineering, vol. 34, pp. 549-554, 2014.
9. M. El-Halwagi and V. Manousiouthakis, "Synthesis of Mass Exchanger Networks," AIChE Journal, vol. 35(8),pp. 1233–1244, 1989.
10. S. A. Papoulias and I. E. Grossman, "A structural optimization approach in process synthesis II: Heat recovery networks," Computer Chemical Engineering, vol. 7(6), pp. 707–721, 1983.

Simultaneous optimization and integration of IGCC plant using novel simplified model

[1,2]M. J. Wang, [1]G. L. Liu, [2]C. W. Hui
[1]Xi'an Jiao Tong Univ., Dep. Chemical Engineering, Xian Ning West Rd. 28, 710049 Xi'an, China; tel. +86 029 82668980, e-mail: guilianliui@mail.xjtu.edu.cn; [2]Hong Kong Univ. of Sci. and Tech. , Dep. Chemical and Biomolecular Engineering, Clear Water Bay, Kowloon, Hong Kong; tel. +852 2358 7137, e-mail: kehui@ust.hk;

As a promising green technology, Integrated Gasification Combined Cycle (IGCC) plant is more efficient and low-pollution electricity generation choice than conventional thermal power plants. Its process is mainly consist of the gasifier, the air separation unit and the combined cycle unit. Going through those key blocks, the fuel, such as coal and biomass, could be converted to syngas and further combusted for power generation. In order to be a viable technology, the bottleneck is located on the high energy production cost, which could be overcame by improving the plant efficiency through process optimization and integration. The opportunity and challenge are given by the complicated configuration and extreme operation conditions, which also demands a flexible and robust IGCC plant simulation to complete deep process analysis. In this work, a novel simplified IGCC mathematical model is proposed considering thermodynamics constraints with energy and mass conservation. The gasification is simulated based on the principle of Gibbs free energy minimization, which holds the good feature of avoiding potential bi-level optimization problem. The proposed gasification model results in syngas composition similar to the experimental data provided in literature. Meanwhile, the cryogenic air separation unit is selected with the double distillation column configuration, which is simulated considering the novel simplification of thermal coupled distillation columns. What's more, the combined cycle unit is simulated with isentropic assumption plus efficiency to complete the process. To validate the novel simplified IGCC model, one base case is built and compared with literature data. Another highlight of this work is the simultaneous optimization and integration methodology instead of conventional sequential one, which gives more guarantee for achieving global optimal solution. If the ideal property assumptions are occupied, a graphical approach could be generated for the simultaneous optimization and integration of IGCC plant. The best material and energy integration scheme in proper operation condition could be selected considered with respect to the overall thermal efficiency of the IGCC plant. The good performance is shown by about 3 % efficiency increment compared with the base case in the case study.

Extraction of green hydrogen at fuel cell quality from mixtures with natural gas

W. Liemberger[1], M. Groß[1], M. Miltner[1], H. Prazak-Reisinger[2], M. Harasek[1]
[1] Inst. of Chem. Eng., Vienna University of Technology, A-1060 Vienna, Getreidemarkt 9/166, Austria, tel. 0043/1/58801-166272, e-mail: werner.liemberger@tuwien.ac.at; [2] OMV Refinery & Marketing GmbH, A-1020 Vienna, Trabrennstr. 6-8

Electrical energy from renewable sources (e.g. solar power, wind) is increasing over the last years. Fluctuating electrical excess energy is a major problem and requires energy storage technologies, which became a key challenge for the energy transition. Very promising solutions are Power-to-Gas concepts.

One solution is that hydrogen (H_2), produced via water hydrolysis, is injected into the natural gas pipeline grid. Currently the H_2 concentration limits in the grid are 4 % (v/v) in Austria and 5 % (v/v) (up to 10 % (v/v)) in Germany. Moreover the maximal pressure can be 60 bar.

This green hydrogen is then stored and transported in the pipelines with high efficiency and low costs (e.g. existing infrastructure). In a further step the hydrogen can be extracted at any desired location along the gas grid. This renewable hydrogen can then be used to supply fueling stations or other applications.

The here presented process extracts hydrogen at fuel cell quality (99.97 % H_2 according to ISO 14678-2:2012) in a multistep process, based on membrane technology and adsorption from mixtures with natural gas (up to 10% (v/v)). The membrane separation step is used for a cost- and energy-efficient pre-enrichment. Subsequently, pressure swing adsorption (PSA) provides the upgrading of the H_2 to the required quality. All remaining residual gases (CH_4, CO_2, N_2, C_2H_4, H_2O, H_2S etc.) are returned to the grid.

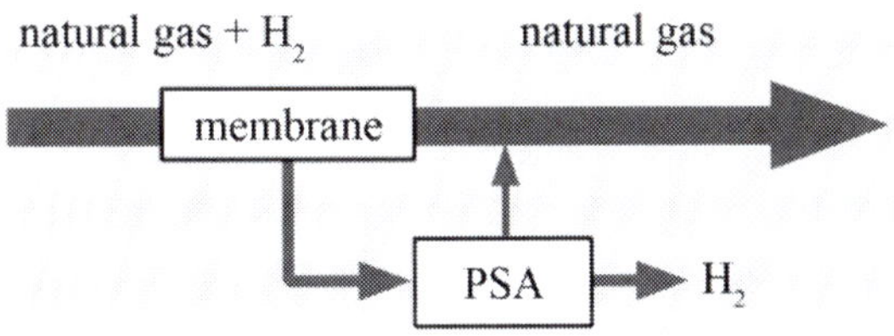

Through the complexity of the combination of two different technologies, process simulation is required for an optimal design. Aspen Plus®, a steady-state process simulation environment was used but had to be extended for membrane and PSA.

The custom membrane model is based on the solution-diffusion approach. It has been validated with experimental data at process conditions (up to 60 bar, various gases and flows). The transient behaviour of the adsorption has been analysed and a steady state model was developed. In a first step single column adsorption simulation (done with Aspen Adsorption®) was setup and validated through various breakthrough experiments (varying pressures, mixtures, flows and bed material). Afterwards a PSA has been simulated with Aspen Adsorption® and was used to build a steady state model for Aspen Plus® that could be integrated into the flowsheet.

This work shows via process simulation the impact of various parameters (pressure levels, stage-cut, adsorbens material, etc.) on the required amount of energy which can further be used for an economic evaluation.

Policy making and models for sustainable building and energy management for settlements

Adrienn Buday-Malik, P. Holndonner, Lajos Vass, K. Matolcsy,
ÉMI Nonprofit Ltd. Liability Company for Quality Control and Innovation in Building,
tel.:+36 13726100 e-mail: amalik@emi.hu; H-2000 Szentendre, DózsaGy. str. 26.

In the current programming period EU supports Hungary's economic and infrastructural cohesion with 12.000 billion(HUF). Efficiency and sustainability of building, infrastructural and energy management are key factors of the success of structural funding policy. There are central initiatives to renew policy instruments and operational tools public bodies applies. Key step towards the sustainability goal is the cooperation with the self-governments who manage at least $1/6^{th}$ of all development projects. ÉMI as a state-owned company engaged also in R&D&I in the building sector, introduced innovative policy models and tools for building and infra management. Cooperation with the self-governments is associated with re-definition of public buildings and infrastructure. Down-to-top sectoral policy making and "SMART" management systems may contribute to higher level of cohesion, which is cornerstone of the "building" of European Union.

SEAP-SECAP – adaptation and mitigation in local policy and strategy making

Sustainable Energy Action Plan (SEAP) or Sustainable Energy and Climate Action Plan (SECAP) are operational documents in which the self-governments define how they intend to reach CO2 reduction and energy efficiency targets by 2020 and 2030. The basis of the action plans is a baseline assessment about the CO2 emission of the cities caused by energy generation, consumption and transport. The action plans define activities and measures to achieve the reduction targets, together with time frames and assigned responsibilities. The energy efficient building renovation including renewable energy generation and intelligent building energy management, the public renewable energy production and utilization in public buildings and the introduction of low carbon emission vehicles into the public transport with smart traffic coordination system are turning up in the local policies and strategies as an instrument for sustainable development, adaptation and mitigation. Cohesion relevance of these documents is tested in frame of the research and to be demonstrated in the study.

NÉER – as an instrument for effective identification of intervention points

Strategies are always measured by actions implemented. ÉMI developed NBEIS/NÉER (National building energy information system) in order to introduce the optimal, measurable and comparable indicator base and establish a decision-supporting tool on the governmental, municipal and – on a longer run - institutional(public bodies, churches, etc.) public buildings. The system requires information about the structure, specification, energy consumption and efficiency, current state of the building and future development plans. ÉMI focuses onthe systemic approach, the sequential improvement and the ability to identify the potential intervention points at buildings and last but not least, the integration of good international practices in the local model. Current R&D&I project aims at a tailor-made and usable tool for public bodies and policy makers facing energy efficiency challenge.

A master plan for the implementation of sustainable enterprise resource planning systems (Part II): development of a roadmap

Abdoulmohammad Gholamzadeh Chofreh[1,2], Feybi Ariani Goni[3], Jiří Jaromír Klemeš[4],
[1]Department of Industrial Engineering, Buein Zahra Technical University, Buein Zahra, Qazvin, Iran; tel. +98 9163028545, e-mail: mohammad_apmie@yahoo.com;
[2]Faculty of Mechanical Engineering, Universiti Teknologi Malaysia, 81310 Johor Bahru, Malaysia
[3]Faculty of Information Science and Technology, Universiti Kebangsaan Malaysia, Bangi, Malaysia; tel. +98 9367120679, e-mail: feybi.ariani@gmail.com;
[4]Faculty of Information Technology and Bionics Pázmány Péter Catholic University, Práter u. 50/a, 1083 Budapest, Hungary; e-mail: klemes.jiri@itk.ppke.hu;

Advanced organizations require implementing a Sustainable Enterprise Resource Planning (S-ERP) system in order to integrate all data, information, and processes across their entire extended value chain. A number of vendors such as Systemanalyse und Programmentwicklung (SAP), Microsoft, and Oracle have developed the system as a holistic solution to support sustainability initiatives. However, the organizations still face the problem in implementing the S-ERP system. There is a lack of a master plan, which shows the stages, dimensions, and steps to assist them to implement this system. In order to solve this issue a master plan has been developed for the implementation of S-ERP system that comprised of a roadmap, a framework, and guidelines. This study focuses on the development of the roadmap that shows the stages intended to implement the S-ERP system. To attain the research objective, this study employed a conceptual research method, which relies primarily on the literature for the development of the roadmap. Various research fields including sustainability, Enterprise Resource Planning (ERP) system, and project management were examined. As a result, the roadmap of the S-ERP system implementation master plan was successfully developed by using the conceptual research method.

Graphical Abstract:

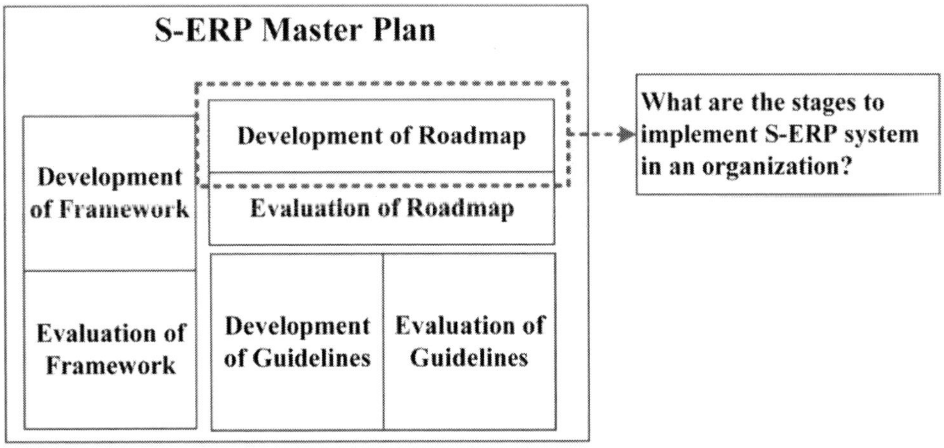

Heat pump integration for total site waste heat recovery

PY Liew[1,2], TG Walmsley[3], JS Lim[2,4]
[1]Department of Environmental Engineering and Green Technology, Malaysia – Japan International Institute of Technology (MJIIT), Universiti Teknologi Malaysia, Jalan Sultan Yahya Petra, 54100 Kuala Lumpur, Malaysia;tel.: +603-22031479, e-mail: pyliew@utm.my; [2]Process Systems Engineering Centre (PROSPECT), Research Institute of Sustainability Environment (RISE), Universiti Teknologi Malaysia, 81310 UTM Johor Bahru, Malaysia; [3]Energy Research Centre, School of Engineering, University of Waikato, Private Bag 3105, Hamilton, New Zealand; [4]Faculty of Chemical and Energy Engineering, Universiti Teknologi Malaysia, 81310 UTM Johor Bahru, Malaysia.

Total Site Heat Integration (TSHI) promotes energy recovery between processes to enhance overall energy efficiency of an industrial complex (Chew et al., 2013).Various industrial waste heat utilisation technologies have been studied to improve the energy efficiency of energy system. Cogeneration has been well established for power generation from high quality industrial waste heat (Sun and Smith, 2014), as well as power recover from steam expansion process. Oluleye et al. (2015)explored the potential for low grade waste heat recovery through Organic Rankine Cycle, Absorption chiller and absorption heat pump using mathematical modelling methodology. District heating has been widely discussed as one of the possible waste heat utilisation(Kapil et al., 2011). In addition, Liew et al. (2015) proposed the integration of district cooling system for utilising industrial low grade waste heat through absorption chillers.

Vapour compression as an open loop heat pump system has good potential to be used to upgrade the waste heat to useful heat in the Total Site system. The effective integration of mechanical heat pump or vapour compressor in TSHI has been discussed by Bagajewicz and Barbaro (2003).Vapour compression systems upgrade low grade waste heat by supplying a low quantity of high pressure steam (thermocompressor) or mechanical work (mechanical-compressor) to generate higher pressure steam, as is common with evaporation systems (Walmsley et al., 2015). The vapour compression system recovers the latent heat content of the industrial waste heat, which reduces cooling demand, decreasing the demand for high quality steam and reducing boiler load.

This paper introduces an effective Total Site targeting methodology to integrate open loop heat pump systems, i.e. vapour compression technologies, into an integrated industrial energy system for enhancing overall site energy efficiency. Industrial waste heat and high quality steam demand are able to be reduced simultaneously though this integration. The energy and cost-benefit of thermocompressor and mechanical-compressor installation are compared through a literature case study.

Reference

Bagajewicz M.J., Barbaro A.F., 2003, On the use of heat pumps in total site heat integration, Computers & Chemical Engineering, 27 (11), 1707-1719.

Chew K.H., Wan Alwi S.R., Klemes J.J., Manan Z.A., 2013, Process modification potentials for total site heat integration, chemical Engineering Transactions, 35, 175-180.

Kapil A., Bulatov I., Smith R., Kim J.K., 2011, Integration of low grade heat with district heating, Chemical Engineering Transactions, 25, 123-128.
Liew P.Y., Wan Alwi S.R., Manan Z.A., Klemeš J.J., Varbanov P.S., 2015, Incorporating district cooling system in total site heat integration, Chemical Engineering Transactions, 45,19-24.
Oluleye G., Jobson M., Smith R., Perry S.J., 2015, Evaluating the potential of process sites for waste heat recovery, Applied Energy, 161, 627-646.
Sun L., Smith R., 2014, A New Steam Turbine Model for Utility System Design and Optimization, Chemical Engineering Transactions, 39, 1399-1405.
Walmsley T.G., Walmsley M.R.W., Neale J.R., Atkins M.J., 2015, Pinch analysis of an industrial milk evaporator with vapour recompression technologies, Chemical Engineering Transactions, 45,7-12.

Application of P-graph techniques for efficient the use of wood processing residues in biorefineries

M. J. Atkins, T. G. Walmsley, B.H.Y. Ong, M. R. W. Walmsley, J.R. Neale
Univ. Waikato, Energy Research Group. Private Bag 3105, 3105 Hamilton, New Zealand; tel. +64 7 8384937, e-mail: matkins@waikato.ac.nz;

It is anticipated that demand for chemicals and fuel derived from sustainably grown biomass will increase over the coming decades. Forest and wood processing residues and waste are likely to become a significant feedstock to large scale biorefineries to produce both renewable fuels and chemicals. Maximising the economic value of these residues whilst simultaneously minimising the environmental impact of the manufactured product is an important task in process and product selection and design. Multiple processing and product pathways exist and it is often unclear what the best options are without detailed assessment or preliminary design.

The P-graph framework was used to examine the economically feasibility of several processing pathways to produce a range of bio-based fuels and chemicals. Promising products and processes were assessed using available cost and conversion/yield data. The availability and cost of wood processing residues and forest residues were based on Geographic Information System (GIS) models for New Zealand. Energy resources, such as renewable electricity, geothermal, and biomass were also included so that efficient and economic use of energy is incorporated into the P-graph.

Particular focus is given to maximising the use of available biomass and other natural resources, in particular geothermal heat. Total site integration methods are applied to the identified economic pathways into the biorefinery and the effect on the utility system and process integration.

One of the approaches for reducing the risk of adopting new conversion technologies is to identify economic synergies with existing processes and infrastructure. Specific integration schemes need to be identified and benefits quantified to assist the successful commercialisation and adoption of these new technologies. These synergies are examples of industrial symbiosis; the sharing of resources, including utilities and services between different co-located production facilities. Total site analysis is also an important analysis tool to maximise the potential integration between existing and new processes. The convergence of other natural resources, such as availability of biomass residues and geothermal heat, provide additional integration options and assist the economics.

A framework for assessing the opportunity for industrial symbiosis, by integrating new biomass processes with existing wood processing cluster will be discussed in this paper.

Innovative catalytic systems for methane steam reforming intensification

V. Palma, A. Ricca, M. Martino, E. Meloni
University of Salerno, Department of Industrial Engineering, via Giovanni Paolo II 132, 84084 Fisciano (SA), ITALY – aricca@unisa.it

Hydrogen is the simplest and the most plentiful element in the universe; despite its simplicity and abundance, hydrogen is not naturally present in molecular form on the Earth, unless combined with other elements. Hydrogen is one of the main constituents of hydrocarbons from which can be extracted by reforming processes. Actually, the main technique to produce hydrogen is methane (or natural gas) steam reforming (SR), a catalytic endothermic process in which the hydrocarbon reacts with steam to produce hydrogen and carbon monoxide. The endothermic nature in such reaction requires that in one hand the process should be carried out at very high temperature (around 850°C), in the other hand a considerable heat flux towards the reaction system is required to sustain the process to achieve high methane conversion[1]. As a result, the requirement of transfer heat from an external source to a high temperature catalytic system determined the main part of operating and fixed costs, therefore resulting a key target in the process intensification direction. In this aim, the minimization of heat transfer resistance in the overall system (heating source – reactor – catalyst – gas phase) appeared the most viable solution. In particular, to achieve a uniform temperature in the whole catalytic volume may increase heat transfer rate. Previous studies demonstrated that highly conductive supports help to flat the radial and axial temperature profiles, so minimizing the heat transfer resistance from the heating medium to the catalytic volume, and obtaining a more uniform catalytic behaviour in the whole volume[2]. In addition, the thermal advantages achieved by using highly thermal conductive honeycomb monolithic catalysts may be further enhanced by forcing the flux to cross the monolithic walls. In this aim the use of Wall-Flow monoliths appears a viable solution towards the methane steam reforming intensification[3]. Aim of this work was to evaluate the improvements in a methane steam reforming process due to the employing of high-conductive monoliths as catalytic supports; in particular, wall-flow configuration effect on system performances were evaluated. It is also demonstrated that rare hearts supports (CeO_2, ZrO_2) were characterized by an improved oxygen mobility[4], resulting in an increased activity towards reforming reactions and in an enhanced resistance towards coke formation. In the present study, structured catalysts for MSR reaction were developed, by optimizing catalytic phase formulation (active phase percentage, support), and adhesion to the mechanical support.

Several nickel based powder catalysts were prepared by wet impregnation method, in order to investigate the Ni loading effect and the support role in the process. Once optimized catalytic formulation, support was modified by adding pseudo-bohemite (as binder for improve adherence of catalyst to the monolithic support): the pseudo-bohemite role in catalytic performances was also investigated. Finally, the selected formulation were supported on monolithic supports, in order to investigate the mechanical support role. Monolithic catalysts were prepared by the impregnation method, as reported in earlier works[3], by a dip impregnation in a slurry, up to reach a washcoat loading of 20%wt, and then the Nickel was deposed on the coated monolith. Finally, some samples were converted in Wall-Flow configuration, by alternately plugging channels in the inlet and outlet sections, so forcing the process gas to cross the porous walls.

Experimental tests were carried out in a tubular lab scale catalytic reactor in isothermal conditions both for the powder and for the monolithic samples. Activity tests on powder formulations (Table 1) evidenced that too high nickel catalysts resulted detrimental for the process, since the reduction of nickel dispersion linked to the reduced specific surface area of the support. On the other hand, ceria supported samples resulted in the higher activity for the higher operating temperature, while ceria-zirconia samples were more active at the lowest temperature, to evidence that in an overall context the former one is the more active, but the low dispersion of active phase may reduce the TOF frequency of the catalyst

Table 1 - Activity tests on powder formulations

	5% Ni / CeO2		10% Ni / CeO2		20% Ni /CeO2		5% Ni / CeO2-ZrO2	
	Conv. CH_4	H_2 %vol	Conv. CH_4	H_2 %vol	Conv. CH_4	H_2 %vol	Conv. CH_4	H_2 %vol
650	38.2%	58.5%	7.9%	20.1%	6.0%	19.1%	51.7%	65.3%
700	63.4%	69.0%	18.6%	40.3%	11.1%	29.8%	66.3%	69.4%
750	87.4%	73.8%	38.0%	58.0%	20.6%	43.9%	76.4%	71.7%
800	95.8%	75.0%	61.9%	68.7%	37.0%	56.8%	79.4%	72.5%

The pseudo-bohemite loading (Figure 1-SX) appeared to reduce catalytic activity, since the effect of ceria was reduced. Anyway, a small amount of pseudo-bohemite slightly increased catalytic performances, due to the sensible increasing of support surface area, and in turn of Nickel dispersion.

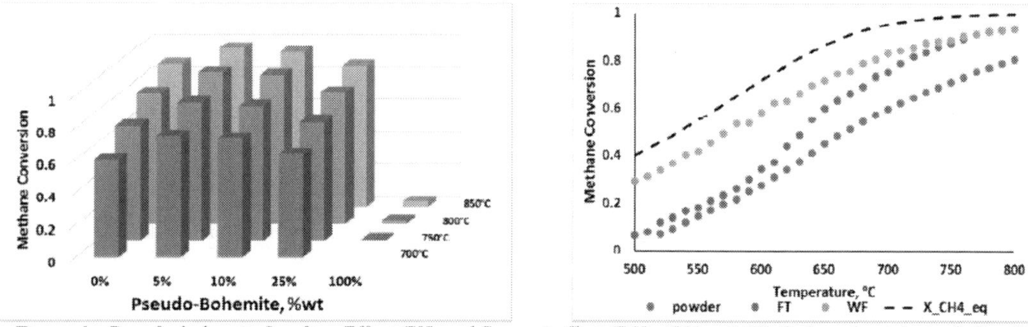

Figure 1 - Pseudo-bohemite Loading Effect (SX) and Support effect (DX) (5%wt Ni/CeO_2 based catalysts, S:C = 3)

Finally the support effect was investigated, by testing both traditional wall-flow and innovative flow-through configurations (Figure 1-DX). As reported, supported catalyst clearly increased catalytic performances in particular at the highest operating temperature, demonstrating that in such extreme conditions heat transfer towards catalytic volume is enhanced by the conductive support. The WF configuration enhancement was more evident in the unfavourable conditions, at the lowest operating temperatures, where reaction kinetics were disadvantaged. In such conditions, the deeper contact between solid and gas phase induced by the wall-flow configuration improved both mass and heat transfer mechanisms between catalyst and reactants, so resulting in an increased reaction rate.

References:

[1] Yan, Y.; Zhang, Z.; Zhang, L.; Wang, X.; Liu, K.; Yang, Z. Int J Hydrogen Ener. 2015, 40, 1886-1893.

[2] Palma, V.; Ricca, A.;Ciambelli, P. Chem Eng J. 2012, 207, 577-586.

[3] Palma, V.; Miccio, M.; Ricca, A.; Meloni, E.; Ciambelli, P. Fuel. 2014, 138, 80-90.

[4] Polychronopoulou, K.; Kalamaras, C.M.; Efstathiou. A.M. Recent Patents on Materials Science 2011, 4, 122-145

Improving sustainability of electric steelworks through process simulations

I. Matino*, V. Colla*, V. Colucci*, P. Lamia*, S. Baragiola[&], C. Di Cecca[#]
*Scuola Superiore Sant'Anna, TeCIP Institute – ICT-COISP, Pisa, Italy,
i.matino@sssup.it, colla@sssup.it
[&]RIVA ACCIAIO S.p.A, Caronno Pertusella, Italy, qualita.milano@rivagroup.com
[#]Politecnico di Milano, Milano, Italy, cosmo.dicecca@polimi.it

The European Steel sector is always committed to investigate solutions for reducing its environmental impact in terms of gas and water emissions, energy consumption and solid waste production. Current technologies and operational practices already allow meeting legal environmental constraints. Nonetheless, sustainable steel production, improving energy and resource efficiency and reducing emissions, is a key objective for this sector, whose social and economic value fully justifies the ongoing and future efforts. Process integration provides viable solutions to improve sustainability of the steel production cycle.

The project entitled *"Environmental Impact Evaluation and Effective Management of Resources in the EAF Steelmaking" (EIRES)*, that is funded by the Research Fund for Coal and Steel, aims at providing electric steelworks with a tool to systematically evaluate and investigate solutions improving the process sustainability. The project includes the development of a process simulation tool for evaluating energy consumptions, atmospheric and water emissions, waste and by-products generations in current or in not-conventional scenarios.

As complex process modifications, before being practically implemented and tested on the field, need a thorough evaluation, a simulation model of the whole electric steelmaking route has been developed, to support steelworks staff in the selection of the process modification which mostly contribute to improve steelwork sustainability while preserving the objectives of high quality production and workers' health and safety protection.

The paper presents such simulation model, which has been developed through the Aspen Plus® simulation suite in order to monitor and predict the impact of each operation unit, including both production units and treatment processes for fumes and wastewater. The models of each sub-process have been tuned and validated through field data and are connected each other through Aspen Simulation Workbook® allowing a more friendly management of the simulation tool using MS Excel® as main interface. MS Excel® also allows the connection of the simulation model to another tool also developed within the EIRES project quantifying different aspects of the environmental impact.

Key words: steelworks modeling; steelmaking sustainability; environmental impact; energy consumption

Gas-phase photocatalytic synthesis of acetaldehyde from ethanol: thermodynamic evaluation and comparison with photocatalytic activity

V. Vaiano, D. Sannino, P. Ciambelli
Department of Industrial Engineering, University of Salerno, Via Giovanni Paolo II, 132, 84084 Fisciano (SA), Italy;
tel. +39 089964006, e-mail: vvaiano@unisa.it;

Photocatalysis is a research area where a wide number of studies and applications have been developed. Heterogeneous photocatalysis is based on the use of semiconductor materials, which are able to induce chemical reactions when excited by solar or artificial light. Large studies are devoted to environmental science for removing pollutants from gaseous streams or wastewater. Recently, an increasing numbers of papers reports about the photocatalysis as suitable method for green organic synthesis. However, all these studies never started from a rigorous estimation of the thermodynamics yield of the photocatalytic reactions. In this view, this work focuses on the thermodynamic evaluations as a preliminary step for the photocatalytic partial oxidation of organic compounds aimed to the synthesis of high valued chemicals. The comparison of thermodynamic data with the experimental ones for the photocatalytic conversion of ethanol to acetaldehyde will be discussed and connected to the kinetics of the photoreaction system.

Thermodynamic analysis has been carried out with the Gaseq program, a Microsoft Windows program with a graphic interface. The basic principle of this program is the minimization of Gibbs free energy, considering chemical equilibrium of perfect gases.

For photocatalytic tests, a specifically developed photocatalysts, based on direct supporting of the active photocatalyst on phosphors particles, were used.

Phosphors (model RL-UV-B-Y; Excitation Wavelength: 365 nm; Emission Wavelength: 440 nm; particles diameter: 5-10 μm, provided by DB Chemic) coated with TiO_2 was prepared by incipient wet impregnation method. Phosphor sample was added to solution of titanium isopropoxide in isopropanol, dried at 80 °C and calcined in air at 400 °C for 3 h. A second impregnation with an aqueous solution of ammonium vanadate was performed to have VO_x/TiO_2/phosphors catalyst. The optimized TiO_2 content was 30 wt % while V_2O_5 nominal loading was 5 wt% based on the TiO_2 loading in the composite sample.

Catalytic tests were carried out feeding 30 L/h (STP) at ethanol concentration in the range 0.2-2 vol % with oxygen/ethanol ratio of 2; temperature and pressure reaction were 100°C and 1 atm, respectively. The contact time was in the range 22-560 ms.

The photocatalytic fluidized bed reactor was illuminated by two UVA-LEDs modules positioned in front of the reactor pyrex windows (light intensity: 90 mW/cm^2). Each module consisted of 40 UV-LEDs emitting at 365 nm. To improve fluidization of photocatalysts, 20 g of glass spheres (Lampugnani) were mixed with samples. The gas composition was continuously measured by continuous CO-CO_2 NDIR analyzer.

Experimental data has shown the influence of the reaction temperature and the contact time in the distribution of the reaction products. Interesting is the control on the products distribution induced by the contact time, which drives towards the thermodynamic expectations, when it increases. The choice of the appropriate contact time in the photocatalytic reaction permits to get the desired product with the highest yield acting as selectivity control.

The potential for digestate thickening in biogas plants and evaluation of possible evaporation methods

M. Vondra, V. Máša, P. Bobák
Brno Univ. of Technology, NETME Centre & Inst. of Process Engineering – VUT ÚPI.
Technická 2, 616 69 Brno, Czech Republic, tel. +420 541144918, email: vondra@upei.fme.vutbr.cz

Biogas plants are up-to-date technologies primarily focused on energy exploitation of different organic waste. Vastly subsidized, biogas plants have become significantly widespread in Europe with more than 14 thousand sites in 2014, reaching almost 8 GW in total electricity installed capacity. Besides biogas, which can be easily utilized for electricity production, operators of the plants must somehow treat several other products of the fermentation process, and liquid digestate is one of them. Liquid digestate contains mostly water and only 3 percent of total solids. Nutrients are sparse but still valuable and the digestate may be used for fertilization. However, the costs of digestate application are rather high and fertilization of agricultural land requires repetitive applications, which increases fuel and time consumption. Other costs of liquid digestate treatment are connected with considerable storage requirements. The same problems are associated with livestock manure management. Additional issue of the biogas plant operation is the utilization of waste heat, as only 15 to 30 percent of the heat produced by the biogas cogeneration is consumed during the fermentation process.

Aforementioned challenges could be theoretically completed at once. The excessive waste heat could provide enough energy for the vaporization of water that is contained in the liquid digestate, thus reducing its volume, increasing nutrients concentration and decreasing associated costs. This paper investigates potential of different evaporation methods, such as multi-stage flash (MSF) evaporation, multi-effect distillation and single effect evaporation for the purpose of liquid digestate thickening. Simple construction, reliable operation and low temperature waste heat utilization are some of the methods' main advantages. Presented considerations are focused on specific heat consumption and other relevant parameters of selected technologies. Presented results shows that the proposed model of MSF evaporator has the best thermal performance.

Problematique approach to analyse barriers in implementing industrial ecology in Philippine industrial parks

M. Promentilla[1], L. Bacudio[2], M. Benjamin[4], A. Chiu[2], K. Yu[3], R. Tan[1], K. Aviso[1]
[1]Center for Engineering and Sustainable Development Research
[2]Industrial Engineering Department
[3]School of Economics
De La Salle University Manila
[4]Chemical Engineering Department, University of Santo Tomas, Philippines
tel. +632 5360223, e-mail: michael.promentilla@dlsu.edu.ph; 2401 Taft Avenue, Malate, Manila, Philippines, 0922

Industrial ecology is recognized as an important framework toward a circular economy wherein industrial systems minimize their environmental burden by mimicking the material cycles and energy cascades found in biological ecosystems. Advocates of such framework in planning of eco-industrial parks suggest that both economic and environmental gains can be attained by transforming the industrial production from a linear to a closed loop system. However, it is imperative first to understand and analyze systematically the barriers in implementing the concept of industrial ecology in industrial parks even at the early planning stage. This work thus proposes a problematique approach to understand and analyse such barriers toward a successful development of eco-industrial parks. A problematique is a term coined by Warfield referring to concepts and tools for a structural model of relationships among members of a set of problems. The problematique is shown to be effective in analyzing the structure that underlies problematic situations, thus increasing the potential for crafting solution through human intervention. An illustrative case study was presented using a methodological framework built from decision making trial and evaluation laboratory (DEMATEL), interpretive structural modeling (ISM) and analytic network process (ANP). Among the identified barriers in an industrial park situated in Philippines, the method reveals the strength and direction of interaction, hierarchy structure, degree of relative importance, and the causal loop mapping to aid stakeholders in systems thinking and problem solving for such complex issues.

Catalytic activities of bimetallic catalysts for low temperature water gas shift reaction

V. Palma[a], M. Martino[a], D. Pisano[a], P. Ciambelli[a].
[a]Univ. of Salerno, Dep. of Industrial Engineering, via G. Paolo II 132, 84084, Fisciano, Italy.
Tel: +39089964147, e-mail: vpalma@unisa.it.

Nowadays hydrogen is mainly produced by Steam Reforming of hydrocarbons, however the composition of the reformate includes also large amounts of carbon monoxide, making it unsuitable in PEM fuel cells applications. The water gas shift (WGS) allows to reduce the CO percentage to less than 1%, in syngas, through a two-stage process, a high temperature stage (HTS), conventionally performed on iron/chromium catalysts over 623 K, and a low temperature stage (LTS), performed on copper/zinc catalysts below 623 K. The WGS reaction is exothermic, therefore, a two-stage process allows to achieve a fast reaction in the HTS and to reach thermodynamic equilibrium in the LTS, unfortunately this kind of configuration requires two different catalytic systems, moreover the Cu-Zn catalysts, used in LTS, are pyrophorics and not suitable in fuel cells applications. From the foregoing consideration it is clear the need to investigate new catalytic systems, active in a wide temperature range, and compatible with the fuel cells. Among the most promising catalysts, for sure, the nobel metals based catalysts guarantee high performance in a wide range of temperatures and are compatible with fuel cells applications. Several studies have been already reported on the use of Pt-based catalysts [1,2] and, among them, for sure the Pt/CeO_2ZrO_2 could be considered the most promising LT-WGS catalyst.

These catalysts show excellent performance, approaching the equilibrium also below 473 K, however are affected by deactivation phenomena, making it necessary to improve the stability; a possible strategy provides the association with other metals.

In this work we studied the catalytic activity of some bimetallic catalysts, containing Platinum in combination with Tin (1%Sn1%Pt/CeO_2ZrO_2), Sodium (1%Na1%Pt/CeO_2ZrO_2) and (1%Mo1%Pt/CeO_2ZrO_2) Molybdenum. The catalysts were prepared by impregnation of ceria-zirconia with aqueous solutions of tin chloride, sodium nitrate, ammonium heptamolybdate and platinum chloride; the resulting catalysts were characterized by XRD, Raman and B.E.T. method, while the WGS experiments were performed at atmospheric pressure, in the temperature range of 423-673 K, with a reacting mixture of 5% CO, 25% H_2O and N_2 balance (500 Ncm^3/min).

References

[1] Panagiotopoulou P., Papavasiliou J., Avgouropoulos G., Ioannides T., Kondarides D.I., 2007, Water-gas shift activity of doped Pt/CeO_2 catalysts, Chemical Engineering Journal 134, 16-22.

[2] Palma V., Pisano D., Martino M., Ricca A., Ciambelli p., 2014, Comparative Studies of Low Temperature Water Gas Shift Reaction over Platinum Based Catalysts, Chemical Engineering Transactions 39, 31-36.

Assisted heat transfer for increased Total Site Heat Integration

T.G. Walmsley[1], M.J. Atkins[1], A.H. Tarighaleslami[1], P.Y. Liew[2]
[1]Energy Research Centre, School of Engineering, University of Waikato, Private Bag 3105, Hamilton, New Zealand; tel. +647 8384937, email: timgw@waikato.ac.nz; [2] Environmental and Green Technology, Malaysia-Japan International Institute of Technology (MJIIT), Universiti Teknologi Malaysia, 54100 Kuala Lumpur, Malaysia

Total Site Heat Integration (TSHI) provides a valuable framework for practical integration of industrial clusters and Locally Integrated Energy Systems (LIES). In general, the method prioritises intra-process heat integration before searching for inter-plant heat integration opportunities via the utility system. The Total Site (TS) method has five main steps: (1) process and stream data specification, (2) process level Pinch Analysis, (3) extraction of excess heating and cooling loads from process Grand Composite Curves, (4) Total Site Profile (TSP) composition, and (5) TS utility consumptions targeting. This paper focuses on step 3 and, specifically, on the removal internal heat recovery pockets, which may be exploited to assist TS heat recovery.

The literature explains the concept of *assisted heat transfer* for TSHI. Bagajewicz and Rodera (2000) first introduced the idea of assisted heat transfer in the context of directly integrating two processes. Their analysis showed that heat integration between two (or more) processes should chiefly occur between the processes' Pinch Temperatures. However, in some presented cases, maximum heat recovery was only achieved by utilising a process heat recovery pocket. Later, Bandyopadhyay et al. (2010) revisited the idea of assisted heat transfer in a more generic TSHI fashion using the utility system for inter-process heat integration. Their study recommended that GCCs are shifted to the utility temperature scale before segments of heat recovery pockets are removed. Segments of pockets on the right hand side of a new intersection point within the pocket were marked for intra-process heat recovery (i.e. removed), while the left hand segments continued to form part of the TSP. Bandyopadhyay et al. (2010) demonstrated that for one example case the modified TSHI method increased TSHI. However, as will be shown in this paper, such an increase in TSHI is not a generally applicable result.

This study investigates how assisted heat transfer can generally apply to increasing TSHI. Analysis results show that assisted heat transfer increases TSHI only when a process heat recovery pocket spans the TS Pinch Region. The maximum increased TSHI can be targeted by comparing each heat recovery pocket to the Site Utility Grand Composite Curve (SUGCC) using background/foreground analysis. Where heat recovery pockets span two steam pressure levels away from the TS Pinch Region, the analysis shows the potential for *assisted shaft work production*. In this case, the source segment of the heat recovery pocket generates steam, which can be utilised in existing steam turbines to extract work. After exiting the turbine, the new steam pressure and temperature should still be sufficiently hot to provide heating to the sink segment of the heat recovery pocket. If a heat recovery pocket falls outside these two situations (assuming direct inter-process integration is disallowed), the entire pockets should be recovered internal to a process.

References

Bagajewicz, M. & Rodera, H., 2000. Energy savings in the total site heat integration across many plants. *Computers & Chemical Engineering*, 24(2–7), pp.1237–1242.

Bandyopadhyay, S., Varghese, J. & Bansal, V., 2010. Targeting for cogeneration potential through total site integration. *Applied Thermal Engineering*, 30(1), pp.6–14.

System synthesis by maximizing sustainability net present value

David Širovnik, [1]Žan Zore, [1]Lidija Čuček, [1]Zorka Novak Pintarič, [1]Zdravko Kravanja
Ulica Anice Kolarič 5, Ptuj, Slovenia, email: david.sirovnik@me.com; [1]Faculty of Chemistry and Chemical Engineering, University of Maribor, Smetanova ulica 17, Maribor, Slovenia

This contribution presents a new concept for multi-objective systems synthesis which is based on a composite sustainability measurement called sustainability net present value (*SNPV*). This is an extension of the recently developed metric sustainability profit (Zore et al., 2016). *SNPV* is a metric composed of economic, environmental and social efficiency, expressed in monetary terms and is defined from the wider macro-economic perspective as well as from an individual company's micro-economic view. By using *SNPV*, it is possible to obtain answer regarding the advisability of a particular investment in terms of sustainability. The presented concept is illustrated by two examples of biofuel supply networks: i) a company's supply network for existing biogas production, and ii) a larger-scale biofuel supply network. Both burdening and unburdening alternatives were considered in the comparative study. Maximal *SNPV* occurs at the appropriate trade-off between economic, environmental and social aspects and yields a solution with alternatives selected in order to provide the highest economic profitability, environmental unburdening and new job opportunities, considering the time value of the money involved.

Acknowledgement

The authors acknowledge financial support from the Slovenian Research Agency (programs P2-0032 and P2-0377, project L2-7633 and PhD research fellowship contract No. 1000-14-0552, activity code 37498) and from the SCOPES joint research project CAPE-EWWR 'Computer Aided Process Engineering Applied to Energy, Water and Waste Reduction During Process Design and Operation'.

Sustainability assessment of chlor-alkali/PVC production chain using system dynamics

X. P. Jia, X. Gan, L. Fan, Z.W. Li, F. Wang
Qingdao University of Science and Technology, School of Environment and Safety, Zhengzhou Rd. 53, 266042 Qingdao, Shandong, China;
tel. +86 53284022617, e-mail: jiaxp@qust.edu.cn.

Sustainability for process industries is an important issue to resolve its high energy consumption, high emission and resource dependency problems. From the perspective of production chains, it is of great significance to investigate the relationships among the environmental, energy, and economic aspects. The aim of this work is to assess the sustainability of the chlor-alkali/PVC production chain.

China is the largest manufacturer and consumer of PVC in the world. China captures about 30% global PVC output and consumption. China has abundant calcium carbide and coal resource, PVC by calcium carbide route captures about 80% of the total output. The chlor-alkali/PVC sector is one of the most energy-intensive and pollution-intensive sectors. The chlor-alkali/PVC production chain of 'coal - coke - calcium carbide - acetylene -polyvinyl chloride-cement' is presented as the case study. The sustainability for various scenarios of production chains using system dynamics is investigated.

First, the "Driving force-Pressure-State-Impact-Response" (DPSIR) framework of the production chain is proposed. Focusing on environmental, energy, and economic aspects, the assessment index system of sustainability is established and each index is determined. After a better understanding of the mechanism of energy consumption, environmental impact, and economic aspects, the corresponding causality loops and system dynamics models are proposed to investigate the interactions of three subsystems.

Five scenarios for long-term goals of production chain are simulated using system dynamics, including base case, resource recovery, energy saving, structural emission reduction, and sustainable production scenarios. The results showed that the sustainable production scenario has the priority to achieve the sustainability of production chains. Compared with the base case scenario, energy consumption per ten thousand yuan of output value of sustainable production scenario is reduced from 78.3 kg coal equivalent to 31.9 kg coal equivalent, COD emission per ten thousand yuan from 0.20 kg to 0.05 kg. This methodology can be used as a decision supporting tool for design, planning, and management of production chains in process industries.

Process design and environmental assessment of plasma-assisted nitrogen fixation incorporating renewables

A. Anastasopoulou [1], S. Butala [1], J. Lang [2], V. Hessel [1], Q. Wang [1]
[1] Department of Chemical Engineering & Chemistry, Eindhoven University of Technology, The Netherlands
[2] Innovation Management, Verfahrenstechnik & Engineering, Evonik Technology & Infrastructure GmbH, Germany

Introduction

Considering the ever increasing environmental concerns about the long-term sustainability of established chemical processes, industrial nitrogen fixation is likely to face great challenges due to its inherent high CO_2 emissions and energy consumption[1]. Based on these facts, it becomes a necessity to embrace a holistic approach towards the advancement of sustainable technologies for the synthesis of fixed nitrogen. Plasma technology has demonstrated promising potential towards the synthesis of nitric oxide and ammonia[2][3]. However, meager attention has been placed on providing a commercial scale process design for the plasma-assisted nitrogen fixation and assessing its environmental performance. This knowledge gap is addressed by this research work which presents an ex-ante design of the nitric acid synthesis process assisted by plasma technology and a life cycle assessment (LCA) study considering the implementation of renewable electricity generation.

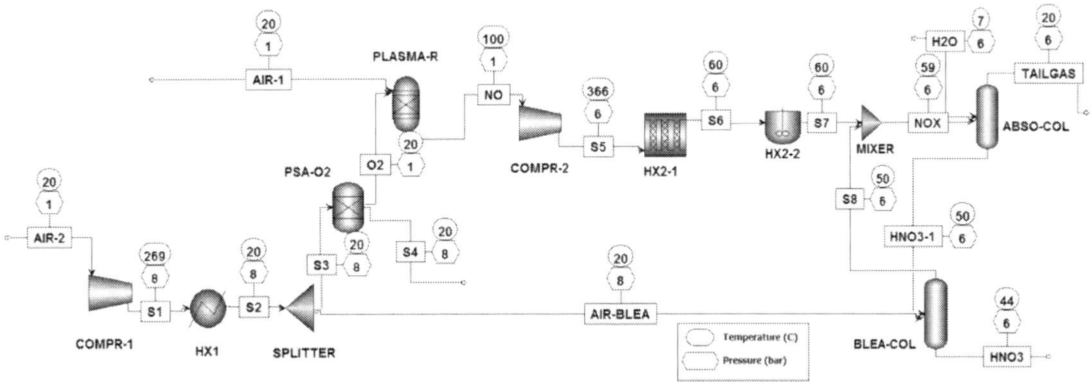

Figure 1:Process flow-sheet of plasma-assisted HNO_3 synthesis

Methodology

As shown in Figure 1, the plasma-assisted nitric acid (HNO_3) production involves: 1) an air separation unit for pure oxygen production 2) a gliding arc plasma reactor where nitric oxide (NO) is synthesized 3) a compression system 4) a system of two heat exchangers where NO gas stream is cooled down and converted simultaneously into nitric dioxide (NO_2) and dinitrogen tetroxide (N_2O_4) and e) an absorption and bleaching column where nitrogen oxides (NO_x) are absorbed by water and bleached with air to produce the final weak HNO_3 solution (60% w.w).

In order to carry out the LCA study, the necessary inventory data has been first obtained from process simulations being conducted in Aspen Plus Software tool with the aid of certain literature data. Different design options based on the NO yield, the power consumption of the plasma reactor, the recycle of the tail gas and the energy recovery in the plasma reactor have been considered and evaluated against certain environmental impact factors.

Results

The base case scenario, employing the design depicted in Figure 1, considers 6% NO yield and 7.7 kWh/kg NO. For the given case the Global Warming Potential (GWP) for the plasma-assisted nitric acid synthesis incorporating different electricity mixes and the conventional production route is depicted in Figure 2. The GWP distribution has been reflected based on three categories: a) the feed which includes the contribution of all upstream activities prior to plasma reactor b) the reaction which involves the impact of the plasma reactor and c) the separation which comprises all downstream activities and their corresponding impact.

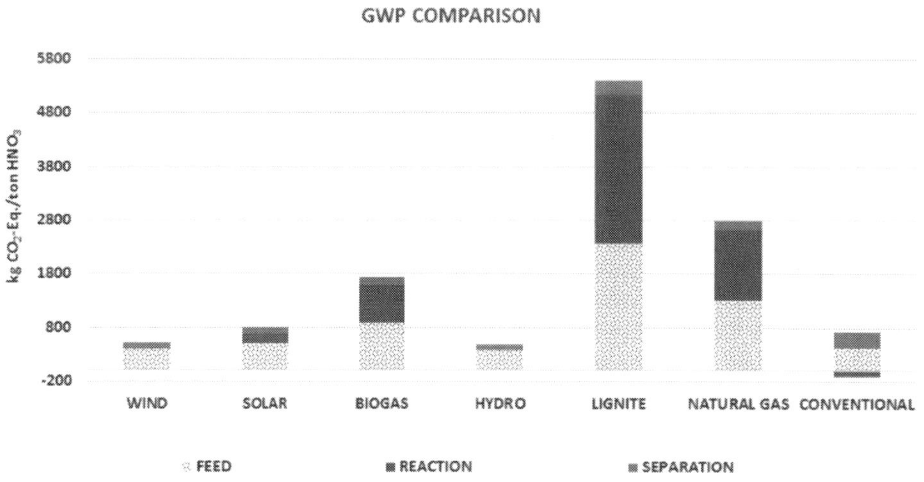

Figure 2. GWP comparison of the plasma - assisted nitric acid process using different energy sources against the conventional process

As it can be deduced, the use of fossil fuels, and more precisely, lignite as a mean of electricity production can impair the GWP of the plasma process by a factor of 9 as compared to the conventional route. On the other hand, the incorporation of renewables-wind, hydro, solar energy and biogas-decrease dramatically the value of the GWP by 68% - 91% as to the lignite electricity form. More precisely, for the wind and hydro energy CO_2 emissions are lower by 15% and 23%, correspondingly, as to the conventional production pathway. Upon such environmental profile, plasma process seems to be a promising sustainable alternative to the synthesis of fixed nitrogen.

References

[1] European Commission, Integrated Pollution Prevention and Control-Reference Document on Best Available Techniques for the Manufacture of Large Volume Inorganic Chemicals - Ammonia , Acids and Fertilisers, I (2007).

[2] A. Anastasopoulou, Q. Wang, V. Hessel, J. Lang, Energy Considerations for Plasma-Assisted N-Fixation Reactions, Processes. 2 (2014) 694–710. doi:10.3390/pr2040694.

[3] B.S. Patil, J.R. Palau, V. Hessel, J. Lang, Q. Wang, Plasma Nitrogen Oxides Synthesis in a Milli-Scale Gliding Arc Reactor: Investigating the Electrical and Process Parameters, Plasma Chem. Plasma Process. (2015) 1–17.

Comparative analysis of the potential environmental impacts generated by the sulphur production plant in a refinery

E. Correa[1], H. Escalante[1], P. Gauthier-Maradei[1], P. Acevedo[2], A. Santis[2]
[1]Universidad Industrial de Santander, Dep. Chemical Engineering, Carrera 27 Calle 9, Bucaramanga, Colombia; tel. +57 (7)344000, e-mail: escala@uis.edu.co;
[2]Universidad Cooperativa de Colombia, Dep. Industrial Engineering, Avenida Caracas No. 37-63, Bogotá, Colombia; Tel +57 (4)4446065, e-mail: paola.acevedop@ucc.edu.co; angelica.santisn@ucc.edu.co

The knowledge and quantification of environmental impacts generated by a production process in a short, medium, and long term, are fundamental aspects to all organizations as an early indication that allows control possible effects of an environmental damage. While it is true that the improvement or strategic investment focus was directed in the past toward the increase of the installed capacity of plants, environmental and social dimensions began to be enacted since the 90s with the Kyoto Protocol. This new situation has caused that governments exercise a greater environmental control over productive sectors, with a more robust environmental legislation and, of course, aligned within the macro trend worldwide.
The sulfur production plant of the refinery studied, is well-know for this environmental requirement. This plant is a source of local air emissions with key components such as sulfur dioxide (SO_2), precursor of "acid rain", and carbon dioxide (CO_2), considered as a greenhouse gas. Recent analytical characterizations of gases emitted by the plant to the atmosphere (580-1600 ppm versus 550 ppm SO_2 admitted by the Colombian legislation), have identified opportunities on technical and operational levels. These opportunities have to do with the improving of the process control, the implementation of online analytical instruments, and to ensure the provision of a stream of fresh air for the oxygen reaction.
Being aware of this problem, the refinery has been proposed within the framework of corporate social responsibility for 2017 to implement a macro environmental project that significantly reduces emissions in the sulfur plant. The project basically involves the design and construction of a new section called the TGTU process (Tail Gas Treatment Unit). This new unit uses surplus gases with the above-described amounts of SO_2 that undergoes the chemical conversion process by hydrogenation to the H_2S gas. Then the H_2S gas is injected to the sulfur production process (Claus process) in order to improve the efficiency of sulfur recovery.
This paper makes an approach to quantify potential environmental impacts induced by the existing and improved process (TGTU) through the methodology of life-cycle assessment (LCA). The LCA is defined by ISO 14040 at the end of the 90s. The results are additional valid arguments within the decision-making criteria about the project and the organization in a general. The LCA's results were estimated by the software SimaPRO from the actual plant data for 2015 and the data process simulation (HYSYS v8.4) for the existing plant and the new section (TGTU). Finally the source of information to develop the model of TGTU was supplied by the licensor of the technology.
Four scenarios were evaluated in total:
Scenario 1- Assessment of current sulfur plant: This scenario seeks to determine the environmental effects of the plant operating with two different production schemes (23 TMD-69TMD). In this scenario no changes were made in the design of the plant, except to enrich the air stream to reaction (O_2) with a specialized gas supplier.

Scenario 2- Assessment of current sulfur plant (without increasing O_2): This scenario raises the same premises as the scenario 1, except that the air to reaction is not enriched with oxygen for the operation at maximum output (69 TMD).
Scenario 3- Assessment of sulfur plant with the built-in TGTU (current production): In this scenario the comparative environmental impacts between the operation of the existing plant and the same plant integrated with the new TGTU section under a scheme of current plant production are evaluated (23 TMD).
Scenario 4- Assessment of sulfur plant with the built-in TGTU (future production): The last scenario, it raises the evaluation of the comparative environmental impact in a scheme of maximum production (69 TMD), with the operation of the existing sulfur plant versus integrated plant with the new TGTU section.

The Table 1 summarizes the results of the LCA for the studied processes. It's possible to identify which implementations of a system including the oxygen enrichment of the air to reaction generates an important 43% reduction in the environmental impacts. It is also noted that to incorporate the TGTU unit into the Claus process represents a further significant reduction (75%) and the same process integrated with the operation at maximum load induces a reduction of up to 80% environmental impacts. The TGTU technology eliminates practically the categories of acidification and respiratory effects, which are generated by acid components of the atmosphere.

Table 1. Comparative summary of scenarios in relation to the current base case

Impact Categories (μPt)	Sulfur Plant	Future Sulfur Plant	% Reduction	Sulfur Plant TGTU	% Reduction	Sulfur Plant TGTU Future	% Reduction
Respiratory Inorganic	574	377	34%	14	98%	6,6	98%
Terrestrial Acidification	5,70E+00	3,70E+00	35%	3,00E-01	95%	0,14	96%
Global Warming	280	113	60%	165	41%	78	31%
Non-renewable energy	31,3	14,3	54%	39,4	-26%	18	-26%
Total	891	508	43%	219	75%	103	80%

As conclusions of the realized work is had that:

- The refinery should incorporate the proposed increase of the concentration of oxygen in the air to reaction of the Claus process, as a measure to obtain significant benefits in the reduction of environmental impacts, in addition to being the basic premise to increase the production of the unit.
- The addition of the new section TGTU to the Claus process allowed to demonstrate significant positive effects in the estimation of the overall environmental impacts, with reductions of up to 75.5% -80%, eliminating practically the categories associated with the acidification of the air resource (respiratory effects and terrestrial acidification).
- The results allow the refinery to justify the implementation of the technology TGTU integrated to the Claus process, not only as an alternative of compliance in environmental legal regulations (emissions from stationary sources), but also as an alternative reduction of the potential impacts that these processes induce on the environment and particularly on the air resource.

Investigation of H_2S and CO_2 removal from gas streams using hollow fibre membrane gas–liquid contactors

Mojtaba Mirfendereski, [1]Toraj Mohammadi
Faculty of Mechanical Engineering and Energy, Shahid Beheshti University, Tehran, Iran; tel. +98 21 73 93 2669, email: m_mirfendereski@sbu.ac.ir
[1]Research Centre for Membrane Separation processes, Faculty of Chemical Engineering, Iran University of Science and Technology, Narmak, Tehran, Iran; tel. +98 21 77240051, email: torajmohammadi@iust.ac.ir

In this work chemical absorption of H_2S and CO_2 from gas streams was carried out in a polypropylene porous asymmetric hollow fibre membrane contactor (HFMC). An aqueous solution containing 0.5 M methyldiethanolamine (MDEA) were used as chemical absorbent solution.
Effects of gas flow rate (360, 420 and 480 nlit/h), liquid flow rate (0.5, 1.5 and 2.5 lit/h), H_2S content (5000, 15000 and 25000 ppm) and CO_2 content (4, 12 and 20 vol.%) on the H_2S and CO_2 outlet concentrations and the separation selectivity were investigated.
The results showed that the removal of H_2S with aqueous solution of MDEA was very high and indicated almost total removal of H_2S. Experimental results also indicated that the membrane contactor was very efficient in the removal of trace H_2S at high gas/liquid flow ratio. The removal of H_2S was almost complete with recovery of higher than 96%.
The selectivity for H_2S was about four order of magnitude higher than that in conventional-packed towers. Using feed gas mixtures containing 5000 ppm H_2S with CO_2 concentrations in the range of 4–20 vol.%, the outlet H_2S concentration of less than 2.0 ppm was attained with less than 2.0 vol.% of CO_2 permeated and absorbed.

Critical reflections on the Chemical Leasing concept

Rodrigo Lozano[a,b,*], Angela Carpenter[b,c], Francisco J. Lozano[d]
[a] Copernicus Institute of Sustainable Development, Utrecht University, The Netherlands [b] Organisational Sustainability, Ltd., Cardiff, United Kingdom [c] School of Earth and Environment, University of Leeds, United Kingdom [d] Tecnológico de Monterrey, Monterrey Campus, Mexico

Chemical Leasing has been developed as a collaborative business model to complement the two main approaches (policy initiatives and scientific/technological) used to foster green chemistry and sustainable chemistry. Chemical Leasing is based on using chemicals more efficiently, reducing waste, and closing the feedback loop more effectively. This is done by shifting the focus away from profit generation, through increased sales, towards a value-added approach by providing a service. Most of the available Chemical Leasing studies have been empirical and descriptive, and generally identify only short-term benefits from process change. This paper provides critical reflections on the Chemical Leasing model based on types of chemicals, green and sustainable chemistry, business models, collaboration, and the chemical leasing cases available. Chemical Leasing offers a more efficient business model alternative to traditional industry practice, bringing economic and environmental benefits to both suppliers and users; however, its use is restricted to some specific types of chemicals (such as solvents and catalysts). The paper proposes a clearer and more precise definition of chemical leasing and argues that chemical leasing needs to be part of a holistic approach, so that the economic, environmental, social, and time dimensions of sustainability are fully addressed.

Total site integration of light hydrocarbons separation process

L. Ulyev[a], M. Vasilyev*[a], A. Maatouk[a], Neven Duic[b] Alisher Khusanov[c]

[a]National Technical University "Kharkiv Polytechnic Institute", 21 Frunze St., 61002 Kharkiv, Ukraine, e-mail: leonid.ulyev@gmail.com
[c]Department of Energy, Power Engineering and Environment, Faculty of Mechanical Engineering and Naval Architecture University of Zagreb, Zagreb, Ivana Lucica 5, 10002 Zagreb, Croatia, e-mail: neven.duic@fsb.hr
[b]M.Auezov South Kazakhstan State University, Tauke hana Avenue, 160012 Shymkent, Kazakhstan, e-mail: husanov@inbox.ru

Ukraine is the largest consumer of hydrocarbons per unit of production in Europe (Ukraine policy review, 2006). The most important point is a reduction of energy consumption in chemical and metallurgical industries as a biggest consumer. This paper deals with energy savings potential of light hydrocarbons separation process. Energy consumption of light hydrocarbons separation process processes typical of Eastern European countries were analysed. Process Integration (PI) was used to perform a preliminary analysis of different units and fulfil the retrofit project of Total Site Integration (TSI). A new heat exchangers network (HEN) were developed and the equipment was calculated with use of the Pinch principles. A complex method was developed and applied to integrate several units at the enterprise site demonstrating the possibility to use heat pumps. Heat pump integration increases heat recovery and offers a solution in order to increase energy savings and project profitability. The estimated payback period for integration Heat Pump of Gas Separation Enterprise is about 127 days and the pathways of plant modernization have been also proposed.

Simultaneous optimization of short term scheduling and heat integration schemes for multipurpose batch plants

L. Sun, X. Zou, H. G. Dong, S. M. Wang
Dalian University of Technology, School of Chemical Engineering, NO. 2 Linggong Road, 116024, Dalian, P.R. China; tel. +86 84986068, e-mail: hgdong@dlut.edu.cn;

Batch processes have been widely used in the production of food, pharmaceutical, fine chemical, biochemical reagents and agricultural chemicals due to its intrinsic flexibility and adaptability. However, energy savings in batch plants were neglected in the past compared to its continous countpart. Fortunately, the emphasis on process sustainability have incentivized academics and industries to develop different energy reduction methodologies and technologies for batch processes in the past two decades, which haved been reviewed by Fernández et al.[1] (2012). To cope with the time-dependent existence of process hot/cold streams which represent the huge challenge of heat integration in batch plants, so most of pinch-based and mathematical approaches published in the early stage explored opportunities for heat integration under predefined production schedule[2], which always lead to suboptimal configurations. Gradually, simultaneous consideration of production scheduling and heat recovery opportunity becomes an attractive subject in batch plants.[3-6]

Barbosa-Póvoa A P F D[3] studied the economic savings in utility requirements while considering the possibility of having direct plant heat integration with associated costs of the involved auxiliary equipment. Majozi[4] proposed a continuous time framework to determine the production schedule that is concomitant with maximum process-process heat transfer and maximum profit. While the continuous Resource-Task Network (RTN) formulation was extended by Chen and Chang[5] to address direct heat integration between two tasks. However, these researches all only consider the potential heat integration candidate between unit to unit (task to task). Recently, Lee and Majozi[6] explored heat integration between streams when they were transferred from a processing unit to another which achieved added benefit of shortened processing time compared to heating and cooling in situ. However, the aforementioned researches ignored heat integration between unit and streams.

As shown in Fig. 1, the potential heat integration schemes for batch plants includes exothermic and endothermic reactor (unit) pairs, hot-cold stream pairs, as well as reactor (unit)-stream, which have not been considerd fully in previous literatures. A new concept of associated tasks has been proposed to describe the heating or cooling task produced by endothermic or exothermic reaction and both the processing tasks i and the associated task k simultaneously exsist in the same period in a reactor where temperature needs to be controled. The batch scheduling formulation is extended from Majozi[4], which is applicable to both multiproduct and multipurpose batch processes. Thus a MINLP model is formulated based on an improved STN shown in Fig. 2 with the objective of maximizing the total profit. At last, A literature example and a case study are used to verify the feasibility and superiority of the methodology.

References

[1] Fernández I, Renedo C J, Pérez S F, et al. A review: Energy recovery in batch processes[J]. Renewable and Sustainable Energy Reviews, 2012, 16(4): 2260-2277.
[2] Vaselenak J A, Grossmann I E, Westerberg A W. Heat integration in batch processing[J]. Industrial & Engineering Chemistry Process Design and Development, 1986, 25(2): 357-366.
[3] Barbosa-Póvoa A P F D, Pinto T, Novais A Q. Optimal design of heat-integrated

multipurpose batch facilities: a mixed-integer mathematical formulation[J]. Computers & Chemical Engineering, 2001, 25(4): 547-559.
[4] Majozi T. Heat integration of multipurpose batch plants using a continuous-time framework[J]. Applied Thermal Engineering, 2006, 26(13): 1369-1377.
[5] Chen C L, Chang C Y. A resource-task network approach for optimal short-term/periodic scheduling and heat integration in multipurpose batch plants[J]. Applied Thermal Engineering, 2009, 29(5): 1195-1208.
[6] Lee J Y, Seid E R, Majozi T. Heat integration of intermittently available continuous streams in multipurpose batch plants[J]. Computers & Chemical Engineering, 2015, 74: 100-114.

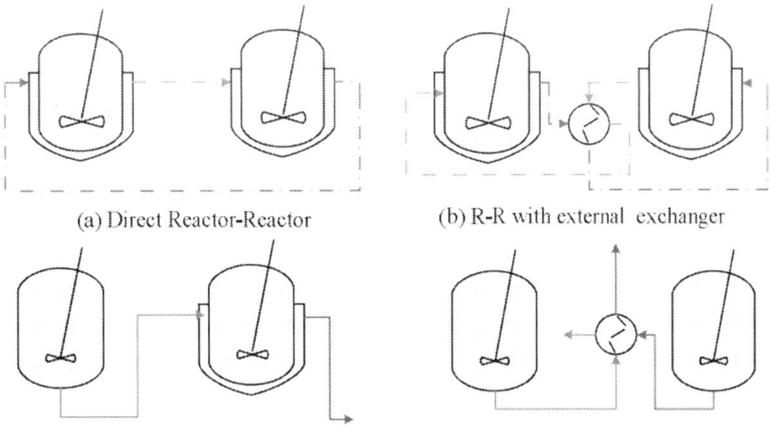

Fig.1 Four types of heat integration schemes for multipurpose batch plants

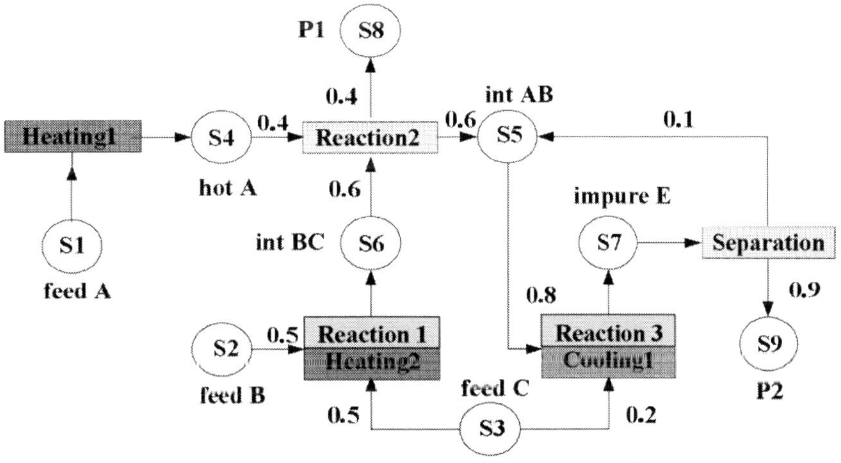

Fig.2 STN involved associated task

A fuzzy association rule mining based approach to identifying SO2 emission concentration patterns of coal-fired boilers

Peng Ran [a], Yase Wang [a], Pei Liu [b], Zheng Li [b].
[a]School of energy & power engineering, North China Electric Power University, Baoding, Hebei Province 071003, China
Tel:+86-13582232866, e-mail: 13582232866@163.com
[b]State Key Lab of Power Systems, Department of Thermal Engineering, Tsinghua University, Beijing 100084, China

Sulphur dioxide emissions control in coal-fired power plants faces significant challenges with increasingly stricter emissions standards. These challenges are further exaggerated by the fact that coal-fired power plants operate strongly irregularly in order to meet the user demand between full-load and part-load conditions, mainly due to requirements from a power grid with the increasing capacity of intermittent renewable power. Acquiring accurate SO2 emissions patterns at these variable loads, thus unstable and unpredictable, are a challenging task. In this paper, we use a fuzzy association rule mining approach, based on real-life operational data, to obtain quantitative SO2 emission concentration patterns of coal-fired boilers. Operational data from a 300MW supercritical power plant over a one-year period are used for analysis purposes. Results show that an optimal operational strategy can be produced by applying the fuzzy association rule mining techniques. A coal-fired power plant, if operated following the optimal operation strategy, can reduce its coal consumption rate by 1.35 g/kWh and SO2 emission concentration by 145 mg/Nm-3, respectively.

Keywords: SO2 emission concentration, coal-fired boilers, operation optimization, energy consumption, fuzzy association rule mining

The potential analysis of renewable sources in the south of Brazil

Osiris Canciglieri Junior*[a], Gerson M. Tiepolo[b], Jair Urbanetz Junior[b,c], Teófilo Miguel de Sousa[d]
[a]GraduateProgram in Production Engineering and Systems – PPGEPS, Pontifical Catholic University of Parana - PUCPR, Imaculada Conceição 115, 80215-901, Curitiba, Paraná, Brazil.
[b]Department of Electrical Engineering, Technological Federal University of Paraná – UTFPR, Sete de Setembro 3165, 80230-901, Curitiba, Paraná, Brazil Program in Production Engineering and Systems – PPGEPS, Pontifical Catholic University of Parana - PUCPR, Imaculada Conceição 115, 80215-901, Curitiba, Paraná, Brazil.
[c]Graduate Program in Civil Engineering, Technological Federal University of Paraná – UTFPR, Sete de Setembro 3165, 80230-901, Curitiba, Paraná, Brazil.
[d]Department of Electrical Engineering, UNESP, Guaratinguetá, SP, Brazil.
*osiris.canciglieri@pucpr.br

One of the key issues for sustainable development is the generation of electricity with more use of renewable sources, and less environmental degradation. Historically the Paraná State of Brazil has been one of the largest producers of electricity in the country, almost entirely from hydropower. However the use of this source in Paraná State is declining due to depletion of water potential and also the pressure of society with respect to environmental, social and economic impacts caused by damming of rivers and flooding of cities and areas to form large reservoirs, hindering increasing its expansion in the electrical matrix. To overcome these limitations, other sources have been researched such as wind, biomass and photovoltaic. The improvement and application of new technologies for generating electricity using an integrated and distributed system may result in a gain on a large scale in the environmental, social and financial aspects. In this context, this paper aims to present an analysis of the potential for electricity generation from renewable sources in the State of Parana - Brazil. The results of this analysis may contribute to development of policies for incentives for research and development projects for the dissemination of renewable sources in the State of Paraná.
Keywords: Renewable Sources, electricity potential, photovoltaic, environment degradation.

Robust model predictive control of heat exchangers in series

J. Oravec, M. Bakošová, A. Mészáros
Slovak University of Technology in Bratislava, Faculty of Chemical and Food Technology, Institute of Information Engineering, Automation and Mathematics, Radlinského 9, SK-812 37, Bratislava, Slovakia;
tel. +421 259 325 364, e-mail: juraj.oravec@stuba.sk.

Recently, more attention has been paid to saving energy and alleviating environmental problems. As the energy prices increase, the energy savings are very important in industry with energy demanding production. To solve this problem, the advanced control strategies have been intensively developed, that can be used in more complicated control problems to improve the process control and to save energy

This study investigates the use of so-called robust model based predictive control (MPC) algorithms for optimal operating of the heat exchangers in series from the stability and economic viewpoints. For the advanced controller design, the influence of uncertain process parameters was taken into account. In order to design the robust MPC, the optimization problem with constraints was formulated in the form of the linear matrix inequalities and then the convex optimization problem was solved using the semidefinite programming. The designed robust MPC strategies were based on the worst-case optimization and on the additional control input saturation.

We investigated a case study devoted to control of heat exchangers in series with two various significant disturbances in the input temperature of the heated medium. Results revealed that the robust MPC improved the control performance and ensured the energy saving during the heat exchangers operation. Although the input temperature of the heated medium varied from 14 °C to 27°C, robust MPC-based methods ensured the more aggressive control action keeping the outlet temperature at the required reference than the well-known LQ optimal control approach. Moreover, the total consumption of the natural gas used for heating was reduced up to 13% in comparison to the LQ optimal control strategy during operation of heat exchangers lasting 1000 s. In practice robust model predictive control may lead to the significant energy savings and reduction the overall costs needed for heating process led in heat exchangers in series.

Acknowledgements
The authors are pleased to acknowledge the financial support of the Scientific Grant Agency VEGA of the Slovak Republic under the grants 1/0973/12 and APVV 0551-11. J. Oravec was also supported by an internal STU grant.

Revamping of heat exchanger network of an Egyptian refinery plant using new temperature driving force (TDF) graphical technique

D. Kamel[1], M. Gadalla[1], F. Ashour[2]

[1]British university in Egypt (BUE), Chemical Eng. Dept., Cairo Ismailia Rd, 11837 El Shrook City; tel.+2 01001819010, e-mail: dina.ahmed@bue.edu.eg; [2]Cairo Univ., Chemical Eng. Dept., Giza.

No one can deny the importance of energy in preserving and developing the level of comfort of the society. The chemical engineering plants, especially crude oil refineries consume huge amount of energy for heating of the raw crude oil before being processed in several units to reach the required product forms as kerosene, gasoline etc.

Revamping or retrofitting of existing chemical engineering plants appears to be an appropriate opportunity for improving and enhancing the existing heat exchanger network's (HEN's) design and performance, revamping is commonly used to modify the existing process for many objectives, such as saving of energy, reducing the environmental emissions and increasing the productivity of the plant. For all these reasons it became important for the researchers to develop and improve the revamping methodologies. Process integration using pinch analysis principals is generally used for revamping of existing heat exchanger networks to maximize the use of the existing hot and cold streams and to minimize the external utilities requirements.

This paper presents a new graphical approach for revamping of existing heat exchanger network, the new graphical approach is based on pinch analysis rules; the HEN is represented on a simple graph, where the cold stream temperatures are plotted on the X-axis while the driving forces for each exchanger are plotted on the Y-axis.

As shown in figure (1), the graph is divided into 5 regions according to Pinch Analysis principals, to easily identify the positions of misallocated coolers, exchangers and heaters

Region 1 is below hot pinch and below cold pinch, therefore optimum region for heat exchangers and cooler.

Region 2 is above the hot pinch and below the cold pinch, therefore non-optimum region for heat exchangers as it opposes the pinch rules (i.e. crossing the pinch).

Region 3 is above the hot pinch and also above the cold pinch, therefore optimum region for heat exchangers and heaters.

Region 4 is below ΔT = Zero line so the presence of any heat exchanger in this region is unfeasible.

Region 5 is below hot pinch and above cold pinch, therefore non-optimum region for heat integration.

This graphical technique describes energy analysis problems in term of the driving force inside the heat exchanger, which is an important factor in the revamping process as the difference in that driving force is involved in the calculation of the area of heat exchangers, and consequently it affects the cost.

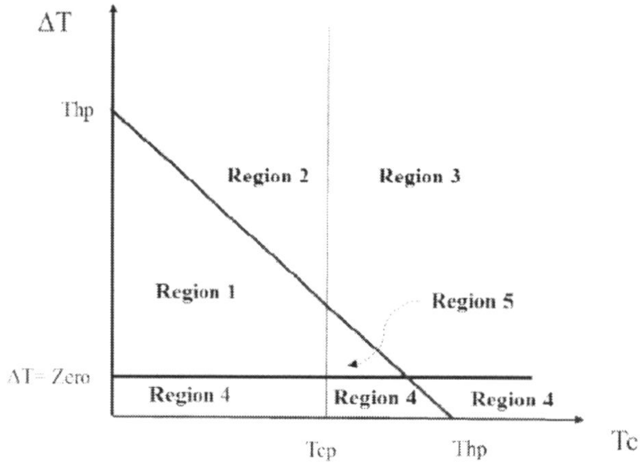

Figure 1: Graphical representation of feasible regions for heat integration

Each exchanger can be represented as a straight line, assuming constant heat capacities within certain temperature ranges, this straight line is drawn between two points, the 1st point is the intersection between the cold supply temperature and the cold end driving force (T_{cs}, ΔT_{ce}) and the second point is the intersection between the cold target temperature and the hot end driving force (T_{ct}, ΔT_{he}) as shown in Figure (2).

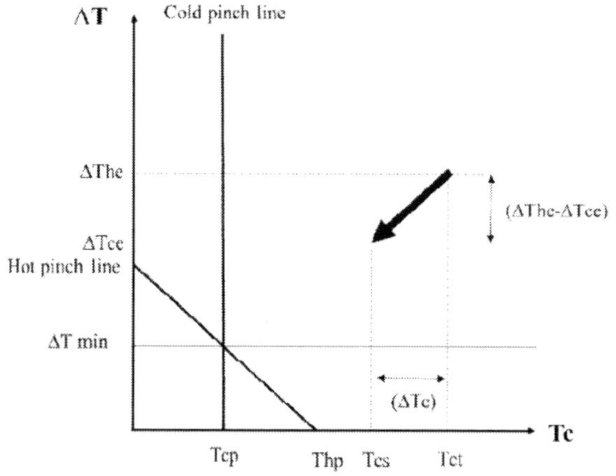

Figure 2: Graphical representation of a single heat exchanger

This graphical approach is applied on an existing HEN in an Egyptian refinery (MIDOR); the raw crude oil is heated in a preheat train of 11 heat exchangers and then enters the furnace, where most of the energy is consumed. The detailed steps of the graphical, conceptual revamping are introduced, and a reduction of 10% of the overall energy consumption is obtained.

Finally, The new technique follows completely the pinch analysis rule, but its advantages is that is it simple as each exchanger can be represented by a straight line, visual and indicative as both the length of the line and its slope has important physical meanings, each exchanger position in the HEN provides its relative area, also the feasible regions for heat integration are easily identified visually.

Warm syngas clean-up processes applied in Synthetic Natural Gas (SNG) production with coal and biomass blending

Po-Chuang Chen, Hsiu-Mei Chiu, Yau-Pin Chyou, [1] Karel Svoboda
Institute of Nuclear Energy Research, Chemistry Division, No. 1000, Wenhua Rd., Jiaan Village, Longtan District, Taoyuan City 32546, Taiwan (R.O.C.).; tel. +886 34711400 x. 5050, e-mail: ypchyou@iner.gov.tw; [1]Institute of chemical process fundamentals of the ASCR, v.v.i., Rozvojová 135, 165 02 Praha 6, Czech Republic.

In Taiwan, the Greenhouse Gas Reduction Act has been issued since 2015, which sets the target to reduce greenhouse gas (GHG) emissions to 50% of the 2005 level by 2050. Natural gas is one of the options to reduce the GHG emissions due to lower CO_2 emission in electricity generation. The value discharged from NGCC (natural gas combined-cycle) is near half of that from coal-fired power plant. The price of natural gas in Taiwan is substantially higher than that of coal, which results in the situation that near 50% of electricity is generated from coal and the capacity factor of NGCC units is relatively low. If cheaper gas fuel could be provided to domestic NGCC units, their capacity factor could be increased; then, the CO_2 emission in power sector could be decreased to help meet the GHG reducing target. Synthetic natural gas (SNG) from solid fuel via gasification is possible to provide a relatively lower price than that of natural gas to NGCC units in Taiwan. The price of SNG from coal has been studied in 2012, and the results showed that the mean price of SNG is US$12.27/MMBtu, while the liquefied natural gas (LNG) is US$14.32/MMBtu [Chang et al., 2012]. It shows the possibility to decrease the CO_2 emission with relatively lower cost of electricity in Taiwan.

The SNG production processes have been built with the commercial chemical process simulator, Pro/II® V8.1.1, to analyse the efficiency improvement with warm gas clean-up processes. The four major blocks, consisted of air separation unit (ASU), gasification island, gas clean-up unit, and methanation processes, were built in a previous study [Chen et al., 2015]. Two different parts in the study, i.e., warm syngas clean-up processes and another kind of biomass that is possibly used in Taiwan, are discussed and compared with the previous study. The warm gas clean-up process is implemented to keep the temperature of syngas in the range of 400°C to increase the available energy which is compared with typical one with lower temperature. A series of sorbents are selected for the processes: e.g., Na_2CO_3-based sorbent to remove HCl and ZnO-based sorbent to deminish sulphur contained in syngas, while MgO-based sorbentfor removal of CO_2 to enhanced the methanation processes. The results show that the efficiency could be increased as warm gas clean-up processes are adopted in the system. Biomass could further reduce the CO_2 emission, due to the advantage of carbon neutral feature. The effect of biomass blended with coal shows the similar trend with previous study, i.e., the system performance is slightly decreased with the blend percentage of biomass.

Reference:
[1] H. W. Chang, C. H. Chen and L. F. Lin, 2012, Techno-economic Analysis on the Use of Substitute Natural Gas (SNG) with Carbon Capture and Storage (CCS), 2012 Taiwan Symposium on Carbon Dioxide Capture, Storage and Utilization, Nov. 25-27, Taiwan.
[2] P. C. Chen, H. M. Chiu, Y. P. Chyou, 2015, Synthetic Natural Gas (SNG) Production via Gasification Process with Blend of Coal and Wood Chip as Feedstock, Chemical Engineering Transactions, Vol. 45, pp. 601-606.

Rotating cylinders for enhanced heat transfer in annealing processes

O. M. H. Fatla [a,b], A.Valera-Medina [a], F. Robinson [c], M. Cichuta [c], Nathan Beynon[c]
[a]School of Physical Sciences and Engineering, Cardiff University, UK, email: FatlaOM@cardiff.ac.uk
[b]Collage of Engineering, Al-Qadisya University, Iraq
[c] COGENT Power, TATA Steels, UK

A particular high temperature annealing process anneals steel coils at 1200°C in a cycle taking several days. This is a batch process in which steel coils are placed under an atmosphere cover over which an electrically heated furnace is used. Heating is performed in a hydrogen atmosphere to protect and chemically remove Sulphur and nitrogen from the steel. The use of fans is impossible due to high temperatures that impact on blade components. By nature of the design of the process and the lack of any convection, the steel and process equipment experience significant temperature differentials resulting in sub optimum properties of the steel and wear / failure of the equipment.

Thus, this project aims to investigate engineering options and challenges to introduce gas circulation into the gap between the heating element and the protective cover thereby exchanging some of the radiant heat transfer to convective heat transfer thereby reducing the temperature differentials in the system and in particular to reduce the local extremely high temperatures which cause the system failure / inefficiencies. Finite volume numerical simulation using ANSYS Fluent has been carried out to examine the effect of one possible solution through the use of rotating cylinders inside the furnaces. The rotating cylinder technique was used to generate recirculation and turbulence inside the furnace cavity, thus improving convection and heat transfer through the entire system. Results demonstrate that fluctuations resulting from the cylinder rotation improve mixing of the fluid and enhance energy transfer.

Process design of acid gas removal units considering various flowsheet configurations and off-shore environments for LNG-FPSO applications

H. Cho, [1]K.-J. Min, J.-K. Kim
Department of Chemical Engineering, Hanyang University, Wangsimni-ro 222, Seongdong-gu, Seoul 133-791, Republic of Korea; tel. +82 (0)2-2220-2331, email: habin88@hanyang.ac.kr, jinkukkim@hanyang.ac.kr;
[1]GS Engineering and Construction, Gran Seoul 33, Jongno-gu, Seoul 110-121, Republic of Korea

The prospects for insufficient energy resource and advancement in process design technologies lead the development of gas processing plants on offshore FPSO (Floating Production Storage & Offloading) platforms. Process technologies of gas processing are well established and widely applied for onshore applications, but, there is still a lack of R&D effort for offshore applications which are influenced by oceanic environment such as ship-motion.

AGRUs (Acid Gas Removal Units) of LNG-FPSO applications consists of two separation columns and under ship-motions, equilibrium is not fully attained, due to the reduction of liquid residence time by weeping and dumping in tray column or liquid mal-distribution in packing column.

Various structural modifications of the column configuration, such as multiple feed or side draw with semi-lean solvent, can be considered in order to improve the separation efficiency and reduce energy requirements. These design options have redistribution effect.

This study aims to assess impacts of ship motions for the performance of AGRUs in LNG-FPSO applications. The results of case studies using commercial process simulators (e.g. AspenPlus®, UniSim® or ProMax®) will be used to demonstrate how oceanic effects can be systematically considered for the process simulation. And in the case study various structural modifications will be evaluated to achieve energy-efficient design of AGRU under oceanic environment.

Acknowledgement
This research was supported by a grant from the GAS Plant R&D Center funded by the Ministry of Land, Infrastructure and Transport (MOLIT) of the Korean government.

INDUS3ES PROJECT- Industrial Energy and Environment Efficiency

H. Sevi,[1] A. Laura, M. Asier
TÜPRAŞ, Turkish Petroleum Refineries Corporation, Petrol Street 25, 41790 Körfez, Kocaeli; tel. +0902623163000, e-mail: sevizeynep.hasdemir@tupras.com.tr[1]
TECNALIA RESEARCH & INNOVATION, Area Anardi 5, 20730 Azpeitia (Spain); tel. +34 946 430 850, e-mail: laura.alonso@tecnalia.com, asier.martinez@tecnalia.com[2]

Large quantities of waste heat are continuously rejected from industries. Most of this waste energy, however, is of low-quality and is not practical or economical to recover it with current technologies. Indus3Es project will develop an innovative Absorption Heat Transformer (AHT) for this purpose, focused on low temperature waste heat recovery especially below 130°C. The Indus3Es System will effectively recover and revalorize almost 50% of the low temperature waste heat, increasing quality of the waste source to the required temperature and reusing it again in the industrial process. The main objective is to develop an economically viable solution for industry, appropriate for existing plants and adaptable to various industrial processes. The developed system will be demonstrated in real environment in TÜPRAŞ, the main refining industry in Turkey, enabling to analyse besides integration aspects, operational and business issues.

TÜPRAŞ is Turkey's largest industrial enterprise with approximately 11 mn ton per year crude processing capacity with a 7.8 Nelson Capacity for İzmit Refinery. Among these complex operations, low grade waste heat management is a complicated issue because utilizing and valuing these sources within operations is difficult due to having low temperature, small capacity and multiplexed waste heat sources. With the purpose of analysing these sources and determining the best recovery with respect to efficiency and profitability, waste heat sources mapping is done by considering heat exchangers using cooling water and air fans for whole refinery. It is determined that for one refinery, there are approximately 250 sources that have duty range between 100-2000 kW. Temperature ranges of these sources are between 70-120°C.

The aim of this study is to determine the real environment application of AHT system within refinery. For this purpose, waste heat sources are analysed and due to availability of sources and suitability of locations, Power Plants are found more suitable for AHT integration. TÜPRAŞ has stated five different possible application cases that have different target and waste heat sources at different temperature levels. All cases are represented in a detailed way in the paper and among these five cases; case 4 is selected as real demonstrator operation. In this case, the steam leaving from the condensate drum which is presently vented to the atmosphere in Power Plant is used as waste heat source for AHT system and the target case is to increase the temperature of boiler feed water entering to the deaerator to decrease steam consumption rate in the deaerator. The procedure of selecting operation for AHT implementation is also represented; critical issues and main barriers from operational side are also discussed in this study. The preliminary configuration of the AHT integrated in the selected case study is presented, as well as a basic simulation of the cycle for the definition of its operation conditions.

A novel freeze drying process by using self-heat recuperation technology

K Bando, Y Kansha, A Tsutsumi
Collaborative Research Center for Energy Engineering, Institute of Industrial Science, The University of Tokyo, 4-6-1, Komaba, Meguro-ku, Tokyo, Japan;
tel. +81 3 5452 6727, e-mail: a-tsu2mi@iis.u-tokyo.ac.jp; fax: +81 3 5454 6728.

Freeze drying technology has been gathering attention for advantage of little loss of taste and aroma. As freezing substances can enhance the rigidity of product, this rigidity to a large extent prevents collapse of the solid matrix remaining, and physical and chemical degradations after drying. Recently, the market of product by freeze drying technology grows larger and larger, because it has come to be used not only for food products but also pharmaceutical products.

However, so far, freeze drying technology has been industrially applied to only high added value products because of large energy consumption, thus it is necessary to save energy. Freeze drying consists of three stages for drying: (1)freezing the water of product, (2)making vacuum, and (3)heating for removing the frozen water as a vapor by sublimation. The required energies for three stages above are the following: (1)amount of sensible heat and latent heat of solidification, cooling, (2)work for vacuum pomp, and (3)amount of latent heat of sublimation, heating. It is reported that these drying stages contain a cooling and a heating processes, and a cooler and a heater works separately. Also these drying stages contain decreasing and recovering pressure process. In these processes, exergy loss of working fluid in drying chanber is occurred by heat emission and pressure release. And that is problem of freeze drying technology.

Then, by introducing the concept of [1]SHR (Self-Heat Recuperation) technology to freeze drying process, the exergy loss can be reduced. SHR is one of exergy recuperation technology, heat are circulated in a feed-effluent heat exchanger of the thermal process by compressing the stream without any heat addition. It is mentioned above that freeze drying technology contains a heater and a cooler in its process. By using SHR technology, the energy consumption of this drying process may be reduced. Additionally, in stage (2) the pressure is decreased by vacuum pomp, and after stage (3) the pressure is increased into ordinary pressure. Work can be recovered in a pressure increasing process, and the work can be recycled for pressure decreasing process. By using work recovery technology, energy consumption of freeze drying technology can be reduced.

In the present paper, we perform process simulations and experiments to estimate the effect of applying SHR technology.

Reference:
1. Yasuki Kansha, Naoki Tsuru, Kazuyoshi Sato, Chihiro Fushimi, Kaoru Shimogawara, Atsushi Tsutsumi, "Self-Heat Recuperation Technology for Energy Saving in Chemical Processes" Industrial and Engineering Chemistry Research 48(16), 7682-7686, 2009

Energy integration studies in gas recovery unit

Parthsarathi Pachori and Amar Nath Samanta
Department of Chemical Engineering, Indian Institute of Technology Kharagpur, Kharagpur 721302, India
amar@che.iitkgp.ernet.in

In this article, heat integration studies are carried out in an industrial gas recovery unit (GRU) consisting of 3 distillations columns, de-ethanizer, depropanizer and debutanizer connected in series. The fundamental approach for heat integration is exchanging heat between the hot process streams and cold process streams. Two major candidates for heat integration are vapor recompression column (VRC) and divided wall column (DWC). Since divided wall column techniques are capital cost intensive, VRC is used to integrate energy in the GRU. In VRC the overhead vapor is compressed to sufficiently high temperature so that required heat can be supplied to the reboiler.

The GRU includes a de-ethanizer, a depropanizer, and a debutanizer with 34, 38 and 28 trays respectively (excluding reboiler and condenser). The feed to de-ethanizer column is given on the 1st tray, the de-ethanizer bottom is fed to 17th tray of depropanizer column and the de-propanizer column bottom is fed to 13th tray of de-ethanizer column. The flowsheet of the base case of GRU is shown in figure 1 and this would be used as a reference to all the heat integrated schemes. The total energy required in this case is 108975.3 MJ/hr with thermodynamic efficiency of de-ethanizer column as 12.63%, depropanizer as 8.26% and debutanizer as 11.47%.

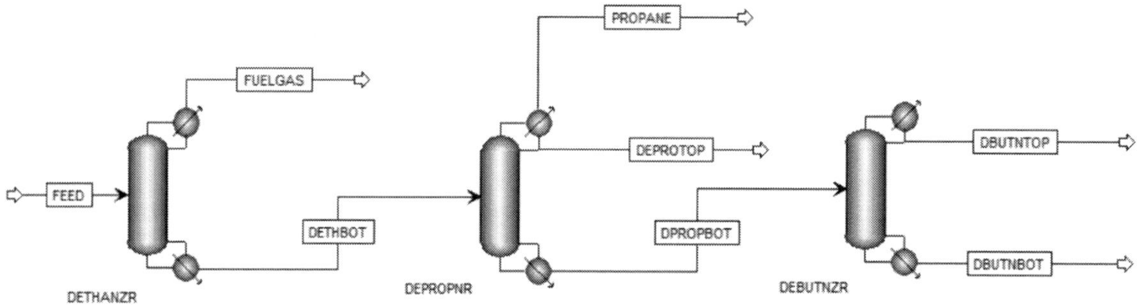

Figure 1. Gas Recovery Unit process diagram

There is a scope of integrating heat in this system using different process hot streams to exchange heat with the process cold streams to reduce the utility load. In view to this, the bottom of all the columns can be used to preheat feed of de-ethanizer column; bottom of second column can be used to preheat its feed or bottom of the third column to preheat the feed of second and third columns. VRC can be applied to all the three columns individually or simultaneously and the equivalent heat can be used as a part of reboiler duty for any of the three columns. In total 27 heat integration schemes, starting from very basic to more complex, are used to find the best scheme for energy savings in GRU. A suitable scheme for energy integration is proposed considering TAC, energy savings and improvement of thermodynamic efficiencies of the columns. A maximum of 28% increase in total energy savings is obtained for the GRU. Energy integration also improves the thermodynamic efficiencies of de-ethanizer from 12.63% to 26.88%. Similar improvements in thermodynamic efficiencies are also observed for de-propanizer and de-butanizer columns. This scheme can also be retrofitted to a conventional gas recovery unit with a payback period of 1.35 years.

Process and utility design and integration for ultra-energy efficient milk powder production

T.G. Walmsley[1], M.J. Atkins[1], J.R. Neale[1], M.R.W. Walmsley[1], M.J.R. Philipp[2], R.H. Peesel[2]
[1]Energy Research Centre, School of Engineering, University of Waikato, Private Bag 3105, Hamilton, New Zealand; tel. +647 8384937, email: timgw@waikato.ac.nz; [2]Universität Kassel, Dep. Umweltgerechte Produkte und Prozesse, Kurt-Wolters-Strasse 3, 34125 Kassel, Germany.

New Zealand produces about 19 billion litres of milk each year. A few percent of the milk production is consumed locally, while the remainder (>95%) is processed into export products, e.g. milk powders, butters, and cheeses. Approximately 75% of New Zealand's milk becomes milk powder for export. Industrial-scale conversion of raw milk into powdered milk is an energy intensive process requiring steam heating, chilling and electrical utilities. The milk powder process has four main processing steps: milk separation, heat treatment, multi-effect evaporation, and spray drying. The two primary dewatering operations – multi-effect evaporation and spray drying – consumes a vast majority of the required energy. In New Zealand, the specific fuel consumption for milk powder factories range from 6,000 to 12,000 MJ/t of powder. The energy use depends on the installed technology, the scale of production, and the boiler fuel efficiency. Given the paramount importance of dairy processing in New Zealand and, particularly, the contribution of milk powder production to export products, reductions in energy and emissions from milk powder processing can have substantial impact on sustainable trade and food processing in New Zealand.

This paper investigates process and utility design and integration pathways for developing an ultra-low energy milk powder plant. To perform the investigation, a model milk powder plant has been implemented in an Excel™ spreadsheet tool and the initial design validated using industrial data. Focus in the study is given to analysing the impact of critical design parameters, new technologies, and novel integrated process designs, on total energy use and emissions. The areas of focus include: (1) new heat treatment/evaporator process design and integration, (2) spray dryer exhaust heat recovery including upgrading, (3) high efficiency poly-generation utility systems, and (4) high concentrate solids (~60%) spray drying. Each of these concepts individually have merit for reducing energy use and emissions. Collectively, the energy and emissions savings are compounded to point where an ultra-low energy plant requires 3,000 MJ/t_p of fuel, equating to an emissions reduction of 50% compared to the current standard industry design. However, it is recognised that the technology for high concentrate solids spray drying still requires further development. Employing the first three strategies lowers fuel use to 3,800 MJ/t_p and a 37% emissions reduction.

Industrial energy supply structures with low primary energy demand and emissions for different countries considering the Energiewende

M. Philipp[1], G. Schumm[1], R. H. Peesel[1], T.G. Walmsley[2], M.J. Atkins[2], J. Hesselbach[1]
[1]Universität Kassel, Dep. Umweltgerechte Produkte und Prozesse, Kurt-Wolters-Straße 3, 34125 Kassel, Germany, tel.: +49561-804-3460, e-mail: philipp@upp-kassel.de
[2]Energy Research Centre, School of Engineering, University of Waikato, Private Bag 3105, Hamilton, New Zealand; tel. +647 8384937, fax: +647 8560115, timgw@waikato.ac.nz

Facing climate change and increasing energy costs, energy efficiency has to be increased at all stages of the energy chain from generation to final consumption. Simultaneously, the advantages of energy efficiency measures have to outweigh their investment costs. For the food processing industry in Germany, the energy costs are already the third biggest impact on the total production costs. Each industry in the different countries of the world has its individual supply structure to convert final energy to useable energy such as steam, process heat, chilled water or cooling water. The impact on costs and environment are highly dependent on the energy source and the utilities used (i.e. Combined Heat and Power (CHP), Heat Pump, Steam boiler). However, the most efficient supply structure in terms of primary energy and CO_2-emissions varies for each country because of the different primary energy factors and emissions for providing electricity.

This study reports the development of an EXCEL™ spreadsheet based tool for analysing the most efficient supply structure in terms of primary energy, carbon emissions and energy costs. The used database is the Global Emissions Model for integrated Systems (GEMIS). The GEMIS database has been used to extract electricity grid emissions factors and energy efficiency ratios.

In the German dairy industry, separate conversion with gas fired steam boiler and cooling with ammonia chillers are the state of the art technologies. It is attractive due to its robustness and low investment costs. The disadvantages are significant heat losses and emissions in the conversion. The other supply scenarios have to compete against this.

The first improvement is done by combined heat and power to provide cogenerated electricity and heat. The cooling is still supplied by ammonia chillers. In countries with high electricity grid emissions factors, the use of decentralised CHP results in savings of primary energy and emissions. In Germany, it is often applied and promoted by engineering consultancies. But the advantages do not occur in all countries and the currently ongoing energy revolution, i.e. Energiewende, has to be taken into account.

High temperature heat pumps are installed in small numbers in the industry. Suitable refrigeration systems are for example two stage ammonia or gas cooled transcritical CO_2 chillers. Speaking in terms of theoretical energy efficiency heat pumps are the best solution because they are closed thermodynamic cycles. This technology seems promising for providing baseloads but it has to be applied correctly in respect to the rules of the Pinch method. The energy and mass balance of the plant must be investigated first to make sure that the hot and cold side, defined by the refrigerant, are also matching the hot and cold region of the Pinch system.

First results show that supply structures such as CHP can be applied in clusters with high emission grid factors, whereas heat pumps are favourable in countries like Norway or France with a large share of nuclear power or renewables. Also, the influence of energy revolution (Energiewende) has a high impact on the efficiency of the conversion paths.

Optimal high-speed passenger transport network with minimal integrated construction and operation energy consumption based on superstructure model

Ruirui Zhang*[a], Pei Liu[a], Chengchuan Zhou[b], Angelo Amorelli[c], Zheng Li[a]
[a]Tsinghua BP Clean Energy Research and Education Centre, Department of Thermal Engineering, Tsinghua University, Beijing 100084, China
[b]Duke University, Nicholas School of the Environment, Box 90328, Durham, NC 27708, United States
[c]BP, Centre of Expertise Applied Chemistry & Physics, Sunbury, UK
NXWHL10@163.com

Economy development and promotion of high-speed transport technologies result in the increase of inter-city high-speed passenger transport. The energy consumption over the lifetime of transport infrastructure, composed of construction and operation energy consumption, is a significant factor that should be considered at the planning stage. In this paper, we present a mixed-integer programming (MIP) model based on superstructure for the energy consumption optimization of inter-city high-speed transport systems, a network design integration of aviation and high-speed railway, accounting both energy consumption during infrastructure construction and energy consumption during subsequent operation processes, to optimize connections between large population centers and connections between modes of transport. Energy consumption during infrastructure construction is obtained related to investment costs using a lifecycle assessment based on the economic input-output method. The shares of construction energy consumption and operation energy consumption during infrastructure lifetime are compared in the case studies of East and Southeast China and the model results show the optimized network structure according to the total energy minimization objective. Optimized transport networks are analyzed in terms of hub-spoke characteristics and city transport functions as sources and sinks. Model results have implications in the energy-saving aspect for actual high-speed railway and civil aviation infrastructure layouts and high-speed transport technology development.

Energy efficiency concepts for batch retort sterilization in the food processing industry

R. H. Peesel[1], M. Philipp[1],G. Schumm[1], J. Hesselbach[1]
[1]Universität Kassel, Dep. Umweltgerechte Produkte und Prozesse, Kurt-Wolters-Strasse 3, 34125 Kassel, Germany, tel.: +49561-804-3460, e-mail: peesel@upp-kassel.de

In 2010, the global industrial sector consumed around 58 PWh. This energy consumption will steadily increase up to 88 PWh in the year 2040. Facing climate change and increasing energy costs, energy efficiency has to be increased at all stages of the energy chain from generation to final consumption. Simultaneously, the advantages of energy efficiency measures have to outweigh their investment costs. For the food processing industry in Germany, energy is the third biggest contributor to total cost. Strong commitment to reduce CO_2-equivalent emissions by the industry is likely required to ensure the achievement of the two-degree-target and to avoid dangerous climate change. The thermal treatment of food with a batch retort sterilization process is one of the main energy consumers in the food processing industry.

Literature review reveals that the present process has been optimized in terms of sterilization temperature, sterilization time, material properties, heating medium and heat transfer coefficient. But, all these optimization actions neglect the energy saving potential of higher initial temperatures of all components included in the process. Also, optimization measures for the cooling process have focused on water recovery. These two neglected areas are the starting point for this study.

The purpose of this study is to highlight the energy saving potentials of a batch retort sterilization process operated by steam and to develop energy efficiency concepts. By using the methodical approach of the onion diagram and the recovery of low grade heat sources, four comprehensive energy efficiency concepts have been developed. Two concepts increase the initial product temperature resulting in lower transient steam consumption rates and a decreased energy demand. The two others transform the waste heat and the energy for product cooling into heat streams for supplying the nearby heat sinks or other processes in the plant.

Results show that the total energy demand can reduce by more than 23 %, and 42 % of the energy injected into the system can be recovered. The scope of the study focused on a single, independent retort to ensure the transferability of the concepts to other case studies. This study also shows, using a practical example, the importance of a comprehensive process understanding and analysis prior to application of Pinch Analysis is needed to ensure that the correct waste heat stream temperatures are defined and all possible heat sinks in the process are considered.

Optimisation of non-isothermal utilities in unified total site targeting method

A.H. Tarighaleslami, T.G. Walmsley, M.J. Atkins, M.R.W. Walmsley, J.R. Neale
Energy Research Centre, School of Engineering, University of Waikato, Private Bag 3105
Hamilton 3240, New Zealand; tel. +64 7 8384937, e-mail: aht5@students.waikato.ac.nz

Total Site Heat Integration is a convenient method for optimal design of large scale utility systems that serve large chemical processes, such as oil and gas refineries, petrochemicals or even lower temperature chemical and process plants. Such systems are required to provide power and heat for processes within the site. However, Total Site Heat Integration of different chemical plants might confront batch, semi-continuous or continuous plants which are clustered into one large site. Excess heat produced from one plant could be transferred to other plants in the cluster using an intermediate fluid which can be isothermal (e.g. steam) or non-isothermal (e.g. hot oil, hot water) fluid. Isothermal fluids address the utilities which its temperature does not change while transferring heat during generation and consumption. On the other hand, non-isothermal utility has significant temperature change during its generation and consumption. As a result, upper and lower temperatures are important considerations for each non-isothermal utility.

In this paper, Total Site optimisation of targeting utility generation and consumption for lower temperature processes, which are mainly use non-isothermal utilities, is presented. The optimisation applies to the recently developed Unified Total Site Targeting method (currently in press). The new method shows that, for low temperature processes where non-isothermal utilities are used, the supply and target temperatures of a utility is an important constraint, while for higher temperature processes, where isothermal utilities are applied, no significant change in targets from conventional Total Site are obtained. Utility supply and target temperatures provide degrees of freedom that may be selected to maximise Total Site heat recovery and/or minimise total operating and capital cost.

Optimal selection of aerobic biological treatment for a petroleum refinery plant

A. Orbecido[2], A. Beltran[2], R. Malenab[2], K. Minano[2], M.Promentilla[1,2]
[1]Center for Engineering and Sustainable Development Research
[2]Chemical Engineering Department
De La Salle University Manila
tel. +6325360223, e-mail: michael.promentilla@dlsu.edu.ph; 2401 Taft Avenue, Malate, Manila, Philippines, 0922

Aerobic biological treatment has been known to be an integral part of a typical wastewater treatment plant that reduces the pollution load of wastewater from either municipality or industry having soluble organic contaminants. For example, the conventional activated sludge process has been widely used in a long time but a variety of biological treatment systems were also being considered in recent years to meet a more stringent discharge standards. In designing such wastewater treatment plant, several criteria have to be taken into account that includes the technical, socio-economic and environmental aspects of the decision problem. This work thus applies a multiple criteria analysis based on Fuzzy AHP for optimal selection of the different aerobic biological treatment technologies. This technique decomposes the decision problem into a hierarchic structure and derives priority weights for the ranking of the alternatives. The decision model also incorporates the ambiguity-type uncertainty when eliciting pairwise comparison judgment from a domain expert. A case study applied to a petroleum refinery plant is presented considering the following alternatives: 1) conventional activated sludge system; 2) cyclic activated sludge system; 3) integrated fixed film activated sludge system; and 4) membrane bioreactor. These wastewater treatment systems were then evaluated and compared with respect to the following criteria: 1) economic criteria such as the capital and operating cost; 2) environmental criteria such as the treated effluent quality, ability to adjust to hydraulic and pollutant loading, ability to cope with oil ingress, and land footprint; 3) technical criteria such as pretreatment and secondary clarifier requirement, reliability and validity of technology, and complexity to operate and control. Sensitivity analysis was also done to understand how the weighting of criteria affects the ranking of the alternatives.

Conceptual design of an autonomous once-through gas-to-liquid process with microchannel Fischer-Tropsch reactors and verified kinetics

M. Ostadi, M. Hillestad[1]
Norwegian University of Science and Technology (NTNU), Department of Chemical Engineering, NO-7491, Trondheim, Norway; [1] corresponding author, e-mail: magne.hillestad@ntnu.no, tel. +4773594122

Converting remote natural gas to liquid fuel is one possible solution to the problem of transporting remote gas to the energy market. However, the high investment cost of gas-to-liquid (GTL) plants prevents large scale exploitation of remote gas reserves. A lean GTL is suggested [1] based on an autothermal reformer with enriched air as oxidant and a once-through Fischer-Tropsch synthesis. In order to maximize the syngas conversion and the production of heavy hydrocarbons, a staged microchannel reactor path with distributed hydrogen feed and product withdraw is proposed. The hydrogen is produced by steam methane reforming in a heat exchange reformer (gas heated reformer), heat integrated with the hot effluent stream from the autothermal reformer. Enriched air is produced with a proven membrane technology and thus cryogenic air separation is avoided. The natural gas is heated by the hot effluent gas from a gas turbine and thus also the fired heater is avoided. The process flow diagram is shown in Figure 1. The process is autonomous in the sense that it is self-sufficient with power and water.

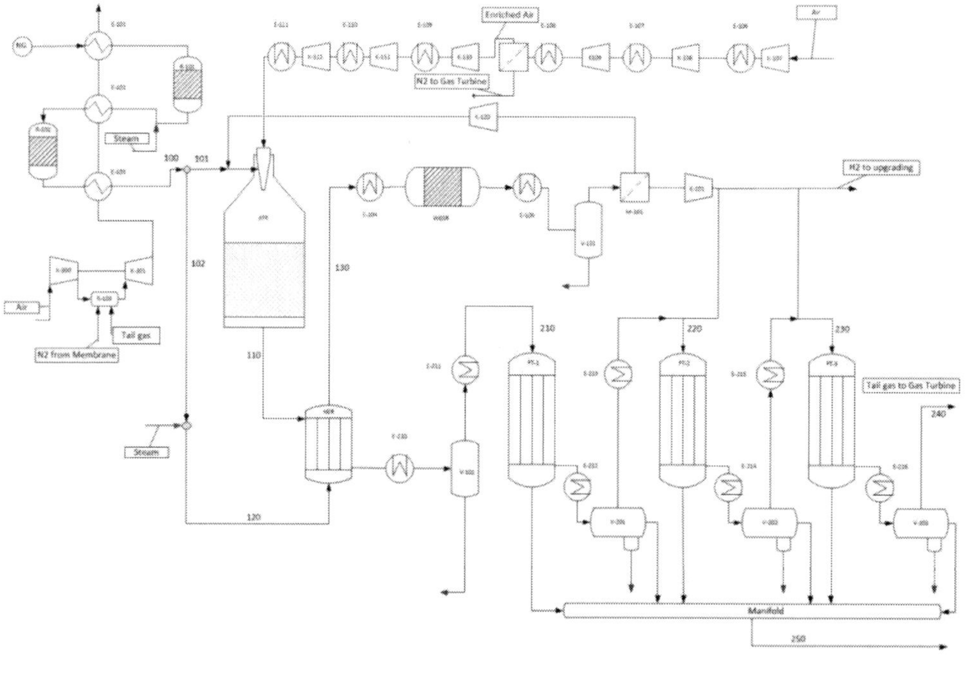

Figure 1: Process flow diagram of the GTL plant.

Here, an alternative kinetic rate model and a model describing the product distribution of the Fischer-Tropsch synthesis are applied [2]. A kinetic rate model, suggested by Ma et al. [3], is verified against data from a microchannel laboratory reactor [4]. In addition, a model describing the product distribution and methane selectivity is developed [2] and fitted to data generated by Yang et al. [4]. The verified kinetic model is implemented in a model of the microchannel reactor with the use of Aspen Custom Modeler (ACM) language. The model is further exported to Aspen Hysys process simulator for simulation and optimization of the entire GTL plant.

The new and verified kinetic model yields higher reaction rates than what was used in our previous investigation [1]. Therefore the carbon and energy efficiencies of the GTL plant are improved. The energy and carbon efficiencies are shown in the Figure below. More importantly, the catalyst volume is only half of what we reported in the previous study and thus the investment cost will be reduced substantially. Since the syngas contains much nitrogen, the cost of the Fischer-Tropsch reactors is the major investment cost. A comparison between our previous investigation [1] and the new highlights the importance of the implemented kinetic model. The new kinetic model along with the product distribution model is specifically suited for microchannel reactor and it gives us a more realistic view of the process in the large scale.

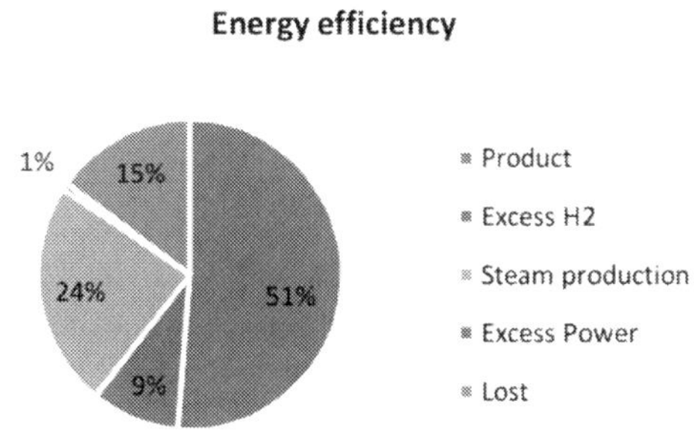

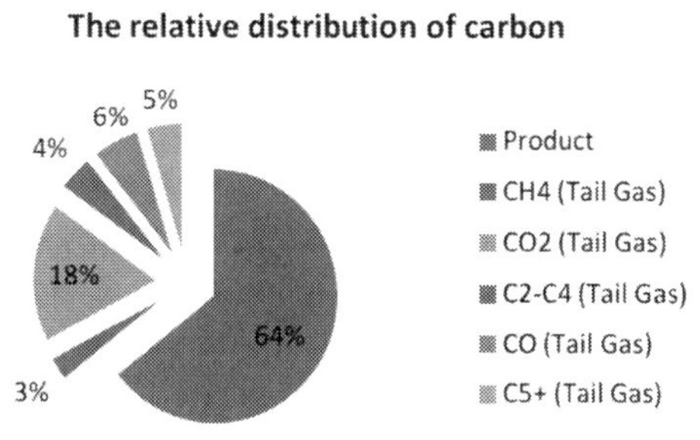

References:

[1] Ostadi M, Dalane K, Rytter E & Hillestad M, Conceptual design of an autonomous once-through gas-to-liquid process - Comparison between fixed bed and microchannel reactors, *Fuel Processing Technology*, Volume **139**, (2015), 186-195.

[2] Ostadi M & Hillestad M, Verification of a macro kinetic model and development of a product distribution model for a commercial Co/Re/γ-Al_2O_3 Fischer-Tropsch catalyst in microchannel reactor, Chisa 2016, Prague.

[3] Ma W, Jacobs G, Sparks DE, Spicer RL, Davis BH, Klettlinger JLS, & Yen CH, Fischer–Tropsch synthesis: Kinetics and water effect study over 25%Co/Al2O3 catalysts, *Catalysis Today*, Volume **228**, (2014), 158-166.

[4] Yang J, Boullosa SE, Myrstad R, Venvik HJ, Pfeifer P & Holmen A, Fischer-Tropsch Synthesis on Co-based Catalysts in a Microchannel Reactor, effect of temperature and pressure on selectivity and stability, *Catalysts, and Catalysis: Advances and Applications*, B. H. Davis, M. L. Occelli, Eds., CRC Press, 2016, p. 223.

Optimal selection of desalination systems using fuzzyAHP and grey relational analysis

R. Eusebio[1,2], A. Orbecido[2], M. Promentilla[1,2]
[1]Center for Engineering and Sustainable Development Research
[2]Chemical Engineering Department
De La Salle University Manila
tel. +632 5360223, e-mail: michael.promentilla@dlsu.edu.ph; 2401 Taft Avenue, Malate, Manila, Philippines, 0922

Water scarcity is an alarming global problem for a growing population with depleting sources of fresh water. Desalination is thus becoming an important solution for water management to address such looming shortage of the municipal water supply. At present, several technologies dominate the desalination industry which can be categorized either as a thermal process such as multi-stage flash distillation or a membrane process such as that of reverse osmosis. New desalination systems are also being developed to make the process more cost-effective and energy efficient. Hence, this work proposes a systematic approach for optimal selection of desalination systems using fuzzy analytic hierarchy process (FAHP) and grey relational analysis (GRA). Fuzzy AHP addresses the vagueness involve in the trade-off of the criteria or attributes used in evaluating the alternatives. On the other hand, the GRA solves the multiple criteria decision problem by aggregating the entire range of performance attribute values for every alternative into a single score in spite of incomplete information. An illustrative case study was presented wherein five desalination systems namely reverse osmosis (RO), combined reverse osmosis and forward osmosis (RO-FO), electrodialysis (ED), multi-stage flash distillation (MFD), and combined forward osmosis and membrane distillation (FO-MD) were evaluated. These desalination systems were compared to each other with respect to energy requirement, land footprint, system efficiency, economic viability, and maturity of technology. Sensitivity analysis was also done to determine the robustness of the modeling results from the variation of weights of the criteria.

An energy hub approach for multiple-plants heat integration

Xiaolu Chen, Chenglin Chang, Yufei Wang*, Xiao Feng
State Key Laboratory of Heavy Oil Processing, China University of Petroleum, Beijing 102249, China
wangyufei@cup.edu.cn

Multi-plants heat integration can be carried out either directly using process streams or indirectly using intermediate fluids. Due to the corresponding characteristics of the heat transfer, direct heat integration can accomplish more energy targets with less number of heat exchangers. While many design methodologies have been developed, most current studies mainly consider energy aspect and the distance factor are always ignored. Since the piping cost for long distance is a major proportion of the total capital cost, this article proposed a mathematical programming methodology for multi-plants heat integration with a centralized energy hub. In this way, streams from different plants can exchange heat in the hub and the piping cost can be largely reduced. Also, the capital governance can be simplified significantly and the networks are often easier to be operated and maintained. A literature example is illustrated to demonstrate the effectiveness of the proposed methodology. Compared with the results of the conventional design, the energy hub approach is a more attractive choice for multi-plants heat integration.

Investigation of heat recovery with pump arounds at a crude oil atmospheric distillation tower

Z. Varga, K. Lóczi
University of Pannonia, MOL Department of Hydrocarbon and Coal Processing,
Egyetem u. 10., H-8200 Veszprém, Hungary;
tel. +36 88624414, e-mail: vargaz@almos.uni-pannon.hu.

Nowadays, energy efficiency improvement is one of the core issues of the hydrocarbon processing industry. Crude oil distillation unit consumes the considerable part of energy being required to produce fuels and petrochemical feedstocks. Even the slightest efficiency improvement can result significant savings in the operating costs as well as contribute to reducing the flue gas emission. The recovery of heat that is used for heating up and partially evaporating the crude oil is inevitable.

This paper summarizes the results of study to be performed to determine the optimal heat recovery of an atmospheric crude oil distillation tower using different pump-around structures.

Model of the atmospheric tower for separating Russian Export Blend type crude oil into naphtha, kerosene, light gas oil, heavy gas oil and atmospheric residue was prepared by Pro II software. Column contained 32 theoretical trays and the side cuts were drawn off from trays 10, 16 and 23, respectively, and feed was introduced to 30th tray.

The following three cases were investigated. Case "A" (base case): column without pump around, it means heat was removed at the top condenser at relatively low temperature level (135°C). Case "B": column with one pump around introduced between tray 17 and 18; the initial temperature of pump around stream was 264°C and the return one was 224°C. Case "C": column with two pump arounds. First is introduced between tray 17 and 18; the initial temperature of pump around stream was 264°C and the return one was 224°C. Second is added between tray 24 and 25; the initial temperature of pump around stream was 297°C and the return one was 277°C. Design constraints were the yields and qualities of the products.

In order to evaluate the investigated cases we prepared energy integration considering all of hot and cold streams of the atmospheric distillation unit. So, the minimum heating and cooling requirement can be estimated for every case.

The most important results of the study are summarized in the table below.

	Case A	Case B	Case C
Duty, kW			
Top condenser	39950	30560	29804
Pump around 1	-	12500	7222
Pump around 2	-	-	6666
Minimum heat requirements, kW			
Cooling, Q_C	23256	14207	13569
Heating, Q_H	68181	58046	57057

Data given in the table above clearly shows that using one or two pump around to the atmospheric tower considerably decreased both the cooling and heating requirements, which results lower fuel consumption, flue gas emission and operating cost.

The required minimum column diameter was also calculated. They were 5180 mm, 4570 mm and 4270 mm for Case "A", "B" and "C", respectively, which shows the installation of pump arounds is not only contributed to decreasing the operation but also the investment costs.

Investigation of fouling in plate heat exchanger at sugar factory

O. Demirskyy[1], P. Kapustenko[1], G. Khavin[2], O. Arsenyeva[1], O. Matsegora[2], S. Kusakov[2], I. Bocharnikov[1]
[1] National Technical University "Kharkiv Polytechnical Institute", Dep. Of Integrated technologies Processes and Apparatuses, 21 Frunze st., Kharkiv, 61002, Ukraine; tel. +380577202223, e-mail: kap@kpi.kharkov.ua;
[2] AO SPIVDRUZHNIST-T LLC, Krasnoznamenny per. 2, off. 19, 61002, Kharkiv, Ukraine; tel. +380577202278, e-mail: o.p.arsenyeva@gmail.com.

Plate heat exchangers are efficient heat transfer equipment and now are widely used in industry. In comparison with shell-and-tube units they provide narrower temperature approach up to 1°C, less metal consumption and high energy saving potential. It makes this equipment more energy efficient and economically preferable. The process of flow movement in PHE is characterized by the highly developed turbulent flow caused by geometrical parameters of the criss-cross channels, formed by the adjacent plates [1]. This affects the hydraulic and thermal performance of the PHE unit. The mathematical simulation of fouling formation process will increase the reliability of this type of equipment and lead to its wider application in industry. In the present paper the dynamic of fouling formation inside the channels of plate heat exchanger is observed.

The fouling formation in heat transfer equipment is the complex process, which is determined by the physical properties of the heat carrier, material of the unit and hydraulic characteristics of the flow. The mathematical model based on the asymptotic behaviour [2] of water fouling is examined. The fouling process supposes the net rate of fouling accumulation as the difference between the fouling deposition rate and the fouling removal rate. The relation for forecasting the fouling resistance dynamics during the time is proposed.

The investigation of precipitation and particulate deposition in purified juice heating PHE for the first stage evaporation, which operates in sugar plant, was examined. In this position M15M plate heat exchanger produced by Alfa Laval is used. The analysis and mathematical simulation of the experimental data are presented. For the juice heaters the content of fouling deposition is mostly the calcium salts as calcium carbonates and sulphates. The parameters of the equation for deposition term estimation were determined for the regarded heat carrier. It allows to determine the deposition term and to simulate the fouling formation in time. The comparison of the experimental data and mathematical calculations showed a good agreement.

The proposed mathematical model enables to predict the fouling formation behaviour in plate purified juice heater and to determine the operation term for the cleaning of this unit. Basing on the observed model, the software, which enables to determine the periods of PHE cleaning during the operation was developed. The comparison of the industrial measurement data with calculation results is presented.

Key words: plate heat exchanger, thin juice heater, precipitation fouling, heat transfer.

Acknowledgement: The support of EC Project DISKNET (FP7-PEOPLE-2011-IRSES-294933) is sincerely acknowledged.

References

1. Arsenyeva O, Kapustenko P, Tovazhnyanskyy L, Khavin G (2013) The influence of plate corrugations geometry on plate heat exchanger performance in specified process conditions. Energy 57: 201-207.

2. Crittenden BD, Yang M, Dong L, Hanson R, Jones J, et al. (2014) Crystallization Fouling With Enhanced Heat Transfer Surfaces. Heat Transfer Engineering 36: 741-749.

A multi-period and multi-objective optimization model of a microgrid system investment model

L. Uy, P. Uy, J. Siy, C. Sy[1]
Industrial Engineering Department, De La Salle University, 2401 Taft Avenue, Manila Philippines, 1004
Tel: +632-5244611 loc 220, Email: charlle.sy@dlsu.edu.ph

There are currently 7 billion people worldwide, and 1.3 billion people of which have limited or totally no access to electricity. The said crisis does not only concern energy shortages; but it is also compounded with environmental issues such as pollution and climate change. An emerging alternative solution to address the said concerns nowadays is through building a Microgrid system. Microgrids (MG) are small-scale, low voltage power supply networks designed to supply electrical load for a small community such as a university campus, a commercial area, a trading state and others (Wang , Wang, Xiao, & Gong, 2014). MGs are also considered as effective means to deal with the issues on increasing demand and shortage of electricity by having its own generating sources which could make the MG system independent from the main grid (Gadonneix, et al., 2013).

This study extends existing mixed-integer linear programming Microgrid investment models by considering multi-period and multi-objective investment setups. The optimization of the Microgrid model provides investors a mix of distribution energy sources that maximizes the Net Present Value (NPV) with consideration of the allowable budget and emission constraint under multiple periods. Through the NPV, the return of investment of the investor would be determined for the period that the model was run. Upon the analysis of the model, it was seen that the model chooses to prioritize the use of Diesel generators until the emission constraint is maxed out. After which, it moves to the utilization of Wind power then Photovoltaic power. The use of electricity from the grid only happens when the demand for electricity is greater than the generating capacity of the system.

In doing the validation and sensitivity analysis, the model was run under different scenarios to identify which factors affect the decision of the model in choosing one source over the other. As a result, it was seen that there are four factors that affect the model's decision: cost, budget, carbon emissions, and useful life. Since the objective of the model is to maximize the NPV of the system, the model would choose to prioritize the least cost among the different Distribution Energy Resources (DER). However, when the emission and useful life enter the picture, the model would integrate the three factors and determine which DER has the least cost for the more output. Through the scenario analysis, it was validated that the model can generate the purchasing pattern given the different parameters. This provides investors the set of equipment that they need to invest on each period based on their limited resources.

Reference:

Gadonneix, P., Sambo, A., Tienan, L., Choudhury, A., Kim, Y., Birnbaum, L., Nadeau, M. (2013). *2013 World Energy Issues Monitor.* London: World Energy Council.

Wang , R., Wang, P., Xiao, G., & Gong, S. (2014). Power demand and supply management in microgrids with uncertainties of renewable energies. *Electrical Power and Energy Systems, 63*, 260-269

Synthesis and design of seawater desalination membrane networks with salt production options

[1]V. Chauhan, [1]S. Y. Alnouri, [1,2]P. Linke,
[1]Department of Chemical Engineering, Texas A&M University, PO Box 23874 Doha, Qatar,
[2]Qatar Energy and Environment Research Institute (QEERI), PO Box 5825, Doha, Qatar

Seawater Reverse Osmosis (SWRO) has become a leading desalination technology, as a result of the global increase in demand for desalinated water. Various Zero Liquid Discharge (ZLD) technologies have been developed, which entail salt production, to combat brine disposal issues. Integrating ZLD options significantly impacts the economics and overall footprints of desalination systems.

Considerable literature is available on the optimization of SWRO membrane networks. In this work, a novel superstructure-based approach has been developed, by building upon earlier work [1-3]. The main motivation of the proposed methodology is to be able to systematically explore different membrane desalination configurations, coupled with salt production technologies. All possible design options have been embedded into the proposed superstructure model, whilst deploying optimization techniques that identify economically optimal solutions. Hence, the method can explore reduced water costs by extracting value from concentrates in the desalination system in the form of salt co-products. This work is the first ever attempt to propose a superstructure-based design approach for this problem.

This work considers SWRO and Nanofiltration (NF) membranes as the primary synthesis units of the membrane desalination network. Contrary to the high rejections of all ions by SWRO membranes, NF membranes offer higher rejections of bivalent ions over monovalent ions and offer potential opportunity to selectively channel streams containing high value ions to salt production operations. In order to capture these differences, detailed membrane module performance has been captured. Furthermore, several Salt Production Processes (SPP) involving desalination brines, which have been proposed and implemented on a commercial scale, have also been considered as a third category for potential synthesis units within the network. Since all synthesis units were modelled to handle the multicomponent composition of water, this enabled the quantification of the different salts extracted within SPP units. Relevant pressure change operations have also been accounted for, and a comprehensive economic objective function was employed.

A case study involving multiple membrane units and SPPs has been utilized to demonstrate the proposed method, using FilmTec SW30 and NF270 by Dow membranes to assemble a 100,000 m3/day production capacity plant. Moreover, the possibility of integrating two different SPP units onto the system were explored. The first SPP option produces calcium carbonate through sodium carbonate softening in a solid contact clarifier, while the second SPP option produces sodium chloride through the sequential pond evaporation and evaporative crystallization. The least cost optimal membrane network designs with salt production have been compared against a base case of SWRO desalination without salt production. The results indicate significant reductions in water costs if salts are co-produced in the desalination system.

[1] S.Y. Alnouri, P. Linke, A Systematic Approach to Optimal Membrane Network Synthesis for Seawater Desalination. Journal of Membrane Science. 417–418, (2012) 96-112.
[2] S.Y. Alnouri, P. Linke, Optimal SWRO Desalination Network Synthesis Using Multiple Water Quality Parameters, Journal of Membrane Science. 444 (2013) 493-512

[3] S.Y. Alnouri, P. Linke, Optimal Seawater Reverse Osmosis Network Design Considering Product Water Boron Specifications. Desalination. 345, (2014) 112-127.

Hybrid-heating-systems for optimized integration of low-temperature-heat and renewable energy

G. Schumm[1], M. Philipp[1], F.Schlosser[1], J. Hesselbach[1], T.G. Walmsley[2], M.J. Atkins[2]
Universität Kassel, Dep. Umweltgerechte Produkte und Prozesse, Kurt-Wolters-Straße 3, 34125 Kassel, Germany, email: schumm@upp-kassel.de; tel. +495618043460.
[2] Energy Research Centre, School of Engineering, University of Waikato, Private Bag 3105, Hamilton, New Zealand; tel. +647 8384937.

The food and beverage industries are significant industrial producers of green-house gas (GHG) emissions. Emission reductions can be achieved by increased energy efficiency and the use of renewable energy to replace fossil fuel use. The main efficiency method within the food and beverage industries is the use of low temperature heat (LTH), i.e. below 100 °C. Sources for LTH include heat recovery from process flows, heat rejection from utility operations (i.e. chillers, combined heat and power, condensing economisers), and renewable energy (i.e. solar thermal). For the import and integration of intermittent renewables in the form of electricity, electric steam generators replacing the coal or natural gas fired steam boilers, which generate high temperature heat (HTH), during supply peaks caused by wind turbines or solar PV, are the-state-of-the-art. The efficient use of LTH and the integration of electricity from renewable energy sources (RE) requires good operational control, flexibility and integration on the demand side.

A hybrid heating system (H^2S) has been developed that can retrofit existing indirect steam heater designs. The idea of the H^2S is to integrate available recovered and renewable LTH and HTH sources into the isolated hot water loop of an indirect steam heater system. It individually provides processes with LTH or HTH controlled by an algorithm based ratio, called the energy recipe (ER). The ER considers both, the optimized use of LTH and the recommended use of HTH during RE-peaks. Retrofitting existing indirect steam heaters into hybrid heating systems, using both heat recovery and steam utility, in the food and beverage industries provides a cost-effective and practical solution for increasing energy efficiency and process demand management. The common setup of an indirect steam heater system consists of an isolated hot water loop connecting a product heater with a steam supplied hot water heater. This set-up is needed to heat sensitive product flows from its supply temperature to its target temperature, thereby avoiding high contact temperatures. Although a H^2S is not the optimal Pinch solution, it is an industrially-acceptable solution, especially for batch and semi-continuous processes, because it can improve process control and operational flexibility, reduce heat exchanger area redundancy, while maintaining product hygiene requirements.

The aim of this paper is to show the functionality, reliability and the benefits of a hybrid heating system. The scope includes the design of control algorithms for the H^2S as well as applications for LTH and RE integration and its cost-benefit in comparison to standard designs. The benefits will be shown by numerical simulations on a semi-continuous whey processing plant programmed with comprehensive MatLab models embedded in Simulink and event based StateFlow controls. Furthermore, a 40 kW_{th} H^2S testing facility, controlled by an industrial PLC, is installed and tested to prove the systems functionality and reliability for different circumstances, which results will be presented as well.

Design and optimization of boil-off gas (BOG) re-liquefaction process for small-scale liquefied natural gas (LNG) propulsion ship

D.H. Kwak, H. Cho, [1]S.H. Park, [1]C.S. Kim, [1]S.J. Seo, J.K. Kim
Department of Chemical Engineering, Hanyang University, Wangsimni-ro 222, Seongdong-gu, Seoul 04763, Republic of Korea; tel. +82 (0)2-2220-2331, email: jinkukkim@hanyang.ac.kr; [1]R&D Center, DongHwa Entec, Gwahaksandan-ro 305beon-gil, Gangseo-gu, Busan, Korea 46742, Republic of Korea.

Recently, the interest for on board propulsion system with LNG has been growing due to the depreciation in LNG price and the regulation of greenhouse gas (GHG) of International Maritime Organization (IMO). LNG propulsion system can be seen as economic option, compared to conventional fuel-using method because LNG price is declined and SOx and NOx treatment is not required.

Typically LNG is contained in cryogenic tank which is below -160℃ so some of boil-off gas (BOG) is generated during voyage. BOG has to be managed with Gas Combustion Unit (GCU) or re-liquefaction process. GCU had been widely used in the past, but recondensation of BOG is preferred in these days for decreasing loss of fuel.

Previous studies performed for the modeling of BOG re-liquefaction process which are mainly applied for large-scale LNG carrier or LNG terminal. However, LNG storage capacity in LNG propulsion ship is relatively small and such small-scale characteristics needs to be carefully addressed. This study aims to design and optimize of BOG re-liquefaction process for LNG propulsion ship. Nitrogen expander cycle is selected for liquefaction process due to the safety issue in onboard system. The refrigeration system is modeled using commercial simulator UniSim Design® and economic evaluation is performed with the result from simulator (e.g. shaft power requirement, area of heat exchanger).

The optimization is also adopted to design cost-effective BOG re-liquefaction process. The optimization model is formulated to find out the optimal configuration and operating condition simultaneously, subject to engineering constraints related to onboard circumstance. And the part-load operation is also studied to consider the variation of BOG flowrate during the operating horizon. The optimization solver in MATLAB® is linked with UniSim Design® to perform the process optimization procedure.

Acknowledgement
This work was supported by the World Class 300 Project (No. S2305678) of the SMBA(Korea).

Bread as a chemical reactor: Triggering the aroma production through chemical kinetics

Davide Papasidero*[a], Alessandro Giorgi[a], Elisa Rocchi[b], Laura Piazza[b], Sauro Pierucci[a], Giulia Bozzano[a], Flavio Manenti[a]
[a]Dipartimento di Chimica, Materiali e Ingegneria Chimica "Giulio Natta", Politecnico di Milano, Piazza Leonardo da Vinci 32, 20133 Milano, Italy
[b]DeFENS, Dept. of Food, Environmental and Nutritional Sciences, Università degli Studi di Milano, Via Mangiagalli 25, 20133 Milano, Italy
davide.papasidero@polimi.it

Recent studies address the structural modifications occurring in the thermal processing of food to interact with the development of aroma and taste compounds. Food sensory science is often related to statistical analysis of panel tests and subjective human responses. From a chemical engineering perspective, food can be treated like a chemical reactor. Bread can be one example of "food chemical reactor", which enormously changes its density (from high to low depending on the leavening process), its aroma (driven from Maillard and other reactions involving carbohydrates, proteins and fats), its color (caramelization and Maillard), its structure (starch gelatinization, gluten hardening), etc. Indeed, transport phenomena and chemical kinetics are consistently involved in this process. A detailed study on the chemical kinetics of bread aroma development can be helpful to achieve an objective quality marker, which can be monitored and controlled dynamically through the process conditions and variables. The identification of the most relevant aroma compounds and the relative production paths description is the key to the kinetic approach. This would be beneficial for industrial purposes: the result of the research is the scientific description of a quantitative-sensorial aspect. This feature can be useful for further integration with physical/phenomenological models and texture models, to achieve an innovative holistic viewpoint for the product development and process control on the industrial scale breadmaking.

Maximization of heat recovery in heat exchangers of a crude oil distillation unit by modification of the existing network

P. Trzcinski,[1]M. Markowski
Department of Process Equipment, Warsaw University of Technology Plock Branch, Poland
Jachowicza 2/4, 09-402 Płock, Poland, tel. 24-2622610; 24-367599; fax 24-2626542; e-mail: trz@pw.plock.pl;
[1]Department of Process Equipment, Warsaw University of Technology Plock Branch, Poland
Jachowicza 2/4, 09-402 Płock, Poland, tel. 24-2622610; 24-367599; fax 24-2626542; e-mail: mark@pw.plock.pl;

The paper presents a method to maximize heat recovery in a HEN as a result of its modification. The suggested method of maximizing heat recovery is based on maximizing its driving force, which is the temperature difference between hot and cold process streams. The calculations have been carried out in three stages:

1. Calculations of the status quo of HEN.
2. Short-cut method calculations of the modified HEN aimed at increasing the probability of finding a global maximum of the regenerated heat, while a constant average was assumed for the value of heat transfer coefficient across the network.
3. Verification calculations of the modified HEN for the variable values of heat transfer coefficients.

The second and the third step of the calculations refer to the network after modification, which involved additional splitting of the cold stream into three streams, and was based on the algorithm of the selection of heat exchangers and process streams by applying the principle of maximizing temperature difference between the hot and cold streams.

The suggested method has been applied to maximize heat recovery in an HEN of a crude oil distillation unit processing crude oil. The calculations have shown that applying the proposed method results in a heat recovery increase by circa 4%.

Keywords: network of heat exchangers, process stream, energy conservation, optimization.

Development of a dynamic fouling model for a heat exchanger

Khayyam Zahid[a], Rajnikant Patel and Iqbal. M. Mujtabaa*
Chemical Engineering Division, School of Engineering, University of Bradford, Bradford, West Yorkshire BD7 1DP, UK. Tel: +44 1274 233645
I.M.Mujtaba@bradford.ac.uk

Fouling is a general term used to describe the deposition of foreign matter on a heat transfer surface. Costs associated with heat exchanger fouling include increased energy consumption, downtime, cleaning, and can also exacerbate problems associated with membranes (Shirazi, 2010). Various models have been proposed in an attempt to model and predict fouling behaviour (Müller-Steinhagen, 2011; Arsenyeva, 2012). In order to improve model reliability, multiple experiments and simulations have been conducted; nevertheless fouling problems within industry are still not accurately predicted leading to non-optimal fouling mitigation. Recent fouling model developments involve using CFD to predict fouling and are in agreement with previously published experimental results and theoretical models (Haghshenasfard, 2015; Pääkkönen, 2016).

The aim of this work is to develop a model for fouling in a heat exchanger which can effectively predict the heat transfer behaviour when a saline fluid is introduced to the system. Discrete rather than dynamic fouling factors are often added to calculate overall heat transfer co-efficient in heat exchangers. This will not be the case in this study as a dynamic fouling factor will be used in order to account for losses in heat transfer efficiency over time. To prove the validity of this model, an experiment has been conducted to verify the model accuracy.

A saline solution was made up using 3 grams of regular table salt per 100 grams of water giving a salt concentration of 30000 ppm (typical seawater salinity). Initial checks were made to ensure all tubing was connected and the equipment was safe to use. The system was set to run in counter-current operation and the hot fluid temperature was set to 40°C (313.16K, typical of last stage temperature of Multistage Flash Desalination process) whilst the cold fluid temperature was set to 20°C (293.16 K, typical of winter seawater temperature). Both hot and cold fluid flow rates were set to 1 litre per minute.

Figures 1 and 2 show a cooling/heating gradient of around 4°C was observed throughout the duration of the experiment. The 0.5°C variance in cooling fluid temperature can be accounted for as the system was turned off overnight and restarted in the morning in order to minimize the risk of damage to the system by leaving the system unattended for an extended period of time. Data logging began when the system reached steady state which took around 90 minutes. The next step in calculating the fouling factor was to observe how the cooling or heating changed over time. Figure 3 shows the relationship between UA_{decay} and time. Figure 4 shows that the thickness of the fouling deposit increases from 0 to around 0.0002mm over 700 minutes which is expected as the flow rates and temperatures were very low.

The development of this fouling model was based on using traditional heat exchanger models and implementing a fouling factor based on the decrease in heat transfer efficiency over time, observed experimentally. The fouling behaviour could be explained by crystallisation and precipitation mechanisms. The experimental data was analysed and a dynamic fouling model for heat transfer was developed from the findings. The following equation is a representation of the overall Kern and Seaton (1959) model developed in this work based on experimental data.

$$U_f = \frac{U_c A}{1 + U_c A\left((0.000004t)\left(1 - e^{-\frac{t}{t_c}}\right)\right)} \tag{16}$$

The model is able to accurately predict the temperature change with an average discrepancy of 1.6°C and 0.4°C for the cold stream and hot stream respectively.

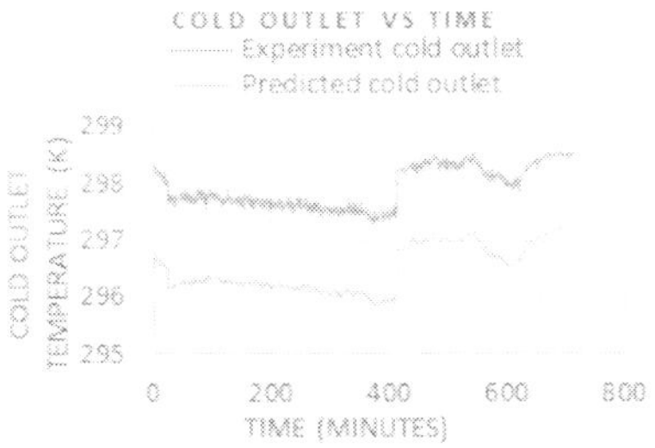

Figure 1: Cold Stream Outlet Temperature (Experiment and Prediction)

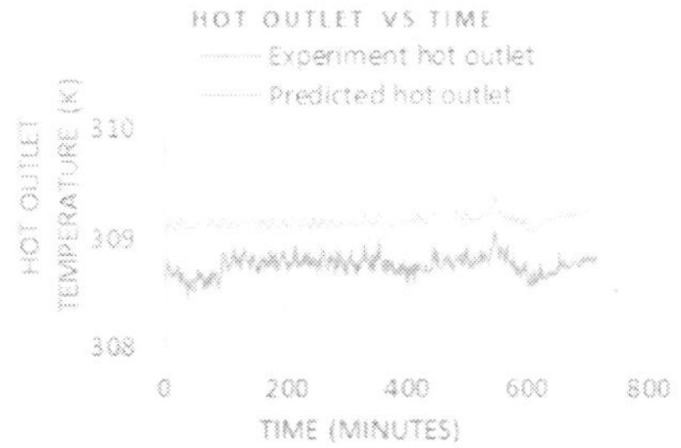

Figure 2: Hot Stream Outlet Temperature (Experiment and Prediction)

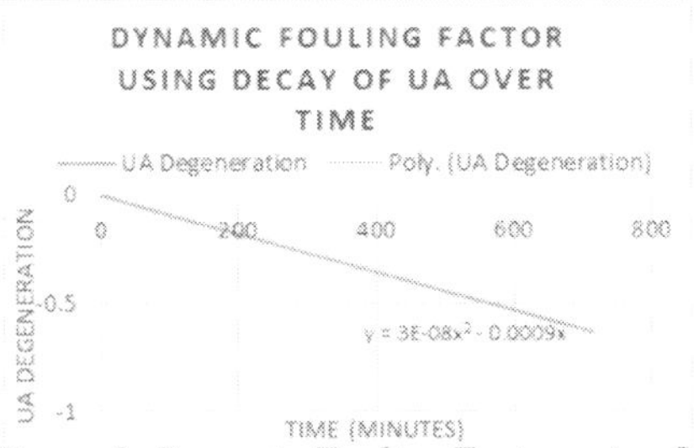

Figure 3: Dynamic Fouling Factor using Heat Transfer Efficiency over Time

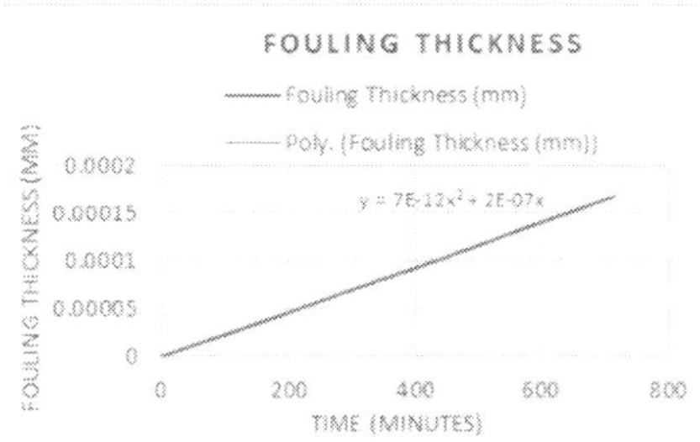

Figure 4: Fouling Thickness against Time

References

Arsenyeva, O.P., Tovazhnyansky, L.L., Kapustenko, P. O., Demirskyy, O.V. (2012) Accounting for thermal resistance of cooling water fouling in plate heat exchangers. Chemical Engineering Transactions. 29. p.1327-1332

Haghshenasfard, M., Hooman, K. (2015). CFD modeling of asphaltene deposition rate from crude oil, Journal of Petroleum Science and Engineering, 128, p. 24-32

Müller-Steinhagen, H., 2011, "Heat Transfer Fouling: 50 Years After the Kern and Seaton Model." Heat Transfer Engineering 32(1) p.1-13.

Pääkkönen, T.M., Ojaniemi, U., Pättikangas, T., Manninen, M., Muurinen E., Keiski, R.L., Simonson C.J. (2016) CFD modelling of CaCO3 crystallization fouling on heat transfer surfaces, International Journal of Heat and Mass Transfer, 97, p. 618-630

Kern, D.Q. and Seaton, R.E. (1959). A Theoretical Analysis of Thermal surface Fouling. Brit. Chem. Eng. 4(5): p. 258-262.

Shirazi, S., Lin, C.J., Chen.D. (2010). Inorganic fouling of pressure-driven membrane processes—a critical review, Desalination 250 (1) p. 236-248.

Tuning parameters of PID controllers for the operation of heat exchangers under fouling conditions

M. Trafczynski, M. Markowski, S. Alabrudzinski, K. Urbaniec
Warsaw University of Technology - Plock Campus. Department of Process Equipment, 2/4 Jachowicza, 09-402 Plock, Poland: tel. +48 24 262 26 10, e-mail: mariant@poczta.fm, fax: +48 24 262 65 42.

It belongs to industrial practice that the operation of a heat exchanger may be affected by fouling which builds up on its heat transfer surface. In a real-life plant, possible consequences of fouling include burning extra fuel to compensate for reduced heat recovery, reducing plant output when the exchangers are cleaned and generating costs of cleaning interventions. Various approaches to the mitigation of fouling effects in industrial heat exchangers and exchanger networks are reported in the literature. On the other hand, although transient states and control of heat exchangers and exchanger networks may also have a detrimental effect on the heat recovery, papers devoted to the role of fouling in control issues are rare.

The present paper is an extension of the work presented at SDEWES2015 conference (M. Trafczynski, M. Markowski, K. Urbaniec, S. Alabrudzinski, The influence of fouling on the dynamic behavior of PID-controlled heat exchangers, Proceedings of the 10th Conference on Sustainable Development of Energy, Water and Environment Systems, SDEWES2015.0847, 1-20, 2015), where changes in the dynamic behavior of shell-and-tube heat exchangers induced by fouling build-up were studied. A mathematical model was proposed of transient heat exchange with the influence of thermal resistance of fouling taken into account. A heat exchanger was considered as an array of cells in which the exchange of heat was modelled using the lumped-parameter approach. Model equations to describe transient states of the heat exchanger were derived from the energy balance of a control volume in which changes in the state of tube-side fluid, tube walls and shell-side fluid were accounted for. As the equations were solved, transfer functions were obtained to describe the relationships between input signals (changes in process parameters at heat exchanger inlet) and responses (changes in temperature at exchanger outlet). Upon linearization of the equations, Laplace transforms were obtained for each exchanger cell. This approach has the advantage of yielding simple analytical relationships which are sufficiently accurate as long as the number of cells defined in the heat exchanger structure is large.

A mathematical model of the entire heat exchanger was implemented and simulations were carried out using Simulink in MATLAB environment. The structure of the dynamic model was visualized to depict transfer functions and signal flows. The model was validated using data records acquired during 3 years of operation of a heat exchanger network coupled with a crude distillation unit. The values of simulated and real temperature at heat exchanger outlet in transient states of the exchanger were compared and found to be in close agreement. The dynamic behavior of the heat exchanger was correctly reproduced for different periods of its uninterrupted operation, that is, no matter if the heat transfer surface was clean or fouled.

Using the validated dynamic model, one can simulate the behavior of a heat exchanger together with its PID control loop and investigate the influence of fouling build-up on specific indexes of quality of operation. A simulation study was carried out on selected PID-controlled exchangers operated in a heat exchanger network coupled with a Crude Distillation Unit (CDU) rated 220 kg crude oil per second. For different periods of uninterrupted exchanger operation, that is, at different values of the thermal resistance of fouling, tuning parameters were determined using the Ziegler-Nichols method. It was found that the optimal values of tuning parameters changed when the thermal resistance of fouling was changed. However, oscillations remained in the characteristics of some controlled exchangers indicating that the new values of tuning parameters determined by the simple Ziegler–Nichols method were ineffective. In the present work, to improve the quality indices of the exchangers control, a more advanced method of the optimization of controller tuning is presented using Signal Constraint in MATLAB/SIMULINK. The importance of this research topic is underscored by the fact that estimated 80% of industrial heat exchangers equipped with PID controllers are operated either in manual mode, or using default values of the tuning parameters of their controllers.

Multiple modifications in stepwise retrofit optimization of heat exchanger networks

N. M. Rohani, E. Ayotte-Sauvé, O. Ashrafi, S. Bédard
Natural Resources Canada, CanmetENERGY, 1615 Lionel-Boulet Blvd., Varennes, QC J3X 1S6, Canada; tel. +1 4506524437, e-mail: navid.rohani@canada.ca; fax: +1 4506525918

This work considers improving the heat recovery of an existing heat exchanger network (HEN) while minimizing the total annualized cost of the project. Sreepathi and Rangaiah (2014) as well as Jezowski (1994) provide literature surveys of HEN retrofit methodologies spanning the period from 1993 to 2014. As described by Furman and Sahinidis (2002) in the HEN synthesis context, approaches to tackle the retrofit problem usually fall in two broad categories: simultaneous approaches (e.g. Ciric and Floudas, 1990; Sorsak and Kravanja 2004; Ponce-Ortega et al., 2008; Nguyen et al., 2010) and stepwise (or sequential) approaches (e.g. Ciric and Floudas, 1989; Yee and Grossmann, 1991; Asante, 1996; Briones and Kokossis, 1996; Athier et al., 1998; Smith et al., 2010).

The focus of this work is on the stepwise approaches which generate, starting from a base case network, a candidate retrofit HEN at each step in order to sequentially build and optimize the HEN topology. More precisely, each step of such approaches consists in two stages. During the first stage, retrofit HEN superstructures are obtained from the previous step by introducing topological modifications which fall in three categories: adding new process-process heat exchangers, relocating existing process-process heat exchangers or splitting process streams and mixing the resulting stream branches. The second stage generates a candidate HEN by minimizing the total annualized cost of the superstructures via numerical optimization.

The network pinch approach, proposed by Asante (1996) and later refined by Smith et al. (2010) as well as Bakhtiari and Bédard (2013), falls into that category. Like simultaneous approaches and stepwise approaches, this method has its share of limitations. For example, it fails when no pinching matches are present in the HEN. Varbanov and Klemes (2000) suggested heuristics to circumvent this problem.

One inherent problem of step-wise approaches may occur when at most one new process-process heat exchanger is added at each step in order to improve heat recovery (e.g. Bakhtiari and Bédard, 2013). More precisely, if in one single step, the new process-process heat exchanger does not form a new cooler-heater path, then it is not possible to reduce the total utility consumption. In such cases, multiple modifications may be necessary in a given step. Consequently, the present study proposes a systematic method for adding multiple process-process heat exchangers in one single step using the notion of *bridge* introduced by Bonhivers et al. (2014). Loosely stated, a bridge is a sequence of process stream temperature intervals which starts from a cooler, alternates between the cold and hot process streams passing through chosen process-process heat exchangers and ends with a heater. Bridges provide a basis for the identification of promising new paths in the network. The present work introduces a superstructure of retrofit options associated with each bridge. Subsequently, an MINLP model minimizing the total annualized cost is solved to obtain the explicit topology of the network. This new approach is illustrated on a milk processing case study where the network pinch (Asante, 1996) fails in finding a retrofit solution. For this example, the diagnosis stage of the network pinch method cannot yield a retrofit project which improves the base case utility consumption. As a solution, multiple modifications in a single step are applied and heat recovery options are identified.

References

[1] B.K. Sreepathi, G.P. Rangaiah. Review of Heat Exchanger Network Retrofitting Methodologies and Their Applications. *Ind Eng Chem Res*, 53 (2014) 11205-11220.
[2] J. Jezowski, Heat-Exchanger Network Grassroot and Retrofit Design - the Review of the State-of-the-Art 0.2. Heat-Exchanger Network Synthesis by Mathematical-Methods and Approaches for Retrofit Design. *Hung J Ind Chem*, 22 (1994) 295-308.
[3] K.C. Furman, N.V. Sahinidis. A Critical Review and Annotated Bibliography for Heat Exchanger Network Synthesis in the 20th Century. *Ind Eng Chem Res*, 41 (2002) 2335-2370.
[4] A.R. Ciric, C.A. Floudas. A mixed Integer Nonlinear Programming Model for Retrofitting Heat-Exchanger Networks. *Ind Eng Chem Res*, 29 (1990) 239-251.
[5] A. Sorsak, Z. Kravanja. MINLP retrofit of heat exchanger networks comprising different exchanger types. *Comput Chem Eng*, 28 (2004) 235-251.
[6] J.M. Ponce-Ortega, A.Jimenez-Gutierrez, I.E. Grossmann. Optimal synthesis of heat exchanger networks involving isothermal process streams. *Comput Chem Eng*, 32 (2008) 1918–1942.
[7] D.Q. Nguyen, A. Barbaro, N. Vipanurat, M.J. Bagajewicz. All-At-Once and Step-Wise Detailed Retrofit of Heat Exchanger Networks Using an MILP Model. *Ind Eng Chem Res*, 49 (2010) 6080-6103.
[8] A.R. Ciric, C.A. Floudas. A Retrofit Approach for Heat Exchanger Networks. *Comput Chem Eng*, 13 (1989) 703-715.
[9] T.F. Yee, I.E. Grossmann. A Screening and Optimization Approach for the Retrofit of Heat-Exchanger Networks. *Ind Eng Chem Res*, 30 (1991) 146-162.
[10] N.D.K. Asante. Automated and Interactive Retrofit Design of Practical Heat Exchanger Networks. University of Manchester Institute of Science and Technology, PhD Dissertation,1996.
[11] V. Briones, A. Kokossis. A new approach for the optimal retrofit of heat exchanger networks. *Comput Chem Eng*, 20-Supplement 1 (1996) S43-S48.
[12] G. Athier, P. Floquet, L. Pibouleau, S. Domenech. A mixed method for retrofitting heat-exchanger networks. *Comput Chem Eng*, 22-Supplement 1 (1998) S505-S511.
[13] R. Smith, M. Jobson, L. Chen. Recent development in the retrofit of heat exchanger networks. Appl Therm Eng, 30 (2010) 2281-2289.
[12] B. Bakhtiari, S. Bedard. Retrofitting heat exchanger networks using a modified network pinch approach. *Appl Therm Eng*, 51 (2013) 973-979.
[14] P.S. Varbanov, J. Klemes. Rules for paths construction for HENs debottlenecking. *Appl Therm Eng*, 20 (2000) 1409-1420.
[15] J.C. Bonhivers, B. Srinivasan, P.R. Stuart. New analysis method to reduce the industrial energy requirements by heat-exchanger network retrofit: Part 1 – Concepts. *Appl Therm Eng*, In Press, Corrected Proof (2014) DOI:10.1016/j.applthermaleng.2014.04.078.

Investigation of the performance of different recuperative cycles for gas turbines/aero engine applications

C.Salpingidou[1], D. Misirlis[3], Z. Vlahostergios[1], S. Donnerhack[2], M. Flouros[2], A. Goulas[1], K.Yakinthos[1]
[1]Laboratory of Fluid Mechanics & Turbomachinery, Department of Mechanical Engineering, Aristotle University of Thessaloniki, Thessaloniki, 54124, e-mail: csalpingidou@eng.auth.gr
[2]MTU Aero Engines AG, Dachauer Strasse 665, Munich, Germany
[3] Technological Educational Institute (TEI) of Central Macedonia, Serres, Greece, e-mail: misirlis@eng.auth.gr, dmissirlis@teiser.gr

This work presents an investigation of the performance of three recuperative cycles for gas turbines, with a particular interest in aero engine applications. In these cycles heat exchangers are integrated in different positions of the gas turbine in order to exploit part of the thermal energy of the hot gas in order to preheat the compressor discharge air before the latter enters the combustion chamber, leading to increased cycle thermal efficiency and reduced fuel consumption. The first configuration under investigation is the conventional recuperative cycle (CR), in which a heat exchanger is placed after the low pressure turbine. In the second configuration, referred to as alternative recuperative cycle (AR), a heat exchanger is placed between the high pressure (HPT) and low pressure turbine (LPT), while in the third configuration, referred to as staged heat recovery (SHR) cycle, two heat exchangers are employed, the primary one between the HPT and LPT turbines and the secondary at the exhaust, downstream the LPT. In the first part of the present work, a parametric conceptual analysis was conducted to investigate the impact of heat exchanger effectiveness and overall pressure ratio (OPR) on cycle performance and identify the optimal operational range for each configuration. This part of the analysis was performed mainly with the use of CAPE-OPEN/COFE and GasTurb11 cycle analysis code. The results show that the CR cycle presents superior performance in relation to the AR cycle for low OPR, while for higher OPR the AR cycle outperforms. In addition, for the SHR cycle, proper selection of the primary and the secondary heat exchangers affect significantly the cycle efficiency. In the second part, this work was focused on the assessment of practical issues regarding the implementation of the AR and the SHR recuperative concepts in gas turbine applications and more specifically in a recuperative aero engine. This presents a major challenge, since the inter-turbine positioning and integration of recuperators should take into consideration issues related with aero engine geometry space constraints and technical limits due to the increased operational temperature of the recuperators. For performing the analysis, the MTU-developed intercooled recuperated (IR) thermodynamic cycle concept, which is presented in figure 1, was used. In this cycle (conventional) recuperation is achieved by placing a system of heat exchangers, figure 2, right after the LPT inside the hot-gas exhaust nozzle. Various recuperator scenarios were examined taking into considerations different axial lengths and tube core arrangements. For the determination of the recuperator inner/outer pressure losses and effectiveness results from previously performed CFD computations were used. The recuperator effectiveness and pressure losses for each scenario were included and assessed in the aero engine CAPE-OPEN/COFE models for the CR, AR and SHR cycles. The analysis was also extended into heat exchanger materials which can cope with the increased temperature range of the inter-turbine positioning. The performance analysis of the three recuperative aero engine cycles showed significant optimization potentials which can be increased by the aero engine design and material selection.

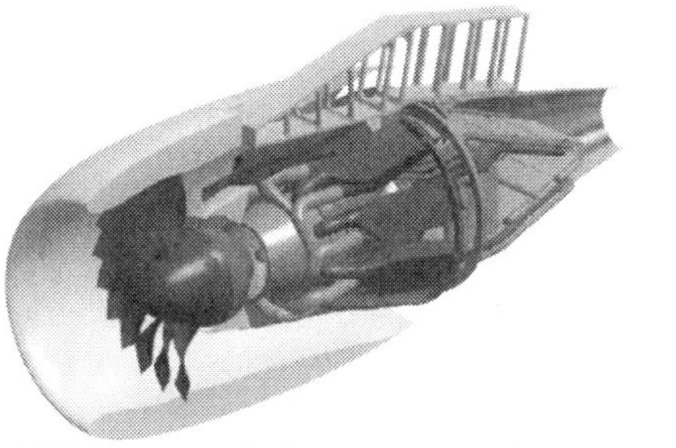

Figure 1. The MTU intercooled recuperated aero engine concept

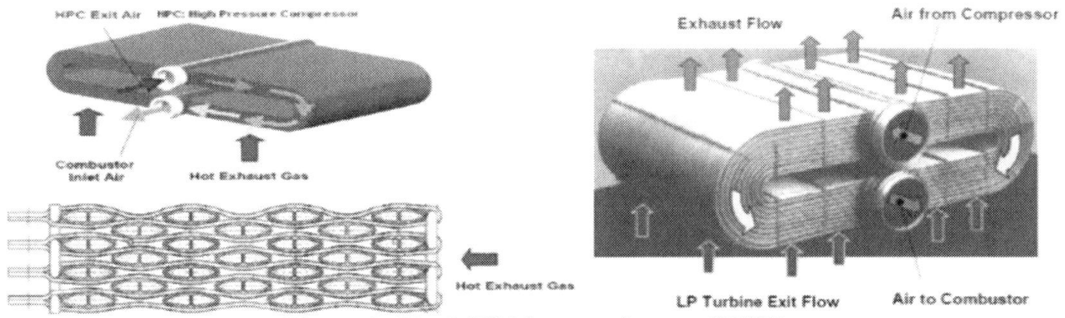

Figure 2. The MTU-heat exchanger (HEX)

References

[1] Dellenback P.A., A Reassessment of the Alternative Regeneration Cycle, J. Eng. for Gas Turbines and Power, Transact.of ASME OCT.2006, Vol.128/783-788

[2] Schonenborn H., Simon B., Ebert E. and Storm P., [2004]. Thermomechanical design of a heat exchanger for a recuperative aero engine. In: Proceedings of ASME Turbo Expo 2004

[3] Goulas A., Donnerhack S., Flouros M., Misirlis D., Vlahostergios Z., Yakinthos K., Thermodynamics Cycle Analysis, Pressure Loss, and Heat Transfer Assessment of a Recuperative System for Aero-Engines, J. Eng. Gas Turbines Power. 2015, 137(4)

Heat transfer and pressure drop in cross-flow welded plate heat exchanger for ammonia synthesis column

L.L. Tovazhnyanskyy[1], P.O. Kapustenko[1], O.Yu. Perevertaylenko[1], G.L. Khavin[2], O.P. Arsenyeva[1], P.Yu. Arsenyev[2]
[1] National Technical University "Kharkiv Polytechnical Institute", Dep. Of Integrated technologies Processes and Apparatuses, 21 Frunze st., Kharkiv, 61002, Ukraine; tel. +380577202223, e-mail: kap@kpi.kharkov.ua;
[2] AO SPIVDRUZHNIST-T LLC, Krasnoznamenny per. 2, off. 19, 61002, Kharkiv, Ukraine; tel. +380577202278, e-mail: o.p.arsenyeva@gmail.com.

Efficient heat recuperation is of primary importance in resolving the problem of efficient energy usage and consequent reduction of fuel consumption and greenhouse gas emissions. Plate Heat Exchanger (PHE) is one of the modern efficient types of compact heat transfer equipment. The principles of the construction and design for different types of PHEs are sufficiently well described in literature, see e.g. Klemes et al. (2015). The conventional type is plate-and-frame PHE, which was initially developed for the food industry and later proved efficient in many other applications. However due to elastomer gaskets the range of plate-and-frame PHE application is limited to pressures up to 25 bar and temperatures up to 180 °C and working fluids friendly to gaskets material. Besides, the cost of the gaskets, especially for severe working conditions, can dramatically increase the cost of heat exchanger as a whole unit. To widen the PHE application range by excluding elastomer gaskets the brazed (BPHE) and welded (WPHE) types of PHE were developed. Here the results of research on the thermal and hydraulic performance of WPHE of special design for high pressures and temperatures are presented.

The construction of investigated WPHE is developed for work in high pressure shell of ammonia synthesis column at pressure up to 32 MPa and temperature up to 520 °C. It consists of the stack of round corrugated plates with diameter 626 mm, which are welded together to form a number of channels for cold and hot streams exchanging heat. The welded collectors of special design are organizing multi pass movement of both streams with overall counter flow. The movement of two streams in one pass block is cross flow. Compare to the flow of streams in conventional plate-and-frame PHE there is significant difference. From hydraulic point of view the stream is entering the channel through almost full cross section, while in channels of plate-and-frame PHE it is entering from distribution collector of small diameter compare to channel width. It causes much smaller local hydraulic resistance at the port zone of WPHE and ensures even flow distribution across channel width. But for the heat transfer cross flow of streams causes the reduction of mean temperature difference compare to counter flow in one pass of plate-and-frame PHE. The overall counter flow in WPHE is making this loss in mean temperature difference smaller, but still this cross flow feature for individual passes should be accounted in correct PHE design. The available literature data are not directly applicable, as the level of fluid mixing across PHE channel is not known.

The heat transfer and pressure drop in WPHE were investigated experimentally using the model consisted of 16 plates representing the block of plates in one pass of WPHE. The thermal and hydraulic characteristics of this model were measured at specially developed test rig with water as test fluid for both streams. The results of experiments were compared with predictions by Equations proposed by Arsenyeva et al. (2012). It confirmed the validity of these equations for PHE channels of different geometry in case of cross flow. It is also estimated the dependence of PHE heat transfer effectiveness (ε) from the number of heat transfer units (NTU) in one pass of WPHE with cross flow of streams. The method for calculation of WPHE with overall counter flow and cross flow inside separate pass is developed.

The validity of the proposed Equations and developed method for WPHE design was confirmed by comparison with the data of tests on WPHE installed in ammonia synthesis column at industrial enterprise of ammonia production. WPHE was installed in existing synthesis column of ammonia unit instead shell-and-tube heat exchanger. The use of WPHE instead shell-and-tube unit enable to cut down the volume occupied by heat exchanger in high pressure shell of ammonia synthesis column and allows increase of the volume of catalyst. It leads to 15% rise of ammonia output.

Key words: plate heat exchanger, ammonia synthesis column, heat transfer, pressure drop.

Acknowledgement: The support of EC Project DISKNET (FP7-PEOPLE-2011-IRSES-294933) is sincerely acknowledged.

References

Klemeš J J, Arsenyeva O, Kapustenko P, Tovazhnyanskyy L. (2015), Compact Heat Exchangers for Energy Transfer Intensification: Low Grade Heat and Fouling Mitigation. CRC Press. Boca Raton, FL, USA.

O.P. Arsenyeva, L.L. Tovazhnyanskyy, P.O. Kapustenko, O.V. Demirskiy (2012), Heat transfer and friction factor in criss-cross flow channels of plate-and-frame heat exchangers, Theor. Found. Chem. Eng. 46 (6) 634-641.

Sequential initialisation technique for synthesising multi-period multiple utilities heat exchanger networks

Adeniyi J. Isafiade*[a], Oludare J. Odejobi[b]
[a]Department of Chemical Engineering, University of Cape Town, South Africa
[b]Department of Chemical Engineering, Obafemi Awolowo University, Ile-Ife, Nigeria.
Aj.isafiade@uct.ac.za

This paper presents a new method for synthesizing heat exchanger networks involving multi-period operations and multiple utilities. Since process parameters, such as streams' flowrate and supply/target temperatures, in multi-period operations usually fluctuate around some nominal conditions, the synthesis models developed for their optimisation need to ensure that the heat load requirement of every stream in every period of operation is satisfied. Solving problems of this nature is not a trivial task due to the presence of multiple periods of operations. The complexity is further compounded in situations where multiple options of utilities are involved in the optimisation task. Hence in this paper, a new synthesis method which involves firstly, solving single period models for each period of operation in the problem, after which the selected matches from each of the single period solution are systematically used to initialise the multi-period superstructure. Contrary to existing methods where utilities are placed or appended only in the first and last stage of the multi-period superstructure model, the newly presented method positions utilities in all stages of the superstructure where process streams of the opposite kind exist. The new approach is applied to two examples and the solutions obtained demonstrate the benefits of the method.

Hydraulic design optimization of multi-scenario cooling water distribution systems

A. L. H. Costa, [1]E. L. R. Tonassi,[2]J. N. M. Souza
Rio de Janeiro State University, Rua São Francisco Xavier, 524, Instituto de Química, Maracanã, Rio de Janeiro, RJ, Brazil; e-mail: andrehc@uerj.br; [1]Petrobras, Avenida República do Chile, 65, Centro, Rio de Janeiro, RJ, Brazil; e-mail: ericadelima@globo.com; [2]National Institute of Industrial Property, Praça Mauá, 7, Centro, Rio de Janeiro, RJ; e-mail: jaimenms@gmail.com

Industrial plants demands the removal of energy from material streams through a set of coolers. Typically, the cold utility employed is represented by cooling water associated to a recirculation system. In this kind of system, cooling water is distributed from the cooling tower to the set of coolers located along the plant. The resultant flow of heated water is returned to the cooling tower. This service demands the construction of a pipe network in order to each heat exchanger to be fed with the established cooling water flow rate. In this context, this paper presents a mathematical programming solution for the hydraulic design of cooling water distribution systems that must operate according to a set of different scenarios. The resultant network must be flexible in order to provide for each scenario the set of flow rates specified by the process operation. The objective function contains the capital costs for piping and pumping and the operational costs associated to the energy consumption in the fluid flow. The main problem constraints involves mechanical energy balances along the network. The utilization of the proposed approach is illustrated through its comparison with the traditional design procedure using a typical example.

A systematic comparison of stagewise/interval-based superstructure approaches for the optimal synthesis of heat exchange networks

Duncan M. Fraser, Michael Short, Joanne Crimes, Oluwatosin S. Azeez, Adeniyi J. Isafiade
Department of Chemical Engineering, University of Cape Town, South Africa
Aj.isafiade@uct.ac.za

This article describes a study in which the various stagewise and interval-based superstructures for the mixed integer non-linear programming (MINLP) optimisation of heat exchanger networks were systematically compared, using exactly the same basis. The effect of using different logarithmic mean temperature difference (LMTD) approximations on the Total Annual Cost of the network and the network structure were examined. The accuracy of the approximations was also analyzed over a wide range of temperature difference ratios. It is concluded that the best approximation, which is the Underwood-Chen approximation, should be used in future mathematical optimisation of Heat Exchanger Network Synthesis problems.

Area-based retrofit and operational optimization of existing crude oil heat exchanger networks

L. M. Ochoa-Estopier,[1] M. Jobson,[1] L. Chen
[1]Centre for Process Integration, School of Chemical Engineering and Analytical Science, The University of Manchester, Sackville Street, Manchester M13 9PL, UK; e-mail: lluvia.ochoa-estopier@manchester.ac.uk,megan.jobson@manchester.ac.uk; Process Integration Limited, Station House, Stamford New Road, Altrincham, Cheshire WA14 1EP, UK; e-mail: lu.chen@processint.com

Retrofit and operational optimisation projects are frequently implemented in crude oil heat exchanger networks (HENs) to reduce operating costs. However, it is a challenge to optimise these HENS because of their complexity and the many practical constraints to be considered (e.g. plant layout, installed area, limited budget, pressure drop limitations, etc.).

This work presents an optimisation approach suitable for industrial crude oil HENs. The approach identifies the retrofit modifications and/or operating conditions that minimise operating and retrofit capital costs. Retrofit modifications considered include adding, deleting and relocating an exchanger, and adding and deleting a stream splitter. Operational optimisation variables include stream split fractions and utilisation of available heat transfer area.

The main feature of this approach is the specification of heat exchangers in terms of heat transfer area. This consideration significantly simplifies the optimisation, while capturing the details of the existing HEN. Temperature-dependent heat capacities are also considered. Practical constraints are implemented to ensure that industrially-relevant solutions can be achieved. These constraints include existing heat transfer area, minimum number of modifications, and forbidden relocations and matches. A case study on an industrial HEN demonstrates how the approach identifies opportunities to reduce energy consumption with minimal structural modifications.

Study on the flow noise mechanism in shell-and-tube heat exchangers based on flow field and sound field synergy principle

Y.P. Cao[a], Y.S. Lin[b], H.B. Ke[b] ,M. Zeng[a], Q.W. Wang*[a]
[a]Key Laboratory of Thermal-fluid Science and Engineering, Ministry of Education, Xi'an Jiaotong University, Xi'an, Shaanxi, 710049, China
[b]Key Lab on Steam Power System, Wuhan Second Ship Des. and Res. Ins, Wuhan, 430025, China.
Corresponding Author: Q.W. Wang, wangqw@mail.xjtu.edu.cn
Reporter: Y.P. Cao, caoyiping12@163.com

Shell-and-Tube Heat Exchangers are widely used in modern industries in national economy as major equipment for implementing technological process. As the noise radiation of pipelines in shell-and-tube heat exchangers can't be ignored, it is the key point to investigating the transfer mechanism of sound energy in pipelines and developing the high efficiency silencers with low penalty of pressure drop. In recent years, scholars have done a lot of research on the shell-and-tube heat exchangers, but few of them are focused on the flow noise mechanism. Since few theories are related to the transfer processes of sound energy from the perspective of matching flow field and sound field synergy principle and the investigation of field synergy principle on the transfer process of sound energy is rarely covered, we study the flow noise mechanism in shell-and-tube heat exchangers based on flow field and sound field synergy principle.

In this paper, we explore the effect of the synergy between flow field and sound field on the transfer process of sound energy and develop the field synergy principle for transfer process of sound energy based on other synergy principle about flow field, temperature field, concentration field and so on, so as to reveal the generation and transfer mechanism of flow noise in pipelines of shell-and-tube heat exchangers. The theoretical analysis and numerical simulation method are coupled to investigate the synergy between flow field and sound field, the flow noise in pipelines and the relationship between field synergy and noise transfer process. The results show that with the increase of synergy between flow and sound field, the wall doing work by the fluid increases, which means sound energy exchange between the wall and the fluid increases.

Keyword: Field synergy principle, flow noise, shell-and-tube heat exchangers, numerical study

A superstructure based optimization framework for batch water network synthesis with multiple wastewater treatment models

S. M. Wang, X. Zou, H. G. Dong*, L. Sun
Dalian University of Technology, School of Chemical Engineering, NO. 2 Linggong Road, 116024, Dalian, P.R. China; tel. +86 84986068,e-mail: hgdong@dlut.edu.cn;

Increasingly requirement of high quality water for designated use and serious environmental crisis calls for efficient water network with effective wastewater treatment facilities to exploit the opportunities for water reuse and remove hazardous material to meet stringent environment legislations.

A number of researches have been proceeded to optimize the water-using network based on various mathematical model with fixed removal ratio for treatment facilities, nevertheless, ignoring that removal ratio is appreciably determined by the cost of wastewater treatment units. To guarantee a true reflection of treatment cost when designing an optimal water network for continuous process, recent works increasingly incorporated wastewater treatment models which consist of reverse osmosis (RO), ultrafiltration(UF) and multimedia filter with the consideration of detail structure and removal ratio of the contaminant as well as recovery ratio of water(Khor, Foo and El-Halwagi et al, 2011; Khor and Chachuat,2012; Baidoo and Majozi, 2015). Besides, Yang and Grossmann (2014) simultaneously synthesized water network with treatment superstructure containing several categories of units so as to achieve a better tradeoff between removal efficiency and cost of the treatment units.

Rigorous treatment model can provide more realistic design, however, the resulting water network model inevitably requires high computational expense which limits the widely industrial application. And incorporating detail structure in water network model mainly aims to exploit the effect of treatment cost on removal ratio that is crucial to water network optimization. Alternatively, for the purpose of model complexity reduction, various level of removal ratio corresponding to different treatment cost can be deduced according to environmental engineering knowledge, which can be introduced to optimize water network in order to consider the impact of discrepancies of treatment systems.

What's more, aforementioned researches pay relatively little attention to water network of batch processes which is more complicated compared to continuous counterpart, as time dimension is intrinsically coupled. In addition, operation mode of treatment units can be broadly categorized into two patterns: continuous and batch with retention time. If treatment units with batch operation mode are involved, such as membrane biological reactor (MBR) and sedimentation tank, schedule is essential for treatment in water network of batch process. And previous literatures on the optimization of batch water network also negelected effect of wastewater treatment system (Gouws, Majozi and Foo, 2010).

Therefore, an improved state-time-space superstructure (STS) (Zhou, Li and Dong, 2009, 2016) was proposed to design water network for batch process, as shown in Fig. 1, which was consisted by four blocks: process operator (PO), distribution network (DN), treatment units (TU), buffer tanks (B). PO represents water using process in truly-batch pattern wherein water is supplied from mixers of DN at the start and discharges wastewater to splitters of DN at the end of its operation, and splitting and mixing of any stream is achieved by corresponding mixers and splitters in distribution network (DN). Furthermore, wastewater from splitters is intercepted by treatment facilities in TU, where various level of removal ratio corresponding to different treatment cost is proposed to substitute for accurate configuration of treatment system. TU is categorized into two types based on operation mode: continuous (TU2) consisted with RO system and adsorption and batch (TU1) comprised by MBR and sedimentation tank, for TU1, there is a schedule as described by time points at time axis. On these bases, the overall MINLP mathematical model can be optimized to cater for less-cost water network configuration. Finally, two case studies are illustrated to verify the feasibility and superiority of the methodology.

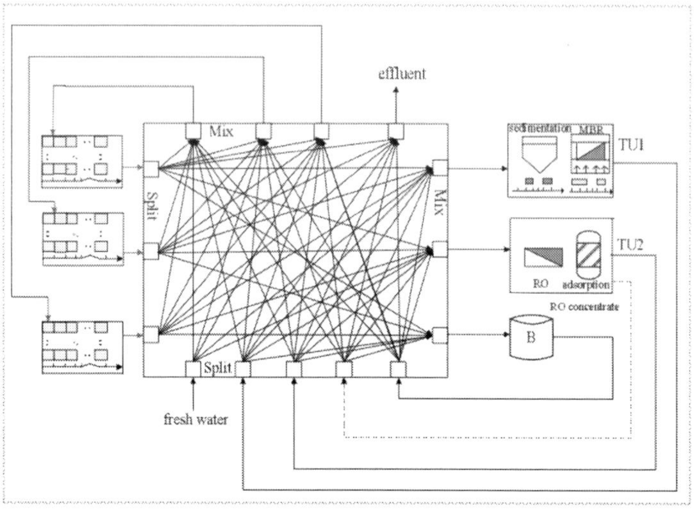

Fig. 1. Improved state- time-space superstructure for batch water network

References

[1] Khor C S, Foo D C Y, El-Halwagi M M, et al. A superstructure optimization approach for membrane separation-based water regeneration network synthesis with detailed nonlinear mechanistic reverse osmosis model[J]. Industrial & Engineering Chemistry Research, 2011, 50(23): 13444-13456.

[2] Khor C S, Chachuat B, Shah N. A superstructure optimization approach for water network synthesis with membrane separation-based regenerators[J]. Computers & Chemical Engineering, 2012, 42: 48-63.

[3] Buabeng-Baidoo E, Majozi T. Effective Synthesis and Optimization Framework for Integrated Water and Membrane Networks: A Focus on Reverse Osmosis Membranes[J]. Industrial & Engineering Chemistry Research, 2015, 54(38): 9394-9406.

[4] Yang L, Salcedo-Diaz R, Grossmann I E. Water Network Optimization with Wastewater Regeneration Models[J]. Industrial & Engineering Chemistry Research, 2014, 53(45): 17680-17695.

[5] Gouws J F, Majozi T, Foo D C Y, et al. Water minimization techniques for batch processes[J]. Industrial & Engineering Chemistry Research, 2010, 49(19): 8877-8893.

[6] Li L J, Zhou R J, Dong H G. State– time– space superstructure-based MINLP formulation for batch water-allocation network design[J]. Industrial & Engineering Chemistry Research, 2009, 49(1): 236-251.

[7] Zhou R J, Li L J, Dong H G. Optimal design of batch mass exchange networks with multipurpose exchange units[J]. Computers & Chemical Engineering, 2016, 84: 536-545.

Performance and kinetic evaluation of an anammox reactor

*P. V. Tikilili[1] and E. M. N. *Chirwa
*University of Pretoria, Department of Chemical Engineering, Pretoria, South Africa.
Tel: +27 12 420 4338, e-mail: phumza2@yahoo.com; [1]Soshanguve East, 172 Tshega street, Pretoria, Fax: +27 12 362 5089

Biological treatment processes are recommended for the treatment of wastewater effluents before discharging into receiving water bodies due to lower operation cost and energy saving . Nitrogen removal is one of the most crucial wastewater treatments required due to its contribution to eutrophication of the water bodies. Various innovative biological nitrogen removal processes such as Single reactor High activity Ammonia Removal Over Nitrite (SHARON), completely autotrophic nitrogen removal over nitrite (Canon) process, De-ammonification and oxygen-limited autotrophic nitrification-denitrification (Oland) process, have been developed. However, anaerobic ammonium oxidation (anammox) process is the latest technique for improved nitrogenous compounds removal from wastewater. This process is carried out by autotrophic bacteria of the type *Planctomycetes* called anammox bacteria which combine ammonium and nitrite to produce nitrogen gas and a small amount of nitrate in anoxic conditions. The anammox process is considered an economical and low energy alternative to the conventional biological nitrogen removal. The application of anammox process has been developed for wastewater treatment using different reactors. However, anammox still has some gaps to be filled with regards to the type of microbial cultures. The types of commonly used microbial cultures for the anammox process are granule-based and attached growth cultures. However, a gap still exists on the use of suspension cultures for anammox process.

In this study, a 5L sequencing batch reactor was successfully operated for the treatment of simulated wastewater using anammox process. The reactor was inoculated with a pre enriched anammox biomass obtained from a wastewater treatment plant. The reactor was operated in two stages differentiated by different nitrogen loading rate (NLR) i.e. 5 $g N \cdot L^{-1} \cdot d^{-1}$ and 6.3 $g \cdot L^{-1} \cdot d^{-1}$. The highest removal efficiency for total nitrogen achieved by the reactor was 93% at the highest nitrogen loading rate of 6.3 $g \cdot L^{-1} \cdot d^{-1}$. Furthermore, a modified Stover Kincannon model and evaluate the performance of the reactor. The modified Stover-Kincannon was more suitable for the description of nitrogen removal in the reactor, with the regression coefficient of R^2 =0.9739.

Valorising of wheat straw by applying combined liquid hot water and ethanol organosolv pretreatment

F. Weinwurm, J. Denner, T. Turk, A. Friedl
Technische Universität Wien, Inst. of chemical engineering, Getreidemarkt 9/166, 1060 Vienna, Austria; tel. +43158801166200, e-mail: anton.friedl@tuwien.ac.at

Lignocellulose based biorefineries, analogous to conventional refineries, aim to produce value-added products from renewable lignocellulosic resources through a series of operations. These products are raw materials for the chemical industry, food and feedstuff production or biofuels. Advantages of lignocellulosic feedstock biorefineries are: (1) their feedstock is often cheap and easily available as a byproduct from agriculture or forestry, (2) they offer a broad range of products. There are different paths for utilizing the biomass. Some use the whole biomass (e.g. biogas or syngas production), others aim to deconstruct it and convert the resulting fragments to produce a variety of chemicals. Lignocellulosic biomass is made up of three major components – cellulose, hemicellulose, and lignin – which have to be separated in order to process them into specific products. The cellulose can be sold as fibres or, like the hemicellulose carbohydrates, processed further to platform and specialty chemicals through biological or chemical conversion. Lignin could be a precursor for phenolic platform chemicals, polymers or specialty chemicals.

There are different treatments available to fractionate the biomass. For example, it can be treated with pressurised water at temperatures from 160 to 240 °C (Liquid hot water treatment), or treated with diluted acid or in alkaline conditions. Also, good fractionation has been accomplished with methods that combine cooking with a sudden pressure release, causing the moisture in the biomass to expand and thereby break up its structure. The organosolv treatment is a promising method to achieve high delignification of the biomass and to obtain relatively high purity lignin, and at the same time retain lower formation of sugar degradation products and cellulose cleavage. Yet, chemically diversified second generation biorefineries are rarely economically successful.

So far, only few authors have combined liquid hot water prehydrolysis with subsequent organosolv treatment of wheat straw. The aim of the presented work was to carry out the two treatments subsequently in order to track the release of carbohydrates and lignin into, and their degradation in, the respective liquid phases. Small reactors (45 mL) were used for both treatments to rapidly heat the mixtures of wheat straw and water, or a water-ethanol solution, to the desired temperature (180 °C). The reactors were quenched 0, 15, 30, 45 or 60 minutes after the temperature was reached and the liquors taken for analysis to track the concentrations of dissolved biomass components. A 1 L reactor was used to prepare larger batches of liquid hot water-treated straw for use as feedstock in the ethanol organosolv treatments.

The liquid samples were analysed for oligomeric and monomeric carbohydrates, their degradation products furfural and 5-hydroxymethylfurfural (HMF), acetic acid and lignin by HPAEC-PAD, HPLC and UV/Vis methods. To compare the treatment conditions between the large and small reactors, the severity factor R0 was used. It represents the combined effect of temperature and time on the biomass and is defined as shown in Equation (1).

$$R0 = \int \exp\left(\frac{T - 100°C}{14.75}\right) dt \qquad (1)$$

The feedstock for the organosolv treatment was prepared by liquid hot water treatment of wheat straw with a target temperature of 180 °C which was reached after 60 minutes. These conditions lead to a R0 of 4713. This severity corresponds to a treatment time of approximately 17 minutes at 180 °C in the small reactors. Temperature-time profiles and the corresponding R0 of both reactor systems are shown in Figure 1a. The total (sum of monomeric and oligomeric) carbohydrate yields were 16.9 % in the large reactor and 11.0 % in the small reactor after 15 minutes (R0: 4295). The different carbohydrate yields could be explained through longer residence time and higher severity in the large reactor, so the use of R0 remains a valid way to compare these two treatments.

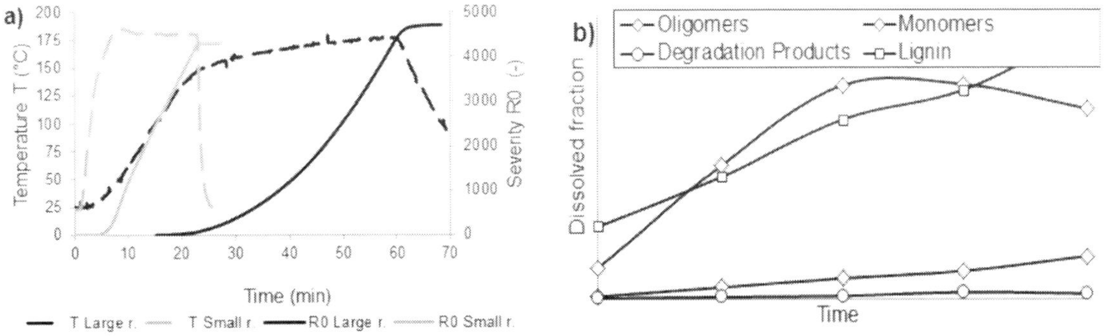

Figure 1: (a) Temperature-time profiles (dashed lines) of small (grey) and large (black) reactors with R0 values (solid lines). (b) Schematic progression of solute concentrations in treatment liquors of a hot water process

The progression of carbohydrates, their degradation products and lignin during liquid hot water treatment is shown in Figure 1b. Mainly oligomeric carbohydrates were released from the straw, which were then broken down to monomers and further degraded to furfural and HMF.

During ethanol organosolv treatment, a similar behaviour was observed with some significant differences. Carbohydrate release was lower by a factor of 2.6, and lignin yield was increased almost threefold after one hour of subsequent ethanol organosolv treatment. UV measurements suggest, that a large portion of carbohydrates formed soluble degradation products, which inflated the lignin measurements at 280 nm.

Uncatalyzed liquid hot water prehydrolysis and subsequent organosolv treatment were carried out successfully and the behaviour of the dissolved components was recorded. In the next steps, the kinetic parameters are calculated and compared with reported values in literature. The optimum treatment parameters for maximum delignification can be chosen accordingly.

Accelerated carbonation of MSWI fly ashes for CO_2 fixation

E. Santos[1], S. Cobo[1], A. Fernández[2], A. Irabien[1]
[1] University of Cantabria, Dept. Chemical & Biomolecular Engineering, Avda. Los Castros s/n, 39005 Santander, Cantabria (Spain); tel. +34 942206749, e-mail: angel.irabien@unican.es
[2]SADER S.A., Camino del Arsenal, 19, 48013 Bilbao, Basque Country (Spain)

Municipal solid waste incineration (MSWI) produces a variety of residues such as bottom ashes, fly ashes and air pollution control residues. Fly ashes correspond to the finest fraction of the remaining residues after thermal treatment being transported by the flue gases beyond the furnace. MSWI fly ashes are considered as hazardous waste because they contain large quantities of chlorides combined with significant concentrations of heavy metals, such as Pb and Zn, which may be leached by water. Consequently, the residues are usually treated by a stabilization/solidification process and are then landfilled [1-2].

Accelerated carbonation is proposed as an alternative technology for MSWI fly ash stabilization since heavy metals could be converted into more stable forms such as oxides or carbonates improving chemical and physical properties of the treated materials. Moreover, carbonation of alkaline residues captures anthropogenic CO_2, identified as one of the main contributors of climate change [3].

The present study focuses on the application of an aqueous phase accelerated carbonation treatment on MSWI fly ashes, aimed at gaining insight on the feasibility of the process at larger scale, identifying also the key factors for its industrial implementation.

Accelerated carbonation experiments were carried out in a 2 l glass reactor by bubbling continuous pure CO_2 in 100 g of ashes at 20ºC. The fly ashes were collected in a MSWI plant located in Zabalgarbi (Spain). Ash samples were oven dried at 105 ºC and then mixed with distilled water at L/S ratio ranging from 0.1 to 8 l/Kg. The compliance leaching test was carried out in accordance to Spanish Standard UNE-EN 12457-4 with L/S=10 l/Kg for 24 h of mixing and then filtered with a 0.45 μm filter paper. Chloride content was determined according to ISO 9297, while the metal concentration (Pb, Zn, Cu, Ni, As, Cd, Sb, Ba, Cr, Se, Mo and Hg) was determined by atomic absorption and ICP-MS. CO_2 uptake was evaluated using thermogravimetric analyses.

Carbonation had a significant influence on Pb and Zn leachability. The Pb release from carbonated ash was reduced by 1-2 orders of magnitude while Zn release was reduced by 1 order of magnitude when comparing with non carbonated MSWI fly ashes, in both cases below the landfill acceptance criteria for inert waste. Release of soluble salts was reduced after carbonation below the landfill acceptance limits for non hazardous waste. The ash showed high reactivity towards CO_2 with a maximum CO_2 uptake of 22%. Finally, progress has been made with a description of the process block diagram, a quantitative material flow analysis and the sizing of the main equipment with the aim of integrating the treated product in the supply chain.

Acknowledgments: Authors would like to thank the financial support from SADER S.A.

References

[1] Baciocchi, R., Costa, G., Di Bartolomeo, E., Polettini, A., Pomi, R.: The effects of accelerated carbonation on CO_2 uptake and metal release from incineration APC residues. Waste Management 29, 2994-3003 (2009).
[2] Cappai, G., Cara, S., Muntoni, A., Piredda, M.: Application of accelerated carbonation on MSW combustion APC residues for metal immobilization and CO_2 sequestration. Journal of Hazardous Materials 207-208, 159-164 (2012).
[3] Ukwattage, N.L., Ranjith, P.G., Yellishetty, M., Bui, H.H., Xu, T.: A laboratory-scale study of the aqueous mineral carbonation of coal fly ash for CO_2 sequestration. Journal of Cleaner Production 103, 665-674 (2015).

Changes in components of tyre-derived oil after sulphur and heteroatom removal using three acids

Natakorn Rujiravorawut[a], Suwannee Missita[b], Sirirat Jitkarnka[a, c, *]
[a]The Petroleum and Petrochemical College, Chulalongkorn University, 254 Chula Soi 12 Wangmai Pathumwan Bangkok 10330, Thailand; tel. (+66)2 2184110, fax: (+66)2 2154459,
[b]TPI Polene Public Company Limited, 299 Moo 5 Mitraparp Rd., T. Tubkwang A. Kangkhoy Saraburi 18260, Thailand; tel. (+66)36399111,fax:(+66)36339228
[c]Center of Excellence on Petrochemical and Materials Technology, Chulalongkorn University Research Building, Chula Soi 12 Wangmai Pathumwan Bangkok 10330, Thailand
*sirirat.j@chula.ac.th

Tyre-derived oil consists of many types of hydrocarbons such as aromatic hydrocarbons and sulphur compounds. Aromatic hydrocarbons can generate gum in the engine, which results in a lower efficiency of the engine and causes black smoke. Moreover, sulphur compounds can also cause air pollution. Conventionally, acid treatment has been adopted by small companies that aimed at reducing aromatics, but it also gives undesired compounds in the tyre-derived oil. It has been reported that sulphuric acid can remove impurity and sulphur compounds in waste lube oil. However, it is suspicious that sulphuric acid used in the acid treatment can also form heteroatom-containing hydrocarbon molecules, such as S compounds, in the oil. So can other acids. In this work, sulphuric acid, nitric acid, and hydrochloric acid were used as chemicals in acid treatment in order to investigate the changes in components in tyre-derived pyrolysis oil. The products were analyzed by using SIMDIST-GC, GC×GC-TOF/MS, and S analyzer. The results showed that the volume of tyre-derived oils decreased about 50%wt together with 50% sludge formation when sulphuric and nitric acids were used. For hydrochloric acid, the volume of tyre-derived oil decreased about 10wt. %. The tyre-derived oil contained 5% gasoline, 25% kerosene, 35% gas oil, LVGO 8% and HVGO 27%. The results from GC×GC-TOF/MS were used to explain the changes in oil components before and after treatments.

Optimizing of air stripper parameters for VOCs removal from groundwater

E. Edreder, [1]A. Imhamed, [2]M. Emtir
National oil corporation, P.O. Box 2655, Tripoli, Libya; tel. +218 21 4836820, eelmahboub@yahoo.com, [1]Sirte oil company, Brega, Libya, [2]Libyan Petroleum Institute, P.O. Box 6431, Tripoli, Libyan; tel. +218214830032, e-mail: memtir@yahoo.com

Groundwater has been important source of fresh water in the past and will continue to be one in the future. This source is threatened by the increase of the number of contaminants resulted from the activities of the chemical and oil industrial activities. The major sources of contaminants are accidental spills, underground storage tanks, waste lagoons, landfills, and septic tanks. There are several water treatment technologies that can be used in the treatment and ultimate removal of these organic chemicals. These treatment technologies include the activated carbon adsorption, packed tower aeration, and the low profile sieve tray towers.

This work provides a design and economic evaluation of the use of packed tower air stripping unit to remove VOCs (Benzene, ethyl-benzene, toluene, and xylene) from groundwater using ASDC program software.

The main objective is determine the packed-tower configuration (diameter and height of packing material), air loading rate and water loading rate that will enable reduction of a given influent concentration of a volatile contaminant to a desired effluent concentration at minimum for specific inflow rate and set of environmental conditions (air and water temperature and pressure).

The operating variables are considered to be stripping factor and pressure drop in the column. First, for a given packing material and a key contaminant component (benzene), the effect of stripping factor and pressure drop on the column design parameters (diameter & height) are investigated for minimum cost optimization. Second, the selection of key contaminant affects the design, operation and cost and this is systematically studied in this work. Third, the sensitivity of three types of packing materials: Super Intalox Saddles, pall-ring1", and nor –Pac 5/8"on the design, operation and cost is studied.

Summarized results indicated that, the nor-Pac 5/8 "packing, benzene as the key design contaminant, pressure drop of 1190 N/m/m and stripping factor of 3 are found to be the optimum air stripper parameters.

Generalisation of the solution of the inverse Richards' problem

M. Vocciante, A.P. Reverberi, V.G. Dovì
Univ. Genova, Dep. of Chemistry and Industrial Chemistry, Via Dodecaneso 31, 16146 Genova tel. +39 010 3536032, e-mail: dovi@istic.unige.it;

In inverse problems defined by models that include partial differential equations, a part of the boundary conditions are unknown and are to be estimated from experimental measurements [1]. We have shown in a previous article [2] that the solution of the inverse Richards' problem can provide useful information on the flux at the bottom of landfills using measurements at the surface only. The possibility of estimating percolation rates at the bottom through the use of surface data can be a useful complement of the information furnished by the vadose measurement system. In addition to pointing to the possible presence of biases of in-situ equipment, this approach makes it possible to use inexpensive mobile equipment to carry out surface measurements. Indeed, the monitoring of leaks in landfill liners can be undermined by the sensors located close to the liner being unable to detect a narrow plume or by sensors located in deeper regions failing to intercept the front of the plume due to the presence of impeding layers that divert the percolating water horizontally. Data misinterpretation can also be the consequence of instrumental malfunctioning, the repair of in-situ equipment being frequently too expensive to be carried out.

In this article, we consider a generalisation that makes it possible to consider the presence of unknown non-linear parameters, such as the effective hydraulic conductivity and the root uptake coefficients. This is accomplished using the method of separation of variables in the resulting estimation problem. Thanks to the linearity of the model, all these conditions can be expressed as linear functions of the unknown lower boundary condition. Otherwise, the relevant non-linear parameters are to be estimated from the data as well.
Using simulated data, we show how the correlation between the linear parameters contained in the unknown lower boundary conditions and the non-linear parameters can reduce the reliability of the monitoring procedure and hence the necessity of limiting the number of the latter.

[1] M. Vocciante, A.P. Reverberi, and V.G. Dovì, A General Method for the Solution of Inverse Problems in Transport Phenomena, Chemical Engineering Transactions 43 (2015), pp. 1615-1620, http://dx.doi.org/10.3303/CET1543270
[2] M. Vocciante, A.P. Reverberi, and V.G. Dovì, Approximate solution of the inverse Richards' problem, Applied Mathematical Modelling (2016), (article in press) http://dx.doi.org/10.1016/j.apm.2015.12.026

Seaweed valorization through agar production: process simulation and economic evaluation

J. D. Martínez-Consuegra[1], A. D. González-Delgado[2] Y. Y. Peralta-Ruiz[3]
[1]Corporacion Universitaria Minuto de Dios-UNIMINUTO, Business Administration Program, Carrera 53 No. 74 - 110, Barranquilla, Colombia.
[2] University of Cartagena, Chemical Engineering Department, Avenida del Consulado #Calle 30 No. 48 – 152, Cartagena, Colombia.
[3]Universidad Libre Seccional Barranquilla, Industrial Engineering Department, Carrera 46 No.48-170, Barranquilla, Colombia.
jairo.martinezc@uniminuto.edu

Seaweed can potentially be used for obtaining valuable products owing to their abundance, diversity and economic value. Since there are several processing alternatives for seaweed valorization, it is important to select those that can contribute to sustainable development of processing topologies under certain criteria. In this paper, Agar production from seaweed is evaluated using process simulation and economic evaluation by varying intermediate product separation alternatives, freeze-thaw (route 1) and evaporation (route 2).

Simulation was performed using UniSim Design Suite under appropriate kinetic and thermodynamic models, economic evaluation was performed taking into account local conditionsand capital and operative costs were calculated, economic feasibility was measured under parameters as Net Present Value, Profitability Index or Payback Period. A sensitivity analysis, was also performed evaluating six scenarios by varying raw material cost and the plant location.

It was found that two routes are feasible from the techno-economic point of view, however, route 2 has a higher profitability due to lower cost fixed capital, operating costs and utilities, presenting a higher contribution to sustainability by lower energy consumption and resources conservation.Finally, sensitivity analysis showed that raw material costs this is the most influential variable in plant profitability, an increase of 50 % the cost of feedstock, the payback period increases 100% and the net present value of the project decreases 50 %.

Development and validation of N-nitrosamine rejection mathematical model using a spiral-wound reverse osmosis process

Mudhar. A. Al-Obaidi, Chakib Kara-Zaïtri and Iqbal. M. Mujtabaa*
Chemical Engineering Division, School of Engineering, University of Bradford, Bradford, West Yorkshire BD7 1DP, UK. Tel: +44 1274 233645
I.M.Mujtaba@bradford.ac.uk

Reverse Osmosis (RO) is a water purification process, used in in water desalination and waste treatment that uses a semipermeable membrane to remove undesirable particles. RO has many immediate advantages including minimum thermal damage, high packing density and lower energy consumption (Reverberi et al., 2014; Fritzmann et al., 2007). Furthermore, spiral-wound RO modules are less susceptible to membrane fouling and are easier to clean (Song, 2002). With the increasing application of RO in the removal of organic compounds, the modelling of membrane separation and mass transport mechanisms constitutes a significant factor in the development of cost effective and optimum design strategies. Various types of models have been proposed to describe the transport phenomena of solvent and solute through the membrane. These models are based on the three-parameter irreversible thermodynamics model (Spiegler and Kedem, 1966), the solution-diffusion model (Lonsdale et al., 1965) and the pore flow models (Jonsson and Boesen, 1975). Successful examples of membrane transport mechanisms based on the Spiegler and Kedem model have been well described in the literature and validated for both wastewater treatment and seawater desalination processes. For example, Ahmad et al. (2007) have developed a model suitable for the multiple solutes system for unsteady-state simulation and prediction of membrane filtration dynamics in terms of permeate flux and concentration of each solute. This model was validated with experimental data derived from pre-treated palm oil mill effluent as a feed using a PVDF hollow fibre membrane module in a pilot plant scale RO system. Whereas, Mane et al. (2009) have studied boron rejection using two commercially spiral-wound modules in pilot-scale and full-scale RO processes by varying pH and pressures. Patroklou et al. (2013) developed a mathematical model of RO for boron rejection and validated the model using experimental data of Mane et al. (2009).

To the best of authors' knowledge, the development of a spiral-wound RO model based on the Spiegler and Kedem concept and its validation of N-nitrosamine compounds have only been conducted by Fujioka et al. (2014). Their model has conidered the variation of operating parameters in *x- axis* with assuming zero permeate pressure. The aim of this paper is to develop a new distributed model with relaxing the proposed assumption of zero pressure at the permeate side. The new idea here is to develop a mathematical steady state model based on the work by Spiegler and Kedem to simulate the rejection of N-nitrosamine compounds using a spiral-wound RO filtration process. The model will account for the spatial variation of all operating parameters along the length of the feed and permeate channels of the module (distributed model). The validation of this model has been corroborated against experimental data derived from the literature. The data includes uncharged N-nitrosamine compounds rejection based on a pilot-scale plant of three cylindrical pressure vessels with spiral-wound elements. The process model developed can be used later to study the variation of solute concentration, pressure, flow rate, solvent and solute fluxes along the length of the feed channel and to investigate the impact of variation in operating conditions on the unit performance.

The model has been validated by comparing the model predictions results with those obtained from actual experimentation for the specific RO membrane. Figure 1 shows the comparison of model rejections of the three selected N-nitrosamines at three different overall permeate fluxes against experimental results. The clear corroboration with experimental data readily shows the suitability of the model to measure the observed rejection data with an accepted error within high operating pressures. Furthermore, Figure 1 shows that the model can be used to simulate the observed data at low operating pressure (low average permeate flux) but with some minor deviation. This might be attributed to the inaccurate estimation of the transport membrane parameters at such pressure.

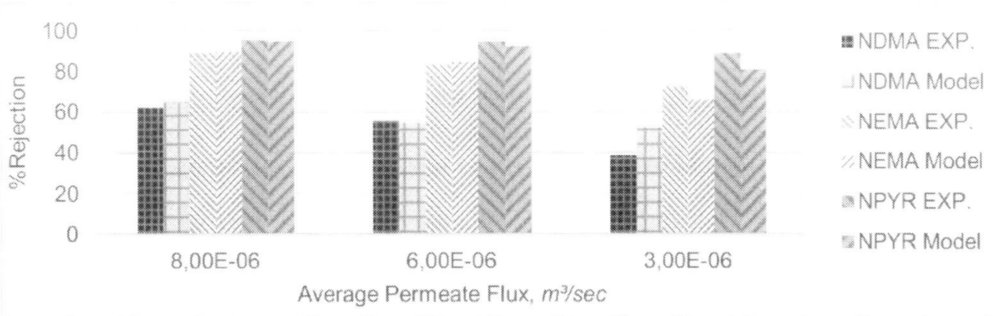

Figure 1: Experimental and model rejections of NDMA, NEMA and NDEA with average permeate flux

References

Ahmad A. L., Chong M. F. and Bhatia S., 2007, Mathematical modelling of multiple solutes system for reverse osmosis process in palm oil mill effluent (POME) treatment. Chemical Engineering Journal, 132, 183-193.

Fritzmann C., Löwenberg J., Wintgens T. and Melin T., 2007, State-of-the-art of reverse osmosis desalination. Desalination, 216, 1-76.

Fujioka T., Khan S. J., Mcdonald J. A., Roux A., Poussade Y., Drewes J. E. and Nghiem L. D., 2014, Modelling the rejection of N-nitrosamines by a spiral-wound reverse osmosis system: Mathematical model development and validation. Journal of Membrane Science, 454, 212-219.

Jonsson G. and Boesen C. E., 1975, Water and solute transport through cellulose acetate reverse osmosis membranes. Desalination, 17, 145-165.

Mane P. P., Park P.-K., Hyung H., Brown J. C. and Kim J.-H., 2009, Modelling boron rejection in pilot- and full-scale reverse osmosis desalination processes. Journal of Membrane Science, 338, 119-127.

Patroklou G., Sassi, K.M. and Mujtaba, I.M., 2013, Simulation of boron rejection by seawater reverse osmosis desalination. Chemical Engineering Transaction, 32, 1873-1878.

Reverberi A., Fabiano B., Cerrato C. and Dovì V., 2014, Concentration polarization in reverse osmposis membranes: Effect of membrane splitting. Chemical Engineering Transaction, 39, 763-768.

Song L., Hong S., Hu J., Ong S. and Ng W., 2002, Simulations of Full-Scale Reverse Osmosis Membrane Process. Journal Of Environmental Engineering, 128, 10, 960.

Spiegler K. S. and Kedem O., 1966, Thermodynamics of hyperfiltration (reverse osmosis): criteria for efficient membranes. Desalination, 1, 311-326.

Modeling of multi-scale uncertainties in biofuel supply chain optimization

D. Yue, F. You
Cornell University, Ithaca, NY 14850

Uncertainties are ubiquitous in a biofuel supply chain, such as in biomass supply, biofuel demand, and biorefinery yield (Awudu & Zhang, 2012). These uncertainties can significantly impact the performance of a biofuel supply chain, thus should be accounted for in supply chain optimization (Lee, 2014). Various approaches have been proposed in the past to address uncertainties in supply chain optimization, including but not limited to stochastic programming, robust optimization, and chance-constraint method (Sahinidis, 2004; Yuan et al., 2016). While most works treated different types of uncertainties equally in a uniform modeling framework, we note that uncertainties should be categorized into strategic and operational scales, and uncertainties at different scales should be treated differently.

As often seen in two-stage optimization problems, strategic decisions are often made "here-and-now", and operational decisions are made in a "wait-and-see" manner (McLean & Li, 2013; You & Grossmann, 2010). We consider strategic uncertainties as those associated with strategic decisions, e.g., biorefinery yield. Once realized, strategic uncertainties remain unchanged for a long period of time and have impacts throughout the lifetime of a project. On the other hand, we consider operational uncertainties as those associated with operational decisions, e.g., supply and demand variations. Operational uncertainties change frequently and often lead to immediate adjustments in operational decisions (Yue et al., 2014). Realizations of both strategic and operational uncertainties are not known exactly "here-and-now". However, strategic uncertainties will be realized once the strategic decisions are implemented and will be treated as constants during the operation phase. In this work, we consider the optimal design and operations of a three-echelon supply chain under strategic and operational uncertainties. The supply chain superstructure consists of a number of biomass suppliers, biofuel customers, and candidate sites for biorefineries. The strategic decisions include location and capacity of the biorefineries to build. The operational decisions include plans for biomass acquisition, biofuel sales, biofuel production, and material transportation and storage. The strategic uncertainty – biorefinery yield – is modeled using multiple scenarios as that in the stochastic programming approach (Gebreslassie et al., 2012). The operational uncertainties – biomass supply and biofuel demand – are modeled using box uncertainty sets as that in the robust optimization approach (Bertsimas & Thiele, 2004; Tong et al., 2014). Note that the operational decisions are dependent on the realization of biorefinery yield, thus leading to a scenario-based robust formulation. The objective is to minimize a special cost function, which accounts for the expectation of multiple scenarios in yield uncertainty and the robustness of operational solutions under supply and demand uncertainties. The model is solved with a combined solution strategy involving the Benders decomposition method and column-and-constraints generation method (Zeng & Zhao, 2013). A case study on optimal design of a potential biofuel supply chain under uncertainty is presented. The solution obtained through the proposed approach outperforms the deterministic solution in an uncertain environment.

References

I. Awudu, J. Zhang, 2012. Uncertainties and sustainability concepts in biofuel supply chain management: A review. Renewable and Sustainable Energy Reviews, 16, 1359-1368.

D. Bertsimas, A. Thiele, 2004. A Robust Optimization Approach to Supply Chain Management. In D. Bienstock & G. Nemhauser (Eds.), Integer Programming and Combinatorial Optimization, 3064, 86-100, Springer Berlin Heidelberg.

B. H. Gebreslassie , Y. Yao, F. You, 2012. Design under uncertainty of hydrocarbon biorefinery supply chains: Multiobjective stochastic programming models, decomposition algorithm, and a Comparison between CVaR and downside risk. AIChE Journal, 58, 2155-2179.

J. H. Lee, 2014. Energy supply planning and supply chain optimization under uncertainty. Journal of Process Control, 24, 323-331.

K. McLean, X. Li, 2013. Robust Scenario Formulations for Strategic Supply Chain Optimization under Uncertainty. Industrial & Engineering Chemistry Research, 52, 5721-5734.

N. V. Sahinidis, 2004. Optimization under uncertainty: state-of-the-art and opportunities. Computers & Chemical Engineering, 28, 971-983.

K. Tong, F. You, G. Rong, 2014. Robust design and operations of hydrocarbon biofuel supply chain integrating with existing petroleum refineries considering unit cost objective. Computers & Chemical Engineering, 68, 128-139.

F. You, I. E. Grossmann, 2010. Integrated multi-echelon supply chain design with inventories under uncertainty: MINLP models, computational strategies. AIChE Journal, 56, 419-440.

Y. Yuan, Z. Li, B. Huang, 2016. Robust optimization under correlated uncertainty: Formulations and computational study. Computers & Chemical Engineering, 85, 58-71.

D. Yue, F. You, S. W. Snyder, 2014. Biomass-to-bioenergy and biofuel supply chain optimization: Overview, key issues and challenges. Computers & Chemical Engineering, 66, 36-56.

B. Zeng, L. Zhao, 2013. Solving two-stage robust optimization problems using a column-and-constraint generation method. Operations Research Letters, 41, 457-461.

Pioneering phase of biomass industry in Malaysia: overview of key risks

Puan, Yatim [a*], Sue Lin, Ngan [b], Hon Loong, Lam [b], Ah Choy, Er [c].
[a] UKM-Graduate School of Business, Universiti Kebangsaan Malaysia, 43600 UKM, Bangi Selangor, Malaysia.
[b] Centre of Excellence for Green Technology, Faculty of Engineering, University of Nottingham Malaysia Campus, Jalan Broga, 43500 Semenyih, Selangor Darul Ehsan, Malaysia
[c] School of Social, Development & Environmental Studies , Faculty of Social Sciences and Humanities, Universiti Kebangsaan Malaysia, 43600 UKM, Bangi Selangor, Malaysia.
[*] Corresponding author e-mail addresses: puan@ukm.edu.my (P. Yatim).

Key words: biomass industry, risk assessment, risk management, financing, feedstock logistics, supply chain

Renewable energy sources have increasingly become important in recent decades not only as ways for improving energy security and mitigating climate change but also help contributing to economic growth and job creation. As the second largest producer of crude palm oil, Malaysia has abundance of biomass residues from palm oil industries and these biomass resources can be converted to bio-energy and bio-chemicals. In addition to biomass from palm oil industries, Malaysia is also rich with other renewable organic matters such as timber waste, rice husk, coconut trunk fiber, sugar cane waste, and municipal wastes. The current pressure for energy efficiency and growing energy demands provide opportunities for increased exploitation of these biomass wastes for power generation. Recognizing the roles of biomass for sustainability and wealth creation, the government has stepped up efforts to promote the growth of the industry. However, despite institutional arrangements, policy frameworks, funding mechanisms and incentives to support the growth of the biomass industry, the industry has not been able to create value along the industry's value chain. While there are many attractive reasons to venture into biomass industry, there are also potential risks. Lack of understanding of risks associated with the industry is often cited as one of the reasons for the industry's slow growth. An industry evolves through life cycle stages and at each stage presents risk factors. This paper attempts to identify several key risks associated with early stage of biomass industry. The findings of the study will help industry players to better understand key risks and improve risk assessments, risk management and mitigation, and ultimately position the industry to be more viable.

Assessment of power generation alternatives through a fuzzy multiobjective mixed integer long-term planning model: case study of non-interconnected areas of Colombia

[1]A. M. Rosso-Cerón, [2]V. Kafarov, [3]Gerardo Latorre and [4]Y. A. Herrera.
[1]Universidad Industrial de Santander, carrera 27 calle 9; tel. +57 7634400 ext. 2603, e-mail: anarosso10@gmail.com; [2]Universidad Industrial de Santander-Bucaramanga-Colombia, carrera 27 calle 9-Bucaramanga-Colombia; tel. +57 7634400, e-mail: kafarov@uis.edu.co; [3]Universidad Industrial de Santander-Bucaramanga-Colombia, carrera 27 calle 9-Bucaramanga-Colombia; tel. +57 7634400, e-mail: glatorre@uis.edu.co; [4]Universidad Autónoma de Bucaramanga, avenida 42 No. 48-11-Bucaramanga-Colombia, +57 76436111, e-mail: yherrera743@unab.edu.co.

Nowadays, energy dependence on fossil fuels (in terms of primary energy consumption representing an average of 35% oil, 27% coal, 23% natural gas, and the rest from alternative energies) triggers significant concerns as they are not renewable energy.

Furthermore, energy generation is necessary for a country development since it is essential for everyday activities such as lighting, using electrical appliances, transport, livelihoods, and education by increasing employment opportunities that lead to economic and social benefits.

In Colombian case, non-interconnected areas are located in isolated places; cover almost 66% of the national territory including about 1,200 settlements, 16 departments, 91 tows and 2 million people. In addition, these zones do not have appropriate access routes (roads and rails), lack industrial and commercial development, public services are limited and undeveloped, and the basic needs of people are unsatisfied.

With the aim of supplying the lack of electricity in non-interconnected areas have been established solutions with diesel generation, which are not sustainable in the long term, since an economical and environmental point of view. In addition initiative for the implementation of renewable energy in non-interconnected areas becomes sustainable option, which means the implementation of projects for development plans in this area.

Therefore, with the goal of planning a sustainable energetic future for supplying the electricity demand until 2024, in two isolated regions (San Andres Island and La Macarena) of non-interconnected areas of Colombia, this work assesses photovoltaic solar, biomass gasification, wind turbines, small hydro, grid extension, and diesel systems trough a fuzzy multiobjective mixed integer lineal programming- MOMILP.

The model has into consideration two objectives: the minimization of the total power system cost (capex, opex, transmission and fuel costs) and CO2 Emissions equivalent (environmental objective) and also design, operational, and efficiency constrains.

In order to test the deterministic model, approximate parameters were obtained from the National Monitoring Center (CNM), Unified Information System (SUI), the Energy Regulation Commission and Gas (CREG), the International Energy Agency (IEA) and NASA for areas of San Andres and La Macarena.

A compromise programming approach is proposed for finding a preferred compromise solution through an interaction between the decision maker and model analyzer, the main goal is to solve a problem with fuzzy multiobjective linear coefficients resulting fuzzy solution in the space defined by its objectives possibility distribution.

Then in this methodology a mathematical support for multi-objective decision-making takes place in a context of uncertainty. Due to unavailability of required data (parameters) over the mid-term decision horizon, those are assumed to be imprecise (fuzzy) in nature, all are defined by their possibility distributions (the semantic interpretation of the degree of membership will handle will be the degree of possibility), that is, we will address the problem of MOMILP with fuzzy parameters using the possibility theory.

Logically this requires working with models based on the estimation of the possible values that can reach the magnitudes involved in it, and therefore, the results will contain a greater or lesser degree of imprecision and uncertainty.

In the case of San Andres Island the results indicate that the environmental objective favours the use of wind systems, while economic objective favours to continue using diesel plants.

On the other hand, in the case of La Macarena, the environmental objective favours the use of biomass systems, while economic leans towards use of small hydro.

Current research focuses on the extension of the proposed model to incorporate uncertainty, and analysis of several alternative scenarios to provide with further insight into the influence of key parameters into the overall power system.

Rapid capital and operating cost estimation for biorefinery processes under development

M. Tsagkari[1,2], A. Kokossis[2], J.L. Couturier[1], J.L. Dubois[1]
[1] Centre de Recherche Rhône Alpes, Arkema, Rue Henri Moissan, 69491, Pierre-Bénite, France, mob. : +33788507095, e-mail : mirela.tsagkari@arkema.com
[2] Dep. Of Process Design and Plant Analysis, School of Chemical Engineering, National Technical University of Athens, 9 Heroon Polytechniou Str., 157 80, Athens, Greece

Quick capital and operating cost estimation at the early stages of a process development is crucial for determining the viability of the project and screening the economic potential of alternative routes. Biorefinery processes employ state-of-the-art technologies and thus, represent high risk business decisions. Therefore, preliminary cost estimation during the R&D stages is necessary to help researchers decide wisely on project continuity. Most references rely on literature information to estimate the costs as historical cost data are proprietary information and are seldom announced. They also draw on various costing techniques and heuristics, which were developed for the needs of the petrochemical and the Oil & Gas Industry and require high level of process detail, which is not available at the early-stages of the process conception.

This study investigates several capital cost estimation methods and reports wide discrepancies in their results, which urge for new developments in the field. The methods rely on "significant process steps" and thermodynamics-based cost estimation relationships and are widely cited in literature, while three well-documented biorefinery processes, representing different types of biorefineries, are used as validation models. The methods' accuracy is evaluated with means of both deterministic and statistical comparison from reported literature estimates and commercial biorefineries' cost data. As the majority was published during the 1970's-80's and is derived by petrochemical processes, their estimates do not account for the technological progress often met within the state-of-the-art biorefinery processes. Therefore, they are considered obsolete and their use is put into dispute.

Fig. 1 illustrates the reported discrepancies for dry corn mill ethanol biorefineries. The authors have been collecting capital cost data for commercial corn-to-ethanol plants (marked as circles). Only three out of the six cost methods fall within the 90% Confidence Intervals of the best regression line (marked as triangles), while the capital estimation reported in literature reference greatly underestimates the plant's final cost (marked as x). This study reports similar figures for chemical and thermochemical biorefineries, as well.

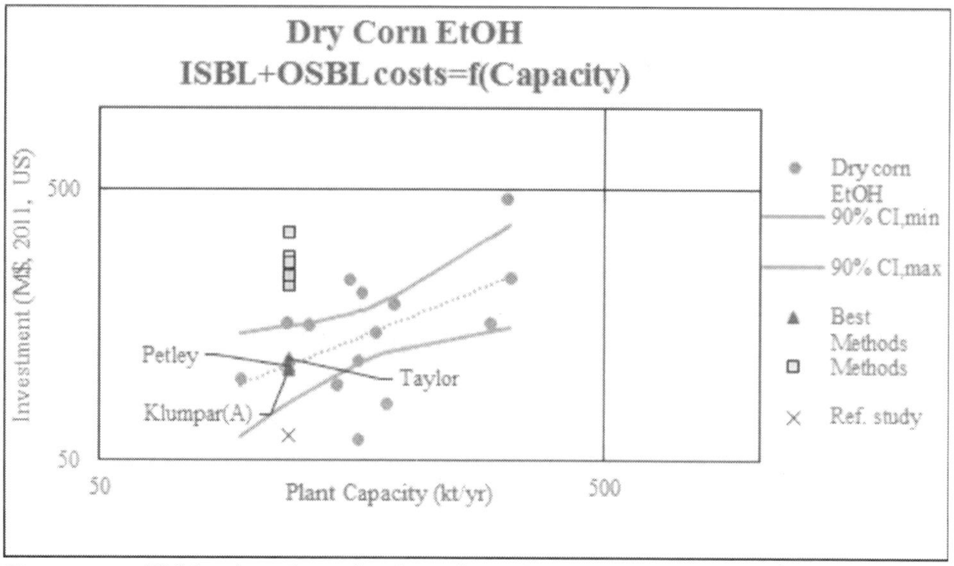

Fig. 1. Dry corn mill bioethanol production plants Investment Costs (M$, 2011,US) vs Plant Capacity (kt/yr)

Therefore, the authors decided to attempt new capital and operating cost estimation methods suitable for biorefineries under development, requiring information available at the start of the process development and within known uncertainty ranges. For the purpose of this work, the authors have been compiling two databases: the first is composed of cost data from 303 commercial cost biorefinery plants and the second consists of detailed flowsheets from 40 discrete biorefinery processes.

Various process parameters and variables were investigated to determine their effect on capital costs. The authors have developed capital cost estimation relationships, either based on curvilinear regression analysis or based on the "significant process step" approach. The uncertainty around the point estimate is determined based on probabilistic range estimating, accounting for both technical and estimation relationship uncertainty.

Furthermore, the authors have developed a new rapid operating cost estimation methodology with uncertainty. The methodology relies on factors requiring minimal user input, i.e. the raw materials cost calculation, for determining the net production cost of the process, employing probabilistic means of determination.

Overview of biomass supply chain optimisation model: HAZOP approach

Chun Hsion. Lim[a,b], Hon Loong. Lam[b]
[a]Crops For the Future, The University of Nottingham Malaysia Campus, Jalan Broga, Semenyih, 43500 Selangor, Malaysia.;
[b]The University of Nottingham Malaysia Campus, Dep. Chemical & Environmental Engineering, Jalan Broga, Semenyih, 43500 Selangor, Malaysia.
Honloong.lam@nottingham.edu.my

Biomass has been one of the potential alternative resources to achieve sustainable system at current stage of development. However, implementation of biomass industry is still a major challenge due to several issues. The main problems are related to various supply chain management limitations such as high logistic cost, low biomass density and fluctuation in availability. Researchers have been working on this matter by introducing biomass supply chain optimisation model to optimise the biomass system. Nevertheless, much improvement of current biomass supply chain optimisation system is expected to fill in the gaps for further improvement. This study gives an overview on current state of biomass supply chain modelling approaches and analyse in detail potential research gaps to improve the feasibility of biomass implementation. A systematic approach is implemented in the study to ensure high quality review. Hazard and Operability Study (HAZOP) is implemented into overview of biomass supply chain for the first time. This approach is generally used in chemical process plant to ensure process safety and operation. The traditional HAZOP methodology is modified into a robust platform for general literature review. Guidewords and deviations are used in brain-storming session as a guideline for hazard identification process. This will ensure a systematic consideration of all potential process hazards within the system. Each identified biomass supply chain process hazards are cross-checked with literature to determine existing rectification proposed by researcher. In other word, existing biomass supply chain optimisation modelling approaches from literature are considered as the safeguard for respective supply chain process hazards. In case of inadequacy of existing approach to cover specific hazards, recommendations are suggested as part of improvement over the current approaches. Based on the overview of biomass supply chain system via HAZOP approach, critical issues in biomass supply chain are identified. This allows higher level of analysis and overview of the system, and act as a platform to enhance existing biomass supply chain optimisation research and development.

Optimization of the future Finnish gas distribution system

M. Mikolajková;[1] F. Pettersson,[1] F. Maréchal,[2] H. Saxén[1]
[1] Åbo Akademi University, Thermal and Flow Engineering Laboratory, Biskopsgatan 8, Åbo 20500, Finland, tel. +358 2 2154915, e-mail: mmikolaj@abo.fi
[2] EPFL Valais Wallis, EPFL IPESE (SCI-STI-FM), Rue de l'Industrie 17, Case postale 440, Sion 1951, Switzerland, tel. +41 21 69 33516, e-mail: francois.marechal@epfl.ch

Natural gas — a fuel with one of the lowest CO_2 emissions among the fuels commonly used in Finland for the heat and electricity production — is supplied to Finland by a pipeline connection from Russia. The existing pipeline connects the southern part of Finland and supplies natural gas to the biggest cities in Finland. Almost 30 TWh of natural gas was supplied to cover the demand [1]. Other possibilities to gas supply would be either to inject locally produced biogas into the pipeline or a prospective injection of re-gasified natural gas from liquid natural gas (LNG) terminals. While biogas injection helps to decrease the emissions, the construction of new LNG terminals and storages along the Finnish coastline is currently seen as a measure to gain more energy independence [2]. A new pipeline construction connecting Finland and Estonia was currently recognized by Finnish and Estonian politicians as an advantageous step to improve energy independence. The future expansion of the natural gas distribution network depends on the selection of an appropriate place for new LNG ports in Finland and their successful integration in the supply network.

The weather conditions have a strong impact on the heat and electricity production and thus on the consumption of natural gas. The existing natural gas network was simulated, considering the seasonal variations and the profiles of the main consumers. The problem was formulated as an optimization problem, making it possible to find the best solutions for given objectives. The model yields information about the mass flows through the pipeline, optimal pressures, compressor duties and the possibility to include prospective new consumers connected to the pipeline. The feasibility of new supply points, e.g., the introduction of re-gasified natural gas from LNG terminals or biogas from potential sites, was also studied. As another option, the flows in a fictitious pipeline connecting Finland and Estonia were simulated under different circumstances, including seasonal and other variation in the demands. An attempt was also made to optimize the use of the pressurized natural gas by energy retrieval using turbines installed in nodes with sufficiently high delivery pressure considering the influence of this on the performance of the entire network.

This work was financially supported by the research program Efficient Energy Use (EFEU) Finland and Graduate School in Chemical Engineering (GSCE).

References

1. Gasum. *Gasum's Year 2014: Gasum Annual Report 2014.* Espoo: Gasum Corporation. , 2015.

2. YLE uutiset. ***Finland and Estonia Confirm LNG Terminal***. YLE. 18.11.2014, 2014 Available from:http://yle.fi/uutiset/finland_and_estonia_confirm_lng_terminal/7635420.

Understanding chemical engineering student's learning of process systems engineering from metacognitive perspectives

M. Promentilla[1,2], K. Aviso[2], R. Lucas[3],
[1]Center for Engineering and Sustainable Development Research
[2]Chemical Engineering Department
[3]Br. Andrew Gonzalez FSC College of Education
De La Salle University Manila
tel. +6325360223, e-mail: michael.promentilla@dlsu.edu.ph; 2401 Taft Avenue, Malate, Manila, Philippines, 0922

Process systems engineering (PSE) in the field of chemical engineering education requires a systems approach to process design and process integration which emphasizes the unity of the processes and considers the interactions between different unit operations and processes. Understanding the behavior and interaction of such systems often times require the evaluation and calculation of multiple mathematical equations which are more easily accomplished in a technology-enhanced learning environment particularly with the use of e-learning and computational tools. The integration of computer aided courses within the curriculum of the undergraduate chemical engineering degree has thus become essential in preparing students for professional practice. In addition, previous works have shown that these discipline specific computer-aided exercises can potentially assist in the development of metacognitive skills. However, the evaluation of the effectiveness of these tools in learning PSE based from metacognitive perspectives has not yet been fully assessed. Early studies on metacognition refer to it as a concept wherein an individual is aware and able to regulate his cognitive activities such as reasoning skills and perception in the learning processes. Traditional lecture delivery for example has not been considered effective for promoting higher order thinking and learning in students. This work thus explores how learning is enhanced through the use of discipline specific problem solving exercises with the aid of various computational tools and software packages. The pilot study involved creating story problems for computer-aided based course in the undergraduate chemical engineering curriculum of a certain private Philippine university. The study utilizes an evaluation instrument based on Metacognitive Activities Inventory (MCAI) to measure the metacognitive skilfulness of the students. This study also presents implications on future research directions in evaluating student's e-learning strategies..

Potential of biomass to resources as an action to mitigate transboundary haze in ASEAN country

[1]S.T. Tan, [1]H. Hashim, [2]A.H.A Rashid, [3]E.K.Y. Wong, [1]J.S. Lim, [1]W.S. Ho, A.B. Jaafar[4]
[1]Process System Engineering Centre (PROSPECT), Research Institute for Sustainable Environment (RISE), Universiti Teknologi Malaysia, Johor, Malaysia;
tel. + 607-5535578, e-mail: haslenda@cheme.utm.my;
[2]Industrial Biotechnology Research Centre, SIRIM Berhad, Shah Alam, Malaysia
[3]STI Strategic Studies Division, Academy of Sciences Malaysia
[4]Ocean Thermal Energy Development Centre, Institute of Future Energy, Universiti Teknologi Malaysia, Johor, Malaysia

For decades, the countries of Southeast Asia, especially Singapore and Malaysia suffer from the attack of transboundary haze caused by forest fires and open burning in Indonesia. Most of the forest burning happened on the Sumatra Island and Kalimantan province in Indonesia, and some remote area in Sabah and Sarawak in Malaysia. The small and medium enterprise and local farmers practice open burning of the surrounding forests and biomass of pulp, paper, and palm oil plantations to clear space for expansion. Given the environmental damage and human health risks, these fires need to stop immediately. One solution to clearing the biomass without fire may be to harvest it and use it for value-added resources.

Conversion of biomass on the field in forest or plantation into value-added product such as power and bioethanol can potentially play a role in an effort to reduce the resulting haze conditions from slash and burn practices. There are competing uses for biomass to resources because of their economic and environmental value for a variety of purposes. Biomass material can be potentially be used to generate power, heat, steam, and bioethanol, which potentially offer high economic returns to the farmers.

To explore the economic potential of biomass-to-resources, two case studies on biomass-to-power and biomass-to-ethanol are incorporated in this study with the suggested feedstock capacity of 2000t of biomass daily. Net present value economic analysis with equity and debt corporate financing method are applied in the case studies to analyse the economic profitable levels of biomass-to-resources.

In the case study of biomass to power, the minimum electricity product cost ranged from \$ 0.23/kWh to \$ 0.19/kWh with variations of equity financing share of 30% to 70%. The minimum product cost is consider high even with the equity financing adoption as compared to the current feed-it-tariff (FiT) incentive of \$ 0.10/kWh, as shown in Figure 1. The case study is repeated with different capacities (2000t/d, 1000t/d, and 500t/d). It can be seen that there is only a marginal reduction in the minimum electricity price (ranged from \$ 0.24/kWh to \$ 0.19/kWh) due to economy-of scale capacity increment. This is due to the high fixed investment cost (approximately \$3000/kW), while the current FiT scheme is relatively low. The low FiT scheme renders the biomass-to-power to be less competitive at the current power industry market.

In the case study of biomass to ethanol, the production cost ranged from \$ 0.64/l to \$ 0.62/l with the movement of equity financing share from 30% to 70% which is higher than the current market ethanol price of \$ 0.58/l, as shown in Figure 2. Figure 2 also shows the variation of ethanol production cost at different capacities and with variation in enzyme costs. The plot demonstrates that economic viability from lower ethanol production cost can be achieved at favourable equity financing ratios, higher capacities (due to economy of scale) and lower enzyme costs

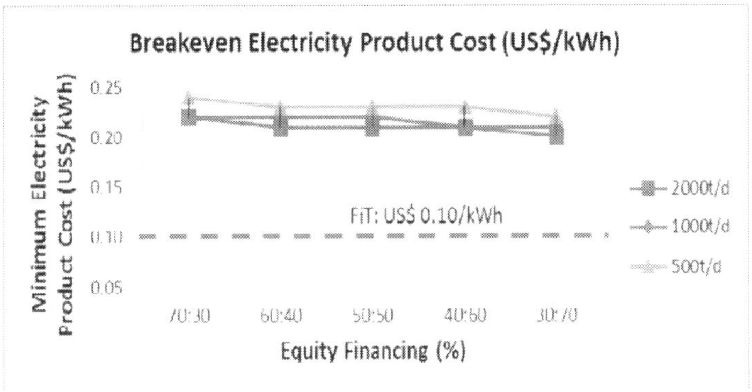

Figure 1 Breakeven of electricity selling price for biomass-to-power in Malaysian context.

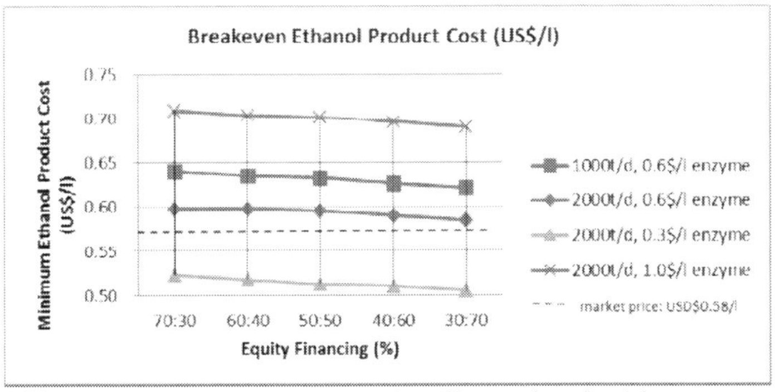

Figure 2 Breakeven of ethanol selling price for biomass-to-ethanol in Malaysian context.

The two case studies reviewed the economic potential of localised biomass-to-power and ethanol in current market. For biomass-to-power, the current FiTscheme is relatively lower than the electricity production cost, rendering the biomass-to-power option less attrative to investors. The rate of FiT scheme in Malaysia was established in year 2011, and is considered not up-to-date on current renewable resources market as various RE resources have been more economically competitative in recent years. In order to promote the utilisation of biomass to power, the current FiT should be reviewed and revised.

The case study of biomass to ethanol, on the other hand, demonstrated a favourable scenario to investors demonstrating that with a financial interest rate of 3%, ethanol production is economically competitive in the current market. Nevertheless, the current interest rate stands at the rate of 5%-8% and with high cost of enzyme in Malaysia, there needs to be some policy and technology intervention to enablea sustainable bioethanol industry in Malaysia.

Although economic return is one of the main causal factors for the regional haze occurrences, it is probably a great motivator for moving away from traditional methods of land clearing which does not yield any economic benefits. As haze episodes may evolve into potentially complex emergencies, the development of an effective technology for biomass utilisation is critical. From the above detailed discussion, turning waste into vaue-added products such as compost, fuel, power, and biochemical will create an economic benefitsand utimately reduce the open burning practices and prevent the haze issue.

Absorption cooling devices with LiBr / H_2O as working media

A. Žužić, [1]V. Filipan
University of Zagreb Faculty of Chemical Engineering and Technology, Marulićev trg 19, HR–10000 Zagreb; tel. +385 995730546, email: azuzic@fkit.hr, zuzic.andreja93@gmail.com;
[1]University of Zagreb Faculty of Chemical Engineering and Technology, Department of Thermodynamics, Mechanical Engineering and Energy, Savska cesta 16, HR–10000 Zagreb; tel/fax: +385 14597109

Nowadays, cooling demand is rapidly increasing in many parts of the world, especially in Mediterranean and moderate climates. Electrically driven compressor chillers are usually used for the refrigeration and space cooling purposes, but their large consummation of electric energy, especially in peak–load periods during the summer, causes some problems in energy systems. This also results with an unwanted increase in consumption of electric energy obtained mainly from fossil fuels. On the other hand, the absorption cooling devices are driven mostly by thermal energy and only a little part of electric energy is used for their operation. Furthermore, solar thermal energy can be used for running the absorption cycles and efficient cooling can be obtained during the periods with largest solar insolation which is in correlation with the system cooling demand. Waste energy from the absorption cycles can be further utilized for the heating of sanitary domestic water. Also, in the periods when there are no needs of cooling, solar thermal energy can be utilized for the space heating of buildings. The market potential for absorption cooling, especially solar cooling, is very large and still under development. These solutions are environmentally friendly because of their lower electricity demand and they also have very low operational costs, which is another advantage. Since the absorption cooling systems are not yet sufficiently recognized in technical practice, an absorption cooling device installed for cooling of a business building in Dubrovnik, Croatia is chosen as the basis for the proposed paper.

This paper contains a short historical review of absorption cooling devices, starting from the first experiments revealing that some liquids are capable to absorb aqueous vapour and some other gases all through the mass produced devices that have become forerunners of many modern absorption systems. The aim of the study is to develop the mathematical model of an absorption cooling process in MATLAB and to test this model by calculating thermodynamic properties of binary mixture LiBr/H_2O using the constant coefficients given by Dong – Seon Kim and Infante Ferreira. The derived model is used for calculation of absorption cooling process in the working conditions of the installed solar cooling plant in Dubrovnik, Croatia. The technical parameters of the system were collected by using the long distance measurement system. The experimental data are then compared to the data calculated by the model. The comparison confirms that the theoretical model can be applied for this case taking into account slight deviations resulting from the simplifications used in the model (particularly from the idealization in the heat transfer processes).

A scheduling model of household appliances with a battery storage system based on charge/discharge characteristics of lithium-ion battery

Feifei Yang, Shaoyong Ke, Xiankun Huang, Yongzhong Liu
Xi'an Jiaotong University, Dept. of Chemical Engineering, Xianning West Rd. 28, 710049 Shaanxi, China;
tel. +86 29 82664752, e-mail: yzliu@mail.xjtu.edu.cn.

Due to over-consumption of fossil fuel and aggravation of greenhouse effects, household application is one of the viable solutions for renewable power system. Lithium-ion batteries are widely used for energy storage and constant energy supply for the excellent electrochemical performances.

This work demonstrates a stand-alone renewable power system combined with battery storage technologies for a household application. An integrated approach is proposed for optimal scheduling of household appliances, coordinated with the battery storage performances based on battery charge and discharge characteristics.

A mixed integer linear programming (MILP) model for scheduling of appliances and battery in household is established to minimize the levelized cost of energy (LCOE) of the system. Battery characteristics, such as charging/discharging rate, depth of discharge, capacity of cell and battery are optimized by considering the lifespan and the LCOE of entire system. The strategies for appliances and battery scheduling in multiple periods are embedded into the model.

Results of the case study show that compared with the results in literature, the LCOE obtained by the proposed model can further be reduced as a result of the optimization of battery characteristics and multi-period scheduling of appliances and battery. The LCOE and lifespan are intensively dependent upon the characteristics of battery charge/discharge.

The results of this work provide deep insights into design and operation of the battery storage system and the corresponding hybrid renewable power system.

Lignin valorization from side-streams produced during agriculture waste pulping and TCF bleaching

J. Fernández-Rodríguez[1], E. Robles[1], O. Gordobil[1], M. González-Alriols[1], J. Labidi[1]
Biorefinery Processes Group, Chemical and Environmental Engineering Department, University of the Basque Country UPV/EHU, Plaza Europa, 1, 20018, Donostia-San Sebastián, Spain
*E-mail: jalel.labidi@ehu.eus, tel.:+34 943017178; fax: +34 943017140

The residual biomass valorization has recently emerged because of its cheaper value and the huge possibilities of conversion that could offer. Traditionally the use of biomass valorization is reduced to energy generation; however biomass presents a chemical composition that allows the transformation of its main macrocomponents (cellulose, hemicelluloses and lignin) into bioproducts which are currently obtained from fossil resources [1].

Blue agave bagasse is one of the most abundant agricultural waste products in western Mexico as represents up to 40 % (on wet weight basis) of the blue agave (*Agave tequilana*) plant, a traditional crop cultivated to produce the distilled spirit called tequila after extracting sugars from the stem, which is then left after being cooked, shredded and milled thus being an undervalued product [2]. Recent works have aimed blue agave bagasse as source for high value-added materials such as nanocellulose [3,4] due to their high amount of α-cellulose (around 50%). These studies have demonstrated the possibility to valorize a residual stream into high value-added products with properties similar to those obtained from neat raw material, but with lower production costs.

Nevertheless, the works focused on the nanocellulose production have traditionally omitted the valorization of the whole biomass fractions, forgetting for instance streams produced during the cellulose purification that are rich in lignin. In this work, the extraction and characterization of lignin obtained as byproduct from a global process of purification of cellulose from blue agave bagasse and leaves is carried out. In this way, the whole valorization of this biomass is studied taking into account the principles of the integrated biorefinery, in which, not only the cellulose valorization is treated, but also the rest of the main fractions of the biomass, in this case, the lignin content; and all this by means of environmental friendly processes.

The working procedure began with a sulfur-free pulping process for the lignin separation. This process consisted in an organosolv process using ethanol-water (70:30 v/v) as white liquor during 90 min at 180 °C in a solid ratio of 1:15. After the separation of the solid fraction from the black liquor, two total chlorine free bleaching stages were effectuated to obtain high purity cellulose for further nanocellulose production. The bleaching stages consisted in a double oxygen treatment with sodium hydroxide (1.5 wt.%). Finally, lignin from the pulping black liquor as well as from the spent bleaching liquors was isolated by selective precipitation to be valorized as phenolic compound source.

Comparison between the obtained lignin from the different stages according to their extraction yield and physico-chemical characterization is carried out. From agave bagasse, higher yields in lignin extraction are achieved than from agave leaves. For example, during the organosolv pulping process the yield was 9.49% for bagasse while for leaves, 5.03% with regard to the initial biomass introduced to the reactor. As it was expected, the lignin extraction yields were reduced in each stage. Also the purity of the lignins, referred to the acid insoluble and soluble lignin, were lower for the bleaching stages, but remaining a level of purity higher than 75% for every sample, a high level taking into account the purity of lignin from other feedstock [5]. The molecular weight distribution of all the samples and the functional chemical groups were evaluated by means of gel permeation chromatography and Fourier transform infrared spectrum, demonstrating in this way the potential of valorization of lignin as part of integrated biorefinery process.

References:

[1] Nunes, L. J. R., Matias, J. C. O., & Catalão, J. P. S. (2016). Wood pellets as a sustainable energy alternative in Portugal. Renewable Energy, 85, 1011-1016.

[2] Kestur GS, Flores-Sahagun THS, Dos Santos LP, Dos Santos J, Mazzaro I, Mikowski A. Characterization of blue agave bagasse fibers of Mexico, Compos Part A-Appl S, 2013;45:153-161. DOI:10.1016/j.compositesa.2012.09.001.

[3] Espino, E., Cakir, M., Domenek, S., Román-Gutiérrez, A. D., Belgacem, N., & Bras, J. (2014). Isolation and characterization of cellulose nanocrystals from industrial by-products of Agave tequilana and barley. Industrial Crops and Products, 62, 552-559.

[4] Robles, E., Urruzola, I., Labidi, J., & Serrano, L. (2015). Surface-modified nano-cellulose as reinforcement in poly (lactic acid) to conform new composites. Industrial Crops and Products, 71, 44-53.

[5] Toledano, A., Serrano, L., & Labidi, J. (2014). Improving base catalyzed lignin depolymerization by avoiding lignin repolymerization. Fuel, 116, 617-624.

Demonstration of biogas methanation combined with membrane-based gas-upgrading in a promising Power-to-Gas-concept

F. Kirchbacher, M. Miltner, [1]M. Lehner, [2]H. Steinmüller, M. Harasek
Institute of Chemical Engineering, Vienna University of Technology, A-1060 Vienna, Getreidemarkt 9/166, Austria; tel: +43 1 58801 166253, e-mail: florian.kirchbacher@tuwien.ac.at, fax: +43 1 58801 15999; [1]Chair for Process Technology and Industrial Environmental Protection, Montanuniversitaet Leoben, A-8700 Leoben, Franz Josef-Strasse 18, Austria; [2]Energieinstitut at the Johannes Kepler University Linz, A-4040 Linz, Altenberger Strasse 69, Austria.

The further development of PV and wind power as renewable energies with their production rate fluctuations both on short- and medium-time-scale result in the necessity of smarter grids and higher energy storage capacities. One very prominent and promising technology for meeting this future electric energy storage demand is the concept of Power-to-Gas. Here, the excess electric energy is converted to hydrogen using alkaline or PEM electrolysis. Most concepts incorporate an immediate subsequent conversion to methane using a local carbon dioxide source and a process of thermo-catalytic or biological methanation. After a final gas upgrading mainly comprising the separation of H_2, CO_2 and H_2O the thereby produced SNG can be fed to the natural gas grid owning a huge potential for energy storage and distribution.

The current work presents the joint research efforts undertaken by the authors in the field of Power-to-Gas-Processes. A process chain consisting of a coupled hydrogen dark fermentation and a biogas fermentation, a thermo-catalytic methanation step and product gas upgrading applying membrane-based gas-permeation is developed and demonstrated on laboratory scale. Fermentation steps deliver a raw biogas mixture containing mainly methane, carbon dioxide, hydrogen and water together with traces of hazardous soubstances like hydrogen sulphide and ammonia. After an initial drying step this gas is compressed and trace component removal is performed by multistage adsorption. Without further treatment this gas is mixed with make-up-hydrogen from electrolysis and fed to the methanation step using nickel as active catalytic material. The gas upgrading consists of an initial drying step by condensation (gas temperature 5 °C) and a subsequent membrane stage separating residual H_2O, H_2 and CO_2. The methane-rich product gas stream is thoroughly analysed and compliance to Austrian gas grid standards is verified (OEVGW G31/G33). The removed gas components are envisaged to be fed back to the mathanation in order to increase methane yields.

The described process chain has been demonstrated on a scale of roughly 0.5 m^3(STP)/h and experimental results will be presented. Special emphasize will be laid on the analysis of the methanation performance considering the already high methane content of the mixed raw biogas. It will be shown that the combination of methanation with membrane-base gas separation technology provides significant advantages for process integration. For this purpose, process simulation will be extensively used for supporting data analysis, basic engineering and process optimisation steps. Modelling and simulation tasks will be accomplished using in-house-modelling algorithms implemented in the well-known commercial software environments ASPEN© Plus© and MATLAB©.

Economic and location analysis of hydrogen extraction facilities for use in a *Power to Gas* framework

M. Groß[1], W. Liemberger[1], M. Miltner[1], H. Prazak-Reisinger[2], M. Harasek[1]
[1] Inst. of Chem. Eng., Vienna University of Technology, A-1060 Vienna, Getreidemarkt 9/166, Austria, tel. 0043/1/58801-166270, e-mail: markus.gross@tuwien.ac.at; [2] OMV Refinery & Marketing GmbH, A-1020 Vienna, Trabrennstraße 6-8

The storage of fluctuating electrical excess energy from renewable sources (e.g. wind, solar power) is a key challenge of our energy transition and has gained significant importance during the last few years. Various Power-to-Gas concepts offer a promising approach in this context. Excess electric energy can be used to produce storable green hydrogen with alkaline or PEM electrolysis. Afterwards, hydrogen can be further converted to methane and stored in the natural gas grid. Contrariwise, hydrogen can also directly be fed to the gas grid up to certain limits (in Austria currently 4 % v/v). Thus, the well-developed existing natural gas grid can act as storage or as transportation device for the renewable hydrogen. In a further step, hydrogen could be extracted at arbitrary locations along the gas grid for decentralized utilization. Thus, it would be possible to supply fueling stations or other applications with delocalized renewable hydrogen (the concept is shown in figure 1).

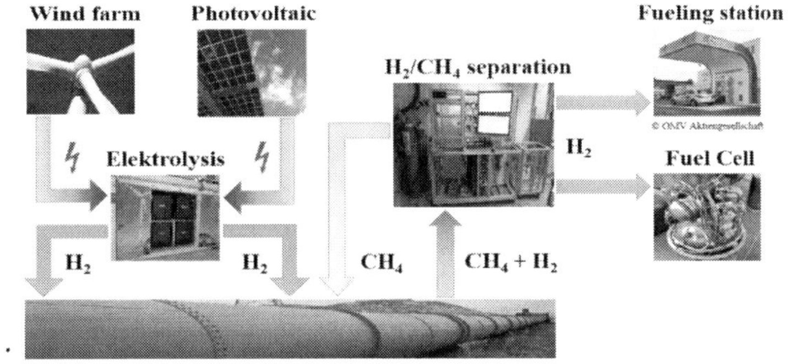

Figure 1: Motivation of the research work

The current work presents an economic analysis of hydrogen extracting facilities. The target of such a facility is the extraction and purification of hydrogen from mixtures with natural gas (1-10 % v/v hydrogen) and providing this hydrogen in fuel cell quality for further utilization. Therefore a process concept is developed, which incorporates a first pre-concentration step applying membrane-based gas permeation and pressure swing adsorption (PSA) as a second process step. The gas quality can be thoroughly analyzed to verify the compliance for the desired application. The developed process must be highly flexible regarding pressure due to different levels in the Austrian gas grid (lowest level 6 bar, highest level up to 60 bar) and hydrogen content in the natural gas. Experimental investigation of the separation process and the individual steps will be performed with pure gases, synthetic gas mixtures and finally with real gases obtained from a real-world hydrogen injection facility (wind2hydrogen project in Auersthal).

After an extensive experimental analysis the power consumption, the efficiency and the flexibility of the extraction facility will be investigated with different process scenarios to determine the total costs and a good operating point of the separation process. Additionally possible economic locations, depending on transportation paths, gas flow and composition of the natural gas mixture, will be considered.

Harnessing geothermal energy resources for sustainable development in Morocco

AE. Barkaoui[1], Y. Zarhloule[1], Varbanor. P.S[2], Klemes. J.J[2]
[1]Laboratory of Hydrogeology-Environment, Faculty of Sciences, University Mohamed Ist, Oujda, Morocco.
[2]Faculty of Information Technology and Bionics, Pázmány Péter Catholic University Práter utca 50/a., H-1083 Budapest, Hungary

Comparing to other energy sources, geothermal energy is a clean, renewable, constant and available worldwide. Due to its reduction capacity in greenhouse gas emissions, this energy is already being used for electricity generation and direct utilization. For Morocco, the use of geothermal energy, among other renewable resources, is primordial in order to break the current dependence on fossil fuel and electricity imports. By 2020, the country is planning to obtain 42% of its power from its own renewable sources. This work presents the status of energy needs and renewable, potential, utilization and the different possibilities of geothermal energy use in Morocco. It also gives a comparison between geothermal energy and the other energy sources regarding environmental issues. It is expected that geothermal energy development will provide considerable environmental benefits by reducing the release of the CO_2 emission into the atmosphere, estimated at around 49.9 million ton in Morocco.

In Morocco, thermal waters are mainly hosted within sedimentary reservoirs, consisting of Liassic limestones with tickness up to 500 m. The geothermal fluid is characterized by a complex deep circulation and it flows up through complex fault systems. The Liassic reservoir of Eastern Morocco belongs to the Atlas domain (Fig.1) which is characterized by a succession of NE-SW horst and graben structures of pre-Miocene age and by Plio-Quaternary basaltic volcanics, produced by crustal thinning. This reservoir is at the origin of twenty-four thermal waters and aquifers, with a temperature ranging from 26 to 54 °C and a maximum flow rate reaching 40 l/s. Some of these hot aquifers, like e.g. Fezouane, near Berkane and Hammam Ben Kachour, at Oujda play an important role in the economy of the area [1,2].

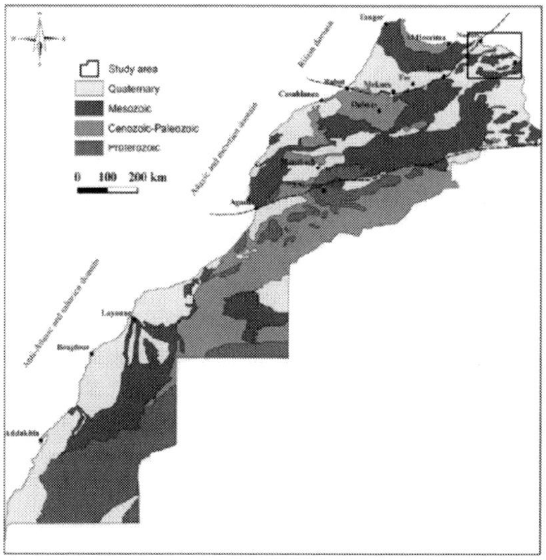

Figure1 .Main structural and geological features of Morocco.

The regional pattern of the heat flow density highlights heat flux increasing northeastward, from less than 60 (north Mauritania) to more than 80-90 mW/m^2 in the eastern Rif, northeastern Morocco, Alboran sea and northwestern Algeria. The distribution of the geothermal gradient [3] highlights the high values in the northeastern part of Morocco, where they can reach 50 °C km^{-1}. This part of the country is also characterized by high residual magnetic anomalies related to widespread Quaternary volcanism.

Many water boreholes were logged within the framework of a geothermal survey conducted in northeastern Morocco to understand better the behavior of the thermal water inside the Liasic geothermal reservoir. Among the recorded thermal profiles the borehole 1624/7 was selected. This hole is characterized by an increase of geothermal gradient at 300 m depth from 29 to 127 °C km-1. At the same depth, lithology changes from clay to dolomite. At about 470 m depth, the temperature is about 50 °C. The shape of the thermal profile argues for a conductive thermal regime both in the upper (clay) and in the lower (carbonate) section of the hole. The dolomitic formation continues until the hole bottom (1042 m depth). By extrapolating the thermal gradient inferred in the lowermost section of the hole, a bottom temperature of about 120 °C is inferred. The compilation of the obtained results with the geological, geochemical and the hydro-geothermal data allowed the construction of the conceptual model.

ACKNOWLEDGMENTS

The authors would like to acknowledge the financial support by the EC and Croatian Ministry of Science Education and Sports projects "INTERGEORES" (NEWFELPRO Grants Agreements 43)

[1] Barkaoui AE., Zarhloule Y., Rimi A., Boughriba M., Verdoya M, Bouri S (2013): Hydrogeochemical investigations of thermal waters in the northeastern part of Morocco. Journal of Environmental Earth Sciences. DOI 10.1007/s12665-013-2582-x

[2] Barkaoui AE., Correia A., Zarhloule Y., Rimi A., Carneiro JF., Boughriba M., Verdoya M (2012) : Reconstruction of remote climate change from borehole temperature measurement in Eastern part of Morocco. Journal of climatic change. DOI 10.1007/s10584-012-0638-7.

[3] Zarhloule, Y., Rimi, A., Boughriba, M., Barkaoui, A.E., Lahrach, A., The geothermal research in Morocco: history of 40 years. In: World Geothermal Congress, 2010.

Investigation of glycerol utilization for methanol production

A. L. H. Costa, P. A. Souza, E. R. A. Lima
Rio de Janeiro State University, Rua São Francisco Xavier, 524, Instituto de Química, Maracanã, Rio de Janeiro, RJ, Brazil; e-mail: andrehc@uerj.br;

Glycerol is a subproduct from the biodiesel production. The increase of the utilization of biodiesel around the world has provided large volumes of this stream available to the market. A possible utilization of glycerol is as raw material for methanol production. In this context, it can be a potential partial substitute of the natural gas in the synthesis gas production. However, the introduction of glycerol brings considerable changes in the mass and energy balances of the methanol plant. Previous works have analysed this impact focused only in the reactional aspects of the steam reforming, without considering the entire system. In this context, the proposed paper presents an analysis of the introduction of glycerol in the methanol production considering all the interconnections among the units in relation to raw material, intermediate products, final products, fuel and utility system. The target is to present an economical comparison of the introduction of glycerol in relation to a plant which operates only with natural gas.

Thermodynamic analysis of a hybrid low-temperature waste heat-compressed air energy storage system

Peng Ran [1], Yase Wang [1], Pei Liu [2], Zheng Li [2], Weidou Ni [2].
[1] North China Electric Power University, School of Energy & Power Engineering.
Yonghuabei Rd. 619, Baoding, Hebei Province 071003, China
Tel: +86-13582232866, e-mail: 13582232866@163.com
[2] State Key Lab of Power Systems, Department of Thermal Engineering, Tsinghua University, Beijing 100084, China

In China, considerable amount of waste heat is emitted from chemical, metallurgical and cement industries, which can be used not only for municipal purposes but also for electricity energy storage if recovered. Compressed air energy storage (CAES) has less geographic restrictions and high energy-capacity, which makes it more attractive. A major challenge with developments of CAES is its capital cost and its configuration complexity. In this paper, a novel hybrid system was proposed on the basis of adiabatic CAES. It can improve the financial return per capital cost and make the configuration simplification by using waste heat. In this new system, which is called hybrid waste heat–compressed air energy storage (WHCAES), thermal energy storage unit is abandon, which makes adiabatic CAESs inexpensive and less complexity, compressed air directly absorb the heat release from waste heat. The mathematical model of the new system is established based on the first and the second law of thermodynamics with respect to their internal status and the efficiency of the storage for one charging and discharging cycle. The waste heat temperature, ambient air temperature and heat exchanger effectiveness are examined to evaluate their effects on the system performance such as power output, energy storage efficiency. Theoretical thermodynamic analyses show that the WHCAES can fully use the waste heat, which is less expensive than an adiabatic CAES system with the same scale of compressors, turbines, and air storage vessel do. And energy storage efficiency of the WHCAES varies between 42% and 56%, depending on the heat exchanger effectiveness and ambient air temperature. Key element to improve the efficiency is to develop high effectiveness heat exchanger. The power out ratio of the WHCAES system to CAES system increases with the waste heat temperature and decreases with the maximum air storage pressure.

Sustainable production of asphalt using biomass as primary process fuel

F. Bühler, T.-V.Nguyen, B. Elmegaard
Technical University of Denmark, Dep. of Mechanical Engineering, Nils Koppels Allé, 2800 Kgs. Lyngby, Denmark;
tel. +45 254102, e-mail: fabuhl@mek.dtu.dk;

Asphalt is the major road paving material in Europe and its production is mainly based on fossil fuels[1]. The possibilities of reducing the fossil energy use have been investigated for many years and the industry has approached this topic from different perspectives. A first one is to optimise the design and operation of the asphalt factories. Another is to develop new additives and mixtures for asphalt, which allow lower production temperatures, while keeping the quality constant. However, asphalt factories still use a great amount of fossil fuels, in particular fuel oil, natural gas and LPG, within the drum dryers, to dry and heat the aggregate before it is mixed with bitumen. New asphalt types are not always accepted by customers, due to uncertainties in the long-term performance. This paper investigates the possibilities of supplying the process fuel for the production of hot-mix asphalt from renewable sources. The focus is on biomass systems, comparing different feedstocks and technologies, as well as how the biomass system can be integrated within the utility system.

Research has focused on the modelling and optimisation of the asphalt plants rotary dryer [2,3]. The energy efficiencies of the stone and asphalt industry are reviewed by Moray et al. [4]. Possible energy efficiency measures are presented in technical reports [1,5–7]. An example of new mixtures is Low-Energy Asphalt (LEA) as presented by Romier et al.[8]. Most relevant to the concept of the current work is a study where the use of hydrogen, originating from wind power, in the rotary dryer is investigated[9]. Biomass feedstocks and conversion technologies are described in detail in the literature [10–12], but articles analysing case studies for the direct integration of biomass resources for the process industry are rare. Most relevant to the current work are the case studies of biomass cogeneration systems for the steelmaking process and animal feed production in Brazil[13,14].

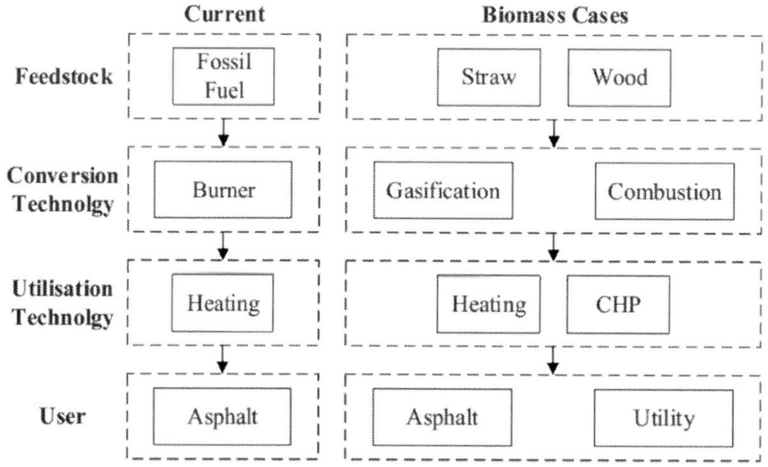

Figure 1: Pathways of the current system and the analysed cases with biomass.

Based on an energy analysis of a hot-mix-asphalt production in a batch mix asphalt factory, which represents the majority of factories in Europe, the potentials and feasibility of utilising biomass as an energy source are investigated. The focus of the analysis is the plants direct-fired rotary dryer, which is the most energy intense process on-site. The integration of gasification and combustion processes (gasifier and furnace), together with a comparison of different feedstocks (wood chips and straw), is investigated. As the asphalt factory produces in batches has short start-up times and is not in operation during cold weather, the options to produce (i) district heat and (ii) combined heat and power are further investigated. The analysed pathways for the current and renewable scenarios are visualised in Figure 1.

For each of the resulting eight cases an economic and thermodynamic analysis is performed to find the best solution. The results show the feasibility of using biomass for process heating in the asphalt factory, with the current legislation and fuel prices. The varying production of asphalt over the seasons and days, as well as short start-up times, cause challenges, which can be however solved with the integration of the production unit to the utility system.

References

[1] EAPA, Environmental Guidelines on Best Available Techniques (BAT) for the Production of Asphalt Paving Mixes, Brussels, Belgium, 2007.

[2] D. Peinado, M. de Vega, N. García-Hernando, C. Marugán-Cruz, Energy and exergy analysis in an asphalt plant's rotary dryer, Appl. Therm. Eng. 31 (2011) 1039–1049.

[3] L. LeGuen, F. Huchet, P. Tamagny, Drying and heating modelling of granular flow: Application to the mix-asphalt processes, J. Appl. Fluid Mech. 4 (2011) 71–80.

[4] S. Moray, N. Throop, J. Seryak, C. Schmidt, C. Fisher, M.D. Antonio, Energy Efficiency Opportunities in the Stone and Asphalt Industry Overview of the Stone Crushing Industry Stone Crushing Process Overview, in: Proc. Twenty-Eighth Ind. Energy Technol. Conf., New Orleans, 2006.

[5] T.J. Young, Energy Conservation at the Plant Makes Sense and Saves Dollars, Hot Mix Asph. Technol. (2008) 28–39.

[6] Carbon & Energy Africa Ltd., Implication of the Proposed Carbon Tax Legislative and Policy Framework on SABITA Members, Cape Town, South Africa, 2011.

[7] O. Stotko, Energy and Related Carbon Emission Reduction Technologies for Hot Mix Asphalt Plants, in: 10th Conf. Asph. Pavements South. Africa - CAPSA, 2011: pp. 1–9.

[8] A. Romier, M. Audeon, J. David, Y. Martineau, F. Olard, Low-Energy Asphalt with Performance of Hot-Mix Asphalt, Transp. Res. Rec. 1962 (2006) 101–112.

[9] K. Gopalakrishnan, S.K. Khaitan, Use of Hydrogen from Renewable Energy Source for Powering Hot-Mix Asphalt Plant, Int. J. Traffic Transp. Eng. 2 (2012) 30–43.

[10] P. McKendry, Energy production from biomass (part 1): Overview of biomass, Bioresour. Technol. 83 (2002) 37–46.

[11] P. Mckendry, Energy production from biomass (part 2): conversion technologies, Bioresour. Technol. 83 (2002) 47–54.

[12] A. Demirbaş, Biomass resource facilities and biomass conversion processing for fuels and chemicals, Energy Convers. Manag. 42 (2001) 1357–1378.

[13] T.L. Oliveira, P.S. Assis, E.M. Leal, J.R. Ilídio, Study of biomass applied to a cogeneration system: A steelmaking industry case, Appl. Therm. Eng. 80 (2015) 269–278.

[14] M.L. de M. Rodrigues, F.R.P. Arrieta, J.R. Sodré, Small scale biomass CHP plant: An assessment for an animal feed industry, Appl. Therm. Eng. 59 (2013) 174–181.

Biofuel production from microalgae using integrated process of solar hydrothermal hydrolysis and microbial conversion

C. Xiao[1,2], Q. Fu[1,2]*, Y. Huang[1,2], X. Zhu[1,2], Q. Liao[1,2]*
[1]Key laboratory of Low-grade Energy Utilization Technologies and Systems, MOE, Chongqing, China; [2]Institute of Engineering Thermophysics, Chongqing University, Chongqing, China; tel.+86 023 65102474, e-mail: fuqian@cqu.edu.cn.

Microalgae are sunlight-driven cell factories that can convert CO_2 to various biofuels. A typical process for biofuel conversion from microalgae usually consists of microalgae cultivation, microalgae biomass drying, and biofuel conversion. The microalgae biomass drying process is energy-intensive, significantly hindering the commercialization of microalgae. Therefore, developing novel energy conversion technologies that directly use wet microalgae biomass is necessary for biofuel production from microalgae. In this regard, microbial fermentation with wet microalgae biomass is of interest. But the hydrolysis rate of the microalgae is relative low due to the complexity of structure and components of the microalgae cell wall, hindering the fermentation process in the next step. Thus it is necessary to pre-treat the microalgae biomass before the fermentation process. Hydrothermal liquefaction is one promising pretreatment technology, which utilizes sub-critical water to convert microalgae biomass to monosaccharide, amino acid, and etc.. However, the high-energy requirement for maintenance of the relative high temperature (usually 150-200°C) and high pressure (usually 1-4 MPa) hinders the commercialization of this technology. In this study, a novel biofuel production process integrated hydrothermal hydrolysis and solar energy was proposed. Solar energy was used as energy input to drive the hydrothermal hydrolysis of microalgae. The hydrolysate of microalgae was used for microbial dark/light fermentation for biogas and biohydrogen production in the following steps. A small demonstration setup was established in lab-scale. The experimental results showed that the hydrolysis process of microalgae biomass could be successfully realized in this integrated system. The biogas production amount of microbial fermentation using the hydrolysate was much higher than that of fermentation directly using microalgae biomass. This system provides an alternative solution for biofuel production from microalgae biomass with less energy requirement.

Application of target-oriented robust optimization on the design of an integrated algal biorefinery

A.T. Ubando[1], C. L. Sy[2]
[1]Mechanical Engineering, De La Salle University, 2401 Taft Avenue, 0922 Manila, Philippines. Tel: (632) 524-4611. Email: aristotle.ubando@dlsu.edu.ph
[2]Industrial Engineering, De La Salle University, 2401 Taft Avenue, 0922 Manila, Philippines. Tel: (632) 524-4611. Email: charlle.sy@dlsu.edu.ph

Integrated algal biorefinery (IAB) offers an efficient way to produce algal biofuels while increasing its economic potential through the production of high-valued co-products. Such bioenergy systems can potentially reduce environmental footprints by reusing wastes materials from other processes. An IAB combines multiple conversion technologies from microalgae to produce an array of products such as biofuels, biochemicals, and bioenergy streams. Design methodologies such as mathematical programming models have been instrumental in the optimal synthesis and design of such complex systems. However, uncertainties may arise where unexpected changes may occur during operations which may result to economic losses or heightened environmental emissions. Uncertainties may be accounted on the design procedure of an IAB through target-oriented robust optimization (TORO) methodology. TORO is a new approach which allows to maximize robustness against uncertainties while satisfying multi-objective targets. A hypothetical but realistic case study of an IAB is used to illustrate the developed methodology on evaluating uncertainties on economic performance and environmental footprint. Results revealed the robust optimal configuration of the IAB based on its targeted economic performance and environmental footprint.

Sustainable milk powder production using enhanced process integration and 100% renewable energy

M.R.W. Walmsley, T.G. Walmsley, M.J. Atkins, J.R. Neale
Energy Research Group, School of Engineering, University of Waikato, Private Bag 3105, Hamilton, New Zealand; tel. +647 8384937, fax: +647 8560115, m.walmsley@waikato.ac.nz

Sustainable food production and processing is a topic of increasing global importance. Traditionally the demand for dairy products per capita has been high in Europe and North America, and low in Asia. However with increasing global connectedness and trade, the demand for dairy in China, India and the Middle East is rapidly rising. Introducing sustainable farming practices and renewable energy supplies into dairy processing is therefore of increasing importance in a carbon emissions and resource constrained world.

California and New Zealand produce similar quantities of milk annually. Most of the milk in California is processed as fluid milk, whereas in New Zealand milk supply far exceeds domestic demand and 75% of the milk is dried into milk powder for export. California also produces milk powders but to a much lesser extent. In terms of renewable energy, California has a favourable supply of solar energy and a highly fertile inner valley region with irrigation supporting intensive farming and potentially biogas production from cow manure. New Zealand, on the other hand, has fertile grazing land, good rainfall, and a large supply of low temperature geothermal heat, considerable waste biomass and a high level of renewable electricity through wind, hydro and geothermal supporting industrial heat pumps. The respective renewable energy, land and water resources give rise to the long-term potential for implementing an entirely sustainable milk powder production industry in these regions.

This paper presents a total site analysis of the requirements to economically integrate renewable energy into an 'ultra-energy efficient milk power factory of the future' in California and New Zealand. The analysis first compares the overall environmental foot print of milk powder production in California and New Zealand, including on-farm and in-factory requirements for land, water, energy and chemicals (including fertiliser use on-farm). Options for supplying in-factory energy needs using 100% renewable energy are then analysed and the most economical options are presented.

In California electrical and process heating and cooling needs are best met using biogas from cow manure for combined heat and power, supplemented by integrated solar thermal-PV during the day. In New Zealand the best option is to use renewable electricity from local wind, hydro and/or geothermal resources, and to use geothermal and heat pumps for process heating up to 150°C and combustion of waste biomass for high temperature process heating up to 210°C. The possibility of upgrading geothermal steam using MVR technology for >150°C heating is also considered. The ultra-efficient milk powder factory presented in the study reduces overall thermal energy demand by 20% compared to current designs, and lowers renewable energy demand by the same amount, which is extremely valuable from a sustainability point of view.

Optimal sizing and design of hybrid renewable energy systems using mathematical programming

J.-Y. Lee, [1] K. B. Aviso, [1] R. R. Tan
Department of Chemical Engineering and Biotechnology, National Taipei University of Technology, 1, Sec. 3, Zhongxiao E. Rd., Taipei 10608, Taiwan, R.O.C.; tel. +886 2 2771 2171 ext. 2524, fax: +886 2 2731 7117, e-mail: juiyuan@ntut.edu.tw; [1] Chemical Engineering Department/Center for Engineering and Sustainable Development Research, De La Salle University, 2401 Taft Avenue, 1004 Manila, Philippines; emails: kathleen.aviso@dlsu.edu.ph (K. B. Aviso), raymond.tan@dlsu.edu.ph (R. R. Tan).

Global concern about the depletion of energy resources, climate change and environmental emissions are among the key drivers of sustainable development. Reducing greenhouse gases (GHGs) emissions and mitigating global warming are becoming socially and economically pressing for nations all over the world. Effective measures to mitigate and adapt to climate change include the use of renewable energy (RE) sources (e.g. solar, wind and biomass), which provide clean and sustainable energy production and serve as a promising alternative to conventional sources. According to "Energy Technology Perspectives 2015" (IEA, 2015), carbon capture and storage (CCS), renewables and end-use energy efficiency are key technologies to achieve the 2°C Scenario (2DS), viz., to limit average global temperature increase to 2°C, and expected to contribute more than 80% of total carbon dioxide (CO_2) emission reductions required in 2050. Furthermore, renewables will be needed mainly in the power sector, where wind and solar photovoltaic (PV) have the potential to provide 22% of annual emissions reduction in 2050 under the 2DS. In addition to their cleanliness and availability, application of RE sources can enhance energy security (with reduced energy dependency) and efficiency in the long run.

A common problem of many RE sources is the dependence on weather and climatic conditions and hence the intermittent and unpredictable nature, which invariably results in a mismatch between the power generation and the load demand. This adversely affects the system performance; moreover, fluctuations in the output of RE sources make it difficult to utilise renewables efficiently in large power networks. Despite the unfavourable features of renewables, integrating two or more complementary sources such as solar and wind can significantly improve the system efficiency and reliability, thereby reducing the dependence on backup energy devices (e.g. batteries, diesel generators and fuel cells). This approach has emerged as a field of study on hybrid renewable energy systems (HRESs), and become increasingly popular in the last decade.

An HRES may be defined as an energy system utilising more than one RE source, such that the electricity generated is more efficient, reliable and cost-effective, offering a better option than single source-based systems (Mahesh and Sandhu, 2015; Sinha and Chandel, 2015). It can work in a grid-connected mode or in isolation, and may incorporate energy storage or a conventional source (e.g. diesel) to further improve the system reliability. Different types of hybrid systems are feasible depending on the local demand and resource availability. In particular, solar and wind are the most promising RE sources widely used for power generation because of their complementary nature. This advantage makes such hybrid systems a popular topic attracting considerable research interest. Typically the size of HRESs ranges from a few to several hundred kilowatts depending on the load. In addition, hybrid systems can be used in urban, rural and remote areas.

The main focus of HRES research has been system sizing and optimisation. This vitally important aspect of HRES design requires efficient methods to ensure reliable power supply and a cost-effective system. Oversizing results in higher system costs, while under-sizing causes insufficient or fluctuating supply of power. In addition, HRESs are widely considered a viable option for electrifying remote areas where grid extension is not feasible or economical and fuel transportation is expensive. Recently, the principles and tools of process integration (PI) have been extended to design and optimise HRESs. There are two general categories of PI techniques, namely pinch analysis and mathematical optimisation. The former sets performance targets ahead of detailed system design, and provides useful insights into the design problem, while the latter is normally preferred for handling complex problems with multiple quality indices or with operational or topological constraints. Essentially, HRES design is analogous to the synthesis of batch resource conservation networks, an active branch of PI. This similarity allows the application of well-established PI approaches and techniques to HRES design and optimisation problems.

In this work, the mathematical model developed by Lee et al. (2014) will be expanded for optimal sizing and design of HRESs, taking into account multiple objectives such as power reliability and system cost, as well as variations in RE supplies and load demands. This calls for the development of a multi-objective, multi-period optimisation framework, which is the main focus of the research. An optimisation model will be developed and applied as a generic decision-support tool to a case study. The model is based on a superstructure that includes all feasible options for power allocation in typical HRESs, and consists of energy balance equations as well as capacity and operational constraints. The objective is to determine the optimal HRES size and configuration, while satisfying the reliability and cost criteria for power supply.

References

International Energy Agency (IEA). Energy Technology Perspectives 2015 - Mobilising Innovation to Accelerate Climate Action. Paris: OECD/IEA; 2015.

Lee J-Y, Chen C-L, Chen H-C. A mathematical technique for hybrid power system design with energy loss considerations. Energy Convers Manage 2014;82:301-7.

Mahesh A, Sandhu KS. Hybrid wind/photovoltaic energy system developments: Critical review and findings. Renew Sustain Energy Rev 2015;52:1135-47.

Sinha S, Chandel SS. Review of recent trends in optimization techniques for solar photovoltaic-wind based hybrid energy systems. Renew Sustain Energy Rev 2015;50:755-69.

Selected methods of advanced biogas upgrading

M. Miltner, [1]A. Makaruk, M. Harasek
Institute of Chemical Engineering, Vienna University of Technology, A-1060 Vienna, Getreidemarkt 9/166, Austria; tel: +43 1 58801 166276, e-mail: martin.miltner@tuwien.ac.at, fax: +43 1 58801 15999; [1]AXIOM Angewandte Prozesstechnik GmbH, A-2483 Ebreichsdorf, Wienerstrasse 114, Halle I-J, Austria.

The upgrading of biogas for the production of biomethane to be injected to the natural gas grid has gained significant importance in recent years. It is often considered to be superior to the production of electricity and district heat with internal combustion engines (CHP) mainly because of the better energy utilisation and the higher flexibility of applying a natural gas substitute. A considerable number of small and industrial scale plants have been commissioned to date representing the portfolio of currently available upgrading technologies. Although supposed to be state-of-the-art, academic research demonstrates that biogas upgrading still provides huge potential for optimisation and development.

High capabilities towards biogas upgrading is ascribed to membrane-based gas-permeation technology due to its flexibility and adaptability. Different membrane materials, membrane module configurations and plant topologies allow for the design of a tailor-made upgrading plant according to the needs of the customer and the application. Furthermore, gas-permeation stands out for low specific energy demand, high process simplicity and plant safety.

To fully exploit the potentials of gas-permeation for biogas upgrading a thorough analysis and design of the process is inevitable. Process simulation has proved to be a valuable tool for this investigation provided that capable and validated model algorithms are utilised. This is especially true during the design of processes incorporating different membrane materials for the selective and individual separation of different biogas constituents like CH_4, CO_2, H_2 and H_2S.

Frequently, the most challenging upgrading step is not the separation of the bulk component CO_2 but the one of certain trace components like H_2S, NH_3 or various hydrocarbons (terpenes, thiophenes, mercaptans). A novel process of short-time-contacted chemical-oxidative scrubbing has recently been commercialized and serves well for the removal of high and strongly fluctuating sulphur contents in the raw biogas. The combination of this technique with an adsorptive precision cleaning step using oxides of iron, zinc or copper facilitates the safe and constant compliance with the strict gas grid regulations even for highly loaded biogas and at reasonable costs.

Current work provides an insight to the authors' research undertaken in the field of biogas upgrading. The application of commercial as well as innovative membrane materials e.g. for the separation of CO_2 and H_2S from methane-rich gas streams is presented by means of numerical and experimental approaches. Additionally, the novel process for desulphurization of carbon dioxide- and methane-rich gaseous energy carriers will be illustrated using results from field-tests of a pilot plant and an industrial scale reference plant. A prospect of optimization potential and future research pathways in the field of biogas upgrading shall be given in a summary.

Comparison of heat exchanger designs using Vipertex 1EHT enhanced heat transfer tubes

David J. Kukulka[1*], Rick Smith[2]
[1] State University of New York College at Buffalo, 1300 Elmwood Avenue, Buffalo, New York, 14222 USA, kukulkdj@buffalostate.edu
[2] Vipertex, 658 Ohio Street, Buffalo New York. USA

A numeric study was performed that compares the performance of heat exchangers using the Vipertex enhanced heat transfer tubes (model 1EHT) to the performance of heat exchangers that use smooth surface tubes. Heat transfer enhancement is an important factor in obtaining energy efficiency improvements in all heat transfer applications. Surface enhancement of the 1EHT tube is accomplished through the use of the primary dimple enhancement and the secondary background pattern made up of petal arrays. Utilization of enhanced heat transfer tubes is an effective method that is utilized in the development of high performance thermal systems. Vipertex™ enhanced surfaces, have been designed and produced through material surface modifications that produce flow optimized heat transfer tubes that increase heat transfer. Current energy demands and the desire to increase efficiencies of systems have prompted the development of enhanced heat transfer surfaces.
Enhanced heat transfer tubes are widely used in many areas (refrigeration, air-conditioning, petro-chemical, chemical, etc.) in order to reduce cost and create a smaller application footprint. A new type of enhanced heat transfer tube has been created; therefore it is important to investigate relevant heat exchanger designs using the Vipertex enhanced surface tube in industrial applications and compare that performance to other tubes. Results include design characteristics and performance predictions using the design simulations produced using HTRI Exchanger Suite (2016). Performance for all cases considered using the Vipertex tube predicted over performance when compared to a smooth tube design. Vipertex tubes produced higher performance and more efficient designs.

Robust controller design for a heat exchanger

A. Vasičkaninová, M. Bakošová, Ľ. Čirka, M. Kalúz
Slovak University of Technology in Bratislava, Faculty of Chemical and Food Technology, Institute of Information Engineering, Automation, and Mathematics, Radlinského 9, 812 37 Bratislava, Slovak Republic; e-mail: anna.vasickaninova@stuba.sk.

The aim is to show the benefits of two advanced control strategies in heat exchanger control with focusing on energy savings. Two robust control techniques were used for controller design: H_∞ control and μ-synthesis.

H_∞ optimal control is a frequency-domain optimization and synthesis theory that was developed in response to the need for a synthesis procedure that explicitly addresses questions of modelling errors. The H_∞ robust controller design theory deals with defined uncertainties, like parametric uncertainty and performance specified as weight filters. The H_∞ theory has some weaknesses; for instance, it does not consider the uncertainty structure. The μ-synthesis theory is a further development of the H_∞ control where the uncertainty structure is considered in the design.

The designed controllers were verified in laboratory conditions on a real-time control of the laboratory heat exchanger in the Multifunction Process Control Teaching System (MPCTS) Armfield PCT 40 (Figure 1). The PCT40 is connected to the computer through the data acquisition and control card, and the MATLAB Real Time Toolbox is used as the control software.

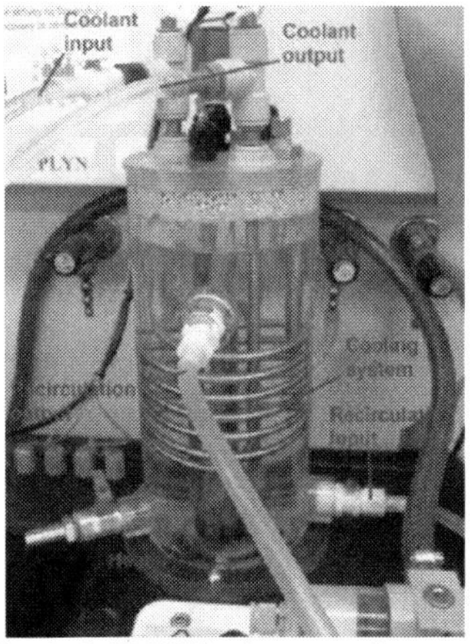

Fig.1. MPCTS - heat exchanger.

The heat exchanger consists of a cylindrical vessel with heated fluid (water) inlet and outlet, the heating element, internal cooling system and a set of sensors. The manipulated variables are the heater power, handled by the solid state relay, and the cooling water flow rate in the cooling coil. The controlled variable is the temperature of the outlet heated stream measured by the temperature sensor. The constant level of heated water in the tank is assured by equal adjustment of the inlet and the outlet water flow rates using two peristaltic pumps

Experimental results confirmed that the robust controllers can effectively control the plant, and illustrated the potentials of the used robust control methods.

A Sankey Diagram approach to quantifying industrial residual energy in China

Z Zhao, T Wang, P Liu, Z Li
State Key Lab of Power Systems, Department of Thermal Engineering, Tsinghua University, Beijing 10084, China; liu_pei@tsinghua.edu.cn

The industrial sector is China's largest energy consumer, accounting for more than 60 percent of China's overall energy consumption. However, a large part of energy in the industrial sector is not used efficiently, and there is great potential for enhancing energy utilization efficiency. In this paper, we provide a quantitative analysis of energy flows in key industrial sectors in China, including iron and steel, electrolytic aluminum, petroleum refining, cement, and eight other sectors. The energy flows are illustrated via Sankey Diagrams, which explicitly indicate stages with huge energy waste thus great potential for improving. Results show that the iron and steel sector has the largest potential for energy recovery, amounting to approximately 300 milliontonnes coal equivalent per year. The energy recovery potential in each sector is further divided into three levels according to technical difficulty and current status, namely full potential, technically recoverable, and already utilized. Based on the results, prioritizing development of some energy recovery technologies is proposed, including coke dry quenching, combined cycle power generation from industrial residual gas, and power generation from low temperature waste heat during cement production, with an expected annual energy saving of 120 milliontonnes coal equivalent.

Hydrogen production by a thermally integrated ATR based fuel processor

V. Palma[1], A. Ricca[1], B. Addeo[2], G. Paolillo[2]
[1] University of Salerno, Department of Industrial Engineering, via Giovanni Paolo II 132, 84084 Fisciano (SA), ITALY – aricca@unisa.it
[2] SOL S.p.A. - 20900 Monza (MB) - ITALY

Nowadays, the combination hydrogen-fuel cell is seen as the most viable solution to counter the antithetic problems of energy growing demand and environmental pollution [1]. The high difficulties in H2 transport and storage make the distributed production a very interesting solution. Since very flexible production as well as quick start-up and shut-down procedure are fundamental features, H2 production by auto-thermal reforming of hydrocarbons results as the optimal solution for the coupling with a fuel cell system. Auto-Thermal Reforming (ATR) is typically considered as the overlap of several chemical reactions, such as fuel partial and total oxidation, steam reforming. The different kinetics of the involved reactions suggest that the oxidation reactions occur as first; the heat developed by fuel oxidation is used to sustain the endothermic steam reforming reaction. Therefore, the choice of an appropriate catalytic system is a fundamental step in the reaction setup. It should show high selectivity to-wards reforming reactions (POX, SR), inhibit side reactions as cracking and coke formation, generate low pressure drop in operating conditions and, finally, it should be characterized by high resistance to mechanical shocks and thermal gradients. Previous studies confirmed the importance of a flat thermal profile along the catalytic bed: in this way, honeycomb and foam monolithic catalysts result as the favourite solutions. Aim of this work was to design and test a thermally integrated fuel processor based on natural gas ATR, aimed to the hydrogen production intensification and to the external duty minimization [2].

A compact fuel processor based on Auto-Thermal Reforming process for natural gas conversion was designed and configured to produce up to 10 Nm3/h of hydrogen. Water and air were delivered to the system, together with the hydrocarbon. The system was provided to 2 catalytic stage: in the ATR module, the hydrocarbons conversion to syngas was obtained, in the WGS module the CO present in the syngas was converted increasing hydrogen content. In the ATR module a 1 L commercial (Johnson Matthey) honeycomb catalyst noble metals based was loaded, while in the WGS 7 L of commercial (Johnson Matthey) pellets catalysts were employed. Due to the very different temperature in the two stages, a compact and optimized heat exchange module were designed and placed between the two reaction volumes. In this way, sensible heat of effluent gas from ATR volume was exploited to pre-heat reactants before feeding to it; moreover, effluent gas were cooled to a temperature consistent to the WGS stage. Such configuration allowed to feed all reactants at room temperature, without external heat exchangers; moreover, the thermal integration was able to reduce system start-up transient times. Preliminary studies were carried out in order to evaluate the reactants rate and the feed ratios effect on system performances.

The process monitoring was assured by evaluating temperature and composition before and after each catalytic stages, in order to have information about catalytic processes performances. Moreover, by monitoring temperature downstream the heat recovery module, the heat transfer efficiency was evaluated. Sample gas analysis was assured by means a Uras-Caldos-Magnos ABB dedicated analyser, optimized for syngas composition analysis.

Preliminary tests were carried out by feeding Natural Gas (NG) or CH_4, air and H_2O, by varying both ATR space velocity and feed ratio, in order to investigate the effect of reactants rates on system performances.

Preliminary tests confirmed that the designed system was characterized by very short start-up times, being able to produce up to 5 Nm^3/h of hydrogen (27%vol) in less than 3 minutes. The transitory phases were strictly related to the overall gas rate, since higher stream rate resulted in faster system response. Despite thermodynamic predictions, due to the thermal integration the system seemed not to be greatly affected by steam content, since too high steam-to-carbon ratio resulted in a decreasing in system temperature. The ATR module well approached the thermodynamic equilibrium, reaching a total methane conversion on all investigated operating conditions. On the other hand, WGS module was able to convert only the 53% of CO, mainly due to the thermodynamic limitations, so also affecting the thermal efficiency that anyway reached very promising values.

Table 1 - ATR and WGS performances in methane processing (ATR GHSV = 15,000 h^{-1})

		H_2O/C	O_2/C	CH_4 Exp. *Th. Eq.*	H_2 Exp. *Th. Eq.*	CO_2 Exp. *Th. Eq.*	CO Exp. *Th. Eq.*	X_CH_4 Exp. *Th. Eq.*
ATR	**Case 1**	0.49	0.6	**2.1%** *1.6%*	**36.8%** *36.1%*	**4.8%** *5.5%*	**13.9%** *12.5%*	**96.7%** *97.6%*
	Case 2	0.60	0.65	**0.6%** *0.7%*	**37.1%** *35.5%*	**4.9%** *5.7%*	**14.3%** *12.3%*	**99.0%** *98.9%*
	Case 3	0.65	0.65	**0.3%** *0.4%*	**37.7%** *36.0%*	**4.8%** *5.5%*	**14.3%** *12.7%*	**99.5%** *99.4%*
	Case 4	0.75	0.65	**0.3%** *0.3%*	**37.7%** *36.4%*	**5.3%** *5.8%*	**13.9%** *12.4%*	**99.5%** *99.6%*

		H_2O/CO	T_{in}, °C	CH_4 Exp. *Th. Eq.*	H_2 Exp. *Th. Eq.*	CO_2 Exp. *Th. Eq.*	CO Exp. *Th. Eq.*	X_CO Exp. *Th. Eq.*
WGS	**Case 1**	0.74	300	**1.9%** *2.0%*	**39.6%** *39.6%*	**9.2%** *12.2%*	**8.4%** *4.4%*	**36.7%** *66.9%*
	Case 2	0.70	327	**0.4%** *0.6%*	**40.3%** *40.1%*	**11.0%** *12.3%*	**6.5%** *4.5%*	**52.1%** *67.1%*
	Case 3	0.70	366	**0.3%** *0.3%*	**41.2%** *40.3%*	**10.3%** *11.7%*	**6.6%** *5.3%*	**51.1%** *61.3%*
	Case 4	0.85	372	**0.3%** *0.3%*	**41.5%** *40.6%*	**11.1%** *12.2%*	**6.2%** *4.8%*	**52.5%** *63.9%*

On the other hand, Natural gas tests evidenced that also ATR stage suffer the "raw" hydrocarbon feed, that partially inhibited the catalyst activity leading to a fuel conversion not higher than 95%. Once again, WGS volume showed a quite low conversion, related to the low kinetic of the selected catalyst in the occurred operating conditions. In addition, the very low steam-to-CO ratio evaluated in the WGS feed didn't exceed 1, so thermodynamically limiting the stage performances.

On the other hand, by increasing ATR GHSV up to 22,500 h^{-1}, the system demonstrated to be able to have a stable production of 10.8 Nm^3/h of hydrogen, reaching a thermal efficiency (LHV based) of 74% despite the non optimal WGS behavior.

In conclusion, the system showed very permising results, matching the target of the production of 10 Nm^3/h of hydrogen with natural gas. Anyway, several aspects may be optimized in order to improve performances, mainly selecting a more active middle-temperature WGS catalyst, able to operate in severe conditions and for temperature below 300°C. Moreover, to recover heat from WGS exhaust stream may allow to feed more steam to the WGS stage, so enhancing system thermodynamic.

References

[1]V. Palma, F. Castaldo, P. Ciambelli, G. Iaquaniello, Clean Technologies and Environmental Policy, 14 (2012) 973-987.

[2] V. Palma, A. Ricca, P. Ciambelli, Int J Hydrogen Energ, 11 (2013) 406-416.

Study of hydrogen sulfide absorption with diethanolamine in methanolic aqueous solutions

F. Amararene, C. Bouallou[1]
MINES ParisTech, PSL-Research University,
CES Centre d'efficacité énergétique des systèmes,
60 Bd Saint Michel, 75006 Paris, France [1]E-mail : chakib.bouallou@mines-paristech.fr

The removal of the sulphur compounds, such as the hydrogen sulfide (H_2S) or the carbonyl sulphide (COS) of natural or synthesis gases is an important operation in the industrial processes. These sulphur species are eliminated because of their high toxicity and their corrosive capacity, but also to reach the specifications imposed by industries of transport of gas and to respect the very strict environmental standards. The sulphur compounds are eliminated by means of various processes, among which that corresponding to the gas absorption in specific liquid solutions. This absorption is carried out by physical, chemical solvents and currently by mixing both in the so called hybrid solvents.

In this work, hydrogen sulfide (H_2S) absorption in a hybrid solvent containing aqueous DiEthanolAmine (DEA) dissolved in methanol have been studied extensively in a modified Lewis cell reactor. Experiments have been achieved in an extended range of temperature, 313-353 K and with two mixtures (50% water, 30% DEA, 20% Methanol), (40% water, 40% DEA, 20% methanol). Through these experiments, three parameters affecting the absorption of hydrogen sulfide (temperature T, loading rate α and mass fraction %) were studied. Absorption is accelerated by raising the temperature while it is penalized by the increase in the loading rate. The absorption rate increases as the mass fraction of DEA varies from 30% to 40% to achieve an optimum loading rate ($\alpha = 0.25$ at T = 333 K and $\alpha = 0.50$ T = 353 K) for which the absorption rate remains practically unchanged.

On-line control of the heat exchanger network under industrial constraints

M. Markowski
Warsaw University of Technology, Department of Process Equipment; Jachowicza 2/4, 09-402 Plock, Poland; e-mail: mark@pw.plock.pl

It is commonly known that the aim of the heat exchanger network (HEN) is maximization of heat recovery. Unfortunately heat recovery is deteriorated by deposits mounting up on the heat transfer surface of the heat exchangers. Moreover the process parameters (temperature, mass flow etc.) are changing with time which also are influencing on heat recovery. Hence, to maximize the heat recovery in real HEN, author proposed a mathematical model of on-line control of HEN. Unfortunately the existing mathematical models of the heat exchanger are inaccurate because heat transfer coefficients are calculated with errors at values ranging from 10 to 40%. To overcome this issues, the author introduced very accurate mathematical model of the heat exchanger. The model is based on industrial measurements of the process stream parameters. These parameters are used for on-line validation of the heat exchanger model. For validated model the results of calculation are very accurate, for the errors of calculations are less than 1%. In this article the proposed model is applied to HEN belonging to selected from industry Crude Distillation Unit (CDU), processing 800 t/h of crude oil. In proposed on-line model the manipulated variables are mass flows of crude oil passing via several branches of the HEN. The results of the HEN simulation show that on-line control of the mass flow thru several branches of the HEN gives opportunity to 1,5 % increase of the total heat recovery.

Magnetically recoverable catalysts for cellulose conversion into glycols

O.V. Manaenkov[1], L.M. Bronstein[2], V.G. Matveeva[1], O.V. Kislitza[1], E.M. Sulman[1], A.I. Sidorov[1], V.Yu. Doluda[1]
[1]Tver State Technical University, Af. Nikitina, 22, Tver, Russia, 170026. Tel./Fax: +7 4822 789317. E-mail: ovman@yandex.ru
[2] Indiana University, Department of Chemistry, 800 E. Kirkwood Av., Bloomington, IN 47405, USA

We used magnetically recoverable Ru catalysts supported on a Fe_3O_4-SiO_2 support that can be separated from the reaction mixture with a magnet. The catalyst was synthesized by a group led by Dr. L. Bronstein. It was shown that the magnetically recoverable catalyst characterized by high selectivity to lower polyols (ethylene and propylene glycol). For addition of Ru species, $Ru(acac)_3$ was dissolved in of THF and mixed with Fe_3O_4-SiO_2. The suspension was stirred overnight in air to allow THF evaporation. The sample was then dried in vacuum at room temperature for a minimum of 2 hours until the sample was entirely dry. The sample was heated in a quartz tube in a tube furnace under argon to 300°C for 3 hours. The catalyst was reduced in hydrogen at ambient pressure during 2 hours at 300 °C.

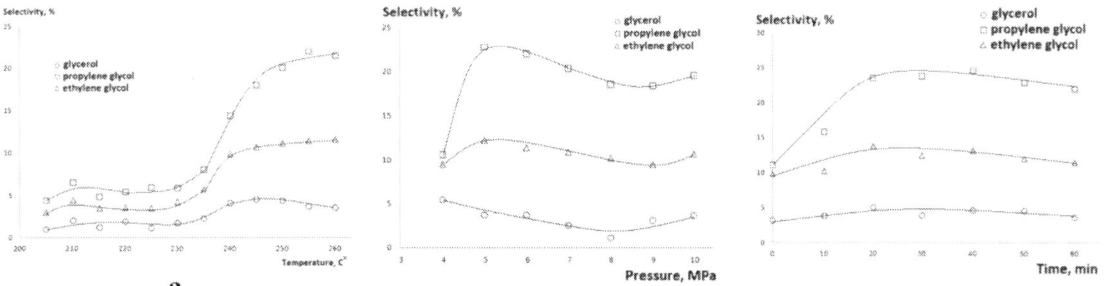

a b c

Fig. 1 Glycerol, propylene glycol and ethylene glycol process selectivity dependence on temperature (a), hydrogen pressure (b) and process time (c): 30 mL water, 600 rpm, 5 % Ru-Fe_3O_4-SiO_2, 0.092 mmol Ru per 1 g cellulose.

Cellulose conversion to polyols was carried out in subcritical water under the following conditions: temperature 205 - 260 °C, hydrogen partial pressure 4 - 106 MPa, process time 10 - 60 min (fig. 1). The experiments were performed in a steel reactor with volume 50 cm^3 (Parr Instrument, USA.). It was shown, that optimal reaction temperature of cellulose hydrolytic hydrogenation process was 255 °C. Optimal hydrogen pressure was found to be 6 MPa. Optimal cellulose hydrolytic hydrogenation time was found to be 50 minutes. The optimal conditions for cellulose hydrolytic hydrogenation over 5 % Ru-Fe_3O_4-SiO_2 catalysts sample was founded: temperature: 245 °C; hydrogen partial pressure: 6 MPa; process time: 50 minutes. The achieved results are: cellulose conversion - 100 %; propylene glycol selectivity ~ 23 %, ethylene glycol selectivity ~ 12 %.

This work was funded by the Russian Science Foundation (Grant 15-08-00455).

Experimental investigation of nanofiltration process for the separation of complex sugar mixtures containing mono- and multivalent salts

Muhammad Ali Samee, Amal Ahmed Elgohary, Michael Harasek, Anton Friedl
Institute of Chemical Engineering, Vienna University of Technology, Getreidemarkt 9/166 A-1060 Vienna, Austria; tel. +43(1)58801-166271; fax: +43(1)58801-15999
muhammad.ali.samee@tuwien.ac.at ; muhammadalisamee@yahoo.com

The energy consumption of sugar manufacturing industry is very high mainly due to the energy demand of multi-stage evaporation. However, alternative processes have been considered to reduce these high operating costs. Reverse osmosis (RO) and nanofiltration (NF) have been examined for the pre-concentration of clarified thin sugar juice to reduce the role of evaporation in the sugar production because evaporation is an energy intensive process [1]. Also, RO/NF can remove water from the aqueous solution at temperature below its boiling point and without phase change. It has positive effect on the quality of the juice because sugar juice is degraded at high temperatures. The clarified thin sugar juice contains 15°Brix (15 w%) of sugar and about 0.25-0.40 w% ionic compounds.

In the present work the effect of presence of single salt and the mixtures of different inorganic electrolytes (NaCl, $MgCl_2$, and Na_2SO_4) during sugar concentration by nanofiltration has been studied experimentally. The experiments were conducted on a lab-scale cross-flow membrane unit OS-MC-01 from Osmota with effective membrane area of 0.008 m^2. Sugar concentration was measured by a digital refractometer DR301-95 from Krüss. The concentrations of dissolved ions in the samples are detectable by conductivity measurements and anions and cations concentration in the feed and permeate sample collected during the experiment were analysed on Dionex 5000+ ion chromatographic equipment. In this work, negatively charged NF membrane (MPF-34) from Koch Membrane was used [2]. It has maximum operating temperature limit of 70 °C, a pressure range of 5-35 bar, and a pH range 0-14. All the experiments were performed at the same process conditions of 32 bar, 60 °C, feed flowrate of 2.5 L/min, sugar feed concentration 15 °Brix and 0.25w% of each electrolyte.

The experiments proved that nanofiltration process is capable of concentrating sugar from 15° Brix to 26° Brix, separating different ions from sugar solution. The evaluation of the data revealed that salt concentration did not affect the retention of the sugar significantly, on contrary the salt rejection was dependent on the sugar concentration and salt concentration as well. The interesting result was that the sugar has a big influence on the rejection of anion and cation and also on the conductivity of electrolytic solution. The presence of sugar in ionic mixture with a maximum valence of two will govern the negative rejection and will force the monovalent ion with the same charge to go through the membrane this will yield a negative rejection for that monovalent ion. For example in a mixture of NaCl, Na_2SO_4 and sugar the negative rejection will be observed for Cl^- as shown in figure 1.

In full manuscript the results of other complex ionic mixtures with sugar will be presented.

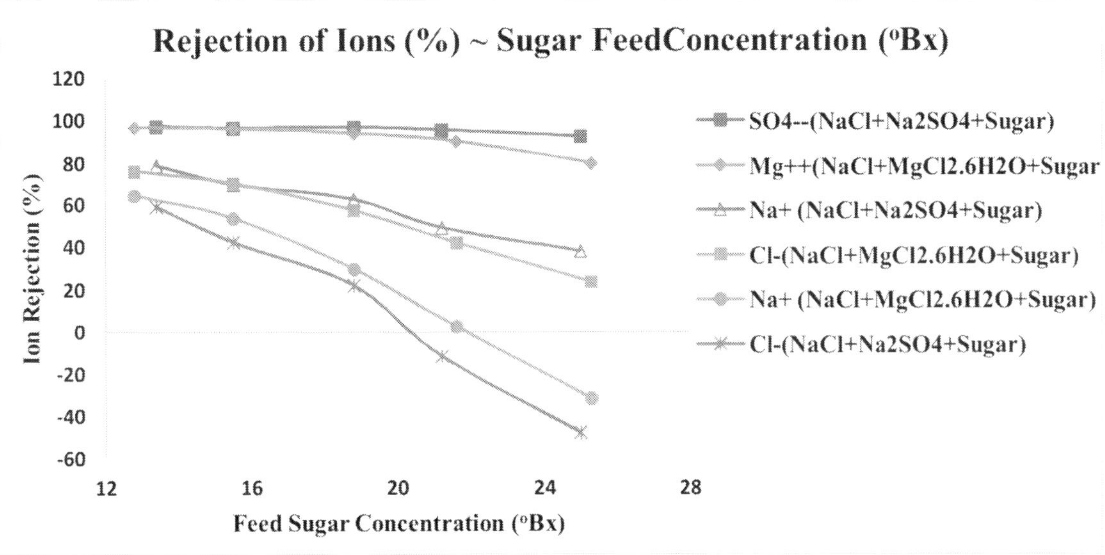

Figure 1: The effect of sugar feed concentration on ionic rejection.

References

[1] S. Gul, M. Harasek, Energy saving in sugar manufacturing through the integration of environmental friendly new membrane processes for thin juice pre-concentration, Applied Thermal Engineering 43 (2012) 128-133.

[2] Koch membrane. http://www.kochmembrane.com/

Visualization study of dynamic behaviours of molten blast furnace slag droplet impacting on different material surfaces

Xian-Yan He [2], Xun Zhu *[1-2], Hong Wang *[1-2], Bing Ding [2], Qiang Liao [1-2], Qian Fu [ab]
[1] Key Laboratory of Low-grade Energy Utilization Technologies and Systems, Chongqing University, Chongqing 400030, China
[2] Institute of Engineering Thermophysics, Chongqing University, Chongqing 400030, China,
Tel: +86-023-65102474, E-mail: zhuxun@cqu.edu.cn, hognwang@cqu.edu.cn

Blast furnace (BF) slag is one of the main by-products in iron and steel industry. In consideration of the tremendous output of BF slag, mature technologies have been developed to make proper use of the resource. The deep value of the waste heat has not been got an effective recovery, although the discharged temperature of BF slag is above 1450°C and the sensible heat is about 1,880,000 kJ per ton slag. At present, the focus was concentrated on developing a dry centrifugal granulation technology which was expected to achieve both heat recovery and comprehensive utilization. It is inevitable that some larger granulated droplets will directly impact onto the chamber wall before solidification because of space restriction of the liquid slag centrifugal granulation unit. In this case, it has a significant negative influence on the operational stability of the waste heat recovery device as the larger molten slag droplet might deposit on the device surface or evenly slag buildup happen. By the way, slag impact surface also exists in boiler and gasifier, and it directly affects the heat transfer efficiency, causes corrosion effect on the water wall, even results in plugging slag.

In the present study, visualization experiments were performed on the dynamic behavior of high temperature slag droplet impact different surface. The experimental parameters, such as the diameter and temperature of the BF slag droplet, the impacting angle and the falling height, were controlled by adjusting the corresponding settings. The surface materials and surface roughness were changed to obtain the effects on the process of molten blast furnace slag droplet impacting solid surface. Meanwhile, the apparent static contact angles on different surfaces were measured by a redesigned sessile drop method, since it's a crucial factor.

The results showed that the apparent static contact angles were obviously different and the range was from 132.7° to 151.4°. Combined with the properties of different substrates, it could be speculated that an adiabatic smooth plate was an ideal material to promote the slag droplet splashing and the stainless steel could reduce the droplet spreading inversely as a heat-absorbent material. Unfortunately, refractory cement was unsuitable to the granulating waste heat recovery unit, as an unavoidable deterioration wall material. Furthermore, the viscous dissipation energy on stainless steel was statistically analyzed and a semiempirical model was developed to predict the maximum spread factor dependent on the initial parameters of slag droplet. The current study proved the model was suitable to the experimental data. The model could guide the design of the heat recovery device.

Keywords: Blast furnace slag; Droplet; Impact; spread

Economic feasibility study on the wastewater treatment plant in fuel companies. A Cuban study case

Y. Busto[1], E. W. Palacios[2], L. M. Rios[3], J. Martin[3], Y. Jiménez[3], H. Perez[4], M. Yera[5]
[1]Technological University of Israel, Francisco Pizarro street and Orellana Avenue E4-142, 170150, Quito, Ecuador; tel. +593 98 4379929, e-mail: yailenb09@gmail.com; [2]Faculty of Odontology, Central University of Ecuador, University Citadel between Bolivia and Salgado street, Quito, Ecuador; tel. +593 99 2728009, e-mail: wsaudades@gmail.com; [3]Study Center of Applied Chemistry, Central University "Marta Abreu" of Las Villas, Highway to Camajuani km $5^{1/2}$, 54830, Villa Clara, Cuba; [4]Fuel Trading Company of Villa Clara. Road to Subfloor 76, 50300, Villa Clara, Cuba; [5]Iberoamerican University of Ecuador, 9 de Octubre and Santa Maria street, Quito, Ecuador; tel. +593 99 4149669, e-mail: miriam.yera62@gmail.com

In the Cuban oil industry, wastewater from the process of reception and distribution of hydrocarbons, contain a wide range of contaminants with high chemical and biological oxygen demand (COD and BOD). However, the most significant pollutants of these effluents are the hydrocarbons, fats and oils due to its high toxicity and ability to spread in thin layers on the water surface which difficult the passage of sunlight, hurting life in such ecosystems.

Two technological alternatives were evaluated from technical, economic and environmental point of view. The first one, considered as short-term solution (rehabilitation of the existing system) and second one (building of a collection pool), considered investment in the medium to long term. Taking into account the equipment acquisition cost as well as the process flow requirements, the total investment cost for both alternatives was estimated. Furthermore, the main dynamic economic indicators: Net Present Value (NPV), Internal Rate of Return (IRR) and Pay Period (PP) were evaluated to develop a comprehensive economic-financial study using the method of "Percentage of Delivered Equipment". In addition, a sensitivity analysis to evaluate the effect of increases or decreases in the price of the process equipment market over the economy of the proposal variant was analysed.

From the economic point of view, alternative 1 reported higher gross annual earnings ($2,139,553) compared to the second variant ($2,131,428). However, considering that the annual income of the company are substantial ($11,791,284 in 2013), the alternative 2 proposes a more effective solution from a technological point of view achieving greater life of the proposed technology and as better operation and maintenance. Moreover, this technological option guarantees a significant reduction on the environmental pollution risk.

Ru-containing catalysts in hydrogenation of D-glucose in flow-type microreactor

V.Yu. Doluda, V.G. Matveeva, E.M. Sulman*, M.E. Grigorev, M.G. Sulman, A.V. Bykov
Tver Technical University, Dep. Biotechnology and Chemistry, A.Nikitina str. 22, 170026, Tver, Russia; tel./fax: +7 4822 789317, e-mail: sulman@online.tver.ru

At present catalytic hydrogenation of D-glucose is an important process of industrial production of D-sorbitol which is widely used in different fields. D-sorbitol can be used as a precursor in chemical and pharmaceutical industry (e.g. for vitamin C production), in polymer production and as sugar substitute. The main hydrogenation catalyst is Ni-Raney which in spite of low cost has many disadvantages such as metal leaching and catalyst deactivation. Ru-containing catalysts are an alternative to common Ni-Raney catalyst hence to decrease their cost it is necessary to decrease the amount of active metal in the catalyst. In this paper the investigation of catalytic D-glucose hydrogenation with use of Ru-containing hypercrosslinked polystyrene in flow-type microreactor settlement is presented.

Continuous hydrogenation of D-glucose was carried out using MN100/Ru-2.85% catalyst in flow-type reactor. The analysis of reaction mixture was done using HPLC.

To study D-glucose catalytic hydrogenation reaction in continuous mode the experiments varying residence time (τ), temperature, hydrogen pressure and glucose initial concentration were carried out. The experiment time was 6 hours that was enough to reach out the steady-state condition.

In all range of experiments the selectivity of D-glucose hydrogenation process reached out 98-99%.

Addition of water in calcium based adsorbent for CO_2 adsorption: Experimental and isotherm modelling studies

[1]N. Azmi, [2]N. A. Rashidi, [3]S. Yusup, [4]K. M. Sabil
[1]Universiti Teknologi PETRONAS, Chemical Engineering Department, 32610 Bandar Seri Iskandar, Perak, Malaysia; tel. +6 05 3688208, e-mail: hudaz7221@gmail.com; [2]Universiti Teknologi PETRONAS, Chemical Engineering Department, 32610 Bandar Seri Iskandar, Perak, Malaysia; tel. +6 05 3688208, e-mail: noradilla.88@gmail.com; [3]Universiti Teknologi PETRONAS, Chemical Engineering Department, 32610 Bandar Seri Iskandar, Perak, Malaysia; tel. +6 05 3688208, e-mail: drsuzana_yusuf@petronas.com.my; [4]Heriot-Watt Universiti Malaysia, Institute of Petroleum Engineering, 62200 Putrajaya, Malaysia; tel. +6 03 889443697, e-mail: K.MSabil@hw.ac.uk.

In this work, a combined process of adsorption and gas hydrate formation has been proposed as an alternative approach for the CO_2 separation from gas streams. The aim of this research work is to study the suitable isotherms for the CO_2 adsorption on dry and wet calcium oxides, via a static volumetric method. The accelerating CO_2 uptake for the wet calcium oxides was observed, due to the formation of CO_2 hydrates and was indicated by an inflection of isotherm at a definite pressure, which resembles the S-shape. The highest CO_2 adsorption capacity at 2°C was attained by the wet calcium oxides, with water ratio (R_w) of 0.67. The experimental data obtained from the volumetric method have been analyzed by various isotherm models such as Langmuir, Freundlich, Redlich-Peterson, Toth, and Sips. The approriateness of these models was evaluated by the non-linear regression method. Based on findings, the Freundlich model adequately describes the CO_2 adsorption onto dry calcium oxides, whilst the Redlich-Peterson is best-fitted for the wet samples.

Keywords: synthesized CaO, CO_2 adsorption, isotherm models, non-linear regression

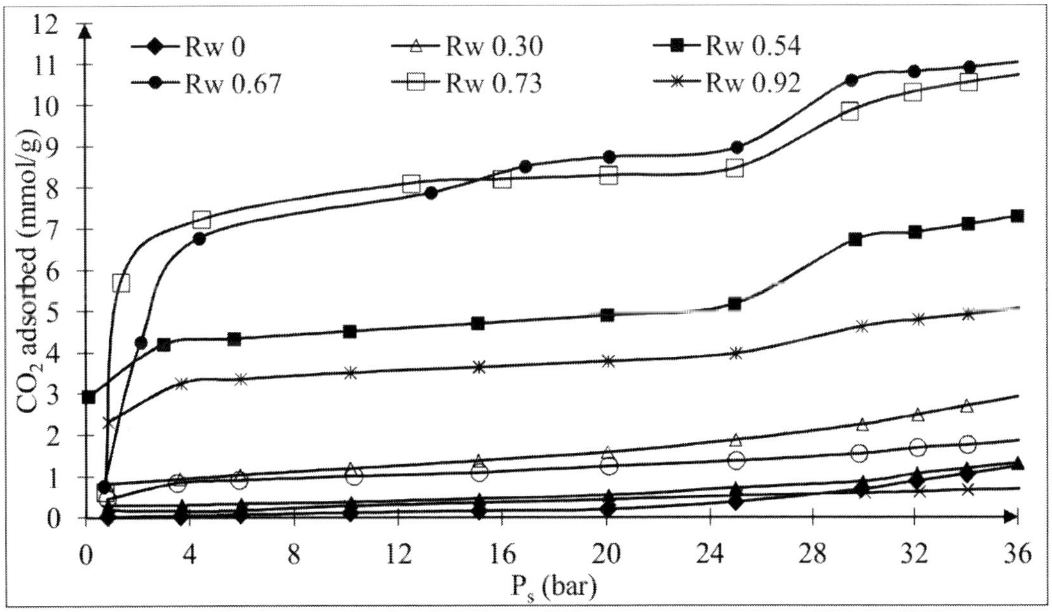

Figure 1 CO_2 uptakes on synthesized CaO at different amount water (R_w) at 2°C

Carbon dioxide capture from model marine diesel engine exhaust by means of K_2CO_3-based sorbents

M. Balsamo, A. Erto, A. Lancia, F. Di Natale
Università degli Studi di Napoli Federico II, Dipartimento di Ingegneria Chimica, dei Materiali e della Produzione Industriale, P.le Tecchio 80, 80125 Naples, Italy;
tel. +39 081 7682236, e-mail: marco.balsamo@unina.it, fax: +39 081 5936936.

Carbon capture and storage (CCS) is considered one of the most efficient strategies for mitigating CO_2 emissions derived from anthropogenic activities. CCS technologies commonly involve the separation of CO_2 from the flue gases, transport of the gas to a suitable storage location and long term isolation from atmosphere. Numerous scientific and industrial activities are currently aimed at the development of high efficiency and low-cost CO_2 purification processes. Nowadays, most studies involve the use of aqueous amine solutions for carbon dioxide scrubbing, which suffer many drawbacks related to the considerable amounts of thermal energy required for absorbent regeneration, solvent degradation and equipment corrosion In 2012, DNV Greece developed a design concept for the application of CCS on-board ships with MEA (Monoethanolammine) scrubbing.

This research group is currently developing a CCS scheme for on board naval installation based on the use of potassium carbonate which converts CO_2 into potassium bicarbonate. The sorbent carbonation reaction can be effective at low temperature and potentially requires less energy for regeneration with respect to MEA scrubbing. The exhaust sorbent can be regenerated in the temperature range 150-300°C to recover CO_2 as an almost pure gas, stored as a liquid in cryogenic tank and disposed, or sold, at docks. Alternatively, the produced potassium bicarbonate could be sold as a value-added product for different applications, such as fire extinguisher or agriculture fertilizer. This solution benefits from the large availability of carbonates and the absence of toxic residuals.

In this paper, we report experimental results on CO_2 capture from a model marine diesel engine exhaust containing 5% CO_2 and 5% H_2O (balance N_2) by means of K_2CO_3 both raw and supported onto porous alumina. CO_2 capture tests were carried out in a fixed bed column integrated in a lab scale unit. The removal efficiency and capture dynamics were investigated in the temperature range 60-105°C. The sorbent regeneration was assessed from 120 to 250°C to define optimal conditions for CO_2 recovery from the spent sorbent.

Novel process for hydrogen production through the sorption enhanced reforming of methane combined with chemical looping combustion

J.R. Fernández, J. C. Abanades
Spanish Research Council, INCAR-CSIC, C/ Francisco Pintado Fe, 26, 33011, Oviedo, Spain; tel. +34 985 11 90 90, e-mail: jramon@incar.csic.es; fax: +34 985 29 76 62

A novel process scheme is analysed for the production of hydrogen from methane, while obtaining a separate stream of CO_2, which is suitable for a subsequent reuse and/or for its permanent geological storage. This system comprises three interconnected fluidized-bed reactors operating at atmospheric pressure. The hydrogen is produced in a sorption enhanced reforming (SER) stage, where the presence of a CaO-based sorbent allows the CO_2 to be removed as it is formed. According to Le Chatelier's principle, the equilibrium is shifted towards hydrogen formation, with the result that higher H_2 production and almost complete conversion of CH_4 and CO are achieved. The main novelty of the process concerns the regeneration of the $CaCO_3$, required to carry out a multicycle operation. The calcination of $CaCO_3$ is a highly endothermic reaction that demands temperatures around 900 ºC in pure CO_2 and at atmospheric pressure. In the proposed process, a redox chemical loop supplies the heat necessary for the calcination. The $CaCO_3$ formed in the SER is put in contact with a very high-temperature stream of iron-based oxygen carrier, which is mainly composed of Fe_2O_3 resulting from the exothermic oxidation of Fe_3O_4 with air in a separate reactor (at 950ºC-1200ºC). The calciner also acts as a fuel reactor, where Fe_2O_3 is reduced again to Fe_3O_4 by feeding a fuel gas (CH_4, H_2, CO), which is converted to CO_2 and steam. Iron-based oxygen carriers are cheap materials and present good stability under repeated oxidation/reduction cycles at high temperatures. Moreover, they are especially suitable for this application because of their high density, which favours their separation by segregation from the CaO and the reforming catalyst (finer and less dense solids) once the calcination/reduction have been carried out. Downstream of the segregator, the more dense iron-rich solids (jetsam) are recycled to the air reactor, while the calcined product and the catalyst (flotsam) are fed into the reformer, so that a continuous double calcium-iron chemical loop is established. The influence of the temperature, steam addition and active contents of calcium and iron-based materials on the H_2 yield and overall energy efficiency has been investigated. A preliminary thermodynamic assessment of the proposed system has been carried out reaching H_2 equivalent efficiencies of up to 81%, which are around 10% points above the efficiency of a reference H_2 production plant based on conventional steam reforming including CO_2 capture with MDEA. Carbon capture efficiencies of around 98% can be obtained, which are significantly higher than those obtained in the reference plant that uses MDEA absorption (around 85%). These results confirm the potential of this novel system for a future development as hydrogen production process with CO_2 capture.

Effect of processing conditions on CO_2 adsorption and mechanical strength of MgO pellets

L. Chen[1,2], Y.D. Ding[1,2]*, X. Zhu[1,2], Q. Liao[1,2], G. Song[1,2],Q. Fu[1,2]
[1]Key laboratory of Low-grade Energy Utilization Technologies and Systems, MOE, Chongqing, China; [2]Institute of Engineering Thermophysics, Chongqing University, Chongqing, China; tel.+86 023 65102474, e-mail: dingyudong@cqu.edu.cn.

MgO is a promising adsorbent for CO_2 capture due to its widely available in nature, low temperature adsorption and water vapor in flue gas can be favorable for the CO_2 adsorption. In the applications, adsorbent was processed to solid particles with a certain shape and a certain hardness firstly, then was used to the large-scale fixed-bed reactor systems. Hence, the processing of MgO pellets is a necessary step, and the processing conditions have an important effect on the hardness and pore structure. In this work, with various calcinations temperature and sodium carboxymethyl cellulose (CMC) mass ratio, MgO pellets were fabricated with a mould by using basic magnesium carbonate pentahydrate as precursors and CMC as binder. The MgO pellet samples was characterized by X-ray diffractometer (XRD), scanning electron microscopy (SEM), specific surface area and porosity analyzer, microcomputer control universal testing machine and thermogravimetric analyzer. The results showed that the MgO pellets possessed a high BET surface area in a range of 81- 102 m^2 g^{-1} and a highly porous structure. The CO_2 adsorption capacity increased first and then decreased with the increasing calcinations temperature from 300 °C to 600 °C, and as well with various CMC mass ratio from 15% to 40%. While the mechanical strength monotonously increased with increasing calcinations temperature and CMC mass ratio. The CMC20%-500 pellets had a desirable adsorption capacity of 24.77 mg g^{-1} CO_2 and an acceptable mechanical strength of 0.388 MPa at 50 °C, 1bar. Moreover, an appropriate amount of carbon inserted in the co-existing MgO particles could prevent the aggregation so as to increase the CO_2 adsorption capacity. The CO_2 adsorption capacity of the pellets was only decreased by 3.6% after 5 cycles. These results indicate that the fabricated pellets used CMC as binder had a stable CO_2 adsorption capacity, desirable mechanical strength and was easy to be scaled up, indicating the promising use for industrial application.

Keywords: MgO pellet; CO_2 adsorption; mechanical strength; calcination temperature; porous structure

An energy efficient top-lit gas-lift bioreactor for algal cultivation and CO_2 mitigation of industrial off-gas

N. Seyed Hosseini[1], H. Shang[1], G. M. Ross[2], and J. A. Scott[1,2]
[1]Bharti School of Engineering, Laurentian University, 935 Ramsey Lake Road, Sudbury, ON, Canada, P3E 2C6; Tel: +1 705 675 1151, Fax: +1 705 675 4862
Email: nseyedhosseini@laurentian.ca (N. Seyed Hosseini) hshang@laurentian.ca (H. Shang) jascott@laurentian.ca (J.A. Scott) ;
[2]Northern Ontario School of Medicine, 935 Ramsey Lake Road, Sudbury, ON, Canada, P3E 2C6; Tel: +1 705 662 7218, Fax: +1 705 675 4862; Email: jascott@laurentian.ca (J.A. Scott); gross@nosm.ca (G.M. Ross)

Specific species of microalgae with high photosynthetic rates and lipid content have the potential to provide a renewable liquid fuel, biodiesel (Chisti, 2007). Furthermore, through microalgae photosynthesis there is opportunity to fix anthropogenic carbon dioxide (CO_2) from industrial point sources to both improve productivity and mitigate greenhouse gas emissions (Seyed Hosseini et al., 2015).

In addition, industrial off-gases released to the environment contain significant amount of waste heat that could be utilized to maintain the temperature of open ponds in cold climate regions (Shang and Scott, 2011).

Among various methods proposed for large-scale cultivation of microalgae (Zhao and Su, 2014) open oval raceways circulated by use of a paddlewheel are currently the most economic option for commercial scale production due to relatively low capital, maintenance and operation costs (Chisti, 2007). They are typically located in regions with warm temperature and also have intensive light intensities, which can cause considerable evaporation losses and possible photoinhibition (Chisti, 2007). These regions are often semi-arid and suffer from lack of freshwater. Whereas regions that are rich in freshwater generally experience seasonal cold climates and consequently are not considered for outdoor algal production unless sources of “free” heat, such as that contained within off-gas can be utilized (Laamanen et al., 2014).

Open commercial raceways typically have large surface areas but only operate at a water depth of 15-30 cm (Zhao and Su, 2014). Providing an appropriate large land space close to a fixed off-gas source may, therefore, prove difficult on an industrial site and distribution piping costs to transfer off-gas to remote algal farms will contribute significantly in the cost of cultivation (Putt et al., 2011). The depth is limited due to restricted light penetration, but if off-gas is to be bubbled through, the shallow depth is likely to lead to reduced CO_2 (and where applicable heat) transfer due to a short residence time. Overall, these limitations have led to low utilization of CO_2 and areal productivity (20 g_{dw} m^{-2} day^{-1}) in traditional raceway ponds.

Alternative designs have been presented to increase productivity and CO_2 capture efficiency in traditional raceways such as: using single or multiple sumps with/without baffles (de Godos et al., 2014), a carbonation column system to circulate the culture through an absorption column (Putt et al., 2011), a carbon supplying device fixed at the bottom of the pond (Su et al., 2008) and an airlift-driven raceway design as a replacement of the current paddlewheel-driven design (Ketheesan and Nirmalakhandan, 2012). Although proposed configurations have provided enhanced CO_2 transfer efficiency, the reported areal productivities were not significantly improved.

A more desirable approach to utilizing CO_2 from off-gas to achieve longer gas-liquid transfer times, as well as providing greater per area (areal) productivity on industrial sites would be to have deeper ponds. In order to make the ponds deeper and avoid the cost of artificial lighting, gas-lift systems could be employed to provide vertical circulation of the microalgae. Gas-lift colums have gained wide acceptance for gas-liquid contacting applications in bioprocessing due to efficient mixing with low stress, high volumetric gas transfer and a lack of microbial growth on the walls (Kumar and Das, 2012).

This paper presents a design for an open bioreactor that uses a gas-lift system to enable deeper open ponds, thereby significantly reducing the operational footprint as well as increasing CO_2 transfer. Growth of *Scenedesmus* sp. in this one-meter deep, top-lit gas-lift bioreactor was evaluated. The results indicated a comparable volumetric biomass productivity (0.06 g_{dw} L^{-1} day^{-1}), but around three times higher productivity per unit area (60.0 g_{dw} m^{-2} day^{-1}) than achieved with traditional raceway ponds (15-25 g_{dw} m^{-2} day^{-1}). The energetic productivity in terms of daily biomass productivity per unit power input (0.41 g_{dw} W^{-1} day^{-1}) obtained was comparable or greater than attained with various photobioreactors (0.1-0.6 g_{dw} W^{-1} day^{-1}). Also, lipid content per dry weight biomass of *Scenedesmus* sp. was increased by 27% with an enhanced level of CO_2 in the feed gas that leads to promote biodiesel production. The algal biomass and lipid productivities have been optimized and validated to improve the economics of biodiesel production.

References

1. Chisti, Y., 2007. Biodiesel from microalgae. Biotechnol. Adv. 25 (3), 294–306.
2. de Godos, I., Mendoza, J.L., Acién, F.G., Molina, E., Banks, C.J., Heaven, S., Rogalla, F., 2014. Evaluation of carbon dioxide mass transfer in raceway reactors for microalgae culture using flue gases. Bioresour. Technol. 153, 307–314.
3. Ketheesan, B., Nirmalakhandan. N., 2012. Feasibility of microalgal cultivation in a pilot-scale airlift-driven raceway reactor. Bioresour. Technol. 108, 196–202.
4. Kumar, K., Das, D., 2012. Growth characteristics of Chlorella Sorokiniana in airlift and bubble column photobioreactors. Bioresour. Technol. 116, 307–313.
5. Laamanen, C.A., Shang, H., Ross, G.M., Scott, J.A., 2014. A model for utilizing industrial off-gas to support microalgae cultivation for biodiesel in cold climates. Energy Convers. Manag. 88, 476–483.
6. Putt, R., Singh, M., Chinnasamy, S., Das, K.C., 2011. An efficient system for carbonation of high-rate algae pond water to enhance CO2 mass transfer. Bioresour. Technol. 102, 3240–3245.
7. Seyed Hosseini, N., Shang, H., Ross, G.M., Scott, J.A., 2015. Microalgae cultivation in a novel top-lit gas-lift open bioreactor. Bioresour. Technol. 192, 432-440.
8. Shang, H., Scott, J.A., 2011. Minimizing leakage from a smelter off-gas system by enhanced control. J. Clean. Prod. 19, 70–77.
9. Su, Z., Kang, R., Shi, S., Cong, W., Cai, Z., 2008. An economical device for carbon supplement in large-scale micro-algae production. Bioprocess Biosyst. Eng. 31, 641–645.
10. Zhao, B., Su. Y., 2014. Process effect of microalgal-carbon dioxide fixation and biomass production: A review. Renew. Sustainable Energy Rev. 31, 121–132.

Participant plants and streams selection for interplant heat integration among three plants

Runrun Song[a], Yufei Wang[a], Xiao Feng*[b]
[a]State Key Laboratory of Heavy Oil Processing, China University of Petroleum, Beijing 102249, China
[b]School of Chemical Engineering & Technology, Xi'an Jiaotong University, Xi'an 710049, China
xfeng@mail.xjtu.edu.cn

Interplant heat integration is an efficient way of energy saving for industry zones. In past few decades, many researches of this topic were proposed, which brings significant energy savings. However, in many previous studies, either all process streams or waste heat source/sink of individual plants were used for integration. This paper introduced a new selection strategy of participant plants and streams before integration, which can reduce the number of participant streams and calculation burden while the energy targets almost keep unchanged. The proposed method can consider both pinch and threshold problems. The parallel structure is selected for integration between three plants, since it can always achieve the largest energy saving than split and series structures and it might be more reliable when considering flow rate or temperature fluctuation in individual plants. A case study is used to show the effectiveness of the strategy.

CO_2-argon-steam oxy-fuel (CARSOXY) combustion for CCS inert gas atmospheres in gas turbines

[1]A. Al-Doboon, [1]A. Valera-Medina, [2]J. Ng, [1]N. Syred

College of Physical Sciences and Engineering, Cardiff University, UK Malaysia Campus, University of Southampton, Malaysia Cardiff School of Engineering, Cardiff University, Queen's Building, The Parade, CF24 3AA, Wales, UK; tel. +44 (0) 7448327557 , e-mail: Al-DoboonAI@cardiff.ac.uk fax: +44(0) 29 2087 4939

Sharp demand for electricity consumption is a serious problem that leads to excessive emissions of CO_2 with its inherent climate change consequences. Oxy-fuel combustion, a technique which uses pure Oxygen with recirculation of CO_2 gas in a combustion process, may provide one path for the solution of the aforementioned issues by decreasing the emissions from power station plants using gas turbine cycles. However, although emissions are reduced with this technology by utilizing CO_2 as a working fluid, carbon dioxide increment lessens the power output and reliability of the turbines used.

Thus, the aim of this work is to investigate how emissions can be reduced without significantly changing efficiency and power output in Advanced Humidified Air Turbine (AHAT) cycles. In this study, 0-D chemical kinetic software (GASEQ) is utilized to generate more than 100 blends of Argon, H_2O and CO_2, aiming at the improvement of thermodynamic properties, reduction of flame temperature and gas recirculation for carbon capture and storage, respectively. An empirical approach combined with the design of experimental models are implemented to investigate the improvement and deterioration in thermodynamic properties such as product temperature (T_o), Specific heat ratio (γ) and Heat capacity (C_p) in order to identify the best blend that will be used in an analytical simulation model. 1-D chemical kinetic software (CHEMKIN - PRO) using GRI-3.0 reaction mechanism is utilized to evaluate through a reaction network the combustion properties and final emissions of the most promising blends. Results indicate there is complete elimination of NOx, with enough CO_2 and H_2O in the flue gases to be sequestrated for further treatment or recirculation. Furthermore, efficiency and power outputs of this AHAT cycle are predicted to be similar to those obtained using conventional fuels and air working fluid conditions.

Numerical investigation of the hydrodynamics in a bubbling fluidized bed with granulated molten slag particles during heat recovery process

Lei Cao [2], Hong Wang [*1-2], Xun Zhu [*1-2], Xian-Yan He [2], Qiang Liao [1-2], Qian Fu [1-2]
[1] Key Laboratory of Low-grade Energy Utilization Technologies and Systems, Chongqing University, Chongqing 400030, China;
[2] Institute of Engineering Thermophysics, Chongqing University, Chongqing 400030, China;
Tel: +86-023-65102474, E-mail: hongwang@cqu.edu.cn, zhuxun@cqu.edu.cn

Abstract:

Steelmaking industry is one of the most energy-intensive industries, accounting for about 5-6% of the world total energy consumption, whereas the waste heat recovery rate is merely 17%[1]. As the main liquid by-products during the steelmaking process, a substantial amount of molten slag, which carrying dramatically high-grade thermal energy is exhausted at the temperature of about 1450-1550 °C is a potential source of energy and raw materials[2]. Currently, the dominant molten slag treatment process is water quench with higher cooling rate, whereafter, vitreous slag is gained for subsequent utilization as an important alternative of Portland cement. Regrettably, this process fails to efficiently recover the considerable thermal energy within the slag and consumes huge amount of water reaches to about 1000-1500 litres per ton slag[3]. Be aimed at heat recovery and water saving, dry granulation heat recovery approach by using air as cooling media sprung up energetically and then, various approaches have been developed. In view of lower heat transfer capability of air, it is imperative to enhance the heat transfer between slag particles and cooling air in all those dry approaches. Due to the characters of higher rate of heat and mass transfer, fluidized bed is identified as a kind of promising device to recover heat from high temperature slag. In order to further understand the rheology in bubbling fluidized bed with granulated molten slag particles, and instruct subsequent trial. It is essential to study the hydrodynamics in a bubbling fluidized bed with granulated molten slag particles which are classified as Geldart D particles [4].

Aiming at this fundamental issue, and to get more inside bed information that is difficult to be obtained by experiment, numerical simulation on hydrodynamics of granulated molten slag particles in a bubbling fluidized bed is conducted to investigate the influence of operation conditions on bed hydrodynamics. A two-fluid Eulerian model with closures from the Kinetic Theory of Granular Flow (KTGF) is applied. Firstly, a comparison between simulation results and Geldart et al.[5] experiments with Geldart D particles in a bubbling fluidized bed was made to validate the accuracy of present model, and a satisfactory level of agreement was achieved. Then, the qualitative flow pattern and the bed pressure drop are obtained with granulated slag. Moreover, the bubble size and concentration at a specific bed height are captured. The results show that the averaged bubble diameter at a specific bed height increases linearly with superficial air velocity increases, and the bubble concentration increases slightly. Furthermore, the axial particle velocity at different bed level have been extracted, the results reveal that the axial particle velocity is nearly symmetric about centre axis of the bed, and the particle velocity is negative at the centre axis of the bed. More important, symmetry vortex formation of slag particles in bubbling fluidized bed can facilitate heat exchange between high temperature slag and cooling air.

Keywords: Granulated molten slag; Bubbling fluidized bed; Computational Fluid Dynamics; Two-fluid model; Hydrodynamics; Bubble characteristics

References:

[1] X. Chunbao Charles, D.-q. Cang, A brief overview of low CO_2 emission technologies for iron and steel making, Journal of Iron and Steel Research, International, 17 (2010) 1-7.
[2] H. Zhang, H. Wang, X. Zhu, Y.-J. Qiu, K. Li, R. Chen, Q. Liao, A review of waste heat recovery technologies towards molten slag in steel industry, Applied Energy, 112 (2013) 956-966.
[3] S. Jahanshahi, Y. Pan, D. Xie, Some fundamental aspects of the dry slag granulation process, in: 9th International Conference on Molten Slags, Fluxes and Salts (Molten12), Beijing, 2012, pp. 27-31.
[4] D. Geldart, Types of gas fluidization, Powder Technology, 7 (1973) 285-292.
[5] D. Geldart, R. Cranfield, The gas fluidisation of large particles, The Chemical Engineering Journal, 3 (1972) 211-231.

Analysis of the influence of turbulence on the heat transfer between spherical particles and planar surfaces

[1]G. Brösigke, [2]A. Herter, [2]M. Rädle, [1]J.-U. Repke
[1]TU Bergakademie Freiberg, Institute of Thermal, Environmental and Natural Products Process Engineering, Leipziger Straße 28, 09596 Freiberg, Germany; tel. +49 3731 39 2283, email: georg.broesigke@tun.tu-freiberg.de; jens-uwe.repke@tun.tu-freiberg.de;
[2]Hochschule Mannheim, PI, John-Deere-Str. 81 A, 68163 Mannheim, Germany
email: a.herter@hs-mannheim.de; m.raedle@hs-mannheim.de

The heat transfer between spherical particles and walls on the one hand and between particles solely on the other hand is relevant in several industrial apparatus, which amongst others include fixed bed reactors, fluidized beds, tube dryers and rotary kilns. The occurring mechanisms are not fully understood jet or rather their different amount of contribution is not quantified satisfactorily. For purposive development and efficient design of mentioned apparatus a fundamental understanding of the occurring mechanisms is crucial.

In a previous work [1] the heat conduction through the gap of gas between a single spherical particle and a planar surface was identified as dominating mechanism for the laminar regime. The investigation was carried out with CFD simulations and the results were validated against both experimental data and a correlation from literature for a static sphere on a planar surface [2].

For a fundamental and detailed description the influence of turbulence is investigated in the next step. The range of transition where the laminar regime changes to a turbulent regime is of special interest in this case in order to identify the source and the conditions for turbulence to be generated for the investigated geometry. In this range a successive superposition of diffusive and convective heat transfer mechanism is expected.

Turbulence models were historically developed to describe fully developed turbulence and therefore fail in the description of the transition regime. Hence the investigation in this work is carried out with direct numerical simulation.

The simulations are carried out with the open source CFD simulation software OpenFOAM®. A customized solver is developed on the basis of the standard solver *chtMultiRegionFoam* in order to meet the special requirements for direct numerical simulation. The mesh is generated with the OpenFOAM® utility *snappyHexMesh*.

On basis of the results gained by the direct numerical simulations a comparison with established turbulence models like the k-ε model or the k-ω-SST model is carried out for fully developed turbulent conditions. First results show qualitatively good agreement in velocity field and dissipation rate for both turbulence models compared to DNS. A slight deviation in the absolute amount of the transport properties is observed as well as a deviation in the point of stall.

This work aims on identifying and quantifying the influence of turbulent flow conditions on the heat transfer.

[1] G. Brösigke, A. Herter, M. Rädle, J.-U. Repke, "Investigations of Heat Transfer Mechanisms between a Moving Sphere and a Static Plate with Computational Fluid Dynamics" 17th Conference on Process Integration, Modelling and Optimisation for Energy Saving and Pollution Reduction, Prague 24.-27.8.2014

[2] E.-U. Schlünder, Chem. Eng. Process. 18: 31-53, 1984

Investigation of the pressure drop of random packed bed adsorbers

B. Haddadi, C. Jordan, H. R. Norouzi , M. Harasek
TU Wien, Institute of Chemical Engineering, Getreidemarkt 9/1662, A-1060 Wien, Austria
tel. +43 1 58801-15924, e-mail: bahram.haddadi.sisakht@tuwien.ac.at

Adsorption is one of the most often used unit operations in chemical engineering. However, due to computational limitations and time constraints simulations are commonly done only with "black box" models or simplified porous media structures. In a new project an OpenFOAM® based CFD solver "adsorpFoam" and the required grid generation work flow using DEM methods for random packings were created to carry out more detailed analysis of adsorption columns. Fully resolved packed beds will be simulated to get more detailed information on gas side maldistribution, bypass streams, near-wall effects (all these factors contribute to the total pressure drop of the packing), local temperature and adsorbate distribution, as well as fluid residence time distributions of random packings in columns.

The current work will summarize the first results of adsorpFoam with respect to the validation using gas flow in small random packings: The code and the procedure have been tested for spheres, monodispersed cylinders and cylindrical particles with size distribution. In all cases good agreement of the experimental results and the CFD simulations have been achieved, the results also compare well to calculations based on Ergun equation. The most critical point is the correct capturing of the void fraction in the packing – which is a strong function of the particle size and particle properties. As an example, the results for an activated carbon packing in a lab scale device are presented:

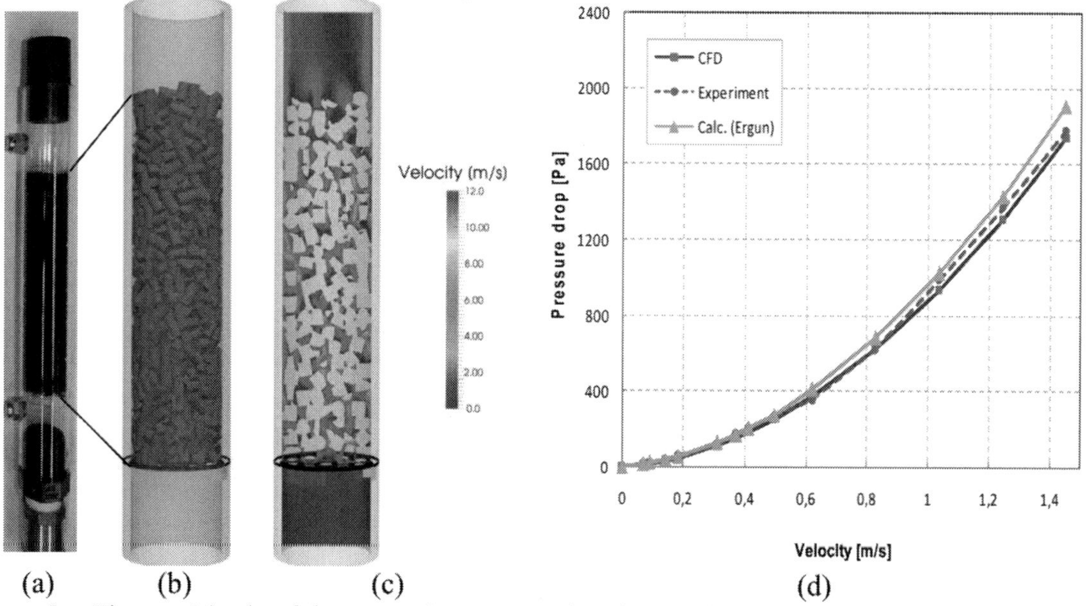

(a) (b) (c) (d)

In Figure (a) the lab setup is presented, Figure (b) shows the equivalent CFD implementation of the random packing, Figure (c) gives the distribution of the velocity magnitude in the symmetry plane and Figure (d) presents the pressure drop of the packing depending on the superficial gas velocity for CFD, experiment and Ergun calculation.

In a next stage of the project multi-component adsorption kinetics will be investigated, which are influenced by the local flow and temperature profile. Compared to common "porous media" CFD approaches realistic and reasonable residence time distributions can be achieved with this method.

Sensitivity analysis of radiative heat loading to tube coil with geometric imperfections

J. Hájek, Z. Jegla
Brno University of Technology, Faculty of Mechanical Engineering, Institute of Process Engineering, Technická 2896/2, 616 69, Brno, Czech Republic; Tel.: +420 541 144 493, e-mail: dohnal@upei.fme.vutbr.cz

The tube coil in real fired heater after some use deviates from the perfect designed geometry. The tubes slightly bend and depending on the quality of the hooks on which they are suspended, they may also shift from their design positions. Such behaviour is not accounted for in the design procedures and it is therefore important to assess the impact of the geometry imperfections on heat loads to the tubes. It has been recently shown in (Jegla et al., 2015) that geometry imperfections may contribute to increased heat flux nonuniformity, but a systematic treatment of the impact of tube position deviations has not been reported so far. Note that CFD combustion simulation reported in that work revealed that the major source of heat flux nonuniformity is the flow pattern, which however depends mostly on operating conditions.

The modelling methodology adopted in the present work is based on idealized flame and refractory wall behaviour, which allows to investigate the radiative heat loading of the tube coil surface. The model conditions are in essence the same as in the landmark work of (Hottel and Sarofim, 1967). Historically the approach to solve the radiative heat transfer equation was to idealise geometries and then apply zonal method. Results were obtained by analytic solution of the heat exchange factor integrals. The only difference here is that we apply modern numerical methods that make it possible to analyse more complex geometries. The geometries analysed are 2-D cuts of a row of tubes, some of which are displaced to simulated the bending and other coil deformations. The main aim of the study is to cover scenarios with one deformed tube in a row of ideally placed ones, some attention is however given also to more complex situations with more deformed tubes.

The results cover an array of geometric configurations that follow from the typical tube coil arrangements used in fired heaters. Namely, the geometry of tube bends (U-turns) has two basic designs, denoted as "short-radius" and " long-radius". These two design options define the two baseline configurations. Both cylindrical and box furnace geometry are considered. In fired heaters with helically shaped tube coil the pitch of tubes may however be arbitrary, so additional consideration is given also to such designs.

Hottel, H. C., and Sarofim, A. F., 1967, Radiative Transfer, McGraw-Hill, New York.

Jegla, Z., Vondál, J., and Hájek, J., 2015, Standards for fired heater design: An assessment based on computational modelling, Applied Thermal Engineering, 89(1), 1068–1078.

Numerical analysis of heat transfer to a biomass fired Stirling engine

C. Jordan, M. Harasek
TU Wien, Institute of Chemical Engineering, Getreidemarkt 9/1662, A-1060 Wien, Austria
tel. +43 1 58801-15924, e-mail: christian.jordan@tuwien.ac.at

In order to reduce net carbon dioxide emissions from generation of electrical energy, alternative methods for thermal power generation have to be investigated. One promising sustainable method is the use of renewable energy carriers like woody biomass (wood pellets). Since transport of biomass also contributes to the carbon footprint of the produced electricity small decentral generation units (e. g. up to 50 kW) are in consideration. In cooperation with Frauscher Thermal Motors GmbH, Hargassner GmbH and Bioenergy 2020+ GmbH a new biomass fired Stirling engine was developed. Besides the combustion chamber itself, the heat transfer from the combustion gas to the Stirling engine working gas at the heat exchanger is a key factor for the success of such concepts.

In this case the heat exchanger consists of a set of radial lamella fin plates which are arranged on tubes containing the working gas. The lamella are exposed to the hot combustion gas at temperatures of up to 1400 °C, heat is transferred via convection and radiation. Using the CFD package Ansys FLUENT® the flow field and convective as well as radiative heat transfer have been simulated. Multiple variants for this heat exchanger structure have been investigated: length and thickness of the lamella, number (spacing) of the lamella as well as dust deposition on the surface of the heat exchanger have been considered. Some of the structures implemented in CFD are shown in Figure 1, a total of about 2 million grid cells have been used for discretization of the geometry.

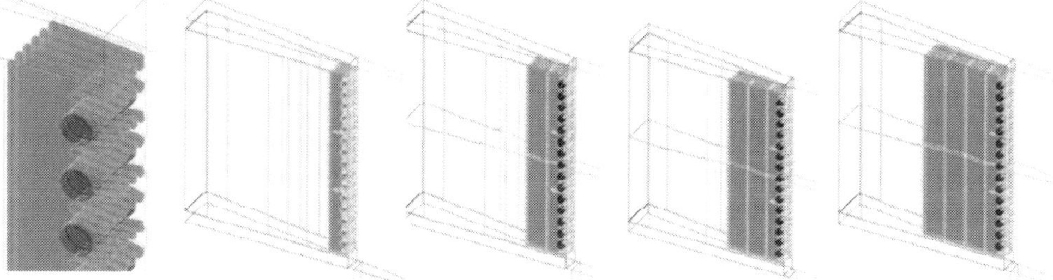

Figure 1: CFD investigation of lamella dimensions: Variation of the length

From the simulations it was found that the design constraints are not very critical – see also Figure 2. The overall heat transfer rate from the hot gas stream to the Stirling engine does not change considerably between the length and thickness variants. Even plain working gas tubes without lamella reach about 90 % of the maximum heat transfer rate. The lamella spacing had little higher influence than fin geometry, but the geometry of the fins can be kept small and simple. Highest influence was attributed to dust and ash deposition from the flue gas onto the heat transfer surface – however, a heat exchanger with larger surface can tolerate thicker deposition layers since more active area for convective heat transfer is available in the system. Nevertheless it is recommended to include a cleaning device for the heat exchanger.

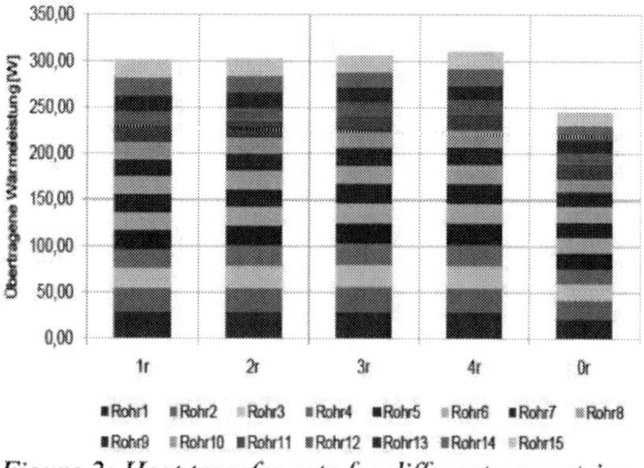

Figure 2: Heat transfer rate for different geometries

Numerical modeling of heat exchangers in gas turbines using CFD computations and thermodynamic cycle analysis tools

C.Salpingidou[1], D. Misirlis[3], Z. Vlahostergios[1], M. Flouros[2], S. Donnerhack[2], K.Yakinthos[1]
[1]Laboratory of Fluid Mechanics & Turbomachinery, Department of Mechanical Engineering, Aristotle University of Thessaloniki, Thessaloniki, 54124, e-mail: csalpingidou@eng.auth.gr
[2]MTU Aero Engines AG, Dachauer Strasse 665, Munich, Germany
[3] Technological Educational Institute (TEI) of Central Macedonia, Serres, Greece, e-mail: misirlis@eng.auth.gr, dmissirlis@teiser.gr

The optimization of energy systems such as gas turbines presents a continuous technological challenge both for economic and environmental reasons. The use of heat exchangers in gas turbines for exhaust gas waste heat recovery can strongly contribute in this direction since it can lead to significant improvement of gas turbine performance through reduction of fuel consumption and pollutant emissions. For proper implementation of a heat exchanger in gas turbines and the maximization of its benefits, engineers should be in the position to assess the most important operational characteristics such as the inner and outer pressure losses and the effectiveness. For this purpose, validated numerical tools are developed since they can provide time- and cost-efficient methods to assess the performance of various heat exchanger designs for gas turbine applications.

The current work deals with the modeling of a state-of-the-art heat exchanger for gas turbine applications with the use of high-fidelity 2D and 3D CFD computations. The derived numerical results were validated against experimental measurements and the heat exchanger characteristics were integrated into the thermodynamic cycle model.

In this paper the analysis was performed in the following steps:

1. Creation of a 2D CFD model of the heat exchanger core outer flow and its validation using experimental measurements
2. Creation of a 3D CFD model of the heat exchanger for both flows (inner/outer) including the effect of heat exchanger material. Computations using the 3D CFD model were performed for a wide range of operational conditions leading to the derivation of heat exchanger characteristics, inner and outer pressure losses and thermal effectiveness as functions of its operating conditions.
3. Development of a 0D thermodynamic cycle model of the gas turbine application and incorporation of the previously derived heat exchanger operational characteristics. Parametric analysis and performance assessment of the gas turbine thermodynamic cycle with the incorporated heat exchanger characteristics and further optimization of the thermodynamic cycle.

This analysis is based on a heat exchanger design which was developed by MTU Aero Engines AG which is presented in figure 1.

The results of this work will demonstrate the potentials of the heat exchanger integration in a gas turbine for significant performance improvement and fuel consumption reduction.

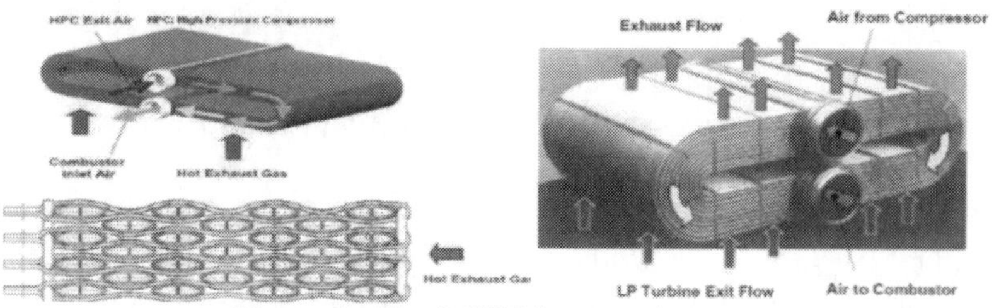

Figure 1. The MTU-heat exchanger

References

[1] D. Missirlis, M. Flouros , K. Yakinthos, Heat transfer and flow field investigation of a heat exchanger for aero engine applications, 2011, Int.J.of Heat and Technology 29 (2) , pp. 57-64

[2] Z. Vlahostergios, D. Missirlis, M. Flouros, C. Albanakis, K. Yakinthos, Effect of turbulence intensity on the pressure drop and heat transfer in a staggered tube bundle heat exchanger, Experimental Thermal and Fluid Science 60 (2015) 75–82

[3] K. Yakinthos, D. Misirlis, Z. Vlahostergios, M. Flouros, S. Donnerhack, A. Goulas, BEST STRATEGIES FOR THE DEVELOPMENT OF A HOLISTIC POROSITY MODEL OF A HEAT EXCHANGER FOR AERO ENGINE APPLICATIONS, GT2015-42408, Proceedings of ASME Turbo Expo 2015: Power for Land, Sea and Air GT2015 June 15-19, 2015, Montreal, Canada

Comments to administrator:

We want to inform you that we prefer to submit this abstract for poster presentation if possible. Even though in the system we might have selected the lecture presentation for this abstract, we prefer and think that this abstract might be more suitable for a poster presentation (in the system we cannot edit at this time the submission selection from 'lecture' to 'poster').

Thus, please consider this abstract for poster presentation.

Fully resolved CFD simulation of a hollow fiber membrane module

M. Harasek, B. Haddadi, M. Miltner, P. Schretter, C. Jordan
TU Wien, Institute of Chemical Engineering, Getreidemarkt 9/1662, 1060 Wien, Austria
tel. +43 1 58801-15925, e-mail: michael.harasek@tuwien.ac.at

The CFD simulation of membrane modules has been covered in literature in multiple ways. However, many known examples are limited to flat membranes, segments of spiral wound modules or single hollow fibers. This is since the extreme aspect ratios and the tiny features of membrane modules require a huge number of mesh cells for sufficient spatial resolution of whole module. But considering the computational resources available today, such validation cases for comparison with experiments can be calculated within reasonable time spans.

As an example the flow and concentration fields in a small hollow fiber membrane module featuring multiple permeate outlet variants will be considered in this investigation. The separation of methane (CH_4), carbon dioxide (CO_2) and oxygen (O_2) by a hollow fiber gas permeation module has been simulated using the solver "LTSmembraneFoam". The solver was derived from the free open-source CFD code OpenFOAM®. The trans-membrane flux at the selective layer was calculated from the partial pressure differences using the solution-diffusion mechanism. The CFD model was validated against 1D process simulation models (Aspen Plus®) and experimental results. The geometry is shown in Figure 1.

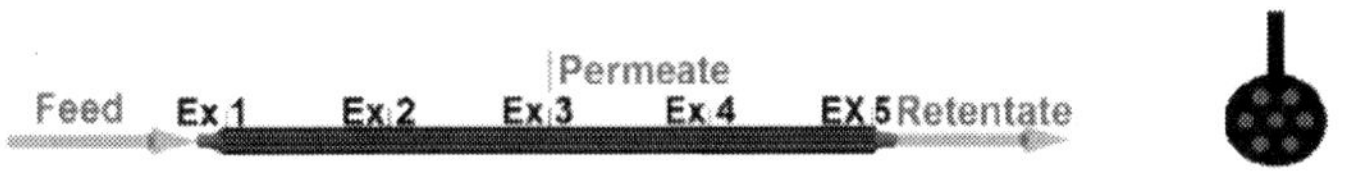

Figure 1. Module geometry with positioning of outlets

All important numbers for characterizing membrane separation units can be calculated from calibrated CFD simulations: stage cut, product recovery, achievable permeate concentration fiber side pressure drop. As an example the mass fraction of carbon dioxide along the simulated membrane module is presented in Figure 2 for all five outlet positions.

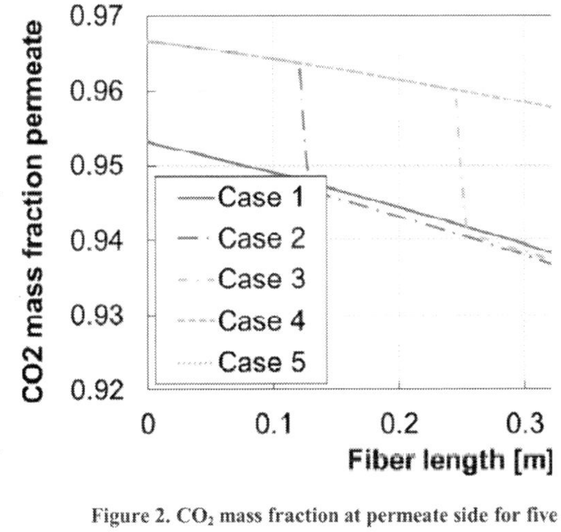

Figure 2. CO_2 mass fraction at permeate side for five different outlet positions

CFD offers substantial benefits to the developers of membrane modules: Since the concentration field and the flow field are fully resolved, the local effect of spacers (mixing promoters) can be considered. It is possible to analyze the influence of concentration polarization and the impact of the module geometry (e.g. position of feed inlet, permeate outlet) and the flow regime (co-/countercurrent, laminar/turbulent). The pressure drop of the module channels can be calculated based on full geometric details instead of only using correlations. Different models for material properties, multi-component diffusion and trans-membrane fluxes can be evaluated. From these detailed calculations new correlations can be derived for use in daily engineering practice to obtain fast and accurate layouts for complex plants.

Velocity measurement downstream of axial guide vane swirl generator

J. Vondál, M. Dohnal, V. Zejda, J. Hájek
Brno University of Technology, Faculty of Mechanical Engineering, Institute of Process Engineering, Technická 2, 61669 Brno, Czech Republic; tel. +420 541144923, e-mail: vondal@fme.vutbr.cz

Swirl-stabilised non-premixed flames are often used in process burners. Their complexity makes them difficult to predict computationally. This work presents measured data as a reliable source to validate numerical simulations. Many recent papers deal with flow predictions incorporating swirl. Several papers also present measured data. None of them, however, introduced measured data for real-size axial guide vane swirl generator as shown on Figure 1. This specific design is frequently utilized in process Low-NO_x burners installed in fired heaters. Thanks to its design it generates complex flow field downstream of the swirler. It can be described as a combination of guide-vane swirl generator and bluff body.

The swirler used in this study blocks 87 % of supply air duct cross-sectional area. The vanes occupy 68.7 % of the duct cross-section. It is capable of generating swirl flow with swirl number ranging from 0.22 up to 0.37 based on the distance from swirler (0.1D up to 1.5D). Presented data consists of particle image velocimetry (PIV) acquired velocity data enhanced with time-statistics and frequency analysis obtained from constant temperature hot-wire anemometry. Figure 2 shows averaged velocity magnitude plotted along the duct length at constant radius of r/R=0.73. Each line was acquired at different angular position. The Figure shows maximum velocity at the position x/D=0.3 with fast decay afterwards caused by intensive transport of axial momentum into angular momentum behind vanes in combination with momentum transport from high velocity region between vanes and duct.

Provided results are intended to be used for validation of numerical predictions. They should be helpful to provide validation especially to widely spread RANS turbulence models. Their importance is further increased in the context of modelling combustion chemistry via the often-used eddy dissipation model, which utilizes the strategy mixed-is-burned and relies on accurate turbulence prediction more than any other chemistry model. The reason is that turbulence is the driving factor for mixing and therefore also for chemistry and heat release. The importance of the turbulence modelling is therefore amplified. So far it is recognized that only with large eddy simulations (LES) are possible successful predictions of in-flame properties. The LES approach is however still too computationally expensive for the simulation of industrial fired heaters due to their huge dimensions (on the order of 10 m) and the need to resolve fine features like gas nozzles with diameters on the order of 1 mm. The only feasible alternative for practical predictions in the present are RANS turbulence models.

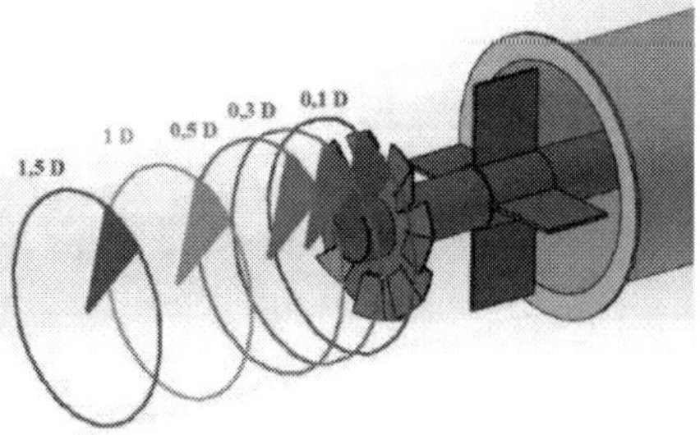

Figure 1 Axial guide vane swirl generator

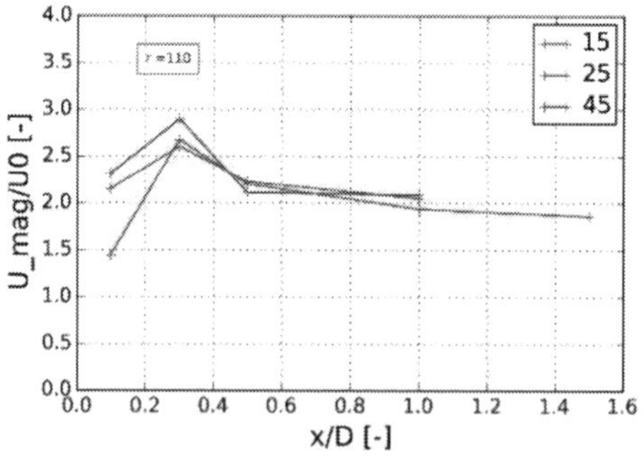

Figure 2 Velocity magnitude in r/R=0.7

Three-dimensional modelling of heat and mass transfer during combustion of low-grade Karaganda coal

A.S. Askarova, S.A. Bolegenova, Sym.A. Bolegenova, V.Ju. Maximov, M. Beketayeva, A.B. Ergaliyeva, Z.Kh. Gabitova, A.O. Nugimanova
al-Farabi Kazakh National Univ., Physical and Technical Dep., al-Farabi av. 71, 050040, Almaty, Kazakhstan;
tel. +77076111586, e-mail: Beketayeva.m@gmail.com.

Thermal power plants are a major source of the harmful impact on the environment. Half of the operating units of the Republic of Kazakhstan operated for over 40 years and require modernization and optimization. In this regard, remains relevant more profound study of the processes of solid fuel combustion, development and implementation of 'clean' technologies to protect the environment and ensure the efficiency of power plant equipment. It is clear that the implementation of direct experimental research directly into the combustion chamber is expensive and not effective, as the results will only make sense for a particular boiler. However, advances in computer technology and methods of numerical simulation allows to carry out computing experiments on the ignition, heat transfer and mechanisms of pulverized coal torch burnout in the boiler furnace without a full-scale experiments.

The study in this paper is focused to research using modern methods of 3D computer modeling processes of heat and mass transfer occurring in the real furnace chambers of industrial power plants during combustion of low-grade Karaganda coal. The data so obtained can be used for calculating the parameters and modes of operation of real fuel systems which helps to find more effective and cleaner ways of exploitation the energy object.

The real combustion process characteristics of low-grade Karaganda coal were determined in two boundary conditions aiming to find the more adequate scheme describing the real technological process. As a result developed the physical, chemical and mathematical models and propose numerical methods for solving the system of nonlinear equations. Also carried out computational experiments on two boundary conditions of the wall temperature (the aerodynamic, heat and mass transfer characteristics – V, T, Q, O_2, CO_2), and studied its effects on the combustion process characteristics. Verification of numerical calculation results with natural experimental data was made and the optimal boundary condition for the temperature on the walls of chamber determined.

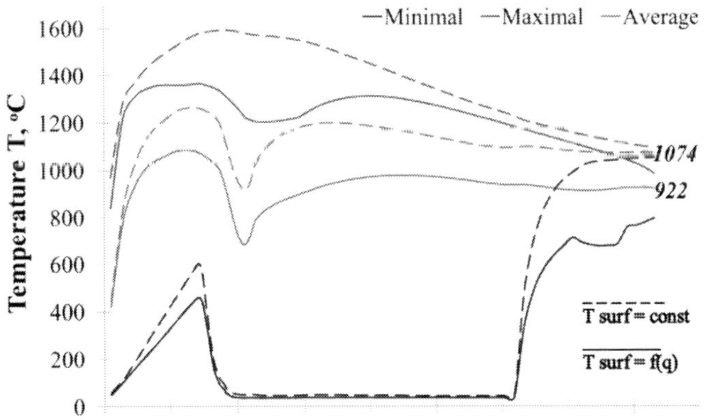

CFD study of fluid flow through regular foams with varying porosity and cell diameter

I. I. Mitrichev, A. V. Jhensa, E. M. Kol'tsova
D. Mendeleyev University of Chemical Technology of Russia, IT department, Miusskaya sq., 9, 125047, Moscow , Russia; tel. +74954952126, e-mail: kolts@muctr.ru

Open-cell foams are a well-known class of materials, which has been extensively used as thermal and noise insulators, particulate filters. However, the research on application of open-cell foams as catalysts has been started only recently [1]. It has been shown that the enhanced flow mixing and heat transfer allow alleviating strong heat- and mass-transfer limitations that is highly important in fast, short-contact time processes.

The fluid flow structure governs the mixing efficiency, so it is necessary to know the former to improve technological processes. For that purpose the mathematical modeling and computer simulations could be employed.

It has been recently demonstrated by Hutter et al. [2] with the use of planar laser induced fluorescence and CFD LES-simulation that the regular, streamwise-periodic foam structure is favorable then commercial random foam to obtain higher mixing efficiency. However, the only regular foam structure studied was based on tetrahedral packing of spheres. We use CFD simulation to perform both RANS and LES simulation of fluid flow through open-cell foam structure on two different packings. These packings are well-known from crystallography BCC and FCC packings.

We assess the boundaries of existence of different flow regimes, and calculate turbulence intensity in the regular foams. We make a comparison between BCC and FCC packed structures based on the values of pressure drop, turbulence intencity, mixing efficiency.

We compare the results of using LES and RANS approaches, the results are close in prediction of pressure drop and mean flow pattern. The analysis of unsteady LES simulation results allow us to find flow pulsations at medium Reynolds numbers.

We consider the influence of foam porosity and cell diameter on the turbulence development and mixing efficiency. For this purpose, a separate series of computational experiments is performed. As a result we obtain the optimal values of regular foam design parameters to obtain maximal mixing efficiency and, consequently, conversion gain in diffusion-limited reactions.

The work is financially supported by RFBR project № 14-07-00960.

1. Groppi, G., Beretta, A., & Tronconu, E. (2011) Structured Catalytic Reactors for Selective Oxidations / In D. Duprez & F. Cavani (Eds.), Handbook of Advanced Methods and Processes in Oxidation Catalysis (pp.943-997), London, Imperial College Press.
2. Hutter, C., Zenklusen, A., Kuhn, S., & von Rohr, P. R. (2011). Large eddy simulation of flow through a streamwise-periodic structure. Chemical Engineering Science, 66(3), 519-529.

CO_2 separation process using circulating fluidized bed based on exergy recuperation

Y. Kansha, M. Ishizuka, H. Mizuno, A. Tsutsumi
Collaborative Research Center for Energy Engineering, Institute of Industrial Science, The University of Tokyo, 4-6-1, Komaba, Meguro-ku Tokyo 153-8505, Japan
tel. +81-3-5452-6720, e-mail: kansha@iis.u-tokyo.ac.jp.

By fossil fuel combustion, large amounts of greenhouse gases (GHGs) are generated and emitted into the atmosphere. The reduction of carbon dioxide (CO_2) emission has become a major target in efforts to suppress global warming. Carbon capture and storage (CCS) has attracted attention in the past decades to reduce greenhouse gas (GHG) emissions. CO_2 separation from gas mixtures requires huge amounts of energy. In fact, CO_2 separation from gas mixtures is one of the most energy intensive process in CCS. Therefore it is necessary to reduce the energy consumption of CO_2 separation for propagation of CCS.

The most widely used method among CO_2 separation is a chemical absorption using amine based solution. In the conventional CO_2 separation by absorption using amine based solution, the thermal energy is supplied from the bottom of regenerator which utilizes reaction heat for regeneration of the solutions and heat of vaporization for stripping. To reduce the supply of the thermal energy, authors have developed CO_2 gas separation processes based on self-heat recuperation (Kishimoto et al. 2011, 2012). Applying the self-heat recuperation to chemical processes, not only the latent heat but also the sensible heat of the process stream can be circulated into the processes without any heat addition (Kansha et al. 2009), leading to minimizing exergy destruction of heat transfer and reduction of the process energy consumption. However, the absorption using solution requires additional energy for separation of liquid/gas, vaporization for stripping.

In this paper, an innovative energy saving CO_2 separation process using chemical absorption was proposed and the energy saving performance of the proposed process was evaluated by simulations and experiments. In this process, metal oxide particles were selected as absorbents not for requiring separation energy of liquid/gas. These particles were fluidized by feed gas mixture for increasing segregation effect in the bed and heat transfer rate from chemical reaction to a heat pump system. Heat of absorption was recuperated and supplied to the heat of desorption through the heat pump by following exergy recuperation. In addition, some of the heat was circulating with absorbents.

References

A. Kishimoto, Y. Kansha, C. Fushimi, A. Tsutsumi (2011) Exergy recuperative CO2 gas separation in post-combustion capture, Industrial and Engineering Chemistry Research, 50, 10128–10135.

A. Kishimoto, Y. Kansha, C. Fushimi, A. Tsutsumi (2012) Exergy recuperative CO2 gas separation in pre-combustion capture, Clean Technologies and Environmental Policy, 14, 465–474.

Y. Kansha, N. Tsuru, K. Sato, C. Fushimi, A. Tsutsumi (2009) Self-heat recuperation technology for energy saving in chemical processes, Industrial and Engineering Chemistry Research, 48, 7682–7686.

Catalytic pyrolysis of volatile tars contained in gaseous products of fast pyrolysis of agricultural waste

Yu.Yu. Kosivtsov, K.V. Chalov, Yu.V. Lugovoy, E.M. Sulman*, A.A. Stepacheva, V.P. Molchanov
Tver Technical University, Dep. Biotechnology and Chemistry, A.Nikitina str. 22, 170026, Tver, Russia; tel./fax: +7 4822 789317, e-mail: sulman@online.tver.ru

The problems of biomass processing for energy production still have no solution on a full scale that is connected with the variety of biomass composition. The difficulties of biomass processing are also determined by the differences in physical chemical characteristics of biomass of different origin. The primary biomass sources are wood, vestures and marsh plants. The raw material composition can change depending on the parts of plants (bark, roots, and branches, stock), storage conditions and biomass harvesting methods. These factors can affect the initial humidity, ash content and as the result heat-generation.

The technologies of fast pyrolysis are applied to pyrofuel production, however taking into account its low operating ability (high oxygen content and corrosion activity) it is necessary to search for the new technological solutions. To generate electrical energy from plant-based biomass waste the fast pyrolysis technologies focused on the increase of the conversion of the initial raw material to the combustible gases with high heat value and low resin content are more preferable.

In this research the gas refining from the tars was carried out by thermocatalytic method. To do this the pyrolysis gases flowed through the heated layer of alumosilicate catalyst impregnated by iron subgroup metal of various concentrations. The use of zeolite catalysts leads to the decrease the tars amount in a pyrolysis gas. The most active catalyst of the tars degradation in a pyrolysis gas for all the types of the studied feedstock is Co-containing catalyst.

The optimal catalyst granule size (1 mm) was chosen depending on the unit capacity according to the gaseous product formation and the system pressure. The use of metal-containing alumosilicate catalyst in the pyrolysis gas refining from the tars results in the change of the gas volume and heat value. Such metal content in the catalyst allows increasing C_1-C_4 hydrocarbons concentration and removing the tars from the combustible gas of the fast pyrolysis of the studied types of biomass.

It was found that catalyst-substrate mass ratio, granules size and surface metal concentration affect the catalyst activity. The experiments showed that for the full removal of the tars from a combustible gas the mass ratio of alumosilicate catalyst-feedstock must be more than 1:20 by weight.

Electrified water sprays generation for gas pollutants emission control

L. Manna[1], F. Di Natale[1], C. Carotenuto[2] and A. Lancia[1]
[1] University of Naples, Department of Chemical, Material and Production Engineering, P.le Tecchio 80, 80125 Napoli, Italy, tel +39-0817682246, e-mail: Francesco.dinatale@unina.it
[2] The Second University of Naples, Department of Information and Industrial Engineering, Via Roma 29, 81031, Aversa, Caserta, Italy

The electrostatic charging of liquid has been used from years in many different fields as injection processes, agricultural spraying of pesticides, electrospinning and also in air pollution control technologies. The integration of electric field to atomization process allows producing sprays with finer droplets size distributions and wider angular dispersions, that are more effective for several gas-liquid processes as absorption or particle capture. However, only some liquids can be effectively be used in electrification processes because of their superficial tension and electrical conductivity. The most of studies conducted in these years were about dielectric liquids as oils or fuels. Indeed, water was rarely studied because of its ionic characteristic that makes the charging process heavy to control. However, our group proved that electrified water sprays are very effective means to remove ultrafine particles and to absorb acid gases and may provide a significant process intensification for gas cleaning technologies.

In our study, we investigated the induction charging of water sprays. Induction was identified as the most useful way to charge at appreciable levels fluids with flow rates of the orders of liters per minutes. This process also allows having a good control of undesired electric discharges phenomena on the operating line.

An experimental set up was designed, constructed and operated to carry out experiments at lab scale. We used a 60 degs hollow cone hydraulic spray nozzle with flow rate of about 2 L/min. An optical analysis was performed to address the breakup length of the spray in absence of electric field. The distances at which the liquid films broke up were used to evaluate the sheet thicknesses both at exit from nozzle and breakup point by using the LISA model.

Two different kind of induction charging electrodes were used: a brass ring of 13 cm diameter and a steel ring 9 cm diameter. The influence of electric field strength was studied by varying both the electrode type and the distance between the nozzle tip and the electrode to evaluate how space charge effects influenced the charging level.

To measure the intensity of droplet charging, the spray current was measured using a faraday cup in the form of a metallic mesh surrounding the nozzle at a certain distance. The ratio between the spray current and the liquid flow rate is the so-called Droplet Charge to Mass Ratio (D-CMR), that is the most useful measure of sprayed droplets charging level.

The experiments revealed that the values D-CMR against the charging potential has a non-monotonic trend with a maximum at a potential named V_{opt} and a zero value for a potential named V_{max}. The tests indicated that the value of V_{opt} is smaller for the smaller ring. The values of the D-CMR are similar for the two nozzles. Moreover, the optimal nozzle placement corresponded to the ring placed at the break-up length distance from the nozzle tip. However, the smaller ring is more prone to wetting and and short circuiting making it less robust than the larger ring.

For $V<V_{opt}$, the Artana model was used to describe data obtained from charging tests. It allowed estimation of an equivalent liquid sheet thickness, that was compared with the LISA model predictions. Although the models assumptions suggested that the two estimate

thickness cannot be quantitatively similar, we expected that the two values were similar. Instead, the equivalent thickness estimated from induction charge calculations were far lower than those predicted by the LISA model. In our opinion, this should indicate that the induction charging phenomena take place on the periphery of the liquid sheet, and that the electric field modified the actual jet break-up dynamics. To this end, future work may benefit for the implementation of optical tests on jet break up and droplet size distribution at different charging potentials.

Targeting of water-using networks of single contaminant involving regeneration reuse/recycling

LIU Jinhao (刘金豪)1, LI Aihong (李爱红)1,2, LIU Zhiyong (刘智勇)3
1 School of Chemical Engineering, Hebei University of Technology, Tianjin 300130, China
2 Department of Chemical Engineering, Chengde Petroleum College, Hebei 067000, China
3 Engineering Research Center of Seawater Utilization Technology of Ministry of Education, School of Marine Science and Engineering, Hebei University of Technology, Tianjin 300130, China

Abstract An effective method is presented for targeting single contaminant water-using networks involving regeneration reuse/recycling. In the targeting procedure, the initial regenerated concentration is first estimated according to the features of water-using processes and regeneration unit. The regeneration pinch corresponding to the initial regenerated concentration is then identified. Finally, the regeneration target is calculated with a few relationships developed from contaminant load balances. If the regeneration pinch does not change, the regeneration target can be obtained by only one step simple calculation. If the regeneration pinch changes, the regeneration targets can be obtained by two-step simple calculation. The results of a few literature examples show that the target of single contaminant water-using networks involving regeneration reuse/recycling can be obtained easily with the proposed method. By comparing the design procedure proposed in this paper with the methods in the literature, it can been seen that: the deeper we understand the features of water-using networks involving regeneration unit, the simpler the design procedure.

Key words Water-using network, Regeneration reuse/recycling, Target, Regeneration pinch

Carbon capture from integrated pulp and board mill

P. Kangas, K. Onarheim[1], V. Hankalin[2], S. Santos[3]
VTT Technical Research Centre of Finland, Biologinkuja 7, 02150 Espoo, tel. +358-20-7224645, e-mail: petteri.kangas@vtt.fi; [1]VTT Technical Research Centre of Finland; [2]ÅF-Consult Ltd, Finland; [3]IEAGHG, United Kindom

Capturing biogenic CO_2 from flue gases emitted by an integrated pulp and board mill could be a feasible option in reducing greenhouse gas emissions and potentially earning additional revenues from "carbon negative" emissions. Modern pulp mills mainly use bio-based raw materials and energy. If biomass is sustainably managed and the emissions from the pulp mill are captured and stored, this could be considered as "carbon negative". With sufficient economic incentives, this could emerge as a new opportunity to the industry. This study evaluates the possibilities of retrofitting CO_2 capture plant to an integrated pulp and board mill. A modern Nordic kraft pulp mill was used as a starting point for the evaluation. The technical performance of the mill was evaluated using Wingems Process Simulation tool. All relevant pulp and board mill units are included in the model. The carbon capture unit is evaluated using Aspen Plus simulation tool which estimates the additional energy demand of the CO_2 capture plant. The CO_2 capture is based on chemical absorption technology using MEA solvent.

In this study, two base cases are defined: (i) Stand-alone pulp mill producing 800,000 air dried tonne (adt) of softwood kraft pulp annually, and (ii) Integrated pulp and board mill producing 740,000 adt of softwood kraft pulp and 400,000 adt 3-ply folding boxboard annually. Both reference mills are energy independent and they export excess electricity to the grid. CO_2 emissions from the mill are 2.1 million metric tonnes per year (MTPY) – 1.6 million MTPY from the recovery boiler, 0.3 million MTPY from the multi-fuel boiler and 0.2 million MTPY from the lime kiln. Only the emissions from the lime kiln are considered partly non-biogenic CO_2 emissions.

For both reference mills with CO_2 capture, several sub-cases were evaluated looking at the performance of capturing CO_2 from different flue gas streams of the recovery boiler, the multi-fuel boiler and the lime kiln. The results of the study indicated that the CCS can be implemented as part of the integrated pulp and board mill. For the standalone pulp mill, the excess energy in form of steam and electricity is sufficient to run the CO_2 capture plant in all cases. However, this should also reduce the amount of electricity exported to the grid. For the integrated pulp and board mill, if CO_2 from the flue gas of the recovery boiler is captured, this would require an additional investment of an auxiliary boiler to fulfil the steam demand of the pulp and board mill with CO_2 capture.

Innovative technology for production of construction materials on the basis of man-made waste of metallurgical industry

Y.M. Fedorchuk [(1)], M.A. Sadenova [(2)*], S.A. Abdulina [(3)]
[(1)] Tomsk Polytechnic University, 30, Lenin avenue, 634050, Tomsk, Russia
[(2)] S. Amanzholov East Kazakhstan State University, 55, Kazakhstan str., 070000 Ust-Kamenogorsk, Kazakhstan
[(3)] D. Serikbayev East Kazakhstan state technical university, 19, Serikbayev str., 070000, Ust-Kamenogorsk, Kazakhstan
tel. +77232255441, e-mail: Sadenova@mail.ru

Technology for producing hydrogen fluoride in any industry (nuclear, chemical, aluminum production) is based on the chemical interaction of sulfuric acid with the mineral fluorite CaF_2, which is dominant in the deposits of fluorspar.

$$CaF_2 + H_2SO_4 = CaSO_4 + 2HF$$

Hydrogen fluoride is a target product that is used for the needs of the relevant industry, but calcium sulfate is a solid waste of this process, it is called in the technical literature fluorine anhydrite (anhydrite - anhydrous calcium sulfate, got in technology of hydrogen fluoride).

Fluorine anhydrite also has the ability to form gypsum, it can be used as a target in the production of building materials including plaster base. Application sulfatkaltsievyh waste justified the development of alternative products based on plaster.

The paper proposes the establishment of production sheathing on the basis of fluorine anhydrite, which are analogous to known plasterboard and fibre reinforced gypsum sheets of the company «Knauf». As raw materials to be used wastes hydrogen fluoride production, that are fluorine anhydrite with various impurities.

As a result, developed a technology for the production of gypsum boards, are defined by their mechanical and physico-chemical properties determine the cost of plates, consider the need for equipment, communications, and raw materials.

Microwave susceptible catalytic diesel particulate filter

V. Palma[1], E. Meloni[1*]
[1]University of Salerno, Department of Industrial Engineering, Via Giovanni Paolo II, 132 - 84084 Fisciano (SA), Italy
*tel. +39089969275, e-mail: emeloni@unisa.it;

The diesel internal combustion engine is one of the most environmentally-friendly vehicle devices, because it emits less carbon dioxide (diesel, 13.4 kg/km; gasoline, 16.6 kg/km) and is more fuel efficient (diesel, 183.1 MJ/100 km; gasoline, 223.5 MJ/100 km) than the universal gasoline stoichiometric engine [1]. However, considerable challenges still exist regarding the emission control of particulate matter (PM), commonly known as soot, expelled by the diesel combustion process. These ultrafine particles, which have aerodynamic diameters of less than 2.5μm ($PM_{2.5}$), are problematic as they can cause serious respiratory diseases, including lung cancer and chronic obstructive pulmonary disease (COPD) [1]. Various filters are commonly used for soot abatement in diesel exhaust, while catalytic oxidation is preferable to intensify burning of soot trapped in filters [2]. The diesel particulate filters (DPFs) are the most important technologies to maintain soot emissions under EU regulations [3], consisting in alternately plugged parallel square channels, so that the exhaust gases are forced to flow through the porous inner walls: in this way the particles are collected on the surface and in the porosity of the channel walls, progressively blocking the pores [3]. Since the pressure loss increases by the formation of a thick soot cake as the PM is accumulated, the DPF needs to be periodically regenerated by burning off the accumulated soot. As the exhaust gas temperature of diesel is typically about 150 °C, this temperature is far too low compared with the minimum combustion temperature of 600 °C needed to burn PM. As a result, some type of system is necessary in order to forcibly regenerate the filter [4]. The oxidation step is promoted by the so-called 'passive' (at temperatures upstream the DPF of ca. 200 – 400 °C) and 'active' regeneration (temperatures upstream the DPF of ca. 550 – 600 °C) [5]. The regeneration of loaded DPF using microwaves belongs to the active systems of regeneration, because, depending on the filter load, the soot is heated up to the temperature for regeneration. The microwave regeneration differs from conventional systems in the point of heat introduction. In conventional systems the heat is transferred to the filter indirectly by heating the exhaust gas [6]. In the microwave regeneration system the heat is coupled directly into the soot: the good dielectric properties of SiC, catalyst and soot, in combination with MW heating and catalytic combustion may result in the effective oxidation of diesel soot at lower temperature and higher reaction rate [7]. The results of our previous deposition and on-line regeneration tests on uncatalysed and Copper-Ferrite catalysed DPF, showed that the K addition to our catalyst formulation and the simultaneous use of microwaves during the regeneration step at lower gas flow rate, allows to reduce the energy supplied and the regeneration time than that required for the uncatalysed filter [7]: in particular we observed the positive effect that NOx had on catalytic activity. Starting by these very promising results, the objectives of this work are to increase the active species load simultaneously modifying the porosimetric characteristics of the support, in order to simultaneously further reduce the PM oxidation temperature and keep low the pressure drop. Furthermore we wanted to verify the feasibility of the MW heating technology by comparing the energy balance of the entire process to the actually employed regeneration technologies.

1. Chanmin Lee, Joo-Il Park, Yong-Gun Shul, Hisahiro Einaga, Yasutake Teraoka, Ag supported on electrospun macro-structure CeO_2 fibrous mats for diesel soot oxidation, Applied Catalysis B: Environmental 174–175 (2015) 185–192.
2. N.V. Vernikovskaya, T.L. Pavlova, V.V. Mokrinskii, D.Yu. Murzin, N.A. Chumakova, A.S. Noskov, Soot particulates abatement in diesel engine exhaust by catalytic oxidation followed their trapping in filters, Chemical Engineering Journal 269 (2015) 416–424.
3. Palma V., Ciambelli P., Meloni E., Optimising the Catalyst Load for Microwave Susceptible Catalysed Dpf, Chemical Engineering Transactions 29, 2012, 637-642
4. Palma V., Ciambelli P., and Meloni E., 2013, Optimal $CuFe_2O_4$ load for MW susceptible catalysed DPF, Chemical Engineering Transactions 35, 2013, 727-732
5. A. Liati, P. Dimopoulos Eggenschwiler, E. Müller Gubler, D. Schreiber, M. Aguirre, Atmospheric Environment 2012, 49, 391-402.
6. I. Imenokhoyev, A. Matthes, H. Gutte. and G. Walter, 2011, Numerical 3D-FEM-simulation made by COMSOL Multiphysics of a microwave assisted cleaning system for a diesel sooty particle filter and its experimental validation, International COMSOL Conference, Ludwigsburg 2011, October 26 – 28.
7. V. Palma, E. Meloni, P. Ciambelli, Optimal loading of a MW susceptible catalysed DPF, Chemical Engineering Transactions 45, 2015, 2035-2040
8. V. Palma, P. Ciambelli, E. Meloni, A. Sin, Fuel 2015, 140, 50 – 61.

When life cycle thinking is necessary for decision making: emerging cleaner technologies in the chlor-alkali industry

M. Margallo, R. Onandía, J. Laso, R. Aldaco, A. Irabien
Department of Chemical and Biomolecular Engineering, University of Cantabria, Av. de los Castros s/n, 39005 Santander, Spain;
Tel. +34 942200931, email: margallom@unican.es

The chlor-alkali industry sector produces chlorine, sodium/potassium hydroxide and hydrogen by the electrolysis of brine. Nowadays, three different electrolysis techniques are applied: mercury, diaphragm, and membrane cell technology. The techniques differ from each other in terms of electrode reaction and electrode materials, and in the way, the produced chlorine is kept separate from sodium hydroxide and hydrogen [1]. Up to the end of the 20th century, the mercury cell technique dominated in Europe, while the diaphragm cell technique dominated in the United States and the membrane cell technique in Japan. However, this pattern has changed and during the period from 1997 to 2012, the share of the mercury and diaphragm cell techniques decreased significantly in the European countries, from 63 % to 26 % and from 24 % to 14 %, respectively, whereas the use of the membrane cell technique increased from 11 % to 59 %. Reasons for the change include the need to replace installations that have reached the end of their service life and environmental concerns over mercury emissions from mercury cell plants [2]. Moreover, the European Commission labels the membrane process as the Best Available Technique (BAT) for the chlor-alkali industry. In this sense, the membrane cell technology has fewer exhausts to the environment and is relatively more efficient in the use of electric power [3]. Nevertheless, the disadvantages of this technology are the high-purity brine required, the low Cl_2 quality, the high steam consumption and great cost of membranes [1].

Despite the fact that the overall energy intensity has been reduced, the issue of energy consumption is still a major matter. A promising approach for reducing the electricity demand of chlor-alkali electrolysis is using oxygen-depolarised cathodes (ODC). ODCs are long known and have been successfully used in chlorine production through electrolysis of hydrogen chloride (HCl). For chlor-alkali electrolysis, a pilot plant using ODCs began operation only recently [4]. The utilisation of ODC in chlor-alkali electrolysis is an integration of an alkaline fuel cell cathode into the membrane electrolysis cell. This lowers the cell voltage by about 1 V at current densities of industrial relevance (e.g. 4 kA/m^2), and thus the electrical energy consumption by about 30 %. The achieved environmental benefit of this technique is a reduction of energy consumption. However, the overall reduction of energy consumption is lower, as some energy is required to produce pure oxygen and because hydrogen is not co-produced, which could otherwise be used in chemical reactions or to produce steam and electricity via combustion or fuel cells [2]. In this sense, the reduced electricity demand does not necessarily imply cleaner chlorine production. According to this, we propose the use of the life cycle assessment (LCA) methodology to determine the environmental performance of the existing electrolysis technologies and to compare it with the ODC technique. The LCA was conducted from cradle to grave taking one ton eq. (1 ton of Cl_2 and 1 ton of NaOH) as functional unit. The study included the salt mining and transport, the brine purification, the electrolysis process, the treatment of products and the waste management (Figure 1). The work included all the mass and energy inputs (energy, fuels, steam, reagents and water) and outputs (products, co-products, waste, air emissions, effluents) of the whole process. The results of the LCA show that the membrane and the ODC technologies have the fewest consumption of energy, materials and water, and therefore, the lowest environmental impacts. However, the comparison of the membrane and ODC technologies depends on the hydrogen valorisation. Hydrogen generated in the membrane cell can be used as fuel substituting natural gas (NG) or to produce electricity via combustion or fuel cells.

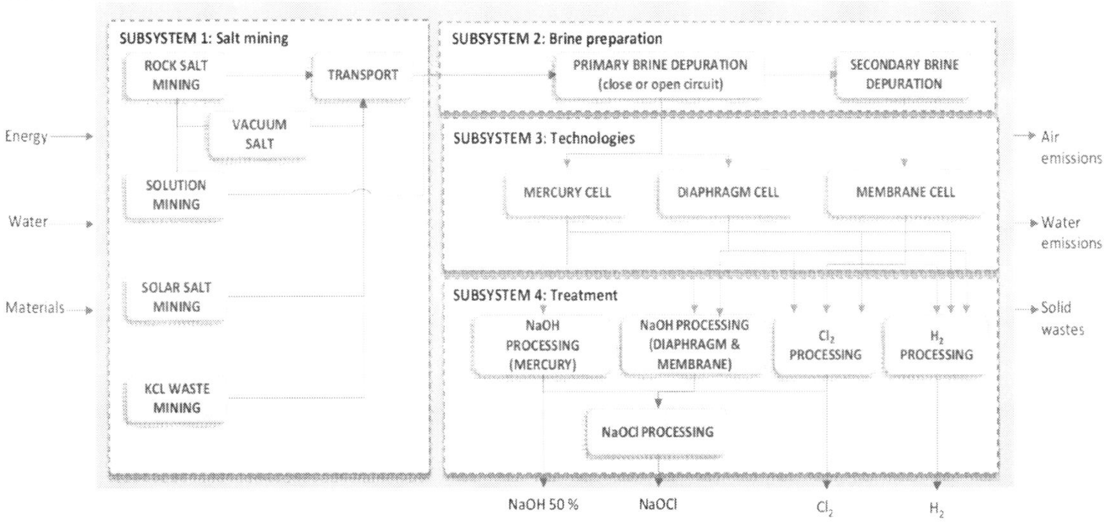

Figure 1. Flow diagram of the gate-to-grave step of the LCA of chlor-alcali process.

This work compared both alternatives and the non-valorisation one with the ODC in terms of consumption of natural resources (NR) and environmental burdens (EB) [5]. The consumption of energy, materials and water in ODC was 4.7 %, 1 % and 10.5 % higher than the NR demand in the membrane without valorisation and with hydrogen valorisation for electricity production. However, the NR demand was substantially reduced in the membrane technology when the hydrogen substituted NG. In this case, the energy, material and water demand of the ODC was a 9.6 %, 14 % and 6 % greater than in the membrane scenario. The results show that the initial premise of a reduction of energy consumption of 30 % using ODC was not fulfilled. The reasons were energy demand to produce oxygen and the inability of hydrogen valorisation. Regarding the environmental burdens, ODC presented the smallest impact in most of air and water categories. Only in global warming (GW) ODC exceeded the impact of the membrane with hydrogen valorisation. Moreover, the ODC impact was very close to those obtained in the membrane without valorisation. According to these results, the ODC yields the lowest environmental, and it has a potential to contribute to cleaner production in the chlor-alkali industry. The unanswered question is the economic feasibility of this technology. Unfortunately, there is not an easy answer. The variables costs of electricity, oxygen, new electrolysers and depolarised cathodes must be evaluated to find this answer.

References

[1] Eurochlor, 2013. An Eco-profile and Environmental Product Declaration of the European Chlor-Alkali Industry Chlorine (The chlor-alkali process). Brussels, Belgium: Eurochlor.

[2] Brinkmann T, Giner G, Schorcht F, Roudier S, Delgado L, 2014. Best Available Techniques (BAT) Reference Document for the Production of Chlor-alkali. Luxembourg: Publications Office of the European Union.

[3] Kiros Y and Bursell M, 2008. Low Energy Consumption in Chlor-alkali Cells Using Oxygen Reduction Electrodes. Int J Electrochem Sci 3, 444-451.

[4] Jung J, Postels S, Bardow A, 2014. Cleaner chlorine production using oxygen depolarized cathodes? A life cycle assessment. J Clean Technol 80, 46-56.

[5] Irabien A, Aldaco R, Dominguez-Ramos A, 2009. Environmental sustainability normalization of industrial processes. Comput Aided Chem Eng, 26, 1105–1109.

Enhanced SO_2 removal by using charged water droplets

F. Di Natale, [1] C. Carotenuto[2], A. Lancia[1]
[1] University of Naples, Department of Chemical, Material and Production Engineering, P.le Tecchio 80, 80125 Napoli, Italy, tel +39-0817682246, e-mail: Francesco.dinatale@unina.it
[2] The Second University of Naples, Department of Information and Industrial Engineering, Via Roma 29, 81031, Aversa (CE), Italy, tel +39-0815010410, email: claudia.carotenuto@unina2.it

In this study we investigate how electrically charged droplets interact with SO_2, a dipolar molecule that is the main acid component in flue gases. This process takes place, for example, in wet electrostatic scrubbers. These units allow the capture of ultrafine and nanometric particles from industrial and engines exhaust gases, and can also provide side-benefits as the capture of acid gases.

There are few studies on absorption of SO_2 by charged droplets and the largest majority of them considered the interactions among a charged water spray and a gas stream containing SO_2. Although providing significant insights on the engineering feasibility of the process, the effectiveness of these experimental procedure in understanding the underpinning physics of the process is limited by the complexities of charged sprays fluid dynamics, that make the analysis of the experiments quite complex. The general finding of these studies is that SO_2 absorption rate significantly increases, even up an order of magnitude, by increasing the charge level of the water spray. However, it is difficult to discriminate how much of this absorption increment is specifically ascribable to interactions among charged droplets and SO_2 and how much is related to an indirect consequence of the presence of the charge that cause a reduction of droplet sizes and a better gas-liquid mixing.

In order to exploit specific droplet-molecule interactions, we developed a new experimental procedure based on the use of an electrospray exerted in dripping mode to produce a stable train of droplets with identical size and charge level. The droplet properties were measured with charging potential varying from 0 to ±3 kV. The electrospray was placed in a closed vessel of about 2 L volume, that is part of a closed loop that included an IR gas analyser. A model gas stream containing nitrogen and 1000 ppm of SO_2 was fed in the loop, the loop was closed and the electrospray was switched on at a given flow rate and voltage. The time course of SO_2 concentration was monitored for about 10 minutes.

To interpret experimental data, we adopted a predictive physical mathematical model for SO_2 absorption that includes both the formation and the falling of the droplets and considers the theoretical models of Matheson and Giardina (1974) to take into account the interfacial interactions of SO_2 with the charged droplet. Different equations for mass transfer rate in the gas and in the liquid domains were considered.

The experimental and modelling results were appreciably consistent and revealed that the SO_2 absorption rate is actually enhanced up to about 15% by the presence of electric charge on the scrubbing droplets. Unfortunately, since the mechanism that control the absorption rate is the mass transfer in the liquid phase, we are not able to stablish the effect of charge-dipole interactions among the droplet and the SO_2 in the gas, which is expected to significantly increase the mass transfer rate in the gas phase.

This result indicated that, compared with a conventional wet scrubber of similar design, a wet electrostatic scrubber can achieve superior SO_2 removal efficiency operating with the same water flow rate or can achieve the same removal efficiency with a lower water flow rate.

Energy efficient co-production of hydrogen and power from brown coal employing direct chemical looping

Muhammad Aziz, Takuya Oda, Atsushi Morihara
Solutions Research Laboratory, Tokyo Institute of Technology
2-12-1 Ookayama, Meguro-ku, Tokyo 152-8550, Japan
Tel: +81-3-5734-3809, Fax: +81-3-5734-3559, e-mail: aziz.m.aa@m.titech.ac.jp

Abstract

As energy source, coal is always interesting due to its wide geographical distribution, great abundance, and low cost. Unfortunately, a conventional utilization of coal leads to some environmental problems, especially its huge CO2 emission. In addition, more than 50% of coal reserves are categorized as low rank coal (brown coal), therefore coal drying becomes a very energy intensive process. Brown coal has characteristics of high moisture content, low calorific value, low sulfur content, etc. Clean and highly efficient energy harvesting from coal is demanded to achieve clean energy harvesting and optimum utilization of brown coal.

Recently, conversion of coal to hydrogen has received great attention from researchers, governments and industries. The motivations behind this phenomena include reduce in environmental impact, increase of energy security, improve the economic development, encouragement in science and research. Hydrogen has high gravimetric energy density compared to gasoline (120 MJ kg-1 than 44 MJ kg-1). Considering wide fuel-utilization including transportation, energy carrier with both volumetric and gravimetric densities is preferable. Hydrogen is considered as secondary clean energy resource which carries the energy and can be converted to useful energy with relatively high efficiency. The utilization of hydrogen includes as fuel in fuel-cell, internal combustion engine, and turbine, and as chemical for ammonia and other hydrogenation

In this study, an integrated system co-producing hydrogen and power from brown coal with high total energy efficiency is proposed and evaluated through process modeling and intensification. The proposed integrated system is composed of four subsequent modules: drying, chemical looping, combined cycle and hydrogenation. To minimize the exergy destruction throughout the integrated system, heat circulation and process integration technologies are employed.

Direct chemical looping is adopted as the main conversion process due to its benefits of fewer units operation, higher hydrogen-to-CO_2 ratio and lower oxygen requirement. In direct chemical looping, the fuel is indirectly oxidized with the oxygen brought by the oxygen carrier, and change the fuel to the desired gases (especially CO_2 and H_2) and heat. As the produced CO_2 in chemical looping is in high purity, CO_2 capture can be bypassed and the produced CO_2 can be delivered or stored for further sequestration. In addition, the produced heat can be utilized as the heat source for power generation. The produced hydrogen is then stored in liquid organic hydrogen carrier (toluene-methylcyclohexane). Figure 1 shows the schematic process flow diagram of the proposed integrated-system for co-producing of hydrogen and power from brown coal.

From process modelling and calculation, the proposed integrated-system shows high total energy efficiency, more than 40%, including drying, CO_2 separation and hydrogen storage.

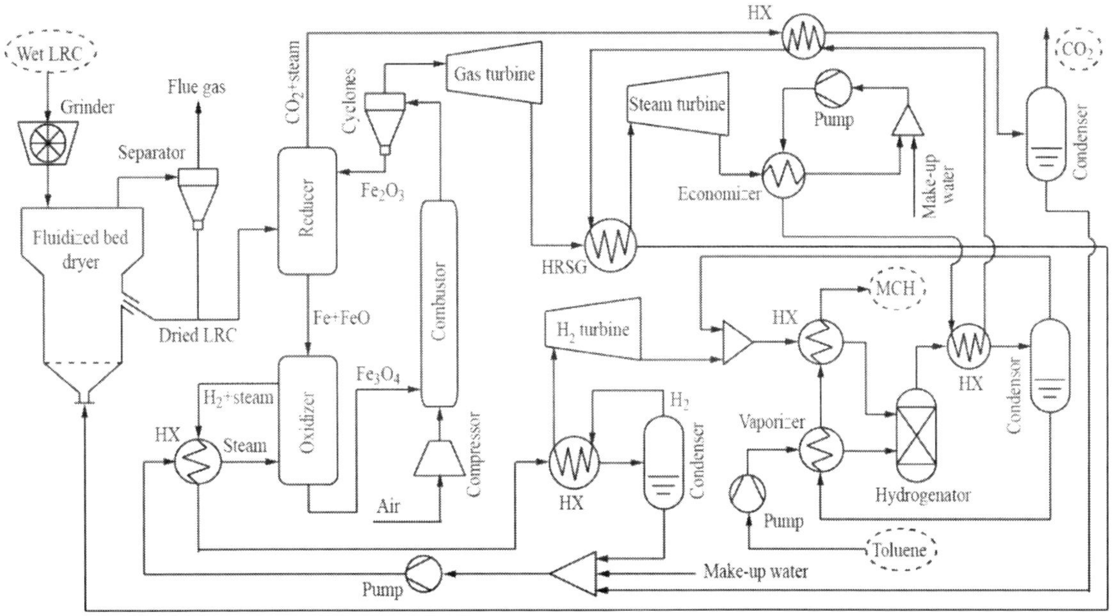

Figure 1 Schematic process flow diagram of the proposed integrated-system for co-producing of hydrogen and power from brown coal

Keywords: brown coal, co-production, hydrogen, chemical looping, energy efficiency, heat circulation, process integration

Supervisory control and monitoring of a hybrid high temperature PEM fuel cell system with Li-Ion batteries and an LPG reformer

C. Ziogou[1], G. Giaouris[2], S. Papadopoulou[3], S.Voutetakis[1]
[1]Chemical Process and Energy Resources Institute (CPERI), Centre for Research and Technology Hellas (CERTH), PO Box 60361, 57001, Thessaloniki, Greece, {cziogou, paris}@cperi.certh.gr
[2]School of Electrical, Electronic and Computer Engineering, Newcastle University, NE1 7RU, England, United Kingdom, damian.giaouris@newcastle.ac.uk
[3] Department of Automation Engineering, Alexander Technological Educational Institute of Thessaloniki, PO Box 141, 54700, Greece, shmira@autom.teithe.gr

This work presents the design and development of a Supervisory Control and Data Acquisition (SCADA) system for an integrated High Temperature Polymer Electrolyte Membrane (HT-PEM) fuel cell that provided power to Li-Ion battery stack. The necessary hydrogen is either from a renewable hydrogen production station or from an LPG reformer system which are both located at CERTH. A Battery Monitoring System (BMS) monitors the Li-Ion stack and it is integrated to the SCADA system. In this work the operation of the hybrid system is evaluated based on information from each subsystem such as battery state of charge and fuel cell temperature. The operation relies on a novel integrated management methodology that incorporates monitoring, evaluation and control features, which are configurable, expandable and flexible according to the specific end-user requirements. The main objective for the integrated system is to be able to operate in a hybrid mode depending on the origin of the hydrogen that can be from a gas reformer or from hydrogen tanks. Another important objective is to be able to charge the battery stacks and to self-sustain the auxiliary subsystems along with the demanded load. Finally the entire operation is automated and the supervisory system protects the lifetime of the involved subsystems (reformer, fuel cell, battery stack). The objective is to preserve the lifetime of the battery while using the optimum amount of fuel either hydrogen from LPG or pure hydrogen.

As Li-Ion batteries are used the charging and discharging functions needs to be carefully monitored. Therefore, the developed BMS is of major importance. The BMS has a protection subsystem of each individual cell of the battery stack. More specifically, the system is developed for a battery stack that consists of 15 Li-ion battery cells of 3.2V/120Ah each, which are connected in series to produce a 48V. In order to protect the battery and the individual cells a state of the art BMS has been designed, implemented and tested. The main requirements for this BMS are the monitoring of the state and the protection from overcharging of each individual cell and at the same time the reduction of the overall cost as much as possible.

The analysis of its operation validates that the developed supervisory control system can manage the thermal and power load in real-time and is able to fulfill the objectives for effective control of the involved subsystems. The flexible and modular design of the system can be applied to a wide range of traction and stationary battery applications since the involved platform is generic and can be adapted upon demand. The supervisory monitoring capabilities enable the optimum utilization of the battery pack and the available fuels (LPG or pure hydrogen). Finally the SCADA system provides informative decisions to the user that monitors the integrated process and enables proactive maintenance operation to ensure that the system will provide the maximum of its capabilities.

Assessment of the exergy performance of an organosolv process using Aspen Plus®

A. Mabrouk, [1]X. Erdocia, [1]M. González Alriols, [1]J. Labidi
National School of Engineers of Gabes, Chemical Engineering Department, Gabès, Tunisia; mabroukaycha@yahoo.fr
[1]University of the Basque Country (UPV/EHU), Chemical and Environmental Engineering Department, Donostia-San Sebastian, Spain; phone: +34-943017178; jalel.labidi@ehu.es

Organosolv fractionation is a successful process to separate lignocellulosic biomass in its main component for the preparation of multiple products including biofuels, chemicals, and materials.

In this study, we investigate the exergetic behaviour of the delignification process of olive tree using ethanol-water as solvent media. The process consisted of several units, starting with a mixer, a heater, a reactor and, finally, a filtration step for the components separation. Exergetic analysis was used as the decision making tool for the evaluation of the global process efficiency and for the identification of the major the sources of irreversibilities. The analysis was carried out with Aspen Plus® program. As a result, the total exergy losses found reached around 98 kW, being the heater the main responsible of this quantity, contributing with about 28.75% of the irreversibility generated in the studied sequence. The obtained information was used as the starting point for the optimization of the process operational conditions for a better energetic behaviour.

TiO_2/zeolite composite on ceramic tiles: influence of the amount of photocatalyst on adsorption and photocatalytic properties

G. Sarno, V. Vaiano, D. Sannino, P. Ciambelli
Department of Industrial Engineering, University of Salerno, Via Giovanni Paolo II, 132, 84084 Fisciano (SA), Italy;
tel. +39 089964006, e-mail: vvaiano@unisa.it;

Photocatalytic functionalization of surfaces involves the conferring of self-cleaning and antibacterial properties due to the spontaneous oxidizing power of semiconductors thin film when irradiated by a light with a suitable range of wavelengths. TiO_2, the most known photocatalyst, has been widely applied for the realization of multifunctional coatings. Recently composites have been deposited as surface layer on ceramic tiles to attribute self-cleaning characteristics. Some papers on TiO_2/zeolites composite in powder form demonstrated that the presence of zeolite enhanced photodegradation process because of the adsorption properties of the zeolite.

In this work, to enhance the self-cleaning properties, TiO_2/Ag-zeolite was supported on ceramic tiles and evaluated the influence on adsorption and photocatalytic properties of the anchored composite with respect to a TiO_2 coating.

The exchanged zeolite was prepared through ionic exchange in two steps, the first conducted at room temperature on Na, K-ferrierite with a solution of NH_4NO_3 1 M while the second ion exchange at room temperature with $AgNO_3$ aqueous solution for 24 h, followed by calcination at 550 °C. Functionalized tiles were prepared by a TiO_2 sol, obtained with titanium isopropoxide, ethanol, acetyl acetone, acetic acid, urea and PEG. In the obtained sol, Ag-zeolite and TRITON X 100, as non-ionic surfactant, were added. The ceramic tiles were obtained via steps-coating technique followed by heat-treatment in air.

Adsorption tests on methylene blue (MB) aqueous solutions were performed at ambient temperature and atmospheric pressure using a properly designed batch recirculating photoreactor in stainless steel. The irradiation of photoreactor was realized by four UV lamps (nominal power 8 W) emitting at 365 nm, placed at the proximity of a pyrex window.

For photocatalytic activity test an aqueous solution with an initial concentration of MB equal to 5 ppm was used.

Characterization results evidenced that the deposition method induced the formation of TiO_2/Ag-zeolite layer, strongly adherent to the tile surface, in which anatase TiO_2 was present. The loading of composite linearly increased with the increase of the coating steps, reaching a value of titanium content on the surface of about 13%.

The presence of zeolite enhanced the adsorption ability of the surface film, which resulted proportional to the increasing of the amount of TiO_2/Ag-zeolite composite on the tile surface.

The photocatalytic performances of the samples functionalized with Ag-zeolite were much higher than those obtained with the samples without zeolite but with the same TiO_2 content. The application of the TiO_2/Ag-zeolite composite resulted in the increasing of the degradation rate of the dye with respect to a simple TiO_2 coating and in enhanced antibacterial properties.

Wheat straw combustion and co-firing for clean heat energy production

I. Barmina, R. Valdmanis, M. Zake
Institute of Physics, University of Latvia, Salaspils, Miera Street 32, LV-2169, tel. +371 29475762, e-mail: mzfi@sal.lv

Because of still increasing interest in environmentally friendly fuel for clean and effective heat energy production and residential heating increases demand for renewable energy resources biomass pellets with most intensive use of wood pellets. Because of limited access to wood residues for production of wood pellets in some regions, using of pellets, which are produced from alternative biomass resources, especially wheat straw is of great interest. The results of previous experimental study have shown [1,2] that the combustion/emission characteristics of wheat straw highly differs from the combustion/emission characteristics of wood, which have higher content of cellulose and of lignin determining, whereas less nitrogen and ash content determining more effective and cleaner heat energy production. With the aim to improve combustion characteristics of wheat straw the detailed experimental study and analysis of the effects of co-combustion of wheat straw with wood pellets and co-firing with gas was carried out.

To assess the effects of wheat straw pellets co-firing with gas and co-combustion with wood pellets on wheat straw thermo chemical conversion, experimental studies were conducted using a batch-size pilot device [1] with integrated biomass gasifier and combustor and total heat output up to 2 kW. To provide co-firing of wheat straw with gas, the propane flame flow was injected radially in the bottom of the combustor, close above the layer of wheat straw pellets. To provide co-combustion of wheat straw with wood pellets, the gasfier was filled with a mixture of wheat straw and wood pellets by varying the mass load of wood pellets in a mixture from 0 up to 100 %. The experimental study of the effects of co-firing and co-combustion on wheat straw thermo chemical conversion includes the joint time-dependent measurements of the main flame characteristics - flame temperature, combustion efficiency, composition of the products and heat output. The results of experimental study have shown that for nearly equal primary and secondary swirling air supply rates in the device and equal values of the additional heat energy supply in a device at co-firing wheat straw with gas and co-combustion of wheat straw with biomass (about 50 %) the most effective improvement of the combustion conditions in the flame reaction zone can be achieved by co-firing wheat straw pellets with gas.

Co-firing of wheat straw pellets with gas results in an enhanced biomass gasification increasing a rate of wheat straw mass loss rate in the gasifier and enhanced flaming combustion of volatiles (CO, H_2) at the primary stage of the thermo chemical conversion of wheat straw ($t < 1000$ s). The enhanced combustion of volatiles results in a pronounced decrease of the average values of the CO mass fraction in the products from 2320 ppm to 1100 ppm and the H_2 mass fraction in the products from 1200 ppm to 540 ppm with a correlating increase of the volume fraction of CO_2 in the products from 6.8 to 9.5 %, products temperature from 480 K to 530 K and combustion efficiency from 74 to 77 % decreasing the air excess in the products by about 50 %. Moreover, the total heat output increases from 0.9 kW to 1.4 kW, decreasing the average value of the NO_x mass fraction in the products from 260 to 210 ppm indicating the formation of cleaner and more effective heat energy production (Figure 1-a).

The less effective and more complicate variation of the gasification/combustion characteristics are observed for the flaming regime of wheat straw at co-combustion with wood pellets, when a decrease of the average values of the CO mass fraction in the products from 88 ppm to 63 ppm and decrease of the H_2 mass fraction in the products from 43 ppm to 35 ppm results in a correlating increase of the volume fraction of CO_2 in the products from 11 to 11.43 %, which can be related to the formation of greenhouse emission. Co-combustion of wheat straw with wood pellets results in an increase of the average value of the products temperature from 490 K to 540 K, decreasing the air excess in the products by about 5 % and mass fraction of the NO_x emission in the products from 23 ppm to 12 ppm. The average value of the flame temperature at flaming combustion stage increases from 1425 K to 1506 K and the average value of total heat output increases from 0.85 kW to 1.1 kW, confirming higher and cleaner heat energy production. It should be noticed that co-combustion of wheat straw with wood pellets with lower content of hemicelluloses and higher lignin content [1] promotes an ignition delay of volatiles at about 200 s by limiting transition from primary smoldering stage to flaming combustion (Figure 1-b).

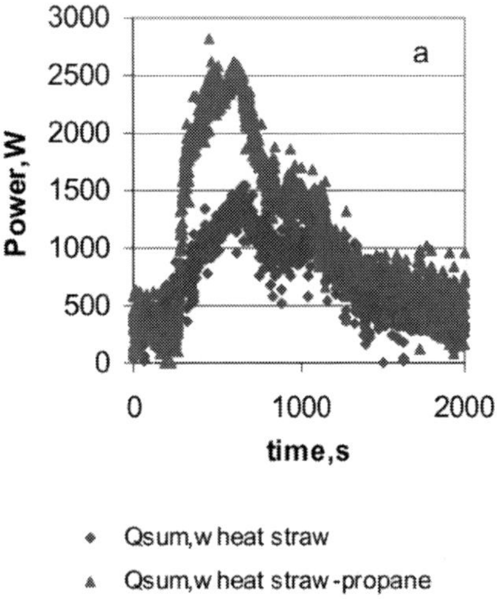

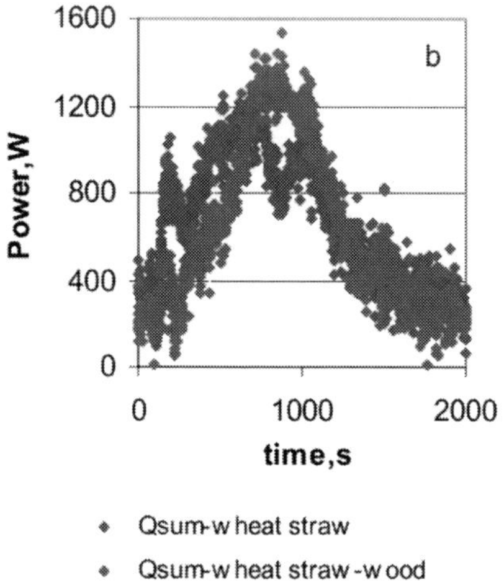

Figure 1. Effect of wheat straw co-firing with gas (a) and co-combustion with wood pellets (b) on the kinetics of heat output.

Considering the effects of wheat straw co-firing with gas and co-combustion with wood pellets it can be concluded that most effective thermo chemical conversion of wheat straw pellets at the flaming combustion stage can be obtained for the regime of co-firing, whereas co-combustion of wheat straw with wood pellets results in cleaner heat energy production determining the enhanced release of greenhouse emission with reduced formation of NO_x emission.

References

1. I. Barmina, A. Lickrastina, M. Zake, A. Arshanitsa, V. Solodovnik, G. Telysheva, 2013, Experimental study of the thermal decomposition and combustion of lignocellulosic biomass pellets, Latvian Journal of Physics and Technical Sciences, N3, 35-48.
2. I. Barmina, A. Lickrastina, M. Zake, A. Arshanitsa, V. Solodovnik, G. Telysheva, 2014, The effect of birch – bark addition on the elemental composition and combustion characteristics of different types of biomass pellets, Chemical Engineering Transactions, 39, 1525-1530.

Modelling of a post-combustion CO_2 capture process using deep belief network

Fei Li[1], Jie Zhang[1], Eni Oko[2], Meihong Wang[2]
[1]School of Chemical Engineering and Advanced Materials, Newcastle University, Newcastle upon Tyne, UK, e-mail: jie.zhang@newcaslte.ac.uk
[2]School of Engineering, University of Hull, Hull HU6 7RX, UK

Some researchers have strongly suggested that deep architectures in neural networks are more efficient than shallow architectures in learning the hidden relationship in a data set. Deep learning has emerged as a very powerful tool in machine learning. Deep learning technique includes two phases: an unsupervised pre-training phase and a supervised back-propagation phase. In the unsupervised pre-training phase, a deep belief network (DBN) is pre-trained as initial weights of the subsequent supervised phase. In the supervised back-propagation phase, the network weights are fine-tuned in a supervised manner. This paper presents a study of modelling post-combustion CO_2 capture process using DBN. DBN with many layers of Restricted Boltzman Machine (RBM) can extract a deep hierarchical representation of training data. In terms of CO_2 capture process, the DBN model predicts CO_2 production rate and CO_2 capture level using the following variables as model inputs: inlet flue gas flow rate, CO_2 concentration in inlet flue gas, pressure of flue gas, temperature of flue gas, lean solvent flow rate, MEA concentration and temperature of lean solvent. A greedy layer-wise unsupervised learning algorithm is introduced to optimize DBN, which can bring better generalization than a single hidden layer neural network. The developed deep architecture network models can then be used in the optimisation of the CO_2 capture process.

NOx emissions by hydrogen and syngas combustion in gas turbine

M. Kanniche
EDF R&D, 6 Quai Watier, 78401 Chatou, France;
tel. +33 130877093, e-mail: Mohamed.kanniche@edf.fr.

Integrated Gasification Combined Cycle (IGCC) is currently one of the most efficiency and clean power production system from coal. It enables the conversion of coal and other solids (biomass, petcoke, solid wastes) or liquid fuels (petroleum residues) to a clean gaseous syngas fuel, while still maintaining NOx and CO emissions targets and high cycle efficiency thank to high efficiency of combined cycles. Pre-combustion methods are the preferred means of capturing CO_2 in IGCC systems. An IGCC plant equipped with precombustion CO_2 capture combined with low emissions of other gases, e.g. NOx, SOx, can be realized through major advancement in IGCC gas turbine technology.

Currently, combustion of syngas and of hydrogen-rich syngas derived from coal in gas turbine is not feasible with low NOx premix systems as it is currently the standard for natural gas combustion.

This paper focuses on pollutant emission performance that could be reached by combustion of syngas and high hydrogen content syngas in industrial burners with premix mode. EDF developed a hybrid approach mixing the turbulent mixing capabilities of CFD tool using pdf approach and the chemical powerful capabilities of homogeneous reactor approach using detailed chemical kinetic models, especially useful for CO and NOx modeling.

The final objective of this approach is to build a simplified tool, which could be applied in industrial gas turbines in order to predict quickly the NOx emission variation versus operating conditions and fuel variations. Preliminary, each combustion chamber has to be characterized within a series of usual variations of ambient conditions and fuel quality.

The calculations of industrial premix burner show a good comparison with Cardiff University measurements reported here versus different equivalence ratios. The comparison is however less accurate for very lean conditions (very low NO emission) where the NO emission is under-predicted. It is well known that chemical reactions scheme GRI 3.0 under-predicts the NO emission for very lean conditions. The results show that NO emissions increase rapidly with increasing equivalence ratio while they increase moderately with increasing pressure. The experiments reveal however a tendency to flashback for higher pressure and for 'base load' operations. Further experimental work is needed to evaluate the flash back propensity to high pressures, which maybe could be minimized by increasing air mass flow rate (decreasing equivalence ratio).

Process simulation and costing study for the FT-liquid fuels production from steam glycerol reforming

Piyapong Hunpinyo[a,c*], Phavanee Narataruksa[b,c]
[a] Division of Chemical and Process Engineering Technology (CPET), Faculty of Engineering and Technology, King Mongkut's University of Technology North Bangkok (Rayong Campus)
[b] Department of Chemical Engineering, Faculty of Engineering, King Mongkut's University of Technology North Bangkok (KMUTNB)
[c] Research and Development Center for Chemical Engineering Unit Operation and Catalyst Design (RCC), STRI Building, Floor 1st and 7th (Room 702), King Mongkut's University of Technology North Bangkok (KMUTNB)
Corresponding author: 19 Moo 11 Tambon Nonglalok, Amphur Bankhai, Rayong 21120, Thailand. Tel.: +66 2 555 2000 Ext 1557, Mobile: +66 80 044 9344; Fax: +66 2 556 1306; Email: piyapong.h@eat.kmutnb.ac.th

There are various outlets for disposal and utilization of the crude glycerol generated in biodiesel plants in Thailand. For large scale biodiesel manufacturers, crude glycerol can be refined into a pure form and then be used in food, pharmaceutical, or cosmetics industries. For small scale manufacturers, however, purification is too expensive to be performed in their manufacturing sites. Their crude glycerol is usually sold to large refineries for upgrading. In recent years, however, with the rapid expansion of biodiesel industry, the market is being flooded with excessive crude glycerol. As a result, biodiesel manufacturers only receive 0.07-0.1 US$/kg for this glycerol. Therefore, manufacturers must seek new, value-added uses for this glycerol.

In order to convert crude glycerol into valued-added products through thermo-chemical methods with an alternative for utilizing this waste stream. This paper focuses on the conceptual design for the production of a promising biofuel source out of waste glycerol to be integrated within a small stand-alone biodiesel plant in the south of Thailand. The biodiesel production plant used in the process model, is supposed to give about 46,000 tons of crude palm oil per year. To produce this amount 14,735 kg/day raw glycerol material is needed. The simulation program proposes a detailed superstructure embedding the alternative technologies involved. The commercial simulators Aspen Plus is used to construct the process modelling, simulations and economic potential are performed to investigate different conditions for an integrated plant. Briefly, glycerol as a waste product from the trans-esterification of small plant oils and animal fats is first reformed using steam reforming. The syngas obtained is cleaned up and its composition is adjusted in terms of the ratio H_2/CO and then fed to the Fischer-Tropsch (FT) reactor. The novelty of this process lies in the use of a Ruthenium-based catalyst for the conversion of syngas to synthetic fuels. The production of FT liquid alkanes is separated while the off-gas as by-product is using for cogeneration with gas engine system to produce the electricity.

The results from the simulations, along with an economic evaluation, appear to be a very promising technology for synthetic fuel production, but still struggling to survive in business for reaching adequate profit. The integrated process is should be as effective as possible and both the secondary steams from pretreatment and steam drying should be used in the reforming and FT plant, to keep the productions costs down. Moreover, there are more channels for further process improvements with the use of systematic process design and optimization methodologies.

Multi-objective synthesis of South Africa's biomass to bioethanol supply network

Mildred Mutenure[1], Lidija Čuček[2], Adeniyi Jide Isafiade[1,*], Zdravko Kravanja[2]
[1] Department of Chemical Engineering, University of Cape Town, South Africa,
[2] Faculty of Chemistry and Chemical Engineering, University of Maribor, 2000 Maribor, Slovenia
aj.isafiade@uct.ac.za

South Africa produces synthetic ethanol and to date there are no commercial bioethanol production plants in operation in South Africa. The government of South Africa has been trying to encourage investments into the biofuel industry (see e.g. DME, 2007). Sugarcane and residues from sugarcane, maize, sorghum, wheat and barley are among the acceptable bioethanol feedstocks in South Africa. As the feedstock cultivation areas are scattered over a wide area and the possible bioethanol feedstocks have low energy density, there is a need to locate the processing facilities in strategic positions such that production and logistic costs and greenhouse gas (GHG) emissions due to transportation are minimized.

To address the above mentioned challenges associated with bioethanol production in South Africa, a mathematical programming approach for the synthesis of optimal bioethanol supply network is developed. It accounts for food and sugar demands, targeted amounts of produced biofuels, various feedstocks and their geographical distributions, tax subsidies, etc. Various processing technologies for ethanol production are considered and different transportation modes are used. Two objectives are considered, economic objective which is measured by annual profit, and environmental objective measured by GHG emissions. The ε-constraint method is used to solve the multi-objective optimisation problem and Pareto analysis is done to check the trade-offs between the objectives. To improve the accuracy of the data used in the model, the study uses geographical information systems software ArcGIS 10.3.1, for data analysis and gathering Google Earth is used, and for optimization GAMS is used.

From the case studies it is obtained that it would be possible to achieve GHG emission savings from use of E10 ethanol-petroleum blend. It would be possible to economically meet the 2 % target penetration of biofuel into the South African national liquid fuel supply.

Acknowledgments
The authors acknowledge the financial support for the NRF project Bioenergy Advanced optimisation, Grant No. 87744, and from the Slovenian Research Agency (programs P2-0032 and P2-0377).

References

Department of Minerals and Energy, 2007, Biofuels Industrial Strategy of the Republic of South Africa, <www.energy.gov.za/files/esources/renewables/biofuels_indus_strat.pdf(2).pdf> Accessed: 15.12.2015

Selective hydrogenation of levulinic acid to gamma-valerolactone using polymer-based Ru-containing catalysts

I.I. Protsenko, A.O. Zaykovskaya, L.Zh. Nikoshvili, V.G. Matveeva, E.M. Sulman
Tver Technical University, Dep. Biotechnology and Chemistry, A.Nikitina str. 22, 170026, Tver, Russia; tel./fax: +7 4822 789317, e-mail: nlinda@science.tver.ru

At present the process of biofuels production on the basis of levulinic acid (LA) is of great importance. LA can be transformed to 2-methyltetrahydrofurane (2-MTHF), which is a fuel additive. Although there is a possibility of direct LA transformation to 2-MTHF, improved yields can be achieved by indirect pathways, which proceed through the production of GVL as an intermediate. 5%-Ru/C is one of the most widespread heterogeneous catalysts of GVL synthesis from LA. However, in order to achieve high conversion of LA, the addition of co-catalysts or the increase of reaction temperature is often necessary. Besides, the key factor influencing the behavior of Ru-containing catalysts of LA hydrogenation is the size and dispersity of Ru nanoparticles (NPs).

Nowadays conformational rigid hypercrosslinked polymeric materials are the most promising organic supports, which are able to control the size and dispersion of catalytically active NPs and to prevent their aggregation and leaching. Despite the fact that recently the catalysts based on hypercrosslinked polymers are widely used in reactions of fine organic synthesis, there are no data on the use of polymer-stabilized Ru-containing NPs in hydrogenation of LA to GVL.

This work is devoted to the investigation of catalytic properties of ruthenium-containing NPs stabilized by hypercrosslinked polystyrene (HPS). Series of catalysts was synthesized at variation of Ru loading and HPS type (HPS bearing amino-groups or non-functionalized HPS). Before the reaction, the catalysts were activated in hydrogen flow at 300°C for 2 h. Hydrogenation of LA was carried out in Parr Series 5000 Multiple Reactor System in aqueous medium at variation of reaction temperature (80-150°C) and hydrogen pressure (10-40 bar). Samples of reaction mixture were analyzed via HPLC method.

Ru/HPS catalysts were shown to allow carrying out the hydrogenation of LA with high yields of GVL (more than 99%) in aqueous medium at mild reaction conditions (100°C and 20 bar of hydrogen partial pressure) without the necessity of co-catalyst addition. Synthesized HPS-based catalysts were shown to be active and selective at the absence of any co-catalysts and can be considered as promising for hydrogenation LA to GVL.

Advanced bio blending components for middle distillates

P. Solymosi, Sz. Tomasek, [1] A Holló. J. Hancsók
University of Pannonia., MOL Department of Hydrocarbon and Coal Processing, Egyetem út 10., H-8201 Veszprém, Hungary; tel. +3688624217, e-mail: solymosip@almos.uni-pannon.hu; [2] MOL Hungarian Oil and GAS Plc., Refining R&D, Olajmunkás utca 2., H-2440, Százhalombatta, Hungary; tel. +368823552817, e-mail: Ahollo@,ol.hu

The consumption of diesel fuel in European Union was almost 200 million tons in 2015. Based on the above mentioned fact, to satisfy the expected 10 energy % renewable ratio in the European Union until 2020, it will be necessary to produce or partially import 20-25 million tons of biodiesel. The problem is not only the available amount of biodiesel but also the 7 volume % blending limit of this component which is described by the EN590:2013 standard. The cause of this is a several unfavourable property of biodiesel (for example: low heat and oxidation stability due to the existence of olefinic double bonds, low cold flow properties, ect.). Accordingly, the application of other renewable fuel blending component is necessary. The use of advanced biofuels is inevitable. The advanced biofuels are made from low-ILUC risk feedstocks regardless from the production technology. One possible way of this is the production of gas oil boiling range straight paraffinic hydrocarbons (C_{16}-C_{22}) via special hydrocracking from waste originated triglycerides or fatty acids. The great potential of this product is that it contains only the best quality components of the diesel fuel. Furthermore, the bio gas oil which was produced from waste or non-feed or food sources can be double counted in the calculation of the renewable ratio. Of course the production of biodiesel from the above mentioned sources is also possible. However, due to the production technology of biodiesel it is feasible in limited extent because of the high sensitivity of the quality of the feedstock.

Based on the above mentioned facts, the aim of our experimental work was the production of high quality motor fuel blending component from non-feed or food sources such as high euric acid rapeseed oil via special hydrocracking. The product was obtained on in-house defined favourable operation parameters, the catalyst was consciously selected ($NiMoP/Al_2O_3$). Based on the results it was concluded that the cold flow properties (cold filtering plugin point) of the products could not fulfil the requirements of the EN590:2013 standard (in case of temperate climate countries the CFPP of the summer grade diesel fuels is max. +5°C, winter grade is max. -20°C). This is because the existence of large amount (35-45 weight %) of high carbon number normal-paraffins (C_{21}-C_{22}), the freezing point of these components is too high (n-C_{21}: 40°C, n-C_{22}: 44°C).

Under the terms of the abovementioned facts we performed additional experiments to improve the cold flow properties of the normal paraffin rich product fractions. One alternative is the isomerization of the products which was obtained in the hydrogenation step; another possible way is adding additive to the hydrogenated product and the fossil diesel fuel blend or the combination of these two methods. (The freezing point of the iso-paraffins which is obtained in the isomerization step is lower (10-30 %) than equal carbon number normal paraffins.) The isomerization experiments were performed on Pt-SAPO-11 and Pt-AlSBA-15 catalysts. The first catalyst had moderate acidity and high Pt content to ensure the high isomerisation and low cracking activity. The investigation of Pt-AlSBA-15 catalyst for this purpose was not reported by our research team. Based on the experimental results it was concluded that in case of Pt-SAPO-11 catalyst the iso-paraffin concentration was higher than 70 weight %, together with this the CFPP was lower than -4°C and -12°C range in favourable process parameters. In case of Pt-AlSABA-15 catalyst we found favourable operation parameter combination ranges where both the cracking reaction occurring in the middle of the molecule and the isomerisation reactions were significant. Based on these results the obtained fractions contained necessarily diesel fuel fraction components and approximately 30 weight % of C_{10}-C_{12} paraffins which are outstanding JET fuel bio-blending components. In the further experimental work we investigated the possibility of the CFPP reduction of the product fraction which obtained in the isomerization step and their blends with fossil diesel fuel by adding MDFI additives. Based on the experimental results it was concluded that the use of MDFI additive in 450 mg/kg amount helps to reduce the CFPP value of the bio-paraffin fossil fuel blend. Applying both isomerization technology and MDFI additive the blending ratio of the produced bio-paraffins may be increased up to 10 weight %. In this way the CFPP requirements of the EN590:2013 standard (both the summer and winter grade requirements of temperate climate) can be fulfilled and can be approached the expected 10 energy % renewable ration by 2020.

JET fuel production from high molecular weight Fischer-Tropsch paraffin mixture

Sz. Tomasek, A. Holló, J. Hancsók
Department of MOL Hydrocarbon and Coal Processing, University of Pannonia, H-8200, Veszprém Egyetem utca 10; tel. +3688624962, fax: +3688624520
e-mail: tomaseksz@almos.uni-pannon.hu

Nowadays the production and utilization of fuels – produced from renewable sources – have come to the front because of the increasing fuel consumption, more stringent environmental regulations (lower sulphur and aromatic content) and crude oil dependency of some countries. The utilization of alternative fuels in transportation is requested by political regulations both in the European Union and around the World. The European Union specifies the use of renewable energy to 10 % in transportation by 2020. In order to reduce the harmful effects significant steps have been also taken in the last years in the aviation sector. In 2009 The European Parliament decided to involve the aviation into the European Emission Trading System (EU ETS) from 2012. In 2009 the International Air Transport Association (IATA) also declared ambitious goals (growing of fuel consumption efficiency by 1.5 % annually by 2020, achieving a carbon neutral growth by 2020, and reducing the CO_2 emission by 50 % relative to 2005 level by 2050). In order to achieve the stated objectives the use of biofuels, which burn cleaner and have lower harmful emission during the entire lifecycle, is also required. Biocomponents of gasoline and diesel fuel are mainly produced nowadays from edible plants which raises many questions. Alternative fuels can be also produced from biomass based synthesis gas by Fischer-Tropsch synthesis. The boiling point of the 40-45 % of Low-Temperature Fischer-Tropsch product is higher than 370 °C, so-called Fischer-Tropsch wax. It is inevitable to transform the less valuable, high molecular weight paraffin product of the Low-Temperature-Fischer-Tropsch synthesis into more valuable products. A possible method could be the hydrocracking since the heavy paraffin product of the aforementioned synthesis is practically free of aromatics and sulphur compounds. For this reason high-quality environmentally friendly fuel can be produced from Fischer-Tropsch waxes through hydrocracking/ hydroisomerization with using expediently chosen catalyst and process parameter combinations.

Based on the foregoing, the aim of our experimental work was to study the hydrocracking of heavy Fischer-Tropsch paraffin mixture on different catalysts (Pt/H-Beta, Pt/H-ZSM5, Pt/H-[BAl]MCM22 zeolites) at various process parameters (temperature: 225-350°C, pressure: 40 bar, liquid hourly space velocity: 1.0 h^{-1}).

In our experimental work we investigated the effects of catalyst composition and process parameters on the yield and quality of liquid products. On the tested catalysts gases (10-48 %) and gasoline (10-60 %) are mainly produced. The C_{21-} product – formed on Pt/H-Beta zeolite – contained 85% isoparaffin hydrocarbon, while on the Pt/H-[BAl]MCM22 and Pt/H-ZSM5 zeolite took place isomerization in smaller degree (iso/n-paraffin ratio: 2-3 and 0.5-0.6 , respectively). In our experiments we could produce fuel components – belonging to different boiling point range – with good yield and we observed shape selective effects and special selectivity as well, due to specific structural of zeolites.

Fatty acid hydrotreatment using hypercrosslinked polystyrene-supported Pd catalysts to produce biofuels

A.A. Stepacheva[1*], V.G. Matveeva[1], E.M. Sulman[1], V.N. Sapunov[2]
[1]Tver Technical University, Dep. Biotechnology and Chemistry, A.Nikitina str. 22, 170026, Tver, Russia; tel./fax: +7 4822 789317, e-mail: a.a.stepacheva@mail.ru
[2]Mendeleev University of Chemical Technology of Russia, Moscow, Russia

Nowadays the biomass conversion is of the great interest. One of the ways to process biomass is the biofuel production. One of the most promising ways for biodiesel synthesis is the biomass hydrofining. Hydrofining includes following processes – hydrodeoxygenation, hydrocracking, hydrogenation – to produce 2nd generation biodiesel in form of linear saturated hydrocarbons or alcohols with carbon number of 15-22.

2nd generation of biodiesel is mixing easily with mineral diesel. Moreover it has higher cetane number, higher energy content, excellent combustion quality, good low-temperature properties, and superior thermal stability, storage stability, and materials compatibility in comparison with both petrol diesel and biodiesel.

In this investigation the hydrotreatment of fatty acids was carried out. The stearic acid (99%, KhimMedService, Russia) was used as the model substrate. n-Dodecane (99.9%, Sigma-Aldrich) was used as a solvent. Palladium supported on hypercrosslinked polystyrene (HPS) of MN270 type (Purolite LTD, UK) was used as a catalyst. The metal loading was varied from 1 up to 5% (wt.). The process was conducted in stainless steel batch reactor PARR-4307 (USA). The following reaction conditions was varied: hydrogen pressure 0.2 – 2.5 MPa, temperature 100 – 260°C, concentration of stearic acid 0.05 – 0.2 M. Liquid samples were taken every hour during the process and analyzed using GC-MS.

It was revealed, that the most effective catalytic system in both processes was the catalyst 1%Pd/HPS-MN-270, which has the high surface area (1300 m^2/g) and average pore size 4-6 nm. This catalyst allows achieving the high products yield in hydrodeoxygenation process (selectivity regarding n-heptadecane was 99% at 100% substrate conversion) as well as in hydrogenation reaction (selectivity regarding stearyl alcohol was 93% at 100% substrate conversion) of stearic acid chosen as model substrate.

Water management in lignocellulosic ethanol production – a case study and comparative analysis from a Swedish perspective

R. Fornell, [1]C-J. Franzén, K. Willquist, A. Petersson
SP Technical Research Institute of Sweden, Box 857, 50115, Borås, Sweden; tel. +46 10 5165014, email: Rickard.fornell@sp.se; [1]Chalmers University of Technology, Göteborg, Sweden

Lignocellulosic ethanol production is on the verge of commercialization, but still there are issues to be resolved for this process. Due to the recalcitrance of lignocellulosic raw materials a harsh pretreatment is most often needed, and this leads to the formation of inhibitory compounds that potentially exacerbate the performance of the following hydrolysis and fermentation reactions. The process is conventionally dilute (suspended solids around 10%) in order for inhibitor concentrations to be kept low, and thus not deteriorate the performance of the bioreactors. This on the other hand leads to large flow rates, high fresh water demand, and increasing costs for waste water treatment. Due to the large flow of waste water the potential for integration with adjacent infrastructures, e.g. CHP plants, industrial plants, pulp mills, might be difficult even if it is shown that these types of integrations are beneficial from a sustainability perspective. Studies on so called High Gravity processes (with suspended solids at 20% or more) have begun to increase. One example of a High Gravity concept is the so called Multifeed approach, developed at Chalmers University of Technology. This concept has the potential to reach very high suspended solids in the fermentor, whilst at the same time decreasing the effect of inhibitory compounds.
This project focuses on the water management of lignocellulosic ethanol processes, and is divided into four parts:

1) A review of inhibitory substances in lignocellulosic ethanol production, including formation, degradation and inhibition in the unit operations included in the ethanol process.
2) Experimental and demo plant analyses of compounds at different positions in an ethanol process using the Multifeed concept with on-site production of flocculating, xylose-fermenting yeast.
3) Application of water pinch analysis with one and three components in order to assess the potential for recycling of water in the Multifeed ethanol process.
4) A comparative analysis of energy and economics between three different conceptual ethanol production processes (dilute, Multifeed, and "conventional" High Gravity fermentation).

The results include a map of potential inhibitors; where they are formed, where they have the most severe inhibitory effect, and where they are degraded in the ethanol process. For the most important inhibitory compounds the analyses made in the project indicate the levels of inhibitory substances at different positions in the studied Multifeed ethanol process, and also highlight by exemplification the importance of taking into account synergistic effects that occur between different inhibitors.

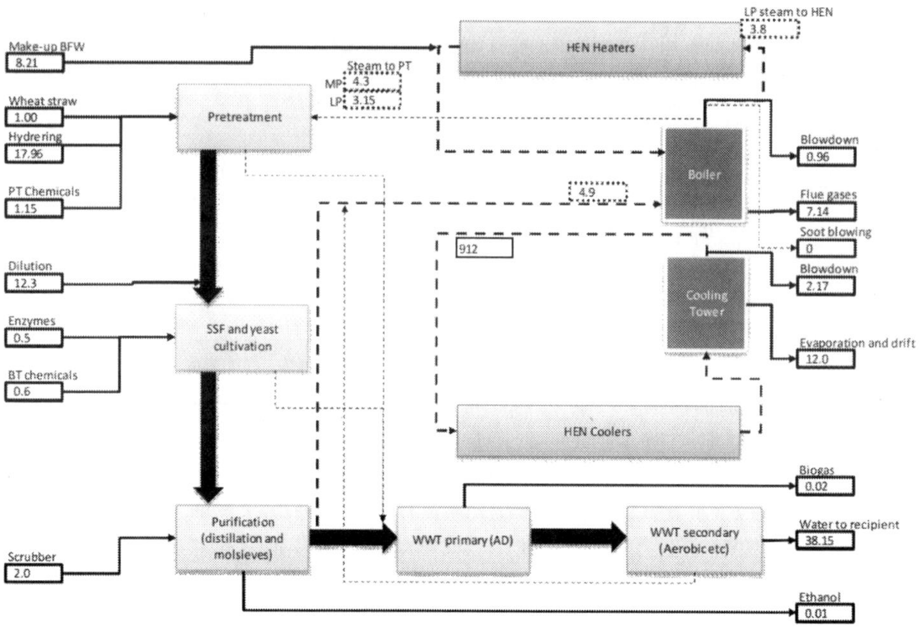

Figure 1. Water inputs and outputs in the studied ethanol process [kg H_2O/s].

Two methods for water pinch analysis, based on literature references, have been applied to the ethanol process (Foo 2013; Zhang, Feng et al. 2013). The results show the difference in water use based on using a one-substance analysis and a three-substance analysis. It can be seen that the one-substance analysis misses potential limits in the process, but also that it is not straightforward to apply water-pinch analysis due to the synergistic effects and a large number of inhibitors being formed and degraded in an ethanol process.

The recycling strategies from the water -pinch analyses have been used in a flowsheet model in order to assess the level of inhibition. The potential accumulation of inhibitory substances due to different recycling strategies was mapped for eight different potentially inhibitory substances.

Based on the flowsheeting models, a techno-economic analysis was done, which concluded that water is an important factor for the economics for the ethanol process, and that an approach enabling higher suspended solids will lead to a potentially higher profitability than the more dilute process. It is however important to again state the necessity to study the effects on ethanol yield of different concepts in more detail, since a potential decrease in yield would reduce the benefits of the higher suspended solids content.

Foo, D. C. Y. (2013). "Process integration for Resource Conservation." CRC press, ISBN: 9781439860489.

Zhang, Q., X. Feng, et al. (2013). "Evolutionary graphical approach for simultaneous targeting and design of resource conservation networks with multiple contaminants." Industrial and Engineering Chemistry Research **52**(3): 1309-1321.

Fuel purpose hydrotreating of free fatty acid by-products and heavy straight run gas oil

D. Sági, A. Holló, [1]G. Varga, J. Hancsók
Pannonia Univ., Dep. MOL Hydrocarbon and Coal Processing, Egyetem str. 10, H-8200 Veszprém, Hungary; tel. +3688624313, e-mail: dsagi@almos.uni-pannon.hu fax: +3688624520; [1]OT Industries, Galvani str. 44, H-1117 Budapest, Hungary, e-mail: VargaGe@otindustries.hu

The European Union (EU) has set a target in the 2009/28/EC directive, which aims that the share of biofuels in transportation have to be at least 10 energy% in 2020. However, only half of this target can be met by first generation biofuels (produced from sugars, oil crops, etc.). For this reason, the EU supports and encourages the development and application of advanced biofuels. The feedstock of these new generation biofuels (wastes, algae, cellulose, etc.) does not endanger and influence the cultivation of food and feed crops. Furthermore, when the renewable share of transportation fuels is accounted, biofuels from waste material and advanced ones will be considered twice. Besides that, after 2020 governments would financially support only advanced biofuels [1].

The research and development of newer generation biofuels for Diesel engines are important, due to the high level gas oil consumption of the EU. Nowadays, the most widely used biocomponent of gas oils is biodiesel, which consists of a mixture of various fatty acid methyl esters (FAME). On the other hand, it is well known that biodiesel has many disadvantages, e.g. bad thermal and oxidative stability, high CFPP value, lower energy content, etc. [2-4]. As a result, the diesel fuel product standard of EU (EN 590:2013) limits the blending rate of biodiesel to 7.0 V/V%. The bio gas oil could be a solution for these problems, which is produced by catalytic hydrogenation of triglycerides and/or fatty acids from different origins (e.g. wastes). Bio gas oil is a mixture of n- and iso-paraffins in gas oil boiling range and they are the most favourable energy source for running Diesel engines [5]. The second generation bio gas oil is an excellent gas oil blending component and it is preferred by the car manufacturers, too [6]. Consequently, the aim of our research work was to produce gas oil blending component with bio gas oil content from waste feedstock.

During our experiments we investigated the fuel purpose hydrogenation of free fatty acid by-products (0, 3, 5, 10, 20%) and heavy straight run gas oil mixtures on a commercial $NiMo/Al_2O_3$ catalyst. The effects of feed compositions and process parameters (temperature, pressure, liquid hourly space velocity (LHSV), hydrogen/feedstock ratio) on the quality and quantity of the main product were investigated. The applied process parameters were the following: P=40-50-60 bar, T=300-380°C, LSHV=1.0-2.0 h^{-1}, hydrogen/feedstock ratio=400 Nm^3/m^3.

Based on the results we concluded that, we produced low sulphur gas oil boiling range products with high yield (96.0-97.5%), from feedstock mixtures with 10-20% free fatty acid content, by applying favourable process parameters (P=40 bar, T=360°C, LSHV=1.0-2.0 h^{-1}, hydrogen/feedstock ratio=400 Nm^3/m^3). The products we obtained by applying the favourable process parameters are suitable for gas oil blending. Nevertheless, even better gas oil blending component could be produced with a further isomerisation step from these products, which would satisfy all the requiements of the diesel fuel standard of the European Union (EN 590:2013).

References

1. ec.europa.eu/energy/en/topics/renewable-energy/biofuels/land-use-change Accessed 04.01.2016
2. Pölczmann Gy., Tóth O., Beck Á., Hancsók J., 2016, Investigation of storage stability of diesel fuels containing biodiesel produced from waste cooking oil, Journal of Cleaner Production, 111, Part A, 85-92.
3. Deshmukh D., Lanjekar R.D., A review of the effect of the biodiesel composition of biodiesel on NOx emission, oxidative stability and cold flow properties, 2016, Renawable and Sustainable Energy Reviews, 54, 1401-1411
4. Moser, B.R., Fuel property enhancement of biodiesel fuels from common and alternative feedstocks via complementary blending, 2016, Renewable Energy, 85, 819-825
5. Hancsók J., Krár M., Magyar Sz., Boda L., Holló A., Kalló D., Investigation of the production of high cetane number biogasoil from pre-hydrogenated vegetable oils over Pt/HZSM-22/Al_2O_3, 2007, Microporous and Mesoporous Materials, 101(1-2), 148-152
6. www.acea.be/uploads/publications/Worldwide_Fuel_Charter_5ed_2013.pdf Accessed 13.01.2016

Energy storage key component ammonia borane: catalytic hydrolysis, and kinetic parameters

Göksel Özkan[1],Tayyibe Şimşek[1], DeryaÖncel Özgür[1], Hatice B. Murathan[1], M. Sena Akkuş[1] ,Gülay Özkan[2]
[1]Department of Chemical Engineering, Gazi University, 06570 Maltepe, Ankara, Turkey
[2]Department of Chemical Engineering, Ankara University, 06100 Tandoğan, Ankara, Turkey;tel.+903125823515, e-mail: gozkan@eng.ankara.edu.tr[1]; Fax: 0312-230 84 34.

Ammonia borane (AB) is a promising material for hydrogen storage because of some prominent features such as solid at room conditions, high chemical stability in neutral aqueous solutions at room temperature, and Its higher theoretical hydrogen content (19.6 wt %) than other hydrogen storage material. It seems to be more practical for mobile and portable applications because of easy releasing H_2 at low temperature between 20^0C and 70 ^0C, less area than the other storage material, its solid or liquid storage form. Hydrogen can be produced from AB vie either a high temperature requirement of thermolysis, catalytic hydrolysis of its aqueous solution which is homogeneous phase or heterogeneous phase or solvolysis route. Ammonium borane with aqueous solution of a suitable catalyst, by hydrolysis Eq. (1) or Eq. (2) given in one reaction takes place, and in theory a total of 3 moles H_2 gas can be obtained. Hydrolysis of AB in the presence of metal catalysts or solid acids releases 3 mol hydrogen from Eq (1) and (2) (at ambient temperature and pressure)

$$NH_3BH_3 + 3H_2O \rightarrow NH_3 + H_3BO_3 + 3H_2 \quad (1)$$
$$NH_3BH_3 + 2H_2O \rightarrow NH_4^+ + BO_2^- + 3H_2 \quad (2)$$

For catalytic hydrolysis of AB, various catalysts and reaction parameters were investigated. Noble metal and alloys catalysts, Such as carbon supported Ru- Co, zeolite confined rhodium, Pt - Co nanoclusters, the colloid cobalt (0) and rhodium (0) catalyst, nano scale Cu/CuO ,polymer supported Ru, Co, Ni, Pd catalyst which are soluble in water. Nano-scale Fe, Co/SiO_2, Pd-TiO_2 supported (Co-Ni-P) catalyst have been tested in the methanolysis of AB. It has been observed that porous supported materials like the different types of carbon material, SiO2, inorganic-based synthetic zeolites, Al_2O_3, metal oxide are used in the study.

In this study, a catalytic hydrolysis of heterogenic phase AB has been experimented with an aim of having high yield of hydrogen production with high efficiency at lower temperatures. Experimental studies has been handled in three phases respectively, synthesize of catalyst, parametric reactions experiments and determining the kinetic parameters. Catalysts has been prepared both with impregnation techniques where as supporting material Al_2O_3, as active material Pt, Co, Sn has been chosen. Some characteristic specifications (SEM, EDS ve BET) has been determined depending on structure, surface area and morphology of synthesized catalyst and seen that active materials has been homogenously spread and cover the surface of support material in nano-scale. In the second phase of study, effect of temperature on hydrogen production from AB has been investigated and amounts of hydrogen productions with efficiency values has been recorded as a result for each experiment. Results points out that %1Pt/Al_2O_3, 5%Co/Al_2O_3 and %2Sn/Al_2O_3 catalysts (at 50^0C) have highest performance with hydrogen production efficiencies as 87%, 75,38%, 34,2% respectively. Seen that Sn has lower performance on Al_2O_3 than Pt and Co active materials. The kinetic parameters of the dehydrogenation of AB are deduced by integral analysis and linear regression methods using exponential and Langmuir-Hinshelwood kinetics model for the reaction rates. The results showed that the reaction with AB dehydrogenation is zeroth order in AB concentration with Co/Al_2O_3 catalysts ($-r_{AB}=k$), first order with Pt/Al_2O_3 and Langmuir-Hinshelwood kinetics model developed based on the experimental Sn/Al_2O_3 data.

Key Words : Catalytic hydrolysis, ammonia borane, Al_2O_3 support, dehydrogenation of AB

The effect of operating conditions and supports on Ni-Cu-Cs three metallic catalyst for steam reforming of ethanol to the production of hydrogen

Göksel Özkan[1],Başak Şahbudak[1], Gülay Özkan[2]
[1]Department of Chemical Engineering, Gazi University, 06570 Maltepe, Ankara, Turkey
[2]Department of Chemical Engineering, Ankara University, 06100 Tandoğan, Ankara, Turkey;tel.+903125823515, e-mail: gozkan@eng.ankara.edu.tr[1]; Fax: 0312-230 84 34.

One of the most indispensable elements of modern societies is sustainable/renewable energy. In this sense, hydrogen is the one of the clean and renewable alternative energy. Hydrogen production from ethanol steam reforming has been used, since ethanol can be obtained from several biomass sources. In addition, ethanol has high hydrogen content, low toxicity and can be easily and safely manipulated and transported. The ethanol steam reforming (SRE) can be represented by the following equations (Eq1-6)

$$C_2H_5OH + 3H_2O \rightarrow 2CO_2 + 6H_2 \quad (1)$$
$$C_2H_5OH + H_2O \rightarrow 2CO + 4H_2 \quad (2)$$
$$C_2H_5OH + 2H_2 \rightarrow 2CH_4 + H_2O \quad (3)$$
$$C_2H_4 + 2H_2O \rightarrow 2CO + 4H_2 \quad (4)$$
$$CH_4 + H_2O \rightarrow 3H_2 + CO \quad (5)$$
$$CO + H_2O \rightarrow CO_2 + H_2 \quad (6)$$

According to Eq(1-6) different types of intermediate products are observed. For that reason we investigated suitable catalyst that is expected to eliminate or minimize the formation of undesired intermediate product. Metal sintering, carbon deposition on the catalysts surface came from ethanol decomposition is the major problem for reducing lifetime of catalyst. Various articles on the design of the catalyst for the SRE have been published. In these articles, the selection of active metals and catalyst support has been analysed as critical factor that allows for obtaining catalysts with high activity, cheap and selectivity. The use of Ni-Cu bimetallic or Ni/Cu/Pd three metallic catalysts are one of the most studied in steam reforming. Other dopants, such as K, La, Li, Ce, and Cs have also been studied. Akiyama et al.(2012) found that Ni/ZrO_2 catalyst doped with Cs yields higher levels of H_2 when compared with the catalyst that is doped with Ce, Li, K, La or Ce. That's why we used Cs. Ozkan et al, (2011) found that the selectivity was found to increase with increase in the amount of Cu and Ni but upper limit for this was discovered to be the catalyst having composition of Cu and Ni between 6 % and 10%. In this study, the amount of Cu and Ni is chosen as 10% and 10% respectively. Ni-Cu-Cs appears interesting but their surface areas are relatively small. For this support material have been used. The supporting materials play an important role in the steam reforming process facilitating the migration of reactive species toward the metal particles, stabilize them and have a great effect on hydrogen selectivity. CeO_2, $Zr0_2$,La_2O_2, MgO supports have been proposed because they favour ethanol dehydrogenation reactions instead of dehydration . However these alkaline oxides have a low surface area such as Ni/ZrO_2 :$12m^2$ /g .Alumina supports are frequently used because of their mechanical and chemical stability under the reaction conditions. But, acidic sites on the support promote ethanol dehydrogenation and lead to formation of ethylene where turn deactivates the catalysts by carbon deposition. So, basic additives that favour water adsorption and mobility of the OH groups on the Al_20_3 surface are added. Alkaline oxide are widely used in reforming formulations to neutralize the acidity of Al_2O_3.Recently, mesoporous materials such as $LiAlO_2$ with well-defined morphologies have attracted much attention. Among many types of mesoporous materials, hollow spheres are of significance for their potential uses in many areas, such as containers for encapsulation, fillers and catalytic supports, because of their low densities, high surface area and excellent optical properties

In this study, The effect of operating conditions and support on the catalytic activity have been investigated by testing the hydrogen generation from ethanol steam reforming in the presence of nickel/copper/caesium supported on active carbon (AC) and $LiAlO_2$.Each catalytic run was carried out at constant feed flow rate, different water/ethanol molar ratios and a reforming temperature from 300 to 700°C. The selectivity values towards hydrogen, hydrogen percentages in product and conversion of ethanol obtained from the reactions were compared. It is concluded that $LiAlO_2$ increased catalysis activity and Ni-Cu-Cs catalyst supported on $LiAlO_2$ had the highest selectivity of hydrogen and hydrogen percentages in product at 600 °C, ratio of 1,85 and 9.

Key words: Steam reforming, Ethanol, Nickel/copper/caesium catalysts

Supersonic separation of natural gas liquids by Twister technology

A. Esmaeili[1]
Saipem Contracting Nigeria Limited, Process Engineering Department; Saipem Base, Port Harcourt, River State, Nigeria, tel.: +989125130090, e-mail: arash.esmaeili2010@gmail.com;

The Twister supersonic separator is a unique combination of known physical processes, combining aero-dynamics, thermodynamics and fluid dynamics to produce an innovative gas conditioning process. Condensation and separation at supersonic velocity is the key for achieving a significant reduction in both capital and operating cost.Twister is a proven gas conditioning technology. Condensation and separation at supersonic velocity provides several unique benefits as the short residence time within the Twister tube prevents hydrate problems, thereby eliminating the use of chemicals and associated regeneration systems. A Twister tube designed for 1 million Sm^3/d (35 MMSCFD) at 100 bar is approximately 2 meters long. The simplicity and reliability of this static device with no rotating parts, operating without chemicals, ensures a simple, environmentally friendly facility, with a high availability, suitable for de-manned operation [1]. Twister can achieve both water and hydrocarbon dew pointing in one unit. The supersonic separation equipment has thermodynamics similar to a turbo expander, combining cyclone gas/liquid separation and re-compression in a compact, tubular device.

Exergy analysis of the air evacuation process from the pressure accumulator of the pneumatic pulsator system

K.J. Wołosz, J. Wernik
Warsaw Univ. Tech., Plock Branch, Dep. Process Equipment, 09-400 Płock, Lukasiewicza 17, Poland; tel. +4824 3675996, e-mail: krzysztof.wolosz@pw.plock.pl.

Pneumatic pulsator systems are made for clearing and unclogging outlets of loose material silos. The system, and its location on a sample silo, is shown in Fig.1. The pulsators utilize energy of compressed air in order to destruct adverse phenomena in loose materials and make the outlets of silos permeable. Main, basic parts of the pulsator are: a head and a pressure accumulator. The head controls the airflow which is driven by the pressure difference between the pressure accumulator and the silo. The most important parameter of the pulsator is a force which can be transmitted onto the loose material bed. This impact force is related to the pressure difference but this relation is not obvious. Many factors and phenomena, such as gas conversion during supersonic airflow for example, make necessity of energy and exergy analyses to be carried out. Analysis of energy flux and calculations of the energy reserves have been reported in previous work [1] by the same authors. This paper proposes a methodology for calculating an irreversibility of a high speed gas conversion with transient airflow conditions taken into account. The aim of this work is to deliver values of energy which is available for conversion to the impact force in order to adverse phenomenon could be destroyed in a silo. The exergy analysis delivers results which make it possible to determine the efficiency of the pneumatic pulsator system.

Fig. 1. A pneumatic pulsator system and its location in a loose material silo with the pressure acumulator zoomed in.

The exergy calculations have been based on results of numerical simulations which have been carried out by using Finite Volume Method which is employed in OpenFOAM® toolbox [2]. The governing equations of the laws of conservation have delivered results which have been validated by using data reported in literature. The validating case considered an outflow from an open end and it has been chosen in order to fulfil both conditions simultaneously: outflow to atmosphere and transient analysis [3].

Except validation of the results, there have also been performed a grid convergence test and an error estimation which show correctness of the numerical model [4].

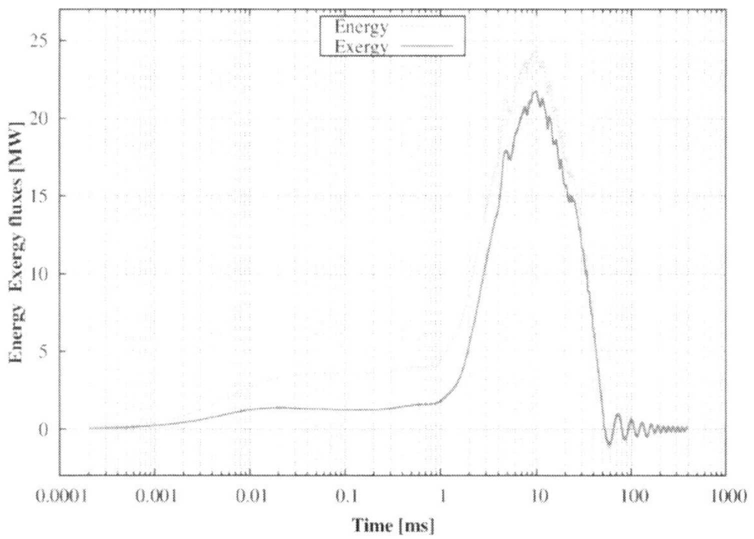

Fig. 2. Energy and exergy fluxes vs. time of the air evacuation process.

By using mean values of flow parameters such as temperature, pressure, velocity and density, the exergy analysis has been carried out. It has been assumed that the flow does not influence the environmental parameters.

The most important period of the process is observed during the first 25 ms from the start, which is why the plot of energy and exergy fluxes in Fig.2 is drawn in semi-logarithmic scale. In order to calculate energy and exergy, the corresponding fluxes are integrated in time and the total values are delivered: energy equals 682.9 kJ, and exergy is 605.2 kJ. The irreversibility of the process is, therefore, 77.7 kJ. This loss occurs in one work cycle of pulsator. The modern pulsators are designed for 1 million and more workcycles. It is easy to calculate that the irreversibility of the gas conversion during the air evacuation can reach values of over 77.7 GJ.

There has been shown that there is a noticeable difference between energy and exergy fluxes. The air evacuation from the pressure accumulator is a gas conversion in critical conditions, therefore, a normal shock is created. These conditions produce losses and energy irreversibility. The exergy is a maximal work which can be transferred to the impact force, therefore, it can be regarded as a reference value for efficiency of the pneumatic pulsator system [5]. The irreversibility of the process is 77.7 kJ but this number doesn't show the rest of losses of the pneumatic pulsator system, which is why the further analyses will be carried out.

References

[1] Wołosz K.J., Wernik J., 2015, The release of energy from the pressure accumulator in the pneumatic pulsator, Chemical Engineering Transactions, 45, 979-984 DOI:10.3303/CET1545164 979

[2] OpenFOAM: The open source CFD toolbox, 2015, www.openfoam.org

[3] Honma, H., Ishihara, M., Yoshimura, T., Maeno, K., Morioka, T., 2003,Interferometric CT measurement ofthree-dimensional flow phenomena on shock waves and vortices, Shock Waves 13, 179–190 DOI:10.1007/s00193-003-0206-1

[4] Jasak H., Gosman A.D., 2003, Element residual error estimate for the finite volume method, Computers & Fluids 32(2), 223–248 DOI:10.1016/S0045-7930(02)00004-X

[5] Szargut J., 2005, Exergy Method: Technical and Ecological Applications, WIT Press, Southampton, UK.

Reaction parameters and an energy optimization for biodiesel production using a supercritical process

[1]N. Farrag, [1,*]M. Gadalla, [2]M. Fouad
[1]The British University in Egypt, Faculty of Engineering, Chemical Engineering Department.
[2]Cairo University, Faculty of Engineering, Chemical Engineering Department.
[*]Mamdouh.Gadalla@Bue.edu.eg

Biodiesel has been proven to be the best reliable alternative for petroleum diesel. Besides, being renewable, it is biodegradable, and non-toxic fuel. This paper aimed to study the production of this green fuel using industrial, competitive techniques; base-catalyzed transesterification and supercritical methanol transesterification. The research involved techniques for reaction parameters optimization and thus embedded optimum values into a simulation-based design procedure for overall energy optimization/integration and emissions reduction.

Literature experimental reaction data for the two manufacturing technologies were taken to conduct an optimization for biodiesel production. The state-of-the-art process flowsheets for the two manufacturing processes were used for the study. This optimization was done for the most affecting parameters on the production processes. The experimental results from the literature for the two techniques were optimally analyzed using parameters analyse carried out using Pareto chart, contour plot methodology, and surface plot methodology. The research study revealed that the key process variable for the base-catalyzed process was the catalyst loading and the optimum conditions for better yield were found to be at 25 °C, 1.3% NaOH and 6:1 methanol to oil molar ratio. On the other hand, for the supercritical methanol process, the most prominent variable was the methanol to oil ratio and the optimum conditions were found to be at 8 minutes and 42:1 methanol to oil ratio. A simplified linear regression analysis was also performed to correlate the biodiesel yield with the process parameters. The optimal process conditions were used to build an ASPEN HYSYS rigorous simulation model for the previous production techniques. The supercritical based-process was chosen for further studies for its better economic performance. As a result, a process optimization was carried out for the reactor volume and the reflux ratios of the distillation columns. The optimum design parameters achieved for the base process were 12 m^3 for the reactor volume, 2 for the methanol recovery column reflux ratio and 3.4 for the biodiesel purification column reflux ratio.

Pinch analysis principles through ASPEN Energy Analyzer software were employed to analyse the energy performance of the overall optimum model obtained from ASPEN HYSYS. The energy targets were calculated for biodiesel production using supercritical methanol approach. Composite curves resulted in 3.4 and 3.7 MW for heating and cooling requirements, respectively compared with 4.4 and 4.7 MW for the original process. Finally, a heat exchanger network was developed to accomplish the energy targets proposed by the composite curves. The resulting integrated process has an improvement over the literature process flowsheet for better energy saving opportunities. The energy consumption of the optimum case has been reduced by 25%, and thus substantial cut in the CO_2 emissions. Utility curves were also generated to determine the loads of hot utilities (i.e. superheated steam) to be produced by the proposed energy integration.

Modelling of coal–biomass blends gasification and power plant revamp alternatives in Egypt's natural gas sector

Dalia Amer[a,*], Mamdouh Gadalla[a], Omar Abdelaziz[b,c], Fatma Ashour[b]
[a]Department of Chemical Engineering, The British University in Egypt, El-Shorouk City, Cairo 11837, Egypt
[b]Department of Chemical Engineering, Cairo University, Giza 12613, Egypt
[c]Department of Chemical Engineering, Lund University, P.O. Box 124, SE-221 00 Lund, Sweden
*Dalia.Amer@bue.edu.eg; Tel.: +20 22 689 0000 (Ext. 1471); fax: +20 22 630 0010

Recently, there has been a growing research interest in the co-gasification of biomass with coal to produce syngas and electricity in a sustainable manner. Co-gasification technology can not only decrease the exploitation of a significant amount of conventional coal resources, and thus lower greenhouse gases (GHG) emissions, but also boost the overall gasification process efficiency. In the present work, a rigorous simulation model of an entrained flow gasifier is developed using the Aspen Plus® software environment. The proposed simulation model is tested for an American coal and the model validation is performed in good agreement with practical data. The feedstocks used in the proposed gasifier model are dry Egyptian coal and a blend of an Egyptian coal and rice straw that is gathered locally. The proposed gasifier model mainly consists of three reactors. The first one is a yield reactor where the coal pyrolysis occurs, the second reactor is a stoichiometric reactor where the gasification reactions arise, and the third reactor is a Gibbs reactor where the water-gas and steam-methane reforming reactions take place. The effect of using a dry coal-rice straw blend (10 wt% rice straw) as feed compared with dry coal on the performance of gasifier is studied. The developed model provides a robust basis for revamping of an existing Egyptian natural gas-based power plant to replace its standard fuel with a coal-rice straw blend, in case of low natural gas supply. The model is further employed to assess different revamping scenarios and alternatives within the natural gas power plant. For a dry blend of (90% Egyptian coal and 10% rice straw), the cold gas efficiency is estimated as 85.7%, while for dry coal it is calculated as 79.61% 7. The revamped Egyptian natural gas power plant decreases the total annualized cost (TAC) by 52.7 % with respect to a new constructed IGCC power plant. Besides, the payback period decreases to 1.24 years rather than 12 years in case of the construction of a new IGCC power plant.

Keywords

Gasification, Coal-rice straw blends, IGCC, Power generation, Process integration

Biodiesel production out of fat derived from environmental sanitation

L.M.Pinotti [1], G. S. B. Salomão[1], L. C. Benevides [1], J. P. Oliveira[2], S. T. A. Cassini [2]
[1] Department of Engineering, Federal University of Espírito Santo, Rodovia BR 101 Norte Km 60 São Mateus – ES – Brazil, tel. +552733121592, e-mail: pinotti2008@hotmail.com
[2] Department of Environmental Engineering, Federal University of Espírito Santo, 29075-910 Vitoria –ES- Brazil

Biodiesel may be cost effective if produced from low cost feedstock such as animal fats, industry grease trap and frying oil, which contains high amount of free fatty acids (FFA). However, when using raw materials with a high content of free fatty acids it is interesting to use enzymes as a catalyst since the alkaline transesterification can generate soap as final product. Fats with high free fatty acid can be converted in biodiesel by esterification/transesterification. In this work fat from university restaurant grease trap (FUR) were used as feedstock for enzymatic synthesis of biodiesel. The reactions were performed using hexane, ethanol and immobilized enzymes lipase from the fungi *Candida antarctica*, *Thermomyces lanuginosa* and *Mucor miehei* with a reaction time of 24 hours. The experiments were performed by using factorial design of type 3^2 to evaluate the best conditions of temperature (30; 40 and 50 ° C) and enzyme concentration (5, 7.5 and 10% w / w fat) for each different enzyme and feedstock. When the enzyme obtained from *C. antarctica* was used, high yield was obtained in all studied conditions and none of the variables was statistically significant. As for the enzymes from the *Thermomyces lanuginosa* and the *Mucor miehei* significant effect of the temperature was reported, higher yield was obtanained with lower temperatures (30-40ºC). The concentration of enzymes only was statistically significant with enzyme from *Mucor miehei,* which had better yield with 10% w/w fat. Still, the lowest yields of biodiesel were obtained when using enzyme from *Mucor miehei*.

Keywords: Biodiesel, lipases, esterification/transesterification, high free fatty acids, fat grease trap.

Bromination of guaiacol and syringol using ionic liquids to obtain bromides

[1]A.Sequeiros, [2] L. Serrano, [1] J. Labidi
[1]Univ. Basque Country, Dep. Chemical and Environmental Engineering, Plaza Europa 1, 20018, Donostia San-Sebastian, Spain; tel. +34 943 017125, e-mail:ane.sequeiros@ehu.es; Laboratoire de Génie des Procédés Papetiers (LGP2), Centre de Recherches sur les Macromolécules Végétales (CERMAV-CNS), 461 rue de la Papeterie, 38402, Saint-Martin-d'Hères Cedex, France.

Brominated organic substrates, particularly aromatics, are of paramount importance as building blocks in organic chemistry as well as in the manufacturing of pharmaceuticals and agrochemicals [1,2]. The replacement of conventionally used toxic elemental bromide and hazardous acidic reagents is one of the mayor challenges in the aromatic bromination [3]. Consequently, the use of ionic liquids is studies as clean alternative to the hazardous reagents. These compounds are usually non-flammable and display non measurable vapor pressure. Also as alternative to the molecular bromide N-bromosuccinimide (NBS) is used, due to its ease of handling, low price, and to the fact that the byproduct, succinimide, can be easily recovered and reconverted back into NBS for its reutilization [2].

The present work aims to study the bromination of guaiacol (G) and syringol (S) in the presence of an ionic liquid. The G and S were used as case study because it is believed that these compounds are highly reactive and their conversion during lignin depolimerization gave products such as catechol, syringyl aldehyde, phenol and *o*-cresol [4,5]. Therefore, the experiments consist in the preparation of the bromination reagent by mixing a shortly microwaved 1-butyl-3-methylimidazolium bromide [Bmim]Br at 150 °C, 50W with NBS followed by adding the aromatic compounds. In the case of the aromatic compounds mixture an equimolecular amount was added. After the reaction was completed liquid-liquid extraction was carried out using two different solvents, diethyl ether and ethyl acetate, with the aim of seeing the change in the reaction yield.

Table 1. Conversion (%) and yield (%) of the resulted products on the bromination with EtAc

Conversion (%) and yield (%) of the resulted products						
Ethyl acetate						
	S		G		S+G	
	Conv (%)	Yield (%)	Conv (%)	Yield (%)	Conv (%)	Yield (%)
G	-		57±6		50±4	
S	46±6		-		60±4	
p-BrG	-			13±3		11±0
o-BrG	-			9±2		7±1
p-BrS		13±2	-			15±1
m-BrS		5±4	-			5±1

In the Table 1 the results of the conversion and yield of each product for the reactions with ethyl acetate are shown. The results showed that the mass balance of this reaction was not ideal as the sum of the product yields was not the same as the reactants conversion. The first reactions showed that the less sterically hindered reactant, the guaiacol (X_G 57%) has higher conversion that syringol (X_S 46%). However, in the reaction with the mix of G and S favored the conversion of X_S 60% over the X_G 50%. Accordingly, the yield of the BrS isomers was higher with the mix of G and S (ΣBrS 20%) than only with S (ΣBrS 18%).

Table 2.Conversion (%) and yield (%) of bromination products with DEE and EtAc and total oil (mg)

	Conversion (%) and yield resulted (%) and the total oil (mg)			
	Diethyl ether		Ethyl acetate	
	Conv (%)	Yield (%)	Conv (%)	Yield (%)
Oil	373.73±35.40		586.60±24.90	
G	57±6		50±4	
S	72±6		60±4	
p-BrG		10±1		11±0
o-BrG		5±0		7±1
p-BrS		15±0		15±1
m-BrS		8±3		5±1
ΣBrG		15		17
ΣBrS		22		20

The table 2 shows the results for the conversion and the yield of the resulting products with diethyl ether (DEE) and ethyl acetate (EtAc). Although the conversion of X_G and X_S was higher with DEE (X_G 57% and X_S 72%) than with EtAc (X_G 50% and X_S 60%) the results showed that in proportion EtAc (ΣBrG 17% and ΣBrS 20%) was able to extract more desired products than the DEE (ΣBrG 15% and ΣBrS 22%).

1.Khosravi K, Kazemi S, Green, mild and efficient bromination of aromatic compounds by HBr promoted by *trans*-3,5-dihydroperoxy-3,5-dimethyl-1,2-dioxolane in waster as solvent. *Chim Chen Lett* **23**: 387-390 (2012).

2.Pavlinac J, Zupan M, Laali KK, Stavber S, Halogenation of organic compounds in ionic liquids. *Tetrahedron* **65**: 5625-5662 (2009).

3.Bora U, Bose G, Chaudhuri MK, Dhar SS, Gopinath R, Khan AT, Patel BK, Regioselective bromination of organic substrates by tetrabutylammoium bromide promoted by V_2O_5-H_2O_2: An environmentally favorable synthetic protocol. *Org Lett* **2**: 247-249 (2000).

4.Wahyudiono, Kanetake T, Sasaki M, Goto M, Decomposition of a lignin model compound under hydrothermal conditions. *Chem Eng Techno* **30**: 1113-1122 (2007).

5.Afifi AI, Chornet E, Thring R, W, Overed, RP, The aryl ether bond reactions with H-donor solvents: guaiacol and tetralin in the presence of catalysts. *Fuel* **75**: 509-516 (1996).

Calculation of the operation parameters of the catalytic converters of the harmful gas impurities

V.E. Ved [1], E.V. Krasnokutskiy [1], M.I. Satayev [2], A.V. Ved [1], A.A. Saipov [2]
[1] National Technical University "Kharkiv Polytechnic Institute", 21 Frunze st., Kharkiv, 61002 Ukraine, e-mail: ved@kpi.kharkov.ua, tel. +38057 7076589; [2] M. Auezov South Kazakhstan State University, 5 avenue Tauke Khan, Shymkent, 160012 Kazakhstan.

In this paper theoretically grounded principles of catalytic converters design for harmful gas emissions purification of various thermal power units.

Experimental studies have shown that the smallest aerodynamic losses in structure of gas converters achieved using airfoils as working elements. This was the basis for use of the geometric structure to create catalytic converters carriers.

The paper considers the catalyst support element which is in the form of a rectangular parallelepiped with a gentle curve in the direction of the main gas flow lines. Support element is an airfoil blade. Workspace converter is filled with rows of blades. Each row of blades is a grating with parallel end walls. A combustible mixture of air and gaseous impurities with known parameters (such as total temperature, total pressure, flow direction, relative mass density of the impurity) enters at the inlet of converter. The atmospheric pressure is given at outlet. It is necessary to define for each row of blades flow parameters: temperature, speed, relative to the mass density of the impurities.

The solution the problem is associated with the development of a mathematical model of heat generation in the catalytic converter.

As a result of structural analysis of dynamic processes occurring in the catalytic converter and decomposition of a complete mathematical model of a catalytic process it is found that a major influence on the process is provided by the intensity of heat release in the combustion process of carbon (soot) and carbon monoxide which are in the mixture. Thus, to describe the process of mixing a two-component gas in this case is sufficient to use the truncated equation derived from Navier-Stokes equations, by dropping the viscous and diffuse members - Euler approach with source members.

On the basis of solution of the problem the algorithm was developed and tested on the example of the flow of harmful gas emissions of incinerators in the inter-blade channels catalyst.

Analysis of the results shows that there is a uniform temperature rise in the direction of gas flow by exothermic neutralization of harmful gases contained in the stream. By the calculation the absolute temperature increment in a grid is obtained and reached 54 °C. The experimentally determined value of raising the temperature of gas during its passage inter-blade channel is 57.5 °C. Thus, a satisfactory agreement between the calculated and experimental research results is obtained.

CALS-based computer-aided support in the chemical industry

A. Bessarabov, T. Stepanova, G. Zaremba, E. Poluboiarinova
R&D Centre "Fine Chemicals", Krasnobogatyrskaya st. 42, 107564 Moscow, Russian Federation; tel. +7 495 9835888, e-mail: bessarabov@nc-mtc.ru

Computer-aided development of specifications, one of the key parts of a product documentation package, requires the use of the state-of-the-art information- management systems. The most advanced computer-aided support system for this is the CALS-technology (Continuous Acquisition and Life cycle Support). This is unified complex based on creation of an integrated information environment supporting all of stages of the production lifecycle. They include stages of marketing studies, engineering, process design, manufacturing, aftersales service and product use, utilization.

Following are some quantitative estimates of effectiveness of CALS implementation in the USA production sector: saving of direct design expenses 10–30%, saving of product development time 40–60%, time saving for introduction of products to the market 25–75%, reduction in defectives and in number of engineering changes 23–73%, saving of expenses for preparation of engineering documentation up to 40%, and saving of expenses for preparation of maintenance documentation up to 30%.

Currently, most Russian enterprises do not realize the importance of implementing, as soon as possible, CALS technologies in industrial production. Should the introduction of CALS in Russian industry be unreasonably delayed, the country may lose external market of knowledge-intensive products and have difficulties in partaking in industrial cooperation.

However, CALS technologies are being implemented actively and successfully mostly at military-industrial enterprises. In chemicals sector, the use of the CALS initiative is described primarily in publications by researchers from the FSUE IREA and R&D Centre "Fine Chemicals" (RDC FC). Over the last 15 years, the two establishments solved a number of most topical problems of the chemicals sector using CALS-specific program complex.

In the nanotechnology realm, the problem of recovery of ultra pure silicon and the assortment of oxide nanopowders in a plasma-chemical reactor under cryogenic quenching was considered. Within the framework of the information-driven CALS project, a master scheme (application protocol), Reference Data for Design (RDFD), for the engineering of an apparatus for the production of nanomaterials was created (Bessarabov, 2014).

The production of high purity materials is one of the priority research areas at IREA and RDC FC. The development of information-driven CALS projects was undertaken along several lines. A set of studies was conducted concerning the engineering of industrial production of highly purified mineral acids. To solve the problem, we devised a CALS project for the production of highly purified mineral acids using RDFD protocol together with the computer-aided quality management CALS system to analytically monitor the range of acids.

Another line of our research in the realm of preparation of highly purified chemical substances is flexible technologies for the preparation of aliphatic hydrocarbons and petroleum spirits of reactive qualification. To address the problem, we also created database patterns for issuing reference data for design and used the CALS project with all the 17 sections of RDFD and the relevant documentation entered in it.

In the area of creating new light-weight heat-protection materials (HPMs), we developed a research-and-information complex consisting of CALS projects targeted at analyzing raw materials, information and patent sources. For instance, in the CALS project analyzing patent sources, patents were arranged into countries-of-patentee groups and according to duration.

Within the framework of organic synthesis, we carried out two work packages. The first one was related to the development of CALS projects in module technology for preparation of a range of ionic liquids (ILs). To study the mechanism of synthesis of ILs containing 1,3-dialkylimidazolium cation, which we use for the synthesis of a range of promising materials, a combined three-stage laboratory-scale plant was devised. The plan includes 6 main process stages. Basing on experimental research, a pilot plant for the synthesis of ILs was developed and CALS project was created for a temporary process guide for preparation of a range of ILs.

Scientific studies under consideration at the second work package were related to the preparation of phosphonoacetic acid (PAA) derivatives. The method for the preparation of 3-methyl PAA ester is based on reaction of phase-transfer PH alkylation in organic phase/solid alkali metal carbonate system. Theoretical and experimental research allowed us to develop an information CALS system structured along three lines: Catalytic Synthesis, Separation of the Target Product, and Product Quality Assessment. Data entered in the subcategories of the system reflect the most important properties of the chemicals and processes.

In the Ecology focus area, a range of works was completed on example of phosphorus-containing waste treatment in Russia and Kazakhstan. An effective method of industrial waste management is to use it for production of road-building material (asphalt concrete). However, because phosphorous slag and phosphogypsum pose health and environmental risks, their compositions and radioactivity levels have to be studied. With this end in view, we entered in the CALS system analytical research data for rare and heavy metals, radioactive elements, etc.

Alongside the issue of industrial phosphorus-containing waste management, we dealt with CALS projects involving the environmental impact of deicing agents (DAs). To handle that problem, we developed a project with all the necessary information entered in it: list of environmental objects subject to DA impact (topsoil, water bodies, vegetation, atmosphere), classes and names of substances occurring is DAs, and quality parameter that were used in the environmental impact assessment. The developed CALS system contains all necessary data on the assessment method, instrument, and output documentation for every quality parameter.

Basing on the ISO 10303 STEP standard, a master scheme (application protocol) for specifications on commercial chemicals was previously developed (Lobanova et al., 2015). In the development of the master scheme, the information structure was based on the interstate standard GOST 2.114-95 prescribing the general rules for the design, description, formatting, coordination, and approval of product specifications. The standard database structure consists of an introductory statement and 8 main upper-level information categories: specifications, safety requirements, environmental monitoring requirements, acceptance rules etc.

Category 5: Testing Methods lays down the techniques, methods and modes of control (tests, measurements, and analyses) of parameters, norms, and characteristic of and requirements to products that should be monitored under the provisions of Category 4: Acceptance Rules. For every control method the following should be established, depending on the particular character of the method: sampling techniques, equipment, materials and chemical reagents, etc.; mode of preparation for the control, method of control, and data processing. All these blocks are subcategories of the CALS system.

References

Bessarabov A (2014) Development and implementation of methods of cybernetics in technologies of chemical reagents and high purity substances. Chemical Engineering Transactions. 39:1003-1008. doi: 10.3303/CET1439168

Lobanova A V, Stepanova T I, Bulatitskii K K, Bessarabov A M (2015) CALS-based Systemic Metrological Studies. Russian Journal of General Chemistry 85(10):2431–2440. doi: 10.1134/S1070363215100382

CO_2 adsorption on polybenzoxazine grafted activated carbon: Effects of amine precursor to synthesize benzoxazine monomer

[1]K. Inkong, [1,2,*]P. Rangsunvigit, [1,2]T. Chaisuwan [3]S. Kulprathipanja
[1]The Petroleum and Petrochemical College, Chulalongkorn University, 254 Soi Chulalongkorn 12, Phayathai Rd., Pathumwan, Bangkok 10330, Thailand; tel. +6622184154, fax +6622154459, *e-mail: Pramoch.r@chula.ac.th; [2]Center of Excellence on Petrochemical and Materials Technology, 7th floor, Chulalongkorn University Research Building, Soi Chulalongkorn 12, Phayathai Rd., Bangkok 10330, Thailand; tel. + 662 218 4148; [3] UOP, A Honeywell Company, Des Plaines, Illinois, U.S.A. 60017; tel. +2245958337, e-mail: Santi.Kulprathipanja@honeywell.com

Carbon dioxide (CO_2) is the major greenhouse gas that contributes to global warming phenomenon. The adsorption process is one of interesting processes to reduce CO_2 because of low energy consumption, non-corrosive problems, and easy regeneration. The amine-functionalized adsorbents have been used to overcome the limitation of classic adsorbents in the CO_2 adsorption field by synergistic effects of physical and chemical adsorption. Polybenzoxazine (PBZ) is a very attractive polymer for the CO_2 adsorption due to more amine functional group. In this work, activated carbon (AC) was functionalized with polybenzoxazine via ring-opening polymerization of benzoxazine monomer synthesized from different amine precursors was investigated for CO_2 adsorption. Benzoxazine monomer (BZ) was synthesized from phenol, paraformaldehyde, and hexamethylenediamie (HMDA) or triethylenetetramine (TETA). The PBZ loading was varied by different benzoxazine monomer solution concentrations from 0.1 to 0.5 g/L. CO_2 adsorption isotherms were investigated at 35°C, 50°C, and 75°C. Adsorbents were characterized by TG-DTA, FT-IR, SEM, and surface area and pore size analyses. The amounts of PBZ grafted on activated carbon were determined by UV-Vis spectroscopy. The results showed that the PBZ grafted activated carbon improved the CO_2 adsorption capacity due to the synergistic effects between physical and chemical adsorption. When the adsorption temperature increased, the CO_2 adsorption capacity of the grafted adsorbent decreased because the physisorption was more dominant than the chemisorption. The decrease in the capacity due to the temperature increase was more pronounced with the unmodified activated carbon. Using TETA resulted in high CO_2 adsorption than using HMDA. It was likely that using TETA to synthesize PBZ containing secondary amine and tertiary amine, whereas only tertiary amine was found in PBZ synthesized by HMDA. The secondary amine can directly react with CO_2, but the tertiary amine has only van der Waals force to attack CO_2. CO_2 adsorbed onto PBZ/AC was completely desorbed by heating at 120°C for 24 hr. There was no significant change in the CO_2 adsorption capacity of the regenerated adsorbents.

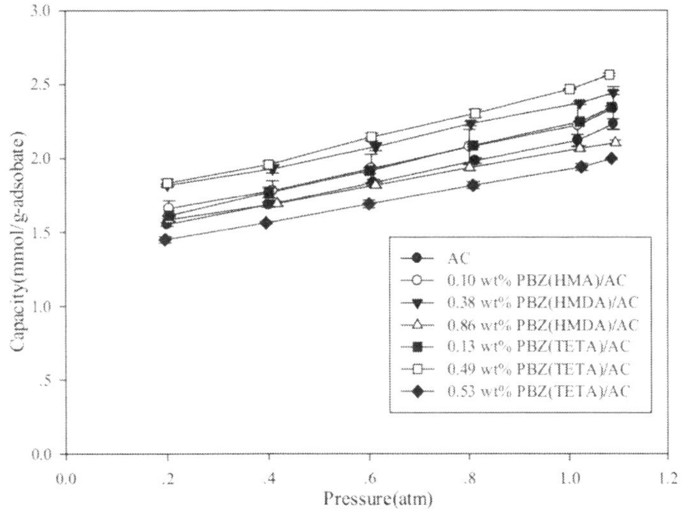

Figure 1 Adsorption isotherms of AC and modified AC at 35°C

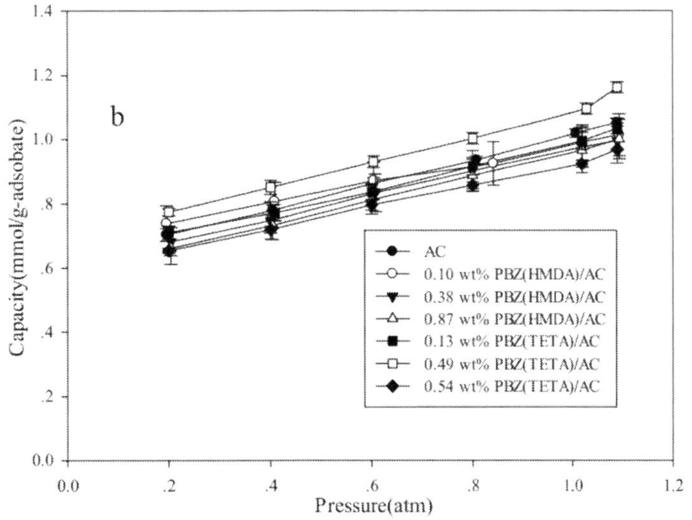

Figure 2 Adsorption isotherms of AC and modified AC at 75°C

Reference

1. Cox, P.M, Betts, R.A., Jones, C.D., Spall, S.A., Totterdell, I.J., Acceleration of global warming due to carbon-cycle feedbacks in a coupled climate model, Nature, 2000, 408, 184-187.

2. Ko, Y.G., Choi, U.S., Shin, S.S., Primary, secondary, and tertiary amines for CO_2 capture: Designing for mesoporous CO_2 adsorbents, Journal of Colloid and Interface Science. 2011, 362, 594-602.

3. Xu, X., Song, C., Andresen, J.M., Miller, B.G., Scaroni, A.W., Preparation and characterization of novel CO2 "molecular basket" adsorbents based on polymer-modified mesoporous molecular sieve MCM-41, Microporous and Mesoporous Materials. 2003, 62, 29-45.

4. Ishida, H., Agag, T., Handbook of Benzoxazine Resins, Elsevier, New York, 2011.

Competitive adsorption of methane and carbon dioxide on different activated carbons

[1]K. Poomisitiporn, [1,2*]P. Rungsunvigit, [1,2]B. Kitiyanan, [3]S. Kulprathipanja
[1] The Petroleum and Petrochemical College, Chulalongkorn University, 254 Soi Chulalongkorn 12, Phayathai Rd., Pathumwan, Bangkok 10330, Thailand; tel. +662 218 4154, fax +6622154459, *e-mail: Pramoch.r@chula.ac.th; [2] Center of Excellence on Petrochemical and Materials Technology, 7th floor, Chulalongkorn University Research Building, Soi Chulalongkorn 12, Phayathai Rd., Bangkok 10330, Thailand; tel. + 662 218 4148 [3] UOP, A Honeywell Company, Des Plaines, Illinois, U.S.A. 60017; tel. +2245958337, e-mail: Santi.Kulprathipanja@honeywell.com

Adsorbed natural gas (ANG) is one of natural gas storage technologies, in which natural gas is adsorbed by a high porosity adsorbent material at relatively low pressure and room temperature with high methane capacity and low cost. Carbon-based materials, like activated carbons, are promising as an ANG storage adsorbent because of high surface area, high porosity, and high volumetric storage capacity. In this research, the adsorption capacity of single and binary component of methane and carbon dioxide was investigated by using activated carbons including coconut shell activated carbon (CSAC), palm shell activated carbon (PSAC), and coal-based activated carbon (CBAC) in a packed bed column at room temperature. The composition ratio of single component of methane and carbon dioxide was fixed at 20 vol% of the total feed gas, and the composition ratios of binary component of methane and carbon dioxide were fixed at 14.6 : 5.4, 10.0 : 10.0, and 5.0 : 15.0 vol% of the total feed gases, respectively. From the result of single component adsorption, it showed that all activated carbons exhibit preferential adsorption for carbon dioxide in relation to methane. In terms of the adsorption capacity of single component, the CSAC showed the highest methane adsorption capacity, 1.49 mmol/g, while the PSAC showed the highest adsorption capacity of carbon dioxide, 5.36 mmol/g. As the result of binary component adsorption, it can be seen that the adsorbed amount increased along with the carbon dioxide composition, while the amount of methane adsorbed decreased, indicating competition for adsorption sites and preferential adsorption of carbon dioxide over methane. In terms of the adsorption capacity of binary component, the methane selectivity is ordered as follows: CSAC > PSAC > CBAC for the whole composition ratios of methane and carbon dioxide. From the overall results, it can be considered that the adsorption capacity and methane selectivity were affected by the physical and chemical properties of activated carbons.

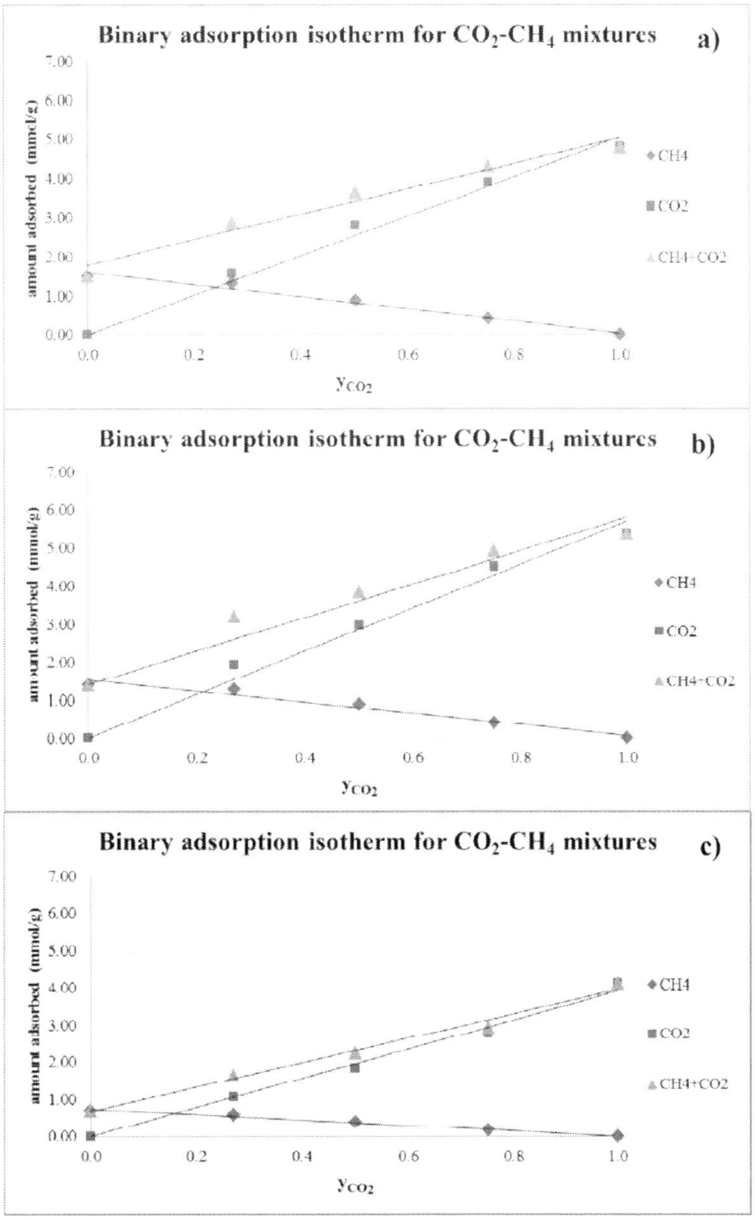

Figure 1 Binary adsorption isotherm for CO_2-CH_4 mixtures at 25 °C and 1 atm of a) CSAC b) PSAC and c) CBAC

Reference

1. Bastos-Neto, M., Canabrava, D.V., Torres, A.E.B., Rodriguez-Castellón, E., Jiménez-López, A., Azevedo, D.C.S. and Cavalcante, C.L. (2007). Effects of textural and surface characteristics of microporous activated carbons on the methane adsorption capacity at high pressures. Applied Surface Science 253(13), 5721-5725.

2. Rios, R.B., Stragliotto, F.M., Peixoto, H.R., Torres, A.E.B., Bastos-Neto, M., Azevedo, D.C.S., and Cavalcante, C.L. (2013). Studies on the adsorption behavior of CO2-CH4 mixtures using activated carbon. Brazilian Journal of Chemical Engineering, 30(4), 939-951.

Characterization of the residue from double alginate extraction from *Sargassum filipendula* seaweed

Camila S. D. Costa, Saulo L. Cardoso, Emily Nishikawa, Melissa G. A. Vieira*,
Meuris G. C. da Silva
State University of Campinas, School of Chemical Engineering, Department of Process and Products Design, Campinas, São Paulo, Brazil
melissagav@feq.unicamp.br

Residue of alginate extraction from *Sargassum filipendula* brown seaweed shows high potential to be applied as bioadsorbent for the removal of heavy metals from contaminated wastewater, since it contains some constituents of the raw algae. Therefore, the waste of the extraction process could be also employed as bioadsorbent, valorizing the residue from alginate extraction and making the metal sorption more economically attractive. In front of this, the alginate extraction was carried out twice to verify if after the double extraction the residue presents the same constituents and the same potential to remove toxic metals. Furthermore, the double extraction was used in order to verify if the alginate extraction yield increases and if it preserves the characteristics of the residue as bioadsorbent. The yield (w/w) of alginate and residue in the first extraction was 24 ± 5.00 % and 52 ± 2.51 %, respectively. With the second extraction, the yield obtained represented 11 ± 1.88 % of alginate and 76 ± 3.84 % of residue. The high amount of residue encourages its use as biosorbent adding value to a waste of the process. The SEM-EDX results indicate that the materials' surface is mainly composed by Si, Na, Mg, Al, S, K, Ca and Fe, being that the presence and reduction of some elements concentration is due the diatoms shells. The spectra obtained from FTIR of first and second residues were similar, indicating that the main functional groups involved in adsorption phenomenon are not totally removed. It suggests that even after extracting more alginate it is possible to use the residue as bioadsorbent.

Determining energy costs for milling solid matter in a ball mills

M. M.Kuznetsova [1],V.E. Ved [1], E.V. Krasnokutskiy [1], M.I. Satayev [2], A.A. Saipov [2], R.F. Altynbekov [2]
[1] National Technical University "Kharkiv Polytechnic Institute", 21 Frunze st., Kharkiv, 61002 Ukraine, e-mail: ved@kpi.kharkov.ua, tel. +38057 7076589; [2] M. Auezov South Kazakhstan State University, 5 avenue Tauke Khan, Shymkent, 160012 Kazakhstan.

In the theory of calculating milling energy costs the most known methods were suggested by Rittinger, Kirpichev and Kick, Bond and Rundquist. But none of them have wide practical application as they take into account only geometrical dimensions of the particles to be ground and the characteristics of the material regardless of the milling method and conditions. All this stands in the way of applying the known milling methods.

We suggest a method for calculating milling energy costs depending on the ratio of the impact and friction forces influence, as well as on the changes in the strength characteristics of the matter during milling.

The effective work used in milling consists of two components: work of impact (A_p) and work of attrition (A_{mp}).

The milling by attrition in a ball mill (according to Taggart) takes about 1.5-2% of the energy, respective of the part of the mass (volume) of the material to be ground. We suggest introducing an energy parameter p that characterizes a share of energy of milling by impact. Consequently the attrition energy share is (1-p). Let's assume that the milling stage – from the initial size of d to a size d_κ, can be carried out in n stages. Then the energy required to fracture a particle of the material equals:

$$A_p = \frac{\sigma_{np}^2 k d^3}{2E} \sum_{i=0}^{n} p^i , \tag{1}$$

where d – the size of the particles (a diameter of a circumscribed sphere of a polyhedron, which is the form that the particles tend to shape), k – coefficient of form, E – elastic modulus (Yung's modulus), Pa, σ_{np} – practical value of ultimate strength of the ground matter particles, Pa.

The work of the attrition force to grind (1-p) a portion of the matter for particles with the initial size of d equals:

$$A_{mp} = \frac{\sigma_m^2 k d^3}{2E} \left(1 - p^n\right), \tag{2}$$

where σ_m – practical value of ultimate strength of the ground matter particles through compression for particles that is determined by Orowan-Kelly equation, Pa.

In order to establish the energy needed for milling in relation to the geometrical size of the particles to be ground, we have to determine the number of stages in each part of this process: The maximum milled extent of the matter is u that equals the ratio of the initial size of the particle to the final size d_k. With each stage every particle will split into a number of lumps. Hence we can claim that

$$a^n = \frac{d^3}{d_\kappa^3} = u^3 . \tag{3}$$

With the maximum value of a=2, we will now determine the number of stages necessary to get the necessary size of the particles that are milled:

$$n = \frac{3\lg u}{\lg 2} \approx 10\lg\frac{d}{d_K}. \quad (4)$$

Thus the dependence of work needed for milling the mass of a matter m_M by the combination of impact and attrition on the initial and the final size of the particles, characteristics of the matter can be determined by the following expression:

$$A = \frac{m_M}{2E\rho}\left(\sigma_{np}^2 \sum_{i=0}^{10\lg\frac{d}{d_K}} p^i + \sigma_T^2\left(1 - p^{10\lg\frac{d}{d_K}}\right)\right). \quad (5)$$

For the first time there has been established the dependence of milling energy costs on the characteristics of the matter to be ground and the milling method in the form of an equation (5). The research also suggests a parameter p that allows setting the relationship of the influence of various milling methods on the milling process. It allowed comparing the energy needed for milling in relation to the shares of impact and attrition fracturing for processes with different dispersion level of the initial matter.

Development of the frequency PID tuning method with filter

K. A. Kozin, A. M. Egorova, S. S.Mikhalevich, [1]F. Manenti, [1]F. Rossi
National Research Tomsk Polytechnic University, Institute of Physics and Technology, Department of Electronics and Automation of Physical Plants, Tomsk 634050, Russia, Lenin st. 2; tel. +79609797171, e-mail: anopt@ya.ru; [1]Politecnico di Milano, Dept. di Chimica, Materiali e Ingegneria Chimica "Giulio Natta", Piazza Leonardo da Vinci 32, 20133 Milano, Italy.

One of the actual research trends of control theory are development of PID tuning methods. It is due to wide spread PID controllers in engineering practice. At the present time there are a lot of different methods that are based on desired time-domain or frequency characteristics. But most of them developed for ideal PID controller, despite of all controllers contain filter for derivative part of PID controller in modern engineering practice. It means, that developed methods for ideal controllers can't guarantee desired characteristics in a real control system (Figure 1). At this paper, authors used the Mikhalevich et al. PID tuning method [1] that is provided desired phase margin and crossover frequency. The aim of research is to find proportional, integral, derivative and filter coefficients of PID controller that will provide desired characteristics (phase margin and crossover frequency) in the system and present the way of applying proposed method in digital control system.

The paper consists of introduction, theoretical research, comparison study of proposed method with known, description of line of centrifugal extractors and development of control system.

The criterions of desired control system that are based on previous research and Aström recommendations [2] are shown in the theoretical part. Then, the analytical approach of proposed PID tuning method is presented. After, comparison proposed method with conventional was done.

Practical approach contains developing the control system of extraction process [3]. Block of centrifugal extractors, as a plant, is used. This block contains 12 centrifugal extractors which are connected in serial line and used for extraction uranium and plutonium. This equipment is a part of Russian Federation large project – closed-loop nuclear fuel cycle. The plant was linearized and digital control system was developed.

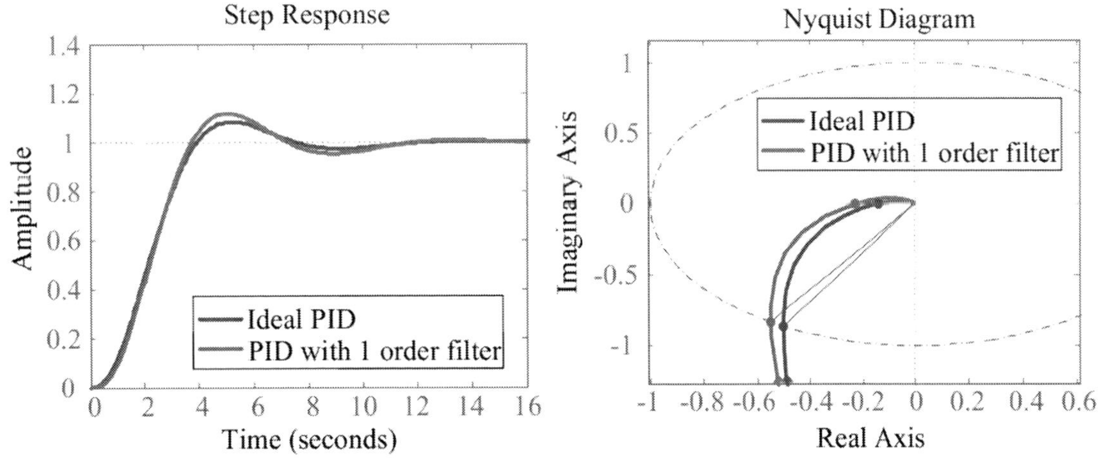

Figure 1 – Comparison frequency and time domain characteristics

References

1. Mikhalevich, S. S., Baydali, S. A., Manenti, F., 2014, Development of a tunable method for PID controller to achieve the desired phase margin, Journal of Process Control, 25, 28-34, DOI: 10.1016/j.jprocont.2014.10.009.

2. Aström, K. J., Hägglund, T., 2005, Advanced PID Control in Research Triangle Park, NC 27709, ISA - Instrumentation, Systems, and Automation Society.

3. Goryunov, A. G., Mikhaylov, V. S., 2012, The automatic control system of a multi-component nonequilibrium extraction process in the pulse column, Journal of Process Control, 22, 1034-1043.

Dimensionless analysis of heat, momentum and mass transfer in a pool fire

L. G. Zárate[a], M. E. Cordero[a], B. Kozanoglu[b] , Alejandro Regalado-Méndez[c]
[a]Ingeniería Química, Universidad Popular Autónoma de Puebla, C.P. 72410. México; tel. + 52 222 229 94 00. [b]Universidad de las Américas-Puebla, Ex hacienda Santa Catarina Martir, 72820. México. [c]Universidad del Mar, Ciudad Universitaria S/N, Puerto Ángel, Oaxaca, 70902, México
*e-mail: luis.zarate@upaep.mx

Statistical analysis about major accidents indicate that fires in the Chemical Industry are the accidents that cause the highest losses and hydrocarbons are the materials to be ignited the first [1]. In this paper, the phenomenon denominated as hydrocarbon pool fire is studied. In this type of fires, the convective mass, momentum and heat transfer are presented simultaneously. In Figure 1, zones with different characteristics are observed and it is evident the presence of concentration, temperature as well as velocity gradients. In this type of scenarios, there are numerous dimensionless numbers which can be used to describe the phenomena [2, 3, 4]. These dimensionless groups provide the advantage of being able to compare behaviors at different scales, where space and time variations are generally present. As a result these dimensionless groups can be used in scaling procedures [4]. In this paper simulations were carried out applying the CFD techniques to the pool fires of different diameters. Figure 2 displays the temperature distribution over the flame in 1.6 s (in K). The results are validated comparing them with previously published experimental data [5]. For the simulation, different models of momentum, heat and mass transfer were coupled with the chemical kinetics, for hydrocarbon fuel combustion, in transient turbulent regime.

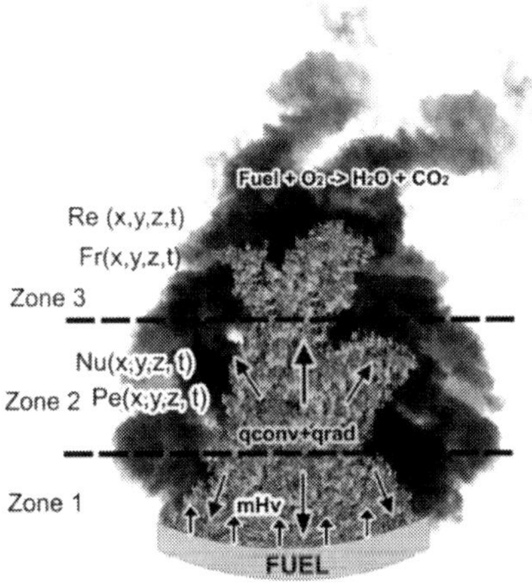

Figure 1. Representation of regions of a pool fire

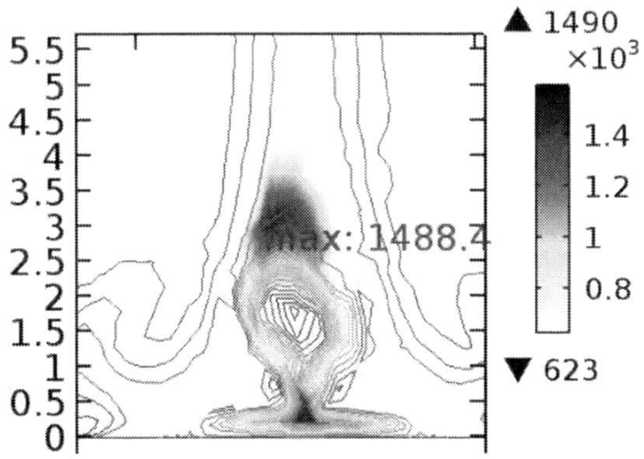

Figure 2. Simulation by CFD of a pool Fire in time 1.6 s.

Dimensionless numbers such as Reynolds, Peclet, Froude and Nusselt, among others, help to describe the characteristic behavior of the flames (geometry, energy transport, kinetic, composition). In the Figures 3 and 4 the behavior of Reynolds and Froude numbers are shown, respectively. In these figures, a sharp increase is observed with Reynolds and Froude numbers at an elevation between 2.5 and 3.0 m, over the curve of 1.6 s. This deflection point corresponds in Figure 2, exactly where the maximum temperature is observed.

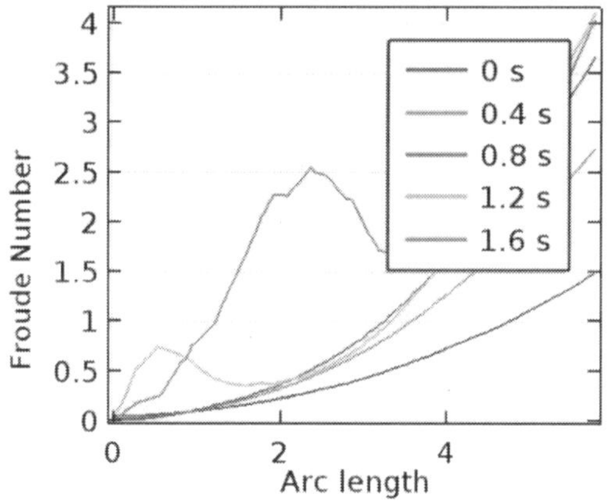

Figure 3. Reynolds number (-) along the central axis (m).

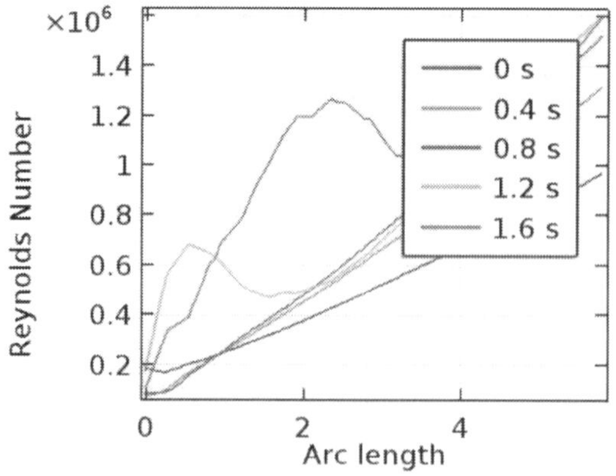

Figure 4. Froude number (-) along the central axis (m).

Figure 5 displays the reaction rate of the fuel in mol/m^{3}s, 1.6 s after the combustion started. Figure 6 shows the dimensionless temperature distribution along the central axis.

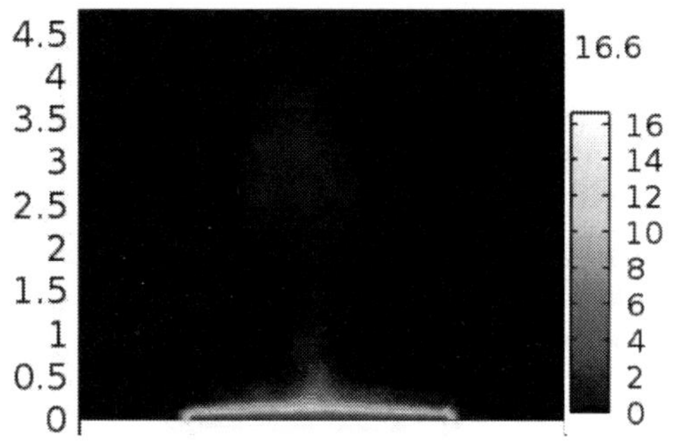

Figure 5. Reaction rate of combustion of a hydrocarbon at time 1.6 s.

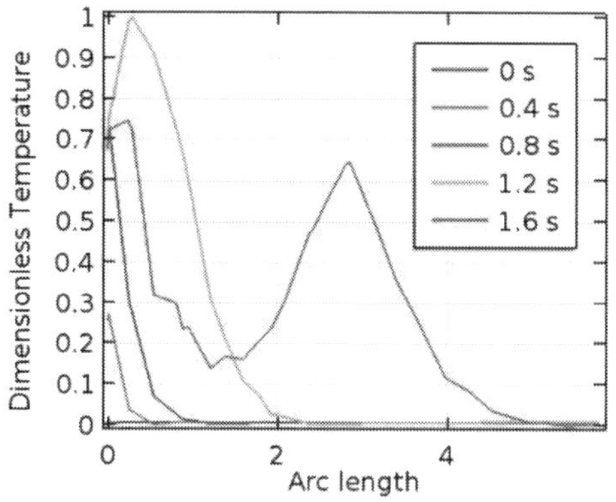

Figure 6. Dimensionless temperature along the central axis in a pool fire.

As it can be seen, at the locations where maximum values or inflection points of the dimensionless numbers are observed, the transfer phenomena occur at a higher rate (mixing, reaction and energy liberation) as happens at 1.6 s, at the elevation between 2.5 and 3.0 m. Simulation through CFD allows to analyze this kind of phenomena at different scales.

References

[1] Lee's loss prevention in the process industries, 2005, 3rd Edition. Edited by Sam Mannan, Oxford, United Kingdom

[2] Dušica J. Pešić, Radoslav H. Kartov, Darko N. Zigar, Valentin Chochev: The impact of wind inertial and fire buoyancy forces on air pollution in street canyon, Facta Universitatis Series: Working and Living Environmental Protection, Vol 11, No2, 2014, Niš

[3] de Ris J. L. Mechanism of buoyant turbulent diffusion flames. Procedia Engineering 62-2013, 13 – 27.

[4] Yang Ch., Thovert J., Debenest. Upscaling of mass and thermal transports in porous media with heterogeneous combustion reactions. International Journal of Heat and Mass Transfer. Volume 84, May 2015, Pages 862–875.

[5] Chatris, J. M., J. Quintela, J. Folch, E. Planas, J. Arnaldos, y J. Casal, "Experimental Study of Burning Rate in Hydrocarbon Pool Fires", Combust. Flame, 126(1-2), 1373-1383, 2001.

Effects of mixed surfactants on methane hydrate formation and dissociation

[1]K. Inkong, [1,*]P. Rangsunvigit, [2]S. Kulprathipanja
[1] The Petroleum and Petrochemical College, Chulalongkorn University, 254 Soi Chulalongkorn 12, Phayathai Rd., Pathumwan, Bangkok 10330, Thailand; tel. +6622184154, fax +6622154459, *e-mail: Pramoch.r@chula.ac.th; [2] UOP, A Honeywell Company, Des Plaines, Illinois, U.S.A. 60017; tel. +2245958337, e-mail: Santi.Kulprathipanja@honeywell.com

Natural gas hydrates are ice-like crystalline compounds and non-stoichiometric compounds formed by hydrogen bonds from water molecules when gaseous molecules such as methane, ethane, propane, nitrogen, and carbon dioxide are trapped inside. Gas hydrates have received much attention not only as a new natural energy resource but also as a new means of natural gas storage and transportation. The ability to store a large amount of natural gas in the hydrate form makes the hydrate technology a good candidate for gas storage. However, slow formation rate and stability of natural gas hydrate have been considered problems hindering the industrial applications. In this work, hydrate promoters including methyl ester sulfonate (MES), sodium dodecyl sulfate (SDS), and mixed MES and SDS were investigated for hydrate formation and dissociation. The formation experiments were conducted in a batch reactor at desired experimental conditions. The dissociation experiment was conducted after the formation with the driving force of 25°C. The results showed that all promoters enhanced the rate of methane formation compared with pure water due to the decrease in the interfacial tension of the gas-liquid phase, resulting in the reduction of diffusion resistance between two phases and the increase in the gas solubility. The temperature profile and the rate of methane hydrate formation of each promoter were different. At the same concentration, the addition of SDS resulted in the faster rate of methane hydrate formation than MES but the gas consumption was higher with MES. Mixed MES – SDS showed the synergetic effects on the rate methane hydrate formation.

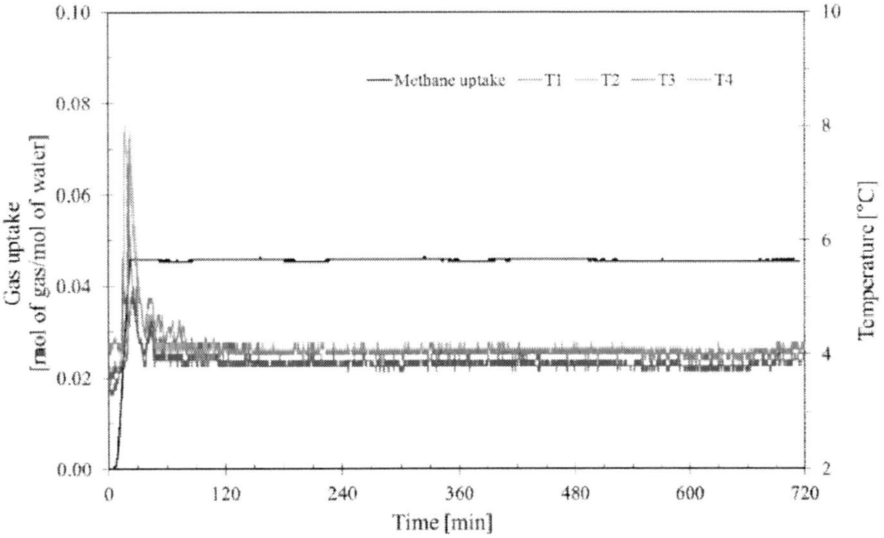

Figure 1 Gas uptake and temperature profiles during the methane hydrate formation of 4 mM SDS at 8 MPa and 4°C

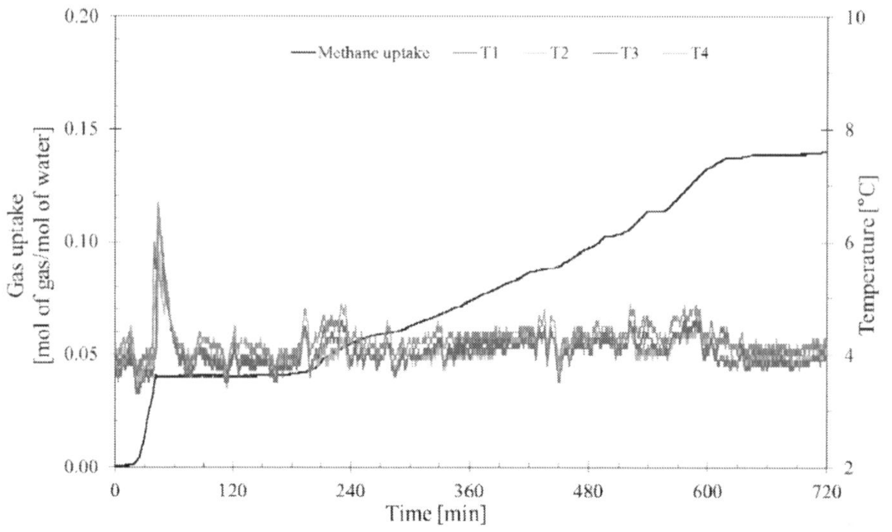

Figure 2 Gas uptake and temperature profiles during the methane hydrate formation of 2 mM MES at 8 MPa and 4°C

Reference

1. Sloan, E.D., and Koh, C.A., Clathrate hydrates of natural gases, 3rd ed., Boca Raton: CRC Press., 2008.
2. Lin, W., Chen, G.L., Sun, C.Y., Guo, X.Q., Wu, Z.K., Liang, M.Y., Chen, L.T., and Yang, L.Y. Effect of surfactant on the formation and dissociation kinetic behavior of methane hydrate. Chemical Engineering Science, 2004, 59, 4449-4455.
3. Ganji, H., Manteghian, M., Zadeh, K.S., Omidkhah, M.R., and Mofrad, H.R., Effect of different surfactants on methane hydrate formation rate, stability and storage capacity, Fuel, 2007, 86, 434-441.
4. Okutani, K., Kuwabara, Y, Mori, Y.H., Surfactant effect on hydrate formation in an unstrirred gas/liquid system: An experiment study using methane and sodium alkyl sulfates, Chemical Engineering Science, 2008, 63, 183-194.

Effects of tetrahydrofuran (THF) and cetyltrimethylammonium bromide (CTAB) on carbon dioxide hydrate formation

[1]S. Limvisitsakul, [1,*]P. Rangsunvigit, [2,**]S. Kulprathipanja, [1]A. Siangsai, [1]K. Inkong
[1]The Petroleum and Petrochemical College 254 Chula Soi 12 Wangmai Pathumwan Bangkok 10330, Thailand; tel. (+66)2 2184110, fax: (+66)2 2154459, *e-mail: Pramoch.r@chula.ac.th
[2]UOP, A Honeywell Company 25 E. Algonquin Rd. Des Plaines, IL 60017; tel. 224-595-8337, **e-mail: Santi.Kulprathipanja@honeywell.com

Fossil fuel is the main resource of the world energy. However, combustion of fossil fuel produces carbon dioxide (CO_2) emission. Carbon capture and storage (CCS) technology has been considered to solve this problem. To increase efficiency of CCS technology, gas hydrates have been investigated as an alternative to Store CO_2.

Gas hydrates are crystalline solids formed from mixtures of water and light gases. Gas hydrate occurs under high pressure and low temperature conditions. However, the utilization of hydrate is restricted because of some disadvantages such as low gas storage density, low formation rate, and long induction time of gas hydrate formation. It has been reported that adding a surfactant can promote gas storage capacity and hydrate formation rate. A surfactant added to natural gas hydrate system can reduce gas-water interfacial tension, which can decrease the inter phase diffusion resistance and increase solubility of gas molecules in solution.

In this research study, carbon dioxide hydrate formation in the presence of different tetrahydrofuran (THF) and cetyltrimethylammonium bromide (CTAB) concentrations was investigated in terms of kinetic and thermodynamic. The experiment was conducted in the quiescent close system. Figure 1 shows the formation of CO_2 hydrate in the presence of 10 mol% THF at 3 MPa and 4°C. The temperature from all thermocouples located at different positions in the crystallizer rises abruptly referring to the hydrate formation and continues to grow until reaches the limit of hydrate formation. In the presence of 10 mol% THF, the gas uptake increases rapidly because of gas hydrate formation. Table 1 shows the experimental conditions along with measured induction time and conversion to hydrate. The results of CO_2 hydrate formation rate in the presence of THF show that both THF and the surfactant can promote the CO_2 hydrate formation rate compared with the pure water system.

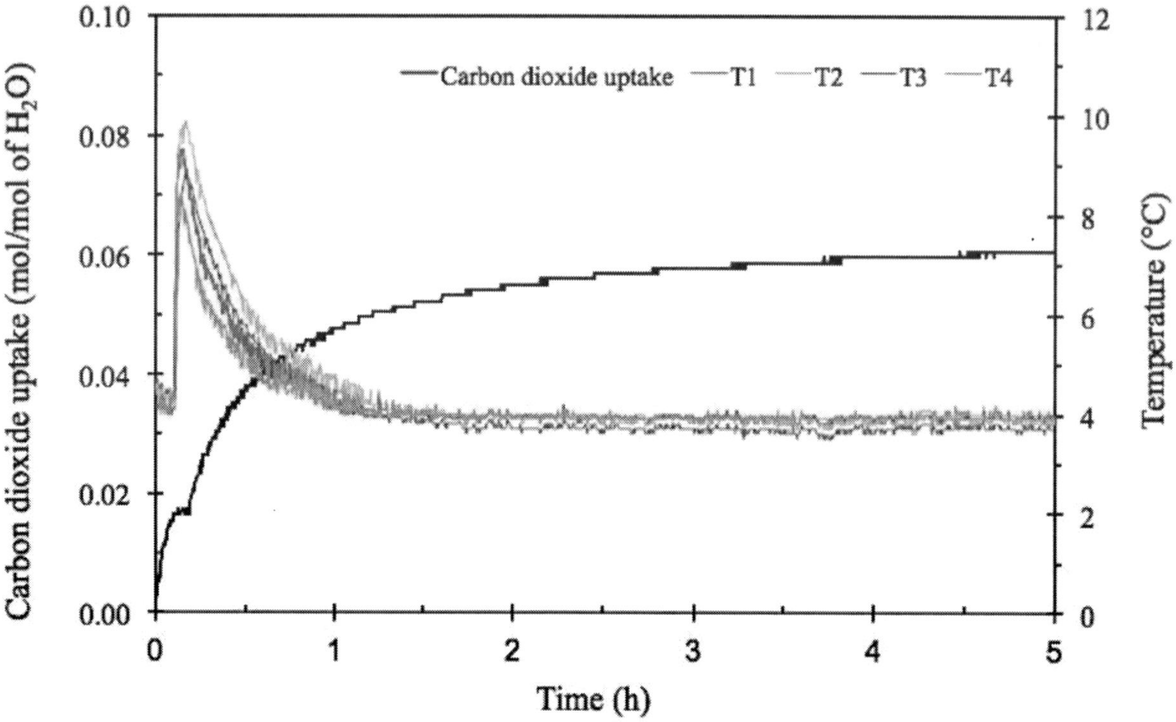

Figure1. Formation of CO_2 hydrate at 3 MPa and 4°C in the presence of 10 mol% THF.

Table1 Experimental conditions along with measured induction time and conversion to hydrate

System	Exp. No.	P(Mpa)	T(K)	%Conversion	Induction time (min)
10 mol% THF	1	3	277.15	53.00%	8.00
	2	3	277.15	42.45%	10.00
	3	3	277.15	40.00%	10.00

Reference:
Demirbas, A. (2010). Methane hydrates as potential energy resource: Part 1–Importance, resource and recovery facilities. Energy Conversion and Management, 51(7), 1547-1561.
Hao, W., Wang, J., and Fan, S. (2008). Evaluation and analysis method for natural gas hydrate storage and transportation processes. Energy Conversion and Management, 49(10), 2546-2553.
Sun, Q., & Kang, Y. T. (2015). Experimental correlation for the formation rate of CO 2 hydrate with THF (tetrahydrofuran) for cooling application. Energy, 91, 712-719.

Effects of tetra-n-butylammonium bromide (TBAB) on methane hydrate formation and dissociation

[1]W. Chanakro, [1,*]P. Rangsunvigit, [2**]S. Kulprathipanja, [1]A. Siangsai, [1]K. Inkong
[1] The Petroleum and Petrochemical College, Chulalongkorn University, 254 Soi Chulalongkorn 12, Phayathai Rd., Pathumwan, Bangkok 10330, Thailand; tel. +662 218 4154, fax +6622154459, *e-mail: Pramoch.r@chula.ac.th; [2]UOP, A Honeywell Company, Des Plaines, Illinois, U.S.A. 60017; tel. +2245958337, **e-mail: Santi.Kulprathipanja@honeywell.com

Methane hydrates have received much attention not only as an alternate fuel energy resource but also as a new means of natural gas storage and transportation. The ability to store large quantities of natural gas in the hydrate form makes it a good alternate for the storage and safe transport of natural gas. However, there are many problems in developing this technology. Hydrate formation is a slow process and needs high formation pressure of methane hydrate. Promoters were discovered to enhance the hydrate formation rate and decrease hydrate formation pressure. Tetra-n-butyl ammonium bromide (TBAB) is a thermodynamic promoter, which can form semiclathrate structure at a much lower pressure and high temperature. In TBAB hydrate crystals, a part of the cage structure is broken to encage the large TBAB molecule. This characteristic of the semiclathrate hydrates leads them to have more gas storage capacity than other thermodynamic promoters. In this work, the effects of TBAB on the thermodynamic and kinetic of methane hydrate formation and dissociation were investigated and compared with pure water system. The experiments of forming TBAB semiclathrate hydrate were performed in batch operation under the desired experimental conditions. The dissociation experiments were performed after the formation with the driving force of 25°C. The results indicated that TBAB enhanced the rate of methane formation and increased the gas consumption compared with pure water system.

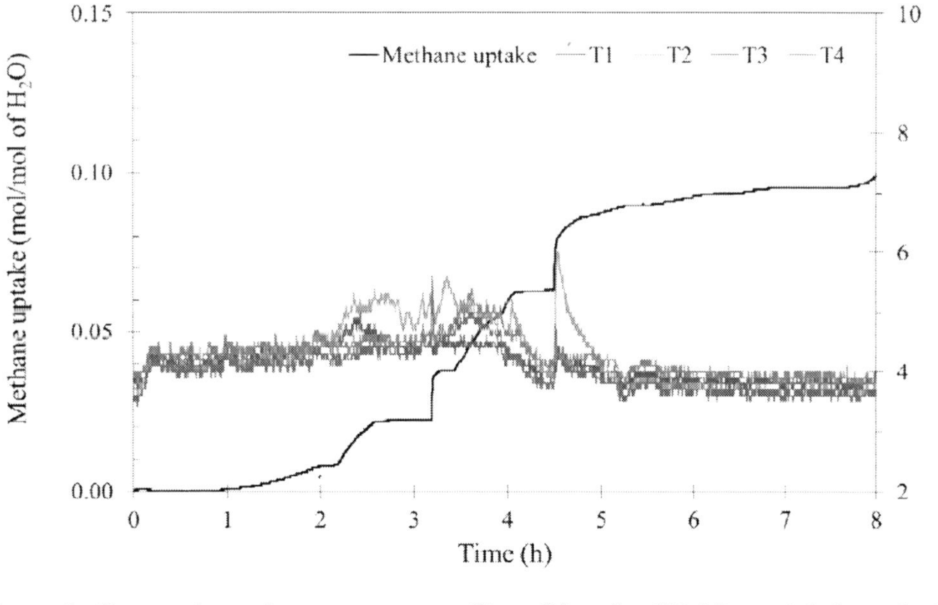

Figure 1. Gas uptake and temperature profiles of forming TBAB semiclathrate hydrate at 4 °C and 0.29 mol% TBAB concentration in batch operation (initial pressure 8 MPa).

References:
1. Demirbas, A. (2010). Methane hydrates as potential energy resource : Part 1 – Importance, resource and recovery facilities. Energy Conversion and Management, 51(10), 1547-1561.
2. Yang, M., Jing, W., Wang, P., Jiang, L., and Song, Y. (2015). Effects of an additive mixture (THF+TBAB) on CO_2 hydrate phase equilibrium. Fluid Phase Equilibria, 401, 27-33.
3. Zhong, D., and Englezos, P. (2012). Methane separation from coal mine methane gas by tetra-n-butyl ammonium bromide semiclathrate hydrate formation. Energy and Fuels, 26(4), 2098-2106.
4. Mohammadi, A.H., Eslamimanesh, A., Belandria, V., and Richon, D. (2011). Phase equilibria of semiclathrate hydrates of CO_2, N_2, CH_4, or H_2+ tetra-n-butylammonium bromide aqueous solution. Journal of Chemical and Engineering Data, 56(10), 3855-3865.

Environment and water treatment

N.Ben Tahar, L.Bentahar G.Otmanine,A.Saoud and H.Mimoun
Faculty of Science
University M'hamed Bougara Boumerdes Algeria
nbentahardz@yahoo.fr

It is often said that water is not necessary to life ... "Water is simply life"
Water is also critical in manufacturing processes in any industry.
Water use in a production unit has been greatly reduced by cutting waste and by technological advances. The restrictions go beyond just saving water they require for recycling in the circuits and fire service and water for extra cooling.
It is from these waters is that life emerged. As as living bodies were transformed into structures more complex and sophisticated, they left the marine environment to reach the earth, drawing water most of their physical constitution. On the ground water that's life. Water is also critical in manufacturing processes in any industry.
Water use in a production unit has been greatly reduced by cutting waste and by technological advances. The restrictions go beyond just saving water they require for recycling in the circuits and fire service and water for extra cooling. In most processes, the water can act either as reactive as a solvent or as an agent used to carry cold or heat. In each specific case of use it is essential to know the water quality. Indeed the natural water that is to say the water in the rough may contain harmful substances which could have adverse consequences on equipment or chemical reactions. So we must recognize all these substances to adjust their content in the water before assigning it to a particular purpose. The legislation also imposes constraints on the content of water discharged into carbon compounds and suspended solids. There are also standard on nitrogen excretion. Subsequently, the water having participated in any manufacturing process must be either recycled or released into the wild. But this water may contain pollutants, so harmful, which may render it unusable without prior treatment or dangerous for the environment. Water should be purified as for recycling for the waste.It is in this context that we have considered the treatment of industrial waste water. We have prepared four types of adsorbents that we used in a laboratory facility for the treatment of waste water from the cheese Boudouaou. Les results are satisfactory.

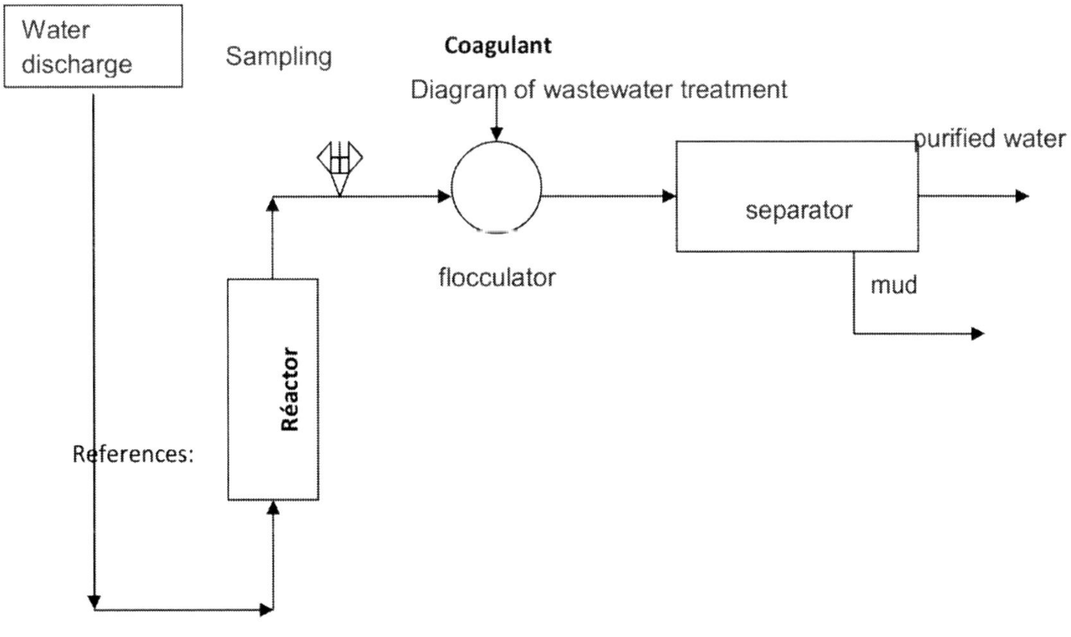

Diagram of wastewater treatment

References:

1/ P.WUITHIER « Raffinage et Génie chimique »Tome II Edition Technip Paris 1972
2/N.Fank KEMMER « Manuel de l'eau » Edition Technique et Documentation 1984
3/J.B ARMANI « Traitement des eaux résiduaires urbaines et industrielles » Paris 1992
4/Mémoire sur le traitement des eaux de Rhoude Nouss, TABTI Hamza FHC 2008

Equilibrium study of cadmium ions adsorption on sericin/alginate particles

Jacqueline O. Lima[a], Mariana F. Ragassi[a], Marcelino L. Gimenes[b], Melissa G. A. Vieira[a], Meuris G. C. da Silva[a*]
[a]School of Chemical Engineering, University of Campinas, UNICAMP, 13083-825, Campinas – SP, Brazil
[b]Department of Chemical Engineering, State University of Maringá, UEM, 87020-900, Maringá – PR, Brazil
meuris@feq.unicamp.br

Cadmium is a pollutant widely found in industrial effluents and its toxicity affects the ecosystem and presents human health risk. The adsorption can be used to remove toxic metals in dilute solutions. Industrial waste, such as sericin, can efficiently replace conventional bioadsorbent materials. Sericin is a protein present in silkworm cocoon and is usually discarded in effluent of spinning processes. Alginate is a polysaccharide used to improve the characteristics of the blend. The objective of this study was to investigate the use of sericin/alginate particles as bioadsorbent to remove cadmium ions (Cd^{2+}) in waste water through adsorption equilibrium study. Equilibrium experiments were conducted to evaluate the influence of adsorption temperature (10, 20, 40 and 60°C), which indicated that the cadmium adsorption capacity increased with decreasing temperature. Adsorption isotherm models, Langmuir, Freundlich, Dubinin-Radushkevich and Temkin were used to analyze the equilibrium data. Cadmium adsorption equilibrium could be described by the Freundlich adsorption model at all the temperatures essayed. The values of the n parameter, which are greater than unity, indicate that Cd^{2+} ions are favorably adsorbed by sericin/alginate particles at the temperature range studied.

Low-temperature steam reforming of raw bio-ethanol over ceria-zirconia supported catalysts

V. Palma, C. Ruocco, A. Ricca
Salerno Univ., Dep. Industrial Engineering, Via Giovanni Paolo II, 132 84084 Fisciano (SA), Italy; tel. +39 089 964027, fax +39 089 964057 email: cruocco@unisa.it

Introduction

The dependency of energy production form fossil fuels is linked to the emission of CO_2 and other pollutants in the atmosphere. As a result, the development of bio-fuels based processes for H_2 production and its use in fuel cells is arousing a great interest. Presently, steam reforming of bio-ethanol, derived from fermentation of agricultural and forest wastes, is considered as a promising route for generation of H_2-rich mixtures. However, the main challenges of such technology are related to catalyst deactivation due to carbon formation [1]. In fact, bioethanol mixtures, besides water and ethanol, may contain several impurities, including higher alcohols, which decrease catalyst resistance towards coking.

In the present work, ethanol steam reforming was investigated over bimetallic Pt-Ni based catalysts supported over ceria-zirconia. Preliminary tests were carried out under pure water/ethanol mixture and the impact of temperature (400-500°C), space velocity (GHSV, 7500-30000 h^{-1}) and H_2O/C_2H_5OH molar ratio (r.a., 3-9) on carbon formation rate was studied. Further, at chosen operative conditions, a mixture containing 1 mol% of the main bioethanol impurities (C_2H_4O, C_3H_7OH, C_4H_9OH, $C_5H_{11}OH$) was fed to the reformer and their effect on catalyst stability was studied at 450°C, 15000 h^{-1} and r.a.=6.

Experimental

3wt%Pt-10wt%Ni/CeO_2-ZrO_2 catalyst was prepared by wet impregnation. The calcined support (CeO_2-ZrO_2 by Rhodia) was added to an aqueous solution containing the Ni salt precursor ($C_4H_6O_4Ni*4H_2O$, Strem) on a heating and stirring plate. Then, the sample was dried at 120°C overnight and calcined at 600°C for 3 h. The Pt deposition was performed by the same procedure and $PtCl_4$ (Strem) was employed as Pt salt precursor The prepared catalysts were characterized by means of several techniques including specific area measurements, X-ray fluorescence and diffraction, temperature programmed reduction.

After reduction in situ (5%H_2 in N_2 stream at 600°C for 1h @10°C/min), the water/ethanol mixture, stored in a tank under nitrogen pressure, was properly diluted with N_2, vaporized and fed to an annular fixed bed reactor, which was placed, in turn, in an electrical oven. 5 heated solenoids allowed the reaction mixture to be sent to the reactor or to the analysis section. All the lines downstream the boiler were heated at 140 °C in order to avoid condensation. The products distribution was monitored through a FT-IR spectrophotometer, paramagnetic and thermoconductivity analysers.

Stability tests without impurities were performed under a 5%C_2H_5OH/30%H_2O/65%N_2 ranging T, GHSV and r.a. respectively between 400-500°C, 7500-30000 h-1 and 3-9. Conversely, the impurities impact (1 mol% with respect to C_2H_5OH) was studied at 450°C, 15000 h^{-1} and r.a.=6. Carbon formation rate (CFR) was evaluated through Temperature Programmed Oxidation (5%O_2/N_2 from room T to 600°C @10°C/min), which allowed the calculation of coke mass (g_C), and starting from the mass of carbon fed during the test ($g_{C,f}$) as well as the time-on-stream (TOS) and the catalytic mass (g_{cat}).

Results & Discussion

During stability tests, carried out at different T, ethanol was completely converted while an increasing trend for H_2O conversion with reaction temperature was observed. However, carbon formation rate rose with a quasi-exponential trend, probably due to the higher extent of CH_4 decomposition reaction leading to H_2 and coke formation. Likewise, the reduction of contact time was responsible for higher CFR values as a consequence of decreased activity towards carbon gasification reaction. However, it's worthwhile noting that coke formation on catalyst surface can be minimized by increasing water/ethanol feed molar ratio. On the basis of such observations, the effect of crude bio-ethanol contaminants on bimetallic catalyst performances was carried out at the following operative conditions: 15000 h^{-1}, 450 °C and r.a.=6. The stability tests with raw mixtures, prepared by adding individually 1 mol % of every impurity, were carried out for 5 h and the results, reported in terms of H_2 selectivity and CFR, are summarized in Figure 1. Total C_2H_5OH conversion was recorded with TOS and a slight increase in H_2 yield with impurities carbon content was observed. Moreover, the results of TPO analysis carried out over the exhaust samples showed slightly higher carbon formation rates with respect to the ones observed without impurities. These findings were due to the reforming contribution of contaminants as well as to the slight increase in carbon selectivity from C_3 to C_5 alcohols: CH_4 decomposition or intermediates polymerization [2] may enhance both H_2 coke deposits formation. The addition of C_2H_4O increased H_2 selectivity with respect to C_3H_7OH co-feeding, despite the lower C-content of this molecule: C_2H_4O is a key intermediate that can undergo decarbonylation to form CH_4 and eventually coke.

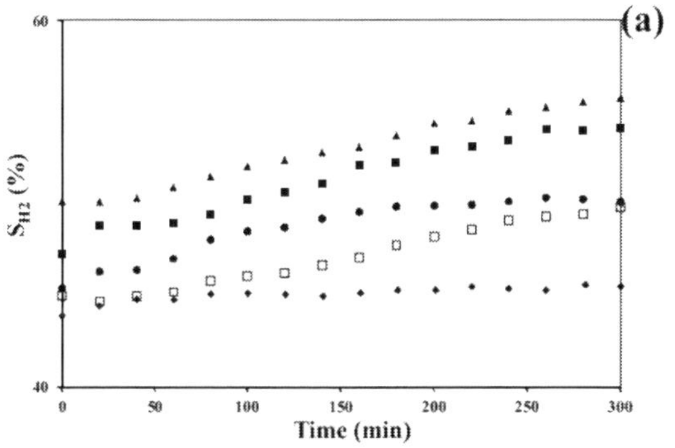

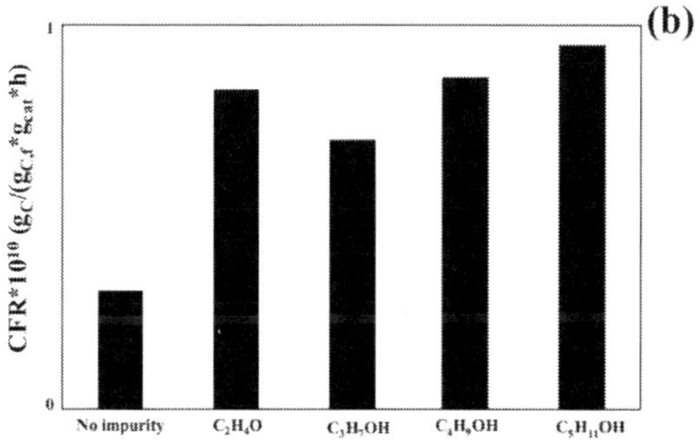

Figure 1. (a) H_2 selectivity and (b) CFR for the different cases :no impurity (diamond), C_3H_7OH (empty square), C_2H_4O (circle), C_4H_9OH (square) and $C_5H_{11}OH$ (triangle); GHSV=15000 h^{-1}, T=450°C.

However, the chosen catalyst showed high resistance to deactivation: after 1 day stability test under a raw bio-ethanol containing all the impurities, C_2H_5OH was still completely converted.

Conclusions

Bio-ethanol steam reforming was studied over a bimetallic catalyst. The effect of operative conditions on catalyst stability was preliminary investigated and the optimal conditions (namely 450°C, 15000 h^{-1} and r.a.=6) for coke minimization were selected. The influence of the main bio-ethanol impurities was also investigated: total C_2H_5OH conversion for 1 day was recorded even under a model bioethanol mixture containing 1 mol % of all the impurities.

References

[1] Kowalik P. et al., 2015, , Catal. Today, 254, 129-134.
[2] Choong C.K. et al., 2011, Appl. Catal. A: Gen., 407, 145-154.

Mathematical model of separation transplutonium and rare earth elements by ion-exchange chromatography method with complexation during elution

Y. A. Chursin, A. I. Gozhimov, S. N. Liventsov, O. V. Shmidt
Tomsk Polytechnic University, 634050, Russia, Tomsk, Lenin Avenue, 30;
tel. 8-913-868-48-10, e-mail: yurychursin@mail.ru;

Spent nuclear fuel contains precious and highly radioactive metals, such as plutonium, americium, curium, europium. Closed nuclear fuel cycle technology allows to distinguish these elements from spent nuclear fuel. Americium and curium are very expensive metals. They are used only in the most important areas of nuclear technology[1]. There are many methods for separation these metals from each other, like extraction methods, pyrochemical methods, chromatographic methods. This paper considers ion-exchange chromatography method with complexation during elution.

General rate ion-exchange chromatography method is based on the difference between distribution coefficients of separated cations in solution. It means at the one hand metal, which is strongly adsorbed by solid phase will slowly move in a chromatographic column, and the other hand, metal, which is weaker adsorbed will move faster. Experiments shows, that distribution coefficients of americium, curium are slightly different and separation by ion-exchange only is non effective.

Injection of chelating agent(pentetic acid) in eluent improved the separation factor. It should be took in consideration in modeling, because it influences the distribution of ions in the column.

Separation processes should be modeling for technology optimization and developing automatic control systems.

In this way, mathematical model should to consider next physical and chemical processes[2, 3]:

- injection and distribution separated components during column load
- ion-exchange between ions in solid and liquid phases
- diffusion in liquid phase (concentration gradient)
- eluent injection
- complexation
- distribution separated components during elution

Developed model allows varying many parameters: composition of input solution, injection period, solid phase composition of chromatographic column, elution starting time, eluent composition. Calculation allows observing distribution of separated components in solid and liquid phases at each time, components concentrations at column output.

The proposed model was checked on the data from article[4] "Separation of curium from americium using composite sorbents and complexing agent solutions" and it has shown its value.

Modeling result shown at figure 1.

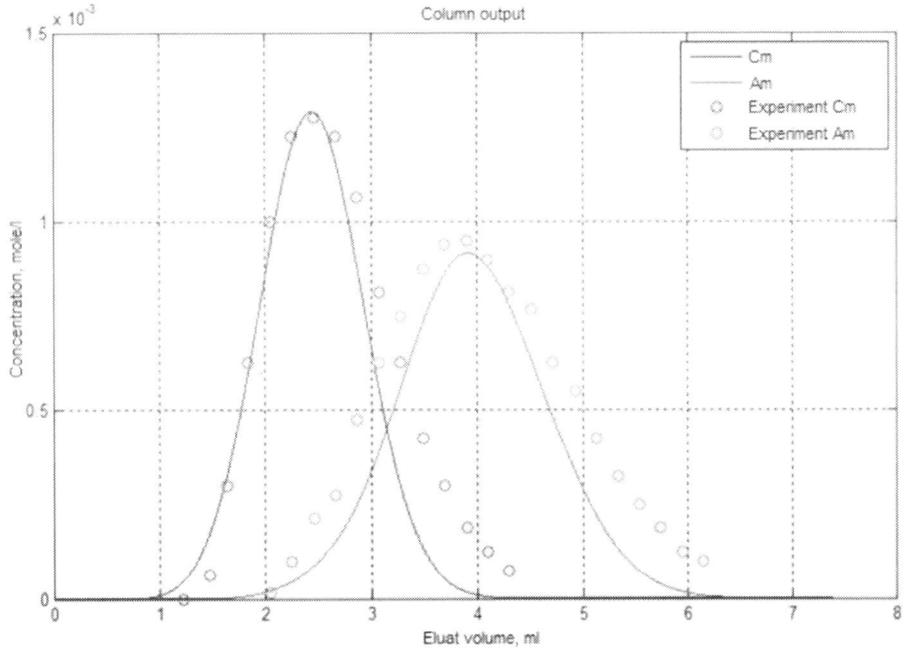

Figure 1 – Modeling results

References:

1. Praktikum po fiziko-himicheskim metodam analiza /I.YA. Gureckij, V.V. Kuznecov, L.B. Kuznecova, E.A. Kuchkarev, O.L. Samorukova, G.G.Svirshchevskaya, I.F. Kolosova, M.B. Ogareva, N.D. Rumyanceva, A.F. ZHukov, L.B. Oganesyan, S.L. Rogatinskaya.. – M.: Himiya, 1987. – 248 p.
2. Kasatkin A.G. Osnovnye processy i apparaty himicheskoj tekhnologii. – M.: Himiya, 1961.
3. Georges G., Atilla F., Dean G. S., Anita M. Katti. 2006. Fundamentals of Preparative and Nonlinear Chromatography.
4. Kamila S˘t'astna´, Jan John, Ferdinand S˘ebesta, Martin Vlk. 2015. J Radioanal Nucl Chem (2015) 304:349–355

Modeling of the electrolyzers cooling system for the fluorine production operator training simulator

O. V. Egorova, N.V. Liventsova, S. N. Liventsov, T. G. Sopelnichenko
National Research Tomsk Polytechnic University, Institute of Physics and Technology, Department of Electronics and Automation of Nuclear Plants, Tomsk 634050, Russia, Lenin st. 2; tel.+79039133867, e-mail: ola_ola81@mail.ru

Fluorine technological production is very dangerous for people and environment because of the use of highly-aggressive components and significant power consumption in case of incorrect control of run planned production suspension caused by operator errors. In such conditions high competence level of operators is vitally important. A operator training simulator (OTS) can be a very effective means of training the operators [1, 2].The key element of OTS is usually presented by a technological system model intended for imitation of processes which occur in processing installations and a system for controlling them during standard and emergency situations[1]. Development of such a model refers to complicated and specific tasks and to find the solution for them by standard tools is rather difficult.

The present work is devoted to mathematical model development of the electrolyzers cooling system for a fluorine production operator training simulator.

The structure, mathematical description of model and results of numeric experiments are presented in the article. The model has a model structure and consists of the modules of process simulation, control system and abnormal situations initiation. The process simulation module performs interactive calculation of mass flow of cooling water in the cooling systems piping net and electrolyzers electrolyte temperature depending on the supply net pressure change, position of control valves placed in the exchanger tube-sheet inputs of electrolyzers, cooling water temperature and temperature and consumption of fluorine hydride supplied into the electrolyzers. The control system module simulates operation of the measuring subsystem and local electrolyzers electrolyte temperature automatic control systems. Mathematical description of the modules contains special variables – "fault indicators". Changing their values the model adjustment becomes possible according to simulation of standard or different emergency modes of the modelled cooling system equipment. "Fault indicators" values presetting is carried out using the emergency situations module, operation of which is based on the mechanism providing probabilistic selection of the simulated situation and its duration time range.

Model performance testing was carried out by comparing the real production experiment data with computational experiment data and expert survey. Testing showed simulation results correlation with the specified requirements at this development stage.

References

1. Cameron D., Clausen C., and Morton W., 2002, in: Braunschweig B., Gani R. (Ed.), Software Architectures and Tools for Computer Aided Process Engeneering (Chapter 1), Elsevier, Amsterdam, the Netherlands.

2. Patle, D.S.; Ahmad, Z.; Rangaiah, G.P., 2014, Operator training simulators in the chemical industry: Review, issues, and future directions, Rev. Chem. Eng., 30, 199–216.

Modelling and optimisation of a crude oil hydrotreating process using neural networks

Wissam A.S. Muhsin, Jie Zhang, Jonathan Lee
School of Chemical Engineering and Advanced Materials, Newcastle University, Newcastle upon Tyne, NE1 7RU, UK; Telephone: +44 (0)191 208 7266; Fax: +44 (0)191 208 5292; e-mail: w.a.muhsin2@newcastle.ac.uk; jie.zhang@newcastle.ac.uk; jonathan.lee@ncl.ac.uk.

This paper presents a study on the data-driven modelling and optimisation of a crude oil hydrotreating process using bootstrap aggregated neural networks. Crude oil can be defined as a complex mixture of hydrocarbons (liquids and gases) which contain many different hydrocarbon compounds with varied appearances and compositions because each oil field has unique specifications of hydrocarbons (Hamadi, 2006).

Most refineries continuously try to improve and upgrade existing operating units or use a new technology in order to meet the environmental regulations concerning the quality and specification of oil products. Changes in operation units are made in response to regulation changes which affect modern refineries (Babich and Moulijn, 2003).Hydrotreating (HDT) is a special process that can be utilised in petroleum refineries to reduce inorganic impurities like sulphur, nitrogen, and oxygen compounds.HDT of crude oil is a new process with challenges that have not been extensively taken into account in the literature, as the conventional process of HDT is conducted for each oil product individually and not for the whole crude oil. Additionally, different process variables should be considered in the HDT process such as charge, pressure, temperature, liquid hourly space velocity (LHSV), and hydrogen to hydrocarbon (H_2/HC) ratio (Nawaf et al., 2015, Jarullah et al., 2011b).Recently, computer and information technology have become increasingly significant in crude oil refineries and industrial processes with the improvement of simulation, modelling, optimisation, and control systems. In this work, a crude oil hydrotreating process was simulated utilising Aspen HYSYS (Version 8.8) to produce simulated HDT process operation data. Also, bootstrap aggregated neural networks (Zhang et al., 1997a, Zhang et al., 1997b) were used to build up an accurate and robust data-driven model for the crude oil hydrotreating process. Process optimization plays a significant role in industrial decision making and is one of the main tools that can be used for obtaining the best plant design, maximising profitability of a plant and minimising its environmental impacts. The target of process optimization is to reduce cost, increase process profits, and process efficiency (Binder et al., 2001, Khalfalla, 2009, Jarullah, 2011a).

A single neural network model was developed first for the purpose of comparison. The network inputs are crude oil flow rate, hydrogen molar flow rate, reactor temperature and pressure. The network outputs are sulphur and nitrogen removal. 150 data samples were produced from the HYSYS simulation of a crude oil hydrotreating process to build a neural network model. These data (150 samples) are divided into three groups: training data (78 samples), testing data (41 samples), and unseen validation data (31 samples). The neural networks were trained by employing the Levenberg-Marquardt training method with early stopping in order to avoid over-fitting in the neural network. The number of hidden neurons was determined by trying a range of hidden neurons and examining their sum of squared errors (SSE) on the testing data. The neural network generating the lowest SSE on the testing data is considered to have suitable number of hidden neurons. Consequently, 30 hidden neurons were used. This study has found that there are some quite large errors on the unseen validation data though model errors on training and testing data appear to be small. This reveals that a single neural network model is not reliable, and therefore a bootstrap aggregated neural network should be considered. The bootstrap aggregated neural network contains 30 single neural networks. The training data for every neural network was acquired via bootstrap re-sampling with replacement of the original training data. The performance of individual neural networks and bootstrap aggregated neural networks with different number of networks was studied in this paper. The major finding was that the bootstrap aggregated neural network models are more accurate and reliable than the individual neural network models.

The developed bootstrap aggregated neural network model was utilised to optimise the HDT of crude oil. The aim of the process optimisation is to maximise sulphur removal in the liquid at the end of reactor effluent subject to various process constraints. The optimisation problem solved in this paper is described as follows:

$$\min J= - \alpha_1 S - \alpha_2 F + \alpha_3 P$$

where J is the objective function, S is the bootstrap aggregated neural networks output, F is the feed flow rate, P is the reactor pressure, α_1, α_2, and $\alpha_3$are weighting factors which are selected based on the relative importance of the corresponding terms and their magnitudes. Large weighting factor was given to the first term in order to maximise the Sulphur recovery. The second term intends to increase the feedstock to maintain high productions. Furthermore, the lowest weighting factor was applied for the third term which aims to decrease the reactor pressure due to the fact that high pressure is not appropriate, as hydrogen partial pressure will not rise because of the physical limitations of high pressure(Jimenez et al., 2005). The sequential quadratic programing (SQP) method implemented via the function "fmincon" in the MATLAB Optimisation Toolbox was used. It can be deduced that when relatively large weighting is applied to the feed flow rate, the product throughput is high with the sacrifice of Sulphur removal. When small weighting is applied to the feed flow rate, the Sulphur removal is the highest but the production throughput is reduced significantly. Thus, there is trade off between Sulphur removal and production rate. The weightings selected will be determined by the particular plant operation objectives.

References

BABICH, I. V. & MOULIJN, J. A. 2003. Science and technology of novel processes for deep desulfurization of oil refinery streams: a review. *Fuel,* 82, 607-631.

BINDER, T., BLANK, L., BOCK, H. G., BULIRSCH, R., DAHMEN, W., DIEHL, M., KRONSEDER, T., MARQUARDT, W., SCHLÖDER, J. & VON STRYK, O. 2001. Introduction to Model Based Optimization of Chemical Processes on Moving Horizons. *In:* GRÖTSCHEL, M., KRUMKE, S. & RAMBAU, J. (eds.) *Online Optimization of Large Scale Systems.* Springer Berlin Heidelberg.

HAMADI, A. S. 2006. Petroleum Refining. Dohuk, Iraq.

JARULLAH, A. T. 2011a. *Kinetic Modelling Simulation And Optimal Operation Of Trickle Bed Reactor For Hydrotreating Of Crude Oil.* PhD Thesis, University of Bradford, UK.

JARULLAH, A. T., MUJTABA, I. M. & WOOD, A. S. 2011b. Kinetic parameter estimation and simulation of trickle-bed reactor for hydrodesulfurization of crude oil. *Chemical Engineering Science,* 66, 859-871.

JIMENEZ, F., NUNEZ, M. AND KAFAROV, V. 2005. *Study and modeling of simultaneous hydrodesulfurization, hydrodenitrogenation and hydrodearomatization on vacuum gas oil hydrotreatment,* in Puigjaner, L. and Espuna, A. (eds.) European Symposium on Computer-Aided Process Engineering-15, 20A and 20B. pp. 619-624.

KHALFALLA, H. A. 2009. *Modelling and optimisation of oxidative desulphurization process for model sulphur compounds and heavy gas oil.* Ph.D. Thesis, University of Bradford, UK.

NAWAF, A. T., GHENI, S. A., JARULLAH, A. T. & MUJTABA, I. M. 2015. Optimal Design of a Trickle Bed Reactor for Light Fuel Oxidative Desulfurization Based on Experiments and Modeling. *Energy & Fuels,* 29, 3366-3376.

ZHANG, J., MARTIN, E. B., MORRIS, A. J. & KIPARISSIDES, C. 1997a. Inferential estimation of polymer quality using stacked neural networks. *Computers & Chemical Engineering,* 21, Supplement, S1025-S1030.

ZHANG, J., MARTIN, E. B., MORRIS, A. J. & KIPARISSIDES, C. 1997b. Prediction of polymer quality in batch polymerisation reactors using neural networks. *Proceedings of the 1997 American Control Conference, Vols 1-6,* 1370-1374.

On modeling the convective mass transfer over moving films interface for the reaction-diffusion systems of the second order

L.M. Musabekova, N.N. Dausheyeva, N.S. Zhumataev, [1]M.A. Jamankarayeva
State University of South Kazakhstan, Tauke Khan, 5, Shymkent, 160012, Kazakhstan; tel. +77052026408, e-mail: mleyla@bk.ru; [1]South Kazakhstan State Pedagogical Institute, Baytursynov str., 13, Shymkent, 160000, Kazakhstan; tel. +77027104891, e-mail: d_madina08@mail.ru.

The film absorption processes occur in many industrial chemical and power enterprises. It is used for purifying gases from toxic admixtures or for extracting the valuable components. However the very limited resources to increase the interfacial mass transfer in the film processes still have been found. Among various causes of those failures, the certain cause, which may be not main but important in our opinion, is the difficulty to solve and to analyze the coupled equations of film motion and mass transfer especially with allowance for the specific peculiarities of the processes at films surfaces.

The paper deals with the methods for approximately analytical solving the non-automodel problems of film absorption, accompanied by the partly reversible reaction of a second order. The basic premise which was used to simplify the approach to creating the model consists in that the chemisorption purification is usually applied in those cases when gases are sparingly soluble, and the process of physical absorption turns out little effective. Therefore, the Peclet number in this case is sufficiently high, and the diffusion boundary layer is very thin. This circumstance gives the opportunity to select such small control parameter as the gradient of surface velocity along the film motion direction, and then we look for the solution of the model in the form of expansion in power series of this parameter.

The main results of the submitted work can be summarized as follows.

1. The methods to obtain approximate analytical solutions of the stationary convective diffusion equation for the coupled problem of isothermal film chemisorption at high Peclet numbers have been worked out.

2. The approximate analytical solutions of the following problems have been obtained: physical absorption of poorly soluble gas; chemisorption with partly reversible reaction both of the first and second orders with specially allowance for the kinetics of surface reaction.

The submitted method allowed also obtaining the approximate analytical solution of the convective diffusion equation taking into account the phenomenon of changing velocity of the film surface induced by the concentration-dependent surface tension.

The obtained results are likely to be of importance to mass transfer modeling and engineering practice in industrial processes and apparatuses design.

Performance evaluation of low-temperature solid oxide fuel cells with SDC–based electrolyte

[1]D. Saebea, [2]S. Authayanun, [3]Y. Patcharavorachot, [4]A. Arpornwichanop
[1]Burapha University, Department of Chemical Engineering, Faculty of Engineering, Long-Hard Bangsan Rd, Chonburi, 20131 Thailand; tel. +66 850562910, e-mail: dangs@eng.buu.ac.th; [2]Srinakharinwirot University, Department of Chemical Engineering, Faculty of Engineering, Rangsit-Nakhon Nayok Rd, Nakorn Nayok, 26120 Thailand; tel. 66 816276456, e-mail: suthida_autt@hotmail.com; [3]King Mongkut's Institute of Technology Ladkrabang, Department of Chemical Engineering, Faculty of Engineering, Chalongkrung Rd, Bangkok, 10520 Thailand; tel. +66 819013479, e-mail: yaneepon.p@hotmail.com; [4]Chulalongkorn University, Department of Chemical Engineering, Faculty of Engineering, Payathai Rd, Bangkok, 10330 Thailand; tel. +66 869713499, e-mail: amornchai.a@chula.ac.th.

Solid oxide fuel cell (SOFC) is regarded as an efficient device for power generation. The SOFC offers many advantages such as a high electrochemical reaction rate, the flexibility of using various fuel types, and the prospect for combined heat and power systems. However, SOFC normally operated at high temperature of 800-1000 °C, resulting in shortened cell lifetimes and increased the fabrication costs. This is obstacle to the commercialization of SOFC technology. Thus, the development of low-temperature SOFCs is interesting research topic. A decrease in the SOFC operating temperature can lower the thermal stress on the SOFC stack. A novel samaria-doped ceria (SDC) is a promising electrolyte for low-temperature SOFCs because its oxygen-ion conductivity is excellent in low temperatures range (500-700 °C). The mathematic model of low-temperature solid oxide fuel cell (SOFC) based on with SDC–based electrolyte is required to study and develop the cell performance. The aims of work are to develop the mathematic model and optimal operating conditions of low-temperature SOFC. In this study, the mathematical model of low-temperature solid oxide fuel cell was simulated by using MATLAB program. The low-temperature solid oxide fuel cell model was validated to ensure the reliability of model. This result indicated that the SOFC model can be used to predict the SOFC behavior. The SOFC model was used to investigate the effects of the operating temperature, pressure, thickness of material and pore of electrode on the cell performance. The simulation results showed that the increase of fuel concentration, operating temperature, pressure and pore of electrode resulted in higher the cell performance. For considering the configuration of SOFC, the cathode-supported SOFC which achieved high power density of 0.7061 kW at temperature of 550°C and current density of 3 Acm^{-2} is suitable for low-temperature SOFC.

Power-to-gas storage optimization through power pinch analysis

A. El Elmi, H. Er-rbib, C. Bouallou[1]
MINES ParisTech, PSL-Research University,
CES Centre d'efficacité énergétique des systèmes,
60 Bd Saint Michel, 75006 Paris, France [1]E-mail : chakib.bouallou@mines-paristech.fr

This work focuses on the optimization of an innovative power storage technology Power - to-Gas using pinch analysis. The concept of Power -to-gas storage is the conversion of power into gas that can be stored in the natural gas network. This concept incurs considerable power losses compared to other storage technologies. Besides, it has a long start time and a long stand-by recovery time that need to be studied. Graphical and numerical approaches have been developed for power optimization. A first a graphical approach is presented. Then a numerical approach will facilitate more targeted responses to the problems experienced with the graphical one. This work acknowledges losses of the rectifier AC (alternating current) to DC (direct current), of the inverter (DC to AC), of the charging and the discharging process.

This work proposes a new approach of Pinch Analysis Method for the optimization of Power-to -gas storage of a hybrid power system using intermittent and renewable energy. Graphical and numerical tools take into consideration new parameters (standby time, start time) and were applied to determine:

- The temporal time ratio of charging, discharging, stand-by time and rest;
- The minimum of outsourced electricity and the maximum of storage capacity.

Different scenarios for energy farms are presented and allow the designers to choose the best alternative for energy systems in a French City. Results show that the best alternative is an onshore wind farm. A sensitive analysis allows us to apprehend the importance of the source's type in the optimization of the storage system.

Preparation and semiconducting properties of Bi_2S_3 films

Z. Grubač, [1]M. Metikoš - Huković
Department of General and Inorganic Chemistry, Faculty of Chemistry and Technology, University of Split, N. Tesle 10, 21000 Split, Croatia, e-mail: grubac@ktf-split.hr;
[1]Department of Electrochemistry, Faculty of Chemical Engineering and Technology, University of Zagreb, P.O. Box 177, 10000 Zagreb, Croatia; e-mail: mmetik@fkit.hr

Solar energy is considered to be the most promising renewable energy source for the future energy demands due to its abundance and inexhaustibility. The major factors to consider in the production of solar cell materials are the bandgap energy matching the solar spectrum and the production cost. Thus, heterojunction solar cells have been getting a lot of attention recently.

Among photovoltaic materials of recent interest, there are the low cost, nontoxic chalcogenides such as ZnS, SnS and Bi_2S_3. Anodically formed sulphide films on metals, in the corresponding electrolyte solution, are compounds with a variable chemical composition for which the exception from the stoichiometric composition is very common. Depending on the character and degree of deviation, the anionic (the excess of metal ions in comparison to the stoichiometric composition) or cationic vacancies will be the main charge carriers during anodic formation and growth of thin surface films. Stabilized films, under the conditions when electronic conductivity prevails, act (have the same characteristics) as *n*- or *p*- type semiconductors [1].

Electrodeposition is a promising method for thin film preparation because of the possibility of large scale production, easy monitoring of the deposition process and easy control over the deposition parameters [2]. Bi_2S_3 in the thin film form is a *n*-type semiconducting material with a layered orthorhombic structure, and its direct and indirect bandgap energies were reported to be 1.25 and 1.7 eV, respectively [3,4]. It is a particularly challenging material because of its absorption coefficient of the order of 10^4 to 10^5 cm^{-1}, reasonable conversion efficiency and stability together with low cost [5].

In this work, the mechanism of formation and growth of sulphide films on polycrystalline bismuth in 0.1 M NaOH solution containing 0.1 M Na_2S was investigated using cyclic voltammetry and electrochemical impedance spectroscopy. Voltammetric measurements offer limited insight into the mechanism of passive film growth. The study of the frequency dependent periodic response of anodically formed semiconducting passive films to *ac* perturbations of potential can provide a more fruitful approach. After the Bi_2S_3 film formation at constant potential, the semiconducting properties of passive film were studied by Mott-Schottky analysis under conditions in which its electronic conductivity prevails. Based on the Brug capacitance calculation method and Harrington-Devine method, the Mott-Schottky analysis was performed [6]. Taking into account the frequency dispersion in Mott-Schottky tests, the semiconducting parameters were determined: the donor density, N_D, the flat-band potential, E_{fb} and the space-charge layer thickness, d_{sc}. Structural, morphological and optical properties of Bi_2S_3 films formed on bismuth are under investigations.

Literature:

1. Z Grubač, M. Metikoš - Huković, J. Electroanal. Chem. 565 (2004) 85.
2. H.M.M.N. Hennayaka, Ho Seong Lee, Thin Solid Films 548 (2013) 86
3. L. M. Peter, J. Electroanal. Chem. 98 (1979) 49.
4. B. Miller, S. Menezes, A. Heller, J. Electroanal. Chem. 94 (1978) 85.

5. V. Balasubramanian, N. Suriyanarayanan, S. Prabahar, Adv. Appl. Sci. Res., 3 (2012) 2369.
6. J. Katić, M. Metikoš - Huković, Acta Chim. Slov. 61 (2014) 350.

Present chemical production simulator model validation method

N.V.Liventsova, O. V. Egorova, S.N. Liventsov, M. O. Blank
National Research Tomsk Polytechnic University, Institute of Physics and Technology, Department of Electronics and Automation of Nuclear Plants, Tomsk 634050, Russia, Lenin Avenue, 2; tel.+79069491359, e-mail: liventsovanina@tpu.ru.

Skills and abilities of an operator and, consequently, validity of managing solutions under real manufacturing conditions depend directly on correct variables behaviour in situations practiced by the operator on the simulator. The suggested approaches in case of unknown input signals are reduced to the following stages: identification of input signals using the quantitative criterion; development of the qualitative assessment technique and model verification with further conclusion about the developed model applicability in the computer simulator. Lately many approaches have been suggested for assessing adequacy and applicability of the simulation model for its use in a computer simulator. However, there is no established theory in this field, thus there is no performance analysis of the suggested approaches.

In case, when the main input and output signals for the developed model are not measured in the production environment, as it is rather challenging to fit equipment with additional sensors, it complicates the use of the direct comparison methods of experimental and simulated data. Manufacturing data obtained during a passive experiment are difficult to simulate on a model due to indefiniteness of output signals. An active experiment on simulation of standard and emergency modes at real manufacture is often impossible. Therefore, the procedure based on expert evaluation method was developed. The procedure consists of two main stages. At the first stage the simulation quality of separate process variables is assessed using a special criterion. It allows verifying the validity of the developed model and identifying its inputs or necessity to adjust any development stage. At these cond stage the model performance in general and its applicability for training purposes with the assistance of manufacture specialists is assessed using a special criterion. The quality of simulation of interconnected measured process variables which allow analyzing cause-and-effect relations is carried out to find the reasons of accident conditions.

The method was tested in the assessment of fluorine production simulation model; the model elements are treated in the work [1].The use of the method pointed out the necessity to adjust the model and as a result the applicability of the model was verified.

References

1. Egorova O. V. , Liventsova N. V. , Efremov E. V. , Grachev M. I. Model of the electrolyzers series power-supply system for the computer trainer-simulator of the fluorine production process control // Procedia Chemistry. - 2014 - Vol. 11. - p. 139-146.

Production of n-butanol in sugarcane biorefineries through ethanol catalysis

M. O. S. Dias [1], L. G. Pereira [2,3], R. Maciel Filho [2,4], A. Bonomi [2,4]
[1] Inst. of Science and Technology, Federal Univ. São Paulo (ICT-UNIFESP), Rua Talim, n° 330 São José dos Campos, SP, Brazil; [2] Laboratório Nacional de Ciência e Tecnologia do Bioetanol (CTBE), Centro Nacional de Pesquisa em Energia e Materiais (CNPEM), Caixa Postal 6192, CEP 13083-970, Campinas, São Paulo, Brazil.; [3] Department of Civil Engineering, University of Toronto, 35 St George St, Toronto, Ontario, Canada; [4] School of Chemical Engineering, Univ. of Campinas (FEQ-UNICAMP), Av. Albert Einstein, 500 , Campinas, Brazil.

Replacement of fossil resources by renewable products is a fundamental step towards a sustainable future. In the chemical industry, sugars and ethanol are envisioned as the most suitable feedstock for production of biobased chemicals, while in the transportation sector ethanol has been increasingly used on a large scale basis in countries like Brazil and the United States. Alternatives for ethanol as biofuel have been investigated, especially in cold climate countries where its use in spark ignition engines is difficult due to low temperatures. n-Butanol may be classified as a drop-in fuel, since its characteristics (higher energy content than ethanol, superior blend stock with gasoline, diesel, biodiesel and ethanol) allow it to be used in gasoline driven engines with little or no modifications. In addition, n-butanol and its derivatives are precursors for several chemical products like paints, solvents and plasticizers, so the replacement of fossil-derived by renewable n-butanol can increase the sustainability of the chemical industry.

Although most of the n-butanol used in the world is derived from oil, industrial production of n-butanol from renewable resources has been envisioned through acetone-butanol-ethanol (ABE) fermentation, a process characterized by low yields and high product (butanol) inhibition by the fermenting microorganism (solventogenic Clostridia strains). Butanol inhibition prevents the use of concentrated sugar solutions, requiring large amounts of water during fermentation - what leads to high stillage generation and high energy consumption during product purification (Mariano et al., 2013). An alternative route that may overcome these issues is based on the use of ethanol produced from conventional alcoholic fermentation using *Saccharomyces Cerevisiae* as feedstock, and its conversion to n-butanol through chemical catalysis. Since ethanol fermentation is a consolidated process at the industry, replacing current ethanol fermentation facilities for ABE fermentation requires costly modifications. An alternative to ABE fermentation, ethanol catalytic conversion may be employed to convert ethanol to n-butanol in existing ethanol production facilities, thus producing a higher quality fuel (or chemical) with less retrofitting than ABE fermentation.

Different catalysts can be used in ethanol conversion to n-butanol, such as hydroxyapatite (Ogo et al., 2011), ruthenium based (Dowson et al., 2013), mixed oxides (Carvalho et al., 2012), among others. The combination of ethanol conversion, n-butanol selectivity and reaction time is crucial for the performance of the process; results available in the literature provide conflicting parameters for ethanol conversion and n-butanol selectivity, which should be maximum, and reaction time, which must be long (several hours) in order to achieve high ethanol conversion. However, long reaction times require high equipment investment and also lead to production of other chemicals, decreasing n-butanol selectivity (Dowson et al., 2013; Tsuchida et al., 2008).

In this study the use of such catalysts was evaluated through process simulation and economic analysis, to assess n-butanol production in a sugarcane biorefinery within the Virtual Sugarcane Biorefinery (VSB) framework, developed at the Brazilian Bioethanol Science and Technology Laboratory (CTBE) of the Brazilian Center for Research in Energy and Materials (CNPEM). Results show that high investment required in the catalyst plant compromises the economic return of the process, so catalysts with low reaction times can be advantageous even though smaller conversions are obtained when compared with longer reaction times.

References

Carvalho, D. L., de Avillez, R. R., Rodrigues, M. T., Borges, L. E. P., Appel, L. G. (2012). Mg and Al mixed oxides and the synthesis of n-butanol from ethanol. Applied Catalysis A: General, 415-416, 96–100.

Dowson, G. R. M., Haddow, M. F., Lee, J., Wingad, R. L., Wass, D. F. (2013). Catalytic Conversion of Ethanol into an Advanced Biofuel: Unprecedented Selectivity for n-Butanol. Angewandte Chemie, 52, 9005–9008.

Mariano, A. P., Dias, M. O. S., Junqueira, T. L., Cunha, M. P., Bonomi, A., Maciel Filho, R. (2013). Utilization of pentoses from sugarcane biomass: techno-economics of biogas vs. butanol production. Bioresource Technology, 142, 390–399.

Ogo, S., Onda, A., Yanagisawa, K. (2011). Selective synthesis of 1-butanol from ethanol over strontium phosphate hydroxyapatite catalysts. Applied Catalysis A: General, 402, 188–195.

Tsuchida, T., Kubo, J., Yoshioka, T., Sakuma, S., Takeguchi, T., & Ueda, W. (2008). Reaction of ethanol over hydroxyapatite affected by Ca/P ratio of catalyst. Journal of Catalysis, 259, 183–189.

Pyrolysis kinetics of two different lead-impregnated olive-industry wastes under nitrogen atmosphere by non-isothermal thermogravimetry

M.A. Martín-Lara, A. Ronda[1], G. Blázquez, A. Pérez, M. Calero
University of Granada, Dep. Chemical Engineering, Avda. Fuentenueva, s/n, 18071 Granada, Spain; tel. +34958244075; e-mail: alirg@ugr.es; fax: +34958248992.

There is a growing interest in renewable and sustainable energy sources. Biomass energy has been around since ancient times when people use to burn wood or coal to heat their homes or prepare food. Huge amount of biomass waste are generated worldwide. Wood still remains the most common source to produce biomass energy. But, apart from wood, other products that are used to create biomass energy include crops, plants, landfills, municipal and industrial waste, trees and agricultural waste. In Mediterranean area highlights olive stone and olive tree pruning. Pyrolysis is an attractive alternative to use of these waste generating appreciable economic value products. Thermogravimetric analysis (TGA) is a widely used analytical technique to observe the thermal decomposition and kinetics of pyrolysis of materials. The main objectives of this word were to conduct a study on the kinetics of pyrolysis of native olive stone and olive tree pruning and lead-loaded olive stone and olive tree pruning after a biosorption process for lead removal from industrial wastewaters. Thermogravimetric analysis was performed at different heating rates (5, 10 and 20 °C/min) under an inert atmosphere of nitrogen. The curves of weight loss (TG) and derived of the weight loss curve (DTG) for metal-loaded samples showed a little increase of the degradation temperature of hemicellulose and cellulose. Also, the general shift to higher temperatures when the heating rate is increased was observed for all samples (native and lead-loaded ones). The kinetics of decomposition of samples was represented by three isoconversional models (Flynn-Wall-Ozawa, kissinguer-and Friedman models) and for an independent parallel reaction model. The activation energy predicted by FWO method for native samples ranged from 100 to 199 kJ/mol and from 80 to 209 kJ/mol for olive stone and olive tree pruning, respectively. Impregnated samples obtained lower or similar activation energies, for example, for FWO method, activation energies ranged from 89 to 146 kJ/mol and from 122 to 209 kJ/mol for lead-loaded olive stone and olive tree pruning, respectively. The mechanistic kinetic model proposed for the pyrolysis of these native and lead-loaded samples could be interpreted considering these waste formed by three independent parts, each one following an independent nth-order reaction. The correlation of the different dynamic pyrolysis tests has been done simultaneously and a unique set of kinetic parameters have been obtained that satisfactorily correlates the different TG runs. Concerning the values of the kinetic parameters obtained applying this mechanistic model in pyrolysis decomposition, activation energy around 120, 239 and 69 kJ/mol and reaction order of 0.86, 1.19 and 3.04 were obtained for each modelled fraction of native olive stone. Reaction orders of native olive tree pruning had high values: 2.52, 3.61 and 2.02 for hemicellulose (fraction 1), cellulose (fraction 2) and lignin (fraction 3), respectively.

Selection of the frequency domain specifications in systems with PID-controller to obtain stability and flexible plant operation

K. A. Kozin, S. S. Mikhalevich, [1]F. Manenti, A.E. Stolpovskiy, [1]F. Rossi
Tomsk Polytechnic University, Dept. of Electronics and Automation of Nuclear Plants, 30 Lenina Ave., 634050 Tomsk, Russia; tel. +7 9610974669, e-mail: kozin@tpu.ru; [1]Politecnico di Milano, Dept. di Chimica, Materiali e Ingegneria Chimica „Giulio Natta", Piazza Leonardo da Vinci 32, 20133 Milano, Italy.

In process control today the most common form of controller is PID. It has a great developmental history and many different representations. Nowadays PID-algorithm is used in most industrial control loops. That is why the problem of tuning such controller is essential. There is a plenty of works, devoted to the tuning of PID-controller, but still there is no simple method, that could furnish specified process characteristics and stability of a control system. The explanation for this fact comprise that PID-controller has three parameters, and one need at least a systematic procedure to find their optimal values.

In this sense the most common PID-controller design methods are quite simple and not enough flexible to provide optimal system performance in specified conditions. Among them are Cohen-Coon method [1] and Ziegler-Nichols method [2], which provide robust performance but poor transient characteristics of the system.

For more than a half of a century large numbers of modifications of these methods have been developed [3]. The most effective of them include variable parameters that depend on expected process performance. As an example of such method consider Bucz et. all modified frequency domain Ziegler-Nichols method with variable weights [4], or Internal Model Control method [5], which allows to achieve acceptable disturbance rejection.

It should be noted that methods, that guarantee system nominal performance or robust stability, are often represented in frequency domain and based on gain and phase margin specifications. However, general analytic expressions for PID – coefficients are still not obtained. In [6] authors offered PI-design algorithm with specified phase margin φ_m and crossover frequency ω_c.

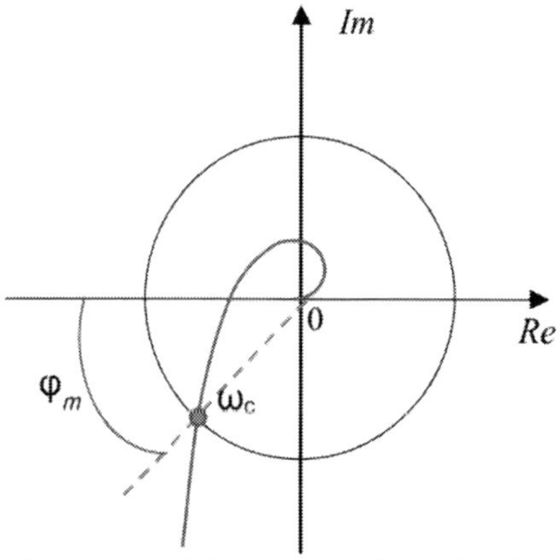

Figure 1 – Considered frequency domain characteristics

Mikhalevich et.al. has modified this method [7] by adding an extra equation, obtained under the condition of minimal overshoot to original system of equations. It allows the parameters of PID-controller, provided specified phase margin and crossover frequency, to be computed.

Still, this method has no any recommendation for choosing the value of crossover frequency for PID parameters computation.

The purpose of this paper is to provide a criterion for choosing values of crossover frequency and phase margin for Mikhalevich et. all PID-tune method, to achieve stabile performance and nominal transient characteristics of a control system. Along with that, a classification of process transfer functions for the method is proposed. This work consider plants without delay time and classical control scheme with unitary feedback. In such systems, robustness and performance are rival characteristics. Therefore, considered criterion can be used to achieve optimal process in specific conditions.

References

1. Cohen G.H. and Coon G.A., Theoretical Consideration of Retarded Control, Trans. ASME, 75, pp. 827-834, 1953.
2. Ziegler J.G. and Nichols N.B., 1942, Optimum Settings for Automatic Controllers, ASME Transactions, 64, 759-768.
3. K.J. Åström, T. Hägglund, Advanced PID Control, ISA Press, Research TrianglePark, USA, 2006.
4. Bucz Š., Veselý V., Kozáková A. and Kozák Š. 2014, A novel PID controller design methodology for specified performance using ultimate plant parameters, IFAC Proceedings Volumes (IFAC-PapersOnline), pp. 4909.
5. D.E. Rivera, M. Morari, S. Skogestad, Internal model control. 4. PID controller design, Ind. Eng. Chem. Process Des. Dev. 25 (1986) 252–265.
6. Hamamci S. E., Tan, N., 2006, Design of PI controllers for achieving time and frequency domain specifications simultaneously, ISA Transactions, 45, 529-543.
7. Mikhalevich, S. S., Baydali, S. A., Manenti, F., 2014, Development of a tunable method for PID controller to achieve the desired phase margin, Journal of Process Control, 25, 28-34.

Steam reforming of toluene over $Ru/SrCO_3$-Al_2O_3 catalysts prepared using an impinging stream micro reactor

H. Iida, A. Fujiyama, S. Tanaka, A. Igarashi, K. Okumura
Department of Applied Chemistry, School of Advanced Engineering, Kogakuin University, 2665-1, Nakano-machi, Hachioji-shi, Tokyo, 192-0015, Japan;

tel. +81-42-628-4620; fax: +81-42-628-4993; e-mail: iida@cc.kogakuin.ac.jp.

1 Introduction

Catalytic steam reforming of hydrocarbons is an important route for hydrogen production for the fuel cells. Kerosene is one of the promising hydrogen sources because of its easy usability and safety. However, kerosene has low reactivity and causes easily carbon deposition on the catalyst surface because it mainly contains alkanes with carbon number of above 10, and aromatic hydrocarbons. In this study, in order to develop a catalyst that has superior catalytic performance for steam reforming of kerosene, steam reforming of toluene as a model of aromatics was performed over $Ru/SrCO_3$-Al_2O_3 catalysts. In addition, in order to improve the catalytic activity of the $Ru/SrCO_3$-Al_2O_3 catalysts, an impinging stream micro reactor[1)] that is desirable for the synthesis of fine particles in liquid phase reaction was applied for an increase in surface area of $SrCO_3$-Al_2O_3 support.

2 Experimental

$SrCO_3$ and $SrCO_3$-Al_2O_3 supports (Sr/Al (molar ratio) = 3.1) were prepared by a conventional precipitation (PR) and an impinging stream method (IM). In the PR method, strontium chloride and sodium carbonate solutions (0.2 mol L^{-1}) were added dropwise to a distilled water or a γ-Al_2O_3 suspension in the beaker at the rate of of 3 mL min^{-1}. The stirring rate was 250 rpm.. In the IM method, stainless steel vessels containing the precursor solutions were pressurized by an N_2 gas. After that, the both solutions were jetted out from nozzles (ID: 500 μm) in a γ-Al_2O_3 suspension in the beaker. the solutions then were impinged at the distance of 2.5 mm from the nozzles (Nozzle distance: 5mm) (Fig. 1). The impact velocity (20 m s^{-1}) was adjusted by controls in the pressure in the vessels. After the centrifugal separation (3000 rpm, 10 min) and the washing were carried out three times, the resultant materials were dried at 353 K for 24 h and calcined at 873 K in an air stream. $Ru/SrCO_3$-Al_2O_3 and $Ru/SrCO_3$ catalysts (Ru: 1.0 wt%) were prepared by a physical mixing with Ru $(PPh_3)_3Cl_2$, thereafter the catalysts were calcined at 873 K, in a stream of N_2[2)]. A commercial Ru/Al_2O_3 catalyst (Ru: 2.0 wt%) was used as a reference catalyst. The reaction was performed at atmospheric pressure using a conventional fixed bed flow reactor. Prior to the reaction, the catalysts were treated in an N_2 stream at 873 K. The reaction conditions for steam reforming of toluene were a reaction temperature of 873 K, an S/C molar ratio of 2.0, a time factor (W/F) of 3.5 or 7.0 g h mol^{-1}.

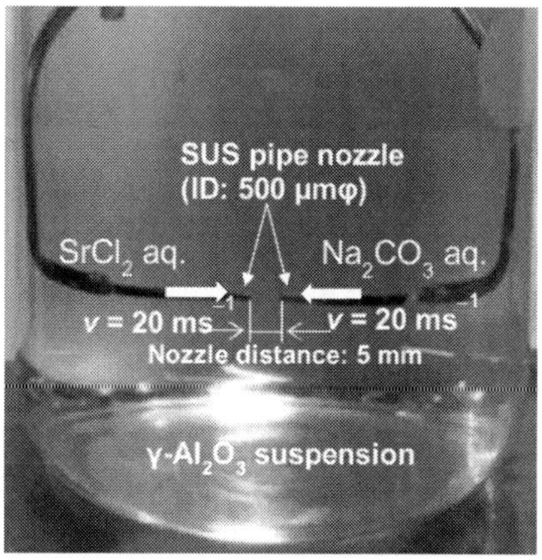

Fig. 1 Impinging stream micro reactor

3 Results and discussion

Table 1 shows the BET surface area and crystallite size of $SrCO_3$ and $SrCO_3$-Al_2O_3 supports prepared by the PR and IM methods. The addition of Al_2O_3 to $SrCO_3$ brought about an increase in BET surface area because it would prevent the sintering of $SrCO_3$ particles under the calcination. The crystallite size of $SrCO_3$ in $SrCO_3$-Al_2O_3 (IM) was smaller than that of $SrCO_3$-Al_2O_3 (PR), whereas the BET surface area of $SrCO_3$-Al_2O_3 (IM) was larger than that of $SrCO_3$-Al_2O_3 (PR). In the IM method, impact of the streams would bring about an rapid mixing of the precursor solutions, resulting in the micronization of $SrCO_3$ crystallite due to the acceleration of formation of many crystal nucleus.

Table 1 BET surface area and crystallite size of $SrCO_3$-Al_2O_3 and $SrCO_3$ supports

Support	Preparation method	BET surface area* / $m^2 g^{-1}$	$SrCO_3$ crystallite size** / nm
$SrCO_3$	PR	0.7	59
$SrCO_3$-Al_2O_3	PR	15	53
$SrCO_3$-Al_2O_3	IM	20	41

* BET single point method.

** $SrCO_3$ (111), Scherrer equation.

Fig. 2 shows the time dependency of catalytic activity of Ru/$SrCO_3$-Al_2O_3 and Ru/$SrCO_3$ catalysts for steam reforming of toluene. Although the toluene conversion of Ru/$SrCO_3$ at a time factor (W/F) of 3.5 g h mol^{-1} was low, the toluene conversions of Ru/$SrCO_3$-Al_2O_3 catalysts at a time factor (W/F) of 3.5 g h mol^{-1} were higher than that of a commercial Ru/Al_2O_3 catalyst at a W/F of 7.0 g h mol^{-1}. This means that the Ru/$SrCO_3$-Al_2O_3 catalysts had extremely superior catalytic activity compared to a commercial Ru/Al_2O_3 catalyst despite low Ru loadings. In addition, the catalytic activity of Ru/$SrCO_3$-Al_2O_3 (IM) was higher than that of Ru/$SrCO_3$-Al_2O_3 (PR) because the surace area of the IM catalyst was higher than that of the PR catalyst. The formation of benzene, which formed by hydrogenating dealkylation ($C_6H_5CH_3 + H_2 \rightarrow C_6H_6 + CH_4$), over the both Ru/$SrCO_3$-$Al_2O_3$ catalysts was a little.

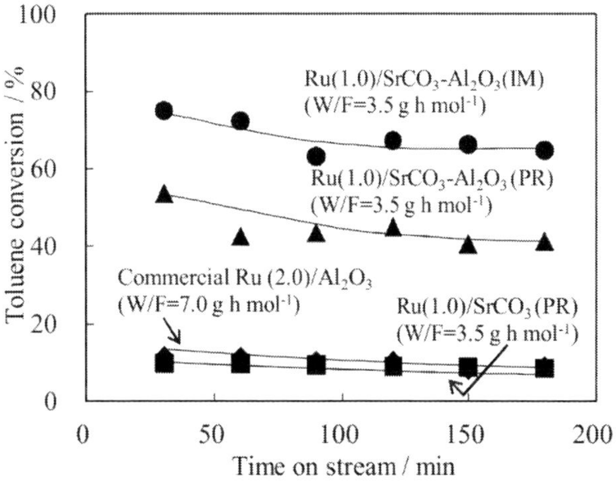

Fig. 2 Time dependency of catalytic activity of Ru/$SrCO_3$-Al_2O_3 and Ru/$SrCO_3$ catalysts for steam reforming of toluene

4 Conclusions

It was revealed that the Ru/$SrCO_3$-Al_2O_3 was effective for steam reforming of toluene, especially the Ru/$SrCO_3$-Al_2O_3 prepared using an impinging stream micro reactor exhibited superior catalytic activity.

References

1) G. Wu, H. Zhou, S. Zhu, *Mater. Lett.*, **61**, 168(2007).
2) H. Iida, K. Noguchi, T. Numa, A. Igarashi, K. Okumura, *Catal. Commun.*, **72**, 101(2015).

Sustainability assessment of technological pathways for biofuel and biochemicals production

M. O. S. Dias [1], L. G. Pereira [2, 3], S. C. Rabelo [2]
[1] Inst. of Science and Technology, Federal Univ. São Paulo (ICT-UNIFESP), Rua Talim, 330, São José dos Campos, Brazil; [2] Laboratório Nacional de Ciência e Tecnologia do Bioetanol (CTBE), Centro Nacional de Pesquisa em Energia e Materiais (CNPEM), Caixa Postal 6192, CEP 13083-970, Campinas, São Paulo, Brazil.; [3] Department of Civil Engineering, University of Toronto, 35 St George St, Toronto, Canada; [4] School of Chemical Engineering, Univ. of Campinas (FEQ-UNICAMP), Av. Albert Einstein, 500 ,Campinas, Brazil.

Biomass and biomass derived feedstock have been increasingly used in the bioenergy and chemical industries for production of renewable products, due to concerns about climate change, fluctuating oil prices and prospects of diminished oil reserves – all of which compromise the sustainability of these industries. Along with biofuels, bioelectricity and biochemicals can promote a shift from a carbon intensive economy to a sustainable biobased one, in which non-food crops, lignocellulosic materials and residues are used as feedstock in efficient conversion processes (Sheldon, 2014).

Evaluating the techno-economic indicators of biorefineries is of upmost importance, since the new routes must be competitive with existing ones based on fossil derived resources (Choi et al., 2015). To be sustainable, biorefineries development must also include analysis of their environmental impacts, through which environmentally-friendly routes can be selected. These analyses, however, are seldom included when new processes are being developed at laboratory scale.

Techno-economic analysis through process modelling and simulation and the subsequent economic analysis can pave the way for the development of sustainable, efficient and financially attractive routes, therefore avoiding experiments that would not be feasible at industrial scale – such as processes based on expensive feedstocks and/or solvents and/or highly energy demanding. Environmental analysis can also indicate the most appropriate characteristics for biorefineries design, selecting routes for further investigation and industrial development based on potential environmental impacts caused by process inputs and outputs

In this study results for techno-economic and environmental analyses of selected technological routes for production of biofuels and biochemicals are provided to demonstrate the need to assess economic and environmental impacts of biorefineries prior to performing laboratory experiments. The Virtual Sugarcane Biorefinery (VSB) framework, developed at the Brazilian Bioethanol Science and Technology Laboratory (CTBE) of the Brazilian Center for Research in Energy and Materials (CNPEM), was used to perform Aspen Plus simulations, economic analysis using electronic spreadsheets and Life Cycle Assessment using Sima Pro. Three case studies will be assessed: integrated first and second generation ethanol production from sugarcane, comparison between production of ethanol and n-butanol in sugarcane biorefineries and the use of different sugarcane hybrids for first generation ethanol production. The conclusions drawn from each case study indicate the directions towards technological development in these areas must be diverted.

If these analyses are carried out in an early stage of the technological development, resources (e.g. solvents, energy, water, feedstock, time) utilization can be optimized and the most suitable routes can be further investigated.

References

Choi, S., Song, C. W., Shin, J. H., Lee, S. Y. (2015). Biorefineries for the production of top building block chemicals and their derivatives. Metabolic Engineering, 28, 223–239.

Sheldon, R. A. (2014). Green and sustainable manufacture of chemicals from biomass: state of the art. Green Chemistry, 16(3), 950–963.

The longitudinal flow of oil and petroleum products in the channels and pipes

Yu.A. Tolchinsky[1], V.E. Ved[1], M.K. Zhantasov[2], V.I. Satayev[2], A.A. Saipov [2]
[1]National Technical University "Kharkiv Polytechnic Institute", 21 Frunze st., Kharkiv, 61002 Ukraine, e-mail: valeriy.e.ved@gmail.com, tel. +38057 7076589;[2] M. Auezov South Kazakhstan State University, 5 avenue Tauke Khan, Shymkent, 160012 Kazakhstan.

The process of heat transfer is considered and the model to describe heat transfer for the transportation of oil and oil products is proposed. The process of heat transfer is considered to be so that takes place in two-phase or three-phase environments. Carrying phase is petroleum or petroleum products. Carried phase (dispersed phase) is water, gas, solid core. The reasons for presence of the solid phase are chemical-physical reactions by type of paraffin forming reaction. The corresponding chemical kinetics of these processes is not considered. Stratal water or the gas is present in the flow in dispersed form. Their distribution has two characteristic scales: macroscopic (about the size of a cross pipe or apparatus) and microscopic. The microscopic scale is determined by the kinetics of emulsification. The curing process and emulsification process are considered to have reached equilibrium. This means that the solid particles are unchanged and set. Equilibrium of emulsification process is achieved by equal velocities emulsification and coagulation. Flowing portion of the flow which located outside the boundaries of the solid core and borders the flow region is considered as a Newtonian with temperature-dependent viscosity. The heat transfer model uses a heat transfer coefficients at the boundaries of the solid core are the flow region. On the boundary of the flow heat transfer coefficients depend on the local flow velocity by the power law with an exponent of one-third and on the boundaries of the solid core with an exponent of one quarter. Two situations are considered. The first situation is when the emulsified phase fills the entire volume of the flow. The second situation is when this phase is concentrated near the boundary flow or solid core. Accounting of multiphase is carried out within multi-phase cell model to the values of viscosity and thermal conductivity. As the result the expressions for these quantities as a function of volume concentration and scale so that the heat transfer coefficients become functions of volume concentrations are obtained. In the case of macroscopic inclusions of the gas phase the contribution of thermal convection inside the inclusion is taken into account. For in such a way determined heat transfer coefficients the equation of convective heat transfer is formulated. This equation is a one-dimensional with the variable of the coordinate along the axis of the tube or apparatus. In this equation the transverse heat transfer is taken into account in the form of surface sources and sinks located at the boundaries of the pipe or the apparatus and the internal borders of the flow region.

The model of intermittency of transient flow of oil

Yu.A. Tolchinsky[1], V.E. Ved[1], M.K. Zhantasov[2], V.I. Satayev[2], A.A. Saipov [2]
[1]National Technical University "Kharkiv Polytechnic Institute", 21 Frunze st., Kharkiv, 61002 Ukraine, e-mail: valeriy.e.ved@gmail.com, tel. +38057 7076589;[2] M. Auezov South Kazakhstan State University, 5 avenue Tauke Khan, Shymkent, 160012 Kazakhstan.

The flow of oil and petroleum products, polymers, organosilicons and other liquids with a viscosity of one order of magnitude greater than the viscosity of water is often carried out in an intermediate hydrodynamic regime. In this regime, the volume occupied by the flow is divided into "laminar" and "turbulent" part so that it can be considered as a similar to two-phase flow in which the role of pseudophase plays laminar and turbulent areas. The model of transient current which contains two main positions is proposed. The first position is that in the flow without borders the intermittency growing through the development of the inherent instability of the flow. Within the Lagrange description of liquid particles in a stream through a sequence of instabilities that are growing, replacing each other and interacting. Then instabilities stabilize and fade. The duration of growth, stabilization and decay depend on the Reynolds number for flow and its value for the boundaries of hydrodynamic regimes - laminar and turbulent as well as on the characteristics of local velocity profile of the liquid particle. The model takes into account the first and second derivatives of the local velocity profile. On the basis of general phenomenological grounds general expressions for the duration of the periods depending on the difference between the values of the flow Reynolds number and hydrodynamic regimes borders, characteristics of flow speed profile and "diffusion" of turbulence are formulated. The essence of the second position is that flow in borders has another source of intermittency. Another source of intermittency is hydrodynamic boundary layer. Because of the greater curvature of the velocity profile near the borders the evolution of instabilities of liquid particles in the boundary layer is growing faster and expressed more strongly. "Boundary" between the boundary layer and the flow core has a fractal character. Between the core flow and boundary layer turbulence is exchanged so that the border of the core flow is a surface source and sink. In particular, some parts of turbulence from the boundary layer fall into the core flow, interacting with portions of the turbulence of the main flow. In the model the interaction of turbulence parts by analogy with the interaction point vortices and dissipative decay of turbulence is considered. The model includes a system of four equations: two equations to determine volume concentrations of laminar and turbulent pseudophases and two equations of like the Navier-Stokes equations to describe the flow pseudo-double-phase fluid with interfacial interaction. The magnitude of this interaction depends on the Reynolds numbers listed above, the first and second derivatives of the flow velocity.

The relationship between the human activities and indoor climate of buildings

P. Kapalo[1], S. Vilcekova[2], F. Domnita[3]
[1]Technical University of Kosice, Dep.of Technical Building Equipment, Vysokoskolska 4, 042 00 Kosice, Slovakia. tel.+421 55 602 4271, e-mail: peter.kapalo@tuke.sk.
[2]Technical University of Kosice, Dep.of Environmental Engineering, Vysokoskolska 4, 042 00 Kosice, Slovakia. tel.+421 55 602 4260, e-mail: silvia.vilcekova@tuke.sk.
[3]Building Services Faculty, Department of Building Services Engineering, Technical University of Cluj-Napoca, 21 Decembrie 1989 Street 128-130, 400 604 Cluj-Napoca, Romania, telephone: +402 64 410 179, e-mail: florin.domnita@insta.utcluj.ro.

The paper describes the partial results of human respiration impact to indoor air quality in space, which was examined. From various other studies it is known, that the maximum concentration of carbon dioxide is determined of 1,000 ppm up to 1,500 ppm in different countries. In the selected room with one occupant person were recorded: the concentration of carbon dioxide, indoor air temperature, relative air humidity, heart-beat intensity (pulse) and blood pressure were recorded. Four measurements with four occupants of different gender, age and weight were performed. The measurements were performed for different human activities – sedentary work with personal computer and stationary bike cycling. The aim of the paper is to research the relation between carbon dioxide concentration and human health.

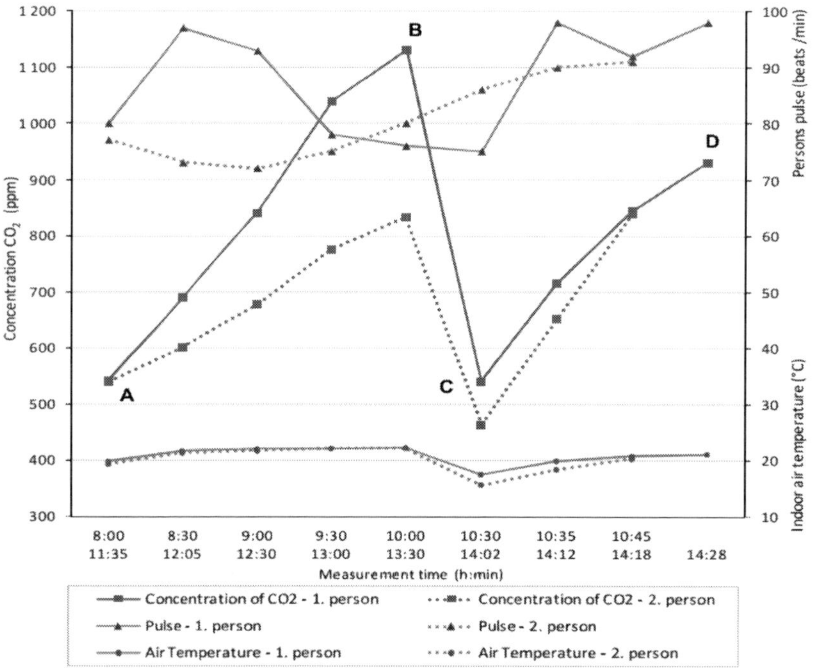

Fig. 1: Carbon dioxide concentration values and heart-beat intensity (pulse) of occupants.

From the analysis it can be concluded that the occupants of a room with increased level of carbon dioxide concentration are making more effort to breathe which can result to earlier fatigue in the workplace and to low productivity of labour.

Studies on liquid flow through the model of the crystallizer segment

W. Suchecki, K. Urbaniec
Płock Campus of Warsaw University of Technology, Department of Process Equipment, Jachowicza 2/4, 09-402 Płock, Poland
tel. +48 24 2622610, fax: +48 24 2626542, e-mail: suchecki@pw.plock.pl

In the sugar industry, in the process of crystallizing the suspension of the afterproduct (crystals III), among others cooling crystallizers with the vertical movement of the tubular cooling elements are utilized. According to the designers and users, in this crystallizer type there is potential for optimization, which could allow for a significant reduction of the apparatus dimensions.

Liquid flow through the model of a single tubular segment (part of the moving cooling element) of the cooling crystallizer was studied. An experimental model of the crystallizer segment was built. The model was utilized for the experimental research on liquid flow, conducted in parallel with the simulation of the segment operation. The method of digital particle image velocimetry was applied to analyse the experimental flow. The experimental results enabled verification of the assumptions of the numerical model based on CFD code Fluent. The finite volume method was applied in the numerical simulation.

The results of experimental research and of numerical simulations were presented to improve understanding of physical conditions that promote or disturb homogenous velocity distribution of the liquid flowing around the segment of the crystallizer model. Additionally, the usefulness of the adopted research methodology in the modeling of flow phenomena and in precise quantitative analysis of the studied flow was confirmed.

Studies on velocity fields around the cavitation vortices generated by the model of a propeller blade

W. Suchecki, K. Urbaniec
Płock Campus of Warsaw University of Technology, Department of Process Equipment, Jachowicza 2/4, 09-402 Płock, Poland
tel. +48 24 2622610, fax +48 24 2626542, e-mail suchecki@pw.plock.pl

The elimination of dangers caused by cavitation phenomena is an important issue in the design of rotary hydraulic machines. These dangers are: cavitation erosion, efficiency decrease as well as vibration and noise. One of the most intensive and dangerous forms of cavitation is vortex cavitation that accompanies the operation of hydraulic machines in which components comprising rotating blades are applied.

Velocity fields around cavitation vortices generated by the model of a propeller blade were experimentally studied in a cavitation tunnel. Flow images were recorded using high-speed camera and processed using digital particle image velocimetry (DPIV) complemented with computer-aided techniques that had been developed for the purpose of this research. These techniques included the removal of image distortions on the basis of calibration mask, determination of instantaneous velocity distributions and removal of air-bubble traces from flow images.

Examples of the results of experimental studies were presented in the form of velocity fields determined in the longitudinal plane as well as in three transverse planes remote from the blade. Instabilities of the cavitating vortex stream and of the local liquid-flow velocity in its surrounding were detected. The impact of the angle of attack of propeller blade on the instability of the vortex stream and the impact of vortex core parameters on the local velocities of the surrounding liquid were determined.

Thermal analysis of the finned channel of a pneumatic pulsator using numerical simulation and thermography

J. Wernik, K.J. Wołosz
Warsaw University of Technology, Plock Branch, Department of Process Equipment
Al. Jachowicza 2/4, 09-402 Plock, Poland
phone +48 24 2622610, e-mail: wernik@pw.plock.pl

The research on heat conduction through the finned channel of pneumatic pulsator was discussed. Rectangular fins are located on the channel.

3-D model of the channel was created. The developed model of 3-D channel of an industrial pulsator has been imported into COMSOL Multiphysics. This program is the environment for modelling and solving scientific and engineering problems based on partial differential equations. As a result, the solution was obtained in the form of temperature distribution on finned channel of pneumatic pulsator and the temperature distribution of the air flowing over the pulsator.

Numerical methods are subject to certain errors, hence it is essential to verify the results.

Real object were subject to thermographic tests in order to verify the correctness of the numerical model. The FLIR SC7600 camera was used. Non-contact diagnostic method based on the observation of infrared radiation has enabled visualization, registration and interpretation of temperature distribution on the surface of the tested channel. The resulting thermal images were compared with the temperature distributions designated using a computer.

The use of thermography in laboratory tests to verify numerical models is possible. Therefore, when designing new devices, both techniques can be used methodically.

Towards carbon neutral combustion. Cofiring solid particles and gaseous fuel

M. M. Fawzi, T. Prodanuks, I.Veidenbergs, D. Blumberga
Riga Technical University, Institute of Energy Systems and Environment, Azenes street 12-1, Riga, Latvia, LV-1048; tel. +371 67089943, e-mail: toms.prodanuks_1@rtu.lv

One of the main obstacles to transfer to carbon neutral energy management is technological transition. An important task is to change fuel used in the boiler furnace from natural gas (fossil fuel) to biomass. The easiest but most expensive solution is the installation of biomass boiler. Other solutions include burner replacement or additional installation, but these options are more complex from the engineering and technical point of view.

By replacing natural gas with wood particles, boiler output is reduced, because value of combustion heat, furnace load and specific combustion parameters are lower, as well as temperature and pollution with nitrogen oxides NOx decrease. It is especially important when energy end user improves energy efficiency and reduces energy consumption. As a result there is excess installed capacity in the boiler house.

Experimental data analysis has been conducted in the research. In the research independent variables that significantly influence fuel combustion when two different phase fuels are combusted simultaneously or when existing fuel is replaced by other fuel have been looked for.

This research is important as natural gas will gradually become replaced by biogas that is carbon neutral. Moreover, biomass cofiring with solid wood particles will reduce the impact on climate change.

Universal multi-functional secondary catalyst carriers for purification of gas emission of thermal power equipments

E.V. Krasnokutskiy [1], B.B. Makhanov [2], V.E. Ved [1], M.I. Satayev [2], A.V. Ponomarenko [1], A.A. Saipov [2]
[1] National Technical University "Kharkiv Polytechnic Institute", 21 Frunze st., Kharkiv, 61002 Ukraine, e-mail: valeriy.e.ved@gmail.com, tel. +38057 7076589; [2] M. Auezov South Kazakhstan State University, 5 avenue Tauke Khan, Shymkent, 160012 Kazakhstan.

In the production units of catalytic converters of exhaust gases of thermal power equipments in totally and internal combustion engines in particular are used in various engineering design solutions on registration of structural features of these units, designed to improve the efficiency of gas emissions purification, such as mutual location and geometry of the branch pipes of input-output exhaust gases, presence of additional branch pipes for supplying auxiliary reagents, using of thermal insulation block of catalytic reduction [3] or the system of its heating, using of various classes of inorganic compounds as a catalyst carrier (ceramic and metal carriers) and varying within wide limits geometry and structure of the catalyst.

The infliction of catalyst on a metal or ceramic surface by known means is associated with a variety of disadvantages, in particular for the reason that can only be used a certain combination of catalyst material and carrier material, so that should be satisfactory and stable adhesion of the catalytic material to the carrier surface (metal or ceramic) the operating conditions of the catalyst. It should be observed corresponding to the ratio between the temperature coefficients of linear expansion (TCLE) in the systems of substrate - catalytic layer, since cyclic temperature changes are occurred thermal stress is directly proportional to the difference of thermal expansion values in the system media is the catalyst that leads to the destruction of the coating. The degradation of the catalytic coating can also be caused by mechanical impact, dust, and other factors.

To solve this problem, a method was developed for controlled formation of catalytically active centers of complex micro-relief on surface of heat-resistant alloys and ceramic materials (primary carriers). It is based an idea in fundamentals that is based on the chemical compositions and new technology of special amorphous or glass-ceramic adhesives (secondary carriers) of the primary media formation, which are applied to the surface by layers, thickness less than 1 μm. Glass-containing coverage of primary carrier (metal or ceramic) is intended for fixation on it's at the glassing temperatures of the secondary carrier layer of catalytically active coating to any value of thermal coefficient of linear expansion (TCLE).

Water recovery and reuse in the fractionation of protein hydrolysate by ultrafiltration and nanofiltration membranes

Ricardo Abejón*[a], Azucena Abejón[a], Marie-Pierre Belleville[b], Aurora Garea[a], Angel Irabien[a], Jose Sanchez-Marcano[b]
[a]Chemical and Biomolecular Engineering Department, Universidad de Cantabria
Av. Los Castros, 39005 Santander, Spain
[b]Institut Européen des Membranes, ENSCM/UM/CNRS - Université de Montpellier
Place Eugène Bataillon, 34095, Montpellier, France
ricardo.abejon@unican.es

The fractionation of a protein hydrolysate obtained from tuna processing by-products by means of a membrane cascade integrating ultrafiltration (UF) and nanofiltration (NF) membranes was proposed in order to separate and purify the protein fraction between 1 and 4 kDa, which is the most interesting for nutraceutical purposes. A simulation model, based on mass balances and empirical equations for describing permeate flux and rejection of protein fractions, was developed and complemented with a simple cost estimation model. The product purity (49.3 %) and the process yield (62.6 %) were independent of the total water consumption of the process, but high water consumptions were required to maintain the total protein content of the stream below upper bounds that assured the absence of membrane clogging. The implementation of a water recovery system, based on an additional tight NF stage, implied improvements in both environmental and economic aspects of the process.

Biomass gasification with dolomite and olivine particles as a bed inventory in presence of catalytic ceramic filter

S. Rapagnà, G. Spinelli
Faculty of Biosciences, Agro-Food and Environmental Technology; University of Teramo, via Balzarini,1 64100 Teramo
Tel +390861266893, e-mail: srapagna@unite.it

The use of biomasses as a renewable energy sources to produce liquid fuels by Fischer-Tropsch synthesis, as well as electrical energy and heat, can contribute significantly to reduce the CO2 emissions to the atmosphere. Nowadays biomasses account for more than 35% of the primary energy consumption in the development countries. Due to the fact that, it is estimated that in the near future, most of the world population will live in these countries, biomasses will be one of the main renewable energy sources of the future. To avoid transportation cost of the biomasses, it is preferable to utilize small transformation plants able to be fuelled with biomasses produced around them. Fluidised bed reactor should be the best solution to transform biomasses having different physical properties in a energy vector, due to its possibility to operate continuously, at high temperatures, and to utilize catalysts into the reactor and outside it to drastically reduce tar, NH3 etc. in the producer gas. The fluidising agent could be air, steam or pure oxygen. Air, that it is necessary to assure temperature level inside the gasifier above 800 °C, gives the production of a gas with low heating value due to the presence of large amount of nitrogen. Oxygen can be the best solution, but the high operating costs associated with its production, made this gas not suitable to be utilised in small biomass gasification plants (Wang and Chen, 2013). The presence of steam into the gasifier is desirable in order to permit the reforming reaction of organic compounds, produced in the devolatilization process of the biomass particles, and for the gasification of char particles inside the gasifier. Tar content in the producer gas can be reduced by utilising olivine particles as a bed inventory. It has been observed that the catalytic acivity of the olivine particles increase with the gasification time due to the formation on the bed particle surface, of a layer of CaO. Compared to fresh olivine, a CaO-coated olivine bed material it is able to decrease tar content in the product gas by 82% (Kirnbauer et al., 2012). In order to eliminate completely the tar in the producer gas by using only one reactor, it has been proposed to utilize catalytic ceramic filters placed inside the fluidised bed gasifier (Rapagnà et al. 2014, D'Orazio A., et al. 2015). The effectiveness of the ceramic filter catalyst to eliminate the tar is strongly dependent on the tar concentration in the product gas. To drastically decrease the amount of tar that reaches the ceramic filter catalyst, dolomite and olivine particles will be utilised as bed inventory. The dolomite fines elutriated from the fluidised bed will deposited on the surface of the ceramic candles. With this arrangement, the dolomite particles will act both as primary catalyst, because is part of the bed inventory, and as a secondary catalyst, because is on the surface of the ceramic filter after the fluidised bed. Results will be presented at the conference.

References

Wang L. and Chen Z., 2013, Applied Thermal Engineering, 59, 278-282.
Kirnbauer F. et al. 2012. Fuel, 95, 553-562.
Rapagnà and Giacomo Spinelli, 2015 Chemical Engineering Transactions, 43, 397-402.
D'Orazio A., et al. 2015, International Journal of Hydrogen Energy, 40, 7282-729.

Microwave pyrolysis of the spruce sawdust for producing high quality syngas

P. Lestinsky, B. Grycova, I. Koutnik, A. Pryszcz
VSB-TU Ostrava, Institute of Environmental Technology, 17. listopadu 15, 70833 Ostrava, Czech Republic;
tel. +420 597 327 331, e-mail: pavel.lestinsky@vsb.cz

Currently, a growing demand for alternative energy sources to replace traditional fossil fuels. Although fossil fuels in the world still enough (still found new reserves or mining techniques – e.g. Shale gas) is needed looking for a solution for future substitute. Wastes from the wood processing industry could be serving as a potential feedstock for energy production, e.g. hydrogen.

Production of energy from wood biomass is as old as humanity itself. In the recent 20 years there has been enormous progress in research of torrefication, pyrolysis and mainly in gasification of the biomass. Products from conventional pyrolysis of the biomass are gas known as syngas, liquid condensate known as bio-oil and char termed as bio-char. Most of the articles dealing with convection pyrolysis of biomass are focused on engaged in the production of bio-oil. Only some works deal with the possibility of producing high-quality syngas (H_2+CO) or hydrogen alone. For this purpose, it seems most appropriate technology microwave pyrolysis.

Microwave irradiation is an alternative heating method and has already been successfully applied to biomass pyrolysis. Compared with convectional heating processes where heat is transferred from the surface to the core of the material through conduction driven by temperature gradient, microwave induce heat at the molecular level by direct conversion of the electromagnetic energy into heat, and therefore, they can provide uniform internal heating for material particles. Microwave induce pyrolysis promote the heterogeneous catalytic reaction and produce greater concentration of syngas and hydrogen in the product gas than convectional pyrolysis. In addition, the instantaneous response if microwave makes it easier for a rapid start-up and shut-down.

The biomass is mixed with a suitable material which significantly absorbs microwave radiation for generation of intensive heating in the reactor. In our case, the spruce sawdust was mixed with char from previous pyrolysis of spruce sawdust. Char from the pyrolysis of sawdust also contains small amounts of ash (up to 5 %), which includes alkali metals and alkaline earth metals. The presence of such Char has a catalytic effect on the pyrolysis process, the cracking of tars and phenolic compounds to the simple gases (H_2 and CO).

In this work we were studied the influence of various parameters of microwave pyrolysis of pine sawdust (microwave power, particle size, the ratio of sawdust and char, pyrolysis duration, etc.) to maximize the yield of hydrogen, or syngas. Experiments were carried out in a conventional microwave oven with power up to 1 kW. It was also studied the influence of parameters on the number of products (gas, liquid and solid), but also the amount of waste water in the liquid condensate.

This work was financially supported by the Ministry of Education, Youth and Sports of the Czech Republic in the "National Feasibility Program I", project LO1208 "Theoretical Aspects of Energetic Treatment of Waste and Environment Protection against Negative Impacts".

Optimization of EDM process for water purification

A. G. Goryunov, I. S. Nadezhdin, [1]F. Manenti, [1]D.Papasidero
Tomsk Polytechnic University, Dept. of Electronics and Automation of Nuclear Plants, Lenina St, 30, Tomsk 634050, Russia; tel.: +7 906 954 26 19, alex1479@tpu.ru; Tomsk Polytechnic University, Dept. of Economics, Lenina St, 30, Tomsk 634050, Russia;
[1]Politecnico di Milano, Dept. di Chimica, Materiali e Ingegneria Chimica „Giulio Natta", Piazza Leonardo da Vinci 32, 20133 Milano, Italy.

In modern world, water purification is main task for population needs, as well as the industrial sewage water purification. Typically, sewage water contains many different harmful impurities. It is adversely affected on humans and environment. Therefore modern water purification plants are organized a multistage purification process. Nowadays, there are known many approaches to water purification: [1]: ion exchange, coagulation, reverse osmosis, distillation, aeration. They are become widely used, but they have several disadvantages. Scientists are developing new technologies to improve the efficiency of water purification plants. Recently, electrical energy is started to use for water purification. One of the method of water purification based on electric energy usage is a method of purification using electrical discharge machining of metal balls in water.

Electrical discharge machining (EDM) process is used in industry for the machining of metal workpieces [2].During the processing of metal balls by electric discharge, erosion products are formed. Erosion products adsorbed ions of harmful impurities contained in the water. This method uses cheap raw materials (metal balls) and allows to clean the water from a wide range of harmful impurities. However, the proposed EDM method has got peculiarities. Typically, electro erosion products are highly dispersed metallic particles. The dimensions of the particles are ranging from 1 to 100 nm. The small metal particles are completely oxidized, and the larger ones are oxidized superficially.

This work proposes optimization of water purification process by EDM method. It is based on developed mathematical model, which is taken into account the influence of electric discharge parameters (discharge voltage, discharge current, pulse-on-time) on concentration of harmful impurities in clean water. The model predictive control (MPC) approach is applied for this reason. The result of optimization is the reducing energy cost.

References

1. Lutchmiah K., Verliefde A.R.D., Roest K., Rietveld L.C., Cornelissen E.R., Forward osmosis for application in wastewater treatment: A review, Water research, 58, 179-197, 2014.
2. Muthuramalingam T., Mohan B., A review on influence of electrical process parameters in EDM process, Archives of civil and mechanical engineering, Vol. 15, 87–94, 2015.

Optimization of uranium crystallization process by using MPC approach

A. O. Ochoa Bique, A. G. Goryunov, [1]F. Manenti
Tomsk Polytechnic University, Dept. of Electronics and Automation of Nuclear Plants, Lenina St, 30, Tomsk 634050, Russia; tel.: +7 923 404 3589, anthonob@tpu.ru; [1]Politecnico di Milano, Dept. di Chimica, Materiali e Ingegneria Chimica „Giulio Natta", Piazza Leonardo da Vinci 32, 20133 Milano, Italy.

Crystallization is a multi-component and multi-phase complex process, which is known as method for separating uranium[1]. Many studies of uranium crystallization are reported in this field[2-4]. They are focused on material's nano-purity achievement. In the framework of "Proryv" project stream a plant for refining of extraction and crystallization treatment process of spent nuclear fuel (SNF) is developed for this reason at the present time in Russia. There needs a computation-intensive for the model simulation of all plant's elements and concurrent operation. There is expected of chemical process simulation and hardware modeling at the same time. The linear crystallizer isa final part of uranium purification there[5]. Automation is implemented at all processing step. The recent interest is improving controllability and operability for safety purposes, but simulation of crystallization process still has many problems[6].

Thus, the model predictive control (MPC) approach was included in the system. It allowed to predicted of controlled variable changes for the time being ahead and provided the best trajectory of the process[7].

This work proposes optimization of crystallization process. It is based on developed mathematical model, which describes the heat and mass transfer processes involved in the process[8]. The result of optimization is the increasing of material's pure state.

References

1. Chikazawa, T., et al., *Batch crystallization of uranyl nitrate.* Journal of Nuclear Science and Technology, 2008. **45**(6): p. 582-587.
2. Homma, S., et al., *Flowsheet study of U-Pu co-crystallization reprocessing system.* Journal of Nuclear Science and Technology, 2008. **45**(6): p. 510-517.
3. Henrich, E., H. Schmieder, and K. Ebert, *Combination of TBP extraction and nitrate crystallization for spent nuclear fuel reprocessing.* Inst. Chem. Eng. Symp. Ser., 1987. **103**: p. 191-205.
4. Shepherd, J.S., et al., *Mathematical modelling of the pre-oxidation of a uranium carbide fuel pellet.* Computers and Chemical Engineering, 2015.
5. Veselov, S., et al., *Mathematic Simulation of Crystallization Refining Process of Spent Nuclear Fuel Reprocessing Desired Products in Linear Crystallizer.* Advanced Materials Research, 2015. **1084**: p. 666-672.
6. Nagy, Z.K., M. Fujiwara, and R.D. Braatz, *Modelling and control of combined cooling and antisolvent crystallization processes.* Journal of Process Control, 2008. **18**(9): p. 856-864.
7. Manenti, F., et al., *Operational optimization of reverse osmosis plant using MPC*, in *Chemical Engineering Transactions*. 2015. p. 247-252.
8. Ochoa Bique, A.O., A.G. Goryunov, and F. Manenti, *A comparison of simulation techniques for uranium crystallization process*, in *Chemical Engineering Transactions*. 2015. p. 793-798.

Production of energy from biomass: near or distant future prospects?

A. G. Goryunov, [1]N.N. Goryunova, [1]A.O. Ogunlana, [2]F. Manenti
Tomsk Polytechnic University, Dept. of Electronics and Automation of Nuclear Plants, Lenina St, 30, Tomsk 634050, Russia; tel.: +7 906 954 26 19, alex1479@tpu.ru; Tomsk Polytechnic University, Dept. of Economics, Lenina St, 30, Tomsk 634050, Russia;
[2]Politecnico di Milano, Dept. di Chimica, Materiali e Ingegneria Chimica „Giulio Natta“, Piazza Leonardo da Vinci 32, 20133 Milano, Italy.

The high cost of energy, limitation/exhaustible and the non-renewable resources lead to the search for alternative energy sources. The types of energy sources that would not hinder the environment and could compete in efficiency and cost effectiveness against the traditional forms of energy. This could be biomass that is currently the sixth in the ranking of supply availability and the fifth in productivity among renewable energy sources. Biomass refers to the total mass of plants and animals present in biogeocenose, in a definite size or level. Biomass is one of the most promising raw materials to satisfy the growing demand for renewable energy and the needs of "green chemistry".

Among the leading countries that use biomass for energy:US, Germany, China, Brazil, Japan. The Research is initiated to assess the prospects of energy production from biomass. The experience of developed European countries in the commercialization project of energy production from biomass is being studied, analyzing the technologies [1-4]. Also being studied is the necessary conditions for the successful implementation of the project, which revealed the possibility of accepting the usage of different technologies in Russia. Applying the tools or instruments of business planning, research was made on the market of biofuels in Europe and on their competitive ability in biomass.

The project researched the possibility of introducing to the market production of energy from waste timber industrial complex and wood-processing. On basis of SWOT-analysis, technologies identify as those external environmental factors and the intrinsic features of the technologies that should be considered when developing strategies of issued product to market.

The study draws conclusions about the competitiveness of the technologies of energy production from biomass among the substitutes in the present economic conditions in Europe and Russia, their demand in the market for energy sources. The authors seek to answer the question about the prospects of these technologies in the future and what needs to be done today for the sustainable use of energy resources.

References

1. Pierucci, S., Ranzi, E., Manenti, F., *Generalized simulation tools for coal, biomass and organic waste gasification processes*. Chemical Engineering Transactions, 45, 1771-1776, 2015.
2. Manenti, F., Leon-Garzon, A.R., Rossi, F., Ravaghi-Ardebili, Z., Pierucci, S., *Solar-supplied Biomass Gasification for the One-step Synthesis of Methanol and Dimethyl Ether*. Chemical Engineering Transactions, 39, 1219-1224, 2014.
3. Ravaghi-Ardebili, Z., Manenti, F.,*Unified modeling and feasibility study of novel green pathway of biomass to methanol/dimethylether*. Applied Energy, 145, 278-294,2015.
4. Ranzi, E., Corbetta, M., Manenti, F., Pierucci, S., *Kinetic Modeling of the Thermal Degradation and Combustion of Biomass*, Chemical Engineering Science, 110, 2-12, 2014.

Sericin and alginate blend as matrix for incorporation of diclofenac sodium

Jacyara M. M. Vidart[a], Mariana Nakashima[a], Thiago L. da Silva[a], Paulo C. P. Rosa[b], Marcelino L. Gimenes[c], Melissa G. A. Vieira[a], Meuris G. C. da Silva[a]*
[a]School of Chemical Engineering, University of Campinas, UNICAMP, 13083-825, Campinas – SP, Brazil
[b]School of Pharmaceutical Sciences, University of Campinas, UNICAMP, 13083-859, Campinas – SP, Brazil
[c]Department of Chemical Engineering, State University of Maringá, UEM, 87020-900, Maringá – PR, Brazil
meuris@feq.unicamp.br

Diclofenac sodium (DS) is a non-steroidal anti-inflammatory widely used clinically and has recurring side effects in gastrointestinal system. In addition, it requires multiple daily doses to maintain its therapeutic drug-blood level. Therefore, the modification of the form of presentation of the DS becomes desirable to improve patient compliance and decrease side effects. Sericin is a globular protein that offers many desirable characteristics for the incorporation of drugs. It is presented in the silkworm cocoons (*Bombyx mori*) and is usually discarded in the wastewater of degumming silk processing. Sodium alginate is a natural polysaccharide extracted from brown seaweed that has an abundant use in drug delivery systems. The blend of sericin and alginate may provide characteristics more suitable for improving the encapsulation. The aim of the current work is to evaluate the content of DS incorporated in sericin/alginate particles. The sericin solution was obtained by degumming process in autoclave (1 kgf/cm^2, 40 min) and the sodium alginate used was of analytical grade (Sigma-Aldrich). Blends of sericin and alginate with DS incorporated were produced in various formulations. These blends were dripped in $CaCl_2$ solution, in the ionic gelation technique, in order to produce the particles. The different formulations used were compared by evaluating the efficiency of incorporation of DS in the sericin/alginate particles. Additionally, the surface morphology was analyzed by SEM - Scanning Electron Microscopy, and the size of the particles was evaluated by OM - Optical Microscopy. The results showed that the efficiency of incorporation reached values in the range of 91.1 ± 2.4 % to 75.5 ± 2.1 %, and the SEM analysis proved the incorporation of the drug in the blend matrix.

SOFC running on steam reforming of biogas: External and internal reforming

S. Authayanun[1], T. Pornjarungsak[2], T. Prukpraipadung[2], D. Saebea[3], A. Arpornwichanop[4]
Y. Patcharavorachot[2]
[1]Srinakharinwirot University, Department of Chemical Engineering, Rangsit Nakhon Nayok Rd., Ongkharak, Nakhon Nayok 26120, Thailand
[2]King Mongkut's Institute of Technology Ladkrabang, Department of Chemical Engineering, Chalongkrung Rd., Ladkrabang, Bangkok 10520, Thailand; tel. +6623298360, e-mail: yaneeporn.pa@kmitl.ac.th
[3]Burapha University, Department of Chemical Engineering, Bangsaen Rd., Muang Chonburi, Chon Buri 20131, Thailand
[4]Chulalongkorn University, Department of Chemical Engineering, Phyathai Rd, Pathumwan, Bangkok 10330, Thailand

Solid oxide fuel cell (SOFC) has attracted considerable interest for distributed power sources due to its high efficiency. The high temperature operation (1073-1273 K) leads to the flexibility of using various fuel types, i.e., methane, methanol and ethanol. In this work, biogas is used as fuel for SOFC because of its availability and renewability. However, it need to convert such fuel into a hydrogen-rich gas required for the electrochemical reaction on the anode side. In general, there are two main approcahes to incorporate a fuel processor in the SOFC operation: external reforming and internal reforming. Each approach has different advantages and disadvantages. This work aims to identify the favourable operating condition of SOFC incorporating with different approaches of reforming based on a thermodynamic calculation.

In this work, two approaches of SOFC operation fuelled by biogas are considered: one where the endothermic steam reforming and electrochemical reaction are operated in different units (external reforming SOFC) and another where biogas is introduced to the SOFC stack and directing reformed on the anode side (internal reforming SOFC). The model of SOFC system is designed and developed using Aspen Plus simulator software. Equilibrium gas composition obtained from steam reforming reaction can be calculated based on Gibbs free energy minimization method. Electrochemical equations taking into account all voltage losses (activation, ohmic and concentration losses) are written in a calculator block by Fortran code. Effect of operating conditions, i.e. temperature (T_{SOFC}), pressure (P_{SOFC}) and steam to carbon molar ratio (S/C) on SOFC performances is examined.

From our preliminary study, it was found that under the same operation conditions (T_{SOFC} = 1,173 K, P_{SOFC} = 3 atm and S/C = 0.5), the internal reforming SOFC has higher electrical efficiency (96.81%) than external reforming SOFC (85.69%). Although this result is expected, the thermal efficiency is further determined to evaluate the operation of SOFC.

Study on the PEDOT:PSS electric properties changes for high power conversion efficiency of organic-inorganic perovskite solar cell

Ki-Hwan Hwang, Hyeon Jin Seo, Yong-Min Lee, Hyun Ji Choi, Yul Hee Lee, Dong In Kim, Sang Hun Nam, Jin-Hyo Boo*
Sungkyunkwan Univ., Dep. chemistry, 2066, SEOBU-RO, JANGAN-GU, SUWON-SI, GYEONG GI-DO, Korea;
tel. +82 312905972, e-mail: nercisus@naver.com, fax: +82 312907075.

Organic-inorganic hybrid perovskite solar cells are one promising candidate for future clean energy sources. The conducting polymer, PEDOT:PSS poly(3,4-ethylene dioxythiophene):poly(styrenesulfonate), is used as hole collecting layer. However, basic research into the effect of the PEDOT:PSS electric properties changes is lacking. In this report we demonstrate that the conductivity of PEDOT:PSS thin films can be enhanced by modifying structures with dipping process in various solution such as polar protic solvents (methanol, ethanol, 2-propanol). And we study on the effect of thermal condition for conductivity enhancement. At the same time, the molecular structures and electrical properties of films were investigated by FE-SEM, Raman spectroscopy and 4-point probe, respectively. To determine power conversion efficiency(PCE) of solar cell, we applied the electric properties changed PEDOT:PSS film on the perovskite solar cells. we used the solar simulator to confirm the effect of electric properties changes on the PCE.

Texture-etched glass substrate for Perovskite solar Cell to reduce the thickness of the thin film

Dong In Kim[1], Sang-Hun Nam[2], Ki-Hwan Hwang[1,2], Yong-Min Lee[1], Jin-Hyo Boo[1,2*]
[1]Department of Chemistry, Sungkyunkwan University, Suwon 440-746, Korea
[2] Institute of Basic Science, Sungkyunkwan University, Suwon 440-746, Korea
[*]Corresponding author e-mail: jhboo@skku.edu

Glass texturing is a sufficient method for changing the surface morphology to enhance a light trapping. In this paper, it is applied on the hybrid inorganic-organic solar cell (perovskite solar cell) for enhancement light trapping and current density. We have wet chemical etched fluorine doped tin oxide(FTO) coated glass substrate using diluted HF solution. The light scattering and trapping properties can be controlled by varying either etchant concentration or etching time. We compared textured perovskite solar cell and non-textured perovskite solar cell. As a result, we reduced absorption layer of textured glass was applied to perovskite solar cell. Its PCE was similar to non-textured was applied to perovskite solar cell due to the light trapping layer effect on short-circuit current density (Jsc) and PCE increasement when we applied it on the Perovskite solar cell.

The influence of fouling build-up in condenser tubes on power generated by a condensing turbine

S. Alabrudzinski, M. Markowski, M. Trafczynski, K. Urbaniec
Warsaw University of Technology, Department of Process Equipment, Plock Campus, Jachowicza 2/4, 09-402 Plock, Poland; tel. +48600218104,
e-mail: slawomir.alabrudzinski@wp.pl

In a thermal power plant steam circuit, steam generated in a high-pressure boiler flows through a turbine coupled with rotary generator and is subsequently condensed in a water-cooled shell-and-tube condenser. The power generated by the turbine is correlated with condenser inlet pressure which is directly influenced by condensation pressure. At a temperature of cooling water not exceeding 35 °C, a typical level of pressure at the outlet of the turbine is below 10 kPa (absolute value). As a rough estimate, an increase in condensation pressure of 1 kPa results in power drop by approximately 1 %. In practical terms, even a minuscule increase in the condensation pressure is followed by economic losses for the plant operator. In order to cover the power demand, it is necessary to operate the steam cycle at reduced efficiency. Consequently, the fuel consumption in the steam boiler is increased, as is the emission of carbon dioxide in the flue gas discharged from the boiler.

The fouling build-up and time-dependent increase in thermal resistance is one of the most crucial phenomena affecting the overall condenser performance. In order to reduce the fouling build-up in tubes, many power-block condensers are equipped with specialized cleaning systems. However, such preventive methods are not fully effective, especially with tube-side fluid being seawater so that typically, the average value of the thermal resistance of fouling is close to 0,0001 m^2K/W. Assuming a typical value of the heat transfer coefficient across clean condenser tubes about 3500 W/m^2K, the influence of the average fouling layer is equivalent to reducing that coefficient to 2600 W/m^2K, that is, by 25 %. A corresponding reduction in the transferred heat stream leads to increased pressure in the condenser and reduced power generated by the turbine.

In this paper, the CFD based estimation method is proposed to investigate the influence of fouling build-up rate in condenser tubes on power generated by turbine. Condenser operation parameters as well as fouling build-up curves are based on literature data for real power-block condensers. The local values of the heat transfer coefficient across condenser tubes are calculated taking account of the fouling thermal resistance along with the convective heat transfer coefficients at shell-side and cooling-water inside the tubes. Bearing in mind that the number of tubes in a typical power-block condenser is very large (for 300 MW condensers may exceed 20,000 tubes in bundle) the continuum (porous zone) approach was adopted. The geometry model of the tube bundle is two-dimensional and the phenomena of heat, mass and momentum transfer were taken into account. The numerical model of condenser operation was implemented using ANSYS Fluent package using its UDF (User Defined Functions) feature to reproduce continuum-type operation of tube bundle zone. The steam turbine power was estimated using Aspen HYSYS package and turbine expander model.

Conclusions and the estimation method for use by the power plant operator were presented.

The effect of time-dependent increase in fouling build-up rate on power generated by steam turbine were discussed.

Cellular automata application for simulation of uranium crystallization process

A. O. Ochoa Bique, D. A. Serikov, A. G. Goryunov, [1]F. Manenti
Tomsk Polytechnic University, Dept. of Electronics and Automation of Nuclear Plants, Lenina St, 30, Tomsk 634050, Russia; tel.: +7 923 404 3589, anthonob@tpu.ru; [1]Politecnico di Milano, Dept. di Chimica, Materiali e Ingegneria Chimica „Giulio Natta", Piazza Leonardo da Vinci 32, 20133 Milano, Italy.

Crystallization is a complex process in the chemical industry and it is known as method for separating uranium[1]. It is important to carry out widely research in this field [2, 3].Development of modern technology for nuclear waste handling is focused on getting of material in the pure state. Aplant for refining of extraction and crystallization treatment process of spent nuclear fuel is developed in Russia for this reason. There is expected model simulation of all plant's parts, whose are included chemical process modeling and hardware modeling at the same time. The simulation of crystallization process still causes many problems [4, 5].

The cellular automata method is proposed as application to simulate of uranium nitrate crystallization process[6]. This approach takes into account the physical laws and it imitates in a digital format allowing specifying complex boundary conditions, considering complex and midline phase transitions and putting forward hypotheses regarding the phase formations and concentration and temperature fields [7]. Therefore, the use of cellular automata method allows monitoring the system.

This work proposes application of cellular automata for simulation of uranium crystallization process and detection of unsteady regimes of its behavior. This approach monitoring a crystal formation is able to achieve the material's nano-purity.

References

1. Lindenberg, C., et al., *Design and optimization of a combined cooling/antisolvent crystallization process.* Crystal Growth and Design, 2009. **9**(2): p. 1124-1136.
2. Henrich, E., H. Schmieder, and K. Ebert, *Combination of TBP extraction and nitrate crystallization for spent nuclear fuel reprocessing.* Inst. Chem. Eng. Symp. Ser., 1987. **103**: p. 191-205.
3. Shepherd, J.S., et al., *Mathematical modelling of the pre-oxidation of a uranium carbide fuel pellet.* Computers and Chemical Engineering, 2015.
4. Lakerveld, R., et al., *The application of a task-based concept for design of innovative industrial crystallizers*, in *Computer Aided Chemical Engineering*. 2009. p. 909-914.
5. Nagy, Z.K., M. Fujiwara, and R.D. Braatz, *Modelling and control of combined cooling and antisolvent crystallization processes.* Journal of Process Control, 2008. **18**(9): p. 856-864.
6. Ochoa Bique, A.O. and A.G. Goryunov, *Simulation of the Uranium Crystallization Process Using Cellular Automata.* Advanced Materials Research, 2015. **1084**: p. 72-76.
7. Toffoli, T. and N.H. Margolus, *Invertible cellular automata: A review.* Physica D: Nonlinear Phenomena, 1990. **45**(1-3): p. 229-253.

Dynamical computer aided model of centrifugal extractors cascade

A. O. Ochoa Bique, E. P. Zelenetskaya, A. G. Goryunov, [1]F. Manenti
Tomsk Polytechnic University, Dept. of Electronics and Automation of Nuclear Plants, Lenina St, 30, Tomsk 634050, Russia; tel.: +7 923 404 3589, anthonob@tpu.ru; [1]Politecnico di Milano, Dept. di Chimica, Materiali e Ingegneria Chimica „Giulio Natta“, Piazza Leonardo da Vinci 32, 20133 Milano, Italy.

High-quality fuel is fundamental to ensuring the effective operation of nuclear plants.

Refinery with extracting technologies is most effective and commonly used method. The extracting process of refinery is referred to as a technology of high responsibility and major hazard. Due to use of components of high reactivity, it is impossible to carry out an experimental research. Therefore, it is necessary to use computer aided models at the development and research stages.

One of the main units of extraction cascade is a centrifugal extractor used for refinery of chemical solutions. The basic technological processes of the centrifugal extractor include mixing the initial solutions in the mixing chamber to ensure mass transfer and separating of the resulting emulsion in the rotor for obtaining the pure chemical compounds. Thus, the combination of the chamber and rotor together constitutes a rigid system of two units, one mixing chamber and the other separating chamber.

The research has shown that the centrifugal extractors cascade is a complex plant. Therefore, the dynamical computer aided model of cascade has been made with the following points in mind:

1. The mass transfer processes in both phases have variable rate, hence, the concentrations of chemical compounds circulating in the system are continuously calculates at different points in each centrifugal extractors. Moreover, values of concentrations are continuously calculates in two different conditions, one is start-up and the other steady state.
2. Since volumes of organic and water phases are variable during whole cascade, the values of the flow rates of solutions are automatically calculates at each modeling time interval.

Furthermore, the computer aided model has been made by taking into account the following considerations:

1 It is possible to change structure of the cascade, hence, the number of centrifugal extractors in the model is variable.

2 The dynamical model is based on some interlocks, such as opposite connection of apparatus, overflow, etc.

3 There are additional modules such as valves, pumps, tanks, etc.

Note that this model is applicable not only for extraction but also for re-extraction processes simulation. Moreover, these two processes could be simulated simultaneously.

By computer aided modeling of different cascade configuration and conditions is possible to solve technological and optimization problems which occur in plants, explore effect of a disturbances that affect systems, carry out analysis, synthesis and design of automatic control systems.

Finding the best available techniques for an environmental sustainable waste management in the fish canning industry

J. Laso[1], M. Margallo[1], J. Celaya[1], P. Fullana[2], C. Gazulla[2], R. Aldaco[1], A. Irabien[1]
[1] Department of Chemical and Biomolecular Engineering, University of Cantabria, Av. de los Castros s/n, 39005 Santander, Spain.
[2] UNESCO Chair in Life Cycle and Climate Change, Escola Superior de Comerç International (ESCI). Universitat Pompeu Fabra (UPF), Passeig Pujades 1, 08003 Barcelona, Spain.

Spain is the top European producer of canned food with more than 343,000 tonnes of product weight produced, valued at 1,500 million [1]. As one of the largest fishing nations in Europe, Spain has a traditional abundant consumption and production of fish. In particular, anchovy is the 5th most popular fish. Although, consumer preferences show a considerable discrepancy depending on the regions. For example, in Cantabria (North Spain), anchovy is the 2nd most preferred fish [2]. In particular, the quality of the Cantabrian canned anchovy is world-renowned and, moreover, it is considered as a "gourmet product" by consumers due to its handmade and traditional manufacture. However, the production of canned anchovy generates a huge amount of solid and liquid wastes (approximately 9,000 tonnes/year) due to manual operations such as removal of heads and filleting. For this reason, it is necessary to design and implement strategies for a sustainable management of the Cantabrian anchovy industry. To reach this objective, in the last years, different directives such as the Directive on Integrated Prevention Pollution and Control (IPPC) [3] have been developed. IPPC proposes the use of the best available techniques (BATs) for those industrial installations included in this Directive. BATs are obtained as a result of a decision making process based on life cycle assessment (LCA), which allows to determine the "hot spots", from raw materials extraction to end of life. However, BATs are usually applied to process rather than to products.

This work proposes a technical procedure focused on the product rather than on the process to determine the BATs of the waste management of the anchovy canning industry.

First, the LCA of a canned anchovy was carried out from cradle to gate (excluding fishing stage) based on data from canning plants of Santoña (Cantabria). The functional unit selected was one can so called "octavillo" (50 g) of Cantabrian anchovy in extra virgin olive oil (EVOO).

Primary data, such as the consumption of electricity, fuels, water and raw materials (salt, brine, EVOO and packaging); and the generation of solid and liquid wastes, were obtained by a questionnaire sent to all of the canning anchovy plants. On the other hand, secondary data such as, the production of packaging, energy, water and raw materials were obtained from databases and literature.

The LCA results indicated that the management of the solid residues was the 90 % of the total impact to air and the 70 % of the total impact to water of all of the gate-to-gate stages. This is due the fact that during the anchovy manufacture about 60 % of the fresh anchovy caught is turned into a waste. Particularly, 96 g of fresh anchovy generates 65 g of solid residues. Heads, spines, guts and rest of anchovy compose these residues that are collected and disposed of in a landfill. The landfilling of the solid residues was the main contributor to the global warming (GW), photochemical ozone formation (POF), aquatic acidification (AqA) and eutrophication (EU) with a percentage of 80 %, 93 %, 98 % and 77 %, respectively. This is due to the emissions of the landfill operation and the waste disposal along the lifetime of the landfill.

In order to reduce the environmental impact of the management of the solid residues from the canning process, several BATs are proposed.
During the canned anchovy process, the solid residues can be divided into two groups: broken anchovies and rest of anchovies from the cutting steps and, on the other hand, heads and spines. These wastes could be used as a raw material for other processes.
Broken anchovies and rest of anchovies could be used to make anchovy paste within the factory.
On the other hand, heads and spines of anchovies could be sent to a flour factory to make fish flour.
The European Commission has promoted the reutilization of waste by means of the circular economy. This concept introduced in several environmental policy initiatives, aims to keep the added value in products for as long as possible and eliminate waste. Figure 1 shows a conceptual diagram illustrating the main phases of the circular economic model applied to the canned anchovy process through the environmental sustainable waste management.

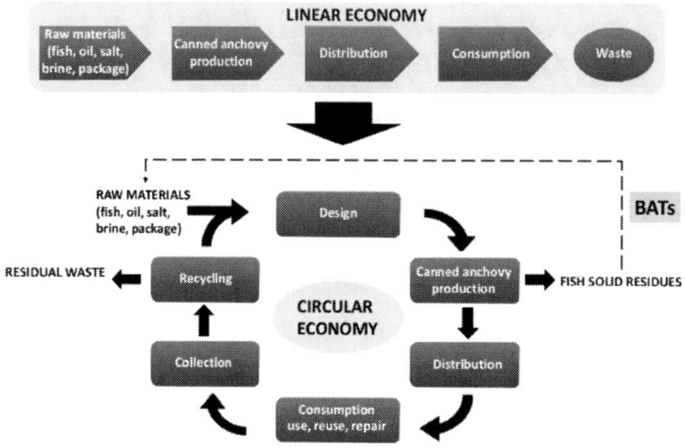

Figure 1. Conceptual diagram of a circular economic model applied to the canned anchovy process.

The use of waste as raw material in the manufacture of anchovy paste and fish flour could improve the environmental performance of the process and reduce the fish losses. Moreover, these measures could increase the economic benefits of the plant, giving economic value to food waste.

Acknowledgements

Authors thank to Ministry of Economy and Competitiveness of Spanish Government for the financial support through the project called GeSAC-Conserva: Sustainable Management of the Cantabrian Anchovies (CTM2013-43539-R). Jara Laso also thanks to the Ministry of Economy and Competitiveness of Spanish Government for the financial support through the research fellowship BES-2014-069368.

References

[1]. Food and Agriculture Organization of the United Nations. Globefish. http://www.globefish.org/the-canned-seafood-sector-in-spain.html. Visited: 27-11-15

[2]. Eurofish. 2012. Overview of the world's anchovy sector and trade possibilities for Georgian anchovy products.

[3]. Directive 2008/1/EC of European Parliament and of the Council of 15 January 2008 concerning integrated pollution prevention and control. Official Journal of the European Union L24, 8-29.

Framework for design optimization: process development of supercritical CO_2 Brayton cycle for coal-fired power plant

Q. Zhao[1,2], M. Mecheri[2], T. Neveux[2], R. Privat[1], JN. Jaubert[1]
[1]LRGP-CNRS, 1 Rue Grandville, 54000 Nancy, France;
[2]EDF R&D, 6 quai Watier 78401 Chatou, France; Tel: +331 30 87 79 68, email: qiao.zhao@edf.fr

Coal is still a key source of energy in the world and coal-fired power plants have always been one of the major sources of electricity. The resurgent interest in enhancement of coal-fired power plants has been driven by restrictive policies on greenhouse gas emissions and the drastic increase of the energy demand.

In this context, power cycle improvement is one of the most frequently foreseen solutions. Both replacement of working fluid and modification of cycle configuration could bring favorable results in enhancement of overall efficiency of coal-fired power plants. Previous preliminary studies based on process modeling showed that the recompression supercritical CO_2 (SC-CO_2) Brayton cycle would be more efficient than the water steam Rankine cycle. However, in order to exploit the industrial feasibility of this SC-CO_2 Brayton cycle, knowledge in including flow scheme synthesis, mass and energy balances, equipment sizing, economic evaluation, process simulation and optimization has been identified as cornerstone.

Moreover, process optimization is a powerful method which can deal with several challenges. It enables to find a global feasible optimum design under consideration of defined criterion as well as trade-offs.

As a part of this overall feasibility study, the Nonlinear Programming (NLP) mono-criterion optimization framework permits to find high performance process architecture of SC-CO_2 Brayton cycle by applying the criterion defined in industrial power generation system. This paper presents a systematic method to find the optimal points of different architectures of SC-CO_2 Brayton cycle applied to coal-fired power plant with respect to industrial constraints.

Firstly, the process modeling within realistic industrial hypotheses has been brought out with the software *ProsimPlus*®. The accuracy and performance of *ProsimPlus*® in terms of energy and mass balance, as well as of thermodynamic equation of state have been benchmarked with previous studies. Secondly, the proposed methodology provides results of optimized process design of a recompression SC-CO_2 Brayton cycle within realistic constraints. In fact, the process variants in this optimization involve both operational and design parameters including pressure ratio, temperature, heat exchange surface and mass flow rate in different components. Furthermore, possible process modifications and/or alternative process architectures of Brayton cycle have been brought out to identify and compare the optimal points of each architecture.

The next step study will focus on process synthesis in order to identify possible and potential processing routes to produce the desired energy efficiency. The methodology of Mixed Integer Nonlinear Programming (MINLP) optimization will be developed.

Gasification of the tar deposits

A. Pryszcz, B. Grycova, I. Koutnik, P. Lestinsky
VSB-TU Ostrava, Institute of Environmental Technologies, 17. listopadu 15, Ostrava - Poruba 708 33;
tel. +420 597 327 315, e-mail: adrian.pryszcz@vsb.cz.

Tar properties vary according to the source of their origin. The major part of tars is formed from the thermal conversion of coal (coal tar), the second part is formed during pyrolysis or gasification of biomass. During the last years, due to efforts to use renewable sources of energy instead of fossil fuel, new technologies for pyrolysis and gasification have been developed, especially for thermal processing of biomass. Tar removal or conversion is considered technologically an extremely challenging problem. Tar can condense or polymerize into more complex structures, which can lead to clogging pipes or heat exchangers. Tar elimination from the gasification product is necessary before additional usage.

Gasification is a process that converts solid or liquid hydrocarbon to obtain synthesis gases. This conversion represents a partial oxidation of the carbon in the feeding material and is generally carried out in the presence of a gasifying carrier, such as air, oxygen, steam or carbon dioxide. This conversion process is influenced by a number of factors, amount of oxidative medium, feedstock composition, temperature, etc.

Tar deposits are taken as a waste and nowadays most of the tar deposits are incinerated therefore it is necessary to search for new methods of using of the tar deposits. Partial oxidation of heavier hydrocarbons or steam reforming of natural gas or naphtha is a widespread method used for producing of syngas gas or hydrogen. This article describes the gasification of tar deposits for production of H_2, CO_2, CO, CH_4. As feedstock were chosen tar deposits from tar storage tanks, after few years because of condensation and polymerization reaction of tars, on the bottom of tar storage tanks a wide layer of tar deposits is created. As gasification medium were used steam and air to maximize the yields of hydrogen or syngas.

This work was financially supported by the Ministry of Education, Youth and Sports of the Czech Republic in the "National Feasibility Program I", project LO1208 "Theoretical Aspects of Energetic Treatment of Waste and Environment Protection against Negative Impacts".

A heterogeneous condensation assisted three-phase bed column to remove submicronic particles

F. La Motta[1], F. Di Natale[1], C.Carotenuto[2], A. Lancia[1]
University of Naples Federico II, Dep. Chemical Engineering, Materials and Industrial Production, P.le Tecchio 80, 80125 Napoli, Italy
tel. +39 081 7682246, e-mail: fdinatal@unina.it; [1] Add Adress
tel. +39 081 7682236, e-mail: francesco.lamotta@unina.it; [1] Add Adress
tel. +39 081 7682236, e-mail: aleserto@unina.it; [1] Add Adress
tel. +39 081 7682243, e-mail: lancia@unina.it; [1] Add Adress

The emission of particulate matter entrained in industrial and vehicles exhaust gases is one of the major health and environmental concerns. The ultrafine particles can remain suspended in the atmosphere for a long time and, once inhaled, they can reach the deepest regions of the lungs and even entering in the circulatory system, causing a wide range of health problems. The traditional particle abatement devices are designed to comply with regulations on particles with sizes above 1 μm, and they are far less efficient in collecting sub-micrometric particles, especially those in the range 0.1-2 μm, called Greenfield gap.

In the last years a couple of new technologies were proposed for improving the particle removal efficiency: wet electrostatic scrubber, based on a strong electrostatic interaction established between particles and droplets, and heterogeneous condensation, based on enlarging the submicronic particles. Indeed, heterogeneous condensation is a promising technique to improve the performances of traditional particle collection devices. This technique consists of the condensation of vapour on the ultrafine particles in order to create a coarser liquid–solid aerosol whose size is larger than the upper limit of the Greenfield Gap. Therefore, the heterogeneous condensation is not a particle collection technique by itself, but a preconditioning technique to be used before other conventional particle collection processes.

In this paper we are reporting the results on an lab-scale concept for a possible unit for submicronic particle removal. The unit is based on a heterogeneous condensation system coupled with a three-phase bed column where the enlarged particles are removed.
In order to investigate the hydrophobicity effect on the removal efficiency two different particle aerosols were tested. Particles consist in simple polystyrene and polystyrene with carboxylic groups on surface, that increase the hydrophilicity.

The aerosol was produced by an aerosol generator, TOPAS 221, equipped with a Laskine nozzle and fed with a solution, where previously were added the particles. The particle laden gas passed through a diffusion dryer to remove water drops and produce a dried aerosol and then entered the heterogeneous condensation unit (HC). This consisted in a glass cylinder, where a distilled water film fall along the tube walls. The water was kept at a controlled temperature by means a thermostatic bath. The aerosol then passed into the three-phase bed columns (3BC), consisting in a glass cylinder high 15 cm, filled with glass beads and cold distilled water for a total height of 9 cm. The gas space- time in the HC and the 3BC units were, respectively, 5,3 s and 13 s.

The aerosol size distribution and concentration were measured using a Laser Aerosol Spectrometer, TSI Nanoscan 3910. Sampling points were available soon before and after the HC-3BC sequence. The particle size distribution of the two sample materials at the HC inlet presents bimodal distribution with the first peak at 25 nm and the second at 140 nm. The first peak was made by the surfactants necessary to avoid coagulation.

The removal efficiency was thus investigated for different film temperature in the HC unit. The benchmark for efficiency improvements related to HC pre-treatment is given by comparison of the removal efficiency at the given temperature THC=50-80°C, with the value of particle removal efficiency with the HC column kept at 30°C, when no condensation occurred.

The experimental results showed that the particle removal efficiency increased by more than 10% for the polystyrene particles with carboxylic groups, by increasing temperature from 50 to 70°C, becoming higher than 95% at 80°C. For the polystyrene particles, the particle capture became very high already at 50°C, when the removal efficiency was close to 90%, reaching 98% at 80°C.

These results showed that higher water film temperature higher is the removal efficiency, pointing out that the heterogeneous condensation takes place enlarging the particles to a size easier to remove.

Cadmium and zinc kinetic adsorption study adsorption onto activated carbon in a single and binary system

F. Di Natale, F. La Motta, A. Erto, A. Lancia
University of Naples Federico II, Dep. Chemical Engineering, Materials and Industrial Production, P.le Tecchio 80, 80125 Napoli, Italy
tel. +39 081 7682236, e-mail: francesco.lamotta@unina.it; University of Naples Federico II, Dep. Chemical Engineering, Materials and Industrial Production, P.le Tecchio 80, 80125 Napoli, Italy

The presence of heavy metals in wastewater is one of the main concerns for environmental protection, due to the intrinsic toxicity and long lifetime of these compounds. Therefore, there is a pressing need to remove them from polluted water, using efficient and cost-effective remediation techniques, which at the same time will not produce toxic residues. Many different technologies have been developed to remove heavy metals from contaminated natural water and wastewater. Among these, adsorption emerged as a highly effective technology applicable both to wastewater treatment and for groundwater remediation. Despite the high number of papers in the literature dealing with heavy metal adsorption, most of the studies were focused on the assessment of single-compound adsorption equilibrium and its dependence on the main process parameters. Conversely, the kinetic aspects of the process were far less studied.

The transfer of knowledge to the design of either industrial devices (e.g. adsorption reactors) or in-situ intervention for groundwater restoration (e.g. permeable adsorbing barrier) requires the knowledge of adsorption process behaviour in experimental conditions next to real case-studies, which are often characterized by the simultaneous presence of different heavy metals. Nevertheless, very few studies dealt on experimental evaluation of adsorption equilibrium and kinetics in solution containing more than one metal.

Water solutions containing cadmium and zinc are typical examples of such systems. Cadmium is unanimously recognized as very dangerous; it is extremely toxic and classified as carcinogenic. Its high toxicity is enhanced by the ability to bioaccumulate in the aquatic ecosystem and it can reach humans through the food chain. Differently, zinc is an essential element for human beings but it can become toxic at high concentrations. The main sources of cadmium and zinc releases in water include metal plating plants, metallurgic industries, battery production and disposal, mine drainage and the leaching from natural rocks where these two metals are often present simultaneously. Therefore, the removal of cadmium by adsorption should often be carried out in presence of zinc .

In this work, an experimental and modelling analysis of cadmium and zinc kinetic adsorption both in mono and in binary systems on a commercial granular activated carbon (Aquacarb 207EATM - provided by Sutcliffe Carbon) is presented. The main object of the work is to clarify the rate controlling mechanisms of cadmium and zinc adsorption and their reciprocal influences.

Adsorption tests were performed in batch and fixed-bed column, at a constant temperature of 20°C in order to individuate the effect of solid-liquid contact type on the adsorption rate. Binary adsorption tests were carried out with different initial concentration ratios of the two analytes (C°_{Cd}:C°_{Zn}).

Furthermore, a dedicated multicomponent adsorption kinetic test was carried out, on activated carbon preloaded with cadmium, on which zinc adsorption kinetics was carried out.

Pseudo-first order, pseudo-second order, Elovich and diffusional models were applied to the experimental data in order to understand the most probable adsorption mechanism.

The experimental results showed that the zinc adsorption kinetic is the same for single-compound and binary systems, well described by a single internal mechanism following the Elovich model. This result is congruent with thermodynamic studies.

On the other hand, the cadmium adsorption kinetic presents substantial differences between the single-compound and binary system behaviours. Indeed, in the single-compound system the adsorption rate showed two control mechanisms: a first one well described by a pore diffusion model and a second one following the Elovich model. Instead in the multicomponent process, a single mechanism control was found, well described by Elovich. Actually, the Elovich energetic parameter was the same for both single and multicomponent adsorption. The carbon selectivity toward zinc is so marked that during the cadmium preloaded adsorption tests, about 34% of the adsorbed cadmium was replaced by the zinc, until a new equilibrium condition arose.

Design methodology and analysis a solid oxide fuel cell-based power plant

C. Mahisanan, [1] S. Authayanun, [2] D. Saebea, [3] Y. Patcharavorachot, [4] A. Arpornwichanop [1]
[1]Computational Process Engineering Research Unit, Chulalongkorn University, Department of Chemical Engineering, Faculty of Engineering, Bangkok, 10330, Thailand; tel. +66 2218 6878, email: amornchai.a@chula.ac.th; [2]Srinakharinwirot University, Department of Chemical Engineering, Faculty of Engineering, Nakorn Nayok 26120, Thailand; [3]Burapha University, Department of Chemical Engineering, Faculty of Engineering, Chonburi 20131, Thailand; [4]King Mongkut's Institute of Technology Ladkrabang, School of Chemical Engineering, Faculty of Engineering, Bangkok, 10520, Thailand.

The solid oxide fuel cell (SOFC) is regarded as a promising technology for power generation due to its high efficiency and low environmental impact. Due to high-temperature operation, SOFC can be integrated with other energy technologies for efficient energy usage. Different designs of SOFC systems are generally performed based on various heuristic methods and decision making. Thus, it is necessary to have a systematic methodology for system design tasks. In this study, a step by step method for design of the SOFC system is proposed. With the proposed design method, a SOFC-based power plant incorporated with an oxyfuel combustion process to reduce CO_2 emission is developed. Equipment selections and all operational conditions are clearly revealed. In addition, the sensitivity analysis of key operating parameters, such as steam to carbon ratio, fuel utilization factor and SOFC temperature, is performed and discussed.

Performance analysis and control structure design for proton exchange membrane fuel cell

N. Chatrattanawet, T. Hakhen, [1]D. Saebea, A. Arpornwichanop
Chulalongkorn University, Computational Process Engineering Research Unit, Department of Chemical Engineering, Faculty of Engineering, Bangkok, 10330, Thailand; tel. +66 869713499, email: amornchai.a@chula.ac.th; [1]Burapha University, Department of Chemical Engineering, Faculty of Engineering, Chonburi, 20131 Thailand; tel. +66 850562910, e-mail: dangs@eng.buu.ac.th.

A proton exchange membrane fuel cell (PEMFC) is regarded as a highly efficient power generation device. The performance of PEMFC depends on various key operating parameters, such as pressure, flow rate and humidity of reactant gases, and cell temperature (60-100 °C). Generally, the cell temperature is known to have an importance on transportation within the PEMFC and its performance. The fuel cell temperature should be limited under a working temperature that does not decompose the material properties of the components. If the fuel cell temperature is too high, the membrane will be damaged. Therefore, the aim of this study is to implement the cell performance analysis and the concept of control structure design for the PEMFC. The steady-state analysis is performed to design the suitable operating condition. The effect of input parameters on cell voltage and cell temperature is analyzed to investigate the dynamic behavior of PEMFC that is important for control design. To obtain an efficient control system, the control structure design of the PEMFC is also considered to find the suitable controlled and manipulated variable. Moreover, the selection of input-output pairings by considering the relative gain array (RGA) as controllability index is applied to control system design. The result showed that the cell voltage and cell temperature depend on the inlet molar flow rates and temperature of hydrogen and air, and operating current density. According to the RGA, the inlet molar flow rates of hydrogen and air are manipulated variables to regulate the cell temperature and partial pressure of hydrogen, respectively.

Environmental impacts assessment of separation of light rare earths from monazite ore

D. Ratthanapra,[1]U. Suwanmanee
Thailand Institute of Nuclear Technology, Research and Development Division, Rangsit Nakhon Nayok St., 29 ongkharak district; tel +662 4019889, [1]Srinakharinwirot Univ., Dep. Chemical Engineering, Rangsit Nakhon Nayok St., 66 ongkharak district; tel.+6626495000, e-mail: unchalee@g.swu.ac.th;

Rare earth elements (REEs) or rare earth metals (REMs) are the group of fifteen lanthanides (atomic numbers 57–71), scandium and yttrium. Lanthanides consist of two groups including the light rare earths, Lanthanum (La), Cerium (Ce), Praseodymium (Pr), Neodymium (Nd), Promethium (Pm), Samarium (Sm), and the heavy rare earths, Europium (Eu), Gadolinium (Gd) Terbium (Tb), Dysprosium (Dy), Holmium (Ho), Erbium (Er), Thulium (Tm), Ytterbium (Yb), Lutetium(Lu). REEs are considered as necessary materials for utilization in numerous applications such as glass polishing, phosphors, lasers, magnets, batteries, high-temperature superconductor, optical glass, as well as catalysts, and its demand and cost are increasing. To meet the ultimate utilization of REES, Life Cycle Assessment (LCA) is the important tool to evaluate and assess its environmental impact. This assessment can also be used for product development for REEs production by evaluating the environmental impact of a product in its whole life cycle. Thus, the objective of this work is to study and assess environmental impacts based on the CML 2 Baseline 2000 method (World 1995) of light REEs production, typically separation methods of CeO_2 production.LCA used for subject to assess the environmental impact of separation methods of cerium oxide (CeO_2) from other REEs via integration with solvent extraction process (method I) and excluded solvent extraction process (method II).The functional unit of this study was selected as one kg of CeO_2. The system boundary include mining and grinding of monazite ore, transportation of raw materials, digestion of monazite ore by alkali method, dissolution process, the separation of U and Th by precipitation, and separation of CeO_2. The data on process was available from a field study conducted in Thailand Institute of Nuclear Technology and applied into SimaPro 7.3. The relative environmental impacts to assess were abiotic depletion, global warming, ozone layer depletion, human toxicity, freshwater aquatic ecotoxicity, marine aquatic ecotoxicity, terrestrial ecotoxicity, photochemical oxidation, acidification, and eutrophication. The results indicated that the separation method of CeO_2 with integrated solvent extraction process (method I) showed better performance than the separation method of CeO_2 without solvent extraction process(method II) because the addition of solvent extraction process can achievable the higher purity of CeO_2 concentration. However, when abiotic depletion, human toxicity, freshwater aquatic ecotoxicity, and photochemical impacts are still relatively high, the production method I should be improved. This LCA results can also be bring helped for product development, scientist and governor plans, and developing the technology performance and process design for valuable products from REEs in Thailand by considering the lower environmental impact of a product in its life cycle.

Variability of energy consumption by applying process integration techniques

Sanchez Razo P., Martinez Patiño P., Picón Nuñez P., Ireta Moreno F., Pizano Martínez A.
[1]Department of Electrical Engineering
[2]Department of Chemical Engineering.
University of Guanajuato. Carr. Salamanca-Valle de Santiago km. 3.5+1.8 km. Comunidad de Palo Blanco, C.P. 36885 Salamanca, Gto. Tel. +52 (464) 64 7 99 40. México.
p.sanchezr.666@gmail.com

The application of process integration techniques in industrial operations has a direct impact on different inputs, for example, water and energy. By making retrofit to an industrial process, not only the main variables are modified, but also affects other process variables.

In this paper, two case studies with techniques to reduce consumption of water and energy are presented. Being water and energy the main inputs and variables; once the heuristic methodology is applied, clearly shows the variability of electricity consumption.

This article highlights the impact of energy and water supplies where electricity varies depending on these inputs; also shows increasing variables to be considered when applying a retrofit study.

The case studies will serve as a reference for the application of process integration techniques where the water and energy (heat) will be involved; and also would include the impact on power consumption.

Bio-hydrogen production from empty palm fruit bunches: Process simulation and economic evaluation of under North-Colombian conditions

D. L. Peréz-Zúñiga[1], E. J. Luna-Barrios[1], Y. Y. Peralta-Ruiz[2], Angel D. González-Delgado[3]
[a]Independent consultant in Chemical Engineering, Anillo vial km 12. Cartagena, Colombia.
[b]Universidad Libre Seccional Barranquilla, Industrial Engineering Department, Carrera 46 No.48-170, Barranquilla, Colombia.
[c]University of Cartagena, Chemical Engineering Department, Avenida del Consulado #Calle 30 No. 48 – 152, Cartagena, Colombia.
luna.1793@hotmail.com

Global energy demand is increasing and hydrogen can contribute to satisfy energy requirements in rural areas, for achieving a sustainable hydrogen production process from energy crops-derived residues, it is required to evaluate alternatives that are currently being used, as well as the potential of novel feedstocks that matches the above-cited properties. Colombia is the fourth largest palm oil producer worldwide, and local research policies are favoring the study of this production chain in order to satisfy specific demands detected for the process, which includes the increase of economic feasibility, this aim can be achieved via generation of valuable co-products as hydrogen from residues obtained in different stages.

This work presents a techno-economic evaluation of the hydrogen production from empty palm fruit bunches under Colombian conditions, using computer-aided tools, stages evaluated were biomass pretreatment, direct gasification, air separation, gas cleaning, shift reaction and hydrogen purification. Economic feasibility was measured taking into account a processing capacity of Colombian north zone with a hydrogen yield of 3989.76 t/yr, with local tax rate, and subsidies available for palm oil crops in Colombia, economic parameters as Net Present Value, Payback Period and profitability index were calculated. Results shows that process presents a high sensitivity to the feedstock cost, In addition, the influence of the tax rate applied to the industrial sector in in Colombia is strong over the profitability of the topology developed, which can discourage the investment in novel processes.

Valorisation of palm oil crop residues through hydrogen production using computer aided process engineering

E. J. Luna-Barrios[1], D. L. Peréz-Zúñiga[1], A. D. González-Delgado[2] Y. Y. Peralta-Ruiz[3],
[1]Independent consultant in Chemical Engineering, Anillo vial km 12. Cartagena, Colombia.
[2]University of Cartagena, Chemical Engineering Department, Avenida del Consulado #Calle 30 No. 48 – 152, Cartagena, Colombia.
[3]Universidad Libre Seccional Barranquilla, Industrial Engineering Department, Carrera 46 No.48-170, Barranquilla, Colombia.
luna.1793@hotmail.com

Nowadays, there are specific needs related to palm oil production chain that must be satisfied in order to achieve sustainability, one of them is the valorization of residual biomass generated during palm oil production. To achieve the above mentioned aim, it is required to evaluate biotechnological alternatives for obtaining high value products, as well as the potential of novel feedstocks that matches the sustainability criterion, computer-aided simulation has been successfully used to evaluate emerging technologies of biomass processing and can help to predict technical behavior of novel technologies for palm residues transformation.

In this work, an evaluation of sustainable bio-hydrogen production from empty palm fruit bunches via gasification of feedstock was developed based on palm production capacity of Colombian north zone where are produced 360,000 t/yr. Simulation takes into account technological limitations and novel equipment performance, special chemical species present in bunches were modeled and process topology was built using a commercial simulation software. A sensitivity analysis was performed by changing relevant operating conditions in the hydrogen production process. Results shows that, with the biotechnological process for hydrogen can be produced 3989.76 t/yr of hydrogen from Empty Palm Fruit Bunches and it is not necessary to use a sulfur removal unit, also it is possible to use Selexol™ for cleaning the hydrogen and retrieve it; Selexol™ flow rate to purify the hydrogen stream was 974,850 kg/h and operating conditions were 10 °C for temperature and 36.2 atm for pressure; using this component is avoided the need to use a desorption tower.

Procedures of non-standard starting engine RD-33

J. Bartoszewicz
Poznan University of Technology, Faculty of Machines and Transportation, Chair of Thermal Engineering, pl. M. Skłodowska-Curie 5, Poznań, 60-965, Poland;
tel. +48616652215, fax: +48616652281, e-mail: jaroslaw.bartoszewicz@put.poznan.pl.

Issues concerning the process of operating the air technology safely are among crucial determinants of proper functioning of any institution using aircrafts as a basis of its activity. Their complexity is a result of many factors generated by accumulation of many systems and their connections in a relatively small space, which is indispensable for proper functioning of the aircraft as a whole. Safety of the technical product means its capability to stay in a state of risk no greater than the risk criterion value in any operating conditions, despite their complexity [8]. Improper functioning of the aircraft elements, sets or functional units being a result of the aircraft damaging cause a threat to the security. Safety issues relative to the air technology are realized on many planes, including among other things a systematic approach to designing the aircraft, that is: using up-to-date designing and technological methods, using modern construction materials of high quality, verification of assumptions during tests with objects and proper operating of this technology.

As far as operators are considered, it means a proper selection of candidates for the profession, their education and training, motivating, keeping them in good physical condition, etc. Second plane of the safety issues is related to assessment of systems at each stage of their existence using methods and valuation criteria useful for the stage under consideration. This article focuses on this approach to the safety problem, and in particular it presents procedures in case of the aircraft failure. It also presents rules of conduct for repairing the breakdowns and some examples. Available measures enabling to detect the reasons of the aircraft failure comprises [3]: technical knowledge, experience of the flying personnel and aircraft maintenance service, details from Flight Data Recorders, technical documentation (manuals, bulletins, diagrams, etc.), tools, control and measurement instruments, and stands for the aircraft diagnosis.

The plane MiG-29 was constructed in the combining wind wings and the fuselage in one bearing surface. Propulsion of the aircraft are two turbo jet engines RD-33. Figure 3 shows the object of analysis: MiG-29 aircraft and one of his engines RD-33.

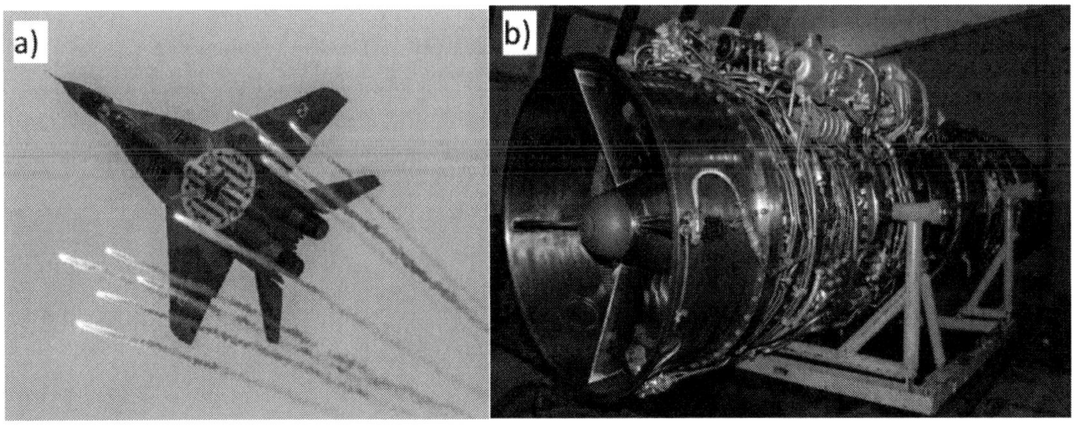

Fig. 1 Object of analysis: a – the plane MiG-29, b – one with Mig-29 engines

In particular, the RD-33 engine consists of [2]:

- low-pressure rotor (LPR) with four-stages axial fan and one-stage axial turbine cooled by air (the nozzle set and turbine blades),
- high-pressure rotor (HPR) with nine-stages axial compressor and one-stage turbine cooled by air (stator vanes of the compressor's first two stages are regulated depending on the angle of attack),
- annular combustion chamber with fuel injectors mounted on the intake manifold,
- afterburner chamber with initial flow mixing upstream of the afterburning manifold,
- subsonic exhaust nozzle with a smooth regulation of critical and exhaust flow,
- accessory gear box with drive to high-pressure rotor (HPR),
- the engine automatic regulation and control systems with its starting system,
- oil system, air intake system, engine parameters control system.

Ground starting of the aircraft power plant is achieved by the turbine starter [1], which drives the engines via the accessory gearbox. The power plant start-up system provides:

- on-ground starting the engines separately in any order,
- starting both engines one after the other,
- cold start-up of the engines and the turbine starter,
- automatic, semi-automatic or manual relight,
- preventive air relight during shooting the missiles or using the aircraft gun.

This article presents methodologies of proceeding being in force in case of detecting a failure in the MiG-29 aircraft. Safe operation of the aircraft equipped with two engines is more difficult than in case of a single-engine aircraft. Presented examples of failures illustrate greater complexity of problems faced by servicing personnel and the pilot when operating the MiG-29 aircraft. They should systematically deepen their knowledge taking part in numerous courses and training courses. Not only does the pilot's and aircraft's safety depend on proper proceeding of servicing personnel, but also minimization of time needed to restore the serviceability of the combat aircraft. Procedures of proceeding in case of any failure during the MiG-29 engine start-up have to be known to the pilot and technical personnel.

[1] Balicki W. and all, Airline turbine engines. Design-Maintenance-Diagnostics, Research Library of the Aviation Institute, Warszawa 2012.
[2] Kowalski M., Issues turbine diagnosis engines using data from the aircraft's flight recorders, Scientific Papers of the Institute of Air Force Technical, Issue No. 31, Warszawa 2012.
[3] Aircraft MiG-29 Piloting Techniques Manual, WLOP 176/94, Poznań 1995.
[4] MiG-29 Aircraft Aerodynamics Practical, WLOP 48/91, Poznań 1992.
[5] Turbojet Engine RD-33 Technical Description and Operation, Vol. 1, Poznań, 1991.
[6] Turbojet Engine RD-33 Technical Description and Operation, Vol. 2, Poznań, 1991.
[7] Staniszewski R., Control of power units, WŁK, Warszawa 1998.
[8] Szczeciński L., Aviation turbine engines, MON, Warszawa, 1965.
[9] Szczeciński L., Twin-rotor, twinflow air turbine engines, WŁK, Warszawa 1971.
[10] Wiatrek R., Engine Theory, WAT, Warszawa 1983.
[11] Żółtowski B., Basics of diagnostics of machines, The Publishing House of the Academy of Agricultural and Technical University, Bydgoszcz 1996.

Use of a holistic-hierarchical approach for improvement of biofuel and energy blocks from microalgae production chain

Y. Y. Peralta-Ruiz[1], J. D. Martinez-Consuegra[2], A. D. González-Delgado[3]
[1]Universidad Libre Seccional Barranquilla, Industrial Engineering Department, Carrera 46 No.48-170, Barranquilla, Colombia.
[2]Corporacion Universitaria Minuto de Dios-UNIMINUTO, Business Administration Program, Carrera 53 No. 74 - 110, Barranquilla, Colombia.
[3]University of Cartagena, Chemical Engineering Department, Avenida del Consulado #Calle 30 No. 48 – 152, Cartagena, Colombia.
jairo.martinezc@uniminuto.edu

Although microalgae biomass can potentially be exploited in the production of biofuels, biotechnological use of biomass of microalgae for biofuel production presents various bottlenecks at different stages to prevent the sustainable use of this energy crop. In this work, the joint use of experimental techniques, characterization, process modeling, sensitivity analysis and process simulation to generate alternatives to improve the supply chain, emphasizing the production of fermentable sugars and oil, both precursors shown third generation biofuels is presented. The hierarchical part of the approach sets experimentation as a tool for selection of raw materials and improving efficiency, and uses modeling and simulation to minimize the use of natural resources allowing to select technologies and processing routes.

The results of the implementation shows in the experimental part an increase in the production of total sugars and cell wall disruption of 75.5 % of extraction efficiency for 120 minutes of hydrolysis, and 92% using a polyfunctional process. While the modeling using MatLab software and process simulation UniSim Design Suite under appropriate kinetic and thermodynamic models, shows that the technology of simultaneous saccharification and co-fermentation allows greater efficiency of ethanol production at large scale.

Valorization of wastewater biomass in North-Colombian region through biofuel production: Simulation and environmental evaluation

Y. Y. Peralta-Ruiz[1], J. D. Martinez-Consuegra[2], A. D. González-Delgado[3]
[1]Universidad Libre Seccional Barranquilla, Industrial Engineering Department, Carrera 46 No.48-170, Barranquilla, Colombia.
[2]Corporacion Universitaria Minuto de Dios-UNIMINUTO, Business Administration Program, Carrera 53 No. 74 - 110, Barranquilla, Colombia.
[3]University of Cartagena, Chemical Engineering Department, Avenida del Consulado #Calle 30 No. 48 – 152, Cartagena, Colombia.
jairo.martinezc@uniminuto.edu

Valorization of agricultural and industrial wastes for the production of high value products has positive impacts from economic and environmental points of view and contributes to sustainable development. however, even the same wastes presents different compositions depending on handling conditions and process topologies must be adapted to local contexts including technological limitations in order to give a more realistic recommendations for the scaling-up of developed technologies In this work, the technical performance of three alternatives for biodiesel production process from wastewater was evaluated using UniSim Design Suite under appropriate kinetic and thermodynamic models.

Lipid extraction step was evaluated under different scenarios, where biomass of microalgae was exposed to three previously developed alternatives for metabolite extraction: hexane-based extraction (HBE), extraction assisted with high-speed homogenization (SHE) and extraction using ethanol/hexane mixture (EHE). Residual wastewater biomass composition was modeled taking into account real information from a Colombian plant and process efficiencies were defined based on experimental results previously obtained by authors. In addition, an esterification stage was included prior to transesterification taking into account high fatty acid content in biomass-derived oil.

Results shows that for a processing capacity of 47 kt/yr, EHE pathway presents the lowest lipid recovery efficiency with a 46 %, and highest energy requirements, the lowest global generated environmental impact (-3.12 PEI/day) and the highest output global environmental impact (5.28 PEI/day), and HBE pathway presented lowest global warming potential and total energy requirements.

Design and simulation of a RF resonant reactor for biochemical reactions

[1]F. Desogus, S. Casu, [2]G. Muntoni, [2]M. B. Lodi, [2]L. Schirru, A. [2]Urru
Cagliari Univ., Dep. Electric and Electronic Engineering, Piazza d'Armi, 09123 Cagliari, Italy; tel. +39 0706755787, e-mail: sergio.casu@diee.unica.it;
[1]Cagliari Univ., Dep. Mechanical Chemical and Material Engineering, Piazza d'Armi, 09123 Cagliari, Italy; tel. +39 0706755079, e-mail: f.desogus@dimcm.unica.it;
[2]Cagliari Univ., Dep. Electric and Electronic Engineering, Piazza d'Armi, 09123 Cagliari, Italy.

Microwaves are known for their ability to enhance chemical processing thanks to their heating capacity. It is rather less well known that microwaves can interact with reacting matter by a different mechanism. In fact, chemical and biochemical reactions are related to electrical forces between the charged parts of the reacting molecules, i.e. electrical interactions at microscopic scale; thus an external electromagnetic field can interact with such reactions [1], and many reports of these interactions were reported in the literature, about both thermal and non-thermal kind [2-4]. However, the real action of microwaves in chemical and biochemical processing has not yet been fully understood, and the mechanism of the radiation effect on enzymes, which are very fragile and temperature sensitive molecules, and whether it is thermal or non-thermal, is still unclear. But, actually, the action of an external electric field modifies the dipolar orientation of molecules and the status of the active sites of the enzyme, enhancing the efficiency and the specificity of enzymatic reactions [5].

Aim of this work is to design and analyze the behaviour of a microwave resonant cavity, to be used as a biochemical (enzyme) reactor, working at the frequency of 2.45 GHz, evaluating the influence of the electromagnetic exposition on the reacting processes, especially on enzyme homogeneous catalyzed reactions performed in a polar medium; in particular the enzymatic hydrolysis of sucrose into glucose and fructose was chosen as a case study.

The reactor was paced inside a resonant cavity, which was designed by Computer Simulation Technology® (CST), a software for the numerical resolution of high frequency electromagnetic problems through Finite Integral Technique (FIT). Utilizing COMSOL Multiphysics®, a software for the numerical simulation of physical problems through Finite Element Method (FEM), the process operation was simulated.

A resonant cavity, which is a particular case of waveguides, is a device used for confining an electromagnetic field inside a spatial region, forcing it to propagate in a fixed direction. Solving Maxwell's equations in the phasor domain, it is possible to describe how the electromagnetic field propagates and distributes along the z-axis of the cylindrical waveguides or transverse to it. It is necessary to make the assumption that the solutions depend on the transverse magnetic modes, so that $H_z=0$ (TM modes). After having determined the waveguide modes, which propagate above the cut-off frequency, the corresponding resonant cavity can be designed knowing the diameter and the resonant frequency, which is 2.45 GHz. It was taken into account that the medium inside the cavity is water and therefore the dielectric constant exhibits a loss behaviour, described by the Debye model [6]; in order to excite a mode independent from the azimuthal coordinate, i.e. TM_{012}, the correct height is 134 mm.

In order to test the effectiveness in the biochemical processing, it was decided to simulate the exposition of a continuous flow of solution, using a reactor made of eight plexiglass tubes allocated into the cavity (Fig. 1). The reactor was centered with respect to the cavity, ensuring

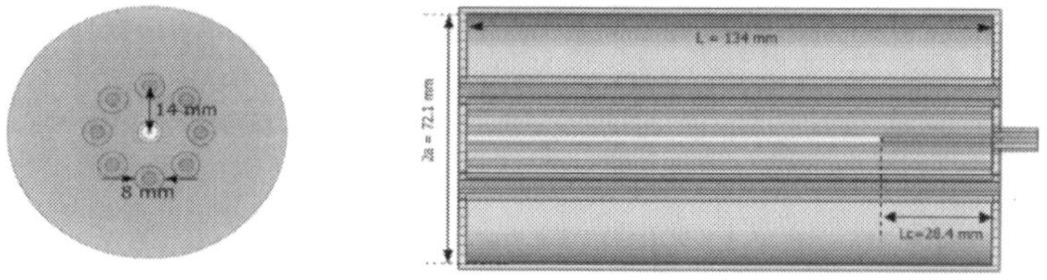

a b

Fig. 1. Sketch of the reactor in the resonant cavity: a) bottom and top view; b) side view.

a uniform absorption of the electromagnetic radiation by the liquid. Moreover, the irradiated liquid is flowing, therefore inhomogeneous microwave exposition is prevented and well mixing conditions are created.

The multiphysic simulation was carried out for different working microwave power in the range of 1÷300 W and for the fluid average axial velocity ranging from $5.0 \cdot 10^{-5}$ to $1.5 \cdot 10^{-1}$ m s^{-1}. The main finding was the fact that no significant deviation from the resonance frequency was registered. But the working conditions of the cavity exert strong influence on the temperature in the system, which is a key parameter in chemical and, above all, biochemical kinetic experiments: it influences the process evolution rate and also can irreversibly damage the enzyme molecules. It Fig. 2 the average temperature values as a function of the mean axial velocity and of the microwave power is reported. Considering that in the sucrose hydrolysis, catalyzed by the invertase enzyme, the temperature should not overcome 60°C [7], since the enzyme loses most of its activity, therefore a useful working region below this temperature threshold was found for P=1 W and velocities in the range $2.0 \cdot 10^{-4} \div 1.5 \cdot 10^{-1}$ m s^{-1}. For increasing power, the available velocity range shrinks (velocity has to be increased and so the process will be carried out with more fluid passages inside the cavity); anyway, the maximum suitable microwave power is 100 W with velocities in the reduced range between $1.0 \cdot 10^{-2}$ and $1.5 \cdot 10^{-1}$ m s^{-1}.

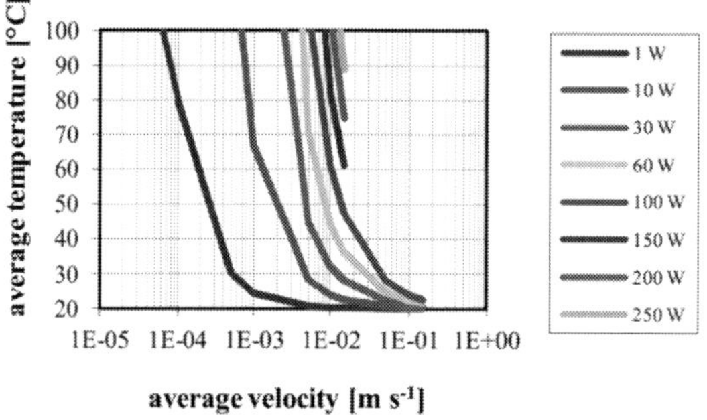

Fig. 2. Outlet average temperature of the fluid for different incident microwave powers and average fluid velocities (inlet temperature fixed at 20°C).

[1] Buchachenko A.L., Frankevich E.L., Chemical generation and reception of radio- and microwaves, VCH Publishers Inc., 1994.
[2] Foster K.R., IEEE *Trans Plasma Sci* 28; 15-23 (2000).
[3] Carta R., Desogus F., *J Environ Chem Eng* 1; 1292-300 (2013).
[4] Carta R., Desogus F., *AIChE J* 56; 1270-78 (2010).
[5] Gong G., Liu D., Huang Y., *Biosyst Eng* 107; 67-73 (2010).
[6] Harrington R.F., Time-harmonic electromagnetic fields, McGraw-Hill, 1961.
[7] Vrábel P., Polakovic M., Stefuca V., Báleš V., *Enzyme Microb Technol* 20; 348-54 (1997).

Experimental determination of the particle dynamics into a rotating tube

F. Desogus, [1]R. Carta
Cagliari Univ., Dep. Mechanical Chemical and Material Engineering, Piazza d'Armi, 09123 Cagliari, Italy; tel. +39 0706755079, e-mail: f.desogus@dimcm.unica.it;
[1]Cagliari Univ., Dep. Mechanical Chemical and Material Engineering, Piazza d'Armi, 09123 Cagliari, Italy; tel. +39 0706755068, e-mail: r.carta@unica.it.

This paper reports the description of a graphic methodology for determining the biomass particle bed characteristics which realize in a cylindrical rotating reactor to be used for the experimental operation of pyrolysis processing at laboratory scale. The apparatus here studied is a quartz tube which can be made rotating with different rotating speed; the tube can work horizontally or slightly inclined with respect to the horizontal plane.

Rotating reactors are a kind of device that is commonly used in some industrial processing applications involving solids, particularly when a strict contact between the solid and a gas phase is needed (mixing, drying, milling, coating, granulation/agglomeration, grinding, calcination, cement manufacturing, etc.). Rotating reactors can be used also in pyrolysis processing, in a wide variety of operating conditions and with many kinds of feed [1-3].

As a consequence of the reaction tube position and rotation, different operating conditions and bed behaviours, i.e. motion regimes, hold-up, residence time and advancing rate can be found; these operating conditions and bed behaviours are also influenced by the biomass characteristics (i.e. kind of material, particle shape and size, water content) and particularly by its flow rate in the axial advancing from the top to the bottom of the tube.

Carrying on a pyrolysis process in a rotary kiln needs a specific knowledge of the behaviour of this kind of system, considering both the solid dynamics which happen inside the tube and the influence of the operating parameters on the product amount and characteristics, and so on the process yield [4]. For these reasons, a process model should take into account all the happening phenomena together (and their interactions): the solid motion dynamics, the heat transfer between the reactor wall and the solid phase, the heat transfer between solid particles, the heat transfer between the solid and the gas phase, the gas phase fluid dynamics, and reaction kinetics. The technique here presented was developed to give an answer to the first issue by a non-invasive way and with high accuracy. In fact, most of the available techniques (magnetic resonance imaging, positron emission particle tracking, fiber optics probes, particle image velocimetry, particle tracking velocimetry, radioactive particle tracking, etc.) are often limited in the possible observations and are not always suitable in the case of ordinary and experimental processing [5].

The apparatus that was used for the present study is a commercial laboratory scale rotating furnace produced by Nabertherm® (rotary tube furnace for continuous processes, model RSR 120-750/11, see Fig. 1a); the tube furnace is a quartz cylinder with a total length of 179 cm, an inner diameter of 10 cm and an outer one of 10.6 cm. The solid particles to be fed to the reactor were at first charged into a filling funnel, from which the particles were carried by a feed screw; at the end of the cylinder, the out coming particles were collected (a sketch of the apparatus is shown in Fig. 2b). The out coming biomass mass flow rate was continuously measured by positioning the collecting recipient on a technical balance (Ohaus®, Adventurer™ Pro AV 4102); the device was connected to a computer by an USB interface and the weight data were collected, at one second intervals, by the Ohaus® Data Acquisition Software.

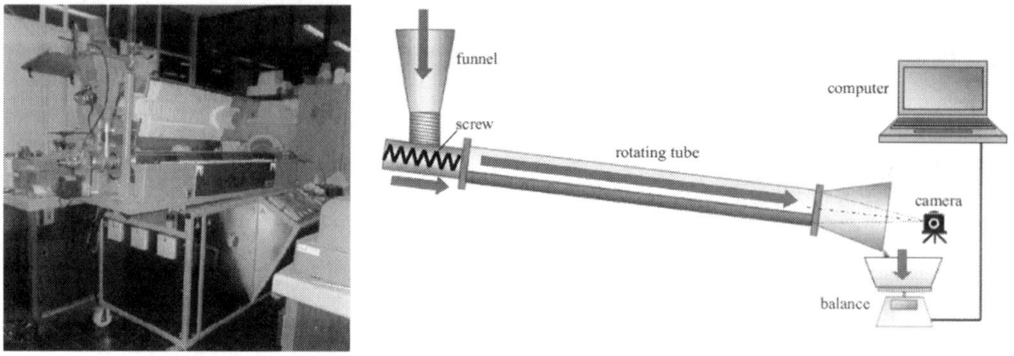

a *b*

Fig. 1. Picture (a) and sketch (b) of the experimental rotating kiln used in the present study.

The methodology is based on pictures of the advancing bed, taken from the bottom (outlet) side. These images are then analysed by means of the perspective geometry rules, and the bed profile can be obtained by using the geometrical relationships which can be defined for the system. Thus, the information on the bed profile so obtained can be used to calculate the parameters of a mathematical model describing the mass transport phenomenon as a function of the three independent variables of the process: the biomass flow rate, the tube slope, and the cylinder rotating rate. In Fig. 2 an example of a graphical bed profile determination is reported.

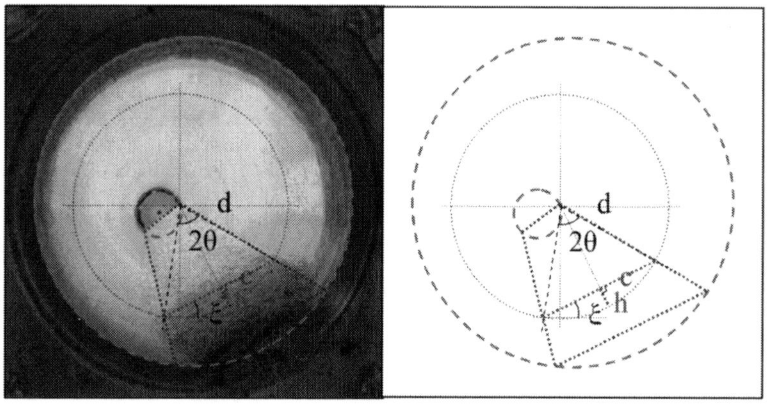

Fig. 2. Example of the graphical procedure to determine the bed characteristics (c: chord; d: diameter; h: bed depth; θ: half of the central angle described by the bed in the transverse plane; ξ: angle between the chord and the horizontal plane).

The methodology here presented will be used in the future development of a whole pyrolysis process model including heat exchange, gas fluid dynamics and pyrolysis reaction kinetics, that are influenced by the bed shape, depth and operating conditions.

[1] Pedroza M.M., Sousa J.F., Vieira G.E.G., Bezerra M.B.D., *J Anal Appl Pyrol* 105; 108-115 (2014).
[2] Murugan S., Gu S., *Renew Sust Energ Rev* 46; 282-295 (2015).
[3] Desogus F., Pili F., Carta R., *Chem Eng Sci* 145; 80-89 (2016).
[4] Descoins N., Dirion J.-L., Howes T., *Chem Eng Process* 44; 315-321 (2005).
[5] Huang A.-H., Kao W.-C., Kuo H.-P., *Adv Powder Technol* 24; 364-72 (2013).

Kinetic modelling of the gas-phase partial oxidation of light hydrocarbons

F. Desogus, [1]R. Carta
Cagliari Univ., Dep. Mechanical Chemical and Material Engineering, Piazza d'Armi, 09123 Cagliari, Italy; tel. +39 0706755079, e-mail: f.desogus@dimcm.unica.it;
[1]Cagliari Univ., Dep. Mechanical Chemical and Material Engineering, Piazza d'Armi, 09123 Cagliari, Italy; tel. +39 0706755068, e-mail: r.carta@unica.it.

During the last decades many studies have been directed to the development of better and better solutions for the efficient use of carbon sources (biomass and organic fuels) to obtain energy, also in the form of electricity, both on large and small scale. Processing liquid fuels can present some disadvantages, both from the technical and economic point of view, and similar (or worse) problems arise when solid fuels are used without pre-treatment, such as untreated biomass [1]. The combustion of gaseous fuels, instead of liquid or solid ones, is normally more efficient and cleaner. Thus, when liquid and solid fuels are available, it could be convenient to convert them to gas before using, instead of directly burning them.

This paper reports the kinetic modelling, with the experimental determination (at 500°C) of parameters, of gas phase reactions between light hydrocarbons (until C_7) and water. The experimental data were obtained with mixtures of heptane and water in different ratios as starting reagents. The aim of the work was to obtain detailed information about kinetics of the gas phase homogeneous chemical reactions which take place during wet (in the presence of water vapour as the oxidizing agent) biomass conversion processing, considering that other experimental studies are, at the moment, being conducted by the authors about biomass pyrolysis processing [2].

The experimental work was carried out by means of a laboratory apparatus (see Fig. 1) consisting in a continuous heated stainless steel reactor (in which the fluid dynamics regime was assumed to be of perfect mixing) placed inside a muffle oven and maintained at a

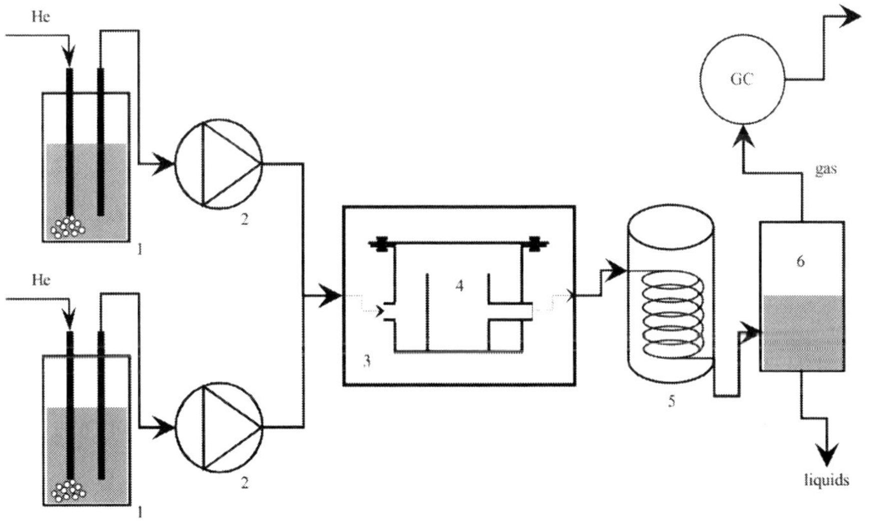

Fig. 1. Scheme of the experimental apparatus (1: reagent reservoirs; 2: multichannel volumetric pump; 3: muffle oven; 4: reactor; 5: cooler-condenser; 6: phase separator; GC: gas chromatograph analysis system).

constant temperature. The pump regulation system made possible to change the reactor inlet volumetric liquid flow rate of each reagent independently, in such a way that also the ratio between the two flow rates could be changed. The out coming gas composition, after cooling and separation of the condensable components, was analysed by an in-line gas chromatograph.

The first tests were conducted by feeding heptane and water. The choice of heptane is due to the following reason: it is one of the linear hydrocarbons, liquid at near to room temperature, and with the smallest molecular weight. Furthermore, it is known that heptane can be subjected to partial oxidation to produce syngas [3], and oxidation and pyrolysis of heptane can be found in many scientific works [4]; also the reaction involving heptane and water has been studied by other researchers [5, 6], and this could be important for comparison purposes.

The reactions to be considered in the kinetic model were chosen according to the literature information [5] and are the following:

$C_7H_{16} + 7\,H_2O \rightarrow 7\,CO + 15\,H_2$ (R1); $C_7H_{16} + 4/3\,H_2 \rightarrow 7/3\,C_3H_8$ (R2);

$C_7H_{16} + 5/2\,H_2 \rightarrow 7/2\,C_2H_6$ (R3); $C_7H_{16} + 6\,H_2 \rightarrow 7\,CH_4$ (R4);

$CH_4 + H_2O \rightleftharpoons 3\,H_2 + CO$ (R5); $CO + H_2O \rightleftharpoons CO_2 + H_2$ (R6);

$C_2H_6 + 2\,H_2O \rightleftharpoons 5\,H_2 + 2\,CO$ (R7); $C_3H_8 + 3\,H_2O \rightleftharpoons 7\,H_2 + 3\,CO$ (R8).

The kinetic constants were numerically determined by the ordinary least squares method, and are reported in Table 1.

	KINETIC CONSTANT		
REACTION	SYMBOL	NUMERICAL VALUE	UNIT
(R1)	k_1	$2.183\cdot10^{-20}$	$mol/(dm^3\cdot min\cdot kPa^8)$
(R2)	k_2	$5.289\cdot10^{-5}$	$mol/(dm^3\cdot min\cdot kPa^{(4/147)})$
(R3)	k_3	$3.127\cdot10^{-5}$	$mol/(dm^3\cdot min\cdot kPa^{(5/98)})$
(R4)	k_4	$8.899\cdot10^{-5}$	$mol/(dm^3\cdot min\cdot kPa^{(1/14)})$
(R5)	k_5	$2.241\cdot10^{-6}$	$mol/(dm^3\cdot min\cdot kPa^2)$
(R6)	k_6	$3.290\cdot10^{-6}$	$mol/(dm^3\cdot min\cdot kPa^2)$
(R7)	k_7	$3.215\cdot10^{-8}$	$mol/(dm^3\cdot min\cdot kPa^3)$
(R8)	k_8	$2.938\cdot10^{-10}$	$mol/(dm^3\cdot min\cdot kPa^4)$

Table 1. Calculated kinetic constants.

The experimental apparatus and procedure, in addition to providing the values of the model parameters here considered, showed to be able for future operations in different conditions, i.e. at different temperatures and with other hydrocarbons and organic compounds as reagents, both pure and in a mixture. The obtained data will be used together with those coming from the parallel studies by the same authors about the wet pyrolysis of biomass for gaseous production and utilization.

[1] Bridgwater A.V., Toft A.J., Brammer J.G., *Renew Sust Energ Rev* 6; 181-246 (2002).
[2] Desogus F., Pili F., Carta R., *Chem Eng Sci* 145; 80-9 (2016).
[3] Al-Hamamre Z., *Int J Hydrogen Energ* 38; 11458-11469 (2013).
[4] Chaos M., Kazakov A., Zhao Z., Dryer F.L., *Int J Chem Kinet* 39; 399-414 (2007).
[5] Abashar M.E.E., *Int J Hydrogen Energ* 38; 861-9 (2013).
[6] Pant K.K., Kunzru D., *Chem Eng J* 87; 219-25 (2002).

Use of microwaves for disinfection of farmland: a feasibility study

[1]F. Desogus, [2]A. Fanti, [2]M. Spanu, [2]M. B. Lodi, S. Casu, [2]G. Mazzarella, [3]B. Bisceglia
Cagliari Univ., Dep. Electric and Electronic Engineering, Piazza d'Armi, 09123 Cagliari, Italy; tel. +39 0706755787, e-mail: sergio.casu@diee.unica.it;
[1]Cagliari Univ., Dep. Mechanical Chemical and Material Engineering, Piazza d'Armi, 09123 Cagliari, Italy; tel. +39 0706755079, e-mail: f.desogus@dimcm.unica.it;
[2]Cagliari Univ., Dep. Electric and Electronic Engineering, Piazza d'Armi, 09123 Cagliari, Italy;
[3]Salerno Univ., Dep. Industrial Engineering, Piazza d'Armi, Via Giovanni Paolo II 132, 84084 Fisciano, Italy.

Recently the scientific community's interest in biological agriculture has significantly increased. "Bio-agriculture" means a precise cultivation method which allows the exclusive use of natural substances, already present in the nature, and excluding chemical products such as manure, herbicides and pesticides. In order to avoid invasive and potentially harmful methods both for cultures and soil health, in recent times the use of electromagnetic waves, in particular microwaves, for soil disinfection from pathogen agents has been taken into account. Microwaves are important because lead to an increase in the rotational energy of molecules, especially water, whose particles possess an electric dipolar moment with angular frequency equal to microwaves. Supposing bacteria and microbes composed mainly of water, the energy transfer, so the temperature increase, can be exploited to inhibit and eliminate them [1].

Soil is a mixture of fragmented minerals with a layer of organic matter and a certain amount of water and air. As better specified in Table 1, three are the main components of soil:

- clay, with particle size below 2 μm;
- silt, with particles in the range 2-20 μm;
- sand, with particle size ranging from 20 μm to 2 mm.

CLAY	SILT	SAND					GRAVEL
		Very Fine	Fine	Medium	Coarse	Very Coarse	
< 0.002	0.002 ÷ 0.05	0.05 ÷ 0.1	0.1 ÷ 0.25	0.25 ÷ 0.5	0.5 ÷ 1.0	1.0 ÷ 2.0	> 2.0

Table 1. Classification of granulometric classes (sizes in millimetres).

The soil texture influences the dielectric properties, in fact, depending on the composition, particular kinds of ground are able to retain different amounts of water: clayey soils can bound more water than sandy ones. So, in this work, main ground configurations are taken into account to better define the interactions between soils and microwaves.

In order to irradiate the soil, we designed a horn antenna with pyramidal base, working at 6 GHz, composed by a waveguide WR159 and aperture dimensions fixed as in Table 2. The radiation system was fed with 1 kW and it exhibited a gain of 10 dB.

a [mm]	b [mm]	c [mm]	a_1 [mm]	b_1 [mm]	c_1 [mm]
40.386	20.193	63.366	71.12	44.45	153.42

Table 2. Dimensions of the designed horn antenna.

The behaviour and response of our antenna was studied and analyzed using a Finite Element Method software for high frequencies electromagnetic problems. The device was placed in an air box with length 1.2 m and width of about 2λ from the aperture. A cube, representing soil, with a side length of 10 cm was placed at a distance of 0.95m from the horn.

Focusing on the soil water content, hence on the volumetric moisture, and on the percentage of sand and clay, the effective dielectric constant of the soil was evaluated comparing different models: the first was proposed by Hallikainen et al. [2] and the second model was developed by Gadani and Vyas [3, 4]. Once the value of dielectric constant was determined, it was possible to observe as the field distribution is affected by the presence of the soil. Hence, we evaluated the average Specific Absorption Rate (SAR) for different kinds of soil at 6 GHz in the presence of our antenna, using the following formula: $SAR = \sigma |\mathbf{E}|^2 / 2\rho$.

Using moisture and texture data for Sardinian cultures from a measure campaign conducted in 2003, we analyzed the cases of dry soil (*F*), soil with a medium moisture (*A*) and wet soil (*G*), with the textures reported in Table 3 and the dielectric constants in Table 4. Supposing the ground as an adiabatic system, and considering that reflected power by soil was 46.8% of the incident one, we performed the thermal analysis with a resulting incident power of 532 W. Hence the SAR distribution along a line is morphologically similar to the electric field distribution inside the soil (see Fig. 1). It is possible to observe that soil with medium moisture is able to absorb a higher amount of energy than dry and wet soils. Then higher temperature in such kind of soil could be reached for a fixed exposition time.

The heat transfer phenomenon related to microwave exposition of soil should be evaluated, in order to assess if the temperature increase necessary to inhibit the microbial activity is achievable and in which kind of soil. With exposition time on the order of 10 min is it possible to reach 50°C of temperature increase, thus leading to an use for real applications.

SOIL	Component fraction			
		SILT		
	SAND	coarse	fine	CLAY
A	0.290	0.354	0.147	0.209
F	0.293	0.365	0.120	0.222
G	0.059	0.809	0.099	0.033

Table 3. Texture of the analyzed soils.

W_v [cm³/cm³]	*A* 0.2940	*F* 0.0935	*G* 0.6557	*A* 0.2940	*F* 0.0935	*G* 0.6557
ε'	14.99	4.860	53.65	14.61	4.913	92.48
ε''	3.364	0.551	12.36	4.057	0.824	30.01
t_δ	0.224	0.113	0.230	0.278	0.168	0.325

Table 4. Dielectric constants for the soils.

	Hallikainen et al. (1985)	Gadani and Vyas (2015)

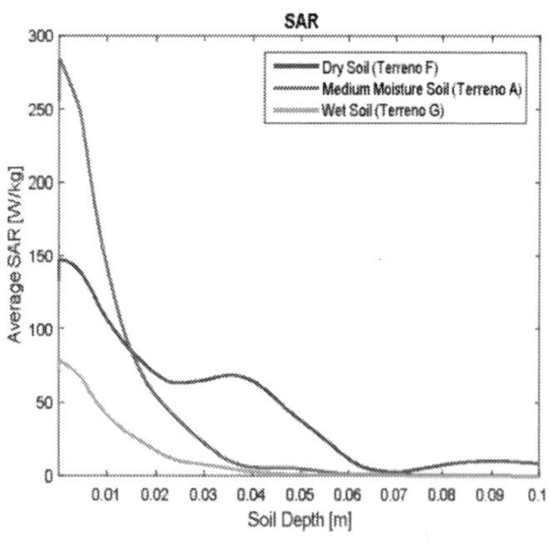

Fig. 1. Volume averaged SAR for the main kinds of soil.

[1] A. S. Komarova, A. A. Likhacheva, D. G. Zvyagintsev, *Moscow Univ Soil Sci Bull* 63, 190-195 (2008).

[2] M. T. Hallikainen, F. T. Ulaby, M. C. Dobson, M. A. El-Rayes, L. K. Wu, , *IEEE Trans Geosci Remote Sens* 1, 25-34 (1985).

[3] D. H. Gadani, A. D. Vyas, *Indian J Pure Appl Phys* 53, 190-198 (2015).

[4] A. R. von Hippel, Dielectric Materials and Applications, Arthur von Hippel editor, 1954.

Crude palm oil production in North Colombia: Process simulation and exergy evaluation

D. Martinez[1], A. Puerta[1], R. Mestre[1], Y. Y. Peralta-Ruiz[2], Angel D. González-Delgado[1]
[1]University of Cartagena, Chemical Engineering Department, Avenida del Consulado #Calle 30 No. 48 – 152, Cartagena, Colombia.
[2]Universidad Libre Seccional Barranquilla, Industrial Engineering Department, Carrera 46 No.48-170, Barranquilla, Colombia.
agonzalezd1@unicartagena.edu.co

The palm production chain in Colombia is facing several technical demands, including the increment of efficiency in the use of energy for the process, for which it is necessary to know its energy sinks and stages susceptible to improvement, from technical and energetic points of view. In this work, a real crude palm oil extraction process located in North Colombian region is analyzed using process simulation and exergy analysis methodologies, in order to quantify the irreversibilities of the productive process. Simulation was performed using specialized software and the effects of operating variables on the process efficiency was evaluated. By exergy analysis were quantified the physical and chemical exergies of the species involved, based on a robust modeled oil composition, the total irreversibilities of the process, the exergy from utilities, exergy from wastes generated during oil extraction and overall exergy efficiency of the process.

Results shows that for a 30 t/h plant, there is a high exergetic content from the solid wastes of the process, like palm fruit bunches, seeds and residual cake, therefore it is proposed an improved topology to optimize the process the incorporation of a palm kernel oil unit, which will help to decrease exergy of wastes.

Big data and Python

J. Buzík, T. Létal, P. Lošák
Brno University of Technology, Faculty of Mechanical Engineering, Institute of Process Engineering, Technická 2, 616 69 Brno, Czech Republic; tel. +420 541144941, e-mail: buzik@upei.fme.vutbr.cz, letal@fme.vutbr.cz, losak.p@fme.vutbr.cz

Turn of 20th and 21th century is distinct with increasing use of computational technologies. At present, these technologies are interweaved in all branches of industry. Thanks to advancements in computational technologies, numerical analyses can be used as a counterpart to traditional design methods in mechanical engineering. Numerical analyses are especially advantageous in non-standard cases, for which there are none or overly simplified analytical methods. On the other hand, numerical analyses are demanding in terms of computational power, expensive software and qualified experts for pre- and post-processing. These disadvantages seriously limit applicability of the numerical analyses. However, there is an alternative that utilizes creating computationally efficient models based on large set of numerical analyses. If applied appropriately, this approach has advantages of analytical as well as numerical methods, therefore the resulting .method can be efficiently applied in practice.

This paper deals with development of simplified methods based on results of computational fluid dynamics (CFD). Using CFD software ANSYS Fluent, influence of cross-flow on tube bundle in superheater chamber is analysed. The goal of the project is reliable evaluation and prediction of high cycle fatigue of the tubes.

CFD presents one of the most computationally demanding methods. In order to obtain good results, fine mesh is usually necessary. This requirement leads to large amount of result data, which have to be consequently processed. There is a lot of general purpose software that could be utilized for particular tasks in the data evaluation, e.g. Hadoop, R, Python with libraries PyDoop and Pandas, etc. However, there is no software that would cover the whole evaluation.

The main content of this paper presents development of special software tool that will be used for efficient processing of CFD result data and finding relations between key input and output parameters. The main features of the software will be graphical user interface, user defined conversion of CFD results to forces acting on tubes and possibility of data reduction using Fourier transformations. The software is being developed mainly using Python programing language. This language is very easy to use – it has less 'boilerplate' code than other languages and the code is very readable. The main disadvantage can be speed, nevertheless, that can be overcome by using specialized libraries for critical tasks.

Thermochemical treatment of rubber seed for biofuels synthesis

A. S. Reshad, V.V. Goud and P. Tiwari
Department of Chemical Engineering, IIT Guwahati, Guwahati, Assam, 781039, India
*Corresponding author. Tel.: +91 361 2582263/2272; fax: +91 361 2582291.
E-mail address:pankaj.tiwari@iitg.ernet.in (PankajTiwari), vvgoud@iitg.ac.in (V. V. Goud)

Rubber seed (RS) comprises of 40-48% shell and 52- 60% kernel by weight of seed. Rubber seed oil (RSO) from kernel was extracted by Soxhlet extraction using various organic solvents. Extracted RSO was used for synthesis of rubber oil methyl ester (ROME). Two steps procedure, esterification with homogeneous catalyst (H_2SO_4) followed by transesterification with heterogeneous ($Ba(OH)_2{\cdot}8H_2O$) was performed using ultrasonic technique. Response surface methodology (RSM) was applied to optimize the process parameters. Maximum conversion RSO(~97%)was achieved at catalyst loading 5.38 %(w/w) with methanol to oil molar ratio 8.09:1 in 13.20 minutes. Physico-chemical properties of RSO and ROME were estimated to measure its suitability for biodiesel production and fuel application respectively. A detailed engine performance and emission characteristic study was conducted with various blends of ROME-Diesel. Further, thermogravimetric analysis (TGA) and reactor pyrolysis of solid by-products such as shell and cake of the RS were conducted to study the pyrolysis temperature range for the production of targeted bio-oil compositions. Further, Waste rubber seed shell (RSS) comprises of 40–48% by weight of seed was used for biofuel and value-added chemicals production through semi-batch pyrolysis reactor. Proximate, ultimate, extractive, thermal behaviors and crystanllite index analyses are important for biomass characterization. The ultimate analysis results revealed that the total organic matter content of RSS is around 96.67wt%. On average 3.53 wt% extractive (hexane and ethanol) were obtained in RSS sample and the crystallinity index was found to be 38.6%. Thermogravimetric analysis (TGA) of RSS was conducted at heating rate of 20°C/min to identify the active pyrolysis zone of temperature. R50 value of RSS at 20°C/min heating rate was found to be 12.72 %/min (376.5°C). The residual weight left after 800°C (0.8 wt%) was an indication of lignin and inorganic minerals presence in RSS. This result was confirmed by proximate and field-emission scanning electron microscopy (FESEM) with energy dispersive X-ray spectroscopy (EDX) analysis of RSS. Effects of pyrolysis temperature (350°C–600°C) and heating rate (10°C/min–40°C/min) on products distribution and heating value of bio-char were investigated. The result showed that maximum liquid yield and carbon-rich bio-char can be obtained at 550°C pyrolysis temperature for heating rate of 30°C/min. With this condition the product distribution was found to be 46.14wt% (liquid), 30.42wt% (bio-char) and 23.44wt% (gas). Bio-char obtained at 550°C pyrolysis temperature has comparable energy content with anthracite (32–34 MJ/kg) solid fuel. Higher calorific value (34.5 MJ/kg), higher fixed carbon, lower ash and moisture content of bio-char produced from RSS pyrolysis can be used as a solid fuel. Liquid products were identified using gas chromatography with mass spectroscopy (GC-MS). Functional group presence also analyzed using Fourier transform infrared (FT-IR) spectroscopy. The physico-chemical-thermal properties of RSS and obtained products (liquid and bio-char) suggest that RSS can be considered as a suitable feedstock for bio-chemical and fuel production

An experimental investigation on seeded granulation of detergent powders

[1]N. Rahmanian, [2]M.H. Halmi, [3]D. Choy, [4]R. Patel, [5]S. Yusup, [6]I.M. Mujtaba
[1]School of Engineering, Faculty of Engineering and Informatics, University of Bradford, Richmond Road, BD7 1DP, UK; tel. +441274234552, e-mail: n.rahmanian@bradford.ac.uk;
[2]Chemical Engineering Department, Universiti Teknologi Petronas, 32610 Seri Iskandar, Perak, Malaysia; tel. +4407401157617, e-mail: m.h.binhalmi@bradford.ac.uk;
[3]Chemical Engineering Department, Universiti Teknologi Petronas, 32610 Seri Iskandar, Perak, Malaysia; tel. +60165618878, e-mail: dinson_ck@hotmail.com;
[4]School of Engineering, Faculty of Engineering and Informatics, University of Bradford, Richmond Road, BD7 1DP, UK; tel. +441274233690, e-mail: r.patel@bradford.ac.uk;
[5]Chemical Engineering Department, Universiti Teknologi Petronas, 32610 Seri Iskandar, Perak, Malaysia; tel. +6053687642, e-mail: drsuzana_yusuf@petronas.com.my;
[6]School of Engineering, Faculty of Engineering and Informatics, University of Bradford, Richmond Road, BD7 1DP, UK; tel. +441274239645, e-mail: i.m.mujtaba@bradford.ac.uk;

Granulation is commonly used to enlarge particle size to impact desirable characteristics and functionality to the granules. Granulation process transforms fine powders into free-flowing, dust-free granules with the presence of liquid binder at certain operating conditions. The main focus of this research is on seeded granulation of detergent powders, a new phenomenon of granulation in which a layer of fine powders surround the coarse particle. This is already proven for calcium carbonate (Rahmanian et al., 2011). Here, detergent granules were produced in a 5 L high shear Cyclomix granulator using different coarse/fine powder ratio (1/3, 1, 3) and different binder ratio of 10%, 20% and 30%. The granules were then characterized for their particle size distribution, strength and structure. It was found that a high percentage (70 wt. %) of granules in the desired size range between 125-1000μm were produced using the powder ratio of 1/3 and a binder content of 10%. Low mean crushing strength (3.0N) with a narrow distribution was obtained using this condition and no seed was found at other ratios. Structure characterization of the detergent granules produced in Cyclomix granulator shows that consistent seeded granules structures are produced under the optimum process and formulation conditions of 1/3 powder ratio with 10% binder.

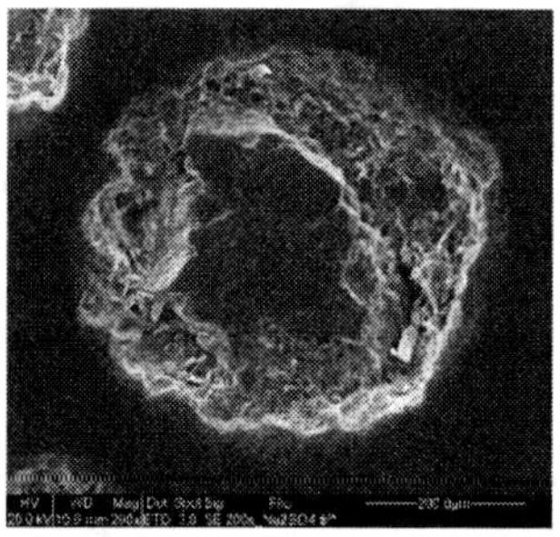

Figure 1: SEM image of a seeded granule 500-600μm at t 10% of binder with the powder ratio of 1/3 (25%F: 75% Coarse).

References:

1) Rahmanian, N., Ghadiri, M., Jia, X., 2011, Seeded granulation. Powder Technology, 206, 53-62.

2) Rahmanian, N., Naji, A., Ghadiri, M., 2011, Effects of process parameters on granules properties produced in a high shear granulator. Chemical Engineering Research and Design, 89(5), 512-518.

Technical and economic impacts on different methods of improving the performance of a gas pipeline

Amar Benmounah and Ammar Chebouba
Department of Transport and Equipment of Hydrocarbons. Faculty of Hydrocarbons and Chemistry. University Mhamed Bougara Boumerdas 35000. Algeria.
amarbenmounah@yahoo.fr, chebouba@yahoo.fr

Today, natural gas represents 24.7% of the global energy consumption. Over 106,000 m^3 of natural gas are consumed every second in the world, (3,350 billion m^3 of gas per year). World production of natural gas in 2012 amounted to 3 193 billion cubic meters.

In a context of growing concern of degradation of the environment, first quality natural gas is its non-polluting nature. For this we propose in this article to improve technical performance gas pipelines existing or future, to today, in order to obtain better performance, lower transportation costs, as well as flexibility of use compared to the competing energies. The Algerian exploitation policy, through the Transport activity of SONATRACH, which manages 16,200 km of pipelines, is to maximize revenues by optimizing the recovery and the means of transport by pipelines. There are several methods of improving the performance of a gas pipeline to know:

(1) increase in the diameter of the pipe, *2)* increase of the discharge pressure; *(3)* doubling of compressor stations; *(4)* cooling of the gas at the beginning of sections; *(5)* improvement of the strength characteristics of metal component pipeline; *(6)* decrease roughness of the pipe; *(7)* monitoring of the integrity of the pipeline; *(8)* improving the capacity by parallel conduct (looping); *(9)* treatment of gas initially; *(10)* Finally it can combine all of these methods together.
Each of these methods has its advantages and disadvantages as feasibility (construction difficulties) and costs generated by the application of the method.

We can enumerate the following cases: the increase in the diameter of the pipeline, whatever limited to 48" or same 56", will greatly reduce the load losses and the number of compressor stations, but will require greater discharge pressures which will increase the temperature of the gas. The latter will not have time to dissipate between stations and therefore will have a negative impact on the flow, hence need to cool the gas. This is not to mention the additional work of civil engineering in relation to the trench, the costs that it induces and the effect on the environment.

The purpose of this article is to list the various problems encountered in the project of a gas pipeline from the technical, organizational and economic point of view. Partial solutions for each case will be offered as well as an optimal solution for improving the performance of a gas pipeline.

Keywords: performance; costs, optimization, heat transfer, reliability.

A case study for developing eco-efficient street lighting system in Saudi Arabia

Khurram Shahzad*[,a], Lidija Čuček[b], Muhammad Sagir[c], Abdul-Sattar Nizami[a],Tariq Iqbal[d], Talal Almeelbi[a], Iqbal M.I. Ismail[a]
[a]Center of Excellence in Environmental Studies, King Abdulaziz University, Jeddah, Kingdom of Saudi Arabia
[b]Faculty of Chemistry and Chemical Engineering, University of Maribor, Maribor, Slovenia
[c]Department of Chemical Engineering, University of Gujrat, Pakistan
[d]Department of Process and Food Engineering, Faculty of Engineering, University of Putra, Malaysia, shahzadkhu@gmail.com

It is now well-known phenomenon that energy efficiency has highest short-term pay out period to decrease overall energy consumption. The replacement of conventional lighting technology with innovative lighting solutions can save up to 40% of lighting energy. The ecological evaluation of street light provision system in King Abdulaziz University (KAU), Jeddah is carried out using Sustainable Process Index (SPI) methodology. This study is carried out selecting three commonly used street illuminating devices i.e. High Pressure Sodium (HPS) lamps, Compact Fluorescent (CF) lamp and Light Emitting Diode (LED). The results show that energy consumption can be decreased by a factor of 1 to 4 by replacing HPS lamp with high efficiency LED lamp. Similarly, environmental assessment results reveal that ecological footprint as well as carbon footprint caused by lighting service can also be lowered by replacing HPS and CF lamps with LED lamps.

Acknowledgement
The authors acknowledge the financial supports from the Ministry of Higher Education, Kingdom of Saudi Arabia (KSA) and from Slovenian Research Agency (program P2-0377).

Case study comparison of bubbling fluidised bed and grate-firing technology based small biomass combined heat and power plants

L. Rummel, A. Paist, E. Latõšov
TallinnUniv. of Technology, Dep. Thermal Engineering, EhitajateRd.5, 19086Tallinn, Estonia;
tel. +3725345 0636, e-mail:leo.rummel@eesti.ee;

The article gives an overview of the first results of a long-term research on work regimes and behaviour of mineral matter in biomass boiler systems. As woody biomass is gaining importance as an energy source for electricity and heat we need to investigate with scientific methodology the main problems that occur with biomass boiler systems.

The results of research done on large power plants and lab scale batch boilers are not applicable in case of small and medium size biomass boilers on which the research has been limited. This study focuses on the rapidly growing small cogeneration field.

Two small-scale combined heat and power (CHP) plant industrial boilers systems are compared based on design and measured operating data. One boiler system uses grate firing and the other one bubbling fluidised bed (BFB) technology. Both CHP plants have capacities of 2 MW electrical and 8 MW thermal. Design parameters and actual operational data of the plants are compared in Table 1 and used main equipment in Table 2.

Table 1. Comparison of design parameters of the two CHP plants

Parameter	Unit	Grate-fired CHP plant		BFB boiler CHP plant	
		Desing	Actual	Design	Actual
Fuel input	MWth	12.6	11.9	12.6	13
Thermal output	MWth	8	-*	8	8.26
Electrical output, gr	MWe	2	2	2	2
Electric output, net	MWe	1.63	1.88	1.73	1.8
Total efficiency**	%	76.4%	-	77.2%	77.4%
* Thermal output was not measured during the performance test.					
** Based on LHV and net electric capacity					
Fuel					
Min LHV	MJ/kg	7.2		6	
Max LHV	MJ/kg	12.6	Avg. 9.6	13	Avg. 7.5
Min moisture	%	35		30	
Max moisture	%	55	Avg. 52.7	60	Avg. 52.9
Min size	mm	1		20	
Max size	mm	250		200	
Boiler					
Steam pressure	bar	51	50.0	25	26.7
Steam temperature	*C	450	-	450	463.1
DH water, supply	*C	90	90	100	100
DH water, return	*C	55	55	55	55
Emissions					
Max dust	mg/Nm3	375	117.6	250	98.39
Max NOx	mg/Nm3	400	148.5	375	199.3
Max CO	mg/Nm3	250	78.5	-	211.8

Table 2. Main equipment used in the two CHP plants

	Grate-fired CHP plant	BFB boiler CHP plant
Furnace	Moving grate	Bubbling fluidised bed (BFB)
Steam cooler	No steam cooler after superheater	Steam cooler after superheater
Flue gas cleaning	Multicyclone	Electrostatic precipitator (ESP)
Cooling	Wet cooling tower	No cooling tower

Based on design data and data collected from performance tests and Supervisory Control and Data Acquisition (SCADA) system, the following conclusions can be made regarding the two CHP plants:

- The net electric output of the grate fired CHP plant is higher and due to this the total efficiency could also be higher. This could be due to the fact that BFB boiler uses fans to keep the bubbling bed in the air and thus consumes more electricity, but also due to the use of ESP.
- The BFB boiler can burn a wider range of biomass fuel by moisture and LHV. The particle size is more limited.
- The BFB boiler has lower dust emissions. This could be due to the use of ESP instead of a multicyclone. Generally the dust emissions of a BFB boiler should be higher due to creation of small particles in the process.
- The NOx and CO emissions of the BFB boiler are higher. This could be due to many reasons. It is possible that the BFB boiler was still being fine-tuned during the testing and these results will be different later.
- Exiting flue gas temperature from the BFB boiler plant was 189*C and from the grate fired boiler it was 148 *C. This shows that the heat from the combustion is collected more efficiently in this grate-fired boiler.

Further research will include comparison of differences in work regimes, long-term efficiency and problems with mineral matter and corrosion with these two case study combined heat and power plants.

Design of water networks with multiple contaminants by concentration potential concepts

Xiao-Yan Fan, Zhi-Yong Liu *
Hebei University of Technology, Tianjin, China
Corresponding author: ZY Liu, Email: liuzhiyong@hebut.edu.cn

It is often difficult to design water networks with multiple contaminants with literature methods. This paper provides a review about the concepts of Concentration Potential and their applications in design of water using networks with multiple contaminants. The values of Concentration Potential of Demands (CPDs) can be used to identify the precedence order of the processes to be performed. The values of Concentration Potential of Sources (CPSs) can be used to determine the reuse order of the source streams. The concepts of Concentration Potential have been successfully applied in design of different types of water using networks with multiple contaminants. It shows that the concepts of Concentration Potential are powerful tools in design of water using networks with multiple contaminants.

Experimental and simulation studies for recovery of residual Assam crude oil: Characterization of rock-fluid and screening of chemicals

P Tiwari*, R Saha, K Anem, R Uppaluri
Department of Chemical Engineering, Indian Institute of Technology Guwahati, Guwahati, Assam, 781039, India
*Corresponding author: Tel.: +91 361 2582263; fax: +91 361 2582291
Email address: pankaj.tiwari@iitg.ernet.in (Pankaj Tiwari)

Surfactants are injected in reservoirs to recover the residual crude oil which cannot be produced by natural pressure difference and/or by water flooding. Surfactant reduces interfacial tension between crude oil and aqueous phase which helps in reducing the trapped from the pore space of the reservoir rock. However, the surfactant when injected reacts with reservoir rock and gets adsorbed due to which the surfactant gets diluted and increases the amount requirement for the process which ultimately increases the cost. The present work focuses on experimental and simulation studies on identifying and characterizing the cost effective combination of alkali-surfactant-polymer for EOR process for Assam crude oil. Assam crude oil of 26° API and viscosity 7.3 Pa.s at 30°C was characterized for various physical and chemical properties including FTIR and NMR analyses. Various surfactants were subjected for their application in enhance oil recovery. Interfacial interaction of Assam oil with different surfactant was studied to observe the reduction in interfacial tension (IFT) which is one of the main objectives in chemical enhanced oil recovery technique. IFT measurements along with other characterization such as thermal stability, phase behaviour, adoption phenomena were performed in detail with various types of surfactant. The selection of appropriate surfactant depends on the rock-fluid nature. Thus the Assam crude oil collected from various wells and rock (core) from different geological conditions were also screened. FTIR and NMR spectroscopy techniques were used to analyse the crude oil samples. Assam rock elemental analysis was performed by EDAX which shows the anionic nature of Assam reservoir rock. Assam rock due to anionic in nature adsorbs more sationic surfactant, medium adsorption for non-ionic and least adsorption for Anionic surfactants. CTAB was found better in reducing IFT compared to other surfactants but its adsorption was significant since it is cationic in nature. Non-ionic surfactant was somewhat better in terms of less adsorption and reducing IFT, whereas anionic surfactant showed lease adsorption but the required reduction in IFT was not achieved. Core flooding experiments were performed with chemical slug prepared with optimum concentrations of alkali, surfactant and polymer. The optimum combination of chemicals and their concentration was established for lowest IFT value and Winser III microemuslion for both with and without brine environment. Conventional (from different geological location of Assam oil field) and synthetic cores of 1.5 inch diameter were characterized for porosity, permeability and mineralogical content of the formation. Core flooding studies were conducted with various optimal ASP chemical solutions to determine the maximum residual oil recovery in addition with the water flooding from Assam oil field. One dimensional mathematical model for chemical flooding was developed for two phase system and oil recovery from secondary (water) and chemical flooding results were compared with experimental findings.

The concept of the test bench to measure the risks of stopping the burner in automatic solid fuel boilers

R. Urbaniak, J. Bartoszewicz, R. Kłosowiak, A. Nygard
Poznan University of Technology, Chair of Thermal Engineering, Piotrowo 3, 60965 Poznan, Poland; tel. +48 6652213, fax: +48 6652281, e-mail: rafal.urbaniak@put.poznan.pl, e-mail: jaroslaw.bartoszewicz@put.poznan.pl, e-mail:robert.klosowiak@put.poznan.pl; e-mail:adam.nygard@doctorate.put.poznan.pl

Modern low-power boilers for solid fuels with automatic fuel delivery systems for combustion chamber, are nowadays increasingly used solution. Their use allows to obtain substantial savings of precise fuel supply to the combustion process while improving greatly environmental combustion process of solid fuels. Unfortunately, along with the use of automated systems supplying fuel to the burner increases the risk of many threats which did not have place for the operation of traditional boilers with manual fuel supply.
The article presents the concept of bench to identify the risks associated with one of the main problems occurring during the operation of automatic solid fuel boilers equipped with screw feeders, this problem is related to the blocking of work.

The burners screw, also known as feeders retort are energy devices, in which fuel from the tank lowers by gravity into the tray (typical auger screw popularly called), which is moved in the direction of the furnace. Figure 2 shows the mode of action of the feeding screw.

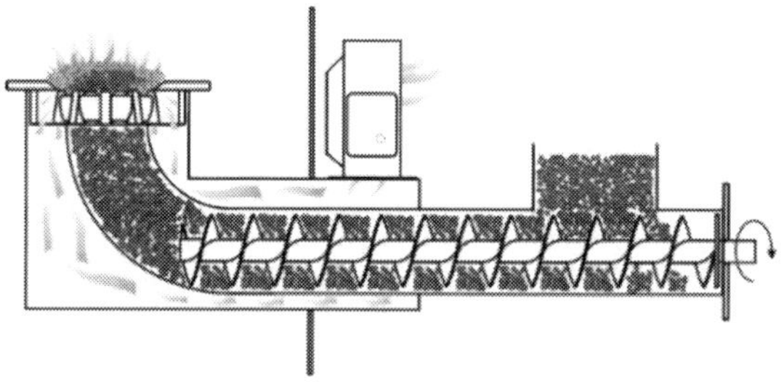

Fig. 1. The method of operation of the retort burner [6]

In order to realize specific functions, designed and performed research station, whose main component is a PC that will allow for the implementation of control algorithms in the LabView environment. Figure 2 shows a block diagram of the automatic system built on the basis of measurement systems manufactured by Adventech, ADAM series [7].

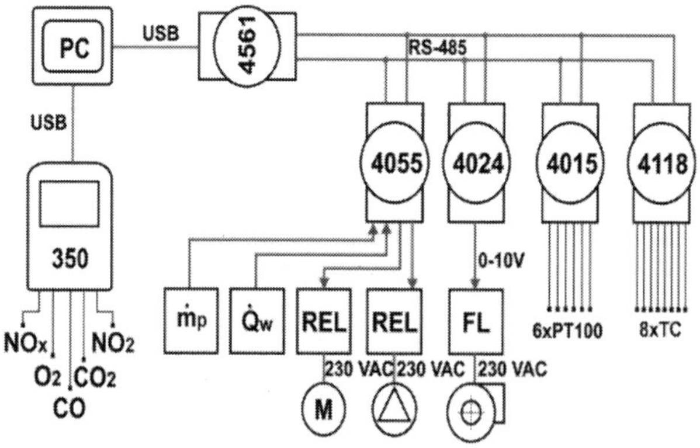

Fig. 2. Block diagram of the measuring system [7]

The main element test stand, was 9 thermocouple. In view of the expected high temperatures, applied type sensors K. Figure 3 shows the arrangement of the probes on the surface of the burner.

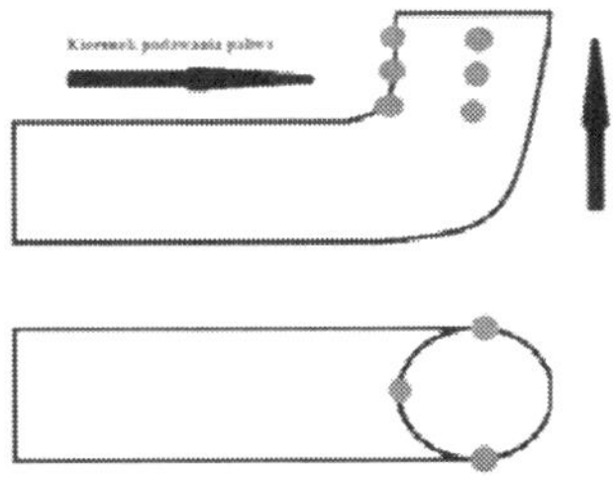

Fig. 3. Distribution of the team thermocouples used for testing the temperatures occurring on the surface of the feeder

Carried out using the position of studies have shown that dangers of blockage of the burner. Figure 4 shows an example of the temperature distribution the feeder during his blocking, recorded with the designed test bench.

Fig. 4. Distribution of temperatures in the tray during normal operation

[1] Czyste ogrzewanie, www. Czysteogrzewanie.pl, dostęp: 2015

[2] Instytut Chemicznej Przeróbki Węgla, „Kryteria emisyjno-energetyczne IChPW na znak bezpieczeństwa ekologicznego", Zabrze, 2006.

[3] Kapitaniak A., Sztraube J., „Poradnik palacza. Budowa i obsługa kotłów rusztowych.", WNT, Warszawa 1991.

[4] Norma PN EN 303-5, „Kotły grzewcze. Część 5: Kotły grzewcze na paliwa stałe z ręcznym i automatycznym zasypem paliwa o mocy nominalnej do 300 kW – Terminologia, wymagania, badania i oznakowanie", PKN

[5] Urbaniak R., Bartoszewicz J., Kłosowiak R., Main Reasons of NOx Emissions From Process of Fuel Cobustion in Low-Power Boilers in Poland, Polish Journal of Environmental Studies, Vol. 24, No. 5 pp ISSN: 2223-2230, 2015.

[6] Zawistowski J., Kotły z podajnikiem, www.zawijan.wordpress.com, dostęp: 2015.

[7] Urbaniak R., Bartoszewicz J., Kłosowiak R., The concept of the test bench for testing automatic biomass boilers, Conference Process Integration, Modelling and Optimisation for Energy Saving and Pollution Reduction 2014, ISBN: 978-80-02-02555-9 , Materiały konferencyjne, Czech Republic, Prague 23-27.08.2014

Modelling and validation with experimental data of the fluid-dynamic behaviour of a spouted bed facility for the gasification of agricultural residues

D. Bove, [1]C. Moliner, [2]M. Curti, [2]G. Rovero, [1]E. Arato, [1]B. Bosio
Free University of Bozen-Bolzano, Faculty of Science and Technology, Piazza Università 5, 39100 Bolzano, Italy; [1]University of Genoa, DICCA, Via Opera Pia 15, 16145 Genova, Italy; [2]Polytechnic of Turin, DISAT, C.so Duca degli Abruzzi 24, 10129, Torino, Italy.

The obtaining of the optimum operational conditions for a given process is the major goal of pilot and, once scaled, industrial units. For the case of fluidised bed reactors, the optimization of the fluid dynamic parameters like gas inlet flow is one of the most important steps for a successful design of the reactor. Experimental trial and error tests provide the most reliable data to evaluate these units even though they are expensive both in time and economic terms. There is, however, a major drawback: the change of one of the parameters of the process may lead to different optimum conditions. It is then when simulation activities play an important role for a reliable design of the reactor. Achieving the right balance between experimental and modeled results will be the key for a good development of a technology.

Within this framework, the spouted bed reactor has been studied. Spouted bed reactors are attracting particular attention for their use in gasification processes because it has been proven a convenient technique for thermo-chemical conversions of biomass through fluidisation processes. This device guarantees an improvement of the gas-solid mass transfer and the minimisation of the energy requirements of the system.

In this work the attention has been focused on the fluid-dynamic behaviour, so that the experimental activity has been carried out in a conical square-based spouted bed facility tested without occurring chemical reactions. Its main dimensions are: side of 200 mm and total high of 2 m. The base has an inclined angle of 60° in order to enhance the recirculation of solids. The air inlet orifice is situated at the center with a diameter of 21 mm.

In order to obtain experimental data, air was used as the gas phase and the particles were PET material. Different initial heights of PET (25 cm, 35 cm and 45 cm) were tested. For each initial quantity of PET, two different air velocities (set velocity and set velocity increased by a 10%) were evaluated.

The obtained experimental data (total pressure and local pressure along the bed of solids) have been used to validate a 3D model developed according to a Computational Fluid Dynamic (CFD) modeling technique. The Eulerian-Eulerian method was applied to predict the gas-solid flow behavior and the kinetic theory of granular flow was incorporated within the method as closure equation. The full Eulerian-Eulerian approach includes the conservation equations of mass and momentum for each phase and turbulence phenomena of the phases.

For this work, the main fluid dynamic parameters of the system were recorded for each case and compared with the values provided by the developed model with the purpose of the optimisation of the inlet gas velocity in order to establish a stable spouting regime.

Subsequently, the validated simulation tool has been used to identify preliminary operating constraints for the experimentation of a pilot plant based on a spouted bed reactor built with the same geometry and dimension of the tested facility and where different agricultural residues (wood chips, prune of apple trees and rice straw) have been gasified.

Impact of hybrid heat transfer enhancement techniques in shell and tube heat exchanger design

E. Shekariana[1], M. R. Jafari Nasr[2] A.H. Tarighaleslami[3], T.G. Walmsley[3], M.J. Atkins[3], N. Sahebjameed[4], M. Alaghehbandan[5]
[1]Department of Chemical Engineering, College of Engineering, Adadshahr Branch, Islamic Azad University, Azadshahr, Iran
[2]Research Institute of Petroleum Industry (RIPI), Tehran, Iran
[3]Energy Research Centre, School of Engineering, University of Waikato, Private Bag 3105 Hamilton 3240, New Zealand; tel. +64 7 8384937, e-mail: aht5@students.waikato.ac.nz
[4]Department of Chemical Engineering, College of Engineering, Quchan Branch, Islamic Azad University, Quchan, Iran
[5]Applied Research Centre for Agricultural Inputs, Karaj, Iran

Shell and Tube Heat Exchangers (STHE) are widely used in various industries such as oil, gas, petrochemical and power plants. Despite the advantages of shell and tube heat exchangers, one of their major problems is low thermal efficiency. Iterative trial and error procedures are commonly applied in heat exchanger design procedures to achieve desirable heat loads subjected to the allowable fluid pressure drops. As a result, to ensure compliance with the allowable pressure drop and achieve a desired duty, heat exchangers are designed larger than actually required.

In recent years a new design algorithm, which could calculate simultaneously desirable heat load with consideration of maximum allowable pressure drops has been used. The method is called the Rapid Design Algorithm (RDA). In this method the results are higher velocities on both the shell and tube sides, higher heat transfer coefficients and the minimum heat transfer area requirement and investment costs. Initially, Kern method has been used to correlate pressure drop, heat transfer coefficient of the fluid, and the heat exchanger area together. Combining pressure drop and basic heat exchange relationships led designers to avoid lengthy trial and error procedure. However, using Kern relations increases shell side errors due to inaccurate relationships. Later, Bell-Delaware equations were used to improve flow pattern in shell side and consequently achieved closer results of thermal-hydraulic analysis to experimental results.

Various techniques have been applied to increase heat transfer rates in HEs and decrease heat and energy losses in process industries. These methods are known as Heat Transfer Enhancement (HTE). HTE can be applied in both shell side and tube side of a shell and tube heat exchanger. STHEs. Helical baffles are more common in shell side, while using tube insert and applying nanoparticles to the service fluid are examples of HTE techniques in tube side. Simultaneous use of RDA design algorithm and HTE techniques may increase the heat transfer charactristics in shell and tube heat exchangers.

The aim of this paper is to develop the required inputs to RDA for three tube-side HTE scenarios for in shell and tube heat exchangers. The three considered scenarios are: Al_2O_3/water nanofluid, twisted tape turbulators, and a hybrid combination of both techniques.

Efficient flow modelling in equipment containing porous elements

V. Turek, D. Fialová, Z. Jegla
Brno Univ. Technol., Fac. Mech. Eng., Inst. Proc. Eng., Technická 2, 61669 Brno, Czech Republic; e-mail: turek@fme.vutbr.cz, tel. +420 541144908, fax: +420 54112372.

The paper discusses a simplified approach to mathematical modelling of flow distribution and pressure drop in process equipment containing porous elements (e.g. matrices of regenerative heat exchangers, tube sheets of dense tube bundles of recuperative heat exchangers, packed beds of catalytic converters, etc.). The modelling approach in question has been developed to be utilized primarily in optimization algorithms and as such it is capable of providing relatively accurate data within rather short times. Still, possible improvements with respect to available solution methods and solution efficiency in general are touched on as well given the intended application.

Acknowledgement

The authors gratefully acknowledge financial support provided by Technology Agency of the Czech Republic within the research project No. TE02000236 "Waste-to-Energy (WtE) Competence Centre".

Novel low-cost upgrading technology to produce biogas enrichment

D. Hidalgo[1,2], F. Corona[1,2], J.M.Martín-Marroquín[1,2], R. Piñero[1,2], G. Antolín[1,2]
[1]CARTIF Centro Tecnológico, Boecillo, Valladolid, 47151, Spain; tel. +34 983546504, e-mail: dolhid@cartif.es.
[2]ITAP Institute, University of Valladolid, Valladolid, 47010, Spain; tel. +34 983423355.

Biomass fuels in the form of wood, crop residues, and animal dung, continue to be the dominant source of cooking energy in India and other developing countries. The National Family Health Survey for 2005-2006, indicated that 71% of households in India rely primarily on solid fuels for cooking. 27% of urban households and 90% of rural households used biomass as their primary cooking fuel in 2005, 69% nationally. These estimates indicate that in 2005 over 770 million Indians living in nearly 160 million households remained primarily dependent on solid fuel for their domestic needs, 97% of which is biomass, the rest coal. Biomass fuels are usually burnt on inefficient, unvented household cooking stoves often in poorly ventilated houses or kitchens. The increased indoor and outdoor pollution is a major effects on public health such as neonatal mortality, respiratory disorders and depression in women. Many consider the alternative is liquefied petroleum gas (LPG) as it is a clean domestic fuel but non-renewable. The production of LPG in India is 170,450 kg as against the demand of 210,456 kg in 2010, and the demand is estimated to grow at 8% per annum. Further LPG is increasingly used in transport sector and government is planning now to reduce subsidy on this product. As a consequence, it is urgent to find an alternative clean biofuel that is renewable and indigenous. Biomethane, also called renewable natural gas, can be a good option.

A major challenge found in the production of renewable natural gas is to remove CO_2 and biogas impurities (mainly H_2S), which are produced during anaerobic digestion of biomass.

Under these circumstances, the BIOFUEL_for_COOKING_AND_CHP project (Figure 1) arises with the aim to contribute to solve the rural energy demand for domestic uses from locally produced biomass and household waste and the demand for electricity and thermal energy derived from renewable sources. One of the main innovations of this project is the development of a biogas upgrading low-cost process based on the combination of chemical scrubbing and low-pressure membrane technologies, that can remove 95-100% of impurities and reduce the CO_2 concentration up to 1-5%, thereby increasing the calorific value of the biogas, and also improving the engines efficacy.

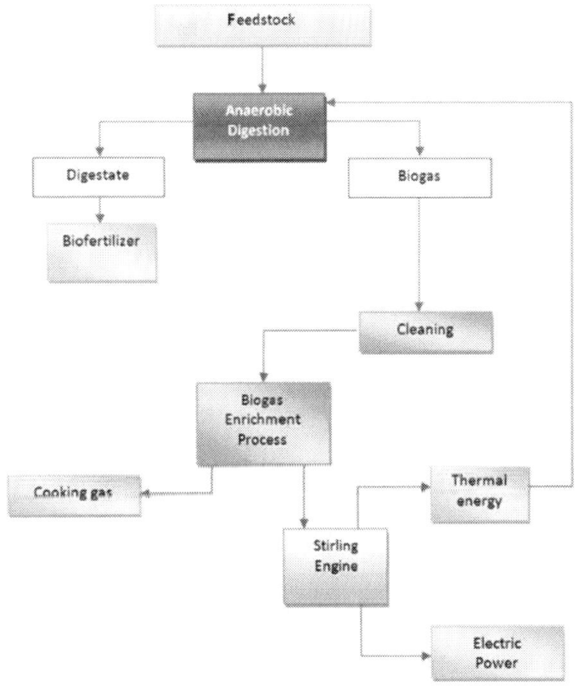

Figure 1. Organic feedstock valorisation process.

How to perform the Kyoto Protocol and the decision of Environmental Forum in Paris

D. Terpugov[1], G Terpugov, G. Kandelaki
1 D. Mendeleyev university of Chemical Technology of Russia, Miusskaya sqr. 9 125047, Moscow, Russia
Tel. +74956836789, e-mail mamont58@mail.ru
D. Mendeleyev university of Chemical Technology of Russia, Miusskaya sqr. 9 125047, Moscow, Russia
Tel. +74956836789, e-mail mamont58@mail.ru
D. Mendeleyev university of Chemical Technology of Russia, Miusskaya sqr. 9 125047, Moscow, Russia

Annotation
It is necessary to increase thermoenergetical efficiency of industrial plants, especially plants of termoenergetical and nuclear complex –that will decrease emissions of greenhouse gases for solving problem of global warming. Some technical ways of solving this problem will be shown in this article.

In June of 2013 the USA president Barak Obama made a statement. His speech was about warming of the climate. Obama told that USA are ready to head the world community for solving the problem of greenhouse gases, and first of all gas emissions of CO_2. If no solution of this common to all mankind problem isn't found the Earth will be involved in **planetary catastrophe.** This conclusion is based on results of ecological forum in Paris in 2015 and Kyoto protocol. According to these documents mankind is to reduce emissions of greenhouse gases (for example reduction of CO_2 up to 26 – 30%) to avoid cathastrophical scenario by 2030.

How we can do it?

At the moment almost 90% of all consumed energy is obtained from organic fuel. Primal energy resources consumption structure in Russia and whole world (in 2000) is shown on fig. 1. More than 1/3 of fuel is burnt in thermal power stations.

Total coefficient efficiency of thermal power stations can be described as complex of coefficients of all stages of process. Change of chemical energy into heat in steam-boilers has efficiency about 88%. Steam turbine changes up to 42% of heat into kinetic energy. Efficiency of generator is high and has efficiency about 98%. So summary of efficiencies is 36% in change from chemical energy to electricity. High-performance thermal stations reach efficiency up to 40%. Even then about 60% of heat of combusted fuel is used for warming environment.

There is only one way to reduce emissions of CO_2 – increasing coefficient efficiency of thermal power stations by using new technologies that make possible trapping and burying (or reusing) of greenhouse gases.

Heat exchanging surface can be regenerated with use of special reagents (acids) and it leads to additional sewage or using drilling method. For acid regeneration expensive alloyed steel is needed. Drilling method takes 4 – 36 hours only for tubes drilling, and even more time is needed for disassembling and reassembling the heat exchanger and to place it into the process line. Of course it is necessary to stop the process or to use the parallel process line. All these operations take time and reduce economic efficiency.

Exfoliating with thermal expansion of tube limescale has thickness 0.1 – 0.2 mm. Comparison of this thickness with known literal data shows that our heat exchangers will **save at least 15% of heat energy of chemical plants, thermo electrical and nuclear power stations.**

Industrial use of our devices will reduce fossil fuel consumption up to 25 – 30 % and reduce greenhouse gases emissions on the same rate. This corresponds with decisions of ecological forum in Paris (December 2015).

Water magnetohydrodynamic generator.

MHD-generator makes straight change of mechanical energy of moving substance into electrical energy (Hall's effect). The working bodies are: combustion products of fossil fuels, штуке gases with addition of alkali metals and their salts, vapors of alkali metals, two-phase mixes of vapors and liquid alkali metals, liquid metals and electrolytes and others. Temperature about 10000 K is needed for making electro conducting working body. Gas enriching with metal vapor (for example alkali) makes possible to lower temperature to 2200 – 3000 K.

Despite wide perspectives and lots of research works MHD generators and MHD-based devices are barely used even today. The problems are high temperature – 2200 – 10000 K and ecological problems as a result of using metals as working body. We have done researches of water energy and they show that water can be used as working body.

Conclusions

These technologies of greenhouse gases trapping and bury and increasing efficiency of thermoelectrical (from 38 – 42% up to 60 – 65%) and nuclear power stations will make possible for mankind to perform Kyoto protocol and decisions of ecological forum in Paris.

The Earth is sick by Venus

D. Terpugov, G Terpugov, O. Katsereva,B. Ksenofontov, R Komlyashov, Z. Vdovenko, S. Ilina, S. Churbanov
D. Mendeleyev university of Chemical Technology of Russia, Miusskaya sqr. 9 125047, Moscow, Russia
Tel. +74956836789, e-mail mamont58@mail.ru
D. Mendeleyev university of Chemical Technology of Russia, Miusskaya sqr. 9 125047, Moscow, Russia
Tel. +74956836789, e-mail mamont58@mail.ru
D. Mendeleyev university of Chemical Technology of Russia, Miusskaya sqr. 9 125047, Moscow, Russia
Tel. +74956836789, e-mail mamont58@mail.ru
Bauman Moscow State University
D. Mendeleyev university of Chemical Technology of Russia, Miusskaya sqr. 9 125047, Moscow, Russia
D. Mendeleyev university of Chemical Technology of Russia, Miusskaya sqr. 9 125047, Moscow, Russia
D. Mendeleyev university of Chemical Technology of Russia, Miusskaya sqr. 9 125047, Moscow, Russia

In June of 2013 the USA president Barak Obama made a statement. His speech was about warming of the climate. Obama told that USA are ready to head the world community for solving the problem of greenhouse gases, and first of all gas emissions of CO_2. If no solution of this common to all mankind problem isn't found the Earth will be involved in **planetary catastrophe.**

We have developed a complex of technologies that are able to solve the problem of the greenhouse gases and cure the Earth from the Venus illness; witch became unsuitable for life in the result of climate warming (greenhouse gases such as CO_2 and other).

Physical and chemical base of these technologies, that will provide saving life on the Earth are shown in this research work.

Keywords: membrane technology, direct osmosis, greenhouse gases – CO_2, SO_2, H_2S, CO_2 trapping from atmosphere, trapping, using and burying of greenhouse gases.

Introduction

Venus and Earth are alike in a few parameters – size, density, acceleration of gravity and total volume of CO_2. CO_2 on Venus is all in the atmosphere, but on the Earth most part of this gas is in connected form. The Ocean contains 60 times more CO_2 than atmosphere.

Solubility of CO_2 in the world Ocean will decrease with the temperature growth and in turn it will lead to increasing temperature growth and more CO_2 outcome. This can lead to warm steam Gulfstream stopping. It can happen in 15 – 20 years and be the beginning of chaos, possible wars, freezing of Europe, world economic collapse and "greenhouse catastrophe".

British scientist Stephen Hawking claims that "mankind will hardly survive in third millennium because Earth will turn into another Venus due to greenhouse effect".

We have developed technologies that make possible:

-removing CO_2 from atmosphere, trapping and disposal CO_2 and SO_2 from gas emissions of thermal power stations (for example)

-obtaining drinking water from air to reduce its concentration in air and its contribution into greenhouse effect.

According to UNESCO crisis of water resources is coming in a whole planet scale and period of 2005 – 2015 years was named the International decennary of actions "Water for life". According to International Health Organization more than 2 billion people suffer from shortage of drinking water in 2004 year. Water consumption increased more than in 7 times, yet population increased in 3 times during XX century. Germany inhabitant expends 127 liters a day, USA – 296, Russia – 278. All over the world process of pollution and wrong use of water resources leads to catastrophic reduction of drinking water reserve.

Drinking water reserve shortage aftermath is grave for all mankind. Lack of drinking water can lead to wars and conflicts, economical and social destabilization of countries. According to UNESCO territory of the USA has status "no sweet-water" and "polluted water".

A potential assessment of renewable energy sources in non-interconnected zones of Colombia, using geographic information system - ArcGIS: study of cases

[1]A. M. Rosso-Cerón, [2]F. Blanco Patiño,[3]J. Araque-Duarte, [4]V. Kafarov.
Universidad Industrial de Santander, carrera 27 calle 9; tel. +57 7634400 ext. 2603, e-mail: anarosso10@gmail.com;
[2]francy.blanco@correo.uis.edu.co,[3]julian.araque2@correo.uis.edu.co, [4]e-mail: kafarov@uis.edu.co

Nowadays, energy dependence on fossil fuels (in terms of primary energy consumption representing an average of 35% oil, 27% coal, 23% natural gas, and the rest from alternative energies) triggers significant concerns as they are not renewable energy.

Furthermore, energy generation is necessary for a country development since it is essential for everyday activities such as lighting, using electrical appliances, transport, livelihoods, and education by increasing employment opportunities that lead to economic and social benefits.

In Colombian case, non-interconnected zones (NIZ) are located in isolated places; cover almost 66% of the national territory including about 1,200 settlements, 16 departments, 91 tows and 2 million people. In addition, these zones do not have appropriate access routes (roads and rails), lack industrial and commercial development, public services are limited and undeveloped, and the basic needs of people are unsatisfied.

Paradoxically, they have great environmental importance with a wide variety of natural resources (water, solar, wind, and biomass) that can be exploited for the production of clean energy, since they take into account population growth, industry, the availability of energy resources, environmental disturbance, and the amount of financial resources for technological, social welfare and cultural development.

Consequently, the aims of this research is assessing energy potential of solar, wind, hydro, and agricultural waste biomass sources and choose the best option for the municipals of three departments with the highest energy demand from NIZ: Meta, Chocó; and Putumayo departments. Potentials are estimated using ArcGIS software and considered social and environmental restrictions like the presence of indigenous reservations, armed groups, and natural parks. It is important to mention that software like METEONORM are not a good alternative to determine potential places of Latin America, since there is a lack of data about meteorological stations and the interpolation of these data do not provide accurate and reliable information. The results show that the water resource is the best choice in Chocó's towns; agricultural waste biomass has the greatest potential in Putumayo's towns, and the most viable option for Meta's towns is the solar resource.

New process development of NGL (Natural Gas Liquid) processes

[1]S. Yoon, [1]J.K. Kim, [2]C.K. Park, [2]B.D. Lee
[1]Department of Chemical Engineering, Hanyang University, Wangsimni-ro 222, Seongdong-gu, Seoul, 04763, Republic of Korea; email: jinkukkim@hanyang.ac.kr
[2]Institute of Chemical & Environmental Process, Jeonjin Entech Co., Ltd, Gadal 1-Ro 38, Gangseo-Gu, Busan, 46729, Republic of Korea;

This presentation addresses a new process development of NGL (natural gas liquids) recovery process for offshore environment. Turbo-expander scheme is adopted for new NGL process for achieving high energy efficiency without compromising hydrocarbon recovery.

In order to carry out process design, the model of NGL process is first built with the aid of process simulator, which is used for evaluating important design variables and understanding design interactions existed in the recovery of hydrocarbon. Optimization framework based on stochastic solver is also built to screen different structural options and to determine the most appropriate operating conditions. The optimization solver is linked with a process simulator, with which computational difficulties can be avoided and high-quality optimal solution can be readily obtained within reasonable computational time.

For process design activities, heat integration method is systematically applied to maximize heat recovery and power recovery in a holistic manner. Also, optimization technique is effectively applied to carry out economic trade-off. The proposed NGL process is evolved from conventional turbo-expander processes, which is resulted from new heat recovery arrangement as well as different pattern of stream splitting and mixing.

Off-shore environment is carefully managed during the design of column internals and column configuration in which ship motion and its impact on mal-distribution of the column is considered.

Acknowledgement
This research was supported from the Technology Innovation (10042424, Floating Production Platform Topside Systems and Equipment Development) funded By the Ministry of Trade, Industry & Energy (MI, Korea).

Equilibrium and thermodynamic studies on adsorption of trivalent chromium by sericin-alginate particles prepared from *Bombyx mori* cocoons

Júlia R. de Andrade[a], Meuris G. C. da Silva[b], Marcelino L. Gimenes[a], Melissa G. A. Vieira*[b]
[a]State University of Maringa, School of Chemical Engineering, Colombo Av., 5790, 87020-900 Maringá, Brazil
[b]University of Campinas, School of Chemical Engineering, Albert Einstein Av., 500, 13083-852 Campinas, Brazil
melissagav@feq.unicamp.br

Particles prepared from the blend between sericin and alginate were used as biosorbent of Cr(III) ions form aqueous solution. In order to show how is the Cr(III) distribution between the liquid phase and the solid phase at the equilibrium, adsorption isotherms were obtained at four different temperatures (20, 30, 40, 60 °C). The isotherm models of Langmuir, Freundlich and Dubinin–Radushkevich were used to test the equilibrium adsorption data. Better fittingswere obtained for both Langmuir and Dubinin–Radushkevich models. Additionally, the thermodynamic parameters enthalpy change, entropy change, and Gibbs free energy change were calculated and the biosorption process was found to be exothermic and spontaneous. The results shown that sericin-alginate particles are a promising adsorbent material for the efficient removal of Cr(III).

Study on the DMSO assisted perovskite deposition with dipping method

K-.H, Hwang, H.J, Seo, Y-.M, Lee, D.I, Kim, S-.H, Nam, J-.H, Boo
Department of chemistry, sungkyunkwan university, suwon-si, 440-746, Korea/+82312905972/+820312907075/nercisus@naver.com

Organic-inorganic hybrid perovskite solar cell is getting a lot of attention as next-generation solar cells. It is important that improvement of the active-layer electrical and optical properties for high power conversion efficiency (PCE) of solar cells. In this work, we study on the morphology and electrical properties of methylammonium lead iodide (MAPbI3) thin films. MAPbI3 was deposited with dipping treatment in diethyl ether. we fabricated solar cells which is consisted with ITO/PEDOT:PSS/MAPbI3/PCBM/Ag. To comparatively study the weight-percent control effects on the resulted electrical and optical properties. The solar cell performance and impedance were investigated with solar-simulator. At the same time, morphology, optical properties and crystalline variations were investigated by field emission scanning electron microscopy (FE-SEM), UV-Vis spectroscopy and XRD respectively.

Numerical simulation and thermographic examination of the heat transfer in a radiator

M. Grabowski, K. Urbaniec, J.Wernik, K.J. Wołosz
Warsaw University of Technology, Plock Branch, Department of Process Equipment
Al. Jachowicza 2/4, 09-402 Plock, Poland
phone +48 24 2622610, e-mail: wernik@pw.plock.pl

Investigations of the heat transfer by conduction and natural convection in a radiator used in electronic equipment were carried out. The radiator consisting of four straight fins made of aluminium is depicted in Fig. 1.
On the basis of a physical model, 3-D geometrical model of the radiator was developed in COMSOL Multiphysics software. Using the finite-element method, numerical simulations were performed to determine temperature distribution at the radiator surface.
In order to validate the numerical model of the heat transfer, an experimental setup comprising an electric heating coil, radiator and thermovision camera FLIR SC7600 was assembled. By varying the voltage, the heat flow supplied by the heating coil was varied. The steady-state temperature distributions at the radiator surface were recorded using the thermovision camera, and parameters of the convective flow field in the surrounding air were measured by anemometry.
Positive results of the validation of the numerical model indicate that thermovision is a suitable experimental tool for this kind of application. Numerical simulation combined with thermographic techniques can be conveniently used in developing new radiator designs.

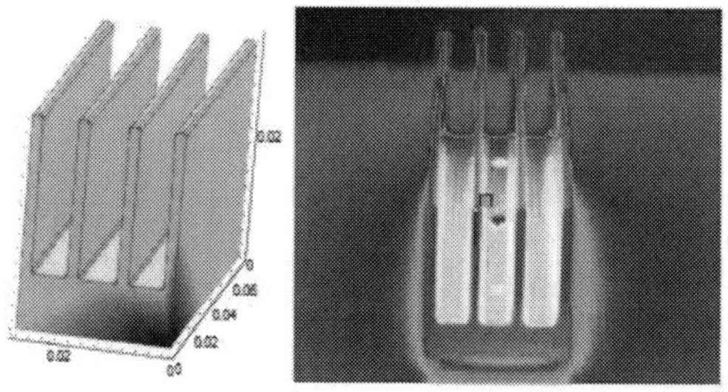

Fig. 1 Radiator geometry and visualisation of the temperature distribution.

Optimization of gas-turbine blades cooling process in the domain of discrete Fourier transform

A. Frąckowiak, M. Ciałkowski
Poznan University of Technology, Chair of Thermal Engineering, Piotrowo 3, 60965 Poznan, Poland; tel. +48 6652213, fax: +48 6652281, e-mail: andrzej.frackowiak@put.poznan.pl,
e-mail: michal.cialkowski@put.poznan.pl;

This paper presents results of calculations concerning determination of temperature distribution on the walls of cooling channels in blades of a gas turbine with one and three channels. To do so, an inverse heat conduction problem with known heat transfer coefficient h (fig.1) and temperature T_o on the outer surface of the blade Γ_{out} was solved:

$$\Omega: \quad \Delta T = 0$$
$$\Gamma_{out}: -k\frac{\partial T}{\partial n} = h\left(T - T_f\right)$$
$$T = T_o$$

where k[W/mK] is the heat conduction coefficient, T_f[K] denotes temperature of the fluid surrounding the region Ω from the outside (T_f = 800K). The distribution of temperature on the Γ_{in} boundary of the multiply-connected domain Ω should be found (the Cauchy problem).

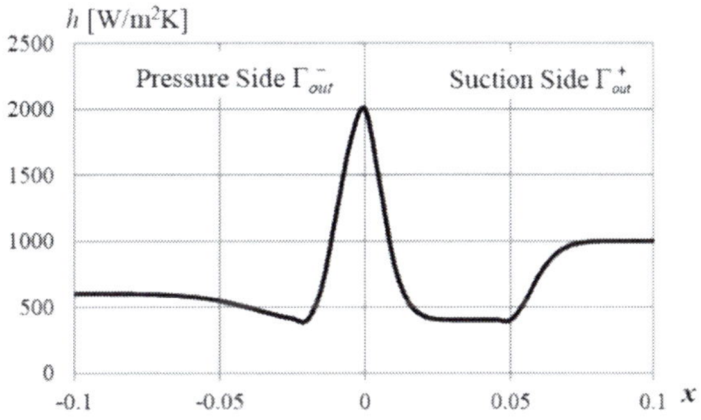

Fig.1. Distribution of the heat transfer coefficient on the outer boundary of the blade

Unknown temperature on the inner boundary of the blade Γ_{in} was assumed in a form of the linear combination of trigonometric functions φ_i so that coefficients c_i standing by those functions were components of discrete Fourier transform, [1]:

$$\Gamma_{in}: \quad T = \sum_{i=0}^{n} c_i \varphi_i$$

Those coefficents were determined from the minimum of quadratic functional in the form

$$J[T] = \frac{1}{2}\int_{\Gamma_{out}} \left(T - T_o\right)^2 ds + \frac{1}{2}\alpha\sum_{i=0}^{n}\left[\frac{i+1}{2}\right]^{2i} c_i^2, \qquad k \ge 0, \quad \alpha > 0$$

Sample results of calculations for a blade with one and three channels are presented in figures 2, 3, 4 and 5.

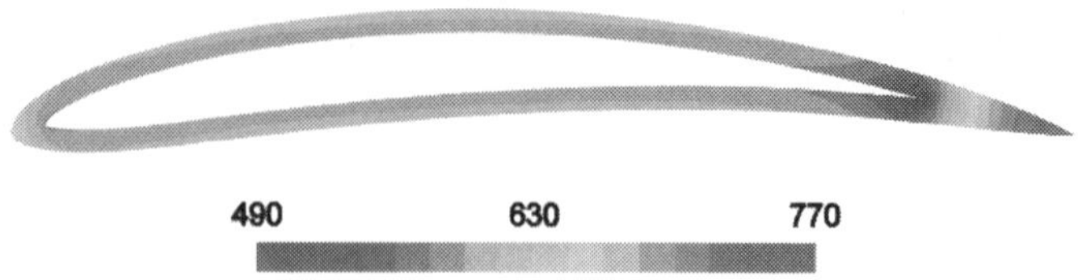

Fig 2. Distribution of temperature in a blade with one cooling channel for $T_o = 600$K, $k = 1$

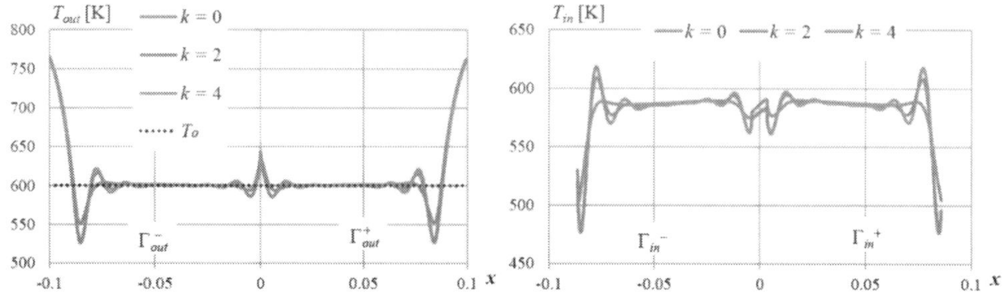

Fig. 3. Distribution of temperature on the outer surface of the blade (left) and on the wall of cooling channel (right); $T_o = 600$K

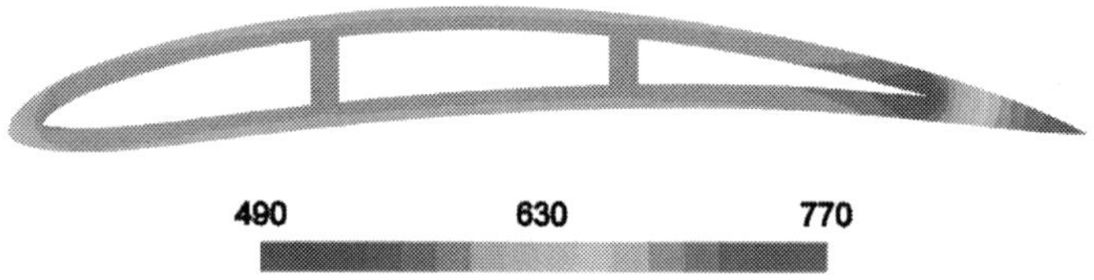

Fig. 4. Distribution of temperature in a blade with three cooling channel for $T_o = 600$K, $k = 1$

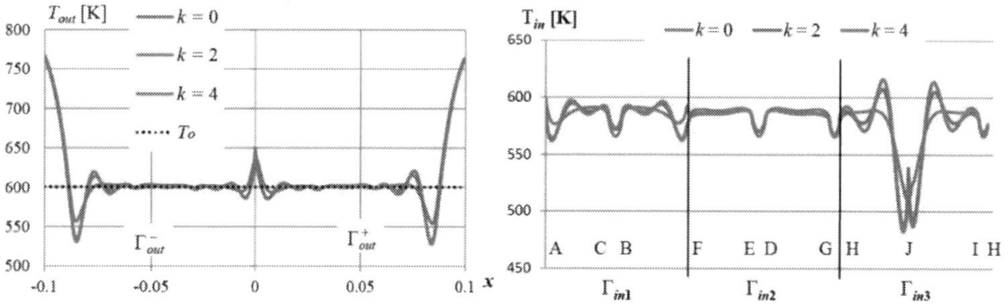

Fig. 5. Distribution of temperature on the outer surface of the blade (left) and on the walls of cooling channels (right) in the blade with three channels; $T_o = 600$K

Optimization of cooling process of the gas-turbine blades in the domain of discrete Fourier trans form in combination with regularization of the solution is characterized by high stability of thermal field. Components responsible for oscillations of the solution were dampened. On the outer surface, in the vicinity of the trailing edge, large derogations from $T_o = 600$K are observed since it is impossible to remove a large amount of heat from this part of the blade through the cooling channel. Increasing the intensity of cooling process in this part of the blade results in significant reduction of temperature on the wall of the channel in the vicinity of the trailing edge.

References

[1] Wróblewska A, Frąckowiak A, Ciałkowski M, *Regularization of the inverse heat conduction problem by the discrete Fourier transform,* Inverse Problems in Science and Engineering, 24(2), pp. 195 – 212, 2016.

H_2S oxidative decomposition for the simultaneous production of sulphur and hydrogen

V. Palma[1], V. Vaiano[1], D. Barba[1], M. Colozzi[2], E. Palo[2], L. Barbato[2], S. Cortese [2]
[1]University of Salerno, Dep. of Industrial Engineering, Via Giovanni Paolo II, 132, 84084 Fisciano (SA), Italy;
tel. +39089964147, e-mail: vpalma@unisa.it
[2]KT Kinetics Technology, Viale Castello Della Magliana 75, 00148 Rome, Italy

Hydrogen is mainly produced from natural gas via methane steam reforming and other established methods for hydrogen production such as partial oxidation of residual oil, gasification of coal and steam-iron process. A very interesting alternative could be the recovery of hydrogen from chemical substances identified as pollutants, such as H_2S.

Hydrogen sulphide is a by-product from sweetening of sour natural gas, hydrodesulphurization of light hydrocarbons, and from upgrading of heavy oils, bitumen and coals.

Hydrogen sulphide is usually removed by the well-known Claus process where H_2S is oxidized to water and elemental sulphur by a two-step reaction; however, due to thermodynamic limitations, H_2S conversion is limited to about 95-97% [1].

As alternative to the production of only sulphur, hydrogen could be also produced by decomposition of H_2S. Many processes have been proposed for the decomposition of H_2S [2]. The most interesting appears to be the thermal decomposition of H_2S. Different works have examined high temperature thermal dissociation of H_2S [3], but no practical industrial application has been realized due to the endothermic nature of the reaction, its equilibrium limitation below temperature of ~ 1300 °C and the facile recombination of H_2 with S_2 during the quenching step [4].

An interesting alternative could be to promote the decomposition reaction of H_2S, with a low concentration of oxygen in the feed stream, in order to obtain simultaneously sulfur, hydrogen and water. In this way, the thermodynamic limitations of thermal decomposition could be overcome, increasing H_2S conversion.

The aim of the work is to study the thermal decomposition of H_2S in the presence of oxygen in temperature range between 700 and 1100°C. The influence of residence time in the reactor has been studied, focusing the attention on the feeding molar ratio (O_2/H_2S) in order to minimize SO_2 selectivity and maximize the hydrogen yield together with a good H_2S removal efficiency.

The experimental results showed that H_2S conversion and H_2 yield increased with the temperature and the O_2/H_2S ratio, even if, the higher oxygen concentration in the feed, favours the total H_2S oxidation reaction enhancing the SO_2 production, thus lowering the H_2 yield. In particular, SO_2 selectivity increased from the 2-15% by varying the O_2/H_2S molar ratio from 0.2 up to 0.35.

The increase of the temperature in the range 700-1100 °C has determined the increase of H_2S conversion and H_2 yield, with a reduction of SO_2 selectivity. The decrease of the SO_2 production with the increase of temperature is likely due to the promotion of Claus reaction that involves the SO_2 consumption, previously produced, with the H_2S residual according to the reaction ($2H_2S + SO_2 \leftrightarrow 3/2\ S_2 + 2H_2O$).

Based on these preliminary results, it was possible to identify the optimal operating conditions, suitable to obtain a high H_2S conversion, a good H_2 yield and a low SO_2 selectivity.

In particular, the O_2/H_2S ratio equal to 0.2, a reaction temperature equal to 1100 °C and a residence time of 150 ms allowed to obtain a H_2S conversion of about 60%, a H_2 yield of 20% and the minimization of the SO_2 selectivity (3%).

Finally, the thermal decomposition of H_2S in the presence of oxygen can be considered a valid alternative to H_2S cracking without oxygen in order to produce simultaneously hydrogen and sulphur.

In particular, the presence of oxygen makes the overall process less expensive in terms of energy duty, since the heat required from the endothermic reactions is provided by the oxidation reactions.

References:

[1] Clark PD, Dowling NI, Huang M. Production of H_2 from catalytic partial oxidation of H_2S in a short-contact time reactor. Catal Commun 2004;5:743-747.

[2] Manenti F, Papasidero D, Frassoldati A, Bozzano G, Pierucci S, Ranzi E. Multi-scale modeling of Claus thermal furnace and waste heat boiler using detailed kinetics. Comput Chem Eng 2013;59:219e25.

[3] Clark PD, Sui R. Toward zero emission of sulfur species: using alumina nanofiber-based adsorbent for Claus tail gas processing. ACS National Meeting Book of Abstracts; 2010.

[4] Mashapa TN, Rademan JD, Van Vuuren MJJ. Catalytic performance and deactivation of precipitated iron catalyst.

Thermodynamic and kinetic based simulation approach to CO_2 and CO methane hydrogenation

S. Sharifian, M. Harasek , M. Miltner
Vienna University of Technology, Institute of Chemical Engineering, Getreidemarkt 9/166, A-1060 Vienna, Austria;
tel. +43 1 58801-166251, email:seyedmehdi.sharifian@tuwien.ac.at.

The trend of carbon dioxide emission mostly reflects energy-related industrial activities over the recent years. Global CO_2 emissions raised up to 35.3 billion tones (Gt) CO_2 in 2013, which is 2% increase compared to 2012. The hydrogenation of Carbon dioxide uses hydrogen as the required high energy material for transformation into more useful fuels or chemicals. This method is a truthful option for utilization of CO_2 and its reduction in atmosphere indeed.

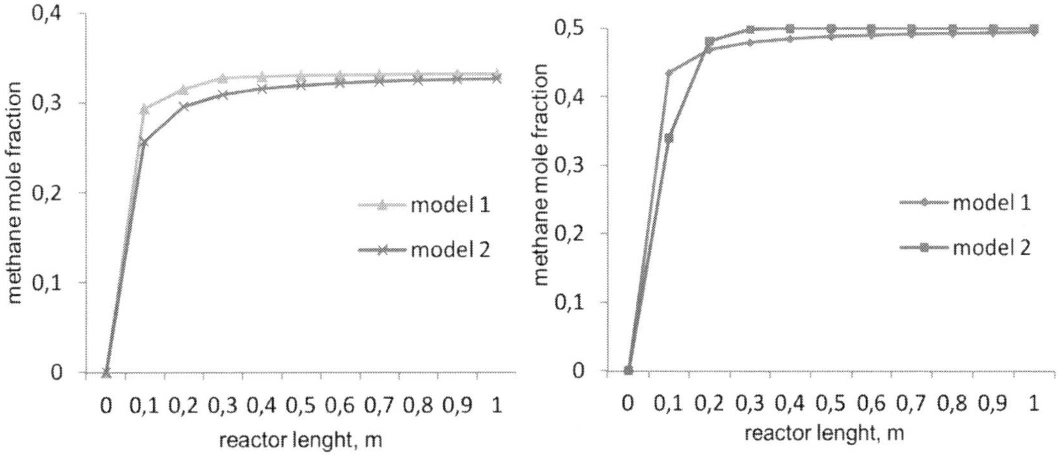

(a) (b)

The current study is based on simulation of CO_2 and CO methanation using Aspen Plus® V8.6. First, different kinetics models which are obtained from literature are used in order to investigate methane mole fraction changes along reactor length. Figure (a) presents CO_2 and figure (b) shows CO hydrogenation. It can be found that methanation of pure CO leads to higher value of methane yield and reactant conversion related to CO_2 at the same operating condition and stociometric feed ratio.

Moreover, a comprehensive thermodynamic simulation will be carried out to express effects of different operating conditions on system performance. It can be seen that both case performances are highly affected by temperature, pressure and feed ratio. Although, CO and CO_2 seem to have similar operation but, they have still differences which will be expressed in full manuscript of this summary.

Another step in this study is validation which is based on comparison of predicted values with three cases from different reactors and operating conditions which are obtained from literature. It will be shown that simulation results have a good agreement with empirical cases and it can be beneficial in reactor design and analysis of methanation process.

Effect of moisture problems in historical buildings and indoor climates

A. Sedlakova, [1]S. Vilcekova, P. Kapalo, [1]E. Kridlova Burdova, L. Tazky
Technical University of Kosice, Institute of Architectural Engineering, Vysokoskolska, 042 00 Kosice, Slovakia;
tel. +421 915 342 777, e-mail: ladislav.tazky@tuke.sk;
Technical University of Kosice, Institute of Environmental Engineering, Vysokoskolska 4, 042 00, Kosice, Slovakia.

In today´s era, historical buildings are frequently having problems with ground moisture. In the times of their construction there were no hydro insulation used to insulate the ground from the constructions. The only solution was to heighten the ground level above the terrain or using air-based insulation systems. After some time, the air channels and ventilated floors were removed (or destroyed) and allowed a free infiltration of a moisture to constructions. Today´s trend is again using the air-based insulation systems to intercept moisture moving, mainly in historical buildings where there is no degradation methods of insulation allowed.

Moisture in constructions has significant effect also on indoor climate when the relative humidity of indoor air rises. High relative humidity creates molds at the wall surfaces, most of them at the lower places, where are the lowest surface temperatures. Historical buildings not commonly used such as small churches, have bigger problems with molds because of poor ventilation.

To raise the quality of indoor air, construction details where moisture problems occur have to be solved. Moisture flow from the ground to the construction has to be intercepted or at least reduce as much as possible. The air channels are the easiest and the less invasive (for the Monument Board). While designing the air channels, materials compatible with historical buildings should be used. Last but not least, environmental load of used materials on nature and society have to be considered.

A new CO_2 absorption data for aqueous solutions of N-methyldiethanolamine + hexylamine

A. Mehassouel[1], R. Derriche[1], C. Bouallou[2]
[1]Ecole Nationale Polytechnique, Laboratoire de valorisation des énergies fossiles, B.P.182, El-Harrach, Alger, Algeria E mail: ammar.mehassouel@g.enp.edu.dz
[2]MINES ParisTech, PSL-Research University,
CES Centre d'efficacité énergétique des systèmes,
60 Bd Saint Michel, 75006 Paris, France E-mail : chakib.bouallou@mines-paristech.fr

This work focuses on the development of a new solvent for CO_2 capture. This new solvent is an aqueous solution with a blend of N-methyldiethanolamine (MDEA) and Hexylamine (HA) as an activator. CO_2 absorption has been investigated between 298 and 333 K using a Lewis cell with a constant interfacial area. New CO_2 absorption data is obtained with several concentrations of MDEA and HA: (MDEA 37 wt% + HA 3 wt%), (MDEA 35 wt% + HA 5 wt%) and (MDEA 33 wt% + HA 7 wt%) to assess the CO_2 absorption flux and the absorption capacity of the new solvent.

The kinetic results are in agreement with a pseudo first order regime of absorption according to film theory for the mixture of MDEA 37 wt% + HA 3 wt%. The activation energy decreased from 43.3 $kJ.mol^{-1}$for MDEA 40 w% to 22.25 $kJ.mol^{-1}$ for the new formulated solvent leading to an energy saving for the regeneration. The addition of small amount of HA leads to a high increase in the CO_2 absorption rates and improves the performance of the process allowing a substantial gain in terms of energy regeneration needs and reducing the cost of capture.

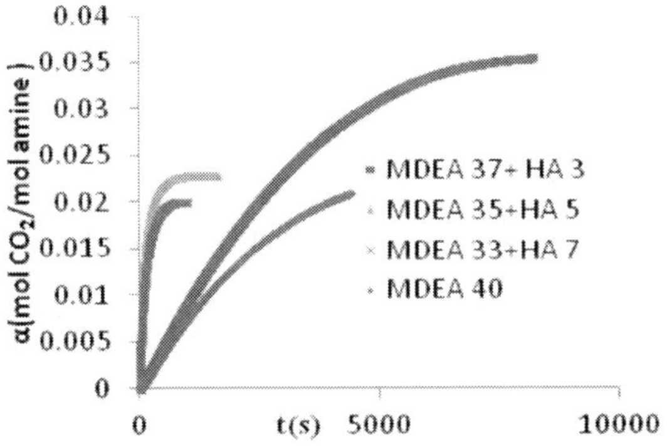

Figure 1: CO_2 loaded versus time for different solvents at 298.15 K

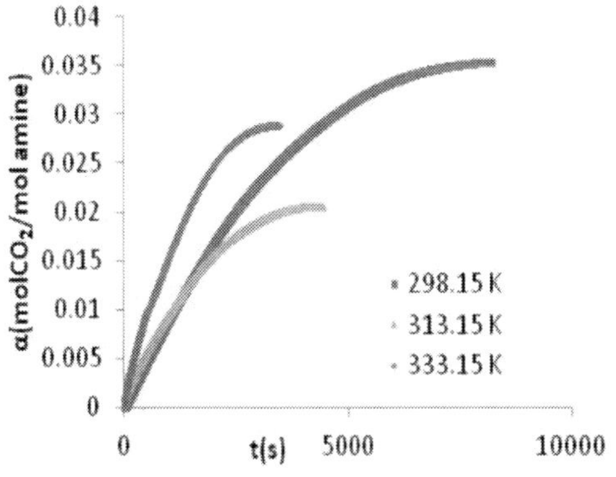

Figure 2: CO_2 loaded versus time for different temperatures and MDEA 37+HA 3 solvent

Characteristics of axial-flow pump measurement

V. Zejda[1], P. Bělohradský[1], L. Houdková[1,2], J. Kundrátek[2], J. Boráň[2], V. Brummer[1]
[1)] Brno Univ. Technology, Fac. Mechanical Eng., Inst. Process Eng., Technická 2896/2, 616 69 Brno, Czech Republic; tel. +420 541144939, e-mail: zejda@upei.fme.vutbr.cz; [2)] Kunst, spol. s r. o., Palackého 1906, 753 01 Hranice, Czech Republic.

This paper deals with a measurement of characteristic of axial-flow pump AMV 400 produced by KUNST Company, which was developed for digesters stirring.

The paper will contain a short review focused on CFD modelling of digesters stirred by similar devices. Then the design of a measuring stand will be described. The measuring stand has been designed with the target of the most geometric similarity to the digester. However the dimensions of real digester and experimental stand cannot be the same, so the differences will be taken into account at the results evaluation.

The main part of the paper will describe the pump characteristic measurement. The pump characteristic will be measured at different operational regimes. The influence of a stirrer rotations and its distance from the surface will be studied.

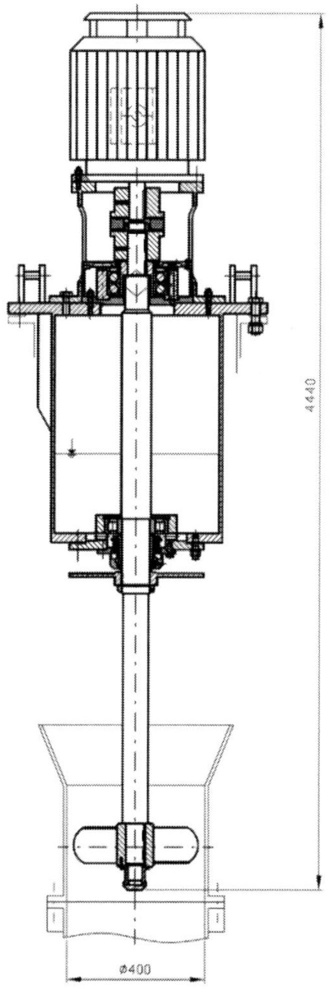

Evaluation of architectural and environmental aspects of passive house

A. Sedlakova, [1]S. Vilcekova, [1]E. Kridlova Burdova, P. Turcsanyi
Technical University of Kosice, Institute of Architectural Engineering, Vysokoskolska 4, 042 00 Kosice, Slovakia;
tel. +421 918 878 509, e-mail: peter.turcsanyi@tuke.sk;
Technical University of Kosice, Institute of Environmental Engineering, Vysokoskolska 4, 042 00, Kosice, Slovakia.

Passive house is a term known very well to engineers and architects all over the world. In these times of minimizing an energy loads and negative emissions, passive houses (or even net-zero houses, positive houses) are one of the best ways to meet European "20-20-20" targets.
The main topic of this paper is to create a complex assessment of a passive house, constructed in Košice, Slovakia. The passive house was built taking in account architectural, environmental and constructional requirements of today's euro codes.
The passive house was evaluated from two points of view. First, software simulations and calculations. The passive house was re-drawn to 3D form and situated to the virtual space of DesignBuilder's software reference year for Košice, Slovakia. It was observed how the passive house performs (energy performance and energy efficiency) throughout a simulated period of one year. Also, environmental indicators were calculated using the Life Cycle Assessment method (LCA), such as embodied energy (EE – energy used for acquiring raw material, manufacture and its transport), CO_2 emissions (ECO_2 – global warming potential GWP) and SO_2 emissions (ESO_2 – acidification potential AP).
Second, in situ measurements were conducted. Measured quantities were evaluated and compared to the software outputs and calculated outcomes. Results of this experiment are displayed in numbers as well as in graphic figures. Our goal was to measure and find out the differences between simulations, calculations and in situ measurements in Košice's region, and how we were able to minimize environmental load on nature and society.

Preliminary design and analysis of regenerative heat exchanger

B. Kilkovský, Z. Jegla
Institute of Process Engineering, Faculty of Mechanical Engineering, Brno University of Technology, Technická 2, 616 69 Brno, Czech Republic;
e-mail: kilkovsky@fme.vutbr.cz.

Heat regenerators transfer heat from one gas to another, with by storage in solids. This type of heat exchanger is used primarily when heat has to be transferred between enormous amounts of gases, when the gases are dirty and liable to plug up the recuperative heat exchangers, or when the gas is too hot or reactive. Regenerative heat exchanger can provide at the same volume much higher heat transfer area and lower friction factor than recuperative heat exchanger. Furthermore, regenerative heat exchanger has in most cases better efficiency and its construction is much simpler.

The calculation of this type heat exchanger is relatively complicated. Moreover, a suitable and adequately accurate calculation method is not freely available. Those are the main reasons why the companies are not interested in this type of heat exchanger. On the grounds of requirements from companies we started to deal with design of regenerative heat exchangers and creation of suitable calculation model that would cover all their requirements.

The various possibilities of regenerative heat exchanger applications will be describe in the paper. For various types of geometries of heat transfer surface their advantages and disadvantages will be described. On the basis of measurements of several types of heat transfer surfaces geometries their possibilities of heat absorption and amount of pressure drop will be compared. The mathematical models for calculation of these heat exchangers will be the main issue of the paper. The possibilities of improvement and extension of the simple calculation model will be also described/introduced. This model can be used for preliminary calculation of regenerator for given purpose.

Simultaneous using of regenerator as heat exchanger and reactor in which, for example, can take place cleaning of waste gas will be discuss.

The possibility of simultaneous use of regenerative heat exchanger as a reactor, where cleaning of waste could take place, will be considered.

Acknowledgement

The authors gratefully acknowledge financial support provided by Technology Agency of the Czech Republic within the research project No. TE02000236 "Waste-to-Energy (WtE) Competence Centre".

Recognition of thermal behaviour of low emission burner using comparative computational simulations

Z. Jegla
Brno University of Technology, Faculty of Mechanical Engineering, Institute of Process Engineering, Technická 2, 616 69 Brno, Czech Republic; e-mail: jegla@fme.vutbr.cz

The paper presents specific developed approach applied for recognition of real thermal behaviour of system of low emission gas fuel burners operated in radiant chamber of vertical cylindrical fired heater with nominal firing capacity of 24 MW placed in crude oil atmospheric distillation unit.

Results of CFD complex numerical computations performed for this vertical cylindrical fired heater to analyse the distribution of heat flux on the tubes of radiant chamber and validated by local heat flux measurements on operated heater are used as real expected values of heat flux distribution. This heat flux distribution profile is searched by comparative way using another type of thermal simulation of radiant chamber - so-called modified plug flow (MPF) model. The matching the desired heat flux profile within the frame of MPF model is performed iteratively based on successively refinement of initially assumed profile of local rate of heat release per unit height of the vertical cylindrical radiant chamber (i.e. fuel burnt profile) adjusted in the sequence of individual virtual height segments from which the vertical cylindrical radiant chamber is build-up using the MPF model.

Fuel burnt profile (for which the heat flux profile along radiant chamber height is the same as the heat flux profile found by CFD simulation) finally obtained by the MPF model then gives important information about real thermal behaviour of low emission burners employed in the radiant chamber of fired heater. This information allows consequently to identify and quantify another true burners thermal characteristics (such as flame length, robustness of flame, radiant chamber height profile of local rate of heat release, etc.) which are necessary and important for accurate thermal design of new radiant chambers (and especially for its inbuilt tube coil heat transfer systems) of fired heaters (for refinery, chemical or petrochemical industry), heat transfer fluids heaters (in pharmaceutical or food-processing industry), power boilers or waste incinerators (in waste-to-energy applications) employing this type and size of low emission gas fuel burners.

Acknowledgement

The authors gratefully acknowledge financial support provided by Technology Agency of the Czech Republic within the research project No. TE02000236 "Waste-to-Energy (WtE) Competence Centre".

The flue gas desulphurization at the first stage of the absorber - O-element

D. Jecha, [1]P. Lestinsky , V. Brummer, J. Jicha
Brno University of Technology, Faculty of Mechanical Engineering, Technicka 2896/2, 616 69 Brno, Czech Republic; tel. +420 541144954, e-mail: jecha@fme.vutbr.cz;
[1]VSB – Technical University of Ostrava, Institut Enviromental Technology, 17. listopadu 15/2172, 708 00 Ostrava-Poruba, Czech Republic; e-mail: pavel.lestinsky@vsb.cz

Processes of thermal decomposition are necessary not only for detoxication, but also for reduction of negative impacts on the environment. At the same time, waste serves as secondary source of energy. However, thermal processing has its drawbacks as well. These include release of emissions into the atmosphere, residual waste and water pollution. Emissions are formed by the polluting substances in the flue gas which are produced by the waste combustion and are of a harmful and toxic nature. Flue gas contains carcinogenic hydrocarbons, carbon monoxide, sulfur dioxide, hydrogen chloride, fine dust particles etc.

In today's industrial practice, it is common that the design of new flue gas cleaning units made by the designing company is based on the experience gained during the implementation of a similar equipment so that the devices are modified according to the specific conditions in the operation.

For effective design of flue gas cleaning devices is however needed to use validated computational techniques and subsequent optimization of the apparatus so that the designed device can work in a real operation with the maximum efficiency. An effective design of wet flue gas cleaning can be done using computational methods developed on the basis of data obtained during experiments carried out on a pilot scale equipment. In this case, experiments on the wet scrubber were used.

Pilot plant is comprised of first stage, i.e. O-element that replaces venturi scrubber that is currently used in practice. O-element is a device used to homogenize the gas-liquid mixture. Upon contact of flue gas and liquid, the hot flue gas is cooled and pollutants are being removed from flue gas. The second stage is designed as a packed column with the structured packing.

By a gradual transition from the planar embodiment of the O-element (the O-element 2R) to the spatial arrangement, the four branched O-element - the O-element 4R originated. The design of equipment for the larger flue gas flow rate is necessary, because the Waste to energy plants incinerating the municipal and hazardous waste commonly produce flue gas with flowrates in the tens of thousands m^3/h.

For such flue gas flow rate would be the originally designed device O-element 2R too space-demanding. Spraying a liquid in the four-branched-O element can be performed in two ways. Firstly, it could be managed in the same manner as in two-branched O-element, i.e. through the orifices in the bottom arcs of the equipment, or by spraying of the liquid around the circumference of the inlet pipe.

Experiments were performed with the two O-4R elements with liquid spray positioned around the perimeter of the inlet pipe and with the liquid spray through the nozzles in the lower arcs of the equipment. The main observed parameter was the SO_2 removal efficiency depending on the L / G ratio. In addition even more important indicator as SO_2 removal efficiency was monitored and it was an O-element pressure loss.

Application of oxide copper chromium catalysts for the purification of exhaust gases

[1]T.S. Baizhumanova, [1]Z.T. Zheksenbaeva, [1]S.A. Tungatarova, [1]R.O. Sarsenova, [2]M.A. Sadenova, [3]S.A. Abdulina
[1]D.V. Sokolsky Institute of Fuel, Catalysis and Electrochemistry, 142, Kunaev str., Almaty, 050010, Republic of Kazakhstan; tel. +77272916632, e-mail: tungatarova58@mail.ru;
[2]S. Amanzholov East Kazakhstan State University 34, 30 Gvardeiskoi divisii str., 070000, Ust-Kamenogorsk, Kazakhstan;
[3]D. Serikbayev East Kazakhstan State Technical University, 19, Serikbayev str., 070000, Ust-Kamenogorsk, Kazakhstan

The problem of chemical safety and sanitary air protection is particularly relevant due to the increase of harmful emissions of industrial enterprises, which have a strong toxic effect. The data on studying the process of deep oxidation of toluene on polyoxide nickel-, copper-, chromium-containing catalysts are presented in this paper. Toluene as the main component of emissions the furniture, cable, footwear and other production has been chosen as model substance of neutralization.

Polyoxide catalysts were prepared by capillary impregnation of alumina modified with cerium by mixed aqueous solutions of metal nitrates by incipient wetness of carrier with subsequent drying under 453 – 473 K (4 – 5 h) and calcination at 873 K (1 – 1.5 h) in air.

Toluene content before and after reaction was analyzed on the Crystal – 2000M chromatograph with a flame ionization detector (24.91 Hz) on a capillary column. Catalytic activity of catalysts was determined in flow installation at deep oxidation of toluene in air at various temperatures (523 – 773 K), space velocities (5 - 15×10^3 h^{-1}) and toluene concentration (320 mg/m^3) in the initial mixture.

Figure 1 presents the data obtained in the oxidation of toluene in synthesized nickel-copper-chromium oxide catalysts at 723 K and space velocity 5×10^3 h^{-1}. The Figure 1 shows that the nickel-copper-chromium-containing polyoxide catalysts of the deep oxidation of toluene by activity are located in the following order: Ni-Cu-Cr/2 % Ce/θ-Al_2O_3 (98.8 %) > Ni-Cu-Cr/θ-Al_2O_3 (93 %) > Ni-Cu/2 % Ce/θ-Al_2O_3 (85 %) > Ni-Cr/2 % Ce/θ-Al_2O_3 (76 %) > Ni/2 % Ce/θ-Al_2O_3 (57 %).

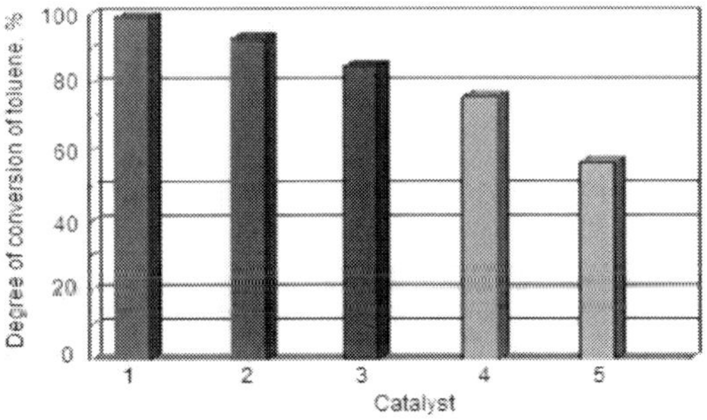

Fig. 1. The oxidation of toluene in air on oxide Ni-Cu-Cr containing catalysts: 1 - NiCuCr/2 % Ce/θ-Al_2O_3; 2 - NiCuCr/θ-Al_2O_3; 3 - NiCu/2 % Ce/θ-Al_2O_3; 4 - NiCr/2 % Ce/θ-Al_2O_3; 5 - Ni/2 % Ce/θ-Al_2O_3

It should be noted that comparison of the activity of multicomponent Ni-Cu-Cr containing catalyst supported on alumina modified by cerium with catalyst supported on alumina without cerium showed that the conversion of toluene on the catalyst Ni-Cu-Cr/2 % Ce/θ-Al_2O_3 significantly higher (98.8 %) than on the Ni-Cu-Cr/θ-Al_2O_3 (93 %).

Effect of process parameters (temperature, space velocity, concentration of toluene in gaseous mixture) on the completeness conversion of toluene (to CO_2 and H_2O) on the Ni-Cu-Cr/2 % Ce/θ-Al_2O_3 polyoxide catalyst was studied. The degree of oxidation of toluene from the variation of space velocities at different temperatures on the Ni-Cu-Cr/2 % Ce/θ-Al_2O_3 catalyst are presented in Table 1. It can be seen that with the increase of space velocity from 5×10^3 to 15×10^3 h^{-1} degree of oxidation of toluene reduces from 98.5 to 89.3 %, respectively.

Table 1: Effect of temperature and space velocity on the degree of conversion of toluene on the Ni-Cu-Cr/2 % Ce/θ-Al_2O_3 catalyst

T, K	Space velocity, $\times10^3$ h^{-1}		
	5	10	15
	(α), %	(α), %	(α), %
523	73.7	81.0	82.1
573	89.5	85.7	83.8
623	94.7	90.5	85.7
673	97.5	91.5	89.3
723	98.5	93.5	89.3
773	98.5	95.2	89.3
Note: The concentration of toluene in the feed - 320 mg/m^3			

Thus, degree of conversion of toluene to CO_2 reaches the 98.5 – 98.8 % on the Ni-Cu-Cr/2 % Ce/θ-Al_2O_3 catalyst at a temperature of 723 – 773 K and space velocity 5×10^3 h^{-1}.

Influence of concentration of toluene in the initial mixture to the efficiency of its conversion on the polyoxide catalysts of various compositions were studied. The data shows that increasing the concentration of toluene from 100 to 320 mg/m^3 in the initial mixture with air leads to a slight decrease in the degree of conversion of toluene on the two component Ni-Cu/2 % Ce/θ-Al_2O_3 and Cu-Cr/2 % Ce/θ-Al_2O_3 catalysts. A noticeable decrease in activity among two component oxide catalysts were found on the nickel-chromium-containing catalyst from 76.6 to 73.0 %. Ni-Cu-Cr/2 % Cr/θ-Al_2O_3 catalyst was the most stable.

Thus, the optimal conditions of deep oxidation of toluene on oxide Ni-, Cu- and Cr-containing catalysts over 2 % Ce/θ-Al_2O_3 were determined. Degree of conversion of toluene reaches 98.5 – 98.8 % on the Ni-Cu-Cr/2 % Ce/θ-Al_2O_3 catalyst at temperatures of 723 – 773 K, GHSV - 5×10^3 h^{-1} and the concentration of toluene in the initial mixture with air – 100 - 320 mg/m^3.

Polyoxide supported Ni-Cu-Cr/2 % Ce/θ-Al_2O_3 catalyst with desired properties for deep oxidation of hydrocarbons - toluene, xylene, styrene, ethyl acetate, butyl acetate, isobutanol, formaldehyde, acetone, ethanol and others who have severe toxic effects on a living organism and flora was developed. It has been shown that the synthesized polyoxide Ni-Cu-Cr catalyst supported on 2 % Ce/θ-Al_2O_3 provides 98.8% toluene conversion to CO_2 at space velocity of 5×10^3 h^{-1}, temperature of 723 – 773 K and 320 mg/m^3 content of toluene in the feed mixture.

The work was supported by the Ministry of Education and Science of the Republic of Kazakhstan (Grant No 0245/GF4, 0246/GF4).

Enhanced utilization of palm oil mill wastes for power generation

Muhammad Aziz[1,*], Tedi Kurniawan[2]
[1] Solutions Research Laboratory, Tokyo Institute of Technology
2-12-1 Ookayama, Meguro-ku, Tokyo 152-8550, Japan
Tel: +81-3-5734-3809, Fax: +81-3-5734-3559, e-mail: aziz.m.aa@m.titech.ac.jp

[2] Faculty of Mechanical Engineering, Universiti Malaysia Pahang
26600 Pekan, Pahang Darul Makmur, Malaysia

Abstract

The production of palm oil and its products has increased rapidly in the last decade due to high demand following the economic growth, especially in developing countries including China, India, and South East Asian countries. It leads to a massive expansion of palm plantation that leads to some environmental problems, especially the waste. The waste treatment in palm oil mills is very important task from both environmental and economic point of views. Palm oil production must be environmentally sustainable if the industries are planning to have an export to some developing countries that are generally demanding strict environmental prerequisites. Hence, the implementation of new environmental friendly technologies are very important and urgently required. Some possible utilization of solid and liquid wastes includes energy-harvesting, industrial material and fertilizer. However, high transportation cost, labor cost and greenhouse gases (GHG) emissions become some barriers in their utilization.

Empty fruit bunch (EFB) and palm oil mill effluent (POME) are the largest part of wastes from palm oil mills. Therefore, an effective utilization of EFB and POME is demanded to be able to increase the economic benefit as well as reducing the environmental problems. In this study, an integrated small-scale energy harvesting (power generation) utilizing EFB and POME with high total energy efficiency is proposed. Therefore, the objective of this study is to propose an innovative process for power generation utilizing EFB and POME from palm oil mills through process modeling and to evaluate the feasibility of the developed process in term of energy efficiency.

The proposed integrated small-scale power generation system utilizing both EFB and POME is shown conceptually in Figure 1. Basically, the system consists of two main modules: EFB gasification and POME digestion modules, and one additional organic Rankine cycle (ORC) module. The solid, dashed and dotted lines represent material, heat and electricity flows, correspondingly.

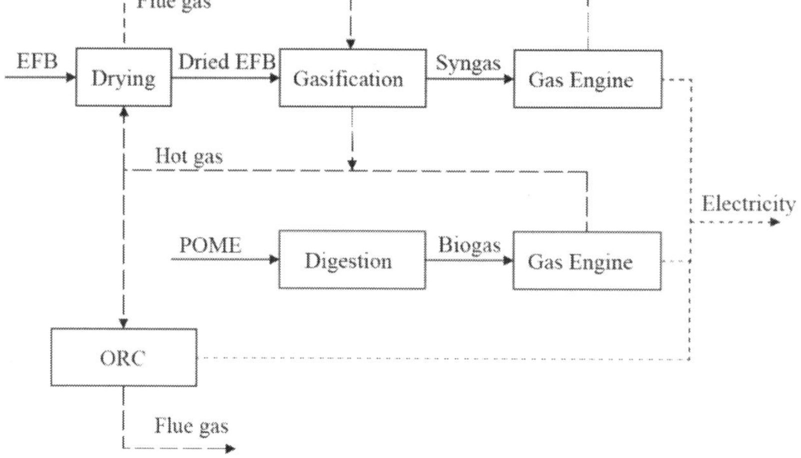

Figure 1 Conceptual process diagram of the proposed integrated-system for power generation from EFB and POME

In process calculation, palm mill having processing capacity of 50 t h^{-1} of fresh fruit bunch is modelled and evaluated. Figure 2 shows the relationship between net generated power and EFB gasification temperature. As the gasification temperature increases, the net generated power increases accordingly, especially the generated power from EFB gasification module. Numerically, the total net generated power increases from 4.16 MW (gasification temperature of 800 °C) to 9.57 MW (gasification temperature of 1000 °C). In addition, it should be noted that the generated power from POME digestion module is not influenced by the gasification temperature and it is constant under same flow rate of FFB in palm oil milling. From POME digestion module, about 1.4 MW of electricity can be generated from produced biogas.

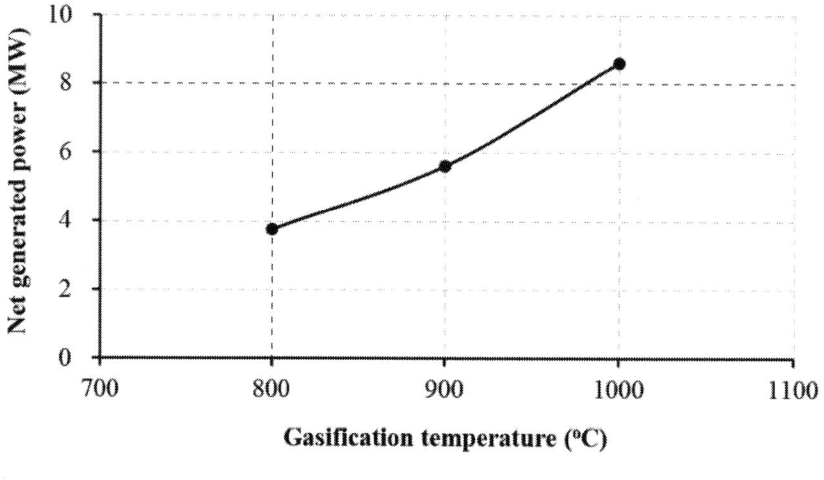

Figure 2 Relationship between net generated power and EFB gasification temperature

Keywords: brown coal, co-production, hydrogen, chemical looping, energy efficiency, heat circulation, process integration

Integration of heat pumps into existing district heating substations

V. Máša, M. Havlásek
Brno Univ. of Technology, NETME Centre & Inst. of Process Engineering – VUT ÚPI.
Technická 2896/2, 616 69 Brno, Czech Republic, tel. +420 541142372, e-mail: masa@fme.vutbr.z

This paper is substantiated with a thorough literary search and presents state of the art in the area of district heating and development trends in air to water heat pumps for heat production. Facts and data from the literary search support the idea that the air to water heat pumps should be more often used as a supplementary source of heat in district heating substations.

Efficiency, or coefficient of performance (COP), is the most important operating characteristic of a heat pump. This paper discusses main factors affecting COP of air to water heat pumps, with required temperature of outlet water and ambient air temperature as the most important ones.

The paper further aims to evaluate potential of integration of air to water heat pumps in existing district heating substations, especially for supply of heating in a building sector. Building sector is a promising area for use of heat pumps with a prospective positive impact on sustainability of the district heating systems. Heat pump is fully integrated in the system and thus efficient for the industrial practice only if the actual return on investment (ROI) is positive. The ROI and COP depend especially on prices of electrical energy, price of heat from district heating and potential state subsidies.

However, integration of heat pumps is associated with certain technical limitations. Temperature of water from heat pumps is lower than the standard temperature of water in district heating systems. The temperature may be though sufficient for heating of potable hot water. If the conditions allow, the heat pumps may significantly enhance district heating, especially during transitional periods of spring and autumn. Top-quality automatic control of this combined heat source is a serious challenge and the paper outlines main requirements on the control system which has to ensure optimum coordination between existing district heating substation and new heat pumps resulting in efficient heat production.

Selective non-catalytic reduction (SNCR) and its efficiency with respect to various combustion parameters

P. Gal, D. Jecha, J. Jicha and P. Stehlik
Brno University of Technology, Institute of Process Engineering, Technicka 2, 616 69, Brno, Czech Republic, tel. +420 5 4114 4955, e-mail: gal@upei.fme.vutbr.cz

Nitrogen oxides, which are formed for example during the combustion process in coal power plants, have negative impact on the environment. It is one of the main component of greenhouse gases and together with sulfur oxides cause acid rains. When reducing the formation and discharge of nitrogen oxides, it is possible to proceed several technological processes. One of the methods that is used in practice for reducing these pollutants, is selective non-catalytic reduction (SNCR). Increasingly stringent emission limits forcing businesses to invest in a system for removing NO_x. SNCR method is one of the cheapest, but sufficiently effective alternative to achieve these levels.

This contribution describes the experimental tests of injecting a reducing agent into the flue gas stream. In this case a reducing agent (with the trade name Satamin) consisting of an aqueous urea solution and an additives. The tests were performed in the boiler K14 at the power plant in Ostrava - Třebovice, CZ. The aim of the tests was to reduce NO_x emissions achieved by primary measures e.g., adjusting the pressure of the boiler unit, low NO_x pulverized coal burners, change media for grinding mill circuits so that the flue gas at the outlet reaches approximately 190 mg/m_N^3, because NO_x emission limit specified by legislation is established to 200 mg/m_N^3.

Efficiency SNCR was determined based on the conversion rate of Satamin and the associated reduction of primary NOx to a final value measured at a chimney. The contribution shows and compares the impact of various factors the concentration of CO in the flue gas stream, the primary value of NO_x, the temperature in the combustion chamber, etc. that affect, positively or negatively, the reaction process of Satamin and NO_x according to the equations:

$$4\ NO + 2\ (NH_2)_2CO + 2\ H_2O + O_2 \rightarrow 4\ N_2 + 6\ H_2O + 2\ CO_2$$
$$6\ NO_2 + 4\ (NH_2)_2CO + 4\ H_2O \rightarrow 7\ N_2 + 12\ H_2O + 4\ CO_2$$

Utilization of bottom ash as catalyst in biomass steam gasification for hydrogen and syn gas production

[1]A. P. Herman, [1]S. Yusup, Z. [1]M. Shahbaz
Biomass Processing Lab, Centre for Biofuel and Biochemical Research, Chemical Engineering Department, University Teknologi PETRONAS, 32610 Seri Iskandar, Perak, Malaysia;
tel. +605 3688225 e-mail: drsuzana_yusuf@petronas.com.my; fax: +605 365 4075.

Bottom ash is a solid waste from thermal power plant and it is usually disposed of into landfills and ash ponds. These disposal methods are not sustainable since new lands need to be acquired as the landfills and ash ponds are fill to its capacity. Bottom ash is also classified as hazardous material and the disposal methods may have detrimental effects on the environmental effect to the area. Hence, more research needs to be done to explore the potential of recycling the bottom ash as more a useful product. The objective of this research is to explore the potential of utilizing bottom ash as catalyst in biomass steam gasification. In this research, bottom ash was used as catalyst in gasification of Palm Kernel Shell (PKS) using Thermo Gravimetric Analyzer coupled with mass spectrometry (TGA/MS). The characterization of the bottom ash shows that the bottom ash contains elements that have been used as catalyst for biomass gasification before such as calcium, iron, aluminum and magnesium. The effects of temperature (650 – 750 ^{o}C) and bottom ash percentage 7% were studied. The produced hydrogen gas content is 36.57 vol % at temperature of 700 ^{o}C while the gas composition of carbon monoxide, carbon dioxide and methane are 24.73, 1.40 and 37.30 vol % respectively. In addition, it is found that the minerals exist in the coal bottom ash help to reduce the formation of carbon dioxide gas while promote the formation of methane in the product gas. In conclusion, the use of bottom ash as catalyst is possible for biomass steam gasification and the product gases composition are comparable with previous researches, however the results need to be validated for bench or pilot scale study.

Dry ice blasting for cleaning - measurement of pellets velocity behind the blasting nozzle

J. Vondál, [1]L. Kotek, J. Hájek
Brno University of Technology, Faculty of Mechanical Engineering, Institute of Process Engineering, Technická 2, 61669 Brno, Czech Republic; tel. +420 541144923, e-mail: vondal@fme.vutbr.cz; [1] Brno University of Technology, Faculty of Mechanical Engineering, Institute of Production Machines, Systems and Robotics, Technická 2, 61669 Brno, Czech Republic.

Dry ice blasting is advanced technology utilized in surface cleaning industry. The technology benefits from soft pellets (hardness similar to chalk) and cooling effect on cleaning surface. Pellets consists of CO_2 in solid form which has sublimation temperature -78.5 °C at atmospheric pressure. Even though it is well recognized technology (Spur, Uhlmann, and Elbing 1999) there is demand on development new blasting nozzles with either increased efficiency or design for specific blasting conditions.

This paper introduces on-site measurement of dry ice pellets velocity exiting nozzle. The measurement is performed with high speed camera Olympus iSPEED2. Measured area was 280 mm x 260 mm while nozzle exit diameter was 8.5 mm. Dry ice pellets at the enter to the system have diameter of 3 mm and length up to 15 mm, which is reduced at the nozzle exit to the maximum of 3 mm in all directions.

In total 8 nozzles were measured at the same conditions. One nozzle diameter was measured with two extensions – 50 mm and 100 mm. Nozzle length is measured from throat to the end. Experiments shows significant dependency of pellet velocity just behind the nozzle on nozzle length (see Figure 1) which is due to momentum transport from air stream to the dry ice pellet.

Numerical simulation was performed to support experimental findings. The results show fast decay of axial velocity behind the nozzle which corresponds to the theoretical values. Air velocity is crucial for momentum transfer to the pellets and its speedup. The longer pellets remains in high velocity region the more momentum is transferred. When the air exits nozzle the velocity decay is similar for many nozzles. The difference is in the length of nozzle downstream of the throat.

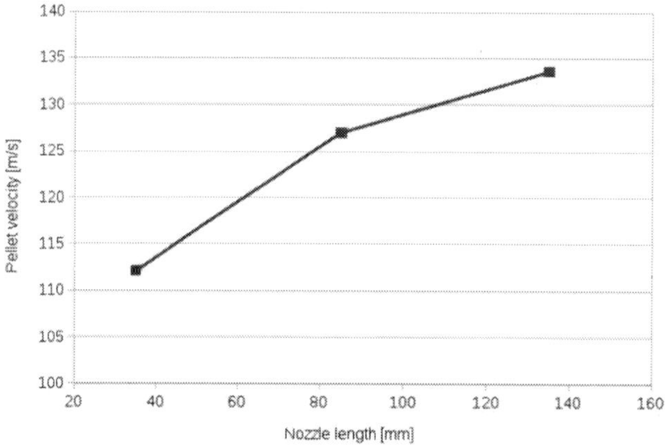

Figure 1: Pellet velocity dependence on nozzle length

Spur, G, E Uhlmann, and F Elbing. 1999. "Dry-Ice Blasting for Cleaning: Process, Optimization and Application." *Wear* 233–235 (December): 402–11. doi:10.1016/S0043-1648(99)00204-5.

Computational analysis of swirling pipe flows

M. Dohnal, J. Hájek
University of Technology, Institute of Process Engineering, Technická 2896/2, 616 69, Brno, Czech Republic; Tel.: +420 541 144 493, e-mail: dohnal@upei.fme.vutbr.cz

This paper is devoted to a numerical study of swirling pipe flows. Swirling pipe flows find application in many practical processes as reviewed in (Mitrofanova, 2003). One example of swirling pipe flow that does not employ twisted tape or any other tube insert is a novel process for calcination of diatomite (Hajek et al., 2015).

The focus in this work is specifically on configurations with swirling flow prepared in controlled laboratory environment, where the inlet conditions closely resemble solid body rotation. Such flow has been studied experimentally e.g. by laser Doppler velocimetry in (Rocklage-Marliani et al., 2003), where the flow has been confined in a straight pipe without any obstructions. Very similar configuration was investigated by the same technique in (Pashtrapanska et al., 2006), where the authors however added a so-called tripping ring close after the swirl generator. The two swirling flows possess many similarities, but there are also notable differences, mainly the presence of large-scale instability generated by the contraction in the second configuration.

Above mentioned experimental results provide a solid database for the validation of CFD simulation of swirling pipe flows. Experiments uncovered significant turbulence anisotropy in swirling pipe flows, and the character of these flows has been analysed in detail. Several computational studies have attempted to predict the flows in the configuration of (Rocklage-Marliani et al., 2003), while no computational studies so far have been performed to predict the configuration of (Pashtrapanska et al., 2006). This work provides a validation study for several often used as well as emerging turbulence models based on the prediction of both these configurations.

One of the focal points of this analysis is the experimentally observed asymmetry and unsteady behaviour in swirling pipe flows. To this end, advanced scale-resolving models with moderate CPU requirements have been used to ascertain their capability to predict the complex features of swirling pipe flows. The validation of such models that promise good accuracy at acceptable CPU costs is of primary importance for engineering CFD practitioners.

Hajek, J., Jurena, T., Baier, B., Bagherzadeh, A., Schnick, T., and Smejkal, Q., 2015, CFD Process Modeling of the Industrial Calcination of Diatomite, Chemie Ingenieur Technik, 87(7), 991–997.

Mitrofanova, O. V., 2003, Hydrodynamics and Heat Transfer in Swirling Flows in Channels with Swirlers (Analytical Review), High Temperature, 41(4), 518–559.

Pashtrapanska, M., Jovanović, J., Lienhart, H., and Durst, F., 2006, Turbulence measurements in a swirling pipe flow, Experiments in Fluids, 41(5), 813–827.

Rocklage-Marliani, G., Schmidts, M., and Ram, V. I. V., 2003, Three-Dimensional Laser-Doppler Velocimeter Measurements in Swirling Turbulent Pipe Flow, Flow, Turbulence and Combustion, 70(1-4), 43–67.

A life cycle inventory model for evaluating transportation impact of a supply chain

A.T. Ubando[1], I. V. Gue[2], J.A. Gonzaga[3], J.B. Biona[4]
[1]Mechanical Engineering, De La Salle University, 2401 Taft Avenue, 0922 Manila, Philippines. Tel: 524-4611 loc. 275. Email: aristotle.ubando@dlsu.edu.ph
[2]Mechanical Engineering, De La Salle University, 2401 Taft Avenue, 0922 Manila, Philippines. Tel: 524-4611 loc. 275. Email: ivan_gue@dlsu.edu.ph
[3]Mechanical Engineering, De La Salle University, 2401 Taft Avenue, 0922 Manila, Philippines. Tel: 524-4611 loc. 275. Email: jeremias.gonzaga@dlsu.edu.ph
[4]Mechanical Engineering, De La Salle University, 2401 Taft Avenue, 0922 Manila, Philippines. Tel: 524-4611 loc. 275. Email: jose.bienvenido.biona@dlsu.edu.ph

The life cycle assessment (LCA) methodology is an important tool in analyzing the environmental implication of a single product flow. It comprises the use of life cycle inventory (LCI) and life cycle impact assessment (LCIA) models to assess environmental impacts. In which, LCI models are databases that accounts for the material flow needed for the specific process and the emissions related to it. One of the important factors in accounting the environmental impact of a product is its transportation through the supply chain. As with most processes, the transportation requires fuel consumption leading to emissions. However, with the complexity of the transportation of a supply chain, it will be difficult to isolate and determine these parameters. As such, this study proposes a LCI model in determining the impact of its transportation in the supply chain. The model will be implemented on the Microsoft Excel program assisted with VBA programming. The model is capable to calculate the emissions and fuel consumption associated with the transportation of the raw materials from three regional levels. The first regional level represents the sink region where the product is produced on-site. The second regional level represents the source region of the input materials. The third regional level represents the source of the source region (second regional level) of the raw materials. The model will account for any product under the Standard International Trade Classification (SITC). A realistic case study in the Philippines is presented as a case study to illustrate the developed model.

A systematic approach for combined reuse of waste heat and CO2 in industrial parks

[1]R. Hassiba, [1,2]D. Al-Mohannadi, [1]S. Alnouri, [1,3]P. Linke,
[1]Department of Chemical Engineering, Texas A&M University at Qatar, PO Box 23874, Doha, Qatar [2]Qatar Foundation Research Division, Doha, Qatar [3]Qatar Energy and Environment Research Institute (QEERI), PO Box 5825, Doha, Qatar

The depletion of natural resources and the increase in greenhouse gases emissions, which constitute mainly from carbon dioxide, has led many policymakers to issue policies to reduce carbon emissions and fuel consumption. However, reducing the energy consumption is constrained by meeting the increase in goods demands governed by the growth in global population. There exist many energy integration techniques that have been applied to conserve natural resources and reduce carbon dioxide emissions without disturbing the production rate. Traditionally, Total Site Analysis is used to find the energy target for an industrial park with a common utility system. Waste heat is used to generate steam that is used as an energy source for the cooler sink streams or is cooled using a cooling utility before discharge. Very recently, carbon integration was introduced to target carbon dioxide emissions and mitigate the effect of greenhouse gases in industrial parks [1].

Opportunities exist to utilize waste heat in carbon integration, to reduce carbon footprint, by systematically designing a low cost carbon-energy integration networks. This work proposes a combined graphical and mathematical method that explores the synergy between Total Site Analysis and Carbon Integration techniques, in order to enhance the overall industrial park efficiency. The proposed method utilizes the unused waste heat from the energy sources to treat, capture and ship the carbon dioxide from the source plant to the sink plant. The result of the optimization is a heat and carbon network to achieve a specified carbon cut at minimum cost. A case study is used to demonstrate the method.

[1] Al-Mohannadi D.M., Linke P., 2015., On the Systematic Carbon Integration of Industrial Parks for Climate Footprint Reduction. Journal of Cleaner Production. (2016), pp. 4053-4064

An initial study on the life cycle assessment of algal biodiesel production in the Philippines

A.T. Ubando[1], I. V. Gue[2], N.A. Toledo[3], S.S. Garibay[4], C.N. Jimenez[5]
[1]Mechanical Engineering, De La Salle University. Addr: 2401 Taft Avenue, 0922 Manila, Philippines. Tel: (02) 524-4611 loc. 275. Email: aristotle.ubando@dlsu.edu.ph
[2]Mechanical Engineering, De La Salle University. Addr: 2401 Taft Avenue, 0922 Manila, Philippines. Tel: (02) 524-4611 loc. 275. Email: ivan_gue@dlsu.edu.ph
[3]Institute of Aquaculture, University of the Philippines Visayas. Addr: Miagao, 5023 Iloilo, Philippines. Tel: (033) 315-8556. Email: ncatph@yahoo.com
[4]Institute of Aquaculture, University of the Philippines Visayas. Addr: Miagao, 5023 Iloilo, Philippines. Tel: (033) 315-8556. Email: sgaribay009@gmail.com
[5]Institute of Aquaculture, University of the Philippines Visayas. Addr: Miagao, 5023 Iloilo, Philippines. Tel: (033) 315-8556. Email: cnjimenez@yahoo.com

Fossil fuels accounts for more than three-fourths of today's global energy consumption. In which, the combustion of these fuels has contributed to the worsening of climate change and of global warming. Not only does fossil fuels are detrimental to the environment, they can be depleted and are not sustainable. As with oil, the world is currently left with a half century worth of supply. In the Philippines, the country is heavily dependent on fossil fuels, as 40% of its energy consumption comes from oil with majority of it being diesel fuel. As such, the country must find a solution to reduce its dependency on fossil fuels. An alternative solution is biofuels due to its reduced greenhouse gas emissions compared fossil fuels. Biofuels production in the Philippines is mandated by the Biofuels Act of 2006 which suggest a certain blend of biofuels in the time horizon. However, with the recent typhoons hitting the country and parasite infestation in the country, feedstocks such as coconut and sugarcane may not produce enough biofuel for the targeted demand of the country in years to come. With this, microalgae is eyed as a potential feedstock for biodiesel production, because of its high oil yield per hectare and fast growth rate. As needed for decision making, an assessment of the viability of the production of algal biofuel must be prepared. Therefore, this study proposes the use of the life cycle assessment (LCA) methodology as an assessment tool for the viability of the production of algal biofuel in the Philippines. This study will focus on the assessment of a cultivation system in the Philippines, which is an open pond cultivation system. An Environmental Design of Industrial Products (EDIP) impact category will be used to assess the algal biofuel production in the Philippines. An uncertainty analysis will be captured to determine the robustness of the assumptions for the model parameters. Results revealed the environmental performance of algal biofuel production in the Philippines compared to other studies.

Comparison of material compositions of roofs in term of environmental and energy performance

Eva Kridlova Burdova[a], Silvia Vilcekova[a], Jaroslav Vojtus[b], Anna Sedlakova[b]
[a]Institute of Environmental Engineering, Faculty of Civil Engineering, Technical University of Košice, Vysokoškolská 4, 042 00 Košice, Slovakia
[b]Institute of Architectural Ingineering, Faculty of Civil Engineering, Technical University of Košice, Vysokoškolská 4, 042 00 Košice, Slovakia
silvia.vilcekova@tuke.sk

In world with limited amount of energy sources and with serious environmental pollution, interest in comparing the environmental embodied impacts of buildings using different structure systems and alternative building materials will be increased. The selection of building materials used in the constructions (floors, walls, roofs, windows, doors, etc.) belongs to one of the most important roles in the phase of building design. This decision has impact on the performance of building with respect to criteria of sustainability. The energy used in extraction, processing and transportation of materials used in building constructions can be significant part of the total energy used over the life cycle of building, particularly in nearly-zero energy buildings. The environmental impacts are expressed by indicators such as embodied energy (EE) from non-renewable resources, embodied $CO_{2.eq}$ emissions (GWP, Global Warming Potential) and embodied SO_{2eq} emissions (AP, Acidification Potential) within system boundary from Cradle to Gate. The aim of analysis is identify the environmental quality of material compositions for designed variants of roof constructions. The final values of assessments are compared by using methods of multi-criteria decision analysis.

Input data validation for complex supply chain models applied to waste management

R. Šomplák[a], L. Zavíralová[a], M. Pavlas[a], V. Nevrlý[b], P. Popela[b]
[a]Institute of Process Engineering, Faculty of Mechanical Engineering, Brno University of Technology Technická 2896/2, Brno, 616 69, Czech Republic;
[b]Institute of Mathematics, Faculty of Mechanical Engineering, Brno University of Technology, Technická 2896/2, 616 69 Brno, Czech Republic.
tel.: +420541144908, e-mail: somplak@upei.fme.vutbr.cz;

This contribution focuses on comprehensive input data analysis prior to the application of complex network flow models. Network flow models are composed of vertices interconnected by edges, the number of which is dependent on particular problem studied. Any quantitative data in vertices may contain inaccurate information. The article is intent on computational system for simulation and forecasting in waste management incomplete data problems. The article introduces enhancements to the mathematical model of previously published tool and its elements. The focus is put on the suitable choice of weights for all input forecasting models and territorial divisions of relevant regions. Another model improvement is based on the support of adding new constraints that link multiple types of waste. The next part of article concentrates on forecasting of waste production by the application of the advanced model. The case study focused on the estimation of trends in data for municipal waste and composition of residual municipal solid waste in the Czech Republic is presented as well.

Transportation cost as an integral part of supply chain optimization in the field of waste management

J. Gregor, [1]R. Šomplák, M. Pavlas
University of Technology, Institute of Process Engineering, Technická 2896/2, 616 69, Brno, Czech Republic; Tel.: +420 541 142 376, e-mail: gregor@upei.fme.vutbr.cz

This paper deals with transportation cost modelling in the field of waste management. Waste transport from producers to the pre-processing facilities and/or plants for final treatment significantly contributes to overall processing cost and should be properly investigated to select the most convenient solution. The main goal of this paper is to describe comprehensive techno-economic model involving several types of transportation systems (e.g. collecting vehicles, transfer stations, intermodal container system). Each transportation system has been analyzed according to its specific parameters, i.e. transport distance, amount of waste transported, etc.

The result of the calculation (e.g. specific cost per ton and kilometer) then serve as inputs for detailed supply chain optimization involving complex networks. Correct implementation of transportation cost with relation to other system key parameters (e.g. distance, capacity of waste) enables getting more reliable output data.

In the paper, authors describe methods of optimization task simplification without losing exactness of real solution. The impact on computational time will be addressed and concrete results related from real industrial case study in the field of waste-to-energy presented.

Numerical modelling of combustion in 1,5MW low-NO_x burner

T. Juřena, P. Bělohradský, P. Skryja, I. Hudák
Brno University of Technology, Faculty of Mechanical Engineering, Institute of Process Engineering, Technická 2, 616 69 Brno, Czech Republic; tel. +420541144943, e-mail: jurena@fme.vutbr.cz.

Results from CFD simulations of swirling non-premixed gas flame using a low-NOx gas burner are presented and compared to experimental data measured during a testing campaign in a semi-industrial water-cooled combustion chamber. The burner is both fuel- and air-staged with maximum capacity of 1,5MW. Numerical simulations are performed with Ansys Fluent software using Reynolds-averaged Navier-Stokes (RANS) turbulence model coupled with eddy-dissipation chemistry model (EDM) and discrete ordinates radiative heat transfer model. This modelling approach is preferable for large-scale combustion applications such as process fired heaters, where employing advanced models (e.g. Large Eddy Simulation) is still computationally demanding.

It is shown, how results are sensitive to the mixing rate constant of EDM, which is employed in a simulation of swirling turbulent diffusion flame. Specifically, temperatures are overestimated with the constant being set to 4 (the default value in Ansys Fluent), which leads to false prediction of NOx formation. Temperature peaks are reduced by lowering the mixing rate constant. An acceptable agreement with measured outlet and peak flame temperatures is achieved, when the constant is set to 0.6. With this value, however, highly increased concentrations of unburned species are reported at the outlet. Predicted values deviate from the measurements about three orders of magnitude. Based on the concentrations of unburned species and oxygen in the flue gas at the outlet, the best results are obtained with the constant 1.2.

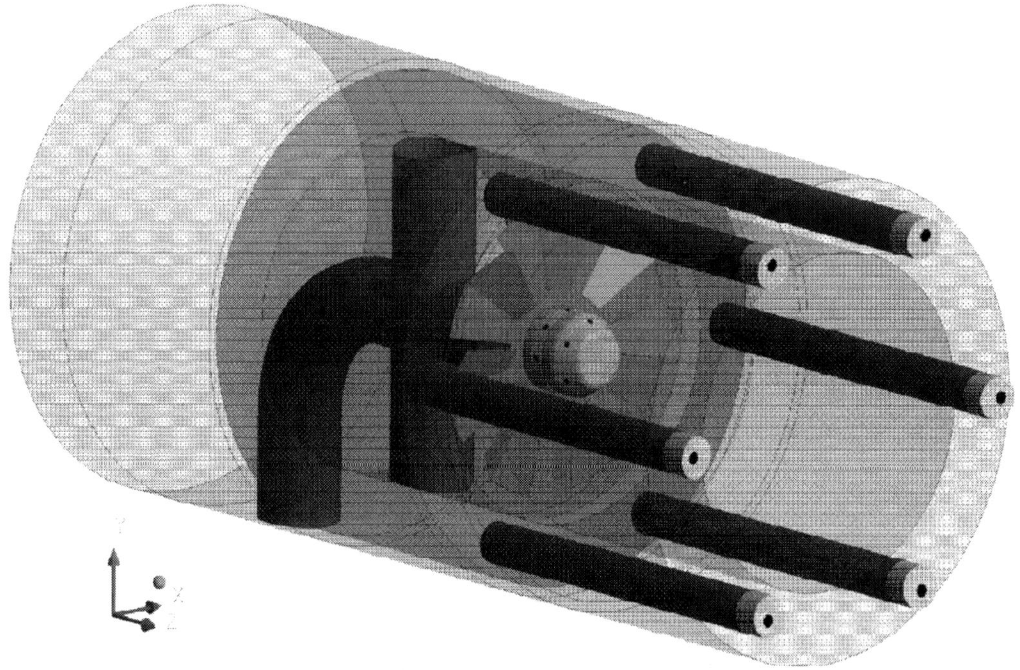

Figure 1 Model of the burner with a primary head, six secondary heads and an axial swirl generator.

The concept of energy recoverable from solid fuel combustion in low-power boilers

R. Urbaniak
Poznan University of Technology, Chair of Thermal Engineering, Piotrowo 3, 60965 Poznan, Poland; tel. +48 6652213, fax: +48 6652281, e-mail: rafal.urbaniak@put.poznan.pl,

In Poland, the main source of heat energy in municipal construction of small power boilers are fired with coal. For several years, the new regulations force their manufacturers of heating devices a significant increase in energy efficiency and reduction of emissions.

However, the combustion process itself is associated with high losses resulting from several factors. The main factor in the fall in energy efficiency is the flue gas losses. This phenomenon is related to a high temperature exhaust gas. Unfortunately, the loss increases with boiler contamination, and by an increase in the generated heat.

Figure 1 shows the course of temperature exhaust gas boiler with a capacity of 15 kW, operating with nominal power, controlled using three different control algorithms.

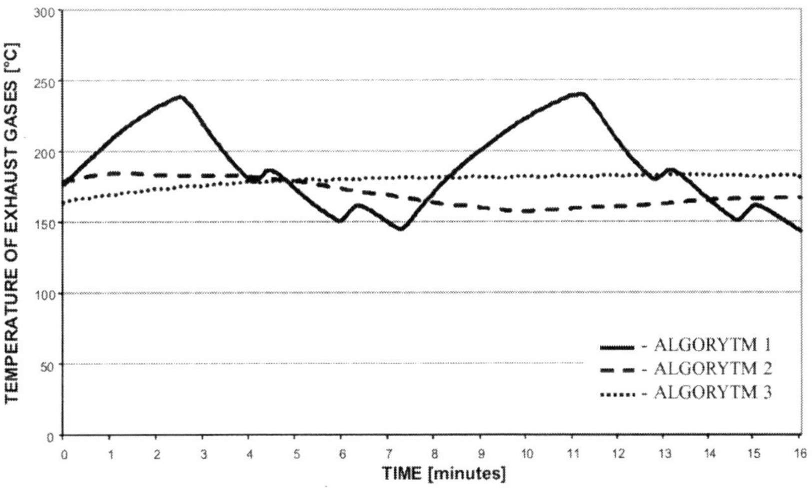

Fig. 1. Diagram temperatures of exhaust gas

As seen very often flue gas temperature exceeds the value of 200 degrees Celsius, which is a great loss of heat energy.

The authors proposed a solution consisting in recovery of excess heat through the use of Peltier cells. Energy recovery system from the exhaust built of Peltier cells uses the Seebeck effect, and thus produce electricity. Conduct research Peltier working in the mode of generating electricity required to build a bench.

Fig. 2. View research stations

The authors of the article were able to reproduce the natural conditions prevailing on the coal-fired boiler. The result of the study was to obtain electricity, which allowed the lighting of the traditional light bulb 10W. Figure 3 shows the effect of work positions in the form of a light bulb.

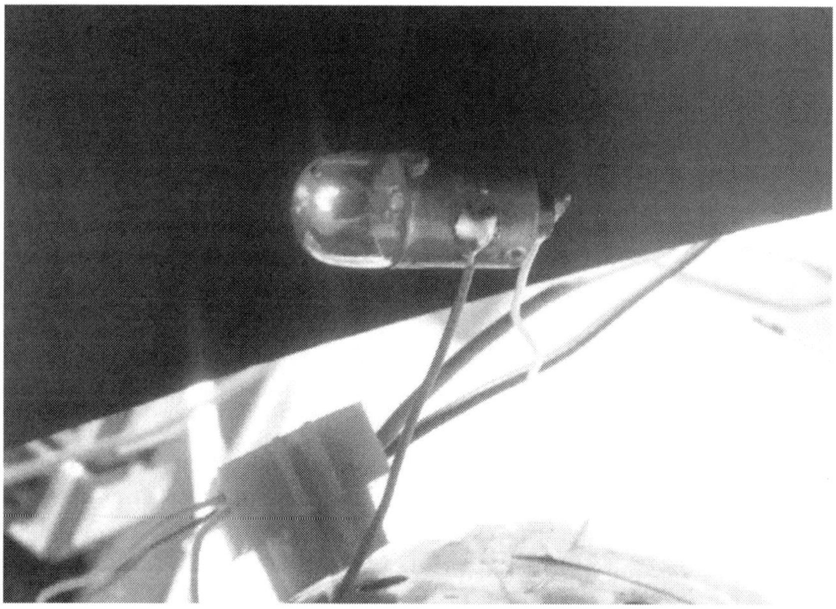

Fig. 3. View glowing electric bulbs

CFD simulations of continuous reactors to study reaction and flow

A. Twerda [1], L.F.G. Geers [1], V. van Pul-Verboom[1]
[1]TNO, P.O. Box 6012, 2600 JA Delft, The Netherlands;
A. Twerda: tel. +31 888 666538, e-mail: aris.twerda@tno.nl;

Process intensification is an increasing trend in chemical engineering and it enables substantial reduction in equipment size, energy consumption and waste production which leads to safer, cost-effective and sustainable technologies. More than in batch reactors detailed information within the whole continuous reactor is required for safe and efficient operation. As measurements are limited, models can provide this information. Especially CFD can give 4D (3 dimensions and time) information of the flow and other quantities, such as temperature and species concentration. This information can be used for trouble shooting, designing equipment or, when the computation speed can be increased sufficiently, even to control the process. In this work, CFD is applied in two particular cases: an OBR and a Helix reactor.

In the Helix, the mixing and reaction of two fluids was modelled. Because the large viscosity contrast between the two fluids, prediction of the flow pattern is not trivial. The CFD model includes laminar single phase flow model with heat transfer and species transport. From the results two important observations could be made: 1) The helical configuration greatly improves the mixing while preventing thermal overshoot and undesired side-product generation. 2) The length of the feeding tube should be shortened as much as possible to avoid inhomogeneous temperature fields hence avoiding the locally high values of degradation product.

In the OBR a crystallization is modelled where the crystals have an important contribution to the viscosity of the solution. In corresponding experiments the reactor was clogged several times due to large amount of crystals in the flow. Here, the standard CFD tool ANSYS Fluent was coupled with the gPROMS software. The kinetics of the crystallization was modelled with a Population Balance Model. The coupled simulations show that the increase in viscosity can be predicted and this technique has the potential to clarify clogging behaviour and estimate the probability of clogging. Currently available experimental data support the potential of the simulations.

Overall we can state that CFD is a valuable tool for acquiring detailed insight in the phenomena occurring in a continuous flow reactor. The two cases showed that it allows predicting the trend of the key processes in the reactor.

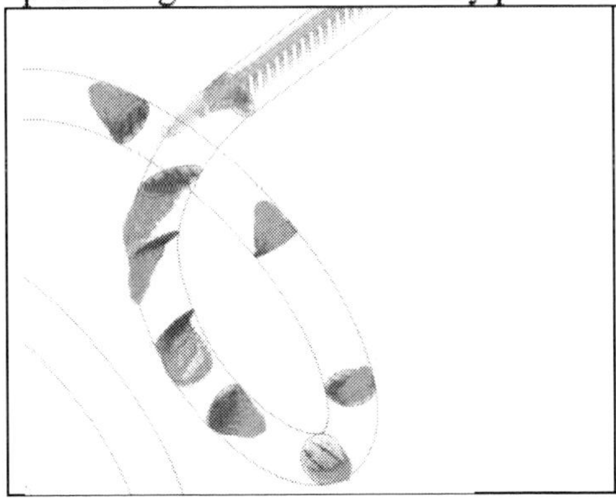

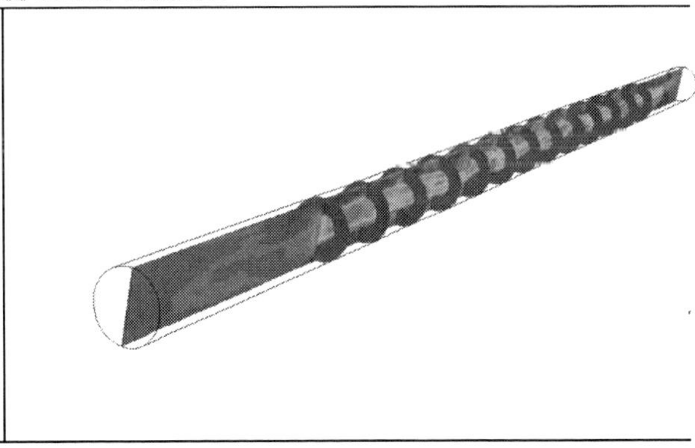

Figure 1 Vector plot and contour plot of the velocity in the Helix and OBR respectively

Development of a high temperature, high pressure bench scale demonstrator for slurry flow reactions

L. Geers[1], R. van Someren, A. van de Runstraat, N. Wennersbusch, A. Twerda, V. van Pul
TNO, Process and Instrument Development, Leeghwaterstraat 44, 2628 CA Delft, The Netherlands; [1]e-mail: leon.geers@tno.nl

Flow chemistry is already well established for bulk chemical production, while pharmaceutical, fine and specialty chemicals are still largely produced in batch facilities. However, the commercial availability of novel continuous flow equipment ranging from micro- to pilot- and plant-scale promises a bright future for flow chemistry in these areas. An ever growing number of technologies are emerging to seamlessly develop continuous processes from lab to production scale, offering huge benefits in terms of development time and process efficiency. Most technologies currently focus on single phase liquid processes, with some attention to liquid-gas processes. In contrast, the application of flow chemistry for processes which employ solids is lagging behind. Especially in the pharmaceutical and specialty chemicals industries, the ratio of heterogeneous to homogeneous reactions is considerable, so there is a wide application field not reaping the benefits of continuous flow.

Over the last decade, TNO has built experience of handling solid particles in continuous flow chemistry. Examples include a Friedel Crafts acylation with solid aluminium chloride utilised in the production of a high value pharmaceutical, the generation of Grignard reagents from solid magnesium, a liquid/gas/solid process for the debromination of a pharmaceutical precursor, and several crystallization processes, all in continuous flow installations. This experience led to the construction of a pilot installation that has been used to produce a higher reaction yield almost twenty times faster than the current batch production unit for a specific process.

On the basis of experience gained in the development of this pilot unit, TNO has designed and constructed a flexible, modular, bench-scale, multi-phase reaction set-up for the quick assessment of the feasibility of solid-liquid and liquid-liquid processes in continuous flow. In parallel, a computational model was developed to determine the operating conditions of the installation based on small scale laboratory experiments. It is also possibe to perform CFD calculations for a more accurate description of the effects of flow phenomena in the reactor on its performance.

Photocatalytic removal of tartrazine dye from aqueous samples on $LaFeO_3/ZnO$ photocatalysts

Vincenzo Vaiano, Giuseppina Iervolino*, Diana Sannino
Department of Industrial Engineering, University of Salerno, Via Giovanni Paolo II, 132, 84084 Fisciano (SA), Italy
giiervolino@unisa.it

Dyes represent a very important group of water pollutants that appear in the effluents of different industries. These contaminants, by preventing the penetration of light, can have negative effects on the aquatic environment, causing problems to the ecosystem, and possibly being carcinogenic and genotoxic on human health. In particular tartrazine is a commonly not-biodegradable dye used in food industries and it, in high amount, could cause allergic and/or intolerance reactions. In the past decade only few effective photocatalysts for organic dye degradation under visible-light irradiation has been reported. Effective photocatalysts, such as ZnO, however, exhibit photodegradation activity only under ultraviolet irradiation. The coupling of ZnO, having a band gap of 3.2 eV, with low band gap photocatalysts could induce the removal of the dyes under visible light. So, in this work, $LaFeO_3$, which has a band gap of about 2.1 eV, was supported on ZnO surface. The aim of this work was to evaluate the efficiency of $LaFeO_3/ZnO$ on the degradation of tartrazine under UV but also under visible light. Furthermore it was evaluated the optimal composition in terms of $LaFeO_3$ amount on ZnO.

Stress analysis of deformed storage tank shell

I. Klučka, R. Nekvasil, J. Buzík
Brno University of Technology, Faculty of Mechanical Engineering, Institute of Process Engineering, Technická 2, 616 69 Brno, Czech Republic; tel. +420 541144941,722768767
e-mail: klucka@fme.vutbr.cz, nekvasil@fme.vutbr.cz, buzik@fme.vutbr.cz

Shell plate material deformations appeared when storage tank operation. Material deformations were significant and may affect the loss of stability for the future operation. The assessment was made on the basis of performed measurements for examination the strength of existing deformations in the current wall thicknesses. Corrective actions and optimal method of operation including regular inspections was proposed. According to the measurements taken were assessed major geometry deformations. These deformations were subjected to a detailed stress analysis and assessed in terms of standards in use.

Modeling and optimization of nitrogen removal processes from natural gas

M.A. Gim, [1]Y.G. Mo, [1]J.T. Chung, J.K. Kim
Department of Chemical Engineering, Hanyang University, Wangsimni-ro 222, Seongdong-gu, Seoul 04763, Republic of Korea; tel. +82 (0)2-2220-2331, email: jinkukkim@hanyang.ac.kr; [1]Research Development Institute, Korea Gas Corporation, Incheonsinhangdae-ro 960, yeonsu-gu, Incheon 21993, Republic of Korea; tel. +82 (0)32-810-4502, Fax: +82 (0)32-810-4502

Traditional natural gas of high quality is now decimated and non-traditional natural gas production has been increasing. Hence, gas production and processing plants are required for processing of natural gas having a high content of nitrogen. Among various options for rejecting nitrogen from the produced natural gas, low-temperature distillation process is widely applied in industries.

This study focuses on modeling and optimization of nitrogen removal process using distillation. The nitrogen removal process is modeled using commercial simulator UniSim Design®. Different flowsheets for low-temperature nitrogen removal process are simulated, and their performance and economics are compared in a systematic manner. The developed process model is validated against literature results. Sensitivity analysis is carried out to understand impacts of key design variables.

The study also aims at developing a process design framework to improve energy efficiency without compromising the separation performance. In order to determine the optimal design configuration and operating conditions for the nitrogen separation process, the optimization technique is applied for minimizing overall operating costs, as an objective function, subject to economic parameters and engineering constraints. Optimization is performed by the stochastic solver in MATLAB®. The case study is presented to illustrate the benefit of modelling and optimization framework proposed in this study, achieving high energy efficiency.

Acknowledgement
This research was respectfully supported by Engineering Development Research Center (EDRC) funded by the Ministry of Trade, Industry & Energy(MOTIE). (No. N0000990).

Optimization of operational strategy for amine-based post-combustion CO_2 capture processes considering part-load performance

S.-Y. Oh, J.-K. Kim
Department of Chemical Engineering, Hanyang University, Wangsimni-ro 222, Seongdong-gu, Seoul 04763, Republic of Korea; tel. +82 222202331, email: jinkukkim@hanyang.ac.kr

Post-combustion CO_2 capture is the most mature technology to reduce CO_2 emissions from flue gas in conventional power plants. Although the power plants usually operate at different working loads in practice, it is typically assumed that the capture system operates at the full working load of the power plant.

This study deals with how the capture plant should be operated under different operating loads and aims to provide a systematic framework for the evaluation of the part-load performance of the CO_2 capture process with power plants. The CO_2 capture process using aqueous monoethanolamine (MEA) is modeled in a commercial simulator UniSim Design®. The part-load power plant models are considered in both natural gas-fired combined cycle and coal-fired plants.

In this work, the part-load performance of CO_2 capture model is validated against literature data, while sensitivity analysis is carried out with a few key design parameters, such as solvent circulation rate and lean loading, at different conditions of working loads for the power plants. The current study especially focuses on developing operational strategy for minimizing regeneration energy and improving process efficiency.

The case study is presented to illustrate an economic impact of the capture process with discontinuous nature at different loads of power plants. Moreover, it will show the effect of the part-load performance on the integrated power plant with the CO_2 capture plant, and with the aid of multi-period modeling approach the most appropriate operating strategies can be systematically obtained.

Acknowledgment:
This research was supported by the Korea Carbon Capture & Sequestration R&D Center through the National Research Foundation of Korea funded by the Ministry of Science, ICT and Future Planning (2014M1A8A1049305).

Process modeling and dynamic simulation of natural gas liquids process

[1]J.S. Oh, [1]S.K. Yoon, [1]H.S. Son, [2]Y.R. Kim, [2]S.M. Park, [1]J.K. Kim,
[1]Department of Chemical Engineering, Hanyang University, Wangsimni-ro 222, Seongdong-gu, Seoul 04763, Republic of Korea; tel. +82(0)2-2220-2331, email: jinkukkim@hanyang.ac.kr
[2]Hyundai Heavy Industries Co., Ltd., 17-10 Mabuk-ro 240beon-gil, Giheung-gu, Yongin-si, Gyeonggi-do 16891, Republic of Korea;

Natural Gas Liquids (NGL) process is applied to recover heavy hydrocarbon components from sale gas for improving economics of gas processing, and turbo-expander is typically employed to facilitate sub-ambient conditions for the separation and to improve energy efficiency. The turbo-expander NGL process is highly integrated between a distillation column, heat exchangers and a turbo-expander and hence maintaining reliable and robust operation is an important issue in industries.

This presentation focuses on process modeling and dynamic simulation of NGL (natural gas liquids) process in order to gain conceptual understanding dynamic behavior and to establish reliable and robust control strategy. Steady-state process modeling is first carried out with UniSim® environment for the NGL process considered, and dynamic simulation and associated with process control schemes are also made.

Dynamic response and safety of NGL processes was evaluated at different feed conditions, with which the most appropriate control schemes was determined. Through a case study, it was confirmed that safety and operability of the process is considerably improved when multiple manipulated variables proposed in this study is applied.

Acknowledgement
This work was supported by the Technology Innovation (10042424, Floating Production Platform Topside Systems and Equipment Development) funded by the Ministry of Trade, industry & Energy (MI, Korea)

Process modeling and optimization of MEG Regeneration Unit (MRU) in FPSO applications

H.S. Son, [1]Y.R. Kim, [1]S.M. Park, J.K. Kim,
Department of Chemical Engineering, Hanyang University, Wangsimni-ro 222, Seongdong-gu, Seoul 04763, Republic of Korea; tel. +82(0)2-2220-2331, email: jinkukkim@hanyang.ac.kr
[1]Hyundai Heavy Industries Co., Ltd., 17-10 Mabuk-ro 240beon-gil, Giheung-gu, Yongin-si, Gyeonggi-do 16891, Republic of Korea;

Monoethyleneglycol (MEG) is widely used for preventing hydrate formation in offshore oil and gas plants, and recovering MEG in a cost-effective manner has been an important subject for decades. Effective operation of MEG regeneration unit requires improved separation performance in distillation column and salt-removal (reclaimer).

This research focuses on modeling and optimization of MEG regeneration unit on Floating Production Storage Offloading (FPSO). The MEG regeneration unit is modeled with commercial simulator Unisim Design® and Aspen Plus, which are used for excluding and including salts, respectively. The model is validated against industrial process data and sensitivity analysis is conducted to evaluate how energy consumption and MEG losses are influenced by heater temperature, distillation column design and other design parameters.

The current study aims at developing a process design framework for improving energy efficiency with advanced separation performance. Reclaimer and pretreatment for removing different kinds of salts are designed to maintain MEG quality as well as maximizing energy efficiency. Process optimization is carried out with MATLAB®, which is strategically linked to the process simulator. With this integrated design tool, economic design configuration and optimal operating conditions for MRU can be systematically determined. A case study will be presented to illustrate how the proposed design framework is applied in practice.

Acknowledgement
This Research was respectfully supported by Engineering Development Research Center (EDRC) funded by the Ministry of Trade, Industry & Energy (MOTIE). (No. N0000990)

Process modelling and optimization of membrane-based post-combustion CO_2 capture for coal-fired power plants

[1]S. Lee, [1]S. Youn, [2]J.H. Lee, [2]J.H. Moon, [2]J.G. Yeo, [1]Y.K. Yeo [1]J.K. Kim
[1]Department of Chemical Engineering, Hanyang University, Wangsimni-ro 222, Seongdong-gu, Seoul 133-791, Republic of Korea; tel. +82 (0)2-2220-2331, email: jinkukkim@hanyang.ac.kr
[2]Korea Institute of Energy Research,152 Gajeong-ro, Yuseong-gu, Deajeon 305-343,Republic of Korea

Carbon dioxide is one of critical green-house gases which cause the global climate change. To allow the transition of the fossil fuel combustion power plant to sustainable electricity supply, CO_2 capture process is required. Membrane processes have been suggested for carbon dioxide capture from flue gas, due to rapid development in membrane materials, low-energy usage, small foot print and easy equipment installation.

We developed single membrane unit model in MATLAB® environment which is validated with mixed-gas experimental results at different feed conditions. The multi-stage membrane processes including recycle stream are simulated, based on validated single membrane model. The reasonable agreement between simulated outputs and experimental results is observed.

In order to implement the membrane process for CO_2 capture for coal-fired power plants in a cost-effective and energy-efficient manner, superstructure approach is applied in the process optimization for finding feasible and economic design of gas separation membrane networks where various design options are systematically screened and economic trade-off between major capital cost (i.e. compressor, vacuum pump, and membrane area etc.) and key operating cost (electricity and steam) are evaluated. The optimization is carried out using stochastic solver based on built-in MATLAB® function GA (genetic algorithms) with minimizing TAC (total annualized cost) per ton of captured CO_2, subject to specified purity and recovery of CO_2.

A case study is presented to verify usefulness of above-mentioned design method and to demonstrate how optimal configuration can be effectively found.

Acknowledgement
This research was supported by the Korea Carbon Capture & Sequestration R&D Center through the National Research Foundation of Korea funded by the Ministry of Science, ICT and Future Planning (2014M1A8A1049305).

Reaction mechanism and kinetics of the mercury solid waste pyrolysis: an isoconversional approach

Y. Busto[1], E. W. Palacios[2], F. M. G. Tack[3], L. M. Peralta[4], M. Yera[5]
[1]Technological University of Israel, Francisco Pizarro street and Orellana Avenue E4-142, 170150, Quito, Ecuador; tel. +593 98 4379929, e-mail: yailenb09@gmail.com; [2]Faculty of Odontology, Central University of Ecuador, University Citadel between Bolivia and Salgado street, Quito, Ecuador; tel. +593 99 2728009, e-mail: wsaudades@gmail.com; [3]Bioengineering Science Faculty, Gent University, Coupure Links 653, 9000B, Gent, Belgium; [4]Department of Chemical Engineering, Faculty of Chemistry and Pharmacy, Central University "Marta Abreu" of Las Villas, Highway to Camajuani km $5^{1/2}$, 54830, Villa Clara, Cuba; [5]Iberoamerican University of Ecuador, 9 de Octubre and Santa Maria streets, Quito, Ecuador; tel. +593 99 4149669, e-mail: miriam.yera62@gmail.com

The present study focuses on the kinetic mechanism of mercury removal from contaminated solid waste generated by chlor-alkali plant using pyrolysis. The isoconversional method of kinetic analysis was used to calculate the kinetic parameters that best describe mercury thermal decomposition in the solid waste. A mechanism involving 6 heterogeneous and homogeneous reactions was proposed to represent the behaviour of mercury compounds in the solid matrix during thermal treatment. The proposed model was compared to nine models previously reported in literatures to elucidate the controlling reaction mechanism. Fitting each of these to the experimental data of thermal decomposition of the mercurial sludge sample, confirmed the hypothesis that not a single mechanism is ruling the process. The D1-diffusion mechanism could be considered the controlling step of the process at high retention times while at low thermal decomposition times (< 15 min) the diffusion mechanism (D1) as well as the third order reaction mechanism (F3) could be controlling the process. Nevertheless, as a first depth-in to the knowledge of this polydisperse and multicomponent system (mercurial sludge), the diffusion mechanism (D1) can be considered the overall controlling stage as an increase of temperature smooths the progress of the chemical reactions involved.

Removal continuous studies of chromium (VI) using sugar cane bagasse

Y. Busto[1], E. W. Palacios[2], I. Aloma[3], L. M. Rios[4], M. Cortez[3], M. Calero[5], M. Yera[6]
[1]Technological University of Israel, Francisco Pizarro street and Orellana Avenue E4-142, 170150, Quito, Ecuador; tel. +593 98 4379929, e-mail: yailenb09@gmail.com; [2]Faculty of Odontology, Central University of Ecuador, University Citadel between Bolivia and Salgado street, Quito, Ecuador; tel. +593 99 2728009, e-mail: wsaudades@gmail.com; [3]Department of Chemical Engineering, Faculty of Chemistry and Pharmacy, Central University "Marta Abreu" of Las Villas, Highway to Camajuani km 51/2, 54830, Villa Clara, Cuba; [4]Study Center of Applied Chemistry, Central University "Marta Abreu" of Las Villas, Highway to Camajuani km $5^{1/2}$, 54830, Villa Clara, Cuba; [5]Department of Chemical Engineering; University of Granada, Granada, Spain; [6]Iberoamerican University of Ecuador, 9 de Octubre and Santa Maria street, Quito, Ecuador; tel. +593 99 4149669, e-mail: miriam.yera62@gmail.com

In this work was studied the behaviour of biosorption process of chromium from aqueous solutions onto modified sugar cane bagasse packed bed columns. The performance of packed beds is described through the concept of the breakthrough curve. The time for breakthrough appearance and the shape of the breakthrough curve are very important characteristics for determining the operation and the dynamic response of a biosorption column. Four mathematical models; Adams–Bohart, Thomas, dose-answers and Yoon–Nelson models were applied to experimental data to predict the breakthrough curves and to determine the characteristic parameters of the column useful for process design. The experimental data show the breakthrough increased meanwhile the feed flow decreased and the packed bed lengths. The breakthrough curve study show the Adams–Bohart model could be used for the description of the initial part of the breakthrough curve. The dose-answers models reproduced the entire breakthrough curve.

Development of titanium-containing catalysts based on the zeolite-bentonite carriers for the conversion of nitrogen oxides

M.A. Sadenova
S. Amanzholov East Kazakhstan State University, 55 Kazakhstan str., 070000 Ust-Kamenogorsk, Kazakhstan. tel. +77232255441, e-mail: Sadenova@mail.ru

It is known that titanium and its compounds possess unique properties and this determines the broad scope of their use. The titanium-containing catalysts have been successfully used for the conversion of toxic components in gas emissions in the process of selective catalytic reduction of nitrogen oxides by ammonia and others. There are many different titanium-containing catalysts which differ in their methods of preparation, structure, form and their application area. So, in Marchenko (2000) were worked out a composition and method of preparing vanadium-titanium catalyst based on oxides of vanadium and titanium (anatase) for reduction of nitrogen oxides by ammonia in the presence of oxygen and was investigated the role of adsorbed forms of identification reactants in the reaction mechanism. In Hey (2009) authors described catalyst carrier comprising kieselguhr and titanium dioxide, characterized by the fact that the ingredients are taken in certain ratio, and the catalyst support is formed in a monolithic structure with a plurality of straight channels. In Isaeva (1999) group of Russian authors have suggested a method of producing a catalyst, comprising a compound of titanium, a stage of impregnating the solid carrier by compound of one or more platinum group metals, wherein the carrier representing aluminum oxide with the finely divided titanium dioxide is used for the preparation of the catalyst.

Synthesis of new systems for the purification of gas emissions from non-traditional sources of raw materials solves two environmental problems: it reduces the amount of solid and pasty waste (sludge) and reduced anthropogenic fumes in the air. For forming ceramic catalyser carrier were prepared based on the powder mixture of zeolite and / or ash and clay and clay content of 20-50%. TiO_2-V_2O_5 was used as active catalyst component. The choice of the titan - vanadium system is caused by two major factors:

- Involvement of technological waste of the Titanium-magnesium plant located in Ust-Kamenogorsk (Ust-Kamenogorsk Titanium-Magnesium Plant) for creation of low-waste resource-saving productions;
- The titan-vanadium systems applied on aluminum oxide in the course of gas purification, which are known with their high efficiency.

The experimental data obtained enabled the identification of the various metals present in test samples, and to perform accurate analysis of the microstructure of the synthesized catalyst systems to predict their properties and preparation of systems with desired properties.

References

Marchenko E.A., 2000, Catalysts and mechanisms of selective catalytic reduction of nitrogen oxides by ammonia, Thesis for the degree of Chemistry PhD Almaty.

Hey Ya.V. (DK), Jorgensen K.S. (DK) Haldor TOPSEE A/S (DK), 2009, Catalyst carrier, a catalyst and process for treating flue gas, Patent №2358801.

Isaeva G.G. (RU), MG Makarenko M.G. (RU), Vinogradov E.Yu. (RU) et. all., 1999, Catalyst for purification of gases from hydrocarbons, nitrogen oxides, carbon monoxide, and a method of its producing, Patent №2135279.

Technological proposal for treating wastewater contaminated with nitro aromatic compounds by simulation

Y. Busto[1], E. W. Palacios[2], L. M. Rios[3], L. M. Peralta[4], M. Yera[5]
[1]Technological University of Israel, Francisco Pizarro street and Orellana Avenue E4-142, 170150, Quito, Ecuador; tel. +593 98 4379929, e-mail: yailenb09@gmail.com; [2]Faculty of Odontology, Central University of Ecuador, University Citadel between Bolivia and Salgado street, Quito, Ecuador; tel. +593 99 2728009, e-mail: wsaudades@gmail.com; [3]Study Center of Applied Chemistry, Central University "Marta Abreu" of Las Villas, Highway to Camajuani km $5^{1/2}$, 54830, Villa Clara, Cuba; [4]Department of Chemical Engineering, Faculty of Chemistry and Pharmacy, Central University "Marta Abreu" of Las Villas, Highway to Camajuani km 51/2, 54830, Villa Clara, Cuba; [5]Iberoamerican University of Ecuador, 9 de Octubre and Santa Maria streets, Quito, Ecuador; tel. +593 99 4149669, e-mail: miriam.yera62@gmail.com

During the process of producing nitro aromatic hazardous compounds are generated wastewater highly polluting and dangerous for the health and workers safety and to the environment.

The present study focuses on carrying out the project of a technology viable in the conditions of the country, for the treatment of wastewater contaminated with nitro aromatic compound (TNR). This compound is dissolved in the wastewater from the production process can form metal salts and very sensitive compounds that accumulate can lead to accidents, hazardous to human health and the environment, which is why it is vitally important disposal. The role of treatment is to achieve the destruction and elimination of nitro aromatic compound wastewater from the synthesis, purification and washing of this product.

In this research, develop methods and mathematical procedures for evaluating and predict the behaviour of complex reactions was needed. An integrated design of semi continuous stirred tank type reactors was achieved in order to minimize the consumption of material resources and to reduce the dumping of waste into the environment. Programs developed by using PSI software, were carried out to obtain the main variable profiles, consistent with the values taken in the experimental runs. Next, the phenomenological models obtained by modelling were validated. Finally, the technological scheme of the proposal process was defined and the main and auxiliary equipment was designed.

Experimental research of flammability and explosion parameters of binary gas mixtures in 1m³

J. Vereš*, J. Skřínský, J. Trávníčková, A. Dalecká
Energy Research Centre, VŠB-TU Ostrava, 17. listopadu 15/2172, 708 33 Ostrava, Czech Republic; e-mail: jan.veres@vsb.cz

LEL, UEL, Pmax, (dP/dT)max, LOC are the most important variables used to characterize fire and explosion hazard of gases. This paper describes validation of a model for predicting the flammability and explosion of binary premixed mixtures of flammable solvents and gas vapours. To confirm the predictive efficiency of the derived parameters, the model was verified by comparing the predicted values with the experimental data for the studied premixed $CH_4/H_2/O_2/N_2$ mixtures. For binary premixed mixtures explosions the specific variant of stainless steel $1 m^3$ sphere was used. With the strengthened $1 m^3$ sphere it is possible to conduct experiments at final pressures not exceeding 20 bar, and initial temperatures, up to 200 °C. A fast acting system for homogenization of the liquid within an adjustable time, called the homogenization time. All parts are temperature controlled and thermally insulated. The development of the explosion in the explosion chamber is monitored independently by two piezo-electrical pressure transducers. The partial pressure method was used for both gas/vapor mixtures preparation. Gases were supplied from compressed gas bottles. The adiabatic explosion pressures, computed by assuming chemical equilibrium within the flame front are examined in comparison with the measured explosion pressures. The influence of initial temperature and heat losses during propagation on explosion pressures and explosion times are discussed for the examined systems. The chemical equilibrium model assumes adiabatic conditions in constant volume, and formation of equilibrium-defined concentrations of post explosion compounds and their expansion due to the temperature rise caused by the liberated heat assuming ideal gas behaviour and ideal condensed phase products. Three thermodynamic models are used to calculate the explosion pressure. The discrepancy between the results of experiment and calculation mirrors the degree of conversion of the initial mixture and the degree the explosion can be considered adiabatic. Efforts were made to identify the main cause of the large discrepancy in two of the combustion regimes. Based on the results, therefore, it appears reasonable to suggest potential application for our model in assessment of fire and explosion hazards, and development of inherently safer designs for chemical processes containing binary mixtures of flammable binary gas mixtures.

Industrial gases supply chain stabilization using model predictive routing and control of the end-user inventories

A. Nikolakopoulos
National Technical University of Athens, Department of Chemical Engineering, Heroon Polytechneiou, 9, Athens, 15780, Greece;
tel. +302107723236, e-mail: nikolako@mail.ntua.gr

This paper proposes a centralized Model Predictive Control strategy as new decision support tool for vehicle routing and inventory control in the distribution of industrial gases via truck deliveries that follows the concept of vendor managed resupply [3]. The dynamical predictive model of the system is based on the multi-period deterministic model of the static problem [5]. Customers' inventory levels are the states of the system and the quantities delivered at each time period to the customers are the control variables. The unknown future demands are considered as the disturbance to the system. A time series decomposition forecasting model is used for future demand predictions and the cost function involves a balanced contribution of the cost of deviating from the safety stock trajectory and the sum of the transportation and inventory costs. The latter serves for examining the effects of averting the stock-out risk to the operating cost from a managerial point of view. An exact linearization of the quadratic term of the objective function guarantees the optimality of solutions.

The basic structure of the adopted strategy [2] is shown in Figure 1. The desired trajectories of safety inventory levels are compared to the predicted outputs of the method and the errors are fed to the optimization model.

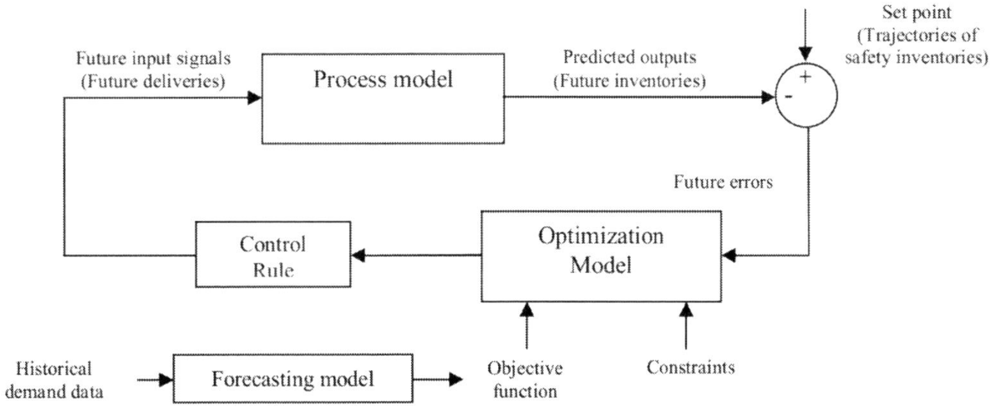

Figure 1 Structure of the Model Predictive strategy for Inventory Routing and Control

The optimization model takes into account the future customers' demands predicted by a forecasting model, the capacity and routing constraints and the cost function, where the future safety inventory tracking and the economic cost error are balanced. The schedule of future deliveries is produced by the optimization model and used as future input signals for the distribution process model which predicts the future inventory levels.

The methodology is illustrated by means of two examples that share the same basic features but differ in terms of the demand profile and the approach adopted for predicting future demand. For both examples a distribution system is considered, which involves 1 supplier, 3 customers, and 3 vehicles. For the first example perfect demand information is assumed for one period ahead, and for the remainder of the horizon, the demand forecast is set equal to the nominal demand. The response of the inventory levels for each customer and two different levels of stock out risk aversion (a = 0.2 and 0.5). In the first case, results (see Figure 2, a = 0.2) show that the safety stock is reached and maintained promptly. For the second case (a = 0.2), the total cost is reduced at the expense of an increased stock-out risk.

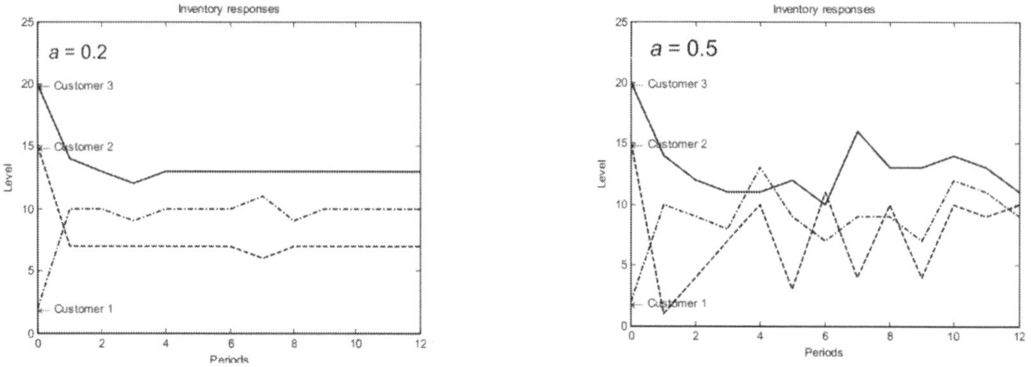

Figure 2 Inventory responses for a = 0.2, and 0.5 for Example 1

For the second example hypothetical historical data are produced for a period of T_h = 120 months and the time series decomposition method [4] was used to build the forecasting model. Figure 3, illustrates the response of the inventory levels. For a = 0.3 (low risk) results show the targeted safety stock trajectory is followed closely by the output levels of model, while for a = 0.5 the total cost is decreased at the cost of an increased stock-out risk.

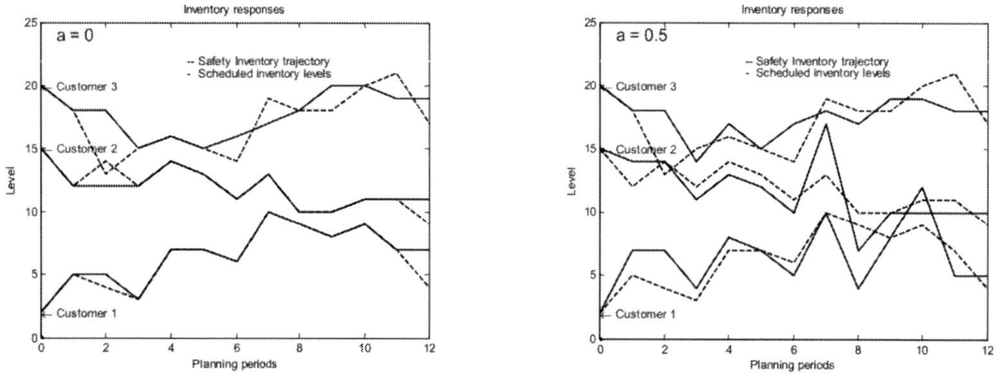

Figure 3 Inventory responses for a = 0.3, and 0.5 for Example 2

Overall, it is shown that by using the proposed MPC strategy, the supply chain can be stabilized in the sense described in [1]. The method offers a useful tool for producing balanced decisions between costs and tracking of the safety inventories while the produced schedules are always feasible and with optimal routing.

References

[1] Amrit R, Rawlings JB, Angeli D (2011) Economic optimization using model predictive control with a terminal cost. Annu Rev Control 35(2): 178-186.

[2] Camacho EF, Bordons C (1999) Model Predictive Control. Springer, Berlin.

[3] Kleywegt AJ, Nori VS, Savelsbergh MWP (2002) The stochastic inventory routing problem with direct deliveries. Transport Sci 36 (1): 94-118.

[4] Montgomery DC, Johnson LA, Gradiner JS (1990) Forecasting and Time Series Analysis. McGraw-Hill, New York.

[5] Nikolakopoulos A (2015) A Metaheuristic Reconstruction Algorithm for Solving Bi-level Vehicle Routing Problems with Backhauls for Army Rapid Fielding. Springer Operations Research/ Computer Science Interfaces Series, 56:141-157.

Quality improvement of middle distillates from thermal decomposition of waste polypropylene

O. Tóth, A. Holló, J. Hancsók
University of Pannonia, MOL Department of Hydrocarbon and Coal Processing, H-8200 Veszprém, Egyetem str. 10., Hungary; tel. +36 88 624962, fax: +36 88 624520
e-mail: totho@almos.uni-pannon.hu

The world production of plastics in the past years was about 250 million tonnes and the corresponding amount in the European Union (EU) increased to 50 million tonnes (~20%). Polyolefins (LDPE, LLDPE, HDPE and PP) are the main plastic materials and they account for 47% of the total plastic demand. As an unavoidable result of this huge demand, the amount of post-consumer plastic waste generated in EU was also high: roughly 25 million tonnes. About 62% of this waste is recovered and 38% is disposed in landfills. The energy recovery and mechanical or feedstock recycling are useful options for the management of plastic wastes. Feedstock recycling of polyolefin plastic wastes is receiving increased attention.

The demand of engine fuels (in Europe the demand for diesel fuel) is continuously growing. Nowadays the amount of used bio-components is increasing, too. One of the diesel biocomponents with excellent application properties is the bio gas oil (the mixture of iso- and n-paraffins). This can be produced from bio-derived materials (vegetable oil, animal fat) and waste-derived materials (used cooking oil, waste animal fats, etc.) too. The mixture of iso- and n-paraffins can also be produced from waste polyolefins in several steps. The gas oil fraction generated in the first step by the thermal decomposition of polyolefins is not yet suitable for engine fuel purposes because of its high olefin content and thus its inappropriate thermal and oxidation stability. Therefore quality improvement is needed for the saturation of olefins and then for the isomerisation of saturated hydrocarbons, which can be performed alone or in mixtures with gas oils. In the latter research field there are no detailed results published.

In our experimental work the quality improvement of blends of crude oil based and waste polyolefin originated gas oil fractions was investigated. The crude derived gas oil fraction had high sulphur and aromatic content. So the aim of our work was not only the quality improvement of the decomposition product with high olefin content, but also the quality improvement of the crude derived gas oil components.

The objective of the first step of the quality improving experiments was the investigation of the catalytic hydrogenation of the thermal decomposition products in 5, 10, 20 and 30% blends with gas oil over transition metal catalyst. In the frame of this work the effect of the process parameters (temperature: 320-360°C, pressure: 40-70 bar, liquid hourly space velocity: 1.0-3.0 h^{-1}, H_2/hydrocarbon ratio: 450 Nm^3/m^3) on the quality and quantity of the liquid products was investigated. The quality improvement experiments were carried out in a 100 cm^3 tubular reactor.

During this quality improvement the paraffinic characteristic of the gas oil has increased because of aliphatic paraffin chains formed by the saturation of olefins and the heteroatom removal reactions. The formed products fulfill the specifications of the valid EU standard (EN 590:2013 Automotive fuels. Diesel. Requirements and test methods), except the cold flow properties. In order to improve this property the isomerisation of n-paraffins to iso-paraffins over a self-developed Pt/SAPO-11 catalyst was investigated, because the freezing points of the iso-paraffins are significantly lower than the n-paraffins (ΔT=10-26°C). During the isomerisation reactions not only the cloud point and cold filtering plugging point (CFPP) of the products but their cetane number also decreased. Thus such process parameters were defined, that caused a large reduction in the cloud point and cold filtering plugging point and a relatively small reduction in the cetane number as a compromise.

It was found that the products with the best properties were the products which contained the highest amount of iso-paraffins with a single methyl branch. The highest increase in the cetane number was achieved by the quality improvement of the 30% thermal decomposition product containing gas oil (9-14 unit, depending on the process parameters). However, the increase in the cloud point and cold filtering plugging point values were the highest in this case (ΔT=6-11°C). During the second quality improving step (the isomerisation), the cetane number of these products decreased with 4-6 units, but the cold filtering plugging point values are significantly decreased (ΔT=8-15°C). In summary with the addition of small amounts of additives high performance diesel gas oil or without the addition of additives excellent gas oil blending component can be produced with the presented catalytic conversions.

Sediment influenced by acid mine drainage in the Smolnik creek – qualitative and quantitative characterization

E. Singovszka, [1]M. Balintova, S. Demcak
Technical University of Kosice, Institute of Environmental Engineering; tel. +421556024154, e-mail: eva.singovska@tuke.sk; [1]Faculty of Civil Engineering,Vysokoskolska 4, 04200 Kosice, Slovakia.

Water and sediment quality monitoring has one of the highest priorities in environmental protection policy. Heavy metals entering into the river can be from natural or anthropogenic sources. Main anthropogenic sources of heavy metal contamination are mining, disposal of untreated and partially treated effluents contain toxic metals, as well as metal chelates from different industries. In Slovak Republic there are some localities with existing acid mine drainage (AMD) generation conditions. The most critical values were observed in the abandoned deposit Smolnik. Waters from the earth surface penetrated the mine and they are enriched with metals and their pH values decreased.

The paper presents comparison of results of quantitative XRF analysis of sediments from the Smolnik creek sampled in uncontaminated part of creek and in the part influenced by AMD. The characterisation of sediments by statistical analysis is presented, too.

Keywords: Sediment, XRF, FTIR, XRD, statistical analysis.

Study of thermal reduction of barium sulphate for barium sulphide preparation

M. Balintova, [1]S. Demcak, M. Holub, P. Pavlikova, A. Estokova
Technical University of Kosice, Institute of Environmental Engineering; tel. +421556024127, e-mail: magdalena.balintova@tuke.sk; [1]Faculty of Civil Engineering, Vysokoskolska 4, 04200 Kosice, Slovakia

Basic methods for sulphate removal processes are reverse osmosis, ion exchange, precipitation by lime, cements, and salts of barium and the biological removal process. Biological sulphate removal is low cost but long term method, on the other hand reverse osmosis and ion exchange are more efficient but much costly. Lime and cement treatment process can be used for neutralization of acid water and partial sulphate removal only to the saturation level i.e. 1 200 mg/l.

The soluble salts of barium are most commonly used for precipitation of sulphate from aquatic acidic solutions to the insoluble product barium sulphate BaSO4. Benefits of precipitation are high sulphate removal efficiency but limitations are toxicity of barium compounds and highly economical costs. For this reason is recycling of BaSO4 to barium sulphide BaS (the precipitating reagent) very important.

The paper deals with study of BaSO4 reduction by carbon to BaS by thermal analysis and infrared spectrometry. DCS analysis shown, that transformation of BaSO4 to BaS was performed at a temperature approx. 800 °C. Thermal analysis and infrared spectra of the product confirmed the change in its composition but process of reduction by carbon was uncompleted and in sample was still present a part of BaSO4. Presence of BaS was confirmed by volatilization of hydrogen sulphide after the mixing of product with distillation water.

Keywords: Barium, regeneration, reduction, thermal analysis.

Pinch analysis of benzol by-product plant

M. Vecer, L. Simkova, I. Koutnik, K. Wichterle, [1]P. Baran
VSB-Technical University of Ostrava, FMME, dept. Chemistry, tr. 17. listopadu 15, 70833, Ostrva-Poruba, Czech Republic; tel. +420 596324230, e-mail: marek.vecer@vsb.cz;
[1]ArcelorMittal Ostrava, 10/C – Corporate Coke Lab, Vratimovska 689, 707 02 Ostrava-Kuncice, Czech Republic; tel. +420 595684269, e-mail: Pavel.Baran@arcelormittal.com;

Benzol by-product plant is ideal process unit with high potential for energy savings. Both heating and cooling processes of material flows are needed here and many heat exchangers are employed for. The material and energetic balance of benzol by-product plant will be presented here.

Results of energy optimization procedure obtained by pinch analysis show few options of relatively simple process changes which might have positive effect on economic aspects of benzol processes. Especially condenser of raw benzol vapours is the unit, where the flow optimization on the cold side can improve process performance significantly, see fig.1.

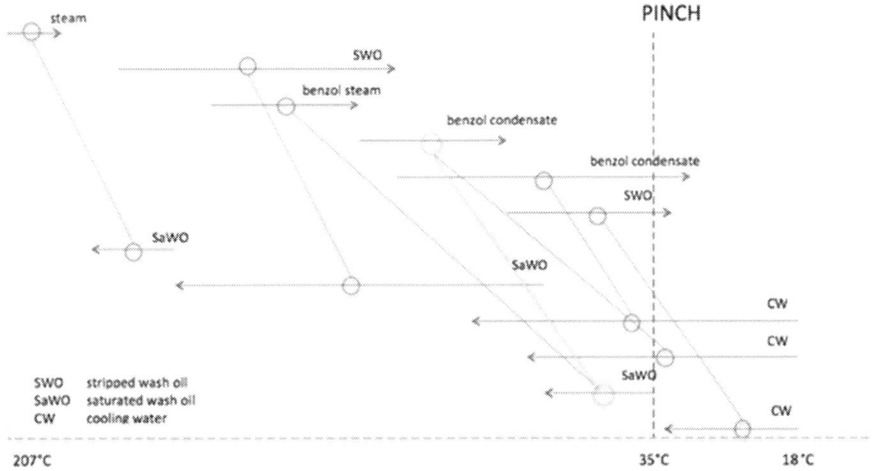

Fig. 1 Grid diagram of benzol by-product plant

Two case studies with modified process setup will be described with respect on material and energy savings. Procedure of pinch analysis has been applied and is reported under the real industrial conditions. Economic evaluation for different model simulations will be provided.

Keyword: pinch analysis, coke plant, heat exchanger, energy, enthalpy balance, material balance

Acknowledgements: This study was supported by "Institute of clean technologies for mining and utilization of raw materials for energy use - Sustainability program" (reg. No. LO1406).

Process of zinc separation from zinc-coated iron scrap

K. Wichterle, M. Vecer, J. Korpas
VSB-Technical University of Ostrava, FMME, Dept. of Chemistry, 70833, Ostrava-Poruba, Czech Republic; tel. +420 596324230, e-mail: marek.vecer@vsb.cz;

Due to increase of steel recycled, content of zinc in airborne particles released from steelworking plants to the off-gas becomes to be important. Until now, large portion of zinc leaves the high-temperature process as a zinc ferrite, which is stable and zinc wining from it is costly. It has been mostly dumped as a waste.

We have realized that during slower heating of scrap to the temperature up to 800°C, first portion of the dust contains valuable zinc oxide which may be separated. Here, unlike to the high-temperature dust, recovery of zinc would be easier.

The dust separation can be, without high investment costs, adapted in tandem oven equipment. Other options are electric arc furnaces employing a preheating section.

Sampling and analysis of the dust from a steelmaking plant, laboratory experiments, and suggestions for the proposed industrial technology modifications are presented.

Keyword: Zinc, Iron, Steel, Scrap, Material recycling

Acknowledgements: Presentation of this study was supported by "Institute of clean technologies for mining and utilization of raw materials for energy use - Sustainability program" (reg. No. LO1406).

Energy efficiency evaluation in a system of brick kilns

R.H. Chavez, [1]J.J. Guadarrama
Instituto Nacional de Investigaciones Nucleares, Carretera Mexico-Toluca s/n, La Marquesa, Ocoyoacac 52750, Mexico, MEXICO; tel. +525553297200 ext. 12654, e-mail: rosahilda.chavez@inin.gob.mx; [1]Engineering School, UAEMex, email: guja1704@gmail.com

The aim of this paper is to determine the thermal efficiency of two coupled brick kilns to usage the residual energy of the combustion gases from the first kiln. The second kiln was determined preheating kiln with the residual energy liberated from the first kiln. The internal dimensions of the two kilns are, radio 1.65m, height 3.3m, capable of accommodating 15000 brick pieces each one, the minimum size of bricks are 0.235m x 0.125m x 0.56m, with weight of 2.5kg and 2.1kg before and after firing, respectively. To ensure that the flue gases have a good natural draft in the second kiln, after yielding its heat, it was determined the rate of flow of flue gases in the ducts that connect both kilns as well as the local temperature and pressure, pressure drop by friction of the gases through the ducts, and the buoyancy of hot gases. The results show the usefulness of coupling two kilns, because it reduces the burning time from 40h to 20h. The thermal efficiency of coupled two kilns was 53% versus 40% in individual operation, achieving a second preheat kiln to 150°C.

Mathematical modelling of the effects of circadian rhythm on microalgal growth in phototrophic and mixotrophic cultures

A. Bose, S. Chakraborty
Department of Chemical Engineering, Indian Institute of Technology Kharagpur, Kharagpur 721302, India.; tel. +913222283930, e-mail: saikat@che.iitkgp.ernet.in; fax: +913222282250.

The production of algae based biofuels has been an important aspect of study to improve the yield and efficiency of such processes. In this study a quantitative model is being proposed to study the effects of circadian rhythms in algal growth processes to optimize the conditions required for maximum biomass growth.

The microalgal species *Chlorella sorokiana* is being studied with respect to its growth in different culture conditions, namely phototrophic and mixotrophic culture medium. The effect of mixing, reactor volume, illumination intensity and frequency has also been studied empirically to define the ideal growth parameters. In order to study the effects of these variables, photobioreactors of varying sizes (0.5 – 2 L) have been installed to experimentally measure the algal growth in phototrophic and mixotrophic conditions, initially inoculated with the culture medium under room temperature conditions. Accumulation of microalgal biomass is measured at fixed intervals of time to quantify the temporal growth rate of the algae, along with the varying parameters of light intensity, period of illumination, and carbon dioxide concentration in the input stream.

A major emphasis lies on the study of algal circadian rhythms which affect the overall growth pattern of the system. To study the effect of illumination cycles on the biomass growth, a Multiple Monod kinetics based growth equation is used in the model. It has been observed that a small fraction of the cells die during the dark period, which has been effectively quantified in the form of 'Night Biomass Loss' as a first order decay function. An accurate estimate of the dissolved carbon dioxide and oxygen concentrations in the liquid phase and absorbed algal phase values have been made, which allows us to measure the amount of carbon dioxide fixed by the algal system.

Our experimental results have been compared with our model simulations predicting the growth of microalgae under different culture conditions. Under 12h light: 12 dark, the highest biomass yield of 1.09 g/L was observed experimentally in 800 ml reactor under phototrophic conditions. The dissolved carbon dioxide during late log phase was in the range of 83 ppm and 86 ppm for 12h light: 12 dark and 24h light regime, respectively. The extent of mixing induced in the system by the input air stream acts as an important controlling factor for growth. For all performed runs, it was found that the system acts in the laminar zone for 2LPM airflow rate. The mathematical model incorporates a light intensity factor, mixing time in the laminar regime and the specific growth rate of the algal species as parameters to provide a generalized study for algal growth in photobioreactor systems. It has been observed that the growth of microalgae is not a continuous process but occurs in a periodic fashion, referred to as circadian rhythms which are best achieved by alternating dark and illuminated periods. Continuous exposure to light for the same time provides less growth than the setup with alternating light and dark exposure, proving the presence of 'fatigue' in the biomass growth which cannot be accounted for.

Effects of microwave output power and sample amount on drying kinetics of samphire

B. Özbek, E. Altun and E. Demirhan
Department of Chemical Engineering, Yıldız Technical University, Davutpaşa Campus Esenler/İstanbul, 34210, Turkey;
tel. +90 212 383 4747, fax: +90 212 383 4725.
e-mail: bozbek@yildiz.edu.tr

Food drying is a traditional method of food preservation, which is also used for the production of special foods and food ingredients [1]. Drying not only affects the water content of the product, but also alters other physical, biological and chemical properties such as enzymatic activity, microbial spoilage, viscosity, hardness, aroma, flavor and palatability of foods [2]. Different drying methods such as hot-air drying, vacuum drying, sun-drying and freeze drying are used in the drying of fruits and vegetables [3]. So that compared with these drying techniques; microwave drying offers opportunities as uniform energy and high thermal conductivity to the inner sides of the material, energy saving, fast start-up, shut-down conditions and high quality of the finish product. It also reduces the drying time and prevents food from decomposing [2-4].

The present study describes the effects of microwave output powers and sample amounts on moisture content, moisture ratio and drying time of samphire. Arçelik (MD-594) programmable domestic microwave oven was used for the drying experiments. To investigate the effect of microwave output power and sample amount on drying kinetics, different output powers; 180, 360, 540, 720 and 900 W; and different amounts of samphire; 25, 50, 75 and 100 g were examined, respectively. Experiments showed that drying time was proportional to microwave output powers and sample amounts for samphire. The mathematical models depending on process conditions were also developed. Also, the minimum energy to evaporate one gram of water was calculated by using the drying constants for each sample.

REFERENCES

[1] Krokida, M.K., Maroulis, Z.B.& Saravacos, G.D. (2001), The effect of the method of drying on the colour of dehydrated products, *International Journal of Food Science and Technology*, 36(1), 53-59.

[2] Barbosa-Canovas, G.V. & Vega-Mercado, H. (1996), *Dehydration of Foods* (1st edition), New York, United States of America.

[3] Ozkan, A.I., Akbudak, B. & Akbudak, N. (2007), Microwave drying characteristiscs of spinach, *Journal of Food Engineering,* 78, 577-583.

[4] Maskan, M. (2000), Microwave/air and microwave finish drying of banana, *Journal of Food Engineering,* 44, 71-78.

Struvite crystallisation in a fluidised bed reactor: modelisation and operation

F. Corona[1,2], D. Hidalgo[1,2], L. Lipsa[1,2], J.M.Martín-Marroquín[1,2] and G. Vilalta[3]
[1]CARTIF Centro Tecnológico, Boecillo, Valladolid, 47151, Spain; tel. +34 983546504, e-mail: fraenc@cartif.es.
[2]ITAP Institute, University of Valladolid, Valladolid, 47010, Spain; tel. +34 983423355.
[3]Thermal Sciences and Fluid Department, Federal University of São João del-Rei (UFSJ), Brazil.

In recent years, the intensification of livestock farms has meant an increase in the quantities of waste products from animals, which have serious environmental problems if they are not properly managed. Conventional approaches to pig manure are its direct application to land. However, these practices result in the loss of nutrients and lead to serious environmental problems. The most common problems directly linked to the disposal of untreated manure are: The formation of odours, ammonia and methane emissions, release of nutrients and pathogens that can affect human health and aesthetic problems. In addition, because of the concentration and the situation of farms, about 30% of manure generated has not agricultural land for application.

The problems described can be converted into an opportunity if it is taken into account that the livestock waste are a renewable resource (by-product) and if they are managed properly, can serve as feedstock for the biofuels industry and fertilisers.

Under these circumstances, the LIFE REVAWASTE project (www.revawaste.eu) arises with the aim to demonstrate a sustainable management of a broad spectrum of wastes (non-recyclable fraction proceeding from waste treatment plants and industrial, together with biomass, livestock and agro-food wastes) in an integrated plant including, among other treatments, anaerobic digestion and pyrolysis. Regarding the element obtained in the anaerobic digestion (digestate), the process does not significantly reduce the concentration of nitrogen or phosphorus. Crystallisation of nitrogen and phosphorus in the form of magnesium phosphate and ammonium hexahydrate ($MgNH_4PO_4{\cdot}6H_2O$) also known as struvite is one of the possible techniques used to remove and / or recover digestate nutrients obtaining a product which can be applied as the basis for high quality organic fertilisers.

In order to study the behaviour of the crystallisation reaction of struvite, a modelisation and simulation of the process was carried out using ANSYS FLUENT Computational Fluid Dynamics (CFD) software. The reaction was simulated in a fluidised bed reactor with a volume of 50 L (2.12 m high and a diameter of 0.215 m). By CFD simulations, the effect of reaction parameters (fluidising air flow, composition of the reaction mixture, reaction time...) in struvite crystallisation has been studied. Different cases has been analysed (modifying only the boundary conditions).

A series of experiments was carried out to verify the validity of the struvite crystallisation model obtained by FLUENT. Different digestate streams have been considered for crystallisation in this paper. Mainly, digestate obtained as a by-product from anaerobic digestion of pig manure and synthetic digestate made from synthetic laboratory reagents with typical real digestate composition, but with the absence of impurities.

Batch experiments were run in a glass fluidised bed reactor. The reactor had the same dimensions as the reactor simulated by FLUENT. The experiments were developed for different levels of the main variables of study of the struvite crystallisation reaction:

nitrogen/phosphorus molar ratio (4, 8, 10), magnesium/ phosphorus molar ratio (1, 1.5, 2), air flow (2, 6, 12 NL/min), reaction time (0.5, 1, 2 h).

The study shows that the model obtaining by FLUENT fits the experimental results. Yield and crystal size increases when magnesium concentration and the reaction time are increased.

Summarising, the results so far show that the crystallisation of struvite in a fluidised bed reactor is a method with good prospects for the valorisation of digestate from anaerobic digestion of agricultural and livestock waste.

Kinetics of microwave-based ionic liquid-mediated catalytic conversion of Ricinus communis to biofuel products

D. Mathur, S. Chakraborty
Department of Chemical Engineering, Indian Institute of Technology Kharagpur, Kharagpur 721302, India.; tel. +913222283930, e-mail: saikat@che.iitkgp.ernet.in; fax: +913222282250.

Lignocellulosic fuels are an important class of second generation biofuels for the future, particularly because they, unlike the first generation cellulosic fuels, do not engender any food vs. fuel debate. The Biomass derived 5-Hydroxy-methylfulral (HMF) has emerged as a key precursor chemical for the production of fine chemicals, liquid fuel and other carbon-based derivatives. The five-member ring compound, 5-Hydroxy-methylfurfural (HMF), is one of the top bio-based platform compounds, which can be easily converted to a novel biofuel molecule 2,5-dimethylfural via selective hydrogenation.

The main objective of this work is to determine the kinetics of a fast, economically viable process for the conversion of lignocelluloses Ricinus communis to HMF – the platform chemical, along with the production of glucose for bioethanol production.

Here, we perform ionic liquid ([BMIM]Cl) based catalytic conversion of *Ricinus communis* (Castor leaves/tree mixtures) under the influence of microwave radiations with $CuSO_4$ as a catalyst in a microwave reactor. The lignocellulosic substrate is first depolymerised into Glucose, which then converts to HMF that further dissociates into levulinc acid and formic acid.

While the conversion of lignocellulose to glucose is a hydrolysis process, the formation of HMF from glucose is a dehydration process, the former requiring the addition of water and the latter necessitating the removal of water. We focus on experimentally finding out the optimal water addition and microwave heating profile (temperature and pressure) for maximizing the HMF yield from *Ricinus communis* in [BMIM]Cl aided with $CuSO_4$. We find an initial increase of glucose concentration with water addition, which decreases above 30% water addition. Similarly, higher conversion of glucose to HMF is observed at a lower percentage of water added since lack of water aids the dehydration process leading to HMF formation. We observe a 6wt. % conversion to HMF for 15% addition of water in 30 mins.

Our kinetic model along with a Power law model (for the dehydration of HMF and conversion of HMF to levulinc acid) and Biphasic Model (for the hydrolysis of cellulose) for quantifying the temporal dynamics of glucose and HMF production is based on our understanding of the crucial role water plays in determining the product distribution. We specify the kinetically favoured conditions for optimal HMF production, and vary the heating profiles in the microwave reactor to obtain an empirical relationship between energy consumption, reaction time and HMF conversion.

The effect of pre-treatment on the lignocellulosic substrate has also been analysed using FESEM (pore structure, size and density), BET (pore-surface analysis), DLS (particle size distribution) and XRD (extent of crystallinity).

The microwave reactor helps align the dipoles in the electromagnetic field created by the combination of ionic liquid and microwave radiation, thus reducing the total reaction time drastically to 30-40 min, while our novel water addition strategy allows us to manipulate the production distribution for maximizing the conversion of lignocelluloses to HMF via glucose.

Effect of Gd, Ce and Ru promoters on $Co/\gamma\text{-}Al_2O_3$ in Fischer-Tropsch synthesis

A. Moradian, F. Bahadoran[1], L. Shirazi, Y. Zamani
Research Institute of Petroleum Industry (RIPI), Gas Research Division;
Tel: +982148252402, e-mail: fbahadoran@yahoo.com; [1]West of Azadi Sport Complex, Tehran, Iran; Fax: +982144739712

Fischer- Tropsch synthesis (FTS) is referring to catalytic process which converts syngas (H_2, CO) to heavier hydrocarbons. FTS catalysts are composed of active metal, support and promoters. Promoters are used to increasing activity and selectivity of the catalyst to heavier hydrocarbons. Addition of rare earth oxide (REO) as a promoter in some study illustrate that the activity and selectivity of the catalyst increased. Huber et al[1] by study on $Gd/Co/SiO_2$ found that the activity increased and methane selectivity decreased. Haibo et al[2] by research on Co/SiO_2 with CeO_2 as a promoter showed that the activity and chain growth probability increased. Effect of Ru and REO on the catalytic performances of FTS is not investigated by detail. Therefore as a main goal of this paper we study on $15Co/0.1Ru/\gamma\text{-}Al_2O_3$ with %6 Gd or Ce as a promoter.

The catalysts were prepared by dry impregnation method and then characterized by XRD, BET, TPR, TEM techniques. Reaction was done in a fixed bed reactor at 210 ºC, atmospheric pressure and $H_2/CO = 2/1$ conditions.

BET analyses of synthesized catalyst are presented in Table 1. Decrease of BET surface area in promoted catalyst is attributed to impregnation of precursor solution on $\gamma\text{-}Al_2O_3$ support. As can be seen in Figure 1, by addition of Ce or Gd and Ru, the Co_3O_4 peak is widened because of smaller cobalt particle size.

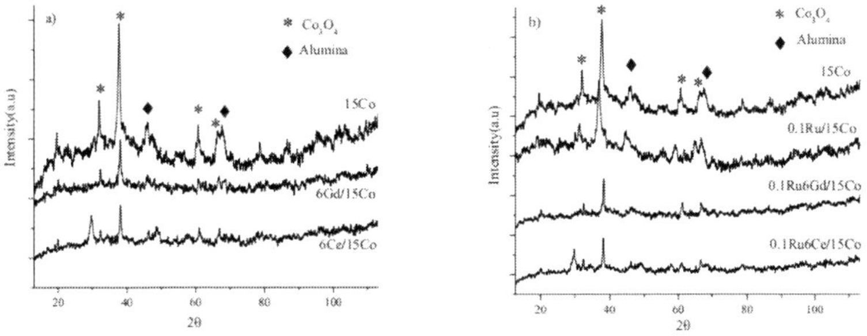

Figure 1. XRD analysis of synthesized catalysts

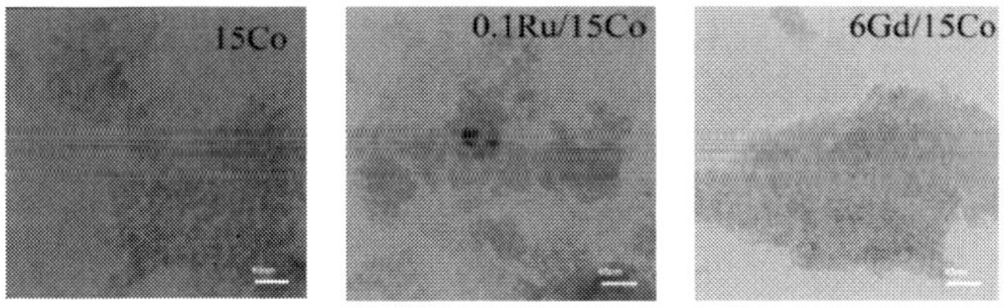

Figure 2. TEM images of synthesized catalysts

The dark area in TEM images, Figure 2, are shows the cobalt particles. As illustrate, by addition of Ru/Gd the cobalt particle size decreased. According to Figure 3, reducibility of Ru promoted catalyst is more than unpromoted one which this effect is due to H_2 spillover. As mentioned before, from XRD calculation results by scherrer equation[3], Ce and Gd reduced particle size which means increasing of Co-alumina interaction and in turn decreases of cobalt reduction. Figure 3.a illustrate that Ce and Gd shifts the reduction peak to higher temperature that confirms this deduction.

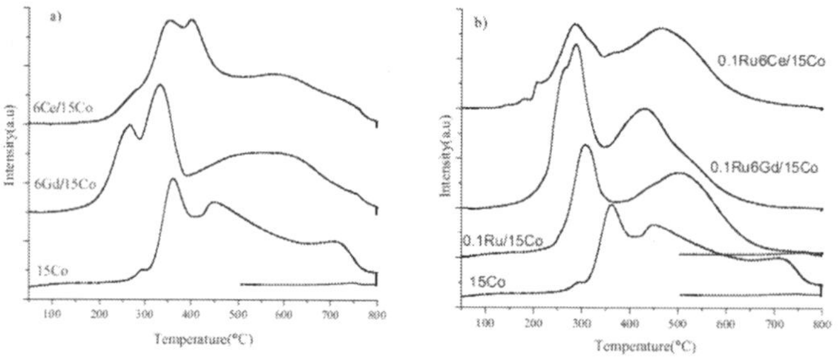

Figure 3. TPR pattern of synthesized catalysts

FTS catalyst activity is influence by two major factors: the dispersion and the reduction of the catalyst. The XRD, TPR and TEM results demonstrate on the decrease of catalyst performance in presence of Gd and Ce. Table 1 show that the catalyst activity (CO_{Rate}) for 0.1Ru/15Co is higher than others which attributed to higher reduction due to H_2 spillover from Ru to cobalt. Moreover, it is observe that the dispersion in presence of Ru is higher than other catalyst which can be increase number of active site and therefore increase catalyst activity. The methane selectivity for 0.1Ru/15Co is higher than other one because of higher reduction of smaller particle size in presence of Ru as a promoter. In presence of 6Gd and 6Ce, the methane selectivity is decreased which can be attributed to Strong Metal Support Interaction(SMSI) effect by Ce and Gd[4]. SMSI effect causes smaller cobalt particle which can not reduce in pre reduction step and results in lower activity and methane selectivity. Ru by H_2 spillover can compensate some of reducibility problem of the catalyst and increases its activity. Also for 6Gd/15Co, 6Ce/15Co and 0.1Ru6Ce/15Co the CO_{Rate} is decreased because of increased number of smaller particle size and therefore decrease catalyst reduction.

Table 1. Fischer- Tropsch synthesis results (210°C, atmospheric pressure, $H_2/CO=2$) , particle size and surface area of catalysts

					Rate(mole/kg.s)		Selectivity						
Catalyst	BET (m^2/g)	Pore Size (Å)	d_{Co3O4} (nm)	%D	$CO_{Rate}(\times 10^5)$		C_1	C_2	C_3	C_4	C_5	C_5^+	α
15Co	103.6	67.7	28.2	4.54	41.8		31	7	13	11	7	26	0.55
0.1Ru/15Co	104.9	65.6	14.1	9.11	97.9		43	7	19	12	7	8	0.51
6Gd/15Co	86.4	62.8	18.7	6.83	25.3		15	6	6	7	7	53	0.64
0.1Ru6Gd/15Co	85.0	64.0	21.6	5.92	45.4		19	5	9	10	9	43	0.65
6Ce/15Co	87.7	57.8	20.1	6.38	12.6		24	7	7	9	8	35	0.59
0.1Ru6Ce/15Co	76.9	63.5	17.6	7.29	31.0		34	7	13	11	8	19	0.56

There is maximum CO_{Rate} for 0.1Ru/15Co and 0.1Ru6Gd/15Co due to higher reduction. More reduction can be attributed to H_2 spillover. In presence of Ru the methane selectivity is increased which related to more number of smaller particle reduction due to H_2 spillover. In presence of Ce and Gd the methane selectivity is decrease because of SMSI effect.

[1] G.W. Huber, S.J.M. Butala, M.L. Lee, C.H. Bartholomew, Gd promotion of Co/SiO_2 Fischer – Tropsch synthesis catalysts, 74 (2001) 45–48.

[2] S. Haibo, S. Shikong, Studies on Promotion Mechanism of CeO_2 in $Co-CeO_2/SiO_2$ Catalyst for Fischer-Tropsch Synthesis Using D_2 Pulse Reaction, Chinese J. Catal.27 (2006) 1096–1100.

[3] T.K. Das, G. Jacobs, P.M. Patterson, W.A. Conner, J. Li, B.H. Davis, Fischer–Tropsch synthesis: characterization and catalytic properties of rhenium promoted cobalt alumina catalysts, Fuel. 82 (2003) 805–815.

[4] C.H. Bartholomew, FUNDAMENTALS OF INDUSTRIAL Second Edition, A JOHN WILEY SONS, INC., Publ. (2006).

Development of a blast furnace model with thermodynamic process depiction by means of the Rist operating diagram

A. Spanlang, [1]W. Wukovits, [2]B. Weiss,
K1-MET GmbH,Stahlstraße 14, 4020 Linz, Austria; tel.+43 732 698975682,e-mail: andreas.spanlang@k1-met.com;[1]TU Wien, Institute of Chemical Engineering, Getreidemarkt 9/166-2, 1060 Vienna, Austria;[2]Primetals Technologies Austria GmbH, Iron- and Steelmaking Technology and Innovation –Direct Reduction, Linz, Austria.

Despite the fact that alternative processes such as Corex® and Finex® have already been established on industrial scale, the blast furnace route continues to be the most important process used to produce pig iron.

Due to its importance, a variety of models of the blast furnace process have been developed in the past decades. In addition to this, a well-established analogue representation of blast furnace operation is given by the Rist operating diagram.

A black box model was implemented in the simulation platform of gPROMS ModelBuilder®based on elemental assignments and empirical component distribution correlations. The main system inputs are iron oxides, coke, additives as well as hot blast and substitute reducing agents in order to produce liquid hot metal, slag and blast furnace gas. The thermodynamic process conditions are analysed in a Rist sub-model enabling the construction of the operating diagram of a blast furnace process. Furthermore, an additional sub-model is used to calculate the adiabatic flame temperature in front of the blast furnace tuyeres.

The modelling concept enables the description of interdependencies between the main blast furnace process, the raceway conditions as well as the overall thermodynamics described by the Rist operating diagram within a single mathematical model. A detailed analysis of the model behaviour under varying inputs was carried out. The achieved simulation results were in good accordance with the reference data thus confirming the applicability of the developed model for simulation of the blast furnace process.

The developed model provides the means for comprehensive simulation of different blast furnace route setups. This allows for further optimisation regarding process efficiency thereby potentially reducing fuel consumption as well as environmental emissions.

Entropy and exergy analysis of different microbial energy conversion pathways

Q. Fu[1,2], G. Cao[1,2], Q. Liao[1,2]*, Y. Huang[1,2], X. Zhu[1,2],
[1]Key laboratory of Low-grade Energy Utilization Technologies and Systems, MOE, Chongqing, China; [2]Institute of Engineering Thermophysics, Chongqing University, Chongqing, China; tel.+86 023 65102474, e-mail: lqzx@cqu.edu.cn.

Entropy and exergy analysis have been widely used in various energy conversion systems and chemical engineering processes. However, few examples were found in microbial energy conversion systems. In this study, we used entropy and exergy analysis to evaluate different microbial energy conversion pathways, including photo fermentation and dark fermentation. Glucose was used as substrate to produce different energy substances (hydrogen, alcohol and methane) in different pathways. The velocity field, temperature field, substrate concentration field and energy substances production rate (R_i /mol·s^{-1}) of the bioreactors were simulated under their optimal operation conditions using a computational fluid dynamic software. Based on the simulation, the chemical exergy ($E_{x,i}$ /J·mol^{-1}, under the ambient temperature of T_0) of the energy substances and total entropy generation rate (S /W·K^{-1}) due to irreversible flow, heat and mass transfer were calculated. In addition, a parameter, the net exergy output ($E_{x,net}=E_{x,i}\cdot R_i-T_0\cdot S$) of the bioreactor, was proposed to evaluate the energy conversion pathways. The results showed that the net exergy output from high to low were dark fermentation for methane, photo fermentation for hydrogen, and dark fermentation for alcohol. In addition, improving the energy substances production rate was the most important since its largest effect on net exergy output. This study will contribute to the optimal choice of different microbial energy conversion pathways.

Keywords: Entropy and exergy analysis; microbial energy conversion; entropy generation rate

Simulation of production process on dimethyl ether from methanol

S. Karacan[1]
[1]Ankara University, Engineering Faculty, Department of Chemical Engineering, Tandoğan 06100, Ankara, TURKEY. Tel:+90(312) 2033523, Fax: +90 (312) 212 15 46 (E-mail: karacan@eng.ankara.edu.tr)

Over the past few years, dimethyl ether (DME) has been given the widespread attention as an environmentally friendly chemical product. Energy is one of the principal concerns of the industrialized world. There is a continuous growth in the energy consumption with the industrial improvements of the developing countries, which results in higher amount of green house gas emissions and elevated energy prices. Moreover, natural gas and petroleum are important raw materials for industry and their consumption rate has increased more than 200 times in the last century. Among them, petroleum is the one having the fastest depletion rate because of its excessive consumption for transportation. Its overuse for transportation (about 57% of total oil consumption) causes problems in the petrochemical industry, in which petroleum is the major raw material. According to the predicted lifetimes of precious raw materials, it is expected that there will be significant fuel shortage problems in the future. Thus, the world is looking for new, cheaper and sustainable solutions; or by other means, alternative energy resources rather than fossil fuels. The fuel alternative of the future should have an easily available, non-petroleum feedstock that enables production and fuel versatility with reduced well to wheel green house gas emissions and enhanced fuel performance. Moreover, the infrastructure for fuel transportation and storage must be well established. Among all the alternative fuel candidates such as hydrogen, methane, methanol, bio-fuels and Fischer Tropsch fuels; ether based fuels, mainly di-methyl ether takes a profound attention. DME serves as a clean, high efficiency compression ignition fuel with low NOx and SOx emissions and reduced particulate matter. It is an environmentally benign fuel and does not possess large concerns with toxicity, infrastructure, transportation and storage as the other alternative fuel options .

DME is produced by dehydrogenation over an industrialized molecular sieve catalyst. The commercial domestic γ-Al2O3 catalyst provided by DME production factory is used in methanol dehydration due to its good catalytic activity. In the is study, simulations were performed using the commercial simulation package ChemCAD (version 6.4). The production process was built by using flow diagram in ChemCAD area. The Plug flow reactor module was used to model the reactions in the fixed-bed reactor. Two distillation columns were used to simulate the distillation process. The rigorous equilibrium stage model, also called SRK model, consists of a set of nonlinear equations to represent material balance, vapor-liquid equilibrium, mole fraction summation and heat balance. Simple HTXR, mixer and pump unit operations were used to set up the process. A comparison between the calculated and measured results has proved that these results are satisfactory. The bed height and the volume of the catalytic bed are calculated aim at 30000 tone per year DME yields and while taking account of high-purity DME production.

Spark plasma sintering simulation of alumina composite modified with carbon nanotubes

N.A. Fedosova[1], E.M. Koltsova[1], E.V. Zharikov[1], I.I. Mitrichev[1], A. S. Shaneva[1].
[1] – Russia, Moscow, Miusskay sq., 9, D. Mendeleyev University of Chemical Technology of Russia, e-mail: kolts@muctr.ru, tel. 8(495)4952126

Spark plasma sintering method is increasingly used for the production of composite ceramic materials. This sintering technique uses pulsed (on-off) directed electric current under low atmospheric pressure to perform high speed consolidation of the powder particles. Spark plasma sintering allows in a short time period obtain porous-free small-grained samples. High reproducibility of results is achieved by the complete controllability of sintering process parameters.

We obtained a high-density ceramic composite based alumina matrix reinforced with carbon nanotubes (20-50 %vol.) sintered by spark plasma sintering. The structure of the resulting material is tightly sintered alumina grains surrounded with carbon nanotubes. Nanotubes are homogeneously distributed in the bulk of composite. The fracture toughness of samples with 50% vol. nanotubes is approximately 3.3 times higher than samples of pure alumina, and the elastic modulus is 2.8 times higher. However, not all temperature conditions yielded nonporous material with improved strength characteristics. In order to determine the optimum parameters for the process of spark plasma sintering method a numerical simulation of the sintering process was carried out. To simulate a change in the porosity of the spark plasma sintering process the balance equation of the pore size (type Liouville equation) was used. The rate of change in pore size depends on the heating rate (for the pulse impact step) and on the maximum holding temperature (for holding step). Kinetic parameters of the rate of change in pore size depends on the amount of carbon nanotubes in the composite.

The sintering process was divided into two independent stages: stage with impact of pulsed electric current and holding stage at the maximum temperature. For each of these stages we made a mathematical description including the factors influencing the process. For the first stage of sintering process such parameter is the heating rate of the compact powder, for the second stage it is the holding temperature. The amount of carbon nanotubes is considered in both stages of the process.

For the numerical solution of the problem two approximation schemes were examined: the scheme "CABARET" and "Z-scheme". These schemes are used in solving problems of continuum mechanics. After investigating these schemes, it was found that "Z-scheme" is advantageous since it is absolutely stable and the scheme "CABARET" conditionally stable at a ratio of $\Delta t / h \leq 1$. Based on this we used "Z-scheme" as the most suitable scheme.

While developing the mathematical model we have introduced additional constants (12 constants). These new parameters are responsible for the specificity of the sintering process in more detail. The numerical values of the constants were found after analytical assessment. It is allowed us to describe a mathematical model more accurately at every stage of the sintering. The relative error on the output characteristic (final porosity) is 5%. Thus the developed mathematical model can be used to calculate the optimal parameters of the sintering process for the ceramic composite to improve its desired characteristics.

This work was supported by a grant RSF 14-19-00522.

Direct air capture using K_2CO_3-based composite sorbent for renewable methane production by Sabatier process

J. V. Veselovskaya, P. D. Parunin, V. S. Derevschikov, A. G. Okunev
Novosibirsk State University, Dep. Physics, Pirogova Str. 2, 630090 Novosibirsk, Russia; tel. +7 9137510292; Boreskov Institute of Catalysis, Lavrentieva Av., 630090 Novosibirsk, Russia; e-mail: jvv@catalysis.ru.

Given depletion of fossil fuels, renewable energy technologies will need to play a bigger role in the future. However, there are still some unresolved problems that prevent these technologies from being economically viable. For instance, large scale integration of wind and solar energy into the electrical grid requires a storage technology for the excess renewable electricity produced by these fluctuating sources.

The electricity generated by renewables can be used for water electrolysis (1) to produce hydrogen:

$$2H_2O = 2H_2 + O_2. \quad (1)$$

Hydrogen is considered to be a perspective fuel for low-carbon economy, but there are still unresolved safety problems regarding its transportation and storage. Therefore, it is reasonable to use the generated hydrogen in Sabatier process (2):

$$CO_2 + 4H_2 = CH_4 + 2H_2O. \quad (2)$$

The synthetic renewable methane, which corresponds to chemically normal natural gas, can be injected into the existing gas network or stored and utilized on demand. This renewable energy storage sequence is also known as power-to-gas.

Renewable methane production by Sabatier process (2) requires a sustainable source of carbon dioxide. It is well known that CO_2 is the major anthropogenic greenhouse gas which contributes to global climate change. Direct CO_2 capture from ambient air (a.k.a. "direct air capture", DAC) offers more flexibility compared to conventional carbon capture from large point sources. Incorporation of the DAC unit with into the renewable energy storage system offers an opportunity to use anthropogenic carbon dioxide as a valuable feedstock for the production of renewable methane which can be utilized in the heating market or as a fuel for transportation.

Carbon dioxide capture from ambient air is more challenging than conventional carbon capture from flue gases because of very low concentration of CO_2 in air (~ 400 ppm) and the necessity to operate in the presence of moisture excess at ambient temperature and pressure. The most developed DAC technologies are based on carbon dioxide absorption by aqueous alkali solutions. The major drawback of this approach is that the regeneration of these solutions is generally multi-stage and energy intensive.

Potassium carbonate is a well-known solid inorganic chemisorbent which reacts with atmospheric CO_2 in the presence of water vapour forming potassium bicarbonate:

$$K_2CO_3 + CO_2 + H_2O = 2KHCO_3. \quad (3)$$

The main problem with using bulk potassium carbonate for CO_2 absorption purposes is low carbonation rate. This problem can be solved by dispersing K_2CO_3 particles inside a porous support material. Composite materials "potassium carbonate inside a porous matrix" are known to be effective sorbents for capturing carbon dioxide from wet flue gases and are regrded as promising materials for DAC.

The main goal of our research is to design the process which will effectively combine DAC using K_2CO_3-based composite sorbent and methane production from captured CO_2 by catalytic reaction (2).

The composite material K_2CO_3/Al_2O_3 was synthesized and studied in temperature swing adsorption (TSA) cycles using the experimental apparatus schematically shown in Figure 1. Each cycle comprised step of carbon dioxide absorption from ambient air, followed by consecutive thermal desorption/methanation step.

It has been shown that the increase in the outer heater temperature from 200 to 325°C during the thermal regeneration step affects the utilization extent of the composite sorbent in the TSA cycle with total CO_2 uptake rising from 1.9 to 4.4 wt. %. Further heater temperature increase up to 400°C has not resulted in any significant change in the total CO_2 uptake by the composite material.

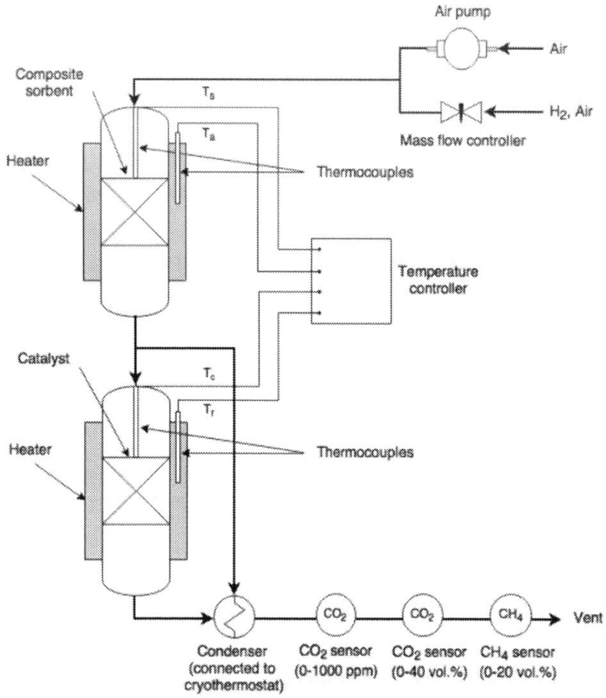

Figure 1: Scheme of the experimental apparatus used for carrying out CO_2 absorption/desorption and CO_2 absorption/methanation experiments

The process combing thermal regeneration of the composite sorbent in hydrogen atmosphere and CO_2 methanation reaction over commercial nickel catalyst NKM-2V was successfully implemented using the catalytic reactor connected to the outlet of the adsorber with the composite sorbent (Figure 1). Possibility to achieve 99% conversion of desorbed CO_2 into CH_4 was demonstrated experimentally using the commercial nickel catalyst heated up to 361-430°C. The further investigation of this process, which has a great potential for carbon-neutral economy, should be focused on optimization of operating conditions and reduction of total energy consumption.

Acknowledgments

The work was performed in the framework of the joint Research and Educational Center for Energy Efficient Catalysis (Novosibirsk State University, Boreskov Institute of Catalysis).

Sorption equilibrium of aromatic pollutants onto dissolved humic acids

Vincenzo Leone[a,b], Dino Musmarra[b,c], Pasquale Iovino[a,b,*], Stefano Salvestrini[a,b], Sante Capasso[b]

[a] Department of Environmental, Biological and Pharmaceutical Sciences and Technologies, Second University of Naples, via Vivaldi 43, 81100 Caserta, Italy.

[b] Environmental Technologies, University Spin Off of Second University of Naples, Via Vivaldi, 43, 81100 Caserta, Italy.

[c] Department of Civil Engineering, Design, Construction and Environment, Second University of Naples, Via Roma 29, 81031 Aversa (CE), Italy.

Humic acids (HA) are a mixture of natural aromatic polymers with high chemical heterogeneity and a broad range of molecular weights; they arise from biological decomposition of plants and other organisms and account for the greater part of organic matter in soil and surface water (Hayer et al., 1998). In this work we investigated the sorption of common organic pollutants onto HA in water solution by equilibrium dialysis experiments using 3.5 kDa cut-off membranes and two fully characterized samples of HA from leonardite and vegetable compost. The dissociation constants and sorption stoichiometries computed from free-ligand versus bound-ligand curves were analyzed in relation to physical-chemical properties of pollutants and HA samples.

References

M.H.B. Hayes, 1998. Humic Substances, in Structure, Properties and Uses, ed. by G. Davies, E. Ghabbour (Royal Society of Chemistry, Cambridge, UK, pp. 1–27.

Sorption of benzene derivatives onto insolubilized humic acids

Vincenzo Leone[a,b], Marco Trifuoggi[d], Dino Musmarra[b,c], Sante Capasso[b], Pasquale Iovino[a,b,*]
[a] Department of Environmental, Biological and Pharmaceutical Sciences and Technologies, Second University of Naples, via Vivaldi 43, 81100 Caserta, Italy.
[b] Environmental Technologies, University Spin Off of Second University of Naples, Via Vivaldi, 43, 81100 Caserta, Italy.
[c] Department of Civil Engineering, Design, Construction and Environment, Second University of Naples, Via Roma 29, 81031 Aversa (CE), Italy.
[d] University of Naples "Federico II", Chemical Sciences Department, 80126 Naples, Italy.

Humic acid (HA) was extracted by basic/acid treatment from leonardite, an oxidation product of lignite associated with near-surface mining and containing up to 90% of HA, and from vegetable compost. Following precipitation as calcium salt, HA was immobilized by heating at 350 °C for 1.5 h (Iovino et al., 2015; Radwan et al., 2015). Immobilized humic acid (Im-HA) was not soluble up to strongly basic pH. Sorption isotherms on Im-HA were determined for historical contaminants (toluene, phenol, o-xylene, benzyl alcohol) and emerging pollutants (diclofenac) with batch equilibrium tests at neutral pH and 25 °C. The experimental data were analyzed using the equations of more common use, i.e. Langmuir, Fowler, bi-Langmuir, Freundlich and BET, and the results interpreted in relation tophysico-chemical properties of sorbates and Im-HA.

References

P. Iovino, V. Leone, S.Salvestrini, S. Capasso, 2015. Sorption of non-ionic organic pollutants onto immobilized humic acids, Desalination and Water Treatment, 1(5), 55-62.
E. K. Radwan, H. H. A. Ghafar, A. S. Moursy, C. H. Langford, A. H. Bedair, G. Achari, 2015. Preparation and characterization of humic acid-carbon hybrid materials as adsorbents for organic micro-pollutants, Environ Sci Pollut Res, 2015, 22, 12035–12049, DOI 10.1007/s11356-015-4468-9.

Metallic affinity of toxic and noble metals by particles produced from sericin, alginate and poly(ethylene glycol)

T. L. Silva[a], V. H. Meinerz[a], J. M. Vidart[a], M. L. Gimenes[b], M. G. A. Vieira[a] and Meuris G. C. Silva[a]
[a]School of Chemical Engineering, State University of Campinas, UNICAMP, 13083-825, Campinas – SP, Brazil
[b]Department of Chemical Engineering, State University of Maringá, UEM, 87020-900, Maringá – PR, Brazil
meuris@feq.unicamp.br

Biosorbents have been used as a potential alternative to the conventional technologies for removal of toxic metals from water and wastewater. The use of waste material as adsorbent offers significant advantages like the low-cost, availability, efficiency, biodegradability, etc. In this study, particles produced from a blend between sericin, alginate and poly(ethylene glycol) (PEG) were used to remove toxic and precious metals from aqueous solutions. Precious metals are found in wastewaters of various industrial processes including metal, ore processing, electronic manufacturing and mining; and the recovery of these metals is profitable to those industrial processes. Sericin is a protein present in cocoon of silkworm silk (*Bombyx mori*) and usually discharged in the effluent of silk production. Alginate is a natural polysaccharide extracted from abundant sources of brown algae (brown seaweed) properties to this material. The addiction of PEG into a blend of sericin-alginate induces chemical modification that can develop better properties to this material. The objective of this work is to evaluate the metallic affinity and adsorption capacity of particles produced by sericin-alginate blend and sericin-alginate-PEG blend. The blends were dripped in aqueous and alcoholic (ethanol) solutions of $CaCl_2$ and $Ca(NO_3)_2$, and then the particles were dried at 100 °C. The simulation of speciation diagrams as functions of pH was performed using Hydra/Medusa software in order to establish the pH to be used in the experiments. The removal of chromium, cadmium, nickel, copper, lead, silver, palladium, platinum and gold ions from aqueous solution by sericin-alginate particles was evaluated through adsorption affinity testes. The results indicated that the presence of PEG improve the adsorption capacity of the biosorbents. For the noble metals the greatest values of reduction and adsorption capacity were obtained to gold (98 – 99 %) and palladium (87 – 89 %), and for the toxic metals it was obtained higher values of reductions to lead (66 – 83 %), cadmium (62 – 79 %), chromium (69 – 74 %) and copper (65 – 74 %).

Simulation of PEMFC with Pt/CNT nanostructured cathode catalyst

I. I. Mitrichev[1], V. A. Vasilenko, A. I. Scherbakov, E. M. Koltsova
D. Mendeleev University of Chemical Technology of Russia, MUCTR, Miusskaya sq. 9, 125047, Moscow, Russia.
tel. +7 495 4952126, e-mail: kolts@muctr.ru.

The technology of proton exchange membrane fuel cells (PEMFC) is one of the most prospective and the fastest developing tendency in power engineering.

Carbon black based platinum catalyst is the most common catalyst for low-temperature PEMFC. Its disadvantage is platinum dissolution in the strongly acidic environment of the polymer electrolyte and corrosion of a carbon support.

In Frumkin Institute of Physical chemistry and Electrochemistry RAS (Moscow, Russia) the effective liquid-phase polyol synthesis method of monoplatinum catalyst on carbon nanotubes (CNT) has been developed. Membrane-electrode assembly (MEA) with usage of catalytic systems synthesized on CNT has been performed.

In this paper we used CFD soft-ware FLUENT ANSYS to simulate phenomena occurring in MEA. Also the mathematical model was added by equations describing decrease of catalysts electrochemical surface area (ECSA) due to coalescence of platinum nanoparticles. With the help of mathematical model we have obtained and compared discharge curves data of two MEA's with usage of 20PtE-TEK (0.42 mg_{Pt}/cm^2) commercial catalyst and catalytic system synthesized on CNT (CNT+15%XC72). The calculated values are in good agreement with the experimental data.

In order to determine ECSA of the cathode catalyst cyclic voltammetry has been taken. Cathodes of two types were subjected to accelerated stress testing in order to determine the characteristics of degradation and electrochemical parameters changes depending on CNT amount in support. Cyclic voltammetry at 65°C, 100% RH, 1 atm and 10 mVs^{-1} scan rate was used to measure the ECSA after 1000, 2000 cycles. In order to determine the stability of the catalysts we have calculated prediction values of ECSA at greater than in the tests conducted number of cycles (5000).

Despite the fact that the initial ECSA of commercial catalyst is three times higher the corresponding value of the synthesized catalyst, after 2000 cycles of testing active catalyst surfaces are equal. Further comparison of the curves shows that at 5000 cycles active surface of commercial catalyst is reducing more than 95%, while reduction of synthesized catalyst ESCA is less than 50%.

Comparison of calculated voltage curves for two catalysts shows that before cycling characteristics of CNT-based catalyst are the same to commercial catalyst in high-voltage area, and in the area of high current density they are inferior to commercial one only by 8-10%. However, after 5000 cycles predicted by mathematical modeling voltage characteristics of the synthesized catalyst are significantly higher than for commercial one. The predicted reduce of discharge characteristics of the synthesized CNT catalyst after 5000 cycles at high current density area is about 10%.

Thus, the results of experimental studies and mathematical modeling has showed that at the tests longer than 2000 cycles nanostructured catalyst synthesized by polyol method is more stable than the commercial one. Synthesized CNT-based catalyst systems show higher stability and can be a good alternative to the commercial catalysts.

This study is supported by state contract with Russian Foundation for Basic Research (RFBR) № 14-29-04026.

Triclosan photolysis removal: a parametric variation study

P. Iovino[1], S. Chianese[2], V. Leone[1], M. Prisciandaro[3], D. Musmarra[2]
[1]Dipartimento di Scienze e Tecnologie Ambientali, Biologiche e Farmaceutiche, Seconda Università degli Studi di Napoli, Via Vivaldi 43, 81100 Caserta, Italy; tel.: +39 0823XXXX, fax: +39 0823XXX, e-mail: pasquale.iovino@unina2.it.
[2]Dipartimento di Ingegneria Civile, Design, Edilizia e Ambiente, Seconda Università degli Studi di Napoli, Real Casa dell'Annunziata, Via Roma 9, 81031 Aversa (CE), Italy; tel.: +39 0815010387, fax: +390815037370, e-mail: simeone.chianese@unina2.it.
[1]Dipartimento di Scienze e Tecnologie Ambientali, Biologiche e Farmaceutiche, Seconda Università degli Studi di Napoli, Via Vivaldi 43, 81100 Caserta, Italy; tel.: +39 0823XXXX, fax: +39 0823XXX, e-mail: vincenzo.leone@unina2.it.
[3]Dipartimento di Ingegneria Industriale e dell'Informazione e di Economia, Università dell'Aquila, viale Giovanni Gronchi 18, 67100 L'Aquila, Italy; tel.: +39 0862434241, fax: +39 0862434203, e-mail: marina.prisciandaro@univaq.it.
[2]Dipartimento di Ingegneria Civile, Design, Edilizia e Ambiente, Seconda Università degli Studi di Napoli, Real Casa dell'Annunziata, Via Roma 9, 81031 Aversa (CE), Italy; tel.: +39 0815010387, fax: +390815037370, e-mail: dino.musmarra@unina2.it.

Several types of compounds, such as pharmaceuticals and personal care products (PPCPs), illegal drugs, etc., are listed among the emerging and micro pollutants. Their increasing presence in the aquatic environment, due the ineffective degradation of traditional wastewater treatments, contributes to its pollution and an effect on water life may be supposed (Rivera-Utrilla et al., 2014).

The effective removal of refractory pollutants to conventional practices requires innovative and aggressive techniques. Photolysis, an Advanced Oxidation Process (AOP), favours the generation of awfully reactive OH radicals,that allow photochemical reactionsfor the degradation of persistentcontaminants (Petrie et al., 2015).

Triclosan(5-chloro-2-(2,4-dichlorophenoxy)phenol), an antimicrobial disinfectant agent, is becoming one of the most commonly detected personal care product in tap and superficial waters, due to its intensive and widespread use (Lapworthet al., 2012).

The photochemical removalof Triclosanwas investigated through a parametric variation study by using a batch reactor of 100 mL equipped with 254 nm fixed UV light at an irradiation intensity of 400 mJ m^{-2}.

The effect of the initial concentration of Triclosan, in the range 0.1-1 nM, and the volume of the exposed solution, in the range 10-30 mL, was investigatedas a function of the treatment time (0-60 min).

Moreover, the removal of Triclosan by heterogeneous catalytic photolysis by TiO_2 was studied. Investigations were carried out by varying TiO_2 concentration in the range 20-100 mg/L and keeping constant Triclosan concentration (0.1mM).

Finally, a reaction mechanism was suggested in order to describe the effects of the parameters investigated and Triclosan degradation by-products were identified and quantified.

Identification and quantification analysis were performed by using a reverse phase high performance liquid chromatography (RP-HPLC system, Shimadzu, Kyoto, Japan) implemented with a Shimadzu DAD detector (SPD-M20A), equipped with Shimadzu LC-20AD.

References

Lapworth, D.J., Baran, N., Stuart, M.E., Ward R.S., 2012. Emerging organic contaminants in groundwater: a review of sources, fate and occurrence. Environ. Pollut. 163, 287–303.

Petrie B., Barden R., Kasprzyk-Hordern B., 2015. A review on emerging contaminants in wastewaters and the environment: Current knowledge, understudied areas and recommendations for future monitoring. Water. Res. 72, 3–27.

Rivera-Utrilla, J., Sánchez-Polo, M., Ferro-García, M.Á., Prados-Joya, G., Ocampo-Pérez, R., 2014. Pharmaceuticals as emerging contaminants and their removal from water. A review.Chemosphere 93(7), 1268-1287.

Measurement of the induction time for nucleation of salt scales on process equipment

M. Prisciandaro[a], G. Mazziotti di Celso[b], D. Karatza[c], A. Lancia[d], D. Musmarra[c]
[a] Univ. of L'Aquila, Dept. DIIIE; viale Giovanni Gronchi 18, 67100 L'Aquila, Italy.
email: marina.prisciandaro@univaq.it
[b] Univ. of Teramo, Faculty of Bioscience, Via C. Lerici, 1, 64023 Mosciano S.A. (TE), Italy.
email: gmazziottidicelso@unite.it
[c] Second Univ. of Naples, Dept. DICDEA; via Roma 29, 81031 Aversa (CE), Italy).
email: karatza@irc.cnr.it; dino.musmarra@unina2.it
[d] Univ. Federico II of Naples, Dept. DICMPI; p.le Tecchio 80, 80125 Naples, Italy.
email: lancia@unina.it

Scale deposition on process equipment surfaces has several disadvantages: in particular, when scales crystallize on heat transfer surfaces, they offer a resistance to the heat flow and can accumulate in pipelines, orifices and other flow passages seriously impeding the process flow. Moreover, calcium carbonate scales, together with calcium sulfate scales, are the major cause of fouling in reverse osmosis membranes, resulting in a continuous decline in desalted water production thus reducing the overall efficiency and increasing operation and maintenance costs. Therefore, from an economic point of view, the formation of these mineral scales (Kirk-Othmer, 2011) is an obstacle to the recovery of potable water from sea or brackish waters as well as to the industrial utilization of many natural waters.

The aim of the paper is to measure the induction time for calcium carbonate precipitation by measuring the induction period for nucleation. Through a well-assessed laser light scattering technique previously devised (Prisciandaro et al., 2009), the induction time is measured and then used to estimate thermodynamic parameters such as the activation energy for nucleation. Experiments are carried with the temperature varying in the range 15 - 35°C. The knowledge of activation energy for nucleation is of major importance for guiding the choise of the most suitable additive for the inhibition of $CaCO_3$ scale formation.

References

Kirk-Othmer Encyclopedia of Chemical Technology, Lancia, A., D. Musmarra, Prisciandaro, M. "Calcium sulfate" 1-22. Fifth Edition 2011, John Wiley & Sons.

Prisciandaro, M., Olivieri, E., Lancia, A., Musmarra, D. Gypsum Scale Control by Nitrilotrimethylenephosphonic Acid. Ind. Eng. Chem. Res. 2009, 48, 10877–10883.

Compression and expansion at the right pinch temperature

Chao Fu*[a], Preben Maurstad Uv[b], Bjørn Nygreen[b], Truls Gundersen[c]
[a]SINTEF Energy Research, Kolbjoern Hejes vei 1A, NO-7491 Trondheim, Norway
[b]Department of Industrial Economics and Technology Management, Norwegian University of Science and Technology, Alfred Getz vei 3, NO-7491 Trondheim, Norway
[c]Department of Energy and Process Engineering, Norwegian University of Science and Technology, Kolbjoern Hejes vei 1A, NO-7491 Trondheim, Norway
fuchao83@hotmail.com

Heat Integration is extended to the topic of Heat and Work Integration when pressure changing equipment such as compressors and expanders are included in heat exchanger network design. The latter topic is much more complex since heat and work have different energy qualities (exergy). Systematic graphical design methodologies have recently been developed for heat and work integration. The methodologies are based on a set of theorems that have been proven based on thermodynamic and mathematical analyses. The theorems show that minimum exergy consumption can be achieved in many cases when compression/expansion starts at the pinch temperature (Pinch Compression/Expansion). The Grand Composite Curve has been used to determine the maximum portions of streams using Pinch Compression/Expansion. However, the pinch temperatures for hot streams (hot pinch) and cold streams (cold pinch) are different. They were not well distinguished in previous studies. Based on a recent mathematical optimization study on the topic of heat and work integration, it is concluded that the pinch identity (hot or cold) should be the same as the identity of the stream at the inlet of compression/expansion. This paper introduces this new insight with an illustrative example. The insight is then verified by thermodynamic and mathematical analyses on various cases. A simple application of the insight to the self-heat recuperation scheme is investigated. An alternative scheme has been proposed. The total work consumption is reduced by 17.7% in a thermal process with butane.

Degradation of tetramethylammonium hydroxide (TMAH) by using advanced oxidation processes

M. Prisciandaro[a], D. Karatza[b], F. Tortora[a], V. Innocenzi[a], A. Lancia[c], D. Musmarra[b]
[a] Univ. of L'Aquila, Dept. DIIIE; viale Giovanni Gronchi 18, 67100 L'Aquila, Italy.
email: marina.prisciandaro@univaq.it; francesco.tortora@graduate.univaq.it; valentina.innocenzi1@ univaq.it
[b] Second Univ. of Naples, Dept. DICDEA; via Roma 29, 81031 Aversa (CE), Italy).
email: karatza@irc.cnr.it; dino.musmarra@unina2.it
[c] Univ. Federico II of Naples, Dept. DICMPI; p.le Tecchio 80, 80125 Naples, Italy.
email: lancia@unina.it

The semiconductor and electronic component manufacturing plants typically generate high-strength wastewaters containing high concentrations of tetramethylammonium hydroxide (TMAH, $(CH_3)_4NOH$) that is recognized as a poisonous, corrosive, slow to biodegrade and eutrophic to aquatic environments. Disposal of TMAH wastewater from an industrial plant is a difficult and costly issue. The recently observed growth of the semiconductor industry and the stringent regulations imposed on industrial and municipal discharges require innovative solutions to treat these wastewaters.

A literature analysis showed that a series of papers describing the treatment of industrial wastewaters contaminated by TMAH has been published. The proposed treatments include separation by using chemical/physical (*e.g.* advanced oxidation processes and adsorption) and biological processes(*e.g.* anaerobic digestion). The treatments so far described are technically complex and costly, because they require a series of chemical and biological processes; therefore the problem of the treatment of micro-electronic wastewaters containing TMAH is still open.

On the other hand, requests in the context of degradation of chemically stable compounds has stimulated the birth and development of a new class of oxidative techniques based upon cavitation in water, which belongs to the class of Advanced Oxidation Processes (AOP). During last two decades, the cavitation has been seen as a source of production of extremely reactive species (·OH radicals, typical of the AOP) thanks to the violent collapses of water vapor bubbles. This aspect makes the cavitation a promising techniques in the field of wastewater and liquid waste treatment, for removal of both biological and chemical contamination. Many review works have been listing several experimental results in the field of wastewater and claiming the advantages of cavitation techniques in this field: experimental data and theoretical analysis are available in literature for the degradation of different kinds of pollutants. Additionally the cavitational techniques (particularly HC) are recognized as a low-cost process enhancer for different kinds of application, thanks to the capacity of focusing energy in a local form.

The idea at the basis of this paper is to test several oxidation processes, among which a sonochemical reactor, already used to degrade pollutants such as hormones and pharmaceutical residues (*see* Capocelli et al., 2012-2014 and Musmarra et al., 2016), to treat a simulated industrial stream containing TMAH.

References

Capocelli, M., Joyce, E., Lancia, A., Mason, T.J., Musmarra, D., Prisciandaro, M., 2012. Sonochemical degradation of estradiols: Incidence of ultrasonic frequency. Chemical Engineering Journal 210, 9-17.

Capocelli, M., Prisciandaro, M., Lancia, A., Musmarra, D., Hydrodynamic Cavitation of p-nitrophenol: a Theoretical and Experimental Insight Chem Eng J (2014) 254: 1-8.

D. Musmarra, M. Prisciandaro, M. Capocelli, D. Karatza, P. Iovino, A. Lancia, Degradation of Ibuprofen by Hydrodynamic Cavitation: reaction pathways and effect of operational parameters, Ultrasonics Sonochemistry (2015) http://dx.doi.org/10.1016/j.ultsonch.2015.09.002

Design of a knowledge-enabled supervisory framework for the detection of abnormal conditions at process pilot plants

C. Ziogou[1], S. Kirnidis[2], D. Ioannidis[2], S. Papadopoulou[3], D. Tzovaras[2], S.Voutetakis[1]
[1]Chemical Process and Energy Resources Institute (CPERI), Centre for Research and Technology Hellas (CERTH), PO Box 60361, 57001, Thessaloniki, Greece, {cziogou, paris}@cperi.certh.gr
[2]Information Technologies Institute (ITI), Centre for Research and Technology Hellas (CERTH), PO Box 60361, 57001, Thessaloniki, Greece, {krinidis, djoannid, Dimitrios.Tzovaras}@iti.gr
[3] Department of Automation Engineering, Alexander Technological Educational Institute of Thessaloniki, PO Box 141, 54700, Greece, shmira@autom.teithe.gr

The objective of this work is to develop a framework able to communicate online with process systems utilizing the existing infrastructure and provide knowledgeable actions to the operators. The integration of data and information derived from complex and highly dynamic processes is of major importance to improve both timely decision making and the improvement of process operation and quality of final products. In general process systems produce each day a large amount of real-time data that provide significant information which currently is analyzed using conventional tools whereas the decision making is partially performed at a later stage using respective tools by the plant operators and the process managers. Therefore, the motivation of this work is mainly the development of a knowledge-based ontology-assisted platform that will enable online decision support procedures using real-time data combined with previously derived information from the process operation. Besides the collection, analysis and deployment of informative actions, the motivation for the development of an advanced knowledge-based platform is the need for smart data visualization combined with diagnosis and alarm notifications capabilities.

This work includes the systematic construction of an ontology that combines components of existing standards such as ISA-95 implemented by a B2MML (Business To Manufacturing Markup Language) XML schema. The functionalities of the developed knowledge-based platform will be demonstrated at an industrial-grade control system of a continuous process. More specifically, a pilot plant that operates at CERTH/CPERI will be used as a demonstration case for the developed framework. The pilot plant is fully automated and is controlled by a supervisory control and data acquisition (SCADA) system which is responsible for the acquisition of the data from the Input/Output (I/O) field network. A set of Human Machine Interfaces (HMIs) represents the status of the each process component.

A Common Information Data Exchange Model (CIDEM) is developed that represent both static (e.g. data related to the reactor operating ranges) and dynamic data (e.g. the temperature of the reactor) of the pilot plant. Events and dynamic data from the process system are stored at a hybrid repository combined with information from the Process and Instrumentation Diagram (P&ID) of the plant. The CIDEM enables the combination of dynamic and static data and provides notification and suggested actions based on a set of predefined rules which are modeled by a Finite State Machine (FSM) structure. An online filtering tool is developed that uses online data from the SCADA system to invoke the specific rule based on the status of the FSM. Online data from the process and information from the data archiving system are combined by the filtering tool. The bidirectional communication is performed using an OPC-enabled middleware whereas the visualization of the resulting information is integrated at the existing interfaces. The developed integrated intelligent notification platform is tested at the process unit at CERTH/CPERI plant floor.

The knowledge-enabled platform provides information related to early warnings of failing experiments or notification for preventive maintenance. An indicative case is the use of the evolution of the temperature profile, or the consumption of power by the heating resistance by observing the controller output besides the input. In the case of a significant deviation it is an indication related to the spatial distribution of the conversion of the reaction in conjunction with the final concentration of the species. The main results of this work demonstrate how a knowledge-based ontology-assisted system can improve the information which is provided to the process operator in an integrated not disruptive way combining both online and historical data from the I/O field and the previously applied actions of the operator.

Strategy for quantification of vented gas emissions from oil storage tanks

M. Samcha[1], A. Boukadi[2]
[1] ETAP, petroleum services department, 4 rue des entrepreneurs Charguia II-2035 Tunisie, tel: +21670838440, fax: +21670838988, e-mail: mariem.samcha@etap.com.tn;
[2] ETAP, department of petroleum services, 4 rue des entrepreneurs Charguia II-2035 Tunisie, tel: +21670838440, fax: +21670838988, e-mail: amel.boukadi@etap.com.tn;

During production, storage and transportation of crude oil, a portion of gas in solution with oil is vented due to temperature or pressure variations.

These gases are released to the atmosphere without being treated or burned raising hazardous air pollution and greenhouse effect.

Evidently, the gas emission represents a loss of considerable energy. But the harmful consequences to the environment are supposed to be of greater importance.

We have quantify three types of emission generated by storage tank: breathing losses, working losses and flash emissions that are estimated by means of several methods.

In this context, we present a review concerning VOC emissions/vented gas in crude oil production and different methods for their quantification.

Various methods have been used to calculate losses flash: Correlation Equation Approach, Emission Factor, Graphic Approch and Simulator Hysys. working and breathing losses emissions were estimated by E&P Tank.

Several results are obtained in this study, such as:

- The majority of VOC emission is concentrated in the flashing emissions.
- The comparison of results obtained proves that Chart is the most overestimated method however VBE and standing are underestimated methods.

In conclusion, on Tunisian fields, it would be interesting to find the adequate way to recover them. This can solve energetic and environmental issue.

Transesterification of waste cooking oil into biodiesel using Aspen HYSYS

S. Karacan[1] , F. Karacan[2], M.T. Çağatay[3]
[1]Ankara University, Engineering Faculty, Department of Chemical Engineering, Tandoğan 06100, Ankara, TURKEY. Tel:+90(312) 2033523, Fax: +90 (312) 212 15 46 (E-mail: karacan@eng.ankara.edu.tr)
[2]Turkish Atomic Energy Authority, Sarayköy Nuclear Research and Training Center, Istanbul Road 30 km., 06983 Saray, Ankara, TURKEY
[3] National Defence Ministry, Ankara District Quality Management Presidency, Etlik 06010, Ankara, Turkey.

Increasing uncertainty about global energy production and supply, environmental concerns due to the use of fossil fuels, and the high price of petroleum products are the major reasons to search for alternatives to petrodiesel. Biodiesel fuel has become more attractive because of its environmental benefits, due to the fact that plants and vegetable oils and animal fats are renewable biomass sources. Biodiesel represents a largely closed carbon dioxide cycle (approximately 78%), as it is derived from renewable biomass sources. Compared to petroleum diesel, biodiesel has lower emission of pollutants, it is biodegradable and enhances the engine lubricity and contributes to sustainability. Biodiesel has a higher cetane number than diesel fuel, no aromatics, no sulfur, and contains 10–11% oxygen by weight. Huge quantities of waste cooking oils and animal fats are available throughout the world, especially in the developed countries. Management of such oils and fats pose a significant challenge because of their disposal problems and possible contamination of the water and land resources. Even though some of this waste cooking oil is used for soap production, a major part of it is discharged into the environment. People around the world use edible oils for cooking, after which the oil is discarded. The amount of heat and water increases the hydrolysis of triglycerides and the percent of free fatty acid (FFA) in the oil. The water and FFA content have a negative influence on the transesterification reaction. The WCO price is two to three times cheaper than vegetable oils, and it also reduces the cost of waste product removal and treatment. Meanwhile, it can significantly decrease the amount of farmland, which is necessary for biodiesel producing corps. The WCO has been categorized by the FFA content to two groups: (a) yellow grease (<15%) and (b) brown grease (>15%). The prices of these types of WCO.

In this study, The process steady-state simulation was performed using ASPEN/HYSYS 8.0 computer software. Biodiesel production from waste cooking oil in the reactive distillation column packed with heteregenous basic CaO catalyst was implemented. The effect of parameters such as inlet temperature, total feed flow rate, molar ratio of methanol to WCO, reflux ratio and reboiler duty on biodiesel mole fraction and conversion were investigated

Evaluation of metal affinity of Ag^{+}, Cd^{2+}, Cr^{3+}, Cu^{2+}, Ni^{2+}, Zn^{2+} and Pb^{2+} in residue of double alginate extraction from *Sargassum filipendula* seaweed

Saulo L. Cardoso, Bárbara P. Moino, Camila S. D. Costa, Meuris G. C. da Silva, Melissa G. A. Vieira *
State University of Campinas, School of Chemical Engineering, Department of Process and Products Design, Campinas, São Paulo, Brazil
melissagav@feq.unicamp.br

The continuous population growth and technological development are contributing factors to the increased contamination of effluents with toxic metals. The removal of such metals from effluents has become an important subject for scientific research. Among the most used methods, adsorption has shown good potential to treat such effluents. In this work we investigated the metals affinity with the residue of alginate extraction from the seaweed *Sargassum filipendula*. The alginate residue has shown good results in the removal of metal contaminated industrial effluent and is an interesting subject of research. The experimental tests were performed using different metal solutions with concentration of $1 mmol.L^{-1}$,at 180 rpm in room temperature 25°C for a period of 24 hours under controlled pH 3.5. The alginate residue can be used as a bioadsorbent in the treatment of metals contaminated effluents with good results. The first and second extractions residue has shown similar ability of ion removal.

Improvement of emission characteristics and thermal efficiency in diesel engine by fueling DGM-diesel blend

M. Saei Moghaddam
Quchan University of Advanced Technology, Dep. Chemical Engineering, Quchan, P.O.Box 94771-67335 I.R.Iran; tel. +985147344002, e-mail: mojtabasaei@qiet.ac.ir;

This research work investigates physico-chemical properties and the exhaust emission of the blended fuel and standard diesel. Diethylene glycol dimethyl ether (DGM) is hopeful alternative fuel as well as fuel additive to reduce smoke emission in heavy-duty diesel engine due to high oxygen content of it. The addition of DGM to the standard diesel fuel caused significant changes in cetane index, viscosity and flash point temperature, but density was not change significantly. Experimental study on exhaust emission was carried out on ECE R-96 8-modes cycle. The result show that in all modes, the average smoke reduction rates of DGM-Diesel fuel blend was 20.5% of that of standard diesel respectively. NO_x emission decreased in modes 5,6 and 7 (maximum torque speed range) more than in modes 1,2,3 and 4 (rated power speed range). By use of DGM, BSHC increased but BSCO decreased. In general low smoke and CO emission was measured with the DGM-Diesel fuel blend.
Keywords: Diesel fuel additive, DGM, Emission improvement.

Physicochemical properties

Physical property tests which are consist of density, viscosity; cetane index and flash point temperature were determined. As shown in Table 4 the density of both fuels which is measured by D 1298 ASTM method were not changed significantly and by addition of DGM to the standard diesel fuel viscosity and flash point temperature were decreased, but cetane index which is measured according to D 976 ASTM method increase for blended fuel.

Table 1
Physical property tests for both fuels

	Standard diesel	Diesel+DGM
Density (Kg/m3)	839	836
Viscosity (mpa.s)	3.1	2.8
Cetane Index	48.2	52
Flash Point Temperature (°F)	158	149

Emissions characteristics

Fig. 1 shows NO_x emission for standard diesel and DGM-Diesel fuel according to ECE R-96 8-modes cycles. The NO_x emission increased with the increase of the load. In the rated power speed condition (modes 1, 2 and 3) the NO_x formation was lower than maximum torque speed. when DGM added to the standard diesel NO_x emission decreased in maximum torque speed range (modes 5,6 and 7) more than in the rated power speed range (modes 1,2,3 and 4), the NO_x reduction rate were 2%, 4.7% and 6.1% of that of standard diesel, in the rated power speed range (modes 1,2,3 and 4) respectively.

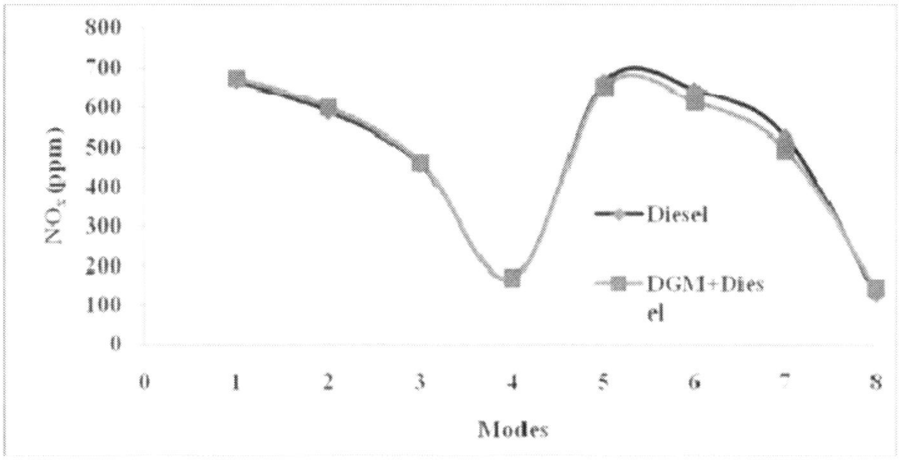

Fig.1. NO_x concentrations for two kinds of fuels

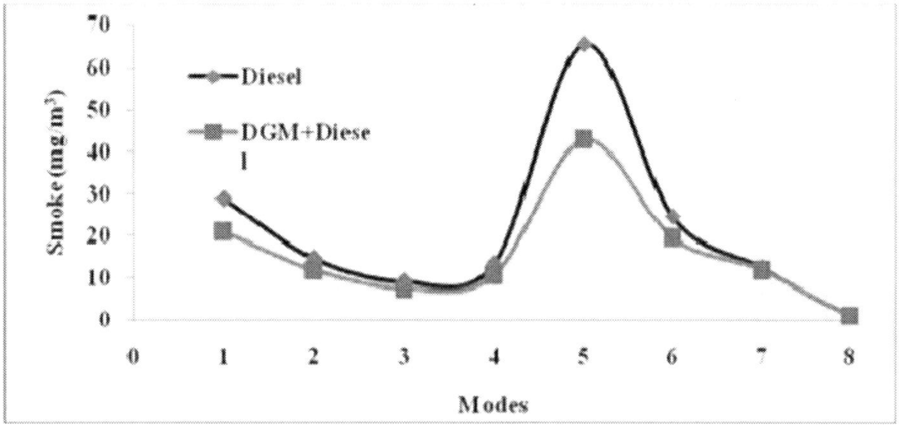

Fig.2. Smoke concentrations for two kinds of fuels

Fig. 2 shows smoke concentration for two kinds of fuels according to ECE R-96 8-modes cycles. In general, the addition of DGM in diesel decreases the smoke concentration; the decrease of smoke emission using DGM can be explained by the enrichment of oxygen content of it. At lower loads smoke concentration of DGM -Diesel blend fuel slightly decrease in comparison with standard diesel fuel. At higher loads, smoke concentration decreases significantly. The smoke concentration decreased in maximum torque speed range (modes 5,6 and 7) more than in the rated power speed range (modes 1,2,3 and 4). In all modes, the average smoke reduction rate of DGM-Diesel fuel blend was 20.5% of that of standard diesel respectively.

New approaches to the synthesis of the catalyst purification of gaseous emissions

M.A. Sadenova [1] *, S.A. Abdulina [2], S. Shagalakova [1]
[1] S. Amanzholov East Kazakhstan State University, 55 Kazakhstan str., 070000 Ust-Kamenogorsk, Kazakhstan
[2] D. Serikbayev East Kazakhstan state technical university, 19, Serikbayev str., 070000, Ust-Kamenogorsk, Kazakhstan
tel. +77232255441, e-mail: Sadenova@mail.ru

The intensive development of advanced technologies in the industry is accompanied by continuous increase in emissions. An mandatory requirement for industrial manufacturing is the cleaning of waste gases from toxic substances to the normative level. A significant contribution to the pollution of the atmosphere is brought of nitrogen oxides (NO_x) and carbon monoxide (CO), appearing in the combustion of fuel. Among the existing methods of the deep cleaning of gas emissions most perspective are catalytic, which are increasingly used because they are quite effective. The basic factors limiting the widespread use of catalytic purification techniques is the high cost, short life, inability to regenerate contacts (in most cases), and technical complications related with the loading / unloading of the contact mass (depending on the construction of reactor and form of catalysts).

Zeolite is a unique natural mineral which has the ability to absorb and permanently retain in its structure particles of different substances. Therefore, in such technological processes where the use of synthetic zeolites disadvantageous, great importance is the use of natural materials. Despite the fact that a high catalytic activity and selectivity of zeolites in many chemical reactions was the cause of intense study on various properties of these catalysts, the nature of the catalytic action of crystalline aluminosilicates yet is studied not enough.

For development of effective, stable and inexpensive catalysts of gas purification will be used the natural sorbents of East Kazakhstan.

The primary purpose of research is reduced to the search and production suitable for long-term exploitation and relatively inexpensive catalysts for cleaning of gases with high activity and selectivity that have sufficient mechanical strength, thermal stability, well-developed geometric surface, stability to influence of aggressive environment and low flow resistance.

The innovative technology explains the choice of natural materials of Kazakh deposits for the synthesis of carriers of catalysts for cleaning gases from toxic components, such as oxides of nitrogen and carbon.

As a result of the use of zeolites in ecological catalysis it proved to be very effective and highly promising. When introduced in the composition or applied to the surface of the catalytically active component was prepared a series of catalysts with high activity and polyfunctional action.

Numerical study of main and retention hole effect on thermal efficiency and pollution emitting in a premix multi-hole flat flame burner

Mohammad Hossein Saberi Moghaddam[1], Mojtaba Saei Moghaddam[2]
[1]Faculty of Mechanical Engineering, Tarbiat Modares University, Tehran, 1456817458, Iran tel.: +982166501452; Email: h.saberi87@gmail.com.
[2] Quchan University of Advanced Technology, Dep. Chemical Engineering, Quchan, P.O.Box 94771-67335 I.R.Iran; tel. +985147344002, e-mail: mojtabasaei@qiet.ac.ir;

Numerical simulation of lean premix methane/air combustion in flat multi-hole burner which is used in condensing boiler, is investigated. Three dimensions domain and two step mechanism for methane combustion is used in this simulation. Presence of distribution mesh leads to uniform flow on the surface of burner and reduces NO_x pollutant emission. Increasing of main hole diameter increases the temperature and thermal efficiency. Although increase of main hole diameter increases thermal efficiency, emits more NO_x pollutant. Radiation is the main heat transfer method in this burner. By Adding retention hole, flame to gets close to surface of burner. Increasing of retention hole diameter, increases thermal efficiency but not sensible.

Key words: Condensing boiler, Multi-hole burner, Numerical simulation, Main and retention hole, Thermal efficiency, NO_x pollutant emission.

Introduction and numerical method

The new generation of boilers which is more efficient than conventional boilers and has less pollutant emission, is designed which is named condensing boiler. Multi-hole flat flame burner is used in condensing boiler which is shown in figure 1. In the present work, the effect of changing the diameter of the main and the retention hole size on temperature, combustion efficiency and NO_x emission are discussed by numerical simulation. Navier-Stokes equations with continuity, energy and species conservation equations and also the equation of state are numerically solved. FLUENT 15 is applied to simulate combustion.

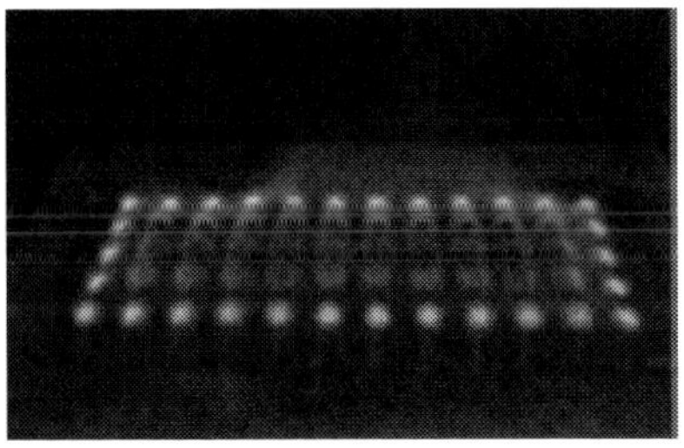

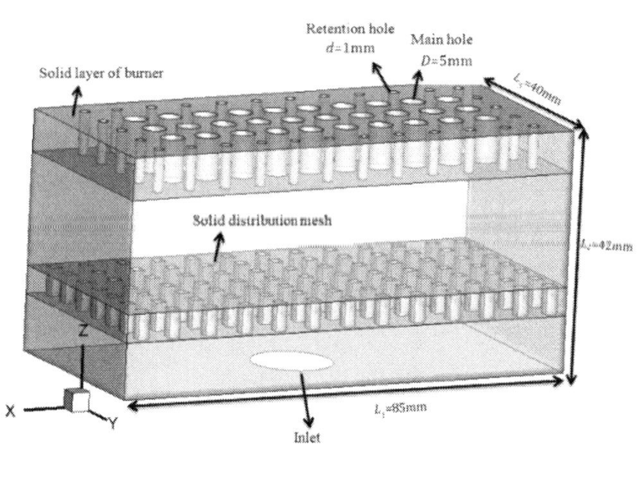

Figure1. Experimental [3] and numerical of flat burner.

Results and discussion

Having a uniform flow is the most characteristic of the multi-hole flat flame burner, which reduces the maximum of temperature. Use of distribution mesh reduces flow fluctuation and reduces 53% NO_x pollutant emission. The main holes have the most important role of production heat in this burner. By increasing main hole diameter, flow mass and releasing combustion energy are increased too. Temperature and NO_x pollutant emitting is increased consequently. The radiation heat transfer is the dominant heat transfer in present work, so that when a main hole diameter equal 5mm, 93% from a total heat transfer is a radiation heat transfer. Presence of retention hole makes the flame to get closer to the surface of burner then flame stability increases. By increasing retention hole size, mass flow of each holes is increased. Then temperature and thermal efficiency is increased, too. It should be noted that increase of temperature and efficiency isn't tangible. According to the required temperature and limit of pollutant emission, main and retention hole diameter equal to 5mm and 1.25mm is chose, subsequently.

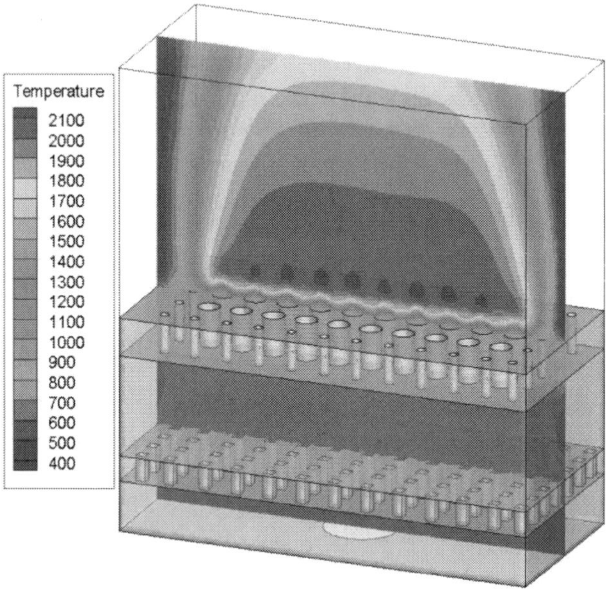

Figure 2. Combustion and flame in multi hole burner.

References

[1] Weber, C., B. Gebhardt, and U. Fahl. "Market transformation for energy efficient technologies—success factors and empirical evidence for gas condensing boilers." Energy 27, no. 3 (2002): 287-315.

[2] Yang, S. W., S. H. Lee, and S. S. Hwang. "Surface flame patterns and stability characteristics in a perforated cordierite burner for fuel reformers." Proceedings of the Institution of Mechanical Engineers, Part A: Journal of Power and Energy 222, no. 1 (2008): 25-30.

[3] Raghavan, Vasudevan. "Experimental Study of Pre-mixed Flames on a Multi-Hole Matrix Burner." International Journal of Integrated Engineering 4, no. 1 (2012).

[4] Lee, Seungro, Sung-Min Kum, and Chang-Eon Lee. "Performances of a heat exchanger and pilot boiler for the development of a condensing gas boiler." Energy 36, no. 7 (2011): 3945-3951.

Numerical simulation of slug flow mass transfer in the pipe with granular layer

D. P. Khramtsov, A. V. Vyazmin, D. A. Nekrasov
Moscow State Univ. Mech. Engin., ul. Staraya Basmannaya 21/4, 105066 Moscow, Russia;
tel. +7 495 2671051, e-mail: info@ttie.ru.

One of the most common flow regimes in the pipes for energy systems, oil and gas production, as well as in microchannels is the slug flow regime of two-phase flow [1]. In such flow a large gas bubble is rising in a pipe or moving under the pressure in the capillary channels, separated by a fluid layer. Mixing of gas and liquid phases, combined with diffusion and low thermal resistance through the thin film between bubble and wall, leads to a substantial intensification of the processes of heat and mass transfer. A satisfactory theory of such flows has not yet been developed. This is due, primarily, to the complexity of such flows. Problems with the theoretical description arise even for the case of motion of a single slug in an inclined pipe [2]. The aim of this work is to develop numerical model of a slug flow with interphase interactions in the pipe filled with granular layer for different inclination angles of the pipe.

Based on the smoothed particle hydrodynamics method a model of the motion of the slug flow with mass transfer phenomena in the pipe was developed. In smoothed particle hydrodynamics method influence of each particle on the properties of nearby particles are computed on the basis of its density and the distance from the particle for which the parameters are calculated [3].

Numerical simulation was performed for slug flow in the pipe filled with spherical particles of different diameters. In order to compare simulation results with experimental data, simulation of slug flow with spheres diameter 10mm (Fig. 1) and 5 mm (Fig. 2) respectively were performed.

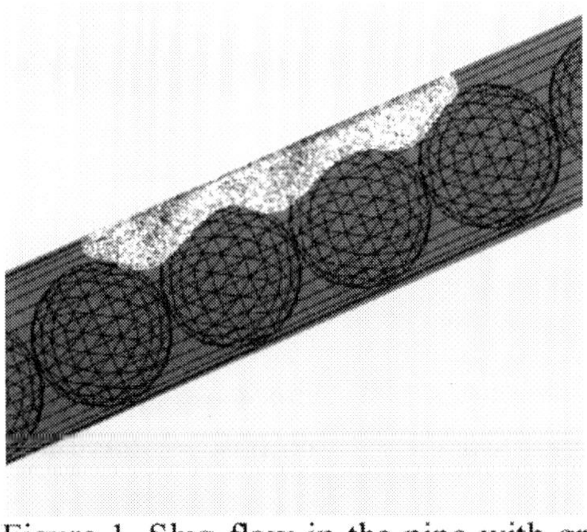

Figure 1. Slug flow in the pipe with granular layer: sphere diameter – 10 mm.

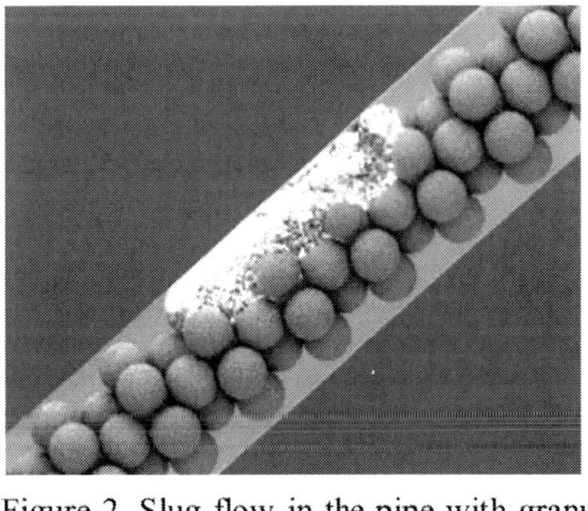

Figure 2. Slug flow in the pipe with granular layer: sphere diameter – 5 mm

The calculations were performed for pipe inclination angles from 10 to 90° with step 5°. To measure mass transfer intensification at different inclination angles an effective dimensionless mass transfer coefficient was used. This coefficient is calculated as a ratio of the volume of the slug at the end of the pipe at a given angle and vertical pipe. The dependence in the experiment and in the numerical modeling showed a direct relationship between the mass transfer intensification and velocity of the slug. The maximum value of the effective mass transfer coefficient is observed for maximum bubble velocity (angle 60°), and the minimum at the lowest inclination angle respectively (Fig. 3).

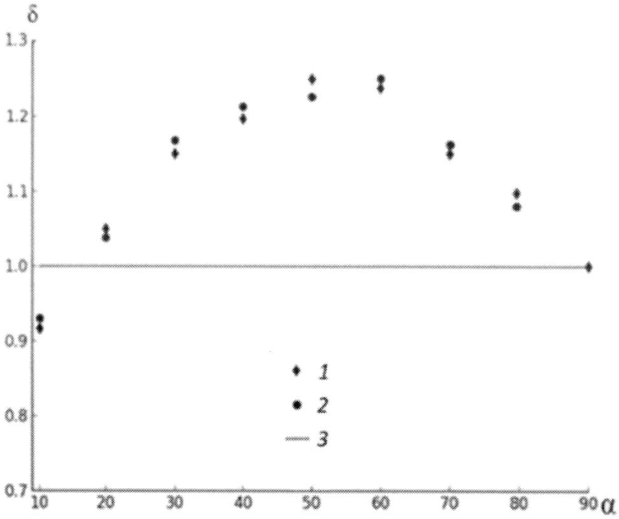

Figure 3. Effective dimensionless mass transfer coefficient depending on the angle of the pipe: 1 – experimental data; 2 – simulation data; 3 – vertical pipe

A numerical study of the slug flow in inclined pipe with mass transfer phenomena was performed. The smoothed particle hydrodynamics method showed its effectiveness in solving hydrodynamics problems that involves interphase interactions in the channels with complex geometry. To assess the intensification of interphase mass transfer an effective mass transfer coefficient was introduced. Calculations showed that the intensity of mass transfer decreases with increasing of slug flow velocity. This effect can be related to the fact that the main contribution to the intensity of mass transfer is due to the time of interphase interaction. The processes of mass transfer are observed not only in macro systems, but also in the microscale systems. Developed model can also be adapted to solve problems of mass transfer processes in porous media and gels.

This work was supported by the Russian Science Foundation (project no. 15-19-00177).

Reference

1. Hessel V., Löwe H., Müller A., Kolb G., 2005, Chemical Micro Process Engineering. Processing and Plants. Wiley VCH Verlag: Weinheim.

2. Pokusaev B. G., Zaitsev A., Zaitsev V. A., 1999, Transport processes in the slug flow regime three-phase media, Teoretical foundations of chemical engineering. vol. 38. p. 595-602 (in Russian).

3. Amada T., Imura M., Yasumuro Y., Manabe Y. Chihara K., 1999, Particle-based fluid simulation on GPU, Proceedings of ACM Workshop on General-purpose Computing on Graphics Processors.

Photocatalysts coupled with luminescent materials to enhance the photocatalytic degradation of organic pollutants in wastewater

V. Vaiano, O. Sacco, D. Sannino, P. Ciambelli
Department of Industrial Engineering, University of Salerno, Via Giovanni Paolo II, 132, 84084 Fisciano (SA), Italy;
tel. +39 089964006, e-mail: vvaiano@unisa.it;

In the last years, heterogeneous photocatalysis has drawn lots of attention as a green method for the removal of pollutants in wastewater. Among different semiconductors, TiO_2 is still the most promising photocatalyst due to being environmentally friendly, since its chemical stability, high oxidative power and low cost. However, it can only be activated by UV light because of the large band gap of about 3.2 eV. It is well known that the UV light just accounts for 5% of solar light, while the residual 45% and 50% of solar light are made up of visible light and near-infrared (NIR) light, respectively. In order to extend the absorption ability of TiO_2 to the visible-light region, many strategies, such as doping with heteroatoms have been developed. Moreover, one of the main limitations for large scale application of photocatalysis is the photon transfer in the photocatalytic reactors.

To improve this feature it is possible to couple the photocatalyst with light emitting phosphorescent particles, substances that exhibit the phenomenon of luminescence.

Phosphors are materials having optically active centres that can absorb an incident radiation, which is re-emitted at shifted specific wavelengths. When phosphors are present together with a suitable photocatalyst, they can catch the radiation coming from external light sources and emit their specific radiation in the core of photoreactor, transferring to the photocatalyst, in its close proximity, additional photons, so improving the photons transfer also in the darker zones of a photoreactor.

Among different inorganic commercial phosphors particles, ZnS and aluminates based materials emit blue visible light when activated with UV light. Thus, to fruitfully capture the light emitted by phosphors, it is necessary to modify TiO_2 photocatalyst to make it active in visible region. In this perspective, TiO_2 has been doped with nitrogen (N-doped TiO_2) for exploiting also of the re-emitted radiation by phosphors.

In this work physical mixtures of N-doped TiO_2 and different phosphors, ZnS-based phosphors (ZSP); $BaMgAl_{10}O_{17}$:Eu (BaMgP); $Sr_{2.90}$ $Eu_{0.03}$ $Dy_{0.07}$ Al_4Si_{11}(SrSiP), having all blue emission, has been tested in the degradation of methylene blue.

N-doped TiO_2 was prepared by sol gel method starting from titanium(IV) isopropoxide, as source of titanium dioxide and ammonia aqueous solutions (30% wt.) as dopant precursor and for the hydrolysis, finally calcined at 450°C until to obtain N/Ti=18.6.

The photocatalysts and the phosphors were characterized by XRD, UV Vis DRS, micro-Raman, specific surface area.

The photocatalytic effect of coupling phosphors and N-doped TiO_2 was studied using methylene blue (MB) as model pollutant showing in all cases, and a beneficial effect on photocatalytic reaction rate if compared with the results obtained in absence of phosphors. In the case of physical mixtures, the optimized relative ratio between the amount of N-doped TiO_2 and phosphors in the dye removal depends on the chemical structure of the phosphors and their harvesting-emitting capacity. The highest MB degradation was obtained coupling N-doped TiO_2 with both BaMgP and SrSiP, being 50% after 15 min of UV irradiation.

Life cycle analysis of microalgae extraction techniques

S. Papadaki, K. Kyriakopoulou, M. Krokida
School of Chemical Engineering, National Technical University of Athens
Zografou Campus, Athens, GR-15700, Greece
Tel.:+30 2107723288, E-mail: spcheng@central.ntua.gr

In this study, multifunctional extracts from Haematococcus pluvialis microalga have been suggested as food additives. H. pluvialis is a green unicellular Chlorophyta alga able to accumulate high levels of astaxanthin (up to 2% of its dry weight). Astaxanthin-rich Haematococcus has already been marketed as a dietary supplement for human consumption. The health benefit of this product is mainly due to its strong antioxidant activity, which is 100 times more than a-tocopherol. Moreover, astaxanthin has an antioxidant action up to 500 times that of vitamin E and is the most stable antioxidant and never turns into a pro-oxidant. In addition, pro-vitamin A, a more effective antioxidant that other carotenoids, decelerate age-related macular degeneration, immunomodulatory effects, etc.. The range of shades from astaxanthin includes red as well as orange. Therefore, it can replace synthetic ones such as Tartrazine, Sunset Yellow etc. Moreover, the significant antioxidant activity that astaxanthin presents can sufficiently compete and replace synthetic antioxidants such as BHA, BHT.

For effective utilization of astaxanthin, the nutrient should be extracted from the microalga. In this study, a comparative analysis between conventional solvent extraction and innovative green extraction methods, using microwaves and ultrasounds, has been conducted. The isolation of the desired natural compounds as well as the environmental impact of each method has been the criteria for the different processes evaluation. A comparative life cycle analysis was carried out, using proper databases and software, in order to evaluate the selected extraction processes' sustainability.

The comparative analysis between conventional solvent extraction and innovative green extraction methods of bioactive compounds from microalgae constitutes the main scientific innovation aspect of the proposed work.

Some tentative results and conclusions are presented below:

- Ultrasound is the preferred method for carotenoids' extraction:
  - High yield
  - Low cost
  - Short time
  - Low environmental impact
- Different solvent systems, ie DMF, achieve high yielding but show severe eco-toxicity.
- Laboratory-scale application with potential scale up practice for nutraceutical, cosmetic and food industries

Optimization of industrial pretreatment of *Spirulina platensis* cyanobacterium

S. Papadaki [1,2], K. Kyriakopoulou [1], M. Stramarkou [1], I. Tzovenis [2], M. Krokida [1]
[1]School of Chemical Engineering, National Technical University of Athens
Zografou Campus, Athens, GR-15700, Greece
[2] MicroPhykos, Ag. Antoniou 24B, Halandri, Athens 15235, Greece
Tel.:+30 2107723288, E-mail: spcheng@central.ntua.gr

Spirulina platensis is a cyanobacterium with known therapeutic properties to diabetes, arthritis, anaemia, cardiovascular diseases and cancer due to its content in antioxidant compounds (phycocyanin) and nutritious compounds (short chain PUFAs and chlorophylls). Phycocyanin is used as a functional ingredient into various food products to enhance their nutritional qualities acting as food colorant, antioxidant and emulsifier. Therefore, it can sufficiently replace or reduce the use of synthetic ones such as Brilliant blue FCF, BHA, BHT etc. Phycocyanin-rich Spirulina has already been marketed as a dietary supplement for human consumption. [1,2]

This study is a laboratory-scale application with potential scale up practice in drying process, a treatment stage that took place directly after the harvesting of the cultivation and shows great effect on the nutritional value of the final product. [3,4] Moreover, the drying stage of microalgal biomass is the most energy consuming stage in the whole value chain of microalgae production and processing, that affects directly the total environmental performance including energy and carbon footprint. The dried methods that studied were: Accelerated Solar Drying and Vacuum Drying.

The dried Spirulina biomass was characterized evaluating spectrophotometrically the pigments content (total carotenoids, b-carotene, chlorophylls, asxtaxanthin, phycocyanin)[5–7]. Moreover, the antioxidant activity of the dried biomass was evaluated through DPPH assay. Finally, a comparative life cycle analysis was carried out, using proper databases and software, in order to evaluate the selected drying processes' sustainability.

References

1. Shimamatsu, H., Hydrobiologia 512,(2004) 39–44 .
2. Nal, M. K. Ü., *J. Food Sci,* .68 (2003) 1144-1148.
3. Tiburcio, P. C., Galvez, F. C. F., Cruz, L. J. & Gavino, V. C., J. Appl. Phycol. 19 (2007) 719–726.
4. Prasetyaningrum, A. & Djaeni, M., Inter. J. of Sci. and Eng., 3 (2012) 1–3.
5. Jeffrey, S. W. & Humphrey, G. F. Biochem. Physiol. Pflanz 167(1975) 191–194.
6. Ben-Amotz, A. & Avron, M. Plant Physiol. 72 (1983) 593–597.
7. Boussiba, S. & Richmond, A. E., Arch. Microbiol. 120 (1979) 155–159.
8. Strati, I. F. & Oreopoulou, V. Food Chem. 129 (2011) 747–752.

The details of heat and mass transfer modeling with time delay

V. G. Sorokin, A. V. Vyazmin, A. I. Zhurov, V. V. Reznik
Moscow State Univ. Mech. Engin., ul. Staraya Basmannaya 21/4, 105066 Moscow, Russia;
tel. +7 495 2671051, e-mail: av1958@list.ru.

The important feature of evolutionary processes, including heat- and mass-transfer processes with chemical conversions, is that, in the general case, the rate of variations in the desired quantities depends not only on the state at the given time point, but also on the entire previous evolution of the process [1]. These systems are called hereditary systems. In the particular case where the state of the system is only determined by a particular time point in the past, rather than the entire evolution of the system, the system is referred to as a delayed feedback system.

Systems with delayed feedback are frequently modelled by reaction–diffusion equations, in which the kinetic function F (the rate of chemical reactions) depends on both the sought concentration function $u = u(x, t)$ and the same function, but with the delayed argument $w = u(x, t - \tau)$. The special case of $F(u, w) = f(w)$ has a simple physical interpretation, i.e., heat- and mass-transfer processes in media with local non-equilibrium have inertial properties, i.e., the system does not react to action instantaneously at the given time point t, as in the classical local equilibrium case, but it reacts by the delay time τ later.

Parabolic equation of heat- and mass-transfer has a physically paradoxical property, i.e., an infinite disturbance propagation rate, which is not observed in nature. Since, when solving non-steady-state heat- and mass-transfer problems, it is necessary to take into account relaxation phenomena associated both with the finiteness of the rate of heat and mass transfer τ_1 and with the finiteness of the times of chemical conversions τ. The following non-linear hyperbolic reaction–diffusion equations with time-varying delay $\tau = \tau(t)$ are analyzed:

$$\tau_1 \frac{\partial^2 u}{\partial t^2} + \frac{\partial u}{\partial t} = a \frac{\partial^2 u}{\partial x^2} + F(u, w), \quad w = u(x, t - \tau(t)), \tag{1}$$

where a - diffusion coefficient, x - longitudinal coordinate. It should be noted that, as a particular case, at $\tau_1 = 0$, considering equation includes parabolic equations with delay.

The numerical solving of various nonlinear partial differential equations and systems of equations with delay and difficulties that arise in this case are described in [2]. In any case, the general disadvantages of numerical methods include: absence universal application when changing the geometric shape of the object, the type of fluid flow, reaction kinetics and inapplicability in the presence of singular points. It should be emphasized that delay substantially complicates the analysis of equations and is a factor that can lead to the instability of the systems being modelled.

The use of exact solutions for numerical simulation of complex equations and systems of equations can be very useful. The term exact solutions with respect to the nonlinear delay of partial differential equations are used in the cases where a solution is expressed as follows:

- The solution can be expressed in terms of elementary functions or can be represented in the closed form (the solution is expressed in terms of indefinite or definite integrals).

- The solution can be expressed in terms of solutions to ordinary differential equations or delay ordinary differential equations (or systems of these equations). Solution methods and various applications of linear and nonlinear ordinary differential equations with delay, which are substantially simpler than nonlinear partial differential equations with delay, are described, e.g., in [3].

- The solution can be expressed in terms of solutions to linear partial differential equations.

Thereby exact solutions to nonlinear differential equations promote the better understanding of the qualitative features of the processes under description (nonuniqueness, spatial localization, blowup regimes, etc.). Formation of exact solutions and transformations database for nonlinear partial differential equations with delay will significantly simplify computing algorithms for numerical simulation. Exact solutions allow a priori understanding of the problems of occurrence of instability in the numerical calculations.

As an example of this approach we now consider equation (1) in the following form:

$$\tau_1 \frac{\partial^2 u}{\partial t^2} + \frac{\partial u}{\partial t} = a \frac{\partial^2 u}{\partial x^2} + uF(w/u), \quad w = u(x, t - \tau(t)). \tag{2}$$

1. Equation (2) yields solution periodic with respect to x,

$$u = [C_1 \cos(\lambda x) + C_2 \sin(\lambda x)]\psi(t),$$

where C_1, C_2, and λ are arbitrary constants and the function $\psi(t)$ is described by the following ordinary differential equation with delay:

$$\tau_1 \psi''(t) + \psi'(t) = -a\lambda^2 \psi(t) + \psi(t) F(\psi(t - \psi(t))/\psi(t)).$$

2. Equation (2) also yields solution of the form

$$u = [C_1 \exp(-\lambda x) + C_2 \exp(\lambda x)]\psi(t),$$

where the function $\psi(t)$ is described by the following ordinary functional-differential equation:

$$\tau_1 \psi''(t) + \psi'(t) = a\lambda^2 \psi(t) + \psi(t) F(\psi(t - \psi(t))/\psi(t)).$$

Equation (1) of the form

$$\tau_1 \frac{\partial^2 u}{\partial t^2} + \frac{\partial u}{\partial t} = a \frac{\partial^2 u}{\partial x^2} + bu + F(u - w), \quad w = u(x, t - \tau(t)).$$

yields solution

$$u = \varphi(x) + \psi(t),$$

where

$$\varphi(x) = \begin{cases} C_1 \cos(\lambda x) + C_2 \sin(\lambda x), & \lambda = \sqrt{b/a} \quad \text{at } b > 0; \\ C_1 \exp(-\lambda x) + C_2 \exp(\lambda x), & \lambda = \sqrt{-b/a} \quad \text{at } b < 0 \end{cases}$$

and the function $\psi(t)$ is described by the following delay differential equation:

$$\tau_1 \psi''(t) + \psi'(t) = b\psi(t) + F(\psi(t) - \psi(t - \psi(t))).$$

This work was supported by the Russian Foundation for Basic Research (project no. 16-08-01252).

Reference

1. Jou D., Casas-Vazquez J., Lebon G., 2010, Extended Irreversible Thermodynamics. 4th ed. Springer. New York.

2. Jackiewicza Z., Zubik-Kowal B., 2006, Spectral collocation and waveform relaxation methods for nonlinear delay partial differential equations, Appl. Numer. Math., 2006, vol. 56, nos. 3–4, p. 433.

3. Driver R.D., 1977, Ordinary and Delay Differential Equations. Springer. New York.

Investigating the energy-water-carbon nexus of mega-scale chemicals production from Appalachian shale gas

C. He, F. You
Cornell University, Ithaca, New York 14850, USA
tel. +1 607-255-1162, e-mail: fy49@cornell.edu

The shale gas boom affects not only the fuel market but also the entire petrochemical industry since most of the shale gas resources in the U.S. are reported to be fairly rich in natural gas liquids (NGLs). (He and You, 2014) The light NGLs such as ethane and propane are very important commodity chemical intermediates in the industrial value chain to produce polymers, plastics, adhesives, etc. With continuous focus on the shale gas-to-chemicals (SGTC) projects, systematic studies on techno-environmental modeling can determine how and where SGTC projects should be developed, as well as the resulting economic and environmental risks. (He and You, 2015) Especially, shale gas producers in the U.S. are in a pessimistic economic climate due to the collapse in oil prices and the low estimated ultimate recovery (EUR). The recent growth in shale gas production has also resulted in increasing concerns with the environmental footprints including climate change, water, etc. For example, quantitatively assessing the leaking data of CH_4 is complex and varies from well to well, potentially leading to serious environmental issues. (Howarth, 2014) In addition, some practical issues influencing project performance should be taken into account as follows. (1) Feedstock supply: the overall performance of an SGTC project are susceptible to the numbers of wet wells (wells with NGLs production) and processing facilities; (2) Project operating time directly affects the absolute environmental footprints, as more wet wells are required as time goes by; (3) Well lifetime indirectly affect the environmental performance, because a longer lifetime of a shale well leads to the reductions of normalized environmental footprints (water, emissions, energy, etc.).

To decipher the true environmental impacts of a mega-scale SGTC project located in Appalachian region, this work systematically addresses techno-environmental modeling and investigates the energy-water-carbon nexus. In order to estimate the numbers of required shale gas wells, wellsites, and gathering systems, detailed feedstock supply and process design models for the multisite distributed gas processing facilities and centralized olefins production plant are established. Then, we conduct a comprehensive investigation on the energy-water-carbon nexus by determining life cycle energy consumption (ηle), life cycle freshwater footprint (ηlw), and life cycle GHG emissions (ηghg). The influences of allocation approach (mass-based allocation and economic value-based allocation) on the environmental impacts are also compared. The results show, based on a mass-based allocation approach, the ηle, ηlw, and ηghg are 13.8–17.2 GJ/t olefins, 3.31–4.28 kg/kg olefins, and 0.83–1.17 kg CO_2-eq/kg olefins, respectively. However, if we use an economic value based allocation method, these values are updated to 37.4–28.7 GJ/t olefins, 7.28–9.42 kg/kg olefins, and 1.80–2.49 kg CO_2-eq/kg olefins, respectively. The values of ηghg indicate that shale gas can be categorized as a low-carbon feedstock based on a mass-based allocation approach, or high-carbon feedstock based on an economic value-based allocation approach.

References

1. C. He, F. You, 2014, Shale gas processing integrated with ethylene production: novel process designs, exergy analysis, and techno-economic analysis, Industrial & Engineering Chemistry Research, 53(28), 11442-11459.
2. C. He, F. You, 2015, Toward more cost-effective and greener chemicals production from shale gas by integrating with bioethanol dehydration: Novel process design and simulation-based optimization, AIChE Journal, 61(4), 1209-1232.
3. R.W. Howarth, 2014, A bridge to nowhere: methane emissions and the greenhouse gas footprint of natural gas, Energy Science & Engineering, 2(2), 47-60.

Decontamination of wastewater using nanocarbon compounds insert in diatomite mesoporous structure

E. Flores,[1] O. Enriquez,[1] J. De la Cruz[1], A. Lópezı, G. Poma[1,2], M. Quintana[1,2]
[1]Engineering and Technological University, Department of Chemical Industrial Engineering, Medrano St. 168, Lima, Peru; tel.+5112305020, e-mail: mquintana@utec.edu.pe;[2]National University of Engineering, School of Physical Engineering, Faculty of Science, Tupac Amaru Av. 210, Lima, Peru; tel. +5114811070

There is a wide variety of techniques for the removal of heavy metals from wastewater. These techniques are useful but some of them are not economically available. Herein, we have synthesized a diatomite earth and graphene compound material. This material was obtained by continuous stirring of diatomite earth in a suspended graphene solution.

Diatomite earth was taken from Piura locality (Peru) and graphene was synthesized in a lab from Merck graphite.

In this research, we carried out a comparative study about the relationship of the diatomite distribution size used, stirring time required to produce the material and graphene concentration of the solution. Finally, we study the behaviour of this compound material to decontamination of heavy metals in wastewater.

Design of a heat recovery steam generator for solid oxide fuel cell-based trigeneration systems

P. Tippawan and A. Arpornwichanop*
Computational Process Engineering Research Unit, Chulalongkorn University, Department of Chemical Engineering, Faculty of Engineering, Bangkok 10330, Thailand; tel. +66 2218 6878, email: amornchai.a@chula.ac.th.

The solid oxide fuel cell-based trigeneration system is regarded as a highly efficient process for heat, power and cooling generation. It can be physically decomposed into three sub-systems, i.e., power system (solid oxide fuel cell, SOFC), heat recovery system (heat recovery steam generator, HRSG) and cooling system (absorption chiller). Due to its operation, the SOFC releases large amounts of thermal energy at relatively high temperatures (873-1273 K). However, a commercially available absorption chiller rated between one to 1000 tons of cooling energy needs input steam at temperature of 423 K. Thus, a large temperature difference occurs between the two crucial elements in the system that strongly affects the overall system performance. The HRSG becomes a primary source of irreversibility and comprehensive understanding is needed to provide guidelines for more efficient use of this energy conversion plant. In this study, a complete strategy for the optimum design of a heat recovery steam generator is outlined. The performance of the heat recovery steam generator is examined by modeling a trigeneration system based on an ethanol-fueled solid oxide fuel cell in Aspen Plus. An ethanol steam reformer operated at temperature of 980 K and steam-to-ethanol ration of 1.8 is chosen to be a fuel processor to produce the hydrogen-rich gas for the SOFC. To maximize the HRSG efficiency, the performance of the HRSG at required different products is investigated in terms of the energy and exergy analyses, with respect to various key operating parameters, such as pressures, mass flow rates and temperature profile on the gas and liquid side. The result shows that to minimize exergy destruction in the HRSG, the temperature difference across the HRSG should be maintained at a minimum and heat recovery steam generators with multiple pressure levels should be employed. The pressure levels of four major components including superheaters, evaporators, economizers and steam drum are calculated and shown at the outstanding performance.

Sawmill waste as source of energy in transportation and power industry

J. Bartoszewicz[1], A. Nygard, R. Kłosowiak, R. Urbaniak
Poznan University of Technology, Faculty of Machines and Transportation, Chair of Thermal Engineering, pl. M. Skłodowska-Curie 5, Poznań, 60-965, Poland;
tel. +48616652215, fax: +48616652281, e-mail: jaroslaw.bartoszewicz@put.poznan.pl.

With growing demand for energy and the European Union directives in force the interest in renewable energy sources utilization also grows. In Poland, assumptions for renewable power industry development were determined in some governmental documents: "Strategy for Renewable Energy Sources Development" (approved by the Parliament on 23 August 2001); "Poland's Energy Policy until 2030" (approved by the Council of Ministers on 10 November 2009) and "Programme for electrical power engineering" (approved by the Council of Ministers on 28 March 2006). A strategic goal of Polish policy is to increase the utilization of renewable energy resources so that a share of this type of energy in gross final energy consumption will reach the minimal level of 15 % in 2020.Report of the Central Statistical Office of Poland from 28 November 2014 presents the percentage share of acquired renewable energy according to its carriers in Poland in 2013. This paper considers the utilization of solid biofuels originating from sawmill waste as the source of energy in transportation and power industry.

Wood designated for energetic purposes may be used in various forms [3, 4]. Most often used forms are as follows:

- sawdust – waste constituting approx. 10% of wood processed in sawmills. It is a product of slicing, milling, etc., in plants with more advanced woodworking. The level of sawdust moisture content differs significantly and is estimated to be from 6÷10% to 45÷65% for sawdust originated directly from freshly-felled trees. For moisture content of 5÷15%, the content of ash is lower than 0%,
- shavings – they are, like sawdust, by-product of wood industry, produced as a result of subsequent stages of wood processing. They are characterized by low moisture content (5÷15%) and contain small amount of contaminants
- wood chips – shredded wood in the form of 5÷50 mm long chips of irregular shapes. Wood waste called wood chips are produced during log processing in wood splitting machines which, according to their characteristics and parameters, process wood into pieces of right size. Wood chips constitute an excellent fuel for boilers,
- bark – valuable product in terms of energy; it is waste of wood processing, constituting from 10 to 15% of the acquired wood weight. It has calorific value of 18.5÷20 MJ/kg, moisture content of 55÷65%, and the content of ash, having slagging tendency, is of 1÷3% of dry weight. Bark is largely processed into sawdust during production processes,
- wood briquettes – have form of cylinder or cube made of dry shredded wood (sawdust, shavings or wood chips) pressed under a high pressure without any adhesive substances. Their calorific value is of 19÷21 GJ/t; moisture content of 6÷8%; and ash content of 0.5÷1% of dry weight,
- pellets – granules of length of a few centimetres and diameter of 6÷25 mm, produced from wood waste, mainly from sawdust and shavings. Pellets are extruded in a rotary press, without any adhesive substances and under high pressure which enables the material to be highly compacted. Pellets may be easily transported, and are practical in terms of storage and operating. They are valuable due to their very low content of ash (0.4÷1% of dry weight). Their calorific value is of 16.5÷17.5 MJ/kg, and moisture content of 7÷12%.

A very important stage in analysing solid fuels, originating from sawmill waste, is calculation of their calorific value. Target product obtained in sawmills from wood waste as a result of wood processing are wood chips. They are shredded wood of a length of approx. 5 to 50 mm. Moisture content for wood chips from fresh pine wood is approx. 40%. If we assume that wood chips are not compacted any way, their volume increases comparing to solid waste. Main factors influencing on biofuels' calorific value are their moisture content and mass density. Results of calorific value calculations for the investigated fuel are summarized in table No. 1. It arises from this table that wood chips of moisture content of approx. 40% have calorific value of approx. 0.68 MWh/m^3, and, in turn, the calorific value of wood chips of moisture content lower than 15 % is 0.91 MWh/m^3.Table No. 2 summarizes demand for heat in separate time intervals for respective compartments. To prepare this summary we had to take into account the year season and the fact that Eberl drier absorbed heat every sixth day.

Tab. 1. Summary of calorific value calculation results

Type of waste	Moisture content in waste	Volume of waste	Volume of non-compacted wood chips	Usable calorific value of wood chips
Wet wood	40 %	555 m3/year	1648 m3/year	0,68 MWh/m3
	40 %	855 m3/year	2890 m3/year	0,68 MWh/m3
Dry wood	< 15 %	45 m3/year	121 m3/year	0,91 MWh/m3

Table No. 4 summarizes results of the fuel mass flow calculations.

	Fuel mass flow [kg/ quarter of year]
Spring / Autumn	366336
Summer	108000
Winter	179712
The need for an entire year	556848
The amount of fuel which has a sawmill	436500÷661500

Example of using sawmill's waste as the source of energy in order to satisfy its demand for thermal energy shows how profitable is to manage this waste properly. This example indicates that utilization of renewable energy sources brings not only ecological benefit but also financial one and as such should not be ignored.

1. Główny Urząd Statystyczny, Energia ze źródeł odnawialnych w 2013 r., Informacje i Opracowania Statystyczne, Warszawa 2014. (in polish)
2. Nowak W., Stachel A., Borsukiewicz-Gozdur A., Zastosowania odnawialnych źródeł energii., Wyd. PolitechnikiSzczecińskiej, Szczecin 2008.(in polish)
3. Sorensen B., Renewable Energy. Elsevier 2011.
4. Urbaniak R., Bartoszewicz J., Kłosowiak R., Main causes of NOx emissions by low-power boilers, Polish Journal of Environmental Studies, Vol. 24, No. 5 (2015), 2223-2230.
5. Dahlquist E., Biomass as Energy Source: Resources, Systems and Applications, Taylor Francis Group, London, 2013.
6. Gnutek Z., Bryczewska-Mazurek A., The thermodynamic analysis of multicycle ORC engine, Energy, 26, 2001, pp. 1075-1082.
7. Angrist S., Direct energy conversion (3td ed.) Boston, Allyn and Bacon, 1976.
8. Berglund M., Börjesson P., Energy efficiency in different types of biomass system, In 12th Europen Biomass Conference, ETA Firenze WIP Munich, 2002, pp. 219-222.

The effect of water temperature in the power consumption on pumps

Pablo Sanchez Razo [(1)], Jesus Martinez Patiño[(1)], Martin Picon Nuñez[(2)], Jose Merced Lozano García[(1)], Carlos Rubio Maya[(3)]
[(1)]Department of Electrical Engineering
[(2)]Department of Chemical Engineering.
University of Guanajuato. Carr. Salamanca-Valle de Santiago Km. 3.5+1.8 Km. Comunidad de Palo Blanco, C.P. 36885 Salamanca, Gto. Tel. +52 (464) 64 7 99 40. México.
[(3)]Faculty of Mechanical Engineering, UMSNH, Morelia, Michoacán, Mexico
p.sanchezr.666@gmail.com

Frequently, motors are subjected to more severe duty cycles than those for which they were designed, such as numerous consecutive starts, humidity, corrosive atmospheres and high temperature, so that the efficiency that the manufacturer offers a new equipment, rarely it obtained due to non-ideal conditions of operation of the equipment in the plant.

Just after the first start of a motor from ambient temperature the heat is stored in the windings, then transferred to the stator laminations, while the heat generated in the rotor and the stator is transferred outside the motor through the air gap. Motor temperature is limited by the temperature endurance of the electrical insulation and mechanical stresses in the cage winding, so the higher the temperature, the engine will be less efficient. The efficiency of an electric motor represents the behavior of this electrical machine for converting electrical power into mechanical power.

In vast sectors of the industry using motors coupled with pumps to move different types of fluids through countless processes, which depending on the conditions of pressure, flow, viscosity and required temperature of the fluid, it affect in one way or another the performance of the pump and therefore the electric motor coupled to the pump. It is intended to evaluate the effect of water temperature of a process, that by thermal conduction transfers heat energy to the pump that moves the water, and its in turn to the electric motor via the mechanical coupling, in order to analyze the consequences for efficiency and power consumption of the electric motor of this temperature gain due to heat conduction into the interior through its power take and other mechanical linkages.

Effective protein separation from microalgae based on simple solvents extraction technology

Phong, W.N.[1], Le, C.F.[2], Show, P.L.[3*], Lam, H.L.[3], Ling, T.C.[1]
[1]Institute of Biological Sciences, Faculty of Science, University of Malaya, 50603 Kuala Lumpur, Malaysia.
[2]School of Pharmacy, Faculty of Science, University of Nottingham Malaysia Campus, Jalan Broga, 43500 Semenyih, Selangor Darul Ehsan, Malaysia.
[3]Department of Chemical and Environmental Engineering, Faculty of Engineering, University of Nottingham Malaysia Campus, Jalan Broga, 43500 Semenyih, Selangor Darul Ehsan, Malaysia. Email: PauLoke.Show@nottingham.edu.my

ABSTRACT
One of the main challenges to extract the proteins from microalgae is the presence of rigid cell wall. This study aims to provide some guidance on the selection of the efficient solvent to facilitate the proteins release during the cell disruption process. The effects of solvent types such as methanol, ethanol, 1-propanol and water in rupturing the microalgal cell wall were studied. It was found that water able to release the highest concentration of proteins from microalgae compared to the other solvents.
Keywords: Microalgae, Cell disruption, Total proteins, Solvents

INTRODUCTION
The mass production of proteins source from microalgae offers several advantages (Solana et al., 2014). Microalgal proteins emerges as the most promising and sustainable alternative to conventional proteins sources such as fishmeal and soybean in aquaculture industry (Roy and Pal, 2014). This is mainly owing to the high protein content (Blackburn and Volkman, 2012; López Barreiro et al., 2013) and nutritional quality of microalgae (Norambuena et al., 2015). Microalgae have the ability to synthesise all types of essential amino acids which is mostly equivalent or even better with that of other higher plant (Spolaore et al., 2006) and their amino acid composition does not significantly affected by changes in environmental conditions (Blackburn and Volkman, 2012). Additionally, microalgae have higher photosynthetic efficiency (López Barreiro et al., 2013), high growth rate, short harvesting cycle (Cheah et al., 2014; Chen et al., 2013; Taher et al., 2014), high disease resistance ability and high biomass density (Roy and Pal, 2014).

However, one of the main difficulties faced in the mass production of proteins commercially is the low recovery of proteins due to the presence of multiple layers of thick cell wall (González-Fernández et al., 2012; Günerken et al., 2015). The development of simple cell disruption method for rapid cell lysis is important to facilitate the release of proteins from microalgae (Safi et al., 2015) in microalgal industry especially in large scale. Therefore, it is vital to choose a suitable solvent for the cell disruption process. This study could provide some guidance on the selection of the efficient solvent to facilitate the proteins release during the cell disruption process of microalgae. The objectives of these studies are to study the effects of different types of solvent in cell disruption process.

Bovine serum albumin (BSA) was used as a standard for the construction of a standard curve. BSA is commonly used as standard because of its high purity and low cost (Sattayasai, 2012). The protein concentration of the microalgae can be measured using the equation generated from the standard curve, $y = 0.0014x$ as shown in Fig. 1.

Fig.2 showed that amongst all the solvents, water used to lyse *C. Tai-03* was found to have the highest proteins concentration with about 3% of protein concentration per dry weight, followed by 1-propanol, ethanol and methanol.

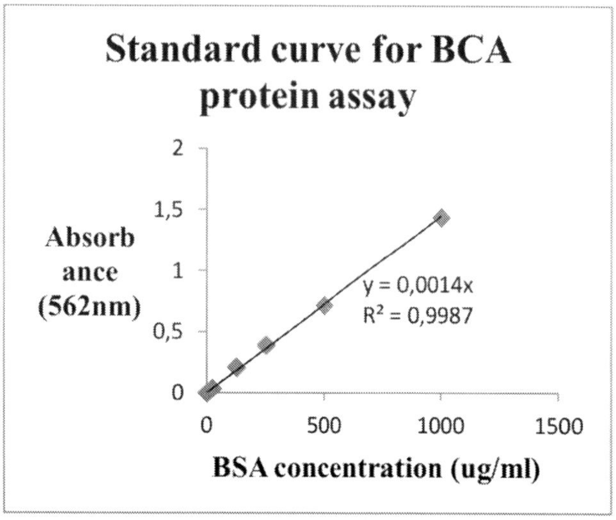

Fig. 1 – Standard curve for BCA protein assay using BSA as the standard.

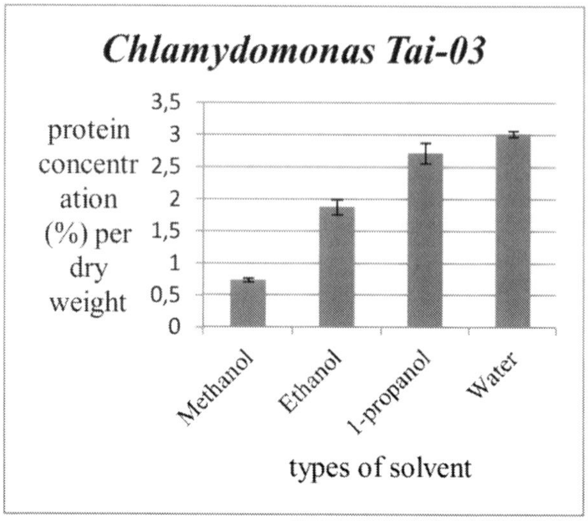

Fig. 2 -- Total proteins content of *Chlamydomonas Tai-03* subjected to different types of solvent. Each bar is an average of three replicates.

CONCLUSIONS

It was found that water could be an excellent solvent for microalgal cell disruption with the indication of higher proteins yield than the other solvents. Further investigation should be done on different strains of microalgae to evaluate the general effectiveness of water in improving the release of proteins yield from microalgae with different composition of cell wall.

Characterization of printed circuit boards of mobile phones discarded in Brazil

L. M. Terena[a], A. F. de Almeida Neto[a], M. L. Gimenes[b], M. G. A. Vieira[a]
[a]University of Campinas, Dept. of Products and Processes Design, Rua Albert Einstein, nº 500, 13083-852, Campinas – São Paulo – Brazil; tel. +55 19 35213928.
[b]State University of Maringá, Dept. of Chemical Engineering, Avenida Colombo, nº 5790, 87020-900, Maringá – Paraná – Brazil, tel. +55 44 3011476.
melissagav@feq.unicamp.br

The characterization of printed circuit boards (PCB) removed from discarded mobile phones in Brazil and an inventory of data of the components in PCB were accomplished in this paper. Data of chemical composition of PCB were obtained from the physicochemical characterization of their components through the analysis of granulometry, scanning electron microscopy with energy dispersive x-ray detector (SEM/EDS), infrared spectroscopy (FT-IR) and inductively coupled plasma optical emission spectrometry (ICP-OES). The analysis allowed a broad view and a highlight for the wide variety of elements and heterogeneity of the material. It can be also highlighted the high amount of copper, lead and other high-value metals in the PCB. This study is relevant for the development of new recycling technologies, once this material has been discarded incorrectly in Brazil. Based on the percentage of mobile phones recycled and disposed in Brazil, it is possible to quantify financial loss resulting from incorrect disposal.

Towards carbon neutral combustion. LCOE analyses of co-firing solid particles and gaseous fuel in Latvia

M. M. Fawzy, T. Prodanuks*, I. Veidenbergs, D. Blumberga
Riga Technical University, Institute of Energy Systems and Environment, Azenes street 12-1, LV-1048, Riga, Latvia, tel +371 29419783, toms.prodanuks_1@rtu.lv

One of the main obstacles to transfer to carbon neutral energy management is technological transition. An important task is to change fuel used in the boiler furnace from natural gas (fossil fuel) to biomass. The easiest but most expensive solution is the installation of biomass boiler. Other solutions include burner replacement or additional installation, but these options are more complex from the engineering and technical point of view.

By replacing natural gas with wood particles, boiler output is reduced, because value of combustion heat, furnace load and specific combustion parameters are lower, as well as temperature and pollution with nitrogen oxides NOx decrease. It is especially important when energy end user improves energy efficiency and reduces energy consumption. As a result there is excess installed capacity in the boiler house.

Feasibility of co-firing is evaluated by application of Levelized Cost of Energy (LCOE) analysis.

This research is important as natural gas will gradually become replaced by biogas that is carbon neutral. Moreover, biomass co-firing with solid wood particles will reduce the impact on climate change

Stand-alone evaluation of the energetic potential of various lignocellulosic residues in Spain

[1]V. Hernández, [2]C.A. García, [3]Eulogio Castro, [3]J.M. Romero-García, [2]C. A. Cardona
[1]Tecnoparque, Servicio Nacional de Aprendizaje SENA, Km 10 vía al Magdalena, Manizales-Colombia
[2]Universidad Nacional de Colombia Sede Manizales, Research Group in Chemical, Catalytic and Biotechnological Processes, Km 9 vía al Aeropuerto, Manizales, Colombia, tel. +57 68879300 ext 55880, e-mail: vhernandezp@unal.edu.co;
[3]Jaen University, Department of Chemical, Environmental and Material Engineering Campus Las Lagunillas, tel. +34 953212163.

The replacement of the fossil raw materials either fully or partially is an objective in many countries, including Spain. It is of special interest to use local biomass such as agricultural, forest, agro-industrial and industrial byproducts, due to their low cost and large availability. Currently, most energy and chemicals are derived from fossil raw materials, although there are several environmental, economic and social concerns related to their extraction and use. These facts have increased the interest in the use of renewable raw materials. Lignocellulosic biomass represents the major renewable source of potentially fermentable carbohydrates and it is mainly composed of cellulose, hemicellulose and lignin. The lignocellulosic complex constitute the principal component of most agricultural, forestry and municipal solid waste and the byproducts of the agroindustry, food industry and other industrial sectors. Its composition, availability and renewability have increased the scientific and academic interest for evaluating the production and recovery of a wide range of value-added products, such as enzymes, reducer sugars, furfural, biofuels, proteins and amino acids, lipids, organic acids, phenols, activated carbon, biopolymers, cosmetics, biosorbents, resins, medicines, bioenergy, biopesticides, biofertilizers, secondary metabolites, biopromotores and surfactants, among others.

Depending on factors such as residue type and amount and the final use of the product obtained by the biomass processing, a wide range of conversion technologies are available. These are classified into thermochemical, biochemical and chemical conversions. Thermochemical processes involve irreversible chemical reactions at high temperatures and different oxidation conditions. Into this classification, pyrolysis, gasification and combustion are included and they differ not only in process parameters but also in the products obtained. The biochemical conversion covers those processes in which the action of microorganisms or enzymes is involved in the production process for obtaining value-added products. The biochemical processes are referred mainly to a) fermentation, where a wide range of products such as ethanol, xylitol, gibberellic acid and compost, among others are produced and b) anaerobic digestion where biofertilizers and biogas are obtained.

In this work, the energetic potential of various Spain lignocellulosic residues (whet straw, rape straw, olive tree pruning and olive stone) was calculated theoretically and compared with the energy content of different fuels (ethanol and biogas) obtained through its processing. The lignocellulosic biomasses were firstly characterized following the NREL laboratory methods. Also its ultimate composition was obtained.

Incorporating non-linear cost functions into P-graph: a wood processing cluster case study

B.H.Y. Ong, T.G. Walmsley, M.J. Atkins, M.R.W. Walmsley, J.R. Neale
Energy Research Centre, School of Engineering, University of Waikato, Private Bag 3105, Hamilton, New Zealand; tel. +647 8384937, email: matkins@waikato.ac.nz;

Process Graphs (P-graph) are bipartite graphs that uses a theoretical approach to solve Process Network Synthesis (PNS) problems. The P-graph framework was first developed to optimise traditional chemical processes to manufacture desired products. P-graph methodology has been applied to different applications (Lam, 2013), for example (i) synthesis of separation networks, (ii) optimal design of steam supply systems, (iii) synthesis of alternative distillation system sequences, (iv) heat exchanger networks, (v) identifying reaction pathways, (vi) supply chain network synthesis, and (vii) risk management.

P-graph applies mixed-integer linear programming (MILP) to optimise process structure models (Heckl et al., 2015). One of the limitations in P-graph is the assumption of linear cost functions; including capital cost, feedstock costs, and product sale price. In P-graph, the capital cost is estimated using a fixed investment cost plus a linear function of plant size (or capital size) that is proportional to input or output material flowrates. However, in process engineering, capital cost estimation normally requires non-linear functions to maintain accuracy across over a broad range of equipment sizes. The cost of wood biomass is also non-linear, being governed by volume and distance. Higher wood volumes increases the average distance travelled to harvest and from which to transport biomass to the industrial site. Product sale price is dependent on the elasticity of its local and/or global market. Some product prices follow linear trends because the volume delivered is well below market demand. Other product prices fall rapidly with volume due to market saturation and over-supply. In the context of wood processing, accounting for non-linear cost/price functions in the application of the P-graph methodology is therefore critical.

The aim of this paper is to incorporate non-linear cost functions into P-graph via a combination of P-graph framework and Mathematical Programming (MP) method. The focus of the study is to use iterative method to approximate the capital cost, feedstock cost, and product price, using a series of linear functions that update with each iteration until convergence is reached. Cost functions are obtained from literature and industrial data sets.

References

Heckl, I., Halász, L., Szlama, A., Cabezas, H., Friedler, F., 2015. Process synthesis involving multi-period operations by the P-graph framework. Computers & Chemical Engineering, 24th European Symposium on Computer Aided Process Engineering(ESCAPE) 83, 157–164. doi:10.1016/j.compchemeng.2015.04.037

Lam, H.L., 2013. Extended P-graph Applications in Supply Chain and Process Network Synthesis. Current Opinion in Chemical Engineering 2, 475–486.

A master plan for the implementation of sustainable enterprise resource planning systems (Part III): evaluation of a roadmap (summary)

Abdoulmohammad Gholamzadeh Chofreh[1,2], Feybi Ariani Goni[3], Jiří Jaromír Klemeš[4],
[1]Department of Industrial Engineering, Buein Zahra Technical University, Buein Zahra, Qazvin, Iran; tel. +98 9163028545, e-mail: mohammad_apmie@yahoo.com;
[2]Faculty of Mechanical Engineering, Universiti Teknologi Malaysia, 81310 Johor Bahru, Malaysia
[3]Faculty of Information Science and Technology, Universiti Kebangsaan Malaysia, Bangi, Malaysia; tel. +98 9367120679, e-mail: feybi.ariani@gmail.com;
[4]Faculty of Information Technology and Bionics Pázmány Péter Catholic University, Práter u. 50/a, 1083 Budapest, Hungary; e-mail: klemes.jiri@itk.ppke.hu;

An implementation of Sustainable Enterprise Resource Planning (S-ERP) systems is challenging that requires multidisciplinary skills, knowledge, and coordinated efforts of all levels in an organization. Related study in S-ERP systems had been done, however, they only highlighted on the imperative of the system and empirical analysis of the system implementation without looking into detail on how to implement this integrated system. This lack of comprehensive plan and direction to guide practitioners in implementing S-ERP system would result in increased sustainability implementation cost and time and the decline in the quality of products and services. This problem motivates the researchers to propose a master plan of the S-ERP system implementation that consists of a roadmap, a framework, and guidelines. A development of S-ERP roadmap has been underscored in the previous paper. For the next phase, the roadmap is then evaluated by using a peer review method, which is the focus of this paper. The evaluation consists of two main processes including data collection and data analysis. In this regard, twelve experts had participated in the peer review. As a result, the developed roadmap needs to be expanded into two roadmaps, namely roadmap towards sustainable enterprise and roadmap towards sustainable integrated enterprise. From the evaluation results, the experts confirmed that the proposed S-ERP implementation roadmap can generally be used for implementing the S-ERP systems in all types of industry. The roadmap could serve as a theoretical base for further research into S-ERP system research, which can give more insight into the system implementation. In addition, this roadmap could serve as a tool for practitioners in implementing S-ERP systems within organizations.

Use of regression discontinuity design for the estimation of the impact of national policies on the development of biomass energy in China

M.S. Dovì
University of St. Andrews , Regulus Hall, Queen's Gardens, St Andrews KY16 9TQ, UK; tel. +44 1334 467065, e-mail: msd2@st-andrews.ac.uk

Regression discontinuity design has become a popular econometric tool for evaluating impact and effectiveness of legislative measures [1]. To evaluate China's national policy for the development of biomass energy sources, I have considered the development of bioethanol production over the last 20 years using data provided by the National Bureau of Statistics of China and the International Energy Agency.

The ratio bioethanol to wheat production has been employed as an instrumental variable in the regression analysis. This choice makes it possible to eliminate endogeneity problems that might turn up, if more general variables are considered.

The resulting regression discontinuity design makes it possible to test the hypothesis of a shock effect by comparing observations close to either side of the cut-off point. Typically, one has to estimate the coefficients $\alpha, \tau, \beta_1, \beta_2$ of the linear regression

$$r = \alpha + \tau D + \beta_1(t - c) + \beta_2 D(t - c) + \varepsilon$$
$$c - h \le t \le c + h$$

In other words, I have used ordinary least squares (OLS), but the data points are limited to an interval $(c-h)-(c+h)$, where c is the cut-off date considered. D is a binary dummy variable equal to zero if $t \le c$ and to one if $t \ge c$. The interval (bandwidth) h is the interval on either side of the cut-off point over which data are considered. It can be reduced provided the significance of the test (which can be derived from the p-values of the estimates) remains sufficient.

The value of τ resulting from the regression indicates the discontinuity estimated at the cut-off date, whereas the coefficients β_1 and β_2 provide the slopes on either side of the cut-off date. The coefficient α represents an uninfluential baseline value.

If the procedure is iterated at each cut-off date corresponding to a major legislative bill, it is possible to estimate the impact of each measure by comparison of the coefficients of the regression analysis and their statistical significance.

[1] David S. Lee and Thomas Lemieux, 2010. "Regression Discontinuity Designs in Economics," Journal of Economic Literature, American Economic Association, 48(2), 281-355

The combustion characteristics of modified plant biomass pellets

I. Barmina[1], R. Valdmanis[1], M. Zake[1], A. Arshanitsa[2], Y. Akishin[2], G. Teysheva[2]
[1]Institute of Physics, University of Latvia, Salaspils, Miera Street 32, LV-2169, tel. +371 29475762, e-mail: mzfi@sal.lv
[2]Institute of Wood Chemistry, Riga, Dzerbenes 21, LV- 1021, tel. +371 67555916, e-mail: ligno@edi.lv

Solid plant biofuel, including wood and agricultural residues, plays a prominent role within the classes of alternative and renewable sources suitable for heat and energy production. However, the low energy content, low density and low hydrophobicity in comparison with solid fossil are the main disadvantages of plant biomass as a source of energy. The pelletization and torrefaction both are recognized as promising technologies for improving fuel properties of plant biomass [1]. The advantages of both pelletization and torrefaction technologies were joined by authors using microwave torrefaction of commercial wood and wheat straw pellets realized in rotated microwave reactor of original construction. The decreasing of particle density of pellets as the result of torrefaction was significantly eliminated by spilling waste cooking and mineral oils on the surface of microwave treated pellets. The detail description of this two-step methods of biomass pellets modification as well the induced transformation of biomass structure are presented elsewhere [2].

The present research aims to investigate the combustion of modified pellets (spruce, aspen, wheat straw) and clarify the impact of each of stages of above mentioned two-step modification of biomass on the processes of its burnout. The composition of combustible volatiles (CO, H_2, CH_4, etc.) obtained as the result of biomass gasification, ignition of volatiles, heat energy production as the result of volatile and char combustion were studied and discussed.

The experimental study of the effect of biomass pellets modification on their gasification/combustion characteristics was carried out using a small-scale batch-size experimental device with heat power up to 2 kW. The experimental device is composed of biomass gasifier and combustor and is developed with the aim to carry out a series of experiments to assess and compare the effect of biomass composition and structure variations on the formation of main gasification/combustion characteristics downstream the swirling flame flow. The diagnostic tools for the complex experimental study of the gasification/combustion characteristics included: the Pt/Pt-Rh thermocouples for the local time-dependent measurements of the flame temperature; an infrared FTIR spectrometer Varian 640-IR for the control of volatiles composition (CO, CH_4, C_2H_2, C_2H_4), a gas-analyzer Testo 350XL to investigate the effect of mw-pretreatment of biomass on emissions composition (CO, CO_2, NO_x). The calorimetric measurements of the cooling water flow were used to estimate the effect of biomass pre-treatment on the heat energy production at different stages of the thermo chemical conversion of activated biomass pellets. The mass consumption rate of biomass pellets (dm/dt) during the thermo chemical conversion was controlled using a test facility that consists of a moving rod supplied with a pointer and allows provide the measurements of biomass weight loss during the gasification of biomass samples.

The results of complex experimental study of gasification/combustion characteristics have shown the effect of mw-induced variations of the main characteristics of biomass pellets (spruce, aspen, wheat straw) on development of the processes of thermal decomposition of biomass pellets, ignition and combustion of volatiles by minimizing the effect of endothermic processes of biomass heating and drying on the main combustion characteristics and formation of the products. The mw assisted torrefaction of biomass pellets results in faster thermal decomposition and gasification of torrefied pellets with faster ignition and more complete combustion of the volatiles, determining higher peak values of CO_2 in the products (Figure 1) determining a correlating increase of the produced heat energy up to 25-30 % with a linear dependence between the field-induced variation of the carbon content in pretreated samples and produced heat energy.

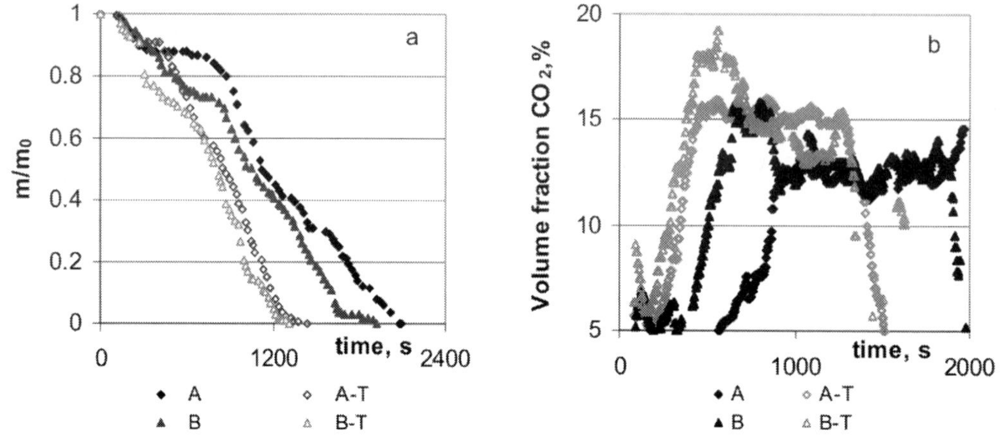

Figure 1. Effect of mw pre-pretreatment on the time-dependent variations of the weight loss of biomass pellets (a) and CO_2 formation (b). A- wheat straw pellets; B- spruce pellets; A-T- mw-pretreatment of wheat straw biomass; B-T- mw-pretreatment of spruce biomass.

The spilling of waste oils (cooking or motors) on the surface of previously mw treated pellets (about 15% from biomass weight) leaded to further acceleration of ignition process. The volatile products of gasification of oil spilled pellets contained higher amount of HC-hydrocarbons, in comparison with gasification products of initial and mw treated pellets. As the results the higher temperature in burnout zone and total heat energy outputs were observed for spilled pellets. The lowest value of air excess coefficient (α) and the highest concentration of CO_2 measured in combustion products of oil spilled pellets gave the evidence that the most completed combustion was achieved. Consequently, the results of the complex experimental measurements allows conclude that proposed modification of plant biomass pellets including their mw torrefaction with further spilling of waste oils is an effective tool to convert a biomass into fuel with improved combustion characteristics.

References

1. A. Uslu, A.P.C. Faaij, P.C.A. Bergman, 2008, Pre-treatment technologies, and their effect on international bioenergy supply chain logistics. Techno-economic evaluation of torrefaction, fast pyrolysis and pelletisation, Energy, 33, 1206–1223.
2. A. Arshanitsa, T. Dizhbite, O. Bikovens, G. Pavlovich, A. Andersone, G. Telysheva, 2016, Effects of microwave treatment on the chemical structure of lignocarbohydrate matrix of softwood and hardwood, Energy Fuels (Web), DOI: 10.1021/acs.energyfuels.5b02462.

Degradation of an organophosphorous insecticide in aqueous medium by electro-Fenton process

Mirvet. Assassi*[a, b], Farid. Madjene [d], Rachida. Maachi[a], Abdeltif. Amrane [c]
[a] Université Houari Boumediene, Institut de Chimie Industrielle, Laboratoire de Génie de la Réaction, Alger 1611, Algérie.
[b] Université Mohamed El Bachir El Ibrahimi Bordj Bou Arreridj, Algérie.
[c] Université Rennes 1, Ecole Nationale Supérieure de Chimie de Rennes, CNRS, UMR 6226, Avenue du Général Leclerc, CS 50837, 35708 Rennes Cedex 7, France.
[d] Unité de Développement des Equipements Solaires, UDES, Centre de Développement des Energies Renouvelables, CDER, 42415, Tipaza, Algérie.
mirvetassassi@yahoo.com

The removal of organophosphorus insecticide phosmet (Imidan) from water has-been Investigated by the electro-Fenton method. The study of various parameters (current, catalyst concentration, temperature, nature of electrolyte and duration of oxygenation) may influence the degradation of phosmet by the electro-Fenton process was carried out: the rise of intensity of the current applied from 60 to 100 mA, increase phosmet elimination rate from 60 to 70%; the maximum performance degradation reaches its with a concentration of Fe^{3+} equal to 0.5mMol.L^{-1} in a concentration range of 0.1 to 0.5 mMol.L^{-1}; The temperature of 20°C seems a suitable temperature for the electro Fenton treatment; an addition of oxygen during the entire experience has increased the degradation yield up to 88.2%.

Forecasting modelling for energetic efficiency in a steelmaking process

S. M. Acosta[1], A. M.O. Sant'Anna[2] O. Canciglieri Junior[3*],
[1]Department of Electronics at Federal University of Technology Paraná (UTFPR), Av. Sete de Setembro, 3165, Curitiba, Brazil;
[2]Mechanical Engineering Department at Federal University of Bahia (UFBA), Rua Aristides Novis, 2, Salvador, Brazil;
[3]Production and System Engineering Graduate Program at Pontifical Catholic University of Parana (PPGEPS/PUCPR), Rua Imaculada Conceicão, 1155, Curitiba, Brazil;
[*]tel. +55 41 3271 2579, e-mail: osiris.canciglieri@pucpr.br[3]; angelo.santanna@pucpr.br[2]

The quality of products and processes is a permanent challenge for industries, and such challenge is no different in steelmaking processes. One of the main problems affecting the quality of steel products is the existence of contaminants in alloy steel, being phosphorus (P) a major contamination element interfering with the steelmaking process. The increased levels of P can severely affect physical integrity of steel bonds, thus threatening the quality of the final product.

This paper proposes a robust approach to model the phosphorus concentration levels in steelmaking process. The proposed approach consists in applying the generalized linear model and artificial neural networks techniques. We used the improved neural network models inspired in the human nervous system for processing the information. The different techniques used for modelling the phosphorus levels investigate the variables that have significant influence on refining process. Based on better predictive model, the increase of phosphorus levels in the final product is related with initial levels of carbon, oxygen, magnesium, manganese oxide and calcium oxide.

The results illustrate the efficiency of the techniques used in the modelling, with emphasis on the adequacy of the predictive models constraints in the refining process. This study presents relevant strategy to model characteristics' of row material based on forecasting strategy to the efficiency of alloys and steel industry.

Keywords

Alloys metal; regression model; neural networks; energy efficiency.

An efficient decomposition algorithm for mixed-integer linear bilevel optimization problems

D. Yue, F. You
Cornell University, Ithaca, NY 14850

Mixed-Integer Bilevel Linear Programming (MIBLP) problems stem from hierarchical decision-making processes and the Stackelberg game, when discrete variables are involved (Bard, 1998). Although many applications in transportation planning, government policy making, revenue management, etc. fit the bilevel optimization framework, the use of MIBLP models in real-world applications is relatively rare (Kalashnikov et al., 2015). This is mainly due to the lack of efficient algorithms for solving MIBLP problems. The complexity of bilevel optimization lies in the property that constraint region of the upper-level program is partially determined by the solutions to a lower-level program (Saharidis et al., 2013). MIBLP problems further complicate the matter because 1) the bilevel feasible region can be nonconvex and disconnected; 2) removing the integrality constraints does not necessarily provide a valid relaxation of the original MIBLP problem; 3) lower-level optimal solutions are not always feasible to the original MIBLP when upper-level connecting constraints are present (i.e., issue of relatively complete response). All these challenges prohibit direct employment of Mixed-Integer Linear Programming (MILP) algorithms for solving MIBLP problems.

After reviewing relevant literature on MIBLP (Xu and Wang, 2014; Zeng and An, 2014) and Mixed-Integer Bilevel Nonlinear Program (MIBNLP) (Kleniati and Adjiman, 2015; Mitsos, 2010; Domínguez and Pistikopoulos, 2010), we discover that existing algorithms for MIBLPs are either subject to simplifying assumptions on the integrity of parameters/variables or restrictions on the presence of upper-level connecting constraints. The goal of this work is to develop an efficient MIBLP algorithm that deals with the least restricted form of MIBLPs. The proposed algorithm requires only two basic assumptions: 1) the MIBLP constraint region is a nonempty compact set; and 2) the inducible region is nonempty. To address this class of MIBLPs, we first reformulate the original MIBLP into an equivalent Generalized Semi-Infinite Program (GSIP), which is further reformulated into an equivalent single-level optimization problem. The issue of relatively complete response is tackled using a projection-based formulation with Fourier-Motzkin elimination and disjunctive programming. Based on this single-level reformulation, a decomposition algorithm is developed that converges to the global optimal solution in finite iterations. The master problem provides a valid lower bound, while two subproblems are used to provide a valid upper bound or to test the feasibility of lower-level solutions. A KKT-condition-based cut is generated using the solutions to the subproblems and added to the master problem at the end of each iteration, so that non-decreasing lower bounds can be obtained successively. Implementation of the algorithm is described along with applications on two examples adapted from the literature and a number of randomly generated numerical examples.

References

J. F. Bard, Practical Bilevel Optimization: Algorithm and Applications, 1998, Kluwer Academic Publishers Dordrecht.

L. F. Domínguez, E. N. Pistikopoulos, 2010, Multiparametric programming based algorithms for pure integer and mixed-integer bilevel programming problems, Computers & Chemical Engineering, 34(12), 2097-2106.

V. V. Kalashnikov, S. Dempe, G. A. Pérez-Valdés, N. I. Kalashnykova, J. F. Camacho-Vallejo, 2015, Bilevel Programming and Applications, Mathematical Problems in Engineering, 2015

P. M. Kleniati, C. S. Adjiman, 2015, A generalization of the Branch-and-Sandwich algorithm: From continuous to mixed-integer nonlinear bilevel problems, Computers & Chemical Engineering, 72, 373-386.

A. Mitsos, 2010, Global solution of nonlinear mixed-integer bilevel programs, Journal of Global Optimization, 47(4), 557-582.

G. D. Saharidis, A. Conejo, and G. Kozanidis, Exact Solution Methodologies for Linear and (Mixed) Integer Bilevel Programming, in Metaheuristics for Bi-level Optimization. 482, 221-245, E.-G. Talbi, Ed., ed: Springer Berlin Heidelberg.

P. Xu, L. Wang, 2014, An exact algorithm for the bilevel mixed integer linear programming problem under three simplifying assumptions, Computers & Operations Research, 41, 309-318.

B. Zeng, Y. An, 2014, Solving Bilevel Mixed Integer Program by Reformulations and Decomposition, Optimization Online.

Diesel engine emission reduction operating with hydrogen produced from ethanol reforming and exhaust gas recirculation

A. C. R. Teixeira, H. F. Lima, L. A. Costa, M. V. C. Gonçalves, J. R. Sodré
Pontifical Catholic University of Minas Gerais, Dep. Mechanical Engineering, Av. Dom José Gaspar 500, 30535-901 Belo Horizonte, MG, Brazil; tel. +55 31 33194911, e-mail: ricardo@pucminas.br.

Exhaust gas recirculation (EGR) is a well known technology used to reduce NO_X emissions from diesel engines, with drawbacks on engine performance. On the other hand, the fast burn speed of hydrogen improves combustion and can counterbalance the awkward effects of EGR. Thus, the aim of this study is to evaluate the performance and emissions from a diesel engine using an ethanol steam reformer to produced hydrogen assisted by an EGR system at different operating conditions. A 50 kW diesel power generator was used in the tests, at full load operation. The EGR system was adapted to the engine, and it was used to both reduce oxides of nitrogen (NO_X) emissions and as a heat source to the steam reformer. The steam reformer was constructed using a high Rhodium concentration catalytic converter, and ethanol fuel was injected in the intake air charge blended with the recirculated exhaust gas at the reformer inlet. Performance and emission parameters were compared for engine operation with and without the fuel reformer and EGR system. The results indicate a reduction in carbon dioxide (CO_2) and oxides of nitrogen (NO_X) emissions, compared with engine operation with the standard configuration.

Distillation pressure influence on Total Site

[1]A. Oslé, [1,2]A.E. Bonet Ruiz, [1]J. Bonet, [2]V. Plesu, [2]P. Iancu, [1]J. Llorens
[1]University of Barcelona, Faculty of Chemistry, Department of Chemical Engineering, Martí i Franqués St. 1, Floor 6, E-08028, Barcelona, Spain; tel. +34 93 4021310, e-mail: bonet@ub.edu; [2] University POLITEHNICA of Bucharest, Centre for Technology Transfer in Process Industries (CTTPI), Gh. Polizu St. 1, Bldg. A, Room A056, RO-011061, Bucharest, Romania; tel. + 40 21 4023808, fax: +40 21 3185900, e-mail: a_bonet@chim.upb.ro.

Distillation is the most used unit operation for separation in chemical industry, as well as being the unit operation consuming the largest amount of energy. Therefore, any improvement in distillation provides great energy savings and decreases the environmental pressure. Coopetition between chemical plants located in a certain area opens new opportunities of energy integration that are not available when they are individually operated. Although a common treatment of residues is a common practice, e.g. wastewater, common hot/cold services are not a usual one. The goal of the present study is to provide a simple tool for a fast evaluation of the advantages that could be derived from a Total Site configuration. The Total Site includes several processes found in the same geographical area, connected through a common central service system, which permits the recovery of heat between the different processes. To avoid any risk, only indirect integration using steam streams between the plants is considered. A hypothetical case study is presented as illustrative example considering five plants with five different mixtures to separate. The mixtures chosen are examples whose distillation sequence has been previously studied and available in open literature. A novel methodology (stacking boxes) is proposed herein, that allows to perform a simple, fast and very visual analysis of possible integrations among all the columns with common heating and cooling services. Possible pressurization of the columns to allow a high degree of heat integration is also considered.

First of all, the minimum energy required when the plants are operated independently is calculated using fully integrated columns. Although the fully integrated columns provide the minimum energy consumption, the use of a column sequence provides higher flexibility for integration. Therefore, the sequences of distillation columns with a high enough distillation sequence efficiency are also considered. Once analysed, the best integrated and sequential cases are chosen according to their distillation sequence efficiency for all of the five processes studied. Infinite/infinite analysis is used to check the feasibility and calculate the streams compositions and flow rates. The minimum heat consumption for each column is determined using a shortcut method based on basic thermodynamic data (entropy and enthalpy) and distillation efficiency.

Alternatives using one single column, sequences of columns or a combination of both are evaluated to minimize the energy requirements. Then the integration of process streams is performed through the common hot and cold services (Total Site). Steam services are produced at different pressure and temperature levels and are then distributed to the site. The integration using the different heating and refrigerating utilities is applied to many columns and chances of integration increase with the number of mixtures to be separated and with the number of temperature levels of the heating services. In addition, the column's pressure is manipulated to fit columns between two levels of the heat services (low and high pressure steam) in order to obtain the maximum efficiency. The new method of stacking boxes proposed in this work is applied to obtain the best system configuration from the energy requirements point of view.

The results for the case study show that modifying the pressure of the distillation columns and integrating them leads to energy savings around 35 % compared to the minimum energy requirements when all the plants are operated independently.

Effect of daylighting on the Optimum Task Performance Zone (OTPZ), energy savings and emissions reduction in office buildings

Y. A. Dodo[1]*, M. Z. Kandar[1], R. Nafida[2], R. Zakaria[3], B. B. Nyakuma[4], A. S Kawuwa[5], F. M. Bashir[1]
[1]Department of Architecture, Faculty of Built Environment, Universiti Teknologi Malaysia, 81310 UTM Skudai, Johor Bahru, Malaysia
[2]Department of Architecture, Centre for the Study of Built Environment in the Malay World (KALAM), Institute for Smart Infrastructure and Innovative construction (IS&IC), Universiti Teknologi Malaysia, 81310 UTM Skudai, Johor Bahru, Malaysia
[3]Construction Technology and Management Centre (CTMC), Faculty of Civil Engineering, Universiti Teknologi Malaysia, 81310 UTM Skudai, Johor Bahru, Malaysia.
[4]Centre of Hydrogen Energy, Institute of Future Energy, Universiti Teknologi Malaysia, 81310 UTM Skudai, Johor Bahru, Malaysia
[5]Department of Architecture, Abubakar Tafawa Balewa University, P.M.B 0248 Bauchi, Bauchi State, Nigeria
*Tel: +6(0)166486703, e-mail: yadodo@utm.my, dyaaminu@yahoo.com

Energy consumption in buildings is a significant contributor to greenhouse gas (GHGs) emissions. Consequently, the incorporation of low cost, sustainable practices can potentially guarantee the paradigm shift towards energy savings and carbon dioxide emissions reduction. The use of natural daylight i.e. "Daylighting" presents a practical opportunity to harness light from the sun into office building designs. This will potentially reduce overdependence on electric lighting which affects users' Works Plane Illuminance (WPI), and improve workplace comfort and performance efficiency in office buildings. However, there have no studies on the effect of Daylighting on WPI and Optimal Occupant Task Performance (OOTP), despite its importance on user experiences in office buildings.

Therefore, the main aim of this study is to examine the effect of daylighting on the Optimum Task Performance Zone (OTPZ), energy savings and emissions reductions in an office building. Consequently, a paper based Post Occupancy Evaluation (POE) test comprising spot measurements and experiments was carried out on 136 occupants in an office space with dimensions; 2900 mm x 4400 mm x 3200 mm and 25 % Window-to-Wall Ratio (WWR).

The results demonstrated that the OTPZ in the office building was characterized by daylight illumination of 300 lux. Furthermore, Work Plane Illuminance (WPI) in the OTPZ ranged from 200 lux to 400 lux while the non-optimum task performance zones (*n*-OTPZ) were found to be unfavorable for occupants performing tasks. The results also indicated that energy savings from determining the OTPZ from daylighting in the office building can lower carbon dioxide (CO_2) emissions by 10 – 20 %. In conclusion, the importance of daylighting in office buildings was highlighted in this study with a view to improving building designs and eliminating hitherto grey areas for construction professionals and energy policy makers in Malaysia.

Keywords: Daylighting, Carbon dioxide, Optimum Task Performance, Energy, Emissions

Effects on energy optimization for process industries: case of variable viscosity

Eid M. Al-Mutairi,
Chemical Engineering Department, King Fahd University of Petroleum and Minerals, Dhahran 31261, Saudi Arabia; tel: +966138607726, email: mutairi@kfupm.edu.sa

Heat Exchanger Network Synthesis (HENS) has been applied in a number of industrial processes for development of optimum heat exchanger network, which guaranteed energy savings and environmental protection by cutting down emissions of hazardous compounds. Most the previous works were based on stringent observation of Pinch Design Method in which most process stream properties were assumed constant. In this work, a new approach which will allow more flexibility in the heat exchanger network design was developed. A novel mathematical modeling approach is proposed to investigate effects of stream properties like heat capacity, viscosity and heat transfer coefficient when they are considered as variable functions of temperature in the shell and tube sides of the heat exchangers in the network. Successful application of this approach will bring forward a more realistic analysis into the design of optimal Heat Exchanger Networks.

Energy optimization for scheduling of water supply system considering pump operations

H. Shi, [1]F. You
Cornell University, Ithaca, NY 14850.

Both water and energy are critical factors for the development of our modern societies [1]. With fast extension of water supply systems, the efficient use of energy in water distribution systems has attracted increasing attention in recent decades. The operation of the water supply systems is expensive because of large amount of water demands in the industry process and the daily life. Optimal scheduling of the water supply systems is to determine the optimal water distribution schedule, which helps obtain remarkable savings by coordinating the pumping with time-varying electricity tariffs [2]. Extensive research efforts have been directed towards constructing the mathematical models which fully represent the characteristics of water supply systems [3]. In those models, nonlinear equations are coupled with binary variables. As a result, the optimization problem is a mixed-integer nonlinear programming (MINLP) and challenging to solve [4].

In this work, we first formulate the integrated problem for scheduling of water supply system with the pump operation into a nonconvex MINLP model. We then propose a novel piecewise outer-approximation algorithm to determine the global optimal solution efficiently. A symbolic reformulation approach is employed to reformulate the original MINLP problem to an equivalent standard form. After choosing a set of initial grid points, we use piece-wise linear functions to underestimate nonconvex terms and tangent lines to underestimate convex terms. The original nonconvex MINLP problem can then be reformulated into a relaxed MILP problem. There are an outer loop and an inner loop in the proposed algorithm. The relaxed MILP problem is in the outer loop. In each iteration, the outer-loop problem is solved first to obtain a solution and passes the values of binary variables into the inner loop (Note that binary variables which are used to formulate the linear underestimations are excluded). In the inner loop, there are an NLP problem and an MILP problem. The NLP problem is obtained by fixing the binary variables in the original MINLP problem to the binary values provided by the outer-loop problem. The inner-loop MILP problem is obtained by fixing the binary variables in the outer-loop MILP problem to the current outer-loop solution. The two inner-loop problems are solved iteratively. The grid points are updated adaptively after the inner-loop MILP problem is solved. The current global optimal solution is checked and updated when the inner-loop NLP problem is solved. The inner loop is terminated until the inner-loop MILP problem is infeasible. An integer cut is then added to the outer-loop problem. The entire algorithm problem is terminated until the outer-loop problem is infeasible. The current optimal solution is regarded as the global optimal solution. Two case studies of real water supply network systems are considered to demonstrate the proposed computational framework and solution method. The small-scale case study shows that our proposed method is much more efficient than solving the problem directly with BARON 14 by reducing the computational time by one order of magnitude (16 CPUs vs. 154 CPUs), and returns the same solution as the direct method. For the large-scale case study, our proposed method returns a global optimal solution with 321.69 CPUs, while BARON 14 fails to return a feasible solution within 10-hour computational time limit.

References

[1] B. Coelho and A. Andrade-Campos, "Efficiency achievement in water supply systems—A review," Renewable and Sustainable Energy Reviews, vol. 30, pp. 59-84, 2014.

[2] B. Ulanicki, J. Kahler, and H. See, "Dynamic optimization approach for solving an optimal scheduling problem in water distribution systems," Journal of Water Resources Planning and Management, vol. 133, pp. 23-32, 2007.

[3] C. D'Ambrosio, A. Lodi, S. Wiese, and C. Bragalli, "Mathematical programming techniques in water network optimization," European Journal of Operational Research, vol. 243, pp. 774-788, 2015.

[4] B. Ghaddar, J. Naoum-Sawaya, A. Kishimoto, N. Taheri, and B. Eck, "A Lagrangian decomposition approach for the pump scheduling problem in water networks," European Journal of Operational Research, vol. 241, pp. 490-501, 2015.

Game theory-based life cycle optimisation framework and application

J. Gao, F. You
Cornell University, Ithaca, NY 14850

An optimally designed supply chain is supposed to reflect the best configuration and operation of all the involved elements to achieve the best target performance. Nowadays, supply chain optimisation is vital to the success of industrial concerns more than ever. Despite various research advances in supply chain optimisation, there are a number of research challenges. These include, but are not limited to, multiobjective challenges and multi-player challenges (Garcia and You, 2015). A very active research topic within the context of multiobjective supply chain design is sustainable supply chain design using the life cycle optimisation approach, which integrates life-cycle assessment (LCA) with supply chain optimisation to demonstrate the trade-off between economic and environmental performances (Yue et al., 2013, Gao and You, 2015). Meanwhile, most of the existing work still view the supply chain from a centralized perspective and integrate the various components of the supply chain into a monolithic model. The decision-maker is assumed to have full control of the entire system so that all the decisions can be implemented successfully. However, in practice the management of a supply chain is normally decentralized. Consequently, the optimal strategy obtained from a cooperative model can be practically unachievable (Von Stackelberg, 2010). Therefore, there is growing interest in both industry and academia to investigate competition among the supply chain players and its impact on supply chain optimisation. Theories such as the leader-follower game can be employed to model such scenarios (Yue and You, 2014). The goal of this work is to propose such a modeling framework that can address the trade-off of multiple objectives within a non-cooperative supply chain.

In this work, we propose a leader-follower game-based life cycle optimisation framework, which incorporates the idea of leader-follower game theory and that of life cycle optimisation. A standard leader-follower game involves two players: a leader and a follower. The leader is recognized to have certain advantages enabling it to take actions first (e.g., the priority of customers in a buyer's market), and then the followers reacts to the leader's decisions rationally. With full knowledge of this supply chain, the leader senses more responsibility and cares about both its own profit and the overall environmental performance of this supply chain. Correspondingly, due to limited information about the whole picture of supply chain, the follower only cares about its own profit. Both leader's decisions and follower's involve design and operational decisions. The resulting problem is formulated as a multiobjective bilevel MILP problem. In the first-stage problem (also known as the leader's problem), there are two objectives minimizing the leader's total cost and the overall GHG emission of this supply chain, respectively. In the second-stage problem (i.e., the follower's problem), the objective is to maximize the follower's profit. This problem is extremely challenging to solve using any off-the-shelf general-purpose solvers. Therefore, we present a novel algorithm that can significantly improve the computational efficiency. In order to demonstrate the proposed modeling framework and solution algorithm, a general supply chain case study is considered. In this case study, the distributors of final products are considered as the leader in this supply chain, who will make design and operational decisions ahead of the follower to optimize not only its total cost, but the overall GHG emission as well. The suppliers and manufacturers are considered as the follower, who makes decisions based on the leader's move and solely cares about maximizing its own profit. According to the optimisation results, the lowest cost achievable for the leader is \$4.7 billion. Correspondingly, the follower's profit is \$1.3 billion, and the overall GHG emission is 24 million tons CO2-eq. On the contrary, the highest cost for the leader is \$6.1 billion, with \$1.5 billion profit for the follower. The corresponding GHG emission is 8 million tons CO2-eq.

Reference

J. Gao and F. You, 2015, Shale Gas Supply Chain Design and Operations toward Better Economic and Life Cycle Environmental Performance: MINLP Model and Global Optimization Algorithm, *ACS Sustainable Chemistry & Engineering*, 3(7), pp. 1282-1291.

D. J. Garcia and F. You, 2015, Supply chain design and optimization: Challenges and opportunities, *Computers & Chemical Engineering*, pp. DOI: 10.1016/j.compchemeng.2015.03.015.

H. Von Stackelberg (2010) Market structure and equilibrium. Springer Science & Business Media.

D. Yue and F. You, 2014, Game-theoretic modeling and optimization of multi-echelon supply chain design and operation under Stackelberg game and market equilibrium, *Computers & Chemical Engineering*, 71(0), pp. 347-361.

Integration of heat exchanger network considering the uncertainty of the reactor

Di Zhang, Guilian Liu
School of Chemical Engineering and Technology, Xi'an Jiaotong University, Xianning Rd. 28, 710049 Xi'an, China; tel: +86 29 82664376, email: guilianliui@mail.xjtu.edu.cn.

Integrating the heat exchanger network (HEN) by pinch technology is an efficient method to decrease the energy consumption. In a HEN, the source and sink streams, and their heating and cooling duties are dictated by the reactor and separation and recycle system. Furthermore, the separation and recycle system is defined by the reactor design. Hence, the reactor design affects the supply/target temperature of one or more source/sink streams, and the integration of the HEN. There is the possibility to reduce the utility further by adjusting the reactor design parameters. To achieve this, it is necessary to consider the uncertainty of the reactor design in the HEN integration.

In this work, a graphical method is proposed to analyze the effect of the reactor's uncertainty. Since the variation of the reactor parameters affects the inlet and outlet streams' target and/or supply temperature, the temperature variation of the corresponding source and sink is accommodated by shifting the corresponding inflection point of the composite curve. Correspondingly, the part of the composite curve below or above the inflection point are shifted to target the heating or cooling utility, and this depends on the relative location of the reactor-related sink and source to the pinch point. Based on this method, rules are proposed for targeting the pinch and minimum heating and cooling utility. When the operating parameters of the reactor change, whether the pinch will change or not can be identified easily according to these rules, as well as the minimum heating and cooling utilities. No complicated calculation and graphs are needed.

Optimal design and risk management of shale gas supply chain

J. Gao, F. You
Cornell University, Ithaca, NY 14850

In recent years, the widespread application of horizontal drilling and hydraulic fracturing technologies has led to a "shale revolution" (EIA, 2015). However, despite the optimistic forecast of shale gas production given by the EIA, a recent report by Post Carbon Institute unveils the fact that the actual future of shale gas may not be as bright as the EIA suggests (Hughes, 2014). From a well-by-well based calculation of shale gas production throughout the U.S., they conclude that the actual profitability of a shale well can be significantly affected by the uncertainty in the estimated ultimate recovery (EUR) and astounding decline rates of production ranging from 60% to 90% in the first three years. Therefore, it is essential to design and operate emerging shale gas supply chains with explicit consideration of EUR uncertainty and actual shale gas production profiles. By reviewing the existing works (Cafaro and Grossmann, 2014, Gao and You, 2015, Yang et al., 2015), we have identified the absence of a comprehensive shale gas supply chain model that considers design and planning decisions under uncertainty. Meanwhile, recently published papers on the life cycle assessment of shale gas highlights the large influence of EUR uncertainty, identified as the most critical source of uncertainty that can significantly influence both the economic and environmental performances of shale gas supply chains. Therefore, the goal of this work is to develop a shale gas supply chain model that not only considers both design and planning decisions, but also properly addresses EUR uncertainty and corresponding risk management.

A two-stage stochastic programming approach was considered to deal with the EUR uncertainty in shale gas supply chain, and the resulting problem is developed as a multiobjective SMILP model (You et al., 2009). The objective was to simultaneously optimize the expected total cost and the financial risk. The latter criterion is measured by conditional value-at-risk (CVaR) and downside risk. All of the design decisions are made in the first-stage problem, including drilling schedule, decisions on construction of processing plants, and corresponding pipeline designs. Meanwhile, recourse operating decisions are made in the second stage corresponding to different EUR sampling data, including the shale gas production, planning on shale gas processing, and distribution to end customers. To solve this computationally challenging problem efficiently, we applied the sample average approximation method and proposed a novel algorithm to take advantage of the model structure. One case study based on the Marcellus shale play was presented to illustrate the applicability of the proposed modeling framework and solution algorithm. According to the optimisation results, the stochastic programming model shows great potential to improve the overall economic performance of shale gas supply chain over a deterministic one: the deterministic model fails to provide feasible solution for more than half of the scenarios. Additionally, the stochastic programming model performs quite well even compared with the perfect information model, of which the total cost is only 9% more. Moreover, after risk management, total expected cost increases but the risk of high-cost scenarios is reduced effectively, and the CVaR management shows its advantage over downside risk management in this specific case study.

Reference

D. C. Cafaro and I. E. Grossmann, 2014, Strategic planning, design, and development of the shale gas supply chain network, *AIChE Journal*, 60(6), pp. 21.

EIA (2015) Annual Energy Outlook 2015 with projections to 2040 U.S. Energy Information Administration.

J. Gao and F. You, 2015, Shale Gas Supply Chain Design and Operations toward Better Economic and Life Cycle Environmental Performance: MINLP Model and Global Optimization Algorithm, *ACS Sustainable Chemistry & Engineering*, 3(7), pp. 1282-1291.

J. D. Hughes (2014) Drilling Deeper: A Reality check on U.S. Government Forecasts for A Lasting Tight Oil & Shale Gas Boom: Post Carbon Institure.

L. Yang, I. E. Grossmann, M. S. Mauter and R. M. Dilmore, 2015, Investment optimization model for freshwater acquisition and wastewater handling in shale gas production, *AIChE Journal*, 61(6), pp. 1770-1782.

F. You, J. M. Wassick and I. E. Grossmann, 2009, Risk management for a global supply chain planning under uncertainty: Models and algorithms, *AIChE Journal*, 55(4), pp. 931-946.

Optimal design of biofuel supply chains using novel hybrid life cycle optimization framework

D. Yue, F. You
Cornell University, Ithaca, NY 14850

Life Cycle Optimization (LCO) has been widely employed as a decision support tool to simultaneously optimize economic and environmental objectives for sustainable design and improvement of systems at various scales, i.e., molecule, process, and supply chain (Yue et al., 2014). As known, LCO is an integrated framework that combines Life Cycle Assessment (LCA) and Multi-Objective Optimization (MOO). To our knowledge, most existing LCO applications are based on the traditional process LCA, where the Life Cycle Inventory (LCI) is derived from a presumed process flow chart. While process LCA provides accurate and detailed information for key processes involved in the system, it suffers from system boundary truncation because it is infeasible to include all relevant processes into the flow chart (Suh and Huppes, 2005). Since a large portion of the life cycle is neglected due to the truncated system boundary, any decisions made in process-based LCO studies are based on incomplete life cycle information thus may not be truly optimal (Hanes and Bakshi, 2015).

The goal of this work is to develop an LCO modeling framework that provides truly optimal decisions based on a complete life cycle system boundary. We, for the first time, seamlessly integrate a hybrid LCA method with MOO in a consistent mathematical programming framework. The proposed framework consists of four parts, namely process model, Input-Output (IO) model, upstream and downstream cutoff models. We complement the truncated process system boundary with the IO model that accounts for exchanges between all industrial sectors in the macroeconomic system. The IO model captures all background activities that are typically neglected by the process model, including mining, grain farming, power generation, financial services, etc. This is achieved by using regional/national IO data coupled with averaged sectoral environmental impact factors (Hendrickson et al., 2006). Extensions to Supply and Use Tables (SUTs) and Multi-Region Input-Output (MRIO) model are also discussed. The upstream cutoff model quantifies the inputs from the IO system to the process system, which requires conversion of physical-unit inputs to equivalent expenditures using market price data. On the other hand, the downstream cutoff model quantifies the outputs from the process system that are consumed in the IO system, which reflects the influence of the system of interest to the entire economy. Unlike the static LCI in LCA studies, the LCI in this integrated hybrid LCO model is flexible. We classify a number of activities in the system and devise continuous and integer variables to represent the decisions in each activity. Such decisions would not only influence the process exchanges but also the exchanges between industrial sectors in the economy and exchanges between the process and IO systems. We demonstrate the proposed model with a real-world application to sustainable design of a nationwide UK advanced biofuel supply chain (Akgul et al., 2012), where decisions such as facility location, technology selection, production plan, transportation and logistics are optimized. We account for 40 processes in the process model and 224 industrial sectors within a bi-region IO model consisting of the UK and Rest-Of-World (ROW) (Wiedmann et al., 2011). LCO results indicate that the total project cost increases from 683 to 913 MM GBP/year when the cap on life cycle greenhouse gas (GHG) emission decreases from 2,003 to 777 kton CO2-equivalent/year. It is remarked that up to

58% of the full life cycle GHG emission can result from the IO system, which otherwise would be ignored if only process LCO is performed.

References

O. Akgul, N. Shah, L. G. Papageorgiou, 2012, Economic optimisation of a UK advanced biofuel supply chain, Biomass and Bioenergy, 41, 57-72.

R. J. Hanes, B. R. Bakshi, 2015, Process to planet: A multiscale modeling framework toward sustainable engineering, AIChE Journal, 61(10), 3332-3352.

C. T. Hendrickso, L. B. Lave, H. S. Matthews, (2006) Environmental life cycle assessment of goods and services: an input-output approach. Washington, DC USA: Resources for the Future.

S. Suh, G. Huppes, 2005, Methods for Life Cycle Inventory of a product, Journal of Cleaner Production, 13(7), 687-697.

T. O. Wiedmann, S. Suh, K. Feng, M. Lenzen, A. Acquaye, K. Scott, J. R. Barrett, 2011, Application of Hybrid Life Cycle Approaches to Emerging Energy Technologies – The Case of Wind Power in the UK, Environmental Science & Technology, 45(13), 5900-5907.

D. Yue, F. You,Snyder S. W., 2014, Biomass-to-bioenergy and biofuel supply chain optimization: Overview, key issues and challenges, Computers & Chemical Engineering, 66, 36-56.

Optimization of process conditions for ash reduction from low grade coal by HF acid using response surface methodology

Sushanta K. Behera[1], Sudipto Chakraborty[1], B. C. Meikap[1,2]
[1]Department of Chemical Engineering Indian Institute Technology (IIT) Kharagpur; West Bengal- 721 302, India; e-mail: skbehera@che.iitkgp.ernet.in;
Tel. +91 -3222 283942, e-mail: sc@che.iitkgp.ernet.in;
Tel. +91 -3222 283958, e-mail: bcmeikap@che.iitkgp.ernet.in
[2]Chemical Engineering, School of Engineering, Howard College, University of Kwazulu-Natal, Durban, South Africa

Summary

This research aims to find out the maximum possible ash reduction from low grade coal by optimizing the operating conditions and solvent concentration by using response surface methodology. The effect of three independent variables; HF concentration (10 to 20% by vol.%), temperature (60-100°C), and time (90-180 min) responsible for ash reduction from low grade coal was investigated by using design of experiment. Quadratic models was developed to correlate the independent variable for removal of ash from low grade coal by applying central composite design (CCD). Experimental data were analyzed by using analysis of variance (ANOVA) and the most influential factor on each experimental design response was identified. Process optimization was done by validating the model to obtain maximum ash reduction from low grade coal.

Experimental set-up and technique

The high ash coal samples was collected from Mahanadi coal field (MCL), Orissa India. The lump coal samples were crushed and sieved to -30 to + 72 British standard sieve (BSS) mesh size particle for present studies. Before start of experiment the sample are dried in hot air oven for 12 hrs at 105 °C. The proximate and ultimate analyses of the samples are presented in Table 1. In this work, coal samples were treated with aqueous hydrofluoric acid (HF) solution at various concentration by varying parameters.

Table 1. Chemical analyses of original coal sample

Coal sample	Proximate analysis, wt %, dry basis			Ultimate analysis wt %, daf basis				
	Volatile matter	Ash	Fixed carbon	Carbon	Hydrogen	Nitrogen	Sulfur	Oxygen (diff.)
Ananta	30.6	39.9	29.5	39.37	3.21	0	0.37	57.05

Experiment arrangement used for optimization of reduction of ash from low grade coal was carried out in 1000 ml of Teflon beaker, stirring and heating is provided by magnetic stirrer hot plate (as shown in **Fig. 1**) and performed solution volume of 100 ml with 10 gm coal sample size. After subsequent leaching, the treated sample washed with distilled water to dilute the HF concentration and filtered than dried in the hot air oven at 90 °C for 2 to 3 hrs. The ash content of raw and leached

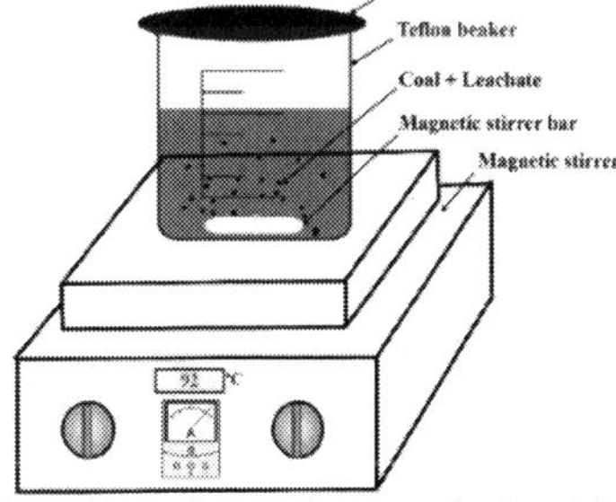

Fig. 1 Experimental set up for leaching of coal.

coal were determined by ASTM D 3174 method.

$$\text{Ash reduction } (\%) = \frac{\text{Raw coal ash} - \text{Leached coal ash}}{\text{Raw coal ash}} \times 100 \qquad (1)$$

Result and Discussion

The statistical software "Design expert (version 8.0.7.1)" has been used for regression analysis of experimental data and draw to response surface plot. The quadratic model has been suggested by the software which was not aliased. The final empirical model interms of coded factor for ash reduction (Y) of coal as in equation (2).

Ash reduction (Y, %) = $74.48 + 3.66\,X_1 + 0.85\,X_2 + 0.43\,X_3 - 0.071\,X_1\,X_2 - 0.064\,X_1\,X_3 - 0.11\,X_2\,X_3 - 0.96\,X_1^2 + 0.33\,X_2^2 + 0.28\,X_3^2$ (2)

To study the combined effect of time and concentration at temperature 80 °C **(Fig. 2),** time and temperature at 15% HF concentration **(Fig. 3),** temperature and concentration at time 135 min on ash reduction of coal **(Fig. 4)** and at evaluate the maximum ash removal at optimizing the conditions **.**

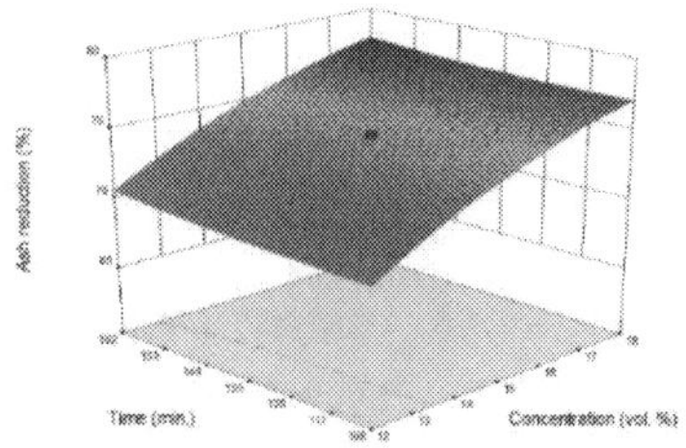

Fig. 2 combined effect of time and concentration on ash reduction at temperature 80 °C

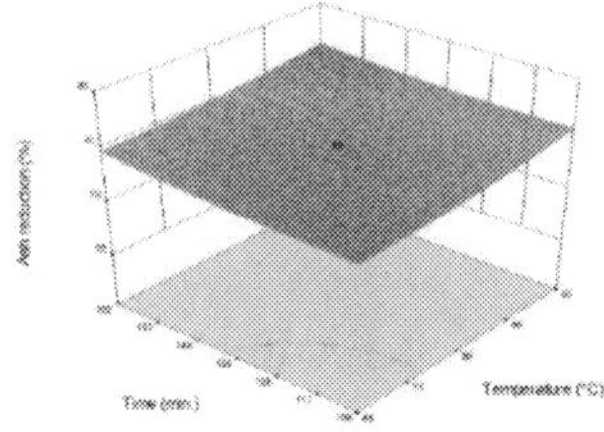

Fig. 3 combined effect of time and temperature on ash reduction at 15% HF concentration.

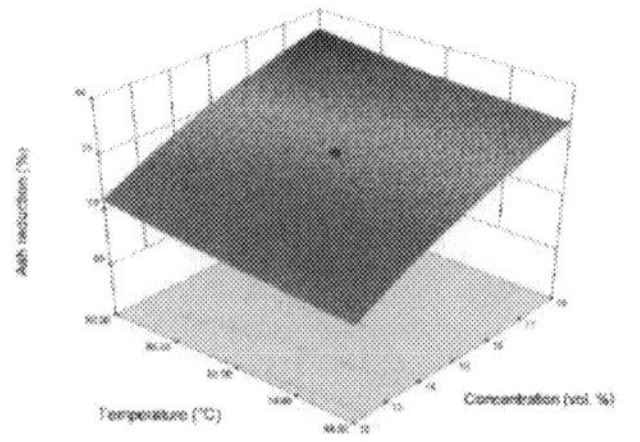

Fig. 4 combined effect of temperature and concentration on ash reduction at time 135

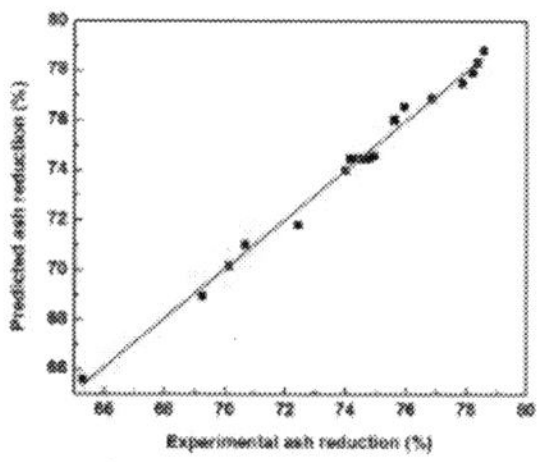

Fig. 5 The actual and predicted plot of coal ash reduction

Conclusion: From response surface modeling the experimental values are satisfied with predicted value (as shown **Fig. 5**) with correlation coefficient of 99%. It was found that from model the predicted ash reduction is direct function of experimental variables. The maximum ash reduction (78.84%) has been found at optimum condition at concentration (18%), temperature (92 °C) and time (162 min).

Reference

1. Sahu, J. N., Acharya, J., Meikap, B. C. Optimization of productions for activated carbons from Tamarind wood by zinc chloride using response surface methodology. *Bioresource Technology*, 101(2010), 1974-1982.
2. Steel, K. M., Patrick, J. W. The production of ultra clean coal by sequential leaching with HF followed by HNO_3. *Fuel*, 82(2003), 1917-1920
3. Ghafari, S., Aziz, H. A., Isa, M. H., Zinatizadeh, A. A. Application of response surface methodology (RSM) to optimize coagulation-flocculation treatment of leachate using poly

aluminum chloride (PAC) and alum. *Journal of Hazardous Materials*, 163(2009), 650-656.

PoPA – SHARPS: A new framework for cost-effective design of hybrid power systems

N.E. Mohammad Rozali[1], S.R. Wan Alwi[2], W.S. Ho[2], Z. A. Manan[2], J.J. Klemeš[3]
Department of Chemical Engineering Department, Universiti Teknologi PETRONAS, 32610 Bandar Seri Iskandar, Perak, Malaysia;
tel. +605 3687582, e-mail: erniza.rozali@petronas.com.my; [2]Process Systems Engineering Centre (PROSPECT), Faculty of Chemical Engineering, Universiti Teknologi Malaysia, 81310 Johor Bahru, Malaysia; [3]Centre for Process Integration and Intensification – CPI[2],Research Institute of Chemical and Process Engineering - MŰKKI, Faculty of Information Technology, University of Pannonia, Egyetem u. 10, H-8200 Veszprém, Hungary.

Development of Power Pinch Analysis (PoPA) for the design and optimisation of Hybrid Power Systems (HPS) is steadily progressing. To date, PoPA has been developed for widespread applications including electricity targeting (Wan Alwi *et al.*, 2012) and allocations (Mohammad Rozali *et al.*, 2013), optimal sizing (Mohammad Rozali *et al.*, 2014), load shifting (Mohammad Rozali *et al.*, 2015a) and HPS storage optimisation (Mohammad Rozali *et al.*, 2015b). Even though PoPA tools are reliable and offer valuable insights for the in-depth analysis of HPS design, the economic aspect still needs more attention.

Systematic Hierarchical Approach for Resilient Process Screening (SHARPS) was introduced by Wan Alwi and Manan (2006), as a cost-screening tool that can lead to a cost-effective network design with the minimum resource utilisation within the desired payback period. Initially, it was developed to economically screen various process changes options, specifically for water network. It was later extended by Lawal *et al.* (2012) to determine a cost-effective carbon emission reduction in buildings. SHARPS has been proven to be an effective cost-screening strategy to enable designers to maximise the potential savings and minimise the investment, ahead of design.

This work presents a new framework for the design of a cost-effective HPS by incorporating PoPA and SHARPS techniques. Minimum electricity targets for HPS powered by renewable and non-renewable sources are first established using PoPA. Afterwards, SHARPS strategy is carried out where the investment versus annual savings (IAS) composite plot covering every power generation technology is plotted. The IAS plot can provide essential quantitative insights for designers to screen all the generation technologies involve, and decide on significant schemes such as using cheaper PV panels, in order to satisfy the desired payback period.

Keywords: Power Pinch Analysis (PoPA), Systematic Hierarchical Approach for Resilient Process Screening (SHARPS), hybrid systems, renewable energy, cost-effective design

References;

Lawal, M., Wan Alwi, S. R., and Manan, Z. A. (2012). A Systematic Method For Cost Effective Carbon Reduction (CECR) In Buildings. *Journal of Applied Sciences*. 12(11), 1186-1190.

Mohammad Rozali, N. E., Wan Alwi, S. R., Abdul Manan, Z., Klemeš, J. J., and Hassan, M. Y. (2014). Optimal sizing of hybrid power systems using power pinch analysis. *Journal of Cleaner Production*. 71, 158-167.

Mohammad Rozali, N. E., Wan Alwi, S. R., Manan, Z. A., and Klemeš, J. J. (2015a). Peak-off-peak load shifting for hybrid power systems based on Power Pinch Analysis. *Energy*. 90, 128-136.

Mohammad Rozali, N. E., Wan Alwi, S. R., Manan, Z. A., Klemeš, J. J., and Hassan, M. Y. (2013). Process integration of hybrid power systems with energy losses considerations. *Energy*. 55, 38-45.

Mohammad Rozali, N. E., Wan Alwi, S. R., Manan, Z. A., Klemeš, J. J., and Hassan, M. Y. (2015b). A process integration approach for design of hybrid power systems with energy storage. *Clean Technologies and Environmental Policy*. 17(7), 2055-2072.

Wan Alwi, S. R., and Manan, Z. A. (2006). SHARPS: A new cost-screening technique to attain cost-effective minimum water network. *AIChE Journal*. 52(11), 3981-3988.

Wan Alwi, S. R., Mohammad Rozali, N. E., Manan, Z. A., and Klemeš, J. J. (2012). A process integration targeting method for hybrid power systems. *Energy*. 44(1), 6-10.

Reduction of carbon footprint for iron and steel industry by Carbon Capture, Utilization and Storage (CCUS) technologies

C.C. Cormos, L. Petrescu, A.M. Cormos
Babes-Bolyai University, Faculty of Chemistry and Chemical Engineering
Arany Janos 11, Postal code: 400028, Cluj-Napoca, Romania
Tel.: +40-264-593833, Fax: +40-264-590818, Email: cormos@chem.ubbcluj.ro

Reducing the greenhouse gas emissions from heat and power sector as well as from other energy-intensive industrial applications (e.g. metallurgy, cement, chemical and petro-chemical processes etc.) is of paramount importance in the modern society. Various carbon capture methods and technologies (e.g. post-, pre- and oxy-combustion) can be applied to reduce the CO_2 emissions [1]. The iron and steel industry is one of the largest industrial sources of CO_2 (about 6% of anthropogenic CO_2 emissions, approx. 1.2 Gt CO_2 / year). Among various carbon capture methods suitable to be implemented in an integrated steel mill several options are evaluated in this paper in post-combustion capture configurations. One capture option is based on gas-liquid absorption using chemical solvents (alkanolamines). Other option is evaluating the calcium looping system used for post-combustion CO_2 capture. As sorbent used in the looping cycle, the natural calcium-based sorbent was evaluated mainly because the easy integration of the spent solid within the steel mill (lime plant). Due to the complexity of integrated steel mill, the analysis was concentrated on the plant sub-systems with the highest CO_2 emissions e.g. power plant, hot stoves etc. which uses the blast & basic oxygen furnace gases to generate the heat and power for steel mill operation. The carbon capture rate is set to be at least 90% relating to the carbon content in the feedstock to be used.

The paper evaluates the conventional size of integrated steel mill with an output of 4 million hot rolled coil (HRC) per year [2]. Mass and energy integration aspects of heat and power plant sub-systems with the carbon capture unit and the influence of various carbon capture options on the overall steel mill performances are discussed. The power plant designs as well as the carbon capture units were modelled and simulated using process flow modelling software (ChemCAD), the simulation results (the mass and energy balances) being used to assess the overall technical and environmental indicators. For comparison reason, a conventional integrated steel mill without carbon capture was also considered. The assessments show that calcium looping system has significant advantages compared not only to benchmark case without capture but also to the gas-liquid absorption CO_2 capture case such as: the higher energy efficiency, lower energy penalty for CO_2 capture, mass and energy integration potential and improved environmental performances.

Keywords: Iron and steel; Carbon Capture, Utilization and Storage (CCUS); Gas-liquid absorption, Calcium looping.

Topic: CO_2 minimisation and mitigation

References:

1. Intergovernmental Panel on Climate Change (IPCC), *Carbon dioxide capture and storage*, Cambridge University Press, UK, 2005.
2. International Energy Agency, Greenhouse Gas R&D Programme (IEAGHG), *Iron and steel CCS study (Techno-economics integrated steel mill)*, Report 2013 / 04, 2013.

Removal of CO_2 in a multistage fluidized bed reactor by monoethanol amine impregnated activated carbon

Dipa Das[a], B. C. Meikap[a,b]
[a]Department of Chemical Engineering, Indian Institute of Technology (IIT) Kharagpur, West Bengal-721302, India.
e-mail: dipadas@che.iitkgp.ernet.in
e-mail: bcmeikap@che.iitkgp.ernet.in, tel:+91-3222-283958; fax:+91-3222-282250
[b]Department of Chemical Engineering,School of Engineering, Howard College,University of Kwazulu-Natal, Durban, South Africa

Summary: The concentration CO_2 in the atmosphere is increasing in a faster rate due to various industrial activities,which causes global warming and climate change. The main contributor for increase of atmospheric CO_2 concentration are fossil fuel combustion for power generation, transport, industry, and domestic use. This trend will rise unless and until we put up a mechanism that will not only curb CO_2 concentration but also put us the development agenda in front. Considering the importance and the seriousness of the problem, increasing attention is being paid to develop processes and equipment for the removal of carbondioxide which can effectively meet the demands of technology and increasingly stringent environmental laws and regulations.The main focus of this paper is based on the capture of CO_2 in a multistage fluidized bed reactor by using amine impregnated activated carbon made from green coconut shell.In this work we developed,designed and fabricated a four staged fluidized bed reactor capable of operating in a continuous regime for the two-phase system at low temperature. The adsorbent used here is activated carbon prepared from green coconut shell and then it has been impregnated with monoethanol amine solution with impregnation ratio 0.6.The effect of the operating parameters such as sorbent flow rate, superficial gas velocity and the weir height on CO_2 removal efficiency in the multistage fluidized bed were investigated.

Experimental setup and procedure: The schematic diagram for four staged fluidized bed reactor was shown in Figure 1. The fluidized bed column was consisted of four stages (0.21m height per stage and 0.095 m internal diameter). Four number of stainless steel made plate (S_1, S_2, S_3, S_4) of 0.002 m thick were used as internal baffles between two stages and each plate was drilled with hole of diameter 0.002 m on a triangular pitch arrangement. Grid plates were covered with fine weir mesh (100 mesh size) with openings smaller than particle size. Down comers (D_1, D_2, D_3, D_4) were made up of perspex cylinder of 0.024 m internal diameter and height of 0.265 m. The weir height was considered to be bed height. The material flows from stage to stage through the down-comer.There were provisions for measuring pressure drop. Gas distributor was there at the bottom of the column. Calibrated rotameter was fitted to measure the air flow rate. A conical hopper was attached at the bottom of the column for storage of the solid. A feeding funnel was there at the top of the column to hold the amine impregnated activated carbon particle and it was attached to the screw feeder. The solid fed to the first stage of the down-

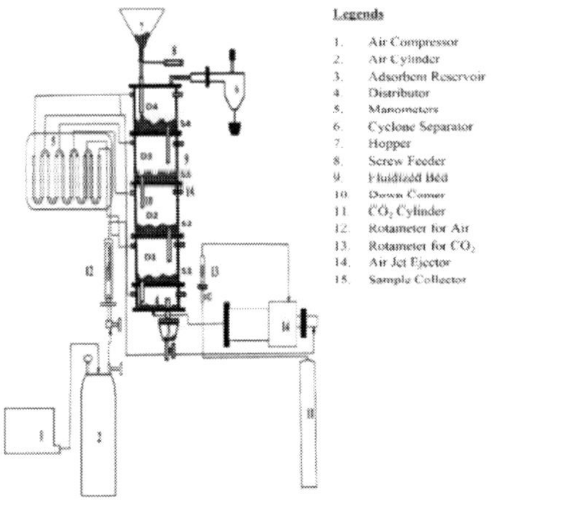

Figure 1..Schematic diagram of the experimental set-up for four stage fluidized bed reactor

comer from the top of the funnel connected to screw feeder and then through Perspex tube. There was provision to feed the air-CO_2 mixture at the bottom of the fluidized bed reactor from the mixing chamber.The gas leaving the top stage was passed through 0.14m diameter cyclone.The experiments were conducted by setting the gas velocity range (0.188-0.353 m/s), solid flow rate range (2.15-4.12 kg/hr) and weir height (30-50 mm). The inlet CO_2 loading concentrations were varied from (3000-20000 ppm). The gas samples were analyzed for CO_2 by the "Orsat Analysis" method. The removal efficiency of CO_2 in each stage can be calculated as:

$$\eta_{CO_2} = \frac{CO_{2\ inlet} - CO_{2\ outlet}}{CO_{2\ inlet}} \times 100 \qquad (1)$$

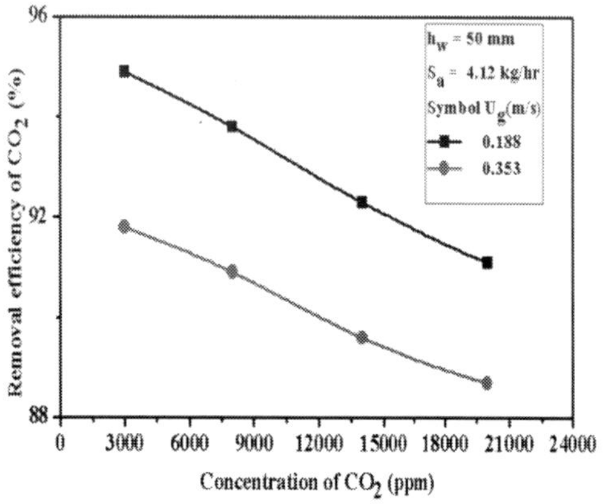

Figure 2. Effect of different gas flow rate on CO_2removal efficieny.

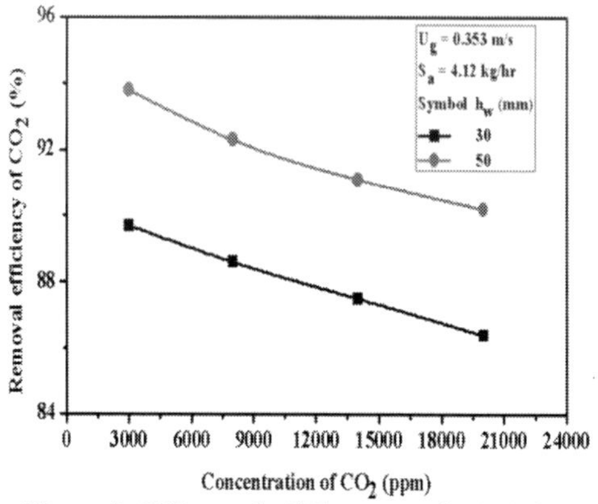

Figure3..Effect of different weir height on CO_2removal efficieny.

Results and discussion: Figure 2. represents that the percentage removal of carbon dioxide decreased with increase in gas velocity due to decrease of solids hold up in the bed there by decreasing the probability of gas-particle interactions.Figure 3. represents that with increase in the weir height the CO_2 removal efficiency increases due to increase in bed volume resulting in more gas solid interaction. Also with increase in inlet CO_2 concentration, the removal efficiency decreases due to formation of the product layer on the adsorbent particles, which results in decrease in sorbent activity.It has been found from the experiment that the maximum removal of CO_2 was 94.9 % under low gas flow rate (0.188 m/s) and high solid flow rate (4.12 kg/hr.) at large weir height of 50 mm.

Conclusion: The result indicates that the removal efficiency of the carbon dioxide was around 95% at a high solid flow rate corresponding to lower gas velocity and at larger weir height at room temperature.

References

1. S. Roy, C.R. Mohanty, Meikap, B. C. Multistage fluidized bed reactor performance characterization for adsorption of carbon dioxide, *Ind. Eng. Chem.Res.*, 48, (2009) 10718-10727.
2. C. R.Mohanty, S.Adapala, B. C.Meikap. Removal of hazardous gaseous pollutants from industrial flue gases by a novel multi-stage fluidized bed desulfurizer,*J. Hazard. Mater.* 165,(2009), 427-434.
3. P.S. Pillay, Y.B.G.Varma. Pressure drop and solid holding time in multistage fluidization,*Powder Technol.* 35(1983) 223–231.

Robust design and synthesis of algae processing network under feedstock and product price uncertainty

J. Gong, F. You
Cornell University, Ithaca, NY, 14850, USA.

With high area productivity, minimum competition with conventional agriculture, low carbon dioxide emissions, and valuable co-products, microalgae provide an excellent feedstock for the production of sustainable biofuel and bioproducts [1]. Much research effort has been devoted to optimal design and synthesis of algal processing networks to identify the optimal pathways based on economic [2] and environmental [3] objectives. The existing models consider deterministic upgrading and manufacturing conditions, while the parameters employed in modelling of these networks are subject to variations resulted from both internal and external sources in practice. An optimal pathway that ignores risks during a decision-making process may lead to suboptimal or even infeasible solutions when uncertainty is inevitably realized. Therefore, it is practical and crucial to tackle uncertainty in processing network optimization models in order to obtain reliable optimal pathways under the presence of various sources of uncertainty. Moreover, return on investment (EOR) is an important economic indicator that shows the efficiency of an investment, but EOR has never been considered an objective function in existing optimization models. It is the aim of this research to obtain the optimal processing route of an algae processing networking under feedstock and product price uncertainty based on EOR.

Based on a comprehensive algae processing network [3], we develop a two-stage adaptive robust mixed-integer fractional programming model to maximize the EOR under feedstock and product price uncertainty. Decisions are made at two stages: (1) a design stage involving the selection of the optimal network configuration; (2) an operational stage pursuing the worst-case operating conditions after the first-stage decisions are made and uncertainty is revealed. The objective function is to maximize the EOR, which is defined as the ratio of net profit divided by total investment. In order to solve the proposed problem, we develop a global optimization algorithm that integrates a parametric algorithm based on root-finding methods to handle the fractional terms in the objective function [4], and a column-and-constraint generation (C&CG) algorithm based on iteratively solving a master problem and a subproblem to tackle multiple inner optimization problems in the model [5]. The optimal processing routes demonstrate higher processing capacities to hedge against feedstock and product price uncertainty, but the use of uncertain budget helps reduce the level of conservatism that is observed in a conventional robust optimization approach.

References

1. R. Davis and A. Aden, "Renewable Diesel from Algal Lipids: An Integrated Baseline for Cost, Emissions, and Resource Potential from a Harmonized Model," Argonne National Laboratory, June 2012.
2. M. Rizwan, J. H. Lee and R. Gani, "Optimal processing pathway for the production of biodiesel from microalgal biomass: A superstructure based approach," *Computers & Chemical Engineering*, vol. 58, pp. 305-314, Nov. 2013.
3. J. Gong and F. Q. You, "Global Optimization for Sustainable Design and Synthesis of Algae Processing Network for CO2 Mitigation and Biofuel Production Using Life Cycle Optimization," *AIChE Journal*, vol. 60, pp. 3195-3210, Sep. 2014.

4. Z. X. Zhong and F. Q. You, "Globally convergent exact and inexact parametric algorithms for solving large-scale mixed-integer fractional programs and applications in process systems engineering," *Computers & Chemical Engineering*, vol. 61, pp. 90-101, Feb. 2014.
5. B. Zeng and L. Zhao, "Solving two-stage robust optimization problems using a column-and-constraint generation method," *Operations Research Letters*, vol. 41, pp. 457-461, Sep. 2013.

Robust design and synthesis of shale gas processing and NGLs recovery processes under uncertain feedstock compositions

J. Gong, F. You
Cornell University, Ithaca, NY, 14850, USA.

Unconventional natural gas from shale, or shale gas, is regarded as an important driver leading the change in North America's energy landscape [1]. Applying mature extraction techniques such as horizontal drilling and hydraulic fracturing, the total natural gas production is predicted to increase by 45% in 2040, majorly boosted by the development of shale gas resources [2]. The fast expansion in upstream shale gas production motivates developing additional facilities to absorb the increasing supplies of shale gas. Raw shale gas primarily consists of methane, natural gas liquids (NGLs), acid gases, nitrogen, and water. In order to transport valuable methane and NGLs products in pipelines for downstream utilization, raw shale gas has to be upgraded in a shale gas processing plant. There are several technologies available to address the separation and upgrading tasks in a shale gas processing plant. However, to the best our knowledge, there is no contribution in the literature to systematically study the overall economic performance considering the best combination of these technologies. In addition, uncertain raw shale gas compositions complicate the design and synthesis of shale gas processes. If a processing plant is designed to process shale gas with fixed compositions, a large amount of off-spec gas will be produced when the real composition deviates from the designed composition. In order to maintain stable supply of qualified gas products, uncertain feedstock compositions must be considered at the design stage. Therefore, it is the aim of this work to obtain the robust design and synthesis of shale gas processing and NGLs recovery processes under uncertain feedstock compositions.

In this work, we propose a comprehensive superstructure of shale gas processing and NGLs recovery for the production of pipeline gas and NGLs from raw shale gas. There are four major sections in the superstructure, namely acid gas removal, dehydration, NGLs recovery, and nitrogen rejection [3-4]. Received raw shale gas is first sent to an amine absorption column to separate carbon dioxide and hydrogen sulfide. The absorbents are regenerated later in a distillation column. For environment protection purposes, the waste sulfur is recovered in a Claus process. Two types of amines absorbents are considered alternatives in this section. In the second section, the acid-gas-free feed undergoes dehydration, which can be achieved by either a glycol absorption process or a condensation process. Next, NGLs are split from methane via a demethanizer column, and constitutes in NGLs are further fractionated through a distillation train. The demethanizer operates under a relatively low temperature, thus requiring a complex utility system. Accordingly, we consider two NGLs recovery process alternatives, which differ with each other in terms of the configuration of the heat exchanger network. Finally, extra nitrogen is split from the methane gas product to increase the higher heating value of the pipeline gas, so that it satisfies the product specifications. Based on the proposed superstructure, we develop a two-stage adaptive robust mixed-integer linear programming model to minimize the total annualized cost. Three groups of constraints are considered, including mass balance constraints, energy balance constraints, and economic evaluation constraints. In the proposed model, integer variables are introduced to model technology selection within the superstructure. The design-stage problem minimizes the annualized investment cost before the uncertainty is revealed; the operational-stage problem minimizes the worst-case annual operating cost, given the design-stage decisions and revealed uncertainty. The optimal solution is compared with results in previous work to show the advantages of the optimal process performance.

References

1. R. A. Kerr, "Natural Gas From Shale Bursts Onto the Scene," *Science*, vol. 328, pp. 1624-1626, June 2010.
2. U. S. EIA, "Annual energy outlook 2015," Washington, DC (DOE/EIA-0383(2015), 2015.
3. C. He and F. You," Toward more cost-effective and greener chemicals production from shale gas by integrating with bioethanol dehydration: Novel process design and simulation-based optimization," *AIChE Journal*, vol. 61, pp. 1209-1232, Apr. 2015.
4. C. He and F. Q. You, "Shale Gas Processing Integrated with Ethylene Production: Novel Process Designs, Exergy Analysis, and Techno-Economic Analysis," *Industrial & Engineering Chemistry Research*, vol. 53, pp. 11442-11459, July 2014.

Sustainable design and operation of supply chains under uncertainty with life cycle optimisation and stochastic programming

J. Gao, F. You
Cornell University, Ithaca, NY 14850

Extensive research in both academia and industry has been done to improve the economic performance for the design and operation of supply chains. In recent years, there have been an increasing awareness of the importance to obtain a sustainable design of supply chains (Garcia and You, 2015). Accordingly, the life cycle optimisation framework that integrates life cycle assessment with multiobjective optimisation is developed as an important tool for sustainable design of supply chains under both economic and environmental concerns (Yue and You, 2013). Meanwhile, another research challenge in supply chain optimisation is known as properly addressing the uncertainties in supply chain design. The techno-economic feasibility of a supply chain is subjected to various sources of uncertainties, which will eventually affect both the economic and environmental performance of the system. Approaches such as stochastic programming and robust optimisation are identified as powerful tools to address the uncertainties emerging in optimisation problems (You and Grossmann). However, by reviewing the existing publications, a gap is found between these two important challenges. Therefore, the goal of this work is to address the sustainable design and operation of supply chains under uncertainty using novel modelling framework integrating both life cycle optimisation and stochastic programming techniques.

In this work, we propose a novel modelling framework for the life cycle optimisation of supply chains under uncertainty, which incorporates both the life cycle optimisation and the stochastic programming techniques. The economic objective is to minimize the expected total cost per functional unit, and the environmental objective is to minimize the expected total GHG emissions per corresponding to each functional unit. The resulting problem is a stochastic mixed-integer linear fractional programming problem (SMILFP), which cannot be solved directly by any off-the-shelf solver. In order to tackle this challenging problem, we propose a novel algorithm integrating the parametric algorithm and a multi-cut L-shaped method. In the outer loop of this algorithm, we apply the parametric algorithm to solve an SMILP subproblem; in the inner loop, we apply the multi-cut L-shaped method to further decompose the SMILP problem. The proposed modelling framework and solution algorithm are illustrated through a state-wide county-level hydrocarbon biofuels supply chain in Illinois, United States. Both the biomass supply uncertainty and demand uncertainty of biofuel products are considered. Major design and operational decisions include the selection of biomass supply regions, construction of biorefineries, selection of processing technologies, production, inventory, and distribution plan, etc. A Pareto-curve is obtained showing the trade-off between the expected cost and GHG emissions per unit gasoline equivalent gallon biofuel under the impacts of uncertainties.

Reference

D. J. Garcia and F. You, 2015, Supply chain design and optimization: Challenges and opportunities, *Computers & Chemical Engineering,* 81, pp. 153-170.

D. Yue, M. A. Kim and F. You, 2013, Design of Sustainable Product Systems and Supply Chains with Life Cycle Optimization Based on Functional Unit: General Modeling Framework, Mixed-Integer Nonlinear Programming Algorithms and Case Study on Hydrocarbon Biofuels, *Acs Sustainable Chemistry & Engineering,* 1(8), pp. 1003-1014.

F. You, J. M. Wassick and I. E. Grossmann, 2009, Risk management for a global supply chain planning under uncertainty: Models and algorithms, *AIChE Journal,* 55(4), pp. 931-946.

Towards the integration of the ethyl (S)-3-hydroxybutyrate production process into a biorefinery concept

R. Hirschmann, N. Borodkin, [1]F. Baganz, V.C. Hass
Hochschule Furtwangen University, Jakob-Kienzle-Str. 17, D-78056 Villingen-Schwenningen, Germany; tel. +49 7720 307 4253, e-mail: volker.hass@hs-furtwangen.de;
[1]University College London, Dept. of Biochemical Engineering, Gower Street, London WC1E, Great Britain.

Ethyl (S)-3-hydroxybutyrate (E3HB) is naturally found as part of kiwi fruit aroma and is an important precursor for the production of several pharmaceuticals. E3HB can be produced in a whole cell biotransformation process from Ethylacetoacetate (EAA), using the yeast *Saccharomyces cerevisiae* as the whole cell biocatalyst.

One way of improving the sustainability of such processes may be their integration into biorefinery concepts. According to the International Energy Agency (Task 42) "Biorefining is the sustainable processing of biomass into a spectrum of marketable products and energy." (IEA 2013). A key property of biorefineries is that the educts are almost completely transformed into products and the amount of non-utilizable products is strictly minimized. Thus, the integration of already existing processes into biorefinery concepts requires research into two directions. First, the process needs to be characterized with respect to its product and side product spectrum under different operational conditions and second, process strategies need to be found minimizing the amount of non-marketable or non-utilizable side products. In this contribution we will focus on the experimental investigation of the E3HB-process under aerobic and anaerobic conditions. In particular anaerobic E3HB production has been poorly characterized so far, although anaerobic processes often exhibit a very promising resource efficiency.

In the course of our research six biotransformation processes have been carried out in 5 to 20 l stirred tank bioreactors, equipped with online probes for temperature, pH and dissolved oxygen tension. The exhaust gas concentrations of O_2 und CO_2 (aerobic conditions) and the volumetric flow rate of CO_2 (anaerobic conditions) were also measured online. Samples were taken to determine concentrations of glucose, biomass, ethanol, EAA and E3HB. EAA and E3HB concentrations were determined by high pressure liquid chromatography (HPLC). Additionally, HPLC was used to determine the concentrations of side products such as organic acids. The six biotransformation runs were performed using the operational conditions given in table 1.

Table 1: E3HB production under aerobic and anaerobic conditions in fed-batch and continous mode

run	mode	runtime h	feed rate h^{-1}	yield $g_{E3HB}\ g_{EAA}^{-1}$	final conc. $g_{E3HB}\ l^{-1}$	production rate $g_{E3HB}\ g_{biomass}\ h^{-1}$
1	aerobic fed-batch	72	0.0267	0.64	12	0.0080
2	anaerobic fed-batch	72	0.0267	0.19	2	0.0041
3	aerobic fed-batch	52	0.1000	0.31	7	0.0120
4	aerobic fed-batch	52	0.0200	0.80	4	0.0061
5	aerobic conti/perfusion	72	0.014 - 0.18 stepwise increasing	0.64	12	0.0060
6	anaerobic conti/perfusion	72	0.014 – 0.18 stepwise increasing	0.46	5	0.0068

Figure 1 illustrates a typical process. After inoculation yeast is grown under aerobic conditions in fed-batch mode in order to maximize biomass yield and minimize ethanol production. After approximately 23 h a biomass concentration of 20 to 25 g l^{-1} is achieved. The feed of glucose solution is reduced and the biotransformation of EAA to E3HB starts by feeding EAA-solution. At this point the E3HB concentration begins to increase, biomass growth is stopped and the ethanol concentration slightly increases as well

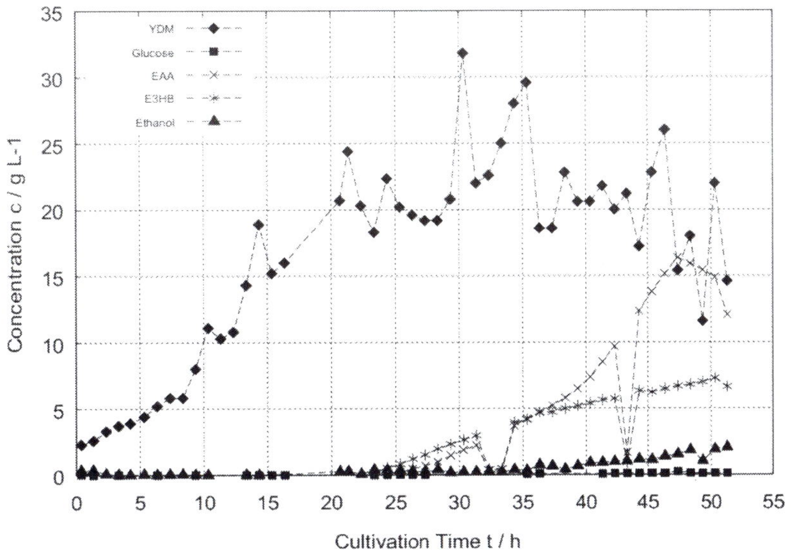

Figure 1: Production of Ethyl (S)-3-hydroxybutyrate in a whole cell biotransformation process

The final product concentrations and yields depend on the operational conditions applied (table 1). In our aerobic experiments a specific E3HB-production rate of $4.1*10^{-3}$ to $1.2*10^{-2}$ g_{E3HB} $g_{biomass}^{-1}$ h^{-1} was observed, which is well within the range of the reported E3HB-production rate of $2*10^{-3}$ to $5.4*10^{-2}$ g_{E3HB} $g_{biomass}^{-1}$ h^{-1} (Sybesma et al. 1998). However, a mass balance analysis of our data indicates that significant amounts of side products were produced. In contrast to the findings of Sybesma et al. (1998) and Rohner et al. (1988) who did not expect E3HB to be produced under anaerobic conditions, we found a specific E3HB-production rate of up to $6.8*10^{-3}$ g_{E3HB} $g_{biomass}^{-1}$ h^{-1}. These findings support the integration the E3HB biotransformation into a biorefinery concept, where the anaerobically produced CO_2 is further utilized (Reule et al. 2015).

Literature

IEA (2013): Task 42: Biorefinery. Available online at http://www.iea-bioenergy.task42-biorefineries.com, checked on 7/9/2015.

Reule, W.; Oppenländer, T.; Hass, V.C.; Hirschmann, R. (2015) (submitted to patent office): Bioraffinerie Verfahren. Patent no. F 2182 - uh / mbr / ed - in assessment.

Rohner, M.; Locher, G.; Sonnleitner B.; Fiechter, A. (1988): Kinetics and modeling of the stereoselective reduction of acetoacetic acid esters by continuously growing cultures of Saccharomyces cerevisiae. In *Journal of Biotechnology* (9), pp. 11–28.

Sybesma, W. F. H.; StraathofA. J. J.; Jongejan J. A.; Pronk, J. T.; Heijnen, J. J. (1998): Reductions of 3-oxo Esters by Baker's Yeast: Current Status. In *Bioanalytics and Biotransformation* (16), pp. 95–134.

Two-stage adjustable robust scheduling against uncertainty for batch processes with general network structure

H. Shi, [1]F. You
Cornell University, Ithaca, NY 14850.

Batch processes have been widely used in the process industry to manufacture all kinds of high-value-added chemicals [1]. Production scheduling is one of the most crucial decisions because of its critical influence on production efficiency and economic performance. Extensive research efforts have been directed towards constructing and solving the deterministic optimization models for batch scheduling [2]. In deterministic models, the parameters used in the production processes are considered to have fixed values. However, this deterministic assumption is not realistic, since uncertainty is inevitable in practice [3]. Conventional robust optimization approaches have been widely used to tackle uncertainty in production scheduling problems [4]. Conventional robust optimization is a modelling framework that only needs moderate information about the uncertainty. However, the optimal solution returned by the conventional robust optimization method takes into account all the realizations of uncertain parameters. As a result, the solution of conventional robust optimization approaches tends to be overly conservative.

In this work, we propose a novel computational framework and solution approaches for two-stage adjustable robust scheduling of batch processes under uncertainty. We first reformulate the deterministic mixed-integer linear programming (MILP) model for batch scheduling problem into a two-stage optimization problem. Symmetric uncertainty sets are then introduced to confine the uncertain parameters, and budgets of uncertainty are used to adjust the degree of conservatism. We then apply both the Benders decomposition algorithm and the column-and-constraint generation (C&CG) algorithm to efficiently solve the resulting two-stage ARO problem, which cannot be tackled directly by any existing optimization solvers. For each solution algorithm, there are a first-stage problem and a second-stage problem which are solved iteratively until the stopping criterion is met. In each iteration of the Benders decomposition algorithm, the second-stage problem provides the master problem with its dual solution, and the master problem generates either an optimality cut or a feasibility cut with the dual solution. In each iteration of the C&CG algorithm, the second-stage problem only provides the master problem with a non-trivial solution, and the master problem generates a new set of second-stage decision variables and a new column of new constraints. Two case studies for batch manufacturing processes are considered to demonstrate the applicability of the proposed computational framework and solution algorithms. The results show that the C&CG algorithm is more computationally efficient than the Benders decomposition algorithm, and the proposed two-stage ARO approach returns 9% higher profits than the conventional robust optimization approach for batch scheduling.

References

[1] C. A. Méndez, J. Cerdá, I. E. Grossmann, I. Harjunkoski, and M. Fahl, "State-of-the-art review of optimization methods for short-term scheduling of batch processes," Computers & Chemical Engineering, vol. 30, pp. 913-946, 2006.

[2] R. E. Burkard and J. Hatzl, "Review, extensions and computational comparison of MILP formulations for scheduling of batch processes," Computers & chemical engineering, vol. 29, pp. 1752-1769, 2005.

[3] Y. Chu, F. You, J. M. Wassick, and A. Agarwal, "Integrated planning and scheduling under production uncertainties: Bi-level model formulation and hybrid solution method," Computers & Chemical Engineering, vol. 72, pp. 255-272, 2015.

[4] Z. Li and M. G. Ierapetritou, "Robust optimization for process scheduling under uncertainty," Industrial & Engineering Chemistry Research, vol. 47, pp. 4148-4157, 2008.

Optimal design and operation of integrated centralized and decentralized energy systems

Liu Wen Hui, Ho Wai Shin*, Lim Jeng Shiun, Haslenda Hashim, Zarina Ab Muis
PROSPECT, Faculty of Chemical and Energy Engineering, Universiti Teknologi Malaysia, Skudai, Johor, Malaysia

Concerns over sustainability of fossil fuels, and increasing awareness for the environment have encouraged countries all over the world to shift from the heavy reliance on fossil fuel to renewable energy (RE) resources for electricity generation. Although implementation of RE has been on the rise, large-scale deployment of RE still remains a challenge, especially in terms of economic and technicality. This paper aim to provide an optimal solution to integrate the current energy system (centralized energy system running on mainly fossil fuels) with the new energy system (the distributed/decentralized energy system) that is economically and technically sound. The analysis is based on a mixed integer programming approach where the model proposed in this study is a time oriented model which is separated into two parts. The first part consist of analysis on a yearly scale where optimal strategy to construct, shutdown, or retrofit, new, old, and existing plant will be revealed. The second part of the model which consist of analysis on an hourly scale will reveal the optimal scheduling and operation of both systems, the interaction between systems will also be defined and optimized. The model will implement a case study from Iskandar Malaysia to demonstrate its capability. The results will aid in development of Iskandar Malaysia at the same time show cases the usefulness of the model as a decision planning and policy making tool.

A novel mathematical method for the simultaneous optimization of the hydrogen network and the hydrogen consuming reactor

Yingjia Wang, Guilian Liu
School of Chemical Engineering and Technology, Xi'an Jiaotong University, Xianning Rd. 28, 710049 Xi'an, China; tel: +86 29 82664376, email: guilianliui@mail.xjtu.edu.cn.

In refinery, fresh hydrogen is an important utility, and its consumption increases rapidly as the processing of the heavier, high-sulfur crude oil increases. The hydrogen consuming reactors are key equipments, and its operating parameters have a great effect on the hydrogen network integration and the minimum fresh hydrogen consumption, as it connecting the hydrogen source and hydrogen sink. The optimization of the reactor operating parameters and hydrogen network should be considered simultaneously.

This paper proposed a new mathematical programming method for the optimization of the hydrogen network. The objective is to increase the economic performance of the hydrogen network and the hydrogen-consuming reactor. In this model, the match between hydrogen source and hydrogen sink, sulfur content of the feed oil, hydrogen consumption and the operating conditions of the reactor considered simultaneously. The latter includes the residence time, pressure, temperature and hydrogen to oil ratio, which influence the hydrogen consumption and the capital cost of the reactor. The quantitative relation between the hydrogen consumption of the reactor and these parameters are employed in the model, instead of directly using the black-box model. Based on this, the overall optimization of the hydrogen network can be achieved easily, with the optimal hydrogen distribution network and the optimal operating conditions of the reactor targeted. The hydrogen network of a refinery is optimized by the proposed method.

Analysis of the influence of permeability in multiphase flow in porous media using Computational Fluid Dynamics (CFD)

A. T. S. Souza, [1]D.da C. Ribeiro, [2]A. P. Meneguelo
Department of Engineering e Technology, Federal University of Espírito Santo – UFES campus São Mateus, Rodovia BR 101 Norte, km 60, Litorâneo, CEP 29.932-540, São Mateus, ES, Brasil; e-mail: alessandra.t.souza@aluno.ufes.com,
[1]daniel.ribeiro@ufes.com.br,[2]ana.meneguelo@ufes.br;

In the petroleum industry, as well as in the others, all the processes should be planned and analyzed seeking the most economical, efficient and with fewer risks to the environment. In this case, the behavior prediction of oil reservoirs is critical to achieving these goals.

Among the properties that characterize an oil reservoir, porosity and permeability are commonly used but they can vary greatly along the porous medium. Thus it is necessary a large set of information, that is organized into a fine mesh, to represent them. However, the greater the amount of information the higher is the computational cost to analyze them (Menezes Filho et al, 2010).

The stored oil is produced through a pressure differential between the reservoir rock and the well. In some cases the rock has sufficient primary energy to maintain oil production, in others not. In cases where it is present in sufficient quantity with the continuous production of the oil, the primary energy dissipates and production decreases.

To maintain a viable production, recovery methods are employed. They can act providing an additional secondary energy or reducing the viscous resistance of the fluid. Analyzing the first option, the recovery method via water injection is the most widely used recovery techniques and mayemploy different arrangements of injection and production wells(Araujo e Neto, 2009).

This article analyzes the influence of the level of information about permeability in the prediction of water and oil flow and breakthrough time, simulating a water injection process in a 2D porous medium saturated of oil, in a geometry of 100 m², in three meshes with different values and configurations of permeability. The contribution of this work is using the Computational Fluid Dynamics (CFD) framework for solving oil reservoir flow which, in general, is deal with specialized reservoir flow simulators.

The simulations were performed on software ANSYS CFX v15.0, which describes the isothermal flow in porous media by the mass conservation equation (1) and an anisotropic version of Darcy's law (2), solving them into discrete volumes defined according the finite volume method based on elements.

$$\frac{\partial}{\partial t}(\emptyset\rho) + \nabla \cdot (\rho \boldsymbol{K} \cdot \boldsymbol{U}) = 0 \tag{1}$$

$$U = -R^{-1}.\nabla p \tag{2}$$

where $\boldsymbol{U}$ is the true velocity, $\emptyset$ is the porosity, ρ is density, $\boldsymbol{K} = (K^{ij})$ is a symmetric rank two tensor, called the area porosity tensor and $\boldsymbol{R} = (R^{ij})$ represents a resistance to flow in the porous medium and it is in general a symmetric positive definite second rank tensor, in order to account for possible anisotropies in the resistance. (ANSYS CFX, 2014).

Results showed that the variation of results among different meshes is significant and can lead to wrong judgments about the efficiency of the method used. CFD results are compared to an open-source reservoir simulator code. This comparison showed that the flow is reproduced qualitatively but final oil recovery factor is overpredicted. The simulation time was also analyzed, the meshes showed differences of up to ten days, which reinforces the idea that the greater the degree of information the higher the computational cost. Thus the head of the project must balance these variables.

Reference

Araújo, M. De V.and Farias Neto, S. R. de. *Estudo numérico da produção de óleo em reservatórios petrolíferos via injeção de água*. In: Congress of Scientific Research in UFCG, IV, Campina Grande-Paraíba, Brasil. 2009.

Menezes Filho, D. de S.; Carvalho, D. K. E. De; Willmersdorf, R. B. and Lyra, P. R. M.. *Técnicas clássicas de transferência de escala para simulações de escoamentos em reservatórios de petróleo altamente heterogêneos.* In: National Congress of Mechanical Engineering, VI, Campina Grande-Paraíba, Brasil. 2010.

ANSYS CFX. Theory Guide v 15, 2014.

Comparative study of the CPC solar collectors

G.O. Prado[1], N.G.Sousa[2]
[1]Federal University of Triângulo Mineiro, Dep. Industrial Engineering, Av. Dr. Randolfo Borges Jr., 1250, Univerdecidade, Uberaba-MG, Brazil;
email:webgop@yahoo.com.br. [2] Dep. Chemical Engineering

Solar energy is one of the main references in the field of renewable energy it is considered clean and practically inexhaustible. The acquisition of this energy source depends, among other factors, of the capture area available. Thus, a more effective interception of the solar rays becomes desirable and feasible using solar concentrators. Such equipment may be mobile, when it has a solar tracking mechanism, or static. Aiming at a system with simplified deployment and use reduced operating resources, the aim of this study was to research and catalog the basic settings of construction and performance of parabolic solar concentrators without tracking, of the CPC type (Compound Parabolic Concentrator). For this, it were identified and analyzed the main characterization parameters of optical, productives, thermals and of functionality, available in the reference area.

Indicators for energy management in cooling systems

Martínez Patiño J. [(1)], Picón-Núñez M. [(2)], Ireta-Moreno F. [(1)], Macías Aguilera F.F[(1)], Rubio-Maya C. [(3)]
[(1)]Department of Electrical Engineering
[(2)]Department of Chemical Engineering.
University of Guanajuato. Carr. Salamanca-Valle de Santiago Km. 3.5+1.8 Km. Comunidad de Palo Blanco, C.P. 36885 Salamanca, Gto. Tel. +52 (464) 64 7 99 40. México.
[(3)]Faculty of Mechanical Engineering, UMSNH, Morelia, Michoacán, Mexico
jesusmp23@ugto.mx

Industrial processes involves different operations where energy is used, either thermal or electrical. To do a good energy management is essential establish indicators that relate directly or indirectly to energy.

There are indications of a global nature or particular inside an industry, even indicators for certain processes. From there the importance to know what type, when, and where the energy is consumed in every industrial process.

A study case is shown in this article where indicators obtained from cooling system are reviewed, including thermal and power electric indicators. The study case will show a way to formulate an energy behavior profile that will improve the energy management in this kind of systems.

Indirect land use change effects of bioconversion technologies

D. J. Garcia, F. You
Cornell University, Ithaca, NY, 14850

Indirect land use change must be a central consideration for bioconversion technologies. Bioconverstion technolgoies, such as corn-to-ethanol technologies, require land not only for the processing facility itself but also indirectly require land for the cultivation of the input bioamss, for example, corn. It is possible that this land could replace or be cultivated in addition to land previously used for food, causing a food versus fuel debate. Often, this indirect land use requirement is not considered in the optimization of bioconverstion processes, processing pathways, or supply chains. However, when it is considered as an objective, the results could be significant, as demonstrated recently by Ferreira-Filho and Horridge in a study in Brazil [1]. Therefore, it is integral to understand the indirect land use requirements of bioconversion technologies to truly and holistically understand the sustainability of biofuels and bioproducts, especially compared to their traditionally produced counterparts.

A comprehensive biomass to bioproducts and biofuels product and process network is optimized over economic and indirect land use criterion to find optimal biomass processing pathways. Hundreds of materials and components are considered as well as 200 bioconversion technologies. Indirect land use change is calculated as the amount of land required to grow, cultivate, and harvest each type of biomass in the network. Various types of biomass are included in the network, such as softwood, sugarcane, corn, and soybeans, and each type of biomass has a unique amount of land required for cultivation. Different yields per land area are considered for each type of biomass. Data for the processing technologies are taken from existing technoeconomic analyses and peer-reviewed research articles. Three objectives of maximization of NPV, maximization of NPV per amount of indirect land required, and the minimization of the indirect land required are considered, resulting in a multiobjective, mixed integer nonlinear fractional program (MINLP). A solution method using NLP subproblems, the parametric algorithm, the ε-constraint method, and a branch-and-refine algorithm using piecewise linear approximations is employed to simplify the model and solve it efficiently and quickly. The results have important environmental and land use change implications for the bioconversion processing sector and is flexible enough to easily accomodate biomass types from different agricultural regions around the world. In this work, representative biomass land requirements for the U.S. state of Illinois are considered.

References

1. J. B. Ferreira-Filho, M. Horridge. "Ethanol Expansion and Indirect Land Use Change in Brazil," Land Use Policy, 36, 595-604, 2014.

Modelling and economic evaluation of Sulcis coal chemical cleaning process

Elisabetta Fois, Antonio Lallai[1], Paola Meloni[1]
Area Impianti di Superficie, Carbosulcis S.p.A., Miniera Monte Sinni, 09010 Gonnesa (CI), Italy, e-mail: fois.elisabetta@carbosulcis.eu; [1]Dipartimento di Ingegneria Meccanica, Chimica e dei Materiali, Università di Cagliari, Piazza d'Armi, 09123 Cagliari (CA), Italy, [1]e-mail: antonio.lallai@dimcm.unica.it; [1]e-mail: meloni@dimcm.unica.it

Low-rank and low quality coal resources are found worldwide, bearing high contents of ashes and sulphur. Efficient and environmentally friendly utilization of low-rank coal has become an important aspect of clean coal technology development. Recent years have seen concerted research efforts to understand the structure and properties of low-rank coals and rapid progress in the development of innovative new technologies for low-rank coal utilization.

Pyritic sulphur and mineral matter are generally removed from coal through gravity concentration or froth-flotation processes, but these methods cannot remove organic sulphur, which in some cases reaches 50% of the total, and requires to be processed through chemical desulphurisation methods, also effective in removing part of inorganic sulphur and mineral matter.

A leaching process in the desulphurization of Sulcis coal was investigated under various leaching conditions. Moreover, this work shows results of the experimental tests which are conducted on Sulcis coal samples by sequential leaching of potassium hydroxide followed by hydrogen peroxide. The tests carried out verify the technical and efficiency of the process. In order to study the mechanism of the leaching treatment, micromorphology of the leached coal was observed by SEM (Scanning Electron Microscope).

Parameters investigated included coal top-size, chemical reagent composition, slurry concentration, reaction temperature, and reaction time. Additional investigations include optimization of coal leaching parameters, product recovery, product stability on both process performance and economics. Finally, we carried out a comparison between the results obtained from the simulation model and the experimental tests results.

The high efficiency of sulphur reduction, on average about 30%, was found to be of special technical relevance for relatively large particle sizes, <5.60 mm, said efficiency being obtained through tests carried out on different particle size classes of leached samples.

It was found that this process enables pyritic sulphur, since it was observed that potassium hydroxide can selectively acts on pyritic sulphur. The sulphur reduction altogether achieved was conveniently associated with a very small reduction in calorific value; this makes the presented process particularly convenient, from both the implementation and the economic points of view. This can be ascribed to the advantageous selection of the type of reagents used and the mild operating conditions, i.e. relatively short times and low temperatures. In fact, it should be noted that the concentrations of these reagents enable the coal matrix to be subjected to a not too aggressive basic attack, hence achieving an economic saving on the reagent, whose cost remarkably affects the process operating costs. At the same time, humic acids and/or humates are extracted from the caustic leaching solution resulting at the end of the process. Thus, this preferred embodiment allows to obtain an alkaline humic extract from Sulcis coal. Surprisingly, a by-product, such as the leaching solution coming from the caustic step of the process, indeed finds a convenient direct or indirect application to commercial agriculture, as fluid extract with a high content of humic substances present therein.

Optimising natural and engineering systems: A foundation for a community-based sustainability modelling framework

D. J. Garcia, F. You
Cornell University, Ithaca, NY, 14850

Sustainability is increasingly an integral goal of process and product design and is a high priority to researchers, policymakers, and businesses around the world. However, it is difficult to specify exactly what sustainable communities quantitatively look like. For example, consider a last remaining empty plot of undeveloped land in a small town. Since this is the last plot of undeveloped land, it currently functions as a final refuge for local wildlife. As tough economic times grip this town, development of the land is proposed to galvanize the economy. The town can build a park, a wildlife reserve, or can allow a factory to expand. Which solution is the most sustainable? The oft-quoted definition of sustainable development as "development that meets the needs of the present without compromising the ability of future generations to meet their own needs" [1] is broadly useful, but it does not provide tools to answer this question. Ultimately, this is an optimisation question, and if quantitative tools can be used to answer it, then we can find a most sustainable solution. We propose a community-based sustainability optimisation modeling framework to address this question.

Communities are modeled as grids. Each grid area is described by a set of relevant parameters (land type, flora and fauna, wind speed distribution, population, types of buildings, water availability, human services provided, pollution types and rates, etc.). Sets of possible projects and policies are given (e.g. urban farming, factory construction, park construction, planting of trees, wildlife preserve, taxes on CO_2 emissions, etc.). Projects and policies are given anticipated capital costs, operating costs, environmental impacts, social impacts, and location of implementation (e.g. rooftop solar can only go on grid areas with buildings), among other key parameters. Each project or policy can affect other project or policies in other areas on the grid (e.g. building a riverfront park will increase the economy of the surrounding neighborhood by increasing the local economic value of water). Ecosystem services (ES), such as air purification, runoff mitigation, noise reduction and recreation, are supplied and demanded by the surrounding environment and the human population, respectively [2]. A most sustainable option can be found by maximizing expected economic and social benefits and minimizing environmental externalities. For example, a community might choose to build a CO_2-emitting factory and plant enough trees to offset the emissions created (creating temporary planting jobs in the meantime).

As this is an enormously broad and ambitious project, we start with a simplified version of the holistic community sustainability model. Environmental phenomena we consider include air-purification and runoff mitigation ESs, effects of human emissions and wastes on the surrounding biosphere (and vice versa), etc. Economic/social aspects include the recreational value of water, numbers of jobs created, GDP per capita of the town, effects of the environment on human health and safety, etc. Some of the multiple objectives include cost minimization, water consumption minimization, greenhouse gas emission minimization, and social benefit maximization. A case study centered on a small town in the region of central Illinois, USA is presented.

References

1. G. Brundtland, M. Khalid, S. Agnelli, S. Al-Athel, B. Chidzero, L. Fadika and V. Hauff et al. "Our Common Future (Brundtland Report)," Oxford University Press, 1987.

2. F. Baró, D. Haase, E. Gómez-Baggethun, and N. Frantzeskaki. "Mismatches between ecosystem services supply and demand in urban areas: A quantitative assessment in five European cities," _Ecological Indicators_, vol. 55, pp. 146-158, 2015.

Screening of reaction-separation processes with recycle streams based on distillation efficiency

[1] J. Bonet, [1,2]A.E. Bonet Ruiz, [2]V. Plesu, [2]P. Iancu, [1]J. Llorens
[1]University of Barcelona, Faculty of Chemistry, Department of Chemical Engineering, Martí i Franqués St. 1, Floor 6, E-08028, Barcelona, Spain; tel. +34 93 4021310, e-mail: bonet@ub.edu; [2] University POLITEHNICA of Bucharest, Centre for Technology Transfer in Process Industries (CTTPI), Gh. Polizu St. 1, Bldg. A, Room A056, RO-011061, Bucharest, Romania; tel. + 40 21 4023808, fax: +40 21 3185900, e-mail: a_bonet@chim.upb.ro.

A new process scheme structure is usually proposed based on heuristics, solutions of similar problems and some experience or intuition. On the other hand, optimization based on design methods provides novel alternatives but their use is limited mainly to the own method developers. Heuristics are still present in chemical engineering education and in chemical engineering practice for preliminary conceptual design, despite of the increasing computing power. A well-known and accepted heuristic for distillation columns is that the optimal reflux is around 1.2 to 1.5 times the minimum reflux. This heuristic is valid independently of energy and steel costs fluctuations due to the sharp increases of reflux or number of stages outside the optimal region and that the overall costs function is quite flat around the minimum. A similar heuristic also exists for the recycle streams, being the optimum around 1.2 times the minimum recycle flow rate. Depending on the catalyst cost and separation difficulty this value becomes higher, e.g. 1.4, or lower, e.g. 1.05. There are some studies about the influence of the recycle streams on the reactor conversion and selectivity, nevertheless the separation requirements are studied in a second step. It is important to notice that most of the energy and investment requirements are related to the separation.

The minimum recycle flow rate is easily determined under the infinite/infinite analysis. The objective function of minimizing the overall distillate flow rates is usually used together with the infinite/infinite analysis. In this way, a value is obtained avoiding the fact that in many cases the heuristics used for distillation column sequencing sometimes are contradictory or disregard some good alternatives. Although robust global optimization methods using powerful and highly efficient optimization algorithms combined with a flow sheet simulator provides accurate results, they become limited by a certain complexity due to the great combinatorial diversity of possible alternatives. Therefore, simple and fast analysis methods are useful for a first alternative screening, before proceeding with rigorous methods. Simplified mathematical models are more suitable in the early stages of design.

A simple parameter for distillation column sequencing in the early stages of conceptual design is the DSE. This parameter calculation requires only streams flow rates and distillation efficiencies that depend on bottoms and distillates temperatures. The DSE provides a measure of the difficulty of the separation from the energy efficiency point of view and it has been used successfully and validated for column sequencing. The goal of the present study is to evaluate the use of the DSE for chemical processes with reaction, separation and recycle streams. The present study combines for this sort of systems the infinite/infinite analysis with the DSE instead of the overall distillate flow rate. The study is in the line that similar methods used for column sequencing are also used for the proposal of the entire process or its retrofitting. However, the DSE value applied to a distillation column train becomes related to the distillation sequence efficiency but when applied to the overall process it becomes even negative. DSE applied to a process is a value useful to compare and rank alternative schemes or operating conditions.

Tackling uncertain feedstock price and demand of product and process networks via multi-stage adaptive robust optimisation

D. J. Garcia, F. You
Cornell University, Ithaca, NY, 14850

Bioconversion processes are exposed to many kinds of uncertainties. For example, biomass feedstock prices fluctuate constantly, and fuel demand levels can also vary. Bioconversion processing pathways should be designed to maximize economic performance while also hedging against these uncertainties. To that end, this work applies an adaptive robust optimization (ARO) approach [1] to formulate a three-stage, robust mixed integer nonlinear programming (MINLP) bioconversion product and process network optimisation model with uncertain feedstock prices and biofuel demands. The objective is to minimise cost. Uncertainty is handled in a multi-stage fashion, with "here-and-now" design (i.e. technology selection) decisions made before biofuel demand uncertainty is revealed. "Wait-and-see" operational decisions are made after revelation of biofuel demand uncertainty. A third stage reveals feedstock price uncertainty and allows calculation of the total annualized cost. A case study is performed for a bioconversion product and process network [2], but the proposed model and solution framework can be applied to a variety of networks.

The model contains eight biomass feedstocks (switchgrass, corn, corn stover, hardwood, softwood, soybean, sugarcane, and microalgae) and hundreds of technologies and materials/compounds. Both bioproducts (e.g. acetone, isobutanol, PHB, etc.) and three types of biofuel (biodiesel, gasoline, and ethanol) are allowed to be produced. Uncertain demand levels for the biofuels must be met. Uncertainty in both feedstock price and biofuel demand are modeled with uncertainty sets and corresponding uncertainty budgets. Off-the-shelf global solvers cannot solve the proposed three-stage optimisation model, so we develop a novel solution algorithm that incorporates a chance-and-constraint algorithm in an outer loop and a branch-and-refine algorithm in an inner loop. The method is shown to solve the three-stage ARO problem both quickly and efficiently. Trade-offs can be identified between more conservative solutions (higher cost) and less conservative solutions (more risk). A series of robust optimal solutions are found for each combination of feedstock price and biofuel demand uncertainty budgets. Smaller uncertainty budgets result in lower total annualized costs, and larger uncertainty budgets result in larger total annualized costs. At maximum budgets of feedstock price and product demand uncertainty, a conventional robust optimal solution is approached, reflecting the most conservative solution and the highest total annualized cost. ARO integration with the bioconversion product and process model allows for the identification of a series of bioconversion pathways depending on how conservative the decision maker wishes to be in the face of uncertainties and can be a vital tool for handling uncertainty.

References

1. D. Bertsimas, E. Litvinov, X. A. Sun, J. Zhao, and T. Zheng. "Adaptive Robust Optimization for the Security Constrained Unit Commitment Problem", _IEEE Transactions on Power Systems_, vol. 28, pp. 52-63, 2013.
2. D. J. Garcia and F. You. "Multiobjective optimization of product and process networks: General modeling framework, efficient global optimization algorithm, and case studies on bioconversion," _AIChE Journal_, vol. 61, pp. 530-554, 2014.

Tackling uncertainty in product and process networks via multi-stage adaptive robust optimisation

J. Gong, D. Garcia, F. You
Cornell University, Ithaca, NY, 14850

Bioconversion processes are exposed to many kinds of uncertainties. For example, biomass feedstock prices fluctuate constantly, and fuel demand levels can also vary. Bioconversion processing pathways should be designed to maximize economic performance while also hedging against these uncertainties. To that end, this work applies an adaptive robust optimization (ARO) approach [1] to formulate a three-stage, robust mixed integer nonlinear programming (MINLP) bioconversion product and process network optimisation model with uncertain feedstock prices and biofuel demands. The objective is to minimise cost. Uncertainty is handled in a multi-stage fashion, with "here-and-now" design (i.e. technology selection) decisions made before biofuel demand uncertainty is revealed. "Wait-and-see" operational decisions are made after revelation of biofuel demand uncertainty. A third stage reveals feedstock price uncertainty and allows calculation of the total annualized cost. A case study is performed for a bioconversion product and process network [2], but the proposed model and solution framework can be applied to a variety of networks.

The model contains eight biomass feedstocks (switchgrass, corn, corn stover, hardwood, softwood, soybean, sugarcane, and microalgae) and hundreds of technologies and materials/compounds. Both bioproducts (e.g. acetone, isobutanol, PHB, etc.) and three types of biofuel (biodiesel, gasoline, and ethanol) are allowed to be produced. Uncertain demand levels for the biofuels must be met. Uncertainty in both feedstock price and biofuel demand are modeled with uncertainty sets and corresponding uncertainty budgets. Off-the-shelf global solvers cannot solve the proposed three-stage optimisation model, so we develop a novel solution algorithm that incorporates a chance-and-constraint algorithm in an outer loop and a branch-and-refine algorithm in an inner loop. The method is shown to solve the three-stage ARO problem both quickly and efficiently. Trade-offs can be identified between more conservative solutions (higher cost) and less conservative solutions (more risk). A series of robust optimal solutions are found for each combination of feedstock price and biofuel demand uncertainty budgets. Smaller uncertainty budgets result in lower total annualized costs, and larger uncertainty budgets result in larger total annualized costs. At maximum budgets of feedstock price and product demand uncertainty, a conventional robust optimal solution is approached, reflecting the most conservative solution and the highest total annualized cost. ARO integration with the bioconversion product and process model allows for the identification of a series of bioconversion pathways depending on how conservative the decision maker wishes to be in the face of uncertainties and can be a vital tool for handling uncertainty.

References

1. D. Bertsimas, E. Litvinov, X. A. Sun, J. Zhao, and T. Zheng. "Adaptive Robust Optimization for the Security Constrained Unit Commitment Problem", _IEEE Transactions on Power Systems_, vol. 28, pp. 52-63, 2013.

2. D. J. Garcia and F. You. "Multiobjective optimization of product and process networks: General modeling framework, efficient global optimization algorithm, and case studies on bioconversion," _AIChE Journal_, vol. 61, pp. 530-554, 2014.

A view to Turkey and the European Union's agricultural mechanization level

A. Ozkan, [1] Y. Dilay[1], H.Kaya[1], O.Gokdogan[2]
Karamanoglu Mehmetbey Univ., Dep.Vocational Hihg School, Dept. Machine Yunus Emre Kampusu 70100 Karaman Turkey;
tel. +905359389119, e-mail: aozkan@kmu.edu.tr;

Technological developments in the global competition are continuously increasing. This development also shows itself in the agricultural sector. This is no doubt an important growing process. Mechanization in the implementation of advanced technology, as well as soil, water, fertilizers, pesticides, etc. by allowing the efficient use of production inputs is an important tool that enables productivity in agriculture. The countries which were developed has dramatically utilized from all of the increase in productivity in agriculture and all of the development on the agriculture and all of the mechanization equipment, depending on the agricultural structure.

The mechanization level can vary depending on according to the technological structure of the country. The reflections of these amendments are extremely natural. In Comparing agricultural mechanization level and structure of Turkey and the EU countries is based on data. These; farmland, average farm size, parcel number, total population, agricultural population, and compared data such as agricultural labor and total agricultural production.

In this study the mechanization levels of Turkey and the European Union were compared. These data with Turkey and the European Union have introduced the current situation. Mechanization level indicators in Turkey are 1.70 kW / ha, 24.10 ha / tractor, tractor 444.00 / 1000 ha, 5.20 equipment / tractors, 4.20 tons / tractor; EU, with the order of 6 kW / ha, 11.30 ha / tractor, 90 tractor / 1000 ha, 11 equipment / tractors, 12 tons / tractor. According to this comparison, the level mechanization of Turkey is considerably lower than the EU average. This case shows that the EU agricultural buildings and businesses is more open to development.

Keywords: European Union, Turkey, mechanization, agricultural structures

Integration of ion exchange process and ultra-filtration on recovery of silica particles from spent tungsten chemical - mechanical polishing slurry

W.J. Yeoh, A. Z. Abdullah
School of Chemical Engineering, Universiti Sains Malaysia, Nibong Tebal, Penang, Malaysia
tel. +60142452515, e-mail: andrew-1018@hotmail.com; 435, Jalan Besar , 34250 Tanjong Piandang, Perak, Malaysia.

Chemical mechanical polishing (CMP) is a process to planarize wafers for microelectronic applications. Tungsten CMP slurries used usually consist of three major components which are oxidizer such as H_2O_2 , catalyst such as ferric nitrate for passivating reaction and abrasive material such as silica for mechanical polishing.

Figure 1: Description on the removal of tungsten film by CMP process.

As shown on figure 1 above, the irregular topography of tungsten film are been remove by CMP process and those film will exist as tungsten oxide in the spent slurry. In order to reduce the cost of manufacturing microchip processor and minimize waste disposal to environment, the silica nanoparticles have to be recovered from the spent slurry which consist of tungsten oxide, silica and ferric nitrate. Hence, this project will be focus on the formulation of the best strategy to achieve good recovery of the silica nanoparticles from the spent slurry to be operated under optimum conditions.

Study results showed that ultrafiltration has the ability to retain the silica particles while let the soluble silica, tungstate and ferric ion to permeate through the membrane. Besides, it also has been found out the selectivity on silica retention is increase when the membrane molecular cut off become smaller. However, the contrary on selectivity and permeate flux has caused the ultrafiltration become inefficient. Besides, fouling is the major concern since they are plenty of particles existed in the system.

In order to enhance the ultrafiltration, the ion exchange process been introduced. By using a strong acid cation resin, it attaches most of the ferric and tungstate ions. Thus, this can enhance the further separation by using ultrafiltration. The ultrafiltration carried out by using polyethersulfone membrane at 100kDa MWCO which aim for reconcentration the slurry and also filter out the silica particles in dissolved form. In this study, data analysis mainly had done by ICP-OES on silica, iron and tungsten concentration, zeta potential, mean particle size and turbidity.

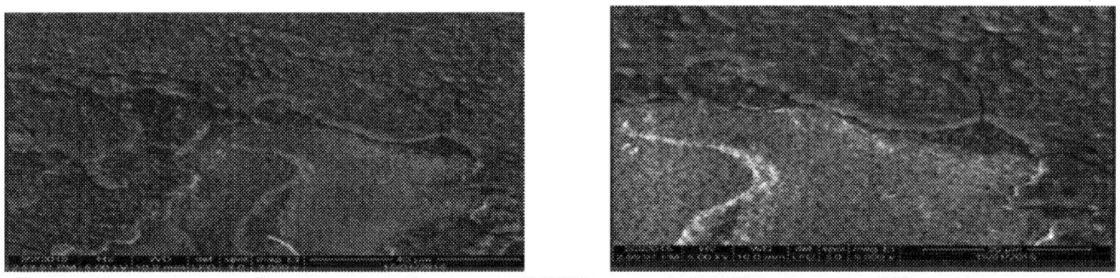

Figure 2: SEM result on membrane surface at 3000x and 5000x magnification.

At the first part of project, two processes carried out separately to study the behaviour and performance of each process. The ultrafiltration process only retained 50% of silica particles when 100kDa membrane was used. 70% of silica particles was retained when 50kDa membrane used, while 80% of silica was retained when 30kDa membrane was used. However, the flux was the lowest at 30kDa. As showed in figure 2, by doing SEM-EDS analysis on the membrane surface, the fouling is mainly caused by the ferric and tungstate ion.

For the second part of project, spent slurry at Ph 2, 4 and 6 has been investigated by using the ion exchange process. The experiment result showed that at most of ferric and tungstate ions are removed from the slurry at pH2. This might be due to dissolution of iron and tungsten into cation at acidic condition. Thus, this can enhance the subsequent ultrafiltration since most of those fouling materials have been removed.

For the third part of project, in order to investigate the performance of the process integration, the ultrafiltration is carried out in series after the ion exchange process. It showed out this can improve the efficiency of the process. The selectivity on the silica retention is high while the flux is acceptable. This was due to the successful removal of tungsten and iron elements by ion exchange process. However, two major scenarios still need to be further investigating which is on regeneration of ion exchange column and the stability of process stream flow rate.

In conclusion, the integration of ion exchange process and ultrafiltration is feasible for recovery of silica particles from the spent tungsten CMP slurry.

Optimization of energy and water consumption in the bioethanol integrated process from molasses and cane bagasse. Colombian case

Valderrama, Claudia[1]; Quintero, Viviana; Viatcheslav, Kafarov.
Universidad Industrial de Santander. Cra 27 calle 9. Tel : 6344000. Bucaramanga, Colombia, e-mail: valderrama.cm@gmail.com[1].

SUMMARY

Colombia is ranking as third in production of ethanol from sugar cane in Latin America in 2014, behind Brazil and Argentina. Currently Colombian sugar industry has 6 distilleries which supply the demand of bioethanol for blending with gasoline by 8% because its usage reduces emissions of greenhouse gases and increase fuel octane, on account of that, it has emerged in recent years as an alternative to fossil fuels in the transport sector.

The raw material for bioethanol production are classified into first and second generation; that refers to the origin of biomass, whether crops such as sugar cane and corn, or lignocellulosic sources such as bagasse or agro-industrial waste and regardless of the source of biomass, generally the key stages of the production process include enzymatic hydrolysis, fermentation with yeast and distilled for purification. In Colombia the bioethanol mostly is produced from sugar cane as feedstock, with a yield of 70 L/ton and the remains bagasse is not exploited as a source of lignocellulose to produce more ethanol.

There are reports form the literature about the amount of water needed to produce one liter of ethanol from corn: 11- 160 liters of water and for the same amount of fuel ethanol from switchgrass is 1.9-4.6 liters of water; these values include the amount required for irrigation.

Being the water the most valuable raw material in the world, justifies any effort focused on reducing consumption, to achieve it, from the viewpoint of process design, it is necessary to perform a sequential steps: the first one is the reduction of energy consumption through energy process integration network (heat exchangers); the second step is to make optimum network design process water, which can be achieved by focusing on the optimization problem of a superstructure formulating the problem of optimal allocation of water as a nonlinear programming problem mixed-integer (NPL).

The purpose of this paper is to establish the framework to the design of a production process of bioethanol using mathematical programming applied to the Integration of process water networks and heat exchanger networks to decrease water and energy consumption in joint production bioethanol from sugar cane and lignocellulosic material (bagasse) for the Colombian case.

REFERENCES

- Ahmetovié, E., & Grossman, I. (2010). Global superstructure optimization for the design of integrated process water networks. American Institute of Chemical Engineers., 434-457.
- ASOCAÑA. (2014). El Sector Azucarero Colombiano, más que azúcar, una fuente de energía renovable para el país. . ASOCAÑA.
- BNDES Y CGEE. (2008). Capítulo 3. Producción de bioetanol. En Bioetanol de caña de azúcar: Energía para el desarrollo sostenible (págs. 69-72). Rio de Janeiro: Editora Senac Rio.
- Federación Nacional de Combustibles. (04 de 09 de 2015). Fedebiocombustibles. Obtenido de http://www.fedebiocombustibles.com/estadistica-produccion-titulo-Alcohol_Carburante_(Etanol).htm
- Moncada, J., Tamayo, J., & Cardona, C. (2014). Integrating frst, second and third generation biorefineries: Incorporating microalgae into the sugarcane biorefinery. Chemical Engineerign Science, 126-140.
- Renewable Fuels Association. (2015). Going Global, 2015 ethanol industry outlook. Washington, DC.: Renewable Fuels Association. Obtenido de http://ethanolrfa.org/pages/World-Fuel-Ethanol-Production
- Soccol, C. e. (2011). Lingocellulosic Bioethanol: Current Status and Future Perspectives. En A. e. Pandey, Biofuels: Alternative feedstocks and conversion process (págs. 101-121). Oxford: ELSEVIER.
- Tomás-Péjo, E., Alvira, P., Ballesteros, M., & Negro, M. (2013). Pretreatment Technologies for Lignocellulose to Bioethanol conversion. En A. e. Pandey, Biofuels: Alternative Feedstocks and Conversion Processes (págs. 149-176). Oxford: ElSevier.
- Walker, G. (2010). Bioethanol feedstocks. En G. Walker, Bioethanol: Science and technology of fuel alcohol. (págs. 31-41). Ventus Publishing ApS.
- Wu, M., Mintz, M., & Wang, M. (2011). CONSUMPTIVE WATER USE IN THE PRODUCTION OF ETHANOL AND PETROLEUM GASOLINE. En M. Wu, & Y. Chiu, Consumptive Water Use in the Production of Ethanol and Petroleum Gasoline — 2011 Update (págs. 7-9). Lemont: Argonne National Laboratory.

The recovery of heat from waste gas at the heating centers

Y. Dilay, [1]., H. Kaya[2]
[1]Karamanoğlu Mehmetbey Univ., Dep. Tec. Voc. H S. 70100 Karaman, Turkey;
Tel. +90 3382262176, e-mail: ydilay@kmu.edu.tr;
[2] Karamanoğlu Mehmetbey Univ., Dep. Tec. Voc. H S. 70100 Karaman, Turkey;
Tel. +90 33822624590, e-mail: hskaya@kmu.edu.tr;

It is possible to saving energy from waste gas which is obtained from heating centers which is working at a high temperature such as with high capacities furnes and some drying systems by benefiting the discarded air for heating the ambient air, for heating water, or using it for processes which have occurred at low temperatures. Because, utilization of waste gas which has a temperature about 180 ^{0}C by increasing the efficient of system, on one hand while it provides the economy of energy and fuel, on the other hand it provides the decreasing of cost and environmental pollution.

In the Export dependent countries for energy sector especially in the cold region, it is very important the recovery of energy from waste gas at high temperatures.

In this study, the recovery systems of waste gas which are used in various sectors in Turkey were analyzed. It was suggested the precautions for these losses on a great scale. In the study the waste of heat which is occurred from chimneys was determined between 15% to 20% and with the recovery systems these rates can be reduced between 5% to 8%.

Effect of temperature on catalytic steam co-gasification of rubber seed shells and plastic HDPE residues

Bridgid Lai Fui Chin[1], Suzana Yusup[2*], Ahmed Al Shoaibi[3], Pravin Kannan[3], Chandrasekar Srinivasakannan[3], Shaharin Anwar Sulaiman[4], Angga Pratama Herman[2]
[1]Department of Chemical Engineering, Faculty of Engineering and Science, Curtin University, Sarawak campus, CDT 250, 98009 Miri Sarawak, Malaysia e-mail: bridgidchin@curtin.edu.my; tel. +6085 443 956.
[2]Biomass Processing Lab, Centre for Biofuel and Biochemical Research, Green Technology MOR, Department of Chemical Engineering, Universiti Teknologi PETRONAS, Tronoh 31750, Perak Darul Ridzuan, Malaysia, e-mail: drsuzana_yusuf@petronas.com.my[*]; tel. +605 3688 208.
[3]Department of Chemical Engineering, The Petroleum Institute, P. O. Box 2533, Abu Dhabi, United Arab Emirates, e-mails: aalshoaibi@pi.ac.ae, pkannan@pi.ac.ae, csrinivasakannan@pi.ac.ae
[4]Department of Mechanical Engineering, Universiti Teknologi PETRONAS, Tronoh 31750, Perak Darul Ridzuan, Malaysia, e-mail: shaharin@petronas.com.my

Co-gasification is one of the alternative means for generating renewable energy source, replacing the diminishing fossil fuels, overcoming the greenhouse gases emissions, and as well increasing development of waste processing technology specifically for plastic and biomass waste. A considerable amount of studies have been published in work related to the gasification behaviour of individual components of plastic and biomass waste. Meanwhile, limited studies have been focused to the binary mixtures of plastic and biomass waste in gasification process. Conversely, no studies on catalytic co-gasification of plastic waste mixtures in a fixed bed gasifier. Fixed bed gasifier is able to provide the benefits of wide temperature distribution, easier-controlled gas quality and minimum particulate carryover compared to fluidized bed gasifier. In the present study, the behaviour of co-gasification of plastic and biomass waste mixtures with the presence of nickel catalyst is investigated. Rubber seed shell is selected as lignocellulosic biomass for gasification feedstock since Malaysia is one of the major rubber growing countries. This results to immediate need to study the potential of biomass rubber seed wastes utilization. High density polyethylene (HDPE) is selected as the plastic waste for gasification process. HDPE is considered to be the most commonly used plastics globally and is said to become the largest used commodity plastic material due to their versatility and wide applications. The consequence of the high demand of HDPE will definitely lead to a production of immense amount of plastic waste which lead to secondary pollution to the environment.

The catalytic co-gasification of rubber seed shell and HDPE mixture is investigated in a bench scale fixed bed system known as high throughput micro reactors (HTMR) system at different reactor temperature of 600-800°C in order to study the effects on performance parameters such as gas and char yield, syngas yield and product gas composition. The operating temperature is selected based on the optimum conditions found from the previous study on the thermogravimetric analysis equipment coupled with the mass spectrometer using these feedstock. The constant process variables used in the HTMR system are rubber seed shells particle size range of 0.100-0.150 mm, HDPE particle size range of 0.355-0.500 mm and percentage of HDPE in the mixture of 20 weight%. The feeding rate of 2g/hr is carried out in this system.

Catalytic decomposition of oil slime in presence of inorganic salts

Yu.Yu. Kosivtsov, K.V. Chalov, Yu.V. Lugovoy, E.M. Sulman*, V.P. Molchanov
Tver Technical University, Dep. Biotechnology and Chemistry, A.Nikitina str. 22, 170026, Tver, Russia; tel./fax: +7 4822 789317, e-mail: sulman@online.tver.ru

The world crude oil production increases every year and at present amounts to 49 billions tons per year. The contribution of the Russian Federation to the global oil recovery budget is about 500 millions tons per year, besides the amount of the produced oil-containing waste can reach up to 10% of the total volume of the produced oil. The oil-containing waste can be obtained at all the stages of processing, storage, transportation and the use of oil. It also requires the vast area for the earthen containers and sludge collectors, resulting in the energy carrier price rise. The thermal treatment resulting in gaseous and liquid fuel formation is considered to be the most economically beneficial. Besides oil waste sensible utilization can be an essential step towards the development of resource-saving technology on the basis of the integrated use of oil raw material.

Compared to other methods of oil-containing waste thermal treatment, catalytic pyrolysis has a number of advantages: (i) rather low process temperatures (400-650 °C), (ii) low sensitivity to the raw material composition and (iii) a closed circuit of processing that comply with the modern requirements of chemical production. Catalytic pyrolysis of a number of organic compounds (wood, polymers, hydrocarbons) has been recently studied with metal chlorides as catalysts. However, to the best of our knowledge, no catalytic studies of metal chlorides in pyrolysis of oil-containing waste have been published.

In the this paper we report on the results of the process of heavy and residual hydrocarbons thermocatalytic processing with metals chlorides KCl, NaCl, $MgCl_2$, $AlCl_3$, $FeCl_2$, $FeCl_3$, $ZnCl_2$, $CoCl_2$ и $NiCl_2$. The pyrolysis process was carried out with an experimental pyrolysis set-up in a temperature range from 450 to 650 °C. The experimental pyrolysis set-up is a metal fixed-bed reactor heated with the electric furnace.

The study of the kinetics of the oil sludge pyrolysis process was carried out using the thermoscales TG 209 F1 (NETZSCH) with different heating rate (1, 5, 10, 15 and 20 °C/min). The oil-containing waste samples with the amount of crude oil 20% (wt.) without a catalyst and with 5% (wt.) $CoCl_2$ were used for the analysis.

Mg_xAl-CO_3 layered double hydroxide as a solid catalyst for esterification of acetic acid with 2-ethylhexanol

J. Kuljiraseth, [1]A. Wangriya, [2]S. Jitkarnka
The petroleum and petrochemical college, Chulalong University, Phayathai Rd. 254, 10330 Bangkok, Thailand; tel. +66 882224676, e-mail: jirayu.kul@hotmail.com;
[1]SCG Chemicals Co Ltd, 1 Siam Cement Rd, Bangsue, Bangkok, 10800, Thailand
[2]e-mail: sirirat.j@chula.ac.th

In a petrochemical plant, a unit in a petrochemical process can produce some acid wastes. Acetic acid is a major component in the acid wastes, which needs to be treated before being released to the environment. However, acetic acid in the wastes can be utilized as a reactant to produce a valuable product via esterification reaction using Mg/Al-type Layered Double Hydroxides (LDHs) as a solid catalyst. For esterification, acetic acid reacts with 2-ethylhexanol, forming 2-ethylhexyl acetate, which can be used as cleaners, paint removers, and a good solvent for nitrocellulose production. The research objectives were to study the effects of Mg/Al ratio in Mg_xAl-CO_3 structure and to investigate different thermal transition behaviors such as physical, chemical, and acid-base properties on the catalytic activity. The catalysts were characterized using BET, FTIR, *In-situ* XRD, XPS, SEM, XRF, STA-FTIR, NH_3-TPD, and CO_2-TPD. The activity of calcined LDHs catalysts was tested on esterification of acetic acid with 2-ethylhexanol to 2-ethylhexyl acetate. The conversion of acetic acid in the solution was enhanced significantly using the LDH as a catalyst, owing to the acid-base properties of the LDH.

Pyrolysis of polymer waste in presence of metal chlorides of iron subgroup

Yu.Yu. Kosivtsov, K.V. Chalov, Yu.V. Lugovoy, E.M. Sulman*, V.P. Molchanov, A.A. Stepacheva
Tver Technical University, Dep. Biotechnology and Chemistry, A.Nikitina str. 22, 170026, Tver, Russia; tel./fax: +7 4822 789317, e-mail: sulman@online.tver.ru

Currently, global stocks of used tires constitute about 25 million tons and the number continues to increase. Currently, share of recycling of used tires by crushing for production of crumb rubber is increasing, though this method is not without drawbacks, as a result of the refining process forms a complex mixture of polymer fibers and rubber crumb (the polymeric cord) which also requires utilization. Currently, share of recycling of used tires by crushing for production of crumb rubber is increasing, though this method is not without drawbacks, as a result of the refining process forms a complex mixture of polymer fibers and rubber crumb (the polymeric cord) which also requires utilization. Among the modern, efficient and flexible forms of recycling of complex mixtures of plastic wastes are particularly important methods of catalytic pyrolysis.

In this work investigation of the influence of the polymeric cord of used tires composition on the course of the catalytic pyrolysis presents. The following iron-subgroup metal chlorides were used as catalysts: nickel chloride; cobalt chloride and iron chloride. The worn tire pyrolysis experiments were conducted at 350–600°C. The experimental pyrolysis system comprised a metal reactor with a fixed substrate layer.

The catalytic effect on polymer cord pyrolysis was estimated from the following characteristics: mass distribution of gaseous, liquid and solid products; the volume and generation rate of gaseous products; composition and heat of combustion of gaseous products; the composition of liquid products; changes in thermal stability of the polymer cord; and specific surface area of the remaining solid.

It has been established that at the collaborative catalytic pyrolysis of rubber and cord in a weight ratio of 1:1, there is an increase of reaction rate and the mass fraction of volatile products, which affects on increasing of efficiency and on reducing of power consumption of the process.

The C_1–C_4 hydrocarbon production achieved by the investigated iron-subgroup metal chlorides decreases in the following order: $Co^{2+} > Ni^{2+} > Fe^{2+}$. The optimum parameters for polymer cord pyrolysis determined from the results of this work are a temperature of 450°C and impregnation with 2 wt% of $CoCl_2$.

Selection of pollution prevention strategies with Target-Oriented Robust Optimization (TORO)

K. Aviso[1], C. Sy[2], R. Tan[1]
[1]Chemical Engineering Department
[2]Industrial Engineering Department
De La Salle University Manila
tel. +632 5244611, e-mail: kathleen.aviso@dlsu.edu.ph; 2401 Taft Avenue, Malate, Manila, Philippines, 0922

The reduction of industrial emissions can be achieved through the use of pollution prevention (PP) measures in addition to pollution control (i.e., end-of-pipe treatment) techniques. In some cases, the simultaneous implementation of multiple PP measures can result in drastic cumulative reductions that can match the performance of pollution control devices. However, performance levels of many such techniques may also be subject to uncertainty. Thus, systematic tools need to be used to plan the implementation of arrays of PP strategies to take into account uncertainties as well as financial constraints. In this work, a nonlinear 0-1 programming model is developed based on the target-oriented robust optimization (TORO) framework. The model is demonstrated using an industrial case study involving the reduction of airborne fluoride emissions from brick kilns; the example considers various PP measures that use mass and heat transfer effects to reduce volatilization of fluoride compounds.

Suzuki cross-coupling in environmentally friendly solvents in a batch mode using hypercrosslinked polystyrene-supported Pd catalysts

N.A. Nemygina[1], L.Zh. Nikoshvili[1,2], M.G. Sulman[1,3], V.G. Matveeva[1], E.M. Sulman[1], L. Kiwi-Minsker[2,4]
[1]Tver Technical University, Dep. Biotechnology and Chemistry, A.Nikitina str. 22, 170026, Tver, Russia; tel./fax: +7 4822 789317, e-mail: nlinda@science.tver.ru
[2]Tver State University, Regional Technological Centre, 170100, Tver, Russia
[3]A.N. Nesmeyanov Institute of Organoelement Compounds of RAS, 119991, Moscow, Russia
[4]Ecole Polytechnique Fédérale de Lausanne, GGRC-ISIC-EPFL, CH-1015, Lausanne, Switzlerand

Suzuki cross-coupling reaction is of great importance as it is widely used in fine organic synthesis for obtaining of active pharmaceutical ingredients, crop protection substances, and polymers. Suzuki cross-coupling is traditionally catalyzed by homogeneous palladium complexes. However, the industrial application of the majority of developed phosphorus- or nitrogen-containing ligands is limited by their high cost. Besides, a common disadvantage of homogeneous catalysts is difficulty of their separation from the reaction mixture and their reuse.

This work is devoted to the synthesis of Pd-containing catalysts based on hypercrosslinked polystyrene (HPS) and study of their catalytic properties in Suzuki cross-coupling of 4-bromoanisole and phenylboronic acid. Polymers as supports are able to provide control over the particle size and dispersion that is the main problem of synthesis of catalytically active metal nanoparticles. Besides, polymers have variety of properties, variation of which allows influencing effectively the processes of nanoparticle formation.

In the present study, a number of HPS-based catalysts was synthesized with different palladium content, type of HPS (non-functionalized or bearing amino groups) and nature of the precursor ($PdCl_2(CH_3CN)_2$, $Pd(OAc)_2$, $PdCl_2$, $Pd(C_6H_5CN)_2Cl_2$). The Suzuki cross-coupling of 4-bromoanisole and phenylboronic acid was carried out in a batch mode at ambient pressure while using low-molecular weight alcohols (*i*-PrOH, EtOH) and water as solvents, and also in $ScCO_2$ medium.

Kinetic study was carried out at ambient pressure at variation of type and concentration of base (Na_2CO_3, NaOH, K_2CO_3), solvent nature and composition, reaction temperature (50-75°C) and gas phase composition. The highest catalyst activity was observed in EtOH-water mixture (5: 1) while using NaOH as a base (the conversion of 4-bromoanisole has reached higher than 98% for less than 1 h of reaction time). Selectivity was higher than 95% with respect to 4-methoxybiphenyl. Thus HPS-supported palladium catalysts were found to be promising (active, selective), simple in preparation and stable for at least four consecutive runs for the Suzuki cross-coupling at mild temperatures along with the absence of the necessity to use phase transfer agents.

Catalytic deoxygenation of simulated bio-oil using $MgAlCO_3$-based catalysts for chemicals production

[1]J.M. C. Malones, [2]A. Wangriya, [1,*]S. Jitkarnka
[1]Chulalongkorn University, Chula Soi 12, Phyathai Road, Pathumwan, Bangkok 10330; tel. +6622184116; email: marvinmalones@gmail.com; University of the Philippines Visayas, Miagao, Iloilo 5023
[2]SCG Chemicals Co. Ltd., Siam Cement Road, Bangsue Bangkok, Thailand 10800; email: aunchanw@scg.co.th
[1,*]Chulalongkorn University, Chula Soi 12, Phyathai Road, Pathumwan, Bangkok 10330; email: sirirat.j@chula.ac.th

The high oxygen content of bio-oil due to the presence of acids, aldehydes, alcohols, ketones, phenol and other oxygenates is the major hindrance for usage. Deoxygenation via catalytic upgrading of bio-oil to fuel-grade oil was revealed to be performed by acid-base active sites created by metal promoters and porous structures in the catalysts, via cracking, dehydration, decarboxylation and decarbonylation. In this work, the use of catalysts derived from layered double hydroxide (LDH) for catalytic deoxygenation of 3 % guaiacol in ethanol, was explored. Guaiacol had been used extensively as a model compound for bio-oil studies because it is a product of lignin pyrolysis, and has a coke-forming tendency. Catalytic testing was done in a U-tube reactor at 400 ^{O}C. $Ni_{0.35}Mg_{2.65}AlCO_3$ and Mg_3AlCO_3 LDH catalysts were initially calcined in-situ at 500 ^{O}C for 5 hours under flowing He. Simulated bio-oil was fed at a rate of 5.1 g/h (WHSV of 2.55 $g/g_{cat}h$), and liquid products were collected every 30 minutes for 3 hours. The catalysts were characterized using XRD, FTIR and Surface Area Analyzer. Gas and liquid products were analyzed using GC with TCD and GC with TOFMS, respectively. XRD and FTIR confirmed successful synthesis of LDH and XRD confirmed that amorphous mixed metal oxides formed from LDHs when calcined at 500 ^{O}C. Catalytic activity testing showed that conversion of guaiacol was high (above 97 %) and the product liquid yield was also high (above 85 %) using calcined LDHs. Hydrocarbons, aromatics, aromatic oxygenates and oxygenates were found in the condensed liquid product while light hydrocarbons were present in the gas product. The partial substitution of Ni to Mg in the layer of the LDH resulted in improved catalytic activity and optimization can be done depending on the desired product.

Direct electricity generation from pre-treated empty fruit bunch using single chamber microbial fuel cell

N. A. Nik Mahmood[1], C. M. Kee[1], N. Amalina[1], T.A. Tuan Abdullah[2], K. A. Ibrahim[2] and N. F. Ghazali[*1] tel. +60735471, email: nazlee@utm.my; 1 Department of Bioprocess and Polymer Engineering, Faculty of Chemical and Energy Engineering, Universiti Teknologi Malaysia, 81310 Skudai, Johor Bahru, MALAYSIA. 2 Department of Chemical Engineering, Faculty of Chemical and Energy Engineering, Universiti Teknologi Malaysia, 81310 Skudai, Johor Bahru, MALAYSIA.

A single chamber microbial fuel cell was operated with empty fruit bunches (EFB) as source of fuel using with mixed microbes as electricigens to produce electricity. The study includes the effect of varying temperature, pH, and substrate or fuel concentration towards the electricity production. The study was designed with 3 different value of temperature (25°C, 37.5 °C and 50°C), pH (5, 7 and 10) and substrate concentration (1.5 g, 5.5g and 7.5g) to analyze power performance. Finally, voltage produced was calculated to obtain power and current and represented as power curves and voltage curves for each parameters and evaluate of its sustainability and productivity on power generation.

Effect of enzyme and substrate loading on oil palm lignocellulose hydrolysis

N. A. Makhtar[1], N.F. Ghazali[1]
[1]Dept of Bioprocess Engineering, Faculty of Chemical and Energy Engineering, Universiti Teknologi Malaysia, 81310 Skudai, Johor, Malaysia ;
tel. +6075535493, e-mail: nazlee@utm.my; fax: +6075588166.

Malaysia as one of the largest contributor of palm oil industry in the world is now facing a great environmental issues regarding the palm oil waste. Empty fruit bunch (EFB) is one of the biomass waste aside from the trunk, frond and fruit mesocarp from oil palm plantation. To be able to convert this waste into products would be a great benefit to the country. For example, bioconversion of oil palm lignocellulose via enzymes to fermentable sugar could be used for the production of bioethanol. However, the bioconversion is challenging due to the cost of enzymes and the interactions between substrate and enzymes. As the study on EFB enzymatic hydrolysis is still limited, we are taking an approach to provide more understanding in this hydrolysis.

There are few factors affecting enzymatic hydrolysis such as pH, temperature, substrate concentration, enzyme concentration and structure of lignocellulose. In order to utilize EFB as the biofuel feedstock, it is important to understand the effect of enzyme concentration and substrate concentration on the bioconversion of EFB. So, in this study, we are investigating the effect of enzyme loading and substrate loading for the maximum conversion of lignocellulose. The results show that as the enzyme loading increased, the rate of reaction and also yield of reducing sugar (RS) and glucose were initially increased and then plateaued. Similar pattern was also applied to reaction rate as the substrate loading increased. However, the yield of RS and glucose decreased as the substrate loading increase. This could be closely related to the lignin interference during hydrolysis. Furthermore, the initial product formation rates were used to determine the kinetic parameters such as maximum rate constant V_{max} and the half saturation constant K_m. From this study, feasible amount of cellulase and EFB substrate could be identified for maximum conversion and facilitate bioethanol production.

Design of integrated solar thermal energy system for multi-period process heat demand

Adeniyi J. Isafiade*[a], Zdravko Kravanja[b], Miloš Bogataj[b]
[a]Department of Chemical Engineering, University of Cape Town, Private Bag X3, Rondebosch 7701, South Africa
[b]Faculty of Chemistry and Chemical Engineering, University of Maribor, 2000 Maribor, Slovenia
aj.isafiade@uct.ac.za

This paper presents a new methodology for integrating renewable energy sources such as solar with chemical process utility demand. In chemical plant operations, the demand for utilities, as well as available process heat loads, are sometimes seasonal or multi-period in nature. Also, the availability of renewable energy such as solar, is time of day dependent as well as seasonal, i.e. time of year dependent. Therefore, in order to fully harness the economic and environmental benefits associated with the use of solar thermal energy, an approach which systematically integrates solar thermal radiation with process heat demand and other sources of utilities in chemical plants, while considering the multi-period profile of both the utility sources and the process streams need to be developed. The approach adopted in this paper, entails the use of the multi-period version of the stage-wise superstructure with an objective function involving a simultaneous minimisation of the annualised investment costs and annual operating cost of the network. The utility sources considered include solar and fossil fuel. The fossil fuel is included as a back-up depending on how much of the process heat demand is satisfied by solar energy. For the solar energy source, concentrated solar thermal collector is considered for both direct and indirect heat integration including thermal energy storage. In the example considered, a reasonable amount of cost savings was obtained when compared to the case without solar thermal integration.

A stochastic fuzzy multi-criteria decision model for optimal selection of low carbon technologies

M. Promentilla[1,2], A. De Serra[2], M. Antonio[2], R. Chuaunsu[2], K. Aviso[1,2], R. Tan[1,2]
[1]Center for Engineering and Sustainable Development Research
[2]Chemical Engineering Department
De La Salle University Manila
tel. +632 5360223, e-mail: michael.promentilla@dlsu.edu.ph; 2401 Taft Avenue, Malate, Manila, Philippines, 0922

Power sectors, as the world's demand for electricity is increasing, are recognized to be as significant contributors of carbon dioxide (CO_2) emissions in a fossil-fuel based economy. Low carbon energy systems are thus being developed, promoted and deployed as part of the solution portfolios to address climate change. However, optimal selection of such systems should take into consideration several criteria such as that of technical, economic, environmental and social aspect of the decision problem. In addition, some emerging technologies may lack accurate information to understand with certainty the behavior of the alternatives with respect to some criteria. The decision maker has also to do tradeoff analysis in prioritizing the alternatives in a complex problem with multiple conflicting criteria. In this work, a stochastic fuzzy Analytic Network Process (ANP) model was thus developed and applied in the prioritization of low carbon energy systems to aid in the decision making process under such uncertainty. This technique decomposes the complex problem into a hierarchic network structure and derives priority weights to rank the alternatives. The decision model incorporated the ambiguity-type uncertainty wherein a calibrated fuzzy scale was used to represent the judgment in pairwise comparisons of alternatives and criteria. Stochastic simulations based on Monte Carlo method were also used for the uncertainty analysis of the priorities derived from the model. An illustrative case study in the Philippines was presented wherein biomass, geothermal, solar, hydro, and wind power were evaluated with respect to tangible criteria such as levelized cost of electricity, carbon footprint, land footprint and water footprints, as well as, intangible criteria such as maturity of technology, social acceptance, and social benefits.

Process design under uncertainties with P-graph methodology

J. Baumgartner, Z. Süle, B. Bertók
University of Pannonia, Department of Computer Science and Systems Technology, Egyetem str. 10., 8200, Veszprém, Hungary
email: baumgartner@dcs.uni-pannon.hu; tel: +3688624000

The structure of process systems in the chemical industry is more complex than simply serial or parallel. Consequently, determining the overall reliability of a chemical process, whether it is capable to produce a desired set of products or not, is theoretically difficult. Kovács et al published a computational method to calculate the structural reliability of a process defined, i.e., what is the probability of remaining at least one subprocess leading to each of the desired products while the risk of failure for any process element is considered.

The presentation of interest introduces a method capable to determine the cost optimal and alternative suboptimal process structures or flowsheets, each of which guarantees a required level of reliability. The proposed method roots in the P-graph approach introduced in the early '90s by Friedler et al, and is to be illustrated by applying it to an actual practical problem of designing a reliable biodiesel supply chain.

Effect of critical point properties of supercritical CO_2 and optimization in discontinuous printed circuit heat exchangers

Wenxiao Chu[a], Ting Ma[a], Min Zeng[a], Yi-tung Chen[b], Qiu-wang Wang*[a]
[a] Key Laboratory of Thermo-Fluid Science and Engineering, MOE, Xi'an Jiaotong University, Xi'an, Shaanxi 710049, China
[b] Department of Mechanical Engineering, University of Nevada, Las Vegas, NV 89154, USA
wangqw@mail.xjtu.edu.cn

As a kind of natural fluids, the carbon dioxide is an environment friendly working medium. In addition, the attractive physical and transport properties of supercritical carbon dioxide (SCO_2) like high specific heat and low viscosity, which is different from normal gases and liquids, are very beneficial in the Brayton Cycle, chemical engineering and refrigeration system. In this paper, the effect of critical point properties of SCO_2 is analysed in a new type of printed circuit heat exchanger (PCHE) with discontinuous airfoil fins (AFFs). The SCO_2 fluid in three regions including precritical state, transcritical state and supercritical state in heat transfer process are analysed. The result shows that there is a strong heat transfer enhancement and an acceptable pressure loss in transcritical state. However, the heat transfer deterioration of SCO_2 is observed after the critical point. Meanwhile, the comprehensive performance is compared by heat transfer rate per unit pump power. Comparing with the rectangular model, it is investigated that the staggered AFF arrangement has better comprehensive heat transfer performance (CHTP) than the parallel one. The study also shows that reducing the flow resistance is the best way to improve the CHTP.

Targeting and design of evacuated-tube solar collector networks

M. Picón-Núñez, G. Martínez-Rodríguez, A.L. Fuentes-Silva
Department of Chemical Engineering, University of Guanajuato, México;
tel. +52 473 73 20006 ext. 8106, e-mail: piconnunez@gmail.com; Noria Ata s/n, Guanajuato, Gto.; fax: +52 473 73 20006 ext. 8101.

In this work the outlet temperature of networks of solar collectors for the supply of thermal energy to low energy intensity industrial processes is determined. Typically a solar collector network exhibits a structure which is a combination of series-parallel arrangement. In design, the number of lines in parallel is determined based on the total mass flow rate required while the number of collectors in series is established by the process thermal requirements. The achievement of the process thermal requirements depends on the thermal performance of the collector technology as well as the availability of the solar radiation. This work shows the development and validation of a thermal model and its use to predict the thermal performance of networks of evacuated tube collectors.

The design of solar collector networks starts by specifying the minimum temperature required; then, the number of collectors that meet the design objective can be obtained. The final outlet temperature of a set of collectors arranged in series is obtained making the outlet temperature of one collector the inlet of the next one. For a given target temperature the number of collectors in series is determined, however for the specified heat load to be achieved a set of collectors in parallel are needed. The thermal model is validated against experimental data. The results show the accuracy of the model and therefore confirm that its use for the prediction of the thermal performance of sets of collectors in series is reliable. Evacuated tube solar collectors are integrated systems that include the storage tank or header. In this work it is shown that the thermal losses in this type of systems is as high as 37.71% of the total energy received. Of these losses, 6.1% is lost in the header and the rest is lost in the tubes by radiation and convection. Three commercial evacuated tube collectors are analyzed. Theoretical prediction of outlet temperature of sets of collectors arranged in series is shown in Fig. 1 and in Fig. 2 the normalized temperature rise shows the similar behaviour of the various commercial technologies.

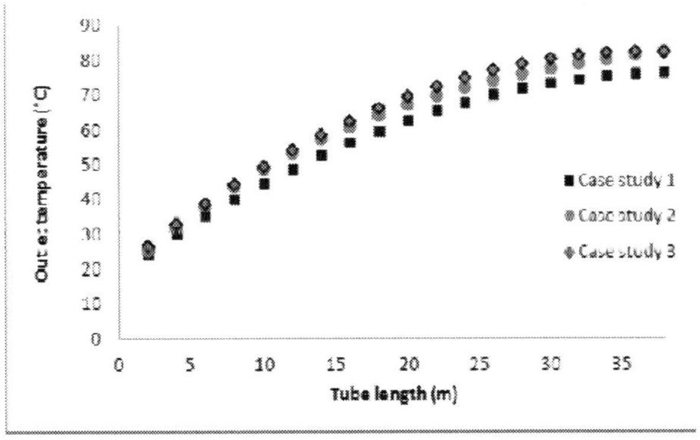

Fig 1: Theoretical prediction of the outlet temperature of solar collectors arranged in series (solar radiation: 806.6 W/m2).

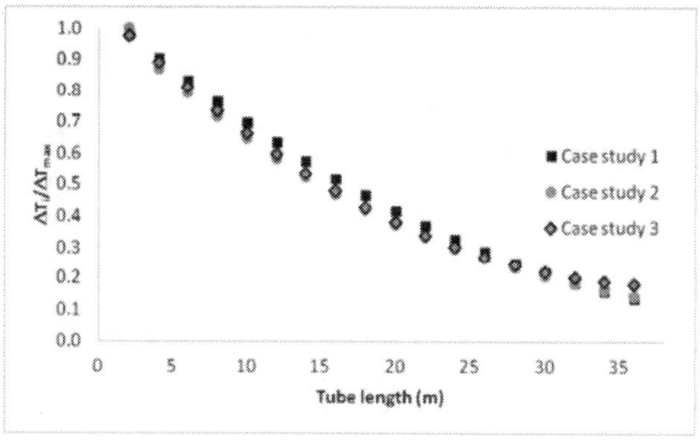

Fig. 2: Normalized temperature rise as a function of the number of solar collectors in series.

An improved graphical approach for visualizing the flexibility of multi-functional energy systems

Raymond R. Tan,[1] Kathleen B. Aviso,[1] Jui-Yuan Lee[2]
[1]Chemical Engineering Department, De La Salle University, 2401 Taft Avenue, 0922 Manila, Philippines; fax: +632 536 4226; [2]Department of Chemical Engineering and Biotechnology, National Taipei University of Technology, Sec. 3, Zhongxiao E. Rd., Taipei 10608 Taiwan, R.O.C.; fax: +886 2 2731 7117.

Multi-functional energy systems such as biorefineries as well as tri- and polygeneration plants enable efficient production of multiple streams by taking advantage of opportunities for heat and mass integration. However, for such systems, the resulting interdependencies among individual process units results makes it necessary to utilize systematic techniques to ensure proper design and operation. For example, the use of geometric representation of feasible operating ranges of trigeneration plants was previously proposed by Lai and Hui (2009) to quantify the capability of such systems to meet variations in product demand.

In this paper, we propose a modified geometric representation of such feasible operating regions for multi-functional energy systems. The modified approach differs from the previous methodology in two key aspects. Firstly, input-output interactions among process units are accounted for, such that the boundaries of the feasible operating regions are not necessarily perpendicular lines (or planes and hyperplanes in the case of higher-dimensional systems); secondly, lower operating limits for partial load operation are also considered, below which process units need to be switched off. The methodology is demonstrated to be useful for quantifying and visualizing operational flexibility via a simple trigeneration case study.

Segregation performance of granulated blast furnace slag particles with wide size distribution in spouted bed for heat recovery

Hua Li[2], Xun Zhu [*1-2], Hong Wang [*1-2], Qian Fu[1-2], Lei Cao[2], Qiang Liao [1-2]
1 Key Laboratory of Low-grade Energy Utilization Technologies and Systems, Chongqing University, Chongqing 400030, China
2 Institute of Engineering Thermophysics, Chongqing University, Chongqing 400030, China, Tel: +86-023-65102474, E-mail: hongwang@cqu.edu.cn zhuxun@cqu.edu.cn

Dry slag granulation has been proposed as alternative to water quenching for its potential in water saving and energy conservation. Typically, it involves with two strategies; (i) turning the molten slag into fine particles by rotary atomizer and (ii) recovering the heat as hot air and stream or converting the heat into chemical energy. Of great concern is the slag particle size distribution which domains the subsequent heat transfer process. However, the size distribution of the as-produced slag particle is often in a wide range of 0-5 mm, resulting in de-fluidization, inhomogeneous residence time, and particles agglomeration in a fluidization bed. To tackle this challenging issue, the present study developed a technique combining the spouted bed and the bubbling bed to enhance the heat recovery from high-temperature slag particles. Spouted bed has prominent capacity to deal with large-size particles, especially for those can't be well fluidized in bubbling fluidized beds. Also, the device can simultaneously segregate a part of fine particles and then transport them into the bubbling fluidized bed by virtue of high fluidizing velocity and baffle. The segregation performance of particles in wide size distribution in spouted bed was experimentally investigated. The effects of fluidizing air velocity, baffle height on the segregation characteristics were identified for the single-run spouted bed. The variations of the segregated particle size (ds), the segregation mass flow rate (Gs) and particle lateral distribution in bubbling fluidized chamber with operational conditions (fluidizing velocities and baffle height) were particularly discussed. The results showed that higher separation mass flow rate (Gs) can be easily obtained at a higher fluidizing velocity and lower baffle height in a static bed height, leading to an increase in the segregation particle size (ds) for coarse particles were separated into the bubbling fluidized bed. On the basis of the operating conditions tested in this study, the fluidizing airflow and baffle height can be adjusted to ameliorate segregation performance.

Keywords: Blast furnace slag; heat recovery; spouted bed; segregation

Determination of optimum production conditions of casting slurry in the manufacture of molten carbonate fuel cell electrodes with the tape casting technique

Gülay Özkan[1],Selim Gemici[1], Erdi Başarır[1], Göksel Özkan[2]
[1]Department of Chemical Engineering, Ankara University, 06100 Tandoğan, Ankara, Turkey;tel.+903125823515,e-mail:gozkan@eng.ankara.edu.tr[1]; Fax: 0312-230 84 34.
[2]Department of Chemical Engineering, Gazi University, 06570 Maltepe, Ankara, Turkey

In this study, Ni anode and $LiAlO_2$ electrolyte green sheets were manufactured for MCFC system. Experimental scale production of green sheets was carried out by using tape casting technique. The method of casting slurry, which is the first stage of preparation, has a direct impact on the final product. Therefore, cast slurries with different ratios of a solvent, a binder, a dispersant, plasticizer loading to the organic compounds were prepared to produce green sheets with the technique of tape casting in order to investigate the viscosity, shear stress, thixotropy and flow characteristics. After sheets were dried, physical properties (mechanical tests) were determined. Mixing method (balling mill and mechanical stirrer) and mixing duration of slurry, mixing weight ratio of binder, solvent, plasticizer, dispersant, Li/K carbonate, Ni,and $LiAlO_2$ powder in the slurry was examined.

Molten Carbonate Fuel Cells (MCFC) are called according to its electrolyte as well as other fuel cells. MCFC has high energy conversion efficiency and operates between 650-700 °C. Usually, it comes at the beginning of the systems used in the local and regional level to high energy needs.

In recent years, fuel cell electrode and electrolyte materials in the production of tape casting technique are known to find the area of intensive use. Nowadays tape casting technique is especially widely used in the production of plate and coating processes such as thin ceramic coating, polymeric coating, porous coating and the metal or alloy coating. Especially, this technique is used in the preparation of membrane, electrodes and electrolyte matrix of the polymeric membrane fuel cell (PEMFC), molten carbonate (MCFC) and solid oxide (SOFC) fuel cell.

In this study, Ni anode and $LiAlO_2$ electrolyte green sheets were manufactured for MCFC system. Experimental scale production of green sheets was carried out by using tape casting technique. The method of casting slurry, which is the first stage of preparation, has a direct impact on the final product. Therefore, cast slurries with different ratios of a solvent, a binder, a dispersant, plasticizer loading to the organic compounds were prepared to produce green sheets with the technique of tape casting in order to investigate the viscosity, shear stress, thixotropy and flow characteristics. After sheets were dried, physical properties (mechanical tests) were determined. Mixing method (balling mill and mechanical stirrer) and mixing duration of slurry, mixing weight ratio of binder, solvent, plasticizer, dispersant, Li/K carbonate, Ni and $LiAlO_2$ powder in the slurry was examined.

Slurry had been mixed in the ball-milling before it was mixed with mechanical stirrer. An amount varying between 32-53 % by weight of nickel powder was used in the anode slurry.

The best weight ratio is 32 % for nickel powder. In addition, 13-52 % by weight of ethanol was used in the anode slurry. The best weight ratio is 41 % for the solvent (ethanol). 1,5 PVB (Polyvinyl butyral) / PEG (Polyethylene glycol) (w/w) was optimized for anode.

22-32,3 % by weight of lithium aluminate ($LiAlO_2$) powder was used in the electrolyte slurry. The best weight ratio is 32 % for lithium aluminate powder. An amount varying between 23-52,3 % by weight of ethanol was used in the electrolyte slurry. The best weight ratio is 45,7 % for the solvent and 4,4:3,8 wt % for K/Li carbonate, 6,2 wt % for PVG, 7,1 wt % for PEG and 0,5 wt % for glycerol was used in the electrolyte mixture.

Keywords: Molten Carbonate Fuel Cell, Tape Casting Technique, Anode, Electrolyte

Effect of experimental conditions on measuring autoignition temperature of Met-OH and Et-OH binary mixtures

J. Vereš*, J. Skřínský, J. Trávníčková, A. Dalecká
Energy Research Centre, VŠB-TU Ostrava, 17. listopadu 15/2172, 708 33 Ostrava, Czech Republic; tel. +420 597 324 948, e-mail: jan.skrinsky@vsb.cz

Auto ignition Temperature (AIT) is an important variable used to characterize the fire and explosion hazard of liquids and must be known for safe handling, storage, and transportation. The measurement of AIT is defined in test methods that are maintained by standardization bodies such as the Energy Institute in the UK, ASTM in the USA, CEN in Europe and ISO internationally. This method, if reported, gives different results. It is found that the measured AIT of 1-pentanol conducted with a flask made of quartz will result in a value higher than the one conducted with a flask made of borosilicate by 10 °C.

Most methods for measuring AIT of liquid chemicals introduce the sample into the apparatus container which is preheated to a specific temperature and AIT is evidenced by the sudden appearance of a flame inside the container and by a sharp rise in the temperature of the gas mixture, but the container shape and container size may vary in different test methods.

This paper describes measurement on AIT of Met-OH and Et-OH Binary Mixtures. We explored effects of flask material, ambient temperature, and ambient humidity on the measured AIT of ethanol using the EN 14522:2005 and ASTM E659-78:2005 methods.

Measurements of the autoignition temperatures were obtained from the AIT-12-1 autoignition tester (SN: 321-OZM-13-01, OZM Research s.r.o., Czech Republic). The experimental set-up consists of an electrically heated crucible furnace capable of attaining a temperature of 600°C or higher, commercial 500-ml borosilicate round-bottom short-necked boiling flask closely wrapped in reflective aluminium foil and fine chromel-alumel thermocouple (36 B and S gage) for measuring the gas temperature inside the flask. The apparatus involved control devices that program the instrument to heat the sample at a specific heating rate within a temperature close to the expected autoignition temperature. AIT temperatures of the pure liquid flammable substances were determined experimentally according to the EN 14522:2005 and ASTM E659-78:2005 standards with an accuracy of 1.5 °C.

AIT of ethanol conducted with a flask made of quartz will result in a value higher than the one conducted with a flask made of borosilicate by 10 °C.

Results reveal that immiscibility in the two liquid phases could be ignored in the measuring of AIT. Potential application for the results concerns the assessment of fire and explosion hazards for chemical processes containing binary immiscible mixtures of Met-OH and Et-OH system.

NOx reduction strategies for combustion of oil-extracted residue

P. W. Li and C. S. Chyang
Chung Yuan Christian University, Department of Chemical Engineering, Chung Pei Rd. 200, Chung Li District, Taoyuan City, Taiwan 32023; tel. +886 32654119, fax: +886 34636242, e-mail: cschyang@gmail.com

Biomass combustion is considered environmental protective because of the carbon neutrality, especially for agricultural waste. *Jatropha curcus* is an oily biomass widely spread in the tropics or in the subtropics and is usually utilized as the feedstock of biodiesel.[1] However, the extracted residue can not be treated as animal feed since there are phorbol ester and curcins.[2] To make the best of the extracted residue of *Jatropha*, the combustion tests were held in a fluidized-bed combustor to observe the performance.

The tests showed that the oil-extracted residue is easy to ignite and the combustion efficiency is above 99%. CO emission is negligible and SO_2 can be captured by inherent metals in the ash favoring desulfurization; NOx concentration, however, shows a unique profile in the combustor, and its emission is way beyond the regulation. Therefore, measures should be taken to improve the NOx emission to meet the requirement.

From the NOx concentration profile, it can be seen that the NOx concentration increases sharply in the freeboard of high temperature. To confirm the effect of freeboard temperature on NOx formation, the freeboard temperature were controlled as a parameter to observe the NOx variation.

On the other hand, it was reported that calcium magnesium acetate could simultaneously reduce SO_2 and NOx.[3] For biomass, the sulfur content is low and the SO_2 emission is negligible, as also verified in this study. So, only acetic acid was used as NOx reducing agent.

In this study, the effects of freeboard temperature and reducing agent (acetic acid) on NOx emission were investigated in a fluidized-bed combustor. The freeboard temperature was set in the range of 700 – 850 °C and the flow rate was between 0 – 30 ml/min. Experimental results revealed that the NOx emission was reduced with the decrease of the freeborad temperature, and the injection of acetic acid could further diminish the NOx emission. However, lower freeboard temperature and an excess of injection would both lead to an increase of CO emission. Under the optimum condition, NOx emission can be controlled to less than 50 ppm.

Reference

1. Yang, C.-Y.; Fang, Z.; Li, B.; Long, Y.-f., Review and prospects of Jatropha biodiesel industry in China. *Renewable and Sustainable Energy Reviews* **2012,** 16, (4), 2178-2190.
2. Prasad, L.; Pradhan, S.; Das, L. M.; Naik, S. N., Experimental assessment of toxic phorbol ester in oil, biodiesel and seed cake of Jatropha curcas and use of biodiesel in diesel engine. *Applied Energy* **2012,** 93, 245-250.
3. Nimmo, W.; Patsias, A. A.; Hampartsoumian, E.; Gibbs, B. M.; Williams, P. T., Simultaneous reduction of NOx and SO2 emissions from coal combustion by calcium magnesium acetate. *Fuel* **2004,** 83, (2), 149-155.

RGB histograms for evaluation of fuel oil stability

A. Vráblík, N. Bringlerová, J.M. Hidalgo-Herrador, R. Černý
Unipetrol Centre of Research and Education, Záluží 1, 436 70 Litvínov, Czech Republic
tel. +420 471122297, e-mail: ales.vrablik@vuanch.cz

Fuel oils are very often used as a marine fuel according to the ISO 8217 for the marine application in the Sulphur Emission Control Area as well as outside this area depending on the total sulphur content.

The marine fuel must meet specific quality parameters according to the relevant standard. The density as well as viscosity are used for basic characterization or categorization to the individual classes. The stability of the marine fuel is very important and frequently discussed parameter. The standardised method for stability is Total Sediment according to ISO 10307. It is quite complicated and time consuming therefore simplified methods for stability assessing are being used - for example drop tests (Spot test, Oliensis test) or different solubility tests (toluene insolubles, heptane insolubles).

The article deals with the development of the new method for fuel oil stability assessment. The method is based on the Spot test according to ASTM D 4740, but the spot is visualised, modified by software and converted into the RGB histograms. Histograms provide the values necessary for calculate so-called V-value parameter. This newly introduced V-value is an objective parameter for evaluation of stability of fuel oils.

In this article V-value parameter is correlated with other properties for evaluation of the fuel oil quality and stability, especially with the most frequently used Total Sediment. It was verified that the new method is a very good and objective tool for the describing of fuel oil stability.

Acknowledgment

This publication is a result of research project solved in a frame of UniCRE centre which infrastructure was supported by the European Regional Development Fund and the state budget of the Czech Republic. The publication was supported by institutional funds (Ministry of Industry and Trade of the Czech Republic).

Simultaneous capture of gaseous SO_2 and HCl with solid lime

G. M. Lin, C. S. Chyang
Chung Yuan Christian University, Department of Chemical Engineering, Chung Pei Rd 200, Chung Li District, Taoyuan City, Taiwan 32023; tel. +886 3 2654119; fax: +886 3 4636242; e-mail: cschyang@gmail.com

In certain cases like incineration of municipal solid waste or co-combustion of fossil fuel and refuse, both SO_2 and HCl usually co-exist in the exhaust gases. The presence of HCl probably indirectly contributes to the formation of the dioxin compounds. Therefore, direct removal of SO_2 and HCl in the furnace is an imperative strategy. From the literature survey, very little were found to study the simultaneous control of SO_2 and HCl at high temperature, suggesting that it still has not been completely explored. Even so, a collective conclusion was drawn that the SO_2 retention efficiency was markedly promoted in the presence of HCl, while the concurrent chlorination reaction was suppressed. The reason for this phenomenon is not clear yet for now, which means more experimental evidences need to be presented and discussed. Thus, the present work was undertaken to elaborate the simultaneous chlorination and sulphation behaviors of the calcined limestone and its interactions under conditions similar to fluidized bed combustors.

Experiments on simultaneous sorption of SO_2 and HCl were carried out in a self-made reactor system, mainly aimed at investigating the behaviors of gas-solid reaction. The fixed-bed reactor made of quartz was designed specially to be capable of handling high-temperature operations of gas-solid reaction. Besides, this reactor already met several necessary criteria as follows: (1) excellent leakproofness, (2) rapid temperature rise for inlet reacting gas, (3) highly efficient cooling for exhaust gas and (4) low space requirement.

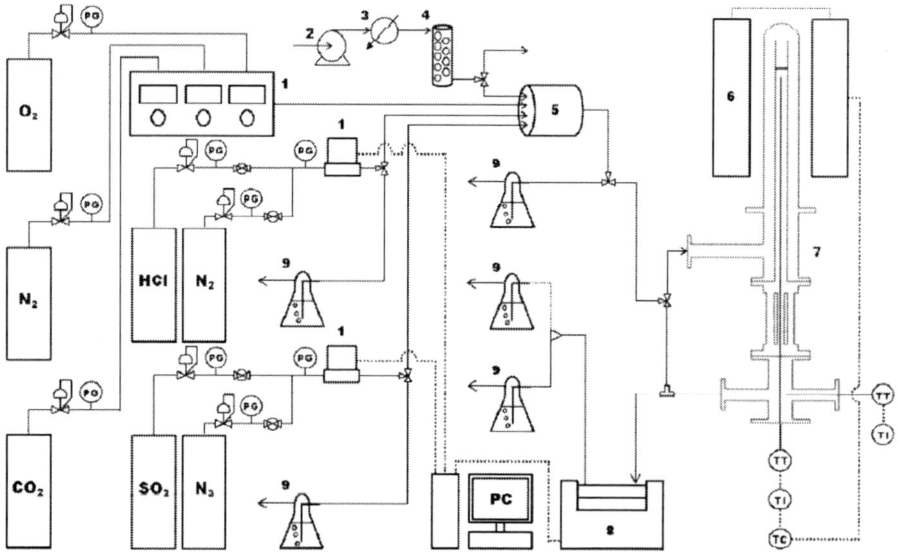

1. Mass Flow Controller
2. Clean Pump
3. Cooler
4. Dryer Column
5. Blending Chamber
6. Electrical Heater
7. Fixed-Bed Reactor
8. FTIR Gas Analyzer
9. NaOH Solution

All the while, the total gas flow rate was maintained at 1000 sccm (i.e. 1000 ml/min at 0 °C and atmospheric pressure); moreover, the composition of CO_2/O_2 in gas mixture was kept constant at 21% (e.g. 6% O_2 and 15% CO_2 in the reacting gas) for all tests. Also considering

the fact that the reactor system was slightly pressurized by the gases supplied from cylinders, and therefore a self-developed method has been successfully applied to minimize the possibility of gas leakage.
Experimental data obtained from fixed-bed operation are presented, showing SO_2 and HCl concentration profiles during the sorption process (also refers to as breakthrough curve). The term "breakthrough curve", which is a plot of effluent concentration as a function of time, was successfully applied to describe the sorption behavior of sulfur and chlorine in terms of the sorbent reactivity. In the present work, tests were done to further consolidate the validity of proposed explanation, and the reason why the sulphation and chlorination of sorbent is promoted and inhibited respectively with the coexistence of SO_2 and HCl was confirmed as well.

Carbon dioxide emissions and utilization opportunities in Pakistan

[1]A. R. Kalair, [2]N. Abas, [3]A. Kalair, [4]N. Khan
[1,]Air University, Islamabad, Pakistan; ali.kalair.ak@gmail.com; [2]Univ. of Gujrat, Dept. of Electrical Engineering, Hafiz Hayat Campus, Gujrat; naeemkalair@uog.edu.pk;
[3,4]COMSATS Insti. Of I.T., Dep. Electrical Engineering, Park Road, Chak Shehzad, Islamabad, Pakistan; tel.+923006490048, nasrullahk@yahoo.com

Abstract: Pakistan is one of the top 10 climate change hit countries. Melting glaciers, flash floods, droughts, heat waves and exploding population have created water, food and energy crises. Acasia and rosewood trees have dried up by scorching heat in most of country. Pakistan is affected by CO2 emissions of upwind neighboring countries under regional warming. Pakistan's annual CO2 emissions were 166.41Mt in 2015. These emissions are attributed to energy (50.7%), agriculture (38.80%), industry (5.8%), wastes (1.8%) and others (2.90%). This paper reports a niche opportunity of utilizing CO2 for domestic and industrial water heating under a carbon capture and storage strategy.

Introduction: Climate may be referred as weather conditions such as air temperatures, precipitations, atmospheric pressures, humidity levels, wind flows, sunshine intensities and cloud covers etc. Any change in weather compared to earlier short or long term observations is called climate change. Short term climate changes continue to occur in environment but long term changes are caused by greenhouse gases (GHG) in atmosphere. Key greenhouse gases include CO2 (76%), CH4 (16%), N2O (6%) and F gases (2%). Out of total 76% CO2 emissions 65% emit during fossil fuels combustions & industrial processes and 11% through forestry and land use patterns. Economic activities leading to GHG emissions include energy (25%), agriculture (24%), industry (21%), transport (14%), energy related activities (10%) and buildings (6%) [Boden, 2015]. Global annual energy consumption has gone beyond 14GTOE and carbon emissions have surpassed 10GtC. Annual GHG emissions have exceeded 53GtCOeq out of which about 40Gt is CO2 [Abas, 2014]. Global simulation analyses show the CO2 emissions, despite concerted efforts for reduction, are likely to increase to 51.88Gt by 2100 which would be a level 52.9% above 2010 levels. Average concentration of greenhouse gases is 430ppm out of which 402ppm is CO2 concentration which is a level never seen in last 800,000 years. The average temperature rise has been recorded to be 1°C from 1880 to 2015 which is halfway down the IPCC target of 2°C by 2100. We must not emit more than 1 Tt of CO2 on top of previously emitted 1.9 Tt to stay under 2°C ceiling which is not possible to without cutting 40-70% emission over 2010 level by 2050. Key drivers of GHG emission include population, gross domestic products (GDP) and global energy systems. Kaya identity points out seven sources and defines global CO2 emission flux (F) to be product of world population, per capita world GDP, energy intensity of world GDP and carbon intensity of energy [Raupach, 2007].

Analysis: Pakistan's overall GHG Emissions (Mt CO2 eq.) are projected to increase from 347 in 2011 to 4621 in 2050. The power sector is one of the major contributors to GHGs in Pakistan. The weighted average baseline emissions factor for power sector in Pakistan is 0.566 tCO2/MWh which for wind, solar and hydropower projects is 0.478 to 0.564 tCO2/MWh. Total installed capacity is 19,735 MW during summer and 14,922 MW during winter season. The most growth in emissions comes from the power sector, but both the industry and transport sectors show steady and significant growth. Per capita CO2 emissions increase from 858 kg/yr to 1650 kg/yr in 2030, but the 2030 value for Pakistan's energy sector remains about 3 times lower than the world average in 2007. The CO2 emissions from the entire energy spectrum of Pakistan are shown in Fig.1

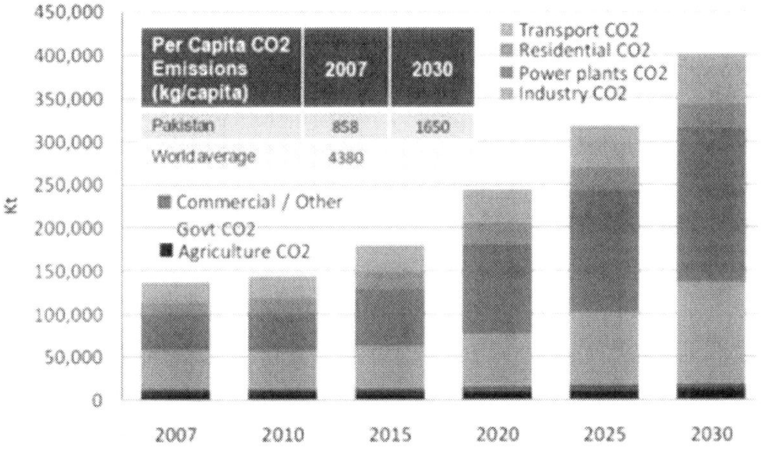

Fig.1 Pakistan's CO2 Emissions by Sector [Mukluks, 2012]

The agriculture and livestock sector is one of the most important sectors of economy in Pakistan. It is responsible for 21% of the GDP, employs 45% of labour force, and earns valuable foreign exchange for the country. Use of feed supplements is expected to reduce CH4 emissions by an average of 23 percent per animal. Global CO2 emissions are 39.6 Gt/y out of which 2.8 GtCO2/y is from cement industries. According World Bank report Pakistan's annual CO2 emissions were 163,177.83 in 2008 which increased to 16, 3753 in 2014 out of which 36.3MtCO2 come from cement industries. During 1990 to 2007 overall green house gas emission have decreased by 15% and in the same era increased to 36% by transport. CO2 emissions from transport sector are 24%. The emissions are caused by sea (10%), rail (25), buses (6%), wheelers (25), passenger cars (45%), trucks (23%), local (5%) and international (7%) air. Mountain glaciers in Himalayan region are melting faster than ever before in known history [Hizer, 2015]. Southeast Asia faces a serious survival threat akin to Collapse of Indus Valley Civilization.

Conclusions: Owing to large scale wind farms and solar parks the net increase in GHG on per capita bases might not increase despite new coal projects yet CCS technology is recommended for current under construction coal power projects in Gudani, Thar and Punjab under $45 billion CPEC program.

References

Boden TA, Marland G, Andres RJ, Global, regional and national fossil duel CO_2 emissions, Carbon Dioxide Information Center, Oak Ridge National laboratory, US DOE, 2015.

IPCC, Working Group III, Fifth Assessment Report, 2010.

N. Abas, N Khan, Carbon Conundrum, Climate Change, CO2 Capture and Consumptions, Journal of CO2 Utilization, Journal of CO2 Utilization, 8, 39-48, 2014.

Raupach MR, Marland G, Cias P, Le Quere C, Canadell JG, Klepper G, Field CB, Global and regional drivers of accelerating CO2 emissions, PNAS, 104, 10288-10293, 2007.

Numerical study on heat transfer and pressure drop performance of an asymmetric fin-and-tube heat exchanger

Z. W. Ren, T. Ma, M. Zeng, Q.W. Wang*
Key Laboratory of Thermo-Fluid Science and Engineering, Ministry of Education, Xi'an Jiaotong University, Xi'an, 710049, China
Corresponding Author: Q.W Wang, wangqw@mail.xjtu.edu.cn
Reporter: Z. W. Ren, renzhiwen2015@stu.xjtu.edu.cn

This research presents a 3-D model of an asymmetric finned tube heat exchanger, and the heat transfer and pressure drop characteristics of its air side are analyzed. Compared to the straight symmetrical fin-and-tube heat exchanger, the heat transfer factor is increased by 13.49% on average when the Reynolds number vary from 547 and 6485. Along with the increase of Reynolds number, the friction factor increases when the Reynolds number is less than 2169. On the contrary, when the Reynolds number is more than 2169, the friction factor decreases, and the decreasing trend becomes larger. The comprehensive performance of the asymmetric fin-and-tube heat exchanger is improved by 14.85% compared to the straight symmetrical fin-and-tube heat exchanger. The influence of inclined angle β under different velocities is also studied. At $15^{\circ} < \beta < 30^{\circ}$, increasing β could significantly strengthen the heat transfer. At $30^{\circ} < \beta < 90^{\circ}$, the increasing trend of heat transfer factor along with increasing of β becomes gradually slow within a larger inlet velocity. At $\beta \approx 30^{\circ}$, the comprehensive performance is the best. The optimum parameters are β=32.11°, a=2.0mm, t=2.0mm, D_{out} =9.0mm, D_{in}=7.0 mm for the laminar flow. The optimum parameters are β=27.81°, a =2.4mm, t=3.4mm, D_{out} =10.8mm, D_{in}=7.9mm for turbulent flow, and its comprehensive performance is increased by 27.65%.

Model predictive control of a membrane reactor for methane steam reforming

A. S. Kyriakides[1,3], S Voutetakis[1], S Papadopoulou[1,2], P Seferlis[1,3]
[1]Chemical Process & Energy Resources Institute (C.P.E.R.I.), CEnter for Research and Technology Hellas (CE.R.T.H.), P.O. Box 60361, 57001, Thermi-Thessaloniki, Greece
[2]Department of Automation Engineering, Alexander Technological Educational Institute of Thessaloniki, P.O. Box 141, 57400 Thessaloniki, Greece
[3]Department of Mechanical Engineering, Aristotle University of Thessaloniki, P.O. Box 484, 54124 Thessaloniki, Greece

In general, methane steam reforming is a conventional method to produce synthesis gas from hydrocarbon fuels. Produced synthesis gas can be used in Fisher-Tropsch and other processes to produce valuable products. Otherwise, hydrogen can be separated from synthesis gas and be used for applications such as fuel cells for cleaner energy production. Significant emphasis has been placed in the study of efficient hydrogen production and purification technologies. The use of a membrane reactor has been proposed, equipped with a Pd based membrane that exhibits an extremely high selectivity of Pd towards hydrogen, in order to improve the efficiency of the process. Hence, high methane conversion can be achieved at much lower reactor temperature than in conventional methane steam reforming reactors, while only hydrogen is simultaneously removed within the reaction zone through the membrane and the subsequent equilibrium shift towards hydrogen production.

The main aim of this study is the design and implementation of an optimal model predictive controller in a plug flow membrane reactor where methane steam reforming and hydrogen purification takes place simultaneously. Therefore, a mathematical model, which can describe the dynamic behavior of the physicochemical phenomena occurring in the reactor is developed. The dynamic, one-dimensional, nonlinear, and pseudo-homogeneous mathematical model of the tubular membrane fixed-bed reactor consists of mass and energy balances where axial gradients of temperature and mass are fully considered. More specifically, the process model takes into consideration the complex reaction mechanism of the reforming reactions, convective and molecular diffusion of the reacting mixture species, diffusion of hydrogen through the membrane, and thermal effects due to reaction and heat exchange with an external heat source. Subsequently, the control problem is defined by appropriately selecting the manipulated and controlled variables, the objective function and the constrains that are imposed in the solution of the optimization problem. The applied algorithm of optimal model predictive control, with constrains enforced at the manipulated variables, employs a linearized version of the rigorous dynamic model around a known system's operating point. The main challenge is to efficiently produce pure hydrogen with maximum methane conversion while operating under conditions that do not favor coke production. The case study of the applied optimal model predictive controller in the non-linear system confirms the controller ability to achieve the desired dynamic behavior both in the case of desired reference point targeting and disturbance compensation.

Energy exchange strategy for autonomous smart-grids with renewable energy sources empowered by an Internet of Things (IOT) architecture

C. Ziogou[1], S. Papadopoulou[2], S.Voutetakis[1]
[1]Chemical Process and Energy Resources Institute (CPERI), Centre for Research and Technology Hellas (CERTH), PO Box 60361, 57001, Thessaloniki, Greece, {cziogou, paris}@cperi.certh.gr
[2] Department of Automation Engineering, Alexander Technological Educational Institute of Thessaloniki, PO Box 141, 54700, Greece, shmira@autom.teithe.gr

This work presents a decision making methodology which is developed for the effective energy exchange between off-grid stations with Renewable Energy Sources (RES) for power generation. When multi-source energy stations (nodes) are connected, a network is formed where the decisions should be made according to the overall supply and demand of energy. Each node needs to exchange information with the monitoring and control station that supervises the operations of the entire network as part of the operations domain of the grid.

The main operation objectives are to exchange power based on demand response strategies, to maximize the usage of available stored power at network level and to utilize the total amount of available renewable energy. The presented methodology consists of a flexible modular decision-making mechanism for networked systems that rely on a multi-agent communication structure using an Internet of Things (IoT) architectural reference model. The rules of operation are represented by propositional logic and the implementation is realized by an agent-based structure, which is executed at each node. The energy sharing is based on a load balancing approach with priorities that are dynamically updated based on the status of each node and the internal handling of its energy resources and local load demand. Although this methodology is shown by a case study which is related to the energy domain, it can be easily extended to other networked systems where exchange of resources is necessary.

The decision making methodology is developed with implementation considerations for the operation of a smart grid that involves three autonomous stations with local energy storage capabilities using a lead-acid battery stack. Furthermore, there are hydrogen generation and long term hydrogen storage options. Each station has a PV array, wind generators and a diesel generator for backup. Also at one of the stations there is a Polymer Electrolyte Membrane (PEM) fuel cell and a PEM electrolyzer for the utilization and production of hydrogen respectively. In order to make this network as self-sustained as possible the main target is to minimize the use of the diesel generator. Therefore, the priority for the energy exchange is firstly to serve the local load demand and secondly to avoid the use of the generator by utilizing the available energy in the network. The protection of the subsystems within each station is also considered as respective indications for the status of the energy storage is available to the supervisory control station. This isolated grid is located at Xanthi, Greece and its operation is monitored using a Supervisory Control and Data Acquisition (SCADA) system while the Machine to Machine (M2M) communication for the decision making is implemented by the IoT enabled architecture using the agent-based middleware. The performance of the involved nodes is explored and the overall operation of the grid is presented whereas the response of each node is monitored online. An industrial grade agent-based IoT architecture can increase the resiliency and flexibility of such networks whereas the dynamic rerouting of energy can significantly improve the robustness and optimum use of the available resources.

Performance improvement of multistage flash-tank heat recovery networks through mathematical modeling and process integration

Krum Semkov[a,b,*], Emma Mooney[a], Michael Connolly[a], Sean Keating[a], Catherine Adley[a]
[a] Department of Chemical and Environmental Science, University of Limerick, Limerick, Ireland
[b] Institute of Chemical Engineering, Bulgarian Academy of Sciences, Sofia, Bulgaria
krum.semkov@ul.ie

Flash Tank heat exchanger systems are usually used with heavy heat transfer conditions such as in the case of high solid content, strong tendencies to scaling and fouling and usually high corrosion aggressive streams. A typical area of application is in the production of alumina by the wide spread Bayer Process (Smith, 2009) known as a very energy intensive technology. This makes the energy efficiency improvement a task of highest importance. For this reason, several approaches have been discussed targeting mainly the betterment of performance of the different stages of the alumina production (Hond et al., 2007).
The present paper aims to improve the efficiency using simultaneously Process Integration (PI) and Combined Temperature and Pressure Management (CTPM) in order to intensify the internal heat exchange. This way the high potential heat (steam) input demand could be reduced and at the same time decreasing the output heat from the system, as was shown in the recently proposed approach (Semkov et al., 2013).
For this purpose the basic mathematical model of the flash tank heat recovery cascades (Korobeinikov et al., 2010) was further developed enhancing its current range of application (Zhelev at al., 2013) by including all important factors such as evaporation, indirect and direct (mixing) condensation, presence of additional input and output streams in a cascade stage, as well as the influence of the non-condensable gases and the superheated state of the vapor. Applying the model as well as the PI and CTPM to a multistage digester cascade, new beneficial feasibilities were discovered. In order to evaluate the overall efficiency improvement numerical experiments were carried out. As a result it was shown that substantial high potential heat (steam) saving accompanied by low grade interim output heat streams reduction can be achieved. As so far some of these outputs have been used for useful purposes (e.g. boiler feed water preheating), an additional saving opportunity through PI appears, replacing them by recovered waste heat from other relatively low grade heat sources e.g. exhaust gas from the calcination process.

Acknowledgement

Authors would like to acknowledge the support of Enterprise Ireland under the Project IP 2013 0223.

References

Hond R., Hiralal I., Rijkeboer Ab, Alumina yield in the Bayer process. Past, present and prospects, 2007, Edited by TMS (The Minerals, Metals & Materials Society).
Korobeinikov A. et al., 2010, Mathematical modelling of internal heat recovery in flash tanks heat exchanger cascades, Proc. 75th European Study Group with Industry, University of Limerick, June 27 - July 2, Limerick, Ireland, 1-12.
Semkov K., Mooney E., Connolly M., Adley C., 2013, Chemical Engineering Transaction, 35 (accepted).
Smith P., 2009, The process of high silica bauxites – Review of existing and potential processes, Hydrometallurgy, 98, 162-176.
Zhelev T., Semkov K., Mooney E., Majozi T., Korobeinikov A., 2013, Heat Transfer Engineering, 34, 24, 1191-1201, DOI: 10.1080/01457632.1013.776888.

An optimization model for the synthesis of decentralized biomass-based polygeneration systems including provision for potable water for remote rural areas

Kathleen B. Aviso,[1] Sheree A. Pagsuyoin,[2] Michelle Almendrala,[1] Florinda T. Bacani,[1] John R. Barajas,[1] Aileen Huelgas-Orbecido,[1] Jana S. Latayan,[3] Michael Angelo B. Promentilla,[1] Joost R. Santos,[4] Raymond R. Tan[1]
[1]De La Salle University, 2401 Taft Avenue, 0922 Manila, Philippines; fax: +632 536 4226; [2]Department of Civil and Environmental Engineering, University of Massachusetts-Lowell, Lowell, MA 01854,USA; fax: +1-978-934-3052; [2]Chemical Engineering Department, [3]Acton-Boxborough Regional High School, Acton, MA 01720, USA; [4]Department of Engineering Management and Systems Engineering, The George Washington University, Washington, DC, USA; fax: +1- 202-994-1249.

Polygeneration systems enable flexible, decentralized production of multiple products with additional gains in efficiency and environmental impact reductions as a result of inherent opportunities for energy and material recovery. In developing countries, such systems are especially promising for remote rural sites, where underutilized biomass residues present a potentially abundant resource for the provision of electricity, fuel for household use (i.e., cooking) and potable water for isolated communities. Process systems engineering (PSE) methods present a useful approach for the rigorous design of such systems, taking into account case-specific supply and demand considerations, as well as technological alternatives.

In this paper, a mixed integer linear programming (MILP) optimization model is proposed for the synthesis of decentralized, biomass-based polygeneration system that incorporates the objective of providing clean drinking water for remote rural areas. The model incorporates as constraints the availability of locally sourced biomass as the main input for the production of electricity, cooking fuel and water for use by geographically isolated communities with limited access to regional imports of such commodities. Technology and process options in the model are selected via Boolean variables. A semi-hypothetical case study based on the requirements of a rural community in the Philippines is solved to demonstrate the use of the optimization model.

Optimization of the passage arrangement for multi-stream spiral-wound heat exchangers using a genetic algorithm

Xiang Li, Gaopeng Zhang, Min Zeng, Qiuwang Wang*
Key Laboratory of Thermal-fluid Science and Engineering, Ministry of Education, Xi'an Jiaotong University, Xi'an, Shaanxi, 710049, China
wangqw@mail.xjtu.edu.cn

The heat exchanger's flexibility is the operation and adjust characteristics of a heat exchanger. The performance of heat exchangers in varied operating conditions is directly affected by their flexibilities. This paper presents the flexible design of one multi-stream spiral-wound heat exchanger in varied operating conditions. Owing to multi-objective and nonlinear characteristics of the optimization, a genetic algorithm (GA) is adopted to solve the problem of the passage arrangement. Variance of the cumulative passage thermal loads is introduced as objective functions, and the sequence number method is adopted as encoding method. The result shows that the flexible design and GAs can be used for the heat exchanger's design and optimization in varied operating conditions.

Supply chain design network model for biofuel and petrochemicals from biowaste

Ismaail A. Emara*[a], Mamdouh Gadalla[b], Fatma Ashour[c]
[a] Master's Student, Chemical Engineering Department, Faculty of Engineering, Cairo University, 12654, Egypt.
[b] Associate Professor, Chemical Engineering Department, Faculty of Engineering, British University in Cairo, 11837, Egypt.
[c] Professor, Head of Chemical Engineering Department, Faculty of Engineering, Cairo University, 12613, Egypt.
ismaailamr@hotmail.com

Supply Chain (SC) is the effective coordination and integration of all activities performed by several business lines and their infrastructures; such as raw material suppliers, manufacturers, distributors and retailers till delivering the products to the customer.
The supply chain facilities location is an important issue which is needed to be solved as it is one of the important pillars in the supply chain design model. Therefore, it is critical for supply chain managing teams and designers to adopt the latest scientific and most effective methods to locate a facility by carrying different scenarios and trade-offs assessment, the used tools are only a decision supporting tools.
This paper concentrates on the development status of a decision-support tool for solving supply chain design including location problem of various facilities, using the MILP mixed integer linear problem solving programming model by constructing a generic analytical algorithm to solve this problem. EXECL SOLVER TOOL, MATLAB SOFTWARE with a GUI (graphical user input interface) programmed by the C# (C SHARP) programming software are used. In the MILP programming model iterative algorithm methods are used, where every iteration would generate a whole different new set of solutions to match some constrains, conditions and service levels added by the programmer. The first solution only matches some previous added criteria, resulting in a new set of solutions in the iteration, until finally an optimal solution rises under dynamic environment.
In this paper, a typical three layers medium term planning mixed integer linear Programme MILP model for supply chain design is presented and applied for chemicals and biofuel production from rice straw biomass. These three stages are commonly: raw materials collecting areas, distribution center and manufacturing plant locations. With the aid of ASPEN HYSYS simulation software and CAPCOST TOOL, Economic feasibility studies are done for designing supply chain design model. Different scenarios are examined which is a good way to price out potential solutions.

Keywords: Mixed Integer Linear Problem, Rice Straw, Biodiesel, Supply Chain Design Model, Economic Feasibility Study, Sustainable Development and Renewable Energy, MATLAB, C SHARP.

Environmental index for palm oil mill

Nabila Farhana Jamaluddin[1], Zarina Ab Muis[1]*, Haslenda Hashim[1], Lim Jeng Shiun[1], , Rahmalan Ahamad[2]
[1]Process System Engineering Centre (PROSPECT),Faculty of Chemical and Energy Engineering
[2]Palm Oil Research Centre (PORC)
Universiti Teknologi Malaysia
Johor Bahru, Skudai, Malaysia
*zarinamuis@utm.my

Palm oil mill has been regarded as a profit making industry for the past decades. Besides revenue from the palm oil production itself, the abundance of biomass could generate high economic return to the palm oils mill by converting it to value added products. However, the palm oil industry currently suffered various criticisms and negative reports from the international non-government organizations on the environmental issue. This paper presents the development and the applicability of environmental index (EI) for palm oil mill in order to enhance the competitiveness of the industry. The methodology will include the index development and identify area of weaknesses (hotspot). The index will be able to identify performance of the mills in Environmental aspect against benchmarks and differentiate every mills performance. By this, the mills can easily identify its hotspot and take necessary steps to improve. The system will also benefit the marketing team to market its products (CPO, PKO etc.) to niche markets that seek uptake from mills with lowest footprints (Carbon, water, energy etc.). The results can be used not only to evaluate the performance of an operating process against standard benchmarking but also to establish the best environmental practices among palm oil industry. It is anticipated that the development of environmental index will be an important instrument for supporting sustainable operation for palm oil mill.

Amino acids recovery from swiftlets (*Aerodramus fuciphagus*) guano employing aqueous two phase system (ATPS)

Hui Yi Leong[1], Pau Loke Show[1,*], Chien Wei Ooi[2], Azizon Abdullah[3]
[1] Food and Pharmaceutical Engineering Research Group, Department of Chemical and Environmental Engineering, Faculty of Engineering, University of Nottingham Malaysia Campus, Jalan Broga, 43500 Semenyih, Selangor Darul Ehsan, Malaysia.
[2] Chemical Engineering Discipline, School of Engineering and Advanced Engineering Platform, School of Engineering, Monash University Malaysia, Jalan Lagoon Selatan, Bandar Sunway, 47500 Selangor Darul Ehsan, Malaysia.
[3] Regional Veterinary Laboratory, Kota Bharu Kubang Kerian, 16150 Kota Bharu, Kelantan.
*Corresponding author: Tel: +603 8924 8605; Fax: +603 8924 8017; Email address: PauLoke.Show@nottingham.edu.my

Swiftlets (*Aerodramus fuciphagus*) guano is a promising source of amino acids. Amino acids are good for skin rejuvenation, and this is due to their powerful effects on repairing and regeneration of cells and tissues as well as involve in building and distribution of collagen. Thereby, this provides a potential alternative against the use of edible bird's nest (EBN) in several industry applications, such as nutraceuticals, cosmetics, skin care products and etc. Aqueous two phase system (ATPS) is claimed to be an effective system in recovery of protein and amino acids. The present study is aimed to effectively recover amino acids from swiftlets guano by employing ATPS. Quantification of amino acids is based on colorimetric assay. The guano sample is provided by the regional veterinary laboratory at Kelantan.

Keywords: *Aerodramus fuciphagus*; amino acids; aqueous two phase system (ATPS); guano

WOOD-BIO-SOL – Biotechnology research network with partners in the Baltic sea region

Bernd Niemeyer, Helena Horn, Stephan Lassen
Helmut-Schmidt-University, University of the Federal Armed Forces Hamburg, Department of Thermodynamics, Holstenhofweg 85, 22043 Hamburg, Germany

The international network WOOD-BIO-SOL was organized to facilitate the process of finding partners for projects in the field of biotechnology. In this context the virtual Bio Technology Institute BTI was installed in order to gather potential cooperation partners from industry located in the Baltic Sea Region, including small and medium-sized enterprises as well as universities and research institutions. The long term goal of the network is the delivery of novel substances such as biomedicals, chemical products and food additives.

AUTHOR INDEX

AUTHOR INDEX

AUTHOR INDEX

AUTHOR INDEX

AUTHOR INDEX

AUTHOR INDEX

AUTHOR INDEX

AUTHOR INDEX

AUTHOR INDEX

AUTHOR INDEX

AUTHOR INDEX

AUTHOR INDEX

AUTHOR INDEX

AUTHOR INDEX

AUTHOR INDEX

AUTHOR INDEX

AUTHOR INDEX

AUTHOR INDEX

AUTHOR INDEX

AUTHOR INDEX

AUTHOR INDEX

AUTHOR INDEX

AUTHOR INDEX

AUTHOR INDEX